THE POCKET OXFORD
GREEK DICTIONARY

The Pocket Oxford Greek Dictionary

Greek–English
English–Greek

A revised edition of
The Oxford Dictionary
of Modern Greek

Compiled by
J. T. PRING

George. Pap.

OXFORD UNIVERSITY PRESS

Oxford University Press, Great Clarendon Street, Oxford OX2 6DP

Oxford New York

Athens Auckland Bangkok Bogota Bombay
Buenos Aires Calcutta Cape Town Dar es Salaam
Delhi Florence Hong Kong Istanbul Karachi
Kuala Lumpur Madras Madrid Melbourne
Mexico City Nairobi Paris Singapore
Taipei Tokyo Toronto Warsaw

and associated companies in
Berlin Ibadan

Oxford is a trade mark of Oxford University Press

Published in the United States by
Oxford University Press Inc., New York

The two parts of The Oxford Dictionary of Modern Greek were
first published separately, Greek–English in 1965 and
English–Greek in 1982. This combined edition was first published
in 1982

British Library Cataloguing in Publication Data

Data available

ISBN 0–19–864196–6 Hbk
ISBN 0–19–864197–4 Pbk
ISBN 0–19–864536–8 Flexi (available in Greece only)

Library of Congress Cataloging in Publication Data

Pring, J. T. (Julian Talbot)
The pocket Oxford Greek dictionary : Greek–English, English–Greek
/ compiled by J. T. Pring.
p. cm.
"A revised edition of The Oxford dictionary of Modern Greek"
ISBN 0–19–864196–6: £8.99.—ISBN 0–19–864197–4 (pbk.)
1. Greek language, Modern—Dictionaries—English. 2. English
language—Dictionaries—Greek, Modern. I. Pring, J. T. (Julian
Talbot). Oxford dictionary of modern Greek. II. Title.
PA1139.E5P78 1995
483'.21—dc20 95–3145 CIP

5 7 9 10 8 6

Typeset by Fotron SA, Athens
Printed in Hong Kong

CONTENTS

ABBREVIATIONS

a., adjective
acc., accusative
adv., adverb
aero, aeronautical
aor., aorist
archit., architecture
biol., biology
bot., botanical
chem., chemistry
conj., conjunction
D., Dutch
dat., dative
E., English
eccl., ecclesiastical
esp., especially
etc., etcetera
F. French
f., feminine
fam., familiar
fig., figurative
G., German
gen., genitive
geom., geometry
Gk., Greek
gram., grammatical
I., Italian
i., intransitive
imper., imperative
int., interjection

joc., jocular
iron., ironic
lit., literally
m., masculine
math., mathematics
mech., mechanics
med., medical
mil., military
mus., music
n., neuter
naut., nautical
neg., negative
nom., nominative
num., numeral
part., participle
pass., passive
pej., pejorative
phys., physics
pl., plural
prep., preposition
pron., pronoun
s., substantive
sing., singulár
sthg., something
subj. subjunctive
T., Turkish
t., transitive
v., verb

PREFACE TO THE REVISED EDITION

In this re-set and revised edition, the old accentual system is replaced by the now widely used Monotonic, with monosyllabic words unmarked, and an acute accent placed on the stressed syllable of other words. Breathings are abolished. The diaeresis keeps its distinctive function.

Both sections have been extended and updated with new material, and the proportion of Greek words given in demotic form has been increased. However, the remarks on Vocabulary set out on pages XI-XVIII of the Introduction are still valid. Although in recent years official use of demotic forms has increased, the great body of older writings, together with dictionaries, encyclopaedias and documents of all kinds, still presents itself unaltered to the researcher and general reader alike; while a degree of formality retains its place in educated speech.

A dictionary intended for general reference must be usable to interpret various kinds of modern Greek, including katharevusa forms; and for a proper understanding of the language one should know one's declensions and conjugations not only in their simpler demotic versions, but also in their fuller and more formal ones.

Any apparent inconsistencies of spelling can be attributed to the fluidity of Greek usage.

J. T. P.

INTRODUCTION

The classical age of Greek literature, under the dominating influence of Athens, was followed by a period of great linguistic change after the conquests of Alexander had carried Greek influence far and wide over the Near East. The Greek which came to be used in common throughout this area, not only by Greeks themselves but also by the native peoples, who learnt it as a second laguage, is known as the Hellenistic Koine, or common Greek. Its main constituent was the Attic-Ionic dialect, which had been perfected as a medium of oratory by Isocrates and Demosthenes, and of philosophical prose by Plato. But this new extension of Greek naturally brought about many changes in vocabulary, grammar, and pronunciation; and various contemporary writings, especially letters preserved in Egyptian papyri, throw enough light on the state of the vernacular to show that it was already developing some of the characteristic features of the present-day language. The age of the Koine, which saw the Roman conquest of Greece and the rise of Christianity, may be set roughly between 300 B.C. and A.D. 300. Outstanding among works representing literary forms of Hellenistic Greek are the History of Polybius and the New Testament. Some academic purists, reacting against the innovations of the Koine, promoted the doctrine of 'Atticism', which set up classical Attic Greek as the only permissible model of prose composition; and this attitude never ceased to affect Greek writing by inhibiting free expression in the natural idiom of the times.

The spirit of Atticism may be said to prevail through the whole Byzantine period from the establishment of Constantinople in 330 until its capture by the Turks in 1453. Although Greek had replaced Latin as the official language of the Eastern Empire by about 600, the development of the vernacular during the following centuries remains largely a matter for speculation, and many writers continued to regard it as unfit for literature. Thus by the middle of the fifteenth century, when western Europe stood poised upon the threshold of the modern world with well-integrated languages and flourishing literatures, Greece was entering upon four hundred years of bondage to the Turks without benefit of either.

Crete, under Venetian administration, held out against the Turks until 1669; and the poetry written in the Cretan dialect of the seventeenth century may be regarded as the beginning of modern demotic literature. Under the Turks, the regional dialects found expression in a body of folk songs reflecting the life of a people deprived of all material and intellectual progress. When Greece won independence in 1830, Athens and the Peloponnese became the political core of the new kingdom, and it is their dialects which form the basis of the standard spoken Greek of today. But the official language of a modern state could not be wrought out of the foklore of a medieval peasantry. Efforts were made to produce a purified form of Greek ('katharevusa') suitable for modern needs. But they became too deeply influenced by the spirit of Atticism; and the problem of finding a natural prose medium supple enough to provide expression in both formal and colloquial terms was not solved to anybody's satisfaction. Now, after a century and a half of independence, Greeks are still frustrated by the 'language question'. But with the spread of education and the growth of journalism and broadcasting, the question begins to solve itself. Demotic and katharevusa cannot be kept apart and a form of Greek is emerging which combines features of both.

THE EVOLUTION OF DEMOTIC

The development of some main features of the modern vernacular is outlined below. It should be borne in mind that they do not represent a coherent or exclusive system; and that other, more conservative, forms are also to be found coexisting with them.

Phonology

After the Hellenistic period the distinguishing power of word-accent lay in its position rather than its pitch; and the former distinctions of vowel-length were lost. Certain words now show a stronger tendency to keep the primary stress in their inflected forms. A ε ι o have remained more or less unchanged in quality; ου was already u before the end of the classical period; and ω merged with o as a mid back vowel after the first century A.D. Aι had merged with ε by the second century A.D., and the second element of αυ ευ ηυ had developed into a labial fricative.

H and ει had both become i by early Byzantine times. Υ and οι were being confused as ü in Hellenistic Greek, and both became i by the tenth century.

Double consonants have been reduced to single, except in a few dialects. The aspirated voiceless stops θ φ χ had changed to fricatives by the fourth century A.D., and initial h (marked by 'rough breathing') had disappeared. By the same date the voiced stops represented by 6 δ γ had become replaced by fricatives. But in certain modern forms the labial and dental stops are still preserved after μ ν, being now written as π τ, e g.: γαμπρός < γαμβρός, δέντρο < δένδρον. The originally voiceless π τ κ following a nasal have changed to voiced stops (which can also occur without the nasal environment, especially in initial position). Among other phonological changes are: (i) Loss of many initial and medial unaccented vowels, including the verbal augment. (ii) Loss of nasals finally and before a continuant consonant, (iii) Dissimilation of voiceless consonant groups, e.g.: φτερό < πτερόν, οχτώ < οκτώ, σκολιό < σχολείον, έκαψα < έκαυσα.

Morphology and Syntax

The dative case, already restricted in New Testament Greek, had fallen out of popular use by the tenth century. Indirect object is now expressed by the accusative (usually after σέ) or genitive. Prepositions govern the accusative. A notable feature of nouns is the extension of a simplified first declension having two terminations in each number, with -ς marking nominative singular of masculines and genitive singular of feminines, and a common plural in -ες, -ων. This includes many nouns adapted from the classical third declension, e.g.: ο έρωτας, η πατρίδα, η 6ρύση. A new plural inflexion -δες is very common, especially in non-classical words. The neuter second declension form in -ι has developed from diminutives in -ιον since the Koine period.

Dual number had become obsolete in later classical Greek. Optative mood, already rare in the Koine, fell out of use in the Byzantine period; and the infinitive became superseded by να with subjunctive. Verbs have lost the old future, perfect, and pluperfect tenses. Futurity is expressed by θα with subjunctive, and conditional mood by θα with imperfect. Compounds of έχω provide perfect, pluperfect, future perfect, and past conditional tenses. Durative and perfective aspect of verbs are based upon present and aorist stems respectively, and this category is now extended to future time. The indeclinable relative pronoun που and active participle in -οντας have been in common use since the Byzantine period.

Other basic features of demotic are: (i) Parataxis in preference to subordination. (ii) Formal redundancy of the types μικρό κοριτσάκι, πιο χειρότερος, ξανακοιμήθηκε πάλι. (iii) Great use of diminutives.

THE GREEK VOCABULARY

The nucleus of the modern vocabulary has been handed down from ancient Greek. It is supplemented by several strata of loan-words, of which the chief are: Latin (from the Hellenistic and Byzantine periods); Italian (from the Venetian and Genoese occupation of Greek lands after 1200); Turkish (from the period 1453-1830); French and, to a lesser degree, English (during the last 150 years). The abundant resources of derivation and composition which Greek possesses have made easy the creation of new words, especially scientific terms, out of the native stock. These include translation-words, such as σιδηρόδ-ρομος (chemin-de-fer), and the reborrowing of Hellenic coinages already current elsewhere, as αεροπλάνον (aeroplane).

As we have noted above, the Greek of daily use is based broadly on *demotic* (or 'common Greek') rather than the more formal *katharevusa* (or 'puristic Greek'), which is by tradition the language of Law, the Church, the official world, and the domains of science and technology. But the terms demotic and katharevusa (D and K hereinafter) have another, separate, meaning. Besides indicating degrees of formality in the manner of expression, they also designate more precise differences arising in morphology and syntax. For instance, a typical D-feature is the conditioned elision of final -ν, and a typical K-feature is the retention of unstressed verbal augment. Dative case forms are exclusively K, as is the genitive absolute construction. Paratactic constructions are typically D. However, this dualism is not fully realized in Greek grammar. When it does occur it may be marked in the primary form of a word but not oblique forms, or vice versa. It is marked for example, in κρύβω (D)/κρύπτω (K) (*I hide*, v.t.), but not in έκρυψα (DK) *(I hid)*, but again in κρύφτηκα (D)/εκρύφθην (K) *(I hid,* v.i.). Of eighteen forms of the definite article (excluding dative), dualism is found in two feminine plurals: οι (D)/αι (K) and τις (D)/τας (K). It is also potential in two accusative singulars: το(ν) (D)/τον (K) and τη(ν) (D)/την (K). The remaining forms of the article are common to D and K. In adjectives where the suffix -ος follows stem-final ǫ, dualism may be shown in feminine forms: δεύτερη (D)/δευτέρα (K) *(second)* etc. The above are but token instances of a widespread and striking feature of modern Greek.

Another function of the terms D and K is to convey the 'feel' that some words acquire from their position on the stylistic scale. Thus θύρα *(door)* is felt to be a puristic word and πόρτα a demotic one. But in many cases such a classification is not feasible. Although, in a general way, D-features are typical of colloquial style and K-features of formal style, the correspondences are not absolute; and K-features have no small place in the composition of colloquial Greek. Moreover there is, historically, a large element of common ground. Many classical words have been transmitted unchanged and remain in general use, the same for all Greek. Such are: άνθρωπος *(man),* σώμα *(body),* θάλασσα *(sea),* μικρός *(small),* και *(and).* Others may have lost currency in daily speech but retain a place in formal or less prosaic styles, e.g. δύναμαι *(can)* τέκνον *(child)* as against common Greek μπορώ, παιδί. Some ancient words, long obsolete, have been revived in officialese, e.g. εκατέρωθεν *(on both sides)* as against κι' από τις δυο μεριές.

Thus the terms D and K have disparate ranges of meaning, and may not always prove adequate signposts for those who explore the paths of Greek. In common parlance *water* is νερό rather than ύδωρ, *red* is κόκκινος rather than ερυθρός, *horse* is άλογο rather than ίππος, and to *fly* is πετώ rather than ίπταμαι, (the first series being D and the second K). But it is also understood that the second series would apply when referring to such things as the *Water Board,* or the *Red Cross,* or *horsepower,* or the *Flying Dutchman.* By the same token, *to wash one's hands* is πλένω τα χέρια μου (D) while *to wash one's hands of it* is νίπτω τας χείρας μου (K). Or again: *with the doors closed* is με κλειστές τις πόρτες (D), but *behind closed doors* is κεκλεισμένων των θυρών (K). All the above items are proper to colloquial usage.

This being so, too much concern with ideas of a 'demotic versus katharevusa' conflict may bring more confusion than clarity to the matter. The dictionary's function is to reflect usage without being committed to either one or other of these philological watchwords: and any preemptive application of D or K labels to headwords would be too suggestive of a dichotomy that belongs rather to the 'language question' than to the language, and too likely to obscure the interplay of those variegated strands of usage from which the fabric of Greek is woven.

This book, then, is designed to meet the need for a compact, up-to-date dictionary suitable both for general reference and the language student. In the Greek-English section much of the less useful material of older dictionaries has been eliminated in favour of a more practical selection of words and idioms in common use, covering the vocabulary of everyday affairs and general literature. Similarly, the aim of the English-Greek section is to give Greek equivalents of English words and phrases with illustrations of their idiomatic usage. In principle the renderings are into colloquial language of everyday use, varying from subtle to commonplace from the polite or profound to the trivial and unbuttoned. Colloquial language does not exclude formal expressions, which are given a place here in due proportion where the context is appropriate.

ARRANGEMENT OF GREEK-ENGLISH SECTION

The headword is in bold type. A swung dash, used in the body of the article, represents either the headword or that part of it preceding a vertical line. Certain derivatives are given under headwords, being shown by the swung dash with termination in bold type. Other derivatives have been treated as separate entries where it seemed more convenient to do so. A shift of accent is usually shown in quotations of the headword: when it is not, it may be deduced from the general rules of accentuation.

It should be clear from the nature of an entry whether it is to be read as a translation of the Greek word, or as a definition or explanation. Thus 'fish' is a translation; but 'sort of fish' is a definition, and implies that an English equivalent word is not found, or is too uncommon to be usefully quoted. Words in Roman type within brackets form part of the translation or definition. 'Bedroom (suite)' indicates that the word means both 'bedroom' and 'bedroom suite'. Words in italic within brackets are explanatory, e.g.: 'diamond *(cards)*', 'demolish *(building)*'. Common idioms are included; and a short phrase may be given to illustrate a particular usage. The derivation of indeclinable loan-words is given in square brackets.

The term *(fam.)* is used, widely but not exhaustively, to show colloquial expressions that might be better avoided in serious or formal utterance. It did not seem advisable to try to assess degrees of informality or acceptability; but the English rendering of a word or phrase may help the reader to judge its stylistic value.

Attention is drawn to two conventions governing the entries for verbs:

(1) γέρνω *v.t. & i.* bend down, tilt; half close *(door); (v.i.)* lean *or* bend over, stoop; lie down; sink *(of star)*.

The above layout indicates that the meanings which precede *(v.i.)* refer to the transitive function of the verb, and those which follow *(v.i.)* refer to its intransitive function.

(2) αποκαλύπτ|ω *v.t.* unveil, uncover, reveal. ~ομαι *v.i.* raise one's hat.

The passive form of an active verb is quoted in the body of the entry if it involves an extension of the range of meaning. In such cases the simple passive inversion of the active meaning is not necessarily excluded by not being mentioned. In the above entry is is understood that αποκαλύπτομαι also means 'be unveiled, uncovered, *or* revealed'.

In this section space has been saved by leaving out many straightforward derivatives and compounds. Thus άτολμος (timid) is given, but not ατολμία (timidity); κρυφά (secretly) and βλέπω (look) are given, but not κρυφοβλέπω (look stealthily).

Among common words usually omitted are:

(1) Feminine equivalents of masculine nouns such as γειτόνισσα (from γείτονας), φοιτήτρια (from φοιτητής), μυλωνού (from μυλωνάς).

(2) Adverbs in -ως and -α formed from adjectives, e.g.: σοφώς (from σοφός), άσχημα (from άσχημος).

(3) Diminutives and augmentatives ending in -άκι, -ίτσα, -ούλης, -αρος, etc.

However, a good many such derivatives can be found by reference to the English-Greek section. Some common prefixes are shown separately, with an indication of their range of meaning, e.g.: καλο-, ξε-, παρα-, σιγο-. For the general rules of derivation and composition a grammar should be consulted. This is also the best source of information on the phonetic, orthographic, and morphological differences *(a)* between puristic and demotic Greek, and *(b)* within the field of demotic itself. Variant spellings of words are so numerous that many alternative forms have had to be omitted from this dictionary. Examples of common types of variant are the following: βιβλίον or βιβλίο, εφημερίς or εφημερίδα, βρύσις or βρύσι or βρύση, δηλώ or δηλώνω, αγαπώμαι or αγαπιέμαι. Often the first letters of words are affected, and their several forms must be sought at widely separated points in the dictionary, e.g.: ολίγος or λίγος, ευρίσκω or βρίσκω, κτίζω or χτίζω. An attempt is made to show (as older dictionaries often do not) when such pairs are not identical in meaning, e.g.: δικαιούμαι and δικαιώνομαι, ρεύμα and ρέμα, αι θέρμαι and οι θέρμες. The problem of variant forms makes it impracticable to give the inflexions of verbs, nouns, and adjectives; but there is an Appendix showing principal parts of a select list of verbs. A full dictionary should deal comprehensively with these matters, and present impartially the puristic and demotic forms of any word liable to such variation. A small dictionary cannot do so, and must refer readers to the grammarian for an explanation of much that is here left unsaid.

ARRANGEMENT OF ENGLISH-GREEK SECTION

Entries are arranged under headwords, which are in bold type, followed by an abbreviation in italic denoting part of speech. Any further indications as to part of speech under the same word are enclosed in brackets. The body of an article consists of the Greek rendering of the headword, with indications of gender, number, etc., in italic. Uses of the headword are illustrated by phrases and idioms in Roman type, with their Greek equivalents. Explanations of or comments on the English and Greek material are given in italic within brackets. Words in Roman type within brackets form part of the material being rendered.

A swung dash is used in the body of an article to represent either the headword or that part of it preceding a vertical line. Derivatives given under headwords are shown by the swung dash with termination in bold type. Some other derivatives are entered as separate headwords, their location being governed by alphabetical order. Where a 'headword' comprises more than one word (e.g. a phrasal verb), a swung dash within the article stands for the whole phrase. Phrasal verbs, when given as separate headwords, are entered immediately after the simple verb, overriding the general alphabetical order. Thus *go* is followed by all its phrasal combinations from *go about* to *go up*, which are then followed by *goad, goal,* etc.

When a word functions as more than one part of speech, or a verb as both transitive and intransitive, these are either taken together as a single entry or given separately, as convenient. Designation of a verb as transitive or intransitive may be omitted if its function is clear in the context of the entry. A few longer entries are divided for convenience into numbered sections.

The problem of handling variant forms in Greek has been alluded to above. Many words occur with variant spellings, as ἔχομε/ἔχουμε, σβήνω/σβύνω, χαλώ/χαλνώ, νταντέλα/δαντέλα, πανί/παννί. There are also variations in the use of accents and iota subscript. The K/D distinction gives rise to other formal differences, among the commonest of which are the following types:

K-form	D-form
το έργον	έργο
το ποτήριον	ποτήρι
το εμπόριον	εμπόριο
ο φύλαξ	φύλακας
ο συγγραφεύς	συγγραφέας
ο ρήτωρ	ρήτορας
ο σωτήρ	σωτήρας
ο χειμών	χειμώνας
ο γείτων	γείτονας

η εικών	εικόνα
η πλάξ	πλάκα
η ομάς	ομάδα
η θέσις	θέση, θέσι
η πατρίς	πατρίδα
η ταυτότης	ταυτότητα
δηλώ	δηλώνω

Examples of such variations will be found in this dictionary. In some entries both K and D forms of a word are given; or a phrase may be varied in different places. Thus *rally round* is given as συσπειρούμαι περί (K) in the Greek-English section and συσπειρώνομαι γύρω από (D) in the English-Greek.

Note on Adverbs. Adverbs are derivable from adjectives in -ος by means of the suffix -α (D) or -ως (K). In general either form may be used colloquially, as βέβαια or βεβαίως *(certainly)*. Some words give only one adverbial form: όμορφα *(nicely)*, κυρίως *(chiefly)*. In certain other cases the two forms have different meanings: ευχάριστα *(pleasantly)*, ευχαρίστως *(with pleasure)*. The general term for well is καλά, but καλώς figures in colloquial idioms like καλώς ήρθες *(welcome!)*. Adjectives of the type πλήρης give adverbs in -ως: πλήρως *(fully)*. The type βαρύς gives -ιά (D) or -έως (K): βαριά or βαρέως *(heavily)*, μακριά *(far)*, ταχέως *(quickly)*. The type ευγνώμων gives -όνως (rare): ευγνωμόνως *(gratefully)*. It is often preferable to avoid an adverbial form in favour of a phrase, e.g. με θρασύτητα rather than θρασέως *(cheekily)*. But no precise rule can be given.

Within an article comment in italic offers the reader various hints about meaning and usage. It might be, for example, a synonym, or a typical subject or object of the verb in question, or a person or thing to which an adjective could be applied. Some meanings are to be inferred from a related entry. Thus under *lighted* (a.) the distinction between αναμμένος and φωτισμένος is inferable from the entry under *light* (v.). Likewise a noun εσφαλμένη ιδέα *(misconception)* can be inferred from the entry under *misconceive* (v.).

Where Greek affords no direct translation of the English, a definition or explanation is attempted, as in *all-rounder, clannish, coroner, cubist, countrified, drop-out, enclave, exoticism, purdah, urbanization, vegetate, be in at the kill, get outside help, have a one-track mind, gild the lily, keep up with the Joneses.* The device of parataxis is useful for items such as *he funked going, it grows on you, live it down, outgrow one's clothes, you did well to come, what grounds have they?* In some cases an idiomatic near-equivalent, or similar proverb, is preferred, as in *be a bad sailor, can't hold a candle to, spare the rod and spoil the child, that's an understatement.*

Certain derivative adjectives and adverbs are best rendered periphrastically, e.g. *allusive, draughty, impressionable, inky, possessive, ideally, proverbially, reluctantly, suspiciously, widely.* Similar treatment of abstract nouns is exemplified in *confrontation, draughtsmanship, inability, inclusion, neighbourliness, retirement.* Some adjectives call for different renderings in respect of (a) things, acts, states, ideas, and (b) persons, e.g. *chivalrous, comatose, English, patriotic.* On the whole, Greek prose uses abstraction and metaphor less boldly than English does; and a figurative turn of phrase often goes better into Greek if recast by means of 'like' or 'as if' or some other softening effect; see *baffle description, be dwarfed by, presidential timber, shoot past, a stone's throw, zigzag down, be at the end of one's tether.*

It is hoped that the selection of examples given will provide readers with many useful grammatical hints, and may help to communicate something of the flavour of Greek idiom.

PRONUNCIATION

The pronunciation of Greek bears a consistent relation to its spelling, so that it need not be separately shown in the body of the dictionary. The phonetic values of the Greek letters are summarized here.

Vowels

	phonetic symbol	nearest English equivalent
η ι υ ει οι υι	i	beat
ε αι	ε	bet
α	a	butt
ο ω	ɔ	bought
ου	u	boot

αυ	af	*cuff*
αυ (before voiced sounds)	av	*love*
ευ	εf	*chef*
ευ (before voiced sounds)	εv	*ever*
ηυ	if	*leaf*
ηυ (before voiced sounds)	iv	*leave*

Note I. The vowel-digraphs included above are pronounced as shown unless the first letter bears an acute accent or the second a diaeresis. In these two cases each letter has its separate value, e.g.:

πολλοί	(pɔli)	ϱολόι	(rɔlɔi)
ἁρπυια	(arpia)	μυϊκός	(miikɔs)
αιτία	(εtia)	αϊτός	(aitɔs)

Note 2. Unstressed i before another vowel is sometimes reduced to a semivowel glide. This is noted in the very rare cases where a distinction of meaning is concerned, e.g.: σκιάζω (-i-) shade, σκιάζω (-j-) frighten.

Consonants

	phonetic symbol	nearest English equivalent
6	v	*over*
γ	γ	(Spanish fue*g*o)
γ (before i and ε)	j	*y*ield
δ	ð	fa*th*er
ζ	z	la*z*y
θ	θ	au*th*or
κ	k	*sk*in
λ	l	*l*eave
μ	m	*m*ay
ν	n	*n*ot
ξ	ks	bo*x*
π	p	*sp*in
ϱ	r	*th*ree
σ, ς	s	*s*ee
σ (before voiced consonants)	z	la*z*y
τ	t	*st*ick
φ	f	*f*at
χ	x	(Scots lo*ch*)
χ (before i and ε)	ç	*h*ew
ψ	ps	ta*ps*

The following consonant-digraphs have special values:

	phonetic symbol	nearest English equivalent
γγ	ŋg	fi*n*ger
γξ	ŋks	sphi*n*x
γχ	ŋx, nç	
γκ (initial)	g	*g*et
γκ (medial)	g, ŋg, ŋk	
μπ (initial)	b	*b*et
μπ (medial)	b, mb, mp	
ντ (initial)	d	*d*o
ντ (medial)	d, nd, nt	
τζ	dz	a*dz*e

Accentuation

Stress-accent coincides with the written accent of a word, and falls upon one of the last three syllables. If the final syllable contains η, ω, ει, or ου, the accent does not fall earlier than the penultimate syllable. (These rules are subject to some modification in demotic usage). The difference between acute, grave, and circumflex accent does not affect pronunciation. In printed Greek the grave accent of final syllables is sometimes replaced by an acute. Fluctuation of usage in this matter is reflected in the dictionary.

———

Among books consulted, I have made the most frequent use of these three: *Lexikon tis Ellinikis Glossis (Proia)*, *Antilexikon* (Th. Vopstandzoglou), and *Lexique Grec moderne Français* (H. Pernot).

I wish to express my thanks to Dr. P. Mackridge, who painstakingly read through the English-Greek manuscript and gave me the benefit of his perceptive criticism and advice on many points. Thanks are also due to all my Greek informants for their kind help; and to the British Academy and the University of London for grants in aid of research. Above all I am indebted to my wife Eleni, without whose collaboration the work could not have been accomplished.

J.T.P.

GREEK-ENGLISH

α- (*privative prefix*) un-, in-, -less.
αβαθής *a.* shallow.
άβαλτος *a.* not placed in position; un-worn, new.
αβανιά *s.f.* false accusation.
αβάντα *s.f.* (*fam.*) profit, advantage; backing.
αβάρετος *a.* indefatigable.
αβαρία *s.f.* jettison; damage (*to ship or cargo*); (*fig.*) κάνω ~ moderate one's pretensions.
αβασάνιστος *a.* not well thought out.
αβάσιμος *a.* groundless.
αβάστακτος *a.* unbearable; uncontrollable.
άβατος *a.* untrodden; inaccessible.
άβαφος *a.* unpainted; not dyed, polished *or* tempered (*of metal*).
άβγαλτος *a.* not having come out *or* off; inexperienced.
αβγατίζω *v.t. & i.* increase.
αβγό *s.n.* egg.
αβγοθήκη *s.f.* egg-cup.
αβγολέμονο *s.n.* soup *or* sauce with eggs and lemon.
αβγοτάραχο *s.n.* roe preserved in wax, botargo.
αβγότσουφλο *s.n.* egg-shell.
αβγουλιέρα *s.f.* egg-cup.
αβγουλωτός *a.* egg-shaped.
αβέβαιος *a.* uncertain.
αβέρτα *adv.* openly.
αβίαστος *a.* unforced, easy.
αβίωτος *a.* impossibly difficult.
αβλαβής *a.* harmless; unharmed; σώος και ~ safe and sound.
αβλεψία *s.f.* oversight.
άβολος *a.* inconvenient, awkward.
άβουλος *a.* irresolute; weak-willed.
άβραστος *a.* uncooked; under-cooked.
αβρός *a.* gracious, delicate, courteous.
άβροχ|ος *a.* dry (*not wetted*); ~οις ποσί without any trouble.
άβυσσος *s.f.* abyss.
αγαθ|ός *a.* good, kind; naïve; τα ~ά possessions.
αγάλια *adv.* slowly, gently.
αγαλλίαση *s.f.* exultation.
άγαλμα *s.n.* statue.
άγαμος *a.* unmarried.
άγαν *adv.* too much. μηδέν ~ nothing in excess.
αγανακτ|ώ *v.i.* be exasperated. ~ηση *s.f.* exasperation.

αγανός *a.* loosely woven.
αγάνωτος *a.* unplated (*metal*).
αγάπη *s.f.* love.
αγαπημένος *a.* dear, loved; favourite.
αγαπητικός *s.m.* lover, sweetheart.
αγαπητός *a.* dear, loved.
αγαπίζω *v.t. & i.* reconcile; become reconciled.
αγαπ|ώ *v.t. & i.* love; όπως ~άτε as you like.
άγαρμπος *a.* unshapely, ill-proportioned, inelegant; uncouth.
αγάς *s.m.* aga; (*fig.*) despot, sybarite.
αγγαρ|εία *s.f.* drudgery; chore; (*mil.*) fatigue (-party). ~εύω *v.t.* impose a task on.
αγγείο *s.n.* vessel; (chamber-) pot; blood-vessel; αισχρόν ~ν rascal.
αγγειοπλαστική *s.f.* pottery.
αγγελία *s.f.* tidings, announcement; small advertisement. ~φόρος *s.m.* messenger.
αγγελικός *a.* angel's; angelic.
αγγέλλω *v.t.* announce.
άγγελμα *s.n.* announcement.
άγγελος *s.m.* angel.
αγγελτήριο *s.n.* note announcing wedding, *etc.*
άγγιγμα *s.n.* touch.
αγγίζω *v.t.* see **εγγίζω.**
άγγιχτος *a.* untouched.
αγγλικός *a.* English.
Άγγλ|ος *s.m.* Englishman. ~ίδα *s.f.* English-woman.
αγγούρι *s.n.* cucumber.
άγδαρτος *a.* not skinned *or* grazed.
αγελάδα *s.f.* cow.
αγελαίος *a.* gregarious (*animal*); (*fig.*) vulgar, commonplace.
αγέλαστος *a.* unsmiling, sullen; not taken in.
αγέλη *s.f.* herd, flock. ~δόν *adv.* in a herd.
αγένεια *s.f.* rudeness.
αγένειος *a.* beardless.
αγενής *a.* ill-mannered.
αγέννητος *a.* unborn; not having given birth.
αγέρας *s.m. see* **αέρας.**
αγέραστος *a.* ageless.
αγέρωχος *a.* haughty.
άγημα *s.n.* detachment; (*naut.*) landing-party.
άγια *adv.* (*fam.*) well; ~ έκανες κι ήρθες you did very vell to come; καλά και ~ that is all very well (*but...*).

αγιάζι *s.n.* frost, cold.

αγιάζω *v.t. & i.* sanctify; asperse; become a saint.

αγίασμα *s.n.* holy water.

αγιασμός *s.m.* (blessing with) holy water.

αγιάτρευτος *n.* incurable; not cured.

αγίνωτος *a.* unripe, not done; impossible.

αγιογδύτης *s.m.* robber of churches; (*fig.*) usurer.

αγιογραφία *s.f.* sacred painting, icon.

αγιόκλημα *s.n.* honeysuckle.

αγιορείτικος *a.* of Mt. Athos.

άγιος *a. & s.m.* holy, sacred; saint.

αγκαζάρω *v.t.* engage (*artist*); book, reserve; invite pressingly.

αγκαζέ 1. *a.* occupied, engaged, booked. 2. *adv.* arm-in-arm. [F. *engagé*].

αγκάθι *s.n.* thorn.

αγκαλά *conj.* although.

αγκάλη *s.f.* 1. embrace; arms, bosom. 2. bay, inlet.

αγκαλιά *s.f.* embrace; arms, bosom, lap; armful; (*adv.*) in one's arms, on one's lap. **~ζω** *v.t.* embrace.

αγκιδ|α *s.f.* splinter, prickle. **~ωτός** *a.* barbed.

αγκινάρα *s.f.* globe artichoke.

αγκίστρι *s.n.* fish-hook.

αγκλίτσα *s.f.* shepherd's crook.

αγκομαχώ *v.i.* pant; be in the throes of death.

αγκύλες *s.f.pl.* square brackets.

αγκύλ|ι *s.n.* prickle, splinter. **~ώνω** *v.t.* prick, sting.

αγκύλωσις *s.f.* (*med.*) anchylosis.

αγκυλωτός *a.* bend.

άγκυρ|α *s.f.* anchor. **~οβολώ** *v.i.* anchor.

αγκώνας *s.m.* elbow; bend.

αγκωνάρι *s.n.* corner; crust.

αγκωνή *s.f.* corner-stone.

αγλαΐζω *v.t.* adorn.

αγλέουρ|ας *s.m.* τρώω τον **~α** gorge oneself.

άγλυκος *a.* unsweetened.

αγναντεύω *v.t.* see from afar.

άγνοια *s.f.* ignorance.

αγνός *a.* pure, chaste.

αγνοώ *v.t.* ignore; not to know.

αγνώμ|ων *a.* ungrateful. **~οσύνη** *s.f.* ingratitude.

αγνώριστος *a.* unrecognizable.

άγνωστος *a.* unknown.

άγονος *a.* sterile, barren; (*fig.*) fruitless; **~** γραμμή unprofitable line (*shipping*).

αγορά *s.f.* market; buying, purchase.

αγοράζω *v.t.* buy.

αγοραίος *a.* market; for hire; (*fig.*) low-class.

αγορανομία *s.f.* market inspectorate.

αγοραπωλησία *s.f.* sale (*transaction*).

αγοραστής *s.m.* buyer.

αγορ|εύω *v.i.* speak in public. **~ευση** *s.f.* speech, address.

αγόρι *s.n.* boy.

αγουρίδα *s.f.* unripe grape.

αγουρόλαδο *s.n.* oil from unripe olives; virgin oil.

άγουρος *a.* unripe.

άγρα *s.f.* hunt, pursuit.

αγράμματ|ος *a.* illiterate. **~οσύνη** *s.f.* illiteracy.

άγραφος *a.* unwritten; **~** χάρτης (*fig.*), worthless document.

αγριάδα *s.f.* wildness, fierceness; couchgrass.

αγριεύω *v.t. & i.* make *or* become wild *or* fierce; scare; become scared.

αγρίμι *s.n.* wild animal, game; (*fig.*) shy, unsociable person.

αγριοκοιτάζω *v.t.* scowl at.

άγριος *a.* wild, fierce.

αγριόχορτο *s.n.* weed.

αγροικία *s.f.* country property; boorishness.

αγροίκος *a.* coarse, boorish.

αγροικώ *v.t. & i.* hear; understand.

αγρόκτημα *s.n.* country property.

αγρονομία *s.f.* rural economy; overseeing of land.

αγρ|ός *s.m.* field. **~ότης** *s.m.* countryman, farmer. **~οτικός** *a.* agrarian, rural.

αγροφυλακή *s.f.* rural police.

αγρυπν|ώ *v.i.* stay awake, keep watch. **~ία** *s.f.* sleeplessness; keeping awake; vigil.

αγυάλιστος *a.* unpolished.

αγυιόπαις *s.m.* street-urchin.

αγύμναστος *a.* untrained.

αγύρευτος *a.* unsought, not in demand.

αγύριστ|ος *a.* not traversed; from which there is no return; not given back; **~ο** κεφάλι stubborn person.

αγύρτης *s.m.* charlatan, quack.

αγχίνους *a.* intelligent.

αγχιστεία *s.f.* relationship by marriage.

αγχόνη *s.f.* gibbet, gallows.

άγχος *s.n.* pathological anxiety, stress.

αγχώδης *a.* stressful.

άγω *v.t. & i.* lead, bring; be in (*year of one's age, etc.*); **~** και φέρω lead by the nose; **~μεν** let us go.

αγωγή *s.f.* upbringing; training; (*med.*) regimen; (*law*) suit, action.

αγώγι *s.n.* load, fare; cost of transport.

αγωγιάτης *s.m.* muleteer.

αγωγός *s.m.* conduit; conductor (*of heat, etc.*); duct, (*electric*) lead.

αγών|ας *s.m.* struggle, exertion, contest, **~ες** (*pl.*) athletic games.

αγωνία *s.f.* agony, anguish; suspense.
αγων|ίζομαι *v.i.* struggle, contend. **~ιστής** *s.m.* contender, fighter.
αγωνιώ *v.i.* be in (death) agony; *(fig.)* struggle, strive.
αδαής *a.* unfamiliar *(not expert).* *(with gen. or σέ)*
αδαμάντινος *a.* diamond; firm, unflinching.
αδάμας *s.m.* diamond.
αδάμαστος *a.* untamed; untameable; indomitable.
αδαμιαί|ος *a.* Adam's; *εν αδαμιαία περιβολή* in the nude.
αδάπανος *a.* costing nothing.
άδεια *s.f.* permission; permit, licence; leave (of absence).
αδειάζω *v.t. & i.* empty; have time to spare; discharge (*gun*).
αδειανός *a.* empty, vacant; unoccupied.
άδειος *a.* empty, vacant.
αδέκαρος *a.* penniless.
αδέκαστος *a.* incorruptible; impartial.
αδελφή *s.f.* sister.
αδέλφια *s.n.pl.* brothers, brother(s) and sister(s).
αδελφικός *a.* fraternal, brotherly.
αδελφός *s.m.* brother.
αδένας *s.m.* gland.
άδενδρος *a.* treeless.
αδέξιος *a.* clumsy, maladroit.
αδερφ- *see* **αδελφ-**.
αδέσποτος *a.* without an owner *or* master.
άδετος *a.* unbound; unmounted, loose.
άδηλος *a.* uncertain, unknown.
αδημονώ *v.i.* be anxiously impatient.
άδης *s.m.* Hades; hell; underworld.
αδηφάγος *a.* gluttonous.
αδιάβαστος *a.* unreadable; unread; unprepared *(in lessons).*
αδιάβατος *a.* impassable.
αδιάβροχ|ος *a.* watertight, waterproof. **~ο** *s.n.* raincoat.
αδιαθεσία *s.f.* indisposition.
αδιάθετος *a.* indisposed *(in health);* not disposed of; *(law)* intestate.
αδιαίρετος *a.* undivided; indivisible.
αδιάκοπος *a.* uninterrupted.
αδιακρισία *s.f.* indiscretion.
αδιά|κριτος *a.* 1. imperceptible. 2. prying. 3. indiscreet. **~κρίτως** *adv.* irrespectively; indiscriminately; importunately.
αδιάλειπτος *a.* incessant.
αδιάλλακτος *a.* intransigent.
αδιάλυτος *a.* indissoluble; not broken up.
αδιάντροπος *a.* shameless.
αδιαπέραστος *a.* impenetrable, impermeable.
αδιάπτωτος *a.* sustained, unfailing.
αδιάρρηκτος *a.* unbreakable; unbroken.

αδιάσπαστος *a.* indissoluble, unbroken.
αδιάφθορος *a.* incorruptible.
αδιαφόρετα *adv.* without profit; to no purpose.
αδιάφορ|ος *a.* indifferent, without interest. **~ία** *s.f.* indifference.
αδιαχώριστος *a.* indivisible; undivided.
αδίδακτος *a.* untaught.
αδιέξοδ|ος *a.* having no outlet; blind *(road).* **~ο** *s.n.* impasse, cul-de-sac.
αδικαιολόγητος *a.* unjustified; unjustifiable.
αδίκημα *s.n.* wrongful act.
αδικία *s.f.* injustice, wrong.
άδικ|ος *a.* unjust, wrong; vain, fruitless. **~ο** *s.n.* injustice, wrong. *έχω* ~ *ο* be wrong. **~α** *adv.* wrongly; in vain.
αδικώ *v.t. & i.* wrong; do wrong.
αδιόρατος *a.* hardly perceptible.
αδιόρθωτος *a.* uncorrected; irreparable; incorrigible.
αδίστακτος *a.* unhesitating.
άδολος *a.* without guile; pure, unadulterated.
αδόξαστ|ος *a.* unsung; (*fam.*) *θα σου αλλάξω τον* ~*ο* I'll give you what for!
αδούλευτος *a.* unworked, uncultivated.
αδράν|εια *s.f.* inertia; inactivity. **~ής** *a.* inert; sluggish.
αδράχνω *v.t.* seize, grasp.
αδράχτι *s.n.* spindle.
αδρός *a.* sizeable, generous.
αδυναμία *s.f.* weakness; addiction; leanness.
αδυνατίζω *v.t. & i.* weaken; make thin; lose weight.
αδύνατ|ος *a.* weak; lean; impossible. **~ώ** *v.i.* be unable.
αδυσώπητος *d.* inexorable.
άδυτ|ος *a.* inaccessible. **~ον** *s.n.* *(eccl.)* sanctuary. **~α** *s.n.pl.* depths.
άδω *v.t. & i.* sing.
αεί *adv.* always.
αειθαλής *a.* evergreen.
αεικίνητον *s.n.* perpetual motion.
αείμνηστος *a.* ever-remembered.
αέναος *a.* perpetual.
αεράμυνα *s.f.* anti-aircraft defence.
αέρ|ας *s.m.* air, wind; mien; (*fam.*) key-money; *πήρε ο νους του* ~*α* he has got above himself; ~*α! battle-cry.*
άεργος *a.* not working.
αερίζω *v.t.* fan, ventilate, air.
αέρινος *a.* light as air, ethereal.
αέρι|ο *s.n.* gas; ~*α* (*pl.*) (*med.*) wind.
αεριούχος *a.* aerated; (*for drinks*) fizzy.
αερίόφως *s.n.* gaslight; (*fam.*) gas.
αερισμός *s.m.* ventilation.
αεριστήρας *s.m.* ventilator, fan.
αεριώδης *a.* gaseous.
αεριωθούμενος *a.* jet-propelled.

αεροβατώ *v.i.* (*fig.*) be in the clouds. be dreaming.
αερογέφυρα *s.f.* airlift.
αεροδρόμιον *s.n.* aerodrome.
αεροκοπανίζω *v.i.* chatter; labour in vain.
αερόλιθος *s.m.* meteorite.
αερολιμήν *s.m.* airport.
αερολογώ *v.i.* talk nonsense.
αεροπειρατία *s.f.* hijack (*of plane*).
αεροπλάν|ο *s.n.* aeroplane. **~οφόρον** *s.n.* (*naut.*) aircraft-carrier.
αεροπορία *s.f.* air-force; air-line.
αεροπορικώς *adv.* by air.
αεροπόρος *s.m.* flyer, airman.
αερόστατο *s.n.* balloon.
αεροστεγής *a.* airtight.
αέτειος *a.* aquiline.
αετός *s.m.* eagle; kite (*toy*).
αέτωμα *s.m.* pediment.
αζευγάρωτος *a.* unpaired; unploughed.
αζημίωτ|ος *a.* without suffering loss; *με το* **~ο** for a consideration.
αζήτητος *a.* unclaimed; not in demand.
αζύγι(α)στος *a.* unweighed; (*fig.*) unconsidered.
άζυμος *a.* unleavened.
άζωτον *s.n.* nitrogen.
άη- (*prefix*) saint.
αηδής *a.* disgusting.
αηδ|ία *s.f.* disgust; (*fam.*) anything disgusting. **~ιάζω** *v.t.* & *i.* disgust; feel disgust. **~ιάστικος** *a.* disgusting.
αηδ|ών *s.f.*, **~όνι** *s.n.* nightingale.
αήρ *s.m.* air.
αήττητος *a.* invincible; unbeatem.
άηχος *a.* noiseless; (*gram.*) surd.
αθανασία *s.f.* immortality.
αθάνατος *a.* immortal, everlasting; *ο ~* American aloe.
αθέατος *a.* invisible; unseen, secret.
άθελα *adv.* unwillingly; unintentionally.
αθέλητος *a.* unintentional; unwilling.
αθέμιτος *a.* illicit.
άθεος *a.* an atheist.
αθεόφοβος *a.* impious.
αθεράπευτος *a.* incurable; uncured.
αθέρας *s.m.* awn; cutting edge; (*fig.*) cream (*best part*).
αθετώ *v.t.* break (*word, etc.*) **~ηση** *s.f.* breach.
αθεώρητος *a.* not officially examined.
άθικτος *a.* intact; untouched.
αθλητ|ής *s.m.* athlete. **~ικός** *a.* athletic, athlete's. **~ισμός** *s.m.* athletics.
άθλιος *a.* miserable; vile; squalid.
άθλος *s.m.* contest, exploit, task.
αθόλωτος *a.* 1. limpid, not turbid. 2. not vaulted.
αθορύβητος *a.* imperturbable.

αθόρυβος *a.* noiseless.
άθραυστος *a.* unbroken; unbreakable.
αθροίζω *v.t.* assemble; add up.
άθροιση *s.f.* addition (*adding up*).
άθροισμα *s.n.* sum total; group (*of people*).
αθρόος *a.* in a body; numerous.
άθυμος *a.* dispirited.
άθυρμα *s.n.* plaything.
αθυρόστομος *a.* garrulous; impertinent.
αθώ|ος *a.* innocent, not guilty; artless. **~ότητα** *s.f.* innocence. **~ώνω** *v.t.* acquit. **~ωση** *s.f.* acquittal.
αϊβασιλιάτικος *a.* of St. Basil's day.
αίγα, αίξ *s.f.* goat.
αίγαγρος *s.m.* chamois.
αιγιαλός *s.m.* sea-shore.
αιγίς *s.f.* shield; aegis.
αίγλη *s.f.* glitter; glory, magnificence.
αιγόκλημα *s.n.* honeysuckle.
αιδέσιμος *a.* reverend.
αιδήμων *a.* modest, bashful.
αιδοίον *s.n.* privy parts.
αιδώς *s.f.* shame, decency.
αιθέριος *a.* ethereal; volatile (*oil*).
αιθήρ *s.m.* ether.
αίθουσα *s.f.* large room, hall.
αιθρία *s.f.* cloudless weather.
αίθριος *a.* clear, serene.
αίλουρ|ος *s.m.* wild cat; member of cat family. **~οειδής** *a.* feline.
αίμ|α *s.n.* blood. **~άσσω** *v.i.* bleed.
αιματηρός *a.* bloody.
αιματοκύλισμα *s.n.* carnage.
αιματοχυσία *s.f.* bloodshed.
αιματώδης *a.* sanguine (*complexion*); like blood.
αιματώνω *v.t.* & *i. see* **ματώνω.**
αιμοβόρος *a.* bloodthirsty.
αιμοδοσία *s.f.* blood-donation.
αιμομιξία *s.f.* incest.
αιμόπτυση *s.f.* spitting of blood.
αιμορραγία *s.f.* haemorrhage.
αιμοσφαίριον *s.n.* blood-corpuscle.
αιμόφυρτος *a.* covered in blood.
αίνιγμα *s.n.* enigma, riddle.
αίνος *s.m.* praise.
άιντε *int.* come on!
αίρ|εση *.f.* heresy, sect; option. **~ετικός** *a.* heretical; sectarian.
αιρετός *a.* elected.
αίρω *v.t.* raise; lift up; raise (*siege*), lift (*embargo*).
αισθάνομαι *v.t.* & *i.* feel (*well, fear, the cold, etc*); sense.
αίσθημα *s.n.* sensation, feeling; **~τα** (*pl.*) feelings (*sympathies*).
αισθηματίας *s.m.* person of responsive feelings.

αισθηματικ|ός *a.* sentimental. **~ότητα** *s.f.* sentimentality.

αισθησιακός *a.* sensual.

αίσθησ|η *s.f.* sense (*bodily faculty*); προκάλεσε **~ιν** it made an impression.

αισθητικός *a.* of the senses; aesthetic; an aesthete; a beauty specialist, **~ή** *s.f.* aesthetics.

αισθητός *a.* perceptible.

αισιόδοξ|ος *a.* optimistic. **~ία** *s.f.* optimism.

αίσιος *a.* auspicious, favourable.

αίσχος *s.n.* shame, infamy.

αισχροκέρδεια *s.f.* illicit gain; profiteering.

αισχρολογία *s.f.* obscene talk.

αισχρός *a.* disgraceful; indecent.

αισχύν|η *s.f.* shame, disgrace. **~ομαι** *v.i.* feel ashamed.

αίτημα *s.n.* demand (*thing asked for*); postulate.

αίτηση *s.f.* application, request.

αιτία *s.f.* cause, reason.

αιτιατική *s.f.* (*gram.*) accusative.

αιτιολογία *s.f.* explanation, giving a reason (*for*).

αίτι|ος *a.* responsible, to blame. **~ον** *s.n.* cause, motive. **~ώμαι** *v.t.* hold to blame.

αϊτός *s.m.* eagle; kite (*toy*).

αιτ|ώ *v.t.* request (*something*). **~ούμαι** *v.t.* beg (*something*).

αίφνης *adv.* 1. suddenly. 2. for example.

αιφνιδιασμός *s.m.* sudden attack, surprise.

αιφνίδιος *a.* sudden, unexpected.

αιχμάλωτ|ος *s.m.* prisoner, captive. **~ίζω** *v.t.* take prisoner; (*fig.*) captivate.

αιχμή *s.f.* point (*of weapon, etc.*); peak. **~ρός** *a.* sharp, pointed.

αιών|ας *s.m.* age, century, era. **~ιος** *a.* eternal. **~όβιος** *a.* centuries old.

αιώρ|α *s.f.* swing. **~ούμαι** *v.i.* swing; (*fig.*) waver.

ακαδημ|ία *s.f.* academy. **~αϊκός** *a.* academic; an academician.

ακαθάριστος *a.* not cleaned *or* peeled; (*fin.*) gross.

ακαθαρσί|α *s.f.* dirt, dirtiness; **~ες** (*pl.*) excrement.

ακάθαρτος *a.* dirty, unclean.

ακάθεκτος *a.* unrestrainable, unbridled.

ακάθιστος *a.* who does not sit down; **~** ύμνος (*eccl.*) Lenten hymn of praise to the Virgin.

ακαθόριστος *a.* undefined; indefinable.

άκαιρος *a.* untimely.

ακακία *s.f.* 1. acacia. 2. guilelessness.

άκακος *a.* harmless, guileless.

ακαλαίσθη|τος *a.* lacking in taste. **~σία** *s.f.* lack of good taste.

ακάλεστος *a.* uninvited.

ακαλλιέργητος *a.* uncultivated; uncultured.

ακάλυπτος *a.* uncovered.

ακαμάτης *a.* lazy.

ακάματος *a.* tireless.

άκαμπτος *a.* inflexible.

ακάμωτος *a.* not done; unripe.

άκανθ|α, ~ος *s.f.* thorn; acanthus.

ακανόνιστος *a.* not settled, not cleared up; disproportioned, irregular.

ακαρεί *a. εν* **~** instantly.

ακαριαίος *a.* instantaneous.

άκαρπος *a.* unfruitful; (*fig.*) fruitless.

ακατάβλητος *a.* 1. indomitable. 2. unpaid (*money*).

ακατάδεκτος *a.* disdainful.

ακαταλαβίστικος *a.* (*fam.*) incomprehensible.

ακατάληπτος *a.* incomprehensible.

ακατάλληλος *a.* unsuitable.

ακαταλόγιστος *a.* not fully responsible (*for one's actions*).

ακαταμάχητος *a.* irrefutable; unconquerable; irresistible.

ακατανόητος *a.* incomprehensible.

ακατάπαυστος *a.* incessant, çontinuous.

ακαταπόνητος *a.* tireless.

ακατάρτιστος *a.* not organized; lacking qualifications *or* knowledge.

ακατάστα|τος *a.* untidy; variable. **~σία** *s.f.* disorder; changeability (*of weather*).

ακατάσχετος *a.* that cannot be stanched *or* stemmed; (*law*) not distrained *or* distrainable; unrestrainable.

ακατάφερτος *a.* who cannot be persuaded; not feasible; unaccomplished.

ακατέβατα *adv.* at a fixed price.

ακατέργαστος *a.* not wrought *or* refined, crude, raw; uncut (*gem*).

ακάτιον *s.n.* skiff.

ακατοίκητος *a.* uninhabited; uninhabi-table.

ακατονόμαστος *a.* nameless.

ακατόρθωτος *a.* not feasible; unaccomplished.

άκαυ(σ)τος *a.* incombustible; unburnt.

ακένωτος *a.* inexhaustible; not emptied.

ακέραιος *a.* whole, intact; honest; **~** αριθμός integer.

ακεφιά *s.f.* low spirits.

ακηδής *a.* negligent.

ακήρυκτος *a.* unproclaimed, undeclared.

ακίνδυνος *a.* not dangerous.

ακίνητ|ος *a.* immovable; real (*estate*). **~ον** *s.n.* item of real estate, building, property. **~ώ** *v.t. & i.* immobilize; remain motionless. **~οποιώ** *v.t.* immobilize.

ακκίζομαι *v.i.* assume coquettish airs.

ακκισμός *s.m.* coquettish air.

ακλείδωτος *a.* unlocked.

ακληρονόμητος *a.* without heirs.

άκληρος *a.* childless; poor; hapless.

ακλήρωτος *a.* unalloted; not drawn (*of lots*); (*mil.*) not yet called up.

άκλιτος *a.* indeclinable.

ακλόνητος *a.* unshakeable, firm.

ακμάζω *v.i.* flourish, prosper; be in one's prime.

ακμαίος *a.* in full vigour.

ακμή *s.f.* point, edge; acme, peak; (*med.*) acne.

άκμων *s.m.* anvil.

ακο|ή *s.f.* hearing; εξ ~ής by hearsay.

ακοίμητος *a.* unable to sleep; vigilant.

ακοινώνητος *a.* 1. unsociable. 2. not having received the sacrament.

ακολασία *s.f.* excess, debauchery.

ακόλαστος *a.* 1. licentious. 2. unpunished.

ακολουθ|ία *s.f.* retinue, suite; coherence; church service; κατ'~ίαν in consequence.

ακόλουθος *s.m.* attendant; page; attaché.

ακόλουθος *a.* next, following; true (*to one's word, etc.*).

ακολουθ|ώ *v.t. & i.* follow; ~εί to be continued.

ακολούθως *adv.* next, afterwards; ως ~ as follows.

ακόμα, ακόμη *adv.* still, yet; more; ~ και even.

ακόν|η *s.f.,* ~ι *s.n.* whetstone. ~ίζω *v.t.* sharpen, whet.

ακόντιο *s.n.* javelin.

άκοπ|ος *a.* not cut; easy, not tiring. ~ως *adv.* without trouble.

ακόρεστος *a.* insatiable.

άκοσμος *a.* unseemly.

ακουμπώ *v.t. & i.* touch; stand, lean; place (*money in bank, etc.*); (*v.i.*) rest, lean (*for support*); put down.

ακούμπωτος *a.* unbuttoned.

ακούραστος *a.* untiring, energetic.

ακούρδιστος *a.* not tuned; not wound up.

ακούσιος *a.* unintentional; against one's will.

ακουστικ|ός *a.* acoustic, auditory. ~ή *s.f.* acoustics. ~ό *s.n.* (*telephone*) receiver; hearing-aid; stethoscope.

ακουστ|ός *a.* audible; famous; (*fam.*) έχω ~ά πως I have heard that...

ακού|ω *v.t. & i.* hear, listen (*to*); obey; ~ω έναν πόνο feel a pain; ~ω μυρωδιά smell a smell; (*fam.*) άκου λέει *phrase of assent.* ~ομαι *v.i.* be well known.

άκρα *s.f. see* **άκρη.**

άκρα *s.n.pl.* limbs.

ακράδαντος *a.* steadfast.

ακραιφνής *a.* pure, sincere.

ακρατ|ής *a.* intemperate. ~εια *s.f.* intemperance; incontinence.

ακράτητος *a.* unrestrainable.

άκρατος *a.* undiluted; (*fig.*) pure.

άκρ|η *s.f.* end, edge, tip, summit; cape; corner; βάζω λεφτά στην ~η put money aside; ~ες μέσες disjointedly.

ακριανός *a. see* **ακρινός.**

ακριβά *adv.* dearly, at a high price.

ακριβαίνω *v.t. & i.* raise price of; become dearer.

ακρίβεια *s.f.* 1. exactness, precision. 2. costliness, expense, high price(s).

ακριβής *a.* exact, correct; punctual.

ακριβοθώρητος *a.* rarely to be seen.

ακριβολόγος *a.* precise in speech.

ακριβός *a.* dear, loved; costly; (*with σε*) sparing of.

ακριβώς *adv.* exactly, precisely.

ακρίδα *s.f.* locust.

ακρινός *a.* at the far end, last.

ακριτικός *a.* of the frontier (ballads).

ακριτομυθία *s.f.* indiscreet talk.

άκριτος *a.* lacking in judgement.

ακροάζομαι *v.t. & i.* listen (to); (*med.*) auscultate.

ακρόαση *s.f.* hearing, audition; audience, interview; auscultation.

ακροατήριο *s.n.* audience, auditorium.

ακροατής *s.m.* listener.

ακρο|βασία *s.f.* acrobatics. ~βάτης *s.m.* acrobat.

ακροβολισμός *s.m.* skirmish.

ακρογιάλι *s.n.* sea-shore.

ακρογωνιαίος *a.* ~ λίθος corner-stone.

ακροθαλασσιά *s.f.* sea-shore.

άκρ|ον *s.n.* end, extremity; των ~ων extremist; φθάνω στα ~α go to extremes.

ακροποδητί *adv.* on tip-toe.

ακρόπολις *s.f.* acropolis, citadel.

άκρ|ος *a.* extreme, complete, utmost; last; εις ~ον extremely. ~ως *adv.* extremely.

ακροσφαλής *a.* precarious.

ακροώμαι *v.t. & i.* listen (to); give audience; auscultate.

ακρωτηριάζω *v.t.* mutilate; perform amputation on.

ακρωτήριο *s.n.* cape, promontory.

ακταιωρός *s.m.f.* coastguard (-vessel).

ακτ|ή *s.f.* shore; ~ές (*pl.*) coast.

ακτήμων *a.* without property, landless.

ακτ|ίνα *s.f.* ray; radius; spoke; ~ίνες (*pl.*) X-rays.

ακτινοβολία *s.f.* radiation, radiance.

ακτινογραφία *s.f.* X-ray photograph(y).

ακτινοθεραπεία *s.f.* radiotherapy.

ακτινοσκοπώ *v.t.* X-ray.

ακτοπλοΐα *s.f.* coastal shipping.

ακυβέρνητος *a.* ungovernable, adrift; without a government.

ακύμαντος *a.* without waves, calm.

άκυρ|ος *a.* invalid, void. **~ώ, ~ώνω** *v.t.* nullify, cancel.

ακωλύτως *adv.* without hindrance.

άκων *a.* against one's will.

αλάβαστρο *s.n.* alabaster.

αλάδωτος *a.* without oil; (*fig.*) without bribery; unbaptized.

αλαζ|ών *s.m.* arrogant. **~ονεία** *s.f.* arrogance.

αλάθευτος *a.* unerring, infallible.

αλάθητ|ος *a.* unerring, infallible. **~ον** *s.n.* infallibility.

αλαλάζω *v.i.* cry, shout (*in joy or triumph*).

αλαλιάζω *v.t. & i.* daze, stupefy; be dazed.

άλαλος *a.* speechless, dumb, agape.

αλαμπουρνέζικ|ος *a.* (*fam.*) bizarre, nonsensical; **~α** *adv.* double Dutch.

αλάνθαστος *a.* with no mistakes; (*fam.*) infallible.

αλάνι *s.n.* vacant plot of ground; street-lounger.

αλάργα *adv.* far off.

άλας *s.n.* salt.

αλάτ|ι *s.n.* salt. **~ιέρα** *s.f.* salt-cellar.

αλατίζω *v.t.* salt.

αλαφιάζ|ω, ~ομαι *v.i.* be startled.

αλαφροΐσκιωτος *a.* who sees the fairies.

αλαφρός *a. see* **ελαφρός.**

αλγεινός *a.* painful, grievous.

άλγος *s.n.* pain, grief.

αλεγράρω *v.t.* liven up.

αλέθω *v.t.* grind (*corn, etc.*).

αλείβω *v.t.* rub, coat, smear, spread.

άλειμμα *s.n.* smearing, greasing; animal fat.

αλείφω *v.t. see* **αλείβω.**

άλειωτος *a. see* **άλιωτος.**

αλέκτωρ *s.m.* cock.

αλεξικέραυνον *s.n.* lightning-conductor.

αλεξίπτωτ|ον *s.n.* parachute. **~ιστής** *s.m.* parachutist.

αλεπού *s.f.* fox.

άλεσμα *s.n.* grinding; grist.

αλέστ|ος *a.* quick, nimble. **~α** *adv.* quickly; *είμαι* **~α** be ready.

αλέτρι *s.n.* plough.

άλ|ευρον, ~εύρι *s.n.* flour.

αλήθ|εια *s.f.* truth; (*adv.*) really, indeed, by the way. **~εύω** *v.i.* come true; be right. **~ής** *a.* true. **~ινός** *a.* true, genuine. **~ώς, ~ινά** *adv.* truly.

αληθοφανής *a.* plausible.

αλησμόνητος *a.* unforgettable; unforgotten.

αλήτης *s.m.* vagabond.

αλί *int.* alas!

αλιάνιστος *a.* not cut up small.

αλιεία *s.f.* fishing, fishery.

αλι|εύς *s.m.* fisherman. **~ευτικός** *a.* of fishing. **~εύω** *v.t. & i.* fish (for).

άλικος *a.* scarlet.

αλίμονο *int.* alas! good gracious! whatever next? **~ σου** woe betide you!

αλίπαστος *a.* salted.

αλισβερίσι *s.n.* commercial dealings.

αλισίβα *s.f.* lye.

αλιτήριος *a.* rascally.

αλιφασκιά *s.f.* sage.

άλιωτος *a.* unmelted, undissolved; not decomposed; that does not wear out.

αλκ|ή *s.f.* bodily strength. **~ιμος** *a.* strong.

αλκο(ο)λικός *a.* alcoholic (*person*); (*fam.*) excessively addicted.

αλκοτέστ *s.n.* breathalyser test.

αλκυόνα *s.f.* kingfisher.

αλλά *conj.* but.

αλλαγή *s.f.* change.

αλλά|ζω *v.t. & i.* change; **~ζει** that alters things. **~ξιά** *s.f.* exchange; set of underwear, suit of clothes.

αλλαντικά *s.n.pl.* kinds of sausage.

αλλάσσω *v.t. & i. see* **αλλάζω.**

αλλαχού *adv.* elsewhere.

αλλεπάλληλος *a.* one on top of another, repeated.

αλλεργία *s.f.* (*med.*) allergy.

αλληγορία *s.f.* allegory.

αλληλεγγ|ύη *s.f.* solidarity, mutual aid. **~υος** *a.* joint.

αλληλένδετος *a.* linked with each other, interdependent.

αλληλογραφ|ία *s.f.* correspondence (*written*). **~ώ** *v.i.* correspond.

αλληλοδιάδοχος *a.* one after another, successive.

αλληλοπαθής *a.* (*gram.*) reciprocal.

αλληλοσυγκρούομαι *v.i.* conflict.

αλληλοτρώγομαι *v.i.* squabble with each other.

αλλήλους *pron. acc. pl.* one another. **~ων** (*gen. pl.*) one another's.

αλληλουχία *s.f.* sequence.

αλλιώς *adv. see* **αλλοιώς.**

αλλοδαπ|ή *s.f.* foreign parts. **~ός** *a.* alien, a foreign subject.

άλλοθεν *adv.* from elsewhere.

άλλοθι *adv.* elsewhere; *n.* alibi.

αλλόθρησκος *a.* of another creed.

αλλοίθωρ|ος *a.* cross-eyed. **~ίζω** *v.i.* squint.

αλλοίμονο *int. see* **αλίμονο.**

αλλοί|ος *a.* different, not the same. **~ώς** *adv.* differently, otherwise.

αλλοι|ώνω *v.t.* alter; falsify. **~ώνομαι** *v.t.* deteriorate, go off. **~ωση** *s.f.* deterioration, falsification.

αλλιώτικος *a.* different; odd (*of character*).

αλλόκοτος *a.* strange, queer, grotesque.

αλλοπρόσαλλος *a.* changeable, fickle.

άλλ|ος *pron. & a.* other, else, rest; next; different; more; **~οι... ~οι** some... others; **~η**

φορά another time; *την ~η φορά* next time; *τίς ~ες* the other day; *χωρίς ~ο* without fail; *εξ ~ου* besides; *~α αντ' ~ων* random talk; *κάθε ~ο* anything but! *~ο τίποτα* nothing but! *κι ~ο* some more; *~α δύο* two more.

άλλοτε *adv.* formerly; another time; *~ μεν... ~ δε* now... now.

αλλοτινός *a.* of former times.

αλλότριος *a.* other people's.

αλλού *adv.* elsewhere; (*fam.*) *~ αυτά* don't tell me that!

αλλό|φρων *a.* frantic, beside oneself. **~φροσύνη** *s.f.* frenzy; violent agitation.

αλλόφωνος *a.* speaking another language.

άλλως *adv.* otherwise; or else; *~ πως* in some other way.

άλλωστε *adv.* besides.

άλμα *s.n.* jump, leap. **~τώδης** *a.* proceeding by leaps and bounds.

άλμη *s.f.* brine.

αλμπάνης *s.m.* farrier.

άλμπουρο *s.n.* mast.

αλμύρα *s.f.* saltness.

αλμυρός *a.* salt; (*fig.*) expensive.

αλόγα *s.f.* mare; (*fam.*) ungainly woman.

αλογάριαστος *a.* beyond calculation; with one's accounts unsettled; spending improvidently.

αλογίσιος *a.* of horses, horsy.

αλόγιστος *a.* lacking judgement; thoughtless; beyond calculation.

άλογ|ο *s.n.* horse; (*chess*) knight. **~όμυγα** *s.f.* horsefly.

άλογος *a.* devoid of reason.

αλόη *s.f.* aloe.

αλοιφή *s.f.* ointment.

αλουμίνιο *s.f.* aluminium.

άλουστος *a.* not having washed.

αλπακάς *s.m.* I. alpaca. 2. metal resembling silver.

άλσος *s.n.* grove, small wooded park.

αλτ *int.* halt!

αλτήρες *s.m.pl.* dumb-bells.

άλτης *s.m.* jumper (*athletic*).

αλτρουισμός *s.m.* altruism.

αλύγιστος *a.* unbending, inflexible.

αλυκή *s.f.* salt-pan.

αλύπητ|ος *a.* pitiless; without sorrow. **~α** *adv.* without pity; unsparingly.

αλυσίδα *s.f.* chain.

άλυσις *s.f.* chain, sequence, succession.

άλυτος *a.* not untied; unsolved; unsolvable; (*fig.*) indissoluble.

αλύτρωτος *a.* unredeemed.

αλυχτώ *v.i.* bark.

άλφα *s.n.* the letter **Α**; (*fig.*) beginning, start.

αλφαβήτα *s.f.* ABC.

αλφάβητ|ο *s.n.* alphabet. **~άριο** *s.n.* primer.

~ικός *a.* alphabetic(al).

αλφάδι *s.n.* level, square, plumb-line.

αλχημεία *s.f.* alchemy.

αλώβητος *a.* intact.

Αλωνάρης *s.m.* July.

αλών|ι *s.n.* threshing-floor. **~ίζω** *v.t. & i.* thresh; (*fig.*) scatter; thrash; (*v.i.*) wander round.

αλώπηξ *s.f.* fox.

άλωση *s.f.* capture, fall.

άμα *adv.* as soon as, when; if; *άμ' έπος άμ' έργον* no sooner said than done.

αμάδα *s.f.* flat stone used in bowls.

αμαζόνα *s.f.* amazon.

αμάθεια *s.f.* ignorance.

αμάθευτος *a.* that has not become known; not used (*to*).

αμαθής *a.* ignorant.

αμάθητος *a.* not learnt (*of lessons*); *see also* **αμάθευτος.**

άμαθος *a.* not used (*to*).

αμάκα *s.f. & adv.* (*fam.*) buckshee.

αμάλ|αγος, ~ακτος *a.* not .pliable; (*fig.*) unrelenting, hard.

αμάν *int.* for mercy's sake!

αμανάτι *s.n.* pawn, security; parcel delivered by private individual.

αμανές *s.m.* love song pleading for compassion. *παίρνω ψηλά τον ~έ* become cocky.

άμ|αξα *s.f.* coach, carriage; horse-cab; (*fam.*) *τά εξ ~άξης* unrestrained abuse. **~αξάς** *s.m.* coachman, driver.

αμάξι *s.n.* horse-cab; (*motor*) car.

αμαξιτός *a.* carriageable.

αμαξοστοιχία *s.f.* train.

αμαξωτός *a.* carriageable.

αμάραντος 1. *a.* unfading. 2. *s.m.* amaranth.

αμαρτάνω *v.i.* sin, err.

αμάρτημα *s.n.* a sin.

αμαρτί|α *s.f.* sin; *έχουμε ~ες ακόμη* we have more troubles in store; *λέω την ~α μου* I make no secret of it; *είναι ~α* (*από τό Θεό*) it is a pity.

αμαρτωλός *a.* sinful; a sinner.

αμασκάλη *s.f.* armpit.

αμαυρ|ός *a.* dim, dark. **~ώνω** *v.t.* darken, obscure; (*fig.*) sully.

αμάχη *s.f.* enmity, hatred.

αμαχητί *adv.* without a fight.

άμαχος *a.* non-combatant.

άμβι|ξ, ~υξ *s.m.* still, alembic.

αμβλύς *a.* blunt, dull; (*geom.*) obtuse. **~ύνω** *v.t.* blunt, take the edge off.

άμβλωση *s.f.* abortion.

αμβροσία *s.f.* ambrosia.

άμβων *s.m.* pulpit, rostrum.

αμ' δέ *int.* not on your life!

άμε *int.* go!

αμέ, αμή *int.* well! of course!

αμείβω *v.t.* reward, recompense.

αμείλικτος *a.* implacable.

αμείωτος *a.* unimpaired.

αμέλγω *v.t.* milk.

αμέλεια *s.f.* negligence, carelessness; laziness (*at lessons*).

αμελέτητ|ος *a.* I. not having prepared (*lesson*); not prepared *or* studied. 2. unmentionable; (*fam., cuisine*) ~α (*s.n.pl.*) mountain oyster.

αμελής *a.* negligent; lazy (*pupil*).

αμελώ *v.t.* neglect.

άμεμπτος *a.* irreproachable.

αμερικανικός *a.* American.

Αμερικάνος *s.m.* Greek from America.

Αμερικανός *s.m.* American.

αμέριμνος *a.* carefree; heedless.

αμέριστος *a.* indivisible; undivided.

αμερό|ληπτος *a.* impartial. ~ληψία *s.f.* impartiality.

άμεσος *a.* direct, immediate.

άμεστος *a.* unripe.

αμέσως *adv.* at once.

αμετάβατος *a.* (*gram.*) intransitive.

αμετάβλητος *a.* unchanged; unalterable.

αμετάκλητος *a.* irrevocable.

αμετα|μέλητος, ~νόητος *a.* unrepentant.

αμετάπειστος *a.* unpersuadable, obstinate.

αμετάπτωτος *a.* undiminished.

αμεταχείριστος *a.* not in use; unused, unworn.

αμέτοχος *a.* not taking part (*in*) (*with gen. or εις*).

αμέτρητος *a.* countless.

άμετρος *a.* measureless.

αμήν *int.* amen!

αμηχαν|ία *s.f.* perplexity, embarrassment. ~ώ *v.i.* be at a loss *or* embarrassed.

αμίαντος 1. *a.* undefiled. 2. *s.m.* asbestos.

αμίλητος *a.* silent; stand-offish; whose favour has not been solicited.

άμιλλ|α *s.f.* rivalry, emulation. ~ώμαι *v.t. & i.* vie with, rival; contend, compete.

αμίμητος *a.* inimitable.

άμισθ|ος *a.* unpaid. ~ί *adv.* without payment.

αμμέ, αμμή *int.* see αμέ.

αμμοκονία *s.f.* mortar, plaster.

άμμ|ος *s.m.f.* sand. ~ώδης *a.* sandy.

αμμουδιά *s.f.* sandy beach.

αμμωνία *s.f.* ammonia.

αμνάδα *s.f.* ewe-lamb.

αμνημόνευτος *a.* unmentioned; immemorial; not commemorated.

αμνήμ|ων *a.* forgetful. ~ονώ *v.i. & t.* be forgetful; forget. ~οσύνη *s.f.* forgetfulness.

αμνησία *s.f.* loss of memory.

αμνησίκακος *a.* not resentful.

αμνηστία *s.f.* amnesty.

αμνός *s.m.* lamb.

αμοιβάδες *s.f.pl.* amoebic dysentery.

αμοιβαίος *a.* mutual, reciprocal.

αμοιβή *s.f.* reward, recompense, fee.

άμοιρος *a.* hapless; not provided with (*with gen.*).

αμολ(λ)|ά(ρ)ω *v.t.* slacken, let loose, release; fly (*kite*). ~ιέμαι *v.i.* hasten, rush.

αμόνι *s.n.* anvil.

αμορτισέρ *s.n.* shock-absorber.

άμορφος *a.* shapeless.

αμόρφωτος *a.* uneducated.

άμουσος *a.* unmusical.

αμπαζούρ *s.n.* lampshade. [F. *abatjour*].

αμπάλωτος *a.* in need of repair.

αμπάρα *s.f.* bolt, bar.

αμπάρι *s.n.* storehouse; hold (*of ship*).

αμπάς *s.m.* coarse woollen cloth, drugget.

άμπ|ελος *s.f.*, ~έλι *s.n.* vine; vineyard.

αμπελουργία *s.f.* viticulture.

αμπελοφάσουλο *s.n.* green bean.

αμπελώνας *s.m.* vineyard.

αμπέχονον *s.n.* soldier's tunic.

άμποτε *adv.* would that!

αμπραγιάζ *s.n.* clutch (*of motor*). [F. *embrayage*].

αμ' πώς *int.* of course!

άμπωτις *s.f.* ebb-tide.

άμυαλος *a.* brainless.

αμυγδαλές *s.f.pl.* tonsils.

αμυγδαλιά *s.f.* almond-tree.

αμύγδαλο *s.n.* almond.

αμυγδαλωτός *a.* almond-shaped; made with almonds.

αμυδρός *a.* dim, faint.

αμύητος *a.* uninitiated.

αμύθητος *a.* untold, fabulous.

άμυλο *s.n.* starch.

άμυν|α *s.f.* defence, resistance. ~ομαι *v.t. & i.* defend (*with gen.*); resist, defend oneself. ~τικός *a.* defensive.

αμυχή *s.f.* scratch.

αμφι- *denotes* round, about, both.

άμφια *s.n.pl.* (*eccl.*) vestments.

αμφιβάλλω *v.i.* doubt.

αμφίβιος *a.* amphibious.

αμφίβολ|ος *a.* doubtful, uncertain. ~ία *s.f.* doubt.

αμφιδέξιος *a.* ambidextrous; dextrous.

αμφίεση *s.f.* dress, attire.

αμφιθέατρ|ο *s.n.* amphitheatre. ~ικώς *adv.* in the shape of an amphitheatre.

αμφιρρ|έπω *v.i.* waver. ~οπος *a.* undecided.

αμφισβητήσιμος *a.* questionable, doubtful.

αμφισβητ|ώ *v.t.* dispute, call in question.

~ούμαι *v.i.* be in dispute. **~ηση** *s.f.* dispute, doubt, question.

αμφίστομος *a.* two-edged.

αμφιταλαντεύομαι *v.i.* waver, be uncertain.

αμφιτρύων *s.m.* host, giver of party.

αμφορεύς *s.m.* amphora, jar.

αμφότεροι *pron.pl.* both.

άμωμος *a.* blameless.

αν *conj.* if, whether; ~ και although.

αν- (*privative prefix*) un-, in-, -less.

ανά *prep.* 1. (*with acc.*) along, over, throughout, across; ~ χείρας in one's hands. 2. (*with nom. & acc.*) (*distributive*) ~ εις one by one; ~ εικοσιτετράωρον every 24 hours; έλαβον ~ τρείς λίρες they received £3 each.

ανα- *denotes* up, back, again, intensification.

ανάβαθος *a.* shallow.

ανάβαθρα *s.f.* ladder.

ανάβαθρο *s.n.* entrance steps.

αναβαίνω *v.i. & t.* come *or* go up; ascend, mount, rise (*in level*); (*v.t.*) go up (*stairs, etc.*).

αναβάλλω *v.t.* postpone, put off.

ανάβαση *s.f.* ascent.

αναβατήρας *s.m.* step (*of vehicle*); lift; hoist.

αναβάτης *s.m.* climber; rider.

αναβιβάζω *v.t. see* **ανεβάζω**.

αναβιώ *v.i.* revive.

αναβλέπω *v.i.* look up; recover one's sight.

αναβλητικός *a.* dilatory, delaying.

αναβλύζω *v.i.* well up.

αναβολεύς *s.m.* stirrup.

αναβολή *s.f.* postponement.

ανάβολος *a.* inconvenient.

αναβρασμός *s.m.* agitation, state of ferment.

αναβροχιά *s.f.* drought.

αναβρύ(ζ)ω *v.i.* well up.

ανάβω *v.t. & i.* light (*fire, etc.*); turn on (*light, etc.*); inflame, provoke; (*v.i.*) light, take fire; get inflamed *or* provoked; break out; go bad.

αναγαλλιάζω *v.i.* exult.

αναγγέλλω *v.t.* announce.

αναγέννηση *s.f.* rebirth; renaissance.

αναγερτός *a.* half-reclining.

αναγινώσκω *v.t.* read.

αναγκάζω *v.t.* compel, oblige.

αναγκαί|ος *a.* necessary. **~ον** *s.n.* w.c.

αναγκασμός *s.m.* compulsion.

αναγκαστικ|ός *a.* compulsory; **~ά έργα** hard labour.

ανάγκη *a.* need, necessity; εν ~ if necessary.

ανάγλυφος *a.* in relief.

αναγνώρισις *s.f.* recognition; admission; reconnaissance.

αναγνωρίζω *v.t.* recognize; admit; reconnoitre.

αναγνωριστικός *a.* of reconnaissance.

ανάγνωση *s.f.* (*act of*) reading.

ανάγνωσμα *s.n.* reading (*matter*).

αναγνωσματάριο *s.n.* school reader.

αναγνωστήρι *s.n.* lectern.

αναγνωστήριο *s.n.* reading-room.

αναγνώστης *s.m.* reader.

αναγνωστικ|ός *a.* concerned with reading. **~ό** *s.n.* primer.

αναγορεύω *v.t.* proclaim (*ruler, etc.*).

αναγούλα *s.f.* feeling of sickness *or* disgust.

αναγράφω *v.t.* inscribe, record, publish.

ανάγ|ω *v.t.* raise; (*math.*) reduce; trace back. **~ομαι** *v.i.* relate to, go back to, date from; put out (*to sea*).

αναγωγεύς *s.m.* (*geom.*) protractor.

αναγωγή *s.f.* reduction, conversion.

ανάγωγος *a.* ill-mannered, badly brought up.

αναδασμός *s.m.* redistribution of land.

αναδάσωσις *s.f.* reafforestation.

αναδεικνύ|ω *v.t.* show to advantage, mark out; elect. **~ομαι** *v.i.* make one's mark.

ανα|δεκτός, ~δεξιμιός *s.m.* godson.

αναδεύω *v.t. & i.* stir; move.

αναδέχομαι *v.t.* undertake; stand sponsor to.

αναδίδω *v.t.* emit, give forth.

αναδίνω *v.i.* give off steam, smell, *etc.*

αναδιπλασιασμός *s.m.* (*gram.*) reduplication.

αναδιφώ *v.t.* search through, scrutinize (*documents*).

αναδουλειά *s.f.* lack of work.

ανάδοχος *s.m. & f.* sponsor, guarantor; godparent.

αναδρομή *s.f.* flashback, back reference; (*mus.*) repeat.

αναδρομικός *a.* retrospective.

αναδύομαι *v.i.* emerge, break surface.

ανάερος *a.* light as air, ethereal; airless.

ανάζερθη *s.f.* (*fam.*) back-handed slap.

αναζητώ *v.t.* search high and low for.

αναζωογονώ *v.t.* revivify.

αναζωπυρώ *v.t.* rekindle.

αναθαρρ|εύω, ~ώ *v.i.* take fresh courage.

ανάθεμα *s.n. & int.* curse; (*eccl.*) interdict.

αναθεματίζω *v.t.* interdict; curse.

αναθέτω *v.t.* charge, entrust (*duty to person*); present (*offering*).

αναθεώρ|ηση *s.f.* revision, review. **~ώ** *v.t.* revise, review.

ανάθημα *s.n.* ex-voto offering.

αναθρώσκω *v.i.* rise (*of smoke*); (*fig.*) be kindled (*of hope*).

αναθυμίαση *s.f.* (emission of) fumes, bad smell.

αναίδεια *s.f.* impudence. **~ής** *a.* impudent.

αναίμακτος *a.* without bloodshed.

αναιμία *s.f.* anaemia.

αναίρεσις *s.f.* refutation; (*law*) cassation; manslaughter; (*mus.*) natural.

αναιρώ *v.t.* refute; revoke.

αναισθησία *s.f.* unconsciousness; (*fig.*) insensibility; heartlessness.

αναισθησιολόγος *s.m.* anaesthetist.

αναισθητικό *s.n.* anaesthetic.

αναισθητοποιώ *v.t.* anaesthetize.

αναίσθητος *a.* unconscious; (*fig.*) indifferent, without feeling.

αναισχυντία *s.f.* shamelessness.

αναίτ|ιος *a.* not responsible, innocent. **~ως** *adv.* without cause.

ανακαινίζω *v.t.* renovate, renew.

ανακαλ|ύπτω *v.t.* discover, find out. **~υψη** *s.f.* discovery.

ανακαλώ *v.t.* recall, revoke, countermand; **~εις την τάξιν** call to order.

ανάκατα *adv.* without fixed order, indiscriminately.

ανακάτεμα *s.n.* mixing; confusion; nausea.

ανακάτευτος *a.* not mixed; not involved.

ανακατεύ|ω *v.t.* mix up, stir; involve; confuse; shake, shuffle. **~ομαι** *v.i.* interfere; feel sick.

ανάκατος *a.* mixed up, muddled.

ανακάτωμα *s.n.* mixing; confusion; nausea.

ανακατώνω *v.t.* see **ανακατεύω.**

ανακάτωση *s.f.* disturbance; nausea.

ανακατωσούρης *s.m.* trouble-maker.

ανακεφαλαιώνω *v.t.* recapitulate.

ανακηρύσσω *v.t.* declare (*winner, etc.*).

ανακινώ *v.t.* stir up; revive (*an old topic*).

ανάκλασις *s.f.* refraction.

ανάκλησις *s.f.* recall; revocation.

ανακλητήριος *a.* of recall.

ανάκλιντρον *s.n.* settee.

ανακοιν|ώνω, ~ώ *v.t.* announce. **~ωθέν** *s.n.* communiqué. **~ωσις** *s.f.* announcement.

ανακόλουθος *a.* inconsequent.

ανακοπή *s.f.* checking; reprieve.

ανακούρκουδα *adv.* in crouching position.

ανακουφ|ίζω *v.t.* relieve, ease. **~ιση** *s.f.* relief.

ανακριβής *a.* inaccurate, inexact.

ανάκρισις *s.f.* interrogation, inquiry.

ανακριτής *s.m.* examining magistrate.

ανάκρουσις *s.f.* recoil (*of gun*); backing water; (*mus.*) performance.

ανακρούω *v.t.* thrust back; **~** *πρύμναν* go astern; (*mus.*) perform.

ανάκτορ|ον *s.n.* palatial residence; **~α** (*pl.*) king's palace.

ανακτώ *v.t.* regain.

ανακύπτω *v.i.* emerge; crop up; recover (*from reverse*).

ανακωχή *s.f.* armistice.

αναλαμβάνω *v.t. & i.* undertake, take up (*again*); (*v.i.*) recover. *ανελήφθη* he ascended into heaven; (*fam.*) he has vanished.

αναλαμπή *s.f.* flash.

ανάλατος *a.* without salt, insipid; lacking charm.

ανάλαφρος *a.* light as air, ethereal.

ανάλγητος *a.* unfeeling.

αναλήθεια *s.f.* untruth.

ανάληψις *s.f.* withdrawal (*of money*); undertaking; resumption; (*eccl.*) ascension.

ανάλλαγος *a.* not having changed one's linen.

αναλλοίωτος *a.* unchanging, fast (*colour*).

αναλογία *s.f.* relation, proportion, ratio.

αναλογίζομαι *v.t.* bethink oneself of.

αναλογικ|ός *a.* proportionate; **~όν εκλογικόν** *σύστημα* proportional representation.

αναλόγιο *s.n.* music-stand; lectern.

ανάλογ|ος *a.* in proportion. **~ο, ~ούν** *s.n.* quota.

αναλογ|ώ *v.i.* correspond; *τι μου ~εί;* what is my share?

αναλόγως *adv.* proportionately, accordingly; (*prep.*) **~** *των περιστάσεων* according to circumstances.

ανάλυ|ση *s.f.* analysis. **~τικός** *a.* analytical; detailed.

αναλύ|ω *v.* analyse; melt; **~ομαι εις** *δάκρυα* dissolve in tears.

αναλφάβητος *a.* illiterate.

αναμαλλιασμένος *a.* dishevelled; worn rough (*of material*).

αναμασώ *v.t.* chew well; (*fig.*) keep on repeating.

ανάμελος *a.* neglectful.

αναμένω *v.t.* wait for, expect.

ανάμεσα *adv.* in between; (*prep.*) **~** *σε* among, between; **~** *από* from between.

αναμεταξύ *adv. & prep.* (*with gen.*) between, among; *στο* **~** in the mean time; **~** *μας* between ourselves.

αναμέτρηση *s.f.* weighing up; confrontation, showdown.

αναμετρώ *v.t.* weight up, take stock of.

αναμιγνύ|ω *v.t.* mix; implicate. **~ομαι** *v.i.* intervene.

ανάμικτος *a.* mixed.

αναμιμνήσκω *v.t.* remind.

αναμίξ *adv.* mixed together.

ανάμιξη *s.f.* mixing; intervention.

άναμμα *s.n.* (*act of*) lighting; inflammation; rotting (*of cereals, etc.*).

αναμμένος *a.* alight; heated; gone bad.

ανάμνηση *s.f.* memory, recollection.

αναμνηστικός *a.* commemorative.

αναμονή *s.f.* waiting, expectation.

αναμόρφωση s.f. reformation.
αναμορφωτήριον s.n. reformatory.
αναμπουμπούλα s.f. bustle, fuss.
αναμφιβόλως adv. indubitably.
αναμφισβήτητος a. indisputable.
ανανάς s.m. pineapple.
ανανδρία s.f. cowardice.
ανανεώνω v.t. renew, renovate.
ανανέωση s.f. renewal.
αναντα πόδοτος a. unrequited; unrepaid; unrepayable.
αναντικατάστατος a. irreplaceable; not replaced.
αναντίρρητος a. incontrovertible.
άναξ s.m. king, sovereign.
ανα|ξαίνω, ~ξέω v.t. (fig.) reopen, rekindle (wound, strife, etc.).
αναξιόλογος a. unworthy of note.
αναξιοπαθής a. suffering undeservedly.
αναξιόπιστος a. untrustworthy.
αναξιοπρεπής a. undignified.
ανάξιος a. unworthy; unfit, incompetent.
ανάπαλιν adv. conversely.
αναπαλλοτρίωτος a. inalienable.
αναπάντεχος a. unexpected.
αναπαράγω v.t. (biol.) reproduce.
αναπαραγωγικός a. reproductive.
αναπαραδιά s.f. (fam.) lack of money.
αναπαράσταση s.f. representation; reconstruction.
αναπαριστάνω v.t. depict, reenact.
ανάπαυλα s.f. respite.
ανάπαυσ|η s.f. rest, repose; ~εις (pl.) comfort.
αναπαυτικός a. comfortable.
αναπαύω v.t. rest, give repose to. ~ομαι v.i. rest, relax, repose. ~θηκε he is at rest (dead).
αναπέμπω v.t. refer back; give forth (smell); offer up (thanks).
αναπεπταμένος a. open (not enclosed); unfurled.
αναπηδώ v.i. jump or spring or shoot up; start, jump.
αναπηρία s.f. infirmity, disability.
ανάπηρος a. disabled, mutilated.
ανάπλαση s.f. remoulding, reforming; recalling to mind.
αναπλάσσω v.t. remould; recall to mind.
αναπλαστικός a. of remoulding or reshaping.
αναπληρώνω v.t. replace; find or act as substitute for.
αναπληρω(μα)τικ|ός a. substitute, deputy; ~ές εκλογές by-election.
αναπληρωτής s.m. deputy, substitute, surrogate.
αναπνευστικός a. respiratory.

αναπνέω v.t. & i. breathe.
αναπνοή s.f. breath, breathing.
ανάποδα adv. the wrong way round, upside down, inside out, awry.
αναπόδεικτος a. unproved; unprovable.
αναπόδεκτος a. unacceptable.
ανάποδη s.f. wrong or reverse side; backhanded slap. παίρνω από την ~ misconstrue.
αναποδι|ά s.f. hitch, reverse; contrariness. ~άζω v.i. become contrary or difficult.
αναποδογυρίζω v.t. & i. turn inside out or upside down.
ανάποδος a. the wrong way round; difficult, perverse.
αναπόδραστος a. unavoidable.
αναπολόγητος a. without a hearing; confounded, at a loss.
αναπολώ v.t. recollect, conjure up.
αναπόσπαστος a. inseparable; indispensable.
αναποφάσιστος a. irresolute.
αναπόφευκτος a. inevitable.
αναπροσαρμογή s.f. readjustment; αυτόματη ~ index-linking.
αναπτερώνω v.t. raise (hopes, morale).
αναπτήρας s.m. (cigarette) lighter.
ανάπτυξη s.f. unfolding; development.
αναπτύσσω v.t. unfold; develop, expound, evolve; deploy (troops), increase (speed).
ανάπτω v.t. & i. see ανάβω.
ανάργυρ|ος a. without money; Αγιοι ~οι SS. Cosmas & Damian.
άναρθρος a. inarticulate; (gram.) without the article.
αναρίθμητος a. countless.
ανάριος a. sparse, not closely set.
αναρμόδιος a. not competent (not authorized officially).
ανάρμοστος a. unseemly.
ανάρπαστος a. snapped up.
ανάρρηση s.f. accession.
αναρριχ|ώμαι v.i. clamber up. ~ητικός a. climbing.
ανάρρωση s.f. convalescence.
αναρτώ v.t. hang up, suspend.
αναρχί|α s.f. anarchy. ~κός a. anarchical; an anarchist.
αναρωτιέμαι v.i. wonder.
ανάσα s.f. breath, breathing; respite.
ανασαίνω v.i. breathe, get one's breath.
ανασηκών|ω v.t. lift slightly; turn, roll or tuck up. ~ομαι (v.i.) sit up.
ανασκάπτω v.t. dig up, excavate; raze.
ανασκαφή s.f. excavation.
ανάσκελα adv. on one's back.
ανασκευάζω v.t. refute.
ανασκοπώ v.t. pass in review, survey.

ανασκουμπώνομαι *v.i.* roll up one's sleeves; prepare for action.

άνασσα *s.f.* queen, sovereign.

ανασταίνω *v.t.* revive; rear (*esp. as foster child*).

ανασταλτικός *a.* delaying, restraining.

ανάσταση *s.f.* resurrection.

ανάστατος *a.* in a stir; in disorder.

αναστατ|ώνω *v.t.* (*fig.*) upset, turn upside down. **~ωση** *s.f.* commotion, upheaval.

αναστέλλω *v.t.* check, stay, inhibit.

αναστεν|άζω *v.t.* sigh. **~αγμός** *s.m.* sigh.

αναστηλών|ω *v.t.* restore (*by rebuilding*); revive. **~ομαι** *v.i.* (*fig.*) rise up.

αναστήλωση *s.f.* (*archaeological*) restoration.

ανάστημα *s.n.* height, stature.

αναστολή *s.f.* check, stay, inhibition.

αναστρέφω *v.t.* invert, reverse.

ανασυγκρότηση *s.f.* rehabilitation (*of resources*).

ανασυνδέω *v.t.* reconnect; (*fig.*) renew (*ties, etc.*).

ανασύρω *v.t.* pull up *or* out, raise, draw forth.

ανασφάλιστος *a.* uninsured.

ανάσχεση *s.f.* containment, curbing.

ανασχηματίζω *v.t.* re-form, reshuffle.

αναταράσσω *v.t.* stir (up), shake.

αναταραχή *s.f.* unrest, commotion.

ανάταση *s.f.* lifting up (*of hands, soul*).

ανατείνω *v.t.* extend, hold up.

ανατέλλω *v.i.* rise (*of sun, etc.*); dawn (*fig.*).

ανατέμνω *v.t.* dissect.

ανατίμηση *s.f.* rise in price, revaluation.

ανατινάζ|ω *v.t.* shake up; blow up. **~ομαι** *v.i.* explode; jump (*from surprise, etc.*).

ανατίναξη *s.f.* explosion, blowing up.

ανατοκισμός *s.m.* compound interest.

ανατολή *s.f.* rising (*of sun, etc.*); east; orient; Levant.

ανατολικ|ός *a.* east, eastern. **~ώς** *adv.* to the east.

Ανατολίτ|ης *s.m.* Oriental; (*esp.*) Greek from Asia Minor. **~ικος** *a.* oriental; (*esp.*) from Asia Minor.

ανατομία *s.f.* anatomy.

ανατρεπτικός *a.* subversive.

ανατρέπ|ω *v.t.* upset, overthrow, tip up. **~ομαι** *v.i.* capsize, overturn.

ανατρέφω *v.t.* rear, bring up.

ανατρέχω *v.i.* refer *or* go back (*to*).

ανατριχι|άζω *v.i.* shiver, shudder, have goose-flesh. **~ιαστικός** *a.* hair-raising. **~ίλα** *s.f.* shiver.

ανατροπή *s.f.* upset, overthrow.

ανατροφή *s.f.* upbringing.

ανάτυπο *s.n.* off-print.

ανατυπώνω *v.t.* reprint.

άναυδος *a.* speechless.

άναυλα *adv.* without paying a fare; (*fam.*) neck-and-crop.

αναφαγιά *s.f.* (*fam.*) not eating enough.

αναφαίνομαι *v.i.* appear, arise.

αναφαίρετος *a.* inalienable.

αναφανδόν *adv.* openly.

αναφέρ|ω *v.t.* mention, quote. **~ομαι** *v.i.* refer, relate.

αναφιλητό *s.n.* see **αναφυλλητό**.

αναφλέγ|ω *v.t.* inflame. **~ομαι** *v.i.* catch fire.

ανάφλεξη *s.f.* combustion; ignition.

αναφορά *s.f.* relation, reference; report; application, petition.

αναφορικός *a.* relative.

αναφουφουλιάζω *v.t.* fluff up *or* out.

αναφυλλητό *s.n.* sob, sobbing.

αναφύομαι *v.i.* (*re*)appear, arise.

αναφωνώ *v.i.* cry out.

αναχαιτίζω *v.t.* check, restrain.

αναχρονισμός *s.m.* anachronism; old-fashioned views.

ανάχωμα *s.n.* mound, bank, dike.

αναχώρηση *s.f.* departure.

αναχωρητής *s.m.* anchorite, hermit.

αναχωρώ *v.i.* set out, leave.

αναψηλάφηση *s.f.* (*law*) retrial.

αναψυκτικά *s.n. pl.* refreshments (*cold drinks, ices*).

αναψυχή *s.f.* recreation, pleasure.

ανδρ- see **αντρ-**.

ανδραγάθημα *s.n.* heroic *or* daring feat.

ανδραγαθία *s.f.* prowess.

ανδράδελφος *s.m.* husband's brother.

ανδράποδο *s.n.* enslaved prisoner; (*fig.*) servile wretch.

ανδρεία *s.f.* bravery.

ανδρείκελο *s.n.* puppet.

ανδρείος *a.* brave.

ανδριάς *s.m.* statue (*of person*).

ανδρικός *a.* male, men's; manly.

ανδρόγυνο *s.n.* married couple.

ανδρούμαι *v.i.* grow to manhood.

ανεβάζω *v.t.* carry *or* lift up; put up (*price*); put on (*play*); (*fig.*) ~ στα ουράνια praise to the skies.

ανεβαίνω *v.t. & i.* see **αναβαίνω**.

ανέβασμα *s.n.* going up, climbing; lifting *or* carrying up; rise (*in price*); production (*of play*).

ανεβοκατεβαίνω *v.t. & i.* go up and down.

ανέγγιχτος *a.* untouched, intact, new.

ανεγείρω *v.t.* raise, erect.

ανέγερση *s.f.* erection, building (*act*); getting up.

ανεγνωρισμένος *a.* recognized.

ανειλημμένος *a.* undertaken.

ανειλικρινής *a.* insincere, dissembling.
ανείπωτος *a.* unutterable; untold (*of feelings*).
ανέκαθεν *adv.* ever, always (*in the past*).
ανεκδιήγητος *a.* beyond description.
ανέκδοτ|ος *a.* unpublished. ~ο *s.n.* anecdote.
ανεκμετάλλευτος *a.* unexploited.
ανεκπλήρωτος *a.* unfulfilled; unfulfillable.
ανεκποίητος *a.* inalienable; not sold.
ανεκτέλεστος *a.* unaccomplished; that cannot be performed.
ανεκτικός *a.* patient, forbearing.
ανεκτίμητος *a.* inexpressible, invaluable.
ανεκτός *a.* bearable; passable.
ανέκφραστος *a.* inexpressible; not expressed; expressionless, inexpressive.
ανελεύθερος *a.* illiberal.
ανελκυστήρ *s.m.* lift, hoist.
ανελλιπ|ής *a.* unfailing; complete. ~ώς *adv.* without missing (*an occasion*).
ανέλπιστος *a.* unexpected.
ανέμη *s.f.* winder (*in spinning*).
ανεμίζω *v.t. & i.* expose to open air; wave in the air; winnow; (*v.i.*) blow about (*of hair, flag, etc.*).
ανεμιστήρας *s.m.* fan, ventilator.
ανεμοβλογιά *s.f.* chicken-pox.
ανεμογκάστρι *s.n.* false pregnancy.
ανεμοδείκτης *s.m.* weathercock.
ανεμοδέρνω *v.i.* be storm-tossed.
ανεμοδούρα *s.f.* weathercock; whirlwind; winder (*in spinning*); wind-swept spot.
ανεμοζάλη *s.f.* hurricane.
ανεμομάζωμα *s.n.* ill-gotten goods.
ανεμόμυλος *s.m.* windmill.
ανεμοπύρωμα *s.n.* erysipelas.
άνεμος *s.m.* wind.
ανεμόσκαλα *s.f.* light ladder.
ανεμοστρόβιλος *s.m.* whirlwind.
ανεμοφράκτης *s.m.* windbreak.
ανεμπόδιστος *a.* unhindered, free.
ανεμώδης *a.* windy.
ανεμώνη *s.f.* anemone.
ανενδοίαστος *a.* unhesitating.
ανένδοτος *a.* unyielding; persistent.
ανενόχλητος *a.* undisturbed.
ανέντιμος *a.* dishonest.
ανεξαίρ|ετος *a.* not excepted *or* exempted. ~έτως *adv.* without exception.
ανεξακρίβωτος *a.* not confirmed.
ανεξάλειπτος *a.* ineffaceable.
ανεξάντλητος *a.* inexhaustible.
ανεξαρτησία *s.f.* independence.
ανεξάρτητος *a.* independent.
ανεξέλεγκτος *a.* not verified; not subject to scrutiny *or* control.
ανεξέλικτος *a.* not fully developed (*in*

personality *or* career)
ανεξερεύνητος *a.* unfathomable; unexplored.
ανεξέταστος *a.* unexamined.
ανεξετάστως *adv.* without inquiry *or* examination.
ανεξήγητος *a.* inexplicable; not explained.
ανεξιθρησκεία *s.f.* religious tolerance.
ανεξίκακος *a.* forgiving.
ανεξίτηλος *a.* indelible.
ανεξιχνίαστος *a.* untraced; untraceable; inscrutable.
ανέξοδος *a.* free, without charge.
ανεξόφλητος *a.* not paid (back) (*of debt or creditor*).
ανεπαίσθητος *a* imperceptible.
ανεπανόρθωτος *a.* irreparable.
ανεπάρκεια *s.f* insufficiency.
ανεπαρκής *a.* insufficient, inadequate.
ανέπαφος *a.* untouched, intact.
ανεπηρέαστος *a.* not affected, uninfluenced.
ανεπίδεκτος *a.* not admitting of (*with gen.*).
ανεπιθύμητος *a.* undesirable; undesired.
ανεπίληπτος *a.* irreproachable.
ανεπίσημος *a.* unofficial, informal.
ανεπιτήδειος *a.* maladroit.
ανεπιτήδευτος *a.* unaffected, simple.
ανεπιτυχής *a.* unsuccessful; ill-chosen.
ανεπιφύλακτος *a.* without reserve.
ανεπρόκοπος *a.* good for nothing.
ανεπτυγμένος *a.* cultivated (*of persons*).
άνεργ|ος *a.* out of work. ~ία *s.f.* unemployment.
ανερεύνητος *a.* unsearched, uninvestigated.
ανερευνώ *v.t.* investigate.
ανερμάτιστος *a.* without ballast; (*fig.*) lacking stability; ~ εις not well versed in.
ανέρχομαι *v.t. & i.* ascend, climb; rise; amount.
ανέρωτος *a.* undiluted.
άνεση *s.f.* relaxation, comfort, ease; με ~η at leisure.
ανεστραμμένος *a.* reversed, upside-down.
ανέτοιμος *a.* unready.
άν|ετος *a.* comfortable; convenient. ~έτως *adv.* in comfort; easily, without constraint.
άνευ *prep.* (*with gen.*) without; εκ των ων ουκ ~ sine qua non.
ανεύθυνος *a.* not responsible.
ανευλαβής *a.* impious; disrespectful.
ανεύρεσις *s.f.* finding.
ανεύρετος *a.* not to be found.
ανευρίσκω *v.t.* find (again), discover.
ανευχάριστος *a.* not satisfied; ungrateful.
ανεφάρμοστος *a.* inapplicable; not fitting; unrealizable.
ανέφελος *a.* cloudless.
ανέφικτος *a.* unrealizable, unattainable.

ανεφοδιάζω *v.t.* provision.
ανεφοδιασμός *s.m.* provisionment.
ανεφοδίαστος *a.* not provisioned.
ανέχεια *s.f.* indigence.
ανέχομαι *v.t.* tolerate, bear.
ανεχόρταγος *a.* insatiable.
ανεψιά *s.f.* niece.
ανεψιός *s.m.* nephew.
ανήθικος *a.* immoral.
άνηθο *s.n.* dill.
ανήκουστος *a.* unheard of.
ανήκω *v.i.* belong.
ανηλεής *a.* pitiless.
ανήλικος *a.* under age; a minor.
ανήλιος *a.* not sunny.
ανήμερα *adv.* on the actual day.
ανήμερος *a.* wild (*beast*).
ανήμπορος *a.* indisposed, ill.
ανήξερ|ος *a.* (*fam.*) κάνω τον ~ο pretend not to know.
ανήρ *s.m.* man.
ανησυχητικός *a.* disturbing.
ανησυχία *s.f.* uneasiness.
ανήσυχος *a.* uneasy, restless.
ανησυχώ *v.t.* & *i.* disquiet; be worried.
ανήφορ|ος *s.m.* uphill slope, acclivity. ~ικός *a.* uphill.
ανθεκτικός *a.* resistant.
ανθηρός *a.* blooming.
άνθηση *s.f.*, ~ισμα *s.n.* flowering, bloom.
ανθίζ|ω *v.i.* flower, blossom; thrive. ~ομαι *v.t.* (*fam.*) see through, tumble to.
ανθίσταμαι *v.i.* resist.
ανθόγαλα *s.n.* cream.
ανθοδέσμη *s.f.* bunch of flowers.
ανθοδοχείο *s.n.* flower-vase.
ανθολογ|ώ *v.i.* select the best specimens. ~ία *s.f.* anthology.
ανθόνερο *s.n.* orange-flower water.
ανθοπώλης *s.m.* florist, flower-seller.
άνθος *s.n.* flower.
ανθός *s.m.* flower; flowering.
ανθόσπαρτος *a.* flower-strewn.
ανθότυρο *s.n.* cream-cheese.
ανθρακεύω *v.i.* coal; (*fig.*) take in supplies.
ανθρακιά *s.f.* glowing fire.
ανθρακικός *a.* carbonic.
ανθρακίτης *s.m.* anthracite; (*naut.*) stoker.
ανθρακωρυχείο *s.n.* coal-mine.
ανθρακωρύχος *s.m.* coal-miner.
άνθραξ *s.m.* coal; carbon; anthrax.
ανθρωπεύω *v.t.* & *i.* civilize, improve; become civilized *or* improved.
ανθρωπιά *s.f.* humaneness; civility; (*fam.*) της ~ς decent.
ανθρώπινος *a.* human.
ανθρωπινός *a.* decent, good enough.
ανθρωπισμός *s.m.* humaneness; human-

ism; (*fam.*) civility.
ανθρωπιστής *s.m.* humanist.
ανθρωποειδής *a* anthropoid.
ανθρωποκτονία *s.f.* homicide.
ανθρωπολόγος *s.m.* anthropologist.
ανθρωπόμορφος *a.* in human shape.
άνθρωπος *s.m.* man, person, human being.
ανθρωπότης *s.f.* mankind, humanity.
ανθρωποφαγία *s.f.* cannibalism.
ανθρωποφοβία *s.f.* misanthropy.
ανθυγιεινός *a.* unhealthy, unhygienic.
ανθυπασπιστής *s.m.* (*mil.*) regimental sergeant-major.
ανθυπολοχαγός *s.m.* (*mil.*) second lieutenant.
ανθυπομοίραρχος *s.m.* (*gendarmerie*) sub-lieutenant.
ανθυποπλοίαρχος *s.m.* (*navy*) sub-lieutenant.
ανθώ *v.i.* blossom, flourish.
ανία *s.f.* boredom. ~ρός *a.* boring.
ανίατος *a.* incurable.
ανίδεος *a.* having no idea; unsuspecting.
ανιδιοτελής *a.* disinterested.
ανιδρύω *v.t.* (re)erect.
ανίερος *a.* impious.
ανικανοποίητος *a.* unsatisfied.
ανίκανος *a.* incompetent, unfit; impotent.
ανίκητος *a.* invincible; unbeaten.
άνιπτος *a.* with hands *or* face unwashed.
ανίσκιωτος *a.* shadeless.
ανισόρροπος *a.* unbalanced.
άνισος *a.* unequal, uneven, unfair.
ανίσταμαι *v.i.* rise (*esp. of resurrection*).
ανιστόρητος *a.* ignorant of history; unfrescoed (*church*).
ανιστορώ *v.t.* recollect, relate; decorate (*with sacred scenes*).
ανίσχυρος *a.* powerless; not valid.
ανίσως 1. *adv.* unevenly. 2. *conj.* if by chance; ~και in case.
ανίχνευση *s.f.* tracking; reconnaissance.
ανιχνεύω *v.t.* track, derect.
ανιψ- *see* ανεψ-.
ανιώ *v.t.* & *i.* bore; be bored.
άνοδος *s.f.* ascent.
ανοησία *s.f.* nonsense, foolishness, folly.
ανόητος *a.* foolish, silly.
ανόθευτος *a.* unadulterated.
άνοιγμα *s.n.* opening; span.
ανοίγ|ω *v.t.* & *i.* open, cut open; cut, dig (*road, door, well, etc.*); (*fam.*) burgle; ~ω το δρόμο widen the road, clear the way (for); ~ω το βήμα μου walk faster; turn on (*light, tap*); let out (*garment*); (*v.i.*) begin; brighten (*of weather*); open up *or* out; come out (*of blossom*); get lighter (*in colour*). ~ομαι *v.i.* (*fig.*) be over-confiding; launch out too ambitiously.
ανοίκειος *a.* unbecoming.

ανοίκιαστος *a.* unlet.
ανοικοδόμητος *a.* not built (on).
ανοικοδομώ *v.t.* (re)build.
ανοικοκύρευτος *a.* untidy; (*man*) not settled down to family life.
ανοικονόμητος *a.* (*fam.*) impossible to deal with.
ανοικτ|ός *a.* open; light in colour; affable, open-handed; *διατηρώ ~ές σχέσεις* have a wide circle of acquaintance; *στα ~ά* on the open sea.
ανοιξιάτικος *a.* of spring.
άνοιξη *s.f.* spring (*season*).
ανοιχτόκαρδος *a.* genial.
ανοιχτομάτης *a.* wide-awake, sharp.
ανοιχτοχέρης *a.* open-handed.
ανομβρία *s.f.* drought.
ανομοιογενής *a.* heterogeneous; uneven.
ανόμοιος *a.* dissimilar, odd.
άνομος *a.* lawless; unlawful.
ανοξείδωτος *a.* stainless, ruskless.
ανόργανος *a.* inorganic.
ανοργάνωτος *a.* unorganized.
ανορεξιά *s.f.* lack of appetite; (*fig.*) half-heartedness.
ανορθογραφία *s.f.* spelling mistake; (*fam.*) solecism, not the thing.
ανορθόδοξος *a.* unorthodox.
ανορθώνω *v.t.* stand upright, raise; restore.
ανοσία *s.f.* immunity (*from disease, etc.*).
ανόσιος *a.* unholy; vile.
ανοσταίνω *v.t. & i.* make *or* become insipid, vapid *or* unattractive.
άνοστος *a.* insipid, vapid, unattractive.
ανούσιος *a. see* άνοστος.
ανοχή *s.f.* patience; tolerance; consent; *οίκος ~ς* brothel.
ανοχύρωτος *a.* unfortified; (*fig.*) unprotected.
αντ- *see* αντι-.
ανταγωνίζομαι *v.t. & i.* rival, conflict (with).
ανταγωνισμός *s.m.* rivalry, conflict.
ανταλλαγή *s.f.* exchange, barter.
αντάλλαγμα *s.n.* thing given in exchange, recompense; *εις ~* in exchange.
ανταλλακτικ|ός *a.* of *or* by exchange. *~ó s.n.* spare part, refill.
ανταλλάξιμ|ος *a.* exchangeable. *~α* (*s.n.pl.*) bonds issued to Gk. refugees from Turkey.
ανταλλάσσω *v.t.* exchange.
αντάμα *adv.* together.
ανταμείβω *v.t.* reward.
αντάμης *a.* brash.
ανταμοιβή *s.f.* reward.
ανταμώνω *v.t.* meet.
αντάμωση *s.f. καλή ~* au revoir.
αντανάκλαση *s.f.* reflection, reverberation.
αντανακλαστικ|ός *a.* reflecting. *~όν s.n.* reflex.

αντανακλώ *v.t. & i.* reflect.
αντάξιος *a.* corresponding in worth.
ανταπεργία *s.f.* lockout.
ανταποδίδω *v.t.* give in return, repay.
ανταποκρίνομαι *v.i.* correspond; respond; come up (*to*).
ανταπόκριση *s.f.* correspondence; response; journalist's dispatch; (*transport*) connection.
ανταποκριτής *s.m.* correspondent.
αντάρα *s.f.* mist; noise, uproar.
ανταρκτικός *a.* antarctic.
ανταρσία *s.f.* rebellion.
αντάρτ|ης *s.m.* rebel, guerrilla. *~ικός a.* rebel. *~ικο s.n.* rebel forces.
αντασφάλεια *s.f.* reinsurance.
ανταύγεια *s.f.* reflection, brightness.
άντε *int.* go! come! come on!
αντέγκληση *s.f.* counter-charge, recrimination.
αντεθνικός *a.* against the national interest.
αντεκδίκηση *s.f.* reprisal.
αντένα *s.f.* aerial.
αντενδείκνυμαι *v.i.* be contra-indicated.
αντεπανάσταση *s.f.* counter-revolution.
αντεπεξέρχομαι *v.i.* cope, manage.
αντεπίθεση *s.f.* counter-attack.
αντεπιστέλλον *a. ~ μέλος* corresponding member.
αντεπιτίθεμαι *v.i.* counter-attack.
αντεραστής *s.m.* rival in love.
άντερο *s.n.* intestine, bowel; (*fam.*) *στριμμένο ~* cantankerous person.
αντέτι *s.n.* custom.
αντέχω *v.i.* resist, endure, hold firm.
αντζούγια *s.f.* anchovy.
αντηλιά *s.f.* strong reflection, glare.
αντηχώ *v.i.* resound; reverberate.
αντί *prep.* (*with gen. or acc.*) instead of, in return for; for (*price*); *~ για* instead of; *~ πληρωμής* by way of payment, for pay; *~ να* (*with verb*) instead of.
αντι- *denotes* opposition, opposite situation, replacement, reciprocation, equivalence, negation, posteriority.
αντιαεροπορικός *a.* anti-aircraft.
αντιαισθητικός *a.* offending good taste, unsightly.
αντιαρματικός *a.* (*mil.*) anti-tank.
αντιβαίνω *v.i.* (*with εις*) go against, be contrary to.
αντίβαρο *s.n.* counter-weight.
αντιβασιλεύς *s.m.* viceroy, regent.
αντιβγαίνω *v.i.* be opposed (*to*); compete.
αντιγραφή *s.f.* copying; (*fig.*) copy.
αντίγραφο *s.n.* copy, transcript.
αντιγράφω *v.t. & i.* copy, copy out, imitate; crib.
αντιδημοτικός *a.* unpopular.

αντίδι *s.n.* endive, chicory.
αντιδιαστολή *s.f.* contradistinction.
αντίδικος *s.m.* opponent (*at law*).
αντίδοτο *s.n.* antidote.
αντίδραση *s.f.* reaction.
αντιδραστήριον *s.n.* (*chem.*) reagent.
αντιδραστικός *a.* reactionary; reactive.
αντιδρώ *v.i.* react; be opposed.
αντίδωρον *s.n.* (*ecl.*) Host.
αντίζηλ|ος *a.* rival. ~ία *s.f.* rivalry.
αντίθεση *s.f.* contrast, opposition.
αντίθετος *a.* contrary, opposite, reverse.
αντιθέτω *v.t.* set against, oppose.
αντίκα *s.f.* antique.
αντικαθιστώ *v.t.* replace; relieve (*sentry, etc.*).
αντικαθρέφτισμα *s.n.* reflection (*of image*).
αντικάμαρα *s.f.* antechamber; (*fam.*) κάνω ~ keep people waiting, keep aloof.
αντικανονικός *a.* irregular; against regulations.
αντικατασκοπεία *s.f.* counter-espionage.
αντικατάσταση *s.f.* replacement.
αντικαταστάτης *s.m.* substitute; successor; understudy.
αντικατοπτρισμός *s.m.* reflection; mirage.
αντίκειμαι *v.i.* be opposed.
αντικειμενικ|ός *a.* objective. ~ότης *s.f.* objectivity.
αντικείμενο *s.n.* thing, object; aim.
αντικλείδι *s.n.* pass-key.
αντικοινωνικός *a.* anti-social.
αντικρούω *v.t.* repulse; refute.
αντίκρυ *adv.* opposite.
αντικρύζω *v.t.* face; behold; meet (*debt, etc.*).
αντικρυνός *a.* opposite, facing.
άντικρυς *adv.* completely, absolutely.
αντίκρυσμα *s.n.* encounter; (*fin.*) cover, security.
αντίκτυπος *s.m.* repercussion.
αντιλαϊκός *a.* against the popular interest; unpopular.
αντίλαλος *s.m.* echo.
αντιλαμβάνομαι *v.t.* & *i.* understand; perceive, notice; be quick in the uptake.
αντιλέγω *v.i.* object, retort; say the opposite.
αντιληπτικός *a.* perceptive.
αντιληπτός *a.* noticed; understandable.
αντίληψη *s.f.* understanding; opinion; perception, notice; quickness of mind; state assistance.
αντιλογία *s.f.* contradiction.
αντιλυσσικός *a.* anti-rabies.
αντιμάμαλο *s.n.* undertow.
αντιμάχομαι *v.t.* fight against; detest.
αντίμετρο *s.n.* countermeasure.
αντιμετωπίζω *v.t.* face, brave.
αντιμέτωπος *a.* face-to-face.

αντιμιλώ *v.i.* answer back.
αντιμισθία *s.f.* reward, fee.
αντιμόνιο *s.n.* antimony.
αντιναύαρχος *s.m.* vice-admiral.
αντινομία *s.f.* contradiction, conflict.
αντίξοος *a.* unfavourable.
αντίο *int.* good-bye. [It. *addio*]
αντιπάθεια *s.f.* dislike, aversion.
αντιπαθητικός *a.* unlikeable, disagreeable.
αντιπαθώ *v.t.* dislike.
αντίπαλος *s.m.* rival, adversary.
αντιπαραβάλλω *v.t.* compare, collate.
αντιπαράτασις *s.f.* cross-examination.
αντιπαρατάσσω *v.t.* array against.
αντιπαρέρχομαι *v.t.* & *i.* pass by *or* over; ignore; escape.
αντιπαροχή *s.f.* giving in exchange (*esp.* of property owner to developer).
αντιπατριωτικός *a.* unpatriotic.
αντιπειθαρχικός *a.* insubordinate.
αντίπερα *adv.* on the farther side.
αντιπερισπασμός *s.m.* diversion, distraction.
αντιπλοίαρχος *s.m.* (*naut.*) commander.
αντίποδ|ας *s.m.* very opposite. ~ες (*pl.*) antipodes.
αντιποιητικός *a.* unpoetical.
αντίποινα *s.n. pl.* reprisals.
αντιπολιτεύομαι *v.i.* be in opposition.
αντιπολίτευσις *s.f.* (political) opposition.
αντίπραξη *s.f.* competition, rivalry.
αντιπρόεδρος *s.m.* vice-president.
αντιπροσωπεία *s.f.* representation; delegation.
αντιπροσωπεύω *v.t.* represent.
αντιπρόσωπος *s.m.* representative, delegate.
αντιρρησίας *s.m.* objector.
αντίρρηση *s.f.* objection.
αντίρροπο *s.n.* counterpoise.
αντίς *prep.* see αντί.
αντισήκωμα *s.n.* compensation; return for favours; (*mil.*) buying-out money.
αντισηπτικός *a.* antiseptic.
αντίσκηνο *s.n.* tent.
αντισμήναρχος *s.m.* (*aero.*) wing-commander.
αντισταθμίζω *v.t.* balance; (*mech.*) compensate, offset.
αντίσταση *s.f.* resistance.
αντιστέκομαι *v.i.* resist.
αντίστιξις *s.f.* (*mus.*) counterpoint.
αντίστοιχος *a.* equivalent, corresponding *ο* ~ μου my opposite number.
αντιστοιχώ *v.i.* be equivalent.
αντιστρατεύομαι *v.t.* conflict with, be bad for.
αντιστράτηγος *s.m.* lieutenant-general.
αντιστρέφω *v.t.* reverse; change.

αντίστροφ|ος *a.* reverse, inverse. **~ως** *adv.* vice versa.

αντισυνταγματάρχης *s.m.* lieutenant-colonel.

αντισυνταγματικός *a.* unconstitutional.

αντισφαίρισις *s.f.* tennis.

αντιτάσσω *v.t.* array (against); oppose; **~** *βίαν εις την βίαν* meet force with force.

αντιτείνω *v.i.* object (*to proposal, etc.*).

αντιτίθεμαι *v.i.* be opposed.

αντίτιμο *s.n.* price, value.

αντιτορπιλικόν *s.n.* (*naut.*) destroyer.

αντίτυπο *s.n.* copy (*of book, etc.*).

αντιφάρμακο *s.n.* antidote.

αντίφα|ση *s.f.* discrepancy, contradiction. **~τικός** *a.* contradictory.

αντιφάσκω *v.i.* contradict oneself.

αντίφωνον *s.n.* antiphon, anthem.

αντίχειρ *s.m.* thumb.

άντληση *s.f.* pumping, drawing of water.

αντλία *s.f.* pump; fire-engine; water-cart.

αντλώ *v.t.* pump, draw off; draw on.

αντοχή *s.f.* resistance; strength.

άντρ- *see* **ανδρ-.**

άντρακλας *s.m.* big man.

άντρας *s.m.* man; husband.

αντρειωμένος *a.* brave, strong.

αντρίκιος *a. see* **ανδρικός.**

αντρογυναίκα *s.f.* masculine woman; brave woman.

αντρόγυνο *s.n.* married couple.

άντρον *s.n.* cave, den.

αντρόπιαστος *a.* without dishonour.

άντυτος *a.* not dressed.

αντωνυμία *s.f.* pronoun.

άνυδρος *a.* waterless.

ανύμφευτος *a.* unmarried.

ανύπανδρος *a.* unmarried.

ανυπάκουος *a.* disobedient.

ανύπαρκτος *a.* non-existent.

ανυπαρξία *s.f.* non-existence; lack.

ανυπέρβλητος *a.* insuperable; unsurpassable.

ανυπερθέτως *adv.* without fail; without delay.

ανυπόδητος *a.* barefooted.

ανυπόκριτος *a.* unfeigned.

ανυπόληπτος *a.* discredited.

ανυπολόγιστος *a.* incalculable.

ανυπόμον|ος *a.* impatient. **~ώ** *v.i.* be impatient. **~ησία** *s.f.* impatience.

ανύποπτος *a.* unsuspecting; not suspect.

ανυπόστατος *a.* groundless; trifling; non-existent.

ανυπότακτος *a.* refractory; (*mil.*) a defaulter.

ανυπόφορος *a.* unbearable.

ανυψών|ω *v.t.* raise; (*fig.*) extol. **~ομαι** (*v.i.*) rise.

άνω *adv.* up, above; **~** *κάτω* in confusion, upset; *η* **~** *Βουλή* Upper Chamber; (*prep. with gen.*) above, over; **~** *ποταμών* incredible.

ανώγι *s.n.* upper storey.

άνωθεν *adv.* from above.

ανώδυνος *a.* painless.

ανωμαλία *s.f.* irregularity, unevenness; anomaly.

ανώμαλος *a.* irregular, uneven; anomalous.

ανώνυμος *a.* anonymous; (*fin.*) **~** *εταιρεία* limited company.

ανώτατος *a.* supreme, highest.

ανώτερ|ος *a.* superior, higher, senior (*in rank*). (*with gen.*) above; **~α** *βία* force majeure; *και εις* **~α** to your continued success!

ανωφελής *a.* fruitless.

ανώφλι *s.n.* lintel.

αξέγνοιαστος *a.* without cares; unconcerned.

άξενος *a.* inhospitable.

άξεστος *a.* rough, uncouth.

αξέχαστος *a.* unforgotten; unforgettable.

αξί|α *s.f.* value, worth, merit; **~ες** (*pl.*) stocks and shares; *κατ'* **~αν** deservedly.

αξιαγάπητος *a.* lovable.

αξιέπαινος *a.* praiseworthy.

αξιέραστος *a.* lovable; charming.

αξίζ|ω *v.t. & i.* be worth, cost; deserve; be of worth; be worth while; **~ει** *τον κόπο* it is worth the trouble; *μου* **~ει** I deserve it.

αξίνα *s.f.* pick.

αξιο- *denotes* worthy of.

αξιοδάκρυτος *a.* deplorable.

αξιοθαύμαστος *a.* wonderful.

αξιοθέατ|ος *a.* worth seeing; *τα* **~α** the sights.

αξιοθρήνητος *a.* lamentable.

αξιοκατάκριτος *a.* reprehensible.

αξιοκρατία *s.f.* meritocracy.

αξιόλογος *a.* worthy of note; distinguished.

αξιόπιστος *a.* worthy of credit, reliable.

αξιόποινος *a.* punishable; deserving punishment.

αξιοποι|ώ *v.t.* develop, exploit. **~ησις** *s.f.* development.

αξιοπρέπεια *s.f.* dignity, correctness, seemliness.

αξιοπρεπής *a.* dignified, correct, seemly.

άξιος *a.* capable; deserving, worthy.

αξιοσημείωτος *a.* noteworthy.

αξιότιμος *a.* estimable.

αξιόχρεος *a.* solvent, reliable.

αξιώ *v.t.* claim, demand; judge worthy.

αξίωμα *s.n.* high office; axiom.

αξιωματικός *s.m.* officer.

αξιωματούχος *s.m.* functionary.

αξιωμένος *a.* capable (*person*).

αξιών|ω *v.t.* judge worthy ~ομαι *v.i.* manage, contrive.

αξίωσn *s.f.* claim, pretension.

αξόδευτος *a.* not spent, used up *or* sold out.

αξύριστος *a.* unshaven.

άξων *s.m.* axis, axle, pivot.

αοίδιμος *a.* of blessed memory.

αοιδός *s.m.f.* singer, bard.

άοκνος *a.* tireless.

αόμματος *a.* blind.

άοπλος *a.* unarmed.

αόρατος *a.* invisible.

αοριστία *s.f.* uncertainty.

αόριστος *a.* vague, indefinite; (*s.m.*) (*gram.*) aorist.

αορτήρ *s.m.* shoulder-strap, sling.

άοσμος *a.* odourless.

απαγγελία *s.f.* recitation; delivery.

απάγκιο *s.n.* spot sheltered from wind.

απαγορευμένος *a.* forbidden.

απαγορ|εύω *v.t.* forbid, prohibit. ~ευσn *s.f.* prohibition.

απαγχονίζω *v.t.* hang (*person*).

απάγω *v.t.* abduct, carry off.

απαγωγή *s.f.* abduction; *εκουσία* ~ elopement; *εις άτοπον* ~ reductio ad absurdum.

απαθανατίζω *v.t.* immortalize.

απάθ|εια *s.f.* impassivity; indifference, apathy. ~ής *a.* impassive; indifferent, apathetic.

απαίδευτος *a.* uneducated, unskilled.

απαίρω *v.i.* set sail.

απαισιόδοξος *a.* pessimistic.

απαίσιος *a.* unpropitious; frightful.

απαίτη|σn *s.f.* claim, demand, requirement. ~τικός *a.* demanding.

απαιτ|ώ *v.t.* demand, claim, require; *τα* ~ούμενα what is needful.

απαλλαγή *s.f.* deliverance, release, discharge.

απαλλαγμένος *a.* absolved (*from*); free (*of*).

απαλλάσσω *v.t.* deliver, release, absolve; relieve (*of duties*).

απαλλοτρι|ώ, ~ώνω, *v.t.* expropriate.

απαλός *a.* soft; gentle.

απάνεμος *a.* sheltered from wind.

απάνθρωπος *a.* inhuman.

άπαντα *s.n.pl.* complete works (*of author*).

απαντέχω *v.t.* await.

απάντημα *s.n.* meeting, encounter.

απάντηση *s.f.* answer.

απαντοχή *s.f.* expectation.

απαντώ *v.t. & i.* answer; meet (*expenses*); (*also pass.*) occur, be found.

απάνω *adv.* up, above, on top; upstairs; ~ ~ superficially; ~ *κάτω* up-and-down,

approximately; *από* ~ (from) above, on top; *έως* ~ to the top; *το* ~ *πάτωμα* the upper floor; ~ *που* at the moment when. (*prep.*) ~ *από* above, over, more than; ~ *σε* upon, in the moment of; *έγραψε το σπίτι* ~ *στο γιο του* he made the house over to his son.

απανωτός *a.* one after another.

άπαξ *adv.* once; inasmuch as.

απαξι|ώ, ~ώνω *v.t. & i.* think unworthy (*of*); disdain (*to*).

απαράβατος *a.* inviolable; inviolate.

απαραβίαστον *s.n.* inviolability.

απαράδεκτος *a.* unacceptable.

απαραίτητος *a.* indispensable.

απαράλλακτος *a.* identical.

απαράμιλλος *a.* unrivalled.

απαράσκευος *a.* unprepared.

απαρατήρητος *a.* unnoticed; unrebuked.

απαρέγκλιτος *a.* steady, unswerving.

απαρέμφατον *s.n.* (*gram.*) infinitive.

απαρέσκεια *s.f.* displeasure.

απαρηγόρητος *a.* inconsolable.

απαριθμώ *v.t.* count, enumerate.

απαρνούμαι *v.t.* renounce, forsake.

απαρτία *s.f.* quorum.

απαρτίζ|ω *v.t.* form, compose, ~ομαι *v.i.* be made up, consist.

άπαρτος *a.* not received; not captured; impregnable.

απαρχαιωμένος *a.* antiquated.

απαρχ|ή *s.f.* outset; ~αί (*pl.*) first-fruits.

άπ|ας *a.* all, whole; *εξ* ~*αντος* without fail; *τα* ~*αντα* complete works.

απάστρευτος *a.* not cleaned, peeled, *or* shelled.

απασχόληση *s.f.* occupation.

απασχολ|ώ *v.t.* occupy; distract attention of; ~*ημένος* busy; preoccupied.

απατεώνας *s.m.* swindler, rogue.

απάτη *s.f.* deceit; illusion.

απατηλός *a.* deceptive, illusory.

απάτητος *a.* untrodden, inaccessible.

άπατος *a.* bottomless.

άπατρις *a.* without a homeland; unpatriotic.

απατ|ώ *v.t.* deceive, cheat. ~ώμαι *v.i.* be mistaken.

απαυδώ *v.i.* tire, become sick (*of*); (*fam.*) make fed up.

άπαυτος *a.* endless.

απαυτ|ός *a.* (*fam.*) so-and-so, what's-his-name. ~ώνω *v.t. & i.* do (*replacing another verb from vagueness or euphemism*).

άπαχος *a.* lacking fat.

απεγνωσμένος *a.* desperate.

απειθαρχ|ώ *v.i.* be insubordinate. ~ία *s.f.* insubordination.

απειθής *a.* disobedient.

απείθεια *s.f.* disobedience.
απειθώ *v.i.* be disobedient.
απεικάζω *v.t.* portray; figure, understand.
απεικονίζ|ω *v.t.* represent, portray. ~ομαι *v.i.* be reflected.
απειλ|ή *s.f.* threat. ~ητικός *a.* threatening.
απειλώ *v.t.* threaten.
απείραχτος *a.* untouched; unharmed.
απειρία *s.f.* 1. inexperience. 2. unlimited number.
άπειρ|ος *a.* 1. inexperienced. 2. countless, infinite. ~ον *s.n.* infinity.
απεκδύομαι *v.t.* divest oneself of, renounce.
απέλαση *s.f.* deportation.
απελευθερ|ώνω *v.t.* set free. ~ωση *s.f.* liberation.
απελπίζ|ω *v.t.* deprive of hope. ~ομαι *v.i.* give up hope.
απελπισία *s.f.* despair; (*fam.*) είναι ~ it is hopeless!
απελπισμένος *a.* in despair.
απελπιστικός *a.* hopeless (*case*).
απέναντι *adv. & prep.* (*with gen.*) opposite; compared with, in relation to; towards (*in part payment*); against (*receipt*).
απεναντίας *adv.* on the contrary.
απένταρος *a.* (*fam.*) without a penny.
απέξω *adv.* (from) outside; by heart. μιλώ ~ ~ για hint at.
απέραντος *a.* boundless.
απέραστος *a.* impassable; unbeatable; not threaded; not entered (*in accounts*); that has not been suffered (*of malady*).
απεργ|ία *s.f.* strike. ~ός *s.m.* striker. ~ώ *v.i.* go on strike.
απερίγραπτος *a.* indescribable.
απεριόριστος *a.* unlimited; not under restraint.
απεριποίητος *a.* neglected, untidy.
απερίσκεπτος *a.* thoughtless, foolish.
απερίσπαστος *a.* with undivided attention.
απέριττος *a.* without superfluity, simple.
απεριφράστως *adv.* straight out, point-blank (*of speech*).
απέρχομαι *v.i.* depart.
απεσταγμένος *a.* distilled.
απεσταλμένος *s.m.* envoy, representative, delegate.
απευθείας *adv.* straight, direct; live.
απευθύν|ω *v.t.* address, send. ~ομαι *v.i.* apply.
απευκταίος *a.* untoward.
απεύχομαι *v.i.* wish not to happen.
απεχθάνομαι *v.t.* detest.
απεχθής *a.* odious. ~εια *s.f.* repugnance.
απέχω *v.i.* be distant; abstain.
απηλιώτης *s.m.* east wind.
απηνής *a.* pitiless, cruel.

απηρχαιωμένος *a.* antiquated.
απηχ|ώ *v.t. & i.* echo; (*fig.*) ~ καλώς produce a good impression. ~ηση *s.f.* (*fig.*) repercussion, effect.
άπιαστος *a.* not caught; impossible to catch; in mint condition.
απίδι *s.n.* pear.
απίθαν|ος *a.* unlikely; (*fam.*) super, ~ότης *s.f.* unlikelihood.
απιθώνω *v.t.* put down; (*fam.*) stick, bung.
απίκο *adv.* (*fam.*) ready to depart. [It. *a picco*]
άπιον *s.n.* pear.
άπιοτος *a.* undrunk; undrinkable; sober.
απίστευτος *a.* unbelievable.
απιστία *s.f.* disbelief; infidelity; disloyalty.
απίστομα *adv.* face downwards.
άπιστος *a.* unbelieving; unfaithful; faithless; disloyal.
απιστώ *v.i.* (*with* εις) disbelieve; deceive; break one's word.
άπλα *s.f.* roominess.
απλά *adv.* simply (*not elaborately*).
απλάγιαστος *a.* not lying down.
απλανής *a.* fixed; expressionless.
άπλαστος *a.* unmoulded; unaffected.
άπλετος *a.* abundant (*of lighting*).
απλήρωτος *a.* 1. unfilled. 2. unpaid.
απλησίαστος *a.* unapproachable.
άπληστος *a.* insatiable, grasping.
απλίκα *s.f.* wall-bracket, sconce.
απλογραφία *s.f.* single-entry book-keeping.
απλοϊκός *a.* naive.
απλοποιώ *v.t.* simplify.
απλ|ός *a.* simple; single. ~ότης *s.f.* simplicity.
απλ|ούς *a.* simple; single. ~ούστατα *adv.* quite simply.
απλοχέρης *a.* open-handed.
απλόχωρος *a.* spacious.
απλυσιά *s.f.* dirtiness.
άπλυτ|ος *a.* unwashed; (*fig.*) τα ~α dirty linen.
άπλωμα *s.n.* roominess; spreading *or* hanging out.
απλών|ω *v.t. & i.* spread, stretch out; hang up (*washing*). ~ομαι *v.i.* extend one's activities.
απλώς *adv.* simply, merely.
απλώστρα *s.f.* clothes-horse.
άπλωτος *a.* not navigable.
απλωτ|ός *a.* outspread. ~ή *s.f.* stroke (*in swimming*).
απνευστί *adv.* in one breath; at a gulp.
άπνοια *s.f.* absence of wind; (*med.*) apnoea.
άπνους *a.* lifeless.

από *prep.* (*with acc. in demotic, gen. in purist Gk.; also with nom.*) 1. (*source, origin*) from, since; ~ αρχής from the beginning; ~ πολλού for a long time (since); ~ μικρός from childhood. 2. (*agent, cause, means, material*) by, from, of, on; πεθαίνω ~ die of; πάσχω ~ suffer from; ζω ~ live on *or* by; ~ χαρτί (*made*) of paper. 3. (*separation, deprivation*) from, of. 4. (*difference, comparison*) from, than. 5. (*distributive, partitive*) πήραν ~ δύο κομμάτια they got 2 pieces each; πάρε ~ εκείνα take one (*or* some) of those. 6. (*content, subject-matter, circumstance*) γεμάτο ~ full of; ανάγκη ~ need of; ξέρω ~ be knowledgeable about; στραβός ~ το ένα μάτι blind in one eye; ~ λεφτά έχει πολλά he is well-off for money. 7. (*route, direction*) along, through, via, past; περνώ ~ το γραφείο I call at the office. 8. (*adverbial phrases*) ~ μέσα on the inside, from inside; ~ δω over here; ~ 'δω κι εμπρός from now on. 9. (*prepositional phrases*) μέσα ~ out of; πριν ~ before.

απο- denotes source, restoration, completion, cessation, posteriority, separation, the contrary.

αποβάθρα *s.f.* landing-stage, gangway, railway platform.

αποβαίνω *v.i.* prove, turn out.

αποβάλλω *v.t.* cast off, dismiss, lose; (*v.i.*) miscarry.

απόβαρον *s.n.* tare (*weight*).

απόβαση *s.f.* disembarkation, landing.

αποβατικός *a.* of *or* for landing.

αποβδόμαδα *adv.* (*fam.*) next week.

αποβιβάζω *v.t.* disembark, unload.

αποβιώ *v.i.* die.

αποβλακών|ω *v.t.* make stupid. ~ομαι *v.i.* become stupid.

αποβλέπω *v.i.* (*with εις*) aim at, aspire to; look to, rely on.

απόβλητα *s.n.pl.* effluent.

απόβλητος *a.* rejected; an outcast.

αποβολή *s.f.* casting off; expulsion; loss; miscarriage; abortion.

αποβραδίς *adv.* yesterday evening; the evening before.

απόβρασμα *s.n.* scum.

απόγειο *s.n.* land-breeze.

απόγειον *s.n.* apogee.

απογειο|ούμαι *v.i.* take off (*in flight*). ~ωσις *s.f.* take-off.

απόγε(υ)μα *s.n.* afternoon. ~τινή *s.f.* afternoon; matinée.

απογίν|ομαι *v.i.* 1. become; τι θα ~ω; what will become of me? 2. get worse and worse. 3. be undone (*annulled*).

απόγνωση *s.f.* desperation.

απογοήτευση *s.f.* disappointment, disillusion.

απογοητεύω *v.t.* disappoint, disillusion.

απόγονος *s.m.* descendant; offspring.

απογραφή *s.f.* census, inventory.

απογυμνώνω *v.t.* strip, plunder; lay bare.

αποδάσωσις *s.f.* deforestation.

αποδεικνύω *v.t.* prove, demonstrate.

αποδεικτικόν *s.n.* certificate.

απόδειξη *s.f.* proof; receipt.

αποδεκατίζω *v.t.* decimate.

αποδέκτης *s.m.* addressee, consignee; (*fin.*) payee.

αποδεκτός *a.* acceptable; accepted.

αποδελτιώνω *v.t.* index (*on cards*).

αποδεσμεύω *v.t.* release, unblock, deregulate.

αποδέχομαι *v.t.* accept; admit.

αποδημητικός *a.* migratory.

αποδημία *s.f.* migration; going *or* living abroad.

απόδημος *a.* expatriate.

αποδιδράσκω *v.i.* escape (*of prisoner*).

αποδίδω *v.t.* give back, return; yield; attribute; render (*pay or perform*), represent.

αποδιοπομπαίος *a.* ~ τράγος scapegoat.

αποδοκιμάζω *v.t.* disapprove; demonstrate against.

απόδοση *s.f.* return; yield; rendering; output; attribution. (*gram.*) apodosis.

αποδοτικός *a.* profitable, productive.

αποδοχ|ή *s.f.* acceptance; acceptation; ~αί (*pl.*) emoluments.

απόδραση *s.f* escape (*of prisoner*).

αποδύομαι *v.i.* undress; (*fig.*) ~ εις τον αγώνα strip for the fray.

αποδυτήριον *s.n.* changing-room.

αποζημιώνω *v.t.* compensate, repay.

αποζημίωση *s.f.* compensation, indemnity, damages.

αποζητώ *v.t.* miss, feel the loss of.

αποζώ *v.i.* scrape a living.

αποθανών *a.* deceased.

αποθαρρυντικός *a.* discouraging.

αποθαρρύνω *v.t.* discourage.

απόθεμα *s.n.* deposit; stock, reserve.

αποθεραπεία *s.f.* complete cure; completion of cure.

αποθετικός *a.* (*gram.*) deponent.

αποθέτω *v.t.* put down; (*fig.*) confide; put aside (*savings*); deposit.

αποθε|ώνω *v.t.* deify; (*fig.*) give a hero's welcome to. ~ωσις *s.f.* apotheosis; (*fig.*) great reception.

αποθηκάριος *s.m.* store-keeper.

αποθηκεύω *v.t.* place in store.

αποθήκη *s.f.* any place for storage; magazine (*of gun*).

αποθνήσκω v.i. die.

αποθρασύνομαι v.i. grow impudent.

αποθυμώ v.t. miss, feel the loss of; feel desire for.

αποικία s.f. colony.

αποικιακ|ός a. colonial. ~ά s.n. pl. groceries.

άποικος s.m. colonist, settler.

Αποκαθήλωσις s.f. Descent from the Cross.

αποκαθιστ|ώ v.t. restore, re-establish; set up, marry (one's children, etc.). ~αμαι v.i. be restored; settle down, marry.

αποκαλύπτ|ω v.t. unveil, uncover, reveal. ~ομαι v.i. raise one's hat.

αποκάλυψις s.f. revelation; Apocalypse.

αποκαλώ v.t. call, name; (fam.) label.

αποκά(μ)νω 1. v.i. get tired. 2. v.t. finish off; τι απέκαμες με το σπίτι; what have you done about the house?

αποκαμωμένος a. exhausted.

αποκάρωμα s.n. torpor (from heat).

αποκατάσταση s.f. restoration, resettlement; marriage.

αποκατινός a. lower, underneath, below.

αποκάτω adv. (from) underneath.

απόκει|μαι v.i. be, repose; εις υμάς ~ται it is up to us.

απόκεντρ|ος a. outlying. ~ωσις s.f. decentralization.

αποκεφαλίζω v.t. decapitate.

αποκηρ|ύσσω v.t. renounce, disavow, disinherit. ~υξις s.f. renunciation; disinheritance.

αποκλεισμός s.m. exclusion; blockade; boycott; lock-out.

αποκλειστικός a. exclusive.

αποκλεί|ω v.t exclude, cut off; ~εται it is an impossibility.

απόκληρος a. disinherited; outcast.

αποκληρώνω v.t. disinherit.

αποκλίνω v.i. lean, incline; diverge.

απόκλισις s.f. deviation, declination.

αποκοιμίζω v.t. put or lull to sleep; (fig.) lull suspicions of.

αποκοιμούμαι v.i. go to sleep.

αποκομίζω v.t. carry away; derive, receive.

απόκομμα s.n. bit; press-cutting; weaning.

αποκοπή s.f. cutting off, amputation; weaning; κατ' ~ν on job-work.

αποκόπτω v.t. cut off; wean.

αποκορύφωμα s.n. acme.

απόκοσμος a. living in solitude.

απόκοτος a. audacious.

αποκούμπι s.n. prop, support.

απόκρεως s.f. carnival.

απόκρημνος a. precipitous.

αποκριές s.f. pl. carnival.

αποκρίνομαι v.i. answer.

απόκριση s.f. answer.

αποκρού|ω v.t. repulse, reject. ~στικός a. repulsive.

αποκρυπτογραφώ v.t. decode.

αποκρύπτω v.t. conceal.

αποκρυσταλλώνω v.t. crystallize.

απόκρυφος a. secret; (eccl.) apocryphal.

απόκτημα s.n. acquisition.

αποκτηνώνω v.t. brutalize.

αποκτώ v.t. gain, obtain, acquire; have (children).

αποκύημα s.n. product (of imagination).

απολαβή s.f. gain; ~αί (pl.) emoluments.

απολαμβάνω v.t. & i. gain, relish, enjoy, feel enjoyment.

απόλαυση s.f. enjoyment.

απολαυστικός a. enjoyable; entertaining; delectable.

απολαύω v.t. gain, get; enjoy, relish; (with gen.) enjoy, experience.

απολείπ|ω v.i be lacking; remain to be done; ~ομαι v.i. be inferior; lag behind.

απολειφάδι s.n. remnant of soap.

απολήγω v.i. lead (to), end up (in).

απολίθωμα s.n. fossil.

απολιθώνω v.t. petrify.

άπολις a. stateless.

απολίτιστος a. uncivilized.

απολλύω v.t. lose; destroy.

απολογία s.f. defence, justification.

απολογισμός s.m. financial statement; account of office, report.

απολογούμαι v.i. defend or justify oneself.

απολύμανση s.f. disinfection.

απολυμαντικό s.n. disinfectant.

απόλυση s.f. release, dismissal.

απολυταρχία s.f. absolutism.

απολυτήριον s.n. certificate of discharge; school leaving-certificate.

απόλυτος a. absolute; cardinal (number).

απολυτός a. loose.

απολυτρώνω v.t. ransom; deliver.

απολύτως adv. absolutely.

απολύ|ω v.t. release, let loose; dismiss; (v.i.) ~σε η εκκλησία church service is over. ~σου hurry!

απολωλώς a. lost; ~ός πρόβατον lost sheep.

απομαθαίνω v.t. 1. learn thoroughly. 2. teach completely. 3. unlearn.

απομακρ|αίνω, ~ύνω v.t. move away. ~υνση s.f. removal, departure. ~υσμένος a. distant.

απόμακρος a. far-off.

απομανθάνω v.t. unlearn.

απομάσσ|ω v.t. wipe off; rub down. ~ομαι v.i. wipe one's hands or face.

απόμαχος a. past active service.

απομεινάρι s.n. remains, left-over.

απομένω v.i. remain, be left; be left

speechless.
απόμερος *a.* out-of-the-way.
απομεσήμερο *s.n.* afternoon.
απομίμηση *s.f.* imitation, fake.
απομνημονεύματα *s.n.pl.* memoirs.
απομνημονεύω *v.t.* memorize.
απομονώνω *v.t.* isolate; insulate.
απομόνωσις *s.f.* isolation, insulation.
απομυζώ *v.t.* suck (*blood, etc.*).
απονέμω *v.t.* award, bestow, deal out.
απονενοημένος *a.* desperate.
απόνερα *s.n.pl.* wake (*of ship*).
απονήρευτος *a.* guileless, innocent.
απονιά *s.f.* unfeelingness.
απονομή *s.f.* award, bestowal.
άπονος *a.* unfeeling.
αποξενώνω *v.t.* alienate, estrange.
απόξεση *s.f.* scraping down.
αποξεχνιέμαι *v.i.* become oblivious (*of time, one's duties, etc.*).
αποξηραίνω *v.t.* dry, drain.
απόξω *adv. see* **απέξω.**
απο|παίδι, ~παιδο *s.n.* neglected *or* disowned child, waif.
αποπαίρνω *v.t.* chide, be sharp with.
αποπάνω *adv.* above, on top.
απόπαπας *s.m.* ex-priest.
απόπατ|ος *s.m.* w.c. **~ώ** *v.i.* go to stool.
απόπειρ|α *s.f.* attempt. **~ώμαι** *v.i.* make an attempt.
αποπέμπω *v.t.* send away, dismiss.
αποπεράτωσις *s.f.* completion.
απόπι(ω)μα *s.n.* heel-tap (*liquor*).
αποπλανώ *v.t.* mislead; seduce.
αποπληξία *s.f.* apoplexy.
αποπληρώνω *v.t.* pay off.
απόπλους *s.m.* sailing, departure.
απόπλυμα *s.n.* dishwater, slops.
αποπλύνω *v.t.* wash well, rinse; (*fig.*) wipe out (*insult, etc.*).
απόπνοια *s.f.* (emission of) smell.
αποποιούμαι *v.t.* refuse, decline.
αποπομπή *s.f.* dismissal.
απόρημα *s.n.* question, problem.
απόρθητος *a.* impregnable.
απορία *s.f.* doubt, uncertainty, perplexity, difficulty; indigence.
άπορος *a.* poor, needy.
απορρέω *v.i.* flow, stem, result (*from*).
απόρρητ|ος *a.* secret; *ο εξ ~ων* confidant.
απορρίμματα *s.n.pl.* rubbish, refuse.
απορρίπτω *v.t.* throw away, cast off; reject, refuse; fail (*examinee*).
απορρίχνω *v.t. & i.* miscarry.
απόρροια *s.f.* consequence.
απορροφ|ώ *v.t.* absorb, take up. **~ητικός** *a.* absorbent. **~ητήρας** *s.m.* extractor.
απορρυπαντικό *s.n.* detergent.

απορώ *v.i.* be at a loss, wonder; be indigent.
αποσαρίδια *s.n.pl.* sweepings.
αποσαφηνίζω *v.t.* elucidate.
απόσβεση *s.f.* extinguishing; erasure; paying *or* writing off (*debt*).
αποσβήνω *v.t. & i.* extinguish, wipe out; be wiped out.
αποσβολώνω *v.t.* confound, abash.
αποσείω *v.t.* shake off.
αποσιωπητικά *s.n.pl.* dots to show omission of word(s).
αποσιωπώ *v.t.* not to mention; hush up.
αποσκευές *s.f.pl.* baggage.
απόσκιο *s.n.* shady spot.
αποσκίρτησις *s.f.* defection.
αποσκοπώ *v.i.* aim, have as object (*to*).
αποσμητικόν *s.n.* deodorant.
αποσοβώ *v.t.* avert, keep off.
απόσπασμα *s.n.* extract, excerpt; detachment.
αποσπερίτης *s.m.* evening star.
αποσπ|ώ *v.t.* tear away, up *or* out; detach. **~ασμένος** *a.* seconded (*to post*).
απόσταγμα *s.n.* distillation (·*product*).
αποστάζω *v.t.* distil.
αποσταίνω *v.i.* get tired.
απόσταξη *s.f.* distilling.
αποστασία *s.f.* revolt, defection, apostasy.
απόστασις *s.f.* distance.
αποστάτης *s.m.* rebel; turncoat.
αποστειρώνω *v.t.* sterilize.
αποστέλλω *v.t.* dispatch; depute.
αποστέργω *v.t.* spurn, disdain.
αποστερ|ώ *v.t.* deprive. *~ούμαι* *v.t.* be deprived of, lose (*with acc. or gen.*).
αποστηθίζω *v.t.* learn *or* say by heart.
απόστημα *s.n.* abscess.
αποστολεύς *s.m.* sender.
αποστολή *s.f.* dispatch, shipment; mission.
απόστολος *s.m.* apostle.
αποστομώνω *v.t.* silence.
αποστράγγισις *s.f.* straining, draining.
αποστρατεία *s.f.* retirement (*of officers*).
αποστράτευσις *s.f.* demobilization.
απόστρατος *s.m.* retired officer.
αποστρέφ|ω *v.t.* avert (*face.*) *~ομαι* *v.t.* detest.
αποστροφή *s.f.* turning away; aversion, abhorrence; apostrophe (*rhetorical*).
απόστροφος *s.f.* (*gram.*) apostrophe.
αποσυμφόρησις *s.f.* decongestion.
αποσύνθεσις *s.f.* decomposition; breakdown.
αποσυντίθεμαι *v.i.* decompose, rot; become disorganized.
αποσύρ|ω *v.t.* withdraw, take away; retract. *~ομαι* *v.i.* withdraw, retire.
αποσχίζομαι *v.i.* separate, secede (*from*).

αποσώνω *v.t.* finish up, finish off.
απότακτος *a.* cashiered, dismissed the service.
αποταμιεύω *v.t.* save up, put by.
απόταξις *s.f.* cashiering, dismissal.
αποτάσσω *v.t.* cashier.
αποτείν|ω *v.t.* address (*words*). ~*ομαι v.i.* address oneself, inquire (*with προς*).
αποτελειώνω *v.t.* finish off.
αποτέλεσις *s.f.* formation.
αποτέλεσμα *s.n.* result, consequence. ~**τικός** *a.* effective.
αποτελ|ώ *v.t.* form, constitute. ~*ούμαι v.i.* consist.
αποτέτοιος *a.* (*fam.*) what's-his-name.
αποτεφρώνω *v.t.* reduce to ashes.
αποτιμώ *v.t.* assess.
αποτινάσσω *v.t.* shake off.
αποτί(ν)ω *v.t.* pay (*debt, penalty, homage*).
απότομος *a.* steep, sheer; sudden, abrupt; brusque.
αποτραβ|ώ *v.t.* withdraw, draw off. ~*ιέμαι v.i.* withdraw.
αποτρέπω *v.t.* avert; dissuade; deter.
αποτριχωτικόν *s.n.* depilatory.
αποτρόπαιος *a.* abominable.
αποτροπή *s.f.* prevention, warding off; dissuasion.
αποτροπιάζομαι *v.t.* abhor.
αποτρύγημα *s.n.* completion of vintage.
αποτρώγω *v.t.* & *i.* eat up; finish eating.
αποτσίγαρο *s.n.* cigarette-end.
αποτυγχάνω *v.i.* fail, be unsuccessful.
αποτυπ|ώνω *v.t.* impress, imprint, stamp. ~*ωμα s.n.* impression, print.
αποτυφλώνω *v.t* blind.
αποτυχαίνω *v.i.* see **αποτυγχάνω.**
αποτυχία *s.f.* failure, lack of success.
απουσ|ία *s.f.* absence; lack. ~*ιάζω v.i.* be absent.
αποφάγια *s.n.pl.* remains of food on plate.
αποφαγωμένος *a.* having finished eating; eaten up.
αποφαίνομαι *v.i.* give an opinion, pronounce, show, be apparent.
αποφασί|ζω *v.t.* & *i.* decide on; make up one's mind; *τον* ~*σαν* they have no hope of his recovery; ~*σμένος* determined; dispaired of (*by doctors*).
απόφασ|η *s.f.* decision, resolve; verdict; *το παίρνω* ~*η* accept the inevitable *or* make up one's mind.
αποφασιστικός *a.* decisive, resolute.
αποφέρω *v.t.* produce, bring in.
αποφεύγω *v.t.* avoid, escape.
αποφθέγγομαι *v.i.* lay down the law.
απόφθεγμα *s.n.* saying.
απόφοιτος *s.m.f.* graduate (*of school or*

university *etc.*).
αποφοιτώ *v.i.* graduate.
αποφορά *s.f.* stink.
αποφόρι *s.n.* cast-off garment.
αποφράς *a.* ~ *ημέρα* unlucky day.
αποφυγή *s.f.* avoidance.
αποφυλακίζω *v.t.* release from prison.
απόφυσις *s.f.* excrescence.
αποχαιρετισμός *s.m.* leave-taking, farewell.
αποχαιρετώ *v.t.* say goodbye to.
αποχαλιν|ώ *v.t.* unbridle, let loose. ~*ούμαι v.i.* lose all restraint.
αποχαυνώνω *v.t.* debilitate, make torpid.
αποχέτευση *s.f.* drainage, drains.
απόχη *s.f.* (*butterfly or fishing*) net.
αποχή *s.f.* abstention; abstinence.
αποχρεμπτικόν *s.n.* expectorant.
αποχρωματίζω *v.t.* take colour out of, fade.
αποχρών *a.* sufficient, justifiable.
απόχρωση *s.f.* 1. fading, discolouring. 2. shade, nuance.
αποχώρηση *s.f.* withdrawal, resignation.
αποχωρητήριο *s.n.* w.c.
αποχωρίζ|ω *v.t.* separate. ~*ομαι v.t.* & *i.* part from; part.
αποχωρισμός *s.n.* separation, parting.
αποχωρώ *v.i.* withdraw, resign.
απόψε *adv.* tonight; during the (*past*) night.
αποψιλώ *v.t.* depilate; (*fig.*) denude, strip.
αποψινός *a.* tonight's.
άποψη *s.f.* (point of) view, aspect.
απόψυξη *s.f. κάνω* ~ defrost.
απραγματοποίητος *a.* unrealizable; not carried out.
άπραγος *a.* inexperienced.
άπρακτος *a.* without achieving one's object; unachieved.
απραξία *s.f.* inactivity; (business) stagnation.
απρέπεια *s.f.* breach of good manners.
απρεπ|ος, ~ήσα. improper, unbecoming.
Απρίλ|ης, ~ιος *s.m.* April.
απρόβλεπτος *a.* unforeseen; unfore-seeable.
απρόθυμος *a.* reluctant.
άπροικ|ος, ~ιστος *a.* without a dowry.
αποκάλυπτος *a.* undisguised, open.
απροκατάληπτος *a.* unbiased.
απρόκλητος *a.* unprovoked.
απρόκοφτος *a.* good-for-nothing.
απρομελέτητος *a.* unpremeditated.
απρονοησία *s.f.* improvidence; thoughtless act.
απρό|οπτος *a.* unforeseen; *εξ* ~*όπτου* unexpectedly; *εκτός* ~*όπτου* barring the unforeseen.
απρόσβλητος *a.* unaffected (*by*); unassailable; not offended.

απροσγείωτος *a.* not having landed (*of aircraft*); (*fig.*) not down-to-earth.

απρόσδεκτος *a.* unacceptable, undesirable.

απροσδιόριστος *a.* indeterminate, indefinite.

απροσδόκητος *a.* unexpected.

απρόσ|εκτος *a.* inattentive. ~εξία *s.f.* inattentiveness.

απρόσιτος *a.* inaccessible; unapproachable; beyond one's means.

απροσ|κάλεστος, ~κλητος *a.* uninvited.

απρόσκοπτος *a.* smooth, without stumbling.

απροσπέλαστος *a. see* **απρόσιτος.**

απροσποίητος *a.* unfeigned.

απροστάτευτος *a.* unprotected.

απροσωπόληπτος *a.* impartial.

απρόσωπος *a.* impersonal.

απροφύλακτος *a.* without precaution; exposed.

απροχώρητον *s.n.* limit, dead end.

άπταιστος *a.* faultless.

άπτερος *a.* wingless.

απτόητος *a.* intrepid, undaunted.

άπτ|ομαι *v.t.* touch (*with gen.*); (*fam.*) μη μου ~ου touchy; fragile, delicate.

απτός *a.* tangible.

απύθμενος *a.* bottomless.

απύλωτ|ος *a.* without gates; (*fig.*) ~ον στόμα one who cannot keep his mouth shut.

απύρετος *a.* without a temperature.

άπω *adv.* η ~ Ανατολή Far East.

απωθώ *v.t.* thrust away, repel, reject.

απώλεια *s.f.* loss; οίκος ~ς house of shame.

απώλητος *a.* unsold; unsaleable.

απών *a.* absent.

απώτατος *a.* farthest.

απώτερος *a.* farther; ~ σκοπός ulterior motive.

άρα *conj.* so, consequently.

άρα *see* **άραγε.**

αρά *s.f.* curse.

Άραβας *s.m.* Arab.

αραβικός *a.* Arabian, Arabic.

αραβόσιτος *s.m.* maize.

αραβούργημα *s.n.* arabesque.

άραγε *particle* I wonder if, can it be that?

άραγμα *s.n.* harbouring, lying at moorings.

αράδα *s.f.* line, row; turn; της ~ς ordinary, common-or-garden; με την ~ in turn. (*adv.*) continuously; ~ ~ in a row.

αραδιάζω *v.t.* arrange in line; (*fig.*) recount, tell.

αράζω *v.t. & i.* moor, tie up, bring *or* come to rest.

αράθυμος *a.* irascible.

αραι|ός *a.* thin, sparse, scattered. ~ά και

πού occasionally.

αραιώνω *v.t. & i.* thin *or* spread out, dilute; make *or* become fewer; rarefy.

αρακάς *s.m.* vetch.

αραλίκι *s.n.* crack, fissure; (*fig.*) opportunity.

αραμπάς *s.m.* cart.

αραξοβόλι *s.n.* anchorage.

αράπης *s.m.* negro; bogyman.

αραπίνα *s.f.* negress; dark girl.

άρατος *a.* (*fam.*) έγινε ~ he vanished.

αράχν|η *s.f.* spider; spider's web. ~ιάζω *v.i.* get covered with cobwebs.

Αρβανίτης *s.m.* Albanian; Greek of Albanian descent.

αρβύλα *s.f.* soldier's boot; (*fam.*) grapevine.

αργά *adv.* slowly; late.

αργάζω *v.t.* tan.

αργαλειός *s.m.* loom.

αργάτης *s.m.* (*fam.*) labourer; windlass.

αργητα *s.f.* delay.

αργία *s.f.* not being at work; holiday, closing (-day); suspension from duty.

άργιλλος *s.f.* clay.

αργοκίνητ|ος *a.* slow-moving; (*fam.*) ~ο καράβι slowcoach.

αργόμισθ|ος *s.m.* one who draws salary without working. ~ία *s.f.* sinecure.

αργοπορία *s.f.* slowness, delay.

αργός *a.* slow, idle, not at work; suspended from duty.

αργόσχολος *a.* doing nothing.

αργότερα *adv.* later on.

αργυραμοιβός *s.m.* money-changer.

αργύρια *s.n.pl.* money.

αργυρολογία *s.f.* collection of money (*esp. by dubious means for one's own ends*).

αργυρ|ος *s.m.* silver. ~ούς *a.* silver.

αργώ *v.i.* be late, lose time; not be working; be closed.

άρδευσις *s.f.* irrigation.

άρδην *adv.* from top to bottom, completely.

αρειμάνιος *a.* warlike, fierce.

Άρειος *a.* ~ Πάγος Areopagus; Supreme Court.

αρεοπαγίτης *s.m.* judge of Supreme Court.

αρέσκεια *s.f.* taste, liking; κατ'~ν at choice.

αρέσκομαι *v.i.* take pleasure (*in*), like (*with* να *or* εις).

αρες-μάρες *s.* (*fam.*) ~ (κουκουνάρες) nonsense.

αρεστός *a.* pleasing, agreeable.

αρέσ|ω *v.i.* please, give pleasure; μου ~ει I like (it).

αρετή *s.f.* virtue.

αρετσίνωτος *a.* unresinated.

αρθριτικ|ός *a.* arthritic. ~ά (*pl.*) arthritis.

αρθρίτις *s.f.* arthritis.

αρθρογράφος *s.m.f.* leader-writer; journalist.

άρθρο *s.n.* article, clause.

αρθρώνω *v.t.* articulate, utter.

άρθρωσις *s.f.* joint, articulation.

αριά *adv.* ~ και πού rarely.

αρίδα *s.f.* (*fam.*) leg.

αρίθμησις *s.f.* counting, numbering.

αριθμητής *s.m.* numerator; adding-machine.

αριθμητικ|ός *a.* numerical, arithmetical. ~ή *s.f.* arithmetic.

αριθμός *s.m.* number, figure.

αριθμ|ώ *v.t.* count, number. ~ούμαι *v.i.* amount (*to*).

άριστα *adv.* excellently.

αριστείον *s.n.* first prize.

αριστερά 1. *s.f.* left hand; left-wing. 2. *adv.* on the left.

αριστερίζω *v.i.* be a leftist.

αριστερός *a.* left; left-handed; left-wing.

αριστεύω *v.i.* win highest mark; be (among) the best.

αριστοκρατία *s.f.* aristocracy.

άριστος *a.* best, excellent.

αριστοτέχνης *s.m.* a master hand.

αριστούργημα *s.n.* masterpiece.

αριστούχος *s.m.f.* winner of highest mark.

αρκετ|ός *a.* enough; a fair amount of. ~ά *adv.* enough, rather, considerably.

αρκούδ|α *s.f.* bear. ~ίζω *v.i.* crawl (*of babies*).

αρκούντως *adv. see* **αρκετά**.

αρκτικός *a.* 1. initial. 2. arctic, northern.

άρκτος *s.f.* bear.

αρκ|ώ *v.i.* suffice, be enough; ~εί να provided that. ~ούμαι *v.i.* confine oneself, be content (*with* να *or* εις).

αρλούμπα *s.f.* foolish talk *or* boast.

άρμα *s.n.* chariot; (*mil.*) ~ μάχης tank.

αρμαθιά *s.f.* bunch, string (*of keys, onions, etc.*).

αρμάρι *s.n.* cupboard, wardrobe.

άρματα *s.n.pl.* arms, weapons.

αρματολός *s.m.* armed Greek having police duties under Turkish occupation.

αρματώνω *v.t.* arm; equip (*ship*).

αρμέγω *v.t.* milk.

Αρμέν|ης, ~ιος *s.m.* Armenian.

αρμενίζω *v.i.* make sail; journey.

αρμένικ|ος *a.* Armenian; (*fam.*) ~η επίσκεψη (unduly) prolonged visit.

άρμεν|ο *s.n.* sail; ship; ~α (*pl.*) rigging.

άρμη *s.f.* brine.

αρμόδι|ος *a.* propitious; proper, competent. ~ότης *s.f.* competency, province.

αρμόζω *v.i.* fit, be suitable.

αρμολόγ|ηση *s.f.*, ~ημα *s.n.* pointing (*of brick work*).

αρμονία *s.f.* harmony.

αρμός *s.m.* joint; gap.

αρμοστής *s.m.* high commissioner.

αρμυρ- *see* **αλμυρ-**.

αρνάκ|ι *s.n.* little lamb; (*fam.*) ~ια (*pl.*) white horses (*waves*).

αρνησίθρησκος *a.* apostate; renegade.

αρνησικυρία *s.f.* veto.

άρνηση *s.f.* refusal; denial; negation.

αρνήσιος *a.* lamb's.

αρνητικός *a.* negative.

αρνί(ον) *s.n.* lamb, mutton.

αρνούμαι *v.t. & i.* refuse; deny.

άρον άρον *adv.* in hot haste.

αρόσιμος *a.* arable.

άροσις *s.f.* ploughing.

αροτρ|ον *s.n.* plough. ~ιώ *v.t.* plough.

αρουρ|α *s.f.* ploughed land; (*fig.*) άχθος ~ης wastrel.

αρουραίος *s.m.* vole.

άρπα *s.f.* harp.

άρπαγας *s.m.* rapacious person.

αρπαγή *s.f.* rapine; seizure, rape.

αρπάγη *s.f.* hook, grapnel.

άρπαγμα *s.n.* rape, plunder; scuffle.

αρπάζω *v.t. & i.* snatch, carry off, seize; catch (*illness*); (*v.i.*) catch fire; catch (*in cooking*). ~ομαι *v.i.* grab hold of (*with* από); get irritated; αρπάχτηκαν they came to blows.

αρπακτικός *a.* rapacious; of prey.

αρπαχτά *adv.* in haste.

αρπάχτρα *s.f.* predatory female.

άρπυια *s.f.* harpy.

αρπώ *v.t. & i. see* **αρπάζω**.

αρραβώνας *s.m.* deposit, earnest; engagement (ring); ~ες (*pl.*) betrothal.

αρραβωνιάζ|ω *v.t.* betroth. ~ομαι *v.i.* get engaged.

αρραβωνιαστικ|ός *s.m.*, ~ιά *s.f.* one's betrothed.

αρράγιστος *a.* uncracked; unbroken; unbreakable.

αρρεναγωγείον *s.n.* boys' school.

άρρεν *s.n.* male child.

αρρενωπός *a.* manly.

άρρηκτος *a.* unbreakable; unbroken.

άρρην *a.* male.

άρρητος *a.* inexpressible.

αρρυθμία *s.f.* irregularity.

αρρύθμιστος *a.* not regulated *or* settled.

αρρωσταίνω *v.t. & i.* make ill; fall ill.

αρρώστι|α *s.f.* illness. ~άρης *a.* sickly, delicate.

άρρωστος *a.* ill.

αρρωστώ *v.t. & i. see* **αρρωσταίνω**.

αρσενικός *a.* male; masculine (*gender*).

άρσις *s.f.* lifting, removal; (*mus.*) upbeat.
αρτέμων *s.m.* (*naut.*) jib.
αρτηρί|α *s.f.* artery. **~ακός** *a.* arterial.
άρτι *adv.* recently, newly.
αρτιγενής *a.* newly born.
αρτιμελής *a.* sound of limb.
άρτ|ιος *a.* whole, intact; even (*number*). **~ον** *s.n.* (*fin.*) par; εις το **~ιον** at face value.
~ίως *adv.* recently; with nothing missing.
αρτοποιείον *s.n.* bakery.
αρτοπωλείον *s.n.* baker's shop.
άρτος *s.m.* bread.
άρτυμα *s.n.* seasoning.
αρτύν|ω *v.t.* season; feed with meat. **~ομαι** *v.i.* break one's fast, eat meat.
αρύομαι *v.t.* draw, derive.
αρχές *s.f. pl.* the authorities.
αρχαΐζω *v.i.* adopt old-fashioned ways *or* expressions.
αρχαϊκός *a.* archaic.
αρχαιοδίφης *s.m.* antiquary.
αρχαιοκαπηλία *s.f.* smuggling of antiques.
αρχαιολόγ|ος *s.m.* archaeologist. **~ία** *s.f.* archaeology.
αρχαί|ος *a.* old, ancient; **~ότερος** senior; **~α** (*s.n.pl.*) antiquities.
αρχαιότητα *s.f.* antiquity; seniority.
αρχαιρεσίαι *s.f. pl.* election of officers.
αρχάριος *a.* beginner.
αρχέγονος *a.* primordial, primitive.
αρχείον *s.n.* archives.
αρχέτυπον *s.n.* archetype.
αρχ|ή *s.f.* beginning, origin; principle; authority; κατ' **~ήν** in principle; κατ' **~άς** in the beginning.
αρχηγείον *s.n.* command headquarters.
αρχηγία *s.f.* command.
αρχηγός *s.m.* commander, leader.
αρχι- *denotes* first, chief, master.
αρχιεπίσκοπος *s.m.* archbishop.
αρχιεργάτης *s.m.* foreman.
αρχιερεύς *s.m.* prelate; chief priest.
αρχίζω *v.t.* & *i.* begin.
αρχικελευστής *s.m.* (*naut.*) chief petty officer.
αρχικός *a.* first, initial.
αρχιμανδρίτης *s.m.* archimandrite.
αρχιμηνιά *s.f.* first day of month.
αρχινώ *v.t.* & *i.* begin.
αρχιπέλαγος *s.n.* archipelago.
αρχιστράτηγος *s.m.* (*mil.*) commander-in-chief.
αρχισυντάκτης *s.m.* editor-in-chief.
αρχιτεκτονική *s.f.* architecture.
αρχιτέκτων *s.m.* architect.
αρχιφύλακας *s.m.* sergeant of city police.
αρχιχρονιά *s.f.* New Year's Day.
άρχομαι *v.i.* & *t.* (*with gen.*) begin.

αρχομανία *s.f.* lust for power.
άρχοντας *s.m.* notable, man of substance; lord.
αρχοντιά *s.f.* distinguished appearance.
αρχοντικό *s.n.* gentleman's house.
αρχοντικός *a.* of distinction, fine.
αρχοντοξεπεσμένος *s.m.* impoverished gentleman.
αρχοντοπιάνομαι *v.i.* give oneself lordly airs.
αρχοντόπουλο *s.n.* son of distinguished or well-to-do family.
αρχοντοχωριάτης *s.m.* vulgar well-to-do person.
αρχύτερα *adv.* earlier; μία ώρα **~** as soon as possible.
άρχω *v.i.* & *t.* (*with gen.*) rule, reign.
άρχων *s.m.* ruler, lord; archon.
αρωγή *s.f.* aid.
άρωμα *s.n.* scent, aroma; flavour (*of ice-cream, etc.*). **~τικός** *a.* scented.
ας 1. *particle* (*introducing wishes, suppositions, etc.*) let, may. 2. (*imperative*) allow, let (alone), leave.
ασάλευτος *a.* stable, not moving.
ασανσέρ *s.n.* lift. (F. *ascenseur*).
άσαρκος *a.* fleshless, skinny.
ασαφ|ής *a.* not clear **~εια** *s.f.* obscurity.
ασβέστης *s.m.* lime.
ασβέστιον *s.n.* calcium.
ασβεστοκονίαμα *s.n.* plaster, mortar.
ασβεστόλιθος *s.m.* limestone.
άσβεστος *a.* *see* **άσβηστος**.
άσβεστος *s.f.* lime.
ασβεστώνω *v.t.* whitewash.
άσβηστος *a.* that is not *or* cannot be extinguished *or* erased.
ασβόλη *s.f.* soot.
ασβός *s.m.* badger.
ασέβ|εια *s.f.* disrespect, impiety. **~ής** *a.* disrespectful, impious.
ασέλγεια *s.f.* licentiousness. **~ής** *a.* licentious.
άσεμνος *a.* obscene, indecent.
ασετυλίνη *s.f.* acetylene.
ασήκωτος *a.* too heavy to lift; still in bed.
ασήμαντος *a.* insignificant.
ασημείωτος *a.* unnoted; unblemished.
ασημένιος *a.* silver.
ασήμι *s.n.* silver.
ασημικά *s.n. pl.* silverware.
ασημοκαπνισμένος *a.* silver-plated.
άσημος *a.* not well-known, obscure.
ασημώνω *v.t.* silver-plate; give a silver coin for luck.
ασηπτικός *a.* aseptic.
ασθένεια *s.f.* illness, weakness, disease.
ασθενής *a.* ill, weak; a patient.

ασθενοφόρον *s.n.* ambulance.
ασθενικός *a.* sickly.
ασθενώ *v.i.* be ill.
άσθμα *s.n.* asthma.
ασθμαίνω *v.i.* pant.
ασίκης *s.m.* (*fam.*) ladies' man; fine fellow.
ασίτευτος *a.* not ripened; not hung (*meat*).
ασιτία *s.f.* under-nourishment.
ασκαρδαμυκτί *adv.* fixedly, without blinking.
ασκέπαστος *a.* uncovered.
ασκεπής *a.* bare-headed.
άσκεπτος *a.* thoughtless.
ασκέρι *s.n.* army, crowd.
ασκημ- *see* **ασχημ-**.
άσκηση *s.f.* exercise, practice.
ασκητής *s.m.* hermit.
ασκητικός *a.* ascetic.
ασκί *s.n.* skin (*for wine, etc.*).
άσκοπος *a.* pointless.
ασκός *s.m.* skin (*for wine, etc.*); sac.
ασκ|ώ *v.t.* exercise, practise (*profession*); exert; carry out. *~ούμαι v.i.* exercise oneself, practise. (*with* εις).
άσμα *s.n.* song.
ασμένως *adv.* gladly.
ασορτί *adv.* to match.
ασουλούπωτος *a.* (*fam.*) badly shaped.
ασπάζομαι *v.t.* kiss, embrace; (*fig.*) adopt.
ασπάλαθος *s.m.* gorse, furze.
ασπασμός *s.m.* kiss, embrace.
ασπαστός *a.* acceptable.
άσπαστος *a.* unbroken, unbreakable.
ασπίδα *s.f.* 1. shield. 2. asp.
άσπιλος *a.* immaculate.
ασπιρίνη *s.f.* aspirin.
ασπίς *s.f.* see **ασπίδα**.
άσπλαχνος *a.* without compassion.
άσπονδος *a.* irreconcilable.
ασπόνδυλος *a.* invertebrate.
ασπούδαστος *a.* uneducated.
ασπράδα *s.f.* whiteness.
ασπράδι *s.n.* white spot; white (*of eye, egg*).
ασπριδερός *a.* whitish.
ασπρίζω *v.t. & i.* whiten, whitewash; become white; appear as a white shape.
ασπρίλα *s.f.* whiteness.
ασπρομάλλης *a.* white-haired.
ασπροπρόσωπος *a.* without loss of credit.
ασπρόρουχα *s.n.pl.* underclothes, linen.
άσπρ|ος *a.* white. *~α s.n.pl.* money.
ασπρόχωμα *s.n.* white clay.
άσσος *s.m.* ace.
αστάθ|εια *s.f.* instability, changeability. *~ής a.* unsteady, changeable.
αστάθμητος *a.* unweighed; (*fig.*) *~ παράγων* imponderable.
αστακός *s.m.* lobster.

αστάρι *s.n.* lining; undercoating.
άστατος *a.* changeable.
αστέγαστος *a.* roofless.
άστεγος *a.* homeless.
αστειεύομαι *v.i.* joke.
αστεΐζομαι *v.i.* joke.
αστείο *s.n.* joke.
αστεί|ος *a.* funny, amusing; trifling (*of things*). *παίρνω στα αστεία* treat as a joke.
αστείρευτος *a.* inexhaustible.
αστεϊσμός *s.m.* joke.
αστέρ|ας *s.m.*, *~ι. s.n.* star.
αστερισμός *s.m.* constellation.
αστερό|εις *a.* starry; *~εσσα s.f.* Stars and Stripes.
αστεροσκοπείον *s.n.* observatory.
αστεφάνωτος *a.* not legally married.
αστήρ *s.m.* star.
αστήρικτος *a.* unsupported; (*fig.*) untenable.
αστιγματισμός *s.m.* astigmatism.
αστικός *a.* urban; civic; bourgeois; civil (*law, state*); middle (*class*).
αστοίχειωτος *a.* not haunted.
αστοιχείωτος *a.* not knowing the rudiments (*of*).
άστοργος *a.* unloving, hard-hearted.
αστός *s.m.* townsman, bourgeois.
αστόχαστος *a.* thoughtless.
άστοχος *a.* that misses the target, unsuccessful, bad.
αστοχώ *v.i.* miss the mark; be wrong; forget.
αστράγαλος *s.m.* ankle.
αστραπή *s.f.* lightning *~δόν adv.* like lightning.
αστραπιαίος *a.* quick as lightning.
αστραπόβροντο *s.n.* thunder and lightning.
αστράπτω *v.i.* lighten; sparkle, gleam.
αστράτευτος *a.* exempt from military service.
αστρικ|ός *a.* stellar. *~ ό s.n.* one's destiny.
άστριφτος *a.* not twisted.
άστρο *s.n.* star.
αστρολόγος *s.m.* astrologer.
αστρονόμ|ος *s.m.* astronomer. *~ία s.f.* astronomy. *~ικός a.* astronomical.
αστροπελέκι *s.n.* thunderbolt.
αστροφεγγιά *s.f.* starlight.
άστρωτος *a.* unmade (*bed*); unlaid (*table*); unpaved; bare (*floor*); not saddled; not broken in.
άστυ *s.n.* city.
αστυνομ|ία *s.f.* police. *~ικός a.* of the police; a member of the police force.
αστυνόμος *s.m.* police inspector.
αστυφιλία *s.f.* drift of population to cities.
αστυφύλακας *s.m.* police constable.
ασυγκίνητος *a.* unmoved.

ασυγκράτητος *a.* unrestrainable.
ασύγκριτος *a.* incomparable.
ασυγύριστος *a.* untidy.
ασυγχρόνιστος *a.* not synchronized; unprogressive.
ασυγχώρητος *a.* unforgiven; unforgivable.
ασυδοσία *s.f.* immunity, exemption; lack of all restraint.
ασύδοτος *a.* immune, exempt; wildly irresponsible.
ασυζήτητ|ος *a.* without being discussed; beyond dispute; out of the question. ~ί *adv.* without discussion.
ασυλία *s.f.* (right of) asylum; (*diplomatic*) immunity; privilege.
ασύλληπτος *a.* not caught; difficult to imagine or understand; elusive (*idea*).
ασυλλόγιστος *a.* thoughtless.
άσυλο *s.n.* place of refuge, asylum.
ασυμβίβαστος *a.* incompatible; not having reached agreement; difficult, obstinate.
ασυμμάζευτος *a.* incorrigible.
ασύμμετρος *a.* disproportionate.
ασυμπάθη|τος ~ιστος *a.* unlikeable; unforgivable.
ασυμπλήρωτος *a.* not completed; not filled; unfillable.
ασυμφιλίωτος *a.* irreconcilable; unreconciled.
ασύμφορος *a.* disadvantageous.
ασυμφώνητος *a.* with whom one has not made an agreement.
ασύμφωνος *a.* incompatible, at variance.
ασυναγώνιστος *a.* defying competition.
ασυναίσθητος *a.* unconscious, involuntary.
ασύνακτος *a.* not gathered.
ασυνάρτητος *a.* incoherent.
ασυνείδητος *a.* unscrupulous; unconscious, unwitting. ~ον *s.n.* the unconscious.
ασυνεπής *a.* inconsistent, inconsequent; not true (*to promise, etc.*).
ασύνετος *a.* foolish, imprudent.
ασυνήθιστος *a.* unusual; unused (*to*).
ασύντακτος *a.* unorganized; not yet composed (*of speech, etc.*); ungrammatical.
ασύρματος *s.m.* wireless.
ασύστατος *a.* unrecommended; unfounded; inconsistent.
ασύστολος *a.* shameless.
άσφαιρος *a.* blank (*shot, etc.*).
ασφάλεια *s.f.* sureness; safety, security; insurance (company); police; fuse; safety-catch.
ασφαλής *a.* sure, safe, secure.
ασφαλίζω *v.t.* secure; insure.
ασφάλιση *s.f.* insurance, assurance.
ασφαλιστήριο *s.n.* insurance policy.
ασφαλιστικός *a.* of safety; of insurance.

ασφάλιστος *a.* 1. not closed. 2. uninsured.
ασφάλιστρ|ο *s.n.* safety-catch; ~α (*pl.*) insurance premium.
άσφαλτος *a.* unerring.
άσφαλτος *s.f.* asphalt; tarred road.
ασφαλώς *adv.* safely; certainly.
ασφόδελος *s.m.* asphodel.
ασφράγιστος *a.* unsealed; unstamped.
ασφυκτιώ *v.i.* suffocate.
ασφυξία *s.f.* suffocation.
άσχ|ετος *a.* irrelevant, unrelated. ~έτως *adv.* independently (*of*).
ασχημάτιστος *a.* not fully formed.
ασχημία *s.f.* ugliness; unseemly act.
ασχημίζω *v.t. & i* make or become ugly.
ασχημονώ *v.i.* commit unseemly acts.
άσχημος *a.* ugly; unseemly; bad.
ασχημοσύνη *s.f.* impropriety.
ασχολία *s.f.* occupation, business.
ασχολίαστος *a.* not annotated; not discussed.
ασχολούμαι *v.i.* be occupied or engaged in (*with με or εις*).
ασώματ|ος *a.* bodiless; των ~ων Michaelmas.
άσωστος *a.* endless; unfinished.
ασωτία *s.f.* dissoluteness, prodigality.
άσωτος *a.* dissolute, prodigal.
αταίριαστος *a.* not matching; incompatible.
άτακτος *a.* irregular; disorderly; naughty.
αταξία *s.f.* irregularity; disorder; naughty act; (*med.*) ataxy.
αταραξία *s.f.* calmness, imperturability.
ατάσθαλος *a.* disorderly, slovenly.
άταφος *a.* unburied.
άτεγκτος *a.* relentless.
άτεκνος *a.* childless.
ατέλεια *s.f.* defect; exemption.
ατελείωτος *a.* endless; unfinished.
ατελεσφόρητος *a.* ineffectual.
ατελεύτητος *a.* endless.
ατελής *a.* incomplete; defective; free of tax.
ατελώνιστος *a.* exempt from duty; not cleared through customs, bonded.
ατενής *a.* intent (*of gaze*).
ατενίζω *v.t.* gaze at.
ατενώς *adv.* fixedly.
ατερμάτιστος *a.* unfinished.
ατέρμων *a.* endless.
άτεχνος *a.* badly done; unskilful.
ατζαμής *s.m.* duffer, unhandy person.
ατζέμ-πιλάφι *s.n.* lamb with rice.
ατημέλητος *a.* unkempt; in dishabille.
άτι *s.n.* stallion, steed.
ατίθασος *a.* untameable, unruly.
ατιμάζω *v.t.* dishonour.
ατίμητος *a.* priceless.
ατιμία *s.f.* dishonour, disgrace; disgraceful

act.

άτιμος *a.* disgraceful, dishonourable.
ατιμωρησία *s.f.* impunity.
ατιμώρητος *a.* unpunished.
ατίμωση *s.f.* disgrace, dishonour (*downfall*).
ατλάζι *s.n.* satin.
άτλας *s.m.* atlas.
ατμάκατος *s.f.* steam-launch.
ατμάμαξα *s.f.* locomotive.
ατμολέβης *s.m.* boiler.
ατμομηχανή *s.f.* steam-engine.
ατμοπλοΐα *s.f.* steam navigation (company).
ατμόπλοιον *s.n.* steamship.
ατμός *s.m.* vapour, steam.
ατμόσφαιρα *s.f.* atmosphere.
άτοκος *a.* sterile; (*fig.*) (*fin.*) without interest.
άτολμος *a.* timid.
ατομικιστής *s.m.* individualist.
ατομικ|ός *a.* personal; individual; atomic. **~ότης** *s.f.* individuality.
άτομο *s.n.* atom; person, individual.
άτονος *a.* languid; colourless; unaccented.
ατονώ *v.i.* flag fall into disuse.
άτοπος *a.* out of place, unbecoming.
ατού *s.n.* trump. (*F.* atout).
ατόφιος *a.* solid (*silver, etc.*); the living image of.
ατράνταχτος *a.* unshakeable; (*fig.*) solid (*wealth*).
ατραπός *s.f.* path.
άτριφτος *a.* not rubbed, scrubbed, grated *or* ground.
άτριχος *a.* hairless.
ατρόμητος *a.* intrepid.
ατροφία *s.f.* atrophy.
ατρόχιστος *a.* unsharpened.
ατρύγητος *a.* unharvested.
άτρωτος *a.* unscathed; invulnerable.
ατσάλ|ι *s.n.* steel. **~ένιος** *a.* of steel. **~ώνω** *v.i.* steel.
άτσαλος *a.* disorderly, untidy; foul (*of speech*).
ατσίγγανος *s.m.* gipsy.
ατσίδα *s.f.* (*fam.*) wide-awake person.
ατύχημα *s.n.* stroke of ill fortune; accident.
ατυχής *a.* unlucky, unfortunate.
ατυχία *s.f.* ill fortune, bad luck.
άτυχος *a.* see **ατυχής.**
ατυχώ *v.i.* suffer bad luck; fail.
αυγερινός *s.m.* morning star.
αυγή *s.f.* dawn.
αυγό *s.n. see* **αβγό.**
Αύγουστος *s.m.* August.
αυθάδ|ης *a.* impudent, impertinent. **~εια** *s.f.* impudence. **~ιάζω** *v.i.* be impertinent.

αυθαίρετος *a.* high-handed, arbitrary.
αυθέντης *s.m.* lord, master.
αυθεντία *s.f.* authority; seigniory.
αυθεντικός *a.* authentic.
αυθημερόν *adv.* on the same day.
αύθις *adv.* once again.
αυθομολογούμενος *a.* self-evident.
αυθόρμητος *a.* spontaneous.
αυθυποβολή *s.f.* auto-suggestion.
αυθωρεί *adv.* forthwith.
αυλαία *s.f.* stage curtain.
αυλάκι *s.n.* groove, furrow, ditch, irrigation trench.
αυλακωτός *a.* grooved, furrowed, rifled, fluted.
αύλαξ *s.f.* see **αυλάκι.**
αυλάρχης *s.m.* lord chamberlain.
αυλή *s.f.* yard; court.
αυλητής *s.m.* piper, flautist.
αυλίζομαι *v.i.* have access from, be reached from (*of entrance to property*).
αυλικός *a.* of the court; a courtier.
αυλόγυρος *s.m.* courtyard wall.
αυλοκόλακας *s.m.* fawner.
αυλός *s.m.* pipe, flute.
άυλος *a.* incorporeal.
αυλωτός *a.* having pipes.
αυνανισμός *s.m.* masturbation.
αυξάν|ω *v.t.* & *i.* increase. **~ομαι** *v.i.* increase.
αύξηση *s.f.* increase; (*gram.*) augment.
αυξομείωσις *s.f.* fluctuation.
αύξων *a.* increasing; (*math.*) ~ αριθμός serial number.
αϋπνία *s.f.* insomnia.
άυπνος *a.* unable to sleep; watchful.
αύρα *s.f.* gentle breeze.
αυριανός *a.* of tomorrow.
αύριο *adv.* tomorrow.
αυστηρός *a.* strict; austere.
αυτ- *denotes* self, same.
αύτανδρος *a.* with all hands.
αυταπάρνησις *s.f.* self-denial.
αυταπάτη *s.f.* self-delusion.
αυτάρεσκος *a.* complacent.
αυτάρκης *a.* self-sufficient; satisfied with little.
αυταρχικός *a.* autocratic.
αυτεξούσι|ος *a.* independent. **~ον** *s.n.* free-will.
αυτεπάγγελτος *a.* self-appointed; ex officio; automatic (*e.g. retirement at age-limit*).
αυτί *s.n.* ear.
αυτοβιογραφία *s.f.* autobiography.
αυτόβουλος *a.* of one's own volition.
αυτόγραφος *a.* autograph.
αυτόδηλος *a.* self-evident.
αυτοδημιούργητος *a.* self-made.

αυτοδιάθεσις s.f. self-determination.
αυτοδίδακτος a. self-taught.
αυτοδικαίως adv. of right.
αυτοδικία s.f. taking law into one's own hands.
αυτοδιοίκησις s.f. self-government.
αυτοθελώς adv. of one's own accord.
αυτόθεν adv. thence.
αυτόθι adv. in the same place; ibidem.
αυτοθυσία s.f. self-sacrifice.
αυτοκαλούμενος a. self-styled.
αυτοκέφαλος a. independent; (eccl.) autocephalous.
αυτοκινητάμαξα s.f. rail motor-car.
αυτοκινητιστής s.m. motorist.
αυτοκίνητ|ος a. self-propelling. ~ο s.n. motorcar. ~όδρομος s.m. motorway.
αυτόκλητος a. self-invited; ~ μάρτυς one who voluteers evidence.
αυτοκράτειρα s.f. empress
αυτοκράτωρ s.m. emperor.
αυτοκρατορ|ία s.f. empire. ~ικός a. imperial.
αυτοκτονία s.f. suicide.
αυτοκτονώ v.i. commit suicide.
αυτοκυβέρνησις s.f. self-government.
αυτοκυριαρχία s.f. self-command.
αυτολεξεί adv. word for word.
αυτόματ|ος a. automatic. ~ον s.n. automaton.
αυτομολ|ώ v.i. desert to the enemy. ~ος s.m. defector.
αυτονόητος a. self-evident.
αυτονομία s.f. autonomy.
αυτόνομος a. autonomous.
αυτοπαθής a. (gram.) reflexive.
αυτοπεποίθηση s.f. self-confidence.
αυτοπροαίρετος a. voluntary.
αυτοπροσώπως adv. in person.
αυτόπτης s.m. eye-witness.
αυτ|ός pron. he; this; ο ~ός the same; η ~ού υψηλότης His Highness.
αυτοστιγμεί adv. immediately.
αυτοσυντήρησις s.f. self-preservation.
αυτοσυντήρητος a. self-supporting.
αυτοσχεδιάζω v.t. improvise.
αυτοσχέδιος a. impromptu, improvised.
αυτοτελής a. self-contained.
αυτού adv. there.
αυτουργός s.m. perpetrator.
αυτούσιος a. in its original state, intact.
αυτοφυής a. indigenous; wild (of plants); native (of metals, etc).
αυτό|φωρος a. (detected) in the very act; επ' ~φώρω red-handed. ~φωρο s.n. (fam.) police-court.
αυτόχειρ s.n. suicide (person).
αυτοχειροτόνητος a. self-appointed.

αυτόχθων a. indigenous (person).
αυτόχρημα adv. purely and simply.
αυτοψία s.f. post-mortem; (magistrate's) investigation on the spot.
αυχ|ήν, ~ένας s.m. nape of neck; (fig) neck, col.
αυχμηρός a. arid.
άφαγ|ος a. who eats insufficient; not having eaten. ~ία s.f. not eating enough.
αφάγωτος a. uneaten; not having eaten; (fig.) intact.
αφαίμαξις s.f. blood-letting; drain (on resources).
αφαίρεσις s.f. taking away; subtraction; abstraction.
αφαιρ|ώ v.t. remove, take away, subtract; steal. ~ούμαι v.i. have one's thoughts elsewhere.
αφαλοκόβω v.t. (fam.) weigh down (with burden); break the back of.
αφαλός s.m. navel.
αφάνα s.f. besom.
αφαν|ής a. invisible; obscure. ~εια s.f. obscurity (being unknown).
αφανίζω v.t. annihilate; harass. ~ομαι v.i. vanish; be destroyed.
αφάνταστος a. unimaginable.
άφαντος a. invisible.
αφαρπάζομαι v.i. flare up (in anger).
αφασία s.f. aphasia.
άφατος a. ineffable.
αφέγγαρος a. moonless.
αφεδρών s.m. anus; w.c.
αφειδής a. lavish.
αφέλεια s.f. simplicity; naïveté; (fam.) fringe (hair).
αφελής a. simple, naïve.
αφέντης s.m. lord, master; proprietor.
αφεντιά s.f. nobility; (fam.) ή ~ σου your honour, you.
αφεντικό s.n. (fam.) boss, governor.
αφερέγγυος a. insolvent.
άφεση s.f. remission, discharge.
αφετηρία s.f. point of departure.
αφέτης s.m. starter (of race).
άφευκτ|ος a. inevitable. ~ως adv. without fail.
αφέψημα s.n. infusion. (tea, herbs, etc.).
αφ|ή s.f. 1. (sense of) touch. 2. lighting; περί λύχνων ~άς at lighting-up time (i.e. at dusk).
αφήγημα s.n. narrative, story.
αφήγη|σις s.f. narration. ~τής s.m. narrator.
αφηγούμαι v.t. narrate.
αφηνιάζω v.i. bolt; take the bit between one's teeth.
αφήνω v.t. let, allow; let alone; let go of, let slip; leave; abandon; entrust; emit.

αφηρημάδα *s.f.* (*act of*) absentmindedness.
αφηρημένος *a.* absent-minded; abstract.
άφθα *s.f.* sore place on gum.
αφθαρσία *s.f.* indestructibility.
άφθαρτος *a.* imperishable.
άφθαστος *a.* unrivalled; unsurpassable.
άφθον|ος *a.* abundant, plenty of. **~ία** *s.f.* abundance. **~ώ** *v.i.* abound.
αφίεμαι *v.i.* relax; be released.
αφιερ|ώνω *v.t.* dedicate, devote. **~ωμα** *s.n.* offering; special feature (*in journal*). **~ωσις** *s.f.* dedication.
αφικνούμαι *v.i.* arrive.
αφιλοκερδής *a.* disinterested.
αφιλότιμος *a.* mean, lacking self-respect; (*joc.*) blighter.
αφίνω *v.t.* see **αφήνω**.
άφιξη *s.f.* arrival.
αφιόνι *s.n.* poppy; opium.
αφιονισμένος *a.* fanatical; stupefied.
αφιππεύω *v.i.* dismount (*from horse*).
αφίσα *s.f.* poster, bill.
αφίσταμαι *v.i.* differ; hold aloof.
αφκιασίδωτος *a.* without make-up; plain.
άφλεκτος *a.* non-inflammable.
αφλογιστία *s.f.* misfire.
άφοβος *a.* fearless.
αφοδεύω *v.i.* go to stool.
αφοδράριστος *a.* unlined.
αφομοι|ώνω *v.t.* assimilate. **~ωσις** *s.f.* assimilation.
αφόντας *conj.* see **αφότου**.
αφοπλίζω *v.t.* disarm; defuse (*lit.*).
αφοπλισμός *s.m.* disarmament.
αφόρετος *a.* unworn, new.
αφόρητος *a.* unbearable.
αφορία *s.f.* scarcity of crops.
αφορίζω *v.t.* excommunicate; delimit.
αφορισμός *s.m.* excommunication; aphorism.
αφορμή *s.f.* occasion, ground.
αφορμίζω *v.i.* fester.
αφορολόγητος *a.* untaxed.
άφορος *a.* unfruitful.
αφορ|ώ *v.t.* relate to, concern; *w.f. όσον ~ά* as regards.
αφοσιούμαι *v.i.* devote oneself.
αφοσίωσις *s.f.* devotion.
αφότου *conj.* since (*time*).
αφού *conj.* since (*causal*); after, when.
αφουγκράζομαι *v.t. & i.* listen (to).
άφρακτος *a.* unfenced.
αφράτος *a.* light and soft; light and crisp (*bread*); plump and white (*flesh*); puffed, billowy.
αφρίζω *v.i.* foam, froth, lather; (*fig.*) rage.
αφρόγαλα *s.n.* cream.
αφροδίσια *s.f.* sexual instinct.

αφροδισιακός *a.* aphrodisiac.
αφροδισιολόγος *s.m.* V.D. specialist.
αφροδίσιος *a.* venereal.
αφρόκρεμα *s.f.* (*fig.*) cream, flower, pick.
αφρόντιστος *a.* neglected; heedless.
αφρός *s.m.* foam, spray, scum; lather; cream.
αφροσύνη *s.f.* foolishness.
αφρούρητος *a.* unguarded.
αφρώδης *a.* frothy, foaming.
άφρων *a.* foolish, thoughtless.
αφτί *s.n.* ear; (*fam.*) δεν ιδρώνει τ' ~ του he doesn't care.
αφτιάζομαι *v.i.* prick up one's ears.
άφτια|στος, ~χτος *a.* not made *or* done; not made up (*of face*).
άφτρα *s.f.* sore place on gum.
αφυδάτωσις *s.f.* dehydration.
αφυπηρετώ *v.i.* complete one's military service.
αφύπνισις *s.f.* awakening.
αφύσικος *a.* unnatural; abnormal.
άφωνος *a.* mute; speechless.
άφωτα *adv.* before dawn.
αφώτιστος *a.* unlit; unenlightened; (*fig.*) unbaptized.
αχ *int.* ah! oh!
αχαΐρευτος *a.* good-for-nothing.
αχάλαστος *a.* not destroyed *or* spoilt; indestructible; not spent *or* changed (*money*).
αχαλίνωτος *a.* unbridled.
αχαμν|ός *a.* lean. **~ά** *s.n.pl.* testicles.
αχανής *a.* vast.
αχαρακτήριστος *a.* (*esp. pej.*) indescribable.
αχάρα|κτος *a.* (*road*) not cut; (*paper*) unruled; (*water*) unrippled. **~γα** *adv.* before dawn.
άχαρις *a.* see **άχαρος**.
αχάριστος *a.* ungrateful; thankless.
άχαρος *a.* ungraceful; giving no pleasure.
αχάτης *s.m.* agate.
άχερο *s.n.* straw, chaff.
αχηβάδα *s.f.* scallop-shell; niche.
άχθος *s.n.* burden.
αχθοφόρος *s.m.* porter.
αχίλλειος *a.* of Achilles.
αχινός *s.m.* sea-urchin.
αχλάδι *s.n.* pear.
αχλύς *s.f.* thin mist; (*fig.*) melancholy.
αχμάκης *s.m.* dolt.
άχνα *s.f.* mist, vapour; (*fam.*) μη βγάλεις ~ do not utter a sound.
αχνάρι *s.n.* footprint; (*fig.*) (paper) pattern.
άχνη *s.f.* mist, vapour; finest powder, sublimate.
αχνίζω *v.t & i.* steam, make misty; give off vapour.

αχνός *s.m.* mist, vapour.
αχνός *a.* pallid.
αχόρταγος *a. see* αχόρταστος.
αχόρταστος *a.* insatiable; unsated.
αχός *s.m.* noise, din.
άχου *int.* oh!
αχούρι *s.n.* stable.
άχραντ|ος *a.* immaculate; ~α μυστήρια sacraments.
αχρείαστος *a.* not needed.
αχρείος *a.* vile, villainous; obscene.
αχρέωτος *a.* not debited; not in debt *or* mortgaged.
αχρησιμοποίητος *a.* unused; unusable.
αχρηστεύ|ω *v.t.* make useless. ~ομαι *v.i.* become useless; fall out of use.
αχρηστία *s.f* disuse.
άχρηστος *a.* useless; not used, waste.
άχρι *prep.* up to, until.
αχρόνι(α)στος *a.* not yet one year old *or* dead.
αχρονολόγητος *a.* undated.
άχρους *a.* colourless; (*fig.*) neutral.
αχρωμάτιστος *a.* not painted; colourless; (*fig.*) neutral.
αχταρμάς *s.m.* transfer from one conveyance to another.
αχτένιστος *a.* unkempt; (*fig.*) not polished up (*of speech, etc.*)
άχτι *s.n.* strong desire (*esp. for vengeance*); τον έχω ~ I bear him a grudge; βγάζω το ~ μου let off steam, satisfy a long-felt desire.
αχτί|δα, ~να *s.f.* ray.
αχτύπητος *a.* unbeaten.
άχυρο *s.n.* straw, chaff.
αχυρών *s.m.* barn.
αχώνευτος *a.* undigested; (*fig.*) unbearable.
αχώριστος *a.* unseparated; inseparable.
άψ|α ~άδα *s.f.* tartness; irritability.
αψεγάδιαστος *a.* faultless.
άψε σβήσε *adv.* in the twinkling of an eye.
άψη(σ)τος *a.* not (sufficiently) cooked *or* baked; (*fig.*) raw.
αψήφιστ|ος *a.* not having voted *or* been voted for; daring; disregarded. ~α *adv.* το πήρα ~α I treated it lightly.
αψηφώ *v.t.* defy, brave, ignore.
αψίδα *s.f.* arch; apse.
αψιδωτός *a.* arched.
αψίκορος *a.* quickly tiring of what one has.
άψιλος *a.* (*fam.*) penniless.
αψιμαχία *s.f.* skirmish.
άψογος *a.* impeccable.
αψύς *a.* hot-tempered; sharp, pungent.
αψυχολόγητος *a.* without psychological insight.
άψυχος *a.* lifeless, spiritless.

αψώνιστος *a.* not bought; not having shopped.
άωρος *a.* unripe; (*fig.*) hasty.
άωτον *s.n.* το άκρον ~ acme, ne plus ultra.

B

βαγένι *s.n.* barrel.
βάγια *s.f.* wet-nurse; nursemaid.
βάγια *s.n.pl.* palm branches.
βαγιά *s.f.* laurel
βαγόνι *s.n.* coach, truck (*railway*).
βάδην *adv.* at walking pace.
βαδίζω *v.i.* walk, march.
βάδισμα *s.n.* gait, walk, walking.
βαζγεστίζω *v.t. & i.* (*fam.*) be fed up (with).
βάζο *s.n.* vase, pot.
βάζω *v.t.* put, apply, place in position; put on (*clothes, radio*); serve with, give (*in shop, at table*); impose (*tax*); hire, employ, get (*person to do sthg.*); ~ τραπέζι lay the table; ~ στην μπάντα lay aside, save up; ~ καλό βαθμό give good marks; ~ στοίχημα wager; ~ φωνές start shouting; ~ αυτί listen; ~ χέρι lay hands (*on*); ~ ένα χέρι lend a hand; ~ χρέος contract a debt; ~ υποψηφιότητα stand for office; ~ πλώρη set out (*for*); ~ με τρόπο put without being noticed; ~ τα δυνατά μου do one's best; τα ~ με quarrel with; ~ μπρος start (off); τον ~ μπροστά *or* πόστα I reprimand him; ~ μέσα imprison; δεν το ~ κάτω I will not admit defeat; το ~ στα πόδια take to one's heels; τον ~ στο πόδι μου I leave him in my place; δεν το βάζει ο νους it is inconceivable; βάλε με το νου σου just imagine; βάλε πως supposing; βάλθηκε να μάθει αγγλικά he set about learning English.
βαθαίνω *v.t. & i.* deepen.
βαθ|έως, ~ιά *adv.* deeply.
βαθμηδόν *adv.* gradually.
βαθμιαίος *a.* gradual.
βαθμίς *s.f.* step, rung; (*fig.*) point on a scale; grade, stage, level.
βαθμολογία *s.f.* marking; graduation; score, rating.
βαθμός *s.m.* degree; rank; mark (*award*).
βαθμούχος *s.m.* (*higher ranking*) officer.
βάθος *s.n.* depth; far end; background; κατά ~ thoroughly; at bottom.
βαθουλαίνω *v.t. & i.* hollow out; sag.
βαθουλός *a.* hollow, sunken.
βαθούλωμα *s.n.* hollow place, depression.
βάθρακ|ας, ~ός *s.m.* frog.
βάθρον *s.n.* base, pedestal; bench.

βαθυ- *denotes* deep.
βαθύνοια *s.f.* sagacity.
βαθύνω *v.t. & i.* deepen.
βαθύπλουτος *a.* very rich.
βαθ|ύς *a.* deep, profound. ~ **ύτης** *s.f.* depth, profundity.
βαθύφωνος *a.* (*mus.*) bass; contralto.
βαίνω *v.i.* go; be on the way (*to*).
βά|ιον *s.n.* palm branch; Κυριακή των ~ῖων Palm Sunday.
βάκιλλος *s.m.* bacillus.
βακτηρία *s.f.* stick, crutch.
βακτηρίδιον *s.n.* bacillus.
βακτηριολογία *s.f.* bacteriology.
βακχεία *s.f.* orgy.
βάκχ|η, ~ίς *s.f.* bacchante.
βακχικός *a.* bacchic.
βαλανίδ|ι *s.n.* acorn. ~ **ιά** *s.f.* oak-tree.
βάλανος *s.f.* acorn.
βαλάντιον *s.n.* purse.
βαλαντώνω *v.i. & t.* be exhausted; exhaust, wear out.
βαλβίδα *s.f.* valve; starting-post.
βαλές *s.m.* footman; knave (*cards*).
βαλί|τζα, ~τσα *s.f.* suitcase.
βαλκανικός *a.* Balkan.
βάλλ|ω *v.t. & i.* fire, shoot, throw; (*fig.*) launch (*verbal*) attack. ~ομαι *v.i.* be fired at; εβλήθη he was hit.
βαλμένος *a.* *1.* worn (*of clothes*). *2.* see **βαλτός**.
βαλς *s.n.* waltz. [F. *valse*].
βάλσαμ|ο *s.n.* balsam, balm. ~ώνω *v.t.* embalm; (*fig.*) console, ease.
βάλσιμο *s.n.* putting (*on*), placing in position.
βάλτος *s.m.* marsh, bog.
βαλτός *a.* set on, instigated.
βαμβακερός *a.* cotton.
βαμβάκι *s.n.* cotton; cotton-wool.
βαμβακοπυρίτιδα *s.f.* gun-cotton.
βάμβαξ *s.m.* cotton; cotton-wool.
βάμμα *s.n.* tincture.
βαμμένος *a.* dyed, painted; (*fig.*) (*fam.*) dyed-in-the-wool.
βάναυσος *a.* rude, rough.
βάνδαλος *s.m.* vandal.
βανίλια *s.f.* vanilla.
βάνω *v.t.* see **βάζω**.
βαπόρι *s.n.* steamer; (*fam.*) έγινε ~ he got furious *or* he was off in a flash.
βαπτίζω *v.t.* baptize, christen; stand sponsor to; (*fam.*) water (*wine, milk*).
βάπτισμα *s.n.* baptism (*sacrament*).
βάπτισις *s.f.* christening.
βαπτιστής *s.m.* baptist.
βαπτιστικ|ός *a.* baptismal. ~όν *s.n.* certificate of baptism.

βάπτω *v.t. & i.* temper (*steel*); dye, paint; polish (*shoes*); (*v.i.*) become stained.
βάραθρο *s.n.* abyss.
βαραίνω *v.t. & i.* weigh down; make heavier; be a burden to, weary; (*v.i.*) weigh, have weight; become heavier; get worse (*in health*); heel over; dip (*of scale*).
βαράω *v.t.* see **βαρώ**.
βαρβαρικός *a.* barbaric.
βάρβαρος *a.* barbarous; (a) barbarian.
βαρβαρότης *s.f.* barbarity.
βαρβάτ|ος *a.* not castrated, ruttish; ~ο άλογο stallion; (*fig.*) competent; considerable.
βαργεστ|ίζω, ~ώ *v.t. & i.* see **βαζγεστίζω**.
βάρδα *int.* (*fam.*) look out!
βάρδια *s.f.* watch, guard; shift.
βάρδος *s.m.* bard.
βαρεία *s.f.* (*gram.*) grave accent.
βαρελάς *s.m.* cooper.
βαρελήσιος *a.* from the barrel.
βαρέλι *s.n.* barrel, cask, drum.
βαρελότο *s.n.* firecracker.
βαρεμένη *a.* pregnant.
βαρετός *a.* boring, burdensome.
βαρέως *adv.* gravely; το φέρω ~ be annoyed *or* upset by it.
βαρήκοος *a.* hard of hearing.
βαριά *adv.* heavily, seriously.
βαρίδι *s.n.* weight, counterweight.
βαρι|έμαι, ~ούμαι *v.i. & t.* be bored (with), be tired (of); not want (to); (*fam.*) δεν ~έσαι never mind about that!
βαριετέ *s.n.* music-hall.
βάριον *s.n.* barium.
βάρκα *s.f.* (*small*) boat, rowing-boat.
βαρκάδα *s.f.* boat-trip.
βαρκάρης *s.m.* boatman.
βαρόμετρο *s.n.* barometer.
βάρος *s.n.* weight, heaviness; burden, responsibility; εις ~ του to his detriment, at his expense.
βαρούλκο *s.n.* windlass.
βαρυ- *denotes* heavy.
βαρύθυμος *a.* dejected.
βαρύν|ω *v.t. & i.* *1.* see **βαραίνω**. *2.* (*v.t.*) be laid to one's charge, fall upon (*of responsibility, etc.*). *3.* ~ομαι see **βαριέμαι**.
βαρυποινίτης *s.m.* long-term prisoner.
βαρύς *a.* heavy; hard, severe, extreme; bad for health; strong (*wine*); deep (*in pitch*); reserved, distant.
βαρυσήμαντος *a.* momentous.
βαρύτης *s.f.* gravity; weight; severity; depth.
βαρύτιμος *a.* costly, valuable.
βαρύτονος I. *a.* bearing grave accent. *2.* *s.m.* (*mus.*) baritone.

βαρύφωνος *a.* deep-voiced.
βαρ|ώ *v.t. & i.* strike, hit; wound, kill (*game*); play (*instrument*); sound, strike (*of bugle, clock, etc.*); (*fam.*) ~*άει η τσέπη του* his pocket is well lined.
βαρώνος *s.m.* baron.
βασάλτης *s.m.* basalt.
βασανίζ|ω *v.t.* examine, go into; torture, torment, rack. ~*ομαι v.i.* have a lot of trouble.
βασανιστήριο *s.n.* rack, torture; torture-chamber.
βάσανο *s.n.* torture; trouble, tribulation; (*fam.*) nuisance; lady-love.
βάσανος *s.f.* scrutiny, inquiry; torture.
βασίζ|ω *v.t.* base. ~*ομαι v.i.* rely.
βασικός *a.* basic, primary.
βασιλεία *s.f.* kingship, royalty; reign.
βασίλειον *s.n.* kingdom, realm.
βασίλεμα *s.n.* setting (*of sun, etc.*).
βασιλεύς *s.m.* king.
βασιλεύω *v.i.* reign; set (*of sun, etc.*); close (*of eyes*). *ζω και* ~ be alive and kicking.
βασιλιάς *s.m.* king.
βασιλική *s.f.* basilica.
βασιλικός *a.* royal; royalist; regal.
βασιλικός *s.m.* basil.
βασίλισσα *s.f.* queen.
βασιλοκτόν|ος *s.m.* regicide (*person*).~*ία s.f.* regicide (*act*).
βασιλομήτωρ *s.f.* queen mother.
βασιλόπαις *s.m.f.* king's son *or* daughter.
βασιλόπιττα *s.f.* New Year's cake.
βασιλό|πουλο *s.n.* prince. ~*πούλα s.f.* princess.
βασιλόφρων *a.* royalist.
βάσιμος *a.* well-founded, sound.
βάσ|η *s.f.* base, basis, foundation; pass-mark; *δίνω* ~*η σε* attach importance to.
βασκαίνω *v.t. & i.* cast evil eye (upon).
βάσκα|μα *s.n.,* ~*νία s.f.* evil eye.
βάσκανος *a.* jealous; ~ *οφθαλμός* evil eye.
βαστάζω *v.t.* carry, support.
βαστ|ώ *v.t. & i.* carry, support; hold, control; keep, have charge of; (*v.i.*) last; hold out, endure; originate (*from*); dare; *δεν του* ~*άει* he does not dare. ~*ιέμαι v.i.* control oneself; maintain one's position *or* condition; ~*ιέμαι καλά* be well *or* well-off.
βατ *s.n.* watt.
βάτα *s.f.* wadding.
βατεύω *v.t.* cover (*of animals*).
βατόμουρο *s.n.* blackberry.
βάτ|ος *s.m.f.,* ~*ο s.n.* bramble.
βατός *a.* passable, fordable.
βάτραχος *s.m.* frog, toad.
βατσίνα *s.f.* vaccine.
βαυκαλίζω *v.t.* rock *or* lull to sleep; (*fig.*)

lull (*with false hopes*).
βαφείο *s.n.* dyers'.
βαφή *s.f.* tempering; dyeing; dye; shoe-polish.
βαφτ- *see* **βαπτ-.**
βάφτ|ιση *s.f.,* ~*ίσια s.n.pl.* christening.
βαφτισιμ|ός *s.m.* godson ~*ά s.f.* god-daughter.
βαφτιστήρι *s.n.* godson.
βάφ|ω *v.t. see* **βάπτω.** ~*ομαι v.i.* make up.
βάψιμο *s.n.* tempering; dyeing; painting; make-up.
βγάζω *v.t.* 1. take, bring, get *or* draw out; take *or* get off, remove, extract; dismiss (*from job*); deduct; ~ *το καπέλο μου* take off one's hat; *έβγαλα ένα δόντι* I had a tooth out; ~ *το φίδι από την τρύπα* pull the chestnuts out of the fire. 2. give forth, produce; ~ *δόντια* cut one's teeth; ~ *σπυριά* come out in spots; *η Ελλάδα βγάζει κρασί* Greece produces wine; *τι σας έβγαλαν;* what refreshment did they offer you? ~ *λεφτά* make money; ~ *λόγο* make a speech; ~ *όνομα* become known; ~ *γλώσσα* answer cheekily; *τον έβγαλαν βουλευτή* he was returned as deputy; *τον έβγαλαν ψεύτη* he was proved a liar; *τον έβγαλαν Κώστα* they named him Costa. 3. work out, make out; *τα έβγαλα σωστά;* did I work it out right? *δεν* ~ *τη γραφή σου* I cannot make out your writing; *τα* ~ *πέρα* manage, succeed. (*v.i.*) *πού βγάζει ο δρόμος;* where does the road lead?
βγαίν|ω *v.i.* go *or* come out; appear, be produced; come off *or* out (*become detached*); diverge, deviate; turn out (*good or bad*); work *or* come out (*of sum, patience game, etc.*); result (*from*); ~*ω δήμαρχος* be elected mayor; ~*ω από το γυμνάσιο* finish one's schooling; *δεν του* ~*ει κανείς στα χαρτιά* none can beat him at cards; *δεν* ~*ει το ύφασμα* the material is not enough; *μου βγήκε σε καλό* it turned out well for me; *βγήκε λάδι* he got off scot free.
βγαλμένος *a.* taken off *or* out; graduated; returned (*of candidate*).
βγάλσιμο *s.n.* extraction, removal; taking *or* getting out *or* off; dislocation.
βδέλλα *s.f.* leech.
βδέλυγμα *s.n.* object of disgust, abomination.
βδελυρός *a.* abominable.
βδομάδα *s.f.* week.
βέβαι|ος *a.* sure, certain. ~*ότητα s.f.* certainty. ~*ως adv.* certainly.
βεβαι|ώνω *v.t.* assure; confirm, affirm. ~*ωση s.f.* assurance; confirmation; certificate. ~*ωτικός a.* confirmatory; affirmative.

βεθαρυμμένος *a.* subject to liabilities; blemished (*record*); heavy (*conscience*).

βέθηλ|ος *a.* profane; sacrilegious. ~ώνω *v.t.* desecrate, sully.

βεθιασμένος *a.* forced, unnatural.

θεγγέρα *s.f.* evening visit.

θεδούρα *s.f.* wooden bucket (*for yogurt*).

θεζίρης *s.m.* vizier.

θέης *s.m.* bey.

θελάδα *s.f.* frock-coat.

θελάζω *v.i.* bleat.

θελανίδι *s.n.* acorn.

Βέλγος *s.m.* a Belgian.

θελέντζα *s.f.* rough blanket.

θεληνεκές *s.n.* range (*of gun*).

θέλο *s.n.* veil.

θελόν|α ~η *s.f.*, ~ι *s.n.* needle, bodkin.

θελονάκι *s.n.* crochet-needle.

θελονιά *s.f.* stitch.

θελονισμός *s.m.* acupuncture.

θέλος *s.n.* arrow, shaft.

θελούδο *s.n.* velvet.

βέλτιστος *a.* best.

βελτί|ων *a.* better. ~ώνω *v.t.* improve. ~ωση *s.f.* improvement.

Βενετία *s.f.* Venice.

θενζινάκατος *s.f.* motor-boat.

θενζίνη *s.f.* petrol, benzine.

θενζινόπλοιο *s.n.* motor-assisted caique.

θεντά|για, ~λια *s.f.* fan.

θεντέτ(τ)α *s.f.* 1. vendetta. 2. star (*actor*).

θεντούζα *s.f.* cupping-glass.

θέρα *s.f.* wedding *or* engagement ring.

θεράντα *s.f.* veranda.

θερβελιά *s.f.* dropping of sheep, rabbits, *etc.*

θερβερίτσα *s.f.* squirrel.

θέργυα *s.f.* stick, rod, (*curtain*) rail.

θερεσέ *adv.* (*fam.*) on credit, on tick; τ'ακούω ~ take it with a grain of salt.

θερίκοκ(κ)ο *s.n.* apricot.

θερμούτ *s.n.* vermouth. [It. *vermuth*].

θερνίκι *s.n.* varnish, polish; (*fig.*) veneer.

θέρος *a.* real, genuine.

θετεράνος *s.m.* veteran.

θέτο *s.n.* veto; (*fam.*) έχω το ~ have the final say.

θήμα *s.n.* step, pace; rostrum; Άγιον~ (*eccl.*) sanctuary.

θηματίζω *v.i.* pace (*up and down*).

θηματοδότης *s.m.* (*med.*) pacemaker.

θήξ, θήχας *s.m.* cough.

θήτα *s.n.* the letter Β.

θήχω *v.i.* cough.

θία *s.f.* force, violence; haste, hurry; μετά ~ς with difficulty.

θιάζ|ω *v.t.* force; rape; press, urge. ~ομαι *v.i.* be in a hurry.

θιαιοπραγία *s.f.* (*bodily*) assault.

θίαι|ος *a.* violent, forcible. ~ότητα *s.f.* violence; ~ότητες (*pl.*) acts of violence.

θιαίως *adv.* violently, forcibly.

θιάσ|η, ~ύνη *s.f.* haste.

θιασμός *s.m.* rape.

θιαστής *s.m.* ravisher.

θιαστικ|ός *a.* in a hurry; pressing, urgent. ~ά *adv.* hurriedly.

θιβλιάριο *s.n.* card, book (*insurance, bank, etc.*).

θιβλικός *a.* biblical.

θιβλιογραφία *s.f.* bibliography.

θιβλιοδέτης *s.m.* bookbinder.

θιβλιοθηκάριος *s.m.* librarian.

θιβλιοθήκη *s.f.* bookcase; library.

θιβλιοκάπηλος *s.m.* publisher of pirated editions.

θιβλιοκρισία *s.f.* book-review.

θιβλίο *s.n.* book.

θιβλιοπώλ|ης *s.m.* bookseller. ~είο *s.n.* bookshop.

θίβλος *s.f.* Bible; χυανή ~ blue book.

θίγλα *s.f.* look-out post.

θιγλίζω *v.t. & i.* keep a look-out (for).

θίδα *s.f.* screw; (*fam.*) είναι *or* έχει ~ he is potty.

θιδάνιο *s.n.* remains of drink in glass.

θιδέλο *s.n.* calf; veal; calfskin.

θιδολόγος *s.m.* screwdriver.

θιδώνω *v.t. & i.* screw.

θίζα *s.f.* visa.

θίζιτα *s.f.* visit; visitor.

θιζόν *s.n.* mink (*fur*).

θίλλα *s.f.* villa.

θίντσι *s.n.* winch.

θιογραφία *s.f.* biography.

θιόλα *s.f.* 1. stock (*plant*). 2. (*mus.*) viola.

θιολέτα *s.f.* violet.

θιολί *s.n.* violin; (*fam.*) (same old) tune *or* game.

θιολι|στής, ~τζής *s.m.* violinist.

θιολόγ|ος *s.m.* biologist. ~ία *s.f.* biology.

θιομηχαν|ία *s.f.* industry, manufacture. ~ικός *a.* industrial.

θιομήχανος *s.m.* industrialist, manufacturer.

θιοπαλαιστής *s.m.* one who struggles to earn a living.

θιοπάλη *s.f.* struggle to earn a living.

θιοπορισμός *s.m.* livelihood.

θιοποριστικός *a.* providing a living.

θί|ος *s.m.* life; διά ~ου for life; (*fam.*) ~ος και πολιτεία one who has had a chequered career.

θιός *s.m. & n.*, ~ό *s.n.* wealth, fortune.

θιοτεχνία *s.f.* light manufacture.

θιοτικ|ός *a.* of living, ~όν επίπεδον standard of living.

θιοχημεία *s.f.* biochemistry.

θίρα *int.* (*naut.*) haul! hoist!
θιταμίνη *s.f.* vitamin.
θιτρίνα *s.f.* shop-window; show-case, cabinet; window-dressing.
θίτσα *s.f.* cane, switch.
θίτσιο *s.n.* bad habit, vice.
θιώ *v.i.* live.
θιώσιμος *a.* viable; livable.
θλαθερός *a.* harmful, injurious, unwholesome.
θλάθη *s.f.* harm, damage, detriment; breakdown.
θλάκας *s.m.* idiot, fool.
θλακεία *s.f.* stupidity; nonsense.
θλακώδης *a.* idiotic, stupid.
θλαμμένος *a.* (*fam.*) cranky.
θλαξ *s.m.f.* idiot, fool.
θλάπτω *v.i. & i.* harm, hurt, injure; wrong.
θλαστάνω *v.i.* sprout, shoot, grow.
θλαστάρι *s.n.* young shoot; scion.
θλαστημ- *see* θλασφημ-.
θλαστήμια *s.f.* curse, oath.
θλάστηση *s.f.* germination; vegetation.
θλαστός *s.m.* shoot; offspring.
θλασφημ|ία *s.f.* blasphemy; cursing. ~ώ *v.t. & i.* blaspheme; curse.
θλάφτω *v.t. & i. see* θλάπτω.
Βλαχία *s.f.* Wallachia.
θλάχικος *a.* Vlach; (*fig.*) uncouth.
Βλάχος *s.m.* a Vlach; (*fig.*) bumpkin.
θλέμμα *s.n.* look, glance, expression.
θλέννα *s.f.* mucus.
θλεννόρροια *s.f.* gonorrhoea.
θλεννώδης *a.* mucous.
θλέπ|ω *v.t. & i.* see; look (after); ~ω και παθαίνω have a hard job (*to*); ~οντας και κάνοντας as the case may be; δε ~ω την ώρα να φύγω I am impatient to leave. *See also* είδα.
θλεφαρίδα *s.f.* eyelash.
θλέφαρο *s.n.* eyelid.
θλέψεις *s.f. pl.* aspirations, designs.
θλήμα *s.n.* missile, projectile.
θλητική *s.f.* ballistics.
θλίτα *s.n.pl.* notchweed.
θλογιά *s.f.* smallpox; disease of vine.
θλογιοκομμένος *a.* pockmarked.
θλογώ *v.t. see* ευλογώ.
θλοσυρός *a.* fierce (*of looks*).
θόας *s.m.* boa (*snake*).
θογγ|ητό *s.n.*, ~ος *s.m.* groan(ing), moan(ing).
θογγώ *v.i.* groan, moan.
θόδι *s.n.* ox; ~α (*pl.*) oxen, cattle.
θοδινό *s.n.* beef.
θόειος *a.* bovine.
θοή *s.f.* confused noise; buzzing, humming, roaring.

θοήθεια *s.f.* aid, assistance, help (*abstract*).
θοήθημα *s.n.* help, aid (*concrete*); reference book.
θοηθητικός *a.* ancillary, auxiliary; favourable (*wind*).
θοηθός *s.m.* help (er), assistant.
θοηθώ *v.t. & i.* help, assist.
θόθρος *s.m.* cesspool.
θόιδι *s.n. see* θόδι.
θολά *s.f.* time (*occasion*).
θολάν *s.n.* 1. steering-wheel. 2. flounce. [F. *volant*]
θολθ|ός *s.m.* bulb; sort of onion; eyeball. ~οειδής *a.* bulbous.
θολεί *v.i.* it is convenient.
θολετός *a.* convenient, feasible.
θολεύ|ω *v.t.* fit in; cope *or* manage with; suit, be convenient to; τα ~ω get along, make do. ~ομαι *v.i. & t.* be comfortable, get settled *or* fixed up; find convenient; make shift.
θολή *s.f.* 1. throwing; firing, fire, shot, 2. comfort, convenience; knack.
θόλι *s.n.* bullet.
θολίδα *s.f.* bullet; meteor; (*naut.*) lead.
θολιδοσκοπώ *v.t.* sound, fathom.
θολικός *a.* convenient, handy; easy to get on with *or* to please.
θόλλεϋ, θόλεϊ *s.n.* volley-ball.
θολτ *s.n.* volt.
θόλτα *s.f.* turn, revolution; stroll, ride; (*fam.*) τα φέρνω ~ manage, make do; evade the issue; τον έφερα ~ I got round him.
θόμθα *s.f.* bomb.
θομθαρδίζω *v.t.* bombard; bomb, shell.
θομθαρδισμός *s.m.* bombardment, bombing.
θομθαρδιστικ|ός *a.* of bombardment. ~όν *s.n.* (*aero.*) bomber.
θομθητής *s.m.* buzzer.
θόμθος *s.m.* buzzing, droning, humming.
θομθυκοτροφία *s.f.* silkworm breeding.
θόμθυξ *s.m.* silkworm cocoon.
θομθώ *v.i.* buzz, drone, hum.
θορά *s.f.* prey, food.
θόρθορος *s.m.* mire.
θορειοανατολικός *a.* north-east.
θορειοδυτικός *a.* north-west.
θόρειος *a.* north, northern, northerly.
θορείως *adv.* northwards, to the north (of).
θοριάς *s.m.* north (wind).
θορινός *a. see* θόρειος.
θορράς *s.m.* north (wind).
θοσκή *s.f.* pasture.
θόσκημα *s.n.* pasturing; pasture-animal.
θοσκοπούλα *s.f.* shepherdess.
θοσκός *s.m.* shepherd.
θοσκοτόπι *s.n.* pasturage.

βόσκω *v.i. & t.* graze, browse; (*fig.*) wander aimlessly; (*v.t.*) pasture; feed on (*of animals*).

βόστρυχος *s.m.* curl.

βοτάν|η *s.f.* ~ι *s.n.* herb, simple.

βοτανίζω *v.t. & i.* weed.

βοτανικ|ός *a.* botanic(al). ~ή *s.f.* botany.

βότανο *s.n.* herb, simple.

βοτανολόγος *s.m.* botanist.

βότρυς *s.m.* bunch of grapes.

βότσαλο *s.n.* pebble.

βουβαίνω *v.t.* make dumb.

βουβ|άλι *s.n.*, **~αλος** *s.m.* buffalo.

βουβαμάρα *s.f.* dumbness.

βουβός *a.* dumb; silent (*film*).

βουβών *s.m.* groin; (*med.*) bubo. **~ικός** *a.* inguinal; bubonic.

βουδιστής *s.m.* Buddhist.

βο(υ)ερός *a.* noisy.

βουή *s.f.*, **~τό** *s.n.* see **βοή.**

βουίζω *v.i.* make a confused noise; buzz, hum, roar.

βούκα *s.f.* mouthful.

βουκέντρ|α *s.f.* ~ι *s.n.*, ~ο, *s.n.* goad.

βουκόλ|ος *s.m.* cowherd. **~ικός** *a.* bucolic.

βούλα *s.f.* seal; spot, dot; papal bull.

βούλευμα *s.n.* (*law*) judicial decision on whether case shall proceed for trial.

βουλεύομαι *v.i.* deliberate.

βουλευτήριον *s.n.* parliament house.

βουλευτής *s.m.* member of Parliament, deputy.

βουλευτικός *a.* parliamentary.

βουλή *s.f.* will, intent; Parliament, Chamber.

βούληση *s.f.* volition, will-power.

βούλιαγμα *s.n.* sinking; collapse.

βουλιάζω *v.t & i.* sink; bring down, bash in; (*fig.*) ruin; (*v.i.*) sink; collapse; (*fig.*) be ruined.

βουλιμία *s.f.* voracity.

βουλκανιζατέρ *s.n.* puncture repair-shop.

βουλλ- *see* **βουλ-.**

βουλοκέρι *s.n.* sealing-wax.

βούλωμα *s.n.* sealing; stopping up, blockage; stopper, cork, plug; filling (*of tooth*).

βουλώνω *v.t. & i.* seal; stop up, block; close, plug, fasten down; fill (*tooth*); brand (*animals*); (*v.i.*) become stopped up.

βούλ|ομαι *v.i.* wish; (*fam.*) *του ~ήθηκε νά* he took it into his head to.

βουλώ *v.t. & i. see* **βουλιάζω.**

βουνήσιος *a.* of the mountains.

βουνό *s.n.* mountain.

βούρδουλας *s.m.* whip.

βούρκος *s.m.* slime, mire, filth.

βουρκώνω *v.i.* brim with tears; threaten rain; get muddy.

βουρλίζ|ω *v.t.* drive mad. **~ομαι** *v.i.* become furious; desire madly.

βούρλο *s.n.* bulrush.

βούρτσ|α *s.f.* brush. **~ίζω** *v.t.* brush.

βους *s.m.f.* ox; *βόες* (*pl.*) oxen, cattle.

βουστάσιον *s.n.* byre; milking-shed.

βούτηγμα *s.n.* dipping, plunging; (*fam.*) pilfering.

βούτημα *s.n.* hard biscuit, *etc.*, for dipping in drink.

βουτηχτής *s.m.* diver; (*fam.*) pilferer.

βουτιά *s.f.* dive, plunge; (*fam.*) pilfering.

βουτσί *s.n.* cask.

βουτυράτος *a.* of *or* like butter.

βούτυρο *s.n.* butter.

βουτ|ώ *v.t. & i.* dip, plunge; (*fam.*) lay hands on, pilfer, pinch; (*v.i.*) dive; *~ηγμένος* covered, loaded (*with*).

βόχα *s.f.* stink.

βοώ *v.t. & i.* shout, cry aloud (for); *see also* **βουΐζω.**

βραβείο *s.n.* prize, award.

βραβεύω *s.n.* reward, give prize to.

βράγχια *s.n.pl.* gills.

βραδάκι *s.n.* early evening.

βραδέως *adv.* slowly.

βραδιά *s.f.* evening, night.

βραδιάζ|ει *v.i.* evening falls. *~ομαι v.i.* be overtaken by nightfall.

βραδιν|ός *a.* evening. **~ό** *s.n.* evening. (meal).

βράδυ 1. *s.n.* evening. 2. *adv.* in the evening.

βραδυ- *denotes* slowness.

βραδυκίνητος *a.* slow-moving.

βραδύνω *v.i.* be slow *or* late.

βραδυπορώ *v.i.* go slowly, lag.

βραδύς *a.* slow.

βραδύτητα *s.f.* slowness, delay.

βράζ|ω *v.i. & i.* boil; seethe, teem; ferment; wheeze; *~ει το αίμα του* he is full of vitality; (*fam.*) *να σε βράσω* expression of vexed disappointment.

βράκα *s.f.* breeches; (*fam.*) trousers, underpants.

βρακί *s.n.* (*fam.*) underpants.

βράση *s.f.* seething, boiling, fermentation.

βράσιμο *s.n.* boiling; wheezing.

βρασμός *s.m.* boiling.

βραστός *a.* boiled.

βραχιόλι *s.n.* bracelet.

βραχίων *s.m.* arm.

βράχν|α, **~άδα** *s.f.* hoarseness.

βραχνάς *s.m.* nightmare.

βραχνιάζω *v.t. & i.* make *or* become hoarse.

βραχνός *a.* hoarse.

βράχος *s.m.* rock.

βραχυ- *denotes* shortness.

βραχυκύκλωμα *s.n.* short circuit.

βραχύνω *v.t.* shorten.

βραχυπρόθεσμος *a.* (*fin.*) short-dated.
βραχ|ύς *a.* short, **~ύτητα** *s.f.* shortness, brevity.
βραχώδης *a.* rocky.
βρε *int* (*fam.*) unceremonious mode of address or cry of surprise, impatience, etc.
βρε(γ)μένος *a.* wet, damp.
βρέξιμο *s.n.* making wet or damp.
βρεσ|ίδι, ~ιμο *s.n.pl.* thing found.
βρετίκ|ια, ά *s.n.pl.* reward for finding.
Βρετταν|ία *s.f.* Britain. **~ικός** *a.* British. **~ός** *s.m.* Briton.
βρεφοκομείο *s.n.* foundling hospital.
βρέφος *s.n.* infant.
βρέχ|ω *v.t. & i.* wet, moisten; (*v.i.*) **~ει** it rains.
βρίζα *s.f.* rye.
βρίζω *v.t. & i.* abuse, revile; swear (at).
βρίθω *v.i.* (*with gen.*) teem (*with*).
βρισι|ά *s.f.* oath, term of abuse; **~ές** (*pl.*) abuse.
βρισίδι *s.n.* string of oaths.
βρίσκ|ω *v.t.* find; hit, strike; obtain; consider, judge. **~ομαι** *v.i.* be, find oneself.
βρογχίτιδα *s.f.* bronchitis.
βρόγχος *s.m.* bronchial tube.
βροντερός *a.* thunderous.
βροντή *s.f.* thunder.
βρόντ|ος *s.m.* loud noise; heavy fall; *στο* **~ο** in vain.
βροντ|ώ *v.i. & t.* thunder, boom, bang; (*v.t.*) bang on; fell, floor; (*fam.*) **~ώ** *κανόνι* go bankrupt; *τα* **~ηξε** *κάτω* he abandoned the job. **~οκοπώ** *v.i.* bang, thump. **~ολογώ** *v.i.* thunder, boom, reverberate.
βροντώδης *a.* thunderous.
βροτός *a.* mortal.
βροχερός *a.* rainy.
βροχή *s.f.* rain, shower.
βρόχι *s.n.* net, snare.
βρόχιν|ος *a.* **~ο** *νερό* rainwater.
βροχοπτώσεις *s.f.pl.* rainfall.
βρόχος *s.m.* noose.
βρυκόλακας *s.m.* vampire.
βρύο *s.n.* moss.
βρύσ|η, ~ις *s.f.* fountain; tap.
βρυχηθμός *s.m.* roar.
βρυχώμαι *v.i.* roar.
βρύω *v.i.* teem (*with*).
βρώμα *s.n.* food.
βρώμα *s.f.* stink; filth; (*fig.*) (*fam.*) bitch, swine.
βρωμερός *a.* dirty; ill-smelling.
βρώμη *s.f.* oats.
βρωμιά *s.f.* filth; (*fig.*) dirty business.
βρωμίζω *v.t. & i.* make or get dirty.
βρώμικος *a.* dirty.
βρώμιος *a.* stinking, gone bad.
βρωμο- *denotes* dirtiness.

βρωμώ *v.i.* 1. stink. 2. **~** *από* be glutted with.
βρώσιμος *a.* eatable.
βύζαγμα *s.n.* suckling; sucking.
βυζαίνω *v.t. & i.* suckle; suck; (*fam.*) milk (*person*).
βυζανιάρικο *s.n.* suckling (*child*).
βυζαντινός *a.* Byzantine.
βυζί *s.n.* breast; nipple (*of gun*).
βυθίζ|ω *v.t.* sink, plunge. **~ομαι** *v.i.* sink; (*fig.*) be abstracted or lethargic.
βύθιση *s.f.* lethargy.
βύθισις *s.f.* sinking.
βύθισμα *s.n.* (*naut.*) draught.
βυθοκόρος *s.f.* dredger.
βυθομετρήσεις *s.f.pl.* soundings.
βυθός *s.m.* bottom, bed (*of sea, etc.*).
βύνη *s.f.* malt.
βύρσα *s.f.* dressed hide.
βυρσοδεψείο *s.n.* tannery.
βύσμα *s.n.* plug, wad.
βύσσιν|ο *s.n.* morello cherry. **~ής** *a.* dark-red.
βυσσοδομώ *v.t. & i.* machinate.
βυτίο *s.n.* barrel, drum; water-cart; tanker.
βωβός *a.* dumb.
βώδι *s.n. see* **βόδι.**
βωλοδέρνω *v.i.* break soil; (*fig.*) struggle, wrestle (*with task*).
βωλοκοπώ *v.t.* harrow.
βώλ|ος *s.m.* clod, lump; **~οι** (*pl.*) marbles (*game*).
βωμολοχία *s.f.* scurrility.
βωμός *s.m.* altar.
βώτριδα *s.f.* moth.

Γ

γαβάθα *s.f.* wooden or earthen bowl.
γαβγίζω *v.t* bark.
γάγγλιον *s.n.* ganglion.
γάβγισμα *s.n.* bark (ing).
γάγγραινα *s.f.* gangrene, canker.
γάζα *s.f.* gauze.
γαζέτα *s.f.* small change.
γαζί *s.n.* stitch, stitching.
γαζία *s.f.* popinac (*sort of acacia*).
γαζώνω *v.t.* stitch (*by machine*); rake (*with fire*).
γαί|α *s.f.* earth; **~αι** (*pl.*) land (*cultivable*).
γαιάνθρακας *s.m.* pit-coal.
γάιδαρος *s.m.* ass, donkey; (*fig.*) boor.
γαϊδουράγκαθο *s.n.* thistle.
γαϊδούρι *s.n. see* **γάιδαρος.**
γαϊδουριά *s.f.* boorishness; piece of rudeness.
γαϊδουροκαλόκαιρο *s.n.* (*fam.*) S. Martin's summer.

γαιοκτήμων *s.m.* owner of cultivable land.
γαϊτανάκι *s.n.* maypole-dance.
γαϊτάνι *s.n.* braid.
γάλα *s.n.* milk; *του ~χτος* suckling (*pig,
etc.*), milk (*chocolate*).
γαλάζιος *a.* blue, azure.
γαλαζοαίματος *a.* blue-blooded.
γαλαζόπετρα *s.f.* turquoise; copper sulphate.
γαλακτερός *a.* made from milk.
γαλακτοκομία *s.f.* dairy-farming.
γαλακτομπούρεκο *s.n.* sort of cake.
γαλάκτωμα *s.n.* emulsion.
γαλανόλευκος *a.* see **κυανόλευκος.**
γαλανός *a.* blue; blue-eyed.
γαλαντόμος *s.m.* generous man.
γαλαξίας *s.m.* Milky Way; milk-tooth.
γαλαρία *s.f.* gallery (*of theatre, mine*).
γαλατάς *s.m.* milkman.
γαλβανίζω *v.t.* galvanize; (*fig.*) rouse to
enthusiasm.
γαλέος *s.m.* lamprey.
γαλέτα *s.f.* hard tack; breadcrumbs.
γαλή *s.f.* cat.
γαλήν|η *s.f.* calm, tranquillity. **~εύω** *v.t. & i.*
make *or* grow calm. **~ιος** *a.* calm.
γαλιάνδρα *s.f.* lark (*bird*).
γαλιφιά *s.f.* flattery.
γαλλικός *a.* French.
γαλλόπουλο *s.n.* young turkey.
γάλλος *s.m.* turkey.
Γάλλος *s.m.* Frenchman.
γαλόνι *s.n.* 1. braid; (*mil.*) stripe. 2. gallon
(-jar).
γαλότσα *s.f.* galosh; clog.
γαλουχώ *v.t.* suckle; (*fig.*) rear.
γαμ|βρός, ~πρός *s.m.* son-in-law; brother-
in-law (*sister's husband*); bridegroom;
(*fig.*) marriageable man.
γαμήλιος *a.* nuptial, wedding.
γάμμα *s.n.* the letter **Γ.**
γάμος *s.m.* marriage, wedding.
γάμπα *s.f.* calf, leg.
γαμψός *a.* hooked.
γανί|λα *s.f.* verdigris; fur. **~άζω** *v.i.* get
furred; feel dry.
γάντζος *s.m.* hook, grapple.
γαντζώνω *v.t.* hook, catch.
γάντι *s.n.* glove.
γανωματάς *s.m.* tinker.
γανώνω *v.t.* plate with tin.
γανωματής, ~τζής, ~ματάς *s.m.* tinker.
γαργαλ|εύω, ~ίζω, ~ώ *v.t.* tickle; (*fig.*)
tempt. **~ιέμαι** *v.i.* be ticklish.
γαργαλιστικός *a.* tempting.
γαργάρα *s.f.* gargle, gargling.
γάργαρος *a.* purling; limpid.
γαρίδα *s.f.* shrimp, prawn.
γαρμπής *s.m.* south-west wind.

γάρμπος *s.n.* line, (*good*) shape.
γαρνίρω *v.t.* garnish, trim.
γαρνιτούρα *s.f.* garnish, trimming.
γαρ(ο)ύφαλλο *s.n.* carnation; clove.
γαστήρ *s.f.* belly.
γάστρα *s.f.* flower-pot.
γαστρικός *a.* gastric.
γαστρονομία *s.f.* gastronomy.
γάτ|α *s.f.* **~ί** *s.n.* **~ος** *s.m.* cat.
γατάκι *s.n.* kitten.
γαυγίζω *v.i.* bark.
γαυριώ *v.i.* exult.
γαύρος *s.m.* sort of whitebait.
γδάρσιμο *s.n.* skinning, barking.
γδέρνω *v.t.* skin, flay; dark; graze; (*fig.*)
fleece.
γδούπος *s.m.* thud.
γδυμνός *a.* see **γυμνός.**
γδύν|ω *v.t.* undress; (*fig.*) rob. **~ομαι** *v.i.* get
undressed.
γδύσιμο *s.n.* undressing; (*fig.*) robbing.
γδυτός *a.* undressed.
γεγονός *s.n.* event; fact.
γεια *s.f.* health; *~ σας or ~ χαρά* greeting; *με
~* good wish to one having something new.
αφήνω ~ bid farewell.
γείσο *s.n.* eaves; peak (*of cap*).
γειτ|νιάζω, ~ονεύω *v.i.* be near together *or*
adjoining.
γειτονιά *s.f.* neighbourhood.
γειτονικός *a.* neighbouring.
γείτ|ων, ~ονας *s.m.* neighbour.
γελάδα *s.f.* cow.
γέλασμα *s.n.* deceit; laughing-stock.
γελαστός *a.* smiling, cheerful.
γελέκι *s.n.* waistcoat.
γέλιο *s.n.* laugh, laughter.
γελοιογραφία *s.f.* caricature; cartoon.
γελοιοποι|ώ *v.t.* make ridiculous. **~ούμαι**
v.i. make oneself ridiculous.
γελοίος *a.* ridiculous.
γελ|ώ *v.i. & t.* laugh; *(with με)* laugh at; cheat,
take in. **~ιέμαι** *v.i.* be deceived *or*
mistaken.
γέλως *s.m.* laugh, laughter.
γελωτοποιός *s.m.* jester.
γέμα *s.n.* (*fam.*) meal; luncheon.
γεμάτος *a.* full, covered (*with*); loaded;
(*fig.*) thick (*in texture*), well-nourished (*of
figure*).
γεμίζω *v.t. & i.* fill, stuff, cover (*with*); load
(*gun*); (*v.i*) become full *or* covered (*with*);
fill out.
γέμιση *s.f.* stuffing, filling (*of bird, cushion,
etc.*); waxing (*of moon*).
γεμιστός *a.* stuffed (*of food*).
γέμω *v.i.* (*with gen.*) be full (*of*).
Γενάρης *s.m.* January.

γεν|εά, ~ιά s.f. race, generation.
γενέθλια s.n.pl. birthday.
γενειάδα s.f. long beard.
γένεσις s.f. genesis, creation.
γενέτειρα s.f. birthplace.
γενετ|ή s.f. εκ ~ής from birth.
γενετήσιος a. generative, sexual.
γενετικός a. genetic.
γέν|ι s.n., ~εια (pl.) beard; αφήνω ~εια grow a beard.
γενικεύω v.t. & i. make general, spread; throw open (discussion); (v.i.) generalize.
γενική s.f. (gram.) genetive.
γενικός a. general.
γενίτσαρος s.m. janissary.
γέννα s.f. childbirth, confinement; (fig.) breed, spawn.
γενναιόδωρος a. generous, liberal.
γενναί|ος a. brave; ~ο δώρο generous gift.
γενναιόφρων a. magnanimous, liberal.
γενναιόψυχος a. brave.
γέννημα s.n. offspring, product; ~ και θρέμμα born and bred; ~τα (pl.) cereals.
γέννηση s.f. birth.
γεννητικός a. genital.
γεννητικότης s.f. birth-rate.
γεννητούρια s.n.pl. (fam.) birth (as social event).
γεννήτρια s.f. (electric) generator.
γεννήτ|ωρ s.m. father; ~ορες (pl.) parents.
γενν|ώ v.t. & i. give birth (to); bear; lay eggs; (fig.) cause, engender. ~ιέμαι v.i. be born.
γενοκτονία s.f. genocide.
γέν|ος s.n. race, tribe; genus; gender; εν ~ει in general.
γερά adv. strongly, hard, steadily.
γεράκι s.n. hawk.
γεράματα s.n.pl. old age.
γεράνι s.n. 1. geranium. 2. sweep (of a well).
γερανός s.m. crane (bird, machine).
γέρας s.n. prize, trophy.
γερατειά s.n.pl. old age.
γέρικος a. old.
γέρνω v.t. & i. bend down, tilt; half close (door); (v.i.) lean or bend over, stoop; lie down; sink (of star).
γερνώ v.t. & i. make or grow old.
γεροδεμένος a. sturdy, strongly built.
γέροντας s.m. old man.
γεροντοκόρη s.f. old maid.
γεροντοπαλήκαρο s.n. old bachelor.
γέρος s.m. old man.
γερός a. healthy, hale, strong, sturdy; whole, undamaged.
γερουσία s.f. senate; (fig.) old fogies.
γερτός a. leaning, bent, aslant.
γέρων s.m. old man.

γεύμα s.n. meal, luncheon. ~τίζω v.i. have lunch.
γεύομαι v.t. (with gen.) taste.
γεύση s.f. taste, flavour; tasting.
γευστικός a. savoury, tasty.
γέφ|υρα s.f., ~ύρι s.n. bridge.
γεω|γραφία s.f. geography. ~γράφος s.m. geographer.
γεω|λογία s.f. geology. ~λόγος s.m. geologist.
γεωμετρ|ία s.f. geometry. ~ικός a. geometric(al).
γεώμηλον s.n. potato.
γεωπονία s.f. (science of) agriculture.
γεωργ|ία s.f. farming; agriculture. ~ός s.m. farmer.
γεωργικός a. agricultural.
γη s.f. earth, ground, land.
γηγενής s.f. indigenous.
γήινος a. terrestrial.
γήπεδο s.n. piece of ground; sports ground.
γήρας s.n. old age.
γηράσκω v.t. & i. make or grow old.
γηρατειά s.f. old age.
γηροκομείον s.n. old people's home.
γητεύω v.t. weave spells over.
για 1. prep. (with acc.) for; because of; instead or on behalf of; about, concerning; γι' αυτό therefore; ~ καλά for certain, definitely; ~ την ώρα for the time being, up to now; ~ δέσιμο crazy; ~ πού whither? ~ το Θεό for God's sake; δεν άφησε βουνό ~ βουνό he left not a single mountain (i.e. unclimbed). 2. adv. as, for; πάει ~ βουλευτής he is standing for Parliament; περνάει ~ ωραία she passes for beautiful; το έχει ~ γούρι he thinks it a lucky omen. 3. conj. ~ να in order to; ~ να μην προσέχης έπεσε κάτω through not taking care you fell down. 4. hortatory particle (with imperative) ~ φαντάσου fancy that! ~ πες μου tell me.
γιαγιά s.f. grandmother.
γιαίνω v.i. heal (of wound).
γιακάς s.m. collar.
γιαλός s.m. sea-shore.
γιαννάκης s.m. (fam.) raw recruit.
γιάντες s.n. wager on pulling wishbone.
γιαούρτι s.n. yogurt.
γιαπί s.n. building or repairs in progress.
γιαρμάς s.m. yellow peach.
γιασεμί s.n. jasmine.
γιατί adv. why.
γιατί adv. because.
γιατρειά s.f. cure.
γιατρεύω v.t. cure, heal.
γιατρικό s.n. medicine.
γιατρός s.m. doctor.

γιατροσόφι *s.n.* nostrum.

γιάτσο *s.n.* (*fam.*) ice-cream.

γιαχνί *s.n.* sort of ragout.

γιγάντειος *a.* gigantic.

γίγας *s.m.* giant.

γιγνώσκω *v.t. see* γινώσκω.

γίδα *s.f.* goat.

γιδοβοσκός *s.m.* goatherd.

γιδοπρόβατα *s.n.pl.* sheep and goats.

γιλέκο *s.n.* waistcoat.

γινάτι *s.n.* obstinacy; spite.

γίν|ομαι *v.i.* become, be done *or* produced; be finished *or* ready; ripen; happen; *τί ~εσαι;* how are you? *τί θα ~ω;* what will become of me? *ο μη γένοιτο* God forbid! *το ~όμενον* (*math.*) product; *~ωμένος* ripe.

γινώσκω *v.t.* know.

γιόκας *s.m.* son (*endearing*).

γιομ- *see* **γεμ-**.

γιορτή *s.f. see* **εορτή**.

γιορτινός *a.* festive. *~ά* Sunday best.

γιος *s.m.* son.

γιουβαρλάκια *s.n.pl.* meat-and-rice balls.

γιουρούσι *s.n.* attack, invasion.

γιουχαΐζω *v.t. & i.* hoot, boo.

γιρλάντα *s.f.* garland.

γιωτ *s.n.* yacht. [D. *yacht*]

γιώτα *s.n.* the letter Ι. *~χής* *s.m.* (*fam.*) motorist.

γκ- *see also* **κ-**.

γκαβός *a.* cross-eyed; blind.

γκάγκαρος *s.m.* (*fam.*) true-born Athenian.

γκάζι *s.n.* gas; paraffin; (*fam.*) *πατώ ~* step on the gas.

γκαζιέρα *s.f.* cooking-stove (*gas or paraffin*).

γκαζόζα *s.f.* bottled lemonade.

γκάιντα *s.f.* bagpipe.

γκαμήλα *s.f.* camel.

γκαράζ *s.n.* garage. [F. *garage*].

γκαρδιακός *a.* cordial; bosom (*friend*).

γκαρίζω *v.i.* bray.

γκαρσόν(ι) *s.n.* waiter; bachelor. [F. *garçon*].

γκαρσονιέρα *s.f.* bachelor's apartment.

γκαστρώνω *v.t.* (*fam.*) get with child; (*fig.*) exasperate.

γκάφα *s.f.* blunder, faux pas.

γκελ *s.n.* bounce; (*fam.*) sex-appeal.

γκέμι *s.n.* bridle.

γκέτα *s.f.* gaiter, puttee.

γκιαούρης *s.m.* giaour.

γκινέα *s.f.* guinea.

γκίνια *s.f.* bad luck.

γκιόσα *s.f.* old she-goat.

γκιουβέτσι *s.n.* (vessel for) meat cooked with Italian pasta.

γκιώνης *s.m.* screech-owl.

γκλίτσα *s.f.* shepherd's crook.

γκορτσιά *s.f.* wild pear.

γκουβερνάντα *s.f.* governess.

γκρέκα *s.f.* meander (*pattern*).

γκρεμίζ|ω *v.t.* cast down (*from height*); demolish; overthrow. *~ομαι v.i.* fall (*from height*); collapse.

γκρέμισμα *s.n.* demolition, overthrow.

γκρεμ(ν)ός *s.m.* abyss, precipice, cliff.

γκρενά *s.n. & a.* colour of garnet. [F. *grenat.*].

γκρί(ζος) *a.* grey.

γκριμάτσα *s.f.* grimace.

γκρίνια *s.f.* nagging, whining.

γκρινιάζω *v.i.* grumble, nag; cry, whimper.

γκρινιάρης *a.* grumbling, nagging.

γλάρος *s.m.* seagull.

γλαρώνω *v.i.* close one's eyes in sleep, drowse; (*of eyes*) close.

γλάστρα *s.f.* flower-pot.

γλαυκός *a.* pale blue.

γλαύξ *s.f.* owl.

γλαφυρός *a.* elegant (*of language*).

γλειφιτζούρι *s.n.* lollipop.

γλείφ|ω *v.t.* lick. *~ομαι v.i.* lick one's lips; *η γάτα ~εται* the cat is washing itself.

γλείψιμο *s.n.* lick, licking.

γλεντζές *s.m.* pleasure-loving person, reveller; rake.

γλέντι *s.n.* merrymaking, party, beanfeast.

γλεντοκοπώ *v.i.* go in for riotous living.

γλεντώ *v.i. & t.* make merry, have a good time; (*v.t.*) dine and wine; enjoy; squander (*money*) on amusement.

γλεύκος *s.n.* must.

γληγ- *see* **γρηγ-**.

γλίνα *s.f.* grease (*left from cooking*).

γλιστερός *a.* slippery.

γλίστρημα *s.n.* slide, slip.

γλιστρώ *v.t. & i.* slip, slide, glide, be slippery; (*fig.*) slip away, escape.

γλίσχρος *a.* niggardly, mean.

γλί|τζα, ~τσα *s.f.* dirt, slime.

γλοιώδης *a.* viscous, sticky; (*fig.*) oily (*person*).

γλόμπος *s.m.* globe (*of light*).

γλουτός *s.m.* buttock.

γλύκ|α, ~άδα *s.f.* sweetness; *~ες* (*pl.*) caresses, tender looks.

γλυκάδ|ι *s.n.* 1. vinegar. 2. *~ια* (*pl.*) sweetbreads.

γλυκαίνω *v.t. & i.* make *or* become sweet *or* mild; alleviate.

γλυκανάλατος *a.* insipid.

γλυκάνισο *s.n.* aniseed.

γλυκατζ|ής *s.m.* *~ού* *s.f.* one with a sweet tooth.

γλυκερίνη *s.f.* glycerine.

γλύκισμα *s.n.* cake, pudding.

γλυκ|ό *s.n.* sweetmeat; preserve; ~ά (*pl.*) pastries, confectionery.
γλυκοαίματος *a.* likeable.
γλυκόζη *s.f.* glucose.
γλυκομίλητος *a.* soft-spoken.
γλυκόξινος *a.* sweetish-sour.
γλυκόπικρος *a.* bitter-sweet.
γλυκόπιοτος *a.* pleasant to drink *or* smoke.
γλυκοπύρουνος *a.* having edible kernel (*esp. apricot*).
γλυκόριζα *s.f.* liquorice.
γλυκ|ός, ~ύς *a.* sweet; mild; *του ~ού νερού* fresh-water, (*fig.*) incompetent; (*woman*) no better than she should be.
γλυκοσαλιάζω *v.i.* trifle flirtatiously.
γλυκοχάρα(γ)μα *s.n.* serene daybreak.
γλυκύτης *s.f.* sweetness; mildness.
γλύπτ|ης *s.m.* sculptor. ~ική *s.f.* (*art of*) sculpture.
γλυ(τ)σίνα *s.f.* wistaria.
γλυτ|ώνω *v.t. & i.* save, deliver, (*v.i*) escape, get away; *φτηνά τη ~ωσε* he got off cheaply.
γλυφίδα *s.f.* chisel.
γλυφός *a.* brackish.
γλύφω *v.t.* carve, sculpt.
γλώσσα *s.f.* 1. tongue; language. 2. sole (*fish*).
γλωσσεύω *v.i.* answer back impertinently.
γλωσσίδι *s.n.* tongue (*of bell, etc.*).
γλωσσικός *a.* lingual; linguistic.
γλωσσοδέτης *s.m.* tongue-twister. *τον έπιασε* ~ he was tongue-tied.
γλωσσοκοπάνα *s.f.* talkative and impertinent woman.
γλωσσολογία *s.f.* philology, linguistics.
γλωσσομαθής *s.m.* linguist.
γλωσσοτρώγω *v.t.* bring bad luck to (*by speaking enviously*).
γλωσσού *s.f.* saucily voluble woman.
γλωσσοφαγιά *s.f.* bringing of bad luck (*by envious talk*).
γνάθος *s.f.* jaw.
γνέθω *v.t.* spin.
γνέφω *v.i.* nod, make a sign.
γνέψιμο *s.n.* nod, sign.
γνήσιος *a.* genuine, true.
γνωματεύω *v.i.* deliver one's opinion.
γνώμη *s.f.* opinion, view.
γνωμικόν *s.n.* maxim, adage.
γνωμοδοτώ *v.i.* deliver one's opinion.
γνώμων *s.m.* T-square, level; (gnomon of) sundial; meter (*for gas, etc.*); (*fig.*) rule, principle.
γνωρίζω *v.t.* know; be *or* get acquainted with; recognize; make known, introduce; ~ *υμίν ότι* I inform you that.
γνωριμία *s.f.* acquaintance; *δεν έχει* ~ he is

unrecognizable.
γνώριμος *a.* known; an acquaintance.
γνώρισμα *s.n.* sign, mark (*abstract*).
γνώση *s.f.* good sense; *βάζω* ~ learn (better) by experience.
γνώσ|ις, ~η *s.f.* knowledge, cognizance; ~εις (*pl.*) knowledge, learning.
γνώστης *s.m.* expert, connoisseur.
γνωστικός *a.* sensible.
γνωστοποιώ *v.t.* notify, give notice of (*to someone*).
γνωστ|ός *a.* known; an acquaintance; *ως* ~όν as is known.
γόβ|α *s.f.* (*woman's*) court shoe. ~άκι *s.n.* slipper.
γογγύζω *v.i.* groan; grumble.
γοερός *a.* loud and poignant.
γόης *s.m.* charmer.
γόησσα *s.f.* enchantress, siren.
γοητεία *s.f.* charm.
γοητευτικός *a.* charming.
γοητεύω *v.t.* charm.
γόητρο *s.n.* prestige.
γομάρι *s.n.* load; donkey-load; donkey.
γόμα *s.f.* gum; india-rubber.
γομφίος *s.m.* molar.
γόμφος *s.m.* peg, pin; joint (*of bones*).
γόμωση *s.f.* loading *or* charge (*of gun*); stuffing.
γόνα *s.n.* (*fam.*) knee.
γονατίζω *v.i. & t.* (make to) kneel; (*fig.*) weigh *or* be weighed down.
γονατιστός *a.* on one's knees.
γόνατο *s.n.* knee.
γον|εύς *s.m.* father; ~είς (*pl.*) parents.
γονικός *a.* parental, inherited.
γονιμοποιώ *v.t.* fecundate.
γόνιμος *a.* fertile, prolific, fruitful.
γονιός *s.m. see* **γονεύς.**
γόνος *s.m.* spawn. (*fig.*) seed, offspring, scion.
γόνυ *s.n.* knee.
γονυκλισία *s.f.* genuflexion.
γονυπετής *a.* on one's knees.
γόος *s.m.* loud cry of pain *or* grief.
γόπα *s.f.* 1. sort of fish. 2. (*fam.*) cigarette-end.
γοργόνα *s.f.* mermaid; (*fig.*) beautiful woman.
γοργός *a.* quick.
γόρδιος *a.* ~ *δεσμός* Gordian knot.
γορίλλας *s.m.* gorilla.
γοτθικός *a.* Gothic.
γούβα *s.f.* cavity, pothole.
γουδ|ί *s.n.* mortar. ~οχέρι *s.n.* pestle.
γούλα *s.f.* gullet; (*fig.*) greed.
γουλί *s.n.* stalk (*of cabbage, etc.*); (*fam.*) bald.

γουλιά *s.f.* drop, mouthful, sip.
γούνα *s.f.* fur.
γουναρικό *s.n.* fur (*garment*).
γούνινος *a.* made of fur.
γούπατο *s.n.* hollow (*in ground*).
γουργουρ|ίζω *v.i.* gurgle, rumble. **~ητό** *s.n.* gurgle, rumble.
γούρι *s.n.* good luck.
γουρλ|ής, ~ίδικος, *a.* bringing good luck.
γουρλομάτης *a.* with bulging eyes.
γουρλώνω *v.t.* open (*eyes*) wide.
γουρλωτός *a.* bulging (*of eyes*).
γούρνα *s.f.* stone basin *or* trough (*for water*).
γουρούν|ι *s.n.* pig. **~α** *s.f.* sow. **~όπουλο** *s.n.* sucking-pig. **~ήσιος** *a.* piggish, pig's.
γουρσούζ|ης *s.m.* unlucky *or* peevish fellow. **~ιά** *s.f.* bad luck. **~ικος** *a.* bringing bad luck.
γουστάρ|ω *v.t.* & *i.* desire, feel like; δεν τον ~ω I do not care for him; ~ω να or μου ~ει να I enjoy.
γούστο *s.n.* taste (*discrimination*); δεν τον κάνω ~ I do not care for him; κάναμε ~ we had an amusing time; κάνω ~ μια μπίρα I feel like a beer; για ~ for fun; δεν έχει ~ there is no fun in that; έχει ~ να it would be a nice thing if (*ironic*).
γουστόζικος *a.* entertaining.
γοφός *s.m.* hip.
γραβάτα *s.f.* necktie.
γραβιέρα *s.f.* gruyère cheese.
γραία *s.f.* old woman.
γραίγος *s.m.* north-east wind.
γράμμα *s.n.* letter. **~τα** (*pl.*) reading and writing.
γραμμάριον *s.n.* gramme.
γραμματ|εύς *s.m.f.* secretary. **~εία** *s.f.* secretariat.
γραμματικ|ός *a.* grammatical; a grammarian; a clerk. **~ή** *s.f.* grammar.
γραμμάτιον *s.n.* bill of exchange.
γραμματοκιβώτιο *s.n.* letter-box.
γραμματοκομιστής *s.m.* postman.
γραμματολογία *s.f.* history of literature.
γραμματόσημο *s.n.* postage-stamp.
γραμμένος *a.* written; (*fig.*) fated; bold (*of features*).
γραμμή *s.f.* line; (*adv.*) straight, in turn; πρώτης ~ς first-class.
γραμμόφωνο *s.n.* gramophone.
γρανάζι *s.n.* gear (*of motor*).
γρανίτα *s.f.* water-ice.
γρανίτης *s.m.* granite.
γραπτ|ός *a.* written. **~ώς** *adv.* in writing.
γραπώνω *v.t.* grab.
γρασάρω *v.t.* grease.
γρασίδι *s.n.* grass.

γρατσουν|ίζω, ~ώ *v.t.* scratch.
γραφειοκρατία *s.f.* bureaucracy, red tape.
γραφείο *s.n.* desk; study; office, bureau.
γραφεύς *s.m.* clerk.
γραφ|ή *s.f.* writing; reading (*of MS.*); Γ ~αι (*pl.*) Scriptures; στο κάτω κάτω της ~ής when all is said and done.
γραφικός *a.* of writing *or* drawing, clerical; (*fig.*) picturesque, graphic.
γραφίς *s.f.* pen.
γραφομηχανή *s.f.* typewriter.
γραφτός *a.* written; (*fig.*) fated.
γράφω *v.t.* & *i.* write; enrol; record; settle (*property*).
γράψιμο *s.n.* writing.
γρήγορ|ος *a.* quick. **~α** *adv.* quickly. **~εύω** *v.i.* hasten.
γρηγορώ *v.i.* watch, be vigilant.
γριά *s.f.* old woman.
γριγρί *s.n.* fishing-boat with lantern.
γρίλ(λ)ι|α *s.f.* grille; slat (*of shutter*). κινητές ~ες louvre.
γριν- ** *see* **γκριν-.
γρίπος *s.m.* drag-net.
γρίππη *s.f.* influenza.
γρίφος *s.m.* riddle.
γροθιά *s.f.* fist; punch.
γροικώ *v.i.* hear.
γρόμπος *s.m.* lump, swelling.
γρό(ν)θος *s.m.* fist; punch.
γρονθοκοπώ *v.t.* punch repeatedly.
γρόσι *s.n.* piastre; ~α (*pl.*) (*fig.*) money.
γρουσ- ** *see* **γουρσ-.
γρυ *adv.* (*fam.*) nothing; not a word.
γρυλλίζω *v.i.* grunt.
γρύλλος *s.m.* cricket; jack (*tool*).
γρυπός *a.* hooked.
γυάλα *s.f.* large glass receptacle.
γυαλάδα *s.f.* shine, gloss.
γυαλένιος *a.* glass.
γυαλί *s.n.* glass (*substance*); mirror; chimney (*of lamp*); ~ά (*pl.*) glasses, spectacles.
γυαλίζω *v.t.* & *i.* polish; shine.
γυαλικά *s.n.pl.* glass-ware.
γυάλινος *a.* glass.
γυάλισμα *s.n.* polishing.
γυαλιστερός *a.* glossy.
γυαλόχαρτο *s.n.* glass-paper.
γυλιός *s.m.* (*mil.*) kitbag.
γυμνάζω *v.t.* train, drill, exercise. **~ομαι** *v.i.* practise, exercise oneself.
γυμνάσια *s.n.pl.* (*mil.*) manoeuvres.
γυμνάσι|ον *s.n.* secondary school. **~άρχης** *s.m.f.* head of such school.
γύμνασμα *s.n.* an exercise.
γυμναστήριον *s.n.* gymnasium.
γυμναστής *s.m.* instructor, trainer.
γυμναστική *s.f.* gymnastics.

γύμνια *s.f.* nakedness.
γυμνισμός *s.m.* nudism.
γυμνιστής *s.m.* nudist.
γυμνόπους *a.* barefoot.
γυμνός *a.* naked, bare.
γυμνοσάλιαγκος *s.m.* slug.
γυμνώνω *v.t.* lay bare, expose; (*fig.*) strip, rob.
γυναίκα *s.f.* woman; wife.
γυναικάδελφος *s.m.* wife's brother.
γυναικάς *s.m.* man who runs after women.
γυναικείος *a.* woman's.
γυναικολόγος *s.m.f.* gynaecologist.
γυναικόπαιδα *s.n.pl.* women and children.
γυναικ|οπρεπής, ~ωτός *a.* effeminate.
γυναικών, ~ίτης *s.m.* women's quarters; harem.
γύναιο *s.n.* hussy, trollop, hag.
γυνή *s.f.* woman; wife.
γύρα *s.f.* walk, stroll; (*fam.*) τα φέρνω ~ make ends meet; βγαίνω στη ~ seek importunately.
γυρεύω *v.t.* seek, look *or* ask for.
γυρίζω *v.t.* & *i.* turn; give back; take *or* escort around; shoot (*film*); (*v.i.*) turn; return; roam; change (*of weather*); change one's view; sink (*of sun*).
γυρίνος *s.m.* tadpole.
γύρις *s.f.* pollen.
γύρισμα *s.n.* turn, turning; return (*giving back*); change; sinking (*of sun*).
γυρισμός *s.m.* return (*coming back*).
γυριστός *a.* curved, winding; turned up *or* down.
γυρνώ *v.t.* & *i.* see γυρίζω.
γυρολόγος *s.m.* pedlar.
γύρος *s.m.* circle, circumference; turn, revolution; hem, brim; hoop (*of barrel*); walk, stroll; round (*of boxing, etc.*); circuit.
γυροφέρνω *v.t.* & *i.* hang around; dwell upon; try to get *or* persuade; τα ~ prevaricate.
γυρτός *a.* bent, leaning; lying down.
γύρω *adv.* round; (*prep.*) ~ σε or από round.
γυφταριό *s.n.* gipsy encampment; (*fig.*) mess, pigsty; riff-raff.
γύφτ|ος *s.m.* gipsy; smith; (*fig.*) miser. ~ιά *s.f.* meanness. ~ισσα *s.f.* gipsy woman.
γυψ *s.m.* vulture.
γύψ|ινος *a.* plaster. ~ωμα *s.n.* plasterwork.
γύψος *s.m.* plaster of Paris.
γων|ία, ~ιά *s.f.* angle, corner; square (*instrument*); fireplace; crust of bread.
γωνι|αίος, ~ακός *a.* corner.
γωνιόλιθος *s.m.* cornerstone.
γωνιόμετρο *s.n.* T-square.
γωνιώδης *a.* angular.

Δ

δα *particle of intensification.*
δάγγειος *s.m.* (*med.*) dengue.
δάγκ|αμα, ~ωμα *s.n.* bite.
δαγκάνα *s.f.* claw, nipper (*lobster's, etc*).
δαγκ|ανιά, ~ωματιά *s.f.* bite.
δαγκ|άνω, ~ώνω *v.t.* & *i.* bite.
δάδα *s.f.* see δας.
δαδί *s.n.* fire-wood (*esp.* pine).
δαίδαλος *a.* & *s.m.* intricate; a labyrinth.
δαίμονας *s.m.* demon, devil.
δαιμον|ίζω *v.t.* enrage. ~ίζομαι *v.i.* be possessed of devils; get enraged; ~ισμένος (*fig.*) mischievous, very clever.
δαιμόνι|ο *s.n.* genius; demon; καινά ~ α revolutionary ideas.
δαιμόνιος *a.* very clever.
δαίμων *s.f.* demon, devil.
δάκνω *v.t.* & *i.* bite.
δακρυγόν|ος *a.* lachrymal; ~α αέρια tear-gas.
δακρύζω *v.i.* see δακρύω.
δάκρυ(ον) *s.n.* tear; (*fig.*) small drop.
δάκρυσμα *s.n.* exuding of drops, oozing, watering.
δακρυσμένος *a.* tearful; oozing.
δακρύω *v.i.* exude drops, ooze, weep; water (*of eyes*).
δακτυλικ|ός *a.* ~ά αποτυπώματα finger-prints (*on record*).
δακτύλιος *s.m.* ring.
δακτυλογράφος *s.m.f.* typist.
δακτυλοδεικτούμενος *a.* in the public eye (*for good or bad*).
δάκτυλος *s.m.* finger; toe; inch; dactyl; (*fig.*) hand, influence.
δαλτωνισμός *s.m.* colour-blindness.
δαμάζω *v.t.* tame.
δαμάλι *s.n.* heifer.
δαμαλισμός *s.n.* vaccination.
δαμάσκηνο *s.m.* plum; ξερό ~ prune.
δαμιζάνα *s.m.* demijohn.
δανδής *s.m.* dandy.
δανείζ|ω *v.t.* lend. ~ομαι *v.t.* borrow.
δανεικ|ός *a.* lent; τα ~ά (*pl.*) money borrowed. ~ά *adv.* on loan.
δάνειο *s.n.* loan.
δανειστής *s.m.* lender.
δαντέλλα *s.f.* lace.
δαντελλωτός *a.* indented (*coastline, etc*).
δαπάν|η *s.f.* expenditure, expense. ~ηρός *a.* costly; extravagant. ~ώ *v.t.* spend, consume.
δάπεδο *s.n.* floor, ground.
δαρ|μός *s.m.*, ~σιμο *s.n.* beating.
δας *s.f.* torch, brand.
δασεία *s.f.* (*gram.*) rough breathing.

δασικός *a.* of forests *or* forestry.
δασκάλ|α, ~ισσα *s.f.* (*primary*) school-mistress; (*pej.*) schoolmarm.
δασκαλεύω *v.t.* coach, prime.
δάσκαλος *s.m.* teacher; (*primary*) school-master; (*pej.*) pedant.
δασμολόγιον *s.n.* tariff.
δασμός *s.m.* duty, tax.
δασολογία *s.f.* forestry.
δασονομία *s.f.* conservancy of forests.
δάσος *s.n.* wood, forest.
δασόφυτος *a.* wooded.
δασ|ύς, ~ός *a.* hairy, bushy.
δασώδης *a.* forested.
δαυλ|ί *s.n.*, ~ός *s.m.* torch, brand.
δαύτος *pron.* this one.
δάφν|η *s.f.* laurel, bay; (*fig.*) δρέπω ~ας triumph, succeed.
δαχτυλήθρα *s.f.* thimble.
δαχτυλιά *s.f.* finger-mark.
δάχτυλο *s.n.* finger; toe.
δαχτυλίδι *s.n.* ring.
δαψιλής *a.* abundant, lavish.
δε *conj.* and; but; on the other hand.
δε(ν) *adv.* not (*before verb*).
δεδηλωμένος *a.* self-confessed.
δεδομέν|ος *a.* given; τα~α (*pl.*) facts, data; ~ου ότι given that.
δέησις *s.f.* prayer.
δει *v.i.* there is need; ολίγου ~ν έπιπτε he nearly fell; πολλού γε και ~ far from it.
δείγμα *s.n.* sample, specimen; token, proof. ~τοληψία *s.f.* sampling.
δεικνύω *v.t. & i.* show, indicate, demon-strate; (*with* να) show how to; point (at); seem, appear.
δείκτης *s.m.* forefinger; indicator; pointer, hand.
δεικτικός *a.* (*gram.*) demonstrative.
δείλ|ι, ~ινό *s.n.* late afternoon.
δειλιάζω, ~ώ *v.i.* falter, flinch.
δειλός *a.* timorous, cowardly.
δείνα(ς) *pron.* ο ~ so-and-so.
δειν|ός *a.* terrible; (*fig.*) clever, expert; τα ~ά (*pl.*) sufferings. ~οπαθώ *v.i.* suffer, have a hard time.
δείπνο *s.n.* supper, dinner.
δεισιδαιμονία *s.f.* superstition.
δείχνω *v.t. & i. see* δεικνύω.
δέκα *num.* ten.
δεκάδα *s.f.* group of ten.
δεκαδικός *a.* decimal.
δεκαεννέα *num.* nineteen.
δεκαέξ(ι) *num.* sixteen.
δεκαεπτά *num.* seventeen.
δεκαετία *s.f.* decade.
δεκάλογος *s.m.* Ten Commandments.
δεκανεύς *s.m.* (*mil.*) corporal.

δεκανίκι *s.n.* crutch.
δεκαοκτώ *num.* eighteen.
δεκαπενθήμερον *s.n.* fortnight.
δεκαπεντασύλλαβος *s.m.* fifteen-syllable line.
δεκαπενταύγουστος *s.n.* 15th August (*feast of Assumption*).
δεκαπέντε *num.* fifteen.
δεκάρα *s.f.* ten-lepta piece. (*fig.*) brass farthing.
δεκάρι *s.n.* ten (*esp. at cards*); ten-drachma piece. ~κο *s.n.* ten-drachma piece.
δεκασμός *s.m.* bribery.
δεκατέσσερεις *a.* fourteen.
δέκατ|ος *a.* tenth; τα ~α (*pl.*) temperature recurrently above normal.
δεκατρείς *a.* thirteen.
Δεκέμβρ|ης, ~ιος *s.m.* December.
δεκοχτούρα *s.f.* Egyptian pigeon.
δέκτης *s.m.* receiver.
δεκτικός *a.* receptive, capable of.
δεκτός *a.* received, accepted; acceptable.
δελεάζω *v.t.* lure, tempt.
δέλεαρ *s.n.* bait.
δελεαστικός *a.* tempting, enticing.
δέλτα *s.n.* the letter Δ; delta.
δελτάριο *s.n.* postcard.
δελτίον *s.n.* bulletin, report; form; card (*ration, identity, etc.*).
δελφίνι *s.n.* dolphin, porpoise; dauphin.
δέμα *s.n.* cord, lace; parcel, bundle.
δεμάτι *s.n.* bunch, sheaf, truss.
δένδρον *s.n.* tree.
δενδροστοιχία *s.f.* avenue of trees.
δέντρ|ο, ~ί *s.n.* tree.
δεντρολίβανο *s.n.* rosemary.
δένω *v.t. & i.* tie (up), bind; bandage; put on (*brake*). assemble (*parts*); set (*jewel*); thicken (*sauce, etc.*); render impotent by spell. (*v.i.*) thicken; form (*of fruit*).
δεξαμενή *s.f.* reservoir; (*naut.*) dock, basin.
δεξαμενόπλοιον *s.n.* tanker.
δεξιά 1. *s.f.* right hand; right wing. 2. *adv.* to the right; propitiously.
δέξιμο *s.n.* welcoming home.
δεξιός *a.* 1. (on the) right; right-hand(ed); right-wing. 2. dextrous.
δεξιοτέχνης *s.m.* virtuoso.
δεξιότητα *s.f.* dexterity.
δεξιώνομαι *v.t.* receive (*as guest*).
δεξίωση *s.f.* reception.
δέομαι 1. *v.t.* (*with gen.*) need. 2. *v.i.* pray.
δέον *s.n.* what is necessary; πλέον του ~τος too much; τα ~τα the needful; respects; ~ να σημειωθή it must be noted.
δεόντως *adv.* properly, duly.
δέος *s.n.* awe; apprehension.
δέρας *s.n.* fleece.

δερβέναγας *s.m.* (*fig.*) tyrant.
δερβένι *s.n.* defile.
δερβίσης *s.m.* dervish.
δέρμα *s.n.* skin; hide, leather.
δερματικός *a.* of the skin.
δερμάτινος *a.* leather.
δερματολόγος *s.m.f.* skin specialist.
δέρ|(ν)ω *v.t.* beat; lash, beat down on (*of elements*). ~νομαι *v.i.* beat one's breast; toss and turn.
δες *imperative of* βλέπω.
δέσιμο *s.n.* tying (up), binding, etc. (*see* δένω); binding (*of book*); (*fam.*) για ~ crazy.
δεσμά *s.n.pl.* fetters; imprisonment.
δεσμεύω *v.t.* bind, make prisoner.
δέσμη *s.f.* bunch, bundle, sheaf; beam (*light*).
δέσμιος *a.* captive.
δεσμός *s.m.* bond, tie; amour.
δεσμώτης *s.m.* prisoner.
δεσπόζω *v.t.* (*with gen.*) dominate, command.
δεσποιν|ίς, ~ ίδα *s.f.* young lady; Miss.
δεσπότης *s.m.* ruler; (*fam.*) bishop.
δεσποτικός *a.* despotic.
Δευτέρα *s.f.* Monday.
δευτερεύων *a.* secondary.
δευτεροβάθμιος *a.* second-grade; (*court*) of Appeal.
δευτερόλεπτο *s.n.* second (*of time*).
δεύτερ|ος *a.* second; εκ ~ου a second time; ~ο πράμα second-rate.
δευτερώνω *v.t. & i.* do *or* happen a second time.
δέχομαι *v.t.* receive, accept.
δη *particle of intensification.*
δήγμα *s.n.* bite, sting.
δήθεν *adv.* as if, supposedly; so-called.
δηκτικός *a.* biting.
δηλαδή *adv.* namely, that is.
δηλητήρι|ο *s.n.* poison, venom. ~άζω *v.t.* poison. ~ώδης *a.* poisonous, venomous.
δηλ|ώ, ~ώνω *v.t.* declare, return; notify; signify.
δήλωση *s.f.* declaration, statement.
δημαγωγός *s.m.* demagogue.
δήμαρχ|ος *s.m.* mayor. ~είον *s.n.* town hall.
δημεύω *v.t.* confiscate.
δημηγορώ *v.i.* speak in public; harangue.
δημητριακά *s.n.pl.* cereals, corn.
δήμιος *s.m.* executioner.
δημιούργημα *s.n.* creation (*thing*).
δημιουργία *s.f.* creation (*act*).
δημιουργικός *a.* creative.
δημιουργός *s.m.* creator.
δημιουργώ *v.t.* create.
δημογραφία *s.f.* population statistics.

δημοδιδάσκαλος *s.m.* primary school-teacher.
δημοκοπία *s.f.* demagogy.
δημοκράτης *s.m.* democrat.
δημοκρατ|ία *s.f.* democracy; republic. ~ικός *a.* democratic, republican.
δημοπρασία *s.f.* auction.
δήμος *s.n.* municipality, borough.
δημοσιά *s.f.* public highway.
δημοσι|εύω *v.t.* publish, promulgate. ~ευση *s.f.* publication. ~ευμα *s.n.* newspaper article.
δημοσιογράφος *s.m.* journalist.
δημόσ|ιος *a.* public; ~ιος υπάλληλος civil servant; το ~ιον the state, the public. ~ία, ίως *adv.* in public.
δημότης *s.m.* citizen of borough.
δημοτικ|ή *s.f.* demotic Greek. ~ιστής *s.m.* partisan of δημοτική.
δημοτικ|ός *a.* municipal; ~ό σχολείο primary school; ~ό τραγούδι folk-song.
δημοτικότητα *s.f.* popularity.
δημοφιλής *a.* popular (*of persons*).
δημοψήφισμα *s.n.* plebiscite, referendum.
δημώδης *a.* popular (*of the people*), folk-.
δηώ *v.t.* ravage.
δι- *denotes* two.
διά *prep.* 1. (*with acc.*) for; because of; about, concerning; ~ τούτο for this reason; ~ τί why? 2. (*with gen.*) through, across; with, by means of, divided by; ~ βίου for life; ~ παντός for ever; ~ της βίας by force; ~ μιας at one go; ~ μακρών at length (*in detail*).
δια- *denotes* passing through; separation; conflict; intensification.
διαβάζω *v.t. & i.* read, study; coach, prime; hear (*person*) his lesson.
διαβάθμισις *s.f.* grading.
διαβαίνω *v.t. & i.* cross, pass.
διαβάλλω *v.t.* traduce, denigrate slyly.
διάβαση *s.f.* crossing, passage; ford, pass.
διάβασμα *s.n.* reading; studying.
διαβασμένος *a.* well-read; schooled.
διαβατήριο *s.n.* passport.
διαβάτης *s.m.* passer-by.
διαβατικός *a.* of passage, passing by; ~ δρόμος frequented street.
διαβατός *a.* negotiable, fordable.
διαβεβαι|ώνω *v.t.* assure. ~ωση *s.f.* assurance.
διάθημα *s.n.* step (*action taken*), démarche.
διαβήτης *s.m.* 1. pair of compasses. 2. (*med.*) diabetes.
διαβιβάζω *v.t.* transmit, convey.
διαβίβαση *s.f.* conveyance (*of message*). ~εις (*mil.*) Signals.
διαβιώ *v.i.* live, pass one's life.

διαβλέπω *v.t.* discern (*below surface*).

διαβόητος *a.* notorious.

διαβολάνθρωπος *s.m.* very shrewd and active person, live wire; (*fig.*) devil.

διαβολεμένος *a.* very shrewd and active; (*fig.*) devilish (*hunger, etc*).

διαβολή *s.f.* sly denigration.

διαβολικός *a.* diabolic(al), devilish.

διάβολ|ος *s.m.* devil (*also as mark of admiration*); ~ου κάλτσα very shrewd and able person.

διαβουλεύσεις *s.f.pl.* deliberations.

διαβούλιο *s.n.* (secret) meeting, plot.

διάβροχος *a.* soaking wet.

διάβρωση *s.f.* corrosion, eating away.

διάγγελμα *s.n.* edict, rescript.

διαγι(γ)νώσκω *v.t.* diagnose.

διαγκωνίζομαι *v.i.* elbow one's way.

διάγνωση *s.f.* diagnosis. τι ~ έκανε; what did he diagnose? διέγνωσε διαβήτη he diagnosed diabetes.

διαγουμίζω *v.t.* pillage; waste.

διάγραμμα *s.n.* diagram, plan.

διαγράφω *v.t.* trace, describe, outline; erase, cross out *or* off.

διάγ|ω *v.i. & t.* live; pass (*time, one's life*); ~ει καλώς he is well.

διαγωγή *s.f.* behaviour, conduct.

διαγωνίζομαι *v.i.* compete.

διαγώνιος *a.* diagonal.

διαγώνισ|μα *s.n.*, ~μός *s.m.* competition (*for award*), examination.

διαδεδομένος *a.* widespread, rife.

διαδέχομαι *v.t.* succeed.

διαδήλωση *s.f.* (*public*) demonstration.

διαδηλωτής *s.m.* demonstrator.

διάδημα *s.n.* diadem, crown.

διαδίδ|ω *v.t.* spread; noise abroad; ~εται ότι it is rumoured that.

διαδικασία *s.f.* procedure; process.

διάδικος *s.m.f.* (*law*) party to action.

διάδοση *s.f.* spreading; rumour.

διαδοχή *s.f.* succession.

διάδοχος *s.m.* successor; crown prince.

διαδραματίζ|ω *v.t.* play (*role*). ~ομαι *v.i.* be enacted (*occur*).

διαδρομή *s.f.* journey, way.

διάδρομος *s.m.* passage, corridor; carpet for this; ~ προσγειώσεως (*aero.*) runway.

διαζευγνύομαι *v.i.* get divorced.

διαζύγιο *s.n.* divorce.

διάζωμα *s.n.* frieze.

διαθέσιμ|ος *a.* available, at one's disposal. ~ότητα *s.f.* availability; suspension from duty.

διάθεσ|η *s.f.* arrangement, disposition; disposal; mood, humour; (*gram.*) voice; ~εις (*pl.*) intentions.

διαθέτ|ω *v.t.* have at one's disposal; dispose, arrange; use; appropriate, devote; bequeath; με ~ει καλά it puts me in a good mood *or* makes me well disposed. *See also* **διατίθεμαι**.

διαθήκη *s.f.* testament; will.

διαθλώ *v.t.* refract.

διαίρεση *s.f.* division; dissension.

διαιρετός *a.* divisible.

διαιρώ *v.t.* divide.

διαίσθηση *s.f.* intuition.

δίαιτα *s.f.* diet.

διαιτησία *s.f.* arbitration.

διαιτητής *s.m.* arbitrator, referee.

διαιτητικός *a.* 1. dietary. 2. of arbitration.

διαιτώμαι *v.i.* eat.

διαιωνίζω *v.t.* perpetuate; protract.

διακαής *a.* ardent.

διακανονίζω *v.t.* regulate, settle.

διάκειμαι *v.i.* be disposed.

διακεκαυμένος *a.* torrid.

διακεκριμένος *a.* distinguished.

διάκενο *s.n.* space, interstice.

διακήρυξις *s.f.* declaration, proclamation.

διακινδυνεύω *v.t. & i.* endanger, risk; run risks.

διακλάδωσις *s.f.* branching, fork; branch (*road, line, etc.*).

διακοίνωσις *s.f.* diplomatic note.

διακονιάρης *s.m.* beggar.

διάκονος *s.m.* deacon.

διακοπή *s.f.* suspension, recess; cutting *or* breaking off; interruption, cut; ~ές (*pl.*) vacation, holidays.

διακόπτης *s.m.* switch, stopcock, main tap.

διακόπτω *v.t.* suspend; cut *or* break off; interrupt.

διακορεύω *v.t.* deflower.

διάκος *s.m.* deacon.

διακόσιοι *a.* two hundred.

διακόσμησις *s.f.* decoration; illumination (*of MSS*).

διάκοσμος *s.m.* decoration (*of room, street, etc.*); décor.

διακοσμώ *v.t.* decorate, adorn.

διακρίν|ω *v.t. & i.* make out, distinguish; differentiate; (*v.i.*) see. ~ομαι *v.i.* be outstanding *or* visible; distinguish oneself.

διάκριση *s.f.* distinction, differentiation; discretion (*judgement, pleasure, choice*); discrimination (*partiality*).

διακριτικόν *s.n.* star, stripe, *etc.*, denoting rank.

διακριτικός *a.* distinguishing; discreet; discretionary.

διακυβεύω *v.t.* risk, stake.

διακύμανσις *s.f.* fluctuation; undulation.

διακωμωδώ *v.t.* ridicule.

διαλαλώ *v.t.* proclaim loudly, cry (*wares*).
διαλαμβάνω *v.t.* deal with, treat of, contain.
διαλανθάνω *v.t. & i.* escape attention of; escape unnoticed.
διαλέγω *v.t.* choose, select.
διάλειμμα *s.n.* interval, break.
διαλείπων *a.* intermittent, occulting.
διαλεκτικός *a.* 1. of dialectic. 2. of dialects.
διάλεκτος *s.f.* dialect.
διάλεξη *s.f.* lecture.
διαλευκάνω *v.t.* elucidate.
διαλεχτός *a.* choice, exceptional.
διαλλαγή *s.f.* settlement (*of difference*).
διαλλακτικός *a.* conciliatory.
διαλογή *s.f.* sorting, scrutiny.
διαλογίζομαι *v.t.* consider, ponder.
διάλογος *s.m.* dialogue.
διάλυμα *s.n.* (*liquid*) solution.
διάλυση *s.f.* solution, dissolution; liquidation, disbanding, breaking off.
διαλυστήρι|α *s.f.*, **~ι** *s.n.* comb.
διαλυτικ|ός *a.* dissolvent; ~*ό στοιχείο* disruptive element. ~**ά** *s.n.pl.* diaeresis.
διαλυτός *a.* soluble.
διαλύω *v.t.* dissolve, liquidate, disband, break off, dispel.
διαμάντ|ι *s.n.* diamond. ~**ικά** *s.n.pl.* jewellery.
διαμαρτύρ|ομαι *v.i.* protest. ~**όμενος** *s.m.* protestant. ~**ία** *s.f.* protest, remonstrance.
διαμαρτυρώ *v.t.* protest (*bill*).
διαμάχη *s.f.* dispute, struggle.
διαμελίζω *v.t.* dismember.
διαμένω *v.i.* reside, sojourn.
διαμέρισμα *s.n.* region, constituency; compartment; apartment, flat.
διαμερισμός *s.m.* apportionment.
διάμεσ|ος *a.* intermediary; in between. ~**ο** *s.n.* interval, space between. *κάνω το* ~ **ο** act as go-between .
διαμετακόμιση *s.f.* transit (*of goods*).
διαμέτρημα *s.n.* bore, calibre.
διάμετρος *s.f.* diameter.
διαμονή *s.f.* (place of) residence, sojourn.
διαμορφώ|νω *v.t.* form, mould. ~**ωση** *s.f.* formation, moulding. ~**ωτικός** *a.* formative.
διαμπάξ *adv.* right through (*of wounds*).
διαμπερής *a.* passing right through (*of wound*).
διαμφισβητώ *v.t.* call in question, contest.
διάνα *s.f.* 1. turkey. 2. bull's-eye. 3. reveille.
διανέμω *v.t.* distribute.
διανόημα *s.n.* thought (*idea*).
διανόηση *s.f.* thought (*process*).
διανοητικός *a.* intellectual; profound.
διάνοι|α *s.f.* intellect; *έχω κατά* ~*α* have in mind.

διανοίγω *v.t.* open up.
διανομεύς *s.m.* distributor; postman.
διανομή *s.f.* distribution.
διανο|ούμαι *v.t. & i.* think (*of*). ~**ούμενος** *s.m.* intellectual.
διάνος *s.m.* turkey.
διανυκτερεύω *v.i.* spend the night; stay up *or* open all night.
διανύω *v.t.* cover (*distance*); be in the course of (*specified period of time*).
διαξιφισμός *s.m.* sword-thrust.
διάολος *s.m. see* **διάβολος.**
διαπασών *s.f.n.* (*mus.*) octave; tuning-fork; (*fam.*) *στή* ~ full-blast.
διαπεραιώνω *v.t.* ferry.
διαπεραστικός *a.* piercing.
διαπερ(ν)ώ *v.t.* pierce; penetrate, pass through.
διαπιστεύω *v.t.* accredit. ~**τήρια** *s.n.pl.* credentials, letters of credence.
διαπιστ|ώνω *v.t.* confirm, note. ~**ωση** *s.f.* confirmation.
διάπλαση *s.f.* formation, conformation.
διάπλατος *a.* wide open.
διαπληκτίζομαι *v.i.* quarrel; exchange blows.
διάπλους *s.m.* (*naut.*) passage, crossing.
διαπνέομαι *v.i.* be animated (*by feeling*).
διαπομπεύω *v.t.* (*fig.*) pillory.
διαπορθμεύω *v.t.* ferry.
διαποτίζω *v.t.* saturate.
διαπραγματεύομαι *v.i. & i.* negotiate; ~ *περί* (*with gen.*) deal with (*topic*).
διαπράττω *v.t.* perpetrate.
διαπρεπής *a.* eminent.
διαπρέπω *v.i.* excel.
διαπύησις *s.f.* suppuration.
διάπυρος *a.* red-hot; (*fig.*) ardent.
διάρθρωσις *s.f.* articulation; structure.
διάρκεια *s.f.* duration.
διαρκής *a.* continuous; permanent.
διαρκώ *v.i.* last.
διαρκώς *adv.* continually, always.
διαρπάζω *v.t.* pillage.
διαρρέω *v.i.* flow through; leak (*of vessel*); leak out (*of liquid, news, etc.*); elapse; fall off (*of following*).
διαρρηγνύω *v.t.* tear, burst *or* break (down, open *or* off); burgle. ~ *τα ιμάτιά μου* protest one's innocence.
διαρρήδην *adv.* flatly.
διαρρήκτης *s.m.* burglar.
διάρρηξη *s.f.* bursting, rupture; burglary.
διαρροή *s.f.* leakage.
διάρροια *s.f.* diarrhoea.
διαρρύθμιση *s.f.* arrangement, disposition.
διασάλευση *s.f.* disturbance; derangement.
διασάφησις *s.f.* clarification; (*customs*)

declaration.
διάσεισn *s.f.* concussion.
διάσελο *s.n.* saddle (*of mountain*), col.
διάσημ|ος *a.* famous. **~α** *s.n.pl.* insignia.
διασκεδάζω *v.t.* & *i.* dispel; entertain, amuse; divert; enjoy oneself.
διασκέδασ|η *s.f.* dispersion, dispelling; entertainment. **~τικός** *a.* amusing.
διασκελίζω *v.t.* step over.
διασκέπτομαι *v.i.* deliberate.
διασκευάζω *v.t.* arrange, adapt (*book, music, etc.*).
διάσκεψη *s.f.* deliberation; conference.
διασκορπίζω *v.t.* scatter, disperse; waste.
διασπαθίζω *v.t.* squander (*wealth*).
διάσπαση *s.f.* fission, splitting; distracting (*of attention*).
διασπείρω *v.t.* disseminate, scatter.
διασπορά *s.f.* dissemination; Dispersion.
διασπώ *v.t.* split; distract (*attention*); break through (*enemy lines*).
διασταλτικός *a.* expanding, dilating.
διάσταση *s.f.* separation, disagreement; dimension. *παίρνω διαστάσεις* grow, intensify (*v.i.*).
διασταυρ|ώνω *v.t.* cross; *~ώθηκα μαζί του* I passed him. **~ωση** *s.f.* crossing.
διαστέλλω *v.t.* distinguish; dilate; expand.
διάστημα *s.n.* space, interval, distance.
διαστημόπλοιον *s.n.* space-ship.
διάστικτος *a.* spotted, dotted.
διαστολή *s.f.* distinction; expansion; (*med.*) diastole; (*mus.*) bar-line.
διαστρεβλώνω *v.t.* twist; distort.
διαστρεμμένος *a.* cantankerous.
διαστρέφω *v.t.* twist; distort; pervert.
διαστροφή *s.f.* perversion; deterioration.
διασυμμαχικός *a.* inter-allied.
διασύρω *v.t.* denigrate.
διασχίζω *v.t.* pass through, traverse.
διασώζω *v.t.* save, preserve.
διάσωση *s.f.* escape, rescue, preservation.
διαταγή *s.f.* order, command.
διάταγμα *s.n.* order (*of executive authority*).
διατάζω *v.t.* order; order about.
διατακτική *s.f.* (*written*) warrant, authority.
διάταξις *s.f.* arrangement; provision, stipulation; *ημερησία* ~ order of day.
διαταράσσω *v.t.* disturb, upset.
διατάσσω *v.t.* arrange; order.
διατεθειμένος *a.* disposed.
διατείν|ω *v.t.* stretch; **~ομαι** *v.i.* assert.
διατελώ *v.i.* (continue to) be; ~ *πρόθυμος* I remain yours truly (*formula ending letter*).
διατέμνω *v.t.* divide, intersect.
διατήρηση *s.f.* maintenance, preservation.
διατηρ|ώ *v.t.* keep, maintain, preserve.

~*ούμαι* *v.i.* keep; remain intact.
διατί *adv.* why.
διατίθεμαι *v.i.* be disposed.
διατιμώ *v.t.* fix price of; tariff.
διατρανώνω *v.t.* say straightforwardly, make quite clear.
διατρέφω *v.t.* keep, support (*family, etc.*).
διατρέχ|ω *v.t.* run through; cover (*distance*); be in the course of (*period, stage, etc.*); ~*ω κίνδυνο* be in danger; *τα* ~*οντα* what is going on; *τα διατρέξαντα* what happened.
διάτρηση *s.f.* perforation, drilling.
διάτρητος *a.* perforated, riddled.
διατριβή *s.f.* sojourn; study, treatise, thesis; diatribe.
διατροφή *s.f.* keep, board; alimony.
διατρυπώ *v.t.* pierce, perforate.
διάττων *s.m.* shooting star.
διατυμπανίζω *v.t.* trumpet, blaze abroad.
διατυπώνω *v.t.* express, formulate, word.
διατύπ|ωση *s.f.* formulation, wording; ~*ώσεις* (*pl.*) formalities.
διαυγ|ής *a.* limpid. **~εια** *s.f.* limpidity, clarity.
δίαυλος *s.m.* narrow channel, strait.
διαφαίνομαι *v.i.* show through, be just visible.
διαφανής *a.* clear, transparent.
διαφεντεύω *v.t.* defend; run, manage.
διαφέρω *v.i.* be different.
διαφεύγ|ω *v.i.* & *t.* escape, slip away; *μου* ~*ει* it escapes me; ~*ω τον κίνδυνο* escape danger.
διαφημ|ίζω *v.t.* make known, advertise, vaunt. **~ιση** *s.f.* advertisement.
διαφθείρω *v.t.* corrupt, seduce, bribe.
διαφθορά *s.f.* corruption.
διαφιλονικώ *v.t.* dispute (*prize, etc.*).
διαφορά *s.f.* difference.
διαφορετικ|ός *a.* different, dissimilar. **~ά** *adv.* otherwise; differently.
διαφορικός *a.* differential.
διάφορο *s.n.* profit, interest.
διάφορος *a.* different, dissimilar, various, mixed.
διάφραγμα *s.n.* diaphragm; partition.
διαφυγή *s.f.* escape; leak; evasion.
διαφυλάσσω *v.t.* preserve intact, keep.
διαφυλετικός *a.* inter-racial.
διαφων|ώ *v.i.* disagree. **~ία** *s.f.* disagreement.
διαφωτίζω *v.t.* solve, clear up; enlighten.
διαχειμάζω *v.i.* winter.
διαχειρίζομαι *v.t.* manage, administer.
διαχείρισις *s.f.* handling, management; supply department.
διαχειριστής *s.m.* manager, administrator.
δια|χέω, **~χύνω** *v.t.* spread, give off; ~*χύνομαι εις* burst into (*applause, etc.*).

διάχυση *s.f.* diffusion; effusive welcome.

διαχυτικός *a.* cordial, effusive.

διάχυτος *a.* diffused, spread about.

διαχωρίζω *v.t.* segregate, separate; dissociate.

διαψεύδω *v.t.* belie, prove false, deny.

διάψευση *s.f.* belying, proving false, denial.

διβάρι *s.n.* hatchery.

δίβουλος *a.* in two minds.

δίγαμος *s.m.f.* bigamist.

δίγλωσσος *a.* bilingual.

δίγνωμος *a.* undecided.

δίδαγμα *s.n.* teaching, lesson; moral.

διδακτέος *a.* (*material*) to be taught.

διδακτικός *a.* of teaching; didactic; educative.

διδακτορία *s.f.* doctorate.

διδακτός *a.* teachable (*of subject*).

δίδακτρα *s.n.pl.* tuition fees.

διδάκτωρ *s.m.f.* doctor (*degree title*).

διδασκαλείον *s.n.* normal school.

διδασκαλία *s.f.* teaching; (*stage*) production.

διδάσκαλος *s.m.* teacher; (*primary*) schoolmaster.

διδάσκω *v.t.* & *i.* teach; produce (*play*).

διδαχή *s.f.* teaching, sermon.

διδόμενον *s.n.* ground (*motive*).

δίδραχμο *s.n.* two-drachma piece.

δίδυμ|ος *a.* twin; ~α (*s.n.pl.*) twins.

δίδω *v.t. see* **δίνω**.

διεγείρω *v.t.* stimulate, rouse.

διέγερση *s.f.* stimulation, excitement.

διεγερτικό *s.n.* stimulant.

διεθνής *a.* international.

διεισδυτικότης *s.f.* insight.

διεισδύω *v.t.* slip in, penetrate.

διεκδικώ *v.t.* lay claim to; contest.

διεκπεραιώνω *v.t.* dispatch, deal with; perform (*task*).

διέλευση *s.f.* crossing, transit.

διελκυστίνδα *s.f.* tug-of-war.

διένεξις *s.f.* dispute.

διενεργώ *v.t.* hold, conduct.

διεξάγω *v.t.* hold, conduct.

διεξοδικός *a.* lengthy, exhaustive.

διέξοδος *s.f.* way out, issue; outlet.

διέπω *v.t.* govern, determine.

διερευνώ *v.t.* investigate.

διερμην|εύς *s.m.f.* interpreter. ~εύω *v.t.* interpret.

διέρχομαι *v.t.* & *i.* pass through, cross; spend (*time*).

διερωτώμαι *v.i.* wonder.

δίεση *s.f.* (*mus.*) sharp.

διεστραμμένος *a.* perverse; perverted.

διετία *s.f.* period of two years.

διευθετώ *v.t.* arrange, settle.

διεύθυνσις *s.f.* address; direction; management.

διευθυντής *s.m.* director, principal; (*mus.*) conductor.

διευθύν|ω *v.t.* direct, administer; steer, drive, pilot; address, direct (*letter, glance, etc.*). ~ομαι *v.i.* (*with* προς) make for.

διευκολύνω *v.t.* facilitate; assist, oblige (*esp. financially*).

διευκρινίζω *v.t.* clarify; clear up; look into.

διευρύνω *v.t.* widen; enlarge.

διεφθαρμένος *a.* corrupt; immoral.

διήγημα *s.n.* story, tale. ~τικός *a.* narrative.

διήγηση *s.f.* narration, narrative.

διηγούμαι *v.t.* & *i.* relate, tell.

διηθώ *v.t.* filter.

διημερεύω *v.i.* spend the day.

διήμερος *a.* lasting two days; two days old.

διηνεκής *a.* continual.

διηπειρωτικός *a.* intercontinental.

διηρημένος *a.* divided.

διθύραμβος *s.m.* dithyramb.

διίσταμαι *v.i.* disagree.

διισχυρίζομαι *v.i.* assert.

δικάζω *v.t.* try, judge.

δικαιοδοσία *s.f.* jurisdiction.

δικαιολογ|ία *s.f.* justification, excuse. ~ώ *v.t.* justify; offer excuses for; side with.

δίκαιον *s.n.* justice, law, right.

δίκαι|ος *a.* just, fair, right; έχω ~ο be right. ~ως *adv.* justly, rightly.

δικαιοσύνη *s.f.* justice.

δικαιούμαι *v.i.* & *t.* (*with* να *or* gen.) be entitled (to).

δικαιούχος *a.* entitled.

δικαίωμα *s.n.* right; ~τα (*pl.*) dues, fees.

δικαιωματικώς *adv.* of right.

δικαιώνω *v.t.* decide in favour of, side with; justify, vindicate.

δικαίωσις *s.f.* justification, vindication.

δίκαννο *s.n.* double-barrelled shot-gun.

δικαστήρι|ον *s.n.* law-court; ~α (*pl.*) (*fam.*) litigation.

δικαστής *s.m.* (*law*) judge.

δικαστικ|ός *a.* judicial; a member of the judiciary. ~ά legal fees.

δικέλλ|α *s.f.*, ~ι *s.n.* two-pronged fork.

δικέφαλος *a.* two-headed.

δίκη *s.f.* law-suit, trial; θεία ~ divine retribution.

δικηγορ|ία *s.f.* legal practice. ~ώ *v.i.* practise law.

δικηγορικ|ός *a.* legal, lawyer's. ~όν γραφείον firm of solicitors. ~ός Σύλλογος the Bar.

δικηγόρος *s.m.f.* lawyer, solicitor, barrister.

δίκην *prep.* (*with gen.*) like.

δίκιο *s.n.* right; έχω ~ be right; παίρνω το ~

του I take his part.

δίκλινο *s.n.* twin-bedded room.

δικλίς *s.f.* valve.

δικογραφία *s.f.* (*law*) documents of a case, brief.

δικολάβος *s.m.* unlicensed lawyer.

δικονομία *s.f.* legal procedure.

δίκοπος *a.* two-edged.

δικός *a.* close, related; ~ *μου* my, mine, my own.

δίκοχο *s.n.* (*mil.*) forage-cap.

δικράνι *s.n.* pitchfork.

δικτάτορ|ας *s.m.* dictator. **~ικός** *a.* dictatorial.

δικτατορία *s.f.* dictatorship.

δίκτυο *s.n.* net; network.

δικτυωτό *s.n.* (*wire*) netting; lattice, grille; duck-board; grid (*of maps*).

δίλημμα *s.n.* dilemma.

διμερής *a.* bipartite.

διμεταλλισμός *s.m.* bimetallism.

δίμηνος *a.* of two months.

διμοιρία *s.f* (*mil.*) platoon.

δίνη *s.f.* whirlpool, whirlwind.

δίνω *v.t.* give, grant; offer (*price*); yield, produce; ~ *και παίρνω* be influential, be the rage; *του* ~ *στα νεύρα* I get on his nerves; ~ *εξετάσεις* take examination; *του* ~ *δίκιο* I take his part.

διό *conj.* for which reason.

διόγκωση *s.f.* swelling.

διόδια *s.n.pl.* toll.

δίοδος *s.f.* passage, pass, crossing; thoroughfare.

διοίκησις *s.f.* administration, command; governing body; province of governor.

διοικητ|ής *s.m.* governor, commandant. **~ικός** *a.* administrative.

διοικώ *v.t.* administer, govern, command.

διόλου *adv.* (not) at all; *όλως* ~ completely.

διομολογήσεις *s.f.pl.* Capitulations.

διονυχίζω *v.t.* examine minutely.

δίοπος *s.m.* (*naut.*) lowest rank of petty officer.

δίοπτρα *s.n.pl.* spectacles.

διόπτρα *s.f.* binoculars.

διορατικός *a.* clear-sighted.

διοργανώνω *v.t.* organize, get up.

διόρθωμα *s.n.* a correction *or* repair.

διορθώνω *v.t.* correct, put right *or* straight; mend; reform; punish.

διόρθωση *s.f.* correction; repair.

διορία *s.f.* time allowed, term, delay.

διορίζω *v.t.* appoint, nominate.

διορύσσω *v.t.* dig through.

διορώ *v.t. see* **διαβλέπω.**

διότι *conj.* because.

διοχετεύω *v.t.* conduct (*fluid*), pipe, carry.

δίπατος *a.* two-storied.

δίπλ|α *s.f.* fold, pleat; ~ες (*pl.*) fritters.

δίπλα *adv.* near, by the side, next door; ~ *σε* near, beside, next to; (*fam.*) *το κόβω* ~ have a nap.

διπλά *adv.* twice as much.

διπλάνο *s.n.* biplane.

διπλανός *a.* near-by, next-door.

διπλαρώνω *v.t. & i.* accost; come alongside.

διπλάσι|ος *a.* double; twice as big. **~άζω** *v.t* double.

δίπλευρος *a.* two-sided.

διπλογραφία *s.f.* double-entry bookkeeping.

διπλοπόδι *adv.* cross-legged.

διπλός *a.* double.

διπλότυπο *s.n.* duplicate (*bill, etc.*).

διπλ|ούς *a.* double; *εις* ~*ούν* in duplicate.

δίπλωμα *s.n.* folding, wrapping; diploma, degree.

διπλωμάτης *s.m.* diplomat; (*fam.*) artful person.

διπλωματ|ία *s.f.* diplomacy; (*fam.*) artfulness. **~ικός** *a.* diplomatic; (*fam.*) artful.

διπλωματούχος *a.* qualified; having diploma *or* degree.

διπλών|ω *v.t.* fold, wrap (up). **~ομαι** *v.i.* double up.

δίποδο *s.n.* biped.

δίπορτ|ος *a.* (*fam.*) *το έχω* ~*ο* have two strings to one's bow.

δίπους *a.* two-legged.

διπρόσωπος *a.* double-faced.

διπυρίτης *s.m.* hard tack.

δις *adv.* twice.

δισάκκι *s.n.* (saddle) bag, haversack.

δισέγγονο *s.n.* great-grandchild.

δισεκατομμύριον *s.n.* thousand million; (*US*) billion.

δίσεκτο *s.n.* leap year.

δίσεχτος *a.* unlucky (*year*).

δισκίο *s.n.* (*med.*) tablet.

δισκοβόλος *s.m.* discus-thrower.

δίσκος *s.m.* discus; disk; tray; collection-plate; gramophone-record.

δισταγμός *s.m.* hesitation.

διστάζω *v.i.* hesitate.

διστακτικός *a.* hesitant.

δίστιχο *s.n.* couplet.

δίστομος *a.* two-mouthed; two-edged.

δισυπόστατος *a.* being in two places at once; showing two sides to one's character.

διττός *a.* double.

διυλίζω *v.t.* filter, distil; (*fig.*) scrutinize, screen.

διυλιστήριον *s.n.* refinery.

διφθερίτις *s.f.* diphtheria.
δίφθογγος *s.f.* diphthong.
διφορούμενος *a.* ambiguous.
δίφραγκο *s.n.* (*fam.*) two-drachma piece.
διχάζω *v.t.* divide, split.
διχάλ|α *s.f.* pitchfork. ~*ωτό*ς *a.* forked.
διχασμός *s.m.* division; disagreement.
διχογνωμία *s.f.* difference of opinion.
διχόνοια *s.f.* dissension.
διχοτόμ|ησις *s.f.* bisection, partition (*act*).
~**ία** *s.f.* dichotomy.
διχοτομώ *v.t.* bisect; partition.
δίχρονος *a.* for two years; two years old; two-stroke (*engine*).
δίχτυ *s.n.* net.
δίχως *prep.* (*with acc.*) without; *με* ~ without; ~ *άλλο* without fail.
δίψα *s.f.* thirst.
διψ|ώ *v.i. & t.* be thirsty; thirst for; ~*ασμένος* thirst.
διωγμός *s.m.* persecution.
διώκτης *s.m.* persecutor.
διώκω *v.t.* persecute; chase, pursue; dismiss, expel; prosecute.
διώνυμον *s.n.* (*math.*) binomial.
διώξιμο *s.n.* dismissal; expulsion.
δίωξις *s.f.* pursuit; prosecution.
διώροφος *a.* two-storied.
διώρυγα *s.f.* canal.
διώχνω *v.t.* send away, dismiss.
δόγης *s.m.* doge.
δόγμα *s.n.* dogma; principle.
δοθιήν *s.m.* boil.
δοιάκι *s.n.* tiller, helm.
δόκανο *s.n.* trap.
δοκάρι *s.n.* beam, girder.
δοκιμάζ|ω *v.t.* test, try on *or* out; taste, sample; attempt; undergo, suffer. ~*ομαι* *v.i.* suffer.
δοκιμασία *s.f.* test; suffering, ordeal.
δοκιμαστικός *a.* trial, tentative; ~ *σωλήν* test-tube.
δοκιμή *s.f.* trial, testing; rehearsal; fitting (*of clothes*); attempt, shot.
δοκίμιο *s.n.* essay; printer's proof.
δόκιμος 1. *a.* of proven worth; classic. 2. *s.m.* (*naut.*) cadet; (*eccl.*) novice.
δοκός *s.f.* beam, girder.
δοκούν *s.n. κατά το* ~ as one sees fit.
δολιεύομαι *v.t.* trick.
δόλιος (-*i*-) *a.* wily.
δόλιος (-*j*-) *a.* poor, unfortunate.
δολιοφθορά *s.f.* sabotage.
δολ(λ)άριον *s.n.* dollar.
δολοπλοκία *s.f.* intrigue.
δόλος *s.m.* bait, trap; guile.
δολοφόν|ος *s.m.* murderer. ~**ία** *s.f.* murder. ~**ώ** *v.t.* murder.

δόλ|ωμα *s.n.* bait, decoy. ~**ώνω** *v.t.* bait.
δομ|ή *s.f.* structure, fabric. ~**ικός** *a.* structural.
δόνησις *s.f.* vibration; (*earthquake*) shock, tremor.
δονκιχωτικός *a.* quixotic.
δόντι *s.n.* tooth, tusk; tine; bit sticking out; cog; *του πονεί το* ~ *γι'αυτή* he is sweet on her; *μιλώ έξω από τα* ~*α* speak straight out *or* severely.
δοντιά *s.f.* tooth-mark.
δον|ώ *v.t.* cause to vibrate, shake. ~*ούμαι v.i.* vibrate.
δόξ|α *s.f.* glory. ~**άζω** *v.t.* glorify, praise; worship, believe in.
δοξάρι *s.n.* (*mus.*) bow.
δοξασία *s.f.* belief.
δοξολογία *s.f.* (*eccl.*) doxology, Te Deum.
δορά *s.f.* skinning; animal's skin.
δορκάς *s.f.* roebuck.
δόρυ *s.n.* spear. ~**φόρος** *s.m.* satellite.
δοσίλογος *a.* answerable; (*fam.*) a quisling.
δόσ|η *s.f.* giving; dose; instalment (*of payment*).
δοσοληψία *s.f.* (*business*) dealing.
δότης *s.m.* donor.
δοτική *s.f.* (*gram.*) dative.
δούγ(ι)α *s.f.* stave (*of barrel*).
δούκας *s.m.* duke.
δουκάτο *s.n.* 1. duchy. 2. ducat.
δούλα *s.f.* maidservant.
δουλεία *s.f.* slavery, bondage.
δουλειά *s.f.* work, job, business; trouble, bother.
δούλεμα *s.n.* running-in (*use*); working up, fashioning; stirring, kneading; digging; leg-pulling.
δουλευτής *s.m.* hard *or* good worker.
δουλεύω *v.i. & i.* work, function; (*v.t.*) work for, serve; work up, fashion; stir, knead; dig (*soil*); pull leg of.
δούλεψη *s.f.* work; pay.
δουλικ|ός *a.* servile. ~**ό** *s.n.* (*fam.*) skivvy.
δουλοπάροικος *s.m.* serf.
δουλοπρεπής *a.* servile.
δούλος *a & s.m.* enslaved; slave, servant.
δούναι *s.n.* (*fin.*) debit; ~*και λαβείν* debit and credit, (*fig.*) dealings, relations.
Δούρειος *a.* ~ *ίππος* Trojan Horse.
δοχείο *s.n.* receptacle, vessel; chamber.
δραγάτης *s.m.* custodian of vineyard.
δραγόνος *s.m.* dragoon.
δραγουμάνος *s.m.* dragoman.
δράκα *s.f.* handful.
δρακόντειος *a.* Draconian.
δράκος *s.m.* ogre; dragon; unchristened baby.
δράκων *s.m.* ogre, dragon; (*fig.*) vigilant

guard.

δράμα *s.n.* drama; (*fig.*) much ado. **~τικός** *a.* dramatic.

δραματολόγιο *s.n.* repertory.

δραματοποιώ *v.t.* dramatize.

δράμι(ον) *s.n.* dram; (*fig.*) small quantity.

δραξ *s.f.* handful.

δραπέτ|ης *s.m.* fugitive, escaped prisoner. **~εύω** *v.i.* escape, get away.

δράσ|ις, ~η *s.f.* activity, action; *άμεσος ~ ις* flying squad.

δρασκελίζω *v.t.* step over; straddle; stride.

δραστήριος *a.* active, diligent; effective.

δραστηριότης *s.f.* activity, effectiveness.

δράστης *s.m.* doer; culprit.

δραστικός *a.* efficacious; drastic.

δράττομαι *v.t.* (*with gen.*) grasp; *~ της ευκαιρίας* take the opportunity.

δραχμή *s.f.* drachma.

δρέπ|ανον, ~άνι *s.n.* sickle.

δρέπω *v.t.* reap. *~ δάφνες* win glory.

δριμ|ύς *a.* sharp, biting, harsh. **~έως** *adv.* severely.

δρομαίος *a.* running.

δρομάς *s.f.* dromedary.

δρομεύς *s.m.* runner.

δρομολόγιο *s.n.* itinerary, route.

δρόμ|ος *s.m.* running, race; speed; journey, distance; way, road, street; *~ο!* off with you!

δροσερός *a.* cool, fresh.

δροσιά *s.f.* cool, freshness; dew.

δροσίζ|ω *v.t. & i.* cool, refresh; get cool (*of weather*). *~ομαι v.i.* cool down; quench one's thirst; (*ironic*) suffer a let-down.

δρόσος *s.f.* dew.

δρυάς *s.f.* dryad.

δρύινος *a.* of oak.

δρυμός *s.m.* forest.

δρυς *s.f.* oak.

δρω *v.i.* act, be active.

δυαδικός *a.* dual; binary.

δυάρι *s.n.* the figure 2; deuce (*at cards*).

δυϊκός *s.m.* (*gram.*) dual.

δύναμαι *v.i.* be able, can.

δυναμικό *s.n.* potential.

δυναμικός *a.* energetic, dynamic.

δύναμ|η *s.f.* strength, power; *~ει* by virtue of; *~εις* (*pl.*) (*mil.*) forces.

δυναμίτιδα *s.f.* dynamite.

δυνάμωμα *s.n.* intensification; increase in strength, loudness, etc.

δυναμώνω *v.t. & i.* strengthen, intensify; (*v.i.*) gain strength *or* force.

δυναμωτικός *a.* tonic, strengthening.

δυναστεία *s.f.* dynasty; rule, reign.

δυναστεύω *v.t.* tyrannize over.

δυνατά *adv.* strongly, hard; well, ably;

loudly, aloud.

δυνατ|ός *a.* strong, powerful, vigorous; good, able; loud; possible; *όσον το ~όν* as far as possible; *βάζω τα ~ά μου* do one's best. **~ότης** *s.f.* possibility.

δύο, δυο *num.* two; *και τα ~* both; *στα ~* in two.

δυόσμος *s.m.* mint (*herb*).

δυσ- *denotes* difficult; bad.

δυσανάβατος *a.* hard to climb.

δυσανάγνωστος *a.* illegible.

δυσανάλογος *a.* disproportionate.

δυσαναπλήρωτος *a.* hard to replace *or* fill.

δυσανασχετώ *v.i.* fret and fume.

δυσαρέσκεια *s.f.* discontent, displeasure; coolness (*between persons*).

δυσάρεστος *a.* unpleasant.

δυσαρεστ|ώ *v.t.* displease. *~ούμαι v.i.* (*with με*) grow cool towards. **~ημένος** dissatirfied.

δυσαρμονία *s.f.* disharmony.

δύσβατος *a.* difficult of passage.

δυσδιάκριτος *a.* hard to discern.

δυσειδής *a.* uncomely.

δυσεντερία *s.f.* dysentery.

δυσεύρετος *a.* hard to find, rare.

δυσθυμία *s.f.* low spirits.

δύση *s.f.* setting (*of sun, etc.*); west; (*fig.*) decline.

δύσκαμπτος *a.* inflexible, stiff.

δυσκίνητος *a.* moving slowly; sluggish.

δυσκοίλι|ος *a.* constipated; constipating. **~ότης** *s.f.* constipation.

δυσκολεύ|ω *v.t. & i.* make difficult; impede; become difficult. *~ομαι v.i.* have difficulty; hesitate.

δυσκολία *s.f.* difficulty.

δύσκολος *a.* difficult; fussy.

δύσληπτος *a.* hard to catch, take, *or* understand.

δυσμαί *s.f.pl.* west; (*fig.*) decline.

δυσμένεια *s.f.* disfavour.

δυσμενής *a.* adverse, unfavourable.

δύσμορφος *a.* uncomely.

δυσνόητος *a.* abstruse.

δυσοίωνος *a.* inauspicious.

δυσοσμία *s.f.* bad smell.

δυσπεψία *s.f.* indigestion.

δύσπιστος *a.* incredulous, mistrustful.

δύσπνοια *s.f.* difficult breathing.

δυσπραγία *s.f.* slump; straitened circumstances.

δυσπρόσιτος *a.* difficult of access.

δυστοκία *s.f.* difficult parturition; (*fig.*) indecision.

δύστροπος *a.* testy, fractious, cantankerous.

δυστύχημα *s.n.* accident, (stroke of) misfortune.

δυστυχία *s.f.* unhappiness; misfortune; poverty.

δυστυχ|ής *a.* unhappy, unfortunate. **~ώς** *adv.* unfortunately.

δύστυχος *a.* unfortunate.

δυστυχ|ώ *v.i.* suffer misfortune *or* poverty; **~ισμένος** *a.* unfortunate.

δυσφημ|ίζω, ~ώ *v.t.* defame, vilify. **~ησις***s.f.* defamation.

δυσφορία *s.f.* malaise, discomfort; unrest; annoyance.

δυσχερ|ής *a.* difficult. **~εια** *s.f.* difficulty, obstacle.

δύσχρηστος *a.* unhandy, awkward (*thing*); rare (*word*).

δυσώδης *a.* foul-smelling.

δύτης *s.m.* diver.

δυτικ|ός *a.* west, western. **~ώς** *adv.* to the west.

δύω *v.i.* set (*of sun, etc.*); (*fig.*) decline.

δυωδία *s.f.* duet.

δώδεκα|α *num.* twelve. **~αριά, ~άδα** *s.f.* dozen. **~ατος** *a.* twelfth.

δωδεκαδάκτυλο *s.n.* (*med.*) duodenum.

δωδεκασύλλαβος *s.m.* twelve-syllable line.

δώμα *s.n.* flat roof; **~τα** (*pl.*) apartments.

δωμάτιο *s.n.* room.

δωρε|ά *s.f.* bequest, donation. **~άν** *adv.* gratis; to no purpose.

δωρητής *s.m.* donor.

δωρ|ίζω, ~ώ *v.t.* donate.

δωρικός *a.* Doric.

δωροδοκ|ία *s.f.* bribery. **~ώ** *v.t.* bribe.

δωροληψία *s.f.* taking of bribes.

δώρον *s.n.* gift, present; ~ άδωρον gift bringing more trouble than profit.

δωσίλογος *a.* see **δοσίλογος**.

δώσ' του *imper.* go to it!

E

ε *int. of summons, interrogation, resignation, etc.*

εαμίτης *s.m.* member of *EAM* (*resistance movement in occupied Greece*).

εάν *conj.* if.

έαρ *s.n.* spring. **~ινός** *a.* of spring, vernal.

εαυτ|ός *pron.* (one)self; ο ~ός μου myself; αφ' ~ού μου of my own accord; ομιλεί καθ' ~όν he talks to himself; καθ' ~ού genuinely, absolutely. ο ~ούλης μου (*fam.*) number one.

έβγα 1. *s.n.* way out; end (*of month, etc.*). 2. *imperative of* **βγαίνω**.

εβδομ|άς, ~άδα *s.f.* week; Μεγάλη ~άδα Holy Week. **~αδιαίος** *a.* week's; weekly.

~αδιάτικο *s.n.* a week's wage.

εβδομή|(κο)ντα *num.* seventy. **~κοστός** *a.* seventieth.

έβδομος *a.* seventh.

έβενος *s.m.f.* ebony.

εβραϊκός *a.* Hebrew, Jewish.

Εβραίος *s.m.* Jew.

έγγαμος *a.* married.

εγγαστρίμυθος *s.m.* ventriloquist.

εγγίζω *v.t. & i.* touch (*upon*); offend, hurt; (*v.i.*) approach.

εγγιστα *adv.* ως ~ approximately.

εγγλέζικος *a.* English.

Εγγλέζος *s.m.* Englishman.

εγγόνι *s.n.* grandchild.

εγγον|ος, ~ός *s.m.* grandson.

εγγράμματος *a.* literate, educated.

εγγραφή *s.f.* enrolment, registration.

έγγραφ|ος *a.* written, in writing. **~ο** *s.n.* document, instrument, paper.

εγγράφω *v.t.* enrol, register.

εγγύη|ση *s.f.* guarantee; bail, security, deposit. **~τής** *s.m.* guarantor, surety.

εγγύς *adv.* near; (*prep. with gen.*) near.

εγγυώμαι *v.t. & i.* guarantee; stand surety *or* vouch for (*with για*).

εγείρ|ω *v.t.* raise, erect; waken, rouse. **~ομαι** *v.i.* get up.

έγερση *s.f.* erection, raising; awakening.

εγερτήριο *s.n.* reveille.

εγκάθετος *a.* hired (*esp. bravo, etc.*).

εγκαθιστ|ώ *v.t.* install, establish, appoint. **~αμαι** *v.i.* settle, establish oneself.

εγκαίνια *s.n.pl.* inauguration.

έγκαιρος *a.* timely, opportune.

εγκαίρως *adv.* in time.

εγκάρδιος *a.* cordial.

εγκάρσιος *a.* transversal.

εγκαρτερώ *v.i.* bear up, persevere.

έγκατα *s.n.pl.* bowels, depths.

εγκατα|λείπω *v.t.* abandon; **~λελειμμένος** deserted.

εγκατάσταση *s.f.* installation; settling.

έγκαυμα *s.n.* burn.

έγκειται *v.i.* εις υμάς ~ η απόφασις the decision rests with you.

εγκέφαλ|ος *s.m.* brain. **~ικός** *a.* cerebral.

έγκλειστος *a.* imprisoned, enclosed.

εγκλείστως *adv.* enclosed (herewith).

εγκλείω *v.t.* confine, enclose.

έγκλημα *s.n.* crime (*criminal act.*) **~τίας** *s.m.* criminal. **~τικός** *a.* criminal. **~τικότης** *s.f.* (*commission of*) crime, delinquency.

εγκλιματίζω *v.t.* acclimatize.

έγκλισις *s.f.* (*gram.*) mood; (*geom.*) inclination.

εγκόλπιο *s.n.* vade-mecum, manual.

εγκόλπιος *a.* worn on the breast; pocket-size.

εγκολπούμαι *v.t.* pocket, appropriate; take up (*protégé, idea*).

εγκοπή *s.f.* incision, notch, slot.

εγκόσμιος *a.* worldly, mundane.

εγκράτεια *s.f.* sobriety, restraint; abstinence.

εγκρίνω *v.t.* approve (of), sanction.

έγκριση *s.f.* approval.

έγκριτος *a.* distinguished, notable.

εγκύκλιος *a.* general (*education*); circular (*letter*).

εγκυκλοπαίδεια *s.f.* encyclopaedia.

εγκυμονώ *v.i. & t.* be pregnant; (*fig.*) be fraught with, bode.

έγκυος *a.* pregnant.

έγκυρος *a.* authoritative; valid.

εγκώμιο *s.n.* eulogy.

έγνοια *s.f.* care, worry.

εγρήγορση *s.f.* vigilance.

εγχείρημα *s.n.* undertaking.

εγχείρηση *s.f.* (*med.*) operation.

εγχειρίδιο *s.n.* manual; dagger.

εγχείριση *s.f.* handing over, delivery.

έγχορδος *a.* (*mus.*) stringed.

έγχρωμος *a.* coloured.

εγχώριος *a.* native, local, of the country.

εγώ *pron.* I; (*s.n.*) ego. **~ισμός** *s.m.* egoism; pride; conceit.

εδάφιον *s.n.* paragraph; verse (*of Bible*).

έδαφος *s.n.* ground, soil; territory. **~ιαίος** *a.* down to the ground. **~ικός** *a.* territorial.

έδεσμα *s.n.* viand, dish.

έδρα *s.f.* seat, see, chair; headquarters; posterior; facet.

εδραιώνω *v.t.* strengthen, consolidate.

εδρεύω *v.i.* sit, reside, have one's headquarters.

εδώ *adv.* here; ~ και δύο χρόνια two years ago.

εδώδιμος *a.* edible; ~α (*s.n.pl.*) provisions.

εδώλιον *s.n.* seat, bench; (*in court*) dock. (*naut.*) thwart.

εθελοντής *s.m.* volunteer.

εθελούσιος *a.* voluntary.

έθιμο *s.n.* custom, tradition. **~οτυπία** *s.f.* etiquette.

εθνάρχης *s.m.* leader of nation.

εθνικισμός *s.m.* nationalism.

εθνικός *a.* national; gentile, heathen. **~οποιώ** *v.t.* nationalize. **~ότης** *s.f.* nationality. **~όφρων** *a.* nationalistic; patriotic.

έθνος *s.n.* nation.

εθνοσυνέλευσις *s.f.* national assembly.

ει *conj.* if; ~ δε if not.

είδα *v.t. aor. of* **βλέπω.** ~ κι απόειδα I got fed up. ~ κι έπαθα I had a hard job.

ειδάλλ|ως, ~ιώς *adv.* if not, otherwise.

ειδεμή *adv.* if not, otherwise.

ειδεχθής *a.* hideous.

ειδήμων *a.* expert in *or* on (*with* σε *or gen.*).

είδησ|ις *s.f.* knowledge, intimation, notice; news; ~εις (*pl.*) news; παίρνω ~η get wind of, notice.

ειδικεύομαι *v.i.* specialize.

ειδικός *a.* special, specific; a specialist.

ειδοποι|ώ *v.t.* inform, notify. **~ησις** *s.f.* notification, notice.

είδ|ος *s.n.* sort, kind; species; article, commodity; ~η (*pl.*) goods, ware.

ειδότες *s.m.pl.* οι ~ those who know.

ειδύλλιο *s.n.* idyll; (*fam.*) romance.

είδωλ|ο *s.n.* idol; image. **~ολάτρης** *s.m.* idolater, pagan.

είθε *particle* would that (*with* να).

ειθισμέν|ος *a.* κατά τα ~α as usual.

είθισται *v.i.* it is the custom.

εικάζω *v.i.* guess, conjecture.

εικασία *s.f.* conjecture.

εικαστικός *a.* 1. conjectural. 2. representative (*of fine arts*).

εική *adv.* ~ και ως έτυχε at random.

εικόνα *s.f.* image, picture; αγία ~ icon.

εικονίζω *v.t.* depict, show.

εικονικός *a.* using imagery; illustrative; fictitious (*sale, etc.*).

εικόνισμα *s.n.* picture; (*esp.*) icon.

εικονογραφ|ώ *v.t.* illustrate. **~ία** *s.f.* illustration.

εικονοκλάστης *s.m.* iconoclast.

εικονοστάσι(ον) *s.n.* shrine; place where icons are hung; (*in church*) screen.

είκοσ|ι *num.* twenty. **~τός** *a.* twentieth. **~άδα** *s.f.* a score.

εικών *s.f.* see **εικόνα.**

ειλικριν|ής *a.* sincere, frank. **~εια** *s.f.* sincerity, frankness.

είλ|ως, ~ωτας *s.m.* helot.

είμαι *v.i.* am.

ειμαρμένη *s.f.* fate, destiny.

ειμή *adv.* except, but (*after negative*).

ειμί *v.i.* am.

είναι *v.i.* is, are; το ~ being, existence.

ειρημένος *a.* aforesaid.

ειρηνεύω *v.t. & i.* pacify, calm; make peace, live in peace.

ειρήνη *s.f.* peace. **~ικός** *a.* peaceful, peaceable, pacific.

ειρηνοδικείον *s.n.* magistrate's court.

ειρκτή *s.f.* prison.

ειρμός *s.m.* train of thought, coherence.

ειρων|εία *s.f.* irony. **~εύομαι** *v.t. & i.* speak ironically (to).

εις, *a. & article m.* one; a, an.

εις, σε *prep.* (*with acc.*) at; in, among; within

(*time*); to, into; on, on to; (*oath*) on, by; στης Μαρίας to *or* at Mary's; ψόφιος στην κούρασή dog-tired; ερωτεύτηκε στα καλά he fell well and truly in love; με πέθανε στη φλυαρία he wore me out with his chatter.

εισαγγελ|εύς *s.m.* (*law*) public prosecutor. **~ία** *s.f.* public prosecutor's office.

εισάγω *v.t.* put in; bring *or* usher in, admit; introduce; import.

εισαγωγ|ή *s.f.* introduction, admission; import(ation); (*mus.*) overture. **~εύς** *s.m.* importer.

εισαγωγικ|ός *a.* introductory; of admission *or* importation. **~ά** *s.n.pl.* inverted commas.

εισβάλλω *v.i.* (*with* εις) invade, burst in; flow into (*of tributary*).

εισβολ|ή *s.f.* invasion, irruption; bout (*of malady*). **~εύς** *s.m.* invader.

εισδύω *v.i.* penetrate, get in.

εισελαύνω *v.i.* (*with* εις) invade; make triumphal entry.

εισέρχ|ομαι *v.i.* (*with* εις) come *or* go in, enter; τα ~όμενα in-tray.

εισέτι *adv.* still; (*not*) yet.

εισέχω *v.i.* go in, be hollow, open inwards.

εισήγη|σις *s.f.* recommendation; report. **~τής** *s.m.* proposer, rapporteur.

εισηγούμαι *v.t.* propose, moot; report on.

εισιτήριο *s.n.* ticket.

εισιτήριος *a.* of entry *or* admission.

εισόδημα *s.n.* revenue, return, income. **~τίας** *s.m.* rentier.

είσοδος *s.f.* entrance, entry; admission.

είσπλους *s.m.* sailing in; mouth (*of harbour, etc.*).

εισπνέω *v.t.* inhale.

εισπράκτορας *s.m.* collector (*of moneys*); conductor (*of vehicle*).

είσπραξ|η *s.f.* collecting (*of moneys*); encashment; **~εις** (*pl.*) takings.

εισπράττω *v.t.* collect (*moneys*); cash.

εισφέρω *v.t.* contribute.

εισφορά *s.f.* contribution.

εισχωρώ *v.i.* penetrate, slip in.

είτα *adv.* then, next.

είτε *adv.* ~ . . . ~ either . . . or, whether . . . or.

εκ, εξ *prep.* (*with gen.*) 1. (*origin*) from. 2. (*composition*) of. 3. (*partitive*) out of, among. 4. (*manner, cause*) ~ της χειρός by the hand; ~ ανάγκης of necessity; ~ νέου anew; ~ ίσου equally.

εκ-, εξ- *denotes* out, off; completely. *See also* **εξε-**.

έκαστ|ος *pron.* each (one); καθ' ~ην every day; τα καθέκαστα the details.

εκάστοτε *adv.* each time; η ~ κυβέρνησις each successive government.

εκάτερ|ος *pron.* each, either (*of two*). **~ωθεν** *adv.* on each side.

εκατόμβη *s.f.* hecatomb.

εκατομμύρι|ο *s.n.* million. **~ούχος** *s.m.* millionaire.

εκατό(ν) *num.* a hundred; τοις *or* τα ~ per cent.

εκατοντάδα *s.f.* (*set of*) one hundred.

εκατονταετηρίδα *s.f.* century; centenary.

εκατονταετής *a.* hundred years'; a centenarian.

εκατονταπλάσιος *a.* a hundredfold.

εκατόνταρχος *s.m.* centurion.

εκατοντούτης *s.m.* centenarian.

εκατοστάρ|ι *s.n.* hundred-drachma note; quarter-oka. **~ικο** *s.n.* hundred-drachma note.

εκατοστ|ός *a.* hundredth. **~όμετρον** *s.n.* centimetre.

εκβαίνω *v.i. see* **βγαίνω**.

εκβάλλω *v.i. & i.* take off; pull *or* have out; eject, expel; give forth, shed; utter; cast up; (*v.i.*) (*of rivers, roads*) flow *or* lead out (*into*).

έκβαση *s.f.* outcome, issue.

εκβιάζω *v.t.* compel; blackmail; extort; force (*passage, etc.*).

εκβολή *s.f.* ejection; mouth (*of river*).

εκβράζω *v.t.* cast up.

εκγυμνάζω *v.t.* train, exercise.

εκδέρω *v.t.* flay, skin; graze.

έκδηλ|ος *a.* clear, apparent. **~ώνω** *v.t.* show, manifest.

εκδίδω *v.t.* issue, publish; pronounce (*judgement*); draw (*bill, etc.*); extradite.

εκδικάζω *v.t.* (*law*) try (*case*).

εκδίκη|ση *s.f.* revenge, vengeance. **~τής** *s.m.* avenger, **~τικός** *a.* revengeful, avenging.

εκδικούμαι *v.t. & i.* avenge; take revenge (on).

εκδιώκω *v.t.* expel, dismiss; (*mil.*) dislodge.

εκδορ|ά *s.f.* skinning; excoriation; abrasion. **~ιον** *s.n.* blistering plaster.

έκδοση *s.f.* issue, publication; edition; version, story; extradition.

εκδότ|ης *s.m.* publisher; drawer (*of bill*). **~ικός** *a.* publishing; issuing.

έκδοτος *a.* addicted; dissolute.

εκδούλευση *s.f.* service, favour.

εκδοχή *s.f.* version, interpretation.

εκδράμω *v.i.* θα ~ *future of* **εξέδραμον**.

εκδρομή *s.f.* excursion, outing.

εκδύω *v.t.* undress.

εκεί *adv.* there; ~ που while, instead of, whereas. **~θεν** *adv.* thence.

εκείνος *pron.* he, that (one).

εκεχειρία *s.f.* truce.

έκθαμβος a. dazzled; enraptured.

εκθειάζω v.t. praise extravagantly.

έκθεμα s.n. exhibit.

έκθεση s.f. putting out, exposure; abandoning; exhibition; statement, report; essay.

εκθέτης s.m. exhibitor.

έκθετ|ος a. exposed. ~**ο** s.n. foundling.

εκθέτω v.t. put out in the open, expose; abandon (child); exhibit; expound, relate; put up to auction; put forward (candidature); compromise. (See **εκτίθεμαι**).

εκθλίβω v.t. squeeze (lemon, etc.); elide.

εκθρονίζω v.t. dethrone.

έκθυμος a. cordial, eager.

εκκαθαρίζω v.t. & i. settle (account, etc.); clarify; conclude; purge, clean up; (v.i.) remain as balance in hand.

εκκαλών s.m. (law) appellant.

εκκεντρικός a. eccentric.

εκκενώ v.t. empty, clear; leave, evacuate; discharge (gun).

εκκένωση s.f. emptying; (electrical) discharge.

εκκίνηση s.f. starting off.

εκκλησία s.f. church.

εκκλησιάζομαι v.i. go to church.

εκκλησίασμα s.n. congregation.

εκκλησιαστικός a. ecclesiastical, religious.

έκκλησις s.f. appeal.

εκκλίνω v.i. deviate.

εκκοκκίζω v.t. remove seeds of; shell; gin.

εκκολάπτω v.t. hatch.

εκκρεμ|ής a. pending, unsettled. ~**ές** s.n. pendulum.

εκκρίνω v.t. secrete (physiologically).

εκκωφαντικός a. deafening.

εκλαϊκεύω v.t. present in popular form.

εκλαμβάνω v.t. take, interpret, construe.

έκλαμπρος a. brilliant, eminent.

εκλέγω v.t. choose, select; elect.

εκλείπω v.i. vanish, be in eclipse; ο εκλιπών the deceased.

έκλειψη s.f. eclipse.

εκλεκτικός a. discriminating (in taste).

εκλεκτ|ός a. choice, select, exceptional; elected; οι ~οί élite.

εκλελυμένος a. lax, dissolute.

εκλέξιμος a. eligible (for office).

εκλιπαρώ v.t. beseech.

εκλογεύς s.m. elector.

εκλογή s.f. choice, selection; ~**ές** (pl.) election(s), ~**ικός** a. electoral; of elections.

έκλυση s.f. release; laxity.

έκλυτος a. dissolute.

εκμαγείο s.n. (plaster) cast.

εκμαιεύω v.t. extract, worm out (truth, etc.).

εκμανθάνω v.t. learn thoroughly, master.

εκμαυλίζω v.t. seduce, corrupt.

εκμεταλλεύομαι v.t. work, exploit.

εκμηδενίζω v.t. annihilate.

εκμισθώ v.t. let out for hire, rent. ~**τής** s.m. lessor.

εκμυζώ v.t. suck out, drain; (fig.) exploit, milk.

εκμυστηρεύομαι v.t. confide (secret).

εκνευρίζω v.t. irritate, annoy.

έκνομος a. unlawful.

εκούσιος a. voluntary.

έκπαγλος a. breath-taking, wondrous.

εκπαίδευσις s.f. education.

εκπαιδευτήριον s.n. school.

εκπαιδευτικός a. educational; a schoolteacher.

εκπαιδεύω v.t. educate, train.

έκπαλαι adv. from time immemorial.

εκπαρθενεύω v.t. deflower.

εκπατρίζω v.t. expatriate.

εκπέμπω v.t. emit, send out; broadcast (by radio).

εκπίπτω v.i. & t. decline in value; ~ θέσεως lose one's position; reduce in value or price; deduct.

έκπληξη s.f. surprise. ~**κτικός** a. surprising.

εκπληρώ v.t. fulfil, carry out.

εκπλήσσω v.t. astonish.

έκπλους s.m. departure (of ship).

εκπνέω v.t. & i. exhale; die; terminate, expire.

εκπνοή s.f. exhalation; expiry.

εκποιώ v.t. sell, dispose of.

εκπολιτίζω v.t. civilize.

εκπομπή s.f. emission; (radio) broadcast.

εκπονώ v.t. produce, achieve (painstakingly); train (for sports).

εκπορεύομαι v.i. come, proceed (from).

εκπορθώ v.t. capture (city, etc.); lay waste.

εκπορνεύω v.t. prostitute.

εκπρόθεσμος a. overdue.

εκπρόσωπ|ος s.m. representative. ~**ώ** v.t. represent, act for.

έκπτωση s.f. decline, fall; loss (of rights, etc.); reduction, rebate, discount.

έκπτωτος a. deposed, dispossessed.

εκπυρσοκρότησις s.f. detonation.

εκρήγνυμαι v.i. explode, burst; erupt; εξερράγη πόλεμος war broke out.

εκρηκτικός a. explosive.

έκρηξη s.f. explosion, eruption, out-break, outburst.

εκριζώ, ~**ώνω** v.t. uproot, pull out, eradicate.

εκροή s.f. outflow, outfall.

έκρυθμος a. abnormal, irregular.

εκσκαφή s.f. excavation.

έκσταση *s.f.* ecstasy, wonderment.
εκστομίζω *v.t.* utter.
εκστρατεία *s.f.* campaign, expedition.
εκσυγχρονίζω *v.t.* update.
εκσφενδονίζω *v.t.* sling, fling; fire (*torpedo, etc.*).
εκτάδην *adv.* (*lying*) at full length.
έκτακτ|ος *a.* temporary, emergency; special, exceptional; excellent. **~ως** *adv.* temporarily; exceptionally.
έκταση *s.f.* extent, expanse; (*gram.*) lengthening.
εκτεθειμένος *a.* exposed; committed; compromised.
εκτείν|ω *v.t.* stretch out, extend, enlarge. **~ομαι** *v.i.* extend.
εκτέλεσις *s.f.* execution; performance, perpetration; fulfilment.
εκτελεστής *s.m.* executor; executant, performer; executioner.
εκτελεστικ|ός *a.* executive; **~όν** *απόσπασμα* firing-squad.
εκτελώ *v.t.* execute; perform, carry out.
εκτελωνίζω *v.t.* clear through customs.
εκτενής *a.* extended, lengthy.
εκτεταμένος *a.* extensive; lengthy.
εκτίθεμαι *v.i.* be exposed *or* exhibited; be committed *or* compromised; **~** (*υποψήφιος*) stand for election. (*See* **εκθέτω**).
εκτιμ|ώ *v.t.* esteem, value; appraise, estimate, **~ησις** *s.f.* esteem; valuation, estimate.
εκτί(ν)ω *v.t.* serve (*sentence*), pay (*penalty*).
εκτομή *s.f.* cutting off; castration.
εκτοξεύω *v.t.* shoot (*arrow*); hurl.
εκτοπ|ίζω *v.t.* displace, dislodge; send into political exile. **~ισμα** *s.n.* displacement (*of fluid*). **~ησις** *s.f.* political exile.
έκτος *a.* sixth.
εκτός *adv. & prep.* outside; except; **~** *της πόλεως* outside the city; **~** *του Γιάννη* except John; **~** *από* except for, apart from; **~** *αν* unless; **~** *εαυτού* beside oneself; **~** *κινδύνου* out of danger.
έκτοτε *adv.* from that time.
εκτραγωδώ *v.t.* dramatize, paint in lurid colours.
εκτραχηλίζομαι *v.i.* become slovenly; let oneself go (*too far*).
εκτραχύνω *v.t.* (*fig.*) worsen (*relations, etc.*).
εκτρέπ|ω *v.t.* divert. **~ομαι** *v.i.* deviate, stray; get drawn into (*dispute, etc.*).
εκτροπή *s.f.* deviation.
έκτροπ|ος *a.* improper; **~α** (*s.n.pl.*) unseemly behaviour.
εκτροχιάζομαι *v.i.* become derailed; (*fig.*) run off the rails.

έκτρωμα *s.n.* abortion, monster.
έκτρωση *s.f.* abortion, miscarriage.
εκτυλίσσ|ω *v.t.* unwind, unwrap. **~ομαι** *v.i.* unfold, happen.
εκτυπώνω *v.t.* emboss; print.
εκτυφλωτικός *a.* blinding.
εκφέρ|ω *v.t.* utter, express; carry to burial. **~ομαι** *v.i.* (*gram.*) be construed.
εκφοβίζω *v.t.* intimidate.
εκφορά *s.f.* funeral.
εκφορτωτής *s.m.* unloader, stevedore.
εκφράζω *v.t.* express, declare.
έκφρασ|η *s.f.* expression. **~τικός** *a.* expressive.
εκφυλίζομαι *v.i.* abate, slacken; degenerate.
εκφυλισμός *s.m.* degeneration, corruption.
έκφυλος *a.* degenerate.
εκφωνη|τής *s.m.,* **~τρια** *s.f.* (*radio*) announcer.
εκφωνώ *v.t.* call (*roll*); deliver (*speech*).
εκχυδαΐζω *v.t.* vulgarize.
εκχύλισμα *s.n.* extract, essence.
εκχωμάτωση *s.f.* clearing away of earth.
εκχωρώ *v.t. & i.* assign, transfer, concede; make way.
εκών *a.* willingly; **~** *άκων* willy-nilly.
έλα *v.i.* come! (*imperative of* **έρχομαι**).
ελαία *s.f.* olive (-tree).
ελαιογραφία *s.f.* oil-painting.
ελαιόλαδο *s.n.* olive-oil.
ελάιο *s.n.* olive-oil; oil.
ελαιοτριβείο *s.n.* olive-press.
ελαιόχρους *a.* olive-coloured.
ελαιόχρωμα *s.n.* oil-paint.
ελαιώδης *a.* oily; containing oil.
ελαιών(ας) *s.m.* olive-grove.
ελασίτης *s.m.* member of *ΕΛΑΣ* (*resistance army in occupied Greece*).
έλασμα *s.n.* metal plate.
ελάσσων *a.* less, lesser; (*mus.*) minor.
ελαστικ|ός *a.* elastic, flexible, resilient. **~ό** *s.n.* rubber; tyre; elastic.
ελ|άτη *s.f.,* **~ατο** *s.n.,* **~ατος** *s.m.* fir.
ελατήριο *s.n.* spring (*of clock, vehicle*); incentive, motive.
ελάττωμα *s.n.* fault, defect. **~τικός** *a.*
ελάττων *a. see* **ελάσσων**.
ελαττώνω *v.t.* lessen, reduce.
ελάττωσις *s.f.* lessening, decrease.
ελαύνω *v.t. & i.* drive; (*v.i.*) ride, press on.
ελ|άφι *s.n.,* **~αφος** *s.m.* deer.
ελαφρόπετρα *s.f.* pumice.
ελαφρ|ός, ~ύς *a.* light, slight; trifling; mild, weak (*tea, etc.*).
ελαφρυντικ|ός *a.* mitigating; **~α** (*s.n.pl.*) extenuating circumstances.
ελαφρώνω *v.t. & i.* lighten, ease, alleviate;

feel eased *or* relieved; get lighter.

ελάχιστ|ος *a.* least, very small; *~ος όρος* minimum; *τουλάχιστον* at least. *~α adv.* very little.

Ελβετ|ός, ~ικός *a.* Swiss.

ελεγείο *s.n.* elegy.

ελεγκτής *s.m.* inspector.

ελεγκτικ|ός *a.* inspectoral; *~όν συνέδριον* Audit Board.

έλεγχος *s.m.* inspection, examination; control; audit; censure, criticism; school report; *~ συνειδήσεως* remorse.

ελέγχω *v.t.* inspect, examine; audit; censure; show, testify.

ελεειν|ός *a.* wretched, vile; pitiable. *~ολογώ v.t.* pity; deplore.

ελεήμ|ων, & i. merciful, compassionate. *~οσύνη s.f.* (giving of) alms.

έλ|εος *s.n.* mercy, charity, compassion; *τα ~έη του Θεού* wealth; *δεν έχω ~εος κρασί* I have not a drop of wine.

ελευθερία *s.f.* liberty, freedom.

ελευθέριος *a.* liberal; generous; loose (*of morals*).

ελεύθερος *a.* free; unmarried.

ελευθερόστομος *a.* outspoken.

ελευθερόφρων *a.* open-minded; free-thinking.

ελευθερώνω *v.t.* set free, release; rid.

έλευσις *s.f.* coming, advent.

ελεφάντειος *a.* elephantine; ivory.

ελεφαντόδους *s.m.* elephant's tusk; ivory.

ελεφαντοστούν *s.n.* ivory.

ελέφας *s.m.* elephant.

ελε|ώ *v.t.* give alms to; show compassion to; *Κύριε ~ησον* Lord have mercy!

ελιά *s.f.* olive (-tree)*;* mole, beauty-spot.

ελιγμός *s.m.* twisting, winding; manoeuvre.

ελικοειδής *a.* spiral, winding.

ελικοκίνητος *a.* propeller-driven.

ελικόπτερο *s.n.* helicopter.

έλιξ *s.m. & f.* coil, spiral; thread (*of screw*); volute (*cochlea*); tendril; propeller.

ελίσσομαι *v.i.* wind, coil; whirl round; perform evolutions.

έλκηθρο *s.n.* sledge.

έλκος *s.n.* ulcer; sore.

ελκυστικός *a.* attractive, tempting.

ελκύω *v.t.* attract, charm.

έλκω *v.t. & i.* draw, pull, tow; *~ την καταγωγήν* be descended (*from*). (*v.i.*) weigh. *έλκομαι προς* gravitate to.

ελλανοδίκης *s.m.* judge (*of contest*).

έλλειμμα *s.n.* deficit, deficiency.

ελλειπτικός *a.* elliptic(al)*;* deficient.

έλλειψη *s.f.* lack, deficiency; ellipse.

Έλλην(ας) *s.m.* Greek.

ελληνικός *a.* Greek.

ελληνισμός *s.m.* the Greek people; Hellenism.

ελληνιστί *adv.* in Greek.

ελλιπής *a.* defective, deficient.

ελλοχεύω *v.i.* lie in ambush.

έλξη *s.f.* pull; traction; attraction, charm; gravitation.

ελονοσία *s.f.* malaria.

έλος *s.n.* marsh, bog.

ελπίζω *v.i. & t.* hope (for), expect.

ελπ|ίς,~ ίδα *s.f.* expectation, hope; *παρ' ~ίδα* contrary to expectation.

ελώδ|ης *a.* marshy; *~εις πυρετοί* malaria.

εμαυτόν *pron.* myself.

εμβαδόν *s.n.* area (*measurement*).

εμβαθύνω *v.i.* delve (*mentally*).

εμβάλλω *v.t.* insert, introduce; *~ εις φόβον* inspire with fear; *~ εις πειρασμόν* lead into temptation.

έμβασμα *s.n.* remittance.

εμβατήριον *s.n.* (*mus.*) march.

έμβλημα *s.n.* emblem, badge, motto.

εμβολή *s.f.* insertion; embolism; (*naut.*) collision, ramming.

εμβολι|άζω *v.t.* graft; inoculate. *~ασμός s.m.* grafting; inoculation, vaccination.

εμβόλιμος *a.* intercalary.

εμβόλιον *s.n.* graft; vaccine.

έμβολον *s.n.* piston; ramrod; ram (*of ship*); sting. (*med.*) blood-clot.

εμβριθής *a.* erudite, serious, profound.

εμβρόντητος *a.* thunderstruck.

έμβρυο *s.n.* embryo, foetus.

εμέ(να) *pron.* me.

εμείς *pron.* we.

εμετ|ος, ~ός *s.m.* vomiting. *~ικός a.* emetic.

εμμανώς *adv.* madly, to distraction.

εμμένω *v.i.* (*with εις*) abide by, adhere to.

έμμεσος *a.* indirect, mediate.

έμμετρος *a.* 1. in moderation. 2. in verse.

έμμηνα *s.n.pl.* menstruation.

εμμίσθος *a.* paid, salaried.

έμμον|ος *a.* persistent, fixed. *~η ιδέα* obsession.

εμός *pron.* my, mine.

έμπα 1. *s.n.* way in; beginning (*of month, etc.*). 2. *imperative of* **μπαίνω**.

εμπάθεια *s.f.* animosity, violent feeling.

εμπαιγμός *s.m.* mockery.

εμπαίζω *v.t.* mock; deceive.

εμπεδώ *v.t.* establish, consolidate.

εμπειρία *s.f.* experience, practice.

εμπειρικός *a.* empiric(al); quack.

εμπειρογνώμων, *s.m.f.* expert.

έμπειρος *a.* experienced, skilled.

εμπεριστατωμένως *adv.* in detail.

εμπιστεύομαι *v.t. & i.* entrust, trust, confide.

εμπιστευτικός *a.* confidential.

έμπιστ|ος *a.* trustworthy. **~οσύνη** *s.f.* trust, confidence.

έμπλαστρον *s.n.* (*med.*) plaster.

έμπλε|ος, ~ως *a.* full, replete.

εμπλοκή *s.f.* engagement; jamming (*of gun*).

εμπλουτίζω *v.t.* enrich; improve quality of.

έμπνευση *s.f.* inspiration.

εμπνέ|ω *v.t.* insufflate; inspire. *~ομαι v.t. & i.* conceive idea of; feel inspired.

εμποδίζω *v.t.* prevent, hinder, obstruct, prohibit.

εμπόδιο *s.n.* hindrance, obstacle.

εμποιώ *v.t.* cause, inspire.

εμπόλεμος *a.* belligerents at war; ~ *δύναμις* war footing; ~ *ζώνη* zone of hostilities.

εμπόρευμα *s.n.* merchandise.

εμπορεύομαι *v.i. & t.* be in commerce; trade *or* deal in; traffic in, exploit.

εμπορικ|ός *a.* commercial, mercantile. **~ο** *s.n.* shop selling textiles; clothier's; haberdasher's.

εμπόριο *s.n.* trade, commerce.

έμπορος *s.m.* merchant, tradesman, vendor.

εμποροϋπάλληλος *s.m.f.* shop-assistant.

εμποτίζω *v.t.* impregnate, imbue.

έμπρακτος *a.* real, actual, put into practice.

εμπρησ|μός *s.m.* arson. **~τής** *s.m.* fire-raiser. **~τικός** *a.* incendiary.

εμπριμέ *s.n.* printed fabric.

εμπρόθεσμος *a.* within prescribed time-limit.

εμπρός *adv. & prep.* forward(s), ahead, in front; ~ *σε* in front of, compared with; ~ *από* before, in front of; ~ *μου* in front of me; *βάζω* ~ start, set in motion, scold; *πηγαίνω* ~ do well, succeed; *το ρολόι πάει* ~ the clock is fast; ~! hallo! (*on telephone*); come in!

έμπροσθεν *adv. & prep.* in front; previously; (*with gen.*) in front of; (*as a.*) (in) front, previous.

εμπρόσθιος *a.* (in) front, fore.

έμπυ|ο *s.n.* pus. **~άζω** *v.i.* suppurate.

εμφαίνω *v.t.* show, indicate.

εμφανής *a.* manifest; prominent.

εμφανίζ|ω *v.t.* show, reveal; (*photography*) develop. **~ομαι** *v.i.* appear, present oneself.

εμφανίσιμος *a.* presentable.

εμφάνιση *s.f.* appearance; presentation; (*photography*) developing.

έμφαση *s.f.* emphasis, stress.

εμφιλοχωρώ *v.i.* creep in (*of errors, etc.*).

έμφοβος *a.* frightened.

εμφορούμαι *v.i.* be animated (*by*).

έμφροντις *a.* pensive, worried.

εμφύλιος *a.* ~ *πόλεμος* civil war.

εμφυσώ *v.t.* insufflate; instil.

εμφυτεύω *v.t.* plant, implant.

έμφυτ|ος *a.* innate. **~ο** *s.n.* instinct.

εμφωλεύω *v.i.* nestle; lurk.

έμψυχ|ος *a.* animate, living. **~ώνω** *v.t.* enliven, encourage.

εν *prep.* (*with dat.*) in; ~ *τούτοις* however. ~ *τάξει* all right.

εν(α) *num. & article n.* one; a, an. *See also* **ένας.**

εναγκαλίζομαι *v.t.* embrace.

ενάγ|ω *v.t.* (*law*) sue; *~ων* plaintiff; *~όμενος* defendant.

εναγωνίως *adv.* anxiously, impatiently.

εναέριος *a.* aerial, by air, overhead.

ενάλιος *a.* found in the sea, marine.

εναλλαγή *s.f.* alternation; exchange.

εναλλάξ *adv.* alternately.

εναλλάσσω *v.t.* alternate.

έναντι *prep.* (*with gen.*) towards (*in part payment*); against (*receipt*).

ενάντια *adv.* contrarily, adversely; ~ *σε* against.

εναντίον *adv. & prep.* (*with gen.*) against.

εναντίον *s.n.* opposite, contrary.

ενάντ|ιος, ~ίος *a.* contrary, adverse; opposed; *απ' ~ίας* on the contrary.

εναντιούμαι *v.t. & i.* (*with gen. or εις*) oppose, thwart.

εναποθέτω *v.t.* place, repose.

εναπόκειται *v.t.* *σε σένα* ~ it rests with you.

εναργής *a.* clear, lucid; evident.

έναρθρος *a.* articulate; (*gram*) with definite article.

εναρκτήριος *a.* inaugural, opening.

εναρμονίζω *v.t.* harmonize.

έναρξη *s.f.* opening, beginning.

ένας *a. & article m.* one; a, an; ~ ~ one by one; ~ *κι* ~ especially good *or* bad; *ο* ~ *τον άλλον* one another.

ενασχόληση *s.f.* occupation.

ένατος *a.* ninth.

ένδακρυς *a.* tearful.

ένδεια *s.f.* want, penury.

ενδεικν|ύω *v.t.* indicate. *~υμαι, ~ύομαι v.i.* be called for *or* necessary.

ενδεικτικ|ός *a.* indicative, symptomatic. **~όν** *s.n.* certificate (*scholastic*).

ένδειξη *s.f.* mark, indication.

ένδεκα *num.* eleven.

ενδελεχής *a.* assiduous.

ενδέχ|εται *v.i.* it is likely, there is a possibility; *~όμενον* eventuality. *~όμενος a.* eventual, possible.

ενδημικός *a.* endemic.

ενδιαιτώμαι *v.i.* (*with εις*) dwell, inhabit.

ενδιάμεσος *a.* in between; intervening.

ενδιατρίβω *v.i.* sojourn; dwell *or* insist (*on*).

ενδιαφέρον *s.n.* interest, concern.
ενδιαφέρ|ω *v.t.* interest, concern. *~ομαι v.i.* be interested (*with για or να*).
ενδιαφέρων *a.* interesting.
ενδίδω *v.i.* give way, give in.
ένδικος *a.* legal.
ενδοιάζω *v.i.* hesitate, be undecided.
ενδοιασμός *s.m.* scruple, compunction.
ενδόμυχος *a.* interior, inward.
ένδ|ον *adv. & prep.* (*with gen.*) within. *~ότερα s.n.pl.* innermost parts; inside story.
ένδοξος *a.* glorious, renowned.
ενδοτικός *a.* ready to make concessions; concessive.
ενδοχώρα *s.f.* hinterland.
ένδυμα *s.n.* garment; dress, attire. *~σία s.f.* suit, dress; attire.
ενδύ|ω. *v.t.* dress *~ομαι v.i. & t.* get dressed; put on.
ενέδρ|α *s.f.* ambush. *~εύω v.t. & i.* (lie in) ambush.
ένεκ|α, ~εν *prep.* (*with gen.*) on account of.
ενενή(κο)ντα *num.* ninety.
ενεός *a.* dumbfounded.
ενέργεια *s.f.* activity, action, operation; effort; energy.
ενεργητικ|ός *a.* energetic; active; (*med.*) laxative. *~όν s.n.* assets; εις το ~όν του to his credit (*morally*). *~ότης s.f.* activeness.
ενεργός *a.* active, operational; effective.
ενεργ|ώ *v.i. & t.* act, take steps; hold, carry out; open bowels of. *~ούμαι v.i.* take place; pass motion of bowels.
ένεση *s.f.* injection.
ενεστώς, *a.* present; (*gram.*) ο ~ present tense.
ενετικός *a.* Venetian.
ενέχυρ|ο *s.n.* pawn, pledge. *~οδανειστής s.m.* pawnbroker.
ενέχ|ω *v.t.* contain. *~ομαι v.i.* be implicated.
ενήλι|ξ, ~ικος *a.* of age, major. *~ικιούμαι v.i.* come of age.
ενήμερ|ος *a.* aware, informed. *~ώνω v.t.* inform; bring up to date.
ένθα *adv.* where.
ενθάδε *adv.* here.
ενθαρρ|ύνω *v.t.* encourage. *~υνση s.f.* encouragement.
ένθεν *adv.* hence; ~ και ~ on both sides.
ένθερμος *a.* warm, fervent.
ενθουσιά|ζω *v.t.* fill with enthusiasm. *~ομαι v.i.* become enthusiastic.
ενθουσιασμός *s.m.* enthusiasm.
ενθουσιαστικός *a.* inspiring enthusiasm.
ενθουσιώδης *a.* enthusiastic.
ενθρόνισης *s.f.* enthronement.
ενθυλακώνω *v.t.* pocket.

ενθυμίζ|ω *v.t.* bring to mind; μου ~ει τον Πέτρο he reminds me of Peter.
ενθύμιο *s.n.* reminder, souvenir.
ενθυμούμαι *v.t.* remember, recall.
ενιαίος *a.* single, unified.
ενιαύσιος *a.* annual; lasting one year.
ενιαυτός *s.m.* year.
ενικός *a.* (*gram.*) singular.
ένιοι *pron.* (*pl.*) some.
ενίοτε *adv.* sometimes.
ενισχ|ύω *v.t.* strengthen, reinforce; confirm. *~υση s.f.* strengthening, re-inforcement.
εννέα *num.* nine. *~κόσιοι a.* nine hundred.
εννιά *num.* nine.
εννιάμερα *s.n.pl.* prayers on ninth day after death.
έννοια (-i-) *s.f.* idea, concept; meaning.
έννοια (-j-) *s.f.* care; worry; ~ σου take care; (*threat*) or don't worry! άλλη ~ δεν έχω I couldn't care less.
έννομος *a.* legal.
εννο|ώ *v.t.* mean; intend; expect, require; understand; notice; *~είται* naturally or it is understood.
ενοικιάζ|ω *v.t.* let, rent, hire (*of either party*); *~εται* 'to let'.
ενοικιαστήριο *s.n.* lease; 'to let' notice.
ενοικιαστής *s.m.* lessee, tenant.
ενοίκι|ο *s.n.* hire; rent. *~οστάσιο s.n.* rent-control.
ένοικος *s.m.* occupant, inmate.
ενόν *s.n.* what is possible; θα φάμε εκ των ~των we shall take pot luck.
ένοπλος *a.* armed.
ενοποίησις *s.f.* unification.
ενοποιώ *v.t.* unify.
ενόρασις *s.f.* intuition.
ενόργανος *a.* organic; instrumental; with instruments.
ενορία *s.f.* parish.
ένορκ|ος *a.* sworn; a juryman. *~ως adv.* on oath.
ενορχήστρωσις *s.f.* orchestration.
ενόσω *adv.* while, as long as.
ενότης *s.f.* unity.
ενοχή *s.f.* guilt, culpability.
ενοχλ|ώ *v.t.* annoy, trouble; molest, pester. *~ηση s.f.* annoyance, trouble. *~ητικός a.* annoying, troublesome.
ένοχ|ος *a.* guilty, culpable. *~οποιώ v.t.* inculpate.
ενσάρκωσις *s.f.* incarnation.
ενσημ|ος *a.* stamped. *~ον s.n.* stamp (*for insurance, etc.*).
ενσκήπτω *v.i.* break out; appear or happen suddenly; burst upon the scene.
ενσπείρω *v.t.* sow, spread (*discord, etc.*).
ενσταλάζω *v.t.* pour drop by drop; instil.

ένστασις s.f. objection; (law) demurrer.
ενστερνίζομαι v.t. embrace, adopt.
ένστικτ|ο s.n. instinct. ~ώδης a. instinctive.
ενσυνείδητος a. conscious, deliberate.
ένσφαιρ|ος a. with bullets; ~ον φυσίγγιον ball cartridge; ~α πυρά live shots.
ενσφράγιστος a. under seal.
ενσωματώνω v.t. embody.
ένταλμα s.n. warrant, writ.
ένταση s.f. stretching, strain; strength, intensity; intensification.
εντατικός a. strong, intensive; aphrodisiac.
ενταύθα adv. here; local (in postal address).
ενταφιάζω v.t. inter.
εντείνω v.t. stretch; intensify, increase; raise (voice).
έντεκα num. eleven.
εντέλεια s.f. perfection.
εντελ|ής a. perfect, complete. ~ώς adv. completely, quite.
εντέλλομαι v.t. & i. order, bid; be ordered.
εντερικ|ός a. intestinal. ~ά (s.n.pl.) diarrhoea.
εντεριώνη s.f. pith.
έντερο s.n. intestine, bowel.
εντεταλμένος a. charged (with), responsible (for); a delegate.
εντεύθεν adv. hence; (with gen.) on this side of.
εντευκτήριον s.n. club, meeting-place, reception-room.
έντεχνος a. skilful, adroit.
έντιμος a. honest, honourable.
έντοκος a. bearing interest.
εντολή s.f. order; mandate; authorization; δέκα ~ές Ten Commandments.
εντομή s.f. incision.
έντομ|ο s.n. insect. ~οκτόνο s.n. insecticide. ~ολόγος s.m. entomologist.
έντονος a. vigorous; strong, vivid, intense.
εντοπίζω v.t. confine, localize.
εντόπιος a. local, native.
εντός adv. & prep. inside, within; ~ της πόλεως inside the city; ~ ολίγου shortly; η ~ επιφάνεια the inner surface.
εντόσθια s.n.pl. entrails.
εντράδα s.f. stewed meat with vegetables.
εντρ- see also ντρ-.
εντριβή s.f. friction, massage.
εντριβής a. versed, experienced.
εντρυφώ v.i. (with εις) take delight in.
έντυπ|ος a. printed. ~ov s.n. printed matter; form.
εντύπωσ|η s.f. impression; sensation. ~ιακός a. impressive.
ενυδρείον s.n. aquarium.
ένυδρος a. aquatic; hydrate.
ενυπόθηκος a. on mortgage.

ενώ conj. while; although.
ενωμοτάρχης s.m. sergeant of gendarmerie.
ενωμοτία s.f. squad.
ενώνω v.t. join, unite.
ενώπιον prep. (with gen.) before, in presence of.
ενωρίς adv. early.
ένωση s.f. union, combination; short circuit.
εξ prep. see εκ.
εξ num. six.
εξαγόμενον s.n. conclusion; (math.) product, result.
εξαγοράζω v.t. avoid by payment, buy off; acquire by payment; redeem, ransom; secure by bribery.
εξαγριώνω v. t. make fierce, infuriate.
εξάγω v. t. take or bring out, extract; export; deduce, infer.
εξαγωγή s.f. export(ation); extraction. ~εύς s.m. exporter.
εξαδέλφη s.f. cousin.
εξάδελφος s.m. cousin.
εξαερίζω v. t. ventilate.
εξαεριστήρας s.m. ventilator.
εξαερώ v.t. vaporize; gasify.
εξαίρεσ|ις, ~η s.f. exception; exemption; extraction; κατ'~ιν by way of exception.
εξαιρετικ|ός a. exceptional; beyond the usual degree; special, excellent. ~ά adv. very, exceptionally.
εξαίρ|ετος a. excellent, specially good. ~ετα adv. very well, famously. ~έτως adv. particularly (of good qualities).
εξαίρω v.t. stress, bring out; sing praises of.
εξαιρώ v.t. except; exempt.
εξαίσιος a. superb.
εξαίφνης adv. suddenly.
εξακολουθώ v.t. & i. continue.
εξακοντίζω v.t. hurl, fling.
εξακόσια a. six hundred.
εξακριβώνω v.t. verify, ascertain.
εξαλείφω v.t. efface, erase, remove.
έξαλλος a. beside oneself.
εξάμβλωμα s.n. abortion, monstrosity.
εξάμετρον s.n. hexameter.
εξάμην|ο s.n., ~ία s.f. six months; rent or pay for such period.
εξαναγκάζω v.t. compel.
εξανδραποδίζω v.t. enslave, subdue.
εξανεμίζομαι v.i. be squandered, go up in smoke.
εξάνθημα s.n. rash (on skin).
εξανίσταμαι v.i. revolt, protest.
εξαντλώ v.t. use up, exhaust.
εξάπαντος adv. for a certainty, without fail.
εξαπατώ v.t. cheat, delude, deceive.
εξαπίνης adv. unawares.
εξάπλωση s.f. spread(ing).

εξαποδώ(ς) *s.m.* (*fam.*) the devil.
εξαπολύω *v.t.* unleash.
εξάπτω *v.t.* excite, inflame.
εξαργυρώνω *v.t.* cash.
εξάρθρωσις *s.f.* dislocation.
έξαρσις *s.f.* exaltation.
εξάρτημα *s.n.* attachment; ~τα (*pl.*) gear, tackle.
εξάρτησις *s.f.* dependence.
εξάρτια *s.n.pl.* (*naut.*) shrouds.
εξάρτυσις *s.f.* (*mil.*) soldier's equipment.
εξαρτώμαι *v.i.* depend.
έξαρχος *s.m.* exarch.
εξασθένησις *s.f.* enfeeblement, weakening.
εξασκ|ώ *v.t.* exercise, drill; discharge (*function*); practise (*profession*); exert (*pressure, etc.*). ~ησις *s.f.* exercise, practice.
εξασφαλίζω *v.t.* make certain of, assure; book, reserve.
εξατμίζ|ω *v.t.* evaporate; (*v.i.*) give off steam. ~ομαι *v.i.* evaporate.
εξάτμισις *s.f.* evaporation, exhaust (*of motor*).
εξαϋλώ *v.t.* render incorporeal; reduce to a shadow.
εξαφαν|ίζω *v.t.* spirit away; wipe out, eliminate. ~ίζομαι *v.i.* disappear; ~ισθέντες (*pl.*) (*mil.*) missing.
έξαφν|ος *a.* sudden, unexpected. ~α *adv.* suddenly; for instance.
εξαχνίζω *v.t.* sublimate.
εξαχρειώνω *v.t.* corrupt, deprave.
έξαψη *s.f.* heat, fit of anger *or* excitement; flushing.
εξεγείρ|ω *v.t.* rouse. ~ομαι *v.i.* rise, revolt.
εξέγερση *s.f.* uprising, revolt.
εξέδρα *s.f.* dais; (*spectators'*) stand; pier.
εξέδραμον *v.i.* (*aorist*) went for an outing.
εξεζητημένος *a.* far-fetched, recherché.
εξε(ι)λιγμένος *a.* developed, advanced.
εξελικτικός *a.* evolutionary; capable of developing.
εξέλιξη *s.f.* evolution, development.
εξελίσσομαι *v.i.* evolve, develop, unfold.
εξελληνίζω *v.t.* hellenize; translate into Greek.
εξεμώ *v.t.* vomit.
εξεπίτηδες *adv.* on purpose.
εξερευνώ *v.t.* explore, investigate. ~ησις *s.f.* exploration, investigation. ~ητής *s.m.* explorer.
εξέρχ|ομαι *v.i.* go *or* come out, leave; τα ~όμενα out-tray.
εξετάζω *v.t.* examine, interrogate.
εξέταση *s.f.* examination, interrogation.
εξεταστής *s.m.* examiner.
εξέταστρα *s.n.pl.* examination fees.

εξευγενίζω *v.t.* ennoble; improve (*stock*).
εξευμενίζω *v.t.* propitiate.
εξευρ|ίσκω *v.t.* manage to find. ~εσις *s.f.* finding.
εξευτελίζω *v.t.* cheapen, lower.
εξευτελιστικός *a.* degrading.
εξέχ|ω *v.i.* stand out. ~ων *a.* prominent.
έξη, έξι *num.* six.
εξήγηση *s.f.* explanation, exposition; translation.
εξηγ|ώ *v.t.* explain, interpret. ~ούμαι *v.i.* make oneself clear, explain.
εξή(κο)ντα *num.* sixty.
εξημερώ *v.t.* tame, civilize.
εξημμένος *a.* heated; conceited.
εξηνταβελόνης *s.m.* miser.
εξηντλημένος *a.* used up, exhausted.
εξής *adv.* ως ~ as follows; τά ~ the following; εις το ~ from now on; και ούτω καθ'~ and so on.
εξιδανικεύω *v.t.* idealize, sublimate.
εξιλασμός *s.m.* propitiation, atonement.
εξιλεών|ω *v.t.* appease. ~ομαι *v.i.* atone.
έξις, ~η *s.f.* habit.
εξίσταμαι *v.i.* be filled with wonder.
εξιστορώ *v.t.* relate in detail.
εξίσωσις *s.f.* equalizing, balancing; (*math.*) equation.
εξιτήριο *s.n.* discharge note.
εξιχνιάζω *v.t.* ferret out, discover.
εξοβελίζω *v.t.* strike out, get rid of.
εξόγκωμα *s.n.* swelling, protuberance.
εξογκώνω *v.t.* swell; exaggerate.
εξόγκωση *s.f.* swelling; exaggeration.
εξοδεύ|ω *v.t.* spend, consume, use (up); sell (*of shops*); waste. ~ομαι *v.i.* be put to expense.
έξοδον *s.n.* expense.
έξοδος *s.f.* going out; way out, exit; sortie; retirement.
εξοικειώνω *v.t.* accustom, familiarize.
εξοικονομ|ώ *v.t.* manage to find; fix up (*person*); meet (*situation*). ~ούμαι *v.i.* make do.
εξοκέλλω *v.i.* run aground; go astray.
εξολοθρεύω *v.t.* exterminate.
εξομαλύνω *v.t.* make level; smooth down.
εξομοιώνω *v.t.* liken, compare.
εξομολόγηση *s.f.* confession.
εξομολογ|ώ *v.t.* confess (*penitent*). ~ούμαι *v.t. & i.* acknowledge, confess.
εξόν *prep.* except; ~ αν unless.
εξοντώνω *v.t.* annihilate.
εξονυχίζω *v.t.* examine minutely.
εξοπλίζω *v.t.* arm.
εξοπλισμός *s.m.* armament, arming; equipment. ηθικός ~ moral principles.
εξοργίζω *v.t.* anger.

εξορ|ία *s.f.* banishment, exile. **~ίζω** *v.t.* banish, exile.

εξορκίζω *v.t.* conjure; exorcise.

εξορμώ *v.i.* rush forth; make a sortie.

εξορύσσω *v.t.* dig up; pluck out.

εξοστρακίζω *v.t.* ostracize; get rid of.

εξουδετερώ *v.t.* neutralize.

εξουθενίζω *v.t.* defeat *or* exhaust utterly.

εξουσία *s.f.* power, authority, right.

εξουσιάζω *v.i. & t.* be in authority (over).

εξουσιοδοτώ *v.t.* authorize, charge.

εξόφθαλμος *a.* with protuberant eyes; clear as daylight.

εξοφλ|ώ *v.t. & i* pay off, discharge (*debt*); (*v.i.*) have done (*with*); be finished, have exhausted one's talent. **~ηση** *s.f.* payment of debt, settlement.

εξοχ|ή *s.f.* 1. country(side), rural spot *or* resort. 2. protrusion; *κατ' ~ήν* pre-minently.

εξοχικός *a.* country, rural.

έξ|οχος *a.* eminent; excellent. **~όχως** *adv.* particularly, very.

εξοχότης *s.f.* excellence; leading light; *η αυτού* ~ His Excellency.

εξτρεμιστής *s.m.* (*violent*) extremist.

εξυβρίζω *v.t.* abuse, vilify.

εξυγιαίνω *v.t.* cleanse; restore to healthy condition.

εξυμνώ *v.t.* glorify, sing praises of.

εξυπακούεται *v.i.* it is understood (*supplied mentally*).

εξυπηρέτ|ησις *s.f.* service, assistance. **~ικός** *a.* helpful, of service.

εξυπηρετώ *v.t.* serve, assist, be of service to.

έξυπν|ος *a.* wide-awake, smart, clever, intelligent. **~άδα** *s.f.* cleverness; wise-crack.

εξυπνώ *v.t. & i.* wake up.

εξυφαίνω *v.t.* weave; (*fig.*) hatch (*plot*).

εξυψώνω *v.t.* elevate; exalt.

έξω *adv. & prep.* out, outside; abroad; *απ'* ~ (from) outside, by heart; ~ *απ'* ~ in a roundabout way; ~ *της πόλεως* outside the city; ~ *από* outside, except for; *μια κι* ~ in one go; ~ *νου* away with cares!

εξώγαμος *a.* illegitimate, bastard.

εξώδικος *a.* unofficial, out-of-court.

έξωθεν *adv.* from outside *or* abroad; outward(ly).

εξώθυρα *s.f.* outside door.

εξωθώ *v.t.* push, drive (*to action*).

εξωκκλήσιον *s.n.* country chapel.

εξώλης *a.* depraved; (*fam.*) ~ *και προώλης* a thoroughly bad lot.

έξωμος *a.* décolleté.

εξωμότης *s.m.* renegade.

εξώπορτα *s.f.* outside door.

εξωραΐζω *v.t.* beautify.

εξώρας *adv.* late.

έξωση *s.f.* expulsion, eviction.

εξώστης *s.m.* balcony.

εξωστρεφής *a.* extrovert(ed).

εξωτερικεύ|ω *v.t.* reveal (*thoughts, feelings*). **~ομαι** *v.i.* reveal one's thoughts *or* feelings.

εξωτερικ|ός *a.* external, exterior; foreign. **~ό** *s.n.* exterior; outward appearance; foreign parts.

εξωτικ|ός *a.* exotic; (*eccl.*) lay. **~ό** *s.n.* supernatural being.

εξωφρενικός *a.* mad, crazy; maddening; inconceivable.

εξωφρενισμός *s.m.* extravagance, eccentricity.

εξώφυλλο *s.n.* cover (*of book*); outside shutter.

εορτάζω *v.t. & i.* celebrate; celebrate one's name-day *or* patronal festival.

εορτάσιμος *a.* observed as a holiday; festive.

εορτή *s.f.* festival, holiday; name-day.

επαγγελία *s.f.* promise.

επαγγέλλομαι *v.t.* promise; be by profession; profess to be.

επάγγελμα *s.n.* profession, trade, calling. **~τικός** *a.* professional.

επαγγελματίας *s.m.* one who plys a trade *or* profession (*not as employee*); professional.

επαγρυπνώ *v.i.* (*with επί*) watch over, look after.

επαγωγ|ή *s.f.* induction. **~ικός** *a.* 1. inductive. 2. charming.

έπαθλον *s.n.* prize, reward.

έπαιν|ος *s.m.* praise; honourable mention. **~ετικός** *a.* complimentary. **~ετός** *a.* praiseworthy. **~ώ** *v.t.* praise, commend.

επαίρομαι *v.i.* be puffed up, swagger.

επαισθητός *a.* perceptible, appreciable.

επαίσχυντος *a.* shameful.

επαίτης *s.m.* beggar.

επαιτώ *v.t. & i.* beg.

επακόλουθ|ος *a.* ensuing. **~α** *s.n.pl.* consequences.

επακολουθώ *v.t. & i.* follow; ensue.

έπακρον *s.n. εις το* ~ in the highest degree.

επάκτιος *a.* coastal.

επαλείφω *v.t.* rub *or* paint (*with medicament*).

επαληθεύω *v.i. & t.* come true; verify.

επάλληλος *a.* successive.

έπαλξ|ις *s.f.* battlement; *επί των ~εων* in the forefront of battle.

επαμφοτερίζω *v.i.* waver; be ambiguous.

επαν(α)- *denotes* again, back, upon.

επάναγκες *s.n.* είναι ~ να it is imperative that...

επανάγω *v.t.* bring back.

επανακάμπτω *v.i.* return.

επανακτώ *v.t.* get back.

επαναλαμβάνω *v.t. & i.* resume; repeat.

επανάληψ|ις, *s.f.* resumption; repetition; *κατ' ~ιν* repeatedly.

επαναπαύομαι *v.i.* rest (*on one's laurels*); rely *or* count (*on*).

επανάσταση *s.f.* revolution.

επαναστάτης *s.m.* insurgent; revolutionary. **~ικός** *a.* revolutionary.

επαναστατώ *v.i. & t.* rebel, revolt; rouse to rebellion; upset.

επανασυνδέω *v.t.* re-establish (*connexion*), join together again.

επαναφέρω *v.t.* bring back, restore; recall (*to mind*).

επανδρώνω *v.t.* man, staff.

επανειλημμέν|ος *a.* repeated. **~ως** *adv.* repeatedly.

επανεκλέγω *v.t.* re-elect.

επανέρχομαι *v.i.* come back.

επάνοδος *s.f.* return.

επανορθ|ώνω *v.t.* redress, rectify. **~ωσις** *s.f.* reparation.

επάνω *adv. see* απάνω.

επανωφόρι *s.n.* overcoat.

επαξίως *adv.* fittingly, deservedly.

επαπειλ|ώ *v.t.* threaten; **~είται** (*v.i.*) there is a threat of.

επάρατος *a.* accursed.

επάργυρος *a.* silver-plated.

επάρκεια *s.f.* sufficiency.

επαρκής *a.* sufficient, adequate. **~ώ** *v.i.* suffice.

έπαρση *s.f.*. hoisting; conceit.

επαρχί|α *s.f.* eparchy (*administrative division*); (*fig.*) the provinces. **~ώτης** *s.m.* provincial. **~ακός, ~ώτικος** *a.* provincial.

έπαυλις *s.f.* villa.

επαύριον *adv. η* ~ the next day, tomorrow.

επαφή *s.f.* touch, contact.

επαχθής *a.* burdensome.

επείγ|ω *v.i.* be urgent. **~ομαι** *v.i.* be in a hurry.

επείγ|ων *a.* urgent. **~όντως** *adv.* urgently, hurriedly.

επειδή *conj.* since, because.

επεισόδιο *s.n.* incident, episode; scene, row.

έπειτα *adv.* then, next; besides; ~ *από* after.

επέκτασις *s.f.* extension; spreading.

επεκτείνω *v.t.* extend; spread.

επεμβαίνω *v.i.* intervene; interfere.

επέμβασις *s.f.* intervention; interference; (*med.*) operation.

επένδυσις *s.f.* facing, lining, covering;

(*fin.*) investment.

επενδύω *v.t.* line, cover; invest.

επενεργώ *v.i.* act, produce an effect (*on*).

επεξεργάζομαι *v.t.* elaborate, work out; process.

επεξεργασία *s.f.* polishing up (*of speech, etc.*); processing (*of materials*).

επεξήγησις *s.f.* elucidation.

επεξηγώ *v.t.* explain in detail.

επέρχομαι *v.i.* come on suddenly, occur; (*with κατά*) attack.

επερώτησις *s.f.* interpellation.

επέτειος *s.f.* anniversary.

επετηρίς *s.f.* anniversary; year-book; (*army, etc.*) list.

επευφημώ *v.t.* cheer, acclaim.

επήκο|ος *a.* εις ~ον πάντων in everyone's hearing.

επηρεάζω *v.t.* influence; have (bad) effect on.

επήρεια *s.f.* influence.

επηρμένος *a.* puffed up with pride.

επί *prep.* 1. (*with acc.*) towards; multiplied by; for (*duration*); ~ πλέον in addition; ~ τα βελτίω for the better. 2. (*with gen.*) on; in the time of, under; ~ τέλους at last; ~ κεφαλής at the head. 3. (*with dat.*) (*occasion, circumstance*) ~ τη ευκαιρία by the way, on the occasion of; ~ κλοπή (*charge*) of theft.

επίατρος *s.m.* (*mil.*) army doctor.

επιβαίνω *v.t.* (*with gen.*) mount; go aboard; (*of animals*) cover.

επιβάλλον *s.n.* air of authority.

επιβάλλ|ω *v.t.* impose, inflict, dictate. **~ομαι** *v.i.* impose one's authority, command respect.

επιβάρυνση *s.f.* (extra) charge.

επιβαρύνω *v.t.* aggravate, burden.

επιβάτ|ης *s.m.* passenger. **~ηγόν** *s.n.* passenger ship. **~ικόν** *s.n.* passenger vehicle *or* ship.

επιβεβαιώνω *v.t.* confirm.

επιβεβλημένος *a.* imperative, bounden.

επιβήτωρ *s.m.* stud-animal; usurper.

επιβιβάζ|ω *v.t.* embark, put aboard. **~ομαι** *v.i.* embark, go aboard.

επιβίωσις *s.f.* survival.

επιβλαβής *a.* injurious.

επιβλέπω *v.t. & i.* supervise, invigilate.

επιβλητικός *a.* imposing.

επιβολή *s.f.* imposition, infliction; authority, sway.

επιβουλεύομαι *v.t.* plot against; wish ill to.

επίβουλος *a.* insidious.

επιβραβεύω *v.t.* award prize to.

επιβραδύνω *v.t.* retard; slow down.

επιγαμία *s.f.* intermarriage.

επίγειος *a.* earthly.
επίγνωση *s.f.* awareness, realization.
επίγονος *s.m.* descendant.
επίγραμμα *s.n.* epigram.
επιγραφ|ή *s.f.* inscription; shop-sign; heading, title (*of book*); address (*of letter*). **~ική** *s.f.* epigraphy.
επιδαψιλεύω *v.t.* lavish.
επιδεικνύω *v.t.* display, exhibit; show off. **~ομαι** *v.i.* show off.
επιδεικτικός *a.* ostentatious; showy.
επιδεινώ, ~ώνω *v.t.* make worse.
επίδειξη *s.f.* display; showing off; (*mil.*) diversion.
επιδεκτικός *a.* susceptible (*of*).
επιδέξιος *a.* adroit.
επιδερμίς *s.f.* epidermis; (*fam.*) complexion.
επίδεσμος *s.m.* bandage.
επιδέχομαι *v.t.* admit of; put up with.
επιδημία *s.f.* epidemic.
επιδίδ|ω *v.t.* hand over, present. **~ομαι** *v.i.* apply oneself.
επιδικάζω *v.t.* (*law*) award.
επιδιορθ|ώνω *v.t.* repair. **~ωση** *s.f.* repair(ing).
επιδιώκω *v.t.* be ambitious of, aim at.
επιδοκιμάζω *v.t.* approve of.
επιδοκιμασία *s.f.* approval.
επίδομα *s.n.* extra payment, allowance.
επίδοξος *a.* presumptive, to be.
επιδόρπια *s.n.pl.* dessert.
επίδοση *s.f.* presentation, delivery; applying oneself; proficiency; record (*in performance*).
επιδότηση *s.f.* subsidy, bounty.
επίδραση *s.f.* effect; influence.
επιδρομή *s.f.* incursion, raid.
επιδρώ *v.i.* (*with επί*) influence, have an effect on.
επειι|κής *a.* lenient. **~εια** *s.f.* leniency.
επίζηλος *a.* enviable.
επιζήμιος *a.* harmful.
επιζητώ *v.t.* be ambitious of, hope for.
επιζώ *v.i.* survive.
επιθανάτιος *a.* death (*bed, agony, etc.*).
επίθεμα *s.n.* compress.
επίθεση *s.f.* affixing, application; attack.
επιθετικός *a.* 1. aggressive, offensive. 2. adjectival.
επίθετο *s.n.* epithet; adjective; surname.
επιθεώρησις *s.f.* inspection; inspectorate; review; revue.
επιθεωρώ *v.t.* inspect, review.
επιθυμία *s.f.* desire.
επιθυμ|ώ *v.t.* desire, wish for; *την~ησα* I miss her. **~ητός** *a.* desirable.
επίκαιρ|ος *a.* timely; well-placed; topical;

τα ~α current events.
επικαλ|ούμαι *v.t.* invoke, appeal to; *~ούμενος* called, nicknamed.
επικαρπία *s.f.* enjoyment; (*law*) usufruct.
επίκειμαι *v.i.* be imminent.
επικερδής *a.* lucrative.
επικεφαλίς *s.f.* heading.
επικήδειος *a.* funeral.
επικήρυξη *s.f.* setting a price on person's head.
επικίνδυνος *a.* dangerous.
επίκληση *s.f.* invocation; nickname, personal epithet.
επικλινής *a.* sloping.
επικοινων|ία *s.f.* intercourse, contact, communication. **~ώ** *v.i.* communicate.
επικός *a.* epic; an epic poet.
επικουρία *s.f.* aid; (*mil.*) relief.
επικράτεια *s.f.* state, land.
επικρατώ *v.i. & t.* prevail, predominate, reign; (*with gen.*) prevail against.
επικρέμαμαι *v.i.* threaten, be imminent.
επικρίνω *v.t.* criticize, censure.
επικριτικός *a.* critical, fault-finding.
επικροτώ *v.t.* approve, applaud.
επίκτητος *a.* acquired.
επικυριαρχία *s.f.* suzerainty.
επικυρώνω *v.t.* ratify, confirm.
επιλαμβάνομαι *v.t.* (*with gen.*) broach (*subject*); address oneself to (*task*).
επιλαχών *s.m.* runner-up (*of candidates*).
επιλεγόμενος *a.* called.
επιλέγω *v.t.* choose, select.
επίλεκτος *a.* picked, outstanding (*of persons*).
επιληψία *s.f.* epilepsy.
επιλήψιμος *a.* reprehensible.
επιλογή *s.f.* selection.
επίλογος *s.m.* epilogue; conclusion.
επίλοιπος *a.* remaining.
επιλοχίας *s.m.* (*mil.*) sergeant-major.
επίμαχος *a.* contested, disputed; controversial.
επιμέλ|εια *s.f.* care, diligence. **~ής** *a.* diligent.
επιμελητεία *s.f.* (*mil.*) commissariat.
επιμελητήριον *s.n.* board, chamber; *Εμπορικόν Ε ~* Chamber of Commerce.
επιμελητής *s.m.* keeper, superintendent; university tutor; editor (of text).
επιμελούμαι *v.t.* look after, care for.
επιμένω *v.t.* persist, insist.
επίμετρον *s.n. εις ~* to boot, for good measure.
επιμήκης *a.* oblong.
επιμιξία *s.f.* inter-marriage; cross-breeding; intercourse.
επιμίσθιον *s.n.* extra pay.

επίμον|ος *a.* persistent; stubborn; pressing, urgent. **~ή** *s.f.* perseverance; obstinacy; isistence.

επίμοχθος *a.* toilsome.

επιμύθιον *s.n.* moral (*of tale*).

επίνειον *s.n.* port serving inland town.

επινίκι|ος *a.* of victory; τα ~α celebration of victory.

επινό|ημα *s.n.* device. **~ητικός** *a.* inventive.

επινοώ *v.t.* invent, devise; fabricate.

επίορκος *s.m.* breaker of oaths.

επιορκώ *v.i.* break one's oath.

επιούσα *s.f.* next day, morrow.

επιούσιος *a.* ~ άρτος daily bread.

επίπαγος *s.m.* crust (*on soft or liquid substance*).

επίπεδο *s.n.* level; (*geom.*) plane.

επίπεδος *a.* plane, level, flat.

επιπίπτω *v.i.* (*with κατά*) fall upon.

επίπλαστος *a.* feigned.

επιπλέω *v.i.* float; (*fig.*) be among the successful.

επίπληξη *s.f.* reprimand.

επιπλήττω *v.t.* reprimand.

επιπλοκή *s.f.* complication.

έπιπλ|ο *s.n.* piece of furniture; ~α (*pl.*) furniture.

επιπλώνω *v.t.* furnish.

επίπλωση *s.f.* furnishing.

επιπόλαιος *a.* superficial.

επίπονος *a.* hard, laborious.

επιπροσθέτως *adv.* in addition.

επίπτωσις *s.f.* effect, repercussion; incidence (*of tax*).

επιρρεπής *a.* inclined, prone (*to*).

επίρρημα *s.n.* adverb.

επιρρίπτω *v.t.* impute.

επιρροή *s.f.* influence.

επισημαίνω *v.t.* stamp; mark, pinpoint.

επισείων *s.m.* pennant.

επίσημ|ος *a.* official; formal; marking a special occasion; οι ~οι official personalities. **~ότης** *s.f.* formality, solemnity.

επίσης *adv.* too, likewise.

επισιτισμός *s.m.* food supply.

επισκεπτήριο *s.n.* visiting-card.

επισκέπτης *s.m.* visitor.

επισκέπτομαι *v.t.* visit.

επισκευ|άζω *v.t.* repair. **~ή** *s.f.* repair(ing).

επίσκεψη *s.f.* visit.

επισκιάζω *v.t.* overshadow, eclipse.

επίσκοπ|ος *s.m.* bishop. **~ή** *s.f.* bishopric; diocese; bishop's palace.

επισκοπώ *v.t.* survey, inspect.

επισμηναγός *s.m.* (*aero.*) squadron-leader.

επισπεύδω *v.t.* hasten, advance date of.

επισταμένως *adv.* with expert attention.

επιστασία *s.f.* supervision.

επιστάτης *s.m.* custodian, janitor, overseer, caretaker.

επιστατώ *v.t. & i.* superintend; be in charge.

επιστεγάζω *v.t.* roof; (*fig.*) crown, cap.

επιστήθιος *a.* ~ φίλος bosom friend.

επιστήμη *s.f.* branch of learning; science.

επιστημονικός *a.* scientific.

επιστήμων *s.m.f.* professionally qualified person; scientist; scholar.

επιστολή *s.f.* letter.

επιστόμιον *s.n.* mouthpiece; valve.

επιστρατεύω *v.t.* call up, mobilize.

επίστρατος *a.* called up, mobilized.

επιστρέφω *v.t. & i.* give back; come back; (*fig.*) rebound.

επιστροφή *s.f.* return.

επίστρωμα *s.n.* covering, layer.

επιστρώνω *v.t.* coat, cover.

επιστύλιον *s.n.* architrave.

επισύρω *v.t.* draw upon *or* attract to oneself.

επισφαλής *a.* precarious.

επισφραγίζω *v.t.* set seal on, confirm; crown, round off.

επισωρεύω *v.t.* pile up, accumulate.

επιταγή *s.f.* order; cheque.

επιτακτικός *a.* imperative.

επίτακτος *a.* requisitioned.

επίταξη *s.f.* requisition.

επιτάσσω *v.t.* enjoin; requisition.

επιτάφιος *a.* funeral; (*s.m.*) Good Friday procession and office.

επιταχύνω *v.t.* increase speed of; advance date of.

επιτείνω *v.t.* intensify.

επιτελάρχης *s.m.* (*mil.*) chief of staff.

επιτελείον *s.n.* chief collaborators *or* staff; (*mil.*) general staff.

επιτελής *s.m.* member of general staff.

επιτέλους *adv.* at last.

επιτετραμμένος *s.m.* chargé d'affaires.

επίτευγμα *s.n.* achievement, success.

επίτευξη *s.f.* attainment, securing.

επιτήδειος *a.* suitable (*for*); clever, smart; deft; crafty.

επίτηδες *adv.* on purpose.

επιτήδευμα *s.n.* trade, occupation; tax on this.

επιτηδεύ|ομαι *v.t. & i.* feign, affect; (*with εις*) be skilled in; ~μένος affected.

επιτήρηση *s.f.* surveillance, observation; invigilation.

επιτηρώ *v.t.* supervise, keep watch on, invigilate.

επιτίθεμαι *v.i. & i.* attack, assault (*with κατά or εναντίον or gen.*).

επίτιμος *a.* honorary.

επιτιμώ *v.t.* reprimand.

επιτόκιον *s.n.* rate of interest.

επίτοκος *a.* about to give birth.
επιτομή *s.f.* summary; abridgement.
επίτομος *a.* short, concise; abridged.
επιτόπιος *a.* local; on the spot.
επιτραπέζιος *a.* for *or* on the table.
επιτρέπω *v.t.* permit.
επιτροπή *s.f.* committee; commission.
επίτροπος *s.m.* guardian, trustee; church-warden; commissioner, agent; commissar.
επιτυγχάνω *v.t. & i.* attain, get; hit, get right; bring off; meet, find (*by chance or design*); be successful.
επιτυχημένος *a.* successful, effective.
επιτυχής *a.* successful.
επιτυχία *s.f.* success.
επιφάνεια *s.f.* surface; *κατ' ~ν* in appearance (*but not reality*). **~κός** *a.* superficial.
επιφανής *a.* distinguished.
Επιφάνια *s.n.pl.* Epiphany.
επίφαση *s.f.* external appearance.
επιφέρω *v.t.* bring about, occasion; *ο ~ν* bearer (*of letter, etc.*).
επίφοβος *a.* dangerous, unsafe.
επιφοίτησις *s.f.* divine inspiration.
επιφορτίζω *v.t.* charge, entrust (*duty to person*).
επιφυλακή *s.f. εν ~* on the alert.
επιφυλακτικός *a.* circumspect.
επιφύλαξη *s.f.* reserve; reservation (*of rights*).
επιφυλάσσω|ω *v.t.* hold in store. **~ομαι** *v.i.* behave with reserve, wait for a suitable occasion.
επιφυλλίδα *s.f.* newspaper serial, literary article.
επιφώνημα *s.n.* exclamation; (*gram.*) interjection.
επιχαρίτως *adv.* graciously.
επίχειρα *s.n.pl.* deserts, just recompense (*of badness*).
επιχείρημα *s.n.* argument; attempt.
επιχειρηματίας *s.m.* businessman, entrepreneur.
επιχειρηματικός *a.* enterprising.
επιχείρησις *s.f.* undertaking, (business) enterprise; (*strategic*) operation.
επιχειρώ *v.t.* undertake; attempt.
επιχορήγησις *s.f.* subvention; civil list.
επίχρισμα *s.n.* coat, veneer.
επιχρίω *v.t.* coat, paint.
επίχρυσος *a.* gilt, gold-plated.
επιχρυσώνω *v.t.* gild.
επιχωματώνω *v.t.* level up, embank.
επιχωριάζω *v.i.* be in use, be found (*in locality*).
επιψηφίζω *v.t.* sanction by vote.
εποίκηση *s.f.* settlement (*abroad*).

εποίκ|ιση *s.f.* **~ισμός** *s.m.* settlement (*at home*).
εποικοδομητικός *a.* edifying.
έπομαι *v.i.* follow.
επό|μενος *a.* following, next; *την ~μένη* on the next day; *ως ήτο ~μενον* as was to be expected. **~μένως** *adv.* consequently.
επονείδιστος *a.* shameful, ignominious.
επονομάζω *v.t.* call, surname.
εποποιία *s.f.* epic (poetry).
εποπτεύω *v.t.* supervise, oversee.
έπος *s.n.* epic poem; *αμ' ~ αμ' έργον* no sooner said than done.
εποστρακίζομαι *v.i.* ricochet.
επουλώνω *v.t.* heal (*wound*).
επουσιώδης *a.* of slight importance, unessential.
εποφθαλμιώ *v.t.* covet, have one's eye on.
εποχή *s.f.* epoch, era, age; season, time; *αφήνω ~* leave one's mark, cause a stir.
έποψη *s.f.* point of view.
επτά *num.* seven. **~κόσιοι** *a.* seven hundred.
επωάζω *v.t.* incubate.
επωδός *s.f.* refrain.
επώδυνος *a.* painful.
επωμίζομαι *v.t.* shoulder (*burden*).
επωμίς *s.f.* epaulette.
επωνυμία *s.f.* nickname, name; title (*of company, etc.*).
επώνυμο *s.n.* surname.
επωφελής *a.* useful, profitable.
επωφελούμαι *v.i.* (*with gen. or από & acc.*) take advantage of.
έρανος *s.m.* fund, collection; contribution.
ερασιτέχνης *s.m.* amateur.
ερασμι(α)κός *a.* Erasmian.
εράσμιος *a.* charming, amiable.
εραστής *s.m.* lover.
εργάζομαι *v.i.* work; function, go.
εργαλείο *s.n.* tool, implement.
εργασία *s.f.* work, employment; workmanship.
εργάσιμος *a.* that can be wrought *or* tilled, *etc.*; working (*hours, day*).
εργαστήριον *s.n.* workshop; studio; laboratory.
εργάτης *s.m.* worker, workman, labourer; (*naut.*) windlass.
εργατιά *s.f.* working class.
εργατικ|ός *a.* of *or* for workers; industrious; a working-class man. **~όν κόμμα** Labour party. **~ά** *s.n.pl.* wages.
εργένης *s.m.* bachelor.
εργοδηγός *s.m.* foreman.
εργοδότης *s.m.* employer.
εργολάβος *s.m.* 1. contractor. 2. beau, lover. 3. kind of macaroon.
εργοληψία *s.f.* (*building, etc.*) contract.

έργ|ον *s.n.* (piece of) work, thing done, task; book, play, film, musical work; ~α (*pl.*) building, *etc.*, operations.
εργοστασιάρχης *s.m.*. factory owner *or* manager.
εργοστάσιο *s.n.* factory, works.
εργοτάξιον *s.n.* construction site.
εργόχειρο *s.n.* needlework.
έρεβος *s.n.* Erebus; pitch darkness.
ερεθίζω *v.t.* irritate, excite.
ερέθισμα *s.n.* irritant; spur, stimulus.
ερεθισμός *s.m.* irritation, excitation.
ερεθιστικός *a.* irritating.
ερείπιο *s.n.* ruin, wreck.
ερειπώνω *v.t.* reduce to ruins.
έρεισμα *s.n.* prop, support.
έρευνα *s.f.* search, research, investigation.
ερευνητής *s.m.* researcher; view-finder.
ερευνώ *v.t.* & *i.* search, examine.
ερήμην *adv.* (*law*) by default.
ερημ|ιά *s.f.* solitude; wilderness. ~ικός *a.* solitary, lonely.
ερημίτης *s.m.* hermit.
ερημοκλήσι *s.n.* isolated country chapel.
ερημονήσι *s.n.* uninhabited island.
έρημος 1. *a.* desolate, deserted, lonely. 2. *s.f.* desert.
ερημώνω *v.t.* lay waste.
ερίζω *v.i.* dispute, quarrel.
Ερινύες *s.f. pl.* Furies.
έρι|ον *s.n.* wool. ~ουργία *s.f.* wool industry. ~ούχον *s.n.* woollen material.
έρις *s.f.* contention, quarrel.
ερίτιμος *a.* valuable; esteemed.
ερίφης *s.n.* (*fam.*) poor devil.
ερίφιο *s.n.* baby kid.
έρμα *s.n.* ballast.
έρμαιο *s.n.* thing tossed by waves; prey, victim.
ερμάριον *s.n.* cupboard, wardrobe.
ερμηνεία *s.f.* interpretation, expounding.
ερμηνευτής *s.m.* interpreter.
ερμηνεύω *v.t.* interpret, expound.
ερμητικός *a.* hermetic.
έρμος *a.* poor, unfortunate, left to one's fate.
ερπετό *s.n.* reptile.
ερπύστρι|α *s.f.* track (*of tractor, tank*). ~οφόρον *s.n.* tracked vehicle.
έρπω *v.i.* crawl.
έρρινος *a.* nasal (*sound*).
ερύθημα *s.n.* blush(ing).
ερυθριώ *v.i.* blush.
ερυθρόδερμος *s.m.* redskin.
ερυθρός *a.* red.
ερυθρωπός *a.* reddish.
έρχ|ομαι *v.i.* come; μου ~εται να I feel a desire to; σας ~εται καλά it suits you; δεν ~εται βολικά it is not convenient; ~εται σε

λογαριασμό it will just about do *or* he is amenable to reason.
ερχόμενος *a.* coming, next.
ερχομός *s.m.* coming, arrival.
ερώμαι *v.t.* (*with gen.*) fall in love with.
ερωμένη *s.f.* mistress, lover.
ερωμένος *s.m.* lover, paramour.
έρ|ως, ~ωτας *s.m.* love, passion.
ερωτεύ|ομαι *v.t.* & *i.* fall in love (with); ~μένος in love.
ερώτημα *s.n.* question, query; problem requiring solution.
ερωτηματικ|ός *a.* interrogative. ~'όν *s.n.* question-mark.
ερωτηματολόγιον *s.n.* questionnaire.
ερώτηση *s.f.* question, inquiry; interrogative sentence.
ερωτιάρ|ης, ~ικος *a.* amorous.
ερωτικ|ός *a.* erotic; amatory; ~ό ποίημα love-poem; ~ό ζεύγος pair of lovers.
ερωτοδουλειά *s.f.* love-affair; intrigue.
ερωτοτροπ|ία *s.f.* flirtation; ~ες (*pl.*) amorous behaviour.
ερωτύλος *s.m.* philanderer.
ερωτώ *v.t.* & *i.* ask, question (*person*); ask (*a question*); inquire.
εσκεμμένος *a.* premeditated.
εσοδεία *s.f.* crop, harvest.
εσοδεύω *v.t.* gather in, harvest.
έσοδο *s.n.* income, revenue.
εσοχή *s.f.* recess, indentation; recessed top floor of building; top flat.
εσπέρα *s.f.* evening.
Εσπερία *s.f.* western Europe.
εσπεριδοειδή *s.n.pl.* citrus fruits.
εσπερινός *a.* evening; (*s.m.*) vespers.
εσπερίς *s.f.* evening party.
έσπερος *s.m.* evening star.
εσπευσμένως *adv.* hastily, in a hurry.
Εσταυρωμένος *s.m.* Christ crucified.
εστί(ν) *v.i.* is.
εστία *s.f.* hearth, fireplace; home; origin, cradle; focus; burner (*of stove*).
εστιατόριο *s.n.* restaurant.
έστω *v.i.* so be it; ~ και even.
εσύ *pron.* thou, you.
εσφαλμένος *a.* erroneous.
εσχάρα *s.f.* grill, gridiron.
εσχατιά *s.f.* furthest point, extremity.
έσχ|ατος *a.* extreme; furthest, last; worst; ~άτη προδοσία high treason. ~άτως *adv.* recently.
έσω *adv.* inside, within; ο ~ κόσμος inner self.
εσώγαμβρος *s.m.* man living with and supported by wife's parents.
εσωκλείω *v.t.* inclose (*in letter*).
εσώρουχα *s.n.pl.* underclothes.

εσωστρεφής *a.* introvert(ed).
εσωτερικό *s.n.* inside, interior.
εσωτερικός *a.* interior, internal; inland, domestic (*not foreign*); intrinsic.
εταζέρα *s.f.* whatnot.
εταίρα *s.f.* courtesan.
εταιρ(ε)ία *s.f.* society, company, firm.
εταίρος *s.m.* companion, associate, fellow, member.
ετερο- *denotes* other, different.
ετερογενής *a.* heterogeneous.
ετεροδικία *s.f.* extraterritorial rights.
ετεροθαλής *a.* with one parent in common. ~ αδελφός stepbrother.
ετερόκλητος *a.* mixed, motley.
έτερ|ος *pron.* (an)other; ο ~ος one *or* the other (*of two*); αφ'~ου besides.
ετερόφωτος *a.* borrowing one's ideas from others.
ετήσι|ος *a.* annual; lasting one year; ~ιες (*s.f.pl.*) etesian winds.
έτι *adv.* still; moreover.
ετικέττα *s.f.* 1. label. 2. etiquette.
ετοιμάζ|ω *v.t.* prepare. ~ομαι *v.i.* get ready.
ετοιμασία *s.f.* preparation.
ετοιμοθάνατος *a.* moribund.
ετοιμόρροπος *a.* tumbledown.
έτοιμ|ος *a.* ready; (*fam.*) τρώω από τα ~α live on capital.
ετοιμότητα *s.f.* readiness (of wit).
έτος *s.n.* year.
έτσι *adv.* so, thus, like this; το δίνουν ~ it is given away free; ~ (να) κουνηθείς if you so much as stir...; ~ κι ~ so so, middling, in any case; ~ ή αλλιώς in any case.
ετσιθελικά *adv.* arbitrarily.
ετυμηγορία *s.f.* verdict.
ετυμολογία *s.f.* etymology.
ευ *adv.* well.
ευ- *denotes* well, easily.
ευαγγέλιον *s.n.* gospel.
Ευαγγελισμός *s.m.* Annunciation.
ευαγής *a.* pure; philanthropic (*of institutions*).
ευάγωγος *a.* well brought up; docile.
ευάερος *a.* well-ventilated.
ευαισθησία *s.f.* sensitiveness.
ευαίσθητος *a.* sensitive.
ευάλωτος *a.* easily captured; corruptible.
ευανάγνωστος *a.* legible.
ευαρέσκεια *s.f.* satisfaction, pleasure.
ευάρεστος *a.* agreeable, pleasant.
ευαρεστούμαι *v.i.* be pleased, deign.
ευάρμοστος *a.* well-matched.
εύγε *int.* bravo!
ευγένεια *s.f.* nobility; politeness, courtesy.
ευγενής, ~ικός *a.* noble; polite, courteous.
εύγευστος *a.* palatable.

ευγλωττία *s.f.* eloquence.
εύγλωττος ·*a.* eloquent.
ευγνωμονώ *v.t.* be grateful to.
ευγνώμ|ων *a.* grateful. ~οσύνη *s.f.* gratitude.
ευδαιμονία *s.f.* felicity; prosperity.
ευδία *s.f.* fine weather.
ευδιάθετος *a.* in good mood; (well) disposed.
ευδιάκριτος *a.* discernible.
εύδιος *a.* fair (*of weather*).
ευδοκιμώ *v.i.* succeed, get on; thrive.
ευδοκώ *v.i.* be pleased, deign.
εύδρομον *s.n.* (*naut.*) cruiser.
ευειδής *a.* good-looking.
εύελπις *a.* hopeful, promising; (*mil.*) a cadet.
ευέξαπτος *a.* irritable.
ευεξία *s.f.* healthy condition, well-being.
ευεργεσία *s.f.* benefaction.
ευεργέτ|ης *s.m.* benefactor. ~ικός *a.* beneficial; beneficent. ~ώ *v.t.* make benefaction to.
ευερέθιστος *a.* irritable.
ευζωία *s.f.* comfortable living.
εύζωνος *s.m.* (*mil.*) evzone (*soldier of kilted Gk. regiment*).
ευήθης *a.* a simpleton.
ευήλιος *a.* sunny (*that gets the sun*).
ευημερώ *v.i.* enjoy prosperity.
εύηχος *a.* melodious.
ευθαλής *a.* luxuriant (*of vegetation*).
ευθανασία *s.f.* easy death; euthanasia.
ευθεί|α *s.f.* straight line; κατ'~αν, απ'~ας (*adv.*) straight, direct.
εύθετος *a.* suitable, proper.
ευθυν- *see* φτην-.
εύθικτος *a.* touchy.
εύθραυστος *a.* fragile, delicate.
εύθρυπτος *a.* friable.
ευθυβολία *s.f.* marksmanship; accuracy (*of firearm*).
ευθυγραμμίζω *v.t.* align, bring into lign.
ευθύγραμμος *a.* rectilinear, straight.
ευθυκρισία *s.f.* right judgement.
ευθυμογράφημα *s.n.* humorous article (*in paper*).
εύθυμ|ος *a.* merry, gay. ~ία *s.f.* merriment, gaiety.
ευθύνη *s.f.* responsibility; έχω την ~ (*with gen.*) be in charge of.
ευθύνομαι *v.i.* be responsible *or* answerable.
ευθύς *adv.* immediately.
ευθύς *a.* straight, upright; honest.
ευθυτενής *a.* erect (*of carriage*).
ευκαιρία *s.f.* opportunity, chance; (*available*) time; (*good*) bargain; σημαία ~ς flag of convenience.

εύκαιρος *a.* opportune; free, available.
ευκάλυπτος *s.m.* eucalyptus.
εύκαμπτος *a.* supple, flexible.
ευκατάστατος *a.* well-to-do.
ευκαταφρόνητος *a.* negligible.
ευκίνητος *a.* nimble, agile.
ευκοίλι|ος *a.* laxative; whose bowels work easily. ~**ότητα** *s.f.* diarrhoea.
ευκολία *s.f.* ease, convenience; service, good turn.
ευκολόπιστος *a.* credulous.
εύκολος *a.* easy; easy-going.
ευκολύν|ω *v.t.* facilitate; help financially. ~*ομαι v.i.* be in a position (*to pay, etc.*).
ευκοσμία *s.f.* decorum, propriety.
εύκρατ|ος, ~ής *a.* temperate.
ευκρινής *a.* clear, distinct.
ευκταίος *a.* to be wished for.
ευκτική *s.f.* (*gram.*) optative.
ευλαβ|ής *a.* devout. ~**ούμαι** *v.t.* show respect for, revere. ~**εια** *s.f.* devoutness.
εύληπτος *a.* easy to catch, understand *or* take.
ευλογημένος *a.* blessed; (*fam.*) *epithet of mild reproach.*
ευλογία *s.f.* blessing, benediction.
ευλογι|ά *s.f.* smallpox. ~**οκομμένος** *a.* pock-marked.
εύλογος *a.* proper, reasonable; plausible.
ευλογώ *v.t.* bless.
ευλύγιστος *a.* supple, flexible.
ευμάρεια *s.f.* luxurious living, opulence.
ευμελής *a.* 1. melodious. 2. well-built (*of persons*).
ευμενής *a.* favourably disposed.
ευμετά|βλητος, ~**θολος** *a.* changeable.
εύμορφος *a.* see **όμορφος**.
ευνή *s.f.* bed.
ευνόητ|ος *a.* understandable; intelligible; αυτό είναι ~*ο* that is understood.
εύνοια *s.f.* favour, goodwill.
ευνοϊκός *a.* favourable, propitious.
ευνοιοκρατία *s.f.* political favouritism.
ευνομία *s.f.* law and order.
ευνομούμενος *a.* well-governed.
ευνοούμενος *a.* favourite.
ευνούχος *s.m.* eunuch.
ευνοώ *v.t.* favour; show favours to.
ευοδ|ώ, ~**ώνω** *v.t.* bring to successful conclusion. ~*ούμαι v.i.* go well, succeed.
ευοίωνος *a.* auspicious.
εύοσμος *a.* fragrant.
ευπάθ|εια *s.f.* sensitiveness; proneness (*to ailment*). ~**ής** *a.* sensitive, delicate.
ευπαρουσίαστος *a.* presentable.
ευπατρίδης *s.m.* patrician.
ευπειθής *a.* obedient.
εύπεπτος *a.* digestible.

εύπιστος *a.* credulous.
εύπλαστος *a.* plastic; shapely (*of body*).
εύπορος *a.* well-off.
ευπρέπεια *s.f.* decorum, propriety.
ευπρεπής *a.* seemly.
ευπρεπίζω *v.t.* tidy, smarten.
ευπρόσβλητος *a.* susceptible (*to ailment*); vulnerable.
ευπρόσδεκτος *a.* acceptable, welcome.
ευπροσήγορος *a.* affable.
ευπρόσιτος *a.* accessible, approachable.
ευπρόσωπος *a.* presentable.
εύρεση *s.f.* finding.
ευρεσιτεχνί|α *s.f.* δίπλωμα ~*ας* patent.
ευρετήριο *s.n.* index, catalogue.
ευρέως *adv.* widely.
εύρημα *s.n.* find; original touch.
ευρίσκ|ω *v.t.* find; hit, strike; obtain; consider, judge. ~*ομαι v.i.* be, find oneself.
ευρύνω *v.t.* widen.
ευρ|ύς *a.* wide, broad, extended. ~**ύτης** *s.f.* breadth.
ευρύχωρος *a.* spacious, roomy.
ευρωπαϊκός *a.* (west) European.
εύρωστος *a.* robust.
εύσαρκος *a.* fat, corpulent.
ευσέβεια *s.f.* piety.
ευσεβής *a.* pious.
ευσπλαγχνία *s.f..* compassion.
ευστάθεια *s.f.* stability, steadiness.
ευσταλής *a.* smart and upstanding.
εύστοχος *a.* well-aimed; to the point, happy.
εύστροφος *a.* agile; quick (*of mind*).
ευσυγκίνητος *a.* easily moved (*emotionally*).
ευσυνείδητος *a.* conscientious, scrupulous.
εύσχημος *a.* decent; specious; without giving offence.
εύσωμος *a.* well-built; large (*of body*).
εύτακτος *a.* orderly, quiet.
ευτέλεια *s.f.* bad quality; baseness.
ευτελής *a.* cheap and nasty; base, mean.
ευτράπελος *a.* humorous.
ευτραφής *a.* well-nourished, plump.
ευτρεπίζω *v.t.* tidy, smarten.
ευτύχημα *s.n.* stroke of good fortune.
ευτυχ|ής *a.* fortunate; happy. ~**ώς** *adv.* luckily, happily.
ευτυχία *s.f.* happiness; good fortune.
ευτυχισμένος *a.* see **ευτυχής**.
ευυπόληπτος *a.* reputable.
ευφάνταστος *a.* over-imaginative.
ευφημισμός *s.m.* praise; euphemism.
εύφημος *a.* ~ μνεία favourable mention.
εύφλεκτος *a.* inflammable.
ευφορία *s.f.* fruitfulness; euphoria.
ευφραδής *a.* eloquent, fluent.
ευφραίνω *v.t.* rejoice, delight.

ευφραντικός *a.* delectable.

ευφρόσυνος *a.* giving joy *or* pleasure.

ευφυ|ής *a.* intelligent; witty. **~ία** *s.f.* intelligence, wit. **~ολόγημα** *s.n.* witticism.

ευφωνία *s.f.* euphony.

εύχαρις *a.* graceful, charming; pleasant.

ευχαρίστηση *s.f.* pleasure, satisfaction.

ευχαριστί|α *s.f.* Eucharist. **~ες** (*pl.*) thanks.

ευχάριστ|ος *a.* pleasant. **~ως** *adv.* with pleasure, gladly.

ευχαριστ|ώ *v.t.* thank; please, gratify. **~ούμαι** *v.i.* be pleased, enjoy oneself; **~ημένος** pleased, satisfied.

ευχέρεια *s.f.* ease, facility; fluency.

ευχερής *a.* fluent.

ευχή *s.f.* prayer, wish; blessing.

εύχομαι *v.t.* wish; give blessing *or* good wishes to.

εύχρηστος *a.* easy to use; in common use (*of word*).

εύχυμος *a.* juicy, succulent.

ευψυχία *s.f.* bravery.

ευώδης *a.* fragrant.

ευωχία *s.f.* feasting.

εφ' *prep. see* επί.

εφαλτήριο *s.n.* vaulting-horse.

εφάμιλλος *a.* (*with gen.*) equal to, a match for.

εφάπαξ *adv.* once only; (in) a lump sum.

εφάπτ|ομαι *v.i.* touch, be in contact. **~ομένη** *s.f.* tangent.

εφαρμογή *s.f.* fit, fitting; application, implementation.

εφαρμόζω *v.t. & i.* fit, apply, put into effect; (*v.i.*) fit.

έφεδρ|ος *s.m.* (*mil.*) reservist. **~εία** *s.f.* (*mil.*) reserve.

εφεκτικός *a.* reserved, cautious.

εφεξής *adv.* henceforth.

έφεσις *s.f.* 1. (*law*) appeal. 2. inclination.

εφετείον *s.n.* Court of Appeal.

εφέτ|ος *adv.* this year. **~ινός** *a.* this year's.

εφεύρεση *s.f.* invention.

εφευρέτης *s.m.* inventor.

εφευρετικός *a.* inventive.

εφευρίσκω *v.t.* invent; contrive.

έφηβος *s.m.* adolescent youth.

εφηβικ|ός *a.* youth's; **~ή ηλικία** adolescence.

εφημερεύω *v.i.* be on duty.

εφημέριος *s.m.* officiating priest.

εφημερ|ίς, ~ίδα *s.f.* newspaper, gazette.

εφήμερος *a.* ephemeral.

εφιάλτης *s.m.* nightmare.

εφίδρωσις *s.f.* sweating.

εφικτός *a.* attainable, possible.

έφιππος *a.* equestrian; a horseman.

εφιστώ *v.t.* **~ την προσοχήν** (*with gen.*) draw (*person's*) attention.

εφόδι|α *s.n.pl.* supplies, equipment; requirements. **~αστής** *s.m.* supplier. **~άζω** *v.t.* supply, equip.

εφοδιοπομπή *s.f.* supply convoy.

έφοδος *s.f.* charge, assault; snap inspection.

εφοπλίζω *v.t.* arm; equip (*ship*).

εφοπλιστής *s.m.* ship-owner.

εφορ(ε)ία *s.f.* supervision; board of governors; revenue department; tax office.

εφορμώ *v.i.* 1. swoop down. 2. be at anchor.

έφορος *s.m.* inspector, director, keeper (*in various public services*).

εφόσον *conj.* provided that.

εφτά *num.* seven. **~κόσιοι** *a.* seven hundred.

εφτάζυμο *s.n.* unleavened bread.

εφταμηνίτης, ~ικος *a.* a seven months' child; underdeveloped physically.

εφτάψυχος *a.* having seven lives.

εχέγγυο *s.n.* pledge, guarantee.

εχέγγυος *a.* solvent; reliable.

έχει(ν) *s.n.* possessions.

εχεμύθεια *s.f.* discretion; *υπό* **~ν** under pledge of secrecy, (*iron.*) inaudibly.

εχθές *adv.* yesterday.

έχθρα *s.f.* enmity; ill-feeling.

εχθρεύομαι *v.t.* hate.

εχθρικός *a.* of the enemy; hostile.

εχθροπραξί|α *s.f.* hostile act; **~ες** (*pl.*) hostilities.

εχθρός *s.m.* enemy.

εχθρότητα *s.f.* hostility.

έχιδνα *s.f.* viper, adder.

εχίνος *s.m.* sea-urchin; hedgehog.

εχτ- *see* εχθ- *or* εκτ-.

έχ|ω *v.t. & i.* have; keep, hold; consider, regard; cost, be worth; **~ει** there is *or* are; *είχε* there was *or* were; **~ει καλώς** all right; *τι* **~εις;** what is the matter with you? *~ω δίκιο* I am right; *το ~ω σε καλό* I think it lucky; *καλύτερα ~ω να* I prefer to; *τα ~ουμε καλά* we are on good terms; *τα ~ει τετρακόσια* he has his head screwed on the right way; *τα ~ω μαζί της* I am annoyed with her *or* I am having an affair with her; *~ει να κάνει* it makes a difference; *~ει ο Θεός* God will provide; *πόσες ~ει ο μήνας;* what day of the month is it? *~ει ως εξής* it goes as follows.

εψές *adv.* yesterday (*evening*).

έψιλον *s.n.* the letter Ε.

εωθιν|ός *a.* morning. **~όν** *s.n.* reveille.

έως 1. *prep.* (*time & place*) until, up to, as far as; (*with acc.*) **~** *την αυγή* until dawn; (*with gen.*) **~** *θανάτου* unto death. 2. *conj.* (*with ον or ότου & να*) until. 3. *adv.* (*with numbers*) about.

Εωσφόρος *s.m.* Lucifer.

Z

ζαβάδα s.f. stupidity.

ζαβλακώνω v.t. stupefy.

ζαβολιά s.f. cheating (esp. at play).

ζαβολιάρης a. mischievous, a cheat.

ζαβ|ός a. crooked, perverse; clumsy, stupid. **~ά** adv. wrongly, awry.

ζαγάρι s.n. hound.

ζακέτα s.f. jacket.

ζάλ|η, ~άδα s.f. giddiness; **~ες, ~άδες** (pl.) fits of giddiness; (fig.) worries.

ζαλίζ|ω v.t. make dizzy or tipsy, confuse, daze. **~ομαι** v.i. become dizzy, tipsy or confused.

ζαμάνια s.n.pl. έχω καιρούς και **~** να τον δω I haven't seen him for ages.

ζαμπάκι s.n. narcissus.

ζαμπόν(ι) s.n. ham.

ζάντα s.f. rim (of wheel).

ζάπλουτος a. very rich.

ζάρα s.f. crease, wrinkle.

ζαργάνα s.f. garfish.

ζαρζαβατικά s.n.pl. vegetables.

ζάρι s.n. die; **~α** (pl.) dice.

ζαρίφης a. (fam.) elegant.

ζαρκάδι s.n. roebuck.

ζαρτιέρα s.f. suspender.

ζάρωμα s.n. creasing, wrinkling; crease, wrinkle. **~τιά** s.f. crease, wrinkle.

ζαρώνω v.t. & i. crease, wrinkle; (fig.) huddle oneself up, cringe, crouch.

ζατρίκι s.n. chess.

ζαφείρι s.n. sapphire.

ζαφορά s.f. saffron.

ζάχαρ|η s.f. sugar. **~άτος** a. sweet (having sugar). **~ένιος** a. sweet; (fig.) honeyed; η **~ένια** (iron.) dear life.

ζαχαριέρα s.f. sugar-bowl.

ζαχαροκάλαμο s.n. sugar-cane.

ζαχαροπλαστείο s.n. confectioner's.

ζαχαρώνω v.t. & i. cover with sugar; (v.i.) candy; (fig.) flirt.

ζαχαρωτά s.n.pl. sweets.

ζεϊμπέκικος s.m. sort of dance.

ζελατίνη s.f. gelatine.

ζελές .s.m. jelly.

ζεματ|ίζω, ~ώ v.t. & i. scald; be very hot; (fam.) rook (charge too much), punish severely. **~ίζομαι** v.i. get a nasty shock. **~ιστός** a. scalding hot.

ζεμπίλι s.n. flat straw basket.

ζενίθ s.n. zenith.

ζερβ|ός a. left-handed; left. **~ά** adv. to the left; (fig.) awry, wrong.

ζέρσεϋ s.n. jersey (material). [E. jersey]

ζέση s.f. boiling; (fig.) ardour.

ζεσταίν|ω v.t. & i. heat, warm; get hot.

~ομαι v.i. get or feel hot.

ζέσταμα s.n. heating (making hot).

ζεστασιά s.f. (right degree of) warmth.

ζέστη s.f. heat; fever; χάνει **~** it is hot.

ζεστ|ός a. hot, warm. **~ό** s.n. hot drink.

ζευγαράκι s.n. (fam.) pair of lovers.

ζευγάρι s.n. pair, couple, brace; yoke (of oxen, land); ploughing **~ίζω** v.i. plough.

ζευγαρώνω v.t. & i. match, mate; (v.i.) mate.

ζευγάς s.m. ploughman.

ζευγνύω v.t. yoke, harness, join; bridge.

ζεύγος s.n. pair, couple, brace.

ζευζέκης a. silly and superficial.

ζεύξη s.f. yoking; bridging.

ζεύω v.t. yoke, harness; put horse, etc., to (cart).

ζέφυρος s.m. zephyr.

ζέψιμο s.n. harnessing.

ζέχνω v.i. stink.

ζέω v.i. boil.

ζήλ(ε)ια s.f. envy, jealousy.

ζηλεμένος a. desirable.

ζηλευτός a. enviable, attractive, excellent.

ζηλεύω v.t. & i. envy; be jealous (of).

ζηλιάρης a. envious, jealous.

ζήλος s.m. zeal.

ζηλότυπος a. jealous.

ζηλοφθονία s.f. (malicious) envy.

ζηλωτής s.m. zealot.

ζημί|α s.f. damage, loss. **~ώνω** v.t. & i. inflict loss on; suffer loss.

ζημιάρης a. mischievous.

ζήση s.f. life; livelihood.

ζήτα s.n. the letter Z.

ζητεία s.f. begging.

ζήτημα s.n. question, problem, issue, matter; είναι **~** it is doubtful.

ζήτηση s.f. demand (for commodity); quest (for truth, etc.).

ζητιάν|ος s.m. beggar. **~εύω** v.t. & i. beg for; go begging.

ζήτω int. long live, hurrah for!

ζητώ v.t. seek, look or ask for; demand; (v.i.) beg. **~ιέμαι** v.i. be in demand.

ζητωκραυγ|ή s.f. cheer, acclamation. **~άζω** v.t. & i. cheer.

ζιζάνιο s.n. tare, weed; (fig.) dissension; mischievous person.

ζόρ|ι s.n. force; trouble, difficulty; με το **~ι** by force. **~ίζω** v.t. force, put pressure on. **~ίζομαι** find it difficult, be pushed, make a special effort.

ζόρικος a. difficult, troublesome.

ζορμπάς s.m. bully.

ζούγκλα s.f. jungle.

ζουζούνι s.n. (small) insect.

ζούλα s.f. στη **~** on the sly.

ζουλεύω v.t. & i. see **ζηλεύω**.

ζούλ|ημα, ~ισμα *s.n.* squeezing.
ζουλώ *v.t.* crush, squeeze, squash, mash.
ζουμ|ί *s.n.* juice; broth. (*fig.*) gist; profit.
  ~ερός *a.* juicy.
ζουμπάς *s.m.* punch (*tool*).
ζουμπούλι *s.n.* hyacinth.
ζουπώ *v.t. see* **ζουλώ.**
ζουριάζω *v.t. & i.* shrink, shrivel.
ζούρλα *s.f.* madness.
ζουρλαίν|ω *v.t.* drive mad. *~ομαι* (*v.i.*) go mad (*about*).
ζουρλομανδύας *s.m.* straitjacket.
ζουρλός *a.* mad; a madcap.
ζοφερός *a.* dark.
ζοχάδ|α *s.f.* (*fam.*, *fig.*) peevishness, peevish person; *~ες* (*pl.*) piles.
ζυγαριά *s.f.* pair of scales.
ζύγι *s.n.* weighing; *με το ~* by weight; *~α* (*pl.*) weights.
ζυγιάζω *v.t. & i. see* **ζυγίζω.**
ζυγίζ|ω *v.t. & i.* weigh; balance; weigh up; arrange (*soldiers*) in line. *~ομαι v.i.* poise, hover.
ζυγ|ός *s.m.* yoke; (beam of) balance; mountain peak; (*mil.*) rank, row. *εφ' ενός ~ού* in single file.
ζυγ|ός *a.* even (*number*); *μονά ~ά* odd or even.
ζυγούρι *s.n.* two-year-old lamb.
ζυγώ *v.i.* (*mil.*) dress.
ζύγωμα *s.n.* 1. approach. 2. cross-bar.
ζυγώνω *v.t. & i.* bring *or* come near.
ζύθ|ος *s.m.* beer. *~οποιία s.f.* brewing.
ζυμάρ|ι *s.n.* dough, paste. *~ικά s.n.pl.* Italian paste (*macaroni, etc.*).
ζύμη *s.f.* leaven; dough; (*fig.*) stuff, quality (*of character*)
ζύμωμα *s.n.* kneading.
ζυμών|ω *v.t. & i.* knead; ferment; (*v.i.*) make bread. *~ομαι v.i.* ferment.
ζύμωση *s.f.* fermentation.
ζω *v.i. & t.* live; (*v.t.*) live through, experience; keep, support (*family*).
ζωάριον *s.n.* small animal.
ζωγραφιά *s.f.* picture.
ζωγραφίζω *v.t.* paint, draw, depict.
ζωγραφική *s.f.* (*art of*) painting.
ζωγραφιστός *a.* painted; (*fig.*) beautiful.
ζωγράφος *s.m.* painter, artist.
ζώδι|ον *s.n.* sign of zodiac; (*fig.*) destiny; *~ακός κύκλος* zodiac.
ζωέμπορος *s.m.* cattle dealer.
ζωή *s.f.* life; vitality; livelihood.
ζωηρεύω *v.t. & i.* make *or* get livelier; get brighter (*of colour*).
ζωηρ|ός *a.* lively, energetic; warm, animated; bright (*of colour*). *~ότης s.f.* vivacity, warmth.

ζωικ|ός *a.* 1. pertaining to life. 2. animal; *~ή μέταξα* pure silk.
ζωμός *s.m.* broth.
ζωνάρι *s.n.* belt, waistband; (*fam.*) *κρεμώ το ~ μου* seek a quarrel.
ζώνη *s.f.* girdle, belt; cordon; zone.
ζωντανεύω *v.t. & i.* revive; bring back *or* come back to life; enliven.
ζωντάνια *s.f.* vitality.
ζωνταν|ός *a.* living, alive; vivid; underdone (*of food*). *~ά s.n.pl.* animals (*esp. cattle*).
ζωντόβολο *s.n.* beast; (*fig.*) dolt.
ζωντοχήρα *s.f.* divorced woman.
ζωντοχήρος *s.m.* divorced man.
ζών|ω *v.t.* gird (on); encircle, surround. *~ομαι v.t.* put on (*sword, lifebelt, etc.*).
ζωογόνος *a.* invigorating.
ζωογονώ *v.t.* invigorate, brace.
ζωοδότης *s.m.* giver of life.
ζωοδόχος *a. ~ πηγή* life-giving spring.
ζωοκλοπή *s.f.* cattle-thieving.
ζωοκομία *s.f.* cattle-raising.
ζωολογ|ία *s.f.* zoology. *~ικός a.* zoological. *~ος s.m.* zoologist.
ζώον *s.n.* animal; (*fig.*) dolt.
ζωοπάζαρο *s.n.* cattle-fair.
ζωοτομία *s.f.* vivisection.
ζωοτροφίαι *s.f. pl.* victuals.
ζωόφιλος *a.* fond of animals.
ζωοφόρος *s.f.* frieze (*of temple*).
ζωπυρώ *v.t.* rekindle.
ζωστήρας *s.m.* belt, baldric; (*med*) shingles.
ζωτικ|ός *a.* vital. *~ότης s.f.* vitality; (*fig.*) vital importance.
ζωύφιον *s.n.* small insect; louse.
ζωώδης *a.* brutish.

Η

η *article f.* the.
ή *conj.* 1. or; *ή...ή* either... or. 2. than.
η *pron. f.* who (*see* **ος**).
ήβη *s.f.* puberty; pubes.
ηβικός *a.* pubic.
ηγεμ|ών *s.m.* prince, sovereign. *~ονία s.f.* hegemony, sovereignty; principality. *~ονικός a.* princely. *~ονεύω v.i.* rule.
ηγεσία *s.f.* leadership.
ηγ|έτης, ~ήτωρ *s.m.* leader.
ηγετικός *a.* leading, of a leader.
ηγούμαι *v.i. & t.* (*with gen.*) lead, be in command (of).
ηγουμέν|η, ~ισσα *s.f.* abbess.
ηγούμενος *s.m.* abbot.
ήγουν *adv.* namely.
ήδη *adv.* already.

ηδον|ή *s.f.* pleasure (*esp. sensual*). **~ικός** *a.* sweet, voluptuous. **~ισμός** *s.m.* hedonism.

ηδονοβλεψίας *s.m.* voyeur, peeping Tom.

ηδύνω *v.t.* sweeten; delight.

ηδυπαθής *a.* sensual; a voluptuary.

ηδύποτο *s.n.* liqueur.

ηδύς *a* sweet. **~έως** *adv.* sweetly.

ηθική *s.f.* ethics, morals; morality.

ηθικολογώ *v.i.* moralize.

ηθικ|όν *s.n.* morale; morality.

ηθικ|ός *a.* ethical, moral; virtuous; **~ός** αυτουργός instigator (*of crime*). **~ότης** *s.f.* morality, virtue.

ηθμός *s.m.* filter.

ηθογραφία *s.f.* description of local customs.

ηθοπλαστικός *a.* (*morally*) uplifting.

ηθοποι|ία *s.f.* acting. **~ός** *s.m.f.* actor, actress.

ήθ|ος *s.n.* character, nature; air, manner; *~η* (*pl.*) manners, habits; morals.

ήκιστα *adv.* very little; not at all.

ηλεκτρίζω *v.t.* electrify. (*fig.*) rouse to enthusiasm.

ηλεκτρικ|ός *a.* electric. **~ό** *s.n.* electricity (*supply*).

ηλεκτρισμός *s.m.* electricity.

ηλεκτροκίνητος *a.* driven by electricity.

ηλεκτρολόγος *s.m.* electrician.

ήλεκτρον *s.n.* amber.

ηλεκτρονική *s.f.* electronics.

ηλεκτρόνιον *s.n.* electron.

ηλεκτροπληξία *s.f.* electric shock.

ηλεκτροφωτίζω *v.t.* light by electricity, lay on electricity to.

ηλιακός *a.* sun's, solar.

ηλίαση *s.f.* sunstroke.

ηλίθιος *a.* idiotic; an idiot.

ηλικία *s.f.* age (*length of life*); (*mil.*) class.

ηλικιούμαι, *v.i.* grow to maturity; grow old.

ηλικιωμένος *a.* elderly.

ηλιοβασίλεμα *s.n.* sunset.

ηλιοθεραπεία *s.f.* sunbathing.

ηλιοκαμένος *a.* sunburnt.

ηλιόλουστος *a.* bathed in sunshine.

ήλι|ος *s.m.* sun; sunflower. **~οστάσιον** *s.n.* solstice. **~όφως** *s.n.* sunlight.

ηλιοτρόπιον *s.n.* heliotrope; litmus.

ήλος *s.m.* nail; corn, callus.

ημαρτημένα *s.n.pl.* errata.

ήμαρτον *v.i.* forgive me! (*see* **αμαρτάνω**).

ημεδαπός *a.* native (*not foreign*).

ημείς *pron.* we.

ημέρα *s.f.* day; της *~ς* laid, caught, *etc.* the same day.

ημερεύω *v.t. & i.* tame, calm; become tame *or* calm.

ημερήσιος *a.* daily; a day's.

ημεροδείκτης *s.m.* almamac.

ημερολόγιο *s.n.* calendar, almanac; diary, log-book.

ημερομηνία *s.f.* date.

ημερομίσθιο *s.n.* day's wage.

ήμερος *a.* domesticated; cultivated (*plant, ground*); tame, gentle, peaceable.

ημερώνω *v.t. & i.* tame; cultivate (*wild plant*); (*fig.*) civilize; (*v.i.*) calm down.

ημέτερος *a.* our, ours.

ημι- *denotes* half.

ημιαργία *s.f.* half-day's holiday.

ημιεπίσημος *a.* semi-official.

ημιθανής *a.* half-dead.

ημίθεος *s.m.* demigod.

ημίκοσμος *s.m.* demi-monde.

ημικρανία *s.f.* migraine.

ημικύκλιο *s.n.* semicircle.

ημιμάθεια *s.f.* smattering of knowledge.

ημίμετρα *s.n.pl.* half-measures.

ημίονος *s.m.f.* mule.

ημιπληγία *s.f.* (*med.*) paralysis of one side.

ημίρρευστος *a.* half-liquid.

ημισέληνος *s.f.* half-moon; crescent.

ήμισ|υς *a.* half; εξ *~είας* half-and-half.

ημισφαίριο *s.n.* hemisphere.

ημιτελής *a.* half-finished; unfinished.

ημιτελικός *a.* semi-final.

ημιτόνιο *s.n.* (*mus.*) semitone.

ημίφωνο *s.n.* (*gram.*) semivowel.

ημίφως *s.n.* dim light; twilight.

ημιχρόνιο *s.n.* half-time (*in games*).

ημίωρο *s.n.* half-hour.

ημπορώ *v.i. see* **μπορώ**.

ηνί|ον *s.n.* rein. **~οχος** *s.m.* charioteer.

ηνωμένος *a.* united.

ηξεύρω *v.t. & i. see* **ξέρω**.

ήπ|αρ *s.n.* liver; *~ατα* (*pl.*) strength; (*fam.*) του κόπηκαν τα *~ατα* he had quite a turn.

ηπατίτις *s.f.* hepatitis.

ήπειρ|ος *s.f.* mainland; continent; Epirus. **~ωτικός** *a.* continental. **~ώτικος** *a.* Epirot.

ήπιος *a.* mild.

ηρέμ|α, **~ως** *adv.* calmly, quietly.

ηρεμ|ος *a.* calm, quiet. **~ία** *s.f.* calm, quiet. **~ώ** *v.i.* be calm, quiet *or* at rest.

ηρωικός *a.* heroic.

ηρωίνη *s.f.* heroin.

ηρώον *s.n.* war memorial.

ήρ|ως, **~ωας** *s.m.* hero. **~ωίς**, **~ωίδα** *s.f.* heroine. **~ωισμός** *s.m.* heroism. **~ωικός** *a.* heroic.

ήσκα *s.f.* tinder.

ησυχάζω *v.i. & t.* grow *or* make calm *or* quiet; rest, sleep; be reassured; be a hermit.

ησυχαστήριο *s.n.* retreat; hermitage.

ήσυχ|ος *a.* quiet, peaceful, free from worry. **~ία** *s.f.* quietness, peace.

ήτα *s.n.* the letter **H**.

ήτοι *adv.* namely.
ήττ|α *s.f.* defeat. ~οπάθεια *s.f.* defeatism.
ήττον *adv.* less; ουχ ~ none the less.
ηττώμαι *v.i.* be defeated.
ηφαιστειογενής *a.* volcanic.
ηφαίστειο *s.n.* volcano.
ηχηρ|ός *a.* resonant, loud; (*gram.*) sonant. ~ότης *s.f.* resonance, sonority.
ηχητικός *a.* producing *or* employing sound.
ηχοβόλισις *s.f.* echo-sounding.
ηχογραφώ *v.t.* record (*on disc, tape*). ~ησις *s.f.* recording.
ηχομόνωσις *s.f.* sound-proofing.
ήχ|ος *s.m.* sound. φράγμα του ~ου sound-barrier.
ηχώ *s.f.* echo.
ηχώ *v.i.* sound, ring.
ηώς *s.f.* dawn.

θα *particle introducing future, conditional, etc., verbs.*
θάβω *v.t.* bury.
θαλάμη *s.f.* chamber (*of gun*).
θαλαμηγός *s.f.* yacht.
θαλαμηπόλος *s.m.f.* manservant; house-maid; steward, stewardess.
θαλαμίσκος *s.m.* cabin; module (*of space-craft*).
θάλαμος *s.m.* room, chamber, ward.
θάλασσα *s.f.* sea; (*fam.*) τα κάνω ~ make a mess of it.
θαλασσιν|ός *a.* of *or* from the sea; a seafarer; ~ά (*s.n.pl.*) shell-fish.
θαλάσσιος *a.* of *or* from the sea, maritime.
θαλασσόβιος *a.* living in *or* frequenting the sea; marine.
θαλασσοδάνειο *s.n.* bottomry; (*fig.*) risky loan.
θαλασσοδέρν|ω *v.t.* buffet. ~ομαι *v.i.* be buffeted by waves *or* (*fig.*) by adversity.
θαλασσοκρατ|ία, ~ορία *s.f.* mastery of the seas.
θαλασσόλυκος *s.m.* sea-dog, old salt.
θαλασσοπνίγομαι *v.i.* suffer hardship at sea; (*fig.*) be in difficulties.
θαλασσοπόρος *s.m.* voyager, navigator.
θαλασσοπούλι *s.n.* sea-bird.
θαλασσώνω *v.t.* τα ~ make a mess of it.
θαλερός *a.* verdant; (*fig.*) youthful.
θάλλω *v.i.* blossom, flourish.
θαλπωρή *s.f.* pleasant warmth; (*fig.*) care, attention.
θάμβος *s.n.* dazzlement; amazement.
θάμνος *s.m.* bush, shrub.

θαμνώδης *a.* bushy, scrubby.
θαμπάδα *s.f.* dullness, mistiness (*of sight, polished surface*).
θαμπός *a.* dim, dull; turbid, clouded.
θάμπωμα *s.n.* dazzlement; amazement; clouding, tarnishing.
θαμπών|ω *v.t. & i.* dazzle; make *or* become dull *or* clouded. ~ομαι *v.i.* be dazzled.
θαμώνας *s.m.* habitué (*of café, etc.*).
θανάσιμος *a.* deadly.
θανατηφόρος *a.* deadly.
θανατικ|ός *a.* capital. ~ό *s.n.* deadly epidemic.
θάνατ|ος *s.m.* death. ~ώνω *v.t.* put to death; (*fig.*) grievously afflict.
θανή *s.f.* death; funeral.
θάπτω *v.t.* bury.
θαρραλέος *a.* bold.
θαρρετός *a.* bold; assured (*not shy*).
θαρρεύω *v.i.* take courage. ~ομαι *v.i.* trust.
θάρρος *s.n.* boldness, courage; assurance.
θαρρ|ώ *v.i. & t.* think, believe; (*v.t.*) τον ~εψα για I took him for.
θαύμα *s.n.* miracle, marvel, wonder.
θαυμάζω *v.t. & i.* wonder at, admire; be surprised.
θαυμάσιος *a.* wonderful, marvellous.
θαυμασμός *s.m.* wonder, admiration.
θαυμαστής *s.m.* admirer.
θαυμαστικ|ός *a.* admiring. ~όν *s.n.* (*gram.*) exclamation mark.
θαυμαστός *a.* admirable, deserving praise.
θαυματοποιός *s.m.* juggler.
θαυματουργός *a.* miracle-working.
θάψιμο *s.n.* burial.
θεά *s.f.* goddess.
θέα *s.f.* sight, view, prospect.
θεαθήναι *v.i. see* θεώμαι.
θέαμα *s.n.* spectacle, show. ~τικός *a.* spectacular. ~τικότητα *s.f.* spectacular quality; TV rating.
θεάρεστος *a.* meritorious (*of good works*).
θεατής *s.m.* spectator.
θεατός *a.* visible.
θεατρικός *a.* for the theatre; theatrical.
θεατρίν|ος *s.m.* actor, ~α *s.f.* actress. ~ίστικος *a.* histrionic.
θέατρο *s.n.* theatre; γίνομαι ~ become a laughing-stock.
θεατρώνης *s.m.* impresario.
θεία *s.f.* aunt.
θειάφι *s.n.* sulphur.
θειικός *a.* sulphuric.
θεϊκός *a.* divine.
θείον *s.n.* sulphur.
θείος *s.m.* uncle.
θείος *a.* divine.
θέλγ|ω *v.t.* charm. ~ητρο *s.n.* charm,

fascination.

θέλημα *s.n.* will (*thing intended*); errand.

θεληματικ|ός *a.* voluntary, intentional. **~ότητα** *s.f.* determination.

θέληση *s.f.* will (*faculty*), volition; will-power; wish (*thing intended*).

θελκτικός *a.* charming, alluring.

θέλ|ω *v.t. & i.* want, need, require; owe; (*v.i.*) be willing; attempt. *~ω να πω* I mean to say; *~εις...~εις* either... or; *~ει δε ~ει* willy-nilly; *λίγο ήθελε να πέσει* he nearly fell.

θέμα *s.n.* subject, topic, theme; (*gram.*) stem; *~τα* (*pl.*) exam paper. *το ~ είναι* the thing is.

θεματοφύλακας *s.m.* depositary, guardian.

θεμέλι|ο *s.n.* foundation, basis. **~ακός** *a.* fundamental. **~ώνω** *v.t.* found.

θέμ|ις *s.f.* justice (*personified*). **~ιτός** *a.* lawful.

θεο- *denotes* 1. of God. 2. utterly.

θεογνωσία *s.f.* religiousness; (*fig.*) *τον έφερα σε* ~ I brought him to his senses.

θεόγυμνος *a.* stark naked.

θεόθεν *adv.* from God.

θεοκατάρατος *a.* accursed.

θεοκρατία *s.f.* theocracy.

θεολόγ|ος *s.m.* thelogian. **~ία** *s.f.* theology. **~ικός** *a.* theological.

θεομηνία *s.f.* natural disaster.

Θεομήτωρ *s.m.* Mother of God.

θεομπαίχτης *s.m.* humbug, rascal.

θεοποιώ *v.t.* deify; (*fig.*) idolize.

θεόρατος *a.* huge (*esp. in height*).

θεός *s.m.* god; *είχαμε θεό* (*fam.*) we were lucky. *δεν έχεις το θεό σου* you're the limit!

θεοσεβής *a.* godly.

θεοσκότεινος *a.* pitch-dark.

θεόστραβος *a.* absolutely blind *or* crooked.

θεότης *s.f.* deity.

Θεοτόκος *s.f.* Mother of God.

θεοφάν(ε)ια *s.n.pl.* Epiphany.

θεοφιλής *a.* dear to God.

θεοφοβούμενος *a.* god-fearing.

θεραπεία *s.f.* cure, treatment, remedy; cultivation (*of arts*).

θεραπευτήριον *s.n.* sanatorium.

θεραπευτικ|ός *a.* curative. **~ή** *s.f.* therapeutics.

θεραπεύω *v.t.* cure, treat, remedy; cultivate (*arts*); satisfy (*need*).

θεράπων *s.m.* servant, devotee; ~ *ιατρός* medical attendant.

θέρετρο *s.n.* summer resort *or* residence.

θεριακλής *s.m.* avid consumer of coffee *or* tobacco.

θεριεύω *v.i.* grow bigger, shoot up.

θερίζω *v.t.* harvest, reap, mow; mow down;

(*fig.*) torment (*of pain, cold, etc.*).

θερινός *a.* of summer; summery.

θεριό *s.n.* wild beast.

θέρ|ισμα *s.n.*, **~ισμός** *s.m.* harvest, reaping, mowing.

θεριστής *s.m.* reaper; June.

θεριστικός *a.* for reaping *or* mowing.

θέρμαι *s.f.pl.* hot springs.

θερμαίν|ω *v.t.* heat, warm. **~ομαι** *v.i.* get a fever.

θέρμανση *s.f.* heating.

θερμαντικός *a.* heating, calorific.

θερμαστής *s.m.* stoker.

θερμάστρα *s.f.* (*heating*) stove.

θέρμ|η *s.f.* fever (*also* **~ες**); (*fig.*) fervour. **~αι** *s.f.pl.* hot springs.

θερμίδα *s.f.* calorie.

θερμόαιμος *a.* warm-blooded; quick to anger.

θερμογόνος *a.* calorific.

θερμοκήπιο *s.n.* hot-house.

θερμοκρασία *s.f.* temperature.

θερμόμετρο *s.n.* thermometer.

θερμός *s.n.* thermos flask.

θερμός *a.* hot, warm; (*fig.*) warm, ardent; expansive.

θερμοσίφωνας *s.m.* water-heater, geyser.

θερμοστάτης *s.m.* thermostat.

θερμότης *s.f.* heat, warmth; (*fig.*) ardour; expansiveness.

θερμοφόρος *s.f.* hot-water bottle.

θέρος *s.m.* harvest.

θέρος *s.n.* summer; harvest.

θεσιθήρας *s.m.* place-seeker.

θέσις *s.f.* position, place, situation; post, employment; seat; room (*available space*); class (*of travel*); thesis; (*mus.*) down-beat.

θέσμιος *a.* statutory.

θεσμός *s.m.* institution, law.

θεσπέσιος *a.* divine, superb.

θεσπίζω *v.t.* decree, enact.

θέσπισμα *s.n.* decree.

θετικισμός *s.m.* positivism.

θετικ|ός *a.* positive; definite, practical, down-to-earth; **~ές** *επιστήμες* natural science.

θετός *a.* adoptive.

θέτω *v.t.* put, set; impose, submit.

θεώμαι *v.t. & i.* look at; be seen; *προς το θεαθήναι* for appearance sake.

θεωρείο *s.n.* (*theatre*) box; (*press, etc.*) gallery.

θεώρημα *s.n.* theorem.

θεώρησις *s.f.* official scrutiny (*of document*); visa.

θεωρητικός *a.* 1. theoretical; a theoretician. 2. of imposing appearance.

θεωρία *s.f.* 1. theory. 2. imposing appearance.

θεωρώ *v.t.* consider, regard; scrutinize (*document*), visa; submit for visa, *etc.*

θηκάρι *s.n.* scabbard, sheath.

θήκη *s.f.* box, case, receptacle.

θηλάζω *v.t. & i.* suckle.

θηλασμός *s.m.* suckling (*process*).

θηλαστικό *s.n.* mammal.

θήλαστρο *s.n.* baby's feeding-bottle.

θηλ(ε)ιά *s.f.* noose, slip-knot, loop; stitch (*in knitting*).

θηλή *s.f.* nipple; papilla.

θήλ|υ, ~υκό *s.n.* female.

θηλυκ|ός *a.* female; feminine. *~ό μυαλό* inventive mind. **~ότητα** *s.f.* femininity.

θηλυκώνω *v.t.* button, do up.

θηλυπρεπής *a.* womanish, effeminate.

θήλυς *a.* female.

θημωνιά *s.f.* stack (*of hay, corn*).

θήρα *s.f.* chase, pursuit; hunting, shooting; game, quarry.

θήραμα *s.n.* game, quarry.

θηρεύω *v.t.* pursue, hunt.

θηριοδαμαστής *s.m.* animal-tamer.

θηρίο *s.n.* wild beast; (*fig.*) monster. *έγινε ~* he got furious.

θηριοτροφείο *s.n.* menagerie.

θηριώδης *a.* ferocious.

θησαυρίζω *v.t. & i.* amass (*riches*); grow rich.

θησαυρός *s.m.* treasure; (*fig.*) store, storehouse; thesaurus.

θησαυροφυλάκιο *s.n.* treasury; strong-room.

θήτα *s.n.* the letter Θ.

θητεία *s.f.* military service; term of office.

θητεύω *v.i.* hold office.

θιασάρχης *s.m.* head of theatrical company.

θίασος *s.m.* theatrical company; troupe.

θιασώτης *s.m.* votary, supporter.

θίγω *v.t.* touch (upon); (*fig.*) offend.

θλάσις *s.f.* breaking; (*med.*) fracture.

θλιβερός *a.* grievous, piteous.

θλίβω *v.t.* squeeze; (*fig.*) afflict, grieve.

θλιμμένος *a.* sad, sorrowful; in mourning.

θλίψη *s.f.* squeezing; (*fig.*) affliction, sorrow, mourning.

θνησιγενής *a.* (*fig.*) still-born.

θνησιμότης *s.f.* death-rate.

θνήσκω *v.i.* die.

θνητ|ός *a.* mortal. **~ότης** *s.f.* mortality.

θόλος *s.m.* dome, cupola, vault.

θολός *a.* dull,'lustreless; blurred, clouded; turbid; confused.

θολ|ώνω *v.t. & i.* make *or* become dull, blurred, turbid *or* confused; *~ωσε το μάτι του* he was seized with anger *or* desire,

etc.

θολωτός *a.* vaulted, domed.

θόρυβος *s.m.* noise, row; sensation, fuss.

θορυβ|ώ *v.i. & t.* make a noise; disturb, worry. *~ούμαι v.i.* get worried.

θορυβώδης *a.* noisy, rowdy.

θούριο *s.n.* martial song.

θράκα *s.f.* embers.

θρανίο *s.n.* (*pupil's*) desk.

θρασομανώ *v.i.* rage; (*of plants*) run riot.

θράσος *s.n.* effrontery, cheek.

θρασύδειλος *a.* blustering but cowardly.

θρασύς *a.* impudent, cheeky.

θρασύτητα *s.f.* insolence, cheek.

θραύση *s.f.* fracture, breaking; havoc.

θραύσμα *s.n.* splinter, fragment.

θραύω *v.t.* break, shatter.

θρεμμένος *a.* well-nourished.

θρεπτικός *a.* nourishing; alimentary.

θρεφτάρι *s.n.* fattened animal.

θρέφω *v.t.* nourish.

θρέψη *s.f.* nutrition.

θρήν|ος *s.m.* lamentation, dirge. **~ητικός** *a.* plaintive, mournful. **~ώ** *v.i. & t.* lament, mourn.

θρηνώδ|ης *a.* plaintive, mournful. **~ία** *s.f.* lament.

θρησκεία *s.f.* religion.

θρήσκευμα *s.n.* faith, creed, denomination.

θρησκευτικός *a.* religious, of religion.

θρησκόληπτος *a.* obsessed with religion.

θρήσκος *a.* religious, god-fearing.

θριαμβεύ|ω *v.i.* triumph. **~της** *s.m.* triumphant one, victor. **~τικός** *a.* triumphal, triumphant.

θρίαμβος *s.m.* triumph.

θρίξ *s.f.* a hair, bristle.

θροΐζω *v.i.* rustle.

θρόισμα *s.n.* rustle.

θρόμβ|ος *s.m.* clot; bead (*of sweat*). **~ωσις** *s.f.* (*med.*) thrombosis.

θρονί *s.n.* seat.

θρονιάζομαι *v.i.* settle comfortably, install oneself.

θρόνος *s.m.* throne.

θρούμπα *s.f.* ripe olive (*fallen from tree*).

θρυαλλίς *s.f.* wick, fuse.

θρύβω *v.t. & i.* crumble.

θρυλείται *v.i.* it is rumoured.

θρύλ|ος *s.m.* legend, tale; rumour. **~ικός** *a.* legendary, heroic.

θρύμμα *s.n.* fragment, crumb. **~τίζω** *v.t.* break into fragments.

θρύψαλο *s.n.* fragment (*of glass, china*).

θυγάτ|ηρ, ~έρα *s.f.* daughter.

θύελλ|α *s.f.* storm. **~ώδης** *a.* stormy, tempestuous.

θυλάκιον *s.n.* pocket.

θύλ|αξ, ~ακος *s.m.* pouch; pocket; sac.

θύμα *s.n.* victim.

θυμάμαι *v.t. & i. see* θυμούμαι.

θυμάρι *s.n.* thyme.

θυμηδία *s.f.* mirth (*esp. ironic*).

θύμηση *s.f.* memory.

θυμητικό *s.n.* memory.

θυμίαμα *s.n.* incense; (*fig.*) adulation.

θυμιατ|ήρι, ~ό *s.n.* censer.

θυμιατίζω *v.t.* cense; (*fig.*) adulate.

θυμίζω *v.t.* remind (*one*) of, recall.

θυμ|ός *s.m.* anger. ~ώδης *a.* irascible.

θυμούμαι *v.t. & i* . remember, recall.

θυμώνω *v.t. & i.* make *or* get angry.

θύρα *s.f.* door.

θυρεοειδής *a.* thyroid.

θυρεός *s.m.* escutcheon.

θυρίδα *s.f.* small window; ticket-office window, guichet; locker, safe-deposit.

θυρόφυλλο *s.n.* leaf of door.

θυρωρός *s.m.f.* door-keeper, hall-porter.

θύσανος *s.m.* tassel, tuft.

θυσί|α *s.f.* sacrifice. ~άζω *v.t.* sacrifice.

θύω *v.t. & i.* sacrifice; (*fig.*) ~ και απολλύω run riot.

θωπεία *s.f.* caress.

θωπεύω *v.t.* caress; (*fig.*) flatter.

θωρακίζω *v.t.* protect with armourplate.

θωρακικός *a.* of the chest.

θώραξ *s.m.* breastplate; thorax.

θωρηκτόν *s.n.* battleship.

θωριά *s.f.* look, appearance; colour.

θωρώ *v.t.* see, look at.

I

ιαίνω *v.t.* cure, heal.

ιαματικός *a.* curative, medicinal.

ίαμβ|ος *s.m.* iambus. ~ικός *a.* iambic.

Ιανουάριος *s.m.* January.

Ιάπωνας *s.m.* a Japanese.

ιαπωνικός *a.* Japanese.

ιάσιμος *a.* curable.

ίασις *s.f.* cure, healing.

ιατρείον *s.n.* surgery, consulting-room, clinic.

ιατρεύω *v.t.* cure, heal.

ιατρική *s.f.* (*science of*) medicine.

ιατρικόν *s.n.* medicine, medicament.

ιατρικός *a.* medical.

ιατροδικαστ|ής *s.m.* (*forensic*) medical expert. ~ική *s.f.* forensic medicine.

ιατρός *s.m.f.* doctor.

ιατροσυμβούλιον *s.n.* consultation between doctors.

ιαχή *s.f.* war-cry.

ιβίσκος *s.m.* hibiscus.

ιδανικ|ός *a.* ideal. ~ό *s.n.* ideal.

ίδε, ιδέ *v.t.* see! lo! (see ιδώ)

ιδέα *s.f.* idea; opinion.

ιδεαλιστής *s.m.* idealist.

ιδεατός *a.* in one's imagination.

ιδεοληψία *s.f.* obsession.

ιδεολογία *s.f.* ideology; utopian idea.

ιδεολόγος *s.m.f.* idealist.

ιδεώδ|ης *a.* ideal. ~ες *s.n.* ideal.

ιδία *adv.* in particular, above all.

ιδιάζ|ων *a.* characteristic, peculiar. ~όντως *adv.* in particular.

ιδιαίτερ|ος *a.* special, particular; ο ~ος private secretary; τα ~α private affairs.

ιδιαιτέρως *adv.* especially; privately.

ιδικός *a. see* δικός.

ιδιοκτησία *s.f.* ownership; property.

ιδιοκτήτης *s.m.* owner, proprietor.

ιδιόκτητος *a.* of one's own, private.

ιδιόμορφος *a.* odd, peculiar.

ιδιοποιούμαι *v.t.* appropriate.

ιδιόρρυθμος *a.* peculiar, eccentric; (*eccl.*) idiorhythmic.

ίδι|ος *a.* (of) one's own; (*one*) self, εγώ ο ~ος I myself; same, ~ος με the same as. ~α *adv.* ~α με the same as, like.

ιδιοσυγκρασία *s.f.* temperament; constitution.

ιδιοτέλ|εια *s.f.* self-interest. ~ής *a.* self-seeking.

ιδιότης *s.f.* property, quality, peculiarity; capacity (*function*).

ιδιότροπος *a.* unusual; fussy.

ιδιοφυ|ΐα *s.f.* talent. ~ής *a.* talented.

ιδιόχειρος *a.* with *or* into one's own hand.

ιδίωμα *s.n.* characteristic; habit; dialect. ~τικός *a.* dialectal.

ιδιω(μα)τισμός *s.m.* idiom (*form of expression*).

ιδίως *adv.* in particular, above all.

ιδιωτεύω *v.i.* live as a private citizen.

ιδιώτ|ης *s.m.* individual, private citizen. ~ικός *a.* private.

ιδού *int.* see! behold!

ίδρυμα *s.n.* foundation, institution (*thing*).

ίδρυσις *s.f.* founding, establishment (*act*).

ιδρυτής *s.m.* founder.

ιδρύω *v.t.* found, establish.

ιδρώ|νω *v.i. & t.* sweat; (*fig.*) toil; cause to sweat; ~μένος sweating.

ιδρ|ώς, ~ώτας *s.m.* sweat. ~ωμα *s.n.* sweating.

ιδώ *v.t.* aorist subjunctive of βλέπω.

ιέραξ *s.m.* falcon, hawk.

ιεραπόστολος *s.m.* missionary.

ιεράρχ|ης *s.m.* prelate. ~ία *s.f.* hierarchy. ~ικός *a.* hierarchical.

ιερατείον *s.n.* clergy.
ιερατικ|ός *a.* priestly; ~ή σχολή seminary.
ιέρεια *s.f.* priestess.
ιερεμιάς *s.f.* jeremiad.
ιερεύς *s.m.* priest.
ιερογλυφικά *s.n.pl.* hieroglyphics.
ιερόδουλος *s.f.* prostitute.
ιερ(ο)εξεταστής *s.m.* inquisitor.
ιεροκήρυξ *s.m.* preacher.
ιερομάρτυς *s.m.f.* holy martyr.
ιερομόναχος *s.m.* monk in holy orders.
ιερόν *s.n.* (*eccl.*) sanctuary.
ιερ|ός *a.* sacred, holy; ~όν οστούν (*anat.*) sacrum.
ιεροσυλία *s.f.* sacrilege.
ιεροτελεστία *s.f.* sacred rite, divine office.
ιερουργώ *v.i.* officiate at church service.
ιεροφάντης *s.m.* hierophant.
ίζημα *s.n.* sediment.
ιησουίτης *s.m.* Jesuit.
ιθαγέν|εια *s.f.* nationality (*by birth*). ~ής *a.* native.
ιθύνω *v.t.* direct, govern, rule.
ικανοποι|ώ *v.t.* satisfy, please. ~ηση *s.f.* satisfaction. ~ητικός *a.* satisfactory, satisfying.
ικαν|ός *a.* capable, competent; sufficient; a fair amount of; (*mil.*) fit. ~ότης *s.f.* capability, competence.
ίκαρος *s.m.* (*aero.*) cadet.
ικεσία *s.f.* entreaty.
ικετεύω *v.t.* entreat.
ικέτης *s.m.* suppliant.
ικμάς *s.f.* sap; (*fig.*) vitality.
ικρίωμα *s.n.* scaffolding; scaffold.
ίκτερος *s.m.* jaundice.
ικτίς *s.f.* animal of weasel tribe.
ιλαρά *s.f.* measles; (*fam.*) βγάζω την ~ feel very hot.
ιλαρ|ός *a.* merry. ~ ότης *s.f.* merriment.
ίλεως *a.* clement.
ίλη *s.f.* (*mil.*) troop of cavalry.
ιλιγγ|ος *s.m.* giddiness, vertigo, ~ιώ *v.i.* become giddy *or* dizzy. ~ιώδης *a.* (*causing one to feel*) giddy *or* dizzy.
ιλύς *s.f.* sediment, mud, slime, silt.
ιμάμης *s.m.* imam (*Moslem priest*); ~ μπαϊλντί dish of aubergines with oil.
ιμάς *s.m.*strap, thong; belt (*in machinery*).
ιματιοθήκη *s.f.* wardrobe; linen-room; cloak-room.
ιμάτιον *s.n.* garment.
ιματιοφυλάκιον *s.n.* cloak-room.
ιματισμός *s.m.* (*outfit of*) clothing.
ιμπεριαλισμός *s.m.* imperialism.
ιμπρεσ(σ)ιονισμός *s.m.* impressionism.
ίνα *s.f.* fibre, filament.

ίνα *conj.* in order to.
ίνδαλμα *s.n.* ideal, idol (*admired person*).
ινδιάνος *s.m.* turkey; red-indian.
ινδικ|ός *a.* Indian; ~ό χοιρίδιο guinea-pig; ~όν κάρυον coconut.
ινδοκάλαμος *s.m.* bamboo.
ινιακός *a.* occipital.
ινστιτούτον *s.n.* institute; ~ καλλονής beauty-parlour.
ίντσα *s.f.* inch.
ινώδης *a.* fibrous, stringy.
ιξώδης *a.* glutinous.
ιοβόλος *a.* venomous.
ίον *s.n.* violet.
ιόνιος *a.* of the Ionian Sea *or* Islands.
ιός *s.m.* venom; (*med.*) virus.
ιουδαίος *s.m.* Jew.
ιούλιος *s.m.* July.
ιούνιος *s.m.* June.
ιόχρους *a.* violet.
ιππασία *s.f.* riding; horsemanship.
ιππεύς *s.m.* rider; cavalryman; knight (*chess*).
ιππεύω *v.i. & t.* ride; mount, sit astride.
ιππικόν *s.n.* cavalry.
ιππικ|ός *a.* of horses. ~οί αγώνες show-jumping.
ιπποδρομίες *s.f.pl.* races.
ιπποδρόμιο *s.n.* racecourse; circus.
ιπποκόμος *s.m.* groom.
ιπποπόταμος *s.m.* hippopotamus.
ίππ|ος *s.m.* horse; δέκα ~ων 10 h.p.
ιπποσκευή *s.f.* harness.
ιππότης *s.m.* knight; (*fig.*) chivalrous man. ~ικός *a.* knightly, chivalrous. ~ισμός *s.m.* chivalry.
ιπποφορβείον *s.n.* stud farm.
ίπτ|αμαι *v.i.* fly. ~αμένη *s.f.* air-hostess. ~μενος *s.m.* flyer, airman.
ιριδίζω *v.i.* be iridescent. ~ισμός *s.m.* iridescence.
ίρις *s.f.* rainbow; iris (*of eye*); iris (*plant*).
ις *s.f.* fibre, filament.
ίσα *adv.* equally, evenly; straight, directly; in a straight line; ~ ~ exactly, precisely; ~ με up to, as far as, until; ήρθαμε ~ ~ we just had enough, were just in time.
ισάδα *s.f.* straightness.
ισάζω *v.t.* straighten; make even.
ίσαλος *s.f.* (*naut.*) water-line.
ίσαμε *prep.* (*with acc.*) up to, as far as, until.
ισάξιος *a.* equal in worth *or* ability.
ισάριθμος *a.* equal in number.
ισημερία *s.f.* equinox.
ισημερ|ινός 1. *a.* equinoctial; equatorial. 2. *s.m.* equator.
ισθμός *s.m.* isthmus.
ίσια *adv. see* **ίσα.**

ίσι|ος *a.* straight, level, upright; honest; equal (*to*), the same (*as*) (*with με*); *το ~ο* justice, right.

ισ(ι)ώνω *v.t.* straighten; make even. **~ωμα** *s.n.* straightening, levelling; level road.

ίσκα *s.f.* tinder.

ίσκιος *s.m.* shade, shadow.

ισλαμικός *a.* Islamic.

ισο- *denotes* equality; level.

ισόβιος *a.* for life, lifelong.

ισοβίτης *s.m.* lifer (*in prison*).

ισόγειο *s.n.* ground floor.

ισοδύναμ|ος *a.* equivalent. **~ώ** *v.i.* be equivalent.

ισοζυγίζ|ω *v.t.* balance, *~ομαι* (*v.i.*) balance, hover.

ισοζύγιον *s.n.* (*fin.*) balance; *εμπορικόν ~* balance of trade.

ισολογισμός *s.m.* balance-sheet.

ίσον *adv.* (*math.*) equals.

ισόπαλ|ος *a.* equally matched; *βγήκαν ~οι* the result was a draw. **~ία** *s.f.* draw, tie.

ισοπεδώνω *v.t.* make level.

ισόπλευρος *a.* equilateral.

ισορροπ|ία *s.f.* balance. **~ώ** *v.i.* & *t.* balance; *~ημένος* well-balanced (*mentally*). **~ος** *a.* balanced.

ίσ|ος *a.* equal (*to*), the same (*as*); *εξ ~ου* equally; *το ~ο* justice, right; *κρατώ το ~ο* sustain accompaniment, (*fig.*) play up (*to*), support.

ισοσκελ|ής *a.* balanced (*of budget*); (*geom.*) isosceles. **~ίζω** *v.t.* balance (*budget*).

ισότητα *s.f.* equality.

ισοτιμία *s.f.* parity.

ισότιμος *a.* equal in rights, rank, *or* value.

ισοϋψής *a.* ~ *καμπύλη* contour-line.

ισοφαρίζω *v.t.* & *i.* counterbalance; come equal, tie.

ισόχρονος *a.* isochronous; simultaneous.

ισοψηφία *s.f.* equal division of votes. **~ώ** *v.i.* obtain same number of votes.

ισπανικός *a.* Spanish.

Ισπανός *s.m.* Spaniard.

ίσταμαι *v.i.* stand.

ιστιοδρομία *s.f.* yacht-racing.

ιστίο *s.n.* sail.

ιστιοπλοΐα *s.f.* sailing.

ιστιοφόρο *s.n.* sailing-vessel.

ιστολογία *s.f.* histology.

ιστόρημα *s.n.* story.

ιστορί|α *s.f.* history; story; *~ες* (*pl.*) (*prolonged*) trouble, bother.

ιστορικ|ός *a.* historical; historic; an historian. **~ό** *s.n.* story, account (*of event*).

ιστοριοδίφης *s.m.* historical researcher.

ιστορώ *v.t.* relate, tell; illustrate (*book, etc.*).

ιστός *s.m.* 1. mast. 2. loom. 3. web; (*biol.*) tissue.

ισχίο *s.n.* hip. *~αλγία* *s.f.* sciatica.

ισχνός *a.* lean, thin, meagre. *~αίνω* *v.t.* & *i.* make *or* become lean, thin *or* meagre.

ισχυρ|ίζομαι *v.i.* assert, claim. **~ισμός** *s.m.* assertion, claim.

ισχυρογνώμων *a.* obstinate.

ισχυρός *a.* strong; intense; loud; (*law*) valid.

ισχύς *s.f.* strength; force, validity; *εν~ύι* in force.

ισχύω *v.i.* have weight *or* validity; be in force. *~ν* (*a.*) valid.

ισ|ώ *v.t.* equalize. *~ούμαι* *v.i.* (*with με*) be equal to.

ίσ|ωμα, ~ώνω *see* **ισιω-**.

ίσως *adv.* perhaps; ~ (*να*) *έρθει* perhaps he will come.

ιταλικός *a.* Italian.

Ιταλ|ός, ~ίδα *s.m.* an Italian.

ιταμός *a.* defiantly impertinent.

ιταμότης *s.f.* insolence.

ιτ|έα, ~ιά *s.f.* willow.

ιχθυοπωλείο *s.n.* fishmonger's.

ιχθυοτροφείο *s.n.* hatchery.

ιχθύς *s.m.* fish.

ιχνηλατώ *v.t.* track down.

ιχνογράφημα *s.n.* sketch.

ιχνογραφία *s.f.* sketching.

ίχνος *s.n.* footprint, track; trace, vestige.

ιώβειος *a.* of Job.

ιωβηλαίον *s.n.* jubilee.

ιώδιο *s.n.* iodine.

ιώμαι *v.i.* be cured.

ιωνικός *a.* Ionic.

ιώτα *s.n.* the letter I.

ιωτακισμός *s.m.* iotacism.

Κ

κάβα *s.f.* wine-cellar; pool (*in cards*).

καβάκι *s.n.* black poplar.

καβαλιέρος *s.m.* escort, partner.

καβάλα *adv.* on horseback; astride; (*fam.*) *ψωνίζω ~* be taken in (*by seller*).

καβαλάρης *s.m.* horseman; (*mus.*) bridge (*of violin*).

καβάλες *s.f.pl.* leap-frog.

καβαλέτο *s.n.* easel.

καβαλικεύω *v.t.* & *i.* ride, mount, sit astride; (*fig.*) dominate.

καβαλίνα *s.f.* dung of horses.

καβάλ|ος *s.m.*, *~ο* *s.n.* fork of trousers.

καβαλώ *v.t.* & *i. see* **καβαλικεύω**.

καβα(ν)τζάρω *v.t.* round, double (*cape*).

καβγ|άς *s.m.* quarrel, row. *~αδίζω* *v.i.*

quarrel. **~ατζής** *s.m.* quarrelsome person.

κάθ|ος *s.m.* cape, headland; (*naut.*) hawser; (*fam.*) παίρνω ~ο get an inkling.

καθούκι *s.n.* shell (*of tortoise, etc.*).

κάθουρας *s.m.* crab; spanner.

καθουρδίζω *v.t.* roast (*beans, etc.*); fry lightly, brown.

καθούρι *s.n.* crab.

καγκάγια *s.f.* (*fam.*) hag.

καγκελάριος *s.m.* chancellor.

κάγκελο *s.n.* banister, rail, railing, grille.

καγκελόπορτα *s.f.* garden gate.

καγκουρό *s.f.* kangaroo.

καγχάζω *v.i.* laugh loudly *or* scornfully.

καδένα *s.f.* (watch) chain.

κάδος *s.m.* bucket, tub.

κάδρο *s.n.* framed picture; frame.

καδρόνι *s.n.* beam (*timber*).

καζάκα *s.f.* loose jacket.

καζανάκι *s.n.* w.c. cistern.

καζάνι *s.n.* boiler; cauldron, copper.

καζίνο *s.n.* casino.

κάζο *s.n.* mishap, flop.

καζούρα *s.f.* κάνω ~ σε tease, rag.

καήλα *s.f.* feeling *or* smell of burning; (*fig.*) ardent desire.

καημέν|ος *a.* burnt; (*fig.*) poor, unfortunate; (*fam.*) ~ε και συ phrase of mild remonstration.

καημός *s.m.* sorrow; yearning.

καθ' see κατά.

καθά(περ) *adv.* as, according to (*with verb*).

καθαίρεσις *s.f.* deposition, dismissal.

καθαίρω *v.t.* cleanse, purge.

καθαιρώ *v.t.* depose, dismiss.

καθάπτομαι *v.t.* (*with gen.*) offend (*honour, etc.*).

καθαρεύουσ|α *s.f.* purified (*i.e.* formal) language. **~ιάνος** *s.m.* user *or* advocate of this.

καθαρίζω *v.t. & i.* clean, clear; peel, shell, gut; burnish; refine; clarify (*explain*); settle (*account*); (*fam.*) clean out, polish off, kill off. (*v.i.*) become clean *or* clear.

καθαριότητα *s.f.* cleanness, cleanliness.

καθάρισμα *s.n.* cleaning, *etc.* (*see* καθαρίζω).

καθαριστήριο *s.n.* (*dry*) cleaners'.

καθαρίστρ(ι)α *s.f.* charwoman.

κάθαρμα *s.n.* refuse, riffraff, scum.

καθαρόαιμος *a.* thoroughbred.

καθαρογράφω *v.t.* make fair copy of.

καθαρ|ός *a.* clean; clear, distinct, plain; straightforward; pure, unmixed; sheer; net; ~ό μυαλό lucid mind; το μέτωπό του είναι ~ό he has nothing to be ashamed of; ~η Εβδομάδα first week of Lent.

καθαρότητα *s.f.* purity; clarity.

καθάρσιον *s.n.* (*med.*) purgative.

κάθαρσις *s.f.* purification; catharsis; quarantine.

καθαρτήριον *s.n.* purgatory.

καθαρτικός *a.* purgative.

καθ|αυτό, ~εαυτού *adv. & a.* real(ly), genuine(ly).

κάθε *a.* every, each; ~ άλλο far from it; το ~ τι everything; ~ πού whenever.

καθέδρ|α *s.f.* (*eccl.*) throne. ~ικός ναός cathedral.

κάθειρξη *s.f.* imprisonment.

καθ|είς, ~ένας *pron.* each one; anyone, the first comer.

καθέκαστα *s.n.pl.* details.

καθέκλα *s.f.* chair.

καθελκύω *v.t.* launch.

καθεστώς *a.* established; το ~ régime; status quo.

κάθετος *a.* vertical; at right-angles.

καθεύδω *v.i.* sleep.

καθηγητής *s.m.* professor; master, teacher.

καθήκι *s.n.* chamber-pot; (*pej.*) scum.

καθήκον *s.n.* duty; ~τα (*pl.*) functions.

καθηλώνω *v.t.* nail *or* pin down; immobilize.

καθηματωμένος *a.* bloodstained.

κάθημαι *v.i. see* κάθομαι.

καθημερ(ι)ν|ός *a.* daily, every day. ~ή *s.f.* working day. ~ό *s.n.* daily bread. ~ώς *adv.* daily.

καθησυχάζω *v.t. & i.* calm, reassure, allay; become calm *or* quiet; be allayed.

καθιδρώς *a.* bathed in sweat.

καθιερώ|νω *v.t.* consecrate; establish; countenance (*a practice*); είναι ~μένο it is customary.

καθίζηση *s.f.* subsidence; (*chem.*) precipitation.

καθίζω *v.t.* seat, place; strand (*ship*).

καθίκι *s.n. see* καθήκι.

καθισιά *s.f.* sitting (*interval of time*).

καθισιό *s.n.* not working, doing nothing.

κάθισμα *s.n.* manner of sitting; seat; subsidence; running aground.

καθίσταμαι *v.i.* become, turn out to be, prove.

καθιστικός *a.* sedentary.

καθιστός *a.* seated.

καθιστώ *v.t.* make, appoint, set up.

καθό *adv.* as, qua.

καθοδηγητής *s.m.* political indoctrinator.

καθοδηγώ *v.t.* give guidance to.

κάθοδος *s.f.* descent, alighting; invasion; (*phys.*) cathode.

καθολικεύω *v.t. & i.* generalize.

καθολικ|ός *a.* universal, general; catholic. ~όν *s.n.* 1. (*eccl.*) nave. 2. ledger.

καθόλου *adv.* in general; (not) at all.

κάθομαι *v.i.* be seated; sit (down); settle, sink; remain; spend time (*doing sthg.*); be without work.

καθομιλουμένη *s.f.* spoken lanuguage, everyday speech.

καθορ|ίζω *v.t.* determine, fix, state precisely. **~ισμός** *s.m.* fixing, settling.

καθοσίωσ|ις *s.f.* consecration; *έγκλημα* **~εως** high treason, outrage.

καθόσον *adv.* (in) so far as; as (*causal*).

καθούμεν|ος *a.* seated; *στα καλά* **~α** out of the blue.

καθρέπτης *s.m.* mirror.

καθρεφτίζ|ω *v.t.* mirror, reflect. **~ομαι** *v.i.* look in the mirror, be reflected.

κάθυγρος *a.* wet, sodden.

καθυστέρηση *s.f.* delay; retardation.

καθυστερ|ώ *v.t. & i.* delay, withhold; be behind *or* late; **~ημένος** backward; **~ούμενα** arrears.

καθώς 1. *conj.* as, when. 2. *adv.* as, such as; **~** *πρέπει* proper, respectable.

και *conj.* and, also, too, even; **~ ... ~** both... and; **~** *να* even if; **~** *να...~* *να μη* whether... or not; *θέλει* **~** *δε θέλει* whether he likes it or not. *Also functions like* **να, ότι, πως, που, διότι.**

καΐκ|ι *s.n.* caique. **~τσής** *s.m.* captain *or* owner of caique.

καΐλα *s.f. see* **καήλα.**

καϊμάκι *s.n.* thick cream; froth on coffee.

καινός *a.* new.

καινοτομία *s.f.* innovation.

καινουργ|ής, ~ιος *a.* new.

καίπερ *conj.* although.

καιρικός *a.* of the weather, atmospheric.

καίριος *a.* timely, telling; mortal (*wound*); vital (*spot*).

καιρ|ός *s.m.* time, period; weather, wind; *με τον* **~ό** eventually; *στον* **~ό** *του* in (its) season.

καιρο|σκοπώ, ~φυλακτώ *v.i.* bide one's time. **~σκόπος** *s.m.f.* opportunist.

καισαρικός *a.* Caesarean.

καϊσί *s.n.* (*sweet-kernelled*) apricot.

καίτοι *conj.* although.

καί|ω *v.t. & i.* burn, shrivel; smart; be hot *or* alight. **~ομαι** *v.i.* be on fire; get burned *or* shrivelled; fuse; *κάηκα* (*fam.*) I got caught out.

κακά *adv.* badly.

κακάο *s.n.* cocoa.

κακαρίζω *v.i.* cackle.

κακαρ|ώνω *v.t. & i.* (*fam.*) perish from cold; *τα* **~ωσε** he kicked the bucket.

κακεντρέχεια *s.f.* malevolence.

κακία *s.f.* vice; malice, ill-will.

κακίζω *v.t.* censure, find fault with.

κακιώνω *v.i.* get angry; become on bad terms.

κακο- *denotes* bad, difficult.

κακοαναθρεμμένος *a.* ill-bred.

κακοβάζω *v.i.* form suspicions; imagine the worst.

κακόβουλος *a.* malevolent.

κακογερνώ *v.i.* suffer unhappy old age; age prematurely; become ugly in old age.

κακογλωσσιά *s.f.* scandalous gossip.

κακόγουστος *a.* in bad taste.

κακοδαιμονία *s.f.* misfortune, distress.

κακοδικία *s.f.* miscarriage of justice.

κακοήθης *a.* evil, immoral; (*med.*) malignant, pernicious.

κακόηχος *a.* ill-sounding.

κακοκαιρία *s.f.* bad weather.

κακοκαρδίζω *v.t. & i.* disappoint, distress; suffer disappointment *or* distress.

κακοκεφαλιά *s.f.* folly, ill-considered action.

κακοκεφιά *s.f.* bad mood.

κακολογία *s.f.* casting aspersions.

κακομαθ|αίνω *v.t. & i.* learn badly; spoil (*child, etc.*); get into bad habits; **~ημένος** ill-behaved, spoiled.

κακομελετώ *v.t. & i.* bring bad luck (on) (*with inauspicious words*).

κακομεταχειρίζομαι *v.t.* ill-treat.

κακό|μοιρος, ~μοίρης *a.* unfortunate, poor (*commiserative*).

κακ|ό(ν) *s.n.* evil; harm; wrong; ill fortune; uproar; *παίρνω από* **~ό** take a dislike to; *το έχω σε* **~ό** I think it a bad omen; *ψάρια* *και* **~ό** fish galore; *τον* **~άκου** in vain; *σκάει απ' το* **~ό** *του* he is bursting with rage.

κακοπαίρνω *v.t.* take amiss; speak roughly to.

κακοπέραση *s.f.* hardship.

κακοπερνώ *v.i.* suffer hardship; have a hard life.

κακοπέφτω *v.i.* make a bad marriage.

κακοπιστία *s.f.* bad faith.

κακοπόδαρος *a.* bringing ill luck.

κακοποι|ός *a.* injurious; a malefactor. **~ώ** *v.t.* ill-treat; rape.

κακο(ρ)ρίζικος *a.* unlucky; (*fam.*) wretched, cussed.

κακός *a.* bad, evil; ill-natured.

κακοσμία *s.f.* bad smell.

κακοσυνηθίζω *v.t. & i.* spoil (*child, etc.*); get into bad habits.

κακοσυσταίνω *v.t.* speak ill of.

κακοτοπιά *s.f.* rough ground; (*fig.*) awkward situation.

κακοτυχία *s.f.* mischance, ill fortune.

κακοτυχίζω *v.t.* express pity for.

κακοτυχώ *v.i.* suffer ill fortune.

κακούργημα *s.n.* felony, crime.
κακουργοδικείον *s.n.* criminal court.
κακούργος 1. *a.* criminal. 2. *s.m.* felon, criminal; (*fig.*) cruel person.
κακουχία *s.f.* hardship.
κακοφαίν|ομαι *v.i.* μου ~εται it offends me, I take exception to it; κακοφανισμένος offended, piqued.
κακοφέρνομαι *v.i.* behave rudely.
κακοφορμίζω *v.t. & i.* (cause to) fester.
κακοφτιαγμένος *a.* badly done *or* made; ill-shaped.
κακόφωνος *a.* cacophonous.
κάκτος *s.f.* cactus.
κακώς *adv.* badly.
κακώσεις *s.f.pl.* injuries, lesions.
καλά *adv.* well; properly; all right; για (τα) ~ seriously, definitely; πάει ~ all right! τα έχω ~ με be on good terms with; σώνει (*or* ναι) και ~ with obstinate insistence; τον κάνω ~ I cure him *or* (*fig.*) I can handle him.
κάλ|αθος *s.m.,* ~άθι *s.n.* basket; ~αθος αχρήστων waste-bin.
καλάι *s.n.* solder.
καλαισθησία *s.f.* good taste.
καλαϊτζής *s.m.* tinker.
καλαμάκι *s.n.* straw (*for drinks*).
καλαμαράς *s.m.* scribe; (*fam.*) penpusher.
καλαμάρι *s.n.* 1. inkstand. 2. cuttlefish.
καλαματιανός *a.* (dance) of Kalamata.
καλάμι *s.n.* reed, cane; fishing-rod; bobbin; shin-bone.
κάλαμος *s.m.* reed, cane; pen.
καλαμοσάκχαρον *s.n.* cane sugar.
καλαμπαλίκι *s.n.* (*fam.*) rowdy crowd; ~α (*pl.*) testicles.
καλαμπόκι *s.n.* maize.
καλαμπούρι *s.n.* pun.
κάλαντα *s.n.pl.* Christmas and New Year carols.
καλαντάρι *s.n.* calendar.
καλαπόδι *s.n.* shoemaker's last.
καλαφατίζω *v.t.* caulk.
καλέμι *s.n.* pen; chisel.
κάλεσμα *s.n.* invitation; ~τα (*pl.*) guests.
καλεσμένος *a.* invited; a guest.
καλή *s.f.* η ~ του (*poetic*) his bride *or* beloved; *see also* **καλός.**
καλημαύκι *s.n.* Orthodox priest's headgear.
καλημέρα *int. & s.f.* good-morning.
καληνύχτα *int. & s.f.* good-night.
καλησπέρα *int. & s.f.* good-afternoon, good-evening.
κάλιο *s.n.* potash.
καλιακούδα *s.f.* crow.
καλιγώνω *v.t.* shoe (*horse*).
καλικάντζαρος *s.m.* goblin.

κάλλια *adv.* better.
καλλίγραμμος *a.* shapely.
καλλιέπεια *s.f.* fine language.
καλλιέργεια *s.f.* cultivation, culture.
καλλιεργήσιμος *a.* cultivable.
καλλιεργητής *s.m.* cultivator.
καλλιεργώ *v.t.* cultivate; foster, develop.
κάλλι|ος *a.* better, superior. ~ο *adv.* better.
καλλιστεί|ο *s.n.* prize for beauty; ~α (*pl.*) beauty contest.
κάλλιστ|ος *a.* most beautiful; best. ~α *adv.* very well.
καλλίτερ- *see* **καλυτερ-.**
καλλιτέχνημα *s.n.* work of art.
καλλιτέχνης *s.m.* artist.
καλλιτεχνία *s.f.* art (*esp. fine arts*); artistry.
καλλιτεχνικός *a.* of art *or* artists; artistic.
καλλονή *s.f.* beauty.
κάλλος *s.n.* beauty.
καλλυντικός *a.* cosmetic.
καλλωπίζω *v.t.* beautify, embellish.
καλμάρω *v.t. & i.* make *or* become calm.
καλντερίμι *s.n.* cobbles, cobbled street.
καλο- *denotes* good, well.
καλοαναθρεμμένος *a.* well-bred.
καλόβαθρο *s.n.* stilt.
καλοβαλμένος *a.* well appointed *or* turned out.
καλοβλέπω *v.i. & t.* have good sight; look well at; look favourably on; covet, ogle.
καλόβολος *a.* easy-going, obliging.
καλόγερος *s.m.* monk; boil; clothes-horse; hat-stand; hopscotch; μπήκε ο ~ στο φαΐ the food is burnt.
καλόγηρος *s.m.* monk.
καλόγνωμος *a.* well-disposed.
καλόγρια *s.f.* nun.
καλοδεμένος *a.* well bound *or* tied; well set (*ring*); well set-up (*man*).
καλοδεχούμενος *a.* welcome; welcoming.
καλοζώ *v.i. & t.* live well; maintain in comfort.
καλοήθης *a.* moral; (*med.*) benign.
καλοθελητής *s.m.* (*ironic*) well-wisher.
καλοθρεμμένος *a.* well-nourished.
καλοκάγαθος *a.* exceptionally good and kind (*person*).
καλοκάθομαι *v.i.* settle firmly and comfortably.
καλοκαίρ|ι *s.n.* summer; summery weather. ~ία *s.f.* fine weather. ~ινός *a.* of summer; summery.
καλοκαμωμένος *a.* well prepared *or* finished; shapely, handsome.
καλοκαρδίζω *v.t. & i.* cheer, gladden; be cheered *or* gladdened.
καλόκαρδος *a.* kindly, jovial.
καλοκοιτάζω *v.t.* look carefully at; look

after well; covet, ogle.
καλομαθ|αίνω *v.t. & i.* learn well; pamper; become pampered; ~*ημένος* well-behaved; pampered.
καλομελετώ *v.t. & i.* bring good luck (to) *(with auspicious words)*.
καλό|μοιρος, ~μοίρης *a.* fortunate.
καλ|ό(ν) *s.n.* good; ~*ά* (*pl.*) benefits, riches; *στο ~ό* so long! *με το ~ό* safely, with God's help; ~*ού κακού or (για) ~ό κακό* for any eventuality; *το ~ό που σου θέλω* take my advice; *παίρνω με το ~ό* wheedle; *τον πήρα από ~ό* I regard him favourably; ~*ό 'νάχεις formula of entreaty*; *σε ~ό σου expression of mild disapproval*; *βάζω τα ~ά μου* put on one's best clothes; *δεν είναι στα ~ά του* he is not in good form *or* health *or* a good mood *or* his right mind.
καλοντυμένος *a.* well-dressed.
καλοπέραση *s.f.* comfort, well-being.
καλοπερνώ *v.i.* have an easy life; have a good time.
καλοπιάνω *v.t.* wheedle, get round.
καλοπόδαρος *a.* bringing good luck.
καλοπροαίρετος *a.* well-intentioned.
καλοριφέρ *s.n.* central-heating (*radiator*); heater. *[F.* calorifère].
καλο(ρ)ρίζικος *a.* lucky, auspicious.
καλόρχεται *v.i.* it fits (*of clothes*); it is welcome.
κάλος *s.m.* corn, callus.
καλ|ός *a.* good; kind; ~*ές τέχνες* fine arts; *ώρα ~ή greeting or farewell*; *μια και ~ή* once and for all; *η ~ή* the right side (*of cloth, etc,*); *τον ξέρω απ' την ~ή του* I know his real character; *του τα είπα από την ~* I gave him a piece of my mind; *ο ~ός της (poetic)* her good man *or* lover; ~*έ familiar mode of address*.
καλοστεκούμενος *a.* well-off.
καλοσυνεύω *v.i.* clear up, brighten.
καλοσυνηθ|ίζω *v.t. & i.* pamper; become pampered; ~*ισμένος* well-behaved, pampered.
καλοσυσταίνω *v.t.* do credit to; speak well of, recommend.
καλόττα *s.f.* crown (*of hat*).
καλότυχ|ος *a.* fortunate. ~*ίζω* *v.t.* envy, consider fortunate.
καλούπ|ι *s.n.* form, mould. ~*ώνω* *v.t.* put into moulds; (*fig.*) deceive.
καλούτσικος *a.* fair, passable, goodish.
καλοφαγάς *s.m.* gourmand.
καλοφαίν|ομαι *v.i. μου ~εται* I am pleased.
καλοφτιασμένος *a.* well made *or* built.
καλπάζω *v.i.* gallop.
καλπάκι *s.n.* fur cap.
καλπασμός *s.m.* gallop.

κάλπη *s.f.* ballot-box; *βάζω* ~ stand for election.
κάλπικος *a.* counterfeit.
καλπονόθευση *s.f.* electoral fraud.
καλπουζάνος *s.m.* counterfeiter; (*fig.*) dishonest person.
κάλτσα *s.f.* sock, stocking; *του διαβόλου* ~ sly fox.
καλτσοδέτα *s.f.* garter.
καλύβ|α, ~η *s.f.,* ~**ι** *s.n.* hut.
κάλυμμα *s.n.* cover, covering; headgear.
κάλυξ *s.m.* (*bot.*) calyx, bud; (*mil.*) cartridge case; (*anat.*) calix.
καλύπτω *v.t.* conceal; cover; defray.
καλυτερεύω *v.t. & i.* improve.
καλύτερ|ος *a.* better. ~**α** *adv.* better.
κάλυψη *s.f.* (act of) covering.
κάλφας *s.m.* foreman mason; apprentice.
καλώ *v.t.* call, name; summon; invite.
καλώδιο *s.n.* rope; (*electric*) cable.
κάλως *s.m.* thick rope, hawser.
καλώς *adv.* well; *έχει* ~ good, agreed; ~ *ωρίσατε* welcome; ~ *τονε* welcome to him; ~ *εχόντων των πραγμάτων* all being well.
καλωσορίζω *v.t.* welcome.
καλωσύνη .*f.* goodness, kindness; fine weather.
κάμα *s.f.* sort of dagger.
καμάκι *s.n.* harpoon.
κάμαρα *s.f.* room.
καμάρα *s.f.* arch, vault.
καμάρι *s.n.* air of pride, jaunty air; object of pride.
καμαριέρα *s.f.* chamber-maid.
καμαρίνι *s.n.* (*stage*) dressing-room.
καμαρότος *s.m.* (*ship's*) steward.
καμαρώνω *v.i. & t.* wear a proud air; take pride (in).
καμαρωτ|ός *a.* 1. arched. 2. with an air of pride. ~**ά** *adv.* proudly, jauntily.
κάματος *s.m.* fatigue; tilling.
καμβάς *s.m.* canvas (*for embroidery*).
καμήλα *s.f.* camel.
καμηλοπάρδαλη *s.f.* giraffe.
καμηλαύκι *s.n. see* **καληαύκι.**
καμινάδα *s.f.* chimney, funnel.
καμινέτο *s.n.* spirit-lamp.
καμινεύω *v.t.* smelt; burn (*in kiln*).
κάμ|ινος *s.f.,* ~**ίνι** *s.n.* furnace, kiln.
καμιόνι *s.n.* lorry.
καμμία *pron. f. see* **κανένας.**
καμμύω *v.t.* half close (*eyes*).
κάμνω *v.t. see* **κάνω.**
καμουτσί(κι) *s.n.* whip.
καμουφλάρω *v.t.* camouflage.
καμπάνα *s.f.* (*church, etc.*) bell; (*fam.*) scolding.
καμπαναριό *s.n.* belfry.

καμπανιά *s.f.* bell-ringing; stroke of bell; (*fam.*) unpleasant hint.

καμπάνια *s.f.* newspaper campaign.

καμπανίτης *s.m.* champagne.

κα(μ)παρ(ν)τίνα *s.f.* gabardine.

κάμπη|η, ~ια *s.f.* caterpillar.

καμπή *s.f.* bend, turn.

καμπίνα *s.f.* cabin.

καμπινές *s.m.* w.c.

καμπίσιος *a.* of the plain.

κάμπος *s.m.* plain, open country.

κάμ|ποσος, ~πόσος *pron.* some, a fair number (of); *κάνω τον ~πόσο* talk big *or* blusteringly.

καμπούρα *s.f.* hump, boss.

καμπούρης *s.m.* a hunchback.

καμπουριάζω *v.t. & i.* hunch, bend; stoop, become bent.

καμπουρωτός *a.* humped, bent, curved.

κάμπτ|ω *v.t. & i.* bend; round (*bend, cape*); (*v.i.*) turn. *~ομαι v.i.* bend, give way; relent; drop (*of prices*).

καμπύλη *s.f.* curve.

καμπύλος *a.* curved, convex, rounded.

καμτσίκι *s.n.* whip.

καμφορά *s.f.* camphor.

κάμψη *s.f.* bending, giving way; lessening, fall.

κάμωμα *s.n.* making; ripening; *~τα* (*pl.*) coquettish airs, affectation of reluctance; antics, exploits (*pej.*).

καμωμένος *a.* done, made; ripe.

καμώνομαι *v.i.* pretend; affect reluctance.

καν *conj.* at least; *ούτε ~* not even; *~ ... ~* either ... or; *~ και* ~ a lot (of).

κανάγιας *s.m.* rascal.

κάνα - δυο *a.* one or two.

κανακάρης *a.* petted, one-and-only; a darling only son.

κανάλι *s.n.* channel, strait.

καναπές *s.m.* sofa, settee.

καναρίνι *s.n.* canary.

κανάτα *s.f.* jug.

κανάτι *s.n.* 1. pitcher. 2. shutter.

κανδήλα *s.f.* votive lamp.

κανείς *pron.* see **κανένας**.

κανέλλα *s.f.* cinnamon.

κανένας *pron.* any, anyone, one; some; (*after neg.*) no, no one; *καμμιά ώρα* about an hour; *καμμιά φορά* some time, sometimes.

κάνθαρος *s.m.* beetle.

κανθός *s.m.* corner of eye.

κανιά *s.n.pl.* (*fam.*) (*spindly*) legs.

κάννι|α, ~η *s.f.* barrel (*of gun*).

κανναβάτσ|α *s.f.*, **~ο** *s.n.* hessian.

κανναβ|ι *s.n.* hemp. **~ούρι** *s.n.* hempseed.

κάνναβις *s.f.* cannabis.

καννίβαλος *s.m.* cannibal.

κανόνας *s.m.* rule (*measure, principle*).

κανόνι *s.n.* cannon; (*fam.*) *ρίχνω or σκάζω ~* go bankrupt; *το σκάζω* ~ play truant.

κανονιά *s.f.* cannon-shot.

κανονίζω *v.t.* arrange, regulate, settle.

κανονικός *a.* regular, ordinary, usual; (*eccl.*) canonical.

κανονιοφόρος *s.f.* gunboat.

κανονισμός *s.m.* settlement, regulation(s).

κάνουλα *s.f.* faucet, tap for barrel.

καντάδα *s.f.* serenade.

καντάρι *s.n.* steelyard; hundredweight.

καντήλα *s.f.* (votive) (*olive-oil*) lamp; blister.

καντήλι *s.n.* small (*olive-oil*) light.

καντίνα *s.f.* canteen (*shop*).

καντράν *s.n.* dial.

κάν|ω *v.t. & i.* 1. do, make, perform; have (*sthg.*) done *or* made; *~ω μαθήματα* have *or* give lessons; *~ω κοστούμι* have a suit made; *~ω μπάνιο* have a bath *or* bathe; *τον ~ω καλά* I cure him *or* (*fig.*) I can handle him; *του ~ω καλό* I do him good; *τον ε~α για* I mistook him for; *τα ~ω θάλασσα* (*or σαλάτα*) make a mess of it; *~ω χωρίς* do without; *~ω πίσω* move back; *δεν ~ω με* I do not get on with; *δεν ~ει* (*να*) it is not right *or* proper (to); *δεν ~ει για* it is not suitable for; *έχει να ~ει για* it makes a difference; *το ίδιο μου ~ει* it is all the same to me; *~ει αέρα* it is windy. 2. produce; *~ω παιδί* have a child; *η Ελλάδα ~ει κρασί* Greece produces wine; *~ω νερά* leak, (*fig.*) hedge. 3. behave, be occupied; spend *or* take time; *~ω σαν τρελλός* (*για*) behave like a madman, be madly keen (on); *~ω τον τρελλό* pretend to be mad; *~ω το γιατρό* be a doctor; *~ω πως* pretend that; *~ω να* make an effort to; *ε~ε καιρό να μας γράψει* he took a long time to write to us; *ε~ε στην Αμερική* he has been in America; *ε~ε γιατρός* he was a doctor; *~ε γρήγορα* make haste! 4. *τι ~ετε;* how are you? *πόσο ~ει;* how much does it cost? *~ει να πάρω κι άλλα λεφτά* there is still some money due to me.

κανών *s.m.* rule (*measure, principle*); (*eccl. & mus.*) canon.

καούρα *s.f.* burning (*sensation*).

κάπα *s.f.* shepherd's cloak.

καπάκι *s.n.* cover, lid, cap.

καπάντζα *s.f.* trap-door; mouse-trap.

καπάρ|ο *s.n.* earnest, deposit. **~ώνω** *v.t.* pay deposit on; book.

καπαρτίνα *s.f.* gabardine.

καπάτσος *a.* resourceful, shrewd.

κάπελας *s.m.* wine-shop keeper.

καπέλλο *s.n.* hat.

καπετάνιος *s.m.* captain (*of ship or irregulars*).

καπηλ|εία *s.f.* exploitation (*for base ends*). **~εύομαι** *v.t.* exploit thus.

καπηλειό *s.n.* wine-shop.

καπηλικός *a* gross, vulgar.

καπίστρι *s.n.* halter.

καπλαμάς *s.m.* veneer.

καπλαντίζω *v.t.* veneer; (*fig.*) put protective cover on (*quilt, book*).

κάπν|α *s.f.* soot. **~ες** fumes.

καπνέμπορος *s.m.* tobacco-merchant.

καπνιά *s.f.* soot.

καπν|ίζω *v.t. & i.* smoke; *του ~ισε να* he took it into his head to.

κάπνισμα *s.n.* smoking.

καπνιστής *s.m.* smoker.

καπνιστός *a.* smoked.

καπνοδόχ|η, ~ος *s.f.* chimney, smoke-stack.

καπνοπώλης *s.m.* tobacconist.

καπν|ός *s.m.* 1. (*pl. ~οι*) smoke; (*pl.*) fumes; (*fig.*) *έγινε ~ός* he vanished. 2. (*pl. τα ~ά*) tobacco, tobacco-plant.

κάποιος *pron.* somebody; some, a certain.

καπόνι *s.n.* capon. (*naut.*) davit.

καπότα *s.f.* shepherd's cloak.

κάποτε *adv.* sometimes; some time; once.

κάπου *adv.* somewhere; *~ ~* now and then; *~ δέκα* about ten.

καπούλια *s.n.pl.* hindquarters.

κάππα *s.n.* the letter **Κ**.

κάππαρη *s.f.* caper (*edible*).

καπρίτσιο *s.n.* caprice; (*mus.*) capriccio.

κάπρος *s.m.* boar.

κάπως *adv.* somewhat.

κάρα *s.f.* head.

καραβάν|α *s.f.* mess-tin; *λόγια της ~ας* nonsense. **~άς** *s.m.* (*pej.*) (boorish) officer.

καραβάνι *s.n.* caravan, company of travellers.

κάραβ|ι *s.n.* ship. **~ιά** *s.f.* shipload.

καραβίδα *s.f.* crayfish.

καραβοκύρης *s.m.* owner *or* skipper of vessel.

καραβόπανο *s.n.* sailcloth.

καραβοτσακισμένος *a.* shipwrecked.

καραγκιόζ|ης *s.m.* (chief character of) shadow-theatre; (*fig.*) comical person. **~λίκι** *s.n.* crudely comical behaviour.

καραδοκώ *v.t.* lie in wait for; be on look-out for.

καρακάξα *s.f.* magpie; (*fam.*) ugly, talkative woman.

καραμέλα *s.f.* sweet, bon-bon; lump-sugar.

καραμούζα *s.f.* toy flute; motor-horn.

καραμπίνα *s.f.* carbine.

καραμπογιά *s.f.* green vitriol; jet-black.

καραμπόλα *s.f.* pile-up (*collision*).

καραντίνα *s.f.* quarantine.

καραούλι *s.n.* watch, guard; sentinel; look-out post.

κεράτι(ο) *s.n.* carat.

καρατομώ *v.t.* behead.

καράφα *s.f.* carafe, decanter.

καράφλα *s.f. see* **φαλάκρα**.

καρβέλι *s.n.* round loaf.

κάρβουν|ο *s.n.* charcoal; coal. **~ιάζω** *v.t. & i.* char. **~ιάρης** *s.m.* charcoal-burner; coal-merchant; coal-man.

κάργα *adv.* (*full*) to the brim; tightly.

κάργια *s.f.* jackdaw.

κάρδαμο *s.n.* cress.

καρδαμώνω *v.i. & t.* become invigorated; invigorate.

καρδερίνα *s.f.* goldfinch.

καρδιά *s.f.* heart; *χάνω ~* take courage; *δε βαστά η ~ μου* I have not the courage *or* the heart (*to*); (*fam.*) *έχω ~ αγγινάρα* have many love affairs.

καρδιακός *a.* cardiac; suffering from heart trouble.

καρδιολόγος *s.m.* heart specialist.

καρδιοχτύπι *s.n.* palpitation; (*fig.*) anxiety.

καρ|έκλα, ~έγλα *s.f.* chair.

καρικατούρα *s.f.* caricature.

καρικώνω *v.t.* darn, mend.

καρίνα *s.f.* keel.

καριοφίλι *s.n.* flint-lock.

καρκινοβατώ *v.i.* (*fig.*) go crabwise, *hence* make no progress.

καρκίνος *s.m.* crab; cancer; palindrome.

καρκινώδης *a.* cancerous.

καρκίνωμα *s.n.* cancerous growth; (*fig.*) affliction, worrying circumstance.

καρμανιόλα *s.f.* guillotine; (*fig.*) danger spot; dishonest card-game; extortionate shop, *etc.*

καρμίρης *s.m.* miser.

καρμπόν *s.n.* carbon-paper.

καρναβάλι *s.n.* carnival.

καρντάν *s.n.* universal joint (*of engine*). [F. *cardan*]

καρότο *s.n.* carrot.

καρ|ούλα *s.f.,* **~ούμπαλο** *s.n.* bump (*on head*).

καρούλι *s.n.* pulley; caster; roller; reel.

καρπαζιά *s.f.* slap on the head.

καρπερός *a.* prolific.

καρπίζ|ω *v.i.*fruit. **~ομαι** *v.t.* enjoy fruits of.

καρπ|ός *s.m.* 1. fruit; *ξηροί ~οί* dried fruit and nuts. 2. wrist.

καρπούζι *s.n.* watermelon.

καρπούμαι *v.t.* enjoy fruits of.

καρποφορώ *v.i.* bear fruit.

καρποφόρος *a.* fruit (*tree*); fruitful; effective; profitable.

καρραγωγεύς *s.m.* cart-driver.

καρρέ 1. *s.n.* foursome (*at cards*); open neck (*of dress*); neck so exposed; hair en brosse; table centre-piece; (*naut.*) ward-room. 2. *a.* square. [F. *carré*].

καρρό *s.n.* diamond (*cards*); check (*pattern*). [F. *carreau*].

κάρρο *s.n.* cart.

καρροτσάκι *s.n.* hand-cart, barrow; pram, push-cart; wheel-chair.

καρσί *adv.* opposite.

κάρτα *s.f.* picture postcard; greeting-card; visiting-card.

καρτέρι *s.n.* ambush.

καρτερία *s.f.* endurance, fortitude.

καρτερικός *a.* steadfast.

καρτερώ *v.i. & t.* show patient endurance; wait (*for*).

κάρτο *s.n.* quarter (*of hour*).

καρυά *s.f.* walnut-tree.

καρύδα *s.f.* coconut.

καρύδι *s.n.* walnut; Adam's apple; κάθε καρυδιάς ~ all sorts and conditions of men.

καρυδιά *s.f.* walnut-tree; walnut (*wood*).

καρύκευμα *s.n.* seasoning, sauce.

καρυοθραύστης *s.m.* nutcracker.

καρυοφύλλι *s.n.* clove.

καρφ|ί *s.n.* nail; ~ιά (*pl.*) pedestrian crossing; ~ί δε μου καίγεται I do not care a pin; κόβω ~ιά shiver with cold; του βάζω τα εφτά ~ιά I tease him.

καρφίτσα *s.f.* pin; brooch.

καρφιτσώνω *v.t.* pin (*on, together*).

κάρφωμα *s.n.* nailing.

καρφ|ώνω *v.t.* nail (up, down, *etc.*); (*fig.*) fix, transfix; του ~ώθηκε η ιδέα he is possessed with the idea; (*fam.*) inform against.

καρχαρίας *s.m.* shark.

κάσσα *s.f.* packing-case; safe; cashdesk; door *or* window frame; (*fam.*) coffin.

κασσέλα *s.f.* wooden chest.

κασσέρι *s.n.* sort of cheese.

κασσίδα *s.f.* (*fam.*) balding, scurvy head.

κάσκα *s.f.* helmet.

κασκαβάλι *s.n.* see **κασέρι**.

κασκέτο *s.n.* cap.

κασμίρι *s.n.* wool suiting material.

κασόνι *s.n.* packing-case.

κασσίτερος *s.m.* tin.

καστανιά *s.f.* 1. chestnut-tree; chestnut (*wood*). 2. collapsible picnic-box.

κάστανο *s.n.* chestnut.

καστανός *a.* chestnut (*colour*).

καστόρι *s.n.* beaver (*fur*); felt (*for hats*); sort of leather; (*fam.*) suede.

κάστρο *s.n.* fortress.

κάστωρ *s.m.* beaver.

κατά *prep.* 1. (*with acc.*) (*place*) towards, by, in the direction of; ~ ξηράν και θάλασσαν by land and sea; καθ' οδόν on the way; ~ κει in that direction; ~ μέρος aside. (*time*) at, about, during; ~ αυτάς these days; καθ' ην στιγμήν at the moment when. (*relation*) according to; ~ τύχη by chance; ~ βάθος at bottom, thoroughly; το κατ' εμέ as for me. (*distribution*) ~ έτος every year, per year; καθ' εκάστην every day. 2. (*with gen.*) against; ~ γης to *or* on the ground; φέρομαι ~ κρημνών be heading for destruction; ~ διαβόλου to the devil.

κατα- *denotes* down; against; very, completely.

κατάβαθα *adv.* deep. τα ~ depths.

καταβαίνω *v.t. & i. see* **κατεβαίνω**.

καταβάλλω *v.t.* lay low, exhaust; pay, put down (*money*); ~ προσπάθειαν make an effort.

καταβαραθρώνω *v.t.* play havoc with, ruin.

κατάβαση *s.f.* descent.

καταβεβλημένος *a.* run-down, done-up.

καταβιβάζω *v.t. see* **κατεβάζω**.

καταβόθρα *s.f.* swallow-hole; covered drain; (*fig.*) insatiable person.

καταβολάδα *s.f.* layer (*horticulture*).

καταβολ|ή *s.f.* payment; από ~ής κόσμου since the world began.

καταβρεκτήρ *s.m.* water-cart.

καταβρέχω *v.t.* sprinkle, spray, soak.

καταβροχθίζω *v.t.* devour.

καταβυθίζω *v.t.* sink (*in liquid*).

καταγγελ|ία *s.f.* (*law*) charge, denunciation; annulment. ~λω *v.t.* charge, denounce.

καταγέλαστος *a.* an object of ridicule.

καταγής *adv.* on *or* to the ground.

καταγίνομαι *v.i.* be busy *or* engaged (*in*).

κάταγμα *s.n.* (*med.*) fracture.

κατάγ|ω *v.t.* win (*victory*). ~ομαι *v.i.* be descended *or* come (*from*).

καταγωγή *s.f.* origin, descent.

καταγώγιο *s.n.* den of thieves; sink of iniquity.

καταδεικνύω *v.t.* prove, demonstrate.

καταδέχομαι *v.t. & i.* deign to accept; take notice of, be nice to (*inferiors*); deign, condescend, permit oneself (*to*).

κατάδηλος *a.* obvious.

καταδίδω *v.t.* betray, give away.

καταδικάζω *v.t.* condemn; sentence.

καταδίκη *s.f.* sentence.

κατάδικος *s.m.f.* convict, prisoner.

καταδιωκτικ|ός *a.* of pursuit. ~όν *s.n.* (*aero.*) fighter plane.

καταδιώκω *v.t.* pursue, chase; persecute.

καταδί|ωξη *s.f.* ~**ωγμός** *s.m.* persecution.

κατάδοση *s.f.* betrayal.

καταδρομή *s.f.* pursuit; (*naut.*) privateering; ~ *της τύχης* persistent ill fortune.

καταδρομικόν *s.n.* (*naut.*) cruiser.

καταδυναστεύω *v.t.* oppress.

καταδύομαι *v.i.* dive.

καταζητώ *v.t.* hunt for, want (*for crime*).

κατάθεση *s.f.* deposit, account; tabling; deposition.

καταθέτω *v.t.* lay (*down*), deposit; lodge, file, table; give (*evidence*); lay down (*arms*).

καταθλιπτικ|ός *a.* oppressive, crushing; ~ή *αντλία* compression pump.

καταιγ|ίς, ~**ίδα** *s.f.* violent storm.

καταιγισμός *s.m.* (*mil.*) concentrated artillery-fire.

καταΐφι *s.n.* sort of oriental cake.

κατακάθ|ι *s.n.* dregs, residue. ~**ίζω** *v.i.* sink, settle, subside.

κατάκαρδα *adv.* seriously, to heart.

κατακέφαλ|α *adv.* on the head; headlong. ~**ιά** *s.f.* blow on the head.

κατακλ|είς *s.f.* conclusion, end; εν ~**είδι** in conclusion.

κατακλίνομαι *v.i.* lie down.

κατακλύζω *v.t.* inundate, flood.

κατακλυσμ|ός *s.m.* deluge; (*fig.*) abundance; *φέρνω τον* ~**ό** make difficulties about trifles.

κατάκοιτος *a.* confined to bed.

κατακόμβη *s.f.* catacomb.

κατάκοπος *a.* tired out, exhausted.

κατακόβ|ω *v.t.* cut deeply *or* in many places *or* on a large scale; cut to pieces. ~**ομαι** *v.i.* get exhausted; make great efforts.

κατακόρυφ|ος *a.* vertical. ~**ο** *s.n.* zenith; (*fig.*) extreme degree.

κατακρατώ *v.t.* withhold unlawfully.

κατακραυγή *s.f.* outcry.

κατακρεουργώ *v.t.* butcher.

κατακρημνίζω *v.t.* hurl down, demolish; (*chem.*) precipitate.

κατακρίνω *v.t.* censure, blame.

κατάκτηση *s.f.* conquest.

κατακτητής *s.m.* conqueror.

κατακτώ *v.t.* conquer, win.

κατακυρώνω *v.t.* award (*judicially*); knock down (*at auction*); give (*contract*).

καταλαβαίνω *v.t. & i.* understand; notice.

καταλαγιάζω *v.t. & i.* calm (down).

καταλαμβάνω *v.t. & i.* seize, possess oneself of; detect (*wrongdoer*); take up (*space*); understand; notice.

καταλείπω *v.t.* leave (behind).

καταλεπτώς *adv.* in detail.

καταλήγω *v.i.* end up (*in, as*); lead (*to*).

κατάληξη *s.f.* termination, end(ing).

καταληπτός *a.* comprehensible.

κατάληψη *s.f.* seizure, occupation; comprehension.

κατάλληλος *a.* suitable.

καταλογίζω *v.t.* impute; charge up.

κατάλογος *s.m.* list; directory; menu; catalogue (*of exhibition, etc.*).

κατάλοιπο *s.n.* residue, remains, left-over.

κατάλυμα *s.n.* billet, lodging.

κατάλυσις *s.f.* abolition; billeting; (*chem.*) catalysis.

καταλύω *v.t. & i.* abolish; be billeted.

καταλώ *v.t.* wear out, use up; digest.

κατάματα *adv.* straight in the eye.

καταμερίζω *v.t.* apportion.

καταμεσής *adv.* right in the middle.

κατάμεστος *a.* very full.

καταμετρώ *v.t.* measure, count; survey (*land*).

κατάμουτρα *adv.* to *or* in one's face.

καταναλ|ίσκω, ~**ώνω** *v.t.* consume, spend, use up.

κατανάλωση *s.f.* consumption, using up.

καταναλωτής *s.m.* consumer.

κατανέμω *v.t.* divide up; allot, assign.

κατανεύω *v.i.* nod assent.

κατανο|ώ *v.t. & i.* fully understand. ~**ηση** *s.f.* understanding.

κατάντ|ημα, ~**ι** *s.n.,* ~**ια** *s.f.* sorry state (*that one is reduced to*).

καταντώ *v.i. & t.* be reduced (*to*), end up (*in, as, by*); reduce, bring (*to*).

κατάνυξη *s.f.* devout concentration.

καταπακτή *s.f.* trap-door.

καταπάτι *s.n.* dregs, residue.

καταπατώ *v.t.* trample on; violate; encroach upon; trespass on.

κατάπαυση *s.f.* cessation, ending.

καταπέλτης *s.m.* catapult; (*fig.*) bolt from the blue.

καταπέτασμα *s.n.* curtain; (*fam.*) *τρώω το* ~ eat one's bellyful.

καταπιάνομαι *v.i.* (*with* με) start work on, take up.

καταπιέζω *v.t.* oppress.

καταπίνω *v.t.* swallow (up).

καταπίπτω *v.i.* fall, collapse; subside; decline.

καταπλακώνω *v.t.* flatten, squash.

κατάπλασμα *s.n.* poultice.

κατάπλατα *adv.* in the middle of the back.

καταπλέω *v.i. & t.* sail in; (*v.t.*) sail down (*river*).

καταπληκτικός *a.* amazing.

κατάπληκτος *a.* amazed.

κατάπληξη *s.f.* amazement.

καταπλήσσω *v.t.* amaze.

κατάπλους *s.m.* sailing in.

καταπνίγω *v.t.* stifle; suppress.

καταπόδι *adv.* close behind, on one's heels.

καταπολεμώ *v.t.* combat.

καταποντίζ|ω *v.t.* sink (*in sea*). ~ομαι *v.i.* sink, founder.

καταπονάω *v.t. & i.* hurt badly; suffer great pain.

καταπον|ώ *v.t.* tire out, exhaust. ~ούμαι *v.i.* get exhausted.

καταπότι *s.n.* pill.

καταπραΰνω *v.t.* assuage, mollify.

κατάπτυστος *a.* despicable.

κατάπτωση *s.f.* collapse, fall; decline.

κατάρα *s.f.* curse.

καταραμένος *a.* cursed; the devil.

καταργώ *v.t.* abolish, abrogate.

καταριέμαι *v.t. & i.* curse.

καταρράκτης *s.m.* waterfall; cataract.

καταρρακώνω *v.t.* tear *or* wear to shreds.

καταρρέω *v.i.* crumble, collapse.

καταρρίπτω *v.t.* fell, shoot down; demolish; beat (*record*).

κατάρ|ρους *s.m.,* ~**ροή** *s.f.* catarrh.

κατάρρυτος *a.* well-watered.

κατάρτι *s.n.* mast.

καταρτίζω *v.t.* form, put together; prepare, equip.

καταρώμαι *v.t. & i.* curse.

κατάσαρκα *adv.* next to the skin.

κατάσβεση *s.f.* quenching, quelling.

κατασκευάζω *v.t.* make, manufacture, construct; concoct.

κατασκεύασμα *s.n.* thing made, construction; concoction.

κατασκευή *s.f.* making, construction; manufacture, make; way a thing is made *or* shaped.

κατασκήνωση *s.f.* camp.

κατασκοπ|ος *s.m.* spy. ~**εία** *s.f.* espionage.

κατασκοτών|ω *v.t.* beat unmercifully. ~ομαι *v.t.* get tired *or* bruised; make a great effort.

κατασπαράσσω *v.t.* tear to pieces.

κάτασπρος *a.* very white.

κατασταλάζω *v.i.* settle (*of sediment*); (*fig.*) reach a decision.

κατασταλτικός *a.* repressive.

κατάσταση *s.f.* state (*of affairs*), condition, situation; status; return (*in accountancy*); list, register; goods, wealth.

καταστατικόν *s.n.* charter, statutes.

καταστέλλω *v.t.* repress, curb.

κατάστηθα *adv.* full on the chest.

κατάστημα *s.n.* establishment, institution, office; shop. ~**τάρχης** *s.m.* shopkeeper.

κατάστικτος *a.* spotted, dotted.

κατάστιχο *s.n.* ledger, register.

καταστολή *s.f.* repression.

καταστρατηγώ *v.t.* get round (*law*).

καταστρεπτικός *a.* destructive.

καταστρέφω *v.t.* destroy, ruin; spoil; deflower.

καταστροφή *s.f.* destruction, ruin; damage; catastrophe.

κατάστρωμα *s.n.* deck.

καταστρώνω *v.t.* draw up, frame.

κατάσχεσις *s.f.* (*law*) attachment, distraint.

κατάσχω *v.t.* sequestrate, distrain on.

κατατάσσ|ω *v.t.* classify, count, rank (*among*). ~ομαι *v.i.* enlist.

κατατομή *s.f.* vertical section; profile.

κατατόπια *s.n.pl.* (*fam.*) ins-and-outs (*of a place*).

κατατοπίζ|ω *v.t.* direct, give guidance to, brief. ~ομαι *v.i.* get one's bearings.

κατατρέχω *v.t.* persecute.

κατατρίβ|ω *v.t.* wear down. ~ομαι *v.i.* waste one's time.

κατατροπώνω *v.t.* rout.

κατατρύχω *v.t.* torment.

καταυλίζομαι *v.i.* bivouac.

καταφανής *a.* evident, clear; conspicuous.

καταφατικός *a.* affirmative.

κατα|φέρνω *v.t.* persuade, win over; beat (*at sthg.*); τα ~φέρνω succeed, manage; μου την ~φερε he played me a trick.

καταφερτζής *s.m.* (*fam.*) one with a knack of succeeding; wangler.

καταφέρ|ω *v.t.* deal (*blow*); μου ~ε μια he fetched me a blow. ~ομαι *v.i.* (*with κατά or εναντίον*) speak against, attack.

καταφεύγω *v.i.* (*with εις*) take refuge in, have recourse to.

καταφθάνω *v.i.* arrive (*unexpectedly*); catch up with.

κατάφορτος *a.* (over) loaded.

καταφρονώ *v.t.* disdain.

καταφυγή *s.f.* refuge (*abstract*); resource.

καταφύγιο *s.n.* (*place of*) refuge, shelter.

κατάφυτος *a.* rich in vegetation.

κατάφωρος *a.* flagrant, obvious (*of misdeed*).

κατάφωτος *a.* brightly lit.

κατάχαμα *adv.* on the ground.

καταχερίζω *v.t.* smack.

καταχθόνιος *a.* infernal; (*fig.*) darkly secretive, deep-laid.

καταχνιά *s.f.* mist.

καταχραστής *s.m.* embezzler.

κατάχρεως *a.* sunk in debt.

κατάχρησ|η *s.f.* excessive use, abuse; embezzlement; ~**εις** (*pl.*) excess (*sensual*).

καταχρηστικώς *adv.* improperly; against

regulations; as an exception.

καταχρώμαι *v.t.* abuse, take undue advantage of; embezzle.

καταχωνιάζω *v.t.* swallow up; hide away.

καταχωρ|ώ, ~ **ίζω** *v.t.* enter (*in register*); insert (*in newspaper*).

καταψηφίζω *v.t.* vote against.

κατάψυξη *s.f.* deep refrigeration.

καταψύω *v.t.* refrigerate

κατεβάζω *v.t.* bring *or* carry *or* take *or* let down; lower; (*fam.*) devour; ~ μια strike a blow; τα ~ pull a long face.

κατεβαίν|ω *v.i. & t.* come *or* go *or* get down; descend; dismount; fall (*in level*); μου ~ει I take it into my head (*to*). (*v.t.*) go down (*street, etc.*).

κατεβασιά *s.f.* flood, downpour; cold in the head; cataract (*of eyes*).

κατέβασμα *s.n.* descent, lowering, taking *or* going down; hernia.

κατεβασμένος *a.* downcast; reduced (*in price*).

κατεβατό *s.n.* page (*paper*).

κατεδαφίζω *v.t.* demolish (*building*).

κατειλημμένος *a.* occupied, taken.

κατεπείγων *a.* very urgent.

κάτεργα *s.n.pl.* galleys (*as punishment*).

κατεργάζομαι *v.t.* work, fashion; plot.

κατεργάρ|ης *s.m.* rascal. ~**ιά** *s.f.* cunning (act), trick, trickery.

κατέρχομαι *v.t. & i.* descend, come down; fall; ~ εις τας εκλογάς take part in elections.

κατεσπευσμένως *adv.* in hot haste.

κατεστημένον *s.n.* the Establishment.

κατεύθυνση *s.f.* direction, course.

κατευθύν|ω *v.t.* direct, turn. ~ομαι *v.i.* turn, go.

κατευνάζω *v.t.* appease, calm, assuage.

κατευόδιο *s.n.* good journey (*wish*).

κατέχω *v.t.* have, hold, occupy; know.

κατεψυγμένος *a.* frozen; frigid (*zone*).

κατηγορηματικός *a.* categorical; (*gram.*) attributive.

κατηγορητήριον *s.n.* (*law*) indictment.

κατηγορία *s.f.* 1. accusation. 2. category.

κατήγορος *s.m.* (*law*) plaintiff; δημόσιος ~ prosecutor.

κατηγορούμεν|ος *a.* accused; (*law*) defendant. ~**ον** *s.n.* (*gram.*) predicate.

κατηγορώ *v.t.* accuse, charge; criticize.

κατηφής *a.* downcast.

κατηφορίζω *v.i.* go downwards, slope down.

κατήφορος *s.m.* downward slope, declivity.

κατηχ|ώ *v.t.* catechize, initiate, admonish. ~**ηση** *s.f.* catechism.

κάτι *pron.* something; some, certain; (*adv.*) somewhat; να ~ μήλα there are (*beautiful*) apples for you!

κάτι *s.n.* fold; strand (*of thread*).

κατισχύω *v.i. & t.* (*with gen.*) prevail (over).

κατι|ών *a.* descending; ~όντες (*s.m.pl.*) descendants.

κατμάς *s.m.* makeweight (*of meat*).

κατοικήσιμος *a.* habitable.

κατοικία *s.f.* habitation, residence.

κατοικίδιος *a.* domestic (*animal*).

κάτοικος *s.m.f.* inhabitant, resident.

κατοικώ *v.i. & t.* (*with acc. or εις*) live, dwell (in), inhabit.

κατολίσθηση *s.f.* landslip.

κατόπιν *adv.* behind; afterwards; ~ από after; τον πήρα το ~ I followed him closely.

κατοπτεύω *v.t.* keep watch on, observe. (*from above*).

κατοπτρίζω *v.t.* see **καθρεφτίζω.**

κάτοπτρον *s.n.* mirror.

κατόρθωμα *s.n.* feat, exploit.

κατορθώνω *v.t.* accomplish, succeed in.

κατορθωτός *a.* feasible.

κάτουρ|ο, ~**λιό** *s.n.* (*fam.*) urine.

κατουρ|ώ *v.i. & t.* make water (on). ~**ιέμαι** *v.i.* desire to make water; wet oneself; (*fam.*) be in a funk.

κατοχή *s.f.* possession, keeping; (*mil.*) occupation.

κάτοχος *s.m.* possessor; ~ της Αγγλικής having a command of English.

κατοχυρώνω *v.t.* fortify; (*fig.*) safeguard.

κάτοψις *s.f.* ground-plan.

κατρακυλώ *v.t. & i.* bring *or* come tumbling down.

κατράμι *s.n.* tar.

κατραπακιά *s.f.* slap on the head.

κατσαβίδι *s.n.* screwdriver.

κατσάβραχα *s.n.pl.* rocky ground.

κατσάδα *s.f.* scolding.

κατσαμάκι *s.n.* affectation of reluctance; evasion.

κατσαρίδα *s.f.* cockroach.

κατσαρόλα *s.f.* saucepan.

κατσαρ|ός *a.* curly. ~**ό** *s.n.* curl.

κατσ(ι)άζω *v.t. & i.* (cause to) waste away.

κατσίβελος *s.m.* gipsy; sloven.

κατσίκα *s.f.* she-goat.

κατσικίσιος *a.* goat's.

κατσίκι *s.n.* kid; goat.

κατσουφιάζω *v.i.* frown, look sullen.

κάτω *adv.* down, below, underneath; on the ground; το ~ when all is said and done; απάνω ~ up-and-down, approximately; από ~ (from) below, underneath; άνω ~ upside down (*in confusion*), upset; η ~ Βουλή Lower Chamber; τα ρίχνω ~ give up, throw in one's hand; βάζω ~ defeat

(*person*); παίρνω την ~ βόλτα get worse;
~ τα χέρια hands off! ~ η κυβέρνηση down
with the government! (*prep. with* από *or*
gen.) below, under, less than; ~ του
μηδενός below zero.
κατώγ|ειο, ~ι *s.n.* (lower) ground-floor.
κάτωθεν *adv.* (from) below.
κατώτατος *a.* lowest; minimum.
κατώτερος *a.* lower; inferior.
κατωφέρεια *s.f.* downward slope.
κατώφλι *s.n.* threshold.
καυγάς *s.m. see* **καβγάς.**
καύκαλο *s.n.* skull; shell (*of crab, etc.*).
καυκί *s.n.* shell (*of crab, etc.*).
καϋμ- *see* **καημ-.**
καύμα *s.n.* heat; burn; κυνικά ~τα dog-days.
καυσαέρια *s.n.pl.* exhaust fumes.
καύσιμ|ος *a.* combustible; ~α (*s.n.pl.*) fuel.
καύση *s.f.* burning; combustion.
καυστήρας *s.m.* burner; blowlamp.
καυστικός *a.* caustic.
καύσωνας *s.m.* heat-wave.
καυτερός *a.* hot, peppery; (*fig.*) scathing.
καυτηριάζω *v.t.* cauterize; brand (*animals*);
(*fig.*) reprobate.
καυτός *a.* burning hot, boiling.
καύχημα *s.n.* boast, object of pride.
καυχηματίας *s.m.* boaster.
καυχησιά *s.f.* boast, boasting.
καυχ|ιέμαι, ~ώμαι *v.i. & t.* boast (of).
καφάσι *s.n.*lattice; crate (*for fruit*).
καφασωτός *a.* trellised, latticed.
καφεκοπτείο *s.n.* shop where coffee is
ground.
καφενείο *s.n.* coffee-house, café.
καφές *s.m.* coffee.
καφετζής *s.m.* café-keeper.
καφετής *a.* coffee-coloured.
καφωδείο *s.n.* café chantant.
καχεκτικός *a.* sickly, stunted.
καχεξία *s.f.* (*med.*) cachexy.
καχύποπτος *a.* mistrustful.
κάψα *s.f.* 1. great heat. 2. crucible.
καψαλίζω *v.t.* singe; toast.
καψερός *a.* (*fam.*) poor, unfortunate.
κάψιμο *s.n.* burning (*act or sensation*); burn.
καψούλες *s.f.pl.* eyelets for shoelaces.
καψ|ούλι, ~ύλλιον *s.n.* percussion-cap;
(*med.*) capsule.
καψώνω *v.i.*feel very hot.
κβάντουμ *s.n.* (*physics*) quantum.
κέδρος *s.f.* cedar.
κείμαι *v.i.* lie; be found.
κείμεν|ος *a.* existing, prevailing (*laws, etc.*).
~ο *s.n.* text.
κειμήλιο *s.n.* precious object, relic,
heirloom.
κείνος *pron. see* **εκείνος.**

κέικ *s.n.* cake. [E. *cake*]
κεκοιμημένοι *s.m. pl.* οι ~ the departed.
κεκορεσμένος *a.* charged, saturated.
κέκτη|μαι *v.t.* have, possess. ~μένα
δικαιώματα vested interests.
κελα(η)δώ *v.i.* sing (*of birds*), warble.
κελαρύζω *v.i.* purl (*of stream*).
κελεπούρι *s.n.* find, bargain.
κέλευσμα *s.n.* command, bidding.
κελευστής *s.m.* (*naut.*) chief petty officer.
κέλης *s.m.* (*riding*) horse; (*naut.*) captain's
launch.
κελλάρι *s.n.* larder.
κελλί *s.n.* cell.
κέλυφος *s.n.* shell (*of nut, egg, etc.*).
κεμέρι *s.n.* money-belt.
κενόδοξος *a.* vain, conceited.
κενό *s.n.* void, empty space; vacuum; gap;
vacancy.
κενός *a.* empty, vacant; vain, idle.
κέντημα *s.n.* sting, prick; needlework,
embroidery.
κεντητός *a.* embroidered.
κεντιά *s.f.* sting, bite, prick; (*fig.*) cutting
remark.
κεντράδι *s.n.* graft.
κεντρί *s.n.* sting (*organ*).
κεντρίζω *v.t.* sting, goad; graft.
κεντρικός *a.* central.
κεντρομόλος *a.* centripetal.
κέντρο *s.n.* 1. sting, prick, goad. 2. centre;
τηλεφωνικό ~ telephone exchange. 3.
place of entertainment *or* refreshment.
κεντρόφυξ *a.* centrifugal.
κέντρωμα *s.n.* sting (*wound*); grafting.
κεντρώνω *v.t.* sting; graft.
κεντώ *v.t. & i.* sting, prick, goad; rouse,
provoke; embroider.
κενώνω *v.t.* empty; serve up (*dish*).
κένωση *s.f.* emptying; evacuation (*of
bowels*); serving up (*of dish*).
κεραία *s.f.* antenna; wire, aerial; trolley (*of
tram*); minus-sign, dash. μέχρι ~ς to the
last detail.
κεραμείον *s.n.* tile-works, pottery.
κεραμίδα *s.f.* tile; (*fam.*) μου ήρθε ~ it was
a bombshell to me.
κεραμίδι *s.n.*tile; peak (*of cap*).
κεραμική *s.f.* ceramics.
κέραμος *s.m.f.* tile.
κέρας *s.n.* horn; wing (*of army, etc.*); ~ της
Αμάλθειας horn of plenty.
κεράσ|ι *s.n.* cherry. **~ένιος** *a.* cherry-like; of
cherry-wood. **~ιά** *s.f.* cherry-tree.
κέρασμα *s.n.* treating (*to drink, etc.*); tip.
κερατάς *s.m.* (*fam.*) cuckold; (*as term of
abuse, also joc.*) bastard.
κερατένιος *a.* (*fam.*) bloody, confounded.

κερατιά *s.f.* carob-tree.

κέρατο *s.n.* horn; (*fam.*) ~ (βερνικωμένο) cross-grained person.

κερατώνω *v.t.* (*fam.*) cuckold.

κεραυνοβόλος *a.* like lightning; sudden and dire; ~ έρωτας love at first sight.

κεραυνοβολώ *v.t.* & *t.* hurl thunder-bolts (at); (*fig.*) dumbfound.

κεραυνός *s.m.* thunderbolt.

κερδ|ίζω, ~αίνω *v.t.* & *i.* earn, gain, win; beat (*competitor*); appear in better light; ~ίζει το βράδυ it looks better at night.

κέρδος *s.n.* gain, profit.

κερδοσκοπώ *v.i.* profiteer; speculate.

κερένιος *a.* waxen.

κερί *s.n.* wax; candle.

κερκίδα *s.f.* shuttle; wedge-shaped tier of seats.

κέρμα *s.n.* fragment; small coin; token. ~τοδέκτης *s.m.* slot-machine.

κερνώ *v.t.* treat, stand, buy (*with double acc.*).

κερώνω *v.t.* & *i.* polish with wax; become pale as wax.

κεσάτι *s.n.* lack of business.

κεσές *s.m.* shallow bowl (*for jam, etc.*).

κετσές *s.m.* coarse felt.

κεφάλα *s.f.* big head.

κεφαλαιοκρατία *s.f.* capitalism.

κεφάλαι|ο *s.n.* 1. (*fin.*) capital; ~α (*pl.*) funds; (*fig.*) asset to community (*person*). 2. chapter.

κεφαλαίον *s.n.* capital letter.

κεφαλαλγία *s.f.* headache.

κεφαλάρι *s.n.* fountainhead; cornerstone; capital (*of column*); bolster; bed head.

κεφάλας *s.m.* big-headed *or* thick-headed person.

κεφαλή *s.f.* head; head, leader; επί ~ής in charge.

κεφάλι *s.n.* head; (*fig.*) brains, driving-force; κάνω τού ~ού μου go one's own way; πέφτω με το ~ apply oneself avidly; βάζω το ~ μου warrant, wager; κατεβάζει το ~ του he is inventive.

κεφαλικός *a.* of the head; capital; ~ φόρος poll-tax.

κεφαλόδεσμος *s.m.* head-scarf; bandeau.

κέφαλος *s.m.* sort of mullet.

κεφαλοτύρι *s.n.* hard cheese (*for grating*).

κεφάτος *a.* in good spirits.

κέφι *s.n.* good mood, gaiety, gusto; στο ~ slightly tipsy, merry; κάνω το ~ μου do as one pleases; κάνω~ make merry.

κεφτές *s.m.* rissole.

κεχρί *s.n.* millet.

κεχριμπάρι *s.n.* amber.

κηδεία *s.f.* funeral.

κηδεμών *s.m.f.* guardian (*of minor*).

κηλεπίδεσμος *s.m.* (*med.*) truss.

κήλη *s.f.* hernia.

κηλίδα *s.f.* stain. ~ιδώνω *v.t.* stain, sully.

κήπος *s.m.* garden.

κηπουρ|ός *s.m.* gardener. ~ική *s.f.* gardening.

κηρήθρα *s.f.* honeycomb.

κηρίο *s.n.* candle; candle-power.

κηροπήγιο *s.n.* candlestick.

κηρός *s.m.* wax.

κήρυγμα *s.n.* preaching, sermon.

κήρ|υξ, ~υκας *s.m.* herald, crier; (*fig.*) partisan, advocate (*of*).

κηρύσσω *v.t.* & *i.* proclaim, declare; preach.

κήτος *s.n.* cetacean; sea-monster.

κηφήνας *s.m.* drone (*bee, also fig.*).

κι *conj. see* και

κιαλάρω *v.t.* look at through binoculars; espy; have one's eye on.

κιάλια *s.n.pl.* binoculars.

κίβδηλος *a.* debased, counterfeit (*of money*); (*fig.*) false.

κιβώτιο *s.n.* large box; ~ ταχυτήτων gear-box.

κιβωτός *s.f.* Ark.

κιγκαλερία *s.f.* hardware.

κιγκλίδωμα *s.n.* balustrade, banisters, railings, iron bars.

κιθάρα *s.f.* guitar.

κιλίμι *s.n.* pileless carpet.

κιλλίβας *s.m.* gun-carriage.

κιλό *s.n.* kilo(gram).

κιμ|άς *s.m.* minced meat; (*fam.*) κάνω ~ά make mincemeat of.

κιμωλία *s.f.* chalk.

κινά *s.f.* henna.

κίναιδος *s.m.* sodomite.

κινδυν|εύω *v.i.* & *t.* be in danger, risk, endanger; ~εψε να πνιγεί he nearly got drowned.

κίνδυνος *s.m.* danger, risk.

κινέζικα *s.n.pl.* Chinese (*language*); (*fam.*) anything unintelligible.

κίνημα *s.n.* movement (*action*); (*fig.*) step, measure; uprising.

κινηματογράφος *s.m.* cinema.

κίνηση *s.f.* movement, motion (*action or abstract*); flow of traffic *or* passers-by; activity (*social, business, etc.*).

κινητήρας *s.m.* motor. ~ιος *a.* motive.

κινητοποιώ *v.t.* mobilize.

κινητ|ός *a.* movable, mobile. ~ά *s.n.pl.* movable property, estate.

κίνητρο *s.n.* motive.

κινίν|η *s.f.*, ~ο *s.n.* quinine; ~ο (*fig.*) anything bitter.

κιν|ώ *v.t.* & *i.* set in movement, work; move, rouse; start (*war, lawsuit*). (*v.i.*) set out.

~ούμαι *v.i.* move; bestir oneself.
κιόλα(ς) *adv.* already; as well.
κιονόκρανο *s.n.* capital (*of column*).
κιονοστοιχία *s.f.* colonnade.
κιόσκι *s.n.* kiosk; pavilion.
κιούγκι *s.n.* drain-pipe.
κιούπι *s.n.* clay storage-jar.
κιρσός *s.m.* varicose vein.
κίσσα *s.f.* magpie.
κισσός *s.m.* ivy.
κιτάπι *s.n.* (*fam.*) notebook.
κιτρινάδα *s.f.* yellowness; paleness.
κιτρινάδι *s.n.* yolk of egg; yellow spot.
κιτριν|ίζω ~ιάζω *v.i. & t.* turn yellow *or* pale.
κίτρινος *a.* yellow; pale.
κίτρον *s.n.* citron.
κιχ *pron.* (*fam.*) δεν βγάζω ~ not utter a sound.
κίων, ~ονας *s.m.* column, pillar.
κλαγγή *s.f.* clank (*of arms*).
κλαδευτήρι *s.n.* pruning-hook.
κλαδεύω *v.t.* prune, lop.
κλαδί *s.n.* branch of tree.
κλάδος *s.m.* branch (*lit. & fig.*); (season of) pruning.
κλαδωτός *a.* with foliage pattern.
κλαί(γ)|ω *v.t. & i.* weep *or* lament (for); pity. **~ομαι** *v.i.* grumble, moan.
κλάκα *s.f.* claque.
κλάμα *s.n.* crying, lamentation.
κλάνω *v.i. & t.* (*fam.*) break wind (at).
κλαουρίζω *v.i.* whimper.
κλαρί *s.n.* branch of tree; (*fig.*) βγαίνω στο ~ take to the hills (*as outlaw*); make one's debut; (*of woman*) adopt free-and-easy mode of life.
κλάσιμο *s.n.* (*fam.*) breaking wind.
κλάση *s.f.* class, age-group.
κλάσμα *s.n.* (*math.*) fraction.
κλασέρ *s.n.* folder, portfolio (*for papers*). [F. *classeur*].
κλασσικός *a.* classic, classical; (*fam.*) monumental (*liar, etc.*).
κλαυθμηρός *a.* plaintive, whimpering.
κλάψ|α *s.f., ~ιμο* *s.n.* crying; grumbling; moaning.
κλαψιάρης *a.* given to crying *or* complaining.
κλαψιάρικος *a.* plaintive, whining.
κλέβω *v.t.* steal; rob; cheat; abduct; κλέφτηκαν they eloped.
κλειδαριά *s.f.* lock.
κλειδαρότρυπα *s.f.* keyhole.
κλειδί *s.n.* key; (*railway*) switch.
κλειδοκύμβαλο *s.n.* piano.
κλειδούχος *s.m.* turnkey; (*railway*) points-man.
κλείδωμα *s.n.* locking.

κλειδώνω *v.t. & i.* lock (up).
κλείδωση *s.f.* (*anat.*) joint.
κλείθρο *s.n.* lock.
κλεινός *a.* renowned.
κλείνω *v.t. & i.* shut, close, ~ το μάτι wink; shut up, put away; turn off (*tap, switch*); stop up, block (*crack, road, view*); conclude, arrange (*treaty, meeting*); book (*seats*); έκλεισε το σπίτι he agreed to take the house; έκλεισε τα πενήντα he has turned fifty. (*v.i.*) shut, close; become blocked; be concluded *or* reached; έκλεισε η πληγή μου my wound has healed; έκλεισε η φωνή μου I have lost my voice.
κλεις *s.f.* see **κλειδί.**
κλείσιμο *s.n.* shutting, closing; conclusion, agreement; healing (*of wound*).
κλεισούρα *s.f.* pass, defile; confinement within doors; frowstiness.
κλειστός *a.* shut, closed.
κλείω *v.t. & i.* see **κλείνω.**
κλέος *s.n.* renown.
κλεπταποδόχος *s.m.f.* receiver of stolen goods.
κλέπτης *s.m.* thief.
κλέπτω *v.t.* see **κλέβω.**
κλεφτά *adv.* hurriedly, furtively.
κλέφτης *s.m.* thief; (*fig.*) klepht (*Gk. irregular fighter against Turks*).
κλέφτικος *a.* 1. of the klephts. 2. thieving (*cat, etc.*).
κλεφτοπόλεμος *s.m.* guerrilla warfare.
κλεφτός *a.* stolen; furtive.
κλεφτουριά *s.f.* the klephts.
κλεφτοφάναρο *s.n.* small torch *or* lantern.
κλεψιά *s.f.* theft.
κλεψιγαμία *s.f.* illicit sexual union.
κλεψιμαίικος *a.* stolen.
κλέψιμο *s.n.* theft.
κλεψιτυπία *s.f.* piracy (*of books*).
κλεψύδρα *s.f.* water-clock.
κλήδονας *s.m.* fortune-telling done on St. John's day.
κλήθρα *s.f.* alder.
κλήμα *s.n.* vine. **~ταριά** *s.f.* climbing vine; vine-arbour. **~τόφυλλο** *s.n.* vine-leaf.
κλήρα *s.f.* child, heir; (*fig.*) luck.
κληρικός *a.* clerical; a clergyman.
κληροδότημα *s.n.* bequest, legacy.
κληρονομία *s.f.* inheritance, heritage.
κληρονομικ|ός *a.* of inheritance; heredita-ry, inherited. **~ότης** *s.f.* heredity.
κληρονόμος *s.m.f.* heir, heiress.
κληρονομώ *v.t.* inherit.
κλήρ|ος *s.m.* 1. lot, portion; fate, destiny; ρίχνω ~ον draw lots. 2. clergy.
κλήρωση *s.f.* election by lot; drawing of lots *or* numbers in lottery.

κληρωτός *a.* conscript.

κλήση *s.f.* summons, call.

κλητεύω *v.t.* serve summons on.

κλητήρ(ας) *s.m.* usher; bailiff.

κλητική *s.f.* (*gram.*) vocative.

κλίβανος *s.m.* oven.

κλίμα *s.f.* clique.

κλίμα *s.n.* climate.

κλιμάκιο *s.n.* detachment, group, party (*of larger organization*).

κλιμακ|ώνω *v.t.* escalate. **~ωσις** *s.f.* escalation.

κλιμακωτός *a.* graduated; stepped.

κλίμακας *s.f.* staircase, ladder; scale.

κλιματισμός *s.m.* air-conditioning.

κλινάμαξα *s.f.* sleeping-car.

κλίνη *s.f.* bed.

κλινήρης *a.* confined to bed.

κλινικ|ός *a.* clinical. **~ή** *s.f.* nursing-home, clinic.

κλινοσκέπασμα *s.n.* bed-covering.

κλίνω *v.t. & i.* lean, bend, bow; bear, tend (*towards*); (*gram.*) inflect, decline, conjugate.

κλίση *s.f.* bending, inclination; list; slope, gradient; change of direction; tendency; (*gram.*) inflexion, declension, conjugation.

κλιτός *a.* inflected.

κλιτύς *s.f.* slope (*of mountain*).

κλίφι *s.n.* pillow-case, mattress-cover.

κλοιός *s.m.* iron ring; (*fig.*) ring (*that closes round quarry*).

κλονίζ|ω *v.t.* shake, rock; unsettle; impair (*health*). **~ομαι** *v.i.* totter; waver, have doubts.

κλονισμός *s.m.* shaking, shock; impairment (*of health*).

κλόουν *s.m.* clown.

κλοπή *s.f.* theft.

κλοπιμαίος *a.* stolen.

κλούβα *s.f.* dog-catcher's van; (*fam.*) black Maria.

κλουβί *s.n.* cage.

κλούβι|ος *a.* addled. **~αίνω** *v.i. & t.* become *or* make addled.

κλυδωνίζομαι *v.i.* be storm-tossed.

κλύσμα *s.n.* enema.

κλωβός *s.m.* cage.

κλωθογυρίζω *v.t. & i.* (*fam.*) hang around; *τα ~* prevaricate.

κλώθω *v.t.* spin.

κλών|ος *s.m.*, **~άρι**, **~ί** *s.n.* branch.

κλώσ|η *s.f.*, **~ιμο** *s.n.* spinning.

κλώσσα *s.f.* brooding hen.

κλωσσόπουλο *s.n.* new-born chick.

κλωσσώ *v.t. & i.* incubate; (*fig.*) be sickening for (*ailment*).

κλωστή *s.f.* thread.

κλωστική *s.f.* spinning.

κλώτσ|ημα *s.n.*, **~ιά** *s.f.* kick.

κλωτσοσκούφι *s.n.* (*fig.*) one who gets kicked around.

κλωτσώ *v.t. & i.* kick; recoil (*of gun*).

κνήμη *s.f.* shin, lower part of leg.

κνίσα *s.f.* fume of roasting meat.

κνώδαλο *s.n.* (*fig.*) nonentity.

κοάζω *v.i.* croak (*of frog*).

κόβ|ω *v.t. & i.* 1. cut (up *or* out)*;* grind (*corn, etc.*); carve (*joint*); pick (*flower*); kill; rob *~ω τα μαλλιά μου* get one's hair cut; *~ω κοστούμι* buy material for suit; (*fig.*) *~ω μονέδα* make money; *~ω δρόμο* haste *or* take short cut; (*fam.*) *~ω εισιτήριο* take *or* issue ticket; *το ~ω δίπλα* have a nap; *το ~ω λάσπη* make off, bolt; *~ει και ράβει* he has influence *or* he talks incessantly; *του έκοψαν ένα μισθό* they fixed him a salary. 2. cut off (*supply*); break off (*relations*); bar (*way, view*); reduce (*wages, etc.*); take away (*appetite*); give up (*smoking, etc.*); *κόψαμε την καλημέρα* we are no longer on speaking terms; *μου έκοψες το αίμα* you gave me a fright. 3. tire, hurt; *με ~ουν τα παπούτσια* my shoes hurt; (*fig.*)*μου ~ει τα χέρια* I feel the loss of it; (*fam.*) *τον έκοψα* I tricked him; *~ το κεφάλι μου* I feel sure (*that*). 4. (*v.i.*) cut; abate (*of wind, fever, etc.*); fade (*of colour*); turn sour; take a short cut; *~ω δεξιά* turn right; (*fam.*) *~ει το κεφάλι του* he is intelligent; *κόψε το αίμα μου* my heart stood still; *δε μου 'κοψε* it did not occur to me (*to*). 5. *~ομαι v.i.* cut oneself; (fig.) take great trouble, put oneself out (*for*); exhaust oneself; *μου κόπηκαν τα ήπατα or τα γόνατα* I felt stunned *or* upset.

κόγχη *s.f.* shell, conch; (*anat.*) eyesocket, conch of ear; (*archit.*) niche.

κογχύλι|η *s.f.*, **~ι** *s.n.* sea-shell.

κόθορνος *s.m.* buskin.

κοιλαίνω *v.t.* hollow out.

κοιλαράς *s.m.* pot-bellied person.

κοιλάδα *s.f.* valley.

κοιλιά *s.f.* belly.

κοιλιόδουλος *s.m.* glutton.

κοιλόπονος *s.m.* belly-ache; pangs of parturition.

κοίλ|ος *a.* hollow, concave, **~ότητα** *s.f.* hollowness; hollow, cavity. **~ωμα** *s.n.* hollow, depression.

κοιμάμαι *v.i. see* **κοιμούμαι.**

κοίμηση *s.f.* (*eccl.*) Assumption.

κοιμητήριο *s.n.* cemetery.

κοιμ|ίζω *v.t.* put to sleep, lull; **~ισμένος** asleep, (*fig.*) dull-witted.

κοιμ|ούμαι, **~ώμαι** *v.i.* sleep, go to bed.

κοινόβι|ον *s.n.* coenobitic monastery.

~ακός *a.* coenobitic.
κοινοβουλευτικός *a.* parliamentary.
κοινοβούλιον *s.n.* parliament.
κοινοκτημοσύνη *s.f.* community of wealth.
κοινολογώ *v.t.* divulge.
κοιν|όν *s.n.* public; **~ά** (*pl.*) public affairs.
κοινοποιώ *v.t.* make known, give notice (of)*;* serve (*summons*).
κοινοπολιτεία *s.f.* commonwealth.
κοινοπραξία *s.f.* cooperative.
κοιν|ός *a.* common; (*held*) in common; ordinary, commonplace; **~ή γνώμη** public opinion; **από ~ού** in common.
κοινότ|ης *s.f.* community; commune (*unit of local government*); **Βουλή των ~ήτων** House of Commons. **~ικός** *a.* of the commune, local.
κοινοτοπία *s.f.* commonplace.
κοινόχρηστος *a.* used in common.
κοινωνία *s.f.* 1. society, social community. 2. Holy Communion.
κοινωνικός *a.* social, sociable.
κοινωνιολογία *s.f.* sociology.
κοινωνώ 1. *v.i.* & *t.* receive Holy Communion *or* administer this to. 2. *v.i.* participate.
κοινωφελής *a.* for the public benefit.
κοίταγμα *s.n.* look; minding; examination.
κοιτάζ|ω *v.t.* look at, examine; take notice of; mind, look after. **~ομαι** *v.i.* look at oneself; be examined (*by doctor*).
κοίτασμα *s.n.* layer, deposit, bed.
κοίτη *s.f.* bed; lair, nest.
κοιτίς *s.f.* cradle.
κοίτομαι *v.i.* be lying down; be bedridden.
κοιτών *s.m.* bedroom.
κοκέτα *s.f.* (*naut.*) berth.
κοκεταρία *s.f.* smartness, stylishness; coquetry.
κοκέτης *a.* smart, stylish (*person*).
κοκκαλιάζω *v.i.* become stiff (*of limbs*).
κοκκαλιάρης *a.* bony.
κόκκαλο *s.n.* bone; ivory (*of piano-key*); shoe-horn; (*fam.*) *έμεινε ~* he was thunderstruck.
κοκκαλώνω *v.i* become stiff; (*fig*) be thunderstruck.
κοκκινάδα *s.f.* redness.
κοκκινάδι *s.n.* red spot; rouge.
κοκκινέλι *s.n.* red resinated wine.
κοκκινίζω *v.t.* & *i.* make *or* become red *or* brown; flush, blush.
κοκκινίλα *s.f.* redness.
κοκκινογούλι *s.n.* beetroot.
κοκκινομάλλης *a.* red-haired.
κόκκινος *a.* red.
κοκκινωπός *a.* reddish.
κόκκος *s.m.* grain, particle; (*coffee*) bean;

bead (*on string*).
κοκκοφοίνικας *s.m.* coconut-palm.
κοκκύτης *s.m.* whooping-cough.
κοκόνα *s.f.* lady; (*iron.*) *η ~* her ladyship.
κόκορας *s.m.* cock.
κοκορέτσι *s.n.* grilled sheep's entrails.
κοκορεύομαι *v.i.* show off like a cock.
κολάζ|ω *v.t.* 1. punish. 2. lessen bad effect of, explain away. 3. tempt. **~ομαι** *v.i.* fall into temptation, do wrong.
κολάι *s.n.* knack.
κολακεία *s.f.* flattery.
κολακευτικός *a.* complimentary.
κολακεύω *v.t.* flatter.
κόλ|αξ, ~ακας *s.m.* flatterer.
κόλαση *s.f.* hell, damnation.
κολασμένος *a.* damned.
κολατσιό *s.n.* snack.
κόλαφος *s.m.* slap in the face.
κολεός *s.m.* sheath, scabbard; (*anat.*) vagina.
κολικός *s.m.* colic.
κολιός *s.m.* sort of mackerel.
κόλλα *s.f.* 1. glue, paste; starch 2. sheet of paper.
κολλάρο *s.n.* collar; flange. (*fam.*) head (*on beer*).
κολλάρ|ω, ~ίζω *v.t.* starch.
κολλέγιο *s.n.* college.
κόλλημα *s.n.* sticking (*on, together*), soldering; piece stuck on.
κολλητήρι *s.n.* solder, soldering iron.
κολλητικός *a.* contagious.
κολλητ|ός *a.* stuck, soldered; close-fitting; contiguous, adjoining. **~ά** *adv.* side-by-side.
κολλητσίδα *s.f.* burr.
κολλίγας *s.m.* share-cropper.
κολ(λ)ιέ *s.n.* necklace. [F. *collier*]
κόλλυβα *s.f.* boiled wheat (*offered at memorial service*).
κολλυβογράμματα *s.n.pl.* smattering of education.
κολ|λώ, ~νώ *v.t.* & *i.* glue, stick, solder (*on, together*); attach, fix; catch *or* give (*malady*). (*v.i.*) stick; fasten *or* attach oneself (*to*); (*fam.*) *δεν ~λάει* that's a likely tale!
κολοβός *a.* crop-tailed; truncated.
κολοκυθάκι *s.n.* courgette.
κολοκύθι *s.n.* vegetable marrow; (*fam.*) *~α* (*pl.*) nonsense.
κολόνα *s.f.* column, pillar; block (*of ice*).
κολόνια *s.f.* eau-de-Cologne.
κολοσσ|ός *s.m.* colossus. **~ιαίος** *a.* colossal.
κολοφών *s.m.* summit, pinnacle.
κολοφώνιο *s.n.* colophony.

κόλπο *s.n.* trick, ruse; knack.

κόλπος *s.m.* bosom; gulf, bay; apoplectic fit.

κολυμβήθρα *s.f.* font.

κολυμβητής *s.m.* swimmer.

κολυμβώ, ~πώ, *v.i.* swim.

κολύμπι *s.n.* swimming, swim.

κομβ|ίον *s.n.* button. **~οδόχη** *s.f.* button-hole.

κομβολόγιον *s.n.* string of beads; rosary; (*fig.*) string, stream (*of lies, people, etc.*).

κόμβος *s.m.* knot (*tied or in wood*); junction; traffic roundabout; (*naut.*) knot.

κόμη *s.f.* (*head of*) hair.

κόμης *s.m.* count, earl.

κομητεία *s.f.* county.

κομήτης *s.m.* comet.

κομίζω *v.t.* bring, bear.

κομιστής *s.m.* bearer.

κόμιστρα *s.n.pl.* transport charges.

κομιτατζής *s.m.* comitadji.

κομιτάτο *s.n.* committee (*esp. revolutionary*).

κόμμα *s.n.* 1. (*political*) party. 2. comma, decimal point.

κομμάρα *s.f.* lassitude.

κομματάρχης *s.m.* (*political*) party-organizer.

κομμάτι *s.n.* piece, bit; (*adv.*) a little, a bit; (*fam.*) *~α να γίνει* never mind! *στα ~α* go to hell!

κομματιάζ|ω *v.t.* break *or* cut *or* tear into pieces. **~ομαι** *v.i.* make great efforts.

κομματίζομαι *v.i.* go in for party politics.

κόμματος *s.m.* (*fam.*) fine bouncing girl *or* woman.

κομμένος *a.* cut, ground; tired out.

κομ(μ)ό *s.n.* chest-of-drawers. [F. *commode*]

κομ(μ)οδίνο *s.n.* night-table.

κομμουνισ|τής *s.m.* communist. **~μός** *s.m.* communism.

κόμμωση *s.f.* coiffure.

κομμωτήριο *s.n.* hairdresser's.

κομπάζω *v.i.* boast.

κομπιάζω *v.i.* choke; (*fig.*) hum and haw, falter.

κομπίνα *v.i.* (*fam.*) scheme; racket.

κομπιναιζόν *s.n.* slip, petticoat. [F. *combinaison*].

κομπλιμέντο *s.n.* compliment.

κομπογιαννίτης *s.m.* quack, charlatan.

κομπόδεμα *s.n.* hoard, nest-egg.

κομπολόγι *s.n.* string of beads.

κόμπος *s.m.* knot (*tied or in wood*); bunion; drop, small quantity; lump in one's throat; *έφτασε ο ~ στο χτένι* things are coming to a head.

κομπόστα *s.f.* stewed fruit.

κομπρέ(σ)σα *s.f.* (*med.*) compress.

κομπριμέ *a.* & *s.n.* compressed; a tablet. [F. *comprimé*]

κομφόρ *s.n.pl.* comfort, conveniences (*in home, etc.*). [F. *confort*]

κομψευόμενος *a.* dandified.

κομψός *a.* smart, elegant.

κομψοτέχνημα *s.n.* bibelot.

κονάκι *s.n.* lodging.

κονδύλ|ιον *s.n.* slate-pencil; item (*of expenditure*).

κόνδυλος *s.m.* knuckle.

κονδυλοφόρος *s.m.* pen-holder.

κονεύω *v.i.* put up, lodge.

κονιάκ *s.n.* brandy. [F. *cognac*]

κονίαμα *s.n.* mortar; plaster.

κόνιδα *s.f.* nit.

κόνικλος *s.m.* rabbit.

κονιορτός *s.m.* dust.

κόνις *s.f.* power, dust.

κονίστρα *s.f.* arena.

κονκάρδα *s.f.* cockade, badge.

κονσέρβα *s.f.* tinned food.

κονσέρτο *s.n.* concert, concerto.

κοντά *adv.* near; nearly; (*prep.*) *~ σε* near, compared with; *~ στα άλλα* in addition; *~ στο νου* it goes without saying.

κονταίνω *v.t.* & *i.* shorten; get shorter.

κοντάκι *s.n.* butt (*of gun*).

κοντακιανός *a.* a shortish person.

κοντάκ|ιον, ~ι *s.n.* (*eccl.*) a hymn.

κοντάρι *s.n.* pole, lance; gaff.

κοντεύ|ω *v.i.* be almost finished *or* ready *or* completed; draw near; be about (*to*); *κόντεψα να χαθώ* I nearly got lost; *~ει μεσημέρι* it is almost midday.

κόντ|ημα, ~εμα *s.n.* shortening.

κοντινός *a.* close, near, near-by; direct (*route*); to happen shortly.

κοντο- *denotes* short, near.

κοντόγεμος *a.* nearly full.

κοντολογής *adv.* briefly.

κοντομάνικος *a.* short-sleeved.

κοντοπίθαρος *a.* a dumpy person.

κόντος *s.n.* shortness.

κοντός *s.m.* pole.

κοντός *a.* short; *~ ψαλμός αλληλούια* that's that!

κοντοστέκ|ω, ~ομαι *v.i.* stop short.

κοντούλα *s.f.* sort of pear.

κοντούλης *a.* of small stature.

κοντόφθαλμος *a.* short-sighted; obtuse.

κόντρα *adv.* counter; in the opposite direction.

κοντραπλακέ *s.n.* plywood.

κοντυλένιος *a.* slim, graceful (*of bodily features*).

κοντύλι *s.n.* see **κονδύλιον.**

κοπάδι *s.n.* flock, herd.

κοπάζω *v.i.* abate.

κοπαν|άω, **~ίζω** *v.t.* beat, pound; (*fig.*) τα ίδια *~άω* keep repeating the same thing; *~ίζω αέρα* talk nonsense; (*fam.*) τα *~άω* drink.

κόπανος *s.m.* pestle, beater; butt (*of gun*); (*fig.*) blockhead.

κοπέλλα *s.f.* young woman, girl.

κοπέλλι *s.n.* lad; apprentice.

κοπετός *s.m.* lamentation.

κοπή *s.f.* cutting.

κόπια *s.f.* copy.

κόπια *s.n.pl.* wages.

κοπι|άζω *v.i.* work hard, toil; take the trouble (*to*); *~άστε μέσα* please come in.

κοπιαστικός *a.* laborious, toilsome.

κοπίδι *s.n.* cutting-tool.

κόπιτσα *s.f.* hook-and-eye fastener.

κόπ|ος *s.m.* pains, trouble; hard work; (*pl.* *~οι* and *~ια*) reward (*for labour*).

κόπρανα *s.n.pl.* excrement.

κοπριά *s.f.* dung, manure.

κοπρίζω *v.t. & i.* manure; defecate (on).

κόπρος *s.f.* dung, manure.

κοπρόσκυλο *s.n.* scavenging mongrel; (*fig.*) wastrel.

κοπρώνας *s.m.* midden, dunghill.

κοπτήρας *s.m.* cutter; incisor.

κοπτική *s.f.* cutting (*in tailoring*).

κόπτ|ω *v.t. see* **κόβω**. *~ομαι* *v.i.* (*fig.*) beat one's breast.

κόπωση *s.f.* fatigue.

κόρα *s.f.* crust.

κόρακας *s.m.* crow.

κορακίστικα *s.n.pl.* secret language; double-dutch.

κορακοζώητος *a.* long-lived (*person*).

κοράλλ|ινος, **~ένιος** *a.* coral.

κοράλλι(ον) *s.n.* coral.

κόραξ *s.m.* crow; hook.

κοράσ|ιον *s.n.*, **~ίς** *s.f.* girl (*not yet adolescent*).

κόρδα *s.f.* string, sinew.

κορδέλλα *s.f.* ribbon, tape (-measure); tapeworm; hairpin-bend.

κορδόνι *s.n.* cord; lanyard; boot-lace; (*adv.*) in succession; *πάει ~* it is going well.

κορδών|ω *v.t.* tighten. *~ομαι* *v.i.* puff oneself out.

κορεννύω *v.t.* stuff, satiate.

κορεσμός *s.m.* satiation.

κόρη *s.f.* maiden, girl; virgin; daughter; pupil (*of eye*).

κορινθιακ|ός *a.* Corinthian; *~ή σταφίδα* (grapes producing) currants.

κοριός *s.m.* bed-bug.

κορίτσι *s.n.* girl; virgin.

κορκός *s.m.* yolk (*of egg*).

κορμί *s.n.* body.

κορμός *s.m.* trunk (*of tree, body, or structure*); sort of cake.

κορμοστασιά *s.f.* carriage, bearing.

κορνάρω *v.i.* sound one's (*motor*) horn.

κορνίζα *s.f.* picture-frame; cornice.

κόρνο *s.n.* (motor-) horn.

κορόιδ|ο *s.n.* laughing-stock; dupe. *~εύω* *v.t.* make fun of, mock; fool.

κορόμηλο *s.n.* plum.

κόρ|ος *s.m.* satiety; *κατά ~ον* to repletion.

κορσές *s.m.* corset.

κορτ|άρω, **~ετζάρω** *v.t. & i.* flirt (with).

κόρτε *s.n.* flirting.

κορυδαλλός *s.m.* skylark.

κορυφαίος *a. & s.m.* leading, pre-eminent; leader of chorus.

κορυφή *s.f.* top of the head; summit, peak; cream (*on milk*); (*fig.*) leading light.

κορυφούμαι *v.i.* reach culmination.

κορύφωμα *s.n.* culmination, height.

κορφή *s.f.* top of the head; summit peak; top shoot (*of plant*).

κόρφος *s.m.* bosom; gulf, bay.

κορώνα *s.f.* crown; *~ γράμματα* heads or tails.

κορώνω *v.t. & i.* heat, inflame; get hot *or* inflamed.

κόσκιν|ο *s.n.* sieve; *~ο από τις σφαίρες* riddled with bullets. *~ίζω* *v.t.* sift.

κοσμάκης *s.m.* common people, man in the street.

κόσμημα *s.n.* ornament; jewel; *~τα* (*pl.*) jewellery.

κοσμήτ|ωρ *s.m.* dean (*of faculty*); steward (*of meeting*).

κοσμικ|ός *a.* 1. cosmic. 2. lay, temporal; worldly. 3. social (*event*), society (*man or woman*); *~ή κίνηση* social activity.

κόσμιος *a.* seemly, modest.

κοσμογυρισμένος *a.* travelled.

κοσμοθεωρία *s.f.* Weltanschauung.

κοσμοϊστορικός *a.* of historic significance.

κοσμοπολιτικός *a.* cosmopolitan.

κόσμ|ος *s.m.* world; people; *όλος ο ~ος* everybody; *του ~ου τα λεφτά* a great deal of money.

κοσμοχαλασιά *s.f.* upheaval; hubbub.

κοσμώ *v.t.* decorate, adorn.

κοστίζω *v.i.* cost; (*fig.*) be a grievous blow.

κόστος *s.n.* cost; cost price.

κοστούμι *s.n.* (*man's*) suit.

κότα *s.f. see* **κόττα**.

κότερο *s.n.* cutter; pleasure-boat.

κοτέτσι *s.n.* chicken-coop.

κοτζάμ *prefix denoting largeness* (*often with contrasting defect*), great big.

κότινος *s.m.* wild olive; (*fig.*) prize.
κοτλέ *a.* corduroy.
κοτολέτα *s.f.* cutlet.
κοτρώνι *s.n.* large stone.
κοτσάνι *s.n.* stem.
κοτσάρω *v.t.* (*fam.*) put, shove, stick; give, land with; impute.
κότσι *s.n.* ankle; bunion; (*fam.*) ~α guts (*courage*); δε βαστούν τα ~α μου it is beyond my strength.
κοτσίδα *s.f.* plait, pigtail.
κοτσονάτος *a.* hale and hearty.
κότσος *s.m.* chignon, bun.
κότσι|υφας *s.m.,* ~ύφι *s.n.* blackbird.
κότ(τ)α *s.f.* hen, chicken. ~όπουλο *s.n.* pullet, chicken.
κοτώ *v.i.* (*fam.*) dare.
κουάκερ *s.n.* (*fam.*) porridge.
κουβαλ|ώ *v.t. & i.* carry, bring; move house. ~ιέμαι *v.i.* pay unexpected *or* unwelcome visit. ~ίδια *s.n.pl.* moving house.
κουβάρι *s.n.* ball (*of wool, etc.*); μαλλιά ~α topsy-turvy, at loggerheads.
κουβαριάζ|ω *v.t.* roll into a ball; crumple; (*fig.*) swindle. ~ομαι *v.i.* curl oneself up; shrink; fall in a heap.
κουβαρίστρα *s.f.* reel, spool.
κουβαρντάς *s.m.* (*fam.*) open-handed person.
κουβάς *s.m.* bucket.
κουβέντ|α *s.f.* talk, conversation, remark. ~ιάζω *v.i. & t.* converse; discuss, criticize.
κουβέρ *s.n.* place laid at table; service charge. [F. *couvert*].
κουβερνάντα *s.f.* governess.
κουβέρνο *s.n.* government.
κουβέρτα *s.f.* blanket; (*naut.*) deck.
κουβερτούρα *s.f.* chocolate icing.
κουβούκλι *s.n.* canopy (*over bier, etc.*).
κουδουνάτος *s.m.* one in fancy dress.
κουδούνι *s.n.* bell; (*fam.*) είναι ~ he is drunk.
κουδουνίζω *v.i.* ring; (*v.t.*) make known.
κουδουνίστρα *s.f.* (*baby's*) rattle.
κουζίνα *s.f.* kitchen; style of cooking; kitchen-stove.
κουζουλός *a.* (*fam.*) dotty.
κουκέτα *s.f.* (*naut.*) berth.
κουκί *s.n.* broad bean; berry, bean.
κουκίδα *s.f.* spot, dot.
κούκλα *s.f.* doll; tailor's dummy; skein (*of wool*); (*head of*) maize; (*fig.*) beautiful child *or* woman.
κούκ(κ)ος *s.m.* 1. cap. 2. cuckoo; (*fam.*) απόμεινε ~ he was left on his own; τρεις κι ο ~ (*iron.*) very few; του κόστισε ο ~ αηδόνι he paid through the nose for it.
κουκουβάγια *s.f.* owl.

κουκούδι *s.n.* pimple.
κουκουές *s.m.* member of *KKE* (*Communist party of Greece*).
κουκούλα *s.f.* hood, cowl; tea-cosy.
κουκουλάρικος *a.* shantung.
κουκούλι *s.n.* silk-worm cocoon.
κουκουλώνω *v.t.* muffle up; bury; hush up; trick *or* force into marriage.
κουκουνάρι *s.n.* pine cone *or* kernel. ~ά *s.f.* stone-pine.
κουκούτσι *s.n.* pip, stone; ούτε ~ not a scrap.
κουλός *a.* one-armed; armless.
κουλούρα *s.f.* ring-shaped loaf; (*fam.*) lifebelt; nought (*no marks*); lavatory-seat.
κουλούρι *s.n.* small ring-shaped biscuit *or* bread.
κουλουριάζω *v.t.* coil, roll up.
κουλτούρα *s.f.* culture.
κουμάντο *s.n.* command, control.
κουμάρι *s.n.* jug.
κουμάσι *s.n.* hen-coop; (*fam.*) καλό ~ a bad lot.
κουμκάν *s.n.* a card game.
κουμπάρα *s.f.* godmother of one's child.
κουμπαράς *s.m.* money-box.
κουμπάρος *s.m.* godfather of one's child; best man (*at wedding*).
κουμπ|ί *s.n.* button, stud, link; switch; του βρήκα το ~ί I found his weak spot. ~ότρυπα *s.f.* button-hole.
κουμπούρα *s.f.* pistol.
κουμπώνω *v.t.* button, do up.
κουμπωτήρι *s.n.* button-hook.
κουνά|θι, ~δι *s.n.* marten.
κουνέλι *s.n.* rabbit.
κούνημα *s.n.* movement, swaying, swinging, rocking.
κούνια *s.f.* cradle; swing; sling.
κουνιάδα *s.f.* sister-in-law (*spouse's sister*).
κουνιάδος *s.m.* brother-in-law (*spouse's brother*).
κουνιστός *a.* rocking, swaying.
κουνούπι *s.n.* mosquito.
κουνουπίδι *s.n.* cauliflower.
κουνουπιέρα *s.f.* mosquito-net.
κουντώ *v.t.* push.
κουν|ώ *v.t. & i.* move, rock, shake, swing; δεν το ~άω από 'δω I won't budge from here. ~ιέμαι *v.i.* move, rock, shake, swing; bestir oneself, get a move on.
κούπα *s.f.* cup, glass; heart (*cards*).
κουπαστή *s.f.* (*naut.*) gunwale. ~ σκάλας banisters.
κουπί *s.n.* oar; κάνω *or* τραβώ ~ row.
κουπόνι *s.n.* remnant (*of cloth*); coupon.
κούρα *s.f.* cure, treatment; doctor's visit.
κουράγιο *s.n.* courage.
κουράδα *s.f.* (*fam.*) turd.

κουράζ|ω *v.t.* tire, weary. **~ομαι** *v.i.* become tired.

κουραμάνα *s.f.* army bread.

κουραμπιές *s.m.* sort of cake; (*fam.*) soldier who dodges front-line service.

κουράντες *s.m.* medical attendant.

κουράρω *v.t.* attend, treat (*of doctor*).

κούραση *s.f.* fatigue.

κουρασμένος *a.* tired.

κουραστικός *a.* tiring, trying.

κουραφέξαλα *s.n.pl.* (*fam.*) nonsense.

κούρβουλο *s.n.* (dried) stem of vine.

κουρδίζω *v.t.* tune; wind up; (*fig.*) provoke, work (*person*) up.

κουρδιστός *a.* clockwork.

κουρέας *s.m.* barber.

κουρείο *s.n.* barber's shop.

κουρελαρία *s.f.* old rags; (*fig.*) ragtag and bobtail.

κουρελ|ής, ~ιάρης *s.m.* ragged person.

κουρέλι *s.n.* rag; (*fam.*) τον *κάνω* ~ I abuse him. **~άζω** *v.t.* tear to shreds.

κούρε(υ)μα *s.n.* haircut; shearing.

κουρεύω *v.t.* cut hair of; shear.

κούρκος *s.m.* turkey.

κουρκούτι *s.n.* batter; pap.

κουρνιάζω *v.i.* perch, roost.

κουρούνα *s.f.* rook.

κούρσα *s.f.* horse-race; errand, trip; racing *or* touring motor-car.

κουρσάρος *s.m.* corsair.

κουρσεύω *v.t.* pillage.

κουρτίνα *s.f.* curtain.

κουσκουσούρης *s.m.* a gossip.

κουσούρι *s.n.* fault, blemish.

κουτάβι *s.n.* puppy, cub; simpleton.

κουτάλι *s.n.* spoon.

κουταμάρα *s.f.* stupidity.

κούτελο *s.n.* forehead.

κουτί *s.n.* box, tin; του ~*ού* tinned, (*fig.*) spick-and-span; μου *'ρθε* ~ it was a godsend to me *or* it fitted me to a T.

κουτορνίθι *s.n.* ninny.

κουτ|ός *a.* stupid, silly. **~ιαίνω** *v.t. & i.* make *or* become stupid.

κουτουλώ *v.t. & i.* butt; nod (*drowsily*).

κουτουπιές *s.m.* instep.

κουτουράδα *s.f.* foolish act.

κουτουρού *adv.* at random, casually.

κούτρα *s.f.* (*fam.*) head.

κουτρουβάλα *s.f.* somersault.

κουτρούλ|ης *a.* bald; του ~η ο γάμος hullabaloo.

κούτσα *s.f.* lameness; (*adv.*) ~ ~ limpingly.

κουτσαβάκης *s.m.* bully, fire-eater.

κουτσαίνω *v.t. & i.* lame; limp.

κουτσαμάρα *s.f.* lameness.

κουτσο- *denotes* deficiency.

κουτσομπολ|ιό *s.n., ~ιά* (*pl.*) gossip.

κουτσοπίνω *v.i.* drink (*alcohol*) slowly.

κουτσ|ός *a.* lame; ~α στραβά after a fashion.

κουτσούβελο *s.n.* (*small*) child.

κουτσουλιά *s.f.* bird's droppings.

κουτσουπιά *s.f.* Judas-tree.

κουτσουρεύω *v.t.* truncate; cut down drastically.

κούτσουρο *s.n.* tree-trunk, log; (*fig.*) blockhead.

κουφαίν|ω *v.t.* deafen. **~ομαι** *v.i.* become deaf.

κουφάλα *s.f.* cavity (*in tree-trunk or tooth*).

κουφαμάρα *s.f.* deafness.

κουφάρι *s.n.* corpse; (*ship's*) hull.

κουφέτο *s.n.* pill; sugared almond.

κουφιαίνω *v.t. & i.* make *or* become hollow.

κουφίζω *v.i.* be hard of hearing.

κούφ|ιος *a.* hollow; decayed, rotten (*tooth, nut, etc.*); muffled (*sound*); (*fig.*) lacking seriousness (*person*); στα ~α without result *or* without noise.

κουφόβρaση *s.f.* dull, sultry weather.

κούφος *a.* lightweight, not serious; empty (*hope, etc.*).

κουφός *a.* deaf.

κούφταλο *s.n.* bent old woman *or* man.

κούφωμα *s.n.* hollow (*in mountainside*); embrasure (*of door or window*).

κόφ|α *s.f., ~ίνι* *s.n.* deep basket, hamper.

κοφτά *adv.* ορθά ~ flatly, straight out.

κοφτερός *a.* sharp (*knife, etc.*).

κοφτήριο *s.n.* (*fam.*) place where one gets swindled.

κόφτρα *s.f.* (*fam.*) gold-digger.

κόφτ|ω *v.t. see* **κόβω**; τι με ~ει; what do I care?

κοχλάδι *s.n.* pebble.

κοχλάζω *v.i.* boil, seethe, bubble up.

κοχλιάριον *s.n.* spoon.

κοχλίας *s.m.* snail; screw; cochlea.

κοχύλι *s.n.* sea-shell.

κόψη *s.f.* cutting edge; cut (*of clothes*).

κοψιά *s.f.* cut (*wound*).

κόψιμ|ο *s.n.* cutting; style, cut; ~ατα (*pl.*) (*fam.*) the gripes.

κοψομεσιάζομαι *v.i.* (*fig.*) break one's back, exhaust oneself.

κραγιόν(ι) *s.n.* crayon; lipstick.

κραδαίνω *v.t.* brandish; vibrate.

κράζω *v.i. & t.* crow, caw; cry out loud; call, summon.

κραιπάλη *s.f.* debauch.

κράμα *s.n.* mixture; alloy.

κράμβη *s.f.* cabbage, kale.

κρανίο *s.n.* cranium, skull.

κράνος *s.n.* helmet.

κράξιμο *s.n.* crowing, cawing.

κρασάτος *a.* wine-coloured; cooked in wine.

κράση *s.f.* (*bodily*) constitution; (*gram.*) crasis.

κρασί *s.n.* wine.

κράσπεδο *s.n.* hem; kerb; foot (*of hill*); edge (*of shore*).

κραταιός *a.* powerful.

κράτει *int.* avast! *κάνε και* ~ go easy!

κράτημα *s.n.* (*act. of*) holding (back).

κρατήρ *s.m.* 1. crater. 2. ancient wine-bowl.

κράτηση *s.f.* detention, arrest; deduction (*from pay*).

κρατητήριον *s.n.* detention-cell.

κρατικοποίησις *s.f.* nationalization.

κρατικός *a.* of *or* by the state.

κράτος *s.n.* power, rule; the state; *κατά* ~ totally (*of defeat*).

κρατούμενος *s.m.* prisoner, detainee.

κρατ|ώ *v.i. & t.* rule, prevail; last, keep (*of time or durability*); come, be descended (*from*); (*v.t.*) have, hold, keep; detain, withhold; book, reserve; ~*ιέμαι καλά* be well off *or* well preserved; *κάνε και λίγο* ~*ει* have a little restraint.

κραυγ|ή *s.f.* shout, cry. ~**άζω** *v.i.* shout, cry out. ~**αλέος** *a.* vociferous; crying.

κραχ *s.n.* (*financial*) crash. [G. *Krach*]

κράχτης *s.m.* decoy, tout.

κρέας *s.n.* flesh, meat.

κρεατόμυγα *s.f.* bluebottle.

κρεβάτι *s.n.* bed.

κρεβατίνα *s.f.* vine-arbour.

κρεβατοκάμαρα *s.f.* bedroom (*suite*).

κρεβατώνω *v.t.* confine to bed.

κρέμα *s.f.* cream.

κρεμάζω *v.t.* hang (up).

κρεμάλα *s.f.* gallows.

κρεμανταλάς *s.m.* lanky fellow.

κρέμασμα *s.n.* hanging (up).

κρεμαστός *a.* hanging; ~*ή γέφυρα* suspension-bridge.

κρεμάστ|ρα *s.f.*, ~**άρι** *s.n.* clothes-hanger; hat-stand.

κρεμμύδι *s.n.* onion; bulb; (*fam.*) turnip (*watch*).

κρεμ|(ν)ώ *v.t. & i.* hang (up)*; -ω* hang (*execute*); (*v.i.*) hang down. ~*ομαι v.i.* hang (*be suspended*). ~*ιέμαι v.i.* hang (*from trapeze, etc.*); hang *or* lean (*out of window, etc.*); lean (*for support*); (*fig.*) place all one's hopes (*in*); ~*άστηκε* he hanged himself *or* was hanged.

κρεοπώλης *s.m.* butcher.

κρεουργώ *v.t.* butcher.

κρεοφάγος *a.* eating meat.

κρεπ *s.n.* crape; crêpe rubber. [F. *crêpe*]

κρεπάρω *v.i.* burst, split.

κρημνίζ|ω *v.t.* cast down (*from height*); demolish; overthrow. ~*ομαι v.i.* fall (*from height*); collapse.

κρημνός *s.m.* precipice, abyss.

κρήνη *s.f.* fountain, spring.

κρηπίδωμα *s.n.* foundation, base; quay; railway platform.

κρησάρα *s.f.* sieve.

κρησφύγετο *s.n.* hiding-place.

κριάρι *s.n.* ram.

κριθαράκι *s.n.* 1. barley-shaped paste (*in soup*). 2. (*med.*) sty.

κριθ|ή *s.f.*, ~**άρι** *s.n.* barley.

κρίκος *s.m.* ring, link; (*lifting*) jack.

κρίμα *s.n. & int.* sin; a pity; *το* ~ *στο λαιμό σου* on your head be it; ~ (*σ*)*τα λεφτά* what a waste of money!

κρίνος *s.m.* lily.

κρίνω *v.t. & i.* judge, deem, consider; decide.

κριός *s.m.* ram; battering-ram.

κρίσιμος *a.* critical, decisive, serious.

κρίση *s.f.* judgement, opinion; critical faculty; piece of criticism; crisis, depression; (*med.*) fit, attack.

κριτήριο *s.n.* criterion.

κριτής *s.m.* judge.

κριτική *s.f.* criticism.

κριτικός *a.* critical (*pertaining to criticism*); a critic.

Κροίσος *s.m.* Croesus; (*fig.*) multimillionaire.

κροκάδι *s.n.* yolk (*of egg*).

κροκόδειλος *s.m.* crocodile.

κρόκος *s.m.* crocus; yolk (*of egg*).

κρομμύδι *s.n.* onion; bulb.

κρόμμυον *s.n.* onion.

κρονόληρος *s.m.* dotard.

κρόσσι *s.n.* fringe.

κρόταλο *s.n.* rattle.

κρόταφος *s.m.* (*anat.*) temple.

κρότος *s.m.* loud noise; (*fig.*) stir.

κρουαζιέρα *s.f.* pleasure-cruise.

κρουνηδόν *adv.* in torrents.

κρουνός *s.m.* spring; tap; (*fig.*) torrent.

κρούση *s.f.* striking; playing (*of instrument*); exploratory attack.

κρούσμα *s.n.* case (*of theft, typhus, etc.*).

κρούστα *s.f.* crust; scab.

κρούσταλλο *s.n.* (*transparent*) ice; icicle.

κρουστός *a.* 1. thickly woven. 2. percussion (*instrument*).

κρούω *v.t.* strike, knock (on); ring (*bell*); play (*instrument*).

κρυάδ|α *s.f.* coldness; ~*ες* (*pl.*) shivers.

κρύβ|ω *v.t.* hide; put away. ~*ομαι v.i.* hide.

κρυμμένος *a.* hidden.

κρύο *s.n.* cold.

κρυολόγημα s.n. a cold.
κρυοπάγημα s.n. frost-bite.
κρύος a. cold, chilly; unresponsive; expressionless, insipid.
κρυπτογραφώ v.i. & t. write in code.
κρύπτω v.t. see κρύβω.
κρυστάλλ|ινος, ~ένιος a. of or like crystal.
κρύσταλλο s.n. crystal (glass); icicle.
κρύσταλλος s.m.f. (chem.) crystal.
κρυφά adv. secretly.
κρύφ|ιος, ~ός a. secret.
κρυφτός a. hidden.
κρύφτω v.t. see κρύβω.
κρύψιμο s.n. (act of) hiding.
κρυψίνους a. secretive.
κρυψώνας s.m. hiding-place.
κρύωμα s.n. cold, chill.
κρυώνω v.t. & i. make, get or feel cold.
κρωγμός s.m. croaking (of birds).
κτέν|ι(ον) s.n. comb; rake.~ίζω v.t. comb. ~ισμα s.n. combing; coiffure.
κτήμα s.n. (landed) property. ~τίας s.m. land-owner.
κτηνίατρος s.m. veterinary surgeon.
κτήνος s.n. beast, brute.
κτηνοτροφία s.f. stock-breeding.
κτηνώδης a. bestial, brutish.
κτήσις s.f. (territorial) possession.
κτητικός a. (gram.) possessive.
κτίζω v.t. build; found (city); wall up.
κτίριον s.n. building.
κτίσιμο s.n. (act of) building.
κτίστης s.m. bricklayer; the Creator.
κτυπ- see χτυπ-.
κύαμος s.m. broad bean.
κυάνιον s.n. cyanide.
κυανόλευκος a. blue-and-white; η ~ the Greek flag.
κυανούς a. blue, azure.
κυβέρνησις s.f. command, government.
κυβερνήτης s.m. commander, governor.
κυβερνητικός a. of or by the government.
κυβερνώ v.t. command, govern.
κυβισμός s.m. cubism.
κύβ|ος s.m. cube; die. ~ικός a. cubic.
κυδώνι s.n. quince; sort of shellfish.
κύηση s.f. gestation.
κυκεώνας s.m. disorder, confusion.
κυκλάμινο s.n. cyclamen.
κυκλικός a. circular; cyclic.
κύκλος s.m. circle; cycle; circuit; ~ εργασιών business turnover; ~ μαθημάτων course of lessons.
κυκλοτερής a. circular.
κυκλοφορ|ώ v.t. & i. circulate. ~ία s.f. circulation; currency; flow of traffic or passers-by.
κύκλωμα s.n. (electric) circuit.

κυκλώνας s.m. cyclone.
κυκλώνω v.t. encircle.
κυκλωτικός a. encircling.
κύκνει|ος a. swan's; ~ον άσμα swan-song.
κύκνος s.m. swan.
κυλικείον s.n. buffet.
κύλινδρος s.m. cylinder, roller.
κυλί|ω v.t. roll (over, along). ~ομαι v.i. roll, wallow. ~ομένη κλίμαξ moving staircase.
κυλότα s.f. breeches; underpants.
κυλ|ώ v.t. & i. roll (over, along); (v.i.) roll along or by. ~ιέμαι v.i. roll, wallow.
κύμα s.n. wave.
κυμαίνομαι v.i. fluctuate, waver.
κυματίζω v.i. wave, flutter.
κυματοθραύστης s.m. breakwater.
κύμβαλον s.n. cymbal.
κύμινο s.n. cummin.
κυνήγη|μα, ~τό s.n. chase, pursuit, search.
κυνήγι s.n. pursuit; hunting, shooting; game, quarry.
κυνηγός s.m. hunter, sportsman.
κυνηγώ v.t. & i. chase, pursue, hunt; go shooting.
κυνικ|ός a. 1. cynical; a cynic. 2. canine; ~ά καύματα dog-days.
κυνισμός s.m. cynicism.
κυνόδους s.m. canine tooth.
κυοφορώ v.i. & t. be pregnant, breed; (fig.) meditate, plan.
κυπαρίσσι s.n. cypress.
κύπελλο s.n. cup.
κυπρίνος s.m. carp.
κύπτω v.t. & i. bend, lean, stoop.
κυρ- (prefix to Christian name or title) mister (in provincial usage).
κυρά 1. f. equivalent of κυρ-. 2. s.f. missus.
κύρης s.m. father; master.
κυρία s.f. lady, mistress; Mrs.; madam.
κυριακάτικ|ος a. of Sunday; ~α (s.n.pl.) Sunday clothes. ~α adv. on Sunday.
Κυριακή s.f. Sunday.
κυριαρχ|ώ v.i. & t. (with gen.) be master of; prevail, be paramount. ~ία s.f. sovereignty, mastery. ~ος a. sovereign, master.
κυριεύω v.t. seize control of, take.
κυριολεκτικώς adv. literally.
κύρι|ος s.m. lord, master; gentleman; mister; ~ε sir.
κύρι|ος a. main, principal, chief; ~ο όνομα proper name; ~ο άρθρο leading article.
κυριότης s.f. (right of) ownership.
κυρίως adv. chiefly, above all; ~ ειπείν properly speaking.
κύρος s.n. weight, authority; validity.
κυρτ|ός a. bent; curved, hooked; convex, bulged; ~ά στοιχεία italics.
κυρώνω v.t. ratify, authenticate, confirm.

κύρωση *s.f.* ratification; penalty, sanction.
κύστις *s.f.* (*anat.*) bladder, cyst.
κυτίον *s.n.* box.
κύτος *s.n.* hold (*of ship*).
κυττάζω *v.t.* see **κοιτάζω.**
κύτταρο *s.n.* (*biol.*) cell.
κυφός *a.* hunchbacked.
κυψέλη *s.f.* bee-hive; ear-wax.
κύων *s.m.f.* dog.
κώδιξ *s.m.* codex; code.
κώδων *s.m.* bell.
κωδωνοκρουσία *s.f.* peal(ing) of bells.
κωδωνοστάσιον *s.n.* belfry.
κωθώνι *s.n.* raw recruit, greenhorn.
κωκ *s.n.* 1. coke. 2. sort of cream bun.
κωλαράς *s.m.* (*fam.*) one broad in the beam.
κωλο- (*fam.*) *epithet of opprobrium.*
κωλομέρι *s.n.* rump, buttock.
κώλον *s.n.* limb; colon (*punctuation*).
κωλοπετσωμένος *a.* wily, sharp, fly.
κώλος *s.m.* (*fam.*) bottom, arse; seat (*of pants*); (*fig.*) bottom; ~ *και βρακί* hand-in-glove.
κωλοφωτιά *s.f.* glow-worm.
κώλυμα *s.n.* obstacle, hindrance.
κωλυσιεργώ *v.i.* be obstructive; filibuster.
κωλύω *v.t.* hinder, prevent.
κωλώνω *v.i.* back, retrace steps.
κώμα *s.n.* coma.
κώμη *s.f.* large village.
κωμικός *a.* comic, funny; a comedian.
κωμόπολις *s.f.* small country town.
κωμωδία *s.f.* comedy; anything comical; *παίζω* ~ play a trick.
κώνειο *s.n.* hemlock.
κώνος *s.m.* cone.
κώνωψ *s.m.* mosquito.
κώπη *s.f.* oar. **~λασία** *s.f.* rowing.
κωφάλαλος *a.* deaf-and-dumb.
κωφεύω *v.i.* turn a deaf ear.
κωφός *a.* deaf.
κώχη *s.f.* corner; crease (*of trousers*).

Λ

λάβα *s.f.* lava.
λαβαίνω *v.t.* see **λαμβάνω.**
λάβαρον *s.n.* banner.
λαβείν *s.n.* (*fin.*) credit (*money owed to one*).
λαβή *s.f.* handle, hilt; hold (*in wrestling*); *δίδω* ~*ν* afford a pretext.
λαβίς *s.f.* tweezers, pincers, tongs, forceps.
λαβομάνο *s.n.* wash-basin.
λάβρα *s.f.* great heat; *φωτιά και* ~ (*fig.*) strong passion, (*fam.*) exorbitant price.

λαβράκι *s.n.* bass (*fish*). scoop (*journalist's*).
λάβρος *a.* vehement, ardent, eager.
λαβύρινθος *s.m.* labyrinth, maze.
λάβωμα *s.n.* wounding; wound. **~τιά** *s.f.* wound.
λαβώνω *v.t.* wound.
λαγάνα *s.f.* unleavened bread (*for first day of Lent*).
λαγαρός *a.* clear, limpid; pure (*gold, etc.*).
λαγήν|α *s.f.,* **~ι** *s.n.* pitcher, jug.
λαγκάδ|ι *s.n.,* **~α, ~ιά** *s.f.* gorge, ravine, valley.
λαγκεμένος *a.* languishing (*look*).
λάγν|ος *a.* lascivious. **~εία** *s.f.* lasciviousness.
λαγοκοιμούμαι *v.i.* sleep with one eye open, nap.
λαγόνες *s.f.pl.* loins.
λαγ|ός *s.m.* hare; (*fig.*) *τάζω* ~*ους με πετραχήλια* make wonderful but empty promises.
λαγούμι *s.n.* underground run, passage *or* drain; mine.
λαγωνικό *s.n.* harrier; hunting-dog; sleuth.
λαδ|άς, ~έμπορος *s.m.* oil merchant.
λαδερό *s.n.* oil-cruet; oil-can.
λαδερός *a.* oily; (*vegetable dish*) cooked with oil.
λαδής *a.* olive-coloured.
λάδι *s.n.* olive-oil; oil; (*fig.*) *του βγάζω το* ~ I drive him hard; *βγήκε* ~ he got off scot-free.
λαδιά *s.f.* 1. oil *or* grease stain. 2. olive crop.
λαδικό *s.n.* oil-cruet; oil-can; (*fig.*) gossiping old woman.
λαδολέμονο *s.n.* oil and lemon dressing.
λαδομπογιά *s.f.* oil-colour paint.
λαδόχαρτο *s.n.* greaseproof paper.
λάδωμα *s.n.* oiling; oil-stain; bribery.
λαδώνω *v.t.* oil; make oily *or* greasy; (*fam.*) bribe.
λαζαρέτο *s.n.* quarantine station.
λάθ|ος *s.n.* error, mistake. **~εύω** *v.i. & t.* be mistaken; mislead.
λαθραί|ος *a.* clandestine, smuggled. **~ως** *adv.* secretly.
λαθρέμπορος *s.m.* smuggler. **~όριο** *s.n.* smuggling.
λαθρεπιβάτης *s.m.* stowaway.
λαθρόβιος *a.* underground (*press*); one with no known means of support.
λαθροθήρας *s.m.* poacher.
λαθροχειρία *s.f.* pilfering.
λαΐδη *s.f.* lady (*of title*).
λαϊκός *a.* popular, of the people; working-class; lay.
λαίλαψ *s.f.* hurricane.

λαίμαργ|ος *a.* gluttonous. **~ία** *s.f.* gluttony.

λαιμητόμος *s.f.* guillotine.

λαιμοδέτης *s.m.* necktie.

λαιμ|ός *s.m.* neck, throat; neckline (*of garment*); τον παίρνω στο ~ό μου I am responsible for his ill fortune.

λακέρδα *s.f.* salted tunny-fish.

λακές *s.m.* lackey.

λάκ(κ)α *s.f.* lacquer.

λάκκα *s.f.* hole (*in ground*); clearing (*in forest*).

λακκάκι *s.n.* small hole; dimple.

λάκκ|ος *s.m.* hole (*in ground*); grave; den; κάτι ~ο έχει η φάβα something queer is afoot.

λακκούβα *s.f.* hole, cavity.

λακτίζω *v.t. & i.* kick.

λακωνικός *a.* laconic.

λαλές *s.m.* poppy.

λάλημα *s.n.* singing, crowing (*of birds*).

λαλιά *s.f.* speech, talk.

λάλος *a.* talkative.

λαλούμενα *s.n.pl.* (*rustic*) fiddles.

λαλώ *v.t. & i.* talk, speak; play (*the drum, fife, etc.*); (*v.i.*) sing (*of birds*); play (*of instruments*).

λάμα *s.f.* 1. thin sheet of metal. 2. blade (*of knife, razor*). 3. llama.

λαμαρίνα *s.f.* sheet-iron; baking-pan.

λάμας *s.m.* llama.

λαμβάνω *v.t.* take; receive, get; ~ την τιμή I have the honour; ~ μέρος take part; ~ χώραν take place.

λάμβδα *s.n.* the letter Λ.

λάμια *s.f.* ogress; glutton.

λάμνω *v.t. & i.* row; fly with slow flaps of wings.

λάμπα *s.f.* lamp; bulb.

λαμπάδα *s.f.* torch; large candle.

λαμπερός *a.* shining.

λαμπίκ|ος *s.m.* still, alembic; (*fig.*) anything clear and transparent. **~άρω** *v.t.* distil; filter.

λαμπιόνι *s.n.* electric bulb.

λαμποκοπώ *v.i.* shine (*from cleanness*).

Λαμπρή *s.f.* Easter.

λαμπρ|ός *a.* bright, brilliant. **~ά** *adv.* admirably. **~ώς** *adv.* brilliantly, splendidly.

λαμπρύνω *v.t.* add lustre to.

λαμπτήρας *s.m.* lamp; bulb.

λαμπυρίδα *s.f.* glow-worm.

λαμπυρίζω *v.i.* twinkle.

λάμπω *v.i.* shine.

λάμψη *s.f.* brilliance; flash.

λανθ|άνω *v.i.* be latent. **~άνομαι** *v.i.* be mistaken; **~ασμένος** wrong.

λανσάρω *v.t.* bring out, launch (*débutante,*

new fashion).

λαντζιέρης *s.m.* scullion.

λαξεύ|ω *v.t.* carve, sculpt. **~τός** *a.* sculptured.

λαογραφία *s.f.* folklore.

λαός *s.m.* people (*race, nation, body of citizens*); common people, populace.

λάου-λάου *adv.* cautiously and cunningly.

λαουτζίκος *s.m.* the common people.

λαούτο *s.n.* lute.

λαοφιλής *a.* popular (*of persons*).

λαπάς *s.m.* rice boiled to pap; poultice; (*fig.*) spineless person.

λαρδί *s.n.* lard.

λάρ|υγγας *s.m.*, **~ύγγι** *s.n.* larynx, throat.

λαρυγγισμός *s.m.* (*mus.*) roulade.

λαρυγγίτ|ις **~ιδα** *s.f.* laryngitis.

λαρυγγολόγος *s.m.* throat specialist.

λάρυγξ *s.m.* larynx.

λασκάρω *v.t. & i.* loosen, slacken (*of rope, etc.*).

λάσπη *s.f.* mud; mortar; ~ η δουλειά μας our efforts have failed; (*fam.*) το 'κοψε ~ he cleared out.

λασπ|ώδης, ~ερός *a.* muddy, soggy.

λασπώνω *v.t. & i.* mortar; make *or* become muddy; become mushy; (*fig.*) make a mess of; be a failure.

λαστιχένιος *a.* made of rubber *or* elastic; (*fig.*) supple.

λάστιχ|ο *s.n.* rubber; elastic; rubber band; india-rubber; tyre; washer; hose; (*boy's*) catapult; ~α (*pl.*) galoshes.

λατέρνα *s.f.* barrel-organ.

λατινικός *a.* Latin.

λατομείο *s.n.* quarry.

λάτρα *s.f.* housework.

λατρεία *s.f.* adoration, worship.

λατρευτός *a.* adored.

λατρεύω *v.t.* adore, worship; look after.

λάτρης *s.m.* devotee, admirer.

λάφυρ|ο *s.n.* booty, spoil. **~αγωγώ** *v.t.* plunder.

λαχαίνω *v.t. & i.* (*with acc.*) meet by chance; (*with gen.*) befall, fall to the lot of; έλαχα εκεί I happened to be there; έλαχε να τον δω I happened to see him.

λαχαναγορά *s.f.* vegetable market.

λαχανιάζω *v.i.* pant.

λαχανικό *s.n.* vegetable.

λάχαν|ο *s.n.* cabbage. **~όκηπος** *s.m.* vegetable-garden. **~οπωλείον** *s.n.* greengrocer's.

λαχείο *s.n.* lottery; (*fig.*) stroke of luck.

λαχνός *s.m.* lot (*drawn*).

λαχτάρα *s.f.* longing, strong desire *or* emotion; shock.

λαχταρώ *v.i. & t.* palpitate; be anxious; long

(for); make anxious.

λέαινα *s.f.* lioness.

λεβάντα *s.f.* lavender.

λεβάντες *s.m.* east wind.

λεβαντίνος *s.m.* levantine.

λεβέντης *s.m.* fine upstanding fellow; brave *or* generous man.

λεβεντιά *s.f.* manliness; upstanding apprearance; generosity.

λέβητας *s.m.* cauldron; boiler.

λεβιέ *s.n.* lever. [F. *levier*]

λεγάμενος *a.* (*fam.*) *o* ~ (*pej.*) you know who, his nibs.

λεγεών, ~α *s.m.f.* legion.

λέ(γ)ω *v.t. & i.* say, tell; recite, sing; call, name; *δε μου ~ει τίποτα* it does not impress me; *~ω να ταξιδέψω* I have in mind to travel; *δεν ~ει να σταματήσει* it shows no sign of stopping; *κοίταξε τι ~ει ο καιρός* look and see what the weather is like; *το 'λεγες αυτό;* would you have imagined it? *δεν σου ~ω* I do not deny it (*but...*); *δεν ~γεται* it is beyond description; *το ~ει η καρδιά του* he is brave; *τι ~ς* fancy that! *~ς και* as if; *που ~ς* well (*resumptive*); *~ει parenthetic phrase to support narrative*; *δε μου ~τε phrase introducing inquiry*; *το ~γειν* eloquence; *θα τα πούμε* we will have a talk; *πες πως κέρδισα* supposing I won; *πες .... πες* either (*because*) ... or (*because*); *τι θα πει τούτο;* what does this mean? *να σου πω* I say, look here (*also as qualified rejoinder or guarded response*).

λεζάντα *s.f.* caption.

λεηλασία *s.f.* pillage.

λεηλατώ *v.t.* pillage.

λεία *s.f.* booty; prey.

λειαίνω *v.t.* make smooth.

λειβάδι *s.n.* pasture (*land*).

λείβομαι *v.i.* be insufficient; be lacking; *μου λείφτηκε μία υάρδα* I was a yard short (*of material*).

λειμών *s.m.* pasture (*land*).

λείξα *s.f.* gluttony.

λείος *a.* smooth.

λείπ|ω *v.i.* be absent *or* away; be missing *or* lacking; run out; *μου ~ει* I miss it *or* feel the lack of it; *μας έλειπες* we missed you; *λίγο έλειψε να πέση* he nearly fell.

λειρί *s.n.* cock's comb.

λειτούργημα *s.n.* office, function.

λειτουργία *s.f.* working, functioning; (*eccl.*) service.

λειτουργιά *s.f.* bread offered for Holy Communion.

λειτουργός *s.f.* functionary, official.

λειτουργ|ώ *v.i.* work, function, go; (*eccl.*)

officiate. *~ιέμαι v.i.* go to church. **~ικός** *a.* functional; liturgical.

λειχήν *s.m.* 1. lichen. 2. (*med.*) herpes.

λείχω *v.t.* lick.

λειψανδρία *s.f.* lack of men.

λείψαν|ο *s.n.* mortal remains; *~α* (*pl.*) remains, debris; relics, bones.

λειψός *a.* deficient.

λειψυδρία *s.f.* shortage of water.

λείωμα *s.n.* pap, pulp; worn-out object.

λειώνω *v.t. & i.* melt, dissolve, reduce to *or* become pulp; wear out, exhaust; get worn out *or* exhausted.

λεκάνη *s.f.* basin, bowl; lavatory-pan; (*anat.*) pelvis.

λεκανοπέδιο *s.n.* basin (*area of land*).

λεκ|ές *s.m.* stain, spot. **~ιάζω** *v.t. & i.* stain; become stained.

λελέκι *s.n.* stork.

λελογισμένος *a.* wise, sensible.

λεμβοδρομία *s.f.* boat-race.

λέμβος *s.f.* small rowing *or* sailing boat.

λεμβούχ|ος *s.m.* boatman. **~ικά** *s.n.pl.* boatman's charges.

λεμονάδα *s.f.* lemonade.

λεμόν|ι *s.n.* lemon. **~ιά** *s.f.* lemon-tree. **~όκουπα** *s.f.* empty half of squeezed lemon.

λεμφατικός *a.* lymphatic.

λέξ|η *s.f.* word; *κατά ~η, ~η προς ~η* word for word.

λεξικ|ό *s.n.* dictionary. **~ογραφία** *s.f.* lexicography.

λεξιλόγιο *s.n.* vocabulary; glossary.

λεοντάρι *s.n.* lion.

λεοντή *s.f.* lion's skin.

λεοντιδεύς *s.m.* lion cub; (*fig.*) dandified youth.

λεοπάρδαλη *s.f.* leopard.

λέπι *s.n.* scale (*of fish, reptile*).

λεπ|ίς, ~ίδα *s.f.*, **~ίδι** *s.n.* (*cutting*) blade.

λέπρα *s.f.* leprosy.

λεπρ|ός *a.* leprous; a leper. **~οκομείο** *s.n.* leper-colony.

λεπτά *s.n.pl.* money.

λεπταίνω *v.t. & i.* make *or* become thin *or* refined.

λεπτεπίλεπτος *a.* very refined *or* delicate.

λεπτοδείκτης *s.m.* minute-hand.

λεπτοκαμωμένος *a.* fine, slender, delicate.

λεπτολογώ *v.t. & i.* examine minutely; split hairs.

λεπτομέρεια *s.f.* detail. **~κός** *a.* (*a matter*) of detail.

λεπτομερ|ής *a.* detailed. **~ώς** *adv.* in detail.

λεπτ|ό *s.n.* 1. minute. 2. lepton (*money*); **~ά** (*pl.*) money.

λεπτ|ός *a.* thin, slim, lean; fine; delicate,

λεπτουργός fragile; light (of cloting); refined, well-mannered; subtle; keen (of senses). ~ότης s.f. delicacy, tact, finesse.

λεπτουργός s.m. cabinet-maker.

λεπτοφυής a. slender, fine.

λεπτύνω see λεπταίνω.

λέρα s.f. dirt; (fam.) dirty dog.

λερωμένος a. dirty, soiled.

λερών|ω v.t. & i. make or become dirty; sully, ~ομαι v.i. get oneself dirty.

λεσβία s.f. lesbian.

λέσι s.n. rotting carcass; stink.

λέσχη s.f. club (institution).

λέτσος s.m. ragamuffin.

λεύγα s.f. league.

λεύκ|α, ~η s.f. poplar.

λευκαίνω v.t. make white, bleach.

λευκοκύτταρον s.n. (biol.) white corpuscle.

λευκαντικό s.n. bleach.

λευκόλιθος s.m. magnesite.

λευκ|ός a. white; blank (paper); (fig.) pure. ~ή βίβλος White Paper.

λευκοσίδηρος s.m. tin-plate.

λευκόχρυσος s.m. platinum.

λεύκωμα s.n. 1. album. 2. white of egg; albumen.

λευτεριά s.f. (fam). freedom; delivery (in childbirth).

λευχαιμία s.f. leukaemia.

λεφτά s.n.pl. money.

λεφτό s.n. minute.

λεχούδι s.n. newly born child.

λεχ|ούσα, ~ώ, ~ώνα s.f. mother of newly born child.

λέω v.t. & i. see λέγω.

λέων s.m. lion.

λεωφορείο s.n. omnibus.

λεωφόρος s.f. avenue, boulevard.

λήγουσα s.f. (gram.) final syllable.

λήγω v.i. terminate, expire; (gram.) end.

λήθαργ|ος s.m. lethargy. ~ικός a. lethargic.

λήθη s.f. oblivion.

λημέρι s.n. den, lair; retreat, refuge; haunt.

λήμμα s.n. headword.

ληνός s.m. wine-press.

ληξιαρχ|είον s.n. register-office. ~ικός a. relating to registry of births, marriages and deaths.

ληξιπρόθεσμος a. (fin.) mature.

λήξη s.f. cessation, expiry.

λήπτης s.m. receiver, recipient.

λησμο|νώ v.t. & i. forget, be forgetful. ~νιά s.f. oblivion. ~σύνη s.f. forgetfulness.

λησταρχείο s.n. nest of brigands.

ληστ|ής s.m. brigand, robber. ~εία s.f. brigandage, robbery. ~εύω v.t. rob.

ληστρικός a. robber's; predatory, extortionate.

λήψη s.f. receipt (act); reception.

λιάζ|ω v.t. put out in the sun. ~ομαι v.i. bask in the sun.

λιακάδα s.f. sunshine.

λιακωτό s.n. sunny balcony or flat roof.

λίαν adv. very, too.

λιανίζ|ω v.t. cut up small. ~ομαι v.i. hover.

λιανικός a. retail.

λιαν|ός a. thin, slim; να μου τα κάνεις ~α explain it clearly to me. ~ά s.n.pl. small change.

λιανοτούφεκο s.n. sporadic rifle-fire.

λιανοτράγουδο s.n. couplet of popular verse.

λιαστ|ός a. dried in the sun; κρασί ~ό wine made from such grapes.

λιβάδι s.n. see λειβάδι.

λιβάν|ι s.n., ~ωτός s.m. incense.

λιβανίζω v.t. cense; (fig.) adulate.

λιβάρι s.n. hatchery.

λίβας s.m. dry, hot S.W. wind.

λίβελλος s.m. libel.

λίβρα s.f. pound (weight).

λιβρέα s.f. livery.

λιγάκι adv. a little.

λίγδα s.f. grease; grease-stain.

λιγνεύω v.t. & i. make or get thin.

λιγνίτης s.m. lignite.

λιγνός a. thin, slender.

λιγοθυμώ v.i. see λιποθυμώ.

λίγο adv. (a) little; ~ πολύ more or less.

λιγό|λογος, ~μίλητος a. of few words.

λίγ|ος a. small in amount; (a) little, (a) few; ~ο ~ο little by little; σε ~ο shortly; παρά ~ο (να) nearly. See ολίγος.

λιγοστεύω v.t. & i. lessen, reduce; grow less.

λιγοστός a. scant, barely sufficient.

λιγότερ|ος a. less, least. ~ο adv. less, least.

λιγουλάκι adv. a bit, a shade.

λιγούρ|α s.f. nausea; feeling of emptiness (in stomach); hungry desire.

λιγουρεύ|ω, ~ομαι v.t. desire greatly (esp. food).

λιγοψυχία s.f. faint-heartedness.

λιγώ|νω v.t. & i. fill with nausea, cloy; make faint; feel cloyed or faint. ~νομαι v.i. feel cloyed or faint; ~μένα μάτια languishing look.

λιθάνθραξ s.m. pit-coal.

λιθάρι s.n. stone.

λιθίασις s.f. (med.) calculus.

λίθινος a. stone.

λιθοβολώ v.t. throw stones at; stone (to death).

λιθογραφία s.f. lithography; lithograph.

λιθόδμητος a. built of stone.

λιθοδομή s.f. stonework, masonrry.

λιθοκόπος *s.m.* stone-breaker.
λιθόκτιστος *a.* built of stone.
λιθοξόος *s.m.* stone-mason.
λίθος *s.m.f.* stone. *λυδία* ~ touchstone.
λιθόστρωτ|ος *a.* paved with stones. **~ο** *s.n.* paved way, pavement.
λικέρ *s.n.* liqueur. [F. *liqueur*]
λίκν|ον *s.n.* cradle. **~ίζω** *v.t.* rock.
λιλά *a.* lilac-coloured. [F. *lilas*]
λιλλιπούτειος *a.* lilliputian.
λίμα *s.f.* 1. file (*tool*); (*fig.*) boring chatter. 2. great hunger.
λιμάζω *v.i. & t.* be famished (for).
λιμάνι *s.n.* port, harbour.
λιμάρης *a.* gluttonous.
λιμάρω *v.t. & i.* file; (*fig.*) (bore with) chatter.
λιμενάρχης *s.m.* harbour-master.
λιμενικός *a.* of port *or* harbour.
λιμήν *s.m.* port, harbour, haven.
λιμνάζω *v.i.* be stagnant; stagnate.
λίμν|η *s.f.* lake. **~ούλα** *s.f.* pond.
λιμνοθάλασσα *s.f.* lagoon.
λιμοκοντόρος *s.m.* (*fam.*) toff, nut.
λιμοκτονώ *v.i.* starve to death.
λιμός *s.m.* famine.
λιμπίζομαι *v.t.* desire greatly; look fondly on.
λινάρι *s.n.* flax.
λινάτσα *s.f.* sacking.
λινέλαιο *s.n.* linseed-oil.
λίνον *s.n.* flax.
λιν|ός *a.* linen. **~ό** *s.n.* linen.
λιο- *denotes* 1. sun. 2. olive.
λιοντάρι *s.n.* lion.
λιοπύρι *s.n.* great heat of sun.
λιοτριβιό *s.n.* oil-press.
λιπαίνω *v.t.* lubricate, oil; manure.
λίπανση *s.f.* lubrication, manuring.
λιπαρός *a.* greasy.
λίπασμα *s.n.* manure, fertilizer.
λιποβαρής *a.* under-weight.
λιποθυμ|ώ *v.i.* faint. **~ία** *s.f.* fainting-fit. **~ος** *a.* in a faint.
λίπος *s.n.* fat, grease; lard.
λιπόσαρκος *a.* spare, skinny.
λιποτάκτης *s.m.* deserter.
λιποψυχία *s.f.* faint-heartedness.
λιπώδης *a.* fatty.
λίπωμα *s.n.* fatty tumour.
λίρα *s.f.* sovereign, pound.
λιρέτα *s.f.* (*Italian*) lira.
λίστα *s.f.* list, menu.
λιτανεία *s.f.* (*eccl.*) procession with icon (*for intercession*).
λιτ|ός *a.* frugal; plain, simple. **~ότης** *s.f.* frugality; simplicity.
λίτρα *s.f.* pound (*weight*); litre.

λιχνίζω *v.t.* winnow.
λιχούδης *a.* greedy (*for food*). **~ιά** *s.f.* greediness; a titbit.
λιώμα *s.n.* pulp; *κάνω* ~ crush, squash; *γίνομαι* ~ get crushed *or* squashed.
λιώνω *see* **λειώνω**.
λοβιτούρα *s.f.* (*fam.*) trick, roguery.
λοβός *s.m.* lobe; pod.
λογαριάζ|ω *v.t. & i.* count up, reckon; regard; take into account; take notice of, pay attention to; intend, expect (*to*). **~ομαι** *v.i.* settle one's account.
λογαριασμός *s.m.* calculation, reckoning, account; bill.
λογάριθμος *s.m.* logarithm.
λογάς *s.m.* chatterbox; maker of empty promises.
λόγγος *s.m.* thicket.
λογή *s.f.* τι ~ής; what sort of? ~ιών ~ιών all sorts of.
λόγια *s.n.pl.* words.
λογιάζω *v.t.* consider, think of.
λογιέμαι *v.i.* be considered.
λογίζομαι *v.t. & i.* take account of; consider oneself; be considered.
λογικεύομαι *v.i.* think sensibly; become sensible.
λογική *s.f.* logic; way of thinking.
λογικ|ό *s.n.,* **~ά** (*pl.*) senses, reason (*sanity*).
λογικός *a.* rational, logical; reasonable.
λόγιον *s.n.* saying, maxim.
λόγι|ος *a.* learned; ~ώτατος a pedant.
λογισμός *s.m.* thought; (*math.*) calculus.
λογιστήριον *s.n.* counting-house; purser's office; accounts dept.
λογιστής *s.m.* accountant, book-keeper; purser.
λογιστική *s.f.* accountancy, book-keeping.
λογογράφος *s.m.* writer, man of letters.
λογοδιάρροια *s.f.* garrulity.
λογοδοτώ *v.i.* render an account.
λογοκλοπία *s.f.* plagiarism.
λογοκρισία *s.f.* censorship.
λογομαχία *s.f.* exchange of abusive words, row.
λογοπαίγνιο *s.n.* play on words.
λόγ|ος *s.m.* 1. speech (*faculty*); speech, address; talk, mention, question (*of*); saying; word (*pl. τα ~ια*); *έχει ~ο* he is a man of his word; *έδωσαν ~ο* they exchanged pledges; *~ου χάριν* for example; *εν ~ω* in question; *άξιος ~ου* worthy of note; *δεν σου πέφτει ~ος* you have no say; *μου έδωσε το ~ο* he called on me to speak; *δεν παίρνει από ~ια* he will not listen to reason; *της πήρα ~ια* I wormed it out of her; *ο ~ος το λέει* in a manner of speaking; *που λέει ο ~ος* as it

were. 2. reason, ground; account, reckoning; ~ω by reason of; ορθός ~ος good sense. κατά μείζονα ~ov with all the more reason; κατ' ουδένα ~ov on no account. 3. (*fam.*) (του) ~ου του *or* σου he, him *or* you, *etc.*

λογοτέχν|ης *s.m.* writer, man of letters. ~**ία** *s.f.* writing; literature.

λογοφέρνω *v.i.* have a row, exchange insults.

λόγχη *s.f.* lance; bayonet.

λοιδορώ *v.t.* revile, mock.

λοιμός *s.m.* pestilence, plague. ~**ώδης** *a.* pestilential, infectious. ~**ωξη** *s.f.* infection.

λοιπόν *conj. & int.* then, so, well (then).

λοιπ|ός *a.* the rest (of); και τα ~ά etcetera; του ~ου in future.

λοίσθι|ος *a.* πνέω τα ~α breathe one's last.

λοκάντα *s.f.* small restaurant.

λόξα *s.f.* 1. slant; piece of material cut on bias. 2. craze, fad.

λοξοδρομώ *v.i.* deviate, shift one's course; (*naut.*) tack.

λοξ|ός *a.* 1. slanting, oblique, at an angle. 2. a faddist. ~**ά** *adv.* slantwise, at an angle.

λόξυγγας *s.m.* hiccup.

λόπια *s.n.pl.* sort of kidney-beans.

λόρδα *s.f.* (*fam.*) great hunger.

λόρδος *s.m.* lord.

λοσιόν *s.f.* lotion (*cosmetic*). [F. *lotion*]

λοστός *s.m.* crowbar.

λοστρόμος *s.m.* (*naut.*) boatswain.

λοταρία *s.f.* lottery.

λούζ|ω *v.t.* wash (*body, hair*); (*fam.*) scold, revile. ~**ομαι** *v.i.* have a bath, wash one's hair.

λουκάνικο *s.n.* sausage.

λουκέτο *s.n.* padlock.

λούκι *s.n.* drainpipe; gutter; bias fold (*dressmaking*).

λουκούλλειος *a.* Lucullan, sumptuous.

λουκουμάς *s.m.* sort of doughnut fried in oil.

λουκούμι *s.n.* Turkish delight (*sweet-meat*); (*fig.*) tender as chicken; μου 'ρθε ~ it was a godsend to me.

λουλάκι *s.n.* indigo, blue (*in laundering*).

λουλούδι *s.n.* flower.

λουμίνι *s.n.* small wick.

λουμπάγκο *s.n.* lumbago.

λούνα παρκ *s.n.* fairground.

λουξ *s.n.* luxury (*a.*) de luxe. [F. *luxe*]

λουρί *s.n.* strap, belt.

λουρίδα *s.f.* strip, belt.

λουσάρ|ω, ίζω *v.t.* (*fam.*) give a touch of luxury to. ~**ομαι** *v.i.* dress up.

λούσιμο *s.n.* washing (*of body, hair*); (*fam.*)

scolding.

λούσ|ο *s.n.* smart clothes; luxury; (*fam.*) άσ' τα ~α come off it!

λουστράρω *v.t.* polish.

λουστρίνι *s.n.* patent-leather; ~α (*pl.*) patent-leather shoes.

λούστρο *s.n.* polish; (*fig.*) smattering.

λούστρος *s.m.* shoeblack.

λουτρικό *s.n.* bath-robe.

λουτρ|ό *s.n.* bath, bathe; bathroom; ~ά (*pl.*) mineral springs, spa, waters; έμεινε στα κρύα του ~ού he was let down.

λουτρόπολις *s.f.* spa (*town*).

λούτσα *s.f.* (*fam.*) γίνομαι ~ get drenched.

λουφάζω *v.i.* lie low, shrink back; (*fam.*) shut up.

λούω *v.t.* wash, bathe (*body*).

λοφίο *s.n.* (*bird's*) crest; (*mil.*) pompon.

λόφ|ος *s.m.* hill. ~**ώδης** *a.* hilly.

λοχαγός *s.m.* (*mil.*) captain.

λοχίας *s.m.* (*mil.*) sergeant.

λόχμη *s.f.* clump of dense undergrowth.

λόχος *s.m.* (*mil.*) company.

λυγαριά *s.f.* osier.

λυγερ|ός *a.* lithe, svelte. ~**ή** *s.f.* damsel, beautiful girl.

λυγ|ίζω, ~ώ *v.t. & i.* bend; σειέμαι και ~ιέμαι assume coquettish airs.

λυγμός *s.m.* sob.

λυγξ *s.m.* lynx.

λυθρίνι *s.n.* sea-bream.

λύκαινα *s.f.* she-wolf.

λυκάνθρωπος *s.m.* werewolf.

λυκαυγές *s.n.* half-light of dawn.

λύκειον *s.n.* secondary school. (*esp. private*).

λύκος *s.m.* wolf; cock (*of gun*).

λυκόσκυλο *s.n.* wolf-hound, alsatian.

λυκόφως *s.n.* dusk.

λυμαίνομαι *v.t.* ravage, prey on.

λύματα *s.n.pl.* sewage, effluent.

λυντσάρω *v.t.* lynch.

λύνω *v.t.* see **λύω**.

λυπάμαι *v.i. & t.* regret, be sorry (for); (*fig.*) be stingy with.

λύπη *s.f.* regret, sorrow, grief; mourning; pity. ~**ρος** *a.* sad, distressing.

λυπημένος *a.* sad, distressed.

λυπούμαι *v.i. & t.* regret, be sorry (for).

λυπώ *v.t.* grieve, sadden.

λύρα *s.f.* lyre. ~**ικός** *a.* lyric, lyrical. ~**ισμός** *s.m.* lyricism.

λύση *s.f.* termination, dissolution (*of partnership, etc.*); solving, settling; solution, answer; denouement; ending.

λυσίκομος *a.* dishevelled.

λύσιμο *s.n.* loosening, untying; solving; dismantling.

λυσιτέλεια s.f utility.
λύσσα s.f. rabies; (fig.) fury, frenzy; (fam.) anything very salty.
λυσσαλέος a. fierce, furious.
λυσσ(ι)άζω v.i. & t. get rabies; (fig.) be mad (with anger, desire); madden; (fam.) thrash; ~ το φαγητό make the food like brine.
λυσσιατρείον s.n. rabies hospital.
λυσσομανώ v.i. rage.
λυσσώ v.i. get rabies; (fig.) be mad (with anger, desire).
λυσσώδης a. furious, fierce.
λύτης s.m. solver.
λυτ|ός a. not tied up, loose; βάζω ~ούς και δεμένους leave no stone unturned.
λύτρα s.n.pl. ransom.
λυτρώνω v.t. ransom; set free.
λυτρωτής s.m. redeemer.
λυχνάρι s.n. see λύχνος.
λυχνία s.f. (any sort of) lamp; (electric) bulb, valve.
λύχνος s.m. small olive-oil lamp.
λύ|ω v.t. loosen, untie, set free; dismantle; solve, settle; terminate, dissolve. ~ομαι v.i. come undone; ~θηκα στα γέλια I split my sides with laughing; ~όμενος collapsible.
λυώνω v.t. & i. see λειώνω.
λώβ|α s.f. leprosy. ~ός a. leprous; a leper.
λωλ|ός a. mad, crazy. ~άδα s.f. madness; mad action. ~αίνω v.t. send crazy.
λωποδύτης s.m. thief, pickpocket.
λωρίον s.n. see λουρί.
λωρίς s.f. see λουρίδα.
λωτός s.m. lotus.

Μ

μα 1.conj. but. 2. particle (with acc.) by; ~ το Θεό by God!
μαβής a. dark blue.
μαγαζ|ί s.n. shop. ~άτορας s.m. shop-keeper.
μαγάρ|α s.f. dirt, dung; (fig.) dirty beast. ~ίζω v.t. & i. pollute, foul; make dirt.
μαγγανεία s.f. sorcery.
μαγγ|άνι, ~ανο s.n. calender, press; vice (tool); loom; wheel (of well, etc.).
μαγγάνιον s.n. manganese.
μαγγανοπήγαδο s.n. wheel-well; (fig.) treadmill.
μάγγας s.m. see μάγκας.
μαγγώνω v.t. crush, squeeze; seize, nab. (v.i.) seize up (of engine).
μαγεία s.f. magic, sorcery; enchantment; anything delightful.

μαγειρείο s.n. kitchen; cook-shop.
μαγείρε(υ)μα s.n. cooking.
μαγειρευτός a. stewed.
μαγειρεύω v.t. & i. cook; prepare food. (fig.) falsify, cook up.
μαγειρικ|ός a. culinary; for cooking. ~ή s.f. cookery (-book).
μαγειρίτσα s.f. soup eaten at Easter.
μά|γειρος, ~γερας s.m. cook.
μαγεμένος a. enchanted.
μαγερειό s.n. kitchen; cook-shop.
μαγεριά s.f. (quantity sufficient for) one cooking.
μαγέρικο s.n. cook-shop.
μαγευτικός a. enchanting, delightful.
μαγεύω v.t. bewitch, enchant, charm.
μάγια s.n.pl. (sorcerer's) spell, charm.
μαγιά s.f. yeast, leaven; (fam.) initial small capital in business enterprise.
μαγιάτικος a. of May.
μαγικός a. magic; (fig.) magical, enchant-ing.
μαγιό s.n. bathing-dress. [F. maillot]
μαγιονέζα s.f. mayonnaise.
μάγισσα s.f. witch, enchantress.
μαγκάλι s.n. brazier.
μάγκ|ας s.m. one who has learnt the seamy side of life; street-urchin; sharper, spiv; clever fellow. ~ικα s.n.pl. slangy talk of μάγκας.
μαγκλαράς s.m. tall, gawky, fellow.
μαγκούρα s.f. crook, staff, cudgel.
μαγκούφης s.m. solitary or wretched person; ownerless animal.
μαγνησία s.f. magnesia.
μαγνήσιον s.n. magnesium.
μαγνήτ|ης s.m. magnet. ~ίζω v.t. magnetize. ~ικός a. magnetic. ~ισμός s.m. magnetism.
μαγνητόφωνο s.n. tape-recorder.
μάγ|ος s.m. magician; ~οι (pl.) Magi.
μαγουλήθρα s.f. mumps.
μάγουλο s.n. cheek.
μαδέρι s.n. joist; scaffold-pole.
μάδημα s.n. plucking, moulting.
μαδώ v.t.& i. pluck; moult, shed hair, leaves, etc.; fall, come out (of hair, leaves, etc.); (fam.) pluck, swindle.
μαέστρος s.m. musical conductor; (fig.) expert.
μάζα s.f. mass.
μάζε|μα s.n. gathering, collecting. ~τα (pl.) trash, riff-raff.
μαζεμένος a. unassuming, well-behaved; crouching, nestling.
μαζεύ|ω v.t. & i. gather, collect; pick up; strike (sail, flag, etc.); take in, reduce compass of; wind (wool) into ball; curb,

restrain; (*v.i.*) shrink (*of cloth*); become purulent. ~*ομαι v.i.* assemble, collect; crouch, nestle; behave with restraint, settle down; ~*εται νωρίς* he goes home early.
μαζί *adv.* together; (*prep.*) ~ *του* with him.
μαζικός *a.* 1. of the masses; mass. 2. held in common.
μαζοχισμός *s.m.* masochism.
μαζώνω *v.t. see* **μαζεύω.**
Μάης *s.m.* May.
μαθαίνω *v.t. & i.* learn; become aware; ascertain; teach; ~ *σε* get used to.
μαθέ(ς) *adv.* certainly, apparently; by any chance?
μαθεύομαι *v.i.* become known.
μάθημα *s.n.* lesson.
μαθηματικ|ά *s.n.pl.,* ~ **ή** *s.f.* mathematics.
μαθηματικός *a.* mathematical; a mathematician.
μαθημένος *a.* learnt; used, accustomed (*to*).
μάθηση *s.f.* learning.
μαθητεία *s.f.* years of study; apprenticeship.
μαθητεύ|ω *v.i.* be a learner *or* apprentice; ~*όμενος* an apprentice, tiro.
μαθητ|ής *s.m.* pupil, schoolboy; disciple. ~**ρια** *s.f.* schoolgirl.
μαθητικ|ός *a.* studious; pupil's; ~*ό θρανίο* school-desk.
μαία *s.f.* midwife.
μαίανδρος *s.m.* meander (*pattern*).
μαιευτήρ *s.m.* obstetrician.
μαιευτήριον *s.n.* lying-in hospital.
μαιευτική *s.f.* obstetrics.
μαϊμού *s.f.* monkey; (*fig.*) ugly *or* cunning person. ~**δίζω** *v.t.* ape, imitate.
μαινά|ς ~**δα** *s.f.* maenad.
μαίνομαι *v.i.* rage.
μαϊντανός *s.m.* parsley; (*fam.*) meddlesome person.
Μάιος *s.m.* May.
μαΐστρος *s.m.* north-west wind.
μαιτρέσσα *s.f.* mistress, concubine.
μακάβριος *a.* macabre.
μακάρι *adv.* would that! ~ *και* even.
μακαρίζω *v.t.* hold fortunate, envy (*without malice*).
μακάριος *a.* blessed; blissful, serene.
μακαρίτης *a.* late, deceased.
μακαρόνια *s.n.pl.* macaroni.
μακελλάρης *s.m.* butcher.
μακελλειό *s.n.* butcher's shop; slaughterhouse; (*fig.*) carnage.
μακέτα *s.f.* sketch, model.
μακιγιάρ|ω *v.t.,* ~**ομαι** *v.i.* (*cosmetic*) make up.
μακρ- *denotes* length.
μακραίνω *v.t. & i.* lengthen, extend; grow

longer *or* taller; drag on.
μακράν *adv.* far, afar.
μάκρεμα *s.n.* lengthening.
μακρηγορώ *v.i.* expatiate.
μακριά *adv.* far, afar.
μακρινάρι *s.n.* long passage; any elongated object.
μακρινός *a.* distant, far-off.
μακριός *a.* long.
μακρόβιος *a.* long-lived.
μακροβούτι *s.n.* under-water swimming.
μακρόθεν *adv.* from afar.
μακρόθυμος *a.* patient, forbearing; forgiving.
μακροπρόθεσμος *a.* long-term.
μάκρος *s.n.* length.
μακρ|ός *a.* long; *διά* ~*ών* at length, in detail. *από* ~*ού* for a long time.
μακροσκελής *a.* (*fig.*) long-winded.
μακρύνω *v.t. & i. see* **μακραίνω.**
μακρύς *a.* long; tall.
μάκτρον *s.n.* towel, napkin.
μαλαγάνας *s.m.* (*fam.*) artful dodger.
μαλ|άζω, ~**άσσω** *v.t.* knead, massage; soften.
μαλάκιον *s.n.* mollusc.
μαλακ|ός *a.* soft; mild. ~**ώνω** *v.t. & i.* soften; get milder.
μαλακτικός *a.* emollient.
μαλάκυνση *s.f.* softening.
μάλαμα *s.n.* gold; (*fig.*) kind-hearted person.
μάλ|η *s.f. υπό* ~*ης* under the arm.
μαλθακός *a.* soft, unused to hardship.
μάλιστα *adv.* especially; indeed; even; yes.
μαλλί *s.n.* wool; hair (*collective*).
μαλλιά *s.n.pl.* hair (*of head*); ~ *κουβάρια* topsy-turvy; *έγιναν* ~ *κουβάρια* they quarrelled.
μαλλιαρή *s.f.* (*fam.*) extreme form of demotic Gk.
μαλλιαρός *a.* hairy; an advocate of *μαλλιαρή*.
μάλλινος *a.* woollen.
μάλλον *adv.* more, rather; to some extent; *κατά το* ~ *ή ήττον* more or less.
μάλωμα *s.n.* scolding; quarrelling.
μαλώνω *v.t. & i.* scold; quarrel.
μαμά *s.f.* mother, mamma.
μάμμη *s.f.* grandmother.
μαμμή *s.f.* midwife.
μαμμόθρεπτος *a.* coddled.
μαμμούθ *s.n.* mammoth.
μαμμούνι *s.n.* insect.
μαμμούνι *s.n.* insect.
μάνα *s.f. see* **μάν(να).**
μανάβ|ης *s.m.* greengrocer. ~**ικο** *s.n.* grenn-grocer's.
μανδ- *see* **μαντ-.**
μανδαρίνος *s.m.* mandarin.

μανδύας *s.m.* cloak; cope.
μανθάνω *v.t. & i. see* μαθαίνω.
μάνι μάνι *adv.* quickly.
μανία *s.f.* madness; rage; passion, mania; *έχω ~ με* be mad on.
μανιάζω *v.i.* become enraged.
μανιακός *a.* maniac(al).
μανιβέλα *s.f.* lever; starting handle.
μανίκ|ι *s.n.* sleeve; handle (*of implement*); (*fam.*) tricky job. ~έτι *s.n.* cuff. ~ετόκουμπο *s.n.* cuff-link.
μανιτάρι *s.n.* mushroom.
μανιώδης *a.* furious; passionately addicted.
μάννα *s.n.* manna.
μάν(ν)α *s.f.* mother; (*fig.*) spring (*water*); original copy; (*fam.*) dab, expert.
μαν(ν)ούλα *s.f.* (*fam.*) mother; *also term of endearment.*
μανόλια *s.f.* magnolia.
μανουάλιον *s.n.* church candelabrum.
μανούβρα *s.f.* manoeuvre; shunting.
μανούρι *s.n.* sort of white cheese.
μανταλάκι *s.n.* clothes-peg.
μάνταλ|ο *s.n.* bolt, latch. ~ώνω *v.t.* bolt, latch.
μαντάρα *s.f.* (*fam.*) wreck, shambles.
μανταρίνι *s.n.* tangerine.
μαντάρ|ω *v.t.* darn. ~ισμα *s.n.* darning.
μαντάτο *s.n.* news. ~φόρος *s.m.* messenger.
μαντεία *s.f.* divination; a prophecy, oracle.
μαντείον *s.n.* oracle (*place*).
μαντέκα *s.f.* wax for moustaches.
μαντέμι *s.n.* cast iron.
μαντεύω *v.t.* divine, guess.
μαντζουράνα *s.f.* marjoram.
μαντήλι *s.n.* kerchief; handkerchief.
μάντης *s.m.* soothsayer.
μαντικ|ός *a.* prophetic. ~ή *s.f.* divination.
μαντινάδα *s.f.* rhyming couplet (*esp. humorous*).
μαντολάτο *s.n.* nougat.
μαντολίνο *s.n.* mandolin.
μάντρα *s.f.* stall, fold, pen, sty; (*enclosing*) wall, fence; storage yard.
μαντράχαλος *s.m.* tall, gawky fellow.
μαντρ|ί *s.n.* pen, fold. ~ίζω *v.t. see* μαντρώνω.
μαντρόσκυλο *s.n.* shepherd's dog.
μαντρώνω *v.t.* pen, fold (*animals*); wall, fence.
μαξιλάρι *s.n.* pillow, cushion.
μαξούλι *s.n.* crop (*esp. grapes*).
μαόνι *s.n.* mahogany.
μαούν|α *s.f.* lighter (*boat*). ~ιέρης *s.m.* lighterman.
μάππα *s.f.* cabbage.
μάππας *s.m.* silly fellow.
μαραγκιάζω *v.i.* shrivel.

μαραγκός *s.m.* joiner; carpenter.
μαράζ|ι *s.n.* pining. ~ώνω *v.i.* pine, languish.
μάραθ(ρ)ο *s.n.* fennel.
μαραίν|ω *v.t.* wither. ~ομαι *v.i.* wither fade.
μαρασμός *s.m.* decline, decay.
μαραφέτι *s.n.* (*fam.*) thing, thingummy.
μαργαρίνη *s.f.* margarine.
μαργαρίτα *s.f.* daisy.
μαργαριτ|άρι *s.n.* pearl. ~ένιος *a.* pearl, pearly.
μαργαρίτης *s.m.* pearl; (*fam.*) howler.
μαργιολιά *s.f.* exercise of wiles, coquettishness.
μαρίδα *s.f.* whitebait; (*fam.*) crowd of small children.
μαριονέτα *s.f.* marionette.
μάρκα *s.f.* brand, trade-mark; monogram; counter, token; (*fam.*) slippery customer.
μαρκάρω *v.t.* mark, stamp, tick off; notice, espy.
μάρμαρ|ο *s.n.* marble; (*fam.*) *έμεινε* ~ο he was dumbfounded. ~ένιος, ~ινος *a.* marble. ~ώνω *v.t. & i.* turn to marble.
μαρμαρυγή *s.f.* gleam, shimmer.
μαρμελάδα *s.f.* jam, marmalade.
μαρξισμός *s.m.* marxism.
μαρούλι *s.n.* lettuce.
μαρς *s.n.* (*mus.*) march. (*imper.*) march!
μαρσιποφόρος *a.* marsupial.
Μάρτ|ης, ~ιος *s.m.* March.
μαρτυρία *s.f.* testimony, evidence.
μαρτυρικός *a.* evidential; martyr's; full of suffering.
μαρτύριο *s.n.* token; martyrdom; torment.
μαρτυρώ *v.t. & i.* testify, give evidence; indicate; inform against; suffer martyrdom.
μάρτυς *s.m.* witness; martyr.
μάς, μας *pron.* us; our.
μασάζ *s.n.* massage.
μασέλα *s.f.* jaw; set of (false) teeth.
μασιά *s.f.* tongs.
μάσκα *s.f.* mask; fancy dress. ~ράς *s.m.* masker; (*fam.*) blighter; scamp.
μασόνος *s.m.* freemason.
μασουλ|ίζω, ~ώ *v.t.* chew, munch.
μασούρι *s.n.* bobbin, spool, reel; stick (*of sulphur, etc.*). (*fam.*) pay-packet.
Μασσαλιώτισσα *s.f.* Marseillaise.
μαστάρι *s.n.* udder.
μαστίγιο *s.n.* horsewhip, switch.
μαστιγώνω *v.t.* whip.
μαστίζω *v.t.* scourge, plague.
μάστιξ *s.f.* whip; scourge.
μαστίχα *s.f.* mastic; sweet *or* liqueur from this.
μάστορ|ας, ~ης *s.m.* (skilled) workman;

handyman; (*fig.*) expert. **~ικός** *a.* craftsman's; masterly.

μαστορεύω *v.t. & i.* make; mend; do a job (*with tools*).

μαστ|ός *s.m.* breast; udder. **~οειδής** *a.* mastoid.

μαστροπός *s.m.* pimp.

μαστροχαλαστής *s.m.* (*iron.*) bungler.

μασχάλη *s.f.* armpit; arm-hole.

μασώ *v.t. & i.* chew; eat.

ματαιόδοξος *a.* conceited.

ματαιοπονώ *v.i.* labour in vain.

μάτ|αιος *a.* vain, fruitless; conceited. **~αίως** *adv.* in vain.

ματαιώνω *v.t.* thwart, prevent; cancel.

ματάκι *s.n.* wink; peep-hole (*in door*).

μάτην *adv.* in vain.

μάτι *s.n.* eye; evil eye; bud (*leaf or shoot*); burner (*of stove*). **~** *της θάλασσας* whirlpool; *αβγό* **~** fried egg; *έχω στο* **~** covet *or* have a down on; *βάζω στο* **~** set one's heart on; *μου μπήκε στο* **~** it tickles my fancy; *θα σου φάω το* **~** I will take revenge on you; *κλείνω* (*or* *κάνω*) *το* **~** wink; *κλείνω τα* **~** *σε* overlook, be tolerant of; *κάνω στραβά* **~** pretend not to see; *βγάζω τα* **~** *μου* ruin oneself; *παίρνω τα* **~** (*or* *των ομματιών*) *μου* go away in despair; *για τα* (*μαύρα*) **~** for politeness' sake; *για τα* **~** *του κόσμου* for appearance' sake; *τα* **~** *σου τέσσερα* keep your eyes skinned! **~** *μου* my dearest!

ματι|ά *s.f.* glance, look; ogle. **~άζω** *v.t.* espy; covet, set one's heart on; take aim at; bewitch, bring bad luck to.

ματίζω *v.t.* splice, add piece on to.

ματογυάλια *s.n.pl.* eyeglasses.

ματόκλαδο *s.n.* eyelash.

ματόφρυδο *s.n.* eyebrow.

ματόφυλλο *s.n.* eyelid.

ματσαράγκα *s.f.* cheating.

ματς *s.n.* (*sport*) match.

ματς μουτς (*onomatopoeia*) (sound of) kisses.

μάτσο *s.n.* bunch, bundle; skein.

ματσούκα *s.f.* club, cudgel.

ματώνω *v.t. & i.* stain with blood; bleed.

μαυρ- *denotes* black:

μαυραγορίτης *s.m.* black-marketeer.

μαυράδι *s.n.* black spot; pupil (*of eye*).

μαυριδερός *a.* darkish.

μαυρ|ίζω *v.t. & i.* darken; blackball; become brown *or* dark; appear darkly, loom; **~ισε** *το μάτι μου* I am driven to desperation (*by poverty, etc.*).

μαυρίλα *s.f.* blackness, darkness; dark colour, tan.

μαυροδάφνη *s.f.* sort of sweet wine.

μαυρομάτης *a.* black-eyed.

μαυροπίνακας *s.m.* blackboard; black-list.

μαυροπούλι *s.n.* starling.

μαύρ|ος *a.* black; dark; negro; diehard; wretched; gloomy; *έκανα* **~** *α μάτια να τον δω* I have not seen him for ages; *τον έκανα* **~** *ο στο ξύλο* I gave him a good hiding; **~** *ο κρασί* red wine.

μαυροφορώ *v.i.* wear mourning.

μαυσωλείον *s.n.* mausoleum.

μάχ|αιρα *s.f.*, **~αίρι** *s.n.* knife; *στα* **~αίρια** at daggers drawn.

μαχαιριά *s.f.* stab.

μαχαιροβγάλτης *s.m.* one quick on the draw; cutthroat.

μαχαιροπήρουνα *s.n.* cutlery.

μαχαιρώνω *v.t.* knife, stab.

μαχαλάς *s.m.* quarter, district.

μάχη *s.f.* battle. **~τής** *s.m.* fighter.

μαχητικ|ός *a.* combative; militant. **~όν** *s.n.* (*aero.*) fighter.

μάχιμος *a.* combatant; in fighting trim.

μαχμουρλής *a.* still heavy with sleep.

μάχ|ομαι *v.i.* fight, contend. (*v.t.*). *με* **~εται** he hates me.

με *prep.* (*with acc.*) 1. (*in company of, among*) with. 2. (*having, characterized by*) with. 3. (*means, instrument, method*) with, by; **~** *τα πόδια* on foot. 4. (*material, content*) of. 5. (*manner, circumstance*) with; **~** *το καλό* God willing; without mishap; **~** *τέτοια χάλια* in such a (*bad*) state. 6. (*rate, measure, price*) **~** *το μήνα* by the month; **~** *κέρδος* at a profit; **~** *τόσα λίγα δεν αγοράζεται* it can not be bought for so little; *δύο* **~** *τρία* 2 by 3 (metres). 7. (*time*) **~** *τον καιρό* in the course of time; **~** *τη σειρά* in turn; *δύο* **~** *τρεις* between 2 and 3 o'clock. 8. (*in regard to*) *τι μ' αυτό;* what of it? *έχω μανία* **~** to be mad on; *σχετικά* **~** regarding. 9. (*despite*) *μ' όλο που* although. 10. (*cause*) *γελώ* **~** laugh at. 11. (*proximity*) *πλάτη* **~** *πλάτη* back to back. 12. (*exchange*) *αλλάζω λίρες* **~** *δραχμές* exchange pounds for drachmas. 13. (*after words denoting similarity, contact, agreement, & opposites*) *παντρεύομαι* **~** get married to; *χωρίζω* **~** part from; *ομοιότητα* **~** resemblance to; *συγγενής* **~** related to; *ανάλογως* **~** in proportion to; *ίδια* **~** *μένα* the same as me.

με *pron.* me.

μεγαθήριο *s.n.* monster.

μέγαιρα *s.f.* shrew, hag.

μεγαλ- *denotes* great, large.

μεγαλείον *s.n.* grandeur, magnificence; **~α** (*pl.*) honours and wealth; (*fam.*) **~ο** first-rate!

μεγαλειότης *s.f.* majesty; *η Αυτού* ~ His Majesty.

μεγαλειώδης *a.* grand, grandiose.

μεγαλέμπορος *s.m.* large-scale (wholesale) merchant.

μεγαλεπήβολος *a.* enterprising on a grand scale.

μεγαληγορία *s.f.* bombast.

μεγαλόδωρος *a.* liberal, generous.

μεγαλοπιάνομαι *v.i.* put on lordly airs.

μεγαλοποιώ *v.t.* exaggerate.

μεγαλοπρεπ|ής *a.* majestic, imposing, grand. ~εια *s.f.* grandeur, imposingness.

μεγάλ|ος *a.* great, grand; large, big; long (*duration*); grown up; elder; ~η Εβδομάδα Holy Week.

μεγαλόσχημος *a.* eminent, important (*person*); one who pretends to be so.

μεγαλόσωμος *a.* heavily built.

μεγαλουργώ *v.i.* accomplish great things.

μεγαλο(υ)σιάνος *s.m.* (*esp. iron.*) big shot.

μεγαλόφρων *a.* magnanimous; conceited.

μεγαλοφυής *a.* endowed with genius. ~ία *s.f.* genius.

μεγαλοφώνως *adv.* aloud.

μεγαλόχαρη *a.* (*the Virgin*) full of grace.

μεγαλόψυχος *a.* magnanimous.

μεγαλύνω *v.t.* glorify, extol.

μεγαλύτερος *a.* greater, bigger, elder.

μεγαλώνω *v.t. & i.* enlarge, make bigger; bring up (*children*); exaggerate; grow in size; grow up.

μέγαρον *s.n.* mansion, official building.

μέγας *a. see* μεγάλος.

μεγάφωνο *s.n.* megaphone; loud-speaker.

μέγγενη *s.f.* vice (*tool*).

μέγεθ|ος *s.n.* size, greatness, extent. ~ύνω *v.t.* magnify, enlarge. ~υνση *s.f.* magnification, enlargement. ~υντικός *a.* magnifying.

μεγιστάν *s.m.* grandee, magnate.

μέγιστ|ος *a.* greatest, largest. ~ον *s.n.* maximum.

μεδούλι *s.n.* marrow (*of bones*).

μέδουσα *s.f.* sort of jellyfish.

μεζ|ές *s.m.* bit of food as appetizer; titbit. *παίρνω στο* ~*έ* make fun of.

μεζούρα *s.f.* tape-measure.

μεθαύριο *adv.* the day after tomorrow.

μέθη *s.f.* intoxication.

μεθοδεύω *v.t.* organize, plan.

μέθοδ|ος *s.f.* method; course, hand-book. ~ικός *a.* methodical.

μεθοκοπώ *v.i.* get drunk.

μεθόριος *a. & s.f.* (of the) frontier. ~ακός *a.* of the frontier.

μεθύσι *s.n.* intoxication.

μεθύσκ|ω *v.t.* intoxicate. ~ομαι *v.i.* get intoxicated.

μεθυσμένος *a.* drunk.

μέθυσος *s.m.* drunkard.

μεθυστικός *a.* intoxicating.

μεθώ *v.i.* become inebriated.

μεθώ *v.t. & i.* intoxicate; get drunk.

μειδι|ώ *v.i.* smile. ~αμα *s.n.* smile.

μείζ|ων *a.* greater; *κατά* ~*ονα λόγον* all the more so; (*mus.*) major.

μειλίχιος *a.* mild (*of manner*).

μειοδοτώ *v.i.* make lowest tender.

μείον *adv.* less; minus.

μειονέκτημα *s.n.* failing, disadvantage, drawback.

μειονεκτικός *a.* disadvantageous.

μειονότης *s.f.* minority.

μειο(νο)ψηφία *s.f.* minority.

μείων *a.* lesser.

μειώνω *v.t.* lessen, reduce; belittle.

μείωση *s.f.* lessening, reduction; loss of moral stature.

μειωτικός *a.* reducing; pejorative, disparaging.

μελαγχολ|ία *s.f.* melancholy, sadness. ~ικός *a.* melancholy, sad. ~ώ *v.i. & t.* feel *or* make sad.

μελάν|η *s.f.*, ~ι *s.n.* ink. ~ιά *s.f.* inkstain; bruise. ~ιάζω *v.t. & i.* bruise; get bruised; turn blue (*with cold*). ~ός *a.* black, bruised.

μελανοδοχείο *s.n.* ink-well, ink-stand.

μέλας *a.* black.

μελάτος *a.* 1. like honey. 2. soft-boiled (*egg*). 3. loose-fitting.

μελαχροινός *a.* dark-complexioned; brunette.

μελαψός *a.* swarthy.

μέλει *v.i. με* ~ I mind it, I care about it.

μελένιος *a.* of *or* like honey.

μελέτη *s.f.* study; practice; treatise; plan, design (*for building, etc.*).

μελετηρός *a.* studious.

μελετώ *v.t. & i.* study; practise; contemplate, have in mind; mention.

μέλημα *s.n.* (*object of*) care *or* concern.

μέλι *s.n.* honey; *μην του* ~*τος* honeymoon.

μελίγγι *s.n.* temple (*of head*).

μελιγγίτης *s.m.* meningitis.

μέλισσα *s.f.* bee.

μελίσσι *s.n.* swarm of bees; (*fig.*) swarm; (*fam.*) *έχω* ~(*α*) keep bees.

μελισσο|κομία, ~τροφία *s.f.* apiculture.

μελιστάλαχτος *a.* honeyed.

μελιτζάνα *s.f.* egg-plant, aubergine.

μελλοθάνατος *a.* condemned to death.

μέλλον *s.n.* future. ~τικός *a.* future.

μελλόνυμφος *a.* about to be married.

μέλλ|ω *v.i.* be about to, be going to; *του* ~*ει* (*or* ~*εται*) *να* he is fated to.

μέλλ|ων *a.* future, elect, to be; *ο ~ων* (*gram.*) future tense.

μελόδραμα *s.n.* opera. **~τικός** *a.* (*fig.*) melodramatic.

μελομακάρο(υ)νο *s.n.* honey-cake.

μελοποιώ *v.t.* set to music.

μέλος *s.n.* 1. limb. 2. member. 3. melody.

μελτέμι *s.n.* etesian wind.

μελωδί|α *s.f.* melody; tune. **~κός** *a.* melodious.

μελώνω *v.t.* spread with honey. (*v.i.*) become set (*of egg*).

μεμβράνη *s.f.* membrane; parchment; stencil.

μεμιάς *adv.* all at once, at one go.

μεμονωμένος *a.* isolated.

μέμφομαι *v.t.* blame, reproach.

μεμψιμοιρώ *v.i.* grumble, complain.

μεν *conj.* on the one hand; *οι ~ ... οι δε* some... others.

μενεξές *s.m.* violet.

μέν|ος *s.n.* ardour, fury; *~εα πνέω* be wrathful.

μέντα *s.f.* peppermint.

μέντιουμ *s.n.* medium (*spiritualistic*). [F. *médium]*

μένω *v.i.* stay, remain; live, reside; be left over; stop, be suspended; fail (*in exam*); be (*of states*); *έμεινα* I was dumbfounded; *έμεινε έγκυος* she is pregnant; *έμεινε ξερός* he was dumbfounded *or* he fell dead; *έμεινε στον τόπο* he died on the spot; *αυτό μόνο έμενε* that is the last straw; *~ από* run out of.

μέρα *s.f.* day.

μεράκ|ι *s.n.* yearning, passion; sorrow (*for thing unaccomplished*); *τα ~ια* sentimental mood; *καμωμένο με ~ι* made with loving care.

μερακλήδικος *a.* done with loving care; fit for a *μερακλής;* charged with passion (*song, etc.*).

μερακλής *s.m.* one who demands and relishes the best; one who loves to do a job well.

μερ|αρχία *s.f.* (*mil.*) division. **~αρχος** *s.m.* divisional commander.

μερδικό *s.n.* share, portion.

μερεμέτι *s.n.* (*household*) repair.

μερί *s.n.* thigh.

μεριά *s.f.* place; part; side.

μεριάζω *v.t. & i.* move aside; make way.

μερίδα *s.f. see* **μερίς.**

μερίδιο *s.n.* share, portion.

μερικεύω *v.t.* particularize.

μερικός *a. & pron.* partial (*not complete*); particular (*not general*); (*pl.*) some, certain.

μέριμν|α *s.f.* care, concern. **~ώ** *v.i.* have care *or* concern (*for*).

μερ|ίς *s.f.* portion; helping; ration; political party; *περάστε το στη ~ίδα μου* charge it to my account.

μέρισμα *s.n.* dividend.

μερισμός *s.m.* allocation, division.

μερμήγκι *s.n.* ant.

μεροδούλι *s.n.* a day's work; wages for this; *~ μεροφάι* from hand to mouth.

μεροκάματο *s.n. see* **μεροδούλι.**

μεροληπτώ *v.i.* take sides; show partiality.

μέρ|ος *s.n.* part; role; party (*to contract*); place; w.c.; *κατά ~ος* aside; *παίρνω το ~ος του* I take his part; *εκ ~ους του* on his behalf.

μέσα *adv.* inside, within; indoors; *τον έβαλαν ~* he was put in prison *or* he was made to lose his money; (*prep.*) *~ σε* in, into, inside; *~ μου* within me.

μεσάζω *v.i.* mediate.

μεσαίος *a.* middle, in between; average.

μεσαί|ων, ~ών *s.m.* middle ages. **~ωνικός** *a.* medieval.

μεσάνυχτα *s.n.pl.* midnight; *έχει ~* he is ignorant *or* in the dark.

μέση *s.f.* middle; waist; *αφήνω στη ~* leave unfinished; *βάζω στη ~* surround; *βγάζω στη ~* bring into the open, reveal; *βγάζω από τη ~* get rid of. *κόβω στη ~* cut in half.

μεσήλιξ *a.* middle-aged.

μεσημβρί|α *s.f.* south; noon. **~νός** *a.* south, southerly; facing south; midday; meridian.

μεσημέρι *s.n.* noon, midday; *μέρα ~* in broad daylight. **~άζω** *v.i.* go on till midday. **~ανός** *a.* midday; *~ανός ύπνος* siesta.

μεσιάζω *v.t.* be half-way through.

μεσιακός *a.* shared, held in common.

μεσιανός *a.* middle, in between.

μεσίστιος *a.* at half-mast.

μεσιτεία *s.f.* intervention, mediation, agency; brokerage; commission.

μεσιτεύω *v.i.* intervene, mediate; act as agent *or* go-between.

μεσίτης *s.m.* broker, (estate) agent; go-between.

μεσιτικ|ός *a.* broker's, agent's; *~ό γραφείο* house-agent's; **~ά** (*s.n.pl.*) brokerage, commission.

μεσόγει|ος *a.* inland; *η ~ος* Mediterranean. **~ακός** *a.* Mediterranean.

μεσοκόβω *v.t.* break the back of.

μεσόκοπος *a.* middle-aged.

μεσολαβ|ώ *v.i.* intervene, mediate; come between (*in space or time*).**~ητής** *s.m.* intermediary

μέσ|ον *1. s.n.* middle; way, means; *~α* (*pl.*) middle; means, resources; (*fam.*) pull, influence. 2. ~ον, ~ω prep. (with gen.) through, via.

μεσοπόλεμος s.m. inter-war years.

μέσος a. middle, mid; medium; average.

μεσοτοιχία s.f. (sharing of) party wall.

μεσουράνησις s.f. zenith, peak.

μεσοφόρι s.n. petticoat.

μεσόφωνος a. mezzo-soprano.

μεστ|ός a. full, well-filled; ripe. ~ώνω v.i. ripen, mature.

μετά 1. adv. afterwards. 2. prep. (with acc.) after; (with gen.) with.

μεταβαίνω v.i. go, proceed.

μεταβάλλω v.t. change, alter.

μετάβασις s.f. going.

μεταβατικός a. transitory; transitional; (gram.) transitive.

μεταβιβάζω v.t. convey; make over.

μεταβλητός a. variable.

μεταβολή s.f. change; about turn.

μεταβολισμός s.m. metabolism.

μετάγγισις s.f. decanting; transfusion.

μεταγενέστερ|ος a. later, subsequent; οι ~οι posterity.

μεταγράφω v.t. transcribe, (law) register.

μεταγωγικ|ός a. of transport. ~όν s.n. (mil.) army service corps; (naut.) transport ship.

μεταδίδω v.t. transmit; spread (infection, etc.).

μετάδο|ση s.f. transmission; spreading (of infection, etc.); αγία ~σις Holy Communion. ~τικός a. contagious; able to expound. ~τικόν s.n. gift of exposition.

μεταθανάτιος a. posthumous.

μετάθεσις s.f. transposition, transfer.

μεταθέτω v.t. transpose; transfer (employee).

μεταίχμιον s.n. (mil.) no-man's-land; (fig.) midway point.

μετακαλώ v.t. call in (expert, esp. from abroad).

μετακινώ v.t. shift, change position of.

μετακομίζω v.t. & i. convey (objects) from one place to another; move house.

μεταλαμβάνω v.t. & t. receive Holy Communion; administer this to; (with gen.) participate in.

μεταλαμπαδεύω v.t. (fig.) carry the torch of, diffuse.

μετάληψη s.f. Holy Communion.

μεταλλείον s.n. mine.

μετάλλευμα s.n. ore.

μεταλλικός a. of or like metal, metallic; mineral.

μετάλλινος a. made of metal.

μετάλλιον s.n. medal.

μέταλλον s.n. metal.

μεταλλουργία s.f. metallurgy.

μεταμελούμαι v.i. be repentant. ~εια s.f. repentance.

μεταμεσημβρινός a. afternoon.

μεταμορφ|ώνω v.t. transform. ~ωσις s.f. transformation; Transfiguration.

μεταμόσχευσις s.f. grafting; (med.) transplant.

μεταμφι|έζω v.t. disguise. ~έζομαι v.i. dress up, wear fancy dress. ~εση s.f. disguise.

μεταμφιεσμένος a. in fancy dress; disguised.

μεταναστ|εύω v.i. migrate, emigrate, immigrate. ~ευσις s.f. migration, emigration, immigration. ~ης s.m. migrant, emigrant, immigrant.

μετάνοια s.f. change of mind; repentance; penitence; prostration.

μετα|νοιώνω, ~νοώ v.i. change one's mind; repent.

μέτ|αξα s.f., ~άξι s.n. silk. ~αξένιος a. silky.

μεταξοσκώληκας s.m. silkworm.

μεταξύ adv. & prep. (with gen.) between; among; ~ μας between ourselves; εν τω ~ in the mean time.

μεταξωτ|ός a. silk. ~ό s.n. silk material.

μεταπείθω v.t. dissuade.

μετάπλαση s.f. renewal (of growth, tissue, etc.).

μεταποιώ v.t. refashion, change style or shape of.

μεταπολεμικός a. post-war.

μεταπολίτευσις s.f. change of political regime.

μετάπτωσις s.f. (sudden) change (of condition).

μεταπωλώ v.t. resell.

μεταρρυθμίζω v.t. reform, reorganize.

μεταστάς a. deceased.

μετάστασις s.f. change-over; (med.) metastasis.

μεταστροφή s.f. change, swing.

μετασχηματιστής s.m. transformer (electrical).

μετατοπ|ίζω v.t. shift. ~ιση s.f. shifting.

μετατρέπω v.t. turn, change, convert.

μετατροπή s.f. alteration, conversion.

μεταφέρω v.t. & i. carry, transport, convey; transfer; carry over (in accounts); move house.

μεταφορ|ά s.f. transport, conveyance; removal; amount carried forward; metaphor. ~ικός a. of transport; metaphorical; τα ~ικά transport charges.

μετα|φράζω v.t. translate. ~φρασις s.f. translation. ~φραστής s.m. translator.

μεταφυτεύω v.t. transplant.

μεταχειρ|ίζομαι v.t. use; treat, behave to

(*person*). **~ισμένος** *a.* used; worn (*not new*); second-hand. **~ιση** *s.f.* use, treatment.

μετεκπαίδευσις *s.f.* further education; post-graduate study.

μετεξεταστέος *a.* referred (*in examination*).

μετέπειτα *adv.* afterwards; οι ~ posterity.

μετέρχομαι *v.t.* practise (*profession*); use (*means*).

μετέχω *v.t.* (*with gen.*) take part in; partake of.

μετέωρο *s.n.* meteor.

μετέωρος *a.* (hanging) in the air; (*fig.*) in suspense; undecided.

μετζεσόλα *s.f.* sole (*of shoe*).

μετοικώ *v.i.* change one's residence; emigrate.

μετόπη *s.f.* metope.

μετόπισθεν *adv.* (*mil.*) τα ~ the rear.

μετοχή *s.f.* (*fin.*) share; (*gram.*) participle.

μετόχι *s.n.* farm belonging to monastery.

μετοχικ|ός *a.* (*fin.*) relating to joint stock; (*gram.*) participial.

μέτοχος *s.m.* participant; shareholder.

μέτρη|μα *s.n.* measuring, counting; ~μό δεν έχουν they are without number.

μετρημένος *a.* limited; temperate, measured.

μετρητής *s.m.* meter.

μετρητ|ός *a.* measurable; τα ~ά cash, ready money; τοις ~οις in cash.

μετριάζω *a.* moderate.

μετρικ|ός *a.* metrical; metric. **~ή** *s.f.* prosody, versification.

μετριοπάθ|εια *s.f.* moderation. **~ής** *a.* moderate.

μέτριος *a.* moderate; medium; mediocre, indifferent.

μετριό|φρων *a.* modest, unassuming. **~φροσύνη** *s.f.* modesty.

μέτρ|ο *s.n.* measure; rule; metre; bar (*music*); moderation; ~α (*pl.*) measurements; measures, steps; παίρνω ~ο (τα ~α) take measurements; λαμβάνω ~α take measures.

μετρ|ώ *v.t.* measure, count; go over (*ground*); pay over (*money*); (*fig.*) calculate, weigh (*words, etc.*); θα ~ηθώ μαζί του I shall match myself against him.

μετωπικός *a.* frontal, head-on.

μέτωπο *s.n.* forehead; (*mil.*) front; έχει ~ προς it faces on to.

μέχρι(ς) *prep.* (*with acc. or gen.*) as far as, up to; until, by; about (*in measuring*); ~ ενός to the last man; ~ τούδε up to now.

μη 1. *adv.* not; don't! οι ~ καπνιστές non-smokers. 2. *conj.* lest; (if) by any chance.

μηδαμινός *a.* of no account, worthless.

μηδέ, μηδείς *see* ουδ.

μηδέν *s.n.* nothing; zero; nought, cipher. **~ικό** *s.n.* nought, cipher.

μηδενίζω *v.t.* give no marks to.

μηδενιστής *s.m.* nihilist.

μήκ|ος *s.n.* length; longitude. **~ύνω** *v.t.* lengthen.

μηκώμαι *v.i.* bellow, moo.

μηλίγγι *s.n. see* μηνίγγι.

μήλ|ο *s.n.* apple; cheekbone. **~ίτης** *s.m.* cider.

μη(ν) *see* μη.

μην, μήνας *s.m.* month.

μηνιαίος *a.* monthly; month's.

μηνιάτικο *s.n.* month's wages *or* rent.

μηνίγγι *s.n.* temple (*of head*). **~ίτις** *s.f.* meningitis.

μήνις *s.f.* wrath. **~ίω** *v.i.* be enraged.

μήνυμα *s.n.* message.

μηνύ|ω *v.t.* (*law*) lodge complaint against, summons. **~υσις** *s.f.* summons. **~υτής** *s.m.* complainant.

μηνώ *v.t. & i.* send word (to *or* of); ~ το γιατρό send for the doctor.

μήπως *conj. & adv.* (I wonder) if; lest. (*also introduces question*).

μηρός *s.m.* thigh.

μηρυκ|άζω, ~ώμαι *v.t. & i.* chew (the cud). **~αστικός** *a.* ruminant.

μήτε *see* ούτε.

μήτ|ηρ, ~έρα *s.f.* mother.

μήτρα *s.f.* womb, uterus; matrix, mould.

μητρικ|ός *a.* 1. maternal; ~ή γλώσσα mother tongue 2. uterine; τα ~ά ailments of the womb.

μητρομανής *a.* nymphomaniac.

μητρόπολ|ις *s.f.* metropolis, capital; cathedral; see of metropolitan bishop. **~ιτικός** *a.* metropolitan.

μητροπολίτης *s.m.* (*eccl.*) metropolitan bishop.

μητρότης *s.f.* motherhood.

μητρ(υ)ιά *s.f.* stepmother.

μητρ(υ)ιός *s.m.* stepfather.

μητρώον *s.n.* register, roll, official list (*of names*).

μητρώος *a.* (inherited) from the mother.

μηχανεύομαι *v.t.* contrive, engineer, think up.

μηχανή *s.f.* machine, machinery; engine, works; (*fam.*) typewriter, camera etc.; μου στησε μια ~ he played me a trick.

μηχάνημα *s.n.* mechanical appliance.

μηχανικ|ός *a.* mechanical; an engineer *or* mechanic *or* architect. **~ή** *s.f.* engineering, mechanics. **~όν** *s.n.* (*mil.*) corps of engineers.

μηχανισμός *s.m.* mechanism.

μηχανοδηγός *s.m.* engine-driver.

μηχανοκίνητος *a.* worked by machinery; (*mil.*) motorized.

μηχανοποιώ *v.t.* mechanize.

μηχανορραφώ *v.i.* hatch plots, intrigue.

μία, μια *num. & article f.* one; a, an; ~ *και or* ~ *που* as, since, seeing that; ~ *και καλή or* ~ *για πάντα* once for all; ~ *εδώ* ~ *εκεί* now here now there; *διά μιας or με μιας* (all) at once, suddenly; *έφαγε* ~ he suffered a blow *or* shock, *etc.*

μιαίνω *v.t.* pollute, defile.

μιαρός *a.* foul, vile.

μίασμα *s.n.* miasma (*fig.*) noxious influence.

μιγάς *s.m.* half-caste; mongrel; hybrid.

μίγμα *s.n.* mixture, blend, alloy, amalgam.

μίζα *s.f.* 1. stake (*money*); rake-off. 2. self-starter.

μιζέρια *s.f.* state of poverty; miserliness.

μικραίνω *v.t. & i.* diminish, make *or* become smaller *or* shorter.

μικράτα *s.n.pl.* childhood.

μίκρεμα *s.n.* shortening.

μικρο- *denotes* small.

μικροαστός *s.m.* petit bourgeois.

μικρόβιο *s.n.* microbe.

μικρογραφία *s.f.* miniature.

μικροδουλειά *s.f.* trifle, small dealing.

μικροκαμωμένος *a.* slightly-built.

μικρόκοσμος *s.m.* 1. microcosm. 2. children.

μικρολόγος *a.* cavilling.

μικρομέγαλος *a.* (*child*) with grown up manners.

μικροπρά(γ)ματα *s.n.pl.* trifles.

μικροπρεπής *a.* mean, petty.

μικρ|ός *a.* small, little; short (*duration*); young; younger; a young servant *or* employee; mean, petty; ~*ού δειν* nearly.

μικροσκόπ|ιο *s.n.* microscope. ~**κός** *a.* microscopic, minute.

μικρόφωνο *s.n.* microphone.

μικρόψυχος *a.* pusillanimous.

μικρύνω *v.t. & i. see* **μικραίνω**.

μικτ|ός *a.* mixed; ~*όν βάρος* gross weight.

μιλιά *s.f.* (*manner of*) speech; *δεν έβγαλε* ~ he did not utter a word; ~ ! not a word!

μίλι(ον) *s.n.* mile.

μιλ|ώ *v.t. & i.* speak; *δε* ~*ιούνται* they are not on speaking terms; ~*ημένος* one whose favour has been solicited; ~*ώ με την τύχη μου* to be smiled on by fortune.

μίμηση *s.f.* imitation.

μιμητ|ής *s.m.* imitator. ~**ικός** *a.* imitative.

μιμόζα *s.f.* mimosa.

μίμος *s.m.* mimic.

μιμούμαι *v.t.* imitate, copy.

μιναρές *s.m.* minaret.

μινιατούρα *s.f.* miniature.

μίνιο *s.n.* red lead.

μινωικός *a.* Minoan.

μίξη *s.f.* mixing.

μις *s.f.* (*fam.*) beauty queen; [E. *Miss*]. *πάω για* ~ compete for such title.

μισαλλόδοξος *a.* bigoted (*esp. in religion*).

μισάνθρωπος *a.* misanthropic.

μισερός *a.* deficient (*physically or mentally*); skimpy.

μισεύω *v.i.* leave one's country.

μισητός *a.* hated, hateful.

μίσθαρνος *a.* mercenary, venal.

μίσθιος *a.* hired, rented.

μισθοδοτώ *v.t.* pay wages to.

μισθολόγιο *s.n.* payroll; rate of pay.

μισθ|ός *s.m.* wages, salary. ~**οφόρος** *s.m.* mercenary.

μισθ|ώ, ~ώνω *v.t.* hire, rent (*from owner*).

μισθωτής *s.m.* hirer, tenant.

μισθωτός *a.* employed; a wage-earner.

μισιάζω *v.t.* halve; be half-way through.

μισ|ό *s.n.* half; *τα* ~*ά* (*pl.*) half, the middle; *στα* ~*ά* half-way, in the middle.

μισο- *denotes* half.

μισογύνης *s.m.* misogynist.

μίσος *s.n.* hate, hatred.

μισός *a.* half; uncompleted; deficient in health; *βλάκας και* ~ a perfect fool. *See* **μισό**.

μισοτιμής *adv.* at half price.

μισότριβη *s.f.* middle-aged woman; (*fig.*) one who has knocked around.

μισοφόρι *see* **μεσοφόρι**.

μισσεύω *see* **μισεύω**.

μισώ *a.* half.

μίσχος *s.m.* stalk; (*fig.*) tall, slim person.

μισώ *v.t.* hate.

μίτος *s.m.* thread, yarn; clew (*of Ariadne*).

μίτρα *s.f.* mitre; cowl (*of chimney*).

μνεία *s.f.* mention; reminding.

μνήμα *s.n.* grave, tomb.

μνημείο *s.n.* monument.

μνημειώδης *a.* monumental; memorable.

μνήμη *s.f.* memory; remembrance.

μνημονεύω *v.t.* mention; commemorate; learn by heart.

μνημονικό *s.n.* memory.

μνημόσυνο *s.n.* requiem; commemorative meeting.

μνήμων *a.* mindful.

μνησίκακ|ος *a.* rancorous. ~**ώ** *v.i.* be resentful, bear a grudge.

μνηστ|εία *s.f.* betrothal. ~**εύω** *v.t.* betroth. ~**εύομαι** *v.i.* become engaged.

μνηστή *s.f.* fiancée.

μνηστήρας *s.m.* fiancé; (*fig.*) suitor, pretender.

μόδ|α *s.f.* fashion; της ~ας in fashion. ~**ίστρα** *s.f.* dressmaker.

μοιάζ|ω *v.t. & i.* resemble (*with acc. or gen.*); look alike; δεν του ~ει να it ill becomes him to.

μοιασ|ίδι, ~ιμο *s.n.* resemblance.

μοίρα *s.f.* 1. fate, lot. 2. squadron. 3. (*geom.*) degree.

μοιράζ|ω *v.t.* share out, distribute; deliver (*of roundsman*); deal (*cards*); ~ω τη διαφορά split the difference. ~ομαι *v.t.* share (*as recipient*).

μοιραί|ος *a.* preordained; inevitable; fatal; ~α γυναίκα femme fatale. ~**ως** *adv.* inevitably.

μοίραρχος *s.m.* captain of gendarmerie; (*naut.*) commodore.

μοιρασιά *s.f.* distribution, sharing.

μοιρολατρία *s.f.* fatalism.

μοιρολόγ|ι *s.n.* dirge. ~**ώ** *v.i. & t.* lament (*the dead*).

μοιχεία *s.f.* adultery.

μοκέτα *s.f.* fitted carpet.

μόκο *adv.* mum's the word.

μολαταύτα *adv.* nevertheless.

μόλις 1. *adv.* just now, only just; hardly; scarcely. 2. *conj.* as soon as.

μολονότι *adv.* although.

μολοσσός *s.m.* large watch-dog.

μόλυβδος *s.m.* lead (*metal*).

μολύβ|ι *s.n.* lead; pencil. ~**ιά** *s.f.* pencil-stroke. ~**ένιος** *a.* lead, leaden.

μόλυνση *s.f.* infection, contamination.

μολύνω *v.t.* infect, contaminate; defile.

μόλυσμα *s.n.* infectious germ *or* ailment.

μομφή *s.f.* blame.

μονάδα *s.f.* unit.

μοναδικός *a.* unique, sole.

μονάζω *v.i.* live in solitude; be a monk.

μονάκριβος *a.* cherished and only.

μοναξιά *s.f.* solitude; loneliness; lonely place.

μονάρχ|ης *s.m.* monarch. ~**ία** *s.f.* monarchy. ~**ικός** *a.* monarchical; a monarchist.

μονάς *s.f.* unit, monad.

μοναστήρι *s.n.* monastery.

μοναστικός *a.* monastic.

μον|άχα, ~αχά *adv.* only.

μοναχικ|ός *a.* 1. monastic. 2. standing alone, isolated. 3. to oneself (*unshared*); δωμάτιο ~ό a room of one's own.

μοναχοπαίδι *s.n.* only child.

μον|άχος, ~αχός *a.* only; alone; ~αχός του by himself *or* on his own; χρυσάφι ~αχό real gold.

μοναχός *s.m.* monk.

μονέδα *s.f.* money; (*fig.*) κόβω ~ coin money.

μονή *s.f.* monastery.

μονήρης *a.* solitary, isolated.

μόνιμ|ος *a.* permanent. ~**οποιώ** *v.t.* make permanent.

μονο- *denotes* single.

μονογενής *a.* only (*child*); one and only.

μονόγραμμα *s.n.* monogram.

μονογραφ|ή *s.f.* initials, paraph. ~**ώ** *v.t.* initial.

μονογραφία *s.f.* monograph.

μονόδρομος *s.m.* one-way street.

μονόζυγο *s.n.* horizontal bar.

μονοιάζω *v.i. & t.* agree, get on well together; make up a quarrel; reconcile.

μονοκατοικία *s.f.* private house.

μονόκλινο *s.n.* single-bed room.

μονοκόμματος *a.* in one piece; rigid; forthright.

μονομανία *s.f.* obsession.

μονομαχία *s.f.* duel.

μονομερής *a.* one-sided; unilateral.

μονομιάς *adv.* all at once; immediately.

μόνο(ν) *adv.* only; ~ που except that; ~ να provided that; ~ να το σκεφθώ at the mere thought of it.

μονόξυλο *s.n.* dug-out canoe.

μονοπάτι *s.n.* footpath.

μονόπατος *a.* one-storied.

μονόπλευρος *a.* one-sided.

μονοπώλ|ιο *s.n.* monopoly. ~**ώ** *v.t.* enjoy monopoly of; monopolize.

μονορρούφι *adv.* straight off, at one go.

μόν|ος *a.* only; alone; ~ος μου by myself, of my own accord; κατά ~ας (*live*) on one's own, (*talk*) to oneself.

μονός *a.* simple (*not compound*); single (*not double*); odd (*not even*).

μονότονος *a.* monotonous.

μονόφθαλμος *a.* one-eyed.

μονόχνωτος *a.* unsociable.

μονόχρωμος *a.* monochrome.

μοντέλο *s.n.* model.

μοντέρνος *a.* modern.

μονύελος *s.m.* monocle.

μον|ώνω *v.t.* isolate; insulate. ~**ωση** *s.f.* insulation. ~**ωτικός** *a.* insulating.

μοριακός *a.* molecular.

μόριο *s.n.* particle; molecule.

μόρτης *s.m.* hooligan, tough.

μορφ|άζω *v.i.* make faces, grimace. ~**ασμός** *s.m.* grimace.

μορφή *s.f.* form; face; appearance; aspect; phase.

μορφολογία *s.f.* morphology.

μορφονιός *s.m.* dandy, coxcomb.

μορφώνω *v.t.* form; train, educate.

μόρφω|ση *s.f.* education, culture. **~τικός** *a.* educative, educational, cultural.

μόστρα *s.f.* (*fam.*) shop-window.

μοσχάρι *s.n.* calf; veal.

μοσχάτο *s.n.* muscatel grape *or* wine.

μοσχοβολώ *v.i.* smell sweetly.

μοσχοκάρυδο *s.n.* nutmeg.

μοσχοκάρφι *s.n.* clove.

μοσχολίβανο *s.n.* frankincense.

μόσχος *s.m.* 1. calf; veal. 2. musk.

μοτοσυκλέτα *s.f.* motor-bicycle.

μού, μου *pron.* me; my.

μουγγός *a.* dumb.

μουγγρίζω *v.i.* bellow, roar, moo, moan.

μουδιάζω *v.t. & i.* numb; be numb.

μουλάρι *s.n.* mule. **~σιος** *a.* mulish.

μούλος *s.m.* bastard.

μουλώ(χ)νω *v.i.* conceal one's thoughts, keep silent; lie low.

μούμια *s.f.* mummy.

μουνουχίζω *v.t.* castrate.

μουντάρω *v.t. & i.* (*fam.*) rush, pounce on.

μούντζ|α *s.f.* opprobrious gesture with open palm. **~ώνω** *v.t.* make μούντζα at; renounce in disgust.

μουντζούρ|α *s.f.* smudge; moral stain. **~ώνω** *v.t.* smudge, blacken.

μουντός *a.* dull, dark, drab.

μουράγιο *s.n.* jetty, breakwater.

μούργα *s.f.* sediment, lees (*of wine, oil*).

μούργος *s.m.* fierce watch-dog; (*fig.*) fierce *or* ugly fellow.

μούρη *s.f.* (*fam.*) face, mug.

μούρλια *s.f.* madness; (*fam.*) είναι ~ it's a dream! (*eulogistic*).

μουρλός *a.* mad, crazy; over-excited.

μουρμουρίζω *v.i. & t.* murmur; grumble. **~α** *s.f.*, **~ητό** *s.n.* murmuring; grumbling.

μουρντάρης *a.* filthy; (*fig.*) a dirty dog.

μούρο *s.n.* mulberry.

μουρουνέλαιο *s.n.* codliver-oil.

μούσα *s.f.* muse.

μουσακάς *s.m.* dish of minced meat, vegetable, *etc.*

μουσαμάς *s.m.* oilcloth, linoleum; raincoat; (*artist's*) canvas.

μουσαφίρης *s.m.* visitor, guest.

μουσείο *s.n.* museum.

μούσι *s.n.* small pointed beard.

μουσικ|ή *s.f.* music; band. **~ός** *a.* musical; a musician.

μούσκεμα *s.n.* soaking; γίνομαι ~ get soaking wet; (*fam.*) τα κάνω ~ make a mess of it.

μουσκεύω *v.t. & i.* soak; get soaked; τα ~ make a mess of it.

μουσκίδι *s.n.* γίνομαι ~ get soaking wet.

μούσκλι *s.n.* moss.

μούσμουλο *s.n.* loquat.

μουσουλμάνος *s.m.* Moslem.

μουσούδι *s.n.* snout.

μουσουργός *s.m.f.* composer.

μουστάκι *s.n.* moustache.

μουσταλευριά *s.f.* jelly made with must.

μουστάρδα *s.f.* mustard.

μουστερής *s.m.* regular customer, buyer.

μούστος *s.m.* must, new wine.

μούτρ|ο *s.n.*, **~α** (*pl.*) face; κάνω ~α adopt an injured air; έχω ~α να have the cheek to; πέφτω με τα ~α apply oneself avidly; ρίχνω τα ~α pipe down; είναι ~ο he's a crook.

μουτρωμένος *a.* sulky.

μούτσος *s.m.* ship's boy.

μουτσούνα *s.f.* snout.

μούχλ|α *s.f.* mould; (*mental*) stagnation. **~ιάζω** *v.i. & t.* moulder, go mouldy; make mouldy.

μοχθηρός *a.* malicious.

μόχθος *s.m.* toil.

μοχλός *s.m.* lever, bar; (*fig.*) promoter, ringleader.

μπ- *see also* **π-**.

μπα *int.* expression of surprise, rejection, negation, or assent.

μπαγαπόντης *s.m.* rogue, trickster.

μπαγάσας *s.m.* rascal.

μπαγιατ|εύω *v.i.* get stale. **~ικος** *a.* stale. **~ίλα** *s.f.* staleness.

μπάγκα *s.f.* (*financial*) bank.

μπαγκάζια *s.n.pl.* baggage.

μπαγκέτα *s.f.* conductor's baton; drumstick; clock (*of stocking*).

μπάγκος *s.m.* bench; counter.

μπαγλαρώνω *v.t.* (*fam.*) nab, collar.

μπάζα *s.f.* trick (*at cards*); profit from sharp practice; δεν πιάνει ~ μπροστά σε it's not a patch on.

μπάζ(ι)α *s.n.pl.* rubble.

μπάζω *v.t. & i.* admit, usher in; (*v.i.*) shrink (*of clothes*).

μπαϊλντίζω *v.i.* swoon.

μπαίν|ω *v.i.* enter, go in; shrink (*of cloth*); ~ω μέσα incur loss (*in business, cards*); ~ω σε μια σειρά settle down; ~ω σε ιδέα become suspicious; μου ~ει η ιδέα it occurs to me; μπήκα στο νόημα I grasped the idea.

μπαϊράκι *s.n.* banner; (*fig.*) σηκώνω ~ revolt.

μπακάλ|ης *s.m.* grocer. **~ικο** *s.n.* grocer's shop.

μπακαλιάρος *s.m.* cod (*esp. dried*).

μπακίρι *s.n.* copper.

μπακλαβάς *s.m.* cake of almonds and honey.

μπαλαντέρ *s.m.* joker (*cards*). [F. *bala-deur*].

μπαλάντζα *s.f.* balance (*for weighing*).

μπαλάσκα *s.f.* cartridge-pouch.

μπαλκόν|ι *s.n.* balcony ~όπορτα *s.f.* french window.

μπάλλα *s.f.* ball; bullet; bale.

μπαλλέτο *s.n.* ballet.

μπαλλόνι *s.n.* balloon.

μπαλ(ν)τάς *s.m.* axe.

μπάλωμα *s.n.* patching, mending; patch.

μπαλών|ω *v.t* patch, mend; (*fig.*) τα ~ω make excuses. ~ομαι *v.i.* find what one wants.

μπάμιες *s.f.pl.* okra, lady's fingers (*vegetable*).

μπαμπάκ|ι *s.n.* cotton; cotton wool. ~ερός *a.* cotton.

μπαμπάς *s.m.* papa.

μπαμπέσης *a.* treacherous.

μπαμπόγρια *s.f.* crone.

μπαμπούλας *s.m.* bogyman.

μπάμπουρας *s.m.* hornet.

μπανάνα *s.f.* banana.

μπανέλα *s.f.* whalebone (stiffener).

μπανιέρα *s.f.* bath (*tub*); bathing-hut.

μπανιερό *s.n.* bathing-dress.

μπαν|ίζω *v.t.* eye, ogle. ~ιστήρι *s.n.* (*fam.*) watching (*esp. through spy-hole*).

μπάνιο *s.n.* bath, bathe; bath (*tub*); bathroom.

μπάντα *s.f.* 1. side; corner (*quiet spot*); βάζω στην ~ put aside (*money*); κάνε στην ~ stand aside. 2. band (*music, robbers*).

μπαντιέρα *s.f.* banner.

μπαξές *s.m.* garden.

μπαούλο *s.n.* trunk, chest.

μπάρα *s.f.* bar.

μπαρκάρω *v.t. & i.* embark.

μπάρμπας *s.m.* (*fam.*) uncle; old man.

μπαρμπέρης *s.m.* barber.

μπαρμπούνι *s.n.* red mullet.

μπαρούτι *s.n.* gunpowder; (*fam.*) έγινε ~ he got furiously angry; βρωμάει ~ there is danger in the air.

μπας και *adv. see* μήπως.

μπάσιμο *s.n.* shrinkage; taking in (*of dress*); entry.

μπασμένος *a.* knowledgeable; in the know; stunted.

μπάσταρδος *s.m.* bastard.

μπαστούνι *s.n.* stick; club (*cards*); τα βρήκε ~α he met with difficulties.

μπατάλικος *a.* large, unwieldy (*person*).

μπατανία *s.f.* woollen blanket.

μπαταρία *s.f.* battery (*electric*).

μπαταριά *s.f.* salvo, volley.

μπατάρω *v.t. & i.* capsize; be quits.

μπατζάκι *s.n.* trouser-leg; (*fam.*) ~α (*pl.*) legs, pins.

μπατζανάκης *s.m.* brother-in-law (*relationship between husbands of sisters*).

μπάτης *s.m.* breeze from the sea.

μπατίρης *a.* broke.

μπάτσ|α *s.f.*, ~ος *s.m.* slap; blow. ~ίζω *v.t.* slap.

μπαφιάζω *v.i.* get sick (*of*).

μπαχαρικό *s.n.* spice.

μπεζ *a.* beige. [F].

μπέ|ης *s.m.* bey; (*fig.*) περνώ ~ικα live in luxury.

μπεκάτσα *s.f.* woodcock.

μπεκιάρης *s.m.* bachelor.

μπεκρής *s.m.* drunkard.

μπεκρουλιάζω *v.i.* tipple.

μπελ|άς *s.m.* nuisance, trouble; βρήκα τον ~ά μου I got involved (*in trouble*). ~αλίδικος *a.* troublesome.

μπελ(ν)τές *s.m.* (*preserved*) tomato purée; jelly (*of fruit*).

μπέμπης *s.m.* (*fam.*) baby.

μπενζίνα *s.f.* petrol; motorboat.

μπέρδεμα *s.n.* confusion, muddle.

μπερδεύ|ω *v.t.* entangle; embroil; involve; confuse; muddle, get wrong; τα ~ω be confused (*in one's words*). ~ομαι *v.i.* get entangled, involved, *or* confused.

μπερμπάντης *s.m.* roisterer, rogue; womanizer.

μπερντές *s.m.* curtain.

μπέσα *s.f.* trustworthiness.

μπετόν *s.n.* concrete. [F. *béton*]

μπετούγια *s.f.* latch.

μπή|γω, ~ζω *v.t.* drive *or* push in, pin on; ~ζω τα γέλια burst out laughing.

μπηχτή *s.f.* (*fig.*) shrewd thrust.

μπιζάρω *v.t.* encore.

μπιζέλι *s.n.* pea.

μπιλιάρδο *s.n.* billiards.

μπιμπελό *s.n.* bibelot [F].

μπιντές *s.m.* bidet.

μπίρα *s.f.* beer.

μπιρμπάντης *s.m.* rascal.

μπισκότο *s.n.* biscuit.

μπίστικος *a.* trusted.

μπίτ(ι) *adv.* (*fam.*) completely; not at all.

μπιφτέκι *s.n.* steak.

μπιχλιμπίδι *s.n.* trinket.

μπλάστρι *s.n.* plaster (*curative*).

μπλε, μπλου *a.* blue. [F. *bleu*]

μπλέκω *v.t. & i.* entangle; involve; get entangled *or* involved.

μπλέξιμο *s.n.* entanglement; amour.

μπλιγούρι *s.n.* groats.

μπλοκάρω *v.t. & i.* block, jam.

μπλούζα *s.f.* blouse.

μπλόφα *s.f.* bluff.

μπογιά *s.f.* paint, dye, colouring-matter; boot-polish.

μπογιαντίζω *v.t.* paint, colour, dye.

μπόγιας *s.m.* executioner; dog-catcher.

μπογιατζής *s.m.* painter, decorator.

μπόγος *s.m.* large, soft bundle (*esp. clothes*).

μποέμικος *a.* bohemian.

μπόι *s.n.* stature; *ρίχνω* ~ grow taller.

μπολ *s.n.* (*small*) bowl.

μπόλι *s.n.* graft; vaccine. ~άζω *v.t.* graft; inoculate.

μπόλικος *a.* abundant, plenty of; loose-fitting.

μπόμπα *s.f.* bomb; cylinder (*of gas*).

μπομπότα *s.f.* maize flour *or* bread.

μποναμάς *s.m.* new year's gift.

μποξάς *s.m.* bundle; shawl.

μπορ *s.n.* brim (*of hat*). [F. *bord*]

μπόρα *s.f.* squall of rain.

μπορετός *a.* possible.

μπορντούρα *s.f.* edge, edging, border.

μπορ|ώ *v.i.* be able, can; ~*ει* it may be, perhaps; *δεν*~*ώ* I cannot *or* I am not well.

μπόσικος *a.* loose, slack; (*fig.*) shallow, unreliable; yielding.

μποστάνι *s.n.* melon-bed.

μπότζι *s.n.* rolling (*of ship*).

μποτίλι|α *s.f.* bottle. ~άρω *v.t.* bottle; (*fig.*) bottle up, block.

μπότα *s.f.* boot.

μπουγάδα *s.f.* (*doing of*) washing.

μπουγάζι *s.n.* narrow passage.

μπουγάτσα *s.f.* confection of flaky pastry.

μπούγιο *s.n.* *κάνω* ~ (*fig.*) make an (*unmerited*) impression.

μπούζι *s.n.* 1. (cold as) ice; (*fig.*) feeble (*of jokes*). 2. mesembrianthemum.

μπουζί *s.n.* sparking-plug. [F. *bougie*]

μπουζούκι *s.n.* sort of mandoline.

μπούκα *s.f.* mouth (*of gun, harbour*).

μπουκάλα *s.f.* large bottle; (*fam.*) *έμεινε* ~ he was left high and dry.

μπουκάλι *s.n.* bottle.

μπουκάρω *v.i.* swarm *or* pour in; enter (*channel, etc.*).

μπουκέτο *s.n.* bouquet.

μπουκιά *s.f.* mouthful; (*fam.*) ~ *και συχώριο* bit of all-right.

μπουκίτσα *s.f.* morsel.

μπούκλα *s.f.* curl.

μπουκοτάρω *v.t.* boycott.

μπούκωμα *s.n.* stuffing up, blockage; bribery.

μπουκώνω *v.t.* stuff with food; (*fig.*) grease the palm of.

μπουλούκι *s.n.* mass, crowd.

μπουλούκος *s.m.* plump dog; (*fam.*) plump person.

μπουμπούκι *s.n.* bud.

μπουμπουνητό *s.n.* rumble of thunder.

μπουνάτσα *s.f.* calm, fine weather at sea.

μπούνια *s.n.pl.* (*naut*) scuppers; (*fig.*) *ως τα* ~ over head and ears.

μπουνιά *s.f.* (blow of) fist.

μπουνταλάς *s.m.* clumsy blockhead.

μπουντρούμι *s.n.* dungeon.

μπούρδα *s.f.* (*fam.*) hot air, nonsense.

μπουρέκι *s.n.* sort of patty *or* cream cake.

μπουρί *s.n.* chimney (*of stove*).

μπουρ|ί, ~ίνι *s.n.* squall of wind; (*fig.*) fit of temper.

μπουρλότο *s.n.* fire-ship; explosive charge. *γίνομαι* ~ flare up.

μπουρμπουλήθρα *s.f.* bubble.

μπουρνέλα *s.f.* mirabelle plum.

μπούρτζι *s.n.* fortified islet.

μπούσουλ|ας *s.m.* compass; (*fig.*) *χάνω τον* ~ *α* be all at sea.

μπουσουλώ *v.i.* go on all fours.

μπούστος *s.m.* bust; bodice.

μπούτι *s.n.* thigh.

μπουφές *s.m.* sideboard; buffet.

μπούφος *s.m.* horned owl; dolt (*fig.*).

μπουχτίζω *v.i. & t.* become satiated; (*fig.*) get fed up; make fed up.

μπόχα *s.f.* stink.

μπράβο *int.* bravo! good!

μπράβος *s.m.* bravo.

μπράτσο *s.n.* arm.

μπρε *int. see* βρε.

μπριάμι *s.n.* sort of ratatouille.

μπριγιαντίνη *s.f.* brilliantine.

μπριζόλα *s.f.* chop, cutlet.

μπρικ 1. *s.n.* (*preserved*) salmon-roe. 2. *a.* tile-red.

μπρίκι *s.n.* 1. pot for boiling coffee. 2. (*naut.*) brig.

μπρος, μπροστά *adv.* forward(s), ahead, in front (*see* εμπρός).

μπροστινός *a.* (in) front.

μπρούμυτα *adv.* face downwards.

μπρούντζ|ος *s.m.* bronze. ~ινος *a.* bronze.

μπρούσκ|ος *a.* ~ο *κρασί* dry wine.

μπύρα *s.f.* beer.

μυ *s.n.* the letter Μ.

μυαλ|ό *s.n.* marrow (*of bone*); brain(*s*); mind, sense; *πήραν τα* ~*ά του αέρα* success has gone to his head; *του πήρε τα* ~*ά* she turned his head.

μυαλωμένος *a.* wise, prudent.

μύγα *s.f.* fly.

μυγιάγγιχτος *a.* touchy.

μυγιάζομαι *v.i.* be touchy.

μύδι *s.n.* mussel.

μυδραλιοβόλον *s.n.* (*mil.*) machine-gun.
μύδρος *s.m.* (*mil.*) shell.
μυελ|ός *s.m.* marrow (*of bone*); brain. *μέχρι ~ού οστέων* to the marrow.
μυζήθρα *s.f.* sort of soft white cheese.
μυζώ *v.t.* suck.
μύη|ση *s.f.* initiation. *~μένος* initiated; knowledgeable.
μύθευμα *s.n.* fabrication.
μυθικός *a.* mythical; legendary, fabulous.
μυθιστόρημα *s.n.* fiction; novel.
μυθιστοριογράφος *s.m.* novelist.
μυθολογία *s.f.* mythology.
μύθος *s.m.* fable; myth; fairy story.
μυθώδης *a.* untrue; fabulous.
μυία, μυίγα *s.f. see* μύγα.
μυικός *a.* muscular.
μυκηθμός *s.m.* bellowing, mooing.
μύκης *s.m.* fungus.
μυκτηρίζω *v.t.* mock.
μυκώμαι *v.i.* bellow, roar, moo.
μύλ|ος *s.m.* mill. *~όπετρα s.f.* millstone.
μυλωθρός *s.m.* miller.
μυλωνάς *s.m.* miller.
μύξ|α *s.f.* (*fam.*) snot. *~ιάρικο s.n.* brat.
μυρ|ίζω *v.t. & i.* smell (of); *δε ~ισα τα δάχτυλά μου* I could not foresee. *~ίζομαι v.t.* smell; (*fig.*) get wind of, suspect.
μύριοι *a.* ten thousand; (*fig.*) countless.
μυρμήγκι *s.n.* ant. *~άζω v.i.* have pins and needles.
μύρμηξ *s.m.* ant.
μυροβόλος *a.* fragrant.
μύρον *s.n.* essential oil; *άγιον ~* chrism.
μυροπωλείο *s.n.* perfumery (*shop*).
μυρτιά *s.f.* myrtle.
μύρτο *s.n.* myrtle-bough.
μυρωδάτος *a.* having a sweet smell.
μυρωδιά *s.f.* smell; perfume; flavour; tit-bit from thing being cooked; *παίρνω ~* get wind of, suspect.
μυρωδικό *s.n.* (*liquid*) perfume; spice.
μυρώνω *v.t.* anoint (*esp. ritually*).
μυς *s.m.* 1. muscle. 2. mouse.
μυσαρός *a.* heinous.
μυσταγωγία *s.f.* initiation.
μύσταξ *s.m.* moustache.
μυστήρι|ον *s.n.* mystery; act of charity; Sacrament. *~ος a.* mysterious. *~ώδης a.* mysterious.
μύστης *s.m.* initiate; (*fig.*) one deeply versed.
μυστικιστής *s.m.* mystic.
μυστικό *s.n.* secret.
μυστικός *a.* secret; secretive; discreet; mystic(al); a secret policeman; *~ δείπνος* Last Supper.
μυστικοσύμβουλος *s.m.* privy councillor;

(*fig.*) confidant.
μυστικότης *s.f.* secrecy.
μυστρί *s.n.* trowel.
μυταράς *s.m.* big-nosed man.
μυτερός *a.* pointed, sharp.
μύτη *s.f.* nose; tip, point; beak.
μύχι|ος *a.* inmost; *τα ~α* inmost part, depths.
μυχός *s.m.* inmost part (*esp. of bay, etc.*).
μυώ *v.t.* initiate.
μυ|ών *s.m.* muscle. *~ώδης a.* muscular.
μύωψ *a.* myopic.
μωαμεθανός *s.m.* a Mohammedan.
μωβ *a.* mauve. [F. *mauve*]
μώλος *s.n.* mole, breakwater.
μώλ|ωψ *s.m.* bruise. *~ωπίζω v.t.* bruise.
μωραίνω *v.t.* make dull *or* stupid.
μωρ|έ, *~ή int.* unceremonious mode of address.
μωρία *s.f.* foolishness.
μωρό *s.n.* baby; ingenuous person. *~λογώ v.i.* talk nonsense.
μωρός *a.* foolish, silly.
μωσαϊκό *s.n.* mosaic.

N

να 1. *particle of subordination* to, that, in order to, let, may, if, *etc.* 2. *prep.* there! there is, there are (*with nom. or acc.*); *να τος or τον* there he is; *να η Μαρία* here's Mary; *να 'μαστε* here we are. 3. *int. of imprecation.*
νάζι *s.n.*, *~α* (*pl.*) coquettish air; *κάνω ~α* affect reluctance. *~άρικος a.* affected.
ναι *adv.* yes; *~ και καλά* with obstinate insistence.
ναϊάς *s.f.* naiad.
νάιλον *s.n.* nylon.
νάμα *s.n.* spring (*water*).
νάνι *s.n.* (*fam.*) sleep (*childish term*).
νάνος *s.m.* dwarf.
νανουρ|ίζω *v.t.* rock *or* lull to sleep. *~ισμα s.n.* rocking, lulling; lullaby.
ναός *s.m.* temple, church.
ναργιλές *s.m.* hookah.
νάρθηξ *s.m.* 1. (*eccl.*) narthex. 2. (*med.*) splint.
ναρκαλιευτικόν *s.n.* (*naut.*) minesweeper.
νάρκη *s.f.* numbness, torpor; electric ray (*fish*); mine (*weapon*).
νάρκισσος *s.m.* narcissus.
ναρκοθέτις *s.f.* (*naut.*) mine-layer.
ναρκ|ώνω *v.t.* make numb, torpid *or* dull. *~ωσις s.f.* torpor; (*med.*) narcosis.
ναρκωτικ|ός *a.* narcotic; *~ά* (*s.n.pl.*) drugs.

νάτριον *s.n.* sodium.
ναυαγιαίρεσις *s.f.* (*naut.*) salvage.
ναυάγιο *s.n.* shipwreck; (*fig.*) wreck.
ναυαγ|ός *s.m.* shipwrecked sailor, castaway.
~ώ *v.i.* be shipwrecked, founder.
ναύαρχ|ος *s.m.* admiral. **~είον** *s.n.* admiralty. **~ίς** *s.f.* flag-ship.
ναύκληρος *s.m.* boatswain.
ναύλα *s.n.pl.* fare (*for any journey*).
νάυλον *s.n.* nylon. [E. *nylon*]
ναύλος *s.m.* freight, passage-money (*esp. by ship*).
ναυλώνω *v.t.* freight, charter (*vessel*).
ναυμαχία *s.f.* sea-battle.
ναυπηγ|είον *s.n.* shipyard. **~ός** *s.m.* shipbuilder.
ναυς *s.f.* ship.
ναυσιπλοΐα *s.f.* shipping; navigation.
ναύσταθμος *s.m.* naval dockyard.
ναυτασφαλιστής *s.m.* marine underwriter.
ναύτης *s.m.* sailor, seaman.
ναυτία(ση) *s.f.* seasickness; nausea.
ναυτική *s.f.* seamanship.
ναυτικόν *s.n.* navy.
ναυτικ|ός *a.* maritime, seafaring; nautical, naval, of shipping; a seafaring man (*sailor, officer, fisherman*).
ναυτιλία *s.f.* shipping; navigation; merchant marine. **~κός** *a.* marine.
ναυτιώ *v.i.* be seasick; feel nausea.
ναυτοδικείον *s.n.* (*law*) court of Admiralty.
ναυτολογώ *v.t.* enlist (*crew*).
ναυτόπαις *s.m.* ship-boy.
ναφθαλίνη *s.f.* mothballs.
νέα 1. *s.f.* girl. 2. *s.n.pl.* news.
νεάζω *v.i.* appear young; remain young in outlook.
νεαν|ίας *s.m.* young man. **~ικός** *a.* youthful.
νεάνις *s.f.* girl.
νεαρός *a.* young.
νέγρος *s.m.* negro.
νειάτα *s.n.pl. see* **νιάτα**.
νέκρα *s.f.* complete absence of noise *or* activity; deadness.
νεκρανάσταση *s.f.* resurrection.
νεκρικός *a.* deathly; of death, funeral.
νεκροθάπτης *s.m.* grave-digger.
νεκροκεφαλή *s.f.* skull.
νεκρολογία *s.f.* obituary.
νεκροπομπός *s.m.* undertaker's mute.
νεκρ|ός *a.* dead; a dead person; **~ά φύσις** still life. **~ό σημείο** neutral gear.
νεκροταφείο *s.n.* cemetery.
νεκροτομείο *s.n.* mortuary.
νεκροφάνεια *s.f.* appearance of death.
νεκροφόρα *s.f.* hearse.
νεκροψία *s.f.* (*med.*) post-mortem.
νεκρώνω *v.t. & i.* deaden; become

deadened; go deathly pale (*from fear*).
νεκρώσιμος *a.* funeral.
νέκταρ *s.n.* nectar.
νέμεσις *s.f.* nemesis.
νέμομαι *v.t.* enjoy use *or* fruits of, hold in usufruct.
νενέ *s.f.* (*fam.*) grandmamma.
νενομισμένος *a.* prescribed, necessary.
νεο- *denotes* new, young.
νεογέννητος *a.* newly born.
νεογνόν *s.n.* newly born baby *or* animal.
νεόδμητος *a.* newly built.
νεοελλαδίτης *s.m.* person from 'new' provinces of Greece (*esp. Macedonia & Thrace*).
νεοελληνικ|ός *a.* of modern Greece. **~ή** *s.f.*, **~ά** *s.n.pl.* modern Greek language.
νεόκτιστος *a.* newly built.
νεολαία *s.f.* youth (*young people*).
νεόνυμφος *a.* newly married.
νεόπλουτος *a.* newly rich.
νέ|ος *a.* new, modern; young; a young man. **~ο** *s.n.* piece of news.
νεοσσός *s.m.* nestling.
νεοσύλλεκτος *s.m.* recruit.
νεοσύστατος *a.* newly established.
νεότης *s.f.* youth.
νεοφερμένος *a.* newly arrived.
νεόφυτος *a.* newly planted; a neophyte.
νεοφώτιστος *a.* newly baptized; (*fig.*) a recent convert (*to ideology*).
νεράιδα *s.f.* sort of fairy; (*fig.*) beautiful woman. *καλή ~* fairy godmother.
νεράντζι *s.n.* bitter orange.
νερ|ό *s.n.* water; rain; *μετάξι με ~ά* watered silk; *χάνω τα ~ά μου* feel lost; *αυτό σηκώνει ~ό* it is not quite true; *ξέρω το μάθημα ~ό* know the lesson by heart; *κάνω ~ά* (*of ship*) leak; (*fig.*) waver; *~α* (*pl.*) grain (*in wood*). *~άκι* glass of water; *το ξέρω ~άκι* I know it by heart.
νερόβραστος *a.* boiled in water; (*fig.*) insipid, ineffectual, limp.
νεροκράτης *s.m.* turncock; water-key.
νερομάν(ν)α *s.f.* spring (*of water*).
νερομπογιά *s.f.* water-colour.
νερομπούλι *s.n.* watery pap.
νερόμυλος *s.m.* water-mill.
νερόπλυμα *s.n.* slops.
νεροποντή *s.f.* violent rainstorm.
νερουλ|ός *a.* watery, thin. **~ιάζω** *v.i.* become watery *or* flabby.
νεροχύτης *s.m.* sink.
νερώνω *v.t.* water (*adulterate*).
νέτα σκέτα *adv.* plainly, frankly.
νετάρω *v.t.* finish, use up.
νέτος *a.* net; (*fam.*) done, finished (*of work*); broke.

νεύμα *s.n.* nod, sign.
νευραλγία *s.f.* neuralgia.
νευραλγικ|ός *a.* (*fig.*) ~ό *σημείο* vital *or* vulnerable spot.
νευριάζω *v.t.* & *i.* make *or* become exasperated.
νευρικός *a.* of the nerves, nervous; highly strung; irritable.
νευρολόγος *s.m.* neurologist.
νεύρ|ο *s.n.* nerve, sinew; (*fig.*) nervous energy; *έχω* ~*α* be in irritable mood; *έχω* ~*άκια* be in a pet.
νευροπάθεια *s.f.* nervous disease.
νευρόσπαστο *s.n.* marionette, puppet.
νευρώδης *a.* sinewy; (*fig.*) vigorous (*of speech*).
νευρώνω *v.i. see* **νευριάζω.**
νευρωτικός *a.* neurotic.
νεύω *v.i.* nod, make a sign.
νεφέλ|η *s.f.* cloud. ~**οειδής** *a.* nebulous; ~**ώδης** *a.* cloudy; (*fig.*) vague, obscure. ~**ωμα** *s.n.* nebula.
νεφελοκοκκυγία *s.f.* cloud-cuckoo-land.
νέφος *s.n.* cloud; smog.
νεφρ|ός *s.m.*, ~**ό** *s.n.* kidney. ~**ικός** *a.* renal.
νέφτι *s.n.* turpentine.
νεωκόρος *s.m.* sacristan.
νεώριο *s.n.* dockyard.
νεωστί *adv.* lately.
νεωτερισμ|ός *s.m.* novelty, innovation, new trend; ~*οί* (*pl.*) fancy goods.
νεώτερ|ος *a.* newer; younger. ~**α** *s.n. pl.* news.
νήμα *s.n.* (cotton) thread.
νηνεμία *s.f.* calm weather.
νηολογώ *v.t.* register (*ships*).
νηοπομπή *s.f.* (*naut.*) convoy.
νηοψία *s.f.* searching of ships (*for contraband of war*).
νηπιαγωγείον *s.n.* kindergarten.
νήπι|ο *s.n.* infant. ~**ακός** *a.* of infants. ~**όθεν** *adv.* from infancy.
νηπιώδης *a.* infantile; (*fig.*) undeveloped, in its infancy.
νηρηίς (και νηρηίδα) *s.f.* nereid.
νησί *s.n.* island. ~**δα** *s.f.* islet; traffic island.
νησιώ|της *s.m.* islander. ~**ικος** *a.* islander's, of the islands.
νήσος *s.f.* island.
νήσσα *s.f.* duck.
νηστεία *s.f.* fasting; lack of food; *τον άφησαν* ~ he was kept in (*at school*).
νηστεύω *v.i.* fast.
νηστήσιμος *a.* lenten (*food*).
νηστικός *a.* not having eaten.
νηφάλιος *a.* sober; calm, collected.
νιανιά *s.f* (*fam.*) pap.
νιαουρίζω *v.i.* mew, miaow.

νιάτα *s.n.pl.* youth.
νίβω *v.t.* wash (*face, hands*).
νίκη *s.f.* victory.
νικητήρια *s.n.pl.* victory celebration.
νικητής *s.m.* victor.
νικηφόρος *a.* victorious.
νικώ *v.t.* & *i.* conquer, defeat, beat; overcome; (*v.i.*) be victorious, win.
νίλα *s.f.* (*fam.*) nasty trick, practical joke.
νιο- *see* **νεο-**.
νιόπαντρος *a.* newly wed.
νιπτήρας *s.m.* wash basin.
νίπτω *v.t. see* **νίβω.**
νισάφι *s.n.* mercy, respite.
νίτρο *s.n.* nitre, saltpetre.
νιφάδα *s.f.* snowflake.
νιώθω *v.t.* understand; notice; feel.
Νοέμβρ|ιος, ~ης *s.m.* November.
νοερός *a.* mental.
νόημα *s.n.* thought (*faculty*); sense, meaning, point; nod, sign.
νοήμ|ων *a.* intelligent. ~**οσύνη** *s.f.* intelligence.
νόηση *s.f.* intellect.
νοητός *a.* intelligible; imaginary, notional.
νοθ|εία, ~ευση *s.f.* adulteration; falsification.
νοθεύω *v.t.* adulterate, falsify.
νόθος *a.* bastard; forged; irregular (*state of affairs*).
νοιάζει *v.t.* *με* ~ I mind, I care.
νοιάζομαι *v.t.* & *i.* look after, care about; be worried.
νοίκ|ι *s.n.* rent. ~**ιάζω** *v.t.* let, rent, hire. ~**άρης** *s.m.* tenant.
νοικοκυρά *s.f.* housewife; landlady; (*fig.*) housewifely woman.
νοικοκυρεύ|ω *v.t.* tidy; set up, provide for. ~**ομαι** settle down.
νοικοκύρης *s.m.* householder; landlord; master; (*fig.*) provident *or* orderly man.
νοικοκυριό *s.n.* household; house-keeping.
νοιώθω *see* **νιώθω.**
νομαδικός *a.* nomadic.
νομάρχ|ης *s.m.* nomarch (*prefect of department*). ~**είον** *s.n.* nomarch's office (*premises*).
νομάς *s.m.* nomad.
νομάτ|οι ~άίοι *s.m.pl.* persons.
νομή *s.f.* pasture, pasturage; (*law*) usufruct.
νομίζω *v.i.* & *t.* think, consider.
νομικ|ός *a.* legal (*of the law*); a member of the legal profession. ~**ά** *s.n.pl.* (*study of*) law. ~**ή** *s.f.* (*science of*) law.
νόμιμ|ος *a.* legal, lawful. ~**ότης** *s.f.* legality; law-abidingness.
νομιμοποιώ *v.t.* legalize, legitimize.
νομιμόφρων *a.* law-abiding; loyal.

νόμισμα *s.n.* coin, specie; currency, money. **~τικός** *a.* monetary.
νομισματολογία *s.f.* numismatics.
νομοθεσία *s.f.* legislation.
νομοθέτ|ης *s.m.* law-giver. **~ώ** *v.i.* make laws.
νομομαθής *s.m.* jurist.
νομομηχανικός *s.m.* county engineer.
νομός *s.m.* department, prefecture.
νόμος *s.m.* law, statute.
νομοσχέδιον *s.n.* parliamentary bill.
νομοταγής *a.* law-abiding.
νονά *s.f.* godmother.
νονός *s.m.* godfather.
νοοτροπία *s.f.* mentality.
νοσηλ|εία *s.f.* nursing (*of patient*). **~εύω** *v.t.* treat (*sick person*).
νόσημα *s.n.* malady.
νοσηρός *a.* unhealthy, morbid.
νοσοκόμ|ος *s.m.*, **~α** *s.f.* nurse. **~είο** *s.n.* hospital.
νόσος *s.f.* malady.
νοσταλγ|ία *s.f.* nostalgia. **~ικός** *a.* nostalgic. **~ώ** *v.i. & t.* feel nostalgic (for).
νοστιμ|άδα, **~ιά** *s.f.* nice flavour, tastiness; piquancy, attractiveness; funniness.
νοστιμ|εύω, **~ίζω** *v.t. & i.* make *or* become tasty *or* attractive. **~εύομαι** *v.t.* find to one's taste, have appetite for.
νόστιμος *a.* tasty, nice; piquant, attractive; funny.
νοσώ *v.i.* be ill.
νότ|α *s.f.* (*diplomatic*) note; (*mus.*) note; **~ες** (*pl.*) (*written*) music.
νοτι|ά *s.f.* south; south wind; humidity. **~άς** *s.m.* south wind.
νοτίζω *v.t. & i.* make *or* become damp.
νότι|ος *a.* south, southern. **~οανατολικός** *a.* south-east(ern). **~οδυτικός** *a.* south-west(ern).
νότος *s.m.* south; south wind.
νουθε|σία *s.f.* admonition. **~τώ** *v.t.* admonish.
νούλα *s.f.* nought; (*fig.*) cipher.
νούμερο *s.n.* number; (*music-hall*) turn; (*fam.*) droll character, card.
νουνεχής *a.* prudent, sensible.
νουνός *s.m. see* **νονός**.
νους *s.m.* intellect, mind; sense, wit; *βάζω με το νου μου* imagine, fancy; *δεν το βάζει ο ~ του* he cannot imagine it; *λέω με το νου μου* say to oneself; *δεν είσαι με το νου σου* you must be mad; (*έχε*) *το νου σου* take care, keep an eye open; *κοντά στο νου* it goes without saying.
νούφαρο *s.n.* water-lily.
νοώ *v.t. & i.* understand; think.
ντ- *see also* **τ-**.

νταβ|άς *s.m.* sort of baking-tin. **~ατζής** *s.m.* (*fam.*) ponce; fence.
νταγιαντίζω *v.t.* bear, endure.
νταγλαράς *s.m.* (*fam.*) lanky fellow.
ντα ής *s.m.* bully, tough.
ντάλα *adv.* **~** *μεσημέρι* at high noon.
ντάλια *s.f.* dahlia.
νταλίκα *s.f.* trailer (*of lorry*).
ντάμα *s.f.* lady, partner; queen (*cards*); draughts.
νταμάρι *s.n.* quarry; (*fig.*) *καλό* **~** of good stock (*physically*).
νταμιτζάνα *s.f.* demijohn.
νταμπλάς *s.m.* apoplexy; (*fig.*) great astonishment.
νταντά *s.f.* (*child's*) nurse.
νταντέλα *s.f.* lace.
νταούλι *s.n.* drum.
ντάρα *s.f.* tare (*weight*).
νταραβέρι *s.n.* (*business*) dealing; trouble (*quarrels*).
νταρί *s.n.* maize.
ντε *int.* *έλα* **~** come on!
ντελάλης *s.m.* public crier.
ντελβές *s.m.* coffee grounds.
ντελικάτος *a.* delicate.
ντεμπραγιάζ *s.n.* (*mech.*) clutch.
ντεπόζιτο *s.n.* tank, cistern.
ντεραπάρω *v.i.* skid.
ντερβ- *see* **δερβ-**.
ντέρτι *s.n.* sorrow, yearning (*esp. lover's*).
ντέφι *s.n.* tambourine.
ντιβάνι *s.n.* divan.
ντιπ *adv. see* **μπιτ**.
ντοκ *s.n.* (*naut.*) dock.
ντοκουμέντο *s.n.* document.
ντοκυμαντέρ *s.n.* documentary.
ντολμάς *s.m.* stuffed vine *or* cabbage leaf.
ντολμές *s.m.* renegade Jew.
ντομάτα *s.f.* tomato.
ντόμπρος *a.* straight, sincere, frank.
ντόπιος *a.* local, native.
ντορβάς *s.m.* peasant's bag; nose-bag.
ντόρος *s.m.* noise row; (*fig.*) sensation.
ντουβάρι *s.n.* wall; (*fig.*) blockhead.
ντούγια *s.f.* stave (*of cask*).
ντουέτο *s.n.* duet.
ντουζίνα *s.f.* dozen.
ντουί *s.n.* (*electric*) socket.
ντουλάπα *s.f.* wardrobe.
ντουλάπι *s.n.* cupboard.
ντουνιάς *s.m.* people, mankind; world.
ντούρος *a.* hard, firm (*in substance*); unbending, erect (*in carriage*).
ντους *s.n.* showerbath. [F. *douche*]
ντρέπομαι *v.i. & t.* be shy; be ashamed (of).
ντροπαλός *a.* shy.
ντροπή *s.f.* shyness; shame.

ντροπιάζω *v.t.* put to shame.
ντυμένος *a.* dressed.
ντύν|ω *v.t.* dress; cover upholster, line. **~ομαι** *v.i.* get dressed.
ντύσιμο *s.n.* dressing; dress, attire.
νυ *s.n.* the letter **Ν**.
νυγμ|ός *s.m.* prick, sting; (*fig.*) hint; **~οί του στομάχου** pangs of hunger.
νυκτ- *see* **νυχτ-**.
νυκτεριν|ός *a.* of night, nocturnal. **~ό** *s.n.* nocturne.
νυκτόβιος *a.* of nocturnal habits.
νύκτωρ *adv.* by night.
νυμφεύω *v.t. & i. see* **παντρεύω**.
νύμφη *s.f.* 1. nymph. 2. larva. 3. *see* **νύφη**.
νυμφίος *s.m.* bridegroom.
νυμφομανής *a.* nymphomaniac.
νυν *adv.* now.
νυξ *s.f.* night.
νύξη *s.f.* prick, sting; (*fig.*) hint, allusion.
νύστ|α *s.f.* **~αγμός** *s.m.* sleepiness.
νυστάζω *v.i.* feel sleepy.
νυσταλέος *a.* sleepy, drowsy.
νυστέρι *s.n.* lancet.
νύφη *s.f.* bride; daughter-in-law; sister-in-law (*brother's wife*).
νυφικός *a.* bridal.
νυφίτσα *s.f.* weasel.
νυφο(μ)πάζαρο *s.n.* evening stroll, promenade.
νυχθημερόν *adv.* day and night.
νύχι *s.n.* nail, claw, talon; hoof; (*fig.*) **~α** (*pl.*) clutches; *περπατώ στα* **~α** *μου* walk on tiptoe; **~** *και κρέας* hand-in-glove.
νυχι|ά *s.f.* scratch. **~άζω** *v.t.* scratch.
νύχτα *s.f.* night; (*adv.*) at night.
νυχτέρι *s.n.* work done at night.
νυχτερίδα *s.f.* bat (*animal*).
νυχτικ|ιά *s.f.* **~ό** *s.n.* nightgown.
νυχτοπούλι *s.n.* night-bird.
νυχτών|ω, **~ομαι** *v.i.* be overtaken by night; **~ει** night falls.
νωθρός *a.* sluggish, indolent.
νωμίτης *s.m.* yoke (*in dressmaking*).
νωπός *a.* fresh (*not stale or dried up*); still damp (*of washing*).
νωρίς *adv.* early.
νώτ|α *s.n.pl.* back; (*mil.*) rear. **~ιαίος** *a.* vertebral **~ιαίος** *μυελός* spinal cord.
νωχελ|ής *a.* nonchalant. **~εια** *s.f.* nonchalance.

Ξ

ξάγναντ|ο *s.n.* view-point. **~εύω** *v.t.* see from afar.

ξαγοράρης *s.m.* confessor.
ξαγορεύ|ω *v.t.* confess (*penitent*). **~ομαι** *v.i.* confess (*to priest*).
ξαγρυπνώ *v.i. & t.* stay awake; keep watch (over).
ξαδέρφη *s.f.* cousin.
ξαδέρφια *s.n.pl.* cousins.
ξάδερφος *s.m.* cousin.
ξαίρω *v.t. & i. see* **ξέρω**.
ξακουσ|μένος, **~τός** *a.* renowned.
ξαλαφρώνω *v.t. & i.* relieve, allay; be relieved; relieve oneself.
ξαλλάζω *v.t.* change (*clothes*).
ξαμολ|ώ *v.t.* unleash. **~ιέμαι** rush.
ξανά *adv.* again.
ξανα- *denotes* again, back.
ξανάβω *v.t. & i.* inflame, irritate; get hot *or* inflamed *or* irritated.
ξαναγεννιέμαι *v.i.* be reborn.
ξαναγυρίζω *v.i. & t.* return back; give back.
ξαναδίνω *v.t.* give back; retake (*exam.*).
ξανακάνω *v.t.* do *or* make again.
ξανακυλώ *v.i.* have a relapse.
ξαναλέω *v.t.* repeat, say again.
ξάναμμα *s.n.* inflammation; heat, irritation.
ξανανιώνω *v.t. & i.* rejuvenate; be rejuvenated.
ξαναπαίρνω *v.t.* take *or* get back; reemploy; buy again; catch (*ailment*) again; recapture; *τον* **~** go to sleep again.
ξαναρχής *adv.* from the beginning.
ξανάρχομαι *v.i.* come again *or* back.
ξανασαίνω *v.i.* rest, relax.
ξανάστροφη *s.f.* wrong *or* reverse side; back-handed slap.
ξανθ|αίνω, **~ίζω** *v.t. & i.* make *or* become lighter (*of hair*).
ξανθ|ός *a.* blond, fair. **~ομάλλης** *a.* fair-haired. **~ωπός** *a.* on the fair side.
ξάνοιγμα *s.n.* becoming lighter *or* brighter; launching out; clearing (*in forest*).
ξανοίγ|ω *v.t. & i.* make *or* become lighter (*in colour*); become brighter (*of weather*); look at. **~ομαι** *v.i.* unbosom oneself; launch out too ambitiously; put out to sea.
ξαντό *s.n.* lint.
ξάπλα *s.f.* (*fam.*) lounging about.
ξάπλωμα *s.n.* lying down; spreading.
ξαπλών|ω *v.t. & i.* spread (out); lie down. **~ομαι** *v.i.* spread; lie down.
ξαπλώστρα *s.f.* deck-chair.
ξαπλωτός *a.* lying down.
ξαποσταίνω *v.i.* rest.
ξαποστέλνω *v.t.* dispatch; (*fam.*) send packing.
ξασπρίζω *v.t. & i.* whiten, bleach; fade.
ξάστερ|ος *a.* starry, clear. **~α** *adv.* clearly, plainly. **~ιά** *s.f.* starry sky.

ξαστόχημα 128 ξενυχτώ

ξαστόχημα s.n. miscalculation.
ξαφν|ιάζω, ~ίζω v.t. surprise, scare. **~ιάζομαι, ~ίζομαι** v.i. be scared or taken aback.
ξαφνικό s.n. surprise (incident); accident, fatality.
ξαφνικ|ός a. sudden, unexpected. **~ά** adv. suddenly.
ξάφνου adv. suddenly.
ξαφρίζω v.t. skim; (fam.) sneak, steal.
ξε- see **εκ** or **εξ.**
ξεβάφω v.t. & i. (cause to) lose colour, fade.
ξεβγάζω v.t. rinse; see off; get rid of; kill; initiate into loose ways.
ξεβρακώνω v.t. remove trousers or pants of; (fig.) show up.
ξεβράκωτος a. trouserless; (fig.) penniless; a sans-culotte.
ξεγδέρνω v.t. flay, skin; scratch, graze.
ξεγελώ v.t. deceive, gull; seduce.
ξεγεννώ v.t. deliver (woman in parturition).
ξε(γ)νοιάζω v.i. become free from care.
ξέ(γ)νοιαστος a. carefree, light-hearted.
ξεγράφω v.t. (fig.) write off.
ξεδιαλέγω v.t. pick out (the best).
ξεδιαλύνω v.t. & i. clear up, unravel; come true.
ξεδιάντροπος a. shameless.
ξεδίνω v.i. unwind, relax.
ξεδιπλώνω v.t. unfold.
ξεδιψώ v.t. & i. quench thirst of; quench one's thirst.
ξεδοντιάρης a. toothless.
ξεθαρρεύω, ~ομαι v.i. grow (too) bold.
ξεθεώνω v.t. (fam.) wear out, exhaust.
ξεθυμαίνω v.i. evaporate; lose strength or pungency; leak (of fumes, gas-pipe, etc.); (fig.) relieve one's feelings.
ξεθωριάζω v.i. fade.
ξεκαθαρίζω v.t. & i. settle (account); clarify; become settled or clarified; (fam.) kill off.
ξεκάνω v.t. dispose of, sell off; (fam.) kill off.
ξεκαρδίζομαι v.t. die of laughing.
ξεκάρφωτος a. unnailed; (fig.) disjointed.
ξεκινώ v.i. start off, set out.
ξεκλέβω v.t. (fig.) steal, snatch (a moment etc.).
ξεκληρίζ|ω v.t. wipe out (family). **~ω, ~ομαι** v.i. die out (of family).
ξεκοιλιάζω v.t. disembowel.
ξεκοκκαλίζω v.t. bone; devour to the bone; (fig.) squander.
ξεκολλώ v.t. & i. unstick; come unstuck.
ξεκομμέν|ος a. isolated; τιμή **~η** fixed price. **~α** adv. bluntly, straight out.
ξεκουμπίζομαι v.i. (fam.) clear out, depart.

ξεκουμπ|ώνω v.t. unbutton. **~ωτος** a. unbuttoned.
ξεκουράζ|ω v.t. rest, relieve, refresh. **~ομαι** v.i. rest, repose.
ξεκουτιαίνω v.t. & i. make or grow softwitted (esp. from old age).
ξεκουφαίνω v.t. deafen.
ξεκόφτω v.t. & i. detach; say straight out; **~** από give up (doing sthg.).
ξελαρυγγίζομαι v.i. shout oneself hoarse.
ξελέω v.t. unsay, take back.
ξελιγώ|νω v.t. cloy; (fig.) tire, exhaust (exp. with waiting); **~μένος** για avid for.
ξελογιάζω v.t. inveigle; turn head of, monopolize interest of.
ξεμαλλι|άζω v.t. pull hair of; **~ασμένος** dishevelled.
ξεμένω v.i. be stranded; **~** από run out of.
ξεμοναχιάζω v.t. draw (person) aside; find (person) alone.
ξεμπαρκάρω v.t. & i. unload, land; (v.i.) land.
ξεμπερδεύω v.t. & i. disentangle, sort out; have done with; kill; (v.i.) have done with it.
ξεμπλέκω v.t. & i. disentangle (oneself); **~** από get rid or free of.
ξεμπουκάρω v.i. come out, issue (as from narrow opening).
ξεμπράτσωτος a. with bare arms.
ξεμυαλίζω v.t. inveigle; seduce; turn head of, monopolize interest of.
ξεμυτίζω v.i. (fig.) show one's nose.
ξεμωραίνομαι v.i. become childish.
ξένα s.n.pl. foreign parts.
ξεναγός s.m.f. (tourists') guide.
ξενηλασία s.f. expulsion of foreigners.
ξενίζ|ω v.t. 1. entertain (guest). 2. surprise, seem strange to. **~ομαι** v.i. be surprised.
ξενικός a. foreign.
ξενιτεύομαι v.i. go to live abroad.
ξενιτιά s.f. (sojourning in) foreign parts.
ξε(ν)νοιάζω v.i. become free from care.
ξενοδουλεύω v.i. go out to work.
ξενοδοχείο s.n. hotel; (fam.) restaurant.
ξενοδόχος s.m. hotel-keeper.
ξενοικιάζομαι v.i. become vacant (of property).
ξένος 1. a. foreign; strange; somebody else's; (with προς) not concerned with. 2. s.m. foreigner, stranger; guest.
ξενοφοβία s.f. xenophobia.
ξενόφωνος a. foreign-speaking.
ξεντύνω v.t. undress.
ξενυχιάζω v.t. trample on toes of.
ξενύχτης s.m. one who stays up late.
ξενυχτώ v.i. & t. stay up late or all night; (v.t.) keep awake at night; sit up with (sick

case).

ξενώνας *s.m.* guest-room; guest-house.

ξεπαγιάζω *v.i. & t.* freeze; get chilblains.

ξεπαραδιάζω *v.t.* leave without a penny.

ξεπαστρεύω *v.t.* clean up; exterminate.

ξεπατώνω *v.t.* wear out *or* remove bottom *or* floor of; tire to death; exterminate.

ξεπεζεύω *v.i.* dismount (*from horse*).

ξεπερ|νώ *v.t.* unthread; surpass; *έλα να ~αστούμε* let us compete together; *~ασμένος* out-of-date, passé.

ξεπεσμός *s.m.* decline, falling off; reduction (*of price*).

ξεπετ|άω *v.t.* flush out; finish (*job*) quickly; (*fig.*) see (*one's children*) fledged. *~ιέμαι v.i.* shoot up (*in stature*); fly *or* jump out suddenly; butt in.

ξεπέφτω *v.i. & t.* fall off, decline in value *or* status; reduce in value *or* price, deduct.

ξεπηδώ *v.i.* fly *or* jump out suddenly.

ξέπλεκος *a.* loose, let down (*of hair*).

ξεπλένω *v.t.* rinse.

ξεπληρώνω *v.t.* pay off (*creditor or debt*).

ξέπλυμα *s.n.* rinsing; (*fig.*) insipid person *or* thing.

ξεποδαριάζω *v.t.* walk the feet off.

ξεπορτίζω *v.i. & t.* go *or* slip out (*of doors*); send (*person*) out.

ξεπούλημα *s.n.* sale, selling off.

ξεπουλώ *v.t. & i.* sell off; (*v.i.*) sell out.

ξεπροβάλλω *v.i.* come into view, pop up.

ξεπροβοδ|ίζω, ~ώ *v.t.* see off.

ξέρα *s.f.* 1. reef (*in sea*). 2. drought.

ξεράδι *s.n.* dead wood; (*fig.*) dry crust; (*fam.*) *κάτω τα ~α σου* hands (*or* feet) off!

ξεραΐλα *s.f.* aridity.

ξεραίν|ω *v.t.* dry (up), parch. *~ομαι v.i.* dry up; die (*of plants*); (*fig.*) *ξεράθηκε* he was dumbfounded; *ξεράθηκε στα γέλια* he died of laughing.

ξερακιανός *a.* lean, spare (*of body*).

ξέρ|ασμα, ~ατό *s.n.* vomit; vomiting.

ξερικός *a.* (*plant*) grown without watering.

ξερνώ *v.t. & i.* vomit, spew.

ξερόβηχας *s.m.* dry cough.

ξεροβήχω *v.i.* clear one's throat.

ξεροβόρι *s.n.* dry, cold north wind.

ξεροβούνι *s.n.* barren mountain.

ξερογλείφομαι *v.i.* lick one's lips (*in anticipation*).

ξεροκαταπίνω *v.i.* gulp (*with embarrassment*).

ξεροκέφαλος *a.* thick-headed; obdurate.

ξεροκοκκαλίζω *v.t.* (*fig.*) nibble away, waste (*money*).

ξερολιθιά *s.f.* rubble-masonry.

ξερονήσι *s.n.* barren *or* desert island.

ξεροπόταμο *s.n.* dry torrent-bed.

ξερ|ός *a.* dry, dried; curt; *~ός μισθός* bare wage. *έμεινε ~ός* he was dumbfounded *or* fell dead; *έπεσε ~ός στον ύπνο* he slept like a log; (*fam.*) *μάζεψε τα ~ά σου* hands (*or* feet) off!

ξεροσταλιάζω *v.i.* become exhausted with standing.

ξεροψωμίζω *v.i.* live on dry bread.

ξε(ρ)ριζώνω *v.t.* uproot, pull off; eradicate.

ξέρω *v.i. & i.* know (how to). *~ από* know about; *~ τα κουμπιά* know the ropes.

ξεσήκωμα *s.n.* revolt.

ξεσηκών|ω *v.t.* rouse; disturb; trace, copy; imitate. *~ομαι* revolt.

ξεσκ|άζω, ~ώ *v.i.* relax (*in recreation*).

ξεσκεπάζω *v.t.* uncover, lay bare; show up.

ξέσκεπος *a.* uncovered; (*fig.*) frank.

ξεσκίζω *v.t.* tear to pieces; (*fig.*) *~ισμένος* profligate.

ξεσκολ|ίζω *v.i.* finish one's schooling; (*fig.*) *~ισμένος* practised.

ξεσκολ|ίζω *v.t.* (*clear of*) dust.

ξεσκονόπανο *s.n.* duster.

ξέσκουρα *adv.* on the surface (*of injuries*).

ξεσκούφωτος *a.* hatless.

ξεσπ|άζω, ~άνω, ~ώ *v.i.* overflow, burst out.

ξεσπαθώνω *v.i.* draw one's sword; (*fig.*) take up cudgels.

ξεστομίζω *v.t.* utter.

ξεστραβώνω *v.t. & i.* straighten; (*fig.*) open eyes of, educate; become straight.

ξεστρώνω *v.t.* take up, remove; clear (*table*); *~ το σπίτι* take up the carpets.

ξέστρωτ|ος *a.* unmade (*bed*); uncarpeted; unpaved; unsaddled; (*fam.*) *~ο γαϊδούρι* extremely rude person.

ξεσυνερίζομαι *v.t.* see **συνερίζομαι**.

ξεσχίζω *v.t.* see **ξεσκίζω**.

ξετινάζω *v.t.* shake (*violently*); leave in weak condition; (*fig.*) reduce to poverty; (*of critic*) maul.

ξετρελλαίνω *v.t.* drive mad.

ξετρυπώνω *v.t. & i.* unearth; remove basting from (*in dressmaking*); pop up.

ξετσιπωσιά *s.f.* (act of) brazen impudence.

ξετσίπωτος *a.* without sense of shame, brazen.

ξετυλίγω *v.t.* unroll, unwrap, unfold.

ξεύρω *v.t. & i.* see **ξέρω**.

ξεφαντώνω *v.i.* make merry.

ξεφεύγω *v.t. & i.* escape, avoid; (*v.i.*) slip out.

ξεφλουδίζω *v.t. & i.* peel; remove bark *or* shell of; (*v.i.*) peel.

ξεφορτ|ώνω *v.t. & i.* unburden, unload. *~ώνομαι v.t.* get rid of; *~ωνέ με or ~ώσου με* leave me in peace!

ξεφτέρι *s.n.* (*fig.*) quick and intelligent person, adept.

ξέφτ|ι *s.n.* loose thread, ravel. **~ίζω, ~ώ** *v.i.* & *t.* become frayed; get past one's prime; pull out (*thread*).

ξεφυλλίζω *v.t.* & *i.* leaf through (*book*); pluck leaves of; shed leaves.

ξεφυσώ *v.i.* let off steam; puff and blow; break wind.

ξεφυτρώνω *v.i.* sprout, spring up.

ξεφωνητό *s.n.* yell, shriek, shout(ing).

ξεφωνίζω *v.i.* shriek, shout.

ξέφωτο *s.n.* glade, clearing.

ξεχαλινάρομαι *v.i.* lose all restraint.

ξεχαρβαλώ|νω *v.t.* make rickety; **~μένος** rickety, disorganized.

ξεχασιάρης *a.* forgetful.

ξεχασμένος *a.* forgetful; forgotten.

ξεχειλίζω *v.t.* & *i.* fill too full; overflow.

ξέχειλος *a.* full to the brim.

ξεχειλώνω *v.i.* become misshapen (at the edge) (*of garments*).

ξεχειμων|ιάζω *v.i.* spend the winter; **~ιασε** the winter is over.

ξεχνώ *v.t.* & *i.* forget.

ξεχρεών|ω *v.t.* pay off (*debt, creditor*); complete payment for. **~ω, ~ομαι** *v.i.* discharge one's debts.

ξεχύν|ω *v.t.* pour forth; break out in (*rash, etc.*). **~ομαι** *v.i.* pour forth, overflow.

ξέχωρα *adv.* separately; **~** *από* apart from, as well as.

ξεχωρίζω *v.t.* & *i.* set apart; single out; distinguish, tell apart; be different, stand out.

ξεχωριστ|ός *a.* separate; exceptional (*pre-eminent*); special (*peculiar*). **~ά** *adv.* separately, apart.

ξεψαχνίζω *v.t.* investigate closely; sound (*person*); (*fam.*) sponge on, soak.

ξεψυχ|ώ *v.i.* & *t.* die; (*fig.*) feel ardent desire; **~ισμένος** *για τσιγάρο* dying for a cigarette. (*v.t.*) worry the life out of.

ξέω *v.t.* scratch, scrape; scratch out.

ξηγ|ιέμαι *v.i. see* **εξηγούμαι**. (*fam.*) **~ημένος** clear, understood, reliable.

ξηλώνω *v.t.* take apart; unmake, unstitch.

ξημεροβραδιάζομαι *v.i.* spend all day long.

ξημέρωμα *s.n.*, **~τα** (*pl.*) daybreak; (*fam.*) staying up all night.

ξημερών|ει *v.i.* day breaks. **~ομαι** *v.i.* wake up *or* find oneself next morning; stay up *or* be kept awake till morning.

ξηρά *s.f.* (*dry*) land.

ξηραίν|ω *v.t.* dry (up), *parch.* **~ομαι** *v.i.* dry up; die (*of plants*).

ξηρασία *s.f.* dryness; drought.

ξηρ|ός *a.* dry, dried; curt; **~οί** *καρποί* dried fruit and nuts.

Ξι *s.n.* the letter Ξ.

ξιδ- *see* **ξυδ-**.

ξίκικος *a.* underweight.

ξινάρι *s.n.* pickaxe.

ξινίζω *v.t.* & *i.* make *or* become sour.

ξινίλα *s.f.* sourness, acidity.

ξινό *s.n.* citric acid.

ξιν|ός *a.* sour, acid; *μου βγήκε* **~ό** it turned out a disappointment after all; *μ' αρέσουν τα* **~ά** have a weakness for amorous adventures.

ξιππ|άζω *v.t.* frighten; astonish, impress. **~άζομαι** *v.i.* become puffed up with pride; **~ασμένος** vainglorious.

ξιφασκία *s.f.* fencing.

ξιφήρης *a.* sword in hand.

ξιφίας *s.m.* swordfish.

ξιφισμός *s.m.* sword-thrust.

ξιφολόγχη *s.f.* bayonet.

ξιφομαχία *s.f.* fencing.

ξίφος *s.n.* sword.

ξιφουλκώ *v.i.* draw one's sword.

ξόανο *s.n.* wooden statue; (*fig.*) dolt.

ξόβεργα *s.f.* lime-twig.

ξοδεύ|ω *v.t.* spend, consume, use (up); sell (*of shops*); waste. **~ομαι** *v.i.* be put to expense.

ξόδεψη *s.f.* sale, demand (*for commodity*).

ξόδι *s.n.* funeral.

ξοδιάζω *v.t. see* **ξοδεύω**.

ξόμπλι *s.n.* pattern for embroidery; (*fig.*) malicious gossip.

ξοπίσω *adv.* (from) behind; again.

ξόρκι *s.n.* spell used in exorcism.

ξουρ|ίζω *v.t.* shave; (*fig.*) bore (*with loquacity*); tell tall stories to; *ο βοριάς μας* **~ισε** the north wind froze us.

ξοφλημένος *a.* done for, a washout.

ξύγκι *s.n.* fat, tallow, grease; groin; *βγάζω από τη μύγα* **~** squeeze blood from a stone.

ξυδάτος *a.* pickled; vinegary.

ξύδι *s.n.* vinegar.

ξυλάδικο *s.n.* timber-yard; firewood-shop.

ξυλάνθρακας *s.m.* charcoal.

ξυλεία *s.f.* timber.

ξυλένιος *a.* wooden.

ξυλεύομαι *v.i.* cut *or* obtain wood.

ξυλιά *s.f.* blow (*with stick*).

ξυλιάζω *v.t.* & *i.* make *or* become stiff (*with cold*) *or* like wood.

ξυλίζω *v.t.* beat, cudgel.

ξύλινος *a.* wooden.

ξύλ|ο *s.n.* (piece of) wood; **~α** (*pl.*) firewood; (*fam.*) *τρώω* **~ο** get a beating; *έπεσε* **~ο** blows were struck; *το* **~ο** *βγήκε από τον παράδεισο* spare the rod and spoil the child.

ξυλογράφημα *s.n.* woodcut.

ξυλοδεσιά *s.f.* wooden frame (*of building*).
ξυλοκάρβουνο *s.n.* charcoal.
ξυλοκέρατο *s.n.* carob.
ξυλοκόπος *s.m.* woodcutter.
ξυλοκοπ|ώ *v.t.* beat, thrash. **~ημα** *s.n.* beating, thrashing.
ξυλοκρέββατο *s.n.* wooden bed; coffin.
ξυλοπόδαρο *s.n.* stilt.
ξυλοστάτης *s.m.* jamb.
ξυλοσχίστης *s.m.* chopper of wood; (*fam.*) one who is no good at his job.
ξυλουργ|ός *s.m.* joiner. **~ική** *s.f.* joinery.
ξυλοφορτώνω *v.t.* beat, thrash.
ξυν- *see* **ξιν-**.
ξύν|ω *v.t.* scratch, scrape, grate; sharpen (*pencil*); scratch out; (*fig.*) sack (*dismiss*). **~ομαι** *v.i.* scratch.
ξυπάζω *v.i. see* ξιπάζω.
ξύπνημα *s.n.* awakening.
ξυπνητήρι *s.n.* alarm-clock.
ξυπν|ητός, ~ιος, ~ός *a.* awake; (*fig.*) wide-awake, smart.
ξυπνώ *v.t. & i.* wake up.
ξυπόλυτος *a.* barefooted.
ξυράφ|ι *s.n.* (*cut-throat*) razor. **~άκι** *s.n.* razor-blade.
ξυρίζ|ω *v.t.* shave. **~ομαι** *v.i.* shave, get shaved.
ξύρισμα *s.n.* shave, shaving.
ξυριστικ|ός *a.* of *or* for shaving; **~ή μηχανή** safety razor.
ξύσιμο *s.n.* scratching, scraping, grating; sharpening (*of pencil*); scratching out.
ξύσματα *s.n.pl.* scrapings.
ξυστ|ός *a.* scraped, grated. **~ά** *adv.* (almost) grazing the surface.
ξύστρα *s.f.* scraper; currycomb; (pencil) sharpener.
ξύω *v.t. see* ξύνω.
ξώδερμα *adv. see* ξυστά.
ξωκκλήσι *s.n.* country chapel.
ξώλαμπρα *adv.* after Easter.
ξωμερίτης *s.m.* stranger.
ξώπετσα *adv. see* ξυστά.
ξωτικιά *s.f.* fairy.
ξωτικό *s.n.* ghost.
ξώφαρσα *adv. see* ξυστά.
ξώφυλλο *s.n.* cover (*of book*); outside shutter.

O

ο *article m.* the.
ο *pron. neuter of* **ος**.
όαση *s.f.* oasis.
οβελίσκος *s.m.* obelisk.

οβελ|ός *s.m.* spit, skewer. **~ίας** *s.m.* spit-roasted lamb.
οβ|ίς ~ίδα *s.f.* shell (*projectile*).
οβολός *s.m.* obol; (*fig.*) modest contribution, mite.
Οβριός *s.m.* (*fam.*) Jew.
ογδόη *s.f.* (*mus.*) octave.
ογδοή(κο)ντα *num.* eighty.
ογδόντα *num.* eighty.
όγδοος *a.* eighth.
ογκανίζω *v.i.* bray.
ογκόλιθος *s.m.* large block of stone.
όγκος *s.m.* mass, bulk, volume; great quantity; (*med.*) swelling, tumour.
ογκούμαι *v.i.* swell.
ογκώδης *a.* massive, bulky.
οδαλίσκη *s.f.* odalisque.
όδε *pron.* this; *μέχρι τούδε* up to now.
οδεύω *v.i.* go, proceed.
οδήγ|ηση *s.f.* driving. **~ητής** *s.m.* leader.
οδηγία *s.f.* guidance, direction, instruction.
οδηγός *s.m.f.* guide; girl-guide; guide-book; driver; directory.
οδηγώ *v.t.* guide, direct, lead; instruct, show how (*to*); drive, steer (*vehicle*).
οδικός *a.* street, road (*plan, etc.*).
οδογέφυρα *s.f.* viaduct.
οδοιπορία *s.f.* journey (*esp. on foot*).
οδοιπορικά *s.n.pl.* journey-money.
οδοιπόρος *s.m.* wayfarer.
οδοιπορώ *v.i.* journey.
οδοκαθαριστής *s.m.* crossing-sweeper.
οδονταλγία *s.f.* toothache.
οδοντίατρος *s.m.f.* dentist.
οδοντόβουρτσα *s.f.* toothbrush.
οδοντο(γ)ιατρός *s.m.f.* dentist.
οδοντογλυφίδα *s.f.* toothpick.
οδοντόπαστα *s.f.* toothpaste.
οδοντοστοιχία *s.f.* set of (false) teeth.
οδοντοφυΐα *s.f.* cutting of teeth.
οδοντόφωνος *a.* (*gram.*) dental.
οδοντωτός *a* having teeth *or* cogs; **~ σιδηρόδρομος** rack-and-pinion railway.
οδοποιία *s.f.* road-construction.
οδ|ός *s.f.* street, thoroughfare, way; (*fig.*) way, path, method; *καθ'~όν* on the way.
οδοστρωτήρας *s.m.* steam-roller..
οδούς *s.m.* tooth, cog.
οδόφραγμα *s.n.* barricade.
οδύν|η *s.f.* pain, grief. **~ηρός** *a.* painful, grievous.
οδυρμός *s.m.* lamentation.
οδύρομαι *v.i.* lament, wail.
οδύσσεια *s.f.* odyssey.
όζον *s.n.* ozone.
όζω *v.i.* stink.
οζώδης *a.* gnarled.
όθεν *adv.* whence; therefore.

οθόνη *s.f.* sheet, cloth; (*cinema*) screen.

Οθωμανός *s.m.* Ottoman, Turk.

οι *article m.f.pl.* the.

οίαξ *s.m.* tiller; helm.

οίδ|α *v.i.* know; *Κύριος ~ε* Lord knows.

οίδημα *s.n.* swelling.

οίηση *s.f.* conceit, presumptuousness.

οίκαδε *adv.* homeward, home.

οικειοθελώς *adv.* of one's own free will.

οικειοποιούμαι *v.t.* appropriate.

οικεί|ος *a.* own, proper; familiar, intimate; *~οι* (*pl.*) relatives, intimates. *~ότητα s.f.* intimacy, familiarity.

οίκημα *s.n.* dwelling, habitation.

οικ|ία *s.f.* house. *~ιακός a.* domestic.

οικισμός *s.m.* settlement, estate.

οικογένει|α *s.f.* family. *~ακός a.* family. *~άρχης s.m.* head of family.

οικοδέσποινα *s.f.* mistress of house, hostess.

οικοδεσπότης *s.m.* master of house, host; landlord.

οικοδομή *s.f.* (*act of*) building; a building (*under construction*).

οικοδόμημα *s.n.* building, edifice.

οικοδόμησις *s.f.* (*act of*) building.

οικοδομώ *v.t.* build.

οίκοθεν *adv.* ~ *εννοείται* it is self- evident.

οικοκύρ|ης *s.m.* householder; landlord; master. *~ική s.f.* domestic science.

οικολογία *s.f.* ecology *~ικός a.* ecological.

οικονομ|ία *s.f.* economy; *~ίες* (*pl.*) savings.

οικονομικ|ός *a.* economic, financial; not expensive; *~ά* (*s.n.pl.*) finances. *~ά adv.* inexpensively.

οικονομολόγ|ος *s.m.* economist; (*fam.*) thrifty person. *~ία s.f.* economics.

οικονόμος *s.m.f.* steward; (*fig.*) thrifty person.

οικονομ|ώ, *~άω v.t.* save, husband (*resources*); find, get hold of (*with some effort*); fix (*someone*) up; *τά ~άω* make ends meet. *~έμαι, ~ιούμαι v.i.* get fixed up.

οικόπεδο *s.n.* building-plot.

οίκ|ος *s.m.* house; business-house; *κατ' ~ον* in the home.

οικόσημον *s.n.* coat of arms.

οικόσιτος *a.* (*animal*) kept in the home; a boarder.

οικοτεχνία *s.f.* cottage industry.

οικότροφ|ος *s.m.f.* boarder. *~είον s.n.* boarding-school, hostel.

οικουμέν|η *s.f.* world, universe. *~ικός a.* oecumenical; *~ική κυβέρνησις* coalition government.

οικουρώ *v.i.* be confined to the house (*by illness*).

οικτ(ε)ίρω *v.t.* pity; despise.

οικτιρμός *s.m.* pity.

οικτίρμων *a.* merciful, compassionate.

οίκτος *s.m.* pity; contempt.

οικτρός *a.* pitiable; wretched, mean.

οίμοι, οϊμέ *int.* alas!

οιμωγή *s.f.* wailing, groaning.

οινοβαρής *a.* intoxicated.

οινολογία *s.f.* oenology.

οινομαγειρείο *s.n.* cheap tavern.

οινόπνευμ|α *s.n.* alcohol. *~ατώδης a.* alcoholic.

οινοποιία *s.f.* wine-making.

οινοποσία *s.f.* (excessive) wine-drinking.

οίνος *s.m.* wine.

οινοχόος *s.m.* cup-bearer.

οίον *adv.* such as.

οιονεί *adv.* as if; quasi.

οίος *pron.* such as, of the kind which.

οιοσδήποτε *pron.* anybody, any; any kind of; whoever, whichever.

οισοφάγος *s.m.* oesophagus.

οίστρ|ος *s.m.* gadfly; (*fig.*) oestrum, inspiration; high spirits. *~ηλατούμαι v.i.* be inspired.

οιων|ός *s.m.* omen, presage. *~οσκόπος s.m.* augur.

οκά *s.f.* oka (*measure of weight*).

οκαζιόν *s.f.* bargain (*in sale*).

οκλαδόν *adv.* squatting, cross-legged.

οκνηρ|ός *a.* lazy. *~ία s.f.* laziness.

οκνός *a.* slow, sluggish; indolent.

οκρίβας *s.m.* easel.

οκτάβα *s.t.* octave.

οκτάγωνος *a.* octagonal.

οκτακόσιοι *a.* eight hundred.

οκτά|πους *s.m.* *~πόδι s.n.* octopus.

οκτώ *num.* eight.

Οκτώβριος *s.m.* October.

όλα *s.n. pl.* everything (*see* **όλος**).

ολάκερος *a.* whole, entire.

ολάνοικτος *a.* wide *or* fully open.

όλβιος *a.* wealthy; blessed, blissful.

ολέθριος *a.* disastrous, ruinous.

όλεθρος *s.m.* calamity, ruin.

ολημερίς *adv.* all day long.

ολίγ- *denotes* little, few.

ολιγάριθμος *a.* few in number.

ολιγαρκής *a.* satisfied with little.

ολιγαρχία *s.f.* oligarchy.

ολιγοδάπανος *a.* thrifty; inexpensive.

ολιγόλογος *a.* taciturn; laconic.

ολίγον *adv.* (a) little.

ολιγόπιστος *a.* unbelieving.

ολίγ|ος *a.* small in amount, (a) little, (a) few; *~ον κατ' ~ον* little by little; *εντός ~ου* soon; *παρ' ~ον* (*να*), *~ου δειν* (*να*) nearly.

ολιγοστεύω *v.t. & i.* lessen, reduce; grow less.

ολιγοστός *a.* scant, barely sufficient.
ολιγοψυχία *s.f.* faint-heartedness.
ολιγωρία *s.f.* negligence, indifference.
ολικός *a.* total.
ολισθαίνω *v.i.* slip.
ολίσθημα *s.n.* slip, slipping; blunder; lapse *(from rectitude).*
ολισθηρός *a.* slippery *(not affording foothold).*
ολκή *s.f.* pull, weight; calibre.
όλμος *s.m. (mil.)* mortar.
ολο- *denotes* all, completely.
όλο *adv.* all *(wholly);* all the time; ~ και περισσότερο more and more:
ολόγεμος, ~γιομος *a.* quite full.
ολόγερος *a.* in sound *or* healthy condition; undamaged.
ολογράφως *adv. (written out)* in full.
ολόγυμνος *a.* stark naked.
ολόγυρα *adv.* all round.
ολοέν(α) *adv.* all the time.
ολοήμερος *a.* lasting all day.
ολόιδιος *a.* exactly the same.
ολόισιος *a.* quite straight.
ολοκαύτωμα *s.n.* thing completely burnt, holocaust; sacrifice, offering.
ολόκληρος *a.* whole, entire; εξ ~ήρου entirely.
ολοκληρ|ώ *v.t.* bring to completion; consummate; *(math.)* integrate. ~ωτικός *a.* complete; totalitarian; *(math.)* integral.
ολολύζω *v.i.* wail, lament.
ολόμαλλος *a.* all wool.
ολομέλεια *s.f.* plenary session.
ολομέταξος *a.* all silk.
ολομόναχος *a.* all alone.
ολονέν *adv.* all the time.
ολονυκτία *s.f.* all-night service *or* revelry.
ολονυχτίς *adv.* all night long.
ολόρθος *a.* upright, erect.
όλος *a.* all, the whole; ~οι everybody; ~α everything; ~οι μας all of us; ~α ~α altogether *(in sum);* ~α κι ~α anything else *(but not that);* ~ος κι ~ος the selfsame *(person);* μ' ~ο που although.
ολοσηρικό *s.n.* satin.
ολοσχερής *a.* utter, total.
ολοταχώς *adv.* at full speed.
ολοτελ|α ~ώς *adv.* completely.
ολότης *s.f.* whole, entirety.
ολούθε *adv.* from all sides.
ολοφάνερος *a.* obvious, evident.
ολόφτυστος *a.* the very spit of.
ολοφύρομαι *v.i.* wail, lament.
ολόφωτος *a.* all lit up.
ολόχαρος *a.* full of joy.
ολοχρονίς *adv.* all the year round.
ολόχρυσος *a.* all gold.

ολόψυχος *a.* wholehearted.
Ολύμπι|α *s.n.pl.* Olympic games. ~άς *s.f.* olympiad. ~ακός *a.* Olympic. ~ονίκης *s.m.* Olympic victor.
ολύμπιος *a.* olympian; godlike.
όλως *adv.* wholly, quite; ~ διόλου completely; ~ υμέτερος yours truly.
ομάδα *s.f.* body, group; team *(racing)* crew; *(mil.)* section.
ομάδ|ι *adv.* together. ~ικός *a.* (in) mass. ~ικώς *adv.* in a body.
ομαλ|ός *a.* level, even, smooth; usual, normal; *(gram.)* regular. ~ότης *s.f.* evenness; normality, order. ~οποίησις *s.f.* normalization.
ομαλύνω *v.t.* smooth, level.
ομάς *s.f. see* ομάδα.
όμβρι|ος *a.* ~α ύδατα rain-water.
ομελέτα *s.f.* omelet.
ομήγυρις *s.f.* assembly, gathering.
ομήλιξ *s.m.f. (person)* of the same age.
ομηρικός *a.* Homeric.
όμηρος *s.m.* hostage.
όμικρον *s.n.* the letter Ο.
ομιλητής *s.m.* speaker.
ομιλητικός *a.* talkative; chatty, friendly.
ομιλία *s.f. (faculty or manner of)* speech; homily, talk, address; converse; mention.
όμιλος *s.m.* group, circle, club.
ομιλώ *v.t. & i.* speak.
ομίχλη *s.f.* fog, mist.
όμμα *s.n.* eye.
ομματοϋάλια *s.n.pl.* spectacles.
ομνύω *v.t. & i.* swear, vow.
ομο- *denotes* same, together.
ομοβροντία *s.f.* salvo, volley.
ομογάλακτος *a.* foster-brother *or* sister.
ομογενής *a.* of common descent; an expatriate Greek.
ομοεθνής *a.* a fellow countryman.
ομοειδής *a.* of the same species.
ομόθρησκος *a.* a co-religionist.
ομόθυμος *a.* unanimous.
ομοιάζω *v.i.* resemble *(with προς);* look alike.
ομοϊδεάτης *a.* of the same way of thinking *(esp. politically).*
ομοιο- *denotes* similarity.
ομοιογεν|ής *a.* homogeneous; even. ~εια *s.f.* homogeneity.
ομοιοκαταληξία *s.f.* rhyme.
ομοιόμορφος *a.* uniform.
ομοιοπαθ|ής *a.* a fellow sufferer. ~ητική *s.f. (med.)* homoeopathy.
όμοι|ος *a.* similar, alike, like, same. ~ότης *s.f.* similarity, resemblance.
ομοίωμα *s.n.* likeness, image.
ομοιωματικά *s.n.pl.* ditto-marks.

ομόκεντρος *a.* concentric.

ομολογία *s.f.* confession, admission; (*fin.*) debenture.

ομόλογον *s.n.* promissory note, bond.

ομολογουμένως *adv.* admittedly, by common consent.

ομολογώ *v.t.* confess, acknowledge.

ομόνοια *s.f.* concord, harmony (*between persons*).

ομοούσιος *a.* consubstantial.

όμορος *a.* with common frontier.

ομόρρυθμος *a.* ~ *εταιρεία* business partnership.

ομορφαίνω *v.t. & i.* embellish; become beautiful.

ομορφάντρας *s.m.* handsome man.

ομορφιά *s.f.* beauty, good looks.

ομορφονιός *s.m.* (*ironic*) foppish youth.

όμορφος *a.* good-looking, handsome, attractive.

ομόσπονδ|ος *a.* confederate. **~ία** *s.f.* federation, confederacy. **~ιακός** *a.* federal.

ομοταξία *s.f.* class (*in natural history*).

ομότιμος *a.* equivalent in rank *or* value; a peer; ~ *καθηγητής* professor emeritus.

ομού *adv.* together.

ομόφρων *a.* like-thinking.

ομοφυλοφιλία *s.f.* homosexuality.

ομοφωνία *s.f.* unanimity.

όμποε *s.n.* oboe.

ομπρέλλα *s.f.* umbrella, sunshade.

ομφαλ|ός *s.m.* navel. ~*ιος λώρος* umbilical cord. **~ικός** *a.* umbilical.

όμφ|αξ *s.m.* unripe grape; ~*ακες εισίν* sour grapes.

ομώνυμος *a.* homonymous. ~ *ρόλος* titlerole.

όμως *adv.* but, nevertheless.

ον *s.n.* being, creature.

όναγρος *s.m.* wild ass, zebra.

όναρ *s.n.* dream.

όνειδος *s.n.* shame, disgrace.

ονειρεύομαι *v.i. & t.* dream (of).

ονειρευτός *a.* dreamlike; dreamed of.

ονειροκρίτης *s.m.* dream-book.

όνειρο *s.n.* dream.

ονειροπολώ *v.i. & t.* day-dream; dream of, desire.

ονειρώδης *a.* like a dream (*in excellence*).

ονηλάτης *s.m.* donkey-driver.

όνομα *s.n.* name; (*gram.*) noun; *βγάζω* ~ acquire a reputation; *έχει το* ~ *του* it is his name-day; ~ *και πράμα* in very truth; ~*τι* by the name of.

ονομάζω *v.t.* name, call; appoint, nominate.

ονομασία *s.f.* naming; nomination; name.

ονομαστί *adv.* by name.

ονομαστικ|ός *a.* nominal; ~*ή εορτή* name-

day. ~*ή* *s.f.* (*gram.*) nominative.

ονομαστός *a.* renowned.

ονοματεπώνυμον *s.n.* Christian name and surname.

ονοματολογία *s.f.* nomenclature.

όνος *s.m.f.* ass.

όντας *part.* being. (*as conj.*) when.

οντότης *s.f.* entity, being; individuality.

όντως *adv.* indeed, really.

όν|υξ *s.m.* 1. nail, claw, talon; *εξ απαλών* ~*ύχων* from earliest childhood. 2. onyx. 3. flange (*of wheel*).

οξαλικός *a.* oxalic.

οξαποδώ(ς) *s.m.* (*fam.*) the devil.

οξεία *s.f.* (*gram.*) acute accent.

οξείδ|ιον *s.n.* oxide. **~ωσις** *s.f.* oxidization.

οξικός *a.* acetic.

όξινος *a.* sour, acid.

όξος *s.n.* vinegar.

οξύ *s.n.* acid.

οξυά *s.f.* beech.

οξυγονοκόλλησις *s.f.* oxy-acetylene welding.

οξυγόνον *s.n.* oxygen.

οξυγονούχ|ος *a.* containing oxygen; ~*ον ύδωρ* peroxide of hydrogen.

οξυδέρκεια *s.f.* keen sight; (*fig.*) discernment.

οξυζενέ *s.n.* peroxide.

οξύθυμος *a.* irascible.

οξύνοια *s.f.* perspicacity.

οξύνους *a.* shrewd.

οξύνω *v.t.* make pointed; sharpen (*wits, etc.*); intensify, exacerbate (*feelings*); (*gram.*) mark with acute accent.

οξύς *a.* sharp, acute, piercing.

οξύτης *s.f.* sharpness, acidity.

οξύτονος *a.* (*gram.*) with acute accent on last syllable.

οξύφωνος *s.m.* (*mus.*) tenor.

όξω *adv. see* **έξω.**

οπαδός *s.m.f.* follower, adherent.

όπερ *pron.* which.

όπερα *s.f.* opera.

οπή *s.f.* hole, opening, cavity.

όπιον *s.n.* opium.

όπισθεν *adv.* behind; *κάνω* ~ reverse, go backwards; (*prep.*) ~ *μου* behind me.

οπίσθι|ος *a.* rear, back; *τα* ~*α* back, hindquarters.

οπισθοβουλία *s.f.* ulterior motive.

οπισθογραφώ *v.t.* endorse (*cheque, etc.*).

οπισθοδρόμ|ησις *s.f.* retrogression. **~ικός** *a.* retrograde; old-fashioned (*in outlook*).

οπισθοφυλακή *s.f.* rearguard.

οπισθοχώρησις *s.f.* retreat, giving way.

οπίσω *adv. see* **πίσω.**

οπλαρχηγός *s.m.* leader of irregular

soldiers.

οπλή *s.f.* hoof.

οπλίζω *v.t.* arm; (*fig.*) strengthen.

όπλισις *s.f.* loading.

οπλισμός *s.m.* armaments; armature (*of magnet*); (*mus.*) key-signature.

οπλίτ|ης *s.m.* infantry soldier. **~αγωγόν** troop-ship.

όπλ|ον *s.n.* weapon; rifle; fighting service; **~α** (*pl.*) arms; τα τρία **~α** the three fighting services. **~οπολυβόλον** *s.n.* light machine-gun.

οπλοστάσιον *s.n.* armoury, arsenal.

οπλοφορία *s.f.* carrying of arms (*by civilians*).

οπόθεν *adv.* whence; **~** διήλθον whichever way I passed.

οποίος *pron.* of what kind; what (*exclamatory*); ο **~** who, which.

όποιος *pron.* whoever, whichever; **~** κι **~** anybody (*the first comer*).

οποιοσδήποτε *pron.* whosoever, whatsoever; any... at all.

οπός *s.m.* sap, juice.

όποτε, ~δήποτε *adv.* whenever.

οπότε *adv.* when (*at which time*); in which case.

όπου *adv.* where, wherever; **~** και να 'ναι at any moment *or* wherever he may be.

ό|που, ~πού *pron.* he who; whom, which.

οπουδήποτε *adv.* wheresoever.

οπτασία *s.f.* a vision.

οπτικ|ός *a.* optical, optic; an optician. **~ή** *s.f.* optics.

οπτ|ός *a.* roast; **~ή** γη terracotta.

οπωρικά *s.n.pl.* fruit.

οπωροπωλείον *s.n.* fruiterer's.

οπωροφόρ|ος *a.* **~α** δένδρα fruit-trees.

όπως 1. *adv.* as, like; **~ ~** somehow or other, by hook or by crook, after a fashion. 2. *conj.* (in order) to.

οπωσδήποτε *adv.* in any case; definitely; **~** σωστά more or less correctly.

οπωσούν *adv.* somewhat, fairly.

όραμα *s.n.* a vision. **~τίζομαι** *v.i.* have visions.

όρασις *s.f.* (*sense of*) sight, vision.

ορατ|ός *a.* visible. **~ότης** *s.f.* visibility.

οργανέτο *s.n.* barrel-organ; (*fig.*) paid supporter, tool.

οργανικός *a.* organic; instrumental.

οργανισμός *s.m.* organism; (*bodily*) system, constitution; organization.

όργανον *s.n.* organ, instrument, agent.

οργανώνω *v.t.* organize.

οργάνωσις *s.f.* organization.

οργανωτής *s.m.* organizer.

οργασμός *s.m.* rut, heat; (*fig.*) feverish

activity.

οργή *s.f.* rage; (*fam.*) να πάρει η **~** damn it! δίνω τόπο στην **~** give way, not insist.

όργια *s.n.pl.* orgies; (*fig.*) corrupt practices. **~άζω** *v.i.* hold orgies; (*fig.*) run riot.

οργίζω *v.t.* anger.

οργίλος *a.* angry; irascible.

οργ(υ)ιά *s.f.* fathom.

οργώ *v.i.* be in rut; (*fig.*) be avid.

όργ|ωμα *s.n.* ploughing. **~ώνω** *v.t.* plough.

ορδή *s.f.* horde.

ορέγομαι *v.t.* feel desire for.

ορειβασία *s.f.* mountaineering.

ορειβάτης *s.m.* climber.

ορειβατικ|ός *a.* of *or* for climbing; **~όν** πυροβολικόν mountain artillery.

ορεινός *a.* of the mountains; mountainous.

ορείχαλκος *s.m.* brass.

ορεκτικ|ός *a.* appetizing; tempting. **~ό** *s.n.* appetizer; hors-d'oeuvre.

ορεξάτος *a.* keen, eager.

όρεξη *s.f.* appetite; desire.

ορεσίβιος *a.* mountain-dwelling.

ορθά *adv.* upright; right, correctly; **~** κοφτά plainly, frankly.

ορθάνοιχτος *a.* wide open.

όρθιος *a.* upright, erect, standing.

ορθογραφία *s.f.* spelling, orthography.

ορθογώνι|ος *a.* rectangular, right-angled. **~ον** *s.n.* rectangle.

ορθόδοξ|ος *a.* orthodox. **~ία** *s.f.* orthodoxy; Orthodox Christianity.

ορθολογισμός *s.m.* rationalism.

ορθοπεδ|ία, ~ική *s.f.* orthopedics.

ορθοποδώ *v.i.* stand erect; (*fig.*) prosper again.

ορθ|ός *a.* upright, erect; straight; right, correct; **~ή** γωνία right-angle. **~ώς** *adv.* right, correctly.

ορθοστασία *s.f.* standing.

ορθότης *s.f.* rightness.

ορθοφροσύνη *s.f.* right thinking.

ορθοφωνία *s.f.* elocution.

όρθρος *s.m.* (*eccl.*) matins.

ορθών|ω *v.t.* raise. **~ομαι** *v.i.* rise up, stand on end; rebel.

οριζόντιος *a.* horizontal.

ορ|ίζω *v.t.* & *i.* bound, delimit; fix, define; rule, control; give orders (to); **~ίστε** invitation to do, say, accept, or contemplate something; **~ίστε;** I beg your pardon? καλώς **~ίσατε** welcome! ας **~ίσει** let him come in; τι **~ίζετε;** what is your pleasure?.

ορίζων, *s.m.* horizon.

όριον *s.n.* boundary, limit.

ορισμένος *a.* fixed, appointed; certain (*unspecified*).

ορισμός *s.m.* fixing, appointing; definition;

order, command.
οριστικ|ός *a.* definite, decisive. **~ή** *s.f.* (*gram.*) indicative.
ορκίζ|ω *v.t.* put on oath. *~ομαι v.i.* swear.
όρκος *s.m.* oath.
ορκωμοσία *s.f.* taking of oath.
ορκωτ|ός *a.* sworn; *~όν δικαστήριο* jury court.
ορμαθός *s.m.* bunch, string (*of keys, onions, etc.*).
ορμέμφυτ|ος *a.* instinctive. **~ον** *s.n.* instinct.
ορμή *s.f.* onrush, violence; impetuousness, passion.
ορμήνεια *s.f.* advice.
ορμηνεύω *v.t.* advise, instruct.
ορμητήριον *s.n.* base, starting-point.
ορμητικός *a.* violent, impetuous.
ορμιά *s.f.* fishing-line.
ορμόνη *s.f.* hormone.
όρμος *s.m.* bay.
ορμώ *v.i.* rush. *~ώμαι v.i.* be motivated (*by*); originate (*from*).
όρνεον *s.n.* bird of prey.
όρνιθα *s.f.* hen.
ορνίθι *s.n.* pullet; bird.
ορνιθοσκαλίσματα *s.n.pl.* (*fig.*) scrawl.
ορνιθοτροφείο *s.n.* poultry-farm.
ορνιθών *s.m.* chicken-coop.
όρνιο *s.n.* bird of prey, vulture; (*fig.*) blockhead.
ορνιός *s.m.* wild fig.
όρνις *s.f.* hen.
ορντινάντσα *s.f.* batman.
οροθεσία *s.f.* demarcation of boundaries.
ορολογία *s.f.* terminology.
οροπέδιον *s.n.* plateau.
όρος *s.n.* mountain.
όρ|ος *s.m.* term (*limit, word*); condition, stipulation; *~οι* (*pl.*) conditions (*terms, circumstances*); *εφ' ~ου ζωής* for life; *φθάνω εις το προς ον ~ον* reach one's destination; *κατά μέσον ~ον* on the average; *ανώτατος ~ος* maximum.
ορ(ρ)ός *s.m.* whey, serum.
οροσειρά *s.f.* mountain-range.
ορόσημο *s.n.* boundary-mark.
οροφή *s.f.* ceiling, roof.
όροφος *s.m.* floor, story.
ορτανσία *s.f.* hydrangea.
ορτσάρω *v.i.* (*naut.*) luff.
όρτ|υξ *s.m.* **~ύκι** *s.n.* quail.
όρυγμα *s.n.* trench, pit.
όρυζα *s.f.* rice.
ορυκτέλαιον *s.n.* petroleum; lubricating oil.
ορυκτ|ός *a.* mineral. **~ολογία** *s.f.* mineralogy. **~όν** *s.n.* mineral.
ορύσσω *v.t.* dig, excavate.
ορυχείον *s.n.* mine, quarry.

ορφανεύω *v.i.* become an orphan.
ορφαν|ός *a.* orphan. **~οτροφείον** *s.n.* orphanage.
ορχήστρα *s.f.* orchestra.
όρχις *s.f.* 1. testicle. 2. orchid.
όρχος *s.m.* (*mil.*) park.
ορχούμαι *v.i.* dance.
ορώ *v.t.* see.
ος *pron.* who, whom, which; *ο εστί* that is to say; *αφ' ης* since (the time when); *δι' ο* for which reason; *καθ' ο* as, qua; *καθ' ην στιγμήν* at the moment when.
οσάκις *adv.* as often as, whenever.
όσ|ιος *a.* holy, blessed; (*fam.*) *~ία Μαρία* person affecting prudishness *or* innocence.
οσμή *s.f.* odour.
όσ|ο(ν) *adv.* as much *or* long as; *~ο. . . τόσο* the more. . . the more; *~ο τό δυνατόν νωρίτερα* as early as possible; *~ο για* as for; *εφ' ~ον* inasmuch as, as long as; *~ον αφορά* as regards; *~ο να 'ρθει* until *or* by the time he comes; *~ο να 'ναι* for all that; *~ο και να βρέχει* however much it may rain.
οσονούπω *adv.* soon, shortly.
όσ|ος *pron.* as much *or* many as, all; *πουλώ ~α ~α* sell at any price.
όσπριο *s.n.* pulse (*legume*).
οστεοφυλάκιον *s.n.* ossuary.
οστεώδης *a.* bony.
όστια *s.f.* 1. (*eccl.*) Host. 2. (*med.*) cachet.
όστις *pron.* who, whoever.
οστούν *s.n.* bone.
οστρακιά *s.f.* scarlet fever.
όστρακ|ο *s.n.* shell; potsherd. **~όδερμος** *a.* crustacean.
όστρια *s.f.* south wind.
οσφραίνομαι *v.t.* smell.
όσφρησις *s.f.* (*sense of*) smell; keen nose.
οσφυαλγία *s.f.* lumbago.
οσφύς *s.f.* waist, loins.
όσχεον *s.n.* scrotum.
όταν *conj.* when.
ότε *conj.* when.
οτέ *adv.* ~ *μεν* . . . ~ *δε* sometimes . . . sometimes.
ότι 1. *conj.* that. 2. *adv.* as soon as; just (*a moment ago*).
ό,τι, ότι *pron.* what, whatever.
οτιδήποτε *pron.* whatsoever; anything at all.
ότου *pron.* *έως* ~ *or μέχρις* ~ until; *αφ'* ~ since.
οτρά *s.f.* needleful of thread.
ου *adv.* not.
ουγγιά *s.f.* ounce.
ούγια *s.f.* selvage.

ουδαμού *adv.* nowhere.
ουδαμώς *adv.* not at all.
ουδέ *adv. & conj.* not even; ~ ... ~ neither .
.. nor.
ουδείς *pron.* no, nobody.
ουδέποτε *adv.* never.
ουδέτερ|ος 1. *pron.* neither. 2. *a.* neutral;
(*gram.*) neuter. **~οποιώ** *v.t.* neutralize.
~ότης *s.f.* neutrality.
ουδόλως *adv.* by no means.
ουζερί *s.n.* place where ouzo, *etc.,* is sold.
ούζο *s.n.* spirit flavoured with aniseed;
cocktail party.
ουίσκι *s.n.* whisky.
ουκ *adv.* not.
ουλαμός *s.m.* (*mil.*) troop.
ουλή *s.f.* scar.
ούλο *s.n.* gum (*of mouth*).
ούρα *s.n.pl.* urine.
ουρά *s.f.* tail; train (*of dress*); queue; (*fam.*)
λεφτά με ~ a lot of money.
ουραγός *s.m.* one who brings up the rear.
ουρανής *a.* sky-blue.
ουράνιον *s.n.* uranium.
ουράνι|ος *a.* heavenly; *~ο τόξο* rainbow. *τα*
~α the skies.
ουρανίσκ|ος *s.m.* palate. **~όφωνος** *a.*
(*gram.*) palatal.
ουρανοκατέβατος *a.* heaven-sent.
ουρανοξύστης *s.m.* skyscraper.
ουραν|ός *s.m.* sky, heaven; canopy. **~όθεν**
adv. from heaven.
ούργιος *a.* muzzy.
ουρητήριο *s.n.* urinal.
ουρί *s.n.* houri.
ουρικός *a.* uric.
ούριος *a.* favourable (*wind*).
ουρλ|ιάζω *v.i.* howl, wail. **~ιασμα** *s.n.*
howl(ing), wail(ing).
ουροδοχείον *s.n.* chamber-pot.
ουρώ *v.i.* urinate.
ους *s.n.* ear.
ουσία *s.f.* substance; essence; flavour.
ουσιαστικ|ός *a.* substantial. **~όν** *s.n.* (
gram.) substantive.
ουσιώδης *a.* essential, vital.
ούτε *adv. & conj.* not even; ~ ... ~ neither
... nor.
ουτιδανός *a.* worthless (*person*).
ουτοπ|ία *s.f.* utopia. **~ικός** *a.* utopian.
ούτος *pron.* he, this (one).
ούτω(ς) *adv.* so, thus; ~ *ώστε* in such a way
as to.
ουχ *adv.* not; ~ *ήττον* none the less.
ουχί *adv.* not.
οφειλέτης *s.m.* debtor.
οφειλή *s.f.* debt.
οφείλ|ω *v.t. & i.* owe; ~ *να πάω* I ought to

go. *~ομαι v.i.* be due (*owed, caused*).
όφελος *s.n.* profit, advantage.
οφθαλμαπάτη *s.f.* optical illusion.
οφθαλμίατρος *s.m.f.* eye-specialist.
οφθαλμοπορνεία *s.f.* lascivious glances.
οφθαλμός *s.m.* eye; bud.
οφθαλμοφανής *a.* obvious.
όφ|ις *s.m.* snake, serpent. **~ιοειδής** *a.* snaky,
serpentine.
οφρύς *s.f.* brow, eyebrow; crest (*of hill*).
οφφίς *s.n.* domestic office. [F. *office*]
όχεντρα *s.f.* viper.
οχετ|ός *s.m.* brain. **~ευση** *s.f.* drainage.
όχημα *s.n.* vehicle, carriage. **~ταγωγόν** *s.n.*
car-ferry.
όχθη *s.f.* bank (*of river*), shore (*of lake*).
όχι *adv. & particle* not; no (*negative*
response).
οχιά *s.f.* viper.
οχλαγωγία *s.f.* disturbance, hubbub.
οχληρός *a.* tiresome.
οχλοβοή *s.f.* disturbance, hubbub.
οχλοκρατία *s.f.* mob-rule.
όχλος *s.m.* populace, mob.
οχτ- *see* **οκτ-**.
οχυρ|ός *a.* strong (*proof against attack*).
~όν *s.n.* strongpoint, fort.
οχυρών|ω *v.t.* fortify. **~ομαι** *v.i.* occupy
strong defensive position; (*fig.*) arm
oneself with excuse *or* pretext.
όφεται *v.i.* *ας* ~ he is to blame.
όψ|η *s.f.* appearance, look; sight; face, right
side, obverse; *εν ~ει* in view; *εξ ~εως* by
sight; *λαμβάνω υπ' ~ιν* take into
consideration.
οψιγενής *a.* posthumous (*child*); late (*in*
maturing).
όψιμος *a.* late (*in maturing*).
οψίπλουτος *a.* nouveau riche.
όψις *see* **όψη**.
οψοφυλάκιον *s.n.* larder.
οψώνιον *s.n.* purchase (*item*).

Π

παγάν|α, ~ιά *s.f.* shooting-party.
παγερός *a.* icy, frigid.
παγετός *s.m.* frost.
παγετώνας *s.m.* glacier.
παγίδ|α *s.f.* trap, snare. **~εύω** *v.t.* trap,
ensnare.
παγίδ|ι *s.n.* rib. **~άκι** *s.n. see* **παϊδάκι**.
πάγιος *a.* fixed, permanent; (*fin.*) consoli-
dated.
παγίς *s.f. see* **παγίδα**.
παγιώνω *v.t.* consolidate.

παγκάρι *s.n.* stall for candles in church.

πάγκος *s.m.* bench.

παγκόσμιος *a.* world-wide.

πάγκρεας *s.n.* (*anat.*) pancreas.

παγόβουνο *s.n.* iceberg.

παγοδρομία *s.f.* skating.

παγοπέδιλο *s.n.* skate.

παγοποιείον *s.n.* ice-factory.

πάγ|ος *s.m.* ice; frost. **~άκι** *s.n.* ice-cube.

παγούρι *s.n.* soldier's water-bottle.

παγώνι *s.n.* peacock.

παγωνιά *s.f.* frost.

παγωνιέρα *s.f.* ice-box.

παγώ|νω *v.t. & i.* freeze; make *or* get cold; **~μένος** frozen; iced. **~μα** *s.n.* freezing.

παγωτό *s.n.* ice-cream.

παζαρ|εύω *v.i. & t.* bargain (about). **~εμα** *s.n.* bargaining.

παζάρι *s.n.* bazaar, market; bargaining.

παθαίν|ω *v.i. & t.* suffer; *καλά να πάθει* it serves him right; *την έπαθε* (*fam.*) he got caught out, he copped it. **~ομαι** *v.i.* be moved, get carried away.

πάθημα *s.n.* setback, misfortune.

πάθηση *s.f.* malady.

παθητικόν *s.n.* (*fin.*) liabilities; (*fig.*) discredit.

παθητικός *a.* passive; charged with passion.

παθιασμένος *a.* ailing; (*fig.*) fanatical.

παθο|λογία *s.f.* pathology. **~λόγος** *s.m.* general practitioner.

πάθ|ος *s.n.* illness; passion; aminosity; vice; **~η** (*pl.*) sufferings, Passion (*of Christ*).

παθών *s.m.* ο ~ the victim.

παιάν *s.m.* paean. **~ίζω** *v.t. & i.* play (*of band*).

παιγνίδι *s.n.* game, sport; toy, plaything; trick; gambling; *~α* (*pl.*) rustic orchestra. **~άρης** *a.* playful; coquettish.

παίγνιον *s.n.* game, sport; toy.

παιγνιόχαρτο *s.n.* playing-card.

παιδαγωγική *s.f.* pedagogy.

παιδάκι *s.n.* little child.

παϊδάκι *s.n.* cutlet.

παιδάρι|ον *s.n.* boy (*about 12 ~ 15*). **~ώδης** *a.* puerile.

παιδεία *s.f.* education; learning.

παίδεμ|α *s.n.*, **~ός** *s.m.* torment.

παιδεραστής *s.m.* paederast.

παιδεύ|ω *v.t.* torment. **~ομαι** *v.i.* strive hard.

παιδί(ον) *s.n.* child; boy; (*fam.*) fellow, chap.

παιδιά *s.f.* game (*play*).

παιδιαρίζω *v.i.* behave like a child.

παιδίατρος *s.m.* (*med.*) paediatrician.

παιδικός *a.* of children, child's; childish.

παιδίσκη *s.f.* girl (*about 12 ~ 15*).

παιδομάζωμα *s.n.* forced levy of children; crowd of children.

παιζογελώ *v.t.* (*fam.*) lead up the garden path.

παίζω *v.i. & t.* play; gamble; *τα ~ με* flirt with; *του τις έπαιξα* I struck him.

παίκτης *s.m.* player; gambler.

παίνεμα *s.n.* praise.

παινεύ|ω *v.t.* praise. **~ομαι** *v.i.* boast.

παίξιμο *s.n.* play, playing; performance; gambling.

παίρν|ω *v.t. & i.* receive, get; take; contain, hold; marry; beat (*at game*); begin (*to*); *τον ~ω* fall asleep; *~ω και δίνω* have influence; *~ω πόδι* make off; *~ω αέρα* become cheeky; *~ω μπρος* start (*of motor*); *~ω από πίσω* follow closely; *~ω στο μεζέ* make a fool of; *~ω απάνω μου* recover (*in health*); *το ~ω απάνω μου* give oneself airs *or* undertake the job; *τον ~ω στο λαιμό μου* I am the cause of his ill fortune; *δεν ~ει από λόγια* he won't listen to reason; *μας ~ει η ώρα* we have time; *πήρε αέρας* the wind has risen; *μου πήρε το κεφάλι* he drove me distracted; *πήραν τα μυαλά του αέρα* he is too big for his boots; *έχω πάρε δώσε με* have contact with; *πάθηκαν* they got married; *μας πήρε η ώρα* we are late.

παις *s.m.f.* child; boy.

παιχνίδι *s.n.* see **παιγνίδι**.

πακέτο *s.n.* packet, package.

πάκο *s.n.* bundle.

πακτωλός *s.m.* (*fig.*) goldmine.

παλαβ|ός *a.* foolhardy, crazy. **~ώνω** *v.t.* drive *or* go mad.

παλάβρα *s.f.* talking big.

πάλαι, ~ά *adv.* of old.

παλαιικός *a.* of the old school.

παλαίμαχος *s.m.* veteran.

παλαιοελλαδίτης *s.m.* native of 'Old Greece'.

παλαιοημερολογίτης *s.m.* follower of Old Calendar.

παλαιόθεν *adv.* from ancient times.

παλαιοπώλης *s.m.* second-hand dealer.

παλαιός *a.* old (*not new*); former.

παλαιστής *s.m.* wrestler.

παλαίστρα *s.f.* arena.

παλαμάκια *s.n.pl.* hand-clapping.

παλαμάρι *s.n.* (*naut.*) cable.

παλάμη *s.f.* palm (*of hand*); span.

παλαμίδα *s.f.* bonito (*fish*).

παλαμίζω *v.t.* caulk.

παλάσκα *s.f.* cartridge-pouch.

παλάτι *s.n.* palace.

παλεύω *v.i.* wrestle, strive.

πάλη *s.f.* wrestling; struggle, conflict.

παληός *a.* see **παλαιός**.

πάλι *adv.* again, once more; on the other hand; back; ~ *καλά* it's not so bad (*as it might have been*); *άλλο* ~ there we go again!

παλιάλογο *s.n.* nag.

παλιάνθρωπος *s.m.* (*fam.*) rascal, bounder, scoundrel.

παλιατζής *s.m.* rag-and-bone man.

παλιάτσα *s.f.* old, worn-out object.

παλιάτσος *s.m.* clown.

παλιγγενεσία *s.f.* regeneration.

πάλιν *adv. see* **πάλι**.

παλινδρομικός *a.* reciprocating (*movement*).

παλινόρθωση *s.f.* restoration (*of régime*).

παλιννόστηση *s.f.* return to one's homeland.

παλινωδώ *v.i.* change one's tune.

παλιο- 1. *denotes* old. 2. *pejorative epithet.*

παλιογυναίκα *s.f.* (*pej.*) woman; tart.

παλιολλαδίτης *s.m.* native of 'Old Greece'.

παλιόπαιδο *s.n.* bad boy.

παλιός *a. see* **παλαιός**.

παλίρροια *s.f.* tide; ebb and flow.

παλιώνω *v.t. & i.* wear out.

παλκοσένικο *s.n.* stage (*theatre*).

παλλαϊκός *a.* of all the people.

παλλακεία *s.f.* concubinage.

πάλλευκος *a.* pure white.

παλ(λ)ηκάρ|ι *s.n.* young man; brave person; bachelor. **~άς** *s.m.* ruffian. **~ιά** *s.f.* bravery.

πάλλω *v.t. & i.* vibrate; throb.

παλμός *s.m.* pulsation, vibration; enthusiasm, feeling; rhythmic spring *or* throw (*in athletics*).

παλούκι *s.n.* stake, post; (*fig.*) difficulty; *του σκοινιού και του* ~*ού* (*fam.*) a thoroughly bad lot.

παλτό *s.n.* overcoat.

παμ- *see* **παν-**.

παμπάλαιος *a.* very old.

πάμπλουτος *a.* very rich.

πάμπολλοι *a.* very many.

παμφάγος *a.* omnivorous; voracious.

πάμφθηνος *a.* dirt cheap.

παμψηφία *s.f.* unanimous vote.

παν- *denotes* all, very.

παν *s.n.* everything; the whole world (*see also* **πας**).

Παναγία *s.f.* the Virgin Mary.

πανάγιος *a.* very holy.

πανάδα *s.f.* brown patch (*on skin*).

πανάκεια *s.f.* panacea.

πανάκριβος *a.* very expensive.

παναμάς *s.m.* panama hat.

πανδαιμόνιο *s.n.* pandemonium.

πανδαισία *s.f.* sumptuous feast.

πάνδημος *a.* of all the people.

πανδοχείο *s.n.* inn.

πανελλήνι|ος *a.* panhellenic. **~ον** *s.n.* the whole of Greece.

πανεπιστήμιο *s.n.* university.

πανέρι *s.n.* wide basket.

πανηγύρι *s.n.* patronal festival; fair; beanfeast; (*fam.*) row, commotion; *είναι για τα* ~*α* he is quite crazy *or* it is no earthly good.

πανηγυρίζω *v.t. & i.* celebrate (*victory*); (*v.i.*) rejoice.

πανηγυρικός 1. *a.* festive, with rejoicing. 2. *s.m.* oration; panegyric.

πανήγυρις *s.f.* patronal festival; fair.

πάνθεον *s.n.* pantheon.

πάνθηρας *s.m.* panther.

πανθομολογούμενος *a.* universally acknowledged.

πανικ|ός *s.m.* panic. **~όβλητος** *a.* panic-stricken.

πανίς *s.f.* fauna.

παν(ν)ί *s.n.* cloth; sail; baby's napkin; ~ *με* ~ without a penny; *ένα* ~ tarred with the same brush.

παννιάζω *v.i.* become pale *or* flabby.

παν(ν)ικά *s.n.pl.* drapery, linen.

παννυχίδα *s.f.* all-night service *or* revelry.

πανομοιότυπο *s.n.* facsimile.

πανοπλία *s.f.* arms, armour.

πανόραμα *s.n.* panorama. **~τικός** *a.* panoramic.

πανούκλα *s.f.* plague; (*fam.*) harridan.

πανούργος *a.* cunning, crafty.

πανσέληνος *s.f.* full moon.

πανσές *s.m.* pansy.

πανσιόν *s.f.* pension, boarding-house.

πανσπερμία *s.f.* mixture of races.

πανστρατιά *s.f.* (mobilization of) all available forces.

πάντα *adv.* always; in any case.

πάντα 1. *s.n.pl.* everything (*see also* **πας**). 2. *s.f. see* **μπάντα**.

πανταλόνι *s.n.* trousers; drawers.

παντατίφ *s.n.* pendant.

πανταχόθεν *adv.* from all parts.

πανταχού *adv.* everywhere.

πανταχούσα *s.f.* encyclical; (*fam.*) long string of reproaches.

παντελής *a.* complete, utter.

παντελόνι *s.n.* trousers; drawers.

παντέρ(η)μος *a.* lone, abandoned.

παντεσπάνι *s.n.* sponge-cake.

παντζάρι *s.n.* beetroot.

παντζούρι *s.n.* shutter.

παντιέρα *s.f.* banner, flag.

παντογνώστης *a.* omniscient.

παντοδύναμος *a.* omnipotent.

παντοειδ|ής *a.* of all kinds. **~ώς** *adv.* in

every way.
παντοκράτ|ωρ *s.m.* lord of all, Almighty.
παντομίμα *s.f.* dumb show.
παντοπώλ|ης *s.m.* grocer. **~είον** *s.n.* grocer's.
πάντοτε *adv.* always.
παντοτινός *a.* eternal.
παντού *adv.* everywhere.
παντόφλα *s.f.* slipper.
παντρειά *s.f.* marriage, wedlock.
παντρεμένος *a.* married.
παντρεύ|ω *v.t.* marry (*of priest, parent, sponsor*). **~ομαι** *v.t. & i.* marry (*spouse*); get married.
παντρολογώ *v.t.* seek a mate for.
πάντως *adv.* in any case, anyhow.
πάνω *adv. see* **απάνω.**
πανωλεθρία *s.f.* heavy loss *or* destruction.
πανώλης *s.f.* plague.
πανώριος *a.* very beautiful.
πανωφόρι *s.n.* overcoat.
παξιμάδα *s.f.* (*fam.*) light woman.
παξιμάδι *s.n.* 1. rusk. 2. nut (*for bolt*).
παπαγαλίζω *v.t.* repeat parrot-fashion.
παπαγάλος *s.m.* 1. parrot. 2. spanner.
παπαδιά *s.f.* priest's wife.
παπαδίστικος *a.* priest's.
παπαδίτσα *s.f.* ladybird.
παπάκι *s.n.* duckling.
παπάρα *s.f.* panada, pap; (*fam.*) dressing-down.
παπαρδέλες *s.f.pl.* twaddle.
παπαρούνα *s.f.* poppy.
πάπας *s.m.* pope.
παπάς *s.m.* priest; king (*cards*); (*fam.*) three-card trick.
παπατρέχας *s.m.* hasty person.
παπί *s.n.* duckling; γίνομαι ~ get soaking wet.
πάπια *s.f.* 1. duck; κάνω την ~ play dumb. 2. bedpan.
παπιγιόν *s.n.* bow-tie. [F. *papillon*]
παπικός *a.* papal.
πάπλωμα *s.n.* cotton quilt.
παπούτσ|ι *s.n.* shoe. **~ής** *s.m.* shoemaker *or* -seller.
παππάς *s.m. see* **παπάς.**
πάππ|ος, ~ούς *s.m.* grandfather.
πάπυρος *s.m.* papyrus.
παρά 1. *conj.* than; but. 2. *prep.* (*with acc.*) near, beside; contrary to, despite; except, less; μία ~ τέταρτο a quarter to one; μέρα ~ μέρα every other day; παρ' αξίαν undeservedly; ~ τρίχα by a hairsbreadth; ~ λίγο (να) nearly; παρ' όλο που even though. (*with gen.*) from, on the part of; by. (*with dat.*) in, among, with, attached to.
πάρα *adv.* (*of intensity*) ~ πέρα farther off; ~

πολύ very *or* too much.
παρα- *denotes* near, beyond, contrary, excess, *etc.*
παραβαίνω *v.t.* break, infringe.
παραβάλλω *v.t.* compare.
παραβάν *s.n.* screen. [F. *paravent*]
παραβαρύνω *v.t. & i.* overload; be a burden to; become overweight, lose one's mobility.
παράβαση *s.f.* infringement.
παραβάτης *s.m.* contravener, transgressor.
παραβγαίνω *v.i. & t.* compete; (*with gen.*) rival.
παραβιάζομαι (-j-) *v.i.* be in too much of a hurry.
παραβιάζω (-i-) *v.t.* force (*entry, door, etc.*), break open; infringe.
παραβλέπω *v.t. & i.* neglect, overlook; turn a blind eye.
παραβολή *s.f.* comparison; parable; (*math.*) parabola.
παράβολο *s.n.* fee, deposit (*in elections, etc.*).
παραβρίσκομαι *v.i.* be present, attend.
παραγάδι *s.n.* trawl-line.
παραγγελί|α *s.f.* instruction, order; message; επί ~α made to order. **~οδόχος** *s.m.* (*commercial*) agent.
παραγγέλλω *v.t. & i.* order, prescribe; send word.
παράγγελμα *s.n.* (word of) command.
παραγεμίζω *v.t. & i.* overfill; become too full; (*cooking*) stuff.
παραγίνομαι *v.i.* become to excess; go too far; get overripe; ~ χοντρός grow too fat.
παραγιός *s.m.* apprentice, boy.
παράγκα *s.f.* wooden hut.
παραγκωνίζω *v.t.* elbow, thrust aside; **~ισμένος** overlooked, left out of things (*at gathering*).
παραγνωρίζω *v.t. & i.* 1. know (*person*) very well. 2. fail to appreciate; mistake identity of; be mistaken (*as to identity*).
παραγραφή *s.f.* (*law*) lapse of right *or* penalty.
παράγραφος *s.m. & f.* paragraph; άλλη ~ a different matter.
παράγω|ω *v.t.* yield, produce. **~ομαι** *v.i.* (*gram.*) be derived.
παραγωγ|ή *s.f.* yield; production; (*gram.*) derivation. **~ικός** *a.* productive. **~ός** *s.m.* producer, grower.
παράγωγο *s.n.* derivative, by-product.
παράγων *s.m.* factor; leading figure.
παραδάκι *s.n.* (*fam.*) money.
παράδειγμα *s.n.* example; ~τος χάριν for example. **~τικός** *a.* exemplary.
παραδειγματίζ|ω *v.t.* make an example of

(*punish*); warn by example. ~*ομαι v.i.* learn by experience; (*with από*) model oneself on.

παράδεισος *s.m.* paradise.

παραδεκτός *a.* accepted; acceptable, admissible.

παραδέρνω *v.i.* toss, struggle. ~**δαρμένη** *s.f.* (*fam.*) belly.

παραδέχομαι *v.t.* admit, acknowledge.

παραδίδ|ω *v.t.* hand over; surrender; give (*lessons*), teach. ~*ομαι v.i.* surrender, give oneself up.

παραδοξολογία *s.f.* paradox ⟨ *figure of speech*).

παράδοξος *a.* strange, odd, paradoxical.

παραδόπιστος *a.* worshipping money.

παράδοσ|η *s.f.* handing over; surrender; teaching; tradition. ~**ιακός** *a.* traditional.

παραδοτέος *a.* due for delivery.

παραδουλεύτρα *s.f.* charwoman.

παραδοχή *s.f.* acceptance.

παραδρομή *s.f.* inadvertence.

παραείμαι *v.i.* be exceedingly *or* too much.

παραέχω *v.t.* have a great deal *or* too much of.

παραζάλη *s.f.* disorder, bustle.

παραθαλάσσιος *a.* by the sea.

παραθερ|ίζω *v.i.* spend the summer. ~**ιστής** *s.m.* summer visitor.

παραθετικός *a.* comparative.

παραθέτω *v.t.* juxtapose; compare; cite, quote; serve, offer (*food, drink*).

παρά|θυρο, ~θύρι *s.n.* window.

παραθυρόφυλλο *s.n.* shutter.

παραίνεση *s.f.* advice.

παραίσθηση *s.f.* hallucination.

παραίτηση *s.f.* (*act of*) resignation.

παραιτ|ώ *v.t.* give up, let go, forsake. ~*ούμαι v.i. & t.* resign, abdicate; (*with gen.*) desist from.

παρακάθημαι *v.i.* attend ⟨*as guest, etc.*).

παρακάθομαι *v.i.* stay too long; eavesdrop.

παράκαιρος *a.* at the wrong time.

παρακαλε(σ)τός *a.* whose favour has to be sought.

παρακάλια *s.n.pl.* entreaties.

παρακαλώ *v.t.* ask, request (*someone*); ~*!* please, don't mention it.

παρακά(μ)ν|ω *v.t.* overdo; ~*ει κρύο* it is very cold.

παρακαμπτήριος *s.f.* diversion (*of route*).

παρακάμπτω *v.t.* get round.

παρακαταθήκη *s.f.* deposit; stock; heritage.

παρακάτ|ω *adv.* farther down. ~**ιανός** *a.* of inferior quality; low-class.

παρακεί *adv. see* **παρέκει.**

παρακείμενος *a.* adjacent; (*gram.*) ο ~ perfect tense.

παρακεντές *s.m.* (*fig.*) parasite.

παρακέντησις *s.f.* (*med.*) puncture, tapping.

παρακινώ *v.t.* incite.

παρακλάδι *s.n.* branch, offshoot.

παράκληση *s.f.* entreaty; (*ritual*) prayer.

παρακμή *s.f.* decline, decadence.

παρακοή *s.f.* disobedience.

παρακολούθηση *s.f.* following, attendance, observation, surveillance.

παρακολουθώ *v.t.* follow, keep up with; monitor, keep watch on; grasp meaning of; attend (*go to*).

παρακόρη *s.f.* adopted daughter; maid.

παρακούω *v.t. & i.* mishear; disobey.

παρακρατικός *a.* illegal (*but connived at by government*).

παρακράτος *s.n.* corruption in politics.

παρακρατώ *v.t. & i.* retain, hold back (*produce*); last too long.

παράκρουση *s.f.* mental derangement.

παράκτιος *a.* coastal.

παραλαβή *s.f.* receipt (*act of receiving*).

παραλαμβάνω *v.t.* take delivery of; take over; take (*partner, etc.*).

παραλείπω *v.t.* omit.

παράλειψη *s.f.* omission.

παραλέω *v.t.* exaggerate.

παραλήγουσα *s.f.* penultimate syllable.

παραλήπτης *s.m.* recipient, addressee.

παραλήρημα *s.n.* delirium.

παραλής *a.* (*fam.*) rich.

παράλια *s.n.pl.* coast.

παραλί|α *s.f.* sea-shore, sea-front. ~**ακός** *a.* by the sea-shore.

παραλλαγή *s.f.* variation, variant.

παράλληλ|ος *a.* parallel. ~**ως** *adv.* correspondingly. ~**ισμός** *s.m.* parallel.

παραλογιάζω *v.t. & i.* drive *or* become mad.

παραλογίζομαι *v.i.* talk nonsense.

παράλογος *a.* unreasonable, absurd.

παραλυσία *s.f.* profligacy; shakiness.

παράλυση *s.f.* paralysis; shakiness.

παραλυτικός *a.* paralytic.

παράλυτος *a.* loose, shaky; paralysed; profligate.

παραλύ|ω *v.t. & i.* make loose *or* shaky; paralyse; become shaky *or* paralysed *or* profligate; ~*μένος* profligate.

παραμάν|α)α *s.f.* 1. wet-nurse, nanny. 2. safety-pin.

παραμελώ *v.t.* neglect.

παραμένω *v.i.* stay, remain, continue to be.

παράμερ|α *adv.* aside, out of the way. ~**ος** *a.* secluded.

παραμερίζω *v.t. & i.* thrust aside; ward off; put *or* get out of the way.

παραμικρός *a.* slightest.

παραμιλώ *v.i.* talk too much; rave; talk in

one's sleep; talk to oneself.

παραμονεύω *v.t.* lie in wait for; spy on.

παραμονή *s.f.* sojourn; eve.

παραμορφώνω *v.t.* disfigure, distort.

παραμπαίν|ω *v.i.* go in too far *or* too often; μου ~ει (*fam.*) he gets my goat.

παραμύθι *s.n.* fairy-tale.

παραμυθία *s.f.* consolation.

παρανόηση *s.f.* misinterpretation.

παράνομ|ος *a.* illegal, illicit. ~ία *s.f.* illegality; unlawful act. ~ώ *v.i.* break the law.

παρανοώ *v.t.* misinterpret.

παράνυμφος *s.m.f.* sponsor (*at wedding*).

παρανυχίδα *s.f.* agnail.

παραξενεύ|ω *v.t. & i.* seem strange to; become fussy. ~ομαι *v.i.* find it strange.

παράξενος *a.* strange, peculiar; fussy.

παραξηλώνω *v.t.* unstitch; (*fam.*) το ~ overdo it.

παραοικονομία *s.f.* black economy.

παραπαίω *v.i.* stagger; (*fig.*) talk disconnectedly.

παραπάν|ω *adv.* higher up; in addition; με το ~ω very much; ~ω από over, more than. ~ήσιος *a.* to spare, superfluous; excess (*charge, etc.*).

παραπατώ *v.i.* take a false step, slip.

παραπείθω *v.t.* persuade (*to wrong course*); dissuade (*from right course*).

παραπέμπω *v.t.* refer; send on; (*law*) commit, impeach. (*naut.*) convoy.

παραπέρα *adv.* farther on.

παραπέτασμα *s.n.* curtain.

παραπετώ *v.t.* cast aside, leave uncared for.

παραπέφτω *v.i.* get mislaid.

παράπηγμα *s.n.* wooden hut; ~τα (*pl.*) hutment.

παραπλανώ *v.t.* lead astray; seduce.

παραπλεύρως *adv.* next door; to the side.

παραπλήσιος *a.* similar.

παραποιώ *v.t.* counterfeit; falsify.

πάρα πολύ *adv.* very much; too much.

παραπομπή *s.f.* reference (*to book, etc.*); referring (*sending on*); committal, impeachment.

παραπονετικός *a.* plaintive.

παραπονιάρ|ης *a.* complaining. ~ικος *a.* plaintive.

παράπον|ο *s.n.* complaint, grievance. ~ιέμαι *v.i.* complain.

παραπόρτι *s.n.* wicket-gate.

παραποτάμιος *a.* riparian.

παραπόταμος *s.m.* tributary (*river*).

παράπτωμα *s.n.* breach, infringement.

παράρτημα *s.n.* branch (*of bank, etc.*); annexe, supplement; special edition (*of paper*).

παρ|άς *s.m.,* ~άδες (*pl.*) (*fam.*) money.

παράσημο *s.n.* decoration, order.

παρασιτικός *a.* parasitic.

παράσιτ|ο *s.n.* parasite; ~α (*pl.*) atmospherics.

παρασιωπώ *v.t.* avoid mentioning.

παρασκευ|άζω *v.t.* prepare. ~ασμα *s.n.* preparation (*substance*).

παρασκευή *s.f.* 1. preparing. 2. Friday.

παρασκήνια *s.n.pl.* wings (*theatre*). ~κός *a.* behind-the-scenes.

παρασπονδώ *v.i.* break an agreement.

παρασταίνω *v.t. see* **παριστάνω.**

παράστασ|η *s.f.* representation, depiction; presence, bearing; performance (*of play, etc.*); (*law*) appearance; ~εις (*pl.*) representations.

παραστάτης *s.m.* 1. one who stands by you. 2. jamb.

παραστατικός *a.* graphic, vivid; ~ χάρτης sketch-map.

παραστέκ|ω, ~ομαι *v.t.* (*with gen.*) stand by; look after.

παράστημα *s.n.* presence, bearing.

παραστράτημα *s.n.* moral lapse.

παρασύνθημα *s.n.* countersign.

παρασύρω *v.t.* sweep away (*by force*); run over; lead on (*into error*).

παράταιρος *a.* odd, not matching.

παράταξ|ις *s.f.* array, order, parade; μάχη εκ ~εως pitched battle; (*political*) party; (*gram.*) parataxis.

παράταση *s.f.* prolongation, extension (*in time*); respite.

παρατάσσω *v.t.* range in line, draw up.

παρατατικός *s.m.* (*gram.*) imperfect tense.

παρατείνω *v.t.* prolong.

παρατεταμένος *a.* prolonged.

παρατήρηση *s.f.* observation; remark; reproof.

παρατηρητ|ής *s.m.* observer. ~ικός *a.* observant; reproving. ~ικότης *s.f.* power of observation.

παρατηρώ *v.t. & i.* observe, remark, notice; reprove.

παράτολμος *a.* rash.

παράτονος *a.* out of tune.

παρατραβώ *v.t. & i.* drag out, prolong; last too long; (*fig.*) go too far.

παρατράγουδο *s.n.* (*untoward*) incident.

παρατρέχω *v.t. & i.* skip, omit; run a race; hasten hither and thither.

παρατσούκλι *s.n.* nickname.

παρατυγχάνω *v.i.* happen to be present.

παρατυπία *s.f.* breach of etiquette.

παρατώ *v.t. see* **παραιτώ.**

πάραυτα *adv.* at once.

παραφέρνω *v.t. & i.* bring too much; see a resemblance in; bear a resemblance.

παραφέρ|ω *v.t.* rouse to passion. ~ομαι *v.i.*

be carried away.
παραφθορά *s.f.* (*linguistic*) corruption.
παραφορά *s.f.* passion.
παραφορτών|ω *v.t.* overload. ~ομαι *v.i.* & *t.* become overloaded; pester.
παράφραση *s.f.* paraphrase.
παράφρ|ων *a.* mad. ~ονώ *v.i.* go mad. ~οσύνη *s.f.* madness.
παραφυάς *s.f.* sucker, shoot; scion.
παραφυλ|άττω, ~άω *v.t.* & *i.* lie in wait (for).
παραφωνία *s.f.* wrong *or* jarring note.
παραχαράκτης *s.m.* counterfeiter.
παραχειμάζω *v.i.* spend the winter.
παραχρήμα *adv.* at once.
παραχώνω *v.t.* bury, hide away; stuff (*into*).
παραχωρ|ώ *v.t.* grant, concede; relinquish. ~ηση *s.f.* concession.
παρδαλ|ός *a.* spotted; many-coloured, gaudy; ~ή γυναίκα light woman.
παρέα *s.f.* company, party; κάνε μου~ keep me company.
πάρεδρος *s.m.* (*law*) assessor.
παρειά *s.f.* cheek.
παρείσακτος *a.* an intruder.
παρεισ|δύω, ~φρέω *v.i.* intrude oneself.
παρεκβαίνω *v.i.* digress.
παρέκβαση *s.f.* digression.
παρέκει *adv.* further on.
παρεκκλήσι(ον) *s.n.* chapel.
παρεκκλίνω *v.i.* deviate; swerve.
παρεκτός *prep.* (*with gen.*) except.
παρεκτροπή *s.f.* deviation; moral aberration.
παρέλαση *s.f.* parade, march-past.
παρελαύνω *v.i.* parade, march past.
παρέλευση *s.f.* lapse, passage (*of time*).
παρελθ|ών *a.* past; το ~όν the past.
παρέλκει *v.i.* it is superfluous.
παρελκύω *v.t.* protract, draw out.
παρεμβαίνω *v.i.* intervene, step in.
παρεμβάλλω *v.t.* interpose.
παρέμβαση *s.f.* intervention.
παρεμπιπτόντως *adv.* incidentally.
παρεμφερής *a.* similar.
παρενέργεια *s.f.* side-effect.
παρεν|θέτω *v.t.* insert. ~θεσις *s.f.* insertion; parenthesis.
παρενοχλώ *v.t.* harass, annoy.
παρεξηγ|ώ *v.t.* misunderstand, take amiss. ~ηση *s.f.* misunderstanding.
παρεπιδημώ *v.i.* be staying temporarily (*abroad*).
παρεπόμενα *s.n.pl.* consequences.
πάρεργον *s.n.* side-line, parergon.
παρερμηνεία *s.f.* wrong interpretation.
παρέρχομαι *v.i.* & *t.* pass, go by (*of time*); come to an end; pass over, skip.

παρεστ|ώς *a.* present; οι ~ώτες those present.
παρευθύς *adv.* immediately.
παρευρίσκομαι *v.i.* be present.
παρεφθαρμένος *a.* debased, corrupt (*linguistically*).
παρέχω *v.t.* afford, provide; occasion.
παρηγορ|ία *s.f.* consolation. ~ώ *v.t.* console. ~ητής *s.m.* comforter.
παρήκοος *a.* disobedient.
παρήλιξ *a.* advanced in years.
παρήχησις *s.f.* alliteration.
παρθένα *s.f.* virgin.
παρθεναγωγείον *s.n.* girls' school.
παρθενί|α *s.f.* virginity. ~κός *a.* virginal; (*fig.*) maiden, first.
παρθένος *s.f.* & *a.* virgin.
Παρθενών *s.m.* Parthenon.
παρίας *s.m.* outcast.
παρίσταμαι *v.i.* be present.
παριστάμενος *s.m.* bystander.
παριστ|άνω, ~ώ *v.t.* represent, depict; perform (*role, play*); pretend to be.
παρκάρ|ω *v.t.* & *i.* park. ~ισμα *s.n.* parking.
πάρκο *s.n.* park.
παρντόν *int.* excuse me! [F. *pardon*]
πάροδ|ος *s.f.* side-street; passing (*of time, illness, etc.*). ~ικός *a.* passing, transient.
παροικία *s.f.* (*foreign*) colony, quarter.
παροιμί|α *s.f.* proverb. ~ώδης *a.* proverbial.
παρομοιάζω *v.t.* & *i.* liken; mistake (*for*); resemble (*with μέ*).
παρόμ|οιος *a.* similar, of the sort. ~οίωσις *s.f.* simile.
παρόν *see* **παρών.**
παρονομασία *s.f.* pun; nickname.
παρονομαστής *s.m.* (*math.*) denominator.
παροξύνω *v.t.* aggravate.
παροξύτονος *a.* accented on penultimate syllable.
παροπλίζω *v.t.* lay up (*ship*).
παρόραμα *s.n.* oversight; erratum.
παροργίζω *v.t.* provoke, anger.
παρόρμησις *s.f.* instigation; urge.
παροτρύνω *v.t.* incite, urge.
παρουσία *s.f.* presence, attendance; δευτέρα ~ second coming.
παρουσιάζ|ω *v.t.* present, produce. ~ομαι *v.i.* appear.
παρουσιάσιμος *a.* presentable.
παρουσιαστικόν *s.n.* (*person's*) appearance, presence.
παροχετεύω *v.t.* divert course of (*water, etc.*); lay on, supply with.
παροχή *s.f.* allocation, contribution, supplying.
παρρησία *s.f.* outspokenness.
πάρσιμο *s.n.* taking; trimming (*reducing*).

παρτέρι *s.n.* flower-bed.

παρτίδ|α *s.f.* lot, batch; (*completed*) game (*of cards, etc.*) **~ες** dealings.

παρτσαδιάζω *v.t.* cut up.

πάρτυ *s.n.* (*social*) party. [E. *party*]

παρυφή *s.f.* hem, selvage; edge.

παρωδία *s.f.* parody.

παρ|ών *a.* present; **το ~όν** the present.

παρωνυχίς *s.f.* agnail; (*fig.*) trifling matter.

παρωπίδες *s.f.pl.* blinkers.

πάρωρα *adv.* late at night.

παρωχημένος *a.* past.

πας *a.* all, every; *διά παντός* for ever; *οι πάντες* everybody; *Αγίων Πάντων* All Saints' Day.

πασαλεί|βω, ~φω *v.t.* smear, daub. **~βομαι** *v.i.* become smeared; get a smattering. **~μμα** *s.n.* daubing; smattering.

πασαμέντο *s.n.* skirting.

πασαπόρτι *s.n.* passport.

πασάρω *v.t.* pass; pass *or* palm off.

πασάς *s.m.* pasha.

πασατέμπο *s.n.* thing done to kill time. **~ς** *s.m.* roasted pumpkin seeds.

πασίγνωστος *a.* well-known.

πασιέντζα *s.f.* patience (*cards*).

πασπαλίζω *v.t.* cover, sprinkle (*with powder, etc.*).

πασπατεύω *v.t.* finger, feel, handle.

πάσσαλος *s.m.* stake, post.

πάσσ|ο *s.n.* step, pace; *με το ~ο* unhurriedly; *πάω ~ο* pass (*at cards, etc.*); *κάνω ~α* assume coquettish airs.

πάστα *s.f.* Italian pasta; a pastry, cake; (*fig.*) stuff, quality (*of character*).

παστέλι *s.n.* sweet of honey and sesame.

παστεριώνω *v.t.* pasteurize.

παστίτσιο *s.n.* baked macaroni.

παστός *a.* salted.

παστουρμάς *s.m.* highly seasoned cured meat.

πάστρ|α *s.f.* cleanness. **~εύω** *v.t.* clean.

παστρικός *a.* clean; (*iron.*) dubious (*character*).

παστώνω *v.t.* salt, cure.

Πάσχ|α *s.n.* Easter. **~άζω** *v.i. & t.* break a fast; eat (as) one's first fruit, *etc.*, of season.

πασχαλιά *s.f.* Easter; lilac.

πασχαλι|άτικος, ~νός *a.* of Easter.

πασχίζω *v.i.* try hard.

πάσχ|ω *v.i. & t.* suffer; *ο ~ων* the patient; *ο παθών* the victim.

πάταγ|ος *s.m.* bang, loud noise; (*fig.*) stir. **~ώδης** *a.* noisy, resounding.

πατάρι *s.n.* loft.

πατάσσω *v.t.* punish severely, crush.

πατάτα *s.f.* potato.

πατατούκα *s.f.* thick rustic overcoat.

πατέντα *s.f.* patent; *βλάκας ~* absolute fool.

πατέρας *s.m.* father.

πατερίτσα *s.f.* pastoral staff; crutch.

πατηκώνω *v.t.* press down *or* in, squash; *την ~* (*fam.*) gorge oneself.

πάτημα *s.n.* step, footstep; footprint; foothold; treading, pressing; pretext.

πατημασιά *s.f.* footprint.

πατήρ *s.m.* father.

πατινάδα *s.f.* serenade.

πατινάρω *v.i.* skate.

πατόκορφα *adv.* from top to toe.

πάτος *s.m.* bottom, base; sole, sock (*of shoe*).

πατούσα *s.f.* sole (*of foot*).

πατριάρχης *s.m.* patriarch.

πατρίδα *s.f.* native country *or* place.

πατρίκιος *a.* patrician.

πατρικός *a.* father's; fatherly.

πάτριος *a.* ancestral; native.

πατρίς *s.f. see* **πατρίδα**.

πατριώτ|ης *s.m.* fellow countryman; patriot; **~ικός** *a.* patriotic. **~ισμός** *s.m.* patriotism.

πατρογονικός *a.* paternal, ancestral.

πατρόν *s.n.* pattern (*dressmaking*). [F. *patron*]

πατροπαράδοτος *a.* traditional; hereditary.

πατρότητα *s.f.* paternity; authorship.

πατρυιός *s.m.* stepfather.

πατρώνυμο *s.n.* father's name.

πατρώος *a.* paternal; ancestral.

πατσαβούρα *s.f.* dishcloth, rag; (*fig.*) rag (*newspaper*); slut.

πατσάς *s.m.* (soup from) tripe.

πάτσ|ι *adv.* (*fam.*) quits. **~ίζω** *v.i.* become quits.

πατσουρός *a.* flat-rosed.

πατ|ώ *v.t. & i.* tread on; press (*bell, pen, etc.*); press down; raid; run over; (*v.i.*) tread, step, touch bottom; *~ω πόδι* set foot, put one's foot down; *~είς με ~ω σε* a crush *or* scrimmage.

πάτωμα *s.n.* floor, ground; story.

πατώνω *v.t. & i.* fix floor *or* bottom to; touch bottom.

παύλα *s.f.* dash (*punctuation*).

παυσίπονο *s.n.* painkiller.

παύσ|ις *s.f.* cessation; dismissal; (*mus.*) rest; **~εις** (*pl.*) vacation.

παύω *v.t. & i.* stop, cease doing; cause to cease; dismiss; (*v.i.*) stop, leave off.

παφλασμός *s.m.* surging *or* swashing noise (*of waves*).

πάφυλας *s.f.* thin sheet of brass *or* tin.

παχαίνω *v.t. & i.* fatten; grow fat.

πάχνη *s.f.* hoar-frost.

παχνί *s.n.* manger.
πάχος *s.n.* thickness, fatness; fat.
παχουλός *a.* plump.
παχύδερμος *a.* thick-skinned; a pachyderm.
παχυλός *a.* gross, crass; fat (*salary*).
παχύνω *v.t. & i.* see **παχαίνω**.
παχύρρευστος *a.* viscous.
παχ|ύς *a.* thick, fat; rich in fat; ~*ιά λόγια* extravagant promises.
παχύσαρκος *a.* obese.
πά|ω *v.i. & t.* go; take (*convey*); ~*ω για* stand or be candidate for; ~*ει* it suits, it is fitting; ~*ει αυτό* it is past *or* done for *or* dead; ~*ει να πει* it means; ~*ει μεσημέρι* it is just on midday; ~*ει και έρχεται* it will do; *τα* ~*με καλά* we get on well; (*fam.*) ~*ει περίπατο* it is a washout *or* done for *or* gone; ~*ω να σκάσω* I am nearly bursting; *σου* ~*ει πολύ* it suits you well (*of dress*) *or* it ill becomes you (*to do sthing.*).
πεδιάδα *s.f.* plain.
πεδικλών|ω *v.t.* hobble; trip up. ~*ομαι v.i.* stumble.
πέδιλο *s.n.* sandal.
πεδιν|ός *a.* of the plain; flat; (*mil.*) ~*ον πυροβολικόν.* field-artillery.
πεδίον *s.n.* plain; ground, field (*of battle, research, etc.*); ~ *Άρεως* parade ground.
πεζή *adv.* on foot.
πεζικόν *s.n.* infantry.
πεζογραφία *s.f.* prose.
πεζοδρόμιο *s.n.* pavement.
πεζοναύτης *s.m.* (*mil.*) marine.
πεζοπορία *s.f.* walking.
πεζός *a.* on foot, pedestrian; in prose; (*fig.*) prosaic.
πεζούλι *s.n.* stone sill, parapet; low wall (*esp. as seat*); terrace (*on hillside*).
παθαίνω *v.i. & t.* die; (*fam.*) be the death of, kill.
πεθαμός *s.m.* death; (*fig.*) purgatory.
πεθερ|ός *s.m.* father-in-law. ~*ά s.f.* mother-in-law. ~*ικά s.n.pl.* in-laws.
πειθαναγκάζω *v.t.* coerce.
πειθαρχ|ία *s.f.* discipline. ~*ικός a.* disciplinary; obedient. ~*ώ v.t.* be obedient.
πειθήνιος *a.* docile.
πείθω *v.t.* persuade.
πειθώ *v.t.* persuasion.
πείνα *s.f.* hunger, starvation. ~*λέος a.* starving.
πειν|ώ *v.i.* be hungry *or* starving; ~*ασμένος* hungry.
πείρα *s.f.* experience, practice.
πείραγμα *s.n.* annoying, teasing; affection (*malady*).
πειρ|άζω *v.t.* vex, annoy; tease; upset, be bad for; disturb, disarrange; tempt;

~*αγμένος* hurt, offended, affected (*in mind*), tainted (*of meat*). (*v.i.*) ~*άζει;* does it matter?
πειρακτικός *a.* vexatious, wounding.
πείραμα *s.n.* experiment, ~*τικός a.* experimental. ~*τίζομαι v.i.* experiment.
πειρασμός *s.m.* temptation.
πειρατής *s.m.* pirate.
πειραχτήρι *s.n.* a tease.
πείσμα *s.n.* obstinacy; *στο* ~ *όλων* in despite of everyone; *το βάζω* ~ *να* firmly resolve to.
πεισματάρης *a.* obstinate, obdurate.
πεισματώδης *a.* dogged, stubborn.
πεισματών|ω *v.t. & i.* spite; make *or* become obstinately resolved. ~*ομαι v.i.* become obstinate.
πείσμων *a.* obstinate.
πειστήριον *s.n.* (*material*) proof.
πειστικός *a.* persuasive, cogent.
πεκούνια *s.n.pl.* (*fam.*) money.
πέλαγ|ος *s.n.* (open) sea. ~*οδρομώ v.i.* sail the seas; (*fig.*) dither. ~*ώνω v.i.* (*fig.*) feel lost *or* swamped.
πελαργός *s.m.* stork.
πελάτ|ης *s.m.* customer, client, patient. ~*εία s.f.* clientele.
πελεκούδι *s.n.* chip, shaving.
πέλεκυς *s.m.* axe.
πελεκώ *v.t.* cut into shape, carve (*wood, stone*); (*fig.*) beat, thrash.
πελιδνός *a.* livid.
πέλμα *s.n.* sole (*of foot, shoe*); sock (*of shoe*); tread (*of tyre*).
πέλος *s.n.* pile, nap (*of cloth*).
πελούζα *s.f.* lawn.
πελτές *s.m.* see **μπελτές**.
πελώριος *a.* huge.
πέμπτ|ος *a.* fifth; ~*η* (*s.f.*) Thursday.
πεμπτουσία *s.f.* quintessence.
πέμπω *v.t.* send.
πενήντα *num.* fifty.
πένης *s.m.* indigent person.
πενθερός *s.m.* see **πεθερός**.
πένθιμος *a.* sad, mournful; mourning, funeral.
πένθ|ος *s.n.* grief; mourning. ~*ώ v.i.* mourn; be in mourning.
πενία *s.f.* indigence.
πενιχρός *a.* meagre, paltry, mean.
πέννα *s.f.* 1. pen. 2. plectrum. 3. penny.
πένομαι *v.i.* be indigent.
πένσα *s.f.* tweezers, forceps; dart (*dress-making*).
πεντ- *denotes* fivefold, extremely.
πεντάγραμμον *s.n.* (*mus.*) stave.
πεντάγωνον *s.n.* pentagon.
πεντακάθαρος *a.* very clean.

πεντακόσιοι *a.* five hundred.
πεντάλι *s.n.* pedal.
πεντάμορφος *a.* very beautiful.
πεντάρα *s.f.* (*fig.*) farthing.
πεντάρι *s.n.* figure 5; five (*cards*).
πέντε *num.* five; τον πήγε ~ ~ (*fam.*) he got the wind up.
πεντήκ|οντα *num.* fifty. **~οστός** *a.* fiftieth.
Πεντηκοστή *s.f.* Whit Sunday.
πεντοζάλης *s.m.* a Cretan dance.
πέος *s.n.* penis.
πεπαιδευμένος *a.* educated.
πεπατημένη *s.f.* the beaten track.
πεπειραμένος *a.* experienced.
πεπεισμένος *a.* convinced, sure.
πεπιεσμένος *a.* ~ χάρτης papier-mâché.
πέπλ|ος *s.m.*, **~ο** *s.n.* veil.
πεποίθηση *s.f.* conviction, confidence.
πεπόνι *s.n.* melon.
πεπραγμένα *s.n.pl.* proceedings (*report*).
πεπρωμένο *s.n.* destiny.
πεπτικός *a.* digestive (*organs*).
πέρα *adv.* over, far, on the other side; εκεί ~ over there; εδώ ~ over here; ~ (για) ~ through and through, completely; ~ από beyond, across; τα βγάζω ~ manage; ~ βρέχει (*fam.*) couldn't care less.
περαιτέρω *adv.* further; για τα ~ for appropriate action.
πέραμα *s.n.* ford; ferry.
πέραν *adv. & prep.* over, on the other side; (*with gen.*) beyond, across.
πέρας *s.n.* end, term, conclusion.
πέραση *s.f.* έχω ~ be legal tender *or* valid; be influential *or* sought after; be frequented (*of road, etc.*).
πέρασμα *s.n.* crossing, passage; threading (*of needle*); passing (*of time, illness*); much frequented spot.
περασμέν|ος *a.* past; το ~ο καλοκαίρι last summer; ~η βελόνα threaded needle.
περαστικ|ός *a.* passing by; transitory; frequented (*road*); ~ά may you get well!
περατώνω *v.t.* finish.
περβάζι *s.n.* door *or* window frame, sill.
περγαμηνή *s.f.* parchment.
πέρδ|ικα *s.f.*, **~ίκι** *s.n.* partridge; είμαι ~ίκι be quite well again; το λέει η ~ικούλα του he is brave *or* bold.
περδικλώνω *v.t. see* **πεδικλώνω**.
πέρδομαι *v.i.* break wind.
περεχύνω *v.t.* pour liquid over, drench.
περήφαν|ος *a.* proud. **~εια** *s.f.* pride. **~εύομαι** *v.i.* grow *or* be proud.
περί *prep.* 1. (*with gen.*) about, concerning; ~ τίνος πρόκειται; what is in question?έχω ~ πολλού think much of. 2. (*with acc.*) round, near; round about, approximately;

(*concerned*) with.
περι- *denotes* 1. around. 2. very.
περιαρπάζω *v.t.* seize; (*fam.*) pour abuse on.
περιαυτολογ|ώ *v.i.* brag. **~ος** *a.* boastful.
περιβάλλ|ω *v.t* surround; clothe. **~ον** *s.n.* surroundings, environment; entourage.
περίβλημα *s.n.* covering, casing.
περιβόητος *a.* famous, notorious.
περιβολή *s.f.* attire, dress; εν αδαμιαία ~ in the nude.
περιβόλ|ι *s.n.* garden, orchard; (*fig.*) είναι ~ι he is an entertaining person. **~άρης** *s.m.* gardener.
περίβολος *s.m.* enclosed ground, yard; surrounding wall; precinct, campus.
περιγελώ *v.t.* mock, cheat.
περίγελως *s.m.* laughing-stock.
περιγιάλι *s.n.* sea-shore.
περίγραμμα *s.n.* outline.
περιγράφω *v.t.* describe. **~ή** *s.f.* description.
περιδεής *a.* fearful, alarmed.
περιδέραιον *s.n.* necklace.
περιδιαβάζω *v.i.* stroll.
περίδρομ|ος *s.m.* (*fam.*) τρώω τον ~ο eat a bellyful.
περιεκτικός *a.* capacious; substantial (*food*); comprehensive and succinct (*speech*).
περιεργάζομαι *v.t.* scrutinize.
περιέργεια *s.f.* curiosity.
περίεργος *a.* curious, inquisitive, strange.
περιέρχομαι *v.t. & i.* travel through, make the round of; come, reach, be reduced (*to*).
περιέχ|ω *v.t* contain. **~όμενο** *s.n.* content; import.
περιζήτητος *a.* sought-after.
περιηγητής *s.m.* tourist.
περιηγούμαι *v.t.* travel through, tour.
περιθάλπω *v.t.* tend, care for.
περίθαλψη *s.f.* care, relief, welfare work.
περιθώριο *s.n.* margin.
περικάλυμμα *s.n.* covering, casing.
περικεφαλαία *s.f.* helmet.
περικλείω *v.t.* enclose, contain.
περικοκλάδα *s.f.* convolvulus; climbing plant.
περικοπή *s.f.* cutting down, reduction; extract (*from book*).
περικόπτω *v.t.* curtail (*expenses, speech, etc.*).
περικόχλιον *s.n.* nut (*for screw*).
περικυκλώνω *v.t.* encircle.
περιλαίμιον *s.n.* (animal's) collar.
περιλάλητος *a.* famous, notorious.
περιλαμβάνω *v.t.* contain, hold; include.
περιληπτικός *a.* comprehensive; concise; (*gram.*) collective.

περίληψη *s.f.* summary, précis.

περιλούζω *v.t.* shower with (*water, abuse*).

περίλυπος *a.* sorrowful.

περιμαζεύω *v.t.* pick up (*scattered objects*); harbour, rescue (*waif or stray*); curb, control.

περιμένω *v.i. & t.* wait; await, expect.

περίμετρος *s.f.* circumference, perimeter.

πέριξ *adv. & prep.* round; τα ~ environs; (*with gen.*) round.

περιοδεία *s.f.* visit, journey (*on business*).

περιοδικ|ός *a.* periodic. **~ό** *s.n.* magazine.

περίοδος *s.f.* period; season.

περίοικοι *s.m.pl.* οι ~ the neighbours.

περίοπτος *a.* conspicuous and admired.

περιορίζω *v.t.* confine, restrict; curb, control; reduce.

περιορισμός *s.m.* confinement, restriction; curbing; reduction; curfew.

περιουσ|ία *s.f.* property, possessions; estate; wealth. ~λος λαός chosen people.

περιοχή *s.f.* region, area, district.

περιπαθής *a.* full of feeling *or* passion.

περιπαίζω *v.t.* mock; dupe.

περίπατ|ος *s.m.* walk, ride, drive, trip; πάω ~ο go for a walk *or* (*fam.*) be a washout *or* done for *or* gone.

περιπατώ *v.i. & t. see* **περπατώ**.

περιπέτει|α *s.f.* adventure, misadventure; ~ες (*pl.*) vicissitudes. **~ώδης** *a.* adventurous (*life, times, etc.*); (*story*) of adventure.

περιπίπτω *v.t.* fall (*into hands, coma, etc.*).

περιπλαν|ώ *v.t.* take *or* send (a long way) round. ~ώμαι *v.i.* wander, roam; lose one's way.

περιπλέκω *v.t.* entagle; complicate, make confused.

περιπλέον *adv.* over and above; το ~ surplus.

περιπλέω *v.t.* sail round.

περιπλοκάδα *s.f. see* **περικοκλάδα**.

περίπλοκ|ος *a.* complicated. **~ή** *s.f.* complication.

περίπλους *s.m.* circumnavigation.

περιπόθητος *a.* much desired.

περιποίη|ση *s.f.* attention, looking after; ~σεις (*pl.*) entertainment, hospitality. **~τικός** *a.* showing attentiveness.

περιποι|ούμαι *v.t.* look after, tend; nurse; show attentiveness to; ~ημένος well turned out; fine (*workmanship*).

περίπολος *s.f.* patrol.

περίπου *adv.* about, roughly.

περίπτερο *s.n.* pavilion, lodge, summerhouse, kiosk.

περίπτυξη *s.f.* embrace.

περίπτωσ|η *s.f.* case, circumstance; εν πάση ~ει anyhow, in any case.

περισκελίς *s.f.* trousers (*esp. of uniform*); pants.

περίσκεψη *s.f.* circumspection.

περισπασμός *s.m.* distraction; worry.

περισπώ *v.t.* distract (*attention*).

περισπωμένη *s.f.* (*gram*) circumflex.

περίσσεια *s.f.* superabundance.

περίσσευμα *s.n.* surplus.

περισσεύ|ω *v.i.* be left over; δε μου ~ει I have none to spare.

περίσσιος *a.* enough and to spare.

περισσ|ός *a.* ως εκ ~ού on top of everything else, into the bargain.

περισσότερ|ος *a.* more; ο ~ος most. **~ο** *adv.* more.

περισταλτικός *a.* restrictive; (*med.*) peristaltic.

περίσταση *s.f.* situation, circumstance, occasion.

περιστατικό *s.n.* event, happening.

περιστέλλω *v.t.* restrict, check, curb.

περιστέρ|ι *s.n.* pigeon, dove. **~εών(ας)** *s.m.* dove-cot.

περιστοιχίζω *v.t.* surround.

περιστολή *s.f.* restriction, curbing.

περιστρέφ|ω *v.t.* rotate. ~ομαι *v.i.* rotate.

περιστροφ|ή *s.f.* rotation; (*fig.*) ~ές (*pl.*) beating about the bush. **~ικός** *a.* revolving.

περίστροφο *s.n.* revolver.

περισυλλέγω *v.t.* pick up (*scattered objects*); harbour, rescue (*waif or stray*).

περισώζω *v.t.* preserve, rescue.

περίτεχνος *a.* finely made; clever; ornate.

περιτοιχίζω *v.t.* build a wall round.

περιτομή *s.f.* circumcision.

περιτριγυρίζω *v.t.* encircle; go round; prowl round; (*fig.*) court.

περίτρομος *a.* terrified.

περιτροπή *s.f.* εκ ~ής by turns.

περιττός *a.* superfluous, needless; ~ αριθμός odd number.

περίττωμα *s.n.* excrement.

περιτύλιγμα *s.n.* wrapping.

περιτυλίσσω *v.t.* wind (*round*); wrap up.

περιφέρει|α *s.f.* periphery, circumference; district; έχω ~ες be broad in the beam. **~ακός** *a.* district, regional. ~ακός δρόμος ring-road.

περιφέρ|ω *v.t.* carry round. ~ομαι *v.i.* wander about.

περίφημος *a.* famous; first-rate.

περιφορά *s.f.* (*religious*) procession.

περίφραγμα *s.n.* fence, hedge.

περίφραξη *s.f.* fencing, enclosure.

περιφραστικός *a.* periphrastic.

περιφρόνη|ση *s.f.* contempt. **~τικός** *a.* contemptuous.

περίφροντις *a.* concerned, taken up (*with*).

περιφρονώ *v.t.* despise, treat scornfully.

περιχύνω *v.t.* splash with, pour over.

περίχωρα *s.n.pl.* environs.

περιώνυμος *a.* renowned, notorious.

περιωπ|ή *s.f.* high standing *or* quality; από ~ής with detachment.

περμανάντ *s.f.* permanent wave.

περνώ *v.t. & i.* 1. (*v.t.*) pass, go past *or* through, pierce; make pass through, thread; send *or* take through *or* across, cross; excel; go through, experience; spend (*time*); transfer; put down (*on bill, etc.*); put on (*clothes*); coat (*with paint*); ~ για mistake for. 2. (*v.i.*) pass, go past; go *or* get through *or* across; be valid *or* accepted; fare, get on; ~ από call at, go via; περάστε please come *or* go in!

περονιάζω *v.t.* pierce.

περονόσπορος *s.m.* mildew (*on plants*).

περούκα *s.f.* wig.

περπάτημα *s.n.* walking; gait.

περπατησιά *s.f.* gait.

περπατώ *v.i. & t.* walk; traverse; take for a walk.

πέρ(υ)σι *adv.* last year. **~νός** *a.* last year's.

πες *imperative of* **λέγω**.

πέσιμο *s.n.* fall, falling.

πεσκέσι *s.n.* gift (*esp. victuals*).

πεσκίρι *s.n.* towel.

πεσμένος *a.* fallen; impaired, deteriorated.

πέστροφα *s.f.* trout.

πετάγομαι *passive of* **πετώ**.

πετάλι *s.n.* pedal.

πεταλίδα *s.f.* limpet.

πέταλ|ο *s.n.* 1. petal. 2. horseshoe; τινάζω τα ~α (*fam.*) kick the bucket.

πεταλούδα *s.f.* butterfly, moth.

πεταλώνω *v.t.* shoe (*horse*).

πέταμα *s.n.* 1.flight. 2. throwing away.

πεταχτ|ός *a.* 1. protruding. 2. sprightly; στα ~α on the wing, hurriedly.

πετειν|ός *s.m.* cock. **~άρι** *s.n.* cockerel.

πετ(ι)μέζι *s.n.* syrup from must.

πέτο *s.n.* lapel.

πετονιά *s.f.* fishing-line.

πετούμενο *s.n.* bird.

πέτρα *s.f.* (precious) stone; (*fig.*) ~ του σκανδάλου cause of the trouble *or* commotion.

πετράδι *s.n.* pebble; precious stone.

πετραχήλι *s.n.* (*eccl.*) stole.

πετρέλαι|ον *s.n.* oil, petroleum. **~οκίνητος** *a.* oil-driven. **~οπηγή** *s.f.* oil-well.

πετρελαιοφόρ|ος *a.* oil-producing. **~ον** *s.n.* tanker.

πετριά *s.f.* blow with a stone; stone's throw; (*fam.*) hint; idée fixe.

πέτρινος *a.* stone.

πετροβολώ *v.t.* stone (*pelt*).

πετρώνω *v.t. & i.* petrify.

πέτσα *s.f.* skin, crust; (*fig.*) shame.

πετσ|ένιος, ~ινος *a.* leather.

πετσέτ|α *s.f.* 1. napkin; towel; **~άκι** *s.n.* paper-napkin. (*fam.*) στρώνω ~α gossip. 2. peseta.

πετσί *s.n.* skin, hide; leather.

πετσοκόβω *v.t.* hack; cut to pieces.

πετυχαίνω *v.t. & i. see* **επιτυγχάνω**.

πετ|ώ 1. *v.i.* fly (off); jump (*for joy*); twitch (*of eyelid*). 2. *v.t.* throw (away); shed, scrap; **~ώ έξω** eject. **~αμένος** thrown away, wasted. 3. **~άγομαι, ~ιέμαι** *v.i.* spring *or* fly up; slip *or* run (*out, round, etc., on errand*); butt in (*verbally*).

πεύκο *s.n.* pine.

πέφτ|ω *v.i.* fall (down *or* off); drop; come off; fall to the lot of; fall (*in date*), occur; decline in force *or* health; **~ω στο κρεβάτι** go to bed; **~ω στη θάλασσα** go in for a bathe; **~ω έξω** run aground, miscalculate; **~ω δίπλα** make up (*to*); **~ω με τα μούτρα σε** fall upon, take up avidly; δε σου **~ει λόγος** you have no say; πού **~ει;** whereabouts is it? έπεσε ξύλο blows were struck.

πεχλιβάνης *s.m.* wrestler, strong man.

πέψη *s.f.* digestion.

πήγα *aorist of* **πάω** *and* **πηγαίνω**.

πηγάδι *s.n.* well.

πηγάζω *v.i.* spring, originate.

πηγαιμός *s.m.* going, way there.

πηγαιν(ο)έλα *s.n.* coming and going.

πηγαινοέρχομαι *v.i.* come and go.

πηγαίνω *v.i. & t.* go; take (*convey*). See **πάω**.

πηγαίος *a.* spring (*water*); (*fig.*) springing naturally *or* effortlessly.

πηγή *s.f.* spring; source.

πηγούνι *s.n.* chin.

πηδάλι|ο *s.n.* rudder; helm; steering-wheel. **~ούχος** *s.m.* helmsman.

πήδημα *s.n.* jump, jumping.

πηδηχτός *a.* springy (*step*).

πηδώ *v.i. & t.* jump, leap, bounce; jump over; omit.

πήζω *v.t. & i.* curdle, thicken.

πηκτ- *see* **πηχτ-**.

πηλάλα *s.f.* running; (*adv.*) at a run.

πηλαλώ *v.i.* run.

πηλήκιο *s.n.* cap (*soldier's or schoolboy's*).

πηλίκον *s.n.* (*math.*) quotient.

πήλινος *a.* clay, earthen.

πηλός *s.m.* clay; mud.

πηλοφόρι *s.n.* hod.

πηνίον *s.n.* spool, (*electric*) coil.

πήξη *s.f.* curdling, congealing.

πήρα *aorist of* **παίρνω**.

πηρούνι *s.n.* fork. (*fam.*) γερό ~ good trencherman.

πήττα *s.f.* sort of cake *or* pie.

πήχη *s.f.* measure of length (2. *ft.*).

πηχτή *s.f.* brawn (*meat*).

πηχτός *a.* congealed; thick.

πήχτρα *s.f.* anything very dense.

πήχυς *s.m.* 1. fore-arm. 2. *see* **πήχη**.

πι *s.n.* the letter **Π**. στο ~ και φι in a trice.

πια *adv.* (*not*) any longer; now, definitely, at last (*often with nuance of impatience*); είναι καιρός ~ it is high time.

πιάνο *s.n.* piano.

πιάν|ω *v.t. & i.* (*v.t.*) 1. catch hold of. 2. catch (*quarry*); αυτός δεν ~εται he is not to be caught. 3. occupy, take possession of; ~ω σπίτι take (*rent*) a house; ~ω τόπο take up space *or* come in useful. 4. afflict, overtake; την έπιασε το κεφάλι της she has a headache; μας έπιασε βροχή we got caught in the rain. 5. affect; με ~ει η θάλασσα I am a bad sailor. 6. begin; ~ω κουβέντα get into conversation; ~ω δουλειά start a (new) job. 7. consider, reckon; αυτό δεν ~εται that does not count. 8. gain, make, get; ~ω τα έξοδά μου recover one's outlay; ~ω νερό draw water; ~ω κρέας put on flesh. (*v.i.*). 9. put in, call (*of ship*). 10. germinate; take effect; stick; (*fig.*) catch on, come off (*of seed, graft, label, play, joke, etc.*). 11. burn, catch (*of food, saucepan, firewood*). 12. come on, break out (*of rain, wind, cold, fire, etc.*). 13. ~ομαι *v.i.* be caught *or* paralysed *or* busy (*doing something else*); quarrel. πιάσου επάνω μου lean on me; πιάστηκε η φωνή του he lost his voice; πιάστηκαν (στα χέρια) they came to blows; πιάστηκε (από λεφτά) he has made his pile.

πιάσιμο *s.n.* holding, catching, touch; taking root; stiffness, paralysis.

πιασμένος *a.* occupied, taken; paralysed.

πιατάκι *s.n.* small plate; saucer.

πιατέλλα *s.f.* large dish.

πιατικά *s.n.pl.* crockery.

πιάτο *s.n.* plate; course (*at meal*).

πιάτσα *s.f.* market (*abstract*); (*fam.*) taxi-rank.

πίδ|αξ *s.m.* jet of water, fountain.

πιέζω *v.t.* press, squeeze, compress; constrain.

πιένα *s.f.* full house.

πίεση *s.f.* pressure; blood-pressure.

πιεστήριο *s.n.* press.

πιέτα *s.f.* pleat.

πίζουλος *a.* niggling (*job*).

πιθαμή *s.f.* span (*of hand*).

πιθανολογία *s.f.* conjecture.

πιθαν|ός *a.* probable, likely. ~ότητα *s.f.* likelihood.

πίθηκος *s.m.* ape.

πίθ|ος *s.m.*, ~άρι *s.n.* large earthen jar.

πίκα *s.f.* 1. pique, offence. 2. spade (*cards*).

πικάντικος *a.* piquant.

πικάρω *s.f.* irritate, pique.

πίκρα *s.f.* bitterness (*taste, grief*).

πικρ|άδα, ~ίλα *s.f.* bitterness (*taste*).

πικραίν|ω *v.t. & i.* make *or* become bitter; (*v.t.*) grieve. ~ομαι *v.i.* be grieved.

πικρία *s.f.* bitterness (*resentment*).

πικρίζω *v.i.* taste bitter.

πικροδάφνη *s.f.* oleander.

πικρ|ός *a.* bitter. ~ότης *s.f.* bitterness.

πικρόχολος *a.* irritable, peevish.

πιλατεύω *v.t.* torment.

πιλάφι *s.n.* pilaf (*rice dish*).

πίλος *s.m.* hat.

πιλότ|ος *s.m.* pilot. ~ίνα *s.f.* pilot-boat.

πινάκιον *s.n.* (*law*) roll, register.

πινακίδα *s.f.* name-plate, number-plate.

πινακλ *s.n.* pinochle (*card game*).

πινακοθήκη *s.f.* picture-gallery.

πίν|αξ, ~ακας *s.m.* list, table; notice-board; blackboard; picture.

πινέζα *s.f.* drawing-pin.

πινέλ|ο *s.n.* (*artist's*) paint-brush. ~ιά *s.f.* brush-stroke.

πίνω *v.t. & i.* drink; (*fam.*) smoke (*cigarette*).

πιο *adv.* more (*with noun, adj., or adv.*).

πιόνι *s.n.* pawn.

πιοτό *s.n.* drinking; drink (*esp. alcoholic*).

πίπα *s.f.* tobacco-pipe; cigarette-holder.

πιπεράτος *a.* peppery, sharp, biting.

πιπέρι *s.n.* pepper (*condiment*), peppercorn.

πιπεριά *s.f.* pepper (*vegetable*); pepper-tree.

πιπιλίζω *v.t.* suck.

πίπτω *v.i.* see **πέφτω**

πισθάγκωνα *adv.* with hands tied behind one's back.

πισίνα *s.f.* swimming-pool.

πισιν|ός *a.* back, rear; ο ~ός backside. κρατώ ~ή allow for any eventuality.

πίσσ|α *s.f.* tar, pitch; ~ες (*pl.*) waste oil (*from ships*). ~ώνω *v.t.* tar.

πισσόχαρτο *s.n.* tarred felt.

πίστα *s.f.* ring, track, dance-floor.

πιστευτός *a.* believable.

πιστεύω *v.t. & i.* believe.

πίστις *s.f.* belief, faith, trust; loyalty; (*fin.*) credit. (*fam.*) μου βγήκε η πίστη (ανάποδα) I had great difficulty.

πιστόλ|ι *s.n.* pistol. ~ιά *s.f.* pistol-shot.

πιστοποιητικόν *s.n.* certificate.

πιστοποιώ *v.t.* certify.

πιστός *a.* faithful, true, loyal.

πιστ|ώνω *v.t.* (*fin.*) credit (with). **~ωση** *s.f.* credit; appropriation. **~ωτής** *s.m.* creditor.

πίσω *adv.* behind; back; again; *το ρολόι πάει* ~ the clock loses *or* is slow; *χάνω* ~ move back; ~ *από* (*prep.*) behind; ~ *μου* behind me.

πίτ|ερο, ~ουρο *s.n.* bran.

πιτσιλάδα *s.f.* freckle.

πιτσιλιά *s.f.* splash (*of mud, etc.*).

πιτσιλίζω, ~ώ *v.t.* splash, sprinkle.

πιτσιρίκος *s.m.* (*fam.*) small boy.

πιτσούνι *s.n.* young pigeon.

πίττα *s.f. see* **πήττα**.

πιτυρίδα *s.f.* scurf.

πιωμένος *a.* drunk.

πλα(γ)ι *s.n.* side.

πλαγιά *s.f.* slope (*of hill*).

πλαγιάζω *v.i. & t.* lie down; put to bed; layer (*plant*); lay (*sthg.*) on its side.

πλαγιαστ|ός *a.* leaning. **~ά** *adv.* (*leaning*) at an angle (*against*).

πλάγ|ιος *a.* side, lateral; oblique; indirect; underhand. **~ιως** *adv.* obliquely; next door.

πλαδαρός *a.* flabby.

πλαζ *s.f.* bathing-beach. [F. *plage*]

πλάθω *v.t.* mould, shape, create.

πλάι *adv.* at the side, next door; ~ ~ side by side; ~ *σε* (*prep.*) next to.

πλαϊνός *a.* adjoining.

πλαίσι|ον *s.n.* frame, framework; chassis. **~ώνω** *v.t.* frame, surround; flank.

πλάκα *s.f.* slab, paving-stone; plaque; (*scholar's*) slate; bar (*soap, etc.*);(*gramophone*) record; (*photographic*) plate; face (*of clock*); *σπάσαμε* ~ we had fun (*at someone's expense*); *έχει* ~ it makes you laugh.

πλακάκι *s.n.* tile (*on floor or wall*); (*fam.*) *τα χάνω* ~ hush it up.

πλακάτ *s.n.* placard.

πλακέ *a.* flat.

πλακί *s.n.* (*fish, vegetables*) cooked with oil and garlic in shallow dish.

πλακόστρωτος *a.* paved, tiled.

πλακούς *s.m.* cake; (*med.*) placenta.

πλάκωμα *s.n.* pressing down; sudden arrival; (*feeling of*) oppressive weight.

πλακώνω *v.t. & i.* flatten, press down; run over; appear *or* happen suddenly; (*fam.*) turn up, blow in.

πλακωτός *a.* flat, flattened.

πλανεύω *v.t.* seduce.

πλάνη *s.f.* 1. error. 2. plane (*tool*).

πλανήτης *s.m.* planet.

πλανίζω *v.t.* plane.

πλανόδιος *a.* itinerant.

πλάνος *a.* alluring, deceitful.

πλαντ|άζω, ~ώ *v.i.* burst, suffocate (*from anger, etc.*).

πλαν|ώ *v.t.* mislead. **~ώμαι** *v.i.* wander; be mistaken.

πλάξ *s.f.* slab, paving-stone; plaque.

πλασάρω *v.t.* sell.

πλασιέ *s.m.* commercial traveller. [F. *placier*]

πλάση *s.f.* moulding; creation.

πλάσμα *s.n.* (ravishing) creature; creation (*of imagination*). (*med.*) plasma.

πλάστης *s.m.* 1. creator. 2. rolling-pin.

πλάστ|ιγξ, ~ιγγα *s.f.* weighing-machine.

πλαστικός *a.* plastic; shapely.

πλαστογραφώ *v.t.* forge; falsify.

πλαστός *a.* false, forged, fictitious.

πλαταγίζω *v.t. & i.* smack (*lips*); flap (*of flag*); lap (*of water*).

πλατ|αίνω, ~ύνω *v.t. & i.* make *or* become wider.

πλάτ|ανος *s.m., ~άνι* *s.n.* plane-tree.

πλατεία *s.f.* square (*of town*); pit (*of theatre*).

πλάτ|η *s.f.* back; (*fam.*) *χάνω ~ες* give clandestine aid; *έχω ~ες* have friends at court.

πλατιά *adv.* widely.

πλάτος *s.n.* width, breadth; latitude.

πλάττω *v.t.* mould, create, fashion

πλατύς *a.* wide, broad.

πλατύσκαλο *s.n.* landing (*of staircase*).

πλάτωμα *s.n.* level spot (*in hills*).

πλέγμα *s.n.* anything of woven or plaited structure; mesh; tissue.

πλειοδοτώ *v.i.* make highest bid.

πλείον *adv.* more.

πλειονό|της *s.f.* majority.

πλειο(νο)ψηφία *s.f.* majority of votes.

πλειοψηφώ *v.i.* receive most votes.

πλειστηριασμός *s.m.* auction.

πλείστ|ος *a.*most, very much; *ως επί το ~ον* for the most part.

πλεκτάνη *s.f.* intrigue.

πλεκτός *a.* knitted, plainted, of wicker.

πλέκω 1.*v.t.* plait, knit. 2. *v.i.* swim.

πλεμόνι *s.n.* lung; lights.

πλέμπα *s.f.* hoi polloi.

πλένω *v.t.* wash.

πλεξ|ίδα, ~ούδα *s.f.* plait, braid.

πλέξιμο *s.n.* knitting; braiding.

πλέον *adv.* more; *επί* ~ in addition; (*not*) any longer; now, definitely, at last (*often with nuance of impatience*).

πλεονάζ|ω *v.i.* be superfluous *or* more numerous. **~ων** redundant.

πλεόνασμα *s.n.* surplus.

πλεονέκτημα *s.n.* advantage, good point,

merit.

πλεονέκτης *a.* greedy, grasping.

πλεονεκτικός *a.* 1. advantageous. 2. grasping.

πλεονεκτώ *v.t.* (*with gen.*) be superior to.

πλεονεξία *s.f.* cupidity.

πλεούμενο *s.n.* ship.

πλερ- *see* **πληρ-**.

πλέριος *a.* full.

πλευρά *s.f.* side; rib.

πλευρίζω *v.i. & t.* (*naut.*) come alongside; (*fig.*) accost.

πλευρικός *a.* side, flanking, lateral.

πλευρίτις *s.f.* pleurisy.

πλευροκοπώ *v.t. & i.* make flanking attack (on).

πλευρό *s.n.* side; rib; flank.

πλεχτό *s.n.* pullover; piece of knitting.

πλέω *v.i.* sail; float; ~ στον πλούτο be rolling in money.

πληβείος *a.* plebeian.

πληγή *s.f.* wound, sore; (*fig.*) pest, plague, affliction.

πληγιάζω *v.t. & i.* make *or* become sore *or* blistered.

πλήγμα *s.n.* blow.

πληγώνω *v.t.* wound; hurt, offend.

πληθ|αίνω, ~ύνω *v.i. & t.* (*v.i.*) multiply; (*v.t.*) augment.

πλήθος *s.n.* crowd; τα ~η the masses.

πληθυντικός *s.m.* (*gram.*) plural.

πληθυσμός *s.m.* population.

πληθώρ|α *s.f.* abundance; (*med.*) plethora. **~ικός** *a.* excessive. **~ισμός** *s.m.* (*fin.*) inflation **~ιστικός** *a.* inflationary.

πληκτικός *a.* boring, dull, gloomy.

πλήκτρο *s.n.* key (*of piano, typewriter*); plectrum; drumstick.

πλημμελειοδικείον *s.n.* (*law*) criminal court (*intermediate*).

πλημμέλημα *s.n.* (*law*) criminal offence (*of medium gravity*).

πλημμελής *a.* faulty, badly done.

πλημμύρα *s.f.* flood.

πλημμυρ|ίζω, ~ώ *v.i. & t.* be in flood, overflow; be flooded; (*v.t.*) flood.

πλην 1. *prep.* (*with gen.*) except, apart from; ~ τούτου besides. 2. *conj.* but. 3. (*math.*) minus.

πλήξη *s.f.* boredom.

πληρεξούσιον *s.n.* (*law*) warrant of attorney.

πληρεξούσιος *s.m.f.* plenipotentiary; representative, proxy.

πλήρης *a.* full; complete.

πληροφορ|ία *s.f.* piece of information; ~ίες (*pl.*) information.

πληροφορ|ώ *v.t.* inform. ~ούμαι *v.i.* learn,

find out.

πληρώ *v.t.* fill; fulfil.

πλήρωμα *s.n.* crew; ~ του χρόνου fullness of time.

πληρωμή *s.f.* payment, reward.

πληρώ|νω *v.t.* pay (for); ~νω τα σπασμένα pay for others' misdeeds; ~μένος paid, suborned.

πληρωτέος *a.* payable.

πληρωτής *s.m.* payer.

πλησιάζω *v.t. & i.* put *or* bring near, approach, draw near

πλησιέστερος *a.* nearer; ο ~ nearest.

πλησίον *adv. & prep.* near; ~ μου near me; ο ~ μου my neighbour; το ~ σχολείον the neighbouring school.

πλησίστιος *a.* with all sails set.

πλησμονή *s.f.* abundance, satiety.

πλήττω *v.t. & i.* strike, wound, afflict; be bored.

πλια *adv. see* **πια**.

πλιάτσικο *s.n.* loot; looting.

πλίθ(ρ)α *s.f.* mud-brick; (*fig.*) stodgy food.

πλινθόκτιστος *a.* built of bricks.

πλίνθος *s.f.* brick.

πλισσές *s.m.* pleating.

πλοηγός *s.m.* pilot.

πλοιάριον *s.n.* small craft.

πλοίαρχος *s.m.* (*naut.*) captain.

πλόιμος *a.* seaworthy.

πλοίο *s.n.* ship.

πλόκαμος *s.m.* tress, lock, plait; tentacle.

πλοκή *s.f.* plot (*of book, play*).

πλουμ|ί, ~ίδι *s.n.* design, ornamentation.

πλους *s.m.* voyage, passage.

πλουσιοπάροχος *a.* lavish.

πλούσιος *a.* rich; sumptuous; plenty of.

πλουταίνω *v.i.* grow rich.

πλουτίζω *v.t. & i.* enrich; grow rich.

πλούτ|ος *s.m. & n.* wealth; richness; τα ~η riches.

πλυντήριο *s.n.* laundry (- room); washing-machine.

πλύνω *v.t.* wash.

πλύσ|η *s.f.*, **~ιμο** *s.n.* (*act of*) washing.

πλυσταριό *s.n.* laundry-room.

πλύστρα *s.f.* washerwoman; washboard.

πλώρη *s.f.* prow.

πλωτάρχης *s.m.* (*naut.*) lieutenant-commander.

πλωτός *a.* navigable; floating.

πνεύμα *s.n.* breath of life, ghost; Άγιον ~ Holy Ghost; spirit; mind; genius; wit; (*gram.*) breathing; άνθρωπος του ~τος cultured person.

πνευματικός 1. *a.* spiritual; mental; intellectual; pneumatic. 2. *s.m.* confessor.

πνευματισμός *s.m.* spiritualism.

πνευματώδης *a.* witty; spirituous.
πνευμονία *s.f.* pneumonia.
πνεύμων *s.f.* lung.
πνευστός *a.* wind (*instrument*).
πνέω *v.i. & t.* blow; ~ τα λοίσθια breathe one's last.
πνιγηρός, *a.* suffocating.
πνιγμός *s.m.* drowning; choking.
πνίγω *v.t.* drown; suffocate; choke, strangle.
πνικτικός *a.* suffocating.
πνίξιμο *s.n.* drowning; choking.
πνοή *s.f.* breath, breathing; inspiration.
ποδάγρα *s.f.* gout.
ποδάρι *s.n.* foot, leg.
ποδαρικό *s.n.* κάνω καλό ~ bring good luck.
ποδηγετώ *v.t.* guide.
ποδήλατο *s.n.* bicycle.
ποδήρης *a.* (*dress*) reaching to the feet.
πόδι *s.n.* foot, leg; στο ~ standing up *or* on the go; με τα ~α on foot; σηκώνω στο ~ cause a commotion among; έμεινα στο ~ του I am acting *or* I acted for him.
ποδιά *s.f.* lap; apron, overall; sill; slope (*of hill*).
ποδίζω *v.i.* (*naut.*) tack; shelter (*in storm*).
ποδοβολητό *s.n.* tramp, thud, clatter (*of feet*).
ποδόγυρος *s.m.* hem; (*fig.*) (*fam.*) skirts, women.
ποδοκρότημα *s.n.* stamping of feet.
ποδοπατώ *v.t.* stamp *or* trample on.
ποδόσφαιρα *s.f.* a football.
ποδόσφαιρο *s.n.* (*game of*) football.
ποδόφρενο *s.n.* footbrake.
πόζα *s.f.* pose; κρατώ ~ be stand-offish.
πόθεν *adv.* whence; το ~ έσχες source of one's wealth.
πόθ|ος *s.m.* desire, longing. ~ώ *v.t.* desire, long for.
ποίημα *s.n.* poem.
ποίηση *s.f.* poetry.
ποιητής *s.m.* poet; maker.
ποιητικός *a.* poetic.
ποικιλία *s.f.* variety, diversity.
ποικίλλω *v.t. & i.* adorn; vary.
ποικίλος *a.* various, variegated, miscellaneous.
ποιμαντορικός *a.* (*eccl.*) pastoral.
ποιμενικός *a.* shepherd's; pastoral.
ποιμήν *s.m.* shepherd; pastor.
ποίμνιον *s.n.* flock, herd.
ποιν|ή *s.f.* penalty, punishment; sentence. ~ικός *a.* penal.
ποιόν *s.n.* quality, character.
ποί|ος, ~ος *pron. & a.* who? which?
ποιότης *s.f.* quality.
ποι|ώ *v.t.* make; do; περί πολλού ~ούμαι

cherish, value.
πόκα *s.f.* poker (*game*).
πολεμικ|ός *a.* of *or* for war; warlike, aggressive. ~ή *s.f.* polemic.~όν *s.n.* (*naut.*) warship.
πολέμιος *a.* hostile; an enemy.
πολεμιστής *s.m.* fighter, warrior.
πόλεμ|ος *s.m.* war. ~οφόδια *s.n.pl.* munitions.
πολεμώ *v.t. & i.* fight (*against*), strive.
πολεοδομία *s.f.* town-planning.
πολικός *a.* polar.
πολιορκ|ία *s.f.* siege. ~ώ *v.t.* besiege.
πόλ|ις, ~η *s.f.* city, town.
πολιτεία *s.f.* polity; state; country; town; attitude, line (*of conduct*); (*fam.*) βίος και ~ chequered career.
πολίτευμα *s.n.* form of government.
πολιτεύομαι *v.i.* go in for politics; handle a situation.
πολιτευόμενος, ~τής *s.m.* one who mixes in politics, politician.
πολίτης *s.m.* citizen; civilian.
Πολίτης *s.m.* a Constantinopolitan.
πολιτική *s.f.* politics; policy.
πολιτικ|ός *a.* civil, civilian; political; a politician; τα ~ά politics; ~ός ανήρ statesman; με ~ά in civilian clothes; ~ός στίχος 15-syllable verse.
πολιτισμένος *a.* civilized, cultured.
πολιτισμός *s.m.* civilization, culture.
πολιτιστικός *a.* cultural.
πολιτογραφώ *v.t.* naturalize (*as citizen*).
πολιτοφυλακή *s.f.* militia.
πολίχνη *s.f.* large village.
πιλλάκις *adv.* many times.
πολλαπλασιάζω *v.t.* multiply; increase.
πολλαπλασιασμός *s.m.* multiplication.
πολλαπλάσιον *s.n.* multiple.
πολλαπλούς *a.* multiple.
πολλοί *a. pl.* many.
πολλοστός *a.* umpteenth.
πόλος *s.m.* pole.
πολτός *s.m.* pap; purée; pulp.
πολύ *adv.* much; very; long; *see* **πολύς.**
πολυ- *denotes* much, very.
πολυάριθμος *a.* numerous.
πολυάσχολος *a.* very busy.
πολυβόλον *s.n.* machine-gun.
πολυγαμία *s.f.* polygamy.
πολύγλωσσος *a.* polyglot.
πολυγράφος *s.m.* duplicator.
πολυέλαιος *s.m.* chandelier.
πολυετής *a.* lasting many years.
πολυθρόνα *s.f.* arm-chair.
πολυκαιρία *s.f.* lapse of time, age.

πολυκατοικία *s.f.* block of flats.
πολυκλινική *s.f.* polyclinic.
πολυκοσμία *s.f.* crowds of people.
πολύκροτον *s.n.* revolver.
πολύκροτος *a.* causing a stir.
πολυκύμαντος *a.* stormy.
πολυλογάς *s.m.* chatterbox.
πολυλογία *s.f.* loquacity.
πολυμάθεια *s.f.* learning, erudition.
πολυμελής *a.* with many members.
πολυμερής *a.* many-sided.
πολυμήχανος *a.* ingenious; wily.
πολύνω *v.t. & i.* increase.
πολύξερος *a.* knowledgeable, a mine of information.
πολυπαθής *a.* who has suffered much.
πολύπειρος *a.* very experienced.
πολυπληθής *a.* numerous, crowded.
πολύπλοκος *a.* involved, complicated.
πολύπους *s.m.* octopus; (*med.*) polypus.
πολυπράγμων *a.* know-all, nosey.
πολ|ύς *a.* much, many, great; long (*time*); *το ~ύ (~ύ)* at the most; *προ ~ού* long since; *ο ~ύς κόσμος* most people.
πολύσαρκος *a.* corpulent.
πολυσήμαντος *a.* with many meanings; momentous.
πολυσύλλαβος *a.* polysyllabic.
πολυσύχναστος *a.* busy, frequented.
πολυτέλ|εια *s.f.* luxury. **~ής** *a.* luxurious.
πολυτεχνείον *s.n.* Polytechnic.
πολύτιμος *a.* valuable, precious.
πολυφαγία *s.f.* gluttony.
πολύφερνος *a.* well-dowered.
πολύφωτο *s.n.* chandelier.
πολυχρονίζω *v.i. & t.* take a long time (over); enjoy a long life; grant or wish long life to.
πολύχρωμος *a.* many-coloured.
πολύωρος *a.* long, prolonged.
πολώνω *v.t.* polarize.
πόμολο *s.n.* door-knob.
πομπή *s.f.* procession; shame, disgrace.
πομπός *s.m.* transmitter.
πομπώδης *a.* pompous.
πονεμένος *a.* see **πονώ**.
πονέντ|ες, ~ης *s.m.* west wind.
πονετικός *a.* sharing sorrows of others.
πονηρεύ|ω *v.t. & i.* make suspicious; become wily. *~ομαι v.i.* become suspicious.
πονηρ|ία *s.f.* cunning. **~ός** *a.* cunning, knowing.
πονόδοντος *s.m.* toothache.
πονοκεφαλιάζω *v.t. & i.* weary, distract; rack one's brains.
πονοκέφαλος *s.m.* headache.
πονόλαιμος *s.m.* sore throat.

πόνος *s.m.* pain, ache; compassion.
πονόψυχος *a.* compassionate.
ποντάρω *v.i. & t.* punt (*bet*); back (*horse, etc.*).
ποντίκ|ι *s.n.*, **~ός** *s.m.* 1. mouse, rat. 2. muscle.
ποντίφηξ *s.m.* pontiff.
ποντοπόρος *a.* sea-going.
πόντ|ος *s.m.* 1. sea. 2. centimetre. 3. point (*in game*). 4. stitch; *μου 'φυγαν ~οι* my stocking is laddered *or* I dropped some stitches. 5. *ρίχνω ~ο* drop a hint.
ποντς *s.n.* punch (*drink*).
πον|ώ *v.i. & t.* feel pain, hurt; (*v.t.*) hurt; be attached to, care for; feel for; *δεν ~άει την τσέπη του* he does not spare his purse. *~εμένος a.* in distress; bad (*injured or ailing*).
ποπός *s.m.* (*fam.*) bottom.
πορδή *s.f.* (*fam.*) fart.
πορεία *s.f.* march; course.
πορεύομαι *v.i.* go, proceed; manage, make do.
πορθμείο *s.n.* ferry.
πορθμός *s.m.* strait.
πορθώ *v.t.* sack, pillage.
πορίζομαι *v.t.* furnish oneself with, get, draw.
πόρισμα *s.n.* inference, conclusion; finding; corollary.
πόρνη *s.f.* prostitute. **~είον** *s.n.* brothel.
πόρ|ος *s.m.* 1. ford. 2. pore. 3. **~οι** (*pl.*) means, resources.
πόρπη *s.f.* buckle, clasp.
πόρρω *adv.* far.
πορσελάνη *s.f.* china.
πόρτα *s.f.* door, gate.
πορτοκάλ|ι *s.n.* orange. **~άδα** *s.f.* orange-ade. **~ής** *a.* orange-coloured. **~ιά** *s.f.* orange-tree.
πορτοφόλ|ι *s.n.* wallet. **~άς** *s.m.* pickpocket.
πορτραίτο *s.n.* portrait.
πορφύρα *s.f.* purple.
πορώδης *a.* porous.
ποσάκις *adv.* how many times.
πόσιμ|ος *a.* fit for drinking; *~ο νερό* drinking-water.
ποσό *s.n.* amount, sum.
πόσο(ν) *adv.* how (much).
πόσος *pron. & a.* how much, how many.
ποσοστόν *s.n.* percentage.
ποσότης *s.f.* quantity.
πόστα *s.f.* post (*mail*); (*fam.*) *τον έβαλα ~* I scolded him.
πόστο *s.n.* strategic position; eligible post.
ποσώς *adv.* (*after neg.*) not at all.
ποταμηδόν *adv.* in torrents.

ποτάμ|ι *s.n.*, **~ός** *s.m.* river.

ποταπός *a.* base, dishonourable.

πότε *adv.* when? ~ ~ now and then; ~ ... ~ sometimes... sometimes.

ποτέ *adv.* once, formerly; ever; (*after neg.*) never.

ποτήρ|ι(ον) *s.n.* a glass; *γερό ~ι* one who likes his liquor.

πότης *s.m.* tippler.

ποτίζω *v.t. & i.* water, give drink to; become damp (*of wall, etc.*).

πότισμα *s.n.* watering.

ποτιστήρι *s.n.* watering-can.

ποτοαπαγόρευση *s.f.* prohibition (*of liquor*).

ποτό *s.n.* drink, beverage.

πότος *s.m.* (heavy) drinking.

που *adv.* where? *για* ~ where are you off to? ~ *να ξέρω* how should I know? ~ *και* ~ occasionally.

πού 1. *pron.* who, whom, which, that; when. 2. *conj.* that.

πουγγί *s.n.* purse.

πούδρα *s.f.* powder (*medicinal, cosmetic*).

πουθενά *adv.* anywhere; nowhere.

πουκαμίσα *s.f.* shift (*garment*).

πουκάμισο *s.n.* shirt; slough (*of snake*).

πουλάδα *s.f.* pullet.

πουλάκι *s.n.* little bird; ~ *μου* term of endearment.

πουλάρι *s.n.* foal.

πουλερικά *s.n.pl.* poultry.

πούλημα *s.n.* sale.

πουλί *s.n.* bird.

Πούλια *s.f.* Pleiades.

πούλιες *s.f.pl.* spangles.

πούλμαν *s.n.* motor-coach. [E. *Pullman*]

πουλώ *v.t.* sell.

πούντα *s.f.* 1. (-nd-) *chill, cold.* 2. (-nt-) point (*of promontory*).

πουντιάζω *v.i. & t.* catch a severe cold; (cause to) feel very cold.

πούπουλ|ο *s.n.* down (*plumage*). *έχω στα ~α v.* feather-bed.

πουρές *s.m.* purée.

πουρί *s.n.* porous stone; scale, fur, tartar.

πουριτανικός *a.* puritanical.

πουρμπουάρ *s.n.* tip. [F. *pourboire*]

πουρνάρι *s.n.* evergreen oak.

πουρνό *s.n.* morning.

πούρο *s.n.* cigar.

πους *s.m.* foot, leg.

πούσι *s.n.* mist.

πούστης *s.m.* (*fam.*) sodomite.

πουτάνα *s.f.* (*fam.*) whore.

πουτίγκα *s.f.* pudding.

πράγμα *s.n.* thing; matter, affair; merchandise, material (*cloth*); ~*τι* in fact, indeed.

πραγματεία *s.f.* treatise.

πραγματεύομαι *v.t.* treat of, deal with (*topic*).

πραγματικ|ός *a.* real, actual. **~ότης** *s.f.* reality.

πραγματιστής, *s.m.* pragmatist.

πραγματογνώμων *s.m.* valuer, assessor.

πραγματοποιώ *v.t.* realize, carry out; fulfil.

πρακτέον *s.n.* what must be done.

πρακτικ|ό *s.n.* written report; ~*ά* (*pl.*) minutes, proceedings.

πρακτικός *a.* practical; without professional training, quack.

πράκτ|ωρ *s.m.* agent. **~ορείο** *s.n.* agency.

πράμα *see* **πράγμα.** *στα* ~*τα* in office (*of government*).

πραμάτεια *s.f.* merchandise.

πραματευτής *s.m.* pedlar.

πραξικόπημα *s.n.* coup d'état.

πράξ|ις *s.f.* deed, act; practical experience; *θέτω εις ~ιν* put into practice; *εν τη ~ει* in practice; registration (*of births, etc.*); (*law*) enactment; (*fin.*) transaction; (*math.*) operation.

πράος *a.* mild, gentle.

πρασιά *s.f.* margin of garden round house.

πρασινάδα *s.f.* green colour; greenery.

πράσιν|ος *a.* green; (*fam.*) ~*α άλογα* nonsense. **~ίζω** *v.t. & i.* make *or* become green.

πράσ|ο *s.n.* leek; (*fam.*) *πιάνω στα* ~*α* catch red-handed.

πρατήριο *s.n.* shop selling single commodity or supplying members of association; ~ *βενζίνης* petrol-station.

πράττω *v.t. & i.* do; act.

πραΰν|ω *v.t.* calm, soothe. **~τικός** *a.* soothing.

πρέζα *s.f.* pinch (*of salt, etc.*)

πρεμούρα *s.f.* anxious eagerness.

πρέπει *v.i.* it is necessary *or* fitting; ~ *να πάω* I must go; *καθώς* ~ respectable, decent, of good manners *or* appearance.

πρέπ|ων *a.* right, fitting, seemly. **~όντως** *adv.* properly, suitably. **~ούμενος** *a.* fitting.

πρεσβεία *s.f.* embassy, legation.

πρεσβεύω *v.t.* profess, believe in.

πρέσβ|υς, ~ευτής *s.m.* ambassador, minister.

πρεσβύτερος *a.* elder.

πρεσβυωπία *s.f.* (*med.*) long-sightedness.

πρέφα *s.f.* a card game; *το πήρε* ~ he got wind of it.

πρήζ|ω *v.t.* (*fig.*) exasperate. **~ομαι** *v.i.* become swollen.

πρηνής *a.* (*lying*) prone.

πρήξιμο *s.n.* swelling; (*fig.*) exasperation.

πρήσκω *v.t. see* **πρήζω.**

πρίγκ|ηψ, ~**ιπας** *s.m.* prince. ~**ίπισσα** *s.f.* princess.

πρίζα *s.f.* (*electric*) socket. *βάζω στην* ~ plug in.

πριμ *s.n.* premium, bonus.

πριν *adv. & prep.* before; ~ *από μένα* before me; ~ (*από*) *ένα μήνα* a month ago; ~ *του πολέμου or* ~ *από τον πόλεμο* before the war.

πρίν|ος *s.m.,* ~**άρι** *s.n.* evergreen oak.

πριόν|ι *s.n.* saw. ~**ίζω** *v.t.* saw. ~**ίδια** *s.n.pl.* sawdust.

πρίσμα *s.n.* prism.

πριτς *int. of mild defiance.*

πριχού *adv.* before.

προ *prep.* (*with gen.*) before, in front of; ~ *ημερών* some days ago *or* before; ~ *παντός or* ~ *πάντων* above all, especially.

προ- *denotes* before.

προάγ|ω *v.t.* advance, promote. ~**ωγή** *s.f.* advancement, promotion. ~**ωγός** *s.m.* pimp.

προαίρεσ|ις *s.f.* intent; *κατά* ~*ιν* optionally.

προαιρετικός *a.* optional; voluntary.

προαίσθη|μα *s.n.,* ~**ση** *s.f.* presentiment.

προαιώνιος *a.* age-long; belonging to the dim past.

προαλείφομαι *v.i.* prepare the ground (*for self-advancement*).

προάλλες *a. τις* ~ the other day.

προανάκρουσμα *s.n.* prelude.

προαναφερθείς *a.* afore-mentioned.

προαπαιτούμενος *a.* prerequisite.

προασπίζω *v.t.* shield, protect.

προάστ(ε)ιον *s.n.* suburb.

προαύλιον *s.n.* forecourt.

πρόβα *s.f.* fitting (*of clothes*); trying out, rehearsal.

προβάδισμα *s.n.* precedence.

προβαίνω *v.i.* advance, proceed; ~ *σε* do, make, carry out.

προβάλλω *v.t. & i.* project, extend; cast (*shadow*); show (*film*); offer (*resistance*); raise (*objection*); put forward (*alibi*); (*v.i.*) appear, come into sight.

προβάρω *v.t.* try on (*clothes*); rehearse.

προβατίνα *s.f.* ewe.

πρόβατο *s.n.* sheep.

προβιβάζω *v.t.* promote (*in rank*).

προβιά *s.f.* sheep's *or* other animal's skin.

πρόβιος *a.* sheep's.

προβλέπω *v.t.* foresee, anticipate; make provision for.

πρόβλεψη *s.f.* forecast; provision (*for*).

πρόβλημα *s.n.* problem. ~**τίζω** *v.t.* puzzle. ~**τίζομαι** *v.i.* speculate, ponder.

προβλήτα *s.f.* jetty.

προβοδίζω *v.t.* see off.

προβολεύς *s.m.* searchlight, headlight; (*cinema*) projector.

προβολή *s.f.* projection, *etc.* (*see* **προβάλλω**).

προβοσκίδα *s.f.* trunk, proboscis.

προγάστωρ *a.* with protuberant belly.

προγενέστερ|ος *a.* of an earlier time; (*pl.*) ~**οι** predecessors.

πρόγευμα *s.n.* breakfast.

προγεφύρωμα *s.n.* bridgehead.

πρόγκα *s.f.* noisy disapproval.

προγκ|άω, ~**ίζω** *v.t. & i.* drive (*animals*) with cries; (*fig.*) shout down, repulse rudely; (*v.i.*) shy, start.

προγνωστικό *s.n.* forecast, tip.

πρόγονος *s.m.* ancestor.

προγονός *s.m.* stepson.

πρόγραμμα *s.n.* programme. ~**τίζω** *v.t.* plan, organize.

προγράφ|ω *v.t.* proscribe. ~**ή** *s.f.* proscription.

προγυμνάζω *v.t.* train; coach (*pupil*).

πρόδηλος *a.* obvious, evident.

προδιαγράφω *v.t.* prearrange.

προδιάθεση *s.f.* predisposition.

προδιαθέτω *v.t.* predispose, influence, prepare.

προδίδω *v.t.* betray; disclose, give away.

προδοσία *s.f.* betrayal; treason.

προδότης *s.m.* betrayer; traitor.

πρόδρομος *s.m.* forerunner, harbinger; (*epithet of John the Baptist*).

προεδρ(ε)ία *s.f.* presidency, chairmanship.

προεδρείον *s.n.* president and his staff.

πρόεδρος *s.m.f.* president, chairman.

προειδοποι|ώ *v.t.* notify in advance; warn. ~**ηση** *s.f.* warning.

προεισαγωγή *s.f.* introduction, preface.

προεκλογικός *a.* pre-election.

προέκταση *s.f.* extension, prolongation (*in space*).

προέλασις *s.f.* (*mil.*) advance.

προελαύνω *v.i.* (*mil.*) advance.

προέλευση *s.f.* provenance.

προεξάρχω *v.i.* stand out (*above others*).

προεξέχω *v.i.* stick out, project.

προεξοφλώ *v.t.* discount, pay off (*debt*) in advance; receive (*pay*) in advance; take for granted, bank on.

προεξοχή *s.f.* projection, ledge.

προεόρτια *s.n.pl.* eve of *or* preparations for festival.

προεργασία *s.f.* preliminary work.

προέρχομαι *v.i.* originate, spring (*from*).

προεστώς *a.* notable, chief.

προετοιμάζω *v.t.* prepare.

προέχω *v.i.* project; excel; come first in

πρόζα *s.f.* prose; θέατρο ~ς prose drama.

προζύμι *s.n.* leaven.

προηγ|ούμαι *v.i.* & *t.* come first; (*with gen.*) precede, come before. ~μένος *a.* advanced, developed.

προηγούμεν|ος *a.* preceding, previous, last. ~ο *s.n.* precedent; έχουν ~α they are on bad terms. ~ως *adv.* previously, before.

προθάλαμος *s.m.* ante-room.

πρόθεση *s.f.* intention; (*gram.*) preposition.

προθεσμία *s.f.* delay, time-limit, term.

προθήκη *s.f.* shop-window; show-case.

προθυμία *s.f.* willingness, readiness.

πρόθυμ|ος *a.* eager, willing, ready; obliging. ~οποιούμαι *v.i.* show readiness, offer.

πρόθυρα *s.n.pl.* parvis, approach; (*fig.*) verge, threshold.

προίκ|α *s.f.* dowry. ~ίζω *v.t.* dower, endow.

προικοθήρας *s.m.* fortune-hunter.

προικιά *s.n.pl.* effects constituting dowry.

προϊόν *s.n.* product; ~τα (*pl.*) produce.

προΐστ|αμαι *v.t.* (*with gen.*) be at the head of; *ο* ~άμενος man in charge, superior.

πρόκα *s.f.* nail, tack.

προκαλύπτω *v.t.* screen, cover.

προκαλώ *v.t.* challenge; provoke, cause.

προκάνω *v.t.* & *i.* catch (up with); be *or* act in time; have enough time (for).

προκάτ *s.n.* prefab.

προκαταβάλλω *v.t.* pay in advance.

προκαταβολ|ή *s.f.* advance payment; deposit. ~ικώς *adv.* in advance.

προκατακλυσμιαίος *a.* antediluvian.

προκαταλαμβάνω *v.t.* occupy beforehand; prepare (*for bad news*); bias.

προκατάληψη *s.f.* bias.

προκαταρκτικός *a.* preliminary.

προκατειλημμένος *a.* biased.

προκάτοχος *s.m.* previous occupier; predecessor.

πρόκει|ται *v.i.* it is a matter *or* question (*of*) (*with* για *or* να); περί τίνος ~ται; what is it about? ~ται να έρθει he is supposed to come; ε~το να στρίψει he was just going to turn; ~μένου να seeing that, if; *το* ~μενον the matter, subject, *or* point.

προκήρυξη *s.f.* proclamation.

πρόκλη|ση *s.f.* challenge; provocation. ~τικός *a.* provocative.

προκόβω *v.i.* thrive, progress, do well.

προκομμένος *a.* industrious, diligent; (*iron.*) *ο* ~ our fine friend.

προκοπ|ή *s.f.* industriousness; success; δεν είναι της ~πής it is no good.

προκριματικός *a.* preliminary (*of contest leading to finals*).

προκρίνω *v.t.* prefer, favour (*plan, solution, etc.*).

πρόκριτος *a.* notable.

προκυμαία *s.f.* quay.

προκύπτω *v.i.* arise, crop up; result, follow.

προ|λαβαίνω, ~λαμβάνω *v.t* & *i.* anticipate, forestall; catch (up with); be *or* act in time; have enough time (for).

προλαλήσας *s.m.* the previous speaker.

προλέγ|ω *v.i.* & *t.* say before; predict. ~όμενα *s.n.pl.* preface.

προλειαίνω *v.t.* (*fig.*) smooth, prepare (*way*).

προλετάριος *a.* proletarian.

προληπτικός *a.* anticipatory, preventive; superstitious.

πρόληψη *s.f.* prevention; superstition.

πρόλογος *s.m.* prologue.

προμάμμη *s.f.* great-grandmother.

προμαντεύω *v.t.* prophesy.

πρόμαχος *s.m.* champion, defender.

προμαχών *s.m.* bastion, bulwark.

προμελέτη *s.f.* premeditation; preliminary study; *εκ* ~ς deliberate(ly).

προμεσημβρία *s.f.* forenoon.

προμήθεια *s.f.* supply, provision; commission (*percentage*).

προμηθευτής *s.m.* supplier, purveyor.

προμηθεύ|ω *v.t.* supply, purvey, ~ομαι *v.t.* get, procure.

προμηνύω *v.t.* portend.

πρόναος *s.m.* church porch.

προνοητικός *a.* having foresight.

πρόνοια *s.f.* foresight, precaution; providence; Υπουργείον ~ς Ministry of Welfare.

προνόμι|ο *s.m.* privilege, prerogative; natural gift; ~ο ευρεσιτεχνίας patent. ~ούχος *a.* privileged. ~ακός *a.* preferential.

προνοώ *v.i.* (*with* περί) see to, provide for.

προξενείον *s.n.* consulate.

προξενεύω *v.t.* propose as marriage partner; (*fig.*) negotiate.

προξενητής *s.m.* go-between (*esp. in arranged marriage*).

προξενιά *s.f.* (*negotiation of*) arranged marriage; (*fig.*) negotiation.

πρόξενος *s.m.* 1. author, cause. 2. consul.

προξενώ *v.t.* cause, bring about.

προοδευτικός *a.* progressive.

προοδεύω *v.i.* progress; develop.

πρόοδος *s.f.* progress; development; progression.

προοίμιο *s.n.* preamble, prelude.

προοιωνίζομαι *v.t.* predict.

προοπτική *s.f.* perspective; prospect in view.

προορατικότης *s.f.* foresight.
προορίζω *v.t.* destine, intend. **~σμός** *s.m.* appointed end, purpose; destination.
προπαγάνδα *s.f.* propaganda.
προπαίδεια *s.f.* primer of arithmetic.
προπαιδευτικός *a.* preparatory (*education*).
πρόπαππος *s.m.* great-grandfather.
προπαραλήγουσα *s.f.* antepenultimate syllable.
προπαραμονή *s.f.* second day before.
προπαρασκευή *s.f.* preparation; coaching.
προπάτορες *s.m.pl.* forefathers.
προπέλλα *s.f.* propeller.
προπέμπω *v.t.*see off.
πρόπερσι *adv.* the year before last.
προπέτασμα *s.n.* screen; ~ καπνού smoke-screen.
προπέτεια *s.f.* impertinence.
προπίνω *v.i.* drink a toast.
πρόπλασμα *s.n.* maquette, model.
προ-πο *s.n.* (*football*) pools.
πρόποδες *s.m.pl.* foot (*of hill*).
προπολεμικός *a.* pre-war.
προπόν|ηση *s.f.* training, coaching. **~ητής** *s.* trainer, coach.
προπορεύομαι *v.i.* get ahead.
πρόποση *s.f.* toast (*drunk*).
προπύλαια *s.n.pl.* propylaeum.
προπύργιον *s.n.* bastion, bulwark.
πρόρρησις *s.f.* prediction.
προς *prep.* 1. (*with acc.*) to, towards (*direction, relation*); for (*use, purpose*); at (*price*); ~ όφελός μου to my advantage; ~ τιμήν του in his honour; ~ το παρόν for the present; ένα ~ ένα one by one; ως ~ εμέ as for me. 2. (*with gen.*) ~ πατρός on the father's side (*of relatives*); ~ Θεού for God's sake! 3. (*with dat.*) ~ τούτοις in addition. 4. (*adv.*) ~ δε in addition.
προσαγορεύω *v.t.* address, salute; call.
προσάγω *v.t.* bring forward, produce.
προσάναμμα *s.n.* tinder, kindling.
προσανατολίζ|ω *v.t.* orientate, direct. **~ομαι** *v.i.* get one's bearings.
προσάπτω *v.t.* impute.
προσαράσσω *v.i.* run aground.
προσαρμογή *s.f.* fitting, fixing; adaptation, adjustment; adaptability.
προσαρμόζω *v.t.* fit, fix; adapt, adjust.
προσάρτημα *s.n.* thing added, annexe.
προσάρτησης *s.f.* annexation.
προσαρτώ *v.t.* annex.
προσβάλλω *v.t.* attack; injure (*health*); offend, affront; jar on; contest (*will, etc.*).
πρόσβαρος *a.* over-weight (*of goods sold*).
πρόσβαση *s.f.* approach.
προσβλητικός *a.* offensive, rude.

προσβολή *s.f.* attack; injury (*to health*); offence, affront; contesting (*of will, etc.*).
προσγει|ούμαι *v.i.* land (*from air*). **~ωμένος** down-to-earth. **~ωση** *s.f.* landing.
προσδίδω *v.t.* lend, impart.
προσδιορίζω *v.t.* determine, fix; ascertain.
προσδοκώ *v.t.* hope for, expect.
προσεγγίζω *v.t. & i.* bring near; approach, draw near; approximate; (*naut.*) put in, call.
προσεκτικός *a.* attentive; careful; circum-spect.
προσέλευσις *s.f.* coming.
προσελκύω *v.t.* attract (*attention*); gain (*supporters, etc.*).
προσέρχομαι *v.i.* attend, present oneself.
προσέτι *adv.* besides, also.
προσευχ|ή *s.f.* prayer. **~ομαι** *v.i.* pray.
προσεχ|ής *a.* next; forthcoming. **~ώς** *adv.* shortly, soon.
προσέχω *v.i. & t.* pay attention (to), notice; be careful (of), mind; look after.
προσηκόντως *adv.* duly, fittingly.
προσ|ηλιακός, ~ήλιος *a.* sunny (*that gets the sun*).
προσήλυτ|ος *s.m.f.* convert. **~ίζω** *v.t.* proselytize, convert.
προσηλ|ώνω *v.t.* nail, fix; **~ωμένος** (*fig.*) attached (*to*), engrossed (*in*).
προσηνής *a.* affable.
πρόσθεσις *s.f.* (*process of*) addition.
πρόσθετ|ος *a.* additional; **~α μαλλιά** postiche.
προσθέτω *v.t.* add.
προσθήκη *s.f.* addition (*thing added*).
πρόσθ|ιος *a.* front, fore. **~ία κολύμβησις** breast-stroke.
προσιτός *a.* accessible; reasonable (*price*).
πρόσκαιρος *a.* temporary.
προσ|καλώ *v.t.* call, summon; invite. **~κεκλημένος** *a.* invited; a guest.
προσκέφαλο *s.n.* pillow, cushion.
προσκήνιο *s.n.* proscenium.
πρόσκληση *s.f.* call, summons; invitation; complimentary ticket.
προσκλητήριον *s.n.* invitation-card; (*mil.*) call, roll-call.
προσκολλ|ώ *v.t.* stick, attach. **~ηση** *s.f.* sticking, attaching; της **~ήσεως** (*one*) who attaches himself unbidden.
προσκομίζω *v.t.* bring forward, produce.
πρόσκομμα *s.n.* obstacle.
πρόσκοπ|ος *s.m.* scout; boy scout. **~ίνα** *s.f.* girl guide.
προσκόπτω *v.i.* stumble; (*with εις*) come up against (*obstacle*).
προσκρούω *v.i.* collide; (*with εις*) strike

against; contravene.

προσκύν|ημα *s.n.* act of worship, respect *or* submission; (place of) pilgrimage. ~ήματα (*pl.*) respects.

προσκυνητής *s.m.* pilgrim.

προσκυνώ *v.t.* worship; pay respects to; submit to.

προσλαμβάνω *v.t.* take on, engage; assume.

προσμένω *v.t.* wait *or* hope for.

πρόσοδ|ος *s.f.* income, revenue. ~οφόρος *a.* fruitful, profitable.

προσοικειούμαι *v.t.* & *i.* gain the support of; (*with* προς) get used to.

προσόν *s.n.* attribute, qualification; advantage.

προσορμίζομαι *v.i.* put into port.

προσοχή *s.f.* attention, care, caution.

προσόψι(ον) *s.n.* towel.

πρόσοψις *s.f.* façade.

προσπάθ|εια *s.f.* attempt, effort. ~ώ *v.i.* try, attempt.

προσπερνώ *v.t.* outdistance; pass.

προσπέφτω *v.i.* (*fig.*) bow down, kowtow.

προσποίησις *s.f.* affectation; dissimulation.

προσποιητός *a.* affected; assumed.

προσποιούμαι *v.t.* feign, put on; pretend; ~ τον αδιάφορο feign indifference. (*v.i.*) dissemble.

προσπορίζομαι *v.t.* procure, gain.

προσταγή *s.f.* order, command.

προστάζω *v.t.* order, command.

προστακτική *s.f.* (*gram.*) imperative.

προστασία *s.f.* protection.

προστατεύ|ω *v.t.* protect; give patronage to. ~όμενος *s.m.* protégé. ~τικός *a.* protective; patronizing.

προστάτης *s.m.* 1. protector. 2. (*anat.*) prostate.

προστίθεμαι *v.i.* be added.

πρόστιμ|ο *s.n.* fine. ~άρω *v.t.* fine.

προστρέχω *v.i.* hasten (*to assistance*); resort.

προστριβή *s.f.* friction.

προστυχαίνω *v.t.* & *i.* make vulgar, lower quality of; become vulgar *or* worse in quality.

προστυχιά *s.f.* (piece of) vulgarity *or* bad manners.

πρόστυχος *a.* vulgar, ill-mannered, caddish; cheap; of bad quality.

προσύμφωνο *s.n.* draft agreement.

προσφά(γ)ι *s.n.* food eaten with bread.

πρόσφατος *a.* fresh, recent.

προσφέρ|ω *v.t.* offer, give, present. ~ομαι *v.i.* offer; be suitable.

προσφεύγω *v.i.* resort.

προσφιλής *a.* dear, loved.

προσφορά *s.f.* offer, bid; tender; offering; ~

και ζήτησις supply and demand; (*eccl.*) bread offered for Holy Communion.

πρόσφορος *a.* suitable, fit.

πρόσφυγας *s.m.* refugee.

προσφυγή *s.f.* recourse, appeal.

προσφυγικός *a.* of *or* for refugees.

προσφυής *a.* suitable, fit.

πρόσφυξ *s.m.f.* refugee.

προσφωνώ *v.t.* address, salute.

πρόσχαρος *a.* cheerful, pleasant.

προσχέδιο *s.n.* draft plan.

πρόσχημα *s.n.* pretext, pretence.

προσχωρώ *v.i.* join, adhere, (*to pact, etc.*).

πρόσω *adv.* ahead.

προσωδία *s.f.* prosody.

προσωνυμία *s.f.* nickname, name.

προσωπάρχης *s.m.* personnel officer.

προσωπείο *s.n.* mask.

προσωπίδα *s.f.* mask.

προσωπικ|ός *a.* personal; ~ά (*s.n. pl.*) private quarrel. ~όν *s.n.* personnel, staff.

προσωπικότητα *s.f.* personality.

προσωπογραφία *s.f.* portrait.

πρόσωπο *s.n.* face; person; part, role.

προσωποποι|ώ *v.t.* personify. ~ηση *s.f.* personification.

προσωρινός *a.* provisional, temporary.

πρότασις *s.f.* proposal, proposition, motion; (*gram.*) sentence, clause.

προτείνω *v.t.* & *i.* extend, hold out, put forward; propose, move, suggest.

προτελευταίος *a.* penultimate.

προτεραία *s.f.* day before.

προτεραιότης *s.f.* priority.

προτέρημα *s.n.* good quality, virtue, advantage.

πρότερ|ος *a.* earlier, previous; εκ των ~ων beforehand, in advance.

προτεσταντισμός *s.m.* protestantism.

προτίθεμαι *v.i.* intend.

προτιμ|ώ *v.t.* prefer. ~ηση *s.f.* preference.

προτιμ|ητέος, ~ότερος *a.* preferable.

προτιν|ός *a.* previous, earlier; τά ~ά the past.

προτομή *s.f.* bust (*sculptured*).

προτού *conj.* & *adv.* before.

προτρέπω *v.t.* urge, instigate.

προτρέχω *v.i.* & *t.* run ahead; (*with gen.*) outrun.

προτροπάδην *adv.* at full speed (*in retreat*).

προτροπή *s.f.* instigation.

πρότυπ|ος *a.* model. ~ον *s.n.* model, example.

προϋπαντώ *v.t.* go to meet (*on arrival*).

προϋπάρχω *v.i.* & *t.* exist previously; (*with gen.*) come before (*in time*).

προϋπο|θέτω *v.t.* presuppose. ~θεσις *s.f.* assumption; prerequisite.

προϋπολογ|ίζω *v.t.* estimate. **~ισμός** *s.m.* estimate; budget.

προύχων *s.m.* notable.

προφαν|ής *a.* obvious, evident. **~ώς** *adv.* obviously.

προφαντός *a.* first *or* early to ripen.

πρόφασ|η *s.f.* pretext, excuse. **~ίζομαι** *v.t.* plead, invoke (*as excuse*).

προφέρ(ν)ω *v.t.* pronounce, utter (*sounds*).

προφήτ|ης *s.m.* prophet. **~εία** *s.f.* prophecy. **~εύω** *v.t.* prophesy. **~ικός** *a.* prophetic.

προφθάνω *v.t.* & *i. see* **προφταίνω**.

προφορ|ά *s.f.* pronunciation. **~ικός** *a.* oral, verbal.

προφτ|αίνω, **~άνω** *v.t.* & *i.* anticipate, forestall; catch (up with); be or act in time; have time *or* money (for); let out (*secret*); *δεν ~αίνω λάδι* I am always needing more oil.

προφυλά(γ)ω *v.t. see* **προφυλάσσω**.

προφυλακή *s.f.* vanguard; outpost.

προφυλακίζω *v.t.* remand in custody.

προφυλακτήρ *s.m.* bumper (*of motor*).

προφυλακτικός *a.* cautious; precautionary, preventive.

προφύλαξις *s.f.* precaution, care.

προφυλάσσ|ω *v.t.* protect, guard. **~ομαι** *v.i.* take precautions.

πρόχειρ|ος *a.* ready to hand; improvised; in draft; *~ο βιβλίο* notebook; *εκ του ~ου* without preparation, extempore.

προχθές *adv.* the day before yesterday; the other day.

πρόχωμα *s.n.* earthwork.

προχωρώ *v.i.* proceed, advance; progress, gain ground.

προωθώ *v.t.* propel; advance, promote.

προώλης *a.* depraved; *see* **εξώλης**.

πρόωρος *a.* premature.

πρύμα *adv.* with fair wind; successfully.

πρύμ(ν)η *s.f.* stern, poop.

πρύταν|ις *s.n.* head of university *or* college; doyen **~εία** *s.f.* office of *πρύτανις*. **~εύω** *v.i.* (*fig.*) rule, prevail.

πρώην *adv.* formerly; *~ πρόεδρος* ex-president.

πρωθυπουργός *s.m.* prime minister.

πρωί 1. *adv.* in the morning, early. 2. *s.n.* morning.

πρωία *s.f.* morning.

πρώιμος *a.* early, premature; precocious (*of plants*).

πρωινός *a.* morning; rising early.

πρωινό *s.n.* 1. morning. 2. breakfast.

πρωκτός *s.m.* anus.

πρώρα *s.f.* prow, bows.

πρώτα *adv.* (at) first; before.

πρωταγωνιστής *s.m.* protagonist.

πρωτάθλημα *s.n.* championship.

πρωταίτιος *s.m.* one mainly responsible, ringleader.

πρωτάκουστος *a.* unheard of.

πρωταπριλιάτικο *s.n.* April fool's trick.

πρωτάρα *s.f.* woman in first child-bed; animal with first litter; tiro.

πρωτάρης *s.m.* tiro.

πρωταρχικός *a.* first in importance.

πρωτεία *s.n.pl.* first place, primacy.

πρωτεργάτης *s.m.* moving spirit.

πρωτεύ|ω *v.i.* be the first, lead. **~ουσα** *s.f.* capital (*city*). **~ων** *a.* primary.

πρωτινός *adv.* of the old school.

πρωτίστως *adv.* first and foremost.

πρωτο- *denotes* first.

πρωτοβάθμιος *a.* of first instance (*court*); of the first rank.

πρωτόβγαλτος *a.* at the beginning of one's career; inexperienced.

πρωτοβουλία *s.f.* initiative.

πρωτοβρόχια *s.n.pl.* first rain (*of autumn*).

πρωτο|γενής **~γέννητος** *a.* first-born.

πρωτόγονος *a.* primitive.

πρωτοδικείον *s.n.* court of first instance.

πρωτοετής *a.* first-year.

πρωτοκαθεδρία *s.f.* place of honour.

πρωτόκολλ|ον *s.n.* 1. register of correspondence. 2. court or diplomatic etiquette. **~ώ** *v.t.* enter (*document*) in register.

πρωτόλεια *s.n.pl.* first-fruits; (*fig.*) juvenilia.

Πρωτομαγιά *s.f.* May Day.

πρωτομηνιά *s.f.* first day of month.

πρωτόπειρος *a.* inexperienced.

πρωτοπορ(ε)ία *s.f.* (*fig.*) vanguard.

πρωτοπόρος *s.m.* pioneer, avant-gardist.

πρώτ|ος *a.* first, foremost; prime (*number*); *~ες ύλες* raw materials; *~ες βοήθειες* first aid. **~α**, **~ον** *adv.* (at) first; before.

πρωτοστατώ *v.i.* play a leading part.

πρωτοσύγκελος *s.m.* senior diocesan official.

πρωτότοκος *a.* first-born.

πρωτότυπ|ος *a.* original; unusual. **~ία** *s.f.* originality. **~ο** *s.n.* original.

πρωτουργός *s.m.* moving spirit.

πρωτοφανής *a.* unprecedented.

Πρωτοχρονιά *s.f.* New Year's Day.

πρωτοψάλτης *s.m.* precentor.

πρωτύτερα *adv.* earlier, previously.

πτ-*see also* **φτ-**.

πταίσμα *s.n.* (*law*) minor criminal offence. **~τοδικείον** *s.n.* lowest criminal court.

πταίω *v.i.* be to blame.

πτέρνα *s.f.* heel.

πτερνιστήρ *s.m.* spur.

πτερόν *s.n.* feather; wing; blade (*of oar,*

propeller).

πτερύγιον *s.n.* fin; aileron; vane.

πτέρυξ *s.f.* wing (*of bird, aeroplane, army, building, political party*).

πτέρωμα *s.n.* plumage.

πτερωτός *a.* winged, feathered.

πτην|όν *s.n.* bird, fowl. **~οτροφία** *s.f.* poultry - farming.

πτήσις *s.f.* flight. **~τικός** *a.* 1. designed for flight. 2. volatile (*oil, etc.*).

πτίλον *s.n.* down.

πτοώ *v.t.* terrify.

πτύελον *s.n.* spittle, sputum.

πτυσσόμενος *a.* folding, collapsible.

πτυχή *s.f.* fold, pleat; wrinkle.

πτυχί|ον *s.n.* degree, diploma. **~ούχος** *a.* qualified, agraduate.

πτύω *v.t. & i.* spit (on).

πτώμα *s.n.* corpse.

πτώσις *s.f.* fall; collapse; decrease; (*gram.*) case.

πτωχ- *see* **~φτωχ-**.

πτωχ|εύω *v.i.* go bankrupt. **~ευσις** *s.f.* bankruptcy.

πτωχοκομείον *s.n.* poor-house.

πυγμαίος *a. & s.m.* pygmy.

πυγμαχ|ία *s.f.* boxing. **~ος** *s.m.* boxer.

πυγμή *s.f.* fist; (*fig.*) drive, determination.

πυγολαμπίδα *s.f.* glow-worm.

πύελος *s.f.* pelvis.

πυθαγόρειος *a.* Pythagorean; ~ *πίναξ* multiplication table.

Πυθία *s.f.* Pythoness; (*iron.*) oracle, prophet.

πυθμήν *s.m.* bottom.

πύθων *s.m.* python.

πυκνοκατοίκητος *a.* densely populated.

πυκν|ός *a.* thick, dense, compact, close together; in quick succession. **~ά, ~ώς** *adv.* thickly, closely.

πυκνότης *s.f.* density, thickness; frequency.

πυκνώνω *v.t. & i.* make *or* grow thicker *or* more extensive *or* more frequent.

πυκνωτής *s.m.* (*electric*) condenser.

πύλη *s.f.* gate (*of city, palace, etc.*); Υψηλή ~ Sublime Porte.

πυλών *s.m.* gateway, portal.

πυναΐζα *s.f.* drawing-pin.

πυξίδα *s.f.* mariner's compass.

πυξ λαξ *adv.* neck-and-crop.

πύξ|ος *s.m.,* **~άρι** *s.n.* box-tree.

πύον *s.n.* pus.

πυρ *s.n.* fire.

πύρα *s.f.* heat of fire; (*fig.*) sensation of burning.

πυρά *s.f.* pyre, stake.

πυρακτώνω *v.t.* make red-hot.

πυραμίδα *s.f.* pyramid.

πύραυλος *s.m.* rocket.

πύραυνον *s.n.* brazier.

πυργίσκος *s.m.* turret.

πυργο|δεσπότης *s.m.* **~δέσποινα** *s.f.* (*now joc.*) master/mistress of the house.

πύργος *s.m.* tower; castle; country-house.

πυρείον *s.n.* match (*for lighting*).

πυρέσσω *v.i.* be feverish.

πυρετ|ός *s.m.* fever, high temperature; (*fig.*) great activity. **~ώδης** *a.* feverish; feverous.

πυρήνας *s.m.* stone, kernel; core; nucleus; cell (*of activists*). **~ικός** *a.* nuclear.

πυρηνέλαιον *s.n.* oil from stones *or* seeds of plants.

πυρίμαχος *a.* fire proof.

πύρινος *a.* burning, fiery.

πυρίτης *s.m.* pyrites.

πυρίτ|ις *s.f.* gunpowder. **~ιδαποθήκη** *s.f.* powder-magazine. **~ιδοποιείον** *s.n.* powder-factory.

πυρκαϊά *s.f.* (*outbreak of*) fire, conflagration.

πυροβολαρχία *s.f.* (*mil.*) battery.

πυροβολείον *s.n.* gun-emplacement.

πυροβολητής *s.m.* gunner.

πυροβολικόν *s.n.* artillery.

πυροβολισμός *s.m.* firing, shot (*of gun*).

πυροβόλ|ος *a.* ~α όπλα firearms. **~ον** *s.n.* (*heavy*) gun.

πυροβολώ *v.i. & t.* fire; shoot at.

πυροδοτ|ώ *v.t.* set off, detonate. **~ηση** *s.f.* detonation.

πυρομαχικά *s.n.pl.* munitions; ammunition.

πυροπαθής *a.* victim of fire.

πυροσβέστης *s.m.* fireman.

πυροσβεστικ|ός *a.* ~ή υπηρεσία fire brigade. ~ή αντλία fire-engine.

πυροστιά *s.f.* trivet.

πυροτέχνημα *s.n.* firework.

πυροφάνι *s.n.* grid for lamp (*in bow of fishing-boat*).

πυρπολητής *s.m.* one who sails a fire-ship.

πυρπολικό *s.n.* fire-ship.

πυρπολώ *v.t.* set on fire.

πυρρός *a.* russet; red-haired.

πυρσός *s.m.* torch, brand; beacon.

πυρώνω *v.t. & i.* make *or* get red hot *or* warm.

πυτζάμα *s.f.* pyjamas.

πυτία *s.f.* rennet.

πυώδης *a.* purulent.

πω *v.t. & i aorist subjunctive of* **λέγω**.

πώγων *s.m.* beard; chin.

πώλησις *s.f.* sale.

πωλ|ητής *s.m.* seller; salesman. **~ήτρια** *s.f.* seller; saleswoman.

πώλος *s.m.* foal, colt.

πωλ|ώ *v.t.* sell; **~είται** for sale.

πώμα *s.n.* stopper, cork, bung, plug.

πω πω *int. or surprise of dismay.*

πωρόλιθος *s.m.* porous stone.

πωρωμένος *a.* (*fig.*) hardened to wrong-doing.

πώρωση *s.f.* loss of sense of right and wrong.

πώς *adv.* how? ~ όχι, why not? ~ ! yes, of course; κάνω ~ και ~ do one's utmost *or* desire strongly.

πώς *adv.* (*enclitic*) somehow, somewhat.

πώς *conj.* that.

Ρ

ραβαῖσι *s.n.* junketing; uproar.

ραβανί *s.n.* sort of cake.

ραβασάκι *s.n.* billet doux.

ραββίνος *s.m.* rabbi.

ραβδί *s.n.* stick, wand.

ραβδ|ίζω *v.t.* beat (*with stick*). ~ισμός *s.m.* beating; blow.

ραβδο|μάντης, ~σκόπος *s.m.* diviner.

ράβδος *s.f.* crook. stick, staff, rod, baton, wand; ingot (*of gold*); rail (*of track*).

ράβδω|σις *s.f.* stripe; flute, groove; ~σεις (*pl.*) fluting, rifling. ~τός *a.* striped; fluted.

ράβ|ω *v.t.* sew (on, up); stitch; make *or* have made (*clothes*); κόβω και ~ω talk unceasingly *or* have influence. ~ομαι *v.i.* have one's clothes made.

ράγα *s.f.* grape; nipple.

ραγάδα *s.f.* chink, crack.

ραγδαίος *a.* rapid, violent, tumultuous, headlong.

ράγια *s.f.* rail (*of railway, etc., track*).

ραγιάς *s.m.* rayah, bondsman.

ραγ|ίζω *v.t. & i.* crack. ~ισμα *s.n.* crack.

ραγού *s.n.* ragout.

ραδιενέργεια *s.f.* radioactivity.

ραδίκι *s.n.* dandelion *or* similar plant.

ραδινός *a.* slender (*of build*).

ραδιογραφία *s.f.* X-ray photograph(y).

ραδιολογία *s.f.* radiology.

ράδιον *s.n.* radium.

ραδιο(τηλε)γράφημα *s.n.* radio-telegram.

ραδιούργ|ος *a.* scheming; an intriguer. ~ία *s.f.* intrigue.

ραδιόφωνο|ο *s.n.* radio, wireless. ~ία *s.f.* broadcasting service. ~ικός *a.* (by) wireless.

ραζακί *s.n.* sort of white grape.

ράθυμος *a.* indolent.

ραιβοσκελής *a.* bow-legged.

ραίνω *v.t.* sprinkle.

ρακένδυτος *a.* dressed in rags.

ρακί *s.n.* sort of spirit.

ράκος *s.n.* rag. (*fig.*) wreck (*person*).

ραλαντί *s.n.* στο ~ in slow motion, ticking over.

ράμμα *s.n.* stitch; έχω ~τα για τη γούνα σου I have a rod in pickle for you.

ραμμένος *a.* sewn.

ραμολ|ίρω *v.i.* become soft-witted. ~ιμέντο *s.n.* soft-witted person.

ράμπα *s.f.* ramp. φώτα της ~ς footlights.

ράμφ|ος *s.n.* beak; burner, jet (*of light*). ~ίζω *v.t.* peck.

ρανίς *s.f.* drop.

ραντεβού *s.n.* appointment, engagement; (*fam.*) lovers keeping tryst. [F. *rendevous*]

ράν|τζο, ~τσο *s.n.* camp-bed.

ραντίζω *v.t.* sprinkle, shower.

ράντισμα *s.n.* spraying.

ραξ *s.f.* grape; nipple.

ραπάνι *s.n.* radish.

ραπίζω *v.t.* slap the face of.

ράπισμα *s.n.* slap.

ράπτης *s.m.* tailor.

ραπτική *s.f.* sewing, dressmaking.

ραπτομηχανή *s.f.* sewing-machine.

ράπτρια *s.f.* dressmaker.

ράπτω *v.t.* sew (on, up), stitch.

ράσο *s.n.* cassock.

ράτσα *s.f.* breed, race; (*fam.*) rascal; από ~ thoroughbred.

ραφείον *s.n.* tailor's shop.

ραφή *s.f.* making (*of clothes*); seam; suture.

ράφι *s.n.* shelf; μένω στο ~ be left on the shelf.

ραφιν|άρω *v.t.* refine. ~άτος *a.* refined, polished.

ραφίς *s.f.* needle.

ράφτ|ης *s.m.* tailor. ~άδικο *s.n.* tailor's shop.

ραφτικά *s.n.pl.* charge for tailoring *or* dressmaking.

ραφτός *a.* sewn.

ράφτρα *s.f.* dressmaker.

ράφτω *v.t.* sew (on, up), stitch.

ραχάτ|ι ~λίκι *s.n.* lazing about. ~ λουκούμ *s.n.* sort of Turkish delight. ~λίδικος *a.* cushy.

ράχη *s.f.* back; spine (*of book*); ridge (*of mountain*).

ραχιτικός *a.* suffering from rickets; malformed.

ραχοκόκκαλο|ο *s.n.*, ~ιά *s.f.* spine.

ράψιμο *s.n.* sewing (on, up); dressmaking, tailoring.

ραψωδία *s.f.* rhapsody.

ρε *int. see* βρε.

ρεαλισ|τής *s.m.* realist. ~μός *s.m.* realism.

ρεβεγιόν *s.n.* Christmas *or* New Year's Eve party.

ρεβέρ *s.n.* turn-up.

ρεβίθι *s.n.* chick-pea.

ρεγάλο *s.n.* present, gift.

ρέγγα *s.f.* herring.

ρέγομαι *v.t.* feel desire for.

ρέγουλ|α *s.f.* order, moderation. **~άρω** *v.t.* regulate, adjust.

ρεδιγκότα *s.f.* redingote.

ρεζέρβα *s.f.* reserve (*stock*), spare (*esp. wheel*).

ρεζές *s.m.* hinge.

ρεζίλεμα *s.n.* humiliation.

ρεζιλεύω *v.t.* make ridiculous; humiliate.

ρεζίλ|ι, ~ίκι *s.n.* (*object of*) ridicule *or* shame; γίνομαι ~ι οϱ παθαίνω ~ίκι become a laughing-stock, be humiliated.

ρείθρο *s.n.* watercourse, gutter.

ρείκι *s.n.* heath, heather.

ρεκλάμ|α *s.f.* advertisement. **~άρω** *v.t.* advertise. **~αδόρος** *s.m.* show-off.

ρέκτης *s.m.* enterprising person.

ρέμα *s.n.*, **~τιά** *s.f.* ravine, torrent (-bed).

ρεμάλι *s.n.* wastrel.

ρεμβ|άζω *v.i.* muse, dream. **~ασμός** *s.m.* reverie. **~ώδης** *a.* dreamy.

ρεμούλα *s.f.* plundering; (*fam.*) graft.

ρεμουλκ- *see* **ρυμουλκ-**.

ρεμπελ|ιό *s.n.* rebellion; idler's existence. **~εύω** *v.i.* be a loafer.

ρέμπελος *a.* idle and ill-disciplined.

ρεμπέτης *s.m.* vagabond.

ρεμπέτικο *s.n.* sort of popular song in oriental style.

ρε(μ)πούμπλικα *s.f.* trilby *or* homburg hat.

ρέντα *s.f.* winning streak.

ρεπάνι *s.n.* radish.

ρεπερτόριο *s.n.* repertory.

ρεπό *s.n.* time off (*between shifts*). [F. *repos*]

ρεπορτάζ *s.n.* report(ing).

ρέπω *v.i.* lean, incline.

ρεσάλτο *s.n.* desperate attempt.

ρεσεψιόν *s.f.* reception desk.

ρέστ|ος *a.* the rest (of). μένω ~ος από run out of. **~α** *s.n.pl.* change (*money*).

ρετάλι *s.n.* remnant (*of cloth*).

ρετιρέ *s.f.* penthouse apartment.

ρετουσάρω *v.t.* retouch.

ρετσέτα *s.f.* prescription, recipe.

ρετσίν|α *s.f.* resin; resinated wine. **~άτο** *s.n.* resinated wine.

ρετσίν|ι *s.n.* resin. **~ιά** *s.f.* (*fam.*) stigma left by false accusation, smear.

ρετσινόλαδο *s.n.* castor-oil.

ρεύμα *s.n.* current; draught.

ρευματ|ισμός *s.m.* rheumatism. **~ικός** *a.* rheumatic.

ρεύομαι *v.i.* belch.

ρευστ|ός *a.* fluid; liquid (*of assets*); unstable. **~ότης** *s.f.* liquidity, fluidity. **~οποιώ** *v.t.* liquefy; turn (*assets*) into cash.

ρεύω *v.i.* & *t.* fall in, crumble; (*fig.*) waste away. (*v.t.*) wear down, exhaust.

ρεφεν|ές *s.m.* quota, share; (με) **~έ** each paying his share.

ρέψιμο *s.n.* 1. belch, belching. 2. ruin, decay.

ρέω *v.i.* flow.

ρήγας *s.m.* king.

ρήγμα *s.n.* breach, crack, split.

ρηθείς *a.* (afore) said.

ρήμα *s.n.* verb.

ρημαγμός *s.m.* devastation, ruin.

ρημάδι *s.n.* ruin; (*fam.*) derogatory epithet; κλείσ' το το~ turn the damn thing off (*e.g. radio*); τά 'καμες ~ό you have made a mess of it. μάζεψε τα ~α σου clear off!

ρημάζω *v.t.* & *i.* ruin, devastate; fall into ruin *or* neglect.

ρηξικέλευθος *a.* pioneering; an innovator.

ρήξις *s.f.* rupture, breaking; conflict.

ρητίν|η *s.f.* resin. **~ίτης** *s.m.* resinated wine.

ρητό *s.n.* saying.

ρητορεύω *v.i.* hold forth, make speeches.

ρητορική *s.f.* oratory.

ρητορικ|ός *a.* rhetorical. **~ή** *s.f.* oratory.

ρητός *a.* explicit, express.

ρήτρα *s.f.* clause.

ρήτ|ωρ *s.m.* orator.

ρηχ|ός *a.* shallow; **~α** (*s.n.pl.*) shallows.

ρίγα *s.f.* (*measuring*) rule; stripe.

ρίγανη *s.f.* origanum (*herb*); (*fam.*) βάλ' του ~ ironic comment on unfulfilled expectations; so much for that!

ριγέ *a.* striped. [F. *rayé*]

ρίγος *s.n.* shiver, shudder.

ριγώ|νω *v.t.* rule *or* make lines. **~τός** *a.* striped, lined.

ριγώ *v.i.* shiver, shudder.

ρίζα *s.f.* root; foot (*of mountain*).

ριζά *s.n.pl.* foot (*of mountain*).

ριζηδόν *adv.* to the roots, radically.

ριζικ|ός *a.* root, original; radical, complete; basic. **~ό** *s.n.* fate, destiny.

ριζοβολώ *v.i.* take root.

ριζοβούνι *s.n.* foot of mountain.

ριζοσπάστ|ης *s.m.* radical (*in opinions*). **~ικός** *a.* radical.

ριζών|ω, ~ομαι *v.i.* become rooted *or* fixed.

ρικνός *a.* wrinkled, warped.

ρίμα *s.f.* rhyme.

ρίνη *s.f.* file (*tool*).

ρινικός *a.* nasal.

ρινίσματα *s.n.pl.* filings.

ρινόκερως *s.m.* rhinoceros.

ριξιά s.f. charge (explosive); shot, throw.

ρίξιμο s.n. throwing; dropping, firing; throw, shot; attack; fortune-telling; cheating.

ριπ|ή s.f. burst (of firing); gust (of wind); εν ~ή οφθαλμού in the twinkling of an eye.

ριπίδιον s.n. (hand) fan.

ρίπτω v.t. throw, cast; overthrow.

ρις s.f. nose.

ρισκάρω v.t. risk.

ρίχν|ω v.t. throw, cast; drop, pour (out); sprinkle; knock or pull down, overthrow; fire (shots); (fam.) cheat; ~ω κανόνι go bankrupt; ~ω γράμμα post a letter; ~ω τη γνώμη μου express one's opinion; ~ω το παιδί have an abortion; ~ω τα χαρτιά tell fortunes from cards; ~ω το καράβι έξω run the ship aground; το ~ω έξω go on the spree, neglect one's responsibilities; το ~ω σε indulge immoderately in (cards, drink, women, etc.); ~ει βροχή it is raining. ~ομαι v.i. & t. throw oneself; (with σε or gen.) attack; make advances to (woman).

ριχτός a. (of clothes) full, not fitted to figure; worn over shoulders, (with arms not in sleeves).

ρίψις s.f. see ρίξιμο.

ριψοκινδυν|εύω v.i. & t. take risks; risk, venture. ~ς a. risky; rash.

ροβολώ v.i. run or roll downhill.

ρόγχος s.m. (bronchial) rattle; death-rattle.

ρόδα s.f. wheel; (fam.) motor-car.

ροδάκινο s.n. peach.

ροδαλός a. rosy (esp. complexion).

ροδάνι s.n. spinning-wheel. η γλώσσα του πάει ~ he's a chatterbox.

ροδέλαιο s.n. attar of roses.

ροδέλα s.f. washer (ring); round slice.

ρόδι s.n. promegranate.

ροδίζω v.i. become rosy; brown (in cooking).

ρόδινος a. rosy; (fig.) rose-coloured.

ροδίτης s.m. sort of pink grape.

ροδοδάφνη s.f. oleander.

ροδοκόκκινος a. ruddy; brown (of cooking food).

ρόδον s.n. rose.

ροδό|στα(γ)μα, ~σταμο s.n. rosewater.

ροδόχρους a. rosy.

ροζ a. pink. [F. rose]

ροζακί s.n. sort of white grape.

ρόζ|ος s.m. knot (in wood); callus. ~ιάρικος a. knotted, gnarled.

ροή s.f. flow, flux, discharge.

ρόιδ|ι, ~ο s.n. pomegranate. τα κάνω ~ο make a mess of it.

ρόκα s.f. 1. distaff. 2. rocket (plant).

ροκάνα s.f. rattle.

ροκάν|ι s.n. plane (tool). ~ίδι s.n. wood-shaving.

ροκανίζω v.t. plane; gnaw; (fig.) nibble at, consume.

ρολάρω v.i. free-wheel.

ρολό s.n. roll (cylindrical); drop-shutter. [F. rouleau]

ρολογάς s.m. watchmaker.

ρολό(γ)ι s.n. clock, watch; meter; (fam.) πάει ~ it goes like clockwork.

ρόλ|ος s.m. roll; role, part; δεν παίζει ~o it makes no difference.

ρομάντζ|α s.f. romantic scenery or setting; enjoyment of these; (mus.) type of song. ~άρω v.i. indulge in reverie.

ρομαντικός a. romantic.

ρομάντζο s.n. romance (story, love-affair).

ρομβία s.f. barrel-organ.

ρόμβος s.m. (goem.) rhombus; peg-top.

ρόμπα s.f. dressing-grown, dress.

ρομφαία s.f. two-edged sword.

ρονρονίζω v.i. purr.

ρόπαλο s.n. club, cudgel.

ροπή s.n. propensity.

ρούβλι s.n. rouble.

ρούζ s.n. rouge. [F. rouge]

ρουθούν|ι s.n. nostril; (fam.) ~ι δεν έμεινε not a soul survived; μου μπήκε στο ~ι he exasperated me (by pestering). ~ίζω v.i. snort.

ρουκέτα s.f. rocket; (fam.) vomiting.

ρουλεμάν s.n. ball-bearings. [F. roulement]

ρουλέτα s.f. roulette.

Ρούμελη s.f. part of Central Greece.

ρούμι s.n. rum.

ρουμπίνι s.n. ruby.

ρούπι s.n. measure of length (about 3 ins.); (fam.) δεν το κουνάω ~ I will not budge.

ρους s.m. flow; (fig.) current.

ρούσσος a. red-haired.

ρουσφέτι s.n. illegitimate political favour.

ρουτίν|α s.f. routine. ~ιέρικος a. humdrum.

ρούφη(γ)μα s.n. (noisy) sipping, sniffing or sucking in.

ρουφηξιά s.n. sip; sniff; puff, pull (at pipe, etc.).

ρουφήχτρα s.f. whirlpool.

ρουφιάνος s.m. pimp; intriguer.

ρο(υ)φώ v.t. sip, sniff or suck in (noisily); absorb.

ρουχαλ- ee ροχαλ-.

ρουχ|ικά s.n.pl., ~ισμός s.m. clothing.

ρούχ|ο s.n. cloth, material; garment. ~α (pl.) clothes, bed-clothes; (fam.) menstrual period; είμαι στα ~α be confined to bed; τρώγεται με τα ~α του he is an inveterate grumbler.

ρόφημα s.n. soup; hot drink.

ροφός s.m. garfish.

ροχαλητό s.n. snore, snoring.

ροχαλ|ίζω v.i. snore. **~σμα** s.n. snore, snoring.

ρόχ|αλο s.n. **~άλα** s.f. gob of spittle.

ρόχθος s.m. roaring (esp. of waves).

ρύ|αξ s.m., **~άκι** s.n. stream, rivulet.

ρύγχος s.n. snout, muzzle; nozzle.

ρύζ|ι s.n. rice. **~όγαλο** s.n. rice-pudding.

ρυθμίζω v.t. regulate, arrange.

ρυθμικός a. rhythmical.

ρυθμιστ|ήρ, ~ής s.m. regulator.

ρυθμός s.m. rhythm, proportion; order, style (of architecture, furniture); pace.

ρύμη s.f. 1. momentum. 2. narrow street.

ρυμοτομία s.f. street-plan.

ρυμούλκα s.f. trailer.

ρυμουλκό s.n. tug-boat; tractor.

ρυμουλκώ v.t. tow; (fig.) lead by the nose.

ρυπαίνω v.t. make dirty; (fig.) defile, sully.

ρύπαν|σις s.f. pollution, soiling. **~τικός** a. polluting.

ρυπαρ|ός a. filthy; vile **~ότης** s.f. filth.

ρύπος s.m. filth; defilement.

ρυτίδ|α s.f. wrinkle; ripple (on water). **~ώνω** v.t. & i. wrinkle, ripple; get wrinkled or rippled.

ρω s.n. the letter P.

ρώγα s.f. grape; nipple.

ρωγμή s.f. crack, fissure.

ρωγοβύζι s.n. baby's feeding-bottle.

ρώθων s.m. nostril.

ρωμαίικ|ος a. of modern Greece; το **~ο** (pej.) Greece (with all her shortcomings). **~α** s.n.pl. Romaic, demotic Greek; μίλα **~α** be explicit!

ρωμαϊκός a. Roman.

Ρωμαίος s.m. a Roman.

ρωμαλέος a. robust, strong.

ρωμανικός a. romanesque; romance (language).

ρωμαντικός a. romantic.

ρώμη s.f. robustness, strength.

Ρωμι|ός s.m. a (modern) Greek. **~οσύνη** s.f. the (modern) Greek people and ethos.

ρώτημα s.n. question, query.

ρωτώ v.t. & i. ask, question (person); ask (something); inquire.

Σ

σ' see σε

σα conj. see σαν.

σάβανο s.n. shroud.

Σάββατ|ο s.n. Saturday; Sabbath. **~οκύριακο** s.n. week-end.

σαβούρ|α s.f. ballast; trash. **~ώνω** v.t. ballast; (την) **~ώνω** (fam.) gorge oneself.

σαγάν|ι, ~άκι s.n. two-handled frying pan; (fam.) **~άκι** dish of eggs & cheese fried in this.

σαγή s.f. harness.

σαγήν|η s.f. fascination. **~εύω** v.t. charm, allure.

σαγκουίνι s.n. blood-orange.

σάγμα s.n. pack-saddle.

σαγόνι s.n. jaw, chin.

σαγονιά s.f. jaw, chin; blow on the jaw; (fam.) έφαγα ~ I was stung (overcharged).

σαθρός a. rotten, decayed; ramshackle.

σαιζόν s.f. season.

σαΐτα s.f. arrow, dart; shuttle.

σακαράκα s.f. (fam.) old motor-car or cycle.

σακάτ|ης s.m. cripple. **~εύω** v.t. cripple, maim.

σάκκα s.f. satchel; game-bag; briefcase.

σακκάκι s.n. jacket.

σακκί s.n. sack; (fam.) βάζω στο ~ dupe.

σακκίδιο s.n. haversack.

σακκοράφα s.f. sack-needle.

σάκκος s.m. sack, sackful; kitbag; jacket.

σακκούλα s.f. sack, bag; paper or plastic bag; pouch.

σακκούλι s.n. sack, bag. **~άζω** v.t. & i. put into sacks; be baggy.

σάκχαρις s.f. see ζάχαρη.

σάκχαρον s.n. (chem.) sugar.

σάλα s.f. drawing-room.

σάλαγ|ο s.n., **~ος** s.m. noise of crowd or herd; herdsman's cry.

σαλαμάνδρα s.f. salamander; slow-combustion stove.

σαλαμούρα s.f. brine.

σαλάτα s.f. salad; τα κάνω ~ make a mess of it.

σαλβάρι s.f. Turkish trousers.

σάλεμα s.n. slight movement, stirring.

σαλέπι s.n. salep.

σαλεύω v.t. & i. move, stir.

σάλι s.n. shawl.

σάλιαγκας s.m. snail.

σαλιάζω v.i. make saliva.

σαλιάρα s.f. bib.

σαλιάρης a. a chatterbox.

σαλιαρ|ίζω v.i. chatter, gossip, drool. **~ισμα** s.n. foolish talk.

σαλιγκάρι s.n. snail.

σάλι|ο s.n. saliva; τρέχουν τα ~α μου my mouth waters.

σαλιώνω v.t. lick, moisten.

σαλόνι s.n. drawing-room (suite).

σάλος s.m. swell (of sea); rolling (of ship); (fig.) commotion.

σαλπάρω v.i. (naut.) weigh anchor.

σάλπιγξ s.f. trumpet, bugle.

σαλπ|ίζω *v.i. & t.* blow trumpet; sound (*call*) on trumpet. **~ισμα** *s.n.* trumpet-call.

σαλτάρω *v.i.* jump.

σαλτιμπάγκος *s.m.* fair-ground acrobat, *etc.*

σάλτος *s.m.* jump.

σάλτσα *s.f.* gravy, sauce.

σαμάρ|ι *s.n.* pack-saddle; coping. **~ώνω** *v.t.* saddle.

σαματάς *s.m.* row (*noise*).

σάματι(ς) *adv.* as if.

σαμιαμίθι *s.n.* small lizard; (*fig.*) puny fellow.

σαμόλαδο *s.n.* sesame oil.

σαμούρι *s.n.* sable.

σαμπάνια *s.f.* champagne.

σαμποτ|άζ *s.n.* sabotage. [F.]. **~άρω** *v.t.* sabotage.

σαμπρέλα *s.f.* inner tube (*of tyre*).

σάμπως *adv.* as if; it seems that; *also introduces question.*

σα(ν) *conj. & adv.* when, whenever, as soon as, if; like, as (if); ~ *να* (it seems) as if; ~τί; what? ~ *ηλίθιος που ήταν τον γέλασαν* being a fool he got cheated; ~ *σήμερα πέρσι* a year ago today.

σανά *s.n.pl.* hay.

σανδάλι(ον) *s.n.* sandal.

σανίδ|α *s.f.*, **~ι** *s.n.* plank, board. **~ώνω** *v.t.* board over.

σανιδόσκαλα *s.f.* gang-plank.

σανίς *s.f.* plank, board.

σανός *s.m.* hay.

σαντιγύ *s.n.* chantilly (*cream, lace*). [F. *Chantilly*]

σαντούρι *s.n.* dulcimer.

σαπίζω *v.t. & i.* rot.

σαπίλα *s.f.* dacay, rottenness; (*fig.*) corruption.

σάπιος *a.* rotten; corrupt.

σαπουνάδα *s.f.* soapy water; lather.

σαπούν|ι *s.n.* soap. **~ίζω** *v.t.* soap.

σαπουνόφουσκα *s.f.* soap-bubble.

σαπρός *a.* rotten; corrupt.

σάπφειρος *s.m.* sapphire.

σαπφικός *a.* sapphic.

σάπων *s.m.* soap.

σάρα *s.f.* (*fam.*) *η ~ και η μάρα* riff-raff.

σαράβαλο *s.n.* (*fam.*) ruin, wreck.

σαράκι *s.n.* woodworm; (*fig.*) canker, secret sorrow, remorse.

σαρακοστ|ή *s.f.* Lent. **~ιανός** *a.* lenten.

σαράντα (*num.*) forty.

σαραντάμερο *s.n.* Advent.

σαρανταποδαρούσα *s.f.* centipede.

σαραντίζω *v.i.* become forty; be forty days old *or* dead *or* delivered of child.

σαράφης *s.m.* money-changer.

σαρδανάπαλος *s.m.* (*fig.*) sybarite.

σαρδέλλα *s.f.* anchovy; sardine; (*fam.*) stripe (*chevron*).

σαρδόνιος *a.* sardonic.

σαρίδι *s.n.* sweepings.

σαρίκι *s.n.* turban.

σάρκ|α *s.f.* flesh. **~ικός** *a.* carnal.

σαρκασ|μός *s.n.* sarcasm. **~τικός** *a.* sarcastic.

σαρκίον *s.n.* (*fam.*) *να σώσει το ~ του* to save his skin.

σαρκοβόρος *a.* carnivorous.

σαρκοφάγος 1. *a.* carnivorous 2. *s.f.* sarcophagus.

σαρκ|ώδης *a.* fleshy, corpulent. **~ωμα** *s.n.* (*med.*) sarcoma.

σαρξ *s.f.* flesh.

σάρπα *s.f.* scarf, shawl.

σάρωθρον *s.n.* broom.

σάρωμα *s.n.* sweeping(*s*); broom.

σαρώνω *v.t.* sweep.

σασ(σ)ί *s.n.* chassis. [F. *chassis*]

σαστίζω *v.t. & i.* disconcert, confuse; get confused.

σατανάς *s.m.* Satan; (*fig.*) sinister *or* very clever person.

σατανικός *a.* satanic(al), diabolical.

σάτιρ|α *s.f.* satire. **~ίζω** *v.t.* satirize. **~ικός** *a.* satirical.

σατίρι *s.n.* butcher's chopper.

σατράπης *s.m.* satrap; despot.

σάτυρος *s.m.* satyr.

σαύρα *s.f.* lizard.

σαφήνεια *s.f.* clarity, distinctness.

σαφής *a.* clear, explicit.

σαχάνι *s.n.* see **σαγάνι.**

σάχης *s.m.* shah.

σάχλα *s.f.* insipidity; silliness. **~μάρα** *s.f.* nonsense. **~μαρίζω** *v.i.* talk nonsense.

σαχλός *a.* insipid; silly.

σάψαλο *s.n.* see **σαράβαλο.**

σβάρνα *s.f.* harrow; (*fam.*) *παίρνω* ~ sweep *or* drag along; make the round of.

σβέλτος *a.* quick, nimble.

σβέρκος *s.m.* nape of neck.

σβήνω *v.t. & i.* extinguish, quench, put out; erase; be quenched *or* erased; go out; (*fig.*) die.

σβήσιμο *s.n.* putting out; erasure.

σβηστός *a.* (put) out, erased.

σβίγγος *s.m.* fritter.

σβουνιά *s.f.* cattle-dung.

σβούρα *s.f.* spinning-top.

σβύνω *v.t. & i.* see **σβήνω.**

σβώλος *s.m.* clod; clot.

σγάρα *s.f.* (*bird's*) crop.

σγουρός *a.* curly, curly-haired.

σε 1. *pron.* you. 2. *prep. see* **εις.**

σέβ|ας *s.n.* **~ασμός** *s.m.* respect, reverence.

σεβ|άσματα, ~η *s.n.pl.* respects.

σεβάσμιος *a.* reverend.

σεβαστός *a.* respected; respectable, not inconsiderable.

σεβντάς *s.m.* love, heart-ache.

σέβομαι *v.t.* respect, revere.

σειρά *s.f.* series, row, line; turn; sequence, order; *με τη* ~ in turn; *της* ~*ς μας* of our (social) class; *της* ~*ς* common, ordinary; *έχει τη* ~ *του* he is comfortably off; *μπαίνω σε* ~ settle down to a steady course.

σειρήνα *s.f.* siren, charmer; hooter; motor-horn.

σειρήτι *s.n.* stripe, chevron (*on sleeve*).

σεις *pron.* you.

σείσιμο *s.n.* shaking, tremor.

σεισμικός *a.* seismic.

σεισμο|παθής, ~πληκτος *a.* stricken by earthquake.

σεισμός *s.m.* earthquake.

σειστός *a.* with swaying gait.

σεί|ω *v.t.* shake. ~έμαι *v.i.* shake; affect undulating gait.

σεκλέτι *s.n.* care, worry.

σελαγίζω *v.i.* shine (*of star, lighthouse*).

σέλας *s.n. βόρειον* ~ northern lights.

σελάχι *s.n.* 1. belt for pistols, *etc.* 2. ray (*fish*).

σελέμ|ης *s.m.* sponger, cadger. ~ίζω *v.t.* cadge.

σελήνη *s.f.* moon.

σεληνιασμός *s.m.* epilepsy.

σεληνόφως *s.n.* moonlight.

σελίνι(ον) *s.n.* shilling.

σέλινο *s.n.* celery.

σελ|ίς, ~ίδα *s.f.* page.

σέλλα *s.f.* saddle.

σεμιγδάλι *s.n.* semolina.

σεμνός *a.* decorous, unassuming, modest.

σεμνοπρεπής *a.* decorous, modest.

σεμνότυφος *a.* prudish.

σέμν|ωμα *s.n.* (*object of*) pride, boast. ~ύνομαι *v.i.* feel proud of, boast about. (*with για*).

σέμπρος *s.m.* share-cropper.

σένα *pron.* you.

σεντέφι *s.n.* mother-of-pearl.

σεντόνι *s.n.* (*bed*) sheet.

σεντούκι *s.n.* linen-chest.

σεξουαλισμός *s.m.* sexual instinct.

Σεπτέμβρ|ης, ~ιος *s.m.* September.

σεπτός *a.* venerated (*of relics, etc.*).

σερβίρω *v.t.* serve (*at table*).

σερβιτόρος *s.m.* waiter.

σερβίτσιο *s.n.* service (*set of plates, etc.*); place-setting.

σεργιάν|ι *s.n.* walk, stroll. ~ίζω *v.t. & i.* take (for) a walk; wander.

σερμαγιά *s.f.* capital, funds.

σερμπέτι *s.n.* sherbet; manure-and-water.

σερνικός *a.* (*fam.*) male.

σέρν|ω *v.t.* draw, pull, drag; lead (*dance*); ~*ω φωνή* cry out; *τι μας έσυρε* how he abused us! *σύρε στο διάολο* go to the devil! ~*ομαι .i.* crawl, drag; drag oneself along; ~*εται γρίππη* there is an epidemic of flu.

σερπετός *a.* sharp, lively, spry.

σέρρα *s.f.* greenhouse.

σέρτικος *a.* strong (*tobacco*).

σεσημασμένος *a.* (*criminal*) whose particulars are known to police.

σέσουλα *s.f.* scoop; (*naut.*) skeet. *με τη* ~ plentifully.

σεφτές *s.m.* first sale of the day.

σήκαλη *s.f.* rye.

σήκωμα *s.n.* lifting; rising, getting up.

σηκών|ω *v.t.* raise, lift up; carry; remove, withdraw; put up with; take; do with, require; *τον* ~*ει το κλίμα* the climate is good for him; ~*ω το τραπέζι* clear away the table; ~*ω στο πόδι* rouse, throw into confusion; ~*ω κεφάλι* get uppish. ~ομαι *v.i.* rise, get up; revolt; *σήκω* get up!

σηκωτός *a.* raised; on a stretcher.

σήμα *s.n.* mark, badge; signal; omen; ~ *κατατεθέν*- trade-mark.

σημαδ|εύω *v.t.* mark; aim at; ~εμένος having physical defects, scarred.

σημάδι *s.n.* mark, scar, birth-mark; target; omen; sing, indication.

σημαδιακός *a.* having physical defects, scarred; exceptional, rare.

σημαδούρα *s.f.* buoy.

σημαία *s.f.* flag, banner.

σημαίν|ω *v.t. & i.* mean, signify; sound, signal (*alarm, etc.*); (*v.i.*) matter, be important; sound, go (*of bells, etc.*); ~ον *πρόσωπον* leading personality.

σημαιοφόρος *s.m.* standard-bearer; (*naut.*) ensign (*lowest rank of officer*).

σήμανση *s.f.* stamping, recording of (criminal's) particulars.

σημαντικός *a.* signifying (*with gen.*); important, noteworthy.

σήμαντρο *s.n.* wood or metal bar serving as church bell; stamp, seal.

σημασία *s.f.* meaning; importance.

σημασιολογία *s.f.* semantics.

σηματο|γράφος, ~φόρος *s.m.* semaphore.

σηματοδοσία *s.f.* signalling.

σηματοδότης *s.m.* traffic-lights.

σημείο *s.n.* sign, indication; point (*place or degree*), stage, pitch; sign, symbol (*maths, etc.*); ~ *του ορίζοντος* point of the compass.

σημείωμα *s.n.* (*written*) note.

σημειωματάρι(ο) *s.n.* notebook, diary.

σημειώ|νω *v.t.* mark; make a note of; take note of, observe; achieve, score. ~*μένος* having physical defects, scarred.

σημείωση *s.f.* (*written*) note.

σημειωτέος *a.* to be noted.

σημειωτ|ός *a. βήμα* ~*όν* mark time!

σήμερ|α, ~ον *adv.* today. ~**ινός** *a.* today's.

σήπομαι *v.i.* rot.

σηπτικός *a.* septic.

σήρα|γξ, ~γγα *s.f.* tunnel.

σηροτροφία *s.f.* production of raw silk.

σησάμι *s.n.* sesame.

σηψαιμία *s.f.* septicaemia.

σήψη *s.f.* rot, decay; (*med.*) sepsis.

σθεναρός *a.* strong, vigorous.

σθένος *s.n.* strength.

σιαγών *s.f.* jaw, chin.

σιάζω *v.t. & i.* settle, put straight; get better; *τάσιαξαν* (*fam.*) they have made it up *or* formed an attachment.

σίαλος *s.m.* saliva.

σιάξιμο *s.n.* straightening, arranging; improvement.

σιάχνω *v.t. & i. see* σιάζω.

σιγά *adv.* quietly, gently, slowly; ~ ~ gradually, carefully.

σιγα|λός, ~νός *a.* quiet, gentle, slow; ~*νό ποτάμι* still waters run deep.

σιγανοπαπαδιά *s.f.* person affecting prudishness *or* innocence.

σιγαστήρ *s.m.* silencer.

σιγή *s.f.* silence.

σιγηλός *a.* silent; taciturn.

σίγμα *s.n.* the letter Σ.

σιγο- *denotes* quietly, slowly.

σιγοβράζω *v.i.* simmer.

σίγουρ|ος *a.* certain, sure, secure, ~**α** *adv.* for certain. ~**άρω** *v.t.* secure, ensure.

σιγώ *v.i.* be silent; subside, abate.

σιδερένιος *a.* iron; (*fig.*) ~ ! may you enjoy robust health!

σιδερικά *s.n.pl.* things made of iron.

σίδερ|ο *s.n.* iron; flat-iron; curling-tongs. ~*α* (*pl.*) ironwork; railway-lines; prison; *έφαγα τα* ~*α* I made a superhuman effort. *στη βράση κολλάει το* ~*ο* strike while the iron is hot.

σιδερ|ώνω *v.t.* iron, press. ~**ωμα** *s.n.* ironing.

σιδηρόδρομ|ος *s.m.* railway. ~**ικός** *a.* of railways; a railwayman. ~**ικώς** *adv.* by rail.

σίδηρος *s.m.* iron.

σιδηροτροχιά *s.f.* railway-track.

σιδηρουργείον *s.n.* forge.

σιδηρουργός *s.m.* blacksmith.

σιδηρούς *a.* iron.

σιδηρούχος *a.* ferruginous.

σίελος *s.m.* saliva.

σικ *a.* chic.

σίκαλη *s.f.* rye.

σιλουέτα *s.f.* silhouette.

σιμά *adv.* near.

σιμιγδάλι *s.n.* semolina.

σιμός *a.* snub-nosed.

σιμώνω *v.t. & i.* approach.

σινάπι *s.n.* mustard.

σινάφι *s.n.* guild, group.

σινδών *s.f.* (*bed*) sheet.

σινεμά *s.n.* cinema.

σινιάλο *s.n.* sign, signal.

σινικός *a.* Chinese.

σιντριβάνι *s.n.* fountain.

σιρίτι *s.n.* braid; stripe (on uniform).

σιρόκος *s.m.* south-east wind.

σιρόπι *s.n.* syrup.

σίτα *s.f.* wire mesh, sieve.

σιταράτος *a.* corn-coloured; (*fig.*) clear, to the point (*of speech*).

σιτάρ|ι *s.n.* wheat. ~**ένιος** *a.* wheaten.

σιτάρκεια *s.f.* self-sufficiency in cereals.

σιτεύω *v.t. & i.* fatten (*livestock*); hang, get tender *or* high (*of meat*).

σιτηρά *s.n.pl.* cereals.

σιτηρέσιον *s.n.* day's ration.

σιτίζω *v.t.* feed.

σιτιστής *s.m.* (*mil.*) quartermaster-sergeant.

σιτοβολών *s.m.* granary.

σιτοδεία *s.f.* failure of wheat crop, famine.

σίτος *s.m.* wheat.

σιφόνι *s.n.* siphon; S-bend pipe.

σίφουνας *s.m.* water-spout, whirlwind.

σίφων *s.m.* siphon; S-bend pipe; water-spout, whirlwind.

σιχαίνομαι *v.t.* loathe, be disgusted by.

σιχαμάρα *s.f.* (object of) disgust.

σιχα|μένος, ~μερός *a.* disgusting.

σιχαμός *s.m.* (object of) disgust.

σιχασιάρης *a.* squeamish.

σιωπή *s.f.* silence.

σιωπηλός *a.* silent, taciturn.

σιωπηρός *a.* tacit.

σιωπώ *v.i.* be silent; stop talking.

σκ- *see* σχ-.

σκάβω *v.t. & i.* dig, carve, hollow out.

σκάγια *s.n.pl.* small shot.

σκάζω *v.t. & i.* burst, crack, chap; exasperate; be exasperated, maddened, choked (*by heat, jealousy, etc.*); (*fam.*) pay, cough up; *το* ~ escape, make off; *σκάσε* shut up! *να σκάσει* damn him!

σκαθάρι *s.n.* 1. beetle. 2. bream.

σκαιός *a.* unmannerly, abrupt, rude.

σκάκι *s.n.* chess.

σκάλα *s.f.* staircase, steps, ladder; stirrup; landing-place, port, harbour; wave (*in hair*); degree; (*mus.*) scale.

σκαλί *s.n.* step, rung; degree.

σκαλίζω *v.t. & i.* hoe, dig over; search, rummage; carve, sculpt; pick (*teeth, nose*).

σκάλισμα *s.n.* hoeing; searching; carving.

σκαλιστήρι *s.n.* hoe.

σκαλιστός *a.* carved, engraved.

σκαλοπάτι *s.n. see* σκαλί.

σκαλτσούνι *s.n.* 1. woollen sock. 2. turnover (*pastry*).

σκαλώνω *v.i.* climb up; (*fig.*) meet with a hitch, get held *or* caught up.

σκαλωσιά *s.f.* scaffolding.

σκαλωτός *a.* having steps; wavy (*of hair*).

σκαμνί *s.n.* stool.

σκαμπάζω *v.t. & i.* (*fam.*) understand, know.

σκαμπανεβάζω *v.i.* (*naut.*) pitch.

σκαμπίλι *s.n.* 1. slap. 2. a card game.

σκαμπρόζικος *a.* risqué.

σκανδάλη *s.f.* trigger.

σκανδαλίζω *v.t.* intrigue, tempt; scandalize. ~ιστικός *a.* tempting.

σκανδαλοθήρας *s.m.* scandalmonger.

σκάνδαλο *s.n.* scandal; βάζω ~α sow dissension.

σκανδαλώδης *a.* scandalous.

σκανταλιάρης, ~ικος *a.* mischievous, naughty; provocative (*sensually*).

σκαντζόχοιρος *s.m.* hedgehog.

σκάνω *v.t. & i. see* σκάζω.

σκαπανεύς *s.m.* sapper, pioneer.

σκαπάνη *s.f.* mattock.

σκαπουλάρω *v.t. & i.* escape.

σκάπτω *v.t. & i. see* σκάβω.

σκάρα *s.f.* grill, grid; grate; rack.

σκαρί *s.n.* (*naut.*) στα ~ά on the stocks; (*of persons*) build; disposition.

σκαρίφημα *s.n.* sketch, rough draft.

σκαρπίνι *s.n.* woman's lace-up shoe.

σκάρτος *a.* defective, dud.

σκαρφαλώνω *v.i.* climb up

σκαρφ|ίζομαι *v.t.* think up, contrive; του ~ίστηκε να he took it into his head to...

σκαρ|ώνω *v.t.* (*fam.*) put in hand, put together; organize, lay on; του ~ωσα μια I played him a trick.

σκάσ|η, ~ίλα *s.f.* chagrin, worry, vexation.

σκασιάρχης *s.m.* truant.

σκάσιμο *s.n.* bursting, cracking; escaping; playing truant; *see also* σκάση.

σκασμός *s.m. see* σκάση. (*int.*) shut up!

σκαστός *a.* resounding; in hard cash; truant.

σκατο- *opprobrious qualification of sthg.*

σκατ|ό *s.n.* (*fam.*) excrement; ~ά (*pl.*) *int.*

of annoyance or defiance. ~ώνω *v.t.* make a mess of.

σκάφανδρον *s.n.* diving apparatus *or* suit.

σκάφη *s.f.* trough; cockle-boat.

σκάφος *s.n.* hull; ship.

σκάψιμο *s.n.* digging.

σκάω *v.t. & i. see* σκάζω.

σκεβρώνω *v.t. & i.* warp, bend; become bent.

σκέλεθρο *s.n.* skeleton.

σκελετός *s.m.* skeleton; framework.

σκέλος *s.n.* leg; side (*of angle, etc.*).

σκεπάζω *v.t.* cover; shelter; conceal.

σκεπάρνι *s.n.* adze.

σκέπασμα *s.n.* covering; cover, lid.

σκεπαστ|ός *a.* covered (in); (*fig.*) covert. ~ά *adv.* covertly.

σκέπη *s.f.* cover; protection.

σκεπή *s.f.* roof.

σκεπτικισ|μός *s.m.* scepticism. ~τής *s.m.* sceptic.

σκεπτικός *a.* thoughtful, pensive.

σκέπτομαι *v.t. & i.* think about; think, reflect.

σκέρτσο|ο *s.n.* charm, coquetry; (*mus.*) scherzo. ~α (*pl.*) coquettish mannerisms.

σκερτσόζ(ικ)ος *a.* coquettish; natty, saucy (*of hat, etc.*).

σκέτος *a.* plain, unmixed; unsweetened.

σκευασία *s.f.* preparation, compound.

σκευοθήκη *s.f.* dresser.

σκεύος *s.n.* utensil, tool, piece of equipment.

σκευοφόρος *s.m.* luggage-van.

σκευοφυλάκιον *s.n.* sacristy.

σκευωρία *s.f.* machination, trick.

σκέψη *s.f.* thought, reflexion.

σκηνή *s.f.* scene; stage; tent.

σκηνικ|ός *a.* of the stage; τα ~ά stage scenery.

σκηνογραφία *s.f.* (designing of) stage scenery.

σκηνο|θεσία *s.f.* stage production. ~θέτης *s.m.* producer.

σκήν|ος, ~ωμα *s.n.* mortal remains.

σκήπτρ|ο *s.n.* sceptre; κατέχω τα ~α bear the palm.

σκήτη *s.f.* small monastery (*dependency of a greater*).

σκιά *s.f.* shade, shadow; ghost.

σκιαγραφώ *v.t.* sketch.

σκιάδι *s.n.* straw hat; parasol.

σκιάζω (-ι-) *v.t.* cast shade upon; conceal (*sun*).

σκιάζ|ω (-ζ-) *v.t.* frighten. ~ομαι *v.t. & i.* be frightened (of).

σκιάχτρο *s.n.* scarecrow.

σκιερός *a.* shady.

σκίζα *s.f.* splinter.

σκίζ|ω *v.t.* split, tear; (*fig.*) ~ω τα ρούχα μου protest one's innocence. ~ομαι *v.i.* make great efforts; protest vehemently.

σκίουρος *s.m.* squirrel.

σκιόφως *s.n.* half-light.

σκιρτώ *v.i.* jump, leap, frisk.

σκίσιμο *s.n.* tear, tearing.

σκιτζής *s.m.* (*fam.*) one who is no good at his job.

σκίτσο *s.n.* sketch.

σκιώδης *a.* shadowy, unsubstantial.

σκλαβιά *s.f.* slavery; (*fig.*) obligation, responsibility.

σκλάβος *s.m.* slave.

σκλαβώνω *v.t.* enslave; (*fig.*) oblige greatly.

σκλήθρα *s.f.* splinter; alder.

σκλήθρος *s.m.* alder.

σκληραγωγώ *v.t.* inure to hardship.

σκληρ|αίνω, ~ύνω *v.t. & i.* harden.

σκληρίζω *v.i.* screech.

σκληρός *a.* hard; severe; unmanageable (*of child*).

σκληροτράχηλος *a.* opinionated.

σκνίπα *s.f.* midge; (*fam.*) very drunk.

σκοινί *s.n.* rope.

σκολειό *s.n.* school.

σκόλη *s.f.* holiday.

σκολιαν|ός *a.* φορώ τα ~ά μου wear one's Sunday best; του έψαλα τα ~ά του I hauled him over the coals.

σκολιός *a.* crooked; difficult (*of persons*).

σκολίωσις *s.f.* (*med.*) scoliosis.

σκολνώ *v.t. & i. see* **σχολάζω**.

σκολόπενδρα *s.f.* centipede.

σκονάκι *s.n.* (*med.*) powder. (*fam.*) crib.

σκόνη *s.f.* dust, powder.

σκονίζω *v.t.* cover with dust.

σκοντάφτω *v.i.* stumble; meet with obstacle.

σκόντο *s.n.* discount.

σκόπελος *s.m.* reef (*above surface*); (*fig.*) stumbling-block.

σκοπευτήριο *s.n.* shooting-range.

σκοπευτής *s.m.* shot (*who shoots*); ελεύθερος ~ sniper, sharpshooter.

σκοπεύω *v.t. & i.* aim at; take aim; intend.

σκοπιά *s.f.* observation post; sentry-box.

σκόπ|ιμος *a.* opportune, expedient; intentional. **~ίμως** *adv.* intentionally.

σκοποβολή *s.f.* target-practice.

σκοπός *s.m.* sentry, sentinel; target; aim, intention; tune.

σκοπώ *v.i.* intend.

σκορ *s.n.* score (*in games*).

σκορβούτον *s.n.* (*med.*) scurvy.

σκορδαλιά *s.f.* sauce of garlic, *etc.*

σκόρδ|ο *s.n.* garlic; ~α (*pl.*) *said to avert evil eye after praise.*

σκόρ|ος *s.m.* moth. **~οφαγωμένος** *a.* moth-eaten.

σκορπίζω *v.t. & i.* scatter, emit; squander; (*v.i.*) disperse, disintegrate.

σκόρπιος *a.* scattered.

σκορπιός *s.m.* 1. scorpion. 2. snapper.

σκορπώ *v.t. & i. see* **σκορπίζω**.

σκοτάδι *s.n.* darkness, dark.

σκοτεινιάζ|ω *v.t. & i.* darken; ~ει it gets dark. ~ομαι *v.i.* be overtaken by darkness.

σκοτεινιά *s.f.* darkness, dark.

σκοτειν|ός *a.* dark, obscure; unlucky, sinister; ήταν ~α it was dark.

σκοτ|ίζω *v.t.* darken, confuse; worry. ~ίζομαι *v.i.* worry, care; (*fam.*) ~ίστηκα I could not care less.

σκότιος *a.* dark, sinister.

σκοτοδίνη *s.f.* giddiness.

σκότος *s.n.* darkness, dark.

σκοτούρα *s.f.* giddiness; (*fig.*) worry, nuisance.

σκοτωμός *s.m.* killing; fag, sweat. (*fig.*) jostling for priority in crowd.

σκοτών|ω *v.t.* kill; harass. ~ομαι *v.i.* hurt oneself badly; (*fig.*) slave away; endeavour with zeal.

σκούζω *v.i.* howl.

σκουλαρίκι *s.n.* ear-ring.

σκουλήκι *s.n.* worm **~ασμένος** *a.* worm-eaten.

σκουμπρί *s.n.* mackerel.

σκουντουφλώ *v.i.* stumble.

σκουντώ *v.t.* push.

σκούπα *s.f.* broom ηλεκτρική ~ vacuum-cleaner.

σκουπιδαριό *s.n.* rubbish-dump.

σκουπίδι *s.n.* rubbish.

σκουπιδιάρης *s.m.* scavenger, dustman.

σκουπίζω *v.t.* sweep, wipe.

σκουπιδοτενεκές *s.m.* dustbin.

σκουραίνω *v.t. & i.* make *or* become dark; get worse.

σκουρι|ά *s.f.* rust. **~άζω** *v.t. & i.* rust.

σκούρκος *s.m.* hornet.

σκούρ|ος *a.* dark; τα βρήκα ~α I met with obstacles. **~α.** *s.n.pl.* shutters.

σκουτί *s.n.* thick woollen cloth.

σκουφ|ί *s.n.,* **~ια** *s.f.,* **~ος** *s.m.* cap. από πού κρατάει η ~ια του; what are his origins?

σκύβαλο *s.n.* siftings of grain; (*fig.*) sweepings.

σκύβω *v.t. & i.* bend, lean, stoop.

σκυθρωπός *a.* sullen, scowling.

σκύλα *s.f.* bitch.

σκυλάκι *s.n.* puppy.

σκυλεύω *v.t.* despoil.

σκυλήσιος *a.* dog's.

σκυλ|ί *s.n.* dog; *έγινε ~ί* he got mad. **~ιάζω** *v.t. & i.* enrange; get enraged.

σκυλοβρίζω *v.t.* revile grossly.

σκυλοβρωμώ *v.i.* stink.

σκυλοκαβγάς *s.m.* dog-fight.

σκυλολόγι *s.n.* canaille, rabble.

σκυλομούρης *a.* dog-faced; (*fig.*) shameless.

σκυλοπνίχτης *s.m.* (*fam.*) unseaworthy boat, tub.

σκύλος *s.m.* dog.

σκυλόψαρο *s.n.* dogfish; shark.

σκύμνος *s.m.* cub.

σκύρα *s.n.pl.* gravel *or* stones for road-making.

σκυταλοδρομία *s.f.* relay-race.

σκυφτός *a.* bent, stooping.

σκύψιμο *s.n.* bending, stooping.

σκωληκόβρωτος *a.* worm-eaten.

σκωληκοειδίτις *s.f.* appendicitis.

σκώληξ *s.m.* worm, larva.

σκώμμα *s.n.* gibe.

σκώπτω *v.t.* mock, jeer at.

σκωρία *s.f.* rust.

σκώτι *s.n.* liver.

σλαβικός *a.* Slav, Slavonic.

σλέπι *s.n.* river barge.

σλιπ *s.n.* briefs.

σμάλτο *s.n.* enamel.

σμαράγδ|ι *s.n.* emerald. **~ένιος** *a.* emerald.

σμάρι *s.n.* swarm.

σμέουρο *s.n.* raspberry.

σμην|αγός *s.m.* (*aero.*) flight-lieutenant. **~αρχος** *s.m.* group-captain.

σμην|ίας *s.m.* (*aero*) sergeant **~ίτης** *s.m.* aircraftman.

σμήνος *s.n.* swarm; (*aero*) flight.

σμίγω *v.t. & i.* mix; meet.

σμίκρυνση *s.f.* reduction in size.

σμίλ|η *s.f.* chisel, scalpel. **~άρι** *s.n.* chisel.

σμίξιμο *s.n.* mixing, meeting.

σμιχτοφρύδης *a.* whose eyebrows meet.

σμόκιν *s.n.* dinner-jacket. [E. *smoking*]

σμπαράλια *d.n.pl.* smithereens.

σμπάρο *s.n.* shot, discharge of gun.

σμύρις *s.f.* emery.

σνομπαρία *s.f.* snobbery; (*fam.*) snob(s).

σοβαρ|ός *a.* serious; *~όν ποσόν* considerable amount. **~ότητα** *s.f.* gravity. **~εύομαι** *v.i.* become serious, talk seriously.

σοβάς *s.m.* plaster.

σοβα(ν)τίζω *v.t.* plaster.

σοβώ *v.i.* be imminent *or* brewing.

σόδα *s.f.* (bicarbonate of) soda; soda-water.

σοδειά *s.f.* crop, harvest.

σόι *s.n.* stock, breed; kind; *από ~* of good family.

σοκάκι *s.n.* (narrow) street.

σοκάρω *v.t.* (*fam.*) shock.

σόκιν *a.* (*fam.*) risqué. [E. *shocking*]

σοκολάτα *s.f.* chocolate.

σόλα *s.f.* sole (*of shoe*).

σόλοικος *a.* ungrammatical; (*fig.*) wrong, out-of-place.

σολομός *s.m.* salmon.

σολομωνική *s.f.* book of magic; mumbo-jumbo.

σομιές *s.m.* springs of bedstead. [F. *sommier*]

σόμπα *s.f.* stove (*for room heating*).

σορόκος *s.m.* south-east wind.

σορόπι *s.n.* syrup. *~α* sloppy sentiments.

σορός *s.f.* coffin; mortal remains.

σος *pron.* your, yours.

σοσιαλιστής *s.m.* socialist.

σου *pron.* your, to you.

σουβάς *s.m .see* σοβάς.

σούβλ|α *s.f.* spit, skewer. **~άκια** *s.n.pl.* skewers of grilled meat.

σουβλερός *a.* sharp, pointed.

σουβλί *s.n.* awl; sharp-pointed object. **~ιά** *s.f.* hole *or* wound from pointed object; twinge.

σουβλίζω *v.t.* impale, pierce; roast on spit; give twinges to; sew clumsily.

σουγιάς *s.m.* penknife.

σούζα *adv.* (*of animals*) *στέκομαι ~* sit up and beg; (*fig.*) fawn.

σούι-γκένερις *a.* in a class of one's own.

σουλατσάρω *v.i.* stroll about.

σουλούπ|ι *s.n.* shape, cut; (*fam.*) cut of one's jib. **~ιάζω**, **~ώνω** *v.t.* improve *or* restore shape of, lick into shape.

σουλτανί *d.n.* fresh *σουλτανίνα*.

σουλτανίνα *s.f.* kind of seedless grape; sultana from this.

σουλτάν|ος *s.m.* Sultan. **~α** *s.f.* Sultana.

σουμάδα *s.f.* orgeat of almonds.

σουξέ *s.n.* success.

σούπα *s.f.* soup.

σούπα-μούπες *s.* chatter about this and that.

σουπιά *s.f.* cuttlefish; (*fig.*) sly person.

σούρα *s.f.* fold, gathering, crease, tuck.

σουραύλι *s.n.* shepherd's pipe.

σουρλουλού *s.f.* gadabout, light woman.

σουρομαδ|ώ *v.t.* pull by the hair. *~ιέμαι* tear one's hair.

σούρουπ|ο, **~ωμα** *s.n.* dusk. **~ώνει** *v.i.* twilight falls. *~ώνομαι* be overtaken by darkness.

σουρτούκης *s.m.* gadabout.

σουρώ|νω *v.t. & i.* strain, filter; gather, pucker; (*v.i.*) crease, shrivel, waste away; *τη ~σε* he is three sheets in the wind; *~μένος* (*fam.*) drunk.

σουρωτήρι *s.n.* strainer, colander.

σουσάμι *s.n.* sesame.

σουσουράδα *s.f.* wagtail; gossip.

σούσουρο *s.n.* rustle; commotion; covert gossip.

σούστα *s.f.* spring (*of seat, etc.*); press-fastener; trap, gig; a folk dance.

σουτ *int.* hush! quiet!

σουτζούκ|ι *s.n.* kind of sausage; sweet of same shape. **~άκια** *s.n.pl.* kind of meatballs.

σουτιέν *s.n.* brassière. [F. *soutien*]

σούφρα *s.f.* wrinkle, crease.

σουφρώνω *v.t.* wrinkle, crease; (*fam.*) pinch, pilfer.

σοφάς *s.m.* sofa.

σοφέρ *see* **σωφέρ** .

σοφία *s.f.* wisdom; erudition.

σοφίζομαι *v.t.* invent.

σοφισ|τεία *s.f.* sophistry. **~τής** *s.m.* sophist.

σοφίτα *s.f.* attic.

σοφολογιότατος *s.m.* (*iron.*) pedant.

σοφός *a.* wise; learned; clever.

σπαγκοραμμένος *a.* miserly.

σπάγκος *s.m.* string; (*fam.*) stingy person.

σπαγκέτο *s.n.* spaghetti.

σπαζοκεφαλιά *s.f.* brain-teaser.

σπάζω *v.t. & i.* break, smash; ~ *το κεφάλι μου* rack one's brains; ~ *στο ξύλο* thrash; *μου έσπασε τη μύτη* I smell something good cooking.

σπαθάτος *a.* sword-bearing; (*fig.*) tall and lithe.

σπάθη *s.f.* sword, sabre.

σπαθ|ί *s.n.* sword; club (*at cards*). *κόβει το ~ί του* what he says goes. **~ιά** *s.f.* stroke *or* scar of sword.

σπανάκι *s.n.* spinach.

σπανίζω *v.i.* be rare *or* scarce.

σπάν|ιος *a.* rare; exceptional. **~ίως** *adv.* rarely, seldom.

σπάν|ις, ~ιότητα *s.f.* rarity, scarcity.

σπανός *a.* hairless, beardless; deficient in hairs, threads, etc.

σπάνω *v.t. & i. see* **σπάζω**.

σπαράγγι *s.n.* asparagus.

σπαραγμός *s.m.* tearing, rending; quivering.

σπαράζω 1. *v.t. see* **σπαράσσω**. 2. *v.i. see* **σπαρταρώ**.

σπαρακτικός *a.* heart-rending.

σπαράσσω *v.t.* tear, rend; (*fig.*) distress.

σπάραχνα *s.n.pl.* gills.

σπάργανα *s.n.pl.* swaddling-clothes.

σπαρμένος *a.* scattered; sown.

σπαρταριστός *a.* freshly caught (*fish*); blooming (*girl*); graphic, vivid; that makes you laugh.

σπαρταρ|ώ, ~ίζω *v.i.* quiver, palpitate; **~ώ** be convulsed (*with merriment*).

σπαρτ|ός *a.* scattered; sown; *τα~ά* crops.

σπασίλας *s.m.* swot.

σπάσιμο *s.n.* breaking; fracture; hernia.

σπασμένος *a.* broken, fractured.

σπασμ|ός *s.m.* spasm. **~ωδικός** *a.* convulsive; **~ωδικά μέτρα** hastily improvised measures.

σπατ|άλη *s.f.* waste, extravagance. **~αλος** *a.* wasteful; extravagant. **~αλώ** *v.t.* waste, squander.

σπάτουλα *s.f.* spatula.

σπείρα *s.f.* coil, spiral; volute, thread (*of screw*); gang.

σπείρω *v.t.* sow.

σπέρμα *s.n.* seed, germ; semen; (*fig.*); offspring.

σπερμολογία *s.f.* gossip.

σπέρνω *v.t. see* **σπείρω**.

σπεύδω *v.i.* hasten.

σπή|λαιον, ~ιο *s.n.*, **~ιά** *s.f.* cave.

σπίθ|α *s.f.* spark. **~οβολώ** *v.i.* sparkle.

σπιθαμ|ή *s.f.* span (*of hand*); ~*ή προς* ~*ή* inch by inch. **~ιαίος** *a.* tiny, dwarfish.

σπιθούρι *s.n.* pimple.

σπιλ(ι)άδα *s.f.* gust of wind.

σπιλ|ώνω *v.t.* sully, stain. **~ωμα** *s.n.* stain, blemish.

σπινθήρ *s.m. see* **σπίθα**.

σπινθηροβόλ|ος *a.* sparkling. **~ώ** *v.i.* sparkle.

σπίνος *s.m.* finch.

σπιούνος *s.m.* spy; calumniator.

σπιρούνι *s.n.* spur.

σπιρτάδα *s.f.* pungency, pungent smell; (*fig.*) quick-wittedness.

σπίρτο *s.n.* alcohol, spirit; match; (*fig.*) ~ *μονάχο* quick-witted person.

σπιρτόζος *a.* witty.

σπιτήσιος *a.* home-made.

σπίτι *s.n.* house, home; firm; *από* ~ of good family.

σπιτικ|ός *a.* home-made; home-loving. **~ό** *s.n.* home.

σπιτονοικοκύρης *s.m.* householder, landlord.

σπιτώνω *v.t.* house; (*fam.*) keep (*woman*).

σπλά(γ)χν|α *s.n.pl.* entrails, bowels; (*fig.*) feelings. **~ο** *s.n.* child, one's own flesh. **~ικός** *a.* compassionate.

σπλην *s.m.*, **~α** *s.f.* (*anat.*) spleen.

σπληνάντερο *s.n.* dish of sheep's intestine with spleen, *etc.*, on spit.

σπληνιάρης *a.* irritable.

σπογγαλιεία *s.f.* sponge-fishing.

σπογγίζω *v.t.* sponge, wipe.

σπόγγ|ος *s.m.* sponge. **~ώδης** *a.* spongy.

σποδός *s.f.* ashes.

σπολλάτη *int.* many thanks (*ironic.*).

σπονδή *s.f.* libation.

σπόνδυλ|ος *s.m.* vertebra; drum (*of columm*). **~ικός** *a.* vertebral; ~*ική στήλη* spine, backbone. **~ωτός** *a.* vertebrate.

σπόντα *s.f.* 1. cushion (*billiards*). 2. hint, dig.

σπορ *s.n.* sport. [E. *sport*]

σπορά *s.f.* sowing; seed; (*fig.*) offspring.

σποράδην *adv.* here and there.

σποραδικός *a.* sporadic, sparse.

σπορέλαιο *s.n.* seed-oil.

σποριάζω *v.i.* seed; run to seed.

σπόρ|ος *s.m.* seed, germ; semen; (*fam.*) ~*ια* (*n.pl.*) seeds (*of edible fruit, vegetables*).

σπουδάζω *v.t. & i.* study; provide with means to study.

σπουδαί|ος *a.* important; of value, outstanding. **~ότης** *s.f.* importance. **~οφανής** *a.* self-important.

σπουδαστ|ής *s.m.*, **~ρια** *s.f.* student.

σπουδή *s.f.* 1. haste, eagerness. 2. study.

σπουργίτης *s.m.* sparrow.

σπρωξ|ιά *s.f.* push, shove. **~ιμο** *s.n.* push(ing); urging.

σπρώχνω *v.t.* push, shove; urge.

σπυρ|ί *s.n.* grain (*of corn*); bead (*of necklace*); grain (*weight*); spot, pimple. **~ωτός** *a.* granular; (*of cooked rice*) not reduced to pap.

σπω *v.t. & i. see* **σπάζω.**

στάβλος *s.m. see* **σταύλος.**

στάγδην *adv.* drop by drop.

σταγονόμετρο *s.n.* dropper; *με το* ~ in driblets.

σταγ|ών, ~όνα *s.f.* drop.

σταδιακ|ός *a.* gradual. **~ώς** *adv.* in stages.

σταδιοδρομία *s.f.* career.

στάδιον *s.n.* stadium; career; stage, phase.

στάζω *v.i. & t.* drip, leak; drip with.

σταθερ|ός *a.* stable, steady, steadfast. **~οποίησις** *s.f.* stabilization. **~ότης** *s.f.* stability.

σταθμά *s.n.pl.* weights; *έχω δύο μέτρα και δύο* ~ apply one standard to others and another to oneself.

σταθμάρχης *s.m.* station-master; station superintendent (*gendarmerie*).

σταθ|μεύω *v.i.* stop, halt, camp, park. **~μευσις** *s.f.* stopping, parking.

στάθμη *s.f.* level, plumb-line; water-level.

σταθμίζω *v.t.* weigh; take level of; weigh consequences of.

σταθμός *s.m.* halt; station; landmark (*in history, etc.*).

στάλα, ~α(γ)ματιά *s.f.* drop.

σταλάζω *v.t. & i.* drop, drip.

σταλακτίτης *s.m.* stalactite.

σταλίζω *v.t. & i.* repose, rest (*of flocks*).

σταλίκι *s.n.* punt-pole.

στάλος *s.m.* (place for) repose of flocks.

σταμάτημα *s.n.* stopping, checking; break.

σταματ|ώ *v.t. & i.* stop, break off, cease activity.

στάμν|α *s.f.*, **~ί** *s.n.* pitcher.

στάμπ|α *s.f.* stamp, imprint. **~άρω** *v.t.* stamp, mark.

στάνη *s.f.* sheepfold.

στανιό *s.n. με το* ~ unwillingly, by force.

σταξ|ιά *s.f.* drop, **~ιμο** *s.n.* dripping.

στάρι *s.n. see* **σιτάρι.**

στασιάζω *v.i.* revolt, mutiny.

στασίδι *s.n.* pew.

στάσιμος *a.* stationary (*not changing*); stagnant; passed over for promotion.

στάσις *s.f.* stop, stopping-place; suspension, stoppage; posture; behaviour, attitude; revolt, mutiny.

στατήρ *s.m.* hundredweight.

στατικ|ός *a.* static. **~ή** *s.f.* statics.

στατιστική *s.f.* statistics.

σταυλάρχης *s.m.* Master of the Horse.

σταύλος *s.m.* stable, cowshed.

σταυροδρόμι *s.n.* crossroads.

σταυροειδής *a.* cruciform.

σταυροκοπιέμαι *v.i.* cross oneself (repeatedly).

σταυρόλεξο *s.n.* crossword puzzle.

σταυροπόδι *adv.* cross-legged (*squatting*).

σταυρ|ός *s.n.* cross, crucifix; *κάνω το ~ό μου* cross oneself; starfish; part of forehead between brows; *του ~ού* feast of (*esp.* Elevation of) Cross (*14 Sept.*).

σταυροφορία *s.f.* crusade.

σταυρώ|νω *v.t.* crucify; place crosswise; make sign of cross over; encounter; tease, badger; exhort; (*fam.*) *δε ~σα δεκάρα* I have not had a penny; (*v.i.*) cross over (*of front of jacket, panel, etc.*).

σταυρωτός *a.* crossed, crosswise; double-breasted.

σταφίδα *s.f.* raisin; *κορινθιακή* ~ currant; grapes *or* vineyard producing these; (*fam.*) *γίνομαι* ~ get drunk *or* shrivelled; *τα μάτια μου έγιναν* ~ I cannot keep my eyes open.

σταφιδιάζω *v.i.* dry up, shrivel.

σταφυλή *s.f.* grapes; uvula.

σταφύλι *s.n.* grapes.

στάχτη *s.f.* ashes.

σταχτής *a.* ashen, grey.

σταχτόνερο *s.n.* lye.

Σταχτοπούτα *s.f.* Cinderella.

στάχυ *s.n.* ear of corn.

σταχ(υ)ολόγημα *s.n.* gleaning; (*fig.*) selection, anthology.

στεατοκήριον *s.n.* tallow-candle.

στεγάζω *v.t.* roof; shelter.

στεγανόπους *a.* web-footed.

στεγανός *a.* air-tight, water-tight.

στέγαση *s.f.* (*act of*) roofing *or* housing.

στέγασμα *s.n.* roof; *see also* **στέγαση**.

στέγη *s.f.* roof; (*fig.*) home.

στεγνοκαθάρισμα *s.n.* dry-cleaning.

στεγν|ός *a.* dry. ~**ώνω** *v.t.* & *i.* dry, get dry. ~**ωτήρας** *s.m.* dryer.

στειλιάρι *s.n.* helve; cudgel; cudgelling.

στείρος *a.* barren, sterile.

στέκα *s.f.* billiard-cue; (*fam.*) tall, thin woman.

στέκι *s.n.* haunt, resort (*in city*); (*trader's*) pitch.

στεκ|άμενος, ~ούμενος *a.* standing, stagnant; in fairly good shape (*of old person*).

στέκομαι *v.i.* stand up, stand still, stop; be, prove to be; happen; μου στάθηκε he stood by me; δε στάθηκε να τη φιλήσω she would not let me kiss her.

στέκ|ω *v.i.* stand up, stand still; δε ~ει it is not fitting; σου ~ει το φόρεμα the dress fits *or* suits you; ~ει καλά he keeps well *or* is well off.

στέλεχ|ος *s.n.* stalk, stem; handle; counterfoil; ~**η** (*pl.*) (*mil.*) cadres.

στέλ|λω, ~νω *v.t.* send.

στέμμα *s.n.* crown.

στέμφυλο *s.n.* residue of grapes after pressing.

στέναγ|μα *s.n., ~μός* *s.m.* sigh(ing), groan (ing).

στενάζω *v.i.* sigh, groan.

στεν|εύω *v.t.* & *i.* reduce width of; hurt by being tight (*of clothes*); get narrow, shrink; ~εψαν τα πράματα things have worsened.

στενογραφία *s.f.* shorthand.

στενοδακτυλογράφος *s.m.f.* shorthand-typist.

στενόκαρδος *a.* inclined to fuss *or* worry.

στενοκέφαλος *a.* narrow-minded.

στενόμακρος *a.* long and narrow.

στενό *s.n.* strait; pass, defile.

στεν|ός *a.* narrow; tight-fitting; close (*friend, etc.*); ~ά τα πράματα (there are) difficulties; περνά ~ά live on a shoestring; βάζω στα ~ά press hard.

στενότης *s.f.* closeness; narrowness; ~ χρήματος shortage of money.

στενοχώρια *s.f.* lack of space; difficulty, worry, vexation; depression.

στενόχωρος *a.* impatient; narrow, cramped, tight; depressing; *see also* **στενόκαρδος**.

στενοχωρ|ώ *v.t.* worry, upset, depress; press hard; ~ημένος upset, hard up.

στενωπός *s.f.* lane.

στέργω *v.i.* consent, agree (*to*) (*with* εις).

στερε|ός *a.* firm, strong, solid; fast (*colour*); ~ά Ελλάς central Greece; ~ά (*s.f.*) mainland.

στερεοποιώ *v.t.* solidify.

στερεότυπος *a.* stereotyped.

στερεύω *v.i.* run dry; cease to run.

στερέωμα *s.n.* strengthening, support, prop; firmament.

στερεών|ω *v.t.* & *i.* strengthen; make fast *or* firm; ~ σε μία δουλειά stick to one job.

στέρ|ησις *s.f.* deprivation; lack; ~ήσεις (*pl.*) privation.

στεριά *s.f.* dry land. ~**νός** *s.m.* landsman, mainlander.

στερλίνα *s.f.* (pound) sterling.

στέρνα *s.f.* cistern.

στέρνον *s.n.* breast (bone).

στερν|ός *a.* later, last; καλά ~ά happy old age.

στερ|ώ *v.t.* deprive (of, with gen.). ~**ούμαι** *v.t.* & *i.* lack, go short of (*with gen.*); live in want.

στέφανα *s.n.pl.* marriage wreaths.

στεφάνη *s.f.* hoop (*of barrel*); rim (*of cup*); crown (*of tooth*).

στεφάνι *s.n.* hoop (*of barrel*); child's hoop; wreath (*esp. in marriage*).

στέφανος *s.m.* hoop (*of barrel*); wreath, crown; (*fig.*) reward.

στεφανών|ω *v.t.* crown; celebrate marriage of. ~**ομαι** *v.t.* & *i.* marry; get married.

στέψις *s.f.* coronation, crowning; marriage ceremony.

στηθαίον *s.n.* parapet, breastwork.

στηθικός *a.* of the chest; consumptive.

στηθόδεσμος *s.m.* brassière.

στήθ|ος *s.n.* chest, breast, bosom; ~**ια** (*pl.*) breasts.

στηθοσκοπώ *v.t.* examine with stethoscope.

στήλη *s.f.* upright slab, pillar; column; electric battery.

στηλιτεύω *v.t.* stigmatize.

στηλώνω *v.t.* shore up, support; give strength to.

στημόνι *s.n.* warp (*weaving*).

στήνω *v.t.* set upright, erect, put up; ~ αυτί prick up one's ears; ~ παγίδα set a trap.

στήριγμα *s.n.* prop, support.

στηρίζ|ω *v.t.* prop, support; base. ~**ομαι** *v.i.* lean (*for support*); be based *or* grounded (*on*).

στήσιμο *s.n.* setting up, mounting.

στητός *a.* firm and erect.

στιβάλι *s.n.* top-boot.

στιβαρός *a.* strong, firm.

στίβος *s.m.* arena (*of stadium*).

στίγμα *s.n.* mark, brand, stigma; point (*in reckoning position*). **~τίζω** *v.t.* blemish; stigmatize.

στιγμή *s.f.* moment; point (*printing*); dot; τελεία ~ full-stop; άνω ~ colon.

στιγμιαίος *a.* instantaneous; momentary.

στιγμιότυπο *s.n.* snapshot.

στικτός *a.* spotted, dotted.

στίλβω *v.i.* shine, gleam.

στιλβώνω *v.t.* polish.

στιλβωτήριον *s.n.* shoe-shine parlour.

στιλπνός *a.* shining, gleaming.

στίξη *s.f.* punctuation.

στιφάδο *s.n.* meat stewed with onions.

στίφος *s.n.* swarm, mass.

στιχομυθία *s.f.* fast-moving dialogue; stichomythia.

στίχ|ος *s.m.* line, verse. **~ουργός** *s.m.* rhymester.

στοά *s.f.* portico; arcade, passage; gallery (*of mine*); masonic lodge.

στοίβ|α *s.f.* pile, heap. **~άζω** *v.t.* pile up; cram, squeeze.

στοιχειόν *s.n.* ghost, spirit.

στοιχειοθεσία *s.f.* type-setting.

στοιχεί|ο *s.n.* element; (*piece of*) evidence (*law*); letter (*of alphabet*); cell (*electric*); ~α (*pl*) rudiments; type (*printing*).

στοιχειώδης *a.* elementary, basic, essential.

στοιχειών|ω *v.i.* become a ghost; become haunted; ~μένος haunted.

στοίχημα *s.n.* bet, wager; βάζω ~ lay a bet. **~τίζω** *v.t.* & *i.* bet.

στοιχίζω *v.i.* cost; cause pain *or* grief.

στοίχος *s.m.* row, file.

στόκος *s.m.* putty.

στολή *s.f.* uniform; costume.

στολίδι *s.n.* piece of finery *or* decoration.

στολίζω *v.t.* adorn, decorate, dress; (*fam.*) abuse.

στολίσκος *s.m.* (*naut.*) flotilla.

στολισμός *s.m.* adornment, decoration.

στόλος *s.m.* navy, fleet.

στόμα *s.n.* mouth; cutting edge.

στόμ|αχος *s.m.*, **~άχι** *s.n.* stomach. **~αχιάζω** *v.i.* get indigestion.

στόμιον *s.n.* orifice, mouth.

στόμφος *s.m.* pompous *or* declamatory way of speaking, rhetoric.

στομώνω *v.t.* & *i.* temper (*metal*); make *or* get blunt.

στοργή *s.f.* love, affection.

στόρι *s.n.* blind.

στουμπίζω *v.t.* pound, crush; stump up.

στουμπώνω *v.t.* & *i.* stuff full, plug; become stuffed full.

στουπί *s.n.* tow; (*fam.*) blind drunk.

στουπόχαρτο *s.n.* blotting-paper.

στουπώνω *v.t.* & *i.* stop up (*hole*); blot; become stopped up.

στουρνάρι *s.n.* flint.

στοχάζομαι *v.t.* & *i.* think, reflect (about).

στόχασ|η *s.f.* considerateness; judgement. **~τικός** *a.* considerate; judicious.

στοχασμός *s.m.* thought.

στόχαστρον *s.n.* foresight (*of gun*).

στόχος *s.m.* target, objective.

στραβά *adv.* awry, askance; wrongly; το 'βαλε ~ he does not care.

στραβισμός *s.m.* squint, squinting.

στραβο- *denotes* crooked, askance, stubborn.

στραβοκάνης *a.* (*fam.*) bandy-legged.

στραβομάρα *s.f.* blindness; (*fig.*) hitch, reverse.

στραβομουτσουνιάζω *v.i.* make a wry face.

στραβόξυλο *s.n.* (*fam.*) contrary *or* cussed fellow.

στραβοπάτημα *s.n.* false step.

στραβ|ός *a.* crooked, twisted; faulty; blind; κάνω ~ά μάτια pretend not to see; τα ~ά (*fam.*) eyes. στα ~ά blindly, at random.

στραβοτιμονιά *s.f.* error in steering; (*fig.*) ill-considered action.

στραβών|ω *v.t.* & *i.* make crooked, spoil; blind; become bent *or* spoiled. ~ομαι *v.i.* become blind.

στραγάλια *s.n.pl.* roasted chick-peas.

στραγγαλίζω *v.t.* strangle.

στραγγίζω *v.t.* & *i.* strain, drain, wring out; become worn out, exhausted.

στραγγιστήρι *s.n.* strainer, colander.

στραγγουλίζω *v.t.* sprain.

στράκ|α *s.f.* crack (*noise*); cracker (*firework*); κάνω ~ες make a sensation.

στραμπουλίζω *v.t.* sprain.

στραπατσάρω *v.t.* damage; batter; humiliate.

στραπάτσο *s.n.* damage, harm.

στράτα *s.f.* way, road.

στρατάρχης *s.m.* field-marshal.

στράτευμα *s.n.* army; ~τα (*pl.*) troops.

στρατεύομαι *v.i.* (be liable to) serve in the army.

στρατεύσιμος *a.* liable to conscription.

στρατηγείον *s.n.* headquarters.

στρατήγημα *s.n.* stratagem.

στρατηγία *s.f.* generalship (*office of general*).

στρατηγικ|ός *a.* strategic. **~ή** *s.f.* strategy.

στρατηγός *s.m.* general.

στρατί *s.n.* road, street.

στρατιά *s.f.* army.

στρατιώτης *s.m.* (private) soldier.

στρατιωτικ|ός *a.* military; ~ός νόμος martial law. **~ό** *s.n.* military service.

στρατοδικείον s.n. court-martial.
στρατοκρατία s.f. stratocracy; militarism.
στρατολογία s.f. recruitment, conscription.
στρατόπεδον s.n. camp.
στρατός s.m. army.
στρατόσφαιρα s.f. stratosphere.
στράτσο s.n. thick wrapping-paper.
στρατών(ας) s.m. barracks.
στράφι adv. πάω ~ go for nothing.
στρεβλ|ός a. crooked, distorted. ~ώνω v.t. twist, distort.
στρείδι s.n. oyster.
στρέμμα s.n. quarter of an acre.
στρέφ|ω v.t. turn; direct. ~ομαι v.i. turn; face about; revolve.
στρεψοδικία s.f. chicanery.
στρίβω v.t. & i. twist, turn; (fam.) το ~ decamp, bunk; του έστριψε (η βίδα) he is cracked.
στρίγ(κ)λα s.f. witch, shrew, hag. ~ιά s.f. shrewishness; screech. ~ίζω v.i. screech.
στριμμένος a. twisted; maliciously perverse.
στατιστική s.f. statistics.
στρίποδο s.n. trestle, tripod, easel.
στριφογυρίζω v.t. & i. whirl round; (v.i.) turn hither and thither; τα ~ prevaricate.
στριφτός a. twisted.
στρίφωμα s.n. hem.
στρίψιμο s.n. twist(ing), turn(ing).
στροβιλίζω v.t. & i. whirl round; waltz.
στρόβιλος s.m. whirlwind, whirlpool; turbine; top; waltz; pine-cone.
στρογγυλ|ός a. round, circular; (fig.) plain (of words). ~εύω v.t. make round. ~οκάθομαι v.i. plant oneself.
στρούγγα s.f. sheepfold; (fig.) herd following blindly.
στρουθί(ον) s.n. sparrow.
στρουθοκάμηλος s.f. ostrich.
στρουμπουλός a. small and plump.
στρόφαλος s.m. handle (for turning).
στροφείον s.n. crank.
στροφεύς s.m. hinge.
στροφή s.f. turn, turning; change of direction; stanza; (mus.) ritornello.
στρόφιγξ s.f. tap, key, hinge.
στρυμώ(χ)νω v.t. squeeze, squash (in crowd); press hard.
στρυφνός a. crabbed; hard (of nut).
στρώμα s.n. layer, stratum; mattress, bed; στο ~ confined to bed.
στρών|ω v.t. spread, lay; cover, pave; ~ω το κρεβάτι make the bed; (v.i.) settle down, find one's feet; work properly; το φόρεμα ~ει καλά the dress fits well. ~ομαι v.i. apply oneself; invite oneself to stay.
στρώση s.f. layer; bedding.

στρωσίδι s.n. carpet; bedding.
στρωτός a. paved; even, regular.
στύβω v.t. squeeze, wring; rack (brains).
στυγερός a. odious, heinous.
στυγνός a. scowling, gloomy.
στυλό s.n. (fam.) fountain-pen. [F. stylo]
στυλοβάτης s.m. stylobate; (fig.) mainstay; pillar (supporter).
στυλογράφος s.m. fountain-pen.
στύλος s.m. column, pillar; post, prop, support.
στυλώνω v.t. shore up, support; give strength to.
στυπτηρία s.f. alum.
στυπτικός a. binding, astringent, styptic.
στυφός a. having astringent taste.
στύψη s.f. alum.
στύψιμο s.n. squeezing, wringing.
στωικός a. stoic (al).
στωμύλος a. talkative, fluent.
συ pron. thou, you.
συβαρίτης s.m. sybarite.
συγ- see συν-.
σύγγαμβρος s.m. see μπατζανάκης.
συγγένεια s.f. relationship, affinity.
συγγεν|ής a. related; a relation, kinsman. ~ολό(γ)ι s.n. all one's relations. ~εύω v.i. be related.
συγγνώμη s.f. pardon.
σύγγραμμα s.n. work of scholarship, treatise.
συγγραφεύς s.m. writer, author.
συγγράφω v.t. write (books, etc.).
σύγκαιρα adv. opportunely; at the same time.
συγκαί|ω, ~ομαι v.i. be chafed.
σύγκαλα s.n.pl. στα ~ μου sane, in normal health.
συγ|καλύπτω v.t. conceal, hush up; ~κεκαλυμμένος veiled.
συγκαλώ v.t. convoke.
συγκατάβασις s.f. condescension.
συγκαταβατικός a. reasonable (price, terms); condescending.
συγκατάθεσις s.f. assent, consent.
συγκαταλέγω v.t. include, count (among).
συγκατανεύω v.i. consent, agree.
συγκατατίθεμαι v.i. assent, consent.
συγκατοικώ v.i. share the same living quarters.
σύγκειμαι v.i. be composed (of).
συγκεκαλυμμένος a. veiled, indirect.
συγκεκριμέν|ος a. concrete. ~ως adv. actually, to be precise, in point of fact.
σύγκελος s.m. official of diocesan staff.
συγκεντρών|ω v.t. gather together, concentrate; centralize. ~ομαι v.i. concentrate.
συγκέντρωσις s.f. gathering, meeting;

party; concentration; centralization.
συγκερασμός *s.m.* compromise (*of opinions*).
συγκεφαλαιώνω *v.t.* recapitulate.
συγκεχυμένος *a.* confused, vague.
συγκιν|ώ *v.t.* move, touch. **~ηση** *s.f.* emotion. **~ητικός** *a.* moving.
σύγκλησις *s.f.* convocation.
σύγκλητος *s.f.* senate (*university & Roman*).
συγκλίνω *v.i.* converge.
συγκλον|ίζω, ~ώ *v.t.* shake, shock.
συγκοινωνία *s.f.* communication(s); public transport.
συγκοινωνώ *v.t.* communicate.
συγκολλώ *v.t.* stick together, weld, solder; *~ σαμπρέλα* mend a puncture.
συγκομιδή *s.f.* harvest.
συγκοπή *s.f.* syncopation; heart-failure.
σύγκορμος *a.* with one's whole body.
συγκρατ|ώ *v.t.* restrain; contain, retain, hold; remember. *~ούμαι* *v.i.* control oneself.
συγκρίνω *v.t.* compare.
σύγκρισ|η *s.f.* comparison. **~τικός** *a.* comparative.
συγκρότ|ημα *s.n.* group, cluster, complex. **~ώ** *v.t.* form, assemble; *~ώ μάχην* give battle.
συγκρ|ούομαι *v.i.* collide; clash. **~ουση** *s.f.* collision, clash.
σύγκρυ|ο *s.n., ~α* (*pl.*) shivering.
συγκυρία *s.f.* coincidence; juncture.
συγυρίζω *v.t.* tidy up; (*fam.*) give (*person*) what for.
συγχαίρω *v.t.* congratulate.
συγχαρητήρια *s.n.pl.* congratulations.
συγχέω *v.t.* get mixed, confuse.
συγχορδία *s.f.* (*mus.*) chord.
σύγχρον|ος *a.* simultaneous; contemporary. **~ίζω** *v.t.* synchronize; bring up to date.
συγχρωτίζομαι *v.i.* associate (*with*).
συγχύζω *v.t.* get mixed; worry, upset, confuse.
σύγχυση *s.f.* confusion, worry; upset (*due to quarrel, etc.*).
συγχων|εύω *v.t.* amalgamate. **~ευση** *s.f.* amalgamation; merger.
συγχώρη|ση *s.f.* forgiveness. **~τός** *a.* forgivable.
συ(γ)χωρ|ώ *v.t.* forgive, pardon; *με ~είτε* excuse me; *~έθηκε* he is dead; *~εμένος* late lamented.
συδ- *see* **συνδ-**.
σύζευξις *s.f.* coupling, union. (*mus.*) slur, tie.
συζήτηση *s.f.* discussion, argument; *ούτε ~* it is quite definite.
συζητ|ώ *v.t. & i.* discuss, debate; raise objections; *δεν ~είται* it is quite definite. **~ήσιμος** *a.* debatable.

συζυγία *s.f.* (*astronomy*) conjunction; (*gram.*) conjugation.
σύζυγ|ος *s.m.f.* spouse, husband, wife. **~ικός** *a.* conjugal.
συζώ *v.i.* cohabit.
σύθαμπο *s.n.* dusk.
συθέμελα *adv.* to the very foundations.
σύκ|ο *s.n.* fig. **~ιά** *s.f.* fig-tree.
συκοφάντ|ης *s.m.* slanderer. **~ώ** *v.t.* slander, defame.
συκώτι *s.n.* liver; *βγάζω τα ~α μου* be as sick as a dog.
σύλησις *s.f.* pillage (*esp. of churches, etc.*).
συλ- *see* **συν-**.
συλλαβ|ή *s.f.* syllable. **~ίζω** *v.t.* read with difficulty.
συλλαλητήριο *s.n.* demonstration, public meeting, rally.
συλλαμβάνω *v.t.* catch, seize, arrest; (*v.t. & i.*) conceive (*child, idea*).
συλλέγω *v.t.* collect; pick.
συλλέκτης *s.m.* collector.
συλλήβδην *adv.* collectively.
σύλληψη *s.f.* capture, arrest; conception.
συλλογή *s.f.* 1. collection; picking. 2. thought, reflexion.
συλλογί|ζομαι *v.t. & i.* think, reflect (about), consider; *~σμένος* pensive.
συλλογικός *a.* collective, joint.
συλλογισμός *s.m.* reflexion; syllogism.
σύλλογος *s.m.* society, association.
συλλυπητήρια *s.n.pl.* condolences.
συλλυπούμαι *v.t.* condole with.
συλφίς *s.f.* sylph.
συλώ *v.t.* pillage (*esp. church, etc.*).
συμ- *see* **συν-**.
συμβαδίζω *v.i.* keep pace; go together.
συμβαίνω *v.i.* happen, occur.
συμβάλλ|ω *v.i.* (*with εις*) contribute to; flow into, unite with. *~ομαι* *v.i.* contract, make an agreement.
συμ|βάν, ~βεβηκός *s.n.* event, happening.
σύμβασις *s.f.* contract, pact; (*pl.*) conventions.
συμβατικός *a.* contractual; conventional.
συμβία *s.f.* wife.
συμβιβά|ζω *v.t.* reconcile; *τα ~σαμε* we made it up. *~ζομαι* *v.i.* reach agreement *or* compromise; be compatible.
συμβιβασμός *s.m.* compromise, settlement.
συμβίωσις *s.f.* living together.
συμβολαιογράφος *s.m.* notary (public).
συμβόλαιον *s.n.* contract.
συμβολή *s.f.* contribution; confluence.
συμβολίζω *v.t.* symbolize.
σύμβολο *s.n.* symbol, emblem; *~ν της πίστεως* Creed.
συμβουλεύ|ω *v.t.* advise. *~ομαι* *v.t.* consult.

~τικός *a.* advisory.
συμβουλη *s.f.* advice.
σύμβουλ|ος *s.m.* counsellor, adviser; councillor. **~ιον** *s.n.* council, board.
συμμαζεύω *v.t.* gather together, tidy; curb, restrain.
σύμμαχ|ος *a.*allied; an ally. **~ία** *s.f.* alliance.
συμμερίζομαι *v.t.* share (*grief, views, etc.*).
συμμετέχω *v.i. & t.* (*with gen.*) take part, have share (in).
συμμετοχή *s.f.* participation.
συμμετρία *s.f.* symmetry, proportion.
συμμετρικός *a.* symmetrical; well-shaped.
συμμορία *s.f.* gang, band.
συμμορίτης *s.m.* gangster.
συμμορφών|ω *v.t.* conform; bring to heel, make compliant; tidy up. **~ομαι** *v.i.* comform, comply.
συμπαγής *a.* compact, solid.
συμπάθεια *s.f.*sympathy; liking, weakness; favourite.
συμπαθής *a.* likeable.
συμπαθητικός *a.* likeable; sympathetic (*nerve, ink*).
συμπάθιο *s.n.* (*fam.*) με το ~ begging your pardon.
συμπαθ|ώ *v.t.* feel sympathy for; have a liking for; με ~άτε excuse me.
συμπαιγνία *s.f.* collusion.
σύμπαν *s.n.* universe; everybody, everybody, everything.
συμπαράσταση *s.f.* solidarity.
συμπατριώτης *s.m.* compatriot.
συμπεθερι|ά *s.f.,* **~ό** *s.n.* relationship by marriage; negotiation of marriage through third party.
συμπέθεροι *s.m.pl.* fathers-in-law of couple (*in relation to each other*); relations by marriage.
συμπεπυκνωμένος *a.* condensed.
συμπερ|αίνω *v.t. & i.* conclude, infer. **~ασμα** *s.n.* conclusion, inference.
συμπεριλαμβάνω *v.t.* include.
συμπερι|φέρομαι *v.i.* behave. **~φορά** *s.f.* behaviour.
σύμπηξις *s.f.* coagulation; formation.
συμπίεσις *s.f.* compression, squeeze.
συμπίπτω *v.i.* coincide, converge; befall, so happen.
σύμπλεγμα *s.n.* complex; cluster, group (*esp. sculpture, etc.*); network; monogram.
συμπλέκ|ω *v.t.* interlace; ~ω τας χείρας clasp one's hands. **~ομαι** *v.i.* come to blows, clash.
συμπλήρωμα *s.n.* complement; supplement. **~τικός** *a.* complementary.
συμπληρώνω *v.t.* complete, finish off; fill (*post, etc.*); fill in (*form*).

συμπλήρωσις *s.f.* completion, filling.
συμπλοκή *s.f.* affray, clash.
σύμπνοια *s.f.* accord.
συμπολιτεία *s.f.* confederation.
συμπολίτευ|σις *s.f.* government party. **~όμενος** *a.* supporting the government.
συμπολίτης *s.m.* fellow citizen.
συμπονώ *v.t.* feel sympathy for.
συμπόσιον *s.n.* banquet.
συμποσούμαι *v.i.* amount (*to*).
συμπράγκαλα *s.n.pl.* belongings.
σύμπραξις *s.f.* co-operation.
συμπτύσσ|ω *v.t.* shorten, abbreviate. **~ομαι** *v.i.* fall back (*of troops*).
σύμπτωμα *s.n.* symptom. **~τικός** *a.* symptomatic; coincidental, chance.
σύμπτωση *s.f.* coincidence.
συμπυκνώ, ~νω *v.t.* condense. **~τήρ, ~τής** *s.m.* condenser.
συμφέρει *v.i. & t.* it is to one's advantage, it pays.
συμφέρον *s.n.* interest, advantage. **~τολόγος** *a.* mercenary. **~τολογία** *s.f.* self-interest.
συμφιλιών|ω *v.t.* reconcile. **~ομαι** make it up.
συμφορά *s.f.* calamity.
συμφόρησις *s.f.* congestion; stroke (*apoplexy*).
σύμφορος *a.* advantageous.
συμφραζόμενα *s.n.pl.* context (*verbal*).
συμφυής *a.* innate.
συμφύρ|ω *v.t.* jumble, confuse. **~ομαι** *v.i.* consort *or* get mixed up (*with*).
σύμφυτος *a.* inherent.
συμφωνητικόν *s.n.* (deed of) contract, agreement.
συμφωνία *s.f.* agreement; stipulation; symphony.
σύμφωνο *s.n.* consonant; pact.
σύμφων|ος *a.* in agreement, in accord; εκ ~ου with one accord. **~α** *adv.* ~α με in accordance with, according to.
συμφωνώ *v.i. & t.* agree; match, correspond; (*v.t.*) agree on; match, engage, hire.
συμφώνως *adv.* accordingly; ~ προς in accordance with, according to.
συμψηφίζω *v.t.* offset, balance.
συν *prep.* (*with dat.*) with; (*math.*) plus; ~ τω χρόνω in the course of time; ~ τοις άλλοις in addition.
συν- *denotes* together, with, completely.
συναγελάζομαι *v.i.* consort, mingle.
συναγερμός *s.m.* alarm, alert; mass-meeting, rally.
συναγρίδα *s.f.* dentex (*sea fish*).
συνάγ|ω *v.t. & i.* bring together, collect; infer; ~εται ότι the inference is that...

συναγωγή *s.f.* (place of) assemblage; collection; synagogue.

συναγωνίζομαι *v.i.* fight together; (*v.t. & i.*) compete (with); rival.

συναγωνιστής *s.m.* brother-in-arms; rival; competitor.

συνάδελφος *s.m.* colleague, confrère.

συνάζω *v.t.* collect, (*people, taxes, etc.*).

συναθροίζω *v.t.* collect, gather (*people, wealth, information; etc.*).

συνάθροισις *s.f.* gathering.

συναινώ *v.t.* consent.

συναίρεσις *s.f.* (*gram.*) contraction.

συναισθάνομαι *v.t.* be conscious of, realize, feel.

συναίσθημα *s.n.* sentiment, feeling; ~τα (*pl.*) emotions. **~τικός** *a.* emotional.

συναίσθηση *s.f.* consciousness, realization, feeling, sense.

συνακόλουθος *a.* consequent; consistent.

συναλλαγή *s.f.* business dealing; ελευθέρα ~ free trade.

συνάλλαγμα *s.n.* (*fin.*) (bill of) exchange; foreign currency; τιμή ~τος rate of exchange. **~τική** *s.f.* bill of exchange.

συναλλάσσομαι *v.i.* deal, do business; mix, associate. (*politics*) wheel and deal.

συνάμα *adv.* at the same time, together.

συναναστρέφομαι *v.t.* associate with.

συναναστροφή *s.f.* relations, company; party.

συναντ|ώ *v.t.* meet. **~ώμαι** *v.i.* meet; be found (*of quotations*). **~ηση** *s.f.* meeting, encounter, match.

συναξάρι(ον) *s.n.* book of saints; (*fig.*) long-winded story; rigmarole.

σύναξις *s.f.* meeting; collection.

συναπάντημα *s.n.* (*esp.* unlucky) meeting.

συναπτός *a.* consecutive, on end.

συνάπτω *v.t.* join (*hands, battle*); append; contract (*marriage, debt*); conclude (*agreement, peace*); ~ σχέσεις make friends.

συναρμολογώ *v.t.* assemble, put together.

συναρπάζω *v.t.* charm, carry away, transport.

συναρπαστικός *a.* thrilling, gripping.

συνάρτησις *s.f.* connexion; coherence.

συνασπισμός *s.m.* coalition.

συναυλία *s.f.* concert.

συναφ|ής *a.* touching, related, similar. **~εια** *s.f.* connexion, relevance.

συνάφι *s.n. see* συνάφι.

συνάχι *s.n.* cold in the head.

σύναψις *s.f.* joining, concluding, *etc.* (see συνάπτω).

συνδαιτυμών *s.m.* fellow diner.

συνδαυλίζω *v.t.* poke (*fire*); (*fig.*) foment.

συνδεδεμένος *a.* (closely) connected; intimate.

σύνδενδρος *a.* wooded.

σύνδεση *s.f.* connexion, link-up, coupling.

σύνδεσμος *s.m.* bond, tie; ligament; league, union; (*mil.*) liaison; (*gram.*) conjuction.

συνδετήρ *s.m.* paper-fastener.

συνδετικός *a.* connecting.

συνδέω *v.t.* join, unite, connect.

συνδιαλέγομαι *v.i.* converse.

συνδιαλλαγή *s.f.* reconciliation; compromise.

συνδιάσκεψις *s.f.* conference.

συνδικαλισμός *s.m.* trade-unionism.

συνδικάτο *s.n.* syndicate; union (*of workers*).

συνδράμω *v.t. & i.* aor. *subj. of* συντρέχω.

συνδρομή *s.f.* help, succour; subscription; conjunction (*of events*). **~τής** *s.m.* subscriber.

σύνδρομον *s.n.* syndrome.

συνδυάζ|ω *v.t.* combine; put together (& *make inference from*); contrive, arrange. **~ομαι** *v.i.* go together, harmonize.

συνδυασμός *s.m.* 1. combination, arrangement; harmonization. 2. (list of) party's candidates at election.

σύνεγγυς *adv.* εκ του ~ very close.

συνεδρι|άζω *v.i.* meet, be in session. **~ασις** *s.f.* session.

συνέδριον *s.n.* congress; ελεγκτικόν ~ Audit Board.

σύνεδρος *s.m.* judge of criminal court; member of a congress.

συνείδησις *s.f.* consciousness (*having one's faculties awake*); awareness; conscience.

συνειδητ|ός *a.* conscious, deliberate. **~οποιώ** *v.t.* realize, become aware of. **~ώς** *adv.* consciously, deliberately.

συνειρμός *s.m.* connexion, coherence; ~ παραστάσεων association of ideas.

συνεισφέρω *v.t. & i.* contribute.

συνεισφορά *s.f.* contribution.

συνεκτικός *a.* strong, resistant; cohesive.

συνέλευσις *s.f.* assembly.

συνεννόηση *s.f.* understanding, agreement; exchange of views; conspiracy.

συνεννοούμαι *v.i.* reach an understanding; exchange views; conspire.

συνενοχή *s.f.* complicity.

συνένοχος *s.m.* accomplice.

συνέντευξη *s.f.* rendezvous; interview.

συνενώ *v.t.* join, unite, combine.

συνεπάγομαι *v.t.* entail, have as consequence.

συνεπαίρνω *v.t.* carry away, overwhelm.

συνεπαρμένος *a.* enraptured.

συνέπεια *s.f.* consequence; consistency.

συνεπ|ής *a.* consistent, true to one's word;

punctual. ~ώς *adv.* consequently.

συνεπτυγμένος *a.* shortened, abridged.

συνεργάζομαι *v.i.* work together, collaborate.

συνεργασία *s.f.* collaboration.

συνεργάτ|ης *s.m.* collaborator; contributor. ~ική *s.f.* co-operative.

συνεργείο *s.n.* workshop; repair shop *or* party; gang, team (*on a job*).

συνεργία *s.f.* complicity; (*med.*) synergy.

σύνεργο *s.n.* implement.

συνεργός *s.m.* accomplice, accessary.

συνεργώ *v.i.* be an accomplice; connive.

συνερίζομαι *v.t.* insist on being even with; keep up petty rivalry with.

συνέρχομαι *v.i.* recover, come to one's senses; come together; ~ εις γάμον contract marriage.

σύνεση *s.f.* good sense.

συνεσταλμένος *a.* shy, reserved.

συνέταιρος *s.m.* partner, associate.

συνετίζω *v.t.* make (*person*) see reason.

συνετός *a.* judicious, sensible.

συνεφέρνω *v.t. & i.* revive, bring to; come to.

συνέχεια *s.f.* continuity; continuation, sequel; (*adv.*) continuously. εισιτήριο ~ς through ticket.

συνεχ|ής *a.* continuous, continual; successive. ~ώς *adv.* continually, in succession.

συνεχίζω *v.t. & i.* continue.

συνέχ|ω *v.t.* possess, seize (*of emotions*); restrain. ~ομαι *v.i.* be adjoining *or* communicating.

συνήγορος *s.m.* advocate, counsel.

συνηγορώ *v.i.* (*with* υπέρ) advocate, speak in favour of.

συνήθει|α *s.f., ~ο* *s.n.* habit, custom.

συνήθ|ης *a.* usual, customary, ordinary. ~ως *adv.* usually.

συνηθ|ίζω *v.t. & i.* accustom; get used to (*with* να *or* σε *or* acc.); be in the habit of; ~ισε να μαγειρεύει she has got used to cooking; ~ίζει να κοιμάται νωρίς he usually goes to bed early; ~ίζεται it is the fashion *or* the usual thing.

συνηθισμένος *a.* used (*to*), accustomed; usual, habitual.

συνημμένος *a.* annexed, enclosed (*of papers*).

σύνθεσις *s.f.* composition; structure; synthesis.

συνθέτης *s.m.* composer.

συνθετικ|ός *a.* component; synthetic. ~ό *s.n.* component.

σύνθετ|ος *a.* compound, composite. ~ο *s.n.* (*gram.*) compound word.

συνθέτω *v.t.* compose.

συνθήκ|η *s.f.* treaty, pact; ~ες (*pl.*) conditions, circumstances.

συνθηκολογώ *v.i.* come to terms; (*mil.*) capitulate.

σύνθημα *s.n.* signal, pass-word. ~τικός *a.* in code.

συνθλίβω *v.t.* squeeze, crush.

συνίζησις *s.f.* subsidence (*of soil*); (*gram.*) fusion of i or e with following vowel.

συνίσταμαι *v.i.* (*with* εκ) consist of.

συνιστώ *v.t.* form, set up (*committee, etc.*), constitute (*be*); introduce (*person*); advise, recommend (*something to somebody*).

συννεφι|ά *s.f.* cloudy weather. ~άζω *v.i.* cloud over.

σύννεφ|ο *s.n.* cloud. ~όκαμα *s.n.* sultry and overcast weather.

συννυφάδα *s.f.* sister-in-law (*relationship between wives of brothers*).

συνοδεύω *v.t.* accompany, escort.

συνοδία *s.f.* suite, escort, convoy, procession; (*mus.*) accompaniment.

συνοδοιπόρος *s.m.f.* fellow traveller.

σύνοδος *s.f.* assembly; synod; session.

συνοδός *s.m.f.* escort; ιπταμένη ~ air-hostess.

συνοικέσιο *s.n.* (negotiation of) arranged marriage.

συνοικία *s.f.* quarter (*of town*). ~ιακός *a.* local, of city quarter.

συνοικισμός *s.m.* settlement (*place where people live*).

σύν|ολο *s.n.* total; εν ~όλω in all. ~ολικός *a.* total.

συνομήλικος *a.* of the same age.

συνομιλητής *s.m.* interlocutor.

συνομιλία *s.f.* conversation.

συνομολογώ *v.t.* conclude (*agreement*).

συνομοσπονδία *s.f.* confederation.

συνομοταξία *s.f.* class (*natural history*).

συνονόματος *a.* a namesake.

συνοπτικός *a.* summary, synoptic.

συνορίζομαι *v.t.* see **συνερίζομαι**.

σύνορ|ο *s.n.* boundary; ~α (*pl.*) frontier. ~εύω *v.i.* have common boundary *or* frontier.

συνουσία *s.f.* coition.

συνοφρυούμαι *v.i.* frown.

συνοχή *s.f.* coherence, cohesion.

συνοψίζω *v.t.* summarize.

σύνοψις *s.f.* summary, synopsis; prayer-book.

συνταγή *s.f.* prescription; recipe.

σύνταγμα *s.n.* constitution; (*mil.*) regiment.

συνταγματάρχης *s.m.* (*mil.*) colonel.

συνταγματικός *a.* constitutional.

συντάκτης *s.m.* editor; author.

συντακτικ|ός *a.* constituent; editorial; syntactic, of syntax. **~όν** *s.n.* (book of) syntax.

συνταξιδιώτης *s.m.* fellow-traveller.

συνταξιοδοτ|ώ *v.t.* pension off. **~ούμαι** retire.

συνταξιούχος *s.m.f.* pensioner.

σύνταξη *s.f.* compilation; writing, drawing up; editorial staff; pension; (*gram.*) construction.

συνταράσσω *v.t.* convulse, shake, trouble.

συντάσσ|ω *v.t.* arrange, compile; write, draw up; (*gram.*) construe. **~ομαι** *v.i.* (*with* μετά) side with; (*gram.*) govern.

συνταυτίζω *v.t.* identify (*associate with*).

συντείνω *v.i.* conduce, play a part.

συντέλεια *s.f.* ~ *του κόσμου* end of the world.

συντέλεση *s.f.* completion.

συντελεστ|ής *s.m.* factor; (*math.*) co-efficient. **~ικός** *a.* contributory.

συντελώ *v.t. & i.* complete, accomplish; play a part, contribute.

συντέμνω *v.t.* shorten.

συντεταγμένη *s.f.* (*math.*) co-ordinate.

συντετριμμένος *a.* contrite, deeply afflicted.

συντεχνία *s.f.* guild.

συντήρη|σις *s.f.* conservation; maintenance, subsistence. **~τικός** *a.* conservative; cautious.

συντηρώ *v.t.* conserve; maintain, support.

συντίθεμαι *v.i.* be composed (*of*).

σύντμησις *s.f.* abbreviation; abridgement.

συντομεύω *v.t.* shorten.

συντομία *s.f.* brevity; saving of space *or* time.

σύντομ|ος *a.* short, brief; concise. **~α** *adv.* briefly; quickly, soon.

συντονίζω *v.t.* co-ordinate.

συντονισμός *s.m.* co-ordination.

σύντονος *a.* intensive, unrelaxing.

συντρέχ|ω *v.t. & i.* aid; (*v.i.*) meet, coincide; contribute; *δε ~ει λόγος* there is no reason.

συντρίβ|ω *v.t.* crush, smash; afflict. **~ή** *s.f.* crushing; contrition.

σύντριμμα *s.n.* fragment; *~τα* (*pl.*) debris.

συντριπτικός *a.* (*fig.*) crushing.

συντροφεύω *v.t. & i.* keep company with; go into partnership.

συντροφ|ιά *s.f.* companionship, company; party; (*adv.*) together; *και ~ία* & Co.

σύντροφος *s.m.f.* companion, comrade; partner.

συντυχαίνω *v.t. & i.* meet by chance; converse with; happen; *συνέτυχε να* it happened that...

συνύπαρξις *s.f.* coexistence.

συνυπάρχω *v.i.* coexist.

συνυφαίνω *v.t.* interweave; (*fig.*) hatch, plot.

συνωθούμαι *v.i.* jostle.

συνωμο|σία *s.f.* conspiracy (*esp. treasonable*). **~τώ** *v.i.* conspire.

συνώνυμ|ος *a.* synonymous. **~ο** *s.n.* synonym.

συνωστισμός *s.m.* jostling, crush.

σύξυλ|ος *a.* a total loss (*of shipwreck*); (*fig.*) *έμεινε ~ος* he was dumb-founded; *μ' άφησε ~ο* he left me in the lurch.

σύρι|γξ, **~γγα** *s.f.* Pan's pipe; syringe.

συρίζω *v.i. & t.* hiss (at), whistle (at).

συριστικός *a.* sibilant.

σύρμα *s.n.* wire. **~τόπλεγμα** *s.n.* wire netting; barbed wire.

συρμός *s.m.* railway train; fashion.

συρόμενος *a.* sliding.

σύρραξις *s.f.* collision, clash.

συρρέω *v.i.* throng, flock.

σύρριζα *adv.* root and branch; very close.

συρροή *s.f.* throng, influx; abundance.

σύρσιμο *s.n.* dragging; creeping.

συρτάρι *s.n.* drawer.

συρτή *s.f.* drag-net.

σύρτης *s.m.* bolt.

σύρτη *s.f.* sandbank.

συρτός 1. *a.* dragging; drawling; sliding (*door*). 2. *s.m.* sort of cyclic dance.

συρφετός *s.m.* mob.

σύρω *v.t. see* **σέρνω**; *σύρ' τα φέρ' τα* comings and goings.

συσκέπτομαι *v.i.* confer, deliberate.

συσκευάζω *v.t.* pack, box, put up.

συσκευασία *s.f.* packing; (*pharmaceutical*) preparation.

συσκευή *s.f.* apparatus.

σύσκεψις *s.f.* conference, consultation.

συσκοτ|ίζω *v.t.* darken, black out; (*fig.*) confuse. **~ισις** *s.f.* black-out.

σύσπασις *s.f.* contraction (*of brow, etc.*).

συσπειρούμαι *v.i.* coil; (*fig.*) (*with* περί) rally round (*leader, etc.*).

συσσίτιο *s.n.* mess, common table; soup-kitchen.

σύσσωμος *a.* united, in a body.

συσσωρεύ|ω *v.t.* accumulate. **~τής** *s.m.* accumulator.

συστάδην *adv.* εκ του ~ hand-to-hand.

συσταίνω *v.t.* introduce (*person*).

συστάς *s.f.* clump (*trees*).

σύστασ|ις *s.f.* composition (*that of which anything is composed*); consistency (*thickness*); formation, setting up; recommendation; *~εις* (*pl.*) references, advice; introductions; *~η* address; *επί ~ει* registered (*letter*).

συστατικ|ός *a.* component, constituent; **~ή** *επιστολή* letter of introduction; **~ά** (*pl.*) ingredients; references.

συστέλλ|ω *v.t.* contract; take in (*sail*). **~ομαι** *v.i.* contract; feel hesitant *or* shy.

σύστημα *s.n.* system, method. **~τικός** *a.* systematic.

συστή|νω *v.t.* see **συνιστώ**. **~μένο** γράμμα registered letter.

συστολή *s.f.* contraction; modesty, shame.

συσφίγγω *v.t.* tighten; constrict.

συσχετίζω *v.t.* correlate, put together.

συφάμελος *a.* with the whole family.

σύφιλις *s.f.* syphilis.

συχαρίκια *s.n.pl.* gift to first bringer of good news.

συχνά, ~κις *adv.* often; ~ πυκνά very often.

συχνάζω *v.i.* (*with* σε) frequent, be habitué of.

συχν|ός *a.* frequent. **~ότης** *s.f.* frequency.

συχωράω *v.t.* see **συγχωρώ**.

συχώριο *s.n.* (*fam.*) forgiveness of sins (*to dead*); μπουκιά και ~ piece of all right.

σύψυχος *a.* body and soul; with all hands.

σφαγείο *s.n.* slaughter-house.

σφαγ|ή *s.f.* slaughter, carnage. **~ιάζω** *v.t.* slaughter; sacrifice; (*fig.*) butcher.

σφάγιο *s.n.* victim, animal (to be) slaughtered.

σφαδάζω *v.i.* writhe, struggle, jerk.

σφάζω *v.t.* slaughter, kill.

σφαίρ|α *s.f.* sphere; ball; bullet. **~ίδιο** *s.n.* small shot; ballot-ball. **~ικός** *a.* spherical.

σφαιριστήριο *s.n.* billiard table *or* room.

σφαλερός *a.* erroneous; unsafe.

σφαλιάρα *s.f.* hard slap in face.

σφαλ|ίζω ~(ν)ώ *v.t.* & *i.* shut in; close. **~ιστός** *a.* shut.

σφάλλ|ω, ~ομαι *v.i.* err, do *or* be wrong; misfire.

σφάλμα *s.n.* wrong act, fault; error; misfire.

σφαντώ *v.i.* look glamorous.

σφάξιμο *s.n.* slaughtering.

σφαχτ|άρι, ~ό *s.n.* animal (to be) slaughtered.

σφάχτης *s.m.* twinge.

σφεν|δόνη, ~τόνα *s.f.* sling.

σφερδούκλι *s.n.* asphodel.

σφετερίζομαι *v.t.* appropriate dishonestly, embezzle.

σφή(γ)κα *s.f.* wasp.

σφήν *s.m., ~α** *s.f.* wedge. **~ώνω** *v.t.* wedge, thrust in between. **~οειδής** *a.* cuneiform.

σφίγγα *s.f.* sphinx.

σφίγγ|ω *v.t.* & *i.* squeeze, clasp; tighten; clench; make *or* get thick *or* hard; become tight, stick; get constipated; ~ω την καρδιά μου master one's grief; έσφιξε το κρύο it

has become intensely cold. ~ομαι *v.i.* make great efforts, strain; squeeze together.

σφιγξ *s.f.* sphinx.

σφίξ|η *s.f.* urgency; urgent desire to relieve oneself; constipation; δεν είναι ~η there is no hurry; έχει ~ες he is in difficulties.

σφίξιμο *s.n.* tightening, squeezing.

σφιχτός *a.* tight, firm, hard; miserly.

σφόδρα *adv.* extremely, strongly.

σφοδρ|ός *a.* strong, violent, intense. **~ότης** *s.f.* violence, intensity.

σφολιάτα *s.f.* flaky pastry.

σφοντύλι *s.n.* flywheel. βλέπω τον ουρανό ~ see stars.

σφουγγάρ|ι *s.n.* sponge. **~άς** *s.m.* spongefisher *or* seller.

σφουγγαρόπανο *s.n.* floor-cloth.

σφουγγάτο *s.n.* omelet.

σφουγγίζω *v.t.* wipe; (*fam.*) sweep the board of.

σφραγίζω *v.t.* seal, stamp; cork *or* close (*bottle, tin, etc.*); fill (*tooth*).

σφραγίς, ~ίδα *s.f.* seal, stamp. **~ιση** *s.f.* sealing, stamping.

σφρίγ|ος *s.n.* youthful exuberance. **~ηλός** *a.* robust, lusty.

σφυγμ|ός *s.m.* pulse; (*fig.*) weak point. **~ομετρώ** *v.t.* take pulse of; sound out, test.

σφύζω *v.i.* pulsate, throb; (*fig.*) be bursting with vitality.

σφύξη *s.f.* pulse, pulsation.

σφύρ|α *s.f., ~ί** *s.n.* hammer. μεταξύ ~ας και άκμονος between devil and deep sea; βγάζω στο ~ί put up to auction. **~ηλατώ** *v.t.* hammer, forge.

σφύριγμα *s.n.* whistling; whistle (*noise*).

σφυρίδα *s.f.* grouper (*fish*).

σφυρίζω *v.i.* hoot, whistle, hiss.

σφυρίχτρα *s.f.* whistle (*instrument*).

σφυροκοπώ *v.t.* hammer; rain blows on.

σφυρόν *s.n.* ankle.

σχάρα *see* **σκάρα**.

σχεδία *s.f.* raft.

σχεδι|άζω *v.t.* draw, sketch, design; plan, intend. **~αστής** *s.m.* designer, draughtsman, **~άγραμμα** *s.n.* sketch, diagram, plan.

σχέδιο *s.n.* sketch, design, plan, draft, project; ~ν πόλεως town-planning authority.

σχεδόν *adv.* almost, nearly.

σχέση *s.f.* relation, connexion.

σχετίζ|ω *v.t.* connect, associate, compare. **~ομαι** *v.i.* be *or* get acquainted.

σχετικ|ά, ~ώς *adv.* relatively; ~α *or* ~ώς με regarding, with reference to.

σχετικ|ός *a.* relative, relevant; on familiar terms. **~ότης** *s.f.* relativity.

σχετλιασμός *s.m.* lamenting, complaining.

σχήμα *s.n.* form, shape; diagram; (*geom.*) figure; format; cloth (*clerical*); salute; gesture; ~ *λόγου* figure of speech.

σχηματ|ίζω *v.t.* form. ~*ομαι* (*v.i.*) form; (*of girl*) acquire womanly figure. ~**ισμός** *s.m.* formation, configuration; (*gram.*) scheme of inflexion.

σχίζα *s.f.* splinter.

σχιζοφρένια *s.f.* schizophrenia.

σχίζ|ω *v.t. & i.* split, cleave; tear. ~*ομαι v.i.* split, tear; fork, divide.

σχίνος *s.f.* lentisk.

σχίσμα *s.n.* crack; schism.

σχισμ|άδα, ~ή *s.f.* crack, fissure, split.

σχιστόλιθος *s.m.* slate, schist.

σχιστός *a.* slit, open.

σχοινίον *s.n.* rope.

σχοινοβάτης *s.m.* rope-dancer, acrobat.

σχοινοτενής *a.* long-winded.

σχολάζω *v.i. & t.* not to be at work; be on vacation; stop work; (*v.t.*) let out (*of school, office, etc.*); dismiss (*from job*).

σχολαστικός *a.* pedantic.

σχολείο *s.n.* school.

σχολή *s.f.* leisure; school (*esp. of university, painting, etc.*).

σχόλη *s.f.* holiday, feast-day.

σχολι|άζω *v.t.* comment on, criticize; annotate, edit. ~**αστής** *s.m.* commentator; editor, scholiast.

σχολικός *a.* of school.

σχόλιο *s.n.* comment; gloss.

σωβινιστής *s.m.* chauvinist.

σώβρακο *s.n.* pants.

σώγαμπρος *s.m. see* εσώγαμβρος.

σώζ|ω *v.t.* save, rescue; preserve. ~*ομαι v.i.* be extant, remain in existence.

σωθικά *s.n.pl.* entrails, bowels.

σωλήν|ας *s.m.* pipe, tube. ~**άριο** *s.n.* small tube (*of medicament, etc.*). ~**ωση** *s.f.* piping.

σώμα *s.n.* body, corps; (*mil.*) ~ στρατού army corps; copy (*of book*).

σωματάρχης *s.m.* (*mil.*) commander of army corps.

σωματεί|ον *s.n.* guild, corporation. ~**ακός** *a.* corporate.

σωματεμπορία *s.f.* (white) slave-traffic.

σωματικός *a.* of the body, bodily, physical; corporal, corporeal.

σωματοφυλακή *s.f.* bodyguard.

σωματώδης *a.* corpulent.

σών|ω *v.t. & i.* save, preserve; use up, run out of; reach (*with hand etc.*); (*v.i.*) reach; be enough; ~*ει και καλά* with obstinate insistence. *να μη σώσει* damn him! ~*ομαι* get used up, run out.

σώος *a.* safe, whole, unharmed.

σωπαίνω *v.i. & t.* stop talking; hush.

σωρεία *s.f.* pile; abundance of.

σωρηδόν *adv.* in great quantity.

σωριάζομαι *v.i.* fall in a heap; collapse.

σωρός *s.m.* heap, pile; lot of.

σωσίας *s.m.* double, living image.

σωσίβιο *s.n.* life-belt *or* jacket.

σώσιμο *s.n.* saving; rescue; finishing up, consuming.

σώσμα *s.n.* last of wine (*in barrel*).

σωστ|ός *a.* proper, right, upright; real, absolute; *με τα* ~*ά του* in his right mind, in earnest; *δύο* ~*ες ώρες* two whole hours. ~**ά** *adv.* correctly, exactly, right(ly).

σώστρα *s.n. pl.* salvage money.

σωτήρ *s.m.* saviour, rescuer. ~**ία** *s.f.* deliverance, rescue, salvation; way out (*of dilemma*). ~**ιος** *a.* saving, redeeming.

σωφ|έρ *s.m.* chauffeur, driver. [F. *chauffeur*]. ~**άρω** *v.t. & i.* drive.

σωφρον|ίζω *v.t.* bring to reason, correct, reform. ~**ιστήριον** *s.n.* reformatory.

σωφροσύνη *s.f.* prudence, sense, moderation.

σώφρων, *a.* prudent, sensible; moderate.

Τ

τα *article & pron. n. pl.* the; them.

ταβάνι *s.n.* ceiling.

ταβατούρι *s.n.* row, noise.

ταβέρνα *s.f.* tavern, eating-house.

τάβλα *s.f.* board; low table; (*fam.*) ~ στο μεθύσι blind drunk.

τάβλι *s.n.* backgammon.

ταγάρι *s.n.* peasant's bag; nosebag.

ταγγ|ός *a.* rancid. ~**ίζω** *v.i.* go rancid.

ταγή *s.f.* fodder.

ταγιέρ *s.n.* woman's suit. [F. *tailleur*]

ταγίζω *v.t.* feed.

τάγμα *s.n.* (*priestly, etc.*) order; (*mil.*) battalion.

ταγματάρχης *s.m.* (*mil.*) major.

ταγμένος *a.* under a vow.

τάδε *pron.* such-and-such; *o* ~(*ς*) so-and-so; ~ *έφη* so (he) spake.

τάζ|ω *v.t.* vow, promise. ~*ομαι v.i.* make a vow.

ταΐζω *v.t.* feed.

ταινία *s.f.* band, strip, ribbon; (*cinema*) film; tape, ribbon (*in machines*); tape-worm.

ταίρι *s.n.* one of a pair; companion, mate; match, equal.

ταιρι|άζω *v.t. & i.* match, assort; (*v.i.*) match, suit, go (*with*); *δεν* ~*άζει* it is not fitting; *τα* ~*άσαμε* we came to an understanding.

ταιριαστός *a.* well-matched.

τάισμα *s.n.* feeding.

τακίμι *s.n.* set (*of implements*); shift (*of workmen*).

τάκος *s.m.* wooden fixing-block; (*fam.*) nice 'piece' (*woman*).

τακούνι *s.n.* heel (*of shoe*).

τάκτ *s.n.* tact. [F. *tact*]

τακτική *s.f.* tactics, policy; method, orderliness.

τακτικ|ός *a.* regular; orderly; ordinal (*number*). **~ά, ~ώς** *adv.* regularly.

τακτοποι|ώ *v.t.* arrange, put in order; settle. **~ησις** *s.f.* arranging, settling.

τακτός *a.* fixed.

ταλαίπωρ|ος *a.* poor, unfortunate. **~ία** *s.f.* hardship, suffering. **~ώ** *v.t.* torment. **~ούμαι** *v.i.* suffer hardship.

ταλαντεύομαι *v.i.* sway, rock; waver.

τάλαντον *s.n.* talent (*aptitude or money*).

τάλας *a.* poor, unfortunate.

ταλέντο *s.n.* (person of) talent.

τάλ(λ)ηρο *s.n.* five-drachma piece.

τάμα *s.n.* vow; ex-voto offering.

ταμάχι *s.n.* insatiable greed.

ταμείον *s.n.* treasury, accounts department; cash-desk, booking-office; pension *or* provident society.

ταμένος *a.* under a vow.

ταμίας *s.m.* cashier, treasurer.

ταμιευτήριον *s.n.* savings bank.

ταμπάκικο *s.n.* tannery.

ταμπάκ|ος *s.m.* snuff. **~ιέρα** *s.f.* snuff-box; cigarette-box *or* case.

ταμπέλα *s.f.* name-plate; registration-plate (*of motor*).

τά(μ)πια *s.f.* rampart.

ταμπλό *s.n.* picture; dash-board; switch-board. [F. *tableau*]

ταμπόν *s.n.* tampon, ink-pad; buffer.

ταμπού *s.n.* taboo.

ταμπουρ|άς *s.m.* sort of guitar. *η κοιλιά μου βαράει ~ά* I am famished.

ταμπούρ|ι *s.n.* (*mil.*) defensive position. **~ώνομαι** *v.i.* occupy such position.

ταμπούρλο *s.n.* side-drum.

τανάλια *s.f.* pliers (*tool*).

τανάπαλιν *adv.* conversely.

τανκ *s.n.* (*mil.*) tank. [E. *tank*]

τανύ|ω, ~ζω *v.t.* stretch. **~ομαι, ~έμαι** *v.i.* stretch oneself; strain (*at stool*). **~τό** *s.n.* straining at stool.

ταξειδ- *see* **ταξιδ-**.

ταξί *s.n.* taxi. **~τζής** *s.m.* taxi-driver. [F. *taxi*]

ταξιάρχης *s.m.* 1. archangel. 2. commander (*of order of chivalry*).

ταξιαρχία *s.f.* (*mil.*) brigade.

ταξίαρχος *s.m.* brigadier.

ταξιδεύω *v.i.* travel.

ταξίδ|ι(ον) *s.n.* journey; **~ια** (*pl.*) travel.

ταξιδιώτης *s.m.* traveller. **~ικός** *a.* of *or* for travelling.

ταξιθέτ|ης *s.m.* **~ρια** *s.f.* theatre-attendant.

ταξικός *a.* class; **~** *αγώνας* class struggle.

ταξίμετρον *s.n.* taximeter.

τάξιμο *s.n.* vow.

ταξινομώ *v.t.* classify.

τάξ|ις *s.f.* 1. order (*sequence, tidiness, calm state*); *εν* **~ει** all right, in order. 2. class, grade; form (*school*); **~εις** (*pl.*) ranks (*of army, etc.*); *πρώτης* **~εως** first class.

τάπα *s.f.* cork, stopper; (*fam.*) **~** *στο μεθύσι* blind drunk.

ταπεινός *a.* humble; mean, base. **~οφροσύνη** *s.f.* humility.

ταπεινών|ω *v.t.* humble, humiliate. **~ομαι** *v.i.* abase oneself; be humbled *or* humiliated.

ταπέτο *s.n.* carpet, rug.

ταπετσαρία *s.f.* wall-covering; upholstery.

τάπ|ης *s.m.* carpet, rug; *ετέθη επί* **~ητος** it came up for discussion. **~ητουργία** *s.f.* manufacture of carpets.

ταπίστομα *adv.* in prone position.

τάραγμα *s.n.* shock.

ταρ|άζω, ~άσσω *v.t.* shake; disturb; punish (*victuals*).

ταραμάς *s.m.* preserved roe.

ταραξίας *s.m.* rowdy person.

ταράτσα *s.f.* flat roof; (*fam.*) *την κάνω* **~** eat a bellyful.

ταραχ|ή *s.f.* disturbance, commotion; upset, shock. **~οποιός** *s.m.* rowdy person. **~ώδης** *a.* stormy, disturbed.

ταρίφα *s.f.* scale of charges.

ταριχεύω *v.t.* embalm; preserve.

ταρσανάς *s.m.* boat-builder's yard.

τάρταρα *s.n.pl.* bowels of the earth.

ταρταρούγα *s.f.* tortoise-shell.

τασάκι *s.n.* ashtray.

τάσι *s.n.* shallow bowl; scale (*of balance*); cymbal.

τάσις *s.f.* stretching (*of arms in drill*); tension (*of gas, electricity*); tendency, inclination.

τάσσ|ω *v.t.* place; marshal; set, appoint. **~ομαι** *v.i.* place oneself; (*fig.*) side (*with*).

ταυ *s.n.* the letter T.

ταύρος *s.m.* bull. **~ομαχία** *s.f.* bull-fight.

ταύτα *pron.n.pl.* these.

ταυτίζ|ω *v.t.* treat as identical, identify. **~ομαι** be identical, identify oneself.

ταυτόσημος *a.* identical in meaning.

ταυτότητα *s.f.* identity (card).

ταυτόχρονος *a.* simultaneous.

ταφή *s.f.* burial.

ταφόπετρα *s.n.* tombstone.

τάφος *s.m.* grave, tomb; (*fig.*) one who never betrays a secret.

τάφρος *s.f.* ditch, trench.

ταφτάς *s.m.* taffeta.

τάχα, ~τε(ς) *adv.* supposedly; as if; I wonder if, can it be that? *μας κάνει τον* ~ *(μου)* he puts on airs.

ταχεία *s.f.* express (*train*).

ταχ|έως *adv.* quickly; soon; *ως ~ιστα* as soon as possible.

ταχιά *adv.* tomorrow.

ταχινή *s.f.* morning.

ταχίνι *s.n.* ground sesame.

ταχύ *adv. & s.n.* (tomorrow) morning.

ταχυβόλο *s.n.* quick-firing gun.

ταχυδακτυλουργία *s.f.* conjuring (trick).

ταχυδρομείον *s.n.* post, mail; post-office.

ταχυδρομικ|ός *a.* postal; of the post-office. **~ά** *s.n.pl.* postage. **~ώς** *adv.* by post.

ταχυδρόμος *s.m.* postman.

ταχυδρομώ *v.t.* post (*letters*).

ταχυκίνητος *a.* fast-moving.

ταχύνους *a.* quick-witted.

ταχύνω *v.t.* quicken.

ταχυπαλμία *s.f.* (*med.*) palpitation.

ταχ|ύς *a.* quick, fast, rapid. **~ύτης** *s.f.* speed, rapidity; gear (*of motor*).

ταψί *s.n.* shallow baking-tin.

ταώς *s.m.* peacock.

τε *conj.* ~...*και* both... and...

τέζα *adv.* stretched, tight; (*fam.*) *έμεινε* ~ he died; ~ *στο μεθύσι* blind drunk.

τεζάρω *v.t.* stretch, tighten. (*fam.*) *τα* ~ die.

τεθλασμέν|ος *a.* broken. **~η** *s.f.* zigzag line.

τεθλιμμένος *a.* afflicted, grief-stricken.

τείνω *v.t. & i.* stretch (out); strain. (*v.i.*) tend; lead (*to*), be aimed (*at*).

τέιον *s.n.* tea.

τείχος *s.n.* wall (*of city, etc.*).

τεκές *s.m.* opium-den.

τεκμήριον *s.n.* token, mark, indication.

τέκνον *s.n.* child.

τεκταίν|ομαι *v.i. & t.* plot, hatch; be hatched. **~όμενα** intrigues.

τέκτων *s.m.* mason; freemason. **~ονισμός** *s.m.* freemasonry.

τελάρο *s.n.* tapestry-frame; stretcher (*for canvas*); frame (*of door, etc.*).

τελεία *s.f.* full-stop; *άνω* ~ semicolon.

τελειοποιώ *v.t.* perfect.

τέλει|ος *a.* perfect, complete. **~α** *adv.* to perfection.

τελειότης *s.f.* perfection.

τελειόφοιτος *a.* final-year student; graduate.

τελείωμ|α *s.n.*, ~**ός** *s.m.* finishing; exhaustion (*of supply*); ~*ό δεν έχει* it is

endless *or* inexhaustible.

τελειώνω *v.t. & i.* exhaust, use up; end, finish; (*fig.*) die.

τελείως *adv.* completely.

τελειωτικός *s.n.*. ultimatum.

τελεσίγραφον *s.n.*. ultimatum.

τελεσίδικος *a.* final (*without appeal*).

τέλεσις *s.f.* celebration (*of rite*); performance (*of deed*).

τελεσφορώ *v.i.* end successfully.

τελετ|ή *s.f.* ceremony, rite. **~άρχης** *s.m.* master of ceremonies. **~ουργώ** *v.i.* officiate.

τελευταί|ος *a.* last; latest; *σαν τον* ~*ο βλάκα* like an utter fool. **~ως** *adv.* lately, recently.

τελευτή *s.f.* end; (*fig.*) death.

τελεύω *v.t. & i.* see **τελειώνω**.

τελεφερίκ *s.n.* cable-car, funicular, ski-lift.

τέλι *s.n.* (*thin*) wire, string (*of banjo, etc.*).

τελικ|ός *a.* final. **~ά** *adv.* in the end.

τέλμα *s.n.* swamp. **~τώνω** *v.i.* get bogged down.

τέλ|ος *s.n.* 1. end; ~*ος πάντων* after all, finally; *επί* ~*ους* at last; (*στο*) ~*ος* at last; *εν* ~*ει* finally. 2. tax.

τελ|ώ *v.t.* perform, do. **~ούμαι** *v.i.* take place. **~ούμενα** events.

τελωνείον *s.n.* customs (house).

τελώνης *s.n.* customs officer.

τεμάχ|ιο *s.n.* piece. **~ίζω** *v.t.* cut in pieces.

τεμεν|άς *s.m.* oriental-style bow; *κάνω* ~*άδες* kowtow.

τέμενος *s.n.* temple, shrine.

τέμν|ω *v.t.* cut; pass through (*of road*); ~*ουσα* (*math.*) secant.

τεμπέλ|ης, ~ικος *a.* lazy. **~ιάζω** *v.i.* loaf. **~ιά** *s.f.* laziness.

τέμπλο *s.n.* iconostasis.

τέναγος *s.n.* fen.

τενεκ|ές *s.m.* tin (*substance*); large can *or* tin; (*fam.*) ignoramus. **~εδένιος** *a.* tin. **~ετζής** *s.m.* tinsmith.

τενόρος *s.m.* tenor.

τέντα *s.f.* tent; awning. (*adv.*) wide open.

τέντζερ|ες, ~ης *s.m.* (*metal*) cooking-pot.

τέντωμα *s.n.* strain, stretching.

τεντ|ώνω *v.t.* stretch (out); open (*door, etc.*) wide; (*fam.*) *τα* ~*ωσε* he died. **~ώνομαι** (*v.i.*) straighten up.

τένων *s.m.* tendon.

τέρας *s.n.* monster; freak, phenomenon; (*fam.*) a terror, a caution.

τεράστιος *a.* huge, enormous.

τερατολογία *s.f.* tall story.

τερατόμορφος *a.* hideous.

τερατούργημα *s.n.* monstrosity; outrageous deed.

τερατώδης *a.* monstrous; ugly; phenomenal.

τερεβινθέλαιον *s.n.* turpentine.

τερετίζω *v.i.* chirp, twitter.

τερηδών *s.f.* woodworm; caries.

τέρμα *s.n.* terminus, end; goal (*sports*). **~τίζω** *v.t. & i.* put an end to, bring to a close. (*v.i.*) finish.

τερματοφύλακας *s.m.* goalkeeper.

τερπνός *a.* agreeable, enjoyable.

τέρπω *v.t.* give pleasure to.

τερτίπι *s.n.* trick.

τέρψις *s.f.* enjoyment.

τέσσαρα *num.* four. **~άκοντα** *num.* forty.

τέσσερ|(ε)ις *a.* four; *με τα ~α* on all fours.

τεταγμένος *a.* placed, fixed; appointed.

τεταμένος *a.* stretched, strained.

τέτανος *s.m.* (*med.*) tetanus.

τεταρταίος *a.* quartan (*fever*).

Τετάρτη *s.f.* Wednesday.

τέταρτο *s.n.* quarter (*of an hour*); quarto; (*mus.*) crotchet.

τέταρτος *a.* fourth.

τετελεσμέν|ος *a.* done; *~ον γεγονός* fait accompli; (*gram.*) *~ος μέλλων* future perfect.

τέτοιος *pron.* such; *ο ~* what's-his-name.

τετράγωνο *s.n.* (*geom.*) square, quadrangle; block (*between streets*).

τετράγων|ος, ~ικός *a.* square; (*fig.*) wellgrounded, logical.

τετράδιο *s.n.* exercise-book.

τετράδυμα *s.n.pl.* quadruplets.

τετράκις *adv.* four times.

τετρακόσι|οι *a.* four hundred; *τα έχει ~α* he has his wits about him.

τετράπαχος *a.* very plump.

τετραπέρατος *a.* very clever *or* cunning.

τετρα|πλάσιος, ~πλούς *a.* fourfold.

τετράποδο *s.n.* quadruped; (*fig.*) ass; brute.

τετράπους *a.* four-footed.

τετρ|άδα *s.f.* set of four; *κατά ~άδας* in fours.

τετράτροχος *a.* four-wheeled.

τετριμμένος *a.* worm; (*fig.*) trite.

τέττιξ *s.m.* cicada.

τεύτλον *s.n.* beetroot.

τεύχος *s.n.* instalment, part, issue.

τέφρα *s.f.* ashes.

τεφρ|ός, ~όχρους *a.* ashen; grey.

τεφτέρι *s.n.* notebook.

τεχνάζομαι *v.t.* contrive, think up, invent.

τέχνασμα *s.n.* artifice, ruse.

τέχν|η *s.f.* art, craft, skill, craftsmanship; skilled occupation; trick, dodge; *καλές ~ές* fine arts.

τεχνηέντως *adv.* adroitly, artfully.

τεχνητός *a.* artificial; forced, feigned.

τεχνική *s.f.* technique.

τεχνικός *a.* technical; skilful; skilled (*in mechanical arts*); a technician.

τεχνίτης *s.m.* craftsman; technician, mechanic; (*fam.*) expert.

τεχνοκρίτης *s.m.* art critic.

τεχνολογία *s.f.* technology; (*gram.*) parsing.

τεχνοτροπία *s.f.* (*artistic*) style.

τέως *adv.* formerly; *~ βουλευτής* ex-M.P.

τζάκι *s.n.* hearth, fireplace; range; *από ~* of distinguished family.

τζαμ|αρία *s.f., ~λίκι** *s.n.* glass panelling *or* partition; space enclosed by this; conservatory.

τζάμι *s.n.* window-pane.

τζαμί *s.n.* mosque.

τζαμόπορτα *s.f.* glazed door.

τζάμπα *adv.* (*fam.*) for nothing (*free, to no purpose*). **~τζής** *s.m.* one who gets things for nothing.

τζαναμπέτης *s.m.* crabbed person, sourpuss.

τζάνερο *s.n.* mirabelle plum.

τζατζίκι *s.n.* relish of yoghurt, cucumber and garlic.

τζελατίνα *s.f.* 1. gelatine. 2. celluloid holder (*for identity card, etc.*).

τζερεμές *s.m.* undeserved fine *or* loss; sluggish man; vicious horse.

τζίβα *s.f.* vegetable fibre.

τζίγκος *s.m.* zinc.

τζιριτζάντζουλες *s.f.pl.* (*fam.*) evasiveness; coquettish airs.

τζίρος *s.m.* (*business*) turnover.

τζίτζικας *s.m.* cicada.

τζιτζιφιά *s.f.* jujube-tree.

τζίφος *s.m.* vain effort.

τζίφρα *s.f.* paraph; initials (*as signature*).

τζόγος *s.m.* card-playing.

τζόκεϋ *s.m.* jockey. [E. *jockey*]

τζούτα *s.f.* jute.

τζουτζές *s.m.* dwarf; (*fig.*) clown.

τζούτζης *s.m.* (*fam.*) cheat, equivocator.

τήβεννος *s.f.* toga, gown.

τηγανητός *a.* fried.

τηγάν|ι *s.n.* frying-pan. **~ίζω** *v.t.* fry. **~ίτα** *s.f.* fritter.

τήκ|ω *v.t.* melt. *~ομαι v.i.* melt; pine, languish.

τηλεβόας *s.m.* loud-hailer.

τηλεβόλον *s.n.* cannon.

τηλεγραφείον *s.n.* telegraph-office.

τηλεγράφημα *s.n.* telegram, cable.

τηλέγραφ|ος *s.m.* telegraph. **~ώ** *v.i. & t.* telegraph, wire.

τηλεοπτικ|ός *a.* of *or* on TV. *μεταδίδω ~ώς* televise.

τηλεόραση *s.f.* television.

τηλεπάθεια *s.f.* telepathy.

τηλεπικοινωνία *s.f.* telecommunication(s).

τηλεσκόπιον *s.n.* telescope.

τηλεφώνημα *s.n.* telephone-call.

τηλεφωνητής *s.m.* telephone-operator.

τηλέφων|ο *s.n.* telephone. **~ώ** *v.i.* & *t.* telephone. **~ικός** *a.* telephone, telephonic. *~ικώς* by telephone.

τήξις *s.f.* melting; pining.

τήρησις *s.f.* observance.

τηρώ *v.t.* keep (up), observe.

τηρ|ώ, ~άω *v.t.* & *i.* look at; take care.

τί *pron.* (*interrogative*) what; (*exclamatory*) what, how.

τι *pron.* something; a little.

τίγκα *adv.* (*fam.*) full to overflowing.

τίγρ|ις, ~η *s.f.* tiger.

τιθασ(σ)εύω *v.t.* tame.

τίθεμαι *v.i.* be put, set *or* imposed.

τίκτω *v.t.* give birth to; bring forth.

τίλιο *s.n.* (infusion of) lime flowers.

τίλλω *v.t.* pull out (*hair, etc.*); card (*wool*).

τιμαλφ|ής *a.* precious; **~ή** (*s.n.pl.*) jewellery.

τιμάριθμος *s.m.* cost-of-living (index).

τιμάριον *s.n.* fief.

τιμή *s.f.* honour; price; value.

τίμημα *s.n.* price; value.

τιμητικός *a.* honorary; doing honour (*to*).

τίμι|ος *a.* honest; honourable; precious (*metal*); *~ος σταυρός or ~ο ξύλο* Holy Cross.

τιμοκατάλογος *s.m.* price-list.

τιμολόγιον *s.n.* invoice; scale of charges.

τιμόνι *s.n.* rudder; tiller, wheel; steering-wheel; handle-bars. **~έρης** *s.m.* steersman.

τιμ|ώ *v.t.* honour, do honour to. **~ώμαι** *v.i.* cost.

τιμωρ|ώ *v.t.* punish. **~ία** *s.f.* punishment.

τίναγμα *s.n.* shaking; jerk, start.

τινάζ|ω *v.t.* shake; hurl; flap (*wings*); flick; *~ω στον αέρα* blow up. **~ομαι** *v.i.* jump (up), start; brush (*something off*) one's clothes.

τίποτ|α, ~ε *pron.* any, anything; (*after neg.*) nothing; *~α!* don't mention it; *άλλο ~ε* very much (so).

τιποτένιος *a.* insignificant; worthless.

τιράντες *s.f.pl.* braces.

τιρμπουσόν *s.n.* corkscrew. [F. *tire-bouchon*]

τις *pron.* who?

τις *pron.* some, someone, one.

τιτάν *s.m.* titan. **~ιος, ~ικός** *a.* titanic.

τίτλ|ος *s.m.* title. *~οι* (*pl.*) securities. **~οφορώ** *v.t.* entitle (*name*).

τμήμα *s.n.* portion, section, segment;

department; police-station.

τμηματάρχης *s.m.* head of department.

τμηματικώς *adv.* piecemeal.

το *article* & *pron. n.* the; it.

τοιούτ|ος *pron.* such; (*fam.*) homosexual. *~οτρόπως* *adv.* thus, in such a way.

τοιχίζω *v.t.* enclose with wall. (*v.i. of ship*) heel.

τοιχογραφία *s.f.* fresco.

τοιχοκόλλησις *s.f.* bill-posting.

τοίχος *s.m.* wall.

τοίχωμα *s.n.* (*inner*) wall (*of vessel, bodily organ*).

τοκετός *s.m.* childbirth.

τοκίζω *v.t.* lend (*money*) at interest.

τοκογλύφος *s.m.* usurer.

τόκ|ος *s.m.* (*fin.*) interest. **~ομερίδιον** *s.n.* dividend coupon.

τόλμη *s.f.* boldness; effrontery. **~ρός** *a.* bold, daring.

τολμώ *v.i.* dare.

τολύπη *s.f.* lump (*of raw wool*); (*fig.*) flake (*of snow*); puff (*of smoke*).

τομάρι *s.n.* hide, skin.

τομάτα *s.f.* tomato.

τομεύς *s.m.* incisor; sector.

τομή *s.f.* cut, incision; section (*diagram*).

τόμος *s.m.* volume (*book*).

τονίζω *v.t.* accent; accentuate, stress; set to music.

τονικός *a.* tonic; tonal.

τονισμός *s.m.* accentuation.

τόννος *s.m.* 1. ton. 2. tunny.

τόν|ος *s.m.* tone; touch, trace, shade (*of sound, colour, etc.*). (*fig.*) force, vigour; *μιλώ με ~ο* speak sharply; *δίνω ~ο σε* liven up; (*gram.*) accent; (*mus.*) tone (*interval*), key.

τον|ώνω *v.t.* invigorate, liven up. **~ωτικός** *a.* bracing, tonic.

τοξεύω *v.t.* shoot with bow.

τοξικ|ός *a.* toxic; poisonous. **~ομανής** *s.m.f.* drug-addict.

τόξ|ο *s.n.* bow (*weapon*); arc, arch, curve; *ουράνιο ~* rainbow; (*mus.*) bow. **~οειδής** *a.* arched, curved. **~ότης** *s.m.* archer.

τοπ(ε)ίον *s.n.* landscape; site, spot.

τόπι *s.n.* (*child's*) ball; bolt (*of cloth*); cannon (-ball).

τοπικ|ός *a.* local. **~ισμός** *s.m.* localism.

τοπιογραφία *s.f.* landscape painting.

τοπογραφία *s.f.* topography.

τοποθεσία *s.f.* situation, site, spot.

τοποθετ|ώ *v.t.* place, put, post; (*fin.*) invest. **~ηση** *s.f.* placing, disposition; investment.

τόπ|ος *s.m.* piece of ground, place, site; country; room (*space*); *επί ~ον* on the spot; *πιάνω ~ο* take up room *or* have

effect *or* come in useful; *έμεινε στον* ~*o* he died on the spot; *αφήνω στον* ~*o* kill instantaneously; *δίνω* ~*o της οργής* give way (*in argument*); (*math.*) locus.
τοπούζι *s.n.* club; (*fig.*) *σαν* ~ lumpishly.
τοπωνυμία *s.f.* place-name.
τορν|εύω *v.t.* turn (*on lathe*); (*fig.*) give elegant shape to. ~**ευτής** *s.m.* turner. ~**ευτός** *a.* well-turned.
τόρνος *s.m.* lathe.
τορπίλλη *s.f.* torpedo. ~**ίζω** *v.t.* torpedo. ~**οβόλον** *s.n.* torpedo-boat.
τος *pron.* he; *να* ~ here he is.
τοσάκις *adv.* so many times.
τόσο(ν) *adv.* so (much); ~ ...*όσο(ν)* as (much)... *as or* both... and; *όχι και* ~ not very much.
τόσ|ος *pron.* of such size *or* quantity; so much, so many; ~*ος δα* very small *or* little; *άλλοι* ~*οι* as many again; *κάθε* ~*o* every so often; *εκατόν* ~*α* a hundred odd; ~ *οι και* ~*οι* many.
τοσούτ|ος *pron.* of such size *or* quantity; *εν* ~*ω* all the same; ~*ω μάλλον* with all the more reason.
τότε *adv.* then (*at that time*); in that case; ~ *που* when.
τουαλέτα *s.f.* toilet; woman's formal dress; dressing-room; dressing-table; cloak-room; lavatory.
τούβλο *s.n.* brick; (*fig.*) blockhead.
τουλάχιστον *adv.* at least.
τούλι *s.n.* tulle.
τουλίπα *s.* tulip.
τουλούμ|ι *s.n.* skin (*receptacle*); *βρέχει με το* ~*ι* it is raining cats and dogs; *τον έκανα* ~*ι* I gave him a good hiding. ~**ίσιος** *a.* (*cheese*) prepared in skin.
τουλούμπα *s.f.* 1. pump. 2. sort of oriental cake.
τουλούπα *s.f. see* **τολύπη.**
τουλ(ου)πάνι *s.n.* muslin (kerchief).
τούμπ|α *s.f.* 1. somersault, fall; (*fam.*) *κάνω* ~*ες* kowtow. 2. mound.
τούμπανο *s.n.* drum; (*fam.*) *το κάνω* ~ blab the secret; *θα σε κάνω* ~ I will give you a hiding.
τουμπάρω *v.t. & i.* overturn; inveigle. (*v.i.*) overturn.
τουμπελέκι *s.n.* small drum.
τουναντίον *adv.* on the contrary.
τουπέ *s.n.* (*fam.*) cheek; boldness. [F. *toupet*]
τουρισμός *s.m.* tourism.
τουρίστ|ας, ~ής *s.m.* tourist.
Τούρκ|α, ~άλα, ~ισσα *s.f.* Turkish woman.
τουρκεύω *v.i.* become a Moslem; (*fig.*) become enraged.

Τουρκία *s.f.* Turkey.
τούρκικ|ος, ~ός *a.* Turkish.
τουρκοκρατία *s.f.* Turkish domination (*of Greece*).
τουρκομερίτης *s.m.* (*Greek*) from Turkey.
Τούρκος *s.m.* Turk; Mohammedan; (*fig.*) *γίνομαι* ~ become enraged.
τούρλα *s.f.* mound; (*adv.*) heaped up.
τουρλού-τουρλού *s.n.* dish of baked vegetables. (*as* a.) all sorts of.
τουρλώ|νω *v.t.* heap up; cause to bulge out. ~**τός** *a.* heaped up; rounded and protruding.
τουρνέ *s.f.* (*theatrical*) tour. [F. *tournée*]
τουρσί *s.n.* pickle.
τούρτα *s.f.* gateau, layer-cake.
τουρτουρίζω *v.i.* shiver.
τουτέστι *adv.* that is, namely.
τούτος *pron.* this (one).
τούφα *s.f.* tuft, clump.
τουφέκ|ι *s.n.* rifle, musket. ~**ιά** *s.f.* gun-shot. ~**ίζω** *v.t.* shoot.
τράβαλα *s.n.pl.* troubles, fuss.
τράβηγμα *s.n.* pulling, pull, tug; drawing (*of liquid*). ~*τα* troubles, scrapes.
τραβηξιά *s.f.* sip; pull (*at pipe, etc.*); *see also* **τράβηγμα.**
τραβ|ώ *v.i. & i.* 1. pull, drag, draw; ~*ώ το δρόμο μου* go one's own way. 2. draw (*weapon, money, liquid, lot, etc.*); (*fam.*) *το* ~*άει* he drinks. 3. absorb, take up (*liquid, exports, etc.*). 4. fire (*shot*), fetch (*blow*). 5. reach speed *or* distance of. 6. attract; *δεν με* ~*άει* it does not appeal to me. 7. call for, want; *ο καιρός* ~*άει πανωφόρι* the weather calls for an overcoat; *η καρδιά μου* ~*άει τσιγάρο* I feel like a cigarette. 8. endure, go through. (*v.i.*) 9. make *or* head for; move on. 10. last, drag on. 11. draw (*of chimney, pipe*). ~*ιέμαι v.i.* withdraw, cease to take part; be in demand (*of goods*); *δεν* ~*ιέται* it is unendurable; (*fam.*) ~*ηγμένος* drunk, (*wood*) seasoned. ~*ηγμένος από τα μαλλιά* far-fetched.
τραγανίζω *v.t.* crunch; (*fig.*) nibble away.
τραγαν|ός *a.* crisp, firm (*to the teeth*). ~**ό** *s.n.* cartilage.
τράγ(ε)ιος *a.* goat's.
τραγέλαφος *s.m.* incongruous situation *or* thing, mix-up.
τραγήσιος *a.* goat's.
τραγί *s.n.* kid.
τραγικός *a.* tragic; a tragic poet.
τράγος *s.m.* goat.
τραγούδημα *s.m.* singing.
τραγούδ|ι *s.n.* song. ~**ιστής** *s.m.* singer. ~**ιστός** *a.* sung; melodious (*voice*). ~**ώ** *v.t.*

& *i.* sing.
τραγωδ|ία *s.f.* tragedy. **~ός** *s.m.f.* tragic actor *or* actress.
τραίνο *s.n.* (*railway*) train.
τρακ *s.n.* stage fright. [F. *trac*]
τρακάρ|ω *v.t.* & *i.* collide (with); meet unexpectedly; (*fam.*) touch (*person*) for (*loan*). **~ισμα** *s.n.* collision; surprise meeting.
τρακατρούκα *s.f.* fire-cracker.
τράκ|ο *s.n.*, **~ος** *s.m.* collision; (*fig.*) attack, blow.
τρακτέρ *s.n.* (*farm*) tractor. [F. *tracteur*]
τραμ, ~βάι *s.n.* tram. [E. *tramway*]
τραμουντάνα *s.f.* north wind.
τράμπα *s.f.* κάνω ~ swap.
τραμπαλίζομαι *v.i.* see-saw; rock up and down.
τραμπούκος *s.m.* (taker of) petty bribe; ruffian.
τραν|ός *a.* clear, plain; large, great, important. **~εύω** *v.i.* grow, increase.
τράνταγμα *s.n.* jolting, shaking; jolt, start.
τραντάζω *v.t.* jolt, shake; fetch (*blow*).
τρανταχτός *a.* (*fig.*) spectacular, on a big scale.
τράπεζα *s.f.* 1. table; αγία ~ altar. 2. (*fin.*) bank.
τραπεζαρία *s.f.* dining-room (suite).
τραπέζι *s.n.* table; στρώνω ~ lay the table; τους είχαμε ~ we had them to dinner.
τραπέζιον *s.n.* (*small*) table; trapeze; (*geom.*) trapezium.
τραπεζίτης *s.m.* 1. banker. 2. molar.
τραπεζιτικός *a.* bank, banking; a bank employee.
τραπεζογραμμάτιο *s.n.* banknote.
τραπεζομάντηλο *s.n.* tablecloth.
τραπεζώνω *v.t.* entertain to dinner.
τράπουλα *s.f.* pack (*of cards*).
τραστ *s.n.* (*fin.*) trust.
τράτα *s.f.* 1. trawl; trawler. 2. a cyclic dance.
τραταμέντ|ο *s.n.* refreshment (*offered to guest*); κάνω τα ~α stand treat.
τρατάρω *v.t.* & *i.* offer *or* pay for (*refreshment*).
τράτο *s.n.* (*sufficient*) time, room, *or* material.
τραυλ|ός *a.* a stammerer. **~ίζω** *v.i.* stammer.
τραύμα *s.n.* wound, injury; trauma. **~τίας** *s.m.* wounded person, casualty. **~τίζω** *v.t.* wound, injure.
τραχανάς *s.m.* (*home-made*) pasta (*for soup*), furmety.
τραχεία *s.f.* windpipe.
τραχέως *adv.* roughly, abruptly.
τραχηλιά *s.f.* collaret, ruff; bib.
τράχηλος *s.m.* neck.

τραχύνω *v.t.* & *i.* roughen; worsen.
τραχύς *a.* rough, hard; tough (*meat*); harsh, abrupt.
τρεις *a.* three; ~ κι ο κούκος very few.
τρέλλα *s.f.* madness; folly; eccentricity; (*fam.*) anything delightful.
τρελλαίν|ω *v.t.* drive mad. **~ομαι** *v.i.* go mad; be driven mad; (*with* για) be mad on.
τρελλοκομείο *s.n.* madhouse.
τρελλός *a.* mad; a lunatic; crazy; mischievous (*child*).
τρεμο|σβήνω, ~φέγγω *v.i.* flicker, glimmer.
τρεμούλ|α *s.f.* trembling, shivering; fright. **~ιάζω** *v.i.* tremble, shiver.
τρέμω *v.i.* tremble, shake, quiver.
τρενάρω *v.t.* & *i.* delay, drag on.
τρέξιμ|ο *s.n.* running; **~ατα** (*pl.*) running here and there, activity.
τρέπ|ω *v.t.* turn, convert; ~ω εις φυγήν put to flight. **~ομαι** *v.i.* turn (*to*).
τρέφω *v.t.* nourish, feed, nurture; grow (*beard*); cherish (*hope*); (*v.i.*) έθρεψε η πληγή the wound has healed.
τρεχάλα *s.f.* running; (*adv.*) at the run.
τρεχάματα *s.n.pl.* running here and there, activity.
τρεχ|άμενος, ~ούμενος *a.* running; current.
τρεχαντήρι *s.n.* small fast sailing-boat.
τρεχάτος *a.* running (*of persons*).
τρέχ|ω *v.i.* & *t.* run; hasten, hurry; go round (*all over the place*); wander; flow; run, leak, τι ~ει; what is going on? (*v.t.*) run, make (*person*) run; η πληγή ~ει πύον the wound is running with pus.
τρέχ|ων *a.* current; τα ~οντα happenings, events.
τρία *num.* three.
τριάδα *s.f.* trinity; trio.
τρίαινα *s.f.* trident.
τριάκοντα *num.* thirty.
τριακόσιοι *a.* three hundred.
τριακοστός *a.* thirtieth.
τριανδρία *s.f.* triumvirate.
τριάντα *num.* thirty.
τριαντάφυλλο *s.n.* rose.
τριάρι *s.n.* figure 3; three (*at cards*).
τριάς *s.f.* trinity; trio.
τριατατικός *a.* post-office worker.
τριβέλι *s.n.* drill, auger; (*fam.*) importunate chatterer.
τριβή *s.f.* friction, rubbing; wear; (*fig.*) practice, experience.
τρίβολος *s.m.* caltrop.
τρίβ|ω *v.t.* rub; polish; grate. **~ομαι** *v.i.* crumble; show signs of wear; (*fig.*) acquire experience.
τρίγλυφ|ον *s.n.* **~ος** *s.m.* triglyph.

τριγμός *s.m.* creak(ing), squeak(ing); grinding (of teth).

τριγυρ|ίζω, ~νώ *v.i. & t.* go *or* wander around. (*v.t.*) go all over (*premises, district*); surround; (*fig.*) run after, lay siege to.

τριγυρίστρα *s.f.* 1. whitlow. 2. gadabout (*woman*).

τριγύρω *adv.* (all) around.

τριγωνικός *a.* triangular.

τριγωνομετρία *s.f.* trigonometry.

τρίγωνο *s.n.* 1. triangle. 2. sort of cake.

τρίγωνος *a.* three-cornered.

τρίδιπλος *a.* triple, threefold.

τρίδυμα *s.n.pl.* triplets.

τριετής *a.* triennial; three years old; third-year.

τρίζω *v.i. & t.* creak, squeak. (*v.t.*) grind (*teeth*); (*fam.*) *του 'τριξα τα δόντια* I spoke to him threateningly.

τριήρης *s.f.* trireme.

τρικαντό *s.n.* three-cornered hat.

τρικλίζω *v.i.* stagger.

τρικλοποδιά *s.f.* tripping up; (*fig.*) piece of trickery.

τρίκλωνος *a.* three-ply (*thread*).

τρικούβερτος *a.* three-decked; (*fig.*) terrific (*jollification or row*).

τρίκυκλο *s.n.* tricycle.

τρικυμ|ία *s.f.* storm (*at sea*). **~ιώδης** *a.* stormy.

τρίκωχο *s.n.* three-cornered hat.

τρίλλια *s.f.* (*mus.*) trill.

τριμελής *a.* with three members.

τριμερής *a.* tripartite.

τρι|μηνία *s.f.,* **~μηνο** *s.n.* quarter (*three months*); a quarter's pay *or* rent.

τρίμην|ος, ~ιαίος *a.* for *or* of three months, quarterly.

τρίμμα *s.n.* fragment, crumb.

τριμμένος *a.* showing signs of wear; polished; grated.

τρίξιμο *s.n. see* **τριγμός**.

τρίπατος *a.* three-storied.

τριπλάσι|ος *a.* triple, threefold. **~άζω** *v.t.* treble.

τριπλ|ός, ~ούς *a.* triple, threefold. *εις ~ούν* in triplicate.

τριποδίζω *v.i.* trot.

τρί|πους *s.m.,* **~ποδο** *s.n.* tripod, trestle, easel; *ως από ~ποδος* ex cathedra.

τρίπτυχο *s.n.* triptych.

τρις *adv.* thrice.

τρισάγιο(ν) *s.n.* Sanctus; a prayer for the dead.

τρίσβαθα *s.n.pl.* depths.

τρισδιάστατος *a.* three-dimensional.

τρισέγγονο *s.n.* great-great-grandchild.

τρισκατάρατος *a.* thrice-cursed; the devil.

τρίστρατο *s.n.* three-cross-roads.

τριτεύω *v.i.* come third.

Τρίτη *s.f.* 1. Tuesday. 2. (*mus.*) third.

τριτογενής *a.* tertiary.

τριτοετής *a.* third-year.

τρίτον 1. *s.n.* third (*portion*). 2. *adv.* thirdly.

τρίτος *a. & s.m.* third; third party.

τριτώνω *v.i. & t.* happen *or* repeat for third time.

τρίφτης *s.m.* grater.

τριφτός *a.* grated, ground.

τριφύλλι *s.n.* clover.

τρίχ|α *s.f.* a hair, bristle; coat, fur (*of animal*); *στην ~α* spick and span; *παρά* (*or από*) *~α να πνιγώ* I was almost drowned. (*fam.*) *~ες* (*pl.*), nonsense, lies.

τριχιά *s.f.* rope.

τριχοειδής *a.* capillary.

τρί|χρους, ~χρωμος *a.* tricolour.

τρίχωμα *s.n.* coat, fur (*of animal*); hair (*on body*).

τριχωτός *a.* hairy.

τρίψιμο *s.n.* rubbing, friction; polishing; grating.

Τριώδιον *s.n.* three weeks of Carnival (*before Lent*).

τρίωρος *a.* lasting three hours.

τρολλές *s.m.* (*overhead*) trolley.

τρόλλεϋ *s.n.* trolley-bus [E. *trolley*]

τρόμαγμα *s.n.* a fright.

τρομάζω *v.t. & i.* frighten; become frightened; (*with να*) have one's work cut out (*to*).

τρομακτικός *a.* frightening; (*fig.*) terrific.

τρομάρα *s.f.* fright, terror; (*fam.*) *~ σου* bless you! (*in surprise or mild protest*).

τρομερός *a.* frightful, terrible; (*fig.*) terrific, amazing.

τρομοκρατ|ώ *v.t.* terrorize. **~ία** *s.f.* terrorism; (reign of) terror. **~ης** *s.m.* terrorist.

τρόμος *s.m.* fright, terror; trembling.

τρόμπα *s.f.* pump; *~ μαρίνα* loud-hailer.

τρομπέτα *s.f.* trumpet.

τρομπόνι *s.n.* trombone; blunderbuss.

τρόπαι|ο *s.n.* trophy; *~α* (*pl.*) triumph. **~ούχος** *a.* triumphant.

τροπάρ|ιον, ~ι *s.n.* (*eccl.*) a hymn; (*fig.*) (*same old*) refrain.

τροπή *s.f.* turn (*change of direction or circumstance*); conversion; solstice.

τροπικ|ός *a.* 1. tropic(al); *οι ~οί* tropics. 2. (*gram.*) of manner.

τρόπις *s.f.* keel.

τροπολογία *s.f.* ·amendment.

τροποποίηση *s.f.* modification.

τρόπ|ος *s.m.* way, manner, method; (*mus.*) mode; *~οι* (*pl.*) manners; *με ~ο* discreetly, surreptitiously; ~ον τινά somewhat; με κάθε ~ο at all costs; ~ος του λέγειν (in a) manner of speaking; έχει τον ~ο του he is well off.

τρούλλος s.m. dome. **~ωτός** a. domed.

τρουμπ- see **τρομπ-**.

τροφαντός a. early (*vegetable*); plump.

τροφή s.f. food, nourishment; board.

τρόφιμα s.n.pl. food, victuals.

τρόφιμος s.m.f. immate, boarder; (*fig.*) nursling.

τροφοδοσία s.f. provisioning (*of army, etc.*).

τροφοδότης s.m. caterer, purveyor. **~ώ** v.t. feed, serve, keep supplied.

τροφός s.f. wet-nurse.

τροχάδην adv. at a run; quickly.

τροχάζω v.i. trot.

τροχαίος 1. s.m. trochee. 2. a. wheeled; ~ο υλικό rolling-stock; ~α κίνηση wheeled traffic; η ~α traffic police.

τροχαλία s.f. pulley.

τροχιά s.f. track, rut; railway track; orbit, trajectory.

τροχίζω v.t. grind (*knife*); (*fig.*) sharpen (*mental faculty*).

τροχιόδρομος s.m. tramway.

τρόχισμα s.n. grinding (*of knives*); (*fig.*) sharpening.

τροχονόμος s.m. traffic-policeman.

τροχοπέδη s.f. brake.

τροχοπέδιλο s.n. roller-skate.

τροχός s.m. wheel; grindstone.

τροχοφόρος a. wheeled. **~ο** s.n. vehicle; paddle-steamer.

τρυγητής s.m. grape-harvester; (*fam.*) September.

τρυγητός s.m. grape-harvest, vintage.

τρυγόνι s.n. turtle-dove.

τρύγος s.m. grape-harvest, vintage.

τρυγώ v.t. gather (*grapes, honey*); (*fig.*) milk, exploit.

τρύπα s.f. hole; κάνω μια~ στο νερό achieve nothing.

τρυπάνι, ~ανον s.n. drill, auger. **~ανίζω** v.t. drill hole in, pierce; (*med.*) trepan.

τρύπημα s.n. perforating, piercing; prick.

τρυπητήρι s.n. awl, punch.

τρυπητός a. perforated, with holes. **~ό** s.n. strainer.

τρύπιος a. in holes.

τρυπιοχέρης a. spendthrift.

τρυπώ v.t. & i. make hole in, pierce; prick. (*v.i.*) prick; go into holes.

τρυπώνω v.t. & i. hide away; baste (*in sewing*). (*v.i.*) hide oneself; squeeze in.

τρυφερός a. tender, affectionate; (*fam.*) ~όν ήμισυ better half.

τρυφερότητα s.f. tenderness, affection; ~τητες (*pl.*) amorous behaviour.

τρυφηλός a. soft, unused to hardship; epicurean.

τρωγαλίζω v.t. gnaw.

τρώγλη s.f. hole, den; hovel. **~οδύτης** s.m. troglodyte.

τρώ(γ)ω v.t. eat, bite; με ~ει it itches; spend, consume, squander; wear out, eat away; worry; nag, importune; beat, get the better of; suffer, undergo; (*fam.*) ~ω ξύλο catch it, get a beating; ~ω τον κόσμο move heaven and earth or search everywhere; ~ω το κεφάλι μου bring disaster on oneself; ~ω τη χυλόπιτ(τ)α be rejected (*of suitor*); δεν ~ει άχυρα he is no fool; τον ~ει η μύτη (or ράχη) του he is asking for trouble; τον~ει η γλώσσα του he is itching to blab. ~ομαι v.i. be eaten or eatable; (*fig.*) be passable; ~ονται they are at each other's throats; φαγώθηκε να he was madly keen to.

τρωικός a. Trojan.

τρωκτικόν s.n. rodent.

τρωτός a. vulnerable. **~ό** s.n. shortcoming.

τσάγαλο s.n. green almond.

τσαγγός a. rancid; (*fam.*) difficult, testy.

τσαγιέρα s.f., **~ερό** s.n. teapot.

τσαγκάρης s.m. shoemaker.

τσαγκαροδευτέρα s.f. (*fam.*) day off.

τσάι s.n. tea.

τσακ s.n. στο ~ on the dot.

τσακάλι s.n. jackal.

τσακίζω 1. v.t. & i. break, smash; (*fig.*) weaken, exhaust. (*v.i.*) be broken (*in health*); abate (*of wind, etc.*) ~ομαι v.i. make great efforts (*to*). 2. v.t. fold.

τσακίρικος a. (*flecked*) blue-green or grey-green (*eyes*).

τσακίρ-κέφι s.n. (*fam.*) slight inebriation.

τσάκιση s.f. crease (*of trousers*).

τσάκισμα s.n. breaking, smashing; ~ατα (*pl.*) coquettish airs; (*fam.*) να πας στα ~ατα go to the devil!

τσακμάκι s.n. tinder-box.

τσάκωμα s.n. catching, arrest; quarrelling. ~τα bickering.

τσακώνω v.t. catch (in the act). ~ομαι v.i. quarrel.

τσαλαβουτώ v.i. splash or flounder about; (*fig.*) do a slovenly job.

τσαλακώνω v.t. crease, crumple.

τσαλαπατώ v.t. trample on.

τσαλίμι s.n. dexterous movement; (*fig.*) coquettish air.

τσάμι s.n. pine-tree.

τσάμικος s.m. sort of cyclic dance.

τσάμπα adv. see **τζάμπα**.

τσαμπί *s.n.* bunch (*of grapes*).

τσαμπουν|ίζω, ~ώ *v.i.* whimper; prate.

τσάμπουρο *s.n.* grape-stalk.

τσανάκι *s.n.* shallow bowl; (*fam.*) scoundrel.

τσάντα *s.f.* handbag; brief-case; shopping-bag; satchel; game-bag.

τσαντίζω *v.t.* vex, rile.

τσαντίρι *s.n.* tent.

τσάπ|α *s.f.,* **~ί** *s.n.* mattock.

τσαπατσούλης *a.* slovenly (*in one's work*).

τσαπέλλα *s.f.* string of dried figs.

τσαρδάκ|α *s.f.,* **~ι** *s.n.* canopy of dried branches (for shade).

τσάρκα *s.f.* (*fig.*) walk, stroll.

τσαρλατάνος *s.m.* cheat, quack.

τσάρος *s.m.* tsar.

τσαρούχι *s.n.* rustic shoe with pompom.

τσάτρα πάτρα *adv.* after a fashion, stumblingly (*esp. of speech*).

τσατσάρα *s.f.* comb.

τσαχπινιά *s.f.* exercise of wiles, coquettishness.

τσεκ *s.n.* cheque. **~άρω** *v.t.* check, tick off. [E. *cheque*]

τσεκούρ|ι *s.n.* axe. **~ώνω** *v.t.* cut with axe; (*fam.*) inflict severe penalty on; plough (*in exam.*).

τσέλιγκας *s.m.* chief shepherd.

τσεμπέρι *s.n.* kerchief.

τσέπ|η *s.f.* pocket; pocketful. **~ώνω** *v.t.* pocket.

τσερβέλο *s.n.* brains, mind.

τσέρι *s.n.* cherry-brandy.

τσέρκι *s.n.* hoop.

τσέτουλα *s.f.* tally; (*fam.*) on tick.

τσευδίζω *v.i.* lisp; have speech defect.

τσευδός *a.* lisping.

τσιγαρίζ|ω *v.t.* fry lightly, brown; (*fig.*) **~ομαι με το ζουμί μου** suffer privation.

τσιγάρ|ο *s.n.* cigarette. **~οθήκη** *s.f.* cigarette-case *or* box. **~όχαρτο** *s.n.* tissue paper.

τσιγγάνος *s.m.* gipsy.

τσιγγούν|ης *a.* stingy; a miser. **~ιά** *s.f.* stinginess. **~εύομαι** *v.t. & i.* skimp, be stingy (with).

τσιγκέλι *s.n.* (meat) hook.

τσιγκλώ *v.t.* goad.

τσίκλα *s.f.* chewing-gum.

τσίκνα *s.f.* smell of burnt meat *or* hair.

τσικν|ίζω *v.t. & i.* burn (*in cooking*); have a burnt smell; (*fam.*) **τα ~ισαν** they had a tiff.

τσικουδιά *s.f.* 1. terebinth. 2. sort of spirit.

τσιλη(μ)πουρδώ *v.i.* gallivant.

τσιλιθήθρα *s.f.* wagtail; (*fig.*) small, thin woman.

τσίμα *adv.* **~ ~** only just; on the very edge.

τσιμέντο *s.n.* cement; (*fam.*) **~ να γίνει** I couldn't care less.

τσιμουδιά *s.f.* (*fam.*) **δεν έβγαλε ~** he kept mum; **~!** not a word!

τσίμπημα *s.n.* prick, pinch, nip; bite, peck; bite of food.

τσιμπιά *s.f.* pinch.

τσιμπίδα *s.f.* tongs, forceps.

τσιμπίδ|ι *s.n.* tweezers; clothes-peg. **~άκι** *s.n.* tweezers; hair-clip.

τσίμπλα *s.f.* (*fam.*) mucus of eyes.

τσιμπολογώ *v.t.* have a bite; pick up, get. **~ήματα** (*s.n.pl.*) pickings, perks.

τσιμπούκι *s.n.* tobacco-pipe.

τσιμπούρι *s.n.* tick (*insect*); (*fig.*) pest.

τσιμπούσι *s.n.* (*fam.*) spread, feast.

τσιμπ|ώ *v.t.* prick, pinch, nip; bite, peck; have a bite of (*food*); (*fam.*) make, pick up, cadge (*small profit or perquisites*); nab. **~ημένος** in love.

τσίνουρο *s.n.* eyelash.

τσινώ *v.i.* kick, get restive.

τσίπα *s.f.* skin, crust, membrane; (*fig.*) (*sense of*) shame.

τσιπούρα *s.f.* gilthead (*sea-fish*).

τσίπουρο *s.n.* sort of spirit.

τσιράκι *s.n.* apprentice, boy; henchman. (*fig.*) **~ του πατέρα του** another edition of his father.

τσιρίζω *v.i.* screech, squall.

τσιριμόνιες *s.f.pl.* assiduous attentiveness.

τσιρίσι *s.n.* glue.

τσιριχτός *a.* screechy.

τσίρκο *s.n.* circus.

τσίρλα *s.f.* (*fam.*) diarrhoea.

τσίρος *s.m.* dried mackerel; (*fig.*) very thin person.

τσιρότο *s.n.* sticking-plaster.

τσίτα *adv.* (*fitting*) very tight (*of clothes*); **~ ~** only just.

τσίτι *s.n.* calico.

τσίτσ|ιδος *a.,~ίδι* *adv.* stark naked.

τσιτσιμπύρι *s.n.* ginger-beer.

τσιτσιρίζω *v.i. & t.* sizzle; (*fig.*) torment slowly.

τσιτ|ώνω *v.t.* stretch; (*fam.*) **τα ~ωσε** he died.

τσιφλίκι *s.n.* large country estate.

τσιφούτης *s.m.* skinflint.

τσίφτης *s.m.* (*fam.*) good sort, trump; adept; smart dresser.

τσίχλα *s.f.* thrush; (*fam.*) very thin person.

τσόκαρο *s.n.* wooden shoe; (*fam.*) vulgar woman.

τσοκολάτα *s.f.* chocolate.

τσολιάς *s.m.* evzone (*soldier of kilted Gk. regiment*).

τσο(μ)πάνης *s.m.* shepherd.

τσόντα|α *s.f.* gusset. **~άρω** *v.t.* join on (a *piece*); (*fig.*) make up (*remainder of sum required*).

τσουβάλι *s.n.* sack, sackful; sacking; (*fam.*) *βάζω στο* ~ trick.

τσουγκράν|α *s.f.* rake. **~ίζω** *v.t.* scratch.

τσουγκρί|ζω *v.t.* knock (*against, together*); clink (*glasses*); (*fig.*) *τα* ~ισαν they fell out.

τσούζω *v.t. & i.* sting, hurt; (make someone smart; (*fam.*) *το* ~ be given to drinking.

τσουκάλι *s.n.* earthen cooking-pot; chamber-pot.

τσουκνίδα *s.f.* nettle.

τσούλα *s.f.* loose-living *or* low-class woman.

τσουλάω *v.i. & t.* slide along; (*v.t.*) drag *or* trail along; push (*vehicle*).

τσουλής *s.m.* untidy person.

τσούλ|ι, ~άκι *s.n.* (door-) mat.

τσουλίστρα *s.f.* slide, chute.

τσουλούφι *s.n.* lock of hair, kiss-curl.

τσούξιμο *s.n.* stinging, smarting; boozing.

τσουράπι *s.n.* woollen sock.

τσουρέκι *s.n.* large bun.

τσούρμο *s.n.* crowd; crew.

τσουρουφλίζω *v.t.* singe.

τσουχτερός *a.* biting, stinging.

τσούχτρα *s.f.* jellyfish.

τσόφλι *s.n.* shell (*of egg, nut*); peel.

τσόχα *s.f.* baize, felt; (*fam.*) slippery customer.

τυγχάνω *v.i. & t.* (happen to) be; *έτυχε να* it happened that; (*with gen.*) enjoy, obtain.

τύλιγμα *s.n.* twisting, winding, coiling, wrapping.

τυλί|γω, ~σσω *v.t.* twist, wind, coil, wrap (up *or* round); (*fam.*) hook (*victim*).

τύλος *s.m.* callus.

τυλ|ώνω *v.t.* fill; (*fam.*) *την* ~ωσε he gorged himself.

τύμβ|ος *s.m.* grave, tomb. **~ωρύχος** *s.m.* grave-robber.

τύμπανο *s.n.* drum; eardrum; tympanum.

τυπικόν *s.n.* (*gram.*) morphology.

τυπικός *a.* formal, conventional; typical.

τυπογράφ|ος *s.m.* printer. **~είον** *s.n.* printing-press. **~ία** *s.f.* printing.

τυποδεμένος *a.* careful of one's appearance.

τυποποίησης *s.f.* standardization.

τύπ|ος *s.m.* imprint, mark; mould; type, model; form, type, kind; (eccentric) character; type, print; the press; (*math.*) formula. ~οι (*pl.*) conventions, rules, form; *για τον* ~*o* for form's sake.

τύπτω *v.t.* hit, strike.

τυπώνω *v.t.* print, imprint.

τύραvv|oς *s.m.* tyrant. **~ία** *s.f.* tyranny; torment, harassment. **~ώ** *v.i. & t.* be a tyrant; torment, harass.

τυρί *s.n.* cheese.

τυρινή *s.f.* last week of Carnival.

τυροκομία *s.f.* cheese-making.

τυρόπητ(τ)α *s.f.* cheese-pie.

τυρός *s.m.* cheese.

τύρφη *s.f.* peat.

τυφεκ- *see* **τουφεκ-**.

τύφλα *s.f.* blindness; (*fam.*) opprobrious gesture with open palm; ~ *στο μεθύσι* blind drunk; ~ *να 'χει αυτό μπροστά στο δικό σου* this one is far inferior to yours.

τυφλόμυγα *s.f.* blind man's buff.

τυφλοπόντικας *s.m.* mole (*animal*).

τυφλ|ός *a.* blind; (*fig.*) ~οίς όμμασι with one's eyes shut.

τυφλοσ(ο)ύρτης *s.m.* foolproof aid *or* instructions; ready-reckoner; crib.

τυφλών|ω *v.t.* blind; (*fam.*) make *τύφλα* at. ~ομαι *v.i.* become blind.

τυφλώττω *v.i.* (*fig.*) be blind *or* indifferent (*to*).

τυφοειδής *a.* ~ *πυρετός* typhoid fever.

τύφος *s.m.* εξανθηματικός ~ typhus; (*κοιλιακός*) ~ typhoid fever.

τυφών, ~ας *s.m.* cyclone.

τυχαίνω *v.i. & t.* (happen to) be; *μου έτυχε* it fell to me; *έτυχε να λείπω* I happened to be away; *τον έτυχα στο δρόμο* I met him in the street.

τυχαί|ος *a.* fortuitous, chance; *δεν είναι* ~*ος άνθρωπος* he is no ordinary man. **~ως** *adv.* by chance.

τυχάρπαστος *a.* one raised by stroke of fortune to power *or* wealth.

τυχεράκιας *s.m.* lucky dog.

τυχερ|ός *a.* fortunate, lucky. *ο* ~*ός* the winner (*in lottery*). ~*ό παιγνίδι* game of chance. ~*ό* *s.n.* fortune, destiny. ~*ά* *s.n.pl.* (*fam.*) casual profits, perks.

τύχη *s.f.* chance, fortune; fate; (good) luck; *στην* ~ at random.

τυχηρός *a. see* **τυχερός**.

τυχοδιώκτης *s.m.* adventurer.

τυχόν *adv.* by any chance; *τα* ~ *έξοδα* incidental expenses.

τυχών *a. ο* ~ the first comer, anybody; *δεν είναι ο* ~ he is no ordinary person.

τύψη *s.f.* pang (*of remorse*), prick (*of conscience*).

τωόντι *adv.* really, in fact.

τώρα *adv.* now, at present; at once; just now; *από* ~ *θα φύγεις;* are you going already?

τωρινός *a.* present, present-day.

Y

ύαινα *s.f.* hyena.
υάκινθος *s.m.* tuberose; jacinth.
υαλ- *see* **γυαλ-**.
υαλοβάμβακας *s.m.* fibreglass.
υαλοπίνακας *s.m.* pane of glass.
ύαλος *s.t.* (pane of) glass.
υαλουργία *s.f.* glass-making.
υαλόφρακτος *a.* glazed.
υαλώδης *a.* glassy; vitreous.
υάρδα *s.f.* yard (*measure*).
ύβος *s.m.* hump.
υβρεολόγιο *s.n.* string of abuse.
υβρίζω *v.t. & i.* abuse, revile, swear (at).
ύβρ|ις *s.f.* insult; oath, term of abuse; ~εις (*pl.*) abuse.
υβριστικός *a.* insulting, abusive.
υγεία, υγειά *s.f.* health.
υγειονομικός *a.* of health, sanitary (*inspector, regulations, etc.*).
υγιαίνω *v.i.* be in good health.
υγιεινή *s.f.* thygiene.
υγιεινός *a.* healthy (*good for health*); hygienic.
υγιής *a.* healthy (*enjoying good health*); (*fig.*) sound.
υγραίνω *v.t.* moisten.
υγρασία *s.f.* moisture, damp(ness).
υγρ|ός *a.* liquid; moist, damp. ~ό *s.n.* liquid, fluid. ~οποιώ *v.t.* liquefy.
υδαρής *a.* aqueous, watery; insipid.
ύδατα *s.n.pl.* waters.
υδαταγωγός *s.m.* water-conduit.
υδατάνθραξ *s.m.* carbohydrate.
υδατογραφία *s.f.* watercolour.
υδατοπτώσεις *s.f.pl.* waterfalls.
υδατόσημον *s.n.* watermark.
υδατοστεγής *a.* water-tight.
υδατοφράκτης *s.m.* dam, weir.
υδραγωγείον *s.n.* aqueduct.
υδραγωγός *s.m.* water-pipe.
υδραντλία *s.f.* water-pump.
υδράργυρος *s.m.* mercury.
υδρατμός *s.m.* vapour.
υδραυλικ|ός *a.* hydraulic; a plumber. ~ή *s.f.* hydraulics. ~ά *s.n.pl.* plumbing.
υδρ|εύομαι *v.i.* be supplied with water. ~ευσις *s.f.* water-supply.
υδρία *s.f.* pitcher.
υδρόβιος *a.* aquatic.
υδρόγειος. *s.f.* terrestrial globe.
υδρογόνον *s.n.* hydrogen.
υδρογραφία *s.f.* hydrography.
υδροηλεκτρικός *a.* hydroelectric.
υδροκέφαλος *a.* hydrocephalous.
υδροκίνητος *a.* water-driven.

υδροκυάνιον *s.n.* prussic acid.
υδρολήπτης *s.m.* water-consumer.
υδρόμυλος *s.m.* water-mill.
υδρονομεύς *s.m.* turncock; water-key.
υδρορρόη *s.f.* gutter (*of roof*).
υδροστάθμη *s.f.* water-level.
υδροστατικός *a.* hydrostatic.
υδροστρόβιλος *s.m.* whirlpool; (*water*) turbine.
υδροσωλήν *s.m.* water-conduit.
υδροφοβία *s.f.* hydrophobia.
υδροφόρος *a.* water-carrying.
υδροφράκτης *s.m.* dam, weir.
υδροχαρής *a.* aquatic (*plant*).
υδροχλωρικός *a.* hydrochloric.
υδρόχρωμα *s.n.* whitewash.
υδρόψυκτος *a.* water-cooled.
ύδρ|ωψ *s.m.*, ~ωπικία *s.f.* dropsy.
ύδωρ *s.n.* water.
υελ- *see* **υαλ-**.
υετός *s.m.* downpour.
υιικός *a.* filial.
υιοθε|σία, ~τηση *s.f.* adoption. ~τώ *v.t.* adopt.
υιός *s.m.* son.
υλακ|ή *s.f.* bark, barking. ~τώ *v.i.* bark.
ύλη *s.f.* matter, substance, material; pus.
υλικός *a.* material, ~όν *s.n.*material, stuff; ingredient.
υλισμός *s.m.* materialism.
υλιστ|ής *s.m.* materialist. ~ικός *a.* materialistic.
υλοποιούμαι *v.i.* materialize.
υλοτομία *s.f.* tree-felling.
υμείς *pron.* you.
υμέναιος *s.m.* wedlock.
υμέτερος *pron.* your, yours.
υμ|ήν, ~ένας *s.m.* (*anat.*) membrane.
ύμν|ος *s.m.* hymn, anthem. ~ητής *s.m.* eulogist, celebrator. ~ώ *v.t.* hymn, celebrate.
υνί *s.n.* ploughshare.
υπαγορ|εύω *v.t.* dictate. ~ευσις *s.f.* dictation; dictate.
υπάγ|ω 1. *v.i.* gò. 2. *v.t.* class, count, bring (*under heading, jurisdiction, etc.*). ~ομαι *v.i.* (*with* εις) belong to, come under.
υπαίθριος *a.* open-air, outdoor.
ύπαιθρ|ος *a.* η~ος the country (*as opposed to towns*); το ~ο the open air.
υπαινιγμός *s.m.* hint.
υπαινίσσομαι *v.t.* hint at.
υπαίτιος *a.* responsible, guilty.
υπακοή *s.f.* obedience.
υπάκουος *a.* obedient.
υπακούω *v.t. & i.* obey.
υπάλληλ|ος *s.m.f.* employee, clerk, shop-assistant; official; δημόσιος ~ος civil servant. ~οκρατία *s.f.* officialdom.

υπαναχωρώ *v.i.* withdraw discreetly; go back on promise.

υπανδρεία *s.f.* marriage, wedlock.

υπανδρεύω *v.t.* see **παντρεύω.**

ύπανδρος *a.* married.

υπάνθρωπος *s.m.* sub-human person.

υπαξιωματικός *s.m.* non-commissioned officer.

Υπαπαντή *s.f.* Candlemas (*2nd Feb.*).

υπαρκτός *a.* existent.

ύπαρξη *s.f.* existence; (a) being.

υπαρχ|ή *s.f.* εξ ~ής from the beginning.

υπαρχηγός *s.m.* second-in-command.

ύπαρχος *s.m.* (*naut.*) second-in-command (*of warship*).

υπάρχ|ω *v.i.* exist, be; ~ει there is; τα ~οντά μου my possessions.

υπασπιστής *s.m.* aide-de-camp; (*mil.*) adjutant.

ύπατος 1. *a.* highest. 2. *s.m.* consul (*Roman or French Republican*).

υπέδαφος *s.n.* subsoil.

υπείκω *v.i.* give way (*to instincts, force, etc.*).

υπεισέρχομαι *v.i.* creep in.

υπεκμισθώνω *v.t.* sublet.

υπεκ|φεύγω *v.t.* & *i.* escape, dodge. ~φυγή *s.f.* escape; evasion, subterfuge.

υπενθυμίζω *v.t.* remind (*person*) of, call to mind.

υπενοικιάζω *v.t.* grant *or* hold sublease of.

υπεξαίρεσις *s.f.* misappropriation.

υπέρ *prep.* 1. (*with acc.*) over, above. 2. (*with gen.*) for, on behalf of; τα ~ και τα κατά pros and cons.

υπερ- *denotes* 1. over, very much. 2. on behalf of.

υπεράγαν *adv.* too (much).

υπεραιμία *s.f.* (*med.*) congestion.

υπεραμύνομαι *v.t.* (*with gen.*) defend, fight for.

υπεράνθρωπος *a.* superhuman; (*s.m.*) a superman.

υπεράνω *prep.* (*with gen.*) over, above.

υπεράριθμος *a.* supernumerary.

υπερασπίζ|ω, ~ομαι *v.t.* defend.

υπεράσπιση *s.f.* defence.

υπεραστικός *a.* long-distance (*telephone, bus-service, etc.*).

υπεραφθονία *s.f.* superabundance.

υπερβαίνω *v.t.* cross, surmount; surpass, exceed, transcend.

υπερβαλλόντως *adv.* exceedingly; too.

υπερβάλλω *v.t.* surpass, exceed; exaggerate.

υπέρβασις *s.f.* exceeding, overstepping.

υπερβατικός *a.* transcendent, transcendental.

υπερβολ|ή *s.f.* excess; exaggeration; (*geom.*) hyperbola. ~ικός *a.* excessive; exaggerated; (given to) exaggerating.

υπέργηρ|ος, ~ως *a.* extremely aged.

υπερδιέγερση *s.f.* over-excitement.

υπερένταση *s.f.* over-strain.

υπερευαίσθητος *a.* over-sensitive.

υπερέχω *v.t.* (*with gen.*) excel, exceed.

υπερήλικος *s.m.f.* elderly person.

υπερήφαν|ος *a.* proud. ~εια *s.f.* pride. ~εύομαι *v.i.* be proud (*of*); grow proud.

υπερηχητικός *a.* supersonic.

υπερθεματίζω *v.i.* bid higher; (*fig.*) go one better.

υπερθετικός *a.* superlative.

υπερθέτω *v.t.* superimpose.

υπερίπταμαι *v.i.* & *t.* (*with gen.*) fly over.

υπερισχύω *v.i.* & *t.* (*with gen.*) prevail (over); overcome, beat.

υπεριώδης *a.* ultra-violet.

υπερκέρασις *s.f.* (*mil.*) outflanking.

υπερκόπωση *s.f.* overwork.

υπέρμαχος *s.m.f.* champion.

υπέρμετρος *a.* excessive.

υπερνικώ *v.t.* overcome.

υπέρογκος *a.* enormous; inordinate.

υπεροπλία *s.f.* superiority in arms.

υπεροπτικός *a.* haughty, proud.

υπέροχ|ος *a.* excellent, superb. ~ή *s.f.* superiority.

υπεροψία *s.f.* haughtiness.

υπερπέραν *s.n.* the beyond, life after death.

υπερπηδώ *v.t.* overleap, surmount.

υπερπόντιος *a.* overseas.

υπέρποτε *adv.* more than ever.

υπερσιτισμός *s.m.* intensive feeding.

υπερσυντέλικος *a.* (*gram.*) pluperfect.

υπέρταση *s.f.* high blood-pressure.

υπέρτατος *a.* highest, supreme.

υπέρτερ|ος *a.* (*with gen.*) superior to. ~ώ *v.t.* excel, surpass.

υπερτιμ|ώ *v.t.* overestimate; raise price of. ~ησις *s.f.* overestimation; rise in price.

υπερτροφία *s.f.* (*med.*) hypertrophy.

υπέρυθρ|ος *a.* reddish; ~οι ακτίνες infra-red rays.

υπερύψηλος *a.* very high.

υπερυψ|ώ *v.t.* make higher; exalt; (*fig.*) και ~ούται more than enough.

υπερφαλάγγισις *s.f.* (*mil.*) outflanking.

υπερφίαλος *a.* overweening.

υπερφυσικός *a.* supernatural; (*fig.*) extraordinary.

υπερώα *s.f.* palate.

υπερωκεάνιο *s.n.* liner.

υπερώο *s.n.* attic, loft; gallery (*of theatre*).

υπερωρία *s.f.* overtime.

υπεύθυνος *a.* responsible, accountable.

υπέχω *v.t* bear, be under; ~ *λόγον* be accountable.

υπήκο|ος 1. *a.* obedient. 2. *s.m.f.* subject (*member of state*). **~ότης** *s.f.* nationality (*by allegiance*).

υπήνεμος *a.* sheltered from wind; (*naut.*) leeward.

υπηρεσία *s.f.* service, duty, function; domestic servant(s); *είμαι της ~ς* be on duty.

υπηρεσιακ|ός *a.* pertaining to one's official duties; *~ή κυβέρνηση* caretaker government.

υπηρέτ|ης *s.m.* servant. **~ρια** *s.f.* maid.

υπηρετώ *v.i. & t.* serve; do one's (*military*) service.

υπναλέος *a.* drowsy, half-asleep.

υπναράς *s.m.* one who loves sleeping.

υπνηλία *s.f.* inordinate desire for sleep.

υπνοβάτης *s.m.* sleepwalker.

ύπνος *s.m.* sleep.

ύπνωσις *s.f.* hypnosis.

υπνωτ|ίζω *v.t.* hypnotize. **~ισμός** *s.m.* hypnotism. **~ικός** *a.* hypnotic.

υπνωτικό *s.n.* soporific (*drug*).

υπνώττω *v.i.* slumber; (*fig.*) be inactive.

υπό *prep.* 1. (*with acc.*) below, beneath, under. 2. (*with gen.*) by (*agent*).

υπ(ο)- *denotes* under, secretly, slightly.

υπόβαθρο *s.n.* pedestal, plinth.

υποβάλλω *v.t.* submit, present; subject, put (*to expense, torture, etc.*); suggest.

υποβαστάζω *v.t.* support, hold up.

υποβιβάζω *v.t.* lower, decrease; degrade, demote; do less than justice to.

υποβλέπω *v.t.* eye (*with distrust or envy*); cast glances at.

υποβλητικός *a.* evocative.

υποβοηθώ *v.t.* support, back up.

υποβολεύς *s.m.* prompter.

υποβολή *s.f.* submission, presentation; suggestion, prompting.

υποβολιμαίος *a.* spurious; not one's own (*opinion, etc.*).

υποβόσκω *v.i.* smoulder.

υποβρύχιο *s.n.* submarine; (*fam.*) sort of sweet.

υποβρύχιος *a.* underwater.

υπογεγραμμένη *s.f.* (*gram.*) iota subscript.

υπόγει|ος *a.* underground. **~ο** *s.n.* basement, cellar.

υπόγλυκος *a.* slightly sweet.

υπογραμμίζω *v.t.* underline.

υπογραμμός *s.m.* *τύπος και* ~ model, exemplary specimen (*of*).

υπογραφή *s.f.* signature, signing.

υπογράφω *v.t.* sign; (*fig.*) approve.

υπογράψας *s.m.* signatory.

υποδαυλίζω *v.t.* poke (*fire*); (*fig.*) foment.

υποδεέστερος *a.* inferior.

υπόδειγμα *s.n.* model, specimen. **~τικός** *a.* exemplary.

υποδεικνύω *v.t.* point out; suggest, propose.

υπόδειξη *s.f.* indication; suggestion.

υποδεκάμετρο *s.n.* decimetre; measuring rule.

υποδέχομαι *v.t.* receive, greet.

υποδηλώ *v.t.* contain a hint of, convey.

υπόδημα *s.n.* boot, shoe. **~τοποιείον** *s.n.* shoemaker's.

υπόδησις *s.f.* (supplying with *or* wearing of) foot-gear.

υποδιαιρ|ώ *v.t.* subdivide. **~εσις** *s.f.* subdivision.

υποδιαστολή *s.f.* (*math.*) decimal point.

υποδιευθυντής *s.m.* assistant director, vice-principal, *etc.*

υπόδικος *a.* (*person*) under trial.

υποδομή *s.f.* infrastructure.

υποδόριος *a.* subcutaneous.

υπόδουλ|ος *a.* enslaved. **~ώνω** *v.t.* subjugate.

υποδοχή *s.f.* reception.

υποδύομαι *v.t.* play role of.

υποζύγιον *s.n.* beast of burden.

υποθάλπω *v.t.* foment; protect, support (*esp. secretly*); pander to.

υπόθεμα *s.n.* (*med.*) suppository.

υπόθεση *s.f.* hypothesis, supposition; matter, affair, business, case; subject, theme.

υποθετικός *a.* hypothetical, supposed; fictitious; (*gram.*) conditional.

υπόθετο *s.n.* suppository.

υποθέτω *v.t. & i.* suppose, presume.

υποθήκ|η *s.f.* mortgage. **~εύω** *v.t.* mortgage. **~οφυλακείον** *s.n.* mortgage registry.

υποκαθιστώ *v.t.* replace (*take place of, exchange for something else*).

υποκατάστασις *s.f.* replacement (*act*).

υποκατάστατον *s.n.* substitute.

υποκατάστημα *s.n.* branch office *or* shop.

υποκάτω *prep.* (*with gen.*) underneath.

υπόκειμαι *v.i.* be subject *or* liable (*to*).

υποκείμεν|ον *s.n.* subject; (*pej.*) person, individual. **~ικός** *a.* subjective.

υποκινώ *v.t.* stir up, incite.

υποκλέπτω *v.t.* steal craftily.

υπο|κλίνομαι *v.i.* bow. **~κλίση** *s.f.* bow.

υποκλοπή *s.f.* theft; (*wire*) tapping, bugging.

υποκόμης *s.m.* viscount.

υποκόπανος *s.m.* butt (*of weapon*).

υποκοριστικόν *s.n.* diminutive.

υπόκοσμος *s.m.* underworld (*social*).

υποκρίνομαι *v.t. & i.* act, play (role of),

personate. (v.i.) dissemble.
υποκρισία s.f. hypocrisy.
υπόκρισις s.f. 1. acting of part. 2. dissembling.
υποκριτής s.m. 1. actor. 2. hypocrite.
υποκριτικός a. hypocritical.
υπόκρουσις s.f. (mus.) accompaniment.
υποκύπτω v.i. bend, give way, succumb.
υπόκωφος a. dull, muffled (sound).
υπόλειμμα s.n. remainder, remnant, relic.
υπολείπομαι v.i. remain over; be inferior, come after.
υπόληψη s.f. regard, esteem.
υπολογίζω v.t. estimate, reckon; take into account, attach importance to. ~**ισμός** s.m. calculation, estimate; εξ ~ισμού with ulterior motive. ~**ιστής** s.m. ηλεκτρονικός ~ιστής computer.
υπόλογος a. responsible, accountable.
υπόλοιπ|ος a. remaining, rest (of). ~**ο** s.n. remainder, balance.
υπολοχαγός s.m. (mil.) lieutenant.
υπομένω v.i. & t. endure.
υπομιμνήσκω v.t. remind (person) of.
υπομισθώνω v.t. hold sublease of.
υπόμνημα s.n. memorandum; ~τα (pl.) commentary (on text).
υπόμνηση s.f. reminder.
υπομοίραρχος s.m. lieutenant of gendarmerie.
υπομον|ή s.f. patience. ~**εύω** v.i. be patient. ~**ητικός** a. patient.
υποναύαρχος s.m. (naut.) rear-admiral.
υπόνοια s.f. suspicion.
υπονομεύω v.t. undermine.
υπόνομος s.m.f. sewer; blasting charge; (mil.) mine.
υπονο|ώ v.t. imply, hint at; ~**είται** it is understood.
υποπίπτω v.i. (with εις) fall into (error, etc.); come to (one's notice).
υποπλοίαρχος s.m. (naut.) lieutenant; first mate.
υποπόδιον s.n. footstool.
υποπροϊόν s.n. by-product.
υποπτεύομαι v.t.& i. suspect.
ύποπτος a. suspect, suspicious.
υποσημείωσις s.f. footnote.
υποσιτισμός s.m. under-nourishment.
υποσκάπτω v.t. undermine.
υποσκελίζω v.t. trip up; supplant.
υποσμηναγός s.m. (aero.) flight-lieutenant.
υποστάθμη s.f. dregs.
υπόστασις s.f. existence; foundation (in fact); (med.) hypostasis.
υποστατικό s.n. landed property.
υπόστεγο s.n. shed, hangar; shelter.
υποστέλλω v.t. strike (flag, etc.); reduce

(speed).
υποστήριγμα s.n. support (that which supports).
υποστηρίζω v.t. support; back; maintain, assert.
υποστήριξη s.f. support; backing.
υποστράτηγος s.m. (mil.) major-general.
υπόστρωμα s.n. 1. substratum. 2. saddle-cloth.
υποσυνείδητον s.n. subconscious.
υπόσχε|ση s.f. promise. ~**τικός** a. promissory.
υπόσχομαι v.t. & i. promise.
υποταγή s.f. submission, obedience.
υποτακτικ|ός a. obedient; a servant. ~**ή** s.f. (gram.) subjunctive.
υπόταση s.f. low blood-pressure.
υποτάσσ|ω v.t. bring into subjection. ~**ομαι** v.i. submit.
υποτείνουσα s.f. (geom.) hypotenuse.
υποτελής a. subordinate, vassal.
υποτίθεμαι v.i. be supposed.
υποτιμ|ώ v.t. underestimate; reduce price of; devalue. ~**ησις** s.f. underestimation; reduction in price; devaluation.
υποτονθορύζω v.t. & i. murmur, hum.
υποτροπή s.f. relapse.
υποτροφία s.f. bursary, scholarship.
υποτυπώδης a. sketchy, rudimentary.
ύπουλος a. underhand, shifty.
υπουργείον s.n. ministry (of state).
υπουργικ|ός a. ministerial; ~**όν συμβούλιον** cabinet.
υπουργία s.f. ministry (term of office).
υπουργός s.m. minister.
υποφαινόμενος a. ο ~ the undersigned; (fam.) yours truly.
υποφερτός a. bearable; passable.
υποφέρω v.t. & i. bear, endure, go through; suffer, feel pain.
υποφώσκω v.i. dawn.
υποχείριος a. under the thumb or power of (with gen.); a stooge.
υποχθόνιος a. subterranean, infernal.
υποχονδρία s.f. hypochondria.
υπόχρεος a. obliged (in duty or gratitude).
υποχρεώνω v.t. oblige (bind or gratify).
υποχρέωση s.f. obligation.
υποχρεωτικός a. obligatory; obliging.
υποχώρηση s.f. retreat, giving way; subsidence.
υποχωρώ v.i. retreat, give way, abate, subside.
υποψήφι|ος a. aspirant; a candidate or applicant. ~**ότης** s.f. candidature, application.
υποψία s.f. suspicion. ~**άζομαι** v.i. suspect.
ύπτιος a. supine, on one's back.

υπώρεια *s.f.* foot of mountain.
ύσκα *s.f.* tinder.
ύστατος *a.* last.
ύστερα *adv.* after(wards). then, later on; furthermore; (*prep.*) ~ *από* after.
υστέρημα *s.n.* one's inadequately small savings.
υστέρησις *s.f.* χρονική ~ time-lag.
υστερ|ία *s.f.*, **~ισμός** *s.m.* hysteria. **~ικός** *a.* hysterical.
υστερ(ι)νός *a.* later, last.
υστεροβουλία *s.f.* ulterior motive.
υστερόγραφον *s.n.* postscript.
ύστερον 1. *adv. see* **ύστερα.** 2. *s.n.* afterbirth.
ύστ|ερος *a.* later; inferior; εκ των ~έρων with hindsight.
υστερώ *v.i.* lag behind; be lacking in; not have (*with gen. or* σε).
υφ- *see* **υπο-**
υφάδι *s.n.* weft.
υφαίνω *v.t.* weave.
ύφαλα *s.n.pl.* part of ship's hull below water-line.
ύφαλος *s.f.* reef, shoal.
ύφανσις *s.f.* weaving; weave.
υφαντός *a* woven.
υφαντουργία *s.f.* textile industry.
υφαρπάζω *v.t.* obtain by trickery.
ύφασμα *s.n.* cloth, material; ~τα (*pl.*) textiles.
ύφεσις *s.f.* abatement; (*barometric*) depression; (*mus.*) flat; détente; recession.
υφή *s.f.* texture.
υφηγητής *s.m.* university lecturer.
υφήλιος *s.f.* earth, globe.
υφίστ|αμαι *v.t. & i.* undergo, suffer. (*v.i.*) exist, be in force. **~άμενος** *s.m.* subordinate.
ύφος *s.n.* style; air, tone.
υφυπουργός *s.m.* under-secretary.
υψηλ|ός *a.* high, tall; (*fig.*) lofty, great; ~ή Πύλη Sublime Porte. **~ότης** *s.f.* Highness.
υψηλόφρων *a.* 1. magnanimous. 2. arrogant.
υψικάμινος *s.f.* blast-furnace.
ύψιλον *s.n.* the letter Υ.
υψίπεδον *s.n.* plateau.
ύψιστος *a.* highest; the All-highest.
υψίφωνος *s.f.* soprano.
υψόμετρον *s.n.* altitude.
ύψ|ος *s.n.* height; εξ ~ους from on high; (*mus.*) pitch. φτάνω στο ~ος draw level with, come abreast of (*with gen.*)
ύψωμα *s.n.* height (*high ground*).
υψωμός *s.m.* raising; rise (*in price*).
υψών|ω *v.t.* raise. **~ομαι** rise.
ύψωσις *s.f.* raising, elevation; rise (*in price*).

Φ

φάβα *s.f.* yellow pea; pease-pudding.
φαγάνα *s.f.* dredger; (*fam.*) insatiable person *or* machine.
φαγάς *s.m.* glutton.
φαγγρί *s.n.* sea-bream.
φαγγρίζω *v.i. see* **φεγγρίζω.**
φαγέδαινα *s.f.* canker.
φαγητό *s.n.* meal, dish, food to eat.
φαγί *s.n.* food to eat.
φαγιάντζα *s.f.* faience; serving-dish.
φαγκότο *s.n.* (*mus.*) bassoon.
φαγοπότι *s.n.* eating and drinking, carousal.
φαγούρα *s.f.* itch, itching.
φάγωμα *s.n.* eating; corrosion, wear.
φαγωμάρα *s.f.* itching; bickering.
φαγωμένος *a.* eaten; worn away; είμαι ~ I have eaten.
φαγώνομαι *v.i.* bicker; get corroded *or* worn.
φαγώσιμ|ος *a.* eatable; ~*a* (*s.n.pl.*) provisions.
φαεινός *a.* brilliant; ηλίου ~ότερον clear as daylight; ~ή ιδέα brainwave.
φαΐ *s.n.* food to eat.
φαιδρ|ός *a.* merry, gay; (*fam.*) laughable. **~ύνω** *v.t.* cheer, gladden.
φαίνομαι *v.i.* appear, be visible; seem, look, be evident; prove, show oneself (*to be*).
φαινομενικώς *adv.* seemingly.
φαινόμεν|ον *s.n.* phenomenon; prodigy; κατά τα ~α to all appearances.
φαιός *a.* grey; darkish.
φάκα *s.f.* mouse-trap.
φάκελος *s.m.* envelope; file, dossier.
φακή *s.f.* lentil(s). αντί πινακίου ~ς for a mess of pottage.
φακίδα *s.f.* freckle.
φακιόλι *s.n.* kerchief.
φακίρης *s.m.* fakir.
φακός *s.m.* lens; magnifier; torch.
φάλαγ|ξ,~γα *s.f.* phalanx; (*mil.*) column.
φάλαινα *s.f.* whale.
φαλάκρ|α *s.f.* baldness; bald patch. **~αίνω** *v.i.* go bald. **~ός** *a.* bald; bare (*of mountain*).
φαλ(λ)ίρω *v.i. & t.* go *or* make bankrupt.
φαλλός *s.m.* phallus.
φαλμπαλάς *s.m. see* **φραμπαλάς.**
φαλτσέτα *s.f.* shoemaker's knife.
φάλτσο *s.n.* wrong note; (*fig.*) mistake. **~άρω** *v.i.* be out of tune; (*fig.*) err.
φαμ|ίλια, ~ελιά *s.f.* family. **~ελίτης** *s.m.* family man.
φάμπρικ|α *s.f.* factory; (*fam.*) scheme, dodge. **~άρω** *v.t.* manufacture; (*fig.*)

fabricate, concoct.

φανάρι *s.n.* 1. lamp, torch, lantern; headlamp; lighthouse; (*fig.*) κρατώ το ~ abet secret lovers. 2. hanging foodsafe. ~α (*pl.*) traffic-lights.

Φαναριώτης *s.m.* Phanariot.

φαναρτζής *s.m.* lamp-maker; tinsmith; carbody repairer.

φανατίζω *v.t.* make fanatical.

φανατικός *a.* fanatical; a fanatic.

φανατισμός *s.m.* fanaticism.

φανέλλα *s.f.* flannel; vest.

φανερ|ός *a.* clear, obvious. ~ώνω *v.t.* reveal, show.

φανός *s.m.* lamp, lantern.

φαντάζομαι *v.t. & i.* imagine, suppose, think; grow conceited.

φαντάζω *v.i.* make an effect, show up well; look glamorous.

φαντάρος *s.m.* foot-soldier.

φαντασία *s.f.* imagination; fantasy; conceit (*pride*); (*mus.*) fantasia.

φαντασιοκόπος *a.* a dreamer.

φαντασιοπληξία *s.f.* extravagant idea; whim, caprice.

φαντασιώδης *a.* imaginary; imaginative; beyond imagination.

φάντασμα *s.n.* ghost.

φαντασμαγορία *s.f.* phatasmagoria.

φαντασμένος *a.* vainglorious.

φανταστικός *a.* imaginary; fantastic.

φανταχτερός *a.* showy, striking.

φάντης *s.m.* knave (*cards*).

φανφαρόνος *a.* braggart.

φάπα *s.f.* slap, cuff.

φάρα *s.f.* (*pej.*) race, breed.

φάρ|αγξ *s.f.* ~άγγι *s.n.* gorge, ravine.

φαράσι *s.n.* dustpan.

φαρδαίνω *v.t. & i.* make *or* become wider; stretch.

φάρδος *s.n.* width, breadth.

φαρδ|ύς *a.* wide, broad; ~ιά-πλατιά sprawling.

φαρέτρα *s.f.* quiver.

φαρί *s.n.* steed.

φαρμακείο *s.n.* chemist's (*fam.*) place where one gets swindled.

φαρμακέμπορος *s.m.* wholesale druggist.

φαρμακερός *a.* poisonous; (*fig.*) biting.

φαρμάκι *s.n.* poison; (*fig.*) anything bitter (*coffee, grief, cold, etc.*).

φαρμακίλα *s.f.* bitter taste; (*fig.*) grief.

φαρμακομύτης *s.m.* venomous person.

φάρμακο *s.n.* drug, medicine.

φαρμακοποιός *s.m.* pharmacist, chemist.

φαρμακώνω *v.t.* poison; (*fig.*) grieve; taste bitter to.

φαρμπαλάς *s.m. see* **φραμπαλάς.**

φάρος *s.m.* lighthouse, beacon.

φάρσα *s.f.* farce; practical joke.

φαρσί *adv.* fluently.

φάρυγ|ξ, ~γας *s.m.* pharynx.

φαρφουρί *s.n.* porcelain.

φασαρία *s.f.* disturbance, to-do; fuss, bother.

φασιανός *s.m.* pheasant.

φασίνα *s.f.* scrubbing, chores of housework.

φάση *s.f.* phase.

φασίσ|τας *s.m.* fascist. ~τικός *a.* fascist. ~μός *s.m.* fascism.

φάσκελο *s.n. see* **μούντζα.**

φασκιά *s.f.* swaddling-band.

φασκόμηλο *s.n.* sage.

φάσκω *v.i.* ~ και αντιφάσκω say now one thing now the opposite.

φάσμα *s.n.* 1. ghost, spectre. 2. spectrum.

φασματοσκόπιον *s.n.* spectroscope.

φασ|όλι, ~ούλι *s.n.* haricot bean.

φασο(υ) λάδα *s.f.* bean soup.

φασο(υ) λάκια *s.f.* green beans.

φασουλής *s.m.* Punch (*puppet*).

φάτνη *s.f.* manger.

φατνίον *s.n.* (*anat.*) alveolus.

φατρία *s.f.* faction. ~άζω *v.i.* form factions.

φάτσα *s.f.* face; front; (*adv.*) face to face.

φαυλόβιος *a.* living a depraved life.

φαυλοκρατία *s.f.* corruption (*esp. in politics*).

φαύλος *a.* depraved; ~ κύκλος vicious circle.

φαφλατάς *s.m.* jabberer.

φαφούτης *a.* toothless.

Φεβρουάριος *s.m.* February.

φεγγάρι *s.n.* moon; ~α-α at certain periods.

φεγγίτης *s.m.* skylight, fanlight.

φέγγος *s.n.* light.

φεγγοβολώ *v.i.* give brilliant light.

φεγγρίζω *v.i.* be semi-transparent; (*fig.*) grow very thin.

φέγγ|ω *v.i.* 1. shine; ~ει it is daylight; έφεξε day dawned; (*fam.*) μου 'φεξε I had unexpected good fortune. 2. *See* **φεγγρίζω.**

φείδομαι *v.t.* (*with gen.*) spare (*abstain from hurting*); be sparing *or* frugal of.

φειδώ *s.f.* thrift. ~λεύομαι *v.i.* be stingy. ~λός *a.* thrifty, sparing; stingy.

φελί *s.n.* finger-shaped piece (*of food*).

φελλός *s.m.* cork; (*fam.*) trivial person.

φέλπα *s.f.* velveteen.

φενάκη *s.f.* wig; (*fig.*) deceit.

φέουδ|ο *s.n.* fief. ~αρχικός *a.* feudal.

φερέγγυος *a.* solvent; reliable.

φερέλπις *a.* promising; (*young*) hopeful.

φερετζές *s.m.* Moslem woman's veil.

φέρετρο *s.n.* coffin.

φερέφωνο *s.n.* (*fig.*) mouthpiece; one who echoes the views of another.

φερμάνι *s.n.* firman (*order from Sultan*).

φερμάρω 1. *v.t.* gaze at. 2. *v.t. & i.* stop (*esp. of vehicle*).

φέρμελη *s.f.* embroidered waistcoat.

φερμένος *a.* arrived, brought.

φερμουάρ *s.n.* (zip) fastener. [F. *fermoir*]

φέρνω *v.t. & i. see* **φέρω**.

φέρσιμο *s.n.* behaviour.

φέρ|ω *v.t. & i.* carry, bear; wear; bring, fetch; bring about, occasion; bring forward, produce; ~ω αντιρρήσεις raise objections; το 'φερε ο λόγος the subject came up; ο λόγος το ~νει in a manner of speaking; ~νω (*with gen. or* σε) be somewhat like, have a suggestion of; φερ' ειπείν for example; ~ε να δούμε let us see. ~ομαι *v.i.* behave; be reputed *or* considered.

φέσι *s.n.* fez; (*fam.*) γίνομαι ~ get drunk, του βάλανε ~ they welshed on him.

φέτα *s.f.* 1. slice. 2. sort of white cheese.

φετίχ *s.n.* fetish.

φέτ|ος *adv.* this year. ~ινός *a.* this year's.

φετφάς *s.m.* arbitrary ruling.

φευ *int.* alas.

φευγ|άλα *s.f.,* ~ιό *s.n.* (*precipitate*) flight.

φευγαλέος *a.* fleeting.

φευγατίζω *v.t.* help to escape.

φευγάτος *a.* gone, departed.

φεύγω *v.i. & t.* leave, depart, go away; escape, fly. (*v.t.*) run away from; όπου φύγει φύγει *said of one who takes to his heels.*

φευκτέος *a.* to be avoided.

φήμ|η *s.f.* rumour; reputation, repute; fame. ~ίζομαι *v.i.* be famous. ~ολογούμαι *v.i.* be rumoured.

φηρίκι *s.n.* sort of apple.

φθά|νω *v.i. & t.* be enough; arrive, draw near; become established *or* recognized; (*with* σε) reach, reach the point (*of*), be reduced (*to*); ~νει να provided that; να μη ~σω on my head be it! λέει ό,τι ~ση he says whatever comes into his head. (*v.t.*) reach, catch up with; come up to, equal; δεν τον έφτασα he was before my time.

φθαρμένος *a.* worn (*worse for wear*).

φθαρτός *a.* perishable.

φθείρ *s.m.f.* louse.

φθείρ|ω *v.t.* damage, spoil, impair; corrupt. ~ομαι *v.i.* wear out; (*fig.*) lose prestige.

φθην- *see* **φτην-**.

φθινόπωρο *s.n.* autumn.

φθίνω *v.i.* wane, draw to a close; decline, waste away.

φθίσ|η *s.f.* consumption. ~ικός *a.* consumptive.

φθόγγ|ος *s.m.* (*vocal*) sound; (*mus.*) note. ~όσημον *s.n.* (*mus.*) (*written*) note.

φθόν|ος *s.m.* (*malicious*) envy. ~ερός *a.* envious. ~ώ *v.t.* be envious of.

φθορά *s.f.* damage, destruction, wear and tear.

φθορεύς *s.m.* destroyer; corrupter.

φθορισμός *s.m.* fluorescence.

φι *s.n.* the letter **Φ**.

φιάλη *s.f.* bottle.

φιγούρ|α *s.f.* (*mus. & dance*) figure; court card; (*fig.*) κάνω ~α cut a (fine) figure. ~άρω *v.i.* figure, appear; cut a fine figure; pose (*as*). ~άτος *a.* stylish, elegant.

φιγουρίνι *s.n.* book of fashion-plates.

φιδές *s.m.* vermicelli.

φίδ|ι *s.n.* snake; βγάζω το ~ι από την τρύπα pull the chestnuts out of the fire. με ζώσανε τα ~ια I got the wind up. ~ίσιος *a.* sinuous. ~ωτός *a.* winding; snaky.

φίλαθλος *a.* interested in sports (*as spectator*).

φιλαλήθης *a.* truthful.

φιλάνθρωπος *s.m.f.* philanthropist.

φιλαράκος *s.m.* (*fam.*) beau; scamp, fine friend.

φιλάργυρος *a.* miserly; a miser.

φιλάρεσκος *a.* coquettish.

φιλαρμονική *s.f.* (*mus.*) band.

φιλαρχία *s.f.* love of power.

φιλάσθενος *a.* delicate, sickly.

φιλαυτία *s.f.* egoism.

φιλέκδικος *a.* vindictive.

φιλελεύθ|ερος *a.* liberty-loving; a Liberal; κόμμα των ~έρων Liberal Party. ~ερισμός *s.m.* liberalism.

φιλέλλην *s.m.f.* philhellene.

φιλ|ενάδα, ~ηνάδα *s.f.* girl *or* woman friend; (*fam.*) mistress.

φίλεργος *a.* fond of work, industrious.

φίλερις *a.* quarrelsome.

φιλές *s.m.* (hair-) net.

φιλέτο *s.n.* 1. fillet (*meat*). 2. piping (*cloth*); narrow band (*decorative*).

φιλεύω *v.t.* (*with double acc.*) give (present) to (person); treat to.

φίλη *s.f.* friend.

φιλήδονος *a.* sensual; a voluptuary.

φίλημα *s.n.* kiss.

φιλήσυχος *a.* peace-loving.

φιλί *s.n.* kiss.

φιλία *s.f.* friendship.

φιλικ|ός *a.* friendly; friend's; member of ~ή Εταιρεία (*secret society for Gk. independence under Turks*).

φιλιππικός *s.m.* philippic.

φίλιππος *s.m.f.* horse-lover; (*fam.*) race-goer.

φιλιστρίνι *s.n.* porthole.

φιλιώνω *v.t. & i.* reconcile (*estranged*

persons); become reconciled, make it up.
φιλμ *s.n.* film (*cinema, photographic*). [E.]
φίλντισι *s.n.* ivory; (*fam.*) mother-of-pearl.
φιλόδικος *a.* litigious.
φιλόδοξ|ος *a.* ambitious. ~**ώ** *v.i. & t.* be ambitious; aspire to.
φιλοδώρημα *s.n.* tip, gratuity.
φιλοζωία *s.f.* love of one's own life; poltroonery.
φιλόκαλος *a.* with a taste for the beautiful.
φιλοκατήγορος *a.* censorious.
φιλοκερδής *a.* greedy of gain.
φιλολογία *s.f.* study of literature. (*fam.*) mere words.
φιλόλογος *s.m.f.* student of literature.
φιλομαθής *a.* eager to learn.
φιλομειδής *a.* smiling, cheerful.
φιλόμουσος *a.* a lover of music or the arts.
φιλόν(ε)ικ|ος *a.* quarrelsome. ~**ία** *s.f.* quarrelling; quarrel. ~**ώ** *v.i.* quarrel.
φιλόξεν|ος *a.* hospitable. ~**ία** *s.f.* hospitality. ~**ώ** *v.t.* entertain, give hospitality to; find room for.
φιλοπαίγμων *a.* given to joking.
φιλοπατρία *s.f.* love of one's country.
φιλοπόλεμος *a.* bellicose.
φιλόπον|ος *a.* fond of work, industrious. ~**ώ** *v.t.* prepare with zealous care.
φιλοπράγμων *a.* meddlesome, inquisitive.
φιλοπρόοδος *a.* progressive.
φίλος *a. & s.m.* dear; friendly; friend.
φιλόσοφ|ος *s.m.f.* philosopher. ~**ία** *s.f.* philosophy. ~**ικός** *a.* philosophic(al).
φιλόστοργος *a.* loving.
φιλοτελισμός *s.m.* philately.
φιλοτέχνημα *s.n.* work of art.
φιλότεχνος *a.* an art-lover.
φιλοτεχνώ *v.t.* create (*a work of art*).
φιλοτιμία *s.f. see* **φιλότιμο.** *χάνω την ανάγκη* ~ make a virtue of necessity.
φιλότιμο *s.n.* one's honour, pride, dignity or face.
φιλότιμος *a.* having a sense of *φιλότιμο*; ambitious to excel; obliging; generous.
φιλοτιμ|ώ *v.t.* put (*person*) on his mettle, shame into. ~*ούμαι v.i.* make it a point of honour.
φιλοφρόνη|μα *s.n.,* ~**ηση** *s.f.* compliment.
φιλόφρων *a.* courteous, attentive.
φίλτατος *a.* dearest.
φίλτρο *s.n.* 1. philtre; (*parental*) love. 2. filter.
φιλύποπτος *a.* mistrustful.
φιλύρα *s.f.* lime-tree.
φιλώ *v.t.* kiss.
φιμ|ώνω *v.t.* muzzle, gag. ~**ωτρο** *s.n.* muzzle.
φινέτσα *s.f.* finesse, delicacy, refinement.

φίνος *a.* fine (*of quality, feeling, manners*).
φιντάν|ι *s.n.* young plant. ~**άκι** *s.n.* (*adolescent*) girl.
φιόγκος *s.m.* bow (*ribbon*).
φιρίκι *s.n.* sort of apple.
φίρμα *s.f.* firm; trade-name; (*fam.*) (one who has made) a reputation.
φίσα *s.f.* slip (*for filing*); chip (*for gambling*); plug (*electric*).
φισέκι *s.n.* cartridge; cartridge-shaped packet of coins.
φισεκλίκι *s.n.* cartridge-belt.
φίσκα *adv.* full to overflowing.
φιστίκι *s.n.* pistachio-nut; ~ *αράπικο* monkey-nut.
φιτίλι *s.n.* wick, fuse; braid, piping; *του βάζω* ~*α* I irritate or inflame him or egg him on.
φκ(ε)ιάνω *v.t. & i. see* **φτιάνω.**
φκυάρι *s.n.* shovel, spade.
φλαμούρι *s.n.* lime-wood; infusion of lime-blossom.
φλάμπουρο *s.n.* pennon.
φλάντζα *s.f.* flange, gasket. ·
φλάουτο *s.n.* flute.
φλασκί *s.n.* gourd.
φλέβα *s.f.* vein; lode; (*fig.*) talent.
Φλεβάρης *s.m.* (*fam.*) February.
φλέγμα *s.n.* phlegm, mucus; (*fig.*) phlegm. ~**τικός** *a.* phlegmatic.
φλεγμονή *s.f.* inflammation.
φλέγ|ομαι *v.i.* burn, be ablaze; ~*ον ζήτημα* burning question.
φλέμα *s.n.* phlegm, mucus.
φλερτ *s.n.* flirtation; person one flirts with. [E. *flirt*]. ~**άρω** *v.i. & t.* flirt (with).
φλεψ *s.f.* vein; lode.
φληναφώ *v.i.* prate.
φλιτζάνι *s.n.* cup; cupful.
φλόγα *s.f.* flame; (*fig.*) fire, passion.
φλογέρα *s.f.* shepherd's pipe.
φλογερός *a.* flaming; (*fig.*) passionate
φλογίζω *v.t.* inflame.
φλογοβόλον *s.n.* (*mil.*) flame-thrower.
φλόγωσις *s.f.* inflammation.
φλοιός *s.m.* peel, rind, skin; shell; bark; crust (*of earth*).
φλοίσβος *s.m.* lapping (*of waves*).
φλοκάτ|α, ~η *s.f.* peasant's thick blanket or cape.
φλόκος *s.m.* (*naut*) jib.
φλομώνω *v.t. & i.* stupefy; (*fig.*) fill with noxious smell or smoke; (*v.i.*) grow pale.
φλοξ *s.f. see* **φλόγα.**
φλοτέρ *s.n.* float, ball-cock.
φλου *a.* vague, in the air, (*plans*); blurred (*picture*).
φλούδ|α *s.f.,* ~**ι** *s.n. see* **φλοιός.**

φλουρί *s.n.* gold florin.

φλύαρ|ος *a.* loquacious. **~ία** *s.f.* loquacity, chatter. **~ώ** *v.i.* chatter, gossip.

φλύκταινα *s.f.* blister; pustule.

φλυτζάνι *s.n. see* **φλιτζάνι**.

φοβάμαι *v.t. & i. see* **φοβούμαι**.

φοβέρα *s.f.* threat.

φοβερίζω *v.t.* threaten.

φοβερός *a.* frightful; terrific, amazing.

φόβητρο *s.n.* scarecrow; bogy.

φοβητσιάρης *a.* timorous.

φοβία *s.f.* phobia.

φοβίζω *v.t.* frighten; threaten.

φόβος *s.m.* fear.

φοβούμαι *v.t. & i.* fear, be afraid (of).

φόδρα *s.f.* lining.

φοινίκι *s.n.* sort of cake.

φοινικικός *a.* Phoenician.

φοίν|ιξ, **~ικας** *s.m.*, **~ικιά** *s.f.* palm-tree; **~ιξ** phoenix.

φοιτ|ώ *v.i.* (*with* εις) frequent; be a student. **~ηση** *s.f.* attendance at a course of study. **~ητής** *s.m.*, **~ήτρια** *s.f.* student. **~ητικός** *a.* students'.

φόλα *s.f.* 1. dog-poison. 2. patch on a shoe.

φολίς *s.f.* scale (*of fish, etc.*).

φον|εύς, **~ιάς** *s.m.* murderer.

φονεύω *v.t.* murder.

φόν|ος *s.m.*, **~ικό** *s.n.* murder.

φόντο *s.n.* bottom, essential character; background; capital (*funds*); (*fig.*) moral *or* intellectual worth.

φόρα 1. *s.f.* way, impetus; force. 2. *adv.* βγάζω (στη *or* στα) **~** (*fig.*) bring into the open.

φορά *s.f.* force; way, impetus; direction (*of wind, current, events*); time (*occasion*); άλλη **~** some other time; άλλη μιά **~** once more; μια **~** once; βλάκας μια **~** a fool and no mistake; εγώ μια **~** I for my part.

φοράδα *s.f.* mare.

φορατζής *s.m.* tax-collector.

φορβάς *s.f.* mare.

φορβή *s.f.* fodder.

φορείον *s.n.* stretcher.

φόρεμα *s.n.* (wearing of) a dress.

φορεσιά *s.f.* suit of clothes; attire.

φορεύς *s.m.* carrier, bearer.

φορητός *a.* portable.

φόρμα *s.f.* 1. form, shape; mould, cake-tin, *etc.*; (*fig.*) σε **~** in good form, fit. 2. form, document. 3. overall.

φοροδιαφυγή *s.f.* tax-evasion.

φορολογ|ία *s.f.* taxation. **~ώ** *v.t.* tax; **~ούμενος** a taxpayer.

φόρος *s.m.* tax, duty.

φόρτε *s.n.* strong point. **στο ~** in full swing, at one's best.

φορτηγ|ός *a.* freight-carrying. **~ό** *s.n.* lorry; cargo-vessel.

φορτίζω *v.t.* charge (*with electricity*).

φορτικός *a.* importunate, pestering.

φορτίο *s.n.* cargo, load, freight; burden.

φορτοεκφορτωτής *s.m.* stevedore.

φόρτος *s.m.* (*fig.*) heavy load.

φορτσάρω *v.t. & i.* force (*lock, voice, etc.*); intensify. (*v.i.*) strengthen (*of wind*).

φόρτωμα *s.n.* loading; load; (*fig.*) burden, trial (*of persons*).

φορτών|ω *v.t. & i.* load (*with*); take on cargo. **~ομαι** *v.t.* pester.

φορτωτική *s.f.* bill of lading.

φορ|ώ *v.t.* wear; put on; του **~εσα** τα παπούτσια I put his shoes on (*for him*); **~εμένα** ρούχα worn clothing.

φουβού *s.f.* charcoal stove; brazier.

φουγάρο *s.n.* tall chimney, funnel.

φουκαράς *s.m.* (*fam.*) poor chap.

φουλ *a. & s.n.* full; full house (*cards*); στο **~** at full speed *or* blast.

φούμαρα *s.n.pl.* big words; castles in the air.

φουμάρ|ω *v.t. & i.* (*fam.*) smoke. τι καπνό **~ει** what sort of man is he?

φούμος *s.m.* soot, lamp-black.

φούντα *s.f.* tassel.

φουντάρω *v.t. & i.* sink (*ship*); (*v.i.*) drop anchor; sink.

φουντούκι *s.n.* hazel-nut.

φουντώνω *v.i. & t.* grow bushy; spread, grow in extent *or* intensity; flare up; (*v.t.*) make (*hair, dress*) bouffant.

φουντωτός *a.* bushy.

φούξια *s.f.* fuchsia.

φούρια *s.f.* impetuous haste.

φούρκ|α *s.f.* 1. gallows. 2. (*indignant*) anger; τον έχω **~α** I am furious with him. **~ίζω** *v.t.* 1. hang. 2. infuriate.

φουρκέτα *s.f.* hairpin.

φούρναρης *s.m.* baker. **~ικο** *s.n.* bakery.

φουρνέλο *s.n.* grid of cooking-stove; blasting-charge.

φουρνιά *s.f.* ovenful; (*fig.*) batch.

φούρνος *s.m.* oven; bakery.

φουρτούνα *s.f.* storm, rough sea; (*fig.*) tribulation.

φούσκα *s.f.* bladder; (*toy*) balloon; blister; bubble.

φουσκάλα *s.f.* blister; bubble.

φουσκί *s.n.* manure.

φουσκοδεντριά *s.f.* rising of sap.

φουσκοθαλασσιά *s.f.* swell (*of sea*).

φούσκωμα *s.n.* inflation, swelling: flatulence; self-importance.

φουσκών|ω *v.t. & i.* swell, inflate, fill with air; pump up (*tyre*); (*fig.*) exaggerate; annoy. (*v.i.*) swell up, become inflated; rise (*of dough*); bulge; puff, pant; (*fig.*) puff oneself up; ~μένος swollen, bulging.

φουσκωτός *a.* curved; puffed, bouffant (*of dress, hair*).

φούστα *s.f.* skirt.

φουστανέλλα *s.f.* kilted skirt (*Gk. national costume*).

φουστάνι *s.n.* (*woman's*) dress.

φουφούλα *s.f.* (hanging portion of) breeches (*Gk. islanders' costume*).

φούχτα *s.f.* palm of hand; handful.

φραγγέλιο *s.n.* whip.

φραγκεύω *v.i.* become a Catholic.

Φραγκ|ιά *s.f.* Western Europe. ~ικος *a.* Frankish; Catholic; West European. ~ος *s.m.* Frank; Catholic.

φράγκο *s.n.* franc; (*fam.*) drachma.

φραγκοκρατία *s.f.* Frankish domination (*of Greece*).

φραγκολεβαντίνος *s.m.* Levantine.

φραγκοπαναγιά *s.f.* (*fam.*) person affecting prudishness *or* innocence.

φραγκόπαπας *s.m.* Catholic priest.

φραγκοστάφυλο *s.n.* redcurrant.

φραγκόσυκο *s.n.* prickly pear.

φραγκοχιώτικα *s.n.pl.* system of writing Gk. with Roman alphabet (*formerly used in Chios, etc.*).

φράγμα *s.n.* barrier; barrage, dam.

φραγμός *s.m.* barrier (*abstract*); (*mil.*) barrage.

φράζω *v.t. & i.* enclose, fence; obstruct, block; stop up; become stopped up.

φρακάρ|ω *v.i.* stick, jam, get blocked. ~ισμα *s.n.* jam, blockage.

φράκο *s.n.* evening-dress suit.

φραμπαλάς *s.m.* furbelow.

φράντζα *s.f.* fringe.

φραντζόλα *s.f.* long loaf.

φράξιμο *s.n.* enclosing; blockage.

φράουλα *s.f.* strawberry; sort of grape.

φράπα *s.f.* shaddock.

φράσ|η *s.f.* phrase. ~εολογία *s.f.* phraseology.

φράχτης *s.m.* fence, hedge.

φρέαρ *s.n.* well.

φρεάτιον *s.n.* well, shaft, manhole.

φρεγ|άς, ~άδα *s.f.* frigate; (*fam.*) fine figure of a woman.

φρέν|α *s.n.pl.*, ~ες *s.f.pl.* reason, wits; έξω ~ών beside oneself (*with anger*); inconceivable.

φρενιάζω *v.t. & i.* enrage; get enraged.

φρενίτις *s.f.* delirium, frenzy; madness.

φρέν|ο *s.n.* brake (*of wheels*). ~άρω *v.i.* brake.

φρενο|βλαβής, ~παθής *a.* mentally deranged.

φρενοκομείον *s.n.* lunatic asylum.

φρενολογία *s.f.* alienism.

φρεσκάδα *s.f.* freshness; coolness.

φρεσκάρω *v.t. & i.* freshen; give a new look to.

φρέσκο *s.n.* 1. fresco. 2. coolness (*of temperature*). 3. (*fam.*) στο ~ in quod.

φρέσκος *a.* fresh, new (*recent, not stale*); cool.

φρικαλέ|ος *a.* frightful. ~ότης *s.f.* atrocity.

φρίκη *s.f.* horror; (*int.*) frightful!

φρικίαση *s.f.* shuddering.

φρικιώ *v.i.* shudder.

φρικ|τός, ~ώδης *a.* horrible, frightful.

φρίσσω *v.i.* ripple; shudder; be horrified.

φρόκαλ|ο *s.n.* rubbish; broom. ~ώ *v.t.* sweep (*with broom*).

φρόνημα *s.n.* morale; ~τα (*pl.*) convictions (*esp. political*).

φρονηματίζω *v.t.* impart prudence *or* moderation to.

φρόνησις *s.f.* prudence, good sense.

φρονιμάδα *s.f.* good behaviour; prudence, good sense.

φρονιμεύω *v.i. & t.* seetle *or* sober down.

φρονιμίτης *s.m.* wisdom-tooth.

φρόνιμ|ος *a.* sensible; well-behaved, good; virtuous. (*int.*) ~α behave!

φροντ|ίζω *v.t. & i.* look after, take care of; (*with για*) see about; θα ~ίσω I will see to it.

φροντ|ίς, ~ίδα *s.f.* care, charge, concern; ~ίδες (*pl.*) things to see to.

φροντιστήριο *s.n.* tutorial establishment; crammer's.

φροντιστής *s.m.* chief steward; property-man.

φρονώ *v.i. & t.* think, believe.

φρούδος *a.* vain, unavailing.

φρουμάρω *v.i.* (*of horses*) snort.

φρουρά *s.f.* guard; garrison.

φρούριο *s.n.* fort, fortress.

φρουρός *s.m.* guard, sentry; (*fig.*) guardian.

φρουρώ *v.t.* guard, watch.

φρούτο *s.n.* fruit.

φρυάττω *v.i.* be enraged.

φρύγανα *s.n.pl.* dry brushwood (*for fire*).

φρυγανιά *s.f.* piece of toast; fried bread with syrup or sugar.

φρύδι *s.n.* eyebrow.

φταίξιμο *s.n.* fault, blame.

φταίχτης *s.m.* person to blame, culprit.

φταίω *v.i.* be to blame.

φτάνω *v.i. & t.* see **φθάνω**.

φτασμένος *a.* a made man.

φτειάνω *v.t. & i.* see **φτιάνω**.

φτελιά *s.f.* elm.

φτενός *a.* thin, slim.

φτέρη *s.f.* fern.

φτέρνα *s.f.* heel.

φτερν|ίζομαι *v.i.* sneeze. **~ισμα** *s.n.* sneeze, sneezing.

φτερό *s.n.* feather; wing; feather-duster; mudguard.

φτερούγ|α *s.f.* wing. **~ίζω** *v.i.* flutter.

φτερωτός *a.* winged; feathered.

φτην|ός *a.* cheap. **~αίνω** *v.t. & i.* make *or* become cheaper. **~ια** *s.f.* cheapness.

φτιά(χ)ν|ω *v.t. & i.* arrange, put right; make, have made; do; τι ~εις; how are you? μου την έφτιαξε he played me a trick; τα ~ω με start an affair with *or* make it up with. (*v.i.*) improve (in appearance). **~ομαι** *v.i.* make up (*one's face*); φτιαγμένος made (up); looking better.

φτιασίδ|ι *s.n.* cosmetic; make-up. **~ώνομαι** *v.i.* make up.

φτιά|σιμο, ~ξιμο *s.n.* putting straight, fixing; making.

φτιαχτός *a.* artificial, affected.

φτου *int.* symbolic of spitting in disgust or to avert evil eye. **~χι απ' την αρχή** back to square one.

φτουρώ *v.i.* go a long way, be economical of use.

φτυάρι *s.n.* shovel, spade.

φτύνω *v.t. & i.* spit (on).

φτυσιά *s.f.* (clot of) spittle.

φτύσιμο *s.n.* spitting.

φτυστός *a.* (*fam.*) ~ ο πατέρας του the very spit of his father.

φτωχαίνω *v.t. & i.* make *or* become poor.

φτώχεια *s.f.* poverty.

φτωχικ|ός *a.* poor (*in quality*); meagre. **~ό** *s.n.* (*fam.*) one's humble home.

φτωχός *a.* poor.

φυγαδεύω *v.t.* help to escape.

φυγάς *s.m.f.* runaway, fugitive.

φυγή *s.f.* flight (*running away*).

φυγόδικος *s.m.f.* (*law*) defaulter.

φυγόκεντρος *a.* centrifugal.

φυγόμαχος *a.* unwilling to fight.

φυγόπονος *a.* unwilling to work; a shirker.

φυγόστρατος *a.* a shirker of military service.

φύκια *s.n.pl.* seaweed.

φυλάγ|ω *v.t.* guard, watch over; protect, look after; keep, set aside; lie in wait for. **~ομαι** *v.i.* be careful, take precautions; **~ομαι από** beware of.

φύλακας *s.m.* keeper, guard, watchman.

φυλακείο *s.n.* guard-house; watchman's hut.

φυλακ|ή *s.f.* prison. **~ίζω** *v.t.* imprison. **~ισμένος** *s.m.* prisoner.

φύλαξ *s.m.* see **φύλακας.**

φυλά(σσ)ω *v.t.* see **φυλάγω.**

φυλαχτό *s.n.* talisman.

φυλή *s.f.* race, tribe. **~αρχος** *s.m.* headman of tribe. **~εϊικός** *a.* racial, tribal.

φυλλάδα *s.f.* booklet.

φυλλάδιο *s.n.* pamphlet; part, instalment (*of book*).

φύλλο *s.n.* leaf; petal; sheet (*paper, metal, pastry*); page; leaf (*of door, table, shutter, etc.*); ~ πορείας marching orders; (*fam.*) playing-card, newspaper.

φυλλοκάρδια *s.n.pl.* depths of one's heart.

φυλλομετρώ *v.t.* leaf through (*book*).

φυλλοξήρα *s.f.* phylloxera.

φυλλορροώ *v.i.* shed leaves; (*fig.*) fade (*of hopes*).

φύλλωμα *s.n.* foliage.

φύλο *s.n.* 1. sex. 2. race.

φυματικός *a.* consumptive.

φυματίωσις *s.f.* tuberculosis.

φύομαι *v.i.* grow, be found (*of plants*).

φύρ|α *s.f.* shrinkage, loss of weight. **~αίνω** *v.i.* shrink, lose weight.

φύραμα *s.n.* paste; (*pej.*) stuff, quality (*of character*).

φύρδην *adv.* ~ μίγδην higgledy-piggledy.

φυρός *a.* deficient in weight; shrunken.

φυσαλ(λ)ίδα *s.f.* bubble; blister.

φυσαρμόνικα *s.f.* accordion; mouth-organ. (*in collision*) telescoped.

φυσέκι see **ΦΙΣΕΚΙ.**

φυσερό *s.n.* bellows; (*fam.*) (*hand*) fan.

φύση *s.f.* see **φύσις.**

φύσημα *s.n.* blowing; puff; (*fig.*) παίρνω ~ depart hurriedly.

φυσίγγι *s.n.* cartridge.

φύσιγξ *s.f.* ampoule (*for serum*).

φυσικ|ά *s.n.pl.*, **~ή** *s.f.* physics.

φυσικό *s.n.* nature, character.

φυσικ|ός *a.* natural; physical; a physicist. **~ά** *adv.* naturally. **~ότης** *s.f.* naturalness.

φυσιογνωμία *s.f.* cast of features, face; appearance; personality (*well-known person*).

φυσιοδίφης *s.m.* naturalist.

φυσιολάτρης *s.m.* lover of nature.

φυσιολογ|ία *s.f.* physiology. **~ικός** *a.* physiological; (*fig.*) normal, healthy.

φύσις *s.f.* nature; **~ει** by nature.

φυσομανώ *v.i.* rage.

φυσ|ώ *v.t. & i.* blow (out). (*v.i.*) puff, blow; **~άει** it is windy; (*fam.*) το ~άει he has money; το ~άει και δεν κρυώνει he cannot get over his chagrin.

φυτεία *s.f.* plantation; vegetation.

φυτ|εύω *v.t.* plant, implant. **~ε(υ)μα** *s.n.* planting. **~ευτός** *a.* cultivated (*of plants*).

φυτικός *a.* vegetable.

φυτοζωώ *v.i.* live penuriously, barely exist;

languish (*of trade, etc.*).
φυτοκομία *s.f.* horticulture.
φυτολογία *s.f.* botany.
φυτό *s.n.* plant.
φύτρα *s.f.* plumule; (*fig.*) lineage.
φυτρώνω *v.i.* grow, come up (*of plants*).
φυτώριο *s.n.* nursery, seed-bed.
φώκια *s.f.* seal (*animal*).
φωλ|εά, ~ιά *s.f.* nest, lair, burrow. **~ιάζω** *v.i.* nestle.
φωνάζω *v.i. & t.* shout, call (out), summon.
φωνακλάς *s.m.* one who talks loudly.
φωνασκώ *v.i.* bawl.
φωνή *s.f.* voice; cry, shout.
φωνήεν *s.n.* vowel.
φωνητικ|ός *a.* vocal; phonetic. **~ή** *s.f.* phonetics.
φωνοληψία *s.f.* recording (*mechanical*).
φωρατής *s.m.* (*radio*) detector.
φωρώμαι *v.i.* be caught in the act; be shown up.
φως *s.n.* light; (*fig.*) (*faculty of*) sight; ~ *φανάρι* clear as daylight. *φώτα* (*pl.*) lights; (*fig.*) knowledge, wisdom; Epiphany.
φωστήρ(ας) *s.m.* (*fig.*) luminary.
φωσφόρος *s.m.* phosphorus.
φωταγωγ|ώ *v.t.* illuminate. **~ός** *s.m.* light-well.
φωταέριο *s.n.* gas.
φωταψία *s.f.* illumination.
φωτεινός *a.* light, bright; lucid.
φωτιά *s.f.* fire; (*smoker's*) light; (*fig.*) great heat; anger; quickness; expensiveness.
φωτίζω *v.i. & t.* shine (on), give light (to), light; enlighten; ~*ει* dawn breaks.
φώτιση *s.f.* enlightenment.
φωτισ|μός *s.m.* lighting. **~τικός** *a.* of lighting.
φωτοαντίγραφο *s.n.* photocopy.
φωτοβολία *s.f.* radiation, shining.
φωτοβολίδα *s.f.* (*mil.*) flare.
φωτογενής *a.* photogenic.
φωτογραφία *s.f.* photograph; photography.
φωτογραφικ|ός *a.* photographic; ~*ή μηχανή* camera. **~ή** *s.f.* photography.
φωτογράφος *s.m.* photographer.
φωτόμετρο *s.n.* light-meter.
φωτοσκίαση *s.f.* shading.
φωτοστέφανος *s.m.* halo.
φωτοτυπία *s.f.* photostat.
φωτοχυσία *s.f.* flood of light.

X

χαβάγια *s.f.* Hawaiian-style song (*to guitar*).

χαβάνι *s.n.* brass mortar.
χαβάς *s.m.* tune.
χαβιάρι *s.n.* caviar.
χαβούζα *s.f.* cistern; rubbish-dump.
χάβρα *s.f.* synagogue; (*fig.*) uproar.
χαγιάτι *s.n.* upper gallery round courtyard.
χαδ- *see* **χαϊδ-**.
χαζεύω *v.i.* gape; idle away one's time.
χάζι *s.n.* pleasure, amusement; *το κάνω* ~ it amuses me.
χαζ|ός *a.* stupid. **~ομάρα** *s.f.* stupidity.
χαϊβάνι *s.n.* beast; (*fig.*) ass.
χάι|δεμα *s.n.* caressing.
χαϊδευτικός *a.* caressing; affectionate.
χαϊδ|εύω *v.t.* caress, stroke, pat; pet, spoil; cajole. **~εύομαι** *v.i.* desire caresses *or* attention; nuzzle, rub (*against*). **~εμένος** spoilt.
χάιδι *s.n.* caress; **~α** (*pl.*) petting, cajolery.
χαϊδιάρης *a.* fond of being petted.
χαϊδιάρικος *a.* caressing; fond of being petted.
χαϊδολογ|ώ *v.t.* caress repeatedly. **~ιέμαι** *v.i. see* **χαϊδεύομαι**.
χαίνω *v.i.* gape (*of wound, abyss, etc.*).
χαιρέκακος *a.* malevolent.
χαιρετ|ίζω, ~ώ *v.t.* greet; salute.
χαιρέτισ|μα, ~μός *s.m.* greeting, salutation. **~μός** (*ceremonial*) salute.
χαίρ|ω *v.i.* be glad; ~*ετε* greeting on arrival *or departure*; ~*ω άκρας υγείας* enjoy the best of health. **~ομαι** *v.i. & t.* be glad; enjoy; *έλα να χαρείς* do please come! *να το* ~*εσαι* I wish you joy of it.
χαίτη *s.f.* mane.
χακί *s.n.* khaki.
χαλάζι *s.n.* hail.
χαλαζίας *s.m.* quartz.
χαλάλι *adv.* (*fam.*) ~ *τα λεφτά* I do not grudge the money; ~ *σου* you are welcome to it, you deserve it. *οι κόποι μου πήγαν* ~ my efforts were not wasted.
χαλαρ|ός *a.* loose, slack. **~ώνω** *v.t.* loosen, relax.
χάλασμα *s.n.* destruction, demolition; a ruin.
χαλασμένος *a.* rotten, gone bad, decayed; broken, out of order; damaged, spoilt; seduced.
χαλασμός *s.m.* destruction; (*fig.*) ~ (*κόσμου*) disaster; great upheaval, storm, crowd *or* excitement.
χαλάστρα *s.f. μου έκαναν* ~ they spoilt my plants *or* cramped my style.
χαλβ|άς *s.m.* halva (*sweetmeat*); (*fam.*) weak-willed person. **~αδόπιττα** *s.f.* sort of nougat.
χαλεπός *a.* difficult.

χαλές *s.m.* privy; (*fam.*) dirty dog.
χάλ|ι *s.n.*, **~ια** (*pl.*) wretched *or* disreputable state.
χαλί *s.n.* carpet, rug.
χαλίκι *s.n.* gravel.
χαλιμ|ά *s.f.* παραμύθια της **~άς** Arabian Nights.
χαλιν|άρι *s.n.*, **~ός** *s.m.* bridle, bit. **-αγωγώ** *v.t.* lead by bridle; (*fig.*) curb.
χαλίφης *s.m.* caliph.
χαλκάς *s.m.* ring, link.
χαλκείο *s.n.* coppersmith's forge; (*fig.*) place where lies are concocted.
χαλκεύω *v.t.* make out of copper; (*fig.*) concoct (*lies, etc.*).
χαλκιάς *s.m.* coppersmith.
χάλκ|ινος, **~ούς** *a.* copper.
χαλκογραφία *s.f.* copperplate engraving.
χαλκομανία *s.f.* transfer (*design*).
χαλκός *s.m.* copper; Εποχή του Χαλκού Bronze Age.
χάλκωμα *s.n.* copper utensil. **~τάς** *s.m.* coppersmith.
χαλνώ *v.t. & i. see* **χαλώ.**
χαλύβδινος *a.* steel.
χαλυβώνω *v.t.* steel; steel-plate.
χάλυ|ψ, **~βας** *s.m.* steel.
χαλ|ώ *v.t. & i.* spoil; break (*put out of order*); break off (*agreement*); spend (*money*); wear out; reduce quality of; pull down (*building*); undo (*coiffure, knitting, etc.*); change (*banknote*); seduce; **~ώ** τον κόσμο kick up a row, move heaven and earth. (*v.i.*) deteriorate, spoil; go bad; get out of order; lose one's looks; **~άει** ο κόσμος there is a great to-do; δεν **~ασε** ο κόσμος things are not so serious.
χαμάδα *s.f.* ripe olive fallen from tree.
χαμαί *adv.* to *or* on the ground.
χαμαιλέων *s.m.* chameleon.
χαμαιτυπείον *s.n.* brothel.
χαμάλ|ης *s.m.* porter; (*fam.*) lout. **~ίκι** *s.n.* toilsome labour.
χαμάμι *s.n.* Turkish bath.
χαμέν|ος *a.* lost; the loser (*at cards*); scatterbrained, good-for-nothing; **~ο** κορμί wastrel; τα έχω **~α** be confused *or* at a loss *or* weak in the head; πήγε **~ο** *or* στα **~α** it was a wasted effort.
χαμερπής *a.* servile, crawling.
χαμηλός *a.* low.
χαμηλοφώνως *adv.* in a low voice.
χαμηλώνω *v.t. & i.* lower, bring down; become lower.
χαμίνι *s.n.* street urchin.
χαμόγελο *s.n.* smile. **~ώ** *v.i.* smile.
χαμό|δεντρο, **~κλαδο** *s.n.* shrub.
χαμο|μήλι, **~μηλο** *s.n.* camomile (tea).

χαμός *s.m.* loss.
χάμου *adv. see* **χάμω.**
χάμουρα *s.n.pl.* harness.
χαμπάρ|ι *s.n.* (piece of) news; παίρνω **~ι** get wind of, notice. **~ίζω** *v.t. & i.* take notice of; (*with* από) know *or* understand about.
χάμω *adv.* to *or* on the ground.
χάνδαξ *s.m.* ditch, trench.
χάνι *s.n.* country inn; caravanserai.
χανούμ(ισσα) *s.f.* Turkish lady.
χανσενικός *s.m.* leper.
χαντάκ|ι *s.n.* ditch, trench. **~ώνω** *v.t.* (*fig.*) ruin (effect of).
χαντζάρι *s.n.* short sword.
χάντρα *s.f.* bead.
χάν|ω *v.t. & i.* lose; miss (*opportunity*); τα **~ω** get confused *or* embarrassed, lose one's head *or* wits; **~ω** τα νερά μου feel a fish out of water. **~ομαι** *v.i.* get lost, lose one's way, disappear; bother, waste one's time; (*fam.*) να χαθείς to hell with you! (*also jocular*); χάσου out of my sight! (*see* **χαμένος**).
χάος *s.n.* chaos.
χάπι *s.n.* pill.
χαρ|ά *s.f.* joy, pleasure; **~ά** θεού a delight; (*fam.*) wedding; μια **~ά** very well, splendidly; γεια **~ά** greeting; **~ά** στο πράμα (*ironic*) that's nothing. στις **~ές** σας to your marriage! παιδική **~ά** children's playground.
χάραγμα *s.n.* engraving, incision; ruling (*of lines*).
χαράδρα *s.f.* ravine.
χαράζ|ω 1. *v.t. see* **χαράσσω.** 2. *v.i.* **~ει** dawn breaks.
χάρακας *s.m.* ruler (*implement*).
χαράκι *s.n.* ruled line (*in copy-book, etc.*).
χαρακιά *s.f.* scratch, mark (*on surface*); groove, rifling; (*ruled*) line.
χαρακτήρ|ας *s.m.* character. **~ίζω** *v.t.* characterize, qualify.
χαρακτηριστικ|ός *a.* characteristic. **~ά** *s.n.pl.* (*person's*) features.
χαράκ|της *s.m.* engraver. **~ική** *s.f.* (*art of*) engraving.
χαράκωμα *s.n.* ruling (*of lines*); trench.
χαρακών|ω *v.t.* rule (*lines*). **~ομαι** *v.i.* take shelter.
χάρ|αμα *s.n.* **~άματα** (*pl.*) dawn.
χαραμάδα *s.f.* chink, crack.
χαραματιά *s.f.* chink, crack; scratch, mark (*on surface*).
χαράμ|ι *adv.* (*fam.*) to no purpose. **~ίζω** *v.t.* waste, spend in vain.
χάραξη *s.f.* engraving, incision; marking out (*of road, etc.*).

χαράσσω *v.t.* engrave; rule (*lines*); score; write, trace; mark out (*road, etc.*).

χαράτσ|ι *s.n.* poll-tax (*under Turks*); (*fig.*) burdensome levy. **~ώνω** *v.t.* (*fam.*) extract contribution from.

χαραυγή *s.f.* daybreak.

χάρβαλο *s.n.* a ruin.

χαρέμι *s.n.* harem.

χάρη *s.f.* charm, attractiveness; good point, advantage; favour, service; gratitude; ~ *για ~ μου* for my sake; *~σε σένα* thanks to you; *λόγου* ~ for example; *αφήνω (μια)* ~ allow for a bit extra; *see* **χάρις**.

χαρίεις *a.* charming.

χαριεντίζομαι *v.i.* be in flirtatious *or* bantering mood; turn on the charm.

χαρίζ|ω *v.t.* give, make a present of; remit (*penalty, debt*); (*fam.*) *δεν ~ω κάστανα* stand no nonsense; (*v.i.*) *σου ~ει* it flatters you (*of dress*). *~ομαι v.i.* (*with σε*) treat with favour *or* indulgence.

χάρ|ις *s.f.* grace, charm; pardon; *~ις εις* thanks to; *~ιν (with gen.)* for (the sake of); *παραδείγματος ~ιν* for example; *προς ~ιν σου* for your sake.

χάρισμα *s.n.* gift; accomplishment; (*adv.*) for nothing, free.

χαριστικ|ός *a.* partial (*biased*); *~ή βολή* coup de grâce.

χαριτόβρυτος *a.* full of charm.

χαριτολογώ *v.i.* say witty things.

χαριτωμένος *a.* charming.

χάρμα *s.n.* (*source of*) delight, joy; ~ *ιδέσθαι* a joy to behold.

χαρμάνι *s.n.* blend.

χαρμόσυνος *a.* glad, joyful (*news, bells, etc.*).

χαροκόπος *a.* pleasure-loving.

χάρος *s.m.* Death.

χαρούμενος *a.* merry, happy.

χαρούπι *s.n.* carob.

χαρταετός *s.m.* kite.

χαρτένιος *a.* paper.

χαρτζιλίκι *s.n.* pocket-money.

χάρτης *s.m.* paper; map; ~ *υγείας* toilet-paper.

χαρτί *s.n.* paper; (*playing*) card; (*fam.*) card-playing; ~ *και καλαμάρι* word for word, in detail.

χάρτινος *a.* paper.

χαρτόδετος *a.* paper-backed.

χαρτοκλέφτης *s.m.* card-sharper.

χαρτοκόπτης *s.m.* paper-knife.

χαρτόμουτρο *s.n.* (*fam.*) addict of cards.

χαρτόνι *s.n.* cardboard.

χαρτονόμισμα *s.n.* paper currency; banknote.

χαρτο|παίκτης *s.m.* gambler. **~παιξία** *s.f.* gambling.

χαρτοπόλεμος *s.m.* (battle of) confetti.

χαρτοπώλης *s.m.* stationer.

χαρτ|ορρίχτρα, **~ού** *s.f.* fortune-teller (*from cards*).

χαρτόσημο *s.n.* stamp (*on document*); stamped paper.

χαρτοφύλακας *s.m.* brief-case.

χαρτοφυλάκιον *s.n.* brief-case; (*fig.*) portfolio (*ministry*).

χαρτωσιά *s.f.* hand, trick (*at cards*); (*fam.*) *δεν πιάνει* ~ *μπροστά σου* he cannot be compared with you.

χαρωπός *a.* cheerful, gay.

χασάπ|ης *s.m.* butcher. **~ικο** *s.n.* butcher's shop. **~ικος** *s.m.* sort of cyclic dance.

χασές *s.m.* cotton fabric.

χάση *s.f.* waning (*of moon*); *στη* ~ *και στη φέξη* every now and then.

χάσιμο *s.n.* loss.

χασίσι *s.n.* hashish.

χάσκω *v.i.* gape (open).

χάσμα *s.n.* chasm; gap, void.

χασμουρητό *s.n.* yawn, yawning.

χασμ|ουριέμαι, **~ώμαι** *v.i.* yawn.

χασμωδία *s.f.* hiatus.

χασομερ|ώ *v.i. & t.* waste time, loiter; cause to waste time. **~ι** *s.n.* delay, idling. **~ης** *a.* dawdler.

χασούρα *s.f.* loss.

χαστούκι *s.n.* slap (*on face*).

χατζής *s.m.* hadji, pilgrim.

χατίρι *s.n.* favour, service; *για το ~ του* for his sake.

χαυλιόδοντας *s.m.* tusk.

χαύνος *a.* languid.

χαφιές *s.m.* (*fam.*) nark, informer.

χάφτω *v.t.* gulp down; (*fig.*) swallow, believe.

χάχαν|ο *s.n.* noisy laugh. **~ίζω** *v.i.* cackle with laughter.

χάχας *s.m.* simpleton.

χα χα χα *int. representing laughter.*

χαχόλικος *a.* (*fam.*) sloppy, ill-fitting (*of clothes*).

χαψιά *s.f.* gulp, mouthful.

χαώδης *a.* chaotic.

χέζ|ω *v.i. & t.* (*fam.*) defecate (on); (*fig.*) consign to the devil. **~ομαι** *v.i.* (*fig.*) be in a funk; *χέστηκα* I could not care less.

χείλι *s.n.* lip.

χειλικός *a.* labial.

χείλος *s.n.* lip; rim (*of vessel*); (*fig.*) brink.

χειμαδιό *s.n.* winter quarters (*for sheep*).

χειμάζομαι *v.i.* suffer hardships (of winter).

χείμαρρος *s.m.* torrent (-bed).

χειμερινός *a.* winter.

χειμών|(ας) *s.m.* winter. **~ιάζει** *v.i.* winter

comes on. **~ιάτικος** *a.* winter; wintry.
χειρ *s.f.* hand.
χειραγωγώ *v.t.* lead by the hand.
χειράμαξα *s.f.* hand-barrow.
χειραφετώ *v.t.* manumit; emancipate.
χειραψία *s.f.* handshake.
χειρίζομαι *v.t.* handle, manage, operate.
χειρισμός *s.m.* handling, management; (*act of*) manoeuvring.
χειριστήρια *s.n.pl.* controls (*of machine*).
χειριστής *s.m.* operator.
χείριστος *a.* worst.
χειροβομβίδα *s.f.* hand-grenade.
χειρόγραφο *s.n.* manuscript.
χειροδικία *s.f.* taking law into one's own hands.
χειροκίνητος *a.* worked by hand.
χειροκροτ|ώ *v.t. & i.* applaud, clap. **~ημα** *s.n.* applause.
χειρόκτιον *s.n.* glove, gauntlet.
χειρολαβή *s.f.* handrail; helve.
χειρόμακτρον *s.n.* towel, napkin.
χειρομαντεία *s.f.* palmistry.
χειρονομ|ία *s.f.* gesture; gesticulation; *να λείπουν οι ~ίες* paws off!
χειροπέδη *s.f.* handcuff.
χειροπιαστός *a.* palpable.
χειροπόδαρα *adv.* hand and foot.
χειροποίητος *a.* hand-made.
χειρότερ|ος *a.* worse; *ο ~ος* worst. **~α** *adv.* worse. **~εύω** *v.t. & i.* make *or* become worse.
χειροτεχνία *s.f.* handicraft.
χειροτονία *s.f.* ordination.
χειρουργικ|ός *a.* surgical. **~ή** *s.f.* surgery.
χειρουργός *s.m.* surgeon.
χειροφίλημα *s.n.* hand-kissing.
χειρόφρενο *s.n.* handbrake.
χειρωνακτικός *a.* manual.
χέλι *s.n.* eel.
χελιδ|ών *s.f.,* **~όνι** *s.n.* swallow.
χελών|α, ~η *s.f.* tortoise, turtle.
χέρι *s.n.* hand; arm; handle; coat (*of paint, etc.*), going-over, treatment (*washing, wiping, etc.*); *~ ~* hand-in-hand; *~ (με) ~* quickly, direct; *βάζω ένα ~* lend a hand; *βάζω ~* lay hands (*on*), fondle (*amorously*); *βάζω στο ~* trick into giving; *από πρώτο ~* at first hand; *έρχομαι στα ~α* come to blows.
χερικό *s.n. κάνω ~* make a start; *βάζω ~* lay hands on, start abusing; *καλό ~* good luck.
χερόβολο *s.n.* sheaf, armful.
χερούλι *s.n.* handle.
χερσαίος *a.* living on land; continental (*climate*).
χερσόνησος *s.f.* peninsula.
χέρσος *a.* uncultivated, fallow.

χέσιμο *s.n.* (*fam.*) defecation; (*fig.*) funk.
χηλή *s.f.* cloven hoof.
χημ|εία *s.f.* chemistry. **~ικός** *a.* chemical; a chemist (*scientist*).
χήνα *s.f.* goose.
χήρ|α *s.f.* widow. **~εύω** *v.i.* be widowed; (*fig.*) be vacant (*of post*).
χήρος *s.m.* widower.
χθες *adv.* yesterday.
χθεσινός *a.* yesterday's; recent.
χθόνιος *a.* infernal (*gods*).
χι *s.n.* the letter **Χ.**
χιαστί *adv.* crosswise.
χίλια *num.* a thousand.
χιλιάρ|α, ~ικη *s.f.* bottle of $3^1/_2$ kilos capacity.
χιλιάρικο *s.n.* thousand-drachma, *etc.*, note.
χιλι|άς, ~άδα *s.f.* thousand.
χιλιαστής *s.m.* Adventist.
χιλιετία *s.f.* millennium.
χιλιόγραμμο *s.n.* kilogram.
χίλι|οι *a.* a thousand; (*fig.*) *~α δύο* a hundred and one.
χιλιόμετρο *s.n.* kilometre.
χιλιοστημόριο *s.n.* thousandth part; insignificant amount.
χιλιοστ|ό, ~όμετρο *s.n.* millimetre.
χιλιοστός *a.* thousandth.
χίμαιρα *s.f.* chimera.
χιμπαντζής *s.m.* chimpanzee.
χινόπωρο *s.n.* autumn.
χιονάτος *a.* show-white.
χιόν|ι *s.n.* snow. **~ιά** *s.f.* snowy weather; sowball. **~ίζει** *v.i.* it snows.
χιονίστρα *s.f.* chilblain.
χιονοδρομία *s.f.* skiing.
χιονόλασπη *s.f.* slush.
χιονόνερο *s.n.* sleet.
χιονοστιβάδα *s.f.* avalanche.
χιούμορ *s.n.* humour. [E.]. **~ιστικός** *a.* humorous.
χιτών *s.m.* robe, tunic; (*anat.*) cornea.
χιτώνιον *s.n.* (*mil.*) tunic.
χιών *s.f.* snow.
χλαίν|α, ~η *s.f.* (*mil.*) greatcoat.
χλαμύς *s.f.* mantle.
χλευ|άζω *v.t.* mock, deride. **~αστικός** *a.* mocking, derisive. **~ασμός** *s.m.* derision.
χλιαίνω *v.t. & i.* make *or* become tepid.
χλιαρός *a.* tepid.
χλιδή *s.f.* luxury.
χλιμιντρ|ίζω, ~ώ *v.i.* neigh, whinny.
χλό|η *s.f.* grass, lawn; verdure. **~ερός** *a.* verdant.
χλωμ|ός *a.* pale, pallid, wan. **~άδα** *s.f.* pallor. **~ιάζω** *v.i.* grow pale.
χλώριον *s.n.* chlorine.
χλωρίς *s.f.* flora.

χλωρός *a.* green, freshly cut; freshly made (*cheese*).
χλωροφόρμιον *s.n.* chloroform.
χλωροφύλλη *s.f.* chlorophyll.
χνότα *s.n.pl.* breath.
χνούδ|ι *s.n.* down, fluff, pile. **~ωτός** *a.* downy, plushy.
χοάνη *s.f.* crucible; funnel, horn.
χόβολη *s.f.* embers.
χοιρίδιο *s.m.* young pig.
χοιρινό *s.n.* pork.
χοιρομέρι *s.n.* ham.
χοίρος *s.m.* pig.
χολέρα *s.f.* cholera.
χολ|ή *s.f.* bile. **~ιάζω** *v.t. & i.* annoy; get annoyed.
χολόλιθος *s.m.* gall-stone.
χολοσκάνω *v.t. & i.* upset, exasperate; get upset *or* exasperated.
χολώ *v.t.* anger.
χονδρέμπορος *s.m.* wholesaler.
χονδρικός *a.* wholesale.
χονδροειδής *a.* coarse, clumsy, boorish.
χόνδρος *s.m.* gristle, cartilage.
χονδρός *a.* thick, coarse, fat; unrefined, vulgar, rude, gross.
χοντραίνω *v.t. & i.* make *or* become thicker *or* fatter; *τα* ~ use offensive language.
χοντροκέφαλος *a.* pig-headed; thick-headed.
χοντροκοπιά *s.f.* clumsy piece of work.
χοντρομπαλάς *s.m.* (*fam.*) very fat man.
χοντρόπετσος *a.* thick-skinned.
χόντρος *s.n.* thickness, fatness.
χοντρός *a.* see **χονδρός**.
χορδ|ή *s.f.* (*mus.*) string; (*anat.*) cord. *θίγω την εναίσθητη* ~*ή* touch the right chord. **~ίζω** *v.t.* tune, wind up.
χορεία *s.f.* group, body.
χορευτ|ής *s.m.,* **~ρια** *s.f.* dancer. **~ικός** *a.* with *or* for dancing.
χορεύω *v.i. & t.* dance; dance with; *τον* ~ *στο ταψί* make him pay dearly; (*fig.*) (*v.i.*) shake, toss, tremble.
χορηγ|ώ *v.t.* provide, grant. **~ία** *s.f.,* **~ημα** *s.n.* grant, subsidy. **~ηση** *s.f.* provision, granting. **~ός** *s.m.* giver, provider.
χορογραφία *s.f.* choreography.
χοροδιδάσκαλος *s.m.* dancing-master.
χοροεσπερίς *s.f.* ball, dance.
χοροπηδώ *v.i.* gambol, jump about.
χορός *s.m.* dance; chorus, choir.
χορταίνω *v.i. & t.* have one's fill (of); get tired of; satisfy, provide abundantly *or* excessively with.
χορτάρι *s.n.* grass. **~άζω** *v.i.* get covered with grass.
χορταρικό *s.n.* vegetable.
χορτασμός *s.m.* satiety.

χορταστικός *a.* filling, substantial.
χορτάτος *a.* satisfied (*having had enough*).
χόρτ|ο *s.n.* grass; **~α** (*pl.*) green vegetables (*esp. dandelions*).
χορτοφάγος *a.* vegetarian.
χορωδία *s.f.* chorus, choir.
χότζας *s.m.* Moslem priest *or* teacher.
χουβαρντάς *s.m.* (*fam.*) liberal *or* generous person.
χουγιάζω *v.t.* drive (*flocks*) by shooing; shout at.
χουζούρι *s.n.* ease, idleness.
χούι *s.n.* (bad) habit; knack.
χουλιάρι *s.n.* spoon; (*fam.*) a gossip.
χουνί *s.n. see* **χωνί**.
χούντα *s.f.* junta.
χουρμάς *s.m.* date (*fruit*).
χους *s.m.* dust, earth.
χούφτ|α *s.f.* palm of hand; handful. **~ιά** *s.f.* handful. **~ιάζω, ~ώνω** *v.t.* grab.
χούφταλο *s.n.* (*fam.*) very old person.
χουχουλίζω *v.t.* blow on (*for warmth*).
χοχλακίζω *v.i.* boil, bubble.
χράμι *s.n.* hand-woven rug *or* blanket.
χρεία *s.f.* necessity, need; (*fam.*) privy.
χρειάζομαι *v.t. & i.* need, require; be necessary; (*fam.*) *τα* ~ become scared.
χρειώδης *a.* necessary.
χρεμετίζω *v.i.* neigh, whinny.
χρεόγραφον *s.n.* (*fin.*) bond, security.
χρεοκοπ|ία *s.f.* bankruptcy. **~ώ** *v.i.* go bankrupt, fail.
χρεολυσία *s.f.* amortization.
χρέος *s.n.* debt; obligation.
χρεοστάσιον *s.n.* moratorium.
χρεών|ω *v.t.* debit. **~ομαι** *v.i.* incur debts.
χρέω|σις *s.f.* debit(ing). **~στικός** *a.* debit.
χρεώστης *s.m.* debtor.
χρεωστώ *v.i. & t. see* **χρωστώ**.
χρήμα *s.n.,* **~τα** (*pl.*) money.
χρηματ|ίζω *v.i.* ε–*ισε δήμαρχος* he served as mayor. **~ίζομαι** *v.i.* take bribes.
χρηματικός *a.* of money, pecuniary.
χρηματιστήριον *s.n.* stock exchange.
χρηματιστής *s.m.* (*fin.*) broker.
χρηματοδοτώ *v.t.* finance.
χρηματοκιβώτιο *s.n.* safe.
χρησιμεύω *v.i.* be of use, serve.
χρησιμοποι|ώ *v.t.* use. **~ησις** *s.f.* use, utilization.
χρήσιμ|ος *a.* useful. **~ότις** *s.f.* use, usefulness.
χρήσ|ις *s.f.* use, enjoyment, usage; *εν* ~*ει* in use; *προς* ~*ιν* for the use (*of*); (*fin.*) financial year.
χρησμός *s.m.* oracle.
χρηστοήθης *a.* moral, upright.
χρηστομάθεια *s.f.* anthology of edifying

passages.
χρηστός *a.* good (*virtuous*).
χρίσμα *s.n.* chrism; anointing; (*fig.*) adoption as party candidate.
χριστιανικός *a.* Christian.
χριστιανισμός *s.m.* Christianity.
χριστιανός *s.m.* Christian; (*fam.*) chap (*esp. in expostulation*).
χριστιανοσύνη *s.f.* Christendom.
Χριστός *s.m.* Christ.
Χριστούγεννα *s.n.pl.* Christmas.
χρίω *v.t.* anoint; plaster.
χροιά *s.f.* complexion, colour, shade.
χρονιά *s.f.* year.
χρονιάζω *v.i.* take ages; become one year old *or* dead.
χρονιάτικο *s.n.* a year's payment.
χρονίζω *v.i.* take a long time, drag on; become chronic.
χρονικογράφος *s.m.* chronicler.
χρονικ|όν *s.n.* chronicle; ~ά (*pl.*) annals.
χρονικός *a.* of time, temporal.
χρόνιος *a.* of ancient origin; chronic.
χρονογράφημα *s.n.* topical comment (*in newspaper*).
χρονογράφος *s.m.* columnist.
χρονολογ|ία *s.f.* date; chronology; era. ~ούμαι *v.i.* date (*from*).
χρονόμετρο *s.n.* chronometer; metronome.
χρονοντούλαπο *s.n.* βάζω στο ~ shelve, pigeon-hole.
χρόν|ος *s.m.* 1. (*pl.* οι ~οι) time (*duration*); (*gram.*) tense, quantity. 2. (*pl.* οι ~οι, τα ~ια) year; προ ~ών years ago; του ~ου next year; ~ια πολλά many happy returns! κακό ~ο να 'χει bad luck to him! πόσω(ν) ~ων είναι how old is he?
χρονοτριβώ *v.i.* waste time, dawdle.
χρυσαλλίς *s.f.* chrysalis.
χρυσάνθεμο *s.n.* chrysanthemum.
χρυσάφ|ι *s.n.* gold. ~ής *a.* the colour of gold. ~ικά *s.n.pl.* jewellery.
χρυσή *s.f.* (*med.*) jaundice.
χρυσίζω *v.i.* shine like gold.
χρυσικός *s.m.* goldsmith.
χρυσίον *s.n.* money.
χρυσοθήρας *s.m.* gold-digger.
χρυσόμαλλ|ος *a.* golden-haired; ~ον δέρας golden fleece.
χρυσός 1. *s.m.* gold. 2. *a.* gold, golden; (*fig.*) good-hearted, lovable.
χρυσούς *a.* gold, golden.
χρυσοχό|ος *s.m.* goldsmith; jeweller. ~είον *s.n.* jeweller's.
χρυσόψαρο *s.n.* goldfish.
χρυσώνω *v.t.* gild.
χρυσωρυχείον *s.n.* gold-mine.
χρώμα *s.n.* colour; paint; suit (*in cards*).

χρωματίζω *v.t.* colour, paint; (*fig.*) give colour *or* expression to; attach (*political*) label to.
χρωματικός *a.* (*mus.*) chromatic.
χρωματισμός *s.m.* colouring, colour.
χρωματιστός *a.* coloured.
χρώμιον *s.n.* chromium.
χρωστήρ *s.m.* paint-brush.
χρωστικός *a.* colouring.
χρωστώ *v.i. & t.* be in debt, owe; be obliged, must; (*fam.*) ~ της Μιχαλούς be dotty.
χταπόδι *s.n.* octopus.
χτένι *s.n.* comb; rake.
χτενίζω *v.t.* comb; dress (*person's*) hair; (*fig.*) polish up (*speech, etc.*).
χτένισμα *s.n.* combing; hairstyle; (*fig.*) polishing up.
χτες *adv.* yesterday.
χτεσινός *a.* yesterday's; recent.
χτήμα *s.n.* (*landed*) property.
χτίζω *v.t.* build; found (*city*); wall in.
χτικιό *s.n.* (*fam.*) tuberculosis; (*fig.*) torment, trial.
χτίριο *s.n.* building.
χτίσιμο *s.n.* (*act of*) building.
χτίστης *s.m.* bricklayer.
χτύπημα *s.n.* blow, stroke, knock; bruise; shock; (*act of*) beating *or* knocking.
χτυπητό *s.n.* roughcast.
χτυπητός *a.* beaten (*egg, etc.*); gaudy, loud; pointed (*hint, etc.*).
χτυποκάρδι *s.n.* throbbing of the heart.
χτύπος *s.m.* knock, beat, tick.
χτυπ|ώ *v.t. & i.* knock (at), beat, strike, ring; clap (*hands*), stamp (*feet*); beat (*time*); beat up (*eggs*); attack; hit, wound; kill (*game*); bid for (*at auction*). (*v.i.*) strike, ring, sound (*of clock, bell, etc.*); beat, throb; knock against something, hurt oneself; ~άει άσχημα it looks *or* sounds bad; ~άει η πόρτα there is a knock at the door; μου ~άει στα νεύρα it gets on my nerves. ~ιέμαι *v.i.* beat one's breast; come to blows.
χυδαΐζω *v.i.* be vulgar; use extreme form of demotic Gk.
χυδαιολογώ *v.i.* speak vulgarly.
χυδαί|ος *a.* vulgar, crude. ~ότης *s.f.* vulgarity.
χυλόπιτ(τ)α *s.f.* sort of macaroni; (*fam.*) τρώω τη ~ be rejected (*of suitor*).
χυλ|ός *s.m.* pap, chyle. ~ώνω *v.i.* form creamy consistency.
χύμα *s.n.* (*used as a. & adv.*) in confusion; loose (*not in package*); on draught; (*fam.*) του τα είπα ~ I told him straight out.
χυμ|ός *s.m.* sap, juice. ~ώδης *a.* juicy.
χυμ|ίζω, ~ώ *v.i.* rush, swoop.

χύν|ω *v.t.* pour (out), shed, spill; cast (*metal*); ~ω εξάνθημα come out in rash; ~ω λάδι στη φωτιά add fuel to the flames. ~ομαι *v.i.* pour (out), spill over; flow out; (*fig.*) rush, swoop.

χύσιμο *s.n.* pouring, shedding, spilling; casting.

χυτήριον *s.n.* foundry.

χυτός *a.* cast (*of metal*); scattered; loose (*of hair*); (*fig.*) well-proportioned (*of limbs*); fitting like a glove (*of clothes*).

χυτοσίδηρος *s.m.* cast iron.

χύτρα *s.f.* earthen *or* metal cooking-pot. ~ ταχύτητος pressure-cooker.

χωλ *s.n.* hall. [E. *hall*]

χωλ|ός *a.* lame. ~**αίνω** *v.i.* be lame, limp; (*fig.*) make slow progress.

χώμα *s.n.* soil, earth; dust; ground. ~**τινος**, ~**τένιος** *a.* earthen, clay.

χώνευση *s.f.* digestion; smelting.

χωνευτήριον *s.n.* crucible, melting-pot.

χωνευτικός *a.* digestive.

χωνευτός *a.* concealed (*of wiring, plumbing*).

χωνεύω *v.t. & i.* digest; smelt; (*fam.*) δεν τον ~ I cannot stand him. (*v.i.*) be half spent *or* decomposed (*of fire, food being cooked, etc.*).

χων|ί *s.n.* funnel, horn. ~**άκι** *s.n.* cornet (*paper bag, ice-cream*).

χών|ω *v.t.* thrust, stuff, stick (*into*); bury; hide. ~ομαι *v.i.* squeeze in; hide; (*fam.*) stick one's nose in.

χώρα *s.f.* country, land; chief town *or* village of locality; (*anat.*) region; λαμβάνω ~ν take place.

χωρατ|ό *s.n.* joke, pleasantry. ~**ατζής** *s.m.* joker. ~**εύω** *v.i.* joke.

χωράφι *s.n.* field.

χωρητικότης *s.f.* capacity; tonnage.

χώρια *adv.* separately, apart; (*prep.*) apart from, not counting.

χωριανός *a.* (fellow) villager.

χωριάτης *s.m.* peasant, countryman; (*fig.*) boor.

χωριάτικος *a.* peasant, country; boorish.

χωρίζ|ω *v.t. & i.* separate, part, divide; divorce. (*v.i.*) get divorced; ~ω με part from. ~ομαι (*v.t.*) part from. (*v.i.*) part.

χωρικ|ό 1. *a.* village, country. 2. *s.m.* peasant, villager. 3. *a.* ~ά ύδατα territorial waters.

χωρι|ό *s.n.* village; (*fam.*) home-town; κάνουμε ~ό we get on well together; γίναμε από δυο ~ά we quarrelled.

χωρίο(ν) *s.n.* 1. village. 2. passage (*in book*).

χωρίς *adv. & prep.* (*with acc.*) without; not counting; ~ άλλο without fail; (*fam.*) με ~ without.

χώρισμα *s.n.* sorting, division (*act*); partition (*structure*), compartment, pigeonhole, cubicle.

χωρισμένος *a.* divided; divorced.

χωρισμός *s.m.* separation, parting; division; sorting; divorce.

χωριστά *adv.* separately, apart; (*prep.*) not counting.

χωριστικός *a.* separatist.

χωριστός *a.* separate, apart.

χωρίστρα *s.f.* parting (*of hair*).

χώρος *s.m.* space, room.

χωροταξία *s.f.* land-planning.

χωροφύλ|ακας *s.m.* gendarme. ~**ακή** *s.f.* gendarmerie.

χωρώ *v.i.* advance.

χωρ|ώ, ~**άω** *v.i. & t.* fit *or* go (*into*), find room enough; δεν ~εί αμφιβολία there is no room for doubt; δεν μου ~άει το καπέλο the hat is too small for me. (*v.t.*) hold, contain, have room for; δεν το ~άει ο νους μου it passes my understanding.

χώσιμο *s.n.* thrusting, sticking (in).

χωστός *a.* sunk, deep-set.

Ψ

ψάθ|α *s.f.* straw, cane; rush mat; (*wide-brimmed*) straw hat; (*fig.*) στην ~α in penury. ~**άκι** *s.n.* (*man's*) boater. ~**ινος** *a.* straw.

ψαλίδα *s.f.* 1. shears, 2. tendril. 3. earwig.

ψαλίδ|ι *s.n.* scissors; curling-tongs. ~**ίζω** *v.t.* cut, clip. ~**ωτός** *a.* swallow-tailed.

ψάλλω *v.t. & i.* sing, chant.

ψαλμ|ός *s.m.* psalm. ~**ωδία** *s.f.* (chanting of) psalms; (*fam.*) grumbling.

ψαλτήριον *s.n.* psalter.

ψάλτης. *s.m.* chorister, singer.

ψαμμίασις *s.f.* (*med.*) gravel.

ψάξιμο *s.n.* searching.

ψαράδικ|ος *a.* fisherman's. ~**ο** *s.n.* fishing-boat; fishmonger's.

ψαραίνω *v.i.* turn grey.

ψαράς *s.m.* fisherman; fishmonger.

ψάρε(υ)μα *s.n.* fishing.

ψαρεύω *v.i. & t.* fish (for); sound, pump (*for information*).

ψαρής *a.* grey-haired; a grey (*horse*).

ψάρι *s.n.* fish.

ψαρόβαρκα *s.f.* fishing-boat.

ψαροκόκκαλο *s.n.* fish-bone; herring-bone (*pattern*).

ψαρόκολλα *s.f.* fish-glue.

ψαρονέφρι *s.n.* tenderloin (*meat*).

ψαρόνι *s.n.* starling.
ψαροπούλα *s.f.* fishing-boat.
ψαρός *a.* grey, grizzled.
ψαύσις *s.f.* touching, feeling (*action*).
ψαύω *v.t.* touch, feel.
ψαχνό *s.n.* lean meat (*without bone*); (*fam.*) έρχομαι στο ~ come to the point. βαράτε στο ~ shoot to kill!
ψάχν|ω *v.t. & i.* search (for). ~ομαι *v.i.* search one's pockets.
ψαχουλεύω *v.i.* search, grope.
ψέγ|ω *v.t.* blame, reprehend. ~άδι *s.n.* fault, defect. ~αδιάζω *v.t.* find fault with.
ψείρ|α *s.f.* louse. ~ιάζω *v.i.* become lousy. ~ίζω *v.t.* rid of lice; (*fam.*) rob.
ψεκ|άζω *v.t.* spray. ~αστήρας *s.m.* spray, vaporizer.
ψελλίζω *v.i. & t.* stammer.
ψέλνω *v.t.* sing (*hymns*); (*fam.*) τον έψαλα I scolded him.
ψέμα *s.n.* lie; (*fam*) τελείωσαν (*or* σώθηκαν) τα ~τα things are beginning to move in earnest; με τα ~τα with very small outlay (*but to good effect*).
ψες *adv.* yesterday.
ψευδαίσθηση *s.f.* illusion.
ψευδάργυρος *s.m.* zinc.
ψευδής *a.* false; artificial, sham.
ψευδίζω *v.i.* lisp; have speech defect.
ψευδολογώ *v.i.* tell lies.
ψεύδομαι *v.i.* lie.
ψευδομάρτ|υς, ~υρας *s.m.f.* false witness. ~υρώ *v.i.* give false evidence.
ψευδορκ|ία *s.f.* perjury. ~ώ *v.i.* commit perjury.
ψεύδος *s.n.* lie, falsehood.
ψευδός *a.* lisping.
ψευδώνυμο *s.n.* pseudonym.
ψεύμα *s.n.* lie.
ψεύ(σ)της *s.m.* liar.
ψευτιά *s.f.* lie, falsehood.
ψευτίζω *v.t. & i.* lower quality of (*goods*); decline in quality.
ψεύτικος *a.* false, sham, artificial; inferior (*in quality*).
ψευτοδουλειά *s.f.* shoddy job.
ψευτο|ζώ, ~περνώ *v.i.* scrape a living.
ψευτοφυλλάδα *s.f.* lying newspaper; (*fig.*). liar.
ψήγμα *s.n.* particle; filings, dust (*of metal*).
ψήκτρα *s.f.* brush.
ψηλαφ|ώ *v.t. & i.* feel, finger; grope (for); feel one's way. ~ητός *a.* palpable; (*fig.*) obvious. ~ιστά *adv.* gropingly.
ψηλομύτης *a.* supercilious.
ψηλ|ός *a.* high, tall; ~ό καπέλο top-hat. ~ά *adv.* high (up).
ψήλωμα *s.n.* high ground, elevation; growing taller.
ψηλώνω *v.t. & i.* make *or* grow higher *or* taller.
ψημένος *a.* cooked, roasted; hardened, seasoned, matured; (*fam.*) in the know.
ψήν|ω *v.t.* bake, roast; cook; (*fig.*) torment; (*fam.*) persuade; τα ψήσανε they are having an affair. ~ομαι *v.i.* (*fig.*) get very hot; ripen; gain experience, become seasoned.
ψησιά *s.f.* potful, panful (*in cooking*).
ψήσιμο *s.n.* baking, roasting; cooking.
ψησταριά *s.f.* apparatus for barbecue.
ψηστικά *s.n.pl.* fee for cooking in public oven.
ψητ|ός *a.* roast, baked, grilled. ~ό *s.n.* roast *or* grilled meat; (*fig.*) heart of the matter.
ψηφιακός *a.* digital.
ψηφιδωτό *s.n.* mosaic.
ψηφίζω *v.i. & t.* vote (for).
ψηφίο *s.n.* (*numerical*) figure; digit; (*alphabetic*) letter.
ψήφισμα *s.n.* resolution (*by vote*).
ψηφοδέλτιο *s.n.* ballot-paper.
ψηφοδόχος *s.f.* ballot-box.
ψηφοθηρώ *v.i.* solicit votes.
ψήφος *s.m.f.* vote; αναλογική ~ proportional representation; μέλαινα ~ black ball.
ψηφοφορία *s.f.* voting; θέτω εις ~ν put to the vote.
ψηφοφόρος *s.m.f.* voter.
ψηφώ *v.t.* heed, respect.
ψι *s.n.* the letter **Ψ**.
ψίαθος *s.m.* straw, cane.
ψίδι *s.n.* vamp (*of shoe*).
ψίθυρ|ος *s.m.* whisper, murmur. ~ίζω *v.i. & t.* whisper, murmur. ~ισμα *s.n.*, ~ισμός *s.m.* whispering, murmuring. ~ιστά *adv.* in a whisper.
ψιλά *s.n.pl.* (small) change; (*fig.*) money.
ψιλαίνω *v.t.* make thinner; raise pitch of (*voice*).
ψιλή *s.f.* (*gram.*) smooth breathing.
ψιλικ|ά *s.n.pl.* haberdashery; small wares. ~ατζήδικο *s.m.* haberdasher's.
ψιλικό *s.n.* (*fam.*) money.
ψιλολογώ *v.t.* examine minutely.
ψιλορωτώ *v.t.* ask (*person*) many details.
ψιλός *a.* thin, fine; ~ά γράμματα small print; shrill; light (*armour*).
ψιλοτραγουδώ *v.i. & t.* hum.
ψιμάρι *s.n.* late-born lamb; (*fig.*) easy prey.
ψιμύθιον *s.n.* make-up (*cosmetic*).
ψιττακ|ός *s.m.* parrot. ~ίζω *v.t.* parrot.
ψίχα *s.f.* crumb (*soft part of loaf*); edible part of nut; pith; (*fig.*) morsel.
ψιχάλα *s.f.* drizzle; rain-drop. ~ίζει *v.i.* it drizzles.

ψίχ|αλο, ~ουλο *s.n.* crumb.
ψιχίον *s.n.* crumb.
ψιψίνα *s.f.* (*fam.*) pussy-cat.
ψιψιρίζω *v.t.* examine minutely.
ψόγος *s.m.* blame, reprehension.
ψουν- *see* ψων-.
ψοφίμι *s.n.* putrefying carcass; ~α (*pl.*) carrion.
ψόφιος *a.* dead (*of animals*); (*fig.*) exhausted; without vitality, spineless; ~ στην κούραση dog-tired.
ψοφολογώ *v.i.* (*fam.*) (*pej.*) sleep.
ψόφος *s.m.* 1. dull noise. 2. death; (*fam.*) biting cold.
ψοφώ *v.i.* die (*of animals*); (*fig.*) ~ για long for, be mad on.
ψυγείο *s.n.* refrigerator; radiator (*of motor*).
ψυκτικός *a.* cooling, refrigerating.
ψυλλιάζ|ω *v.i.* & *t.* get fleas; make suspicious. ~ομαι get suspicious.
ψύλλ|ος *s.m.* flea; (*fam.*) για ~ου πήδημα at the least provocation, for the merest trifle. καλιγώνει τον ~ο there are no flies on him.
ψύξη *s.f.* refrigeration; chill.
ψυχαγωγ|ία *s.f.* recreation, diversion. ~ικός *a.* recreational.
ψυχανάλυσις *s.f.* psycho-analysis.
ψυχαρισμός *s.m.* linguistic doctrine of Psycharis (1854-1929), *viz.* advocacy of extreme demotic Gk.
ψυχή *s.f.* soul; heart; energy, spirit, courage; butterfly; μου έβγαλε την~ he exasperated *or* exhausted me.
ψυχίατρ|ος *s.m.* psychiatrist. ~είον *s.n.* mental hospital. ~ική *s.f.* psychiatry.
ψυχικάρης *a.* compassionate.
ψυχικό *s.n.* (*act of*) charity; alms.
ψυχικ|ός *a.* psychical; psychic; ~ή διάθεση humour; ~ός κόσμος one's moral resources; ~ή οδύνη (*law*) mental distress (*as ground for damages*).
ψυχογιός *s.m.* adopted son; boy (*servant*).
ψυχοκόρη *s.f.* adopted daughter.
ψυχολόγ|ος *s.m.* psychologist. ~ία *s.f.* psychology. ~ικός *a.* psychological.
ψυχολογώ *v.t.* read soul *or* mind of; ~ημένος done with psychological insight.
ψυχομαχώ *v.i.* be in the throes of death.
ψυχοπαθής *a.* a psychopath.
ψυχοπαίδι *s.n.* adopted child.
ψυχοπόνια *s.f.* compassion.
ψυχόρμητον *s.n.* instinct.
ψυχορραγώ *v.i.* be in the throes of death.
ψύχος *s.n.* cold.
ψυχοσάββατο *s.n.* All Souls' Day.
ψυχοσύνθεσις *s.f.* one's psychological make-up.

ψύχρα *s.f.* chilly weather.
ψύχραιμ|ος *a.* cool, composed. ~ία *s.f.* sangfroid, composure.
ψυχραίν|ω *v.i.* & *t.* get cold (*of weather*); (*v.t.*) cool, make cold. ~ομαι *v.i.* cool off (*of feelings*).
ψυχρόαιμος *a.* cold-blooded (*fish, etc.*).
ψυχρολουσία *s.f.* cold shower *or* douche; (*fam.*) dressing-down.
ψυχρ|ός *a.* cold. ~ότης *s.f.* coldness.
ψυχρούλα *s.f.* freshness (*of weather*).
ψύχω *v.t.* freeze, cool, make cold.
ψυχωμένος *a.* mettlesome.
ψύχωσις *s.f.* psychosis; (*fam.*) mania, complex.
ψυχωφελής *a.* edifying.
ψωμ|άς *s.m.* baker. ~άδικο *s.n.* bakery.
ψωμ|ί *s.n.* bread; loaf. ~άκι *s.n.* roll *or* piece of bread.
ψωμοζήτης *s.m.* mendicant.
ψωμοζώ *v.i.* eke out one's existence.
ψωμοτύρι *s.n.* bread and cheese.
ψωμώνω *v.i.* develop well, fill out.
ψωνίζω *v.t.* & *i.* buy; do shopping; (*fam.*) get hold of, pick up (*person*); catch (*cold, etc.*); την ~ become queer *or* crazy.
ψώνι|ο *s.n.* thing bought; ~α (*pl.*) purchases, shopping; (*fam.*) έχω ~ο have a craze (*for*); είναι ~ο he is a queer fish.
ψώρα *s.f.* scabies, mange; (*fig.*) a pest (*person*). ~λέος *a.* mangy; wretched.
ψωριάζω *v.i.* become mangy.
ψωριάρης *a.* mangy; (*fig.*) skinny, poverty-stricken, beggarly.
ψωρίασις *s.f.* (*med.*) psoriasis.
ψωριώ *v.i.* have scabies *or* mange.
ψωροκώσταινα *s.f.* (*fam.*) poverty-stricken Greece.
ψωροπερηφάνεια *s.f.* shabby gentility.

Ω

ω *int.* of surprise, alarm, etc.
ωάριον *s.n.* ovum.
ώδε *adv.* 1. thus. 2. here.
ωδείον *s.n.* school of music.
ωδή *s.f.* ode.
ωδικ|ός *a.* singing. ~ή *s.f.* singing (lesson).
ωδίνες *s.f.pl.* pangs of childbirth.
ωθ|ώ *v.t.* push, impel. ~ηση *s.f.* pushing, impulsion.
ωιμέ(να) *int.* alas!
ωκεανός *s.m.* ocean.
ωκύπους *a.* fleet-footed.
ωλένη *s.f.* (*anat.*) forearm.
ωμέγα *s.n.* the letter Ω.

ωμοπλάτη *s.f.* shoulder-blade.

ώμορφος *a. see* όμορφος.

ώμός *s.m.* shoulder.

ωμός *a.* 1. raw. 2. hard, cruel. 3. slow, spineless (*person*).

ωμότητα *s.f.* cruelty; ~ητες (*pl.*) atrocities.

ων *v.i.* (*participle*) being.

ώνια *s.n.pl.* provisions, goods.

ωοειδής *a.* oval.

ωοθήκη *s.f.* ovary.

ωόν *s.n.* egg.

ωοτόκος *a.* oviparous.

ώρ|α *s.f.* hour; time; με την ~α by the hour; ~α με την ~α at any moment; ~ες ~ες now and then; με τις ~ες for hours; ήρθε στην (*or* με την) ~α του he came punctually *or* (*ironic*) late; είναι με τις ~ες του he has his moods; είναι στην ~α της she is about to give birth; (*πάνω*) στην ~α at the very moment, just at the right (*or* wrong) moment; της ~ας fresh, (cooked) to order; της κακιάς ~ας of poor quality.

ωραία *adv.* beautifully, well; (*int.*) good, very well.

ωραιοπαθής *a.* in love with beauty.

ωραί|ος *a.* beautiful, handsome; fine, good. ~ότης *s.f.* beauty.

ωράριο *s.n.* (*employee's*) hours of work.

ωριαίος *a.* hourly; lasting an hour.

ωριμάζω *v.i.* ripen, mature.

ωρίμανσις *s.f.* ripening.

ώριμ|ος *a.* ripe, mature. ~ότης *s.f.* ripeness, maturity.

ώριος *a.* beautiful.

ωρισμέν|ος *a.* fixed, determined; certain (*unspecified*). ~ως *adv.* definitely.

ωροδείκτης *s.m.* hour-hand.

ωρολόγιον *s.n.* clock, watch; ηλιακόν ~ sundial; time-table (*of work*); (*eccl.*) breviary.

ωρολογοποιός *s.m.* watchmaker.

ωροσκόπιο *s.n.* horoscope.

ωρυγή *s.f.* howl, howling.

ωρύομαι *v.i.* howl.

ώς 1. *prep.* (*with acc.*) until, up to, as far as. 2. *conj.* ~ ότου (*να*) until. 3. *adv.* (*with numbers*) about.

ως 1. *adv.* as, like, such as. 2. *conj.* as, while, as soon as. 3. *see* ώς.

ωσάν *adv.* like, as, as if.

ωσαύτως *adv.* also, likewise.

ωσότου *conj.* until, by the time.

ώσπερ *adv.* like, as.

ώσπου *conj.* until, by the time.

ώστε *conj.* so, accordingly, consequently; that (*as result*); ούτως ~ in such a way as to.

ωστόσο *adv.* yet, however; all the same.

ωτακουστ|ώ *v.i.* eavesdrop. ~ής *s.m.* eavesdropper.

ωτίον *s.n.* ear.

ωτομοτρίς *s.f.* rail motor-car. [F. *automotrice*]

ωτορινολαρυγγολόγος *s.m.* ear-nose-and-throat specialist.

ωτοστόπ *s.n.* hitch-hiking.

ωφέλεια *s.f.* benefit, good, profit (*general*); είδα ~ I benefited.

ωφέλημα *s.n.* benefit, good, profit (*particular*).

ωφελιμισ|μός *s.m.* utilitarianism. ~τικός *a.* utilitarian.

ωφέλιμος *a.* beneficial, useful, of service.

ωφελ|ώ *v.t.* benefit, do good to, help. ~ούμαι *v.i.* benefit, profit.

ωχ, ώχου *int. of pain or distress.*

ώχρα *s.f.* ochre.

ωχριώ *v.i.* become pale; (*fig.*) pale (*to insignificance*).

ωχρ|ός *a.* pale, pallid; (*fig.*) faint, indistinct. ~ότης *s.f.* pallor.

ENGLISH–GREEK

a, an *article* ένας *m.*, μία *f.*, ένα *n.* I have ~ house έχω (ένα) σπίτι. I haven't got ~ house δεν έχω σπίτι. she is ~ nurse είναι νοσοκόμα he was ~ friend of mine ήταν φίλος μου. half ~ hour μισή ώρα. twice ~ week δύο φορές την εβδομάδα. £1 ~ metre μία λίρα το μέτρο. what ~ lovely day! τι ωραία μέρα!

aback *adv.* take ~ ξαφνιάζω.

abacus *s.* άβαξ. *m.*

abandon *v.* εγκαταλείπω. **~ed** (*deserted*) εγκαταλελειμμένος, (*depraved*) ακόλαστος. **~ment** *s.* εγκατάλειψη *f.*

abase *v.* ταπεινώνω. **~ment** *s.* ταπείνωση *f.*

abash *v.* feel **~ed** τα χάνω, ντρέπομαι.

abate *v.i.* κοπάζω. (*v.t.*) ελαττώνω, (*law*) ακυρώ. **~ment** *s.* ελάττωση *f.*, ύφεση *f.*

abattoir *s.* σφαγείο *n.*

abbess *s.* ηγουμένη *f.*

abbey *s.* μοναστήρι *n.*, αββαείο *n.*

abbot *s.* ηγούμενος *m.*

abbreviat|e *v.* συντομεύω, συντέμνω. **~ion** *s.* συντόμευση *f.*, σύντμηση *f.* (*form, of word*) συντετμημένη λέξη.

ABC *s.* αλφαβήτα *f.* (*fig.*, *rudiments*) στοιχειώδεις βάσεις

abdicat|e *v.t.* παραιτώ. (*v.i.*) παραιτούμαι. **~ion** *s.* παραίτηση *f.*

abdom|en *s.* κοιλιά *f.* **~inal** *a.* κοιλιακός.

abduct *v.* απάγω. **~ion** *s.* απαγωγή *f.* **~or** *s.* απαγωγέας *m.*

abed *adv.* στο κρεββάτι.

aberration *s.* παρεκτροπή *f.*

abet *v.* υποβοηθώ.

abeyance *s.* be in ~ (*custom, law*) δεν ισχύω, (*question*) εκκρεμώ.

abhor *v.* αποτροπιάζομαι. **~rence** *s.* αποτροπιασμός *m.* **~rent** *a.* αποτροπιαστικός.

abide *v.* (*dwell*) διαμένω, (*last*) διαρκώ, (*bear*) ανέχομαι. I can't ~ him (*fam.*) δεν τον χωνεύω. ~ by (*keep*) τηρώ, ~ by the consequences υφίσταμαι τις συνέπειες.

abiding *a.* διαρκής, μόνιμος.

ability *s.* (*power*) δυνατότης *f.* (*competence*) ικανότης *f.* (*talent*) ταλέντο *n.*

abject *a.* άθλιος, (*servile*) δουλοπρεπής.

abjure *v.* απαρνούμαι.

ablative *s.* (*gram.*) αφαιρετική *f.*

ablaze *a.* φλεγόμενος, (*fig.*) λάμπων. be ~ φλέγομαι. the garden is ~ with colour ο

κήπος είναι πνιγμένος στο χρώμα.

able *a.* (*capable*) ικανός, άξιος. be ~ μπορώ, δύναμαι. **~-bodied** *a.* εύρωστος.

ablution *s.* perform one's ~s νίπτομαι.

abnegation *s.* αυταπάρνηση *f.*

abnormal *a.* αφύσικος, ανώμαλος. **~ity** *s.* ανωμαλία *f.* **~ly** *adv.* αφύσικα. the weather is ~ly hot κάνει αφύσικη ζέστη.

aboard *adv. & prep.* ~ the ship επί του πλοίου, μέσα στο πλοίο. go ~ επιβιβάζομαι. go ~ the ship επιβαίνω του πλοίου, μπαίνω στο πλοίο.

abode *s.* κατοικία *f.*

abol|ish *v.* καταργώ. **~ition** *s.* κατάργηση *f.*

abominab|le *a.* βδελυρός. (*fam., very nasty*) απαίσιος. **~ly** *adv.* απαίσια.

abominat|e *v.* απεχθάνομαι. **~ion** *s.* (*feeling*) απέχθεια *f.* (*thing*) *use* abominable.

aborigin|al *a.* αυτόχθων. **~es** *s.* αυτόχθονες *m.pl.*

abort *v.i.* αποβάλλω, κάνω αποβολή. **~ion** *s.* (*act*) έκτρωση *f.* (*fig.*) (*ugly person*) έκτρωμα *n.*, (*person or thing*) εξάμβλωμα *n.* **~ive** *a.* be ~ive αποτυγχάνω.

abound *v.i.* (*be plentiful*) αφθονώ. ~ in (*teem with*) βρίθω (with gen.), είμαι γεμάτος (*with acc. or* από *& acc.*).

about *prep. & adv.* 1. (*prep.*) (*in, on*) γύρω σε. the books were scattered ~ the room τα βιβλία ήταν σκορπισμένα γύρω στο δωμάτιο. have you any money ~ you? έχεις λεφτά απάνω σου; (*concerning*) για (*with acc.*), περί (*with gen.*). what is it ~ ? περί τίνος πρόκειται; how ~ a drink? τι θα έλεγες για ένα ποτό; yes, you can go, but what ~ us? καλά, σεις μπορείτε να πάτε, αλλά εμείς; 2. (*adv.*) (*approximately*) περίπου, πάνω κάτω. he is ~ my age είναι περίπου (*or* πάνω κάτω) στην ηλικία μου. (*of time*) ~ 10 o'clock γύρω στις δέκα, κατά (*or* περί) τις δέκα. it's ~ time καιρός είναι. 3. be ~ to μέλλω να (*formal*). I was ~ to ring you when you turned up πήγαινα να σου τηλεφωνήσω όταν παρουσιάστηκες. just as he was ~ to become a minister he died πάνω που θα γινότανε υπουργός πέθανε. 4. (*here and there*) γύρω, leave rubbish ~ αφήνω σκουπίδια γύρω. the news is going ~ κυκλοφορούν τα νέα. there is a lot of flu ~ σέρνεται γρίπη. 5. come ~ (*v.i.*) συμβαίνω. bring ~ (*v.t.*)

προξενώ. ~ turn (s. & int.) μεταβολή. f.

above adv. & prep. 1. (adv.) απάνω, (ε)πάνω, the floor ~ το επάνω πάτωμα. the ~ statement η ανωτέρω (or παραπάνω) δήλωση. as was stated ~ ως ανεφέρθη ανωτέρω. from ~ άνωθεν. ~- mentioned προαναφερθείς. 2. (prep.) πάνω από (with acc.), υπεράνω (with gen.). ~ all πάνω απ' όλα, προ παντός, προ πάντων. he is ~ all suspicion είναι υπεράνω πάσης υποψίας. ~ average άνω του μετρίου. he got ~ himself (fam.) πήραν τα μυαλά του αέρα.

above-board a. τίμιος.

abras|ion s. εκδορά f. **~ive** a. λειαντικός, (fig.) τραχύς.

abreast adv. δίπλα-δίπλα. (naut.) in line ~ κατά μετωπικήν γραμμήν. keep ~ of the times είμαι συγχρονισμένος.

abridge v. συντέμνω. **~ment** s. (of book) συντετμημένη έκδοσις.

abroad adv. έξω, στο εξωτερικό. go (to live) ~ ξενιτεύομαι, αποδημώ. spread or noise ~ διαδίδω.

abrogat|e v. ακυρώ. **~ion** s. ακύρωση f.

abrupt a. απότομος. **~ly** adv. απότομα. **~ness** s. (of manner) σκαιότητα f.

abscess s. απόστημα n.

abscond v. φεύγω κρυφά, (fam.) το σκάω. (law) φυγοδικώ.

absence s. (being away) απουσία f. (lack) έλλειψη f. ~ of mind αφηρημάδα f. leave of ~ άδεια f.

absent a. απών. be ~ (away) απουσιάζω, (lacking) λείπω. **~ee** s. απών, απουσιάζων a.

absent-minded α. αφηρημένος. **~ness** s. αφηρημάδα f.

absolute a. απόλυτος. **~ly** adv. απολύτως.

absolution s. άφεσιν αμαρτιών.

absolutism s. απολυταρχία f.

absolve v. απαλλάσσω.

absorb v. απορροφώ. **~ent** a. απορροφητικός. **~ing** a. ενδιαφέρων.

absorption s. απορρόφηση f.

abstain v. ~ from απέχω (with gen.), αποφεύγω (with acc. or να).

abstemious a. (person) εγκρατής, (in eating) λιτοδίαιτος. (meal) λιτός. **~ly** adv. λιτά.

abstention s. αποχή f.

abstinence s. εγκράτεια f., αποχή f. (with από).

abstract a. αφηρημένος.

abstract s. περίληψη f.

abstract v. βγάζω. (filch) σουφρώνω.

abstracted a. αφηρημένος, become ~ βυθίζομαι.

abstraction s. (idea) αφηρημένη έννοια.

(state of mind) αφηρημάδα f.

abstruse a. δυσνόητος.

absurd a. παράλογος, γελοίος. **~ity** s. ανοησία f. **~ly** adv. γελοίως.

abundance s. αφθονία f.

abundant a. άφθονος. **~ly** adv. άφθονα.

abuse v. (υ)βρίζω. (misuse) καταχρώμαι.

abuse s. βρισιές, ύβρεις f.pl. (misuse) κατάχρηση f.

abusive a. υβριστικός. **~ly** adv. υβριστικά.

abut v. ~ on (of land) συνορεύω με, (of building) ακουμπώ σε.

abysmal a. (fam.) απερίγραπτος. **~ly** adv. απεριγράπτως.

abyss s. άβυσσος f.

acacia s. ακακία f., γαζία f.

academic a., **~ian** s. ακαδημαϊκός.

academy s. ακαδημία f.

acanthus s. άκανθα f. (archit.) άκανθος f.

accede v.i. ~ to (join) προσχωρώ εις, εντάσσομαι εις (assent) συναινώ εις. ~ to throne ανεβαίνω στο θρόνο.

accelerat|e v.t. επιταχύνω. (v.i.) επιταχύνομαι. **~ion** s. επιτάχυνση f. **~or** s. επιταχυντής m. (fam.) γκάζι n.

accent s. (stress, mark) τόνος m. (pronunciation) προφορά f. (v.) τονίζω. **~ual** a. τονικός.

accentuat|e v. τονίζω. **~ion** s. τονισμός m.

accept v. (απο)δέχομαι. be ~ed (agreed to) γίνομαι δεκτός.

acceptable a. (welcome) ευπρόσδεκτος, (allowable) παραδεκτός.

accept|ance s. **~ation** s. αποδοχή f.

accepted a. (usual) παραδεδεγμένος, (admitted) δεκτός.

access s. πρόσβαση f. (entrance) είσοδος f. (free use) ελευθέρα χρήσις. give ~ to (place) οδηγώ σε. obtain ~ to (person) γίνομαι δεκτός από. enjoy free ~ έχω το ελεύθερο (with σε or να). difficult of ~ δυσπρόσιτος. (attack) παροξυσμός m.

accessary s. συνεργός m.

accessible a. (place) προσιτός, (person) ευπρόσιτος.

accession s. (joining) προσχώρηση f., ένταξη f. (to throne) ανάρρηση f. (thing added) προσθήκη f.

accessory a. συμπληρωματικός. (s., adjunct) εξάρτημα n., (women's) αξεσουάρ n.

accidence s. (gram.) τυπικόν n.

accident s. (chance event) σύμπτωση f. (mishap) ατύχημα n., (serious) δυστύχημα n. (collision) σύγκρουση f. by ~ κατά τύχην.

accidental a. τυχαίος. **~ly** adv. τυχαίως, κατά τύχην.

accidented a. ανώμαλος.

acclaim ν. ζητωκραυγάζω. (*declare*) ανακηρύσσω. (*approve of*) επιδοκιμάζω.

acclamation s. ζητωκραυγές *f.pl.*

acclimatiz|e ν. εγκλιματίζω. **~ation** s. εγκλιμάτιση *f.*

accommodat|e ν. (*hold*) παίρνω, (*put*, *lodge*) βάζω, τακτοποιώ. (*put up*) βολεύω, φιλοξενώ, (*fix up*) εξοικονομώ, (*oblige*) διευκολύνω. (*adapt*) προσαρμόζω. **~ing** *a.* βολικός.

accommodation s. (*lodging*) κατάλυμα *n.*, they found no ~ δεν βρήκαν πού να μείνουν. (*adaptation*) προσαρμογή *f.* (*compromise*) συμβιβασμός *m.*

accompany ν. συνοδεύω.

accompan|iment s. (*consequence*) επακόλουθο *n.* (*trimming*) γαρνιτούρα *f.* (*mus.*, *etc.*) συνοδεία *f.* **~ist** s. ακομπανιατέρ *m.*

accomplice s. συνεργός *m.*

acomplish ν. εκτελώ, (*bring off*) επιτυγχάνω.

accomplished *a.* (*socially*) με πολλά χαρίσματα. an ~ pianist σπουδαίος πιανίστας. ~ fact τετελεσμένο γεγονός.

accomplishment s. (*doing*) εκτέλεση *f.* (*feat*) κατόρθωμα *n.* (*gift*) χάρισμα *n.*

accord ν.*t.* (*give*) παρέχω. (*v.i.*) (*agree*) συμφωνώ.

accord s. ομοφωνία *f.* with one ~ ομοφώνως. of his own ~ αφ' εαυτού, μόνος του.

accordance s. in ~ with σύμφωνα με.

according *adv.* ~ as καθόσον. ~ to κατά, σύμφωνα με (*with acc.*).

accordingly *adv.* (*as may be required*) αναλόγως. (*therefore*) επομένως.

accordion s. ακορντεόν *n.*, (*fam.*) φυσαρμόνικα *f.*

accost ν. πλευρίζω, διπλαρώνω.

account s. (*fin.*) λογαριασμός *m.*, keep ~s κρατώ βιβλία. put it on my ~ γράψτε το. (*estimation*) take ~ of υπολογίζω, παίρνω υπόψη. it is of no ~ δεν έχει σημασία. (*reason*) λόγος *m.* on ~ of λόγω *or* ένεκα (*with gen.*), για *or* διά (*with acc.*). on that ~ γι' αυτό, on no ~ επ' ουδενί λόγω. (*profit*) on my own ~ για λογαριασμό μου. turn to good ~ επωφελούμαι (*with gen. or* από). (*description*) περιγραφή *f.*

account ν. (*deem*) θεωρώ, λογαριάζω. ~ for (*answer for*) δίδω λόγον (*with gen.*), (*explain*) εξηγώ.

accountable *a.* υπεύθυνος, υπόλογος.

accountan|cy s. λογιστική *f.* **~t** s. λογιστής *m.*

accoutrements s. εξοπλισμός *m.*

accredit ν. διαπιστεύω, **~ed** *a.* (*person*) διαπεπιστευμένος, (*idea*) παραδεδεγμένος.

accretion s. επαύξηση *f.*

accrue ν. προκύπτω.

accumulat|e ν.*t.* συσσωρεύω. (*v.i.*) συσσωρεύομαι. **~ion** s. (*act*) συσσώρευση *f.* (*mass*) σωρός *m.* **~or** s. συσσωρευτής *m.*

accuracy s. ακρίβεια *f.*

accurate *a.* ακριβής. **~ly** *adv.* ακριβώς.

accursed *a.* καταραμένος.

accusation s. κατηγορία *f.*

accusative s. (*gram.*) αιτιατική *f.*

accuse ν. κατηγορώ. **~d** s. (*law*) κατηγορούμενος *m.* **~r** s. κατηγορών *m.*, (*esp. law*) κατήγορος *m.*

accustom ν.*t.* συνηθίζω (*also* be *or* get **~ed** *or* ~ oneself to). **~ed** *a.* συνηθισμένος.

ace s. άσσος *m.* he was (*or* is) within an ~ of winning παρά τρίχα να κερδίσει.

acerbity s. δριμύτητα *m.*

acetic *a.* οξικός.

acetylene s. ασετυλίνη *f.*

ach|e s. πόνος *m.* (*v.*) πονώ. be **~ing for** λαχταρώ (*with acc. or* να).

achieve ν. κατορθώνω, πετυχαίνω. **~ment** s. (*success*) κατόρθωμα *n.* (*fulfilment*) επίτευξη *f.* (*thing achieved*) επίτευγμα *n.*

Achilles s. ~' heel η αχίλλειος πτέρνα.

acid s. οξύ *n.* citric ~ ξινό *n.* (*a.*) ξινός. **~ity** *a.* οξύτητα *f.*, ξινίλα *f.*

acknowledge ν. (*recognize*) αναγνωρίζω, (*admit*) παραδέχομαι. ~ receipt βεβαιώ λήψιν. **~ment** s. αναγνώριση *f.* (*reply*) απάντηση *f.*

acme s. ακμή *f.*, αποκορύφωμα *n.*

acne s. ακμή *f.*

acolyte s. (*assistant*) βοηθός *m.*

acorn s. βελανίδι *n.*

acoustic *a.* ακουστικός. **~s** s. ακουστική *f.*

acquaint ν. ειδοποιώ. be *or* get **~ed with** (*person*) γνωρίζω. I am **~ed** with the case έχω γνώσιν της υποθέσεως. they got **~ed** γνωρίστηκαν.

acquaintance s. γνωριμία *f.* (*knowledge*) γνώση *f.* an ~ of mine γνωστός (*or* γνώριμος *or* γνωριμία) μου. make ~ of *see* acquaint.

acquiesce ν. συγκατατίθεμαι. **~nce** s. συγκατάθεση *f.*

acquire ν. αποκτώ. **~d** *a.* επίκτητος. it is an ~d taste πρέπει να το συνηθίσει κανείς. **~ment** s. (*skill*) προσόν *n.*

acquisition s. (*getting*) απόκτηση *f.* (*thing got*) απόκτημα *n.*

acquisitive *a.* φιλοκτήμων.

acquit ν. αθωώνω. he **~ted** himself bravely φέρθηκε γενναία. **~tal** s. αθώωση *f.*

acre s. τέσσερα στρέμματα. (*fig.*) **~s** στρέμματα *n.pl.*

acrid *a.* δριμύς.

acrimon|y s. δριμύτητα f. **~ious** a. δριμύς.f
acrobat s. ακροβάτης m. **~ic** a. ακρο-
βατικός. **~ics** s. ακροβασία f.
acropolis s. ακρόπολη f.
across prep. & adv. 1. (beyond) πέρα από
(with acc.), πέραν (with gen.). ~ the river
the forest begins πέρα από το ποτάμι (or
πέραν του ποταμού) το δάσος αρχίζει. he
swam ~ the river πέρασε το ποτάμι
κολυμπώντας. a bridge ~ the Thames
γεφύρι πάνω στον Τάμεση. the house ~
the street το απέναντι σπίτι. 2. (adv.)
απέναντι. send (or take) ~ (v.t.) περνώ
απέναντι. go ~ (v.t.&i.) περνώ, διαβαίνω.
I came ~ a very interesting book έπεσε στα
χέρια μου ένα πολύ ενδιαφέρον βιβλίο. I
came ~ somebody I hadn't seen for a long
time βρέθηκε μπροστά μου κάποιος που
είχα να δω πολύν καιρό.
acrostic s. ακροστιχίδα f.
act s. πράξη f. ~ of Parliament νομοθέτημα
n. (music-hall, etc.) νούμερο n. as I was in
the ~ of phoning you απάνω που σου
τηλεφωνούσα. (caught) in the ~ (a.) αυτό-
φωρος, (adv.) επ' αυτοφώρω, στα πράσα.
act v.i. (take action) πράττω, ενεργώ.
(function, go) λειτουργώ, δουλεύω.
(pretend) προσποιούμαι, υποκρίνομαι. ~
on (affect) επιδρώ (or επενεργώ) επί (with
gen.). ~ as (deputize for) αναπληρώνω.
(v.t.) (a play, role) παίζω, παριστάνω, (a
role) υποδύομαι, υποκρίνομαι.
acting s. (performance) παίξιμο n. (art,
profession) θέατρο n. (a.) (as deputy)
αναπληρωματικός.
action s. (general) ενέργεια f., δράση f.
(deed) πράξη f., έργο n. (battle) μάχη f.
(bodily movements) κινήσεις f.pl. (of
machine) μηχανισμός m. (of drug) ενέργεια f. man of ~ άνθρωπος της δράσης.
line of ~ γραμμή f., τακτική f. go into ~
(act) ενεργώ, (set to work) καταπιάνομαι
με, (join, battle) συνάπτω μάχην. put into
~ βάζω σε ενέργεια, θέτω εις πράξιν. put
out of ~ αχρηστεύω, (damage) προκαλώ
βλάβη σε, (mil.) θέτω εκτός μάχης. what ~
will the government take? τι μέτρα θα
πάρει (or πώς θα ενεργήσει) η κυβέρνηση;
(law) bring an ~ against κάνω αγωγή
εναντίον (with gen.).
actionable a. (law) υποκείμενος εις ποι-
νικήν αγωγήν.
activate v. δραστηριοποιώ.
active a. ενεργητικός, δραστήριος. be ~ or
take ~ part in ασχολούμαι ενεργώς με. he
was ~ during the revolution έδρασε κατά
την επανάσταση. (gram.) ~ voice ενερ-
γητική φωνή.

activit|y s. δραστηριότητα f., ενέργεια f.
(social, business) κίνηση f. **~ies**
απασχολήσεις f. pl.
act|or s. ηθοποιός m., θεατρίνος m. **~ress**
s. ηθοποιός f., θεατρίνα f.
actual a. πραγματικός. **~ly** adv. πραγματικά, συγκεκριμένως.
actuary s. εμπειρογνώμων ασφαλίσεων.
actuate v. (machine) κινώ, (person) κινώ,
ωθώ.
acuity s. οξύτητα f.
acumen s. οξύνοια f.
acupuncture s. βελονισμός m.
acute a. οξύς. (clear-sighted) διορατικός,
(great) μεγάλος, έντονος, (clever) έξυπνος. ~ angle οξεία γωνία. ~ accent
οξεία f.
AD μετά Χριστόν.
adage s. παροιμία f.
Adam s. ~'s apple μήλο του Αδάμ.
adamant a. άκαμπτος.
adapt v. προσαρμόζω, (book, music) διασκευάζω. **~ation** s. προσαρμογή f.
διασκευή f. **~or** s. προσαρμογέας m.
adaptab|le a. (person) προσαρμόσιμος.
~ility s. ικανότητα προσαρμογής.
add v. προσθέτω. ~ up αθροίζω. ~ up to
ανέρχομαι εις. ~ to (increase) αυξάνω.
adder s. οχιά f.
addict s. drug ~ τοξικομανής. coffee or
tobacco ~ θεριακλής m. **~ed** a. be ~ed to
έχω μανία με. ~ed to drink έκδοτος εις την
μέθην. **~ion** s. μανία f. drug ~ion τοξι-
κομανία f.
addition s. (process) πρόσθεση f. (thing
added) προσθήκη f. in ~ επί πλέον, κοντά
στα άλλα.
additional a. πρόσθετος. **~ly** adv. επιπλέον.
addled a. κλούβιος. make or become ~
κλουβιαίνω.
address s. (speech) προσφώνηση f.
(petition) αίτηση f. (residence) διεύθυνση f.
address v. (words) αποτείνω. (deliver ~ to)
προσφωνώ. (speak to) αποτείνομαι προς
(with acc.). ~ oneself to (task) επιδίδομαι
εις. (send) απευθύνω. **~ee** s. παραλήπτης m.
adept a. ειδήμων.
adequa|cy s. επάρκεια f. **~ate** a. επαρκής,
αρκετός, (suitable) κατάλληλος. **~ately**
adv. αρκετά, καταλλήλως.
adhere v. (stick) κολλώ. ~ to (abide by)
εμμένω εις.
adheren|ce s. εμμονή f. **~t** s. οπαδός m.f.
adhes|ion s. προσκόλληση f. **~ive** s.
κολλητική ουσία.
adjacent a. παρακείμενος.
adjective s. επίθετο n.
adjoin v.i. συνέχομαι. (v.t.) είμαι πλάι σε.

~ing *a.* συνεχόμενος, παρακείμενος.

adjourn *v.t.* διακόπτω. (*v.i.*) διακόπτομαι. (*go elsewhere*) αποσύρομαι. **~ment** *s.* διακοπή *f.*

adjudge *v.* (*judge*) κρίνω, (*award*) απονέμω.

adjudicat|e *v.i.* (*give judgement*) βγάζω απόφαση. **~ion** *s.* απόφαση *f.* **~or** *s.* κριτής *m.*

adjunct *s.* εξάρτημα *n.*

adjure *v.* εξορκίζω.

adjust *v.* (*adapt*) προσαρμόζω. (*set right*) σιάζω, τακτοποιώ, (*clock*) ρυθμίζω, (*seat*) κανονίζω. **~able** *a.* ρυθμιζόμενος. it is ~able κανονίζεται. **~ment** *s.* προσαρμογή *f.* σιάξιμο *n.* τακτοποίηση *f.* ρύθμιση *f.* κανονισμός *m.*

adjutant *s.* υπασπιστής *m.*

ad lib. *adv.* κατ' αρέσκειαν. (*v., fam.*) αυτοσχεδιάζω.

administer *v.* (*manager*) διαχειρίζομαι, (*govern*) διοικώ, (*dispense*) απονέμω. ~ oath to ορκίζω. ~ medicine to δίνω φάρμακο σε.

administrat|ion *s.* διαχείριση *f.*, διοίκηση *f.*, απονομή *f.* (*Government*) κυβέρνηση *f.* **~ive** *a.* διοικητικός. **~or** *s.* διαχειριστής *m.*

admirab|le *a.* θαυμαστός. **~ly** *adv.* θαυμάσια, μια χαρά.

admiral *s.* ναύαρχος *m.* **~ty** *s.* ναυαρχείο *n.*

admiration *s.* θαυμασμός *m.*

admir|e *v.* θαυμάζω. **~er** *s.* θαυμαστής *m.*, θαυμάστρια *f.* **~ing** *a.* θαυμαστικός. **~ingly** *adv.* με θαυμασμό.

admissible *a.* παραδεκτός.

admission *s.* (*acknowledgement*) αναγνώριση *f.* (*entry*) είσοδος *f.*, no ~ απαγορεύεται η είσοδος.

admit *v.* (*acknowledge*) αναγνωρίζω, παραδέχομαι. ~ of επιδέχομαι. (*let in*) βάζω μέσα. be ~ted (*gain entry*) μπαίνω, εισάγομαι, γίνομαι δεκτός. **~tedly** *adv.* ομολογουμένως.

admon|ish *v.* νουθετώ. **~ition** *s.* νουθεσία *f.*

ad nauseam *adv.* κατά κόρον.

ado *s.* φασαρία *f.*

adolescen|ce *s.* εφηβική ηλικία. **~t** *a.* εφηβικός. (*s.*) έφηβος *m.*

adopt *v.* υιοθετώ. **~ion** *s.* υιοθεσία *f.* **~ive** *a.* θετός.

ador|e *v.* λατρεύω. **~ed** λατρευτός. **~able** *a.* αξιολάτρευτος. **~ation** *s.* λατρεία *f.*

adorn *v.* στολίζω. **~ment** *s.* στολισμός *m.*

adrift *a.* ακυβέρνητος. (*fig.*) turn ~ αποδιώχνω.

adroit *a.* επιδέξιος.

adulation *s.* θυμίαμα *n.*

adult *a.&s.* ενήλιξ, ενήλικος.

adulterat|e *v.* νοθεύω. **~ed** *a.* νοθευμένος. **~ion** *s.* νόθευση *f.*

adulter|er *s.* μοιχός *m.* **~ess** *s.* μοιχαλίδα *f.* **~y** *s.* μοιχεία *f.* commit ~y μοιχεύω.

advance *s.* πρόοδος *f.* (*mil.*) προέλαση *f.* (*increase*) αύξηση *f.* (*loan*) δάνειο *n.* make ~s to κάνω προτάσεις σε. in ~ (*ahead*) εμπρός, (*beforehand*) από πριν, (*paid*) προκαταβολικώς. in ~ of (*time*) πριν από, (*place*) εμπρός από.

advance *v.i.* προχωρώ, (*mil.*) προελαύνω. (*v.t.*) (*suggest*) υποβάλλω, (*promote*) προάγω, (*increase*) αυξάνω, (*lend*) δανείζω. ~ (*date of*) επιταχύνω.

advanced *a.* (*modern*) προηγμένος. (*of studies*) ανώτερος. (*in age, development*) προχωρημένος. (*mil.*) προκεχωρημένος.

advancement *s.* προαγωγή *f.*

advantage *s.* (*superior quality*) πλεονέκτημα *n.* (*interest*) συμφέρον *n.* (*profit*) όφελος *n.*, ωφέλεια *f.*, it is to my ~ είναι προς όφελός μου, με συμφέρει. take ~ of επωφελούμαι (*with gen.*), (*exploit*) εκμεταλλεύομαι. have the ~ (*over person*) βρίσκομαι σε πλεονεκτική θέση. to the best ~ επί το συμφερότερον.

advantageous *a.* πλεονεκτικός, επωφελής. **~ly** *adv.* επωφελώς.

advent *s.* άφιξη *f.* (*eccl.*) σαραντάμερο *n.*

adventitious *a.* τυχαίος.

adventur|e *s.* περιπέτεια *f.* **~er**, **~ess** *s.* τυχοδιώκτης *m.f.*

adventurous *a.* (*life, action, tale*) περιπετειώδης. (*person*) τολμηρός, (*taking risks*) ριψοκίνδυνος.

adverb *s.* επίρρημα *n.*

adversary *s.* αντίπαλος *m.*

advers|e *a.* (*conditions*) δυσμενής, αντίξοος. (*weather*) δυσμενής, ενάντιος. (*criticism*) δυσμενής. **~ely** *adv.* ενάντια, δυσμενώς. **~ity** *s.* κακοτυχία *f.*

advertise *v.* διαφημίζω, (*fam.*) ρεκλαμάρω. **~ment** *s.* διαφήμιση *f.* ρεκλάμα *f.* (*small*) αγγελία *f.*

advice *s.* συμβουλή *f.* (*notice*) ειδοποίηση *f.* I seek his ~ τον συμβουλεύομαι.

advisable *a.* I don't think it ~ for you to go there δεν θα συνιστούσα να πας εκεί.

advise *v.t.* (*person*) συμβουλεύω, (*course of action*) συνιστώ. (*inform*) ειδοποιώ. you would be well ~d to... θα κάνατε καλά νά.

advis|er *s.* σύμβουλος *m.* **~ory** *a.* συμβουλευτικός.

advoca|te *s.* συνήγορος *m.* (*v.*) συνηγορώ υπέρ (*with gen.*) **~cy** *s.* συνηγορία *f.*

aeon *s.* αιώνας *m.*

aerated *a.* αεριούχος.

aerial *a.* εναέριος. (*s.*) κεραία *f.*, αντένα *f.*

aerodrome s. αεροδρόμιο n.
aerodynamics s. αεροδυναμική f.
aeronaut s. αεροναύτης m. ~**ics** s. αεροναυτική f.
aeroplane s. αεροπλάνο n.
aesthet|e s. εστέτ m. ~**ic** a. αισθητικός. ~**ics** s. αισθητική f.
afar adv. μακριά.
affable a. προσηνής.
affair s. υπόθεση f., πράγμα n. δουλειά f. they are having an ~ τα έχουν, έχουν σχέσεις. Ministry of Home A~s Υπουργείον Εσωτερικών. in the present state of ~s όπως έχουν τα πράγματα.
affect v. επηρεάζω. επιδρώ (with επί & gen.). (move) συγκινώ. (upset or ~ health of) πειράζω. (feign) προσποιούμαι.
affectation s. προσποίηση f.
affect|ed a. (pretended) προσποιητός. (in manner) επιτηδευμένος. be ~ed έχω προσποιητούς τρόπους. ~**ing** a. συγκινητικός.
affection s. αγάπη f., συμπάθεια f. (malady) πάθηση f.
affectionate a. τρυφερός, στοργικός.
affidavit s. ένορκος κατάθεση.
affiliat|e v. (attach) συνδέω. ~**ion** s. σύνδεση f.
affinity s. συγγένεια f.
affirm v. βεβαιώνω. ~**ation** s. βεβαίωση f. ~**ative** a. καταφατικός.
affix v. επιθέτω.
afflict v. θλίβω. ~**ion** s. (sorrow) θλίψη f., πίκρα f. (misfortune, blow) πλήγμα n.
affluen|t a. πλούσιος. ~**ce** s. πλούτη n.pl., αφθονία f.
afford v. (give) παρέχω. I can't ~ it δεν έχω τα απαιτούμενα χρήματα. I can't ~ to vote πο δεν είμαι σε θέση να ψηφίσω όχι.
afforestation s. δάσωση f.
affray s. συμπλοκή f.
affront s. προσβολή f. (v.) προσβάλλω.
afield adv. μακριά.
aflame a. φλεγόμενος.
afloat adv. (at sea) στη θάλασσα. stay ~ κρατιέμαι στην επιφάνεια.
afoot adv. (on foot) πεζή. (fig.) what's ~? τι τρέχει; there's mischief ~ (fam.) κάποιο λάκκο έχει η φάβα.
aforesaid a. προαναφερθείς.
a fortiori adv. κατά μείζονα λόγον, τοσούτω μάλλον.
afraid a. φοβισμένος. be ~ (of) φοβούμαι. I'm ~ it will rain φοβούμαι μη (or μήπως or πως θα) βρέξει. I was ~ he wouldn't come φοβήθηκα μη δεν (or μήπως δεν or πως δεν θα) έρθει. I was ~ it might turn out bad for me φοβήθηκα μήπως μου έβγαινε σε κακό. I'm ~ he's lost the way (possibly)

φοβούμαι πως (or μήπως or μην) έχασε το δρόμο. I'm ~ he's not at home (regrettably) λυπούμαι αλλά δεν είναι σπίτι.
afresh adv. ξανά, εκ νέου.
African a. αφρικανικός. (person) Αφρικανός m. Αφρικάνα f.
after prep. μετά, ύστερα από (with acc.). one ~ another απανωτός, ο ένας πίσω από τον άλλο. the day ~ tomorrow μεθαύριο. ~ all τέλος πάντων. ~ all, he's only a child είναι παιδί επιτέλους. I was right ~ all τελικά είχα δίκιο.
after conj. αφού, μετά που.
after adv. μετά, ύστερα, κατόπιν.
after-effect s. συνέπεια f.
aftermath s. παρεπόμενα n.pl.
afternoon s. απόγευμα n. good ~ καλησπέρα. (a.) απογευματινός.
afterthought s. μεταγενέστερη σκέψη.
afterwards adv. μετά, ύστερα, κατόπιν.
again adv. πάλι, ξανά. marry ~ ξαναπαντρεύομαι. ~ and ~ επανειλημμένως. as much ~ άλλος τόσος.
against prep. (opposition) κατά, εναντίον (with gen.). (proximity) κοντά σε. lean ~ ακουμπώ σε. I knocked ~ a chair χτύπησα πάνω σε μία καρέκλα. ~ one's will ακουσίως. ~ receipt έναντι αποδείξεως. ~ a rainy day για ώρα ανάγκης.
agape adv. με ανοικτό το στόμα.
age s. ηλικία f. (old ~) γήρας n., γεράματα n.pl. come of ~ ενηλικιούμαι. of the same ~ ομήλικος, under ~ ανήλικος. (period) εποχή f., αιώνας m. middle ~s μεσαίωνας m. I haven't seen him for ~s έκανα μαύρα μάτια να τον δω. ~s ago προ πολλού καιρού.
age v.t. & i. γερνώ, γηράσκω.
aged a. (very old) πολύ ηλικιωμένος. middle-~ μεσόκοπος, μεσήλικος. a boy ~ ten αγόρι δέκα ετών.
ageless a. (eternal) αιώνιος. (always young) αγέραστος, αειθαλής.
agency s. πρακτορείο n. (instrumentality) μεσιτεία f., μεσολάβηση f.
agenda s. ημερήσια διάταξη.
agent s. πράκτορας m., μεσίτης m. (commercial) παραγγελιοδόχος m. (representative) αντιπρόσωπος m. ~ provocateur προβοκάτορας m. ~'s μεσιτικός.
agglomeration s. ανομοιογενής σωρός.
aggrandizement s. αύξηση προσωπικού πλούτου και ισχύος.
aggravat|e v. επιδεινώνω. (annoy) εκνευρίζω. ~**ing** a. εκνευριστικός. ~**ion** s. επιδείνωση f. εκνευρισμός m.
aggregate a. συνολικός. (s.) σύνολο n. in the ~ εν συνόλω. (v.t.) αθροίζω.
aggression s. επίθεση f.

aggressive *a.* επιθετικός. **~ness** *s.* επιθετικότητα *f.*

aggressor *s.* ο επιτιθέμενος.

aggrieved *a.* πικραμένος.

aghast *a.* κατατρομαγμένος. I was ~ when I heard it έφριξα όταν το άκουσα.

agil|e *a.* εύστροφος, σβέλτος. **~ity** *s.* ευστροφία *f.*

agitat|e *v.t.* ταράζω. (*v.i.*) (*campaign*) κινούμαι. **~ed** *a.* ταραγμένος. **~ion** *s.* ταραχή *f.* (*campaign*) εκστρατεία *f.* **~or** *s.* ταραξίας *m.*

aglow *a.* (*with fire*) φλογισμένος. be ~ λάμπω.

agnostic *s.* αγνωστικιστής *m.* **~ism** *s.* αγνωστικισμός *m.*

ago *adv.* long ~ προ πολλού. some days ~ προ ημερών. a month ~ προ ενός μηνός *or* πριν (από) ένα μήνα *or* εδώ κι ένα μήνα. how long ~ ? πριν πόσο καιρό;

agog *a.* & *adv.* be (all) ~ βράζω από αναμονή.

agon|y *s.* αγωνία *f.* be in **~y** αγωνιώ. **~izing** *a.* (*decision*) αγωνιώδης, (*pain*) τρομερός.

agrarian *a.* αγροτικός.

agree *v.* συμφωνώ. ~ on terms συμφωνώ τους όρους. he ~d to do it δέχτηκε να το κάνει. he ~d to take the house έκλεισε το σπίτι. ~ to (*a proposal*) συναινώ εις. it ~s with me μου κάνει καλό. it doesn't ~ with me με πειράζει.

agreeab|le *a.* (*pleasant*) ευχάριστος. (*in agreement*) σύμφωνος. is that ~le to you? συμφωνείτε σ' αυτό; **~ly** *adv.* ευχάριστα.

agreement *s.* συμφωνία *f.* (*pact*) σύμβαση *f.* reach an ~ συμβιβάζομαι. in ~ σύμφωνος. in ~ with (*adv.*) σύμφωνα με.

agricultur|e *s.* γεωγία *f.* **~al** *a.* γεωργικός.

aground *adv.* run ~ (*v.t.*) ρίχνω έξω. (*v.i.*) εξοκέλλω.

ague *s.* πυρετός *m.*, θέρμες *f.pl.*

ahead *adv.* εμπρός, μπρος, μπροστά. ~ of εμπρός (*etc.*) από. get ~ of προσπερνώ.

aid *s.* βοήθεια *f.*, (*specific*) βοήθημα *n.* in ~ of υπέρ (*with gen.*). (*v.*) βοηθώ.

aide-de-camp *s.* υπασπιστής *m.*

Aids *s.* (*disease*) έιντς *n.*

ail *v.* what ~s him? τι έχει; **~ing** *a.* άρρωστος. **~ment** *s.* αρρώστια *f.*

aim *v.* ~ at (*target*) σκοπεύω, σημαδεύω. he ~ed the gun at me, έστρεψε το όπλο εναντίον μου. (*a blow*) καταφέρω. (*intend*) σκοπεύω (*with να*).

aim *s.* (*act of ~ing*) σκόπευση *f.* (*purpose*) σκοπός *m.* miss one's ~ αστοχώ.

aimless *a.* άσκοπος. **~ly** *adv.* ασκόπως.

air *s.* αέρας *m.*, αήρ *m.* by ~ αεροπορικώς. in the ~ (*uncertain*) φλου. (*tune*) σκοπός *m.*

(*manner*) ύφος *n.* put on ~s το παίρνω απάνω μου.

air *a.* (*base, raid etc.*) αεροπορικός. ~ force πολεμική αεροπορία *f.* ~ gun αεροβόλο *n.* ~ hostess αεροσυνοδός *f.* ~ terminal αερολιμένας *m.*

air *v.t.* αερίζω. (*show off*) επιδεικνύω, (*express*) εκφράζω.

air-conditioning *s.* κλιματισμός *m.*

air-cooled *a.* αερόψυκτος.

aircraft *s.* αεροσκάφος *n.* **~-carrier** *s.* αεροπλανοφόρο *n.*

airlift *s.* αερογέφυρα *f.*

airline *s.* (*company*) αεροπορική εταιρεία, (*route*) γραμμή *f.*

airmail *s.* by ~ αεροπορικώς.

airman *s.* αεροπόρος *m.*

airport *s.* αερολιμένας *m.*, αεροδρόμιο *n.*

airtight *a.* αεροστεγής.

airy *a.* ευάερος. (*fig., carefree*) ξέγνοιαστος.

aisle *s.* διάδρομος *m.*

ajar *adv.* leave the door ~ αφήνω την πόρτα μισάνοιχτη.

akimbo *adv.* with arms ~ με τα χέρια στη μέση.

akin *a.* συγγενής (*with* με).

alabaster *s.* αλάβαστρο *n.*

alacrity *s.* προθυμία *f.*

à la mode *adv.* της μόδας.

alarm *s.* συναγερμός *m.* (*fear*) τρόμος *m.* ~ clock ξυπνητήρι *n.*

alarm *v.* (*also* feel ~) τρομάζω, ανησυχώ. **~ing** *a.* ανησυχητικός. **~ist** *s.* που όλο προβλέπει καταστροφές.

alas *int.* αλίμονο.

albatross *s.* άλμπατρος *m.*

albino *s.* αλμπίνος *m.*

album *s.* λεύκωμα *n.*, άλμπουμ *n.*

alchem|y *s.* αλχημεία *f.* **~ist** *s.* αλχημιστής *m.*

alcohol *s.* οινόπνευμα *n.* **~ic** *a.* (*liquor*) οινοπνευματώδης, (*person*) αλκοολικός. **~ism** *s.* αλκοολισμός *m.*

alcove *s.* (*for statue*) αχηβάδα *f.* (*for bed, etc.*) εσοχή *f.*

alder *s.* κλήθρα *f.*

ale *s.* μπίρα *f.* ~ house καπηλειό *n.*

alert *s.* συναγερμός *m.* on the ~ σε επιφυλακή. be on the ~ for έχω το νου μου για (*v.*) προειδοποιώ.

alert *a.* άγρυπνος, ξύπνιος. **~ness** *s.* επαγρύπνηση *f.*

algebra *s.* άλγεβρα *f.* **~ic** *a.* αλγεβρικός.

alibi *s.* άλλοθι *n.*

alien *a.* (*thing*) ξένος, (*person*) αλλοδαπός. **~ate** *v.* αποξενώνω. **~ism** *s.* φρενολογία *f.* **~ist** *s.* φρενολόγος *m.*

alight *a.* (*on fire*) φλεγόμενος. (*lamp, stove, etc.*) αναμμένος. (*illuminated*) φωτισμένος. (*fig., face, etc.*) λάμπων.

alight v. (get down) κατεβαίνω, (settle) κάθομαι.

align v. ευθυγραμμίζω. ~**ment** s. ευθυγράμμιση f.

alike a. όμοιος. they look ~ μοιάζουν. (adv.) το ίδιο.

alimony s. διατροφή f.

alive a. ζωντανός. be ~ ζω, ευρίσκομαι εν ζωή. (fig.) be ~ with βρίθω (with gen.). be ~ to αντιλαμβάνομαι.

alkali s. αλκάλιο, n. ~**ne** a. αλκαλικός.

all s. (everything) το παν, τα πάντα.

all a. & pron. όλος, πας. ~ (the) children όλα τα παιδιά. ~ of you όλοι σας. he ate it ~ (or ~ of it) τό 'φαγε όλο. it's ~ gone (used up) τελείωσε. ~ day όλη μέρα. he grumbles ~ the time όλο (or ολοένα) γκρινιάζει. ~ the time I was sitting there όσο καθόμουν εκεί. ~ the year (round) όλο το χρόνο. ~ night long ολονυχτίς. with ~ speed με πλήρη ταχύτητα. ~ those present όλοι (or πάντες) οι παρόντες. in ~ probability κατά πάσαν πιθανότητα. for ~ his money (in spite of) παρ' όλα τα λεφτά του. ~ (that) you told me όλα όσα μού είπατε. ~ I ask is this δεν ζητώ παρά τούτο. that's ~ αυτά είναι όλα. for ~ I care πολύ που με νοιάζει. ~ in ~ (or when ~'s said and done) στο κάτω κάτω.

all adv. ~ right καλά, εντάξει. ~ in (inclusive) όλα μαζί, (exhausted) εξαντλημένος. ~ the same (yet) παρ' όλα αυτά. (not) at ~ καθόλου. ~ the better τόσο το καλύτερο. ~ over Europe παντού στην Ευρώπη. it's ~ over mud είναι γεμάτο λάσπη. I'm ~ wet μούσκεψα. it's ~ crooked είναι όλο στραβά. that's very well (but...) καλά και άγια. it's ~ one (or the same) to me το ίδιο μού κάνει. ~ lit up ολόφωτος, ~ gold ολόχρυσος.

Allah s. Αλλάχ m.

allay v. (fear) καθησυχάζω, (pain) καταπραΰνω.

allege v. ισχυρίζομαι. the ~ed thief ο υποτιθέμενος (or δήθεν) κλέφτης. ~**ation** s. ισχυρισμός m.

allegiance s. πίστη f.

allegor|y s. αλληγορία f. ~**ical** a. αλληγορικός.

allerg|y s. αλλεργία f. ~**ic** a. αλλεργικός.

alleviat|e v. ανακουφίζω. ~**ion** s. ανακούφιση f.

alley s. στενοσόκακο n. blind ~ αδιέξοδο n.

alliance s. δεσμός m. (of states) συμμαχία f. (by marriage) συμπεθεριά f.

allied a. (of allies) συμμαχικός. (related) συγγενής. see ally iv.

alliteration s. παρήχηση f.

allocat|e v. (duties) καθορίζω, (funds) διαθέτω, χορηγώ. ~**ion** s. (act) καθορισμός m.

χορήγηση f. (share) μερίδιο n.

allot v. see allocate. ~**ment** s. μικρό κομμάτι γης ενοικιαζόμενο προς καλλιέργεια.

allow v. (let) αφήνω, επιτρέπω. not ~ed δεν επιτρέπεται. (own) παραδέχομαι. (give) δίνω. (grant funds) χορηγώ. the appeal was ~ed η προσφυγή έγινε δεκτή. ~ for (take into account) παίρνω υπόψη.

allowance s. (grant), επίδομα n. (deduction) έκπτωση f. make ~(s) for λαμβάνω υπ' όψιν.

alloy s. κράμα n.

all-powerful a. παντοδύναμος.

all-rounder s. με πολλά ενδιαφέροντα και πολλές ικανότητες.

allude v. ~ to αναφέρομαι εις.

allur|e v. δελεάζω. ~**ing** a. δελεαστικός.

allus|ion s. υποδήλωση f., νύξη f. make ~ion to αναφέρομαι εις. ~**ive** a. (style, etc.) με πλάγιες αναφορές.

alluvial a. προσχωματικός.

ally s. συνδέω. be allied to (by marriage etc.) συνδέομαι με. (s.) σύμμαχος m.

almanac s. ημερολόγιο n., καζαμίας m.

almighty a. παντοδύναμος. (fam.) ~ row τρικούβερτος καβγάς.

almond s. αμύγδαλο n. (a.) made of ~s or ~-shaped αμυγδαλωτός. ~-**tree** s. αμυγδαλιά f.

almost adv. σχεδόν, see nearly.

alms, ~-giving s. ελεημοσύνη f.

aloe s. αλόη f. American ~ αθάνατος m.

aloft adv. ψηλά.

alone a. μόνος, μοναχός. they left him ~ (by himself) τον άφησαν μόνο, (in peace) τον άφησαν ήσυχο. I did it ~ (unaided) το έκανα μόνος μου. let (sthg.) ~ αφήνω. (adv.) μόνο(ν).

along prep. από (with acc.). I was going ~ Stadium street περνούσα από την οδό Σταδίου. (all) ~ the river κατά μήκος του ποταμού.

along adv. (on, onward) εμπρός. move ~ προχωρώ. it's coming ~ well προχωράει. we get ~ well τα πάμε καλά. farther ~ παραπέρα. all ~ μαζί με. all ~ από την αρχή.

alongside adv. δίπλα. (prep.) δίπλα σε. come ~ (v.i.) διπλαρώνω, πλευρίζω (with σε).

aloof a. (of manner) αδιάφορος. (adv.) (apart) χωριστά. hold ~ μένω εις απόστασιν.

aloud adv. δυνατά.

alpha s. άλφα n. A~ and Omega το άλφα και το ωμέγα.

alphabet s. αλφάβητο n. ~**ic(al)** a. αλφαβητικός. in ~ical order κατ' αλφαβητική σειρά.

Alps s. Άλπεις f.pl. **~ine** a. άλπειος.

already adv. κιόλας, ήδη. are you leaving ~? από τώρα θα φύγεις;

also adv. και, επίσης. not only... but also; όχι μόνο... αλλά και. (fam.) ~-ran αποτυχημένος.

altar s. βωμός m. (in Christian churches) αγία τράπεζα.

alter v.t. & i. αλλάζω. (v.t.) μεταβάλλω, (clothes) μεταποιώ. that ~s matters αλλάζει.

alteration s. αλλαγή f. (of clothes) μεταποίηση f.

altercation s. λογομαχία f.

alternat|e v.t. εναλλάσσω. (v.i.) εναλλάσσομαι. **~ing** current εναλλασσόμενο ρεύμα. **~ion** s. εναλλαγή f.

alternate a. εναλλασσόμενος. on ~ days μέρα παρά μέρα. **~ly** adv. εναλλάξ, μια ο ένας μια ο άλλος.

alternative a. άλλος. (s.) κάτι άλλο, άλλη λύση, άλλη εκλογή. **~ly** adv. ή (conj.).

although conj. μολονότι, αν και, παρ' όλο που, μ' όλο που.

altitude s. υψόμετρον n.

altogether adv. (wholly) όλως διόλου, πέρα για πέρα, τελείως. I don't ~ agree δεν συμφωνώ απολύτως. (on the whole) εν τω συνόλω, συνολικά.

altru|ism s. αλτρουισμός m. **~ist** s. αλτρουιστής m. **~istic** a. αλτρουιστικός.

aluminium s. αλουμίνιο n.

always adv. πάντα, πάντοτε. he is ~ grumbling όλο (or ολοένα or διαρκώς) γκρινιάζει.

am v. I ~ είμαι.

amalgam s. αμάλγαμα n.

amalgamat|e v.t. συγχωνεύω. (v.i.) συγχωνεύομαι. **~ion** s. συγχώνευση f.

amanuensis s. βοηθός συγγραφέα.

amass v. συσσωρεύω.

amateur s. ερασιτέχνης m. **~ish** a. (thing) κακότεχνος, (person) του γλυκού νερού.

amatory a. ερωτικός.

amaz|e v. καταπλήσσω. I was ~ed έμεινα κατάπληκτος. **~ement** s. κατάπληξη f., θάμβος n. **~ing** a. καταπληκτικός.

Amazon s. αμαζόνα f.

ambassador s. πρέσβης m., πρεσβευτής m.

amber s. κεχριμπάρι n., ήλεκτρο n.

ambidextrous a. αμφιδέξιος.

ambience s. ατμόσφαιρα f.

ambigu|ity s. αμφιβολία f. **~ous** a. διφορούμενος, αμφίβολος.

ambit s. (range) έκταση f., σφαίρα f.

ambiti|on s. φιλοδοξία f. **~ous** a. φιλόδοξος, be **~ous** φιλοδοξώ.

ambivalent a. διφορούμενος.

amble v. (stroll) σεργιανίζω.

ambrosia s. αμβροσία f.

ambulance s. ασθενοφόρο n.

ambush s. ενέδρα f. (v.t.) ενεδρεύω (also lie in ~). lay an ~ στήνω ενέδρα.

ameliorat|e v. βελτιώνω. **~ion** s. βελτίωση f.

amen int. αμήν.

amenable a. (tractable) ευάγωγος, (to law) υπο(τε)ταγμένος. ~ to reason λογικός.

amend v. (correct) διορθώνω, (alter in detail) τροποποιώ. **~ment** s. τροπολογία f.

amends s. αποζημίωση f. make ~ to or for αποζημιώνω.

ameni|ty s. (pleasant surroundings) πολιτισμός m. (of character, climate, etc.) γλυκύτητα f. **~ies** s. (comforts) βολές f.pl., ευκολίες f.pl.

American a. αμερικανικός. (person) Αμερικανός m., Αμερικανίδα f.

amethyst s. αμέθυστος m.

amiable a. προσηνής, ευγενικός.

amicabl|e a. φιλικός. **~y** adv. φιλικά.

amid, **~st** prep. ανάμεσα σε.

amiss adv. κακώς, στραβά. I take it ~ μου κακοφαίνεται. (a.) στραβός.

amity s. φιλικές σχέσεις.

ammonia s. αμμωνία f.

ammunition s. πυρομαχικά n.pl. (fig., for argument) όπλα n.pl.

amnesia s. αμνησία f.

amnesty s. αμνηστία f. (v.) αμνηστεύω.

amoeba s. αμοιβάς f.

amok adv. he ran ~ τον έπιασε αμόκ.

among, **~st** prep. (location) ανάμεσα σε, μέσα σε (with acc.), (division, distribution) μεταξύ (with gen.). ~ the crowd ανάμεσα στο πλήθος. ~ the trees μέσα στα δέντρα. they came out from ~ the trees βγήκαν μέσα από τα δέντρα. ~ other things μεταξύ άλλων. not one ~ them κανένας απ' αυτούς. he fell ~ thieves έπεσε στα χέρια κλεφτών. (the property) was divided ~ the five children μοιράστηκε ανάμεσα στα πέντε παιδιά.

amoral a. χωρίς συνείδηση της ηθικής. **~ism** s. αμοραλισμός m.

amorous a. (person) ερωτύλος, (behaviour) ερωτικός.

amorphous a. άμορφος.

amortization s. χρεολυσία f.

amount s. (sum) ποσό n. (quantity) ποσότητα f. he has any ~ of money (fam.) έχει λεφτά με ουρά.

amount v. (add up to) ανέρχομαι εις. (fig.) it ~s to this, that... σημαίνει ότι.

amphibi|an s. αμφίβιο n. **~ous** a. αμφίβιος.

amphitheatre s. αμφιθέατρο n. (sited or built) like an ~ αμφιθεατρικώς.
amphora s. αμφορεύς m.
ampl|e a. άφθονος. **~y** adv. άφθονα.
amplif|y v. (sound) ενισχύω. (story) διευρύνω. **~ication** s. ενίσχυση f. διεύρυνση f. **~ier** s. ενισχυτής m.
ampoule s. αμπούλα f.
amputat|e v. ακρωτηριάζω. **~ion** s. ακρωτηριασμός m.
amulet s. φυλαχτό n.
amuse v. διασκεδάζω (also ~ oneself or be ~d). I was ~d at his stories διασκέδασα (or έκανα γούστο) με τις ιστορίες του.
amus|ement s. διασκέδαση f. for ~ement για γούστο. **~ing** a. διασκεδαστικός, γουστόζικος.
an see a.
anachronis|m s. αναχρονισμός m. **~tic** a. αναχρονιστικός.
anaemi|a s. αναιμία f. **~c** a. αναιμικός.
anaesthesia s. αναισθησία f.
anaesthet|ic s. αναισθητικό n. **~ist** s. αναισθησιολόγος m. **~ize** v. αναισθητοποιώ.
anagram s. αναγραμματισμός m.
analogous a. παρόμοιος.
analogue s. αντίστοιχο n. ~ computer αναλογικός υπολογιστής.
analogy s. παραλληλισμός m. draw an ~ between A and B παραλληλίζω το A με το B.
analy|se v. αναλύω. **~sis** s. ανάλυση f. **~tic(al)** a. αναλυτικός.
anarch|y s. αναρχία f. **~ic(al)** a. αναρχικός. **~ism** s. αναρχισμός m. **~ist** s. αναρχικός m.
anathema s. it is ~ to me (fam.) το σιχαίνομαι σαν τις αμαρτίες μου.
anatom|y s. ανατομία f. **~ic(al)** a. ανατομικός. **~ist** s. ανατόμος m.
ancest|or s. πρόγονος m. **~ral** a. προγονικός. **~ry** s. γενεαλογική σειρά.
anchor s. άγκυρα f. (v.i.) αγκυροβολώ. **~age** s. αγκυροβόλιο n.
anchorite s. αναχωρητής m.
anchovy s. αντζούγια f.
ancient a. αρχαίος.
ancillary a. βοηθητικός.
and conj. και. ten ~ a half δεκάμισυ. one hundred ~ ten εκατόν δέκα. for hours ~ hours ώρες ολόκληρες. now ~ then κάθε τόσο. women ~ children γυναικόπαιδα n.pl. night ~ day μέρα νύχτα. it is nice ~ warm in here κάνει ωραία ζέστη εδώ μέσα. come ~ see me έλα να με δεις. try ~ come προσπάθησε να' ρθεις.
anecdote s. ανέκδοτο n.

anemone n. ανεμώνη f.
anew adv. εκ νέου, ξανά.
angel s. άγγελος m. **~ic(al)** a. αγγελικός.
anger s. θυμός m. (v.t.) θυμώνω, εξοργίζω.
angle s. γωνία f. (point of view) άποψη f.
angle v. ψαρεύω. **~r** s. ψαράς m.
anglicize v. εξαγγλίζω.
angling s. ψάρεμα n.
angr|y a. θυμωμένος. get ~ θυμώνω. **~ily** adv. θυμωμένα.
anguish s. αγωνία f.
angular a. γωνιώδης.
animal s. ζώο n. (a.) ζωικός, (carnal) σαρκικός.
animat|e a. έμψυχος. (v.t.) εμψυχώνω. **~ed** a. ζωηρός. **~ion** s. ζωηρότητα f.
animosity s. έχθρα f.
aniseed s. γλυκάνισο n.
ankle s. αστράγαλος m.
annal|s s. χρονικά n.pl. **~ist** s. χρονικογράφος m.
annex v. προσαρτώ. **~ation** s. προσάρτηση f. **~e** s. προσάρτημα n., παράρτημα n.
annihilat|e v. εκμηδενίζω. **~ion** s. εκμηδένιση f.
anniversary s. επέτειος f.
Anno Domini adv. μετά Χριστόν. (fam., advancing years) η πάροδος της ηλικίας.
annotat|e v. σχολιάζω. **~or** s. σχολιαστής m.
announce v. αναγγέλλω. **~ment** s. αγγελία f. **~r** s. (radio) εκφωνητής m.
annoy v. ενοχλώ, εκνευρίζω. **~ance** s. ενόχληση. f. **~ing** a. ενοχλητικός.
annual a. ετήσιος. (s., year-book) επετηρίδα f. **~ly** adv. ετησίως, κάθε χρόνο.
annuity s. ετήσια πρόσοδος.
annul v. ακυρώνω. **~ment** s. ακύρωση f.
Annunciation s. (eccl.) Ευαγγελισμός m.
anodyne a. παυσίπονος. (s.) παυσίπονο n.
anoint v. χρίω. **~ing** s. χρίσμα n.
anomal|y s. ανωμαλία f. **~ous** a. ανώμαλος.
anonym|ous a. ανώνυμος. **~ity** s. ανωνυμία f.
anorexia s. ανορεξία f.
another pron. & a. (different) άλλος. ~ time μια άλλη φορά. I've read this book, give me ~ διάβασα τούτο το βιβλίο, δώστε μου ένα άλλο. (one more) would you like ~ cup? θέλετε άλλο (or ακόμα) ένα φλιτζάνι; one ~ see one. one after ~ see after.
answer v.i. απαντώ, αποκρίνομαι. (v.t.) he ~ed my letter απάντησε στο γράμμα μου. (solve) λύω. it doesn't ~ my purpose δεν μου κάνει. ~ back (rudely) αντιμιλώ. ~ for ευθύνομαι για. ~ to (treatment) αντιδρώ σε, (description) συμφωνώ με. ~ to the name of... ακούω στο όνομα. **~ing** machine αυτόματος τηλεφωνητής.

answer s. απάντηση f., απόκριση f. (solution) λύση f. ~**able** a. υπόλογος.
ant s. μυρμήγκι n. ~**hill** s. μυρμηγκοφωλιά f.
antagonis|m s. εχθρότητα f. ~**t** s. αντίπαλος m. ~**tic** a. εχθρικός. be ~tic to εναντιούμαι εις.
antagonize v. you ~ him τον διαθέτεις εχθρικώς.
antarctic a. ανταρκτικός.
antecedent a. προηγούμενος. his ~s το παρελθόν του, (forbears) οι πρόγονοί του.
antechamber s. προθάλαμος m.
antedate v. προχρονολογώ. (precede) προηγούμαι (with gen.).
antediluvian a. προκατακλυσμιαίος.
antenna s. κεραία f.
antepenultimate a. ~ syllable προπαραλήγουσα f.
anterior a. προηγούμενος.
ante-room s. προθάλαμος m.
anthem s. ύμνος m. (eccl.) αντίφωνον n.
anthology s. ανθολογία f. make an ~ of ανθολογώ.
anthracite s. ανθρακίτης m.
anthrax s. άνθρακας m.
anthropoid a. ανθρωποειδής.
anthropolog|y s. ανθρωπολογία f. ~**ist** s. ανθρωπολόγος m.
anthropomorphous a. ανθρωπόμορφος.
anti-aircraft a. αντιαεροπορικός. ~ defence αεράμυνα f.
antibiotic a. αντιβιοτικός.
anti-body s. αντίσωμα n.
anticipate v. (expect) περιμένω, προσδοκώ. (forestall) προλαβαίνω.
anticipation s. (expectation) προσδοκία f.
anticlimax s. απογοητευτική μετάπτωση.
antics s. κόλπα, καμώματα n.pl.
antidote s. αντίδοτο n.
antimony s. αντιμόνιο n.
antinuclear a. αντιπυρηνικός.
antipath|y s. αντιπάθεια f. ~**etic** a. be ~etic το έχω αντιπάθεια σε.
antipodes s. αντίποδες m.pl.
antiquar|ian a. ειδικευμένος στις αντίκες. ~**y** s. αρχαιοδίφης m.
antiquated a. απαρχαιωμένος.
antique a. παλαιός. (s.) αντίκα f.
antiquit|y s. αρχαιότητα f. ~**ies** s. αρχαία n.pl., αρχαιότητες f.pl.
anti-rabies a. αντιλυσσικός. ~ clinic λυσσιατρείο n.
anti-semitism s. αντισημιτισμός m.
antiseptic a. αντισηπτικός.
antisocial a. αντικοινωνικός.
anti-tank a. (mil.) αντιαρματικός.
antithesis s. (contrast) αντίθεση f. (the

opposite) αντίθετο n.
antler s. κέρας n.
anus s. πρωκτός m.
anvil s. αμόνι n.
anxiety s. ανησυχία f. (with impatience) αδημονία f. (desire) επιθυμία f.
anxious a. (uneasy) ανήσυχος. feel ~ αδημονώ, λαχταρώ. be ~ (eager) to επιθυμώ να. ~**ly** adv. με ανησυχία, με αδημονία.
any a. & pron. 1. (corresponding to partitive article some). did you buy ~ cheese? αγόρασες τυρί; there isn't ~ δεν έχει. (for a, an) there isn't ~ road δεν έχει δρόμο. 2. (one or other, some) if ~ other person comes αν έρθει κανένα άλλος. do you know of ~ house to let? μήπως ξέρεις να νοικιάζεται κανένα σπίτι; if you find ~ tomatoes αν βρεις τίποτα ντομάτες (or καμιά ντομάτα). will ~ of these do? σας κάνει τίποτα (or κανένα) απ' αυτά; in case of ~ complaint διά παν παράπονον. 3. (none) there isn't ~ hope δεν υπάρχει καμιά ελπίδα. there aren't ~ (emphatic) δεν υπάρχουν καθόλου. I don't like ~ of these δεν μου αρέσει κανένα (or τίποτα) απ' αυτά. not on ~ account επ' ουδενί λόγω. 4. (no matter which) it can be had from ~ kiosk πουλιέται στο κάθε περίπτερο. ~ of these ties will suit me οποιαδήποτε απ' αυτές τις γραβάτες μού κάνει. come ~ time you like ελάτε ό,τι (or οποιαδήποτε) ώρα θέλετε. if at ~ time you need me αν καμιά φορά με χρειαστείς. 5. in ~ case οπωσδήποτε. at ~ rate τουλάχιστον.
any adv. he's not working ~ longer δεν εργάζεται πια. is that ~ better? είναι καθόλου (or κάπως) καλύτερα; it isn't ~ good (doesn't serve) δεν κάνει, (is valueless) δεν αξίζει τίποτα.
anybody n. 1. κανείς, κανένας. as soon as ~ comes μόλις έρθει κανείς. I can't see ~ δεν βλέπω κανέναν. did you see ~? είδες κανέναν; 2. (everybody, the first comer) ο καθένας, οποιοσδήποτε, (fam.) όποιος κι όποιος. ~ can do it ο καθένας μπορεί να το κάνει. he doesn't keep company with just ~ δεν κάνει παρέα με τον καθένα (or με οποιονδήποτε or με όποιον κι όποιον). it's not a place where just ~ goes εκεί δεν πάει όποιος κι όποιος. 3. ~ interested should apply to the management πας ενδιαφερόμενος να αποτανθή εις την διεύθυνσιν. ~ desiring information οι επιθυμούντες πληροφορίας.
anyhow adv. (in any case) οπωσδήποτε, (by any means) με κανένα τρόπο, (negli-

gently) όπως όπως.
anyone *see* anybody.
anything *n.* τίποτα. if ~ happens αν συμβεί τίποτα. he didn't say ~ δεν είπε τίποτα. she hasn't ~ to wear δεν έχει τι να φορέσει. (*no matter what*) το καθετί, οτιδήποτε, he'll eat ~ αυτός τρώει το καθετί (*or* οτιδήποτε). ~ you may say ό,τι και να πεις. he is ~ but stupid είναι κάθε άλλο από κουτός. (*adv.*) is he ~ like his father? μοιάζει καθόλου με τον πατέρα του;
anyway *adv.* οπωσδήποτε.
anywhere *adv.* πουθενά. are you going ~ θα πάτε πουθενά; I couldn't see it ~ δεν το είδα πουθενά. I haven't got ~ to sleep δεν έχω πού να κοιμηθώ. (*no matter where*) οπουδήποτε, ~ you like οπουδήποτε θέλεις.
aorist *s.* (*gram.*) αόριστος *m.*
apace *adv.* με γοργό ρυθμό.
apache *s.* απάχης *m.*
apart *adv.* χωριστά, χώρια, ξέχωρα. they live ~ ζουν χωριστά. with one's feet ~ με τα πόδια ανοιχτά. I can't tell them ~ δεν τους ξεχωρίζω. it came ~ in my hands έμεινε στα χέρια μου. how far ~ are the two houses? πόσο απέχει το ένα σπίτι από το άλλο; set ~ βάζω στην μπάντα (*or* κατά μέρος). take ~ (*to pieces*) λύνω. ~ from (*prep.*) χώρια, χωριστά, εκτός (*all with* από & *acc.*), πλην (*with gen.*), ~ from the fact that εκτός του ότι.
apartment *s.* (*room*) δωμάτιο *n.* (*flat*) διαμέρισμα *n.*
apath|y *s.* (*lack of emotion*) απάθεια *f.* (*lack of interest*) αδιαφορία *f.* **~etic** *a.* απαθής, αδιάφορος. **~etically** *adv.* με απάθεια, με αδιαφορία.
ape *s.* πίθηκος *m.* (*v.*) πιθηκίζω.
aperient *a.* ευκοίλιος. (*s.*) καθαρτικό *n.*
aperitif *s.* απεριτίφ *n.*
aperture *s.* οπή *f.*
apex *s.* (*lit. & fig.*) κορυφή *f.* (*only fig.*) κορύφωμα *n.*
aphorism *s.* αφορισμός *m.*
aphrodisiac *s.* αφροδισιακό *n.*
apiary *s.* μελισσοκομείο *n.*
apiece *adv.* they got 10 drachmas ~ πήραν από δέκα δραχμές. they cost 10 drachmas ~ κοστίζουν 10 δραχμές το ένα.
aplomb *s.* αυτοκυριαρχία *f.*
apocalypse *s.* αποκάλυψη *f.*
apocryphal *a.* πλαστός.
apogee *s.* απόγειο *n.*
apologetic *a.* he was ~ ζήτησε συγγνώμη. he wrote me an ~ letter μου έγραψε ζητώντας συγγνώμη.
apolog|ia *s.* απολογία *f.* **~ist** *s.* απολογητής *m.*

apologize *v.* ζητώ συγγνώμη.
apology *s.* make an ~ ζητώ συγγνώμη. (*defence*) δικαιολογία *f.*
apopl|ectic *a.* αποπληκτικός. **~exy** *s.* αποπληξία *f.*
apost|asy *s.* αποστασία *f.* **~ate** *s.* αποστάτης *m.*
a posteriori *adv.* εκ των υστέρων.
apost|le *s.* απόστολος *m.* **~olic** *a.* αποστολικός.
apostrophe *s.* (*gram.*) απόστροφος *m.* (*rhetorical*) αποστροφή *f.*
apothecary *s.* φαρμακοποιός *m.*
apotheosis *s.* αποθέωση *f.*
appal *v.* προκαλώ φρίκη σε. be ~led φρίττω. **~ling** *a.* φρικτός, φρικώδης. **~lingly** *adv.* φρικτά.
apparatus *s.* συσκευή *f.*
apparel *s.* αμφίεση *f.*
apparent *a.* (*clear*) προφανής, φανερός, (*seeming*) φαινομενικός. **~ly** *adv.* καθώς φαίνεται.
apparition *s.* οπτασία *f.*
appeal *s.* (*request*) έκκληση *f.* (*law*) έφεση *f.* (*attraction*) έλξη *f.* she has a lot of ~ είναι πολύ ελκυστική.
appeal *v.i.* (*ask*) κάνω έκκληση. (*law*) κάνω έφεση. ~ to (*move*) συγκινώ, (*attract*) ελκύω.
appealing *a.* (*moving*) συγκινητικός, (*attractive*) ελκυστικός.
appear *v.* (*come into view, present oneself*) φαίνομαι, εμφανίζομαι, παρουσιάζομαι. (*seem*) φαίνομαι. (*of advocate for client*) παρίσταμαι. (*of actor*) εμφανίζομαι. (*of book*) βγαίνω.
appearance *s.* εμφάνιση *f.* (*person's presence*) παρουσιαστικό *n.* (*outward* ~) εξωτερικό *n.* (*look*) όψη *f.*, he has the ~ of a sick man έχει όψη αρρώστου. to all ~s κατά τα φαινόμενα. for the sake of ~ για τα μάτια του κόσμου, προς το θεαθήναι. keep up ~s σώζω τα προσχήματα.
appease *v.* κατευνάζω, (*satisfy*) ικανοποιώ. **~ment** *s.* κατευνασμός *m.*
appeliant *s.* (*law*) εκκαλών *m.*
appellation *s.* ονομασία *f.*
append *v.* επισυνάπτω. **~age** *s.* προσάρτημα *n.* **~ed** *a.* συνημμένος.
appendicitis *s.* σκωληκοειδίτις *f.*
appendix *s.* (*of book*) παράρτημα *n.*
appertain *v.* ανήκω.
appetite *s.* όρεξη *f.* lack of ~ ανορεξία *f.*
appetiz|er *s.* (*to eat*) ορεκτικό *n.* **~ing** *a.* ορεκτικός.
applau|d *v.* επευφημώ, (*praise*) επαινώ. **~se** *s.* επευφημία *f.*
apple *s.* μήλο *n.* ~ of discord μήλον της

έριδος. it is the ~ of my eye το προσέχω ως κόρην οφθαλμού. **~-tree** s. μηλιά f.

appliance s. (tool) σύνεργο n. (mechanical) μηχάνημα n. see apparatus.

applicable a. εφαρμόσιμος. it is ~ to me? εφαρμόζεται σε μένα;

applicant s. υποψήφιος m.

application s. (using) εφαρμογή f. (putting) επίθεση f. (request) αίτηση f. (diligence) επιμέλεια f.

applied a. εφαρμοσμένος.

apply v.t. (put) βάζω. (put into use) εφαρμόζω. ~ oneself to (task) επιδίδομαι σε. (v.i.) (be valid) ισχύω. (make an application) κάνω αίτηση, ~ for a post ζητώ θέση. ~ to (address oneself) απευθύνομαι σε. (concern) αφορώ (v.t.).

appoint v. (person) διορίζω, (time, task, etc.) καθορίζω. well ~ed καλοβαλμένος.

appointment s. (of person) διορισμός m. (post) θέση f. (rendezvous) συνέντευξη f., ραντεβού n.

apportion v. μοιράζω, διανέμω. ~ blame κατανέμω τις ευθύνες. **~ment** s. μοιρασιά f., διανομή f.

apposite a. (suitable) κατάλληλος, (well put) εύστοχος.

apposition s. (gram.) παράθεση f.

appraise v. εκτιμώ. **~al** s. εκτίμηση f.

appreciable a. αισθητός. **~ly** adv. αισθητώς.

appreciate v.t. (value) εκτιμώ, (enjoy) απολαμβάνω, (understand) αντιλαμβάνομαι. I ~ what he did for me τον ευγνωμονώ για όσα έκανε για μένα. (v.i.) it has ~ed in value ανέβηκε η τιμή του.

appreciation s. εκτίμηση f. (enjoyment) απόλαυση f. (rise in value) ανατίμηση f. show one's ~ δείχνω την ευγνωμοσύνη μου.

appreciative a. he had an ~ audience το ακροατήριο τον υποδέχτηκε ευνοϊκά.

apprehend v. (understand) αντιλαμβάνομαι, (arrest) συλλαμβάνω.

apprehension s. (worry) ανησυχία f. (arrest) σύλληψη f. **~ive** a. ανήσυχος.

apprentice s. μαθητευόμενος m. (v.) be ~d μαθητεύω. **~ship** s. μαθητεία f.

apprise v. ~ (person) of πληροφορώ για.

approach v.t. & i. πλησιάζω, ζυγώνω. winter ~es πλησιάζει (or κοντεύει) ο χειμώνας.

approach s. (act) προσπέλαση f. (entrance) είσοδος f. (way, path, lit. & fig.) δρόμος m. the ~ of spring το πλησίασμα της άνοιξης. difficult of ~ δυσπρόσιτος. he ~ κάνω προτάσεις. his ~ to the matter ο τρόπος με το οποίο αντιμετωπίζει την υπόθεση. ~

road οδός προσεγγίσεως.

approachable a. ευπρόσιτος.

approbation s. έγκριση f., επιδοκιμασία f.

appropriate a. κατάλληλος. **~ly** adv. καταλλήλως.

appropriate v. (take) οικειοποιούμαι, (devote) προορίζω. **~ion** s. οικειοποίηση f. προορισμός m.

approval s. έγκριση f., επιδοκιμασία f. on ~ επί δοκιμή.

approve v. εγκρίνω. ~e of εγκρίνω, επιδοκιμάζω. **~ing** a. επιδοκιμαστικός.

approximate a. the ~ number ο κατά προσέγγισιν αριθμός. **~ly** adv. κατά προσέγγισιν, περίπου.

approximate v.i. ~ to πλησιάζω, προσεγγίζω. **~ion** s. προσέγγιση f.

appurtenance s. εξάρτηση f.

apricot s. βερίκοκο n. **~-tree** s. βερικοκιά f.

April s. Απρίλιος, Απρίλης m. ~ fool's day prank πρωταπριλιάτικο ψέμα.

a priori adv. εκ των προτέρων.

apron s. ποδιά f. he is tied to her ~-strings είναι κολλημένος στα φουστάνια της.

apse s. αψίς, αψίδα f.

apt a. (to the point) εύστοχος. an ~ pupil μαθητής με αντίληψη. he is ~ to catch cold έχει προδιάθεση στα κρυολογήματα. **~ly** adv. καλά, καλώς.

aptitude s. ικανότητα f.

aquarium s. ενυδρείο n.

aquatic a. υδρόβιος, (sport) θαλάσσιος.

aqueduct s. υδραγωγείο n.

aquiline a. αέτειος, (nose) γρυπός.

Arab a. αραβικός. (person) Άραβας m.

arabesque s. αραβούργημα n. (mus.) αραμπέσκ n.

Arabian, **~ic** a. αραβικός. ~ic (language) αραβική f.

arable a. καλλιεργήσιμος.

arbiter s. (judge) κριτής m. (lord) κύριος m.

arbitrary a. (despotic) δεσποτικός, (highhanded) αυθαίρετος. **~ily** adv. αυθαιρέτως, (fam.) με το έτσι θέλω. (at random) στην τύχη.

arbitrate v.i. ενεργώ διαιτησίαν. **~ion** s. διαιτησία f. submit (case) for ~ion υποβάλλω εις διαιτησίαν. **~or** s. διαιτητής m.

arbour s. σκιάδα f. vine ~ κληματαριά f.

arc s. τόξο n. **~-light** s. λυχνία βολταϊκού τόξου.

arcade s. στοά f.

arch s. αψίδα f., καμάρα f., τόξο n.

arch v.i. (form an ~) σχηματίζω αψίδα. (v.t.) (make like an ~) καμπουριάζω.

arch a. τσαχπίνης.

archaeological a. αρχαιολογικός. **~ist** s. αρχαιολόγος m. **~y** s. αρχαιολογία f.

archa|ic *a.* αρχαϊκός. ~ism *s.* αρχαϊσμός *m.*
archangel *s.* αρχάγγελος *m.*
archbishop *s.* αρχιεπίσκοπος *m.* ~ric *s.* αρχιεπισκοπή *f.*
archduke *s.* αρχιδούκας *m.*
arched *a.* καμαρωτός, τοξοειδής.
archer *s.* τοξότης *m.* ~y *s.* τοξοβολία *f.*
archetype *s.* αρχέτυπο *n.*
archimandrite *s.* αρχιμανδρίτης *m.*
archipelago *s.* αρχιπέλαγος *n.*
architect *s.* αρχιτέκτων *m.* ~ural *a.* αρχιτεκτονικός. ~ure *s.* αρχιτεκτονική *f.*
architrave *s.* επιστύλιο *n.*
archives *s.* αρχείο *n.*
arctic *a.* αρκτικός.
ardent *a.* θερμός, διακαής. ~ly *adv.* θερμώς.
ardour *s.* θέρμη *f.,* ζέση *f.* (*zeal*) ζήλος *m.*
arduous *a.* σκληρός, κοπιαστικός.
are *v.* we ~ είμαστε, είμεθα. you ~ είστε, είσθε. they ~ είναι. as things ~ όπως έχουν τα πράγματα.
area *s.* (*measurement*) εμβαδόν *n.* (*region*) περιοχή *f.*
arena *s.* παλαίστρα *f.*
Areopagus *s.* Άρειος Πάγος *m.*
Argentine *a.* Αργεντινός.
argue *v.* (*maintain*) υποστηρίζω, (*discuss*) συζητώ (*also* ~ about). (*prove*) δεικνύω. don't ~! μη συζητάς *or* μην αντιλέγεις.
argument *s.* (*discussion*) συζήτηση *f.* (*reason*) επιχείρημα *n.* (*plot*) υπόθεση *f.*
argumentative *a.* ~ person αντιρρησίας *m.,* (*fam.*) πνεύμα αντιλογίας.
aria *s.* (*mus.*) άρια *f.*
arid *a.* ξηρός, (*waterless*) άνυδρος. ~ity *s.* ξεραΐλα *f.*
aright *adv.* καλά, καλώς.
arise *v.* (*come about*) προκύπτω, παρουσιάζομαι, γίνομαι. (*get up*) σηκώνομαι.
aristocra|cy *s.* αριστοκρατία *f.* ~t *s.* αριστοκράτης *m.* ~tic *a.* αριστοκρατικός.
arithmetic *s.* αριθμητική *f.* ~al *a.* αριθμητικός.
ark *s.* κιβωτός *f.*
arm *s.* χέρι *n.,* μπράτσο *n.,* with open ~s με ανοικτές αγκάλες. she fell into his ~s έπεσε στην αγκαλιά του. keep at ~'s length κρατώ σε απόσταση. ~-in-~ αγκαζέ. *see* arms.
arm *v.t.* οπλίζω, εξοπλίζω, (*v.i.*) οπλίζομαι. ~ed οπλισμένος. ~ed forces ένοπλοι δυνάμεις.
armada *s.* αρμάδα *f.*
armament *s.* (*arming*) εξοπλισμός *m.* ~s οπλισμός *m.*
arm-chair *s.* πολυθρόνα *f.*
Armenian *a.* αρμενικός. (*person*) Αρμένης *m.,* Αρμένισσα *f.*

armful *s.* αγκαλιά *f.*
arm-hole *s.* μασχάλη *f.*
armistice *s.* ανακωχή *f.*
armour *s.* suit of ~ πανοπλία *f.* ~ed *a.* (*mil.*) τεθωρακισμένος. ~-plate *s.* χαλύβδινα ελάσματα.
armoury *s.* οπλοστάσιον *n.*
armpit *s.* μασχάλη *f.*
arms *s.* όπλα *n.pl.* take up ~ παίρνω τα όπλα. lay down ~ καταθέτω τα όπλα. be up in ~ επαναστατώ.
army *s.* στρατός *m.* ~ corps σώμα στρατού. (*fig., horde*) ασκέρι *n.* serve in ~ στρατεύομαι.
aroma *s.* άρωμα *n.* ~tic *a.* αρωματικός.
around *adv.* γύρω, all ~ γύρω γύρω. somewhere ~ here κάπου εδώ γύρω. (*prep.*) γύρω από, περί (*with acc.*). walk ~ the room κόβω βόλτες μέσα στο δωμάτιο. walk ~ the house φέρνω γύρω το σπίτι.
arouse *v.* εγείρω, (*pity, etc.*) κινώ, προκαλώ, (*waken*) ξυπνώ.
arraign *v.* κατηγορώ.
arrange *v.* (*agree on, fix*) συμφωνώ, κανονίζω. (*set in order, put right*) τακτοποιώ, σιάζω. (*place in position*) τοποθετώ. (*settle matter*) κανονίζω, τακτοποιώ. (*mus.*) διασκευάζω. (*see* adjust).
arrangement *s.* τακτοποίηση *f.,* σιάξιμο *n.,* τοποθέτηση *f.* (*agreement*) συμφωνία *f.* (*mus.*) διασκευή *f.* make ~s κανονίζω.
arrant *a.* ~ liar αρχιψεύταρος *m.* ~ fool θεόκουτος *m.* ~ nonsense κολοκύθια με τη ρίγανη.
array *v.* (*marshal*) παρατάσσω, (*clothe*) περιβάλλω. ~ed (*clothed*) περιβεβλημένος. (*s.*) (*order*) παράταξη *f.* (*clothes*) περιβολή *f.* (*display*) έκθεση *f.*
arrear *s.* ~s of rent καθυστερούμενα ενοίκια. be in ~ with taxes καθυστερώ φόρους.
arrest *v.* (*seize*) συλλαμβάνω, (*stop*) σταματώ. it ~ed my attention τράβηξε την προσοχή μου. (*s.*) σύλληψη *f.* under ~ υπό κράτησιν. ~ing *a.* εντυπωσιακός.
arriv|e *v.* φθάνω, αφικνούμαι. ~al *s.* άφιξη *f.* ~iste *s.* αριβίστας *m.*
arrogan|ce *s.* αλαζονεία *f.* ~t *a.* αλαζών.
arrogate *v.* ~ to oneself διεκδικώ παρ' αξίαν.
arrow *s.* βέλος *n.,* σαΐτα *f.*
arsenal *s.* οπλοστάσιο *n.*
arson *s.* εμπρησμός *m.*
art *s.* (*~istic creativeness*) καλλιτεχνία *f.* (*skill, branch of ~*) τέχνη *f.* he devoted himself to ~ αφοσιώθηκε στην καλλιτεχνία. work of ~ έργο τέχνης, καλλιτέχνημα *n.* fine ~s καλές τέχνες. ~s (*ruses*) κόλπα *n.pl.* Faculty of A~s Φιλοσοφική Σχολή.

artefact s. χειροποίητο αντικείμενο.

arter|ial a. αρτηριακός. **~y** s. αρτηρία f.

artesian a. ~ well αρτεσιανόν φρέαρ.

artful a. πονηρός. **~ly** adv. πονηρά.

arthrit|is s. αρθρίτις f. **~ic** a. αρθριτικός.

artichoke s. (globe) αγκινάρα f.

article s. (thing) πράγμα n. (item) είδος n., ~s of clothing είδη ρουχισμού. (in press & document & gram.) άρθρο n.

articled a. μαθητευόμενος, (to solicitor) ασκούμενος.

articulat|e v. αρθρώνω. (a.) έναρθρος. (of person) be **~e** έχω ευχέρεια λόγου. **~ed** a. αρθρωτός. **~ion** s. άρθρωση f.

artifice s. τέχνασμα n. **~r** s. τεχνίτης m.

artificial a. τεχνητός, (fam.) ψεύτικος. **~ity** s. επιτήδευση f. **~ly** adv. τεχνητώς.

artillery s. πυροβολικόν n.

artisan s. τεχνίτης m.

artist s. καλλιτέχνης m. (painter) ζωγράφος m. **~e** s. αρτίστας m., αρτίστα f.

artistic a. (artist's, of ~ merit) καλλιτεχνικός. (with good taste) καλαίσθητος. **~ally** adv. καλλιτεχνικώς.

artistry s. καλλιτεχνία f.

artless a. αφελής, αθώος. **~ly** adv. αφελώς.

Aryan a. ιαπετικός.

as adv. & conj. 1. ~ ... ~ (comparative) τόσο(ς) ... όσο(ς) (και) or εξίσου ... με. we haven't had ~ many bathes this year ~ last δεν κάναμε τόσα (or τόσο πολλά) μπάνια φέτος όσα (or όσο) πέρσι. does Mary talk ~ quickly ~ you? μιλάει η Μαίρη τόσο γρήγορα όσο εσύ or εξίσου γρήγορα με σένα or γρήγορα σαν και σένα or γρήγορα όπως (κι) εσύ or γρήγορα όσο (κι) εσύ; (simile) he is ~ strong ~ a lion είναι δυνατός σα λιοντάρι. it's ~ sweet ~ honey είναι γλυκό μέλι. (various) come ~ soon ~ possible ελάτε όσο το δυνατό νωρίτερα or όσο πιο νωρίς μπορείτε. ~ soon ~ we arrived μόλις (or άμα) φτάσαμε. ~ long (or much) ~ you like όσο θέλετε. I can eat fish ~ long ~ it isn't frozen τρώω ψάρι, αρκεί να μην είναι κατεψυγμένο. she brought figs ~ well ~ grapes έφερε και σταφύλι και σύκα. let's go ~ far ~ the shops πάμε ως (or μέχρι) τα μαγαζιά. ~ far ~ I know απ' ό,τι ξέρω. ~ far ~ I'm concerned όσον αφορά εμένα. ~ far ~ the eye can see ώσπου φτάνει το μάτι. 2. (while) καθώς, ενώ. (because) αφού, επειδή, μια που. rich ~ he is (being rich) όντας πλούσιος or σαν πλούσιος που είναι, (although rich) παρ' όλο που είναι πλούσιος. so ~ to (purpose) για να, (effect) ώστε να. 3. (various) ~ you wish όπως θέλεις. do ~ I told you κάνε όπως σου είπα. I regard him ~ a brother

τον έχω σαν αδελφό. speaking ~ a father μιλώντας σαν (or ως) πατέρας. such influence ~ I possess όση επιρροή διαθέτω. he came dressed ~ Pierrot ήρθε ντυμένος πιερότος. ~ for me όσο για μένα. ~ it were σα να λέμε. ~ it is στην πραγματικότητα. ~ if, ~ though σα να.

asbestos s. αμίαντος m.

ascend v. ανεβαίνω. ~ the stairs ανεβαίνω τη σκάλα. ~ the throne ανεβαίνω στο θρόνο.

ascendan|cy s. have ~cy over κυριαρχώ (with gen.). **~t** s. be in the ~t κυριαρχώ.

Ascension s. (eccl.) Ανάληψη f.

ascent s. άνοδος f., ανάβαση f., ανέβασμα n. (upward slope) ανήφορος m.

ascertain v.t. εξακριβώνω. (v.i.) πληροφορούμαι.

ascetic a. & s. ασκητικός. **~ism** s. ασκητισμός m.

ascrib|e v. αποδίδω. it is ~able to αποδίδεται σε.

aseptic a. ασηπτικός.

ash, ~es s. στάχτη f., τέφρα f. reduce to ~es αποτεφρώνω. A~ Wednesday Καθαρή Τετάρτη. **~tray** s. τασάκι n.

ashen a. σταχτής, (with fear, etc.) λευκός ως σουδάριον.

ashore adv. στην ξηρά. go ~ αποβιβάζομαι. (of ship) run ~ εξοκέλλω.

Asi|an, ~atic a. ασιατικός. (person) Ασιάτης m.

aside adv. κατά μέρος. put ~ (save) βάζω στη μπάντα. set ~ (nullify) ακυρώνω. move or thrust ~ παραμερίζω (v.t. & i.). (s.) παρατήρησις κατ' ιδίαν.

asinine a. (fig.) ηλίθιος.

ask v. (enquire) (ε)ρωτώ. (~ for, seek) ζητώ. (request person) παρακαλώ. (invite) (προσ) καλώ. he ~ed me what I wanted με ρώτησε τι θέλω. he ~ed for me με ζήτησε. he ~ed me for money μου ζήτησε λεφτά. he ~ed to see me ζήτησε να με δει. he ~ed me to see him με παρακάλεσε να τον 'δω. he ~ed me to tea με (προσ) κάλεσε για τσάι. he ~ed me a question μου έκανε μια ερώτηση.

askance adv. look ~ at κοιτάζω καχύποπτα.

askew adv. λοξά, στραβά.

aslant adv. λοξά.

asleep a. κοιμισμένος. be ~ κοιμάμαι. fall ~ αποκοιμούμαι, με παίρνει ο ύπνος.

asparagus s. σπαράγγι n.

aspect s. (look) όψη f., (point of view) άποψη f. the house has a north ~ το σπίτι βλέπει προς το βοριά. (gram.) ποιόν ενεργείας.

asperity s. τραχύτητα f. speak with ~ μιλώ

με τραχύ τόνο.

asperse v. (slander) κακολογώ. (besprinkle) ραντίζω.

aspersion s. cast ~s on κακολογώ. casting of ~s κακολογία f.

asphalt s. άσφαλτος f.

asphodel s. ασφόδελος m., (fam.) σφερδούκλι n.

asphyxi|a s. ασφυξία f. ~ate v.t. πνίγω. be ~ated αφυκτιώ. ~ation s. πνίξιμο n.

aspirant s. υποψήφιος m.

aspirate v. (gram.) δασύνω. (s.) δασύ σύμφωνο.

aspiration s. (wish) φιλοδοξία f. ~s βλέψεις f.pl.

aspir|e v. ~e το φιλοδοξώ (with acc. or να). (covet) αποβλέπω σε. he is an ~ing actor φιλοδοξεί να γίνει ηθοποιός.

aspirin s. ασπιρίνη f.

ass s. γάιδαρος m. (person) βλάκας m.

assail v. επιτίθεμαι κατά (with gen.). (fig.) be ~ed with κατέχομαι από. ~ant s. ο επιτιθέμενος.

assassin s. δολοφόνος m. ~ate v. δολοφονώ. ~ation s. δολοφονία f.

assault s. (general) επίθεση f. (bodily) βιαιοπραγία f. (rape) βιασμός m. (mil.) έφοδος f. it was taken by ~ κατελήφθη εξ εφόδου.

assault v. επιτίθεμαι κατά (with gen.). (rape) βιάζω.

assemblage s. συγκέντρωση f., συνέλευση f.

assemble v.t. (gather) συναθροίζω, συγκεντρώνω. (put together) συναρμολογώ. (v.i.) συναθροίζομαι, συγκεντρώνομαι.

assembly s. συνέλευση f. ~ line χώρος συναρμολογήσεως.

assent s. συγκατάθεση f. (v.) συγκατατίθεμαι.

assert v. (declare) ισχυρίζομαι. ~ one's right διεκδικώ το δικαίωμά μου. ~ one's authority επιβάλλομαι.

assert|ion s. (statement) ισχυρισμός m. ~ive a. δογματικός.

assess v. (value) εκτιμώ, (fix) ορίζω. ~ment s. εκτίμηση f. ~or s. (law) πάρεδρος m.

asset s. (good thing) αγαθό n. ~s (possessions) αγαθά n.pl., (fin.) ενεργητικό n. (fam.) it's a great ~ having the telephone είναι μεγάλο πράγμα να έχεις τηλέφωνο.

assidu|ity s. ενδελέχεια f. ~ous a. ενδελεχής. ~ously adv. ενδελεχώς.

assign v. (make over) εκχωρώ. (fix) ορίζω. (give) παραχωρώ. (duty to person) αναθέτω, I was ~ed to look after the children μου ανέθεσαν την επίβλεψη των παιδιών.

assignation s. ραντεβού n.

assignment s. (mission) αποστολή f. (task) καθήκον n.

assimilat|e v. αφομοιώνω. ~ion s. αφομοίωση f.

assist v. βοηθώ. ~ance s. βοήθεια f. ~ant s. βοηθός m. (shop) υπάλληλος m.f.

assizes s. συνεδριάσεις δικαστηρίων.

associate v.t. & i. (link) συνδέω. ~ (mix) with συναναστρέφομαι. ~ oneself or be ~d with έχω να κάνω με.

associat|e s. συνέταιρος m. ~ion s. (link) σχέση f. (partnership) συνεταιρισμός m. ~ion of ideas συνειρμός παραστάσεων.

assort|ed a. διάφορος, ποικίλος. ~ment s. ποικιλία f.

assuage v. ανακουφίζω.

assum|e v. (appearance) προσλαμβάνω. (undertake) αναλαμβάνω. (feign) προσποιούμαι. (suppose) υποθέτω, ~ing that... αν υποτεθή ότι. ~ed a. προσποιητός.

assumption s. (supposition) υπόθεση f. (eccl.) A~ of the Virgin Mary Κοίμησις της Θεοτόκου.

assurance s. (statement) βεβαίωση f. (certainty) βεβαιότητα f. (confidence) αυτοπεποίθηση f. (fin.) ασφάλιση f.

assure v. (tell) βεβαιώνω, (make sure of) εξασφαλίζω. ~d a. βέβαιος. ~dly adv. βεβαίως.

asterisk s. αστερίσκος m.

astern adv. όπισθεν, go ~ ανακρούω όπισθεν.

asthma s. άσθμα n.

astir adv. (up) στο πόδι. (a., excited) ξεσηκωμένος.

astonish v. εκπλήσσω. ~ed a. έκπληκτος. ~ing a. εκπληκτικός. ~ment s. έκπληξη f.

astound v. καταπλήσσω. ~ed a. κατάπληκτος, εμβρόντητος. ~ing a. καταπληκτικός.

astray adv. lead ~ παραπλανώ. go ~ παραπλανώμαι, (get mislaid) παραπέφτω.

astride adv. καβάλα.

astringent a. (to skin) στυπτικός, (to taste) στυφός. (severe) δριμύς.

astrolog|er s. αστρολόγος m. ~y s. αστρολογία f.

astronaut s. αστροναύτης m.

astronom|er s. αστρονόμος m. ~ic(al) a. αστρονομικός. ~y s. αστρονομία f.

astute a. ξύπνιος, ανοιχτομάτης. ~ness s. εξυπνάδα f.

asunder adv. χωριστά, (into pieces) σε κομμάτια.

asylum *s.* άσυλο *n.* lunatic ~ φρενοκομείο *n.*

at *prep.* σε, εις. ~ home (στο) σπίτι. ~ my uncle's στου θείου μου. ~ two o'clock στις δύο, ~ Easter το Πάσχα, ~ night τη νύχτα. ~ that time εκείνη την εποχή. ~ his expense (*detriment*) εις βάρος του, ~ the company's expense με έξοδα της εταιρείας. ~ the first opportunity με την πρώτη ευκαιρία. ~ a profit με κέρδος. I laughed ~ his jokes γέλασα με τα αστεία του. get in ~ the window μπαίνω από το παράθυρο. call ~ the office περνώ από το γραφείο. ~ once αμέσως, all ~ once ξαφνικά. ~ one go μονορούφι. ~ hand κοντά. ~ a run τροχάδην, τρεχάλα. (not) ~ all καθόλου. ~ that (*thereupon*) οπότε, πάνω σ' αυτό. *there was only one hotel* and a poor one ~ that ούτε κι αυτό της προκοπής. they are selling ~ 10 drachmas each πουλιούνται δέκα δραχμές το ένα. what is he ~? (*doing*) με τι ασχολείται; (*up to*) τι μαγειρεύει; he is hard ~ it εργάζεται εντατικά. he is ~ it again πάλι ξανάρχισε τα ίδια. she is always on ~ him όλο τα έχει μαζί του.

atavism *s.* αταβισμός *m.*

athe|ism *s.* αθεϊσμός *m.* **~ist** *s.* άθεος *m.* **~istic** *a.* αθεϊστικός.

Athenian *a.* αθηναϊκός. (*person*) Αθηναίος *m.*, Αθηναία *f.*

athirst *a.* διψασμένος.

athlet|e *s.* αθλητής *m.* **~ic** *a.* (*also* ~e's) αθλητικός. **~ics** *s.* αθλητισμός *m.*

atlas *s.* άτλας *m.*

atmospher|e *s.* ατμόσφαιρα *f.* **~ics** *s.* παράσιτα *n.pl.*

atom *s.* άτομο *n.* **~ic** *a.* ατομικός.

atone *v.* ~ for (*put right*) επανορθώνω, (*expiate*) εξιλεώνομαι για. **~ment** *s.* (*eccl.*) εξιλέωση *f.*

atroci|ous *a.* φρικαλέος (*fam.*) φοβερός. **~ties** *s.* ωμότητες *f.pl.* **~ty** *s.* (*fam., awful object*) εξάμβλωμα *n.*

atrophy *s.* ατροφία *f.* (*v.*) ατροφώ.

attach *v.* προσκολλώ. (*tie*) δένω. (*law*) κατάσχω. ~ importance το δίδω σημασίαν εις. be ~ed to (*linked with*) συνδέομαι με, (*fond of*) αγαπώ. ~ oneself το προσκολλώμαι εις, κολλάω σε.

attachment *s.* (*act*) προσκόλληση *f.* (*love*) αγάπη *f.* (*linking*) σύνδεση *f.* (*accessory*) εξάρτημα *n.* (*law*) κατάσχεση *f.*

attaché *s.* ακόλουθος *m.* **~-case** *s.* χαρτοφύλακας *m.*

attack *v.t.* επιτίθεμαι κατά (*with gen.*), ρίχνομαι σε. (*not physically*) προσβάλλω. (*criticize*) χτυπώ. (*s.*) επίθεση *f.* (*fit*) κρίση *f.* **~er** *s.* ο επιτιθέμενος.

attain *v.* (*gain*) επιτυγχάνω. ~ to φθάνω εις.

~able *a.* εφικτός.

attainment *s.* (*achieving*) επίτευξη *f.* (*talent*) ταλέντο *n.*

attar *s.* ~ of roses ροδέλαιο *n.*

attempt *v.i.* (*try*) προσπαθώ, πασχίζω. (*v.t.*) (*undertake*) επιχειρώ. (*s.*) απόπειρα *f.* they made an ~ on his life αποπειράθηκαν να τον δολοφονήσουν.

attend *v.i.* (*be present*) παρίσταμαι, (*pay attention*) προσέχω. (*v.t.*) (*escort*) συνοδεύω. (*patient*) κουράρω. (*classes, etc.*) παρακολουθώ. it is ~ed by danger είναι επικίνδυνο. ~ to (*listen, etc.*) προσέχω, (*see to sthg.*) φροντίζω για, (*show attentions to*) περιποιούμαι, (*serve*) υπηρετώ.

attendance *s.* (*presence*) παρουσία *f.* (*at school, etc.*) φοίτηση *f.* (*medical*) παρακολούθηση *f.* (*service*) υπηρεσία *f.*, be in ~ είμαι της υπηρεσίας. there was a good ~ είχε πολύ κόσμο.

attendant *a.* επακόλουθος (*s.*) (*employee*) υπάλληλος *m.f.* (*of car-park, etc.*) φύλακας *m.* (*of personage*) ακόλουθος *m.* medical ~ θεράπων ιατρός.

attention *s.* προσοχή *f.* pay ~ προσέχω. don't pay any ~ to him μην του δίνεις σημασία. (*care*) περιποίηση *f.* stand to ~ στέκομαι προσοχή.

attentive *a.* (*heedful*) προσεκτικός, (*polite*) περιποιητικός. **~ly** *adv.* προσεκτικά, περιποιητικά.

attenuate *v.* (*weaken*) εξασθενίζω. **~d** *a.* (*reduced*) μειωμένος, (*tapering*) κοντυλένιος.

attest *v.t.* (*show*) αποδεικνύω, (*put on oath*) ορκίζω. (*v.i.*) (*swear*) ορκίζομαι. ~ to μαρτυρώ.

attic *s.* σοφίτα *f.*

attire *s.* περιβολή *f.* **~d** *a.* ντυμένος.

attitud|e *s.* (*of body, mind*) στάση *f.* (*of mind*) θέση *f.* strike an ~e ποζάρω. **~inize** *v.* ποζάρω.

attorney *s.* (*lawyer*) δικηγόρος *m.* (*proxy*) πληρεξούσιος *m.* warrant of ~ πληρεξούσιον *n.*

attract *v.* (*draw*) προσελκύω. (*charm*) ελκύω, τραβώ.

attraction *s.* έλξη *f.* **~s** (*pleasures*) τέρψεις *f.pl.*, (*charms*) θέλγητρα *n.pl.* she has great ~ for him τον τραβάει πολύ.

attractive *a.* ελκυστικός. **~ness** *s.* ελκυστικότητα *f.*

attribute *s.* ιδιότητα *f.* (*symbol*) έμβλημα *n.*

attribut|e *v.* αποδίδω. **~able** *a.* it is ~able το αποδίδεται σε.

attrition *s.* κατατριβή *f.*

attuned *a.* προσαρμοσμένος.

aubergine *s.* μελιτζάνα *f.*

auburn *a.* καστανόξανθος.

auction *s.* πλειστηριασμός *m.* put up to ~ βγάζω σε πλειστηριασμό.

audaci|ous *a.* τολμηρός, θρασύς. ~ty *s.* τόλμη *f.*, θράσος *n.*

audible *a.* he was scarcely ~ μόλις ακουγόταν.

audience *s.* (*listeners*) ακροατήριο *n.* (*interview*) ακρόαση *f.*

audio-visual *a* οπτικοακουστικός.

audit *v.* ελέγχω. (*s.*) A~ Board ελεγκτικόν συνέδριον. ~or *s.* (*of accounts*) ελεγκτής *m.* (*hearer*) ακροατής *m.*

audition *s.* οντισιόν *f.* have an ~ πάω για οντισιόν.

auditorium *s.* αίθουσα *f.*

auditory *a.* ακουστικός.

aught *s.* for ~ I know πολύ που ξέρω. for ~ I care πολύ που με νοιάζει.

augment *v.* αυξάνω (*s.*, *gram.*) αύξηση *f.* ~ation *s.* αύξηση *f.*

augur *v.* προοιωνίζομαι. it ~s well είναι καλός οιωνός. ~y *s.* (*omen*) οιωνός *m.*

August *s.* Αύγουστος *m.*

august *a.* αξιοσέβαστος.

aunt *s.* θεία *f.*

aura *s.* ατμόσφαιρα *f.*

au revoir *int.* καλή αντάμωση.

aurora *s.* ~ borealis βόρειον σέλας.

auspices *s.* under the ~ of υπό την αιγίδα (*with gen.*).

auspicious *a.* αίσιος, ευοίωνος. ~ly *adv.* αισίως.

auster|e *a.* αυστηρός, (*plain*) λιτός. ~ely *adv.* αυστηρά, λιτά. ~ity *s.* αυστηρότητα *f.* λιτότητα *f.*

Australian *a.* αυστραλιανός. (*person*) Αυστραλός *m.*, Αυστραλέζα *f.*

Austrian *a.* αυστριακός.

autarchy *s.* αυταρχία *f.*

autarky *s.* αυτάρκεια *f.*

authentic *a.* αυθεντικός. ~ate *v.* επιβεβαιώνω. ~ity *s.* αυθεντικότητα *f.*

author *s.* (*writer*) συγγραφεύς *m.f.* (*creator*) δημιουργός *m.*

authoritarian *a.* απολυτόφρων.

authoritative *a.* (*commanding*) επιτακτικός, (*valid*) έγκυρος.

authorit|y *s.* (*power*) εξουσία *f.* (*delegated*) εξουσιοδότηση *f.* (*weight, validity*) κύρος *n.* (*an expert*) αυθεντία *f.* the ~ies οι αρχές. (*be*) under the ~y of υπό τας διαταγάς, (*act*) on the ~y of κατ' εντολήν (*both with gen.*).

authorize *v.* (*sanction*) εγκρίνω, (*give authority to*) εξουσιοδοτώ, (*permit*) επιτρέπω.

authorization *s.* έγκριση *f.* εξουσιοδότηση *f.*

(*permission*) άδεια *f.*

autobiography *s.* αυτοβιογραφία *f.*

autocrat *s.*, ~ic *a.* αυταρχικός.

autograph *s.* αυτόγραφο *n.*

automat|ic *a.* αυτόματος. ~ically *adv.* αυτομάτως. ~ion *s.* αυτοματισμός *m.* ~on *s.* αυτόματο *n.*

automobile *s.* αυτοκίνητο *n.*

autonom|ous *a.* αυτόνομος. ~y *s.* αυτονομία *f.*

autopilot *s.* αυτόματος πιλότος.

autopsy *s.* (*post-mortem*) νεκροτομία *f.*

auto-suggestion *s.* αυθυποβολή *f.*

autumn *s.* φθινόπωρο *n.* ~al *a.* φθινοπωρινός.

auxiliary *a.* βοηθητικός.

avail *v.t.* (*help*) ωφελώ. ~ oneself of επωφελούμαι (*with gen.*). (*s.*) it is of no ~ δεν ωφελεί.

available *a.* (*at disposal*) διαθέσιμος. nothing was ~ in the shops τίποτα δεν βρισκόταν στα μαγαζιά.

avalanche *s.* χιονοστιβάδα *f.* (*fig.*) βροχή *f.*

avant-gard|e *s.* πρωτοπορία. *f.* ~ist *s.* πρωτοπόρος *m.*

avaric|e *s.* φιλαργυρία *f.* ~ious *a.* φιλάργυρος.

aveng|e *v.t.* (*person, act*) (*also* ~e oneself on) εκδικούμαι. ~er *s.* εκδικητής *m.* ~ing *a.* εκδικητικός.

avenue *s.* λεωφόρος *f.* (*of trees*) δενδροστοιχία *f.* (*fig.*) δρόμος *m.*

aver *v.* διατείνομαι.

average *a.* μέσος. (*s.*) μέσος όρος, on the ~ κατά μέσον όρον. (*v.*) (*come to*) ανέρχομαι κατά μέσον όρον εις.

avers|e *a.* be ~e το έχω αντιπάθεια σε. ~ion *s.* αντιπάθεια *f.*

avert *v.t.* (*eyes, etc.*) αποστρέφω, (*ward off*) αποτρέπω.

aviary *s.* πτηνοτροφείον *n.*

aviat|ion *s.* αεροπορία *f.* ~or *s.* αεροπόρος *m.*

avid *a.* άπληστος. ~ity *s.* απληστία *f.* ~ly *adv.* απλήστως.

avocado *s.* αβοκάντο *n.*

avoid *v.* αποφεύγω. ~ance *s.* αποφυγή *f.*

avow *v.* ομολογώ. ~al *s.* ομολογία *f.* ~edly *adv.* ομολογουμένως.

await *v.* περιμένω, αναμένω.

awake *a.* ξύπνιος. stay ~ αγρυπνώ. be ~ to έχω συναίσθηση (*with gen.*).

awake, ~n *v.t.* & *i.* ξυπνώ. (*fig.*) (*v.t., rouse*) εγείρω, (*v.i.*) ~ to (*realize*) καταλαβαίνω. ~ning *s.* αφύπνιση *f.*

award *v.* απονέμω. (*grant*) χορηγώ. (*law*) επιδικάζω. (*s.*) απονομή *f.* χορήγηση *f.* (*prize*) βραβείο *n.* (*law*) επιδίκασις *f.*

award-winning *a.* βραβευμένος.
aware *a.* (*informed*) ενήμερος. be ~ of γνωρίζω. **see awake. ~ness** *s.* συνείδηση *f.*
awash *a.* πλημμυρισμένος.
away *adv.* be ~ (*from home, etc.*) λείπω. it is 10 minutes ~ from here απέχει δέκα λεπτά από δω. the house is ~ from the road το σπίτι είναι μακριά από το δρόμο. go *or* get ~ φεύγω, gone ~ φευγάτος. throw ~ πετώ. far and ~ κατά πολύ. explain (*sthg.*) ~ δίνω εξηγήσεις για. look ~ κοιτάζω αλλού. work ~ (*v.i.*) εργάζομαι εντατικά. *see* die, take ~, *etc.*
awl|e *s.* δέος *n.* **~ful** *a.* τρομερός. **~fully** *adv.* τρομερά.
awhile *adv.* για λίγο.
awkward *a.* (*inconvenient*) άβολος, (*maladroit*) αδέξιος, (*embarrassing*) δύσκολος. **~ness** *s.* αβολιά *f.* αδεξιότητα *f.* δυσκολία *f.*
awl *s.* σουβλί *n.*
awning *s.* τέντα *f.*
awry *adv.* (*crookedly*) στραβά. go ~ πηγαίνω ανάποδα.
axe *s.* τσεκούρι *n.* have an ~ to grind έχω δικά μου συμφέροντα. (*v.t.*) (*expenditure*) περικόπτω, (*staff*) απολύω.
axiom *s.* αξίωμα *n.* **~atic** *a.* αυταπόδεικτος.
axis *s.* άξων *m.*
axle *s.* άξων *m.*
aye *adv.* (*yes*) ναι, (*always*) πάντα.
azalea *s.* αζαλέα *f.*
azure *a.* γαλάζιος, κυανούς.

B

baa *v.* βελάζω. **~ing** *s.* βέλασμα *n.*
babble *v.i.* φλυαρώ, (*of water*) κελαρύζω. (*s.*) φλυαρία *f.* κελάρυσμα *n.* **~r** *s.* φλύαρος *a.*
babe *s.* νήπιο *n.*, μωρό *n.*
babel *s.* βαβυλωνία *f.*
baboon *s.* πίθηκος *m.*
baby *s.* μωρό *n.*, (*fam.*) μικρό *n.* (*fig.*) I was left holding the ~ μου φόρτωσαν άθελά μου την ευθύνη. (*a.*) μικρός. **~hood** *s.* νηπιακή ηλικία. **~ish** *a.* μωρουδίστικος.
baby-sit *v.i.* φυλάω το μωρό εν απουσία των γονιών.
bacchanal *a.* βακχικός. (*s.*) βακχίς *f.* **~ia** *s.* βακχικόν όργιον.
bacchante *s.* βάκχη *f.*, βακχίς *f.*
bacchic *a.* βακχικός.
bachelor *a.* άγαμος (*s.*) εργένης *m.* old ~ γεροντοπαλίκαρο *n.*
bacillus *s.* βακτηρίδιον *n.*
back *s.* (*of person, animal, chair, etc.*) ράχη *f.*, πλάτη *f.* (*of person, animal*) νώτα *n.pl.* (

far end) βάθος *n.* (*hinder part, reverse side*) το πίσω μέρος. ~ to ~ πλάτη με πλάτη. put his ~ up τον πικάρω. I turn my ~ on him του γυρίζω τις πλάτες. at the ~ of (*behind*) πίσω από. on one's ~ ανάσκελα.
back *v.i.* (*go ~wards*) κάνω όπισθεν. ~ water σιάρω. ~ out αποσύρομαι. ~ down (*fig.*) ανακρούω πρύμναν. ~ on to συνορεύω με. (*v.t.*) (*support*) υποστηρίζω, (*horse*) ποντάρω σε.
back *a.* πισινός, οπίσθιος. the ~ room το πισινό (*or* πίσω) δωμάτιο. take a ~ seat μπαίνω στο περιθώριο. ~ number παλαιό φύλλο, (*fig.*) be a ~ number έχω περάσει στο περιθώριο.
back *adv.* (ο)πίσω. move ~ κάνω πίσω. go *or* come *or* give ~ επιστρέφω, γυρίζω (πίσω). get ~ (*regain*) ξαναπαίρνω. go ~ to (*date from*) ανάγομαι εις. go ~ on one's word πατώ το λόγο μου. get one's own ~ εκδικούμαι. ~ to front ανάποδα.
backache *s.* have a ~ πονάει η πλάτη μου.
back-bencher *s.* απλός βουλευτής.
backbit|e *v.* κακολογώ. **~ing** *s.* κακολογία *f.*
backbone *s.* ραχοκοκαλιά *f.* (*also fig., main support*) σπονδυλική στήλη. (*fig., spirit*) ψυχή *f.*
backchat *s.* αντιμιλιά *f.* indulge in ~ γλωσσεύω, αντιμιλώ.
backer *s.* υποστηρικτής *m.* (*of horse*) πονταδόρος *m.*
backfire *s.* επιστροφή φλογός. (*v.*) (*fam., of engine*) κλωτσώ. (*fig.*) his attempt ~d η απόπειρα ξέσπασε στο κεφάλι του.
backgammon *s.* τάβλι *n.*
background *s.* φόντο *n.*, βάθος *n.*
backhanded *a.* (*equivocal*) διφορούμενος. ~ slap *or* stroke ανάποδη *f.*
backing *s.* υποστήριξη *f.*
backlash *s.* (*fig.*) αντίδραση *f.*
backlog *s.* συσσωρευθείσα εργασία.
backside *s.* πισινός *m.*
backslide *v.* ξαναπαίρνω στραβό δρόμο.
backstage *a.* παρασκηνιακός. (*adv.*) στα παρασκήνια.
backstair|s *s.* σκάλα υπηρεσίας. (*a.*) ~(s) ύπουλος. ~(s) influence τα μέσα.
backward *a.* (*retarded*) καθυστερημένος, (*in season*) όψιμος, (*shy*) ντροπαλός. (*towards the back*) προς τα πίσω.
backwards *adv.* προς τα πίσω, όπισθεν. go ~ κάνω όπισθεν. (*the wrong way round*) ανάποδα. go ~ and forwards πηγαινοέρχομαι. fall over ~ πέφτω ανάσκελα.
backwash *s.* κύματα προκαλούμενα από διερχόμενο πλοίο. (*fig.*) επακόλουθα *n. pl.*
backwater *s.* (*fig.*) in a ~ μακριά από το κύριο ρεύμα.

backwoods s. απομακρυσμένη περιοχή.
bacon s. μπέικον n.
bacteri|a s. βακτήρια n.pl. **~ology** s. βακτηριολογία f.
bad a. άσχημος. (esp. morally ~) κακός (also of luck, omen, temper, technique). (harmful) βλαβερός, it's ~ for you βλάπτει. (decayed) χαλασμένος, go ~ χαλώ. feel ~ αισθάνομαι άσχημα. have a ~ throat πονάει ο λαιμός μου. a ~ lot (fam.) εξώλης και προώλης. ~ luck to him! κακό χρόνο να 'χει! too ~ ! τι κρίμα! not ~ όχι κι άσχημος.
badge s. σήμα n.
badger s. ασβός m.
badger v.t. ενοχλώ, φορτώνομαι, (fam.) τρώω.
badly adv. άσχημα, κακώς, (hurt, beaten, etc.) σοβαρώς. need (sthg.) ~ έχω μεγάλη ανάγκη από. I want it ~ το θέλω πολύ. (of project) it turned out ~ ήταν αποτυχία. be ~ off (poor) απορώ χρημάτων. ~ made κακοφτιαγμένος.
bad-tempered a. (person) ανάποδος, ζόρικος, γρουσούζης.
baffle v.t. (thwart person) εμποδίζω. (prevent action) ματαιώνω. (perplex) ξενίζω. it ~s description δεν περιγράφεται.
bag s. (hand, shopping, travelling) τσάντα f. (paper, plastic) σακούλα f. (valise) βαλίτσα f. (large ~ or sack) σάκκος m.
bag v. (game) χτυπώ. (fam.) (steal) σουφρώνω, (secure) παίρνω.
baggage s. αποσκευές f.pl.
baggy a. σακουλιασμένος.
bagpipe s. γκάιδα f.
bail s. εγγύησις f. release on ~ απολύω επί εγγυήσει. go ~ for εγγυώμαι δια (with acc.).
bailiff s. (law) κλητήρας m. (land-agent) επιστάτης m.
bairn s. παιδί n.
bait s. δόλωμα n. (v.) δολώνω, (tease) πειράζω.
baize s. τσόχα f.
bake v. ψήνω **~d** a. ψητός.
bake-house s. φούρνος m., ψωμάδικο n.
baker s. ψωμάς m. **~y** s. see bake-house.
balance s. (scales) ζυγαριά f. (steadiness) ισορροπία f. (remainder) υπόλοιπο n. ~ of trade εμπορικόν ισοζύγιον. ~ of power ισορροπία των δυνάμεων.
balance v.t. & i. (make or remain steady) ισορροπώ. (v.t.) (weigh up) ζυγίζω. (match, make up for) αντισταθμίζω. (budget) ισοσκελίζω. well ~d ισορροπημένος.
balance-sheet s. ισολογισμός m.
balcony s. μπαλκόνι n. (of theatre) εξώστης m.

bald a. φαλακρός, go ~ φαλακραίνω. (style) ξηρός. **~ly** adv. ξηρώς. **~ness** s. (also ~ patch) φαλάκρα f.
balderdash s. ανοησίες f.pl.
baie s. μπάλα f.
bale v. ~ out (aero.) πέφτω με αλεξίπτωτο.
baleful a. βλοσυρός.
balk v.t. (obstruct) εμποδίζω. (v.i.) (shy) σκιάζομαι. ~ at δειλιάζω μπροστά σε.
Balkan a. βαλκανικός.
ball s. σφαίρα f. (in games) μπάλα f. (child's) τόπι n. (wool, etc.) κουβάρι n. (dance) χορός m. roll up into a ~ (v.i.) γίνομαι κουβάρι. (fig.) set the ~ rolling δίνω την πρώτη ώθηση.
ballad s. μπαλάντα f.
ballast s. σαβούρα f., έρμα n. (v.) σαβουρώνω, ερματίζω.
ball-bearing s. ρουλεμάν n.
ballet s. μπαλέτο n.
ballistic a. βλητικός.
balloon s. αερόστατο n. (toy) μπαλόνι n., φούσκα f.
ballot s. ψηφοφορία f. put to the ~ θέτω εις ψηφοφορίαν. (v.) ψηφίζω. **~-box** s. ψηφοδόχος f., κάλπη f.
ballyhoo s. (fam., publicity) θορυβώδης διαφήμιση.
balm s. βάλσαμο n. **~y** a. μυρωμένος, αβρός. (breeze) ηδύπνους.
balsam s. βάλσαμο n.
balustrade s. κιγκλίδωμα n.
bamboo s. ινδοκάλαμος m., μπαμπού n.
ban v. απαγορεύω. (s.) απαγόρευση f.
banal a. τετριμμένος.
banana s. μπανάνα f.
band s. (strip) λουρίδα f. (rubber) λαστιχάκι n. (mus. & design) μπάντα f. (gang) συμμορία f.
band v.i. (unite) ενώνομαι.
bandage s. επίδεσμος m. (v.) επιδένω. **~d** δεμένος.
bandit s. ληστής m. **~ry** s. ληστεία f.
bandolier s. φυσεκλίκι n.
bandwagon s. climb on the ~ (fig.) πάω με το ρεύμα.
band|y v. ~y words λογομαχώ. his name is ~ied about τον κουτσομπολεύουν.
bandy-legged a. στραβοκάνης.
bane s. κατάρα f. the ~ of my life ο εφιάλτης μου. **~ful** a. βλαβερός.
bang s. & int. (blow) χτύπημα n. (noise) βρόντος m. go with a ~ έχω επιτυχία. ~ go (my chances, etc.) πάνε στα κομμάτια. ~! μπουμ!
bang v.t. & i. βροντώ. he ~ed (on) the door βρόντησε την πόρτα.
bangle s. βραχιόλι n.

banish v. εξορίζω, (idea, etc.) βγάζω, διώχνω. **~ment** s. εξορία f.

banister s. κάγκελο n. ~ rail κουπαστή της σκάλας.

bank s. (of river) όχθη f. (sloping ground) πλαγιά f. (heap, clump) στοίβα f. (of earth) ανάχωμα n.

bank v.t. ~ up (stream, etc.) προσχώνω, (form into pile) συσσωρεύω, (v.i.) συσσωρεύομαι. ~ up the fire σκεπάζω τη φωτιά.

bank v.i. (also be ~ed) (of car, road) γέρνω. **~ed** a. επικλινής.

bank s. (fin.) τράπεζα f. ~ employee τραπεζιτικός m. ~ holiday αργία f. (a.) (banker's, of banking) τραπεζιτικός. (v.t.) (deposit) καταθέτω. (v.i.) ~ on βασίζομαι σε. **~book** s. βιβλιάριο n. **~er** s. τραπεζίτης m. **~note** s. χαρτονόμισμα n.

bankrupt a. χρεοκοπημένος. go ~ χρεοκοπώ, φαλίρω. **~cy** s. χρεοκοπία f.

banner s. μπαϊράκι n., (μ)παντιέρα f.

banquet s. δείπνο n., συμπόσιο n. (v.) συμποσιάζομαι.

banter s. άκακο πείραγμα.

bapt|ism s. βάπτισμα n. certificate of ~ism βαπτιστικόν n. **~ismal** a. βαπτιστικός. **~ist** s. βαπτιστής m. **~ize** v. βαπτίζω.

bar s. (gold) ράβδος f. (soap, etc.) πλάκα f. (bolt of door) αμπάρα f. (mus.) μέτρο n. (for drinks) μπαρ n. (fig., obstacle) εμπόδιο n. ~s (of window, prison) σίδερα n.pl. (legal profession) δικηγορικόν επάγγελμα.

bar v. (door) αμπαρώνω. (close) κλείνω, (prevent) εμποδίζω, (exclude, forbid) αποκλείω, απαγορεύω.

bar, ~ring prep. εκτός, πλην (with gen.), εκτός από (with acc.).

barb s. αντίστροφος αιχμή αγκίστρου. (fig.) βέλος n. **~ed** a. (fig., of words) αιχμηρός. ~ed wire ακιδωτό σύρμα, (fence) ακιδωτό συρματόπλεγμα.

barbar|ian a. & s. βάρβαρος. **~ic** a. βαρβαρικός. **~ism** s. βαρβαρισμός m. **~ity** s. βαρβαρότητα f. **~ous** a. (savage) άγριος, (cruel) ωμός, (awful) απαίσιος.

barbecue s. ψησταριά f. (v.t.) ψήνω στη σούβλα.

barber s. κουρέας m. ~'s shop κουρείο n.

bard s. αοιδός m. βάρδος m.

bare a. γυμνός, (mountain) φαλακρός. (empty) άδειος. make a ~ living φυτοζωώ. (v.t.) (also lay ~) αποκαλύπτω, (show) δείχνω. **~ness** s. γυμνότητα f.

barefaced a. αναιδής.

barefoot a. ξιπόλητος.

bareheaded a. ασκεπής, ξεκαπέλωτος.

barely adv. (scarcely) μόλις. (poorly) φτωχά. (simply) απλά, απέριττα.

bargain s. (agreement) συμφωνία f. (good purchase) κελεπούρι n., ευκαιρία f. (fam.) into the ~ επιπλέον.

bargain v.i. κάνω παζάρια. (v.i. & t.) (also ~ about) παζαρεύω. (fam.) I didn't ~ for it δεν το περίμενα. **~ing** s. παζάρεμα n., παζάρι n.

barge s. (river) σλέπι n. (lighter) μαούνα f.

barge v. ~ into or against χτυπώ. ~ in πλακώνω, (verbally) πετιέμαι.

baritone s. βαρύτονος m.

bark s. (of tree) φλοιός m. (v.t.) (strip) ξεφλουδίζω. (graze) γδέρνω.

bark v.i. (of dog) γαυγίζω. (s.) γαύγισμα n.

bark s. (vessel) μπάρκο n.

barley s. κριθάρι n.

barmy a. (fam.) τρελλός.

barn s. αχυρώνας m.

barnacle s. όστρακο n.

barometer s. βαρόμετρο n.

baron s. βαρώνος m.

barracks s. στρατώνας m.

barrage s. (dam) φράγμα n., υδροφράκτης m. (mil.) φραγμός πυρός.

barrel s. βαρέλι n. (of gun) κάννη f. **~organ** s. λατέρνα f.

barren a. στείρος, άγονος. **~ness** s. στειρότητα f. (only of land) αφορία f.

barricade s. οδόφραγμα n. (v.) (street) φράζω, (door) αμπαρώνω. ~ oneself αμπαρώνομαι, κλείνομαι.

barrier s. φράγμα n. (fig.) εμπόδιο n.

barrister s. δικηγόρος m.f.

barrow s. καροτσάκι n.

barter v. ανταλλάσσω. (s.) ανταλλαγή f.

basalt s. βασάλτης m.

base a. άτιμος, ποταπός. (metal) ευτελής, (coin) κίβδηλος.

base s. βάση f. (v.) βασίζω. **~less** a. αβάσιμος.

basement s. υπόγειο n.

bash v. (hit) χτυπώ. (of object) get ~ed (in) βουλιάζω. (s.) χτύπημα n. (fam.) have a ~ κάνω μια προσπάθεια.

bashful a. ντροπαλός.

basic a. βασικός, θεμελιώδης. **~ally** adv. βασικά.

basil s. βασιλικός m.

basin s. λεκάνη f. wash-hand ~ νιπτήρας m. (geog.) λεκανοπέδιο n.

basis s. βάση f., θεμέλιο n.

bask v.i. λιάζομαι. (fig.) ~ in (enjoy) απολαμβάνω.

basket s. καλάθι n., (big, deep) κόφα f., κοφίνι n. (open, flat) πανέρι n.

bass a. & s. (mus.) βαθύφωνος, μπάσος.

bass s. (fish) λαβράκι n.

bassoon s. φαγκότο n.
bastard a. εξώγαμος, νόθος. (s.) (fam., term of abuse) κερατάς m.
bastion s. προπύργιον n.
bat s. (animal) νυχτερίδα f. (implement) ρόπαλο n. (fig.) off one's own ~ μόνος μου.
batch s. φουρνιά f.
bated a. with ~ breath με αγωνία.
bath s. λουτρό n., μπάνιο n. have a ~ κάνω μπάνιο, λούζομαι. she is ~ing the baby κάνει μπάνιο (σ)το μωρό. Turkish ~ χαμάμ(ι) n. ~-**tub** s. μπανιέρα f.
bath|e v.t. πλένω, λούω, (v.i.) κάνω μπάνιο. ~**er** s. λουόμενος m. ~**ing** s. κολύμπι n. ~**ing-dress** s. μαγιό n.
bathos s. απογοητευτική μετάπτωση ύφους (ομιλίας ή γραπτού λόγου).
bathroom s. λουτρό n., μπάνιο n.
batman s. ορντινάντσα f.
baton s. (mus.) μπαγκέτα f. (mil.) ράβδος f.
battalion s. τάγμα n.
batten v.i. ~ on παχαίνω εις βάρος (with gen.).
batten s. σανίδα f. (v.) ~ down ασφαλίζω.
batter s. (cookery) κουρκούτι n.
batter v. βροντοκτυπώ, κοπανώ. ~**ed** a. χτυπημένος, στραπατσαρισμένος. ~ing ram s. κριός m.
battery s. (electric) μπαταρία f. (mil.) πυροβολαρχία f.
battle s. μάχη f. (v.) (also ~ against) πολεμώ. ~-**cry** s. ιαχή f. ~-**field** s. πεδίον μάχης. ~-**ship** s. θωρηκτόν n.
battlement s. έπαλξη f.
batty a. (fam.) παλαβός.
bauble s. μπιχλιμπίδι n.
baulk see balk.
bauxite s. βωξίτης m.
Bavarian a. βαυαρικός. (person) Βαυαρός.
bawdy a. άσεμνος. ~ talk ομιλία γεμάτη αισχρολογίες. ~-**house** s. οίκος ανοχής.
bawl v. κραυγάζω.
bay s. (of sea) κόλπος m. (laurel) δάφνη f. (recess) εσοχή f. ~-**window** s. προεξέχον παράθυρο.
bay v. ουρλιάζω. (s.) (fig.) at ~ κολλημένος στον τοίχο.
bayonet s. ξιφολόγχη f.
bazaar s. αγορά f.
BBC βρεττανική ραδιοφωνία.
BC προ Χριστού.
be v. είμαι. (exist) υπάρχω, (become) γίνομαι, (be situated) βρίσκομαι. so ~ it έστω. it was not to ~ ήταν μοιραίο να μη γίνει. it is a thing to ~ avoided πρέπει να αποφευχθεί. it is not to ~ borne δεν υποφέρεται. greatly to ~ desired λίαν επιθυμητός.

the amount to ~ paid το πληρωτέον ποσόν. (notice of sale) to ~ sold πωλείται. he was not to ~ seen δεν εθεάθη. it is to ~ published next year θα (or πρόκειται να) εκδοθεί του χρόνου. when will you ~ back? πότε θα γυρίσεις; or πότε θα είσαι πίσω; that is the ~-all and end-all (of the matter), αυτό είναι το παν. (a.) to-~ μέλλων, the bride to-~ η μέλλουσα νύφη. would-~ υποψήφιος. see also been, being.
beach s. παραλία f. (bathing) πλαζ f. (v.t.) ανελκύω.
beachcomber s. λευκός αλήτης στα νησιά του Ειρηνικού.
beacon s. πυρσός m. (naut.) φάρος m.
bead s. χάντρα f. (drop) σταγόνα f. string of ~s κομπολόι n. ~-**y** a. ~-y eyes μάτια σαν χάντρες.
beading s. κορδόνι n.
beak s. ράμφος m.
beaker s. κύπελλο n.
beam s. (structural) δοκός f., δοκάρι n. (of balance) ζυγός m. (naut.) ζυγόν n. (ray) ακτίνα f., (electronic) δέσμη f. (smile) χαμόγελο n. (fam.) on one's ~ ends απένταρος. be broad in the ~ έχω περιφέρειες. (v.i.) ακτινοβολώ, λάμπω.
beans s. (broad) κουκκιά n.pl. (green) φασο(υ)λάκια n.pl. (haricot) φασόλια n.pl. (kidney) λόπια n.pl. (coffee) κόκκοι m.pl. bean soup φασο(υ)λάδα f. (fig.) full of ~ κεφάτος. (fam.) he spilt the ~ του ξέφυγε το μυστικό.
bear s. άρκτος f., αρκούδα f.
bear v.t. & i. (child) γεννώ. (carry) φέρ(ν)ω, βαστάζω, βαστώ. (lean) ακουμπώ, (proceed) τραβώ, (produce) κάνω, (endure) υποφέρω. (tolerate) ανέχομαι. I can't ~ him (fam.) δεν τον χωνεύω. ~ oneself φέρομαι. ~ in mind έχω υπ' όψιν. bring to ~ εξασκώ. please ~ with me να έχετε υπομονή.
bear down v. (crush) συντρίβω. ~ on (approach) πλησιάζω, (attack) επιτίθεμαι εναντίον (with gen.).
bear in v. it was borne in on him that... αντιλήφθηκε ότι.
bear out v.t. επιβεβαιώνω.
bear up v.i. αντέχω.
bearable a. υποφερτός.
beard v. αντιμετωπίζω.
beard s. γένια n.pl., (long) γενειάδα f. (small pointed) μούσι n. grow a ~ αφήνω γένια. ~**less** a. αγένειος.
bearer s. (of letter) ο κομίζων.
bearing s. (mech.) τριβεύς m., κουζινέτο n.
bearing s. (manner) παράστημα n. have a ~ on έχω σχέση με. ~**s** s. προσανατολισμός

m. get one's ~s προσανατολίζομαι, κατατοπίζομαι.

beast *s. (lit. & fig.)* ζώο *n.*, κτήνος *n.* ~ of burden υποζύγιο *n.* **~ly** *a.* σιχαμερός, αηδής.

beat *v.t. & i. (strike, pulsate)* χτυπώ. *(v.t.) (thrash)* ξυλοκοπώ. *(excel)* περνώ. *(defeat)* νικώ, be ~en ηττώμαι. ~ time κρατώ το χρόνο. ~ down *(price)* κατεβάζω. ~ down on δέρνω. ~ off αποκρούω. ~ up *(eggs)* χτυπώ, *(person)* ξυλοκοπώ.

beat *s. (stroke)* χτύπος *m.*, παλμός *m.* go on one's ~ κάνω την περιπολία μου.

beaten *a. (metal)* χτυπητός, σφυρήλατος. *(egg)* χτυπητός. keep to the ~ track ακολουθώ την πεπατημένην. *(defeated)* ηττημένος.

beatif|ic *a.* πανευτυχής. **~y** *v.t.* ανακηρύσσω εις άγιον.

beating *s.* χτύπημα *n. (thrashing)* δαρμός *m.* he got a ~ *(fam.)* έφαγε ξύλο.

beatitude *s.* μακαριότης *f.*

beau *s.* δανδής *m.*

beautiful *a.* ωραίος, όμορφος. become ~ ομορφαίνω. **~ly** *adv.* ωραία, όμορφα.

beautify *v.* ομορφαίνω, εξωραΐζω.

beauty *s.* ομορφιά *f.*, ωραιότητα *f.*, κάλλος *n. (beautiful woman)* καλλονή *f. (fine specimen)* αριστούργημα *n.*, όνειρο *n.* ~ contest καλλιστεία *n.pl.* ~ queen *(fam.)* μις *f.* ~ parlour ινστιτούτο καλλονής.

beaver *s.* κάστόρι *n.*

becalm *v.* be ~ed πέφτω σε νηνεμία.

because *conj.* γιατί, διότι, επειδή. *(prep.)* ~ of λόγω, εξ αιτίας *(with gen.)*, για *(with acc.)*.

beck *s.* be at his ~ and call είμαι στις διαταγές του.

beckon *v.i.* γνέφω, κάνω νόημα *(with* σε).

become *v.i.* γίνομαι. *(often by* -ίζω, -εύω, -αίνω *or pass.)* ~ white ασπρίζω, ~ better καλυτερεύω, ~ dearer ακριβαίνω, ~ blind στραβώνομαι. ~ of απογίνομαι, what became of your plan? τι απέγινε το σχέδιό σου;

become *v.t. (suit)* αρμόζω, πρέπει. such behaviour does not ~ you δεν σου αρμόζει *(or* πρέπει) τέτοια συμπεριφορά. *(of dress)* πηγαίνω, πάω. the hat ~s you σου πάει το καπέλλο.

becoming *a. (proper)* πρέπων. *(looking well) see* become *(v.t.)*.

bed *s.* κρεββάτι *n.*, κλίνη *f.* confined to one's ~ κλινήρης. *(base)* βάση *f. (of flowers)* παρτέρι *n. (of river)* κοίτη *f.* **~bug** *s.* κοριός *m.* **~-clothes**, **~ding** *s.* κλινοσκεπάσματα *n.pl.* **~-pan** *s.* πάπια *f.* **~ridden** *a.* κατάκοιτος. **~-room** *s.* κρεβατοκάμαρα

f. **~-sitter** *s.* δωμάτιο νοικιασμένο για διαμονή. **~-spread** *s.* κουβρ-λί *n.*

bedeck *v.* καλλωπίζω.

bedevil *v.* μπερδεύω, τυραννώ.

bedlam *s.* τρελλοκομείο *n.*

bedraggled *a.* λασπωμένος.

bee *s.* μέλισσα *f.* swarm of ~s μελίσσι *n.* keep ~s έχω μελίσσι(α). *(fam.)* make a ~ line for τραβώ κατευθείαν για. **~hive** *s.* κυψέλη *f.*

beech *s.* οξυά *f.*

beef *s.* βοδινό *n.*

been *v.* where have you ~? πού ήσουν; have you ~ there? έχετε πάει εκεί; it's ~ sold πουλήθηκε *or* έχει πουληθεί. has the postman ~? ήρθε *(or* πέρασε) ο ταχυδρόμος; what have you ~ reading? τι διάβαζες; I've ~ here about a year βρίσκομαι εδώ κάπου ένα χρόνο. I haven't ~ here since April δεν έχω έρθει εδώ από τον Απρίλιο. I haven't ~ there for six months έχω να πάω έξι μήνες εκεί. he's ~ and got married πήγε και παντρεύτηκε. he's a has-~ αυτός πάει πια.

beer *s.* μπίρα *f.*, ζύθος *m.* `*(fig.)* small ~ ασήμαντο πράγμα.

beet, **~root** *s.* παντζάρι *n.*, τεύτλον *n.* sugar ~ σακχαρότευτλον *n.*

beetle *s.* κάνθαρος *m.*, σκαθάρι *n.*

befall *v. (happen)* συμβαίνω. *(happen to)* what befell him? τι του συνέβη; τι του έτυχε;

befit *v.* it ~s you σου αρμόζει. **~ting** *a.* πρέπων. **~tingly** *adv.* πρεπόντως.

before *adv. (earlier)* πριν, πρωτύτερα. *(in front)* μπροστά, εμπρός. *(formerly)* προηγουμένως, πριν. long ~ προ πολλού. has he been here ~? έχει ξανάρθει εδώ; I had never seen him ~ δεν τον είχα ξαναδεί. on the day ~ την προηγούμενη *(μέρα).

before *conj.* πριν (να), προτού *(with subj.)* ~ I leave home I will phone you πριν (να) φύγω από το σπίτι θα σου τηλεφωνήσω. *(by the time that)* it was past twelve ~ they left ώσπου να φύγουν ήταν περασμένες δώδεκα. on the day ~ he departed την προηγουμένη της αναχωρήσεώς του.

before *prep.* πριν από *(with acc.)* προ *(with gen.).* ~ long σε λίγο. *(in presence of)* μπροστά σε *(with acc.)*, ενώπιον *(with gen.).*

beforehand *adv.* εκ των προτέρων.

befriend *v. (help)* βοηθώ.

beg *v.i. & t. (~ alms)* ζητιανεύω. *(ask)* ζητώ, παρακαλώ. *(entreat)* ικετεύω, I ~ to differ θα μου επιτρέψετε να διαφωνήσω. I ~ to inform you λαμβάνω την τιμήν να σας πληροφορήσω. it's going ~ging, δεν το θέλει κανείς. ~ the question θεωρώ ως αποδεδειγμένον το αποδεικτέον.

beget *v.* γεννώ.

beggar *s.* ζητιάνος *m.*, διακονιάρης *m.* (*fam.*) poor ~ φουκαράς *m.* **~ly** *a.* πενιχρός. **~y** *s.* διακονιά *f.*

begin *v.* & *i.* αρχίζω, αρχινώ. (*v.t.*, *work, conversation, etc.*) πιάνω. ~ again ξαναρχίζω. to ~ with πρώτα πρώτα. **~ner** *s.* αρχάριος *m.*

beginning *s.* αρχή *f.* (*opening*) έναρξη *f.* in *or* at the ~ στην αρχή, κατ' αρχάς.

begone *int.* φύγε, άπαγε.

begrudge *v.* ζηλεύω. I don't ~ him his holiday δεν τον ζηλεύω που έχει διακοπές.

beguile *v.* (*cheat*) εξαπατώ. (*amuse*) διασκεδάζω, we ~d the journey with talk διασκεδάσαμε στο δρόμο κουβεντιάζοντας. ~ the time σκοτώνω την ώρα μου.

behalf *s.* on ~ of υπέρ (*with gen.*), (*in the name of*) εκ μέρους, εξ ονόματος, για λογαριασμό (*all with gen.*).

behave *v.* φέρομαι, συμπεριφέρομαι. ~ yourself! φρόνιμα! (*function*) λειτουργώ. well ~d φρόνιμος.

behaviour *s.* συμπεριφορά *f.*, διαγωγή *f.* **~ism** *s.* ψυχολογία της συμπεριφοράς.

behead *v.* αποκεφαλίζω.

behest *s.* διαταγή *f.*

behind *adv.* (από) πίσω, όπισθεν. stay *or* be left ~ μένω πίσω. be ~ (*in arrear*) καθυστερώ.

behind *prep.* (από) πίσω από (*with acc.*). ~ the times ασυγχρόνιστος.

behind *s.* πισινός *m.*

behindhand *a.* καθυστερημένος.

behold *v.* βλέπω. ~! ιδού! a joy to ~ χάρμα ιδέσθαι.

beholden *a.* υπόχρεος.

behove *v.* it ~s you to do it πρέπει να το κάνεις.

beige *a.* & *s.* μπεζ.

being *s.* (*existence*) είναι *n.*, ύπαρξη *f.* (*creature*) ον *n.* human ~ ανθρώπινον ον. come into ~ γίνομαι. in ~ υπάρχων. for the time ~ προς το παρόν. (*v.*) (*participle*) όντας (*indeclinable*). the house is ~ built το σπίτι χτίζεται. that ~ so ούτως εχόντων των πραγμάτων. he entered without ~ noticed μπήκε χωρίς να γίνει αντιληπτός. by ~ patient he achieved his aim πέτυχε το σκοπό του με το να είναι υπομονετικός. ~ very tall is a disadvantage το να είναι κανείς πολύ ψηλός είναι μειονέκτημα.

belabour *v.* ξυλοκοπώ.

belated *a.* καθυστερημένος, (*overtaken by night*) νυχτωμένος. **~ly** *adv.* αργά.

belch *v.i.* ρεύομαι. (*v.t.*, *send forth*) βγάζω.

(*s.*) ρέψιμο *n.*

beleaguer *v.* πολιορκώ.

belfry *s.* καμπαναριό *n.*

belie *v.* διαψεύδω.

belief *s.* (*faith*) πίστη *f.* (*opinion*) γνώμη *f.* beyond ~ απίστευτος. to the best of my ~ απ' ό,τι ξέρω. folk ~s λαϊκές δοξασίες.

believable *a.* πιστευτός.

believe *v.* πιστεύω. I ~ he's gone νομίζω (*or* μου φαίνεται) πως έχει φύγει. make ~ that κάνω πως. ~r *s.* πιστός *a.*

belittle *v.* (*undervalue*) υποτιμώ, (*disparage*) κριτικάρω.

bell *s.* κουδούνι *n.* (*church*) καμπάνα *f.* alarm ~ κώδων κινδύνου. he rang the ~ *or* the ~ rang χτύπησε το κουδούνι. ringing of ~s κωδωνοκρουσία *f.* **~-shaped** κωδωνοειδής.

belle *s.* καλλονή *f.*

bellicose *a.* φιλοπόλεμος.

belligeren|cy *s.* εμπόλεμος κατάστασις. **~t** *a.* εμπόλεμος.

bellow *v.* μουγγρίζω. (*s.*) μούγγρισμα *n.*

bellows *s.* φυσερό *n.*

belly *s.* κοιλιά *f.* eat one's ~ful τρώω το καταπέτασμα. (*v.*) φουσκώνω. **~-ache** *s.* κοιλόπονος *m.*

belong *v.* ανήκω. ~ under (*be classified*) υπάγομαι εις. **~ings** *s.* υπάρχοντα *n.pl.*

beloved *a.* αγαπημένος.

below *adv.* (από) κάτω.

below *prep.* κάτω από, υπό (*with acc.*), κάτω (*with gen.*). ~ the surface υπό την επιφάνειαν. ~ zero κάτω του μηδενός, υπό το μηδέν.

belt *s.* ζώνη *f.* (*mech.*) ιμάς *m.* (*v.t.*) ζώνω, (*thrash*) δέρνω.

bemoan *v.* θρηνώ.

bemuse *v.* ζαλίζω.

bench *s.* (*park, work*) πάγκος *m.* (*schoolroom*) θρανίο *n.* (*parliamentary*) εδώλιο *n.* (*fig., lawcourt*) δικαστήριο *n.*

bend *v.t.* κάμπτω, λυγίζω. (*head*) σκύβω, (*eyes, steps*) στρέφω. (*v.i.*) (*stoop*) σκύβω. (*lean, give way, also fig.*) κάμπτομαι, λυγίζω. (*change direction*) στρίβω. ~ down σκύβω. on ~ed knee γονατιστός. be bent on είμαι αποφασισμένος (*with να or* για).

bend *s.* καμπή *f.*, στροφή *f.*

beneath (*adv.*) (από) κάτω. (*prep.*) (από) κάτω από, υπό (*with acc.*) ~ consideration αναξιόλογος. it is ~ you (*unworthy*) δεν στέκει σε σένα.

benediction *s.* ευλογία *f.*

benefac|tion *s.* ευεργεσία *f.* make ~ion ευεργετώ. **~or** *s.* ευεργέτης *m.* **~ress** *s.* ευεργέτις *f.*

benefice s. θέσις εφημερίου με κατοικία και απολαβές.

benefic|ence s. αγαθοεργία f. ~**ent** a. αγαθοεργός. ~**ial** a. ωφέλιμος. ~**iary** s. (one entitled) ο δικαιούχος.

benefit v.t. ωφελώ. (v.i.) ωφελούμαι. (s.) ωφέλεια f., όφελος n. I got much ~ είδα μεγάλη ωφέλεια. ~ performance ευεργετική παράσταση.

benevolen|t a. (person, sentiment) φιλάνθρωπος, (institution, purpose) φιλανθρωπικός. ~**ce** s. φιλανθρωπία f.

benighted a. (backward) θεοσκότεινος. get ~ (overtaken by dark) νυχτώνομαι.

benign a. πράος, (med.) καλοήθης.

bent s. κλίση f. to the top of one's ~ όσο τραβάει η καρδιά μου.

bent a. στραβός, κεκαμμένος, λυγισμένος. become ~ στραβώνω. (body) κυρτός, (head) σκυμμένος.

benumb v. ναρκώνω, ξυλιάζω.

benzine s. βενζίνη f.

beque|ath v. κληροδοτώ. ~**st** s. κληροδότημα n.

bereave v. ~ (person) of στερώ (with gen.). be ~d (or bereft) of χάνω, στερούμαι. the ~d relatives οι τεθλιμμένοι συγγενείς. ~**ment** s. απώλεια f.

bereft a. στερημένος.

beret s. μπερές m.

berry s. καρπός m. see mul~, etc.

berserk a. go ~ εκτραχηλίζομαι.

berth v.t. & i. αράζω. (s.) (mooring) αγκυροβόλιο n. (at wharf) θέση f. (situation) θέση f. (sleeping-place) κουκέτα f. give a wide ~ to αποφεύγω.

beseech v. ικετεύω. ~**ing** a. παρακλητικός, ικετευτικός.

beseem see befit.

beset v. περικυκλώνω. his ~ting sin το μεγάλο ψεγάδι του.

beside prep. (next to) δίπλα or πλάι σε, (compared with) εν συγκρίσει με, (apart from) εκτός από (all with acc.). ~ the point εκτός θέματος. be ~ oneself (with anger) είμαι έξω φρενών, (with joy) πετώ από τη χαρά μου.

besides adv. (moreover) εξ άλλου, (as well) επί πλέον. (prep.) (as well as) εκτός από.

besiege v. πολιορκώ.

besmear v. αλείφω.

besmirch v. λερώνω, (fig.) κηλιδώνω.

besom s. σκούπα από κλαδιά.

besot v. αποβλακώνω.

bespatter v. πιτσιλίζω.

bespeak v. (order) παραγγέλλω, (engage) κρατώ, (imply) δείχνω.

bespoke a. επί παραγγελία.

besprinkle v. ραντίζω.

best a. ο καλύτερος. do one's ~ βάζω τα δυνατά μου. make the ~ of it το παίρνω απόφαση. be at one's ~ (mood) είμαι στις καλές μου. ~ man κουμπάρος m.

best adv. καλύτερα. the ~ dressed man ο πιο καλοντυμένος.

best v.t. βάζω κάτω, (fam.) τρώω.

bestial a. κτηνώδης. ~**ity** s. κτηνωδία f.

bestow v. απονέμω.

bestrew v. (scatter) σκορπίζω. the street was ~n with flowers γέμισε ο δρόμος λουλούδια.

bestride v. καβαλλικεύω. (fig.) ~ the scene (be dominating figure) κατέχω κυριαρχική θέση.

bet s. στοίχημα n. (v.) στοιχηματίζω, βάζω στοίχημα.

betake v. ~ oneself to (go) κατευθύνομαι προς (with acc.), (have recourse) καταφεύγω εις.

bethink v. ~ oneself of αναλογίζομαι.

betide v. συμβαίνω. woe ~ you! ανάθεμά σε! αλίμονό σου!

betoken v. σημαίνω, (portend) προμηνύω.

betray v. προδίδω. ~**al** s. προδοσία f. ~**er** s. προδότης m.

betroth v. αρραβωνιάζω, μνηστεύω. ~**al** s. αρραβώνες m.pl., μνηστεία f. ~**ed** s. μνηστήρας m., μνηστή f.

better a. & n. καλύτερος. my ~s οι ανώτεροί μου. ~ half (wife) τρυφερόν ήμισυ. for the ~ προς το καλύτερον. for ~ for worse και για τις καλές και για τις κακές μέρες. get ~ καλυτερεύω, (regain health) αναλαμβάνω. get the ~ of (outwit) βάζω κάτω, (defeat) νικώ.

better adv. καλύτερα. ~ and ~ όλο και καλύτερα. you will be all the ~ for it θα σου κάνει καλό. the sooner the ~ όσο γρηγορότερα τόσο καλύτερα. you had ~ (not) go there καλύτερα να (μην) πας εκεί. think ~ of it αλλάζω γνώμη.

better v.t. (improve) βελτιώνω, (excel) ξεπερνώ. ~ oneself (socially/financially) βελτιώνω την κοινωνική/οικονομική θέση μου. ~**ment** s. βελτίωση f.

between adv. ανάμεσα, (ανα)μεταξύ.

between prep. ανάμεσα σε (with acc.) μεταξύ (with gen.), ~ ourselves μεταξύ μας. ~ 2 and 3 o'clock δύο με τρεις. (he was looking at me) from ~ the curtains ανάμεσα από τις κουρτίνες.

bevelled a. μπιζουτέ.

beverage s. ποτό n.

bevy s. ομάδα f.

bewail v. θρηνώ.

beware v. (also ~ of) προσέχω, έχω το νου

μου (σε), φυλάγομαι (από).

bewilder v. (also be ~ed) σαστίζω.

bewitch v. μαγεύω. **~ing** a. μαγευτικός.

bey s. μπέης m.

beyond adv. πέρα, πέραν. (prep.) πέρα από (with acc.), πέραν (with gen.). be or go ~ (exceed) υπερβαίνω. it's ~ me δεν το καταλαβαίνω.

bias v. προκαταλαμβάνω. he is ~sed είναι προκατειλημμένος. (s.) (prejudice) προκατάληψη f. cut on the ~ κόβω λοξά. **~-fold** λούκι n.

bib s. σαλιάρα f.

bibelot s. μπιμπελό n. (indeclinable).

Bibl|e s. Βίβλος f. **~ical** a. βιβλικός.

biblio|graphy s. βιβλιογραφία f. **~phile** s. βιβλιόφιλος m.

bibulous a. ~ person μπεκρής m., μπεκρού f.

biceps s. δικέφαλος βραχιόνιος μυς, (fam.) μυς του μπράτσου.

bicker v. μαλώνω.

bicycle s. ποδήλατο n. ride a ~ κάνω ποδήλατο.

bid v. (tell, say) λέω. (invite) καλώ. (offer price) προσφέρω. ~ for (lot at auction) χτυπώ, ~ highest πλειοδοτώ. (at cards) δηλώνω.

bid s. (offer) προσφορά f. (effort) προσπάθεια f. make a ~ for freedom προσπαθώ να δραπετεύσω.

bidder s. ο προσφέρων. highest ~ πλειοδότης m.

bidding s. (at auction) οι προσφορές f.pl. I do his ~ τον ακούω, εκτελώ τις διαταγές του.

bide v. ~ one's time καιροσκοπώ.

bidet s. μπιντές m.

biennial a. (lasting 2 years) διετής, (every 2 years) ανά διετίαν. **~ly** adv. κάθε δύο έτη, ανά διετίαν.

bier s. βάθρο φερέτρου.

bifocal a. διεστιακός.

bifurcat|e v. διακλαδώνω. **~ion** s. διακλάδωση f.

big a. μεγάλος. ~ with child έγκυος. grow or make **~ger** μεγαλώνω. ~ words παχιά λόγια. (fam.) talk ~ κάνω τον καμπόσο. he is too ~ for his boots πήρε ο νους του αέρα. (by suffix) ~ man άντρακλας m. ~ woman γυναικάρα f. ~ dog σκύλαρος m. ~ house σπιταρώνα f., etc.

bigam|ist s., **~ous** a. δίγαμος. **~y** s. διγαμία f.

bigot s., **~ed** a. μισαλλόδοξος. **~ry** s. μισαλλοδοξία f.

bigwig s. μεγαλουσιάνος m.

bike s. ποδήλατο n.

bilateral a. διμερής.

bile s. χολή f.

bilge s. (fam., nonsense) ανοησίες f.pl. **~-water** s. σεντίνα f.

bilingual a. δίγλωσσος. **~ism** s. διγλωσσία f.

bilious a. have a ~ attack κάνω εμετό. (peevish) πικρόχολος.

bilk v. κλέβω.

bill s. (account) λογαριασμός m. (parliamentary) νομοσχέδιο n. (banknote) χαρτονόμισμα n. (poster) τοιχοκόλλημα n. (bird's) ράμφος n. ~ of exchange συναλλαγματική f. ~ of fare κατάλογος m. ~ of lading φορτωτική f. **~-hook** s. κλαδευτήρι n. **~-posting**, **~-sticking** s. τοιχοκόλληση f.

billet s. (lodging) κατάλυμα n. (v.t.) στρατωνίζω. be ~ed καταλύω. **~ing** s. κατάλυση f.

billet-doux s. ραβασάκι n.

billiards s. μπιλιάρδο n.

billion s. (UK) τρισεκατομμύριο n., (USA) δισεκατομμύριο n.

billow s. κύμα n. (v.) κυματίζω. **~y** a. κυματοειδής.

billy-goat s. τράγος m.

bin s. δοχείο n. litter ~ κάλαθος αχρήστων, (fam.) καλάθι n.

binary s. δυαδικός.

bind v. δένω. (oblige) υποχρεώνω, δεσμεύω, see bound a.

binding a. δεσμευτικός. (s.) δέσιμο n.

binge s. (fam.) γλέντι n. go on a ~ γλεντώ.

binoculars s. κιάλια n.pl.

binomial s. διώνυμο n.

biochem|ical a., **~ist** s. βιοχημικός. **~istry** s. βιοχημεία f.

biograph|er s. βιογράφος m. **~ical** a. βιογραφικός, **~y** s. βιογραφία f.

biolog|ical a. βιολογικός. **~ist** s. βιολόγος, m., **~y** s. βιολογία f.

biopsy s. βιοψία f.

bipartisan a. δικομματικός.

bipartite a. διμερής.

biped n. δίπους a.

biplane s. δίπλανο n.

birch s. (tree) σημύδα f. (rod) βέργα f. (v.) ραβδίζω με βέργα.

bird s. πουλί n., πτηνό n. ~ of passage αποδημητικό πτηνό, (fig.) περαστικός επισκέπτης. ~'s eye view θεά εξ ύψους. kill two ~s with one stone μ' ένα σμπάρο δυο τρυγόνια. a ~ in the hand... κάλλιο ένα και στο χέρι.

biro s. στυλό διαρκείας.

birth s. γέννηση f. give ~ (to) γεννώ. from or by ~ εκ γενετής. **~-control** s. πρόληψη γεννήσεων. **~-rate** s. ποσοστό των γεννήσεων.

birthday s. γενέθλια n.pl.

birthplace s. γενέτειρα f.

birthright s. πρωτοτόκια n.pl.

biscuit s. μπισκότο n. ship's ~ γαλέτα f. (fam.) take the ~ see cake s.

bisect n. διχοτομώ. ~ion s. διχοτόμησις f.

bishop s. επίσκοπος m. ~ric s. επισκοπή f.

bit s. (piece) κομμάτι n. a ~ of (some) λίγος a. a good ~ of money κάμποσα (or αρκετά) λεφτά. a ~ (adv.) (rather) λιγάκι, κομμάτι. not a ~ καθόλου. ~ by ~ κομμάτι κομμάτι. fall to ~s καταρρέω. break or tear to ~s κομματιάζω. (horse's) χαλινάρι n. take the ~ between one's teeth (lit. & fig.) αφηνιάζω, see brace.

bitch s. σκύλα f. ~y a. κακεντρεχής.

bite v. δαγκώνω, (of insects) τσιμπώ. (take bait) τσιμπώ. (fam.) the ~r bit πήγε για μαλλί και βγήκε κουρεμένος.

bite s. (act, wound) δάγκωμα n. (wound), mouthful) δαγκανιά f. (insect's) τσίμπημα n. have a ~ of food τσιμπώ κάτι.

biting a. δηκτικός, φαρμακερός.

bitter a. πικρός, (cold) δριμύς, (hate, enemy) άσπονδος, (person) πικραμένος, (mordant) δηκτικός. (disappointment) σκληρός. make or become ~ πικραίνω. ~ly adv. πικρά, σκληρά. ~ness a. (feeling) πικρία f. (taste) πίκρα f., πικρίλα f.

bitter-sweet a. γλυκόπικρος.

bitumen s. άσφαλτος f.

bivouac s. καταυλισμός m. (v.) καταυλίζομαι.

bizarre a. αλαμπουρνέζικος.

blab v.i. ακριτομυθώ. (v.t.) ~ the secret (fam.) το κάνω τούμπανο.

black a. μαύρος, jet ~ κατάμαυρος. make or become ~ μαυρίζω. (fig.) ~ sheep κακό κουμάσι. ~ Maria κλούβα f. ~-and-blue μελανιασμένος, in ~ and white (drawn) με μελάνι, (in writing) γραπτώς. things are looking ~ σκούρα τα πράγματα. I got ~ looks με αγριοκοίταξαν.

blackball v. μαυρίζω.

blackberry s. βατόμουρο n.

blackbird s. κότσυφας m.

blackboard s. μαυροπίνακας m.

blacken v.t. & i. μαυρίζω. (v.t.) (defame) δυσφημίζω, (sully) αμαυρώνω.

black-eyed a. μαυρομάτης.

blackguard s. παλιάνθρωπος m. ~ly a. βρώμικος.

black-haired a. μαυρομάλλης.

blacking s. μαύρο βερνίκι.

blackleg s. απεργοσπάστης m..

blacklist v.t. γράφω στο μαυροπίνακα.

blackmail v. εκβιάζω. (s.) εκβιασμός m. ~er s. εκβιαστής m.

black-marketeer s. μαυραγορίτης m.

blackness s. (also ~ colour or spot) μαυρίλα f.

blackout s. (no lights) συσκότιση f. (of person) λιποθυμία f.

blacksmith s. σιδηρουργός m.

bladder s. κύστη f. (of ball) σαμπρέλα f.

blade s. (of knife) λεπίδα f. (of safety razor) ξυραφάκι n., λάμα f. (of oar) πλάτη f. (of grass, corn) φύλλο n.

blah s. (fam.) αερολογίες f.pl.

blame v. μέμφομαι. be to ~ φταίω.

blame s. (censure) μομφή f. (fault) φταίξιμο n. ~less a. άμεμπτος. ~worthy a. αξιόμεμπτος.

blanch v.i. (grow pale) χλομιάζω. (v.t.) (make white) ασπρίζω, (almonds) ξεφλουδίζω.

bland a. πράος, απαλός. ~ly adv. he ~ly announced είχε το τουπέ να αναγγείλει.

blandishment s. καλόπιασμα n.

blank s. κενό n. draw a ~ δεν επιτυγχάνω.

blank a. κενός. (paper) λευκός. (expressionless) ανέκφραστος. (cheque) ανοικτός. (wall, window) τυφλός. (shot) άσφαιρος. (unrhymed) ανομοιοκατάληκτος. ~ refusal ωμή άρνηση. ~ly adv. ανεκφράστως, απολύτως.

blanket s. κουβέρτα f. (hand-woven) χράμι n. (v.) (fig.) σκεπάζω.

blar|e v. βουίζω. ~ing a. στη διαπασών.

blasé a. μπλαζέ.

blasphem|e v. βλασφημώ. ~y s. βλασφημία f.

blast v. (ruin) ρημάζω, (blow up) ανατινάζω, (burn, shrivel) καίω, (int., fam.) ~! να πάρει ο διάολος! ~ed a. (fam.) παλιο-, βρωμο-.

blast s. (wind) ριπή f. (explosion) έκρηξη f. blow a ~ (on whistle, etc.) σφυρίζω. at full ~ (of sound) στη διαπασών, (of works, engine) φουλ. ~-furnace s. υψικάμινος f.

blatant a. (deed) κατάφορος, (person) ξετσίπωτος. ~ly adv. καταφώρως, αδιάντροπα.

blaze v.i. (burn) φλέγομαι, (be brilliant) φεγγοβολώ. ~ up (in anger) ξεσπώ. ~ away (shoot) πυροβολώ συνεχώς, (fig.) εργάζομαι πυρετωδώς. (v.t.) ~ a trail ανοίγω δρόμο. ~ abroad διατυμπανίζω.

blaze s. (flame) φλόγα f. (conflagration) πυρκαγιά f. (outburst) ξέσπασμα n. ~ of publicity το φως της δημοσιότητος.

blazon s. οικόσημον n. (v.) (decorate) διακοσμώ, (proclaim) διακηρύσσω.

bleach v.t. & i. ξασπρίζω. (v.t.) (linen) λευκαίνω, (hair) ξανοίγω. (s.) λευκαντικό n.

bleak a. (exposed) εκτεθειμένος. (desolate) έρημος. (cheerless) ψυχραντικός.

bleary *a.* τσιμπλιάρης.

bleat *v.* βελάζω. (*s.*) βέλασμα *n.*

bleed *v.i.* ματώνω. (*v.t.*) they bled him του πήραν αίμα, (*also fig.*) του έκαναν αφαίμαξη. **~ing** *s.* μάτωμα *n.* αφαίμαξη *f.*

blemish *s.* ελάττωμα *n.* (*v.*) χαλώ.

blench *v.* ξαφνιάζομαι.

blend *s.* χαρμάνι *n.* (*v.t.*) (*mix*) συμμιγνύω. (*v.i.*) (*go well together*) ταιριάζω.

bless *v.* ευλογώ. be ~ed with απολαμβάνω. **~ed** *a.* ευλογημένος, (*epithet of holy man*) όσιος. (*in Beatitudes*) μακάριος. **~ing** *s.* ευλογία *f.*

blight *s.* σκουριά *f.* (*v.*) χαλώ, καταστρέφω. **~er** *s.* (*fam.*) μασκαράς *m.*

blind *s.* (*of window*) στορ *n.* (*stratagem*) κόλπο *n.*

blind *a.* τυφλός, στραβός. become ~ τυφλώνομαι, στραβώνομαι. be ~ to δεν βλέπω. turn a ~ eye κάνω στραβά μάτια. ~ alley αδιέξοδο *n.* ~ man's buff τυφλόμυγα *f.* ~ drunk τύφλα στο μεθύσι.

blind *v.t.* στραβώνω, τυφλώνω. **~ing** *a.* εκτυφλωτικός. **~ly** *adv.* τυφλώς. **~ness** *s.* τύφλα *f.*, στραβομάρα *f.*

blindfold *s.* με δεμένα τα μάτια.

blink *v.i.* ανοιγοκλείνω τα μάτια μου. (*of distant lights*) τρεμοσβήνω. (*as warning signal*) αναβοσβήνω.

bliss *s.* ευδαιμονία *f.* **~ful** *a.* πανευτυχής. (*delightful*) τερπνότατος.

blister *s.* φουσκάλα *f.* (*v.i.*) φουσκαλιάζω. **~ing** *a.* δριμύς.

blithe *a.* χαρούμενος.

blizzard *s.* χιονοστρόβιλος *m.*

bloated *a.* φουσκωμένος.

bloater *s.* καπνιστή ρέγγα.

blob *s.* σταγόνα *f.* (*of colour*) βούλα *f.*

bloc *s.* μπλοκ *n.*

block *s.* (*piece*) τεμάχιο *n.* (*log*) κούτσουρο *n.* (*unhewn stone*) ογκόλιθος *m.* (*cut stone*) κυβόλιθος *m.* (*ice*) κολόνα *f.* (*buildings*) τετράγωνο *n.* (*flats*) πολυκατοικία *f.* (*pulley*) τροχαλία *f.* (*obstruction*) εμπόδιο *n.* (*traffic jam*) μποτιλιάρισμα *n.*

block *v.* (*hinder*) εμποδίζω, (*road, etc.*) φράζω. (*stop up, also* become ~ed) βουλώνω. **~age** *s.* βούλωμα *n.*, απόφραξη *f.*

blockade *s.* αποκλεισμός *m.* (*v.*) μπλοκάρω.

blockhead *s.* κούτσουρο *n.*, τούβλο *n.*

blond(e) *a.* ξανθός.

blood *s.* αίμα *n.* bad ~ έχθρα *f.* in cold ~ εν ψυχρώ. his ~ is up ανάψανε τα αίματά του. ~ red αιματόχρους. covered in ~ αιμόφυρτος. **~curdling** *a.* φρικιαστικός. **~donation** *s.* αιμοδοσία *f.* **~feud** *s.* βεντέτα *f.* **~group** *s.* ομάδα αίματος *f.*

~less *a.* αναίμακτος. **~letting** *s.* αφαίμαξη *f.* **~orange** *s.* σαγκουίνι *n.* **~pressure** *s.* πίεση *f.* **~shed** *s.* αιματοχυσία *f.* **~shot** *a.* κόκκινος. **~stained** *a.*. ματωμένος. **~thirsty** *a.* αιμοβόρος. **~vessel** *s.* αγγείο *n.*

bloody *a.* αιματηρός, (*bleeding*) ματωμένος. (*fam.*) βρωμο-, παλιο-. ~ swine βρωμόσκυλο *n.* **~minded** *a.* στραβόξυλο *n.*

bloom *s.* (*flower*) άνθος *n.* (*flowering*) άνθηση *f.* (*on fruit*) χνούδι *n.* (*prime*) ακμή *f.* in ~ ανθισμένος. in full ~ εις πλήρη ακμήν.

bloom *v.* ανθίζω. **~ing** *a.* (*girl*) δροσερός. (*old person*) ανθηρός.

bloomer *s.* (*fam.*) γκάφα *f.* **~s** *s.* βράκα *f.*

blossom *s. see* bloom. (*v.*) ανθίζω, (*come out*) ανοίγω.

blot *v.* (*stain*) κηλιδώνω, (*dry ink*) στουπώνω. ~ out (*erase*) εξαλείφω, (*hide*) καλύπτω. (*s.*) (*lit. & fig.*) κηλίδα *f.* (*ink*) μελανιά *f.*

blotch *s.* λεκές *m.* (*on skin*) πανάδα *f.*

blotting-paper *s.* στουπόχαρτο *n.*

blouse *s.* μπλούζα *f.*

blow *s.* (*stroke*) χτύπημα *n.* (*with fist*) μπουνιά *f.* (*shock*) πλήγμα *n.* it was a great ~ to him του κόστισε πάρα πολύ. come to ~s έρχομαι στα χέρια. at one ~ διά μιας.

blow *v.t. & i.* φυσώ (*v.i.*) (*of flag,hair, etc. in wind*) ανεμίζω. (*of trumpet*) ηχώ. (*of whistle, siren*) σφυρίζω. (*of fuse*) καίομαι. ~ hot and cold ταλαντεύομαι. the door blew open ο αέρας άνοιξε βιαίως την πόρτα. dust blew in our faces ο αέρας μας έφερε σκόνη στο πρόσωπό μας. it blew away *or* off το πήρε ο άνεμος. it blew down το έρριξε ο άνεμος. ~ in (*turn up*) πλακώνω. ~ over (*cease*) κοπάζω. ~ out (*v.t. & i.*) σβήνω. ~ up (*v.t.*) (*inflate*) φουσκώνω, (*scold*) λούζω, (*shatter*) τινάζω στον αέρα, (*v.i., explode*) τινάζομαι.

blow-out *s.* (*fam.*) I had a ~ (*food*) έφαγα τον περίδρομο, (*a puncture*) μ' έπιασε λάστιχο.

blowzy *a.* a ~ female μία τσούλα.

blubber *v.* (*cry*) κλαίω με πολλά δάκρυα.

blubber *s.* λίπος φάλαινας.

bludgeon *v.* ξυλίζω. (*s.*) ρόπαλο *n.*

blue *a.* μπλε, (*azure*) γαλάζιος, κυανούς. (*esp. eyes*) γαλανός. (*dark*) μαβής. (*fam.*) feel ~ βαρυθυμώ. ~ jokes σόκιν *n.pl.* out of the ~ στα καλά καθούμενα. **~and-white** *s.* (*Gk. flag*) κυανόλευκος *f,* **~blooded** *a.* γαλαζοαίματος. **~book** *s.* κυανή βίβλος. **~bottle** *s.* κρεατόμυγα *f.* **~collar worker** *s.* εργάτης βιομηχανίας *m.* **~print** *s.* σχέδιο *n.* ~**stocking** *s.* λογία *f.*

bluff *s.* (*cliff*) κρημνός *m.* (*a.*) (*abrupt*) απότομος, (*frank*) ντόμπρος.

bluff *s.* (*deception*) μπλόφα *f.* (*v.i.*) μπλοφάρω.

blunder *s.* γκάφα *f.* (*v.*) κάνω γκάφα. ~**er** *s.* γκαφαδόρος *m.*

blunderbuss *s.* τρομπόνι *n.*

blunt *v.* αμβλύνω, (*also* get ~) στομώνω.

blunt *a.* (*not sharp*) αμβλύς, στομωμένος. it is ~ δεν κόβει. make *or* get ~ στομώνω. ~ refusal ωμή άρνηση. (*person*) ντόμπρος, απότομος. ~**ly** *adv.* ντόμπρα, απότομα.

blur *v.* (*make dim*) (*also* become ~red) θολώνω ~**red** *a.* θολός, αμυδρός, αόριστος.

blurt *v.* ~ it out μου ξεφεύγει.

blush *v.* κοκκινίζω. (*s.*) κοκκίνισμα *n.* at first ~ εκ πρώτης όψεως.

bluster *v.* (*of wind*) μαίνομαι, (*of person*) απειλώ θεούς και δαίμονες. (*s.*) (*of wind*) φυσομανητό *n.* (*bombast*) μεγαλαυχία *f.* ~**er** *s.* φανφαρόνος *m.* ~**y** *a.* θυελλώδης.

boa *s.* (*snake*) βόας *m.*

boar *s.* (*wild*) κάπρος *m.*, αγριογούρουνο *n.* (*domestic*) χοίρος *m.*

board *s.* (*plank*) σανίδι *n.* (*black* ~, *notice* ~) πίνακας *m.* (*council*) συμβούλιον *n.* (*committee*) επιστροπή *f.* (*food*) τροφή *f.* ~ and lodging οικοτροφία *f.* on ~ ship επί του πλοίου. go on ~ επιβιβάζομαι (*with* σε). above ~ (*a.*) τίμιος, (*adv.*) τίμια. across the ~ γενικά. sweep the ~ κερδίζω τα πάντα.

board *v.* (*put* ~s *over*) σανιδώνω. (*feed*) ταΐζω, (*be fed*) οικοσιτώ. (*go on* ~) επιβαίνω (*with gen*). ~**er** *s.* οικότροφος *m.f.*, (*at school*) εσωτερικός *m.* ~**ing-house** *s.* πανσιόν *f.* ~**ing-school** *s.* σχολείο με εσωτερικά παιδιά.

boast *v.i.* κομπάζω, καυχιέμαι. (*v.t.*) our town ~s a fine park η πόλη μας είναι περήφανη για το ωραίο πάρκο της. (*s.*) καυχησιά *f.* (*thing one is proud of*) καύχημα *n.* ~**ing** *s.* κομπασμος *m.*

boastful *a.* καυχησιάρης. ~**ly** *adv.* με κομπασμό. ~**ness** *s.* κομπασμός *m.*

boat *s.* βάρκα *f.* (*skiff, lifeboat, etc.*) λέμβος *f.* (*large*) καράβι *n.* we are all in the same ~ όλοι βράζουμε στο ίδιο καζάνι. burn one's ~s (*fig.*) παίρνω αμετάκλητη απόφαση. (*v.*) go ~ing πηγαίνω βαρκάδα. ~**hook** *s.* κοντάρι *n.*, γάντζος *m.* ~**man** *s.* βαρκάρης, λεμβούχος *m.* ~**race** *s.* λεμβοδρομία *f.* ~**swain** *s.* λοστρόμος *m.*

boater *s.* (*hat*) ψαθάκι *n.*

bob *v.i.* ~ up and down χοροπηδώ. (*v.t.*) ~bed hair τα μαλλιά κομμένα κοντά.

bobbin *s.* μασούρι *n.*

bode *v.t. & i.* προμηνύω.

bodice *s.* μπούστος *m.*

bodiless *a.* ασώματος.

bodily *a.* σωματικός. (*adv.*) (*all together*) όλοι μαζί, (*by force*) διά της βίας.

bodkin *s.* βελόνα *f.*

body *s.* (*human, animal*) σώμα *n.*, κορμί *n.* (*corpse*) πτώμα *n.* (*group, institution, thing*) σώμα *n.* (*team*) ομάδα *f.* (*mass*) όγκος *m.* (*of vehicle*) αμάξωμα *n.*, καροσερί *f.* (*main part*) κύριο μέρος. (*substantial character*) ουσία *f.* ~ politic κράτος *n.* in a ~ ομαδικώς. ~ and soul (*adv.*) ψυχή τε και σώματι. ~**building** *s.* γυμναστική αναπτύξεως του σώματος. ~**guard** *s.* σωματοφυλακή *f.*, (*one man*) σωματοφύλακας *m.* ~**work** *s.* καροσερί *n.*

Boeotian *a. & s.* Βοιωτός.

bog *s.* έλος *n.*, βάλτος *m.* get ~ged down (*fig.*) σκαλώνω. ~**gy** *a.* ελώδης.

boggle *v.i.* δειλιάζω.

bogus *a.* ψεύτικος.

bogy(man) *s.* μπαμπούλας *m.*

bohemian *s.* βοημός, μποέμ *m.* (*a.*) μποέμικος.

boil *s.* δοθιήν *n.* (*fam.*) καλόγερος *m.*

boil *v.* βράζω, come to the ~ παίρνω βράση. ~ over ξεχειλίζω. ~ away εξατμίζομαι. (*fig.*) it ~s down to this εν περιλήψει έχει ως εξής. ~**ed** *a.* βραστός.

boiler *s.* καζάνι *n.*, (ατμο)λέβητας *m.*

boiling *a.* (*liquid*) βράζων. (*fig.*) (*very hot*) καυτός.

boiling *s.* βρασμός *m.*, βράσιμο *n.* ~ point σημείο ζέσεως.

boisterous *a.* θορυβώδης.

bold *a.* τολμηρός. (*vivid*) έντονος, (*distinct*) ευκρινής. (*type*) μαύρος. make ~ τολμώ. ~ as brass αναιδής. ~**ly** *adv.* τολμηρώς. ευκρινώς ~**ness** *s.* τόλμη *f.*

bole *s.* κορμός *m.*

bollard *s.* δέστρα *f.*

bolster *s.* κεφαλάρι *n.* (*v.*) ~ (*up*) υποστηρίζω.

bolt *s.* (*on door*) σύρτης *m.* (*with nut*) μπουλόνι *n.* (*cloth*) τόπι *n.* like a ~ from the blue αιφνιδιαστικώς. (*v.t.*) (*door*) μανταλώνω.

bolt *v.i.* (*run off*) το σκάω, (*of horse*) αφηνιάζω, (*v.t., swallow*) χάφτω.

bolt *adv.* ~ upright ολόρθος.

bomb *s.* βόμβα *f.*

bomb, ~**ard** *v.* βομβαρδίζω. ~**ardment** *s.* βομβαρδισμός *m.* ~**er** *s.* (*aero.*) βομβαρδιστικόν *n.*

bombast *s.* κομπορρημοσύνη *f.* ~**ic** *a.* κομπαστικός.

bombshell s. it came as a ~ to me μου ήρθε κεραμίδα.

bona fide a. σοβαρός, πραγματικός, καλής πίστεως.

bonbon s. καραμέλα f.

bond s. (tie) δεσμός m. ~s (fetters) δεσμά n.pl. (contract) συμβόλαιο n. (debenture) ομολογία f. (promissory note) ομόλογον n. in ~ υπό διαμετακόμιση.

bond v.t. (bind) δένω.

bondage s. δουλεία f.

bone a. κοκαλένιος, από κόκαλο. ~less a. ακόκαλος.

bone s. κόκκαλο n., οστούν n. ~s (remains) οστά n.pl. ~ of contention μήλον της έριδος. make no ~s (about) δεν διστάζω (να). I have a ~ to pick with you έχω παράπονα μαζί σου. (v.) ξεκοκκαλίζω. **~-dry** a. κατάξερος. **~-headed** a. κουτός. **~-idle** a. αρχιτεμπέλαρος.

bone-shaker s. (fam.) σακαράκα f.

bonfire s. φωτιά f.

bonnet s. σκούφια f.

bonny a. όμορφος.

bonus s. δώρο n. (also on shares, etc.) έκτακτο μέρισμα.

bony a. κοκκαλιάρης.

boo v. γιουχαΐζω. (s.) γιούχα n. **~ing** s. γιουχάισμα n.

booby s. βλάκας m. **~-trap** s. παγίδα f.

book v.t. (enter) καταγράφω, (reserve) κρατώ, κλείνω, (performer) αγκαζάρω (v.i., take ticket) βγάζω εισιτήριο.

book s. βιβλίο n. (bank-~, etc.) βιβλιάριο n. ~s (accounts) κατάστιχα n.pl. I bring him to ~ του ζητώ ευθύνες. I am in his good ~s με καλοβλέπει. it doesn't suit my ~ δεν μου κάνει. **~-case** s. βιβλιοθήκη f. **~-ing-office** s. ταμείο n. **~-keeper** s. λογιστής m. **~-keeping** s. λογιστική f. **~-let** s. φυλλάδα f. **~-maker** s. πράκτορας στοιχημάτων στις κούρσες. **~-seller** s. βιβλιοπώλης m. **~-shelf** s. βιβλιοθήκη f. **~-shop** s. βιβλιοπωλείο n. **~-worm** s. (fig.) βιβλιοφάγος m.f.

bookbind|ery s. βιβλιοδετείον n. **~-ing** s. βιβλιοδεσία f.

bookie s. see bookmaker.

boom s. (barrier) ζεύγμα n. (noise) βρόντος m. (trade) οικονομικός οργασμός. (v.i.) (sound) βροντώ, (thrive) ευδοκιμώ.

boon s. (favour) χατίρι n. (sthg. useful) ευλογία f.

boor s., **~-ish** a. αγροίκος. **~-ishness** s. αγροικία f.

boost v.t. (lift) σηκώνω, (increase) αυξάνω, (publicize) διαφημίζω, (advance) προωθώ, (strengthen) ενισχύω. (s.) σπρώξιμο

n. διαφήμιση f. ενίσχυση f. **~-er** s. (mech.) ενισχυτής m. (dose) συμπληρωματική δόση.

boot s. μπότα f. (soldier's) αρβύλα f. top-~ στιβάλι n. (of car) πορτ-μπαγκάζ n. he got the ~ (fam.) τον έδιωξαν. (adv.) to ~ εις επίμετρον. (v.) (kick) κλοτσώ. ~ out πετώ έξω.

bootblack s. λούστρος m.

booth s. μπάγκος m. (telephone, etc.) θάλαμος m.

bootlace s. κορδόνι n.

bootlegging s. λαθρεμπόριο οινοπνευματωδών ποτών.

booty s. λάφυρα n.pl. λεία f.

booz|e v. (fam.) το τσούζω. (s.) πιοτό n. **~-ing** s. μεθοκόπι n.

border v. (adjoin) συνορεύω με, γειτνιάζω με. (lie along edge of) βρίσκομαι στην παρυφή (with gen.) ~ed with lace με μπορντούρα από δαντέλα.

border s. (edge) παρυφή f, (edging) μπορντούρα f. (frontier) σύνορα n.pl. (a.) (of the frontier) παραμεθόριος.

borderland s. παραμεθόριος περιοχή. (fig.) μεταίχμιον n.

borderline s. σύνορο n. (fig.) ~ case αμφισβητήσιμη υπόθεση. it is on the ~ βρίσκεται στη διαχωριστική γραμμή.

bore s. (calibre) ολκή f. (nuisance) μπελάς m.

bor|e v. (hole, tunnel etc.) ανοίγω. (for oil, etc.) κάνω γεωτρήσεις. (weary) κουράζω. be ~ed βαριέμαι, πλήττω. he ~es me με κάνει και πλήττω. **~-edom** s. πλήξη f., ανία f. **~-ing** a. πληκτικός, ανιαρός.

born a. γεννημένος. be ~ γεννιέμαι. ~ and bred γέννημα και θρέμμα.

borough s. δήμος m. ~ council δημοτικόν συμβούλιον.

borrow v. δανείζομαι. **~-ed** a. δανεικός. **~-er** s. ο δανειζόμενος.

bosh s. (fam.) ανοησίες f.pl.

bosom s. στήθος n. (fig.) in the ~ of his family στους κόλπους της οικογενείας του. ~ friend επιστήθιος φίλος.

boss s. (protuberance) προεξοχή f. (on shield) επίσημο n.

boss v.t. (bungle, miss) δεν πετυχαίνω. (s.) make a ~ shot τα κάνω σαλάτα. **~-eyed** a. στραβός.

boss s. (fam., master) αφεντικό n. (v.t.) (control, run) διευθύνω. ~ (person) around εξουσιάζω. who is the ~ here? ποιος εξουσιάζει εδώ; **~-y** a. εξουσιαστικός.

botan|ic(al) a. βοτανικός. **~-ist** s. βοτανολόγος m. **~-y** s. βοτανική f.

botch v. (*repair*) μπαλώνω. (*bungle*) τα κάνω σαλάτα. a. ~ed job τσαπατσούλικη δουλειά.

both a. & pron. και οι δύο, αμφότεροι. ~ of us και οι δυο μας. on ~ sides και από τις δύο μεριές. ~... and τόσο... όσο και. you can't have it ~ ways ή το ένα ή το άλλο.

bother v.t. & i. (*give trouble to*) ενοχλώ. (*take trouble*) κοπιάζω, σκοτίζομαι, κάνω τον κόπο (να). I can't be ~ed (*to*) βαριέμαι (να). ~! να πάρει η οργή! (*s.*) μπελάς m. ~**some** a. ενοχλητικός.

bottle s. μπουκάλι n., φιάλη f. (*v.*) εμφιαλώνω, μποτιλιάρω. ~ up (*block*) μποτιλιάρω, (*repress*) συγκρατώ. ~**neck** s. μποτιλιάρισμα n.

bottom s. (*of sea, well*) βυθός m., πάτος m., πυθμήν m. (*of hole, garden, street*) βάθος n. (*of stairs*) βάση f. (*of page*) κάτω μέρος, (*of receptacle, list, class, & general*) πάτος m. (*buttocks*) πισινός m., (*fam.*) ποπός m. at ~ κατά βάθος. touch ~ πατώ, (*fig.*) φθάνω στο κατώτερο σημείο. get to the ~ of εξηγώ. be at the ~ of είμαι πίσω από. knock the ~ out of ανατρέπω.

bottom a. (*lowest*) ο κατώτερος, (*last*) τελευταίος. he came ~ ήρθε πάτος.

bottomless a. απύθμενος.

bough s. κλάδος m.

bought a. (*lit.*) αγοραστός, (*bribed*) αγορασμένος.

boulder s. κοτρόνι n.

boulevard s. λεωφόρος f.

bounc|e v.i. πηδώ. (*s.*) πήδημα n. (*swagger*) κομπασμός m. (*fam.*, *of ball*) γκελ n. ~**er** s. μπράβος m. ~**ing** a. ζωηρός. ~ing girl (*fam.*) κόμματος m.

bound v.i. (*leap*) πηδώ, (*esp. with emotion*) σκιρτώ. (*s.*) άλμα n. by leaps and ~s αλματωδώς.

bound v.t. (*limit*) ορίζω, περιορίζω. be ~ed by περιβάλλομαι από. (*s.*) όριο n. beyond (*or within*) the ~s of πέραν (*or εντός*) των ορίων (*with gen.*). his effrontery knows no ~s το θράσος του δεν έχει όρια. it is out of ~s απαγορεύεται.

bound a. (*of book*) δεμένος. (*connected*) be ~ up with έχω σχέση με. (*occupied*) be ~ up in απασχολούμαι με. (*going*) be ~ for προορίζομαι για. (*obliged*) I am ~ to go είμαι υποχρεωμένος να πάω. (*sure*) he is ~ to be late ασφαλώς θα αργήσει.

boundary s. όριο n., σύνορο n. ~ mark ορόσημο n.

bounden a. ~ duty επιβεβλημένον καθήκον.

bounder s. (*fam.*) παλιάνθρωπος m.

boundless a. απέραντος.

bount|eous, ~**iful** a. γενναιόδωρος, ~**y** s.

γενναιοδωρία f. (*reward*) αμοιβή f.

bouquet s. μπουκέτο n. (*of wine*) άρωμα n.

bourgeois s. αστός m. (*fam.*) μπουρζουάς m. (a.) (*ideas, etc.*) μικροαστικός. ~**ie** s. αστική τάξη.

bout s. (*fit*) κρίση f. he had a ~ of drinking το έρριξε στο ποτό.

bovine a. βόειος.

bow v.i. (*lit. & fig.*) υποκλίνομαι. (*v.t. & i.*) (*bend*) λυγίζω, γέρνω. (*v.t.*) (*one's head*) σκύβω, (*one's knee*) κλίνω. ~ down (*oppress*) καταθλίβω. be ~ed (*with age*) έχω γείρει. (*s.*) (*salute*) υπόκλιση f.

bow s. (*of ship*) πρώρα f., πλώρη f.

bow s. (*weapon & mus.*) τόξο n. (*mus.*) δοξάρι n. have two strings to one's ~ το έχω δίπορτο, (*knot*) φιόγκος m. ~**legged** a. στραβοκάνης.

bowdlerize v. αποκόπτω τα ακατάλληλα χωρία.

bowel s. έντερο n. ~s (*entrails, compassion*) σπλάχνα n.pl. (*depths*) έγκατα n.pl. pass motion of ~s ενεργούμαι.

bower s. (*arbour*) σκιάδα f.

bowl s. (*basin*) λεκάνη f., (*small*) μπολ n. (*of ceiling light*) πλαφονιέρα f.

bowl v.t. & i. κυλώ. (*fig.*) ~ (*person*) over καταπλήσσω. ~ along (*v.i.*) τρέχω. (*s.*) (*ball*) μπάλα f. ~s παιχνίδι της μπάλας.

bowler s. ~ hat μπομπέ m., σκληρό n.

bowman s. τοξότης m.

bowsprit s. πρόβολος m., μπομπρέσο n.

box s. κουτί n. (*chest, also gear ~, letter ~, etc.*) κιβώτιο n. (*theatre*) θεωρείο n., ~office ταμείο n. (*on ear*) χαστούκι n. (*shrub*) πυξάρι n. (*v.t.*) βάζω σε κουτί. ~ in *or* up (*confine*) περιορίζω.

box v.i. πυγμαχώ. ~**er** s. πυγμάχος m. ~**ing** s. πυγμαχία f., μποξ n, ~ing ring ρινγκ n.

boy s. αγόρι n., (*little*) αγοράκι n. ~s' (*school*) αρρένων. ~**hood** s. παιδική ηλικία. ~**ish** a. αγορίστικος, σαν αγοριού.

boycott s. μποϋκοτάζ n. (*v.*) μποϋκοτάρω.

bra s. (*fam.*) σουτιέν n.

brace s. (*support*) στήριγμα n. (*pair*) ζευγάρι n. (*tool*) ~ and bit χειροδράπανο n. ~**s** s. τιράντες f.pl.

brac|e v. (*tighten*) σφίγγω, (*support*) στηρίζω. (*invigorate*) ζωογονώ, τονώνω. ~**ing** a. τονωτικός.

bracelet s. βραχιόλι n.

bracken s. φτέρη f.

bracket s. κονσόλα f., φουρούσι n. (*gram.*) ~s παρένθεση f. square ~s αγκύλες f.pl. in ~s εν παρενθέσει. (*v.*) (*couple*) ενώνω.

brackish a. γλυφός.

bradawl s. σουβλί n.

brag v. καυχιέμαι. ~**gart** s. καυχησιάρης m.

~ging s. καυχησιά f.
braid s. (*trimming*) γαλόνι n., γαϊτάνι n. (*tress*) πλεξούδα f. (*v.*) πλέκω.
brain s. (*lit. & fig.*) μυαλό n., (*lit.*) εγκέφαλος m. **~s** (*intellect*) μυαλό n. (*as eaten or blown out*) μυαλά n.pl. he's got it on the ~ του έχει γίνει έμμονη ιδέα. rack one's ~s σπάζω το κεφάλι μου. he's a first-class ~ είναι εγκέφαλος (*v.t.*) (*kill person with blow*) του ανοίγω το κεφάλι. **~less** a. ηλίθιος. **~-washing** s. πλύση εγκεφάλου. **~-wave** s. φαεινή ιδέα. **~y** a. έξυπνος.
braise v.t. ψήνω στην κατσαρόλα.
brake s. (*of wheels*) τροχοπέδη f. (*lit. & fig.*) φρένο n. (*v.*) φρενάρω.
bramble s. βάτος m.
bran s. πίτουρο n.
branch s. (*of tree*) κλάδος m., κλαρί n., κλαδί n. (*subdivision*) κλάδος m. (*of army*) όπλο n. (~ *line or road*) διακλάδωση f. (~ *office*) υποκατάστημα n.
branch v.i. (*diverge*) διακλαδίζομαι. ~ out (*expand*) επεκτείνομαι.
brand s. (*torch*) δαυλός m. (*stigma*) στίγμα n. (*make*) μάρκα f. (*v.*) (*mark*) σημαδεύω, (*fig.*) στιγματίζω. **~-new** a. κατακαίνουργος.
brandish v. κραδαίνω.
brandy s. κονιάκ n.
brash a. αυθάδης, αντάμης.
brass s. ορείχαλκος m., (*fam.*) μπρούντζος m. (*fig.*) (*effrontery*) αυθάδεια f., (*money*) παράδες m.pl. (*mus.*) the ~ τα χάλκινα. (*fam.*) top ~ ανώτεροι αξιωματικοί.
brass a. ορειχάλκινος, (*fam.*) μπρούντζινος. get down to ~ tacks έρχομαι στο ψαχνό. **~y** . (*fig.*) χτυπητός και χυδαίος.
brassière s. σουτιέν n.
brat s. μυξιάρικο n.
bravado s. αποκοτιά f.
brave v. αψηφώ.
brave a. ανδρείος, γενναίος. ~ person παλληκάρι n. **~ly** adv. γενναία, παληκαρίσια. **~ry** s. ανδρεία f.
bravo s. μπράβος m. (*int.*) μπράβο!
brawl s. καβγάς m. (*v.*) καβγαδίζω. **~er** s. καβγατζής m.
brawn s. (*meat*) πηχτή f. (*strength*) ρώμη f. **~y** a. νευρώδης.
bray v. γκαρίζω. (*s.*) γκάρισμα n.
brazen a. ορειχάλκινος. (*fig.*) ξεδιάντροπος. (*v.*) ~ it out προσπαθώ να ξεμπλέξω από μία δύσκολη θέση με θρασύτητα.
brazier s. μαγκάλι n.
breach s. (*hole*) ρήγμα n. (*rupture*) ρήξη f. (*of rule*) παράβαση f. (*of agreement*) αθέτηση f. ~ of the peace διατάραξη της ησυχίας. (*v.*) κάνω ρήγμα σε, παραβαίνω.

bread s. ψωμί n., άρτος n. ~ and cheese ψωμοτύρι n. earn one's ~ βγάζω το ψωμί μου. **~-crumbs** γαλέτα f. **~-winner** προστάτης οικογένειας.
breadth s. πλάτος n., ευρύτητα f., φάρδος n.
break s. (*interruption, pause*) διακοπή f. (*interval*) διάλειμμα n. (*gap*) κενό n. (*opening*) άνοιγμα n. (*rupture*) ρήξη f. there's a ~ in the wire έσπασε το σύρμα. ~ of day χαράματα n.pl.
break v.t. (*lit. & fig.*) σπάζω, (*put out of order*) χαλώ. (*a rule*) παραβαίνω, an agreement) αθετώ. (*a journey*) διακόπτω. ~ the news (*of*) αναγγέλλω με τρόπο (ότι). ~ one's word πατώ το λόγο μου. ~ the force of (*the wind*) κόβω, (*an attack*) αναχαιτίζω.
break v.i. (*lit. & fig.*) σπάζω, (*go wrong*) χαλώ. day ~s χαράζει. ~ even έρχομαι ίσα ίσα. ~ loose ξεφεύγω. ~ with (*person*) τα χαλώ με. ~ away from (*group*) ξεκόβω από.
break down v.t. (*resistance*) κάμπτω, (*analyse*) αναλύω. (*v.i.*) (*collapse*) καταρρέω. (*of vehicle*) παθαίνω βλάβη, χαλώ. (*come to nothing*) αποτυχαίνω.
breakdown s. (*nervous*) νευρικός κλονισμός. (*stoppage*) διακοπή f. (*of vehicle*) βλάβη f. (*analysis*) ανάλυση f.
break in v.t. (*accustom*) εξοικειώνω, (*a horse*) δαμάζω. (*v.i.*) (*of thieves*) κάνω διάρρηξη. (*interrupt*) somebody broke in κάποιος πετάχτηκε.
break-in s. διάρρηξη f.
break into v. (*laughter*) ξεσπάζω σε, (*premises*) διαρρηγνύω.
break off v.t. (*detach*) κόβω, (*a betrothal*) διαλύω, (*relations*) διακόπτω. (*v.i.*) (*come off*) σπάζω, (*stop talking*) διακόπτω.
break out v.i. (*escape*) δραπετεύω. (*of fire, war*) εκρήγνυμαι. (*exclaim*) ξεσπάζω. ~ in spots βγάζω σπυριά.
break-out s. δραπέτευση f.
break through v.t. (*mil.*) διασπώ. the sun broke through the clouds ο ήλιος ξεπρόβαλε μέσα από τα σύννεφα. (*make hole in*) τρυπώ. (*v.i.*) (*advance*) ανοίγω δρόμο.
break-through s. (*mil.*) διάσπασις f. (*success*) επίτευγμα n.
break up v.t. (*divide*) διαιρώ. (*dissolve, disperse, demolish*) διαλύω. (*v.i.*) (*of school*) διακόπτω. (*be dissolved, etc.*) διαλύομαι.
break-up s. διάλυση f.
breakable a. εύθραυστος.
breakage s. σπάσιμο n.
breaker s. μεγάλο κύμα.
breakfast s. πρωινό n.

breakneck *s.* at ~ speed με ιλιγγιώδη ταχύτητα.

breakwater *s.* κυματοθραύστης *m.*

breast *s.* (*chest*) στήθος *n.*, (*pl.* στήθη). (*woman's*) στήθος *n.*, (*pl.* στήθη *or* στήθια), (*fam.*) βυζί *n.* make a clean ~ of ομολογώ. (*v.t.*) (*face*) αντιμετωπίζω, (*climb*) α-νεβαίνω σε. **--bone** *s.* στέρνο *n.* **--plate** *s.* θώρακας *m.* **-work** *s.* πρόχωμα *n.*

breath *s.* πνοή *f.*, αναπνοή *f.*, ανάσα *f.* all in one ~ απνευστί. be out of ~ λαχανιάζω. it took my ~ away μου 'κοψε την ανάσα. speak under one's ~ μιλώ μες' στο στόμα μου. hold one's ~ κρατώ την αναπνοή μου. not a ~ of suspicion ούτε ίχνος υποψίας. there isn't a ~ of wind δεν φυσάει τίποτε. ~ test έλεγχος πνοής.

breathe *v.t. & i.* αναπνέω. (*v.i.*) (*also get one's breath*) ανασαίνω. ~ one's last πνέω τα λοίσθια. don't ~ a word μη βγάλεις κιχ (*or* άχνα).

breather *s.* ανάπαυλα *f.*

breathing *s.* αναπνοή *f.* **--space** ανάπαυλα *f.*

breathless *a.* (*panting*) λαχανιασμένος, (*tense*) αγωνιώδης.

breech *s.* (*of gun*) θάλαμη. *f.* **-es** *s.* κυλότα *f.* βράκα *f.*

breed *s.* ράτσα *f.* (*v.t.*) (*livestock*) τρέφω, (*give birth to, lit. & fig.*) γεννώ. (*bring up*) ανατρέφω. (*v.i.*) πολλαπλασιάζομαι. **-ing** *s.* ανατροφή *f*, of good ~ing με καλή ανατροφή.

breeze *s.* αύρα *f.*, αεράκι *n.* sea-~ε μπάτης *m.* **-y** *a.* ~y day μέρα με αύρα. this is a ~y spot εδώ πάντα φυσάει. (*person*) ζωηρός, χωρίς τύπους.

breviary *s.* σύνοψη *f.*

brevity *s.* συντομία *f.* (*of speech only*) βραχυλογία *f.*

brew *v.* (*prepare*) παρασκευάζω. (*fig.*) be ~ing επικρέμαμαι. **-ery** *s.* ζυθοποιείον *n.* **-ing** *s.* ζυθοποιία *f.*

bribe *s.* δωροδόκημα *n.* (*fam.*) petty ~ τραμπούκο *n.* take ~s δωροδοκούμαι, χρηματίζομαι. (*v.t.*) δωροδοκώ.

bribery *s.* δωροδοκία *f.* (*taking of bribes*) δωροληψία *f.*

brick *s.* τούβλο *n.*, πλίνθος *f.* (*sunbaked*) πλίθ(ρ)α *f.* built of ~ πλινθόκτιστος. (*fam.*) drop a ~ κάνω γκάφα. he's a ~ είναι εντάξει άνθρωπος. **-work** *s.* πλινθοδομή. *f.* **-yard** *s.* τουβλάδικο *n.*

bride *s.* νύφη *f.* **-al** *a.* νυφικός. **-egroom** *s.* γαμπρός *m.*

bridge *s.* γέφυρα *f.*, γεφύρι *n.* (*cards*) μπριτζ *n.* (*v.*) γεφυρώνω. **-head** *s.* προγεφύρωμα *n.*

bridle *s.* χαλινάρι *n.* (*v.t.*) (*horse*) χαλινώνω, (*temper*) χαλιναγωγώ. (*v.i., show resentment*) τινάζω το κεφάλι μου θιγμένος. **--path** *s.* μονοπάτι για ιππείς.

brief *a.* σύντομος. in ~ *see* briefly. **-ly** *adv.* εν συντομία, με λίγα λόγια.

brief *s.* (*law*) δικογραφία *f.*, φάκελλος *m.* (*fig.*) hold no ~ for δεν πρόσκειμαι εις. (*v.t.*) (*law*) αναθέτω υπόθεσιν εις. (*inform*) ενημερώνω. **-ing** *s.* ενημέρωση *f.*

brief-case *s.* τσάντα *f.*, χαρτοφύλακας *m.*

brier *s.* αγριοτριανταφυλλιά *f.*

brig *s.* (*naut.*) μπρίκι *n.*

brigad|e *s.* (*mil.*) ταξιαρχία *f.* **-ier** *s.* ταξίαρχος *m.*

brigand *s.* ληστής *m.* **-age** *s.* ληστεία *f.*

bright *a.* (*full of light*) φωτεινός, (*shining* ~ly) λαμπερός, (*colour*) ζωηρός, (*cheerful*) γελαστός, (*clever*) έξυπνος. ~ idea φαεινή ιδέα. **-en** *v.t. & i.* ζωηρεύω. **-ly** *adv.* λαμπερά, ζωηρά.

brightness *s.* λαμπεράδα *f.* (*cleverness*) εξυπνάδα *f.*

brilliance *s.* λάμψη *f.* (*talent*) ιδιοφυΐα *f.*

brilliant *a.* (*light*) πολύ λαμπερός. give ~ light φεγγοβολώ. (*fig.*) λαμπρός. **-ly** *adv.* λαμπρώς. (*very well*) λαμπρά. shine ~ly φεγγοβολώ.

brilliantine *s.* μπριγιαντίνη *f.*

brim *s.* χείλος *n.* (*of hat*) μπορ *n.*, γύρος *m.* full to the ~ ξέχειλος.

brim *v.* ~ over ξεχειλίζω. **-ming over** ξεχειλισμένος. **--full** *a.* γεμάτος.

brimstone *s.* θείο *n.*

brin|e *s.* άλμη *f.*, σαλαμούρα *f.* **-y** *a.* αλμυρός. (*s., fam.*) θάλασσα *f.*

bring *v.t.* φέρνω. ~ oneself (*to do sthg.*) αποφασίζω. ~ about επιφέρω. ~ back (*restore*) επαναφέρω, (*return*) φέρνω πίσω, (*call to mind*) θυμίζω. ~ down (*régime, temperature*) ρίχνω, (*prices*) κατεβάζω, (*aircraft*) ρίχνω, καταρρίπτω. ~ forward (*adduce*) παρουσιάζω, (*accounts*) μεταφέρω. ~ in (*yield*) αποφέρω, (*introduce*) εισάγω. ~ off πετυχαίνω. ~ on (*cause*) προκαλώ. ~ out βγάζω, (*show*) δείχνω. ~ round (*revive*) συνεφέρνω, (*persuade*) καταφέρνω. ~ under (*category*) υπάγω εις. ~ up (*child*) ανατρέφω, (*mention*) αναφέρω, (*vomit*) κάνω εμετό. ~ to bear εξασκώ. ~ to an end θέτω τέρμα σε. ~ to light φέρω εις φως.

brink *s.* χείλος *n.*

brisk *a.* ζωηρός. business is ~ η αγορά έχει κίνηση. **-ly** *adv.* ζωηρά. **-ness** *s.* ζωηρότητα *f.*

bristl|e *s.* τρίχα *f.* (*hog's*) γουρουνότριχα *f.* **-y** *a.* άγριος, σκληρός. (*touchy*) εύθικτος.

bristle *v.i.* (*stand up*) σηκώνομαι. ~ with anger εξαγριώνομαι. ~ with (*be full of*) βρίθω (*with gen.*).
British *a.* βρεττανικός, (*person*) Βρεττανός.
brittle *a.* εύθραυστος.
broach *v.* ανοίγω.
broad *a.* φαρδύς, πλατύς, ευρύς. (*coarse*) χοντρός. in ~ outline σε γενικές γραμμές. in ~ daylight μέρα μεσημέρι. it is 50 metres ~ έχει 50 μέτρα πλάτος. have a ~ Scotch accent μιλώ με έντονη σκωτσέζικη προφορά. ~**ly** *adv.* (*speaking*) γενικά.
broadcast *s.* εκπομπή *f.* (*v.*) εκπέμπω, μεταδίδω. ~ing service ραδιοφωνία *f.* ~ing station ραδιοφωνικός σταθμός.
broaden *v.t. & i.* φαρδαίνω, πλαταίνω. (*v.t.*) (*esp. fig.*) διευρύνω, ανοίγω.
broad-minded *a.* ευρύνους. ~**ness** *s.* ευρύνοια *f.*
broadside *s.* (*fig.*) (*attack*) επίθεση *f.* ~ on από την πλευρά.
brocaded *a.* χρυσούφαντος.
brochure *s.* φυλλάδιο *n.*
broil *v.* ψήνω στη σχάρα. ~ed της σχάρας.
broke *a.* (*fam.*) απένταρος. go ~ ρίχνω κανόνι.
broken *a.* σπασμένος, (*out of order*) χαλασμένος. ~ health κλονισμένη υγεία. he was ~-hearted τσάκισε η καρδιά του. ~ in (*fig.*) εξοικειωμένος.
broker *s.* (*middleman*) μεσίτης *m.* (*fin.*) χρηματιστής *m.* ~**age** *s.* μεσιτεία *f.*
bronch|ial *a.* βρογχικός. ~ial tube βρόγχος *m.* ~**itis** *s.* βρογχίτιδα *f.*
bronze *s.* μπρούντζος *m.* (*a.*) μπρούντζινος. B~ Age εποχή του χαλκού.
brooch *s.* καρφίτσα *f.*
brood *s.* (*birds*) κλωσόπουλα *n.pl.* (*fig.*) φάρα *f.* (*v.i.*) (*of hen*) κλωσσώ. (*fig., of person*) κάθομαι και σκέπτομαι και στενοχωριέμαι. ~**y** *a.* (*fig.*) μελαγχολικός.
brook *s.* ρυάκι *n.*
brook *v.* ανέχομαι.
broom *s.* σκούπα *f.* (*fig.*) new ~ νέος προϊστάμενος με υπερβολικό ζήλο. ~**stick** *s.* σκουπόξυλο *n.*
broth *s.* ζουμί *n.*
brothel *s.* πορνείο *n.*
brother *s.* αδελφός *m.* ~s and sisters αδέλφια *n.pl.* ~-in-law (*spouse's* ~) κουνιάδος *m.*, (*sister's husband*) γαμπρός *m.*, (*wife's sister's husband*) μπατζανάκης *m.* ~**hood** *s.* αδελφότητα *f.* ~**ly** *a.* αδελφικός.
brow *s.* (*eyebrow*) φρύδι *n.* (*forehead*) μέτωπο *n.* (*of hill*) κορυφή *f.*
browbeat *v.* εκφοβίζω.
brown *a.* καφέ, (*eyes, hair*) καστανός, (*bread*) μαύρος. ~ paper χαρτί περι-

τυλίγματος. (*v.i. & t.*) (*also* get ~) μαυρίζω, (*in oven*) ροδοκοκκινίζω. (*fam.*) ~ed off βαριεστημένος.
browse *v.* (*lit.*) βόσκω. (*fig.*) ~ among (*books*) ξεφυλλίζω.
bruise *s.* μελανιά *f.* (*v.*) (*also* get ~d) μελανιάζω. ~**r** *s.* σκληρός πυγμάχος.
brunette *s.* μελαχροινή *f.*
brunt *s.* bear the ~ υφίσταμαι το κύριο βάρος.
brush *s.* βούρτσα *f.* (*skirmish*) αψιμαχία *f.*
brush *v.* βουρτσίζω. ~ against περνώ ξυστά από. ~ away (*flies*) διώχνω, (*tears*) σκουπίζω. ~ up ξεσκονίζω. ~ (*dust, etc. off*) one's clothes τινάζομαι.
brushwood *s.* χαμόκλαδα *n.pl.*
brusque *a.* απότομος. ~**ly** *adv.* απότομα.
brutal *a.* (*cruel*) ωμός. ~**ity** *s.* ωμότητα *f.* ~**ize** *v.* αποκτηνώνω. ~**ly** *adv.* ανηλεώς.
brut|e *s.* κτήνος *n.* ~**ish** *a.* κτηνώδης. ~**ishly** *adv.* κτηνωδώς. ~**ishness** *s.* κτηνωδία *f.*
bubble *s.* φουσκάλα *f.* (*v.*) κοχλάζω, (*effervesce*) αφρίζω. ~ over ξεχειλίζω. ~ up αναβλύζω.
buccaneer *s.* κουρσάρος *m.* (*fig.*) ασυνείδητος τυχοδιώκτης.
buck *s.* (*dandy*) δανδής *m.* (*male*) αρσενικός *a.* (*fam.*) pass the ~ φορτώνω την ευθύνη σε άλλον.
buck *v.i.* (*of horse*) χοροπηδώ. ~ up (*v.t.*) (*encourage*) εμψυχώνω, (*revive*) αναζωογονώ, (*v.i.*) εμψυχώνομαι, (*hasten*) κάνω γρήγορα. (*int.*) ~ up! κουνήσου.
bucket *s.* κουβάς *m.*
buckle *s.* αγκράφα *f.* (*v.t.*) κουμπώνω. ~ on ζώνομαι. (*v.i.*) (*crumple*) κάμπτομαι.
bucolic *a.* βουκολικός.
bud *s.* (*flower*) μπουμπούκι *n.* (*shoot*) μάτι *n.* (*v.i.*) μπουμπουκιάζω. (*fig.*) ~ding poet εκκολαπτόμενος ποιητής.
Buddhist *s.* βουδιστής *m.*
budge *v.t. & i.* κουνάω. he won't ~ from here δεν το κουνάει από 'δω.
budget *s.* προϋπολογισμός *m.* (*v.*) ~ for κάνω προϋπολογισμό για.
buffalo *s.* βουβάλι *n.*
buffer *s.* προφυλακτήρ *m.* ~ state μικρό κράτος μεταξύ δύο μεγάλων.
buffet *s.* (*collation*) μπουφές *m.* (*bar*) κυλικείο *n.*
buffet *v.* χτυπώ. be ~ed (*by wind, fortune*) ανεμοδέρνω, (*by waves, fortune*) θαλασσοδέρνω. (*s.*) χτύπημα *n.*
buffoon *s.* καραγκιόζης *m.* ~**ery** *s.* καραγκιοζιλίκια *n.pl.*
bug *s.* κοριός *m.*
bugbear *s.* εφιάλτης *m.*
bugging *s.* (*fam.*) παγίδευση *f.* ~ of phone-

calls υποκλοπή τηλεφωνημάτων.

buggy s. αμαξάκι n.

bugle s. σάλπιγγα f.

build s. κατασκευή f., καμωσιά f.

build v. κτίζω. ~ up (create) δημιουργώ, (increase) αυξάνω. ~ on (rely) βασίζομαι σε. well built καλοφτιαγμένος. built-up area οικοδομημένη περιοχή.

builder s. κτίστης m. (contractor) οικοδόμος m. ~'s yard μάντρα f.

building s. κτίριο n. (official or imposing) μέγαρο n. (under construction) οικοδομή f. (act of ~) οικοδόμηση f. ~ site γιαπί n. ~ plot οικόπεδο n. ~ society οικοδομικός συνεταιρισμός.

built-in a. εντοιχισμένος.

bulb s. βολβός m. (lamp) λάμπα f. ~ous a. βολβοειδής.

bulge v. φουσκώνω. (s.) φούσκωμα n.

bulk s. όγκος m. in ~ χονδρικώς. the ~ of (most) το μεγαλύτερο μέρος. ~y a. ογκώδης.

bull s. ταύρος m. hit the ~s-eye βρίσκω τη διάνα. (fig.) take the ~ by the horns αντιμετωπίζω αποφασιστικά μια δύσκολη θέση.

bulldozer s. μπουλντόζα f.

bullet s. σφαίρα f. ~proof a. αλεξίσφαιρος.

bulletin s. δελτίον n.

bullfight s. ταυρομαχία. f. ~er s. ταυρομάχος m.

bullion s. gold ~ χρυσός εις ράβδους.

bullock s. βόδι n.

bully s. θρασύδειλος a. (ruffian) ψευτονταής m. (v.) εκφοβίζω. ~ing s. εκφοβισμός m.

bulrush s. βούρλο n.

bulwark s. προπύργιον n.

bump s. (blow) χτύπημα n. (jolt) τράνταγμα n. (on head) καρούμπαλο n.

bump v. (proceed with jolts) τραντάζομαι. ~ against or into χτυπώ, πέφτω απάνω σε. I ~ed my head on the door χτύπησα το κεφάλι μου στην πόρτα. (fam.) ~ off ξεκάνω.

bumper s. (fender) προφυλακτήρ m. (glass) γεμάτο ποτήρι. (a.) έξτρα.

bumpkin s. βλάχος m.

bumptious a. αυτάρεσκος.

bumpy a. a ~ road δρόμος με πολλές γούβες. we had a ~ ride τρανταχτήκαμε στο δρόμο.

bun s. τσουρέκι n.

bunch s. (grapes) τσαμπί n. (keys) αρμαθός m. (flowers) μάτσο n. (people) ομάδα f.

bunch v. ~ed up στριμωγμένος.

bundle s. δέμα n. (clothes, etc.) μπόγος m. (papers, etc.) μάτσο n. (v.) (thrust) χώνω.

bung s. τάπα f.

bung v. (fam.) (thrust) χώνω. ~ed up βουλωμένος.

bungalow s. μονόροφο σπίτι.

bungle v. he ~d it τα 'κανε θάλασσα (or σαλάτα). ~r s. μαστροχαλαστής m.

bunion s. κότσι n.

bunk v. (fam., run off) το κόβω λάσπη.

bunk s. (bed) κουκέτα f.

bunk(um) s. (fam.) κολοκύθια n.pl.

buoy s. σημαδούρα f. (v.) ~ up (fig.) αναπτερώνω. ~ant a. cork is ~ant ο φελλός επιπλέει. (fig., in disposition) αμέριμνος. with a ~ant step με ελαστικό βήμα.

burden s. βάρος n., φορτίο n. (tonnage) χωρητικότητα f. (theme) θέμα n. (refrain) επωδός f.

burden v. φορτώνω, επιβαρύνω. ~some a. βαρετός, επαχθής.

bureau s. γραφείο n.

bureaucra|cy s. γραφειοκρατία f. ~t s. γραφειοκράτης m. ~tic a. γραφειοκρατικός.

burgeon v. βλαστάνω.

burgess s. δημότης m.

burg|lar s. διαρρήκτης m. ~lary s. διάρρηξη f. ~le v. διαρρηγνύω.

burial s. θάψιμο n., ταφή f.

burke v. ~ the issue τα φέρνω βόλτα.

burlesque s. παρωδία f. (v.) παρωδώ.

burly a. ρωμαλέος, γεροδεμένος.

burn v.t. καίω. (v.i.) (be alight or hot) καίω. (be on fire, get scorched or spoilt) καίομαι. (be combustible) καίομαι. be ~ing (of person, feel hot) καίομαι, (with emotion) φλέγομαι. get ~t to ashes αποτεφρώνομαι. ~ one's fingers (fig.) καίομαι. he is burnt out (creatively) εξόφλησε.

burn s. έγκαυμα n. (stream) ρυάκι n.

burner s. (of stove) εστία f., μάτι n. (jet) μπεκ n.

burning s. κάψιμο n. (sensation) καούρα f.

burning a. (alight) αναμμένος, (hot) καυτός, (ardent) διακαής. ~ question φλέγον ζήτημα.

burnish v. στιλβώνω.

burnt a. καμένος.

burr s. (plant) κολλητσίδα f.

burrow s. τρύπα f. (v.) σκάβω. ~ into (fig.) ερευνώ.

bursar s. ταμίας m. ~y s. (grant) υποτροφία f.

burst v.t. σκάζω. (v.i.) σκάζω, κρεπάρω. (bubble, bud) σκάζω, (boil) ανοίγω. (bomb, boiler) εκρήγνυμαι, (storm) ξεσπάζω. (be over-full) ξεχειλίζω. I am nearly ~ing πάω να σκάσω. be ~ing with laughter σκάζω στα γέλια. ~ out crying ξεσπάζω σε κλάματα. the door ~ open η πόρτα άνοιξε με βία. he is ~ing to tell τον τρώει η γλώσσα του. the mountains ~ into view τα βουνά ξανοίχτηκαν μπροστά μας. we had a ~ tyre μας έπιασε λάστιχο. ~ in ορμώ μέσα. ~ upon the scene ενσκήπτω. ~ through see break.

burst s. (anger) ξέσπασμα n. (firing) ριπή f. there was a ~ of flame ξεπετάχτηκε μια φλόγα. there was a ~ of applause ξέσπασαν σε χειροκροτήματα.

bursting s. σκάσιμο n.

burly v. θάβω, (thrust, hide) χώνω. ~ied in thought βυθισμενος σε σκέψεις.

bus s. λεωφορείο n. miss the ~ (fig.) χάνω την ευκαιρία. ~ station σταθμός λεωφορείων.

bush s. θάμνος m. ~y a. θαμνώδης, (brows, etc.) δασύς.

bushel s. μόδιον n. (fig.) ~s of πολλά.

busily adv. με δραστηριότητα.

business s. 1. (matter, affair) υπόθεση f. (duty, work, concern) δουλειά f. mind your own ~ να κοιτάς τη δουλειά σου. send (person) about his ~ ξαποστέλνω. 2. (commercial enterprise: particular) επιχείρηση f., (in general) επιχειρήσεις pl., (fam.) δουλειές pl. on ~ για δουλειές. (trade) εμπόριο n. be in ~ εμπορεύομαι. ~ dealings συναλλαγές. f.pl. 3. (right) you have no ~ (to) δεν έχετε κανένα δικαίωμα (να).

business a. εμπορικός. ~-like a. μεθοδικός ~-man s. επιχειρηματίας m.

buskin s. κόθορνος m.

bust s. (woman's) μπούστος m. (sculptured) προτομή f.

bust v.t. & i. (fam.) σπάζω. go ~ ρίχνω κανόνι.

bustle v. ~ about έχω τρεχάματα. (s.) κίνηση f., πηγαινέλα n.

busy a. (occupied) απασχολημένος. be ~ with καταγίνομαι με. keep (person) ~ απασχολώ. (street, etc.) περαστικός. I had a ~ day η μέρα μου ήταν γεμάτη. the shops are ~ τα μαγαζιά έχουν κίνηση. a ~ man πολυάσχολος άνθρωπος.

busybody s. be a ~ χώνω τη μύτη μου παντού.

but conj., adv. & prep. αλλά, μα, παρά. last ~ one προτελευταίος. last ~ three τρίτος προ του τελευταίου. next-door ~ one παραδιπλανός. all ~ one όλοι εκτός ενός. he cannot ~ agree δεν μπορεί παρά να συμφωνήσει. we can ~ try μπορούμε τουλάχιστον να επιχειρήσουμε. it takes ~ an hour παίρνει μόνο μία ώρα. not a day passes ~ they have a quarrel δεν περνάει

μέρα που να μη μαλώσουν. it is all ~ finished σχεδόν τελείωσε. we all ~ lost the train παρά λίγο να χάσουμε το τραίνο. ~ for your help... αν δεν ήσουν εσύ... I've heard nothing ~ good of him δεν άκουσα παρά (μόνον) επαίνους γι' αυτόν. she eats nothing ~ fruit δεν τρώει (τίποτ') άλλο από φρούτα or δεν τρώει παρά (μόνο) φρούτα. he is anything ~ stupid κάθε άλλο παρά βλάκας είναι. (I can agree to) anything ~ that όλα κι όλα μα όχι αυτό.

butcher s. χασάπης m. ~'s shop χασάπικο n., κρεοπωλείον n. (v.) κατακρεουργώ. ~y s. (carnage) μακελειό n.

butler s. μαιτρ ντ' οτέλ m., μπάτλερ m.

butt s. (cask) βαρέλι n. (of gun) (υπο)κόπανος m., κοντάκι n. the ~s σκοπευτήριον n. (target, fig.) στόχος m. (of cigarette) αποτσίγαρο n.

butt v. κουτουλώ. ~ in πετάγομαι.

butter s. βούτυρο n. (v.) αλείβω με βούτυρο. ~ up κολακεύω. ~y a. σαν βούτυρο.

buttercup s. νεραγκούλα f.

butterfly s. πεταλούδα f.

buttery s. αποθήκη τροφίμων.

buttock s. γλουτός m.

button s. κουμπί n. (v.) κουμπώνω.

buttonhole s. κουμπότρυπα f. (flower) μπουτονιέρα f. (v.) (fam.) he ~d me με στρίμωξε και άρχισε το βιολί του.

buttress s. αντέρεισμα n. (v.) στηρίζω.

buxom a. ~ woman κόμματος m.

buy v. αγοράζω. ~ back ξαναγοράζω. ~ off or out εξαγοράζω. ~ oneself out of the army εξαγοράζω τη στρατιωτική μου θητεία. ~er s. αγοραστής m. ~ing s. αγορά f.

buzz, ~ing s. βόμβος m., βουητό n.

buzz v. βομβώ, βουΐζω. (fam.) ~ off το σκάω. ~er s. βομβητής m.

by prep. 1. (agent) από (with acc.), υπό (with gen.). (near) κοντά σε. stand ~ (person) παρατέκομαι (with gen.). (past) see past. (time) ~ moonlight με το φεγγάρι, ~ night νύχτα, ~ tomorrow ως (or μέχρι) αύριο. ~ the time he arrives ώσπου (or μέχρι) να φτάσει. (measure) multiplied ~ επί (with acc.), divided ~ διά (with gen.), ~ the kilo με το κιλό. (means, manner) ~ gas με γκάζι, ~ ship με βαπόρι, ~ post ταχυδρομικώς, ~ the arm από το μπράτσο, ~ heart απ' έξω. (in accordance with) σύμφωνα με. 2. (with gerund) you caught cold ~ going out without an overcoat για να βγεις (or με το να βγεις or βγαίνοντας) έξω χωρίς παλτό κρυολόγησες. 3. (various) ~ sight εξ όψεως, ~ profession εξ επαγγέλματος, ~ force διά της βίας, ~ sea διά θαλάσσης, ~ land and sea κατά ξηράν και θάλασσαν, ~ mistake κατά λάθος, ~ God! μα το Θεό!

by *adv. see* aside, near, past. ~ and ~ ύστερα. ~ the ~ (*or* way) αλήθεια. ~ and large γενικά.

bye-bye *s.* (*fam.*) go to ~ κάνω νάνι. (*int.*) αντίο.

by-election *s.* αναπληρωματικές εκλογές.

bygone *a.* ~ days τα παλιά χρόνια. let ~s be ~s περασμένα ξεχασμένα.

by-law *s.* δημοτική απόφασις.

by-pass *s.* (*road*) περιφερικός *m.* (*v.*) παρακάμπτω, (*fig., dodge*) καταστρατηγώ.

by-product *s.* υποπροϊόν *n.*

byre *s.* βουστάσιο *n.*

by-road *s.* δρομάκι *n.*

bystander *s.* παριστάμενος *m.*

by-street *s.* πάροδος *f.*

by-way *s.* (*fig.*) ~s of history ιστορικά γεγονότα δευτερευούσης σημασίας.

by-word *s.* he is a ~ for stinginess η τσιγγουνιά του είναι παροιμιώδης.

Byzantine *a.* βυζαντινός.

C

cab *s.* (*horse*) άμαξα *f.* (*taxi*) ταξί *n.* ~**man** *s.* αμαξάς *m.*

cabal *s.* φατρία *f.*, κλίκα *f.*

cabaret *s.* καμπαρέ *n.*

cabbage *s.* λάχανο *n.*

cabin *s.* (*hut*) καλύβα *f.* (*ship's*) καμπίνα *f.* ~**ed** *a.* στριμωγμένος.

cabinet *s.* (*display*) βιτρίνα *f.* (*storage*) ντουλάπι *n.* (*political*) στενόν υπουργικόν συμβούλιον. ~-**maker** *s.* καλλιτέχνης επιπλοποιός.

cable *s.* (*rope, wire*) καλώδιο *n.* (*message*) τηλεγράφημα *n.* (*v.*) τηλεγραφώ. ~-**car** *s.* τελεφερίκ *n.* ~-**television** *s.* καλωδιακή τηλεόραση.

cabotage *s.* ακτοπλοΐα *f.*

cache *s.* κρυψώνας *m.*

cachet *s.* (*distinction*) κασέ *n.*

cackle *v.* κακαρίζω. (*s.*) κακάρισμα *n.*

cacophon|y *s.* κακοφωνία *f.* ~**ous** *a.* κακόφωνος.

cactus *s.* κάκτος *f.*

cad *s.* παλιοτόμαρο *n.*, μούτρο *n.* ~**dish** *a.* βρωμερός.

cadaverous *a.* κατάχλωμος.

cadence *s.* (*mus.*) πτώση *f.* (*of voice*) διακύμανση *f.*

cadet *s.* (*naut.*) δόκιμος *m.* (*mil.*) εύελπις *m.* (*aero.*) ίκαρος *m.* (*younger son*) νεότερος γιος.

cadge *v.* σελεμίζω. ~**r** *s.* σελέμης *m.*

cadre *s.* (*mil.*) στέλεχος *m.*

caesura *s.* τομή *f.*

café *s.* καφενείο *n.*

cage *s.* κλουβί *n.* (*v.*) εγκλωβίζω.

cagey *a.* επιφυλακτικός.

caique *s.* καΐκι *n.*

cairn *s.* σωρός λίθων.

cajole *v.* καλοπιάνω. ~**ry** *s.* καλόπιασμα *n.*, χάδια *n.pl.*

cake *s.* (*general*) γλύκισμα *n.* (*slab*) κεκ, κέικ *n.* (*layer*) τούρτα *f.* (*pastry*) πάστα *f.* (*soap, etc.*) πλάκα *f.* (*fam.*) they sold like hot ~s έγιναν ανάρπαστα. it takes the ~ αυτό είναι το άκρον άωτον.

caked *a.* (*hardened*) πηγμένος. (*coated*) γεμάτος, πασαλειμμένος.

calamit|y *s.* συμφορά *f.* ~**ous** *a.* ολέθριος.

calcium *s.* ασβέστιο *n.*

calculat|e *v.* υπολογίζω. ~**ed** *a.* μελετημένος. ~**ing** *a.* συμφεροντολόγος. ~**ion** *s.* υπολογισμός *m.* ~**or** *s.* αριθμομηχανή *f.*

calculus *s.* (*math.*) λογισμός *m.*

calendar *s.* ημερολόγιο *n.*

calf *s.* (*animal*) μοσχάρι *n.* (*leather*) βιδέλο *n.* (*of leg*) γάμπα *f.*

calibre *s.* διαμέτρημα *n.* (*fig.*) αξία *f.*

calico *s.* τσίτι *n.*

caliph *s.* χαλίφης *m.*

call *s.* (*shout*) φωνή *f.* (*summons*) φωνή *f.*, κλήση *f.* (*bird's*) κραυγή *f.* (*trumpet*) σάλπισμα *n.* (*phone*) τηλεφώνημα *n.* (*visit*) επίσκεψη *f.* pay a ~ κάνω επίσκεψη. on ~ διαθέσιμος. there is no ~ δεν υπάρχει λόγος.

call *v.i.* (*shout, cry*) φωνάζω. (*visit*) περνώ. (*v.t.*) (*name*) καλώ, ονομάζω, λέω, (*only persons*) βγάζω. (*by epithet, nickname*) αποκαλώ. (*summon*) καλώ, (*convoke*) συγκαλώ, (*send for*) φωνάζω, (*waken*) ξυπνώ, (*consider*) θεωρώ. I ~ your attention το εφιστώ την προσοχή σας εις. ~-**box** *s.* τηλεφωνικός θάλαμος.

call at *v.* περνώ από, (*of ship*) πιάνω σε.

call for *v.* (*demand*) ζητώ, (*need*) απαιτώ. I will ~ you θα περάσω να σε πάρω.

call forth *v.* προκαλώ.

call in *v.t.* (*send for*) καλώ, (*withdraw*) αποσύρω.

call off *v.* ματαιώνω.

call on (**upon**) *v.* (*require*) καλώ, (*appeal to*) ζητώ από. he called on me (*as visitor*) πέρασε να με δει.

call out *v.i.* φωνάζω. (*v.t.*) καλώ.

call up *v.t.* (*mil.*) καλώ υπό τα όπλα. (*phone to*) τηλεφωνώ σε.

call-up *s.* κλήση υπό τα όπλα.

caller *s.* επισκέπτης *m.*

calling s. επάγγελμα n.

callipers s. διαβήτης m.

callous a. άπονος, σκληρός. ~ly adv. άπονα.
~ness s. απονιά f., σκληρότητα f.

callow a. άβγαλτος.

callus s. κάλος m.

calm v.t. & i. (also ~ down) ηρεμώ, ησυχάζω, καλμάρω. (s.) γαλήνη f., ηρεμία f.
(weather) νηνεμία f. (a.) γαλήνιος, ήρεμος.
(composed) ψύχραιμος. ~ly adv. ήρεμα, με
ψυχραιμία.

calor|ie s. θερμίδα f. ~ific a. θερμικός.

calumn|y s. συκοφαντία f. ~iate v. συκο-
φαντώ. ~iator s. συκοφάντης m.

Calvary s. Γολγοθάς m.

camber s. καμπυλότητα f.

cambric s. βατίστα f.

camel s. γκαμήλα f.

camellia s. καμέλια f.

cameo s. καμέα f.

camera s. φωτογραφική μηχανή. in ~ κεκλεισμένων των θυρών. ~man s. εικο-
νολήπτης m.

camomile s. χαμομήλι n.

camouflage v. καμουφλάρω. (s.) καμου-
φλάζ n.

camp s. στρατόπεδο n. (holiday, etc.)
κατασκήνωση f. (v.) στρατοπεδεύω. (pitch
tent) στήνω σκηνή. go ~ing κάνω κάμ-
πινγκ. ~ing site κατασκήνωση f.

camp-bed s. ράντζο n.

camp-follower s. (fig.) κολλητσίδα f.

campaign s. εκστρατεία f. (press) καμπάνια f. (v.) εκστρατεύω, κάνω εκστρατεία.
κάνω καμπάνια.

campanile s. καμπαναριό n.

camphor s. καμφορά f.

cam-shaft s. εκκεντροφόρος άξων.

can s. (jug) κανάτι n. (tin, large) τενεκές m.,
(small) κουτί n. (of food) κονσέρβα f. (v.t.)
βάζω στο κουτί. ~ning industry κονσερ-
βοποιία f. ~ned a. του κουτιού. ~opener
s. ανοιχτήρι n.

can v. (be able) μπορώ να. ~ you see it? το
βλέπεις; I ~'t hear you δεν σ' ακούω. it ~'t
be done δεν γίνεται. you ~'t imagine δεν
φαντάζεστε, δεν μπορείτε να φαντασθείτε.

Canadian a. καναδικός, (person) Καναδός.

canal s. διώρυγα f. ~ize v. διοχετεύω.

canard s. ψευδής είδηση.

canary s. καναρίνι n.

cancel v.t. ακυρώνω, ματαιώνω. (cross out)
διαγράφω. ~ out (math.) απαλείφω. they ~
(each other) out εξουδετερώνονται.
~lation s. ακύρωση f., ματαίωση f.

cancer s. καρκίνος m. ~ous a. καρκινώδης,
(fig.) σαν καρκίνωμα.

candelabrum s. πολυέλαιος m., πολύφωτο n.

candid a. ειλικρινής, ντόμπρος. ~ly adv.
ειλικρινά.

candidat|e s. υποψήφιος a. ~ure s. υποψη-
φιότητα f.

candle s. κερί n., (large) λαμπάδα f. (fig.)
can't hold a ~ to δεν πιάνει χαρτωσιά
μπροστά σε. ~-stick s. καντηλέρι n.,
κηροπήγιον n.

candour s. ειλικρίνεια f.

candy s. κάντιο n. (sweets) καραμέλες f.pl.

cane s. κάλαμος m. (switch) βέργα f. ~ sugar
καλαμοσάκχαρο n. (a.) καλαθένιος.

cane v. δέρνω με βέργα. get ~d τρώω ξύλο.

canine a. κυνικός. ~ tooth κυνόδοντας m.

canister s. κουτί n.

canker s. γάγγραινα f., σαράκι n.

cannabis s. χασίς n.

cannibal s. καννίβαλος m., ανθρωποφάγος
m. ~ism s. ανθρωποφαγία f.

cannon s. κανόνι n. τηλεβόλο n. ~ade s.
κανονιοβολισμός m. ~-ball s. μπάλλα f.

canny a. προσεκτικός.

canoe s. κανό n. (dug-out) μονόξυλο n.

canon s. κανών m. (a.) κανονικός. ~ical a.
κανονικός. ~ize v. ανακηρύσσω άγιον.

cant v.t. & i. (tilt) γέρνω. (s.) κλίση f.

cant s. (hypocrisy) υποκρισίες f.pl.
(thieves') αργκό f. ~ing a. υποκριτικός.

cantankerous a. στρυφνός, τζαναμπέτης.

canteen s. καντίνα f. (mess tin) καραβάνα f.
(water-bottle) παγούρι n. ~ of cutlery
σερβίτσιο μαχαιροπήρουναν εντός ειδικού επίπλου.

canter s. μικρός καλπασμός. preliminary ~
(fig.) δοκιμαστική διαδρομή.

canticle s. ύμνος m.

cantilever s. πρόβολος m., φουρούσι n.

canton s. καντόνι n. ~ment s. επισταθμία f.

cantor s. πρωτοψάλτης m.

canvas m. κανναβάτσο n. (artist's) καμβάς
m. under ~ (camping) σε σκηνές, (sailing)
με σηκωμένα πανιά.

canvass v.t. (discuss) συζητώ. (v.i.) (solicit
votes) ψηφοθηρώ, (seek custom) αναζητώ
πελατεία.

canyon s. φαράγγι n.

cap s. σκούφος m. (peaked) κασκέτο n.,
πηλίκιο n. bathing ~ σκούφια f. (cover)
κάλυμμα n., καπάκι n. (fig.) set one's ~ at
βάζω στο μάτι. (v.) σκεπάζω. (outdo)
ξεπερνώ.

capability s. ικανότητα f.

capab|le a. ικανός, άξιος. (thing) ~le of
improvement επιδεκτικός βελτιώσεως.
~ly adv. με ικανότητα.

capacious a. ευρύχωρος.

capacity s. (content) χωρητικότητα f.
(ability) ικανότητα f. (understanding)

αντίληψη f. with a seating ~ of 100 με εκατό θέσεις. in my ~ as a doctor υπό την ιδιότητά μου ως ιατρού. in a private ~ όχι επαγγελματικώς.

caparisoned a. στολισμένος.

cape s. (garment) κάπα f. (geog.) ακρωτήριον n.

caper s. (edible) κάπαρη f.

caper v. χοροπηδώ. (s.) χοροπήδημα n.

capital s. (city) πρωτεύουσα f. (letter) κεφαλαίο n. (archit.) κιονόκρανο n. (fig.) κεφάλαιο n. make ~ out of εκμεταλλεύομαι.

capital a. (chief) κεφαλαιώδης, (punishment) θανατικός. (first rate) πρώτης τάξεως. (int.) θαυμάσια! **-ly** adv. θαυμάσια, μια χαρά.

capital|ism s. κεφαλαιοκρατία f. **~ist** s. κεφαλαιοκράτης m. (a.) κεφαλαιοκρατικός.

capitalize v. (fin.) κεφαλαιοποιώ. ~ (on) (profit by) επωφελούμαι (with gen.).

capitation s. ~ tax κεφαλικός φόρος.

capitulat|e v. συνθηκολογώ. **~ion** s. συνθηκολόγηση f. C~ions διομολογήσεις f.pl.

capon s. καπόνι n.

capric|e s. καπρίτσιο n. **~ious** a. καπριτσιόζος. **~iously** adv. καπριτσιόζικα.

capsize v.t. & i. μπατάρω.

capstan s. αργάτης m.

capsule s. (med.) καψούλα f. (of bottle) πώμα n.

captain s. (naut.) πλοίαρχος m., (fam.) καπετάνιος m. (mil.) λοχαγός m. (of team) αρχηγός m. (of irregulars) καπετάνιος m.

caption s. επικεφαλίδα f., τίτλος m. (of picture) λεζάντα f.

captious a. γκρινιάρης.

captivat|e v. γοητεύω. **~ing** s. γοητευτικός.

captiv|e a. & s. αιχμάλωτος, δέσμιος, υπόδουλος. **~ity** s. αιχμαλωσία f.

captor s. ο συλλαμβάνων.

capture v. (men, beasts, ideas) συλλαμβάνω, (cities) κυριεύω, καταλαμβάνω, (hearts) κατακτώ. (s.) σύλληψη f. κατάκτηση f. (of cities) πάρσιμο n.

car s. (motor) αυτοκίνητο n., αμάξι n. ~ crash καραμπόλα f. ~ park πάρκινγκ n.

carafe s. καράφα f.

caramel s. καμένη ζάχαρη, (sweet) καραμέλα f.

caravan s. (company) καραβάνι n. (cart) τροχόσπιτο n. ~ site κάμπιγκ n. **~serai** s. χάνι n.

caraway s. είδος κύμινου.

carbine s. καραμπίνα f.

carbohydrates s. αμυλούχες τροφές.

carbon s. άνθραξ m. ~ paper καρμπόν n. ~

copy αντίγραφο με καρμπόν. **~ic** a. ανθρακικός.

carbuncle s. (boil) δοθιήν m.

carburettor s. καρμπυρατέρ n.

carcass s. ψοφίμι n. (meat) σφάγιο n.

card s. (fam., odd person) τύπος m.

card s. (substance) χαρτόνι n. (filing) καρτέλα f. (post) δελτάριο n., (picture) κάρτα f. (visiting) κάρτα f. (ration, identity, etc.) δελτίο n. (playing) τραπουλόχαρτο n. play ~s παίζω χαρτιά. put one's ~s on the table παίζω με ανοιχτά χαρτιά. it's on the ~s είναι πιθανό.

cardboard s. χαρτόνι n. (a.) από χαρτόνι.

cardigan s. κάρντιγκαν n., πλεκτό γιλέκο με μανίκια.

cardinal s. (eccl.) καρδινάλιος m. (a.) κύριος, (number) απόλυτος.

card-index s. ευρετήριο καρτελών.

cardiolog|y s. καρδιολογία f. **~ist** s. καρδιολόγος m.

card-sharper s. χαρτοκλέφτης m.

care s. (attention) προσοχή f. (charge) φροντίδα f. (worry) έγνοια f. take ~ (of) (be careful, mind) προσέχω, (see to) φροντίζω για, (look after, keep safely) φυλάω. take ~! το νου σου! πρόσεχε! (as threat) έννοια σου!

care v.i. (be concerned, mind) νοιάζομαι. I don't ~ δεν με νοιάζει. little he ~s or he couldn't ~less πολύ που τον νοιάζει, σκοτίστηκε. (be interested) ενδιαφέρομαι. (like) I don't ~ for it δεν μου αρέσει. would you ~ to...? θα σας άρεσε να...; ~ for (look after) φροντίζω, περιποιούμαι. the child was well ~d for το φροντίσανε καλά το παιδί. the garden looks well ~d for ο κήπος φαίνεται περιποιημένος.

careen v. καρενάρω.

career s. σταδιοδρομία f., καριέρα f.

career v. τρέχω ορμητικά.

carefree a. ξέγνοιαστος, αμέριμνος.

careful a. προσεκτικός. be ~ (of) προσέχω. **~ly** adv. προσεκτικά, με προσοχή.

careless a. απρόσεκτος, αμελής. be ~ (of) δεν προσέχω, αμελώ. **~ly** adv. απρόσεκτα, με αμέλεια. **~ness** s. απροσεξία f., αμέλεια f.

caress v. χαϊδεύω, θωπεύω (s.) χάδι n., θωπεία f. **~ing** a. χαϊδευτικός, θωπευτικός.

caretaker s. επιστάτης m. (door-keeper) θυρωρός m. ~ government υπηρεσιακή κυβέρνησις.

careworn a. ταλαιπωρημένος.

car-ferry s. οχηματαγωγό n., φέριμποτ n.

cargo s. φορτίο n. **~-vesel** s. φορτηγό n.

caricature s. χαρικατούρα f.

carnage *s.* μακελειό *n.*

carnal *a.* σαρκικός.

carnation *s.* γαρίφαλο *n.*

carnival *s.* καρναβάλι *n.* (*before Lent*) αποκριές *f.pl.*

carnivorous *a.* σαρκοβόρος.

carob *s.* χαρούπι *n.* (*tree*) χαρουπιά *f.*

carol *v.* κελαϊδώ (*s.*) ~(s) κάλαντα *n.pl.*

carouse *v.* μεθοκοπώ.

carp *s.* κυπρίνος *m.*

carp *v.* ~ (at) κατακρίνω. **~ing** *a.* γκρινιάρικος. (*s.*) γκρίνια *f.*

carpent|er *s.* μαραγκός *m.* **~ry** *s.* ξυλουργική *f.*

carpet *s.* χαλί *n.*, τάπης *m.* fitted ~ μοκέτα *f.* take up the ~s ξεστρώνω. (*fig.*) put on the ~ (*censure*) επιπλήττω, παρατηρώ. (*v.*) ντύνω με χαλί, (*fig.*, *cover*) στρώνω.

carriage *s.* (*vehicle*) όχημα *n.* (*railway*) βαγόνι *n.* (*transport*) μεταφορά *f.*, (*charges*) μεταφορικά *n.pl.* (*bearing*) παράστημα *n.*

carrier *s.* μεταφορεύς *m.* (*of germs etc.*) φορεύς *m.* ~ pigeon ταχυδρομική περιστερά. **~bag** σακούλα *f.*

carrion *s.* ψοφίμι *n.*

carrot *s.* καρότο *n.*

carry *v.t.* (*bring*) φέρω, (*take*, *convey*) πηγαίνω, (*transport*) μεταφέρω, μετακομίζω, κουβαλώ. (*accommodate*) παίρνω. (*be holding*) κρατώ, κουβαλώ, βαστάζω. (*money*) κρατώ, (*a mark, etc.*) φέρω. (*support*) βαστάζω, φέρω. (*entail*) συνεπάγομαι. (*capture*) καταλαμβάνω. (*math.*) ~ five πέντε τα κρατούμενα. ~ (*joke, etc.*) too far παρατραβώ. ~ the day νικώ. ~ one's point επιβάλλω την άποψή μου. be carried (*voted*) εγκρίνομαι, (*held up*) υποστηρίζομαι. ~ oneself well έχω καλό παράστημα. (*v.i.*) (*reach*) φθάνω. his voice carries well η φωνή του ακούγεται καλά. **~cot** *s.* πορτ-μπεμπέ *n.*

carry away *v.* (*win*) αποκομίζω. get carried away παρασύρομαι.

carry forward *v.* (*fin.*) μεταφέρω. carried forward εις μεταφοράν.

carry off *v.* (*abduct*) απάγω, κλέβω, (*win*) αποκομίζω. carry it off (*succeed*) τα βγάζω πέρα.

carry on *v.* (*continue*) συνεχίζω, εξακολουθώ. (*conduct*) διεξάγω. (*behave*) φέρομαι, συμπεριφέρομαι. (*make scenes*) κάνω σκηνές. he is carrying on with a married woman τα έχει με μία παντρεμένη.

carry out *v.* (*do*) εκτελώ, εφαρμόζω, πραγματοποιώ.

carry through *v.* (*complete*) φέρω εις πέρας, (*help*) βοηθώ.

cart *s.* κάρρο *n.* (*v.*) κουβαλώ.

cartilage *s.* χόνδρος *m.*

cartography *s.* χαρτογραφία *f.*

carton *s.* κουτί *n.*

cartoon *s.* γελοιογραφία *f.*

cartridge *s.* φυσίγγι *n.*

carv|e *v.* (*meat*) κόβω, (*engrave*) χαράζω, (*sculpt*) σκαλίζω, γλύφω. **~ing** *s.* γλυπτό *n.*

caryatid *s.* Καρυάτις *f.*

cascade *s.* καταρράκτης *m.* (*v.*) τρέχω καταρρακτωδώς.

case *s.* (*circumstance*) περίπτωση *f.* (*affair*) υπόθεση *f.* it is a ~ of πρόκειται για. it is often the ~ συμβαίνει συχνά. it is not the ~ δεν είναι έτσι. in that ~ εν τοιαύτη περιπτώσει. in any ~ εν πάση περιπτώσει, οπωσδήποτε. such being the ~ οπότε, ούτως εχόντων των πραγμάτων. just in ~ για καλό και για κακό. in ~ it does not happen σε περίπτωση που δεν (θα) συμβεί. in ~ it rains μη τυχόν και βρέξει in ~ of fire σε περίπτωση πυρκαγιάς. (*of disease, theft*) κρούσμα *n.* (*patient*) ασθενής *m.* (*gram.*) πτώσις *f.* have a strong ~ έχω πιθανότητες να κερδίσω. make (out) one's ~ δικαιολογούμαι. make (out) a ~ (*for, against*) παραθέτω επιχειρήματα (υπέρ, κατά *with gen.*) ~ history ιστορικό *n.*

case *s.* (*packing*) κασόνι *n.* (*glass*) βιτρίνα *f.* (*protective cover*) θήκη *f.* (*type*) lower ~ μικρά γράμματα, upper ~ κεφαλαία *n.pl.*

casement *s.* κορνίζα *f.* ~ window παράθυρο με μεντεσέδες.

cash *s.* μετρητά *n.pl.* (*money in general*) λεφτά *n.pl.* in ~ τοις μετρητοίς. (*v.*) εξαργυρώνω. (*fig.*) ~ in on επωφελούμαι (*with gen.*).

cashier *s.* ταμίας *m.*

cashier *v* (*mil.*) αποτάσσω.

cashmere *s.* υπερεκλεκτό είδος μαλλιού.

casing *s.* περίβλημα *n.*

casino *s.* καζίνο *n.*

cask *s.* βαρέλι *n.*

casket *s.* κασετίνα *f.*

casserole *s.* γκιουβέτσι *n.*

cassette *s.* κασέτα *f.* ~ deck κασετόφωνο *n.*

cassock *s.* ράσο *n.*

cast *s.* (*throw*) ριξιά *f.* (*mould*) καλούπι *n.* (*type*) τύπος *m.* (*of play*) οι ηθοποιοί. plaster ~ εκμαγείο *n.*

cast *v.* (*throw, also lots, light, anchor, vote*) ρίχνω, (*shed*) βγάζω, (*metal*) χύνω. ~ an eye ρίχνω μια ματιά. ~ about ψάχνω. ~ one's mind back στρέφω τη σκέψη μου. he is ~ as Macbeth θα υποδυθεί τον Μάκβεθ. be ~ in the form of έχω το σχήμα (*with gen.*). be ~ down αποθαρρύνομαι. be ~

away (naut.) ναυαγώ. ~ing vote αποφασιστική ψήφος. ~ iron χυτοσίδηρος m. ~-iron a. (fig.) (case) ακλόνητος, (constitution) σιδερένιος.

cast off v. (abandon) αποβάλλω, (disinherit) αποκηρύσσω. (rope) αμολάρω. cast-off clothing παλιά ρούχα.

cast up v. (reckon) υπολογίζω. (wash ashore) εκβράζω.

castanets s. καστανιέτες f.pl.

castaway s. ναυαγός m.

caste s. κάστα f. lose ~ ξεπέφτω.

castigate v. (verbally) επιτιμώ αυστηρά. (thrash) τιμωρώ με ξύλο.

castle s. πύργος m. (fort) κάστρο n.

castor s. (wheel) καρούλι n. sugar ~ ζαχαριέρα f. ~ oil ρετσινόλαδο n. ~ sugar ψιλή ζάχαρη.

castrat|e v. ευνουχίζω. ~ion s. ευνουχισμός m.

casual a. (by chance) τυχαίος, (careless) αδιάφορος. ~ labourer μεροκαματιάρης m. ~ly adv. αδιάφορα.

casualt|y s. δυστύχημα n. (victim) θύμα n. ~ies (losses) απώλειες f.pl.

casuist s. καζουιστής m.

cat s. γάτα f. tom ~ γάτος m. wild ~ or member of the ~ family αίλουρος m. (pej., woman) στρίγγλα f. be like a ~ on hot bricks κάθομαι στα κάρβουνα. they live a ~-and-dog life ζούνε σαν το σκύλο με τη γάτα.

cataclysm s. κατακλυσμός m. ~ic a. κατακλυσμιαίος.

catacomb s. κατακόμβη f.

catalogue s. κατάλογος m.

catalyst s. καταλύτης m.

catapult s. καταπέλτης m. (boy's) σφεντόνα f. (v.t.) εκτοξεύω.

cataract s. καταρράκτης m.

catarrh s. κατάρρους m.

catastroph|e s. καταστροφή f. ~ic a. ολέθριος.

catcall s. σφύριγμα n.

catch s. (of ball) πιάσιμο n. (fastening) μάνταλο n. with a ~ in one's voice με σπασμένη φωνή. we had a good ~ (of fish) πιάσαμε πολλά ψάρια. there's a ~ in it έχει κάτι το ύποπτο.

catch v.t. (seize, grasp) (also ~ hold of) πιάνω. (meaning, fugitive) συλλαμβάνω, (illness) κολλώ, αρπάζω. (be in time for) προφταίνω. it caught my eye (or attention) τράβηξε την προσοχή μου. we got caught in the rain μας έπιασε η βροχή. he caught his breath πιάστηκε η αναπνοή του. I caught my sleeve (on a nail, etc.) πιάστηκε το μανίκι μου. I caught my hand

(in the door) μάγγωσα το χέρι μου. he caught me a blow μου κατέφερε μια. (fam.) you'll ~ it! θα φας ξύλο! (v.i.) (burn) αρπάζω, πιάνω.

catch on v.i. (succeed) πιάνω, (understand) μπαίνω.

catch out v.t. (surprise) τσακώνω.

catch up v. ~ (with) προφταίνω.

catching a. κολλητικός.

catchment s. ~ area λεκάνη απορροής.

catchword s. σύνθημα n.

catchy a. (tune) που σ' ελκύει.

catech|ize v. κατηχώ. ~ism s. κατήχηση f.

categorical a. κατηγορηματικός ~ly adv. κατηγορηματικώς.

categor|y s. κατηγορία. f. ~ize v. (classify) ταξινομώ, (distinguish) διακρίνω.

cater v. ~ for (feed) τροφοδοτώ. (fig.) προβλέπω για. ~er s. προμηθευτής τροφίμων. ~ing s. which firm is doing the ~ing for the reception? ποιο κατάστημα ανέλαβε τον μπουφέ της δεξιώσεως;

caterpillar s. κάμπια f. (of vehicle) ερπύστρια f.

caterwaul v. ουρλιάζω.

catgut s. χορδή f.

cathar|sis s. κάθαρση f. ~tic a. καθαρτικός.

cathedral s. μητρόπολη f., καθεδρικός ναός.

catholic a. καθολικός.

cat's-paw s. όργανο n.

cattle s. βόδια n.pl. (pej., people in bondage) ραγιάδες m.pl.

catty a. κακεντρεχής, κακός.

caucus s. επιτροπή κόμματος.

cauldron s. καζάνι n.

cauliflower s. κουνουπίδι n.

caulk v. καλαφατίζω.

cause s. αιτία f., αίτιον n. (reason) λόγος m. (purpose) σκοπός m. make common ~ ενεργώ από κοινού. (v.) προξενώ, προκαλώ, δημιουργώ. (be responsible for) γίνομαι αίτιος για. (make) κάνω.

causeway s. υπερυψωμένος δρόμος.

caustic a. καυστικός.

cauterize v. καυτηριάζω.

caution s. επιφυλακτικότητα f., προσοχή f. (reprimand) επίπληξη f. (warning) προειδοποίηση f. (v.) επιπλήττω, προειδοποιώ.

cautious a. προσεκτικός, επιφυλακτικός.

cavalcade s. παρέλαση f.

cavalier s. ιππότης m. (a.) αγενής.

cavalry s. ιππικόν n.

cave s. σπηλιά f., σπήλαιο n. ~ man άνθρωπος των σπηλαίων (v.) ~ in καταρρέω, (fig.) υποτάσσομαι.

cavern s. άντρο n. ~ous a. (sound) σπηλαιώδης, (*eyes*) βαθουλός, (*mouth*) σαν πηγάδι.

caviar *s.* χαβιάρι *n.* (*red*) μπριχ *n.*

cavil *v.* μικρολογώ. **~ling** *a.* μικρολόγος.

cavity *s.* κοιλότης *f.* (*in tree, tooth*) κουφάλα *f.*

cavort *v.* χοροπηδώ.

caw *v.* κρώζω. **~ing** *s.* κρωγμός *m.*, κρώξιμο *n.*

CD Διπλωματικόν Σώμα *n.*

cease *v.* παύω. **~fire** *s.* κατάπαυση πυρός.

ceaseless *a.* ακατάπαυστος. **~ly** *adv.* ακατάπαυστα.

cedar *s.* κέδρος *m.*

cede *v.* εκχωρώ.

ceiling *s.* ταβάνι *n.* (*also fig.*) οροφή *f.*

celebrate *v.t.* (*praise*) υμνώ, (*honour*) τιμώ, (*rejoice over*) πανηγυρίζω. (*an anniversary*) εορτάζω. (*perform*) τελώ. **~d** *a.* ονομαστός, (*persons only*) διάσημος.

celebration *s.* (*of rite*) τέλεσις *f.* (*of feast, event*) εορτασμός *m.* πανηγυρισμός *m.*

celebrity *s.* διασημότητα *f.*

celery *s.* σέλινο *n.*

celestial *a.* ουράνιος.

celiba|cy *s.* αγαμία *f.* **~te** *a.* άγαμος.

cell *s.* κελλί, κελλίον *n.* (*biol.*) κύτταρο *n.* (*electric*) στοιχείο *n.* (*of activists*) πυρήνας *m.*

cellar *s.* υπόγειο *n.* (*wine*) κάβα *f.*

cello *s.* βιολοντσέλο *n.*

cellophane *s.* σελοφάν *n.*

cellular *a.* κυτταρώδης.

Celtic *a.* κελτικός, (*person*) Κέλτης.

cement *s.* τσιμέντο *n.* (*v.*) τσιμεντάρω, (*fig.*) παγιώνω.

cemetery *s.* νεκροταφείο *n.*

censer *s.* θυμιατό *n.*

censor *s.* λογοκριτής *m.* (*v.*) λογοκρίνω. **~ship** *s.* λογοκρισία *f.*

censorious *a.* φιλοκατήγορος.

censure *v.* επικρίνω. (*s.*) επίκριση *f.*

census *s.* απογραφή *f.*

cent *s.* σεντ *n.* (*fig.*) πεντάρα *f.* per ~ τοις εκατόν.

centaur *s.* κένταυρος *m.*

centenarian *s.* εκατοντούτης *m.*

centenary *s.* εκατονταετηρίδα *f.*

centigrade *a.* Κελσίου.

centimetre *s.* εκατοστόμετρο *n.* (*fam.*) πόντος *m.*

centipede *s.* σαρανταποδαρούσα *f.*

central *a.* κεντρικός. **~ly** *adv.* κεντρικώς.

centraliz|e *v.* συγκεντρώνω. **~ation** *s.* συγκέντρωση *f.*

centre *s.* κέντρο *n.* (*of infection, etc.*) εστία *f.* (*politician*) of the C~ κεντρώος.

centrifugal *a.* φυγόκεντρος.

centripetal *a.* κεντρομόλος.

centurion *s.* εκατόνταρχος *m.*

century *s.* αιώνας *m.*

ceramic *a.* κεραμικός. (*s.*) **~s** (*art*) κεραμική *f.* (*objects*) κεραμικά *n.pl.*

Cerberus *s.* Κέρβερος *m.*

cereal *s.* δημητριακό *n.*

cerebral *a.* εγκεφαλικός.

ceremonial *a.* επίσημος. (*s.*) (*ritual*) ιεροτελεστία *f.*

ceremon|y *s.* τελετή *f.* (*formal behaviour*) επισημότητα *f.*, τύποι *m.pl.* stand on ~y είμαι των τύπων. Master of C~ies τελετάρχης *m.*

certain *a.* (*unspecified*) a ~ gentleman κάποιος κύριος. of a ~ age κάποιας ηλικίας. ~ doctors think μερικοί (*or* ορισμένοι) γιατροί νομίζουν. on ~ conditions υπό ορισμένους όρους. at ~ times σε ορισμένα διαστήματα. to a ~ extent μέχρις ενός σημείου, μέχρι τινός.

certain *a.* (*convinced*) σίγουρος, βέβαιος. (*reliable*) σίγουρος, ασφαλής. for ~ μετά βεβαιότητος. make ~ (*about, that*) βεβαιώνομαι. make ~ of (*seats, supplies, etc.*) εξασφαλίζω. he is ~ to know the answer εκείνος ασφαλώς (*or* σίγουρα) θα ξέρει τη λύση. **~ly** *adv.* σίγουρα, βέβαια, βεβαίως, ασφαλώς.

certainty *s.* βεβαιότητα *f.* bet on a ~ στοιχηματίζω στα σίγουρα. for a ~ εκτός πάσης αμφιβολίας.

certificate *s.* πιστοποιητικόν *n.* (*diploma*) πτυχίον *n.* **~d** *a.* πτυχιούχος.

certif|y *v.* πιστοποιώ, βεβαιώνω. **~ication** *s.* πιστοποίηση *f.* **~ied** *a.* (*as correct*) επικυρωμένος.

certitude *s.* βεβαιότητα *f.*

cessation *s.* παύση *f.*

cess|pit, ~pool *s.* βόθρος *m.*

cetacean *a.* κητώδης. (*s.*) κήτος *n.*

chafe *v.t.* τρίβω, γδέρνω. (*v.i.*) τρίβομαι, (*fig.*) εκνευρίζομαι.

chaff *v.* πειράζω. (*s.*) πειράγματα *n.pl.*

chaff *s.* (*winnowed*) ανεμίδια *n.pl.* (*straw*) άχυρο *n.*

chagrin *s.* στενοχώρια *f.* **~ed** *a.* στενοχωρημένος.

chain *s.* αλυσίδα *f.* (*watch*) καδένα *f.* **~s** (*fetters*) δεσμά *n.pl.* in **~s** (*captive*) δέσμιος. (*n.*) αλυσοδένω.

chair *s.* καρέκλα *f.*, (*of office*) έδρα *f.*, take the ~ προεδρεύω, κατέχω την έδρα.

chairman *s.* πρόεδρος *m.f.* **~ship** *s.* προεδρία *f.*

chalice *s.* (*eccl.*) δισκοπότηρον *n.*

chalk *s.* κιμωλία *f.* (*geology*) ασβεστόλιθος

m. (*fam.*) by a long ~ κατά πολύ, (*after neg.*) με κανένα τρόπο.
challenge *v.* προκαλώ, (*dispute*) αμφισβητώ. (*s.*) πρόκλησηf.
chamber *s.* αίθουσα f. (*anatomy, etc.*) θάλαμος *m.* (*of gun*) θαλάμη f. (*of Parliament*) Βουλή f. ~ music μουσική δωματίου.
chamberlain *s.* αρχιθαλαμηπόλος *m.*
chambermaid *s.* καμαριέραf.
chamber-pot *s.* καθίκι *n.*
chameleon *s.* χαμαιλέων *m.*
chamois *s.* αίγαγρος *m.* (*leather*) σαμουά *n.*
champ *v.* μασουλίζω. (*the bit*) δαγκώνω, (*be impatient*) ανυπομονώ.
champagne *s.* σαμπάνια f.
champion *s.* (*defender*) υπέρμαχος *m.f.* (*winner*) πρωταθλητής *m.* (*v.*) μάχομαι υπέρ (*with gen.*). **~ship** *s.* (*defence*) υποστήριξη f. (*contest*) πρωτάθλημα *n.*
chance *s.* (*fortune, accident*) τύχη f. by ~ τυχαίως, κατά τύχη, κατά σύμπτωση. (*probability*) πιθανότητα f. (*opportunity*) ευκαιρία f. do you by any ~ know? μήπως ξέρετε κατά τύχην; take a ~ να αφήνω στην τύχη. game of ~ τυχερό παιχνίδι. (*a.*) τυχαίος.
chance *v.t.* (*risk*) διακινδυνεύω. (*v.i.*) I ~d to be there έτυχε να είμαι εκεί, έλαχα εκεί. ~ upon βρίσκω κατά τύχην.
chancellor *s.* καγκελάριος *m.*
chancy *a.* a ~ business επιχείρηση αμφιβόλου εκβάσεως.
chandelier *s.* πολύφωτο *n.*
chandler *s.* ship ~ προμηθευτής πλοίων.
change *s.* αλλαγή f. (*money due back*) ρέστα *n.pl.* small ~ ψιλά. *n.pl.* ~ of clothes αλλαξιά f. ring the ~s αλλάζω τη σειρά. for a ~ χάριν αλλαγής.
change *v.t. & i.* αλλάζω. (*banknote into coins*) χαλώ, κάνω ψιλά. (*convert, alter*) (*v.t.*) μεταβάλλω, μετατρέπω, (*v.i.*) μεταβάλλομαι, μετατρέπομαι. ~ one's mind αλλάζω γνώμη, μετανοώ. ~ one's tune αλλάζω τροπάρι. ~ hands αλλάζω χέρια.
changeab|le *a.* ευμετάβλητος. **~ility** *s.* αστάθεια f.
channel *s.* (*sea*) πορθμός *m.* English C~ Μάγχη f. (*for irrigation, etc.*) αυλάκι *n.* (*way*) δρόμος *m.* (*medium*) μέσον *n.* (*for trade*) διέξοδος f. (*TV*) κανάλι *n.* (*v.*) (*direct*) διοχετεύω, (*groove*) αυλακώνω.
chant *v.* ψάλλω. (*s.*) ψάλσιμο *n.*, ψαλμωδία f.
cha|os *s.* χάος *n.* **~otic** *a.* χαώδης.
chap *v.* (*also get* ~ped) σκάζω.
chap *s.* (*fam.*) παιδί *n.*
chapel *s.* παρεκκλήσι *n.*

chaperon *s.* συνοδός f. (*v.*) συνοδεύω.
chaplain *s.* ιερεύς *m.*
chapter *s.* (*of book*) κεφάλαιο *n.* give ~ and verse αναφέρω λεπτομερειακώς τις πηγές μου.
char *v.* μισοκαίω.
character *s.* χαρακτήρας *m.* (*person*) πρόσωπο *n.* (*eccentric*) τύπος *m.* **~istic** *a.* χαρακτηριστικός. (*s.*) χαρακτηριστικό *n.*
characteriz|e *v.* χαρακτηρίζω. **~ation** *s.* χαρακτηρισμός *m.*
charade *s.* συλλαβόγριφος *m.*
charcoal *s.* ξυλοκάρβουνο *n.* **~-burner** *s.* καρβουνιάρης *m.*
charge *v.i.* (*rush*) επιτίθεμαι, ορμώ (*with* εναντίον *& gen.*). (*v.t.*) (*fill*) γεμίζω, (*with electricity*) φορτίζω. (*ask as price*) παίρνω, ζητώ. (*debit*) χρεώνω. (*adjure*) εξορκίζω. (*accuse*) κατηγορώ. he was ~d with theft κατηγορήθη επί κλοπή. (*entrust*) he was ~d with an important mission του ανετέθη μία σπουδαία αποστολή.
charge *s.* (*attack*) έφοδος f., επίθεση f. (*of gun*) γόμωση f. (*accusation*) κατηγορία f. (*liability*) βάρος *n.* (*price*) τιμή f. is there any ~? πληρώνει κανείς; free of ~ δωρεάν. (*fee*) αμοιβή f. list of ~s τιμολόγιο *n.* (*care*) take ~ of αναλαμβάνω. they left the baby in ~ of a neighbour άφησαν το μωρό υπό τη φροντίδα μίας γειτόνισσας. who is in ~ here? ποιος είναι υπεύθυνος (*or* επί κεφαλής) εδώ;
chargé d'affaires *s.* επιτετραμμένος *m.*
chariot *s.* άρμα *n.* **~eer** *s.* ηνίοχος *m.*
charitable *a.* ελεήμων, (*lenient*) επιεικής, (*philanthropic*) φιλανθρωπικός. ~ institution ευαγές ίδρυμα.
charity *s.* (*love*) αγάπη f. (*leniency*) επιείκεια f. (*almsgiving*) ελεημοσύνη f. give money to ~ δίνω για φιλανθρωπικούς σκοπούς.
charlatan *s.* κομπογιαννίτης *m.*, τσαρλατάνος *m.*
charm *s.* χάρη f., (*stronger*) γοητεία f., θέλγητρο *n.* (*spell*) μάγια *n.pl.* (*trinket*) φυλαχτό *n.* (*v.*) γοητεύω, θέλγω, μαγεύω. **~er** *s.* γόης *m.*, γόησσα f. **~ing** *a.* χαριτωμένος, (*stronger*) γοητευτικός, θελκτικός.
charnel-house *s.* οστεοφυλάκιον *n.*
chart *s.* χάρτης *m.* (*statistical*) παραστατικός χάρτης. (*v.*) χαρτογραφώ.
charter *s.* καταστατικός χάρτης *m.* (*hire*) ναύλωση f. (*v.*) (*hire*) ναυλώνω. ~ed accountant ορκωτός λογιστής.
charwoman *s.* παραδουλεύτραf.
chary *a.* επιφυλακτικός.

chase v. κυνηγώ. ~ away διώχνω. (s.) κυνήγι n. give ~ to καταδιώκω.

chasm s. χάσμα n.

chassis s. πλαίσιο n., σασί n.

chaste a. αγνός.

chasten v. φρονηματίζω. in ~ed mood με πεσμένα τα φτερά.

chastise v. τιμωρώ. ~ment s. τιμωρία f.

chastity s. αγνότητα f.

chat s. κουβέντα n. we had a ~ πιάσαμε το κουβεντολόγι. (v.t.) ~ up καταφέρνω. ~ty a. ομιλητικός.

chattel s. κινητό n.

chatter v. φλυαρώ, (of teeth) χτυπώ. (s.) φλυαρία f. ~box s. φλύαρος a.

chauffeur s. οδηγός m., σοφέρ m.

chauvinism s. σοβινισμός m.

cheap a. φτηνός, on the ~ φτηνά. (poor) ευτελής, (mean, petty) μικροπρεπής. ~ jack γυρολόγος, m. (adv.) (also ~ly) φτηνά. he got off ~ly φτηνά τη γλύτωσε. ~ness s. φτήνια f.

cheapen v. φτηναίνω, (fig.) εξευτελίζω.

cheat v.t. εξαπατώ, γελώ, ξεγελώ. (v.i.) (at cards, etc.) κλέβω, (crib) αντιγράφω. (s.) απατεώνας m. ~ing s. απάτη f., κλέψιμο n. (in games) ζαβολιές f.pl.

check v. (examine) ελέγχω. (restrain) αναχαιτίζω, συγκρατώ. (list, etc.) τσεκάρω, ελέγχω. ~ up on ελέγχω. ~ in φτάνω. ~ out φεύγω.

check s. (control) έλεγχος m. (stoppage) εμπόδιο n. keep a ~ on ελέγχω. hold in ~ συγκρατώ. ~-up s. εξέταση f., (fam.) τσεκάπ n.

check s. & a. (pattern) καρρό n.

checkmate v. (chess) κάνω ματ. (fig.) φέρω εις αδιέξοδον.

cheek s. μάγουλο n. (effrontery) θράσος n. ~y a. θρασύς. be ~y βγάζω γλώσσα.

cheer v.t. (acclaim) επευφημώ, ζητωκραυγάζω. (gladden) δίνω κουράγιο σε. (v.i.) ~ up παίρνω κουράγιο. ~ up! κουράγιο! ~ing a. ενθαρρυντικός.

cheer s. (acclamation) ζητωκραυγή f. be of good ~ παίρνω κουράγιο. make good ~ καλοτρώγω.

cheerful a. (person) κεφάτος, (willing) πρόθυμος, (atmosphere) χαρούμενος. ~ly adv. με κέφι, πρόθυμα. ~ness s. ευδιαθεσία f.

cheerless a. καταθλιπτικός.

cheery a. πρόσχαρος.

cheese s. τυρί n. ~ pie τυρόπιτα f. ~-paring a. τσιγγούνης.

chef s. αρχιμάγειρος m.

chemical a. χημικός (s.) χημική ουσία.

chemise s. πουκαμίσα f.

chemist s. χημικός. m. (druggist) φαρμακοποιός m. ~'s shop φαρμακείο n. ~ry s. χημεία f.

cheque s. επιταγή f., τσεκ n.

chequered a. (with light & shade) ημισκιασμένος. (fig.) ~ career βίος και πολιτεία.

cherish v. περιποιούμαι με αγάπη, (hope, etc.) τρέφω.

cherry s. κεράσι n. (tree) κερασιά f. (a.) κερασένιος. ~-brandy s. τσέρι n.

cherub s. αγγελούδι n. ~ic a. χερουβικός.

chess s. σκάκι n.

chest s. (of body) στήθος n. (box) κασέλα f. ~ of drawers κομό n.

chestnut s. κάστανο n. horse-~ αγριοκάστανο n. (tree) (αγριο)καστανιά f. (a.) καστανός.

chevron s. γαλόνι n., σειρήτι n.

chew v. μασώ, μασουλώ. ~ the cud μηρυκώμαι, (fig., ponder) σκέπτομαι. ~ing-gum s. τσίκλα f.

chiaroscuro s. φωτοσκίαση f.

chic a. κομψός. (s.) κομψότητα f.

chicanery s. στρεψοδικία f.

chick s. νεοσσός m.

chicken s. κότα f., κοτόπουλο n. ~pox s. ανεμοβλογιά f.

chick-pea s. ρεβίθι n. (roasted) στραγάλι n.

chicory s. (endive) αντίδι n.

chide v. μαλώνω.

chief a. κύριος. (in rank) πρώτος, αρχι-. (s.) αρχηγός m. (superior) προϊστάμενος m. (mil.) ~ of staff επιτελάρχης m. ~ly adv. κυρίως.

chieftain s. αρχηγός m. (of band) οπλαρχηγός m.

chilblain s. χιονίστρα f.

child s. τέκνο n., παιδί n., μικρό n. ~'s play παιγνιδάκι n. with ~ έγκυος. ~'s, ~rens' a. παιδικός.

child|bed, ~birth s. τοκετός m. in ~birth στη γέννα.

childhood s. παιδική ηλικία.

childish a. παιδικός, (pej.) παιδαριώδης.

childless a. άτεκνος.

childlike a. αφελής.

chill s. (coldness) ψυχρότητα f., κρυάδα f. (illness) κρυολόγημα n. take the ~ off κόβω την κρυάδα. (v.) παγώνω. ~ (or be ~ed) to the bone ξυλιάζω. (a.) ψυχρός. ~y a. ψυχρός. it's ~y κάνει ψύχρα.

chime v.i. ηχώ. ~ the hour χτυπώ την ώρα. ~ in συμφωνώ. ~ in with συμφωνώ με. (s.) (of clock) κουδούνι n. ~s (of bells) καμπάνες f.pl.

chimera s. χίμαιρα f.

chimney s. καπνοδόχος f., καμινάδα f.,

(*tall*) φουγάρο *n*. ~ corner γωνιά του τζακιού.

chimpanzee *s*. χιμπαντζής *m*.

chin *s*. πηγούνι *n*. (*fig*.) keep one's ~ up δεν αποθαρρύνομαι, δεν το βάζω κάτω.

china *s*. πορσελάνη *f*. (*crockery*) πιατικά. *n. pl*. (*a*.) από πορσελάνη.

Chinese *a*. κινέζικος, (*person*) Κινέζος.

chink *s*. (*gap*) χαραμάδα *f*. (*sound*) κουδούνισμα *n*. (*v.i.*) κουδουνίζω.

chip *s*. (*fracture*) τσάκισμα *n*. (*fragment of marble, wood*) πελεκούδι *n*. ~s (*potatoes*) πατάτες τηγανητές. he has a ~ on his shoulder είναι εύθικτος, δεν μπορεί να σκεφθεί αμερόληπτα. ~ off the old block γιος του πατέρα του. (*v*.) (*cut*) πελεκώ. ~ped (*damaged*) χτυπημένος, τσακισμένος. ~ in πετάγομαι.

chiropody *s*. θεραπεία κάλων και άλλων παθήσεων του ποδιού.

chirp *v*. τερετίζω. ~y *a*. καλόκεφος.

chisel *s*. σμίλη. *n*. cold ~ καλέμι *n*. (*v*.) σμιλεύω.

chit *s*. (*note*) σημείωμα *n*. ~ of a girl κοριτσόπουλο *n*.

chit-chat *s*. ψιλοκουβέντα *f*.

chivalr|y *s*. ιπποτισμός *m*. ~ous *a*. (*deed*) ιπποτικός, (*man*) ιππότης *m*.

chivvy *v*. κυνηγώ.

chlorine *s*. χλώριο *n*.

chloroform *s*. χλωροφόρμιο *n*.

chock *s*. τάκος *m*.

chock|-a-block, ~-full *a*. παραγεμισμένος.

chocolate *s*. σοκολάτα *f*. (*a*.) σοκολατένιος.

choice *s*. εκλογή *f*. (*variety*) ποικιλία *f*. (*preference*) προτίμηση *f*. have no ~ δεν έχω περιθώριο εκλογής, δεν μπορώ να κάνω αλλιώς. (*a*.) εκλεκτός, διαλεχτός.

choir *s*. χορωδία *f*.

choke *v.t*. πνίγω, (*v.i*.) πνίγομαι. (*v.t*., *block*) (*also* ~ up) φράζω, βουλώνω. ~ down καταπίνω. ~ off αποθαρρύνω.

choking *a*. πνικτικός. (*s*.) πνίξιμο *n*.

choler *s*. οργή *f*. ~ic *a*. οργίλος.

cholera *s*. χολέρα *f*.

choos|e *v*. διαλέγω, εκλέγω. do as one ~es κάνω όπως μου αρέσει. ~y *a*. εκλεκτικός, δύσκολος.

chop *v*. (*cut*) κόβω, (*with axe*) τσεκουρώνω. ~ to pieces κατακόβω. ~ up small ψιλοκόβω. ~ off αποκόπτω. ~ and change όλο αλλάζω. (*s*.) (*lamb*) παϊδάκι *n*. (*pork, veal*) μπριζόλα *f*. (*blow*) χτύπημα *n*. ~per *s*. μπαλτάς *m*.

choppy *a*. the sea is ~ έχει κυματάκι.

chop-sticks *s*. ξυλαράκια *n.pl*.

choral *a*. (*composition*) χορωδιακός, (*with* ~ *accompaniment*) με χορωδία.

chord *s*. (*mus*.) συγχορδία *f*. (*fig*. *string*) χορδή *f*. it strikes a ~ (*in memory*) κάτι μου φέρνει στο νου.

chore *s*. αγγαρεία *f*.

choreography *s*. χορογραφία *f*.

chorister *s*. ψάλτης *m*.

chorus *s*. χορός *m*. (*of song*) ρεφρέν *n*. a ~ of protest έντονες διαμαρτυρίες.

chosen *a*. εκλεκτός. ~ people περιούσιος λαός.

christen *v*. βαπτίζω. ~ing *s*. βάπτισις *f*., βαφτίσια *n.pl*.

Christendom *s*. χριστιανοσύνη *f*.

Christian *a*. χριστιανικός, (*person*) χριστιανός. ~ name βαπτιστικό (*or* μικρό) όνομα. ~ity *s*. χριστιανισμός *m*.

Christmas *s*. Χριστούγεννα *n.pl*. Father C~ Αϊ-Βασίλης. (*a*.) χριστουγεννιάτικος.

chromatic *a*. χρωματικός.

chromium *s*. χρώμιο *n*.

chronic *a*. χρόνιος.

chronicle *s*. χρονικό *n*. (*v*.) καταγράφω, εξιστορώ. ~r *s*. χρονικογράφος *m*.

chronolog|y *s*. χρονολογία *f*. ~ical *a*. χρονολογικός.

chronometer *s*. χρονόμετρο *n*.

chrysalis *s*. χρυσαλλίδα *f*.

chrysanthemum *s*. χρυσάνθεμο *n*.

chubby *a*. στρουμπουλός.

chuck *v*. πετώ, ρίχνω. ~ (up) (*abandon*) παρατώ, αφήνω. ~ out (*person*) βγάζω έξω, (*thing*) πετώ. ~er out μπράβος *m*.

chuckle *v*. γελώ συγκρατημένα.

chug *v*. κινούμαι ξεφυσώντας.

chum *s*. φίλος *m*. ~ up πιάνω φιλίες. ~my *a*. φιλικός. be ~my with έχω φιλίες με.

chump *s*. (*log fool*) κούτσουρο *n*. (*meat*) από μπούτι. (*fam*.) off one's ~ τρελλός.

chunk *s*. κομματάρα *f*. ~y *a*. χοντρός.

church *s*. εκκλησία *f*. (*building only*) ναός *m*. (*fig*.) enter the ~ περιβάλλομαι το σχήμα. ~ service λειτουργία *f*. ~warden *s*. επίτροπος *m*.

churlish *a*. κακότροπος, στρυφνός.

churn *s*. καρδάρα *f*.

chute *s*. (*slide*) τσουλήθρα *f*.

cicada *s*. τζίτζικας *m*.

cicatrice *s*. ουλή *f*.

cicerone *s*. ξεναγός *m*.

cider *s*. μηλίτης *m*.

cigar *s*. πούρο *n*.

cigarette *s*. τσιγάρο *n*. ~-case *s*. ταμπακιέρα *f*. ~-end *s*. αποτσίγαρο *n*. ~-holder *s*. πίπα *f*.

cinder *s*. ~s στάχτες *f.pl*. it was burnt to a ~ (*food*) έγινε κάρβουνο, (*house*) απανθρακώθηκε.

Cinderella *s*. Σταχτοπούτα *f*.

cinema s. κινηματογράφος m.
cinnamon s. κανέλλα f.
cipher s. (nought) μηδενικό n. (numeral) αριθμός m. (code) κρυπτογραφία f. telegram in ~ κρυπτογραφικό τηλεγράφημα (v.) κρυπτογραφώ.
circle s. κύκλος m. (theatre) εξώστης m. come full ~ διαγράφω πλήρη κύκλο. in a ~ σε σχήμα κύκλου, κυκλικά. (v.i.) διαγράφω κύκλους. (v.t.) κάνω το γύρο (with gen.).
circuit s. (journey round) γύρος m., κύκλος m. (district) περιφέρεια f. (electric) κύκλωμα n. short ~ βραχυκύκλωμα n. ~ous a. take a ~ous route κάνω κύκλο. by ~ous means διά πλαγίων μέσων.
circular a. κυκλικός. (s.) (letter) εγκύκλιος f. (leaflet) διαφημιστικό έντυπο. ~ize v. they were ~ized τους εστάλησαν έντυπα.
circulat|e v.t. & i. κυκλοφορώ. ~ion s. κυκλοφορία f. ~ory a. κυκλοφορικός.
circumcis|e v. περιτέμνω. ~ion s. περιτομή f.
circumference s. περιφέρεια f.
circumflex s. (gram.) περισπωμένη f.
circumlocution s. περίφραση f.
circumnavigation s. περίπλους m.
circumscribe v. περιορίζω.
circumspect a. προσεκτικός. ~ion s. περίσκεψη f.
circumstance s. (case) περίπτωση f. (event) περιστατικό n. ~s (situation) περιστάσεις f. pl., (conditions) συνθήκες f.pl. financial ~s οικονομική κατάσταση. be in reduced ~s τα έχω στενά. in no ~s σε καμιά περίπτωση. under the ~s ούτως εχόντων των πραγμάτων.
circumstantial a. (detailed) λεπτομερής, εμπεριστατωμένος. ~ evidence περιστατική ένδειξις.
circumvent v. παρακάμπτω. (law, etc.) καταστρατηγώ.
circus s. τσίρκο n.
cistern s. δεξαμενή f., στέρνα f. (tank) ντεπόζιτο n. (w.c.) καζανάκι n.
citadel s. ακρόπολη f.
cit|e v. (quote) παραθέτω. ~ation s. (mention) μνεία f.
citizen s. πολίτης m. (inhabitant) κάτοικος m. (subject) υπήκοος m.f. (burgess) δημότης m. ~ship s. υπηκοότητα f.
citric a. κιτρικός. ~ acid ξινό n.
citron s. κίτρο n.
citrus s. ~ fruit εσπεριδοειδή n.pl., (fam.) ξινά m.pl.
city s. πόλις f. (great) μεγαλούπολις f. C ~ (of London) Σίτυ n.
civic a. (citizen's) πολιτικός, (municipal)

δημοτικός.
civil a. (civilian) πολιτικός, (polite) ευγενικός. ~ law αστικόν δίκαιον. ~ list βασιλική χορηγία. ~ servant δημόσιος υπάλληλος. ~ service δημόσιες υπηρεσίες. ~ war εμφύλιος πόλεμος. ~ly adv. ευγενικά.
civilian a. πολιτικός. (s.) πολίτης m.
civility s. ευγένεια f.
civiliz|e v. εκπολιτίζω. ~ation s. πολιτισμός m. ~ed a. πολιτισμένος.
clack s. κλικ-κλακ n.
clad a. ντυμένος.
claim v.t. (demand) απαιτώ, αξιώ, (lay ~ to) διεκδικώ, (seek) ζητώ. (v.i.) (assert, profess) ισχυρίζομαι. (s.) (demand) αξίωση f., διεκδίκηση f. (right) δικαίωμα n. (assertion) ισχυρισμός m. lay ~ to διεκδικώ. make a ~ (insurance, etc.) ζητώ αποζημίωση. ~ant s. διεκδικητής m.
clairvoyance s. μαντική ικανότητα.
clamber v. σκαρφαλώνω.
clammy a. υγρός, (sweaty) ιδρωμένος.
clam|our s. φωνές, κραυγές f.pl. (v.) φωνάζω, κραυγάζω. ~orous a. κραυγαλέος.
clamp v.t. σφίγγω, συνδέω. ~ down on χτυπώ, απαγορεύω. (s.) (tool) νταβίδι n.
clan s. πατριά f.
clandestine s. λαθραίος.
clang v.i. (of bells) κουδουνίζω, (of gates, etc.) ηχώ, βροντώ. (s.) κουδούνισμα n., μεταλλικός ήχος. ~er s. (fam.) γκάφα f.
clank s. μεταλλικός ήχος, (of arms) κλαγγή f.
clannish a. δεμένος με το σόι μου.
clap v.t. (strike) χτυπώ. (fam., put, stick) κολλώ. (v.t. & i.) (applaud) χειροκροτώ. (s.) (of thunder) βροντή f. ~ping s. χειροκροτήματα n.pl.
claptrap s. ανοησίες f.pl.
claque s. κλάκα f.
claret s. μπορντό n.
clarif|y v.t. (make plain) αποσαφηνίζω, διευκρινίζω. ~ication s. αποσαφήνιση f., διευκρίνιση f.
clarinet s. κλαρίνο n.
clarion s. σάλπιγγα f. ~ call σάλπισμα n.
clarity s. διαύγεια f. (of meaning only) σαφήνεια f.
clash v.t. (beat) χτυπώ. (v.i.) (resound) αντηχώ, (conflict) συγκρούομαι, (of colours) δεν ταιριάζω. (s.) (noise) μεταλλικός ήχος, (of arms) κλαγγή f. (conflict) σύγκρουση f. (affray) συμπλοκή f.
clasp v. σφίγγω. (s.) (buckle) πόρπη f. (of handbag, etc.) σούστα f. (squeeze) σφίξιμο n.
class s. (grade, form) τάξις f. (social)

κοινωνική τάξη f. (sort) είδος n. (fam., style) στυλ n. (biol. & mil.) κλάσις f. (lesson) μάθημα n. first ~ πρώτης τάξεως. ~ distinction κοινωνική διάκρισις. ~ struggle or war ταξικός αγώνας. (v.) κατατάσσω.

classic a. κλασσικός. ~s κλασσικές σπουδές. ~al a. κλασσικός.

classif|y v. ταξινομώ. ~ication s. ταξινόμησις f.

classy a. (fam.) στυλάτος, κλάσεως.

clatter s. θόρυβος m. (v.t.) χτυπώ. (v.i.) ~ along περνώ με θόρυβο.

clause s. (proviso) ρήτρα f. (article) άρθρον n. (gram.) πρόταση f.

claustrophobia s. κλειστοφοβία f.

claw s. νύχι n. (of shellfish) δαγκάνα f. (v.) νυχιάζω, ξεσκίζω με τα νύχια.

clay s. άργιλος f., πηλός m. (a.) (soil) αργιλώδης, (made of ~) πήλινος.

clean a. καθαρός, (seemly) ευπρεπής, (not indecent) αθώος. with ~ lines με λιτές γραμμές. come ~ ομολογώ τα πάντα. (adv.) πέρα για πέρα, όλωσδιόλου. (v.t. & i.) (also ~ up, out) καθαρίζω.

clean, ~ing s. καθάρισμα n.

cleaner s. (woman) καθαρίστρια f. dry ~'s καθαριστήριο n.

clean|liness, ~ness s. καθαριότητα f., πάστρα f.

cleanse v. καθαρίζω, (morally) εξαγνίζω.

clear a. (transparent) διαυγής, διαφανής. (weather, sky) αίθριος, καθαρός, (day) φωτεινός, (night) ξάστερος. (voice, tone, view, photo, case, mind, conscience, profit) καθαρός. (sound) ευκρινής, (image, meaning) σαφής. (evident) καταφανής, φανερός, (certain) βέβαιος. (unobstructed) ελεύθερος. make oneself ~ γίνομαι σαφής. I am not ~ about it δεν έχω σαφή ιδέα του θέματος. ~ of (free) απαλλαγμένος από, the ship was ~ of the harbour το πλοίο είχε βγει από το λιμάνι.

clear adv. καθαρά, ευκρινώς. keep or steer ~ of αποφεύγω. get ~ of (free) απαλλάσσομαι από, (away from) φεύγω από. stand ~ στέκομαι σε απόσταση, stand ~ of στέκομαι μακριά από.

clear v.t. (make tidy) καθαρίζω, (empty, unblock) αδειάζω, ελευθερώνω, (acquit) αθωώνω. (jump over) υπερπηδώ, (get round) περνώ ξυστά από. (a profit) καθαρίζω. (goods through customs) εκτελωνίζω. (free, rid) απαλλάσσω (with από). ~ oneself απαλλάσσομαι. ~ the table (after meal) σηκώνω το τραπέζι. ~ the table of books καθαρίζω το τραπέζι από βιβλία or παίρνω τα βιβλία από το τρα-

πέζι. ~ a way or passage ανοίγω δρόμο. (v.i., of weather) καθαρίζω.

clear away v.t. μαζεύω, (after meal) σηκώνω το τραπέζι. (v.i.) (clouds) καθαρίζω, (mist) διαλύομαι.

clear off v.t. (debt) εξοφλώ. (v.i.) (fam.) το σκάω.

clear out v.t. (empty) αδειάζω. (v.i.) (fam.) το σκάω.

clear up v.t. (mess) καθαρίζω, (mystery) ξεδιαλύνω. (v.i.) (weather) καθαρίζω, ξανοίγω.

clearance s. (of departing ship) άδεια απόπλου, (customs) εκτελωνισμός m. (clearing up) καθάρισμα n.

clearing s. (in weather) ξάνοιγμα n. (in forest) ξέφωτο n. ~-house γραφείον συμψηφισμών.

clearness s. see clarity.

cleav|e v. (split) σχίζω, (cling) προσκολλώμαι. ~-age s. σχίσμα n.

clef s. κλειδί n.

cleft s. σχισμή f. (a.) σχιστός.

clematis s. αγράμπελη f.

clemen|cy s. επιείκεια f. ~t a. επιεικής, (weather) μαλακός.

clench v. σφίγγω.

clergy s. κλήρος m. ~man s. κληρικός, ιερωμένος m.

cleric s., ~al a. κληρικός.

clerk s. γραφεύς m. (bank, etc.) υπάλληλος m. (of court, council) γραμματεύς m. (eccl.) κληρικός m.

clever a. έξυπνος, (adroit) επιδέξιος, επιτήδειος, (witty) ευφυής. ~ness s. εξυπνάδα f. δεξιότητα f. ευφυΐα f.

clew s. (of Ariadne) μίτος m.

cliché s. κοινοτοπία f.

click v.t. (tongue) πλαταγίζω, (heels) χτυπώ. (v.i.) κάνω κλικ.

client s. πελάτης m. ~ele s. πελατεία f.

cliff s. απότομος βράχος, γκρεμός m. ~s βραχώδης ακτή.

climate s. κλίμα n.

climax s. (απο)κορύφωμα n. reach a ~ φθάνω στο κατακόρυφο.

climb v.i. (also ~ up) ανεβαίνω, (of road) ανηφορίζω, (fig.) αναρριχώμαι. (v.t.) ανεβαίνω or σκαρφαλώνω (with σε). ~ down (v.t.) κατεβαίνω, (v.i.) (fig.) υποχωρώ, κάνω νερά. ~er s. ορειβάτης m.

climbing s. ορειβασία f. (a.) (equipment, club) ορειβατικός, (plant) αναρριχητικός.

clime s. (fig.) τόπος m.

clinch v. (a deal) κλείνω. (s.) αγκάλιασμα n. (wrestling) μεσολαβή f.

cling v.i. ~ to (ideas) εμμένω εις, (ideas, family, etc.) είμαι (προσ)κολλημένος σε.

(*for support*) είμαι πιασμένος από. (*hold tightly*) κρατώ σφιχτά, αρπάζομαι από. (*fit closely*) κολλώ σε. we clung together μείναμε κολλημένοι ο ένας στον άλλον. ~ing *a.* κολλητός.

clinic *s.* κλινική *s.* ~al *a.* κλινικός.

clink *v.i.* κουδουνίζω. (*v.t.*) (*glasses*) τσουγκρίζω.

clink *s.* (*fam.*, *prison*) φρέσκο *n.*

clip *v.* ψαλιδίζω, (*hair*) κουρεύω, (*ticket*) τρυπώ, (*fasten*) συνδέω. (*s.*) (*shearing*) κούρεμα *n.* (*blow*) φάπα *f.* (*fastener*) συνδετήρας *m.* hair-~ τσιμπιδάκι *n.*

clippers *s.* κουρευτική μηχανή.

clipping *s.* (*from paper*) απόκομμα *n.*

clique *s.* κλίκα *f.*

cloak *s.* κάπα *f.*, μπέρτα *f.* (*fig.*) μανδύας *m.*, (*of darkness*) πέπλο *n.* (*v.*) (*cover*) σκεπάζω, (*fig.*) καλύπτω. ~**room** *s.* ιματιοφυλάκιον *n.*, γκαρνταρόμπα *f.* (*w.c.*) τουαλέττα *f.*

clock *s.* ωρολόγιο *n.*, ρολόι *n.* (*of stocking*) μπαγκέτα *f.* (*v.*) ~ in *or* out χτυπώ καρτέλα. ~**wise** *a.* δεξιόστροφος. ~**work** *a.* κουρδιστός. go like ~work πάω ρολόι.

clod *s.* σβώλος *m.* (*fig.*) (*also* ~-*hopper*) χοντράνθρωπος *m.*

clog *s.* τσόκαρο *n.*

clog *v.t.* (*block*) φράζω, βουλλώνω, (*with mud*, *etc.*) γεμίζω.

cloister *s.* περιστύλιο *n.* (*fig.*) μοναστήρι *n.* live a ~ed life ζω αποτραβηγμένος.

close *a.* (*kinship*) κοντινός, στενός, (*friend*, *connexion*) στενός. (*weather*) βαρύς. (*scrutiny*) επιμελημένος. (*secretive*) κλειστός, (*stingy*) σφιχτός, (*strict*) αυστηρός. at ~ quarters από κοντά. ~ combat μάχη εκ του συστάδην. ~ resemblance μεγάλη ομοιότητα. with ~ attention επισταμένως. a ~ contest σχεδόν ισόπαλος αγώνας. keep a ~ watch on παρακολουθώ προσεκτικά. he had a ~ shave (*fig.*) φτηνά τη γλύτωσε. they stood ~ together στάθηκαν κοντά ο ένας στον άλλον.

close *adv.* (*also* ~ by) κοντά, πλησίον. (*prep.*) ~ to *or* by κοντά σε (*with acc.*), πλησίον (*with gen.*).

close *v.t.* & *i.* κλείνω, (*end*) τελειώνω. ~ ranks πυκνώνω τους ζυγούς. ~ down κλείνω. ~ in (*on quarry*) περισφίγγω τον κλοιό (*get shorter*) μικραίνω. ~ up (*gap*, *etc.*) κλείνω. ~ with (*opponent*) πιάνομαι με, (*over bargain*) συμφωνώ με.

close *s.* (*end*) τέλος *n.* bring to a ~ θέτω τέρμα εις.

closed *a.* κλειστός. behind ~ doors κεκλεισμένων των θυρών. with one's eyes ~ με

κλειστά τα μάτια.

closely *adv.* (*near*) από κοντά, (*tightly*) σφιχτά, (*thickly*) πυκνά, (*attentively*) προσεκτικά. (*in resemblance*) πολύ.

closet *s.* (*room*) μικρό δωμάτιο, (*store*) αποθήκη *f.* (*w.c.*) αποχωρητήριο *n.* (*v.*) be ~ed κλείνομαι.

closing *s.* κλείσιμο *n.* (*being shut for holiday*, *etc.*) αργία *f.* (*a.*) τελευταίος.

closure *s.* see closing.

clot *v.i.* πήζω, (*of blood*) θρομβούμαι. (*s.*) θρόμβος *m.*

cloth *s.* (*material*) ύφασμα *n.* (*piece of* ~) πανί *n.* (*fig.*, *clerical*) σχήμα *n.* floor-~ σφουγγαρόπανο *n.*

clothe *v.* ενδύω, ντύνω. ~d ντυμένος.

clothes *s.* ενδύματα *n.pl.*, ρούχα *n.pl.* ~-**brush** *s.* βούρτσα των ρούχων. ~-**horse** *s.* απλώστρα *f.* ~-**peg** *s.* μανταλάκι *n.*

clothier *s.* υφασματέμπορος *m.* ~'s shop εμπορικό *n.*

clothing *s.* (*outfit*) ρουχισμός *m.* (*clothes*) ρούχα *n.pl.*

cloud *s.* νέφος *n.*, σύννεφο *n.* (*fig.*) be under a ~ είμαι υπό δυσμένειαν. ~**burst** *s.* νεροποντή *f.*

cloud *v.i.* (*of sky*, *face*, *also* ~ over) συννεφιάζω. (*v.i.* & *t.*) (*of eyes*, *liquid*) θολώνω, (*dull*) θαμπώνω. (*v.t.*, *spoil*) χαλώ. ~**ed** *a.* θαμπός, θολωμένος. ~**y** *a.* νεφελώδης. ~y weather συννεφιά *f.*

cloud-cuckoo-land *s.* νεφελοκοκκυγία *f.*

clout *s.* (*blow*) καρπαζιά *f.* (*rag*) κουρέλι *n.*

clove *s.* γαρίφαλο *n.*

cloven *a.* ~ hoof δίχηλος οπλή.

clover *s.* τριφύλλι *n.* (*fig.*) live in ~ περνώ ζωή και κότα.

clown *s.* παλιάτσος *m.*, (*circus*) κλόουν *m.* (*boor*) βλάχος *m.* (*v.*) κάνω τον καραγκιόζη. ~**ish** *a.* χοντρός, γελοίος.

cloy *v.t.* & *i.* (*also* become ~ed) μπουχτίζω. it ~s the appetite κόβει την όρεξη. ~**ing** *a.* μπουχτιστικός.

club *s.* (*stick*) ρόπαλο *n.* (*institution*) όμιλος *m.*, λέσχη *f.* (*cards*) σπαθί *n.* (*v.t.*) (*hit*) χτυπώ. (*v.i.*) ~ together κάνω ρεφενέ. ~**bable** *a.* κοινωνικός.

cluck *v.* κακαρίζω.

clue *s.* ίχνος *n.*, στοιχείο *n.* not have a ~ (*fam.*) δεν έχω ιδέα.

clump *s.* τούφα *f.* (*trees*) συστάδα *f.* (*v.i.*) περπατώ βαριά.

clums|y *a.* (*person*) αδέξιος, (*object*) χονδροειδής. ~**y** piece of work χοντροκοπιά *f.* ~**iness** *s.* αδεξιότητα *f.* (*of shape*) έλλειψη κομψότητας.

cluster *s.* (*people*, *houses*, *etc.*) ομάδα *f.* (*curls*) τούφα *f.* (*trees*, *growing flowers*) συστάδα f. (grapes) τσαμπί n. (v.i.) μαζεύομαι, συγκεντρώνομαι.

clutch v. κρατώ σφιχτά, (grab) πιάνομαι από. (s.) (mech.) αμπραγιάζ n. fall into the ~es of πέφτω στα νύχια (with gen.).

clutter v. γεμίζω. ~ed γεμάτος.

coach s. (carriage) άμαξα f. (railway) βαγόνι n. (motor) πούλμαν n. ~**man** s. αμαξάς m.

coach v. προγυμνάζω, προπονώ. (s.) (teacher) προγυμναστής m. (in sports) προπονητής m. ~**ing** s. προγύμναση f. προπόνηση f.

coadjutor s. βοηθός m.

coagulat|e v. πήζω. ~**ion** s. πήξη f.

coal s. κάρβουνο n. (fig.) haul over the ~s κατσαδιάζω. carry ~s to Newcastle κομίζω γλαύκα εις Αθήνας. (v.i.) ανθρακεύω. ~-**mine** s. ανθρακωρυχείο n.

coalesc|e v. ενώνομαι. ~**ence** s. ένωση f.

coalition s. συνασπισμός m.

coarse a. χονδρός, χοντρός. (language) χυδαίος. ~**ly** adv. χοντρά, ~ly cut χοντροκομμένος. ~**n** v.t. & i. σκληραίνω, τραχύνω, (of looks) αγριεύω. ~**ness** s. τραχύτητα f. (of manners) χυδαιότητα f.

coast s. ακτές f.pl., παράλια n.pl. Blue C~ κυανή ακτή. (fig.) the ~ is clear το πεδίον είναι ελεύθερον. (v.) παραπλέω. ~**er** s. (naut.) ακτοπλοϊκό n. ~**wise** adv. παραλιακώς.

coast v.i. (on cycle, etc.) ρολάρω. (fig.) προχωρώ χωρίς να καταβάλλω μεγάλη προσπάθεια.

coastal a. παράκτιος, παραλιακός. ~ shipping ακτοπλοΐα f. (fort, guns, etc.) επάκτιος.

coastguard s. ακτοφυλακή f.

coastline s. παραλιακή γραμμή.

coat s. (over~) παλτό n., πανωφόρι n. (man's jacket) σακκάκι n., (woman's) ζακέτα f. (animal's) τρίχωμα n. (layer) στρώμα n. (of paint) χέρι n. ~ of arms οικόσημον n. (v.) επιχρίω, (cover, fill) σκεπάζω, γεμίζω.

coat-hanger s. κρεμάστρα f.

coating s. επίστρωμα n.

coax v. καταφέρνω, (wheedle) καλοπιάνω. ~**ingly** adv. με καλοπιάσματα.

cob s. (nut) φουντούκι n. corn~ κούκλα f.

cobble, ~-**stone** s. (also ~d path) καλντερίμι n.

cobble v. μπαλώνω. ~**r** s. μπαλωματής m.

cobra s. κόμπρα f.

cobweb s. αράχνη f.

cocaine s. κοκαΐνη f.

cochineal s. κοχενίλη f.

cochlea s. κοχλίας m.

cock s. πετεινός m., κόκορας m. (tap) κάνουλα f. (of gun) λύκος m. ~-and-bull story παραμύθι n. (v.) (raise) σηκώνω. ~ one's eye ρίχνω μια ματιά. ~ed hat τρικαντό n.

cockade s. κονκάρδα f.

cock-a-hoop a. be ~ αγαλλιώ.

cock-crow s. χαράματα n.pl.

cockerel s. πετεινάρι n.

cock-eyed a. στραβός.

cockle s. (shellfish) κυδώνι n. ~-shell (boat) καρυδότσουφλο n.

cockney s. λαϊκός τύπος του Λονδίνου. (speech) λαϊκή προφορά του Λονδίνου.

cockpit s. στίβος κοκορομαχιών, (fig.) πεδίον πολλών μαχών. (aero.) θάλαμος διακυβερνήσεως.

cockroach s. κατσαρίδα f.

cockscomb s. λειρί n.

cocksure a. he's a ~ fellow πιστεύει πως τα ξέρει όλα.

cocktail s. κοκτέιλ n.

cocky a. see cocksure.

cocoa s. κακάο n.

coconut s. καρύδα f. ~ palm κοκκοφοίνικας m.

cocoon s. κουκούλι n. (v.) κουκουλώνω.

cocotte s. κοκότα f.

cod s. μπακαλιάρος m. ~-liver oil μουρουνόλαδο n.

coddle v. χαϊδεύω, κανακεύω.

cod|e s. κώδιξ m. (secret writing) κρυπτογραφία f. write in ~e κρυπτογραφώ. ~e name συνθηματικόν όνομα. ~**ing** s. κρυπτογράφηση f.

codex s. κώδιξ m.

codicil s. κωδίκελλος m.

codify v. κωδικοποιώ.

co-education s. μικτή εκπαίδευση.

coefficient s. συντελεστής m.

coequal a. ισότιμος.

coerc|e v. εξαναγκάζω, ζορίζω. ~**ion** s. εξαναγκασμός m.

coeval a. (of same epoch) της αυτής εποχής, (of same duration) της αυτής διαρκείας.

coexist v. συνυπάρχω. ~**ence** s. συνύπαρξη f. ~**ent** a. συνυπάρχων.

coextensive a. (in space) ίσης εκτάσεως, (in time) ίσης διαρκείας (both with με).

coffee s. καφές m. ~-**bar** s. καφενείο-μπαρ n. ~-**coloured** a. καφετής. ~-**house** s. καφενείο n. ~-**pot** s. καφετιέρα f. ~-**table** s. τραπεζάκι n.

coffer s. χρηματοκιβώτιο n.

coffin s. φέρετρο n.

cog s. δόντι n. (fig.) ~ in the machine ο πέμπτος τροχός της αμάξης. ~-**wheel** οδοντωτός τροχός, γρανάζι n.

cogent *a.* σοβαρός, ισχυρός.

cogitat|e *v.* συλλογίζομαι. ~ion *s.* συλλογισμός *m.*

cognate *a.* συγγενής.

cognition *s.* αντίληψη *f.*

cogniz|ance *s.* γνώση *f.*, take ~ance λαμβάνω γνώσιν. ~ant *a.* εν γνώσει.

cohabit *v.* συζώ.

coher|e *v.* έχω συνοχή. ~ence *s.* συνοχή *f.*, συνειρμός *m.* ~ent *a.* (*consistent*) συνεπής, (*clear*) καθαρός.

cohes|ion *s.* συνοχή *f.* ~ive *a.* συνεκτικός.

cohort *s.* (*fig.*) ομάδα *f.*

coiffure *s.* χτένισμα *n.*

coil *v.t.* κουλουριάζω. (*v.i.*) κουλουριάζομαι, συσπειρώνομαι. (*s.*) σπείρα *f.* (*electric*) πηνίο *n.*

coin *s.* κέρμα *n.*, νόμισμα *n.* (*v.*) κόβω. (*fig.*) ~ money κόβω μονέδα. (*concoct*) κατασκευάζω. ~age *s.* (*currency*) νόμισμα *n.* (*coined word*) νεολογισμός *m.* ~er *s.* παραχαράκτης *m.*

coincid|e *v.* συμπίπτω. ~ence *s.* σύμπτωση *f.* ~ental *a.* συμπτωματικός. ~entally *adv.* κατά σύμπτωσιν.

coition *s.* συνουσία *f.*

coke *s.* κοκ *n.*

col *s.* διάσελο *n.*

colander *s.* σουρωτήρι *n.*

cold *a.* κρύος, ψυχρός. (*temperament, mood*) ψυχρός. feel ~ κρυώνω. (*weather*) it is ~ κάνει κρύο. get *or* make ~ (*lit. & fig.*) κρυώνω, ψυχραίνω. in ~ blood εν ψυχρώ. get ~ feet (*fig.*) τα χρειάζομαι. pour ~ water on αποδοκιμάζω.

cold *s.* κρύο *n.*, ψύχος *n.* (*illness*) κρυολόγημα *n.*, (*in nose*) συνάχι *n.* catch ~ κρυολογώ, κρυώνω. have α ~ είμαι κρυολογημένος *or* συναχωμένος.

cold-blooded *a.* (*fish, etc.*) ψυχρόαιμος, (*cruel*) ανηλεής, (*premeditated*) προμελετημένος. *see* cold *a.*

cold-shoulder *v.t.* γυρίζω τις πλάτες σε.

cold-storage *s.* in ~ στο ψυγείο.

colic *s.* (*med.*) κωλικός *m.*

collaborat|e *v.* συνεργάζομαι. ~ion *s.* συνεργασία *f.* ~or *s.* συνεργάτης *m.* (*with enemy*) δοσίλογος *m.*

collaps|e *v.i.* καταρρέω, σωριάζομαι. (*s.*) κατάρρευση *f.* ~ible *a.* (*folding*) πτυσσόμενος.

collar *s.* κολλάρο *n.* (*of coat*) γιακάς *m.* (*animal's*) περιλαίμιο *n.*

collar *v.* (*seize*) αρπάζω, (*steal*) σουφρώνω.

collate *v.* παραβάλλω.

collateral *a.* (*kinship*) πλάγιος, (*security*) επιβοηθητικός.

collation *s.* (*comparison*) παραβολή *f.*

(*meal*) κολατσιό *n.*

colleague *s.* συνάδελφος *m.f.*

collect *v.t.* μαζεύω, (*contributions, people, signatures*) συνάζω, (*revenue*) εισπράττω, (*specimens*) συλλέγω, (*one's thoughts, strength*) συγκεντρώνω. (*come and get*) παίρνω. (*v.i.*) μαζεύομαι, (*of people only*) συγκεντρώνομαι, συναθροίζομαι. (*for charity*) κάνω έρανο. ~ed *a.* (*cool*) ψύχραιμος. ~ed works τα άπαντα.

collection *s.* (*of revenue*) είσπραξις *f.* (*of specimens*) συλλογή *f.* (*pile*) σωρός *m.* (*for charity*) έρανος *m.* (*receipt of goods*) παραλαβή *f.*

collective *a.* συλλογικός.

collector *s.* συλλέκτης *m.* (*of taxes*) εισπράκτωρ *m.*

college *s.* κολλέγιο *n.*, σχολή *f.*

collide *v.* συγκρούομαι.

collier *s.* (*miner*) ανθρακωρύχος *m.* (*ship*) καρβουνιάρικο *n.* ~y *s.* ανθρακωρυχείο *n.*

collision *s.* σύγκρουση *f.*

colloquial *a.* ~ speech καθομιλουμένη γλώσσα.

colloquy *s.* συνομιλία *f.*

collusion *s.* συμπαιγνία *f.*

colon *s.* (*anatomy*) κόλον *n.* (*mark*) δύο τελείες.

colonel *s.* συνταγματάρχης *m.*

colonial *a.* αποικιακός. ~ism *s.* αποικιοκρατία *f.*

colonist *s.* άποικος *m.*

coloniz|e *v.* αποικίζω. ~ation *s.* αποίκιση *f.*

colonnade *s.* κιονοστοιχία *f.*

colony *s.* αποικία *f.* (*of foreign residents, etc.*) παροικία *f.* (*biol.*) κοινωνία *f.*

coloss|us *s.* κολοσσός *m.* ~al *a.* κολοσσιαίος.

colour *s.* χρώμα *n.* ~ scheme συνδυασμός των χρωμάτων. feel off ~ δεν είμαι στα καλά μου. ~s (*flag*) σημαία *f.*, join the ~s κατατάσσομαι στο στρατό. come off with flying ~s σημειώνω λαμπρή επιτυχία. show one's true ~s αποκαλύπτομαι αυτός που είμαι. ~ TV χρωματιστή τηλεόραση.

colour *v.t.* χρωματίζω, (*dye, paint*) βάφω, (*fig., influence*) επηρεάζω. (*v.i.*) (*turn red*) κοκκινίζω. ~ed χρωματιστός.

colour-bar *s.* φυλετική διάκριση.

colour-blind *a.* be ~ δεν διακρίνω τα χρώματα. ~ness *s.* δαλτωνισμός *m.*

colourful *a.* γεμάτος χρώμα.

colouring *s.* (*matter, complexion*) χρώμα *n.* (*artist's use of colour*) χρώματα *n.pl.*

colourless *a.* άχρωμος. (*chem.*) άχρους.

colt *s.* πουλάρι *n.*

column *s.* (*archit.*) κίων *m.*, κολόνα *f.*, στήλη *f.* (*of spine, air, print*) στήλη *f.* (*mil.*)

φάλαγγα f. ~ist s. χρονογράφος m.
coma s. κώμα n. in a ~ σε κωματώδη κατάσταση. ~tose a. (state) κωματώδης.
comb s. χτένι n., τσατσάρα f. (cock's) λειρί n. (v.) (lit. & fig.) χτενίζω. ~ one's hair χτενίζομαι.
combat s. αγώνας m. single ~ μονομαχία f. (v.i.) αγωνίζομαι. (v.i. & t.) πολεμώ. (a.) μάχιμος. ~ant a. μάχιμος. (s.) μαχόμενος m. ~ive a. μαχητικός.
combe s. (valley) αυλών m.
combination s. συνδυασμός m.
combine v.t. συνδυάζω. (v.i.) συνενώνομαι. ~d συνδυασμένος.
combustible a. καύσιμος.
combustion s. καύση f.
come v. έρχομαι, (arrive) φτάνω. ~! έλα (sing.), ελάτε (pl.). (become) γίνομαι, ~ undone λύνομαι. (occur, be found) βρίσκομαι, (quotation) συναντώμαι. ~ from (be descended) κατάγομαι από, (spring) προέρχομαι από. that ~s of disobeying αυτά είναι τα αποτελέσματα της ανυπακοής. ~ to (throne, total) ανέρχομαι εις. it ~s to the same thing καταλήγει στο ίδιο συμπέρασμα. I have ~ to realize έφτασα στο συμπέρασμα. it came to nothing δεν βγήκε τίποτα απ' αυτό. he won't ~ to much δεν θα κάνει πολλά. if it ~s to that αν τα πράγματα έχουν έτσι. ~ to pass συμβαίνω. in years to ~ στο μέλλον. ~ what may ό,τι και να συμβεί. ~ of age ενηλικιώνομαι. ~ and go πηγαινοέρχομαι.
come about v.i. γίνομαι, συμβαίνω.
come across v.t. (find) βρίσκω, (meet) συναντώ. (v.i.) see come over.
come along v.i. (progress) προχωρώ, πάω καλά. ~! άντε, βιάσου!
come away v.i. (get detached) ξεκολλώ, (depart) φεύγω.
come back v.i. επανέρχομαι, επιστρέφω, (to memory) ξαναέρχομαι. stage a come-back ξαναβγαίνω στο προσκήνιο.
come before v.t. (precede) προηγούμαι (with gen.).
come between v.i. (intervene) παρεμβαίνω. (in time or space) μεσολαβώ. (v.t.) (separate) χωρίζω.
come by v.i. (pass) περνώ. (v.t., get) βρίσκω, αποχτώ.
come down v.t. & i. κατεβαίνω. (v.i.) (in status) ξεπέφτω, (fall) πέφτω. (from former times) έχω διατηρηθεί. ~ to earth (lit. & fig.) προσγειώνομαι. ~ on λαμβάνω αυστηρά μέτρα εναντίον (with gen.), επιπλήττω, (rebuke) κατσαδιάζω.
come-down s. ταπείνωση f., ξεπεσούρα f.

come forward v.i. (offer) προσφέρομαι.
come-hither a. (fam.) προκλητικός.
come in v.i. (enter) μπαίνω, ~! εμπρός. (participate) συμμετέχω, (have part to play) παίζω ρόλο, (take office) έρχομαι στα πράγματα, (start being used) εισάγομαι, (into fashion) γίνομαι της μόδας, (in a race) έρχομαι. (of income) he has a lot coming in from rents εισπράττει πολλά από ενοίκια. ~ useful πιάνω τόπο. ~ for (undergo) υφίσταμαι, (get) παίρνω. ~ on μετέχω εις.
come into v.t. (inherit) κληρονομώ.
come off v.i. βγαίνω. (succeed) πετυχαίνω, (happen) γίνομαι, (become fact) πραγματοποιούμαι. he came off well βγήκε κερδισμένος. ~ it! άσ' τα λούσα.
come on v.i. (progress) προχωρώ, πάω καλά, (make good) προκόβω. (start) rain came on or it came on to rain έπιασε βροχή, night comes on νυχτώνει. the lights came on (in theatre, etc.) άναψαν τα φώτα, (after power-cut) ήρθε το φως. I feel a cold coming on θα με πιάσει (or θα μου κατέβει) συνάχι. when does the case ~? πότε δικάζεται η υπόθεση; when does the play ~? πότε θα ανέβει (or θα παιχθεί) το έργο;~! άντε, μπρος.
come out v.i. βγαίνω, (of blossom) ανοίγω. (make début) βγαίνω στον κόσμο. (be revealed) αποκαλύπτομαι, γίνομαι γνωστός. ~ in spots βγάζω σπυριά.
come over v.i. (be heard) ακούομαι. the message didn't ~ to the audience το ακροατήριο δεν κατάλαβε. (feel) ~ faint με πιάνει (or μου έρχεται) ζαλάδα. what has ~ you? τι σ' έπιασε;
come round v.i. (agree) πείθομαι, (recover) συνέρχομαι. (call at house) περνώ.
come through v.t. (experience) περνώ. (v.i.) (survive) τη γλυτώνω, τη σκαπουλάρω. (of news) γίνομαι γνωστός.
come to v.i. (also ~ oneself, ~ one's senses) συνέρχομαι.
come under v.t. (be classed) υπάγομαι εις.
come up v.t. & i. ανεβαίνω. (v.i.) (of plants) βγαίνω. (arise) προκύπτω, (for discussion) συζητούμαι, (for trial) δικάζομαι. ~ against αντιμετωπίζω. ~ to (approach) πλησιάζω, (reach as far as) φτάνω μέχρι, (satisfy) ικανοποιώ. ~ with (draw level) προφταίνω, (find) βρίσκω.
come upon v.t. (find) βρίσκω, (meet) συναντώ.
comed|y s. κωμωδία f. ~ian s. κωμικός m.
comely a. όμορφος.
comer s. late ~s καθυστερημένοι m.pl. take on all ~s μετριέμαι με όποιον τύχει. the

first ~ (*anybody*) ο πρώτος τυχών.
comet *s.* κομήτης *m.*
comfort *s.* (*consolation*) παρηγορία *f.*
(*ease*) άνεση *f.* with all ~s με όλες τις
ανέσεις, με όλα τα κομφόρ. that is a great
~ (*fam.*) είναι μεγάλο πράγμα. cold ~
μαύρη παρηγοριά. (*v.*) παρηγορώ, (*ease*)
ανακουφίζω.
comfortab|le *a.* άνετος, (*chair, etc.*)
αναπαυτικός. feel ~le αισθάνομαι άνετα.
~**ly** *adv.* άνετα, αναπαυτικά. be ~ly off έχω
οικονομική άνεση.
comforter *s.* (Job's) ~ (Ιώβειος) παρηγο-
ρητής *m.*
comforting *a.* παρηγορητικός, ανακου-
φιστικός.
comfortless *a.* χωρίς ανέσεις.
comic, ~al *a.* κωμικός.
coming *a.* ερχόμενος, (*to be*) μέλλων. the ~
generation η μέλλουσα γενεά. a ~ man
άνθρωπος με μέλλον.
coming *s.* ερχομός *m.* έλευση *f.* Second C~
δευτέρα παρουσία. (*introduction*) εμφάνιση *f.* ~(s) and going(s) πηγαινέλα *n.*
comma *s.* κόμμα *n.* inverted ~s εισαγωγικά
n.pl.
command *v.* (*order*) διατάζω, προστάζω.
(*be in* ~ *of*) (*army*) διοικώ, (*ship*) κυβερνώ. (*be master of*) κυριαρχώ. (*have at
disposal*) διαθέτω. (*overlook*) δεσπόζω
(*with gen.*). ~ respect επιβάλλομαι.
command *s.* (*order*) διαταγή *f.*, προσταγή *f.*
(*of army, etc.*) διοίκηση *f.* take ~
αναλαμβάνω τη διοίκηση. in ~ επί κεφαλής. under his ~ υπό τας διαταγάς του. at
my ~ (*disposal*) στη διάθεσή μου. (*of sea,
etc.*) κυριαρχία *f.* have ~ of (*languages,
etc.*) κατέχω. have ~ over oneself είμαι
κύριος των παθών μου.
commandant *s.* διοικητής *m.* (*of garrison*)
φρούραρχος *m.*
commandeer *v.* επιτάσσω.
commander *s.* αρχηγός *m.* (*of order of
chivalry*) ταξιάρχης *m.* (*naut.*) αντιπλοίαρχος *m.* ~-in-chief (*mil.*) αρχιστράτηγος *m.*
commanding *a.* (*site*) δεσπόζων, (*of
person*) επιβλητικός. ~ officer αξιωματικός επικεφαλής.
commandment *s.* εντολή *f.*
commando *s.* στρατιώτης καταδρομών,
(*fam.*) κομάντο *m.*
commemorat|e *v.* εορτάζω. ~**ion** *s.*
εορτασμός *m.* ~**ive** *a.* αναμνηστικός.
commence *v.* αρχίζω. ~**ment** *s.* έναρξη *f.*
commend *v.* (*praise*) επαινώ, (*recommend*) συνιστώ. (*entrust*) εμπιστεύομαι. it
does not ~ itself to me δεν το εγκρίνω.

~**able** *a.* αξιέπαινος. ~**ation** *s.* επιδοκιμασία *f.*
commensurable *a.* (*number*) σύμμετρος.
be ~ έχω κοινόν μέτρον.
commensurate *a.* ανάλογος.
comment *s.* σχόλιο *n.*, παρατήρηση *f.* (*v.*)
(*also* ~ on) σχολιάζω. ~**ator** *s.* σχολιαστής
m.
commentary *s.* (*written*) υπομνήματα *n.pl.*
(*spoken*) σχόλια *n.pl.*
commerc|e *s.* εμπόριο *n.* Chamber of C~e
εμπορικόν επιμελητήριον. ~**ial** *a.* εμπορικός. ~ial traveller πλασιέ *m.* (*s.*) διαφή-
μηση *f.*
commiserate *v.* ~ **with** συλλυπούμαι.
commissar *s.* κομισάριος *m.* ~**iat** *s.* (*mil.*)
επιμελητεία *f.*
commission *s.* (*order*) εντολή *f.* (*to artist,
etc.*) παραγγελία *f.* (*body*) επιτροπή *f.*
(*payment*) προμήθεια *f.* ~ **agent** παραγγελιοδόχος *m.* (*mil.*) get one's ~ ονομά-
ζομαι αξιωματικός.
commission *v.* (*person*) δίνω εντολή σέ,
(*thing*) δίνω παραγγελία για. I ~ed him to
find me a house του ανέθεσα να μου βρει
σπίτι.
commissionaire *s.* θυρωρός με λιβρέα.
commissioner *s.* επίτροπος *m.* High C~
Ύπατος Αρμοστής.
commit *v.* (*entrust*) εμπιστεύομαι, (*consign*)
παραπέμπω, (*bind*) δεσμεύω, (*perpetrate*)
διαπράττω. ~ to memory αποστηθίζω,
μαθαίνω απ' έξω. ~**ment** *s.* υποχρέωση *f.*
~**tal** *s.* παραπομπή *f.*
committee *s.* επιτροπή *f.*
commodious *a.* ευρύχωρος.
commodity *s.* προϊόν *n.*, είδος εμπορίου.
commodore *s.* (*naut.*) (*commander of
squadron*) μοίραρχος *m.*
common *s.* (*land*) εξοχικός κοινοτικός
χώρος. out of the ~ ασυνήθιστος. in ~ από
κοινού, we have many things in ~ έχουμε
πολλά κοινά σημεία.
common *a.* κοινός, (*used jointly*) κοινόχρηστος, (*usual*) συνηθισμένος, (*vulgar*)
χυδαίος. ~ ground κοινά σημεία. ~**ly** *adv.*
κοινώς, συνήθως.
commoner *s.* κοινός αστός.
commonplace *a.* τετριμμένος. (*s.*) κοινοτοπία *f.*
common-room *s.* (*senior*) αίθουσα καθη-
γητών, (*junior*) αίθουσα φοιτητών.
commons *s.* αστική τάξη. House of C~
Βουλή των Κοινοτήτων. be on short ~ δεν
έχω αρκετή τροφή.
commonwealth *s.* κοινοπολιτεία *f.*
commotion *s.* φασαρία *f.*, αναστάτωση *f.*
communal *a.* (*used jointly*) κοινόχρηστος.

commune s. (district) κοινότης f. (group living communally) κοινόβιο n. (of 1871) Κομμούνα f.

commune v. επικοινωνώ. ~ with nature γίνομαι ένα με τη φύση.

communicate v.t. μεταβιβάζω. (v.i.) επικοινωνώ. (receive Communion) μεταλαμβάνω.

communication s. (transmission) μεταβίβαση f. (message) μήνυμα n. (contact) (also ~s) επικοινωνία f.

communicative a. ομιλητικός.

communion s. επικοινωνία f. hold ~ see commune. Holy C~ Θεία Κοινωνία.

communiqué s. ανακοινωθέν n.

commun|ism s. κομμουνισμός m. ~ist s. κομμουνιστής m. (a., ideas, etc.) κομμουνιστικός.

community s. (having in common) κοινότητα f. (the public) κοινό n. (colony) παροικία f., (biol.) κοινωνία f.

commute v.t. (change) μετατρέπω. (v.i.) πηγαίνω τακτικά στη δουλειά μου με τη συγκοινωνία.

compact s. (agreement) σύμβαση f. (lady's) πουδριέρα f.

compact a. (dense) συμπαγής, (style) λιτός, (convenient) βολικός. ~ disc ψηφιακός δίσκος. ~ly adv. λιτά, βολικά.

companion s. σύντροφος m.f. (one of a pair) ταίρι n. ~able a. ευχάριστος. ~ship s. συντροφιά f.

company s. συντροφιά f. (group at party, outing, etc.) παρέα f. (business) εταιρεία f. (theatre) θίασος m. (mil.) λόχος m. keep ~ with κάνω παρέα με. part ~ with χωρίζω με. he is good ~ κάνει καλή παρέα.

comparable a. παρόμοιος, παρεμφερής. they are not ~ δεν συγκρίνονται.

comparative a. σχετικός, (studies) συγκριτικός. (s., gram.) συγκριτικός m. ~ly adv. σχετικά, συγκριτικώς.

compare v.t. συγκρίνω, (liken) παρομοιάζω (with προς & acc.). (v.i.) (bear comparison) συγκρίνομαι. ~d to εν συγκρίσει με.

comparison s. σύγκριση f. in ~ with εν συγκρίσει με. (gram.) degrees of ~ παραθετικά n.pl.

compartment s. διαμέρισμα n. (of drawer, handbag) χώρισμα n.

compass s. (range) έκταση f. (instrument) πυξίδα f., μπούσουλας m. points of the ~ σημεία του ορίζοντος. (pair of) ~es διαβήτης m.

compass v. (surround) περικυκλώνω, (grasp mentally) συλλαμβάνω, (engineer) μηχανεύομαι.

compassion s. ευσπλαχνία f. take ~ on ευσπλαχνίζομαι. ~ate a. (ευ)σπλαχνικός.

compatible a. σύμφωνος. be ~ συμβιβάζομαι, συμφωνώ.

compatriot s. συμπατριώτης m.

compel v. αναγκάζω, (enforce) επιβάλλω. ~ling a. επιτακτικός.

compendious a. επίτομος.

compensat|e v.t. (person for loss) αποζημιώνω. (mech.) αντισταθμίζω. (v.i.) ~e for (deficiency) αναπληρώνω. ~ion s. αποζημίωση f.

compet|e v. (also ~e with) συναγωνίζομαι. ~ition s. συναγωνισμός m. ~itive a. συναγωνιστικός. ~itor s. συναγωνιζόμενος m.

compet|ence s. ικανότητα f. (official capacity) αρμοδιότητα f. ~ent a. ικανός, (properly qualified) αρμόδιος. ~ently adv. με ικανότητα.

compil|e v. συντάσσω. ~ation s. σύνταξη f. ~er s. συντάκτης m.

complacen|t a. αυτάρεσκος, ευχαριστημένος από τον εαυτό. μου. ~cy s. αυταρέσκεια f.

complain v. παραπονούμαι. ~t s. παράπονο n. lodge a ~t υποβάλλω παράπονον. (malady) ασθένεια f.

complaisant a. υποχρεωτικός.

complement v. συμπληρώνω. (s.) συμπλήρωμα n. ~ary a. συμπληρωματικός.

complete v. (make whole) ολοκληρώνω, συμπληρώνω. (a form, etc.) συμπληρώνω. (finish) (απο)τελειώνω, (απο)περατώνω.

complete a. τέλειος, πλήρης. (intact) ακέραιος, (entire) ολόκληρος. he is a ~ stranger to me μου είναι τελείως άγνωστος. ~ly adv. τελείως, πλήρως, εντελώς.

completion s. συμπλήρωση f., αποπεράτωση f.

complex a. περίπλοκος. (s.) σύμπλεγμα n. (fam., mental) κόμπλεξ n. ~ity s. περιπλοκή f. a matter of some ~ity μία περιπεπλεγμένη υπόθεσις.

complexion s. χρώμα n., (fam.) δέρμα n. (fig.) όψη f., μορφή f.

compli|ant a. υπάκουος. ~ance s. συμμόρφωση f. in ~ance with σύμφωνα με.

complicat|e v. περιπλέκω. ~ed a. περίπλοκος. ~ion s. περιπλοκή f., μπέρδεμα n. (med.) επιπλοκή f.

complicity s. συνενοχή f.

compliment s. φιλοφρόνημα n. (honour) τιμή f. pay ~s κάνω κομπλιμέντα. my ~s to your mother τα σέβη μου στη μητέρα σας. (v.) συγχαίρω.

complimentary *a.* φιλοφρονητικός, εγκωμιαστικός. (*free*) δωρεάν. ~ ticket πρόσκληση *f.*

comply *v.* συμμορφώνομαι.

component *a.* συστατικός. (*s.*) συστατικό *n.* (*part of appliance*) εξάρτημα *n.*

comport *v.* ~ oneself φέρομαι. ~ with ταιριάζω με. **~ment** *s.* συμπεριφορά *f.*

compose *v.* (*form*) απαρτίζω, αποτελώ, (*write*) γράφω, συγγράφω, (*music*) συνθέτω, (*set up type*) στοιχειοθετώ, (*a dispute*) ρυθμίζω. (*get under control*) συγκεντρώνω, ~ oneself ησυχάζω. be ~ed of απαρτίζομαι *or* αποτελούμαι από, συνίσταμαι εκ (*with gen.*).

composed *a.* ήρεμος.

composer *s.* συνθέτης *m.*

composite *a.* σύνθετος.

composition *s.* σύνθεση *f.* (*of substance*) σύσταση *f.* (*essay*) έκθεση *f.* (*psychological*) ψυχοσύνθεση *f.*

compositor *s.* στοιχειοθέτης *m.*

compost *s.* φουσκί *n.*

composure *s.* ηρεμία *f.*

compound *a.* σύνθετος. ~ interest ανατοκισμός *m.* (*s.*) (*gram.*) σύνθετο *n.* (*chem.*) ένωση *f.* (*enclosure*) περίβολος *m.*

compound *v.t.* (*make*) ετοιμάζω, κάνω, παρασκευάζω. (*v.i.*) (*come to terms*) συμβιβάζομαι.

comprehend *v.* (*grasp*) κατανοώ. (*include*) περιλαμβάνω.

comprehens|ion *s.* κατανόηση *f.* **~ible** *a.* καταληπτός. **~ive** *a.* περιληπτικός.

compress *s.* κομπρέσα *f.*

compress *v.* συμπιέζω, (*ideas*) συμπυκνώνω. **~ed** *a.* συμπιεσμένος. **~ion** *s.* συμπίεση *f.* συμπύκνωση *f.* **~or** *s.* συμπιεστής *m.*

comprise *v.* (*include*) περιλαμβάνω, (*consist of*) αποτελούμαι από.

compromise *s.* συμβιβασμός *m.* (*v.i.*) (*agree*) συμβιβάζομαι. (*v.t.*) (*expose*) εκθέτω, be ~d εκτίθεμαι.

compuls|ion *s.* αναγκασμός *m.* **~ive** *a.* a ~ive liar παθολογικός ψεύτης. **~ory** *a.* αναγκαστικός, υποχρεωτικός.

compunction *s.* ενδοιασμός *m.*, τύψεις *f.pl.*

compute *v.* υπολογίζω. **~r** *s.* ηλεκτρονικός υπολογιστής *m.*

comrade *s.* σύντροφος *m.* **~ship** *s.* αδελφοσύνη *f.*

con *v.* (*study*) μελετώ, (*fam., cheat*) εξαπατώ. **~-man** *s.* απατεώνας *m.*

concatenation *s.* αλληλουχία *f.*

concave *a.* κοίλος.

conceal *v.* αποκρύπτω. ~ed κρυμμένος. **~ment** *s.* απόκρυψη *f.* in ~ment κρυμμένος.

concede *v.* (*admit*) παραδέχομαι, (*grant*) παραχωρώ.

conceit *s.* έπαρση *f.*, οίηση *f.* (*of style*) επιτήδευση ύφους.

conceited *a.* ματαιόδοξος, επηρμένος, (*fam.*) φουσκωμένος γάλλος. he is ~ (*fam.*) έχει ιδέα για τον εαυτό του, το έχει πάρει απάνω του.

conceivab|le *a.* νοητός. by every ~le means με κάθε δυνατόν μέσον. **~ly** *adv.* πιθανώς. not ~ly με κανένα τρόπο.

conceive *v.* συλλαμβάνω. I can't ~ why... δεν μπορεί να συλλάβει ο νους μου γιατί *or* δεν μπορώ να διανοηθώ γιατί. I have ~d a dislike for it άρχισε να μη μου αρέσει.

concentrate *v.t.* συγκεντρώνω. (*chem.*) συμπυκνώνω. (*v.i.*) συγκεντρώνομαι.

concentration *s.* συγκέντρωση *f.* ~ camp στρατόπεδο συγκεντρώσεως.

concentric *a.* ομόκεντρος.

concept *s.* έννοια *f.* **~ion** *s.* σύλληψη *f.* (*idea*) ιδέα *f.*

concern *v.* (*relate to*) αφορώ. (*interest*) ενδιαφέρω, all those ~ed οι ενδιαφερόμενοι. as far as the government is ~ed όσον αφορά την κυβέρνησιν. be ~ed in *or* with (*occupied*) ασχολούμαι με, (*be about*) αναφέρομαι σε. be ~ed about (*anxious*) ανησυχώ για, it ~s me (*worries*) με ανησυχεί.

concern *s.* (*interest*) ενδιαφέρον *n.* (*care*) φροντίδα *f.* (*worry*) ανησυχία *f.* (*matter*) υπόθεση *f.*, δουλειά *f.* (*undertaking*) επιχείρηση *f.* it is a going ~ βρίσκεται εν πλήρει λειτουργία. it's no ~ of his δεν είναι δική του δουλειά.

concerned *a.* (*worried*) ανήσυχος.

concerning *prep.* για, σχετικώς προς (*with acc.*), περί (*with gen.*), όσον αφορά.

concert *s.* (*mus.*) συναυλία *f.* in ~ εκ συμφώνου.

concerted *a.* take ~ action ενεργώ από κοινού.

concertina *s.* (*fam.*) φυσαρμόνικα *f.*

concerto *s.* κοντσέρτο *n.*

concession *s.* (*giving way*) παραχώρηση *f.* (*of right*) εκχώρηση *f.* (*right*) δικαίωμα *n.*

concierge *s.* θυρωρός *m.f.*

conciliat|e *v.* (*calm*) κατευνάζω, (*reconcile*) συμφιλιώνω. **~ion** *s.* κατευνασμός *m.* συμφιλίωση *f.* **~ory** *a.* κατευναστικός, συμφιλιωτικός.

concise *a.* επίτομος, σύντομος. **~ly** *adv.* με συντομία. **~ness** *s.* συντομία *f.*

conclave *s.* (*fig.*) σύσκεψη *f.*

conclud|e *v.t.* (*end*) τελειώνω, τερματίζω.

(*an agreement*) συνάπτω. (*v.i.*) τελειώνω, τερματίζομαι. (*infer*) συμπεραίνω. **~ing** *a.* (*final*) τελικός.

conclusion *s.* (*end*) τέλος *n.*, τέρμα *n.* (*inference*) συμπέρασμα *n.* come to the ~ φτάνω στο συμπέρασμα. a foregone ~ γνωστό εκ των προτέρων. in ~ εν κατακλείδι.

conclusive *a.* αδιαμφισβήτητος.

concoct *v.* κατασκευάζω. **~ion** *s.* κατασκεύασμα *n.*

concomitant *a.* συνακόλουθος.

concord *s.* ομόνοια *f.* **~ant** *a.* σύμφωνος.

concourse *s.* (*throng*) συρροή *f.*

concrete *a.* (*not abstract*) συγκεκριμένος.

concrete *s.* μπετόν *n.* (*a.*) από μπετόν.

concubine *s.* παλλακίδα *f.*

concur *v.* (*of persons*) συμφωνώ, (*of things*) συμπίπτω. **~rence** *s.* συμφωνία *f.* σύμπτωση *f.* **~rent** *a.* ταυτόχρονος. **~rently** *adv.* ταυτοχρόνως.

concussion *s.* διάσειση *f.*

condemn *v.* καταδικάζω. the **~ed** man (*to death*) ο μελλοθάνατος. **~ation** *s.* καταδίκη *f.*

condens|e *v.t.* συμπυκνώνω. (*v.i.*) συμπυκνώνομαι. **~ed** συμπεπυκνωμένος. **~ation** *s.* συμπύκνωση *f.* **~er** *s.* συμπυκνωτής *m.*

condescen|d *v.* καταδέχομαι. **~ding** *a.* (*pej.*) καταδεκτικός σε σημείο περιφρονήσεως. **~sion** *s.* (*pej.*) συγκατάβαση σε σημείο περιφρονήσεως.

condiment *s.* καρύκευμα *n.*

condition *s.* (*state*) κατάσταση *f.* (*term*) όρος *m.* **~s** (*circumstances*) συνθήκες *f.pl.*

condition *v.* (*regulate*) ρυθμίζω, (*accustom*) συνηθίζω. be **~ed** by εξαρτώμαι από. well **~ed** σε καλή κατάσταση.

conditional *a.* υπό όρους. (*gram.*) υποθετικός. be **~** on προϋποθέτω. **~ly** *adv.* υπό όρους.

condole *v.* **~** with συλλυπούμαι. **~nces** *s.* συλλυπητήρια *n.pl.*

condone *v.* παραβλέπω.

conduce *v.* **~** to συμβάλλω εις.

conduct *v.* (*manage*) διαχειρίζομαι, (*hold*) διεξάγω, (*escort*) οδηγώ. (*mus.*) διευθύνω. **~** oneself συμπεριφέρομαι.

conduct *s.* διαχείριση *f.* διεξαγωγή *f.* (*behaviour*) συμπεριφορά *f.*, διαγωγή *f.*

conductor *s.* (*mus.*) διευθυντής *m.* (*bus*) εισπράκτορας *m.* (*phys.*) αγωγός *m.*

conduit *s.* αγωγός *m.*

cone *s.* κώνος *m.* (*of tree*) κουκουνάρι *n.*

coney *s.* κουνέλι *n.*

confabulation *s.* συνομιλία *f.*

confection *s.* (*sweetmeat*) γλυκό *n.* (*dress*) κομψό ρούχο. **~er's** *s.* ζαχαροπλαστείο *n.*

~ery *s.* είδη ζαχαροπλαστικής.

confeder|ate *a. & s.* ομόσπονδος, (*accomplice*) συνεργός *m.* **~acy**, **~ation** *s.* ομοσπονδία *f.*

confer *v.t.* (*bestow*) παρέχω, απονέμω. (*v.i.*) (*deliberate*) συσκέπτομαι. **~ence** *s.* διάσκεψη *f.* **~ment** *s.* απονομή *f.*

confess *v.t.* ομολογώ. (*eccl.*) (*v.i.*) εξομολογούμαι. **~or** *s.* εξομολογητής *m.*

confession *s.* ομολογία *f.* (*eccl.*) εξομολόγηση *f.* go to ~ πάω να εξομολογηθώ.

confetti *s.* χαρτοπόλεμος *m.*

confidant *s.* έμπιστος φίλος *m.*

confide *v.t.* (*secret*) εκμυστηρεύομαι, (*entrust*) εμπιστεύομαι. (*v.i.*) **~** in (*trust*) εμπιστεύομαι, έχω εμπιστοσύνη σε.

confidence *s.* (*trust*) εμπιστοσύνη *f.* (*assurance*) πεποίθηση *f.* (*secret*) μυστικό *n.* **~** trick απάτη *f.* in **~** υπό εχεμύθειαν. take into one's **~** ανοίγω την καρδιά μου σε.

confident *a.* βέβαιος. be **~** έχω πεποίθηση. **~ly** *adv.* με βεβαιότητα.

confidential *a.* εμπιστευτικός. **~ly** *adv.* εμπιστευτικώς.

configuration *s.* διαμόρφωση *f.*

confine *v.* (*limit*) περιορίζω, (*enclose*) εγκλείω, (*shut up*) κλείνω. **~d** *a.* (*in space*) στενός, κλειστός. (*to house*) κλεισμένος. (*of woman*) be **~d** γεννώ.

confinement *s.* περιορισμός *m.* (*birth*) τοκετός *m.* solitary **~** απομόνωση *f.*

confines *s.* όρια *n.pl.*

confirm *v.* επιβεβαιώνω. **~ation** *s.* επιβεβαίωση *f.* **~ed** *a.* (*inveterate*) αμετάπειστος.

confiscat|e *v.* κατάσχω. **~ion** *s.* κατάσχεση *f.*

conflagration *s.* μεγάλη πυρκαγιά *f.*

conflict *s.* διαμάχη *f.*, σύγκρουση *f.* (*v.*) συγκρούομαι. **~ing** *a.* συγκρουόμενος.

confluence *s.* συμβολή *f.*

conform *v.i.* συμμορφώνομαι. **~able** *a.* σύμφωνος.

conformation *s.* διαμόρφωση *f.*

conformity *s.* συμφωνία *f.*, προσαρμογή *f.* in **~** with συμφώνως προς.

confound *v.* (*mix*) μπερδεύω, συγχέω, (*amaze*) καταπλήσσω. (*fam.*) **~** it! να πάρει η οργή. **~ed** *a.* (*fam.*) διαβολεμένος, παλιο-. **~edly** *adv.* διαβολεμένα.

confront *v.* (*also* be **~ed** with) αντιμετωπίζω. **~ation** *s.* αντιμετώπιση *f.* have a **~ation** έρχομαι αντιμέτωπος.

confuse *v.* (*mix*) μπερδεύω, συγχέω. (*disconcert, also* get **~d**) σαστίζω.

confused *a.* (*mixed up*) μπερδεμένος, (*obscure*) ασαφής, (*dim*) θαμπός.

confusion s. (*mixing*) σύγχυσηf. (*disorder*) ανακατωσούραf. (*embarrassment*) σαστισμάραf., σύγχυσηf.
confute v. αντικρούω.
congeal v. πήζω. ~ed a. πηχτός.
congenial a. ευχάριστος.
congenital a., ~ly adv. εκ γενετής.
congested a. ~ traffic πυκνή κυκλοφορία. (*overpopulated*) πυκνοκατοικημένος.
congestion s. συμφόρηση f. (*med.*) υπεραιμία f.
conglomeration s. σύμπηγμα n.
congratulat|e v. συγχαίρω. ~ions s. συγχαρητήρια n.pl.
congregat|e v.i. συναθροίζομαι. ~ion s. (*eccl.*) εκκλησίασμα n.
congress s. συνέδριο n. (*USA*) Κογκρέσο n.
conifer s. κωνοφόρο n. ~ous a. κωνοφόρος.
conjectur|e s. εικασία f. (v.) εικάζω. ~al a. εικαστικός.
conjointly adv. από κοινού.
conjugal a. συζυγικός.
conjugat|e v. (*gram.*) κλίνω. ~ion s. συζυγία f.
conjunction s. (*gram.*) σύνδεσμος m. in ~ with (*persons*) από κοινού με.
conjuncture s. συγκυρία f.
conjure v. (*appeal to*) εξορκίζω. (*produce*) παρουσιάζω, βγάζω. ~ up επικαλούμαι, (*bring back*) επαναφέρω.
conjur|ing s. (*also* ~ing trick) ταχυδακτυλουργία f. ~or s. ταχυδακτυλουργός m.
conk s. (*fam.*) μύτη f. (v.) it has ~ed out πάει περίπατο.
connect v.t. συνδέω. ~ed a. συνδεδεμένος, (*related*) συγγενής, (*adjoining*) συνεχόμενος. be well ~ed έχω γερές πλάτες, έχω υψηλές σχέσεις.
connection s. (*act*) σύνδεση f. (*union*) ένωση f. (*relation*) σχέση f. in ~ with εν σχέσει με. run in ~ with (*of transport*) έχω ανταπόκριση με.
conniv|e v. συνεργώ, κάνω στραβά μάτια. ~e at παραβλέπω. ~ance s. with his ~ance με την ανοχή του.
connoisseur s. γνώστης m.
connot|e v. υποδηλώ. ~ation s. υποδήλωση f.
connubial a. συζυγικός.
conquer v. νικώ, (*country, hearts*) κατακτώ. ~ing a. νικηφόρος. ~or s. νικητής m. κατακτητής m.
conquest s. κατάκτηση f.
conscience s. συνείδηση. f. in all ~ ειλικρινώς. ~less or without ~ ασυνείδητος.
conscientious a. ευσυνείδητος. ~ objector αντιρρησίας συνειδήσεως. ~ly adv. ευσυνειδήτως. ~ness s. ευσυνειδησία f.
conscious a. be ~ (*with faculties awake*) έχω τις αισθήσεις μου, (*aware*) έχω συνείδηση, αντιλαμβάνω, νιώθω. (*of action, feeling*) συνειδητός. ~ly adv. συνειδητώς.
consciousness s. (*awareness*) συνείδησηf. (*realization*) επίγνωση f. (*feeling*) συναίσθησηf. lose ~ χάνω τις αισθήσεις μου.
conscript v. στρατολογώ. (s.) κληρωτός m. ~ion s. στρατολογία f.
consecrat|e v. καθιερώνω, (*anoint*) χρίω. ~ion s. καθιέρωσηf. χρίσιςf.
consecutive a. συνεχής, συναπτός. ~ly adv. συναπτώς.
consensus a. γενική γνώμη.
consent v. συγκατατίθεμαι, συναινώ. (s.) συγκατάθεση f., συναίνεση f. age of ~ ενηλικιότης f. (*divorce*) by mutual ~ κοινή συναινέσει.
consequence s. συνέπεια f. in ~ of συνεπεία (*with gen.*) (*importance*) person of ~ σημαίνον πρόσωπον. it is of no ~ δεν έχει σημασία.
consequent a. επακόλουθος. ~ly adv. επομένως, συνεπώς.
consequential a. σπουδαιοφανής.
conservancy s. υπηρεσία συντηρήσεως. ~ of forests δασονομία f.
conservation s. διατήρησηf.
conservat|ive a. συντηρητικός. ~ism s. συντηρητικότητα f.
conservatoire s. ωδείον n.
conservatory s. σέρραf., θερμοκήπιο n.
conserve v. διατηρώ.
consider v. (*think about*) σκέπτομαι, (*study*) μελετώ, εξετάζω. (*take account of*) λαμβάνω υπ' όψιν, υπολογίζω. (*be of opinion*) νομίζω. (*regard*) θεωρώ, έχω. I ~ it my duty το θεωρώ καθήκον μου. I ~ it good luck το έχω για γούρι. all things ~ed σε τελευταία ανάλυση.
considerab|le a. (*noteworthy*) σημαντικός, υπολογίσιμος. (*no little*) αρκετός. ~ly adv. αρκετά.
considerate a. στοχαστικός.
consideration s. (*thoughtfulness*) στόχαση f. (*thought*) σκέψη f. (*study*) μελέτη f., εξέταση f. (*factor*) πράγμα n. under ~ υπό εξέτασιν. take into ~ λαμβάνω υπ' όψιν, λογαριάζω. for a ~ (*money*) αντί πληρωμής.
considering prep. όταν λάβει κανείς υπ' όψιν (*with acc. or* ότι), εάν ληφθεί υπ' όψιν (*with nom. or* ότι). δεδομένου (*with* ότι). he did quite well ~ τα κατάφερε καλά αναλόγως.
consign v. (*despatch*) αποστέλλω, (*hand over*) παραδίδω, (*relegate*) εξαποστέλνω. ~ment s (*despatch*) αποστολή f. (*load*)

φορτίο n.

consist v. ~ of αποτελούμαι από, συνίσταμαι εκ (*with gen.*). **~ency** s.(*thickness*) σύσταση f. (*being the same*) συνέπεια f.

consistent a. συνεπής. ~ with σύμφωνος μέ. **~ly** adv. με συνέπεια.

console v. παρηγορώ. **~ation** s. παρηγοριά f.

consolidat|e v. (*strengthen*) σταθεροποιώ, (*unify*) ενοποιώ. **~ed** a. πάγιος. **~ion** s. σταθεροποίηση f., ενοποίηση f.

consommé s. κονσομέ n.

conson|ant a. σύμφωνος. (s.) σύμφωνο n. **~ance** s. συμφωνία f.

consort s. σύζυγος m.f. (v.) ~ with (*be compatible*) συμβιβάζομαι μέ, (*mix with*) συναναστρέφομαι, (*pej.*) συμφύρομαι μέ.

consortium s. κονσόρτιουμ n.

conspectus s. σύνοψη f.

conspicuous a. (*easily seen*) ευδιάκριτος, εμφανής. (*talent, courage, etc.*) διακεκριμένος, εξέχων. play a ~ part παίζω σημαντικό ρόλο, διακρίνομαι. be *or* make oneself ~ ξεχωρίζω. a ~ colour χρώμα που ξεχωρίζει. **~ly** adv. εμφανώς.

conspir|e v. συνωμοτώ. **~acy** s. συνωμοσία f. **~ator** s. συνωμότης m.

constab|le s. αστυφύλαξ m. **~ulary** s. αστυνομία f.

constancy s. (*stability*) σταθερότης f. (*fidelity*) πίστη f.

constant a. (*stable*) σταθερός, (*true*) πιστός, (*recurring*) συνεχής. **~ly** adv. συνεχώς, όλο. (*see* continually).

constellation s. αστερισμός m.

consternation s. κατακεραύνωση f. filled with ~ κατάαραγμένος.

constipat|ed, ~ing a. δυσκοίλιος. **~ion** s. δυσκοιλιότητα f.

constituency s. εκλογική περιφέρεια.

constituent a. συστατικός. (s.) συστατικό n. (*elector*) ψηφοφόρος m.

constitute v. (*appoint*) διορίζω, (*set up*) συγκροτώ, (*form, be*) αποτελώ. be so ~d as to είμαι καμωμένος έτσι που.

constitution s. (*political*) σύνταγμα n. (*bodily*) οργανισμός m., κράση f. (*composition*) σύσταση f. (*of society, etc.*) καταστατικόν n.

constitutional a. συνταγματικός. (s.) (*fam., walk*) περίπατος m. **~ly** adv. συμφώνως προς το σύνταγμα.

constrain v. αναγκάζω. **~ed** a. βεβιασμένος, αμήχανος. **~t** s. (*restriction*) περιορισμός m. (*awkwardness*) αμηχανία f.

constrict v. σφίγγω, συσφίγγω. **~ion** s. σφίξιμο n., σύσφιξη f. **~or** s. συσφιγκτήρ m.

construct v. (*building, triangle*) κατασκευάζω. (*make*) φτιάνω, (*assemble*)

συναρμολογώ, (*plot of play, novel*) πλέκω. (*gram.*) συντάσσω. well ~ed καλοφτιαγμένος. **~or** s. κατασκευαστής m.

construction s. (*making*) κατασκευή f. (*thing made*) κατασκεύασμα n. (*gram.*) σύνταξη f. (*fig.*) put a wrong ~ on παρεξηγώ. put another ~ on δίδω άλλην ερμηνείαν εις.

constructive a. θετικός. **~ly** adv. θετικά.

construe v. (*interpret*) εκλαμβάνω. (*gram.*) (*analyse*) αναλύω, (*combine*) συντάσσω.

consul s. πρόξενος m. (*Roman, French Republican*) ύπατος m. **~ar** a. προξενικός. **~ate** s. προξενείον m.

consult v.t. συμβουλεύομαι. (v.i.) συσκέπτομαι.

consultant s. εμπειρογνώμων m. (*med.*) ειδικός ιατρός.

consultation s. σύσκεψη f. (*with doctor*) επίσκεψη f. doctors' ~ ιατροσυμβούλιο n.

consultative a. συμβουλευτικός.

consulting-room s. (*med.*) ιατρείο n.

consume v. (*use*) καταναλίσκω, (*use up*) ξοδεύω, τρώω. it was ~d by fire έγινε παρανάλωμα του πυρός. (*fig.*) be ~d by (*envy, etc.*) κατατρώγομαι από.

consumer s. καταναλωτής m. ~ goods καταναλωτικά αγαθά.

consummate a. τέλειος.

consummat|e v. ολοκληρώνω. **~ion** s. ολοκλήρωση f.

consumption s. κατανάλωση f. (*med.*) φυματίωση f., φθίση f.

consumptive a. φυματικός, φθισικός.

contact s. επαφή f. (*communication*) επικοινωνία f. make (*or* come into) ~ with έρχομαι σε επαφή με, (*touch*) αγγίζω. be in ~ (*touching*) εφάπτομαι. (v.t.) (*communicate with*) επικοινωνώ με. **~lens** s. φακός επαφής m.

contagi|on s. μετάδοση f. (*also fig.*) μόλυνση f. **~ous** a. μεταδοτικός.

contain v. (*have, hold*) περιέχω, περιλαμβάνω. (*have room for*) χωράω. (*hold back*) συγκρατώ, αναχαιτίζω.

container s. κιβώτιον n.

containment s. συγκράτηση f., ανάσχεση f.

contaminat|e v. μολύνω. **~ion** s. μόλυνση f.

contemplate v. (*look at*) κοιτάζω, (*reflect on*) συλλογίζομαι, (*have in view*) σκέπτομαι (*with* νά).

contemplat|ion s. στοχασμός m., σκέψη f. **~ive** a. συλλογισμένος.

contemporaneous a. σύγχρονος.

contemporary a. σύγχρονος, (*of today*) σημερινός.

contempt s. περιφρόνηση f., καταφρόνηση

f. hold in ~ περιφρονώ. ~ of court ασέβεια προς το δικαστήριον.

contemptib|le *a.* άξιος περιφρονήσεως, αξιοκαταφρόνητος. **~ly** *adv.* ελεεινά.

contemptuous *a.* περιφρονητικός. **~ly** *adv.* με περιφρόνηση.

contend *v.* (*fight*) αγωνίζομαι, (*compete*) διαγωνίζομαι. (*assert*) ισχυρίζομαι. **~er** *s.* (*fighter*) αγωνιστής *m.* (*competitor*) διαγωνιζόμενος *m.*

content *a.* ικανοποιημένος. be ~ (*with, to*) ικανοποιούμαι (*with* με, να), αρκούμαι (*with* σε, να).

content *v.* ικανοποιώ. ~ oneself *see* content *a.*

content *s.* (*satisfaction*) ικανοποίηση *f.*

content *s.* (*that contained*) περιεχόμενο *n.* (*capacity*) περιεκτικότητα *f.*

contented *a. see* content *a.*

contention *s.* (*dispute*) φιλονικία *f.* (*assertion*) ισχυρισμός *m.* bone of ~ μήλον της έριδος.

contentious *a.* (*person*) φιλόνικος, φίλερις, (*topic*) αμφισβητούμενος.

contentment *s.* ευδαιμονία *f.*

contest *s.* (*struggle*) αγώνας *m.* (*competition*) διαγωνισμός *m.*

contest *v.t.* (*challenge*) αμφισβητώ, (*a will, etc.*) προσβάλλω. (*lay claim to*) διεκδικώ. (*v.i.*) (*quarrel*) φιλονικώ. **~ant** *s.* διεκδικητής *m.* (*competition*) διαγωνιζόμενος *m.*

context *s.* (*of words*) συμφραζόμενα *n.pl.* (*fig., framework*) πλαίσιο *n.*

contiguous *a.* συνεχόμενος, κολλητός. be ~ συνέχομαι.

continent *s.* ήπειρος *f.* (*fam., mainland Europe*) ηπειρωτική Ευρώπη. **~al** *a.* ηπειρωτικός.

continen|t *a.* εγκρατής. **~ce** *s.* εγκράτεια *f.*

contingency *s.* ενδεχόμενο *n.*, απρόοπτο γεγονός.

contingent *a.* be ~ on εξαρτώμαι από. (*s.*) (*mil.*) δύναμη *f.*

continual *a.* συνεχής, επανειλημμένος. **~ly** *adv.* συνεχώς, επανειλημμένως. he is ~ly interrupting me όλο με διακόπτει.

continuance *s.* (*of action*) συνέχιση *f.* (*of person*) παραμονή *f.*

continuation *s.* (*part that follows*) συνέχεια *f.* (*prolongation*) παράταση *f.* (*resumption*) επανάληψη *f.*

continue *v.t. & i.* εξακολουθώ. (*v.t.*) συνεχίζω. (*v.i.*) συνεχίζομαι. ~ to be εξακολουθώ (*or* συνεχίζω) να είμαι.

continuity *s.* συνειρμός *m.*

continuous *a.* συνεχής, αδιάκοπος. **~ly** *adv.* συνεχώς, αδιάκοπα.

contort *v.* συσπώ. **~ed** συσπασμένος. **~ions**

s. go through ~ions στρεβλώνω το κορμί μου.

contour *s.* ισούψής καμπύλη. ~ map υψομετρικός χάρτης. (*outline*) περίγραμμα *n.*

contraband *s.* λαθρεμπόριο *n.* (*goods*) λαθραία εμπορεύματα.

contracept|ion *s.* πρόληψις εγκυμοσύνης. **~ive** *s.* αντισυλληπτικό *n.*

contract *s.* συμβόλαιο *n.* (*for construction work*) εργολημψία *f.*

contract *v.i.* (*undertake*) κάνω συμβόλαιο, συμφωνώ. (*v.t.*) (*marriage, debts*) συνάπτω, (*habit*) αποκτώ, (*illness*) παθαίνω, προσβάλλομαι υπό (*with gen.*). **~or** *s.* εργολάβος *m.*

contract *v.t.* (*draw together*) συστέλλω. (*v.i.*) συστέλλομαι. ~ one's brow συνοφρυούμαι. **~ed** *a.* (*gram.*) συνηρημένος. **~ion** *s.* συστολή *f.* (*gram.*) συναίρεση *f.* **~ual** *a.* συμβατικός.

contradict *v.* διαψεύδω, (*say the opposite*) αντιλέγω. ~ oneself αντιφάσκω. **~ion** *s.* διάψευση *f.* αντιλογία *f.* αντίφαση *f.* **~ory** *a.* αντιφατικός.

contradistinction *s.* αντιδιαστολή *f.*

contralto *a.* βαθύφωνος, κοντράλτο.

contraption *s.* πράμα *f.*, κατασκεύασμα *n.* (*pej.*) κολοκύθι *n.*

contrapuntal *a.* αντιστικτικός.

contrariness *s.* αναποδιά *f.*

contrary *a.* αντίθετος, ενάντιος. (*perverse*) ανάποδος. ~ το αντίθετος με (*with acc.*), (*adv.*) αντιθέτως προς.

contrary *s.* εναντίον *n.*, αντίθετο *n.* on the ~ απεναντίας. το the ~ αντιθέτως.

contrast *s.* αντίθεση *f.* in ~ το εν αντιθέσει μέ. (*v.t.*) αντιπαραβάλλω, συγκρίνω. (*v.i.*) ~ with κάνω αντίθεση με. **~ing** *a.* αντίθετος.

contraven|e *v.* (*infringe*) παραβαίνω, (*conflict with*) αντιβαίνω εις. **~tion** *s.* παράβαση *f.*

contretemps *s.* κάζο *n.*, αναποδιά *f.*

contribut|e *v.t.* συνεισφέρω. (*v.i.*) ~e to (*play a part in*) συμβάλλω εις, συντελώ εις, (*write for*) συνεργάζομαι σε. **~ory** *a.* συντελεστικός.

contribut|ion *s.* (*thing given*) συνεισφορά *f.* (*part played*) συμβολή *f.* (*article*) άρθρο *n.*, συνεργασία *f.* **~or** *s.* συνεισφέρων *m.* (*to journal*) συνεργάτης *m.*

contrit|e *a.* συντετριμμένος. **~ion** *s.* συντριβή *f.*

contriv|e *v.* (*devise*) επινοώ, εφευρίσκω, (*a plot*) μηχανεύομαι, (*manage*) καταφέρνω. **~ance** *s.* εφεύρεση *f.*

control *v.* (*have power over*) εξουσιάζω, (*check, keep under* ~) ελέγχω, συγκρατώ.

(direct) διευθύνω, (*govern*) κυβερνώ.
control s. (*power*) εξουσία f. (*restraint, supervision*) έλεγχος m. (*direction*) διεύθυνση f. be in ~ of έχω υπό την εξουσία (*or* τον έλεγχο *or* τη διεύθυνσή) μου. lose ~ of oneself χάνω τον έλεγχο του εαυτού μου. ~ room/tower θάλαμος/ πύργος ελέγχου.
controls s. (*restraints*) περιορισμοί m.pl. (*of machine*) σύστημα ελέγχου.
controller s. (*of funds*) διαχειριστής m. (*director*) διευθυντής m.
controversly s. αντιγνωμία f., συζήτηση f. **~ial** a. (*topic*) αμφισβητήσιμος, (*speech*) επίμαχος. a ~ial character πολυσυζητούμενο πρόσωπο.
controvert v. αμφισβητώ, διαψεύδω.
contumacy s. ισχυρογνωμοσύνη f.
contumely s. περιφρόνηση f.
contusion s. μωλωπισμός m.
conundrum s. αίνιγμα n.
conurbation s. αστικό συγκρότημα.
convalesclence s. ανάρρωση f. **~ent** a. αναρρωνύων. be ~ent βρίσκομαι εν αναρρώσει.
convector s. αερόθερμο n.
convene v.t. συγκαλώ. (*v.i.*) συνέρχομαι.
convenience s. ευκολία f, άνεση f. at your ~ όποτε (*or* όπως) σας διευκολύνει, με την άνεσή σας. it is a great ~ είναι μεγάλο πράμα. marriage of ~ γάμος από συμφέρον. flag of ~ σημαία ευκαιρίας. (*w.c.*) αποχωρητήριο n.
convenient a. βολικός. it is ~ for me μου έρχεται εύκολο (*or* βολικό), με βολεύει. find a ~ opportunity βρίσκω την κατάλληλη ευκαιρία.
convent s. μονή γυναικών.
convention s. (*meeting*) συνέδριο n. (*diplomatic*) σύμβασις f. **~s** (*rules of behaviour*) τύποι m.pl., κοινωνικές συμβάσεις. **~al** a. συμβατικός, τυπικός.
converge v. συγκλίνω.
conversant a. be ~ with γνωρίζω από, έχω γνώση (*with gen.*).
conversation s. κουβέντα f., συνομιλία f. **~ai** a. (*he spoke*) in a ~al manner σα να συζητούσε.
converse v. κουβεντιάζω, συνομιλώ.
converse s. (*opposite*) αντίστροφο n. **~ly** adv. αντιστρόφως.
conversion s. μετατροπή f. (*to faith*) προσηλυτισμός m. (*change of faith*) αλλαξοπιστία f. (*embezzlement*) σφετερισμός m.
convert v.t. μετατρέπω, (*to faith*) προσηλυτίζω, (*embezzle*) σφετερίζομαι.
convert s. (*to faith*) προσήλυτος m. (*to*

ideology) νεοφώτιστος a.
convertible a. (*fin.*) μετατρέψιμος it is ~ μετατρέπεται.
convex a. κυρτός.
convey v. (*transport*) μεταφέρω, (*message*) διαβιβάζω, (*idea, feelings*) εκφράζω, αποδίδω. (*law*) μεταβιβάζω.
conveyance s. (*vehicle*) μεταφορικόν μέσον, όχημα n. (*carrying*) μεταφορά f. διαβίβαση f. (*law*) μεταβίβαση f.
convincle v. πείθω. **~ed** πεπεισμένος. **~ing** a. πειστικός.
convict v. αποδεικνύω ένοχον. (*s.*) κατάδικος m. **~ion** s. (*law*) καταδίκη f. (*belief*) πεποίθηση f. carry ~ion είμαι πειστικός.
convivial a. γεμάτος κέφι. **~ity** s. (*mood*) κέφι n. (*merrymaking*) γλέντι n.
convolke v. συγκαλώ. **~cation** s. σύγκληση f.
convoluted a. συνεστραμμένος, στριφτός. (*fig.*) (*difficult*) στριμμένος. (*elaborate*) πολύπλοκος.
convolvulus s. περικοκλάδα f.
convoy v. συνοδεύω. (*s.*) (*ships*) νηοπομπή f. (*supplies*) εφοδιοπομπή f. (*column*) φάλαγγα f. (*escort*) συνοδεία f.
convulsle v. συνταράσσω, (*med.*) συσπώ. be ~ed with laughter σκάω στα γέλια. **~ion** s. σπασμός m. (*seismic, political*) αναταραχή f. **~ive** a. σπασμωδικός.
coo v. κουκουρίζω. (*fig.*) μιλώ χαϊδευτικά. bill and ~ γλυκοσαλιάζω. **~ing** s. κουκούρισμα n.
cook v.t. & i. μαγειρεύω. (*of food, v.i.*) μαγειρεύομαι, ψήνομαι. (*falsify*) μαγειρεύω, ~ up μαγειρεύω. **~shop** s. μαγειρειό n.
cook s. μάγειρος m., μάγερας m., μαγείρισσα f.
cooker s. (*stove*) κουζίνα f.
cookery s. (*also* ~ book) μαγειρική f.
cookie s. (*Scots*) κουλουράκι n., (*USA*) μπισκότο n.
cooking s. μαγείρεμα n. (*a.*) ~ utensils μαγειρικά σκεύη.
cool a. δροσερός, (*calm*) ήρεμος, (*indifferent*) ψυχρός, χλιαρός, (*impudent*) αναιδής. it is ~ weather έχει ψυχρούλα. keep ~ (*fig.*) διατηρώ την ψυχραιμία μου.
cool v.t. δροσίζω. (*v.i.*) ~ (down, off) (*weather*) δροσίζω, (*excited person*) ηρεμώ, καλμάρω, (*friendship, etc.*) κρυώνω, ψυχραίνομαι. **~ing** a. δροσιστικός.
cool-headed a. ψύχραιμος.
coolly adv. ψυχρός, (*cheekily*) αναιδώς.
coolness s. δροσιά f. (*calm*) ψυχραιμία f. (*cheek*) αναίδεια f.
coomb s. (*valley*) αυλών m.
coop s. κουμάσι n. (*v.*) ~ up (*fig.*) κλείνω.

co-op s. συνεργατική f.
cooper s. βαρελάς m.
cooperat|e v. συνεργάζομαι. ~ion s. συνεργασία f., σύμπραξη f. ~ive a. συνεργάσιμος. (s.) συνεργατική f.
co-opt v. εισδέχομαι.
coordinat|e v. (ideas) συντονίζω, (movements) συνδυάζω. ~ion s. συντονισμός m. συνδυασμός m.
cop v. (fam.) (nab) τσιμπώ. he ~ped it την έπαθε. (s.) (fam.) αστυνομικός m.
cope v. ~ with κάνω με, τα καταφέρνω με, τα βγάζω πέρα με, (face) αντιμετωπίζω.
coping s. archit. σαμάρι n. ~-stone (fig.) αποκορύφωμα n.
copious a. άφθονος, πλούσιος. ~ly adv. άφθονα.
copper s. χαλκός m., μπακίρι n. (fam.) αστυνομικός m. (a.) χάλκινος.
copperplate a. ~ engraving χαλκογραφία f. (fig.) ~ writing καλλιγραφία f.
coppersmith s. χαλκωματάς m. μπακιρτζής m.
coppice, copse s. δασύλλιον n.
Coptic a. κοπτικός.
copulat|e v. συνουσιάζομαι. ~ion s. συνουσία f.
copy v. αντιγράφω. (s.) αντίγραφο n. (of book) αντίτυπο n. (of paper) φύλλο n.
copyright s. κοπυράιτ n., πνευματική ιδιοκτησία.
coquetry s. φιλαρέσκεια f., κοκεταρία f.
coquett|e s. φλερτατζού f. ~ish a. φιλάρεσκος.
coral s. κοράλλι(ον) n. (a.) κοράλινος.
cord s. κορδόνι n. (vocal) χορδή f. (umbilical) λώρος m. (v.) δένω.
cordial a. εγκάρδιος, (dislike) έντονος. ~ity s. εγκαρδιότητα f. ~ly adv. εγκαρδίως. I ~ly dislike him μου είναι πολύ αντιπαθητικός.
cordon s. (of police, etc.) κλοιός m. (v.) ~ off αποκλείω.
corduroy s. βελούδο κοτλέ.
core s. πυρήνας m. (fig.) to the ~ μέχρι μυελού οστέων, ως το κόκαλο. (v.) ξεκουκουτσιάζω.
co-religionist s. ομόθρησκος m.
co-respondent s. συγκατηγορούμενος m.
Corinthian a. κορινθιακός.
cork s. φελλός m. (a.) από φελλό. (v.) βουλώνω. (fig.) ~ up καταπνίγω. ~screw s. τιρμπουσόν n.
corm s. βολβός m.
corn s. (cereals) σιτηρά n.pl., δημητριακά n.pl. (wheat) σίτος m., σιτάρι n. (oats) βρώμη f. (maize) καλαμπόκι n., αραβόσιτος m.

corn s. (on foot) κάλος m.
cornea s. κερατοειδής χιτών.
corner s. γωνία f. out of the ~ of one's eye με την άκρη του ματιού. just round the ~ εδώ παρακάτω. (fig.) turn the ~ καβατζάρω. (monopoly) μονοπώλιο n. (a.) γωνιαίος, γωνιακός.
corner v.t. (drive into ~) στριμώχνω στη γωνία, κολλώ στον τοίχο. (monopolize) μονοπωλώ.
cornerstone s. αγκωνάρι n., γωνιόλιθος m. (fig.) ακρογωνιαίος λίθος.
cornet s. (mus.) κορνέτα f. (ice-cream) χωνάκι n.
cornice s. κορνίζα f.
cornucopia s. κέρας της Αμαλθείας.
corollary s. πόρισμα n.
coronary a. (anat.) στεφανιαίος.
coronation s. στέψη f.
coroner s. υπάλληλος με ανακριτικά καθήκοντα σε περιπτώσεις βιαίου θανάτου.
coronet s. διάδημα n.
corporal a. σωματικός. (s., mil.) δεκανεύς m.
corporat|e a. σωματειακός, (joint) συλλογικός. ~ion s. (company) εταιρεία f. (civic) δημοτικόν συμβούλιον.
corporeal a. σωματικός.
corps s. σώμα n. (mil.) ~ commander σωματάρχης m.
corpse s. πτώμα n.
corpul|ent a. σωματώδης. ~ence s. παχυσαρκία. f.
corpuscle s. αιμοσφαίριο n.
correct a. ακριβής, ορθός, σωστός. (manners, dress, etc.) καθώς πρέπει. ~ly adv. ορθώς, σωστά. ~ness s. ακρίβεια f., ορθότητα f.
correct v. διορθώνω. ~ion s. διόρθωση f. ~ive a. διορθωτικός.
correlat|e v.t. συσχετίζω. ~ion s. συσχέτιση f.
correspond v. (in value, degree) αντιστοιχώ (with προς & acc.). (agree, be consistent) ανταποκρίνομαι (with σε), συμφωνώ (with με). (write) αλληλογραφώ (with με).
correspond|ence s. αντιστοιχία f. συμφωνία f., σχέση f. αλληλογραφία f. ~ent s. επιστολογράφος m.f. (of journal) ανταποκριτής m.
corresponding a. αντίστοιχος. ~ly adv. αναλόγως.
corridor s. διάδρομος m.
corroborat|e v. επιβεβαιώνω. ~ion s. επιβεβαίωση f.
corro|de v.t. διαβιβρώσκω, τρώγω. ~sion s. διάβρωση f., φάγωμα n. ~sive a. διαβρωτικός.
corrugated a. (metal) αυλακωτός, (paper)

κυματοειδής.

corrupt ν. διαφθείρω. (a.) διεφθαρμένος, σάπιος. (*dishonest*) ανέντιμος. (*linguistically*) παρεφθαρμένος. **~er** s. φθορεύς m.

corruption s. διαφθορά f., σαπίλα f. (*bribery*) δωροδοκία f.

corsair s. κουρσάρος m.

corset s. κορσές m.

cortège s. πομπή f.

coruscate ν. σπιθοβολώ.

cosh s. μαγκούρα f., ρόπαλο n. (ν.) χτυπώ με ρόπαλο.

cosily adv. αναπαυτικά.

cosmetic s. καλλυντικό n.

cosmic a. κοσμικός.

cosmopolitan a. κοσμοπολιτικός, (*person*) κοσμοπολίτης m.

cosset ν. παραχαϊδεύω.

cost s. κόστος n. (*price*) τιμή f. **~s** (*expenses*) έξοδα n.pl. ~ of living κόστος ζωής n. to my ~ εις βάρος μου. at all ~s πάση θυσία. at great ~ (*expense*) αντί μεγάλης δαπάνης. at what ~! (*suffering*) τι μου κόστισε!

cost v.i. κοστίζω, στοιχίζω. (v.t.) (*assess price of*) κοστολογώ.

costive a. δυσκοίλιος.

costly a. ακριβός.

costume s. ενδυμασία f.

cosy a. αναπαυτικός.

cot s. κρεβατάκι n.

coterie s. κύκλος m.

cottage s. εξοχικό σπιτάκι. ~ cheese άσπρο ανάλατο τυρί. ~ industry οικοτεχνία f.

cotton s. βαμβάκι n. βάμβαξ m. (a.) μπαμπακερός, βαμβακερός. **~-wool** s. μπαμπάκι n.

couch s. ανάκλιντρο n.

couch v.t. (*express*) διατυπώνω.

couchette s. κουκέτα f.

cough ν. βήχω. ~ up (*phlegm*) φτύνω, (*money*) σκάζω. (s.) βήχας m.

could ν. she ~ go tomorrow θα μπορούσε να πάει αύριο. I ~n't find it δεν μπορούσα να το βρω. If only I ~ find the time να μπορούσα μόνο να 'βρισκα τον καιρό. he ~n't swim δεν ήξερε κολύμπι. It ~n't be heard δεν ακουγόταν.

council s. συμβούλιο n. ~house εργατική πολυκατοικία. **~lor** s. σύμβουλος m.

counsel s. συμβουλή f. take ~ συσκέπτομαι. keep one's ~ κρύβω τις προθέσεις μου. (*law*) συνήγορος m.

counsel ν. συμβουλεύω. **~lor** s. σύμβουλος m.

count s. (*title*) κόμης m.

count s. (*counting*) μέτρημα n. lose ~ χάνω

το λογαριασμό. on all ~s από πάσης απόψεως. (*law*) κεφάλαιον κατηγορίας.

count v.t. (*add up*) μετρώ, (*reckon, consider*) θεωρώ, λογαριάζω. (v.i.) (*be taken into account*) υπολογίζομαι, έχω σημασία.

count in v.t. (*include*) λογαριάζω.

count on ν. βασίζομαι σε, υπολογίζω σε.

count out ν. (*one by one*) μετρώ ένα ένα. (*the House*) was counted out ανεβλήθη η συνεδρίασις ελλείψει απαρτίας, (*of boxer*) έπαθε νοκάουτ. count me out μη με υπολογίζετε.

count-down s. αντίστροφος μέτρησις.

countenance s. όψη f., έκφραση f. keep one's ~ μένω ατάραχος. lose ~ τα χάνω. put out of ~ σαστίζω. (ν.) ανέχομαι, εγκρίνω.

counter s. (*in shop*) πάγκος m. (*fig.*) under the ~ κρυφά. (*disc*) μάρκα f. (*meter*) μετρητής m.

counter ν. αποκρούω. (*prep.*) ~ to εναντίον (*with gen.*).

counteract ν. εξουδετερώνω. **~ion** s. εξουδετέρωση f.

counter-attack ν. αντεπιτίθεμαι. (s.) αντεπίθεση f.

counter-attraction s. be a ~ το συναγωνίζομαι.

counterbalance ν. αντισταθμίζω.

counterblast s. βιαία ανταπάντηση.

countercharge s. αντικατηγορία f.

counter-espionage s. αντικατασκοπία f.

counterfeit a. προσποιητός, (*money*) κίβδηλος. (v.) (*feign*) προσποιούμαι, (*forge*) παραποιώ, (*money*) παραχαράσσω. **~er** s. παραχαράκτης m.

counterfoil s. στέλεχος n.

countermand ν. ακυρώνω.

counter-offensive s. αντεπίθεση f.

counterpane s. κουβρ-λί n.

counterpart s. αντίτυπο n. αντίστοιχο n. (*person*) αντίστοιχος m.

counterplot s. αντιστρατήγημα n. (*of play*) δευτερεύουσα πλοκή.

counterpoint s. (*mus.*) αντίστιξη f., κοντραπούντο n.

counterpoise s. αντιστάθμισμα n.

counter-productive a. αντιπαραγωγικός.

counter-revolution s. αντεπανάσταση f. **~ary** s. αντεπαναστατικός.

countersign ν. προσυπογράφω. (s.) παρασύνθημα n.

countess s. κόμησσα f.

counting-house s. λογιστήριο n.

countless a. αμέτρητος, αναρίθμητος.

countrified a. μέσα στην εξοχή. (*person*) με γούστα και συνήθειες της εξοχής.

country s. (*land*) χώρα f., τόπος m. (*native land*) πατρίδα f. (*rural parts*) εξοχή f., (*opposed to town*) ύπαιθρος f.

country a. εξοχικός. ~ house πύργος m.

countryman s. άνθρωπος που ζει στην εξοχή. fellow ~ (συμ)πατριώτης m.

countryside s. εξοχή f.

county s. κομητεία f.

coup s. επιτυχία f. ~ d'état πραξικόπημα n. ~ de grâce χαριστική βολή.

couple s. ζεύγος n., ζευγάρι n. married ~ αντρόγυνο n. a ~ of times κάνα δύο φορές.

couple v.t. συνδέω. (v.i.) ζευγαρώνομαι. ~ing s. σύνδεση f. ζευγάρωμα n.

couplet s. δίστιχο n.

coupon s. δελτίο n., κουπόνι n.

courage s. θάρρος n., κουράγιο n., ~ous a. θαρραλέος, γενναίος. ~ously adv. θαρραλέως, με κουράγιο.

courgette s. κολοκυθάκι n.

courier s. αγγελιαφόρος m. (*escort*) συνοδός m.f.

course s. (*march, way, advance*) πορεία f., δρόμος m. (*direction*) κατεύθυνση f. (*of study*) κουρ n. (*of lectures*) κύκλος m. (*of treatment*) θεραπεία f. (*of injections*) σειρά f. (*at meal*) πιάτο n., (*archit.*) στοίχος m. of ~ φυσικά, βεβαίως. as a matter of ~ φυσικώ τω λόγω, είναι αυτονόητο ότι. in due ~ εν ευθέτω χρόνω. in the ~ of time με τον καιρό, με την πάροδο του χρόνου. in the ~ of the week εντός της εβδομάδος, κατά το διάστημα της εβδομάδος. in the ~ of conversation κατά τη διάρκεια της συνομιλίας. in the ordinary ~ of events κανονικώς. the illness will run its ~ η αρρώστια θα κάνει τον κύκλο της. justice will take its ~ η δικαιοσύνη θα ακολουθήσει το δρόμο της. be off one's ~ παρεκκλίνω της πορείας μου. in ~ of construction υπό κατασκευήν. what ~ (of action) shall we take? πώς θα ενεργήσουμε; or τι μέτρα θα λάβουμε; the right ~ ο σωστός δρόμος.

course v.t. κυνηγώ. (v.i.) τρέχω.

court s. (*yard, King's*) αυλή f. (*law*) δικαστήριον n. criminal ~ κακουργιοδικείον n. ~ of appeal εφετείον n. (*tennis*) γήπεδο n. ~ shoe γόβα f.

court v. (*lit. & fig.*) ερωτοτροπώ με. (*favours, etc.*) επιδιώκω. ~ disaster πάω γυρεύοντας την καταστροφή μου. ~ing couple ερωτικό ζευγάρι. (*s.*) pay ~ to επιδιώκω την εύνοια (*with gen.*). ~ship s. ερωτοτροπία f.

courteous a. ευγενικός. ~ly adv. ευγενικά.

courtesan s. εταίρα f.

courtesy s. ευγένεια f. ~ies φιλοφρονήματα

n.pl.

courtier s. αυλικός m.

courtly a. με αριστοκρατικό ύφος.

court-martial s. στρατοδικείον n. be ~led δικάζομαι από στρατοδικείον.

courtyard s. αυλή f.

cousin s. εξάδελφος, ξάδερφος m. εξαδέλφη, ξαδέρφη f. ~s ξαδέρφια n.pl.

cove s. λιμανάκι n.

covenant s. συμφωνητικόν n. (v.) συμβάλλομαι.

cover v. σκεπάζω. (*protect, conceal, suffice for, deal with*) καλύπτω. (*aim gun at*) σημαδεύω. (*travel a distance*) διανύω, κάνω. ~ with (*fill*) γεμίζω, ~ed with γεμάτος (από). ~ up (*conceal*) συγκαλύπτω.

cover s. σκέπασμα n., κάλυμμα n. (*shelter, protection*) κάλυψη f. (*refuge*) καταφύγιο n. (*lid*) καπάκι n. (*of book*) εξώφυλλο n. take ~ καλύπτομαι, κρύβομαι, under ~ of (*protection*) καλυπτόμενος από, (*pretence*) υπό το πρόσχημα (*with gen.*). ~ charge κουβέρ n.

coverage s. κάλυψη f. have a wide ~ καλύπτω πολλά.

covering s. κάλυμμα n. (a.) ~ fire προστατευτικά πυρά.

coverlet s. κουβρ-λί n.

covert a. (*glance*) κρυφός, (*threat*) συγκεκαλυμμένος.

covet v. ζηλεύω, (*long for*) λιμπίζομαι. ~ed περιπόθητος.

covetous a. πλεονέκτης. ~ness s. πλεονεξία f.

cow s. αγελάδα f. (*female of species*) θηλυκός a.

cow v. εκφοβίζω.

coward s., ~ly a. δειλός, άνανδρος. ~ice s. δειλία f., ανανδρία f.

cower v. ζαρώνω από φόβο.

cowherd s. βουκόλος m.

cowl s. κουκούλα f.

coxcomb s. μορφονιός m.

coxswain s. πηδαλιούχος m.

coy a. ντροπαλός.

cozen v. (*inveigle*) τυλίγω. (*person out of money*) he ~ed me out of a tenner (*fam.*) με έβαλε στο χέρι και του'δωσα ένα χιλιάρικο.

crab s. καρκίνος m., καβούρι n.

crabbed a. στρυφνός.

crack s. (*in cup, dish*) ράγισμα n. (*in wall, etc.*) ρωγμή f., χαραμάδα f., σκάσιμο n. (*noise*) κρακ n., (*of whip*) στράκα f. (*joke*) εξυπνάδα f. he got a ~ on the head έφαγε μια στο κεφάλι.

crack v.t. & i. ραγίζω, (*nuts*) σπάζω. (v.t.)

(*head*) σπάζω, (*whip*) κροταλίζω. ~ jokes
λέω αστεία. (*v.i.*) (*of paint, skin*) σκάζω, (*of voice*) σπάζω. (*make noise*) κάνω κρακ. get ~ing! άντε μπρος! ~ up (*v.t.*) εκθειάζω, ρεκλαμάρω, (*v.i.*) καταρρέω. ~ down on χτυπώ, απαγορεύω.

crack *a.* (*expert*) άσσος *m.* a ~ shot άσσος στο σημάδι. (*choice*) επίλεκτος.

cracked *a.* ραγισμένος. (*fam., mad*) τρελλός, παλαβός.

crackle *v.* (*fire*) τριζοβολώ, (*guns*) κροταλίζω. (*s.*) τρίξιμο *n*, κροτάλισμα, *n.*

crackling *s.* ξεραμένη πέτσα ψητού γουρουνόπουλου.

crackpot *s. & a.* παλαβός.

cracksman *a.* διαρρήκτης *m.*

cradle *s.* κούνια *f.*, (*also fig.*) λίκνον *n.* (*v.*) κρατώ στην αγκαλιά μου. (*fig., put*) τοποθετώ.

craft *s.* (*art, skill*) τέχνη *f.* (*cunning*) πονηριά *f.* (*trade*) επάγγελμα *n.* (*guild*) σινάφι *n.* (*boat*) σκάφος *n.*, (*pl.*) σκάφη.

craftsman *s.* τεχνίτης *m.* ~**ship** *s.* τέχνη *f.*

craft|y *a.* πονηρός. ~**ily** *adv.* πονηρά.

crag *s.* βράχος *m.* ~**gy** *a.* βραχώδης, (*face*) χαραγμένος.

cram *v.t.* (*thrust*) χώνω, (*fill*) γεμίζω. ~ with food μπουκώνω, he ~med all the bread into his mouth μπούκωσε όλο το ψωμί. (*v.i.*) they ~med (themselves) into the bus στριμώχθηκαν μέσα στο λεωφορείο. (*v.i., prepare for exam*) προγυμνάζομαι.

cram-full *a.* φίσκα, τίγκα (*adv.*).

crammer's *s.* φροντιστήριο *n.*

cramp *s.* (*of muscles*) γράμπα *f.* (*v.*) εμποδίζω. ~**ed** *a.* (*confined*) στριμωγμένος. (*style*) στρυφνός.

crane *s.* (*bird, machine*) γερανός *m.* (*v.*) (*also* ~ one's neck) τεντώνω το λαιμό μου.

crank *s.* (*handle*) μανιβέλα *f.* (*v.*) βάζω μπρος.

crank *s.* (*eccentricity*) μανία *f.*, βίδα *f.* (*person*) λόξας *m.*, εκκεντρικός *a.* ~**y** *a.* be ~y έχω βίδα, είμαι λοξός.

cranny *s.* χαραμάδα *f.*, άνοιγμα *n.*

crape *s.* κρεπ *n.*

crapulous *s.* (*person*) μέθυσος, μπεκρής.

crash *v.i.* (*collapse*) καταρρέω, γκρεμίζομαι, (*fall*) πέφτω. (*of business*) φαλίρω, (*of plane*) συντρίβομαι. ~ into πέφτω απάνω σε, (*of car*) τρακάρω με. (*v.t.*) (*one's fist, stick, etc.*) βροντώ, χτυπώ, (*a car*) τρακάρω.

crash *s.* (*noise*) βρόντος *m.*, πάταγος *m.* (*fin.*) κραχ *n.* (*of plane*) πτώση *f.* (*collision*) σύγκρουση *f.*, τρακάρισμα *n.* ~ programme εντατικό πρόγραμμα.

crass *a.* (*person*) θεόκουτος. (*complete*)

πλήρης, απύθμενος.

crate *s.* κασόνι *n.* (*for fruit*) καφάσι *n.*

crater *s.* κρατήρας *m.*

cravat *s.* γραβάτα *f.*

crav|e *v.* (*beg for*) he ~ed my forgiveness με εκλιπάρησε να τον συγχωρήσω. ~e for (*desire*) λαχταρώ. ~**ing** *s.* λαχτάρα *f.*

craven *a.* δειλός.

crawl *v.* σέρνομαι, έρπω, (*of infant*) μπουσουλώ, (*of vehicle*) προχωρώ αργά, (*abase oneself*) προσκυνώ. (*fam.*) ~ing with γεμάτος από. (*s.*) (*swimming*) κρόουλ *n.*

crayfish *s.* καραβίδα *f.*

crayon *s.* κραγιόνι *n.*

craz|e *v.* τρελλαίνω. (*s.*) μανία *f.*, λόξα *f.* ~**y** *a.* τρελλός. be ~y about τρελλαίνομαι για, έχω μανία με.

creak *v.* τρίζω. (*s.*) τρίξιμο *n.* ~**y** *a.* ~y stairs σκάλες που τρίζουν.

cream *s.* κρέμα *f.* (*of milk & fig.*) αφρόγαλα *n.* (*thick*) καϊμάκι *n.* (*whipped*) σαντιγύ *f.* (*a.*) (*colour*) κρεμ. (*v.t.*) ~ (off) παίρνω την αφρόκρεμα.

creamy *a.* (*rich*) παχύς, (*complexion*) βελουδένιος, (*voice*) απαλός.

crease *v.* τσαλακώνω, ζαρώνω. (*s.*) τσαλάκωμα *n.*, ζαρωματιά *f.* (*of trousers*) τσάκιση *f.*, κώχη *f.*

creat|e *v.* δημιουργώ, (*world, man*) πλάθω. (*cause*) προκαλώ. (*give rank to*) κάνω. ~**ive** *a.* δημιουργικός.

creation *s.* (*act*) δημιουργία *f.* (*thing*) δημιούργημα *n.* (*the Universe*) πλάση *f.*

creator *s.* δημιουργός *m.* (*God*) Πλάστης *m.*

creature *s.* πλάσμα *n.* (*being*) ον *n.* (*animal*) ζώο *n.* (*tool*) ενεργούμενο *n.*, τσιράκι *n.*

crèche *s.* βρεφικός σταθμός.

credence *s.* πίστη *f.* give ~ to δίδω πίστιν εις. letters of ~ διαπιστευτήρια *n.pl.*

credentials *s.* (*paper*) συστατική επιστολή. have good ~ χαίρω καλής φήμης.

credib|le *a.* πιστευτός. ~**ility** *s.* αξιοπιστία *f.* ~**ly** *adv.* be ~ly informed πληροφορούμαι από αξιόπιστη πηγή.

credit *s.* (*credence*) πίστη *f.* (*good name*) όνομα *n.* (*honour*) τιμή *f.* give ~ for (*approve*) αναγνωρίζω. he's cleverer than I gave him ~ for είναι πιο έξυπνος απ' ό,τι τον θεωρούσα. it does you ~ σε τιμά. he did me ~ με έβγαλε ασπροπρόσωπο. he emerged without loss of ~ βγήκε ασπροπρόσωπος. he acquitted himself with ~ τα 'βγαλε πέρα καλά. (*fin.*) (*advance*) πίστωση *f.* (*balance*) ενεργητικό *n.* (*solvency*) φερεγγυότητα *f.* his ~ is good είναι φερέγγυος. on ~ επί πιστώσει. to his ~ (*morally*) προς τιμήν του, (*of his doing*) εις το ενεργητικόν του. take the ~

for it αποκομίζω τον έπαινο.
credit *a.* (*also* of ~) (*fin.*) πιστωτικός. ~ card πιστωτική κάρτα.
credit *v.* (*believe*) πιστεύω. I did not ~ him with so much sense δεν τον θεωρούσα τόσο λογικό. he is ~ed with great eloquence του αποδίδεται μεγάλη ευγλωττία. (*fin.*) πιστώνω.
creditab|le *a.* αξιέπαινος. **~ly** *adv.* καλά.
creditor *s.* πιστωτής *m.*
credo *s.* my ~ το πιστεύω μου.
credul|ous *a.* εύπιστος. **~ity** *s.* ευπιστία *f.*
creed *s.* (*eccl.*) σύμβολον της πίστεως, το πιστεύω. (*credo*) πιστεύω *n.* (*religion*) θρήσκευμα *n.*
creek *s.* κολπίσκος *m.*
creep *v.* έρπω, σέρνομαι. ~ in (*to room*) μπαίνω συρτά, (*to hole*) χώνομαι, (*penetrate*) εισχωρώ. ~ away *or* out φεύγω αθόρυβα. ~ over (*possess*) καταλαμβάνω. ~ up (*rise*) ανεβαίνω σιγά σιγά, (*approach*) πλησιάζω αθόρυβα.
creeper *s.* αναρριχητικό φυτό.
creep|s *s.* (*fam.*) have the ~s ανατριχιάζω. **~y** *a.* ανατριχιαστικός.
cremat|e *v.* καίω. **~ion** *s.* αποτέφρωση *f.*
crêpe *s.* κρεπ *n.*
crescent *s.* ημισέληνος *f.*, μισοφέγγαρο *n.*
cress *s.* κάρδαμο *n.*
crest *s.* (*bird's*) λοφιά *f.* (*of helmet*) λόφος *m.* (*heraldic*) επίσημον θυρεού. (*of hill*) οφρύς *f.* (*of wave*) κορυφή *f.* **~-fallen** *a.* με πεσμένα φτερά.
cretinous *a.* (*fam.*) ηλίθιος.
cretonne *s.* κρετόν *n.*
crevasse *s.* ρωγμή παγετώνος.
crevice *s.* ρωγμή *f.*
crew *s.* πλήρωμα *n.* (*pej., lot, gang*) παλιοπαρέα *f.*
crib *s.* (*manger*) φάτνη *f.* (*cradle*) λίκνο *n.* (*cabin*) καλύβα *f.* (*v.*) (*confine*) περιορίζω. (*copy*) αντιγράφω.
crick *s.* I've got a ~ in my neck (*or* back) πιάστηκε ο λαιμός μου (*or* η μέση μου).
cricket *s.* γρύλλος *m.*
crier *s.* ντελάλης *m.*
crime *s.* έγκλημα *n.* (*commission of* ~s) εγκληματικότητα *f.*
criminal *a.* (*act*) εγκληματικός. (*of crime*) ποινικός. (*s.*) εγκληματίας *m.*
crimp *v.* (*hair*) κατσαρώνω, (*cloth*) πλισάρω.
crimson *a.* βαθυέρυθρος, (*cardinal's*) πορφυρούς. (*fam.*) κόκκινος.
cringe *v.* ζαρώνω, (*fig.*) έρπω. he did not ~ δεν φοβήθηκε.
crinkle *v.t. & i.* ζαρώνω. (*v.t.*) τσαλακώνω.
crinoline *s.* κρινολίνο *n.*

cripple *v.* καθιστώ ανάπηρον, σακατεύω. (*also fig.*) παραλύω. (*immobilize*) ακινητοποιώ.
cripple *s.*, **~d** *a.* ανάπηρος.
crisis *s.* κρίση *f.*
crisp *a.* τραγανός, ξεροψημένος, (*hair*) σγουρός. (*air*) ξερός και δροσερός. (*manner*) κοφτός, ξερός. **~s** *s.* (*potato*) τσιπς *n.pl.*
criss-cross *a. & adv.* με διασταυρούμενες γραμμές.
criterion *s.* κριτήριο *n.*
critic *s.* (*judge*) κριτής *m.* (*reviewer*) κριτικός *m.* art ~ τεχνοκρίτης *m.*, music ~ μουσικοκριτικός *m.*
critical *a.* (*of criticism*) κριτικός. ~ mind κριτικό πνεύμα. (*demanding*) απαιτητικός. (*fault-finding*) επικριτικός, αυστηρός. (*crucial*) κρίσιμος. **~ly** *adv.* αυστηρά, κρισίμως.
criticism *s.* κριτική *f.*
criticize *v.* (*judge*) κρίνω. (*find fault with*) επικρίνω, κριτικάρω.
critique *s.* (*review*) κριτικό άρθρο.
croak *v.* (*frog*) κοάζω, (*bird*) κρώζω. (*fig.*) γκρινιάζω. (*s.*) κοασμός *m.* κρώξιμο *n.* **~y** *a.* βραχνός.
crock *s.* (*jar*) στάμνα *f.* (*for cooking*) τσουκάλι *n.* (*fam., ruin*) σαράβαλο *n.* **~ery** *s.* πιατικά *n.pl.*
crocodile *s.* κροκόδειλος *m.*
crocus *s.* κρόκος *m.*
Croesus *m.* Κροίσος *m.*
crofter *s.* μικροκαλλιεργητής *m.*
crone *s.* μπαμπόγρια *f.*
crony *s.* παλιόφιλος *m.* a ~ of mine ένας από την παρέα μου.
crook *s.* (*bishop's*) ράβδος *f.* (*shepherd's*) γκλίτσα *f.* (*bend*) καμπή *f.* (*swindler*) απατεώνας *m.*, (*fam.*) μούτρο *n.* (*v.*) κάμπτω.
crooked *a.* (*bent*) στραβός, ~ street σκολιά οδός. (*dishonest*) ανέντιμος, (*act*) κατεργάρικος.
croon *v.* σιγοτραγουδώ.
crop *s.* σοδειά *f.*, (*yield*) παραγωγή *f.* **~s** (*sown*) σπαρτά *n.pl.*, (*harvested*) σοδειά *f.*, συγκομιδή *f.* (*of rumours, etc.*) σωρός *m.* (*bird's*) γούλα *f.* (*haircut*) κούρεμα *n.*
crop *v.t.* (*graze on*) βόσκω. (*sow*) σπέρνω. (*cut off*) κόβω, (*hair*) κουρεύω. (*v.i.*) our apples ~ped well είχαμε καλή παραγωγή από μήλα. ~ up προκύπτω.
cropper *s.* (*fam.*) come a ~ (*fall*) κουτρουβαλιάζομαι, (*fail*) σπάζω τα μούτρα μου.
croquette *s.* κροκέτα *f.*
cross *s.* σταυρός *m.* make sign of ~ κάνω το

σταυρό μου. (*hybrid*) διασταύρωση *f.* (*cut*) on the ~ λοξά.
cross *s.* (*transversal*) διαγώνιος. (*angry*) θυμωμένος, be ~ θυμώνω. we are at ~ purposes έχουμε παρεξηγηθεί. **~ly** *adv.* θυμωμένα.
cross *v.t.* (*legs, etc.*) σταυρώνω, (*swords*) διασταυρώνω. (*go across*) (*desert, ocean*) διασχίζω, (*river, bridge*) διαβαίνω, περνώ. ~ the road περνώ απέναντι. it ~ed my mind πέρασε από το μυαλό μου, μου κατέβηκε. ~ oneself σταυροκοπιέμαι. our letters ~ed τα γράμματά μας διασταυρώθηκαν. ~ off *or* out διαγράφω. **~ed** *a.* σταυρωτός.
cross-examine *v.* εξετάζω κατ' αντιπαράστασιν.
cross-eyed *a.* αλλήθωρος.
cross-fire *s.* διασταυρούμενα πυρά.
cross-grained *a.* ~ person στραβόξυλο *n.*
crossing *s.* πέρασμα *n.*, διάβαση *f.* (*of ways*) διασταύρωση *f.*
cross-legged *adv.* σταυροπόδι.
cross-question *v. see* cross-examine.
cross-reference *s.* παραπομπή *f.*
crossroads *s.* σταυροδρόμι *n.* (*fig.*) αποφασιστικό σημείο.
cross-section *s.* εγκαρσία τομή, (*fig.*) αντιπροσωπευτικό δείγμα.
crosswise *adv.* σταυρωτά.
crossword *s.* σταυρόλεξο *n.*
crotch *s.* (*of trousers*) καβάλλος *m.*
crotchet *s.* (*mus.*) τέταρτο *n.* (*fad*) λόξα *f.* **~y** *a.* ιδιότροπος.
crouch *v.* (*huddle*) ζαρώνω, (*squat*) κάθομαι ανακούρκουδα, (*for spring*) συσπειρούμαι.
croupier *s.* κρουπιέρης *m.*
crow *v.* λαλώ. (*fig.*) ~ over one's success θριαμβολογώ για την επιτυχία μου.
crow *s.* κουρούνα *f.* (*fig.*) as the ~ flies σε ευθεία γραμμή.
crowbar *s.* λοστός *m.*
crowd *s.* πλήθος *n.* great ~s κοσμοσυρροή *f.* (*fam., set*) κύκλος *m.*
crowd *v.i.* (*collect*) συναθροίζομαι, (*into*) στριμώχνομαι (μέσα σε). (*v.t.*) (*fill*) γεμίζω, (*thrust*) χώνω. get ~ed out μένω απ' έξω. **~ed** *a.* γεμάτος.
crown *s.* (*king's*) στέμμα *n.* (*martyr's, victor's*) στέφανος *m.* (*of tooth, coin*) κορόνα *f.* (*of hill, head*) κορυφή *f.*
crown *v.* (*king*) στέφω, (*victor*) στεφανώνω (*pass.* στέφομαι). be ~ed with success στέφομαι υπό επιτυχίας. to ~ all κοντά σε όλα τ' άλλα. (*the hill*) is ~ed with a chapel έχει εκκλησάκι στην κορυφή του.
crowned *a.* εστεμμένος.
crowning *s.* στέψη *f.* (*a.*) ~ touch κορύφωμα

n. the ~ folly το κορύφωμα της βλακείας.
crozier *s.* ποιμαντορική ράβδος.
crucial *a.* κρίσιμος.
crucible *s.* χωνευτήριο *n.*
crucifix *s.* Εσταυρωμένος *m.* **~ion** *s.* σταύρωση *f.*
cruciform *a.* σταυροειδής.
crucify *v.* σταυρώνω.
crude *a.* (*material*) ακατέργαστος, ~ oil ακάθαρτο (*or* αργό) πετρέλαιο. (*primitive*) πρωτόγονος, (*behaviour*) χοντρός, (*workman, work*) άτεχνος, κακότεχνος. ~ bit of work χοντροκοπιά *f.* **~ly** *adv.* χοντρά, άτεχνα, κακοτέχνως.
crud|eness, ~ity *s.* χοντράδα *f.* κακοτεχνία *f.*
cruel *a.* σκληρός, άσπλαχνος. **~ly** *adv.* σκληρά. **~ty** *s.* σκληρότητα *f.*, ασπλαχνία *f.*
cruet *s.* λάδι και ξύδι.
cruise κρουαζιέρα *f.* (*v.*) (*go on* ~) κάνω κρουαζιέρα. (*of vehicle*) πηγαίνω με μέτρια ταχύτητα, (*patrol*) περιπολώ.
cruiser *s.* (*naut.*) καταδρομικόν *n.*
crumb *s.* ψίχουλο *n.* (*soft part of loaf*) ψίχα *f.*
crumble *v.t.* θρυμματίζω, τρίβω. (*v.i.*) θρυμματίζομαι, τρίβομαι, (*of power, buildings*) καταρρέω. ~ into dust γίνομαι σκόνη, κονιορτοποιούμαι.
crumbly *a.* εύθρυπτος.
crumple *v.t. & i.* (*crease*) ζαρώνω, τσαλακώνω. (*fall down*) σωριάζομαι, (*of resistance*) διαλύομαι. (*get squashed*) συνθλίβομαι, γίνομαι φυσαρμόνικα (*or* πήτα).
crunch *v.* τραγανίζω. (*s.*) (*fam.*) when it comes στην κρίσιμη στιγμή.
crusade *s.* σταυροφορία *f.* (*v.*) (*fig.*) κάνω σταυροφορία. **~r** *s.* σταυροφόρος *m.*
crush *v.t.* συνθλίβω, ζουλώ, (*also fig.*) συντρίβω. (*cram*) στριμώχνω. (*crumple*) τσαλακώνω. (*v.i.*) (*get crumpled*) τσαλακώνομαι. ~ into (*room, etc.*) στριμώχνομαι μέσα σε.
crush *s.* (*crowd*) συνωστισμός *m.* (*fam.*) have a ~ on είμαι τσιμπημένος με.
crushing *a.* συντριπτικός.
crust *s.* (*of loaf*) κόρα *f.* I prefer a ~ προτιμώ μία γωνιά. (*of pie*) κρούστα *f.* (*on liquids*) τσίπα *f.* (*earth's*) φλοιός *m.* (*fig.*) not even a ~ of bread ούτε ένα ξεροκόμματο.
crustacean *a.* οστρακόδερμος.
crusty *a.* ξεροψημένος, (*person*) απότομος.
crutch *s.* δεκανίκι *n.*, πατερίτσα *f.*
crux *s.* this is the ~ of the problem εδώ είναι ο κόμπος.
cry *v.* κραυγή *f.*, φωνή *f.* (*watchword*) σύνθημα *n.* (*tears*) have a ~ κλαίω. be a far ~ from απέχω πολύ από. he fled with the police in full ~ το 'σκασε με την αστυνομία

κατάποδι.

cry *v.i.* (*also* ~ out) κραυγάζω, φωνάζω. (*weep*) κλαίω. (*v.t.*) (*wares*) διαλαλώ, (*news*) αναγγέλλω. ~ down εκφράζομαι υποτιμητικά για. ~ up διαφημίζω. ~ off (*v.i.*) αποσύρομαι.

crying *s.* κλάμα *n.*, κλάματα *n.pl.*, κλάψιμο *n.* start ~ βάζω τα κλάματα. (*a.*) (*need*) κατεπείγων.

crypt *s.* κρύπτη *f.*

cryptic *a.* (*obscure*) δυσνόητος, (*enigmatic*) αινιγματικός.

crystal *s.* (*chem.*) κρύσταλλος *m.f.* (*glass*) κρύσταλλο *n.*

crystal *a.* κρυσταλλένιος. **~line** *a.* κρυστάλλινος. **~lize** *v.i.* αποκρυσταλλούμαι.

cub *s.* μικρό *n.*, νεογνό *n.* (*unmannerly youth*) γαϊδούρι *n.*

cubby-hole *s.* τρύπα *f.*

cub|e *s.* κύβος *m.* **~ic** *a.* κυβικός.

cubicle *s.* χώρισμα *n.*

cub|ism *s.* κυβισμός *m.* **~ist** *s.* καλλιτέχνης ασχολούμενος με κυβισμό.

cuckold *s.* κερατάς *m.* (*v.*) κερατώνω.

cuckoo *s.* κούκος *m.*

cucumber *s.* αγγούρι *n.*

cud *s.* chew the ~ μηρυκάζω, (*fig.*) συλλογίζομαι.

cuddl|e *v.t.* αγκαλιάζω. (*v.i.*) ~e up to κουλουριάζομαι στην αγκαλιά (*with gen.*). **~y** *a.* που προκαλεί το χάδι.

cudgel *s.* ρόπαλο *n.* (*fig.*) take up the ~s ξεσπαθώνω (*with* υπέρ & *gen.*). (*v.*) ξυλοκοπώ. ~ one's brains σπάω το κεφάλι μου.

cue *s.* (*billiards*) στέκα *f.* (*signal*) σύνθημα *n.* take one's ~ from ακολουθώ.

cuff *s.* (*blow*) μπάτσος *m.* (*v.*) μπατσίζω.

cuff *s.* (*of sleeve*) μανικέτι *n.* (*fig.*) off the ~ εκ του προχείρου. **~-links** *s.* μανικετόκουμπα *n.pl.*

cul-de-sac *s.* αδιέξοδο *n.*

culinary *a.* μαγειρικός.

cull *v.* (*pick*) δρέπω, (*choose*) επιλέγω.

culminat|e *v.* κορυφούμαι. **~ion** *s.* κορύφωμα *n.*

culpab|le *a.* ένοχος. **~ility** *s.* ενοχή *f.*

culprit *s.* ένοχος *a.*

cult *s.* λατρεία *f.* (*fad*) the newest ~ ο τελευταίος συρμός.

cultivable *a.* καλλιεργήσιμος.

cultivat|e *v.* καλλιεργώ. **~ed** *a.* καλλιεργημένος. **~ion** *s.* καλλιέργεια *f.* **~or** *s.* καλλιεργητής *m.*

cultur|e *s.* πολιτισμός *m.*, κουλτούρα *f.* (*growing*) καλλιέργεια *f.* **~al** *a.* πολιτιστικός, μορφωτικός. ~al ties πνευματικοί δεσμοί. **~ed** *a.* πολιτισμένος, καλλιερ-

γημένος.

culvert *s.* οχετός *m.* (*conduit*) αγωγός *m.*

cumb|ersome, **~rous** *a.* βαρύς, άβολος.

cumulative *a.* σωρευτικός.

cuneiform *a.* σφηνοειδής.

cunning *a.* πονηρός, πανούργος. (*s.*) πονηρία *f.*, κατεργαριά *f.*

cup *v.* (*bleed*) βάζω βεντούζες σε. ~ one's hands to drink water πίνω νερό με τη χούφτα μου.

cup *s.* φλιτζάνι *n.* (*trophy, chalice*) κύπελλο *n.* (*of wine, sorrow*) ποτήριον *n.* **~-bearer** *s.* οινοχόος *m.* **~ful** *s.* φλιτζάνι *n.*

cupboard *s.* ντουλάπι *n.* (*fig.*) ~ love αγάπες από συμφέρον.

cupidity *s.* πλεονεξία *f.*

cupola *s.* τρούλλος *m.*

cupping-glass *s.* βεντούζα *f.*

curate *s.* βοηθός εφημερίου.

curative *a.* θεραπευτικός. (*s.*) θεραπευτικό μέσο.

curator *s.* διευθυντής *m.*

curb *s.* χαλινός *m.* (*v.*) (*fig.*) χαλιναγωγώ, συγκρατώ.

curdle *v.t.* & *i.* πήζω. (*v.i.*) κόβω, (*fig., of blood*) παγώνω.

curds *s.* ξινόγαλα *n.*, κομμένο γάλα.

cure *v.* (*person, illness*) θεραπεύω, γιατρεύω, (*correct*) διορθώνω, (*preserve*) διατηρώ. (*s.*) (*remedy*) θεραπεία *f.*, γιατρειά *f.* (*treatment*) κούρα *f.*, θεραπευτική αγωγή *f.*

curfew *s.* (*mil.*) απαγόρευση της κυκλοφορίας.

curio *s.* περίεργο έργο τέχνης.

curiosity *s.* περιέργεια *f.* (*thing*) κάτι το αξιοπερίεργο.

curious *a.* περίεργος, (*strange only*) παράξενος. **~ly** *adv.* περίεργα. ~ly shaped με περίεργο σχήμα. ~ly enough περιέργως.

curl *s.* (*hair*) βόστρυχος *m.*, μπούκλα *f.*, κατσαρό *n.* (*smoke*) τολύπη *f.* (*bend*) στροφή *f.*

curl *v.t.* & *i.* (*hair*) κατσαρώνω. (*wind*) (*v.t.*) τυλίγω, (*v.i.*) τυλίγομαι, (*of river, road*) στρίβω. ~ one's lip στραβώνω το στόμα μου περιφρονητικά. ~ up (*v.i.*) κουλουριάζομαι, (*of paper, etc.*) στρίβω.

curling-tongs *s.* ψαλίδι *n.*

curly *a.* (*hair*) κατσαρός, σγουρός. (*winding*) ελικοειδής. **~haired** *a.* σγουρομάλλης.

curmudgeon *s.* (*miser*) σπαγγοραμμένος *a.* (*churl*) στρυφνός *a.*

currant *s.* κορινθιακή σταφίδα.

currency *s.* (*money*) νόμισμα *n.* foreign ~ συνάλλαγμα *n.* (*use*) χρήση *f.* give ~ to θέτω εις κυκλοφορία. gain ~ ξαπλώνω.

current *s.* ρεύμα *n.* (*fig., of events*) πορεία *f.*

current *a.* be ~ κυκλοφορώ. during the ~

month κατά τον τρέχοντα μήνα, κατά το μήνα που διατρέχουμε. ~ rate *(of exchange)* τρέχουσα τιμή. ~ events τα επίκαιρα. ~ issue *(of paper)* τελευταίο φύλλο. ~ account τρεχούμενος λογαριασμός.

currently *adv.* *(now)* σήμερον, τώρα.

curriculum *s.* πρόγραμμα μαθημάτων. ~ vitae βιογραφικό σημείωμα.

curry *v.* κάρρι *n.*

curry *v.* ~ favour with κολακεύω, καλοπιάνω.

curse *s.* κατάρα *f.* *(oath)* βλαστήμια *f.* *(v.t.)* καταριέμαι, αναθεματίζω. *(v.i., swear)* βλαστημώ, βρίζω. **~d** *a.* καταραμένος.

cursing *s.* βλαστήμιες *f.pl.,* βρισιές *f.pl.*

cursive *a.* συνεχής.

cursory *a.* βιαστικός. give a ~ glance to κοιτάζω στα πεταχτά.

curt *a.* απότομος. **~ly** *adv.* απότομα, κοφτά.

curtail *v.* *(journey)* συντομεύω, *(speech, expenses)* περικόπτω. **~ment** *s.* συντόμευση *f.* περικοπή *f.*

curtain *s.* κουρτίνα *f.* *(stage)* αυλαία *f.* *(fig.)* παραπέτασμα *n.* *(v.t.)* βάζω κουρτίνες σε. ~ off χωρίζω με κουρτίνα.

curtsey *s.* υπόκλιση *f.* *(v.)* υποκλίνομαι.

curve *s.* καμπύλη *f.* *(in road)* στροφή *f.,* καμπή *f.* *(v.i.)* κάνω καμπύλη. **~d** *a.* καμπύλος.

cushion *s.* μαξιλλάρι *n.* *(v, fig.)* προστατεύω.

cushy *a.* ~ job ραχατλίδικη δουλειά.

cussed *a.* ~ person στραβόξυλο *n.* **~ness** *s.* στραβοξυλιά *f.*

custard *s.* κρέμα *f.*

custodian *s.* φύλακας *m.* *(fig., of traditions, etc.)* θεματοφύλακας *m.*

custody *s.* *(keeping)* φύλαξη *f.* *(of children)* κηδεμονία *f.* take into ~ *(arrest)* συλλαμβάνω. detain in ~ προφυλακίζω.

custom *s.* *(tradition)* έθιμο *n.* *(habit)* συνήθεια *f.* *(patronage)* πελατεία *f.* ~-made *or* built επί παραγγελία.

customar|y *a.* συνήθης, συνηθισμένος. **~ily** *adv.* συνήθως.

customer *s.* πελάτης *m.* *(fam., person)* τύπος *m.*

custom-house *s.* τελωνείο *n.*

customs *s.* *(duties)* δασμοί *m.pl.* C~ *(service)* τελωνείον *n.* C~ official τελωνειακός υπάλληλος.

cut *s.* κόψιμο *n.* *(incision)* τομή *f.* *(excision)* περικοπή *f.* *(with whip)* καμτσικιά *f.* *(with sword)* σπαθιά *f.* ~ and thrust *(fig.)* διαξιφισμοί *m.pl.* short ~ *see* short. a ~ above ανώτερος *(with από or gen.).*

cut *a.* κομμένος. ~ and dried έτοιμος, στερεότυπος.

cut *v.t. & i.* κόβω. ~ in two κόβω στα δύο. it ~s easily *(of knife)* κόβει εύκολα, *(of meat)* κόβεται εύκολα, ~ one's teeth βγάζω δόντια. have one's hair ~ κόβω τα μαλλιά μου. *(lawn, hair)* κουρεύω. *(a hole, road, etc.)* ανοίγω. *(expenditure)* περικόπτω. *(of wind, cold)* ξουρίζω. *(offend)* θίγω, αγγίζω. *(not be present)* κάνω σκασιαρχείο. ~ and run το κόβω λάσπη. ~ short *see* curtail. ~ corners *(fig.)* κάνω τσαπατσούλικη δουλειά. ~ it fine προλαμβάνω μετά βίας. ~ a *(fine)* figure κάνω φιγούρα. it ~s both ways είναι δίκοπο μαχαίρι. he ~ me *(dead)* έκανε πως δεν με είδε. he doesn't ~ any ice with me *(impress)* δεν μου γεμίζει το μάτι. his views don't ~ much ice ο λόγος του δεν έχει πέραση.

cut across *v.t.* *(traverse)* διασχίζω, *(clash with)* συγκρούομαι με, *(transcend)* υπερβαίνω.

cut back *v.t.* *(prune)* κλαδεύω, *(reduce)* περικόπτω.

cut in *v.i.* *(interrupt)* διακόπτω, *(of car)* κόβω αντικανονικά.

cut off *v.t.* κόβω. *(intercept)* αποκόπτω, *(isolate)* απομονώνω, *(disinherit)* αποκληρώνω, *(interrupt, disconnect)* διακόπτω.

cut out *v.t.* *(with scissors)* κόβω, *(a rival)* εκτοπίζω, *(take out)* βγάζω, κόβω, *(stop)* ~ smoking κόβω το τσιγάρο. cut it out! κόφ' το! be ~ for είμαι κομμένος και ραμμένος για. he had his work ~ to find it είδε και έπαθε για να το βρεί. *(v.i., stop)* σταματώ.

cut up *v.t.* κόβω. *(in pieces)* τεμαχίζω, *(meat)* λιανίζω. *(v.i.)* *(of cloth)* κόβομαι. ~ rough θυμώνω. be ~ στενοχωριέμαι, πικραίνομαι.

cut up *a.* *(distressed)* στενοχωρημένος, πικραμένος, φαρμακωμένος.

cute *a.* *(sharp)* έξυπνος *(attractive)* χαριτωμένος, όμορφος. a ~ little kitten ένα γατάκι γλύκα.

cuticle *s.* πετσάκια των νυχιών.

cutlass *s.* ναυτικό σπαθί.

cutlery *s.* μαχαιροπίρουνα *n.pl.*

cutlet *s.* κοτολέτα *f.* *(lamb)* παϊδάκι *n.*

cutter *s.* *(naut.)* κότερο *n.*

cut-throat *s.* μαχαιροβγάλτης *m.* *(razor)* ξυράφι *n.*

cutting *s.* *(act)* κόψιμο *n.* *(from newspaper, cloth)* απόκομμα *n.* *(of plant)* ξεμασκαλίδι *n.* *(of railway)* όρυγμα *n.*

cutting *a.* *(remark)* δηκτικός, *(also wind)* τσουχτερός.

cuttlefish *s.* σουπιά *f.*

cyanide *s.* κυανιούχον άλας.

cybernetics s. κυβερνητική f.
cyclamen s. κυκλάμινο n.
cycl|e s. κύκλος m. (bicycle) ποδήλατο n.
(v.) κάνω ποδήλατο, (go by bike) πάω με
ποδήλατο. ~ic a. κυκλικός. ~ing s. ποδη-
λασία f. ~ist s. ποδηλάτης m.
cyclone s. κυκλώνας m.
cyclop|s s. κύκλωψ m. ~ean a. κυκλώπειος.
cyclostyle v. πολυγραφώ.
cygnet s. μικρός κύκνος.
cylind|er s. κύλινδρος m. ~rical a. κυλιν-
δρικός.
cymbal s. κύμβαλον n., τάσι n.
cynic s., ~al a. κυνικός. ~ism s. κυνισμός m.
cynosure s. επίκεντρον προσοχής.
cypress s. κυπαρίσσι n.
Cypriot a. κυπριακός. (person) Κύπριος
m., Κυπρία f.
Cyrillic a. κυριλλικός.
cyst s. κύστη f.
czar s. τσάρος m.
Czech a. τσεχικός. (person) Τσέχος.

D

dab v.t. (put) επιθέτω, βάζω. (one's eyes)
σκουπίζω. ~ with powder πουδράρω.
~ with paint επαλείφω με χρώμα. (s.) it
needs a ~ of paint θέλει ένα χέρι λαδομπογιά. there's a ~ of paint on my coat το
παλτό μου αλείφτηκε μπογιά.
dab s. (fam., expert) μάνα f.
dabble v.i. (in water) πλατσουρίζω. (fig.)
ανακατεύομαι (with με).
dactyl s. δάκτυλος m.
dad, ~dy s. μπαμπάς m.
dado s. πασαμέντο n.
daffodil s. κίτρινος νάρκισσος.
daft a. τρελλός. don't be ~ δεν είσαι με τα
καλά σου.
dagger s. εγχειρίδιο n. look ~s αγριοκοιτάζω. at ~s drawn στα μαχαίρια.
dago s. (pej.) νοτιοαμερικανός m.
dahlia s. ντάλια f.
daily s. καθημερινός, ημερήσιος. ~ bread
επιούσιος άρτος (s.) (fam.) (paper) ημερήσια εφημερίδα. (woman) παραδουλεύτρα f. (adv.) καθημερινός, ημερησίως.
daint|y a. (fine, delicate) κομψός, λεπτός,
λεπτοκαμωμένος. (fastidious) δύσκολος.
(tasty) νόστιμος. ~ies λιχουδιές f.pl. ~ily
adv. κομψά, όμορφα. (carefully) με
προσοχή.
dairy s. γαλακτοκομείο n. (shop) γαλακτο-
πωλείο n. ~ produce προϊόντα γάλακτος
n.pl. ~-farming s. γαλακτοκομία f.

dais s. εξέδρα f.
daisy s. μαργαρίτα f.
dale s. κοιλάδα f.
dall|iance s. (delay) χασομέρι n. (trifling)
ερωτοτροπίες f.pl. ~y v.i. (delay) χασομερώ. (trifle) ερωτοτροπώ.
dam v. φράσσω. (fig.) θέτω φραγμόν εις.
(s.) φράγμα n.
dam s. (animal's) μάνα f.
damage s. ζημιά f. (items of ~) ζημιές f.pl.
(to ship, cargo) αβαρία f. ~s (law)
αποζημίωση f. (v.) προξενώ ζημιά σέ. get
~d παθαίνω ζημιά. (a.) ~d χαλασμένος.
damask s. δαμάσκο n.
dame s. (fam.) κυρά f. old ~ γριά f.
damn v.t. καταδικάζω. (critically) επικρίνω.
(curse) διαολοστέλνω. ~able a. απαίσιος.
~ation s. καταδίκη f. (int.) see damn (int.).
~ed a. καταδικασμένος, (in hell) κολασμένος.
damn int. να πάρει ο διάολος!
damn a. & s. in this ~(ed) weather μ' αυτό
το βρωμόκαιρο (or παλιόκαιρο). I don't
care a ~ σκοτίστηκα, δεν δίνω πεντάρα.
damp s. (also ~ness) υγρασία f. (a.) υγρός,
νοτισμένος become ~ (of wall, etc.) ποτίζω.
damp v. βρέχω, υγραίνω. (fire, sound)
μετριάζω την ένταση (with gen.). (fig.)
~ his enthusiasm μαραίνω (or βάζω πάγο
σ') τον ενθουσιασμό μου.
dampen v. see damp v.
damper s. (mus.) σουρντίνα f. (of chimney)
καπνοσύρτης m. (on spirits) ψυχρολουσία f.
damsel s. νεαρά κόρη.
damson s. είδος δαμάσκηνου.
danc|e s. (also ~ing) χορός m. she led me a
~e μου έβγαλε την ψυχή ανάποδα. (a.)
χορευτικός.
dance v. χορεύω. ~ attendance on ακολουθώ
υποτακτικά. ~r s. χορευτής m. χορεύτρια f.
dandelion s. αγριοραδίκι n.
dandle v. χορεύω.
dandruff s. πιτυρίδα f.
dand|y s. δανδής m. ~ified a. κομψευ-
όμενος.
danger s. κίνδυνος m. be in ~ (of) κινδυνεύω (να). ~ous a. επικίνδυνος. ~ously
adv. επικινδύνως.
dangle v.t. κρεμώ. ~ one's legs αφήνω τα
πόδια μου να κρέμονται. ~ prospects
before δελεάζω. (v.i.) κρέμομαι. ~ after
(person) περιτριγυρίζω, (also thing)
κυνηγώ.
dank a. υγρός.
dapper a. ~ little man καλοβαλμένο ανθρωπάκι.
dappled a. παρδαλός, με βούλλες.

dare *v.t.* (*challenge*) προκαλώ. (*v.i.*) τολμώ, κοτώ. (*fam.*) if you ~ αν σου κοτάει, αν σου βαστάει. I ~ say you are right (*without reservation*) πιθανώς έχεις δίκιο *or* μπορεί να έχεις δίκιο. he had his merits, I ~ say, but... δε λέω, ήταν άνθρωπος αξίας, μα.

daredevil *s.* απόκοτος *a.*

daring *s.* τόλμη *f.* αποκοτιά *f.* (*a.*) τολμηρός, απόκοτος.

dark *a.* (*day, room, thoughts*) σκοτεινός. (*colour, shade*) σκούρος. (*complexion*) μελαχροινός, (*swarthy*) μελαψός. make or get ~ σκοτεινιάζω, σκουραίνω. it was ~ (*not daylight*) ήταν σκοτεινά. things are looking ~ η κατάστασις είναι ζοφερή. keep it ~ το κρατώ μυστικό. ~ green (*a.*) βαθυπράσινος.

dark, ~ness *s.* σκότος *n.*, σκοτάδι *n.*, σκοτεινιά *f.* ~ colour σκούρο χρώμα, be in the ~ βρίσκομαι στο σκοτάδι. before ~ πριν να σκοτεινιάσει (*or* να βραδιάσει).

darken *v.t. & i.* (*in colour*) κάνω (*or* γίνομαι) πιο σκούρο, σκουραίνω. (*complexion*) μαυρίζω. (*room, sky, countenance, situation*) σκοτεινιάζω.

darkly *adv.* σκοτεινά, (*with threats*) απειλητικά.

darling *a.* αγαπημένος, χαριτωμένος. (*as endearment*) αγάπη μου.

darn *v.* καρικώνω, μαντάρω. (*s.*) (*also* ~ing) καρίκωμα *n.*, μαντάρισμα *n.*

dart *s.* βέλος *n.* (*in dressmaking*) πένσα *f.*

dart *v.t.* ρίχνω (*v.i.*) (*run*) τρέχω. ~ up *or* out ξεπετιέμαι.

dash *v.t.* (*throw*) ρίχνω, πετώ, (*strike*) χτυπώ. (*hopes, etc.*). συντρίβω. ~ (*v.i., rush*) ορμώ. (*int.*) ~! να πάρη η οργή.

dash *s.* (*rush*) ορμή *f.* (*energy*) ζωντάνια *f.* μπρίο *n.* cut a ~ κάνω φιγούρα. (*in writing*) παύλα *f.* (*fam.*) (*small amount*) μια ιδέα.

dashboard *s.* ταμπλό *n.*

dashing *a.* (*outfit*) φιγουράτος, εντυπωσιακός. ~ young man παλικάρι *n.*

dastardly *a.* απαίσιος.

data *s.* δεδομένα *n.pl.*, στοιχεία *n.pl.* ~ bank βάση δεδομένων.

date *s.* (*fruit*) χουρμάς *m.*

date *s.* (*chronological*) χρονολογία *f.* (*day of month*) ημερομηνία *f.* what ~ is it today? πόσες του μηνός έχουμε σήμερα; (*time, period*) εποχή *f.* (*appointment*) ραντεβού *n.* to ~ μέχρι σήμερον. out-of-~ (*timetable, etc.*) μη ενημερωμένος, (*idea, etc.*) ξεπερασμένος. up-to-~ ενημερωμένος. bring up-to-~ ενημερώνω.

date *v.t.* χρονολογώ. (*v.i.*) ~ back to *or* from

χρονολογούμαι από. be ~ed φαίνομαι απαρχαιωμένος.

dative *s.* δοτική *f.*

daub *v.* πασαλείβω.

daughter *s.* κόρη *f.* ~-in-law *s.* νύφη *f.*

daunt *v.* αποθαρρύνω. (*scare*) εκφοβίζω. ~ing *a.* δύσκολος. ~less *a.* ατρόμητος.

davit *s.* επωτίς *f.*, καπόνι *n.*

dawdl|e *v.* χασομερώ. ~er *s.* χασομέρης *a.* ~ing *s.* χασομέρι *n.*

dawn *s.* αυγή *f.*, ξημέρωμα *n.*, χάραμα *n.*, χαράματα *n.pl.* (*fig.*) αρχή *f.* (*v.*) day ~s χαράζει, ξημερώνει (*fig., begin*) αρχίζω. it ~s on me that καταλαβαίνω ότι...

day *s.* ημέρα *f.*, μέρα *f.* ~ of the month ημερομηνία *f.*, in the old ~s τον παλιό καιρό. in those ~s εκείνη την εποχή, win the ~ νικώ, κερδίζω. my ~ off η ελεύθερη μέρα μου. let's call it a ~ ας μείνουμε εδώ. it's had its ~ έφαγε τα ψωμιά του. he was before my ~ δεν τον έφτασα. all ~ long ολημερίς. ~ and night νυχθημερόν, μέρανύχτα. per ~ *or* by the ~ την ημέρα, ημερησίως. this ~ week σήμερα οκτώ. the other ~ προχθές, τις προάλλες, προ ημερών. every other ~ μέρα παρά μέρα. ~ after ~ κάθε μέρα. ~ by ~ *or* any ~ now μέρα με τη μέρα. from ~ to ~ από τη μία μέρα στην άλλη. a ~-to-~ existence μεροδούλι μεροφάι. (on) the ~ before την προηγουμένη. the ~ before the wedding την παραμονή του γάμου. the ~ before yesterday προχθές. the ~ after την επομένη, the ~ after tomorrow μεθαύριο. ~ tripper εκδρομεύς *m.*

daybreak *s.* *see* dawn.

daydream *v.* ονειροπολώ.

daylight *s.* φως της ημέρας. (*fig.*) see ~ αρχίζω να καταλαβαίνω.

daytime *s.* in the ~ την ημέρα, όσο διαρκεί το φως της ημέρας.

daze *v.* ζαλίζω, αποβλακώνω.

dazzl|e *v.* θαμπώνω. ~ed *a.* (*enraptured*) έκθαμβος. ~ing *a.* (*lit. & fig.*) εκθαμβωτικός, (*only of light*) εκτυφλωτικός.

deacon *s.* διάκ(ον)ος *m.*

dead *a.* νεκρός, πεθαμένος. (*animal*) ψόφιος. (*plant*) ξερός. (*silence*) νεκρικός. (*inactive*) νεκρός. (*numb*) μουδιασμένος. (*muffled*) υπόκωφος, πνιγμένος. fall ~ πέρτω ξερός. go ~ (*numb*) μουδιάζω, (*not work*) δεν λειτουργώ. ~ end αδιέξοδο *n.* ~ heat ισοπαλία *f.* ~ loss πλήρη αποτυχία. ~ drunk στουπί (*or* σκνίπα) στο μεθύσι. ~ tired (*or* ~-beat) ψόφιος στην κούραση. it has become a ~ letter περιέπεσεν εις αχρηστίαν. make a ~ set at

επιτίθεμαι εναντίον (*with gen.*). in the ~ of
winter στην καρδιά του χειμώνα.
dead *adv.* (*entirely*) απολύτως, (*exactly*)
ακριβώς.
deaden *v.* (*sound*) καταπνίγω, (*pain*)
αμβλύνω, νεκρώνω.
deadline *s.* what's the ~? πότε τελειώνει η
προθεσμία;
deadlock *s.* αδιέξοδο *n.*
deadly *a.* θανάσιμος.
deadpan *a.* ανέκφραστος.
deaf *a.* κουφός. ~ and dumb κωφάλαλος.
become ~ κουφαίνομαι. turn a ~ ear
κωφεύω. **~en** *v.* ξεκουφαίνω. **~ening** *a.*
εκκωφαντικός. **~ness** *s.* κουφαμάρα *f.*
deal *s.* (*wood*) άσπρο ξύλο.
deal *v.t.* (*distribute, also* ~ out) μοιράζω. (*a
blow*) καταφέρω. (*v.i.*) ~ at (*shop*) ψωνίζω
από. ~ in (*wares*) εμπορεύομαι. ~ with
(*handle*) χειρίζομαι, (*settle*) κανονίζω, (*be
occupied with*) ασχολούμαι με, (*touch on*)
θίγω, αναφέρομαι εις, (*be about, come in
contact with*) έχω να κάνω με. ~ harshly
with μεταχειρίζομαι σκληρά.
deal *s.* (*agreement*) συμφωνία *f.* (*piece of
business*) δουλειά. *f.* it's a ~! σύμφωνοι.
it's my ~ (*cards*) εγώ μοιράζω (*or* κάνω).
deal *s.* (*quantity*) a good (*or* great) ~ (of)
(*as a.*) πολύς, (*adv.*) πολύ.
dealer *s.* έμπορος *m.*, εμπορευόμενος *m.*
dealings *s.* δοσοληψίες *f.pl.* (*fam.*) have ~
έχω πάρε-δώσε, έχω νταραβέρια. (*fin.*)
αγοραπωλησίες *f.pl.*, συναλλαγές *f.pl.*
dean *s.* (*of faculty*) κοσμήτωρ *m.*
dear *a.* (*costly*) ακριβός, δαπανηρός. get ~er
ακριβαίνω.
dear *a.* (*loved*) ακριβός, αγαπητός, προσφιλής. my ~ αγαπητέ μου, αγάπη μου, my
~est (πολυ)αγαπημένε μου. ~ me! μη μου
λες! ~ me no! όχι δά! oh ~ ωχ, Θεέ μου.
dear, ~ly *adv.* (*at high cost*) ακριβά, (*much*)
πολύ.
dearness *s.* (*costliness*) ακρίβεια *f.* her ~ to
me το να μου είναι προσφιλής.
dearth *s.* έλλειψη *f.*
death *s.* θάνατος *m.* D~ (*personified*) χάρος
m. at ~'s door ετοιμοθάνατος. be in the
throes of ~ ψυχορραγώ. put to ~ θανατώνω. be bored to ~ πλήττω θανάσιμα.
(*fam.*) be the ~ of πεθαίνω. be in at the ~
παρίσταμαι στην τελική έκβαση. ~ penalty
θανατική ποινή. ~ sentence καταδίκη εις
θάνατον.
death-duty *s.* φόρος κληρονομίας.
deathless *a.* αθάνατος.
deathly *a.* νεκρικός.
death-rate *s.* θνησιμότητα *f.*
débâcle *s.* ξαφνική κατάρρευση.

debar *v.* αποκλείω.
debag *v.* (*fam.*) ξεβρακώνω.
debark *v.t. & i.* ξεμπαρκάρω.
debase *v.* εξευτελίζω. **~d** *a.* (*morally*)
ποταπός. **~ment** *s.* εξευτελισμός *m.*
debat|e *v.* συζητώ. (*s*) συζήτηση *f.* **~able** *a.*
συζητήσιμος.
debauch *s.* όργιο *n.* (*esp. drinking*)
κραιπάλη *f.* (*v.*) διαφθείρω, εκμαυλίζω.
~ed *a.* διεφθαρμένος. **~ery** *s.* ακολασία *f.*,
όργια *n.pl.*
debenture *s.* ομολογία *f.*
debilit|y *s.* αδυναμία *f.* **~ate** *v.* εξασθενίζω.
debit *s.* δούναι *n.*, παθητικόν *n.* ~ balance
χρεωστικόν υπόλοιπον. (*v.*) χρεώνω.
debonair *a.* πρόσχαρος.
debouch *v.i.* (*river, road*) εκβάλλω, (*troops*)
ξεχύνομαι.
debris *s.* συντρίμματα *n.pl.*
debt *s.* χρέος *n.* incur ~s χρεώνομαι, κάνω
χρέη. be in ~ είμαι χρεωμένος. **~or** *s.*
οφειλέτης *m.*
debunk *v.* (*fam.*) ξεφουσκώνω.
debut *s.* ντεμπούτο *n.* **~ante** *s.* δεσποινίδα
που βγαίνει στον κόσμο.
decade *s.* δεκαετία *f.*
decad|ence *s.* παρακμή *f.* **~ent** *a.* become
~ent παρακμάζω. a ~ent age εποχή πα-
ρακμής.
decalogue *s.* δεκάλογος *m.*
decamp *v.* το σκάω.
decant *v.t.* μεταγγίζω. (*fam., unload*)
ξεφορτώνω. **~er** *s.* καράφα *f.*
decapitate *v.* αποκεφαλίζω.
decay *v.* (*decompose*) αποσυντίθεμαι, (*of
tooth, wood, etc.*) σαπίζω, (*of beauty*)
χαλώ, μαραίνομαι. (*decline*) ξεπέφτω,
παρακμάζω. (*s.*) αποσύνθεση *f.*, σάπισμα
n., χάλασμα *n.*, παρακμή *f.*
deceased *a.* αποθανών. (*s.*) the ~ ο μακαρίτης.
deceit *s.* απάτη *f.* **~ful** *a.* απατηλός.
deceive *v.* απατώ, ξεγελώ, παραπλανώ.
December *s.* Δεκέμβριος, Δεκέμβρης *m.*
decency *s.* (*propriety*) ευπρέπεια *f.*
(*civility*) ανθρωπιά *f.* common ~ στοιχειώδης ευγένεια.
decent *a.* (*proper*) ευπρεπής, ~ people
καθώς πρέπει άνθρωποι, a ~ fellow καλός
τύπος. (*of quality*) της ανθρωπιάς. quite ~
καλούτσικος. **~ly** *adv.* ευπρεπώς. (*well*)
καλά.
decentraliz|e *v.* αποκεντρώνω. **~ation** *s.*
διοικητική αποκέντρωση *f.*
decept|ion *s.* εξαπάτηση *f.* **~ive** *a.* it is ~ive
σε γελάει.
decide *v.* αποφασίζω. **~d** *a.* (*determined*)
αποφασισμένος, (*definite*) αναμφι-

σβήτητος, καθαρός. **~dly** adv. αποφασιστικά, αναμφισβητήτως.
deciduous a. φυλλοβόλος.
decimal a. δεκαδικός. ~ point κόμμα n., υποδιαστολή f.
decimat|e v. αποδεκατίζω. **~ion** s. αποδεκατισμός m.
decipher v. αποκρυπτογραφώ. (fam., make out) βγάζω. **~ment** s. αποκρυπτογράφηση f.
decisive a. αποφασιστικός. **~ly** adv. αποφασιστικά. **~ness** s. αποφασιστικότητα f.
deck v. στολίζω.
deck s. κατάστρωμα n., (of bus, etc.) όροφος m. (cards) τράπουλα f. clear the ~s προετοιμάζομαι για δράση. **~-chair** s. ξαπλώ(σ)τρα f.
declaim v. απαγγέλλω με στόμφο. ~ against καταφέρομαι εναντίον (with gen.).
declamat|ion s. στομφώδης απαγγελία. **~ory** a. στομφώδης.
declar|e v.t. (proclaim) κηρύσσω. (goods, etc.) δηλώνω. (v.i.) (assert) ισχυρίζομαι. **~ation** s. κήρυξη f., δήλωση f., ισχυρισμός m.
declension s. (gram.) κλίση f.
declinable a. κλιτός.
decline v.i. (in health, etc.) εξασθενίζω, (in power, etc.) παρακμάζω, πέφτω. (v.t., gram.) κλίνω. (v.t. & i., refuse) αρνούμαι.
decline s. (physical, moral) κατάπτωση f., (power) παρακμή f., (activity) κάμψη f.
declivity s. κατωφέρεια f., κατήφορος m.
declutch v. ντεμπραγιάρω.
decoction s. αφέψημα n.
decode v. αποκρυπτογραφώ.
décolleté a. ντεκολτέ.
decolonize v. αποαποικιοποιώ.
decompos|e v.i. (rot) αποσυντίθεμαι, σαπίζω. (dissolve) διαλύομαι **~ition** s. αποσύνθεση f., διάλυση f.
decompression s. αποσυμπίεση f.
deconsecrate v. καταργώ.
decontaminate v. απολυμαίνω.
decontrol v.t. αίρω τον έλεγχο επί (with gen.).
décor s. διάκοσμος m., ντεκόρ n.
decorate v. διακοσμώ, (paint) βάφω. (with medal, etc.) παρασημοφορώ.
decoration s. διακόσμηση f. **~s** (festive) διάκοσμος m., (with flags) σημαιοστολισμός m. (honour) παράσημο n.
decorative a. διακοσμητικός. (fam., smart) κομψός.
decorator s. interior ~ διακοσμητής m. we've got the ~s in έχουμε βαφίματα.
decorous a. κόσμιος, ευπρεπής.

decorum s. ευπρέπεια f.
decoy s. κράχτης m., (trap) παγίδα f., (bait) δέλεαρ n. (v.) παρασύρω, δελεάζω.
decrease v.t. μειώνω, ελαττώνω. (v.i.) μειούμαι, ελαττώνομαι, πέφτω. (s.) μείωση f., ελάττωση f., πτώση f.
decree v. θεσπίζω, διατάζω. (s.) θέσπισμα n., διάταγμα n. (law) απόφαση f.
decrepit a. ερείπιο n.
decry v. κρίνω δυσμενώς.
dedicat|e v. αφιερώνω. **~ion** s. αφιέρωση f., (devotion) αφοσίωση f.
deduce v. συνάγω, βγάζω το συμπέρασμα (ότι).
deduct v. αφαιρώ, (withhold) κρατώ.
deduction s. (inference) συμπέρασμα n., (subtraction) αφαίρεση f., (from pay) κράτηση f.
deed s. (act) πράξη f. brave or daring ~ ανδραγάθημα n. **~s** (law) τίτλοι m.pl.
deem v. θεωρώ.
deep a. βαθύς, it is ten metres ~ έχει βάθος δέκα μέτρα. ~ in debt βουτηγμένος στα χρέη, ~ in thought βυθισμένος σε σκέψεις. (secretive) κρυψίνους. (great) μεγάλος. (serious) σοβαρός. (absorbed) απορροφημένος. in ~ water (fig.) σε δύσκολη θέση. go in off the ~ end γίνομαι έξω φρενών. (adv.) βαθιά.
deep s. (sea) βυθός m.
deep-dyed a. βαμμένος.
deepen v.t. & i. βαθαίνω. (intensify) (v.t.) εντείνω, (v.i.) εντείνομαι.
deep-freeze s. κατάψυξη f.
deep-laid a. καταχθόνιος.
deeply adv. βαθιά.
deep-rooted, ~-seated a. βαθιά ριζωμένος.
deep-sea a. ανοικτού πελάγους.
deer s. έλαφος f., ελάφι n.
deescalate v.t. αποκλιμακώνω.
deface v. προξενώ ζημιές σε. (make illegible) εξαλείφω.
de facto a. & adv. ντε φάκτο.
defalcation s. κατάχρηση f.
defam|e s. δυσφημώ. **~ation** s. δυσφήμηση f. **~atory** a. δυσφημιστικός.
default v. φυγοδικώ, αθετώ το λόγο μου. (fin.) δεν πληρώνω τα χρέη μου. (s.) αθέτηση f. by ~ ερήμην, in ~ of ελλείψει (with gen.). **~er** s. φυγόδικος m. (mil.) ένοχος παραπτώματος.
defeat v. νικώ, be ~ed νικώμαι, ηττώμαι. (thwart) ματαιώνω. (s.) ήττα f., ματαίωση f.
defeat|ism s. ηττοπάθεια f. **~ist** a. ηττοπαθής.
defecat|e v. αποπατώ, ενεργούμαι. **~ion** s.

αποπάτηση *f.*

defect *s.* ελάττωμα *n.* ~**ive** *a.* ελαττωματικός.

defect *v.* αποστατώ. ~**ion** *s.* αποστασία *f.* ~**or** *s.* αποστάτης *m.*

defence *s.* (*protection*) προστασία *f.* (*justification*) δικαιολογία *f.* (*law*) υπεράσπιση *f.* (*mil.*) άμυνα *f.*, ~s (*works*) αμυντικά έργα. (*a.*) αμυντικός. ~**less** *a.* ανυπεράσπιστος, (*unarmed*) άοπλος.

defend *v.* προστατεύω, υπερασπίζω, υπεραμύνομαι (*with gen.*). (*justify*) δικαιολογώ. ~ oneself αμύνομαι, δικαιολογούμαι.

defendant *s.* εναγόμενος *m.*, (*accused*) κατηγορούμενος *m.*

defender *s.* υπερασπιστής *m.*

defensible *a.* (*in argument*) υποστηρίξιμος. (*in war*) που επιδέχεται υπεράσπιση.

defensive *a.* αμυντικός. be on the ~ ευρίσκομαι εν αμύνη.

defer *v.t.* (*put off*) αναβάλλω. (*v.i.*) (*yield*) ενδίδω.

deferen|ce *s.* σεβασμός *m.* ~**tial** *a.* be ~tial δείχνω σεβασμό. ~**tially** *adv.* με σεβασμό.

deferment *s.* αναβολή *f.*

defiance *s.* περιφρόνηση *f.*, πρόκληση *f.* in ~ of (*of person*) περιφρονώντας (*with acc.*), (*of law*) κατά παράβασιν (*with gen.*).

defiant *a.* προκλητικός, απειθής. ~**ly** *adv.* προκλητικά.

deficiency *s.* ανεπάρκεια *f.*, (*deficit*) έλλειμμα *n.* mental ~ διανοητική καθυστέρηση *f.*

deficient *a.* (*faulty*) ελλιπής, ελαττωματικός. be ~ in μου λείπει, στερούμαι (*with gen.*). mentally ~ διανοητικά καθυστερημένος.

deficit *s.* έλλειμμα *n.*

defile *s.* στενό *n.*, δερβένι *n.*, κλεισούρα *f.*

defile *v.* ρυπαίνω, (*profane*) βεβηλώνω. ~**ment** *s.* ρύπανση *f.*, βεβήλωση *f.*

define *v.* (*determine*) προσδιορίζω, καθορίζω. (*make clear in outline*) χαράσσω. (*give meaning of*) δίδω τον ορισμόν (*with gen.*).

definite *a.* (*article, answer*) οριστικός, (*time, place*) προσδιορισμένος, (*sure*) σίγουρος. (*distinct*) καθαρός. ~**ly** *adv.* σίγουρα, δίχως άλλο.

definition *s.* ορισμός *m.* (*clarity*) καθαρότητα *f.*

definitive *a.* οριστικός.

deflat|e *v.* ξεφουσκώνω. I ~ed him του έκοψα τον αέρα. ~**ion** *s.* (*fin.*) αντιπληθωρισμός *m.*

deflect *v.t.* εκτρέπω. be ~ed παρεκκλίνω, (*of bullet*) εποστρακίζομαι. ~**ion** *s.* εκτροπή *f.*

deflower *v.* διακορεύω.

deforest *v.* αποδασώνω, αποψιλώνω.

deform *v.* παραμορφώνω. ~**ity** *s.* παραμόρφωση *f.*

defraud *v.* εξαπατώ, ξεγελώ. they ~ed him of his money του έκλεψαν τα λεφτά του.

defray *v.* καλύπτω.

defrost *v.* ξεπαγώνω. ~**ing** *s.* (*of fridge*) απόψυξη *f.*

deft *a.* επιδέξιος.

defunct *a.* be ~ (*of usage*) ατονώ. the ~ ο εκλιπών.

defuse *v.* αφοπλίζω. (*fig.*) καθιστώ ακίνδυνον.

defy *v.* (*ignore*) αψηφώ, δεν λογαριάζω. (*despise*) περιφρονώ. (*challenge*) προκαλώ. (*not admit of*) δεν επιδέχομαι.

degener|ate *a.* έκφυλος. (*v.*) εκφυλίζομαι. ~ate into καταντώ. ~**ation**, ~**acy** *s.* εκφυλισμός *m.*

degrad|e *v.* εξευτελίζω. ~**ation** *s.* εξευτελισμός *m.*, ξεπεσμός *m.* ~**ing** *a.* εξευτελιστικός.

degree *s.* βαθμός *m.* (*social position*) κοινωνική θέση. (*diploma*) δίπλωμα, πτυχίο *n.* (*geom.*) μοίρα *f.* to a ~ εξαιρετικά, to some ~ μέχρις ενός σημείου, by ~s βαθμηδόν, σιγά σιγά.

dehydrat|e *v.* αφυδατώνω. ~**ion** *s.* αφυδάτωση *f.*

de-ice *v.* αποπαγώνω.

deif|y *v.* θεοποιώ. ~**ication** *s.* θεοποίηση *f.*

deign *v.i.* καταδέχομαι. not ~ απαξιώ.

deity *s.* θεότητα *f.* (*God*) Θεός *m.*

deject|ed *a.* άθυμος, αποθαρρημένος. ~**ion** *s.* αποθάρρυνση *f.*

de jure *adv.* ντε γιούρε.

delay *v.t.* & *i.* καθυστερώ, (*put off*) αναβάλλω. (*s.*) καθυστέρηση *f.*, αναβολή *f.*

delect|able *a.* υπέροχος, (*fam.*) μούρλια. ~**ation** *s.* τέρψη *f.*, απόλαυση *f.*

delegate *v.t.* (*depute person*) αποστέλλω, δίδω εντολήν εις. (*entrust task*) αναθέτω. ~ authority αναθέτω υπεύθυνες εργασίες στους υφισταμένους μου.

delegat|e *s.* απεσταλμένος *m.* ~**ion** *s.* αντιπροσωπεία *f.*

delet|e *v.* σβήνω, διαγράφω. ~**ion** *s.* διαγραφή *f.*

deleterious *a.* βλαβερός, επιβλαβής.

deliberate *a.* (*intentional*) σκόπιμος, προμελετημένος, εσκεμμένος. (*careful*) προσεκτικός. ~**ly** *adv.* σκοπίμως, επίτηδες, εκ προμελέτης, εσκεμμένως, προσεκτικά.

deliberat|e *v.i.* (*alone*) διαλογίζομαι, (*together*) συσκέπτομαι. ~**ion** *s.* σκέψη *f.*, συζήτηση *f.*, with ~ion προσεκτικά, (*slowly*) αργά. ~**ive** *a.* συμβουλευτικός.

delicacy s. λεπτότητα f. (*food*) λιχουδιά f.
delicate a. λεπτός, ντελικάτος. (*easily
broken*) εύθραυστος. (*of colour*) απαλός.
(*sensitive*) ευαίσθητος, ευπαθής. **~ly** adv.
λεπτά, απαλά.
delicatessen s. ορεκτικά n.pl.
delicious a. νοστιμότατος, υπέροχος.
delight s. (*pleasure*) τέρψη f., απόλαυση f.
(*joy*) χαρά f. Turkish ~ λουκούμι n.
delight v.t. (*please*) τέρπω, (*charm*)
γοητεύω, (*give joy to*) δίνω μεγάλη χαρά
σε. (*v.i.*) ~ in (*doing sthg.*) μου δίνει μεγάλη
ευχαρίστηση (*with* να). be ~ed χαίρομαι
(*or* ευχαριστιέμαι) πολύ.
delightful a. γοητευτικός, υπέροχος. it's ~!
(*fam.*) είναι τρέλα (*or* όνειρο *or* μούρλια).
it's ~ to be here είναι απόλαυση για μένα
να βρίσκομαι εδώ. **~ly** adv. υπέροχα. it's
~ly cool here έχει μια απολαυστική
δροσιά εδώ.
delimitation s. οροθεσία f.
delineat|e v. απεικονίζω. **~ion** s. απει-
κόνιση f.
delinqu|ency s. ροπή προς το έγκλημα.
~ent a. be ~ent παραμελώ τα καθήκοντά
μου, αψηφώ τους ηθικούς νόμους.
delirious a. be ~ παραληρώ. (*fam.*) be ~
with joy είμαι εξάλλος από τη χαρά μου.
delirium s. παραλήρημα n. (*only lit.*)
παραμιλητό n.
deliver v.t. (*save, rid*) σώζω, απαλλάσσω,
(*send*) στέλνω, (*bring*) φέρνω, (*give*)
δίνω, (*transmit*) διαβιβάζω, (*utter*)
απαγγέλλω, (*a lecture*) κάνω, (*a blow*)
καταφέρω, (*letters*) μοιράζω, διανέμω. ~
up παραδίδω. ~ oneself of (*opinion*)
εκφράζω. be ~ed of (*child*) γεννώ.
deliverance s. απαλλαγή f. (*rescue*) σωτηρία f.
deliverer s. σωτήρας m.
delivery s. (*handing over*) παράδοση f. (*of
letters*) διανομή f. (*birth*) γέννα f.
(*speaker's*) απαγγελία f.
dell s. μικρή δασώδης κοιλάδα.
delouse v. ξεψειριάζω, ψειρίζω.
delta s. δέλτα n.
delude v. εξαπατώ, ξεγελώ. ~ oneself αυτα-
πατώμαι, ξεγελώ τον εαυτό μου.
deluge s. κατακλυσμός m. (*v.t.*) κατακλύζω.
delusion s. αυταπάτη f., εσφαλμένη ιδέα.
suffer from ~s υποφέρω από παραι-
σθήσεις.
delusive a. απατηλός.
de luxe a. πολυτελείας.
delve v.i. (*dig*) σκάβω. ~ into ανασκαλεύω,
σκαλίζω, (*only papers*) αναδιφώ, (*bag,
etc.*) ψάχνω σε.
demagog|ue s. δημαγωγός m. **~ic** a.

δημαγωγικός. **~y** s. δημαγωγία f.
demand s. (*request*) απαίτηση f., (*claim*)
αξίωση f. (*for goods or services*) ζήτηση f.
be in ~ έχω ζήτηση, in great ~ περιζήτητος.
on ~ εις πρώτην ζήτησιν. he makes too
many ~s on my patience καταχράται (*or*
κάνει κατάχρηση) της υπομονής μου. he
makes many ~s on me ζητάει πολλά από
μένα. the strikers' ~s τα αιτήματα των
απεργών.
demand v. (*request*) απαιτώ, (*claim*) αξιώ,
(*need*) χρειάζομαι. **~ing** a. απαιτητικός.
demarcation s. οροθεσία f. (*separation*)
διαχωρισμός m.
démarche s. διάβημα n.
demean v. ~ oneself ταπεινώνομαι. **~ing** a.
ταπεινωτικός.
demeanour s. συμπεριφορά f.
demented a. παράφρων.
demerit s. μειονέκτημα n.
demi- *prefix* ημι-.
demigod s. ημίθεος m.
demijohn s. νταμιτζάνα f.
demilitarize v. αποστρατικοποιώ.
demimonde s. ημίκοσμος m.
demise s. θάνατος m.
demobiliz|e v. αποστρατεύω. **~ation** s.
αποστράτευση f.
democracy s. δημοκρατία f.
democrat s. δημοκράτης m. (*in politics*)
δημοκρατικός a. **~ic** a. δημοκρατικός.
~ically adv. δημοκρατικώς. **~ize** v.
εκδημοκρατικοποιώ.
demographic a. δημογραφικός.
demol|ish v. γκρεμίζω, (*only building*) κα-
τεδαφίζω. **~ition** s. κατεδάφιση f.
demon s. (*spirit*) δαίμονας m. (*monster*)
τέρας n. (*fig.*) (*forceful person*) θηρίο n.
demonstrable a. ευαπόδεικτος, δυνάμενος
να αποδειχθεί
demonstrate v.t. (*prove*) αποδεικνύω,
(*explain*) επιδεικνύω, δείχνω. (*v.i.*) (*protest*) διαδηλώνω.
demonstration s. επίδειξη f., (*protest*)
διαδήλωση f.
demonstrative a. εκδηλωτικός. (*gram.*)
δεικτικός.
demonstrator s. (*protester*) διαδηλωτής m.
demoralization s. (*defeatism*) ηττοπάθεια
f. (*degeneration*) εκφυλισμός m.
demoralize v. (*corrupt*) χαλώ, (*hurt morale
of*) σπάω το ηθικό (*with gen.*). become ~d
χάνω το ηθικό μου.
demot|e v. υποβιβάζω. **~ion** s. υποβιβασμός
m..
demotic a. δημοτικός, ~ Greek δημοτική f.
~ally adv. δημοτικά. **~ist** s. δημοτικιστής m.
demur v. φέρω αντιρρήσεις. (s.) without ~

χωρίς αντιρρήσεις.
demure *a.* κόσμιος, (*coy*) σεμνοφανής.
(*fam.*) ~ miss φραγκοπαναγιά *f.*
den *s.* (*beasts'*) φωλιά *f.*, (*robbers'*) άντρο *n.*
(*secret resort*) λημέρι *n.*, κρησφύγετο *n.*
(*opium*) τεκές *m.* (*work room*) καταφύγιο *n.*
denationalize *v.* αποκρατικοποιώ.
denial *s.* άρνηση *f.* (*of rumour*) διάψευση *f.*
denigrat|e *v.* δυσφημώ. **~ion** *s.* δυσφήμηση *f.*
denim *a.* ντένιμ. **~s** *s.* τζην *n.*
denizen *s.* κάτοικος *m.* the ~s of the jungle τα ζώα της ζούγκλας.
denominat|e *v.* καλώ, ονομάζω. **~ion** *s.* (*sect*) αίρεση *f.* (*money*) αξία *f.* **~or** *s.* (*math.*) παρονομαστής *m.*
denote *v.* δείχνω, σημαίνω.
dénouement *s.* λύση *f.*
denounce *v.* καταγγέλλω.
dens|e *a.* πυκνός, (*stupid*) χοντροκέφαλος. **~ely** *adv.* πυκνώς, ~ely populated πυκνοκατοικημένος. **~ity** *s.* πυκνότητα *f.*
dent *s.* χτύπημα *n.* (*v.*) χτυπώ.
dental *a.* (*of teeth*) οδοντικός, (*of dentistry*) οδοντιατρικός.
dentist *s.* οδοντογιατρός, οδοντίατρος *m.f.* **~ry** *s.* οδοντιατρική *f.*
denture *s.* οδοντοστοιχία *f.*, μασέλλα *f.*
denude *v.* απογυμνώνω.
denunciation *s.* καταγγελία *f.*
den|y *v.t.* (*guilt, charge, fact*) αρνούμαι, (*rumour, etc.*) διαψεύδω. he was ~ied entry του αρνήθηκαν την είσοδο. he ~ies himself nothing δεν αρνείται τίποτα στον εαυτό του.
deodorant *s.* αποσμητικό *n.*
depart *v.* φεύγω, (*set out*) αναχωρώ, ξεκινώ. ~ from (*abandon*) εγκαταλείπω. the ~ed (*dead*) οι αποθανόντες.
department *s.* (*governmental*) υπηρεσία *f.*, διεύθυνση *f.* (*ministry*) υπουργείο *n.* (*of university*) τμήμα *n.* (*prefecture*) νομός *m.* (*area of activity*) σφαίρα *f.*, ειδικότητα *f.* (*of shop, etc.*) τμήμα *n.* ~ store μεγάλο κατάστημα.
departmental *a.* υπηρεσιακός.
departure *s.* (*setting out*) αναχώρηση *f.* (*deviation*) παρέκκλιση *f.* (*change*) αλλαγή. *f.* new ~ (*fig.*) νέος προσανατολισμός. make new ~s ανοίγω νέους ορίζοντες. ~ lounge αίθουσα αναχωρήσεων.
depend *v.* (*be contingent*) εξαρτώμαι (*with* από), it all ~s εξαρτάται. (*rely*) βασίζομαι, στηρίζομαι (*with* σε). ~ upon it! να είστε βέβαιος.
dependable *s.* (*person, news*) αξιόπιστος. is (*person*) ~ ? μπορεί να βασιστεί κανείς σε...;

depend|ant, ~ent *s.* προστατευόμενο μέλος της οικογένειας.
depend|ent *a.* be ~ent *see* depend. (*gram.*) εξαρτημένος. **~ence** *s.* εξάρτηση *f.* **~ency** *s.* χώρα υπό ξένην εξάρτησιν.
depict *v.* παριστάνω. **~ion** *s.* παράσταση *f.*
depilatory *s.* αποτριχωτικό *n.*
depleted *a.* (*empty*) άδειος, (*less*) μειωμένος. become ~ αδειάζω, λιγοστεύω.
deplorab|le *a.* αξιοθρήνητος, ελεεινός, οικτρός. **~ly** *adv.* ελεεινά, φοβερά.
deplore *v.t.* (*regret*) λυπούμαι για, (*protest against*) διαμαρτύρομαι κατά (*with gen.*).
deploy *v.* αναπτύσσω. **~ment** *s.* ανάπτυξη *f.*
deponent *a.* (*gram.*) αποθετικός.
depopulation *s.* μείωση του πληθυσμού.
deport *v.* (*expel*) απελαύνω. ~ oneself (*behave*) φέρομαι. **~ation** *s.* απέλαση *f.* **~ment** *s.* συμπεριφορά *f.*
depose *v.* (*remove*) εκθρονίζω. (*state*) καταθέτω.
deposit *v.* (*lay, put*) βάζω, αποθέτω, (*leave*) αφήνω, (*in bank*) καταθέτω, (*pay as earnest*) προκαταβάλλω. (*s.*) κατάθεση *f.* προκαταβολή *f.*, καπάρο *n.* (*mud, etc.*) ίζημα *n.* (*of mineral*) κοίτασμα *n.*
deposition *s.* (*of Christ*) αποκαθήλωση *f.* (*dethronement*) εκθρόνιση *f.* (*statement*) κατάθεση *f.*
depository *s.* αποθήκη *f.*
depot *s.* (*goods*) αποθήκη *f.* (*bus*) γκαράζ *n.*
deprav|e *v.* διαφθείρω. **~ed** *a.* διεφθαρμένος. **~ity** *s.* διαφθορά *f.*
deprecat|e *v.* είμαι εναντίον (*with gen.*). **~ion** *s.* αποδοκιμασία *f.* **~ory** *a.* αποδοκιμαστικός.
depreciat|e *v.t.* υποτιμώ. (*v.i.*). πέφτω **~ion** *s.* υποτίμηση *f.* πτώση *f.* **~ory** *a.* υποτιμητικός.
depredation(s) *s.* ληστεία *f.*
depress *v.* (*press down*) πατώ, (*make gloomy*) στενοχωρώ, καταθλίβω. **~ing** *a.* στενόχωρος, καταθλιπτικός. **~ion** *s.* (*gloom*) κατάθλιψη *f.* (*barometric, economic*) ύφεση *f.* (*hollow*) βαθούλωμα *n.*, γούβα *f.*
deprivation *s.* στέρηση *f.*
deprive *v.* στερώ. he was ~d of his pension τον στέρησαν της συντάξεώς του *or* του στέρησαν τη σύνταξή του. he was ~d of his livelihood στερήθηκε τα προς το ζην.
depth *s.* βάθος *n.* it is ten metres in ~ έχει βάθος δέκα μέτρων. in the ~ of winter στην καρδιά του χειμώνα. ~s of the earth τα έγκατα της γης. get out of one's ~ χάνω τα νερά μου.
deputation *s.* αντιπροσωπία *f.*
depute *v.* εξουσιοδοτώ. I ~d him to... του

ανέθεσα να.
deput|ize v.i. ~ize for αντιπροσωπεύω, αναπληρώνω. ~y s. αντιπρόσωπος m. αναπληρωτής m.
derail v. εκτροχιάζω. ~ment s. εκτροχίαση f.
derange v. (upset) αναστατώνω. he is ~d το μυαλό του έχει διασαλευθεί. ~ment s. διασάλευση f.
derelict a. εγκαταλελειμμένος. ~ion s. παράλειψη f.
derequisition v. αίρω την επίταξιν (with gen.).
deride v. κοροϊδεύω, χλευάζω.
derision s. εμπαιγμός m. χλευασμός m. hold in ~ κοροϊδεύω.
deris|ive, ~**ory** a. (mocking) χλευαστικός, (ridiculous) εξευτελιστικός, γελοίος. ~**ively** adv. κοροϊδευτικά.
derivation s. παραγωγή f.
derivative s. παράγωγο n. (a.) (of creative work) it is ~ δεν διακρίνεται για την πρωτοτυπία του.
derive v. (obtain) αποκομίζω. be ~d παράγομαι.
derogatory a. μειωτικός.
derrick s. (crane) γερανός m. (of well) ικρίωμα γεωτρήσεως.
dervish s. δερβίσης m.
desalinization s. αφαλάτωση f.
descant v. ~ on εξυμνώ.
descend v.i. & t. κατεβαίνω, κατέρχομαι. ~ upon (attack) επιπίπτω κατά (with gen.). (v.i.) (fall) πέφτω, (lower oneself) καταντώ. be ~ed from κατάγομαι από (or εκ).
descendant s. απόγονος m.
descent s. (going down) κατέβασμα n., κάθοδος f. (slope) κατωφέρεια f. (attack) επίθεση f. (ancestry) καταγωγή f.
describe v. περιγράφω.
descript|ion s. περιγραφή f. ~**ive** a. περιγραφικός.
descry v. διακρίνω.
desecrat|e v. βεβηλώνω. ~**ion** s. βεβήλωση f.
desert s. έρημος f.
desert, ~**ed** a. έρημος, εγκαταλειμμένος.
desert v.t. εγκαταλείπω, αφήνω. (v.i.) (mil.) λιποτακτώ. ~**er** s. λιποτάκτης m. ~**ion** s. εγκατάλειψη f. λιποταξία f.
deserts s. ό,τι μου αξίζει.
deserve v. δικαιούμαι (with να or gen.), he ~s punishment του αξίζει τιμωρία (or να τιμωρηθεί). ~**dly** adv. δικαίως.
deserving a. άξιος.
desiccat|e v. αποξηραίνω. ~**ion** s. αποξήρανση f.
desiderate v. επιθυμώ.
design s. σχέδιο n. (intent) σκοπός m., by ~

σκοπίμως. have ~s on έχω βλέψεις επί, (for evil) έχω κακές προθέσεις επί (with gen.).
design v.t. σχεδιάζω, (destine) προορίζω. (v.i.) (intend) σκοπεύω. ~**edly** adv. σκοπίμως.
designat|e v. (name) ονομάζω, (fix) διορίζω. (a.) νεοεκλεγείς. ~**ion** s. ονομασία f. διορισμός m.
designer s. σχεδιαστής m., (stage) σκηνογράφος m.
designing a. ραδιούργος.
desirab|le a. επιθυμητός. it is very ~le that θα ήτο ευκταίον (or ευχής έργον) να. ~**ility** s. επιθυμητόν n.
desire v. επιθυμώ, (crave) ποθώ, (request) παρακαλώ. (s.) επιθυμία f., πόθος m. have no ~ δεν έχω καμιά όρεξη (with να or για).
desirous a. be ~ (of) επιθυμώ.
desist v.i. παύω, σταματώ.
desk s. γραφείο n., (school) θρανίο n., (teacher's) έδρα f. ~**top** a. επιτραπέζιος.
desolate a. έρημος, παντέρημος. (v.) be ~d (grieved) είμαι κατασυντετριμμένος.
desolation s. (of place) ερήμωση f., (of person) συντριβή της καρδιάς.
despair s. απελπισία f. be filled with ~ με πιάνει απελπισία. in ~ απελπισμένος.
despair v. ~ of απελπίζομαι (with για or πως). his life is ~ed of τον έχουν αποφασισμένο. ~**ing** a. απελπισμένος. ~**ingly** adv. απελπισμένα.
despatch v. see dispatch.
desperado s. παλικαράς m., μπράβος m.
desperate a. απεγνωσμένος, (violent) αποφασισμένος για όλα. (grave, extreme) απελπιστικός, φοβερός. (remedy) της απελπισίας. ~**ly** adv. απεγνωσμένα, φοβερά.
despicab|le a. άξιος περιφρονήσεως. ~**ly** adv. αισχρά.
despise v. περιφρονώ, καταφρονώ.
despite prep. (also in ~ of) ~ his age παρά την ηλικία του. ~ our having been away so long παρ' όλο που λείψαμε τόσο καιρό.
desp|oil v. ληστεύω, απογυμνώνω. ~**oliation** s. λήστευση f.
despond v. αποθαρρύνομαι. ~**ency** s. αποθάρρυνση f. ~**ent** a. αποθαρρημένος.
despot s. τύραννος m., δεσποτικός a. (fam.) σατράπης m. ~**ic** a. δεσποτικός. ~**ically** adv. δεσποτικά. ~**ism** s. δεσποτισμός m.
dessert s. επιδόρπια n.pl. ~-**spoon** s. κουτάλι της κομπόστας.
destination s. προορισμός m.
destine v. προορίζω. he was ~d to die poor ήταν της μοίρας του (or γραφτό του) να

πεθάνη φτωχός.

destiny *s.* πεπρωμένον *n.*, μοίρα *f. see* destine.

destitut|e *a.* άπορος, απέναταρος. be ~e of στερούμαι (*with gen.*). ~**ion** *s.* ένδεια *f.*, πενία *f.*

destroy *v.* καταστρέφω, (*wipe out*) αφανίζω, (*kill*) σκοτώνω.

destroyer *s.* καταστροφεύς *m.* (*naut.*) αντιτορπιλλικόν *n.*

destruction *s.* καταστροφή *f.* (*wiping out*) αφανισμός *f.* (*damage*) καταστροφές, ζημίες *f.pl.*

destructive *a.* καταστρεπτικός. ~ child παιδί που καταστρέφει τα πάντα.

desultory *a.* ξεκάρφωτος, χωρίς σύστημα.

detach *v.* βγάζω, αποσυνδέω, (*men*) αποσπώ. ~**ed** *a.* (*impartial*) αμερόληπτος. ~ed house σπίτι χωρίς μεσοτοιχία. get ~ed (*come off*) βγαίνω.

detachment *s.* αμεροληψία *f.* (*lack of interest*) αδιαφορία *f.* (*mil.*) απόσπασμα *n.*

detail *s.* λεπτομέρεια *f.* in ~ λεπτομερώς. relate the ~s (*of event*) διηγούμαι τα καθέκαστα. (*v.*) (*describe*) εκθέτω λεπτομερώς. (*order*) διατάσσω. ~**ed** *a.* λεπτομερής.

detain *v.* κρατώ, (*delay*) καθυστερώ. ~ in custody προφυλακίζω. ~**ee** *s.* κρατούμενος *m.*

detect *v.* ανακαλύπτω, (*distinguish*) διακρίνω. ~**ion** *s.* ανακάλυψη *f.*, ανίχνευση *f.*

detective *s.* ντέτεκτιβ *m.* ~ story αστυνομικό μυθιστόρημα.

detector *s.* (*electrical*) ανιχνευτής *m.*

détente *s.* ύφεση *f.*

detention *s.* κράτηση *f.* (*school*) he got an hour's ~ έμεινε μία ώρα τιμωρία.

deter *v.* αποτρέπω. ~**rent** *a.* αποτρεπτικός.

detergent *s.* απορρυπαντικόν *n.*

deteriorat|e *v.i.* (*spoil*) χαλώ, αλλοιώνομαι, (*get worse*) χειροτερεύω. ~**ion** *s.* αλλοίωση *f.*, χειροτέρευση *f.*

determinate *a.* ορισμένος.

determination *s.* (*fixing*) καθορισμός *m.* (*resoluteness*) αποφασιστικότητα *f.* (*decision*) απόφαση *f.*

determine *v.* (*fix.*) καθορίζω, (*verify*) εξακριβώνω, (*decide*) αποφασίζω, (*make person do sthg.*) κάνω, πείθω. (*end*) (*v.t.*) διαλύω, (*v.i.*) λήγω.

determined *a.* (*resolved on sthg.*) αποφασισμένος, make ~ efforts ενεργώ αποφασιστικώς.

detest *v.* απεχθάνομαι, μισώ. ~**able** *a.* απεχθής, αντιπαθέστατος. ~**ably** *adv.* μέχρις αηδίας. ~**ation** *s.* απέχθεια *f.*, μίσος *n.*

dethrone *v.* εκθρονίζω. ~**ment** *s.* εκθρόνιση *f.*

detonat|e *v.t.* πυροδοτώ. (*v.i.*) εκπυρσοκροτώ. ~**ion** *s.* εκπυρσοκρότηση *f.* (*bang*) έκρηξη *f.* ~**or** *s.* πυροκροτητής *m.*

detour *s.* make a ~ βγαίνω (*or* αποκλίνω) από το δρόμο μου.

detract *v.* ~ from μειώνω. ~**ion** *s.* επίκριση *f.* ~**or** *s.* επικριτής *m.*

detriment *s.* βλάβη *f.* to his ~ (*materially*) προς βλάβην του, (*morally*) εις βάρος του. ~**al** *a.* επιβλαβής. ~**ally** *adv.* επιβλαβώς.

detritus *s.* συντρίμματα *n.pl.*

deuce *s.* (*fam.*) διάβολος *m.* (*cards*) δυάρι *n.* ~**d** *a.* (*fam.*) διαβολεμένος.

devalu|e *v.* υποτιμώ. ~**ation** *s.* υποτίμηση *f.*

devastat|e *v.* καταστρέφω, ρημάζω. (*emotionally*) συγκλονίζω. ~**ing** *a.* καταστρεπτικός, (*remark*) συντριπτικός, (*beauty*) εκθαμβωτικός. ~**ion** *s.* καταστροφή *f.*, ρήμαγμα *n.*

develop *v.t.* (*body, faculties, talent, idea, heat*) αναπτύσσω, (*natural resources*) αξιοποιώ, (*film*) εμφανίζω, (*spread, broaden*) επεκτείνω. he ~ed an ulcer του παρουσιάστηκε έλκος. (*v.i.*) (*grow*) αναπτύσσομαι, μεγαλώνω, (*unfold, of events*) εξελίσσομαι, (*take shape*) σχηματίζομαι, (*spread*) επεκτείνομαι. ~**ing** country αναπτυσσόμενη χώρα *f.*

developing *s.* (*of film*) εμφάνιση *f.*

development *s.* ανάπτυξη *f.*, αξιοποίηση *f.*, επέκταση *f.*, εξέλιξη *f.*, σχηματισμός *m.*

deviat|e *v.* παρεκκλίνω, αποκλίνω. ~**ion** *s.* παρέκκλιση *f.*, απόκλιση *f.* (*altered route*) παρακαμπτήριος *f.*

device *s.* (*invention*) επινόημα *n.* (*trick*) τέχνασμα *n.*, κόλπο *n.* (*apparatus*) συσκευή *f.* (*emblem*) έμβλημα *n.* leave him to his own ~s άφησέ τον να κάνει ό,τι του αρέσει.

devil *s.* διάβολος *m.* poor ~ φουκαράς *m.* between the ~ and the deep sea μπρος γκρεμός και πίσω ρέμα. ~**ish** *a.* (*of hunger, cold*) διαβολεμένος, (*wicked*) διαβολικός.

devil-may-care *a.* ξένοιαστος, αψήφιστος.

devil|ment, ~**ry** *s.* (*devil's work*) διαβολοδουλειά *f.* (*mischief*) διαβολιά *f.* (*daring*) παρατολμία *f.*

devious *a.* (*means*) πλάγιος, (*sly*) πονηρός. take a ~ route κάνω κύκλο.

devise *v.* επινοώ, εφευρίσκω, μηχανεύομαι.

devoid *a.* ~ of χωρίς (*with acc.*), he is ~ of common sense του λείπει ο κοινός νους.

devolution *s.* (*decentralization*) αποκέντρωση *f.* ~ of power μεταβίβαση εξουσιών.

devolve *v.t.* μεταβιβάζω. (*v.i.*) περιέρχομαι. all the work ~d upon me όλη η δουλειά έπεσε στις πλάτες μου.

devote v. αφιερώνω, ~ oneself αφοσιώνομαι. ~d a. αφοσιωμένος. ~dly adv. με αφοσίωση.

devotee s. λάτρης m. (*follower*) οπαδός m., πιστός a.

devotion s. αφοσίωση f. ~s προσευχές f.pl.

devour v. καταβροχθίζω, (*fig.*) (κατα)τρώγω.

devout a. ευσεβής, (*sincere*) ειλικρινής, θερμός. ~ly adv. ευσεβώς, θερμώς. ~ness s. ευσέβεια f.

dew s. δροσιά f. ~-drop s. δροσοσταλίδα f. ~y a. δροσόλουστος.

dexter|ous a. επιδέξιος. ~ity s. επιδεξιότητα f.

diabet|es s. διαβήτης m. ~ic a. διαβητικός.

diabolic, ~al a. σατανικός.

diadem s. διάδημα n.

diagnos|e v. κάνω διάγνωση. what did he ~e? τι διάγνωση έκανε; he ~ed TB διέγνωσε φυματίωση. ~is s. διάγνωση f.

diagonal a. διαγώνιος. (s.) διαγώνιος f. ~ly adv. διαγωνίως.

diagram s. σχεδιάγραμμα n. ~matically adv. σχηματικώς.

dial s. καντράν n. (v.) (*phone*) παίρνω, καλώ. ~ling code κωδικός αριθμός m.

dialect s. διάλεκτος f. ~al a. διαλεκτικός.

dialectic s. διαλεκτική f.

dialogue s. διάλογος m.

diamet|er s. διάμετρος f. ~rically adv. εκ διαμέτρου, διαμετρικά.

diamond s. διαμάντι n., αδάμας m. (*cards*) καρρό n. rough ~ (*fig. of person*) άξεστος αλλά καλός. (a.) αδαμάντινος. ~-shaped a. ρομβοειδής, (*fam.*) μπακλαβαδωτός.

diapason s. διαπασών f.n.

diaper s. (*baby's*) πάνα f., πανιά n.pl.

diaphanous a. διαφανής.

diaphragm s. διάφραγμα n.

diarrhoea s. διάρροια f.

diary s. ημερολόγιο n. (*pocket*) ατζέντα f.

diaspora s. διασπορά f.

diatribe s. διατριβή f.

dice s. ζάρια n.pl. (v.) (*play* ~) παίζω ζάρια, (*fig.*) ριψοκινδυνεύω. (*cut up*) λιανίζω. ~y a. παρακινδυνευμένος.

dichotomy s. (*bot.*) διχοτόμηση f. (*division*) διαχωρισμός m.

dickens s. (*fam.*) διάβολος m. the ~ of a... διαβολεμένος.

dick(e)y s. (*shirt-front*) πλαστρόν n. (a.) (*shaky*) ασταθής, feel a bit ~ δεν είμαι στα καλά μου. ~-bird s. πουλάκι n.

dictaphone s. ντικταφόν n.

dictate v. υπαγορεύω. ~ to (*order about*) επιβάλλομαι εις. ~s s. προσταγές f.pl.

dictation s. υπαγόρευση f. from ~ καθ'

υπαγόρευσιν, do ~ (*exercise*) κάνω ορθογραφία.

dictator s. δικτάτορας m. ~ial a. δικτατορικός. ~ship s. δικτατορία f.

diction s. (*style*) ύφος n. (*enunciation*) απαγγελία f.

dictionary s. λεξικό n.

dictum s. ρητό n.

didactic a. διδακτικός.

diddle v. ξεγελώ, he ~d me out of a tenner μου βούτηξε ένα χιλιάρικο.

did see do.

die v. (*person*) πεθαίνω, (*animal*) ψοφώ, (*plant*) μαραίνομαι. (*fig.*) σβήνω. never say ~! κουράγιο! a custom that ~s hard έθιμο που δεν χάνεται εύκολα. he ~d of laughing πέθανε (*or* ξεράθηκε *or* ψόφησε *or* έσκασε) στα γέλια. be dying to (*desire*) ψοφώ να, έχω σφοδρή επιθυμία να, he is dying to tell τον τρώει η γλώσσα του, he is dying to get away δεν βλέπει την ώρα να φύγει, he is dying for a cigarette είναι ξεψυχισμένος για τσιγάρο. ~ away *or* down (*of storm, row*) κοπάζω, (*of breeze*) πέφτω, (*of fire, interest*) σβήνω, (*of sound*) σβήνω, χάνομαι. ~ out εκλείπω, (*of fire*) σβήνω. ~ off πεθαίνει ένας-ένας. *see* dying.

die s. κύβος m., ζάρι n. the ~ is cast ερρίφθη ο κύβος. (*for engraving*) μήτρα f. ~-casting s. χύσιμο υπό πίεση.

diehard a. αδιάλλακτος, (*only politically*) βαμμένος.

diesel s. & a. ντήζελ.

diet s. (*political, nutritional*) δίαιτα f. (v.) κάνω δίαιτα. ~etics s. διαιτητική f. ~ician s. διαιτολόγος m.

differ v. (*be unlike*) διαφέρω. (*not agree*) we ~ διαφέρουμε, οι γνώμες μας διαφέρουν.

difference s. διαφορά f. that makes a ~ αυτό αλλάζει τα πράγματα. it makes no ~ to me το ίδιο μου κάνει. does it make any ~? αλλάζει τίποτα; split the ~ μοιράζω τη διαφορά. make a ~ (*draw distinction*) κάνω διάκριση.

different a. (*general*) διαφορετικός, (*other*) άλλος, (*various*) διάφοροι (*pl.*). that's a ~ matter (*or* story) αυτό είναι άλλη υπόθεση (*or* παράγραφος). a ~ kind of car άλλο είδος (*or* άλλου είδους) αυτοκίνητο. they sell ~ kinds of merchandise πουλάνε διαφόρων ειδών εμπορεύματα. they have ~ tastes έχουν διαφορετικά γούστα, διαφέρουν τα γούστα τους. ~ly adv. διαφορετικά.

differential a. διαφορικός. (s.) διαφορικό n.

differentiate v.t. (*see or show to be different*) διακρίνω, ξεχωρίζω. (v.i.) ~ between (*treat differently*) κάνω διακρίσεις μεταξύ (*with gen.*).

difficult *a.* (*thing, person*) δύσκολος, (*person*) σκολιός. (*of achievement*) δυσχερής. make *or* become δυσκολεύω.

difficult|y *s.* δυσκολία *f.* δυσχέρεια *f.* have ~y δυσκολεύομαι. make ~ies φέρω δυσκολίες.

diffid|ence *s.* διστακτικότητα *f.* ~ent *a.* διστακτικός. ~ently *adv.* διστακτικά.

diffraction *s.* διάθλαση *f.*

diffuse *a.* (*style*) πολύλογος, φλύαρος.

diffus|e *v.* διαχέω. ~ed *a.* διάχυτος. ~ion *s.* διάχυση *f.*

dig *v.* σκάβω, (*a hole, well, etc.*) ανοίγω, (*thrust*) χώνω, (*prod*) σπρώχνω, (*excavate*) ανασκάπτω. ~ (oneself) in (*shelter*) οχυρώνομαι. ~ one's heels (*or* toes) in μένω ακλόνητος. ~ into the food πέφτω με τα μούτρα στο φαΐ. ~ over (*garden*) σκαλίζω, σκάβω. ~ out βγάζω, (*find*) ξετρυπώνω. ~ up (*plant*) βγάζω, the tree must be dug up το δέντρο πρέπει να βγει.

dig *s.* (*archaeological*) ανασκαφές *f.pl.* (*nudge*) σκουντιά *f.* (*hint*) that was a ~ at me αυτό ήταν πετριά για μένα.

digest *v.* χωνεύω. (*s.*) περίληψη *f.* ~ible *a.* εύπεπτος. ~ive *a.* πεπτικός.

digestion *s.* πέψη *f.*, χώνευση *f.*, χώνεψη *f.*, good for the ~ χωνευτικός.

digger *s.* (*mech.*) εκσκαφεύς *m.*

diggings *s.* live in ~ μένω σε νοικιασμένο δωμάτιο.

digit *s.* ψηφίο *n.* ~al *a.* ψηφιακός.

dignif|y *v.* εξευγενίζω, (*joc.*) εξωραΐζω. ~ied *a.* αξιοπρεπής.

dignitary *s.* αξιωματούχος *m.*

dignity *s.* αξιοπρέπεια *f.* stand on one's ~ παίρνω ύφος. it is beneath my ~ είναι κατώτερον της αξιοπρεπείας μου.

digress *v.* φεύγω από το θέμα μου. ~ion *s.* παρέκβαση *f.*

digs *s.* *see* diggings.

dike *s.* ανάχωμα *n.* (*ditch*) τάφρος *m.*

dilapidated *a.* σαραβαλιασμένος, (*building*) ετοιμόρροπος.

dilat|e *v.i.* (*expand*) διαστέλλομαι. ~e upon επεκτείνομαι επί (*with gen.*). ~ion *s.* διαστολή *f.*

dilatory *a.* αναβλητικός.

dilemma *s.* δίλημμα *n.*

dilettante *s.* ντιλετάντης *m.* (*a.*) ντιλετάντικος.

dilig|ence *s.* επιμέλεια *f.* ~ent *a.* επιμελής. ~ently *adv.* επιμελώς.

dill *s.* άνηθο *n.*

dilly-dally *v.* χρονοτριβώ.

dilut|e *v.* διαλύω, αραιώνω, (*wine*) νερώνω.

~e(d) *a.* διαλελυμένος, αραιωμένος, (*wine*) νερωμένος. ~ion *s.* (*act*) διάλυση *f.* (*product*) διάλυμα *n.*

dim *v.t.* (*lights*) χαμηλώνω, (*outshine*) σκιάζω. (*v.i.*) (*grow* ~) θολώνω, αδυνατίζω, αμβλύνομαι.

dim *a.* (*light*) θαμπός, αμυδρός, (*eyes, sight*) θολός, αδύνατος, (*memory, wits*) αμβλύς, θαμπός, αδύνατος. (*vague*) αόριστος. take a ~ view of αποδοκιμάζω.

dimension *s.* διάσταση *f.* three-~al τρισδιάστατος.

diminish *v.t. & i.* μικραίνω, λιγοστεύω. (*v.t.*) μειώνω, ελαττώνω. (*v.i.*) μειώνομαι, ελαττώνομαι.

diminution *s.* μείωση *f.*, ελάττωση *f.*, σμίκρυνση *f.*, λιγόστεμα *n.*

diminutive *a.* μικρούλικος, τόσος δα, τοσοδούλης. (*s.*) (*gram.*) υποκοριστικόν *n.*

dimly *adv.* θολά, θαμπά, αμυδρώς, αορίστως.

dimness *s.* αμυδρότητα *f.*, αοριστία *f.*

dimple *s.* λακκάκι *n.*

dim-witted *a.* χαζός.

din *s.* αντάρα *f.*, βουητό *n.*, σαματάς *m.* (*v.i.*) βουΐζω, αντηχώ. (*v.t.*) he ~ned it into me μου 'φαγε τ' αυτιά.

dine *v.i.* δειπνώ.

dinghy *s.* βαρκάκι *n.*

dingy *a.* μαυρισμένος, βρώμικος.

dining-car *s.* βαγκόν-ρεστοράν *n.*

dining-room *s.* τραπεζαρία *f.*

dinky *a.* (*fam.*) κομψούλικος.

dinner *s.* δείπνο *n.* we had him to ~ yesterday τον είχαμε τραπέζι χθες το βράδυ. at ~-time (*lunch*) το μεσημεράκι. ~-jacket *s.* σμόκιν *n.*

dint *s.* by ~ of χάρις εις, χάρη σε.

diocese *s.* επισκοπή *f.*

dip *v.t.* βουτώ, (*flag*) κατεβάζω, (*headlights*) χαμηλώνω. (*v.i.*) (*go lower*) βουτώ, (*of balance*) γέρνω, (*of road, etc.*) κατηφορίζω. ~ into (*a book*) ξεφυλλίζω, ~ into one's pocket ξοδεύω.

dip *s.* βούτηγμα *n.* (*in ground*) κατηφοριά *f.* (*in sea*) βουτιά *f.*

diphtheria *s.* διφθερίτιδα *f.*

diphthong *s.* δίφθογγος *f.*

diploma *s.* δίπλωμα *n.* holder of ~ διπλωματούχος *m.f.*

diplomacy *s.* διπλωματία *f.*

diplomat, ~ist *s.* διπλωμάτης *m.* ~ic *a.* διπλωματικός, (*person*) διπλωμάτης *m.*

dipsomaniac *s.* διψομανής *a.*

diptych *s.* δίπτυχον *n.*

dire *a.* τρομερός, (*need*) έσχατος.

direct *a.* (*not indirect*) άμεσος, (*straight*) κατ' ευθείαν, ίσιος, (*frank*) ντόμπρος,

ευθύς. in ~ contact with σε άμεση επαφή με. the most ~ route ο πιο σύντομος δρόμος. (*orders, etc.*) ~ from London απευθείας από το Λονδίνο.
direct *v.* (*show way to*) οδηγώ, (*turn*) στρέφω, (*address*) απευθύνω, (*order*) διατάσσω, (*manage*) διευθύνω.
direction *s.* (*course*) κατεύθυνση *f.* sense of ~ αίσθηση προσανατολισμού. (*management*) διεύθυνση *f.* (*guidance*) καθοδήγηση *f.* (*order*) διαταγή *f.* ~s (*instructions*) οδηγίες *f.pl.*
directive *s.* κατευθυντήριος γραμμή.
directly *adv.* (*straight*) κατ' ευθείαν, (*at once*) αμέσως. ~ opposite ακριβώς απέναντι. (*conj., as soon as*) αμέσως μόλις.
director *s.* διευθυντής *m.* ~y *s.* οδηγός *m.*
dirge *s.* θρήνος *m.,* μοιρολόι *n.*
dirt *s.* βρώμα *f.,* ακαθαρσία *f.* (*earth*) χώμα *n.* (*fig.*) throw ~ at συκοφαντώ. treat (*person*) like ~ μεταχειρίζομαι σα σκουπίδι. show the ~ λερώνω εύκολα. ~ road χωματόδρομος *m.* ~ cheap πάμφθηνος, τζάμπα.
dirty *a.* βρώμικος, ακάθαρτος, (*body, clothes*) λερωμένος. get ~ (*of clothes*) λερώνω, (*of persons*) λερώνομαι. give (*person*) a ~ look αγριοκοιτάζω. ~ business (*or trick*) βρωμοδουλειά *f.* ~ dog βρωμόσκυλο *n.*
dirty *v.t. & i.* λερώνω, βρωμίζω.
dis- *prefix* α-, αν-, απο-, αφ-, εκ-, εξ-, ξε-.
disability *s.* (*disqualification*) ανικανότητα *f.* (*disablement*) αναπηρία *f.*
disable *v.* καθιστώ ανάπηρον, (*prevent*) εμποδίζω (*with* να). ~d *s.* ανάπηρος. (*ship*) που έχει αχρηστευθεί. ~ment *s.* αναπηρία *f.*
disabuse *v.* I ~d him of the idea τον έβγαλα από την πλάνη.
disadvantage *s.* μειονέκτημα *n.* it turned out to my ~ η υπόθεσις βγήκε προς ζημίαν μου. ~ous *a.* μειονεκτικός.
disaffect|ed *a.* δυσαρεστημένος. ~ion *s.* δυσαρέσκεια *f.,* δυσφορία *f.*
disafforest *v.* αποψιλώνω, αποδασώνω.
disagree *v.* διαφωνώ, δεν συμφωνώ. ~ with (*upset*) πειράζω. ~able *a.* δυσάρεστος, αντιπαθητικός. ~ment *s.* διαφορά *f.,* διαφωνία *f.*
disallow *v.* απορρίπτω.
disappear *v.* εξαφανίζομαι, χάνομαι. ~ance *s.* εξαφάνιση *f.*
disappoint *v.* απογοητεύω, (*hopes*) διαψεύδω. ~ing *a.* απογοητευτικός. ~ment *s.* απογοήτευση *f.*
disapprobation *s.* αποδοκιμασία *f.*
disapprov|e *v.* (*also* ~e of) αποδοκιμάζω, δεν εγκρίνω. ~al *s.* αποδοκιμασία *f.* ~ing

a. αποδοκιμαστικός. ~ingly *adv.* με αποδοκιμασία.
disarm *v.i.* αφοπλίζομαι. (*v.t.*) (*lit. & fig.*) αφοπλίζω. ~ament *s.* αφοπλισμός *m.* ~ing *a.* αφοπλιστικός.
disarrange *v.* αναστατώνω, κάνω άνωκάτω.
disarray *s.* in ~ αναστατωμένος, άνω-κάτω, (*half dressed*) μισοντυμένος.
disassociate *v.t.* see dissociate.
disast|er *s.* συμφορά *f.* καταστροφή *f.* (*failure*) φιάσκο *n.* ~rous *a.* καταστροφικός, ολέθριος, φιάσκο *n.* ~rously *adv.* καταστροφικώς.
disavow *v.* απαρνούμαι. ~al *s.* απάρνηση *f.*
disband *v.t.* διαλύω. (*v.i.*) διαλύομαι. ~ment *s.* διάλυση *f.*
disbel|ief *s.* δυσπιστία *f.* (*in God*) απιστία *f.* ~ieve *v.* δεν πιστεύω.
disburse *v.* δαπανώ.
disc *s.* δίσκο *m.*
discard *v.* απορρίπτω, πετώ.
discern *v.* διακρίνω, (*by insight*) διαβλέπω. ~ible *a.* be ~ible διακρίνομαι. ~ing *a.* διορατικός, (*in taste*) εκλεκτικός. ~ment *s.* διορατικότης *f.*
discharge *v.t.* (*give forth*) βγάζω, (*unload*) εκφορτώνω, (*let fly*) εκτοξεύω. (*send away*) απολύω, (*accused person*) απαλλάσσω. (*perform*) εκτελώ, (*pay off*) εξοφλώ. (*fire*) he ~d his gun at the lion άδειασε το όπλο του απάνω στο λιοντάρι. (*flow out*) the river ~s... το ποτάμι εκβάλλει.
discharge *s.* εκφόρτωση *f.* εκτόξευση *f.* απόλυση *f.* απαλλαγή *f.* εκτέλεση *f.* εξόφληση *f.* (*secretion*) έκκριση *f.* (*electric*) εκκένωση *f.* ~ papers απολυτήριο *n.*
disciple *s.* μαθητής *m.*
disciplinarian *s.* he is a ~ απαιτεί αυστηρή πειθαρχία.
disciplinary *a.* πειθαρχικός.
discipline *s.* πειθαρχία *f.* (*branch*) κλάδος *m.* (*v.*) επιβάλλω πειθαρχία εις, (*punish*) τιμωρώ.
disclaim *v.* αρνούμαι. ~er *s.* (*of responsibility*) άρνηση ευθύνης. (*of right*) παραίτηση *f.*
disclos|e *v.* αποκαλύπτω. ~ure *s.* αποκάλυψη *f.*
disco *s.* ντίσκο *n.*
discobolus *s.* δισκοβόλος *m.*
discol|our *v.t. & i.* ξεβάφω. (*v.i.*) ξεθωριάζω. ~oration *s.* αποχρωματισμός *m.*
discomfit *v.* be ~ed βρίσκομαι σε αμηχανία. ~ure *s.* αμηχανία *f.*
discomfort *s.* (*hardship*) ταλαιπωρία *f.* (*lack of amenities*) έλλειψη ανέσεων.

(bodily) δυσφορία f., στενοχώρια f.
discommode v. ενοχλώ.
disconcert v. αναστατώνω, σαστίζω. I
found his behaviour ~ing η συμπεριφορά
του με τάραξε.
disconnect v. αποσυνδέω, κόβω. ~ed a.
αποσυνδεδεμένος, κομμένος, (speech)
ασυνάρτητος. ~edly adv. ξεκάρφωτα,
ασυνάρτητα.
disconsolate a. απαρηγόρητος, (disappointed) απογοητευμένος.
discontent s. δυσαρέσκεια f., δυσφορία f.
(grievance) παράπονο n. ~ed a. δυσαρεστημένος.
discontinue v. σταματώ, διακόπτω. ~ance
s. διακοπή f. ~ous a. διακεκομμένος,
ασυνεχής.
discord s. δυσαρμονία f. (strife) έριδα f.,
διχόνοια f. (mus.) διαφωνία f. ~ant a.
κακόφωνος.
discotheque s. ντισκοτέκ f.
discount s. έκπτωση f. σκόντο n. at a ~ με
έκπτωση. (fig.) it is at a ~ (not wanted) δεν
ζητιέται, (not valued) δεν εκτιμάται.
discount v.t. (lessen effect of) μειώνω,
(disbelieve) δεν δίνω πίστη σε.
discountenance v. δεν εγκρίνω.
discourage v. αποθαρρύνω, (persuade not
to) αποτρέπω. ~ement s. αποθάρρυνση f.
~ing a. αποθαρρυντικός.
discourse s.ομιλία f., λόγος m. (converse)
συνομιλία f. (v.) ομιλώ, συνομιλώ.
discourteous a. αγενής. ~esy s. αγένεια f.
discover v. ανακαλύπτω. ~er s. ο ανακαλύψας. ~y s., ανακάλυψη f.
discredit v.t. (not believe) δεν δίνω πίστη
σε, (old theory, etc.) θέτω υπό αμφισβήτησιν. (put to shame) ντροπιάζω. (s.)
ντροπή f. fall into ~ χάνω την υπόληψή
μου. ~able a. ντροπιαστικός, αναξιοπρεπής.
discreet a. διακριτικός, εχέμυθος. ~ly adv.
διακριτικά.
discrepancy s. αντίφαση f., διαφορά f.
discretion s. (tact) διακριτικότητα f.,
εχεμύθεια f., (prudence) φρόνηση f.
(judgement, pleasure) διάκριση f. ~ary a.
διακριτικός.
discriminate v.i. (in favour of or against)
κάνω διακρίσεις (with υπέρ or εις βάρος
& gen.). (perceive difference in) ~ between A and B ξεχωρίζω τον Α από τον Β.
discriminating a. εκλεκτικός.
discrimination s. (judgement) κρίση f.
(partiality) διάκριση f.
discursive a. πολύλογος και ασυνάρτητος.
discus s. δίσκος m.
discuss v. συζητώ, κουβεντιάζω. ~ion s.

συζήτηση f., κουβέντα f.
disdain v. περιφρονώ. (to do sthg.) απαξιώ
(να). (s.) περιφρόνηση f. ~ful a. περιφρονητικός.
disease s. (general) ασθένεια f., αρρώστια
f., νόσος f. (malady) νόσημα n., αρρώστια
f. ~d a. άρρωστος, (organ) προσβεβλημένος.
disembark v.t. αποβιβάζω. (v.i.) αποβιβάζομαι. (v.t. & i.) ξεμπαρκάρω. ~ation s.
αποβίβαση f., ξεμπαρκάρισμα n.
disembodied a. άυλος.
disembowel v. ξεκοιλιάζω.
disenchantment s. απογοήτευση f.
disengage v.t. αποσυνδέω. (withdraw)
αποσύρω, (troops) απαγκιστρώνω. ~d a.
(free) ελεύθερος. ~ment s. (mil.) απαγκίστρωση f.
disentangle v.t. ξεμπερδεύω, (also ~
oneself) ξεμπλέκω. ~ment s. ξέμπλεγμα n.
disestablish v. (eccl.) χωρίζω από το
κράτος.
diseuse s. ντιζέζα f.
disfavour s. δυσμένεια f.
disfigure v. (persons) παραμορφώνω,
(objects) προξενώ ζημιές σε, (landscape)
ασχημίζω. ~ment s. παραμόρφωση f.
(scar, etc.) σημάδι n.
disfranchise v. στερώ πολιτικών δικαιωμάτων.
disgorge v. ξερνώ.
disgrace s. (shame) ντροπή f., αίσχος n. in
~ υπό δυσμένειαν. (downfall) πτώση f.,
ταπείνωση f. (v.) ντροπιάζω.
disgraceful a. αισχρός, φοβερός. it is ~
είναι ντροπή (or αίσχος). ~ly adv. αισχρά,
φοβερά.
disgruntled a. μουτρωμένος.
disguise s. μεταμφίεση f. (v.) μεταμορφώνω, (hide) κρύβω. ~ oneself μεταμφιέζομαι.
disgust s. αηδία f., σιχαμάρα f. (v.t.)
αηδιάζω. be ~ed by αηδιάζω με, σιχαίνομαι. ~ing a. αηδιαστικός, σιχαμερός,
σιχαμένος. it is ~ing είναι αηδία.
dish s. πιατέλα f. (food) πιάτο n. wash the
~es πλένω τα πιάτα. (fam., pretty woman)
μπουκιά και συχώριο.
dish v. ~ out μοιράζω, ~ up σερβίρω. (fam.)
I ~ed him του την έφερα.
dishabille s. in ~ ατημέλητος, μισοντυμένος.
dishcloth s. πιατόπανο n.
disharmony s. δυσαρμονία f.
dishearten v. αποκαρδιώνω. ~ing a. αποκαρδιωτικός.
dishevelled a. ξεμαλλιασμένος, ατημέλητος.

dishonest *a.* ανέντιμος. he is ~ δεν έχει τιμιότητα. ~**ly** *adv.* ανέντιμα, όχι τίμια. ~**y** *s.* έλλειψη εντιμότητας.

dishonour *s.* ατιμία *f.*, ντροπή *f.* (*v.*) ατιμάζω, ντροπιάζω. ~**able** *a.* άτιμος, αισχρός. ~**ably** *adv.* άτιμα, αισχρά.

dishwasher *s.* πλυντήριο πιάτων.

dishwater *s.* απόπλυμα *n.* (*fig.*) like ~ νερομπούλι *n.*

disillusion *v.* απογοητεύω. be ~ed βγαίνω από την πλάνη μου. ~**ment** *s.* απογοήτευση *f.*

disinclin|ed *a.* απρόθυμος. ~**ation** *s.* απροθυμία *f.*

disinfect *v.* απολυμαίνω. ~**ant** *s.* απολυμαντικό *n.* ~**ion** *s.* απολύμανση *f.*

disingenuous *a.* ανειλικρινής.

disinherit *v.* αποκληρώνω.

disintegrat|e *v.i.* διαλύομαι. ~**ion** *s.* διάλυση *f.*

disinter *v.* ξεθάβω.

disinterested *a.* αντιδιοτελής.

disjointed *a.* ασυνάρτητος.

disjunctive *a.* (*gram.*) διαζευκτικός.

disk *s.* δίσκος *m.*

dislike *v.* αντιπαθώ, δεν μου αρέσει. (*s.*) αντιπάθεια *f.*

dislocat|e *v.* εξαρθρώνω, βγάζω. ~**ion** *s.* εξάρθρωση *f.*

dislodge *v.* εκτοπίζω, βγάζω.

disloyal *a.* άπιστος, (*of troops, political followers*) μη νομιμόφρων. ~**ty** *s.* απιστία *f.* έλλειψη νομιμοφροσύνης.

dismal *a.* (*sad*) μελαγχολικός, (*dreary*) καταθλιπτικός, άθλιος.

dismantle *v.t.* (*take to pieces*) ξηλώνω, λύω, (*take down*) διαλύω. (*ship*) ξαρματώνω, παροπλίζω.

dismay *v.* τρομάζω. (*s.*) τρομάρα *f.*

dismember *v.* διαμελίζω. ~**ment** *s.* διαμελισμός *m.*

dismiss *v.* απολύω, διώχνω, (*from mind*) βγάζω. ~**al** *s.* απόλυση *f.*, διώξιμο *n.*

dismount *v.i.* κατεβαίνω. (*from horse*) ξεπεζεύω.

disobedi|ence *s.* ανυπακοή *f.*, απείθεια *f.* ~**ent** *a.* ανυπάκουος, απειθής. be ~ent απειθώ.

disobey *v.t.* παρακούω, (*infringe*) παραβαίνω.

disobliging *a.* αγενής, μη εξυπηρετικός.

disorder *s.* ακαταστασία *f.*, αταξία *f.* (*commotion*) ταραχές *f.pl.* (*med.*) πάθηση *f.* ~**ed** *a.* ταραγμένος.

disorderly *a.* (*untidy*) ακατάστατος, (*unruly*) θορυβώδης. ~ house οίκος ανοχής.

disorganiz|e *v.* αναστατώνω. ~**ation** *s.* αναστάτωση *f.*

disorient(at)ed *a.* become ~ μπερδεύομαι, χάνω τον μπούσουλα.

disown *v.* αποκηρύσσω, δεν αναγνωρίζω.

disparage *v.* υποτιμώ. ~**ment** *s.* υποτίμηση *f.*

dispar|ate *a.* ανόμοιος, διάφορος, ~**ity** *s.* ανομοιότητα *f.* διαφορά *f.*

dispassionate *a.* αμερόληπτος. ~**ly** *adv.* αμεροληπτως.

dispatch *v.* αποστέλλω, (*business*) διεκπεραιώνω. (*kill*) αποτελειώνω.

dispatch *s.* (*sending*) αποστολή *f.* (*speed*) with ~ γρήγορα. (*journalist's*) ανταπόκριση *f.* ~ box *or* case χαρτοφύλακας *m.* ~**er** *s.* αποστολεύς *m.*

dispel *v.* διαλύω, διασκορπίζω.

dispensary *s.* φαρμακείο *n.*

dispensation *s.* (*exemption*) απαλλαγή *f.* (*of providence*) θέλημα *n.* (*distribution*) απονομή *f.*

dispense *v.* απονέμω, (*medicine*) παρασκευάζω, (*prescription*) εκτελώ. ~ with κάνω χωρίς.

dispers|e *v.t.* διασκορπίζω. (*v.i.*) διασκορπίζομαι. ~**al** *s.* διασκορπισμός *m.* ~**ion** *s.* (*of light*) ανάλυση *f.* (*of Jews, Gks.*) διασπορά *f.*

dispirited *a.* άθυμος.

displace *v.* (*take place of*) εκτοπίζω, (*shift*) μετακινώ. ~d persons εκτοπισμένα άτομα. ~**ment** *s.* (*naut.*) εκτόπισμα *n.*

display *v.* (*show, show off*) επιδεικνύω, (*exhibit*) εκθέτω, (*allow to appear*) δείχνω. (*feelings*) εκδηλώνω.

display *s.* (*ostentation*) επίδειξη *f.* (*exhibition, show*) έκθεση *f.* (*of emotion*) εκδήλωση *f.*

displeas|e *v.* δυσαρεστώ, δεν αρέσω σε. ~**ing** *a.* δυσάρεστος. ~**ure** *s.* δυσαρέσκεια *f.*

disport *v.* ~ oneself διασκεδάζω, (*frolic*) χοροπηδώ.

disposable *a.* (*available*) διαθέσιμος, (*made of paper*) χάρτινος.

disposal *s.* διάθεση *f.* have at one's ~ διαθέτω, έχω στη διάθεσή μου. have for ~ διαθέτω, έχω για πούλημα. ~ of sewage αποχέτευση *f.* ~ of rubbish συγκέντρωση και πέταμα σκουπιδιών. waste ~ unit σκουπιδοφάγος *m.*

dispose *v.* (*arrange*) διατάσσω, (*incline*) προδιαθέτω. feel ~d έχω διάθεση. well ~d (*favourable*) ευνοϊκά διατεθειμένος, (*in good mood*) ευδιάθετος. ~ of (*get rid*) ξεφορτώνομαι, (*throw away*) πετώ, (*settle, deal with*) κανονίζω, (*sell*) πουλώ.

disposition *s.* (*arrangement*) διάταξη *f.*, (*temperament, control over*) διάθεση *f.*

dispossess *v.* (*oust*) εκδιώκω. he was ~ed of the house του αφαίρεσαν το σπίτι.

disproportionate *a.* δυσανάλογος. **~ly** *adv.* δυσαναλόγως.

disprove *v.t.* αποδεικνύω ως ανακριβές.

disputable *a.* αμφισβητήσιμος.

disputat|ion *s.* συζήτηση *f.* **~ious** *a.* φιλόνικος.

dispute *v.i.* (*quarrel*) μαλώνω. (*v.t.*) (*challenge*) διαμφισβητώ, (*strive for*) διαφιλονικώ, διαμφισβητώ. (*resist*) αντίσταμαι εις. (*s.*) συζήτηση *f.* it is beyond ~ είναι εκτός αμφισβητήσεως, ούτε συζήτηση.

disqualif|y *v.* (*render ineligible*) καθιστώ ακατάλληλον, (*prevent*) εμποδίζω, (*a player*) αποκλείω. **~ication** *s.* (*being ~ied*) αποκλεισμός *m.* (*cause of this*) αιτία αποκλεισμού.

disquiet *v.t.* (*also* be ~ed) ανησυχώ. (*s.*) ανησυχία *f.* **~ing** *a.* ανησυχητικός.

disregard *v.* αγνοώ, δεν δίνω σημασία σε.

disrepair *s.* be in ~ θέλω επισκευή.

disreputable *a.* (*person*) ύποπτος, ανυπόληπτος. (*haunt*) ύποπτος, κακόφημος. (*shabby*) ελεεινός, ~ overcoat παλιοπαλτό *n.*

disrepute *s.* κακή φήμη. fall into ~ βγάζω κακό όνομα.

disrespect *s.* έλλειψη σεβασμού. **~ful** *a.* ασεβής. **~fully** *adv.* ασεβώς.

disrobe *v.i.* γδύνομαι.

disrupt *v.* (*break up*) διαλύω, (*interrupt*) διακόπτω. **~ion** *s.* διάλυση *f.*, διακοπή *f.* **~ive** *a.* διαλυτικός.

dissatisf|y *v.* δεν ικανοποιώ. **~ied** *a.* δυσαρεστημένος. **~action** *s.* απαρέσκεια *f.*

dissect *v.* (*biol.*) ανατέμνω, (*fig.*) εξετάζω. **~ion** *s.* ανατομή *f.* εξέταση *f.*

dissemble *v.t.* κρύβω. (*v.i.*) προσποιούμαι, υποκρίνομαι. **~r** *s.* υποκριτής *m.*

disseminat|e *v.* διασπείρω, διαδίδω. **~ion** *s.* διάδοση *f.*

dissension *s.* διάσταση *f.*, διαίρεση *f.*

dissent *v.* διαφωνώ. (*s.*) διαφωνία *f.* **~ient** *a.* διαφωνών.

dissertation *s.* διατριβή *f.*

disservice *s.* be *or* do a ~ το βλάπτω.

dissid|ence *s.* διαφωνία *f.* **~ent** *a.* διαφωνών, διστάμενος.

dissimilar *a.* ανόμοιος. **~ity** *s.* ανομοιότητα *f.*

dissimilation *s.* (*gram.*) ανομοίωση *f.*

dissimulat|e *v.* *see* dissemble. **~ion** *s.* προσποίηση *f.*, υπόκριση *f.*

dissipat|e *v.t.* (*squander*) σπαταλώ, (*scatter*) διασκορπίζω. (*v.i.*) διασκορπίζομαι. **~ed** *a.* άσωτος. **~ion** *s.* ασωτία *f.*

dissociat|e *v.* χωρίζω. ~e oneself from αρνούμαι κάθε σχέση με. **~ion** *s.* χωρισμός *m.* (*chem.*) διάσταση *f.*

dissolute *a.* άσωτος.

dissolution *s.* διάλυση *f.*

dissolve *v.t.* διαλύω. (*v.i.*) διαλύομαι, (*into tears*) αναλύομαι.

disson|ance *s.* δυσαρμονία *f.* **~ant** *a.* κακόφωνος.

dissua|de *v.* μεταπείθω, αποτρέπω. **~sion** *s.* αποτροπή *f.*

distaff *s.* ρόκα *f.* on the ~ side από τη μητρική πλευρά.

distance *s.* απόσταση *f.* at ten metres ~ σε απόσταση δέκα μέτρων. in the ~ μακριά, στο βάθος. it is within walking ~ μπορεί κανείς να πάει με τα πόδια. keep one's ~ τηρώ τις αποστάσεις. keep (*person*) at a ~ κρατώ σε απόσταση.

distant *a.* (*place, journey, etc.*) μακρινός, (*aloof*) ψυχρός. there is a ~ view of the sea η θάλασσα διακρίνεται από μακριά. the ~ future το απώτερον μέλλον. **~ly** *adv.* (*coldly*) ψυχρώς. we are ~ly related έχουμε μακρινή συγγένεια.

distaste *s.* αντιπάθεια *f.* **~ful** *a.* δυσάρεστος.

distemper *s.* (*paint*) ασβεστόχρωμα *n.* (*n.*) ασβεστώνω.

distemper *s.* (*illness*) αρρώστια *f.* (*animals'*) μόρβα *f.*

distend *v.* φουσκώνω.

distich *s.* δίστιχο *n.*

distil *v.t.* αποστάζω, διυλίζω. (*v.i.*) (*drip*) στάζω. **~lation** *s.* (*process*) απόσταξη *f.* (*product*) απόσταγμα *n.* **~led** *a.* απεσταγμένος. **~lery** *s.* οινοπνευματοποιείον *n.*

distinct *a.* (*clear to senses*) ευδιάκριτος, ξεκάθαρος, (*to mind*) σαφής, ξάστερος. (*noticeable*) αισθητός. (*different*) διαφορετικός. (*apart*) χωριστός, as ~ from εν αντιθέσει προς (*with acc.*).

distinction *s.* (*differentiation, high quality*) διάκριση *f.* (*difference*) διαφορά *f.* (*of appearance*) αρχοντιά *f.* **~s** (*awards*) τιμητικές διακρίσεις. of ~ διακεχριμένος, (*esp. socially*) περιωπής. without ~ αδιακρίτως.

distinctive *a.* (*mark, sign*) διακριτικός, (*characteristic*) χαρακτηριστικός. **~ly** *adv.* καθαρά.

distinctly *adv.* (*clearly*) ξεκάθαρα, καθαρά. (*without doubt*) αναμφισβητήτως.

distinctness *s.* ευκρίνεια *f.*, καθαρότητα *f.*

distinguish *v.* διακρίνω, ξεχωρίζω. ~ oneself διακρίνομαι.

distinguishable *a.* ευδιάκριτος. it is barely ~ μόλις διακρίνεται.

distinguished *a.* διαπρεπής, διακεχριμένος. (*in appearance*) φίνος, αρχοντικός.

distort *v.* διαστρεβλώνω, παραμορφώνω.

distract   296   **do**

~ion s. διαστρέβλωσηf., παραμόρφωσηf.

distract v. (divert from worry) διασκεδάζω.
(drive mad) τρελλαίνω, I am ~ed by the
noise ο θόρυβος μου έχει πάρει το κεφάλι.
~ attention of αποσπώ την προσοχή (with
gen.).

distracted a. be ~ τα έχω χαμένα. ~ly adv.
σαν τρελλός (or τρελλήf.).

distraction s. (amusement) διασκέδασηf.
(being anxious) ανησυχία f. (frenzy)
τρέλλα f. love to ~ αγαπώ μέχρι τρέλλας.
drive to ~ κάνω έξω φρενών, τρελλαίνω.

distrain v. κάνω κατάσχεση. ~t s. κατάσχεσηf.

distraught a. έξαλλος, αλλόφρων.

distress s. (trouble) στενοχώριαf. (sorrow)
λύπη f. (exhaustion) εξάντληση f. (indigence) πενία f. (danger) κίνδυνος m. (v.)
στενοχωρώ. ~ed a. στενοχωρημένος, εξαντλημένος. ~ing a. οδυνηρός, θλιβερός.

distribute v. μοιράζω, διανέμω, (spread
about) απλώνω. ~ion s. διανομή f.
(classificatory) κατανομή f. ~or s. διανομεύς m.

district s. περιφέρεια f. (postal) τομεύς m.
(region) περιοχήf. (a.) περιφερειακός.

distrust v. δεν εμπιστεύομαι. (s.) δυσπιστία
f. ~ful a. δύσπιστος.

disturb v. (intrude on) ενοχλώ, (displace)
αναστατώνω, πειράζω, (make uneasy)
ανησυχώ. (peace, quiet) διαταράσσω.
~ance s. (of person) ενόχληση f. (of
things) αναστάτωμα n. (of peace) φα-
σαρίαf., ταραχήf. ~ing a. ανησυχητικός.

disunion s. διχασμός m. ~ite v. διχάζω.
~ity s. διχασμός m.

disuse s. αχρηστία f. fall into ~ περιπίπτω
εις αχρηστίαν, αχρηστεύομαι. ~d a.
αχρηστευμένος.

disyllablic a. δισύλλαβος. ~le s. δισύλλαβος λέξις.

ditch s. χαντάκι n. die in the last ~
ανθίσταμαι μέχρις εσχάτων.

ditch v.t. (a car) ρίχνω στο χαντάκι. ~ an
aeroplane προσθαλασσώνομαι αναγκα-
στικώς. (fam., dismiss) he got ~ed του
έδωσαν τα παπούτσια στο χέρι.

dither v. αμφιταλαντεύομαι. (s.) be in a ~ τα
έχω χαμένα.

dithyramb s. διθύραμβος m.

ditto s.n. το ίδιο. (marks) ομοιωματικά
s.n.pl.

ditty s. τραγουδάκι n.

divagation s. παρέκβασηf.

divan s. ντιβάνι n.

dive v.i. βουτώ, (naut.) καταδύομαι. (fig.)
(rush, thrust oneself) ορμώ, χώνομαι. (s.)
βουτιά f., κατάδυση f. (aero.) vertical ~
κάθετος εφόρμησις. (low resort) υπόγα f.

~r s. βουτηχτής m.

diverge v. αποκλίνω, απομακρύνομαι,
βγαίνω. ~ence s. απόκλιση f. ~ent a.
αντίθετος, διαφορετικός.

divers a. διάφοροι pl.

diverse a. ποικίλος. ~ify v. ποικίλλω. ~ity
s. ποικιλίαf.

diversion s. (turning aside) διοχέτευση f.
(altered route) παρακαμπτήριος f. (amusement) διασκέδασηf., (distraction & mil.)
αντιπερισπασμός m.

divert v. (turn aside) διοχετεύω, (amuse)
διασκεδάζω. ~ attention αποσπώ την
προσοχή. ~ing a. διασκεδαστικός.

divest v. I ~ him of the right του αφαιρώ το
δικαίωμα. ~ oneself of απεκδύομαι (with
gen.).

divide v.t. (into component parts) διαιρώ,
(share out) μοιράζω, (take share of)
μοιράζομαι, (separate) χωρίζω, διαχωρίζω. ~e by 1,000 διαιρώ διά του χίλια.
(v.i.) χωρίζομαι. ~ed a. διηρημένος,
χωρισμένος. ~ing a. διαχωριστικός.

dividend s. μέρισμα n. ~ coupon τοκο-
μερίδιον n.

divination s. μαντικήf.

divine a. (lit. & fig.) θείος, (fig.) θεϊκός
θεσπέσιος. (s.) (theologist) θεολόγος m.
~ly adv. ~ly beautiful (woman) όμορφη
σαν θεά, she signs ~ly τραγουδάει θε-
σπέσια (or υπέροχα).

divine v. μαντεύω. ~er s. ραβδοσκόπος m.
~ing s. ραβδοσκοπίαf.

diving s. κατάδυση f., βουτιές f.pl. ~
apparatus σκάφανδρο m.

divinity s. θεότης f. (theology) θεολογία f.

divisible a. διαιρετός. ~ility s. διαιρετόν n.

division s. (into components) διαίρεση f.
(sharing) μοιρασιά f., καταμερισμός m.
(separation) (δια)χωρισμός m. (disagreement) διχασμός m. (grade) βαθμός m.
(mil.) μεραρχία f. ~al commander
μέραρχος m.

divisive a. ~ measures μέτρα που προκα-
λούν διχόνοια.

divorce s. διαζύγιο n., (also fig.) χωρισμός
m. (v.t.) χωρίζω. get ~d χωρίζω, διαζευγνύομαι, παίρνω διαζύγιο.

divorced a. διεζευγμένος. (fam.) ζωντοχήρος m., ζωντοχήραf.

divulge v. αποκαλύπτω.

dizzly a. (height, speed) ιλιγγιώδης, (of
person) ζαλισμένος. make ~y ζαλίζω, get
~y ζαλίζομαι. ~iness s. ζαλάδαf.

do v. (anomalous) ~ you like it? σας αρέσει;
no I don't όχι, δεν μου αρέσει. did he go
home? πήγε σπίτι; yes he did ναι, πήγε. but
you did tell me! μα μου το είπες.

(*question-tags*) ~ don't you? did/didn't they? *etc*. έτσι; δεν είν' έτσι; ε; (*as response to statement*) α έτσι; αλήθεια; (*other responses*) (*who said so?*) I did εγώ. (*he speaks French*) so ~ I κι εγώ. (*he likes French*) so ~ I κι εμένα. nor ~/did I ούτε εγώ, ούτε εμένα. I feel better than I did yesterday αισθάνομαι καλύτερα από χθες (*or* απ' ό,τι χθες). they left before we did έφυγαν πριν από μας. ~ pay attention! για πρόσεχε σε παρακαλώ. ~ please come! έλα, να χαρείς (*or* να ζήσεις). don't! μη! don't listen μην ακούς. you don't say! μη μου πεις!

do *v.t.* (*work, lessons, accounts, portrait, favour, etc.*) κάνω, (*music, a play*) παίζω, (*writing*) γράφω. ~ one's hair χτενίζω τα μαλλιά μου, χτενίζομαι. ~ the cooking μαγειρεύω. ~ the washing up πλένω τα πιάτα. be done (*happen*) γίνομαι, (*finished*) τελειώνω, it can't be done δεν γίνεται, what is done can't be undone ό,τι έγινε δεν ξεγίνεται. no sooner said than done αμ' έπος αμ' έργον. the meat isn't done (*not finished cooking*) το κρέας δεν ψήθηκε ακόμα, (*is underdone*) δεν είναι ψημένο. my work is never done η δουλειά μου δεν τελειώνει ποτέ. ~ one's best βάζω τα δυνατά μου. ~ good (to) ωφελώ. ~ harm (to) βλάπτω. (*suit, satisfy*) will this ~ (for) you? αυτό σας κάνει; they did us very well μας περιποιήθηκαν πολύ. he does himself well ζει καλά. what do you ~ with yourself all day? πώς περνάς τη μέρα σου; it isn't done (*is not the right thing*) δεν επιτρέπεται. (*cheat*) I've been done με γέλασαν. well done! μπράβο!

do *v.i.* (*fare, prosper*) how do you ~ ? χαίρω πολύ. it/he is ~ing well (*of undertaking*) πάει καλά, (*thriving in career, soil*) ευδοκιμεί. (*be suitable, proper*) it won't ~ δεν κάνει, this will ~ excellently αυτό θα κάνει θαυμάσια. that will ~ (*is enough*) αρκετά, φτάνει. nothing ~ing δεν γίνεται τίποτα. make ~ (*get along*) τα φέρνω βόλτα, τα βολεύω, βολεύομαι, (*make shift*) I must make ~ with my old coat πρέπει να περάσω με το παλιό μου παλτό, (*be satisfied*) αρκούμαι (*with* σε). have to ~ with έχω σχέση με, έχω να κάνω με. I could ~ with some more money χρειάζομαι κι άλλα λεφτά. I could ~ with a beer γουστάρω (*or* κάνω γούστο) μια μπίρα. ~ without κάνω χωρίς. have done with τελειώνω με. have you done? τελείωσες;

do *s.* (*fam.*) (*party*) πάρτυ *n.*, γλέντι *n.* (*letdown*) κοροϊδία *f.* ~s-and-dont's κανόνες συμπεριφοράς.

do away *v.* ~ with (*abolish*) καταργώ, (*throw out*) πετώ, (*get rid of*) ξεφορτώνομαι, (*kill*) καθαρίζω.
do by *v.* he does well by his employees μεταχειρίζεται καλά τους υπαλλήλους του. he was hard done by τον αδίκησαν.
do down *v.* (*cheat*) εξαπατώ.
do for *v.* it is done for (*no good*) πάει αυτό *or* πάει περίπατο. done for (*of person*) (*ruined*) κατεστραμμένος, χαμένος, (*exhausted*) αποκαμωμένος, (*not expected to live*) καταδικασμένος.
do in *v.* (*kill*) καθαρίζω. I feel done in είμαι εξαντλημένος.
do out *v.* (*tidy*) σιάζω. (*deprive*) he did me out of the money μου έφαγε τα λεφτά. their visit did me out of my siesta η επίσκεψή τους με έκανε να χάσω τον ύπνο μου.
do up *v.* (*button*) κουμπώνω, (*repaint*) βάφω, (*repair*) φτιάνω, (*wrap*) τυλίγω, (*tie*) δένω. I feel done up αισθάνομαι χάλια.
docile *a.* ήρεμος, πειθήνιος. ~ity *s.* ευπείθεια *f.*
dock *s.* (*prisoner's*) εδώλιο *n.*
dock *v.t.* (*cut*) κόβω, (*wages*) κάνω κράτηση από.
dock *v.i.* (*naut.*) προσορμίζομαι, αράζω. (*s.*) (*repair*) δεξαμενή *f.*, (*loading*) νεωδόχος *m.*, ντοκ *n.* ~er *s.* λιμενεργάτης *m.*
docket *s.* (*label*) ετικέτα *f.* (*voucher*) απόδειξη *f.*
dockyard *s.* νεώριο *n.* (*naval*) ναύσταθμος *m.*
doctor *s.* γιατρός, ιατρός *m.* (*degree title*) διδάκτωρ *m.* (*v.*) (*treat*) κουράρω, (*neuter*) ευνουχίζω, (*falsify*) μαγειρεύω, (*dope*) ρίχνω ναρκωτικό σε.
doctrine *s.* δόγμα *n.* ~aire, ~al *a.* δογματικός.
document *s.* έγγραφο *n.* (*v.*) τεκμηριώνω. ~ary *a.* έγγραφος. ~ary film ντοκυμανταίρ *n.*
dodder *v.* περπατώ με κλονιζόμενο βήμα. ~ing, ~y *a.* ξεμωραμένος.
dodge *v.t.* (*elude*) ξεφεύγω. (*v.i.*) (*slip*) πετάγομαι.
dodge *s.* κόλπο *n.* ~er *s.* artful ~er μάρκα *f.* ~y *a.* αμφίβολος, δύσκολος.
dodo *s.* dead as the ~ απαρχαιωμένος.
doe *s.* (*deer*) ελαφίνα *f.* (*rabbit*) κουνέλα *f.* (*hare*) λαγίνα *f.*
doer *s.* (*perpetrator*) δράστης *m.* (*man of deeds*) άνθρωπος των έργων.
doff *v.* βγάζω.
dog *v.* κυνηγώ, ακολουθώ κατά πόδας.
dog *s.* σκύλος *m.*, σκυλί *n.*, κύων *m.* (*fig.*) lucky ~ τυχεράκιας *m.* gay ~ γλεντζές *m.* sly ~ κατεργάρης *a.*, dirty ~ βρωμόσκυλο

n. go to the ~s πάω κατά διαβόλου. let
sleeping ~s lie δεν θίγω τα κακώς κείμενα.
not have a ~'s chance δεν έχω την
παραμικρή ελπίδα. ~'s σκυλίσιος. **~days**
s. κυνικά καύματα. **~eared** *a.* (*book*) με
σελίδες τσακισμένες. **~fight** *s.* σκυλοκαβγάς *m.* **~fish** *s.* σκυλόψαρο *n.* **~sbody**
s. κλωτσοσκούφι *n.* **~star** *s.* Σείριος *m.*
~tired *a.* ψόφιος στην κούραση.
dogged *a.* επίμονος. **~ly** *adv.* επιμόνως.
~ness *s.* επιμονή *f.*
doggerel *s.* στίχοι της κακιάς ώρας.
doggo *adv.* lie ~ λουφάζω.
dogma *s.* δόγμα *n.* **~tic** *a.* δογματικός.
~tism *s.* δογματισμός *m.* **~tize** *v.* δογ-
ματίζω.
do-gooder *s.* αφελής φιλάνθρωπος.
doing *s.* (*action*) that was your ~ εσύ το
έκανες. tell me about your recent ~s πες
μου με τι ασχολείσαι τελευταία. **~s** *s.*
(*fam.*, *thing*) give me the ~s δώσ' μου το
αποτέτοιο.
doldrums *s.* ζώνη των νηνεμιών. (*fig.*) the
market is in the ~ υπάρχει απραξία στην
αγορά. he is in the ~ είναι στις κακές του.
dole *s.* (*charity*) βοήθημα *n.* be on the ~
παίρνω επίδομα ανεργίας. (*v.t.*) ~ out
μοιράζω.
doleful *a.* μελαγχολικός.
doll *s.* κούκλα *f.* (*v.t.*) ~ up στολίζω.
dollar *s.* δολλάριο *n.*
dollop *s.* κομμάτι *n.*
dolorous *a.* μελαγχολικός, λυπηρός.
dolphin *s.* δελφίνι *n.*
dolt *s.* μπούφος *m.*
domain *s.* βασίλειο *n.* (*fig.*) σφαίρα *f.*,
τομέας *m.*
dome *s.* θόλος *m.*, τρούλλος *m.* **~d** *a.*
θολωτός.
domestic *a.* (*of family*) οικογενειακός, (*of
household*) οικιακός. (*not foreign*) (*product*) εγχώριος, (*affairs, airline*) εσωτερικός. (*animal*) κατοικίδιος. ~ science
οικοκυρική *f.* ~ servant υπηρέτης *m.*,
υπηρέτρια *f.*
domesticate *v.* (*animal*) εξημερώνω,
(*plant*) εγκλιματίζω. **~d** *a.* (*man*) νοικοκύρης *m.* (*woman*) νοικοκυρά *f.*
domesticity *s.* αγάπη προς την οικογενειακή ζωή.
domicile *s.* διαμονή *f.*, τόπος διαμονής. (*v.*)
be ~d διαμένω.
dominant *a.* (*prevailing*) κυριαρχών, (*of
heights & mus.*) δεσπόζων.
dominat|e *v.* (*overlook*) δεσπόζω (*with
gen.*). (*control*) κυριαρχώ (*with gen. or
πάνω σε*). **~ion** *s.* κυριαρχία *f.*
domineer *v.i.* ~ over φέρομαι αυταρχικώς

σε. she ~s over him τον κάνει ό,τι θέλει,
τον έχει υποχείριο. **~ing** *a.* κυριαρχικός,
αυταρχικός.
dominion *s.* κυριαρχία *f.*, εξουσία *f.* ~s
(*possessions*) κτήσεις *f.pl.*
domino *s.* (*dress*) ντόμινο *n.* ~s (*game*)
ντόμινο *n.*
don *s.* καθηγητής πανεπιστημίου.
don *v.* φορώ.
donat|e *v.* δωρίζω. ~ **ion** *s.* δωρεά *f.*
done *v.* see do.
donkey *s.* γάιδαρος *m.*, γαϊδούρι *n.* (*fig.*,
fool) χαζός *a.* I haven't been there for ~'s
years χρόνια και ζαμάνια έχω να πάω εκεί.
~work *s.* αγγαρίες *f.pl.*
donnish *a.* (*fig.*) σχολαστικός.
donor *s.* δωρητής *m.*
doodle *v.i.* τραβώ γραμμές αφηρημένα.
doom *s.* μοίρα *f.* (*destruction*) καταστροφή
f. (*v.*) καταδικάζω. **~sday** *s.* till ~sday
μέχρι δευτέρας παρουσίας.
door *s.* πόρτα *f.*, θύρα *f.* front ~ εξώπορτα *f.*
behind closed ~s κεκλεισμένων των
θυρών. lie at the ~ of βαρύνω. show the ~
to διώχνω. next ~ στο διπλανό σπίτι. next
~ to πλάι σε, δίπλα σε. three ~s away τρία
σπίτια πιο πέρα. out of ~s έξω, στο
ύπαιθρο. out-of-~ (*a.*) see outdoor.
~keeper *s.* θυρωρός *m.*
doornail *s.* dead as a ~ ψόφιος.
doorstep *s.* κατώφλι *n.*
doorway *s.* (*entrance*) είσοδος *f.* in a ~ κάτω
από μία πόρτα.
dope *s.* ναρκωτικό *n.* (*fam.*) (*information*)
πληροφορία *f.*, (*v.t.*) ρίχνω ναρκωτικό σε,
(*horse*) ντοπάρω. **~y** *a.* σαν ναρκωμένος.
doric *a.* δωρικός.
dormant *a.* (*in abeyance*) εν αχρηστία,
(*latent*) λανθάνων. (*in winter*) εν νάρκη.
dormitory *s.* υπνωτήριο *n.*
dose *s.* δόση *f.* (*v.*) he was ~d with quinine
τον πότισαν κινίνο.
doss *v.* ~ (down) κοιμάμαι. **~house** *s.*
άσυλο αστέγων.
dossier *s.* φάκελλος *m.*
dot *s.* κουκκίδα *f.*, στιγμή *f.* on the ~ ακριβώς
στην ώρα. (*v.*) (*place* ~ *on*) βάζω στιγμή
σε. fields ~ted with poppies χωράφια με
σκόρπιες παπαρούνες (*or διάσπαρτα με
παπαρούνες*).
dot|e *v.* ~e on (*thing*) τρελλαίνομαι για,
(*person*) κάνω σαν τρελλός για. **~age** *s.*
ξεμώραμα *n.* **~ard** *s.* ξεμωραμένος γέρος.
become a ~ard ξεμωραίνομαι.
dotted *a.* πουαντιγέ. see dot.
dotty *a.* παλαβός. he is ~ χρωστάει της
Μιχαλούς.
double *a.* διπλός.

double *adv.* διπλά, (*twice as much*) διπλάσια, διπλάσιος από, see ~ τα βλέπω διπλά, bent ~ διπλωμένος στα δύο. (*we gave a lot for our house*) but theirs cost ~ μα το δικό τους κόστισε διπλάσια. he gets ~ my salary παίρνει το διπλάσιο μισθό από μένα (*or* δύο φορές όσο εγώ). Olympus is ~ the height of Parnes ο Όλυμπος είναι διπλάσιος στο ύψος από την Πάρνηθα.

double *s.* (*person*) σωσίας *m.* (*running*) at the ~ τροχάδην, τρέχοντας.

double *v.i.* (*become* ~ *in size, etc.*) διπλασιάζομαι. (*v.t.*) (*make* ~) διπλασιάζω, (*fold*) διπλώνω, (*go round*) κάμπτω, καβαντζάρω, (*clench*) σφίγγω. ~ back (*v.i.*) γυρίζω πίσω. ~ up (*v.i.*) διπλώνομαι στα δύο.

double-barrelled *a.* ~ shot-gun δίκαννο *n.*

double-bass *s.* κοντραμπάσσο *n.*

double-breasted *a.* σταυρωτός.

double-cross *v.* προδίδω, εξαπατώ.

double-dealing *s.* δι(πλο)προσωπία *f.*

double-decker *a.* διώροφος.

double-dutch *s.* αλαμπουρνέζικα, κινέζικα *n.pl.*

double-dyed *a.* του αισχίστου είδους.

double-edged *a.* δίκοπος.

double-entry *a.* ~ bookkeeping διπλογραφία *f.*

double-faced *a.* διπρόσωπος.

double-quick *adv.* γρήγορα-γρήγορα, (στο) άψε-σβήσε.

double-talk *s.* παραπλανητικές κουβέντες.

double-width *a.* (*cloth*) διπλόφαρδος.

doubly *adv.* διπλά, you must make ~ sure πρέπει να είστε υπερβέβαιοι, what makes it ~ important εκείνο που του δίνει αυξημένη σπουδαιότητα.

doubt *s.* αμφιβολία *f.* cast ~ on αμφισβητώ. when in ~ σε περίπτωση αμφιβολίας. there is no (room for) ~ δεν χωράει αμφιβολία. no ~, without ~ αναμφιβόλως, ασφαλώς.

doubt *v.i.* αμφιβάλλω. (*v.t.*) αμφιβάλλω για. I ~ his word δεν τον πιστεύω. ~ing Thomas άπιστος Θωμάς.

doubtful *a.* (*of things*) αμφίβολος, αβέβαιος, (*shady*) ύποπτος. be ~ (*of persons*) αμφιβάλλω, έχω αμφιβολίες. **~ly** *adv.* (*he replied*) ~ly σαν να είχε αμφιβολίες.

doubtless *adv.* αναμφιβόλως.

douche *s.* ντους *n.* cold ~ (*lit. & fig.*) ψυχρολουσία *f.*

dough *s.* ζύμη *f.*, ζυμάρι *n.* **~nut** *s.* είδος λουκουμά.

doughty *a.* ανδρείος, ηρωικός. ~ deed ανδραγάθημα *n.*

dour *a.* αυστηρός.

douse *v.* (*drench*) μουσκεύω, (*extinguish*) σβήνω.

dove *s.* περιστέρι *n.* (*also fig.*) περιστερά *f.* **~cote** *s.* περιστεριώνας *m.* **~tail** *s.* χελιδονουρά *f.* (*v.i.*) (*fig.*) συμφωνώ, συνδυάζομαι.

dowager *s.* χήρα ευγενούς. ~ duchess χήρα δούκισσα. (*fam.*) αρχοντική ηλικιωμένη κυρία.

dowdy *a.* (*clothes*) άχαρος, (*person*) άχαρα ντυμένος.

dower *v.* προικίζω. (*s.*) μερίδιο χήρας.

down *s.* (*plumage*) πούπουλο *n.* (*hair*) χνούδι *n.*

down *prep.* go ~ the river κατεβαίνω το ποτάμι. go ~ the mine κατεβαίνω στο ορυχείο. fall ~ a hole πέφτω σε ένα λάκκο. fall ~ the stairs πέφτω από τις σκάλες. tears were running ~ her face τα δάκρυα κυλούσαν στο πρόσωπό της.

down *adv.* κάτω, (*on the ground*) χάμω, (*low*) χαμηλά. ~ to here ως εδώ. ~ to the present day ως τα σήμερα. ~ to her waist μέχρι τη μέση της. from the head ~ από το κεφάλι και κάτω. cash ~ μετρητά. ~ and-out στο δρόμο, στην ψάθα. ~-at-heel κουρελιασμένος. ~-in-the-mouth αποκαρδιωμένος. ~-to-earth προσγειωμένος. ~ with slavery! κάτω η σκλαβιά. go ~ with (*illness*) πέφτω άρρωστος με, αρρωσταίνω από. be ~ on δείχνομαι αμείλικτος εναντίον (*with gen.*) κυνηγώ. be ~ on one's luck έχω ατυχίες. get ~ to business (*or* to brass tacks) έρχομαι εις το προκείμενο. come ~ in the world ξεπέφτω. he's not ~ yet δεν κατέβηκε ακόμα. you are ~ to speak tomorrow είσαι στο πρόγραμμα να μιλήσεις αύριο. it suits me ~ to the ground μου έρχεται μια χαρά. *see also* bend, come, fall, go, knock, lie, put, run, shout, take, *etc.*

down *a.* ~ payment τοις μετρητοίς.

down *v.t.* (*fam.*) ρίχνω, κατεβάζω, ~ tools τα βροντάω κάτω.

down *s.* (*fam.*) I have a ~ on him τον έχω στο μάτι, τα έχω μαζί του.

downcast *a.* άθυμος. with eyes ~ με χαμηλωμένα τα μάτια.

downfall *s.* πτώση *f.*, καταστροφή *f.*

downgrade *v.* υποβιβάζω. (*s.*) be on the ~ παίρνω την κάτω βόλτα.

downhearted *a.* αποκαρδιωμένος.

downhill *a.* κατηφορικός. (*adv.*) go ~ κατηφορίζω, κατεβαίνω, (*fig.*) παίρνω την κάτω βόλτα.

down-market *a.* κατωτέρας κατηγορίας.

downpour *s.* νεροποντή *f.*

downright *a.* (*blunt*) ντόμπρος, (*veritable*) πραγματικός, σωστός, a ~ idiot βλάκας μία φορά. *(adv.)* πραγματικά.

downs *s.* λοφώδης έκταση.

downstairs *adv.* κάτω, go ~ κατεβαίνω τη σκάλα, in a ~ room σε ένα από τα κάτω δωμάτια.

downstream *adv.* *(movement)* με το ρεύμα, *(situation)* πιο κάτω.

downtown *adv.* στο κέντρο. *(a.)* κεντρικός.

downtrodden *a.* καταπιεσμένος.

downward *a.* ~ slope κατήφορος *m.* be on the ~ path παίρνω τον κατήφορο. a ~ movement μία κίνηση προς τα κάτω. ~ (**s**) *adv.* προς τα κάτω.

downy *a.* (*face, plant*) χνουδωτός, (*pillow*) πουπουλένιος. (*fam., sly*) πονηρός.

dowry *s.* προίκα *f.*

dowse *v.* ραβδοκοπώ. ~**r** *s.* ραβδοσκόπος *m.*

doyen *s.* *(of diplomatic corps)* πρύτανις *m.* *(leading figure)* κορυφαίος *a.*

doze *v.* μισοκοιμούμαι. have a ~ τον παίρνω.

dozen *s.* ντουτζίνα *f.* about a ~ καμιά δωδεκαριά. I saw ~s είδα ένα σωρό.

drab *s.* τσούλα *f.*

drab *a.* *(scene, life)* άχρωμος, πληκτικός, *(colour)* μουντός. ~**ness** *s.* *(of scene)* μονοτονία *f.* *(of life)* ανιαρότης *f.*

drachma *s.* δραχμή *f.*

draconian *a.* δρακόντειος.

draft *s.* *(of plan)* προσχέδιο *n.* *(of MS)* πρόχειρο *n.* *(detachment)* απόσπασμα *(n. fin.)* επιταγή *f.* *(v.)* προσχεδιάζω, σχεδιάζω το πρόχειρο *(with gen.)*, αποσπώ.

drag *v.t.* τραβώ, σέρνω. ~ the bottom *(of lake, etc.)* ερευνώ το βυθό. ~ one's feet (*fig.*) χασομερώ. ~ in *(introduce)* φέρνω, εισάγω. ~ out (*prolong*) παρατραβώ. *(v.i.)* ~ on (*proceed slowly*) σέρνομαι, *(of life)* τραβώ, κρατώ.

drag *s.* *(net)* γρίπος *m.* *(impediment)* εμπόδιο *n.* *(on wheel)* τροχοπέδη *f.*

draggled *a.* λασπωμένος.

dragoman *s.* δραγουμάνος *m.*

dragon *s.* δράκων *m.* ~**fly** *s.* λιβελούλη *f.*

dragoon *s.* δραγόνος *m.* *(v.)* εξαναγκάζω.

drain *s.* οχετός *m.* *(small drop)* σταλίτσα *f.*, γουλιά *f.* *(on resources)* αφαίμαξη *f.* (*fig.*) it's down the ~ πήγε χαράμι.

drain *v.t.* αποχετεύω, *(dry, reclaim)* αποξηραίνω, *(leave to dry)* στραγγίζω, *(empty)* αδειάζω. (*fig.*) *(exhaust)* στραγγίζω, εξαντλώ. *(v.i.)* αποχετεύομαι, στραγγίζω.

drainage *s.* αποχέτευση *f.* *(a.)* αποχετευτικός.

drainpipe *s.* σωλήνας αποχετεύσεως.

dram *s.* δράμι *n.* *(small amount)* pour me a ~ βάλε μου ένα δαχτυλάκι.

drama *s.* *(lit. & fig.)* δράμα *n.* *(~tic art)* θέατρο *n.* ~**tist** *s.* δραματουργός *m.* ~**tize** *v.* δραματοποιώ.

dramatic *a.* δραματικός. ~**ally** *adv.* με δραματικό τρόπο.

drape *v.t.* *(cover)* επενδύω, καλύπτω, *(arrange)* τακτοποιώ. ~ oneself τυλίγομαι.

draper *s.* ~'s *(shop)* κατάστημα νεωτερισμών και ψιλικών. ~**y** *s.* νεωτερισμοί *m.pl.* *(in sculpture, etc.)* πτυχές *f.pl.*

drastic *a.* *(strong)* δραστικός, *(stern)* αυστηρός, *(extraordinary)* πρωτοφανής. ~**ally** *adv.* δραστικά.

draught *s.* *(air)* ρεύμα *n.* *(of drink)* ρουφηξιά *f.*, at a ~ μονορούφι. *(of chimney)* τράβηγμα *n.* *(of ship)* βύθισμα *n.* ~ animal υποζύγιο *n.* ~ beer μπίρα του βαρελιού. ~**y** *a.* ~y room δωμάτιο με ρεύματα.

draughts *s.* ντάμα *f.*

draughtsman *s.* σχεδιαστής *m.* a good ~ δυνατός στο σχέδιο. ~**ship** *s.* it suffers from bad ~ship υστερεί στο σχέδιο.

draw *s.* *(in lottery)* κλήρωση *f.* *(in game)* ισοπαλία *f.* *(centre of interest)* επίκεντρο ενδιαφέροντος, *(success)* επιτυχία *f.*

draw *v.* *(with pen, etc.)* σχεδιάζω, *(a line)* τραβώ. *(depict)* απεικονίζω, ζωγραφίζω.

draw *v.t.* (*pull, haul*) τραβώ, έλκω, σέρνω. *(extract)* βγάζω. *(curtain, sword, cork, card, lots, etc.)* τραβώ. *(vehicle)* τραβώ, έλκω. *(nail, tooth, moral, conclusion)* βγάζω, εξάγω. *(liquid)* τραβώ, αντλώ. *(bow)* τεντώνω, *(bill, cheque)* εκδίδω, *(cash, strength)* παίρνω, *(tears, applause)* αποσπώ, *(support)* προσελκύω. *(a fowl)* καθαρίζω. *(attract)* τραβώ, προσελκύω, I feel ~n to her με ελκύει. ~ breath παίρνω αναπνοή, ~ one's last breath πνέω τα λοίσθια. ~ inspiration εμπνέομαι. ~ a distinction κάνω διάκριση. I ~ the line at that δεν το ανέχομαι. ~ it mild! μην τα παραλές. he drew my attention to the danger μου επέστησε την προσοχή επί του κινδύνου. *(v.i.)* *(of chimney, pipe)* τραβώ. *(be equal)* they drew βγήκαν ισόπαλοι, ήσθαν ισοπαλία.

draw aside *v.t.* παίρνω κατά μέρος. *(v.i.)* παραμερίζω.

draw away *v.i.* απομακρύνομαι.

draw back *v.t.* *(one's hand, etc.)* αποτραβώ, *(curtains)* ανοίγω. *(v.i.)* αποσύρομαι.

draw in *v.t.* *(claws)* μαζεύω. *(v.i.)* (*get shorter*) μικραίνω. *(of train)* μπαίνω στο σταθμό.

draw level *v.* ~ with προφταίνω.

draw near *v.i.* *(be not far off)* κοντεύω, πλησιάζω. *(v.t.)* (*get close to*) πλησιάζω.

draw on v. (use) (experience, source) καταφεύγω εις, (savings) αντλώ από. (allure) δελεάζω, (to wrongdoing) παρασύρω. (get nearer) πλησιάζω, κοντεύω. (proceed) προχωρώ.

draw out v.t. (money) αποσύρω, (prolong) παρατείνω. (encourage) I drew him out τον έκανα να μιλήσει. (v.i.) (get longer) μεγαλώνω. long drawn out (story, etc.) παρατραβηγμένος, (winter, suffering) παρατεταμένος.

draw up v.t. (range) παρατάσσω, (compile) συντάσσω, (a boat) τραβώ έξω. draw oneself up τεντώνομαι. (v.i.) (stop) σταματώ.

drawback s. μειονέκτημα n.

drawbridge s. κινητή γέφυρα.

drawer s. συρτάρι n. chest of ~s κομό n.

drawers s. (man's) σώβρακο n., (woman's) κυλότα f., βρακί n.

drawing s. (art) σχέδιο n. (as school subject) ιχνογραφία f. (sketch, plan) σχέδιο n. he is fond of ~ του αρέσει να σχεδιάζει.

drawing-pin s. πινέζα f.

drawing-room s. σαλόνι n.

drawl v.i. σέρνω τη φωνή μου. (s.) συρτή φωνή.

drawn a. (sword) γυμνός, (face) τραβηγμένος, (game) the game was ~ βγήκαν ισόπαλοι.

dray s. κάρρο n.

dread s. τρόμος m. (v.) τρέμω, φοβούμαι.

dreadful a. φοβερός. **~ly** adv. φοβερά.

dream s. όνειρο n. (unrealizable) ονειροπόλημα n. have a ~ βλέπω όνειρο. **~like** a. ονειρώδης. **~y** a. (person) ονειροπόλος m.f. (music, etc.) απαλός, (dim) θολός.

dream v. (also ~ about) ονειρεύομαι, (daydream) ονειροπολώ. (suppose, imagine) φαντάζομαι. I shouldn't ~ of it ούτε θα μπορούσα να το διανοηθώ. ~ up επινοώ, σκαρφίζομαι. **~er** s. ονειροπόλος m.f.

dreary a. καταθλιπτικός, ανιαρός, πληκτικός.

dredge v. βυθοκορώ. ~ up γριπίζω. **~r** s. βυθοκόρος f., φαγάνα f.

dregs s. κατακάθι n., κατακάθια pl. υποστάθμη f.

drench v. μουσκεύω. get ~ed μουσκεύω, γίνομαι μούσκεμα (or λούτσα). he ~ed the salad with oil πλημμύρισε τη σαλάτα με λάδι.

dress s. (attire) ενδυμασία f., (way of ~ing) ντύσιμο n. (woman's) φουστάνι, φόρεμα n., (formal) τουαλέττα f. full ~ (uniform) μεγάλη στολή. ~ coat or suit φράκο n. fancy ~ αποκριάτικα ρούχα, fancy ~ ball

χορός μεταμφιεσμένων. ~ circle εξώστης m. ~ rehearsal γενική δοκιμή.

dress v.t. ενδύω, ντύνω, (food) ετοιμάζω, (wound) επιδένω, (wood, stone) πελεκώ, (hair) χτενίζω. (adorn) στολίζω. (v.i.) (also get ~ed) ντύνομαι. (mil.) ζυγώ, right ~! ζυγείτε από δεξιά.

dresser s. (furniture) μπουφές με ράφια.

dressing s. (getting dressed) ντύσιμο n. (med.) (bandage) επίδεσμος m., (putting it on) επίδεση f. (for salad) λαδολέμονο n., λαδόξυλο n. (of textiles) κολλάρισμα n. (manure) λίπασμα n.

dressing-down s. κατσάδα f.

dressing-gown s. ρόμπα f.

dressing-room s. τουαλέττα f., (stage) καμαρίνι n.

dressing-table s. τουαλέττα f.

dressmaker s. μοδίστρα f., ράφτρα f. **~ing** s. ραπτική f.

dressy a. κομψός. (woman) κοκέτα, (garment) κοκέτικος.

dribble v.i. βγάζω σάλια, (drip) στάζω. he ~s τρέχουν τα σάλια του.

driblet s. in ~s με το σταγονόμετρο.

dried a. (fruits, etc.) ξερός, (liquids) σκόνη f.

drier s. see dryer.

drift v. (be carried) παρασύρομαι, (wander) περιφέρομαι ασκόπως. (s.) (movement) κίνηση f. (course) πορεία f. (of snow) στιβάδα f. (meaning) έννοια f.

drifter, **~-net** s. ανεμότρατα f. **~-wood** s. εκβρασμένα συντρίμματα ξύλου.

drill s. (tool) τρυπάνι n. (v.) (make hole in) τρυπώ, τρυπανίζω. ~ a hole ανοίγω τρύπα.

drill s. (mil. & gym) άσκηση f. (fig.) what's the ~? τι κάνουμε τώρα; (v.t.) γυμνάζω. (v.i.) γυμνάζομαι.

drill s. (cloth) ντρίλι n.

drily adv. ξερά, κάπως ειρωνικά.

drink v. πίνω. ~ in ρουφώ. (s.) (beverage) ποτό n. he has taken to ~ το 'χει ρίξει στο πιοτό. **~er** s. (habitual) πότης m. (patron of bar, etc.) ο πίνων.

drinkable a. πόσιμος. it is not ~ δεν πίνεται.

drinking s. πιοτό n. **~-water** s. πόσιμο νερό.

drip v.t. & i. στάζω. (s.) στάλα f., σταγόνα f. **~ping** s. στάξιμο n. (fat) λίπος n. (a.) ~ping (wet) μούσκεμα, παπί.

drip-dry a. που δεν χρειάζεται σιδέρωμα.

drive v.i. (be driver) οδηγώ, (be conveyed) πηγαίνω με αυτοκίνητο (by car) or με αμαξάκι (by carriage). he let ~e at me μου έδωσε μία. what are you ~ing at; τι υπονοείς; (v.t.) (beasts, vehicle) οδηγώ. (convey) πηγαίνω, (activate) κινώ, (thrust) σπρώχνω, ρίχνω, (expel) διώχνω, απωθώ, (*compel*) αναγκάζω. (*force to work hard*) ξεθεώνω, βγάζω το λάδι (*with gen.*). (*a road, tunnel*) διανοίγω, (*a nail*) καρφώνω, (*a screw*) βιδώνω. ~e a hard bargain διαπραγματεύομαι σκληρά. ~e mad τρελλαίνω. ~e into a corner στριμώχνω. I drove it home to him του το έβαλα μέσα στο κεφάλι, τον έκανα να καταλάβει.

drive *s.* (*excursion*) περίπατος *m*, βόλτα *f.*, (*by car*) αυτοκινητάδα *f.*, (*by carriage*) αμαξάδα *f.* (*course covered*) διαδρομή *f.* an hour's ~ μία ώρα με το αυτοκίνητο. *mech.*) κίνηση *f.* (*energy*) δραστηριότητα *f.* (*campaign*) καμπάνια *f.*

drivel *v.* μωρολογώ. ~ling idiot βλακόμετρο *n.* (*s.*) σαλιαρίσματα *n.pl.*

driver *s.* οδηγός *m.* (*of coach, carriage*) αμαξάς *m.* (*of car*) σοφέρ *m.*

driving *s.* οδήγηση *f.*, σοφάρισμα *n.* have ~ lessons μαθαίνω να οδηγώ. ~ licence άδεια οδηγήσεως. ~ test εξετάσεις για άδεια οδηγού.

drizzle *s. & v.* ψιχάλα *f.* it ~s ψιχαλίζει, ψιλοβρέχει.

droll *a.* κωμικός, αστείος. ~ery *s.* αστειότητα *f.*

drone *s.* κηφήν *m.* (*fig.*) κηφήνας *m.* (*hum*) βόμβος *m.* (*of voices*) μουρμουρητό *n.* (*v.*) βομβώ, μουρμουρίζω.

drool *v.* σαλιαρίζω.

droop *v.i.* (*of head*) γέρνω, (*of flowers*) γέρνω, μαραίνομαι. his spirits ~ed έχασε το ηθικό του.

drop *s.* σταγών *f.*, σταγόνα *f.*, στάλα *f.* not a ~ ούτε (μία) στάλα. (*fall*) πτώση *f.* at the ~ of a hat αμέσως, χωρίς πολλά παρακάλια. a hundred-foot ~ γκρεμός ύψους εκατό ποδιών. he's had a ~ too much το έχει τραβήξει λίγο.

drop *v.i.* πέφτω. (*fall in* ~s) στάζω. (*of person, collapse*) σωριάζομαι, ~ dead μένω στον τόπο. (*of voice*) χαμηλώνω. he let ~ a remark πέταξε μία κουβέντα.

drop *v.t.* (*let fall*) I ~ped the book μου έπεσε το βιβλίο. I ~ped a stitch μου έφυγε ένας πόντος. (*passenger, parcel, message*) αφήνω. (*men, supplies from air*) ρίπτω. (*bomb, anchor, hint*) ρίχνω. (*voice*) χαμηλώνω, κατεβάζω. (*knock down*) ρίχνω κάτω. (*give up*) εγκαταλείπω, παρατώ, αφήνω. ~ smoking κόβω το τσιγάρο. (*cease to be friends with*) κόβω (*or* διακόπτω) σχέσεις με. (*exclude from office*) βγάζω, αφήνω έξω. (*not pronounce*) δεν προφέρω. ~ me a line γράψε μου δυο λόγια. ~ it! άσ' το, σταμάτα.

drop away *v.i.* (*scatter*) σκορπίζω, (*get*

less) πέφτω, λιγοστεύω.
drop back, behind *v.i.* μένω πίσω.
drop in *v.i.* I will ~ this evening θα περάσω απόψε.
drop off *v.i.* (*come off*) βγαίνω, (*sleep*) τον παίρνω. see also drop away.
drop on *v.* (*deal severely with*) κυνηγώ, δείχνομαι αμείλικτος εναντίον (*with gen.*).
drop out *v.i.* αποσύρομαι, (*of ranks*) μένω πίσω. (*abandon studies*) εγκαταλείπω τις σπουδές μου. (*socially*) αρνούμαι να ζήσω σύμφωνα με το κατεστημένο.
drop-out *s.* ο αρνούμενος να ζήσει σύμφωνα με το κατεστημένο.
dropper *s.* σταγονόμετρο *n.*
droppings *s.* (*of candle*) σταξίματα *n.pl.* (*sheep's, rabbit's*) βερβελιά *f.* (*cow's*) σβουνιά *f.* (*horse's*) (γ)καβαλίνα *f.* (*bird's*) κουτσουλιά *f.* (*all also in pl.*).
dropsy *s.* υδρωπικία *f.*
dross *s.* (*fig.*) παλιόπραμα *n.*, σκουπίδι *n.*
drought *s.* ανομβρία *f.*, ξηρασία *f.*
drove *s.* αγέλη *f.* in ~s αγεληδόν. ~r *s.* βοσκός *m.*
drown *v.t.* πνίγω. (*v.i.*) (*also* get ~ed) πνίγομαι. ~ing ~s. πνιγμός *m.*, πνίξιμο *n.*
drowse *v.* μισοκοιμούμαι. ~ily *adv.* νυσταγμένα. ~iness *s.* νύστα *f.* ~y *a.* μισοκοιμισμένος, νυσταλέος.
drubbing *s.* give (*person*) a ~ ξυλοκοπώ.
drudge *s.* είλωτας *m.* (*v.*) μοχθώ. ~ry *s.* μόχθος *m.*, αγγαρεία *f.*
drug *s.* φάρμακο *n.* (*narcotic*) ναρκωτικό *n.* it is a ~ on the market δεν έχει ζήτηση. ~ addict τοξικομανής *m.* (*v.*) they ~ged him τον πότισαν ναρκωτικό. they ~ged his food του έβαλαν ναρκωτικό στο φαΐ του.
drugget *s.* φτηνό κάλυμμα του πατώματος.
druggist *s.* φαρμακοποιός *m.* ~'s *s.* φαρμακείο *n.*
drugstore *s.* φαρμακείο *n.*
druid *s.* δρυΐδης *m.*
drum *s.* τύμπανο *n.*, τούμπανο *n.* (*of cable*) καρούλι *n.* (*barrel*) βαρέλι *n.* (*of column*) σπόνδυλος *m.* beating of ~s τυμπανοκρουσία *f.*
drum *v.* (*also* play ~) κρούω τύμπανον. (*fig., beat*) χτυπώ, (*with fingers*) παίζω ταμπούρλο με τα δάχτυλα., ~ up συγκεντρώνω. I ~med it into him το έχωσα στο μυαλό του. ~mer *s.* τυμπανιστής *m.*
drumfire *s.* (*mil.*) πυρ φραγμού.
drunk *a.* μεθυσμένος. get ~ μεθώ. ~ard *s.* μέθυσος *m.*, μπεκρής *m.*
drunken *a.* μεθυσμένος. ~ orgy όργιον μέθης. ~ brawl καβγάς μεθυσμένων. ~ness *s.* μέθη *f.*, μεθύσι *n.*

dry a. (*body, hair, clothes, etc.*) στεγνός. (*skin, mouth, soil, climate, vegetation, battery*) ξερός, (*wine*) μπρούσκος, (*manner, book, facts*) ξερός. (*thirsty*) διψασμένος, feel ~ διψώ. run ~ (*river, well* & *fig.*) στερεύω. go ~ (*ban liquor*) εφαρμόζω ποταπαγόρευσιν. ~ as a bone κατάξερος. ~ cleaning στεγνό καθάρισμα. ~ dock ξηρά δεξαμενή. ~ land ξηρά f. ~ nurse νταντά f. ~ rot σήψη ξυλείας κτιρίων, (*fig.*) λανθάνουσα σήψη. ~ walling ξερολιθιά f.

dry v.t. στεγνώνω, ξεραίνω, (*drain*) αποξηραίνω. (*v.i.*) (*also* get ~) στεγνώνω, ξεραίνομαι. ~ up (*fam.*) το βουλώνω.

dryad s. δρυάς f.

dryer s. στεγνωτήριο n. (*for hair*) σεσουάρ n.

dryness s. ξηρότητα f. (*of earth*) ξηρασία f.

dual a. (*control, ownership*) διπλός. (*gram.*) δυϊκός. **~-purpose** a. διπλής χρήσεως.

dub v. (*fam., call*) βαφτίζω, βγάζω. (*film*) ντουμπλάφω.

dubious a. (*mistrustful*) δύσπιστος. (*causing doubt*) αμφίβολος, (*suspect*) ύποπτος. I am ~ αμφιβάλλω, έχω αμφιβολίες. **~ly** adv. δύσπιστα, σα να είχα αμφιβολίες.

ducal a. δουκικός.

ducat s. δουκάτο n.

duchess s. δούκισσα f.

duchy s. δουκάτο n.

duck s. πάπια f., νήσσα f. (*fam., attractive child, etc.*) κούκλα f., (*score of nil*) μηδέν n. lame ~ (*person*) άνθρωπος ανάπηρος, (*firm*) επιχείρηση που χωλαίνει.

duck v.i. σκύβω. (*v.t.*) (*put under water*) βουτώ. (*fam., avoid*) αποφεύγω. **~ing** s. he got a ~ ing τον βούτηξαν στο νερό.

duckboards s. σανίδωμα προς αποφυγήν λασπών.

duckling s. παπάκι n.

duct s. αγωγός m. (*anat.*) πόρος m. **~ile** a. όλκιμος, (*fig.*) μαλακός.

dud a. & s. (*thing*) σκάρτος, (*person*) μηδενικό n., ντενεκές m.

dudgeon s. in (high) ~ φουρκισμένος.

duds s. (*fam., clothes*) κουρέλια n.pl.

due a. (*payable*) οφειλόμενος, πληρωτέος. (*necessary, proper*) δέων, απαιτούμενος. (*just*) δίκαιος. (*expected*) the train is ~ in at any moment now το τραίνο αναμένεται όπου νά 'ναι. the performance is ~ to start at ten η παράσταση πρόκειται να αρχίσει στις δέκα. (*owed or caused*) it is ~ to the fact that... οφείλεται στο ότι. in ~ form κανονικώς. (*adv.*) ~ to (*because of*) εξαιτίας, λόγω (*with gen.*). (*of points of compass*) κατευθείαν.

due s. οφειλόμενον n. to give the devil his ~ να πούμε και του στραβού το δίκιο. **~s** δικαιώματα n.pl., τέλη n.pl.

duel, **~ling** s. μονομαχία. fight a ~ μονομαχώ. **~list** s. μονομάχος m.

duet. δυωδία f., ντουέτο n.

duffer s. ατζαμής m.

dug s. μαστάρι n.

dug-out s. αμπρί n. ~ canoe μονόξυλο n.

duke s. δούκας m. **~dom** s. δουκάτο n.

dulcimer s. σαντούρι n.

dull a. (*obtuse*) βραδύνους. (*depressed*) άκεφος, feel ~ πλήττω. (*bored*) βαριεστημένος, (*boring*) πληκτικός. (*colour, weather*) μουντός, (*sound*) υπόκωφος, (*market*) αδρανής. (*not shiny*) θαμπός, (*not sharp*) αμβλύς.

dull v.t. & i. θαμπώνω. (*v.t.*) (*blunt*) αμβλύνω, (*lessen*) ελαφρώνω. (*v.i.*) (*cloud over*) συννεφιάζω.

dullard s. βραδύνους a.

dullness s. (*of wits*) βραδύνοια f. (*being bored*) βαριεστιμάρα f. (*monotony*) μονοτονία f.

dully adv. (*not brightly*) θαμπά, (*in dull mood*) βαριεστημένα.

duly adv. δεόντως, κανονικώς.

dumb a. βουβός. he was struck ~ έμεινε άναυδος. become ~ βουβαίνομαι. ~ show παντομίμα f. (*silly*) χαζός. **~ly** adv. σιωπηλά. **~ness** s. βουβαμάρα f. (*silliness*) χαζομάρα f.

dumb-bells s. αλτήρες m.pl.

dumbfound v. κατακεραυνώνω. be **~ed** μένω εμβρόντητος (*or* ενεός).

dummy s. ομοίωμα n. (*tailor's*) κούκλα f. (*baby's*) πιπίλα f. (*at bridge*) μορ. (*a.*) τεχνητός, ψεύτικος. ~ run δοκιμή f.

dump v. πετώ, απορρίπτω, ξεφορτώνω. (*s.*) (*supplies*) αποθήκη f. (*rubbish*) σκουπιδότοπος m. (*pej., place*) παλιότοπος m. **~ing** s. (*trade*) ντάμπινγκ n.

dumps s. (*fam.*) be (down) in the ~ είμαι στις μαύρες μου.

dumpy a. κοντόχοντρος.

dun a. γκρι-καφέ.

dun v. πιέζω (*s.*) πιεστικός δανειστής.

dunce s. κούτσουρο n.

dunderhead v. ντουβάρι n.

dune s. αμμόλοφος m.

dung s. κοπριά f. *see* droppings.

dungarees s. εργατική φόρμα.

dungeon s. μπουντρούμι n.

dunk v. βουτώ.

duodecimal a. δωδεκαδικός.

duodenal a. δωδεκαδακτυλικός.

dupe v. κοροϊδεύω, πιάνω κορόιδο. (*s.*) κορόιδο n.

duple *a.* διμερής.

duplicate *a.* (*same*) όμοιος. we have ~ keys έχουμε από τα ίδια κλειδιά. in ~ εις διπλούν. (*s.*) (*of bill, etc.*) διπλότυπο *n.* (*similar specimen*) ακριβές αντίγραφο.

duplicat|e *v.t.* (*repeat*) επαναλαμβάνω, (*make copies of*) βγάζω αντίγραφα (*with gen.*) **~ion** *s.* επανάληψη *f.* **~or** *s.* πολύγραφος *m.*

duplicity *s.* διπροσωπία *f.*

durab|le *a.* ανθεκτικός, γερός, αντοχής. **~ility** *s.* ανθεκτικότητα *f.*, αντοχή *f.*

duration *s.* διάρκεια *f.* for the ~ of the war (*looking back*) όσο διήρκεσε *or* (*looking forward*) όσο διαρκέσει ο πόλεμος.

duress *s.* under ~ υπό πίεσιν.

during *prep.* κατά (*with acc.*), κατά το διάστημα *or* κατά τη διάρκεια (*with gen.*).

dusk *s.* σούρουπο *n.*, λυκόφως *n.* **~y** *a.* σκοτεινός, (*in colour*) μελαψός.

dust *s.* σκόνη *f.*, κονιορτός *m.* throw ~ in his eyes ρίχνω στάχτη στα μάτια του. bite the ~ πέφτω ηττημένος. shake the ~ off one's feet ρίχνω μαύρη πέτρα πίσω μου. they raised the ~ *or* had a ~-up έγινε καβγάς, χάλασαν το κόσμο. **~bin** *s.* σκουπιδοτενεκές *m.* **~cart** *s.* αυτοκίνητο των σκουπιδιών. **~jacket** *s.* κάλυμμα βιβλίου. **~man** *s.* σκουπιδιάρης *m.* **~pan** *s.* φαράσι *n.*

dust *v.t.* ξεσκονίζω, (*powder*) πουδράρω. **~er** *s.* ξεσκονόπανο *n.* **~ing** *s.* ξεσκόνισμα *n.*

dusty *a.* γεμάτος σκόνη, (*object*) σκονισμένος, get ~ γεμίζω σκόνη, σκονίζομαι. (*fam.*) not so ~ αρκετά καλός.

Dutch *a.* ολλανδικός. (*person*) Ολλανδός *m.*, Ολλανδέζα *f.* (*fam.*) go ~, have a ~ treat πληρώνω με ρεφενέ. double ~ αλαμπουρνέζικα *n.pl.*

dutiable *a.* φορολογήσιμος.

dutiful *a.* υπάκουος, πιστός, ευπειθής. **~ly** *adv.* ευπειθώς.

duty *s.* καθήκον *n.* (*obligation*) υποχρέωση *f.* (*function, service*) υπηρεσία *f.* be on ~ είμαι της υπηρεσίας, off ~ εκτός υπηρεσίας. be in ~ bound έχω υποχρέωση. do ~ for (*be used as*) χρησιμεύω για. present one's ~ (*respects*) υποβάλλω τα σέβη μου.

duty *s.* (*tax*) δασμός *m.*, φόρος *m.* free of ~ (*a.*) ατελής, αφορολόγητος, (*adv.*) ατελώς.

dwarf *s. & a.* νάνος *m.* (*v.*) our house is ~ed by the block of flats next door το σπίτι μας φαίνεται σαν νάνος μπροστά στη διπλανή πολυκατοικία.

dwell *v.* (*live*) ζω, διαμένω, κατοικώ. ~ on (*treat at length*) επιμένω σε, ενδιατρίβω εις, (*stress*) τονίζω, (*hold note, etc.*) κρατώ. **~er** *s.* town-~ers κάτοικοι πόλεων. **~ing** *s.* σπίτι *n.*, κατοικία *f.*

dwindle *v.* ελαττώνομαι, λιγοστεύω, πέφτω. (*waste away*) φθίνω.

dye *s.* βαφή, *f.* μπογιά *f.*, (*fig.*) of the deepest ~ του αισχίστου είδους. (*v.*) βάφω. (*fig.*) ~d-in-the-wool βαμμένος. **~ing** *s.* βαφή *f.*, βάψιμο *n.*

dyer's *s.* βαφείο *n.*

dying *s.* θάνατος *m.* (*a.*) (*on point of death*) ετοιμοθάνατος, (*last*) τελευταίος. (~ *out*) εκλείπων, a ~ art μία τέχνη που εκλείπει.

dyke *s.* ανάχωμα *n.* (*ditch*) τάφρος *m.*

dynam|ic *a.* δυναμικός. **~ics** *s.* δυναμική *f.* **~ism** *s.* δυναμισμός *m.*

dynamite *s.* δυναμίτης *f.*

dynamo *s.* γεννήτρια *f.*, δυναμό *n.*

dynast|y *s.* δυναστεία *f.* **~ic** *a.* δυναστικός.

dysentery *s.* δυσεντερία *f.*

dyslexia *s.* δυσλεξία *f.*

dyspep|sia *s.* δυσπεψία *f.* **~tic** *a.* δυσπεπτικός.

E

each *a.* κάθε, ~ time κάθε φορά. (*pron.*) ~ one ο καθείς, ο καθένας, ~ of us ο καθένας μας, ~ in turn ο καθένας με τη σειρά του. they cost 10 drachmas ~ κοστίζουν δέκα δραχμές το ένα (*or* το καθένα). I gave them 10 drachmas ~ τους έδωσα από δέκα δραχμές. ~ and every εις έκαστος. ~ (*of two*) εκάτερος, on ~ side και από τη μια και από την άλλη πλευρά, εκατέρωθεν. they help ~ other βοηθάει ο ένας τον άλλο *or* βοηθούνται μεταξύ τους *or* αλληλοβοηθούνται. we see ~ other often βλεπόμαστε συχνά.

eager *a.* (*to please, etc.*) πρόθυμος, ολοπρόθυμος. be ~ ανυπομονώ, επιθυμώ. he is ~ (*anxious*) to leave δεν βλέπει την ώρα να φύγει. **~ly** *adv.* προθύμως, με ανυπομονησία. **~ness** *s.* προθυμία *f.*, ανυπομονησία *f.*

eagle *s.* αετός *m.*

ear *s.* αυτί *n.* (*of corn*) στάχι *n.* give (*or* lend an) ~ ακούω. up to the ~s in work πνιγμένος στη δουλειά. set them by the ~s σπέρνω ζιζάνια ανάμεσά τους. keep an ~ to the ground έχω το αυτί μου τεντωμένο για το τι λέγεται και γίνεται. lend a favourable ~ τείνω ευήκοον ους. he got a thick ~ έφαγε μια καρπαζιά. **~-drum** *s.* τύμπανο *n.* **~-phones** *s.* ακουστικά κεφαλής. **~-wax** *s.* κυψέλη *f.*

earl *s.* κόμης *m.* **~dom** *s.* κομητεία *f.*

ear|ly *a.* (*in the day*) πρωινός, (*in season*) πρώιμος, (*in good time*) έγκαιρος. from an

~y age από πολύ νωρίς. we had an ~y meal φάγαμε νωρίς. it is ~y closing day τα μαγαζιά είναι κλειστά το απόγευμα. (adv.) (ε)νωρίς, (in the morning) πρωί, (in season) πρώιμα. ~y in May στις αρχές του Μάη. ~ier adv. νωρίτερα, (in the day) πιο πρωί. ~iest a. at the ~iest opportunity όσο το δυνατόν νωρίτερα. my ~iest memories οι πιο παλιές μου αναμνήσεις.
earmark v. προορίζω.
earn v. κερδίζω. ~ings s. κέρδη n.pl.
earnest a. (serious) σοβαρός, (ardent) θερμός. ~ly adv. σοβαρώς, θερμώς.
earnest s. (money) καπάρο n. (example) δείγμα n.
earring s. σκουλαρίκι n.
earshot s. within (or out of) ~ εντός (or εκτός) ακτίνος ακοής.
earth s. γη f. (soil) χώμα n. (electric) προσγείωση f. down to ~ προσγειωμένος. the biggest on ~ ο μεγαλύτερος του κόσμου. run (person) to ~ ανακαλύπτω. why on ~ did you do that? γιατί στην οργή το 'κανες αυτό; the ~'s sphere γήινη σφαίρα.
earthen a. (made of earth) χωματένιος. (made of clay) πήλινος. ~ware s. (collective) πήλινα σκεύη. (a.) πήλινος.
earthly a. γήινος, επίγειος. (fam.) that is no ~ use δεν χρησιμεύει σε τίποτα. for no ~ reason για κανένα λόγο. have no ~ chance δεν έχω την παραμικρή ελπίδα.
earthquake s. σεισμός m.
earthwork s. πρόχωμα n.
earthy a. it has an ~ smell μυρίζει χώμα. (fig.) χοντρός, χοντροκομμένος.
ease s. ευκολία f., ευχέρεια f. (comfort) άνεση f. feel at ~ νιώθω άνετα. he put me at my ~ με έκανε να νιώθω άνετα.
ease v.t. (relieve) ανακουφίζω, (assist) διευκολύνω, (loosen) χαλαρώνω. (v.i.) ~ (off) (become less) κόβω, πέφτω.
easel s. καβαλέτο n., ορχίβας m.
easily adv. εύκολα, άνετα.
east s. ανατολή f. to the ~ of Athens ανατολικώς των Αθηνών. Near (or Far) E~ Εγγύς (or Άπω) Ανατολή.
east, ~erly, ~ern, ~ward a. ανατολικός. ~ wind λεβάντες m., απηλιώτης m. ~wards adv. προς ανατολάς.
Easter s. Πάσχα n., Λαμπρή f. (a.) πασχαλινός.
easy a. εύκολος, ευχερής, (comfortable) άνετος. (free from worry) ήσυχος. (not tiring) ~ work ξεκούραστη δουλειά. have an ~ conscience έχω τη συνείδησή μου ήσυχη (or αναπαυμένη). (the room) is ~ to heat θερμαίνεται εύκολα. (adv.) take

things ~ ζω λιγότερο εντατικά. ~ does it! σιγά σιγά! ~ on the wine! σιγά το κρασί! ~-chair s. πολυθρόνα f. ~-going a. καλόβολος.
eat v. τρώ(γ)ω, μασώ. he had ~en είχε φάει or ήταν φαγωμένος. (fig.) ~ one's words παίρνω πίσω τα λόγια μου. ~-able a. φαγώσιμος. it isn't ~able δεν τρώγεται. ~er s. meat-~er κρεοφάγος a.
eating s. το τρώγειν. he likes ~ του αρέσει να τρώει. ~ and drinking φαγοπότι n.
eau-de-Cologne s. κολόνια f.
eaves s. γείσο n.
eavesdrop v.i. ωτακουστώ, (also ~ on) κρυφακούω. ~per s. ωτακουστής m.
ebb s. (tide) άμπωτις f. (fig.) παρακμή f., on the ~ σε κάμψη. (v.) (of tide) πέφτω. (fig.) παρακμάζω, (of life, daylight) σβήνω.
ebony s. έβενος m.f.
ebullient a. be ~ ξεχειλίζω από ενθουσιασμό.
eccentric a. εκκεντρικός. ~ity s. εκκεντρικότητα f.
ecclesiastic s. κληρικός m. ~al a. εκκλησιαστικός.
echelon s. (mil.) κλιμάκιον n. (of civil service, etc.) βαθμίς f.
echo s. ηχώ f., αντίλαλος m. (v.i.) αντηχώ, αντιλαλώ. (v.t.) απηχώ. (fig., repeat) επαναλαμβάνω.
eclectic a. be ~ in one's tastes δεν έχω περιορισμένα γούστα.
eclipse s. έκλειψη f., there is an ~ of the moon το φεγγάρι κάνει έκλειψη. (v.t., fig.) επισκιάζω.
ecolog|y s. οικολογία f. ~ical a. οικολογικός.
economic a. οικονομικός. ~s s. οικονομολογία f., οικονομικά n.pl., οικονομική f.
economical a. (thing) οικονομικός, (person) οικονόμος m.f. ~ly adv. (inexpensively) οικονομικά, (from an economic point of view) οικονομικώς.
economist s. οικονομολόγος m.f.
economize v.i. κάνω οικονομίες.
economy s. οικονομία f. (fig., of style) λιτότητα f. black ~ παραοικονομία f.
ecsta|sy s. έκσταση f. ~tic a. εκστατικός.
ecumenical a. οικουμενικός.
eczema s. έκζεμα n.
eddy s. δίνη f., στρόβιλος m. (v.) στροβιλίζομαι.
Eden s. Εδέμ f.
edge s. άκρη f. (of blade) κόψη f. (rim) χείλος n. at the very ~ άκρη-άκρη. take the ~ off (knife) στομώνω, (appetite) κόβω. I have the ~ on him τον έχω ξεπεράσει. on ~ εκνευρισμένος.

edge v.t. (push slowly) σπρώχνω σιγά σιγά. ~ one's way forward προχωρώ βαθμιαίως. (of material, etc.) ~d with yellow με κίτρινη μπορντούρα. (of hat, border, etc.) it is ~d with flowers έχει μπορντούρα από λουλούδια.

edgeways adv. πλαγίως.

edging s. μπορντούρα f. (only of cloth) ρέλι n., τελείωμα n.

edgy a. νευρικός.

edible a. φαγώσιμος. it is not ~ δεν τρώγεται.

edict s. διάταγμα n.

edification s. ηθική εξύψωση. for your ~ προς γνώσιν σας.

edifice s. οικοδόμημα n.

edify v. εξυψώνω ηθικώς. ~ing a. εποικοδομητικός.

edit v. (prepare for publication) ετοιμάζω προς έκδοσιν. he ~s the "Times". είναι συντάκτης των Τάιμς. (critically) επιμελούμαι (with gen.). who ~ed the work? ποιος είχε την επιμέλειαν της εκδόσεως; (a book) ~ed by J. Vlachos υπό την επιμέλεια του Ι. Βλάχου. ~ion s. έκδοση f.

editor s. (of paper) συντάκτης m. (of critical edition) επιμελητής m. ~ial a. συντακτικός. ~ial staff σύνταξη f. (s.) κύριον άρθρον.

educat|e v. εκπαιδεύω, μορφώνω. ~ed a. μορφωμένος. ~or s. εκπαιδευτικός m.

education s. παιδεία f., εκπαίδευση f., μόρφωση f. ~al a. εκπαιδευτικός, μορφωτικός. ~alist s. ειδικός στα εκπαιδευτικά ζητήματα.

eel s. χέλι n.

eerie a. μυστηριώδης.

efface v. εξαλείφω. ~ oneself μένω εθελοντικώς παράμερα.

effect s. (result) αποτέλεσμα n. (influence) επίδραση f. (consequence) συνέπεια f. (impression) εντύπωση f., (in art) εφφέ n. have an ~ on επηρεάζω. put into ~ or give ~ to εφαρμόζω. come into ~ τίθεμαι εν ισχύι. take ~ (bring results) φέρνω αποτέλεσμα. in ~ ουσιαστικά. to no ~ χωρίς αποτέλεσμα. to the same ~ με το ίδιο νόημα.

effect v. πραγματοποιώ, επιτυγχάνω. they ~ed an entry κατόρθωσαν να μπουν μέσα. ~ an insurance policy κάνω ασφάλεια.

effective a. αποτελεσματικός, (striking) πετυχημένος, (actual) πραγματικός, (operative) ισχύων, (of armed forces) μάχιμος. ~ly adv. αποτελεσματικώς, πετυχημένα, πραγματικώς, ~ness s. αποτελεσματικότητα f. επιτυχία f.

effectual a. τελεσφόρος.

effectuate v.t. see effect.

effemin|ate a. θηλυπρεπής. ~acy s. θηλυπρέπεια f.

effervesc|e v. αφρίζω. (fig.) (of person) ξεχειλίζω από ζωή. ~ence s. άφρισμα n. ~ent a. (drink) αεριούχος.

effete a. εξαντλημένος.

efficac|y s. αποτελεσματικότης f. ~ious a. αποτελεσματικός.

effici|ent a. (person) ικανός, άξιος, (thing) αποτελεσματικός, (both) αποδοτικός. ~ency s. ικανότητα f., αποδοτικότητα f. ~ently adv. με ικανότητα, αποτελεσματικώς.

effigy s. ομοίωμα n.

effluent s. εκροή f. (sewage, etc.) απόβλητα n.pl., λύματα n.pl.

effort s. (attempt) προσπάθεια f. (trouble) κόπος m.

effortless a. αβίαστος. ~ly adv. ακόπως.

effrontery s. αυθάδεια f.

effusive a. υπερβολικά διαχυτικός. ~ness s. υπερβολική διαχυτικότητα.

egalitarian s. & a. υπέρμαχος της κοινωνικής ισότητος.

egg s. αβγό n. bad ~ (lit.) κλούβιο αβγό, (fig.) παλιάνθρωπος m. put all one's ~s in one basket διακυβεύω τα πάντα σε μια επιχείρηση. ~-cup s. αβγουλιέρα f., αβγοθήκη f. ~-head s. (fam.) διανοούμενος m. ~-plant s. μελιτζάνα f. ~-shaped f. ωοειδής. ~-shell s. αβγότσουφλο n.

egg v. ~ on παρακινώ, σπρώχνω.

ego s. εγώ n. ~-centric a. εγωκεντρικός.

ego|ism s. εγωισμός m. ~ist s. εγωιστής m. ~istic a. εγωιστικός.

egot|ism s. περιαυτολογία. f. ~ist s. περιαυτολόγος a.

egregious a. an ~ fool βλάκας με πατέντα.

egress s. έξοδος f.

Egyptian a. αιγυπτιακός. (person) Αιγύπτιος.

eh int. ε.

eight num. οκτώ, οχτώ, ~ hundred οκτακόσιοι a. ~h a. όγδοος.

eighteen num. δεκαοκτώ. ~th a. δέκατος όγδοος.

eight|y num. ογδόντα. ~ieth a. ογδοηκοστός.

either adv. & a. ~ ... or ή ... ή, είτε ... είτε. (he doesn't like it) and nor do I ~ ούτε εγώ. ~ (of two) εκάτερος. ~ of them will suit me και τα δύο μου κάνουν. you can have ~ of these books μπορείς να πάρεις όποιο από τα δύο βιβλία προτιμάς. on ~ side εκατέρωθεν, και από τις δύο πλευρές.

ejaculat|e v. φωνάζω. ~ion s. φωνή f.

eject v.t. εκβάλλω, βγάζω. (person) πετώ

έξω. (*from machine*) εκτινάσσω. ~**ion** *s.*
(*of person*) εκδίωξη *f.* (*from machine*)
εκτίναξη *f.*

eke *v.* ~out συμπληρώνω. ~ out of living
απoζώ.

elaborat|e *a.* (*complicated*) πολύπλοκος,
(*detailed*) λεπτομερής. (*of thing made*)
πολυδουλεμένος. (*v.t.*) επεξεργάζομαι,
εκπονώ. ~**ely** *adv.* (*in detail*) λεπτομερώς.
~**ion** *s.* (*working out*) εκπόνηση *f.* (*in art*)
βαριά διακόσμηση.

elapse *v.* παρέρχομαι, περνώ.

elastic *a.* ελαστικός, (*made of ~*) λαστιχένιος. (*s.*) λάστιχο *n.* ~**ity** *s.* ελαστικότητα *f.*

elat|e *v.* κατενθουσιάζω. ~**ed** *a.* feel ~ed
πετάω από τη χαρά μου. ~**ion** *s.* ενθουσιασμός *m.*

elbow *s.* αγκώνας *m.* (*fam.*) ~ grease γερό
καθάρισμα και γυάλισμα. I haven't got ~
room είμαι στριμωγμένος (*v.t.*) παραγκωνίζω. ~ one's way διαγκωνίζομαι.

elder *a.* μεγαλύτερος. ~ statesman παλαίμαχος πολιτικός. (*s.*) my ~s οι μεγαλύ-
τεροί μου. (*church*) πρεσβύτερος *m.* ~**ly** *a.*
ηλικιωμένος.

eldest *a.* ο μεγαλύτερος.

elect *a.* (*best*) εκλεκτός. the mayor ~ ο
μέλλων δήμαρχος.

elect *v.t.* εκλέγω. (*v.i.*, *decide*) αποφασίζω.
~**ed** *a.* αιρετός.

election *s.* εκλογή *f.*, εκλογές *f.pl.* ~**eer** *v.i.*
ψηφοθηρώ. ~**eering** *s.* ψηφοθηρία *f.*

elector *s.* εκλογεύς *m.* ~**al** *a.* εκλογικός.
~**ate** *s.* οι ψηφοφόροι.

electric *a.* ηλεκτρικός. ~ atmosphere (*fig.*)
ηλεκτρισμένη ατμόσφαιρα. ~ lighting ηλε-
κτροφωτισμός *m.* ~ shock ηλεκτροπληξία
f. I got an ~ shock (*fam.*) με χτύπησε το
ρεύμα. ~ blanket/chair ηλεκτρική κου-
βέρτα/καρέκλα.

electrical *a.* ηλεκτρικός, (*fig.*, *tense*)
ηλεκτρισμένος. ~ engineering ηλεκτρομη-
χανολογία *f.* ~ engineer ηλεκτρολόγος
μηχανικός. ~**ly** *adv.* με ηλεκτρισμό.

electrician *s.* ηλεκτρολόγος *m.*

electricity *s.* ηλεκτρισμός *m.* the ~ is cut off
(*fam.*) κόπηκε το ηλεκτρικό (*or* το ρεύμα).
~ bill λογαριασμός του ηλεκτρικού.

electrif|y *v.* (*railway*, *etc.*) εξηλεκτρίζω,
(*audience*) ηλεκτρίζω, γαλβανίζω.
~**ication** *s.* εξηλεκτρισμός *m.* ~**ying** *a.* he
gave an ~ying performance ηλέκτρισε το
ακροατήριο.

electrocut|e *v.* θανατώνω διά ηλεκτρο-
πληξίας. ~**ion** *s.* ηλεκτροπληξία *f.*

electron *s.* ηλεκτρόνιο *n.* ~**ics** *s.* ηλεκτρονική *f.*

electroplated *a.* επάργυρος.

elegance *s.* κομψότητα *f.* (*of language*)
γλαφυρότητα *f.* (*of line*, *fam.*) γάρμπος *n.*

elegant *a.* κομψός, (*of style*) γλαφυρός.
(*gesture*, *appearance*) φιγουράτος. ~**ly**
adv. κομψά, με κομψότητα, με γλαφυρότητα.

elegy|y ελεγείο *n.* ~**iac** *a.* ελεγειακός.

element *s.* στοιχείο *n.* ~**ary** *a.* στοιχειώδης.
~ary school δημοτικόν σχολείον.

elephant *s.* ελέφαντας *m.* (*fig.*) white ~
ογκώδες και άχρηστο αντικείμενο. ~**ine** *a.*
(*unwieldy*) ελεφαντοειδής.

elevat|e *v.* ανυψώνω, (*promote*) προάγω.
~**ed** *a.* (*style*, *etc.*) υψηλός, (*railway*)
εναέριος. ~**ing** *a.* ηθοπλαστικός. ~**or** *s.*
ανελκυστήρ *m.*

elevation *s.* (*raising*) ανύψωση *f.* (*height*)
ύψος *n.* (*promotion*) προαγωγή *f.* (*archit.*)
πρόσοψη *f.* (*high ground*) ύψωμα *n.*

eleven *num.* ένδεκα, έντεκα. ~**th** *a.* ενδέ-
κατος.

elf *s.* (*sprite*) σκανταλιάρικο αγερικό.
(*child*) σκανταλιάρικο παιδί. ~**in** *a.*
γοητευτικά σκανταλιάρης.

elicit *v.* (*response*) προκαλώ, (*the truth*)
αποσπώ.

elide *v.* εκθλίβω.

eligible *a.* κατάλληλος, έχων τα προσόντα.
be ~ for pension δικαιούμαι συντάξεως.
(*fam.*) ~ young man καλός γαμπρός.

eliminat|e *v.* βγάζω, αποκλείω. (*from body*)
αποβάλλω. (*math.*) εξαλείφω. ~**ion** *s.*
αποκλεισμός *m.* αποβολή *f.*

elision *s.* έκθλιψη *f.*

élite *s.* οι εκλεκτοί.

elixir *s.* ελιξήριον *n.*

ellip|se *s.* έλλειψη *f.* ~**tical** *a.* ελλειπτικός.

elm *s.* φτελιά *f.*

elocution *s.* ορθοφωνία *f.*

elongat|e *v.* επιμηκύνω. ~**ion** *s.* επιμήκυνση *f.*

elope *v.* he ~d with her την έκλεψε. they ~d
κλέφτηκαν, αλληλοαπήχθησαν. ~**ment** *s.*
εκουσία απαγωγή.

eloqu|ence *s.* ευφράδεια, *f.* ευγλωττία *f.*
~**ent** *a.* ευφραδής, εύγλωττος. ~**ently** *adv.*
με ευφράδεια.

else *adv.* someone ~ κάποιος άλλος,
nothing ~ τίποτ' άλλο, where ~? πού
αλλού; how ~? πώς αλλιώς; or ~ ειδάλλως,
ειδεμή, αλλιώς. ~**where** *adv.* (κάπου αλλού).

elucidat|e *v.* αποσαφηνίζω. ~**ion** *s.* αποσα-
φήνιση *f.*

elu|de *v.* ξεφεύγω από. it ~des me μου
διαφεύγει. ~**sive** *a.* (*of fugitive*) άπιαστος,
(*of idea*) ασύλληπτος. be ~sive (*hard to
find*) δεν βρίσκομαι εύκολα.

emaciated *a.* κάτισχνος. become ~ (*fam.*) σουρώνω.

emanate *v.* προέρχομαι, πηγάζω.

emancipat|e *v.* χειραφετώ. **~ion** *s.* χειραφέτηση *f.*

emasculat|e *v.* ευνουχίζω. **~ion** *s.* ευνουχισμός *m.*

embalm *v.* βαλσαμώνω, ταριχεύω. **~ment** *s.* βαλσάμωμα *n.*

embankment *s.* επιχωμάτωση *f.*

embargo *s.* εμπάργκο *n.*

embark *v.t.* επιβιβάζω, παίρνω. (*v.i.*) επιβιβάζομαι. (*v.t. & i.*) μπαρκάρω. ~ on (*begin*) αρχίζω. **~ation** *s.* επιβίβαση *f.*

embarrass *v.* στενοχωρώ, φέρνω σε δύσκολη θέση. be ~ed στενοχωριέμαι, βρίσκομαι σε δύσκολη θέση. **~ing** *a.* στενόχωρος, άσχημος, (*question*) που φέρνει σε αμηχανία. **~ment** *s.* στενοχώρια *f.* αμηχανία *f.*

embassy *s.* πρεσβεία *f.*

embattled *a.* έτοιμος προς μάχην.

embedded *a.* σφηνωμένος, μπηγμένος.

embellish *v.* εξωραΐζω, στολίζω, καλλωπίζω. **~ment** *s.* εξωραϊσμός *m.*, στολισμός *m.*, καλλωπισμός *m.*

embers *s.* θράκα *f.*

embezzle *v.* καταχρώμαι. **~ment** *s.* κατάχρηση *f.* **~r** *s.* καταχραστής *m.*

embitter *v.* πικραίνω. **~ed** *a.* πικραμένος.

emblem *s.* έμβλημα *n.*

embod|y *v.* (*include*) περιλαμβάνω, (*give tangible form to*) ενσαρκώνω. (*render, express*) αποδίδω. **~iment** *s.* ενσάρκωση *f.*, προσωποποίηση *f.*

embolden *v.* ενθαρρύνω.

embossed *a.* ανάγλυφος, (*print*) έκτυπος.

embrace *v.* αγκαλιάζω, εναγκαλίζομαι. (*adopt*) ασπάζομαι. (*include*) περιλαμβάνω. (*s.*) αγκαλιά *f.*, αγκάλιασμα *n.*, εναγκαλισμός *m.*

embrasure *s.* (*for gun*) πολεμίστρα *f.* (*in room*) κούφωμα *n.*

embroider *v.* κεντώ, (*fig.*) στολίζω. **~ed** *a.* κεντητός. **~y** *s.* κέντημα *n.*

embroil *v.* (*also* get ~ed) μπλέκω.

embryo *s.* έμβρυο *n.* **~nic** *a.* εμβρυώδης.

emend *v.* διορθώνω. **~ation** *s.* διόρθωση *f.*

emerald *s.* σμαράγδι *n.* (*a.*) σμαραγδένιος.

emerg|e *v.* (*from water*) αναδύομαι, (*come out*) βγαίνω, (*come to light*) ανακύπτω, (*crop up, result*) προκύπτω. **~ence** *s.* εμφάνιση *f.* **~ent** *a.* ~ent countries αναπτυσσόμενες χώρες.

emergency *s.* κρίσιμη περίσταση. state of ~ κατάσταση ανάγκης. ~ exit έξοδος κινδύνου.

emery *s.* σμύριδα *f.* **~paper** *s.* σμυριδό-

χαρτο *n.*

emetic *a.* εμετικός.

emigr|ant *s.* μετανάστης *m.* **~ate** *v.* μεταναστεύω. **~ation** *s.* μετανάστευση *f.*

émigré *s.* εμιγκρές *m.*

emin|ence *s.* (*height*) ύψος *n.* (*high ground*) ύψωμα *n.* (*distinction*) διασημότητα *f.* **~ent** *a.* διάσημος, διαπρεπής. **~ently** *adv.* εξαιρετικά, απολύτως.

emir *s.* εμίρης *m.*

emissary *s.* απεσταλμένος *m.*

emission *s.* εκπομπή *f.*

emit *v.* εκπέμπω.

emoluments *s.* απολαβές *f.pl.*, αποδοχές *f.pl.*

emotion *s.* συγκίνηση *f.* the ~s αισθήματα, συναισθήματα *n.pl.*

emotional *a.* συναισθηματικός, (*easily moved*) ευσυγκίνητος, (*moving*) συγκινητικός. **~ly** *adv.* με συγκίνηση. **~ly** unstable συναισθηματικά ασταθής.

emotive *a.* συγκινητικός.

empathy *s.* βαθιά κατανόηση.

emperor *s.* αυτοκράτωρ *m.*

emphas|is *s.* έμφαση *f.* **~ize** *v.* τονίζω.

emphatic *a.* εμφατικός. **~ally** *adv.* εμφατικώς, με έμφαση. **~ally** agree συμφωνώ απολύτως.

empire *s.* αυτοκρατορία *f.*

empiric, **~al** *a.* εμπειρικός. **~ally** *adv.* εμπειρικώς. **~ism** *s.* εμπειρισμός *m.* **~ist** *s.* εμπειρικός *m.*

employ *v.* (*use*) χρησιμοποιώ, (*keep busy, give work to*) απασχολώ. be ~ed in (*doing sthg.*) ασχολούμαι με, (*a bank, etc.*) δουλεύω σε. (*s.*) be in the ~ of εργάζομαι για. **~ee** *s.* υπάλληλος *m.f.* **~er** *s.* εργοδότης *m.*

employment *s.* χρησιμοποίηση *f.* απασχόληση *f.*, ασχολία *f.* (*work*) εργασία *f.* (*see* employ). ~ agency γραφείον ευρέσεως εργασίας.

emporium *s.* (*market*) αγορά *f.* (*shop*) κατάστημα *n.*

empower *v.* (*enable*) επιτρέπω (*with* σε *or* gen.). (*authorize*) εξουσιοδοτώ.

empress *s.* αυτοκράτειρα *f.*

emptiness *s.* κενό *n.*

empty *a.* άδειος, αδειανός, (*vain, void*) κενός. **~-handed** *a.* με άδεια χέρια. **~-headed** *a.* άμυαλος.

empty *v.t. & i.* αδειάζω. (*v.t.*) εκκενώνω. (*v.i.*) (*of river*) εκβάλλω.

emulat|e *v.t.* αμιλλώμαι, συναγωνίζομαι. παραβγαίνω με. **~ion** *s.* άμιλλα *f.* in ~ion of αμιλλώμενος.

emulous *a.* be ~ of (*person*) *see* emulate, (*honours, etc.*) επιζητώ.

emulsion s. γαλάκτωμα n.

enable v. επιτρέπω (with σε or gen.).

enact v. θεσπίζω, (play) παίζω, παριστάνω. **~ment** s. θέσπισις f. παίξιμο n.

enamel s. σμάλτο n. (v.) σμαλτώνω. (a.) (also ~led) εμαγιέ.

enamoured a. ερωτευμένος. become ~ of ερωτεύομαι.

encamp v. στρατοπεδεύω. **~ment** s. κατασκήνωση f.

encase v. εγκιβωτίζω, (fig.) περικαλύπτω.

enchain v. αλυσοδένω, (fig.) αιχμαλωτίζω.

enchant v. μαγεύω, γοητεύω. **~ing** a. μαγευτικός. **~ment** s. μαγεία f. **~ress** s. μάγισσα f., γόησσα f.

encircle v. περικυκλώνω, ζώνω. **~lement** s. περικύκλωση f. **~ling** a. κυκλωτικός.

enclave s. έδαφος (territory) or κρατίδιο (state) περικλεισμένο μέσα σε ξένη χώρα.

enclose v. (shut in) εγκλείω, (surround) περικλείω, (fence) περιφράσσω, (in letter) εσωκλείω.

enclosure s. (fencing) περίφραξη f. (ground) περίβολος m. (in letter) εσώκλειστο n., συνημμένο n.

encomium s. εγκώμιο n.

encompass v. περιστοιχίζω.

encore int. μπις. (v.) μπιζάρω. (s.) μπις, αγκόρ n.

encounter v. αντιμετωπίζω, συναντώ. (s.) συνάντηση f.

encourage v. ενθαρρύνω. **~ement** s. ενθάρρυνση f. **~ing** a. ενθαρρυντικός.

encroach v. ~ on καταπατώ. **~ment** s. καταπάτηση f.

encumber v. εμποδίζω, (παρα)φορτώνω. **~rance** s. εμπόδιο n., βάρος n.

encyclopaedia s. εγκυκλοπαιδεία f.

end s. (extremity) άκρη f. at the ~ of the street στην άκρη του δρόμου. (finish) τέρμα n., πέρας n., τέλος n. bring to an ~ θέτω τέρμα εις, φέρω εις πέρας. the ~s of the earth τα πέρατα της γης. the ~ of the world η συντέλεια του κόσμου. make both ~s meet τα οικονομάω, τα φέρνω βόλτα. keep one's ~ up δεν τα βάζω κάτω. think no ~ of θαυμάζω πολύ, έχω μεγάλη ιδέα για. (object) σκοπός m. appointed ~ προορισμός m. on ~ όρθιο, for hours on ~ επί ώρες ολόκληρες. in the ~ (στο) τέλος, τελικά, εν τέλει.

end a. ακρινός. the ~ house το ακρινό σπίτι.

end v.t. & i. τελειώνω, τερματίζω. (v.i.) (cease, expire) λήγω. ~ up (as, by, in) καταλήγω, he ~ed up as a doorkeeper κατέληξε κλητήρας. (gram.) ~ in λήγω (or τελειώνω) σε.

endanger v. (δια)κινδυνεύω, θέτω εις κίνδυνον.

endear v. ~ (person) to καθιστώ προσφιλή εις. ~ oneself to γίνομαι αγαπητός σε. **~ing** a. αξιαγάπητος. **~ments** s. γλυκόλογα n.pl.

endeavour v. προσπαθώ, πασχίζω. (s.) προσπάθεια f.

endemic a. ενδημικός.

ending s. τέλος n. (gram.) κατάληξη f.

endive s. αντίδι n.

endless a. ατελείωτος, που δεν τελειώνει ποτέ. (patience, etc.) απεριόριστος, που δεν έχει όρια. go to ~ trouble τσακίζομαι, σκοτώνομαι, χαλάω τον κόσμο.

endorse v. οπισθογραφώ, (confirm) επιδοκιμάζω. **~ment** s. οπισθογράφηση f. επιδοκιμασία f.

endow v. (with money) κληροδοτώ. (fig.) προικίζω. **~ment** s. κληροδότημα n. (act) κληροδότηση f. (fig.) χάρισμα n.

endue v. δίνω σε. **~ed** with προικισμένος με.

endurance s. αντοχή f. (patience) καρτερία f. past ~ ανυπόφορος.

endure v.i. (last) αντέχω, βαστώ, κρατώ. (v.t.) (bear, undergo) υποφέρω, αντέχω. (stand up to) αντέχω σε. **~able** a. υποφερτός. **~ing** a. διαρκής, μόνιμος.

enema s. κλύσμα n.

enemy s. εχθρός m. (a.) εχθρικός.

energy s. (phys.) ενέργεια f. (person's) δύναμη f., ενεργητικότητα f. **~etic** a. ενεργητικός. **~etically** adv. με ενεργητικότητα.

enervate v. εξασθενίζω, αποχαυνώνω. **~ing** a. αποχαυνωτικός.

enfant terrible s. τρομερό παιδί.

enfeeble v. εξασθενίζω.

enfold v. αγκαλιάζω, τυλίγω.

enforce v. επιβάλλω, εφαρμόζω αναγκαστικά. **~ment** s. επιβολή f., εφαρμογή f.

enfranchise v. παρέχω πολιτικά δικαιώματα σε.

engage v.t. (hire) συμφωνώ, νοικιάζω, (book) πιάνω, κλείνω, (a performer) αγκαζάρω, (staff) προσλαμβάνω. (attention) προσελκύω. (the enemy) συμπλέκομαι με. ~ first gear βάζω πρώτη. he ~d me in conversation με έπιασε στις κουβέντες. (v.i.) (undertake) αναλαμβάνω, υπόσχομαι. (guarantee) εγγυώμαι. ~ in (occupy oneself) ασχολούμαι με. be ~d in είμαι απασχολημένος με or καταγίνομαι με. get ~d αρραβωνιάζομαι.

engaged a. (busy) απασχολημένος, (betrothed) αρραβωνιασμένος, (taken) κατειλημμένος, πιασμένος. the (phone) number is ~ η γραμμή είναι κατειλημμένη.

engagement s. (of staff) πρόσληψη f. (with enemy) συμπλοκή f. (promise) υποχρέωση f. (betrothal) αρραβώνας m., αρραβώνες m.pl. (meeting) ραντεβού n. have a previous ~ έχω μία ανειλημμένη υποχρέωση.

engaging a. γοητευτικός.

engender v. γεννώ.

engine s. μηχανή f. ~-**driver** μηχανοδηγός m.

engineer s. μηχανικός m. (mil.) E~s Μηχανικόν n. (v.t.) (arrange) μηχανεύομαι. ~**ing** s. μηχανική f.

English a. αγγλικός, εγγλέζικος. (person) Άγγλος m., Αγγλίδα f. (s.) (language) αγγλικά, εγγλέζικα n.pl. in ~ αγγλικά, αγγλιστί. the ~ (people) οι Άγγλοι. ~**man** s. Άγγλος, Εγγλέζος m. ~**woman** s. Αγγλίδα, Εγγλέζα f.

engraft v. (fig.) εμφυτεύω.

engrav|e v. χαράσσω. ~**er** s. χαράκτης m. ~**ing** s. (art) χαρακτική f. (print) γκραβούρα f.

engross v. (absorb) απορροφώ.

engulf v. καταπίνω.

enhance v. (pleasure, beauty) προσθέτω σε, (value) ανεβάζω.

enigma s. αίνιγμα n. ~**tic** a. αινιγματικός.

enjoin v.t. (impose) επιβάλλω, (counsel) συνιστώ.

enjoy v.t. απολαμβάνω, χαίρομαι. (use or fruits of) καρπούμαι, νέμομαι. (health, respect, etc.) απολαύω, χαίρω (both with gen.). (doing sthg.) μου αρέσει να. I ~ reading poetry απολαμβάνω διαβάζοντας ποίηση. ~ oneself ευχαριστιέμαι, διασκεδάζω. when I retire I shall sit back and ~ myself όταν θα πάρω τη σύνταξή μου θα κάθομαι και θα απολαμβάνω.

enjoyabl|e a. απολαυστικός, ευχάριστος. ~**y** adv. ευχάριστα.

enjoyment s. απόλαυση f. (recreation) ψυχαγωγία f.

enlarge v.t. μεγεθύνω, (widen) διευρύνω, (extend) επεκτείνω. become ~d (med.) διογκούμαι. (v.i.) ~ upon επεκτείνομαι επί (with gen.). ~**ment** s. μεγέθυνση f., επέκταση f., διόγκωση f.

enlighten v. διαφωτίζω. ~**ed** a. φωτισμένος. ~**ment** s. διαφώτιση f. (philosophy of reason) διαφωτισμός m.

enlist v.t. (men) στρατολογώ, (support) επιτυγχάνω, εξασφαλίζω. (v.i.) κατατάσσομαι. ~**ment** s. κατάταξη f.

enliven v. ζωηρεύω.

en masse adv. ομαδικώς.

enmesh v. τυλίγω, μπερδεύω.

enmity s. έχθρα f.

ennoble v. εξευγενίζω.

ennui s. ανία f., πλήξη f.

enormity s. (of crime) το τερατώδες.

enormous a. τεράστιος, πελώριος. ~**ly** adv. τεραστίως, (fam.) καταπληκτικά, πάρα πολύ. ~**ness** s. τεράστιο μέγεθος.

enough a. αρκετός. adv. αρκετά. be ~ φθάνω, αρκώ. I have had ~ of him τον έχω βαρεθεί.

enquire v. see inquire.

enrage v. εξαγριώνω. (fam.) he got ~d έγινε σκυλί.

enrapture v. συναρπάζω, ξετρελαίνω, καταγοητεύω.

enrich v.t. εμπλουτίζω. ~ oneself πλουτίζω. ~**ment** s. εμπλουτισμός m.

enrol v.t. εγγράφω. (v.i.) εγγράφομαι, (mil.) κατατάσσομαι. ~**ment** s. εγγραφή f.

en route adv. στο δρόμο. ~ to καθ' οδόν προς (with acc.).

ensconce v. τοποθετώ. ~ oneself στρώνομαι.

ensemble s. σύνολο n.

enshrine v. φυλάσσω.

enshroud v. σκεπάζω.

ensign s. (flag) σημαία f. (officer) σημαιοφόρος m.

enslave v. σκλαβώνω, υποδουλώνω, εξανδραποδίζω. ~**d** a. υπόδουλος, σκλαβωμένος. ~**ment** s. σκλάβωμα n. υποδούλωση f.

ensnare v. παγιδεύω.

ensu|e v. επακολουθώ. ~**ing** a. (resultant) επακόλουθος. (next) επόμενος, ακόλουθος.

ensure v.t. (make sure of) εξασφαλίζω. (guarantee) εγγυώμαι (with ότι or για). (v.i.) (against) ασφαλίζομαι.

entail v.t. (have as consequence) συνεπάγομαι, (necessitate) θέλω. (law) αφήνω ως καταπίστευμα. (s.) κληρονομία υπό τον όρον αναπαλλοτριώτου.

entangle v. μπλέκω, μπερδεύω. get ~d μπλέκω, μπερδεύομαι. ~**ment** s. μπλέξιμο n.

enter v.t. & i. (go or come in) μπαίνω, εισέρχομαι. (a profession) ακολουθώ. he ~ed the room μπήκε στο δωμάτιο. (inscribe, insert) καταχωρώ. (enrol) (v.t.) εγγράφω, (v.i.) εγγράφομαι, γράφομαι. ~ for (v.i.) (race, competition) δηλώνω συμμετοχή σε. ~ into (relations, contract) συνάπτω. ~ into details μπαίνω σε λεπτομέρειες. that does not ~ into the matter αυτό είναι άσχετο. ~ upon αρχίζω, αναλαμβάνω.

enterpris|e s. επιχείρηση f. (initiative) πρωτοβουλία f. ~**ing** a. επιχειρηματικός.

entertain v.t. φιλοξενώ, ξενίζω. we ~ed

them to dinner τους είχαμε τραπέζι. (*amuse, also* be ~ed) διασκεδάζω, (*recreate*) ψυχαγωγώ, (*hopes, feelings*) τρέφω, έχω, (*a proposal*) μελετώ. (*v.i.*) they ~ a lot δέχονται συχνά.
entertaining *a.* διασκεδαστικός, γουστόζικος. **~ly** *adv.* διασκεδαστικά.
entertainment *s.* φιλοξενία *f.* διασκέδαση *f.* ψυχαγωγία *f.* (*show*) θέαμα *n.*
enthral *v.* σκλαβώνω, συναρπάζω. **~ling** *a.* συναρπαστικός.
enthrone *v.* ενθρονίζω. **~ment** *s.* ενθρόνιση *f.*
enthuse *v.i.* ενθουσιάζομαι.
enthusiasm *s.* ενθουσιασμός *m.* fill with ~ ενθουσιάζω. inspiring ~ ενθουσιαστικός.
enthusiast *s.* he is a gardening ~ είναι ενθουσιώδης κηπουρός, έχει μανία με την κηπουρική.
enthusiastic *a.* ενθουσιώδης. **~ally** *adv.* ενθουσιωδώς, με ενθουσιασμό.
entic|e *v.* δελεάζω, (*lead astray*) παρασύρω. **~ement** *s.* δέλεασμα *n.* **~ing** *a.* δελεαστικός.
entire *a.* ολόκληρος. **~ly** *adv.* ολότελα, εξ εξολοκλήρου. **~ty** *s.* ολότητα *f.*
entitle *v.* (*call*) τιτλοφορώ. (*allow*) επιτρέπω (*with σε or gen.*). be ~d to δικαιούμαι (*with να or gen. or acc.*). person ~d ο δικαιούχος. **~ment** *s.* δικαίωμα *n.*
entity *s.* οντότης *f.*
entomb *v.* ενταφιάζω. **~ment** *s.* ενταφιασμός *m.*
entomolog|ist *s.* εντομολόγος *m.f.* **~y** *s.* εντομολογία *f.* **~ical** *a.* εντομολογικός.
entourage *s.* περιβάλλον *s.*, κύκλος *m.*
entrails *s.* σωθικά, εντόσθια, σπλάχνα *n.pl.*
entrance *s.* είσοδος *f.* ~ examination εισαγωγικές εξετάσεις.
entranc|e *v.* μαγεύω, καταγοητεύω. **~ing** *a.* μαγευτικός, θελκτικός.
entrant *s.* υποψήφιος *m.*
entreat *v.* ικετεύω, εκλιπαρώ. **~y** *s.* παράκληση *f.*, ικεσία *f.* **~ingly** *adv.* ικευτικά.
entrench *v.* περιχαρακώνω. ~ oneself οχυρώνομαι. ~ on καταπατώ. **~ment** *s.* περιχαράκωμα *n.*
entrepreneur *s.* επιχειρηματίας *m.*
entresol *s.* ημιώροφος *m.*
entrust *v.* (*sthg. to person*) εμπιστεύομαι.
entry *s.* (*entrance*) είσοδος *f.* (*in ledger*) καταχώρηση *f.* (*in notebook*) σημείωμα *n.* (*in dictionary*) λήμμα *n.* (*for contest*) δήλωση συμμετοχής. ~ form έντυπο αιτήσεων. ~ visa βίζα εισόδου.
entwine *v.* περιτυλίσσω.
enumerat|e *v.* απαριθμώ. **~ion** *s.* απαρίθμηση *f.*
enunciat|e *v.* δηλώνω, διατυπώνω, (*pro-*

nounce) προσφέρω. **~ion** *s.* δήλωση *f.* προσφορά *f.*
envelop *v.* καλύπτω, περιβάλλω. **~e** *s.* φάκελος *m.*
envenom *v.* δηλητηριάζω, πικραίνω.
enviable *a.* επίζηλος, ζηλευτός.
envious *a.* ζηλιάρης, φθονερός. **~ly** *adv.* με ζήλια.
environ *v.* περιβάλλω. **~s** *s.* περίχωρα *n. pl.*, τα πέριξ. **~ment** *s.* περιβάλλον *n.*
envisage *v.* (*face*) αντιμετωπίζω, (*imagine*) φαντάζομαι.
envoy *s.* απεσταλμένος *m.*
envy *s.* ζήλια *f.* (*malicious*) φθόνος *m.*, ζηλοφθονία *f.* (*v.*) ζηλεύω, φθονώ. I ~ him his money τον ζηλεύω για τα λεφτά του.
eparchy *s.* επαρχία *f.*
epaulette *s.* επωμίς *f.*
ephemeral *a.* εφήμερος.
epic *a.* επικός. ~ poet επικός *m.* (*s.*). (*poem, story*) έπος *n.*
epicure *s.* καλοφαγάς *m.* **~an** *a.* (*philosophy*) επικούρειος,· (*meal*) εκλεκτής ποιότητος, (*person*) τρυφηλός.
epidemic *a.* επιδημικός. (*s.*) επιδημία *f.*
epigram *s.* επίγραμμα *n.* **~matic** *a.* επιγραμματικός.
epigraphy *s.* επιγραφική *f.*
epilep|sy *s.* επιληψία *f.*, σεληνιασμός *m.* **~tic** *a.* επιληπτικός.
epilogue *s.* επίλογος *m.*
Epiphany *s.* Θεοφάνεια *n.pl.*
episcopal *a.* επισκοπικός.
episod|e *s.* επεισόδιο *n.* **~ic** *a.* επεισοδιακός.
epistle *s.* επιστολή *f.*
epitaph *s.* επιτάφιον επίγραμμα *n.*
epithet *s.* επωνυμία *f.*
epitom|e *s.* επιτομή *f.* **~ize** *v.* συνοψίζω. (*fig. represent in miniature*) παρουσιάζω σε μικρογραφία.
epoch *s.* εποχή *f.* **~-making** *a.* κοσμοϊστορικός.
equable *a.* (*temper*) ήρεμος, (*climate*) εύκρατος.
equal *a.* ίσος, όμοιος (*with* με). (*in ability*) ισάξιος, (*in number*) ισάριθμος, (*in rights, rank, value*) ισότιμος. (~*ly matched*) ισόπαλος. on ~ terms επί ίσοις όροις. with ~ politeness με την ίδια ευγένεια. ~ to the occasion αντάξιος των περιστάσεων. ~ to (*having the ability*) άξιος (*with* για *or* να), not feel ~ to δεν έχω το κουράγιο (*with* για *or* για να). (*s.*) (*peer*) ταίρι *n.* he has no ~ δεν έχει το ταίρι του. one's ~s οι όμοιοί μου. **~ly** *adv.* εξίσου, την ίδια. **equal** *v.* (*in amount*) ισοφαρίζω. (*be as good as*) είμαι ίσος με, (*come up to*) φθάνω, no

one can ~ him κανείς δεν τον φθάνει, δεν έχει τον όμοιό του (or το ταίρι του), δεν του βγαίνει κανείς. (math.) ισούμαι προς, see also equals.

equality s. ισότητα f.

equalize v.t. εξισώνω. (v.i.) ισοφαρίζω. **~r** s. (sport) ισοφάρισμα n.

equals s. (math. symbol) ισον n. 2 + 2= 4 δύο και δύο ίσον τέσσερα.

equanimity s. αταραξία f.

equat|e v. εξισώνω. **~ion** s. εξίσωση f.

equator s. ισημερινός m. **~ial** a. ισημερινός.

equestrian a. ιππικός, (on horseback) έφιππος.

equidistant a. βρισκόμενος σε ίση απόσταση.

equilateral a. ισόπλευρος.

equilibrium s. ισορροπία f.

equino|x s. ισημερία f. **~ctial** a. ισημερινός.

equip v. εφοδιάζω. (ship) αρματώνω, εφοπλίζω.

equipment s. εφοδιασμός m. (ship's) αρμάτωμα n., εφοπλισμός m. (supplies) εφόδια n.pl. (gear) εξαρτήματα n.pl. (soldier's) εξάρτυση f. (technical) εξοπλισμός m.

equitabl|e a. δίκαιος. **~y** adv. δικαίως.

equity s. δικαιοσύνη f.

equival|ent a. (of same value, power, etc.) ισότιμος, ισοδύναμος, (corresponding) αντίστοιχος, be ~ent to ισοδυναμώ, αντιστοιχώ (both with προς & acc.). (s.) ισότιμο n., ισοδύναμο n., αντίστοιχο n. **~ence** s. ισοτιμία f., ισοδυναμία f.

equivoc|al a. διφορούμενος, (dubious) αμφίβολος. **~ation** s. υπεκφυγή f.

era s. εποχή f.

eradicat|e v. ξερ(ρ)ιζώνω, εκριζώ. **~ion** s. ξερ(ρ)ίζωμα n.

eras|e v. σβήνω, εξαλείφω. **~ure** s. σβήσιμο n., εξάλειψη f.

erect a. ολόρθος, στητός. (bearing) ευθυτενής. become ~ ορθώνομαι, stand ~ στέκομαι όρθιος. (adv.) όρθια.

erect v. ανεγείρω. **~ion** s. (act of building) ανέγερση f. (thing built) κτίριο n.

ero|de v. διαβρώσκω, κατατρώγω. **~sion** s. διάβρωση f.

erotic a. ερωτικός. **~ism** s. ερωτισμός m.

err v. σφάλλω, απατώμαι, λαθεύω, κάνω λάθος. **~ant** a. the ~ant party ο ένοχος.

errand s. θέλημα n. I must do some ~s έχω να κάνω μερικές δουλειές. on a fool's ~ άδικα.

erratic a. (changeable) άστατος. the train service is ~ τα τραίνα δεν είναι τακτικά. his visits are ~ δεν είναι τακτικοί στις επισκέψεις του. **~ally** adv. άτακτα, χωρίς

σύστημα.

erratum s. παρόραμα n.

erroneous a. λανθασμένος, εσφαλμένος. **~ly** adv. εσφαλμένα, (by mistake) κατά λάθος.

error s. λάθος n., σφάλμα n. in ~ κατά λάθος, make an ~ of judgement πέφτω έξω.

erudit|e a. πολυμαθής. **~ion** s. πολυμάθεια f.

erupt v.i. εκρήγνυμαι, (fig.) ξεσπάω. **~ion** s. έκρηξη f., (fig.) ξέσπασμα n.

escalat|e v.t. εντείνω, (fam.) κλιμακώνω. (v.i.) εντείνομαι, κλιμακώνομαι. **~ion** s. ένταση f., κλιμάκωση f. **~or** s. κυλιομένη κλίμαξ.

escapade s. περιπέτεια f., τρέλλα f.

escape v.t. & i. ξεφεύγω, γλυτώνω. (v.i.) (of prisoner) δραπετεύω, (fam.) το σκάω. he narrowly ~d drowning παρά λίγο να πνιγεί. it ~d his notice διέφυγε την προσοχή του. it ~d me (I forgot it or did not notice it) μου διέφυγε. (a groan) ~d his lips ξέφυγε από τα χείλη του.

escape s. διαφυγή f. (of prisoner) δραπέτευση f. he had a narrow ~ φτηνά τη γλίτωσε.

escarpment s. γκρεμός m.

eschew v. αποφεύγω.

escort s. (person) συνοδός m.f. (company) συνοδεία f. (lady's) καβαλιέρος m. (v.) συνοδεύω.

Eskimo s. Εσκιμώος m.

esoteric a. δυσνόητος για τους πολλούς.

especial a. ιδιαίτερος. **~ly** adv. ιδιαιτέρως.

espionage s. κατασκοπεία f.

espous|e v. (fig.) ασπάζομαι. **~al** s. (adoption) υιοθέτηση f.

espy v. I ~ her την παίρνει το μάτι μου.

essay s. δοκίμιο n. (v.t., test) δοκιμάζω. (v.i., attempt) προσπαθώ.

essence s. ουσία f. (extract) εκχύλισμα n., απόσταγμα n.

essential a. ουσιώδης, (indispensable) απαραίτητος. (s) the ~s of the case τα πλέον ουσιώδη της υποθέσεως. the ~s (equipment) τα απαραίτητα. **~ly** adv. κατά βάθος.

establish v. (settle) εγκαθιστώ, (found) ιδρύω, (determine) προσδιορίζω. (one's innocence) αποδεικνύω, (a custom) καθιερώνω.

establishment s. (act) εγκατάσταση f. ίδρυση f. (thing) ίδρυμα n. κατάστημα n. the E~ το κατεστημένον. (staff) προσωπικό n. (of facts) προσδιορισμός m. (of innocence) απόδειξη f. (of custom) καθιέρωση f.

estate s. (landed property) κτήμα n.,

αγρόκτημα *n.*, υποστατικό *n.* housing ~ οικισμός *m.* (*assets*) περιουσία *f.* real ~ ακίνητα *n.pl.* movable ~ κινητά *n.pl.* ~ *agent* κτηματομεσίτης *m.* man's ~ ανδρική ηλικία.

esteem *v.* (*value*) εκτιμώ, (*consider*) θεωρώ. (*s*) υπόληψη *f.*

estimable *a.* αξιότιμος.

estimate *v.* (*reckon*) υπολογίζω, (*for work to be done*) προϋπολογίζω. (*s*). (*reckoning*) υπολογισμός *m.* (*idea*) form an ~ of σχηματίζω γνώμη για. (*for work*) submit an ~ υποβάλλω προϋπολογισμό.

estimation *s.* εκτίμηση *f.*, κρίση *f.*

estrange *v.* αποξενώνω. **~ment** *s.* αποξένωση *f.*

estuary *s.* εκβολή *f.*

etc. κτλ.

etcetera και τα λοιπά, και άλλα. (*s.*) the ~s τα διάφορα.

etch *v.* χαράσσω. **~ing** *s.* χαλκογραφία *f.*

eternal *a.* αιώνιος. **~ly** *adv.* αιωνίως.

eternity *s.* αιωνιότητα *f.*

ether *s.* αιθέρας *m.* **~eal** *a.* αιθέριος.

ethic|s *s.* ηθική *f.* (*moral aspect of sthg.*) ηθική πλευρά. **~al** *a.* ηθικός. **~ally** *adv.* ηθικώς.

ethn|ic *a.* εθνικός. **~ic** minority εθνικότητες *f.pl.* **~ology** *s.* εθνολογία *f.*

etiquette *s.* εθιμοτυπία *f.*, ετικέττα *f.* breach of ~ παρατυπία *f.*

etymology *s.* ετυμολογία *f.*

eucalyptus *s.* ευκάλυπτος *m.*

Eucharist *s.* Ευχαριστία *f.*

eugenics *s.* ευγονισμός *m.*

eulog|ize *v.* εγκωμιάζω. **~y** *s.* εγκώμιο *n.*

eunuch *s.* ευνούχος *m.*

euphemis|m *s.* ευφημισμός *m.* **~tic** *a.* ευφημιστικός.

euphon|ic, ~ious *a.* ευφωνικός. **~y** *s.* ευφωνία *f.*

euphoria *s.* αίσθημα ευφορίας.

European *a.* ευρωπαϊκός, (*fam.*) φράγκικος. (*person*) Ευρωπαίος.

euthanasia *s.* ευθανασία *f.*

evacuat|e *v.t.* εκκενώνω. ~e bowels ενεργούμαι, αποπατώ. **~ion** *s.* εκκένωση *f.* αποπάτηση *f.*

evade *v.* (*person, duty*) αποφεύγω, (*a blow*) ξεφεύγω, (*arrest, attention*) διαφεύγω. ~ the question (*or payment*) αποφεύγω να απαντήσω (*or* να πληρώσω).

evaluat|e *v.* εκτιμώ. **~ion** *s.* εκτίμηση *f.*

evanescent *a.* φευγαλέος.

evangel|ical *a.* ευαγγελικός. **~ist** *s.* ευαγγελιστής *m.*

evaporat|e *v.t.* εξατμίζω. (*v.i.*) εξατμίζομαι. ~ed milk γάλα εβαπορέ. **~ion** *s.* εξάτμιση *f.*

evasion *s.* (*avoidance*) αποφυγή *f.* (*subterfuge*) υπεκφυγή *f.* tax ~ φοροδιαφυγή *f.*

evasive *a.* give an ~ answer απαντώ με υπεκφυγές. ~ action ελιγμός διαφυγής. **~ly** *adv.* με υπεκφυγές.

eve *s.* παραμονή *f.*, προτεραία *f.*

even *a.* (*equal*) ισ(ι)ος, (*regular, steady*) κανονικός. (*surface*) ομαλός, (*colour, temperature, development*) ομοιόμορφος, (*temper*) ήρεμος, (*number*) ζυγός. (*quits*) πάτσι (*adv.*). (*~ly matched*) ισόπαλος. break ~ είμαι στα λεφτά μου. get ~ with εκδικούμαι.

even *v.* ισ(ι)ώνω, ισ(ι)άζω. ~ (*things*) up (*restore balance of*) φτιάνω, διορθώνω.

even *adv.* ακόμα και. (*with comparative*) John is ~ taller than you ο Γιάννης είναι ακόμα πιο ψηλός κι από σένα. ~ if *or* though (έστω) και να, ακόμα κι αν. not ~ ούτε και, ούτε καν. ~ so ακόμα κι έτσι.

even-handed *a.* αμερόληπτος.

evening *s.* βράδυ *n.*, βραδιά *f.* (*early ~*) βραδάκι *n.* in the ~ (το) βράδυ, tomorrow ~ αύριο (το) βράδυ. good ~ καλησπέρα. ~ falls βραδιάζει.

evening *a.* βραδινός. ~ dress βραδινό ένδυμα. ~ meal βραδινό *n.* ~ paper απογευματινή εφημερίδα. ~ party εσπερίδα *f.* ~ star έσπερος *m.*, αποσπερίτης *m.* ~ class βραδινό μάθημα *n.*

evenly *adv.* ομαλά, (*equally*) ίσα.

evenness *s.* ομαλότητα *f.*

event *s.* γεγονός *n.*, συμβάν *n.*, συμβεβηκός *n.*, περιστατικό *n.* in the ~ of war σε περίπτωση πολέμου. at all ~s πάντως, εν πάση περιπτώσει. in any ~ ό,τι και να συμβεί.

eventful *a.* περιπετειώδης. I have had an ~ day η μέρα μου ήταν γεμάτη γεγονότα.

eventual *a.* τελικός. **~ity** *s.* ενδεχόμενο *n.* **~ly** *adv.* τελικά.

ever *adv.* (*always*) πάντοτε, πάντα. for ~ για πάντα, he is for ~ grumbling όλο (*or* διαρκώς) γκρινιάζει. as good as ~ όπως πάντοτε καλός. ~ since (*adv.*) έκτοτε, από τότε κι έπειτα, (*conj.*) από τότε που. he gets ~ more crotchety γίνεται όλο και πιο ανάποδος. (*at any time*) ποτέ. has it ~ happened? συνέβη ποτέ; better than ~ καλύτερα από ποτέ, υπέρποτε καλά, no one ~ listens to me ποτέ δεν μ' ακούει κανείς. if ~ he comes αν έρθει ποτέ. hardly ~ σχεδόν ποτέ. (*fam.*) ~ so (much) πάρα πολύ.

evergreen *a.* αειθαλής.

everlasting *a.* αιώνιος. **~ly** *adv.* διαρκώς, αιωνίως.

evermore *adv.* for ~ για πάντα.

ever-remembered *a.* αείμνηστος.

every *a*. κάθε, πας. ~ day κάθε μέρα, καθ' εκάστην. ~ so often κάθε τόσο. ~ other day μέρα παρά μέρα. ~ one of them όλοι τους. ~ man ο καθένας, έκαστος. in ~ way από κάθε άποψη.

everybody *pron*. ο καθένας, όλοι, όλος ο κόσμος. ~ who wishes όποιος θέλει, όσοι επιθυμούν, οι επιθυμούντες.

everyday *a*. καθημερινός.

everyone *pron. see* everybody.

everything *pron*. το κάθε τι, όλα, τα πάντα. ~ you do το κάθετι που κάνεις *or* όλα όσα κάνεις *or* ό,τι και να κάνεις. money is not ~ τα λεφτά δεν είναι το παν.

everywhere *adv*. παντού. ~ you go όπου και να πας.

evict *v*. εκδιώκω, *(tenant)* κάνω έξωση σε. **~ion** *s*. έξωση *f*.

evidence *s*. *(testimony)* μαρτυρία *f*. *(indication)* ένδειξη *f*. piece of ~ ενδεικτικό σημείο. be ~ of μαρτυρώ. give ~ *(in court)* μαρτυρώ, καταθέτω. be in ~ φαίνομαι, θεώμαι, είμαι θεατός.

evident *a*. προφανής, καταφανής. **~ly** *adv*. προφανώς.

evil *a*. κακός, κακοήθης. ~ eye κακό μάτι, βάσκανος οφθαλμός. *(s.)* κακό *n*. *(sin)* αμαρτία *f*. **~doer** *s*. κακοποιός *m*.

evince *v*. δείχνω.

evocative *a*. υποβλητικός.

evoke *v.t*. *(call up image)* ξαναφέρνω στο νου. *(produce response)* προκαλώ.

evolution *s*. εξέλιξη *f*. **~ary** *a*. εξελικτικός.

evolve *v.t*. αναπτύσσω. *(v.i.)* εξελίσσομαι.

ewe *s*. προβατίνα *f*. **~**-lamb αμνάδα *f*.

ewer *s*. κανάτα *f*.

ex- πρώην.

exacerbate *v*. *(person)* ερεθίζω, *(make worse)* επιδεινώνω.

exact *a*. ακριβής. **~ly** *adv*. ακριβώς. **~ness** *s*. ακρίβεια *f*.

exact *v*. απαιτώ. **~ing** *a*. απαιτητικός, *(task)* δύσκολος.

exaggerat|e *v*. υπερβάλλω, μεγαλοποιώ, εξογκώνω. **~ed** *a*. υπερβολικός. **~ion** *s*. υπερβολή *f*., εξόγκωση *f*.

exalt *v*. *(raise)* εξυψώνω. *(extol)* εκθειάζω. **~ation** *s*. *(of spirits)* έξαρση *f*. **~ed** *a*. υψηλός.

exam *s*. εξετάσεις *f.pl*. sit for ~ δίνω εξετάσεις. ~ fees εξέταστρα *n.pl*.

examination *s*. εξέταση *f*., έλεγχος *m*. *(law, interrogation)* ανάκριση *f. see also* exam.

examine *v*. εξετάζω. *(inspect)* ελέγχω, *(minutely)* εξονυχίζω. *(law, interrogate)* ανακρίνω. **~r** *s*. εξεταστής *m*.

example *s*. παράδειγμα *n*. *(of excellence)* υπόδειγμα *n*. for ~ παραδείγματος *(or*

λόγου*)* χάριν. make an ~ of παραδειγματίζω.

exasperat|e *v*. νευριάζω *(also* get ~ed). *(fam.)* he ~es me μου δίνει στα νεύρα. **~ing** *a*. εκνευριστικός. **~ion** *s*. εκνευρισμός *m*.

excavat|e *v*. ανασκάπτω. **~ion** *s*. ανασκαφή *f*. **~or** *s*. *(machine)* εκσκαφέας *m*. who was the ~or? *(of archaeological site)* ποιος έκανε τις ανασκαφές;

exceed *v.t*. υπερβαίνω, ξεπερνώ. **~ingly** *adv*. πάρα πολύ, υπερβολικά. he is ~ingly stingy παραείναι τσιγγούνης.

excel *v.t*. υπερέχω *(with gen.)*, υπερτερώ. *(v.i.)* διαπρέπω, διακρίνομαι.

excellen|ce *s*. εξαιρετική ποιότητα. **~cy** *s*. His E~cy η αυτού εξοχότης.

excellent *a*. υπέροχος, άριστος, εξαίρετος, έξοχος, έκτακτος. **~ly** *adv*. υπέροχα, άριστα, εξαίρετα.

except *prep. & conj*. εκτός, πλην *(with gen.)*. εκτός από, παρά *(with* να). he does nothing ~ sleep δεν κάνει άλλο παρά να κοιμάται. I like the house, ~ that it is too dear μου αρέσει το σπίτι, μόνο που είναι πολύ ακριβό. I learnt nothing ~ that he had got married δεν έμαθα τίποτε άλλο εκτός του ότι παντρεύτηκε.

except *v*. εξαιρώ. present company ~ed οι παρόντες εξαιρούνται. **~ing** *prep. & conj. see* except.

exception *s*. εξαίρεση *f*. by way of ~ κατ' εξαίρεσιν. without ~ ανεξαιρέτως. take ~ φέρνω αντιρρήσεις, *(be offended)* θίγομαι.

exceptional *a*. εξαιρετικός, *(unusual)* ασυνήθης. **~ly** *adv*. εξαιρετικά.

excerpt *s*. απόσπασμα *n*.

excess *s*. υπερβολή *f*. *(sensual)* καταχρήσεις *f.pl*. an ~ of caution υπερβολική προσοχή. in ~ of *(more than)* πάνω από. in ~ καθ' υπερβολήν.

excess *a*. ~ weight επίπλέον βάρος. ~ fare πρόσθετο εισιτήριο.

excessive *a*. υπερβολικός, υπέρμετρος. **~ly** *adv*. υπερβολικά, καθ' υπερβολήν.

exchange *v*. *(blows, glances)* ανταλλάσσω, *(seats, etc.)* αλλάζω. *(I took it back to the shop and)* they ~d it for a new one μου το άλλαξαν με ένα άλλο.

exchange *s*. ανταλλαγή *f*. in ~ σε αντάλλαγμα. *(fin.)* συνάλλαγμα *n*. rate of ~ τιμή συναλλάγματος. stock ~ χρηματιστήριο *n*. bill of ~ συναλλαγματική *f*. telephone ~ τηλεφωνικό κέντρο.

exchequer *s*. δημόσιο ταμείο *n*., θησαυροφυλάκιο *n*.

excise *s*. φόρος παραγωγής.

excis|e *v*. κόγω, βγάζω. **~ion** *s*. περικοπή *f*.

excitable *a.* be ~ εξάπτομαι εύκολα, *etc.* (*see* excite).

excite *v.* (*cause*) προκαλώ, (*inflame*) εξάπτω, ερεθίζω, (*pleasurably*) συναρπάζω. get ~d (*inflamed*) εξάπτομαι, (*expectantly*) ξεσηκώνομαι, (*pleasurably*) ενθουσιάζομαι.

excited *a.* (*emotionally*) συγκινημένος, (*expectantly*) ξεσηκωμένος, (*pleasurably*) ενθουσιασμένος. ~**ly** *adv.* με θέρμη, με έξαψη.

excitement *s.* συγκίνηση *f.* (*joy*) χαρά *f.* (*commotion*) αναστάτωση *f.*, φασαρία *f.*

exciting *a.* συναρπαστικός.

excla|im *v.* αναφωνώ. ~**mation** *s.* αναφώνημα *n.* (*gram.*) επιφώνημα *n.* ~mation mark θαυμαστικό *n.*

exclu|de *v.* αποκλείω. ~**sion** *s.* αποκλεισμός *m.* to the ~sion of all others εξαιρουμένων όλων των άλλων.

exclusive *a.* αποκλειστικός. (*club, etc.*) για τους λίγους εκλεκτούς. ~ of εκτός από ~**ly** *adv.* αποκλειστικά.

excommunicat|e *v.* αφορίζω. ~**ion** *s.* αφορισμός *m.*

excoriate *v.* γδέρνω. (*fig., censure*) επικρίνω δριμύτατα.

excrement *s.* περιττώματα, κόπρανα *n.pl.*

excrescence *s.* εξάμβλωμα *n.*

excret|e *v.* απεκκρίνω. ~**a** *s.* περιττώματα *n.pl.* ~**ion** *s.* απέκκρισις *f.*

excruciating *a.* φρικτός. ~**ly** *adv.* φρικτά.

exculpate *v.* αθωώνω.

excursion *s.* εκδρομή *f.* (*a.*) ~ train εκδρομικό τραίνο. ~**ist** *s.* εκδρομέας *m.*

excus|e *v.* (*justify*) δικαιολογώ, (*exempt*) εξαιρώ, (*pardon*) συγχωρώ. ~e me με συγχωρείτε. ~**able** *a.* that is ~able δικαιολογείται αυτό. ~**ably** *adv.* δικαιολογημένα.

excuse *s.* δικαιολογία *f.* (*pretext*) πρόφαση *f.* make ~s δικαιολογούμαι.

execrab|le *a.* απαίσιος, φρικτός. ~**ly** *adv.* απαίσια, φρικτά.

execrat|e *v.i.* καταριέμαι. (*v.t., hate*) απεχθάνομαι. ~**ion** *s.* κατάρα *f.* απέχθεια *f.*

execut|e *v.* εκτελώ. (*law*) (*a deed*) επικυρώ, (*a will*) εκτελώ. ~**ant**, ~**or** *s.* εκτελεστής *m.*

execution *s.* εκτέλεση *f.* ~**er** *s.* δήμιος *m.*

executive *a.* εκτελεστικός. (*s.*) (*official*) ανώτερος υπάλληλος.

exemplary *a.* υποδειγματικός.

exemplify *v.* (*be example of*) είμαι παράδειγμα, (*give example of*) δίδω παράδειγμα (*both with gen.*).

exempt *v.* εξαιρώ, απαλλάσσω. (*a.*) απαλλαγμένος (*with gen.*) ~**ion** *s.* απαλλαγή *f.*, εξαίρεση *f.*

exercise *s.* άσκηση *f.*, εξάσκηση *f.* (*physical* ~*s*) γυμναστική *f.* (*task*) γύμνασμα *n.* ~ book τετράδιο *n.* (*mil.*) ~s γυμνάσια *n.pl.*

exercise *v.t.* ασκώ, εξασκώ. (*train*) γυμνάζω. (*perplex*) απασχολώ. (*v.i.*) (*take* ~) γυμνάζομαι.

exert *v.* (*use*) ασκώ. ~ oneself κοπιάζω, καταβάλλω προσπάθειες.

exertion *s.* (*use*) άσκηση *f.* (*trouble*) κόπος *m.* (*attempt*) προσπάθεια *f.*

exhal|e *v.t.* αναδίδω, βγάζω. (*v.i.*) (*breathe out*) εκπνέω. ~**ation** *s.* (*effluvium*) αναθυμίαση *f.* (*breath*) εκπνοή *f.*

exhaust *s.* (*of motor*) εξάτμιση *f.*

exhaust *v.* εξαντλώ, ~ one's talent εξοφλώ. ~**ed** *a.* εξαντλημένος. ~**ing** *a.* εξαντλητικός. ~**ion** *s.* εξάντληση *f.*

exhaustive *a.* εξαντλητικός, εξονυχιστικός. ~**ly** *adv.* εξονυχιστικώς.

exhibit *v.* επιδεικνύω, (*put on show*) εκθέτω. (*s.*) έκθεμα *n.* ~**or** *s.* εκθέτης *m.*

exhibition *s.* επίδειξη *f.* (*show*) έκθεση *f.* make an ~ of oneself γελοιοποιούμαι. ~**ism** *s.* (*showing off*) φιλεπιδειξία *f.* ~**ist** *s.* φιγουρατζής *m.*

exhilarat|e *v.* ζωογονώ. ~**ing** *a.* ζωογόνος, τονωτικός. ~**ion** *s.* αίσθημα ευεξίας.

exhort *v.* παραινώ. ~**ation** *s.* παραίνεση *f.*

exhum|e *v.* εκθάπτω. ~**ation** *s.* εκταφή *f.*

exigen|t *a.* απαιτητικός, (*urgent*) επείγον. ~**ce**, ~**cy** *s.* (*need*) ανάγκη *f.* ~**cies** απαιτήσεις *f.pl.*

exile *s.* εξορία *f.* (*person*) εξόριστος *m.f.* (*v.*) εξορίζω. (*only of political* ~) εκτοπίζω.

exist *v.* υπάρχω, υφίσταμαι. (*live*) ζω. ~**ent** *a.* υπαρκτός. ~**ing** *a* υπάρχων, υφιστάμενος.

existence *s.* ύπαρξη *f.* (*being*) υπόσταση *f.* come into ~ πρωτοπαρουσιάζομαι, πρωτοεμφανίζομαι, κάνω την εμφάνισή μου.

existential|ism *s.* υπαρξισμός *m.* ~**ist** *s.* υπαρξιστής *m.*

exit *s.* έξοδος *f.* ~ visa βίζα εξόδου.

exodus *s.* έξοδος *f.*

ex officio *a.* αυτεπάγγελτος.

exonerat|e *v.* αθωώνω, απαλλάσσω. ~**ion** *s.* απαλλαγή *f.*

exorbitant *a.* υπερβολικός. ~ price (*fam.*) φωτιά και λάβρα. ~**ly** *adv.* υπερβολικά.

exorc|ism *s.* εξορκισμός *m.* ~**ize** *v.* εξορκίζω.

exotic *a.* εξωτικός. ~**ism** *s.* εισαγωγή εξωτικών στοιχείων στην τέχνη.

expand *v.t.* (*substance*) διαστέλλω, (*activity*) επεκτείνω, (*story*) αναπτύσσω. (*v.i.*) (*of substance*) διαστέλλομαι, (*swell*) φουσκώνω. (*of activity*) επεκτείνομαι, (*of*

flower) ανοίγω. *(of person)* ξανοίγομαι. *(of river, lake)* διευρύνομαι. *(of city).* απλώνω.
expanse *s.* έκταση *f.*
expansion *s. (of substance)* διαστολή *f. (of activity, territory, etc.)* επέκταση *f. (of story, etc.)* ανάπτυξη *f.*
expansive *a. (wide)* ευρύς. *(of substance)* διασταλτικός. *(of person)* become more ~ γίνομαι λιγότερο συγκρατημένος, ξεκου-μπώνομαι.
expatiate *v.* μακρηγορώ.
expatriate *v.t.* εκπατρίζω. ~ oneself εκπα-τρίζομαι, αποδημώ. *(a. & s.)* απόδημος.
expatriation *s.* εκπατρισμός *m.,* αποδημία *f.*
expect *v.* περιμένω, αναμένω, *(hope for)* προσδοκώ. *(suppose)* φαντάζομαι, υποθέτω.
expect|ant *a.* προσδοκών. *(mother)* εγκυ-μονούσα. **~antly** *adv.* με προσδοκία. **~ancy** *s.* προσδοκία *f.*
expectation *s.* προσδοκία *f.* contrary to ~ παρά προσδοκίαν. ~ of life πιθανή διάρκεια ζωής.
expector|ant *a.* αποχρεμπτικός. **~ate** *v.* φτύνω.
expedi|ent *a.* σκόπιμος. *(s.)* μέσον *n. (re-course)* διέξοδος *f.* **~ency** *s.* σκοπιμότητα *f.*
expedite *v.* επισπεύδω.
expedition *s.* αποστολή *f. (mil.)* εκστρατεία *f. (speed)* ταχύτητα *f.*
expeditious *a.* ταχύς. **~ly** *adv.* ταχέως.
expel *v.* διώχνω, εκδιώκω, εκβάλλω, βγάζω. *(from school)* αποβάλλω.
expend *v.* ξοδεύω. **~able** *a. (funds)* αναλώσιμος. *(men, supplies, etc.)* δυνά-μενος να θυσιαστεί.
expenditure *s. (spending of money)* έξοδα *n.pl.* δαπάνες *f.pl.* ~ of time δαπάνη χρόνου.
expense *s.* δαπάνη *f.,* έξοδο *n.* at the public ~ δημοσία δαπάνη. at my ~ με δικά μου έξοδα, *(to my detriment)* εις βάρος μου. ~s έξοδα *n.pl.*
expensive *a.* ακριβός, δαπανηρός. **~ly** *adv.* ακριβά.
experience *s.* πείρα *f. (knowledge gained)* εμπειρία *f.* I had a strange ~ κάτι παράξενο μου συνέβη *or* είχα μία παράξενη εμπειρία. *(v.) (undergo)* δοκιμάζω, υφίσταμαι, περνώ. **~d** *a.* πεπειραμένος, έμπειρος.
experiment *s.* πείραμα *n. (v.)* πειραμα-τίζομαι. **~al** *a.* πειραματικός. **~ally** *adv.* πειραματικώς, δοκιμαστικώς.
expert *a.* πεπειραμένος, έμπειρος. get ~ advice παίρνω γνώμη ειδικού. *(s.)* ειδικός

m. (professional adviser) εμπειρογνώμο-νας *m.* **~ise** *s. (skill)* (επι)δεξιότης *f. (knowledge)* ειδικές γνώσεις.
expiat|e *v.* εκτίω ποινή για, εξιλεώνομαι για **~ion** *s.* εξιλέωση *f.*
expir|e *v.* λήγω, εκπνέω. **~ation, ~y** *s.* λήξη *f.*
explain *v.t.* εξηγώ, *(justify)* δικαιολογώ. *(v.i.) (make oneself clear)* εξηγούμαι.
explanat|ion *s.* εξήγηση *f.* δικαιολογία *f.* **~ory** *a.* επεξηγηματικός.
expletive *s.* βλαστήμια *f.*
explicit *a. (of order, statement) (definite)* ρητός, συγκεκριμένος, *(clear)* σαφής, ξεκάθαρος, *(frank)* ξέσκεπος. *(of person)* σαφής. be ~! πες το ξεκάθαρα. **~ly** *adv.* ρητώς, συγκεκριμένα, σαφώς, ξεκάθαρα, ξέσκεπα. **~ness** *s.* σαφήνεια *f.*
explode *v.t.* ανατινάζω. *(a theory)* ανατρέ-πω. *(v.i.)* ανατινάζομαι, εκρήγνυμαι. *(with anger, etc.)* ξεσπάω.
exploit *s.* κατόρθωμα *n.,* άθλος *m.*
exploit *v.* εκμεταλλεύομαι, *(develop)* αξιοποιώ. **~ation** *s.* εκμετάλλευση *f.* αξιοποίηση *f.*
explora|tion *s.* εξερεύνηση *f.* **~ory** *a.* εξερευνητικός.
explor|e *v.* εξερευνώ. **~er** *s.* εξερευνητής *m.*
explos|ion *s (act)* ανατίναξη *f. (event)* έκρηξη *f. (of anger)* ξέσπασμα *n.* **~ive** *a.* εκρηκτικός. *(s.)* εκρηκτική ύλη.
exponent *s.* ερμηνευτής *m. (math.)* εκθέτης *m.*
export *v.* εξάγω. *(s.)* εξαγωγή. *f.* **~er** *s.* εξαγωγεύς *m.*
expose *v.* εκθέτω, *(show up)* αποκαλύπτω. ~ oneself εκτίθεμαι. **~d** *a.* εκτεθειμένος.
exposition *s.* ανάπτυξη *f. (exhibition)* έκθεση *f.*
expostulat|e *v.* διαμαρτύρομαι. **~ion** *s.* διαμαρτυρία *f.*
exposure *s.* έκθεση *f. (showing up)* αποκάλυψη *f. (photographic)* πόζα *f.*
expound *v.* αναπτύσσω.
express *v.* εκφράζω, *(formulate)* διατυ-πώνω. *(squeeze out)* εκθλίβω. ~ oneself εκφράζομαι.
express *a. (explicit)* ρητός, σαφής. *(fast)* train ταχεία *f.* ~ letter κατεπείγον γράμμα.
expression *s.* έκφραση *f. (formulation, wording)* διατύπωση *f.* **~less** *a.* ανέκ-φραστος.
expressive *a.* εκφραστικός. **~ly** *adv.* εκ-φραστικώς. **~ness** *s.* εκφραστικότητα *f.*
expressly *adv.* ρητώς, σαφώς. *(on purpose)* επίτηδες, ειδικώς.
expressway *s. (US)* αυτοκινητόδρομος *m.*
expropriat|e *v. (land)* απαλλοτριώνω. **~ion**

(*of medium quality*) καλούτσικος. (*of price*) λογικός, (*weather*) αίθριος, (*wind*) ούριος. a ~ amount of αρκετός. ~ (*specious*) words παχιά λόγια. it isn't ~ δεν είναι σωστό (*or* εντάξει). make a ~ copy of καθαρογραφώ. (*adv.*) bid ~ το φαίνομαι σα να.

fairly *adv.* (*justly*) δικαίως, (*honestly*) τίμια, (*of degree*) αρκετά. ~ and squarely (*exactly*) ακριβώς, (*frankly*) ντόμπρα. ~ well καλούτσικα. (*fam.*) it ~ took my breath away μου 'κοψε κυριολεκτικά την ανάσα.

fair-haired *a.* ξανθομάλλης.

fairness *s.* (*honesty*) εντιμότητα *f.* (*impartiality*) αμεροληψία *f.* in ~ to him για να είμαστε δίκαιοι απέναντί του.

fair-spoken *a.* πειστικός.

fairway *s.* δίοδος *f.*

fairy *s.* νεράιδα *f.* ~-**tale** *s.* παραμύθι *n.* (*a.*) παραμυθένιος.

fait-accompli *s.* τετελεσμένον γεγονός.

faith *s.* (*belief*) πίστη *f.* (*religion*) θρήσκευμα *n.* (*trust*) εμπιστοσύνη *f.* in good ~ καλή τη πίστει.

faithful *a.* πιστός, (*exact*) ακριβής. ~**ly** *adv.* πιστά, με ακρίβεια. (*of promising*) ειλικρινά. yours ~**ly** μετά τιμής.

faithless *a.* άπιστος. ~**ness** *s.* δολιότητα *f.*

fake *a.* & *s.* ψεύτικος. (*v.t.*) παραποιώ.

falcon *s.* γεράκι *n.*

fall *s.* πτώση *f.* (*tumble*) πέσιμο *n.* (*autumn*) φθινόπωρο *n.* ~s (*water*) καταρράκτης *m.*

fall *v.i.* (*also* ~ down, over) πέφτω. ~ to bits καταρρέω. ~**en** *a.* πεσμένος, (*deposed*) έκπτωτος. the ~ en (*in battle*) οι πεσόντες.

fall away *n.* (*defect*) αποστατώ, αποσκιρτώ.

fall back *n.* αποσύρομαι. ~ on καταφεύγω σε.

fall behind *v.i.* μένω πίσω, (*be late*) καθυστερώ.

fall for *v.* ενθουσιάζομαι για. he fell for it (*was tricked*) τυλίχτηκε.

fall in *v.i.* (*mil.*) παρατάσσομαι.

fall into *v.* (*hands, coma, sin, etc.*) περιπίπτω εις.

fall off *v.i.* (*decline*) ξεπέφτω.

fall out *v.* (*of hair, etc.*) πέφτω, (*happen*) συμβαίνω, (*quarrel*) τσακώνομαι. (*mil.*) λύω τους ζυγούς.

fall-out *s.* (*nuclear*) διαρροή *f.*

fall over *v.* πέφτω κάτω. ~ the cliff πέφτω από το βράχο. I fell over a stone σκόνταψα σε μια πέτρα και έπεσα. ~ oneself *or* ~ backwards (*do utmost*) τσακίζομαι.

fall through *v.i.* (*fail*) ματαιώνομαι.

fall to *v.* (*begin*) αρχίζω. (*food, work*) πέφτω με τα μούτρα (*with* σε).

fall under *v.* (*category*) υπάγομαι εις. ~ the spell of καταγοητεύομαι από.

fall upon *v.* (*of responsibility*) βαρύνω (*v.t.*). (*attack*) επιτίθεμαι εις.

fall within *v.* (*belong to*) ανήκω σε.

fallacy *s.* πλάνη *f.* ~**ious** *a.* εσφαλμένος.

fallible *a.* man is ~ ο άνθρωπος δεν είναι αλάνθαστος.

fallow *a.* χέρσος.

false *a.* ψευδής, ψεύτικος, (*erroneous*) εσφαλμένος, (*faithless*) άπιστος ~ alarm ψεύτικος συναγερμός. ~ note παραφωνία *f.* ~ step γκάφα *f.* play (*person*) ~ απατώ. give ~ evidence ψευδομαρτυρώ. by ~ pretences δι' απάτης. ~**hood** *s.* (*lie*) ψεύδος *n.*, ψευτιά *f.*

falsify *v.* παραποιώ.

falsity *s.* (*wrongness*) ανακρίβεια *f.*, (*treachery*) δολιότητα *f.*

falter *v.* (*in speech*) κομπιάζω (*waver*) διστάζω.

fame *s.* φήμη *f.* ~**d** *a.* φημισμένος.

familiar *a.* (*intimate*) οικείος, (*known*) γνωστός. be ~ with ξέρω, be on ~ terms έχω οικειότητα. be too ~ (*cheeky*) παίρνω οικειότητα ~**ity** *s.* (*intimacy*) οικειότητα *f.* (*knowledge*) γνώση *f.* ~**ize** *v.* εξοικειώνω.

family *s.* οικογένεια *f.* ~ man οικογενειάρχης *m.* ~ tree γενεαλογικό δέντρο *n.* (*a.*) οικογενειακός.

famine *s.* λιμός *m.*

famish *v.* be ~ed *or* ~ing πεινώ λυσσωδώς, πεθαίνω της πείνας.

famous *a.* (*person*) διάσημος, (*thing*) ξακουστός. (*notorious*) περιβόητος. ~**ly** *adv.* (*fam.*) μια χαρά.

fan *s.* ανεμιστήρας *m.* (*hand*) βεντάλια *f.* (*admirer*) λάτρης *m.* ~ club όμιλος θαυμαστών *m.* (*v.*) φυσώ, ~ oneself φυσιέμαι. ~ the flames (*fig.*) ρίχνω λάδι στη φωτιά.

fanatic *s.*, ~**al** *a.* φανατικός. make ~al φανατίζω. ~**ism** *s.* φανατισμός *m.*

fancier *s.* ειδικός (*with* σε).

fanciful *a.* φαντασιώδης, παράδοξος.

fancy *s.* (*imagination*) φαντασία *f.* (*idea*) ιδέα *f.* (*whim*) καπρίτσιο *n.* take a ~ to συμπαθώ. it took my ~ με είλκυσε.

fancy *v.* (*imagine*) φαντάζομαι, ~y that! για φαντάσου! he ~ies himself as νομίζει πως είναι. (*desire*) γουστάρω (*v.t.*). I don't ~y it δεν μου αρέσει.

fancy *a.* (*not plain*) πολυτελείας. (*of design*) φανταιζί. ~ goods νεωτερισμοί *m.pl.* in ~ dress μεταμφιεσμένος.

fanfare *s.* φανφάρα *f.*

fang *s.* κυνόδους *m.* (*snake's*) φαρμακερό δόντι.

fanlight s. φεγγίτης m.
fantasia s. φαντασία f.
fantastic a. (strange) φανταστικός, (wonderful) υπέροχος, (absurd) εξωφρενικός. ~ally adv. υπέροχα, εξωφρενικά.
fantasy s. (imagination) φαντασία f. (false idea) φαντασίωση f.
far adv. (distance) μακριά, (degree) πολύ. ~ and away, by ~ κατά πολύ. so ~ ως τώρα, για την ώρα. in so ~ as εφόσον. ~ from it! κάθε άλλο! it is ~ from true πολύ απέχει από την αλήθεια. go too ~ (fig.) το παρακάνω. as ~ as the station μέχρι το σταθμό. as ~ as I know απ' ό,τι ξέρω. as ~ as I'm concerned όσον αφορά εμένα. (so) ~ from admiring it I don't like it όχι μόνο δεν το θαυμάζω, δε μ' αρέσει.
far a. (also ~-off, ~-away) μακρινός. few and ~ between αραιοί. F~ East Άπω Ανατολή. at the ~ end στην άλλη άκρη. a ~-away look ονειροπόλο βλέμμα.
farc|e s., ~ical a. φάρσα f.
fare v.i. (get on) τα πάω, περνώ. (be fed) τρώγω. how did it ~ with you? πώς τα πήγες;
fare s. (cost of travel) ναύλα n.pl. (in bus, etc.) εισιτήριο n. (in taxi) what is the ~ ? πόσα γράφει το ρολόι; (passenger) πελάτης m. (food) τροφή f. bill of ~ κατάλογος m.
farewell s. αποχαιρετισμός m. bid (person) ~ αποχαιρετώ. ~! χαίρε! αντίο! (a.) αποχαιρετιστήριος.
far-fetched a. παρατραβηγμένος.
far-flung a. εκτεταμένος.
far-gone a. πολύ προχωρημένος.
farm s. αγρόκτημα n., φάρμα f. (v.t.) (land) καλλιεργώ. ~ out εκμισθώνω. ~er s. γεωργός m. ~yard s. αυλή φάρμας.
farrago s. συνονθύλευμα n.
far-reaching a. ευρείας εκτάσεως.
far|-seeing, ~-**sighted** a. προνοητικός.
farther adv. πιο μακριά, μακρύτερα. ~ on πιο πέρα. ~ back πιο πίσω. (a.) απώτερος. at the ~ end στην άλλη άκρη.
farthest a. απώτατος, ο πιο μακρινός (adv.) πιο μακριά, μακρύτερα.
fascinat|e v. (allure) σαγηνεύω. (interest greatly) γοητεύω. ~ing a. σαγηνευτικός, γοητευτικός. ~ion s. σαγήνη f. γοητεία f. have a ~ion for γοητεύω.
fasc|ism s. φασισμός m. ~ist s. φασίστας m. (a.) φασιστικός.
fashion s. (way) τρόπος m. (custom, style) μόδα f., in ~ της μόδας. after a ~ κατά κάποιο τρόπο, όπως όπως (book of) ~ plates φιγουρίνι n. (v.) (of clay) πλάθω, (wood, stone) σκαλίζω. (create) δημι-

ουργώ.
fashionab|le a. (person, resort) κοσμικός, (clothes, resort) της μόδας. ~ly adv. με την τελευταία λέξη της μόδας.
fast v. νηστεύω. (s.) νηστεία f.
fast a. (firm) στερεός, make ~ στερεώνω, (tie up) δένω. (of colour) ανεξίτηλος. (quick) ταχύς, γρήγορος. (of clock) it is ~ πάει μπρος. play ~ and loose with κάνω κατάχρηση (with gen.), (person's affections) παίζω με.
fast adv. (firmly) στερεά, (tightly) σφιχτά. (quickly) γρήγορα. be ~ asleep κοιμάμαι βαθιά. it is raining ~ βρέχει δυνατά.
fasten v.t. (secure) ασφαλίζω, στερεώνω. (affix) κολλώ, (tie up) δένω, (tighten) σφίγγω. ~ together συνδέω. (v.t. & i.) (shut) κλείνω, (do up) κουμπώνω. ~ upon (pretext, etc.) αρπάζω. he always ~ s on me όλο τα βάζει μαζί μου (or με μένα).
fastener s. (press) σούστα f. (hook & eye) κόπιτσα f. (zip) φερμουάρ n. (paper) συνδετήρ m.
fastening s. (of window, etc.) ασφάλεια f. (of bag) κλείσιμο n. (of garment) κούμπωμα n.
fastidious a. (of fine taste) εκλεκτικός, (fussy) δύσκολος, σχολαστικός.
fasting s. νηστεία f.
fastness s. οχυρό n.
fat a. παχύς, χοντρός. (greasy) λιπαρός. (salary) παχυλός. (fam.) a ~ lot he knows! τα πολλά που ξέρει.
fat s. λίπος n., πάχος n. (fig.) the ~ is in the fire προμηνύεται φασαρία. live off the ~ of the land περνώ ζωή και κότα. ~ness s. πάχος n.
fatal a. (causing failure, appointed by destiny) μοιραίος. (causing death) θανατηφόρος. ~ly adv. θανασίμως.
fatal|ism s. μοιρολατρία f. ~ist s. μοιρολάτρης m. ~ity s. (event) συμφορά f. (person killed) θύμα n.
fate s. (destiny) μοίρα f., πεπρωμένο n. (one's personal ~) ριζικό n. (what may happen) τύχη f. they left him to his ~ τον άφησαν στην τύχη του. he met a tragic ~ είχε τραγικό τέλος. it was ~d that ήταν μοιραίο (or πεπρωμένο or γραφτό) να.
fathead s. ~ed a. μάπας m., κουτορνίθι n.
father s. πατέρας m. (eccl.) πατήρ m. (v.) (lit. & fig.) γεννώ. ~ on (to) (of child, idea, book) αποδίδω την πατρότητα εις, (of responsibility) αποδίδω την ευθύνη εις ~hood s. πατρότητα f. ~-**in-law** m. ~**land** s. πατρίδα f. ~**ly** a. ~'**s** a. πατρικός.
fathom s. οργ(υ)ιά f. (v.) κατανοώ. ~**less** a.

απύθμενος.
fatigu|e s. κόπωση f. (mil.) αγγαρεία f. (v.) κουράζω. **~ing** a. κουραστικός.
fatt|en v. παχαίνω, (cattle) σιτεύω. **~ed** calf ο μόσχος ο σιτευτός.
fatu|ous a. βλακώδης. (purposeless) μάταιος. **~ity** s. ανοησία f.
faucet s. κάνουλα f.
fault s. (defect) ελάττωμα n. (blame) φταίξιμο n. (wong act) σφάλμα n. be at ~ (to blame) φταίω. find ~ with (person) επικρίνω. she finds ~ with everything όλα τα βρίσκει στραβά. (in rock) γεωτεκτονικό ρήγμα. (v.) βρίσκω ελάττωμα σε. **~less** a. άψογος. **~y** a. ελαττωματικός.
faun s. φαύνος m.
fauna s. πανίδα f.
faux pas s. γκάφα f.
favour s. (approval, goodwill) εύνοια f. (pariality) μεροληψία f. (kind act) χάρη f., χατίρι n., (political) ρουσφέτι n. (badge) κονκάρδα f. in ~ of υπέρ (with gen.) in our ~ προς όφελός μας.
favour v. ευνοώ, προτιμώ. ~ (person) with (oblige) τιμώ με.
favourab|le a. ευνοϊκός, ευμενής. (wind) ούριος, (weather) αίθριος. **~ly** adv. ευνοϊκά, ευμενώς.
favourit|e a. & s. ευνοούμενος, (best liked) (πιο) αγαπημένος. **~ism** s. ευνοιοκρατία f.
fawn s. ελαφάκι n. (a.) (colour) κιτρινόφαιος.
fawn v. ~ on κάνω τούμπες μπροστά σε. **~ing** a. δουλοπρεπής.
fax-machine s. τηλεομοιοτυπικό μηχάνημα.
fear s. φόβος m. no ~! δεν υπάρχει φόβος. I did not go in for ~ of disturbing him δεν μπήκα μέσα από φόβο μη τυχόν (or μήπως) τον ενοχλήσω. (v.) φοβάμαι, φοβούμαι.
fearful a. (afraid) ανήσυχος, awful) φοβερός. be ~ φοβάμαι. **~ly** adv. (with fear) φοβισμένα, (awfully) φοβερά.
fearless a. άφοβος. **~ly** adv. άφοβα. **~ness** s. αφοβία f.
fearsome a. τρομακτικός.
feasib|le a. εφικτός. **~ility** s. εφικτό n.
feast s. (eccl.) εορτή f. (banquet, lit. & fig.) πανδαισία f. (v.t.) (regale) γλεντώ, χορταίνω. ~ one's eyes on δεν χορταίνει το μάτι μου να βλέπει. (v.i.) ευωχούμαι. **~-day** s. εορτή f. **~ing** s. ευωχία f.
feat s. κατόρθωμα n.
feather s. φτερό n. ~ in one's cap επίτευγμα n. show the white ~ δειλιάζω. birds of a ~ flock together όμοιος ομοίω. in high ~ κεφάτος (v.) ~ one's nest κάνω την μπάζα

μου. **~ed** a. πτερωτός.
feather-brained a. κοκκορόμυαλος.
feather-duster s. φτερό n.
feather-weight s. βάρος φτερού, (fig.) ασήμαντος a.
feathery a. (light & soft) αφράτος.
feature s. χαρακτηριστικό n. (element) στοιχείο n. (article in journal) ειδικό άρθρο (v.i.) παίζω ρόλο. (v.t.) παρουσιάζω. **~less** a. μονότονος.
February s. Φεβρουάριος m., Φλεβάρης m.
feckless a. αχαΐρευτος.
fecund a. γόνιμος. **~ity** s. γονιμότητα f.
fed part. get ~ up (with) βαριεστίζω.
feder|al a. ομοσπονδιακός. **~ate** v.i. σχηματίζω ομοσπονδία. **~ation** s. ομοσπονδία f.
fee s. (professional) αμοιβή f. tuition ~ δίδακτρα n.pl. exam ~ εξέταστρα n.pl.
feeb|le a. αδύνατος, αδύναμος. (joke) σαχλός. **~ly** adv. άτονα, χωρίς δύναμη. **~-le-minded** a. διανοητικά καθυστερημένος.
feed v.t. ταΐζω, (nourish) τρέφω, (keep supplied) τροφοδοτώ. (v.i.) ~ on τρέφομαι με. (s.) (animals') ζωοτροφή f. have a ~ τρώω. (mech.) τροφοδότηση f. **~ing** s. τάισμα n., τροφοδότηση f.
feedback s. (technical) ανάδραση f. (response) ανταπόκριση f.
feel v.t. & i. (by touch) ψαύω, πιάνω. ~ how heavy it is! πιάσε να δεις πόσο βαρύ είναι. ~ one's way περπατώ ψηλαφώντας. ~ (search) for ψάχνω για. (have emotion, sensation) αισθάνομαι, νιώθω. it ~s as if φαίνεται σα να. how does it ~? πώς σου φαίνεται; it ~s rough είναι τραχύ στην αφή. I don't ~ myself (up to form) δεν είμαι ο εαυτός μου. I ~ the heat με πειράζει η ζέστη. I ~ for him τον συμπονώ. ~ equal (or up) to αισθάνομαι ικανός να (or για). ~ like (be in mood for) έχω διάθεση (or όρεξη) για, γουστάρω. (think) νομίζω.
feel s. I can tell by the ~ of it καταλαβαίνω από την αφή του (or το πιάσιμό του).
feeler s. (antenna) κεραία f. (fig.) put out ~s κάνω βολιδοσκοπήσεις, βολιδοσκοπώ.
feeling s. (touch) I've lost the ~ in my right leg έχασα την αίσθηση του δεξιού ποδιού μου. (awareness, emotion) αίσθημα n. person of ~ άνθρωπος με αισθήματα. he does not show much ~ είναι αναίσθητος. hurt (person's) ~s προσβάλλω τα αισθήματά του. I have no hard ~s δεν κρατώ κακία. ~ runs high τα πνεύματα είναι εξημμένα. (presentiment) προαίσθημα n. (opinion) γνώμη f.
feign v. προσποιούμαι. ~ indifference

προσποιούμαι τον αδιάφορο. **~ed** *a*. προσποιητός.

feint *s*. ψευδεπίθεση *f*.

felicitat|e *v*. συγχαίρω. **~ion(s)** *s*. συγχαρητήρια *n.pl*.

felicit|ous *a*. εύστοχος. **~y** *s*. (*happiness*) ευδαιμονία *f*.

feline *a*. αιλουροειδής. ~ grace χάρις αιλούρου.

fell *v*. ρίχνω κάτω. ~ing of trees υλοτομία *f*.

fell *a*. δεινός, (*disease*) επάρατος.

fell *s*. (*hide*) τομάρι *n*. (*hill*) λόφος *m*.

fellow *s*. (*comrade*) σύντροφος *m*. (*equal*) όμοιος *m*. (*one of a pair*) ταίρι *n*. (*chap*) παιδί *n*. (*of college, etc*.) εταίρος *m*. **~-citizen** *s*. συμπολίτης *m*. **~-country-man** *s*. ομοεθνής *m*. **~-traveller** *s*. συνταξιδιώτης *m*. (*also politically*) συνοδοιπόρος *m*. **~-sufferer** *s*. ομοιοπαθής *m*.

fellowship *s*. συντροφιά *f*.

felon *s*. κακούργος *m*. **~ious** *a*. εγκληματικός. **~y** *s*. κακούργημα *n*.

felt *s*. τσόχα *f*. (*for hats*) φετο *n*.

female *a*. θηλυκός. (*s*.) θήλυ *n*. (*fam*.) θηλυκό *n*.

feminin|e *a*. (*gram*.) θηλυκός. (*charm, curiosity, etc*.) γυναικείος. she is very ~e είναι πραγματικό θηλυκό (*or* πολύ γυναίκα). **~ity** *s*. θηλυκότητα *f*.

femin|ism *s*. φεμινισμός *m*. **~ist** *s*. φεμινιστής *m*., φεμινίστρια *f*.

fen *s*. βάλτος *m*.

fenc|e *s*. φράκτης *m*. (*receiver*) κλεπταποδόχος *m*. (*fig*.) sit on the ~e καιροσκοπώ. (*v*.) (*put* ~e *round*) περιφράσσω. **~ing** *s*. περίφραξη *f*.

fenc|e *v*. (*fight*) ξιφομαχώ. (*fig*.) αποφεύγω. **~ing** *s*. ξιφομαχία *f*.

fend *v*. ~ off αποκρούω. ~ for oneself τα βγάζω πέρα μόνος μου.

fender *s*. προφυλακτήρ *m*. (*naut*.) στρωμάτσο *n*.

fennel *s*. μάραθο *n*.

ferment *s*. (*leaven*) μαγιά *f*. (*fig*.) state of ~ αναβρασμός *m*. (*v.t*.) ζυμώνω (*v.i*.) ζυμώνομαι. **~ation** *s*. ζύμωση *f*.

fern *s*. φτέρη *f*. **~y** *a*. γεμάτος φτέρες.

feroci|ous *a*. θηριώδης. **~ously** *adv*. θηριωδώς, άγρια. **~ty** *s*. θηριωδία *f*.

ferret *s*. νυφίτσα *f*. (*v*.) (*rummage*) σκαλίζω, ~ out ανακαλύπτω.

ferro-concrete *s*. μπετόν αρμέ *n*.

ferrous *a*. σιδηρούχος.

ferry *s*. πορθμείο *n*. (*v.t.*) διαπορθμεύω. (*v.i*.) διαπορθμεύομαι. **~-boat** *s*. φερυμπότ *n*.

fertil|e *a*. γόνιμος. **~ity** *s*. γονιμότητα *f*.

fertiliz|e *v*. γονιμοποιώ, (*soil*) λιπαίνω. **~ation** *s*. γονιμοποίηση *f*. λίπανση *f*. **~er** *s*. χημικό λίπασμα.

ferven|t *a*. (*love*) θερμός, (*support*) ένθερμος. **~cy** *s*. θέρμη *f*. **~tly** *adv*. θερμά, με θέρμη.

fervid *a*. φλογερός, παθιασμένος. **~ly** *adv*. φλογερά, με πάθος.

fervour *s*. θέρμη *f*., πάθος *n*.

fester *v*. μολύνομαι, αφορμίζω. (*fig*., *of insult, etc*.) it ~s in my mind με καίει. **~ing** *a*. μολυσμένος, αφορμισμένος.

festiv|al *s*. (*feast*) εορτή *f*. (*of music, etc*.) φεστιβάλ *n*. **~e** *a*. εορταστικός, (*music, bells*) χαρμόσυνος. **~ity** *s*. εορτασμός *m*., γλέντι *n*. **~ies** εορταστικές εκδηλώσεις.

festoon *s*. γιρλάντα *f*. (*v*.) στολίζω.

fetch *v*. παίρνω, φέρνω. I ~ed him from the station τον πήρα από το σταθμό. ~ me the book φέρε μου το βιβλίο. she makes him ~ and carry τον έχει σαν υπηρέτη. (*a price*) πιάνω, (*a blow*) καταφέρω. ~ up (*arrive*) καταλήγω.

fetching *a*. χαριτωμένος.

fête *s*. πανηγύρι *n*. (*v*.) αποθεώνω.

fetid *a*. βρώμικος, δύσοσμος. **~ness** *s*. βρώμα *f*.

fetish *s*. φετίχ *n*. (*fig*.) make a ~ of έχω μανία με.

fetter *v*. δένω, δεσμεύω. (*s*.) (*shackle*) πεδούκλι *n*. **~s** (*lit. & fig*.) δεσμά *n.pl*.

fettle *s*. in good ~ σε καλή φόρμα.

feud *s*. έχθρα *f*., βεντέτα *f*.

feudal *a*. φεουδαρχικός.

fever *s*. πυρετός *m*. **~ish** *a*. πυρετώδης. **~ishly** *adv*. πυρετωδώς.

few *a*. λίγοι, ολίγοι. a good ~ αρκετοί, not a ~ ουκ ολίγοι. **~er** *a*. λιγότεροι. he ate no ~er than five cakes έφαγε πέντε πάστες, αν αγαπάς (*or* παρακαλώ).

fey *a*. αλαφροΐσκιωτος.

fez *s*. φέσι *n*.

fianc|é *s*. μνηστήρας, αρραβωνιαστικός *m*. **~ée** *s*. μνηστή, αρραβωνιαστικιά *f*.

fiasco *s*. φιάσκο *n*.

fiat *s*. φετφάς .

fib *s*. ψεματάκι *n*.

fibre *s*. ίνα *f*. vegetable ~ φυτική ίνα. moral ~ ηθική υπόσταση. **~-glass** *s*. υαλομβάμβαξ *m*.

fibrous *a*. ινώδης.

fickle *a*. ασταθής, άστατος, (*person only*) άπιστος. **~ness** *s*. αστάθεια *f*.

fiction *s*. (*fabrication*) αποκύημα της φαντασίας. (*novels*) μυθιστορήματα *n.pl*. (*law*) legal ~ πλάσμα δικαίου. **~al** *a*. φαντασιώδης.

fictitious *a*. πλαστός. (*law, of sale, etc*.) εικονικός. (*imaginary*) φανταστικός.

fiddle *s*. βιολί *n*. (*fig*.) play second ~ παίζω

δευτερεύοντα ρόλο. (v.) παίζω βιολί. **~r** s. βιολιστής m.

fiddle v. (*waste time*) χασομερώ. (*falsify*) μαγειρεύω. ~ with σκαλίζω, πασπατεύω. (s.) (*fam., sharp practice*) κομπίνα f. be on the ~ κάνω κομπίνες.

fidelity s. πίστη f. (*accuracy*) ακρίβεια f.

fidget (v.i.) δεν κάθομαι ήσυχα. stop ~ing! κάτσε ήσυχα. (*get nervous*) νευριάζω. ~ with παίζω με, πασπατεύω. (s.) (*person*) νευρόσπαστο n. have the ~s see fidget v.

fief s. φέουδο n.

field s. χωράφι n., αγρός m. (*of vision, activity, battle*) πεδίο n. **~-artillery** s. πεδινόν πυροβολικόν. **~-day** s. ημέρα γυμνασίων. (*fig.*) μία πολύ γεμάτη μέρα. **~-events** s. υπαίθρια σπορ. **~-glasses** s. κιάλια n.pl. **~-marshal** s. στρατάρχης m. **~-work** s. (*research*) επιτόπιος επιστημονική έρευνα. (*mil.*) πρόχειρο οχύρωμα.

fiend s. (*fig.*) τέρας n. **~ish** a. σατανικός. **~ishly** adv. σατανικά. (*fam.*) it is ~ishly cold κάνει διαβολεμένο κρύο.

fierce a. άγριος, (*struggle*) σκληρός, λυσσώδης, (*heat, cold*) δυνατός. **~ly** adv. άγρια, σκληρά, λυσσωδώς, δυνατά.

fiery a. φλογερός, (*speech*) πύρινος, (*temper*) παράφορος.

fife s. φλογέρα f.

fifteen num. δεκαπέντε. 15-syllable line (*verse*) δεκαπεντασύλλαβος m. **~th** a. δέκατος πέμπτος.

fifth a. πέμπτος ~ column πέμπτη φάλαγξ.

fiftly num. πενήντα. go ~y-~y πάω μισά-μισά. **~ieth** a. πεντηκοστός.

fig s. σύκο n. (*tree*) συκιά f. wild ~ ορνιός m. (*fam.*) not care a ~ δεν δίνω δεκάρα.

fight v.i. μάχομαι, (*of dogs, etc.*) μαλώνω. (*struggle*) αγωνίζομαι, (*wrestle*) παλεύω. (v.i. & t.) πολεμώ. (v.t.) (*a fire, illness, proposal*) καταπολεμώ. ~ down υπερνικώ. ~ off (*repel*) απωθώ. ~ shy of αποφεύγω. (s.) μάχη f., αγώνας m., πάλη f. (*row*) καβγάς m.

fighter s. πολεμιστής m., αγωνιστής m. (*aero.*) καταδιωκτικόν n.

fighting a. (*combative*) μαχητικός, (*forces, etc.*) μάχιμος. (s.) μάχη f. street ~ οδομαχίες f.pl. there was hard ~ έγινε σκληρή μάχη.

figment s. ~ of the imagination πλάσμα της φαντασίας.

figurative a. μεταφορικός.

figure s. (*number*) ψηφίο n. (*diagram & geom.*) σχήμα n. (*amount*) ποσό n. (*dance, mus.*) φιγούρα f. (*human form*) ανθρώπινη μορφή. (*bodily*) have a good ~ είμαι σιλουέτα, έχω καλό κορμί. fine ~ of a man

(*or* woman) λεβέντης m., λεβέντισσα f. cut a fine ~ κάνω ωραία φιγούρα. cut a sorry ~ κάνω θλιβερή εντύπωση. leading ~ σημαίνον πρόσωπον. ~ of speech σχήμα λόγου. (v.) (*appear*) εμφανίζομαι, (*calculate*) υπολογίζω.

figure-head s. (*naut.*) γοργόνα f. (*fig.*) προσωπικότης κατέχουσα υψηλόν αξίωμα κατ' όνομα.

filbert s. φουντούκι n.

filch v. βουτώ.

file s. (*tool*) λίμα f. (v.t.) λιμάρω.

file s. (*for papers*) φάκελλος m. (v.t.) (*in folders*) κατατάσσω, (*in archives*) βάζω στο αρχείο. (*submit*) υποβάλλω.

file s. (*row*) στοίχος m., γραμμή f. in single ~ ο ένας πίσω από τον άλλο, (*mil.*) εφ' ενός ζυγού. (v.i.) ~ in μπαίνουμε ένας-ένας.

filial a. υιικός.

filibuster v.i. κωλυσιεργώ.

filings s. ρινίσματα n.pl.

fill v.t. & i. (*make or become full*) γεμίζω. (a tooth) σφραγίζω, (a vacancy) συμπληρώνω. ~ up (a form) συμπληρώνω, (a vessel) γεμίζω. ~ out (*enlarge*) γεμίζω, φουσκώνω. it ~s the bill είναι ό,τι χρειάζεται.

fill s. have one's ~ of μπουχτίζω από. eat one's ~ τρώγω καλά, χορταίνω.

filling s. γέμιση f. (*in tooth*) σφράγισμα n. ~ station πρατήριο βενζίνης.

fillip s. (*with finger*) μικρό τίναγμα με το δάχτυλο. (*fig.*) give a ~ to (*stimulate*) τονώνω, κεντρίζω, (*speed up*) επιταχύνω.

filly s. φοραδίτσα f. (*fam., girl*) mettlesome ~ ζωντανό κοριτσόπουλο.

film s. (*haze*) θαμπάδα f. (*layer*) στρώμα n. (*photo.*) φιλμ n. (*cinema*) ταινία f., έργο n. (v.t.) (*make ~ of*) γυρίζω, κινηματογραφώ. **~-star** s. βεντέτα της οθόνης.

filter s. φίλτρο n. (v.t.) διυλίζω, φιλτράρω. (v.i.) ~ through (*leak*) διεισδύω, (*of coffee, light*) περνώ, (*of news*) διαρρέω. **~ing** s. διύλιση f., φιλτράρισμα n.

filth, ~iness s. βρωμιά f. **~y** a. βρωμερός, ρυπαρός.

fin s. πτερύγιον n.

final a. (*in end position*) τελικός. (*last possible, e.g. edition, day, word*) τελευταίος. (*definite*) οριστικός. (*unalterable*) τελεσίδικος. (s.) ~s (*sport*) τελικά n.pl. (*exams*) πτυχιακές εξετάσεις. **~ly** adv. τελικά, εν τέλει, (*definitely*) οριστικά.

finale s. φινάλε n.

final|lity s. οριστικότης f. with (an air of) ~ity με ύφος τελειωτικό. **~ize** v. οριστικοποιώ.

financ|e s. (*also* ~es) οικονομικά n.pl. (v.) χρηματοδοτώ. **~ial** a. οικονομικός, (*pe-*

cuniary) χρηματικός. ~ier s. κεφαλαιούχος m.

finch s. σπίνος m.

find v.t. βρίσκω. (provide) προμηθεύω, all found με όλα τα έξοδα. ~ (person) guilty κηρύσσω ένοχο. ~ out (learn) μαθαίνω, (uncover, discover) ανακαλύπτω. how did it ~ its way into my pocket? πώς βρέθηκε στην τσέπη μου; (v.i.) (give verdict) αποφαίνομαι. ~er s. ~er will be rewarded ο ευρών αμειφθήσεται.

finding s. ανεύρεση f. (conclusion of committee, etc.) πόρισμα n. (law) απόφασις f.

fine s. πρόστιμο n. (v.) βάζω πρόστιμο σε.

fine s. in ~ εν συντομία.

fine a. (pleasing) ωραίος, (excellent) εκλεκτός, (thin, keen) λεπτός, (delicate, refined) λεπτός, φίνος. (of particles) ψιλός, (of gold) καθαρός, (of weather) ωραίος. ~ arts καλές τέχνες.

fine adv. (well) πολύ καλά, μια χαρά. (thinly, delicately) (also ~ly) ψιλά, λεπτά. ~ly chopped, cut up ~ ψιλοκομμένος. ~ly cut (features) λεπτοκαμωμένος. ~ly made (furniture, etc.) ωραία καμωμένος. cut it ~ τα καταφέρνω ίσα-ίσα.

finery s. στολίδια n.pl.

finesse s. λεπτότητα f., φινέτσα f.

finger s. δάχτυλο n., δάκτυλος m. have at one's ~ tips παίζω στα δάχτυλα. she twists him round her little ~ τον παίζει στα δάχτυλά της. have a ~ in every pie έχω παντού το δάχτυλό μου. keep one's ~s crossed χτυπώ ξύλο. put one's ~ on προσδιορίζω ακριβώς. (v.) ψηλαφώ, πασπατεύω. ~-mark s. δαχτυλιά f. ~nail s. νύχι n. ~print s. δακτυλικό αποτύπωμα. ~stall s. δακτυλήθρα f.

finicky a. (person) ιδιότροπος. ~ piece of work ψιλοδουλειά f.

finish v.t. & i. τελειώνω. ~ up (v.t.) τελειώνω, (v.i.) καταλήγω. ~ off (v.t.) (consume, complete) τελειώνω, (kill) ξεκάνω, καθαρίζω. he's ~ed! (done for) πάει κι αυτός. (s.) (end) τέλος n. (of workmanship) τελείωμα n., λεπτότητα εργασίας.

finite a. πεπερασμένος.

fir s. έλατο n. ~cone s. κουκουνάρι n.

fire s. φωτιά f. (outbreak) πυρκαγιά f. (passion) φλόγα f. (fervour) θέρμη f. be on ~ καίομαι, catch ~ πιάνω φωτιά, set ~ to βάζω φωτιά σε. (mil.) βολή f., πυρ n. be under ~ βάλλομαι, under enemy ~ υπό τα εχθρικά πυρά. open ~ ανοίγω πυρ. by ~ and sword διά πυρός και σιδήρου.

fir|e v.i. ~e (at) πυροβολώ (κατά with gen.). guns are ~ing τα κανόνια ρίχνουν. ~e

away! δος του! (v.t.) (shots) τραβώ, ρίχνω. (rocket, arrow) εκτοξεύω, (torpedo) εκσφενδονίζω. ~e a revolver at him του ρίχνω με περίστροφο, του ρίχνω πιστολιές. ~e questions at him τον βομβαρδίζω με ερωτήσεις. (set on ~e) πυρπολώ, (inspire) εμπνέω, (bake) ψήνω. (fam.) (dismiss) διώχνω.

fire-alarm s. σειρήνα κινδύνου.

fire-arms s. πυροβόλα όπλα.

fire-brand s. δαυλί n. (fig.) ταραχοποιός m.

fire-break s. ζώνη προστασίας εναντίον πυρκαγιάς.

fire-brigade s. πυροσβεστική υπηρεσία.

fire|-engine s. πυροσβεστική αντλία. ~man s. πυροσβέστης m.

fire-escape s. έξοδος κινδύνου.

fireplace s. τζάκι n.

fireproof a. πυρίμαχος.

fire-raising s. εμπρησμός m.

fire-ship s. πυρπολικό n.

firewood s. (kindling) προσάναμμα n. (logs) καυσόξυλα n.pl.

firework s. πυροτέχνημα n.

firing s. πυροβολισμός m. ~ squad εκτελεστικό απόσπασμα.

firm s. φίρμα f., εταιρεία f. ~ of solicitors δικηγορικόν γραφείον.

firm a. (steady) σταθερός, στερεός, (hard) σφιχτός, (strict) αυστηρός, (resolute) αποφασιστικός. (adv.) stand ~ (on feet, base) στέκομαι καλά, (on principles) μένω ακλόνητος. hold ~ κρατιέμαι γερά. ~ness s. αποφασιστικότητα f.

firmly adv. σταθερά, αυστηρά, γερά. be ~ resolved to... το έχω πάρει απόφαση να. ~ believe (that) είμαι πεπεισμένος (ότι).

first a. πρώτος, (initial) αρχικός. ~ aid πρώτες βοήθειες. ~ night (of play) πρεμιέρα n. ~ rains (of autumn) πρωτοβρόχια n.pl. at ~ hand από πρώτο χέρι, at ~ sight εκ πρώτης όψεως, at the ~ attempt με την πρώτη. ~ thing tomorrow αύριο πρωί πρωί. he was the ~ to understand πρώτος αυτός (or ήταν ο πρώτος που) κατάλαβε. the very ~ ο πρώτος πρώτος, ο αρχικός. (s.) at ~ στην αρχή, from the ~ από την αρχή.

first adv. πρώτα, (for the ~ time) για πρώτη φορά. ~ of all πρώτα πρώτα, πρώτα απ' όλα. ~ come ~ served προηγούνται οι πρώτοι. ~ly adv. πρώτον (μεν).

fiscal a. οικονομικός.

fish s. ψάρι n. like a ~ out of water σαν ψάρι έξω από το νερό. pretty kettle of ~ μπλέξιμο n. have other ~ to fry έχω άλλες ασχολίες. queer ~ αλλόκοτος άνθρωπος. (v.) (also ~ for) ψαρεύω. ~ up or out βγάζω.

fish-bone s. ψαροκόκκαλο n.

fisher s. αλιεύς m. ~**man** s. ψαράς m. ~**y** s. αλιεία f. (place) ψαρότοπος m.

fish-glue s. ψαρόκολλα f.

fish-hook s. αγκίστρι n.

fishing s. ψάρεμα n., αλιεία f. (a.) αλιευτικός. ~**boat** s. ψαράδικο n. ~**net** s. δίχτυ n. (seine) τράτα f. ~**rod** s. καλάμι n.

fishmonger s. ψαράς m. ~'s shop ψαράδικο n., ιχθυοπωλείον n.

fishy a. ~ smell or taste ψαρίλα f. (fig.) ύποπτος.

fissure s. ρωγμή f.

fist s. γροθιά f.

fit s. (attack) κρίση f., παροξυσμός m. have a ~ παθαίνω κρίση, (fam.) he'll have a ~! θα μείνει. (of anger, etc.) ξέσπασμα n. (of activity) έκρηξη f. by ~s and starts χωρίς σύστημα.

fit a. (suitable) κατάλληλος, (competent) ικανός, (worthy) άξιος. be ~ (well) είμαι καλά. it is not ~ to eat δεν τρώγεται. see ~ (to) κρίνω σκόπιμο (να). he is as ~ as a fiddle είναι περίκαλα στην υγεία του. (s.) (of garment) it is a good ~ έρχεται καλά.

fit v. (go into) ταιριάζω σε, χωράω σε. (the key) ~s the lock ταιριάζει στην κλειδαριά. (the book) doesn't ~ my pocket δεν χωράει στην τσέπη μου. (of garment) it ~s you σου έρχεται, σου πάει, σου εφαρμόζει. (apply) προσαρμόζω, βάζω, (equip) εφοδιάζω, (train) γυμνάζω. ~ **in** v.i. (tally, agree) συμφωνώ. ~ **on** v.t. (a garment) προβάρω. ~ **out, up** v.t. εφοδιάζω, (a ship) αρματώνω. ~ **together** v.t. συναρμολογώ.

fitful a. διακεκομμένος. ~**ly** adv. διακεκομμένα.

fitly adv. καταλλήλως, πρεπόντως, όπως πρέπει (or έπρεπε).

fitness s. καταλληλότητα f. ικανότητα f. (health) υγεία f.

fitted a. be ~ for είμαι ικανός (with για or να).

fitter s. (mech.) εφαρμοστής m.

fitting a. (proper) πρέπων, αρμόζων. it is not ~ (that) δεν κάνει (να). ~**ly** adv. πρεπόντως, όπως πρέπει (or έπρεπε).

fitting s. (tailor's) πρόβα f., δοκιμή f. ~**s** s. εξαρτήματα n.pl.

five num. πέντε. ~ hundred πεντακόσιοι a.

fix v. (place) τοποθετώ, (make fast) στερεώνω, (one's eyes, etc.) καρφώνω. (put right) φτιάνω, (appoint, arrange) ορίζω, (decide) αποφασίζω. ~ (person) up βολεύω, get ~ed up βολεύομαι. why do you ~ on me? γιατί τα βάζεις με μένα; (s.) be in a ~ βρίσκομαι σε δύσκολη θέση.

fixed a. (price, etc.) ορισμένος, (perma-

nent) μόνιμος, (immobile) σταθερός, (of property) ακίνητος. (idea) έμμονος. ~**ly** adv. (gaze) ατενώς.

fixture s. (household) εξάρτημα n. (fam.) he's a ~ here έχει κολλήσει εδώ. (sports) αθλητικό γεγονός.

fizz v. αφρίζω. ~**y** a. αφρώδης.

fizzle v. συρίζω. ~ out ξεθυμαίνω.

flabbergast v. I was ~ed έμεινα κόκκαλο.

flabb|y a. πλαδαρός. (fam., spineless) λαπάς m. ~**iness** s. πλαδαρότητα f.

flaccid a. πλαδαρός. ~**ity** s. πλαδαρότητα f.

flag v.i. πέφτω, εξασθενώ, μειώνομαι. (of plants) μαραίνομαι.

flag s. σημαία f. ~ of convenience σημαία ευκαιρίας. (v.) the streets were ~ged οι δρόμοι ήταν σημαιοστολισμένοι. ~**ship** s. ναυαρχίς f. ~**staff** s. κοντάρι n.

flagellat|e v. μαστιγώνω. ~**ion** s. μαστίγωση f.

flagon s. καράφα f.

flagrant a. κατάφορος. ~**ly** adv. it is ~ly unjust είναι κατάφορος αδικία.

flagstone s. πλάκα f.

flail v.t. κοπανώ. (v.i.) χτυπιέμαι.

flair s. ~ for languages επίδοση στις γλώσσες. he has a ~ for bargains μυρίζεται τις ευκαιρίες. she has ~ έχει φλαιρ.

flak|e s. (snow) νιφάδα f. (loose bit of paint, etc.) φλούδα f. (v.i.) (also ~e off) ξεφλουδίζω. ~**y** a. ~y pastry ζύμη σφολιάτα.

flamboyant a. (person) φιγουράτος, (colour) φανταχτερός. ~**ly** adv. φανταχτερά.

flame s. φλόγα f. go up in ~s αναφλέγομαι. (fig., sweetheart) έρωτας m. (v.i.) (burn) φλέγομαι.

flaming a. (on fire) φλεγόμενος, (fiery) φλογερός, (very red) κατακόκκινος.

flammable a. εύφλεκτος.

flange s. φλάντζα f.

flank v. πλαισιώνω. the road is ~ed by trees ο δρόμος έχει δέντρα και από τις δύο πλευρές.

flank s. (of body) λαγών f. (of hill) πλαγιά f. (of building) πλευρά f. (mil.) πλευρόν n.

flanking a. πλευρικός. make a ~ attack πλευροκοπώ.

flannel s. φανέλλα f. (a.) φανελλένιος.

flap v.t. χτυπώ. (v.i.) πλαταγίζω, (flutter) φτερουγίζω. (s.) (blow) χτύπημα n. (of wing) φτερούγισμα n. (of sail, etc.) πλατάγισμα n. (of pocket) καπάκι n. (of table) πτυσσόμενο φύλλο. (fuss) ταραχή f. get in a ~ αναστατώνομαι.

flare v.i. (widen) ανοίγω. (blaze, also ~ up) φουντώνω, (in anger) αφαρπάζομαι. (s.) (widening) άνοιγμα n. (mil.) φωτοβολίδα f. ~**up** s. φούντωμα n.

flash *v.i.* αστράφτω. the lighthouse ~es ο φάρος εκπέμπει αναλαμπές. (*pass rapidly*) περνώ αστραπιαίως. (*v.t.*) ~ the torch on it το φωτίζω με το φακό.
flash *s.* αναλαμπή *f.*, λάμψη *f.* (*lightning*) αστραπή *f.* in a ~ αστραπιαίως. be a ~ in the pan βγαίνω τζίφος. ~-**back** *s.* αναδρομή στο παρελθόν. ~**light.** (*photo.*) φλας *n.* ~**y** *a.* χτυπητός.
flask *s.* φιάλη *f.*
flat *s.* (*level ground*) πεδινή περιοχή, (*apartment*) διαμέρισμα *n.* (*mus.*) ύφεση *f.*
flat *a.* (*level*) επίπεδος, (*horizontal*) οριζόντιος, (*in shape*) πλακωτός, πλακέ. (*below pitch*) χαμηλός, (*insipid*) ανούσιος, (*downright*) κατηγορηματικός. (*omelet, etc.*) πεσμένος, (*battery*) άδειος, (*paint*) θαμπός, ματ. ~ chest στήθος πλάκα. ~ nose πλακουτσή μύτη. ~ rate ενιαίο μισθολόγιο. have a ~ tyre μ' έπιασε λάστιχο. fall ~ (*fail*) αποτυγχάνω, fall ~ on one's back πέφτω ανάσκελα. that's ~! ούτε συζήτηση. I knocked him ~ τον έριξα κάτω.
flat *adv.* (*plainly, frankly*) ορθά κοφτά, (*exactly*) ακριβώς, (*of refusal*) κατηγορηματικώς. it won't lie ~ δεν στρώνει. go ~ out (*to achieve sthg.*) κάνω τα αδύνατα δυνατά. ~**ly** *adv.* κατηγορηματικώς.
flatten *v.* (*level*) ισοπεδώνω, (*squash*) πλακώνω, (*make lie flat*) στρώνω, ισιώνω.
flatter *v.* κολακεύω. ~**er** *s.* κόλακας *m.* ~**ing** *a.* κολακευτικός. ~**y** *s.* κολακεία *f.*
flatulence *s.* αέρια *n.pl.*
flaunt *v.t.* επιδεικνύω. (*v.i.*) (*of flag*) κυματίζω, (*of person*) επιδεικνύομαι.
flavour *s.* γεύση *f.* (*of wine, tea, ice*) άρωμα *n.* (*v.*) ~ed with αρωματισμένος με. ~**ing** *s.* καρύκευμα *n.* ~**less** *a.* άνοστος.
flaw *s.* ελάττωμα *n.* (*crack*) ρωγμή *f.* ~**less** *a.* τέλειος.
flax *s.* λινάρι *n.*
flay *v.* γδέρνω.
flea *s.* ψύλλος *m.* (*fig.*) it's a mere ~-bite δεν είναι τίποτα.
fleck *v.* πιτσιλίζω. ~ed διάστικτος. waves ~ed with foam αφρισμένα κυματάκια. (*s.*) πιτσιλιά *f.*, κηλίδα *f.*, στίγμα *n.*
fledge *v.* they are not ~d δεν έχουν βγάλει όλα τα φτερά τους. fully ~d (*fig.*) φτασμένος. ~**ling** *s.* ξεπεταρούδι *n.*
flee *v.i.* τρέπομαι εις φυγήν. (*v.t.*) φεύγω από, (*shun*) αποφεύγω.
fleece *s.* μαλλί *n.* Golden F~e χρυσόμαλλον δέρας. (*v.*) (*swindle*) γδέρνω. ~**y** *a.* σαν μαλλί.
fleet *s.* στόλος *m.*
fleet *a.* ~ of foot, ~-footed γοργοπόδαρος. ~**ing** *a.* φευγαλέος. ~**ly** *adv.* γοργά.

flesh *s.* σάρκα *f.* put on ~ παίρνω κρέας. in the ~ ζωντανός. ~ and blood η ανθρώπινη φύση. my own ~ and blood το αίμα μου. the spirit is willing but the ~ is weak το μεν πνεύμα πρόθυμον η δε σαρξ ασθενής. ~**ly** *a.* σαρκικός. ~~**pots** *s.* τρυφηλός βίος. ~**y** *a.* σαρκώδης.
flex *s.* καλώδιο *n.* (*v.*) κάμπτω. ~**ible** *a.* εύκαμπτος. ~**ibility** *s.* ευλυγισία *f.* ευελιξία *f.*
flick *v.t.* (*strike*) χτυπώ. (*a whip, speck of dust, etc.*) τινάζω, (*an insect*) διώχνω. (*propel with a ~*) πετώ. (*s.*) τίναγμα *n.*
flicker *v.i.* (*of light*) τρεμοσβήνω, (*of eyelid*) παίζω, (*of smile*) φτερουγίζω. (*s.*) (*of light*) τρεμούλιασμα *n.* (*of eyelid*) παίξιμο *n.* ~ of hope αμυδρή ελπίδα.
flight *s.* (*running away*) φυγή *f.* put to ~ τρέπω εις φυγήν. (*escape*) διαφυγή *f.* (*in air*) πτήση *f.* (*of fancy*) πέταγμα *n.* (*group of birds, planes*) σμήνος *m.* ~ of stairs σειρά σκαλοπατιών. ~**y** *a.* ~y girl τρελλοκόριτσο *n.*
flimsy *a.* (*thin*) λεπτός, (*weak*) αδύνατος.
flinch *v.* δειλιάζω.
fling *v.t.* ρίχνω. (*v.i.*) (*rush*) ορμώ. (*s.*) ρίξιμο *n.* have one's ~ το ρίχνω έξω.
flint *s.* (*of lighter*) τσακμακόπετρα *f.* ~~**lock** *s.* καρυοφύλλι *n.* ~**y** *a.* σκληρός.
flip *v.* & *s.* see flick.
flipp|ant *a.* επιπόλαιος. ~**ancy** *s.* επιπολαιότητα *f.*
flipper . πτερύγιον *n.*
flirt *v.i.* κάνω κόρτε, (*also fig.*) ερωτοτροπώ. (*lit. only*) (·*also* ~ with) φλερτάρω. (*v.t.*) (*a fan*) παίζω με. (*s., also* ~*atious*) φλερτατζού *f.* ~**ation,** ~**ing** *s.* φλερτ, κόρτε *n.*
flit *v.* (*fly*) πετώ, (*pass*) περνώ φευγαλέα. (*s., fam.*) do a ~ το σκάω.
float *s.* (*support*) φλοττέρ *n.* (*fisherman's*) φελλός *m.* (*on wheels*) άρμα *n.*
float *v.i.* (*be on surface*) επιπλέω, (*drift*) πλέω. (*v.t.*) (*lift*) σηκώνω, (*move*) κινώ. (*fin.*) (*a company*) ιδρύω, (*a loan*) εκδίδω. ~**ing** *a.* (*that* ~*s*) πλωτός, (*variable*) κυμαινόμενος.
flock *s.* κοπάδι *n.* (*of small birds*) σμήνος *n.* (*eccl.*) ποίμνιον *n.* (*v.*) (*throng*) συρρέω, ~ together συναθροίζομαι, μαζεύομαι.
flog *v.* μαστιγώνω. (*fam.*) πουλώ. ~ a dead horse ματαιοπονώ. ~**ging** *s.* μαστίγωμα *n.*
flood *s.* πλημμύρα *f.* (*v.t.* & *i.*) πλημμυρίζω. ~**gate** *s.* open the ~gates for all kinds of abuses ανοίγω το δρόμο για κάθε είδους παρανομία.
flood|light *s.* (*lamp*) προβολεύς *m.* (*v.*) φωταγωγώ. ~**lit** *a.* φωταγωγημένος.
floor *s.* πάτωμα *n.* take the ~ (*to speak*)

flop 327 foal

flop παίρνω το λόγο, (*of dancers*) μπαίνω στην πίστα. wipe the ~ with κατατροπώνω. (*v.*) (*knock down*) ρίχνω κάτω, (*confound*) αποστομώνω. **~board** *s.* σανίδα *f.* **~ing** *s.* πάτωμα *n.*

flop *v.i.* (*move*) κουνιέμαι αδέξια, (*drop*) σωριάζομαι. (*fail*) αποτυγχάνω. (*s.*) αποτυχία *f.* **~py** *a.* μαλακός. ~py disc δισκέτα *f.*

flora *s.* χλωρίς *f.*

floral *a.* με λουλούδια.

florid *a.* (*colour*) κατακόκκινος, (*style*) περίκομψος.

florin *s.* (*gold*) φλουρί *n.*

florist *s.* ανθοπώλης *m.* ~'s shop ανθοπωλείο *n.*

flotilla *s.* στολίσκος *m.*

flotsam *s.* έκβρασμα *n.*

flounce *s.* βολάν *n.* (*v.i.*) τινάζομαι. ~ out αποσύρομαι με θυμό.

flounder *v.i.* τσαλαβουτώ, (*in speech*) κομπιάζω.

flour *s.* αλεύρι *n.* **~y** *a.* από (*or* σαν) αλεύρι.

flourish *v.i.* (*of arts, trade*) ανθώ, (*of plants*) ευδοκιμώ. (*in health*) είμαι καλά. (*be in one's prime*) ακμάζω. (*v.t.*) κραδαίνω.

flourish *s.* (*ornament*) ποίκιλμα *n.* (*gesture*) εντυπωσιακή χειρονομία. (*of trumpets*) φανφάρα *f.*

flout *v.* αψηφώ.

flow *v.* ρέω, (*of tears, traffic*) κυλώ. ~ into (*of river*) εκβάλλω εις. (*s.*) ροή *f.* (*tide*) πλημμυρίδα *f.* (*copious supply*) πλημμύρα *f.*

flower *s.* λουλούδι *n.*, (*also fig.*) άνθος *n.* in ~ ανθισμένος. (*v.*) ανθίζω, (*fig.*) ανθώ. **~bed** *s.* παρτέρι *n.* **~pot** *s.* γλάστρα *f.* **~y** *a.* (*fig.*) ~y language ωραίες φράσεις.

flowing *a.* (*clothes, hair, lines*) χυτός, (*style*) ρέων. ~ beard μακριά γένια. be ~ with πλημμυρίζω από.

flu *s.* γρίππη *f.*

fluctuat|e *v.* διακυμαίνομαι. **~ion** *s.* διακύμανση *f.*

flue *s.* μπουρί *n.*

fluen|t *a.* (*speech*) ευχερής, (*person*) be ~t έχω ευχέρεια, μιλώ με ευχέρεια. **~cy** *s.* ευχέρεια *f.* **~tly** *adv.* ευχερώς.

fluff *s.* χνούδι *n.* (*v.t.*) φουσκώνω. **~y** *a.* χνουδωτός. ~y hair αφράτα (*or* φουσκωτά) μαλλιά.

fluid *a.* ρευστός. (*s.*) υγρό *n.* **~ity** *s.* ρευστότητα *f.*

fluke *s.* (*of anchor*) νύχι *n.* (*stroke of chance*) εύνοια της τύχης, by a ~ τυχαίως.

flume *s.* τεχνητό αυλάκι για ύδρευση.

flummox *v.* αποστομώνω.

flunkey *s.* λακές *m.*

fluorescent *a.* φθορίζων. ~ lamp λαμπτήρας φθορισμού.

flurr|y *s.* (*of wind & rain*) μπουρίνι *n.* (*excitement*) αναστάτωση *f.* get ~ied αναστατώνομαι.

flush *v.i.* (*become red*) κοκκινίζω, (*flow*) ξεχύνομαι. (*v.t.*) (*cleanse*) ξεπλένω με καταιονισμό. ~ the w.c. τραβώ το καζανάκι. (*put up*) ξεπετώ (*drive out*) διώχνω. be ~ed (*with anger*) γίνομαι κατακόκκινος (από θυμό).

flush *s.* (*also* **~ing**) κοκκίνισμα *n.*, έξαψη *f.* in the ~ of victory στη μέθη της νίκης.

flush *a.* (*level*) στην ίδια επιφάνεια, (*with money*) γεμάτος λεφτά.

fluster *v.* σαστίζω, αναστατώνω.

flute *s.* φλάουτο *n.* (*ancient*) αυλός *m.*

flut|ed *a.* ραβδωτός. **~ing** *s.* ραβδώσεις *f.pl.*

flutter *v.i.* (*of winged creature*) πεταρίζω, φτερουγίζω, (*of flag*) κυματίζω, (*of heart*) χτυπώ. (*v.t.*) (*wings*) χτυπώ, (*handkerchief*) κουνώ. (*upset*) αναστατώνω, ~ the dovecotes δημιουργώ θόρυβο. (*s.*) (*of wings*) φτερούγισμα *n.* (*of eyelid*) παίξιμο *n.* be in a ~ είμαι άνω-κάτω. have a ~ δοκιμάζω την τύχη μου.

fluvial *a.* ποτάμιος.

flux *s.* ροή *f.* state of ~ ρευστή κατάσταση.

fly *a.* (*knowing*) ξύπνιος.

fly *s.* μύγα, μυία *f.* ~ in the ointment (*fig.*) παραφωνία *f.* there are no flies on him δεν χάφτει μύγες. **~blown** *a.* μυγοχεσμένος. **~swatter** *s.* μυγοσκοτώστρα *f.*

fly *v.t.* (*a plane*) οδηγώ, κυβερνώ. (*traverse*) περνώ. (*convey*) μεταφέρω αεροπορικώς. (*a kite*) πετώ, (*a flag*) υψώνω. (*v.i.*) πετώ, ίπταμαι. (*travel by air*) ταξιδεύω αεροπορικώς. (*rush*) ορμώ, (*flee*) τρέπομαι εις φυγήν. (*of flag, hair, etc.*) ανεμίζω. let (*sthg.*) ~ εκτοξεύω. (*let*) ~ at επιτίθεμαι (*with* κατά & *gen.*). ~ away φεύγω. ~ in the face of αψηφώ, (*be contrary*) είμαι αντίθετος προς. ~ into a rage εξοργίζομαι. ~ off the handle γίνομαι θηρίο. ~ high ψηλαρμενίζω, έχω φιλοδοξίες. ~ open ανοίγω απότομα. send ~ing (*expel*) πετώ έξω, (*upset*) ανατρέπω.

flyer *s.* αεροπόρος *m.*

flying *s.* (*act of* ~) πτήση *f.* night ~ νυκτερινή πτήση. he has taken up ~ πάει για πιλότος. he is mad about ~ έχει μανία με αεροπλάνα.

flying *a.* (*that flies*) ιπτάμενος. ~ squad άμεσος δράσις, ~ visit βιαστική επίσκεψη. take a ~ jump πηδώ με φόρα. come off with ~ colours σημειώνω λαμπρή επιτυχία.

fly-wheel *s.* σφόνδυλος *m.*

foal *s.* πουλάρι *n.*

foam s. αφρός m. ~ rubber αφρολέξ n. (v.) αφρίζω.

fob v. ~ (sthg.) off on (person) πασάρω σε. he ~bed me off with excuses μου πάσαρε (or με ξεγέλασε με) διάφορες δικαιολογίες.

focus s. εοτία f., in ~ ρυθμισμένος, (of photo) καθαρός. (of attention) κέντρο n. (v.t.) (adjust) ρυθμίζω, (concentrate) συγκεντρώνω.

fodder s. φορβή f. (fig., of humans) βορά f.

foe s. εχθρός m.

foetus s. έμβρυο n.

fog s. ομίχλη f. **~ged** a. θαμπός. **~gy** a. ομιχλώδης. it is ~gy (weather) έχει ομίχλη. **~lamps** s. φώτα ομίχλης.

fogey s., **~ish** a. με περιορισμένες αντιλήψεις. old ~s (fam.) γερουσία f.

foible s. αδυναμία f.

foil s. (wrapping) αλουμινόχαρτο n. be a ~ to τονίζω, δείχνω.

foil s. (weapon) ξίφος n.

foil v. (a plan) ματαιώνω, (a person) εμποδίζω, be ~ed αποτυγχάνω.

foist v. πασάρω.

fold s. (sheep) στάνη f. (fig., bosom of Church) κόλποι της Εκκλησίας.

fold v. διπλώνω. ~ one's arms σταυρώνω τα χέρια (μου). (s.) δίπλα f., πτυχή f. **~er** s. φάκελλος m. **~ing** a. πτυσσόμενος.

foliage s. φύλλωμα n.

folk, ~s s. άνθρωποι m.pl.

folk a. (popular) λαϊκός. **~-dance** s. λαϊκός χορός. **~lore** s. λαογραφία f.

follow v. ακολουθώ, (understand, observe) παρακολουθώ. (a fugitive) καταδιώκω. (as result) έπομαι, προκύπτω, (signify) σημαίνω. ~ up (exploit) εκμεταλλεύομαι. ~ suit κάνω τα ίδια. as ~s ως εξής.

follower s. οπαδός m. **~ing** s. (supporters) οπαδοί m.pl. (votaries) θιασώτες m.pl.

following a. (next) επόμενος. the ~ (items) τα εξής, τα επόμενα, τα κατωτέρω, τα ακόλουθα. (prep.) ~ the minister's decision κατόπιν της αποφάσεως του υπουργού.

folly s. (also act of ~) τρέλλα f.

foment v. (fig.) υποδαυλίζω.

fond a. (loving) τρυφερός. be ~ of (person) συμπαθώ, μου είναι αγαπητός, (thing) μου αρέσει, αγαπώ. **~ly** adv. τρυφερά. he ~ly believes έχει την απλοϊκότητα να πιστεύει. **~ness** s. συμπάθεια f., αγάπη f. (weakness) αδυναμία f.

fondle v. χαϊδεύω.

font s. κολυμβήθρα f.

food s. (nourishment, lit. & fig.) τροφή f. (victuals) τρόφιμα n.pl. (prepared) φαΐ n., φαγητό n. **~stuff** s. είδος διατροφής.

fool s. βλάκας m. make a ~ of γελοιοποιώ. live in a ~'s paradise ζω στα σύννεφα. (v.t.) (trick) ξεγελώ. (v.i.) (meddle) παίζω. ~ around σαχλαμαρίζω.

foolery s. σαχλαμάρες f. pl.

foolhardy a. παράτολμος. **~iness** s. παρατολμία f.

foolish a. ανόητος. **~ly** adv. ανόητα. **~ness** s. ανοησία f.

foolproof a. (plan) αλάνθαστος. (of implement) it is ~ κανείς δεν μπορεί να το χαλάσει.

foot s. πόδι n. (metrical) πους m. (infantry) πεζικό n., ~-soldier οπλίτης m. on ~ με τα πόδια. set on ~ βάζω μπροστά. rise to one's feet σηκώνομαι όρθιος. be on one's feet (busy) είμαι στο πόδι. find one's feet αποκτώ πείρα. put one's ~ down πατώ πόδι. put one's ~ in it κάνω γκάφα. trample under ~ τσαλαπατώ. sweep (person) off his feet συνεπαίρνω. (of hill) πρόποδες m.pl. (of page) τέλος n. (of bed) πόδια n.pl. (v.) ~ it περπατώ. ~ the bill πληρώνω.

foot-and-mouth a. αφθώδης.

football s. (game) ποδόσφαιρο n., (ball) μπάλλα f.

foothills s. πρόποδες m.pl.

foothold s. πάτημα n. get a firm ~ (fig.) εγκαθίσταμαι γερά.

footing s. (foothold) πάτημα n. (fig.) on an equal ~ υπό ίσους όρους, on a war ~ σε κατάσταση πολέμου. put on a firm ~ σταθεροποιώ.

footlights s. φώτα της ράμπας.

footling a. ασήμαντος.

footman s. λακές m.

footmark s. ίχνος n.

footnote s. υποσημείωση f.

footpath s. μονοπάτι n.

footsore a. I am ~ ξεποδαριάστηκα.

footstep s. πάτημα n.

footstool s. σκαμνάκι n.

fop s., **~pish** a. δανδής m.

for prep. 1. (general) για (with acc.), ~ me για μένα, he's leaving ~ Paris φεύγει για το Παρίσι. ask, look, wait, pay ~ (use v. with direct object), he asked ~ money ζήτησε λεφτά. go ~ a walk πάω περίπατο, see ~ yourself να δεις και μόνος σου. weep ~ joy κλαίω από τη χαρά μου. wait ~ hours περιμένω επί ώρες. ~ my part όσο για μένα. what is it ~? σε τι χρησιμεύει; I haven't seen him ~ a week έχω να τον δω μία εβδομάδα. he's been absent ~ two days είναι δύο μέρες που λείπει. they won't be back ~ a month δεν θα γυρίσουν πριν από ένα μήνα. I shall wait ~ him to come θα τον περιμένω να έρθει. it is right ~ you to go το σωστό είναι να πας. it is too late ~ you to leave είναι πολύ αργά για να φύγεις. ~ all I care πολύ που με νοιάζει. ~ all that (*in spite of that*) παρ' όλα αυτά, (*although*) παρ' όλο που. 2. (*in favour of*) υπέρ (*with gen.*). I'm ~ staying on είμαι υπέρ του να μείνουμε. 3. (*exchange, price*) με (*with acc.*). exchange it ~ a new one το ανταλλάσσω μ' ένα καινούργιο. I got it ~ ten pounds το πήρα (με) δέκα λίρες. 4. (*purpose*) προς (*with acc.*). ~ your sake προς χάριν σας. ~ the convenience of the public προς εξυπηρέτησιν του κοινού.

for *conj.* (*because*) επειδή, γιατί, διότι.

forage *s.* ζωοτροφές *f.pl.* (*v.*) ψάχνω.

foray *s.* επιδρομή *f.*

forbear *s.* πρόγονος *m.*

forbear *v.i.* αποφεύγω (*with* να). I could not ~ to protest δεν μπορούσα να μη διαμαρτυρηθώ. **~ance** *s.* υπομονή *f.*

forbid *v.* απαγορεύω (*with acc. of thing, σε or gen. of person*). God ~! Θεός φυλάξοι! ο μη γένοιτο! **~den** fruit απαγορευμένος καρπός. **~ding** *a.* (*severe*) αυστηρός, (*threatening*) απειλητικός, (*wild*) άγριος.

force *v.* (*oblige*) αναγκάζω, (*impose*) επιβάλλω, (*thrust*) χώνω, (*press*) πιέζω, (*shove*) σπρώχνω. (*a door*) παραβιάζω, (*one's voice*) φορτσάρω. ~ one's way in εισέρχομαι διά της βίας. I ~d his hand τον ζόρισα να πάρει απόφαση. **~d** *a.* (*unnatural*) βεβιασμένος.

force *s.* (*strength*) δύναμη *f.* (*violence*) βία *f.* (*mil.*) ~s δυνάμεις *f.pl.* by ~ διά της βίας, με το ζόρι. in ~ (*valid*) εν ισχύι, be in ~ ισχύω. the family turned up in ~ η οικογένεια παρουσιάστηκε εν σώματι. join ~s ενώνομαι. through ~ of circumstances λόγω των περιστάσεων. by sheer ~ of habit από απλή συνήθεια. ~ **majeure** *s.* ανωτέρα βία.

forceful *a.* (*person*) δυναμικός, (*argument*) πειστικός. **~ly** *adv.* με δύναμη, πειστικά.

forceps *s.* λαβίς *f.*

forc|ible *a.* (*violent*) βίαιος, (*persuasive*) πειστικός. **~ibly** *adv.* βιαίως, με το ζόρι, πειστικά.

ford *s.* πέραμα *n.* (*v.*) περνώ από το πέραμα. **~able** *a.* διαβατός.

fore *a.* πρόσθιος, μπροστινός. (*naut.*) πλωριός. (*adv.*) ~ and aft πρύμα-πλώρα. (*s.*) to the ~ εις το προσκήνιον.

forearm *s.* πήχυς *m.*

forebod|e *v.* προοιωνίζομαι. **~ing** *s.* προαίσθημα *n.*

forecast *s.* πρόγνωση *f.* (*v.*) προβλέπω.

foreclosure *s.* κατάσχεσις *f.*

forefathers *s.* πρόγονοι *m.pl.*

forefinger *s.* δείκτης *m.*

forefront *s.* πρώτη γραμμή.

forego|ing *a.* προαναφερθείς. **~ne** *a.* a ~ne conclusion γνωστόν εκ των προτέρων.

foreground *s.* πρώτο πλάνο.

forehead *s.* μέτωπο *n.*

foreign *a.* ξένος, (*affairs, trade*) εξωτερικός. ~ subject αλλοδαπός *m.* ~ parts το εξωτερικό. ~-speaking ξενόφωνος. **~er** *s.* ξένος *m.*, ξένη *f.*

foreman *s.* αρχιεργάτης *m.*

foremost *a.* πρώτος, (*leading*) κορυφαίος. (*adv.*) μπροστά. first and ~ πρώτον και κύριον.

forename *s.* βαφτιστικό όνομα, μικρό όνομα.

forensic *a.* δικανικός. ~ medicine ιατροδικαστική *f.*

forerunner *s.* πρόδρομος *m.*

foresee *v.* προβλέπω. in the ~able future όσο μπορεί να προβλέψει κανείς.

foreshadow *v.* προμηνύω.

foreshore *s.* ακτή *f.*

foreshortening *s.* αλλαγή διαστάσεων λόγω προοπτικής.

foresight *s.* πρόνοια *f.*, προνοητικότητα *f.* (*of gun*) στόχαστρο *n.*

forest *s.* δάσος *n.* **~ed** *a.* δασώδης. **~er** *s.* δασοκόμος *m.* **~ry** *s.* δασολογία *f.*

forestall *v.* προλαβαίνω.

foretaste *s.* have a ~ of δοκιμάζω προκαταβολικώς.

foretell *v.* προλέγω.

forethought *s.* προνοητικότητα *f.*

foreword *s.* πρόλογος *m.*

forfeit *v.* χάνω. (*s.*) (*penalty*) τίμημα *n.* (*in games*) τιμωρία *f.* **~ure** *s.* απώλεια *f.*

forgather *v.* συνέρχομαι, μαζεύομαι.

forge *s.* σιδηρουργείο *n.*

forge *v.t.* (*metal & fig.*) σφυρηλατώ, (*counterfeit*) πλαστογραφώ. (*v.i.*) ~ ahead πάω μπρος, προπορεύομαι. **~d** *a.* πλαστός. **~ry** *s.* (*act*) πλαστογραφία *f.* (*thing*) πλαστό *a.*

forget *v.* ξεχνώ, λησμονώ. ~ oneself (*act badly*) ξεχνώ τη θέση μου. **~-ful** *a.* ξεχασιάρης. **~fulness** *s.* λησμοσύνη *f.*

forgiv|e *v.* συγχωρώ, (*remit*) χαρίζω. **~able** *a.* συγχωρητέος. **~eness** *s.* συγχώρηση *f.* **~ing** *a.* ανεξίκακος.

forgo *v.* παραιτούμαι (*with gen.*), κάνω χωρίς.

fork *s.* (*digging*) δικέλλα *f.* (*table*) πηρούνι *n.* (*of trousers*) καβάλος *m.* (*in road*) διακλάδωση *f.* (*v.i.*) διακλαδίζομαι. (*v.t., fam.*) ~ out πληρώνω. **~ed** *a.* διχαλωτός.

forlorn *a.* έρμος. ~ hope αμυδρά ελπίδα.

form *s.* (*shape, appearance*) μορφή *f.* (*kind*) είδος *n.* (*manner*) τρόπος *m.* (*gram.*) τύπος *m.* (*conventions*) τύποι *m.pl* (*printed* *paper)* έντυπο *n.* *(bench)* μπάγκος *m.* *(of school)* τάξη *f.* *(mould)* καλούπι *n.* be in good ~ είμαι σε φόρμα. for ~'s sake για τον τύπο.

form *v.t.* *(organize, give shape to)* σχηματίζω. *(develop, mould)* διαμορφώνω, διαπλάθω. *(make)* φτιάνω. *(be, constitute)* αποτελώ. *(an alliance)* συνάπτω, *(a plan)* καταστρώνω. *(v.i.)* *(take shape)* σχηματίζομαι, διαμορφώνομαι. *(be produced)* γίνομαι.

formal *a.* *(conventional)* τυπικός, *(official)* επίσημος. **~ly** *adv.* τυπικώς, επισήμως.

formalit|y *s.* τυπικότητα *f.* a mere ~y απλός τύπος. **~ies** τυπικώσεις *f.pl.*

format *s.* σχήμα *n.*

formation *s.* σχηματισμός *m.* *(of body, character, rocks)* διάπλαση *f.*, διαμόρφωση *f.*

formative *a.* διαμορφωτικός.

former *a.* *(earlier)* παλαιός, *(of ~ times)* αλλοτινός. *(not latter)* ο πρώτος. the ~ ... the latter ο μεν... ο δε. ~ spouse πρώην σύζυγος. **~ly** *adv.* άλλοτε.

formidab|le *a.* τρομερός. **~ly** *adv.* τρομερά.

formless *a.* άμορφος.

formula *s.* τύπος *m.*, φόρμουλα *f.* **~te** *v.* διατυπώνω. **~tion** *s.* διατύπωση *f.*

fornication *s.* κλεψιγαμία *f.* *(law)* συνουσία *f.*

forsake *v.* εγκαταλείπω.

for|swear *v.* απαρνούμαι. ~swear oneself επιορκώ. **~sworn** *s.* επίορκος.

fort *s.* φρούριο *n.*

forth *adv.* εμπρός. go back and ~ πηγαινοέρχομαι. from this time ~ από εδώ κι εμπρός. and so ~ και ούτω καθ' εξής. come ~ βγαίνω. bring *or* put ~ βγάζω.

forthcoming *a.* *(imminent)* επικείμενος, προσεχής. *(helpful)* πρόθυμος. the money was not ~ δεν βρέθηκαν τα χρήματα.

forthright *a.* ντόμπρος.

forthwith *adv.* πάραυτα, αμέσως.

fortif|y *v.* οχυρώνω. *(fig., strengthen)* δυναμώνω, στυλώνω. **~ication** *s.* οχύρωση *f.*

fortissimo *adv.* *(fam.)* στη διαπασών.

fortitude *s.* καρτερία *f.*

fortnight *s.* δεκαπενθήμερον *n.* a ~ *(from or ago)* today *(σαν)* σήμερα δεκαπέντε. **~ly** *a.* δεκαπενθήμερος.

fortress *s.* φρούριο *n.*

fortuitous *a.* τυχαίος. **~ly** *adv.* τυχαίως.

fortunate *a.* *(person)* τυχερός, *(choice, etc.)* ευτυχής. we had a ~ escape σωθήκαμε ως εκ θαύματος. **~ly** *adv.* ευτυχώς.

fortune *s.* *(chance, luck)* τύχη *f.* *(wealth)* περιουσία *f.* **~-hunter** *s.* προικοθήρας *m.* **~-teller** *s.* *(from cards)* χαρτορίχτρα *f.* *(from palm)* χειρομάντισσα *f.*

fort|y *num.* σαράντα. have ~y winks παίρνω έναν υπνάκο. **~ieth** *a.* τεσσαρακοστός.

forum *s.* τόπος όπου συζητούνται δημόσια θέματα.

forward *a.* *(in front)* πρόσθιος, μπροστινός. *(early in development)* *(plants)* πρώιμος, *(children)* πρόωρος, *(prompt)* πρόθυμος, *(pert)* βέβαιος για τον εαυτό μου.

forward, ~s *adv.* εμπρός, μπρος, μπροστά. go ~ προχωρώ. *see* forth.

fossil *s.* απολίθωμα *n.* **~ized** *a.* απολιθωμένος.

foster *v.* τρέφω, *(cultivate)* καλλιεργώ. *(a.)* *(adoptive)* θετός. **~child** *s.* ψυχοπαίδι *n.*

foul *v.t.* *(pollute)* ρυπαίνω. *(get tangled with)* μπερδεύομαι με, *(collide with)* συγκρούομαι με.

foul *a.* ρυπαρός, βρωμερός, *(language)* αισχρός, *(atrocious)* απαίσιος. ~ play δόλιο τέχνασμα, *(crime)* έγκλημα *n.* fall ~ of έχω τραβήγματα με. **~ly** *adv.* αισχρά, *(of violent attack)* θηριωδώς.

found *v.* ιδρύω, *(base)* στηρίζω. **~er** *s.* ιδρυτής *m.*

foundation *s.* *(act)* ίδρυση *f.*, *(thing)* ίδρυμα *n.*, *(base, basis)* θεμέλιο *n.*

founder *v.i.* *(naut. & fig.)* ναυαγώ.

foundling *s.* έκθετο *n.* ~ hospital βρεφοκομείο *n.*

fountain *s.* σιντριβάνι *n.*, πίδακας *m.* **~-head** *s.* πηγή *f.* **~-pen** *s.* στυλό *n.*

four *num.* τέσσερα *(a.)* τέσσερ(ε)ις. on all ~s με τα τέσσερα. in ~s κατά τετράδας. ~ hundred τετρακόσιοι. **~fold** *a.* τετραπλάσιος. **~th** *a.* τέταρτος.

four-footed *a.* τετράπους.

four-square *a.* σταθερός.

fourteen *num.* δεκατέσσερα. *(a.)* δεκατέσσερ(ε)ις. **~th** *a.* δέκατος τέταρτος.

fowl *s.* *(bird)* πτηνό *n.* *(chicken)* κότα *f.* wild ~ αγριοπούλια *n.pl.*

fox *s.* αλεπού *f.* *(v.)* *(fam.)* ξεγελώ. **~y** *a.* *(fam.)* πονηρός.

fracas *s.* καβγάς *m.*

fraction *s.* *(bit)* κομματάκι *n.* *(math.)* κλάσμα *n.* *(fam.)* just a ~ μία ιδέα.

fractious *a.* δύστροπος. **~ness** *s.* δυστροπία *f.*

fracture *s.* θραύση *f.* *(med.)* κάταγμα *n.* *(crack)* ρωγμή *f.* **~d** *a.* σπασμένος.

fragil|e *a.* εύθραυστος, *(person)* λεπτεπίλεπτος, *(insecure)* ανασφαλής. **~ity** *s.* εύθραυστον *n.* ανασφάλεια *f.*

fragment *s.* *(bit)* κομμάτι *n.* *(of bomb)* θραύσμα *n.* *(of glass, china)* θρύψαλο *n.* *(v.i.)* θραύομαι, θρυμματίζομαι. **~ary** *a.* αποσπασματικός.

fragran|t *a.* ευώδης. **~ce** *s.* ευωδία *f.*

frail *a.* (*weak*) αδύνατος, (*in health*) λεπτής
υγείας. (*insecure*) ανασφαλής. **~ness, ~ty**
s. αδυναμία *f.* ανασφάλεια *f.*

frame *v.t.* (*shape, build*) διαμορφώνω,
(*formulate*) διατυπώνω, (*devise*) σκαρώνω. be ~d for (*suited*) είμαι φτιαγμένος
για. (*a picture, etc.*) πλαισιώνω.
(*fam.*) he was ~d του σκηνοθέτησαν
κατηγορία. (*v.i.*) (*develop*) αναπτύσσομαι.

frame *s.* (*of structure, spectacles*) σκελετός
m. (*picture*) κορνίζα *f.* (*door, window*)
κάσα *f.* (*body*) κορμί *n.* ~ of mind ψυχική
διάθεση. **~work** *s.* πλαίσιο *n.*

franc *s.* φράγκο *n.*

franchise *s.* δικαίωμα ψήφου.

frank *a.* ειλικρινής, ίσιος, ντόμπρος. **~ly**
adv. ειλικρινώς. **~ness** *s.* ειλικρίνεια *f.*

frankincense *s.* λιβάνι *n.*

Frankish *a.* φράγκικος, (*person*) Φράγκος.

frantic *a.* (*person*) τρελλός, έξαλλος,
(*applause*) φρενιτιώδης. (*desperate*)
απεγνωσμένος. it drives me ~ με κάνει έξω
φρενών. **~ally** *adv.* φρενιτιωδώς, μέχρις
απελπισίας. (*fam., very*) απίστευτα.

fratern|al *a.* αδελφικός. **~ity** *s.* (*group*)
αδελφότητα *f.* (*feeling*) αδελφοσύνη *f.*
~ize *v.* έχω φιλικές σχέσεις. (*of troops,
etc.*) συναδελφώνομαι.

fratricide *s.* αδελφοκτονία *f.*

fraud *s.* απάτη *f.* (*person*) απατεώνας *m.* by
~ με δόλο, με απάτη. **~ulent** *a.* (*promise,
deal*) απατηλός. **~ulent person** απατεώνας *m.*

fraught *a.* γεμάτος. (*fam.*) ανησυχητικός.

fray *s.* αγώνας *m.*

fray *v.i.* (*also* become ~ed) ξεφτίζω,
τρίβομαι. my nerves are ~ed τα νεύρα μου
είναι σμπαραλιασμένα (*or* είναι σμπαράλια).

freak *s.* (*natural*) τέρας *m.* (*man-made*) τερατούργημα *n.* (*queer person*) αλλόκοτος
άνθρωπος. (*a.*) ~ storm θύελλα πρωτο-
φανούς εντάσεως. **~ish** *a.* τερατώδης,
(*queer*) αλλόκοτος.

freckle *s.* πιτσιλάδα *f.*, φακίδα *f.* **~d** *a.* με
φακίδες *f.*

free *v.t.* (*liberate*) ελευθερώνω, (*rid*)
απαλλάσσω. ~ oneself from απαλλάσ-
σομαι από, γλυτώνω από.

free *a.* (*not restricted*) ελεύθερος, (*with
money, advice*) σπάταλος (*with* σε). (*too
familiar in speech*) αθυρόστομος. (*gratis*)
δωρεάν (*adv.*). ~ from *or* of απαλλαγμένος
από, χωρίς. have a ~ hand έχω ελευθερία
δράσεως. make ~ with (*person*) παίρνω
θάρρος με, (*thing*) βάζω χέρι σε.

free-and-easy *a.* χωρίς τύπους.

freebooter *s.* πειρατής *m.*

freedom *s.* ελευθερία *f.*

freelance *a.* ανεξάρτητος.

freely *adv.* ελεύθερα.

freemason *s.* τέκτονας *m.* μασόνος *m.* **~ry**
s. τεκτονισμός *m.* (*fig.*) αλληλεγγύη *f.*

free-wheel *v.* ρολάρω. (*fig.*) ενεργώ χωρίς
να καταβάλλω μεγάλη προσπάθεια.

freez|e *v.t. & i.* παγώνω. (*fam., of persons,
make or feel very cold*) ξεπαγιάζω, πουντιάζω. I am ~ing πούντιασα. it is ~ing
(*weather*) κάνει παγωνιά. (*v.t.*) (*of food*)
καταψύχω, (*of prices*) καθηλώνω. (*s.*)
(*frost*) παγωνιά *f.* deep ~e κατάψυξη *f.*
~er *s.* ψυκτικός θάλαμος.

freezing *s.* πάγωμα *n.* ~ point σημείον
πήξεως. (*a.*) (*icy*) (*wind, look*) παγερός,
(*path, hand, object*) παγωμένος. (*for* ~)
ψυκτικός. *see* frozen.

freight *v.* (*charter*) ναυλώνω, (*load*) φορτώνω. (*s.*) (*charter*) ναύλωση *f.* (*load*)
φορτίο *n.* (*charge*) ναύλος *m.* **~er** *s.* φορτηγό *n.*

French *a.* γαλλικός. (*person*) Γάλλος *m.*,
Γαλλίδα *f.* ~ window μπαλκονόπορτα *f.* ~
bean φραγκοφάσουλο *n.* ~ fries πατάτες
τηγανητές. take ~ leave στρίβω αλά γαλ-
λικά.

frenetic *a.* έξαλλος. **~ally** *adv.* φρενιτιωδώς.

frenz|y *s.* φρενίτιδα *f.* **~ied** *a.* φρενια-
σμένος, φρενήρης. become ~ied φρενιάζω.

frequency *s.* συχνότητα *f.*

frequent *a.* συχνός, (*usual, common*) συνηθισμένος, κοινός. a ~ visitor τακτικός
επισκέπτης. (*v.t.*) (*cafés, theatres, etc.*)
συχνάζω σε, (*company*) συναναστρέφομαι. a ~ed road περαστικός δρόμος. **~ly**
adv. συχνά.

fresco *s.* τοιχογραφία *f.*

fresh *a.* (*not stale*) φρέσκος, νωπός. (*new*)
καινούργιος, νέος. (*air*) καθαρός, (*water*)
γλυκός, (*complexion, breeze*) δροσερός.
he is ~ from college μόλις αποφοίτησε.
(*fam.*) get ~ (*cheeky*) παίρνω αέρα. feel ~
(*rested*) αισθάνομαι ξεκούραστος.

fresh, ~ly *adv.* άρτι, προσφάτως, νεο-,
φρεσκο-. ~(ly) baked φρεσκοψημένος, ~ly
arrived (*person*) νεοφερμένος, άρτι
αφιχθείς.

freshen *v.t.* (*revive*) φρεσκάρω. (*v.i., of
weather*) δροσίζω. **~ing** *s.* (*brightening
up*) φρεσκάρισμα *n.*

freshness *s.* φρεσκάδα *f.* he has a ~ of
approach to his subject εξετάζει το θέμα
υπό νέο πρίσμα.

fret *v.t.* τρώγω. (*v.i.*) τρώγομαι. **~ful** *a.*
γκρινιάρης.

fretwork *s.* διάτρητος ξυλοκοπτική.

friable *a.* εύθρυπτος.

friar *s.* καλόγερος *m.*

fricative *a.* προστριβόμενος.

friction *s.* τριβή *f.* (*disagreement*) προστριβή *f.*

Friday *s.* Παρασκευή *f.* Good ~ Μεγάλη Παρασκευή. Man ~ Παρασκευάς *m.*

fridge *s.* ψυγείο *n.*

fried *a.* τηγανητός.

friend *s.* φίλος *m.*, φίλη *f.* make ~s again (*v.i.*) φιλιώνω. they made ~s έπιασαν φιλίες. a ~'s house φιλικό σπίτι.

friendl|y *a.* φιλικός. **~iness** *s.* φιλικότητα *f.*

friendship *s.* φιλία *f.*

frieze *s.* (*in room*) κορνίζα *f.* (*of building*) διάζωμα *n.* (*sculptured*) ζωφόρος *f.*

frigate *s.* φρεγάδα *f.*

fright *s.* τρόμος *m.* I got a ~ πήρα μια τρομάρα. he looked a ~ έμοιαζε σα σκιάχτρο.

frighten *v.* τρομάζω, φοβίζω. be ~ed (of) φοβούμαι. become ~ed με πιάνει φόβος. **~ed** *a.* τρομαγμένος. **~ing** *a.* τρομακτικός.

frightful *a.* τρομερός. **~ly** *adv.* τρομερά.

frigid *a.* ψυχρός. **~ity** *s.* ψυχρότητα *f.*

frill *s.* φρίλι *n.*, βολάν *n.* (*fig.*) στολίδι *n.* without ~s χωρίς καλλωπισμούς. **~y** *a.* με φρίλια.

fringe *s.* φράντζα *f.*, κρόσσι *n.* (*of hair*) αφέλεια *f.* (*outskirts*) περίχωρα *n.pl.* (*of group, party*) περιθώριο *n.* (*of forest, etc.*) άκρη *f.*, παρυφή *f.* ~ benefits πρόσθετα οφέλη (εργαζομένου).

frippery *s.* μπιχλιμπίδια *n.pl.*

frisk *v.i.* σκιρτώ. (*v.t.*) (*fam., search*) ψάχνω. (*s.*) κούνημα *n.* **~y** *a.* ζωηρός.

fritter *s.* είδος τηγανίτας. (*v.t.*) ~ away σκορπίζω, (*one's time*) χάνω.

frivol|ity *s.* επιπολαιότητα. *f.* **~ous** *a.* επιπόλαιος. **~ously** *adv.* επιπόλαια.

frizzle *v.i.* (*in cooking*) τσιτσιρίζω. (*v.t. & i.*) (*of hair*) κατσαρώνω.

fro *adv.* to and ~ πέρα δώθε. go to and ~ πηγαινοέρχομαι.

frock *s.* φουστάνι *n.* (*child's*) φουστανάκι *n.* **~-coat** *s.* ρεντιγκότα *f.*

frog *s.* βάτραχος *m.* **~march** *v.t.* they ~marched him τον κουβάλησαν σπρώχνοντας.

frolic *v.* (*of animals*) χοροπηδώ, (*of people*) διασκεδάζω με χαρούμενα παιχνίδια. (*s.*) παιχνίδι *n.* **~some** *a.* παιχνιδιάρικος.

from *prep.* από (*with acc.*), εξ, εκ (*with gen.*). ~ spite από κακία, ~ experience εκ πείρας, ~ earliest childhood εξ απαλών ονύχων. ~ now (*or* here) on από 'δω και πέρα. a week ~ today σήμερα οκτώ. tell him ~ me πες του εκ μέρους μου. (*painted*) ~ nature (*or* life) εκ του

φυσικού. we parted ~ our friends χωρίσαμε με τους φίλους μας. it prevented me ~ coming μ' εμπόδισε να 'ρθω. ~ 3 till 4 o'clock τρεις με τέσσερις. ~ being first in the class he has ended up bottom από πρώτος στην τάξη κατάντησε τελευταίος. he's got fat ~ too much eating πάχυνε από το πολύ φαΐ. I caught cold ~ not putting on an overcoat για να μη φορέσω παλτό άρπαξα κρυολόγημα.

frond *s.* φύλλο *n.*

front *s.* (*fore part*) μπροστινό μέρος, (*of building*) πρόσοψη *f.*, φάτσα *f.* sea-~ παραλία *f.* (*boldness*) θράσος *n.* (*mil.*) μέτωπο *n.* on the political ~ στην πολιτική αρένα. (*disguise, bluff*) βιτρίνα *f.*

front *adv. & prep.* in ~ μπροστά, εμπρός, έμπροσθεν. in ~ of (*ahead of*) μπροστά από, (*facing, in presence of*) μπροστά σε, ενώπιον (*with gen.*).

front *a.* εμπρόσθιος, μπροστινός. ~ seat μπροστινό κάθισμα, ~ row *or* rank πρώτη σειρά, ~ entrance κυρία είσοδος. ~ organization μετωπική οργάνωση.

front *v.t.* (*face*) βλέπω σε. ~ed with stone με πέτρινη πρόσοψη.

frontage *s.* πρόσοψη *f.*

frontal *a.* (*of forehead*) μετωπιαίος, (*archit.*) της προσόψεως. ~ attack μετωπική επίθεση.

frontier *s.* σύνορα *n.pl.*, μεθόριος *f.* (*a.*) (*town, people*) παραμεθόριος, (*matters*) μεθοριακός.

frontispiece *s.* προμετωπίδα *f.*

frost *s.* παγωνιά *f.* hoar ~ πάχνη *f.*, αγιάζι *n.* (*fam.*) it was a ~ απέτυχε. **~ed** *a.* (*opaque*) θαμπός. become ~ed (over) σκεπάζομαι με παγωνιά.

frost|-bite *s.* κρυοπάγημα *n.* his fingers got ~-bitten έπαθε κρυοπαγήματα στα δάχτυλα. the plants are ~-bitten τα φυτά είναι καμμένα από την παγωνιά.

frosty *a.* (*cold, lit. & fig.*) ψυχρός. *see also* frosted.

froth *s.* αφρός *m.* (*on coffee*) καϊμάκι *n.* (*v.*) αφρίζω. **~y** *a.* αφρώδης.

frown *v.* συνοφρυούμαι, κατσουφιάζω. **~ing** *a.*, **~ingly** *adv.* (*also* with a ~) συνοφρυωμένος, κατσουφιασμένος.

frowsty *a.* που μυρίζει κλεισούρα.

frowzy *a.* (*squalid*) βρώμικος, *see also* frowsty.

frozen *a.* παγωμένος. I am ~ πάγωσα. deep-~ κατεψυγμένος.

fructify *v.t.* γονιμοποιώ. (*v.i.*) καρποφορώ.

frugal *a.* λιτός. (*of person*) οικονόμος *s.m.f.* **~ity** *s.* λιτότητα *f.* (*thrift*) οικονομία *f.* **~ly** *adv.* λιτά, οικονομικά.

fruit s. (*lit. & fig.*) καρπός m. (*edible*) φρούτο n., (*collective*) φρούτα n.pl., οπωρικά n.pl. ~**juice** χυμός m. bear ~ καρποφορώ. ~**tree** οπωροφόρο δέντρο.

fruitful a. καρποφόρος, γόνιμος. ~**ness** s. γονιμότητα f.

fruition s. come to ~ αποφέρω καρπούς.

fruitless a. (*fig.*) άκαρπος, μάταιος. ~**ly** adv. ματαίως.

fruity a. it has a ~ smell μυρίζει φρούτο. (*mature*) ώριμος. (*suggestive*) σκαμπρόζικος.

frump s., ~**ish** a. κάρρο n.

frustrat|e v. (*a plan*) ματαιώνω, (*a person*) εμποδίζω. ~**ed** a. ανικανοποίητος. ~**ion** s. απογοήτευση f. (*of expectations*) ματαίωση f.

fry v.t. τηγανίζω. (s.) small ~ (*people of no importance*) ασημαντότητες f.pl., (*children*) μαρίδα f.

frying s. τηγάνισμα n. ~**pan** s. τηγάνι n.

fuchsia s. φούξια f.

fuddle v. ζαλίζω. (s.) in a ~ ζαλισμένος.

fuddy-duddy s. γεροπαράξενος m.

fudge v. ψευτοκάνω. (s., *nonsense*) κολοκύθια n.pl.

fuel s. καύσιμα n.pl. ~ tank (*of car*) ντεπόζιτο βενζίνης. add ~ to the fire ρίχνω λάδι στη φωτιά. (v.i.) (*with coal*) ανθρακεύω, (*with petrol*) παίρνω βενζίνη.

fug s. αποπνικτική ατμόσφαιρα λόγω κλεισούρας.

fugitive a. φευγαλέος. (s.) φυγάς m., δραπέτης m.

fulfil v. (*carry out*) εκτελώ, εκπληρώνω, (*realize*) πραγματοποιώ. ~**ment** s. εκτέλεση f. πραγματοποίηση f.

full a. (*filled*) γεμάτος, πλήρης. (*complete*) πλήρης. ~ of γεμάτος (*with acc.*). ~ fare ολόκληρο εισιτήριο. ~ figure γεμάτο κορμί. ~ house γεμάτη αίθουσα. ~ moon πανσέληνος f. ~ stop τελεία f. ~ trousers φαρδύ παντελόνι. to the ~ πλήρως. (*payment*) in ~ εις το ακέραιον. ~ up γεμάτος, ~ to overflowing ξέχειλος. in ~ awareness εν πλήρη συνειδήσει, at ~ speed ολοταχώς. ~**y** adv. πλήρως, τελείως. (*write*) more ~y εκτενέστερα.

full-blooded a. δυνατός, ρωμαλέος.

full-blown a. (*flower*) ολάνοιχτος, (*fig.*) τέλειος.

full-dress a. επίσημος. ~ uniform μεγάλη στολή.

full-face a. (*portrait*) ανφάς.

full-grown a. (*person*) ενήλικος, (*tree, animal*) σε πλήρη ανάπτυξη.

full-length a. (*portrait*) ολόσωμος.

full-scale a. σε φυσικό μέγεθος.

full-time a. τακτικός.

fuller s. ~'s earth σαπουνόχωμα f.

fulminate v. εξακοντίζω μύδρους.

fulsome a. υπερβολικός.

fumble v. ψαχουλεύω.

fume s. (*smoke*) καπνός m. (*of roast meat*) κνίσα f. ~s (*noxious*) αναθυμιάσεις f.pl. (*petrol*) καυσαέρια n.pl. (v.i.) (*chafe*) εκνευρίζομαι.

fumigat|e v. απολυμαίνω με κάπνισμα. ~**ion** s. απολύμανση f.

fun s. διασκέδαση f. have ~ διασκεδάζω. in or for ~ στ' αστεία. make ~ of κοροϊδεύω.

function s. (*official duty*) λειτούργημα n., καθήκον n. (*purpose of sthg.*) σκοπός m., λειτουργία, δουλειά f. (*ceremony*) τελετή f. (*reception*) δεξίωση f. (*math.*) συνάρτηση f. (v.i.) λειτουργώ.

functional a. (*of disorders*) λειτουργικός, (*practical*) πρακτικός.

functionary s. αξιωματούχος m.

fund s. (*stock, supply*) απόθεμα n. (*money collected*) έρανος m. (*for pensions, etc.*) ταμείο n. ~s χρήματα n.pl:, (*stock*) χρεόγραφα n.pl.

fundamental a. θεμελιώδης. (s.) ~s βασικές αρχές. ~**ly** adv. βασικά.

funeral s. κηδεία f. (a.) (*march, bell*) πένθιμος. ~ service νεκρώσιμος ακολουθία, ~ oration επικήδειος λόγος, ~ procession νεκρική πομπή.

funereal a. πένθιμος, (*pace*) αργός.

fungus s. μύκης m.

funicular s. σιδηροδρομικά οχήματα ελκόμενα με καλώδια, (*fam., incorrectly*) τελεφερίκ n.

funk s. & v. (*fam.*) he was in a ~ τα χρειάστηκε. he ~ed going δείλιασε και δεν πήγε.

funnel s. (*utensil*) χωνί n. (*of engine*) φουγάρο n., καπνοδόχος f.

funn|y a. (*amusing*) αστείος, (*peculiar*) παράξενος, περίεργος. ~y business ύποπτη δουλειά. ~**ily** adv. αστεία, παράξενα. ~ily enough κατά περίεργη σύμπτωση, περίεργως.

fur s. (*on animal*) τρίχωμα n. (*dressed*) γούνα f. (*as article of attire*) γουναρικό n. (*in kettle, etc.*) πουρί n. (a.) (*made of ~*) γούνινος.

furbelow s. φραμπαλάς m.

furbish v. (*polish*) γυαλίζω, (*fig.*) φρεσκάρω.

furious a. μανιώδης, λυσσώδης. become ~ γίνομαι θηρίο (or έξω φρενών). ~**ly** adv. μανιωδώς, λυσσωδώς, (*of pace*) δαιμονισμένα.

furl v. διπλώνω.

furlough s. άδεια f.

furnace s. κάμινος f., καμίνι n.

furnish v.t. (a person) εφοδιάζω, (a house) επιπλώνω. (make available) (supplies) προμηθεύω, (opportunity, evidence, etc.) παρέχω, δίνω. ~ing(s) s. επίπλωση f.

furniture s. έπιπλα n.pl. piece of ~ έπιπλο n.

furore s. cause a ~ ξετρελλαίνω (or χαλάω) τον κόσμο.

furred a. (animal) τριχωτός, (tongue) άσπρος. the pipes are ~ οι σωλήνες έχουν πιάσει πουρί.

furrier s. γουναράς m.

furrow s. (ploughed) αυλάκι n. (wrinkle) αυλακιά f., ρυτίδα f. (v.) αυλακώνω.

furry a. τριχωτός, (like fur) σαν γούνα.

further adv. πιο μακριά. ~ on πιο πέρα, ~ back πιο πίσω. (in addition) επί πλέον, (more) περισσότερο. (a.) (more distant) απώτερος. (additional) περισσότερος, κι άλλος, νεώτερος, επι πλέον. for ~ action (in office, etc.) διά τα περαιτέρω.

further v. προάγω, προωθώ. ~ance s. προαγωγή f.

furthermore adv. επι πλέον.

furthermost a. ο πιο μακρινός.

furthest a. απώτατος, ο πιο μακρινός, (adv.) πιο μακρυά.

furtive a. (act) κρυφός, κλεφτός. (person) ύπουλος. ~ly adv. κρυφά.

fur|y s. μανία f., λύσσα f., παραφορά f. (of storm) μανία f. be in a ~y είμαι έξω φρενών. F~ies Εριννύες f.pl.

furze s. ασπάλαθος m.

fuse v.t. & i. (melt) λειώνω, (v.t.) (blend) συγχωνεύω. the lights have ~d κάηκε η ασφάλεια. (s.) (for explosion) φιτίλι n. (electric) ασφάλεια f.

fuselage s. άτρακτος f.

fusillade s. τουφεκίδι n.

fusion s. συγχώνευση f.

fuss s. φασαρία f. make a ~ of περιποιούμαι. (v.i.) (worry) ανησυχώ, (make difficulties) φέρνω τον κατακλυσμό. (v.t.) ενοχλώ.

fuss|y a. (person) τιτίζης, ιδιότροπος. (of style) παραφορτωμένος. ~iness s. ιδιοτροπία f.

fustian s. (fig., bombast) μεγάλα κούφια λόγια.

fusty a. μουχλιασμένος.

futil|e a. μάταιος, (person) ανίκανος. ~ity s. ματαιότητα s.

future s. μέλλον n. (gram.) μέλλων m. (a.) μελλοντικός. my ~ husband ο μέλλων σύζυγός μου.

futuristic a. φουτουριστικός.

fuzzy a. (blurred) θαμπός. ~ hair αράπικα σγουρά μαλλιά.

G

gab s. (fam.) have the gift of the ~ έχω λέγειν.

gabardine s. (γ)καμπαρντίνα f.

gabble v. μιλώ (or λέω) γρήγορα-γρήγορα. (s.) γρήγορη και ακατάληπτη ομιλία.

gable s. αέτωμα n.

gad v.i. (also ~ about) γυρίζω γλεντοκοπώντας. ~about s. be a ~about see gad.

gadfly s. οίστρος m.

gadget s. μικροεφεύρεση f. σύνεργο n., μαραφέτι n.

gaff s. (fisherman's) γάντζος m. (fam.) blow the ~ φανερώνω το μυστικό.

gaffe s. γκάφα f.

gaffer s. μπάρμπας m.

gag s. φίμωτρο n. (joke) αστείο n. (v.t.) φιμώνω. (v.i., make jokes) κάνω αστεία.

gaga a. ξεμωραμένος.

gage s. τεκμήριο n. εγγύηση f. (v.) δίδω ως εγγύησιν.

gaiety s. ευθυμία f., κέφι n.

gaily adv. με κέφι. ~ dressed ντυμένος με χαρούμενα (or ζωηρά) χρώματα.

gain v. (profit, ground, esteem, etc.) κερδίζω, (one's object) πετυχαίνω. ~ strength αποκτώ δύναμη, (after setback) ανακτώ δύναμη. I have ~ed 2 kilos (weight) πήρα (or έβαλα) δύο κιλά. he ~ed recognition by his work το έργο του του απέφερε φήμη. (one's destination) φθάνω εις. ~ on (quarry) πλησιάζω, (pursuer) αφήνω πίσω. ~ (person) over παίρνω με το μέρος μου. (s.) (profit) κέρδος n. (increase) αύξηση f.

gainer s. ο κερδισμένος.

gainful a. επικερδής.

gainsay v. αμφισβητώ, αρνούμαι.

gait s. περπατησιά f., βάδισμα n.

gaiter s. γκέτα f.

gala s. γκαλά, f., εορτασμός m.

galaxy s. γαλαξίας m.

gale s. θύελλα f.

gall s. (bile, also fig.) χολή f. (fam., impudence) θράσος n. ~-bladder s. χοληδόχος κύστις.

gall s. (sore) ερεθισμός από τριβή. (v.) ερεθίζω. ~ing a. εκνευριστικός.

gallant a. (brave) ανδρείος, γενναίος. (fine) ωραίος, περήφανος, λεβέντικος. (ladies' man) γαλάντης m. ~ly adv. γενναίως, περήφανα. ~ry s. ανδρεία f. (chivalry) ιπποτισμός m.

galleon s. γαλιόνι n.

gallery s. (portico, arcade) στοά f. (balcony) εξώστης m. (corridor) διάδρομος m.

(picture--) πινακοθήκη f. (showroom) γκαλερί f. (of mine) στοά f. (of theatre) υπερώο n., (fam.) γαλαρία f. (press, visitors') θεωρείο n.

galley s. γαλέρα f. (ship's) μαγειρείον n. ~s (prison) κάτεργα n.pl. ~ slave κατάδικος των γαλερών.

galley-proof s. ασελιδοποίητο δοκίμιο.

Gallic a. γαλλικός. ~ism s. γαλλισμός m.

gallipot s. βαζάκι για αλοιφή.

gallivant v.i. τσιληπουρδώ.

gallon s. γαλόνι n.

gallop v. καλπάζω. (s.) καλπασμός m.

gallows s. κρεμάλα f.

galore adv. εν αφθονία, και κακό, κι άλλο τίποτα.

galoshes s. γαλότσες f.pl.

galvanize v. γαλβανίζω.

gambit s. (fig.) μανούβρα f.

gamble v. παίζω, ~le away χάνω στα τυχερά παιγνίδια. (s.) it's a ~le είναι καθαρό παιγνίδι (or ζήτημα τύχης). ~ler s. (χαρτο)παίκτης m. ~ling s. παιγνίδι n., τζόγος m., χαρτοπαιξία f.

gambol v. σκιρτώ. (s.) (also ~ling) σκίρτημα n.

game s. παιγνίδι n. (round) παρτίδα f. (jest) we were only having a ~ τα κάναμε για πλάκα. ~s (contests) αγώνες m.pl., (sports) σπορ n.pl. (fig.) play the ~ παίζω τίμιο παιγνίδι, δεν κάνω ζαβολιές. the ~ is up το παιγνίδι χάθηκε. give the ~ away προδίδω το μυστικό. make ~ of κοροϊδεύω, he's having a ~ with us μας κοροϊδεύει. (trick) κόλπο n. he's at the same old ~ ξανάρχισε τα δικά του.

game s. (hunted) θήραμα n. (also as food) κυνήγι n. (fig.) fair ~ εκτεθειμένος εις κατακρίσεις.

game a. έτοιμος, πρόθυμος. ~ly adv. με κουράγιο.

gamekeeper s. φύλακας αγροκτήματος για θηράματα.

gaming s. παιγνίδι n.

gammon s. χοιρομέρι παστό όχι ψημένο. (fam.) σαχλαμάρες f.pl.

gamut s. the whole ~ όλη η κλίμακα.

gander s. χήνα m.

gang s. (of workmen) ομάδα f., συνεργείο n. (of criminals) συμμορία f., σπείρα f. (fam., company) συντροφιά f., παρέα f. the old ~ η παλιοπαρέα. (v.) (fam.) ~ up on συνενώνομαι εναντίον (with gen.). ~ up with πάω μαζί με.

gangling a. ψηλόλιγνος.

gangplank s. σανιδόσκαλα f.

gangrene s. γάγγραινα f.

gangster s. συμμορίτης m., γκάγκστερ m.

gangway s. διάδρομος m. (naut.) διαβάθρα f.

gaol s. φυλακή f. (v.) φυλακίζω. ~er s. δεσμοφύλακας m.

gap s. κενόν n., χάσμα n. (in fence, etc.) άνοιγμα n. (deficiency) έλλειψη f. (difference) διαφορά f.

gape v.i. (be wide open) χάσκω, (of wound, abyss) χαίνω. ~ at κοιτάζω με ανοικτό στόμα. (yawn) χασμουριέμαι.

garage s. γκαράζ n. (v.) βάζω στο γκαράζ.

garb s. ενδυμασία f., στολή f. (v.) ντύνω. ~ed ντυμένος.

garbage s. σκουπίδια n.pl. ~ can σκουπιδοτενεκές m.

garble n. (distort) διαστρεβλώνω, (mix up) μπερδεύω.

garden s. κήπος m., περιβόλι n. kitchen ~ λαχανόκηπος m. (fam.) lead up the ~ path παραπλανώ. (v.) ασχολούμαι με τον κήπο. ~ing s. κηπουρική f. ~-suburb s. κηπούπολη f.

gardener s. κηπουρός m. be an enthusiastic ~ έχω μανία με την κηπουρική.

gargantuan a. τεράστιος.

gargle s. γαργάρα f. (v.) κάνω γαργάρα.

gargoyle s. τερατόμορφον στόμιον υδρορρόης.

garish a. φανταχτερός, χτυπητός.

garland s. γιρλάντα f. (v.) στολίζω με γιρλάντες.

garlic s. σκόρδο n. ~ sauce σκορδαλιά f.

garment s. ρούχο n.

garner v. μαζεύω.

garnet s. & a. γκρενά n.

garnish v. γαρνίρω. (s.) γαρνιτούρα f.

garret s. σοφίτα f.

garrison s. φρουρά f. (v.t.) (troops) εγκαθιστώ, (city) εγκαθιστώ φρουράν εις.

garrotte v. στραγγαλίζω. (s.) στραγγάλη f.

garrulous a. φλύαρος. ~ity s. φλυαρία f.

garter s. καλτσοδέτα f. (Order) Περικνημίς f.

gas s. αέριο n. (domestic) γκάζι n. (petrol) βενζίνη f. (fam.) step on the ~ πατώ γκάζι. (v.t.) δηλητηριάζω δι' αερίου. (v.i.) φλυαρώ. ~bag s. (fam.) φαρλατάς m. ~eous a. αεριώδης. ~-fire s. θερμάστρα γκαζιού. ~light s. αεριόφως n.

gash s. κόψιμο n., εγκοπή f. (v.) κόβω.

gasket s. παρέμβυσμα n., φλάντζα f.

gasoline s. βενζίνη f.

gasp, ~ing s. κομμένη ανάσα. at one's last ~ στα τελευταία μου, (fig.) εξαντλημένος. (v.) κόβεται η αναπνοή μου, λαχανιάζω. ~er s. (fam.) τσιγάρο n.

gassy a. αεριούχος.

gastric a. γαστρικός. ~itis s. γαστρίτιδα f. ~onomic a. γαστρονομικός.

gate s. (of city, palace, & fig.) πύλη f. (of garden, field) καγκελόπορτα f. (of courtyard) αυλόπορτα f. (way in) είσοδος f. (barrier) φράγμα n. ~post s. παραστάτης m. ~way s. είσοδος f. (fig.) πύλη f.

gateau s. τούρτα f.

gate-crash v. μπαίνω απρόσκλητος (σε). ~er s. απρόσκλητος a.

gather v.t. (flowers, people, belongings, etc.) μαζεύω, (crops) συγκομίζω, (revenue) εισπράττω, (experience, impressions) αποκτώ. ~ speed αναπτύσσω ταχύτητα. ~ together συγκεντρώνω, (what is scattered) περιμαζεύω. (sewing) κάνω σούρες σε. (infer) I ~ that... έχω την εντύπωση ότι. (v.i.) μαζεύομαι, συναθροίζομαι. ~ed a. (sewing) σουρωτός, (brow) συνοφρυωμένος. ~ing s. συγκέντρωση f.

gauche a. κοινωνικώς αδέξιος.

gaudy a. χτυπητός.

gauge s. (criterion) κριτήριο n. (thickness) πάχος n. (instrument) μετρητής m., δείκτης m. take the ~ of μετρώ, ζυγίζω. broad ~ line γραμμή μεγάλου πλάτους. (v.) μετρώ, (fig.) εκτιμώ, ζυγίζω.

gaunt a. αποστεωμένος, (place) γυμνός.

gauntlet s. γάντι n. (fig.) throw down the ~ ρίχνω το γάντι. run the ~ of the critics αντιμετωπίζω δυσμενή κριτική.

gauze s. γάζα f. (a.) από γάζα.

gavel s. σφυρί n.

gawky a. ~ fellow μαντράχαλος m.

gay a. χαρούμενος, εύθυμος, (colours, etc.) ζωηρός. lead a ~ life γλεντοκοπώ. (fam., homosexual) ομοφυλόφιλος.

gaze v. ~ at κοιτάζω επιμόνως, ατενίζω. (s.) βλέμμα n.

gazelle s. γκαζέλα f.

gazette s. εφημερίδα f.

gazetteer s. γεωγραφικό λεξικό.

GB Μεγάλη Βρεττανία.

gear s. (equipment) εξοπλισμός m. σύνεργα n.pl. (clothes) ρούχα n.pl. (goods) είδη n.pl. (apparatus) συσκευή f. (of motor) ταχύτητα f. engage first ~ βάζω πρώτη. is the car in ~ or out of ~ ? είναι η ταχύτητα μέσα ή δεν είναι; (v.t.) (adapt) προσαρμόζω. ~-box v.t. κιβώτιο ταχυτήτων. ~-lever s. μοχλός ταχυτήτων. ~-wheel s. γρανάζι n.

gelatine s. ζελατίνη f.

geld v. μουνουχίζω.

gelignite s. ζελινίτις f.

gem s. πολύτιμος λίθος, (also fig.) κόσμημα n. (fig., of persons) διαμάντι n. (fam., funny joke) αμίμητο αστείο.

gendarmerie s. χωροφυλακή. f.

gender s. γένος n.

gene s. (biol.) γονίδιο n.

genealog|y s. γενεαλογία f. ~ical a. γενεαλογικός.

general a. γενικός. in ~ use γενικής χρήσεως. in ~ γενικώς. the ~ public το ευρύ κοινό. ~ post (fig.) ανασχηματισμός m. (med.) ~ practitioner ιατρός παθολόγος. (s.) (mil.) στρατηγός m.

generalit|y s. the ~y of οι περισσότεροι. ~ies γενικότητες f.pl.

generaliz|e v. γενικεύω. ~ation s. γενίκευση f.

generally adv. (in general) γενικώς, γενικά. (usually) κατά κανόνα, συνήθως.

generat|e v. γεννώ, (heat, current, etc.) παράγω. ~ion s. γενεά f. παραγωγή f. ~ive s. (of procreation) γενετήσιος, (productive) παραγωγικός. ~or s. γεννήτρια f.

generic a. γενικός.

gener|ous a. γενναιόδωρος, ανοιχτοχέρης, (magnanimous) μεγαλόψυχος, (ample) γενναίος. ~ously adv. γενναιόδωρα. ~osity s. γενναιοδωρία f., απλοχεριά f. μεγαλοψυχία f.

genesis s. γένεση f.

genetic a. γενετικός. ~s s. γενετική f.

genial a. πρόσχαρος, εγκάρδιος. (mild) ήπιος. ~ity s. εγκαρδιότητα f. ~ly adv. εγκάρδια.

genital a. γεννητικός. ~s s. γεννητικά όργανα.

genitive a. (gram.) γενική f.

genius s. μεγαλοφυΐα f. (spirit) πνεύμα n. evil ~ κακός δαίμονας. of ~ (as a.) μεγαλοφυής. have a ~ for έχω ταλέντο σε.

genocide s. γενοκτονία f.

genre s. (painting) με σκηνές από την καθημερινή ζωή.

gent|eel a. (iron.) καθώς πρέπει. ~ility s. ευγένεια f., αρχοντιά f. (iron.) ψευτοαριστοκρατικότητα f.

gentile s. ο μη Ιουδαίος.

gent|le a. (by birth) ευγενής, (not rough) ήμερος, μειλίχιος, (manners) αβρός. (breeze, touch) απαλός, (reproof) μαλακός, ήπιος, (animal) ήρεμος. ~y adv. αβρά, απαλά, μαλακά, ήπιως.

gentleman s. κύριος m. ~ly a. κύριος, καθώς πρέπει, (bearing) αρχοντικός. in a ~ly way σαν κύριος.

gentleness s. αβρότητα f. απαλότητα f. ηπιότητα f.

gentlewoman s. κυρία f.

gentry s. τάξη των αρχόντων, αρχοντολόι n.

genuflect v. κλίνω το γόνυ. ~ion s. γονυκλισία f.

genuine a. γνήσιος, αληθινός. (serious) σοβαρός, (sincere) ειλικρινής. ~ly adv. αληθινά, ειλικρινά.

genus s. γένος n.

geograph|y s. γεωγραφία f. **~er** s. γεω-
γράφος m. **~ic, ~ical** a. γεωγραφικός.
geolog|y s. γεωλογία f. **~ical** a. γεωλογικός.
~ist s. γεωλόγος m.
geometr|y s. γεωμετρία f. **~ic, ~ical** a.
γεωμετρικός.
geophysics s. γεωφυσική f.
geopolitics s. γεωπολιτική f.
geranium s. γεράνι(ον) n.
geriatrics s. γηριατρική f.
germ s. (biol. & fig.) σπέρμα n. (microbe)
μικρόβιο n. **~icide** s. μικροβιοκτόνο n.
German a. γερμανικός. (person) Γερμανός
m., Γερμανίδα f. (language) γερμανικά
n.pl.
germane a. be ~ to έχω σχέση με.
germinat|e v.i. βλαστάνω. (v.t.) γονιμο-
ποιώ. (v.i. & t., fig.) γεννώ. **~ion** s.
βλάστηση f.
gerrymander v. ~ an election νοθεύω εκλογές. **~ing** s. απάτη f.
gerund s. (gram.) γερούνδιον n.
gestation s. κύηση f.
gesticulat|e v. κάνω χειρονομίες. **~ion** s.
χειρονομία f.
gesture s. (lit. & fig.) χειρονομία f. (nod,
sign) νεύμα n., νόημα n. (v.) χειρονομώ,
γνέφω, κάνω νόημα.
get v.i. (become) γίνομαι, ~ rich γίνομαι
πλούσιος, ~ angry θυμώνω, ~ married
παντρεύομαι, ~ dressed ντύνομαι, ~
broken σπάζω. he got trapped in the lift
κλείστηκε στο ασανσέρ, we got caught in
the rain μας έπιασε η βροχή. ~ into routine
μπαίνω στη ρουτίνα, ~ into conversation
πιάνω κουβέντα, ~ going βάζω μπρος. it
~s hot in July πιάνει ζέστη τον Ιούλιο, I'm
~ting hungry άρχισα να πεινώ. (come)
they got to know each other γνωρι-
στήκανε, ~ to know (learn) μαθαίνω, he
got to like it in the end συνήθισε και στο
τέλος του άρεσε. (go) πηγαίνω, (arrive) ~
here/there φτάνω. ~ home γυρίζω σπίτι, ~
nowhere δεν καταλήγω πουθενά. have got
(possess) έχω. have got to (must) I've got
to leave πρέπει να φύγω, has it got to be
done now? πρέπει να γίνει τώρα; no, it
hasn't got to be όχι δεν είναι ανάγκη. see
also better, cold, drunk, late, ready, tired,
used, well, etc.
get v.t. (acquire) αποκτώ, (obtain) παίρνω,
(earn, gain, win) κερδίζω, (fetch) φέρνω,
(receive) λαμβάνω, παίρνω. (an illness)
κολλώ, αρπάζω, (a bus, train) παίρνω.
(make, cause) κάνω, I can't ~ the door (to)
shut δεν μπορώ να κάνω να κλείσει η
πόρτα, (of persons) they got him to speak
to the boss τον έβαλαν να μιλήσει στον

προϊστάμενο, ~ (woman) with child
καθιστώ έγκυον. ~ one's hair cut κόβω τα
μαλλιά μου, ~ a suit made ράβω κοστούμι,
~ it dry-cleaned το στέλνω για στεγνό
καθάρισμα, I must ~ the roof mended
πρέπει να φέρω άνθρωπο να επισκευάσει
τη στέγη. (convey) I got him to the station
in time τον πήγα στο σταθμό εγκαίρως.
(enjoy, suffer) you ~ a good view from
here έχει ωραία θέα από 'δω, our house ~s
the sun το σπίτι μας το βλέπει ο ήλιος. I
got my feet wet βράχηκαν τα πόδια μου,
we got a puncture μας έπιασε λάστιχο. he
got 6 months in jail έφαγε (or του
κοπάνησαν) έξι μήνες φυλακή. (find)
βρίσκω, ~ the right answer (to sum, etc.)
βρίσκω τη λύση, you can't ~ it in the shops
δεν βρίσκεται (or δεν υπάρχει) στα
μαγαζιά. I don't ~ it (see the point) δεν
μπήκα, I've got it! το βρήκα. (put) ~
(sthg.) in on βάζω, ~ (sthg.) out/off βγάζω.
(pass or make sthg. go) through, over, etc.
περνώ. (see also below).
get about v.i. κυκλοφορώ.
get across v.i. περνώ, (fig.) γίνομαι
αντιληπτός. (v.t.) I got it across to him τον
έκανα να καταλάβει.
get along v.i. (fare) πηγαίνω, πάω. we ~
well τα πάμε καλά. (depart) φεύγω. (fam.)
~ with you! άντε ρε!
get at v.t. (reach) φτάνω, (find) βρίσκω,
(approach) πλησιάζω, (bribe) δωροδοκώ.
what is he getting at? τι υπονοεί;
get-at-able a. is the place ~? είναι εύκολο
να πάει κανείς εκεί; he's not ~ in the
mornings δεν είναι εύκολο να τον πλη-
σιάσει κανείς τα πρωινά.
get away v.i. (escape) ξεφεύγω, το σκάω,
(depart) φεύγω, ~ with it (act successfully)
τα καταφέρνω, ~ with (steal) κλέβω.
(fam.) ~ with you! άντε ρε!
getaway s. δραπέτευση f.
get back v.i. επιστρέφω, γυρίζω. (v.t.) I
never got back the book I lent him δεν μου
επέστρεψε το βιβλίο που του δάνεισα. get
one's own back παίρνω πίσω το αίμα μου.
get by v. περνώ, (fig.) περνώ, τα βολεύω.
get down v.t. κατεβάζω, (swallow) καταπίνω. it gets me down με εκνευρίζει. (v.i.)
κατεβαίνω, ~ to work στρώνομαι στη
δουλειά.
get in v.i. (enter) μπαίνω (μέσα), (find
room) χώνομαι, μπαίνω, (return home)
γυρίζω, (arrive) φτάνω, (be elected)
εκλέγομαι, ~ with συνδέομαι με. (v.t.)
(insert) βάζω, χώνω (μέσα). (obtain supply
of) παίρνω, (summon) φωνάζω, φέρνω,
(collect) μαζεύω. (bring inside) παίρνω

μέσα. (*find time for*) προφταίνω να. (*find room for*) χωρώ. I couldn't get a word in δεν με άφησαν να πω λέξη.

get into *v.* μπαίνω σε, (*clothes*) φορώ, (*habits*) αποκτώ. ~ trouble έχω μπλεξίματα. ~ bed πέφτω στο κρεββάτι.

get off *v.t.* (*remove*) βγάζω, (*dispatch*) στέλνω, (*save, avoid*) γλυτώνω. (*v.i.*) (*alight*) κατεβαίνω, βγαίνω, (*depart*) ξεκινώ. (*escape*) γλυτώνω, he got off lightly φτηνά τη γλύτωσε. ~ with (*start affair*) τα φτιάνω με.

get on *v.t.* (*mount*) ανεβαίνω σε, they ~ my nerves μου δίνουν στα νεύρα. (*v.i.*) (*fare*) πηγαίνω, πάω, we ~ well τα πάμε καλά. (*advance*) προχωρώ, be getting on (*in age*) μεγαλώνω, it's getting on for midday πλησιάζει (*or* κοντεύει) μεσημέρι. (*continue*) ~ with the job εξακολουθώ τη δουλειά μου.

get out *v.t.* (*remove, publish*) βγάζω, (*extract*) αποσπώ. (*v.i.*) (*come out, alight*) βγαίνω, (*become known*) did it ~? έγινε γνωστό; (*go away*) φεύγω, το σκάω. ~ of (*avoid*) αποφεύγω, (*a habit*) αποβάλλω, ~ of practice ξεσυνηθίζω.

get over *v.t.* (*fence, etc.*) υπερπηδώ, (*fig., surmount*) ξεπερνώ. (*conquer*) κατανικώ, (*recover from*) συνέρχομαι από. let's get it over quickly ας το τελειώσουμε γρήγορα. he can't ~ his chagrin το φυσάει και δεν κρυώνει.

get round *v.t.* (*dodge, go round*) παρακάμπτω, (*circumvent*) καταστρατηγώ. (*persuade*) φέρνω βόλτα, καταφέρνω. (*v.i.*) ~ to (*doing sthg.*) βρίσκω την ευκαιρία να.

get through *v.t.* (*fence, etc.*) περνώ από, (*a test*) περνώ, (*a job*) τελειώνω, (*get sthg. accepted*) περνώ. (*v.i.*) (*pass, be accepted*) περνώ. ~ to (*reach*) φτάνω (σε), (*on phone*) συνδέομαι με.

get under *v.t.* (*subdue*) καταστέλλω.

get up *v.i.* (*rise*) σηκώνομαι, (*mount*) ανεβαίνω. (*v.t.*) (*raise*) σηκώνω, (*climb*) ανεβαίνω σε, (*dress*) ντύνω, (*organize*) οργανώνω. (*do, tidy, arrange*) φτιάνω, (*merchandise*) συσκευάζω. (*study*) μελετώ, (*prepare a topic*) προετοιμάζω. she was got up as Columbine ήταν ντυμένη Κολομπίνα. ~ to (*tricks, etc.*) μαγειρεύω, σκαρώνω.

get-up *s.* ντύσιμο *n.*, εμφάνιση *f.* (*disguise*) μεταμφίεση *f.*

gew-gaw *s.* μπιχλιμπίδι *n.*

geyser *s.* (*natural*) θερμοπίδακας *m.* (*heater*) θερμοσίφωνας *m.*

ghastly *a.* (*awful*) φρικτός, (*livid*) πελιδνός,

~ pallor νεκρική ωχρότητα.

gherkin *s.* (pickled) ~ αγγουράκι *n.* (τουρσί).

ghetto *s.* γκέτο *n.*

ghost *s.* (*spectre*) φάντασμα *n.*, στοιχειό *n.* (*spirit*) πνεύμα *n.*, Holy G~ Άγιον Πνεύμα. give up the ~ ξεψυχώ, I haven't the ~ of an idea δεν έχω την παραμικρή ιδέα. (*v.t., fam.*) συγγράφω (βιβλίο, ομιλία) για άλλον επί πληρωμή. ~**ly** *a.* (*spectral*) που μοιάζει με φάντασμα, (*spiritual*) πνευματικός.

ghoul *s.* τέρας *n.* ~**ish** *a.* (*idea, action, etc.*) μακάβριος.

giant *s.* γίγας *m.* (*a.*) γιγαντιαίος.

giaour *s.* γκιαούρης *m.*

gibber *v.* παραληρώ. ~**ish** *s.* αλαμπουρνέζικα *n.pl.*

gibbet *s.* κρεμάλα *f.*

gibe *s.* σκώμμα *n.* (*v.*) ~ at σκώπτω.

giblets *s.* εντόσθια *n.pl.*

gidd|y *a.* (*height, pace*) ιλιγγιώδης, (*person*) ζαλισμένος, make ~y ζαλίζω, feel ~y ζαλίζομαι. (*flighty*) αλαφρόμυαλος, τρελλός *f.* ~**ily** *adv.* ιλιγγιωδώς. ~**iness** *s.* ζάλη *f.*, ζαλάδα *f.*, ίλιγγος *m.*

gift *s.* δώρο *n.* (*accomplishment*) χάρισμα *n.* (*talent*) ταλέντο *n.* ~**ed** *a.* προικισμένος.

gigantic *a.* γιγαντιαίος.

giggle *v.* γελώ πνιχτά. (*s.*) πνιχτό νευρικό γέλιο.

gigolo *s.* ζιγκολό *n.*

Gilbertian *a.* κωμικός και εξωφρενικός, άνω ποταμών.

gild *v.* (επι)χρυσώνω. (*fig.*) ~ the pill χρυσώνω το χάπι. ~ the lily προσπαθώ να καλλωπίσω κάτι που είναι τέλειο. ~ed youth χρυσή νεολαία.

gills *s.* βράγχια *n.pl.*, σπάραχνα *n.pl.*

gilt *a.* χρυσός, επιχρυσωμένος. (*s.*) επιχρύσωση *f.* it took the ~ off the gingerbread αυτό χάλασε την υπόθεση. ~~**edged** *a.* (*fin.*) ασφαλής.

gimcrack *s.* παλιοπράμα *n.*

gimlet *s.* τρυβέλι *n.*

gimmick *s.* (*fam.*) τέχνασμα *n.*, κόλπο *n.*

gin *s.* (*trap*) παγίδα *f.* (*liquor*) τζιν *n.*

ginger *a.* (*hair*) πυρός. ~~**haired** *a.* κοκκινομάλλης.

ginger *s.* ζιγγίβερις *f.* (*fam., fig.*) ζωντάνια *f.* (*v.*) ~ up δραστηριοποιώ, τονώνω, δίνω ζωή σε. ~~**beer** *s.* τζιτζιμπίρα *f.* ~~**group** *s.* ομάδα ακτιβιστών.

gingerly *adv.* προσεκτικά.

gipsy *s.* αθίγγανος, γύφτος *m.* τσιγγάνα, γύφτισσα *f.* (*a.*) (*also* ~'s) τσιγγάνικος, γύφτικος.

giraffe *s.* καμηλοπάρδαλη *f.*

gird *v.* ~ on ζώνομαι, ~ up one's loins (*fig.*) ανασκουμπώνομαι. ~ at σκώπτω. (*encircle*) περιβάλλω, sea-girt περιβαλλόμενος από τη θάλασσα.
girder *s.* δοκός *f.*
girdle *s.* ζώνη *f.* (*corset*) κορσές *m.* (*v.*) περιζώνω.
girl *s.* κορίτσι *n.*, κοπέλλα *f.* νέα, νεαρή *f.* ~ friend φιλενάδα *f.* ~hood *s.* (*youth*) νιάτα *n.pl.* ~ish *a.* κοριτσίστικος.
girth *s.* περιφέρεια *f.*
gist *s.* ουσία *f.*
give *v.t.* δίδω, δίνω, (*offer*) προσφέρω, (*make present of*) χαρίζω, δωρίζω, (*cause, create*) προκαλώ, δημιουργώ. ~ (*person*) a ring κάνω ένα τηλεφώνημα σε, ~ a cry βγάζω ένα ξεφωνητό, ~ battle συγκροτώ μάχην, ~ birth to γεννώ. I'll ~ it to you! θα σε δείρω. ~ away (*bestow*) χαρίζω, (*betray*) προδίδω. ~ lessons παραδίδω μαθήματα.
give *v.i.* (*bend*) κάμπτομαι, λυγίζω, (*of substance*) έχω ελαστικότητα. ~ way (*yield*) υποχωρώ, (*collapse*) καταρρέω, βουλιάζω. ~ way to (*succumb*) υποκύπτω εις, (*passions*) παρασύρομαι από, (*be succeeded by*) αντικαθίσταμαι από. his grief gave way to joy η λύπη του μετεβλήθη εις χαράν, η λύπη του έγινε χαρά.
give *s.* ελαστικότητα *f.* ~ and take αμοιβαίες υποχωρήσεις.
give back *v.t.* επιστρέφω, ξαναδίνω.
give forth *v.t.* βγάζω, αναδίδω, (*sounds*) εκπέμπω.
give in *v.i.* (*yield*) υποχωρώ, ενδίδω. (*v.t.*) (*hand in*) παραδίδω.
give off *v.t.* αναδίδω.
give on *v.* (*also* ~ to) (*face*) βλέπω σε (*or* προς).
give out *v.t.* (*distribute*) μοιράζω, διανέμω, (*announce*) αναγγέλλω. (*v.i.*) (*end*) τελειώνω, εξαντλούμαι.
give over *v.* (*stop*) παύω. the flower garden has been given over to vegetables ο ανθόκηπος μετετράπη εις λαχανόκηπον, τον ανθόκηπον τον κάναμε λαχανόκηπο.
give up *v.t.* (*abandon*) παρατώ, εγκαταλείπω, (*surrender*) παραδίδω, (*a habit*) κόβω, (*write off*) ξεγράφω. (*the doctors*) have given him up (*as incurable*) τον έχουν καταδικάσει, τον έχουν αποφασισμένο. (*devote*) αφιερώνω. give oneself up (*surrender*) παραδίδομαι, (*devote oneself*) αφοσιώνομαι, δίνομαι, (*indulge immoderately in*) το ρίχνω σε. (*v.i.*) I won't ~! δεν το βάζω κάτω.
given *a.* (*specified*) δεδομένος. ~ that...

δεδομένου ότι, ~ the opportunity δοθείσης της ευκαιρίας, αν δοθεί η ευκαιρία. (*inclined*) be ~ το έχω τάση προς, ρέπω προς.
giver *s.* ο δίδων.
glacial *a.* παγετώδης.
glacier *s.* παγετώνας *m.*
glad *a.* be *or* feel ~ χαίρομαι, είμαι ευχαριστημένος. (*willing*) πρόθυμος. (*of news, sound*) χαρμόσυνος. I should be ~ to know θα επιθυμούσα (*or* θα ήθελα) να μάθω. I should be ~ to help you πολύ ευχαρίστως (*or* με χαρά μου) θα σας βοηθούσα. ~den *v.t.* χαροποιώ. my heart is ~dened ανοίγει η καρδιά μου. ~ly *adv.* ευχαρίστως, με χαρά. ~ness *s.* χαρά *f.*
glade *s.* ξέφωτο *n.*
gladiator *s.* μονομάχος *m.*
gladiolus *s.* γλαδιόλα *f.*
glam|our *s.* (*allure*) σαγήνη *f.* ~orous *a.* σαγηνευτικός, με ψεύτικη αίγλη. ~orize *v.* δίνω ψεύτικη αίγλη σε.
glanc|e *s.* βλέμμα *n.*, ματιά *f.* (*of light*) λάμψη *f.* at a ~e με μια ματιά. (*v.*) (*gleam*) λάμπω. ~e at ρίχνω ένα βλέμμα (*or* μια ματιά) σε. ~e off (*be deflected*) γλιστρώ και ξεφεύγω (από). ~ing *a.* πλάγιος.
gland *s.* αδένας *m.* ~ular *a.* αδενικός.
glar|e *v.* (*stare, also* ~e at) αγριοκοιτάζω. (*be dazzling*) λάμπω εκτυφλωτικά. (*s.*) (*stare*) άγριο βλέμμα, (*light*) εκτυφλωτικό φως ~ing *a.* (*dazzling*) εκτυφλωτικός, (*gaudy*) χτυπητός. (*gross*) καταφανής.
glass *s.* (*substance*) γυαλί *n.*, ύαλος *f.* (~ware) γυαλικά *n.pl.* (*vessel*) ποτήρι *n.* (*mirror*) καθρέφτης *m.* (*barometer*) βαρόμετρο *n.* (*lens*) φακός *m.* ~es (*spectacles*) γυαλιά *n.pl.* (*binoculars*) κιάλια *n.pl.*
glass *a.* γυαλένιος, γυάλινος, υάλινος. ~ case βιτρίνα *f.* ~ jar γυάλα *f.* ~house *s.* σέρα *f.* ~paper *s.* γυαλόχαρτο *n.*
glassy *a.* (*vitreous, lustreless*) υαλώδης. (*shiny*) στιλπνός, (*slippery*) γλιστερός. a ~ sea θάλασσα γυαλί (*or* λάδι).
glaucoma *s.* γλαύκωμα *n.*
glaucous *a.* γλαυκοπράσινος. (*bloomy*) χνοώδης.
glaz|e *v.i.* (*grow dull*) θαμπώνω. (*v.t.*) (*fit panes to*) βάζω τζάμια σε, (*coat surface of*) υαλοβερνικώνω. (*fig., conceal*) ~e over συγκαλύπτω. (*s.*) (*of pottery*) υαλοβερνίκωμα *n.* ~ed *a.* (*with glass*) τζαμωτός, υαλόφρακτος, (*of picture*) με τζάμι. (*pottery*) με υαλώδες επίχρισμα. ~ier *s.* τζαμτζής, τζαμάς *m.*
gleam *v.* λάμπω, γυαλίζω (*of subdued light*) φέγγω, (*with cleanness*) αστράφτω. (*s.*)

λάμψη f. (fig., of hope, etc.) αχτίδα f. ~ing
a. γυαλιστερός, αστραφτερός.
glean v. σταχυολογώ, μαζεύω. ~ings s.
σταχυολογήματα n.pl.
glee s. αγαλλίαση f., χαρά f. (mus.) καντάδα
f. ~ful a. γεμάτος χαρά. ~fully adv. με
χαρά.
glen s. λαγκάδι n.
glib a. (facile) ευχερής, (plausible) ανειλι-
κρινώς γλυκομίλητος. ~ly adv. ευχερώς.
glide v.i. (also ~ along, by) γλιστρώ, περνώ
αθόρυβα. (of time) κυλώ. ~r s. (aero.)
ανεμοπλάνο n.
glimmer v. τρεμοσβήνω. (s.) μαρμαρυγή f.
(fig., of hope) αχτίδα f. not a ~ of interest
ούτε το παραμικρό ενδιαφέρον.
glimpse s. I caught a ~ of him τον πήρε το
μάτι μου. let me have a ~ of your paper για
να ρίξω μια ματιά στην εφημερίδα σου.
(v.) see above.
glint v. λάμπω, αστράφτω. (s.) ανταύγεια f.,
λάμψη f.
glisten v. γυαλίζω. ~ing a. γυαλιστερός.
glitter v. αστράφτω, σπιθοβολώ. (s.) λάμψη
f. (fig.) αίγλη f. ~ing a. αστραφτερός,
(fig.) λαμπρός.
gloaming s. σούρουπο n., λυκόφως n.
gloat v. ~ over (with avarice) χαίρομαι
άπληστα με, (with malice) χαίρομαι από
κακεντρέχεια με. ~ing a. θριαμβευτικός.
global a. παγκόσμιος.
globe s. (sphere) σφαίρα f. terrestrial ~
υδρόγειος σφαίρα. (light) γλόμπος m.
~-trotter s. κοσμογυρισμένος a.
globul|e s. σταγονίδιο n. blood ~e αιμο-
σφαίριο n. ~ar a. σφαιρικός.
gloom s. σκοτάδι n. (fig.) κατάθλιψη f. ~y
a. (dark) σκοτεινός, (depressing) βαρύς,
καταθλιπτικός. ~ily adv. μελαγχολικά.
glorif|y v. εξυμνώ, εκθειάζω, ~ication s.
εξύμνηση f.
glor|y s. δόξα f. (v.) ~y in υπερηφανεύομαι
για. ~ious a. ένδοξος, λαμπρός. (wonderful) υπέροχος, θαυμάσιος.
gloss s. (shine) γυαλάδα f. (fig., of respectability, etc.) επίφαση f. (comment) σχόλιο
n. (v.) (comment on) σχολιάζω. ~ over
συγκαλύπτω. ~y a. στιλπνός, (paper)
γκλασέ.
glossary s. γλωσσάριο n.
glove s. γάντι n. hand-in-~ κώλος και βρακί.
glow v. λάμπω. (s.) λάμψη f. (reflected light)
ανταύγεια f. (bodily) ξάναμμα n. ~ing a.
πυρωμένος, κόκκινος, (rosy) ρόδινος.
(enthusiastic) θερμός, ενθουσιώδης, in
~ing colours με ενθουσιασμό. ~-worm s.
πυγολαμπίδα f.
glower v. ~ at αγριοκοιτάζω. ~ingly adv.

άγρια.
glucose s. γλυκόζη f.
glue v. κόλλα f. (v.) (lit. & fig.) κολλώ.
~ together συγκολλώ.
glum a. κατσουφιασμένος, κακόκεφος. ~ly
adv. κακόκεφα.
glut v.t. (market) πλημμυρίζω, (appetite)
χορταίνω. ~ oneself, be ~ted (with) χορταίνω (s.) υπεραφθονία f. ~ted a.
χορτάτος.
glutinous a. γλοιώδης.
glutton s., ~ous a. λαίμαργος, (πολυ)φαγάς
m. (fig.) a ~ for work αχόρταγος στη
δουλειά. (fam.) a ~ for punishment καρ-
παζοεισπράκτορας m. ~y s. λαιμαργία f.
glycerine s. γλυκερίνη f.
GMT ώρα Γκρήνουιτς.
gnarled a. ροζιάρικος.
gnash v. τρίζω.
gnat s. σκνίπα f.
gnaw v. ροκανίζω, (fig.) βασανίζω. ~ing of
hunger νυγμοί της πείνας, λιγούρα f.
gnome s. νάνος m.
go v.i. 1. (move, proceed) πηγαίνω, πάω,
(depart) φεύγω, (pass, go by) περνώ,
(find room, fit) χωρώ. ~ for a walk πάω
περίπατο, ~ for a bathe πάω για μπάνιο, ~
shopping πάω για ψώνια. ~ to bed πάω για
ύπνο (or να κοιμηθώ). ~ and see who it is
πήγαινε να δεις ποιος είναι. he went and
got married πήγε και παντρεύτηκε. it
won't ~ in my bag δεν χωράει μεσ' στην
τσάντα μου. keep the conversation ~ing
κρατώ την κουβέντα. ~ one's own way
κάνω του κεφαλιού μου. as things ~ έτσι
που ειναι τα πράγματα. get ~ing ξεκινώ,
αρχίζω. who ~es there? τις ει; 2. (extend,
reach) εκτείνομαι εις, φτάνω σε. ~ far
(succeed) προκόβω, πάω μπρος, a pound
used to ~ a long way αγόραζες πολύ με μία
λίρα, he ~es too far το παρακάνει, that's
~ing too far αυτό πάει πολύ. he has two
months to ~ έχει ακόμα δύο μήνες
(μπροστά του). ~ one better (than him) τον
ξεπερνώ. 3. (become, be) γίνομαι. ~ wrong
or bad χαλώ, ~ red κοκκινίζω, ~ to sleep
με παίρνει ο ύπνος, ~ on strike απεργώ, ~
mad τρελλαίνομαι, ~ scot free βγαίνω
λάδι, ~ hungry πεινώ, they are ~ing cheap
πουλιούνται φτηνά, they are ~ing begging
θα πάνε χαμένα (or άδικα). 4. (function,
work) δουλεύω, λειτουργώ. ~ easy! σιγά,
αγάλι' αγάλια! ~ easy on the wine με
οικονομία το κρασί! ~ it το ρίχνω έξω, ~
it! εμπρός, δωσ' του! ~ it alone τα βγάζω
πέρα μόνος μου. it ~es without saying
κοντά στο νου (κι η γνώση). what he says
~es ό,τι πει είναι νόμος. 5. (give sound,

sign) the phone went χτύπησε το τηλέφωνο, it went bang έκανε μπουμ, he went like that έκανε έτσι. the story ~es that... λέγεται (*or* θρυλείται) ότι. it ~es as follows έχει ως εξής. 6. (*fail, collapse*) the light bulb has gone κάηκε η λάμπα. my wallet's gone χάθηκε (*or* πάει *or* έκανε φτερά) το πορτοφόλι μου. the wine's all gone τέλειωσε (*or* πάει) το κρασί. it's gone for a burton πάει περίπατο. (*get broken*) σπάζω, (*be abolished*) καταργούμαι, φεύγω. 7. (*release*) let ~ (of) αφήνω. let oneself ~ (*be merry*) το ρίχνω έξω, (*burst out*) ξεσπώ, (*lose morale*) εγκαταλείπω τον εαυτό μου. *see also* going, gone.

go *s.* (*fam.*) (*animation*) ζωντάνια *f.* on the ~ στο πόδι, all the ~ της μόδας, all at one ~ μονομιάς, at the first ~ με την πρώτη. it's no ~ δεν γίνεται, it's a ~! σύμφωνοι! have a ~ κάνω μια προσπάθεια. from the word ~ από την αρχή.

go about *v.i.* (*circulate*) κυκλοφορώ. (*v.t.*) (*deal with, do*) καταπιάνομαι με, κάνω.

go after *v.t.* (*chase*) κυνηγώ, (*a job*) επιδιώκω.

go against *v.t.* (*oppose*) αντιτάσσομαι εις. (*affect adversely*) ζημιώνω, έρχομαι ανάποδα σε. (*be contrary to*) προσκρούω εις.

go ahead *v.i.* (*advance*) προχωρώ, (*do well*) προοδεύω.

go-ahead *a.* προοδευτικός. (*s.*) he got the ~ πήρε άδεια να προχωρήσει.

go along *v.i.* (*proceed*) προχωρώ. ~ with (*accompany*) συνοδεύω, (*agree*) συμφωνώ με, I ~ with him πάω με τα νερά του. (*v.t.*) (*the street, etc.*) περνώ (από).

go at *v.* (*attack*) ρίχνομαι σε.

go back *v.* (*return*) γυρίζω πίσω. ~ on (*break*) παραβαίνω, (*desert*) εγκαταλείπω. ~ to (*be descended from*) κατάγομαι, (*date from*) χρονολογούμαι από.

go before *v.t.* (*precede*) προηγούμαι (*with gen.*), (*appear before*) παρουσιάζομαι ενώπιον (*with gen.*).

go beyond *v.t.* (*orders*) υπερβαίνω, (*bounds*) ξεπερνώ.

go by *v.* (*pass*) περνώ. ~ the church περνώ μπρος από την εκκλησία. (*be guided by*) στηρίζομαι από, (*judge by*) κρίνω από. he goes by the name of Smith φέρει το όνομα Σμιθ.

go-by *s.* give (*person*) the ~ (*ignore*) αγνοώ, (*leave out*) αφήνω έξω.

go down *v.i.* κατεβαίνω, (*of ship*) βυθίζομαι, (*of sun*) δύω, (*of wind, prices, curtain*) πέφτω, (*of swelling, fever*) υποχωρώ. (*be defeated*) ηττώμαι, (*be accepted*) γίνομαι δεκτός, ~ well αρέσω, (*be noted*) σημειώνομαι. ~ to (*reach*) φθάνω μέχρι. ~ with (*illness*) αρρωσταίνω με. ~ in history περνώ στην ιστορία, μένω αξέχαστος.

go for *v.t.* (*to fetch*) πάω να φέρω, (*attack*) ρίχνομαι σε. it went for nothing (*was in vain*) πήγε άδικα. that goes for me too κι εγώ (*or* κι εμένα).

go in *v.t. & i.* μπαίνω (*with* σε), (*go indoors*) πάω μέσα. the key won't ~ το κλειδί δεν μπαίνει. the joint won't ~ the oven το κρέας δεν χωράει στο φούρνο. (*of sun, moon*) κρύβομαι. ~ for (*hobby*) ασχολούμαι με, (*exam*) δίνω, (*contest*) συμμετέχω εις. ~ for politics πολιτεύομαι, αναμιγνύομαι εις την πολιτικήν. ~ for medicine σπουδάζω γιατρός, ~ for teaching γίνομαι εκπαιδευτικός, ~ for the civil service πάω για δημόσιος υπάλληλος. I don't ~ for that δεν είναι του γούστου μου.

go into *v.t.* (*enter*) ~ the room μπαίνω (μέσα) στο δωμάτιο. ~ the army γίνομαι στρατιωτικός, ~ business γίνομαι επιχειρηματίας. ~ mourning πενθώ, ~ fits of laughter ξεσπώ σε γέλια. (*investigate*) εξετάζω.

go off *v.i.* (*go away*) φεύγω, (*start*) ξεκινώ, (*abscond*) το σκάω. (*deteriorate*) χαλώ, (*in performance*) δεν είμαι σε φόρμα. (*of gun*) βαρώ, (*of pistol*) παίρνω (φωτιά). ~ to sleep αποκοιμούμαι. it went off well πήγε καλά. (*v.t.*) I've gone off it δεν μου αρέσει πια.

go-off *s.* at the first ~ με την πρώτη.

go on *v.i.* (*advance*) προχωρώ, (*continue, also* ~ with) συνεχίζω, εξακολουθώ. (*behave*) φέρομαι, (*argue*) συζητώ, (*complain, nag*) παραπονιέμαι, γκρινιάζω. (*fit*) μπαίνω, χωράω. (*happen*) συμβαίνω, γίνομαι, it's been going on for years αυτό κρατάει χρόνια τώρα. he is going on for 50 κοντεύει πενήντα. ~ to ~ with για την ώρα. ~ (with you)! ασ' τα αυτά! (*v.t.*) (*rely on*) στηρίζομαι σε.

go out *v.i.* βγαίνω, (*to functions*) βγαίνω έξω. (*be extinguished*) σβήνω. (*end*) τελειώνω, (*of fashion*) περνώ, πέφτω. the colour went out of her cheeks έχασε το χρώμα της. a call went out (*appeal*) έγινε έκκληση.

go over *v.t.* (*cross*) περνώ, (*do again*) ξανακάνω, (*scrutinize*) εξετάζω, ελέγχω, (*revise*) διορθώνω. (*see premises*) πάω να επισκεφθώ. (*v.i.*) (*make effect*) κάνω καλή εντύπωση. ~ to (*join*) προσχωρώ εις. ~ from gas to electricity αλλάζω από γκάζι σε ηλεκτρισμό.

go round *v.t.* περνώ γύρω από, (*get past*) παρακάμπτω. (*visit gallery, etc.*) επισκέ-

πτομαι, ~ the world κάνω το γύρο του
κόσμου. (v.i.) (revolve) γυρίζω, περιστρέφομαι, (circulate) κυκλοφορώ. (be
enough) φτάνω για όλους.
go through v.t. (pass through) περνώ μέσα
από, (traverse) διασχίζω, (pierce) διατρυπώ. (scrutinize) εξετάζω, (search)
ερευνώ, ψάχνω. (study) μελετώ, (discuss)
συζητώ. (perform) εκτελώ. (endure) τραβώ. (use up) ξοδεύω. (v.i.) (be concluded)
γίνομαι, (be approved) περνώ. ~ with it
προχωρώ ως το τέλος.
go under v.i. (sink) βυθίζομαι, (succumb)
υποκύπτω, (fail) αποτυγχάνω.
go up v.t. & i. ανεβαίνω, (the stairs) τη
σκάλα, (the mountain) στο βουνό. (v.i.)
(explode) ανατινάζομαι. ~ in flames
παίρνω φωτιά. ~ to (approach) πλησιάζω.
I'll ~ to ten pounds for it θα ανέβω μέχρι
δέκα λίρες.
goad s. βουκέντρα f. (fig.) κίνητρο n. (v.)
(lit. & fig.) κεντρίζω. (fig.) παρακινώ,
ωθώ.
goal s. σκοπός m., στόχος m. (football)
τέρμα n., γκολ n. ~-keeper s. τερματοφύλακας m.
go-as-you-please a. με την άνεσή σου.
goat s. αιξ f., κατσίκα f., γίδα f. (he-~)
τράγος m. ~herd s. γιδοβοσκός m. ~'s a.
(milk, etc.) κατσικίσιος.
gobbet s. κομμάτι n.
gobble v.t. καταβροχθίζω. (v.i., of turkey)
γλουγλουκίζω. ~dygook s. (fam.) ακατα-
λαβίστικη φρασεολογία.
go-between s. μεσολαβητής m., ο μεσάζων.
goblet s. κύπελλο n.
goblin s. καλικάντζαρος m.
go-cart s. καροτσάκι n.
god s. θεός m., θεότης f. G~ knows ένας θεός
ξέρει, Κύριος οίδε. G~ willing Θεού
θέλοντος, G~ forbid Θεός φυλάξοι, by G~
για το Θεό, in G~'s name για το όνομα του
Θεού, for G~'s sake προς Θεού, thank G~
δόξα τω Θεώ, δόξα σοι ο Θεός. ~fearing
a. θεοφοβούμενος. ~less a. άθεος. ~like
a. θεϊκός. ~ly a. θεοσεβής.
god|child s. βαφτιστήρι n. ~daughter s.
βαφτισιμιά f., αναδεξιμιά f. ~father s.
νονός m. ~mother s. νονά f. fairy ~mother
καλή νεράιδα. ~parent s. ανάδοχος m.f.
~son s. βαφτισιμιός m., αναδεξιμιός m.
goddess s. θεά f.
godforsaken a. άθλιος. what a ~ hole! τι
βρωμότοπος!
godhead s. θεότητα m.
godsend s. it was a ~ to me μου ήρθε από
τον ουρανό.
god-speed s. wish ~ to κατευοδώνω.

go-getter s. καταφερτζής m.
goggle v.i. γουρλώνω τα μάτια. ~-eyed a.
γουρλομάτης.
goggles s. χοντρά γυαλιά.
going s. πηγαιμός m. coming(s) and ~(s)
πηγαινέλα n. hard ~ δύσκολος δρόμος.
while the ~ is good πριν παρουσιαστούν
εμπόδια. (v.) it is ~ to rain θα βρέξει, he
was ~ to be a doctor επρόκειτο να γίνει (or
ήταν να γίνει or θα γινότανε) γιατρός. be
still ~ strong κρατιέμαι καλά. see also go.
going-over s. (scrutiny) εξέταση f., έλεγχος
m. (revision) διόρθωση f. (tidying) τακτοποίηση f. (ransacking) λεηλάτηση f.
goings-on s. (pej.) έκτροπα n.pl.
gold s. χρυσός m., χρυσάφι n., μάλαμα n.
(colour) χρυσό n. heart of ~ χρυσή (or
μαλαματένια) καρδιά. (a.) (in substance)
χρυσός, (in colour) χρυσαφένιος, χρυ-
σαφής.
gold-digger s. χρυσοθήρας m. (fig., of woman) κόφτρα f.
golden a. see gold. (age, calf, Horde, Horn,
number, rule) χρυσούς.
goldfinch s. καρδερίνα f.
goldfish s. χρυσόψαρο n.
gold-mine s. χρυσωρυχείο n.
goldsmith s. χρυσοχόος m.
golf s. γκολφ n.
golliwog s. κούκλα αραπίνα.
gondol|a s. γόνδολα f. ~ier s. γονδολιέρης m.
gone a. far ~ προχωρημένος. he's ~ on Mary
πονεί το δόντι του για τη Μαρία. I'm
rather ~ on this picture με ελκύει αυτός ο
πίνακας. it's ~ twelve o'clock είναι
περασμένες δώδεκα. see also go.
goner s. (fam.) he's a ~ πάει αυτός.
gong s. γκογκ n.
gonorrhoea s. βλεννόρροια f.
good a. καλός, (person only) αγαθός, (well-
behaved) φρόνιμος. it is ~ for me μου κάνει καλό. it is ~ (valid) for a month ισχύει
για ένα μήνα. a ~ way (distance) αρκετά
μακριά, a ~ time (duration) κάμποση ώρα,
in ~ time εγκαίρως. a ~ deal of αρκετός, a
~ many (or few) αρκετοί. as ~ as new σχεδόν καινούργιος. is it ~ to eat? τρώγεται;
it's a ~ thing (or job) we didn't go καλά (or
ευτυχώς) που δεν πήγαμε. so ~ of you (to
write, etc.) καλωσύνη σας, πολύ ευγενικό
εκ μέρους σας. have a ~ time περνώ καλά,
(as wish) καλή διασκέδαση! make ~ (v.t.) (
promises) εκπληρώνω, (damage) αποζημιώνω για, (fill holes in) βουλλώνω,
σπατουλάρω. make ~ one's losses ανακτώ
τα χαμένα. make ~ one's escape κατορθώνω να δραπετεύσω. make ~ (v.i.)
προκόβω, ευδοκιμώ.

good s. καλό n. (*advantage, profit*) όφελος n. for ~ για πάντα, for the ~ of για το καλό (*or* προς όφελος) (*wth gen.*). it does me ~ μου κάνει καλό, με ωφελεί. he is up to no ~ κάτι κακό σκαρώνει. it's no ~ (*worthless*) δεν αξίζει τίποτα, (*profitless*) δεν ωφελεί. what's the ~ of it? σε τι θα ωφελήσει αυτό; *or* τι όφελος θα βγει απ' αυτό;

goods s. (*possessions*) αγαθά n.pl. (*merchandise*) εμπορεύματα n.pl. leather ~ δερμάτινα είδη, worldly ~ περιουσία f.

goodbye int. αντίο, γεια σας. say ~ to αποχαιρετώ.

good-for-nothing a. αχαΐρευτος.

good-humoured a. πρόσχαρος, ανοιχτόκαρδος.

goodish a. καλούτσικος. a ~ way from here αρκετά μακριά από 'δω.

good-looking a. ωραίος.

goodly a. (*big*) μεγάλος.

good-natured a. καλοκάγαθος, μαλακός.

goodness s. καλοσύνη f. (*nutritional value*) θρεπτική ουσία. my ~! for ~ sake! για όνομα του Θεού! thank ~! δόξα να 'χει ο Κύριος!

good-tempered a. πρόσχαρος, (*animal*) ήμερος.

goodwill s. φιλική διάθεση. (*trade*) αέρας m., πελατεία f.

goody s. (*edible*) λιχουδιά f.

goody-goody a. με αυτάρεσκη σεμνοπρέπεια.

gooey a. γλοιώδης.

goose s. χήνα f. (*fam., fig.*) κουτορνίθι n. ~ that lays the golden eggs όρνις χρυσοτόκος. (*fam.*) I'll cook his ~ for him θα τον θάψω. **~-flesh** s. ανατριχίλα f. **~-step** s. βάδισμα της χήνας.

Gordian a. ~ knot γόρδιος δεσμός.

gore a. αίμα n.

gore v. διαπερνώ με κέρατα.

gorge s. (*ravine*) φαράγγι n. (*gullet*) my ~ rises αηδιάζω.

gorge v.i. τρώω το καταπέτασμα. (*v.t.*) καταβροχθίζω. **~d** a. χορτάτος.

gorgeous a. (*dazzling*) εκθαμβωτικός. (*excellent*) υπέροχος, θαύμα n. (*fam.*) we had ~ weather είχαμε καιρό μαγεία, we had a ~ time περάσαμε υπέροχα.

Gorgon s. (*fig.*) τερατώδης γυναίκα.

gorilla s. γορίλλας m.

gormandize v. see gorge.

gormless a. κουτός.

gorse s. ασπάλαθος m.

gory a. αιμόφυρτος.

gosh int. Θεέ μου!

gosling s. χηνάκι n.

go-slow s. λευκή απεργία.

gospel s. ευαγγέλιον n.

gossamer s. & a. (like) ~ αραχνοΰφαντος.

gossip s. (*talk*) κουβεντολόι n. (*scandal*) κουτσομπολιό n. (*person*) κουτσομπόλης m. κουτσομπόλα f. (*v.*)(*also* ~ about) κουτσομπολεύω. (*chat*) κουβεντιάζω.

got see get.

Gothic a. γοτθικός.

gouge s. σκαρπέλο n. (*v.*) ~ out βγάζω.

gourd s. κολοκύθα f. (*as receptacle*) φλασκί n.

gourmet s. καλοφαγάς m.

gout s. ουρική αρθρίτιδα, ποδάγρα f.

govern v. (*rule*) κυβερνώ, (*a province*) διοικώ, (*master*) κυριαρχώ (*with gen.*). (*determine*) ρυθμίζω, κανονίζω. (*gram.*) συντάσσομαι με. **~ance** s. διακυβέρνηση f. **~ing** a. διευθύνων, ~ing body διοικητικό συμβούλιο.

governess s. γκουβερνάντα f.

government s. κυβέρνησις f. form of ~ πολίτευμα n. G~ House κυβερνείον n. (*a., also* ~al) κυβερνητικός.

governor s. (*of colony*) κυβερνήτης m. (*of bank, military*) διοικητής m. (*of prison*) διευθυντής m. (*of school, hospital*) μέλος της εφορίας. (*mech.*) ρυθμιστής m. (*fam.*) the ~ (*dad*) ο γέρος μου, (*boss*) το αφεντικό. **~-general** s. γενικός διοικητής.

gown s. τουαλέττα f. (*official robe*) τήβεννος f. (*house* ~) ρόμπα f. **~ed** a. ντυμένος.

GP παθολόγος m.

grab v. (*also* ~ at) αρπάζω. ~ hold of αρπάζομαι από. (*snatch for oneself*) βουτώ. (*s.*) make a ~ at αρπάζω. (*mech.*) αρπάγη f. **~ber** s., **~bing** a. αρπακτικός.

grace s. (*charm, elegance, mercy*) χάρις f. ~s (*goddesses*). Χάριτες f.pl., (*social*) χαρίσματα n.pl. (*delay*) προθεσμία f. be in the good ~s of έχω την εύνοια (*with gen.*). with a bad ~ απρόθυμα. he had the ~ to apologize ευδόκησε να ζητήσει συγγνώμη. (*the Virgin*) full of ~ Μεγαλόχαρη. his G~ (*eccl.*) η Αυτού Μακαριότης. (*v.*) (*adorn*) κοσμώ, (*honour*) τιμώ.

graceful a. κομψός, αρμονικός, γεμάτος χάρη. (*esp. of body*) καλλίγραμμος. (*act*) ευγενικός. **~ly** adv. με χάρη, ευγενικά.

graceless a. (*lacking grace*) άχαρος, (*shameless*) αδιάντροπος.

gracious a. (*polite*) ευγενής, (*condescending*) καταδεκτικός. (*Sovereign*) χαριτόβρυτος. (*int.*) good ~! Θεέ μου! μη χειρότερα! **~ly** adv. με καταδεκτικότητα. **~ness** s. καταδεκτικότητα f.

gradation s. βαθμιαία αλλαγή.

grad|e s. (*step, rank*) βαθμός m. (*quality*) ποιότητα f. (*school form*) τάξη f. (*slope*) κλίση f. be on the up ~e ανεβαίνω, πάω πάνω, make the ~e πετυχαίνω. (v.) διαβαθμίζω. **~ing** s. διαβάθμιση f.

gradient s. κλίση f.

gradual a. βαθμιαίος. **~ly** adv. βαθμηδόν, σιγά-σιγά.

graduat|e v.t. (mark in degrees) βαθμολογώ, (apportion on scale) κλιμακώνω. (v.i.) (get degree) αποφοιτώ, παίρνω το δίπλωμά μου. (s.) απόφοιτος m.f. **~ed** a. βαθμολογημένος. κλιμακωτός. **~ion** s. βαθμολόγηση f. κλιμάκωση f. αποφοίτηση f.

graffiti s. συνθήματα στους τοίχους.

graft s. εμβόλιον n. μπόλι n. (med.) μόσχευμα n. (influence) ρουσφετολογία f. (v.) εμβολιάζω, μπολιάζω. (med.) μεταμοσχεύω. (use influence) ρουσφετολογώ. **~ing** s. εμβολιασμός m. μπόλιασμα n. μεταμόσχευση f.

grain s. (wheat, corn) σιτηρά n.pl. (single seed) κόκκος m. σπυρί n. (of salt, sand & fig.) κόκκος m. (weight) σπυρί n. (texture) υφή f. (pattern in wood) νερά n.pl. (cut) with the ~ στα ίσια, against the ~ στο κόντρα. (fig.) it goes against the ~ for me (to do that) είναι ενάντια στις αρχές μου.

grammar s. γραμματική f. **~-school** s. γυμνάσιο n. **~ian** s. γραμματικός m.

grammatical a. γραμματικός. **~ly** adv. γραμματικώς.

gramme s. γραμμάριο n.

gramophone s. γραμμόφωνο n.

granary s. σιτοβολών m.

grand a. (in titles) μέγας. (most important) μεγάλος, (splendid) μεγαλοπρεπής, (impressive) επιβλητικός, (puffed up) επηρμένος. (fam., good) περίφημος. ~ piano πιάνο με ουρά. **~ly** adv. μεγαλοπρεπώς.

grand|child s. εγγόνι n. **~daughter** s. εγγονή f. **~father** s. παππούς m. **~mother** s. μάμμη f., γιαγιά f. **~son** s. εγγονός m.

grandee s. μεγιστάν m.

grandeur s. (of Alps, Rome, etc.) μεγαλείο n. (of occasion) μεγαλοπρέπεια f.

grandiloquent a. (language) στομφώδης, φουσκωμένος.

grandiose a. μεγαλειώδης, (plans) μεγαλεπήβολος.

grandstand s. εξέδρα f.

grange s. (barn) σιτοβολών m. (house) εξοχικό σπίτι με φάρμα.

granite s. γρανίτης m.

granny s. (fam.) γιαγιά f.

grant v.t. (give) δίνω, (a wish) ικανοποιώ, (a right) παραχωρώ, (a diploma) απονέμω, (funds) χορηγώ. (admit) αναγνωρίζω, παραδέχομαι. the request was ~ed η αίτηση έγινε δεκτή. (s.) επιχορήγηση

f., επίδομα n. **~ing** s. χορήγηση f.

granted a. ~ that... δεδομένου ότι. take it for ~ that... το θεωρώ ως βέβαιον ότι.

granul|ar a. κοκκώδης. **~ated** a. ~ated sugar χοντρή ζάχαρη.

grape s. ρώγα f. **~s** σταφύλι n., σταφύλια pl. sour ~s! όμφακες εισίν. **~-hyacinth** s. μούσκαρι n. **~-shot** s. βολιδοθήκη f. **~-vine** s. κλήμα n. (fig.) I heard it on the ~-vine η αρβύλα το λέει.

grapefruit s. γκρέιπ-φρουτ n.

graph s. διάγραμμα n. **~ic** a. γραφικός, παραστατικός. **~ically** adv. γραφικά.

grapnel s. αρπάγη f.

grappl|e v.t. γραπώνω. (v.i.) ~e with (person) αρπάζω, γραπώνω, (problem) καταπιάνομαι με. **~ing-iron** s. αρπάγη f.

grasp v.t. πιάνω, αρπάζω, (take firm hold of) πιάνομαι από. ~ the opportunity δράττομαι της ευκαιρίας. (understand) συλλαμβάνω, αντιλαμβάνομαι. (s.) (action) πιάσιμο n. he has a powerful ~ (with hands) έχει δυνατά χέρια. (fig., grip, power) χέρι n. (understanding) αντίληψη f. it is beyond my ~ δεν μπορώ να το καταλάβω. **~ing** a. αρπαχτικός.

grass s. χορτάρι n., γρασίδι n., χλόη f. overgrown with ~ χορταριασμένος. she's a ~ widow ο σύζυγός της απουσιάζει προσωρινός. (fig.) ~ roots λαϊκόν φρόνημα. put out to ~ βόσκω, (fig.) θέτω εις σύνταξιν. **~hopper** s. ακρίδα f. **~y** a. σκεπασμένος με γρασίδι.

grate s. εσχάρα f., τζάκι n.

grate v.t. (rub) τρίβω, ξύνω, (one's teeth) τρίζω. (v.i.) (creak, squeal) τρίζω. ~ on (fig.) ερεθίζω, it ~s on my nerves μου δίνει στα νεύρα. **~r** s. τρίφτης m.

grateful a. ευγνώμων, (pleasant) ευχάριστος. **~ly** adv. με ευγνωμοσύνη, ευχάριστα.

gratif|y v. ικανοποιώ. **~ication** s. ικανοποίηση f. **~ying** a. ικανοποιητικός.

grating s. (at window) καφασωτό n. (in pavement) εσχάρα f.

gratis adv. δωρεάν, (fam.) τζάμπα.

gratitude s. ευγνωμοσύνη f.

gratuitous a. (free) δωρεάν. (uncalled-for) αδικαιολόγητος. **~ly** adv. άνευ λόγου.

gratuity s. (tip) φιλοδώρημα n. (bounty) επίδομα n.

grave v. χαράσσω. ~n image είδωλο n.

grave s. τάφος m. **~stone** s. ταφόπετρα f.

grave a. & s. (gram.) ~ accent βαρεία f.

grave a. σοβαρός. **~ly** adv. σοβαρά.

gravel s. χαλίκι n. (v.t.) στρώνω με χαλίκι. (fam., nonplus) φλομώνω. **~led** a. (path) χαλικοστρωμένος.

gravid *a.* έγκυος.
gravitat|e *v.* έλκομαι. ~ion *s.* έλξη *f.*
gravity *s.* (*weight*) βαρύτης *f.* force of ~ έλξις της βαρύτητος. (*chem.*) specific ~ ειδικόν βάρος. (*seriousness*) σοβαρότης *f.*
gravy *s.* σάλτσα *f.*
graz|e *v.t. & i.* (*pasture*) βόσκω. (*v.t.*) (*bark*) γδέρνω, (*touch in passing*) περνώ ξυστά από. ~ing *s.* ~ing (land) βοσκοτόπι *n.*
grease *s.* λίπος *n.* (*lubricant*) γράσο *n.* (*dirt*) λίγδα *f.* (*v.*) λιπαίνω, γρασάρω. (*fig.*) ~ the palm of λαδώνω. ~proof *a.* ~proof paper λαδόχαρτο *n.*
greas|y *a.* λιπαρός, (*slippery*) γλιστερός, (*dirty*) λιγδιασμένος, (*unctuous*) γλοιώδης. ~iness *s.* λιπαρότητα *f.*
great *a.* μεγάλος. a ~ many πάρα πολλοί, a ~ deal (*adv.*) πάρα πολύ. Alexander the G~ ο Μέγας Αλέξανδρος. he is a ~ reader διαβάζει πολύ. he's ~ at mending things είναι μοναδικός στο να διορθώνει το καθετί. he's a ~ one for dropping bricks δεν τον φτάνει κανείς στις γκάφες. the ~ thing is to act in time το σπουδαιότερο πράγμα είναι να ενεργήσεις εγκαίρως. (*fam., fine*) περίφημος, we had a ~ time περάσαμε περίφημα. (*fam., pej.*) a ~ big lout κοτζάμ γάιδαρος. ~ly *adv.* πολύ.
greatcoat *s.* πανωφόρι *n.* (*mil.*) χλαίνη *f.*
greater *a.* μεγαλύτερος. G~ London η μείζων περιοχή του Λονδίνου.
great|-granddaughter *s.* δισέγγονη *f.* ~-grandfather *s.* πρόπαππος *m.* ~-grandmother *s.* προμάμμη *f.* ~grandson *s.* δισέγγονος *m.*
great-hearted *a.* μεγαλόψυχος.
greatness *s.* (*size*) μέγεθος *n.* (*eminence*) μεγαλείο *n.*
Grecian *a.* ελληνικός.
greed, ~iness *s.* πλεονεξία *f.* (*for food*) λαιμαργία *f.* ~y *a.* πλεονέκτης, λαίμαργος, κοιλιόδουλος *m.* ~ily *adv.* άπληστα, λαίμαργα.
Greek *a.* ελληνικός, (*person*) Έλλην(ας), Ρωμιός *m.*, Ελληνίς, ~ίδα, Ρωμιά *f.* (*s.*) (*language*) ελληνικά *n.pl.* (*fam.*) it's all ~ to me είναι κινέζικα για μένα. ~ key pattern μαίανδρος *m.*, γκρέκα *f.*
green *a.* πράσινος, (*unripe*) άγουρος, (*novice*) άψητος, (*gullible*) χαζός, (*fresh, alive*) θαλερός, ζωντανός. make *or* become ~ πρασινίζω. (*fam.*) get the ~ light παίρνω άδεια να προχωρήσω. (*s.*) (*colour*) πράσινο *n.* (*grassy spot*) πελούζα *f.* ~s (*edible*) πράσινα χορταρικά. ~ery *s.* πρασινάδα *f.*
greengage *s.* πράσινο δαμάσκηνο.
greengrocer *s.* μανάβης *m.* ~'s shop

μανάβικο *n.*, οπωροπωλείον *n.*
greenhorn *s.* πρωτάρης *m.*
greenhouse *s.* θερμοκήπιο *n.*
greenwood *s.* (*fig.*) take to the ~ βγαίνω στο κλαρί.
greet *v.* (*salute*) χαιρετώ, (*receive*) υποδέχομαι. (*fig.*) strange sounds ~ed my ears παράξενοι ήχοι έφτασαν στα αυτιά μου. ~ing *s.* χαιρετισμός *m.*
gregarious *a.* αγελαίος.
Gregorian *a.* γρηγοριανός.
grenad|e *s.* χειροβομβίδα *f.* ~ier *s.* γρεναδιέρος *m.*
grey *a.* γκρίζος, σταχτής, (*hair*) ψαρός. ~ matter φαιά ουσία. become ~ ασπρίζω. (*s. horse*) ψαρής *m.* ~beard *s.* μπάρμπας *m.* ~-haired *a.* ψαρός, ψαρομάλλης. ~hound *s.* λαγωνικό *n.*
grid *s.* εσχάρα *f.* (*of maps*) δικτυωτό *n.* ~iron *s.* σχάρα *f.*
grief *s.* λύπη *f.* come to ~ αποτυγχάνω, την παθαίνω. ~-stricken *a.* συντετριμμένος.
grievance *s.* παράπονο *n.*
grieve *v.i.* με τρώει η λύπη. ~ for (*lament loss of*) θρηνώ. (*v.t.*) λυπώ, θλίβω. his conduct ~s me η συμπεριφορά του με λυπεί (*or* με κάνει να λυπούμαι *or* μου προξενεί λύπη). I am ~d by your news λυπούμαι για τα νέα σου.
grievous *a.* οδυνηρός, σοβαρός, βαρύς. ~ly *adv.* βαριά.
grill *s.* σχάρα *f.* (*v.*) ψήνω στη σχάρα, (*fig.*) ανακρίνω αυστηρά. ~ed *a.* (*food*) στη σχάρα.
grille *s.* γρίλλια *f.*
grim *a.* (*fierce, desolate*) άγριος, (*relentless*) σκληρός, (*awful*) φρικτός. like ~ death απελπισμένα, μέχρις εσχάτων.
grimace *s.* μορφασμός *m.* (*v.*) κάνω γριμάτσες.
grim|e *s.* βρώμα *f.* ~y *a.* βρώμικος. make *or* become ~y βρωμίζω.
grin *s.* χαμόγελο *n.* (*v.*) χαμογελώ. (*fig.*) ~ and bear it το παίρνω φιλοσοφικά (*or* στωικά).
grind *v.t.* (*in mill*) αλέθω, κόβω, (*in mortar*) κοπανίζω. (*knives*) τροχίζω, (*teeth*) τρίζω. (*crush*) συντρίβω, (*thrust*) χώνω. (*fig.*) (*oppress*) καταπιέζω. (*turn*) γυρίζω. (*v.i.*) (*scrape*) τρίζω, (*toil*) μοχθώ. (*s.*) (*drudgery*) αγγαρεία *f.* (*fam.*) the daily ~ το μαγγανοπήγαδο. ~er (*mill*) μύλος *m.* (*knife-~er*) ακονιστήρι *n.* (*man who ~s knives*) τροχιστής *m.* ~ing *s.* άλεσμα *n.* κοπάνισμα *n.* τρόχισμα *n.* τρίξιμο *n.* καταπίεση *f.* (*a., oppressive*) συντριπτικός.
grindstone *s.* ακονόπετρα *f.* (*fig.*) keep one's nose to the ~ δεν σηκώνω κεφάλι

από τη δουλειά. he keeps them with their noses to the ~ δεν τους αφήνει να πάρουν ανάσα.

grip v. (*seize*) πιάνω, (*clasp*) σφίγγω, (*hold on to*) πιάνομαι από. (*compel attention of*) καθηλώνω.

grip s. (*action*) πιάσιμο n., σφίξιμο n. (*clutches*) νύχια n.pl. have a strong ~ έχω δυνατά χέρια. have a firm ~ of the situation είμαι κύριος της καταστάσεως. have a good ~ of (*understand*) κατέχω καλά. be in the ~ of election fever έχω καταληφθεί από προεκλογικό πυρετό. come to ~s with (*fight*) συμπλέκομαι με, get to ~s with (*problem*) καταπιάνομαι με. lose one's ~ (*lit.*) γλιστρώ, (*fig.*) κάμπτομαι. get a ~ on oneself συνέρχομαι.

gripe v.i. γκρινιάζω. (s.) ~s κοιλόπονος m.

gripping a. συναρπαστικός, συγκινητικός.

grisly a. φρικιαστικός.

grist s. άλεσμα n. (*fig.*) it's all ~ to his mill ωφελείται από όλα αυτά.

grist|le s. χόνδρος m. **~ly** a. it is ~ly έχει χόνδρους.

grit s. άμμος m.f., ψιλό χαλίκι. (*in one's eye*) κόκκος σκόνης, σκουπιδάκι n. (*pluck*) ψυχή f. he's got ~ είναι ψυχωμένος άνθρωπος. (v.t.) (*one's teeth*) σφίγγω. **~ty** a. αμμώδης.

groan v.i. βογγώ. (*fig.*) ~ under the yoke στενάζω υπό τον ζυγόν. (s.) (*also* ~ing) βόγγος m., βογγητό n.

grocer s. μπακάλης m. ~'s shop μπακάλικο n., παντοπωλείον n. **~ies** s. είδη μπακαλικής.

grog s. γκρογκ n.

grogg|ly a. ασταθής. **~iness** s. αστάθεια f.

groin s. βουβών m.

groom s. ιπποκόμος m. (*bride~*) γαμπρός m. (v.) (*tend*) περιποιούμαι, (*fig., prepare*) προετοιμάζω. well ~ed περιποιημένος. **~sman** s. παράνυμφος m., κουμπάρος m.

groove s. αυλάκι n. (*fig.*) get into a ~ αποτελματώνομαι. **~d** a. αυλακωτός.

grope v. (*blindly*) ψηλαφώ, (*in pocket, etc.*) ψάχνω. ~ one's way πηγαίνω ψηλαφητά.

gross s. δώδεκα δωδεκάδες.

gross v.t. (*bring in*) αποφέρω.

gross a. (*total*) συνολικός, (*not net*) ακαθάριστος. ~ weight μικτό βάρος. in (the) ~ χονδρικά, συνολικά.

gross a. (*fat, coarse, rude*) χονδρός, (*person only*) άξεστος. (*flagrant*) κατάφανής, (*crass*) πλήρης. (*vegetation*) οργιώδης. **~ness** s. χοντράδα f.

grossly adv. (*exaggerated, ignorant*) τελείως, πέρα για πέρα. (*insulting*) εις

άκρον.

grotesque a. (*ludicrous*) γελοίος, (*strange*) αλλόκοτος. (*of situation*) τραγελαφικός. (*in art*) γκροτέσκ. **~ly** adv. γελοία, αλλόκοτα.

grotto s. σπηλιά f.

grouch v.i. γκρινιάζω.

ground a. αλεσμένος, κομμένος, τριμμένος.

ground v.t. (*a ship*) καθίζω, ρίχνω έξω. (*base*) στηρίζω. (*instruct*) see grounding. (v.i.) (*of ship*) προσαράσσω, καθίζω.

ground s. (*earth*) γη f., έδαφος n. to or on the ~ χάμω, κατά γης, above ~ στην επιφάνεια. piece of ~ κομμάτι γης. (*sports, parade*) γήπεδο n. (*battle*) πεδίο n. (*hunting, etc.*) περιοχή f. ~s (*of house*) πάρκο n., κήποι m.pl. (*background*) φόντο n. common ~ κοινά σημεία. fall to the ~ αποτυγχάνω, get off the ~ πραγματοποιούμαι. gain (*or* lose) ~ κερδίζω (*or* χάνω) έδαφος, give ~ υποχωρώ. shift one's ~ αλλάζω τακτική, stand one's ~ κρατώ γερά. be on firm ~ πατώ γερά. keep one's feet on the ~ είμαι προσγειωμένος. break new ~ ανοίγω νέους δρόμους. go to ~ (*fig.*) κρύβομαι, run to ~ ξετρυπώνω. cover much ~ (*travel*) καλύπτω μεγάλο διάστημα, (*deal with*) διατρέχω ευρύ πεδίον. it suits me down to the ~ μου έρχεται κουτί (*or* περίφημα). ~ floor ισόγειο n. ~ nuts φιστίκια αράπικα.

ground s. (*reason*) λόγος m., αιτία f. ~ for divorce λόγος διαζυγίου, on ~s of illness για λόγους υγείας. what ~s have they for claiming such an amount? πού στηρίζονται και διεκδικούν ένα τέτοιο ποσό; **~less** a. αβάσιμος, αδικαιολόγητος.

grounding s. get a good ~ in παίρνω γερές βάσεις σε.

ground-plan s. κάτοψη f.

ground-rent s. ενοίκιο γης.

grounds s. (*dregs*) κατακάθι n. (*coffee*) ντελβές m.

groundwork s. βάση m.

group s. ομάδα f. (*companies, buildings*) συγκρότημα n. (*people in company*) παρέα f., ομάδα f., όμιλος m., γκρουπ n. (*statuary*) σύμπλεγμα n. in the same age ~ της αυτής ηλικίας. (v.i.) συγκεντρώνω, διατάσσω, (*classify*) ταξινομώ. **~ing** s. συγκέντωση f. διάταξη f.

group-captain s. (*aero.*) σμήναρχος m.

grouse v.i. γκρινιάζω. (s.) παράπονο n.

grouts s. κατακάθια n.pl.

grove s. άλσος n.

grovel v. (*fig.*) έρπω. **~ling** a. χαμερπής.

grow v.i. (*of plants*) φυτρώνω, φύομαι. (*get bigger*) μεγαλώνω, (*increase*) αυξάνω, (*develop*) αναπτύσσομαι, (*spring*) προέρχομαι. ~ into (become) γίνομαι, ~ up μεγαλώνω. ~ out of (habit) ξεπερνώ, he's ~n out of his clothes δεν του χωρούν πια τα ρούχα του. it grew dark σκοτείνιασε. it ~s on you συνηθίζεις και σου αρέσει. ~ accustomed (to) συνηθίζω, ~ old γερνώ. (v.t.) (cultivate) καλλιεργώ. ~ a beard αφήνω γένια. ~er s. καλλιεργητής m. ~ing s. καλλιέργεια f.

growl v. γρυλλίζω.

grown-up a. & s. μεγάλος, (mature) ώριμος.

growth s. (increase) αύξηση f. (development) ανάπτυξη f. (cultivation, yield) παραγωγή f. there is a thick ~ of weeds έχουν φυτρώσει πολλά αγριόχορτα. he has three days' ~ of beard έχει να ξυριστεί τρεις μέρες. (med.) όγκος m.

groyne s. κυματοθραύστης m.

grub s. σκουλήκι n. (fam., food) φαγητό n.

grub v.i. σκαλίζω. ~ up ξεριζώνω.

grubbly a. βρώμικος. ~iness s. βρώμα f.

grudge s. I bear him a ~ του κρατώ κακία. (v.) I ~ having to pay for it μου κακοφαίνεται που έχω να το πληρώσω. I don't ~ him his success δεν ζηλεύω την επιτυχία του.

grudging a. (stingy) φειδωλός. be ~ of one's money είμαι σφιχτοχέρης. ~ly adv. με το ζόρι.

gruel s. κουρκούτι n.

gruelling a. εξαντλητικός. (s.) they gave him a ~ τον ξεθέωσαν, του έβγαλαν την πίστη ανάποδα.

gruesome a. φρικιαστικός, μακάβριος.

gruff a. (manner) απότομος, (voice) τραχύς.

grumblle v. παραπονιέμαι, γκρινιάζω. ~ler s., ~ling a. γκρινιάρης. ~ling s. γκρίνια f.

grumply a. κατσούφης, μουτρωμένος. ~iness s. γκρίνια f.

grundyism s. σεμνοτυφία f.

grunt v. γρυλλίζω. (s.) γρύλισμα n.

gruyère s. γραβιέρα f.

guarantlee v. εγγυώμαι (with ότι or για). ~eed εγγυημένος. (s.) εγγύηση f. ~or s. εγγυητής m.

guard s. (body of men) φρουρά f. (man) φρουρός m. (posture) άμυνα f. keep ~ (over) φρουρώ. stand ~ φυλάω σκοπός. be on one's ~ έχω το νου μου. be caught off one's ~ καταλαμβάνομαι εξ απροόπτου. (protective device) προφυλακτήρ m.

guard v.t. φυλά(γ)ω, (mil.) φρουρώ. (protect) προστατεύω. (v.i.) ~ against φυλάγομαι από, προφυλάσσομαι από. ~ed a. επιφυλακτικός. ~edly adv. επιφυλακτικά.

guardian s. φύλακας m. (protector) προστάτης m. (of minor) κηδεμών m. ~ship s.

προστασία f. κηδεμονία f.

guerrilla s. αντάρτης m. (a.) ανταρτικος. ~ war(fare) ανταρτοπόλεμος m.

guess v. μαντεύω, (calculate) υπολογίζω. (think) I ~ it's going to rain θαρρώ πως θα βρέξει. I ~ so έτσι λέω. (s.) that was a good ~ of yours καλά το μάντεψες. ~ing, ~work s. εικασία f. by ~work κατ' εικασίαν.

guest s. φιλοξενούμενος m., προσκεκλημένος m. (at hotel) πελάτης m. ~house s. ξενώνας m., πανσιόν f.

guffaw (v.) χάχανο n. χαχανίζω.

guichet s. θυρίδα f.

guidance s. καθοδήγηση f.

guide s. οδηγός m.f. (tourists') ξεναγός m.f. (adviser) σύμβουλος m. (example) παράδειγμα n. (girl-~) προσκοπίνα f. (v.) (conduct) οδηγώ, (instruct) καθοδηγώ, (direct) κατευθύνω, (advise) συμβουλεύω. (show round) ξεναγώ. be ~d by ακολουθώ. ~book s. οδηγός m. ~lines s. κατευθυντήριες γραμμές.

guild s. συντεχνία f., σινάφι n., σωματείο n.

guile s. δόλος m. by ~ με δόλο. ~ful a. δόλιος. ~fully adv. δολίως. ~less a. απονήρευτος. ~lessly adv. αφελώς.

guillotine s. λαιμητόμος f. (v.) καρατομώ.

guilt s. ενοχή f. ~less a. αθώος.

guiltly a. ένοχος. be found ~y κηρύσσομαι ένοχος. plead not ~y αρνούμαι την ενοχή μου. ~ily adv. με ύφος ενόχου, ένοχα.

guinea s. γκινέα f.

guinea-pig s. ινδικό χοιρίδιο, (fig.) πειραματόζωο n.

guise s. (form) he appeared in the ~ of a beggar ενεφανίσθη ως ζητιάνος. (fig.) under the ~ of friendship προσποιούμενος τον φίλο.

guitar s. κιθάρα f.

gulf s. (of sea) κόλπος m. (fig., gap) χάσμα n. G~ Stream Κόλπιον Ρεύμα.

gull s. (bird) γλάρος m.

gull v. (fam.) εξαπατώ. ~ible a. μωρόπιστος.

gullet s. οισοφάγος m. (throat) γούλα f.

gully s. ρείθρο n.

gulp v.t. χάφτω, καταπίνω. ~ down or back (suppress) καταπίνω, πνίγω. (v.i.) (from embarrassment) ξεροκαταπίνω. (s.) χαψιά f. at one ~ μονορούφι.

gum s. (of mouth) ούλο n. ~boil s. παρουλίτιδα f.

gum v. κολλώ. ~med up (eyes) τσιμπλιασμένος. (s.) (sticky) γόμμα f. ~ mastic μαστίχα f. (chewing) τσίκλα f., μαστίχα f. (secretion of eyes) (fam.) τσίμπλα f. ~ boots ψηλές μπότες από καουτσούκ. (fam.) up a ~ tree σε δύσκολη

θέση. **~my** *a.* γεμάτος γόμα (*or* ρετσίνι).
gum *s.* (*fam., int.*) by ~ για το Θεό!
gumption *s.* πρακτικό μυαλό.
gun *s.* (*heavy*) πυροβόλο *n.* (*firearm*) όπλο *n.* spray-~ πιστόλι *n.* (*fig.*) stick to one's ~s δεν υποχωρώ. (*see* rifle, revolver, cannon, *etc.*) (*v.*) (*fam.*) be ~ning for κυνηγώ, ~ down φονεύω με όπλο. **~boat** *s.* κανονιοφόρος *f.* **~fire** *s.* πυροβολισμοί *m.pl.* **~man** *s.* γκάγκστερ *m.* **~point** *s.* at ~point υπό την απειλήν όπλου. **~powder** *s.* μπαρούτι *n.,* πυρίτιδα *f.* **~smith** *s.* οπλοποιός *m.*
gunner *s.* πυροβολητής *m.* **~y** *s.* πυροβολική *f.*
gun-running *s.* λαθρεμπόριο όπλων.
gunshot *s.* τουφεκιά *f.* out of ~ εκτός βολής.
gunwale *s.* κουπαστή *f.*
gurgle *v.* γουργουρίζω, κάνω γλου-γλου. (*s.*) γουργουρητό *n.*
gush *v.* ξεχύνομαι, πηδώ. (*fig.*) μιλώ με υπερβολική διαχυτικότητα. (*s.*) (*flow*) ορμητική ροή. **~ing** *a.* (*fig.*) υπερβολικά διαχυτικός.
gusset *s.* τσόντα *f.*
gust *s.* (*wind*) σπιλιάδα *f.* (*rain*) μπόρα *f.* (*fig., outburst*) ξέσπασμα *n.* **~y** *a.* με δυνατό αέρα.
gusto *s.* όρεξη *f.,* κέφι *n.*
gut *v.* (*fish, etc.*) καθαρίζω. be ~ted by fire κατακαίομαι.
gut *s.* έντερο *n.* (*narow passage*) στενό *n.* ~s (*substance*) ουσία *f.* (*fam., bowels*) έντερα *n.pl.* (*spirit*) θάρρος *n.* he has ~s το λέει η καρδιά του. have no ~s είμαι λαπάς. **~less** *a.* ψόφιος, λαπάς *m.*
gutter *v.* (*of candle*) στάζω.
gutter *s.* (*of roof*) υδρορρόη *f.* (*of street*) ρείθρο *n.* (*fig.*) βούρκος *m.* ~ press κίτρινος τύπος. **~snipe** *s.* αλήτης *m.,* γουρούνι *n.*
guttural *a.* τραχύς. (*gram.*) λαρυγγικός.
guvnor *s.* (*fam.*) *see* governor.
guy *s.* (*rope*) σχοινίον στηρίξεως. (*grotesque figure*) καραγκιόζης *m.* (*chap*) τύπος *m.* (*v.*) κοροϊδεύω.
guzzle *v.* (*food*) χάφτω, (*drink*) ρουφώ.
gym *s.* (*place*) γυμναστήριο *n.* (*practice*) γυμναστική *f.* **~nasium** *s.* γυμναστήριο *n.*
gymnast *s.* γυμναστής *m.* **~ic** *a.* γυμναστικός, **~ics** *s.* γυμναστική *f.*
gynaecology *s.* γυναικολογία *f.* **~ical** *a.* γυναικολογικός. **~ist** *s.* γυναικολόγος *m.*
gyp *v.* (*fam.*) κλέβω, γδύνω.
grizzl|ed, ~ly *a.* ψαρός, γκριζομάλλης. ~ly bear φαιά άρκτος.
gypsum *s.* γύψος *m.*
gypsy *s. see* gipsy.

gyrat|e *v.* περιστρέφομαι. **~ion** *s.* περιστροφή *f.*
gyroscop|e *s.* γυροσκόπιο *n.* **~ic** *a.* γυροσκοπικός.

H

haberdasher *s.* ψιλικατζής *m.* **~y** *s.* ψιλικατζήδικα *n.pl.*
habiliments *s.* ενδυμασία *f.*
habit *s.* συνήθεια *f.* form a ~ αποκτώ συνήθεια. from force of ~ από συνήθεια. be in the ~ of συνηθίζω να. (*dress*) ενδυμασία *f.*
habitable *a.* κατοικήσιμος.
habitat *s.* φυσικό περιβάλλον. **~ion** *s.* (*abode*) κατοικία *f.* (*inhabiting*) κατοίκηση *f.*
habitual *a.* συνηθισμένος. a ~ liar καθ' έξιν ψεύτης. **~ly** *adv.* από συνήθεια.
habituate *v.t.* (*also* ~ oneself) συνηθίζω.
habitude *s.* συνήθεια *f.*
habitué *s.* θαμώνας *m.*
hack *v.t.* πετσοκόβω, (*kick*) κλοτσώ, (*s.*) (*cut*) κοψιά *f.* (*kick*) κλοτσιά *f.*
hack *s.* (*horse*) άλογο *n.* (*fig., pen-pusher*) καλαμαράς *m.* (*a.*) της ρουτίνας. **~ing** *s.* ιππασία *f.* (*a.*) ~ing cough ξερόβηχας *m.*
hackles *s.* with ~ up αγριεμένος.
hackney *a.* (*carriage*) αγοραίος.
hackneyed *a.* τετριμμένος.
had *v. see* have.
Hades *s.* άδης *m.*
haemophilia *s.* αιμοφιλία *f.*
haemorrhage *s.* αιμορραγία *f.*
haemorrhoids *s.* αιμορροΐδες *f.pl.*
haft *s.* λαβή *f.*
hag *s.* μέγαιρα *f.* (*old*) μπαμπόγρια *f.* **~-ridden** *a.* βασανισμένος.
haggard *a.* τσακισμένος.
haggl|e *v.i.* κάνω παζάρια. (*also* ~e over) παζαρεύω. **~ing** *s.* παζάρεμα *n.*
hagiographer *s.* συγγραφεύς βίων αγίων.
hail *s.* χαλάζι *n.* (*v.*) it is ~ing πέφτει χαλάζι.
hail *v.* (*greet*) χαιρετίζω, (*summon*) φωνάζω. within ~ing distance εις απόστασιν φωνής. ~ from προέρχομαι από.
hair *s.* (*single*) τρίχα *f.* (*collective*) (*of head*) μαλλί *n.,* μαλλιά *pl.* (*on body*) τρίχες *f.pl.* (*animal's*) μαλλί *n.,* τρίχωμα *n.* do one's ~ χτενίζομαι, get one's ~ cut κόβω τα μαλλιά μου. tear one's ~ τραβώ τα μαλλιά μου. my ~ stood on end σηκώθηκαν οι τρίχες του κεφαλιού μου. keep your ~ on! μην εξάπτεσαι. he didn't turn a ~ έμεινε ατάραχος. let one's ~ down (*fig.*) αποβάλλω τους

τύπους. split ~s λεπτολογώ. **~brush** s. βούρτσα των μαλλιών. **~cut** s. κούρεμα n. **~do** s. χτένισμα n. **~dresser** s. κομμωτής m., κομμώτρια f. **~dresser's** s. κομμωτήριο n. **~grip** s. τσιμπιδάκι n. **~pin** s. φουρκέτα f. **~pin** bends κορδέλλες f.pl. **~raising** a. ανατριχιαστικός. **~'s-breadth** s. he escaped drowing by a ~'s-breadth παρά τρίχα να πνιγεί. **~shirt** s. τρίχινος χιτώνας. **~style** s. κόμμωση f. **~y** a. τριχωτός, μαλλιαρός.

halcyon a. ~ days αλκυονίδες ημέρες.

hale a. ~ (and hearty) κοτσονάτος.

hale v.t. σύρω.

half a. & s. μισός a., μισό n., ήμισυ n. cut in ~ κόβω στα δύο (or στη μέση). ~ the people were foreigners ο μισός κόσμος ήταν ξένοι. ~ (of) my day was wasted η μισή μου μέρα πήγε χαμένη. ~ an hour μισή ώρα. three and a ~ hours τρεις και μισή or τρεισήμισυ ώρες. it is ~ past three η ώρα είναι τρεισήμισι. we will go halves θα τα μοιρασθούμε μισά-μισά. I don't do things by halves δεν μου αρέσει να κάνω μισές δουλειές.

half adv. μισο-, ημι-. **~dressed** μισοντυμένος, **~finished** ημιτελής. ~ way along the road στα μισά του δρόμου, ~ way through the piece στα μισά του έργου. **~and-~** μισό μισό, **~yearly** κάθε έξι μήνες. it's ~ the size of theirs είναι το μισό από το δικό τους. this isn't ~ as good τούτο είναι πολύ κατώτερο. (fam.) not ~ ! πολύ. he didn't ~ catch it την έπαθε για καλά.

half-baked a. (fig.) ανόητος.

half-breed s. μιγάς m.f.

half-brother s. ετεροθαλής αδελφός.

half-caste s. μιγάς m.f.

half-hearted a. χλιαρός. **~ly** adv. με μισή καρδιά.

half-holiday s. ημιαργία f.

half-mast s. at ~ μεσίστιος.

half-moon s. ημισέληνος f.

half-pay s. on ~ σε διαθεσιμότητα.

half-seas-over a. (fam.) μισομεθυσμένος.

half-time s. ημιχρόνιο n.

half-witted a. χαζός.

hall s. (entrance) χωλ n. (large room) αίθουσα f. (country house) πύργος m.

hallmark s. σφραγίδα f.

hallo int. (greeting) γεια σου. (summons) έ! (surprise) μπα! (on telephone) εμπρός.

hallow v. αγιάζω. **~ed** a. ιερός.

Hallowe'en s. παραμονή των Αγίων Πάντων.

hallucination s. παραίσθηση f.

halo s. φωτοστέφανος m.

halt v.t. & i. σταματώ. ~! αλτ! (s.) (stop) στάση f. (station) σταθμός m. call or come to a ~ σταματώ.

halt a. (lame) χωλός. (v.i.) (limp) κουτσαίνω, (hesitate) διστάζω. **~ing** a. διστακτικός.

halter s. καπίστρι n. (noose) βρόχος m.

halve v.t. (divide) χωρίζω στη μέση, (reduce) περικόπτω στο ήμισυ.

ham s. ζαμπόν n. **~fisted, ~handed** a. αδέξιος.

ham s. ερασιτέχνης. (v.) το παρακάνω.

hamlet s. χωριουδάκι n.

hammer s. σφυρί n. (fig.) come under the ~ βγαίνω στο σφυρί. (fam.) ~ and tongs με μανία. (v.) κοπανώ, χτυπώ. ~ out (a plan) εκπονώ. **~and-sickle** s. σφυροδρέπανο n.

hammock s. αμάκ n.

hammy a. (fam.) υπερβολικός.

hamper s. σκεπαστό κοφίνι.

hamper v.t. παρεμποδίζω.

hamstring v.t. (fig.) παραλύω.

hand v.t. (give) δίνω. he ~ed her out of the carriage τη βοήθησε να κατέβει από την άμαξα. (fam.) I've got to ~ it to him τον αναγνωρίζω. ~ down or on μεταδίδω. ~ in (petition) υποβάλλω, (exam paper) παραδίδω. ~ over παραδίδω, ~ out διανέμω, ~ round προσφέρω.

hand s. χέρι n. (of clock) δείκτης m. (worker) εργάτης m. (crew member) μέλος του πληρώματος. (fig.) (influence) δάκτυλος m. on the one ~ από τη μία (μεριά), αφ' ενός. on the other ~ από την άλλη (μεριά), αφ' ετέρου. an old ~ παλιά καραμπίνα, he's an old ~ at this game είναι πεπειραμένος σε τέτοιες δουλειές. she's a good ~ at embroidery είναι πολύ καλή στο κέντημα. give or lend a ~ βοηθώ, take in ~ αναλαμβάνω. take (or have) a ~ in ανακατεύομαι σε. have a lot on one's ~s, have one's ~s full είμαι πολύ απασχολημένος. turn one's ~ to επιδίδομαι εις. put one's ~ on βρίσκω. lay ~s on (seize) αρπάζω. keep one's ~ in διατηρώ τη δεξιότητά μου. play into his ~s παίζω το παιχνίδι του. get out of ~ γίνομαι εκτός ελέγχου. it has come to ~ παρελήφθη. win ~s down νικώ χωρίς κόπο. the matter in ~ το υπό συζήτησιν θέμα. have an hour in ~ έχω μια ώρα ακόμα. be on ~ είμαι διαθέσιμος. (bound) ~ and foot χειροπόδαρα. (sink) with all ~s αύτανδρος a. she waits on him ~ and foot του τα δίνει όλα στο χέρι. from ~ to mouth μεροδούλι μεροφάι. ~ in ~ χέρι-χέρι. ~ in glove κώλος και βρακί. (fight) ~ to ~ εκ του σύνεγγυς. (close) at ~ κοντά. by ~ (made) με το χέρι, (delivered) ιδιοχείρως. ~s off! κάτω τα χέρια. ~s up! ψηλά τα χέρια. ready to ~ πρόχειρος. ~**bag** s. τσάντα f. ~**bill** s. φέιγβολάν n. ~**book** s. εγχειρίδιο n. ~**brake** s. χειρόφρενο n. ~**cart** s. χειράμαξα f. ~**cuff** s. χειροπέδη f. (v.) βάζω χειροπέδες σε. ~**ful** s. φούχτα f. (fig., child) ζωηρός (a.). ~**kerchief** s. μαντήλι n. ~**luggage** s. φορητές αποσκευές. ~**made** a. χειροποίητος, καμωμένο με το χέρι. ~**maid(en)** s. υπηρέτρια f. ~**picked** a. επίλεκτος. ~**set** s. ακουστικό n. ~**shake** s. χειραψία f. ~**writing** s. γραφικός χαρακτήρας.

handicap s. εμπόδιο n., μειονέκτημα n. (sport) ισοζύγιασμός m. (v.) παρεμποδίζω, (sport) βάζω χάντικαπ σε.

handicraft s. χειροτεχνία f.

handiwork s. my ~ έργον των χειρών μου. (iron.) that's his ~! αυτό είναι κατόρθωμά του.

handl|e s. (of tool) μανίκι n. (of weapon, cutlery) λαβή f. (of cup, door) χερούλι n. (knob) πόμολο n. starting ~e μανιβέλα f. (v.) (manage, use) χειρίζομαι, (touch) πιάνω, ψαύω, (treat) μεταχειρίζομαι, (trade in) εμπορεύομαι. ~**er** s. χειριστής m. ~**ing** s. χειρισμός m., μεταχείριση f. ~**ebar(s)** s. τιμόνι n.

handout s. (alms) ελεημοσύνη f. (statement) ανακοίνωση f. (notes) σημειώσεις f.pl. (leaflet) έντυπος διαφήμιση.

handsome a. ωραίος, (gift) γενναίος, (sum) σεβαστός. ~**ly** adv. με γενναιοδωρία.

handy a. (ready) πρόχειρος, (convenient) βολικός, (clever) επιδέξιος. (fam.) come in ~ πιάνω τόπο. ~**man** s. άνθρωπος για όλες τις δουλειές.

hang v.t. κρεμώ, (person) απαγχονίζω. (one's head) σκύβω. ~ out (flag) βγάζω, (washing) απλώνω. ~up κρεμώ, (phone) κλείνω. (v.i.) κρέμομαι. ~ about or around (wait) περιμένω, (loiter) τριγυρίζω. ~ back μένω πίσω. ~ down (below proper level) κρεμώ. ~ fire χρονίζω. ~ out (live) κάθομαι, (frequent) συχνάζω. ~ out of the window κρεμιέμαι έξω από το παράθυρο. ~ on (not give up) κρατιέμαι, (wait) περιμένω. ~ on to (keep) κρατώ, (persist in) εμμένω εις, ~ on to me! (for support) πιάσου επάνω μου. ~ together (be united) είμαστε ενωμένοι, (be consistent) συμφωνούμε. ~ upon his lips κρέμομαι από τα χείλια του. ~ it! να πάρει ο διάολος. (s.) (fam.) get the ~of καταλαβαίνω. not care a ~ δεν δίνω δεκάρα. ~**dog** a. with a ~dog look σαν δαρμένο σκυλί. ~**er** s. κρεμάστρα f. ~**er-on** s. παρακεντές m., κολλητσίδα f.

~**man** s. δήμιος m. ~**nail** s. παρωνυχίδα f.

hanging s. κρέμασμα n. ~s κουρτίνες f.pl. (a.) κρεμαστός.

hangover s. αδιαθεσία μετά από μεθύσι. (fig.) a ~ from pre-war days κατάλοιπο της προπολεμικής εποχής.

hank s. (skein) κούκλα f.

hanker v. ~ for επιθυμώ. ~**ing** s. επιθυμία f.

hanky-panky s. απάτη f.

hap s. τύχη f. (v.) τυχαίνω.

ha'penny s. μισή πένα.

haphazard a. τυχαίος. ~**ly** adv. στην τύχη.

hapless a. ατυχής.

happen v. (occur) συμβαίνω, γίνομαι. (of mischance) what ~ed to him? τί έπαθε; (chance) τυχαίνω, I ~ed to be there έτυχα (or έλαχα) εκεί, I ~ed to be away έτυχε (or έλαχε) να λείπω. ~ on (find) βρίσκω κατά τύχη. ~**ing** s. συμβάν n., γεγονός n.

happ|y a. ευτυχής, (persons only) ευτυχισμένος. (pleased) ευχαριστημένος, (merry) χαρούμενος. (apt) επιτυχημένος. ~y Easter! καλό Πάσχα. ~**ily** adv. ευτυχισμένα, (luckily) ευτυχώς, ~**iness** s. ευτυχία f. ~**y-go-lucky** a. ξέ(γ)νοιαστος.

harangue s. κραυγαλέα εκφώνηση λόγου. (v.) βγάζω κραυγαλέο λόγο (σε).

harass v. (attack) παρενοχλώ, (trouble) βασανίζω. ~**ment** s. παρενόχληση f.

harbinger s. προάγγελος m.f.

harbour s. λιμάνι n. (fig.) καταφύγιο n. (v.t.) (shelter) παρέχω άσυλο εις. (thoughts, etc.) τρέφω. (contain) έχω. ~**master** s. λιμενάρχης m.

hard a. (not soft, severe) σκληρός, (not easy) δύσκολος, (strong) δυνατός, βαρύς, (intensive) σύντονος. be ~ on him τον μεταχειρίζομαι σκληρά, είμαι αυστηρός απέναντί του. he finds it ~ to wake up ξυπνάει δύσκολα, δυσκολεύεται να ξυπνήσει. have a ~ job (to) βλέπω και παθαίνω (να). try one's ~est βάζω τα δυνατά μου. ~ and fast αμετάβλητος, ~ of hearing βαρήκοος. ~ luck γκίνια f. (also βy δυσ-, δυσκολο-), ~ to understand δυσνόητος, ~ to believe δυσκολοπίστευτος. ~**board** s. πεπιεσμένο ξύλο, νοβοπάν n. ~**bound** a. δεμένος. ~**headed** a. πρακτικός. ~**hearted** a. σκληρόκαρδος. ~**ness** s. σκληρότητα f. ~**ware** s. είδη κιγκαλερίας. (computers) κυρίως μηχανήματα.

hard adv. (work, fight) σκληρά, (pull, rain) δυνατά, (beg, stare) επίμονα. ~ by κοντά. be ~ at it εργάζομαι εντατικά. be ~ pressed πιέζομαι. be ~ put to it δυσκολεύομαι. be up δεν έχω λεφτά, be ~ up for δυσκολεύομαι να βρω, follow ~ on the heels of ακολουθώ κατά πόδας. ~**bitten**

a. σκληραγωγημένος. **~-boiled** *a. (egg)* σφιχτός, (*person)* σκληρός.
harden *v.t. & i.* σκληραίνω. (*v.t.*) (*inure)* σκληραγωγώ. ~ed to crime πωρωμένος. ~ed steel βαμμένο σίδερο.
hardihood *s.* θράσος *n.*
hardiness *s.* τόλμη *f.* σκληραγωγία *f.* ανθεκτικότητα *f.* (*see* hardy).
hardly *adv.* μόλις, σχεδόν. we've ~ got time μόλις που έχουμε καιρό. I ~ know him δεν τον γνωρίζω σχεδόν. I need ~ tell you περιττόν να σας πω. they can ~ have arrived yet δεν μπορεί να έφτασαν ακόμα. I have ~ enough to pay the bill έχω μόλις και μετά βίας να πληρώσω το λογαριασμό. I have ~ begun μόλις άρχισα. *see* scarcely.
hardy *a. (bold)* τολμηρός, (*robust)* σκληραγωγημένος, (*of plants)* ανθεκτικός.
hare *s.* λαγός *m.* **~-brained** *a.* ανόητος.
harem *s.* χαρέμι *n.*
haricot-bean *s.* φασόλι *n.*
hark *v.* ~! άκου! *(iron.)* ~ at him! άκουσον-άκουσον! ~ back το επανέρχομαι σε.
harlequin *s.* αρλεκίνος *n.*
harlot *s.* πόρνη *f.* **~ry** *s.* πορνεία *f.*
harm *s.* κακό *n. (damage)* ζημία *f.* do ~ (to) βλάπτω. come to (*or* suffer) no ~ δεν παθαίνω τίποτα. there's no ~ in asking δεν βλάπτει σε τίποτα αν (*or* να) ρωτήσουμε. there's no ~ in him δεν έχει κακία μέσα του. out of ~'s way εκτός κινδύνου. (*v.*) βλάπτω. **~ful** *a.* βλαβερός, επιβλαβής, επιζήμιος. **~less** *a. (drug, experiment)* αβλαβής, (*person, animal, joke)* άκακος. **~lessly** *adv.* χωρίς ζημία.
harmon|ic, ~ious *a.* αρμονικός. **~iously** *adv.* αρμονικά.
harmonium *s.* αρμόνιο *n.*
harmon|y *s.* αρμονία *f.* **~ize** *v.t. (mus. & fig.)* εναρμονίζω. (*v.i.*) (*agree)* συμφωνώ.
harness *s. (beast's)* σαγή *f.*, χάμουρα *n.pl* (*fig., straps)* λουριά *n.pl.* in ~ δουλεύοντας. (*v.*) ζεύω. (*fig. utilize for power)* τιθασσεύω.
harp *s.* άρπα *f.*
harp *v.* ~ on αναμασώ, επαναλαμβάνω διαρκώς.
harpoon *s.* καμάκι *n.* (*v.*) καμακώνω.
harpy *s.* άρπυια *f.*
harridan *s.* μέγαιρα *f.*
harrow *s.* σβάρνα *f.* (*v*). σβαρνίζω, (*fig.)* σπαράζω. **~ing** *a.* σπαρακτικός.
harry *v.* παρενοχλώ, (*pester)* κυνηγώ.
harsh *a. (severe)* σκληρός, (*rough)* τραχύς. **~ly** *adv.* σκληρά. **~ness** *s.* σκληρότητα *f.* τραχύτητα *f.*
harum-scarum *a.* ελαφρόμυαλος.
harvest *s. (crops)* συγκομιδή *f.*, σοδειά *f.*

(vintage) τρύγος *m. (season)* θέρος *m.*, θερισμός *m.* (*v.*) θερίζω, τρυγώ. **~er** *s.* θεριστής *m.* (*machine)* θεριστική μηχανή.
has-been *s. (fam.)* ξοφλημένος *α.*
hash *s.* ξαναμαγειρεμένο κρέας. (*fig.)* make a ~ of it τα κάνω σαλάτα. (*v.*) λιανίζω.
hashish *s.* χασίσι *n.*
hassock *s.* μαξιλλάρι γονυκλισίας.
haste *s.* σπουδή *f.*, βία *f.*, βιασύνη *f.* be in ~ βιάζομαι, επείγομαι. make ~ σπεύδω, κάνω γρήγορα, βιάζομαι. make ~! κάνε γρήγορα, βιάσου! (*v.*) *see* hasten.
hasten *v.t. (an event)* επισπεύδω, (*an activity)* επιταχύνω. they ~ed him away τον έβγαλαν άρον-άρον έξω. (*v.i.*) σπεύδω, κάνω γρήγορα.
hast|y *a.* βιαστικός. **~y-tempered** οξύθυμος. **~ily** *adv.* βιαστικά, εσπευσμένως. **~iness** *s.* βιασύνη *f.*
hat *s.* καπέλλο *n.*, πίλος *m.* ~ in hand (*fig.)* παρακαλώντας. raise one's ~ αποκαλύπτομαι. (*fig.)* I take off my ~ to him του βγάζω το καπέλλο, τον αναγνωρίζω. pass round the ~ κάνω έρανο. keep it under one's ~ το κρατώ μυστικό. talk through one's ~ λέω μπούρδες. at the drop of a ~ χωρίς παρακάλια. a bad ~ παλιάνθρωπος *m.* old ~ ξεπερασμένος *a.* my ~! Κύριε ελέησον! I'll eat my ~ (if) να μου τρυπήσεις τη μύτη (αν).
hatch *v.t.* εκκολάπτω, (*fig.)* (*contrive)* μηχανεύομαι. (*v.i.*) εκκολάπτομαι. **~ery** *s.* λιβάρι *n.*
hatchet *s.* μπαλ(ν)τάς *m.* (*fig.)* bury the ~ συμφιλιώνομαι.
hate *v.* μισώ, σιχαίνομαι. I ~ him τον μισώ. I ~ quarrelling (*or* rice-pudding) σιχαίνομαι τους καβγάδες (*or* το ριζόγαλο). I ~ bothering you λυπούμαι που σας ενοχλώ. (*s.*) μίσος *n.* **~ful** *a.* (*person)* μισητός, (*job, weather)* σιχαμερός.
hatpin *s.* καρφίτσα του καπέλου.
hatred *s.* μίσος *n.*
hatstand *s.* καλόγερος *m.*
hatter *s.* καπελλάς *m.*
haught|y *a.* (*person)* υπερόπτης, (*behaviour)* υπεροπτικός. **~iness** *s.* υπεροψία *f.*
haul *v.t.* έλκω, τραβώ, σύρω. (*tow)* ρυμουλκώ. ~ down στέλλω. (*s.*) (*act)* τράβηγμα *n. (of fish)* ψαριά *f.* (*spoils)* λεία *f.* **~age** *s.* μεταφορά *f.* (*charges)* μεταφορικά *n.pl.* **~ier** *s.* εργολάβος μεταφορών.
haunch *s.* μπούτι *n.* on one's ~es ανακούρκουδα.
haunt *v.* (*pursue)* καταδιώκω, (*possess)* κατέχω. ~ed στοιχειωμένος. it is ~ed by pickpockets συχνάζεται από λωποδύτες.

(s.) (in town) στέκι n. (in country) λημέρι n. it is a ~ of nightingales είναι γεμάτο αηδόνια.

have anomalous v. you ~ been there, ~n't you? έχετε πάει εκεί, έτσι; yes, I ~ ναι, έχω πάει.

have v.t. (possess) έχω. (a walk, bath, child, go) κάνω. (tea, one's medicine) παίρνω. ~ a fight μαλώνω, ~ lunch γευματίζω, ~ a tooth out βγάζω ένα δόντι, ~ a suit made ράβω κοστούμι. ~ (sthg.) done see get (v.t.). we had a puncture μας έπιασε λάστιχο. I'll ~ that one θα πάρω εκείνο. I won't ~ it (allow) δεν το επιτρέπω. what did you ~ to eat? τι φάγατε; we had them to dinner τους είχαμε τραπέζι. let me ~ it back δωσ' μου το πίσω. has he had measles? έβγαλε ιλαρά; I've been had (tricked) μου την έφτιαξαν. ~ it your own way! όπως αγαπάς. let 'em ~ it! δώσ' του! ~ it out with εξηγούμαι με. ~ to do with (relate to) έχω σχέση με, (mix with) έχω σχέσεις με. rumour has it that... διαδίδεται ότι. (v.i.) ~ to (must) I shall ~ to leave tomorrow πρέπει να φύγω αύριο. they had to get up early έπρεπε να ξυπνήσουν νωρίς. you don't ~ to stay δεν είσαι υποχρεωμένος (or αναγκασμένος) να μείνεις.

have on v.t. (be wearing) φορώ, what did she ~ ? τι φορούσε; (be doing) I have nothing on tonight είμαι ελεύθερος απόψε. (tease) δουλεύω.

have up v.t. be had up (get summons) παίρνω κλήση, (get arrested) συλλαμβάνομαι.

haven s. λιμάνι n. (fig.) καταφύγιο n.

haversack s. σακκίδιο n. (soldier's) γυλιός m.

haves s. (fam.) ~ and have nots οι έχοντες και οι μη έχοντες.

havoc s. θραύση f. καταστροφή f. play ~ with κάνω καταστροφή σε.

hawk s. γεράκι n.

hawk v.t. πουλώ από πόρτα σε πόρτα. (news) διαδίδω. ~er s. γυρολόγος m.

hawser s. παλαμάρι n.

hay s. σανός m. (fig.) make ~ of it τα κάνω σαλάτα. make ~ while the sun shines επωφελούμαι της περιστάσεως. ~cock s. θημωνιά f. ~fever s. πυρετός του χόρτου. ~stack s. θημωνιά f.

haywire a. (fam.) go ~ πάω στραβά.

hazard s. (chance) τύχη f. (risk) κίνδυνος m. (v.) διακινδυνεύω, παίζω. ~ous a. επικίνδυνος.

haze s. καταχνιά f. (fig.) σύγχυση f. ~y a. αμυδρός, (fig.) αβέβαιος.

hazel s. (nut.) φουντούκι n. (tree) φου-

ντουκιά f. (a.) ~ eyes ανοικτά καστανά μάτια.

he pron. αυτός, εκείνος. there ~ is! να τος, να τον.

head s. κεφάλι n., κεφαλή f. (of cattle) κεφάλια n.pl. (of water, steam) στήλη f. (of nail, tool) κεφάλι n. (of list, page) κορυφή f. (of state, church, procession) κεφαλή f. (leader) αρχηγός m. (director) διευθυντής m. (of department) προϊστάμενος m. (promontory) ακρωτήριο n. bed ~ κεφαλάρι n. ~ of stairs κεφαλόσκαλο n. (on beer) κολλάρο n. per ~ κατ' άτομό. ~s or tails κορόνα γράμματα. from ~ to foot από την κορυφή ως τα νύχια. ~ first με το κεφάλι. at the ~ (of) (in front, in charge) επικεφαλής (with gen.). go ~ over heels κουτρουβαλώ. come to a ~ φτάνω σε κρίσιμο σημείο, (of a boil) ωριμάζω. things are coming to a ~ έφτασε ο κόμπος στο χτένι. have a good ~ for έχω μυαλό για. give (person) his ~ δίδω ελευθερίαν εις. keep one's ~ διατηρώ την ψυχραιμία μου. keep one's ~ above water επιπλέω, κατοσθώνω να σωθώ. lose one's ~ τα χάνω. he's off his ~ του έστριψε, τρελλάθηκε. he's not right in the ~ δεν τα έχει σωστά. turn the ~ of ξεμυαλίζω. it went to his ~ τον χτύπησε στο κεφάλι, πήραν τα μυαλά του αέρα. he's got it into his ~ (that) του μπήκε στο μυαλό (ότι), he took it into his ~ (to) του κατέβηκε or του κάπνισε (να). the idea never entered my ~ ούτε μου πέρασε η ιδέα. I can't make ~ or tail of it δεν μπορώ να βγάλω άκρη.

head v.t. (govern) διευθύνω. (lead) ηγούμαι, είμαι επικεφαλής (both with gen.). ~ the list, poll, etc. βγαίνω πρώτος. ~ off κόβω το δρόμο (with gen.), (fig.) αποτρέπω. (v.i.) ~ for τραβώ για, κατευθύνομαι προς.

head a. πρώτος. ~ gardener αρχικηπουρός m. ~ waiter μαιτρ m. ~ post-office κεντρικό ταχυδρομείο. ~ache s. πονοκέφαλος m. ~band s. κεφαλόδεσμος m. ~dress s. μεταμφίεση της κεφαλής. ~er s. (dive) βουτιά f. ~gear s. κάλυμμα της κεφαλής. ~ing s. ~line s. επικεφαλίδα f., τίτλος m. ~lamp s. προβολεύς m. ~long a. ορμητικός. (adv.) με το κεφάλι. ~man s. φύλαρχος m. ~master s. διευθυντής m. (secondary) γυμνασιάρχης m. ~mistress s. διευθύντρια f. (secondary) γυμνασιάρχης f. ~on a. μετωπικός. (adv.) κατά μέτωπον. ~phones s. ακουστικά κεφαλής. ~quarters s. (of firm) έδρα f. (operational) στρατηγείο n. ~sman s. δήμιος m. ~strong a. ισχυρογνώμων. ~way s. πρόο-

δος *f.* make ~**way** προοδεύω. ~-**wind** *s.*
αέρας κόντρα. ~**word** *s.* λήμμα *n.* ~**y** *a.*
μεθυστικός.
heal *v.t.* θεραπεύω, γιατρεύω. (*v.i.*) επουλώνομαι, γιαίνω. ~**er** *s.* θεραπευτής *m.*
~**ing** *s.* θεραπεία *f.* (*a.*) θεραπευτικός.
health *s.* υγεία *f.* (*administrative domain*)
υγιεινή *f.* be in bad ~ δεν είμαι καλά στην
υγεία μου. enjoy the best of ~ χαίρω άκρας
υγείας. (*a.*) (*officer, regulations, etc.*)
υγειονομικός, ~ certificate πιστοποιητικόν υγείας. ~ centre υγειονομικό κέντρο.
~**y** *a.* (*in good ~ & fig.*) υγιής, (*good for*
~) υγιεινός. have a ~y appetite τρώω με
όρεξη.
heap *s.* σωρός *m.*, στοίβα *f.* (*fam.*) I've got
~s έχω ένα σωρό. (*v.t.*) στοιβάζω, (*collect*)
μαζεύω. ~ with (*load*) φορτώνω, γεμίζω.
hear *v.* ακούω. (*law, try*) δικάζω. I have ~d
of it το έχω ακούσει, το έχω ακουστά. ~
from παίρνω νέα από. ~ out ακούω μέχρι
τέλους. I won't ~ of it δεν θέλω να ακούσω
λέξη για αυτό. he could hardly be ~d μόλις
ακουγότανε. ~ ~! μπράβο! ~**er** *s.* ακροατής
m. ~**say** *s.* φήμη *f.* from ~say εξ ακοής.
hearing *s.* (*sense*) ακοή *f.* be hard of ~
βαρυακούω. within ~ εις ακουστήν απόστασιν, in my ~ μπροστά μου. (*audience*)
ακρόαση *f.* gain a ~ γίνομαι δεκτός σε
ακρόαση. be condemned without a ~
καταδικάζομαι αναπολόγητος. (*law*) the
case is due for ~ shortly η υπόθεση θα
δικαστεί προσεχώς. ~-**aid** *s.* ακουστικό *n.*
hearken *v.* ακούω.
hearse *s.* νεκροφόρα *f.*
heart *s.* καρδιά *f.* (*fig. essence*) ουσία *f.*
after my own ~ της αρεσκείας μου. in my
~ of ~s στα κατάβαθα της καρδιάς μου. to
my ~'s content όσο τραβάει η καρδιά μου.
I have it at ~ το 'χω βάλει μεσ' την καρδιά
μου. I have set my ~ on it έχει κολλήσει η
καρδιά μου σ' αυτό. ~ and soul ολοψύχως,
at ~ κατά βάθος, by ~ απ' έξη, learn by ~
αποστηθίζω. take ~ παίρνω κουράγιο,
take to ~ παίρνω κατάκαρδα, take (*person*) to one's ~ έχω στην καρδιά μου. lose
~ αποθαρρύνομαι, lose one's ~ to ερωτεύομαι, wear one's ~ on one's sleeve
δείχνω τα αισθήματά μου. it was enough
to break one's ~ ήταν να σπαράζει η
καρδιά του ανθρώπου. (*at cards*) κούπα *f.*
(*a.*) (*of the ~*) καρδιακός. ~-**ache** *s.* καημός
m., μαράζι *n.* ~-**attack** *s.* συγκοπή *f.*
~-**beat** *s.* παλμός *m.* ~-**breaking** *a.* σπαρακτικός. ~-**broken** *a.* περίλυπος μέχρι θανάτου. ~-**burn** *s.* καούρα *f.* ~-**burning** *a.*
πικρία *f.* ~-**en** *v.* εμψυχώνω. ~-**ening** *a.*
ενθαρρυντικός. ~-**failure** *s.* συγκοπή *f.*

~-**felt** *a.* βαθύς. ~-**less** *a.* άσπλαχνος.
~-**lessness** *s.* σκληρότητα *f.* ~-**rending** *a.*
σπαρακτικός. ~-**searching** *s.* βαθιά
αυτοεξέταση. ~-**strings** *s.* play on his
~strings θίγω τις εναίσθητες χορδές του.
~-**throb** *s.* (*fam.*) βάσανο *n.* ~-**to**-~ *a.*
χωρίς προσχήματα.
hearth *s.* τζάκι *n.* (*fig.*) ~ and home οικογενειακή εστία.
heart|y *a.* (*cordial*) εγκάρδιος, θερμός,
(*hale*) σφριγηλός, (*support, etc.*) ένθερμος. a ~y eater γερό πηρούνι. ~**ily** *adv.*
θερμά, (*with zest*) με όρεξη. laugh ~ily
γελώ με την καρδιά μου. I am ~ily sick of
it το έχω βαρεθεί μέχρις αηδίας.
heat *s.* ζέστη *f.* (*phys.*) θερμότης *f.* (*fervour*)
θέρμη *f.* (*anger, excitement*) έξαψη *f.*,
παραφορά *f.* (*fever*) πυρετός *m.* (*rut*)
οργασμός *m.* (*in sports*) προκριματικός
αγώνας. ~ wave κύμα καύσωνος. (*v.t.*)
ζεστάινω, θερμαίνω. ~**er** *s.* (*stove*) θερμάστρα *f.* (*of water*) θερμοσίφωνας *m.*
heated *a.* (*with emotion, wine*) ξαναμμένος,
(*angry*) οργισμένος. get ~ εξάπτομαι,
ξανάβω. ~**ly** *adv.* οργισμένα.
heath *s.* (*place*) χερσότοπος *m.* (*plant*)
ρείκι *n.*
heathen *a.* ειδωλολάτρης. (*fig.*) βάρβαρος.
heather *s.* ρείκι *n.*
heating *s.* (*system, provision of heat*)
θέρμανση *f.* (*making sthg. hot*) ζέσταμα *n.*
(*a.*) θερμαντικός.
heave *v.t.* (*throw*) ρίχνω, (*raise*) σηκώνω,
(*haul*) τραβώ. ~ a sigh βγάζω στεναγμό,
αναστενάζω. (*v.i.*) (*of ship, etc.*) ταλαντεύομαι, (*gently*) λικνίζομαι. her bosom
~d with emotion κοντανάσαινε από τη
συγκίνηση. ~ to σταματώ. (*s.*) τράβηγμα *n.*
heaven *s.* ουρανός *m.* move ~ and earth
χαλώ τον κόσμο. ~ knows! Κύριος οίδε!
good ~s! Κύριε ελέησον! for ~'s sake για
όνομα του Θεού, προς Θεού. ~**ly** *a.*
ουράνιος. (*fam.*) we had a ~ly time
περάσαμε μούρλια (*or* όνειρο).
heav|y *a.* βαρύς, (*difficult*) δύσκολος,
(*tiring*) κουραστικός, (*big*) μεγάλος,
(*strong, intense*) δυνατός. make ~y
weather (*fig.*) δυσκολεύομαι. be a ~y
drinker πίνω πολύ. find it ~y going
προχωρώ με δυσκολία. time hangs ~y ο
χρόνος περνάει αργά. ~**ily** *adv.* βαρέως,
(*very, much*) πολύ, (*strongly*) δυνατά.
~**iness** *s.* βάρος *n.* ~**y-handed** *a.* καταπιεστικός, (*awkard*) αδέξιος. ~**y-laden** *a.*
βαρυφορτωμένος. ~**yweight** *a. & s.*
βαρέων βαρών.
Hebrew *a.* εβραϊκός. (*person*) Εβραίος.
heckl|e *v.* παρενοχλώ. ~**ing** *s.* παρενόχληση *f.*

hectare s. εκτάριον n.

hectic a. (flushed) αναμμένος. (fig.) πυρετώδης, γεμάτος δραστηριότητα.

hector v.t. φέρνομαι δεσποτικά σε. ~ing a. δεσποτικός.

hedge s. φράχτης από θάμνους. (v.t.) περιφράσσω. (v.i.) απαντώ με υπεκφυγές, (waver) κάνω νερά. (v.t.) ~ one's bets ποντάρω διπλά.

hedgehog s. σκαντζόχοιρος m.

hedon|ism s. ηδονισμός m. ~ist s. ηδονιστής m.

heed v. (also pay ~ to) προσέχω. (s.) προσοχή f. ~ful a. προσεκτικός. ~less a. απρόσεκτος, be ~less of δεν προσέχω. ~lessly adv. απρόσεκτα.

heel s. φτέρνα f. (of shoe) τακούνι n. at one's ~s (close) κατά πόδας. come to ~ πειθαρχώ. take to one's ~s το βάζω στα πόδια. down at ~ κουρελής. (fam.) well ~ed πλούσιος.

heel v.i. γέρνω.

hefty a. (big) μεγάλος, (strong) δυνατός. (of object, bulky) ογκώδης.

hegemony s. ηγεμονία f.

heifer s. δαμαλίδα f.

height s. ύψος n. (stature) ανάστημα n. (high ground) ύψωμα n. (acme) ακμή f. the ~ of fashion η τελευταία λέξη της μόδας.

heighten v.t. (lift) υψώνω, (increase) αυξάνω, (intensify) επιτείνω.

heinous a. στιγερός.

heir, ~ess s. κληρονόμος m.f. be ~ το κληρονομώ. ~ to throne διάδοχος m.f. ~loom s. κειμήλιο n.

helicopter s. ελικόπτερον n.

helium s. ήλιον n.

hell s. κόλαση f. (fam.) raise ~ χαλάω τον κόσμο. to ~ with it! στο διάολο. be ~-bent on (doing sthg.) θέλω σώνει και καλά να. ~ish a. διαβολικός, (fam.) διαβολεμένος.

Hellen|e s. Έλλην(ας) m., Ελληνίδα f. ~ic a. ελληνικός. (person) see H~e. ~ism s. ελληνισμός m. ~ize v.t. εξελληνίζω.

Hellenist s. ελληνιστής m. ~ic ελληνιστικός.

hello int. see hallo.

helm s. πηδάλιο n. ~sman s. πηδαλιούχος m.

helmet s. κράνος n., κάσκα f. (warrior's) περικεφαλαία f. crash ~ κάσκα f.

helot s. είλωτας m.

help v. βοηθώ. (remedy, avoid) it can't be ~ed δεν γίνεται αλλιώς, I couldn't agreeing with him δεν μπορούσα να μη συμφωνήσω μαζί του. don't stay longer than you can ~ να μείνεις όσο το δυνατόν λιγότερο. (serve food) I ~ him to potatoes του βάζω

(or σερβίρω) πατάτες. ~ oneself to (take) παίρνω μόνος μου, (steal) βουτώ.

help s. βοήθεια f. be of ~ (to) βοηθώ. (helper) βοηθός m.f. ~er s. βοηθός m.f. ~ful a. εξυπηρετικός, χρήσιμος. ~ing s. μερίδα f. (a.) give a ~ing hand δίδω χείρα βοηθείας. ~less a. (powerless) ανίσχυρος, (incapacitated) αδύναμος, ανήμπορος. be ~less (unable) αδυνατώ. ~mate s. σύντροφος m.f.

helter-skelter adv. προτροπάδην.

hem s. στρίφωμα n. (v.) στριφώνω. (fig.) ~ in περικλείω, περικυκλώνω.

hemisphere s. ημισφαίριο n.

hemlock s. κώνειο n.

hemp s. καννάβι n. ~en a. από καννάβι.

hen s. κόττα f. (fam.) ~ party γυναικοπαρέα f.

hence adv. (from here) από 'δω, (from now) από τώρα, (as a result) εκ τούτου, γι' αυτό. ~forth, ~forward adv. εφεξής.

henchman s. οπαδός m.

henna s. χένα f.

henpecked a. he is ~ τον σέρνει από τη μύτη η γυναίκα του.

hepatitis s. ηπατίτιδα f.

heptagon s. επτάγωνο n.

her pron. I saw ~ την είδα, for ~ για εκείνη, ~ house το σπίτι της, I gave it to ~ της το έδωσα. ~self see self.

herald s. κήρυκας m. (fig.) προάγγελος m. (v.) προαγγέλλω. ~ry s. εραλδική f.

herb s. βότανο n. ~age s. βοσκή f. ~al a. από βότανα.

herbaceous a. (border) με πολυετή φυτά.

herbivorous a. χορτοφάγος.

herculean a. ηράκλειος.

herd s. αγέλη f. vulgar ~ αγελαίον πλήθος, ~ instinct αγελαίον ένστικτον. (v.t.) (tend) φυλάω, (fig.) ~ together μαζεύω κοπαδιαστά. ~sman s. βοσκός m., βουκόλος m.

here adv. εδώ, ενταύθα. ~ we are! να μας. ~ you are (offering) ορίστε. over ~ από 'δω, προς τα εδώ. ~ lies (buried) ενθάδε κείται. ~abouts adv. εδώ γύρω. ~after adv. στο μέλλον. ~by adv. διά του παρόντος. ~tofore adv. μέχρι τούδε. ~upon adv. κατόπιν τούτου. ~with adv. με το παρόν, διά της παρούσης.

heredit|ary a. κληρονομικός. ~y s. κληρονομικότητα f.

here|sy s. αίρεση f. ~tic s., ~tical a. αιρετικός.

heritage s. κληρονομία f.

hermetically adv. ερμητικώς.

hermit s. ερημίτης m. αναχωρητής m. ~age s. ερημητήριο n.

hernia s. κήλη f.

hero s. ήρωας m. **~ine** s. ηρωίδα f. **ism** s. ηρωισμός m.
heroic a. ηρωικός. **~s** (fam.) μεγάλα λόγια. **~ally** adv. ηρωικά.
heroin s. ηρωίνη f.
herring s. ρέγγα f. (fig.) red ~ παραπλανητικό επιχείρημα. **~-bone** s. (pattern) ψαροκόκκαλο n.
hers pron. δικός της.
hesit|ant a. διστακτικός. **~ancy** s. διστακτικότητα f.
hesitat|e v. διστάζω. **~ion** s. δισταγμός m.
hessian s. κανναβάτσο n.
heterodox a. ετερόδοξος.
heterogeneous a. ανομοιογενής.
heterosexual a. ετεροφυλόφιλος.
hew v. κόβω, πελεκώ.
hexagon s. εξάγωνο n.
hexameter s. εξάμετρο n.
heyday s. μεσουράνημα n.
hiatus s. χασμωδία f.
hibernation s. χειμερία νάρκη.
hibiscus s. ιβίσκος m.
hiccup s. λόξυγγας m. have the ~s έχω λόξυγγα.
hidden a. κρυφός, κρυμμένος.
hide s. τομάρι n., δορά f. (leather) δέρμα n. **~bound** a. (fig.) στενόμυαλος.
hide v.t. κρύβω. (v.i.) κρύβομαι. **~-and-seek** s. κρυφτό n. ~ **out** s. κρησφύγετο n.
hideous a. ειδεχθής, απαίσιος. **~ly** adv. φρικτά. **~ness** s. ειδέχθεια f.
hiding s. be (or go) into ~ κρύβομαι. (fam.) (beating) ξύλο n., ξυλοκόπημα n. get a ~ τρώω ξύλο. **~-place** s. κρυψώνας m.
hierarchy s. ιεραρχία f.
hieroglyphics s. ιερογλυφικά n.pl.
hi-fi a. υψηλής πιστότητος.
higgledy-piggledy adv. φύρδην-μίγδην.
high a. ψηλός, (elevated, exalted) υψηλός. it is 10 metres ~ έχει δέκα μέτρα ύψος. (fam., drunk) μεθυσμένος. (meat) πολύ σιτεμένος. (wind) δυνατός. H~ Court Ανώτατον Δικαστήριον, H~ Commissioner Ύπατος Αρμοστής. ~ jump άλμα εις ύψος. ~ life μεγάλη ζωή, ~ living καλοπέραση f. at ~ noon μεσ' στο καταμεσήμερο, ντάλα μεσημέρι. ~ opinion καλή ιδέα. ~ praise μεγάλοι έπαινοι. ~ school γυμνάσιο n. ~ seas διεθνή ύδατα. ~ spirits κέφι n. ~ spots αξιοθέατα n.pl. ~ tide φουσκονεριά f., (see tide). ~ treason εσχάτη προδοσία. ~ words θυμωμένα λόγια. it is ~ time είναι καιρός πια (with να). have a ~ (old) time το ρίχνω έξω, ξεφαντώνω. have a ~ colour έχω κόκκινο πρόσωπο. be left ~ and dry μένω στα κρύα του λουτρού. **~-born** a. υψηλής καταγωγής. **~brow** a.

σοβαρός, υψηλού επιπέδου. **~-class** a. ανωτέρας ποιότητος. **~er** a. (in rank) ανώτερος. **~est** a. ανώτατος, ύψιστος. **~-falutin** a. σπουδαιοφανής. **~-flown** a. υπερβολικός. **~-frequency** a. υψηλής συχνότητος. **~-handed** a. αυταρχικός. **~-hat** a. περιφρονητικός. **~-lands** s. ορεινή περιοχή. **~-light** v. (fig.) υπογραμμίζω, εξαίρω. (s.) κορύφωμα n. **~-minded** a. υψηλόφρων. **~-ness** s. υψηλότης f. **~-pitched** a. ψιλός. **~-rise** a. υψηλός. **~-spirited** a. γενναίος, ψυχωμένος, (lively) ζωηρός. **~-tech** a. υψηλής τεχνολογίας. **~-water mark** s. ανώτατη στάθμη ύδατος. **~way** s. δημοσιά f. **~wayman** s. ληστής m.
high adv. (also ~ up) ψηλά. ~ and low παντού, on ~ στον ουρανό.
highly adv. (much, very) πολύ he is ~ paid πληρώνεται πολύ καλά, I think ~ of him τον εκτιμώ πολύ, he is ~ thought of χαίρει μεγάλης εκτιμήσεως. **~-coloured** a. (fig.) υπερβολικός. **~-strung** a. υπερευαίσθητος.
hijack v. κλέβω. **~er** s. (of plane) αεροπειρατής m.
hik|e, ~ing s. πεζοπορία f. **~er** s. πεζοπόρος m.
hilari|ous a. κεφάτος, εύθυμος. **~ty** s. ιλαρότητα f.
hill s. λόφος m. **~ock** s. λοφίσκος m. **~side** s. πλαγιά f. **~y** a. (country) λοφώδης, (road) πάνω σε λόφους.
hilt s. λαβή f. (fig.) up to the ~ πέρα για πέρα.
him pron. I saw ~ τον είδα, I gave ~ a book του έδωσα ένα βιβλίο, for ~ για εκείνον. **~self** pron. see self.
hind s. (deer) ελαφίνα f.
hind a. οπίσθιος, πισινός. **~most** a. τελευταίος. **~quarters** s. καπούλια n.pl. **~sight** s. by ~sight εκ των υστέρων.
hind|er v. εμποδίζω. **~rance** s. εμπόδιο n.
hinge s. μεντεσές m., ρεζές m. (v.i.) (fig.) ~ upon εξαρτώμαι από.
hint s. υπαινιγμός m. (trace) ίχνος n. (advice) συμβουλή f. drop a ~ ρίχνω ένα πόντο, he took the ~ μήπκε με την πρώτη. (v.) ~ at υπονοώ, υπαινίσσομαι.
hinterland s. ενδοχώρα f.
hip s. ισχίον n., γοφός m.
hippopotamus s. ιπποπόταμος m.
hire v. μισθώνω, νοικιάζω, (an employee) προσλαμβάνω. ~ out εκμισθώνω. (s.) μίσθωση f., ενοικίαση f. πρόσληψη f. (pay) αμοιβή f. **~ling** s. πληρωμένος a. **~-purchase** s. on ~ purchase με δόσεις.
hirsute a. μαλλιαρός.
his pron. ~ book το βιβλίο του, is it ~? είναι

δικό του;
hiss *v.* (*also* ~ at) σφυρίζω. (*s.*) σφύριγμα *n.*
histology *s.* ιστολογία *f.*
histor|y *s.* ιστορία *f.* **~ian** *s.* ιστορικός *m.*
~ic, ~ical *a.* ιστορικός. **~ically** *adv.* ιστορικώς.
histrionic *a.* θεατρινίστικος. **~s** *s.* θεατρινίστικα καμώματα.
hit *v.* χτυπώ, (*target*) βρίσκω. you've ~ it (*succeeded*) το πέτυχες, he was ~ (*by missile*) εβλήθη, χτυπήθηκε. ~ one's head on a post χτυπώ το κεφάλι μου σε μία κολόνα. they don't ~ it off δεν τα πάνε καλά. ~ on βρίσκω κατά τύχην. ~ out at επιτίθεμαι κατά (*with gen.*). ~ hard πλήττω. (*s.*) χτύπημα *n.* (*success*) επιτυχία *f.*
hitch *s.* εμπόδιο *n.* it went off without a ~ πήγε ομαλά, πήγε ρολόι.
hitch *v.* (*attach*) προσδένω. ~ up (*raise*) σηκώνω. (*fam.*) ~ a ride κάνω ωτοστόπ. **~-hiking** *s.* ωτοστόπ *n.*
hither *adv.* (προς τα) εδώ. **~to** *adv.* μέχρι τούδε.
hit-or-miss *a.* στην τύχη.
hive *s.* κυψέλη *f.* (*fig.*) ~ off (*v.t.*) αποχωρίζω, (*v.i.*) αποχωρίζομαι.
hoard *s.* θησαυρός *m.* (*v.*) μαζεύω και κρύβω.
hoarding *s.* (*billboard*) ψηλός ξύλινος φράκτης (για διαφημίσεις).
hoarfrost *s.* πάχνη *f.*
hoarse *a.* βραχνός. **~ness** *s.* βραχνάδα *f.*
hoary *a.* άσπρος, (*fig.*) πανάρχαιος.
hoax *s.* φάρσα *f.* (*v.t.*) παίζω φάρσα σε.
hob *s.* μάτι κουζίνας.
hobble *v.i.* (*limp*) κουτσαίνω. (*v.t.*) (*bind*) πεδικλώνω. (*s.*) πέδικλο *n.*
hobbledehoy *s.* μαντράχαλος *m.*
hobby *s.* χόμπι *n.* **~-horse** *s.* (*fam.*) βίδα *f.*, μανία *f.*
hobgoblin *s.* καλλικάντζαρος *m.*
hobnailed *a.* με καρφιά.
hobnob *v.* ~ with συναναστρέφομαι.
hobo *s.* αλήτης *m.*
hocus-pocus *a.* απάτη *f.*
hod *s.* πηλοφόρι *n.*
hoe *s.* σκαλιστήρι *n.* (*v.*) σκαλίζω.
hog *s.* χοίρος *m.* (*fig.*) go the whole ~ φτάνω στα άκρα. **~-wash** *s.* (*fam.*) μπούρδες *f.pl.*
hoist *v.* υψώνω, ανεβάζω. (*s.*) τράβηγμα *n.* (*machine*) αναβατήρας *m.*
hold *v.t.* κρατώ, (*possess*) έχω, (*support*) βαστώ, (*contain*) περιέχω, (*have room for*) χωράω, παίρνω. (*restrain*) συγκρατώ, (*consider*) θεωρώ. (*a post*) κατέχω, (*a meeting*) κάνω, οργανώνω, be held (*take place*) γίνομαι. ~ water (*be logical*) στέκο-

μαι. ~ one's own (*or* one's ground) κρατώ γερά. ~ one's peace (*or* one's tongue) σωπαίνω. ~ one's hand (*not act*) καθυστερώ τη δράση μου. they were ~ing hands κρατιούνταν από το χέρι. (*v.i.*) (*endure*) αντέχω, (*last out*) κρατώ. ~ fast, firm, tight κρατιέμαι. ~ (good) (*be valid*) ισχύω. ~ forth (*talk*) μακρηγορώ.
hold back *v.t.* (*restrain*) συγκρατώ, (*hinder*) εμποδίζω, (*hide*) κρύβω. (*v.i.*) επιφυλάσσομαι, διστάζω.
hold by *v.t.* (*stick to*) παραμένω πιστός σε.
hold down *v.t.* (*repress*) καταπιέζω, (*a job*) κρατώ.
hold in *v.t.* συγκρατώ.
hold off *v.t.* κρατώ σε απόσταση. (*v.i.*) περιμένω σε απόσταση, απέχω. the rain held off δεν έβρεξε.
hold on *v.i.* (*resist*) ανθίσταμαι. ~ on! (*stop*) στάσου! ~ to (*keep securely*) κρατώ καλά, (*for support*) πιάνομαι από. (*v.t.*) (*keep in place*) κρατώ.
hold out *v.t.* (προ)τείνω, (*give*) δίνω. (*v.i.*) (*resist*) αντέχω, (*insist*) επιμένω, (*contain oneself*) κρατιέμαι.
hold over *v.t.* (*defer*) αναβάλλω.
hold to *v.* (*stick to*) κρατώ, παραμένω πιστός σε. I held him to his promise τον υποχρέωσα να τηρήσει την υπόσχεσή του.
hold up *v.t.* (*raise*) σηκώνω, (*as example*) παρουσιάζω, (*interrupt*) διακόπτω, (*delay*) καθυστερώ, (*rob*) ληστεύω.
hold-up *s.* διακοπή *f.* (*robbery*) ληστεία *f.*
hold with *v.* εγκρίνω (*usually neg.*).
hole *s.* τρύπα *f.*, οπή *f.* make a ~ in ανοίγω τρύπα σε, τρυπώ. go into ~s τρυπώ. in ~s τρύπιος. pick ~s in βρίσκω τρωτά σε. be in a ~ (*fig.*) βρίσκομαι σε δύσκολη θέση. **~-and-corner** *a.* μυστικός.
holiday *s.* (*public*) αργία *f.* (*feast-day*) εορτή *f.* (*leave*) άδεια *f.* ~s διακοπές *f.pl.* **~-maker** *s.* (*in summer*) παραθεριστής *m.* (*excursionist*) εκδρομεύς *m.*
holiness *s.* αγιότητα *f.*
hollo, ~a *v.i.* φωνάζω.
hollow *a.* κούφιος, κοίλος, (*sunken*) βαθουλός. (*of sound*) σπηλαιώδης. (*false*) απατηλός, (*fig., empty*) κενός. (*s.*) (*depression*) βαθούλωμα *n.* (*of tree, tooth*) κουφάλα *f.* (*v.t.*) βαθουλώνω, ~ out σκάβω. (*adv.*) (*fam.*) beat (*person*) ~ νικώ κατά κράτος.
holly *s.* λιόπρινο *n.*
holm-oak *s.* πρινάρι *n.*
holocaust *s.* ολοκαύτωμα *n.*
holy *a.* άγιος. (*Gospel, Office, Synod, ground, war*) ιερός. ~ man άγιος *m.* ~ water αγιασμός *m.* H~ Land οι Άγιοι Τόποι, H~ Week Μεγάλη Εβδομάδα, H~

of Holies Τα Άγια των Αγίων.

homage s. do (or pay) ~ δηλώ υποτέλειαν, (fig.) αποτίω φόρον τιμής.

home s. (house) σπίτι n. (country) τόπος m., πατρίδα f. give a ~ to (accommodate) φιλοξενώ, feel at ~ αισθάνομαι άνετα, make oneself at ~ φέρομαι σαν στο σπίτι μου. be at ~ (receive visits) δέχομαι. Sailors' H~ Οίκος του Ναύτου. (nursing, etc.) κλινική f. ~ team ντόπιοι. **~coming** s. γυρισμός m. (of emigrant) παλιννόστηση f. **~land** s. πατρίδα f. **~ly** a. (simple) απλός, (warm) ζεστός, (like ~) σπιτικός. **~made** a. σπιτίσιος, σπιτικός. **~sick** a. feel ~sick έχω νοσταλγία. **~spun** a. (fig.) απλός. **~stead** s. αγρόκτημα n. **~ward(s)** a. & adv. προς το σπίτι. **~work** s. (formal) κατ' οίκον εργασία, (fam.) μαθήματα n.pl.

home a. (of household) οικιακός, του σπιτιού. (of country: products) εγχώριος, ντόπιος, (affairs) εσωτερικός. (of family) οικογενειακός. ~ town γενέτειρα f. ~ rule αυτοδιοίκηση f. ~ team or side ντόπιοι n.pl. H~ Office Υπουργείον Εσωτερικών.

homicide s. ανθρωποκτονία f.

homily s. κήρυγμα n.

homoeopathy s. ομοιοπαθητική f.

homogene|ous a. ομοιογενής, **~ity** s. ομοιογένεια f.

homograph s. ομόγραφος λέξις.

homophone s. ομώνυμος λέξις.

homosexual a. & s. ομοφυλόφιλος.

hone s. ακόνι n. (v.) ακονίζω.

honest a. τίμιος, έντιμος. (frank) ειλικρινής. **~ly** adv. τίμια, ειλικρινά. **~y** s. τιμότητα f., εντιμότητα f.

honey s. μέλι n. of or like ~ μελένιος. **~comb** s. κηρήθρα f. **~ed** a. μελιστάλακτος, όλο μέλι. **~moon** s. μήνας του μέλιτος. **~suckle** s. αγιόκλημα n.

honorarium s. αμοιβή f.

honor|ary a. (person) επίτιμος, (degree, etc.) τιμητικός. **~ific** a. τιμητικός.

honour v.t. τιμώ. (a bill, etc.) πληρώνω, εξοφλώ.

honour s. τιμή f. (esteem) εκτίμηση f. (uprightness) εντιμότητα f., τιμιότητα f. (mark of distinction) τιμητική διάκριση f. affair of ~ μονομαχία f. guard of ~ τιμητική φρουρά, word of ~ λόγος τιμής, maid of ~ δεσποινίς επί των τιμών. make it a point of ~ φιλοτιμούμαι. be in ~ bound έχω ηθική υποχρέωση. do the ~s περιποιούμαι τους καλεσμένους. in ~ of προς τιμήν (with gen.). **~able** a. έντιμος, τίμιος. **~ably** adv. τίμια.

hood s. κουκούλα f. **~ed** a. με κουκούλα.

hoodwink v.t. πιάνω κορόιδο.

hoof s. οπλή f., νύχι n.

hoo-ha s. φασαρία f.

hook s. (for hanging up, grappling) γάντζος m. (small) γαντζάκι n. (fish) αγκίστρι n. (meat) τσιγκέλι n. (reaping) δρεπάνι n. ~-and-eye fastener κόπιτσα. by ~ or by crook οπωσδήποτε, on one's own ~ μόνος μου. sling one's ~ στρίβω, ~ line and sinker τελείως. (v.t.) γαντζώνω, (fasten dress, etc.) κουμπώνω. (fig.) (inveigle) τυλίγω, βουτώ. get ~ed on κυριεύομαι από το πάθος (with gen.). ~ it στρίβω. **~ed** a. γαμψός. **~nosed** a. με γαμψή μύτη.

hookah s. ναργιλές m.

hooligan s. ταραχοποιός m.

hoop s. στεφάνη f., τσέρκι n. put (person) through the ~ υποβάλλω εις δοκιμασίαν.

hoopoe s. έποψ m., τσαλαπετεινός m.

hooray int. μπράβο, ζήτω.

hoot v.i. (of owl) σκούζω, (of motor) κορνάρω, (of siren) σφυρίζω. (v.t. & i.) (jeer) γιουχαΐζω. (s.) (also ~ing) σκούξιμο n. κορνάρισμα n. γιουχαΐσμός m. **~er** s. σειρήνα f.

hop v. πηδώ (στο ένα πόδι). (fam.) ~ it στρίβω. (s.) πήδημα n. (fam., dance) χορός m. you caught me on the ~ με βρήκες απροετοίμαστο.

hope s. ελπίς, ελπίδα f. (v.) (also ~ for) ελπίζω, (expect) περιμένω.

hopeful a. be ~ έχω ελπίδα, ελπίζω. (optimistic) αισιόδοξος, (causing hope) ελπιδοφόρος, the situation seems more ~ η κατάσταση παρέχει μεγαλύτερες ελπίδες. (fam.) young ~ φερέλπις νεανίας. **~ly** adv. έχοντας ελπίδες, αισιόδοξα.

hopeless a. (despairing) απελπισμένος, (despaired of) απελπιστικός. (fam.) it (or he) is ~ (or a ~ case) είναι απελπισία. **~ly** adv. απελπισμένα. **~ness** s. απελπιστική κατάσταση.

hoplite s. οπλίτης m.

hopper s. (in mill) χωνί n.

hop-scotch s. κουτσό n., καλόγερος m.

horde s. ορδή f.

horizon s. ορίζοντας m. **~tal** a. οριζόντιος. **~tally** adv. οριζοντίως.

hormone s. ορμόνη f.

horn n. (animal's) κέρας n., κέρατο n. (substance) κέρατο n. (mus.) κόρνο n. (motor) κόρνο n., κλάξον n. sound one's ~ κορνάρω, shoe-~ κόκκαλο n. (snail's) αντένα. f. draw in one's ~s υποχωρώ. on the ~s of a dilemma προ διλήμματος. ~ of plenty κέρας της Αμαλθείας. Golden H~ Κεράτειος Κόλπος, Χρυσούν Κέρας. (a.) κοκάλινος, από κόκκαλο. **~ed** a. κερασφόρος, με κέρατα. **~y** a. κερατοειδής,

(*gnarled*) ϱοζιάϱικος.

hornet s. σκούϱκος m. (*fig.*) stir up a ~'s nest δημιουϱγώ φασαϱίες για τον εαυτό μου.

hornpipe s. χοϱός ναυτών.

horoscope s. ωϱοσκόπιο n.

horrendous a. ανατϱιχιαστικός.

horrib|le a. φοβεϱός, φϱικτός. **~ly** adv. φοβεϱά, φϱικτά.

horrid a. αποκϱουστικός, αντιπαθέστατος.

horrific a. φϱικιαστικός, φϱικαλέος.

horrify v. πϱοκαλώ φϱίκη σε. **~ing** a. φϱικιαστικός, φϱικαλέος.

horror s. φϱίκη f. I have a ~ of it μου πϱοκαλεί φϱίκη. **~-struck** a. be ~-struck φϱίσσω.

hors-d'oeuvre s. οϱεκτικό n.

horse s. άλογο n., ίππος m. (*cavalry*) ιππικό n. (*vaulting*) εφαλτήϱιο n. (*clothes*) καλόγεϱος m. white ~s (*on waves*) αφνάκια n.pl. ~ racing κούϱσες f.pl. **~-back** s. on ~back έφιππος a., καβάλλα adv. **~-chestnut** s. αγϱιοκαστανιά f. **~-flesh** s. (*to eat*) αλογίσιο κϱέας, (*horses*) άλογα n.pl. **~-fly** s. αλογόμυγα f. **~-hair** s. από τϱίχες αλόγου. **~-man** s. καβαλλάϱης m., ιππεύς m. **~-play** s. θοϱυβώδες παιχνίδι. **~-power** s. ιπποδύναμη f. 12 ~-power δώδεκα ίππων. **~-radish** s. χϱάνο n. **~-sense** s. κοινός νους. **~-shoe** s. πέταλο n. **~-whip** v. μαστιγώνω. (s.) καμτσίκι n. **~-woman** s. ιππεύτϱια f.

horticultur|e s. φυτοκομία f. **~ist** s. φυτοκόμος m.

hos|e s. (*apparel*) κάλτσες f.pl. **~iery** s. ανδϱικά είδη.

hose, ~pipe s. λάστιχο n., σωλήνας ποτίσματος. (v.t.) καταβϱέχω με το λάστιχο.

hospice s. (*travellers'*) ξενώνας m. (*for sick*) άσυλο n.

hospitab|le a. φιλόξενος. **~ly** adv. φιλόξενα.

hospital s. νοσοκομείο n.

hospitality s. φιλοξενία f.

host s. οικοδεσπότης m. (*of inn*) ξενοδόχος m. (*eccl.*) Η~ όστια f. (*biol.*) ξενιστής m. (*multitude*) πλήθος n. (*army*) στρατιά f.

hostage s. όμηϱος m.

hostel s. οικοτϱοφείο n. (*of university*) οίκος φοιτητού. youth ~ ξενώνας νεότητας.

hostess s. οικοδέσποινα f. air ~ ιπτάμενη f., αεϱοσυνοδός f.

hostil|e a. εχθϱικός. **~ity** s. εχθϱικότητα f. **~ities** εχθϱοπϱαξίες f.pl.

hot a. ζεστός, θεϱμός, (*peppery*) καυτεϱός, ~ springs θεϱμαί πηγαί. it is ~ (*weather*) κάνει ζέστη. get (*or* feel) ~ ζεσταίνομαι. I

went ~ all over έπαθα έξαψη. (*fam.*) get ~ under the collar ανάβω, εξάπτομαι, με πιάνει θυμός. ~ on the scent επί τα ίχνη. get into ~ water έχω μπλεξίματα. blow ~ and cold αλλάζω συνεχώς γνώμη. he is ~ on Bach είναι ειδικός στον Μπαχ. (*fig.*) ~ air μπούϱδες f.pl. **~-bed** s. (*fig.*) εστία f. **~-blooded** a. θεϱμόαιμος. **~-foot** adv. ολοταχώς. **~-head** s., **~-headed** a. αυθόϱμητος. **~-house** s. θεϱμοκήπιο n. **~-ly** adv. (*angrily*) θυμωμένα, (*with drive*) με πάθος. **~-plate** s. (*of stove*) μάτι n. **~-pot** s. ϱαγκού n. **~-tempered** a. ευέξαπτος. **~-water-bottle** s. θεϱμοφόϱα f.

hot v.t. ~ up ξαναζεσταίνω. (v.i.) ~ up ξαναζεσταίνομαι. (*fig.*) things are ~ting up παϱουσιάζεται κίνηση.

hotch-potch s. συνονθύλευμα n.

hotel s. ξενοδοχείο n. **~ier** s. ξενοδόχος m.

hound s. λαγωνικό n. (v.) καταδιώκω.

hour s. ώϱα f. for ~s επί ώϱες, με τις ώϱες. ~ hand (*of clock*) ωϱοδείκτης m. **~-glass** s. αμμόμετϱο n. **~-ly** a. ωϱιαίος. (*adv.*) κάθε ώϱα, (*by the ~*) με την ώϱα.

houri s. ουϱί n.

house s. σπίτι n. (*business, royal, God's*) οίκος m. (*of Parliament*) Βουλή f. private ~ μονοκατοικία f. keep ~ κϱατώ το σπίτι, move ~ μετακομίζω, set up ~ εγκαθίσταμαι. full ~ πλήϱης αίθουσα, ~ of cards χάρτινος πύϱγος. it brought down the ~ χάλασε ο κόσμος. we get on like a ~ on fire τα πάμε μια χαϱά. (v.) στεγάζω, (*put up*) φιλοξενώ, (*store*) αποθηκεύω. **~-agent** s. κτηματομεσίτης m. **~-boat** s. πλωτό σπίτι. **~-breaker** s. διαϱϱήκτης m. **~-keeper** s. οικονόμος m.f. **~-keeping** s. νοικοκυϱιό n. **~-maid** s. υπηϱέτϱια f. **~-warming** s. εγκαίνια του σπιτιού. **~-wife** s., **~-wifely** a. νοικοκυϱά f. **~-work** s. δουλειές του σπιτιού.

household s. οίκος m., σπιτικό n. (a.) οικιακός, του σπιτιού. ~ gods εφέστιοι θεοί. ~ word πασίγνωστος a. **~er** s. νοικοκύϱης m.

housing s. στέγαση f. ~ question στεγαστικό ζήτημα.

hovel s. τϱώγλη f.

hover v. ζυγίζομαι, (*waver*) ταλαντεύομαι.

how adv. πώς, ~ much πόσο, (a.) πόσος. ~ far πόσο μακϱιά, ~ old πόσων χϱονών. ~ wide is it? τι (*or* πόσο) φάϱδος έχει; ~ often do the trains run? κάθε πόσο έχει τϱαίνο; ~ long did you stay there? πόσον καιϱό μείνατε εκεί; ~ long does it take? πόση ώϱα θέλει; ~ nice! τι ωϱαία!

however adv. (*nevertheless*) όμως. ~ much it may rain όσο και να βϱέχει. (*in whatever* *manner*) με οποιονδήποτε τρόπο. ~ you please όπως σας αρέσει.

howitzer *s.* οβιδοβόλον *n.*

howl *v.* ουρλιάζω, ωρύομαι. (*s.*) (*also* ~ing) ουρλιασμα *n.*

hoyden *s.* τρελλοκόριτσο *n.*

HP ιπποδύναμη *f.*

hub *s.* πλήμνη *f.* (*fig.*) κέντρο *n.*

hubbub *s.* οχλοβοή *f.*, χάβρα *f.*

huckster *s.* γυρολόγος *m.*

huddle *v.i.* (*also* be ~d) στριμώχνομαι, σφίγγομαι, (*nestle*) κουλουριάζομαι. be ~d (*of village, house*) είμαι χωμένος. (*s.*) (*pile*) σωρός *m.* go into a ~ συσκέπτομαι κρυφά.

hue *s.* χρώμα *n.* ~ and cry (*pursuit*) γενική καταδίωξη, (*outcry*) κατακραυγή *f.*

huff *s.* in a ~ προσβεβλημένος. ~**y** *a.* μυγιάγγιχτος.

hug *v.* σφίγγω στην αγκαλιά μου, (*fig.*, *cling to*) προσκολλώμαι εις, (*of garment*) σφίγγω, (*the coast, etc.*) περνώ σύρριζα σε. (*s.*) σφιχταγκάλιασμα *n.*

huge *a.* πελώριος, τεράστιος, (*esp. in height*) θεόρατος. ~**ly** *adv.* λίαν, πάρα πολύ.

hulk *s.* (*pej.*) (*ship*) σαπιοκάραβο *n.* (*person, also* ~ing *a.*) μπατάλικος.

hull *s.* (*ship's*) σκάφος *n.*, κουφάρι *n.*

hull *s.* (*shuck*) φλοιός *m.* (*v.*) ξεφλουδίζω.

hullabaloo *s.* χάβρα *f.*, σαματάς *m.* cause a ~ χαλάω τον κόσμο.

hullo *int. see* hallo.

hum *v.* βουίζω, βομβώ, (*of person*) μουρμουρίζω. ~ and haw τα μασώ. (*the office, etc.*) is ~ming with activity σφύζει από δραστηριότητα.

human *a.* ανθρώπινος. ~ being ανθρώπινον ον.

humane *a.* γεμάτος ανθρωπιά, (*classical*) κλασσικός. ~**ly** *adv.* με ανθρωπιά.

human|ism *s.* ανθρωπισμός *m.* ~**ist** *a.* ανθρωπιστής *m.*

humanitarianism *s.* φιλανθρωπισμός *m.*

humanit|y *s.* (*mankind*) ανθρωπότητα *f.* (*humaneness*) ανθρωπιά *f.* ~**ies** (*classics*) κλασσικές σπουδές.

humanize *v.* εξανθρωπίζω.

humb|le *a.* ταπεινός. (*v.*) ταπεινώνω. ~**ly** *adv.* ταπεινά. ~**ly** born ταπεινής καταγωγής.

humbug *s.* (*sham*) μπούρδες *f.pl.*, κοροϊδία *f.* (*person*) τσαρλατάνος *m.*

humdrum *a.* μονότονος, της αράδας.

humid *a.* υγρός. ~**ity** *s.* υγρασία *f.*

humiliat|e *v.* ταπεινώνω. ~**ion** *s.* ταπείνωση *f.*

humility *s.* ταπεινοφροσύνη *f.*

humorist *s.* χιουμορίστας *m.*

humorous *a.* αστείος. (*drawing, etc.*) χιουμοριστικός. he is very ~ έχει πολύ χιούμορ. ~**ly** *adv.* με χιούμορ.

humour *s.* (*temper*) διάθεση *f.* be in good ~ είμαι στις καλές μου, έχω κέφι. (*being funny*) χιούμορ. sense of ~ αίσθηση του χιούμορ. (*v.t.*, *a person*) του κάνω τα κέφια, πάω με τα νερά του.

hump *s.* ύβος *m.*, καμπούρα *f.* (*bad mood*) have the ~ είμαι κακόκεφος. ~**ed** *a.* καμπουρωτός.

hunch *v.* καμπουριάζω. (*s.*) (*premonition*) προαίσθηση *f.* ~**back** *s.* καμπούρης *a.*

hundred *num.* εκατό(ν). ~s εκατοντάδες *f.pl.* two ~ (*a.*) διακόσιοι. ~-drachma note εκατοστάρι(κο) *n.* (*fig.*) a ~ and one χίλια δύο. ~**th** *a.* εκατοστός. ~**weight** *s.* στατήρ *m.*

hunger *s.* πείνα *f.* (*fig.*) δίψα *f.* (*v.*) ~ for (*fig.*) διψώ για.

hungry *a.* πεινασμένος, (*fig.*) διψασμένος (για). be *or* feel ~ πεινώ.

hunk *s.* κομματάρα *f.*

hunt *v.t.* & *i.* κυνηγώ. (*search, also* ~ for) ψάχνω, γυρεύω, αναζητώ. ~ down καταδιώκω. (*s.*) κυνήγι *n.*, θήρα *f.* (*search*) ψάξιμο *n.* ~**er** *s.* κυνηγός *m.* ~**ress** *s.* κυνηγός *f.* (*Diana*) Κυνηγέτις *f.*

hunting *s.* κυνήγι *n.*, θήρα *f.* (*a.*) κυνηγετικός. ~ ground κυνηγότοπος *m.*, (*fig.*) good (*or* happy) ~ ground παράδεισος *m.*

huntsman *s.* κυνηγός *m.*

hurdle *s.* καλαμωτή *f.* (*in race & fig.*) εμπόδιο *n.*

hurdy-gurdy *s.* ρομβία *f.*

hurl *v.* ρίχνω, πετώ, τινάζω, (*also fig.*) εκτοξεύω. ~ oneself ρίχνομαι, ορμώ.

hurly-burly *s.* φασαρία *f.*

hurrah *int.* ζήτω, μπράβο.

hurricane *s.* λαίλαψ *f.*

hurried *a.* βιαστικός. ~**ly** *adv.* βιαστικά, εσπευσμένως.

hurry *s.* σπουδή *f.*, βία *f.*, βιασύνη *f.* be in a ~ βιάζομαι, επείγομαι. (*see* hurried).

hurry *v.t.* (*also* ~ along, on, up) (*an event*) επισκεύδω, (*an activity*) επιταχύνω. (*send quickly*) στέλνω βιαστικά. ~ him up τον κάνω να βιαστεί. they hurried him away τον έβγαλαν άρον-άρον έξω. (*v.i.*) (*also* ~ up) σπεύδω, κάνω γρήγορα, βιάζομαι. ~ up! κάνε γρήγορα, βιάσου! ~ away *or* off φεύγω βιαστικά.

hurt *v.t.* & *i.* (*feel or cause pain*) my foot ~s (me) (με) πονάει το πόδι μου. he ~ himself χτύπησε, he ~ his foot χτύπησε το πόδι του. (*harm*) βλάπτω, (*offend*) θίγω. (*trouble, matter*) πειράζω. be ~ (*in*

accident) πληγώνομαι. (come to harm) the
plant won't ~ if it's left in the sun το φυτό
δεν θα πάθει τίποτα αν μείνει στον ήλιο.
(s.) βλάβη f., ζημία f., κακό n. (a.) (injured)
πληγωμένος, (offended) πειραγμένος. ~ful
a. βλαβερός, επιβλαβής. (to feelings)
πειρακτικός.
hurtle v.i. ορμώ, (through air) εκσφενδο-
νίζομαι, (down) κατρακυλώ.
husband s. άνδρας m., σύζυγος m. (v.t.)
οικονομώ.
husband|man s. καλλιεργητής m. ~ry s.
γεωργία f. (fig.) οικονομία f.
hush v.t. & i. (stop talking) σωπαίνω, (make
or become quiet) ησυχάζω. ~ up
αποσιωπώ. in ~ed tones χαμηλοφώνως.
(s.) σιωπή f. (of night) σιγαλιά f. (int.)
σιωπή! (a.) ~~~ μυστικός. ~-money s.
δωροδοκία f.
husk s. φλοιός m. (v.) ξεφλουδίζω.
husk|y a. (hoarse) βραχνός, (strong) γερο-
δεμένος. ~iness s. βραχνάδα f. ρωμα-
λεότητα f.
hussy s. αναιδής a.
hustings s. προεκλογική κίνηση. (platform)
εξέδρα f.
hustle v.t. (push) σπρώχνω, (hasten) επισπεύδω. (v.i.) βιάζομαι. (s.) βιασύνη f.
hut s. καλύβα f. (army, etc.) παράπηγμα n.
~ment s. παραπήγματα n.pl.
hutch s. κουμάσι για κουνέλια.
hyacinth s. ζουμπούλι n.
hybrid a. νοθογενής (s.) διασταύρωση f.
hydra s. ύδρα f.
hydrangea s. ορτανσία f.
hydrant s. υδροστόμιο n.
hydraulic a. υδραυλικός.
hydrochloric a. υδροχλωρικός.
hydro-electric a. υδροηλεκτρικός.
hydrogen s. υδρογόνο n.
hydrophobia s. υδροφοβία f.
hyena s. ύαινα f.
hygien|e s. υγιεινή f. ~ic a. υγιεινός.
~ically adv. υγιεινώς.
hymn s. ύμνος m. (v.) υμνώ.
hyper- υπερ-.
hyperbole s. υπερβολή f.
hyper-critical a. υπερβολικά αυστηρός.
hyper-sensitive a. υπερευαίσθητος.
hypertension s. υπέρταση f. (fam.) πίεση f.
hyphen s. ενωτικό n.
hypno|sis s. ύπνωση f. ~tic a. (drug)
υπνωτικός, (gaze) που σε υπνωτίζει.
~tism s. υπνωτισμός m. ~tist s. υπνω-
τιστής m. ~tize v. υπνωτίζω.
hypochondria s. υποχονδρία f. ~c s.
υποχονδριακός m.
hypocri|sy s. υποκρισία f. ~te s. υποκριτής

m. ~tical a. υποκριτικός.
hypodermic a. υποδόριος.
hypotenuse s. (math.) υποτείνουσα f.
hypothe|sis s. υπόθεση f. ~tical a. υποθε-
τικός.
hyster|ia s. υστερία f. ~ical a. υστερικός.
~ically adv. υστερικά. ~ics s. have ~ics
με πιάνει κρίση νεύρων.

I

I pron. εγώ.
iambic a. ιαμβικός.
Iberian a. ιβηρικός.
ice s. πάγος m. (confection) παγωτό n. ~ age
περίοδος των παγετώνων. (fig.) cut no ~
with δεν εντυπωσιάζω. skate on thin ~
εισέρχομαι εις επικίνδυννον έδαφος. (v.t. &
i.) (also ~ over) παγώνω. ~d παγωμένος.
~berg s. παγόβουνο n. ~box s. παγωνιέρα
f. ~cream s. παγωτό n. ~cube s. παγάκι
n. ~rink s. πίστα παγοδρομιών. ~
skating s. παγοδρομίες f.pl.
Icelandic a. ιολανδικός.
icicle s. παγοκρύσταλλος m.
icily adv. ψυχρώς.
icing s. γκλασάρισμα n.
icon s. (αγία) εικών f., εικόνισμα n.
iconocl|asm s. εικονοκλασία f. ~ast s.
εικονοκλάστης m. ~astic a. εικονοκλα-
στικός.
icy a. (attitude) παγερός, (thing) παγωμένος.
Id ταυτότητα αναγνώρισης.
idea s. ιδέα f. (opinion) γνώμη f.
ideal a. ιδανικός, ιδεώδης. (s.) ιδανικό n.,
ιδεώδες n. ~ly adv. ~ly I should like to...
το ιδεώδες θα ήταν να.
ideal|ism s. ιδεαλισμός m. ~ist s. ιδεα-
λιστής m. ~istic a. ιδεαλιστικός.
identical a. ο ίδιος.
identification s. διαπίστωση ταυτότητος.
~ paper(s) ταυτότητα f.
identify v.t. (establish identity of) διαπιστώνω την ταυτότητα (with gen.). (recognize) αναγνωρίζω. (treat as identical)
ταυτίζω, ~ oneself with ταυτίζομαι με.
identity s. (also ~ card) ταυτότητα f.
ideolog|y s. ιδεολογία f. ~ical a. ιδεολογι-
κός.
idiocy s. ηλιθιότητα f., βλακεία f.
idiom s. (language) ιδίωμα n. (peculiar
form) ιδιω(μα)τισμός m. ~atic a. ιδιω-
ματικός.
idiosyncra|sy s. (fam.) ιδιομορφία f. ~tic
a. ιδιόμορφος.

idiot *s.*, **~ic** *a.* ηλίθιος, ανόητος. **~ically** *adv.* ηλίθια, ανόητα.

idle *a.* (*lazy*) οκνηρός, τεμπέλης. (*not working*) αργός, (*machine*) σταματημένος, (*capital*) νεκρός. (*worthless*) μάταιος. **~ness** *s.* τεμπελιά *f.*

idle *v.i.* (*be* ~) τεμπελιάζω, χασομερώ. (*v.t.*) ~ away χάνω. **~r** *s.* χασομέρης *a.*

idly *adv.* (*slowly*) αργά, με το πάσσο μου. (*vainly*) ματαίως. (*doing nothing*) με σταυρωμένα χέρια.

idol *s.* είδωλο *n.* (*admired person*) ίνδαλμα *n.* **~ize** *v.* θαυμάζω μέχρις ειδωλολατρίας, she ~izes him τον έχει ίνδαλμα.

idolat|er *s.* ειδωλολάτρης *m.* **~rous** *a.* ειδωλολατρικός. **~ry** *s.* ειδωλολατρία *f.*

idyll *s.* ειδύλλιο *n.* **~ic** *a.* ειδυλλιακός.

if *conj.* αν, εάν. even ~ και να, ακόμα κι αν. as ~ σα να (*with indicative*). ~ only the weather would improve! μακάρι να έφτιαχνε ο καιρός. ~ only it wasn't so far! να μην ήταν μόνο τόσο μακριά. I should like to go ~ only for a few days θα ήθελα να πάω έστω και για λίγες μέρες. (*I wouldn't do it*) - not even ~ you paid me ακόμα και να με πλήρωνες.

igloo *s.* ιγκλού *n.*

ignit|e *v.t. & i.* ανάβω. **~ion** *s.* ανάφλεξη *f.* ~ion key κλειδί μίζας.

ignob|le *a.* επαίσχυντος, ντροπιαστικός. **~ly** *adv.* αισχρά.

ignomin|y *s.* ατίμωση *f.* **~ious** *a.* ατιμωτικός. **~iously** *adv.* κατά επαίσχυντον τρόπον.

ignoramus *s.* αμαθής *a.*, ζώον *n.*

ignor|ance *s.* αμάθεια *f.* (*unawareness*) άγνοια *f.* **~ant** *a.* (*pej.*) αμαθής. (*unaware*) ανίδεος, be ~ant of (*not know*) αγνοώ.

ignore *v.* (*danger, etc.*) αψηφώ, (*overlook*) παραβλέπω. (*take no notice of, snub*) αγνοώ.

ikon *s. see* icon.

ilex *s.* πρινάρι *n.*

ill *a.* (*sick*) άρρωστος, ασθενής. be ~ ασθενώ, fall *or* make ~ αρρωσταίνω. (*bad*) κακός. ~ feeling έχθρα *f.* ~ fortune *or* luck κακοτυχία *f.* (*s.*) κακό *n.* **~s** (*misfortunes*) κακά *n.pl.*, κακοτυχίες *f.pl.*(adv.) κακώς. ~ at ease στενοχωρημένος. **~advised** *a.* ασύνετος. **~assorted** *a.* αταίριαστος. **~behaved** *a.* ανάγωγος, (*child*) κακομαθημένος. **~bred** *a.* κακοαναθρεμμένος. **~breeding** *s.* κακή ανατροφή. **~disposed** *a.* κακώς διατεθειμένος. **~fated** *a.* (*person*) κακότυχος, δύσμοιρος, (*action*) μοιραίος. **~favoured** *a.* δυσειδής. **~gotten** *a.* παράνομος. **~humoured** *a.* κακότροπος. **~mannered** *a.* ανάγωγος.

~judged *a.* ασύνετος, (*indiscreet*) αδιάκριτος. **~matched** *a.* αταίριαστος. **~natured** *a.* κακός. **~omened** *a. see* ill-fated. **~starred** *a. see* ill-fated. **~tempered** *a.* κακότροπος. **~timed** *a.* άκαιρος.

ill-|treat, **~use** *v.* κακομεταχειρίζομαι, **~use**, **~treatment** *s.* κακομεταχείριση *f.*

illegal *a.* παράνομος. **~ity** *s.* παρανομία *f.* **~ly** *adv.* παρανόμως.

illegible *a.* δυσανάγνωστος.

illegitimacy *s.* (*irregularity*) ανορθοδοξία *f.* (*of birth*) on account of his ~ λόγω του ότι είναι νόθος.

illegitimate *a.* ανορθόδοξος, (*by birth*) εξώγαμος.

illiberal *a.* ανελεύθερος.

illicit *a.* αθέμιτος. **~ly** *adv.* αθεμίτως.

illiter|ate *a.* αγράμματος. **~acy** *s.* αναλφαβητισμός *m.*

illness *s.* αρρώστ(ε)ια *f.*, ασθένεια *f.*

illogical *a.* he is ~ δεν είναι λογικός. **~ly** *adv.* ενάντια στη λογική.

illuminat|e *v.* (*light*) φωτίζω, (*festively*) φωταγωγώ. (*fig. make clear*) διαφωτίζω. (*decorate*) διακοσμώ. **~ion** *s.* (*festive*) φωταψία *f.* (*enlightenment*) φώτιση *f.*

illusion *s.* ψευδαίσθηση *f.* (*false idea*) he is under the ~ that ζει με την αυταπάτη ότι. (*hallucination*) παραίσθηση *f.* optical ~ οπτική απάτη.

illusory *a.* απατηλός.

illustrat|e *v.t.* ιστορώ. (*a scene or book*) εικονογραφώ, (*by example*) επεξηγώ. **~ion** *s.* (*picture*) εικόνα *f.* (*act*) εικονογράφηση *f.* (*example*) παράδειγμα *n.* **~ive** *a.* επεξηγηματικός.

illustrious *a.* ένδοξος.

image *s.* ομοίωμα *n.*, είδωλο *n.* (*also mental*) εικόνα *f.* the very ~ of his father φτυστός ο πατέρας του. **~ry** *s.* παραστατικόν ύφος.

imaginable *a.* διανοητός. the greatest mistake ~ το μεγαλύτερο λάθος που μπορείς να φανταστείς.

imaginary *s.* φανταστικός.

imagination *s.* φαντασία *f.*

imaginative *a.* be ~ έχω φαντασία. over-~ ευφάνταστος.

imagine *v.* φαντάζομαι.

imam *s.* ιμάμης *m.*

imbalance *s.* δυσαναλογία *f.*

imbecil|e *a. & s.* ηλίθιος. **~ity** *s.* ηλιθιότητα *f.*

imbibe *v.* απορροφώ.

imbroglio *s.* ανακατωσούρα *f.*

imbue *v.* εμποδίζω.

IMF Διεθνές Νομισματικόν Ταμείον.

imitat|e *v.* (απο)μιμούμαι. **~ion** *s.* απομί-

μησηf. in ~ion of κατά μίμησιν (with gen.).
(a.) ψεύτικος, ιμιτασιόν. ~ive a. μιμητικός. ~or s. μιμητής m.
immaculate a. άψογος. (eccl.) I~ Conception Άμωμος Σύλληψης. ~ly adv. άψογα.
immanent a. ενυπάρχων.
immaterial a. άυλος. (not important) it is ~ δεν έχει σημασία.
immatur|e a. ανώριμος. ~ity s. ανωριμότητα f.
immeasurab|le a. άμετρος. ~ly adv. ασυγκρίτως.
immediate a. άμεσος. ~ly adv. (at once) αμέσως. (conj., as soon as) μόλις.
immemorial a. (customs) προαιώνιος, (trees, etc.) αιωνόβιος. from time ~ από αμνημονεύτων χρόνων.
immens|e a. (in extent) απέραντος, (in bulk) τεράστιος. ~ely adv. πάρα πολύ. ~ity s. μέγεθος n..
immers|e v. βουτώ, βυθίζω. ~ed in thought βυθισμενος σε σκέψεις. ~ion s. βούτηγμα n.
immigr|ant s. μετανάστης m. ~ate v. έρχομαι ως μετανάστης. ~ation s. μετανάστευση f.
imminent a. επικείμενος. be ~ επίκειμαι.
immobil|e a. ακίνητος. ~ity s. ακινησία f. ~ize v. ακινητοποιώ.
immoderate a. υπερβολικός. ~ly adv. υπερβολικά.
immodest a. άσεμνος. ~ly adv. άσεμνα. ~y ʿs. έλλειψη σεμνότητας.
immolat|e v. θυσιάζω. ~ion s. θυσία f.
immoral a. ανήθικος, (dissolute) έκλυτος. ~ity s. ανηθικότητα f. (debauchery) ακολασία f.
immortal a. αθάνατος. ~ity s. αθανασία f. ~ize v. απαθανατίζω.
immovab|le a. ακίνητος, (of purpose) ακλόνητος. ~ly adv. (fixed) μονίμως.
immun|e a. απρόσβλητος, (exempt) απαλλαγμένος. (med.) be ~e έχω ανοσία. ~ity s. (exemption) απαλλαγή f. (diplomatic) ασυλία f. (med.) ανοσία f.
immuniz|e v. ανοσοποιώ. ~ation s. ανοσοποίηση f.
immure v. εγκλείω.
immutable a. αμετάβλητος.
imp s. (joc.) διαβολάκι n.
impact s. πρόσκρουση f. (effect) επίδραση f.
impair v. βλάπτω, (reduce) μειώνω.
impale v. παλουκώνω.
impalpable a. (to mind) ασύλληπτος, (to senses) ανεπαίσθητος.
imparisyllabic a. περιττοσύλλαβος.
impart v. (give) δίδω, δίνω, (convey) μεταδίδω.

impartial a. αμερόληπτος. ~ity s. αμεροληψία f. ~ly adv. αμεροκήπτως.
impassable a. αδιάβατος.
impassioned a. φλογερός, (esp. of love) περιπαθής.
impassiv|e a. απαθής. ~ely adv. με απάθεια. ~ity s. απάθεια f.
impatience s. ανυπομονησία f.
impatient a. ανυπόμονος, be or get ~ ανυπομονώ, (anxiously) αδημονώ. be ~ of δεν ανέχομαι. ~ly adv. ανυπομόνως.
impeach v. (motives) αμφισβητώ. (person for high crimes) παραπέμπω με κατηγορία. ~ment s. παραπομπή f.
impeccab|le a. άψογος. ~ly adv. άψογα.
impecunious a. απένταρος.
impede v. εμποδίζω.
impediment s. εμπόδιο n. have an ~ in one's speech δυσκολεύομαι στην ομιλία. ~a s. αποσκευές f. pl.
impel v. ωθώ, (provoke) παρακινώ.
impending a. επικείμενος. be ~ επίκειμαι.
impenetrab|le a. αδιαπέραστος. ~ility s. (phys.) αδιαχώρητον n.
impenitent a. αμετανόητος.
imperative a. επιτακτικός, (duty, etc.) επιβεβλημένος. (s., gram.) προστακτική f.
imperceptib|le a. ανεπαίσθητος. ~ly adv. ανεπαισθήτως.
imperfect a. ατελής, ελαττωματικός. (badly done) πλημμελής. (s., gram.) παρατατικός m. ~ly adv. ελαττωματικά. ~ion s. ατέλεια f. (fault) ελάττωμα n.
imperial a. αυτοκρατορικός. ~ism s. ιμπεριαλισμός m. ~ist s. ιμπεριαλιστής m. ~istic a. ιμπεριαλιστικός.
imperil v. εκθέτω εις κίνδυνον.
imperious a. αυταρχικός.
imperishable a. άφθαρτος.
impermanent a. παροδικός.
impermeable a. αδιάβροχος.
impersonal a. απρόσωπος. ~ly adv. απροσώπως, απρόσωπα.
impersonat|e v. υποδύομαι, (mimic) μιμούμαι. ~ion s. μίμηση f. ~or s. a good ~or καλός μίμος.
impertin|ence s. αναίδεια f. ~ent a. αναιδής. ~ently adv. αναιδώς.
imperturbab|le a. ατάραχος. ~ility s. αταραξία f.
impervious a. (person) ανεπηρέαστος, (material) αδιαπέραστος, (to air, water) στεγανός.
impetu|ous a. ορμητικός. ~ously adv. ορμητικώς. ~osity s. ορμητικότητα f.
impetus s. (momentum) φόρα f. (thrust) ώθηση f.
impiety s. ανευλάβεια f.

impinge v. (on senses) γίνομαι αισθητός (with σε). (fam., encroach on) καταπατώ (with acc.).
impious a. ανευλαβής.
impish a. σκανταλιάρης.
implacable a. αμείλικτος.
implant v. εμφυτεύω.
implausible a. απίθανος.
implement s. σύνεργο n., εργαλείο n.
implement v. εφαρμόζω. ~ation s. εφαρμογή f.
implicat|e v. αναμιγνύω, μπλέκω. ~ion s. (involvement) ανάμιξη f. (import) έννοια f., σημασία f.
implicit a, (implied) υπονοούμενος, it is ~ υπονοείται. (absolute) πλήρης, απόλυτος. ~ly adv. πλήρως, απόλυτα.
implore v. εκλιπαρώ.
imply v. υπονοώ, (presuppose) προϋποθέτω.
impolite a. αγενής. ~ness s. αγένεια f.
impolitic a. ασύμφορος.
imponderable a. αστάθμητος. (s.) ~s αστάθμητοι παράγοντες.
import v. (bring in) εισάγω. (mean) σημαίνω. (s.) εισαγωγή f. σημασία f., έννοια f. ~ation s. εισαγωγή f. ~er s. εισαγωγέας m.
import|ant a. σημαντικός, σπουδαίος. it is ~ant έχει σημασία. ~ance s. σημασία f., σπουδαιότητα f. person of ~ance σημαίνον πρόσωπον.
importun|e v. φορτώνομαι. ~ate a. φορτικός. ~ity s. φορτικότητα f.
impose v.t. επιβάλλω. he ~d himself on our company μας κόλλησε. (v.i.) ~ on (people) εκμεταλλεύομαι, (kindness) καταχρώμαι (both with acc.).
imposing a. επιβλητικός.
imposition s. (act) επιβολή f. (burden) βάρος n. (trick) απάτη f. (punishment) τιμωρία f.
impossibility s. it is an ~ είναι αδύνατο. owing to the ~ of λόγω του ότι είναι αδύνατο να.
impossib|le a. αδύνατος. (fam.) he is ~le! είναι ανοικονόμητος (or ανυπόφορος). ~ly adv. (fam.) αφάνταστα.
impost s. δασμός m.
impost|or s. απατεώνας m. ~ure s. απάτη f.
impot|ent a. ανίκανος. ~ence s. ανικανότητα f.
impound v.t. κατάσχω.
impoverish v. (also become ~ed) φτωχαίνω. ~ed gentlewoman ξεπεσμένη αρχόντισσα.
impracticable a. ανεφάρμοστος, (road) αδιάβατος.
impractical a. it is ~ δεν είναι πρακτικό.

imprecation s. αναθεματισμός m.
impregnable a. απόρθητος.
impregnate v. γονιμοποιώ, (saturate) διαποτίζω, (imbue) εμποτίζω.
impress v. εντυπωσιάζω, κάνω εντύπωση σε. I ~ed upon him that... του τόνισα ότι. (s.) σφραγίδα f.
impression s. εντύπωση f. (of book) έκδοση f. ~able a. που εντυπωσιάζεται εύκολα.
impression|ism s. ιμπρεσιονισμός m. ~ist s. ιμπρεσιονιστής m. ~istic a. όχι εμπεριστατωμένος.
impressive a. εντυπωσιακός.
imprint v. (απο)τυπώνω, (on the mind) εντυπώνω. (s.) αποτύπωμα n.
imprison v. φυλακίζω. ~ment s. φυλάκιση f.
improbab|le a. απίθανος. ~ility s. απιθανότητα f.
impromptu a. πρόχειρος, αυτοσχέδιος. (adv.) εκ του προχείρου.
improper a. (unseemly) απρεπής. (inaccurate) λανθασμένος, όχι σωστός. ~ use κατάχρηση f. ~ly adv. απρεπώς, όχι σωστά, (of linguistic usage) καταχρηστικώς.
impropriety s. απρέπεια f.
improv|e v.t. βελτιώνω. (v.t. & i.) καλυτερεύω. (develop, exploit) εκμεταλλεύομαι, ~e on (surpass) ξεπερνώ. ~ing a. (morally) ηθοπλαστικός.
improvement s. βελτίωση f., καλυτέρευση f. it is an ~ on είναι καλύτερο από, ξεπερνάει.
improvid|ent a. απρονόητος. ~ence s. απρονοησία f.
improvis|e v. αυτοσχεδιάζω. ~ed a. αυτοσχέδιος. ~ation s. αυτοσχεδιασμός m.
imprud|ence s. έλλειψη συνέσεως. ~ent a. ασύνετος. ~ently adv. ασύνετα.
impud|ence s. θρασύτητα f., θράσος n., αναίδεια f. ~ent a. θρασύς, αναιδής. ~ently adv. αναιδώς.
impugn v. αμφισβητώ.
impulse s. (thrust) ώθηση f. (urge) παρόρμηση f. I had an ~ to μου ήρθε να. on ~ αυθόρμητα. on the ~ of the moment με την έμπνευση της στιγμής.
impulsive a. αυθόρμητος. ~ly adv. αυθόρμητα.
impunity s. ατιμωρησία f. with ~ ατιμωρητί.
impur|e a. ακάθαρτος, (unchaste) ασελγής, (adulterated) νοθευμένος. ~ity s. ακαθαρσία f. (moral) ασέλγεια f.
imput|e v. αποδίδω, καταλογίζω. ~ation s. κατηγορία f., καταλογισμός m.
in adv. μέσα. be ~ (at home) είμαι (στο) σπίτι, get ~ (arrive) φτάνω, come ~! εμπρός! narrow trousers are ~ τα στενά παντελόνια είναι της μόδας, my luck is ~

η τύχη με ευνόησε, whatever government is ~ οποιαδήποτε κυβέρνηση και να είναι στην αρχή (*or* στα πράματα). in order to keep one's hand (*or* eye) ~ για να μην ξεχνώ την τέχνη μου. all ~ (*inclusive*) όλα μαζί, (*exhausted*) εξαντλημένος. day ~ day out μέρα μπαίνει μέρα βγαίνει, go ~ and out μπαινοβγαίνω. ~s and outs τα μέσα και τα έξω. we are ~ for a difficult time έχουμε να περάσουμε (*or* μας περιμένουν) δύσκολες ώρες. be ~ on (*take part in*) συμμετέχω (*with gen.*). be ~ with (*on good terms*) τα έχω καλά με.

in *prep.* 1. (*place, time*) σέ, εις, μέσα σε, *etc.* ~ the room (μέσα) στο δωμάτιο, ~ this direction προς τα εδώ. ~ two years σε δύο χρόνια, ~ spring την άνοιξη, ~ April τον Απρίλιο, ~ the evening το βράδυ. we arrived ~ good time φτάσαμε εγκαίρως (*or* με την ώρα μας *or* στην ώρα μας). fruit ~ season φρούτα της εποχής. ~ the time *or* reign of επί (*with gen.*). 2. (*manner, condition*) say it ~ Greek πες το ελληνικά. what's that ~ Greek? πώς λέγεται αυτό στα ελληνικά; he said it ~ fun το είπε στα αστεία. ~ fashion στη μόδα *or* της μόδας. she was dressed ~ black ήταν ντυμένη στα μαύρα. the girl ~ black το κορίτσι με τα μαύρα ~ pencil με μολύβι, ~ a loud voice με δυνατή φωνή. I'm not going out ~ this weather δεν βγαίνω έξω με τέτοιο καιρό. they flocked ~ thousands συνέρρεαν κατά χιλιάδες. (*walk*) ~ twos δύο-δύο, (*break*) ~ two τα δύο. ~ order (*correct*) εν τάξει, ~ use εν χρήσει, ~ all εν συνόλω. ~ case of need εν (*or* στην) ανάγκη, ~ cash τοις μετρητοίς. ~ a hurry βιαστικός (*a.*), βιαστικά (*adv.*). ~ a rage οργισμένος, ~ secret κρυφά, ~ love ερωτευμένος. 3. (*respect*) be interested ~ ενδιαφέρομαι για. the best ~ the world ο καλύτερος του κόσμου (*or* στον κόσμο). blind ~ one eye στραβός από το ένα μάτι. the way ~ which he spoke ο τρόπος που (*or* με τον οποίο) μιλούσε. it is 10 metres ~ diameter έχει διάμετρο δέκα μέτρα. he doesn't see it ~ this light δεν το βλέπει υπό αυτό το πρίσμα. a country rich ~ tradition μία χώρα πλούσια σε παραδόσεις. the thing ~ itself το πράγμα αυτό καθ' εαυτό. ~ that, ~ so far as καθόσον, εφ' όσον.

inability *s.* he informed me of his ~ to be present με πληροφόρησε ότι δεν του ήταν δυνατό να παρευρεθεί.

inaccessible *a.* απρόσιτος.

inaccur|ate *a.* ανακριβής. **~acy** *s.* ανακρίβεια *f.*

inact|ion *s.* απραξία *f.* **~ive** *a.* (*not func-* *tioning*) αδρανής, εν αδρανεία. be **~ive** αδρανώ. (*indolent*) νωθρός.

inadequ|ate *a.* ανεπαρκής. **~acy** *s.* ανεπάρκεια *f.*

inadmissible *a.* απαράδεκτος.

inadvert|ence *s.* απροσεξία *f.* **~ent** *a.* (*person*) απρόσεκτος, (*act*) αθέλητος. **~ently** *adv.* άθελά μου.

inalienable *a.* αναπαλλοτρίωτος.

inan|e *a.* ανόητος. **~ity** *s.* ανοησία *f.*

inanimate *a.* (*dead*) νεκρός. (*lifeless, lit. &* *fig.*) άψυχος.

inanition *s.* ασιτία *f.*

inapplicable *a.* ανεφάρμοστος.

inappropriate *a.* ακατάλληλος.

inapt *a.* αδέξιος. (*not suitable*) ακατάλληλος.

inarticulate *a.* (*speech*) άναρθρος. (*person*) (*from surprise*) άναυδος. he is ~ δεν μπορεί να εκφρασθεί καλά.

inartistic *a.* όχι καλλιτεχνικός.

inasmuch *adv.* ~ as εφ' όσον.

inattent|ion *s.* απροσεξία *f.* **~ive** *a.* απρόσεκτος, (*neglectful*) αμελής.

inaudible *a.* be ~ δεν ακούομαι.

inaugural *a.* εναρκτήριος.

inaugurat|e *v.* εγκαινιάζω. **~ion** *s.* εγκαίνια *n.pl.*

inauspicious *a.* δυσοίωνος.

inborn *a.* έμφυτος.

inbred *a.* έμφυτος. (*of kinship*) they were ~ παντρεύονταν μέσα στην ίδια τους την οικογένεια.

inbreeding *s.* ενδογαμία *f.*

incalculable *a.* (*untold*) ανυπολόγιστος, (*unpredictable*) που δεν μπορεί να προβλεφθεί.

incandescent *a.* πυρακτωμένος.

incantation *s.* μαγική επίκληση.

incapable *a.* ανίκανος. (*not admitting of*) ανεπίδεκτος (*with gen.*).

incapacit|y *s.* ανικανότητα *f.* **~ate** *v.* καθιστώ ανίκανον, αχρηστεύω.

incarcerat|e *v.* φυλακίζω. **~ion** *s.* φυλάκιση *f.*

incarnat|e *v.* ενσαρκώνω. (*a.*) ενσαρκωμένος. **~ion** *s.* ενσάρκωση *f.*

incautious *a.* απρόσεκτος.

incendiary *a.* εμπρηστικός.

incense *s.* λιβάνι *n.*

incense *v.* εξοργίζω.

incentive *s.* ελατήριο *n.*, κίνητρο *n.*

inception *s.* έναρξη *f.*

incessant *a.* ακατάπαυστος. **~ly** *adv.* ακατάπαυστα.

incest *s.* αιμομιξία *f.* **~uous** *a.* αιμομιχτικός.

inch *s.* ίντσα *f.* ~ by ~ σπιθαμή προς

σπιθαμή. (v.) ~ forward προχωρώ σπιθαμή προς σπιθαμή.

incidence s. (occurrence) συχνότητα εμφανίσεως, (of taxes, etc.) επίπτωση f. (phys.) πρόσπτωσις f.

incident s. επεισόδιο n., συμβάν n.

incidental a. (chance) τυχαίος. ~ to που είναι το ακολούθημα (with gen.). ~ly adv. παρεμπιπτόντως. (to introduce remark) αλήθεια.

incinerate v. αποτεφρώνω.

incipient a. πρωτοεμφανιζόμενος.

incis|e v. χαράσσω, εγχαράσσω. ~ion s. τομή f., εγκοπή f. ~ive a. κοφτερός, οξύς.

incite v. υποκινώ. ~ment s. υποκίνηση f.

incivility s. αγένεια f.

inclem|ent a. σκληρός, (weather) κακός. ~ency s. σκληρότητα f.

inclination s. (leaning, lit. & fig.) κλίση f. (disposition) διάθεση f.(tendency) τάση f.

incline v.t. & i. (lean, bend) κλίνω. (v.i.) (tend) τείνω, έχω τάση. I am ~d to think τείνω να πιστέψω. I am ~d to prefer the first plan αποκλίνω υπέρ του πρώτου σχεδίου. (be disposed) be favourably ~d διάκειμαι ευμενώς, είμαι καλά διατεθειμένος. I feel ~d to drop the whole business έτσι μούρχεται να εγκαταλείψω όλη την υπόθεση. (s.) κλίση f.

include v. (συμ)περιλαμβάνω.

including prep. μαζί με. ~ the tip συμπεριλαμβανομένου και του φιλοδωρήματος, και το φιλοδώρημα μαζί.

inclusion s. a good feature of this dictionary is the ~ of many new technical terms αυτό το λεξικό έχει το καλό ότι περιλαμβάνει πολλούς νέους τεχνικούς όρους.

inclusive a. & adv. it cost thirty pounds ~ κόστισε τριάντα λίρες όλα μαζί.

incognito a., adv., s.n. ινκόγνιτο.

incoher|ent a. ασυνάρτητος. ~ence s. ασυναρτησία f.

incombustible a. άκαυστος, άφλεκτος.

income s. εισόδημα n. ~ tax φόρος εισοδήματος.

incoming a. εισερχόμενος.

incommensurate a. δυσανάλογος.

incommode v. ενοχλώ, βάζω σε κόπο.

incommunicado a. σε απομόνωση.

incomparab|le a. ασύγκριτος. ~ly adv. ασυγκρίτως.

incompatible a. (thing, state) ασυμβίβαστος, (persons) αταίριαστος.

incompetence s. ανικανότητα f.

incompetent a. ανίκανος, (work, worker) άτεχνος. be ~ to δεν είμαι ικανός να. ~ly adv. όχι ικανοποιητικά, άτεχνα.

incomplete a. (not finished) ημιτελής,

(partial) μερικός, (with part missing) ελλιπής.

incomprehensible a. ακατανόητος.

inconceivab|le a. (beyond the limits of thought) ασύλληπτος, (very strange) αδιανόητος. (fam.) απίστευτος. ~ly adv. αφάνταστα, απίστευτα.

inconclusive a. όχι αποφασιστικός.

incongruous a. γελοίος.

inconsequent a. ασυνεπής.

inconsiderable a. ασήμαντος.

inconsiderate a. αστόχαστος. ~ness s. αστοχασιά f.

inconsist|ent a. ασυνεπής. ~ency s. ασυνέπεια f.

inconsolable a. απαρηγόρητος.

inconspicuous a. (unnoticed) απαρατήρητος. be ~ (not easily seen) δεν διακρίνομαι εύκολα.

inconst|ant a. ασταθής, άστατος. (person only) άπιστος. ~ancy s. αστάθεια f. απιστία f.

incontestable a. αναμφισβήτητος.

incontin|ent a. ακρατής. ~ence s. ακράτεια f.

incontrovertible a. αναμφισβήτητος.

incovenience s. ενόχληση f., μπελάς m. (v.t.) ενοχλώ, βάζω σε μπελά.

inconvenient a. άβολος. at an ~ time σε ακατάλληλη ώρα. it is ~ for me to see him now δεν με βολεύει (or δεν μου είναι βολικό) να τον δω τώρα.

incorporat|e v.t. ενσωματώνω. ~ion s. ενσωμάτωση f.

incorporeal a. άυλος.

incorrect a. (inaccurate) ανακριβής, (mistaken) λανθασμένος. (of manners) απρεπής. ~ly adv. ανακριβώς, λανθασμένα. (of manners) όχι σύμφωνα με τους τύπους.

incorrigible a. αδιόρθωτος.

incorruptible a. αδιάφθορος.

increas|e v.t. & i. αυξάνω. (s.) αύξηση f. ~ingly adv. όλο και περισσότερο.

incredib|le a. απίστευτος. ~ly adv. απίστευτα.

incredul|ous a. δύσπιστος. ~ity s. δυσπιστία f.

increment s. προσαύξηση f.

incriminat|e v.t. ενοχοποιώ. ~ing a. ενοχοποιητικός. ~ion s. ενοχοποίηση f.

incubat|e v. επωάζω, εκκολάπτω. ~ion s. επώαση f. ~or s. εκκολαπτήριο n.

incubus s. (fig.) εφιάλτης m.

inculcate v. εμφυτεύω.

inculpate v. ενοχοποιώ.

incumbent a. it is ~ on us είναι καθήκον μας. (s.) κάτοχος m.

incur v.t. (debt) συνάπτω, (lose) υφίσταμαι,

(*hostility*) επισύρω.
incurab|le *a.* ανίατος, (*incorrigible*) αδιόρθωτος. **~ly** *adv.* ~ly lazy αδιόρθωτος τεμπέλης.
incurious *a.* αδιάφορος.
incursion *s.* επιδρομή *f.*
indebted *a.* υπόχρεος, υποχρεωμένος.
indecen|t *a.* άσεμνος, αισχρός. **~cy** *s.* αισχρότητα *f.*
indecipherable *a.* δυσανάγνωστος.
indecis|ion *s.* αναποφασιστικότητα *f.* **~ive** *a.* αναποφάσιστος.
indeclinable *a.* άκλιτος.
indecor|ous *a.* απρεπής. **~um** *s.* απρέπεια *f.*
indeed *adv.* πράγματι, μάλιστα, στ' αλήθεια. (*int.*) αλήθεια!
indefatigable *a.* ακούραστος, άοκνος.
indefensible *a.* (*town*) ευάλωτος, (*action, etc.*) αδικαιολόγητος.
indefinable *a.* απροσδιόριστος.
indefinite *a.* αόριστος. **~ly** *adv.* (*time*) επ' αόριστον.
indelible *a.* ανεξίτηλος.
indelic|ate *a.* χοντρός. **~acy** *s.* χοντράδα *f.*
indemni|fy *v.* αποζημιώνω. **~ty** *s.* (*security*) εγγύηση *f.* (*compensation*) αποζημίωση *f.*
indent *v.t.* (*a line*) αρχίζω πιο μέσα. (*v.i.*) ~ for (*order*) παραγγέλλω, **~ed** *a.* οδοντωτός, (*coastline*) δαντελωτός.
indenture *s.* σύμβαση μαθητείας.
independ|ence *s.* ανεξαρτησία *f.* **~ent** *a.* ανεξάρτητος. **~ently** *adv.* ανεξαρτήτως.
indescribab|le *a.* απερίγραπτος. **~ly** *adv.* απερίγραπτα.
indestructible *a.* ακατάλυτος.
indeterminate *a.* ακαθόριστος.
index *s.* (*pointer*) δείκτης *m.* (*list*) ευρετήριο *n.* (*v.t.*) (*an item*) βάζω στον κατάλογο.
Indian *a.* ινδικός. (*person*) Ινδός, (*American*) Ινδιάνος. ~ corn καλαμπόκι *n.* ~ summer γαϊδουροκαλόκαιρο *n.*
india-rubber *s.* γόμμα *f.*, λάστιχο *n.*
indicat|e *v.* δείχνω, ενδεικνύω. **~ion** *s.* ένδειξη *f.* **~or** *s.* δείκτης *m.*
indicative *a.* ενδεικτικός. (*s., gram.*) οριστική *f.*
indict *v.* κατηγορώ. **~ment** *s.* κατηγορία *f.*
indifference *s.* αδιαφορία *f.*
indifferent *a.* αδιάφορος, (*poor*) μέτριος. **~ly** *adv.* (*poorly*) μέτρια, έτσι κι έτσι. (*just the same*) το ίδιο.
indig|ence *s.* ένδεια *f.* **~ent** *a.* ενδεής.
indigenous *a.* αυτόχθων.
indigest|ible *a.* δύσπεπτος. **~ion** *s.* δυσπεψία *f.*, βαρυστομαχιά *f.*
indign|ant *a.* αγανακτισμένος. **~antly** *adv.* με αγανάκτηση. **~ation** *s.* αγανάκτηση *s.*
indignity *s.* προσβολή *f.*

indigo *s.* λουλάκι *n.*
indirect *a.* έμμεσος, πλάγιος. **~ly** *adv.* εμμέσως.
indiscipline *s.* απειθαρχία *f.*
indiscreet *a.* (*ill-judged*) αδιάκριτος, (*blabbing secrets*) ακριτόμυθος.
indiscretion *s.* αδιακρισία *f.* (*blabbing*) ακριτομυθία *f.*
indiscriminate *a.* (*person*) όχι εκλεκτικός. deal ~ blows μοιράζω αδιακρίτως γροθιές. **~ly** *adv.* αδιακρίτως.
indispensable *a.* απαραίτητος.
indisposed *a.* (*unwell*) αδιάθετος. (*unwilling*) he is ~ δεν είναι διατεθειμένος.
indisposition *s.* αδιαθεσία *f.* (*disinclination*) απροθυμία *f.*
indisputab|le *a.* αναμφισβήτητος. **~ly** *adv.* αναμφισβήτως.
indissolub|le *a.* αδιάρρηκτος. **~ly** *adv.* αδιαρρήκτως.
indistinct *a.* (*to senses*) δυσδιάκριτος, (*to mind*) συγκεχυμένος, (*to both*) αμυδρός. **~ly** *adv.* αμυδρώς. speak ~ly δεν μιλώ καθαρά.
indistinguishable *a.* (*to senses*) δυσδιάκριτος. they are ~ (*from each other*) δεν ξεχωρίζουν.
individual *a.* ατομικός, (*characteristic*) ιδιαίτερος, δικός μου. (*s.*) άτομο *n.* **~ly** *adv.* (*separately*) χωριστά. he spoke to each child ~ly μίλησε με κάθε παιδί προσωπικά (*or* ένα ένα).
individual|ist *s.* ατομικιστής *m.* **~ity** *s.* (*separateness*) ατομικότητα *f.* (*character*) προσωπικότητα *f.*
indivisib|le *a.* αδιαίρετος. **~ility** *s.* αδιαίρετον *n.*
indoctrinat|e *v.* εμποτίζω, κατηχώ. **~ion** *s.* εμποτισμός *m.*, κατήχηση *f.* **~or** *s.* καθοδηγητής *m.*
Indo-European *a.* ινδοευρωπαϊκός.
indol|ence *s.* νωθρότητα *f.* **~ent** *a.* νωθρός.
indomitable *a.* αδάμαστος.
indoor *a.* του σπιτιού, (*sports*) κλειστού χώρου. ~s *adv.* μέσα (στο σπίτι).
indubitab|le *a.* αναμφίβολος. **~ly** *adv.* αναμφιβόλως.
induce *v.* (*persuade*) πείθω, (*bring on*) προκαλώ. **~ment** *s.* κίνητρο *n.*, δέλεαρ *n.*
induction *s.* (*logic, electric*) επαγωγή *f.*
indulge *v.t.* (*person*) κάνω τα χατίρια (*with gen.*), (*desires*) ικανοποιώ. (*v.i.*) ~ in απολαμβάνω (*with acc.*).
indulg|ence *s.* (*enjoyment*) απόλαυση *f.* (*leniency*) επιείκεια *f.* **~ent** *a.* επιεικής. be ~ent to χαρίζομαι σε.
industrial *a.* βιομηχανικός. ~ relations βιομηχανικές σχέσεις. **~ist** *s.* βιομήχανος *m.*

industrializ|e v. εκβιομηχανίζω. **~ation** s. εκβιομηχάνιση f.
industrious a. φιλόπονος, εργατικός.
industry s. βιομηχανία f. (diligence) φιλοπονία f., εργατικότητα f.
inebriat|e v.t. (also get ~ed) μεθώ. **~ed** μεθυσμένος. (s.) μέθυσος m. **~ion** s. μέθη f.
inedible a. it is ~ δεν τρώγεται.
ineffab|le a. ανείπωτος. **~ly** adv. ανείπωτα.
ineffect|ive, ~ual a. (without desired effect) ατελεσφόρητος. see inefficient.
inefficient a. (person) όχι ικανός, (thing) όχι αποτελεσματικός. **~ency** s. ανικανότητα f. ανεπάρκεια f.
inelegant a. άκομψος.
ineligible a. (unsuitable) ακατάλληλος, (for election) όχι εκλέξιμος.
inept a. άτοπος.
inequality s. ανισότητα f. (of surface) ανωμαλία f.
inequit|able a. άδικος. **~y** s. αδικία f.
ineradicable a. ανεκρίζωτος.
inert a. αδρανής. **~ia** s. αδράνεια f.
inescapable a. αναπόφευκτος.
inessential a. επουσιώδης.
inestimable a. ανεκτίμητος.
inevitab|le a. μοιραίος. **~ly** adv. μοιραίως.
inexact a. ανακριβής. **~itude** s. ανακρίβεια f.
inexcusable a. ασυγχώρητος.
inexhaustible a. ανεξάντλητος.
inexorable a. αδυσώπητος.
inexpedient a. ασύμφορος.
inexpensive a. φτηνός, οικονομικός. **~ly** adv. οικονομικά.
inexperience s. απειρία f. **~d** a. άπειρος.
inexpert a. αδέξιος.
inexplicable a. ανεξήγητος.
inexpressible a. ανέκφραστος.
inextinguishable a. άσβεστος.
inextricably adv. I am ~ involved είμαι ανακατωμένος έτσι που δεν μπορώ να ξεμπλέξω.
infallib|le a. αλάθητος. **~ility** s. αλάθητον n.
infam|y s. όνειδος n., αισχύνη f. **~ous** a. βδελυρός.
infancy s. νηπιακή ηλικία. in one's (or its) ~ νηπιώδης, σε νηπιώδη κατάσταση.
infant s. νήπιο n., βρέφος n. ~ school νηπιαγωγείο n. ~ prodigy παιδί-θαύμα. (a.) νηπιακός.
infantile a. (of infants) παιδικός, (puerile) παιδαριώδης.
infantry s. πεζικόν n. **~man** s. φαντάρος m., οπλίτης m.
infatuat|e v. ξετρελλαίνω, **~ed** ξετρελλαμένος. **~ion** s. τρελλό πάθος.
infect v. μολύνω, **~ed** μολυσμένος. **~ion** s. μόλυνση f.

infectious a. (disease, humour) μεταδοτικός, κολλητικός. (person) don't go near him — he's ~ μην πας κοντά του—, θα κολλήσεις.
infer v. συμπεραίνω. **~ence** s. συμπέρασμα n.
inferior a. κατώτερος. this one is far ~ to yours τύφλα να 'χει αυτό μπροστά στο δικό σου. **~ity** s. κατωτερότητα f. **~ity complex** σύμπλεγμα κατωτερότητος.
infernal a. (regions) καταχθόνιος. (fam., wretched) σιχαμένος, (hellish) διαβολεμένος. **~ly** adv. (fam.) it's **~ly cold** κάνει ένα κρύο του διαβόλου.
inferno s. κόλαση f.
infertil|e a. (land) άγονος, (female) στείρος. **~ity** s. στειρότητα f.
infest v. γεμίζω. **~ed** with γεμάτος από.
infidel a. άπιστος. **~ity** s. απιστία f.
infiltrat|e v. (go into) διεισδύω εις. **~ion** s. διείσδυση f.
infinite a. άπειρος. **~ly** adv. απείρως. **~simal** a. απειροελάχιστος.
infinitive s. (gram.) απαρέμφατον n.
infinity s. άπειρον n.
infirm a. (step) ασταθής, (person) αδύναμος. **~ity** s. αδυναμία f. (of old age) αναπηρία f.
infirmary s. νοσοκομείο n.
inflame v. (excite) εξάπτω. get **~d** εξάπτομαι, (red) ερεθίζομαι.
inflamm|able a. εύφλεκτος. **~ation** s. φλεγμονή f. **~atory** a. εμπρηστικός.
inflat|e v. φουσκώνω. **~ion** s. φούσκωμα n. (fin.) πληθωρισμός m. **~ionary** a. πληθωριστικός.
inflect v. (voice) χρωματίζω. (gram.) κλίνω.
inflexib|le a. άκαμπτος. **~ly** adv. ακάμπτως. **~ility** s. ακαμψία f.
inflexion s. (of voice) διακύμανση f. (gram.) κλίση f.
inflict v. (penalty, one's presence) επιβάλλω, (suffering) προκαλώ. ~ wounds on πληγώνω. **~ion** s. (imposition) επιβολή f. (trouble) ενόχληση f., πληγή f.
influence v. επηρεάζω. επιδρώ επί (with gen.). (s.) επιρροή f., επίδραση f. (fam., string-pulling) μέσα n.pl.
influential a. be ~ έχω επιρροή, έχω μέσα. (play a part) παίζω ρόλο, επιδρώ ευνοϊκά.
influenza s. γρίππη f.
influx s. εισροή f.
inform v. πληροφορώ. ~ against καταγγέλλω. keep (person) **~ed** κρατώ ενήμερο. **~er** s. καταδότης m. (fam.) χαφιές m.
informat|ion s. πληροφορίες f.pl. piece of **~ion** πληροφορία f. **~ive** a. (book, talk, etc.) διαφωτιστικός. he was very **~ive** μου

είπε πολλά.
infra-dig *a.* (*fam.*) αναξιοπρεπής.
infra-red *a.* ~ rays υπέρυθροι ακτίνες.
infrastructure *s.* υποδομή *f.*
infrequent *a.* σπάνιος. **~ly** *adv.* σπανίως.
infringe *v.* (*break*) παραβαίνω, (*encroach on*) καταπατώ. **~ment** *s.* παράβαση *f.* καταπάτηση *f.*
infuriate *v.* φουρκίζω. get ~d (*fam.*) γίνομαι βαπόρι.
infus|e *v.* ενσταλάζω. **~ion** *s.* ενστάλαξη *f.* (*of herbs, tea*) αφέψημα *n.*
ingen|ious *a.* (*person*) πολυμήχανος, (*device*) έξυπνος. **~uity** *s.* (*of person*) εφευρετικότητα *f.* (*of device*) εξυπνάδα *f.*
ingenuous *a.* αφελής. **~ness** *s.* αφέλεια *f.*
ingle-nook *s.* γωνία του τζακιού.
inglorious *a.* άδοξος.
ingoing *a.* εισερχόμενος.
ingot *s.* ράβδος *f.*
ingrained *a.* ριζωμένος.
ingratiat|e *v.* ~e oneself with καλοπιάνω. **~ing** *a.* κολακευτικός.
ingratitude *s.* αγνωμοσύνη *f.*
ingredient *s.* συστατικό *n.*
ingress *s.* είσοδος *f.*
inhabit *v.* κατοικώ. **~able** *a.* κατοικήσιμος. **~ant** *s.* κάτοικος *m.f.*
inhale *v.* εισπνέω, (*smoke*) ρουφώ.
inherent *a.* ενυπάρχων, (*vested*) ανήκων. **~ly** *adv.* εκ φύσεως.
inherit *v.* κληρονομώ. **~ed** *a.* κληρονομικός. **~or** *s.* κληρονόμος *m.f.*
inheritance *s.* κληρονομία *f.* by ~ κληρονομικώς.
inhibit *v.* αναστέλλω. **~ed** *a.* (*person*). με ψυχολογικές αναστολές. **~ion** *s.* ψυχολογική αναστολή.
inhospitable *a.* αφιλόξενος.
inhuman *a.* απάνθρωπος. **~ity** *s.* απανθρωπιά *f.*
inimical *a.* εχθρικός, (*harmful*) επιβλαβής.
inimitable *a.* αμίμητος.
iniquit|ous *a.* εις το έπακρον άδικος. (*fam.*) εξωφρενικός. **~y** *s.* μεγάλη αδικία.
initial *a.* πρώτος, αρχικός. **~ly** *adv.* αρχικώς.
initial *v.* μονογραφώ. **~s** *s.* αρχικά *n.pl.* (*as signature*) μονογραφή *f.*
initiate *v.* (*start*) αρχίζω, (*into mysteries, etc.*) μυώ. ~d (με)μυημένος. (*s.*) μύστης *m.*
initiation *s.* (*starting*) έναρξη *f.* (*of persons*) μύηση *f.*
initiative *s.* πρωτοβουλία *f.*
inject *v.* εγχέω, (*fig.*) εμφυσώ. (*med.*) ~ morphia κάνω ένεση μορφίνης. **~ion** *s.* ένεση *f.* have *or* give an ~ion κάνω ένεση.
injudicious *a.* αούνετος.
injunction *s.* εντολή *f.*

injure *v.* (*wound*) τραυματίζω, (*harm*) βλάπτω, (*wrong*) αδικώ, (*offend*) θίγω.
injur|y *s.* (*wound*) τραύμα *n.* (*harm*) βλάβη *f.,* ζημία *f.* (*wrong*) αδικία *f.* **~ious** *a.* επιβλαβής, (*to interests*) επιζήμιος.
injustice *s.* αδικία *f.* do ~ to αδικώ.
ink *s.* μελάνι *n.* **~-pot**, **~-well** *s.* μελανοδοχείο *n.* (*v.*) **~-stain** *s.* μελανιά *f.* **~-y** *a.* ~y darkness πίσσα σκοτάδι, ~y fingers δάχτυλα όλο μελάνια.
inkling *s.* have an ~ of έχω μία ιδέα για.
inland *a.* (*away from sea*) μεσόγειος. (*domestic*) εσωτερικός.
in-laws *s.* πεθερικά *n.pl.*
inlay *s.* (*wood*) μαρκετερί *n.*
inlet *s.* (*of sea*) κολπίσκος *m.*
inmate *s.* τρόφιμος *m.*
inmost *a.* βαθύτερος. ~ parts (*depths*) ενδότερα *n.pl.*
inn *s.* πανδοχείο *n.,* χάνι *n.* **~keeper** *s.* πανδοχεύς *m.*
innards *s.* (*fam.*) εντόσθια *n.pl.*
innate *a.* έμφυτος.
inner *a.* εσωτερικός. ~ circle στενός κύκλος. ~ tube σαμπρέλα *f.* **~most** *a. see* inmost.
innings *s.* σειρά *f.*
innoc|ent *a.* αθώος, (*harmless*) αβλαβής, (*simple*) αφελής. **~ently** *adv.* αθώως, αφελώς. **~ence** *s.* αθωότητα *f.* αφέλεια *f.*
innocuous *a.* αβλαβής.
innovat|e *v.* νεωτερίζω. **~ion** *s.* νεωτερισμός *m.* **~or** *s.* νεωτεριστής *m.*
innuendo *s.* υπαινιγμός *m.,* υπονοούμενο *n.*
innumerable *a.* αναρίθμητος.
inoculat|e *v.* εμβολιάζω. **~ion** *s.* εμβολιασμός *m.*
inoffensive *a.* άκακος.
inoperative *a.* be ~ δεν ισχύω.
inopportune *a.* παράκαιρος.
inordinate *a.* υπερβολικός, υπέρογκος.
inorganic *a.* ανόργανος.
in-patient *s.* εσωτερικός *a.*
input *s.* είσοδος *f.,* εισαγωγή *f.*
inquest *s.* ανάκριση *f.,* έρευνα *f.*
inquir|e *v.* ρωτώ, ζητώ πληροφορίες. ~e into κάνω έρευνες για. **~er** *s.* ο ζητών πληροφορίες. **~ing** *a.* (*curious*) περίεργος, (*questioning*) ερωτηματικός.
inquiry *s.* (*question*) ερώτηση *f.* (*investigation*) ανάκριση *f.,* έρευνα *f.*
Inquisition *s.* (*eccl.*) Ιερά Εξέτασις.
inquisitive *a.* περίεργος, (*nosey*) αδιάκριτος. **~ness** *s.* περιέργεια *f.*
inquisitor *s.* ιεροεξεταστής *m.* **~ial** *a.* ιεροεξεταστικός.
inroad *s.* (*attack*) επιδρομή *f.* make ~s on (*time, resources*) τρώω.
inrush *s.* εισροή *f.*

insalubrious a. ανθυγιεινός.

insan|e a. παράφρων, τρελλός. become ~e παραφρονώ. **~ity** s. παραφροσύνη f.

insanitary a. ανθυγιεινός.

insatiab|le a. ακόρεστος, άπληστος. **~ility** s. απληστία f.

insatiate a. ακόρεστος.

inscribe v. (write) γράφω, (carve) χαράσσω, (dedicate) αφιερώνω.

inscription s. (entry) εγγραφή f. (on stone, etc.) επιγραφή f. (dedication) αφιέρωση f.

inscrutable a. ανεξιχνίαστος.

insect s. έντομο n., (tiny) ζωύφιο n. **~icide** s. εντομοκτόνο n.

insecur|e a. (not certain) όχι σίγουρος, (weak) επισφαλής. **~ity** s. έλλειψη σιγουριάς, ανασφάλεια f.

inseminat|e v. γονιμοποιώ. **~ion** s. γονιμοποίηση f.

insensate a. παράλογος.

insensibility s. αναισθησία f.

insensib|le a. (unconscious, unfeeling) αναίσθητος, (not knowing) αγνοών, (imperceptible) ανεπαίσθητος. **~ly** adv. ανεπαίσθητα.

insensitiv|e a. (aesthetically) χωρίς ευαισθησία, (unfeeling) αναίσθητος. **~ity** s. έλλειψη ευαισθησίας, αναισθησία f.

inseparable a. αχώριστος.

insert v. εισάγω, βάζω. (in newspaper) καταχωρίζω. **~ion** s. εισαγωγή f., βάλσιμο n. καταχώριση f.

inset s. (photo, table, etc.) εντός πλαισίου.

inshore a. παράκτιος. (adv.) προς την ακτή.

inside adv. μέσα. ~ out το μέσα έξω, ανάποδα. turn ~ out (room, etc.) κάνω άνω κάτω, (garment) γυρίζω μέσα έξω. know ~ out ξέρω απέξω κι ανακατωτά.

inside prep. εντός (with gen.), μέσα σε (with acc.) ~ my coat κάτω από το παλτό μου. from ~ the house από μέσα από το σπίτι.

inside s. εσωτερικό n., μέσα n. the ~ of the road το μέσα μέρος του δρόμου. on the ~ από μέσα. (fam., belly) κοιλιά f., (bowels) άντερα n.pl.

insider s. (με)μυημένος a.

insidious a. δόλιος, ύπουλος.

insight s. διεισδυτικότης f. get an ~ into παίρνω μια ιδέα για.

insignia s. εμβλήματα n.pl.

insignific|ant a. ασήμαντος. **~ance** s. ασημαντότης f.

insincer|e a. ανειλικρινής. **~ity** s. ανειλικρίνεια f.

insinuat|e v. (hint) υπαινίσσομαι. ~e oneself χώνομαι, γλιστρώ. **~ion** s. υπαινιγμός m.

insipid a. άνοστος **~ity** s. ανοστιά f.

insist v. επιμένω. ~ on (demand) απαιτώ. **~ence** s. επιμονή f. **~ent** a. επίμονος. **~ently** adv. επίμονα.

insol|ent a. θρασύς. **~ence** s. θρασύτητα f. **~ently** adv. αναιδώς.

insoluble a. (substance) αδιάλυτος, (problem) άλυτος.

insolv|ent a. αφερέγγυος. **~ency** s. αφερεγγυότης f.

insomnia s. αϋπνία f.

insouci|ant a. αμέριμνος. **~ance** s. αμεριμνησία f.

inspect v. επιθεωρώ, ελέγχω. (look at) εξετάζω. **~ion** s. επιθεώρηση f. έλεγχος m. εξέταση f.

inspector s. (school, police) επιθεωρητής m. (ticket) ελεγκτής m. (tax) έφορος m. **~ate** s. επιθεώρηση f.

inspiration s. έμπνευση f.

inspir|e v. εμπνέω. it ~es him with fear του εμπνέει φόβο. **~ed** a. εμπνευσμένος. **~ing** a. it is ~ing σε εμπνέει.

instability s. αστάθεια f.

install v. εγκαθιστώ. **~ation** s. εγκατάσταση f.

instalment s. (of payment) δόση f. (of publication) τεύχος n. (of story) συνέχεια f.

instance s. (example) παράδειγμα n. (case) περίπτωση f. for ~ παραδείγματος χάριν. in the first ~ κατά πρώτον. court of first ~ πρωτοδικείον n. at the ~ of τη αιτήσει (with gen.). (v.t.) αναφέρω.

instant s. στιγμή f.

instant a. άμεσος, (food) στιγμιαίος. (urgent) επείγων. (date) of the 3rd ~ της τρίτης τρέχοντος. **~ly** adv. αμέσως.

instantaneous a. στιγμιαίος, ακαριαίος. **~ly** adv. ακαριαίως.

instead adv. he was ill so I came ~ ο ίδιος ήταν άρρωστος και στη θέση του ήρθα εγώ. ~ of (with s.) αντί για, (with v.) αντί να. ~ of studying he sits doing nothing δεν μελετάει παρά κάθεται και χαζεύει.

instep s. κουτουπιές m., ταρσός m.

instigat|e v. υποκινώ. **~ion** s. υποκίνηση f. **~or** s. υποκινητής m.

instil v. ενσταλάζω.

instinct a. (charged) εμποτισμένος, γεμάτος.

instinct s. ένστικτο n. (as psychological term) ορμέμφυτο n. **~ive** a. ενστικτώδης. **~ively** adv. ενστικτωδώς.

institute v. (found) ιδρύω, (set up) συνιστώ, (start) προβαίνω εις. (a custom, rules) καθιερώνω. (s.) ινστιτούτο n., ίδρυμα n.

institution s. (setting up) καθιέρωση f. (starting) έναρξη f. (custom, regular thing) θεσμός m. (society) ίδρυμα n.

instruct v. (teach) διδάσκω, (direct) δίδω

εντολήν εις, (*inform*) πληροφορώ. ~or *s.*
δάσκαλος *m.*. εκπαιδευτής *m.*
instruction *s.* διδασκαλία *f.* (*order*) εντολή
f. ~s (*guidance*) οδηγίες *f.pl.* ~al *a.* μορφωτικός.
instructive *a.* διαφωτιστικός.
instrument *s.* όργανο *n.* (*surgical*) εργαλείο *n.*
instrumental *a.* (*mus.*) ενόργανος. be ~ in
συντελώ εις. ~ist *s.* ο παίζων μουσικό
όργανο. ~ity *s.* through the ~ity of με τη
μεσολάβηση (*with gen.*).
insubordinat|e *a.* απειθάρχητος. ~ion *s.*
απειθαρχία *f.*
insufferable *a.* αφόρητος, ανυπόφορος.
insuffici|ency *s.* ανεπάρκεια *f.* ~ent *a.*
ανεπαρκής.
insular *a.* (*fig.*) περιορισμένων αντιλήψεων. ~ity *s.* στενότητα αντιλήψεων.
insulat|e *v.* μονώνω. ~ing *a.* μονωτικός.
~ion *s.* μόνωση *f.*
insult *v.* προσβάλλω. (*s.*) προσβολή *f.* ~ing
a. προσβλητικός.
insuperable *a.* ανυπέρβλητος.
insupportable *a.* αφόρητος, ανυπόφορος.
insurance *s.* ασφάλεια *f.* (*act of insuring*)
ασφάλιση *f.* ~ policy ασφαλιστήριο *n.* ~
premium ασφάλιστρα *n.pl.* take out ~
ασφαλίζομαι. (*a.*) ασφαλιστικός.
insure *v.t.* ασφαλίζω. (*v.i.*) ασφαλίζομαι. ~r
s. ασφαλιστής *m.*
insurgent *a.* επαναστατικός. (*s.*) αντάρτης *m.*
insurmountable *a.* ανυπέρβλητος.
insurrection *s.* εξέγερση *f.*
intact *a.* άθικτος, ακέραιος.
intake *s.* εισαγωγή *f.*
intangible *a.* ακαθόριστος, που δεν συλ-
λαμβάνεται. ~ assets άυλα αγαθά.
integer *s.* ακέραιος αριθμός.
integral *a.* ~ part αναπόσπαστο μέρος.
(*math.*) ολοκληρωτικός.
integrat|e *v.* ολοκληρώνω. ~ed circuit
ολοκληρωμένο κύκλωμα. ~ion *s.* ολο-
κλήρωση *f.*
integrity *s.* ακεραιότητα *f.*
intellect *s.* διάνοια *f.* ~ual *a.* διανοητικός,
πνευματικός. (*s.*) διανοούμενος *m.*, άνθρωπος του πνεύματος.
intelligence *s.* νοημοσύνη *f.*, ευφυΐα *f.*
(*fam.*) εξυπνάδα *f.* (*information*) πληροφορίες *f.pl.*
intelligent *a.* νοήμων, ευφυής. (*fam.*)
έξυπνος. ~ly *adv.* ευφυώς, (*sensibly*)
λογικώς.
intelligentsia *s.* ιντελιγκέντσια *f.*
intelligib|le *a.* νοητός, καταληπτός, ευνό-
ητος. ~ly *adv.* σαφώς. ~ility *s.* σαφήνεια *f.*
(*distinctness*) ευκρίνεια *f.*
intemper|ance *s.* ακράτεια *f.* ~ate *a.*

ακρατής, υπερβολικός. ~ately *adv.*
υπερβολικά.
intend *v.t.* (*destine*) προορίζω, (*mean*)
εννοώ. (*v.i.*, *purpose*) έχω σκοπό, σκοπεύω, θέλω.
intended *a.* (*to be*) μελλοντικός, (*on
purpose*) σκόπιμος. (*s.*) (*future spouse*)
μνηστήρας *m.*, μνηστή *f.*
intense *a.* (*vivid*) έντονος, (*strong*) δυνατός, (*violent*) σφοδρός, (*charged*)
ηλεκτρισμένος. ~ly *adv.* εντόνως, (*very*)
τρομερά. it was ~ly cold έκανε τρομερό
κρύο.
intens|ify *v.* εντείνω. ~ity *s.* ένταση *f.*
(*violence*) σφοδρότητα *f.* ~ive *a.* εντατικός. ~ively *adv.* εντατικά.
intent *a.* (*engrossed*) προσηλωμένος,
(*determined*) αποφασισμένος. (*of thought,
gaze*) έντονος. ~ly *adv.* εντόνως.
intent *s.* πρόθεση *f.* to all ~s and purposes
κατ᾽ ουσίαν.
intention *s.* πρόθεση *f.*, σκοπός *m.* well ~ed
με καλές προθέσεις, καλοπροαίρετος.
intentional *a.* σκόπιμος. ~ly *adv.* σκοπίμως,
επίτηδες.
inter *v.* ενταφιάζω. ~ment *s.* ενταφιασμός *m.*
interact *v.i.* αλληλεπιδρώ. ~ion *s.* αλλη-
λεπίδραση *f.*
interbreed *v.t.* διασταυρώνω. (*v.i.*) δια-
σταυρώνομαι.
intercede *v.* επεμβαίνω, μεσολαβώ.
intercept *v.* (*arrest*) συλλαμβάνω, (*the
enemy*) αναχαιτίζω, (*a message*) υποκλέπτω, (*a letter, messenger*) σταματώ.
interception *s.* σύλληψη *f.* αναχαίτιση *f.*
υποκλοπή *f.* σταμάτημα *n.*
intercession *s.* μεσολάβηση *f.*
interchange *v.* ανταλλάσσω. (*s.*) ανταλλαγή *f.* ~able *a.* ανταλλάξιμος.
intercom *s.* εσωτερικό τηλέφωνο.
intercommunal *a.* διακοινοτικός.
intercommunicat|e *v.* αλληλεπικοινωνώ.
~ion *s.* αλληλεπικοινωνία *f.*
intercontinental *a.* διηπειρωτικός.
intercourse *s.* επικοινωνία *f.*, σχέσεις *f.pl.*
sexual ~ συνουσία *f.*
interdepend|ent *a.* αλληλένδετος. ~ence
αλληλεξάρτηση *f.*
interdict *s.* απαγόρευση *f.*
interest *s.* ενδιαφέρον *n.* (*advantage*) συμφέρον *n.* (*influence*) επιρροή *f.* have an ~
in (*business*) μετέχω εις. (*fin.*) τόκος *m.*,
with ~ (*fig.*) με το παραπάνω. rate of ~
επιτόκιον *n.*
interest *v.* ενδιαφέρω. ~ed ενδιαφερόμενος
be ~ed in ενδιαφέρομαι (*with να or για*).
interesting *a.* ενδιαφέρων. (*fam.*) in an ~
condition σε ενδιαφέρουσα. ~ly *adv.* he

interface 371 intrigue

talked ~ly on that topic η ομιλία του επάνω σ' εκείνο το θέμα ήταν ενδιαφέρουσα.
interface s. (computer) υποδοχή f.
interfer|e v.i. επεμβαίνω, ανακατεύομαι. ~e with (obstruct) εμποδίζω, (tamper) πειράζω. ~ence s. επέμβαση f., ανάμιξη f. (radio) παράσιτα n.pl. ~ing a. που χώνει τη μύτη του παντού.
interim a. προσωρινός. (s.) in the ~ εν τω μεταξύ.
interior a. εσωτερικός, (feelings) ενδόμυχος. (s.) εσωτερικό n. Minister of I~ Υπουργός Εσωτερικών.
interject v. παρεμβάλλω. ~ion s. (gram.) επιφώνημα n.
interlac|e v.t. συμπλέκω. (v.i.) συμπλέκομαι. ~ing branches κλαδιά πλεγμένα το ένα μεσ' στο άλλο.
interlard v. ποικίλλω.
interlinked a. αλληλένδετος.
interlock v.t. συνδέω. (v.i.) συνδέομαι.
interlocutor s. συνομιλητής m.
interloper s. παρείσακτος a.
interlude s. διάλειμμα n.
intermarr|iage s. επιγαμία f. ~y v.i. they ~y παντρεύονται μεταξύ τους.
intermediary s. μεσολαβητής m.
intermin|able a. ατελείωτος. ~ably adv. ατελείωτα.
intermingl|e v.t. αναμιγνύω. (v.i.) αναμιγνύομαι. ~ling s. ανάμιξη f.
intermission s. διάλειμμα n.
intermittent a. διαλείπων. (occasional) περιοδικός, σποραδικός. ~ly adv. κατά περιόδους, σποραδικά.
intern v. θέτω υπό περιορισμόν. ~ee s. κρατούμενος m. ~ment s. περιορισμός m.
intern(e) s. εσωτερικός a.
internal a. εσωτερικός. ~ly adv. εσωτερικώς.
international a. διεθνής. ~ly adv. διεθνώς. ~ize v. διεθνοποιώ.
internecine a. ~ struggle or war αλληλοσφαγή f.
interpellation s. επερώτηση f.
interplay s. αμοιβαία επίδραση.
interpolat|e v. παρεισάγω. ~ion s. (sthg. added) παρεμβολή f.
interpose v.t. (a veto, etc.) προβάλλω. (v.i.) παρεμβαίνω.
interpret v.t. ερμηνεύω. (v.i.) διερμηνεύω. ~ation s. ερμηνεία f. ~ing s. διερμηνεία f.
interpreter s. (expounder, performer) ερμηνευτής m., ερμηνεύτρια f. (translator) διερμηνεύς m.f.
interregnum s. μεσοβασιλεία f.
interrogat|e v. ανακρίνω. ~ion s. ανάκριση f. ~or s. ο ανακρίνων.

interrogat|ive, ~ory a. ερωτηματικός.
interrupt v. διακόπτω. ~er s. ο διακόπτων. ~ion s. διακοπή f.
intersect v.t. τέμνω. (v.i.) τέμνομαι. ~ion s. διασταύρωση f.
intersperse v. (diversify) ποικίλλω. ~d with trees με σκόρπια δέντρα.
interstices s. μικρά διάκενα.
intertwined a. περιπλεγμένος.
interval s. (break) διάλειμμα n. (space) διάστημα n.
interven|e v.i. (occur, mediate) μεσολαβώ, (interfere, mediate) επεμβαίνω. ~tion s. μεσολάβηση f. επέμβαση f.
interview s. συνέντευξη f. (v.) (of reporter) παίρνω συνέντευξη από. (of employer, etc.) βλέπω.
inter|weave v. συνυφαίνω. ~woven a. συνυφασμένος.
intestate a. αδιάθετος.
intestin|e s. έντερο n. ~al a. εντερικός.
intimacy s. οικειότητα f., στενές σχέσεις. (fam., coition) συνουσία f.
intimat|e v. γνωστοποιώ. ~ion s. γνωστοποίηση f.
intimate a. (close) στενός, (familiar) οικείος, (private) απόκρυφος. (of knowledge) βαθύς. be ~ with συνδέομαι στενά με. (s.) ~s οικείοι m.pl. ~ly adv. (closely) στενά, (inside out) απέξω κι ανακατωτά.
intimidat|e v. εκφοβίζω. ~ion s. εκφοβισμός m.
into prep. σε, εις, μέσα σε (all with acc.). come ~ (inherit) κληρονομώ.
intolerab|le a. ανυπόφορος. ~ly adv. ανυπόφορα.
intoler|ance a. έλλειψη ανεκτικότητας. ~ant a. be ~ant δεν έχω ανεκτικότητα.
inton|ation s. (of voice) διακύμανση της φωνής. (eccl.) ψάλσιμο n. ~e v. ψάλλω.
intoxic|ant s. οινοπνευματώδες ποτό. ~ate v. (also get ~ated) μεθώ. ~ated μεθυσμένος.
intoxicat|ing a. μεθυστικός. ~ion s. μέθη f.
intractable a. δύσκολος, (child) σκληρός, (material) δυσκολομεταχείριστος.
intransig|ence s. αδιαλλαξία f. ~ent a. αδιάλλακτος.
intransitive a. (gram.) αμετάβατος.
in-tray s. εισερχόμενα n.pl.
intrepid a. ατρόμητος. ~ity s. παληκαριά f.
intricac|y s. (of design) λεπτότητα f. ~ies of the law οι δαίδαλοι του νόμου.
intricate a. περίπλοκος, μπερδεμένος. (of design) λεπτοδουλεμένος.
intrigu|e s. μηχανορραφία f. (amour) μπλέξιμο n. (v.i.) (plot) μηχανορραφώ. (v.t.) (interest) it ~es me με βάζει σε περιέργεια.

~ing *a.* μυστηριώδης.

intrinsic *a.* ουσιαστικός. **~ally** *adv.* ουσιαστικά.

introduce *v.* (*persons*) συνιστώ. (*bring in, insert*) εισάγω.

introduct|ion *s.* σύστασηf. εισαγωγήf. letter of ~ion συστατική επιστολή. **~ory** *a.* εισαγωγικός.

introspect|ion *s.* ενδοσκόπηση f. **~ive** *a.* ενδοσκοπικός.

introvert *s.* εσωστρεφής *a.*

intrude *v.i.* (*of person*) έρχομαι σε ακατάλληλη ώρα, γίνομαι ενοχλητικός. ~ on ταράσσω, ενοχλώ. **~r** *s.* παρείσακτος, απρόσκλητος *a.*

intrus|ion *s.* ενόχλησηf. see intrude. **~ive** *a.* παρείσακτος, (*gram.*) παρεισφρητικός.

intuition *s.* διαίσθησηf., ενόρασηf.

intuitive *a.*, **~ly** *adv.* εκ διαισθήσεως. be ~ έχω διαίσθηση.

inundat|e *v.* κατακλύζω, πλημμυρίζω. get ~ed with κατακλύζομαι *or* πλημμυρίζω από. **~ion** *s.* κατακλυσμός *m.*, πλημμύρα f.

inured *a.* ψημένος.

invade *v.* εισβάλλω εις, (*disturb*) ταράσσω, (*encroach on*) καταπατώ. **~r** *s.* εισβολεύς *m.*

invalid *s.* ασθενής, άρρωστος *a.* (*v.*) (*mil.*) ~ out αποστρατεύω λόγω αναπηρίας.

invalid *a.* (*not valid*) άκυρος, ανίσχυρος. **~ate** *v.* ακυρώ. **~ation** *s.* ακύρωσηf.

invaluable *a.* ανεκτίμητος.

invariab|le *a.* αμετάβλητος, (*regular*) τακτικός. **~ly** *adv.* πάντα, τακτικά.

invasion *s.* εισβολήf.

invective *s.* (*abuse*) ύβρεις f.pl. (*violent attack*) φιλιππικός *m.*

inveigh *v.i.* ~ against καταφέρομαι εναντίον (*with gen.*).

inveigle *v.* παραπείθω.

invent *v.* εφευρίσκω, (*fabricate*) επινοώ, πλάθω. **~ion** *s.* εφεύρεση f. (*false tale*) μύθευμα *n.* **~ive** *a.* επινοητικός, ~ive mind θηλυκό μυαλό. **~or** *s.* εφευρέτης *m.*

inventory *s.* απογραφήf.

inverse *a.* αντίστροφος. in ~ ratio κατ' αντίστροφον λόγον. **~ly** *adv.* αντιστρόφως.

inversion *s.* αντιστροφή f. (*mus., chem.*) αναστροφήf.

invert *v.* αναστρέφω. (*turn upside down*) αναποδογυρίζω. **~ed** *a.* αναστραμμένος. ~ed commas εισαγωγικά *n.pl.*

invertebrate *a.* ασπόνδυλος.

invest *v.t.* (*besiege*) πολιορκώ. (*fin.*) τοποθετώ. (*fam.*) ~ in (*buy*) αγοράζω. ~ (*person, thing*) with (*quality*) προσδίδω σε, (*authority*) αναθέτω σε. **~ment** *s.* (*fin.*) επένδυσηf. (*also fig.*) τοποθέτησηf. **~or** *s.*

μέτοχος *m.*

investigat|e *v.* ερευνώ, εξετάζω. **~ion** *s.* έρευνα f., εξέταση f., under ~ion υπό εξέτασιν. **~or** *s.* ερευνητής *m.*

investiture *s.* τελετή απονομής αξιώματος.

inveterate *a.* (*person*) μανιώδης, αδιόρθωτος. (*deep-rooted*) βαθιά ριζωμένος.

invidious *a.* προσβλητικός, (*task*) δυσάρεστος.

invigilat|e *v.i.* επιβλέπω. **~or** *s.* ο επιβλέπων, επιτηρητής *m.*

invigorat|e *v.* αναζωογονώ. **~ing** *a.* αναζωογονητικός. **~ion** *s.* αναζωογόνηση f.

invincib|le *a.* αήττητος, ακαταμάχητος. **~ly** *adv.* αήττητα.

inviolab|le *a.* απαράβατος. **~ility** *s.* απαραβίαστον *n.*

inviolate *a.* απαραβίαστος, άθικτος.

invisib|le *a.* αόρατος. (*of person who does not appear*) αθέατος. (*exports, etc.*) άδηλος. ~le ink συμπαθητική μελάνη. it is ~le δεν διακρίνεται. **~ly** *adv.* έτσι που να μη διακρίνεται.

invit|e *v.* (*προσ*)καλώ, (*encourage*) προκαλώ. **~ing** *a.* ελκυστικός. **~ation** *s.* πρόσκληση f.

invocation *s.* επίκλησηf.

invoice *s.* τιμολόγιο *n.*

invoke *v.* επικαλούμαι.

involuntary *a.* αθέλετος.

involve *v.t.* (*entail*) συνεπάγομαι, (*require*) θέλω, (*implicate*) αναμιγνύω, μπλέκω. get ~d αναμιγνύομαι, μπλέκω. **~d** *a.* (*complicated*) περιπεπλεγμένος. **~ment** *s.* ανάμιξηf.

invulnerable *a.* άτρωτος.

inward *a.* εσωτερικός, ενδόμυχος. (*incoming*) εισερχόμενος, (*towards the inside*) προς τα μέσα. ~ voyage ταξίδι του γυρισμού. **~(s)** *adv.* προς τα μέσα (*or* τα έσω). **~ly** *adv.* μέσα μου, κρυφά.

iodine *s.* ιώδιο *n.*

Ionic *s.* ιωνικός.

iota *s.* γιώτα *n.* (*fig., jot*) ίχνος *n.*

IOU *s.* γραμμάτιο *n.*

IQ *s.* διανοητικό πηλίκο *n.*

irascible *a.* οξύθυμος.

irate *a.* οργισμένος.

ire *s.* οργήf.

iridesc|ent *a.* be ~ent ιριδίζω. **~ence** *s.* ιριδισμός *m.*

iris *s.* (*eye, plant*) ίριδα f.

Irish *a.* ιρλανδικός. (*person*) Ιρλανδός *m.*, Ιρλανδέζα f.

irk *v.* ενοχλώ. **~some** *a.* ενοχλητικός.

iron *s.* σίδηρος *m.*, σίδερο *n.* (*things made of* ~) σιδερικά *n.pl.* (*flat-*~) σίδερο *n.* **~s** (*fetters*) σίδερα *n.pl.* strike while the ~ is hot στη βράση κολλάει το σίδερο. have too many ~s in the fire κυνηγώ πολλούς λαγούς.

iron *a.* σιδερένιος. (*curtain, Chancellor, etc.*) σιδηρούς.

iron *v.* σιδερώνω. ~ out (*fig.*) εξομαλύνω. ~**ing** *s.* σιδέρωμα *n.*

ironic, ~al *a.* ειρωνικός. ~**ally** *adv.* ειρωνικά.

ironmonger *s.* έμπορος ειδών κιγκαλερίας. ~**y** *s.* είδη κιγκαλερίας.

ironwork *s.* σιδερικά *n.pl.* ~**s** *s.* σιδηρουργείο *n.*

irony *s.* ειρωνεία *f.*

irradiate *v.* φωτίζω.

irrational *a.* (*devoid of reason*) άλογος, (*against reason*) παράλογος. ~**ly** *adv.* παράλογα.

irreconcilable *a.* (*persons*) άσπονδος, (*ideas*) ασυμβίβαστος.

irrecoverable *a.* (*expense*) αγύριστος.

irredeemable *a.* αδιόρθωτος, (*fin.*) μη αποσβεννύμενος.

irredentism *s.* αλυτρωτισμός *m.*

irreducible *a.* the ~ minimum το ελαχιστότατο.

irrefutable *a.* ακαταμάχητος.

irregular *a.* (*against rules, practice*) αντικανονικός. (*uneven & gram.*) ανώμαλος. (*disorderly & mil.*) άτακτος. (*shape, features, intervals*) ακανόνιστος. (*pulse*) άρρυθμος. be ~ in one's attendance δεν πηγαίνω τακτικά.

irregular|ity *s.* ανωμαλία *f.*, αταξία *f.*, αρρυθμία. *f.* ~**ly** *adv.* αντικανονικά, ανώμαλα, άτακτα, άρρυθμα.

irrelevant *a.* άσχετος. it is ~ δεν έχει σχέση.

irreligious *a.* άθρησκος.

irremediable *a.* αδιόρθωτος.

irreparable *a.* ανεπανόρθωτος.

irreplaceable *a.* αναντικατάστατος.

irrepressible *a.* ακατάσχετος.

irreproachable *a.* άμεμπτος.

irresistible *a.* (*force, argument*) ακαταμάχητος, (*delightful*) τρομερά ελκυστικός.

irresolute *a.* αναποφάσιστος.

irrespective *a.* (*as adv.*) ~ of ανεξάρτητα από, ανεξαρτήτως (*with gen.*).

irresponsib|le *a.* (*person*) που δεν έχει το αίσθημα της ευθύνης. ~**ility** *s.* επιπολαιότητα *f.*

irretrievable *a.* ανεπανόρθωτος.

irrever|ent *a.* ανευλαβής. ~**ence** *s.* ανευλάβεια *f.*

irreversible *a.* αμετάκλητος.

irrevocab|le *a.* αμετάκλητος. ~**ly** *adv.* αμετάκλητα.

irrigat|e *v.* αρδεύω. ~**ion** *s.* άρδευση *f.*

irritab|le *a.* ευερέθιστος. be ~le νευριάζω εύκολα. ~**ly** *adv.* νευριασμένα.

irritant *a.* ερεθιστικός.

irritate *v.t.* (*body, temper*) ερεθίζω, (*temper*) εκνευρίζω. get ~d ερεθίζομαι, νευριάζω.

irritat|ing *a.* ερεθιστικός, εκνευριστικός. ~**ion** *s.* ερεθισμός *m.* εκνευρισμός *m.*

irruption *s.* εισβολή *f.*

is *v.* είναι.

Islam *s.* Ισλάμ *n.* ~**ic** *a.* ισλαμικός.

island *s.* νησί *n.*, νήσος *f.* (*traffic-~*) νησίδα *f.* desert ~ ερημονήσι *n.* (*a.*) νησιώτικος. ~**er** *s.* νησιώτης *m.*

isle *s.* νησί *n.* ~**t** *s.* νησάκι *n.*

isolat|e *v.* απομονώνω. ~**ed** *a.* απομονωμένος. ~**ion** *s.* απομόνωση *f.*

isolation|ism *s.* απομονωτισμός *m.* ~**ist** *s.* οπαδός του απομονωτισμού.

issue *v.i.* (*come out*) βγαίνω, εξέρχομαι, (*flow out*) εκρέω, (*result from*) προέρχομαι (από), (*end in*) καταλήγω (εις). (*v.t.*) (*give forth*) βγάζω, (*distribute*) διανέμω, (*publish, circulate*) εκδίδω.

issue *s.* (*way out*) διέξοδο *f.* (*outflow*) εκροή *f.* (*outcome*) έκβαση *f.* (*off-spring*) τέκνα *n.pl.* (*distribution*) διανομή *f.* (*publication*) έκδοση *f.* (*number, instalment*) τεύχος *n.*, αριθμός *m.* (*of newspaper*) φύλλο *n.* (*question*) ζήτημα, θέμα *n.* at ~ υπό συζήτησιν. take *or* join ~ συζητώ.

isthmus *s.* ισθμός *m.*

it *pron.* αυτό *n.* I can see ~ το βλέπω. ~**self** *pron. see* self.

italics *s.* κυρτά στοιχεία.

itch *s.* φαγούρα *f.* (*fig., desire*) πόθος *m.* (*v.*) I am ~ing έχω φαγούρα. my neck ~es με τρώει ο λαιμός μου. (*fig.*) he is ~ing to tell (*a secret*) τον τρώει η γλώσσα του. he is ~ing to get away δεν βλέπει την ώρα να φύγει.

item *s.* (*of performance*) νούμερο *n.* (*of expenditure*) κονδύλι(ον) *n.* (*article, goods*) είδος *n.* (*piece*) κομμάτι *n.* (*thing*) πράμα *n.*, αντικείμενο *n.* (*job*) δουλειά *f.* (*of news*) είδηση *f.* (*on agenda*) θέμα *n.*

itemize *v.* καταγράφω αναλυτικά.

iterat|e *v.* επαναλαμβάνω. ~**ion** *s.* επανάληψη *f.*

itinerant *a.* πλανόδιος.

itinerary *s.* δρομολόγιο *n.*

its *pron.* I like ~ colour μου αρέσει το χρώμα του.

ivory *s.* ελεφαντοστούν *n.*, φίλντισι *n.*

ivory *a.* (*made of* ~) από ελεφαντόδοντα. (*like* ~) φιλντισένιος (*fig.*) living in an ~ tower αποτραβηγμένος σ' ένα δικό μου κόσμο.

ivy *s.* κισσός *m.*

J

jab v.t. (thrust) μπήγω. (give a ~ to) σκουντώ. (s.) σκούντημα n. (fam., injection) ένεση f.

jabber v.i. αεροκοπανίζω. (s.) φλυαρία f.

jack s. (lifting) γρύλος m. (at cards) βαλές m., φάντης m. ~ of all trades πολυτεχνίτης m. (v.t.) ~ up σηκώνω με γρύλλο.

jackal s. τσακάλι n.

jackanapes s. μορφονιός m.

jackass s. (fig.) κουτεντές m.

jackdaw s. κάργια f.

jacket s. σακκάκι n. (cover) κάλυμμα n.

jack-in-office s. μικρομανδαρίνος m.

jack-in-the-box s. διαβολάκι n.

jack-knife s. σουγιάς m .(v.i., fam.) γίνομαι φυσαρμόνικα.

jackpot s. (fam.) hit the ~ πιάνω την καλή.

Jacobin s. Ιακωβίνος m.

jade s. (stone) νεφρίτης m. (horse) παλιά-λογο n. (woman) κατεργάρα f.

jaded a. (person) αηδιασμένος, (appetite) κομμένος.

jag v.t. σχίζω. ~ged a. μυτερός.

jail s. φυλακή f. (v.) φυλακίζω. ~er s. δεσμοφύλακας m.

jam s. (edible) μαρμελάδα f. (fam.) money for ~ εύκολο κέρδος.

jam v.t. (thrust) χώνω, (wedge) σφηνώνω, (squeeze) στρυμώχνω. the streets were ~med with people οι δρόμοι ήταν πήχτρα από κόσμο. (v.i.) (also get ~med) φρακάρω, (of works) παθαίνω εμπλοκή.

jam s. (congestion) συμφόρηση f. (traffic) μποτιλιάρισμα n. (in machinery) εμπλοκή f. (difficulty) μπλέξιμο n. ~ming s. (radio) παρεμβολή παρασίτων.

jamb s. παραστάτης m.

jam-packed a. τίγκα adv.

jangle v.i. κουδουνίζω δυσάρεστα. (v.t.) (keys, etc.) παίζω. ~d nerves τεντωμένα νεύρα.

janissary s. γενίτσαρος m.

janitor s. θυρωρός m.

January s. Ιανουάριος, Γενάρης m.

japan s. λάκα f.

Japanese a. ιαπωνικός. (person) Ιάπωνας m., Ιαπωνέζα f.

jape s. αστείο n.

jar v.t. (strike) χτυπώ, (jolt, shake, upset) τραντάζω, κλονίζω. (v.i.) ~ (on) χτυπώ άσχημα (σε). ~ with δεν ταιριάζω με. ~ring note παραφωνία f. (s.) τράνταγμα n.

jar s. (vessel for oil, etc.) πιθάρι n. (glass) γυάλα f. (small) βάζο n.

jargon s. κορακίστικα n.pl.

jasmine s. γιασεμί n.

jaundice s. ίκτερος m. (v.) take a ~d view τα βλέπω όλα μαύρα.

jaunt s. εκδρομή f.

jaunt|y a. καμαρωτός. at a ~y angle λοξοβαλμένος. ~ily adv. καμαρωτά.

javelin s. ακόντιο n.

jaw s. σαγόνι n. (fig.) from the ~s of death από του χάρου τα δόντια. (fam., talk) λίμα f. (v.i.) λιμάρω. ~-bone s. γνάθος f.

jazz s. τζαζ f. (v., fam.) ~ up ζωηρεύω. ~y a. φανταχτερός.

jealous a. ζηλιάρης. be ~ (of.) ζηλεύω. ~y s. ζήλια f.

jeans s. τζην n.

jeer v. (also ~ at) (mock) κοροϊδεύω, (boo) γιουχαΐζω. ~ing s. κοροϊδία f. γιουχάισμα n.

jejune a. ξηρός, κούφιος.

jelly s. ζελέ n. ~-fish s. μέδουσα f., τσούχτρα f.

jeopard|y s. κίνδυνος m. be in ~y κιν-δυνεύω. ~ize v.t. διακινδυνεύω.

jerk s. τίναγμα n. (twitch) σπασμός m. physical ~s γυμναστικές ασκήσεις. (v.t.) τινάζω, τραβώ απότομα. (v.i.) τραντά-ζομαι.

jerk|y a. σπασμωδικός. ~ily adv. σπασμω-δικά, με τραντάγματα.

jerry-built a. ψευτοκτισμένος.

jersey s. (sailor's, sportsman's) φανέλλα f. (material) ζέρσεϊ n.

jest s. αστεϊσμός m. (v.) αστεΐζομαι. ~er s. γελωτοποιός m.

jesuitical a. ιησουιτικός.

jet s. (of fountain) πίδακας m. (flame) γλώσσα f. (of burner) μπεκ n. (aero.) τζετ n. ~-propelled a. αεριωθούμενος. ~-set s. τζετ-σετ n.

jet s. (mineral) γαγάτης m. ~-black a. κατάμαυρος.

jetsam s. έκβρασμα n.

jettison v. ρίχνω στη θάλασσα, (fig.) εγκαταλείπω (s.) αβαρία f.

jetty s. προβλήτα f.

jewel s. κόσμημα n. (fig.) θησαυρός m. ~ler s. κοσμηματοπώλης m. ~ry s. κοσμήματα n.pl.

Jewish a. εβραϊκός, εβραίικος. (person) Εβραίος, Ιουδαίος.

jib s. (naut.) φλόκος m. (fig.) cut of one's ~ σουλούπι n.

jib v.i. (of horse) κωλώνω. (fig.) αντι-τίθεμαι, αρνούμαι.

jiffy s. (fam.) in a ~ στο πι και φι, αμέσως.

jig s. (mus.) ζιγκ n. (v.i.) χοροπηδώ.

jiggery-pokery s. ματσαράγκες f.pl.

jilt v. εγκαταλείπω.

jingle v.i. & t. κουδουνίζω. (s.) κουδούνισμα n.

jingoism s. σωβινισμός m.
jinks s. high ~s γλέντι τρελλό.
jinx s. (person) γρουσούζης m., put a ~ on ματιάζω.
jitter|s s. have the ~s κατέχομαι από φόβο και νευρικότητα. **~y** a. φοβισμένος.
Job s. patience of ~ ιώβειος υπομονή.
job s. δουλειά f. (post) θέση f. (household repair) μερεμέτι n. odd ~ μικροδουλειά f. do a ~ (with tools) μαστορεύω. that's a good ~! έτσι μπράβο! it's a good ~ that... καλά που. it's quite a ~ είναι ολόκληρη επιχείρηση. I had a ~ to find the house ταλαιπωρήθηκα (or είδα κι έπαθα or μου βγήκε η πίστη ανάποδα) να βρω το σπίτι.
jobbery s. ρουσφετολογία f.
job-work s. εργασία κατ' αποκοπήν.
jockey s. τζόκεϋ m. (v.) μανουβράρω.
jocose a. φιλοπαίγμων.
jocular a. εύθυμος.
jog v.t. σπρώχνω, (memory) φρεσκάρω. (v.i.) ~ along προχωρώ σιγά-σιγά, (make do) τα φέρνω βόλτα.
joggle v.t. ταρακουνώ. (v.i.) ταρακουνιέμαι.
join v.t. (unite) ενώνω, (tie, connect) συνδέω, (splice) ματίζω, (meet) συναντώ. (a party) προσχωρώ εις, (a club) γίνομαι μέλος (with gen.), (the ladies) πάω να βρω. (~ the company of) κάνω παρέα με. (merge with) ενώνομαι με. ~ on a piece (of material) κάνω τσόντα. they ~ed hands πιάστηκαν από τα χέρια. (v.i.) ενώνομαι, (meet) συναντώμαι. ~ in παίρνω μέρος (σε). ~up κατατάσσομαι, ~ with συνεργάζομαι με. (s.) ένωση f.
joiner s. ξυλουργός m. **~y** s. ξυλουργική f.
joint s. άρθρωση f. (anat.) κλείδωση f. (woodwork) ένωση f. (masonry) αρμός m. ~ of meat κομμάτι κρέας για ψητό. out of ~ εξαρθρωμένος. **~ed** a. αρθρωτός, με αρθρώσεις.
joint a. κοινός, ομαδικός, (operations, staff, etc.) συν(δε)δυασμένος. ~ owner συνιδιοκτήτης m. ~-stock company μετοχική εταιρεία. **~ly** adv. από κοινού.
joist s. πατόξυλο n., καδρόνι n.
jok|e s. αστείο n. practical ~e φάρσα f. (v.) αστειεύομαι. **~er** s. (at cards) μπαλαντέρ m. **~ingly** adv. στα αστεία.
jollification s. γλέντι n.
jollity s. ευθυμία f.
jolly a. (person) κεφάτος, πρόσχαρος, (occasion) χαρούμενος, ευχάριστος. (adv., fam.) πολύ. (v.t.) καλοπιάνω.
jolt v.t. τραντάζω, (fig., shock) ξαφνιάζω. (v.i.) τραντάζομαι. (s.) τράνταγμα n., ξάφνιασμα n.
Jonah s. (fig.) γρουσούζης m.

jost|le v.t. σπρώχνω, στρυμώχνω. get ~led συνωστίζομαι. **~ling** s. συνωστισμός m., στρίμωγμα n.
jot s. not a ~ ούτε ίχνος.
jot v.t. ~ down σημειώνω. **~tings** s. σημειώματα n.pl.
journal s. ημερολόγιο n. (paper) εφημερίδα f. (periodical) περιοδικό n. **~ism** s. δημοσιογραφία f. **~ist** s. δημοσιογράφος m.f. **~istic** a. δημοσιογραφικός.
journey s. ταξίδι n. three hours' ~ διαδρομή τριών ωρών. (v.) ταξιδεύω.
journeyman s. μεροκαματιάρης m.
joust s. κονταροχτύπημα n.
Jove s. Δίας m. (int.) by ~ it's cold! πω πω τι κρύο! by ~ you're right! έχεις δίκιο μα την αλήθεια!
jowl s. σαγόνι n.
joy s. χαρά f. **~-ride** s. αυτοκινητάδα f.
joy|ful, ~ous a. καταχαρούμενος, (news, bells) χαρμόσυνος. **~fully, ~ously** adv. με χαρά.
jubil|ant a. be ~ant πανηγυρίζω, αγαλλιώ. **~ation** s. πανηγυρισμός m., αγαλλίαση f.
jubilee s. ιωβηλαίον n. diamond ~ εξηκοστή επέτειος.
Judaism s. ιουδαϊσμός m.
Judas s. Ιούδας m. **~-tree** s. κουτσουπιά f.
judge s. κριτής m. (law) δικαστής m. (v.) κρίνω, (law) δικάζω. (reckon) υπολογίζω, (consider) θεωρώ.
judgement s. (faculty, opinion) κρίση f. (law) απόφαση f. (view) γνώμη f. (divine retribution) θεία δίκη. Last J~ ημέρα της Κρίσεως.
judicature s. δικαιοσύνη f. (judges) δικαστικόν σώμα.
judicial a. (of courts) δικαστικός, (of mind) έχων ευθυκρισία.
judiciary s. δικαστικόν σώμα.
judicious a. γνωστικός. **~ly** adv. γνωστικά.
judo s. τζούντο n.
jug s. κουμάρι n., κανάτι n. (large) κανάτα f. (fam.) in ~ στο φρέσκο. **~ged** a. (cooked) στη στάμνα.
juggle v.i. κάνω ταχυδακτυλουργίες. **~r** s. ταχυδακτυλουργός m.
juic|e s. χυμός m. (of body) υγρόν n. **~y** a. χυμώδης, ζουμερός. (scandalous) γαργαλιστικός.
jujube-tree s. τζιτζιφιά f.
Julian a. ιουλιανός.
July s. Ιούλιος m.
jumble v.t. ανακατώνω. (s.) ανακατωσούρα f.
jump v.i. & t. πηδώ, (rise) ανεβαίνω. (be startled) ξαφνιάζομαι, you made me ~ με ξάφνιασες. ~ the rails εκτροχιάζομαι, ~ the fence πηδώ πάνω από το φράχτη. ~

about χοροπηδώ, ~ up πετάγομαι όρθιος.
he ~ed out at me πετάχτηκε μπροστά μου.
he ~ed at it (*of opportunity*) το δέχτηκε με
ενθουσιασμό.

jump *s.* πήδημα *n.* (*rise*) άνοδος *f.* high (*or
long*) ~ άλμα εις ύψος (*or* εις μήκος). **~er**
s. (*who* **~s**) άλτης *m.* (*garment*) καζάκα *f.*
(*knitted*) τρικό *n.*, πλεχτό *n.* **~y** *a.*
νευρικός.

junction *s.* (*union*) ένωση *f.*, σύνδεση *f.*
(*place*) διακλάδωση *f.*, κόμβος *m.*

juncture *s.* at the present ~ στο σημείο που
βρίσκεται η υπόθεση.

June *s.* Ιούνιος *m.*

jungle *s.* ζούγκλα *f.*

junior *a. & s.* νεώτερος, (*in rank*) κατώτερος.

junk *s.* παλιάτσες *f.pl.* **~-shop** *s.* παλια-
τζήδικο *n.*

junketing *s.* γλεντοκόπι *n.*

Junoesque *a.* με ηγεμονικό παράστημα.

junta *s.* χούντα *f.*

juridical *a.* δικανικός.

jurisdiction *s.* δικαιοδοσία *f.*

jurisprudence *s.* νομική *f.*

jurist *s.* νομομαθής *m.*

juror *s.* ένορκος *m.*

jury *s.* (*law*) ένορκοι *m.pl.* (*of contest*)
κριτική επιτροπή. **~man** *s.* ένορκος *m.*

just *adv.* (*exactly*) ακριβώς. ~ as it was έτσι
όπως ήταν. ~ as I was leaving home καθώς
(*or* την ώρα που *or* πάνω που) έφευγα απ'
το σπίτι. it is ~ as good as mine είναι εξ
ίσου καλό με το δικό μου. only ~ μόλις. we
~ had enough (*money, etc.*) ήρθαμε ίσα-
ίσα. ~ now (*a moment ago*) τώρα, μόλις,
προ ολίγου. (*merely*) απλώς, did you come
~ for that? ήρθες μόνο και μόνο για αυτό;
it is ~ lovely είναι πραγματικά ωραίο. ~
wait a moment για περίμενε λίγο.

just *a.* δίκαιος, (*right*) ορθός, σωστός. **~ly**
adv. δικαίως, δίκαια, ορθώς.

justice *s.* δικαιοσύνη *f.* to do him ~ (*we must
admit that...*) για να είμαστε δίκαιοι
απέναντί του. do oneself ~ φαίνομαι
αντάξιος του εαυτού μου. do ~ to the meal
τιμώ δεόντως το φαγητό. J~ of the Peace
ειρηνοδίκης *m.*

justif|y *v.* (*vindicate*) δικαιώνω, (*excuse,
defend*) δικαιολογώ. the end ~ies the
means ο σκοπός αγιάζει τα μέσα. **~ication**
s. δικαίωση *f.* δικαιολογία *f.*

jut *v.i.* (*also* ~ out) προεξέχω.

jute *s.* τζούτα, γιούτα *f.*

juvenile *a.* (*youthful*) νεανικός, (*child-
rens'*) παιδικός, (*childish*) παιδαριώδης.
(*s.*) παιδί *n.* (*minor*) ανήλικος *a.*

juxtapos|e *v.t.* αντιπαραθέτω. **~ition** *s.*
αντιπαράθεση *f.* in ~ition πλάι-πλάι.

K

kale *s.* κατσαρό λάχανο.

kaleidoscop|e *s.* καλειδοσκόπιο *n.* **~ic** *a.*
σαν καλειδοσκόπιο.

kangaroo *s.* καγκουρώ *n.*

kebab *s.* σουβλάκια *n.pl.*

keel *s.* καρίνα *f.*, τρόπις *f.* on an even ~
σταθερός. (*v.*) ~ over αναποδογυρίζω.

keen *a.* (*edge*) κοφτερός, (*wind*) τσουχτερός. (*mind, senses*) οξύς. (*eager*) ενθουσιώδης. he is ~ on her της έχει αδυναμία. I
am a ~ fisherman μου αρέσει πολύ το
ψάρεμα. **~ly** *adv.* (*hard*) σκληρά, (*eagerly*)
ζωηρά. **~ness** *s.* (*enthusiasm*) ενθουσιασμός *m.* (*of senses*) οξύτητα *f.* (*of blade*)
αιχμηρότητα *f.*

keep *s.* earn one's ~ βγάζω το ψωμί μου. for
~s για πάντα. (*of castle*) ακροπύργιο *n.*

keep *v.t.* (*have, run, maintain, have in
stock*) έχω. (*protect, put aside*) φυλάγω.
(*prevent*) εμποδίζω. (*continue to have,
cause to do or be*) κρατώ. (*a shop, a
promise*) κρατώ. ~ the books *or* accounts
κρατώ τα βιβλία. it ~s me going με κρατάει
στη ζωή. it ~s me busy με απασχολεί. ~ him
waiting τον έχω και περιμένει. ~ apart
(*prevent from meeting*) κρατώ χω-
ρισμένους, (*separate*) χωρίζω.

keep *v.i.* (*remain good*) διατηρούμαι. butter
doesn't ~ in hot weather το βούτυρο δεν
διατηρείται με τη ζέστη. (*continue to be*) ~
cool (*fig.*) μένω ψύχραιμος, ~ quiet
κάθομαι ήσυχος. is he ~ing well? είναι
καλά από υγεία; (*continue doing*) he ~s
ringing up τηλεφωνεί συνέχεια. she ~s
scolding me όλο με μαλώνει. ~ going
εξακολουθώ να εργάζομαι (*of person*) *or*
να λειτουργώ (*of institution, machine*).

keep at *v.* ~ it επιμένω. keep him at it δεν
τον αφήνω να ησυχάσει.

keep away *v.t.* κρατώ μακριά. (*v.i.*)
κρατιέμαι μακριά, δεν πάω κοντά.

keep back *v.t.* (*emotion*) συγκρατώ. (*hide*)
κρύβω, (*withhold*) κρατώ. (*v.i.*) κρατιέμαι
μακριά, μένω πίσω.

keep down *v.t.* (*oppress, curb*) καταστέλλω,
(*limit*) περιορίζω. (*one's head*) κρατώ
χαμηλά. (*v.i.*) (*hide*) κρύβομαι.

keep from *v.t.* (*prevent*) εμποδίζω. (*hide*)
he kept the truth from me μου έκρυψε την
αλήθεια. (*v.i.*) I couldn't ~ laughing δεν
μπορούσα να κρατηθώ και να μη γελάσω.

keep in *v.t.* (*detain*) κρατώ μέσα. he got kept
in (*at school*) έμεινε τιμωρία. to keep my
hand in για να μη χάσω τη φόρμα μου.
(*v.i.*) ~ with διατηρώ φιλικές σχέσεις με.

keep off v.t. (avert) αποτρέπω, εμποδίζω. keep the rain off (oneself) προφυλάσσομαι από τη βροχή. (v.i.) (abstain from) κόβω, αποφεύγω. (not go near) κρατιέμαι μακριά από. (not mention) δεν θίγω. (not happen) if the rain keeps off αν δεν πιάσει βροχή. ~! μακριά!

keep on v.t. (not take off) δεν βγάζω. (keep in place) κρατώ στη θέση του, συγκρατώ. (fam.) keep your hair (or shirt) on! μην εξάπτεσαι. (v.i.) (continue to) εξακολουθώ να. it keeps on raining βρέχει συνέχεια. ~ at (nag) τρώω, γίνομαι φόρτωμα σε.

keep out v.t. κρατώ έξω. ~ the cold προφυλάσσομαι από το κρύο. (v.i.) ~ of (avoid) αποφεύγω, (not enter) δεν μπαίνω μέσα σε, (not get involved in) δεν ανακατεύομαι σε.

keep to v.i. (observe) τηρώ, (follow) ακολουθώ, (limit oneself to) περιορίζομαι σε. ~ the left κρατώ την αριστερά. (v.t.) (restrict) περιορίζω. keep (oneself) to oneself αποφεύγω τις παρέες.

keep under v.t. (subject peoples) καταπιέζω, (fire) καταστέλλω.

keep up v.t. (custom, courage, contact) διατηρώ. (property) συντηρώ. (keep doing) συνεχίζω, εξακολουθώ. ~ one's singing εξακολουθώ το τραγούδι. keep one's end up δεν το βάζω κάτω. ~ appearances κρατώ τα προσχήματα. keep your chin (or pecker) up! κουράγιο! (v.i.) ~ with (do as well as) φτάνω, (in speed) προφταίνω, (both) εξακολουθώ. (stay in contact) εξακολουθώ να έχω σχέσεις με. ~ with the Joneses κοιτάζω να μην πέσω μικρότερος από τους γείτονες.

keeper s. (guard) φύλακας m. (owner) ιδιοκτήτης m. (overseer) επιστάτης m.

keeping s. (observance) τήρηση f. (charge) in my ~ υπό την φύλαξίν μου. (of pets, etc.) διατήρηση f. it is worth ~ αξίζει να φυλαχτεί. (harmony) in ~ with σύμφωνος με. be in ~ συμφωνώ, ταιριάζω.

keepsake s. ενθύμιο n.

keg s. βαρελάκι n.

ken s. it is beyond my ~ ξεπερνά τις γνώσεις μου.

kennel s. σπιτάκι σκύλου. ~s κυνοτροφείον n.

kerb s. κράσπεδο πεζοδρομίου.

kerchief s. τσεμπέρι n., μαντήλι n.

kernel s. πυρήνας m. (of nut) ψίχα f.

kerosene s. πετρέλαιο n.

kestrel s. κιρκινέζι n.

ketch s. καΐκι n.

kettle s. δοχείο με ράμφος για βράσιμο νερού. (fam.) pretty ~ of fish! ωραίο μπλέξιμο! a different ~ of fish άλλη παράγραφος. ~-drum s. τύμπανο n.

key s. κλειδί n., κλείς f. (of piano, typewriter) πλήκτρον n. (mus.) τόνος, κλειδί n., (fig., tone, style) τόνος m. pass- ~ αντικλείδι n. ~ money αέρας m. ~board s. κλαβιέ n. ~hole s. κλειδαρότρυπα f. ~note s. (mus.) τονική f. (fig.) κεντρική ιδέα. ~stone s. κλείς αψίδος. (fig.) βάση f. ~word s. λέξη κλειδί.

key a. (basic) βασικός, (vital) ζωτικός. ~ man κλειδί n. (v.) ~ed up σε κατάσταση υπερεντεταμένης αναμονής.

khaki a. & s. χακί.

khan s. (inn) χάνι n.

kick v. κλωτσώ, λακτίζω. ~ one's heels χάνω τον καιρό μου. ~ up one's heels το ρίχνω έξω. ~ against the pricks αντιστέκομαι ματαίως. ~ up a row κάνω φασαρία (or καβγά). (fam.) ~ off αρχίζω. ~ the bucket τα τινάζω.

kick s. κλωτσιά f., λάκτισμα n. (fam.) get a ~ out of it μου δίνει ευχαρίστηση. for ~s (excitement) για τη συγκίνηση, (fun) για την πλάκα.

kickshaw s. (toy) μπιχλιμπίδι n. (dish) λιχουδιά f.

kid s. (goat) κατσικάκι n., τραγί n., (newborn) ερίφιο n. (child) παιδάκι n.

kid v.t. κοροϊδεύω. are you ~ding? αστειεύεσαι;

kid a. ~ gloves γάντια από σεβρό. (fig.) handle (person) with ~ gloves φέρομαι με το γάντι σε, (thing) μεταχειρίζομαι προσεκτικά.

kidnap v. απάγω. ~per s. απαγωγεύς m. ~ping s. απαγωγή f.

kidney s. νεφρό n., νεφρός m. ~-shaped σε σχήμα νεφρού. ~-bean s. φασόλι n.

kill v. σκοτώνω, (plants) καίω. ~ off ξεκαθαρίζω, ξεχάνω. ~ two birds with one stone μ' ένα σμπάρο δυο τρυγόνια. my corn is ~ing me ο κάλος μου με πεθαίνει (or με αφανίζει). she was dressed up to ~ το φόρεμά της έκαψε καρδιές.

kill s. θανάτωση f., σκότωμα n., be in at the ~ παρευρίσκομαι στην τελική θριαμβευτική έκβαση του αγώνος.

killer s. δολοφόνος m.

killing s. σκοτωμός m. he made a ~ out of it του απέφερε μεγάλο κέρδος.

killing a. (fam.) (funny) αστείος, (exhausting) εξουθενωτικός.

killjoy s. he is a ~ σου χαλάει το κέφι.

kiln s. κάμινος f.

kilo s. κιλό n. ~gram s. χιλιόγραμμο n. ~metre s. χιλιόμετρο n. ~watt s. κιλοβάτ n.

kilt s. πλισέ φούστα σκωτσέζικη.

kin s. συγγενείς m.pl. next of ~ πλησιέστερος

συγγενής.

kind s. (sort) είδος n. something of the ~ κάτι τέτοιο. nothing of the ~! (not at all) καθόλου. what ~ of person? τι είδους (or τι λογής) άνθρωπος; τι είδος ανθρώπου; all ~s of books κάθε είδους (or κάθε λογής or λογιώ-λογιώ or όλων των ειδών) βιβλία. I had a ~ of premonition είχα κάποια προαίσθηση. pay in ~ πληρώνω εις είδος.

kind a. καλόκαρδος, καλός, αγαθός. (act, letter, etc.) ευγενικός. ~-hearted person χρυσός άνθρωπος, μάλαμα n. ~ness s. καλωσύνη f.

kindergarten s. νηπιαγωγείον n.

kind|le v.t. & i. ανάβω. (v.t.) (provoke) προκαλώ, κινώ. ~ling s. προσάναμμα n.

kindly adv. ευγενικά. will you ~ tell me? έχετε την καλωσύνη να μου πήτε; (a.) see kind a.

kindred s. & a. (persons) συγγενείς m.pl. (a., similar) συγγενικός. ~ spirit αδελφή ψυχή.

kine s. αγελάδες (f.pl.).

king s. βασιλεύς, βασιλιάς m. (at cards) ρήγας, παπάς m. ~'s son βασιλόπουλο n. ~ly a. βασιλικός. ~pin s. (fig.) κινητήριος μοχλός. ~ship s. βασιλεία f.

kingdom s. βασίλειον n. (fam.) ~ come ο άλλος κόσμος.

kingfisher s. αλκυών f.

kink s. στρίψιμο n. (mental) βίδα f. ~y a (fam.) εκκεντρικός.

kinship s. συγγένεια f.

kins|man, ~woman s. συγγενής m.f.

kiosk s. κιόσκι n., περίπτερον n.

kipper s. καπνιστή ρέγγα.

kirk s. εκκλησία f.

kiss s. φιλί n., φίλημα n. (v.t.) φιλώ, ασπάζομαι. (v.i.) φιλιέμαι.

kit s. (soldier's) εξάρτυση f., σκευή f. (tools, equipment) σύνεργα n.pl. ~bag s. γυλιός m.

kitchen s. κουζίνα f., μαγειρείον n. ~-garden s. λαχανόκηπος m.

kite s. (toy) αϊτός m., χαρταετός m. (fig.) fly a ~ κάνω βολιδοσκοπήσεις.

kith s. ~ and kin φίλοι και συγγενείς.

kitten s. γατάκι n., γατούλα f.

kitty s. (joint fund) ταμείο n.

kleptomania s. κλεπτομανία f. ~c s. κλεπτομανής a.

knack s. (ability) ταλέντο n. (trick) κόλπο n.

knapsack s. σακκίδιο n.

knav|e s. απατεώνας m. (cards) βαλές, φάντης m. ~ery s. απάτη f. ~ish a. (person) απατεώνας m. ~ish tricks κατεργαρίες (f.pl.).

knead v. ζυμώνω, (rub) μαλάζω. ~ing s. ζύμωμα n.

knee s. γόνατο n. bow the ~ κλίνω το γόνυ. ~-deep a. μέχρι το γόνατο.

kneel v. γονατίζω. ~ing γονατιστός a.

knell s. πένθιμη κωδωνοκρουσία. (fig.) προμήνυμα συμφοράς.

knickers s. (boys') κοντά παντελόνια. (women's) κυλότα f., βρακί n.

knick-knack s. μπιχλιμπίδι n.

knife s. μαχαίρι n. (fig.) get one's ~ into καταδιώκω. (v.) μαχαιρώνω. ~-grinder s. ακονιστής m.

knight s. ιππότης f. ~ly a. ιπποτικός.

knit v.t. πλέκω, (brow) ζαρώνω. (v.t. & i.) (join) δένω. ~ted a. πλεκτός. ~ting s. πλέξιμο n.

knob s. (handle) πόμολο n. (switch) κουμπί n. (piece) κομμάτι n. (protuberance) εξόγκωμα n.

knobbly a. (knee) κοκκαλιάρικος, (stick) με ρόζους.

knock s. χτύπημα n., χτύπος m. (loss) ζημιά f. (reverse) αναποδιά f. there was a ~ at the door χτύπησε η πόρτα.

knock v.t. χτυπώ, κρούω. (sthg. off or out of) πετώ, ρίχνω. ~ a hole in ανοίγω μία τρύπα σε. I ~ed (my leg, head, etc.) against the door χτύπησα (or έπεσα) πάνω στην πόρτα. ~ at the door χτυπώ την πόρτα.

knock about v.t. (ill-use) κακομεταχειρίζομαι, (damage) στραπατσάρω. (v.i.) he has knocked about a lot έχει γυρίσει παντού, (lived a full life) έχει ζήσει τη ζωή του.

knock back v.t. (fam., drink) πίνω.

knock down v.t. ρίχνω κάτω, (buildings) γκρεμίζω. (at auction) κατακυρώνω.

knock-down a., at ~ prices με χτυπημένες τιμές.

knock off v.t. (make, do) σκαρώνω. (deduct) κόβω. (fam.) I'll knock his head off θα του σπάσω το κεφάλι. (v.i.) (stop work) σταματώ, σχολάω.

knock out v.t. (boxing) ρίχνω νοκάουτ. get knocked out βγαίνω νοκάουτ. (fig.) he was knocked out (by news) έμεινε κόκαλο. (defeat) νικώ, (put out of action) αχρηστεύω.

knock-out s. νοκάουτ n.

knock up v.t. (make) σκαρώνω, (waken) ξυπνώ, (exhaust) εξαντλώ.

knocker s. χτυπητήρι n.

knock-kneed a. στραβοκάνης.

knoll s. λοφίσκος m.

knot v.t. (tie) δένω. ~ted hair μπλεγμένα μαλλιά.

knot s. κόμπος m. (also naut.) κόμβος m. (in wood) ρόζος m. Gordian ~ γόρδιος δεσμός. ~ty a. (wood) ροζιάρικος, (problem) δύσκολος.

know *v.t. & i.* ξέρω, γνωρίζω. (*experience, make acquaintance of*) γνωρίζω. (*learn*) μαθαίνω. he ~s English ξέρει αγγλικά, γνωρίζει τα αγγλικά. he has ~n better days έχει γνωρίσει καλύτερες μέρες. I got to ~ him last year τον γνώρισα πέρσι. I got to ~ (of) it by chance το έμαθα τυχαίως. do you ~ how to do it? ξέρεις πώς να το κάνεις; do you ~ of a good tailor? ξέρεις έναν καλό ράφτη; do you ~ anything about plumbing? ξέρεις (τίποτα) από υδραυλικά; I ~ nothing about their plans δεν ξέρω τίποτα για τα σχέδιά τους. I ~ all about that τα ξέρω όλα για αυτό. you ought to have ~n better έπρεπε να έχεις περισσότερη γνώση. he made himself ~n to me μου συστήθηκε ο ίδιος. his insolence ~s no bounds η αυθάδειά του δεν γνωρίζει όρια. as far as I ~ απ' ό,τι ξέρω. not that I ~ of καθ' όσον ξέρω, όχι. there's no ~ing κανείς δεν ξέρει. let (*person*) ~ πληροφορώ. ~ by heart ~ ξέρω απ' έξω. be in the ~ είμαι μέσα (*or* μπασμένος). ~ thyself! γνώθι σαυτόν.
know-all *s.* (*iron.*) he is a ~ τα ξέρει όλα.
know-how *s.* (*fam.*) απαιτούμενες γνώσεις.
knowing *a.* (*smart*) έξυπνος, (*sly*) πονηρός. ~**ly** *adv.* (*slyly*) πονηρά. (*consciously*) εν γνώσει μου.
knowledge *s.* (*knowing*) γνώση *f.* (*what a person knows*) γνώσεις (*f.pl.*) have no ~ of αγνοώ, δεν ξέρω. it came to my ~ περιήλθε εις γνώσιν μου, έφτασε στα αυτιά μου, πληροφορήθηκα. without his father's ~ χωρίς να το ξέρει ο πατέρας του, εν αγνοία του πατρός του. have a good ~ of history έχω καλές γνώσεις ιστορίας. to the best of my ~ απ' ό,τι ξέρω.
knowledgeable *a.* be ~ έχω καλές γνώσεις, είμαι καλώς πληροφορημένος, είμαι μπασμένος.
known *a.* γνωστός. make ~ γνωστοποιώ. well-~ πασίγνωστος. as is (well) ~ ως γνωστόν.
knuckle *s.* κλείδωσις του δακτύλου.
knuckle *v.* he ~d down to the job στρώθηκε στη δουλειά. ~ under το βάζω κάτω.
Ko νοκάουτ *n.*
Koran *s.* Κοράνι *n.*
kowtow *v.* (*fig.*) κάνω τούμπες.
kudos *s.* δόξα *f.*

L

lab *s.* εργαστήριο *n.*
label *s.* ετικέττα *f.* (*fig.*) χαρακτηρισμός *m.*

(*v.t.*) κολλώ ετικέττα σε, χαρακτηρίζω.
labial *a.* χειλικός.
laboratory *s.* εργαστήριο *n.*
laborious *a.* (*task*) κοπιαστικός, (*style*) δύσκολος. ~**ly** *adv.* με κόπο.
labour *s.* (*work*) εργασία *f.*, δουλειά *f.* (*toil*) κόπος *m.*, μόχθος *m.* (*workers*) εργάτες *m.pl.*, εργατικά χέρια. (*of Hercules*) άθλος *m.* ~s of childbirth ωδίνες του τοκετού, in ~ επίτοκος. hard ~ καταναγκαστικά έργα. Ministry of L~ Υπουργείον Εργασίας. L~ Party Εργατικόν Κόμμα. (*a.*) (*also* member of ~ party) εργατικός.
labour *v.i.* κουράζομαι, κοπιάζω, μοχθώ. (*of ship*) παραδέρνω. ~ in vain ματαιοπονώ. he ~s under the idea that... του έχει καρφωθεί η ιδέα ότι. (*v.t.*) (*insist on*) επεκτείνομαι επί (*with gen.*). ~**ed** *a.* δύσκολος. ~**er** *s.* εργάτης *m.*
labyrinth *s.* λαβύρινθος *m.*
lace *s.* δαντέλα *f.* (*of shoe*) κορδόνι *n.* (*v.t.*) (*fasten*) δένω.
lacerat|e *v.* ξεσχίζω, σπαράσσω. ~**ion** *s.* ξέσχισμα *n.*
lachrymose *a.* κλαψιάρης.
lack *s.* έλλειψη *f.* for ~ of food ελλείψει τροφής. I feel the ~ of it μου λείπει. (*v.t.*) I ~ food στερούμαι τροφής, δεν έχω τροφή, μου λείπει τροφή. (*v.i.*) be ~ing (*missing, absent*) λείπω, δεν υπάρχω. be ~ing in (*personal quality*) υστερώ σε.
lackadaisical *a.* (*person*) νωθρός, αδιάφορος.
lackey *s.* λακές *m.*
laconic *a.* λακωνικός.
lacquer *s.* βερνίκι *n.*, λάκα *f.* (*v.*) βερνικώνω με λάκα.
lad *s.* αγόρι *n.*, παιδί *n.*
ladder *s.* κινητή σκάλα *f.* (*fig.*) κλίμακα *f.* rope ~ ανεμόσκαλα *f.* (*v.i.*) my stocking is ~ed μου έφυγαν πόντοι.
laden *a.* φορτωμένος.
la-di-da *a.* σνομπ.
lading *s.* bill of ~ φορτωτική *f.*
lady *s.* κυρία *f.* (*of title*) λαίδη *f.* Our L~ η Παναγία. ~'s man γαλάντης *m.* ~**bird** παπαδίτσα *f.* ~**killer** *s.* καρδιοκατακτητής *m.* ~**like** *a.* κυρία *f.* ~like behaviour συμπεριφορά κυρίας.
lag *v.i.* (*be slow*) βραδυπορώ, καθυστερώ, μένω πίσω. (*s.*) καθυστέρηση *f.* time ~ χρονική υστέρηση. ~**gard** *a. & s.* βραδυκίνητος. (*fam.*) αργοκίνητο καράβι.
lagging *s.* θερμομονωτικό περίβλημα *n.*
lagoon *s.* λιμνοθάλασσα *f.*
lair *s.* φωλιά *f.*
laird *s.* Σκωτσέζος κτηματίας.
laity *s.* λαϊκοί *m.pl.*

lake s. λίμνη f.

lama s. λάμα(ς) m.

lamb s. αρνάκι n., (also eccl.) αμνός m.

lambent a. λάμπων, (wit) σπινθηροβόλος.

lame a. κουτσός, χωλός, (excuse) αδύνατος. be ~ κουτσαίνω, χωλαίνω. (v.t.) κουτσαίνω. **~ly** adv. αδύνατα, όχι πειστικά.

lament v.i. οδύρομαι, (v.t. & i.) θρηνώ, (the dead) μοιρολογώ. (s.) (also ~ation) οδυρμός m., θρήνος m. **~able** a. αξιοθρήνητος. **~ed** a. πολυθρήνητος.

laminated a. φυλλωτός. (wood) κόντρα-πλακέ.

lamp s. λάμπα f., λαμπτήρ m. (with olive-oil) λύχνος m., λυχνάρι n. (votive) καντήλα f. (of vehicle, etc.) φανός m., φανάρι n. **~-post** s. στύλος m. **~-shade** s. αμπαζούρ n.

lampoon s. σάτιρα f. (v.) σατιρίζω.

lance s. λόγχη f. **~-corporal** s. υποδεκανεύς m.

lance v.t. (prick) ανοίγω. **~t** s. νυστέρι n.

land s. γη f. (dry ~) ξηρά f. (country) χώρα f. (estate) κτήμα n. by ~ κατά ξηράν. (breeze) off the ~ απόγειος. **~ed** a. ~ed property κτηματική περιουσία, ~ed proprietor γαιοκτήμων m.

land v.t. & i. (from ship) ξεμπαρκάρω, (v.t.) αποβιβάζω. (v.i.) αποβιβάζομαι, (aero.) προσγειώνομαι. (fam.) he ~ed the job κατόρθωσε να πάρει τη θέση. he got ~ed in jail βρέθηκε στη φυλακή. I ~ed him a blow του κατάφερα μια.

landing s. (act) αποβίβαση f. (aero.) προσγείωση f. (of stairs) πλατύσκαλο n. **~-craft** s. αποβατικόν σκάφος. **~-stage** s. αποβάθρα f.

land|lady s. σπιτονοικοκυρά f. **~lord** s. σπιτονοικοκύρης m. (of inn) ξενοδόχος m.

landlocked a. περιβαλλόμενος από ξηράν.

landmark s. σημείο προσανατολισμού, (fig.) σταθμός m.

landmine s. νάρκη ξηράς.

landowner s. κτηματίας m. (of cultivable land) γαιοκτήμων m.

landscape s. τοπ(ε)ίο n.

landslide s. κατολίσθηση f.

landsman s. στεριανός a.

lane s. δρομάκι n. (of traffic) λουρίδα f.

language s. γλώσσα f. (a.) γλωσσικός.

languid a. χαύνος, νωθρός, άτονος.

languish v. μαραίνομαι, λειώνω. ~ing looks λιγωμένα μάτια.

languor s. νωθρότητα f., ατονία f. **~ous** a. λιγωμένος.

lanky a. (limbs) ξερακιανός. ~ fellow κρεμανταλάς m.

lantern s. φανός m., φανάρι n.

lap s. αγκαλιά f., ποδιά f. (knees) γόνατα n.pl. (stage) στάδιο n. (of race) γύρος m. live in the ~ of luxury κολυμπώ στη χλιδή.

lap v.t. (wrap) περιτυλίγω. (drink) ρουφώ, (fig.) ~ up χάφτω. (v.i.) (of waves) φλοισβίζω. **~ping** s. (of water) φλοίσβος m.

lapel s. πέτο n.

lapse v.i. (die out) χάνομαι, (of privilege) παραγράφομαι. (backslide) ολισθαίνω, (into coma, etc.) περιπίπτω, (of time) παρέρχομαι. (s.) (mistake) λάθος n. (moral) ολίσθημα n. (of time) πάροδος f. ~ of memory διάλειψη μνήμης.

larceny s. κλοπή f.

lard s. λαρδί n.

larder s. κελλάρι n.

large a. μεγάλος. at ~ (free) ελεύθερος, (in general) εν γένει. **~ly** adv. κατά μέγα μέρος. **~-minded** a. γενναιόφρων. ~scale a. μεγάλης κλίμακος.

largesse s. μοίρασμα δώρων με απλοχεριά.

lark s. (bird) κορυδαλλός m. (fun) αστείο n. we had a ~ κάναμε πλάκα, διασκεδάσαμε.

larva s. κάμπια f.

laryn|x s. λάρυγγας m. **~gitis** s. λαρυγγίτις f.

lascivious a. λάγνος. **~ness** s. λαγνεία f.

laser s. λέιζερ n.

lash v.t. (tie) δένω, (whip) μαστιγώνω, (tail, etc.) χτυπώ. (v.i.) ~ out at επιτίθεμαι βιαίως κατά (with gen.). (s.) (stroke of whip) καμτσικιά f. (of eye) βλεφαρίδα f., ματόκλαδο n., τσίνουρο n.

lass s. κορίτσι n.

lassitude s. ατονία f.

lasso s. λάσσο n.

last s. (shoemaker's) καλαπόδι n. stick to one's ~ περιορίζομαι στις δουλειές που ξέρω.

last a. & s. (final, conclusive) τελικός, τελευταίος, (most recent) τελευταίος. (of dates, seasons) περασμένος. it happened ~ Tuesday έγινε την περασμένη Τρίτη. ~ year πέρσι. ~ night (during the night) απόψε, (yesterday evening) χθες το βράδυ. ~ but one προτελευταίος. L~ Supper Μυστικός Δείπνος. breathe one's ~ πνέω τα λοίσθια. at ~ (στο) τέλος, επιτέλους. to the ~ μέχρι τέλους. the ~ of my money τα τελευταία μου λεφτά. I hope we've seen the ~ of him ελπίζω να μην τον ξαναδούμε. we shall never hear the ~ of it θα βαρεθούμε να ακούμε για αυτό.

last adv. για τελευταία φορά, τελευταία. he spoke ~ μίλησε τελευταίος. **~ly** adv. τέλος, τελικά.

last v. κρατώ, διαρκώ, βαστώ. he won't ~ out the night δεν θα βγάλει τη νύχτα. the coat will ~ we out till spring το παλτό θα μου κρατήσει ως την άνοιξη. **~ing** .*a*. διαρκής, μόνιμος.

latch *s*. μπετούγια *f*. (*v.t*.) κλείνω. (*fam*.) ~ on to (*understand*) καταλαβαίνω. **~-key** *s*. κλειδί *n*.

late *a*. (*of maturing fruit, etc*.) όψιμος, (*deceased*) μακαρίτης. (*delayed: use v. or adv*.) be ~ αργώ, καθυστερώ. they are ~ άργησαν. the train was an hour ~ το τρένο καθυστέρησε μία ώρα (*or* είχε μία ώρα καθυστέρηση). it is (*getting*) ~ είναι αργά. in the ~ afternoon αργά το απόγευμα. in ~ March κατά τα τέλη Μαρτίου. of ~ *see* lately. **~r** *a*. μεταγενέστερος.

late *adv*. αργά. **~r** (on) αργότερα. **~ly** *adv*. τελευταίως, προσφάτως.

latent *a*. λανθάνων. be ~ λανθάνω.

lateral *a*. πλάγιος. **~ly** *adv*. πλαγίως.

latest *a*. τελευταίος. at the ~ το αργότερο.

lath *s*. πήχυς *m*.

lathe *s*. τόρνος *m*. turn on the ~ τορνεύω.

lather *s*. (*soapy*) σαπουνάδα *f*. (*sweaty*) αφρός *m*. (*v.t*.) σαπουνίζω. (*v.i*.) (*of soap*) πιάνω, (*of horse*) αφρίζω, (*of water*) κάνω σαπουνάδα.

Latin *a*. λατινικός. (*s*.) λατινικά *n.pl*.

latitude *s*. (*geographic*) πλάτος *n*. **~s** περιοχές *f.pl*. (*freedom*) ελευθερία *f*. (*margin*) περιθώριο *n*.

latrine *s*. αποχωρητήριο *n*.

latter *a*. τελευταίος. (*s*.) the ~ ο δεύτερος.

lattice *s*. δικτυωτό *n*., καφασωτό *n*.

laud *v*. εξαιρώ. **~able** *a*. αξιέπαινος. **~atory** *a*. επαινετικός.

laugh *v.i*. γελώ, (*mock*) κοροϊδεύω. I ~ at his jokes γελώ με τα αστεία του. ~ it off το ρίχνω στο αστείο. ~ one's head off. ξεκαρδίζομαι στα γέλια. (*s*.) γέλιο *n*., γέλως *m*. I had the ~ of him του την έφερα, τον κανόνισα. **~able** *a*. (*funny*) αστείος, (*absurd*) γελοίος, για γέλια.

laughing-stock *s*. περίγελος *m*., κορόιδο *n*.

laughter *s*. γέλιο *n*. be bursting with ~ σκάω στα γέλια.

launch *v.t*. (*a ship*) καθελκύω, (*a fashion*) λανσάρω, (*a rocket*) εκτοξεύω. (*set going*) βάζω εμπρός, προωθώ. (*v.i*.) ~ out (*into activity*) μπαίνω, (*ambitiously*) ξανοίγομαι. (*s*.) (*act*) καθέλκυση *f*. (*boat*) άκατος *f*.

laund|er *v*. πλένω και σιδερώνω. **~ress** *s*. πλύστρα *f*.

laundry *s*. (*place*) πλυντήριο *n*., πλυσταριό *n*. (*clothes*) ρούχα για πλύσιμο.

laureate *a*. δαφνοστεφής.

laurel *s*. δάφνη *f*. look to one's ~s φροντίζω να μη χάσω την υπεροχή μου.

lava *s*. λάβα *f*.

lavatory *s*. (*wash-basin*) νιπτήρας *m*. (*w.c*.) καμπινές *m*., μέρος *n*., αποχωρητήριο *n*., απόπατος *m*. **~ pan** *s*. λεκάνη *f*.

lavender *s*. λεβάντα *f*.

lavish *a*. σπάταλος, αφειδής, (*sumptuous*) πλούσιος. (*v*.) επιδαψιλεύω. **~ly** *adv*. αφειδώς, πλούσια.

law *s*. (*general*) νόμος *m*. (*jurisprudence*) νομική *f*. (*legal studies*) νομικά *n.pl*. make ~s. νομοθετώ. go to ~ πάω στα δικαστήρια. lay down the ~ αποφαίνομαι δογματικώς. take the ~ into one's own hands καταφεύγω σε χειροδικία. be a ~ unto oneself εφαρμόζω δικούς μου νόμους. civil (*or* criminal *or* maritime) ~ αστικόν (*or* ποινικόν *or* ναυτικόν) δίκαιον. **~-abiding** *a*. νομοταγής, νομιμόφρων. **~-breaker** *s*. παραβάτης του νόμου. **~-court** *s*. δικαστήριο *n*. **~ful** *a*. νόμιμος. **~-giver** *s*. νομοθέτης. **~-less** *a*. άνομος. **~-lessness** *s*. ανομία *f*. **~-suit** *s*. δίκη *f*. **~yer** *s*. δικηγόρος *m.f*. **~yer's** *a*. δικηγορικός.

lawn *s*. πελούζα *f*.

lax *a*. (*loose*) χαλαρός, (*negligent*) αμελής. **~ative** *a*. καθαρτικός. **~ity** *s*. χαλαρότητα *f*., αμέλεια *f*.

lay *s*. (*song*) τραγούδι *n*.

lay *a*. λαϊκός. ~ figure ανδρείκελο *n*.

lay *v.t*. (*put*) βάζω, (*rest, lean*) ακουμπώ, (*place in position*) τοποθετώ, (*impose*) επιβάλλω, (*allay*) κατευνάζω, (*a ghost*) εξορκίζω, (*eggs*) γεννώ, (*a carpet, floor*) στρώνω, (*bets*) βάζω, (*a trap*) στήνω, (*plans*) καταστρώνω, (*table, plates, etc*.) βάζω, (*a case, facts*) εκθέτω, (*information, wreath*) καταθέτω. ~ concrete ρίχνω πλάκα. the scene is laid in England η σκηνή γίνεται στην Αγγλία. ~ bare αποκαλύπτω, ~ low ρίχνω κάτω, ~ waste ερημώνω. ~ oneself open εκτίθεμαι. ~ hands on (*find*) βρίσκω, (*seize*) αρπάζω, (*assault*) σηκώνω χέρι σε. be laid to the charge of βαρύνω. ~ the dust κάνω να κατακαθίσει η σκόνη.

lay about *v.i*. ~ one μοιράζω χτυπήματα αδιακρίτως.

layabout *s*. αλήτης *m*.

lay aside *v.t*. (*keep*) βάζω στην μπάντα, (*give up*) αφήνω, εγκαταλείπω, παραμερίζω.

lay by *v.t*. βάζω στην μπάντα.

lay-by *s*. πάρκινγκ *n*.

lay down *v.t*. (*put down*) βάζω κάτω, (*relinquish*) παραιτούμαι από, (*arms*) καταθέτω. (*sacrifice*) θυσιάζω. (*rules, etc*.) καθορίζω, (*see also* law). (*a ship*) βάζω στα σκαριά.

lay in *v.t.* εφοδιάζομαι με, προμηθεύομαι.
lay off *v.t.* (*dismiss*) απολύω, (*discontinue*) κόβω. (*v.i.*) σταματώ. ~! (*fam.*) μαχριά!
lay on *v.t.* (*water, gas, etc.*) βάζω, (*organize*) οργανώνω, (*fam.*) lay it on (thick) τα φουσκώνω, τα παραλέω.
lay out *v.t.* (*prepare*) ετοιμάζω, (*exhibit*) εκθέτω, (*spend*) ξοδεύω, (*plan*) σχεδιάζω, (*knock down*) ρίχνω κάτω, (*spread*) απλώνω. lay oneself out καταβάλλω κάθε προσπάθεια.
layout *s.* σχέδιο *n.*, διάταξη *f.*, διαρρύθμιση *f.*
lay up *v.t.* (*store*) εναποθηκεύω. (*a ship*) παροπλίζω. the vessel is laid up το πλοίο είναι δεμένο. (*a car*) κλείνω στο γκαράζ. laid up (*of person*) κρεββατωμένος, κλινήρης.
layer *s.* στρώμα *n.* (*of plant*) καταβολάδα *f.*
layman *s.* (*not clergy*) λαϊκός *a.* (*not expert*) ο μη ειδήμων.
laz|e *v.i.* τεμπελιάζω. ~**iness** *s.* τεμπελιά *f.* ~**y** *a.* τεμπέλης. ~**ybones** *s.* τεμπελχανάς *m.*
L-driver *s.* νέος οδηγός.
lead *s.* (*metal*) μόλυβδος *m.* μολύβι *n.* (*naut.*) βολίδα. *f.* ~**free** *a.* αμόλυβδος.
lead, ~**en** *a.* μολύβδινος, μολυβένιος. ~en (*fig.*) βαρύς, σκούρος.
lead *v.t. & i.* (*of guide, road, action, etc.*) οδηγώ, (*bring*) φέρνω. (*be ~er of*) ηγούμαι (*with gen.*). (*excel*) υπερέχω (*with gen.*), (*come first*) προηγούμαι, (*at cards*) παίζω πρώτος. ~ a life διάγω βίον. ~ astray (*mislead*) γελώ, (*into wrong*) παρασύρω, παραπλανώ. ~ by the nose σέρνω από τη μύτη. ~ off (*begin*) αρχίζω. ~ on (*tempt*) βάζω σε πειρασμό. ~ up to καταλήγω σε.
lead *s.* (*example*) παράδειγμα *n.* (*superiority*) υπεροχή *f.* take (*or be in*) the ~ προηγούμαι, είμαι πρώτος. which country has the ~ in technology? ποια χώρα έχει την πρωτοπορία στην τεχνολογία; take the ~ (*in campaign, revolt*) πρωτοστατώ. (*conduit*) αγωγός *m.* (*dog's*) λουρί *n.*
leader *s.* ηγέτης *m.*, αρχηγός *m.* (*of movement, revolt*) πρωτουργός *m.*, πρωτοστάτης *m.* (*in paper*) κύριο άρθρο. ~**ship** *s.* ηγεσία *f.* (*faculty*) ηγετικές ικανότητες.
leading *a.* (*first*) πρώτος, (*chief*) κύριος, (*in excellence*) κορυφαίος. ~ light εξοχότητα *f.* ~ lady πρωταγωνίστρια *f.* ~ article κύριο άρθρο, ~question παρατειστική ερώτηση. play a ~ part παίζω πρωτεύοντα ρόλο.
leaf *s.* φύλλο *n.* (*fig.*) take a ~ from (*person's*) book αντιγράφω. turn over a new ~ διορθώνομαι. (*v.*) ~ through φυλλομετρώ, ξεφυλλίζω. ~**less** *a.* άφυλλος.

~**y** *a.* πυκνόφυλλος, γεμάτος φύλλα.
leaflet *s.* έντυπο *n.*, φυλλάδιο *n.*
league *s.* (*distance*) λεύγα *f.*, τρία μίλια.
league *s.* (*compact*) σύνδεσμος *m.* L~ of Nations Κοινωνία των Εθνών. in ~ with συνεννοημένος με.
leak *v.i.* (*of liquid, tap, utensil, roof, etc.*) τρέχω, στάζω. (*of ship*) κάνω νερά. ~ out (*of gas, news*) διαρρέω. (*v.t.*, *information*) αποκαλύπτω.
leak, ~**age** *s.* διαρροή *f.*, διαφυγή *f.* (*loss*) απώλεια *f.* (*hole*) τρύπα *f.*
lean *a.* ισχνός, (*meat*) άπαχος. (*s.*) ψαχνό *n.* ~**ness** *s.* ισχνότητα *f.*
lean *v.t. & i.* κλίνω, (*against sthg.*) ακουμπώ. (*v.i.*) (*for support*) στηρίζομαι, (*be tilted or bent*) γέρνω, (*stoop*) σκύβω. (*tend*) ρέπω, κλίνω (*with προς*). ~**ing** *s.* τάση *f.*
leap *v.i. & t.* (*also* ~ over) πηδώ. ~ up (*also of heart*) αναπηδώ. ~ at (*offer*) δέχομαι με ενθουσιασμό. the dog ~t at me το σκυλί όρμησε επάνω μου. (*s.*) πήδημα *n.* by ~s and bounds αλματωδώς. ~**frog** *s.* καβάλλες *f.pl.* ~**year** *s.* δίσεκτο έτος.
learn *v.t. & i.* μαθαίνω. (*get news of*) μαθαίνω (*with για or ότι*), πληροφορούμαι (*with acc. or ότι*). ~ by heart μαθαίνω απ' έξω, αποστηθίζω. (*teach*) μαθαίνω. ~**ed** *a.* σοφός, πολυμαθής, (*of writings*) λόγιος. ~**er** *s.* μαθητής *m.* (*novice*) μαθητευόμενος. he's a ~er μαθαίνει. ~er driver μαθητευόμενος οδηγός. ~**ing** *s.* μάθηση *f.*, γνώσεις *f.pl.*
lease *s.* μίσθωση *f.* (*fam., as contract*) συμβόλαιο *n.* (*fig.*) take a new ~ of life ξαναλώνω. (*v.t.*) (*of tenant*) μισθώνω, (*of owner*) εκμισθώνω, (*of both*) (ε)νοικιάζω. ~**hold** *a.* με μακροχρόνια μίσθωση. ~**holder** *s.* κάτοχος ιδιοκτησίας με μακροχρόνια μίσθωση.
leash *s.* λουρί *n.* (*fig.*) hold in ~ χαλιναγωγώ. strain at the ~ δεν βλέπω την ώρα να ελευθερωθώ.
least *a.* ελάχιστος, ο λιγότερος. (*slightest*) παραμικρός. not in the ~ καθόλου. at ~ τουλάχιστον. to say the ~ για να μην πούμε τίποτα χειρότερο. (*adv.*) λιγότερο. I don't want to offend anybody, ~ of all Mary δεν θέλω να προσβάλω κανέναν και κυρίως τη Μαρία.
leather *s.* δέρμα *n.*, πετσί *n.* (*a.*) δερμάτινος, πέτσινος. ~**y** *a.* σαν πετσί.
leave *s.* άδεια *f.* take ~ of (*person, etc.*) αποχαιρετώ. take ~ of one's senses τρελαίνομαι. take French ~ φεύγω χωρίς άδεια. without so much as by your ~ με το έτσι θέλω.

leave *v.t.* αφήνω, (*give up*) παρα(ι)τώ. ~ me
alone αφήστε με ήσυχο. ~ behind δεν
παίρνω μαζί μου, (*forget*) ξεχνώ, (*outdistance*) αφήνω πίσω, ξεπερνώ. ~ off
(*stop*) παύω, σταματώ (*with acc. or* να). ~
out παραλείπω. be left (απο)μένω, be left
over περισσεύω. (*v.i., depart*) φεύγω.

leaven *s.* προζύμι *n.* (*v.*) (*fig.*) ελαφρύνω.

leavings *s.* απομεινάρια *n.pl.*

lecher *s.,* ~ous *a.* λάγνος. ~y *s.* λαγνεία *f.*

lecture *s.* διάλεξη *f.* give a ~ κάνω διάλεξη.
(*fam., scolding*) κατσάδα *f.* (*v.i.*) ~ on
(*teach*) διδάσκω. (*v.t.*) (*scold*) κατσαδιάζω. ~r *s.* ομιλητής *m.* (*university*) υφηγητής *m.*

ledge *s.* (*sill*) περβάζι *n.*, ποδιά *f.* (*of rock
in sea*) ξέρα *f.* (*projection*) προεξοχή *f.*

ledger *s.* κατάστιχο *n.*, καθολικό *n.*

lee *n. see* leeward.

leech *s.* βδέλλα *f.*

leek *s.* πράσο *n.*

leer *s.* πονηρό βλέμμα. (*v.*) ~ at κοιτάζω
πονηρά. ~y *a.* πονηρός.

lees *s.* κατακάθι *n.* to the ~ μέχρι τρυγός.

leeward *a.* υπήνεμος. (*adv.*) to ~ υπηνέμως.

leeway *s.* (*fig., margin*) περιθώριο *n.* make
up ~ (*fig.*) καλύπτω καθυστέρηση.

left *a.* αριστερός. (*s.*) (*hand, wing*) αριστερά
f. on *or* to the ~ (*of*) αριστερά (*with gen.*).
~~handed *a.* αριστερός, αριστερόχειρας,
ζερβοχέρης. ~ist *s.* αριστερίζων *m.*
~wards *adv.* προς τα αριστερά.

leg *s.* πόδι *n.*, σκέλος *n.* (*fam.*) γάμπα *f.* (*of
meat*) μπούτι *n.* (*of trousers*) μπατζάκι *n.*
pull the ~ of δουλεύω. on one's last ~s στα
τελευταία μου.

legacy *s.* κληρονομιά *f.*

legal *a.* νομικός, (*lawful*) νόμιμος. ~ly *adv.*
νομικώς, νομίμως. ~ity *s.* νομιμότητα *f.*

legaliz|e *v.* νομιμοποιώ. ~ation *s.* νομι-
μοποίηση *f.*

legate *s.* (*papal*) νούντσιος *m.*

legatee *s.* κληρονόμος *m.*

legation *s.* πρεσβεία *f.*

legend *s.* (*inscription*) επιγραφή *f.* (*caption*)
λεζάντα *f.* (*tale*) θρύλος *m.* ~ary *a.*
θρυλικός.

legerdemain *s.* ταχυδακτυλουργία *f.*

legib|le *a.* ευανάγνωστος. ~ility *s.* ευανά-
γνωστον *n.*

legion *s.* λεγεών *m.f.* (*fig.*) they are ~ είναι
αναρίθμητοι.

legislat|e *v.* νομοθετώ. ~ion *s.* νομοθεσία *f.*
~ive *a.* νομοθετικός. ~or *s.* νομοθέτης *m.*
~ure *s.* νομοθετικό σώμα.

legitim|ate *a.* νόμιμος, (*proper*) εύλογος.
(*child*) γνήσιος. ~acy *s.* νομιμότητα *f.*
γνησιότητα *f.* (*rightness*) ορθότητα *f.* ~ize

v. νομιμοποιώ.

legum|e *s.* όσπριο *n.* ~inous *a.* οσπριο-
ειδής.

leisure *s.* ελεύθερες ώρες. be at ~ έχω ελεύθερο καιρό. at one's ~ με την ησυχία μου.
~ly *a.* αβίαστος, άνετος, χωρίς βιασύνη. at
a ~ly pace χωρίς να βιάζομαι.

lemon *s.* λεμόνι *n.* ~ tree λεμονιά *f.* ~ade *s.*
λεμονάδα *f.* ~~squeezer *s.* λεμονο-
στύφτης *m.*

lend *v.* δανείζω, (*impart*) προσδίδω. it
doesn't ~ itself to... δεν ειναι κατάλληλο
για. ~ a hand δίδω χείρα βοηθείας, δίνω
ένα χέρι. ~ a favourable ear τείνω ευήκοον
ους. ~er *s.* δανειστής *m.*

lending-library *s.* δανειστική βιβλιοθήκη.

length *s.* μάκρος *n.*, μήκος *n.* (*duration*)
διάρκεια *f.*, χρονικό διάστημα *n.*, μάκρος
n. (*of book*) μέγεθος *n.* (*piece*) τεμάχιο *n.*,
dress ~ κομμάτι για φόρεμα. be a metre in
~ έχω μήκος ένα μέτρο. go to any ~s
μετέρχομαι παν μέσον, μεταχειρίζομαι
κάθε μέσον. at ~ (*finally*) επιτέλους, (*in
detail*) διά μακρών. keep (*person*) at
arm's ~ κρατώ σε απόσταση. (*lying*) at full
~ φαρδύς-πλατύς.

lengthen *v.t. & i.* μακραίνω. (*v.t.*) (επι)-
μηκύνω. (*v.i., of days*) μεγαλώνω. ~ing *s.*
μάκρεμα *n.* (*extension in space*) επέκταση
f., (επι)μήκυνση *f.*

lengthways *adv.* κατά μήκος.

lengthy *a.* εκτενής, εκτεταμένος.

lenien|t *a.* επιεικής. ~cy *s.* επιείκεια *f.*

lens *s.* φακός *m.*

Lent *s.* Σαρακοστή *f.* ~en *a.* σαρακο-
στιανός.

lentil *s.* φακή *f.*

lentisk *s.* μαστιχιά *f.*

leonine *a.* σα λιοντάρι.

leopard *s.* λεοπάρδαλη *f.*

leper *s.* λεπρός *a.* ~~colony *s.* λεπροκομείο
n.

lepr|osy *s.* λέπρα *f.* ~ous *a.* λεπρός.

lesbian *s.* λεσβία *f.*

lese-majesty *s.* έγκλημα καθοσιώσεως.

lesion *s.* (*med.*) βλάβη *f.*

less *a.* ολιγότερος. (*adv.*) λιγότερο, ολιγώ-
τερο. none the ~ παρ' όλα αυτά. it was
nothing ~ than a miracle ήταν στ' αλήθεια
θαύμα. no ~ than (*at least*) τουλάχιστον.
no ~! (*if you please!*) αν αγαπάτε.

less *adv.* (*minus*) μείον.

lessee *s.* μισθωτής *m.*

lessen *v.t. & i.* λιγοστεύω. (*v.t.*) μειώνω.

lesser *a.* μικρότερος.

lesson *s.* μάθημα *n.* let it be a ~ to you να
σου γίνει μάθημα.

lessor *s.* εκμισθωτής *m.*

lest *conj.* μη(ν), μη τυχόν, μήπως.

let *s.* without ~ or hindrance ελευθέρως.

let *v.t.* (*grant for rent*) (ε)νοικιάζω, εκμισθώνω. "to ~" ενοικιάζεται, (*such notice*) ενοικιαστήριο *n.* (*s.*) (*also* ~ting) νοίκιασμα *n.*

let *v. imper.* ~'s go for a walk πάμε (ένα) περίπατο; ~'s see για να δούμε. ~ me think στάσου να σκεφθώ. ~ him do what he likes ας κάνει ό,τι θέλει. don't ~'s quarrel ας μη μαλώνουμε. ~ there be no mistake δε θέλω παρεξηγήσεις.

let *v.t.* (*allow, enable*) αφήνω, επιτρέπω σε, ~ (*sthg.*) go *or* slip αφήνω. ~ (*person*) know ειδοποιώ. ~ me alone! αφήστε με ήσυχο. ~ it alone! (*don't touch*) ασ' το, (*don't interfere*) μην ανακατεύεσαι. ~ well alone! μη θίγετε τα καλώς κείμενα. ~ alone (*not to mention*) (για) να μην αναφέρω. ~ blood κάνω αφαίμαξη.

let down *v.t.* (*lower*) κατεβάζω, (*hair*) λύνω, αφήνω κάτω, (*garment*) μακραίνω, (*disappoint*) αφήνω στα κρύα του λουτρού.

let-down *s.* απογοήτευση *f.*

let in *v.t.* (*a visitor, the wet*) μπάζω, (*a garment*) στενεύω. let oneself in (*to house*) ανοίγω την πόρτα. I let myself in (*or* got ~) for a lawsuit βρέθηκα μπλεγμένος με δικαστήρια. I was ~ for £1,000 μπήκα μέσα χίλιες λίρες.

let into *v.t.* I let him into the secret τον έμπασα στο μυστικό. ~ a wall (*of plaque, etc.*) εντοιχίζω. let a piece into a garment φαρδαίνω ένα ρούχο προσθέτοντας τσόντα.

let off *v.t.* (*not punish*) συγχωρώ, (*release*) απαλλάσσω. I got ~ lightly φτηνά τη γλύτωσα. he ~ the gun έριξε με το όπλο. they let him off the fine του χάρισαν το πρόστιμο.

let on *v.i.* ~ about (*tell*) ξεφουρνίζω, μαρτυρώ. don't ~! μη σου ξεφύγει.

let out *v.t.* (*from custody*) απολύω. (*a visitor*) συνοδεύω στην πόρτα. (*a garment*) φαρδαίνω, (*the cat, air, etc.*) βγάζω, (*a rope*) αμολάω, (*a cry, etc.*) αφήνω. ~ the bath-water αδειάζω το μπάνιο. (*for hire*) νοικιάζω. let me out! αφήστε με να βγω. he ~ the secret, he let the cat out of the bag του ξέφυγε το μυστικό. (*v.i.*) ~ at (*attack*) ρίχνομαι εναντίον (*with gen.*).

let-out *s.* ευκαιρία για ξεγλίστρημα.

let up *v.i.* (*slacken*) λασκάρω, (*stop*) σταματώ, (*of rain, etc.*) κόβω.

let-up *s.* διακοπή *f.*

lethal *a.* θανατηφόρος, (*weapon*) φονικός.

lethargy *s.* (*med.*) λήθαργος *m.* (*sluggishness*) νωθρότητα *f.* (*sleepiness*) υπνηλία *f.* **~ic** *a.* ληθαργικός, νωθρός.

letter *s.* (*symbol*) γράμμα *n.*, στοιχείο *n.* (*message*) γράμμα *n.*, επιστολή *f.* to the ~ κατά γράμμα. man of ~s άνθρωπος των γραμμάτων. **~box** *s.* γραμματοκιβώτιο *n.* **~ed** *a.* (*person*) διαβασμένος. **~ing** *s.* γράμματα *n.pl.*

lettuce *s.* μαρούλι *n.*

levant *v.i.* (*fam.*) το σκάω (χωρίς να πληρώσω).

Levant *s.* ανατολική Μεσόγειος. **~ine** *a.* (φραγκο)λεβαντίνος.

level *s.* επίπεδο *n.* (*of water*) στάθμη *f.* (*instrument*) αλφάδι *n.*, στάθμη *f.* on a ~ with (*in height*) στο ίδιο ύψος με, (*in value*) στο αυτό επίπεδο με. (*fam.*) on the ~ εν τάξει.

level *a.* (*flat, straight*) επίπεδος, ισ(ι)ος, ομαλός, (*equal*) ισόπαλος. ~ spoonful κοφτή κουταλιά. ~ crossing ισόπεδος διάβαση. he drew ~ with me με έφτασε, he drew ~ with the church (*came abreast of*) έφτασε στο ύψος της εκκλησίας.**~headed** *a.* ισορροπημένος, λογικός.

level *v.* ισοπεδώνω, εξομαλύνω. (*raze*) κατεδαφίζω, ρίχνω. **~ling** *s.* ισοπέδωση *f.* κατεδάφιση *f.*

lever *s.* μοχλός *m.*, λεβιέ *n.* **~age** *s.* (*fig.*) bring ~age to bear on βρίσκω τρόπο να επηρεάσω.

leviathan *s.* (*fig.*) μεγαθήριο *n.*

levity *s.* ελαφρότητα *f.*

levy *s.* (*of money*) επιβολή φορολογίας. (*of troops*) στρατολογία *f.* (*v.*) επιβάλλω, στρατολογώ.

lewd *a.* ασελγής. **~ness** *s.* ασέλγεια *f.*

lexicographer *s.* λεξικογράφος *m.f.*

lexicon *s.* λεξικό *n.*

liabilit|y *s.* (*responsibility*) ευθύνη *f.* (*obligation*) υποχρέωση *f.* (*tendency*) τάση *f.* **~ies** (*fin.*) παθητικό *n.* limited ~y company εταιρεία περιορισμένης ευθύνης.

liable *a.* (*responsible*) υπεύθυνος, (*subject*) υποκείμενος. be ~ (*to*) υπόκειμαι (εις), έχω τάση (να).

liaison *s.* (*amour*) δεσμός *m.* (*mil.*) σύνδεσμος *m.*

liar *s.* ψεύτης *m.*, ψεύτρα *f.*

libation *s.* σπονδή *f.*

libel *s.* λίβελλος *m.* (*as legal offence*) δυσφήμηση *f.* (*v.t.*) δυσφημώ. **~lous** *a.* δυσφημιστικός.

liberal *a.* φιλελεύθερος, (*generous*) γενναιόδωρος, (*abundant*) άφθονος,(*~-minded*) ελευθερόφρων. **~ly** *adv.* γενναιόδωρα, άφθονα. **~ity** *s.* γενναιοδωρία *f.* (*of mind*) ελευθεροφροσύνη f.

liberal|ism s. φιλελευθερισμός m. **~ize** v. φιλελευθεροποιώ.

liberat|e v. απελευθερώνω. **~ion** s. απελευθέρωση f. **~or** s. απελευθερωτής m.

libertine s. έκλυτος a.

libert|y s. ελευθερία f. take the ~y of λαμβάνω το θάρρος να. take ~ies παίρνω θάρρος, (with text) παρερμηνεύω (v.t.).

libidinous a. λάγνος.

librar|y s. βιβλιοθήκη f. **~ian** s. βιβλιοθηκάριος m.

licence s. άδεια f. (misuse of freedom) κατάχρηση ελευθερίας, ασυδοσία f.

license v.t. (a person) χορηγώ άδεια σε.

licentious a. ακόλαστος.

lichen s. λειχήν m.

lick v. γλείφω, λείχω. (fam., excel) τρώω. ~ one's lips γλείφομαι, ~ the dust τρώω χώμα. ~ (person) into shape στρώνω. (s.) (also ~ing) γλείψιμο n. (fam.) get a ~ing τρώω ξύλο.

lid s. κάλυμμα n., καπάκι n.

lie s. ψέμα n., ψεύδος n., ψευτιά f. give the ~ to διαψεύδω. (v.i.) ψεύδομαι, λέω ψέματα.

lie v.i. (be) είμαι, (stay) μένω, (be situated) κείμαι, ευρίσκομαι, βρίσκομαι. (be spread out) απλώνομαι. ~ heavy on βαραίνω. ~ down ξαπλώνω, ξαπλώνομαι, πλαγιάζω. he is lying down πλαγιάζει, είναι πλαγιασμένος (or ξαπλωμένος). it ~s with you (to decide) σε σένα απόκειται. see how the land ~s ερευνώ το έδαφος.

liege s. υποτελής a.

lieu s. in ~ of αντί (with gen.).

lieutenant s. (naut.) υποπλοίαρχος m. (mil.) υπολοχαγός m. **~-commander** s. πλωτάρχης m. **~-colonel** s. αντισυνταγματάρχης m. **~-general** s. αντιστράτηγος m.

life s. ζωή f., βίος m. for ~ διά βίου, εφ' όρου ζωής. bring or come to ~ ζωντανεύω. have the time of one's ~ περνώ ωραία όσο ποτέ άλλοτε. true to ~ παρμένος από τη ζωή. as large as ~ με σάρκα κι οστά. not on your ~! αμ δε! ~ imprisonment ισόβια δεσμά. **~-belt** s. σωσίβιο n. **~-boat** s. ναυαγοσωστική λέμβος. **~-giving** a. ζωογόνος. **~-less** a. νεκρός, (fig.) ψόφιος. **~-like** a. σαν ζωντανός. **~-size(d)** a. σε φυσικό μέγεθος. **~-time** s. ζωή f.

lift v.t. σηκώνω, (a ban) αίρω. (fam., steal) ξαφρίζω. (v.i.) (of fog, etc.) διαλύομαι. **~ing** s. σήκωμα n.

lift s. (elevator) ασανσέρ n., ανελκυστήρας m. (fam.) I gave him a ~ τον πήρα στο αμάξι. give a ~ to (spirits) αναπτερώνω.

ligament s. σύνδεσμος m.

ligature s. (med.) απολίνωσις f.

light s. φως n. (for smoker) φωτιά f. (window) παράθυρο n. ~ and shade κιάροσκούρο n. come to ~ ανακαλύπτομαι. throw (new) ~ on ρίχνω (νέο) φως σε. see in a different ~ βλέπω υπό άλλο πρίσμα. see in a favourable ~ βλέπω με ευνοϊκό μάτι. in the ~ of what you have said παίρνοντας υπ' όψιν όσα είπατε. according to his ~s σύμφωνα με τις ικανότητές του. traffic ~s φανάρια (της τροχαίας).

light v.t. (set burning) ανάβω, (give light to) φωτίζω. ~ up (as decoration) φωταγωγώ. (v.i.) (come alight) ανάβω. ~ up (shine) λάμπω. (find) ~ upon βρίσκω κατά τύχην. **~ed** a. αναμμένος, φωτισμένος.

light a. (well lit) φωτεινός, (of colour) ανοικτός, (not heavy) ελαφρός. make ~ of αψηφώ, περιφρονώ. make ~ work of κάνω χωρίς κόπο. (adv.) (tread, sleep) ελαφρά. (travel) με λίγες αποσκευές. **~-fingered** a. be ~-fingered δεν έχω καθαρό χέρι. **~-headed** a. ζαλισμένος. **~-hearted** a. ξέγνοιαστος. **~-house** s. φάρος m. **~ly** adv. ελαφρώς, (dismissively) περιφρονητικά. he got off ~ly φτηνά τη γλύτωσε. **~ness** s. ελαφρότητα f. **~-weight** s. (fig., person) μετριότητα f. **~-well** s. φωταγωγός m. **~-year** s. έτος φωτός.

lighten v.t. & i. (make or get bright) φωτίζω. (get lighter in colour) ανοίγω. (make or get less heavy) ελαφραίνω, ελαφρύνω. (v.i.) (there is lightning) it ~s αστράφτει.

lighter s. (boat) μαούνα f. (for cigar, etc.) αναπτήρας m.

lightning s. αστραπή f. like ~ (a.) αστραπιαίος, (adv.) αστραπιαίως. **~-conductor** s. αλεξικέραυνο n.

lights s. (organ) πλεμόνι n.

lignite s. λιγνίτης m.

like v.t. & i. (a person) συμπαθώ, μου αρέσει. I ~ beer μου αρέσει η μπίρα. my sister didn't ~ it δεν άρεσε της αδελφής (or στην αδελφή) μου. I should ~ to go θα ήθελα να πάω. would you ~ an ouzo? (fam.) πίνεις ένα ούζο; I ~ that one best προτιμώ εκείνο. if you ~ αν θέλεις. **~able** a. συμπαθητικός.

like s. (equal, match) ταίρι n. we shall never see the ~ of it again ποτέ δεν θα δούμε παρόμοιο (or τέτοιο) πράγμα. (fam.) the ~s of us άνθρωποι σαν κι εμάς. and the ~ και τα παρόμοια. ~s and dislikes γούστα n.pl.

like a. & adv. (alike) όμοιος, παρόμοιος. be ~ μοιάζω (with με or gen.), he's ~ his brother μοιάζει του αδελφού (or με τον

αδελφό) του. they are ~ each other
μοιάζουν. I feel ~ a cigarette η καρδιά μου
τραβάει τσιγαράκι. I don't feel ~ working
δεν έχω διάθεση για δουλειά. something ~
this κάτι τέτοιο. what's he ~ ? τι είδους
άνθρωπος είναι; what's your apple ~? πώς
είναι το μήλο σου; what does she look ~?
πώς είναι; it looks ~ rain το πάει για
βροχή. he looks ~ winning (σα να) φαίνεται πως θα κερδίσει. it looks ~ it (is
probable) έτσι φαίνεται. (general) what
did it seem (or feel, sound, taste, etc.) ~?
πώς ήταν;
like prep. σαν, όπως. he thinks ~ me
σκέπτεται όπως εγώ (or σαν κι εμένα). he
behaves ~ a madman κάνει σαν τρελλός. ~
this έτσι. (fam.) ~ anything πολύ.
likelihood s. πιθανότητα f.
likely a. πιθανός. are you ~ to go there?
είναι πιθανό να πας εκεί; it is ~ (that)
ενδέχεται (να).
like-minded a. be ~ έχω την ίδια γνώμη.
liken v. παρομοιάζω.
likeness s. ομοιότητα f. (portrait) πορτραίτο n. it is a good ~ μοιάζει πολύ.
likewise adv. παρομοίως. (conj., also)
επίσης.
liking s. (for people, animals) συμπάθεια f.
be to one's ~ είναι της αρεσκείας μου. see
like v.
lilac s. πασχαλιά f. (a.) λιλά.
lilliputian a. λιλλιπούτειος.
lilting a. λικνιστικός.
lily s. κρίνος m., κρίνο n.
limb s. μέλος n. (bough) κλαδί n. (fig.) out
on a ~ εκτεθειμένος.
limber a. ευλύγιστος. (v.) ~ up (v.t.) γυμνάζω, (v.i.) γυμνάζω τους μυς μου.
limbo s. be in ~ έχω περιπέσει εις λήθην.
lime s. ασβέστης m., άσβεστος f. ~**stone** s.
ασβεστόλιθος m. ~**twig** s. ξόβεργα f.
limelight s. προβολεύς m. (fig.) in the ~ στο
φως της δημοσιότητος.
limit s. όριο n. that is the ~ ως εδώ και μη
παρέκει. know no ~s δεν έχω όρια. you're
the ~! δεν έχεις τον θεό σου. (v.t.) περιορίζω. ~**less** a. απεριόριστος.
limitation s. περιορισμός m. his ~s τα όρια
των δυνατοτήτων του.
limp a. πλαδαρός, (of plants, etc.) μαραμένος.
limp v. κουτσαίνω, χωλαίνω.
limpet s. πεταλίδα f. ~**mine** s. νάρκη
βεντούζα.
limpid a. λαγαρός. ~**ity** s. διαύγεια f.
linchpin s. παραξόνιο n. (fig.) κεντρικός
μοχλός.
line s. γραμμή f. (company) εταιρεία f. (row)

αράδα f., σειρά f. toe the ~ υπακούω. ~ of
march κατεύθυνση f. range in ~ παρατάσσω. (verse) στίχος m. (rope) σχοινί n.
(fishing) ορμιά f. (on face) ρυτίδα f.
(goods) είδος n. drop a ~ (write) στέλνω
δυο λόγια. hard ~s ατυχία f. I draw the ~ at
that δεν φτάνω ως εκεί, we must draw a ~
somewhere πρέπει να τεθεί ένα όριο. ~s
(method) τρόπος m., σύστημα n. on the ~s
of σύμφωνα με. on English ~s κατά το
αγγλικό σύστημα. (read) between the ~s
ανάμεσα στις γραμμές. what's his ~? τι
δουλειά κάνει; he's in the building ~
ασχολείται με οικοδομικές εργασίες. it's
not in my ~ δεν είναι της ειδικότητός μου.
come into ~ with συμμορφώνομαι (or
ευθυγραμμίζομαι) με. in ~ with σύμφωνα
με. get a ~ on (fam.) μαθαίνω για.
line v. (clothes, curtains) φοδράρω, (box,
drawer) στρώνω. (fig., fill) γεμίζω. see
lined.
lineage s. καταγωγή f.
lineaments s. χαρακτηριστικά n.pl.
linear a. γραμμικός.
lined a. (clothes) φοδραρισμένος. (drawer)
στρωμένος. (face, paper) χαρακωμένος.
(with wood panelling) με επένδυση ξύλου.
well ~ (full) γεμάτος.
linen s. λινό n. (~ articles) ασπρόρουχα n.pl.
(fig.) wash one's dirty ~ in public βγάζω
τα άπλυτά μου στη φόρα. (a.) λινός.
liner s. υπερωκεάνιο n.
linger v. (stay) μένω, (dawdle) χρονοτριβώ,
(drag on) παρατραβώ, δεν λέω να
τελειώσω. ~**ing** a. (slow) αργός, (long
drawn out) μακρόσυρτος.
linguist s. γλωσσομαθής m. (philologist)
γλωσσολόγος m. ~**ic** a. γλωσσικός. ~**ics** s.
γλωσσολογία s.
lining s. (clothes) φόδρα f. (other) επένδυση
f.
link s. κρίκος m. cuff ~s μανικετόκουμπα
n.pl. (v.t.) συνδέω. they ~ed arms πιάστηκαν αγκαζέ. ~**ing** a. συνδετικός.
linoleum s. μουσαμάς m.
linseed-oil s. λινέλαιο n.
lint s. ξαντό n.
lintel s. υπέρθυρο n.
lion s. λιοντάρι n., λέων m. ~'s share μερίδα
του λέοντος. ~'s skin λεοντή f. ~**ess** s.
λέαινα f. ~**ize** v.t. περιποιούμαι ως
διασημότητα.
lip s. χείλι n., χείλος n. ~**service** s. κούφιος
έπαινος. ~**stick** s. κραγιόν n.
liquefy v.t. ρευστοποιώ. ~**action** s. ρευστο-
ποίηση f.
liqueur s. λικέρ n., ηδύποτον n.
liquid a. ρευστός, υγρός. (s.) υγρό n. ~**ate**

v.t. διαλύω, (*debt*) εξοφλώ. (*kill off*) καθαρίζω. **~ation** *s.* (*fin.*) go into ~ation χρεοκοπώ. **~ity** *s.* ρευστότητα *f.*

liquor *s.* (*liquid*) υγρό *n.* (*alcoholic*) οινοπνευματώδη ποτά. (*indulgence in* ~) πιοτό *n.*

liquorice *s.* γλυκόρριζα *f.*

lira *s.* (*money*) λιρέττα *f.*

lisp *v.* τσευδίζω. **~ing** *a.* τσευδός.

list *s.* κατάλογος *m.* (*v.t.*) (κατα)γράφω.

list *v.i.* (*of ship*) γέρνω (*s.*) κλίση *f.*

listen *v.* (*also* ~ to) ακούω. he won't ~ to reason δεν παίρνει από λόγια. **~er** *s.* ακροατής *m.*

listless *a.* άτονος.

lists *s.* enter the ~ συμμετέχω στον αγώνα.

litany *s.* παρακλήσεις *f.pl.*

literal *a.* κυριολεκτικός. (*translation*) λέξιν προς λέξιν. **~ly** *adv.* κυριολεκτικά. don't take it ~ly μην τα παίρνεις τοις μετρητοίς.

literary *a.* λογοτεχνικός, φιλολογικός. ~ man (*writer*) λογοτέχνης *m.*, (*scholar*) λόγιος *m.*

literate *a.* εγγράμματος.

literature *s.* λογοτεχνία *f.* (*study of* ~) φιλολογία *f.*

lithe *a.* λυγερός, εύκαμπτος **~ness** *s.* ευκαμψία *f.*

lithograph, ~y *s.* λιθογραφία *f.*

litigant *s.* διάδικος *m.f.*

litigat|e *v.* καταφεύγω στα δικαστήρια. **~ion** *s.* προσφυγή σε δικαστήριο, δίκη *f.*

litigious *a.* φιλόδικος.

litre *s.* λίτρο *n.*

litter *s.* (*couch*) φορείο *n.* (*new-born animals*) γέννα *f.*, μικρά *n.pl.* (*rubbish*) σκουπίδια *n.pl.* (*untidy state*) ακαταστασία *f.* (*beasts' bedding*) στρωματιά *f.*, άχυρο *n.*

litter *v.t.* they ~ed the street with rubbish λέρωσαν το δρομο με σκορπισμένα σκουπίδια.

little *a.* (*size*) μικρός, (*quantity*) (*also* a ~) (ο)λίγος. (*also by suffix*) ~ house σπιτάκι *n.* ~ John Γιαννάκης *m.* ~ cat γατίτσα, γατούλα *f.* it makes ~ difference (*whether we stay or go*) είναι το ίδιο. no ~ ουκ ολίγος.

little *s.* ολίγον *n.*, λίγο *n.* I got ~ out of it δεν απεκόμισα πολλά. I see ~ of him δεν τον βλέπω συχνά. from what ~ I know από τα λίγα που ξέρω. ~ or nothing σχεδόν τίποτα. he says ~ μιλάει λίγο. he is satisfied with ~ ικανοποιείται με λίγα.

little *adv.* λίγο, ολίγον. ~ by ~ λίγο λίγο. very ~ ελάχιστα, πολύ λίγο. it is ~ better than robbery είναι καθαρή κλεψιά. ~ does he suspect ούτε που βάζει ο νους του. a ~

(*somewhat*) λίγο, λιγάκι, κάπως. not a ~ ουκ ολίγον. let's wait a ~ ας περιμένουμε λίγο (*or* λιγάκι). I'm a ~ better είμαι λίγο (*or* κάπως) καλύτερα. would you like a ~ more soup? θέλετε ακόμα λίγη σούπα;

live *a.* ζωντανός. ~ coals αναμμένα κάρβουνα. ~ shots ένσφαιρα πυρά, ~ wire (*fig.*) δυναμικό πρόσωπο. (*adv., of transmission*) απευθείας.

live *v.i.* ζω, (*dwell*) μένω, κατοικώ, κάθομαι. ~ well καλοζώ, ~ like a lord περνώ μπέικα. long ~ England ζήτω η Αγγλία. ~ with (*cohabit*) συζώ με, (*endure*) υποφέρω, ανέχομαι. ~ on (*survive*) επιζώ. ~ on one's pay ζω από το μισθό μου. he ~s on milk τρέφεται με γάλα. ~ on capital τρώω το κεφάλαιό μου. ~ by one's wits ζω από κομπίνες. ~ up to one's principles δεν υπολείπομαι των αρχών μου. (*v.t.*) (*experience*) ~ an easy life ζω βίον άνετον. we ~d through the occupation ζήσαμε την κατοχή. ~ down the scandal επιζώ και κάνω να ξεχαστεί το σκάνδαλο.

livelihood *s.* πόροι ζωής, μέσα συντηρήσεως, τα προς το ζην.

livelong *a.* the ~ day ολημερίς (*adv.*).

live|ly *a.* ζωηρός. **~liness** *s.* ζωηρότητα *f.*

liven *v.i. & i.* ζωηρεύω.

liver *s.* συκώτι *n.*, ήπαρ *n.*

livery *s.* (*dress*) λιβρέα *f.*

livestock *s.* ζώα κτηνοτροφίας.

livid *a.* πελιδνός. (*fam., furious*) πράσινος από το θυμό μου.

living *a.* ζων, ζωντανός. (*animate*) έμψυχος.

living *s.* earn one's ~ κερδίζω τη ζωή μου. scrape a ~ αποζώ. he makes a ~ by trade ζει εμπορευόμενος. easy ~ καλοζωία *f.* good ~ καλοφαγία *f.* standard of ~ βιοτικό επίπεδο. ~ is cheap η ζωή είναι φθηνή. what does he do for a ~? τι εργασία κάνει; (*eccl.*) *see* benefice. **~-room** *s.* σαλόνι *n.*

lizard *s.* σαύρα *f.*

llama *s.* λάμα *f.*

lo! *int.* ιδού!

load *s.* φορτίο *n.* (*amount carried*) φόρτωμα *n.* (*fig., burden*) βάρος *n.* (*fam.*) ~s (of) ένα σωρό. (*v.t. & i*) φορτώνω. (*v.t.*) (*gun*) γεμίζω. **~ing** *s.* φόρτωση *f.* φόρτωμα *n.*

loaf *s.* ψωμί *n.* (*round, square*) καρβέλι *n.* (*long*) φραντζόλα *f.* (*ring-shaped*) κουλούρα *f.*

loaf *v.i.* χασομερώ, χαζεύω. **~er** *s.* χασομέρης *a.*

loan *s.* δάνειο *n.* on ~ δανεικά (*adv.*). (*v.*) δανείζω.

lo(a)th *a.* απρόθυμος. nothing ~ πρόθυμος.

loath|e *v.* σιχαίνομαι, αποστρέφομαι. **~ing** *s.* σιχαμάρα *f.* **~some** *a.* σιχαμερός.

lobby *s.* (*entrance*) είσοδος *f.* (*ante-room*) προθάλαμος *m.* (*political*) ομάδα υποστηρικτών ορισμένων συμφερόντων που εξασκεί πίεση στα κυβερνητικά παρασκήνια. (*v.t.*) ~ one's M.P. μιλώ στο βουλευτή μου.

lobster *s.* αστακός *m.*

local *a.* τοπικός. (*born or produced* ~*ly*) ντόπιος. the ~ butcher ο χασάπης της γειτονιάς. ~ colour τοπικό χρώμα. ~ government τοπική αυτοδιοίκηση. **~ly** *adv.* στη γειτονιά. **~ism** *s.* τοπικισμός *m.* **~ize** *v.* (*restrict*) εντοπίζω.

locality *s.* (*position*) τοποθεσία *f.* (*district*) περιοχή *f.*, γειτονιά *f.* (*fam.*) have a good bump of ~ έχω καλό προσανατολισμό.

locat|e *v.t.* (*place*) τοποθετώ, (*find*) βρίσκω. be ~ed βρίσκομαι. **~ion** *s.* τοποθεσία *f.* **~ive** *s.* (*gram.*) τοπική *f.*

loch *s.* λίμνη *f.*

lock *s.* (*of door*) κλειδαριά *f.*, κλείθρο *n.* (*on canal*) υδατοφράκτης *m.* (*of hair*) μπούκλα. *f.* (*grip*) λαβή *f.* (*v.t.* & *i.*) (*also* ~ up) κλειδώνω. (*grip*) σφίγγω. **~er** *s.* ντουλαπάκι *n.*

locket *s.* μενταγιόν *n.*

locksmith *s.* κλειδαράς *m.*

locomot|ion *s.* μετακίνηση *f.* **~ive** *a.* κινητήριος. (*s*) μηχανή *f.*

locum *s.* ~ (tenens) αναπληρωτής ιατρός. *or* ιερεύς.

locus *s.* (*math.*) τόπος *m.*

locust *s.* ακρίδα *f.* ~ bean χαρούπι *n.* ~ tree χαρουπιά *f.*

lode *s.* φλέβα *f.* **~star** *s.* πολικός αστέρας. **~stone** *s.* μαγνήτης *m.*

lodge *s.* (*porter's*) θυρωρείο *n.* (*hunting*) περίπτερο *n.* (*masonic*) στοά *f.*

lodge *v.t.* (*house*) στεγάζω, φιλοξενώ. (*stick*) κολλώ, (*deposit*) καταθέτω, (*submit*) υποβάλλω. (*v.i.*) (*stay*) μένω, (*stick*) κολλώ. **~r** *s.* νοικάρης *m.*, νοικάρισσα *f.* take ~rs νοικιάζω δωμάτια.

lodging *s.* κατάλυμα *n.* ~s νοικιασμένα δωμάτια. let ~s νοικιάζω δωμάτια.

loft *s.* υπερώο *n.*, σοφίτα *f.*

loft|y *a.* υψηλός, (*disdainful*) ακατάδεκτος. **~ily** *adv.* περιφρονητικά. **~iness** *s.* (*of sentiments*) ευγένεια *f.* (*disdain*) ακαταδεξία *f.*

log *s.* κούτσουρο *n.* (*ship's*) παρκέτα *f.* (*v.*) (*enter*) καταγράφω. **~book** *s.* ημερολόγιο *n.*

logarithm *s.* λογάριθμος *m.*

loggerheads *s.* they are at ~ είναι μαλωμένοι.

logic *s.* λογική *f.* **~al** *a.* λογικός. **~ally** *adv.* λογικά.

logistics *s.* διοικητική μέριμνα.

logo *s.* λογότυπο *n.*

loins *s.* οσφύς *f.*, λαγόνες *f.pl.*

loiter *v.* χασομερώ. **~er** *s.* χασομέρης *a.*

loll *v.i.* (*hang*) κρέμομαι, (*lean*) ακουμπώ, (*lie*) είμαι ξαπλωμένος.

lone *a.* μοναχικός, μόνος. play a ~ hand ενεργώ χωρίς βοήθεια.

lone|ly *a.* (*place*) απομονωμένος. feel ~ly αισθάνομαι μοναξιά. have a ~ly life είμαι πολύ μόνος. **~liness** *s.* μοναξιά *f.*

lonesome *a. see* lonely.

long *v.* ~ for λαχταρώ. ~ to λαχταρώ να, το έχω καημό να. **~ing** *s.* λαχτάρα *f.*, καημός *m.* **~ingly** *adv.* με λαχτάρα.

long *a.* μακρύς, (*time*) πολύς, (*book*) μεγάλος. a ~ journey ταξίδι μεγάλης διάρκειας. it was a ~ operation η εγχείρηση διήρκεσε πολύ. how ~ is your holiday? πόσο θα διαρκέσουν οι διακοπές σας; it is 20 metres ~ έχει μήκος είκοσι μέτρα. in the ~ run τελικά. he won't be ~ δεν θα αργήσει. **~distance** *a.* (*bus, phonecall*) υπεραστικός. **~haired** *a.* μακρυμάλλης. **~hand** *s.* in ~hand με το χέρι. **~legged** *a.* μακροκάνης. **~range** *a.* (*gun*) μεγάλου βεληνεκούς, (*plane*) μεγάλης ακτίνος δράσεως. **~term** *a.* μακράς διαρκείας(*fin.*) μακροπρόθεσμος. **~winded** *a.* (*speech*) μακροσκελής, (*person*) φλύαρος.

long *s.* it won't take ~ δεν θα πάρει πολλή ώρα. will you be away for ~? θα λείψετε καιρό; before ~ εντός ολίγου.

long *adv.* πολύ, ~ ago προ πολλού, not ~ ago τώρα τελευταία. be ~ (*in doing sthg.*) αργώ (*with* να). how ~ is it since you saw them? πόσον καιρό έχεις να τους δεις; as (*or* so) ~ as (*provided that*) αρκεί να, υπό την προϋπόθεσιν ότι. as ~ as you like όσο καιρό θέλεις. **~drawn-out** *a.* παρατεταμένος. **~lived** *a.* μακρόβιος. **~playing** *a.* μακράς διαρκείας. **~standing** *a.* παλαιός, υφιστάμενος από μακρού.

longer *adv.* I'm not staying any ~ δεν θα μείνω άλλο (*or* περισσότερο). he's no ~ in charge of the department δεν είναι πια επικεφαλής του τμήματος. how much ~ will you be? πόση ώρα θα κάνεις ακόμα;

longevity *s.* μακροζωία *f.*

longitude *s.* μήκος *n.*

longshoreman *s.* λιμενεργάτης *m.*

longways *adv.* κατά μήκος.

look *s.* (*glance*) βλέμμα *n.*, ματιά *f.* (*appearance*) όψη *f.*, εμφάνιση *f.* take (*or* have) a ~ ρίχνω μια ματιά. have good ~s είμαι όμορφος. black ~s εχθρικά βλέμματα. by the ~ of things όπως φαίνεται. his ~s have improved η όψη (*or* η εμφάνισή) του έχει φτιάξει. she has lost her ~s έχει χαλάσει.

look v. (also ~ at) κοιτάζω. (appear, seem) φαίνομαι. ~ like (resemble) μοιάζω (see like). it ~s as if it's going to rain σάμπως θα βρέξει. ~ here! για κοίταξε, κοίταξε να δεις! ~ sharp! κουνήσου! she ~s her best in blue την κολακεύουν τα μπλε. he ~s himself again έχει συνέλθει. he ~s the part του ταιριάζει ο ρόλος.

look about v.i. ψάχνω. ~ one κοιτάζω γύρω μου.

look after v.t. (tend) φροντίζω, περιποιούμαι, κοιτάζω, προσέχω. (see about) φροντίζω για.

look back v.i. ~ at or on (recollect) αναπολώ, ξαναθυμούμαι. he never looked back διαρκώς προόδευε.

look down v.i. ~ on (despise) περιφρονώ.

look for v.t. (seek) ζητώ, γυρεύω, ψάχνω για. (expect) ζητώ, περιμένω να βρω. be looking for trouble πάω γυρεύοντας.

look forward v.i. ~ to (with pleasure) προσμένω με χαρά, ανυπομονώ να, περιμένω ανυπόμονα. ~ to (await) περιμένω, (foresee) προβλέπω.

look in v.i. (pay a call) περνώ. ~ at the chemist's περνώ από το φαρμακείο. I will ~ on you tomorrow θα περάσω να σε δω αύριο.

look-in s. he won't get a ~ δεν έχει ελπίδα επιτυχίας.

look into v.t. (investigate) εξετάζω.

look on (v.i.) (be onlooker) παρακολουθώ. ~ to (of aspect) βλέπω προς. (v.t.) (regard) θεωρώ.

look out v.i. (be on watch) προσέχω, έχω το νου μου. ~ on or over (of aspect) βλέπω προς. ~! (threat) έννοια σου! (take care) πρόσεχε! (v.t.) (select) διαλέγω.

look-out s. (post) σκοπιά f. (sentinel) σκοπός m. be on the ~ for έχω τα μάτια μου ανοιχτά για, (lie in wait for) παραμονεύω.

look round v. (turn one's head) γυρίζω το κεφάλι μου. (survey the scene) κοιτάζω γύρω, ερευνώ.

look up v.t. (a word, etc.) ψάχνω να βρω, (a friend) περνώ να δω, επισκέπτομαι. ~ to (respect) σέβομαι. (v.i.) σηκώνω τα μάτια μου. (get better) πάω προς το καλύτερο.

looking-glass s. καθρέφτης m.

loom s. αργαλειός m.

loom v.i. ξεπροβάλλω, (dimly) διαφαίνομαι, (darkly) μαυρίζω. ~ large (in one's thoughts, etc.) κατέχω μεγάλη θέση.

loon, ~y s. & a. (fam.) τρελλός.

loop s. θηλιά f. (v.) προσδένω με θηλειά. (aero.) ~ the ~ κάνω λούπινγκ.

loophole s. πολεμίστρα f. (fig.) παρα-

θυράκι n.

loose a. (not tied up) (dog) λυτός, (hair) λυτός, χυτός, (not tight) χαλαρός, (unstable) ασταθής. (scattered) σκόρπιος, (not in package) χύμα (adv.). (vague) ασαφής, (disjointed) ξεκάρφωτος, (not literal) ελεύθερος. (on the) ~ (at large) ελεύθερος, (morally) ελευθερίων ηθών. ~ money ψιλά n.pl. it is ~ (tooth, brick) κουνάει, (floorboard, handle) παίζει, (garment, shoe) είναι φαρδύς. work ~ ξεβιδώνομαι, χαλαρώνομαι. get ~ (escape) ξεφεύγω. let ~ (dog) λύνω, (an attack) εξαπολύω. be at a ~ end δεν έχω τι να κάνω. go on the ~ ξεφαντώνω. have a screw ~ έχω βίδα.

loosely adv. χαλαρά, (vaguely) ασαφώς. ~ translated κατά ελευθέραν μετάφρασιν.

loosen v.t. & i. λασκάρω. (v.t.) χαλαρώνω, (belt, screw) ξεσφίγγω. ~ up (muscles) χαλαρώνω.

loot s. λάφυρα n.pl. λεία f., πλιάτσικο n. (v.) λαφυραγωγώ, λεηλατώ. ~ing s. λεηλασία f., πλιάτσικο n.

lop v. κλαδεύω.

lope v. προχωρώ πηδώντας.

lopsided a. it's ~ δεν είναι ίσιος.

loquacious a. φλύαρος, πολυλογάς. ~ity s. φλυαρία f., πολυλογία f.

loquat s. μούσμουλο n.

lord s. λόρδος m. (God) Κύριος m. (feudal, etc.) αφέντης m. (v.) ~ it φέρομαι δεσποτικά.

lordly a. (haughty) υπεροπτικός, (splendid) μεγαλοπρεπής.

lordship s. his ~ η εξοχότης του.

lore s. (traditions) παραδόσεις f.pl.

lorry s. φορτηγό n.

lose v. χάνω, get lost χάνομαι. ~ sight of χάνω από τα μάτια μου, (fig.) we mustn't ~ sight of that δεν πρέπει να μας διαφεύγει. ~ one's head χάνω το μυαλό μου. lost in thought βυθισμένος σε σκέψεις.

loser s. χαμένος m. he is a bad ~ το φέρνει βαρέως όταν χάνει.

loss s. απώλεια f., χάσιμο n. (material) ζημία f. sell at a ~ πουλώ με ζημία. cut one's ~es περιορίζω τις ζημίες μου. be at a ~ for words δεν βρίσκω (or δυσκολεύομαι να βρω) τα κατάλληλα λόγια.

lot s. (thing drawn) λαχνός m., κλήρος m. cast or draw ~s βάζω κλήρο. (one's fate) μοίρα f. it fell to my ~ μου έπεσε να... throw in one's ~ with συμμερίζομαι την τύχη (with gen.). (fam.) a bad ~ κάθαρμα n., χαμένος m.

lot s. (for building) οικόπεδο n. (batch, consignment) παραλαβή f., παρτίδα f. (piece, item) κομμάτι n.

lot s. & a. (quantity) ~s or a ~ (of) πολύς, ένα σωρό. the ~ of you όλοι σας. that's the ~ αυτά είναι όλα. it's a ~ of nonsense είναι όλο βλακείες. (adv.) a ~ πολύ. a ~ I care! πολύ που με νοιάζει.

loth a. see loath.

lotion s. λοσιόν f.

lottery s. λαχείο n.

lotus s. λωτός m. (fig.) be a ~-eater απολαμβάνω μια γαλήνια ζωή τεμπελιάζοντας.

loud a. δυνατός, (colour) χτυπητός, (behaviour) χυδαίος. (adv.) (out) ~ δυνατά. ~ly adv. δυνατά, χτυπητά. ~-mouthed a. φωνακλάς m. ~ness s. δύναμη f. ~speaker s. μεγάφωνο n.

lounge s. σαλόνι n. (v.i.) (idle) τεμπελιάζω, (lean) ακουμπώ, (recline) είμαι ξαπλωμένος. ~-suit s. κοστούμι n.

lous|e s. ψείρα f. ~-y a. ψειριασμένος. (fig.) (of bad quality) ελεεινός, χάλια. (as abusive epithet) βρωμο-.

lout s. γουρούνι n.

lovable a. αξιαγάπητος.

love v. αγαπώ. I ~ singing μου αρέσει πολύ το τραγούδι I should ~ to go θα μου άρεσε πολύ να πάω.

love s. αγάπη f. (tender solicitude) στοργή f., (physical) έρωτας m. fall in ~ (with) ερωτεύομαι. in ~ ερωτευμένος. make ~ to κάνω κόρτε σε. not for ~ or money με κανένα τρόπο. there is no ~ lost between them δεν χωνεύονται. ~-affair s. ειδύλλιο n. ~-letter s. ερωτικό γράμμα. ~-lorn, ~-sick a. που μαραζώνει από έρωτα. ~-match s. γάμος από έρωτα. ~-story s. ρομάντζο n.

love|ly a. (beautiful) πολύ όμορφος, ωραιότατος. (delightful) υπέροχος, πολύ ωραίος, θαυμάσιος. ~liness s. ομορφιά f., ωραιότητα f.

lover s. εραστής m. (of things, animals) φίλος m. (of person) αγαπητικός m. ~s ερωτευμένοι m. pl. pair of ~s ζευγαράκι n. (paramour) εραστής m., ερωμένος m.

low v. (of cattle) μυκώμαι.

low a. (position, sound, price) χαμηλός, (birth, station) ταπεινός, (quality) ευτελής, (speed, number, visibility) μικρός, (comedy) χοντρός, (company) πρόστυχος. in ~ spirits κακόκεφος. run ~ εξαντλούμαι, lay ~ ρίχνω κάτω, lie ~ παραμένω κρυμμένος. (adv.) χαμηλά. ~-born a. ταπεινής καταγωγής. ~-brow s. με κοινά γούστα. ~-down s. (fam.) get the ~-down πληροφορούμαι. ~-fat a. ~-fat diet χαμηλά λιπαρά. ~-key(ed) a. συγκρατημένος. ~-land(s) s. κάμπος m.

lower a. κατώτερος, (sound, height) χαμη-

λότερος. the ~ floors τα κάτω πατώματα. (adv.) πιο κάτω, χαμηλότερα.

lower v. χαμηλώνω, (let down) κατεβάζω, (lessen, degrade) μειώνω. ~ing s. χαμήλωμα n., κατέβασμα n., μείωση f.

lowly a. ταπεινός.

loyal a. (of subjects, friends) πιστός, (of troops, political followers, etc.) νομιμόφρων. (devoted) αφοσιωμένος. ~ly adv. πιστά. ~ist s. κυβερνητικός a. ~ty s. πίστη f. νομιμοφροσύνη f.

lozenge s. παστίλια για το λαιμό. ~-shaped a. ρομβοειδής.

Ltd. ΕΠΕ, Α.Ε.

lubricant s. λιπαντικό n.

lubricat|e v. λιπαίνω. ~ion s. λίπανση f.

lubric|ious a. ασελγής. ~ity s. ασέλγεια f.

lucid a. διαυγής, σαφής, καθαρός. ~ interval φωτεινό διάλειμμα. ~ity s. διαύγεια f.

Lucifer s. Εωσφόρος m.

luck s. τύχη f. in ~ τυχερός. good ~ (wish) καλή επιτυχία, (omen of this) γούρι n. bad ~ ατυχία, κακοτυχία f., (omen of this) γουρσουζιά f. worse ~! δυστυχώς, good ~ in your new home! καλορίζικο το καινούργιο σπίτι. bad ~ to him κακό χρόνο να 'χει! ~less a. άτυχης.

luck|y a. τυχερός. (of person bringing ~) γουρλής, (of thing) γουρλήδικος. ~ily adv. ευτυχώς.

lucrative a. επικερδής.

lucre s. (pej.) κέρδος n., χρήμα n.

Lucullan a. λουκούλλειος.

ludicrous a. γελοίος.

lug v. (drag) τραβώ, (carry) κουβαλώ.

luggage s. αποσκευές f.pl. ~-van s. σκευοφόρος f.

lugubrious a. μελαγχολικός.

lukewarm a. χλιαρός.

lull s. there was a ~ in the conversation η κουβέντα έπεσε.

lull v.t. (calm) καθησυχάζω. (to sleep) νανουρίζω, (delude) βαυκαλίζω, (both) αποκοιμίζω. ~aby s. νανούρισμα n.

lumbago s. οσφυαλγία f., λουμπάγκο n.

lumber s. (wood) ξυλεία f. (useless objects) παλιοέπιπλα n.pl. (v.t.) φορτώνω. (v.i.) (move heavily) μετακινούμαι βαριά.

luminous a. φωτεινός.

lump s. (piece) κομμάτι n. (of earth, clay, etc.) βώλος, σβώλος m. (on head) καρούμπαλο n. (on body) εξόγκωμα n., βώλος m. (in throat) κόμπος m. ~ sum εφάπαξ (adv. as s.n.).

lump v.t. ~ together βάζω μαζί. (fam.) ~ it το καταπίνω. ~-y a. σβωλιασμένος. become ~y σβωλιάζω.

lunacy s. φρενοβλάβεια f. (also fig.) τρέλλα f.

lunar a. σεληνιακός.

lunatic s. φρενοβλαβής, παράφρων a. ~ asylum φρενοκομείο n. (a.) τρελλός.

lunch, ~eon s. μεσημεριανό (γεύμα) n. (at) ~-time το μεσημεράκι.

lung s. πνεύμων m.

lunge s. απότομο χτύπημα, (with sword) ξιφισμός m. (v.) ~ at χτυπώ απότομα, εφορμώ κατά (with gen.).

lurch v.i. τρικλίζω, (of ship) μποτζάρω. (s.) τρίκλισμα n. μπότζι n. leave (person) in the ~ αφήνω μπουκάλα.

lure s. δέλεαρ n. (v.) δελεάζω, (lead on) παρασύρω.

lurid a. (red) φλογερός, (glaring) χτυπητός, (hair-raising) ανατριχιαστικός.

lurk v. κρύβομαι.

luscious a. (juicy, lit. & fig.) ζουμερός, (music, art) αισθησιακός.

lush a. οργιώδης, πλούσιος.

lust s. σαρκική επιθυμία, (also fig.) πόθος m. (v.) ~ for ποθώ. ~ful a. λάγνος.

lustily adv. δυνατά.

lust|re s. (shine) γυαλάδα f., λάμψη f. (glory) αίγλη f., ~rous a. γυαλιστερός, (eyes) λαμπερός.

lusty a. (person) ρωμαλέος, (action) δυνατός.

lute s. λαούτο n.

luxe s. de ~ πολυτελείας.

luxuriant a. οργιώδης, πλούσιος, άφθονος. ~ly adv. άφθονα.

luxuriate v. ~ in απολαμβάνω.

luxurious a. πολυτελής. ~ly adv. πολυτελώς.

luxury s. (state, item) πολυτέλεια f. ~ goods είδη πολυτελείας. (pampered living) ευμάρεια f. in the lap of ~ στη χλιδή.

lye s. αλισίβα f.

lying s. (telling lies) το ψεύδεσθαι. (a.) a ~ scoundrel παλιάνθρωπος ψεύτης.

lying s. (reclining) ~ down ξάπλωμα n. (a.) ~ down πλαγιαστός, ξαπλωμένος. ~in s. τοκετός m. ~in-state s. λαϊκό προσκύνημα.

lymphatic a. λυμφατικός.

lynch v.t. λυντσάρω.

lynx s. λυγξ m.

lyre s. λύρα f.

lyric, ~al a. λυρικός. (s.) ~s (words of song) στίχοι m.pl.

M

MA μάστερς.

ma s. (fam.) μαμά f.

ma'am s. κυρία f.

macabre a. μακάβριος.

macadam s. μακαντάμ n. ~ized a. σκυροστρωμένος.

macaroni s. μακαρόνια n.pl.

macaroon s. αμυγδαλωτό n.

mace s. εμβληματική ράβδος.

machiavellian a. μακιαβελλικός.

machinations s. μηχανορραφίες f.pl.

machine s. μηχανή f. (of party, etc.) μηχανισμός m. (v.t.) (sew) ράβω στη μηχανή. ~gun s. πολυβόλο n. ~made a. μηχανοποίητος. ~tool s. μηχανικό εργαλείο.

machinery s. (machines) μηχανές f.pl. (works) μηχανισμός m.

machismo s. επίδειξη ανδρισμού.

mackerel s. σκουμπρί n., κολιός m.

mackintosh s. αδιάβροχο n.

mad a. τρελλός, drive ~ τρελλαίνω, go ~ τρελλαίνομαι, be ~ on τρελλαίνομαι για. like ~ σαν τρελός. ~ dog λυσσασμένο σκυλί. (angry) θυμωμένος. ~cap s. τρελλάρας m. ~den v. τρελλαίνω. it's ~dening είναι να τρελλαθείς. ~ly adv. μέχρι τρέλλας, σαν τρελλός. ~ness s. τρέλλα f.

madam s. κυρία f. (of bordello) τσατσά f.

made part. καμωμένος, φτιαγμένος, (manufactured) κατασκευασμένος. he is a ~ man η επιτυχία του είναι εξασφαλισμένη. he is ~ for the job είναι κομμένος και ραμμένος για τη δουλειά. ~-to-measure επί παραγγελία. see also make v.

Madonna s. Παναγία f.

madrigal s. μαδριγάλιο n.

Maecenas s. Μαικήνας m.

maelstrom s. δίνη f.

maenad s. μαινάς f.

maestro s. μαέστρος m.

mafia s. μαφία f.

magazine s. (journal) περιοδικό n. (powder) πυριτιδαποθήκη f. (of gun) γεμιστήρ m.

maggot s. σκουλήκι n. ~y a. σκουληκιασμένος.

magic s. μαγεία f. like ~ ως διά μαγείας. (a.) (also ~al) μαγικός. ~ian s. μάγος m. (entertainer) θαυματοποιός m.

magisterial a. αυστηρός.

magistrate s. ειρηνοδίκης m. examining ~ ανακριτής m.

magnanim|ous a. μεγαλόψυχος. ~ity s. μεγαλοψυχία f.

magnate s. μεγιστάν m.

magnesia s. μαγνησία f.

magnesium s. μαγνήσιον n.

magnet s. μαγνήτης m. ~ic a. μαγνητικός.

~ism s. μαγνητισμός m. **~ize** v. μαγνητίζω.

magneto s. μαγνιατό n.

magnific|ence s. μεγαλείο n. **~ent** a. μεγαλοπρεπής. (fam.) θαύμα n. **~ently** adv. μεγαλοπρεπώς. (fam.) θαυμάσια.

magni|fy v. μεγεθύνω, (exaggerate) μεγαλοποιώ, (extol) μεγαλύνω. **~fying** glass μεγεθυντικός φακός. **~fication** s. μεγέθυνση f.

magniloquent a. μεγαλόστομος.

magpie s. καρακάξα f., κίσσα f.

mahogany s. μαόνι n.

maid s. (girl) κόρη f. (servant) υπηρεσία f. old ~ γεροντοκόρη f.

maiden s. κόρη f., παρθένος f. (a.) παρθενικός, ~ name πατρικό όνομα. **~head** s. παρθενιά f.

mail s. (armour) πανοπλία f. (fig.) ~ed fist σιδηρά πυγμή.

mail s. (post) ταχυδρομείο n. (letters) αλληλογραφία f., γράμματα n.pl. (a.) ταχυδρομικός. ~ order ταχυδρομική παραγγελία. (v.) στέλνω ταχυδρομικώς.

maim v. ακρωτηριάζω.

main a. (chief) κύριος. the ~ chance προσωπικό συμφέρον. by ~ force με το ζόρι. (s.) in the ~ κατά το πλείστον. **~ly** adv. κυρίως.

main s. (pipe) κεντρικός αγωγός. ~s (system) δίκτυο n. ~ water νερό της βρύσης. (sea) πέλαγος n.

mainland s. στερεά f.

mainspring s. κύριο ελατήριο, (fig.) κίνητρο n.

mainstay s. (fig.) στυλοβάτης m.

mainstream s. (fig.) επικρατούσα τάση.

maintain v.t. (keep) διατηρώ, (support, keep in condition) συντηρώ. (v.i.) (assert) ισχυρίζομαι, λέω.

maintenance s. συντήρηση f.

maisonnette s. διαμέρισμα με δύο ορόφους.

maize s. καλαμπόκι n., αραβόσιτος m.

majestic a. μεγαλοπρεπής. **~ally** adv. μεγαλοπρεπώς.

majesty s. μεγαλείο n. His M~ η Αυτού Μεγαλειότης.

major a. (greater) μεγαλύτερος, (chief) κυριότερος, (higher) ανώτερος. (premise & mus.) μείζων. ~ operation (enterprise) σοβαρό εγχείρημα, (med.) σοβαρή εγχείρηση.

major s. (mil.) ταγματάρχης m. **~-general** υποστράτηγος m.

majority s. (of votes) πλειο(νο)ψηφία f. (of people, things) πλειονότητα f. (manhood) reach one's ~ ενηλικιώνομαι.

make s. (way sthg. is made) κατασκευή f. (brand, type) μάρκα f., τύπος m. be on the ~ είμαι τυχοθήρας.

make v.t. & i. (general) κάνω, (construct) κάνω, φτιάνω. (earn) κερδίζω. ~ the beds στρώνω τα κρεβάτια. ~ into (convert) μετατρέπω σε. have a suit made ράβω ένα κοστούμι. be made γίνομαι. what is it made of? από τι είναι (καμωμένο); wine is made from grapes το κρασί γίνεται από σταφύλια. what do you ~ of it? πώς σας φαίνεται; ~ it a rule to έχω ως κανόνα. ~ a success of πετυχαίνω σε. ~ a habit of συνηθίζω να. ~ light of δεν δίνω σημασία σε. ~ sure (or certain) of επιβεβαιώνω. ~ the best of it το παίρνω απόφαση. I can't ~ sense of it δεν μπορώ να βγάλω νόημα. ~ it known that καθιστώ γνωστό ότι. I made him angry τον θύμωσα, τον έκανα να θυμώσει. what time do you ~ it? τι ώρα λέει το ρολόι σου; ~ public δημοσιεύω. ~ good (v.i., succeed) πετυχαίνω, (v.t., rectify) επανορθώνω, (compensate for) αποζημιώνω για. ~ do τα φέρνω βόλτα. ~ do with αρχκίομαι εις. ~ it (succeed) τα καταφέρνω, (get there) φθάνω εγκαίρως. ~ oneself ridiculous γίνομαι γελοίος. ~ oneself useful γίνομαι χρήσιμος. ~ oneself scarce εξαφανίζομαι. ~ believe that... κάνω πως, (imagine) φαντάζομαι ότι. see also made.

make away with v.t. βγάζω από τη μέση, (kill) καθαρίζω.

make for v.t. (proceed to) κατευθύνομαι προς, τραβώ για. (contribute to) συμβάλλω εις, βοηθώ (σε).

make off v.i. στρίβω, το σκάω ~ with κλέβω.

make out v.t. (a bill, etc.) κάνω. (discern) ~ διακρίνω (understand) καταλαβαίνω, βγάζω. ~ a case for αποδεικνύω ότι. (v.i.) (assert) ισχυρίζομαι. (get on) how are you making out? πώς τα πας;

make over v.t. (convert) μετατρέπω. (property) he made over the house to his son έγραψε το σπίτι στο γιο του.

make up v.t. (complete) συμπληρώνω. (invent) κατασκευάζω, πλάθω. (constitute) αποτελώ, απαρτίζω. (prepare) ετοιμάζω, (a prescription) εκτελώ. (cloth for garment) ράβω. (one's face) βάφω. make it up συμφιλιώνομαι. (v.i.) (use cosmetics) βάφομαι, φτιάνομαι. ~ to κολακεύω, καλοπιάνω. ~ for αποζημιώνω για. ~ for lost time ανακτώ το χαμένο χρόνο.

make-up s. (cosmetic) μακιγιάζ n. (composition) σύσταση f. (temperament) ψυχοσύνθεση f.

make-believe s. φαντασία f. land of ~ χώρα ονείρου. (pej.) προσποίηση f.

maker s. κατασκευαστής m. Μ~ Πλάστης m.

makeshift a. πρόχειρος.

makeweight s. συμπλήρωμα βάρους, (of meat) κατμάς m.

making s. history in the ~ ιστορία εν τω γίγνεσθαι. his troubles are none of my ~ εγώ δεν φταίω για τα βάσανά του. it will be the ~ of him θα του βάλει μυαλό, θα τον δημιουργήσει. it has the ~s of... έχει ό,τι χρειάζεται για να γίνει. they do a lot of music-~ παίζουν πολλή μουσική.

maladjusted a. απροσάρμοστος.

maladroit a. ανεπιτήδειος.

maladministration s. κακή διαχείριση.

malady s. νόσος f., πάθηση f.

malaise s. δυσφορία f.

malaria s. ελονοσία f.

malcontent s. δυσαρεστημένος a.

male a. & s. αρσενικός, άρρην. (man's, of men) ανδρικός. ~ nurse νοσοκόμος m. ~ sex ανδρικό φύλο.

malediction s. κατάρα f.

malefactor s. κακοποιός m.

malevol|ent a. κακεντρεχής. **~ence** s. κακεντρέχεια f.

malform|ed a. δύσμορφος. **~ation** s. δυσμορφία f.

malic|e s. κακία f., κακοβουλία f. **~ious** a. κακός, κακόβουλος. **~iously** adv. με κακία.

malign v. κακολογώ. (a.) βλαβερός.

malignant a. κακεντρεχής. (med.) κακοήθης.

malinger v. προσποιούμαι (or κάνω) τον άρρωστο.

mallard s. αγριόπαπια f.

malleable a. μαλακός, εύπλαστος.

mallet s. ξύλινη σφύρα.

mallow s. μολόχα f.

malmsey s. δυνατό γλυκό κρασί.

malnutrition s. υποσιτισμός m.

malodorous a. δύσοσμος.

malpractice s. κατάχρηση εξουσίας.

malt s. βύνη f.

maltreat v. κακομεταχειρίζομαι. **~ment** s. κακομεταχείριση f.

malversation s. κατάχρηση χρημάτων.

mamma s. μαμά f.

mammal s. θηλαστικό n.

mammoth s. μαμούθ n.

man s. (male) άνδρας, άντρας m. (human) άνθρωπος m. ~ in the street ο μέσος άνθρωπος. best ~ κουμπάρος m. no ~'s land ουδετέρη ζώνη. ~ of the world άνθρωπος του κόσμου. ~ about town κοσμικός τύπος. inner ~ (belly) κοιλιά f. to

a ~ (all) μέχρις ενός. see also man's. **~eater** s. ανθρωποφάγος a. **~handle** v. (move) μεταφέρω με τα χέρια, (ill-treat) κακομεταχειρίζομαι. **~hole** s. ανθρωποθυρίδα f. **~hunt** s. ανθρωποκυνηγητό n. **~made** a. τεχνητός. **~of-war** s. (naut.) πολεμικό n. **~power** s. (men) άνδρες m. shortage of ~power έλλειψη εργατικών χεριών. **~slaughter** s. ανθρωποκτονία f. see also man's.

man v.t. (provide men for) επανδρώνω, (stand to) παίρνω θέση σε.

manacles s. χειροπέδες f.pl.

manage v.t. (direct) διαχειρίζομαι, διευθύνω, (handle) χειρίζομαι. (v.i.) (succeed) καταφέρνω, κατορθώνω. how did you ~? πώς τα κατάφερες; I can't ~ without your help δεν μπορώ να τα βγάλω πέρα χωρίς τη βοήθειά σου.

manageable a. (thing) ευκολομεταχείριστος, (person) εύκολος.

management s. (direction, directors) διαχείριση f., διεύθυνση f. (handling) χειρισμός m.

manager s. διευθυντής m. **~ess** s. διευθύντρια f. **~ial** a. των μάνατζερ.

mandarin s. μανδαρίνος m.

mandat|e s. εντολή f. **~ory** a. υποχρεωτικός.

mandible s. σιαγών f.

mandoline s. μαντολίνο n.

mane s. χαίτη f.

manful a. αντρίκιος. **~ly** adv. αντρίκια.

manganese s. μαγγάνιο n.

mangle s. ψώρα f. **~y** a. ψωριάρης.

manger s. φάτνη f.

mangle s. μάγγανο n.

mangle v. κατακρεουργώ. (fig., spoil) σκοτώνω.

manhood s. ανδρική ηλικία. (men) άνδρες m.pl. (quality) ανδρισμός m.

manila s. μανία f. **~ac, ~acal** a. & s. μανιακός.

manicure s. μανικιούρ n. (v.) ~ one's nails φτιάνω (or κάνω) τα νύχια μου.

manifest a. έκδηλος. (v.) εκδηλώνω. (s.) (naut.) δηλωτικό n. **~ation** s. εκδήλωση f. **~ly** adv. καταφανώς.

manifesto s. διακήρυξη f., μανιφέστο n.

manifold a. πολλοί και διάφοροι.

manikin s. ανθρωπάκι n.

manipulat|e χειρίζομαι, (influence) επηρεάζω. **~ion** s. χειρισμός m.

mankind s. ανθρωπότητα f.

man|ly a. αντρίκιος, λεβέντικος. **~liness** s. ανδρισμός m., λεβεντιά f.

manna s. μάννα n.

mannequin s. μαννεκέν n. (dummy) κούκλα f.

manner s. (*way*) τρόπος *m.* (*sort*) είδος *n.* (*artistic style*) τεχνοτροπία *f.* ~s (*behaviour*) τρόποι *m.pl.*, συμπεριφορά *f.*, (*customs*) συνήθειες *f.pl.* all ~ of κάθε λογής (*or* είδους). in a ~ of speaking σα να πούμε.

mannered *a.* επιτηδευμένος. ill-~ κακότροπος, well-~ καλότροπος.

mannerism s. ιδιομορφία *f.* (*art*) μανιερισμός *m.*

mannish *a.* ανδροπρεπής.

manoeuvre s. μανούβρα *f.*, ελιγμός *m.* ~s (*mil.*, *naut.*) ασκήσεις *f.pl.* (*v.*) μανουβράρω.

manor s. φέουδο *n.*

man's *a.* (*consider sthg.*) from a ~ point of view από ανδρική πλευρά. men's clothes ανδρικά ρούχα. men's changing-room αποδυτήριον ανδρών.

mansion s. μέγαρο *n.*, μεγαλοπρεπές οίκημα.

mantelpiece s. πλαίσιο του τζακιού.

mantle s. μανδύας *m.* (*gas*) αμίαντο *n.* (*v.*) σκεπάζω.

manual s. εγχειρίδιο *n.*, οδηγός *m.*

manual *a.* (*work*) χειρωνακτικός. (*operated by hand*) χειροκίνητος. ~ly *adv.* με το χέρι.

manufacture *v.* κατασκευάζω. (*fam.*, *invent*) τεχνάζομαι. (*s.*) κατασκευή *f.* (*industry*) βιομηχανία *f.* ~s (*goods*) βιομηχανικά είδη. ~r *s.* βιομήχανος *m.*, κατασκευαστής *m.*

manure s. κοπριά *f.* (*v.t.*) λιπαίνω.

manuscript s. χειρόγραφο *n.* (*a.*) (in) ~ χειρόγραφος.

many *a.* πολλοί. how ~ πόσοι. so ~ τόσοι. a great ~ πάρα πολλοί. a good ~ αρκετοί. as ~ as you like όσα θέλεις.

many-coloured *a.* πολύχρωμος.

many-sided *a.* (*question*) πολυμερής, πολύπλευρος. (*person*) με πολλές ικανότητες.

map s. χάρτης *m.* off the ~ απόμερος. (*v.t.*) χαρτογραφώ. ~ out (*fig.*) κανονίζω, σχεδιάζω.

maple s. σφένδαμνος *f.*

maquis s. (*wartime Resistance*) αντίσταση *f.*

mar *v.* χαλώ.

marathon s. μαραθώνιος *m.*

marauder s. ληστής *m.*

marble s. μάρμαρο *n.* ~s (*game*) βόλοι *m.pl.* (*a.*) μαρμάρινος, μαρμαρένιος.

March s. Μάρτιος, Μάρτης *m.*

march *v.i.* βαδίζω, κάνω πορεία. ~ past παρελαύνω. (*v.t.*) they ~ed him off to the police-station τον πήγαν στο τμήμα.

march s. πορεία *f.* ~ past παρέλαση *f.* steal a ~ on προσλαμβάνω. (*mus.*) εμβατήριο *n.*

marches s. σύνορα *n.pl.*

mare s. φοράδα *f.* (*fig.*) ~'s nest τζίφος *m.*

margarine s. μαργαρίνη *f.*

margin s. περιθώριο *n.* (*of lake*) όχθη *f.* (*of road*) άκρη *f.* ~al *a.* (*small*) μικρός, (*in the* ~) στο περιθώριο. ~ally *adv.* κάπως.

marine *a.* (*that grows in sea*) ενάλιος. (*of shipping*) ναυτικός. ~ artist θαλασσογράφος *m.*

marine s. (*soldier*) πεζοναύτης *m.* merchant *or* mercantile ~ ναυτιλία *f.*, εμπορικό ναυτικό.

mariner s. ναύτης *m.* master ~ πλοίαρχος του εμπορικού ναυτικού.

marionette s. μαριονέτα *f.*

marital *a.* συζυγικός.

maritime *a.* (*power*) θαλάσσιος. (*law*) ναυτικός. (*province*) παραθαλάσσιος.

marjoram s. μαντζουράνα *f.*

mark s. σημάδι *n.* trade ~ σήμα κατατεθέν. (*indication*) ένδειξη *f.*, σημάδι *n.* (*trace*) ίχνος *n.* (*target*) στόχος *m.*, σημάδι *n.* hit the ~ βρίσκω το στόχο, miss the ~ αστοχώ. beside (*or* wide of) the ~ άστοχος. (*at school*) βαθμός *m.* post-~ σφραγίδα. *f.* punctuation ~s σημεία της στίξεως. make one's ~ διακρίνομαι. not feel up to the ~ δεν αισθάνομαι καλά. it doesn't come up to the ~ δεν είναι ό,τι πρέπει.

mark *v.* (*make* ~s *on*) σημαδεύω, κάνω σημάδια σε. (*indicate*) σημειώνω. (*observe*) προσέχω. (*give* ~s *to*) βαθμολογώ. ~ off ξεχωρίζω. ~ out (*trace*) χαράσσω, (*destine*) προορίζω. ~ time κάνω βήμα σημειωτόν, (*fig.*) δεν προχωρώ.

marked *a.* (*clear*) καταφανής, φανερός. ~ man (*suspect*) σταματαρισμένος, (*promising*) με καλές προοπτικές για το μέλλον. ~ly *adv.* καταφανώς.

marker s. (*indicator*) σημάδι *n.*

market s. αγορά *f.* put (*or* come) on the ~ βγάζω (*or* βγαίνω) στην αγορά. ~ forces / research τάσεις / έρευνα της αγοράς. play the ~ παίζω στο χρηματιστήριο. ~ value εμπορική αξία. street ~ λαϊκή αγορά. open ~ ελεύθερη αγορά. (*v.t.*) (*sell*) πουλώ. ~able *a.* εμπορεύσιμος.

marking s. (*at school*) βαθμολογία *f.* (*animals', etc.*) διακριτικός χρωματισμός.

marksman s. σκοπευτής *m.*

marmalade s. μαρμελάδα νεράντζι.

marmoreal *a.* μαρμαρένιος.

maroon *v.* εγκαταλείπω. (*s.*) (*signal*) συναγερμός *m.* (*a.*) (*colour*) μαρόν.

marquee s. μεγάλη σκηνή.

marquetry s. μαρκετερί *n.*

marquis s. μαρκήσιος *m.*

marriage s. γάμος *m.* give in ~ παντρεύω. ~ of convenience γάμος από συμφέρον, arranged ~ γάμος από συνοικέσιο. (relation) by ~ εξ αγχιστείας. ~able a. της παντρειάς.

married a. έγγαμος, παντρεμένος.

marrow s. (of bone) μυελός m., μεδούλι n. (vegetable) κολοκύθι n. (fig.) to the ~ μέχρι το κόκκαλο. (essence) ουσία f.

marry v.t. & i. (also get married) παντρεύομαι. (v.t.) (of priest, sponsor) παντρεύω. (fig.) (match) ταιριάζω, (splice) ματίζω.

Marseillaise s. Μασσαλιώτις f.

marsh s. βάλτος m., έλος n. ~y a. βαλτώδης.

marshal s. (of ceremonies) τελετάρχης m. (mil.) στρατάρχης m.

marshal v. τακτοποιώ, (mil.) παρατάσσω. (conduct) οδηγώ. ~ling yard σταθμός συνθέσεως συρμών.

mart s. αγορά f.

martial a. πολεμικός. ~ law στρατιωτικός νόμος.

martinet s. πολύ αυστηρός.

martyr s. μάρτυρας m. ~'s μαρτυρικός. be a ~ to headaches βασανίζομαι από πονοκεφάλους. (v.) be ~ed μαρτυρώ. ~dom s. μαρτύριο n. those who suffered ~dom οι μαρτυρήσαντες.

marvel v. (also ~ at) θαυμάζω, μου προκαλεί έκπληξη. (s.) θαύμα n. ~lous a. θαυμάσιος. ~lously adv. θαυμάσια.

marxist a. (ideas) μαρξιστικός. (s.) μαρξιστής m.

mascot s. μασκότ f.

masculin|e a. αρρενωπός. (gram.) αρσενικός. ~ity s. αρρενοπρέπεια f.

mash v. πολτοποιώ. ~ed potato πατάτες πουρέ.

mask s. προσωπίδα f., μάσκα f. (fig.) προσωπείο n. (v.t.) καλύπτω. ~ed ball μπαλ μασκέ.

masoch|ism s. μαζοχισμός m. ~ist s. μαζοχιστής m.

mason s. χτίστης m. (also freemason) τέκτονας m. ~ic a. τεκτονικός, μασονικός. ~ry s. (of building) λιθοδομή f.

masquerade v.i. ~ as εμφανίζομαι ως. (s.) (fig.) υπόκριση f.

mass s. (eccl.) θεία λειτουργία.

mass s. (bulk) όγκος m., μάζα f. (heap) σωρός m. (crowd) πλήθος n. (abundance) αφθονία f. ~es of πάρα πολλοί. a ~ of (full of) γεμάτος από. the ~es οι μάζες. (v.t.) συγκεντρώνω. (v.i.) συγκεντρώνομαι, μαζεύομαι.

mass a. μαζικός. ~ production μαζική παραγωγή. ~ produced της σειράς. ~ media μέσα μαζικής ενημερώσεως.

massacre v. κατασφάζω. (s.) σφαγή f.

massage v.t. κάνω μασάζ σε.

massive a. ογκώδης, (features) βαρύς. (substantial) δυνατός.

mast s. ιστός m., κατάρτι n. at half ~ μεσίστιος.

master m. κύριος m. (ship's) καπετάνιος m. (boss) αφέντης m., αφεντικό n. (school) καθηγητής m., (δι)δάσκαλος m. (expert) άσσος m. ~ of house οικοδεσπότης m. old ~ μεγάλος ζωγράφος της Αναγεννήσεως. be one's own ~ είμαι ανεξάρτητος. be ~ of one's trade είμαι κάτοχος της τέχνης μου.

master v. (difficulties) υπερνικώ. (a subject, trade) γίνομαι κάτοχος (with gen.) (hold in check) κυριαρχώ (with gen.), συγκρατώ. (get possession of) κυριεύω.

master a. ~ hand αριστοτέχνης m. ~ key αντικλείδι n. ~ mind εγκέφαλος m. ~ stroke εμπνευσμένο στρατήγημα. ~ switch γενικός διακόπτης. ~ craftsman (artist) αριστοτέχνης m.

masterful a. αυταρχικός.

masterly a. (work) αριστοτεχνικός.

masterpiece s. αριστούργημα n.

mastery s. κυριαρχία f. (skill) δεξιοτεχνία f. (of subject) βαθιά γνώση.

mastic s. μαστίχα f.

masticat|e v. μασώ. ~ion s. μάσημα n.

masturbation s. αυνανισμός m.

mat a. ματ.

mat s. (door) ψάθα (or τσουλάκι) των ποδιών. (table) σου-πλα n. (fam.) on the ~ υπό κατηγορίαν.

mat v. see matted.

match s. (for lighting) σπίρτο n. ~box s. σπιρτόκουτο n. ~wood s. it was broken to ~wood έγινε κομματάκια.

match s. (game) ματς n.

match s. (equal) ταίρι n. meet one's ~ βρίσκω το ταίρι μου. he has not his ~ είναι εκτός συναγωνισμού. he is no ~ for them δεν τα βγάζει πέρα μαζί τους. they are a perfect ~ ταιριάζουν περίφημα.

match s. (marriage) γάμος m. they made a ~ of it παντρεύτηκαν. he is a good ~ (eligible) είναι καλός γαμπρός. ~maker s. προξενήτρα f.

match v.t. & i. ταιριάζω. none can ~ him δεν έχει το ταίρι του. handbag and gloves to ~ τσάντα και γάντια ασορτί. ~less a. απαράμιλλος, που δεν έχει το ταίρι του.

mat|e s. σύντροφος m.f. (of animal, etc.) ταίρι n. (naut.) first ~ ε υποπλοίαρχος m. (v.t.) ζευγαρώνω. (v.i.) ζευγαρώνομαι. ~ing s. ζευγάρωμα n.

material s. υλικό n., ύλη f. raw ~s πρώτες ύλες. writing ~s γραφική ύλη. building ~s υλικά οικοδομών. (cloth) ύφασμα n.

material *a.* υλικός, (*essential*) ουσιώδης, (*relevant*) σχέσιν έχων. **~ly** *adv.* ουσιωδώς.

material|ism *s.* υλισμός *m.* **~ist** *s.* υλιστής *m.* **~istic** *a.* υλιστικός.

materialize *v.* πραγματοποιούμαι.

matern|al *a.* μητρικός. **~ ity** *s.* μητρότητα *f.* **~ity** hospital μαιευτήριο *n.*

matey *a.* φιλικός.

mathemat|ical *a.* μαθηματικός. **~ician** *s.* μαθηματικός. **~ics** *s.* μαθηματικά *n.pl.*

matinée *s.* απογευματινή *f.*

matins *s.* (*eccl.*) όρθρος *m.*

matriarch *s.* γυναίκα αρχηγός οικογένειας. **~y** *s.* μητριαρχία *f.*

matriculat|e *v.i.* εγγράφομαι σε πανεπιστήμιο *n.* **~ion** *s.* εγγραφή σε πανεπιστήμιο.

matrimon|y *s.* κοινωνία γάμου. **~ial** *a.* συζυγικός.

matrix *s.* μήτρα *f.*, καλούπι *n.*

matron *s.* κυρία *f.* (*of hospital*) προϊσταμένη *f.* (*of school*) οικονόμος *f.* **~ly** *a.* (*figure*) ευτραφής.

matted *a.* πυκνός και μπερδεμένος.

matter *s.* (*affair*) υπόθεση *f.*, δουλειά *f.*, θέμα *n.*, πράγμα *n.* (*question*) ζήτημα *n.* (*substance*) ύλη *f.*, ουσία *f.* (*pus*) πύον *n.* printed ~ έντυπα *n.pl.* how do ~s stand? πώς είναι τα πράγματα; as a ~ of course φυσικά. as a ~ of fact το γεγονός είναι, εδώ που τα λέμε. for that ~ και μάλιστα. (*importance*) no ~ δεν πειράζει. no ~ how difficult it is όσο δύσκολο κι αν είναι. no ~ what you say ό,τι και να πεις. (*wrong*) what's the ~? τι συμβαίνει; what's the ~ with you? τι σου συμβαίνει; τι έχεις; there is something the ~ with my watch κάτι έχει (*or* κάτι δεν πάει καλά με) το ρολόι μου.

matter *v.* it ~s πειράζει, έχει σημασία.

matter-of-fact *a.* θετικός, προσγειωμένος.

matting *s.* ψαθί *n.*

mattock *s.* τσάπα *f.*

mattress *s.* στρώμα *n.*

matur|e *v.i.* ωριμάζω, (*of bill*) λήγω. (*a.*) ώριμος. **~ity** *s.* ωριμότητα *f.*

maudlin *a.* κλαψιάρης.

maul *v.* τραυματίζω. (*fig., of reviewer*) ξετινάζω.

mausoleum *s.* μαυσωλείον *n.*

mauve *a.* μωβ.

maverick *s.* (*fig.*) ανυπότακτος *a.*

maw *s.* στομάχι *n.* (*jaws*) σιαγόνες *f.pl.*

mawkish *a.* ανούσιος.

maxim *s.* γνωμικό *n.*

maximize *v.* αυξάνω στον ανώτατο βαθμό.

maximum *a.* ανώτατος. (*s.*) μάξιμουμ *n.*, ανώτατο όριο.

May *s.* Μάιος, Μάης *m.* **~ day** πρωτομαγιά *f.*

may *v.* you ~ be right μπορεί να έχετε δίκιο. they ~ not have come μπορεί (*or* ενδέχεται) να μην έχουν έρθει. we ~ as well begin μπορούμε να αρχίσουμε, δεν αρχίζουμε; ~ I sit down? μπορώ (*or* επιτρέπεται) να καθήσω; ~ all go well with you εύχομαι να σας πάνε όλα καλά. *see also* might *v.*

maybe *adv.* ίσως.

mayor *s.* δήμαρχος *m.*

maypole *s.* ~ dance γαϊτανάκι *n.*

maze *s.* λαβύρινθος *m.*

me *pron.* (ε)μένα, με. for ~ για μένα, he saw ~ με είδε.

meadow *s.* λειβάδι *n.*

meagre *a.* ισχνός, πενιχρός. **~ly** *adv.* φτωχικά. **~ness** *s.* πενιχρότητα *f.*

meal *s.* φαγητό *n.*, γεύμα *n.* (*flour*) αλεύρι *n.*

mealy-mouthed *a.* be ~ αποφεύγω να μιλώ ντόμπρα.

mean *a.* (*poor*) φτωχικός, ταπεινός, χαμηλός. (*unworthy*) ανάξιος, (*stingy*) τσιγγούνης. **~ly** *adv.* πενιχρός, άσχημα, κακώς. **~ness** *s.* τσιγγουνιά *f.*

mean *a.* (*average*) μέσος. (*s.*) μέσος όρος.

mean *v.* (*signify*) σημαίνω, (*of persons only*) εννοώ. what does this ~? τι σημαίνει (*or* τι θα πει) αυτό; (*intend*) σκοπεύω, θέλω. ~ well έχω καλές προθέσεις. I didn't ~ to offend him δεν ήθελα να τον προσβάλω. ~ business δεν αστειεύομαι. I ~t it for a joke τα είπα γι' αστείο. (*destine*) he wasn't ~t for politics δεν ήταν κομμένος για πολιτικός.

meander *v.* περιπλανώμαι, (*of river*) ελίσσομαι.

meaning *s.* σημασία, έννοια *f.*

means *s.* μέσον *n.*, μέσα *n.pl.* (*resources*) πόροι *m.pl.* he has ~ (*money*) έχει τον τρόπο του. beyond my ~ εκτός των δυνάμεών μου. he is by no ~ a fool δεν είναι καθόλου κουτός. it can't be done by any ~ δεν γίνεται με κανένα τρόπο. by all ~ (*willingly*) ευχαρίστως. (*he got in*) by ~ of a ladder με τη βοήθεια μιας σκάλας. this event was the ~ of our reconciliation αυτό έγινε η αιτία να συμφιλιωθούμε.

mean|time, ~while *adv.* στο (*or* εν τω) μεταξύ.

measles *s.* ιλαρά *f.*

measly *a.* (*fam.*) μίζερος.

measurab|le *a.* within ~le distance κοντά. **~ly** *adv.* αισθητά.

measure *s.* μέτρον *n.* tape ~ μεζούρα *f.* get the ~ of (*fig.*) αναμετρώ. beyond ~ αφάνταστα. in large ~ κατά μέγα μέρος. for good ~ επιπλέον, από πάνω. made to ~

επί μέτρω.
measure v. μετρώ, (weigh up) αναμετρώ. ~ one's strength αναμετρούμαι. ~ up to (correspond with) ανταποκρίνομαι σε, (qualify for) έχω τα απαιτούμενα προσόντα για. ~d a. μετρημένος, μελετημένος.
measureless a. (number) άμετρος, (space) απέραντος.
measurement s. μέτρηση f. ~s μέτρα n. pl.
meat s. κρέας n. ~ ball κεφτές m. ~less a. ~less day ημέρα ακρεοφαγίας. ~-safe s. (hanging) φανάρι n. ~y a. (fig.) γεμάτος ουσία.
mechanic s. μηχανικός m., τεχνίτης m.
mechanical a. μηχανικός. ~ly adv. μηχανικώς.
mechanics s. (science) μηχανική f. (how sthg. works) μηχανισμός m.
mechan|ism s. μηχανισμός m. ~ize v. μηχανοποιώ. ~ization s. μηχανοποίηση f.
medal s. μετάλλιο n.
medallion s. μενταγιόν n.
meddle v.i. ανακατεύομαι. ~r s., ~some a. πολυπράγμων.
media s. (means) μέσα n.pl. see mass a.
medial a. μεσαίος.
median a. μέσος, (εν)διάμεσος.
mediat|e v. μεσολαβώ. ~ion s. μεσολάβηση f. ~or s. μεσολαβητής m., ο μεσάζων.
medical a. ιατρικός. ~ student φοιτητής της ιατρικής. ~ly adv. ιατρικώς, από ιατρικής απόψεως.
medicament s. φάρμακο n.
medicat|ed a. φαρμακευτικός. ~ion s. it needs ~ion θέλει φάρμακα.
medicinal a. ιαματικός.
medicine s. (physic) γιατρικό n., φάρμακο n. (science) ιατρική f. (fam.) take one's ~ like a man δέχομαι την τιμωρία μου γενναία. get a dose of one's own ~ πληρώνομαι με το ίδιο νόμισμα. ~-man s. μάγος-ιατρός m.
medieval a. μεσαιωνικός.
mediocr|e a. μέτριος. ~ity s. μετριότητα f.
meditat|e v.i. (think) συλλογίζομαι. (v.t.) (plan) μελετώ, σχεδιάζω. ~ion s. στοχασμός m. ~ive a. συλλογισμένος.
medium s. (means) μέσο n. happy ~ ορθόν μέτρον. (spiritualist) μέντιουμ n. (a.) μέσος. of ~ height μετρίου αναστήματος. ~ wave μεσαίο κύμα.
medlar s. μούσμουλο n.
medley s. σύμφυρμα n. (colours) ανακάτωμα n. (writing, music) συνονθύλευμα n.
meek a. (submissive) υποτακτικός, (gentle) πράος.
meet a. (right) αρμόζων.
meet v.t. συναντώ, ανταμώνω. (make

acquaintance of) γνωρίζω. have you met my sister? έχετε γνωριστεί με την αδελφή μου; (satisfy) ικανοποιώ, συμμορφώνομαι προς. it ~s the case είναι κατάλληλο. (of bargaining) I met him half-way μοιραστήκαμε τη διαφορά. (v.i.) συναντιέμαι, (in session) συνεδριάζω. (become joined) ενώνομαι. make both ends ~ τα φέρνω βόλτα. ~ with (find) βρίσκω, συναντώ, (suffer) υφίσταμαι, παθαίνω. he met with an accident του έτυχε (or του συνέβη) ένα δυστύχημα. ~ with losses έχω ζημιές.
meeting s. (individuals) συνάντηση f. (group) συνέλευση f. (political) συγκέντρωση f.
megalomania s. μεγαλομανία f.
megaphone s. τηλεβόας m., χωνί n.
melancholy s. μελαγχολία f. (a.) μελαγχολικός.
mêlée s. συμπλοκή f.
mellifluous a. μελίρρυτος.
mellow a. (ripe) ώριμος, (soft) απαλός. (wine) παλαιός. (v.) ωριμάζω.
melodious a. γλυκός.
melodrama s. μελόδραμα n. ~tic a. μελοδραματικός.
melod|y s. μελωδία f. ~ic a. μελωδικός.
melon s. πεπόνι n. water ~ καρπούζι n.
melt v.t. & i. λειώνω, (fig.) μαλακώνω. (v.i., of metal) τήκομαι. ~ down (v.t.) λειώνω, τήκω. ~ away (v.i.) διαλύομαι, χάνομαι.
melting a. (look) τρυφερός, (tune) γλυκός. ~-point s. σημείο τήξεως. ~-pot s. χωνευτήριον n.
member s. μέλος n. ~ship s. (members) μέλη n.pl. total ~ship ολομέλεια f.
membrane s. μεμβράνη f.
memento s. ενθύμιο n.
memoir s. μελέτη f. ~s απομνημονεύματα n.pl.
memorable a. αλησμόνητος. (of historic occasion) μνημειώδης.
memorandum s. υπόμνημα n.
memorial a. αναμνηστικός. (s.) (monument) μνημείο n.
memorize v. αποστηθίζω.
memory s. μνήμη f. (recollection) ανάμνηση f. from ~ από μνήμης. in ~ of εις μνήμην (or εις ανάμνησιν) (with gen.). within living ~ στην εποχή μας. it escaped my ~ μου διέφυγε.
menac|e v. απειλώ. (s.) απειλή f. ~ing a. απειλητικός.
menagerie s. θηριοτροφείο n.
mend v.t. επισκευάζω, φτιάνω, επιδιορθώνω. (patch) μπαλώνω. ~ one's ways διορθώνομαι. (v.i.) βελτιώνομαι,

φτιάνω.

mendac|ious *a.* ψευδής, αναληθής. (*person*) ψευδολόγος. **~ity** *s.* ψευδολογίες *f.pl.*

mendicant *s.* ζητιάνος *m.*

menial *a.* ~ work χονδρές δουλειές. (*s.*) υπηρέτης *m.*

meningitis *s.* μηνιγγίτιδα *f.*

menopause *s.* εμμηνόπαυση *f.*

men|ses *s.* έμμηνα *n.pl.* **~struation** *s.* εμμηνόρροια *f.*

mensuration *s.* μέτρηση *f.*

mental *a.* διανοητικός. (*fam., mad*) τρελλός. ~ hospital ψυχιατρείο *n.* ~ arithmetic υπολογισμοί με το μυαλό. **~ly** *adv.* διανοητικώς.

mentality *s.* νοοτροπία *f.*

mention *v.* αναφέρω. don't ~ it! παρακαλώ. (*s.*) make no ~ of δεν αναφέρω. honourable ~ εύφημος μνεία.

mentor *s.* μέντωρ *m.*

menu *s.* κατάλογος φαγητών.

mercantile *a.* εμπορικός. *see* marine.

mercenary *a.* συμφεροντολόγος. (*s.*) μισθοφόρος *m.*

merchandise *s.* εμπόρευμα *n.*, εμπορεύματα *pl.*

merchant *s.* έμπορος *m.* (*a.*) εμπορικός. **~man** *s.* εμπορικό πλοίο.

merciful *a.* επιεικής, (*eccl.*) ελεήμων. **~ly** *adv.* επιεικώς. (*fam., luckily*) ευτυχώς.

merciless *a.* ανηλεής. **~ly** *adv.* ανηλεώς.

mercurial *a.* (*fig.*) ζωηρός, σπιρτόζος.

mercury *s.* υδράργυρος *m.*

mercy *s.* επιείκεια *f.* (*eccl.*) έλεος *n.* at the ~ of the waves εις το έλεος των κυμάτων. Lord have ~ Κύριε ελέησον.

mere *s.* λίμνη *f.*

mere *a.* απλός. a ~ trifle ένα τίποτα. the ~ thought of it makes me laugh και μόνο που το σκέπτομαι γελάω. it was the ~st accident ήταν τελείως τυχαίο. a ~ glance will tell you και μία απλή ματιά θα σας δείξει. **~ly** *adv.* μόνο, απλώς.

meretricious *a.* επιδεικτικός.

merge *v.t.* συγχωνεύω. (*v.i.*) συγχωνεύομαι. **~r** *s.* συγχώνευσις *f.*

meridian *s.* μεσημβρινός *m.*

meringue *s.* μαρέγκα *f.*

merit *s.* αξία *f.* (*v.t.*) αξίζω. his appointment was well ~ed πήρε τη θέση επαξίως (*or* με την αξία του).

meritocracy *s.* αξιοκρατία *f.*

meritorious *a.* αξιέπαινος.

mermaid *s.* γοργόνα *f.*

merr|y *a.* εύθυμος, κεφάτος. make ~y γλεντώ. **~ily** *adv.* χαρούμενα, εύθυμα. **~iment** *s.* ευθυμία *f.*, κέφι *n.* **~ymaking** *s.* χαρές και πανηγύρια.

merry-go-round *s.* αλογάκια *n.pl.*

mesh *v.t.* (*catch*) πιάνω. (*v.i.*) (*mech.*) εμπλέκομαι. (*s.*) πλέγμα *n.* (*single space*) θηλιά *f.* (*mech.*) εμπλοκή *f.* **~es** δίκτυα *n.pl.*

mesmer|ism *s.* μεσμερισμός *m.* **~ize** *v.* υπνωτίζω.

mess *s.* (*common table*) συσσίτιο *n.* (*v.i.*) συσσιτώ.

mess *s.* (*disorder*) ακαταστασία *f.* (*dirt*) ακαθαρσία *f.* (*trouble*) be in a ~ έχω τράβαλα. make a ~ of it τα κάνω θάλασσα (*or* σαλάτα).

mess *v.i.* ~ about (*idle*) χάνω την ώρα μου, χασομερώ. ~ about with (*tinker*) πασπατεύω, πειράζω. (*v.t.*). ~ up (*muddle*) κάνω άνω κάτω, (*dirty*) λερώνω, (*spoil*) χαλ(ν)ώ, (*an object*) βγάζω τα μάτια (*with gen.*).

message *s.* μήνυμα *n.*, παραγγελία *f.* (*written*) σημείωμα *n.* send a ~ to μηνώ, ειδοποιώ.

messenger *s.* αγγελιαφόρος *m.*

messy *a.* (*untidy*) ακατάστατος, (*dirty*) βρώμικος. ~ work δουλειά που κάνει ακαταστασία (*or* που λερώνει).

metabolism *s.* μεταβολισμός *m.*

metal *s.* μέταλλο *n.* **~s** (*railway*) ράγιες *f.pl.* (*a.*) μετάλλινος. **~led** *a.* (*road*) σκυροστρωμένος. **~lic** *a.* μεταλλικός. **~lurgy** *s.* μεταλλουργία *f.*

metamorphosis *s.* μεταμόρφωση *f.*

metaphor *s.* μεταφορά *f.* **~ical** *a.* μεταφορικός.

metaphysic|s *s.* μεταφυσική *f.* **~al** *a.* μεταφυσικός.

mete *v.* ~ out απονέμω, (*punishment*) επιβάλλω.

meteor *s.* μετέωρο *n.* **~ic** *a.* μετεωρικός. **~ite** *s.* αερόλιθος *m.*

meteorology *s.* μετεωρολογία *f.*

meter *s.* μετρητής *m.*, ρολόι *n.*

method *s.* μέθοδος *f.* **~ical** *a.* μεθοδικός. **~ically** *adv.* μεθοδικώς.

Methodism *s.* μεθοδισμός *m.*

meticulous *a.* πολύ ακριβής και προσεκτικός. **~ly** *adv.* με μεγάλη ακρίβεια στις λεπτομέρειες.

metre *s.* μέτρο *n.*

metric, ~al *a.* μετρικός. **~ally** *adv.* μετρικώς.

metro *s.* μετρό *n.*

metropol|is *s.* μητρόπολη *f.* **~itan** *a.* μητροπολιτικός, της μητροπόλεως. (*s., eccl.*) μητροπολίτης *m.*

mettle *s.* καρδιά *f.*, ψυχή *f.* (*of horse*) σφρίγος *n.* be on one's ~ φιλοτιμούμαι. show one's ~ δείχνω ποιος είμαι. put

(*person*) on his ~ παρακινώ το φιλότιμό του.
mettlesome *a.* ψυχωμένος.
mew *v.* νιαουρίζω.
mews *s.* στάβλοι αρχοντόσπιτων.
mezzo-soprano *a.* μεσόφωνος.
miaow *v.* νιαουρίζω. (*s.*) νιαούρισμα *n.*
miasma *s.* μίασμα *n.*
Michaelmas *s.* εορτή του Αγίου Μιχαήλ.
microbe *s.* μικρόβιο *s.*
microchip *s.* μικροκύκλωμα *n.*
microcosm *s.* μικρόκοσμος *m.*
microphone *s.* μικρόφωνο *n.*
microscop|e *s.* μικροσκόπιο *n.* **~ic** *a.* μικροσκοπικός.
microwave *a.* ~ oven φούρνος μικροκυμάτων.
mid *a.* in ~ May στα μέσα του Μαΐου. in ~ week μεσοβδόμαδα, in ~ air στον αέρα.
midday *s.* μεσημέρι *n.*(*a.*) μεσημεριανός.
midden *s.* σωρός σκουπιδιών.
middle *s.* μέση *f.* in the ~ of the street στη μέση του δρόμου. in the ~ of March στα μέσα του Μαρτίου. he phoned me when I was in the ~ of washing up με πήρε την ώρα ακριβώς που έπλενα τα πιάτα.
middle *a.* μεσαίος, μέσος. ~ class μεσαία τάξη. ~ age μέση ηλικία, ~-aged μεσήλιξ, μεσόκοπος. M~ Ages Μεσαίωνας *m.* ~-of-the-road policy μετριοπαθής πολιτική.
middleman *s.* τα ενδιάμεσα χέρια.
middling *a.* μέτριος.
midge *s.* σκνίπα *f.*
midget *s.* νάνος *m.*(*a.*) μίνι.
midland *a.* κεντρικός.
midnight *s.* μεσάνυχτα *n.pl.* (*a.*) μεσονύκτιος. burn the ~ oil ξενυχτώ εργαζόμενος διανοητικώς.
midriff *s.* διάφραγμα *n.* (*fam.*) στομάχι *n.*
midshipman *s.* δόκιμος *m.*
midst *s.* in the ~ of (*activity*) πάνω που, (*place*) μέσα σε. in our ~ ανάμεσά μας.
midsummer *s.* καρδιά του καλοκαιριού. ~ day του Άϊ-Γιαννιού.
midway *adv.* εις το μέσον της διαδρομής.
midwife *s.* μαία *f.*, μαμμή *f.* **~ry** *s.* μαιευτική *f.*
mien *s.* όψη *f.*
might *v.* he ~ have been saved θα μπορούσε να είχε σωθεί. we ~ never have met μπορεί να μην είχαμε γνωριστεί ποτέ. (*they worked hard*) so that their children ~ have a better life για να ζήσουν καλύτερα τα παιδιά τους. *see also* may.
might *s.* δύναμις *f.*
mighty *a.* πανίσχυρος, (*huge*) πελώριος, (*vast*) απέραντος. (*adv., fam.*) πολύ.
migraine *s.* ημικρανία *f.*
migrant *s.* μετανάστης *m.*

migrat|e *v.* μεταναστεύω. **~ion** *s.* μετανάστευση *f.*, αποδημία *f.* **~ory** *a.* μεταναστευτικός, αποδημητικός.
mild *a.* (*lenient*) επιεικής, (*slight*) ελαφρός. (*weather, person*) ήπιος, μαλακός. (*penalty, flavour, drink*) ελαφρός. draw it ~! μην τα παραλές.
mildly *adv.* μαλακά, ελαφρώς. to put it ~ για να μην πούμε τίποτα χειρότερο.
mildew *s.* μούχλα *f.* (*bot.*) περονόσπορος *m.* become ~ed μουχλιάζω.
mile *s.* μίλι *n.* (*fam.*) ~s better εκατό φορές καλύτερος.
mileage *s.* απόσταση σε μίλια. what's your ~? πόσα μίλια έχεις κάνει;
milestone *s.* πέτρινος δείκτης που σημειώνει τα μίλια. (*fig.*) σταθμός *m*, ορόσημον *n.*
milieu *s.* περιβάλλον *n.*
milit|ant *a.* αγωνιστικός. (*s.*) μαχητικός *a.* **~ancy** *s.* αγωνιστικότης *f.*
militarism *s.* μιλιταρισμός *m.*
military *a.* στρατιωτικός. ~ government στρατοκρατία *f.*
militate *v.* ~ against επενεργώ δυσμενώς επί (*with gen.*).
militia *s.* πολιτοφυλακή *f.*
milk *s.* γάλα *n.* it's no use crying over spilt ~ ό,τι έγινε έγινε. ~ and water (*fig.*) ανούσιος, νερόβραστος. (*v.t.*) (*lit. & fig.*) αρμέγω. **~maid** *s.* κοπέλλα που αρμέγει τις αγελάδες. **~man** *s.* γαλατάς *m.* **~sop** *s.* μαμμόθρεπτος *a.* **~y** *a.* γαλακτώδης. M~y Way Γαλαξίας *m.*
mill *s.* (*to grind, pump*) μύλος *m.* (*factory*) εργοστάσιο *n.* (*fig.*) go through the ~ υφίσταμαι δοκιμασίες. run-of-the-mill της αράδας. (*v.t.*) (*grind*) αλέθω. (*v.i.*) they ~ around *or* about πηγαινοέρχονται σα μυρμήγκια. **~er** *s.* μυλωνάς *m.* **~-pond** *s.* (*fig.*) sea like a ~pond θάλασσα λάδι (*or* γυαλί). **~-stone** *s.* μυλόπετρα *f.* (*fig.*) βάρος *n.* **~-wheel** *s.* τροχός του υδρόμυλου.
millennium *s.* χιλιετία *f.*
millet *s.* κεχρί *n.*
millimetre *s.* χιλιοστόμετρο *n.*
milliner *s.* καπελλού *f.* **~y** *s.* γυναικεία καπέλα.
million *s.* εκατομμύριο *n.* **~aire** *s.* εκατομμυριούχος *m.*
mime *s.* μιμική *f.* (*v.t.*) εκφράζω με μιμική.
mimic *a.* πλαστός. (*s.*) μίμος *m.* (*v.t.*) μιμούμαι.
mimicry *s.* μιμητική ικανότητα. (*in nature*) μιμητισμός *m.*
minaret *s.* μιναρές *m.*
minatory *a.* απειλητικός.

mince v.t. κάνω κιμά. not ~ one's words δεν μασώ τα λόγια μου.

mincemeat s. (meat) κιμάς m. (fig.) I made ~ of him τον έκανα κιμά.

mince-pie s. χριστουγεννιάτικο σκαλτσούνι.

mincing a. ~ machine κρεατομηχανή f. ~ gait κουνήματα n.pl.

mind s. νους m., μυαλό n. presence of ~ ετοιμότητα f., (coolness) ψυχραιμία f. absence of ~ αφηρημάδα f. to my ~ κατά τη γνώμη μου. bear or keep in ~ θυμάμαι, έχω υπ' όψιν. call to ~ (remember) θυμάμαι, (remind of) θυμίζω. change one's ~ αλλάζω γνώμη, μετανοώ. give one's ~ to στρέφω την προσοχή μου σε, have a (good) ~ to λέω να. make up one's ~ αποφασίζω. set one's ~ on βάζω στο μυαλό μου να. keep one's ~ on προσέχω. take one's ~ off αποσπώ τη σκέψη μου από. (do sthg. to) take one's mind off one's worries ξεδίνω. I've got him on my ~ με απασχολεί. he knows his own ~ ξέρει τι θέλει. be out of one's ~ or not be in one's right ~ δεν είμαι στα καλά μου. be in two ~s αμφιταλαντεύομαι. be of the same ~ συμφωνώ. I gave him a piece of my ~ του τα είπα έξω από τα δόντια. great ~s think alike τα μεγάλα πνεύματα συναντώνται.

mind v.t. & i. (take care, look after) προσέχω, κοιτάζω. ~ you don't fall πρόσεχε να μην πέσεις. ~ out! πρόσεχε! who is ~ing the baby? ποιος προσέχει (or κοιτάζει) το μωρό; ~ one's own business κοιτάω τη δουλειά μου. ~ one's P's and Q's προσέχω τη συμπεριφορά μου. (int.) ~ you... ξέρεις. (worry, object) I don't ~ (it) δεν με πειράζει (or νοιάζει). never ~! έννοια σου, δεν πειράζει. don't ~ about that μην ανησυχείς για αυτό. would you ~? (doing sthg.) θα είχατε την καλωσύνη να ...; (if) θα σας ενοχλούσε αν...; I shouldn't ~ a cigar θα κάπνιζα ευχαρίστως (or δεν θα έλεγα όχι για) ένα πούρο.

minded a. διατεθειμένος. be mathematically ~ έχω μαθηματικό μυαλό. (fam.) be fashion~ δίνω πολλή σημασία στη μόδα.

mindful a. be ~ of προσέχω, έχω υπ' όψιν.

mine pron. δικός μου.

mine v.t. (dig up) εξορύσσω. (lay ~es in) ναρκοθετώ, ποντίζω νάρκες σε.

mine s. (coal. etc.) ορυχείο n., μεταλλείο n. (explosive) νάρκη f. (tunnel) υπόνομος m.f. ~field s. ναρκοπέδιο n. ~layer s. (naut.) ναρκοθέτης f. ~sweeper s. (naut.) ναρκαλιευτικόν n.

miner s. μεταλλωρύχος m. (coal) ανθρακωρύχος m.

mineral a. ορυκτός. ~ water μεταλλικό νερό.

(s.) ορυκτό n. ~ogy s. ορυκτολογία f.

mingl|e v.t. αναμιγνύω, ενώνω. (v.i.) αναμιγνύομαι, ενώνομαι. ~ing s. ανάμιξη f.

miniature s. μινιατούρα f. (a.) μικρού σχήματος, εν μικρογραφία.

minimal a. ελάχιστος. ~ly adv. ελάχιστα.

minimize v. μειώνω στο ελάχιστο.

minimum a. ελάχιστος. (s.) μίνιμουμ n., κατώτατο όριο.

minion s. τσιράκι n.

minister s. υπουργός m. Prime M~ πρωθυπουργός m. (eccl.) πάστωρ m. (v.i.) ~ to υπηρετώ, φροντίζω. ~ial a. υπουργικός.

ministration s. παροχή βοήθειας, περιποίηση f.

ministry s. (department) υπουργείον n. (body) υπουργικόν συμβούλιον. (eccl.) enter the ~ γίνομαι παπάς.

mink s. βιζόν n.

Minoan a. μινωικός.

minor a. (lesser) ελάσσων. (small) μικρός, (secondary) δευτερεύων, (operation) δευτερεύουσης σημασίας. (mus.) ελάσσων. (s.) (not of age) ανήλικος a.

minority s. (of votes) μειο(νο)ψηφία f. (of people, things) μειονότης f.

minstrel s. αοιδός m.

mint s. (herb) δυόσμος m.

mint s. (for coining) νομισματοκοπείον n (fig.) a ~ of money πολλά λεφτά. in ~ condition ολοκαίνουργ(ι)ος. (v.t.) ~ money κόπτω νομίσματα, (fam., make a lot) κόβω μονέδα.

minuet s. μενουέτο n.

minute s. λεπτό n. (moment) στιγμή f. the ~ he comes αμέσως μόλις έρθει. (note) σημείωμα n. ~s (of meeting) πρακτικά n.pl.

minute a. μικροσκοπικός. (detailed) λεπτομερής. ~ly adv. λεπτομερώς.

minutiae s. μικρολεπτομέρειες f.pl.

minx s. τσαχπίνα f., ζωηρό κορίτσι.

mirac|le s. θαύμα n. ~ulous a. θαυματουργός. have a ~ulous escape σώζομαι ως εκ θαύματος. (incredible) καταπληκτικός.

mirage s. αντικατοπτρισμός m.

mir|e s. λάσπη f., ~y a. λασπώδης, (boots, etc.) λασπωμένος.

mirror s. καθρέπτης m. (v.) καθρεφτίζω.

mirth s. φαιδρότητα f. ~ful a. φαιδρός.

misadventure s. ατύχημα n.

misalliance s. αταίριαστος γάμος.

misanthrop|e s., ~ic a. μισάνθρωπος.

misapply v. (funds) καταχρώμαι.

misapprehen|d v. παρεξηγώ. ~sion s. παρεξήγηση f.

misappropriat|e v. (funds) καταχρώμαι. ~ion s. κατάχρηση f.

misbehav|e v. φέρομαι απρεπώς. (of child) κάνω αταξίες. **~iour** s. κακή συμπεριφορά f., αταξίες f.pl.

miscalculat|e v.i. πέφτω έξω. **~ion** s. κακός υπολογισμός m.

miscarriage s. αποβολή f. ~ of justice κακοδικία f.

miscarry v.i. (of woman) αποβάλλω. (fail) αποτυγχάνω, ναυαγώ. (of letter) χάνομαι.

miscast a. ακατάλληλος.

miscellan|y s. ποικιλία f., συνονθύλευμα n. **~eous** a. ποικίλος, διάφορος.

mischance s. ατυχία f. by ~ από κακή σύμπτωση.

mischief s. (damage) βλάβη f. (harm) κακό n. do ~ to βλάπτω, ζημιώνω. make ~ βάζω λόγια. get up to ~ κάνω ανοησίες. (fun) σκανταλιά f. (slyness) πονηριά f. **~-maker** s. ανακατωσούρης m.

mischievous a. (bad) άσχημος, κακός. (playful) σκανταλιάρης, άτακτος. (sly) πονηρός. **~ly** adv. με κακή πρόθεση. (slyly) πονηρά.

misconceive v.t. έχω εσφαλμένη ιδέα για.

misconduct v.t. διαχειρίζομαι κακώς. ~ oneself φέρομαι απρεπώς. (s.) κακή συμπεριφορά, παράπτωμα n.

miscon|strue v. παρερμηνεύω. **~struction** s. παρερμηνεία f.

miscount v. κάνω λάθος στο μέτρημα.

miscreant s. κάθαρμα n.

misdeed s. αδίκημα n.

misdemeanour s. πταίσμα n.

misdirect v. (person, energies) δίνω λανθασμένη κατεύθυνση σε, (letter) βάζω λάθος διεύθυνση σε.

miser s., **~ly** a. φιλάργυρος, τσιγγούνης.

miserab|le a. (bad) άθλιος, αξιοθρήνητος. (unhappy) δυστυχισμένος. (mean) μίζερος, ελεεινός, της κακιάς ώρας. **~ly** adv. άθλια, δυστυχισμένα, ελεεινά.

misery s. αθλιότητα f. δυστυχία f. μιζέρια f.

misfire v.i. (of gun) παθαίνω αφλογιστία. (of engine) δεν παίρνω. (fig.) πέφτω στο κενό.

misfit s. (person) απροσάρμοστος a. (garment) it is a ~ δεν εφαρμόζει καλά.

misfortune s. ατυχία f., κακή τύχη.

misgiving s. (apprehension) κακό προαίσθημα, (doubt) ενδοιασμός m.

misgovern v. κακοδιοικώ. **~ment** s. κακοδιοίκηση f.

mishandle v. χειρίζομαι κακώς.

mishap s. αναποδιά f.

misinform v. δίνω λανθασμένη πληροφορία σε.

misinterpret v. παρερμηνεύω. **~ation** s. παρερμηνεία f.

misjudge v.t. πέφτω έξω (or γελιέμαι) στην κρίση μου για.

mislay v. I have mislaid it δεν ξέρω πού το έβαλα.

mislead v. παραπλανώ. **~ing** a. παραπλανητικός.

mismanage v. διαχειρίζομαι κακώς. **~ment** s. κακή διαχείριση.

misnomer s. αταίριαστη ονομασία.

misogynist s. μισογύνης m.

misplaced a. (trust, etc.) αδικαιολόγητος.

misprint s. τυπογραφικό λάθος.

mispron|ounce v. προφέρω εσφαλμένως. **~unciation** s. εσφαλμένη προφορά.

misquot|e v. παραθέτω εσφαλμένως. **~ation** s. λάθος στην παραπομπή.

misrepresent v. παραποιώ. **~ation** s. παραποίηση f.

misrule s. αναρχία f.

miss s. (title) δεσποινίς f.

miss s. (failure to hit) ξαστόχημα n., άστοχη βολή. give (sthg.) a ~ παραλείπω. score a near ~ παρά λίγο να πετύχω, (see also nearly).

miss v.t. (fail to get, find, etc.) δεν πετυχαίνω, χάνω. you haven't ~ed anything δεν έχασες τίποτα. ~ one's aim αστοχώ. ~ the point δεν μπαίνω στο νόημα. ~ out (omit) παραλείπω. ~ out on χάνω. (fail to perform) I ~ed my lesson έχασα (or έλειψα από) το μάθημά μου. he never ~es going to see his mother each week δεν παραλείπει να επισκέπτεται τη μαμά του κάθε εβδομάδα. he narrowly ~ed being killed παρά λίγο να σκοτωθεί (see also nearly). (notice absence of) I ~ them μου λείπουν, μου έλειψαν. I never ~ed my purse till I got home αντιλήφθηκα ότι έλειπε το πορτοφόλι μου μόνον όταν έφτασα σπίτι μου.

missal s. σύνοψις f.

misshapen . κακοφτιαγμένος, στραβωμένος.

missile s. βλήμα n.

missing a. (lost) χαμένος. (soldier, etc.) αγνοούμενος. be ~ λείπω.

mission s. αποστολή f. **~ary** s. ιεραπόστολος m.

missis s. κυρία f.

missive s. επιστολή f.

misspelt a. ανορθόγραφος.

misspend a. χαραμίζω. he had a misspent youth τα νιάτα του πήγαν άδικα.

misstatement s. ανακρίβεια f.

mist s. ομίχλη f., καταχνιά f. (v.t. & i.) ~ (over) (of eyes, mirror, etc.) θολώνω, θαμπώνω. (of scenery, v.i.) σκεπάζομαι από ομίχλη.

mistake s. λάθος n., σφάλμα n. by ~ κατά

λάθος. make a ~ κάνω λάθος, σφάλλομαι. it's hot and no ~! ζέστη μια φορά! (or κάνει ζέστη, όχι αστεία).

mistake v.t. (misinterpret) παρανοώ. I mistook him for someone else τον παραγνώρισα. I mistook him for his brother τον πήρα για τον αδελφό του.

mistaken a. εσφαλμένος, λανθασμένος. be ~ κάνω λάθος, σφάλλομαι, απατώμαι. **~ly** adv. εσφαλμένως.

mister s. κύριος m.

mistimed a. άκαιρος.

mistletoe s. γκι n.

mistranslat|e v. μεταφράζω λάθος. **~ion** s. λανθασμένη μετάφραση.

mistress s. κυρία f. (of house) οικοδέσποινα f. (school) καθηγήτρια f., δασκάλισσα f. (paramour) ερωμένη f., μαιτρέσσα f.

mistrial s. δικαστικό λάθος.

mistrust v. δεν έχω εμπιστοσύνη σε. (s.) έλλειψη εμπιστοσύνης. **~ful** a. δύσπιστος. **~fully** adv. με δυσπιστία.

misunderstand v. (person) παρεξηγώ, (thing) παρανοώ. (both) δεν καταλαβαίνω καλά. **~ing** s. παρεξήγηση f.

misuse s. κακή χρήση. (of funds, power) κατάχρηση f. (v.) κάνω κακή χρήση (with gen.), καταχρώμαι. (ill-treat) κακομεταχειρίζομαι.

mite s. (offering) οβολός m. (tiny child, animal) τόσος δα, τοσούτσικος a.

mitigat|e v. μειώνω. (pain, anger) καταπραΰνω, καλμάρω. (penalty) μετριάζω. **~ing** circumstances ελαφρυντικές περιστάσεις. **~ion** s. μείωση f. μετρίαση f.

mitre s. (eccl.) μίτρα f.

mitten s. γάντι χωρίς δάκτυλα.

mix v.t. ανακατεύω, ανακατώνω, αναμιγνύω. ~ (up) (confuse) μπερδεύω. get ~ed up (involved) ανακατεύομαι, (in mind) μπερδεύομαι. (v.i.) (go well together) ταιριάζω. ~ with (associate) συναναστρέφομαι (with acc.).

mixed a. μιχτός. (assorted) ανάμιχτος, διάφορος ~ (up) (confused) ανακατεμένος, μπερδεμένος. ~ motives ποικίλα ελατήρια.

mixer s. (concrete) μπετονιέρα f., (food) μίξερ n. good ~ κοινωνικός άνθρωπος.

mixing s. (stirring) ανακάτωμα n. (contact) ανάμιξη f.

mixture s. μίγμα f.

mix-up s. σαλατοποίηση f., ανακατωσούρα f.

moan v. στενάζω, βογγώ. (fam., grumble) γκρινιάζω. (s.) (also ~ing) στεναγμός m., βογγητό n. γκρίνια f.

moat s. τάφρος m.

mob s. όχλος m. ~ rule οχλοκρατία f. (v.) πολιορκώ, συνωστίζομαι γύρω από.

mobil|e a. κινητός, (features) εκφραστικός. **~ity** s. ευκινησία f.

mobiliz|e v. κινητοποιώ, επιστρατεύω. **~ation** s. κινητοποίηση f. επιστράτευση f.

mock v. κοροϊδεύω, χλευάζω. (a.) ψεύτικος. **~ery** s. κοροϊδία f., χλεύη f. (travesty) παρωδία f. **~ing** a. κοροϊδευτικός. **~-up** s. πρόχειρο ομοίωμα.

modal a. (gram.) εγκλιτικός.

mode s. (way & mus.) τρόπος m. (fashion) μόδα f.

model s. (example) υπόδειγμα n., πρότυπο n. (representation, sthg. for copying, mannequin, dress, car) μοντέλο n.

model a. (exemplary) υποδειγματικός, πρότυπος. ~ pupil πρότυπο μαθητού, ~ boat ομοίωμα πλοίου.

model v. (mould) πλάθω. ~ oneself on παίρνω ως υπόδειγμα. well ~led (features) εύγραμμος, λαξευτός.

moderate v.t. μετριάζω, περιορίζω. (v.i.) μετριάζομαι, κόβω.

moderate a. μέτριος, (in opinions) μετριοπαθής. (temperate) μετρημένος. (price) λογικός. **~ly** adv. μετρίως, μέτρια.

moderation s. μετριοπάθεια f. in ~ με μέτρο, εν μέτρω.

modern a. σύγχρονος, μοντέρνος, νεότερος. ~ Greek (a.) νεοελληνικός. **~ize** v. εκσυγχρονίζω, εκμοντερνίζω. **~ization** s. εκσυγχρονισμός m.

modest a. (chaste) σεμνός, (simple) απλός, απέριττος. (unassuming) μετριόφρων. (of means) μέτριος. **~ly** adv. με σεμνότητα, απλά, απέριττα.

modesty s. σεμνότητα f. μετριοφροσύνη f.

modicum s. a ~ of λίγος.

modif|y v. (change) τροποποιώ. (lessen) μετριάζω. **~ication** s. τροποποίηση f. μετριασμός m.

modish a. της μόδας.

modulat|e v.t. (radio) διαμορφώνω. (one's voice) χρωματίζω. **~ion** s. (mus.) μετατροπία f. (of voice) χρωματισμός m., διακύμανση f.

modus s. ~ operandi τρόπος ενεργείας. ~ vivendi προσωρινός διακανονισμός.

Mohammedan s. Μωαμεθανός m.

moist a. υγρός. **~en** v. υγραίνω. **~ure** s. υγρασία f.

molar s. τραπεζίτης m.

mole s. (on skin) ελιά f.

mole s. (animal) τυφλοπόντικας m.

mole s. (breakwater) μώλος m.

molecule s. μόριο n.

molest v. παρενοχλώ. (ill-treat) κακοποιώ.

mollify v. μαλακώνω.
mollusc s. μαλάκιο n.
mollycoddle v. παραχαϊδεύω. (s.) μαμμόθρεφτος a.
molten a. λειωμένος.
moment s. στιγμή f. at any ~ όπου να 'ναι. the ~ he arrives αμέσως μόλις φτάσει. (importance) σημασία f.
momentar|y a. (brief) στιγμιαίος. (at every moment) από στιγμή σε στιγμή. ~ily adv. για μία στιγμή.
momentous a. βαρυσήμαντος.
momentum s. κεκτημένη ταχύτητα. gain ~ παίρνω φόρα.
monarch s. μονάρχης m. ~ic a., ~ist s. μοναρχικός. ~ism s. μοναρχισμός m. ~y s. μοναρχία f.
monast|ery s. μοναστήρι n., μονή f. ~ic a. μοναστικός, (of monks) μοναχικός.
Monday s. Δευτέρα f.
monetary a. νομισματικός.
money s. χρήμα n., χρήματα pl., λεφτά n.pl. (currency) νόμισμα n. ready ~ μετρητά n.pl. ~-box s. κουμπαράς m. ~-changer s. σαράφης m. ~ed a. ~ed man λεφτάς m. ~-lender s. τοκογλύφος m., τοκιστής m. ~-order s. ταχυδρομική επιταγή.
mongrel s. σκύλος όχι εκλεκτής ράτσας.
monitor v. ελέγχω, παρακολουθώ. (s.) (listener) ο παρακολουθών, (device) ανιχνευτής m.
monk s. μοναχός m., καλόγερος m.
monkey s. μαϊμού f. make a ~ of ρεζιλεύω. (fam.) get one's ~ up φουρκίζομαι. ~ business κατεργαριές f.pl. (v.) ~ with πειράζω, πασπατεύω. ~-nut s. αράπικο φιστίκι.
monocle s. μονόκλ n.
monogamy s. μονογαμία f.
monogram s. μονόγραμμα n.
monograph s. μονογραφία f.
monolithic a. μονολιθικός.
monologue s. μονόλογος m.
monopol|y s. μονοπώλιο n. ~ize v. μονοπωλώ.
monosyllable s. μονοσύλλαβη λέξη.
monotone s. speak in a ~ μιλώ χωρίς αλλαγή του ύψους του τόνου.
monoton|y s. μονοτονία f. ~ous a. μονότονος. ~ously adv. μονότονα.
monsoon s. μουσσώνας m.
monster s. τέρας n. (a.) τεράστιος.
monstrosity s. τερατούργημα n.
monstrous a. (huge) τεράστιος, (also atrocious) τερατώδης. ~ly adv. τερατωδώς.
month s. μήνας m., μην m. ~ly a. μηνιαίος. ~ly wage μηνιάτικο n. (adv.) μηνιαίως, κατά μήνα.

monument s. μνημείο n. ~al a. μνημειώδης. ~al mason μαρμαράς m.
moo v. μουγκανίζω. (s.) μουγκανητό n.
mooch v. ~ about αλητεύω.
mood s. διάθεση f. good ~ κέφι n. bad ~ κακοκεφιά f. he's in a good ~ είναι στα κέφια (or στις καλές) του. be in the ~ for έχω διάθεση (or κέφι or όρεξη) για. (gram.) έγκλισις f. ~y a. κατσούφης, κακόκεφος.
moon v.i. ~ about χαζεύω. ~y a. χαζός.
moon s. φεγγάρι n., σελήνη f. full ~ πανσέληνος f. (fam.) promise (person) the ~ τάζω λαγούς με πετραχήλια. once in a blue ~ στη χάση και στη φέξη. ~less a. αφέγγαρος. ~lit a. φεγγαρόλουστος. ~shine s. (fam.) παραμύθια n.pl. ~struck a. νεραϊδοπαρμένος.
moonlight s. σεληνόφως n. by ~ με το φεγγάρι, one ~ night μια βραδιά με φεγγάρι. (v.i.) (fam.) κάνω κι άλλη δουλειά εκτός από την κανονική μου για να ζήσω.
moor, ~land s. ρεικότοπος m.
moor v.t. & i. αράζω. (v.t.) δένω. ~ings s. αγκυροβόλιο n.
Moorish a. μαυριτανικός.
moot v. ανακινώ. (a.) αμφισβητήσιμος.
mop v. (also ~ up) σφουγγίζω, σφουγγαρίζω. (absorb) απορροφώ. (fig.) ~ up (rout) κατατροπώνω.
mop s. σφουγγαρίστρα f. (fam., hair) πυκνά μαλλιά.
mope v.i. με κατέχει ακεφιά και μελαγχολία.
moral a. ηθικός. ~ resources ψυχικός κόσμος. ~ support συμπαράστασηf. (s., of tale) επιμύθιο n. ~ize v. ηθικολογώ. ~ly adv. ηθικώς.
morale s. ηθικό n.
morality s. ηθική f. ~ play θρησκευτικό δράμα.
morals s. ηθική f., ήθη n.pl., ηθικές αρχές. without ~ ανήθικος.
morass s. τέλμα n.
moratorium s. χρεωστάσιον n.
morbid a. αρρωστημένος, νοσηρός, (med.) παθολογικός.
mordant a. δηκτικός.
more a. & pron. (in measure, degree) περισσότερος, (additional) άλλος. it needs ~ effort θέλει περισσότερη προσπάθεια. two ~ books άλλα (or ακόμα) δύο βιβλία. bring some ~ wine φέρε κι άλλο (or ακόμα λίγο) κρασί. there isn't any ~ δεν έχει άλλο. I have no ~ money left δεν μου έχουν μείνει πια άλλα λεφτά. it's a pity we don't see ~ of each other κρίμα που δεν βλε-

πόμαστε περισσότερο (or πιο συχνά). we saw no ~ of them δεν τους ξαναείδαμε.

more adv. περισσότερο, πιο πολύ. ~ powerful δυνατότερος, πιο δυνατός. ~ easily πιο εύκολα. ~ and ~ όλο και περισσότερο. ~ or less πάνω-κάτω, λίγο-πολύ, κατά το μάλλον ή ήττον. once ~ άλλη (or ακόμα) μια φορά. what's ~ και μάλιστα. all the ~ so (because...) τοσούτω μάλλον. the ~ you try the easier it gets όσο πιο πολύ προσπαθεί κανείς τόσο ευκολότερο γίνεται. he's had ~ than two glasses of wine έχει πιει πάνω από δύο ποτήρια κρασί. he worries ~ than I do στενοχωριέται περισσότερο από μένα. (with neg.) πια. be no ~ δεν υπάρχω πια. we don't see them any ~ δεν τους βλέπουμε πια. there were not ~ than ten of them δεν ήταν πάνω από δέκα. she's little ~ than a servant λίγο απέχει από το να είναι υπηρέτρια. no ~ do I ούτε εγώ. nothing ~ τίποτ' άλλο.

morello s. βύσσινο n.

moreover adv. εξ άλλου, άλλωστε, επί πλέον.

mores s. ήθη n.pl.

morganatic a. μοργανατικός.

morgue s. νεκροτομείο n.

moribund a. ετοιμοθάνατος.

morning s. πρωί n., πρωινό n. good ~! καλημέρα. this ~ σήμερα το πρωί. first thing in the ~ πρωί-πρωί. (a.) πρωινός.

moron s., **~ic** a. (fam.) βλακόμετρο n.

morose a. κατσούφης.

morph|ia, ~ine s. μορφίνη f.

morphology s. μορφολογία f.

morrow s. what has the ~ in store for us? τι μας επιφυλάσσει η αύριον (or το αύριο);

morse a. ~ code μορσικός κώδιξ.

morsel s. μπουκιά f., κομματάκι n.

mortal a. (man) θνητός. (injury) θανατηφόρος, θανάσιμος. (sin, enemy, hatred) θανάσιμος. (fam.) every ~ thing τα πάντα. **~ly** adv. θανάσιμα.

mortality s. (mortal condition) θνητότητα f. (death rate) θνησιμότητα f.

mortar s. (substance) κονίασμα n., λάσπη f. (bowl) γουδί n. (mil.) όλμος m.

mortgage s. (to raise money) υποθήκη f. (to buy house) στεγαστικό δάνειο. (v.) υποθηκεύω.

mortification s. ντρόπιασμα n. (med.) γάγγραινα f.

mortif|y v.t. feel ~ied ντροπιάζομαι. ~y the flesh αυτοτιμωρούμαι προς εξαγνισμόν. (v.i., med.) γαγγραινιάζω.

mortuary s. νεκροτομείο n.

mosaic s. ψηφιδωτό n., μωσαϊκό n.

Moslem a. μουσουλμανικός. (person) Μουσουλμάνος m.

mosque s. τζαμί n.

mosquito s. κουνούπι n. **~-net** s. κουνουπιέρα f.

moss s. μούσκλι n. **~y** a. χορταριασμένος.

most a. & s. ο περισσότερος, ο πιο πολύς. the ~ beautiful woman η πιο ωραία (or η ωραιότερη) γυναίκα. ~ people ο πολύς κόσμος, οι περισσότεροι άνθρωποι. at ~ το (πολύ) πολύ. (for) ~ of the year το μεγαλύτερο μέρος του χρόνου. for the ~ part ως επί το πλείστον. make the ~ of (present favourably) παρουσιάζω όσο μπορώ καλύτερα, (benefit from) επωφελούμαι (with gen.).

most adv. περισσότερο (από κάθε τι άλλο). ~ often ως επί το πλείστον. ~ of all πάνω απ' όλα, προ παντός, προ πάντων. (very) πάρα πολύ.

mostly adv. (mainly) κυρίως, κατά το πλείστον. (usually) συνήθως.

mote s. μόριο σκόνης. (fig., in brother's eye) κάρφος n.

moth s. πεταλουδίτσα (της νύχτας) f. (in clothes) σκώρος m. **~-balls** s. ναφθαλίνη f. **~-eaten** a. σκωροφαγωμένος.

mother s. μητέρα f. (fam.) μάνα f. M~ of God Θεομήτωρ f. (a.) μητρικός. ~ country πατρίδα f. (v.t.) περιποιούμαι σαν μητέρα. **~-hood** s. μητρότητα f. **~-in-law** s. πεθερά f. **~-of-pearl** s. σιντέφι n., (fam.) φίλντισι n. **~-less** a. ορφανός από μητέρα.

motherly a. (love) μητρικός. (woman) που ξεχειλίζει από μητρική καλωσύνη.

motif s. μοτίφ n. (mus.) μοτίβο n.

motion s. (movement) κίνηση f. set in ~ θέτω εις κίνησιν. (proposal) πρόταση f. pass a ~ (of bowels) ενεργούμαι. **~less** a. ακίνητος.

motion v. (make sign) γνέφω, κάνω νόημα.

motivat|e v. be ~ed by έχω ως κίνητρο. **~ion** s. κίνητρα n.pl.

motive s. κίνητρο n., ελατήριο n. (a.) κινητήριος.

motley a. (colour) πολύχρωμος, (crowd, etc.) ανακατεμένος. (s.) στολή γελωτοποιού.

motor s. μηχανή f., κινητήρας m. (car) αυτοκίνητο n. **~-boat** s. βενζινάκατος f., μπενζίνα f. **~-coach** s. πούλμαν n. **~-cycle** s. μοτοσυκλέτα f. **~-ing** s. το οδηγείν, αυτοκινητισμός m. **~-ist** s. αυτοκινητιστής m. **~-ized** a. μηχανοκίνητος. **~-way** s. αυτοκινητόδρομος m.

mottled a. διάστικτος, (skin) με λεκέδες.

motto s. ρητό n. (maxim) αρχή f.

mould v. φορμάρω, (clay) πλάθω,

(*character*) διαπλάθω.

mould *s.* (*form*) φόρμα *f.*, καλούπι *n.* (*earth*) χώμα *n.* (*fungus*) μούχλα *f.* **-y** *a.* μουχλιασμένος. (*fam.*, *wretched*) της κακιάς ώρας.

moulder *v.i.* αποσυντίθεμαι. (*bones*) λειώνω, (*leaves*) σαπίζω. (*walls*) μουχλιάζω, καταρρέω.

moulding *s.* κορνίζα *f.*

moult *v.* μαδώ.

mound *s.* σωρός *m.* (*burial*) τύμβος *m.*

mount *v.t.* ~ the stairs ανεβαίνω τη σκάλα. ~ the ladder ανεβαίνω στη σκάλα. (*horse*) καβαλλικεύω, (*a jewel*) δένω, (*a play*) ανεβάζω. (*v.i*) (*also* ~ up) ανεβαίνω.

mount *s.* όρος *n.* Mt Olympus ο Όλυμπος. (*of picture*) χαρτόνι *n.* (*horse*) άλογο *n.* (*base*) βάση *f.* (*gun-carriage*) κιλλίβας *m.* (*of jewel*) δέσιμο *n.*

mountain *s.* βουνό *n.*, όρος *n.* (*a.*) βουνίσιος, ορεινός, **-eer** *s.* ορειβάτης *m.* **-ous** *a.* ορεινός, (*huge*) σαν βουνό.

mountebank *s.* τσαρλατάνος *m.*, κομπογιαννίτης *m.*

mourn *v.t.* & *i.* πενθώ, κλαίω. **-er** *s.* πενθών, (τε)θλιμμένος *a.* **-ful** *a.* πένθιμος, λυπηρός.

mourning *s.* πένθος *n.* be in ~ μαυροφορώ. (*a.*) πένθιμος.

mouse *s.* ποντικός *m.*, ποντίκι *n.* **--trap** *s.* φάκα *f.*

moustache *s.* μουστάκι *n.*

mousy *a.* (*timid*) ντροπαλός.

mouth *v.t.* απαγγέλλω στομφωδώς.

mouth *s.* στόμα *n.* (*of bottle*) στόμιο *n.* (*of river*) εκβολή *f.* (*of cannon, harbour*) μπούκα *f.* (*of cave, harbour*) είσοδος *f.* by word of ~ προφορικώς, δια ζώσης. down in the ~ αποκαρδιωμένος. **-ful** *s.* μπουκιά *f.* **--organ** *s.* φυσαρμόνικα *f.* **-piece** *s.* επιστόμιο *n.* (*fig.*) φερέφωνο *n.*

movable *a.* κινητός.

move *v.t.* κινώ, κουνώ. (*shift*) μετακινώ, (*transport*) μεταφέρω. (*affect*) συγκινώ. (*impel*) ωθώ. (*cause to change attitude*) κλονίζω. (*transfer to new post*) μεταθέτω. ~ a resolution υποβάλλω πρόταση. ~ one's chair nearer πάω την καρέκλα μου πιο κοντά. (*v.i.*) κινούμαι. (*stir, be unstable*) κουνώ, κουνιέμαι. (*shift*) μετακινούμαι. (*circulate*) κυκλοφορώ. (*change residence*) μετακομίζω. (*go*) πηγαίνω. ~ forward, along, on προχωρώ. ~ back κάνω πίσω. ~ away απομακρύνομαι. ~ in (*settle*) εγκαθίσταμαι. ~ out (*leave*) φεύγω, (*of house*) μετακομίζω. ~ up (*make room*) κάνω θέση, (*rise*) ανεβαίνω.

move *s.* κίνηση *f.* (*action*) ενέργεια *f.* (*step*)

βήμα *n.* (*initiative*) πρωτοβουλία *f.* on the ~ εν κινήσει. make a ~ (*act*) κινούμαι, (*leave*) ξεκινώ. get a ~ on! κουνήσου.

movement *s.* κίνηση *f.* (*political, etc.*) κίνημα *n.* (*mechanism*) μηχανισμός *m.* (*mus.*) μέρος *n.* (*of bowels*) κένωση *f.*

mover *s.* prime ~ κινητήρια δύναμις.

movie *s.* ταινία *f.* ~s κινηματογράφος *m.*

moving *a.* συγκινητικός.

mow *v.t.* (*grass*) κουρεύω, (*hay*) θερίζω. ~ down (*fig.*) θερίζω.

MP βουλευτής *m.*

much *a.* πολύς. how ~ πόσος. so ~ τόσος. as ~ as you like όσο θέλεις. he's not ~ of a singer δεν είναι σπουδαίος τραγουδιστής. think ~ of έχω σε μεγάλη εκτίμηση. make ~ of (*exaggerate*) αποδίδω μεγάλη σημασία σε, (*fuss over*) περιποιούμαι. I couldn't make ~ of it δεν κατάλαβα καλά.

much *adv.* πολύ. very *or* too ~ πάρα πολύ. ~ to my surprise προς μεγάλη μου έκπληξη. ~ as I should like to go παρ' όλο που θα ήθελα να πάω. ~ the same σχεδόν ο ίδιος. (*fam.*) ~ of a ~ness σχεδόν τα ίδια.

mucilage *s.* κόλλα *f.*

muck *s.* (*dung*) κοπριά. *f.* (*filth*) βρώμα *f.* (*v.t.*) ~ up χαλ(ν)ώ, κάνω μούσκεμα, (*dirty*) λερώνω. (*v.i.*) ~ about κάνω την ώρα μου. ~ about with πειράζω. **--raking** *s.* σκανδαλοθηρία *f.* **-y** *a.* βρώμικος.

muc|us *s.* βλέννα *f.* (*nasal, fam.*) μύξα *f.* (*of eyes, fam.*) τσίμπλα *f.* **-ous** *a.* βλεννώδης.

mud *s.* λάσπη *f.* throw ~ at (*fig.*) συκοφαντώ. his name is ~ είναι υπό δυσμένειαν. **--brick** *s.* πλίθ(ο)α *f.* **-guard** *s.* φτερό *n.*

muddle *s.* (*disorder*) ακαταστασία *f.* in a ~ μπερδεμένος, (*dazed*) ζαλισμένος. (*v.t.*) μπερδεύω, ανακατώνω. (*v.i.*) ~ through τα κουτσοβολεύω. **-r** *s.* be a ~r (*or* ~-headed) δεν κόβει το μυαλό μου.

muddy *a.* (*road*) λασπώδης, (*shoes, etc.*) λασπωμένος.

muff *s.* (*for hands*) μανσόν *n.* (*duffer*) ατζαμής *a.* (*v.*) ~ it αποτυγχάνω.

muffin *s.* ψημένο τσουρεκάκι.

muffle *v.* (*wrap*) κουκουλώνω, τυλίγω. ~d voice πνιχτή φωνή. **-r** *s.* κασκόλ *n.*

mufti *s.* in ~ με πολιτικά.

mug *s.* (*cup*) κούπα *f.* (*fam., face*) μούτρο *n.*, (*fool*) κορόιδο *n.* ~'s game πεταμένος κόπος.

mug *v.* ~ up μελετώ εντατικά.

mug *v.* (*rob*) επιτίθεμαι και ληστεύω. **-ging** *s.* επίθεση με πρόθεση ληστείας.

muggy *a.* αποπνικτικός.

mulberry *s.* μούρο *n.* (*tree*) μουριά *f.*

mulatto *s.* μιγάς *m.f.*

mul|e s. ημίονος *m.f.* μουλάρι *n.* ~eteer *s.* αγωγιάτης *m.* ~ish *a.* ξεροκέφαλος.
mull *v.* (*wine*) ζεσταίνω. ~ over (*ponder*) συλλογίζομαι.
mullet *s.* κέφαλος *m.* red ~ μπαρμπούνι *n.*
mullion *s.* χώρισμα παραθύρου.
multifarious *a.* πολλών ειδών.
multimillionaire *s.* πολυεκατομμυριούχος *m.*
multinational *a.* πολυεθνικός.
multiple *a.* πολλαπλός. (*s.*) πολλαπλάσιο *n.*
multiplication *s.* πολλαπλασιασμός *m.*
multiplicity *s.* πολλαπλότητα *f.* a ~ of duties πλήθος καθηκόντων.
multiply *v.t.* πολλαπλασιάζω. (*v.i.*) πολλαπλασιάζομαι.
multi-storey *a.* πολυώροφος *a.*
multitud|e *s.* πλήθος *n.* ~inous *a.* πολυπληθής.
mum *s.* (*fam.*) μαμά *f.*
mum *a.* (*silent*) keep ~ το βουλλώνω. ~'s the word! μιλιά! τσιμουδιά! μόκο!
mumble *v.i.* τρώω τα λόγια μου.
mumbo jumbo *s.* ακατάληπτη γλώσσα. (*fraud*) απάτη *f.*
mummery *s.* παντομίμα *f.* (*pej.*) καραγκιοζλίκι *n.*
mumm|y *s.* μούμια *f.* (*fam.*, *mother*) μαμά *f.* ~ify *n.* ταριχεύω.
mumps *s.* μαγουλάδες *f.pl.*
munch *v.* μασουλίζω.
mundane *a.* εγκόσμιος (*dull*) ~ job δουλειά ρουτίνα.
municipal *a.* δημοτικός, του δήμου. ~ity δήμος *m.*
munific|ent *a.* γενναιόδωρος. ~ence *s.* γενναιοδωρία *f.*
muniments *s.* τίτλοι *m.pl.*
munitions *s.* πολεμοφόδια *n.pl.*
mural *s.* τοιχογραφία *f.*
murder *s.* φόνος *m.*, δολοφονία *f.* (*v.*) φονεύω, δολοφονώ. ~er *s.* φονιάς *m.*, δολοφόνος *m.* ~ess *s.* φόνισσα *f.*
murderous *a.* (*weapon, etc.*) φονικός. with ~ intent με δολοφονικές προθέσεις. he gave me a ~ glance με κοίταξε μ' ένα βλέμμα σα να με μαχαίρωνε.
murky *a.* σκοτεινός, μαύρος.
murmur *v.* μουρμουρίζω. (*s.*) μουρμουρητό *n.* without a ~ χωρίς παράπονο.
muscatel *s.* μοσχάτο *n.*
muscle *s.* μυς *m.*, ποντίκι *n.* (*fig.*) ρώμη *f.* (*v.*) (*fam.*) ~ in on (*person's business*) εκτοπίζω κάποιον και του παίρνω τις δουλειές.
muscular *a.* (*of muscles*) μυϊκός, (*strong*) μυώδης.
muse *s.* μούσα *f.*

muse *v.* ρεμβάζω, συλλογίζομαι.
museum *s.* μουσείο *n.*
mushroom *s.* μανιτάρι *n.* (*v.i.*, *spread*) ξεφυτρώνω παντού.
mushy *a.* (*soft*) it's gone all ~ έγινε νιανιά (*or* πολτός). (*fig.*) γλυκανάλατος.
music *s.* μουσική *f.* set to ~ μελοποιώ. ~-hall *s.* βαριετέ *n.* ~-stand *s.* αναλόγιο *n.*
musical *a.* μουσικός, (*melodious*) μελωδικός, is he ~? του αρέσει η μουσική; ~ly *adv.* μουσικώς, μελωδικά.
musician *s.* μουσικός *m.f.* ~ly *a.* μουσικός, γεμάτος μουσικότητα. ~ship *s.* μουσικότητα *f.*
musk *s.* μόσχος *m.*
musket *s.* μουσκέτο *n.*
muslin *s.* μουσελίνα *f.*
mussel *s.* μύδι *n.*
must *s.* (*juice*) γλεύκος *n.*, μούστος *m.*
must *v.* I ~ go there πρέπει να πάω εκεί. they ~ have arrived θα έχουν (*or* πρέπει να έχουν) φτάσει. it ~ have been raining all night θα έβρεχε (*or* πρέπει να έβρεχε) όλη νύχτα. I ~ say ομολογώ ότι...
mustard *s.* μουστάρδα *f.* ~ seed σιναπόσπορος *m.*
muster *v.t.* συγκεντρώνω. (*v.i.*) συγκεντρώνομαι. (*s.*) συγκέντρωση *f.* it will pass ~ θα περάσει, θα κάνει τη δουλειά του.
musty *a.* μουχλιασμένος. it smells ~ μυρίζει μούχλα.
mutab|le *a.* ασταθής. ~ility *s.* αστάθεια *f.*
mutation *s.* μετάλλαξη *f.* (*gram.*) vowel ~ μεταφωνία *f.*
mute *a.* βουβός, αμίλητος (*also gram.*) άφωνος. ~ly *adv.* (*use a.*).
mutilat|e *v.* ακρωτηριάζω. ~ion *s.* ακρωτηριασμός *m.*
mutin|y *s.* στάση *f.* (*v.i.*) κάνω στάση, στασιάζω. ~eer *s.* στασιαστής *m.* ~ous *a.* (*action*) στασιαστικός, (*man*) στασιαστής *m.*
mutter *v.* μουρμουρίζω. ~ing *s.* μουρμούρα *f.*
mutton *s.* αρνί *n.*, αρνίσιο κρέας. dead as ~ τελείως νεκρός. ~-headed *a.* κουτός.
mutual *a.* (*reciprocal*) αμοιβαίος, (*in common*) κοινός. ~ aid αλληλεγγύη *f.* ~ly *adv.* αμοιβαίως.
muzzle *s.* (*snout*) ρύγχος *n.*, μουσούδι *n.* (*appliance*) φίμωτρο *n.* (*of gun*) μπούκα *f.* (*v.*) φιμώνω.
muzzy *a.* (*dazed*) ούργιος, ζαλισμένος, (*blurred*) θολός.
Mycenean *a.* μυκηναϊκός.
my *pron.* ~ house το σπίτι μου. oh ~! πω πω!
myop|ia *s.* μυωπία *f.* ~ic *a.* μύωψ.
myriad *s.* μυριάδα *f.*
myrmidons *s.* (*fig.*) τσιράκια *n.pl.*
myrtle *s.* μυρτιά *f.*

myself *see* self.

mysterious *a.* μυστηριώδης. ~ person μυστήριος άνθρωπος. **~ly** *adv.* μυστηριωδώς.

mystery *s.* μυστήριο *n.* it's a ~! μυστήριο πράγμα!

mystic *s.* μυστικιστής *m.*

mystic, ~al *a.* μυστικός. **~ism** *s.* μυστικισμός *m.*

mystif|y *v.* I was ~ied by his reply έμεινα έκπληκτος με την απάντησή του. **~ication** *s.* έκπληξη *f.*

mystique *s.* (*of institution*) μυστηριακή ακτινοβολία. (*of art*) μυστικό *n.*

myth *s.* μύθος *m.* (*sthg. untrue*) μύθευμα *n.* **~ical** *a.* μυθικός.

mytholog|y *s.* μυθολογία *f.* **~ical** *a.* μυθολογικός.

N

nab *v.* (*fam., catch*) τσιμπώ, τσακώνω.

nadir *s.* ναδίρ *n.*

nag *s.* (*pej.*) παλιάλογο *n.*

nag *v.t. & i.* γκρινιάζω. (*v.i.*) τρώω, (*of pain*) τυραννώ. **~ger** *s.* γκρινιάρης *m.*, γκρινιάρα *f.* **~ging** *s.* μουρμούρα *f.*, γκρίνια *f.*

naiad *s.* ναϊάς *f.*

nail *s.* (*on body*) νύχι *n.*, όνυξ *m.* (*implement*) καρφί *n.*, πρόκα *f.*, ήλος *m.* hit the ~ on the head πετυχαίνω διάνα. hard as ~s πολύ σκληρός. on the ~ αμέσως. **~-brush** *s.* βούρτσα των νυχιών.

nail *v.t.* (*also* ~ up *or* down) καρφώνω. (*fig.*) (*one's eyes*) προσηλώνω, (*a person*) καθηλώνω, (*a lie*) ξεσκεπάζω.

naive *a.* αγαθός, αφελής. **~ty** *s.* αφέλεια *f.*

naked *a.* γυμνός. with the ~ eye διά γυμνού οφθαλμού. **~ness** *s.* γύμνια *f.*

namby-pamby *a.* γλυκανάλατος.

name *s.* όνομα *n.*, ονομασία *f.* (*surname*) επώνυμο, επίθετο *n.* Christian ~ μικρό όνομα. full ~ ονοματεπώνυμο *n.* by the ~ of Peter ονόματι Πέτρος. in the ~ of the law εν ονόματι του νόμου. he has made a ~ for himself έβγαλε όνομα. what's his ~? πώς ονομάζεται; πώς τον λένε; ~ day ονομαστική εορτή. it is my ~ day έχω το όνομά μου. call (*person*) ~s βρίζω. (*v.*) ονομάζω, (*set, appoint*) ορίζω.

nameless *a.* (*unknown*) άγνωστος, (*awful*) ακατονόμαστος.

namely *adv.* δηλαδή, ήτοι.

nameplate *s.* ταμπέλα *f.*

namesake *s.* συνονόματος *a.*

nanny *s.* νταντά *f.*

nanny-goat *s.* κατσίκα *f.* (*old*) γκιόσα *f.*

nap *s.* (*of cloth*) πέλος *n.*

nap *s.* (*sleep*) have a ~ παίρνω έναν υπνάκο. (*fig.*) he was caught ~ping τον έπιασαν στον ύπνο.

nape *s.* ~ of neck σβέρκος *m.*, αυχένας *m.*

naphtha *s.* νάφθα *f.*

napkin *s.* πετσέτα *f.* (*baby's*) πάνα *f.*

nappy *s.* πάνα *f.*

narcissus *s.* νάρκισσος *m.*

narcotic *a.* ναρκωτικός. (*s.*) ναρκωτικό *n.*

nark *s.* (*fam.*) χαφιές *m.*

narrat|e *v.* αφηγούμαι. **~or** *s.* αφηγητής *m.* **-ion** *s.* *see* narrative.

narrative *s.* (*general*) αφήγηση *f.* (*particular*) αφήγημα *n.* (*a.*) αφηγηματικός.

narrow *a.* στενός, (*limited*) περιορισμένος. ~ gauge (*a.*) στενής γραμμής. I had a ~ escape (*or* squeak) φτηνά τη γλύτωσα (*s.*) **~s** στενά *n.pl.* (*v.i.*) στενεύω. **~ly** *adv.* μόλις. *see* nearly. **~-minded** *a.* στενόμυαλος. **~-mindedness** *s.* στενομυαλιά *f.* **~ness** *s.* στενότητα *f.*

nasal *a.* ρινικός. (*also gram.*) έρρινος.

nascent *a.* εν τω γίγνεσθαι.

nastiness *s.* (*of food, weather*) χάλι *n.* (*of person*) κακιά *f.*

nast|y *a.* άσχημος, (*dirty*) βρωμερός, (*difficult*) δύσκολος. (*in temper*) κακός. it smells ~y μυρίζει άσχημα. ~y piece of work (*fam., person*) κάθαρμα *n.*, βρωμάνθρωπος *m.* **~ily** *adv.* κακώς.

nation *s.* έθνος *n.* leader of the ~ εθνάρχης *m.* **~-wide** *a.* πανεθνικός.

national *a.* εθνικός. ~ assembly εθνοσυνέλευση *f.* ~ debt δημόσιο χρέος. (*s.*) (*subject*) υπήκοος *m.f.* **~ly** *adv.* εθνικώς, καθ' όλην την χώρα.

nationalism *s.* εθνικισμός *m.*

nationalist *s.* εθνικιστής *m.* **~ic** *a.* εθνικόφρων.

nationality *s.* (*by race*) εθνικότητα *f.* (*by allegiance*) υπηκοότητα *f.* (*by birth*) ιθαγένεια *f.*

nationaliz|e *v.* εθνικοποιώ. **~ation** *s.* εθνικοποίηση *f.*

native *a. & s.* ιθαγενής. a ~ of London Λονδρέζος γέννημα και θρέμμα. speak like a ~ μιλώ σα να έχω γεννηθεί στη χώρα. ~ land πατρίδα *f.* ~ language μητρική γλώσσα. (*a.*) (*local*) εγχώριος, ντόπιος. (*natural*) έμφυτος. go ~ οικειοποιούμαι τον τρόπο ζωής των ντόπιων.

nativity *s.* γέννηση του Χριστού.

natty *a.* κομψός, (*hat*) σκερτζόζικος, (*device, gadget*) έξυπνος.

natural *a.* φυσικός, (*innate*) έμφυτος. (*with innate ability*) γεννημένος ~ child εξώγαμο παιδί. ~ resources φυσικοί πόροι. **~ly** adv. φυσικά, (by nature) εκ φύσεως. **~ness** s. φυσικότητα f.

naturalist s. φυσιοδίφης m.

naturalize v. πολιτογραφώ. **~ation** s. πολιτογράφησις f.

nature s. φύση f. (sort) είδος n. good ~ καλωσύνη f. in a state of ~ γυμνός, by ~ εκ φύσεως. something in the ~ of a vase κάτι σαν βάζο, ένα είδος βάζου.

naught s. μηδέν n., τίποτα n. set at ~ αψηφώ.

naughty a. (child) ζωηρός, άτακτος, σκανταλιάρικος. (improper) σόκιν. ~y boy! (joc.) κακό παιδί. be ~y κάνω αταξίες. **~iness** s. αταξία f.

nausea a. ναυτία f. (fig.) αηδία f. **~ate** v. it ~ates me μου φέρνει αναγούλα. **~ous** a. αηδιαστικός.

nautical a. ναυτικός.

naval a. ναυτικός. ~ battle ναυμαχία f. ~ dockyard ναύσταθμος m. ~ officer αξιωματικός του ναυτικού.

nave s. κεντρικόν κλίτος.

navel s. ομφαλός m. ~ orange ομφαλοφόρο πορτοκάλι.

navigable a. (river) πλωτός, (boat) πλόιμος.

navigate v.t. (ship) κυβερνώ, (seas) διαπλέω. (fig.) πιλοτάρω. **~ion** s. (science) ναυτική f. (shipping) ναυτιλία f. (voyaging) ναυσιπλοΐα f. **~or** s. (voyager) θαλασσοπόρος m. (officer) αξιωματικός πορείας.

navvy s. σκαφτιάς m.

navy s. ναυτικόν n. ~ blue μπλε μαρέν.

nay adv. όχι. (or rather) ή μάλλον.

near adv. & prep. κοντά, πλησίον. ~ me κοντά μου, ~ the house κοντά στο σπίτι. ~ here εδώ κοντά. come ~ to (in performance) φθάνω, πλησιάζω. Christmas is drawing ~ κοντεύουν τα Χριστούγεννα. we are getting ~ the station κοντεύουμε στο σταθμό. I am getting ~ the end of the book βρίσκομαι σχεδόν στο τέλος του βιβλίου, κοντεύω να τελειώσω το βιβλίο. we've nowhere ~ finished πολύ απέχουμε από το να τελειώνουμε. ~ by κοντά. (v.) πλησιάζω. see also above.

near a. κοντινός. ~ relative στενός συγγενής. N~ East εγγύς Ανατολή. ~ side (of vehicle) πλευρά προς το πεζοδρόμιο. the ~ future το εγγύς μέλλον. **-ly** adv. be ~ly related έχω στενή συγγένεια.

nearby a. κοντινός.

nearest a. κοντινότερος, πλησιέστερος. (most direct) συντομότερος.

nearly adv. (almost) σχεδόν. not ~ καθόλου. the building is ~ finished το κτίριο είναι

σχεδόν τελειωμένο. I am ~ fifty κοντεύω (or πλησιάζω) τα πενήντα. it's ~ three o'clock κοντεύουν τρεις. we are ~ there (or ready or finished) κοντεύουμε. it is ~ time to start κοντεύει η ώρα να (or που θα) ξεκινήσουμε. he ~ choked κόντεψε (or λίγο έλειψε or παρά λίγο) να πνιγεί, λίγο ακόμα και πνιγόταν, ολίγου (or μικρού) δείν επνίγετο. it is not ~ enough δεν φθάνει καθόλου.

neat a. (tidy) τακτικός, (well shaped) καλλίγραμμος, (felicitous) εύστοχος, επιτυχημένος. (in dress) καλοβαλμένος. (drink) σκέτος. **-ly** adv. τακτικά, εύστοχα. **~ness** s. (order) τάξη f. (of appearance) περιποιημένη εμφάνιση.

nebula s. νεφέλωμα n. **~ous** a. νεφελώδης, (fig.) ακαθόριστος.

necessarily adv. κατ' ανάγκην, απαραιτήτως.

necessary a. αναγκαίος, απαραίτητος. it is ~ χρειάζεται. if ~ εν ανάγκη.

necessity s. ανάγκη f. make a virtue of ~y κάνω την ανάγκη φιλοτιμία. **~ies** τα αναγκαία, τα απαραίτητα. **~ate** v. απαιτώ, επιβάλλω, χρειάζομαι. (entail) συνεπάγομαι. **~ous** a. άπορος.

neck s. λαιμός m. ~ and ~ ισόπαλοι. ~ or nothing όλα για όλα. break one's ~ σκοτώνομαι. stick one's ~ out εκτίθεμαι. he got it in the ~ την έπαθε. he was ejected ~ and crop τον έδιωξαν άναυλα. he was breathing down my ~ (following) με ακολουθούσε κατά πόδας, (watching) στεκόταν πάνω από το κεφάλι μου. **-lace** s. περιδέραιο n. κολιέ n. **~line** s. ντεκολτέ n. **~-tie** s. γραβάτα f., λαιμοδέτης m.

nectar s. νέκταρ n.

née a. (Mrs Jones) ~ Smith το γένος Σμιθ.

need s. ανάγκη f., χρεία f. be in ~ of χρειάζομαι, έχω ανάγκη από. there is no ~ of δεν χρειάζεται (with να or nom.), δεν υπάρχει ανάγκη (with να or από). in case of ~ σε περίπτωση ανάγκης, χρείας τυχούσης.

need v. χρειάζομαι (with να or acc.), έχω ανάγκη (with να or από). I ~ it το χρειάζομαι, μου χρειάζεται, το έχω ανάγκη. ~ I go? χρειάζεται (or πρέπει) να πάω; it ~s repairing χρειάζεται (or θέλει) διόρθωμα. it ~ not be done yet δεν χρειάζεται (or δεν είναι ανάγκη) να γίνει ακόμα. you ~ not have said so δεν χρειαζόταν (or δεν ήταν ανάγκη) να το πεις. I didn't ~ to go (so I didn't) δεν χρειάστηκε να πάω.

needful a. what is ~ ό,τι χρειάζεται, τα απαιτούμενα. (fam., money) το παραδάκι.

needle s. βελόνα f. look for a ~ in a haystack ζητώ ψύλλους στα άχυρα. (v.t.) (goad) κεντώ, τσιγκλώ. ~work s. εργόχειρο n. (sewing) ράψιμο n.

needless a. άχρηστος, περιττός. ~ to say περιττόν να λεχθεί. ~ly adv. χωρίς λόγο.

needy a. άπορος.

ne'er-do-well s. ανεπρόκοπος a.

nefarious a. κακοήθης.

negation s. άρνηση f.

negative v.t. (proposal) απορρίπτω, (theory) ανατρέπω.

negative a. αρνητικός. (s.) (photo) αρνητικό n. ~ly adv. (also in the ~) αρνητικά.

neglect v. παραμελώ. (s.) (leaving undone) παραμέληση f. state of ~ κατάσταση εγκατάλειψης. ~ful a. αμελής.

neglig|ent a. αμελής, απρόσεκτος. ~ence s. αμέλεια f., απροσεξία f. ~ible a. ασήμαντος, αμελητέος.

negotiable a. (transferable) μεταβιβάσιμος, (passable) βατός. (of terms) διαπραγματεύσιμος.

negotiat|e v.t. (also ~e about) διαπραγματεύομαι. (pass) περνώ. (fin.) μεταβιβάζω. ~ions s. διαπραγματεύσεις f.pl. ~or s. διαπραγματευτής m.

negress s. αραπίνα f., νέγρα f., μαύρη f.

negro s. αράπης m., νέγρος m., μαύρος m. (a.) (of things) των νέγρων, των μαύρων.

neigh v. χλιμιντρώ. (s.) χλιμίντρισμα n.

neighbour s. γείτονας m., γειτόνισσα f. my ~ (fellow man) ο πλησίον μου. (v.) ~ (upon) γειτονεύω, γειτνιάζω. ~hood s. γειτονιά f. ~ing s., γειτονικός.

neighbour|ly a. (act) φιλικός, (person) καλός γείτονας. ~liness s. φιλική διάθεση προς τους γείτονες.

neither adv. ούτε. ~ do I ούτε εγώ. ~... nor ούτε... ούτε. (pron.) ούτε ο ένας ούτε ο άλλος. ~ of the two κανένας απ' τους δύο.

nemesis s. νέμεση f.

neo-Hellenist s. νεοελληνιστής m.

neolithic a. νεολιθικός.

neologism s. νεολογισμός m.

neon s. ~ light λάμπα με νέον.

neophyte s. νεοφώτιστος a.

nephew s. ανεψιός m. ~s and nieces ανίψια n.pl.

nepotism s. νεπωτισμός m.

nerve s. νεύρο n. (courage) θάρρος n. (cheek) θράσος n. it gets on my ~s μου δίνει στα νεύρα, με εκνευρίζει. fit of ~s νευρική κρίση. in a state of ~s σε κατάσταση νευρικής ταραχής. he lost his ~ έχασε το θάρρος του. (v.) ~ oneself

nest s. φωλιά f. feather one's ~ κάνω την μπάζα μου. (v.i.) φωλιάζω, (build ~) χτίζω φωλιά. ~egg s. (fig.) κομπόδεμα n. ~ling s. νεοσσός m.

nestle v. φωλιάζω, μαζεύομαι. the house ~s in a hollow το σπίτι είναι φωλιασμένο σε μία κοιλάδα.

net a. (fin.) καθαρός.

net s. δίχτυ n. δίχτυο n. (tennis, hair) φιλές m. (tulle) τούλι n.

net v.t. (catch) πιάνω (στο δίχτυ). (cover) σκεπάζω με δίχτυωτό. (gain) κερδίζω.

nether a. ~ world ο κάτω κόσμος.

netting s. δίχτυωτό n. wire ~ συρματόπλεγμα n.

nettle s. τσουκνίδα f. (fig.) grasp the ~ αρπάζω τον ταύρο από τα κέρατα. (v.) ερεθίζω.

network s. δίχτυον n.

neuralgia s. νευραλγία f.

neurasthenia s. νευρασθένεια f.

neuritis s. νευρίτιδα f.

neurolog|y s. νευρολογία f. ~ist s. νευρολόγος m.

neur|osis s. νεύρωση f. ~otic a. νευρωτικός.

neuter a. ουδέτερος. (v.t.) ευνουχίζω.

neutral a. ουδέτερος. ~ gear νεκρό σημείο. ~ity s. ουδετερότητα f.

neutraliz|e v. εξουδετερώνω. ~ation s. εξουδετέρωση f.

never adv. ποτέ, ουδέποτε. ~ in my life ποτέ μου. it ~ occurred to me ούτε μου πέρασε από το νου. ~ mind! δεν πειράζει, έννοια σου! well I ~! για φαντάσου! (fam.) on the ~ ~ (hire purchase) με δόσεις. ~-ending a. ατελείωτος. ~-to-be-forgotten a. αλησμόνητο.

nevertheless adv. όμως, εν τούτοις, παρ' όλα αυτά.

new a. νέος, καινούργιος. (fresh) φρέσκος. N~ World νέος κόσμος. N~ Testament Καινή Διαθήκη. N~ Year νέον έτος, καινούργιος χρόνος. N~ Year's Day Πρωτοχρονιά f. ~ suit καινούργιο κοστούμι, ~ bread φρέσκο ψωμί. ~est fashion τελευταία λέξη της μόδας. ~-born a. νεογέννητος. ~comer s. νεοφερμένος, καινούργιος a. ~fangled a.

οπλίζομαι με θάρρος. ~-racking a. τρομακτικός.

nervous a. νευρικός, (timid) φοβητσιάρης. (sinewy, vigorous) νευρώδης. ~ disease νευροπάθεια f. ~ breakdown νευρικός κλονισμός. ~ly adv. νευρικά, (timidly) φοβισμένα. ~ness s. νευρικότητα f. (timidity) ατολμία f.

nervy a. νευρικός.

καινοφανής. **~-laid** *a.* φρέσκος, της ημέρας.

newly *adv.* άρτι, νεωστί, προσφάτως. ~ built νεόκτιστος, νεόδμητος. ~ painted φρεσκοβαμμένος. ~ married νεόνυμφος. (*fam.*) ~-weds νιόπαντροι *m.pl.* ~ published book άρτι εκδοθέν βιβλίον.

newness *a.* το καινούργιο. (*inexperience*) απειρία *f.*

news *s.* νέα *n.pl.* (*bulletin*) ειδήσεις *f.pl.* piece of ~ ένα νέο.

newsagent's *s.* πρακτορείο εφημερίδων.

newspaper *s.* εφημερίδα *f.* **~-seller** *s.* εφημεριδοπώλης *m.*

next *a.* άλλος, επόμενος. ~ day την άλλη μέρα, την επομένη. ~ Monday την επόμενη Δευτέρα, ~ Monday week την άλλη Δευτέρα. ~ door (*adv.*) δίπλα. **~-door** *a.* πλαϊνός. **~-of-kin** *s.* ο πιο κοντινός συγγενής.

next *adv. & prep.* έπειτα, μετά, ύστερα, κατόπιν, ακολούθως. what ~! τι άλλο ακόμα; what comes ~? τι ακολουθεί; τι έρχεται μετά; who's ~? ποιος έχει σειρά; when ~ (*or* ~ time) you come this way όταν θα ξαναπεράσεις από 'δω. (*prep.*) ~ to (*also* ~ door to) δίπλα σε, πλάι σε, (*in order of choice*) μετά από, ύστερα από. live on ~ to nothing ζω με το τίποτα. she eats ~ to nothing δεν τρώει σχεδόν τίποτα.

NHS ΙΚΑ.

nib *s.* πεννάκι *n.*

nibble *v.* ροκανίζω, (*at bait*) τσιμπώ.

nibs *s.* (*iron.*) his ~ η αφεντιά του.

nice *a.* καλός, όμορφος, ωραίος. (*person*) καλός, συμπαθητικός, ευγενικός. (*fine, subtle*) λεπτός, (*fastidious*) εκλεπτυσμένος, (*fussy*) δύσκολος. we had a ~ time ευχαριστηθήκαμε, περάσαμε ωραία. **~ly** *adv.* όμορφα, ωραία.

nicet|y *s.* ακρίβεια *f.* to a ~y στην εντέλεια. **~ies** λεπτομέρειες *f.pl.*

niche *s.* κόγχη *f.* (*fam., position*) βολική θέση.

nick *s.* (*cut*) μικρό κόψιμο. in the ~ of time πάνω στην ώρα. (*fam., prison*) in the ~ στο φρέσκο. (*v.*) (*cut*) κόβω. (*fam., steal, arrest*) τσιμπώ.

nickel *s.* νικέλιο *n.*

nickname *n.* παρατσούκλι *n.*

nicotine *s.* νικοτίνη *f.*

niece *s.* ανεψιά *f.*

nifty *a.* (*fam.*) (*smart*) κομψός, (*quick*) σβέλτος.

niggardly *a.* (*person*) τσιγγούνης, (*amount*) μίζερος.

nigger *s.* αράπης *m.* (*fig.*) ~ in the woodpile μελανό σημείο.

niggling *a.* (*fam.*) ασήμαντος, (*job*) πίζουλος.

night *s.* νύχτα *f.* ~ and day μέρα νύχτα. spend the ~, stay up *or* open all ~ διανυκτερεύω. be overtaken by ~ νυχτώνομαι. ~ falls νυχτώνει. by ~ νύχτα. (*a.*) νυχτερινός. **~-bird** *s.* νυχτοπούλι *n.* **~-dress, ~-gown** *s.* νυχτικό *n.* **~-fall** *s.* σούρουπο *n.* **~ly** *adv.* κάθε βράδυ, κάθε νύχτα. (*a.*) νυχτερινός. **~-mare** *s.* εφιάλτης *m.*, βραχνάς *m.* **~-school** *s.* φροντιστήριο *n.* **~-work** *s.* νυχτέρι *n.*

nightingale *s.* αηδόνι *n.*

nihil|ism *s.* μηδενισμός *m.* **~ist** *s.* μηδενιστής *m.*

nil *s.* μηδέν *n.*

nimb|le *a.* σβέλτος. (*mentally*) εύστροφος. **~ly** *adv.* σβέλτα.

nincompoop *s.* κουτορνίθι *n.*

nine *num.* εννέα, εννιά. dressed up to the ~s ντυμένος στην τρίχα. ~ hundred εννιακόσιοι *a.* **~-pins** *s.* τσούνια *n.pl.*

nineteen *num.* δεκαεννέα. **~th** *a.* δέκατος ένατος.

ninet|y *num.* ενενήντα. **~ieth** *a.* ενενηκοστός.

ninny *s.* κουτορνίθι *n.*

ninth *a.* ένατος.

nip *v.* τσιμπώ. ~ in the bud καταπνίγω εν τη γενέσει. I'll ~ round to the baker's θα πεταχτώ στο φούρνο. (*s.*) τσίμπημα *n.*

nipper *s.* (*claw*) δαγκάνα *f.* (*fam., child*) πιτσιρίκος *m.* ~s (*tool*) τσιμπίδα *f.*

nipple *s.* ρώγα *f.*, θηλή *f.*

nitr|e *s.* νίτρο *n.* **~ic** *a.* νιτρικός.

nitrogen *s.* άζωτο *n.*

no (*negative response*) όχι. (*a.*) κανένας. ~ buses come this way από 'δω δεν περνάει κανένα λεωφορείο. there's ~ water δεν έχει νερό. there's ~ knowing δεν ξέρει κανείς. there's ~ going back now τώρα πια δεν χωρεί υποχώρηση. he's ~ better (*in health*) δεν είναι καλύτερα. ~ admittance! απαγορεύεται η είσοδος. (*I'm going out*) rain or ~ rain βρέχει ξεβρέχει, είτε βρέχει είτε όχι. ~ mans land ουδετέρα ζώνη.

nob *s.* (*fam.*) αριστοκράτης *m.* (*head*) κεφάλα *f.*

nobble *v.t.* (*fam.*) (*acquire*) βουτώ. (*person*) ψήνω. (horse) κάνω να χάση.

nobility *s.* ευγένεια *f.* (*of character*) υψηλοφροσύνη *f.* (*splendour*) μεγαλοπρέπεια *f.* the ~ οι ευγενείς.

nob|le *a.* (*rank*) ευγενής, (*character*) υψηλόφρων, (*deed*) γενναίος. (*splendid*) μεγαλοπρεπής. **~leman** *s.* ευγενής *m.* **~ly** *adv.* υψηλοφρόνως, γενναίως, μεγαλοπρεπώς.

nobody *pron.* κανείς, κανένας. (*s.*) μηδενικό *n.*, τιποτένιος άνθρωπος.

nocturn|e *s.* νυκτερινό *n.* **~al** *a.* νυκτερινός. (*of ~al habits*) νυκτόβιος.

nod *v.i.* γνέφω, νεύω, (*drowsily*) κουτουλώ. (*in assent*) κάνω ναι με το κεφάλι. have a ~ding acquaintance with (*a subject*) ξέρω λίγο από, έχω μία ιδέα από. (*s.*) γνέψιμο *n.*, νόημα *n.*

noise *s.* (*particular or general*) θόρυβος *m.* (*particular & loud*) κρότος *m.* (*sound*) ήχος *m.* make a ~ θορυβώ, (*fig.*) κάνω κρότο. (*v.t.*) ~ abroad διαδίδω.

noiseless *a.* αθόρυβος. **~ly** *adv.* αθόρυβα.

noisome *a.* αηδής, (*noxious*) βλαβερός, (*ill-smelling*) βρώμιος.

nois|y *a.* θορυβώδης. **~ily** *adv.* θορυβωδώς.

nomad *s.* νομάς *m.* **~ic** *a.* νομαδικός.

nomenclature *s.* ονοματολογία *f.*

nominal *a.* κατ' όνομα, ονομαστικός. **~ly** *adv.* κατ' όνομα, ονομαστικώς.

nominat|e *v.* (*appoint*) διορίζω, (*propose*) υποδεικνύω. **~ion** *s.* διορισμός *m.* υπόδειξη *f.* (*as candidate*) χρίσμα *n.*

nominative *s.* (*gram.*) ονομαστική *f.*

nominee *s.* ο υποδειχθείς υποψήφιος.

non- *prefix* μη. **~-appearance** η μη εμφάνιση. **~-smokers** οι μη καπνίζοντες. *also by* α(ν)-, **~-existent** ανύπαρκτος.

non-aggression *s.* ~ pact συμφωνία μη επιθέσεως.

nonce *s.* for the ~ για την ώρα.

nonchal|ant *a.* ψύχραιμος. **~ance** *s.* ψυχραιμία *f.*

non-combatant *a.* άμαχος.

non-commissioned *a.* ~ officer υπαξιωματικός.

non-committal *a.* επιφυλακτικός.

nonconform|ist *a. & s.* ανορθόδοξος, ανεξάρτητος. **~ity** *s.* ανεξαρτησία *f.*

nondescript *a.* ακαθόριστος.

none *pron. & adv.* (*no one*) κανείς, κανένας, (*not any*) καθόλου. ~ of them κανένας απ' αυτούς, ~ of us κανένας μας. (*he has money but*) I have ~ εγώ δεν έχω καθόλου. ~ other than ο ίδιος. ~ but the best μόνο το καλύτερο. ~ of your tricks! άσε τα αστεία! **~-the-less** παρ' όλα αυτά.

nonentity *s.* μηδενικό *n.*

non-inflammable *a.* άφλεκτος.

non-intervention *s.* η μη επέμβαση.

nonplus *v.t.* (*also* be ~sed) σαστίζω.

nonsense *s.* ανοησία *f.*, μπούρδες *f.pl.* κολοκύθια *n.pl.* stand no ~ δεν χαρίζω κάστανα.

non sequitur *s.* παράλογο επιχείρημα.

non-starter *s.* χωρίς πιθανότητα επιτυχίας.

non-stop *adv.* συνεχώς, χωρίς διακοπή. (*a.*) (*vehicle*) χωρίς σταθμούς.

nook *s.* γωνιά *f.*

noon *s.* μεσημέρι *n.* at high ~ ντάλα μεσημέρι. **~day** *a.* μεσημβρινός.

no-one *pron.* κανείς, κανένας.

noose *s.* βρόχος *m.*, θηλιά *f.*

nor *conj.* ούτε.

norm *s.* το κανονικό.

normal *a.* κανονικός. (*natural*) φυσιολογικός. ~ school διδασκαλείον *n.* **~ity** *s.* το κανονικό. **~ly** *adv.* κανονικά.

north *s.* βορράς, βοριάς *m.* to the ~ of Athens προς βορράν (*or* βορείως) των Αθηνών. ~ wind βοριάς *m.*

north, ~ern, ~erly *a.* βόρειος, βορεινός.

north-east *a.* βορειοανατολικός. ~ wind γραίγος *m.*

northward(s) *adv.* βορείως.

north-west *a.* βορειοδυτικός. ~ wind μαΐστρος *m.*

Norwegian *a.* νορβηγικός. (*person*) Νορβηγός.

nose *v.t.* (*smell*) μυρίζομαι. ~ out ξετρυπώνω. (*v.i.*) ~ into χώνω τη μύτη μου σε. ~ forward προχωρώ σιγά σιγά.

nose *s.* μύτη *f.* show one's ~ ξεμυτίζω. lead by the ~ σέρνω από τη μύτη. I paid through the ~ for it μου κόστισε ο κούκκος αηδόνι. under his very ~ μεσ' στη μύτη του. turn up one's ~ at περιφρονώ. look down one's ~ at κοιτάζω περιφρονητικά. **~-bag** *s.* ντορβάς *m.* **~-dive** *s.* (*fig.*) βουτιά *f.* **~-gay** *s.* μπουκέτο *n.* **~y** *a.* πολυπράγμων.

nostalg|ia *s.* νοσταλγία *f.* **~ic** *a* νοσταλγικός. I feel ~ic (for home) νοσταλγώ (το σπίτι μου).

nostril *s.* ρουθούνι *n.*

nostrum *s.* γιατροσόφι *n.*, πανάκεια *f.*

not *adv.* δεν, μη(ν), όχι. 1. (*negativing verbs*) he isn't here δεν είναι εδώ. if he doesn't come αν δεν έρθει. (*tag questions*) don't they? won't it? couldn't you? *etc.* ε; δεν είν' έτσι; (*after* ας, να) let's ~ waste time ας μη χασομερούμε. I told him ~ to leave του είπα να μη φύγει. ~ to mention (για) να μην αναφέρω. (*with participle*) ~ knowing what had occurred μη ξέροντας τα διατρέξαντα. those ~ having tickets οι μη έχοντες εισιτήρια. 2. (*not negativing verbs*) ~ here όχι εδώ. if ~ αν όχι. whether... or ~ είτε... είτε όχι. will you come or ~? θα έρθεις ναι ή όχι; it may or it may ~ μπορεί ναι μπορεί όχι. I hope ~ ελπίζω όχι. ~ at all καθόλου. ~ even ούτε. ~ even John would do that κι ο Γιάννης ακόμα δεν θα τό 'κανε αυτό. 3. (*commands*) don't! μη! don't laugh μη γελάς. ~ on the grass! μη στο γρασίδι! ~ a word!

μιλιά!

notab|le *a.* αξιόλογος. (*s.*) (*important person*) διασημότητα *f.*, σημαίνον πρόσωπον. ~**iy** *adv.* ιδιαιτέρως.

notary *s.* συμβολαιογράφος *m.*

notation *s.* σημειογραφία *f.*

notch *s.* εγκοπή *f.* (*v.*) (*fam.*) ~ up σημειώνω.

note *s.* (*letter*) σημείωμα *n.* (*written record*) σημείωση *f.* (*diplomatic*) νότα *f.* of ~ σπουδαίος. make a ~ of σημειώνω, κρατώ σημείωση (*with gen. or* για). (*mus.*) νότα *f.*, φθόγγος *m.* (*tone of voice*) τόνος *m.* jarring ~ παραφωνία *f.* (*heed*) take ~ (of) προσέχω. compare ~s ανταλλάσσω από- ψεις. (*money*) χαρτονόμισμα *n.* ~**book** *s.* σημειωματάριο *n.* ~**case** *s.* πορτοφόλι *n.*

note *v.* σημειώνω, λαμβάνω υπ' όψιν. it is to be ~d that... σημειωτέον ότι. ~**d** *a.* ονομαστός. ~**worthy** *a.* αξιοσημείωτος.

nothing *s.* τίποτα, (*zero*) μηδέν *n.* next to ~ σχεδόν τίποτα. ~ but! κι άλλο τίποτα! for ~ (*free*) τζάμπα, (*in vain*) άδικα. our efforts went for ~ (*or* came to ~) οι κόποι μας πήγαν χαμένοι. ~ doing! δεν γίνεται. that has ~ to do with it αυτό δεν έχει καμία σχέση. have ~ to do with him! μην έχεις νταραβέρια μαζί του. there's ~ like home- made bread τίποτα δεν φτάνει το σπιτήσιο ψωμί. he's ~ like his brother δεν μοιάζει καθόλου με τον αδελφό του. he is ~like as clever as she is δεν τη φτάνει καθόλου στην εξυπνάδα.

notice *s.* (*advice*) ειδοποίηση *f.* (*attention*) προσοχή *f.* (*announcement*) αγγελία *f.* take no ~ of it! μην του δίνεις προσοχή (*or* σημασία). I gave them ~ of my arrival τους ειδοποίησα για την αφιξή μου. I gave him ~ (*of dismissal*) του κοινοποίησα από- λυση, τον ειδοποίησα ότι απολύεται. at short ~ σε σύντομο χρονικό διάστημα. (*review*) κριτική *f.* ~**board** πίνακας ανακοινώσεων.

notice *v.* προσέχω, παρατηρώ, αντιλαμ- βάνομαι.

noticeab|le *a.* (*perceptible*) αισθητός, (*evident*) καταφανής. be ~le (*stand out*) ξεχωρίζω. ~**ly** *adv.* αισθητώς.

notif|y *v.t.* (*person*) ειδοποιώ. (*declare*) δηλώνω. ~**iable** *a.* δηλωτέος. ~**ication** *s.* ειδοποίηση *f.* δήλωση *f.*

notion *s.* ιδέα *f.* ~**al** *a.* υποθετικός.

notori|ety *s.* κακή φήμη. ~**ous** *a.* διαβόητος.

notwithstanding *conj.* παρ' όλο που, μ' όλο που, μολονότι. (*adv.*) μολαταύτα. (*prep.*) παρά (*with acc.*).

nougat *s.* μαντολάτο *n.*

nought *s.* (*figure O*) μηδέν *n.* (*nothing*) τίποτα *n.* set at ~ αψηφώ.

noun *s.* όνομα *n.*, ουσιαστικό *n.*

nourish *v.* τρέφω. ~**ing** *a.* θρεπτικός. ~**ment** *s.* τροφή *f.*

nous *s.* (*fam.*) κοινός νους.

nouveau riche *a.* (*person*) νεόπλουτος, (*thing*) νεοπλουτίστικος. ~ ways νεοπλου- τισμός *m.*

novel *a.* πρωτότυπος, καινοφανής. (*s.*) μυθιστόρημα *n.* ~**ist** *s.* μυθιστοριογράφος *m.f.*

novelty *s.* (*strangeness*) το καινοφανές. (*innovation*) καινοτομία *f.* (*new invention*) νέα εφεύρεση. (*novel or fancy item*) νεωτερισμός *m.* this is quite a ~ αυτό είναι κάτι καινούργιο.

November *s.* Νοέμβριος, Νοέμβρης *m.*

novice *s.* πρωτάρης *m.* (*eccl.*) δόκιμος μοναχός.

now *adv.* τώρα. up to ~ έως τώρα. from ~ on από τώρα κι εμπρός. ~ and then πού και πού. just ~ μόλις τώρα. ~ then! λοιπόν. ~ this way ~ that μια από 'δω, μια από 'κει. (*as conj.*) ~ (*that*) he's gone τώρα που έφυγε.

nowadays *adv.* σήμερα, τη σημερινή εποχή.

nowhere *adv.* πουθενά. ~ near as good κατά πολύ υποδεέστερος. *see also* nothing like.

noxious *a.* βλαβερός.

nozzle *s.* στόμιο *n.*

nuance *s.* απόχρωση *f.*

nubile *a.* της παντρειάς.

nucle|us *s.* πυρήνας *m.* ~**ar** *a.* πυρηνικός. ~ar waste πυρηνικά κατάλοιπα.

nud|e *a.* γυμνός. ~**ist** *s.* γυμνιστής *m.* ~**ity** *s.* γύμνια *f.*

nugget *s.* ~ of gold ακατέργαστος βώλος χρυσού.

nuisance *s.* μπελάς *m.* he is a ~ είναι ενοχλητικός.

null *a.* (*void*) άκυρος. ~**ify** *v.* ακυρώνω. ~**ity** *s.* ακυρότητα *f.*

numb *a.* μουδιασμένος. get *or* be ~ μουδιάζω. my fingers are ~ μούδιασαν τα δάχτυλά μου. (*v.t.*) μουδιάζω. ~**ness** *s.* μούδιασμα *n.*

number *s.* αριθμός *m.* a ~ of μερικοί, αρκετοί. any ~ of πολλοί. one of their ~ ένας απ' αυτούς. twenty in ~ είκοσι τον αριθμόν. without ~ αμέτρητος. (*item, act*) νούμερο *n.* (*of journal*) τεύχος *n.* (*fam.*) his ~ is up πάει αυτός. (*fam.*) look after ~ one απάνω απ' όλα ο εαυτούλης μου. ~**less** *a.* αμέτρητος, αναρίθμητος. ~**plate** *s.* πινακίδα *f.*

number *v.t.* αριθμώ. (*include*) συγκα- ταλέγω. (*v.i.*) (*amount to*) ανέρχομαι σε.

numeral *s.* αριθμός *m.*

numerator *s.* αριθμητής *m.*

numerical *a.* αριθμητικός. **~ly** *adv.* αριθμητικώς.
numerous *a.* πολυάριθμος.
numismatics *s.* νομισματική *f.*
nun *s.* καλογραία, καλόγρια *f.* **~nery** *s.* μονή καλογραιών.
nuptial *a.* γαμήλιος.
nurse *s.* (*for the sick*) νοσοκόμος *m.f.*, νοσοκόμα *f.* (*child's*) νταντά *f.* (*wet-~*) παραμάννα *f.*, (*also fig.*) τροφός *f.*
nurse *v.t.* (*patients*) νοσηλεύω, (*suckle*) θηλάζω, (*clasp*) κρατώ στην αγκαλιά μου. (*fig.*) τρέφω. **~ling** *s.* βυζανιάρικο *n.*
nursery *s.* δωμάτιο των παιδιών. (*of plants & fig.*) φυτώριο *n.* **~** school νηπιαγωγείο *n.* **~man** *s.* ανθοκόμος *m.*
nurture *s.* ανατροφή *f.* (*v.*) ανατρέφω. (*fig.*) τρέφω. he was ~d in the School of Paris γαλουχήθηκε από τη Σχολή του Παρισιού.
nut *s.* (*edible*) ξηρός καρπός. (*see* walnut, *etc.*). (*implement*) παξιμάδι *n.* (*fam.*) a hard ~ δύσκολο πρόβλημα. (*fam., head*) κεφάλι *n.* **~s** (*mad*) τρελλός.
nutcrackers *s.* καρυοθραύστης *m.*
nutmeg *s.* μοσχοκάρυδο *n.*
nutriment *s.* τροφή *f.*
nutrition *s.* θρέψη *f.* **~al** *a.* διαιτητικός.
nutri|ious, ~ive *a.* θρεπτικός.
nutshell *s.* καρυδότσουφλο *n.* (*fig.*) in a ~ με λίγα λόγια, εν συντομία.
nutty *a.* από καρύδια, σαν καρύδι.
nuzzle *v.t.* χαϊδεύομαι επάνω σε.
nylon *s.* νάιλον *n.*
nymph *s.* νύμφη *f.*
nymphomaniac *a.* & *s.* νυμφομανής.

O

O *int.* (*with vocative*) ω.
oaf *s.* γουρούνι *n.*
oak *s.* δρυς *f.*, βαλανιδιά *f.* (*a.*) (*also ~en*) δρύινος.
oakum *s.* στουπί *n.*
OAP συνταξιούχος γήρατος.
oar *s.* κουπί *n.* (*fig.*) put one's ~ in ανακατεύομαι. **~sman** *s.* κωπηλάτης *m.*
oasis *s.* όαση *f.*
oast-house *s.* κλίβανος λυκίσκου.
oaten *a.* από βρώμη.
oath *s.* όρκος *m.* put on ~ ορκίζω, take an ~ ορκίζομαι. break one's ~ επιορκώ. on ~ (*sworn*) ένορκος, (*adv.*) ενόρκως. (*swear-word*) βλαστήμια *f.* string of ~s βρισίδι *n.*
oatmeal *s.* αλεύρι από βρώμη.
oats *s.* βρώμη *f.* sow one's wild ~ γλεντώ τα νιάτα μου.

obdurate *a.* πεισματάρης.
obedi|ent *a.* υπάκουος, ευπειθής. **~ence** *s.* υπακοή *f.*, ευπείθεια *f.* **~ently** *adv.* ευπειθώς.
obeisance *s.* (*bow*) υπόκλιση *f.* do ~ to προσκυνώ.
obelisk *s.* οβελίσκος *m.*
obes|e *a.* παχύσαρκος. **~ity** *s.* παχυσαρκία *f.*
obey *v.t.* υπακούω, (*the law*) σέβομαι. (*comply with*) συμμορφώνομαι προς.
obituary *s.* νεκρολογία *f.*
object *s.* (*of pity, comment, etc. & gram.*) αντικείμενο *n.* (*thing*) αντικείμενο, πράγμα *n.* ~ lesson παράδειγμα *n.* (*purpose*) σκοπός *m.* (*pej.*) what an ~ he looks! είναι γελοιογραφία. distance no ~ δεν ενδιαφέρει η απόσταση.
object *v.t.* (*be opposed*) αντιτίθεμαι, έχω αντίθετη γνώμη, φέρω αντίρρηση. ~ to δεν εγκρίνω, είμαι εναντίον (*with gen.*). ~ that (*sthg. is the case*) παρατηρώ ότι.
objection *s.* αντίρρηση *f.* raise an ~ φέρω αντίρρηση. he took ~ to my tie δεν του άρεσε η γραβάτα μου. (*difficulty, snag*) εμπόδιο *n.*
objectionab|le *a.* (*unacceptable*) απαράδεκτος, (*nasty*) δυσάρεστος. (*person*) αντιπαθητικός. **~ly** *adv.* απρεπώς, άσχημα.
objectiv|e *a.* αντικειμενικός. (*s.*) αντικειμενικός σκοπός. **~ely** *adv.* αντικειμενικά. **~ity** *s.* αντικειμενικότητα *f.*
objector *s.* αντιρρησίας *m.*, ο αντιτιθέμενος, ο ενάντιος.
obligat|e *v.* υποχρεώνω. **~ory** *a.* υποχρεωτικός.
obligation *s.* υποχρέωση *f.* be under an ~ είμαι υποχρεωμένος.
oblige *v.t.* (*require, do favour to*) υποχρεώνω. be ~d είμαι υποχρεωμένος. could you ~ me by...? θα είχατε την καλωσύνη (*or* θα μου κάνατε τη χάρη) να... much ~d! καλωσύνη σας. ~ (*person*) greatly σκλαβώνω.
obliging *a.* υποχρεωτικός, πρόθυμος. **~ly** *adv.* πρόθυμα.
oblique *a.* λοξός, πλάγιος. **~ly** *adv.* λοξά, πλαγίως. **~ness** *s.* λοξότητα *f.*
obliquity *s.* (*moral*) υπουλότητα *f.*
obliterat|e *v.* εξαλείφω, σβήνω. **~ion** *s.* εξάλειξη *f.*, σβήσιμο *n.*
oblivi|on *s.* λήθη *f.*, λησμονιά *f.* **~ous** *a.* ~ous of one's surroundings μη έχων συναίσθηση του περιβάλλοντος.
oblong *a.* επιμήκης.
obloquy *s.* δυσφήμηση *f.*
obnoxious *a.* απεχθής, δυσάρεστος.
oboe *s.* όμποε *n.*
obscen|e *a.* αισχρός. **~ity** *s.* αισχρότητα *f.*

obscurant|ism s. σκοταδισμός m. ~ist s. σκοταδιστής m.

obscur|e a. σκοτεινός. (not well known) άσημος, άγνωστος. (v.t.) σκιάζω, σκεπάζω. ~ity s. σκοτεινότητα f., σκοτάδι n. ασημότητα f., αφάνεια f.

obsequies s. κηδεία f.

obsequious a. κόλακας m.

observable a. αισθητός.

observance s. (keeping up) τήρηση f. ~s (rites) τύποι m.pl.

observant a. παρατηρητικός.

observation s. (watching) παρατήρηση f. (being observant) παρατηρητικότητα f. (remark) παρατήρηση f., σχόλιο n. keep under ~ παρακολουθώ.

observatory s. αστεροσκοπείον n.

observe v. (keep) τηρώ. (notice, remark) παρατηρώ. (keep watch on) παρακολουθώ. (see) βλέπω, προσέχω. ~r s. παρατηρητής m.

obsess v. κατέχω. ~ion s. ιδεοληψία f. ~ive a. έμμονος.

obsolescent a. που τείνει να εκλείψει.

obsolete a. πεπαλαιωμένος.

obstacle s. εμπόδιο n.

obstetr|ician s. μαιευτήρ m. ~ics s. μαιευτική f.

obstinacy s. πείσμα n.

obstinate a. πείσμων, πεισματάρης. ~ly adv. επιμόνως, με πείσμα.

obstreperous a. ζόρικος, βάναυσος.

obstruct v. εμποδίζω. (road, way) κλείνω, (pipe) βουλλώνω.

obstruction s. (act) παρεμπόδιση f., παρακώλυση f. (thing) εμπόδιο n. (of pipe) βούλλωμα n.

obstructive a. be ~ ακολουθώ τακτική παρακωλύσεως.

obtain v.t. (acquire) αποκτώ, (procure) βρίσκω, προμηθεύομαι. (v.i.) (be in observance) κρατώ. ~able a. be ~able βρίσκομαι.

obtrude v.t. επιβάλλω. (v.i.) γίνομαι φορτικός.

obtrusive a. ενοχλητικός, φορτικός.

obtuse a. αμβλύς. ~ness s. αμβλύτητα f.

obviate v.t. αποτρέπω, παρακάμπτω, προλαμβάνω.

obvious a. προφανής, φανερός. ~ly adv. προφανώς.

occasion s. (cause) αφορμή f., λόγος m. (juncture, opportunity) ευκαιρία f. (circumstance) περίπτωση, περίσταση f. for every ~ για κάθε περίπτωση. as ~ demands αναλόγως της περιστάσεως. if ~ arises αν παραστεί ανάγκη. equal to the ~ αντάξιος των περιστάσεων. on the ~ of επί

τη ευκαιρία (with gen.). (v.t.) προξενώ, προκαλώ.

occasional a. (sporadic) σποραδικός. (not frequent) αραιός, κατά διαστήματα. he pays us ~ visits μας έρχεται πού και πού. ~ly adv. see above.

occident s. δύση f. ~al a. δυτικός.

occult a. απόκρυφος.

occulting a. (light) εκλείπων.

occupant s. (of premises) ένοικος m. (of seat) κάτοχος m.

occupation s. (holding, possession) κατοχή f. (what one does) απασχόληση f. (calling) επάγγελμα n. ~al a. επαγγελματικός.

occupier s. see occupant.

occup|y v. (position, country) κατέχω, (seize) καταλαμβάνω, (inhabit) κατοικώ, (person's attention) απασχολώ, (take up, fill) παίρνω. it ~ies a lot of space παίρνει πολύ τόπο.

occur v. (happen) συμβαίνω, γίνομαι. (be found) βρίσκομαι, απαντώμαι. it ~red to me μου πέρασε η ιδέα, μου πέρασε από το μυαλό.

occurrence s. (event) γεγονός n., συμβάν n., περιστατικό n. it is of rare ~ συμβαίνει σπάνια.

ocean s. ωκεανός m. ~-going, beyond the ~ υπερωκεάνιος. ~ic a. ωκεάνιος.

ochre s. ώχρα f.

o'clock adv. at 3 o'clock στις τρεις η ώρα.

octagon s. οκτάγωνο n.

octave s. (mus.) οκτάβα f.

October s. Οκτώβριος, Οχτώβρης m.

octogenarian s. ογδοντάρης m.

octopus s. οκτάπους m. χταπόδι n.

ocular a. I had ~ proof of it το διαπίστωσα με τα μάτια μου.

oculist s. οφθαλμίατρος m.

odd a. (number) μονός, περιττός. twenty ~ είκοσι και κάτι. ~ man out ο αποκλεισθείς. (occasional) ~ jobs μικροδουλειές f.pl. at ~ moments κάπου κάπου. (not matching) παράταιρος. (strange) παράξενος, περίεργος.

oddity s. (strangeness) παράξενο n. (funny habit) παραξενιά f. (funny thing) κάτι το παράξενο.

oddly adv. παράξενα. ~ enough το παράξενο είναι ότι...

oddment s. απομεινάρι n. (remnant of cloth) ρετάλι n.

odds s. (likelihood) πιθανότητες f.pl. (betting) στοιχήματα n.pl. the ~ are in their favour οι πιθανότητες είναι υπέρ αυτών. what are the ~ on the favourite? τι στοιχήματα έχει το φαβορί; fight against heavy ~ μάχομαι εναντίον ισχυροτέρων δυνάμεων. it makes no ~ δεν έχει καμιά σημασία. they are at ~ είναι τσακωμένοι. ~ and ends διάφορα μικροπράματα.

ode s. ωδή f.

odi|ous a. απεχθής, αντιπαθέστατος. **~um** s. αντιπάθεια f.

odorous a. εύοσμος.

odour s. μυρωδιά f., οσμή f. he is in bad ~ with the government η κυβέρνηση δεν τον βλέπει με καλό μάτι. **~less** a. άοσμος.

odyssey s. οδύσσεια f.

of prep. the city ~ Athens η πόλις των Αθηνών. on the 1st ~ July την πρώτη Ιουλίου. the corner ~ the street η γωνιά του δρόμου. a boy ~ ten ένα αγόρι δέκα χρόνων. what sort ~ person is he? τι είδος άνθρωπος (or ανθρώπου) είναι; a box ~ matches ένα κουτί σπίρτα. the room was full ~ people το δωμάτιο ήταν γεμάτο κόσμο. I have need ~ you σε έχω ανάγκη, (with you stressed) έχω ανάγκη από σένα. it is made ~ wood είναι (καμωμένο) από ξύλο. what did he die ~? από τι πέθανε; a lot ~ money πολλά λεφτά. a heart ~ gold χρυσή καρδιά. a friend ~ mine ένας φίλος μου. the south ~ England η νότιος Αγγλία. to the west ~ London δυτικώς του Λονδίνου. what do you think ~ Peter? τι ιδέα έχεις για τον Πέτρο; a jug ~ hot water μία κανάτα με ζεστό νερό. how kind ~ you! πολύ ευγενικό εκ μέρους σας. what ~ it? και τι μ' αυτό; free ~ charge δωρεάν.

off adv. ~ with you, be ~! φύγε, φεύγα! I must be ~ πρέπει να φύγω. I had a day ~ είχα μια ελεύθερη μέρα (or μία μέρα άδεια). the milk is ~ (or has gone ~) το γάλα ξίνισε. the switch is ~ ο διακόπτης είναι κλειστός. the light is ~ το φως είναι σβηστό. the marriage is ~ ο γάμος διαλύθηκε. the brake is ~ το φρένο δεν είναι βαλμένο. he had his coat ~ είχε βγαλμένο το σακκάκι. be well ~ βαστιέμαι καλά. be badly ~ έχω φτώχειες. he is badly ~ for company του λείπει η συντροφιά. how are you ~ for money? πώς τα πας από λεφτά; that remark was a bit ~ η κουβέντα δεν ήταν πολύ εντάξει. 5 miles ~ σε απόσταση πέντε μιλίων. ~ and on κάθε τόσο.

off prep. he took it ~ the table το πήρε από το τραπέζι. two miles ~ Cape Sounion, δύο μίλια έξω από το Σούνιο. just ~ the main road δύο βήματα από τον κεντρικό δρόμο. I'm ~ smoking έκοψα το τσιγάρο. I'm ~ my food δεν απολαμβάνω το φαγητό μου. be ~ colour δεν είμαι στα καλά μου.

off a. on the ~ chance στην τύχη. the ~ side (of vehicle) η αντίθετη προς το πεζοδρόμιο πλευρά. an ~ day (when things go badly) μία ανάποδη μέρα.

offal s. (rubbish) απορρίμματα n.pl. (meat) εντόσθια n.pl.

off-beat a. (fam.) ανορθόδοξος, εξεζητημένος.

off-duty a. see duty.

offence s. (infringement) παράβαση f. (wrongful deed) αδίκημα n. (hurting) προσβολή f. give ~ το προσβάλλω. take ~ προσβάλλομαι, θίγομαι. no ~ taken μηδέν παρεξήγηση. (attack) επίθεση f.

offend v. (hurt) προσβάλλω, θίγω. ~ against παραβαίνω. be ~ed προσβάλλομαι, θίγομαι.

offender s. (guilty person) υπαίτιος a., φταίχτης m. (contravener) παραβάτης m.

offensive a. (offending) προσβλητικός, (aggressive) επιθετικός, (disagreeable) δυσάρεστος. (s.) επίθεση f. take the ~ επιτίθεμαι.

offer v.t. προσφέρω, (one's hand) τείνω, (resistance) προβάλλω. (v.i.) προσφέρομαι, (occur) παρουσιάζομαι, (propose) προτείνω.

offer s. προσφορά f. (proposal) πρόταση f. **~ing** s. αφιέρωμα f. (contribution) έρανος m.

off-hand a. (abrupt) απότομος. (adv.) προχείρως.

office s. (post) θέση f., (high) αξίωμα n. take ~ αναλαμβάνω υπηρεσία. be in ~ (of government) είμαι στην αρχή. (duties, functions) καθήκοντα n.pl. (room) γραφείο n. (premises) γραφεία pl. ~ block τετράγωνο γραφείων. (Ministry) υπουργείον n. domestic ~s βοηθητικοί χώροι. good ~s εξυπηρέτηση f.

officer s. (armed forces) αξιωματικός m. (employee) υπάλληλος m. (of society) μέλος της επιτροπής.

official s. υπάλληλος m., λειτουργός m. (high) αξιωματούχος m.

official a. επίσημος. (pertaining to ~ duties) υπηρεσιακός. **~dom** s. (fam.) γραφειοκρατία f. **~ly** adv. επισήμως.

officiate v. ~ as chairman εκτελώ τα χρέη του προέδρου. (eccl.) ιερουργώ, χοροστατώ.

officious a. (too eager) ενοχλητικά πρόθυμος. (obstructive) επιμένων σε γραφειοκρατικές λεπτομέρειες. **~ness** s. υπέρμετρος ζήλος.

offing s. (fig.) in the ~ επικείμενος.

off-licence s. άδεια πωλήσεως οινοπνευματωδών ποτών προς κατανάλωσιν εκτός του καταστήματος.

off-print s. ανάτυπο n.

off-putting a. απωθητικός.

offset v. αντισταθμίζω.

offshoot s. παρακλάδι n.

off-shore a. (island, etc.) κοντά στη στεριά. ~ wind απόγειο n., στεριανός αέρας.

offspring s. βλαστός m., γόνος m. (fam., children) παιδιά n.pl.

off-white a. υπόλευκος.

oft adv. συχνά.

often adv. συχνά. how ~ have I told you? πόσες φορές σου είπα; how ~ do the trains run? κάθε πόσο έχει τραίνο; every so ~ κάθε τόσο. as ~ as not συνήθως.

ogle v. γλυκοκοιτάζω. (s.) ματιά f.

ogre s. δράκος m.

oh int. ωχ, άχου.

oil s. (edible) λάδι n., έλαιον n. (lubricating) λάδι n. (mineral) ορυκτέλαιον, πετρέλαιον n. ~s (paint) ελαιοχρώματα n.pl. strike ~ (fig.) πλουτίζω. (dish) cooked with ~ λαδερός. (v.t.) λαδώνω. **~can** s. λαδωτήρι n. **~cloth** s. μουσαμάς m. **~driven** a. πετρελαιοκίνητος. **~lamp** s. (with olive-~) λυχνάρι n. (with paraffin) λάμπα του πετρελαίου. **~paint** s. ελαιόχρωμα n., λαδομπογιά f. **~painting** s. ελαιογραφία f. he is no ~painting είναι o τέρας ασχήμιας. **~press** s. ελαιοπιεστήριον n. λιοτριβιό n. **~rig** s. εξέδρα αντλήσεως πετρελαίου. **~skins** s. μουσαμαδιά f. **~tanker** s. δεξαμενόπλοιο n. **~well** s.πετρελαιοπηγή f.

oily a. (substance) ελαιώδης, λαδερός. (covered in oil) λαδωμένος, από λάδια. (unctuous) γλοιώδης.

ointment s. αλοιφή f.

OK see **okay.**

okay adv. καλά, εντάξει. (v.t.) εγκρίνω.

okra s. μπάμιες f.pl.

old a. παλαιός, παλιός. (ancient) αρχαίος. (animal, tree) γέρικος. (person) ηλικιωμένος. ~ woman γραία, γριά f. ~ man γέροντας, γέρος m., (gaffer) μπάρμπας m. how ~ are you? πόσων χρόνων είστε; how ~ is the house? πόσο παλιό είναι το σπίτι; grow ~ γερνώ, γηράσκω. ~ age γεράματα n.pl., γήρας n. ~ age pension σύνταξη γήρατος. an ~ hand πεπειραμένος. ~ hat (fam.) ξεπερασμένος. ~ people's home γηροκομείο n. in the ~ days τον παλιό καιρό. any ~ thing οποιοδήποτε πράμα. **~fashioned** a. του παλιού καιρού, (person) παλαιών αρχών, (clothes) ντεμοντέ. **~maidish** a. γεροντοκορίστικος.

olden a. παλαιός.

oleaginous a. ελαιώδης.

oleander s. πικροδάφνη f.

oligarch s., **~ic** a. ολιγαρχικός.

olive s. (also~-tree) ελαία, ελιά f. **~branch** s. κλάδος ελαίας. **~grove** s. ελαιών(ας)

m. **~press** s. see oil-press.

olympian a. ολύμπιος.

Olympic a. Ολυμπιακός. ~ Games Ολυμπιακοί Αγώνες, (fam.) Ολύμπια n.pl. ~ victor Ολυμπιονίκης m.f.

omelet s. ομελέτα f.

omen s. οιωνός m. of good ~ ευοίωνος.

ominous a. δυσοίωνος. **~ly** adv. απειλητικά.

omission s. παράλειψη f.

omit v. παραλείπω.

omnibus s. λεωφορείο n. (a.) ~ volume μεγάλος τόμος με διάφορα έργα.

omnipot|ent a. παντοδύναμος. **~ence** s. παντοδυναμία f.

omnisci|ent a. παντογνώστης m. **~ence** s. παντογνωσία f.

omnivorous a. παμφάγος.

on adv. (forward) εμπρός, μπρος. (continuity) walk ~ προχωρώ. work ~ εξακολουθώ να εργάζομαι. he talked ~ and ~ μιλούσε και μιλούσε χωρίς σταματημό. later ~ αργότερα. and so ~ και ούτω καθ' εξής. what's ~? (at theatre) τι παίζει; have nothing ~ (be naked) δεν φορώ τίποτα, (be at leisure) είμαι ελεύθερος. he had his coat ~ φορούσε το παλτό του. is the milk ~? (to boil) είναι το γάλα πάνω στη φωτιά; the light is ~ το φως είναι αναμμένο. the tap is ~ η βρύση είναι ανοικτή. the brake is ~ το φρένο είναι βαλμένο. the strike is still ~ εξακολουθεί ακόμα η απεργία. the party's ~ το πάρτι θα γίνει. it's not ~ (fam.) δεν γίνεται.

on a. ~ side (of vehicle) η προς το πεζοδρόμιο πλευρά. (of switch, etc.) in the «~» position στο « ανοικτός».

on prep. 1. (place) σε, εις (with acc.), επί (with gen.), ~ the table στο τραπέζι, επί της τραπέζης. ~ the radio στο ραδιόφωνο. ~ my right στα δεξιά μου. a village ~ Pelion ένα χωριό στο Πήλιο. I got ~ to the roof ανέβηκα στη στέγη. I got ~ to him ~ the phone τον πήρα στο τηλέφωνο. ~ the spot (there) επί τόπου. ~ earth επί της γης. 2. (date) ~ Tuesday την Τρίτη. ~ the twentieth (of month) στις είκοσι. ~ a day in spring μία μέρα της άνοιξης. ~ my birthday στα γενέθλιά μου. 3. (concerning) για (with acc.), ~ business για δουλειές. a lecture ~ Plato μία διάλεξη για τον Πλάτωνα. an expert ~ painting ειδικός στη ζωγραφική. be mad ~ έχω μανία με. 4. (circumstance) επί (with dat.), ~ a charge of theft επί κλοπή. ~ no account επ' ουδενί λόγω. ~ the occasion of your birthday επί τη ευκαιρία των γενεθλίων σας. ~ the advice of my lawyer κατά συμβουλή του δικηγόρου μου. ~ my arrival home όταν έφτασα σπίτι. ~ his retirement όταν πήρε τη σύνταξή του. ~ the coming of spring με τον ερχομό της άνοιξης. ~ turning my head (*I saw...*) στρέφοντας το κεφάλι μου. 5. (*various*) be ~ sale πουλιέμαι. be ~ fire καίομαι. be ~ time είμαι στην ώρα μου. be ~ trial δικάζομαι. be ~ leave έχω άδεια. be away ~ holiday λείπω σε διακοπές. ~ purpose επίτηδες. ~ the sly (στα) κρυφά. ~ the cheap με λίγα λεφτά.

once *adv.* (*one time*) μία φορά, άπαξ. (*formerly*) μία φορά κάποτε. ~ upon a time μία φορά κι έναν καιρό. at ~ αμέσως. ~ and for all μια για πάντα, μια και καλή. ~ in a while κάπου-κάπου. ~ more ακόμα μια φορά. ~ or twice μία δύο φορές. all at ~ (*suddenly*) ξαφνικά, (*together*) όλοι μαζί, (*at one go*) μονομιάς. (*conj.*) ~ they find out μόλις (*or* μια και) το μάθουν.

once-over *s.* (*fam.*) give (*sthg.*) the ~ ρίχνω μια ματιά σε.

oncoming *s.* προσέγγιση *f.*

one *num.* ένα. the number ~ topic of conversation το υπό αριθμόν ένα θέμα συζητήσεως. *see also* number.

one *a., s. & pron.* ένας *m.*, μία *f.*, ένα *n.* ~ of my friends ένας από τους φίλους μου. ~ by ~ *or* ~ at a time ένας-ένας. ~ or two κάνα-δυο, μερικοί. as ~ man σαν ένας άνθρωπος. they love ~ another αγαπιούνται, αγαπάει ο ένας τον άλλον. for ~ thing πρώτα-πρώτα. ~ and all όλοι. it is all ~ to me το ίδιο μου κάνει. be at ~ with συμφωνώ με. ~ in a million ένας στο εκατομμύριο. be ~ up (*have advantage*) προηγούμαι. this ~ τούτος, that ~ εκείνος. absent ~s οι απόντες, little ~s τα μικρά. the ~ with gold teeth εκείνος με τα χρυσά δόντια. like ~ possessed σαν τρελλός. the ~s who survived εκείνοι που επέζησαν, οι επιζήσαντες. the ~ who wins όποιος κερδίσει, ο κερδισμένος. he's not ~ to be taken in δεν είναι κανένας που θα πιανόταν κορόιδο. ~ must be careful what ~ says πρέπει κανείς να προσέχει τα λόγια του. I for ~ disagree εγώ προσωπικά διαφωνώ. what with ~ thing and another με το ένα και με το άλλο. the English are ~ thing, the Greeks are another άλλο οι Άγγλοι, άλλο οι Έλληνες. ~ after another απανωτός, αλλεπάλληλος.

one-armed *a.* μονόχειρ.

one-eyed *a.* μονόφθαλμος.

one-legged *a.* μονοπόδαρος.

onerous *a.* βαρύς.

oneself *pron. see* self.

one-sided *a.* μονομερής, μονόπλευρος.

one-time *a.* πρώην (*adv.*)

one-track *a.* he has a ~ mind το μυαλό του στρέφεται διαρκώς στο ίδιο θέμα.

one-way *a.* ~ ticket απλό εισιτήριο. ~ street μονόδρομος *m.*

onion *s.* κρεμμύδι *n.* (*fam.*) know one's ~s ξέρω τη δουλειά μου.

onlooker *s.* θεατής *m.*

only *a.* μόνος. ~ child μοναχοπαίδι *n.* ~ son μοναχογιός *m.* one and ~ μοναδικός.

only *adv.* μόνο(ν), μονάχα. we were ~ just in time μόλις που προφτάσαμε. I shall be ~ too pleased to help you με πολλή ευχαρίστηση θα σας βοηθήσω. if ~ I could! να μπορούσα μόνο. ~ yesterday χθες ακόμα. *see also* if.

only *conj.* we like it, ~ it's too expensive μας αρέσει, μόνο που είναι πολύ ακριβό.

onomatopoeia *s.* ονοματοποιία *f.*

onrush *s.* ορμή *f.*

onset *s.* (*attack*) επίθεση *f.* (*start*) αρχή *f.* (*of malady*) εισβολή *f.*

onshore *a.* ~ breeze θαλάσσια αύρα, μπάτης *m.*

onslaught *s.* σφοδρά επίθεση.

onto *prep. see* on.

onus *s.* ευθύνη *f.*, βάρος *n.*

onward *a. & adv.*, ~s *adv.* (προς τα) εμπρός.

ooze *v.i.* (*leak*) διαρρέω, (*drip*) στάζω. (*v.t.*, *drip with*) στάζω. (*s.*) βούρκος *m.*

opacity *s.* αδιαφάνεια *f.*

opal *s.* οπάλι *n.*

opaque *a.* αδιαφανής.

open *v.t. & i.* ανοίγω, (*begin*) αρχίζω. ~ out (*v.t.*, *spread*) ανοίγω, ξεδιπλώνω, (*v.i.*) (*of bud*) ανοίγω, (*of person*) ξανοίγομαι, (*of view*) ανοίγομαι, απλώνομαι. ~ up (*v.t.*) ανοίγω, (*set in being*) βάζω μπρος. ~ on to βγάζω σε.

open *a.* ανοικτός. (*frank*) ειλικρινής, (*free*) ελεύθερος. (*unsettled*) εκκρεμής. (*undisguised*) απροκάλυπτος. ~ secret κοινό μυστικό. with ~ arms με ανοικτές αγκάλες. ~ country ύπαιθρος *f.* in the ~ (air) στο ύπαιθρο. on the ~ sea στα ανοικτά. come out into the ~ αποκαλύπτω τις προθέσεις μου. be ~ to offers δέχομαι προσφορές. lay oneself ~ to εκτίθεμαι σε. keep an ~ mind δεν είμαι προκατειλημμένος. ~-air *a.* υπαίθριος. ~-ended *a.* απεριόριστος. ~-eyed *a.* ανοιχτομάτης. ~-handed *a.* ανοιχτοχέρης. ~-hearted *a.* ανοιχτόκαρδος. ~-minded *a.* απροκάλυπτος. ~-mouthed *a.* με ανοιχτό το στόμα. ~-plan *a.* με λίγους εσωτερικούς τοίχους.

opener *s.* (*for can, etc.*) ανοιχτήρι *n.*

opening *a.* (*first*) πρώτος. (*s.*) (*gap*) άνοιγμα *n.*, οπή *f.* (*beginning*) έναρξη, αρχή *f.* (*opportunity*) ευκαιρία *f.*

openly 418 orchestra

openly *adv.* φανερά, απροκάλυπτα.

openness *s.* ειλικρίνεια *f.*

opera *s.* όπερα *f.,* μελόδραμα *n.* ~-tic *a.* του μελοδράματος, της όπερας. ~-glasses *s.* κιάλια *n.pl.*

operable *a.* (*workable*) που δουλεύει. (*med.*) εγχειρήσιμος.

operate *v.i.* (*be active*) δρω, (*be in force*) ισχύω, (*function, go*) εργάζομαι, λειτουργώ. (*med.*) ~ on εγχειρίζω. (*v.t.*) (*an appliance, machine*) χειρίζομαι, (*cause mechanical part to move*) κινώ, θέτω εις κίνησιν. (*have, run*) έχω.

operating-theatre *s.* χειρουργείο *n.*

operation *s.* (*working*) λειτουργία *f.* (*strategic*) επιχείρηση *f.* (*med.*) εγχείριση *f.* ~al *a.* (*of operations*) επιχειρήσεων. (*ready for action*) εις κατάστασιν δράσεως.

operative *a.* (*working*) εν λειτουργία. (*essential*) ουσιώδης. (*s., worker*) τεχνίτης *m.*

operator *s.* χειριστής *m.* (*phone*) τηλεφωνητής *m.,* τηλεφωνήτρια *f.*

operetta *s.* οπερέττα *f.*

ophthalm|ia *s.* οφθαλμία *f.* ~ic *a.* οφθαλμικός.

opiate *s.* ναρκωτικό *n.*

opine *v.* φρονώ.

opinion *s.* γνώμη *f.* in my ~ κατά τη γνώμη μου. what is your ~ of them? τι γνώμη έχετε για αυτούς; have a good ~ of έχω καλή ιδέα για, (*esteem*) εκτιμώ. ~ poll σφυγμομέτρηση της κοινής γνώμης.

opinionated *a.* ισχυρογνώμων.

opium *s.* όπιο *n.*

opponent *s.* αντίπαλος *m.*

opportune *a.* (*event*) επίκαιρος, (*time*) κατάλληλος. ~ly *adv.* εγκαίρως, πάνω στην ώρα.

opportun|ism *s.* καιροσκοπία *f.* ~ist *s.* καιροσκόπος *m.f.*

opportunity *s.* ευκαιρία *f.*

oppose *v.* (*put forward in opposition*) αντιτάσσω. (*be against*) αντιτίθεμαι εις, εναντιώνομαι σε. (*combat*) καταπολεμώ. they were ~d by superior enemy forces βρέθηκαν αντιμέτωποι με ισχυρότερες εχθρικές δυνάμεις.

opposed *a.* (*opposite*) αντίθετος. be ~ to *see* oppose. as ~ to κατ' αντίθεσιν προς (*with acc.*).

opposite *a.* (*contrary*) αντίθετος. they went in ~ directions πήραν αντίθετους δρόμους. (*facing*) αντικρυνός. on the ~ bank (*of river*) στην αντικρυνή (*or* στην απέναντι) όχθή. (*corresponding*) my ~ number ο αντίστοιχός μου. (*s.*) αντίθετο *n.*

opposite *adv. & prep.* απέναντι. the house ~ το απέναντι (*or* το αντικρυνό) σπίτι. ~

the church απέναντι από την εκκλησία.

opposition *s.* (*contrast*) αντίθεση *f.* (*resistance*) αντίσταση *f.* (*political party*) αντιπολίτευση *f.,* be in ~ αντιπολιτεύομαι.

oppress *v.* (*keep under*) καταδυναστεύω, καταπιέζω. (*weigh down*) πιέζω, καταθλίβω. ~ion *s.* καταπίεση *f.* ~ive *a.* (*hard, unjust*) καταπιεστικός, (*atmosphere*) καταθλιπτικός.

opprobri|um *s.* αίσχος *n.* ~ous *a.* υβριστικός.

opt *v.i.* ~ for διαλέγω. ~ out of αποσύρομαι από.

optative *s.* (*gram.*) ευκτική *f.*

optic, ~al *a.* οπτικός. ~al illusion οφθαλμαπάτη *f.* ~ian *s.* οπτικός *m.* ~s *s.* οπτική *f.*

optimism *s.* αισιοδοξία *f.*

optimist *s.,* ~ic *a.* αισιόδοξος. ~ically *adv.* με αισιοδοξία.

optimum *a.* καλύτερος, βέλτιστος. ~ conditions οι πιο ευνοϊκές συνθήκες.

option *s.* (*choice*) εκλογή *f.* (*right of choice*) δικαίωμα εκλογής. (*commercial*) οψιόν, πριμ *n.* with ~ of purchase με δικαίωμα αγοράς. I have no ~ but to... δεν μπορώ παρά να. ~al *a.* προαιρετικός.

opul|ence *s.* πλούτος *n.* ~ent *a.* πλούσιος.

or *conj.* ή, (*after neg.*) ούτε. either... ~ ή... ή, είτε... είτε. whether... ~ είτε... είτε. ~ else (*otherwise*) ειδάλλως, ειδεμή. five ~ six πέντε-έξι. in a week ~ so σε καμιά εβδομάδα περίπου. rain ~ no rain I shall go βρέχει ξεβρέχει θα πάω. whether he likes it ~ not he must eat it θέλει δε θέλει θα το φάει.

orac|le *s.* (*place*) μαντείον *n.* (*response*) χρησμός *m.* (*iron., person*) Πυθία *f.* ~ular *a.* μαντικός.

oral *a.* προφορικός. (*s., exam.*) προφορικά *n.pl.* ~ly *adv.* (*spoken*) διά ζώσης, προφορικώς, (*taken*) διά του στόματος.

orange *s.* πορτοκάλι *n.* ~-tree πορτοκαλιά *f.*

orange *a.* (*colour*) πορτοκαλής. (*s.*) πορτοκαλί *n.*

orangeade *a.* πορτοκαλάδα *f.*

orang-outang *s.* οραγκουτάγκος *m.*

oration *s.* αγόρευση *f.*

orator *s.* ρήτορας *m.* ~ical *a.* ρητορικός. ~y *s.* ρητορική *f.*

oratorio *s.* ορατόριο *n.*

orb *s.* σφαίρα *f.*

orbit *s.* τροχιά *f.* (*v.t.*) περιστρέφομαι περί (*with acc.*).

orchard *s.* κήπος οπωροφόρων δένδρων, περιβόλι *n.*

orchestr|a *s.* ορχήστρα *f.* ~al *a.* ορχηστικός. ~ate *v.* ενορχηστρώνω.

orchid s. ορχιδέα f.
ordain v. (order) ορίζω, διατάσσω. (eccl.) χειροτονώ.
ordeal s. δοκιμασία f.
order s. (calm or proper state, system of things) τάξη f. (condition) κατάσταση f. (biol. & social class) τάξη f. (priestly, of chivalry) τάγμα n. (decoration) παράσημο n. (archit.) ρυθμός m. (sequence) σειρά f. (command) διαταγή f., εντολή f. (for goods, service) παραγγελία f. (money) επιταγή f. put or set in ~ τακτοποιώ, βάζω τάξη σε. in ~ (correct) εν τάξει, is everything in ~? είναι όλα εν τάξει; is it in ~? (permitted) επιτρέπεται; in good (or working) ~ σε καλή κατάσταση. out of ~ (not working) χαλασμένος, (against rules) εκτός κανονισμού. get out of ~ χαλώ, δεν είμαι εν τάξει. keep ~ τηρώ την τάξη. she keeps her children in ~ έχει τα παιδιά της πειθαρχημένα. call to ~ επαναφέρω εις την τάξιν. the usual ~ of things η συνήθης κατάσταση πραγμάτων. (qualities) of a high ~ υψηλού επιπέδου. take (holy) ~s περιβάλλομαι το ιερατικόν σχήμα. in ~ of seniority κατά σειράν αρχαιότητος. give ~s διατάσσω. the ~ of the day ημερήσια διάταξη, (fam., in fashion) της μόδας. by ~ of κατά διαταγήν (with gen.). (made) to ~ επί παραγγελία. (fam.) a tall ~ δύσκολη δουλειά. (purpose) in ~ to (or that) για να, διά να.
order v. (command) διατάσσω. (goods, etc.) παραγγέλλω. (arrange) τακτοποιώ, κανονίζω. he ~ed his men to advance διέταξε τους άνδρες του να προχωρήσουν. he ~ed the prisoner to be brought before him διέταξε να προσκομισθεί ο αιχμάλωτος (or να προσκομίσουν τον αιχμάλωτο) μπροστά του. he was ~ed to leave the country διετάχθη να φύγει από τη χώρα. ~ people about δίνω διαταγές.
orderliness s. τάξη f.
orderly a. τακτικός, μεθοδικός. in an ~ way τακτικά, με τάξη.
orderly s. (mil.) ορντινάντσα f., στρατιώτης-υπηρέτης m. ~ officer αξιωματικός υπηρεσίας.
ordinal a. (number) τακτικός.
ordinance s. διάταξη f.
ordinary a. (usual) συνήθης, συνηθισμένος. (plain) κοινός. in the ~ way συνήθως. out of the ~ ασυνήθιστος, διαφορετικός.
ordination s. χειροτονία f.
ordnance s. (guns) πυροβολικόν n. (department) υπηρεσία υλικού πολέμου. O~ Survey χαρτογραφική υπηρεσία.
ordure s. ακαθαρσίες f.pl.

ore s. μετάλλευμα n.
organ s. όργανο n. ~ grinder λατερνατζής m.
organic a. οργανικός. **~ally** adv. οργανικώς, (fundamentally) κατά βάσιν.
organism s. οργανισμός m.
organization s. (organizing, being organized, organized body, e.g. trade union) οργάνωση f. (system or service, e.g.) United Nations, Tourist) οργανισμός m.
organiz|e v. οργανώνω. **~er** s. οργανωτής m.
orgasm s. οργασμός m.
org|y s. όργιο n. hold ~ies οργιάζω.
orient s. ανατολή f. (v.t.) προσανατολίζω. ~al a. της Άπω Ανατολής.
orientat|e v. προσανατολίζω. **~ion** s. προσανατολισμός m.
orifice s. στόμιο n.
origin s. (first cause) αρχική αιτία, (provenance) προέλευση f. (descent) καταγωγή f.
original a. (first) αρχικός. (not copied, creative) πρωτότυπος. (s.) πρωτότυπο n. **~ity** s. πρωτοτυπία f. **~ly** adv. στην αρχή, με πρωτοτυπία.
originate v.t. (cause) γεννώ, προκαλώ. (think of) επινοώ, συλλαμβάνω. (v.i.) προέρχομαι, ξεκινώ.
originator s. who was the ~ of this scheme? ποιος συνέλαβε αυτό το σχέδιο;
ornament s. (also ~ation) διακόσμηση f. (vase, etc.) κομψοτέχνημα n. (piece of decoration, also fig.) κόσμημα n. (v.t.) διακοσμώ, στολίζω, ποικίλλω.
ornamental a. διακοσμητικός.
ornate a. βαριά (or πλούσια) διακοσμημένος. (verbally) περίκομψος.
ornithology s. ορνιθολογία f.
orphan s. ορφανός a. be ~ed, become an ~ ορφανεύω. **~age** s. (condition) ορφάνια f. (home) ορφανοτροφείο n.
orthodox a. ορθόδοξος. **~y** s. ορθοδοξία f.
orthograph|y s. ορθογραφία f. **~ic** a. ορθογραφικός.
orthopedic a. ορθοπεδικός. **~s** s. ορθοπεδική, ορθοπεδία f.
oscillat|e v.i. ταλαντεύομαι. **~ion** s. (phys.) ταλάντωσις f.
osier s. λυγαριά f.
oss|ify v.i. αποστεούμαι. **~uary** s. οστεοφυλάκιο n.
ostensibly adv. δήθεν.
ostentat|ion s. επίδειξη f. **~ious** a. επιδεικτικός.
osteopath s. χειροπράκτωρ m. **~y** s. χειροπρακτική f.
ostrac|ize v. εξοστρακίζω. **~ism** s. εξοστρακισμός m.
ostrich s. στρουθοκάμηλος f.

other *a. & pron. (different)* άλλος, διαφορετικός. *(additional)* άλλος. every ~ day μέρα παρά μέρα. the ~ day τις προάλλες. some day or ~ κάποια μέρα. ~ things being equal αν επιτρέπουν οι περιστάσεις. on the ~ hand από την άλλη (μεριά). have they any ~ children? έχουν άλλα παιδιά;

otherwise *adv.* διαφορετικά, αλλιώς. *(in other respects)* κατά τα άλλα.

otiose *a.* περιττός.

otter *s.* ενυδρίς *f.*

Ottoman *a.* οθωμανικός. *(person)* Οθωμανός.

ottoman *s.* κασέλλα-ντιβάνι *f.*

ought *v.* I ~ to go and see him πρέπει *(or* θα έπρεπε*)* να πάω να τον δω. I ~ to have bought it έπρεπε να το είχα αγοράσει. I ~ not to have bought it δεν έπρεπε να το αγοράσω. fish ~ to be eaten fresh το ψάρι πρέπει να τρώγεται φρέσκο.

ounce *s.* ουγκιά *f.*

our *pron.* ~ house το σπίτι μας. ~ own, ~s δικός μας. **~selves** *see* self.

oust *v.* εκδιώκω.

out *adv.* έξω. ~ with it! λέγε! he is ~ *(not at home)* λείπει, είναι έξω. be ~ *(in calculation)* πέφτω έξω, I wasn't far ~ δεν έπεσα πολύ έξω. have a tooth ~ βγάζω ένα δόντι. have it ~ *(with)* εξηγούμαι, I had it ~ with my brother εξηγηθήκαμε με τον αδελφό μου. he is ~ *(or is going all ~)* for the governorship έχει βαλθεί να γίνει κυβερνήτης. take, get *or* put ~ βγάζω. come, go *or* turn ~ βγαίνω. it is ~ *(of light)* είναι σβησμένο, *(of secret)* βγήκε στη μέση, *(of sun)* λάμπει. the trees are ~ in bloom τα δέντρα έχουν ανθίσει. an ~ and ~ liar ένας ψεύτης πέρα για πέρα. he was ~ and away the best πέρα κατά πολύ καλύτερος.

out of *prep.* από, έξω από, μέσα από, εκτός. drink ~ the bottle πίνω από το μπουκάλι. look ~ the window κοιτάζω έξω από το παράθυρο. take it ~ the box το παίρνω μέσα από το κουτί. from ~ the sea μέσα από τη θάλασσα. ~ Athens εκτός Αθηνών. ~ danger εκτός κινδύνου. ~ season εκτός εποχής. ~ stock *or* print εξαντλημένος. *(made)* ~ bits of wood από κομμάτια ξύλα. ~ curiosity από περιέργεια. ~ bed σηκωμένος. one person ~ ten ο ένας στους δέκα. feel ~ it νιώθω παραμελημένος. get ~ the way φεύγω από τη μέση. put ~ the way βγάζω από τη μέση. he is ~ a job δεν έχει δουλειά. we are *(or* we have run*)* ~ petrol μείναμε από βενζίνη.

outback *s.* ενδοχώρα της Αυστραλίας.

outbid *v.t.* I ~ him ξεπέρασα την προσφορά του.

outboard *a. (motor)* εξωλέμβιος.

outbreak *s. (of violence)* έκρηξη *f. (of disease)* εκδήλωση *f.*

outbuilding *s.* υπόστεγο *n.*

outburst *s.* ξέσπασμα *n.*

outcast *s.* απόβλητος *a.*

outclass *v.t.* υπερτερώ.

outcome *s.* έκβαση *f.*

outcry *s.* κατακραυγή *f.*

outdated *a.* ξεπερασμένος.

outdistance *v.* ξεπερνώ.

outdo *v.* ξεπερνώ. ·

outdoor *a. (activities)* υπαίθριος, της υπαίθρου. *(clothes)* για έξω.

outdoors *adv.* έξω.

outer *a.* εξωτερικός, έξω. **~most** *a.(furthest)* πιο μακρινός, ακραίος. the ~most layer το επάνω επάνω στρώμα.

outfit *s. (clothes)* ιματισμός *m. (tools, etc.)* σύνεργα *n.pl. (camping, etc.)* εξοπλισμός *m.* **~ters** *s.* men's ~ters κατάστημα ανδρικών ειδών.

outflank *v.* υπερφαλαγγίζω.·

outflow *s.* εκροή *f.*

outgeneral *v.* υπερτερώ στη στρατηγική.

outgoing *a. (person)* αποχωρών, *(papers)* εξερχόμενος. *(fam, sociable)* κοινωνικός. *(s.)* ~s έξοδα *n.pl.*

outgrow *v. (a habit, in stature)* ξεπερνώ. ~ one's clothes μεγαλώνω και τα ρούχα μου πέφτουν μικρά.

outhouse *s.* υπόστεγο *n.*

outing *s.* εκδρομή *f.*

outlandish *a.* παράξενος.

outlast *v.t.* διαρκώ περισσότερο από. *(of person)* επιζώ *(with gen.).*

outlaw *s.* επικηρυγμένος *a.,* εκτός νόμου. *(v.)* επικηρύσσω, θέτω εκτός νόμου.

outlay *s.* έξοδα *n.pl.*

outlet *s.* εκροή *f. (fig.)* διέξοδος *f.*

outline *s.* περίγραμμα *n. (summary)* περίληψη *f.* in ~ περιληπτικώς, εν περιλήψει.

outlive *v.* he ~d the war επέζησε του πολέμου. he ~d his brothers έζησε περισσότερο από τους αδελφούς του. he won't ~ the night δεν θα βγάλει τη νύχτα. it has ~d its usefulness έπαψε να είναι χρήσιμο.

outlook *s. (view)* θέα *f. (mental)* αντίληψη *f. (prospect)* προοπτική *f.,* πρόβλεψη *f.*

outlying *a.* απόμερος.

outmanoeuvre *v. see* outgeneral.

outmoded *a.* ξεπερασμένος.

outnumber *v.,* they ~ed us ήταν περισσότεροι από μας.

out-of-date *a.* ξεπερασμένος.

out-of-door *a.* υπαίθριος, της υπαίθρου. ~s *adv.* έξω.

out-of-the-way *a.* (*place*) απόμερος. ~ information πληροφορίες έξω από τις κοινές.

out-of-work *a.* άνεργος.

out-patient *s.* εξωτερικός ασθενής.

outpost *s.* προφυλακή *f.*

outpouring *s.* ξεχείλισμα *n.*

output *s.* (*produce*) παραγωγή *f.* (*energy, etc.*) απόδοση *f.*

outrage *s.* (*offence*) προσβολή *f.* (*scandal*) αίσχος *n.*, σκάνδαλο *n.* (*v.t.*) προσβάλλω, σκανδαλίζω.

outrageous *a.* αισχρός, σκανδαλώδης. **~ly** *adv.* σκανδαλωδώς. (*fam.*) τρομερά.

outrider *s.* έφιππος συνοδός.

outright *a.* (*frank*) ντόμπρος, (*sheer*) καθαρός, πέρα για πέρα.

outright *adv.* (*straight off*) αμέσως. (*point-blank*) κατηγορηματικώς. the shot killed him ~ η σφαίρα τον άφησε στον τόπο.

outrun *v.t.* (*competitors, allotted time*) ξεπερνώ. (*fig.*) προτρέχω (*with gen.*).

outset *s.* αρχή *f.*

outshine *v.* επισκιάζω, υπερέχω (*with gen.*).

outside *adv.* έξω. (*prep.*) εκτός (*with gen.*), έξω από.

outside *a.* εξωτερικός. an ~ chance αμυδρή πιθανότητα. get ~ help καταφεύγω σε ξένο χέρι. (*s.*) εξωτερικό *n.*

outsider *s.* (*one from outside*) ξένος *m.* (*pej.*) εκτός του κατεστημένου κύκλου. (*horse*) χωρίς πιθανότητες.

outskirts *s.* περίχωρα *n.pl.*

outsmart *v.* ξεγελώ.

outspoken *a.* ντόμπρος. **~ly** *adv.* ντόμπρα. **~ness** *s.* ντομπροσύνη *f.*

outspread *a.* απλωμένος.

outstanding *a.* (*eminent*) διαπρεπής. (*qualities*) ξεχωριστός. (*special*) εξαιρετικός. (*noteworthy*) σημαντικός. (*unsettled*) εκκρεμής. be ~ (*of object, person*) ξεχωρίζω, εξέχω. **~ly** *adv.* εξαιρετικά.

outstay *v.* he ~ed his welcome παρέμεινε μέχρι φορτικότητος.

outstretched *a.* (*body*) ξαπλωμένος. (*arms*) απλωμένος, τεντωμένος.

outstrip *v.* ξεπερνώ.

out-tray *s.* ξεερχόμενα *n.pl.*

out-turned *a.* στραμμένος προς τα έξω.

outvote *v.t.* (*reject proposal*) καταψηφίζω, (*beat at polls*) πλειοψηφώ (*with gen.*).

outward *a.* (*outer*) εξωτερικός. (*direction*) προς τα έξω. ~ journey πηγαιμός *m.* **~ly** *adv.* εξωτερικώς, φαινομενικώς.

outward(s) *adv.* προς τα έξω.

outwear *v.* (*last longer than*) κρατώ περισσότερο από. I have outworn these shoes πάλιωσαν τα παπούτσια μου. *see*

outworn.

outweigh *v.t.* ζυγίζω περισσότερο από. (*fig.*) υπερτερώ (*with gen.*).

outwit *v.* ξεγελώ.

outwork *s.* προκεχωρημένον οχυρόν.

outworn *a.* (*fig.*) ξεφτισμένος.

ouzo *s.* ούζο *n.*

oval *a.* ωοειδής.

ovary *s.* ωοθήκη *f.*

ovation *s.* be given an ~ επευφημούμαι.

oven *s.* φούρνος *m.* (*small furnace*) κλίβανος *m.* **~ful** *s.* φουρνιά *f.* **~proof** *a.* άκαυστος.

over *adv.* ~ here εδώ πέρα. ~ there εκεί πέρα. ~ and ~ επανειλημμένος. (all) ~ again ξανά. it's ~ (*finished, stopped*) τελείωσε, (*past*) πέρασε. get, go *or* come ~ περνώ, διαβαίνω. come ~ and see us πέρασε από το σπίτι να μας δεις. fall ~ πέφτω κάτω. brim *or* boil ~ ξεχειλίζω. turn ~ (*the page, in bed*) γυρίζω. have a bit (left) ~ μου έχει μείνει κάτι. my shoes are all ~ mud τα παπούτσια μου είναι γεμάτα λάσπες. I was trembling all ~ έτρεμα σύσσωμος (*or* όλος). that's George all ~ τέτοιος είναι ο Γιώργος.

over *prep.* (*on top of*) πάνω σε. (*above, more than*) πάνω από. ~ and above εκτός από, επιπλέον. (*going across*) a bridge ~the Thames γεφύρι πάνω από τον Τάμεση. (*extent*) ~ a wide area σε μία μεγάλη περιοχή. all ~ the world σ' όλο τον κόσμο. all ~ the place παντού. (*duration*) ~ the past six months στο (*or* κατά το) τελευταίο εξάμηνο. (*about*) they quarrelled ~ money μάλωσαν για λεφτά. we had trouble ~ that affair είχαμε φασαρίες μ' εκείνη (*or* πάνω σ' εκείνη) την υπόθεση.

over- υπερ-, παρα-.

overall *a.* συνολικός, (*adv.*) συνολικά.

overall *s.* ποδιά *f.* **~s** φόρμα *f.*

overawe *v.* επιβάλλομαι σε.

overbalance *v.i.* χάνω την ισορροπία μου. (*v.t.*) ανατρέπω.

overbearing *a.* αυταρχικός.

overblown *a.* (*bloom*) πολύ ανοιγμένος. (*fig., inflated*) φουσκωμένος.

overboard *adv.* στη θάλασσα. throw ~ (*fig.*) εγκαταλείπω.

overburden *v.* παραφορτώνω.

overcast *a.* συννεφιασμένος.

overcharge *v.* he ~d me μου φούσκωσε το λογαριασμό, μου πήρε παραπάνω (λεφτά).

overcoat *s.* παλτό *n.*, πανωφόρι *n.*

overcome *v.* υπερνικώ. (*fig.*) be ~ by (*temptation*) υποκύπτω σε, (*emotion*) συγκινούμαι από. he was ~ by fear κατατρόμαξε. he was ~ by fatigue ήταν ψόφιος στην κούραση. he was ~ by the heat τον είχε αποκάμει η ζέστη. he is ~ by grief τον έχει καταβάλει η λύπη.

overcrowd ν. παραφορτώνω.

overdo ν. παρακάνω.

overdue a. (*payment*) εκπρόθεσμος. be ~ (*late*) έχω καθυστέρηση. (*of reforms, etc.*) it was long ~ έπρεπε να είχε γίνει προ πολλού.

overestimate ν. υπερεκτιμώ.

overflow ν. ξεχειλίζω. (*s.*) ~ pipe σωλήν υπερχειλίσεως.

overgrown a. (*covered*) σκεπασμένος. (*too tall*) he is ~ έχει πρόωρη ανάπτυξη.

overhang ν.t. κρέμομαι πάνω από. **~ing** a. που κρέμεται στο κενό. (*fig.*) επικρεμάμενος.

overhaul ν. (*examine*) εξετάζω λεπτομερώς. (*catch up with*) προφταίνω, (*pass*) ξεπερνώ.

overhaul, ~ing s. λεπτομερής εξέταση. (*repairs*) επισκευή f.

overhead a. (*cable, etc.*) εναέριος. (*adv.*) πάνω από το κεφάλι μας (*or* τους, *etc.*) the flat ~ το από πάνω διαμέρισμα. (*s.*) ~s (*expenses*) γενικά έξοδα.

overhear ν. ακούω τυχαίως. I ~d them saying that... πήρε το αφτί μου να λένε ότι.

overjoyed a. καταχαρούμενος.

overkill s. εξόντωση f.

overland adv. & a. κατά ξηράν, διά ξηράς.

overlap ν.i. (*of visits, duties*) συμπίπτω. (ν.t.) (*of objects*) καβαλλικεύω, σκεπάζω μερικώς.

overlay ν.t. επιστρώνω. (s.) επίστρωμα n.

overleaf adv. όπισθεν.

overlie ν.t. σκεπάζω.

overload ν. παραφορτώνω. (s.) (*electric*) υπερφόρτιση f.

overlook ν. (*have view over*) βλέπω προς. (*supervise*) επιβλέπω. (*condone*) παραβλέπω. (*not notice*) I ~ed it μου διέφυγε.

overlord s. φεουδάρχης m.

overmastering a. ακαταμάχητος.

overnight adv. (*for the night*) τη νύχτα. (*in the night*) κατά τη νύχτα. (*in one night*) σε μια νύχτα. (*fig., of sudden development*) από τη μια μέρα στην άλλη. I put them to soak ~ τα έβαλα να μουσκέψουν αποβραδίς. (*a.*) βραδινός.

overplay ν. ~ one's hand (*fig.*) ενεργώ τολμηρά υπερεκτιμώντας τις δυνατότητές μου.

overpower ν. (*in war*) κατανικώ. (*in hand-to-hand struggle*) καταβάλλω. (*subdue*) δαμάζω. **~ing** a. (*force, desire*) ακαταμάχητος. (*heat, fumes*) αποπνικτικός.

overrate ν. υπερτιμώ. be ~d έχω υπέρ το δέον επαινεθεί.

overreach ν. ξεγελώ. ~ oneself το παρακάνω στην προσπάθειά μου και αποτυγχάνω.

overreact ν.i. υπεραντιδρώ.

override ν. (*prevail over*) υπερισχύω (*with gen.*). (*set aside*) ανατρέπω. (*go against*) παραβαίνω. of ~ing importance κεφαλαιώδους σημασίας.

overrule ν. (*an act, decision*) ανατρέπω. (*law*) ακυρώ.

overrun ν.t. κατακλύζω, (*allotted time*) ξεπερνώ. we are ~ with mice μας έχουν κατακλύσει τα ποντίκια. the country is ~ by bandits. τη χώρα τη λυμαίνονται ληστές.

oversea(s) a. (*expedition, possessions*) υπερπόντιος. (*trade, etc.*) εξωτερικός. (*adv.*) στο εξωτερικό.

oversee ν. επιστατώ. **~r** s. επιστάτης m.

overshadow ν. επισκιάζω.

overshoot ν. υπερβαίνω.

oversight s. παράλειψη f. by ~ εκ παραδρομής.

oversleep ν. I overslept με πήρε ο ύπνος.

overstate ν. μεγαλοποιώ. **~ment** s. υπερβολή f.

overstay f. *see* outstay.

overstep ν. ~ the mark υπερβαίνω τα εσκαμμένα.

over-strain s. υπερένταση f.

overstrung a. (*nervous*) με τεντωμένα τα νεύρα.

overt a. έκδηλος, φανερός. **~ly** adv. στα φανερά, εκδήλως.

overtake ν.t. (*pass*) ξεπερνώ, (*of cars*) προσπερνώ. (*of disaster*) πλήττω. we were ~n by night μας έπιασε (*or* μας βρήκε) η νύχτα, νυχτωθήκαμε.

overtax ν. ~ one's strength υπερεντείνω τις δυνάμεις μου.

overthrow ν. ανατρέπω, ρίχνω. (s.) ανατροπή f.

overtime s. υπερωρία f. work ~ κάνω υπερωρίες.

overtone s. (*fig.*) κάποιος τόνος.

overture s. (*offer*) πρόταση f. (*mus.*) εισαγωγή f.

overturn ν. & i. αναποδογυρίζω.

overweening a. υπερφίαλος.

overweight a. βαρύτερος του κανονικού.

overwhelm ν. (*flood*) κατακλύζω, (*with kindness*) κατασκλαβώνω. be ~ed (*by enemy, grief, disaster*) συντρίβομαι, (*with work*) πνίγομαι. **~ing** a. συντριπτικός. **~ingly** adv. συντριπτικά, πέρα για πέρα.

overwork ν.t. παρακουράζω, (*person only*) παραφορτώνω με δουλειά. (*use too often*) κάνω κατάχρηση (with gen.). (v.i.) παρακουράζομαι. (s.) υπερκόπωση f.

overwrought a. be ~ βρίσκομαι σε υπερδιέγερση.

ov|um s. ωάριον n. **~iparous** a. ωοτόκος.

owe v. οφείλω, χρωστώ. I ~ him a grudge του κρατώ κακία, έχω πίκα μαζί του.

owing a. there is £10 ~ οφείλονται δέκα λίρες. (prep.) ~ to λόγω, εξ αιτίας (both with gen.).

owl s. κουκουβάγια f., γλαυξ f.

own a. one's ~ δικός μου (or του, etc.). for reasons of one's ~ για προσωπικούς λόγους. a room of her ~ ένα δωμάτιο καταδικό της (or όλο δικό της). on one's ~ μόνος μου. hold one's ~ δεν υποχωρώ, αμύνομαι. come into one's ~ (be recognized) αναγνωρίζομαι, (show merits) δείχνω τις ικανότητές μου. get one's ~ back παίρνω πίσω το αίμα μου. with my ~ eyes με τα ίδια μου τα μάτια.

own v. έχω, κατέχω, είμαι κάτοχος (with gen.). who ~s the house? σε ποιον ανήκει το σπίτι; ποιος είναι ο ιδιοκτήτης του σπιτιού; (recognize, admit) αναγνωρίζω. (admit, also ~ up) ομολογώ.

owner s. ιδιοκτήτης m., ιδιοκτήτρια f., κάτοχος m.f. **~ship** s. ιδιοκτησία f., κυριότητα f.

ox s. βόδι n. **~en** βόδια pl.

oxide s. οξείδιο n.

oxidiz|e v.t. οξειδώνω. **~ation** s. οξείδωση f.

oxy-acetylene a. ~ welding οξυγονοκόλληση f.

oxygen s. οξυγόνο n.

oyster s. στρείδι n.

ozone s. όζον n.

P

P s. mind one's ~'s and Q's προσέχω τη συμπεριφορά μου.

pa s. μπαμπάς m.

pace s. (step) βήμα n. (gait) βηματισμός m. (speed) ταχύτητα f., ρυθμός m. quicken the ~ επιταχύνω το βήμα. keep ~ συμβαδίζω. set the ~ καθορίζω το ρυθμό. put (person) through his ~s δοκιμάζω. at a good ~ με γοργό ρυθμό.

pace v. ~ up and down βηματίζω. ~ out μετρώ με βήματα.

pacemaker s. (med.) βηματοδότης m.

pacific a. ειρηνικός. **~ation** s. ειρήνευση f.

pacif|ism s. ειρηνισμός m. **~ist** s. ειρηνιστής m. (a.) ειρηνιστικός. **~y** v. ειρηνεύω.

pack s. (bundle) δέμα n. (knapsack) σακκίδιο n. (hounds) κυνηγετική αγέλη. ~ of lies σωρός ψέματα. ~ of cards τράπουλα f. (packet) πακέτο n.

pack v.t. (merchandise) συσκευάζω. (fill) γεμίζω, (cram) στοιβάζω. ~ one's bags φτιάχνω τις βαλίτσες μου. ~ off, send ~ing ξαποστέλνω. (v.i.) (squeeze into) χώνομαι. ~ up, ~ it in σταματώ. it has ~ed up χάλασε.

package s. δέμα n. (v.) συσκευάζω. ~ deal λύση-πακέτο.

pack-animal s. υποζύγιο n.

packed a. γεμάτος.

packet s. πακέτο n. (postal) δέμα n.

packing s. συσκευασία f., πακετάρισμα n. **~-case** κασόνι n.

pact s. συνθήκη f., σύμφωνο n.

pad s. μαξιλλαράκι n. (paper) μπλοκ n. (for inking) ταμπόν n. (v.t.) (put padding into) βάζω βάτα σε, (fig.) παραγεμίζω. (v.i.) ~ along αλαφροπατώ.

padding s. βάτα f. (fig.) παραγέμισμα n.

paddle s. κουπί του κανό. (v.) (row) κωπηλατώ, (walk in water) τσαλαβουτώ στο νερό.

paddock s. λειβάδι για άλογα.

padlock s. λουκέτο n. (v.) κλείνω με λουκέτο.

paean s. παιάν m.

paed- see ped-.

pagan a. ειδωλολατρικός. (s.) ειδωλολάτρης m. **~ism** s. ειδωλολατρία f.

page s. (paper) σελίδα f.

page s. (boy) νεαρός ακόλουθος. (v.t., call) καλώ.

pageant s. θεαματική ιστορική παράσταση. **~ry** s. θεαματικότητα f.

paid a. πληρωμένος. put ~ to (fig.) χαντακώνω.

pail s. κουβάς m., κάδος m.

pain s. πόνος m. have a ~ in one's leg με (or μου) πονάει το πόδι μου. (fam.) he's a ~ in the neck μου έχει γίνει στενός κορσές. does he feel much ~? πονεί πολύ; on ~ of επί ποινή (with gen.).

pain v. it ~s me to see him πονώ (or υποφέρω) να τον βλέπω. your attitude ~s me η στάση σου με στενοχωρεί. **~ed** στενοχωρημένος.

painful a. οδυνηρός, (sad) θλιβερός. it is ~ πονεί. **~ly** adv. με κόπο. (fam., very) φοβερά.

pain-killer s. παυσίπονο n.

painless a. ανώδυνος. **~ly** adv. ανώδυνα.

pains s. κόπος m. take great ~ καταβάλλω μεγάλους κόπους (or μεγάλες προσπάθειες).

painstaking a. (work) κοπιαστικός, (per-

son) επιμελής.

paint *s.* χρώμα *n.*, μπογιά *f.*

paint *v.* βάφω, μπογια(ν)τίζω. (*a picture*) ζωγραφίζω. (*fig., depict*) απεικονίζω, (*describe*) περιγράφω. (*med.*) κάνω επάλειψη. ~ the town red ξεφαντώνω.

paintbrush *s.* βούρτσα *f.* (*artist's*) πινέλο *n.*

painter *s.* (*decorator*) μπογιατζής *m.*, ελαιοχρωματιστής *m.* (*artist*) ζωγράφος *m.f.*

painting *s.* βάψιμο *n.* (*art*) ζωγραφική *f.* (*picture*) ζωγραφικός πίνακας.

pair *s.* ζεύγος *n.*, ζευγάρι *n.* ~ of scissors ψαλίδι *n.* ~ of trousers παντελόνι *n.* in ~s κατά ζεύγη. (*v.t.*) (*mate*) ζευγαρώνω. ~ off (*v.i.*) ζευγαρώνω.

pal *s.* φίλος *m.* (*v.*) ~ up with πιάνω φιλίες με.

palace *s.* παλάτι *n.* (*royal only*) ανάκτορα *n.pl.*

palaeography *s.* παλαιογραφία *f.*

palaeolithic *a.* παλαιολιθικός.

palaeontology *s.* παλαιοντολογία *f.*

palatable *a.* εύγεστος, (*fig.*) ευκολοχώνευτος.

palatal *a.* (*gram.*) ουρανισκόφωνος.

palate *s.* ουρανίσκος *m.* (*fig.*) have a good ~ είμαι γνώστης.

palatial *a.* σαν παλάτι.

palaver *s.* συζήτηση *f.*

pale *s.* (*stake*) πάσσαλος *m.* (*fig.*) beyond the ~ ανυπόφορος, (*of impossible person*) ανοικονόμητος.

pale *a.* χλομός, (*also fig.*) ωχρός. (*colour*) ανοιχτός. (*complexion only*) χίτρινος. (*v.i.*) (*become ~*) χλωμιάζω, κιτρινίζω. (*also ~* into insignificance) ωχριώ. ~**ly** *adv.* ωχρά. ~**ness** *s.* χλωμάδα *f.*,ωχρότητα *f.*

palette *s.* παλέτα *f.*

palindrome *s.* καρκίνος *m.*

paling *s.* φράκτης από πασσάλους.

palisade *s.* πασσαλωτός φράκτης.

pall *s.* υφασμάτινο επικάλυμμα φερέτρου. (*fig., mantle*) πέπλος *m.* ~ bearer ο ακολουθών την σορόν.

pall *v.* it ~s on me το βαριέμαι.

pall|et, ~**iasse** *s.* (*bed*) αχυρόστρωμα *n.*

palliat|e *v.* ελαφρύνω. ~**ive** *s.* καταπραϋντικό *n.*

pallid *a.* ωχρός. ~**ly** *adv.* ωχρά. ~**ness** *s.* ωχρότητα *f.*

pally *a.* φιλικός.

palm *s.* (*tree*) φοινικιά *f.* (*eccl.*) ~ branches βάγια *n.pl.* P~ Sunday Κυριακή των Βαΐων. (*fig.*) bear the ~ είμαι υπεράνω όλων.

palm *s.* (*of hand*) παλάμη *f.* (*fam.*) grease the ~ of λαδώνω. (*v.*) ~ off πασάρω. ~**istry** *s.* χειρομαντεία *f.*

palmy *a.* in his (*or* its) ~ days στις δόξες του.

palpab|le *a.* ψηλαφητός, (*clear*) φανερός. ~**ly** *adv.* ψηλαφητώς, φανερά.

palpitat|e *v.* σπαρταρώ. ~**ion** *s.* χτυποκάρδι *n.*

pals|ied *a.* παραλυτικός. ~**y** *s.* παράλυση *f.*

paltry *a.* τιποτένιος, μηδαμινός.

pamper *v.* παραχαϊδεύω.

pamphlet *s.* φυλλάδιο *n.*

pan *s.* (*saucepan*) κατσαρόλα *f.* (*pot*) χύτρα *f.*, τέντζερης *m.*, (*frying*) τηγάνι *n.* (*baking*) ταψί *n.* (*w.c.*) λεκάνη *f.*

pan *v.i.* (*for gold*) πλένω. (*fig.*) ~ out well έχω καλή έκβαση.

panacea *s.* πανάκεια *f.*

panache *s.* with ~ με αέρα.

pancake *s.* είδος τηγανίτας.

pandemonium *s.* πανδαιμόνιο *n.*

pander *s.* μαστροπός *m.* (*v., fig.*) ~ to υποθάλπω.

pane *s.* τζάμι *n.*

panegyric *s.* πανηγυρικός *m.*

panel *s.* (*group*) ομάδα *f.* (*list*) κατάλογος *m.* control ~ ταμπλό *n.* (*of door*) ταμπλάς *m.* (*of ceiling*) φάτνωμα *n.* ~**led** *a.* ντυμένος με ξύλο. ~**ling** *s.* (*wood*) ξυλεπένδυση *f.*

pang *s.* (*pain*) πόνος *m.* ~s of childbirth ωδίνες *f. pl.* ~s of remorse τύψεις *f. pl.*

panic *s.* πανικός *m.* (*v.i.*) πανικοβάλλομαι. ~-**stricken** *a.* πανικόβλητος.

panjandrum *s.* (*joc.*) μεγάλη προσωπικότητα.

pannier *s.* καλάθι *n.*

panoply *s.* πανοπλία *f.*

panoram|a *s.* πανόραμα *n.* ~**ic** *a.* πανοραμικός.

pansy *s.* πανσές *m.*

pant *v.* λαχανιάζω, ασθμαίνω. ~ for ποθώ. ~**ing** λαχανιασμένος.

pantheism *s.* πανθεϊσμός *m.*

panther *s.* πάνθηρας *m.*

pantomime *s.* (*dumb show*) παντομίμα *f.*

pantry *s.* οφρίς *n.* (*larder*) κελλάρι *n.*

pants *s.* (*men's*) σώβρακο *n.* (*women's*) παντελόνι *n.*, κυλότα *f.*

pap *s.* (*nipple*) θηλή *f.* (*food*) χυλός *m.*

papa *s.* μπαμπάς *m.*

pap|acy *s.* (*system*) παπισμός *m.* ~**al** *a.* παπικός.

paper *s.* χάρτης *m.*, χαρτί *n.* (*news*) εφημερίδα *f.* (*exam*) θέματα γραπτής εξετάσεως. ~s (*documents*) χαρτιά *n.pl.* (*a.*) χάρτινος, χαρτένιος. (*v.*) (*wall*) επενδύω με ταπετσαρία. ~ over καλύπτω. ~**back** *s.* βιβλίο τσέπης. ~-**clip** *s.* συνδετήρ *m.* ~-**knife** *s.* χαρτοκόπτης *m.* ~-**mill** *s.* εργοστάσιο χαρτοποιίας. ~-**weight** *s.* πρες-παπιέ *n.*

paperwork *s.* ρουτινιέρικες δουλειές του γραφείου.

papier-maché s. πεπιεσμένος χάρτης.
papist s. παπιστής m.
papyrus s. πάπυρος m.
par s. (fin.) ισοτιμία f. above ~ άνω του αρτίου. (fig.) feel below ~ δεν αισθάνομαι πολύ καλά. on a ~ with ισάξιος με.
parable s. παραβολή f.
parabola s. παραβολή f.
parachut|e s. αλεξίπτωτο n. **~ist** s. αλεξιπτωτιστής m.
parade s. (mil.) παράταξη f. (march past) παρέλαση f. ~ ground πεδίον ασκήσεων. (showing off) επίδειξη f. (v.i.) παρατάσσομαι, παρελαύνω. (v.t.) κάνω επίδειξη (with gen.), επιδεικνύω.
paradigm s. παράδειγμα n.
paradise s. παράδεισος m. fool's ~ απατηλή ευδαιμονία. bird of ~ παραδείσιον πτηνόν.
paradox s. παράδοξο n. (figure of speech) παραδοξολογία f. **~ical** a. παράδοξος. **~ically** adv. παραδόξως.
paraffin s. (fuel) φωτιστικό πετρέλαιο. (med.) παραφινέλαιο n.
paragon s. ~ of virtue υπόδειγμα αρετής.
paragraph s. παράγραφος f.
parallel a. & s. παράλληλος. draw a ~ between κάνω παραλληλισμό μεταξύ (with gen.). without ~ άνευ προηγουμένου.
paraly|se v. (also become ~sed) παραλύω. **~sed** παράλυτος. **~sis** s. παράλυση f. **~tic** a. παραλυτικός.
paramount a. of ~ importance υψίστης σημασίας. of ~ necessity υπερτάτης ανάγκης. duty is ~ το καθήκον είναι υπεράνω όλων.
paramour s. ερωμένος m. ερωμένη f.
paranoi|a s. παράνοια f. **~ac** a. παρανοϊκός.
paranormal a. εκτός της ακτίνος των επιστημονικών γνώσεων.
parapet s. στηθαίο n., παραπέτο n.
paraphernalia s. καλαμπαλίκια n.pl.
paraphrase v. παραφράζω. (s.) παράφραση f.
parasite s. παράσιτο n.
parasol s. παρασόλι n.
paratrooper s. αλεξιπτωτιστής m.
parboil v.t. μισοβράζω.
parcel s. δέμα n. (v.) ~ out διαμοιράζω.
parch v. ξεραίνω. ~ed ξεραμένος. I feel ~ed ξεράθηκε το στόμα μου.
parchment s. περγαμηνή f.
pardon v. συγχωρώ. (s.) συγγνώμη f. beg ~ ζητώ συγγνώμη, I beg your ~ με συγχωρείτε. (for crime) χάρη f. grant a ~ to απονέμω χάριν εις.
pardonab|le a. δικαιολογημένος. **~ly** adv. δικαιολογημένα.
pare v. κόβω, (peel) ξεφλουδίζω.
parent s. γονιός m. (pl.) γονιοί, γονείς.

parentage s. of unknown ~ αγνώστων γονέων. of humble ~ από ταπεινή οικογένεια.
parental a. πατρικός.
parenthesis s. παρένθεση f. in ~ εν παρενθέσει.
parenthetic a. παρενθετικός. **~ally** adv. παρενθετικώς.
par excellence adv. κατ' εξοχήν.
pariah s. παρίας m.
parings s. αποκόμματα n.pl. (skin) φλούδες f.pl.
parish s. ενορία f. **~ioner** s. ενορίτης m.
parity s. ισότητα f.
park s. κήπος m., πάρκο n. car ~ χώρος σταθμεύσεως αυτοκινήτων. (v.i.) (of car) σταθμεύω. (v.t. & i.) παρκάρω. **~ing** s. στάθμευση f. **~-meter** s. παρκόμετρο n. **~-ticket** s. κλήση f.
parlance s. γλώσσα f., ομιλία f.
parley v. διαπραγματεύομαι. (s.) διαπραγμάτευση f.
parliament s. κοινοβούλιο n., βουλή f. **~ary** a. κοινοβουλευτικός.
parlour s. σαλόνι n.
parlous a. επικίνδυνος.
parochial a. ενοριακός, (fig.) περιορισμένος. **~ism** s. περιορισμένα ενδιαφέροντα.
parody s. παρωδία f. (v.) παρωδώ.
parole s. on ~ επί λόγω.
paroxysm s. παροξυσμός m.
parquet s. παρκέτο n., παρκέ n.
parricide s. (act) πατροκτονία f. (person) πατροκτόνος m.f.
parrot s. παπαγάλος m. (v.i.) παπαγαλίζω.
parry v. αποκρούω.
parse v. τεχνολογώ.
parsimon|ious a. τσιγγούνης. **~y** s. τσιγγουνιά f.
parsley s. μαϊντανός m.
parson s. εφημέριος m. **~age** s. σπίτι του εφημερίου.
part s. μέρος n. (bit) κομμάτι n. (role) ρόλος m. (instalment) τεύχος n. spare ~ ανταλλακτικό n. in ~(s) εν μέρει, μερικώς. in great ~ κατά μέγα μέρος. for the most ~ ως επί το πλείστον. for my ~ όσο για μένα. on the ~ of εκ μέρους (with gen.). take ~ λαμβάνω μέρος. I take his ~ παίρνω το μέρος του. take it in good ~ το παίρνω από την καλή του πλευρά. three ~s (full, etc.) κατά τα τρία τέταρτα. (man) of many ~s με πολλές ικανότητες.
part v.t. & i. (separate) χωρίζω. ~ from (leave) χωρίζω με, (only after meeting) χωρίζομαι από. ~ company χωρίζω, (disagree) δεν συμφωνώ. ~ with αποχωρίζομαι.

partake v. ~ of συμμετέχω εις, (eat) παίρνω, τρώγω.

Parthian a. ~ shot πάρθιον βέλος.

partial a. (in part) μερικός, (biased) μεροληπτικός. (fond) he is ~ to lobster του αρέσει ο αστακός, he is ~ to blondes έχει αδυναμία στις ξανθές. ~**ly** adv. μερικώς, εν μέρει. ~**ity** s. (bias) μεροληψία f. (liking) αδυναμία f., συμπάθεια f.

participant s. συμμετέχων m.

participat|e v. συμμετέχω. ~**ion** s. συμμετοχή f.

participle s. μετοχή f.

particle s. (chem.) σωματίδιο n., μόριο n. (dust, sand, truth) κόκκος m. (evidence) ίχνος n. (gram.) μόριο n. he doesn't take a ~ of interest δεν δείχνει ούτε το παραμικρό ενδιαφέρον.

particoloured a. ποικιλόχρωμος.

particular a. ειδικός, ιδιαίτερος, συγκεκριμένος. (exact) ακριβής, (careful) προσεκτικός, (demanding) απαιτητικός, (hard to please) δύσκολος. for a ~ reason για ένα ειδικό λόγο. this is a ~ interest of mine ενδιαφέρομαι ειδικώς για αυτό. I'm not ~ δεν έχω ιδιαίτερη προτίμηση. nothing (in) ~ τίποτα το ιδιαίτερο. in this ~ case σε αυτή τη συγκεκριμένη περίπτωση. I am not referring to any ~ person δεν αναφέρομαι σε κανένα συγκεκριμένο πρόσωπο. a ~ friend of mine ένας από τους καλούς φίλους μου.

particular|ize v.i. (go into details) επεκτείνομαι σε λεπτομέρειες. ~**ly** adv. ειδικώς, ιδιαιτέρως, συγκεκριμένως.

particulars s. (details) λεπτομέρειες f.pl., (personal) στοιχεία n.pl. (information) πληροφορίες f.pl. (what happened) τα καθέκαστα n.pl.

parting s. χωρισμός m. (hair) χωρίστρα f.

partisan s. οπαδός m. (guerrilla) παρτιζάνος m. in a ~ spirit με μεροληψία.

partition s. (act) διχοτόμηση f. (of Poland) διαμελισμός m. (thing) χώρισμα n. (v.t.) διχοτομώ. ~ off χωρίζω.

partly adv. μερικώς, εν μέρει.

partner s. (companion) σύντροφος m.f. (collaborator) συνεργάτης m.f. (business) συνέταιρος m.f. (dance) καβαλιέρος m., ντάμα f. (cards, game) συμπαίκτης m. (v.t.) (dance) χορεύω με, (game) παίζω με.

partnership s. (bond) σύνδεσμος m. (collaboration) συνεργασία f. go into (business) ~ συνεταιρίζομαι, ιδρύω εταιρεία.

partridge s. πέρδικα f.

part-time a. have a ~ job εργάζομαι απασχολούμενος μερικώς.

parturition s. τοκετός m.

party s. (political) κόμμα n. (group) ομάδα f. (company) παρέα f. (reception) δεξίωση f., πάρτυ n. evening ~ εσπερίδα f. third ~ τρίτος m. parties (to contract) οι συμβαλλόμενοι, (to lawsuit) οι διάδικοι. be a ~ to συμμετέχω εις.

party-wall s. μεσοτοιχία f.

parvenu s. νεόπλουτος m.

pasha s. πασάς m.

pass s. (mountain) δίοδος f., στενό n. (permit) άδεια f., πάσσο n. (state of affairs) κατάσταση f. come to ~ συμβαίνω. come to such a ~ that... φθάνω σε τέτοιο σημείο που. make a ~ at ρίχνομαι σε.

pass v.t. (go past) περνώ, (leave behind) προσπερνώ, (approve) εγκρίνω, (spend time) περνώ, (give, hand) δίνω, (place, send, put) περνώ. (a bill, resolution) ψηφίζω, περνώ, εγκρίνω. (a candidate) περνώ. ~ the exam περνώ στις εξετάσεις. ~ sentence on καταδικάζω. ~ judgement εκδίδω απόφασιν, (non-legal) κρίνω. ~ the hat round βγάζω δίσκο. ~ a remark κάνω παρατήρηση. ~ water ουρώ.

pass v.i. (go past, be successful or approved) περνώ. she ~es for beautiful περνάει για ωραία. (at cards, etc.) πάω πάσσο.

pass away v. (cease) περνώ, (die) πεθαίνω.

pass by v.t. περνώ (κοντά) από, (ignore) αγνοώ. (v.i.) περνώ.

pass off v.t. (palm off) πασσάρω. (of indiscretion) ~ it off τα μπαλώνω. pass oneself off as an expert παριστάνω τον ειδικό. (v.i.) (cease) περνώ. (take place) γίνομαι, λαμβάνω χώραν.

pass on v.t. (news, disease) μεταδίδω, (order, message) διαβιβάζω, (tradition) μεταβιβάζω. (v.i., die) πεθαίνω.

pass out v.i. (of cadet) αποφοιτώ, (faint) λυποθυμώ.

pass through v.t. (traverse) διασχίζω, περνώ από. pass a thread through the needle περνώ μία κλωστή στη βελόνα. ~ a phase περνώ μία φάση.

pass up v.t. (neglect) αφήνω να μου ξεφύγει.

passab|le a. (road) διαβατός, (fair) υποφερτός, μέτριος. ~**ly** adv. καλούτσικα.

passage s. (act, place) διάβαση f., πέρασμα n. (of time) πέρασμα n. (journey) ταξίδι n. (way through) δρόμος m. (corridor) διάδρομος m. (alley) στενό n. (of book) χωρίο n. (of words) κουβγένια f., (hostile) (also ~ of arms) διαξιφισμός m.

pass-book s. βιβλιάριο καταθέσεων.

passenger s. επιβάτης m. (a.) επιβατικός. ~ ship επιβατηγό n.

passer-by s. διαβάτης m., περαστικός a.

passing *a.* περαστικός. (*s.*) πέρασμα *n.* (*disappearance*) εξαφάνιση *f.*

passion *s.* πάθος *n.* (*rage*) οργή *f.* have a ~ for έχω πάθος με. get in a ~ εξοργίζομαι. charged with ~ παθητικός. P~ (*of Christ*) τα Πάθη.

passionate *a.* (*ardent*) θερμός, (*love, patriotism, appeal*) φλογερός, (*love, anger*) παράφορος, (*voice, music*) παθητικός, (*embrace*) θερμός, περιπαθής. **~ly** *adv.* με πάθος.

passive *a.* (*person*) απαθής. (*attitude & gram.*) παθητικός. **~ly** *adv.* απαθώς.

pass-key *s.* πασπαρτού *n.*

passport *s.* διαβατήριο *n.*

password *s.* σύνθημα *n.*

past *prep.* (*beyond*) πέρα από. go ~ (*v.i.*) περνώ. go ~ the church περνώ έξω από την εκκλησία. it is ~ ten o'clock είναι περασμένες δέκα. it is ten ~ two είναι δύο και δέκα. I am ~ caring έπαψα να ενδιαφέρομαι. the coat is ~ repair(ing) το παλτό δεν σηκώνει πια άλλη επισκευή. I wouldn't put it ~ him to... τον θεωρώ ικανό να. be ~ one's prime έχω περάσει τα καλύτερα χρόνια μου. ~ endurance ανυπόφορος.

past *a.* περασμένος. for some time ~ από καιρό. (*s.*) παρελθόν *n.*

pasta *s.* ζυμαρικά *n.pl.*

paste *s.* (*for pastry*) ζυμάρι *n.* (*sticking*) κόλλα *f.* (*v.*) κολλώ.

pastel *s.* παστέλ *n.*

pasteurized *a.* παστεριωμένος.

pastiche *s.* απομίμηση *f.*

pastille *s.* παστίλια *f.*

pastime *s.* διασκέδαση *f.*

pastor *s.* πάστωρ *m.* **~al** *a.* (*eccl.*) ποιμαντορικός. (*scene*) βουκολικός.

pastry *s.* πάστα *f.*, σφολιάτα *f.*, ζυμάρι *n.* (*very thin*) φύλλο *n.* (*a cake*) πάστα *f.*, γλυκό *n.* **~cook** *s.* ζαχαροπλάστης *m.*

pasture *s.* (*land*) λιβάδι *n.*, βοσκοτόπι *n.* (*also herbage*) βοσκή *f.* (*v.*) βόσκω.

pasty *s.* κρεατόπηττα *f.*

pat *v.* χτυπώ ελαφρά. (*fig.*) pat (*person*) on the back επιδοκιμάζω. (*s.*) ελαφρό χτύπημα.

pat *adv.* στο τσακ, αμέσως. stand ~ δεν αλλάζω γνώμη.

patch *v.* μπαλώνω. ~ up επιδιορθώνω προχείρως. (*fig.*) ~ up a quarrel συμφιλιώνομαι.

patch *s.* (*for mending*) μπάλωμα *n.* (*plaster*) λευκοπλάστης *m.* (*spot, area*) σημείο *n.* (*piece, bit*) κομμάτι *n.* (*stain*) λεκές *m.* (*fam.*) not be a ~ on δεν πιάνω μπάζα μπροστά σε.

patchy *a.* (*uneven*) ανομοιογενής, (*blotchy*) με λεκέδες.

patent *s.* προνόμιο ευρεσιτεχνίας, πατέντα *f.* (*v.*) πατεντάρω.

patent *a.* (*plain*) προφανής. ~ leather λουστρίνι *n.* (*special*) ειδικός, (*clever*) πρωτότυπος. **~ly** *adv.* προφανώς.

paterfamilias *s.* οικογενειάρχης *m.*

paternal *a.* πατρικός. **~ism** *s.* πατερναλισμός *m.* **~istic** *a.* πατερναλιστικός.

paternity *s.* πατρότητα *f.*

path *s.* μονοπάτι *n.* (*pavement*) πεζοδρόμιο *n.* (*of star, bullet*) τροχιά *f.* (*way, also fig.*) δρόμος *m.* the beaten ~ πεπατημένη *f.*

pathetic *a.* αξιολύπητος, οικτρός. his state was ~ ήταν να τον κλαις.

pathology *s.* παθολογία *f.*

pathos *s.* πάθος *n.*

pathway *s.* δρόμος *m.* (*pavement*) πεζοδρόμιο *n.*

patience *s.* υπομονή *f.* ~ of Job Ιώβειος υπομονή. be out of ~ έχω χάσει την υπομονή μου. (*cards*) πασιέντζα *f.*

patient *s.* ασθενής *m.f.* (*client*) πελάτης *m.*

patient *a.* υπομονετικός. **~ly** *adv.* υπομονετικά.

patois *s.* τοπική διάλεκτος.

patriarch *s.* σεβάσμιος πρεσβύτης. (*eccl.*) πατριάρχης *m.* **~al** *a.* πατριαρχικός.

patrician *a. & s.* πατρίκιος.

patricide *see* parricide.

patrimony *s.* κληρονομία *f.*

patriot *s.* πατριώτης *m.* **~ic** *a.* πατριωτικός, (*person*) πατριώτης *m.* **~ism** *s.* πατριωτισμός *m.*

patrol *s.* (*activity*) περιπολία *f.* (*party*) περίπολος *f.* (*v.i.*) περιπολώ.

patron *s.* προστάτης *m.* (*rich*) ~ of arts Μαικήνας *m.* ~ saint προστάτης, άγιος, (*of city*) πολιούχος *a.* (*customer*) πελάτης *m.*, (*of café, etc.*) θαμών *m.*

patronage *s.* προστασία *f.* (*custom*) πελατεία *f.*

patronize *v.* (*support*) υποστηρίζω, (*go regularly to*) πηγαίνω τακτικά σε. (*condescend to*) μεταχειρίζομαι με προσβλητικά προστατευτικό ύφος.

patter (*talk*) φλυαρία *f.* (*noise*) αλαφροχτύπημα *n.* (*v.*) φλυαρώ, αλαφροχτυπώ.

pattern *s.* (*model*) πρότυπο *n.*, υπόδειγμα *n.* (*sample*) δείγμα *n.* (*design*) σχέδιο *n.* (*dress*) πατρόν *n.* (*v.*) σχεδιάζω.

paucity *s.* σπάνις *f.*

paunch *s.* κοιλιά *f.* (*fat*) κοιλάρα *f.* **~y** *a.* κοιλαράς *m.*

pauper *s.* άπορος *a.*

pause *s.* διακοπή *f.* (*mus.*) παύση *f.* (*v.*) σταματώ.

pave v. πλακοστρώνω. (fig.) ~ the way ανοίγω (or προλειαίνω) το δρόμο.
paved a. (slabs, tiles) πλακόστρωτος, (stones) λιθόστρωτος.
pavement s. πεζοδρόμιο n.
pavilion s. περίπτερο n.
paving-stone s. πλάκα f.
paw s. πόδι n. (fam.) ~s off! μάζεψε τα ξερά σου! (v.t.) (handle) πασπατεύω. ~ the ground χτυπώ το έδαφος.
pawn s. (chess & fig.) πιόνι n. (thing pledged) ενέχυρο n. (v.) ~ one's watch βάζω το ρολόι μου ενέχυρο. ~broker s. ενεχυροδανειστής m.
pay s. αμοιβή f., μισθός m.
pay v.t. & i. (also ~ for) πληρώνω. ~ a visit κάνω μία επίσκεψη. ~ attention προσέχω. ~ down (in part) προκαταβάλλω. ~ back or off (debt, creditor) ξεπληρώνω, εξοφλώ. I shall ~ him out (or back) for it (punish) θα του το πληρώσω, θα μου το πληρώσει. he didn't ~ me for it δεν μου το πλήρωσε, δεν με πλήρωσε για αυτό. he paid through the nose for it του κόστισε ο κούκος αηδόνι.
pay in v.t. (deposit) καταθέτω.
pay off v.i. (succeed) επιτυγχάνω. (v.t.) (dismiss) απολύω, (a debt, creditor) see pay.
pay-off s. (fam.) ώρα της πληρωμής.
pay out v.t. (money) καταβάλλω, (rope) αμολάω, (a person) εκδικούμαι.
payable a. πληρωτέος.
payee s. δικαιούχος m. (of draft) αποδέκτης m.
payment s. πληρωμή f.
pay-phone s. τηλέφωνο για το κοινό.
pay-roll s. μισθολόγιο n. be on the ~ of μισθοδοτούμαι από.
PC προσωπικός ηλεκτρονικός υπολογιστής.
pea s. μπιζέλι n. ~s αρακάς m. sweet ~ μοσχομπίζελο n.
peace s. ειρήνη f. (quiet) ησυχία f. (order) τάξη f. hold one's ~ σωπαίνω. keep the ~ τηρώ την τάξιν. make ~ (of belligerents) συνάπτω ειρήνη. make one's ~ συμφιλιώνομαι, τα φτιάνω. he gives me no ~ δεν με αφήνει ήσυχο.
peaceable a. ειρηνικός.
peaceful a. (not violent) ειρηνικός, (calm) ήρεμος, γαλήνιος. ~ly adv. ειρηνικά, ήρεμα.
peace-keeping a. ειρηνευτικός.
peach s. ροδάκινο n. (yellow) γιαρμάς m. (tree) ροδακινιά f (fam., sthg. excellent) όνειρο n. a ~ of a girl κορίτσι να το πιείς στο ποτήρι.
peacock s. παγώνι n.
peak s. (top) κορυφή f. (of achievement) κορύφωμα n. (of intensity) αιχμή f. ~

period περίοδος αιχμής. (of cap) γείσο n.
peal s. (of bells) κωδωνοκρουσία f. ~ of laughter ξέσπασμα γέλιων. ~ of thunder μπουμπουνητό n. (v.i.) καμπανίζω, ηχώ.
peanut s. αράπικο φιστίκι.
pear s. αχλάδι n. (tree) αχλαδιά f. wild ~ γκορτσιά f.
pearl s. μαργαρίτης m., μαργαριτάρι n. (a.) (also ~y) μαργαριταρένιος.
peasant χωρικός, χωριάτης m., χωριάτισσα f. (a) χωριάτικος. ~ry s. χωριικοί m.pl.
pease-pudding s. φάβα f.
peat s. τύρφη f.
pebb|le s. βότσαλο n. ~ly a. με βότσαλα.
peccadillo s. μικροπαράπτωμα n.
peck s. (fam.) ~ of troubles πολλά βάσανα.
peck v. τσιμπώ. (s.) τσίμπημα n.
peculation s. κατάχρηση f.
peculiar a. (own, special) ιδιαίτερος, ειδικός. (strange) παράξενος, περίεργος, αλλόκοτος. ~ly adv. ιδιαιτέρως, περίεργα, αλλόκοτα.
peculiarity s. (special feature) ιδιότητα f. (strangeness) περίεργον n. (eccentricity) παραξενιά f.
pecuniary a. χρηματικός.
pedagogic, ~al a. παιδαγωγικός.
pedal s. πετάλι n. (mus.) πεντάλ n.
pedant s., ~ic a. σχολαστικός. ~ically adv. σχολαστικά.
peddle v.t. πουλώ στη γύρα. ~r s. γυρολόγος m., πραματευτής m.
pederasty s. παιδεραστία f.
pedestal s. βάθρο n.
pedestrian s. & a. πεζός m. ~ crossing διάβαση πεζών.
pediatrics s. παιδιατρική f.
pedigree s. γενεαλογικόν δένδρον. ~ dog σκύλος ράτσας.
pediment s. αέτωμα n.
pedlar s. γυρολόγος m., πραματευτής m.
peek v. ~ at κρυφοκοιτάζω.
peel s. φλοιός m., φλούδι n. (v.t. & i.) (fruit, walls, skin) ξεφλουδίζω, (fruit) καθαρίζω. ~ off (unstick) ξεκολλώ.
peep s. λαθραίο βλέμμα. (v.) ~ at κρυφοκοιτάζω. ~ out of βγαίνω (or ξεπετιέμαι or ξεμυτίζω) μέσα από. ~ing Tom ηδονοβλεψίας m. ~~hole s. οπή παρατηρήσεως, (in front door) ματάκι n.
peer v. ~ at κοιτάζω προσεκτικά.
peer s. (one's equal) ταίρι n. (lord) ομότιμος a., λόρδος m.
peerage s. (rank) αξίωμα λόρδου. (body) λόρδοι m.pl.
peeress s. γυναίκα ομότιμος, (peer's wife) σύζυγος λόρδου.
peerless a. απαράμιλλος.

peeve v. (also be ~d) νευριάζω.
peevish a. γουρσούζης. ~ness s. γκρίνια f.
peg s. γόμφος m., ξυλόπροκα f. (stake) παλούκι n. (clothes) μανταλάκι n. ~s (for hanging) κρεμαστάρι n. off the ~ (ready made) έτοιμος. take (person) down a ~ κόβω τον αέρα σε. (fig., ground, pretext) αφορμή f.
peg v. στερεώνω με γόμφους (or με παλούκια). (clothes) πιάνω, (prices) καθηλώνω. ~ down (fig., restrict) περιορίζω. ~ away εργάζομαι επιμόνως. ~ out (die) πεθαίνω.
pejorative a. υποτιμητικός.
pelf s. παραδάκι n.
pelican s. πελεκάνος m.
pellet s. (pill) δισκίο n. (shot) σκάγι n. (ball) βωλαράκι n.
pell-mell adv. βιαστικά.
pellucid a. διαυγής, λαγαρός.
pelt s. δορά f., τομάρι n.
pelt v.t. (shower) βομβαρδίζω. he was ~ed with stones του έρριξαν πέτρες. (v.i.) it ~ed with rain έβρεξε καταρρακτωδώς. (s.) at full ~ ολοταχώς, δρομαίως.
pen s. (for writing) πέννα f. (v.) γράφω.
pen s. (animals') μάνδρα f., μαντρί n. (v., lit. & fig.) μαντρώνω.
penal a. ποινικός. ~ servitude καταναγκαστικά έργα.
penalize v. τιμωρώ. (fig.) be ~d φέρομαι εις μειονεκτικήν θέσιν.
penalty s. ποινή f. pay the ~ τιμωρούμαι. on ~ of επί ποινή (with gen.).
penance s. αυτοτιμωρία ως μετάνοια.
pence s. πέννες f.pl.
pencil s. μολύβι n. (v.) γράφω με μολύβι. delicately ~led καλογραμμένος.
pendant s. (of necklace) παντατίφ n. (of chandelier) κρεμαστό κρύσταλλος.
pending a. the decision is ~ η απόφαση αναμένεται. (prep.) (until) εν αναμονή (with gen.), (during) κατά (with acc.).
pendulous a. κρεμασμένος, κρεμαστός.
pendulum s. εκκρεμές n.
penetrable a. διαπερατός.
penetrate v. διαπερνώ, διεισδύω εις. (also of idea) εισχωρώ εις. (pierce) διατρυπώ, (imbue) εμποτίζω.
penetrating a. (voice, glance, cold) διαπεραστικός. (mind) οξύς, διορατικός.
penetration s. διείσδυση f., εισχώρηση f. (mental) διορατικότητα f.
penguin s. πιγκουΐνος m.
penicillin s. πενικιλλίνη f.
peninsula s. χερσόνησος f.
penis s. πέος n.
penitence s. μετάνοια f.

penitent a. be ~ μετανοώ. (words, etc.), γεμάτος μετάνοια. (s.) μετανοών a. ~ial a. της μετανοίας.
penitentiary s. φυλακή f.
penknife s. σουγιάς m.
penmanship s. καλλιγραφία f.
pen-name s. ψευδώνυμο n.
pennant s. επισείων m.
penniless a. απένταρος.
penny s. πέννα f. not a ~ ούτε μία πεντάρα. a pretty ~ πολλά λεφτά. turn an honest ~ βγάζω τίμια το ψωμί μου. the ~ has dropped (I understand) μπήκα. I'm not a ~ the wiser δεν κατάλαβα τίποτα. (fam.) spend a ~ πάω στο μέρος.
pension s. (lodging) πανσιόν f.
pension s. (pay) σύνταξη f. ~er s. συνταξιούχος m.f.
pensive a. συλλογισμένος. ~ly adv. συλλογισμένα.
pent a. κλεισμένος. ~ up συγκρατημένος.
pentagon s. πεντάγωνο n.
Pentecost s. Πεντηκοστή f.
penthouse s. (shed) υπόστεγο n. (top flat) ρετιρέ n.
penultimate a. προτελευταίος. (gram.) ~ syllable παραλήγουσα f.
penurious a. φτωχός. ~y a. ένδεια f.
people s. (nation, populace) λαός m. (persons) άνθρωποι m.pl., (in general) κόσμος m. my ~ (family, followers) οι δικοί μου. young ~ οι νέοι, η νεολαία. the common ~ κοσμάκης m. what will ~ say? τι θα πει ο κόσμος; ~'s (party, etc.) λαϊκός. (v.t.) (inhabit) κατοικώ, (fill) γεμίζω.
pep s. ζωντάνια f. ~ talk ενθαρρυντικά λόγια. (v.t.) ~ up ζωντανεύω.
pepper s. (condiment) πιπέρι n. (vegetable, tree) πιπεριά f. ~ pot πιπεριέρα f. (v.t.) (fig., throw) ρίχνω, (cover) γεμίζω.
peppermint s. μέντα f.
peppery a. πιπεράτος, (person) ευέξαπτος.
peptic a. (system) πεπτικός.
per prep. (by, via) διά (with gen.). ~ annum κατ' έτος. ~ cent τοις εκατό. ~ capita ανά κεφαλήν. ~ kilo το κιλό. as ~ κατά, συμφώνως προς (both with acc.).
peradventure adv. ίσως. (after if, lest) τυχόν.
perambulate v. περιφέρομαι. ~ion s. περιοδεία f., βόλτα f. ~or s. καρροτσάκι n.
perceive v. (see) διακρίνω, (understand) αντιλαμβάνομαι.
percentage s. ποσοστό n.
perceptible a. αισθητός, αντιληπτός. ~ly adv. αισθητώς.
perception s. αντίληψη f. organs of ~

αισθητήρια όργανα.
perceptive *a.* αντιληπτικός. ~**ly** *adv.* με αντιληπτικότητα. ~**ness** *s.* αντιληπτικότητα *f.*
perch *s.* κούρνια *f.* (*v.i.*) κάθομαι. (*v.t.*) βάζω. a house ~ed on a rock σπίτι σκαρφαλωμένο σ' ένα βράχο.
perchance *adv.* ίσως. (*after* if) τυχόν, κατά τύχην.
percolat|e *v.t. & i.* περνώ. (*of news*) ~ through διοχετεύομαι. ~**or** *s.* καφετιέρα με φίλτρο.
percussion *s.* επίκρουση *f.* (*mus.*) κρουστά όργανα.
perdition *s.* όλεθρος *m.*
peregrination *s.* περιπλάνηση *f.*
peremptory *a.* κατηγορηματικός.
perennial *a.* αιώνιος, (*plant*) πολυετής.
perfect *a.* τέλειος, (*full*) πλήρης, (*excellent*) θαυμάσιος. (*utter*) a ~ fool σωστός βλάκας. (*gram.*) παρακείμενος *m.*, future ~ τετελεσμένος μέλλων. (*v.*) τελειοποιώ.
perfection *s.* (*state*) τελειότητα *f.* (*act*) τελειοποίηση *f.*
perfectly *adv.* (*to perfection*) τέλεια, (*completely*) τελείως.
perfid|ious *a.* δόλιος, άπιστος. ~**y** *s.* δολιότητα *f.*, απιστία *f.*
perforat|e *v.* διατρυπώ. ~**ed** *a.* διάτρητος, (*paper*) με τρύπες. ~**ion** *s.* διάτρηση *f.* (*in paper*) τρύπες *f.pl.*
perforce *adv.* κατ' ανάγκην.
perform *v.* (*carry out*) εκτελώ. (*play*) παίζω, παριστάνω.
performance *s.* εκτέλεση *f.*, παράσταση *f.* (*functioning*) απόδοση *f.* (*in games, etc.*) επίδοση *f.* (*fam., business*) υπόθεση *f.*
performer *s.* εκτελεστής *m.* (*actor*) ηθοποιός *m.f.*
perfume *s.* άρωμα *n.*, μυρωδιά *f.* (*v.*) αρωματίζω. be ~d with violets μοσχοβολώ μενεξέ.
perfunctory *a.* τυπικός, (*showing no interest*) αδιάφορος.
pergola *s.* κρεββατίνα *f.*
perhaps *adv.* ίσως, πιθανώς.
peril *s.* κίνδυνος *m.* ~**ous** *a.* επικίνδυνος.
perimeter *s.* περίμετρος *f.*
period *s.* περίοδος *f.* (*age, era*) εποχή *f.* (*costume, etc.*) of the ~ της εποχής. ~**ic** *a.* περιοδικός. ~**ically** *adv.* περιοδικώς.
periodical *s.* περιοδικό *n.*
peripatetic *a.* περιπατητικός.
peripher|al *a.* περιφερικός, περιφερειακός. ~**y** *s.* περιφέρεια *f.*
periphras|is *s.* περίφραση *f.* ~**tic** *a.* περιφραστικός.
periscop|e *s.* περισκόπιο *n.* ~**ic** *a.* περι-

σκοπικός.
perish *v.* (*die*) πεθαίνω, (*in natural disaster*) χάνομαι. (*wear out*) φθείρομαι. (*fam.*) be ~ing (*or* ~ed with) cold πεθαίνω από το κρύο. it is ~ing cold κάνει ψόφο. ~ the thought! Θεός φυλάξοι!
perishable *a.* (*beauty, etc.*) φθαρτός, (*goods*) αλλοιώσιμος, που χαλάει.
perjure *v.* ~ oneself ψευδορκώ.
perjur|ed *a.*, ~**er** *s.* ψεύδορκος. ~**y** *s.* ψευδορκία *f.*
perk *v.t.* ~ up (*lift*) σηκώνω, (*v.t. & i.*) (*revive*) ξαναζωντανεύω.
perks *s.* (*fam.*) τυχερά *n.pl.*
perky *a.* ζωηρός, κεφάτος. (*hat*) σκερτσόζικος.
perm *s.* (*fam.*) περμανάντ *f.*
permanen|ce *s.* μονιμότητα *f.* ~**cy** *s.* (*job, etc.*) μόνιμος *a.*
permanent *a.* μόνιμος, (*exhibition*) διαρκής. ~**ly** *adv.* μονίμως.
permeable *a.* διαπερατός.
permeate *v.* διαπερνώ, διεισδύω εις. (*fill*) γεμίζω.
permissible *a.* επιτρεπτός. is it ~? επιτρέπεται;
permission *s.* άδεια *f.*
permissive *a.* (*laws*) επιτρεπτικός. ~ society ανεκτική κοινωνία. ~**ness** *s.* ανεκτικότητα *f.*
permutation *s.* μετάθεση *f.*
pernicious *a.* βλαβερός.
pernickety *a.* (*fussy*) δύσκολος, (*ticklish*) λεπτός.
peroration *s.* κατακλείς της αγορεύσεως.
peroxide *s.* υπεροξείδιον *n.* ~ of hydrogen οξυγονούχον ύδωρ, (*fam.*) οξυζενέ *n.*
perpendicular *a.* κάθετος.
perpetratle *v.* διαπράττω. ~**ion** *s.* διάπραξη *f.* ~**or** *s.* δράστης *m.*
perpetual *a.* διαρκής, αέναος, αδιάκοπος. ~ motion αεικίνητον *n.* ~**ly** *adv.* διαρκώς.
perpetuate *v.* διαιωνίζω.
perpetuity *s.* in ~ εις το διηνεκές.
perplex *v.* μπερδεύω, σαστίζω. ~**ed** *a.* αμήχανος. ~**ity** *s.* αμηχανία *f.*
perquisites *s.* τυχερά *n.pl.*
persecut|e *v.* καταδιώκω. ~**ion** *s.* διωγμός *n.* ~ion complex μανία καταδιώξεως. ~**or** *s.* διώκτης *m.*
persever|ance *s.* επιμονή *f.* ~**ing** *a.* επίμονος. ~**e** *v.* επιμένω.
Persian *a.* περσικός. (*person*) Πέρσης *m.*, Περσίδα *f.*
persiflage *s.* αστεϊσμοί *m.pl.*
persist *v.* (*not give up*) επιμένω. (*continue to be*) εξακολουθώ.
persistent *a.* επίμονος, εξακολουθητικός.

~ly συνεχώς.

person s. πρόσωπον n., άνθρωπος m., άτομο n. in ~ αυτοπροσώπως. any ~s wishing to take part οι επιθυμούντες να λάβουν μέρος.

persona s. ~ non grata ανεπιθύμητο πρόσωπο.

personable a. εμφανίσιμος.

personage s. πρόσωπον n., προσωπικότητα f.

personal a. (affairs, use, appearance, remarks, pronoun) προσωπικός. (rights, servant) ατομικός. **~ity** s. προσωπικότητα f.

personally adv. (for oneself) προσωπικώς, (in person) αυτοπροσώπως.

person|ification s. προσωποποίηση f. **~ify** v. προσωποποιώ.

personnel n. προσωπικό n.

perspective s. προοπτική f. (fig.) see it in its right ~ το βλέπω από τη σωστή του όψη.

perspicac|ious a. διορατικός. **~ity** s. διορατικότητα f.

perspicu|ity s. σαφήνεια f. **~ous** a. σαφής.

perspir|ation s. (sweating) ίδρωμα n. (sweat) ιδρώτας m. **~e** v. ιδρώνω.

persuade v. πείθω, (coax) καταφέρνω. **~d** πεπεισμένος.

persuasion s. πειθώ f. (conviction) πεποίθηση f. (religion) θρήσκευμα n.

persuasive a. πειστικός. **~ly** adv. με πειστικότητα.

pert a. (person) προπετής.

pertain v. ~ το αφορώ.

pertinac|ious a. επίμονος. **~ity** s. επιμονή f.

pertinent a. σχετικός, σχέσιν έχων.

perturb v. ανησυχώ. **~ation** s. ανησυχία f.

perus|al s. ανάγνωση f. εξέταση f. **~e** v. διαβάζω, εξετάζω.

pervade v. (prevail in) επικρατώ εις, (fill) γεμίζω.

pervasive a. που επικρατεί παντού.

perverse a. ανάποδος. ~ person (fam.) κέρατο n. **~ness** s. αναποδιά f.

perversion s. διαστροφή f.

perversity s. αναποδιά f.

pervert s. διεστραμμένος a.

pervert v.t. διαστρέφω, (morally) διαφθείρω. **~ed** a. (twisted) στραβός, (sexually) διεστραμμένος.

pessimism s. απαισιοδοξία f.

pessimist s., **~ic** a. απαισιόδοξος.

pest s. (nuisance) πληγή f. see plague.

pester v. ενοχλώ. (fam.) he **~ed** me μου έγινε τσιμπούρι.

pestilence s. λοιμός m., πανώλης f.

pestilent, ~ial a. λοιμώδης, (fam.) σιχαμένος.

pestle s. κόπανος m., γουδοχέρι n.

pet v. χαϊδεύω, κανακεύω.

pet s. (animal) κατοικίδιο ζώο. (s. & a.) (favourite) αγαπημένος, (spoilt) χαϊδεμένος, κανακάρης. my ~! χρυσό μου! ~ name χαϊδευτικό όνομα. my ~ aversion η αντιπάθειά μου.

pet s. (temper) be in a ~ έχω νευράκια.

petal s. πέταλο n.

petard s. he was hoist with his own ~ πιάστηκε στην παγίδα που είχε στήσει για άλλους.

peter v. ~ out σβήνω, χάνομαι.

petit a. ~ bourgeois μικροαστός m. ~ four πτιφούρ n.

petite a. μικροκαμωμένη.

petition s. αίτηση f. (v.i.) αιτούμαι, υποβάλλω αίτηση. **~er** s. αιτών m., αιτούσα f.

petrel s. (fig.) he is a stormy ~ η παρουσία του προοιωνίζεται έριδα.

petri|faction s. απολίθωση f. **~fy** v.t. πετρώνω, απολιθώνω. (v.i.) πετρώνω, απολιθώνομαι.

petrochemicals s. πετροχημικά n.pl.

petrol s. βενζίνη f. ~ station πρατήριο βενζίνης. ~ gauge δείκτης βενζίνης. ~ tank ρεζερβουάρ.

petroleum s. πετρέλαιο n.

petticoat s. μεσοφόρι, μισοφόρι n. ~ government γυναικοκρατία f.

pettifogging a. λεπτολόγος, στρεψόδικος.

pettiness s. μικροπρέπεια f.

pettish see peevish.

petty a. ασήμαντος, μικρο-. (person) μικροπρεπής. ~ cash μικροέξοδα n.pl. (naut.) ~ officer υπαξιωματικός m. chief ~ officer κελευστής m.

petul|ance s. νεύρα n.pl. **~ant** a. νευρικός.

pew s. στασίδι n.

phalanx s. φάλαγγα f.

phall|ic a. φαλλικός. **~us** s. φαλλός m.

Phanariot s. Φαναριώτης m.

phantasm s. ψευδαίσθηση f. **~agoria** s. φαντασμαγορία f.

phantom s. φάντασμα n.

Pharaoh s. Φαραώ m.

pharisaic, ~al a. φαρισαϊκός.

pharisee s. φαρισαίος m.

pharmaceutical a. φαρμακευτικός.

pharmacology s. φαρμακολογία f.

pharmacopoeia s. φαρμακοποιία f.

pharmacy s. (dispensing) φαρμακευτική f. (shop) φαρμακείο n.

pharynx s. φάρυγξ m.

phase s. φάση f. (v.) ~ in (or out) εισάγω (or καταργώ) σταδιακώς.

pheasant s. φασιανός m.

phenomenal a. φαινομενικός. (fam.) (prodigious) καταπληκτικός. **~ly** adv. καταπληκτικά.

phenomenon *s.* φαινόμενο *n.*
phial *s.* φιαλίδιο *n.*
philander *v.* φλερτάρω. ~er *s.* κορτάκιας *m.*
philanthrop|ic *a.* φιλανθρωπικός. ~ist *s.* φιλάνθρωπος *m.* ~y *s.* φιλανθρωπία *f.*
philatel|ist *s.* φιλοτελιστής *m.* ~y *s.* φιλοτελισμός *m.*
philhellen|e *s.* φιλέλλην *m.* ~ic *a.* φιλελληνικός. ~ism *s.* φιλελληνισμός *m.*
philippic *s.* φιλιππικός *m.*
philistine *s.* φιλισταίος *m.*
philolog|ical *a.* γλωσσολογικός. ~ist *s.* γλωσσολόγος *m.* ~y *s.* γλωσσολογία *f.*
philosopher *s.* φιλόσοφος *m.*
philosophical *a.* φιλοσοφικός, (*resigned*) φιλόσοφος *m.* ~ly *adv.* φιλοσοφικά, σάν φιλόσοφος.
philosoph|ize *v.* κάνω το φιλόσοφο. ~y *s.* φιλοσοφία *f.*
philtre *s.* φίλτρο *n.*
phlegm *s.* (*mucus*) φλέμα *n.* (*calm*) φλέγμα *n.* ~atic *a.* φλεγματικός.
phobia *s.* φοβία *f.*
Phoenician *a.* φοινικικός, (*person*) Φοίνικας.
phoenix *s.* φοίνικας *m.*
phone *s.* τηλέφωνο *n.* (*v.i.*) τηλεφωνώ. (*v.t.*) παίρνω στο τηλέφωνο. (*a.*) ~ book/box/ number τηλεφωνικός κατάλογος / θάλαμος / αριθμός. ~ call τηλεφώνημα.
phonetic *a.* φωνητικός. ~ian *s.* φωνητικός επιστήμων. ~s *s.* φωνητική *f.*
phoney *a.* (*fam.*) ψεύτικος, δήθεν.
phonic *a.* φθογγικός.
phonograph *s.* γραμμόφωνο *n.*
phosphate *s.* φωσφορικόν άλας.
phosphoresc|ence *s.* φωσφορισμός *m.* ~ent *a.* φωσφορίζων.
phosphorus *s.* φωσφόρος *m.*
photo *s.* φωτογραφία *f.*
photocopy *s.* φωτοαντίγραφο *n.*
photogenic *a.* φωτογενής.
photograph *s.* φωτογραφία *f.* ~er *s.* φωτογράφος *m.* ~ic *a.* φωτογραφικός. ~y *s.* φωτογραφία *f.*
photostat *s.* φωτοτυπία *f.*
phrase *s.* (*expression*) έκφραση *f.* (*gram. & mus.*) φράση *f.* (*v.*) εκφράζω.
phrase-book *s.* εγχειρίδιο φράσεων ξένης γλώσσας.
phraseology *s.* φρασεολογία *f.*
phrasing *s.* (*mus.*) φρασάρισμα *n.*
phrenetic *a.* φρενήρης.
phut *adv.* (*fam.*) go ~ χαλώ.
phylloxera *s.* φυλλοξήρα *f.*
physic *s.* (*medicine*) γιατρικό *n.* (*v.*) δίνω γιατρικό σε.
physical *a.* φυσικός, (*bodily*) σωματικός. ~

education (*or* jerks) γυμναστική *f.* ~ly *adv.* φυσικώς, σωματικώς.
physician *s.* ιατρός παθολόγος.
physic|s *s.* φυσική *f.* ~ist *s.* φυσικός *m.*
physiognomy *s.* φυσιογνωμία *f.*
physiolog|ical *a.* φυσιολογικός. ~ist *s.* φυσιολόγος *m.* ~y *s.* φυσιολογία *f.*
physiotherap|ist *s.* φυσιοθεραπευτής *m.* ~y *s.* φυσιοθεραπεία *f.*
physique *s.* σωματική διάπλαση, κράση *f.*
pianist *s.* πιανίστας *m.*
piano *s.* πιάνο *n.* grand ~ πιάνο με ουρά. ~forte *s.* κλειδοκύμβαλο *n.*
piastre *s.* γρόσι *n.*
picaresque *a.* (*story*) με περιπέτειες τυχοδιωκτών.
pick *s.* (*also* ~-axe) αξίνα *f.*
pick *s.* (*best part*) αφρόκρεμα *f.*, ό,τι εκλεκτό. take one's ~ διαλέγω το καλύτερο.
pick *v.t.* (*cut, pluck*) κόβω, μαζεύω. (*peck*) τσιμπώ. (*select*) διαλέγω, (*take, remove*) βγάζω. (*a bone*) γλείφω, (*one's teeth, nose, a lock*) σκαλίζω. ~ holes in βρίσκω τρωτά σε. ~ a quarrel στήνω καβγά. ~ one's way βαδίζω προσεκτικά. ~ and choose διαλέγω και παίρνω μόνο το καλύτερο. I had my pocket ~ed μου βούτηξαν το πορτοφόλι. I've got a bone to ~ with you θα μου δώσεις λόγο.
pick at *v.t.* (*eat*) τσιμπώ χωρίς όρεξη.
pick off *v.t.* (*kill*) σκοτώνω ένα ένα, (*remove*) βγάζω.
pick out *v.t.* (*choose*) διαλέγω, (*distinguish*) ξεχωρίζω, (*remove*) βγάζω, (*underline*) τονίζω.
pick up *v.t.* (*sthg. dropped, laid down*) μαζεύω. (*lift*) σηκώνω, παίρνω στα χέρια μου, (*get*) βρίσκω, (*learn*) μαθαίνω, (*unearth*) ξετρυπώνω. (*collect, call for*) παίρνω. (*make casual acquaintance of*) ψωνίζω. (*a habit*) αποκτώ. (*v.i.*) (*get better, of person*) συνέρχομαι, αναλαμβάνω, παίρνω επάνω μου. (*of conditions*) βελτιώνομαι.
pickaback *adv.* στην πλάτη.
picked *a.* (*choice*) διαλεχτός, επίλεκτος.
picker *s.* συλλέκτης *m.*
picket *s.* (*mil.*) προφυλακή *f.* (*strikers'*) σκοποί *m.pl.* (*v.t.*) βάζω σκοπούς μπροστά σε.
picking *s.* (*of fruit, etc.*) συγκομιδή *f.* ~s τσιμπολογήματα *n.pl.*
pickle *s.* (*edible*) τουρσί *n.* (*brine*) άλμη *f.* in a ~ (*fam.*) μπλεγμένος. I have a rod in ~ for him έχω ράμματα για τη γούνα του. (*v.t.*) κάνω τουρσί.
pick-me-up *s.* τονωτικό *n.*
pickpocket *s.* πορτοφολάς *m.*
pick-up *s.* (*audio*) πικάπ *n.* (*fam., person*)

ψώνιο *n.* ~ **van** ημιφορτηγό.

picnic *s.* πικνίκ *n.* it's no ~ δεν είναι παίξε-γέλασε. (*v.*) πάω για πικνίκ.

pictorial *a.*, ~**ly** *adv.* με εικόνες.

picture *s.* εικών, εικόνα *f.* (*on wall*) πίνακας *m.* (*film*) ταινία *f.*, ~s κινηματογράφος *m.* (*fig.*, *beautiful scene*) ζωγραφιά *f.* the ~ of health η προσωποποίηση της υγείας. put (*person*) in the ~ ενημερώνω. ~ gallery πινακοθήκη *f.* ~ postcard κάρτα *f.* (*v.t.*) απεικονίζω. ~ to oneself φαντάζομαι.

picturesque *a.* γραφικός. ~**ly** *adv.* γραφικά. ~**ness** *s.* γραφικότητα *f.*

piddling *a.* τιποτένιος.

pidgin *s.* (*fam.*) it's not my ~ δεν είναι της αρμοδιότητός μου. ~ English παρεφθαρμένα αγγλικά της Άπω Ανατολής.

pie *s.* (*cheese, meat*) πίττα *f.* fruit ~ γλύκισμα με φρούτο και σφολιάτα. have a finger in the ~ είμαι ανακατεμένος. eat humble ~ ρίχνω τα αφτιά μου. easy as ~ πολύ εύκολο.

piebald *a.* με άσπρες και μαύρες κηλίδες.

piece *s.* κομμάτι *n.*, τεμάχιο *n.* ~ of advice συμβουλή *f.* ~ of furniture έπιπλο *n.* 10-drachma ~ δεκάρικο *n.* nice ~ (*fam.*, *woman*) κόμματος *m.* they are all of a ~ μοιάζουν, ταιριάζουν. take to ~s ξηλώνω. pull or tear to ~s κομματιάζω. come to ~s διαλύομαι. break to ~s (*v.i.*) γίνομαι κομμάτια. go to ~s (*fig.*) καταρρέω, διαλύομαι. I gave him a ~ of my mind του τα είπα από την καλή. (*v.t.*) ~ together συναρμολογώ.

piecemeal *adv.* κομμάτι- κομμάτι, τμηματικώς.

pied *a.* ποικιλόχρωμος.

pier *s.* (*landing*) αποβάθρα *f.* (*of arch*) αψιδοστάτης *m.*, ποδαρικό *n.*

pierce *v.* διατρυπώ. (*penetrate*) διεισδύω εις, διαπερνώ, (*the heart*) σχίζω. ~**ing** *a.* διαπεραστικός.

pierrot *s.* πιερότος *m.*

piety *s.* ευσέβεια *f.* (*duty*) σέβας *n.*

piffle *s.* ανοησίες *f.pl.* ~**ling** *a.* ασήμαντος.

pig *s.* χοίρος *m.*, γουρούνι *n.*

pigeon *s.* περιστέρι *n.* (*young*) πιτσούνι *n.*

pigeon-hole *s.* θυρίδα *f.* (*v.*) ταξινομώ, (*put aside*) βάζω στο χρονοντούλαπο.

piggish *a.* (*greedy*) λαίμαργος, (*dirty*) βρώμικος.

piggy *s.* γουρουνάκι *n.* ~ bank κουμπαράς *m.*

pigheaded *a.* ξεροκέφαλος.

pig-iron *s.* χυτοσίδηρος *m.*

pigment *s.* χρωστική *f.* ~**ation** *s.* χρωματισμός *m.*

pigmy *a. & s.* πυγμαίος.

pike *s.* (*spear*) κοντάρι *n.* (*toll*) διόδια *n.pl.*

pikestaff *s.* plain as a ~ ολοφάνερος.

pilaf *s.* πιλάφι *n.*

pilaster *s.* πιλάστρι *n.*

pile *s.* (*beam*) πάσσαλος *m.* (*of carpet*) πέλος *n.*

pile *s.* (*heap*) σωρός *m.*, στοίβα *f.* (*electric*) στήλη *f.* (*funeral*) πυρά *f.* make one's ~ κάνω την μπάζα μου.

pile *v.t.* (*heap*) (*also* ~ up) στοιβάζω. (*load*) φορτώνω. ~ up (*v.t.*) (*accumulate*) συσσωρεύω, (*money*) μαζεύω, (*v.i.*) συσσωρεύομαι, μαζεύομαι, (*collide*) τρακάρω, (*run aground*) εξοκέλλω. ~ into στριμώχνομαι μέσα σε. ~ it on τα παραλέω.

piles *s.* αιμορροΐδες *f.pl.*

pilfer *v.* κλέβω, βουτώ. ~**er** *s.* κλέφτης *m.* ~**ing** *s.* κλέψιμο *n.*

pilgrim *s.* προσκυνητής *m.* ~**age** *s.* προσκύνημα *n.*

pill *s.* χάπι *n.*

pillage *v.* λεηλατώ. (*s.*) λεηλασία *f.*

pillar *s.* κίων *m.*, στύλος *m.*, κολόνα *f.* (*fig.*) στύλος, στυλοβάτης *m..*, be driven from ~ to post δεν μ' αφήνουν σε χλωρό κλαδί. ~**box** *s.* γραμματοκιβώτιο *n.*

pillion *s.* ride ~ (*horse*) κάθομαι πισωκάπουλα, (*motorcycle*) κάθομαι από πίσω.

pillory *v.* (*fig.*) διαπομπεύω.

pillow *s.* μαξιλλάρι *n.*, προσκέφαλο *n.* (*v.*) ακουμπώ. ~**case** *s.* μαξιλλαροθήκη *f.*, κλίφι *n.*

pilot *s.* (*naut.*) πιλότος *m.*, πλοηγός *m.* (*aero.*) πιλότος *m.* (*fig.*, *guide*) οδηγός *m.* (*a.*, *experimental*) δοκιμαστικός. (*v.*) οδηγώ. ~**boat** *s.* πιλοτίνα *f.* ~**burner** *s.* οδηγός καυστήρας *m.*

pimp *s.* μαστροπός *m.* (*v.*) μαστροπεύω.

pimple *s.* σπυρί *n.* get ~es βγάζω σπυριά. ~**y** *a.* σπυριασμένος.

pin *s.* καρφίτσα *f.* I don't care a ~ καρφί δεν μου καίγεται. get ~s and needles μουδιάζω. for two ~s I'd punch his nose μόλις κρατιέμαι να μην του σπάσω τα μούτρα.

pin *v.* καρφιτσώνω, πιάνω με καρφίτσα, (*one's hopes*) στηρίζω. ~ (*blame*) on φορτώνω σε. ~ down (*immobilize*) καθηλώνω, (*beneath wreckage*) πλακώνω. he won't be ~ned down δεν θέλει να δεσμευθεί.

pinafore *s.* ποδιά *f.*

pincers *s.* λαβίδα *f.* (*lobster's*) δαγκάνες *f.pl.*

pinch *s.* τσιμπιά *f.* (*of salt, etc.*) πρέζα *f.* at a ~ στην ανάγκη. feel the ~ τα έχω στενά.

pinch *v.t.* τσιμπώ, (*one's finger*) μαγκώνω. (*of shoe*) πιάνω, (*fig.*) that's where the shoe ~es εκεί είναι η σφίξη. (*v.i.*) ~ and scrape σφίγγω το ζωνάρι μου.

pin-cushion s. μαξιλλάρι για καρφίτσες.
pine s. πεύκο n.
pine v. λειώνω, μαραζώνω. ~ for (desire) λαχταρώ.
pineapple s. ανανάς m.
pine-cone s. κουκουνάρι n.
ping-pong s. πιγκ-πογκ n.
pinion s. (wing) φτερούγα f. (v., bind) δένω.
pink a. ροζ. (s.) (flower) γαρύφαλλο n. in the ~ of condition σε άριστη κατάσταση he's in the ~ είναι υγιέστατος.
pink v.i. (of engine) χτυπώ.
pin-money s. χαρτζιλίκι n.
pinnace s. άκατος f. (steam) ατμάκατος f.
pinnacle s. (peak) κορυφή f. (on roof) βέλος n. (of fame) κολοφών m.
pinpoint v. επισημαίνω επακριβώς.
pinprick s. μικροενόχληση f.
pint s. πίντα f.
pioneer s. πρωτοπόρος m. (mil.) σκαπανεύς m. (v.i.) καινοτομώ. (v.t.) εισάγω. ~ing a. ρηξικέλευθος.
pious a. ευσεβής, θρήσκος. (dutiful) φερόμενος μετά σεβασμού. ~ly adv. ευσεβώς.
pip s. (seed) κουκούτσι n. (fam.) have the ~ είμαι στις κακές μου.
pipe s. σωλήνας m. (mus.) αυλός m. (bucolic) φλογέρα f., (Pan's) σύριγξ f. (whistle, note) σφύριγμα n. (smoker's) πίπα f., τσιμπούκι n.
pipe v.t. (convey liquid) διοχετεύω. (play) παίζω. (v.t.) (whistle) σφυρίζω. ~ up πετάγομαι και λέω. ~ down μαζεύομαι.
piped a. (edged) με φιτίλι. ~ water τρεχούμενο νερό.
pipeline s. (oil) πετρελαιαγωγός m. (fig.) in the ~ (supplies) καθ' οδόν, (plans) υπό επεξεργασίαν.
piper s. αυλητής m. (fig.) you must pay the ~ το ήθελες – θα το πληρώσεις.
piping s. (edging) φιτίλι n. (system of pipes) σωληνώσεις f.pl. six foot of ~ σωλήνας μήκους έξι ποδιών. (a.) ~ hot καυτός.
piqu|ancy s. νοστιμάδα f. ~ant a. πικάντικος.
pique s. πίκα f. (v.t.) (stir) κεντρίζω. ~d κακοφανισμένος.
piqué a. πικέ.
piracy s. πειρατεία f.
pirat|e s. πειρατής m. (a.) πειρατικός. (v.) κλέβω, (books) ανατυπώνω παρανόμως. ~ical a. πειρατικός.
pistachio s. φιστίκι n. (tree) φιστικιά f.
pistol s. πιστόλι n.
piston s. έμβολο n., πιστόνι n.
pit s. λάκκος m. (mine) ορυχείο n. (scar) σημάδι n.
pit v. (set against) βάζω εναντίον (with

gen.). ~ oneself against ανταγωνίζομαι. be ~ted against έχω ως αντίπαλο.
pitch s. (tar) πίσσα f. (for game) γήπεδο n. (trader's) στέκι n. (degree) βαθμός m. (slope) κλίση f. (of tone) ύψος n., ~ black κατάμαυρος, ~ dark θεοσκότεινος.
pitch v.t. (throw) ρίχνω, (set up) στήνω, (relate) λέω. (v.i.) (of ship) σκαμπανεβάζω. ~ on διαλέγω. ~ into (work) ρίχνομαι σε, (food) πέφτω με τα μούτρα σε, (opponent) επιτίθεμαι εις. ~ed battle μάχη εκ παρατάξεως.
pitcher s. στάμνα f.
pitchfork s. δικράνι n. (v.) χώνω, βάζω.
piteous a. οικτρός, θλιβερός. ~ly adv. θλιβερά.
pitfall s. παγίδα f.
pith s. ψίχα f. (essence) ουσία f. ~ily adv. αποφθεγματικώς. ~y a. νευρώδης, λιτός και ζουμερός.
pithead s. είσοδος ορυχείου.
pitiab|le a. ελεεινός, ~ly adv. ελεεινά.
pitiful a. (state) ελεεινός, (feeling pity) σπλαχνικός. ~ly adv. ελεεινά, σπλαχνικά.
pitiless a. ανήλεος. ~ly adv. ανηλεώς.
pittance s. γλίσχρος μισθός. for a mere ~ για ένα κομμάτι ψωμί.
pitted a. (scarred) σημαδεμένος.
pity s. οίκτος m., έλεος n. have ~ on λυπάμαι. a ~ (regrettable) κρίμα n. it's a ~ you didn't see him κρίμα που δεν τον είδατε.
pity v.t. λυπάμαι. he is to be pitied είναι αξιολύπητος, είναι να τον λυπάσαι.
pivot s. άξονας m. (v.i.) περιστρέφομαι. ~al a. (fig.) βασικός.
placard s. πλακάτ n.
placate v. εξευμενίζω.
place s. μέρος n. (position, rank) θέση f. (site, locality, spot) τόπος n. all over the ~ παντού. in the first ~ πρώτα- πρώτα. out of ~ (remark) εκτός τόπου, (dress) αταίριαστος. take ~ γίνομαι, λαμβάνω χώραν. take the ~ of αντικαθιστώ. in ~ of αντί (with gen.), αντί για (with acc.).
place v. βάζω, τοποθετώ, (an order) δίδω. (classify) κατατάσσω. you know how I am ~d ξέρετε την κατάστασή μου.
placid a. ήρεμος, πράος. ~ity s. ηρεμία f., πραότητα f.
plage s. πλαζ f.
plagiar|ism s. λογοκλοπία f. ~ist s. λογοκλόπος m. ~ize v. κλέβω.
plague s. πανώλης f. (annoyance) πληγή f. (v.) βασανίζω.
plaid s. σκωτσέζικο ύφασμα.
plain s. πεδιάδα f., κάμπος m.
plain a. (evident) ξεκάθαρος, ολοφάνερος. (comprehensible) καθαρός. (simple, or-

dinary) απλός, (unadorned) απέριττος. (with nothing added) σκέτος. (in looks) όχι ωραίος, του σωρού. ~ truth καθαρή αλήθεια. ~ dealing εντιμότητα f. it was ~ sailing δεν υπήρχαν δυσκολίες. ~-spoken ντόμπρος.

plainly adv. (clearly) ολοφάνερα, καθαρά. (simply) απλά, (frankly) ντόμπρα. speak ~! μίλα ρωμαίικα.

plainness s. (clarity) σαφήνεια f. (simplicity) απλότητα f. (frankness) ειλικρίνεια f.

plainsong s. γρηγοριανόν μέλος.

plaint s. παράπονο n. (law) αγωγή f. **~iff** s. ενάγων m.

plaintive a. παραπονετικός, (music) κλαψιάρικος.

plait s. πλεξούδα f. (v.) κάνω πλεξούδα.

plan v.t. σχεδιάζω, μελετώ. (arrange) σχεδιάζω, προγραμματίζω. (v.i., intend) προτίθεμαι, λέω. ~ned (intentional) προμελετημένος. (s.) σχέδιο n., σχεδιόγραμμα n. (arrangements) σχέδιο n. draw up the ~s (of building, etc.) κάνω τη μελέτη. **~ning** s. προγραμματισμός m.

plane s. (level) επίπεδο n., πλάνο n. (aero.) αεροπλάνο n.

plane s. (tool) ροκάνι n. (v.) ροκανίζω.

plane s. (tree) πλάτανος m.

planet s. πλανήτης m.

plangent a. κλαψιάρικος.

plank s. σανίδα f. (v.) (fam.) ~ down πετώ κάτω.

plant s. φυτό n. (installations) εγκαταστάσεις f.pl. (fam., swindle) μηχανή f. (a.) (of plants) φυτικός.

plant v. φυτεύω. (place) βάζω, (a blow) καταφέρω, (a sharp instrument) καρφώνω. (conceal) κρύβω. ~ oneself (take up position) στέκομαι, (comfortably) θρονιάζομαι, στρώνομαι.

plantation s. φυτεία f.

planter s. καλλιεργητής m.

plaque s. πλάκα f.

plash s. (of fountain) κελάρυσμα n. (v.) κελαρύζω.

plasma s. πλάσμα n.

plaster s. (for walls) σοβάς m. ~ of Paris γύψος m. ~ cast γύψινο εκμαγείο. ~ board γυψοσανίδα f. (med.) έμπλαστρο n. sticking ~ λευκοπλάστης m.

plaster v.t. (cover with ~) σοβαντίζω. (cover) σκεπάζω, γεμίζω. (stick) κολλώ. (fam.) ~ed (drunk) σουρωμένος. **~er** s. σοβατζής m.

plastic a. (malleable) εύπλαστος, (arts, surgery & synthetic) πλαστικός. (s.) πλαστικό n. **~ity** s. πλαστικότητα f.

plate s. (for food) πιάτο n. (silverware)

ασημικά n.pl. brass ~ πινακίδα f. (of metal) έλασμα n., φύλλο n. dental ~ μασέλλα f. (illustration) εικόνα f. (engraving, photo) πλάκα f.

plate v. (with armour) θωρακίζω, (with gold) επιχρυσώνω, (with silver) επαργυρώνω. ~d επάργυρος.

plateau s. οροπέδιο n.

platform s. (speaker's) βήμα n. (railway) αποβάθρα f.

platinum s. πλατίνα f.

platitud|e s. κοινοτοπία f. **~inous** a. τετριμμένος.

Platonic a. πλατωνικός.

platoon s. (mil.) διμοιρία f.

platter s. πιατέλλα f.

plaudits s. χειροκροτήματα n.pl.

plausible a. αληθοφανής. he is ~ τα λόγια του είναι αληθοφανή.

play v. παίζω. (of sunlight) παιχνιδίζω, (of fountain) λειτουργώ. ~ a trick on σκαρώνω μία φάρσα σε. ~ the fool σαχλαμαρίζω. ~ the man φέρομαι σαν άνδρας. ~ fair φέρομαι τίμια. ~ the game δεν κάνω ζαβολιές. ~ (person) false προδίδω. ~ second fiddle παίζω δευτερεύοντα ρόλο. ~ it cool ενεργώ με ψυχραιμία. ~ fast and loose with κάνω κατάχρηση (with gen.). ~ into the hands of one's opponent παίζω το παιχνίδι του αντιπάλου μου.

play s. (recreation, games) παιχνίδι n. (style, quality of playing) παίξιμο n. ~ on words λογοπαίγνιο n. bring (or come) into ~ θέτω (or τίθεμαι) εις ενέργεια. give more ~ to (rope) λασκάρω, (imagination, etc.) αφήνω πιο ελεύθερο. (drama) (θεατρικό) έργο.

play down v.t. μειώνω τη σπουδαιότητα (or τη σοβαρότητα) (with gen.). (v.i.) I ~ to him περιορίζομαι εις το διανοητικό του επίπεδο.

play off v.t. play one off against the other κάνω να διάκεινται εχθρικώς ο ένας προς τον άλλον, παρακινώ την εχθρικότητα του ενός προς τον άλλον.

play on v.t. (exploit) εκμεταλλεύομαι.

play out v. played out (exhausted) εξαντλημένος, (out of date) ξεπερασμένος.

play up v.t. (exaggerate) υπερβάλλω. (torment) ταλαιπωρώ, she plays him up τον σέρνει από τη μύτη. (v.i.) ~ to κολακεύω.

play-acting s. (fig.) προσποίηση f.

player s. παίκτης m. (actor) ηθοποιός m.

playful a. παιχνιδιάρης.

playground s., (school) αυλή f. (public) παιδική χαρά. (fig.) τόπος αναψυχής.

playhouse s. θέατρο n.

playing s. (*performance*) παίξιμο n. ~-**card** s. παιγνιόχαρτο n.

playmate s. σύντροφος m.f.

plaything s. παιχνίδι n. (*fig.*) άθυρμα n.

playwright s. θεατρικός συγγραφεύς.

PLC AE.

plea s. έκκληση f. (*excuse*) πρόφαση f.

plead v. κάνω έκκληση, (*offer as excuse*) προφασίζομαι. (*law*) υποστηρίζω. ~ guilty (*or* not guilty) ομολογώ (*or* αρνούμαι) την ενοχή μου. ~**ing** a. παρακλητικός.

pleasant a. ευχάριστος, (*person*) ευγενικός, συμπαθητικός. ~**ly** adv. ευχάριστα, ευγενικά.

pleasantry s. αστείο n.

please v. (*gratify*) ευχαριστώ, δίνω ευχαρίστηση σε. (*be to liking of*) αρέσω σε. ~ yourself, do as you ~ κάνε ό,τι θέλεις (*or* όπως σου αρέσει). (if you) ~ (*request*) παρακαλώ. if you ~! (*iron.*) αν αγαπάτε! be ~d (*satisfied*) είμαι ευχαριστημένος, (*glad*) χαίρομαι.

pleasing a. ευχάριστος.

pleasur|able a. ευχάριστος. ~**ably** adv. ευχάριστα.

pleasure s. ευχαρίστηση f. (*sensual*) ηδονή f. take ~ in βρίσκω ευχαρίστηση σε. with ~ ευχαρίστως.

pleat s. πιέτα f. (*v.*) κάνω πιέτες σε, πλισσάρω. ~**ing** s. πλισσές m.

plebeian a. πληβείος.

plebiscite s. δημοψήφισμα n.

plectrum s. πέννα f.

pledge s. (*promise*) υπόσχεση f. (*guarantee*) εγγύηση f. (*token*) τεκμήριο n. (v.) (*pawn*) βάζω ενέχυρο. ~ oneself υπόσχομαι.

Pleiades s. Πούλια f.

plenary a. πλήρης, in ~ session εν ολομελεία.

plenipotentiary a. & s. πληρεξούσιος.

plenteous a. άφθονος. ~**ness** s. αφθονία f.

plentiful a. άφθονος. ~**ly** adv. άφθονα.

plenty s. αφθονία f. in ~ εν αφθονία. ~ of άφθονος, μπόλικος.

pleonastic a. πλεοναστικός.

plethora s. υπεραφθονία f.

pleurisy s. πλευρίτις f.

pliab|ility s. ευκαμψία f. ~**le** a. εύκαμπτος, (*person*) ενδοτικός.

pliant see pliable.

pliers s. τανάλια f.

plight s. (*difficulty*) δύσκολη θέση, (*sorry state*) χάλι n.

plight v. ~ oneself (*or* one's word) υπόσχομαι.

plimsolls s. πάνινα παπούτσια με λαστιχένιες σόλες.

plinth s. βάση f.

plod v. σέρνομαι. (*fig.*) ~ on μοχθώ.

plonk s. (*thud*) γδούπος m. (*v.t.*) ~ down πετώ κάτω.

plop v.i. ~ down πέφτω κάτω.

plosive a. (*gram.*) στιγμιαίος.

plot s. (*of land*) κομμάτι n. (*for building*) οικόπεδο n. (*of play*) πλοκή f., υπόθεση f. (*conspiracy*) συνωμοσία f.

plot v.t. (*mark out*) χαράσσω, (*plan*) σχεδιάζω, (*reckon*) υπολογίζω. (*v.i.*) (*conspire*) συνωμοτώ. ~**ter** s. συνωμότης m.

plough s. άροτρον n., αλέτρι n. (*v.i.*) ζευγαρώνω. (*v.t.*) οργώνω. ~ one's way (*fig.*) προχωρώ με κόπο. (*in exam*) κόβω. ~**man** s. ζευγάς m. ~**share** s. υνί n.

plover s. βροχοπούλι n.

ploy s. ελιγμός m.

pluck s. κουράγιο n. ~**y** a. γενναίος, θαρραλέος.

pluck v. (*bird & fig.*) μαδώ, (*pick*) κόβω, (*pull, twitch*) τραβώ. ~ out βγάζω. ~ up courage παίρνω κουράγιο. (*in exam*) κόβω.

plug s. βούλωμα n., τάπα f. (*electric*) φίσα f. (*of motor*) μπουζί n. pull the ~ τραβώ το καζανάκι.

plug v.t. βουλώνω, ταπώνω. (*advertise*) διαφημίζω. ~ in βάζω στην πρίζα. ~ away μοχθώ.

plum s. δαμάσκηνο n., κορόμηλο n. (*mirabell*) μπουρνέλα f. (*fam., best job*) καλύτερη θέση.

plumage s. πτέρωμα n.

plumb v. βολιδοσκοπώ. (s.) βαρίδι n. ~-line, νήμα της στάθμης. out of ~ κεκλιμένος. (*adv.*) ακριβώς.

plumb|er s. υδραυλικός m. ~**ing** s. υδραυλικά n.pl.

plume s. φτερό n. (*of helmet*) λοφίο n. (*of smoke*) τολύπη f. borrowed ~s δανεικά στολίδια. (v.) ~ oneself καμαρώνω.

plummet s. βαρίδι n. (v.) πέφτω σα βολίδα.

plump a. παχουλός, (*person only*) στρουμπουλός.

plump v.t. (*make round*) αναφουφουλιάζω, (*put down*) πετώ κάτω. (*v.i.*) ~ for διαλέγω, παίρνω απόφαση για.

plunder v. λεηλατώ. (s.) (*act*) λεηλασία f., (*things*) λάφυρα n.pl., λεία f., (*both*) πλιάτσικο n.

plunge v.t. βυθίζω, (*only into liquid*) βουτώ. (*v.i.*) βυθίζομαι, βουτώ. (*fall abruptly*) καταπίπτω.

plunge s. βουτιά f. take the ~ παίρνω σοβαρή απόφαση.

plunger s. έμβολο n.

pluperfect a. (*gram.*) υπερσυντέλικος.

plural *s.* (*gram.*) πληθυντικός *m.*
plus (*math.*) συν.
plush *a.* (*fam.*) (*also* ~y) πολυτελείας.
plutocracy *s.* πλουτοκρατία *f.*
plutocrat *s.* πλουτοκράτης *m.* ~**ic** *a.* πλουτοκρατικός.
plutonium *s.* πλουτώνιο *n.*
ply *s.* three-~ wool τρίκλωνο μαλλί. ~-wood κοντραπλακέ *n.*
ply *v.t.* (*wield*) χειρίζομαι. ~ the oar κωπηλατώ. (*a trade*) ασκώ. ~ with questions ταράζω στις ερωτήσεις. he plied me with food μου έδωσε να φάω πολύ. (*v.i.*) κάνω τη διαδρομή.
pneumatic *a.* (*worked by compressed air*) πεπιεσμένου αέρος, (*springy*) ελαστικός. ~ drill κομπρεσέρ *n.*
pneumonia *s.* πνευμονία *f.*
poach *v.* κυνηγώ παρανόμως. (*trespass on*) καταπατώ.
poached *a.* ποσέ.
poacher *s.* λαθροθήρας *m.*
pocket *s.* τσέπη *f.* be in ~ βγαίνω κερδισμένος. be out of ~ βγαίνω χαμένος. (*under eyes*) σακκούλα *f.* (*of air*) κενό *n.* (*mil.*) θύλακος *m.* (*a.*) της τσέπης. (*v.t.*) τσεπώνω. ~ one's pride το καταπίνω. ~-**book** *s.* πορτοφόλι *n.* ~-**knife** *s.* σουγιάς της τσέπης. ~-**money** *s.* χαρτζιλίκι *n.*
pock-marked *a.* βλογιοκομμένος.
pod *s.* λοβός *m.*, λουβί *n.*
podgy *a.* κοντόχοντρος.
podium *s.* εξέδρα *f.*
poem *s.* ποίημα *n.*
poet *s.* ποιητής *m.*
poetic, ~al *a.* ποιητικός. ~**ally** *adv.* με ποίηση.
poetry *s.* ποίηση *f.*
pogrom *s.* πογκρόμ *n.*
poignant *a.* οδυνηρός, σπαρακτικός.
point *s.* (*in space, development, measurement*) σημείο *n.* (*moment*) στιγμή *f.* (*in scoring*) βαθμός *m.* (*theme, matter*) θέμα *n.*, ζήτημα *n.* (*item, particular*) σημείο *n.* (*of weapon, knife*) αιχμή *f.* (*of needle, pencil, umbrella, land*) μύτη *f.* (*of star*) ακτίνα *f.* good ~ (*merit*) προσόν *n.* ~ of view άποψη *f.* boiling ~ σημείον βρασμού. get the ~ αντιλαμβάνομαι. beside the ~ εκτός θέματος. to the ~ επί του θέματος. up to a ~ μέχρις ενός σημείου. when it came to the ~ όταν ήρθε η ώρα. in ~ of fact στην πραγματικότητα. what is the ~ of (doing)? τι ωφελεί να...; I can't see the ~ of δεν βλέπω για ποιο λόγο να. make a ~ of θεωρώ απαραίτητο να. make it a ~ of honour to φιλοτιμούμαι να. I was on the ~ of phoning you ό,τι πήγαινα να σου

τηλεφωνήσω. on the ~ of death ετοιμοθάνατος. on the ~ of collapse (*structure*) ετοιμόρροπος. it is not my strong ~ δεν διακρίνομαι για αυτό.
point *v.* ~ to *or* at (*indicate*) δείχνω. ~ out υποδεικνύω, (*stress*) τονίζω.
point-blank *adv.* (*range*) εκ του σύνεγγυς, (*straight out*) ορθά-κοφτά.
pointed *a.* μυτερός, (*clear*) σαφής. ~ remark καμπανιά *f.* ~**ly** *adv.* σαφώς.
pointer *s.* δείχτης *m.* (*indication*) ένδειξη *f.*
points *s.* (*railway*) κλειδί *n.*
poise *s.* (*balance*) ισορροπία *f.* (*of body*) στάση *f.* (*ease*) άνεση *f.* (*v.t.*) (*hold*) κρατώ. (*v.i.*) (*hover, of bird*) ζυγίζομαι. ~**d** *a.* (*ready*) έτοιμος, (*at ease*) με άνεση.
poison *s.* δηλητήριο *n.* (*v.*) δηλητηριάζω. ~**ing** *s.* δηλητηρίαση *f.* ~**ous** *a.* δηλητηριώδης. (*fig.*) απεχθής.
poke *s.* (*sack*) buy a pig in a ~ αγοράζω γουρούνι στο σακκί. (*thrust*) σπρωξιά *f.*
poke *v.t.* (*sthg. into sthg.*) χώνω, (*nudge*) σκουντώ, (*fire, etc.*) σκαλίζω. ~ fun at κοροϊδεύω. ~ about (*v.i.*) (*potter*) χασομερώ, (*rummage*) σκαλίζω. ~ out (*v.t.*) βγάζω, (*v.i.*) ξεπροβάλλω.
poker *s.* (*game*) πόκερ *n.*
poky *a.* ~ place *or* room τρύπα *f.*
polar *a.* πολικός. ~**ity** *s.* πολικότητα *f.*
polariz|ation *s.* πόλωση *f.* ~**e** *v.t.* πολώ. (*fig.*) διαιρώ σε δύο αντίθετα στρατόπεδα.
pole *s.* (*point*) πόλος *m.* ~s apart εκ διαμέτρου αντίθετοι.
pole *s.* (*post*) κοντάρι *n.*, στύλος *m.* (*punt*) σταλίκι *n.* (*jumping*) κοντός *m.* (*fam.*) be up the ~ έχω μπλέξει άσχημα.
pole-axe *v.* σφάζω.
polemic *s.* πολεμική *f.* ~**al** *a.* εριστικός.
police *s.* αστυνομία *f.* ~ constable αστυφύλαξ *m.* ~ inspector αστυνόμος *m.* ~ court αυτόφωρο *n.* ~ station τμήμα *n.* (*a.*) αστυνομικός. (*v.t.*) τηρώ υπό έλεγχον. ~**man** *s.* αστυνομικός *m.*
policy *s.* πολιτική *f.* (*tactic*) τακτική *f.* (*insurance*) ασφαλιστήριο *n.*
polio *s.* πολιομυελίτιδα *f.*
polish *v.* γυαλίζω, στιβώνω, λουστράρω. (*fig., refine*) εκλεπτύνω. ~ up (*speech, etc.*) χτενίζω. ~ off (*finish*) τελειώνω γρήγορα, (*food*) ξεκαθαρίζω, (*kill*) καθαρίζω, (*put paid to*) χαντακώνω. ~ floors κάνω παρκέ.
polish *s.* (*effect*) γυαλάδα *f.* (*substance*) λούστρο *n.*, βερνίκι *n.* (*fig.*) ραφινάρισμα *n.* ~**ed** *a.* γυαλισμένος. (*fig.*) ραφιναρισμένος.
polite *a.* ευγενικός. ~ society καλός κόσμος. ~**ly** *adv.* ευγενικά. ~**ness** *s.* ευγένεια *f.*

politic *a.* (*expedient*) σκόπιμος. body ~ πολιτεία *f.*
political *a.* πολιτικός. **~ly** *adv.* πολιτικώς.
politician *s.* πολιτευτής *m.*, πολιτικός *m.*, πολιτευόμενος *m.*
politics *s.* πολιτική *f.*, πολιτικά *n.pl.*
poll *s.* (*head*) κεφάλι *n.* ~-tax κεφαλικός φόρος. *v.* (*trees*) κλαδεύω.
poll *s.* (*voting*) ψηφοφορία *f.* go to the ~s πάω να ψηφίσω. head the ~ έρχομαι πρώτος στις εκλογές. (*survey of opinion*) σφυγμομέτρηση της κοινής γνώμης. (*v.i.*, *vote*) ψηφίζω. (*v.t.*, *receive votes*) συγκεντρώνω.
pollen *s.* γύρη *f.*
pollinat|e *v.* επικονιάζω. **~ion** *s.* επικονίαση *f.*
pollut|e *v.* μολύνω, ρυπαίνω. **~ed** μολυσμένος. **~ion** *s.* μόλυνση *f.*, ρύπανση *f.*
polonaise *s.* πολωνέζα *f.*
poltergeist *s.* θορυβοποιό δαιμόνιο.
poltroon *s.* δειλός *a.*
polychrome *a.* πολύχρωμος.
polygam|ous *a.* πολύγαμος. **~y** *s.* πολυγαμία *f.*
polyglot *a.* πολύγλωσσος.
polygon *s.* πολύγωνο *n.*
polymath *s.* πολυμαθής *a.*
polyp, ~us *s.* πολύπους *m.*
polyphony *s.* πολυφωνία *f.*
polysyllabic *a.* πολυσύλλαβος.
polytechnic *s.* πολυτεχνείο *n.*
polytheism *s.* πολυθεΐα *f.*
pomegranate *s.* ρόδι *n.* (*tree*) ροδιά *f.*
pommel *s.* (*sword*) λαβή *f.* (*saddle*) μπροστάρι *n.*
pomp *s.* μεγαλείο *n.* **~ous** *a.* πομπώδης.
pompon *s.* φούντα *f.*
ponce *s.* νταβατζής *m.*
pond *s.* λιμνούλα *f.*
ponder *v.t. & i.* συλλογίζομαι. (*v.t.*) ζυγίζω.
ponderous *a.* βαρύς. **~ly** *adv.* βαρειά.
poniard *s.* εγχειρίδιο *n.*
pontiff *s.* Supreme P~ Μέγας Ποντίφιξ.
pontifical *a.* ποντιφικός. (*fig.*) με ύφος ποντίφικος.
pontificate *v.* αποφαίνομαι δογματικώς.
pontoon *s.* πάκτων *m.* ~ bridge πλωτή γέφυρα.
pony *s.* πόνεϋ *m.*
poodle *s.* κανίς *m.*
pooh-pooh *v.* εκδηλώνω περιφρόνηση για.
pool *s.* (*water*) λιμνούλα *f.* (*swimming*) πισίνα *f.*
pool *s.* (*at cards*) ποτ *n.* (*business*) κοινοπραξία *f.* (*common supply*) κοινόχρηστος υπηρεσία. (*v.t.*) συγκεντρώνω σε κοινό ταμείο.
poop *s.* πρύμνη *f.*

poor *a.* φτωχός, (*unfortunate*) κακομοίρης, καημένος. (*health*) λεπτός. (*bad*) κακός. (*weak, unskilled*) αδύνατος. ~ quality (*goods*) δεύτερο πράμα. there was a ~ attendance οι παρόντες ήταν λίγοι.
poorly *a.* αδιάθετος. (*adv.*) (*badly*) άσχημα, (*not richly*) φτωχικά. ~ lighted κακοφωτισμένος.
pop *s.* (*noise*) παφ *n.* (*drink*) γκαζόζα *f.*
pop *s. & a.* (*art, etc.*) ποπ *n.*
pop *v.i.* (*go ~*) κάνω παφ. (*v.t., put*) βάζω, ~ out βγάζω. ~ the question κάνω πρόταση γάμου. (*v.i.*) (*into bed*) πέφτω. ~ out, round, *etc.*, (*on errand*) πετάγομαι, πετιέμαι. ~ up (*spring*) πετάγομαι. ~ off (*leave*) φεύγω, (*die*) πεθαίνω. his eyes ~ped out γούρλωσε τα μάτια του.
popcorn *s.* ψημένο καλαμπόκι.
pope *s.* πάπας *m.* **~ry** *s.* παπισμός *m.*
pop-eyed *a.* γουρλομάτης *m.*
popinjay *s.* μορφονιός *m.*
poplar *s.* λεύκη *f.*
poplin *s.* ποπλίνα *f.*
poppet *s.* κούκλα *f.*
poppy *s.* παπαρούνα *f.*
poppycock *s.* (*fam.*) μπούρδες *f.pl.*
populace *s.* λαός *m.*
popular *a.* (*of the people*) λαϊκός, (*common*) κοινός, (*frequented*) πολυσύχναστος. (*in fashion*) της μόδας, (*favoured product*) της προτιμήσεως. (*liked*) (*person*) δημοφιλής, (*pastime, resort*) λαοφιλής. **~ly** *adv.* γενικώς.
popularity *s.* δημοτικότητα *f.*
popularize *v.* (*simplify*) εκλαϊκεύω, (*promote*) προωθώ.
populat|e *v.* κατοικώ. **~ion** *s.* πληθυσμός *m.* ~ion statistics δημογραφία *f.*
populist *a.* λαϊκός.
populous *a.* πυκνοκατοικημένος.
porcelain *s.* πορσελάνη *f.*
porch *s.* σκεπαστή είσοδος.
porcupine *s.* ακανθόχοιρος *m.*
pore *s.* πόρος *m.*
pore *v.* ~ over είμαι σκυμμένος απάνω σε.
pork *s.* χοιρινό *n.*
pornograph|ic *a.* πορνογραφικός. **~y** *s.* πορνογραφία *f.*
por|osity *s.* πορώδες *n.* **~ous** *a.* πορώδης.
porpoise *s.* φώκαινα *f.*
porridge *s.* (*fam.*) κουάκερ *n.*
port *s.* (*sea*) λιμάνι *n.*, λιμήν *m.* (*a.*) λιμενικός.
port *s. & a.* (*side*) αριστερά *f.* to ~ αριστερά.
portable *a.* φορητός.
portal *s.* πυλών *m.*
portcullis *s.* καταρρακτή *f.*
portend *v.* προμηνύω.

portent s. οιωνός m. (*prodigy*) θαύμα n.
~ous a. (*ominous*) δυσοίωνος, (*pompous*) πομπώδης.
porter s. αχθοφόρος m., χαμάλης m. (*hall-~*) θυρωρός m. **~age** s. (*charge*) αχθοφορικά n.pl.
portfolio s. χαρτοφυλάκιο n.
porthole s. φιλιστρίνι n.
portico s. πρόστοον n.
portion s. (*share*) μερίδιο n. (*dowry*) προίκα f. (*part*) μέρος n. (*of food*) μερίδα f. (*fate*) μοίρα f. (*v.t.*) μοιράζω, (*dower*) προικίζω.
portly a. σωματώδης.
portmanteau s. βαλίτσα f.
portrait s. προσωπογραφία f. πορτραίτο n.
portray v. απεικονίζω, δείχνω. **~al** s. απεικόνιση f.
Portuguese a. πορτογαλικός. (*person*) Πορτογάλος m., Πορτογαλίδα f.
pose v.t. (*question*) θέτω, (*problem*) δημιουργώ. (*place*) τοποθετώ. (*v.i.*) (*sit, put on airs*) ποζάρω. ~ as (*pretend to be*) παριστάνω. (s.) πόζα f.
poser s. δύσκολη ερώτηση.
poseur s. be a ~ παίρνω πόζες.
posh a. (*fam.*) (*things*) πολυτελείας, κομψός. (*people*) σπουδαίος.
position s. (*place, job*) θέση f. (*of house, city, etc.*) τοποθεσία f. (*circumstances*) θέση f., κατάσταση f. (*attitude*) στάση f. (*v.*) τοποθετώ.
positive a. (*not negative*) θετικός, (*certain*) βέβαιος, (*explicit*) ρητός. (*real, out-and-out*) πραγματικός. (*cocksure*) δογματικός. **~ly** adv. θετικώς, απολύτως, πραγματικώς.
positivism s. θετικισμός m.
posse s. απόσπασμα n.
possess v. κατέχω. ~ oneself of (*obtain*) αποκτώ, (*seize*) καταλαμβάνω. be ~ed of έχω. ~ed (*of devils*) δαιμονισμένος.
possession s. (*keeping*) κατοχή f. take ~ of γίνομαι κάτοχος (*with gen.*). be in ~ of έχω, κατέχω, έχω στην κατοχή μου. (*thing owned*) απόκτημα n. my ~s τα υπάρχοντά μου. (*territorial*) κτήση f.
possessive a. (*gram.*) κτητικός. she is a ~ mother επιβάλλεται πλέον του δέοντος στα παιδιά της.
possibility s. δυνατότητα f. (*chance*) ενδεχόμενο n. there is a ~ that ενδέχεται (*or* υπάρχει πιθανότητα) να.
possib|le a. δυνατός, μπορετός. a ~le solution μία πιθανή λύση. ~le survivors οι ενδεχομένως επιζώντες. as many as ~le όσο το δυνατόν περισσότεροι. **~ly** adv. (*perhaps*) ενδεχομένως, πιθανώς. I can't ~ly μου είναι αδύνατο.

post s. (*stake*) πάσσαλος m. (*lamp. etc.*) κολόνα f., στύλος m. (*position, job*) θέση f. (*station*) σταθμός m. (*mil.*) forward ~ εμπροσθοφυλακή f.
post s. (*mail*) ταχυδρομείο n. has the ~ come? πέρασε ο ταχυδρόμος; ~ office ταχυδρομείο n. **~code** ταχυδρομικός τομεύς.
post v.t. (*place*) τοποθετώ, (*on walls*) τοιχοκολλώ. (*letters*) ταχυδρομώ, ρίχνω. ~ up (*ledger*) ενημερώνω. keep (*person*) ~ed κρατώ ενήμερο. (*v.i., hasten*) σπεύδω.
post- *prefix* μετα-.
postage s. ταχυδρομικά τέλη. what is the ~? πόσο είναι το γραμματόσημο; ~ stamp γραμματόσημο n.
postal a. ταχυδρομικός. ~ order ταχυδρομική επιταγή.
postbox s. γραμματοκιβώτιο n.
postcard s. ταχυδρομικό δελτάριο, (*picture*) κάρτα f.
post-chaise s. ταχυδρομική άμαξα.
post-date v.t. μεταχρονολογώ.
poster s. αφίσσα f.
poste restante s. πόστ ρεστάντ n.
posterior a. (*place*) οπίσθιος, (*time*) μεταγενέστερος. (s.) οπίσθια n.pl.
posterity s. οι μεταγενέστεροι.
postern s. παραπόρτι n.
postgraduate a. μεταπτυχιακός.
post-haste adv. ολοταχώς.
posthumous a. ~ works έργα εκδοθέντα μεταθανατίως.
postman s. ταχυδρόμος m.
post-mortem s. (*med.*) νεκροψία f.
postmark s. σφραγίδα f.
postmaster s. διευθυντής ταχυδρομείου. P~ General Υπουργός Ταχυδρομείων.
postpone v. αναβάλλω. **~ment** s. αναβολή f.
postprandial a. επιδόρπιος.
postscript s. υστερόγραφο n.
postulate s. αξίωμα n., προϋπόθεση f. (*v.*) αξιώ, θέτω ως προϋπόθεσιν.
posture s. στάση f. (*v.*) παίρνω πόζες.
posy s. μπουκέτο n.
pot s. (*general*) δοχείο n. (*chamber*) καθήκι n. (*cooking*) χύτρα f., (*earthen*) τσουκάλι n., (*metal*) τέντζερης m. (*flower*) γλάστρα f. take ~ luck τρώω ό,τι βρεθεί. ~s of money λεφτά με ουρά. go to ~ πάω κατά διαβόλου.
pot v. (*plants*) βάζω σε γλάστρα. (*shoot*) χτυπώ. **~ted** (*meat, fish*) διατηρημένος σε βάζο, (*abridged*) συνεπτυγμένος.
potable a. πόσιμος.
potash s. ποτάσα f.
potations s. πιοτί n., τσούξιμο n.
potato s. πατάτα f.

pot-bellied a. κοιλαράς m., κοιλαρού f.

pot-boiler s. έργον μικρής καλλιτεχνικής αξίας γινόμενον προς βιοπορισμόν.

poten|cy s. δύναμη f. ~t a. δυνατός, ισχυρός.

potentate s. ηγεμών m.

potential a. (latent) λανθάνων, (possible) ενδεχόμενος, πιθανός. (s.) δυναμικό n. ~ity s. δυνατότητα f. ~ly adv. ενδεχομένως, πιθανώς.

pot-hole s. γούβα f.

pot-house s. καπηλειό n.

potion s. (philtre) φίλτρο n.

potsherd s. όστρακο n.

pottage s. σούπα f. for a mess of ~ αντί πινακίου φακής.

potter v.i. πηγαινοέρχομαι χασομερώντας.

potter s. αγγειοπλάστης m.

pottery s. (art) αγγειοπλαστική f., κεραμική f. (products) είδη κεραμικής.

potty a. (insignificant) ασήμαντος, (mad) τρελλός.

pouch s. σακκούλα f. (bag) τσαντάκι n. (marsupial's) μάρσιπος m.

poulterer s. έμπορος πουλερικών.

poultice s. κατάπλασμα n.

poultry s. πουλερικά n.pl.

pounce v. ~ on ορμώ επάνω σε. (s.) πήδημα n.

pound s. (money) λίρα f. (weight) λίβρα f. (enclosure) μάντρα f.

pound v.t. κοπανώ. (v.t. & i) χτυπώ.

pour v.t. χύνω, ρίχνω ~ out (drinks) βάζω, σερβίρω. (v.i.) χύνομαι. ~ out or forth ξεχύνομαι. (run) τρέχω, ρέω. (throng, flock) συρρέω. it's ~ing with rain ρίχνει γενναία βροχή. ~ cold water on υποδέχομαι χωρίς ενθουσιασμό.

pout v. σουρώνω τα χείλια μου.

poverty s. φτώχεια f., ένδεια f., πενία f. ~-stricken a. (person) πενόμενος, (home, etc.) άθλιος.

powder v.t. (reduce to ~) κάνω σκόνη, κονιορτοποιώ. (one's face) πουδράρω. (v.i.) (become ~) γίνομαι σκόνη, κονιορτοποιούμαι. ~ed milk γάλα σκόνη.

powder s. σκόνη f. (cosmetic) πούδρα f. (gun) μπαρούτι n. ~-magazine s. πυριτιδαποθήκη f.

power s. δύναμη f. (moral force, validity) ισχύς f. (authority, sway) εξουσία f. (of lens, etc.) ισχύς f. ~s (mental) ικανότητες f.pl., (bodily) δυνάμεις f.pl., (competency) αρμοδιότητες f.pl. the ~s that be οι υφιστάμενες αρχές. in ~ (political) στην εξουσία, στα πράγματα. high ~ed (engine) μεγάλης ισχύος, (person) δυναμικός. (law) ~ of attorney πληρεξουσιότης f. ~ cut διακοπή ρεύματος. ~ politics πολιτική

ισχύος. ~-driven a. μηχανοκίνητος. ~-house s. (fig.) εστία δυναμισμού. ~less a. ανίσχυρος. ~-station s. εργοστάσιο παραγωγής ηλεκτρισμού.

powerful a. δυνατός, ισχυρός. ~ly adv. δυνατά, ισχυρώς.

pow-wow s. (joc.) συζήτηση f.

pox s. σύφιλη f.

PR δημόσιες σχέσεις.

practical a. πρακτικός, θετικός. ~ joke φάρσα f. ~ly adv. θετικά, (almost) σχεδόν.

practice s. (not theory) πράξη f. (habitual action) συνήθεια f. (exercise) άσκηση f., εξάσκηση f. (patients, clients) πελατεία f. in ~ στην πράξη. put into ~ θέτω εις την πράξιν, εφαρμόζω. it is the ~ συνηθίζεται. make a ~ of συνηθίζω να. be out of ~ μου λείπει η άσκηση. sharp ~ βαγαποντιά f.

practise v. (apply) εφαρμόζω, (follow) ακολουθώ. (profession) ασκώ. (exercise oneself) ασκούμαι. ~ the piano ασκούμαι στο πιάνο.

practised a. έμπειρος, πεπειραμένος.

practitioner s. (med.) general ~ παθολόγος m.

praetorian a. πραιτωριανός.

pragmat|ic a. πρακτικός. ~ism s. πραγματισμός m.

prairie s. πεδιάδα f., κάμπος m.

praise v. επαινώ, (God) δοξάζω. (s.) έπαινος m. sing ~s of εξυμνώ.

praiseworthy a. αξιέπαινος.

pram s. καρροτσάκι n.

prance v. χοροπηδώ. (of horse) ορθώνομαι στα πίσω πόδια.

prank s. σκανταλιάρικο παιγνίδι.

prate v. σαλιαρίζω.

prattle v. φλυαρώ.

prawn s. γαρίδα f.

pray v.i. προσεύχομαι. (v.t.) (beg person) παρακαλώ, εκλιπαρώ.

prayer s. δέηση f., προσευχή f. ~-book s. προσευχητάριο n.

pre- prefix προ-.

preach v. κηρύσσω, (fig.) κάνω κήρυγμα. ~er s. ιεροκήρυκας m.

preamble s. προοίμιο n.

prearranged a. προσχεδιασμένος.

precarious a. (chancy) επισφαλής, (perilous) επικίνδυνος.

precaution s. προφύλαξη f. take ~s λαμβάνω τα μέτρα μου.

preced|e v. προηγούμαι (with gen.). ~ing προηγούμενος. ~ence s. προβάδισμα n. take ~ence προηγούμαι. ~ent s. προηγούμενο n.

precept s. αρχή f., κανόνας m. ~or s. διδάσκαλος m.

precinct, ~s s. (area) περιοχή f., (boun

daries) όρια *n.pl.* (*of church, school, etc.*) περίβολος *m.*

preciosity *s.* επιτήδευση *f.*

precious *a.* (*valuable*) πολύτιμος, (*affected*) επιτηδευμένος. (*fam.*) we have ~ little time μας μένει πολύ λίγος καιρός. (*iron.*) you and your ~ theories! εσύ με τις περίφημες θεωρίες σου!

precipice *s.* γκρεμός *m.*

precipitate *v.t.* (*throw*) γκρεμίζω, ρίχνω. (*hasten*) επισπεύδω. (*chem.*) κατακρημνίζω. (*s.*) (*chem.*) καθίζημα *n.* (*a.*) βιαστικός. **~ly** *adv.* βιαστικά.

precipitation *s.* (*chem.*) καθίζηση *f.* (*haste*) βία *f.*

precipitous *a.* απόκρημνος. **~ly** *adv.* απότομα.

précis *s.* περίληψη *f.*

precise *a.* ακριβής. **~ly** *adv.* ακριβώς.

precision *s.* ακρίβεια *f.*

preclude *v.* αποκλείω.

precocious *a.* (*plant*) πρώιμος, (*child*) με πρόωρη πνευματική ανάπτυξη.

preconception *s.* προκατάληψη *f.*

precursor *s.* πρόδρομος *m.*

predator *s.* αρπακτικό ζώο. **~y** *a.* αρπακτικός.

predecease *v.t.* πεθαίνω πριν από.

predecessor *s.* προηγούμενος *a.* (*previous holder*) προκάτοχος *m.*

predestin|ation *s.* προορισμός *m.* **~e** *v.* προορίζω.

predetermine *v.* προκαθορίζω.

predicament *s.* δύσκολη θέση.

predicate *v.* ισχυρίζομαι. (*s., gram.*) κατηγόρημα *n.*

predict *v.* προβλέπω, προλέγω. **~able** *a.* που μπορεί να προβλεφθεί. **~ion** *s.* πρόβλεψη *f.*

predilection *s.* προτίμηση *f.*

predispose *v.* προδιαθέτω. be **~d** (*mentally*) είμαι προδιατεθειμένος, (*physically*) έχω προδιάθεση.

predominance *s.* επικράτηση *f.*

predominant *a.* επικρατών. **~ly** *adv.* κυρίως.

predominate *v.* επικρατώ.

preeminent *a.* be ~ (over) υπερέχω (*with gen.*), **~ly** *adv.* κυρίως, κατά κύριο λόγο.

preempt *v.* αποκτώ πρώτος. **~ive** *a.* προληπτικός.

preen *v.t.* (*of bird*) σιάζω. (*of person*) ~ oneself (*tidy*) φτιάχνομαι, (*be proud*) υπερηφανεύομαι.

preexist *v.i.* προϋπάρχω. **~ence** *s.* προϋπαρξη *f.*

prefab *s.* προκάτ *n.*

prefabricated *a.* προκατασκευασμένος.

preface *s.* πρόλογος *m.* (*v.*) προλογίζω.

prefatory *a.* προεισαγωγικός.

prefect *s.* (*governor*) νομάρχης *m.* (*school*) επιμελητής *m.* **~ure** *s.* (*area*) νομός *m.* (*office*) νομαρχία *f.*

prefer *v.* προτιμώ, (*submit*) υποβάλλω.

preferab|le *a.* προτιμότερος, προτιμητέος. **~ly** *adv.* προτιμότερα, κατά προτίμησιν.

prefer|ence *s.* προτίμηση *f.* **~ential** *a.* προνομιακός.

preferment *s.* προαγωγή *f.*

prefigure *v.* προεικονίζω.

prefix *v.* (*to book, etc.*) προτάσσω. (*s.*) πρόθεμα *n.*

pregnancy *s.* εγκυμοσύνη *f.*

pregnant *a.* έγκυος, (*significant*) βαρυσήμαντος. (*fig.*) be ~ with (*fraught*) εγκυμονώ.

prehensile *a.* συλληπτήριος.

prehistor|ic *a.* προϊστορικός. **~y** *s.* προϊστορία *f.*

prejudge *v.* προδικάζω.

prejudice *s.* (*bias*) προκατάληψη *f.* (*law*) to the ~ of προς ζημίαν (*with gen.*). without ~ to με πάσαν επιφύλαξιν (*with gen.*).

prejudic|e *v.* (*influence*) προδιαθέτω δυσμενώς, (*harm*) παραβλάπτω. **~ed** προκατειλημμένος. **~ial** *a.* επιζήμιος, (*or use v.*).

prelate *s.* ανώτερος κληρικός ιεράρχης *m.*

preliminary *a.* προκαταρκτικός, (*remarks*) προεισαγωγικός, (*contest*) προκριματικός.

prelude *s.* προοίμιο *n.* (*mus.*) πρελούντιο *n.* (*v.t.*) (*precede*) προηγούμαι (*with gen.*) (*preface*) προλογίζω.

premarital *a.* προγαμιαίος.

premature *a.* πρόωρος. **~ly** *adv.* πρόωρως.

premeditate *v.* προμελετώ **~d** (*crime*) εκ προμελέτης.

premier *a.* πρώτος. (*s.*) πρωθυπουργός *m.* **~ship** *s.* πρωθυπουργία *f.*

première *s.* πρεμιέρα *f.*

premise *s.* πρόταση *f.* (*v.*) αναφέρω.

premises *s.* κτίριο *n.* on the ~ εντός του κτιρίου.

premium *s.* πρίμ *n.* (*insurance*) ασφάλιστρα *n.pl.* (*fee, reward*) αμοιβή *f.* put a ~ on επιβραβεύω. petrol is at a ~ (*fig.*) υπάρχει μεγάλη ζήτηση βενζίνης.

premonition *s.* προαίσθημα *n.*

preoccupation *s.* (*care*) έννοια *f.* (*absorption*) απορρόφηση *f.*

preoccupied *a.* απασχολημένος, (*distrait*) αφηρημένος.

preordain *n.* προορίζω.

prep *s. see* homework.

preparation *s.* (*act*) ετοιμασία *f.*, προετοιμασία *f.*, προπαρασκευή *f.* (*product*) παρασκεύασμα *n.* be in ~ ετοιμάζομαι. *see* prepare.

preparatory *a.* προπαρασκευαστικός. ~ to προ (*with gen.*).

prepare *v.t.* (*breakfast, room, etc.*) ετοιμάζω, (*for action*) προετοιμάζω. (*v.i.*) ετοιμάζομαι, προετοιμάζομαι.

prepared *a.* (*ready to cope*) προετοιμασμένος, (*willing*) διατεθειμένος, (*made ready*) έτοιμος. (*of product*) παρασκευασμένος. ~ **ness** *n.* ετοιμότης *f.*

prepay *v.* προπληρώνω.

preponder|ance *s.* υπεροχή *f.* **~antly** *adv.* σε μεγαλύτερο βαθμό. **~ate** *v.i.* υπερέχω, επικρατώ.

preposition *s.* (*gram.*) πρόθεση *f.*

prepossess *v.* I was ~ed by his manners με εντυπωσίασαν οι τρόποι του.

preposterous *a.* εξωφρενικός.

prerequisite *s.* be a ~ προαπαιτούμαι.

prerogative *s.* προνόμιο *n.*

presage *s.* προμήνυμα *n.* (*v.*) προμηνύω.

prescribe *v.* (*lay down*) καθορίζω, (*medicine*) δίνω.

prescription *s.* συνταγή *f.*

prescriptive *a.* καθοδηγητικός.

presence *s.* παρουσία *f.* (*appearance*) παρουσιαστικό *n.* ~ of mind (*wits*) ετοιμότητα *f.*, (*coolness*) ψυχραιμία *f.* in the ~ of παρουσία (*with gen.*).

present *a.* παρών, those ~ οι παρόντες. the ~ writer ο γράφων. (*current*) τωρινός, σημερινός. at the ~ time στη σημερινή εποχή. the ~ mayor ο τωρινός (*or* ο νυν) δήμαρχος.

present *s.* (*time*) παρόν *n.* (*gram.*) ενεστώς *m.* at ~ τώρα, σήμερον. for the ~ προς το παρόν, για την ώρα. up to the ~ έως τώρα.

present *s.* (*gift*) δώρο *n.* make a ~ of χαρίζω.

present *v.t.* (*offer*) προσφέρω, (*submit*) υποβάλλω, (*introduce, exhibit*) παρουσιάζω. (*a weapon*) προτείνω. ~ arms παρουσιάζω όπλα. ~ oneself παρουσιάζομαι.

presentable *a.* εμφανίσιμος, παρουσιάσιμος.

presentation *s.* (*manner of presenting*) παρουσίαση *f.* (*thing offered*) προσφορά *f.* (*payable*) on ~ επί τη εμφανίσει.

presentiment *s.* προαίσθηση *f.*

presently *adv.* (*soon*) σε λίγο, προσεχώς, εντός ολίγου.

preservation *s.* διατήρηση *f.* in a good state of ~ καλώς διατηρημένος. (*escape from harm*) διάσωση *f.* (*protection*) προστασία *f.*

preservative *s.* συντηρητικό *n.*

preserv|e *n.* (*keep in being, from loss or*

decay) διασώζω, (*in good condition*) διατηρώ, συντηρώ. (*protect*) προστατεύω. (*s.*) (*jam*) μαρμελάδα *f.* (*fig.*) σφαίρα δικαιοδοσίας. (*for game, etc.*) περιοχή προστασίας. **~er** *s.* σωτήρας *m.*

preside *v.* (*as chairman*) προεδρεύω. ~ over (*as head*) προΐσταμαι (*with gen.*).

presidency *s.* προεδρία *f.*

president *s.* πρόεδρος *m.f.* **~ial** *a.* προεδρικός.

press *v.t.* πιέζω, (*a button*) πατώ. (*squeeze*) σφίγγω, (*thrust*) χώνω, (*stick, fasten*) κολλώ, (*iron*) σιδερώνω. be ~ed for time μου λείπει καιρός. (*v.i.*) (*insist*) επιμένω. ~ on pen (*or* nerve) πιέζω την πέννα (*or* το νεύρο). ~ down on (*afflict*) καταπιέζω. ~ on προχωρώ. ~ on with συνεχίζω.

press *s.* (*urgency*) φόρτος *m.* (*of crowd*) συνωστισμός *m.* (*squeeze*) σφίξιμο *n.* (*machine*) πιεστήριο *n.* (*newspapers*) τύπος *m.* printing-~ τυπογραφείο *n.*, in the ~ υπό εκτύπωσιν. ~ conference διάσκεψη τύπου.

press-fastener *s.* σούστα *f.*

pressing *a.* (*need, task*) επείγων, (*attitude*) επίμονος, (*invitation*) θερμός. **~ly** *adv.* με επιμονή.

pressure *s.* πίεση *f.* put ~ on πιέζω. ~ gauge μανόμετρο *n.* ~ cooker χύτρα ταχύτητος.

pressurized *a.* με τεχνητή ατμοσφαιρική πίεση.

prestig|e *s.* γόητρον *n.* **~ious** *a.* που προσδίδει γόητρον.

presumably *adv.* πιθανώς.

presume *v.* (*suppose*) υποθέτω, (*venture*) τολμώ. (*take liberties*) ξεθαρρεύομαι. ~ upon καταχρώμαι (*with gen.*).

presumption *s.* υπόθεση *f.* (*boldness*) θράσος *n.*

presumptive *a.* κατά συμπερασμόν. (*heir*) πιθανός.

presumptuous *a.* θρασύς.

presuppos|e *v.* προϋποθέτω. **~ition** *s.* προϋπόθεση *f.*

pretence *s.* πρόσχημα *n.* (*claim*) φιλοδοξία *f.* by false ~s δι' απάτης.

pretend *v.* προσποιούμαι. ~ that κάνω πως. ~ to be mad κάνω τον τρελλό. ~ to (*claim*) έχω αξιώσεις για (*or* ότι). **~ed** *a.* προσποιητός. **~er** *s.* (*claimant*) μνηστήρ *m.*

pretension *s.* (*claim*) αξίωση *f.* see also pretentiousness.

pretentious *a.* με αδικαιολόγητες αξιώσεις, (*showy*) επιδεικτικός. **~ness** *s.* αδικαιολόγητες αξιώσεις.

preterite *a.* (*gram.*) αόριστος.

preternatural *a.* υπερφυσικός.

pretext *s.* πρόφαση *f.*, αφορμή *f.*

prettification s. φτηνή διακόσμηση.
prettily adv. όμορφα.
prettiness s. ομορφιά f.
pretty a. νόστιμος, όμορφος, ωραίος. (iron.) ωραίος. (adv.) (fairly) αρκετά, (almost) ~ well σχεδόν. (fam.) be sitting ~ τα 'χω βολέψει καλά.
prevail v. επικρατώ. ~ against υπερισχύω, επικρατώ (with gen.). ~ upon πείθω, καταφέρνω. ~**ing** a. (wind) επικρατών, (idea) κρατών.
prevall|ence s. επικράτηση f. ~**ent** a. διαδεδομένος.
prevaricat|e v. τα στριφογυρίζω. ~**ion** s. υπεκφυγή f., στριφογύρισμα n.
prevent v. εμποδίζω, (avert) αποτρέπω, προλαμβάνω.
prevention s. πρόληψη f. (of war, etc.) προληπτικά μέτρα. (hygienic) προφύλαξη f. ~ of cruelty to animals προστασία ζώων.
preventive a. προληπτικός, προφυλακτικός.
preview s. προκαταρκτική παρουσίαση.
previous a. προηγούμενος. ~ to πριν από. ~**ly** adv. πριν, πρωτύτερα, προηγουμένως.
pre-war a. προπολεμικός.
prey s. (as food) βορά f. (also fig.) θύμα n. (bird, beast) of ~ αρπακτικός. (v.) ~ on (of animals) καταδιώκω. (plunder) λυμαίνομαι, (torment) βασανίζω.
price s. τιμή f. cost ~ κόστος n. at any ~ πάση θυσία. not at any ~ με κανένα τρόπο. put a ~ on the head of επικηρύσσω. beyond ~ ανεκτίμητος. what ~ ...? (iron.) τι να πει κανείς για...; ~-cut μείωση τιμών. (v.) ορίζω την τιμή (with gen.). be ~d at (cost) τιμώμαι, κοστίζω.
priceless a. ανεκτίμητος. (fam.) (joke) ξεκαρδιστικός, (situation) ανεκδιήγητος, (event, person) κωμωδία f.
price-list s. τιμοκατάλογος m.
pricey a. αλμυρός.
prick v. κεντώ, τσιμπώ. (pierce) τρυπώ. my conscience ~s me με τύπτει η συνείδησίς μου. ~ up one's ears τεντώνω τα αφτιά μου. (s.) κέντημα n. (of conscience) τύψη f.
prickle s. αγκάθι n. (v.) τσιμπώ.
prickly a. (plant) γεμάτος αγκάθια. have a ~ feeling (oneself) μυρμηγκιάζω. (fig., touchy) ευερέθιστος. ~ pear φραγκόσυκο n.
pride s. περηφάνια f. (proper) ~ (self-respect) φιλότιμο n. (air or object of ~) καμάρι n. false ~ ματαιοδοξία f. take ~ in υπερηφανεύομαι για, καμαρώνω (για). take ~ of place έχω εξέχουσα θέση. (v.) ~ oneself on, see pride s.
priest s. ιερεύς m. παπάς m. ~'s wife παπαδιά f. ~**ess** s. ιέρεια f.

priestly a. ιερατικός.
prig s. ~**gish** a. αυτάρεσκος και σχολαστικός.
prim a. (prudish) σεμνότυφος, (stiff) σφιγμένος. ~**ly** adv. αυστηρά.
primacy s. πρωτεία n.pl.
prima facie a. & adv. εκ πρώτης όψεως.
primal a. πρώτος, κύριος.
primar|y a. πρώτος, πρωταρχικός, κύριος, (colour) βασικός. ~**y** school δημοτικό σχολείο. of ~y importance πρωτίστης σημασίας. ~**ily** adv. κυρίως.
primate s. (eccl.) αρχιεπίσκοπος m. (zoology) ~s πρωτεύοντα n.pl.
prime s. ακμή f. ~ of life άνθος της ηλικίας.
prime a. (chief) κύριος, (best) εκλεκτός. (necessity) πρωταρχικός, (number) πρώτος. ~ mover (fig.) πρωτεργάτης m. P~ Minister πρωθυπουργός m.
prime v. (gun, pump) γεμίζω, (instruct) δασκαλεύω, (with drink) ποτίζω. (with paint, etc.) ασταρώνω.
primer s. (book) αλφαβητάριο n. (paint) αστάρι n.
primeval a. αρχέγονος.
priming s. (of gun) γόμωση f. (of surface) αστάρωμα n. (of person) δασκάλεμα n.
primitive a. πρωτόγονος.
primogeniture s. rights of ~ πρωτοτόκια n.pl.
primordial a. αρχέγονος.
primrose s. παναγίτσα f. (fig.) take the ~ path το ρίχνω στις απολαύσεις.
primula s. πρίμουλα f.
prince s. πρίγκιπας m. (king's son only) βασιλόπουλο n. (of Peace, Darkness) Άρχων m. (fig., magnate) μεγιστάν n. ~**ly** a. πριγκιπικός, (lavish) πλούσιος.
princess s. πριγκίπισσα f. (king's daughter only) βασιλοπούλα f.
principal s. (head) διευθυντής m. (actor) πρωταγωνιστής m. (capital sum) κεφάλαιο n. (a.) κύριος. ~**ity** s. πριγκιπάτο n. ~**ly** adv. κυρίως.
principle s. αρχή f. in ~ κατ' αρχήν. on ~ εκ πεποιθήσεως. ~d με αρχές. moral ~s ηθικός εξοπλισμός.
prink v. ~ (oneself up) καλλωπίζομαι.
print s. (trace, mark) αποτύπωμα n. (letters) στοιχεία n.pl. (copy) αντίτυπο n. (engraving) γκραβούρα f. (fabric) εμπριμέ n. in ~ τυπωμένος, εκδοθείς, out of ~ εξαντλημένος.
print v. τυπώνω, εκτυπώνω. (impress & fig.) αποτυπώνω. ~**ed** a. τυπωμένος, έντυπος. ~**er** s. τυπογράφος m.
printing s. (art) τυπογραφία f. (act) εκτύπωση f. ~ press τυπογραφείο n.

prior *a.* πρότερος. have a ~ claim έχω προτεραιότητα, προηγούμαι. ~ to προ (*with gen.*). **~ity** *s.* προτεραιότητα *f.*

prior *s.* (*eccl.*) ηγούμενος *n.* **~ess** *s.* ηγουμένη *f.*

prise *v. see* prize (*force*).

prism *s.* πρίσμα *n.* **~atic** *a.* πρισματικός.

prison *s.* φυλακή *f.*

prisoner *s.* φυλακισμένος *m.*, κρατούμενος *m.* (*esp. fig.*) δεσμώτης *m.* ~ of war αιχμάλωτος *m.* take ~ αιχμαλωτίζω.

pristine *a.* παλαιός, αρχικός.

privacy *s.* (*peace*) ησυχία *f.* they don't respect my ~ δεν με αφήνουν ήσυχο.

private *a.* ιδιωτικός, (*individual*) ατομικός. (*secret*) μυστικός, (*confidential*) ιδιαίτερος. ~ citizen απλός πολίτης. ~ affairs προσωπικές υποθέσεις. ~ house μονοκατοικία *f.* ~ parts γεννητικά όργανα. ~ secretary ιδιαιτέρα *f.* by ~ arrangement δι' ιδιωτικής συμφωνίας. in ~ ιδιαιτέρως, (*of wedding, etc.*) σε στενό οικογενειακό κύκλο.

privateer *s.* κουρσάρικο *n.*

privately *adv.* ιδιαιτέρως, σε στενό κύκλο.

privation *s.* στέρηση *f.* (*destitution*) στερήσεις *f.pl.*

privilege *s.* προνόμιο *n.* **~d** *a.* προνομιούχος.

privily *adv.* κρυφά.

privy *a.* be ~ to έχω γνώση (*with gen.*). P~ Council Ανακτοβούλιον *n.* P~ Purse βασιλική χορηγία.

privy *s.* απόπατος *m.*

prize *s.* (*award*) βραβείο *n.*, έπαθλο *n.* (*booty*) λάφυρο *n.*, λεία *f.* (*sthg. valued*) αγαθόν *n.*

prize *a.* βραβευθείς (*fam.*) ~ idiot βλάκας με περικεφαλαία.

prize *v.* (*value*) εκτιμώ. (*force*) ~ open ανοίγω διά της βίας.

pro *s.* επαγγελματίας *m.*

pro *prep.* υπέρ (*with gen.*), ~s and cons τα υπέρ και τα κατά. ~ tem προσωρινά. ~-English αγγλόφιλος, υπέρ της Αγγλίας.

probability *s.* πιθανότητα *f.*

probab|le *a.* πιθανός. **~ly** *adv.* πιθανώς.

probate *s.* επικύρωση *f.* grant *or* take out ~ of επικυρώνω.

probation *s.* δοκιμασία *f.* **~ary** period περίοδος δοκιμασίας. **~er** *s.* δόκιμος *a.*

probe *s.* (*instrument*) μήλη *f.* (*v.*) εξετάζω.

probity *s.* εντιμότητα *f.*

problem *s.* πρόβλημα *n.* (*issue*) ζήτημα *n.* **~atic** *a.* προβληματικός.

procedure *s.* διαδικασία *f.*

proceed *v.* (*advance*) προχωρώ, (*take steps*) προβαίνω. ~ from προέρχομαι από. ~ with συνεχίζω.

proceeding *s.* (*activity*) ενέργεια *f.* (*act*) πράξη *f.* ~s (*events*) συμβάντα *n.pl.*, (*report*) πεπραγμένα *n.pl.* take legal ~s πάω στα δικαστήρια.

proceeds *s.* εισπράξεις *f.pl.*

process *s.* (*functioning*) λειτουργία *f.* (*procedure*) διαδικασία *f.* (*course*) πορεία *f.* (*of manufacture, operating*) μέθοδος *f.*, τρόπος *m.* it is a slow ~ δεν γίνεται γρήγορα. in ~ of construction υπό κατασκευήν. in the ~ of time με την πάροδο του χρόνου.

process *v.t.* (*materials*) επεξεργάζομαι. (*v.i.*, *go by*) παρελαύνω. **~ing** *s.* επεξεργασία *f.* **~ion** *s.* παρέλαση *f.*

prociaim *v.* (*views*) διακηρύσσω, (*king, etc.*) ανακηρύσσω. (*reveal*) αποκαλύπτω.

proclamation *s.* διακήρυξη, ανακήρυξη *f.*

proclivity *s.* τάση *f.*

proconsul *s.* (*fig.*) κυβερνήτης αποικίας.

procrastinat|e *v.* αναβάλλω. **~ion** *s.* αναβλητικότητα *f.*

procur|e *v.* προμηθεύομαι, βρίσκω. **~er** *s.* μαστροπός *m.*

prod *v.* σκουντώ, σπρώχνω. (*s.*) σπρώξιμο *n.*

prodigal *a.* άσωτος, σπάταλος. ~ son άσωτος υιός. **~ity** *s.* (*plenty*) αφθονία *f.* (*waste*) σπατάλη *f.*

prodigious *a.* (*wonderful*) καταπληκτικός, (*huge*) τεράστιος. **~ly** *adv.* καταπληκτικά.

prodigy *s.* θαύμα *n.*

produc|e *v.* (*present*) παρουσιάζω, βγάζω. (*cause*) προκαλώ, δημιουργώ. (*crops, etc.*) παράγω, βγάζω, κάνω. (*return, yield*) αποδίδω. (*bring forth, create*) παράγω. (*s.*) προϊόντα *n.pl.* **~er** *s.* παραγωγός *m.* (*stage*) σκηνοθέτης *m.*

product *s.* προϊόν *n.* (*math.*) γινόμενον *n.*

production *s.* παραγωγή *f.* (*stage*) σκηνοθεσία *f.* (*work*) έργο *n.*

productiv|e *a.* παραγωγικός. **~ity** *s.* παραγωγικότητα *f.*

profanation *s.* βεβήλωση *f.*

profane *v.* βεβηλώνω. (*a.*) (*not sacred*) κοσμικός, (*impious*) ανίερος. (*blasphemer*) βλάσφημος.

profanity *s.* ανιερότητα *f.* (*words*) βλαστήμιες *f.pl.*

profess *v.t.* (*believe in*) πρεσβεύω, (*teach*) διδάσκω. (*v.i.*) (*declare*) δηλώνω, (*claim*) φιλοδοξώ. **~ed** *a.* (*acknowledged*) δεδηλωμένος, (*alleged*) δήθεν.

profession *s.* (*calling*) επάγγελμα *n.* (*declaration*) ομολογία *f.*

professional *a.* (*matter*) επαγγελματικός. ~ man επαγγελματίας *m.* ~ writer εξ επαγγέλματος συγγραφεύς. ~ player επαγγελ-

ματίας παίκτης. **~ism** s. (quality) επαγ-
γελματική ικανότητα. **~ly** adv. επαγγελματικώς.
professor s. καθηγητής m. καθηγήτρια f.
proffer v. προσφέρω.
profici|ency s. ικανότητα f. **~ent** a. ικανός,
εντριβής. **~ent** in English κάτοχος της
αγγλικής.
profile s. κατατομή f., προφίλ n.
profit s. (gain) κέρδος n. (benefit) όφελος
n. make a ~ αποκομίζω κέρδος. **~** margin
περιθώριο κέρδους. (v.t.) ωφελώ, χρησιμεύω σε. (v.i.) κερδίζω, ωφελούμαι.
profitab|le a. επικερδής, επωφελής. **~ly** adv.
επικερδώς, επωφελώς.
profiteer v.i. κερδοσκοπώ. (s.) κερδοσκό-
πος m. **~ing** f. κερδοσκοπία f.
profitless a. ανωφελής.
profit-sharing a. ~ scheme σχέδιο συμμετοχής.
proflig|acy s. ανηθικότητα f., ασωτεία f.
~ate a. ανήθικος, άσωτος.
profound a. βαθύς, (erudite) εμβριθής. **~ly**
adv. βαθύτατα, (absolutely) τελείως.
profundity s. βαθύτητα f. εμβρίθεια f.
profuse a. άφθονος, υπερβολικός. be ~ in
είμαι σπάταλος σε, (apologies, etc.)
αναλύομαι εις. **~ly** adv. άφθονα, υπερβολικά.
profusion s. αφθονία f.
progenitor s. πρόγονος m.
progeny s. απόγονοι m.pl.
prognosis s. πρόγνωση f.
prognostic s. προγνωστικό n. **~ate** v.
προβλέπω, προμαντεύω. **~ation** s. πρόβλεψη f.
programme s. πρόγραμμα n. (v.) προγραμματίζω.
progress s. πρόοδος f. (course) πορεία f.
(of army) προέλαση f. make ~ σημειώνω
πρόοδο. be in ~ γίνομαι, (negotiations)
διεξάγομαι, (works) εκτελούμαι.
progress v. προοδεύω, προχωρώ. (army)
προελαύνω. **~ion** s. κίνηση f. (math.)
πρόοδος f. **~ive** a. προοδευτικός.
prohibit v. απαγορεύω (with acc. of thing,
σε of person). (prevent) εμποδίζω. **~ion** s.
απαγόρευση f. (of drink) ποτοαπαγόρευση
f. **~ive** a. απρόσιτος **~ory** a. απαγορευ-
τικός.
project v.t. (plan) σχεδιάζω. (hurl) ρίπτω,
(cast, extend) προβάλλω. (make known,
put across) διαφημίζω, προβάλλω. (v.i.,
stick out) προεξέχω. (s.) σχέδιο n.
projection s. (act & geom.) προβολή f.
(thing) προεξοχή f.
projector s. προβολεύς f.
proletar|ian a. προλετάριος. **~iat** s. προ-

λεταριάτο n.
proliferat|e v.i. πολλαπλασιάζομαι. **~ion** s.
πολλαπλασιασμός m.
prolific a. γόνιμος, παραγωγικός.
prolix a. μακρήγορος. **~ity** s. μακρηγορία f.
prologue s. πρόλογος m.
prolong v. παρατείνω, (a line) προεκτείνω.
~ed παρατεταμένος. **~ation** παράταση f.
προέκταση f.
promenade s. (walk, place) περίπατος m.
(v.i.) κάνω βόλτα. (v.t.) (display) επιδεικνύω.
prominence s. give ~ to αποδίδω σημασίαν
εις. come into ~ (thing) αποκτώ σημασία,
(person) γίνομαι γνωστός.
prominent a. (bones, etc.) προεξέχων. (on
ground) εμφανής, δεσπόζων. (leading)
εξέχων, (important) σημαντικός.
prominently adv. figure ~ (play large part)
παίζω σημαντικό ρόλο, (be easily seen)
διακρίνομαι εύκολα.
promiscu|ity s. (esp.) ελεύθερες σεξουα-
λικές σχέσεις. **~ous** a. (casual) τυχαίος,
(mixed) ανακατωμένος. **~ously** adv. στην
τύχη.
promise s. υπόσχεση f. land of ~ γη της
επαγγελίας. show ~ υπόσχομαι πολλά. (v.)
υπόσχομαι, (foretell) προμηνύω. he ~s
well υπόσχεται πολλά.
promising a. the future looks ~ το μέλλον
προμηνύεται καλό.
promissory a. ~ note χρεωστικό ομόλογο.
promontory s. ακρωτήριο n.
promot|e v. προάγω, προωθώ. (in rank)
προβιβάζω. **~ion** s. προαγωγή f., προ-
ώθηση f. προβιβασμός m.
prompt a. γρήγορος, πρόθυμος, άμεσος. (v.)
ωθώ, κάνω, εμπνέω. (actor) υποβάλλω εις.
~er s. υποβολεύς m. **~ing** s. υποβολή f.
promptitude s. ταχύτητα f., προθυμία f.
promptly adv. αμέσως.
promulgat|e v. (law) δημοσιεύω, (ideas)
διαδίδω. **~ion** s. δημοσίευση f. διάδοση f.
prone adv. (face down) μπρούμυτα. (a.)
(liable) υποκείμενος. be ~ to υπόκειμαι εις
(or να). **~ness** s. τάση f.
prong s. δόντι n.
pronoun s. αντωνυμία f.
pronounce v.t. (articulate) προφέρω,
(judgement) εκδίδω. (declare) κηρύσσω.
(v.i.) ~ on αποφαίνομαι επί (with gen.). **~d**
a. έντονος, αισθητός. **~ment** s. δήλωση f.
pronunciation s. προφορά f.
proof s. απόδειξη f. (printer's) δοκίμιο n.
proof a. ~ against ανθεκτικός σε.
prop s. στήριγμα n. (v.) στηρίζω.
propaganda s. προπαγάνδα f.
propagat|e v. αναπαράγω, (fig.) διαδίδω.

propel   446   prototype

~**ion** s. αναπαραγωγή f. διάδοση f.

propel v. προωθώ. ~**lant** s. προωθητικό n.

propeller s. έλικας f., προπέλλα f.

propensity s. ροπή f.

proper a. (right) σωστός, ορθός. (seemly) πρέπων, (suitable) κατάλληλος. (competent, of officials) αρμόδιος. (respectable) καθώς πρέπει. in the ~ way σωστά, όπως πρέπει. the ~ man (for post) ο ενδεδειγμένος. a ~ fool σωστός βλάκας. ~ name κύριο όνομα. Greece ~ η κυρίως Ελλάδα.

properly adv. σωστά, καλά, πρεπόντως, όπως πρέπει. ~ speaking κυρίως ειπείν. we got ~ beaten μας νίκησαν κατά κράτος.

property s. (esp. immovable) ιδιοκτησία f. (building) ακίνητο n. (land) κτήμα n. (wealth) περιουσία f. (effects) υπάρχοντα n.pl., αγαθά n.pl. (attribute) ιδιότητα f. common ~ (known to all) κοινό μυστικό.

prophecy s. προφητεία f.

prophesy v. προφητεύω.

prophet s. προφήτης m. ~**ic** a. προφητικός. ~**ically** adv. προφητικά.

prophyl|actic a. προφυλακτικός. ~**axis** s. προφύλαξη f.

propinquity s. εγγύτητα f.

propitiat|e v. εξευμενίζω. ~**ion** s. εξευμενισμός m. ~**ory** a. εξευμενιστικός.

propitious a. ευνοϊκός. ~**ly** adv. ευνοϊκά.

proportion s. (part) μέρος n. (relation) αναλογία f. in ~ ανάλογος. out of ~ δυσανάλογος. in ~ to ανάλογα με, αναλόγως (with gen.). well ~ed με καλές αναλογίες.

proportional a. ανάλογος ~ representation αναλογικό εκλογικό σύστημα. ~**ly** adv. αναλόγως.

proportionate a. ανάλογος. ~**ly** adv. αναλόγως.

proposal s. πρόταση f.

propose v. προτείνω, (a motion, measure) εισηγούμαι. (intend) σκοπεύω, προτίθεμαι. ~ a toast κάνω πρόποση. ~ marriage κάνω πρόταση γάμου.

proposition s. πρόταση f. (fam., matter) υπόθεση f., δουλειά f.

propound v. εισηγούμαι.

proprietary a. (owner's) του ιδιοκτήτη. (of ownership) ιδιοκτησίας. ~ medicine σπεσιαλιτέ n.

proprietor s. ιδιοκτήτης m.

propriet|y s. (rightness) ορθότητα f. (decorum) ευπρέπεια f. ~**ies** τύποι m.pl.

propuls|ion s. προώθηση f. ~**ive** a. προωθητικός.

prorogue v. διακόπτω.

prosaic a. πεζός. ~**ally** adv. πεζά.

proscenium s. προσκήνιο n.

proscr|ibe v. προγράφω. ~**iption** s. προγραφή f.

prose s. πεζογραφία f., πεζός λόγος. (as a.) πεζός. ~ writer πεζογράφος m.

prosecute v. (continue) συνεχίζω, (at law) διώκω.

prosecution s. (at law) δίωξη f. (prosecuting side) κατηγορία f. (of duties) άσκηση f.

prosecutor s. κατήγορος m. Public P~ εισαγγελεύς m.

proselyt|e s. προσήλυτος m.f. ~**ize** v. προσηλυτίζω.

prosody s. μετρική f.

prospect s. (view) θέα f. (of outcome) προοπτική f. (hope) ελπίδα f. there is no ~ δεν υπάρχει πιθανότητα. have brilliant ~s έχω λαμπρό μέλλον μπροστά μου. in ~ εν όψει.

prospect v. ερευνώ.

prospective a. (to be) μέλλων, (likely) πιθανός, (would-be) υποψήφιος. (development, works) μελλοντικός.

prospectus s. φυλλάδιο n., μπροσούρα f.

prosper v. προκόβω, πάω καλά, ευδοκιμώ. ~**ity** s. ευημερία f.

prosperous a. επιτυχημένος, ευημερών. be ~ ευημερώ.

prostate s. (anat.) προστάτης m.

prostitut|e s. πόρνη f. (v.) εκπορνεύω. ~e oneself πορνεύομαι, πουλιέμαι. ~**ion** s. πορνεία f.

prostrate a. (face down) πεσμένος μπρούμυτα. (overcome) τσακισμένος, εξαντλημένος.

prostrate v. ρίχνω χάμω, (overcome) τσακίζω, εξαντλώ. ~ oneself πέφτω μπρούμυτα, (kowtow) προσπέφτω.

prostration s. εξάντληση f., κατάπτωση f.

prosy a. πεζός.

protagonist s. πρωταγωνιστής m.

protect v. (care for) προστατεύω, (guard) προφυλάσσω.

protection s. προστασία f. προφύλαξη f. ~**ism** s. προστατευτισμός m.

protective a. προστατευτικός, προφυλακτικός.

protector s. προστάτης m. (device) προφυλακτήρας m.

protégé a. & n. προστατευόμενος m.

protein s. πρωτεΐνη f.

protest v.i. (object) διαμαρτύρομαι, (assert) ισχυρίζομαι (with ότι). (s.) διαμαρτυρία f.

protestant a. & n. διαμαρτυρόμενος. ~**ism** s. προτεσταντισμός m.

protestation s. βεβαίωση f.

protocol s. πρωτόκολλον n., εθιμοτυπία f.

prototype s. πρωτότυπο n.

protract *v.* παρατείνω. ~ed παρατεταμένος.

protractor *s.* (*geom.*) αναγωγεύς *m.*

protrud|e *v.t.* βγάζω έξω. (*v.i.*) βγαίνω έξω, προεξέχω. ~ing *a.* (*eyes*) γουρλωμένος, (*teeth, ears*) πεταχτός.

protrusion *s.* προεξοχή *f.*

protuber|ance *s.* εξόγκωμα *n.* ~ant *a.* προεξέχων.

proud *a.* περήφανος. be ~ (of) υπερηφανεύομαι (για), καμαρώνω. (*glorious*) ένδοξος, (*in bearing*) καμαρωτός, (*disdainful*) ακατάδεκτος. be ~ to... θεωρώ τιμή μου να. (*fam.*) do (*person*) ~ περιποιούμαι ιδιαιτέρως. ~ly *adv.* με περηφάνια, καμαρωτά.

prove *v.t.* (*show*) αποδεικνύω, (*verify*) επαληθεύω, (*test*) δοκιμάζω. (*v.i.*) (*turn out to be*) αποδεικνύομαι, φαίνομαι. it ~d useful to me μου φάνηκε χρήσιμο.

proven *a.* αποδεδειγμένος.

provenance *s.* προέλευση *f.*

provender *s.* τροφή *f.*

proverb *s.* παροιμία *f.* ~ial *a.* παροιμιώδης ~ially *adv.* they are ~ially brave η ανδρεία τους είναι παροιμιώδης.

provide *v.* παρέχω, (*equip*) εφοδιάζω. ~ for (*make provision*) προβλέπω, εξασφαλίζω, (*support*) συντηρώ. ~ against λαμβάνω τα μέτρα μου εναντίον (*with gen.*).

provided *conj.* ~ (that) αρκεί να, υπό τον όρον ότι.

providence *s.* πρόνοια *f.*

provident *a.* οικονόμος. ~ society ταμείον πρόνοιας. ~ly *adv.* προνοητικά.

providential *a.* have a ~ escape σώζομαι ως εκ θαύματος.

provider *s.* ο παρέχων.

province *s.* (*general*) επαρχία *f.*, χώρα *f.* (*administrative region of Greece*) διαμέρισμα *n.* the ~s (*fig.*) η επαρχία. (*fig., area*) σφαίρα *f.* (*competency*) αρμοδιότητα *f.*

provincial *a.* επαρχιακός. (*fig., unrefined*) επαρχιώτικος, (*person*) επαρχιώτης *m.* ~ism *s.* επαρχιωτισμός *m.*

provision *s.* παροχή *f.*, εφοδιασμός *m.* ~s (*food*) τρόφιμα *n.pl.*

provisional *a.* προσωρινός. ~ly *adv.* προσωρινώς.

provisionment *s.* ανεφοδιασμός *m.*

proviso *s.* όρος *m.* with the ~ that υπό τον όρον ότι.

provocation *a.* (*act*) πρόκληση *f.* (*annoyance*) ενόχληση *f.* at the least ~ με το τίποτα, για ψύλλου πήδημα.

provocative *a.* προκλητικός.

provok|e *v.* (*cause*) προκαλώ, (*make*) κάνω. (*vex*) ενοχλώ. ~ing *a.* ενοχλητικός.

provost *s.* (*of college*) διευθυντής *m.* (*Scottish mayor*) δήμαρχος *m.*

prow *s.* πλώρη *f.*

prowess *s.* (*valour*) ανδρεία *f.* (*skill*) ικανότητα *f.*

prowl *v.* τριγυρίζω με ύποπτους σκοπούς.

proximity *s.* εγγύτης *f.* in the ~ (of) κοντά (σε).

proxy *s.* (*person*) πληρεξούσιος *m.* by ~ διά πληρεξουσίου.

prude *s.* σεμνότυφος *a.*

prudence *s.* σωφροσύνη *f.*, φρόνηση *f.*

prudent *a.* σώφρων, γνωστικός. ~ly *adv.* με γνώση, με φρόνηση.

prud|ery *s.* σεμνοτυφία *f.* ~ish *a.* σεμνότυφος.

prune *s.* δαμάσκηνο *n.*

prun|e *v.* κλαδεύω, (*fig.*) ψαλιδίζω. ~ing *s.* κλάδεμα *n.* ψαλίδισμα *n.* ~ing-hook *s.* κλαδευτήρι *n.*

pruri|ence *s.* λαγνεία *f.* ~ent *a.* λάγνος.

pry *v.* ~ into χώνω τη μύτη μου σε. ~ing *a.* αδιάκριτος.

PS ΥΓ.

psalm *s.* ψαλμός *m.*

psalter *s.* ψαλτήριον *n.*

pseudo- ψευδο-.

pseudonym *s.* ψευδώνυμο *n.*

psyche *s.* ψυχή *f.*

psychiatr|ist *s.* ψυχίατρος *m.* ~y *s.* ψυχιατρική *f.*

psychic, ~al *a.* ψυχικός.

psychoanaly|se *v.* κάνω ψυχανάλυση σε. ~sis *s.* ψυχανάλυση *f.* ~st *s.* ψυχαναλυτής *m.*

psychological *a.* ψυχολογικός. ~ly *adv.* ψυχολογικώς.

psycholog|ist *s.* ψυχολόγος *m.* ~y *s.* ψυχολογία *f.*

psychopath *s.*, ~ic *a.* (*person*) ψυχοπαθής, (*condition*) ψυχοπαθητικός.

psychosis *s.* ψύχωση *f.*

ptomaine *s.* πτωμαΐνη *f.*

pub *s.* ταβέρνα *f.*

puberty *s.* ήβη *f.*

pubic *a.* ηβικός.

public *a.* δημόσιος, (*state-controlled*) κρατικός. ~ holiday εορτή *f.* ~ house ταβέρνα *f.* ~ opinion κοινή γνώμη. ~ school (*UK*) ιδιωτικό σχολείο μέσης εκπαίδευσης. ~ spirit μέριμνα για το κοινό συμφέρον. (*s.*) the ~ το κοινό. in ~ δημοσία.

publican *s.* ταβερνιάρης *m.*

publication *s.* (*making known*) δημοσίευση *f.* (*of book*) έκδοση *f.* (*book*) βιβλίο *n.*

publicist *s.* πολιτικός σχολιαστής.

public|ity *s.* (*being known*) δημοσιότητα *f.* (*advertising*) διαφήμιση *f.* ~ize *v.* δίδω δημοσιότητα σε, διαφημίζω.

publish v. (of publisher) εκδίδω, (of writer, also make known) δημοσιεύω. ~ing house εκδοτικός οίκος. ~**er** s. εκδότης m.

puce a. κασταvέρυθρος.

pucker v. ζαρώνω.

puckish a. σκανταλιάρικος.

pudding s. πουτίγκα f.

puddle s. (pool) λακκούβα f. (v.i.) (dabble) τσαλαβουτώ.

pudgy a. κοντόχοντρος.

puer|ile a. παιδαριώδης. ~**ility** s. (act, utterance) παιδαριωδία f. (quality) παιδαριώδης νοοτροπία.

puff s. φύσημα n. (of smoke) τολύπη f. (of wind) ριπή f. (at cigar, etc.) ρουφηξιά f. (praise) ρεκλάμα f.

puff v.i. (pant) (also ~ and blow) ξεφυσώ, ασθμαίνω, φουσκώνω, (of engine) βγάζω καπνό. (v.t.) (give out) βγάζω, (swell out) φουσκώνω. (cigar, etc.) τραβώ μια ρουφηξιά από. (praise) ρεκλαμάρω. ~ oneself up φουσκώνω.

puffed a. (hair, dress) φουσκωτός, (panting) λαχανιασμένος. ~ up (proud) φουσκωμένος.

puffy a. (swollen) πρησμένος.

pugil|ism s. πυγμαχία f. ~**ist** s. πυγμάχος m. ~**istic** a. πυγμαχικός.

pugnacious a. με εριστική διάθεση. ~**ly** adv. βιαίως.

pugnacity s. εριστική διάθεση.

puke v. ξερνώ.

pukka a. γνήσιος.

puling a. κλαψιάρικος.

pull s. τράβηγμα n. (at pipe, etc.) ρουφηξιά f. (attraction) έλξη f. (influence) μέσα n.pl. (advantage) πλεονέκτημα n.

pull v.t. & i τραβώ. (v.t.) σέρνω, σύρω. ~ to pieces (fig.) κάνω σκόνη. ~ (person's) leg πειράζω. ~ one's weight βάζω τα δυνατά μου. ~ a long face κατεβάζω τα μούτρα. ~ strings βάζω μέσα. have strings to ~ έχω μέσα.

pull about v.t. τραβολογώ.

pull ahead v.i. αποσπώμαι.

pull at v.t. τραβώ. (pipe, bottle) τραβώ μια ρουφηξιά από.

pull back v.i. αποσύρομαι.

pull down v.t. (lower) κατεβάζω, (demolish) γκρεμίζω, (overthrow) ανατρέπω, (weaken) τσακίζω.

pull in v.t. (crowd, nets) τραβώ, (money) μαζεύω. (v.i.) (arrive) φτάνω, (to side) τραβώ στην άκρη.

pull off v.t. (remove), βγάζω, (achieve) πετυχαίνω.

pull on v.t. (gloves, etc.) βάζω.

pull out v.t. (extract) βγάζω, (open) ανοίγω.

(v.i.) (start off) ξεκινώ. (in driving) βγαίνω από τη γραμμή. ~ of αποχωρώ από.

pull over v.i. (in driving) τραβώ στην άκρη του δρόμου.

pullover s. πουλόβερ n.

pull round v.t. (revive) συνεφέρνω. (v.i.) συνέρχομαι.

pull through v.t. & i. γλυτώνω.

pull together v.t. (rally) συγκεντρώνω. (revive) συνεφέρνω, (of nourishment) στυλώνω. (v.i.) συμφωνώ, συνεργάζομαι. pull oneself together ανακτώ το ηθικό μου, μαζεύομαι.

pull up v.t. (raise) σηκώνω, ανεβάζω, (uproot) ξεριζώνω. (v.t. & i) (stop) σταματώ.

pullet s. κοτόπουλο n., πουλάδα f.

pulley s. τροχαλία f., καρούλι n.

pullulate v.i. πολλαπλασιάζομαι. ~ with βρίθω (with gen.).

pulmonary a. πνευμονικός.

pulp s. πολτός m. (of fruit) σάρκα f. (v.) πολτοποιώ, κάνω λείωμα. ~**ing** s. πολτοποίηση f.

pulpit s. άμβων m.

pulsat|e v. (beat) πάλλω, κτυπώ, (fig., be vibrant) σφύζω. ~**ing** a. παλμικός. ~**ion** s. παλμός m.

pulse s. (beat) σφυγμός m. (v.i.) σφύζω, κτυπώ.

pulse s. (edible) όσπρια n.pl.

pulverize v. κονιοποιώ, (fig.) κάνω σκόνη.

pumice s. ελαφρόπετρα f.

pummel v. γρονθοκοπώ.

pump v.t. αντλώ, τρομπάρω. ~ out βγάζω, αδειάζω. ~ up φουσκώνω. (fig.) (question) ψαρεύω, (instil) εμφυσώ. (s.) αντλία f., τρόμπα f. (for inflating) τρόμπα f.

pump s. (woman's shoe) γόβα f. (man's) λουστρίνι n.

pumpkin s. κολοκύθα f.

pun s. λογοπαίγνιο n.

punch s. (tool) ζουμπάς m. (small) τρυπητήρι n. (v.) τρυπώ, ανοίγω τρύπες σε.

punch s. (blow) γροθιά f. (fig., vigour) δύναμη f. not pull one's ~es μιλάω έξω από τα δόντια. (v.) δίνω γροθιά σε, (repeatedly) γρονθοκοπώ.

punch s. (drink) πόντς n.

Punch s. Φασουλής m. as pleased as ~ κατευχαριστημένος.

punctili|o s. τύποι m.pl. ~**ous** a. be ~ous προσέχω τους τύπους.

punctual a. ακριβής. ~**ity** s. ακρίβεια f. ~**ly** adv. εγκαίρως, στην ώρα μου.

punctuat|e v. (interrupt) διακόπτω. ~**ion** s. στίξη f. (marks) σημεία στίξεως.

puncture s. τρύπημα n. (med.) παρακέντηση f. (fam.) I got a ~ (in tyre) μ' έπιασε λάστιχο. (ν.) τρυπώ.

pundit s. αυθεντία f.

pungency s. δριμύτητα f.

pungent a. δριμύς, καυστικός, καυτερός. ~ smell σπιρτάδα f.

punish v. τιμωρώ, (maltreat) κακομεταχειρίζω. (fam.) they ~ed the victuals ταράξανε τα φαγιά. ~**able** a. αξιόποινος. ~**ment** s. τιμωρία f.

punitive a. προς τιμωρίαν. ~ taxation πολύ βαριά φορολογία.

punt s. λέμβος ποταμού προωθούμενη με σταλίκι.

punt v. (bet) ποντάρω. ~**er** s. πονταδόρος m.

puny a. καχεκτικός. (fig.) ασήμαντος.

pup s. κουτάβι n. (fam.) be sold a ~ πιάνομαι κορόιδο.

pupil s. μαθητής m., μαθήτρια f. (of eye) κόρη f.

puppet s. ανδρείκελο n. ~ show κουκλοθέατρο n.

puppy s. σκυλάκι n. (fig.) ομορφονιός m.

purblind a. (fig.) βλάκας.

purchase s. αγορά f. ~s ψώνια n.pl. (hold) get a (good) ~ on αδράχνω, πιάνομαι γερά από. (v.) αγοράζω. ~**r** s. αγοραστής m.

purdah s. be in ~ (of women) κρατούμαι περιορισμένη μακριά από τα ανδρικά μάτια.

pure a. (clean, unmixed) καθαρός. (unadulterated, chaste) αγνός. (mere, also ~ and simple) καθαρός, σκέτος. (science) αφηρημένος. ~**ly** adv. απλώς, καθαρώς.

purée s. πολτός m. (tomato) μπελτές m. (potato, etc.) πουρές m.

purgative s. καθαρτικό n.

purgatory s. καθαρτήριον n. (fig.) μαρτύριο n.

purge v. (clean) καθαρίζω. (rid) εκκαθαρίζω, (get rid of) αποβάλλω, (sins) αποπλύνω. (s.) εκκαθάριση f.

purification s. καθάρισμα n. (feast of Candlemas, 2nd Feb.) Υπαπαντή f.

purify v. καθαρίζω.

purist s. εκλεκτικός a. (advocate of puristic Gk.) καθαρευουσιάνος m. ~**ic** a. ~ic Greek καθαρεύουσα f.

puritan s. ~**ical** a. (person) πουριτανός m. (idea, etc.) πουριτανικός.

purity s. καθαρότητα f., αγνότητα f.

purl v. κελαρίζω.

purloin v. κλέβω.

purple s. (fig.) πορφύρα f. (colour) πορφυρούν χρώμα. (a.) βαθυκόκκινος. get ~ in the face κοκκινίζω. (fig.) ~ patch παθητικό απόσπασμα.

purport v.i. (to be) φιλοδοξώ. (s.) έννοια f.

purpose s. σκοπός m. (resolution) αποφασιστικότητα f. on ~ σκοπίμως, επίτηδες. to the ~ χρήσιμος, σχετικός. to good ~ αποτελεσματικά. to no ~ άδικα, αδίκως, εις μάτην. (v.) σκοπεύω. ~**fully** adv. με αποφασιστικότητα. ~**ly** adv. σκοπίμως, επίτηδες.

purr v. ρονρονίζω, (of engine) βουίζω. (s.) ρονρόνισμα n. βουή f.

purse s. (lady's) πορτοφόλι n. (also fig.) πουγγί n. hold the ~-strings βαστώ το ταμείο.

purser s. λογιστής m.

pursuance s. in ~ of (accordance) συμφώνως προς, (execution) εις εκτέλεσιν (with gen.).

pursue v. (chase) κυνηγώ, καταδιώκω. (go on with) συνεχίζω, (aim at) επιδιώκω. ~**r** s. διώκτης m.

pursuit s. (chasing) καταδίωξη f. (of happiness, wealth) κυνήγι n., επιδίωξη f. (occupation) εργασία f., απασχόληση f. they went in hot ~ of him τον πήραν κυνήγι.

purulent a. πυώδης.

purvey v. προμηθεύω. ~**or** s. προμηθευτής m. ~**ance** s. προμήθεια f.

pus s. πύον n., έμπυο n.

push s. σπρωξιά f., σπρώξιμο n. (attack) επίθεση f. give the bell a ~ χτυπώ το κουδούνι. have plenty of ~ είμαι αποφασισμένος να επιτύχω. (fam.) get the ~ παύομαι, at a ~ στην ανάγκη.

push v. σπρώχνω (insert) χώνω. (exert pressure on) πιέζω. (button) πατώ, (claims) διεκδικώ, (wares) διαφημίζω, προωθώ, (drugs) πλασάρω. ~ oneself forward προβάλλομαι. ~ one's way through the crowd περνώ παραμερίζοντας το πλήθος. be ~ed for time μου λείπει καιρός.

push along v.i. I must ~ πρέπει να πηγαίνω.

push around v.t. σέρνω από 'δω κι από 'κει.

push aside v.t. παραμερίζω.

push away, back v.t. απωθώ.

push forward v.i. προχωρώ. (v.t.) ~ oneself forward (αυτο)προβάλλομαι.

push off v.i. (go) στρίβω.

push on v.i. προχωρώ. ~ with συνεχίζω. (v.t.) προωθώ, σπρώχνω.

push over v.t. ρίχνω κάτω, ανατρέπω.

push-over s. (fam.) εύκολη επιτυχία.

push through v.t. & i. περνώ.

push up v.t. ανεβάζω.

push-cart s. χειράμαξα f.

pusher s. αρριβίστας m.

pushful a. be ~ προβάλλομαι αναιδώς.

pushing s. σπρώξιμο n. (a.) see pushful.

pusillanimous a. δειλός.

puss s. γατούλα f.

pussy-cat s. ψιψίνα f.

put v. (general) βάζω. (say, express) λέω, εκφράζω, (explain, set out) εκθέτω, (estimate) υπολογίζω. (invest) τοποθετώ. ~ right or straight or in order κανονίζω, τακτοποιώ. ~ a question θέτω μία ερώτηση. ~ an end to θέτω τέρμα σε. ~ one in mind of θυμίζω. ~ one's foot in it κάνω μια γκάφα. ~ in hand βάζω μπροστά, καταπιάνομαι με. ~ on trial δικάζω. ~ to shame ντροπιάζω. ~ to death θανατώνω. ~ to the test δοκιμάζω. ~ to sleep (child) κοιμίζω, (animal) δίνω κάτι να ψοφήσει. be hard ~ to it δυσκολεύομαι. I ~ it to him that του υπέδειξα ότι.

put about v.t. (circulate) διαδίδω, (trouble) ενοχλώ. (v.i., of ship) ορτσάρω.

put across v.t. (convey) μεταδίδω.

put aside v.t. (save) βάζω στην μπάντα, (on one side) βάζω κατά μέρος. (leave) αφήνω.

put away v.t. κρύβω, βάζω στη θέση του. (give up) εγκαταλείπω, (confine) εγκλείω. (kill) καθαρίζω.

put back v.t. βάζω πίσω, (delay) καθυστερώ. (v.i., of ship) επαναπλέω.

put by v.t. βάζω στην μπάντα.

put down v.t. αφήνω κάτω. (window, passenger) κατεβάζω, (money) καταβάλλω, (umbrella) κλείνω. (suppress) καταστέλλω, (humble) ταπεινώνω. (write) γράφω, σημειώνω. (attribute) αποδίδω. ~ one's foot down πατώ πόδι. (v.i.) (of plane) προσγειώνομαι.

put forth v.t. βγάζω, (exert) καταβάλλω.

put forward v.t. (submit) υποβάλλω, (propose) προτείνω. (clock, date) βάζω μπρος.

put in v.t. βάζω (μέσα). (do) κάνω, (submit) υποβάλλω. ~ a good word λέω μια καλή κουβέντα. ~ for (post) εκθέτω υποψηφιότητα για. (v.i., of ship) ~ at πιάνω σε.

put off v.t. (postpone) αναβάλλω, (get rid of) βγάζω, (distract) εκνευρίζω, (with excuse) ξεφορτώνομαι. (dissuade) αποτρέπω, (offend) φέρνω αηδία σε. (make hostile) προδιαθέτω δυσμενώς. (v.i.) ξεκινώ.

put on v.t. βάζω. (clothes) φορώ, βάζω. (light, gas) ανάβω, (clock) βάζω μπρος. (a play) ανεβάζω. (weight, airs) παίρνω, (speed) αυξάνω. (add) προσθέτω, (feign) προσποιούμαι.

put out v.t. βγάζω (έξω). (light, fire) σβήνω. (extend) απλώνω. (dislocate) εξαρθρώνω. (vex) πειράζω, στενοχωρώ. (in order) to put him out of his pain για να ανακουφισθεί από τους πόνους. (v.i., of ship) ανάγομαι.

put over v.t. (convey) μεταδίδω.

put through v.t. (effect) φέρω εις πέρας, (connect) συνδέω. put (person) through it υποβάλλω σε σκληρή δοκιμασία, του βγάζω το λάδι.

put together v.t. (parts) συναρμολογώ. (thoughts) συγκεντρώνω, (things) μαζεύω.

put up v.t. ανεβάζω. (hair, hands, game) σηκώνω, (building) ανεγείρω, (tent) στήνω, (umbrella) ανοίγω, (money) βάζω. (guest) φιλοξενώ. (resistance) προβάλλω. (proposal) προτείνω. (pack) συσκευάζω. (get ready) ετοιμάζω. (to auction) βγάζω. ~ for sale πουλώ. (instigate) βάζω. put (persons's) back up πικάρω. (v.i.) (lodge) μένω. ~ with (be content) αρκούμαι εις (endure) ανέχομαι. ~ for (election) υποβάλλω υποψηφιότητα για. a put-up job απάτη f.

put upon v.t. (exploit) εκμεταλλεύομαι.

putative a. υποτιθέμενος.

putre|fy v.i. σαπίζω, σήπομαι. ~fied σάπιος. ~faction s. σήψη f.

putresc|ent a. σάπιος. ~ence s. σαπίλα f.

putrid a. σάπιος. (fam.) βρωμερός.

puttee s. γκέτα f.

putty s. στόκος m.

puzzle s. αίνιγμα n., πρόβλημα n. (game) γρίφος m. (v.t.) μπερδεύω, σαστίζω. ~ out (solve) ξεδιαλύνω. (v.i.) σπάζω το κεφάλι μου. ~ment s. απορία f.

puzzling a. (difficult) δύσκολος. his conduct is ~ δεν μπορώ να καταλάβω τη συμπεριφορά του.

pygmy a. & s. πυγμαίος.

pyjamas s. πιτζάμα f.

pylon s. στύλος m.

pyramid s. πυραμίδα f.

pyre s. πυρά f.

pyrotechnics s. (fig.) εντυπωσιακό ρητορικό ύφος.

Pyrrhic a. ~ victory πύρρειος νίκη.

python s. πύθων m.

Pythoness s. Πυθία f.

pyx s. (eccl.) αρτοφόριον n.

Q

Q s. (fam.) on the Q.T. κρυφά.

qua conj. ως.

quack v. κάνω κουά-κουά.

quack s. αγύρτης m., κομπογιαννίτης m. (doctor) πρακτικός ιατρός. ~ remedy γιατροσόφι n. ~ery s. αγυρτεία f.

quad s. 1. (of college) αυλή f. 2. see quadruplet.

quadrang|le s. (geom.) τετράπλευρο n. (of college) αυλή f. **~ular** a. τετράπλευρος.
quadratic a. δευτεροβάθμιος.
quadrille s. καντρίλιες f.pl.
quadruped s. τετράποδο n.
quadruple a. τετραπλάσιος. (v.t.) τετραπλασιάζω. (v.i.) τετραπλασιάζομαι.
quadruplet s. τετράδυμο n.
quaff v. πίνω με μεγάλες ρουφηξιές.
quagmire s. βάλτος m.
quail s. ορτύκι n.
quail v. δειλιάζω, (of courage) λυγίζω.
quaint a. (picturesque) γραφικός, (unusual) περίεργος. **~ly** adv. γραφικά, με περίεργο τρόπο.
quake v. τρέμω, (shiver) τρεμουλιάζω.
Quaker s. κουάκερος m. ~ oats κουάκερ n.
qualification s. (reservation) επιφύλαξη f. (description) χαρακτηρισμός m. **~s** προσόντα n.pl.
qualif|y v.t. (limit) περιορίζω, (describe) χαρακτηρίζω. (give the right) it does not ~y him to... δεν του δίνει το δικαίωμα να. be ~ied (for post) έχω τα προσόντα. I am not ~ied (to give opinion, etc.) δεν είμαι σε θέση (να). (v.i., have the right) δικαιούμαι (with να).
qualified a. (limited) περιορισμένος. give ~ approval (to) εγκρίνω με επιφυλάξεις. (having diploma) διπλωματούχος, πτυχιούχος.
qualitative a. ποιοτικός.
quality s. (kind) ποιότητα f. (attribute) ιδιότητα f. (ability) ικανότητα f. (worth) αξία f. good ~ προτέρημα n. bad ~ ελάττωμα n.
qualm s. ενδοιασμός m. **~s** of conscience τύψεις f.pl. (sick feeling) αναγούλα f.
quandary s. δίλημμα n.
quantitative a. ποσοτικός.
quantit|y s. ποσότητα f. **~ies** of πολλοί, αφθονία f. (with gen.). unknown ~y (math.) άγνωστο μέγεθος, (fig.) αστάθμητος παράγων.
quantum s. (phys.) κβάντουμ n. ~ theory κβαντική θεωρία f.
quarantine s. καραντίνα f.
quarrel v. μαλώνω, καβγαδίζω, τσακώνομαι. we have ~led είμαστε μαλωμένοι (or τσακωμένοι). ~ with τα βάζω με, (disagree) διαφωνώ με, (complain of) παραπονούμαι για.
quarrel s. μάλωμα n., καβγάς m. **~some** a. φιλόνικος, καβγατζής.
quarry s. (hunted beast) θήραμα n., κυνήγι n. (human) ο (κατα)διωκόμενος.
quarry s. (for stone) λατομείο n., νταμάρι n. (fig., source) πηγή f. (v.t.) (dig out)

ορύσσω, βγάζω, (fig.) ψάχνω. **~ing** s. εξόρυξη, λατόμηση f.
quart s. τέταρτο του γαλονιού.
quarter s. (fourth part) τέταρτο n. (time) a ~ to one μία παρά τέταρτο. a ~ past one μία και τέταρτο. (of year) τρίμηνο n., τριμηνία f. (source) πηγή f. (district) συνοικία f. (colony of residents) παροικία f. in every ~ of the globe σε όλα τα μέρη του κόσμου. from every ~ από παντού. in high ~s στους ανωτέρους κύκλους. at close ~s κοντά, (fighting) εκ του συστάδην. (mercy) they gave no ~ δεν εφείσθησαν ουδενός. ask for ~ ζητώ χάρη.
quarter v.t. (lodge) βρίσκω κατάλυμα για. **~s** s. (residence) κατοικία f. (for crew, servants) διαμερίσματα n.pl. (billet, lodging) κατάλυμα n.
quarter|deck s. (fig.) οι αξιωματικοί. **~master** s. (mil.) αξιωματικός επιμελητείας.
quarterly a. τριμηνιαίος.
quartet s. κουαρτέτο n.
quartz s. χαλαζίας m.
quash v. (annul) ακυρώ, (suppress) καταπνίγω.
quasi- prefix οιονεί.
quatrain s. τετράστιχο n.
quaver s. (mus.) όγδοο n.
quaver v. (of voice) τρεμουλιάζω. **~ing** a. τρεμουλιαστός.
quay s. προκυμαία f.
queasy s. feel ~, have a ~ stomach αισθάνομαι αναγούλα.
queen s. βασίλισσα f. (at cards) ντάμα f. ~ mother βασιλομήτωρ f. (v., fam.) ~ it κάνω τη βασίλισσα. **~ly** a. βασιλικός.
queer a. (strange) αλλόκοτος, παράξενος. (unwell) αδιάθετος. he's a bit ~ in the head είναι ελαφρώς βλαμμένος. (fam.) be in Q~ street έχω χρέη, έχω μπλεξίματα. (s., fam.) ομοφυλόφιλος a. (v.t., spoil) χαλώ. they ~ed my pitch μου έκαναν χαλάστρα. **~ly** adv. παράξενα. **~ness** s. παραξενιά f.
quell v. καταπνίγω.
quench v. σβήνω. **~less** a. άσβηστος.
querulous a. παραπονιάρης. **~ly** adv. παραπονιάρικα. **~ness** s. γκρίνια f.
query s. ερώτημα n. (question-mark) ερωτηματικό n. (v.i., ask) ρωτώ. (v.t., call in question) αμφισβητώ.
quest s. αναζήτηση f. (v.) ~ for αναζητώ.
question s. ερώτημα n., (also gram.) ερώτηση f. (matter, problem) ζήτημα n. (doubt) αμφιβολία f. the case in ~ η προκειμένη υπόθεση. out of the ~ εκτός συζητήσεως, αδύνατος. without (or beyond) ~ αναμφισβητήτως, there is no ~

of it (*it is certain*) ούτε συζήτηση. it is out of the ~ αποκλείεται.
question *v.* (*interrogate*) ανακρίνω, εξετάζω. (*express doubt about*) αμφιβάλλω για, αμφισβητώ. (*v.i., wonder*) διερωτώμαι. **~er** *s.* ο ερωτών. **~ing** *a.* ερωτηματικός.
questionable *a.* (*statement*) αμφισβητήσιμος. of ~ value αμφιβόλου αξίας.
questionnaire *s.* ερωτηματολόγιο *n.*
queue *s.* ουρά *f.* (*v.*) ~ (up) κάνω ουρά.
quibbl|e, **~ing** *s.* (*evasion*) υπεκφυγή *f.* **~e** *v.i.* (*split hairs*) ψιλολογώ για ασήμαντα πράγματα. **~er** *s.* ψιλολόγος αντιρρησίας.
quick *a.* (*rapid*) γρήγορος, ταχύς, (*nimble*) σβέλτος. at a ~ pace με γοργό ρυθμό. be ~! βιάσου, κάνε γρήγορα! (*hurried*) βιαστικός. (*lively, intelligent*) έξυπνος. he has a ~ mind (*or* eye) κόβει το μυαλό του (*or* το μάτι του). **~ly** *adv.* γρήγορα. **~ness** *s.* γρηγοράδα *f.*, ταχύτητα *f.* (*of mind*) αντίληψη *f.*
quick *s.* cut (*person*) to the ~ θίγω καιρίως.
quicken *v.t.* (*pace*) επιταχύνω, (*stir, liven*) κεντρίζω. (*v.i.*) επιταχύνομαι, ζωντανεύω.
quicksand *s.* κινούμενη άμμος.
quickset *a.* ~ hedge φράχτης από θάμνους.
quicksilver *s.* υδράργυρος *m.*
quick-tempered *a.* be ~ θυμώνω με το τίποτα.
quick-witted *a.* ξύπνιος. ~ person σπίρτο μοναχό. **~ness** *s.* ετοιμότητα *f.*
quid *s.* (*tobacco*) κομμάτι καπνού για μάσημα. (*fam., pound*) λίρα *f.*
quiescent *a.* ακίνητος.
quiet *a.* ήσυχος, (*noiseless*) αθόρυβος, (*not speaking*) αμίλητος, (*simple, not gaudy*) απλός, on the ~ κρυφά, keep it ~ το κρατώ μυστικό. (*s.*) ησυχία, ηρεμία, γαλήνη *f.* **~ly** *adv.* ήσυχα, αθόρυβα, απλά.
quiet, **~en** *v.t. & i.* ησυχάζω, καλμάρω.
quiet|ness, **~ude** *s.* ηρεμία *f.*, γαλήνη *f.*
quiff *s.* αφέλεια *f.*
quill *s.* (*feather*) φτερό *n.* (*pen*) πέννα *f.* (*of animal*) αγκάθι *n.*
quilt *s.* πάπλωμα *n.* **~ed** *a.* καπιτονέ.
quin *s.* (*fam.*) πεντάδυμο *n.*
quince *s.* κυδώνι *n.* (*tree*) κυδωνιά *f.*
quinine *s.* κινίνη *f.*
quintessence *s.* πεμπτουσία *f.*
quintet *s.* κουϊντέτο *n.*
quintuplet *s.* πεντάδυμο *n.*
quip *s.* αστεϊσμός *m.* (*v.*) αστειεύομαι.
quire *s.* δεσμίδα είκοσι τεσσάρων φύλλων χαρτιού γραψίματος.
quirk *s.* εκκεντρικότητα *f.*
quisling *s.* (*fam.*) δοσίλογος *m.*
quit *v.t.* (*leave*) αφήνω, φεύγω από. (*stop*) σταματώ. (*v.i., depart*) φεύγω. (*a.*) be ~ of

ξεφορτώνομαι.
quite *adv.* (*completely*) τελείως, εντελώς, όλως διόλου. (*rather, fairly*) αρκετά, μάλλον. ~ (so)! ακριβώς. not ~ όχι ακριβώς. (*indeed, truly*) όντως, πραγματικά. ~ evident ολοφάνερος, ~ straight ολόισιος, ~ white κάτασπρος. I am ~ well again (*after illness*) έγινα περδίκι.
quits *a.* we are ~ είμαστε πάτσι. I'll be ~ with you θα σ' εκδικηθώ.
quittance *s.* εξόφληση *f.*
quiver *s.* (*archer's*) φαρέτρα *f.*
quiver *v.* τρέμω, (*of eyelid*) παίζω. (*s.*) τρεμούλα *f.*
qui vive *s.* on the ~ εν επιφυλακή.
quixot|ic *a.* δονκιχωτικός. **~ry** *s.* δονκιχωτισμός *m.*
quiz *v.* (*tease*) πειράζω, (*look at*) κοιτάζω ειρωνικά, (*question*) εξετάζω. (*s.*) σειρά ερωτήσεων. **~zical** *a.* ειρωνικός.
quoit *s.* κρίκος *m.* ~s παιγνίδι με κρίκους.
quondam *a.* πρώην *adv.*
quorum *s.* απαρτία *f.*
quota *s.* αναλογούν *n.*, ποσοστό *n.* (*permitted number*) επιτρεπόμενος αριθμός.
quotation *s.* (*passage*) απόσπασμα *n.* (*estimate*) προσφορά *f.* ~ marks εισαγωγικά *n.pl.* (*fin.*) τρέχουσα τιμή.
quote *v.* (*a passage*) παραθέτω, (*mention*) αναφέρω. (*a price*) δίνω.
quoth *v.* είπα, είπε.
quotidian *a.* καθημερινός.
quotient *s.* (*math.*) πηλίκον *n.*

R

rabbi *s.* ραββίνος *m.*
rabbit *s.* κουνέλι *n.*
rabble *s.* συρφετός *m.*, όχλος *m.* **~-rousing** *a.* δημαγωγικός.
rabid *a.* λυσσασμένος. (*fig.*) φανατικός. (*politically*) βαμμένος.
rabies *s.* λύσσα *f.* ~ clinic λυσσιατρείο *n.*
race *s.* (*breed*) φυλή *f.*, ράτσα *f.* human ~ ανθρώπινο γένος. (*a.*) φυλετικός.
race *s.* (*on foot*) αγώνας δρόμου, (*horse, car*) κούρσα *f.* the ~s ιπποδρομίες *f.pl.* obstacle ~ δρόμος μετ' εμποδίων. (*fig., effort to be in time*) αγώνας *m.*
race *v.i.* συναγωνίζομαι στην ταχύτητα, (*go fast*) τρέχω. ~course *s.* ιπποδρόμιο *n.* **~horse** *s.* άλογο κούρσας.
racial *a.* φυλετικός. **~ism** *s.* φυλετισμός *m.* **~ist** *s.* ρατσιστής *m.*
racing *s.* κούρσες *f.pl.*
rack *s.* (*holder*) ράφι *n.*, κρεμάστρα *f.*, θήκη

f. **~railway** *s.* οδοντωτός σιδηρόδρομος.

rack *s.* (*torture*) βασανιστήριο *n.* go to ~ and ruin ρημάζω, πάω κατά διαβόλου. (*v.t.*) βασανίζω. ~ one's brains σπάζω το κεφάλι μου.

racket *s.* (*din*) θόρυβος *m.* (*fuss*) φασαρία *f.* (*roguery*) λοβιτούρα *f.,* κομπίνα *f.* (*of gangs*) γκαγκστερική επιχείρηση. (*tennis*) ρακέτα *f.*

racy *a.* ζωηρός, πικάντικος.

radar *s.* ραντάρ *n.*

raddled *a.* φτιασιδωμένος.

radi|ant *a.* be ~ant (*with*) ακτινοβολώ (από). **~ance** *s.* λάμψη *f.*

radiat|e *v.t.* εκπέμπω, διαχέω. (*v.i.*) (*of light*) ακτινοβολώ. (*of streets, etc.*) εκτείνομαι ακτινοειδώς. **~ion** *s.* ακτινοβολία *f.*

radical *a.* ριζικός, (*of opinions*) ριζοσπαστικός. (*s.*) ριζοσπάστης *m.* **~ly** *adv.* ριζικώς.

radio *s.* (*wireless*) ασύρματος *m.* (*set*) ραδιόφωνο *n.* (*a.*) ραδιοφωνικός. **~-active** *a.* ραδιενεργός. **~-gram(ophone)** *s.* ραδιογραμμόφωνο *n.* **~logy** *s.* ακτινολογία *f.* **~telegram** *s.* ραδιο(τηλε)γράφημα *n.* **~therapy** *s.* ακτινοθεραπεία *f.*

radish *s.* ραπανάκι *n.*

radium *s.* ράδιο *n.*

radius *s.* ακτίνα *f.*

raffish *a.* άσωτος, έκδοτος.

raffle *s.* λοταρία *f.* (*v.t.*) βάζω στη λοταρία.

raft *s.* σχεδία *f.*

rafter *s.* καδρόνι *n.*

rag *s.* κουρέλι *n.,* ράκος *n.* (*fam., newspaper*) πατσαβούρα *f.* (*fam.*) glad ~s γιορτινά *n.pl.*

rag *v.t.* (*tease*) πειράζω. (*v.i.*) κάνω φάρσες. (*s.*) (*lark*) φάρσα *f.*

ragamuffin *s.* κουρελής *m.*

rag-bag *s.* (*fig.*) συνονθύλευμα *n.*

rage *s.* οργή *f.* get in a ~ εξοργίζομαι, (*fam.*) γίνομαι βαπόρι (*or* έξω φρενών). it's all the ~ χαλάει κόσμο. (*v.i.*) μανιάζω, λυσσομανώ.

ragged *a.* (*material*) κουρελιασμένος, (*person*) ρακένδυτος, κουρελής. (*uneven*) ανώμαλος.

raging *a.* μανιασμένος, (*pain*) φρικτός.

ragtag *s.* (*fam.*) ~ and bobtail η σάρα και η μάρα.

raid *s.* επιδρομή *f.* (*sudden*) αιφνιδιασμός *m.* (*v.t.*) κάνω επιδρομή κατά (*with gen.*). (*rob*) ληστεύω. **~er** *s.* επιδρομεύς *m.* ληστής *m.*

rail *s.* κάγκελο *n.* (*for curtain*) βέργα *f.* (*hand-~*) κουπαστή *f.* (*of track*) ράβδος *f.,* ράγια *f.* by ~ σιδηροδρομικώς. run off the ~s (*lit. & fig.*) εκτροχιάζομαι.

rail *v.t.* (*enclose*) περιφράσσω. **~ed** off περιφραγμένος. **~ing(s)** *s.* κιγκλίδωμα *n.*

rail *v.i.* ~ at τα βάζω με, βρίζω. **~ing** *s.* φωνές *f.pl.,* γκρίνια *f.*

raillery *s.* πείραγμα *n.*

rail|road, **~way** *s.* σιδηρόδρομος *m.* (*a.*) (*also* ~wayman) σιδηροδρομικός.

raiment *s.* ενδυμασία *f.*

rain *v.i.* it ~s βρέχει. (*fall like* ~) πέφτει βροχή. (*v.t.*) ~ gifts upon προσφέρω άφθονα δώρα σε. it ~s cats and dogs βρέχει με το τουλούμι.

rain *s.* βροχή *f.* **~bow** *s.* ουράνιο τόξο. colours of the ~bow χρώματα της ίριδος. **~coat** *s.* αδιάβροχο *n.* **~fall** *s.* βροχόπτωση *f.*

rainwater *s.* όμβρια ύδατα, (*as opposed to tapwater*) βρόχινο νερό. ~ pipe υδρορροή *f.*

rainy *a.* βροχερός. (*fig.*) for a ~ day για τις δύσκολες μέρες.

raise *v.* (*head, eyes, dust, anchor*) σηκώνω. (*building, question, point*) εγείρω. (*glass, flag, voice*) υψώνω. (*price, morale, window*) ανεβάζω. (*hat*) βγάζω, (*siege*) αίρω, (*wreck*) ανελκύω, (*loan*) δανείζομαι, (*objection*) προβάλλω. (*family*) μεγαλώνω, (*cattle, etc.*) εκτρέφω, (*crops*) καλλιεργώ. (*in rank*) προάγω, (*increase*) αυξάνω, (*provoke*) προκαλώ. (*find, obtain*) πορίζομαι, βρίσκω. ~ Cain *or* hell *or* the roof χαλώ τον κόσμο.

raisin *s.* σταφίδα *f.*

rake *s.* (*implement*) τσουγκράνα *f.* (*slope*) κλίση *f.* (*roué*) παραλυμένος *a.* **~d** *a.* (*sloping*) επικλινής.

rake *v.t.* τσουγκρανίζω, (*with shots*) γαζώνω, (*scan*) περισκοπώ. (*command view of*) δεσπόζω (*with gen.*). (*fig.*) ~ in μαζεύω, ~ up σκαλίζω, ξεθάβω.

rakish *a.* (*ship*) αεροδυναμικός. (*hat*) at a ~ angle λοξοφορεμένος. (*dissolute*) παραλυμένος.

rally *s.* συναγερμός *m.* (*demonstration*) συλλαλητήριο *n.* (*motor*) ράλλυ *n.*

rally *v.t.* (*troops, etc.*) ανασυντάσσω. ~ one's spirits αναθαρρεύω. ~ one's strength ανακτώ δυνάμεις. (*chaff*) πειράζω. (*v.i.*) ανασυντάσσομαι. (*recover*) συνέρχομαι. they ~ round him συσπειρώνονται γύρω του.

ram *s.* κριάρι *n.* (*battering* ~) κριός *m.* (*ship's*) έμβολο *n.*

ram *v.t.* (*thrust*) μπήγω, χώνω, (*compress*) συμπιέζω. (*collide with*) συγκρούομαι με.

ramble *v.* (*stroll*) σεργιανίζω, (*roam*) γυρίζω. (*straggle*) απλώνομαι ακατάστατα, (*in speech*) πολυλογώ. (*s.*) ~s (*travels*) περιπλανήσεις *f.pl.*

rambling *a.* (*speech*) ασυνάρτητος, (*build-*

ing) λαβυρινθοειδής.

ramification *s.* διακλάδωση *f.*

ramp *s.* (*incline*) ράμπα *f.* (*swindle*) απάτη *f.*

ramp *v.i.* (*act boisterously*) τρεχοβολώ.

rampage *v.i.* αποχαλινώνομαι.

rampant *a.* (*heraldry*) ορθός. be ~ (*vegetation, vice*) οργιάζω, (*disease*) έχω φουντώσει.

ramparts *s.* επάλξεις *f.pl.*

ramrod *s.* εμβολεύς *m.* erect as a ~ ντούρος.

ramshackle *a.* ξεχαρβαλωμένος.

rancid *a.* ταγγός.

ranc|our *s.* μνησικακία *f.* **~orous** *a.* μνησίκακος.

random *a.* τυχαίος. at ~ στην τύχη, στα κουτουρού.

randy *a.* (*noisy*) θορυβώδης, (*lustful*) λάγνος.

range *v.t.* (*set in line*) παρατάσσω. ~ oneself with συντάσσομαι με. (*v.i.*) (*wander*) περιφέρομαι, (*extend*) εκτείνομαι. (*vary*) ποικίλλω, κυμαίνομαι. **~r** *s.* δασοφύλακας *m.*

range *s.* (*line, row*) σειρά *f.* (*choice, variety*) ποικιλία *f.*, εκλογή *f.* (*extent*) πεδίο *n.*, έκταση *f.* firing ~ πεδίο βολής. (*of gun*) βεληνεκές *n.* (*radius*) ακτίνα *f.* at a ~ of 10 miles σε ακτίνα δέκα μιλίων. at long ~ από μεγάλη απόσταση. long-~ (*plane*) μακράς ακτίνος δράσεως, (*gun*) μεγάλου βεληνεκούς. **~finder** *s.* τηλέμετρον *n.*

rank *s.* (*line*) γραμμή *f.* (*of soldiers*) ζυγός *m.* ~s τάξεις *f.pl.* ~ and file απλοί στρατιώτες, (*fig.*) κοσμάκης *m.* (*grade*) βαθμός *m.* (*station*) τάξη *f.* (*v.t.*) κατατάσσω. (*v.i., be counted*) κατατάσσομαι, θεωρούμαι.

rank *a.* (*in growth*) θεριεμένος, (*in smell*) βρωμερός. (*absolute*) σωστός, καθαρός.

rankle *v.i.* αφήνω πικρία.

ransack *v.* (*search*) κάνω άνω κάτω. (*plunder*) λεηλατώ.

ransom *s.* λύτρα *n.pl.* hold to ~ ζητώ λύτρα για (*v.*) λυτρώνω.

rant *v.* μεγαληγορώ.

rap *s.* χτύπημα *f.* not give a ~ δεν δίνω δεκάρα. (*v.*) χτυπώ.

rapaci|ous *a.* αρπακτικός. **~ty** *s.* αρπακτικότητα *f.*

rape *s.* (*abduction*) αρπαγή *f.* (*violation*) βιασμός *m.* (*v.*) βιάζω.

rapid *a.* ταχύς, γρήγορος. ~s καταρράκτης *m.* **~ity** *s.* ταχύτητα *f.* **~ly** *adv.* ταχέως, γρήγορα.

rapier *s.* ξίφος *n.* (*fig.*) ~ thrust πνευματώδης αντεπίθεση.

rapine *s.* διαρπαγή *f.*

rapport *s.* ψυχική επαφή.

rapporteur *s.* εισηγητής *m.*

rapprochement *s.* προσέγγιση *f.*

rapt *a.* (*absorbed*) απορροφημένος, (*attention*) τεταμένος.

raptur|e *s.* έκσταση *f.* **~ous** *a.* εκστατικός.

rare *a.* σπάνιος, (*thin*) αραιός. **~ly** *adv.* σπανίως.

rare|ly *v.t. & i.* αραιώνω. **~action** *s.* αραίωση *f.*

rarity *s.* σπανιότητα *f.* (*rare thing*) σπάνιο πράγμα.

rascal *s.* (*rogue*) αλιτήριος *a.* (*scamp*) μασκαράς *m.*

rash *s.* εξάνθημα *n.*

rash *a.* παράτολμος. **~ness** *s.* παρατολμία *f.*

rasher *s.* φέτα μπέικον.

rasp *s.* ράσπα *f.* (*v.*) λιμάρω, (*fig.*) ερεθίζω. **~ing** *a.* (*voice*) τραχύς.

raspberry *s.* σμέουρο *n.*

rat *s.* ποντικός *m.* (*fig.*) smell a ~ μυρίζομαι κάτι ύποπτο. (*v.*) ~ on προδίδω.

ratchet *s.* οδοντωτός τροχός.

rate *s.* (*speed*) ταχύτητα *f.* (*of growth, production*) ρυθμός *m.* (*cost*) τιμή *f.* (*percentage*) ποσοστό *n.* (*scale of charges*) τιμολόγιο *n.* birth ~ ποσοστό γεννήσεων. first ~ πρώτης τάξεως. at any ~ οπωσδήποτε. ~s δημοτικός φόρος. **~able** *a.* φορολογήσιμος.

rat|e *v.t.* (*reckon*) υπολογίζω, (*consider*) θεωρώ. (*v.i., rank*) θεωρούμαι. **~ing** *s.* εκτίμηση *f.*

rat|e *v.t.* (*scold*) κατσαδιάζω. **~ing** *s.* κατσάδα *f.*

rather *adv.* μάλλον. ~ good μάλλον (or αρκετά) καλός. ~ tired κάπως κουρασμένος. I would ~ resign than do what you ask θα προτιμούσα να παραιτηθώ παρά να κάνω αυτό που μου ζητάτε. would you ~ have studied medicine? θα προτιμούσες να είχες σπουδάσει ιατρική;

ratif|y *v.* (επι)κυρώνω. **~ication** *s.* (επι)κύρωση *f.*

ratio *s.* αναλογία *f.*

ration *s.* μερίδα *f.* day's ~ σιτηρέσιο *n.* ~s τρόφιμα *n.pl.* (*v.*) κατανέμω με δελτίο. (*restrict*) περιορίζω.

rational *a.* λογικός. **~ly** *adv.* λογικώς.

rational|ism *s.* ορθολογισμός *m.* **~ist** *s.* ορθολογιστής *m.*

rationaliz|e *v.* (*behaviour*) αιτιολογώ. (*industry, etc.*) οργανώνω πιο λογικά. **~ation** *s.* λογική εξήγηση, ορθολογιστική οργάνωση.

rattle *s.* κρόταλο *n.*, ροκάνα *f.* (*baby's*) κουδουνίστρα *f.* (*noise*) κροτάλισμα *n.* death-~ ρόγχος *m.*

rattle *v.* (*make or cause to make a noise*) χτυπώ, (*frighten*) τρομοκρατώ. ~ off (*recite or play quickly*) απαγγέλλω (or εκτελώ) τροχάδην.

rattlesnake s. κροταλίας m.

raucous a. βραχνός.

ravage v. ερημώνω, καταστρέφω. (the body) σακατεύω. (plunder) λεηλατώ. (s.) ~s καταστροφές f.pl. (of time) φθορές f.pl.

rav|e v. μαίνομαι. ~e about (praise) εκθειάζω. ~ing a. παράφρων. (s.) τρελλά λόγια.

ravel v. (fray) ξεφτίζω. ~ out ξεμπερδεύω.

raven s. κόρακας m. (a., black) κατάμαυρος.

ravenous a. λιμασμένος.

ravish v. (carry off) απάγω, αρπάζω, (rape) βιάζω. (enrapture) καταγοητεύω. ~ing a. (woman) πεντάμορφος. (music, etc.) που σε καταγοητεύει.

raw a. ωμός. (unworked) ακατέργαστος, ~ material πρώτη ύλη. (unskilled) άβγαλτος. ~ recruit (fam.) Γιαννάκης m. (wound) ανοιχτός, (weather) ψυχρός. ~ deal άδικη μεταχείριση. touch him on the ~ θίγω το ευαίσθητο σημείο του.

ray s. ακτίς f., αχτίδα f.

rayah s. ραγιάς m.

rayon s. ρεγιόν n.

raze v. ισοπεδώνω.

razor s. ξυράφι n. (safety) ξυριστική μηχανή. on the ~'s edge στην κόψη του ξυραφιού. ~-blade s. λάμα f., ξυραφάκι n.

RC ρωμαιοκαθολικός.

re prep. αναφορικώς προς (with acc.).

re- prefix ανα-, ξανα-, επαν-. (as adv.) ξανά, εκ νέου.

reach v.t. (arrive at) φθάνω σε. (touch with hand, etc.) φθάνω. ~ out (one's hand) for απλώνω το χέρι μου για. can you ~ me the cigarettes? μπορείς να μου πιάσεις τα τσιγάρα; (v.i.) φθάνω, as far as the eye can ~ όσο (or ώσπου) φτάνει το μάτι.

reach s. within ~ (near) κοντά, (accessible) προσιτός. out of ~ (far) μακρυά, (inaccessible) απρόσιτος. (of river) έκταση f.

react v. αντιδρώ. ~ion s. αντίδραση f. ~ionary a. & s. αντιδραστικός.

read v. διαβάζω, (for degree) σπουδάζω. it ~s as follows λέει τα εξής. ~ between the lines διαβάζω ανάμεσα στις γραμμές. you ~ too much into it του δίνεις αδικαιολόγητη ερμηνεία. ~ up μελετώ. well-~ διαβασμένος, well-~ in history δυνατός στην ιστορία.

reader s. αναγνώστης m. (university teacher) καθηγητής m. (textbook) αναγνωστικό n.

readily adv. (easily) εύκολα, (willingly) πρόθυμα.

readiness s. ετοιμότητα f. (willingness) προθυμία f.

reading s. διάβασμα n. (of parliamentary bill & as lesson) ανάγνωσις f. (matter) ανάγνωσμα n. (interpretation) ερμηνεία f. (of MS) γραφή f. (of instrument) ένδειξη f. ~-room s. αναγνωστήριο n.

readjust v.t. επαναρυθμίζω, αναπροσαρμόζω. ~ment s. αναπροσαρμογή f.

ready a. έτοιμος. ~ money μετρητά n.pl. make or get ~ (v.t.) (dinner, room, etc.) ετοιμάζω, (for action) προετοιμάζω, (v.i.) ετοιμάζομαι, προετοιμάζομαι. ~-made a. έτοιμος. ~-reckoner s. τυφλοσούρτης m.

reaffirm v. δηλώνω ξανά.

reafforestation s. αναδάσωση f.

real a. πραγματικός. ~ estate ακίνητη περιουσία. ~ly adv. πραγματικά. ~ly? αλήθεια; ~ity s. πραγματικότητα f.

real|ism s. ρεαλισμός m. ~ist s. ρεαλιστής m. ~istic a. ρεαλιστικός.

realiz|e v. (plans, hopes) πραγματοποιώ, (a price) πιάνω, (understand) συνειδητοποιώ. ~ation s. πραγματοποίηση f. (understanding) αντίληψη f.

realm s. βασίλειον n. (fig.) σφαίρα f., τομέας m.

reams s. ~ of ένας σωρός.

reap v. θερίζω. ~ the fruits (of) δρέπω τους καρπούς. ~er s. (human) θεριστής m. (machine) θεριστική μηχανή.

reappear v. επανεμφανίζομαι, ξαναπαρουσιάζομαι.

rear v.t. (set up) εγείρω, (lift) σηκώνω. (bring up) ανατρέφω, (animals) εκτρέφω. (v.i.) (of horse) ανορθώνομαι στα πισινά μου πόδια.

rear s. (back part) οπίσθιο (or πίσω) μέρος. at or in the ~ (of) πίσω (of). bring up the ~ έρχομαι τελευταίος. (a.) the ~ door η πίσω πόρτα. ~-admiral s. υποναύαρχος m. ~-guard s. οπισθοφυλακή f. ~guard action μάχη οπισθοφυλακής.

rearm v.t. επανεξοπλίζω. (v.i.) επανεξοπλίζομαι. ~ament s. επανεξοπλισμός m.

rearrange v. αναδιαρρυθμίζω, τακτοποιώ διαφορετικά. (plans) τροποποιώ. ~ment s. αναδιαρρύθμιση f. τροποποίηση f., αλλαγή f.

reason s. λόγος m. (cause) αιτία f. (sense) λογική f. (sanity) λογικά n.pl. (intellect) νόηση f. it stands to ~ είναι κατάδηλον. he won't listen to ~ δεν παίρνει από λόγια. by ~ of εξ αιτίας (with gen.). he doesn't like driving at night, and with good ~ δε θέλει να οδηγεί νύχτα, και δικαιολογημένα (or με το δίκιο του). with all the more ~ κατά μείζονα λόγον.

reason v.i. (think) σκέπτομαι, συλλογίζομαι. ~ with (person) προσπαθώ να λογικέψω. ~ that (*maintain*) υποστηρίζω ότι. (*v.t.*) ~ out βρίσκω τη λύση. **~ing** *s.* τρόπος της σκέψεως.

reasonab|le *a.* λογικός. **~ly** *adv.* λογικά.

reassur|e *v.* καθησυχάζω. **~ance** *s.* καθησύχαση *f.* **~ing** *a.* καθησυχαστικός.

rebate *s.* έκπτωση *f.*

rebel *s.* επαναστάτης *m.*, αντάρτης *m.* (*v.*) επαναστατώ.

rebell|ion *s.* αντάρσία *f.* **~ious** *a.* (*child, lock of hair*) ανυπότακτος. ~ious troops στρατεύματα των στασιαστών.

rebirth *s.* αναγέννηση *f.*

rebound *v.* (*of ball*) κάνω γκελ. (*fig.*) επιστρέφω. (*s.*) (*fig.*) on the ~ από αντίδραση.

rebuff *v.* αποκρούω. (*s.*) απότομη άρνηση.

rebuke *s.* παρατήρηση *f.* (*v.t.*) κάνω παρατήρηση σε.

rebut *v.* διαψεύδω. **~tal** *s.* διάψευση *f.*

recalcitrant *a.* απείθαρχος.

recall *v.* (*summon back*) ανακαλώ. (*remember*) θυμάμαι, ανακαλώ στη μνήμη. (*remind one of*) (ξανα)θυμίζω. (*s.*) (*summons*) ανάκληση *f.* (*fig.*) those years are beyond ~ εκείνα τα χρόνια πάνε πια και δεν ξαναγυρίζουν.

recant *v.t.* αποκηρύσσω. **~ation** *s.* αποκήρυξη *f.*

recapitulat|e *v.* ανακεφαλαιώνω. **~ion** *s.* ανακεφαλαίωση *f.*

recapture *v.* ξαναπιάνω, (*fig.*) ξαναζωντανεύω.

recede *v.* υποχωρώ.

receipt *s.* (*act*) παραλαβή *f.*, λήψη *f.* (*statement*) απόδειξη *f.* (*recipe*) συνταγή *f.* ~s (*takings*) εισπράξεις *f.pl.*

receive *v.* (*take delivery of*) παραλαμβάνω. (*a blow, guests*) δέχομαι. (*greet, lit. & fig.*) υποδέχομαι. (*get*) παίρνω, λαμβάνω. (*undergo*) υφίσταμαι. **~d** *a.* (*usual*) παραδεδεγμένος.

receiver *s.* παραλήπτης *m.* (*part of apparatus*) δέκτης *m.*, (*of phone*) ακουστικό *n.* (*of stolen goods*) κλεπταποδόχος *m.*

recent *a.* πρόσφατος. **~ly** *adv.* προσφάτως.

receptacle *s.* δοχείο *n.*

reception *s.* υποδοχή *f.* (*party*) δεξίωση *f.* (*of hotel*) ρεσεψιόν *f.* (*of signals*) λήψη *f.*

receptive *a.* (*person*) με ανοικτό μυαλό. be ~ to δέχομαι εύκολα.

recess *s.* (*break*) διακοπή *f.* (*niche*) κόγχη *f.* (*place set back*) εσοχή *f.* **~es** (*of mind, etc.*) μύχια *n.pl.*

recession *s.* (*economic*) κάμψη *f.*

recherché *a.* εξεζητημένος.

recipe *s.* συνταγή *f.*

recipient *s.* παραλήπτης *m.*

reciprocal *a.* αμοιβαίος. (*gram.*) αλληλοπαθής. **~ly** *adv.* αμοιβαίως.

reciprocat|e *v.* ανταποδίδω. **~ing** *a.* (*mech.*) παλινδρομικός. **~ion** *s.* ανταπόδοση *f.*

recital *s.* (*tale*) εξιστόρηση *f.* (*mus.*) ρεσιτάλ *n.*

recitation *s.* απαγγελία *f.*

recite *v.* (*tell*) εξιστορώ. (*declaim*) απαγγέλλω.

reckless *a.* απερίσκεπτος. **~ly** *adv.* απερίσκεπτα. **~ness** *s.* απερισκεψία *f.*

reckon *v.* λογαριάζω, υπολογίζω. (*think*) νομίζω. ~ on βασίζομαι σε.

reckoning *s.* (*calculation*) υπολογισμός *m.* (*bill*) λογαριασμός *m.* his day of ~ η ημέρα που θα δώσει λόγο.

reclaim *v.* (*waste land*) εκχερσώνω, (*drain*) αποξηραίνω. (*persons*) επαναφέρω στον ίσιο δρόμο.

reclamation *s.* (*of land*) εγγειοβελτίωσις *f.* (*moral*) μεταβολή επί το καλύτερον.

reclin|e *v.i.* ξαπλώνω. ~ing seat ανακλινόμενο κάθισμα.

recluse *s.* αποτραβηγμένος από τον κόσμο.

recognition *s.* αναγνώριση *f.*

recognizance *s.* υποχρέωση *f.*

recogniz|e *v.* αναγνωρίζω. **~able** *a.* he is not ~able δεν τον αναγνωρίζει κανείς.

recoil *v.i.* (*gun*) κλωτσώ. (*person*) κάνω πίσω. (*rebound*) επιστρέφω. (*s.*) ανάκρουση *f.*

recollect *v.* αναπολώ, θυμάμαι. **~ion** *s.* ανάμνηση *f.*

recommend *v.t.* συνιστώ, (*entrust*) εμπιστεύομαι.

recommendation *s.* σύσταση *f.* ~s (*of committee, etc.*) υποδείξεις *f.pl.* (*good point*) προσόν *n.* letter of ~ συστατική επιστολή.

recompense *v.* ανταμείβω. (*s.*) ανταμοιβή *f.*

reconcil|e *v.* συμβιβάζω. become ~ed (*make it up*) συμφιλιώνομαι, (*submit to*) υποτάσσομαι εις. **~iation** *s.* συμφιλίωση *f.*

recondite *a.* βαθύς.

recondition *v.* επισκευάζω.

reconnaissance *s.* αναγνώρισις *f.*

reconnoitre *v.* ανιχνεύω, κάνω αναγνώριση.

reconstruct *v.* (*building*) ανοικοδομώ, (*events*) κάνω αναπαράσταση (*with gen.*). **~ion** *s.* ανοικοδόμηση *f.* αναπαράσταση *f.*

record *s.* (*account*) αναγραφή *f.* (*mention*) μνεία *f.* (*disc*) δίσκος *m.*, πλάκα *f.* (*achievement*) ρεκόρ *n.* R~s αρχεία *n.pl.* (*personal file*) φάκελος *m.*, (*police* ~) ποινικό μητρώο. have a bad ~ έχω κακό ιστορικό. off the ~ ανεπισήμως.

record *v.* αναγράφω, (*relate*) αναφέρω. (*on*

disc, tape) ηχογραφώ. **~ing** *s.* εγγραφή *f.*, ηχογράφηση *f.*

record-player *s.* πικάπ *n.*

recount *v.* (*tell*) αφηγούμαι, (*count again*) ξαναμετρώ. (*s., at election*) νέα καταμέτρηση.

recoup *v.* (*person*) αποζημιώνω. ~ one's losses βγάζω τα χαμένα.

recourse *s.* (*expedient*) διέξοδος *f.* have ~ to *see* resort *v.*

recover *v.t.* ανακτώ, ξαναβρίσκω. (*v.i.*) συνέρχομαι. **~y** *s.* ανάκτηση *f.* (*of health, etc.*) αποκατάσταση *f.*

recreation *s.* ψυχαγωγία *f.* **~al** *a.* ψυχαγωγικός.

recrimination *s.* αλληλοκατηγορία *f.*

recrudescence *s.* νέο ξέσπασμα.

recruit *s.* (*mil.*) νεοσύλλεκτος *m.* (*fig.*) νέος οπαδός *m.* (*v.*) στρατολογώ. **~ment** *s.* στρατολογία *f.*

rectangl|e *s.* ορθογώνιο *n.* **~ular** *a.* ορθογώνιος.

rectif|y *v.* διορθώνω. **~ication** *s.* διόρθωση *f.*

rectilinear *a.* ευθύγραμμος.

rectitude *s.* ευθύτης *f.*

rector *s.* (*of university*) πρύτανις *m.* (*priest*) εφημέριος *m.*

rectum *s.* ορθό *n.*

recumbent *a.* ξαπλωμένος, πλαγιασμένος.

recuperat|e *v.i.* αναρρωνύω. **~ion** *s.* ανάρρωση *f.*

recur *v.* (*happen again*) επαναλαμβάνομαι, (*to mind*) επανέρχομαι. **~rence** *s.* επανάληψη *f.* **~ent** *a.* επαναλαμβανόμενος. (*med.*) υπόστροφος.

red *a.* κόκκινος, (*Cross, Sea, etc.*) ερυθρός. in the ~ χρεωμένος. paint the town ~ το ρίχνω έξω. see ~ γίνομαι θηρίο. go or become ~ κοκκινίζω. ~ carpet (*fam.*) επίσημη υποδοχή. ~ tape γραφειοκρατία *f.* ~ lead μίνιο *n.* **~den** *v.* κοκκινίζω. **~dish** *a.* κοκκινωπός. **~-haired** *a.* κοκκινομάλλης. **~-handed** *a.* επ' αυτοφώρω. **~-hot** *a.* πυρακτωμένος, (*fig.*) ένθερμος. **~-letter** *a.* **~-**letter day αλησμόνητη μέρα. **~skin** *s.* ερυθρόδερμος *m.*

redeem *v.* (*pay off*) εξοφλώ, (*fulfil*) εκπληρώ. (*save*) λυτρώνω, σώζω. (*make up for*) αντισταθμίζω. **~er** *s.* λυτρωτής *m.*

redemption *s.* λύτρωση *f.* past ~ ανεπανόρθωτος.

redo *v.t.* ξαναφτιάνω, ξανακάνω.

redolent *a.* be ~ of μυρίζω, αποπνέω, (*fig.*) θυμίζω.

redouble *v.* διπλασιάζω.

redoubtable *a.* τρομερός.

redound *v.i.* συμβάλλω.

redress *v.* (*a wrong*) επανορθώνω, (*the*

balance) αποκαθιστώ. (*s.*) ικανοποίηση *f.*

reduce *v.t.* μειώνω, ελαττώνω. (*subdue*) υποτάσσω. (*math.*) ανάγω. be ~d to (*in circumstance*) καταντώ. ~ (*person*) to (*make, compel*) φέρω σε, αναγκάζω να. (*v.i.*) (*fam.*) κάνω δίαιτα.

reduction *s.* μείωση *f.* (*of price*) έκπτωση *f.*

redund|ant *a.* (*worker*) υπεράριθμος, (*word*) περιττός. **~ancy** *s.* περίσσευμα *n.* ~ancy payment αποζημίωση λόγω απολύσεως.

reduplication *s.* αναδιπλασιασμός *m.*

reecho *v.* αντιλαλώ.

reed *s.* κάλαμος *m.*, καλάμι *n.*

reef *s.* (*below surface*) ύφαλος *f.* (*above*) σκόπελος *m.* (*both*) ξέρα *f.*

reek *v.* (*also* ~ of) βρωμώ. (*s.*) μπόχα *f.*, βρώμα *f.*

reel *s.* (*dance*) σκωτσέζικος χορός.

reel *s.* καρούλι *n.* (*of thread*) κουβαρίστρα *f.*

reel *v.t.* (*wind*) τυλίγω. ~ off (*recite*) αραδιάζω. (*v.i., stagger*) παραπαίω, (*fig.*) κλονίζομαι.

reelect *v.t.* επανεκλέγω.

refectory *s.* τραπεζαρία *f.* (*monastery*) τράπεζα *f.*

refer *v.t.* (*send on*) παραπέμπω, (*ascribe*) αποδίδω. (*v.i.*) ~ to (*allude*) αναφέρομαι εις, (*resort*) προσφεύγω εις. **~ee** *s.* διαιτητής *m.*

reference *s.* παραπομπή *f.* αναφορά *f.* (*mention*) μνεία *f.* (*testimonial*) σύσταση *f.* (*connection*) σχέση *f.* terms of ~ αρμοδιότητα *f.* with ~ to σχετικά με. ~ book βοήθημα *n.*

referendum *s.* δημοψήφισμα *n.*

refill *s.* ανταλλακτικό *n.*

refine *v.* ραφινάρω, (*manners*) εκλεπτύνω, (*oil*) διυλίζω. **~d** *a.* ραφινάτος. **~ment** *s.* φινέτσα *f.*, λεπτότητα *f.* **~ry** *s.* διυλιστήριο *n.*

refit *s.* (*naut.*) επισκευή *f.*

reflect *v.t.* αντανακλώ, (*of mirror & fig.*) καθρεφτίζω. (*v.i.*) (*also* ~ about) συλλογίζομαι. ~ on (*badly*) ζημιώνω.

reflection *s.* αντανάκλαση *f.* (*of images*) αντικαθρέφτισμα *n.* (*image*) εικόνα *f.* (*thought*) σκέψη *f.* (*discredit*) it is a ~ on his honour θίγει την τιμή του.

reflective *a.* στοχαστικός.

reflex *a.* ανακλαστικός. (*s.*) ανακλαστικό *n.*

reflexive *a.* (*gram.*) αυτοπαθής.

reform *v.t.* (*laws, conditions*) μεταρρυθμίζω. (*morally*) she has **~ed** him τον έχει κάνει άλλο άνθρωπο. (*v.i.*) he has **~ed** έχει γίνει άλλος άνθρωπος, έχει διορθωθεί. (*s.*) μεταρρύθμιση *f.*

reformation *s.* μεταρρύθμιση *f.*

reformatory *s.* αναμορφωτήριο *n.*

refract v. διαθλώ. **~ion** s. διάθλαση f.

refractory a. ανυπότακτος.

refrain s. επωδός f.

refrain v. ~ from αποφεύγω (*with acc. or* να).

refresh v. (*repose*) ξεκουράζω, (*cool*) δροσίζω. (*freshen up*) φρεσκάρω, ~ oneself (*with food, drink*) παίρνω κάτι. **~ing** a. δροσιστικός, (*welcome*) ευπρόσδεκτος.

refreshment s. ξεκούρασμα n. take ~ τρώω κάτι. **~s** (*cold drinks, ices*) αναψυκτικά n.pl. (*meal*) ελαφρό γεύμα.

refrigerat|ion s. ψύξη f. (*deep*) κατάψυξη f. **~ed** a. κατεψυγμένος. **~or** s. ψυγείο n.

refuel v.t. ανεφοδιάζω. (v.i.) ανεφοδιάζομαι. **~ling** s. ανεφοδιασμός m.

refugee s. (*abstract*) καταφυγή f. (*place of* ~) καταφύγιο n. (*both*) άσυλο n. take ~ in καταφεύγω σε.

refuges v. πρόσφυγας m.

refund v. επιστρέφω. (s.) get a ~ of παίρνω πίσω.

refuse s. απορρίμματα n.pl. ~ collector σκουπιδιάρης.

refus|e v. αρνούμαι, (*reject*) απορρίπτω. **~al** s. άρνηση f. απόρριψη f. (*right of deciding*) δικαίωμα εκλογής.

refut|e v. διαψεύδω, αντικρούω. **~ation** s. διάψευση f.

regain v. ανακτώ, ξαναβρίσκω. (*get back to*) γυρίζω σε.

regal a. βασιλοπρεπής.

regale v. (*with wine, etc.*) περιποιούμαι, (*with stories, etc.*) ψυχαγωγώ. ~ oneself ευωχούμαι.

regalia s. εμβλήματα n.pl.

regard v. (*consider*) θεωρώ, έχω. (*heed*) προσέχω, (*concern*) αφορώ, (*look at*) κοιτάζω. as ~s όσον αφορά.

regard s. (*esteem*) υπόληψη f., εκτίμηση f. hold in high ~ έχω σε μεγάλη υπόληψη. pay ~ to δίνω σημασία σε, προσέχω, λογαριάζω. have ~ for (*proprieties, feelings*) σέβομαι. have no ~ for αδιαφορώ για. (*kind*) ~s χαιρετίσματα n.pl. with ~ to το όσον αφορά. **~ing** prep. σχετικά με.

regardless a. ~ of danger αψηφώντας τους κινδύνους, ~ of consequences αδιαφορώντας για τις συνέπειες.

regatta s. λεμβοδρομίες f.pl.

regency s. αντιβασιλεία f.

regenerat|e v. αναγεννώ. **~ion** s. αναγέννηση f.

regent s. αντιβασιλεύς m.

regicide s. (*act*) βασιλοκτονία f. (*person*) βασιλοκτόνος m.

régime s. (*form of government*) πολίτευμα n. (*prevailing, established*) καθεστώς n. (*fam.*) see regimen.

regimen s. (*med.*) αγωγή f. (*diet*) δίαιτα f.

regiment s. (*mil.*) σύνταγμα n. **~al** a. του συντάγματος.

regiment v. υποβάλλω σε αυστηρή πειθαρχία. **~ation** s. αυστηρή οργάνωση.

region s. περιοχή f. in the ~ of (*about*) περίπου. **~al** a. τοπικός.

register s. (*official list*) μητρώον n. (*of correspondence*) πρωτόκολλο n. (*of voters*) κατάλογος m. (*of voice, etc.*) έκταση f.

register v.t. (*enter*) καταχωρώ, (*enrol*) εγγράφω, (*show*) δηλώνω. (v.i.) (*enrol oneself*) εγγράφομαι. ~ed letter συστημένο γράμμα. (*with police, etc.*) δηλώνομαι.

registration s. καταχώριση f. εγγραφή f. (*of birth, etc.*) δήλωση f.

registrar s. ληξίαρχος m.

registry, ~-office s. ληξιαρχείον n.

regress v. οπισθοδρομώ. **~ion** s. οπισθοδρόμηση f. **~ive** a. οπισθοδρομικός.

regret v.t. λυπούμαι για, (*repent*) μετανοώ. do you ~ not having had a child? το 'χεις καημό που δεν έκανες παιδί; (s.) λύπη f. **~fully** adv. με λύπη.

regrettab|le a. δυσάρεστος, λυπηρός. **~ly** adv. δυσάρεστα.

regular a. (*consistent, habitual*) τακτικός, (*proper, usual*) κανονικός. (*even*) ομαλός. (*customary*) συνηθισμένος. (*fam., thorough*) σωστός. **~ity** s. (*of habit*) τακτικότητα f. (*of features*) κανονικότητα f. (*punctuality*) ακρίβεια f. **~ly** adv. τακτικά, κανονικά.

regularize v. τακτοποιώ.

regulat|e v. κανονίζω, ρυθμίζω. **~ion** s. (*adjustment*) ρύθμιση f. (*rule*) κανονισμός m. (a.) κανονικός. **~or** s. ρυθμιστής m.

regurgitate v.t. εξεμώ. (v.i.) αναρρέω.

rehabilitat|e v. αποκαθιστώ. **~ion** s. αποκατάσταση f. (*of lands, resources*) ανασυγκρότηση f.

rehash v. ξαναμαγειρεύω. (s.) αναμασήματα n.pl.

rehears|e v. κάνω δοκιμή, (*tell*) εξιστορώ. **~al** s. δοκιμή f., πρόβα f.

reign v. βασιλεύω. (s.) βασιλεία f.

reimburse v. (*money*) επιστρέφω, (*person*) αποζημιώνω.

rein s. ηνίο n. (*fig.*) give free ~ to αφήνω ελεύθερο. give ~ to one's anger αφήνω το θυμό μου να ξεσπάσει. keep a tight ~ σφίγγω τα λουριά (*with* σε). (v.) ~ in χαλιναγωγώ.

reincarnation s. μετενσάρκωση f.

reinforce v. ενισχύω. **~d** concrete μπετόν αρμέ. **~ment** s. ενίσχυση f.

reinstate v. αποκαθιστώ.
reiterate v. επαναλαμβάνω.
reject v. απορρίπτω. **~ion** s. απόρριψη f.
rejoice v.t. προξενώ χαρά σε. (v.i.)
αναγαλλιάζω. **~ing** s. αναγάλλια f.,
ξέσπασμα χαράς.
rejoin v. (come back to) επανέρχομαι σε.
(answer) ανταπαντώ. **~der** s. απάντηση f.
rejuvenat|e v. ξανανιώνω. **~ion** s. ξανά-
νιωμα n.
relapse v. ξανακυλώ. (s.) υποτροπή f.
relate v.t. (tell) αφηγούμαι, (connect) συνδέω. (v.i., refer) αναφέρομαι, έχω σχέση.
related a. σχετικός, (in family) συγγενής. be
~ to συγγενεύω με.
relation s. (connection) σχέση f. (kinship)
συγγένεια f. (kinsman) συγγενής m. **~s**
(kin) συγγενολόι n. have ~s (dealings)
with έχω πάρε-δώσε με. **~ship** s. σχέση f.
συγγένεια f.
relative a. σχετικός. (gram.) αναφορικός.
(s.) συγγενής. **~ly** adv. σχετικά.
relativity s. σχετικότητα f.
relax v.t. χαλαρώνω, (esp. rope, etc.)
λασκάρω. (v.i.) (of muscles, severity)
χαλαρώνομαι. (of person) αφήνω τον
εαυτό μου, (take it easy) ξεκουράζομαι,
αναπαύομαι. **~ed** (calm) ήρεμος. **~ing** a.
(climate) αποχαυνωτικός.
relaxation s. χαλάρωση f. ξεκούραση f.
ανάπαυση f. (recreation) ψυχαγωγία f.
relay v. αναμεταδίδω. (s.) (radio) ανα-
μετάδοση f. (gang) βάρδια f. ~ race
σκυταλοδρομία f.
release v. (deliver) απαλλάσσω, (let loose)
απολύω, αμολάω, (let go of) αφήνω.
(make available) θέτω εις κυκλοφορίαν.
(s.) απαλλαγή f. (esp. of prisoner)
απόλυση f. ~ button κουμπί αποσυμ-
πλοκής.
relegate v. υποβιβάζω.
relent v. κάμπτομαι. **~less** a. αδυσώπητος,
ανένδοτος. **~lessly** adv. ανηλεώς, ανεν-
δότως.
relevan|t a. σχετικός. **~ce** s. σχέση f.
reliab|le a. (person, news) αξιόπιστος,
(machine, service) δοκιμασμένος, καλός.
~ility s. αξιοπιστία f.
reli|ance s. εμπιστοσύνη f. **~ant** a. be ~ant
on βασίζομαι σε, έχω πεποίθησιν εις.
relic s. (eccl.) λείψανο n. (fig.) υπόλειμμα n.
relief s. ανακούφιση f. (help) βοήθεια f.
(mil.) επικουρία f.pl. (variety) ποικιλία f.
(welfare work) περίθαλψη f. (replacement)
αντικατάσταση f. (archit.) ανάγλυφο n.
(fig.) bring into ~ εξαίρω, τονίζω.
relieve v. (pain, person) ανακουφίζω,
ξαλαφρώνω. (person of burden, anxiety)

απαλλάσσω. (the guard, watch) αντικαθιστώ. (monotony) σπάω, ποικίλλω.
(help) βοηθώ. (dismiss from post) απολύω.
(make stand out) τονίζω. ~ one's feelings
ξεθυμαίνω. ~ oneself ανακουφίζομαι, ξαλαφρώνω.
religion s. θρησκεία f.
religious a. θρησκευτικός, (person) θρήσκος. **~ly** adv. (fam.) με θρησκευτική
ακρίβεια.
relinquish v. (a hope, custom) εγκαταλείπω.
(a rope) παρατώ. (a post, attempt)
παραιτούμαι (with gen.). (hand over)
παραδίδω. ~ one's hold of (fig.) αφήνω
από τα χέρια μου.
reliquary s. λειψανοθήκη f.
relish v. απολαμβάνω, νοστιμεύομαι. I don't
~ (the idea, etc.) δεν μου πολυαρέσει. (s.)
(zest) όρεξη f. (piquancy) νοστιμιά f. (tit-
bit) μεζές m.
reluctance s. απροθυμία f.
reluctant a. απρόθυμος. **~ly** adv. με το ζόρι,
με το στανιό.
rely v. βασίζομαι, στηρίζομαι (both with σε).
remain v. μένω. (be left)΄ απομένω, υπολείπομαι. (continue to be) παραμένω.
~der s. υπόλοιπο n. **~ing** a. υπόλοιπος.
remains s. (ruins) ερείπια n.pl. (mortal)
λείψανα n.pl. (food on plate) αποφάγια
n.pl. the ~ of the money τα υπόλοιπα
λεφτά.
remake v. ξαναφτιάνω.
remand v. προφυλακίζω.
remark v. παρατηρώ. (s.) παρατήρηση f.
~able a. αξιοσημείωτος. **~ably** adv. εξαιρετικά, πολύ.
remedy s. θεραπεία f., γιατρειά f. (redress)
αποζημίωση f. (medicine) φάρμακο n. (v.)
(put right) διορθώνω, (cure) γιατρεύω.
remember v. θυμάμαι. ~ me to your brother
τα σέβη μου (or χαιρετίσματα) στον
αδελφό σας.
remembrance s. ανάμνηση f.
remind v. θυμίζω. it ~s one of Italy θυμίζει
την Ιταλία. ~ me to write θύμισέ μου να
γράψω. **~er** s. υπόμνηση f.
reminiscence s. ανάμνηση f.
reminiscent a. be ~ of θυμίζω.
remiss a. αμελής.
remission s. (of sins) άφεση f. (of fees)
απαλλαγή f. (abatement) ύφεση f.
remit v. (debt, penalty) χαρίζω, (lessen)
μειώνω, (send) εμβάζω. **~tance** s. έμβασμα n.
remnant s. υπόλειμμα n. (of cloth) ρετάλι n.
(of food) απομεινάρι n.
remonstr|ance s. διαμαρτυρία f. **~ate** v.
διαμαρτύρομαι.

remorse s. μεταμέλεια f. ~ful a. μετα-
νοητικός, (person) μετανιωμένος. ~less
a. αμείλικτος.

remote a. απόμακρος, μακρινός. (out-of-
the-way) απόκεντρος, (dim) αμυδρός. he
hasn't got the ~st chance, δεν έχει (ούτε)
την παραμικρή ελπίδα.

removal s. (taking away) αφαίρεση f.,
βγάλσιμο n. (moving house) μετακόμιση f.
remove v.t. αφαιρώ, βγάζω. (furniture)
μετακομίζω.
remunerat|e v. ανταμείβω. ~ion s. αντα-
μοιβή f.

renaissance s. αναγέννηση f.

rend v. σχίζω.

render v.t. (tribute) αποδίδω, (in exchange)
ανταποδίδω. (offer, give) προσφέρω,
(perform, express) αποδίδω. ~ (down)
(melt) λειώνω. (make) καθιστώ. ~ing s.
απόδοση f.

rendezvous s. ραντεβού n.

renegade s. αρνησίθρησκος m. (turncoat)
αποστάτης m.

renew v. ανανεώνω. ~al s. ανανέωση f.

renounce v. (disavow) αποκηρύσσω, (give
up) απαρνούμαι.

renovat|e v. ανακαινίζω. ~ion s. ανακαί-
νιση f.

renown s. φήμη f. ~ed a. φημισμένος. be
~ed for φημίζομαι για.

rent s. (tear) σχίσιμο n.

rent s. (money) ενοίκιον, νοίκι n., μίσθωμα
n. (v.t.) (of tenant) μισθώνω, (of owner)
εκμισθώνω, (of both) (ε)νοικιάζω.
~control s. ενοικιοστάσιον n.

rental s. ενοίκιον n., μίσθωμα n.

rentier s. εισοδηματίας m.

renunciation s. (disavowal) αποκήρυξη f.
(giving up) απάρνηση f.

reopen v.t. & i. ξανανοίγω, (begin again)
ξαναρχίζω.

reorganize v.t. αναδιοργανώνω.

repair v.t. (mend) επισκευάζω, (error, etc.)
επανορθώνω. (v.t., go) μεταβαίνω.

repair s. επισκευή f. in good ~ σε καλή
κατάσταση. beyond ~ (adv.) ανεπανόρ-
θωτα. ~ shop or party συνεργείο n.

reparation s. επανόρθωση f.

repartee s. πνευματώδης απάντηση.

repast s. γεύμα n.

repatriat|e v. επαναπατρίζω. ~ion s. επανα-
πατρισμός m.

repay v. (debt, creditor) ξεπληρώνω, (re-
quite) ανταποδίδω. ~ment s. (of debt)
εξόφληση f. (return) επιστροφή f. (re-
quital) ανταπόδοση f.

repeal v. ακυρώ. (s.) ακύρωση f.

repeat v.t. επαναλαμβάνω. (recite) απαγ-

γέλλω. (s.) επανάληψη f. ~ed a. επανειλημμένος. ~edly adv. επανειλημμένως.

repel v. αποκρούω. ~lent a. αποκρουστικός.

repent v. μετανοώ, μετανοιώνω. ~ance s.
μετάνοια f. ~ant a. μετανοών, μετανοιω-
μένος.

repercussion s. αντίκτυπος m.

repert|oire, ~ory s. ρεπερτόριο n.

repetit|ion s. επανάληψη f. ~ious a.
μονότονος. ~ive a. επαναληπτικός.

repine v. στενοχωρούμαι.

replace v. (put back) αποκαθιστώ, ξα-
ναβάζω. (fill place of) αντικαθιστώ,
αναπληρώνω.

replacement s. (person) αναπληρωτής m.
(act) αντικατάσταση f. (spare part)
ανταλλακτικό n. I will get a ~ for it θα το
αντικαταστήσω.

replenish v. ξαναγεμίζω, συμπληρώνω.

replet|e a. γεμάτος. ~ion s. κορεσμός m.

replica s. πανομοιότυπο n.

reply v. απαντώ. (s.) απάντηση f.

report v.t. & i. αναφέρω. (submit a ~) υποβάλλω έκθεση. it is ~ed (that) μεταδίδεται,
αναφέρεται (ότι). (v.i.) (present oneself)
παρουσιάζομαι.

report s. αναφορά f., έκθεση f. (school)
έλεγχος m.f. (piece of news) πληροφορία
f. (rumour) διάδοση f. (repute) όνομα n.,
φήμη f. (bang) κρότος m. ~age s. ρεπορτάζ n.

reporter s. δημοσιογράφος m., ρεπόρτερ m.

repose v.t. (lay) ακουμπώ, (give rest to)
αναπαύω, (base) στηρίζω. (v.i.) (enjoy
rest) αναπαύομαι, (be based) στηρίζομαι.
(be, lie) βρίσκομαι. (s.) (rest) ανάπαυση f.
(calm) γαλήνη f., ηρεμία f. ~ful a.
γαλήνιος, ήρεμος.

repository s. αποθήκη f. (fig., of information, etc.) θησαυρός m.

reprehen|d v. κατακρίνω. ~sible a. αξιο-
κατάκριτος.

represent v. (depict) παριστάνω, αναπαριστώ, απεικονίζω. (act for, typify)
εκπροσωπώ, αντιπροσωπεύω. (assert)
διατείνομαι, (point out) τονίζω.

representation s. αναπαράσταση f. απεικόνιση f. αντιπροσώπευση f. ~s παρα-
στάσεις f.pl.

representative s. εκπρόσωπος m., αντιπρόσωπος m. (a.) αντιπροσωπευτικός (or
use v.).

repress v. καταστέλλω. ~ion s. κατάπνιξη
f. (psychology) απώθηση f. ~ive a.
κατασπιεστικός.

reprieve s. αναβολή f. (v.) παρέχω αναβολή
σε.

reprimand v. επιτιμώ. (s.) επιτίμησηf.
reprisal s. αντεκδίκησηf. ~s αντίποινα n.pl.
take ~s καταφεύγω εις αντίποινα.
reproach v. κατηγορώ. (s.) κατηγορία f.,
παράπονο n. (shame) ντροπή f. beyond ~
άψογος. ~ful a. παραπονετικός.
reprobate s. ανήθικος a. (v.) αποδοκιμάζω.
reproduce v.t. (copy) αντιγράφω, (represent) αναπαριστώ. (biol.) αναπαράγω,
(v.i.) πολλαπλασιάζομαι.
reproduct|ion s. αντίγραφο n. αναπα-
ράσταση f. αναπαραγωγή f. ~ive a.
αναπαραγωγικός.
reproof s. παρατήρησηf.
reprove v. κάνω παρατηρήσεις σε.
reptile s. ερπετό n.
republic s. δημοκρατία f. ~an a. & s. δημοκρατικός.
repudiat|e v. (disavow) απαρνούμαι, (dis-
own) αποκηρύσσω. ~ion s. απάρνηση f.
αποκήρυξηf.
repugn|ance s. απέχθειαf. ~ant a. απεχθής.
repuls|e v. απωθώ. (s.) απώθησηf. ~ion s.
αποστροφήf. ~ive a. αποκρουστικός.
reputable a. ευυπόληπτος.
reputation s. όνομα n.,υπόληψηf., make a
~ βγάζω όνομα.
repute s. όνομα (v.) be ~d to be θεωρούμαι.
~d a. θεωρούμενος, υποθετικός. ~dly adv.
υποθετικώς.
request v.t. (person) παρακαλώ, ζητώ από.
(thing) ζητώ. (s.) αίτησηf. at the ~ of κατ'
αίτησιν or τη αιτήσει (with gen.). be in ~
έχω ζήτηση, είμαι περιζήτητος.
requiem s. μνημόσυνο n.
require v. (need) (also be ~d) χρειάζομαι.
(demand) απαιτώ. ~d a. απαιτούμενος.
requirement s. απαίτηση f. ~s (things
needed) απαιτούμενα n.pl.
requisite a. απαιτούμενος.
requisition s. επίταξηf. (v.) επιτάσσω. ~ed
a. επίτακτος.
requit|e v. (repay) ανταποδίδω, (avenge)
εκδικούμαι. ~al s. ανταπόδοσηf. εκδίκησηf.
rescind v. ακυρώ.
rescue v. διασώζω, γλυτώνω. (s.) διάσωση
f. ~r s. σωτήρας m.
research s. έρευνα f. (v.) ερευνώ, κάνω
έρευνα σε.
resembl|e v. ομοιάζω προς, μοιάζω (με).
they ~e each other μοιάζουν. ~ance s.
ομοιότηταf.
resent v. I ~ it μου κακοφαίνεται. ~ful a.
κακοφανισμένος. ~ment s. κακοφανι-
σμός m.
reservation s. (of rights, mental) επιφύλαξη f. (of seats) make a ~ κρατώ θέση.
(area) περιοχήf.

reserve v. (keep) κρατώ, (hold in store)
επιφυλάσσω. (seats) κρατώ.
reserve s. (store) απόθεμα n. (limitation,
condition) επιφύλαξη f. (caution, restraint) επιφυλακτικότητα f. (mil.) εφεδρεία f. have in ~ έχω κατά μέρος, έχω
ρεζέρβα. (a.) (mil.) εφεδρικός, έφεδρος.
reserved a. (in manner) συγκρατημένος,
(cautious) επιφυλακτικός. (taken) κρατη-
μένος.
reservist s. έφεδρος m.
reservoir s. δεξαμενήf.
resettle v.t. μετοικίζω. (v.i.) μετοικώ.
~ment s. μετοίκησηf.
reshuffle v. (fig.) ανασχηματίζω.
reside v. διαμένω. ~ in (belong to) ανήκω σε.
residence s. διαμονήf. (house) κατοικία f.
resident a. (permanent) μόνιμος, (internal) εσωτερικός. (s.) κάτοικος m. f. ~ial a.
με κατοικίες.
residu|e s. υπόλειμμα n. (dregs) κατακάθι
n. ~al a. υπολειμματικός.
resign v.t. (hand over) παραχωρώ. ~ oneself
υποτάσσομαι. ~ oneself to it το παίρνω
απόφαση. (v.i.) παραιτούμαι. ~ed a. be
~ed to it το έχω πάρει απόφαση.
resignation s. (act) παραίτηση f. with ~
αδιαμαρτύρητα.
resili|ent a. ελαστικός. ~ence s. ελαστι-
κότητα f.
resin s. ρητίνη f., ρετσίνι n. ~ated wine
ρετσίνα f. ~ous a. ρητινώδης.
resist v.t. ανθίσταμαι, αντιστέκομαι (both
with εις). ~ant a. ανθεκτικός. ~less a.
ακαταμάχητος.
resistance s. αντίσταση f. (strength)
ανθεκτικότητα f. take the line of least ~
κοιτάζω να βρω τον ευκολότερο τρόπο.
resolute a. αποφασιστικός, (bold) άφοβος.
~ly adv. αποφασιστικά, άφοβα. ~ness s.
αποφασιστικότητα f.
resolution s. (decision) απόφαση f. (by
vote) ψήφισμα n. (of problem) λύση f.
(being resolute) αποφασιστικότητα f.
resolve v.i. (decide) αποφασίζω. (v.t.)
(solve, end) λύω, (break up) διαλύω,
(convert) μετατρέπω. (s.) απόφασηf. ~d a.
αποφασισμένος.
resonan|t a. ηχηρός. ~ce s. αντήχηση f.
(loudness) ηχηρότητα f.
resonator s. συνηχητής m.
resort v. ~ to (course) καταφεύγω εις,
(person) προσφεύγω εις.
resort s. (recourse) προσφυγήf. it was our
last ~ ήταν το τελευταίο μας καταφύγιο. in
the last ~ ελλείψει άλλου μέσου. summer ~
θέρετρο n. (haunt) στέκι n.
resound v. αντηχώ. ~ing a. ηχηρός.

resource *s.* (*inventiveness*) επινοητικότητα *f.* ~s πόροι *m.pl.* μέσα *n.pl.* ~**ful** *a.* επινοητικός. he is ~ful κόβει το μυαλό του.

respect *v.* (*honour*) εκτιμώ. (*concern*) έχω σχέση με.

respect *s.* εκτίμηση *f.*, σεβασμός *m.* pay one's ~s υποβάλλω τα σέβη μου. in certain ~s από ορισμένες απόψεις. in ~ of σχετικά με, όσον αφορά.

respectability *s.* ευπρέπεια *f.* καθωσπρεπισμός *m.*

respectab|le *a.* (*honest*) έντιμος, (*decent*) της προκοπής. (*person*) καθώς πρέπει. a ~le sum σεβαστό ποσό. ~**ly** *adv.* ευπρεπώς.

respectful *a.*, ~**ly** *adv.* με σεβασμό.

respecting *prep.* σχετικά με.

respective *a.* we sat down in our ~ seats καθίσαμε ο καθένας στη θέση του. ~**ly** *adv.* αντιστοίχως.

respirat|ion *s.* αναπνοή *f.* ~**or** *s.* αναπνευστήρ *m.* ~**ory** *a.* αναπνευστικός.

respite *s.* (*rest*) ανάπαυλα *f.* (*delay*) αναβολή *f.*

resplendent *a.* απαστράπτων.

respond *v.* (*answer*) αποκρίνομαι, απαντώ. (*to feelings*) ανταποκρίνομαι, αντιδρώ θετικά. (*react*) αντιδρώ. ~ to (*obey*) ακούω σε.

respons|e *s.* (*answer*) απάντηση *f.* (*to feelings*) ανταπόκριση *f.* (*reaction*) αντίδραση *f.* ~**ive** *a.* use respond.

responsib|le *a.* υπεύθυνος, (*trustworthy*) εμπιστοσύνης. ~**ility** *s.* ευθύνη *f.* ~**ly** *adv.* με ευθύνη, υπεύθυνα.

rest *v.t.* (*put*) βάζω. (*lean*) ακουμπώ, (*support*) υποστηρίζω. (*give repose to*) αναπαύω, ξεκουράζω. (*v.i.*) (*be*) είμαι. (*lean*) ακουμπώ, (*be supported*) στηρίζομαι. (*repose*) αναπαύομαι, ξεκουράζομαι. (*lie*) κείμαι, (*remain*) (παρα)μένω. it ~s with you σε σας εναπόκειται. ~ upon (*of glance, etc.*) πέφτω απάνω σε.

rest *s.* (*repose*) ανάπαυση *f.* (*between shifts*) ρεπό *n.* have a ~ ξεκουράζομαι. set at ~ καθησυχάζω. be at ~ (*not worried*) ησυχάζω. come to ~ σταματώ. (*support*) στήριγμα *n.* (*mus.*) παύση *f.*

rest *s.* (*remainder*) υπόλοιπο *n.* the ~ of the fruit τα υπόλοιπα φρούτα. for the ~ of the time τον υπόλοιπο καιρό. the ~ (*others*) οι υπόλοιποι, οι άλλοι.

restaurant *s.* εστιατόριο *n.*

restful *a.* ~ holiday ξεκούραστες διακοπές. a ~ colour χρώμα που ξεκουράζει τα μάτια.

restitution *s.* αποκατάσταση *f.* make ~ παρέχω αποζημίωση.

restive *a.* ατίθασος.

restless *a.* ανήσυχος.

restoration *s.* αποκατάσταση *f.* (*of régime*) παλινόρθωση *f.* (*archaeological*) αναστήλωση *f.*

restorative *a.* δυναμωτικός.

restore *v.* αποκαθιστώ. (*repair*) επισκευάζω. (*a monument*) αναστηλώνω. (*bring back*) επαναφέρω. (*return*) επιστρέφω.

restrain *v.* συγκρατώ. ~ oneself κρατιέμαι, μαζεύομαι.

restraint *s.* (*moderation*) μετριοπάθεια *f.* (*restriction*) περιορισμός *m.* with ~ συγκρατημένα. without ~ (*freely*) ελεύθερα. we admired their ~ θαυμάσαμε το πόσο συγκρατημένοι ήταν. put a ~ on (*limit*) περιορίζω.

restrict *v.* περιορίζω. ~**ion** *s.* περιορισμός *m.* ~**ive** *a.* περιοριστικός.

result *v.* ~ in (*of the cause*) έχω ως αποτέλεσμα (*with* να *or* acc.). ~ from (*of the effect*) απορρέω, προκύπτω, προέρχομαι (*all with* από).

result *s.* αποτέλεσμα *n.* ~**ant** *a.* προερχόμενος, επακόλουθος.

resume *v.* (*go on with*) επαναλαμβάνω. (*regain*) ανακτώ. (*recapitulate*) συγκεφαλαιώνω. ~ one's seat επιστρέφω στη θέση μου.

résumé *s.* περίληψη *f.*

resumption *s.* επανάληψη *f.*

resurgence *s.* αναζωπύρωση *f.*

resurrect *v.* (*fig.*) ξεθάβω. (*lit.*) be ~ed ανίσταμαι. ~**ion** *s.* ανάσταση *f.*

resuscitat|e *v.* ξαναζωντανεύω. ~**ion** *s.* ξαναζωντάνεμα *n.*

retail *a.* λιανικός. (*v.t.*) (*sell*) πουλώ λιανικώς. (*tell*) αφηγούμαι. ~**er** *s.* έμπορος λιανικής πωλήσεως.

retain *v.* (*keep*) κρατώ, (*keep in place*) συγκρατώ. (*preserve*) διατηρώ. (*a lawyer*) βάζω. ~**er** *s.* (*servant*) υπηρέτης *m.* (*fee*) προκαταβολή *f.*

retaliat|e *v.* καταφεύγω εις αντίποινα. ~**ion** *s.* αντεκδίκηση *f.*

retard *v.* καθυστερώ. ~**ed** *a.* καθυστερημένος.

retch *v.i.* αναγουλιάζω.

retent|ion *s.* συγκράτηση *f.* ~**ive** *a.* (*memory*) στεγανός, (*soil*) που κρατάει το νερό.

retic|ent *a.* επιφυλακτικός. ~**ence** *s.* επιφυλακτικότητα *f.*

retinue *s.* ακολουθία *f.*

retire *v.i.* αποτραβιέμαι, αποσύρομαι. (*on pension*) παίρνω σύνταξη. ~**d** *a.* συνταξιούχος. ~d officer απόστρατος *m.*

retirement *s.* συνταξιοδότηση *f.* (*of officers*) αποστρατεία *f.* (*seclusion*) in ~ απομονωμένος.

retiring *a.* ντροπαλός, μαζεμένος.
retort *v.* απαντώ. (*s.*) απάντηση *f.*
retouch *v.* ρετουσάρω.
retrace *v.* (*mentally*) ξαναθυμούμαι. ~ one's steps ξαναγυρίζω πίσω.
retract *v.t.* (*take back*) παίρνω πίσω, (*pull in*) μαζεύω.
retreat *v.* υποχωρώ, οπισθοχωρώ. (*s.*) υποχώρηση, οπισθοχώρηση *f.* (*place of refuge*) καταφύγιο *n.*
retrench *v.i.* περικόπτω τα έξοδά μου.
~**ment** *s.* περικοπή εξόδων.
retrial *s.* αναψηλάφησις *f.*
retribution *s.* τιμωρία *f.*, εκδίκηση *f.*
retrieve *v.* (*get back*) επανακτώ, (*save*) διασώζω.
retrograde *a.* οπισθοδρομικός.
retrospect *s.* look at in ~ ανασκοπώ. in ~ ανασκοπώντας τα περασμένα. ~**ive** *a.* (*law, etc.*) αναδρομικός. ~**ively** *adv.* αναδρομικώς.
return *v.* (*give back*) επιστρέφω, (*come back*) γυρίζω, επιστρέφω. (*elect*) βγάζω, (*declare*) δηλώνω.
return *s.* επιστροφή *f.* γυρισμός *m.*, επάνοδος *f.* (*yield*) απόδοση *f.* (*declaration*) δήλωση *f.* (*report*) έκθεση *f.*, (*ticket*) μετ' επιστροφής, many happy ~s χρόνια πολλά. in ~ (*exchange*) εις αντάλλαγμα.
reunion *s.* (*after absence*) ξανασμίξιμο *n.* (*gathering*) συγκέντρωση *f.*
reunite *v.* be ~d ξανασμίγω.
revalue *v.* επανεκτιμώ.
revamp *v.* (*patch up*) μπαλώνω. (*freshen up*) ξαναφρεσκάρω.
reveal *v.* αποκαλύπτω.
reveille *s.* εγερτήριο *n.*
revel *v.* (*make merry*) ξεφαντώνω. ~ in (*fig.*) απολαμβάνω. (*s.*) διασκέδαση *f.* ~**ry** *s.* ξεφάντωμα *n.*
revelation *s.* αποκάλυψη *f.*
revenge *s.* εκδίκηση *f.* take ~ (on) εκδικούμαι. have one's ~ παίρνω το αίμα μου πίσω.
revenge *v.* (*also* ~ oneself on) εκδικούμαι. ~**ful** *a.* εκδικητικός.
revenue *s.* εισόδημα *n.* (*public*) έσοδα *n.pl.* ~ department εφορία *f.*
reverberat|e *v.* αντηχώ. ~**ion** *s.* αντήχηση *f.*
rever|e *v.* ευλαβούμαι. ~**end** *a.* σεβάσμιος. ~**ence** *s.* ευλάβεια *f.* ~**ent** *a.* ευλαβής. ~**ently** *adv.* ευλαβικά.
reverie *s.* ρεμβασμός *m.*
revers *s.* ρεβέρ *n.*
reversal *s.* αντιστροφή *f.* αναποδογύρισμα *n.* (*of character, policy, etc.*) μεταστροφή *f.* (*change*) ριζική μεταβολή *f.*
reverse *v.t.* αντιστρέφω. (*change*) μετα-

βάλλω ριζικά. (*annul*) ανατρέπω. (*v.i.*) (*in driving*) κάνω όπισθεν.
reverse *s.* (*opposite*) αντίθετο *n.* (*other side*) ανάποδη *f.* (*setback*) ατύχημα *n.* (*a.*) αντίστροφος. ~ gear όπισθεν *n.*
reversible *a.* (*mechanism*) αναστρεφόμενος, (*cloth*) ντουμπλ-φας.
reversion *s.* επάνοδος *f.*
revert *v.* επανέρχομαι.
review *s.* (*survey*) ανασκόπηση *f.* (*revision*) αναθεώρηση *f.* (*journal, inspection*) επιθεώρηση *f.* (*critical*) κριτική *f.* (*v.*) ανασκοπώ, αναθεωρώ, επιθεωρώ, κρίνω.
revile *v.* εξυβρίζω.
revis|e *v.* (*review*) αναθεωρώ, (*correct*) διορθώνω. ~**ion** *s.* αναθεώρηση *f.* διόρθωση *f.* ~**ionist** *s.* ρεβιζιονιστής *m.*
revival *s.* αναζωογόνηση *f.* (*rebirth*) αναγέννηση *f.* (*religious*) αφύπνιση *f.*
revive *v.t.* αναζωογονώ, (*bring to*) συνεφέρνω. (*freshen*) φρεσκάρω, (*bring back*) επαναφέρω. (*v.i.*) ξαναζωντανεύω, (*come to*) συνέρχομαι.
revivify *v.* αναζωογονώ.
rev|oke *v.t.* ανακαλώ. ~**ocation** *s.* ανάκληση *f.*
revolt *s.* εξέγερση *f.*, ξεσήκωμα *n.* (*armed*) ανταρσία *f.* (*v.i.*) εξεγείρομαι, επαναστατώ, ξεσηκώνομαι. (*v.t., sicken*) αηδιάζω.
revolting *a.* (*in revolt*) ξεσηκωμένος. (*sickening*) αηδιαστικός, (*abhorrent*) αποκρουστικός.
revolution *s.* επανάσταση *f.* (*turn*) (περι)στροφή *f.* ~**ize** *v.* αλλάζω ριζικά.
revolutionary *a.* επαναστατικός. ~ ideas καινά δαιμόνια. (*s.*) επαναστάτης *m.*
revolv|e *v.t.* περιστρέφω, (*in mind*) σταθμίζω. (*v.i.*) περιστρέφομαι. ~**ing** *a.* περιστροφικός, περιστρεφόμενος.
revolver *s.* περίστροφο *n.*
revue *s.* επιθεώρηση *f.*
revulsion *s.* (*reaction*) μεταστροφή *f.* (*disgust*) αηδία *f.*
reward *v.* ανταμείβω. (*s.*) ανταμοιβή *f.*
rhapsod|y *s.* ραψωδία *f.* ~**ic** *a.* ραψωδικός. ~**ize** *v.* ~ize over εξυμνώ.
rhetoric *s.* ρητορική *f.* (*pej.*) στόμφος *m.* ~**al** *a.* ρητορικός, στομφώδης.
rheumat|ic *a.* ρευματικός. be ~ic υποφέρω από ρευματισμούς. ~**ism** *s.* ρευματισμός *m.*
rhinoceros *s.* ρινόκερως *m.*
rhombus *s.* ρόμβος *m.*
rhym|e *s.* ομοιοκαταληξία *f.* without ~e or reason χωρίς κανένα λόγο. (*v.i.*) ριμάρω. ~**ed, ~ing** *a.* ομοιοκατάληκτος.
rhythm *s.* ρυθμός *m.* ~**ic(al)** *a.* ρυθμικός.
rib *s.* πλευρό *n.*

ribald *a.* χοροϊδευτιχός. **~ry** *s.* χοροϊδία *f.*

ribbon *s.* χορδέλλα *f.* (*of order*) ταινία *f.* tear to ~s χάνω χουρέλια.

rice *s.* ρύζι *n.*

rich *a.* πλούσιος, (*soil*) εύφορος, (*food*) παχύς. (*luxurious*) πολυτελής, grow *or* make rich πλουτίζω. **~es** *s.* πλούτη *n.pl.*

richly *adv.* πλούσια. he ~ deserved his punishment έλαβε τα επίχειρα της κακίας του.

richness *s.* πλούτος *m.* (*luxury*) πολυτέλεια *f.*

rick *s.* θημωνιά *f.*

rick *v.* στραμπουλίζω.

rick|ets *s.* ραχίτιδα *f.* **~ety** *a.* (*person*) ραχιτικός, (*object*) ξεχαρβαλωμένος, (*régime*) ετοιμόρροπος.

ricochet *s.* εποστρακισμός *m.* (*v.*) εποστρακίζομαι.

rid *v.* απαλλάσσω. get ~ of απαλλάσσομαι από, ξεφορτώνομαι, (*dismiss*) διώχνω, (*throw away*) πετώ. **~dance** *s.* good ~ance! καλά ξεχουμπίδια.

riddle *s.* αίνιγμα *n.*

riddle *v.* χάνω κόσκινο.

ride *v.t.* καβαλλικεύω. ~ a bicycle πηγαίνω με ποδήλατο. (*v.i.*) (*go riding*) χάνω ιππασία. (*be conveyed*) πηγαίνω. ~ up ανεβαίνω.

ride *s.* (*jaunt*) περίπατος *m.*, βόλτα *f.* take for a ~ (*deceive*) χοροϊδεύω.

rider *s.* ιππεύς *m.*

ridge *s.* (*of hill*) ράχη *f.* (*of furrow*) πτυχή *f.* (*of roof*) χορφιάς *m.*

ridicu|le *s.* εμπαιγμός *m.*, χοροϊδία *f.* (*v.*) περιγελώ, χοροϊδεύω. **~ous** *a.* γελοίος.

riding *s.* ιππασία *f.*

rife *a.* διαδεδομένος. ~ with γεμάτος.

riff-raff *s.* η σάρα και η μάρα.

rifle *v.* (*search*) ψάχνω,(*rob*) λεηλατώ.

rifle *s.* τουφέκι *n.* **~-range** *s.* σκοπευτήριο *n.* **~-shot** *s.* τουφεκιά *f.*

riffing *s.* ραβδώσεις *f.pl.*

rift *s.* σχισμή *f.*, ρωγμή *f.* (*fig.*) σχίσμα *n.*

rig *v.* (*ship*) εξοπλίζω. (*falsify*) φτιάνω, (*ballot*) νοθεύω. ~ up στήνω. ~ out ντύνω.

rig *s.* (*of ship*) εξαρτισμός *m.* (*fam.*, *dress*) ντύσιμο *n.*, in full ~ με μεγάλη στολή.

rigging *s.* (*ship's*) (ε)ξάρτια *n.pl.* (*of ballot*) νόθευση *f.*

right *a.* (*not left*) δεξιός. ~ hand *or* side δεξιά *f.* (*adv.*) (*also* on *or* to the ~) δεξιά. **~-handed** *a.* δεξιός. **~-wing** *a.* δεξιός. (*s.*) δεξιά *f.*

right *s.* (*justice*) δίκαιο(ν) *n.* (*what is proper*) ορθόν *n.* (*entitlement*) δικαίωμα *n.* set *or* put to ~s επανορθώνω.

right *v.* επανορθώνω. (*correct position of*) ισορροπώ, ξαναφέρνω σε ισορροπία.

right *a.* (*just*) δίκαιος, (*correct, proper*) σωστός, ορθός. (*suitable*) κατάλληλος, (*exact*) ακριβής. be ~ (*of person*) έχω δίκιο. put ~ διορθώνω. do what is ~ χάνω ό,τι πρέπει. just at the ~ moment πάνω στην ώρα. the ~ side (*of cloth, etc.*) η καλή. (*fam.*) get on the ~ side of καλοπιάνω. in one's ~ mind με τα σωστά μου. be all ~ είμαι εν τάξει. (*fam.*) a bit of all-~ μπουκιά και συχώριο. ~ angle ορθή γωνία.

right *adv.* (*correctly*) σωστά, (*exactly*) ακριβώς. ~ round γύρω-γύρω, ολόγυρα. ~ in the middle καταμεσής. ~ away αμέσως. it serves you ~ καλά να πάθεις. all ~ (*correctly*) σωστά, (*duly*) καλά, εν τάξει. **~-angled** *a.* ορθογώνιος. **~-thinking** *a.* ορθόφρων.

right *int.* (*also* all ~, ~ oh, ~ you are) καλά, εν τάξει.

righteous *a.* δίκαιος. **~ly** *adv.* δικαίως.

rightful *a.* νόμιμος. **~ly** *adv.* νομίμως.

rightly *adv.* δικαίως, σωστά. he ~ refused με το δίκιο του αρνήθηκε.

rigid *a.* άκαμπτος. **~ity** *s.* ακαμψία *f.*

rigmarole *s.* (*nonsense*) χουραφέξαλα *n.pl.* (*long story*) συναξάρι *n.*

rigorous *a.* (*strict*) αυστηρός, (*harsh*) σκληρός.

rigour *s.* αυστηρότητα *f.* σκληρότητα *f.*

rile *v.* εκνευρίζω.

rill *s.* ρυάκι *n.*

rim *s.* (*of vessel, crater*) χείλος *n.* (*of wheel*) ζάντα *f.* ~s (*of spectacles*) σκελετός *m.*

rime *s.* (*frost*) πάχνη *f.*

rind *s.* (*fruit*) φλούδι *n.* (*bacon*) πέτσα *f.*

ring *s.* κρίκος *m.*, δακτύλιος *m.* (*circle*) κύκλος *m.* (*for finger*) δαχτυλίδι *n.* (*halo*) φωτοστέφανος *m.* (*sound*) ήχος *m.* (*round fugitive*) κλοιός *m.* (*for sports*) ριγκ *n.* (*gang*) σπείρα *f.* there is a ~ at the bell χτυπάει το χουδούνι. ~ road περιφερειακός *m.*

ring *v.t.* (*bell, etc.*) χτυπώ, κρούω. (*surround*) περικυκλώνω. (*fig.*) it ~s a bell θυμίζει κάτι. ~ in αναγγέλλω. (*v.i.*) χτυπώ, χουδουνίζω. (*resound*) ηχώ. ~ off κλείνω. (*v.t. & i.*) ~ up παίρνω στο τηλέφωνο.

ringing *a.* ηχηρός.

ringing *s.* χουδώνισμα *n.* (*in ears*) βούισμα *n.* ~ of bells κωδωνοκρουσία *f.*

ringleader *s.* αρχηγός *m.*, πρωταίτιος *m.*

ringlet *s.* κατσαρό *n.*

rink *s.* πίστα *f.*

rinse *v.* ξεπλένω. (*s.*) ξέπλυμα *n.*

riot *s.* οχλαγωγία *f.*, ταραχές *f.pl.* (*of colour, etc.*) όργιο *n.* run ~ αποχαλινώνομαι, (*of plants*) οργιάζω. (*v.*) στασιάζω. **~er** *s.* ταραξίας *m.* **~ing** *s.* see riot. **~ous** *a.*

θορυβώδης, (in revolt) ξεσηκωμένος. (of party, binge) τρικούβερτος.

rip v. (tear) σχίζω. ~ off or out αποσπώ, βγάζω. ~ open ανοίγω, (disembowel) ξεκοιλιάζω. (v.i.) let ~ (attack) δίνω ένα τράκο. I let ~ at him του 'δωσα και κατάλαβε. let her ~ (drive fast & fig.) πατώ γκάζι.

rip s. (tear) σχίσιμο n. (rake) παραλυμένος a. ~-off s. (fam.) απάτη f.

riparian a. παραποτάμιος.

ripe a. ώριμος. ~ness s. ωριμότητα f.

ripen v. ωριμάζω. ~ing s. ωρίμανση f.

riposte s. (fig.) εύστοχη απάντηση.

ripple s. ρυτίδα f. (of sound) κελάρυσμα f. (of hair) κατσάρωμα n.

ripple v.t. (ruffle) ρυτιδώνω. (v.i.) (purl) κελαρύζω. ~ing s. (of surface) ρυτίδωση f. (noise) κελάρυσμα n.

rise v. (get up) σηκώνομαι, (of sun, etc.) ανατέλλω. (from dead) ανθίσταμαι. (go up) ανεβαίνω. (of ground) ανηφορίζω, (of building) υψώνομαι. (in degree, rank) ανέρχομαι. (in revolt) εξεγείρομαι. (have origin) πηγάζω. (get louder, stronger) δυναμώνω. (adjourn) διακόπτω. ~ to (bait) τσιμπώ.

rise s. (upward slope) ανηφοριά f. (high ground) ύψωμα n. (upward progress) άνοδος f. (increase) αύξηση f. ~ in price ανατίμηση f. give ~ to προκαλώ. take a ~ out of δουλεύω.

rising s. (of sun, etc.) ανατολή f. (revolt) εξέγερση f.

risk v. (one's life) διακινδυνεύω, (doing sthg.) κινδυνεύω να. (stake) παίζω.

risk s. κίνδυνος m. one who takes ~s ριψοκίνδυνος. at the ~ of με κίνδυνο να. at one's own ~ υπό ιδίαν ευθύνην. ~-y a. παρακινδυνευμένος, ριψοκίνδυνος.

risqué a. τολμηρός, σόκιν.

rissole s. κεφτές m.

rite s. τελετή f.

ritual a. τυπικός, (customary) καθιερωμένος. (s.) (abstract) τυπικό n., (a ceremony) ιεροτελεστία f. ~ism s. προσήλωση στις ιεροτελεστίες.

rival a. αντίζηλος. (s.) αντίπαλος m. (v.t.) συναγωνίζομαι, αμιλλώμαι. ~ry s. αντιζηλία f. άμιλλα f.

river s. ποτάμι n. (also fig.) ποταμός m. ~side s. ακροποταμιά f.

rivulet s. ρυάκι n.

road s. δρόμος m., οδός f. (a.) οδικός. ~works έργα οδοποιίας. ~side s. άκρη του δρόμου. ~-making s. οδοποιία f.

road|s, ~ stead s. αγκυροβόλιο n.

roam v. περιπλανώμαι. ~er s. πλάνη a. ~ing

s. περιπλάνηση f.

roar v. βρυχώμαι. (fig.) (of people) ουρλιάζω, (of elements) μουγγρίζω. ~ with laughter σκάω στα γέλια. do a ~ing trade κάνω χρυσές δουλειές.

roast v. ψήνω, (coffee, etc.) καβουρντίζω. (a.) ψητός, στο φούρνο. (s.) ψητό n. ~ing s. ψήσιμο n. καβούρντισμα n.

rob v. ληστεύω, (deprive) αποστερώ. ~ber s. ληστής m. ~bery s. ληστεία f.

robe s. (woman's) ρόμπα f. ~s (of office) στολή f., (eccl.) άμφια n.pl. (v.t.) ενδύω, περιβάλλω. (v.i.) ενδύομαι. ~d a. ντυμένος.

robin s. κοκκινολαίμης m.

robot s. ρομπότ n.

robust a. εύρωστος. ~ly adv. γερά. ~ness s. ευρωστία f.

rock s. βράχος m. run on the ~s (lit. & fig.) ναυαγώ. (fam.) on the ~s (broke) απένταρος, (with ice) με παγάκια.

rock v.t. (gently) κουνώ, λικνίζω. (violently) σείω, κλονίζω. (v.i.) κουνιέμαι, σείομαι, κλονίζομαι. ~ing-chair s. κουνιστή πολυθρόνα.

rock-bottom s. touch ~ πέφτω πολύ χαμηλά, φτάνω στο κατώτατο σημείο. (a.) κατώτατος.

rocket s. πύραυλος m. (v.i.) πηγαίνω στα ουράνια.

rocky a. βραχώδης.

rod s. ράβδος f. (fishing) καλάμι n. (curtain, stair) βέργα f.

rodent s. τρωκτικό n.

roe s. (beast) ζαρκάδι n. (edible) hard ~ αβγά ψαριού, soft ~ γάλα ψαριού.

rogue s. κατεργάρης m. ~ery s. κατεργαριά f. ~ish a. κατεργάρικος, (mischievous) τσαχπίνικος.

roisterer s. γλεντζές m.

role s. ρόλος m.

roll s. (paper) ρολό n. (cloth) τόπι n. (bread) ψωμάκι n. (list) κατάλογος m. (thunder) βροντή f. (drum) τυμπανοκρουσία f. (of ship) μπότζι n. walk with a ~ σειέμαι.

roll v.t. (along or over) κυλώ. (pastry) ανοίγω, (cigarette) στρίβω. ~ up (paper, umbrella, etc.) κάνω ρολό, (wrap) τυλίγω, (coil) κουλουριάζω, (into a ball) κουβαριάζω. ~ oneself up (in blanket, etc.) τυλίγομαι. ~ up one's sleeves ανασκουμπώνομαι. (v.i.) (down, along or by) κυλώ, (wallow) κυλιέμαι. (walk with a ~) σειέμαι. (of ship) κάνω μπότζι. (of thunder) βροντώ. ~ up (coil) κουλουριάζομαι, (of blind, etc.) ανεβαίνω, (fam., arrive) καταφθάνω.

roll-call s. προσκλητήριο n.

roller s. (garden) κύλινδρος m. (for hair) μπικουτί n. ~-**skate** s. τροχοπέδιλο n.

rollicking a. ζωηρός. have a ~ time γλεντοκοπώ.

rolling a. κυλιόμενος. (country) see undulating. be ~ in money κολυμπώ στα λεφτά. (fig.) ~ stone ανεπρόκοπος a. ~-**pin** s. πλάστης m. ~-**stock** s. τροχαίο υλικό σιδηροδρόμου.

roly-poly a. στρουμπουλός.

Romaic s. (demotic Gk.) ρωμαίικα n.pl.

Roman a. ρωμαϊκός. (person) Ρωμαίος m.

Roman Catholic a. & s. ρωμαιοκαθολικός.

romance a. (language) ρωμανικός.

romance s. (story) ρομάντζο n. (love-affair) ειδύλλιο n. (mus.) ρωμάντζα f. (romantic quality) ρωμαντική ατμό-σφαιρα, κάτι το ρωμαντικό.

Romanesque a. ρωμανικός.

romantic a. ρωμαντικός. ~**ism** s. ρωμαντισμός m.

romp v. παίζω. (s.) παιγνίδι n.

rood s. (eccl.) Εσταυρωμένος m.

roof s. στέγη f. (cover) σκεπή f. flat ~ ταράτσα f. (of cave) θόλος m. ~ of mouth ουρανίσκος m. (v.) στεγάζω.

rook s. (bird) κουρούνα f.

rook v. (fam., overcharge) γδέρνω, μαδώ.

room s. (space) χώρος m. make ~ κάνω τόπο. there is not ~ for us δεν χωράμε, δεν έχει τόπο για μας. is there ~ for us in the car? μας χωράει το αυτοκίνητο;

room s. (in house) δωμάτιο n., κάμαρα f. (large) αίθουσα f. ~s (lodging) διαμέρισμα n. ~**y** a. ευρύχωρος.

roost s. κούρνια f. (fig.) rule the ~ κάνω κουμάντο. (v.) κουρνιάζω.

root s. ρίζα f. ~ and branch σύρριζα. (v.t.) become ~ed ριζώνω. ~ out ξερριζώνω. (fix) καρφώνω. (v.i., search) ψάχνω.

rope s. σχοινί n. (fig.) give ~ (to) δίνω ελευθερία κινήσεων. know the ~s ξέρω τα κουμπιά.

rope v. ~ (together) δένω με σχοινί. ~ in (fig.) επιστρατεύω. ~**y** a. (fam.) της κακιάς ώρας.

rosary s. κομποσκοίνι n.

rose s. ρόδο n., τριαντάφυλλο n. ~-**tree** τριανταφυλλιά f. (sprinkler) τρυπητό n. ~-**petal** s. ροδόφυλλο n. ~-**water** s. ροδόσταμο n.

rose a. ροδόχρους, ροζ. ~-**coloured** (lit. & fig.) ρόδινος.

rosemary s. δεντρολίβανο n.

rosette s. ροζέτα f. (archit.) ρόδαξ m.

rosin s. ρητίνη f.

roster s. κατάλογος m.

rostrum s. βήμα n.

rosy a. ρόδινος.

rot v.t. & i. σαπίζω. (s.) σαπίλα f. (fam.) σαχλαμάρες f.pl.

rotate v.t. περιστρέφω. (v.i.) περιστρέφομαι. ~**ion** s. in ~ion εκ περιτροπής.

rote s. by ~ απ' έξω.

rotten a. σάπιος, (fam.) χάλια. ~**ness** s. σαπίλα f.

rotund a. στρογγυλός, (voice) ηχηρός.

roué s. ακόλαστος a.

rouge s. κοκκινάδι n.

rough a. (violent) βίαιος, (manner, voice) τραχύς, (road) ανώμαλος, (sea) κυματώδης. (hard) σκληρός, (uncouth) άξεστος, (unrefined) χονδρός, (unwrought) ακατέργαστος. (improvised) πρόχειρος, (difficult) δύσκολος. ~ and ready χοντροκαμωμένος, πρόχειρος. ~**ly** adv. βιαίως, τραχιά, πρόχειρα. (about) περίπου.

rough v. ~ up (hair) ανακατώνω, (maltreat) κακοποιώ. ~ in (or out) σχεδιάζω πρόχειρα. ~ it περνώ χωρίς ανέσεις.

rough adv. (violently) βιαίως. sleep ~ κοιμάμαι όπου να 'ναι. cut up ~ θυμώνω.

rough-and-tumble s. συνωστισμός m.

roughen v.t. & i. τραχύνω, αγριεύω.

rough-hewn a. χοντροπελεκημένος.

rough-house s. καβγάς m.

roughness s. τραχύτητα f. (violence) βία f.

roulette s. ρουλέτα f.

round a. (shape, words, sum) στρογγυλός, (in a circle) κυκλικός. ~ trip ταξίδι με επιστροφή. at a ~ pace με γοργό βήμα. ~**ness** s. καμπυλότητα f.

round s. (slice) φέτα f. (series) σειρά f. (cycle) κύκλος m. (of visits, boxing, etc.) γύρος m. daily ~ ρουτίνα f., (drudgery) μαγγανοπήγαδο n. (cartridge) φυσίγγι n. fire ten ~s πυροβολώ δέκα φορές. do (or go or make) the ~s (of) κάνω το γύρο (with gen.), (of doctor) κάνω σειρά επισκέψεων. in the ~ (sculpture) ολόγλυφος.

round adv. γύρω. all or right ~ ολόγυρα, γύρω-γύρω. turn or go ~ γυρίζω. I'll go (or pop or slip) ~ to the grocer's θα πεταχτώ στον μπακάλη. go ~ (deviously) πάω γύρω, (be enough) φτάνω για όλους. come ~ συνέρχομαι. hand (sthg.) ~ περνώ γύρω.

round prep. γύρω σε or από (with acc.). go ~ the shops κάνω το γύρο των μαγαζιών. I'll show you ~ the garden θα σας δείξω τον κήπο. ~ about noon γύρω στις δώδεκα. get ~ (v.t.) (pass) παρακάμπτω, (persuade) φέρνω βόλτα.

round v.t. (make ~) στρογγυλεύω. (get ~) παρακάμπτω. ~ off στρογγυλεύω. ~ up μαζεύω, (*arrest*) συλλαμβάνω. (*v.i.*) ~ on *or* upon ρίχνομαι σε.

roundabout *s.* (*at fair*) αλογάκια *n.pl.* (*traffic*) κόμβος *m.* (*a.*) say it in a ~ way τα λέω περιφραστικώς (*or* από σπόντα).

rounded *a.* στρογγυλεμένος, καμπύλος.

roundly *adv.* (*tell*) ρητά, (*curse*) από την καλή.

round-up *s.* συγκέντρωση *f.* (*arrest*) σύλληψη *f.*

rousle *v.t.* (*waken*) ξυπνώ, (*excite*) εξεγείρω. ~e oneself κουνιέμαι. **~ing** *a.* ζωηρός.

rout *v.* (*defeat*) κατατροπώνω. ~ out ξετρυπώνω. (*s.*) κατατρόπωση *f.*

route *s.* δρόμος *m.* the quickest ~ ο πιο σύντομος δρόμος. (*of procession*) διαδρομή *f.* (*itinerary*) what ~ will you take? τι δρομολόγιο θα ακολουθήσετε; the nearest bus ~ η πιο κοντινή γραμμή λεωφορείων.

routine *s.* ρουτίνα *f.* (*a.*) συνηθισμένος.

rovle *v.* περιπλανώμαι. have a ~ing eye γλυκογυρίζω τα μάτια. ~ing commission αποστολή που συνεπάγεται περιοδεία. **~er** *s.* περιπλανώμενος *a.*

row *s.* (*line*) σειρά *f.* all in a ~ αράδα αράδα.

row *s.* (*quarrel*) καβγάς *m.* (*noise*) θόρυβος *m.* (*fuss*) φασαρία *f.* (*v.i.*) καβγαδίζω.

row *v.i.* (*with oars*) κάνω κουπί. **~ing** *s.* κωπηλασία *f.*

rowdy *a.* θορυβώδης.

royal *a.* βασιλικός. **~ist** *a.* βασιλόφρων, βασιλικός.

royallty *s.* (*quality*) βασιλεία *f.* (*persons*) μέλη της βασιλικής οικογένειας. ~ties (*author's*) συγγραφικά δικαιώματα.

rub *v.t.* τρίβω. (*v.i.*) τρίβομαι. ~ out σβήνω. ~ it in το κοπανάω. ~ along τα καταφέρνω, (*agree*) τα πάω καλά. ~ shoulders with έρχομαι σε επαφή με.

rubber *s.* λάστιχο *n.*, καουτσούκ *n.* (*india-* ~) γόμα *f.*, λάστιχο *n.*

rubber *a.* λαστιχένιος. ~ band λαστιχάκι *n.*, ~ stamp σφραγίδα *f.* **~y** *a.* σαν λάστιχο.

rubbing *s.* τρίψιμο *n.*

rubbish *s.* σκουπίδια *n.pl.* (*nonsense*) ανοησίες *f.pl.* **~ dump** σκουπιδαριό *n.* ~y *a.* τιποτένιος.

rubble *s.* μπάζ(ι)α *n.pl.*

rubicund *a.* κοκκινοπρόσωπος.

ruby *s.* ρουμπίνι *n.*

ruck *s.* δίπλα *f.*

rucksack *s.* σάκκος *m.*

ructions *s.* φασαρία *f.*

rudder *s.* τιμόνι *n.* **~less** *a.* ακυβέρνητος.

ruddy *a.* κόκκινος.

rude *a.* (*roughly made*) χοντροφτιαγμένος.

(*impolite*) αγενής. (*unpolished*) αγροίκος. (*sudden, jarring*) βίαιος, απότομος. **~ly** *adv.* (*not skilfully*) άτεχνα, (*not politely*) με αγένεια, (*suddenly*) βίαια, απότομα, **~ness** *s.* αγένεια *f.*

rudiment|s *s.* στοιχεία *n.pl.* **~ary** *a.* στοιχειώδης, (*sketchy*) υποτυπώδης.

rue *v.t.* μετανοώ για. **~ful** *a.* αποθαρρημένος.

ruffian *s.* κακούργος *m.* (*hired*) μπράβος *m.*

ruffle *v.* (*disturb*) διαταράσσω. ~ hair of αναμαλλιάζω. (*annoy*) πειράζω.

rug *s.* χαλί *n.*, τάπητας *m.* (*wrap*) κουβέρτα *f.*

rugged *a.* άγριος. (*of face*) χαρακωμένος, οργωμένος. (*hard*) σκληρός.

ruin *s.* (*destruction*) καταστροφή *f.* (*of building*) ρημάδι *n.*, (*also fig.*) ερείπιο *n.* **~ation** *s.* καταστροφή *f.*

ruin *v.* καταστρέφω, ρημάζω. **~ed** *a.* κατεστραμμένος, (*building*) ερειπωμένος. **~ous** *a.* καταστρεπτικός, ολέθριος.

rule *s.* (*principle*) αρχή *f.* (*what is laid down*) κανόνας *m.* (*sway*) εξουσία *f.* (*measure*) μέτρο *n.* as a ~ κατά κανόνα, against the ~s αντικανονικός. ~ of thumb εμπειρικός κανόνας.

rule *v.* (*also ~ over*) κυβερνώ. (*hold sway*) κυριαρχώ. (*lay down*) διατάσσω. ~ out αποκλείω. (*lines*) χαρακώνω.

ruler *s.* (*lord*) άρχοντας *m.* the ~s of Europe οι κυβερνώντες την Ευρώπη. (*implement*) χάρακας *m.*

ruling *s.* απόφαση *f.*

ruling *a.* ~ class ιθύνουσα τάξη. ~ passion κυριαρχούν πάθος.

rum *s.* ρούμι *n.*

rum *a.* (*event*) παράξενος, (*person*) μυστήριος.

Rumanian *a.* ρουμανικός. (*person*) Ρουμάνος *m.*, Ρουμανίδα *f.*

rumble *v.i.* (*thunder*) μπουμπουνίζω, (*belly*) γουργουρίζω. ~ past *or* by περνώ με πάταγο. (*s.*) μπουμπουνητό *n.* γουργούρισμα *n.* (*of waggon, etc.*) πάταγος *m.*

ruminant *s.* μηρυκαστικόν *n.*

ruminatle *v.* (*chew cud*) μηρυκάζω. (*ponder*) συλλογίζομαι. **~ive** *a.* συλλογισμένος.

rummage *v.* σκαλίζω.

rummy *s.* (*cards*) ραμί *n.*

rumour *s.* φήμη *f.* (*v.*) it is ~ed φημολογείται, διαδίδεται.

rump *s.* καπούλια *n.pl.* (*fig.*) υπόλειμμα *n.* ~ steak κόντρα φιλέτο.

rumpus *s.* (*fuss*) φασαρία *f.* (*noise*) σαματάς *m.*

run *s.* τρέξιμο *n.* go for a ~ πάω να τρέξω.

(trip) βόλτα f., γύρος m. an hour's ~ μίας ώρας διαδρομή. (series) σειρά f. (demand) ζήτηση f. (mus.) σκάλα f., (vocal) λαρυγγισμός m. chicken ~ κοτέτσι n. have the ~ of the house έχω το ελεύθερο στο σπίτι, έχω το σπίτι στη διάθεσή μου. have a good ~ for one's money έχω αντάλλαγμα για τους κόπους μου. the usual ~ of events η συνηθισμένη ρουτίνα. out of the common ~ άνω του μετρίου. have a long ~ κρατώ πολύ in the long ~ τελικά. ~-of-the-mill της αράδας. be on the ~ (busy) είμαι στο πόδι, (fugitive) κρύβομαι. at a ~ τρέχοντας.

run v.i. (go fast, flow, drip, leak, etc.) τρέχω, (extend) εκτείνομαι, (pass, go) περνώ. (of life, events) κυλώ. (of machinery) δουλεύω. (of buses, etc.) λειτουργώ. (of play) παίζομαι. (of colour) βγάζω. (of words) it ~s as follows έχει ως εξής. (seek election) θέτω υποψηφιότητα, (see ~ for). (of stitches) see ladder. be ~ning with είμαι γεμάτος από. it ~s in the family υπάρχει στην οικογένεια. the tune is ~ning in my head ο σκοπός γυρίζει στο μυαλό μου. it has six months to ~ ισχύει για (or λήγει μετά) έξι μήνες. how often do the trains ~? κάθε πότε έχει τραίνο; ~ before the wind ουριοδρομώ. also ran αποτυχών.

run v.t. (manage) διαχειρίζομαι. (a business) διευθύνω, (a house) κρατώ, (a car) συντηρώ. (convey) πηγαίνω, (put) βάζω, (pass) περνώ. (operate bus service, etc.) θέτω σε κυκλοφορία. (a blockade) διασπώ. ~ (person) close ακολουθώ κατά πόδα. ~ risk of κινδυνεύω να. ~ to earth ξετρυπώνω. the illness will ~ its course η ασθένεια θα κάνει την πορεία της.

run across v.t. (meet) πέφτω επάνω σε. the roads ~ the plain οι δρόμοι διασχίζουν την πεδιάδα.

run after v.t. κυνηγώ.

run away v.i. φεύγω, (flee) τρέπομαι εις φυγήν, (fam.) το σκάω, don't ~ with the idea μη σου περάσει από το μυαλό. he lets his imagination ~ with him παρασύρεται από τη φαντασία του.

runaway s. φυγάς m. (a.) ~ horse αφηνιασμένο άλογο.

run down v.t. (denigrate) κακολογώ. (collide with) συγκρούομαι με. (v.i.) (of clock) ξεκουρδίζομαι, (of battery) αδειάζω. feel ~ είμαι εξαντλημένος.

run for v. ~ Parliament πάω για βουλευτής.

run in v.t. (engine, etc.) στρώνω, (arrest) συλλαμβάνω.

run into v. (meet, collide with) πέφτω επάνω σε. (amount to) ανέρχομαι εις.

run off v.i. see run away. ~ with (steal) κλέβω. (v.t.) (empty) αδειάζω, (print) τραβώ, (write) γράφω στα πεταχτά.

run on v.i. (talk) he will ~ η γλώσσα του πάει ροδάνι. (continue) συνεχίζομαι.

run out v.i. the bread has ~ το ψωμί τελείωσε. we have ~ of bread μείναμε από ψωμί. ~ on εγκαταλείπω.

run over v.t. (recapitulate) ανακεφαλαιώνω. (knock down) πατώ. (v.i.) (overflow) ξεχειλίζω.

run through v.t. (lit. & examine) διατρέχω. (review) ανασκοπώ, (rehearse) κάνω δοκιμή. (use up) σπαταλώ. a shiver runs through my body ρίγος διατρέχει το σώμα μου. (pierce) διατρυπώ.

run-through s. ανασκόπηση f., δοκιμή f.

run to v. (afford) I can't ~ that η τσέπη μου δεν το επιτρέπει. ~ seed ξεπέφτω, φαίνομαι παραμελημένος.

run up v.t. (raise) υψώνω, (put together) σκαρώνω. ~ bills χρεώνομαι. (v.i.) ~ against (person) πέφτω επάνω σε, (problem) αντιμετωπίζω.

run-up s. παραμονές f.pl.

runner s. δρομεύς m. (carpet) διάδρομος m. (of sledge, etc.) ολισθητήρ m.

runner-up s. (in exam) ο επιλαχών, (in contest) ο δεύτερος.

running s. τρέξιμο n. (in race) δρόμος m. (functioning) λειτουργία f. (management) διεύθυνση f. be in the ~ έχω ελπίδες.

running a. ~ water τρεχούμενο νερό. ~ commentary περιγραφή (αγώνος) επί τόπου. ~ fire συνεχές πυρ. three days ~ τρεις επανωτές μέρες.

runt s. (person) χοντοστούλης a.

runway s. διάδρομος προσγειώσεως.

rupture s. ρήξη f. (med.) κήλη f. (v.t.) διαρρηγνύω.

rural a. (not urban) αγροτικός, (countrified) εξοχικός.

ruse s. τέχνασμα n., κόλπο n.

rush s. (plant) βούρλο n. (of baskets) ψάθα f.

rush v.i. ορμώ, χυμώ. (v.t.) (send quickly) αποστέλλω εν μεγάλη σπουδή. they ~ed him to hospital έσπευσαν να τον πάνε στο νοσοκομείο. (capture) καταλαμβάνω εξ εφόδου. ~ (sthg.) through περνώ βιαστικά. (press, urge) βιάζω. be ~ed είμαι πνιγμένος στη δουλειά.

rush s. (onrush) ορμή f. the ~ of modern life ο βιαστικός ρυθμός της σημερινής ζωής. there was a ~ for tickets τα εισιτήρια έγιναν ανάρπαστα. I had a ~ to get here on time σκοτώθηκα (or τσακίστηκα) να φτάσω εγκαίρως. why such a ~? γιατί τόση βία; ~ hour ώρα αιχμής.

rusk s. παξιμάδι n.

russet a. κατακόκκινος.

Russian a. ρωσσικός. (person) Ρώσσος m. Ρωσσίδα f.

rust s. σκουριά f. (v.) σκουριάζω. **~less** a. ανοξείδωτος. **~y** a. σκουριασμένος.

rustic a. χωριάτικος, (décor) ρουστίκ. (s.) χωριάτης m.

rustle s. θρόισμα n. (v.) θροΐζω.

rustler s. ζωοκλέφτης m.

rut s. (in ground) αυλακιά f. (fig.) επαχθής ρουτίνα, get into a ~ αποτελματώνομαι.

rut s. (sexual) βαρβατίλα f. **~tish** a. βαρβάτος.

ruthless a. ανηλεής, αλύπητος, **~ly** adv. ανηλεώς, αλύπητα.

rye s. σίκαλη f.

S

sabb|ath s. σάββατο n. (Sunday) Κυριακή f. **~atical** a. σαββατικός.

sable a. μαύρος. (s.) (animal) σαμούρι n.

sabot s. τσόκαρο n.

sabot|age s. δολιοφθορά f., σαμποτάζ n. (v.) σαμποτάρω. **~eur** s. δολιοφθορεύς m.

sabre s. σπάθη f.

saccharin s. σακχαρίνη f.

sacerdotal a. ιερατικός.

sack s. (large) τσουβάλι n., σακκί n. (small) σακκούλι n. **~cloth** s. σακκόπανο n. in ~cloth and ashes εν σάκκω και σποδώ.

sack v. λεηλατώ. (fam., dismiss) απολύω.

sacrament s. μυστήριον n.

sacred a. ιερός, (writings, etc.) θρησκευτικός. ~ cow (iron.) κάτι το ιερό και όσιο.

sacrifice s. θυσία f. (v.) θυσιάζω.

sacrileg|e s. ιεροσυλία f. **~ious** a. βέβηλος.

sacristan s. νεωκόρος m.

sacrosanct a. απαραβίαστος.

sad a. (person) λυπημένος, (thing) λυπηρός. (deplorable) ελεεινός. **~den** v. λυπώ. **~ly** adv. λυπημένα. **~ness** s. λύπη f., θλίψη f.

saddle s. σέλλα f., σαμάρι n. (v.) σελλώνω. (burden) φορτώνω. **~bag** s. δισάκι n.

sad|ism s. σαδισμός m. **~ist** s. σαδιστής m. **~istic** a. σαδιστικός.

safe s. (for money) χρηματοκιβώτιον n. (for food) φανάρι n. **~deposit** s. (locker) θυρίδα f.

safe a. (harmless) αβλαβής, (of toy, animal, etc.) ακίνδυνος. (unharmed) σώος. (cautious) προσεκτικός. (secure, sure) ασφαλής. ~ and sound σώος και αβλαβής. **~conduct** s. πάσσο n. **~guard** v. εξασφα-

λίζω, (s.) εγγύηση f. **~keeping** s. (custody) διαφύλαξις f. for ~keepingγια ασφάλεια. **~ly** adv. ασφαλώς. we arrived ~ly φτάσαμε αισίως.

safety s. ασφάλεια f. (a.) ασφαλείας. **~belt** s. ζώνη ασφαλείας. **~catch** s. ασφάλεια f. **~pin** s. παραμάννα f.

saffron s. ζαφορά f.

sag v. (give way) βουλιάζω, (droop) κρέμομαι. **~ging** a. βουλιαγμένος, κρεμασμένος.

saga s. έπος n.

sagac|ious a. αγχίνους. **~ity** s. αγχίνοια f.

sage a. & s. σοφός. **~ly** adv. σοφώς.

sage s. (herb) φασκόμηλο n.

sail s. πανί n., ιστίο n. (of mill) πανί n. set ~ αποπλέω, κάνω πανιά. **~cloth** s. καραβόπανο n.

sail v.i. πλέω, (set ~) κάνω πανιά. (v.t.) (traverse) διαπλέω. (be in command of) κυβερνώ. ~ up (river) αναπλέω, ~ down καταπλέω, ~ round περιπλέω. ~ into (fam., attack) ρίχνομαι σε.

sailing s. (sport) ιστιοπλοΐα f. (departure) απόπλους m. **~vessel** s. ιστιοφόρο n.

sailor s. ναύτης m. (seafaring man) ναυτικός m. (fig.) be a bad (or good) ~ (δε) με πιάνει η θάλασσα.

saint s. & a. άγιος m., αγία f. **~hood** s. αγιοσύνη f. **~ly** a. άγιος.

sake s. for the ~ of χάριν (with gen.), για (να). for your ~ προς χάριν σας, για το χατίρι σου.

salac|ious a. λάγνος. **~ity** s. λαγνεία f.

salad s. σαλάτα f. (fam.) ~ days νεανικά χρόνια.

salamander s. σαλαμάνδρα f.

salami s. σαλάμι n.

salar|y s. μισθός m. **~ied** a. μισθωτός.

sale s. πώληση f. (clearance) there is a ~ on έχει εκπτώσεις, (auction) πλειστηριασμός m. it is for ~ πωλείται, πουλιέται.

salep s. σαλέπι n.

sales|man s. πωλητής m. **~woman** s. πωλήτρια f.

salient a. εξέχων.

saline a. αλατούχος.

saliv|a s. σάλιο n., σάλια pl. **~ate** v. σαλιάζω.

sallow a. κιτρινιάρης.

sally s. εξόρμηση f. (of wit) ευφυολόγημα n. (v.) ~ out (make a ~) εξορμώ, (fam., set out) ξεκινώ.

salmon s. σολομός m.

saloon s. σαλόνι n.

salt s. αλάτι n. (pl. άλατα). (fam.) old ~ θαλασσόλυκος m. take it with a grain of ~ δεν τα χάφτω. (v.) αλατίζω. **~cellar** s.

αλατιέρα *f.* **~pan** *s.* αλυκή *f.*
salt, ~y *a.* αλμυρός. **~iness** *s.* αλμύρα *f.*
saltpetre *s.* νίτρο *n.*
salubrious *a.* υγιεινός.
salutary *a.* ωφέλιμος.
salut|e *s.* χαιρετισμός *m.* (*v.*) χαιρετίζω, χαιρετώ. **~ation** *s.* χαιρετισμός *m.*
salvage *s.* (*naut.*) ναυαγιαιρεσία *f.* (*payment*) σώστρα *n.pl.* (*a., vessel, etc.*) ναυαγοσωστικός. (*v.*) διασώζω. **~d** goods διασωθέντα πράγματα.
salvation *s.* σωτηρία *f.*
salve *s.* αλοιφή *f.* (*v.*) (*fig.*) ~ one's conscience ελαφρύνω τη συνείδησή μου. (*salvage*) διασώζω.
salvo *s.* ομοβροντία *f.*
same *a. & pron.* ο ίδιος. it's all the ~ to me το ίδιο μου κάνει. the ~ to you! επίσης. at the ~ time (*all together*) όλοι μαζί, (*simultaneously*) συγχρόνως, (*yet*) μολαταύτα. all the ~ κι όμως. the ~ people (whom) we saw yesterday οι ίδιοι άνθρωποι που είδαμε χθες. he goes to the ~ school as I do πάει στο ίδιο σχολείο με μένα. I caught the ~ bus as last time πήρα το ίδιο λεωφορείο όπως και την άλλη φορά. it's the ~ old story (*fam.*) τα αυτά τοις αυτοίς.
sameness *s.* ομοιομορφία *f.*
samovar *s.* σαμοβάρι *n.*
sample *s.* δείγμα *n.* (*v.*) δοκιμάζω.
sanatorium *s.* σανατόριον *n.*
sanctify *v.* καθιερώνω.
sanctimonious *a.* (*person*) φαρισαίος *m.* (*fam.*) he is a ~ fellow κάνει τον άγιο Ονούφριο.
sanction *v.* επιτρέπω, εγκρίνω. (*s.*) (*approval, penalty*) κύρωση *f.*
sanctity *s.* ιερότητα *f.*
sanctuary *s.* (*eccl.*) ιερόν *n.* (*asylum*) άσυλο *n.*
sanctum *s.* (*fam.*) ιδιαίτερο δωμάτιο.
sand *s.* άμμος *m.f.* stretch of ~ (*by sea*) αμμουδιά *f.* **~bag** *s.* αμμόσακος *m.* **~bank** *s.* σύρτη *f.* **~dune** *s.* αμμόλοφος *m.* **~paper** *s.* γυαλόχαρτο *m.* **~stone** *s.* αμμόπετρα *f.* **~y** *a.* αμμώδης.
sandal *s.* πέδιλο *n.*
sandwich *s.* σάντουιτς *n.* (*v.*) (*squeeze*) σφίγγω, στριμώχνω.
sane *a.* υγιής τον νουν. (*sensible*) λογικός.
sang-froid *s.* ψυχραιμία *f.*
sanguine *a.* (*hopeful*) αισιόδοξος, (*redfaced*) ερυθροπρόσωπος.
sanitary *a.* (*of sanitation*) υγειονομικός, (*clean*) καθαρός, υγιεινός.
sanitation *s.* (*protection of health*) υγειονομικά μέτρα. (*drainage*) αποχέ-

τευση *f.* (*hygiene*) υγιεινή *f.*
sanity *s.* πνευματική υγεία. lose one's ~ χάνω τα λογικά μου. (*sense*) λογική *f.*
Santa Claus *s.* ο Άϊ-Βασίλης.
sap *s.* (*juice*) οπός *m.* **~less** *a.* άζουμος. **~ling** *s.* νεοφύτευτο δενδρύλλιο. **~py** *a.* οπώδης.
sap *s.* (*tunnel*) λαγούμι *n.* (*v.*) υπονομεύω. **~per** *s.* (*mil.*) σκαπανεύς *m.*
Sapphic *a.* σαπφικός.
sapphire *s.* σάπφειρος *m.*, ζαφείρι *n.* (*a.*) ζαφειρένιος.
Saracen *s.* Σαρακηνός *m.*
sarc|asm *s.* σαρκασμός *m.* **~astic** *a.* σαρκαστικός.
sarcophagus *s.* σαρκοφάγος *f.*
sardine *s.* σαρδέλλα *f.*
sardonic *a.* σαρδόνιος.
sartorial *a.* ραπτικός. ~ elegance φροντισμένο ντύσιμο.
sash *s.* ζώνη *f.*
sash-window *s.* συρόμενο παράθυρο (καθέτως).
Satan *s.* σατανάς *m.* **~ic** *a.* σατανικός.
satchel *s.* τσάντα *f.*
satellite *s.* δορυφόρος *m.* (*underling*) τσιράκι *n.*
sate *v. see* satiate.
satiat|e *v.* χορταίνω. **~ed** χορτάτος, κορεσμένος. I am **~ed** χόρτασα.
satiety *s.* χορτασμός *m.*, κορεσμός *m.* to ~ κατά κόρον.
satin *s.* ατλάζι *n.*, σατέν *n.*
satir|e *s.* σάτιρα *f.* **~ical** *a.* σατιρικός. **~ize** *v.* σατιρίζω.
satisfact|ion *s.* ικανοποίηση *f.* **~ory** *a.* ικανοποιητικός.
satis|fy *v.* ικανοποιώ, (*convince*) πείθω. I am **~fied** with little χορταίνω με λίγα. **~fying** *a.* ικανοποιητικός, (*filling*) χορταστικός.
satrap *s.* σατράπης *m.*
saturat|e *v.* (*also get* **~ed**) μουσκεύω, διαποτίζω. **~ion** *s.* διαπότιση *f.* **~ion point** σημείο κορεσμού.
Saturday *s.* Σάββατο *n.*
saturnine *a.* κατσούφης.
satyr *s.* σάτυρος *m.*
sauce *s.* σάλτσα *f.* (*cheek*) αυθάδεια *f.*
saucepan *s.* κατσαρόλα *f.*
saucer *s.* πιατάκι *n.*
saucy *a.* αυθάδης, (*fam., smart*) σκερτσόζικος.
saunter *v.* σεργιανίζω. (*s.*) σεργιάνι *n.*
sausage *s.* λουκάνικο *n.* (*in general*) αλλαντικά *n.pl.*
savage *a.* άγριος. **~ly** *adv.* άγρια. **~ness** *s.* αγριότητα *f.*

savant s. σοφός m.

save v.t. (rescue, preserve) σώζω, γλυτώνω. (keep) φυλάω, (husband) οικονομώ. ~ up (money) βάζω στην μπάντα. (expend less time) κερδίζω. (avoid) αποφεύγω, γλυτώνω. (relieve) απαλλάσσω. they ~d trouble and expense by doing it themselves γλυτώσανε και κόπο και έξοδα κάνοντάς το μόνοι τους. it will ~ you having to cook θα σε απαλλάξει από το μαγείρεμα. ~ one's skin σώζω το σαρκίον μου. (v.i.) ~ up κάνω οικονομίες.

sav|e, ~ing prep. εκτός από, εκτός (with gen.). ~e that... εκτός του ότι.

saving a. ~ clause επιφύλαξη f. he has the ~ grace of... έχει τουλάχιστον το καλό ότι.

saving s. οικονομία f. ~s οικονομίες f. pl. ~s bank ταμιευτήριον n.

saviour s. σωτήρ m.

savoir-faire s. τακτ n.

savour s. γεύση f. (fig.) νοστιμιά f. (v.) (relish) απολαμβάνω. (fig.) ~ of μυρίζω. ~less a. άνοστος.

savour|y a. γευστικός, (not sweet) πικάντικος. ~iness s. νοστιμάδα f.

saw s. (saying) λόγιον n.

saw s. (tool) πριόνι n. (v.) πριονίζω, κόβω. ~dust s. πριονίδια n.pl. ~mill s. πριονιστήρι n. ~yer s. πριονιστής m.

saxophone s. σαξόφωνο n.

say v. λέγω, λέω. I ~! (for attention) να σου πω, (surprise) μη μου πεις! let us ~ ας πούμε. I dare ~ κατά πάσαν πιθανότητα. I must ~ (own) ομολογώ. ~ (supposing) they refused πες πως αρνήθηκαν. not to ~ για να μην πω. to ~ nothing of χωρίς να αναφέρουμε. that goes without ~ing αυτό να λέγεται, κοντά στο νου. there's no ~ing δεν μπορεί να πει κανείς. he looked at me as much as to ~ με κοίταξε σα να 'λεγε. (s.) you have no ~ (in the matter) δεν σου πέφτει λόγος.

saying s. ρητόν n.

scab s. (crust) κά(ρ)καδο n. (itch) ψώρα f. ~by a. ψωραλέος.

scabbard s. θηκάρι n.

scabies s. ψώρα f.

scabrous s. (fig.) σκαμπρόζικος.

scaffold s. ικρίωμα n. ~ing s. σκαλωσιά f.

scald v. (also be ~ing hot) ζεματίζω. (s.) ζεμάτισμα n. ~ing a. ζεματιστός.

scale s. (of fish) λέπι n. (fur) πουρί n. (v.t.) (fish) ξελεπίζω, (teeth) καθαρίζω. (v.i.) ~ off ξεφλουδίζομαι.

scale v.t. (climb) σκαρφαλώνω επάνω σε. (fig.) ανέρχομαι εις.

scale s. (graded system) κλίμακα f. (v.) ~ up αυξάνω, ~ down μειώνω.

scale s. (of balance) δίσκος m. ~s ζυγαριά f. turn the ~(s) (fig.) κλίνω την πλάστιγγα, (weigh) ζυγίζω. hold the ~s even κρίνω δίκαια.

scallop s. (fish) χτένι n. ~s (ornament) φεστόνι n. ~ed a. με οδοντώσεις.

scallywag s. παλιόπαιδο n.

scalp s. δέρμα του κεφαλιού.

scaly a. λεπιδωτός.

scamp s. παλιόπαιδο n.

scamp v. ~ one's work τσαλαβουτώ.

scamper v. τρέχω.

scan v.t. (look at) κοιτάζω, εξετάζω. (glance at) ρίχνω μια ματιά σε. (mechanically) ανιχνεύω. (v.i.) (of verse) έχω καλό μέτρο.

scandal s. σκάνδαλο n. (gossip) κουτσομπολιό n. ~monger s. κουτσομπόλης m., σκανδαλοθήρας m.

scandal|ize v. σκανδαλίζω. ~ous a. σκανδαλώδης, (tongue, etc.) κακός. ~ously adv. σκανδαλωδώς.

Scandinavian a. σκανδιναυικός (person) Σκανδιναυός.

scansion s. μετρική ανάλυση.

scant a. λιγοστός. pay ~ attention δεν δίνω αρκετή προσοχή. be ~ of breath μου κόβεται η αναπνοή. (v.) φείδομαι (with gen.).

scant|y a. ανεπαρκής. ~ily adv. ανεπαρκώς. ~iness s. ανεπάρκεια f.

scapegoat s. αποδιοπομπαίος τράγος.

scapegrace s. τρελλόπαιδο n.

scar s. σημάδι n., ουλή f. (v.t.) αφήνω σημάδια σε. ~red a. σημαδεμένος.

scarce a. (rare) σπάνιος. be ~ (rare, not enough) σπανίζω, δεν βρίσκομαι. make oneself ~ εξαφανίζομαι.

scarcely adv. μόλις, σχεδόν. I ~ had time to eat μόλις που πρόφτασα να φάω. you will ~ believe it ούτε και θα το πιστέψεις. ~ ever σχεδόν ποτέ. she can ~ have done that δεν μπορεί να το έκανε αυτό. ~ anybody knew him σχεδόν κανείς δεν τον ήξερε. I had ~ got out of the house when it started raining δεν είχα βγει καλά καλά από το σπίτι και άρχισε να βρέχει. see hardly.

scare v. ξαφνιάζω, τρομάζω. (s.) (alarm) πανικός m. ~crow s. σκιάχτρο n.

scarf s. (big) σάρπα f. (small) κασκόλ n.

scarify v. (fig.) καυτηριάζω.

scarlet a. άλικος, κατακόκκινος. ~ fever οστρακιά f.

scathing a. δριμύς. ~ly adv. δριμέως.

scatter v. (δια)σκορπίζω. ~ed a. σκόρπιος. ~brained a. κοκορόμυαλος.

scavenger a. σκουπιδιάρης m.

scenario s. σενάριο n.

scene s. σκηνή f. come on the ~ παρουσιάζομαι. change of ~ αλλαγή περιβάλλοντος. behind the ~s στα παρασκήνια.behind-the-scenes παρασκηνιακός.

scenery s. (natural) φύση f. (stage) σκηνικά n.pl.

scenic a. (natural) φυσικός, (stage) σκηνογραφικός.

scent s. μυρωδιά f. (perfume) άρωμα n. (animal's sense of smell) όσφρηση f. (track) ίχνη n.pl. on the ~ επί τα ίχνη. put off the ~ παραπλανώ.

scent v. (perfume) αρωματίζω. (discern) μυρίζομαι. ~ed a. (sweet-smelling) αρωματικός.

sceptic s. ~al a. σκεπτικιστής m. ~ism s. σκεπτικισμός m.

sceptre s. σκήπτρο n.

schedule s. πρόγραμμα n.

scheme s. (arrangement) διάταξη f. (of colours) συνδυασμός m. (plan) σχέδιο n., (dishonest) φάμπρικα f., κομπίνα f. (v.i.) μηχανορραφώ. ~er s., ~ing a. μηχανορράφος, δολοπλόκος.

schism s. σχίσμα n. ~atic a. σχισματικός.

schizophren|ia s. σχιζοφρένεια f. ~ic a. σχιζοφρενής.

scholar s. επιστήμων m.f. ~ly a. (of work) επιστημονικός.

scholarship s. (learning) επιστήμη f. (scholarly ability) επιστημοσύνη f. (award) υποτροφία f.

scholastic a. (of schools) εκπαιδευτικός, (dry) σχολαστικός.

school s. σχολείο n. (of university, arts, vocational) σχολή f. of the old ~ της παλαιάς σχολής. come (or let) out of ~ σχολιάζω (v.) (teach) μαθαίνω, (curb) χαλιναγωγώ.

school a. (book, doctor, etc.) σχολικός. ~ days μαθητικά χρόνια. ~ desk μαθητικό θρανίο. ~ leaver απόφοιτος γυμνασίου. ~boy s. μαθητής m. ~fellow s. συμμαθητής m., συμμαθήτρια f., ~girl s. μαθήτρια f. ~ing s. εκπαίδευση f. ~master s. (primary) δάσκαλος, (secondary) καθηγητής m. ~mistress s. δασκάλα, καθηγήτρια f.

schooner s. σκούνα f.

sciatica s. ισχιαλγία f.

science s. επιστήμη f. natural or physical ~ θετικές επιστήμες. social ~ κοινωνιολογία f.

scientific a. επιστημονικός. ~ally adv. επιστημονικά.

scientist s. θετικός επιστήμων.

scintillat|e v. σπινθηροβολώ. ~ing a. σπινθηροβόλος.

scion s. βλαστός m.

scissors s. ψαλίδι n.

scoff v. (also ~ at) κοροϊδεύω. ~ingly adv. κοροϊδευτικά.

scoff v. (devour) καταβροχθίζω.

scold v. μαλώνω, κατσαδιάζω. I gave him a ~ing (fam.) τον έψαλα. (s.) στρίγγλα f.

sconce s. απλίκα f. (fam., head) κούτρα f.

scone s. μικρό αφράτο ψωμάκι.

scoop s. σέσουλα f. (journalist's) λαβράκι n.(v.) ~ up μαζεύω, (excavate) σκάβω.

scoot v. τρέχω. ~er s. motor ~er βέσπα f.

scope s. (prospect, outlet) προοπτική f. (range of activity) πεδίον δράσεως. beyond his ~ εκτός των δυνατοτήτων του. outside the ~ of the committee εκτός της αρμοδιότητος της επιτροπής. it falls within the ~ of the undertaking εμπίπτει εντός του πλαισίου της επιχειρήσεως.

scorch v. καίω, (in ironing) τσιρώνω. be ~ing hot ζεματώ, see singe.

score s. (twenty) εικοσαριά f. ~s of πάμπολλοι.

score v. (win) κερδίζω. ~ a goal πετυχαίνω γκολ. (v.i.) ~ off βάζω κάτω. (s.) (points won) σκορ n. (debt) settle old ~s παίρνω το αίμα μου πίσω.

score s. (mus.) νότες f.pl. (conductor's) παρτιτούρα f.

score v. (cut, rule) χαράζω. ~ out σβήνω. (s.) (scratch) χαρακιά f.

scorn v. περιφρονώ. (s.) περιφρόνηση f. ~ful a. περιφρονητικός. ~fully adv. περιφρονητικά.

scorpion s. σκορπιός m.

scot s. go ~ free βγαίνω λάδι.

Scot|ch, ~tish a. σκωτσέζικος. (person) Σκωτσέζος m., Σκωτσέζα f.

scoundrel s. παλιάνθρωπος m.

scour v. (rub) τρίβω, (search through) τριγυρίζω, (dig) σκάβω. ~ing s. (rubbing) τρίψιμο n.

scourge s. μάστιγα f. (v.) μαστιγώνω.

scout s. πρόσκοπος m. (v.) ~ around for ψάχνω για.

scout v. (reject) απορρίπτω.

scowl v. (also ~ at) αγριοκοιτάζω. (s.) άγριο βλέμμα.

scraggy a. κοκκαλιάρης, πετσί και κόκκαλο.

scramble v.i. σκαρφαλώνω, (jostle) συνωστίζομαι. (v.t.) (a message) μπερδεύω. ~d eggs αβγά χτυπητά στο τηγάνι.

scrap s. (bit) κομματάκι n. (iron) παλιοσίδερα n.pl. ~s (uneaten food) αποφάγια n.pl. there's not a ~ left δεν έμεινε ίχνος. they had a ~ (fight) ήρθαν στα χέρια.

scrap v.t. (throw away) πετώ. (v.i., quarrel)

καβγαδίζω.

scrape v. (rub) ξύνω, (dig) σκάβω, (graze) γδέρνω. ~ off or out βγάζω. ~ up or together μαζεύω. ~ along, ~ a living περνώ στενά, φυτοζωώ. ~ through (a narrow place) περνώ ξυστά, (just succeed) μόλις τα καταφέρνω.

scrape s. (escapade, iron.) κατόρθωμα n. get into a ~ βρίσκω τον μπελά μου.

scraper s. ξύστρα f.

scrap-heap s. throw on the ~ πετώ στα σκουπίδια.

scraping s. (act) ξύσιμο n. ~s ξύσματα n.pl. there was bowing and ~ δούλεψε υπόκλιση.

scrappy a. (ill organized) ακατάρτιστος.

scratch v.t. γδέρνω, ξύνω. (one's body) ξύνω. (of cat, bramble) γρατσουνίζω. ~ oneself ξύνομαι. (scrape the ground) σκαλίζω. ~ out ξύνω.

scratch s. γδάρσιμο n. (by cat, bramble) γρατσουνιά f. start from ~ αρχίζω από το μηδέν. come up to ~ ανταποκρίνομαι επαξίως.

scratch a. πρόχειρος. we had a ~ meal φάγαμε εκ των ενόντων.

scratchy a. (bad quality) της κακιάς ώρας. (that scratches) που γρατσουνίζει.

scrawl s. ορνιθοσκαλίσματα n.pl. (v.t.) γράφω βιαστικά.

scrawny a. πετσί και κόκκαλο. (person only) ξερακιανός.

scream v. ξεφωνίζω. ~ with laughter ξεκαρδίζομαι στα γέλια. (s.) ξεφωνητό n. (fam.) a perfect ~ κάτι το ξεκαρδιστικό.

screech v. σκληρίζω. (s.) στριγγλιά f.

screen s. προπέτασμα n. (furniture) παραβάν n. (eccl.) εικονοστάσιο n. (cinema) οθόνη f. ~ play ταινία f. (sieve) κόσκινο n.

screen v. (hide) προκαλύπτω, κρύβω. (protect) προστατεύω. (sift) κοσκινίζω. ~ off χωρίζω με παραβάν. (investigate) he is being ~ed ερευνούν το παρελθόν του.

screw s. βίδα f. (propeller) έλικας f. (pay) μισθός m. ~ of paper χωνάκι n. turn of the ~ σφίξιμο της βίδας. he has a ~ loose (fam.) του έστριψε η βίδα. put the ~s on (person) ασκώ πίεση σε.

screw v.t. (also ~ down, in, on) βιδώνω. ~ up (tighten) σφίγγω, (wrinkle) ζαρώνω. ~ up one's courage κάνω κουράγιο. have one's head ~ed on the right way τα έχω τετρακόσια. ~ed (fam.) μεθυσμένος. (v.i.) βιδώνω.

screwdriver s. κατσαβίδι n.

screwy a. (fam.) παλαβός.

scribble v. (write hastily) γράφω στο γό-

νατο. (s.) ορνιθοσκαλίσματα n.pl. ~r s. (iron.) μουντζουροχάρτης m.

scribe s. καλαμαράς m.

scrimmage s. συμπλοκή f. (v.i.) συνωστίζομαι.

scrip s. (fin.) σκριπ n.

script s. (writing) γραφή f. (text) γραπτά n.pl.

scripture s. Αγία Γραφή.

scrofulous a. χοιραδικός. (fig.) βρώμικος.

scroll s. κύλινδρος m. (ornament) έλικας f.

scrotum s. όσχεο n.

scroungle v. (steal) βουτώ, (cadge) διακονεύω. ~er s. σελέμης m. be a ~er ζω με την αμάκα.

scrub v. τρίβω, σφουγγαρίζω. ~bing s. τρίψιμο n., σφουγγάρισμα n.

scrub s. (brushwood) χαμόκλαδα n.pl. ~by a. (ground) θαμνώδης, (chin) αξύριστος.

scruff s. by the ~ of the neck από το σβέρκο.

scruffly a. βρώμικος. ~iness s. βρώμικη κατάσταση.

scrum s. συνωστισμός m.

scruple s. ηθικός δισταγμός, ηθικός ενδοιασμός. (v.) διστάζω.

scrupulous a. (person) ευσυνείδητος, (honesty) σχολαστικός, υπερβολικός. ~ly adv. (very) εις άκρον. ~ness s. ευσυνειδησία f.

scrutin|ize v. εξονυχίζω, εξελέγχω, εξετάζω λεπτομερώς. ~y s. λεπτομερής εξέταση.

scud v. τρέχω, (naut.) ουριοδρομώ.

scuffle s. συμπλοκή f. (v.) συμπλέκομαι.

scull v. κουπί n.

scullion s. λαντζιέρης m.

sculpt, ~ure v. γλύφω, σκαλίζω, λαξεύω. (make statue of) κάνω το άγαλμα (with gen.). ~ured a. γλυπτός.

sculptor s. γλύπτης m.

sculptur|e s. (art) γλυπτική f. (piece) γλυπτό n. ~al a. γλυπτικός.

scum s. (lit.) αφρός m. (lit. & fig.) απόβρασμα n.

scupper v. (also be ~ed, lit & fig.) βουλιάζω. (s.) ~s μπούνια n.pl.

scurf s. πιτυρίδα f.

scurril|ous a. (person) βωμολόχος, (words) υβριστικός. ~ity s. βωμολοχία f.

scurry v. τρέχω.

scurvy s. (med.) σκορβούτο n. (a.) ~y trick βρωμοδουλειά f. ~y fellow βρωμόμουτρο n. ~ily adv. αισχρά.

scuttle v.t. (sink) βουλιάζω. (v.i.) (run) τρέχω, (make off) το σκάω.

scythe s. δρεπάνι n. (v.) θερίζω.

sea s. θάλασσα f. Aegean S~ Αιγαίον Πέλαγος. Black S~ Εύξεινος Πόντος, Μαύρη Θάλασσα. heavy ~ θαλασσοταραχή f. calm ~ μπουνάτσα f. put out to ~ ανοίγομαι εις το πέλαγος, αποπλέω. (journey) by ~ διά θαλάσσης. (situated) by the ~ παραθαλάσσιος. on the open ~ στα ανοικτά. beyond the ~s υπερπόντιος. under the ~ υποβρύχιος. command of the ~s θαλασσοκρατία f. (fig.) feel all at ~ αμηχανώ.

sea a. θαλάσσιος, θαλασσινός. **~-breeze** s. μπάτης m. **~board** s. ακτή f. **~-captain** s. καπετάνιος m. **~-dog** s. θαλασσόλυκος m. **~faring** a. ναυτικός, θαλασσινός. **~-fight** s. ναυμαχία f. **~-front** s. παραλία f. **~-girt** a. περιβρεχόμενος από θάλασσα. **~-going** a. ποντοπόρος. **~-green** a. γαλαζοπράσινος. **~gull** s. γλάρος m. **~-legs** s. find one's ~legs συνηθίζω στην κίνηση του πλοίου. **~-level** s. στάθμη της θάλασσας. **~-loving** a. θαλασσοχαρής. **~-man** s. ναύτης m. good ~man καλός ναυτικός. **~manship** s. ναυτική τέχνη. **~-monster** s. κήτος n. **~-plane** s. υδροπλάνο n. **~-power** s. ναυτική δύναμη. **~-scape** s. θαλασσογραφία f. **~-shell** s. κοχύλι n. **~-shore** s. παραλία f., γιαλός m., ακρογιάλι n. by or along the ~-shore παραλιακός. **~-sick** a. feel or be ~-sick με έχει πιάσει η θάλασσα. **~-sickness** s. ναυτία f. **~-side** a. παραθαλάσσιος, παραλιακός. (s.) to (or at) the ~side (κοντά) στη θάλασσα. **~-urchin** s. αχινός m. **~-weed** s. φύκια n.pl. **~-worthiness** s. πλοϊμότητα f. **~-worthy** a. πλόιμος.

seal s. (animal) φώκια f.

seal s. (on document) σφραγίδα f. under ~ ενσφράγιστος. (v.) σφραγίζω, (stop up) βουλλώνω. ~ off αποκλείω. his fate is ~ed η τύχη του εκρίθη. **~ing-wax** s. βουλοκέρι n.

seam s. ραφή f. (in rock) φλέβα f. ~ed (marked) σημαδεμένος. **~stress** s. ράφτρα f.

seamy a. (fig.) ~ side κακή πλευρά.

sear v. καίω, (cauterize) καυτηριάζω. (fig.) σκληραίνω.

search v.t. & i. ψάχνω. ~ high and low for αναζητώ. (conduct ~ of) ερευνώ. (s.) αναζήτηση f., έρευνα f. (of ships) νηοψία f. **~er** s. ερευνητής m.

searching s. ψάξιμο n. (a.) (detailed) λεπτομερής, (penetrating) διαπεραστικός.

searchlight s. προβολεύς m.

season v.t. (flavour) αρτύνω, (mature) ωριμάζω, (temper) μετριάζω. **~ed** a. (experienced) ψημένος, (wood) τραβηγμένος. **~ing** s. άρτυμα n.

season s. εποχή f. (social) σαιζόν f. (space of time) διάστημα n. in ~ στην εποχή του, (timely) επίκαιρος. **~al** a. εποχιακός.

seasonable a. (timely) επίκαιρος, (weather) της εποχής.

season-ticket s. διαρκές εισιτήριο.

seat s. κάθισμα n. (bench) παγκάκι n. (place in vehicle, theatre, etc.) θέση f. (of office-holder, institution) έδρα f. (of trouble, disease) εστία f. (of body, garment) πισινός m. country ~ πύργος m.

seat v. (place) καθίζω, (have room for) παίρνω, χωρώ. ~ oneself κάθομαι. **~ed** a. καθιστός. **~ing** s. θέσεις f.pl.

secede v. αποχωρώ.

secesssion s. αποχώρηση f.

seclude v. απομονώνω. **~d** a. (quiet) ήσυχος, (out-of-the-way) αποτραβηγμένος.

seclusion s. (quiet) ησυχία f. (isolation) απομόνωση f.

second a. δεύτερος. on March 2nd στις δύο Μαρτίου. ~ name επώνυμο n. he is ~ to none δεν του βγαίνει κανείς. come off ~ best βγαίνω χαμένος. play ~ fiddle παίζω δευτερεύοντα ρόλο. have ~ thoughts μετανοιώνω. on ~ thoughts κατόπιν ωριμωτέρας σκέψεως. ~ in command υπαρχηγός m., (naut.) ύπαρχος m. in one's ~ childhood ξεμωραμένος. he has ~ sight μπορεί να προβλέπει το μέλλον. **~ly** adv. δεύτερον.

second s. (time) δευτερόλεπτο n. (in duel) μάρτυς m. (in boxing) βοηθός m.

second v.t. (a proposal) υποστηρίζω. (an officer, official) αποσπώ. ~ed αποσπασμένος. **~ment** s. απόσπαση f.

secondary a. δευτερεύων. ~ education μέση εκπαίδευση.

second-hand a. μεταχειρισμένος, (also at ~) από δεύτερο χέρι. ~ dealer παλαιοπώλης m.

second-rate a. (article) δεύτερο πράμα, (person) δεύτερης σειράς, μετριότητα f.

secrecy s. μυστικότητα f.

secret a. μυστικός, (hidden) κρυφός. ~ police ασφάλεια f. **~ly** adv. μυστικά, κρυφά.

secret s. μυστικό n. open ~ κοινό μυστικό. make no ~ of δεν κρύβω. be in the ~ είμαι μπασμένος στο μυστικό. in ~ see ~ly.

secretariat s. γραμματεία f.

secretary s. γραμματεύς m.f. private ~ ιδιαίτερος m., ιδιαιτέρα f. ~ general γενικός γραμματεύς. S~ of State υπουργός m.

secret|e v. (hide) κρύβω (med.) εκκρίνω. **~ion** s. έκκριση f.

secretive a. κρυψίνους. **~ness** s. κρυψίνοια f.

sect s. αίρεση f. **~arian** a. αιρετικός.

section s. (part) τμήμα n. (geom.) τομή f.

~al *a.* (*fitted together*) λυόμενος. **~al** interests ιδιαίτερα συμφέροντα.

sector *s.* τομεύς *m.*

secular *a.* κοσμικός. (*music*) μη εκκλησιαστικός.

secure *a.* ασφαλής. (*guaranteed*) εξασφαλισμένος. **~ly** *adv.* ασφαλώς.

secure *v.* (*make firm or safe*) ασφαλίζω, (*loan, etc.*) εξασφαλίζω. (*obtain*) εξασφαλίζω.

securit|y *a.* ασφάλεια *f.* (*guarantee*) εγγύηση *f.* stand **~y** for εγγυώμαι για. **~ies** (*fin.*) τίτλοι *m.pl.*

sedate *a.* σοβαρός, μετρημένος.

sedative *a.* ηρεμιστικός.

sedentary *a.* καθιστικός.

sediment *s.* ίζημα *n.*

sediti|on *s.* στάση *f.* **~ous** *a.* στασιαστικός.

seduc|e *v.* αποπλανώ. **~tion** *s.* αποπλάνηση *f.* (*charm*) σαγήνη *f.* **~tive** *a.* σαγηνευτικός.

sedulous *a.* ακούραστος.

see *v.* βλέπω. (*understand*) καταλαβαίνω. (*escort*) συνοδεύω. I saw her home την πήγα σπίτι. **~** for oneself διαπιστώνω ο ίδιος. there was nothing to be **~n** τίποτα δεν φαινόταν. I saw him come out of the house τον είδα να βγαίνει (*or* που έβγαινε) από το σπίτι. **~** to *or* about (*deal with*) κοιτάζω *or* φροντίζω για. **~** (to it) that... (*make sure*) κοιτάζω *or* προσέχω να. it reamins to be **~n** θα δούμε. **~ing** that... μια και, έχοντας υπ' όψιν ότι. **~** (*person*) off ξεπροβοδίζω. **~** (*person*) through (*aid*) (παρα)στέκομαι (*with* σε *or* gen.). **~** (*business*) through φέρω εις πέρας. **~** through (*a fraud*) μυρίζομαι.

see *s.* (*eccl.*) επισκοπή *f.*, έδρα *f.*

seed *s.* σπόρος *m.* (*offspring*) σπέρμα *n.* have gone to **~** (*fig.*) έχω κατάντια. (*v.i.*) (*make* **~s**) σποριάζω. **~ling** *s.* φιντάνι από σπόρο.

seedy *a.* (*ill*) αδιάθετος, (*shabby*) άθλιος.

seek *v.* ζητώ, (*try*) προσπαθώ, (*have as ambition*) επιδιώκω. sought after περιζήτητος.

seem *v.* φαίνομαι. we **~** to have arrived φαίνεται πως φτάσαμε.

seeming *a.* φαινομενικός. **~ly** *adv.* φαινομενικά.

seem|ly *a.* καθώς πρέπει, ευπρεπής. **~liness** *s.* ευπρέπεια *f.*

seep *v.* διαρρέω. **~age** *s.* διαρροή *f.*

seer *s.* μάντις *m.*

see-saw *s.* τραμπάλα *f.* (*v.*) τραμπαλίζομαι.

seethe *v.* βράζω.

segment *s.* τμήμα *n.* (*v.*) διαιρώ σε τμήματα.

segregat|e *v.* διαχωρίζω, (*isolate*) απομο-

νώνω. **~ion** *s.* διαχωρισμός *m.* απομόνωση *f.*

seine *s.* τράτα *f.*

seism|ic *a.* σεισμικός. **~ology** *s.* σεισμολογία *f.*

seize *v.t.* (*grab*) αρπάζω. (*take possession of*) καταλαμβάνω, (*impound*) κατάσχω. (*v.i.*) **~** up μαγγώνω.

seizure *s.* αρπαγή *f.* (*capture*) σύλληψη *f.* (*distraint*) κατάσχεση *f.* (*med.*) προσβολή *f.*

seldom *adv.* σπανίως.

select *v.* επιλέγω, διαλέγω. (*a.*) επίλεκτος, διαλεχτός. **~ion** *s.* (*choosing*) επιλογή *f.* (**~ed** *examples*) εκλογή *f.* **~or** *s.* επιλογεύς *m.*

selectiv|e *a.* (*choosy*) εκλεκτικός. (*mil.*) **~e** service ειδικές κατηγορίες. **~ity** *s.* (*of radio*) διαχωριστικότητα *f.*

self, selves *pron.* 1. (*intensive*) ο ίδιος. I my**~** εγώ ο ίδιος, we our**~** εμείς οι ίδιοι. he did it by him**~** (*unaided*) το 'κανε (από) μόνος του. she lives by her**~** ζει μοναχή της. he's his old **~** ξαναέγινε ο παλιός. 2. (*reflexive*) (i) τον εαυτό (*with* gen. *personal pron.*), they wanted it for them**~** το ήθελαν για τον εαυτό τους. take care of your**~** πρόσεχε τον εαυτό σου. she said to her**~** είπε μέσα της (*or* με το νου της). he talks to him**~** μιλάει με τον εαυτό του, ομιλεί καθ' εαυτόν. between our**~** μεταξύ μας. (ii) (*by passive v.*) defend one**~** αμύνομαι, cross one**~** σταυροκοπιέμαι. aren't you ashamed of your**~**? δεν ντρέπεσαι; (*by active v.*) I enjoyed my**~** διασκέδασα, he hurt him**~** χτύπησε. (iii) (*by* αυτο-.) admire one**~** αυτοθαυμάζομαι.

self- αυτ-, αυτο-.

self-centred *a.* εγωκεντρικός.

self-confidence *s.* αυτοπεποίθηση *f.*

self-conscious *a.* (*shy*) ντροπαλός.

self-contained *a.* αυτοτελής.

self-control *s.* αυτοκυριαρχία *f.*

self-deception *s.* αυταπάτη *f.*

self-defence *s.* in **~** εν αμύνη.

self-denial *s.* αυταπάρνηση *f.*

self-determination *s.* αυτοδιάθεση *f.*

self-discipline *s.* αυτοπειθαρχία *f.*

self-employed *a.* επαγγελματίας *m.*

self-esteem *s.* αυταρέσκεια *f.*

self-evident *a.* αυτόδηλος.

self-government *s.* αυτοδιοίκηση *f.*

self-important *a.* σπουδαιοφανής.

self-indulgent *a.* ρέπων προς τάς απολαύσεις.

self-interest *s.* ιδιοτέλεια *f.*, συμφεροντολογία *f.*

selfish *a.* εγωιστής *m.*, εγωίστρια *f.* **~ness** *s.* εγωισμός *m.*

self-knowledge *s.* το γνώθι σαυτόν.

selfless *a.* ανιδιοτελής.
self-made *a.* αυτοδημιούργητος.
self-possessed *a.* (*bold*) θαρρετός, (*cool*) ψύχραιμος.
self-preservation *s.* αυτοσυντήρηση *f.*
self-reliant *a.* be ~ στηρίζομαι στον εαυτό μου.
self-respect *s.* αυτοσεβασμός *m.*
self-righteous *a.* φαρισαίος.
self-sacrifice *s.* αυτοθυσία *f.*
self-satisfied *a.* αυτάρεσκος.
self-seeking *a.* συμφεροντολόγος. (*s.*) συμφεροντολογία *f.*, εγωισμός *m.*
self-service *s.* αυτοεξυπηρέτηση *f.*
self-starter *s.* (*mech.*) μίζα *f.*
self-styled *a.* αυτοκαλούμενος.
self-sufficient *a.* αυτάρκης.
self-supporting *a.* αυτοσυντήρητος.
self-taught *a.* αυτοδίδακτος.
self-willed *a.* ισχυρογνώμων.
sell *v.t.* πουλώ. ~ off ξεπουλώ. "to be sold" (*notice*) πωλείται. we are sold out ξεπουλήσαμε. everything is sold out όλα εξαντλήθηκαν. I've been sold (*cheated*) μου τη σκάσανε. he got sold up (*by distraint*) του τα βγάλανε στο σφυρί. (*v.i.*) (*be sold*) πουλιέμαι. **-er** *s.* πωλητής *m.*
sell *s.* (*fam., let-down*) απογοήτευση *f.*
selvage *s.* ούγια *f.*
semantics *s.* σημασιολογία *f.*
semblance *s.* εμφάνιση *f.*
semen *s.* σπέρμα *n.*
semicirc‖le *s.* ημικύκλιο *n.* **~ular** *a.* ημικύκλιος.
semicolon *s.* άνω τελεία.
seminal *a.* (*idea, etc.*) γονιμοποιός.
semi-official *a.* ημιεπίσημος.
Semitic *a.* σημιτικός.
semitone *s.* ημιτόνιο *n.*
semivowel *s.* ημίφωνο *n.*
senat‖e *s.* (*university, Roman*) σύγκλητος *f.* (*upper house*) γερουσία *f.* **~or** *s.* γερουσιαστής *m.*
send *v.* στέλνω. ~ (*person*) packing ξαποστέλνω. ~ away διώχνω. ~ down κατεβάζω, (*student*) αποβάλλω. ~ for (*summon*) καλώ, φωνάζω. ~ forth *or* out βγάζω. ~ up ανεβάζω, (*mock*) κοροϊδεύω. ~ on (*letters*) διαβιβάζω. **~-off** *s.* ξεπροβόδισμα *n.*
sender *s.* αποστολεύς *m.*
senil‖e *a.* ξεμωραμένος. **~ity** *s.* γεροντική άνοια.
senior *a.* & *s.* (*in years*) μεγαλύτερος, (*in rank*) ανώτερος. **~ity** *s.* (*of rank*) αρχαιότητα *f.*
sensation *s. see* feeling. ~ of cold αίσθηση του κρύου. I had a ~ of falling είχα την

εντύπωση πως έπεφτα. (*stir*) make a ~ προκαλώ αίσθηση, κάνω ντόρο. go in search of new ~s επιδιώκω νέες συγκινήσεις.
sensational *a.* εντυπωσιακός, (*trial, scandal*) πολύκροτος. (*journal, writer*) που επιδιώκει να κάνει θόρυβο.
sense *s.* (*bodily faculty*) αίσθηση *f.* the five ~s οι πέντε αισθήσεις. (*right mind*) he has taken leave of his ~s δεν είναι στα λογικά του. come to one's ~s συνέρχομαι. ~ of beauty αίσθηση του ωραίου. ~ of injustice αίσθημα της αδικίας. ~ of duty συνείδηση του καθήκοντος. (*practical wisdom*) λογική *f.* talk ~ μιλώ λογικά. there's no ~ in it δεν έχει νόημα. (*meaning*) έννοια *f.*, σημασία *f.* the word has many ~s η λέξη έχει πολλές έννοιες (*or* σημασίες). I can't make ~ of it δεν βγάζω νόημα. (*v.*) αισθάνομαι.
senseless *a.* (*foolish*) ανόητος, (*unconscious*) αναίσθητος.
sensibility *s.* αισθαντικότης *f.*, συναισθηματικός κόσμος.
sensib‖le *a.* (*reasonable*) λογικός, (*person only*) γνωστικός, (*gift, clothes*) πρακτικός. (*perceptible*) αισθητός. be ~le of έχω επίγνωση (*with gen.*). **~ly** *adv.* λογικά, πρακτικά, αισθητά.
sensitive *a.* (*mentally*) ευαίσθητος, (*physically*) ευπαθής. (*of instrument*) λεπτός. **~ly** *adv.* με ευαισθησία. **~ness** *s. see* sensitivity. (*of instrument*) λεπτότητα *f.*
sensitivity *s.* ευαισθησία *f.* (*bodily*) ευπάθεια *f.*
sensual *a.* αισθησιακός, (*person*) φιλήδονος. **~ity** *s.* ηδυπάθεια *f.*
sensuous *a.* αισθησιακός.
sentence *s.* (*law*) καταδίκη *f.* (*gram.*) πρόταση *f.* (*v.*) (*law*) καταδικάζω.
sententious *a.* (*person*) ηθικολόγος *m.* **~ly** *adv.* speak ~ly ηθικολογώ. **~ness** *s.* ηθικολογία *f.*
sentient *a.* ~ beings όντα με αισθήσεις και αντιδράσεις.
sentiment *s.* (*feeling*) αίσθημα *n.* (*collectively*) συναισθήματα *n.pl.* ~s (*views*) απόψεις *f.pl.*
sentimental *a.* αισθηματικός, συναισθηματικός. **~ity** *s.* (συν)αισθηματικότητα *f.*
sentinel *s.* φρουρός *m.*
sentry *s.* φρουρός *m.* **~-box** *s.* σκοπιά *f.*
separable *a.* που χωρίζεται.
separate *v.t.* & *i.* χωρίζω.
separat‖e *a.* χωριστός. (*private, of one's own*) ιδιαίτερος. **~ely** *adv.* χωριστά, χώρια. **~ion** *s.* χωρισμός *m.*
sepsis *s.* σήψη *f.*

September s. Σεπτέμβριος, Σεπτέμβρης m.

septic a. σηπτικός.

septuagenarian s. εβδομηντάρης m.

sepulch|re s. τάφος m. **~ral** a. (voice) σπηλαιώδης.

sequel s. συνέχεια f.

sequence s. διαδοχή f., σειρά f.

sequestered a. αποτραβηγμένος, ήσυχος.

sequestrat|e v. κατάσχω. **~ion** s. κατάσχεση f.

sequin s. (coin) τσεκίνι n. (ornament) πούλια f.

seraglio s. σεράι n.

Serbian a. σερβικός. (person) Σέρβος.

serenade s. σερενάτα f.

seren|e a. γαλήνιος. **~ity** s. γαλήνη f.

serf s. δουλοπάροικος m. **~dom** s. δουλοπαροικία f.

sergeant s. (mil.) λοχίας m. (aero.) σμηνίας m. (gendarmerie) νωματάρχης m. **~major** s. επιλοχίας m.

serial a. ~ number αύξων αριθμός. in ~ form σε συνέχειες. (s.) (newspaper) επιφυλλίδα f. (T.V.) σήριαλ n.

series s. σειρά f.

serious a. σοβαρός. grow ~ (of person) σοβαρεύομαι. **~ly** adv. σοβαρά. take it ~ly το παίρνω στα σοβαρά. **~ness** s. σοβαρότητα f.

sermon s. κήρυγμα n. **~ize** v.i. κάνω κήρυγμα.

serpent s. φίδι n., όφις m. **~ine** a. φιδωτός, οφιοειδής.

serrated a. πριονωτός, οδοντωτός.

serried a. πυκνός.

serum s. ορ(ρ)ός m.

servant s. υπηρέτης m., υπηρέτρια f. (fam.) υπηρεσία f. civil ~ δημόσιος υπάλληλος.

serve v.t. & i. (as servant, official, soldier, etc.) υπηρετώ. (v.t.) (provide with facilities) εξυπηρετώ. (treat, use) μεταχειρίζομαι, it ~s him right καλά να πάθει. (~ food) σερβίρω. (law) ~ a sentence εκτίω ποινή, ~ a summons κοινοποιώ κλήση. (v.i.) (be suitable) χρησιμεύω, κάνω. it didn't ~ δεν έκανε. (at tennis) σερβίρω.

service s. (woking, branch of employment) υπηρεσία f. enter domestic ~ πάω υπηρέτης. (act of assistance) εξυπηρέτηση f. do (person) a ~ εξυπηρετώ. of ~ (a.) εξυπηρετικός. at your ~ στη διάθεσή σας. the three fighting ~s τα τρία όπλα. on active ~ εν ενεργεία. military ~ θητεία f. (church) λειτουργία f. (set of dishes) σερβίτσιο n. (upkeep of car, etc.) συντήρηση f., (fam.) σέρβις n. (tip) φιλοδώρημα n. (transport) a frequent bus ~ πυκνή συγκοινωνία με το λεωφορείο.

(a.) υπηρεσιακός. ~ **area** τομέας υποστήριξης.

servic|e v.t. (or have ~ed) κάνω σέρβις σε. it needs ~ing θέλει σέρβις.

serviceable a. πρακτικός.

serviette s. πετσέτα f.

servil|e a. δουλοπρεπής. **~ity** s. δουλοπρέπεια f.

serving s. (act) σερβίρισμα n. (portion) μερίδα f.

servitude s. δουλεία f.

sesame s. σουσάμι n. (fig.) open ~ κάτι που σου ανοίγει όλες τις πόρτες.

session s. (meeting) συνεδρίαση f. (term, period) σύνοδος f. be in or go into ~ συνεδριάζω.

set s. (of tools) τακίμι n. (of stamps, novels, etc.) σειρά f. toilet ~ σειρά ειδών τουαλέτας. wireless ~ ραδιόφωνο n. (people) smart ~ κοινωνικοί κύκλοι. I don't belong to their ~ δεν είμαι του κόσμου τους. (of person's head, dress, etc.) κόψη f. make a dead ~ at (attack) see set on.

set a. (arranged, fixed) ωρισμένος, καθωρισμένος. (permanent) μόνιμος. all ~ (ready) έτοιμος. be ~ upon (keen to) είμαι αποφασισμένος να.

set v.i. (sun) βασιλεύω, δύω. (limb, fruit, jelly) δένω, (harden) πήζω. (of tide, opinion) στρέφομαι, (of body, character) σχηματίζομαι. he ~ to work βάλθηκε στη δουλειά.

set v.t. (put) βάζω. he ~ them laughing τους έκανε να γελάσουν. (instruct or cause to do sthg.) βάζω. (regulate) ρυθμίζω. (conditions, exam) ορίζω. (example) δίνω, (trap, scene) στήνω, (teeth) σφίγγω, (bone) βάζω στη θέση του. (jewel) δένω, (table) στρώνω. ~ sail αποπλέω. ~ free ελευθερώνω. ~ to music μελοποιώ. ~ to rights επανορθώνω. ~ in order τακτοποιώ. ~ at rest καθησυχάζω. ~ a price on (head of) επικηρύσσω. ~ (up) type στοιχειοθετώ.

set about v.t. (start) αρχίζω. (attack) see set on.

set aside v.t. (money, etc.) βάζω κατά μέρος (or στην μπάντα). (disregard) παραμερίζω, (annul) ακυρώ.

set back v.t. (hinder) εμποδίζω.

set-back s. αναποδιά f.

set down v.t. (passenger) κατεβάζω, (write) γράφω, (attribute) αποδίδω, (consider) θεωρώ.

set forth v.t. εκθέτω. (v.i.) ξεκινώ.

set in v.i. αρχίζω.

set off v.i. ξεκινώ. (v.t.) (explode) ανάβω, (show up) δείχνω. (cause person to do *sthg.*) κάνω να.

set on *v.* (*attack*) επιτίθεμαι εναντίον (*with gen.*). (*instigate*) βάζω.

set out *v.i.* ξεκινώ. (*v.t., present*) εκθέτω, (*arrange*) τακτοποιώ, απλώνω.

set to *v.i.* (*get going: past tense only*) he ~ (*work*) βάλθηκε στη δουλειά, (*eating*) στρώθηκε στο φαΐ.

set-to *s.* have a ~ τσακώνομαι.

set up *v.t.* (*erect*) στήνω, (*provide for*) αποκαθιστώ. (*start*) αρχίζω. ~ house ανοίγω σπίτι. set oneself up as an expert παριστάνω τον ειδικό. (*v.i.*) ~ as a doctor γίνομαι γιατρός. well ~ (*a.*) καλοφτιαγμένος.

set-up *s.* κατάσταση πραγμάτων.

settee *s.* καναπές *m.*

setting *s.* (*environment*) περιβάλλον *n.* (*of sun*) δύση *f.* (*of jewel*) δέσιμο *n.* (*of hair*) μιζαμπλί *n.*

settle *v.i.* (*take up residence*) εγκαθίσταμαι, (*alight*) κάθομαι. (*sink*) κατακαθίζω. (*decide*) αποφασίζω. ~ for (*accept*) δέχομαι. ~ in τακτοποιούμαι. ~ up τακτοποιώ το λογαριασμό. (*v.t.*) (*colonize*) αποικίζω. (*a dispute, affair, bill*) κανονίζω. (*the dust*) κάνω να καθίσει. (*make comfortable*) βολεύω. (*property on person*) γράφω σε.

settle down *v.i.* (*become serious*) στρώνω. (*to task*) στρώνομαι. (*marry, etc.*) νοικοκυρεύομαι. (*get into routine*) συνηθίζω. (*in armchair, etc.*) βολεύομαι. (*grow calm*) καταλαγιάζω. (*become quiet*) ησυχάζω.

settled *a.* (*fixed*) σταθερός, (*paid*) εξοφλημένος.

settlement *s.* (*agreement*) συμφωνία *f.* (*of debt*) εξόφληση *f.* (*colonizing*) αποίκηση *f.* (*colony*) αποικία *f.* (*of refugees: process*) αποκατάσταση *f.*, (*place*) συνοικισμός *m.* marriage ~ προικοσύμφωνο *n.*

settler *s.* άποικος *m.*

seven *num.* επτά, εφτά. ~ hundred επτακόσιοι. **-th** *a.* έβδομος.

seventeen *num.* δεκαεπτά. **-th** *a.* δέκατος έβδομος.

sevently *num.* εβδομήντα. **-ieth** *a.* εβδομηκοστός.

sever *v.t.* (*cut*) κόβω, (*divide*) χωρίζω. (*v.i.*) σπάζω.

several *a.* μερικοί. ~ times κάμποσες φορές. ~ of us μερικοί από μας. they gave their ~ opinions ο καθένας έδωσε τη γνώμη του.

severe *a.* (*person, style*) αυστηρός, (*injury*) σοβαρός, (*weather*) δριμύς. **-ly** *adv.* αυστηρά, σοβαρά.

severity *s.* αυστηρότητα *f.* σοβαρότητα *f.*

δριμύτητα *f.*

sew *v.* (*also* ~ on *or* up) ράβω. **-ing** *s.* (*art*) ραπτική *f.* (*act*) ράψιμο *n.* ~ing machine ραπτομηχανή *f.*

sew|age *s.* νερά των υπονόμων, λύματα *n.pl.* **-er** *s.* υπόνομος *m.f.*

sewn *a.* ραμμένος.

sex *s.* φύλο *n.* **-ual** *a.* γενετήσιος. **-uality** *s.* σεξουαλισμός *m.* **-y** *a.* προκλητικά ελκυστικός.

sextant *s.* εξάς *m.*

sexton *s.* νεωκόρος *m.*

shabby *a.* (*object*) φθαρμένος, (*person*) φτωχοντυμένος. (*behaviour*) πρόστυχος, αισχρός.

shack *s.* καλύβι *n.*

shackle *v.* δένω. (*fig.*) δεσμεύω. **-s** *s.* δεσμά *n.pl.*

shade *s.* σκιά *f.*, ίσκιος *m.* (*of lamp*) αμπαζούρ *n.* put in the ~ (*fig.*) επισκιάζω. (*gradation of colour, opinion, etc.*) απόχρωση *f.* (*ghost*) σκιά *f.* the ~s (*underworld*) άδης *m.* (*fam.*) a ~ better λιγάκι καλύτερα.

shad|e *v.t.* σκιάζω. (*v.i.*) ~e off into πηγαίνω προς (*with acc.*). **-ing** *s.* φωτοσκίαση *f.*

shaded *a.* (*shady*) σκιερός. (*drawing*) με φωτοσκιάσεις.

shadow *s.* σκιά *f.*, ίσκιος *m.* (*of doubt, etc.*) ίχνος *n.* (*fig.*) be under a ~ είμαι υπό δυσμένειαν. (*v.t.*) παρακολουθώ. **--theatre** *s.* καραγκιόζης *m.* **-y** *a.* σκιώδης.

shady *a.* (*giving or in shade*) σκιερός. (*suspect*) ύποπτος. ~ deal (*fam.*) λοβιτούρα *f.*

shaft *s.* (*of spear*) ξύλο *n.* (*of tool*) μανίκι *n.* (*of cart*) τιμόνι *n.* (*of column*) κορμός *m.* (*of well, mine*) φρέαρ *n.* (*of light*) ακτίνα *f.* (*mech.*) άξων *m.* (*arrow & fig.*) βέλος *n.*

shaggy *a.* δασύς και ατίθασος. (*dog*) μαλλιαρός.

shah *s.* σάχης *m.*

shake *v.t.* σείω, κουνώ. (*mop, rug, etc.*) τινάζω, (*bottle*) ταράσσω, κουνώ. (*jolt*) τραντάζω. (*rock, weaken*) κλονίζω. (*shock*) συγκλονίζω. ~ oneself τινάζομαι. ~ hands κάνω χειραψία. ~ one's head (*denial*) γνέφω όχι. ~ off απαλλάσσομαι από, αποτινάσσω. ~ up ανακατεύω, (*rouse*) κουνώ, (*a pillow*) αναφουφουλιάζω. (*v.i.*) κουνιέμαι, (*tremble*) τρέμω. (*be jolted*) τραντάζομαι. (*in explosion, etc.*) σείομαι.

shake *s.* κούνημα *n.* (*of hand*) χειραψία *f.*

shaky *a.* ασταθής.

shall *v.* I ~ write to him θα του γράψω. ~ I write to him? να του γράψω; ~ we go for a walk? πάμε ένα περίπατο; you ~ have your wish θα γίνει το κέφι σου.

shallow *a.* (*lit. & fig.*) ρηχός. (*fig.*) επιπόλαιος, κούφιος. (*s.*) ~s ρηχά *n.pl.*

sham *v.* ~ sickness προσποιούμαι τον άρρωστο. he is ~ming υποκρίνεται.

sham, ~ming *s.* προσποίηση *f.* (*a.*) προσποιητός, ψεύτικος.

shamble *v.* σέρνομαι.

shambles *s.* μακελλειό *n.* (*fig., mess*) κυκεώνας *m.*

shame *s.* (*feeling*) ντροπή *f.*, αιδώς *f.* (*infamy*) αίσχος *n.* feel ~ ντρέπομαι. put to ~, bring ~ on ντροπιάζω. ~ on you! ντροπή σου, δεν ντρέπεσαι! (*v.t.*) ντροπιάζω. be ~d (*into doing sthg.*) φιλοτιμούμαι. ~-faced *a.* ντροπιασμένος. ~ful *a.* ντροπιαστικός, επαίσχυντος. ~fully *adv.* αισχρώς.

shameless *a.* ξεδιάντροπος. ~ly *adv.* αναιδώς. ~ness *s.* ξεδιαντροπιά *f.*

shampoo *v.* λούζω. (*s.*) λούσιμο *n.* (*substance*) σαμπουάν *n.*

shamrock *s.* τριφύλλι *n.*

shank *s.* (*stem*) στέλεχος *n.* (*of anchor*) άτρακτος *m.f.* (*fam.*) ~s (*legs*) κανιά *n.pl.*

shanty *s.* παράγκα *f.* ~-town *s.* παραγκούπολη *f.*

shape *s.* (*configuration*) σχήμα *n.* (*form, guise*) μορφή *f.* in the ~ of a triangle με σχήμα τριγώνου, in human ~ με ανθρώπινη μορφή. give ~ to διαμορφώνω. take ~ διαμορφώνομαι. knock into ~ σουλουπώνω. get out of ~ ξεφορμάρομαι. (*fig.*) be in good ~ πάω καλά.

shape *v.t.* διαμορφώνω, (*mould*) πλάθω. (*v.i.*) ~ well (*fig.*) πάω καλά, προοδεύω.

shapeless *a.* άμορφος. (*ill-shaped*) ασουλούπωτος.

shapely *a.* κομψός, τορνευτός. (*esp. body*) καλλίγραμμος.

share *v.t.* ~ (out) (*as distributor*) μοιράζω. ~ (in) (*as recipient*) μοιράζομαι. (*a person's grief, views, etc.*) συμμερίζομαι. (*have a ~ in*) συμμετέχω.

share *s.* (*portion*) μερίδιο *n.* lion's ~ μερίδα του λέοντος. fair ~ το δικαιούμενον. it fell to my ~ περιήλθε εις το μερίδιόν μου, μου έπεσε. go ~s in μοιράζομαι. *see* share *v.*

share *s.* (*fin.*) μετοχή *f.* ~-holder *s.* μέτοχος *m.f.*

share-cropper *s.* σέμπρος *m.*, κολλήγας *m.* **share-out** *s.* μοιρασιά *f.*

shark *s.* καρχαρίας *m.*

sharp *a.* (*cutting*) κοφτερός. (*pointed*) μυτερός. (*clear-cut*) καθαρός. (*abrupt*) απότομος. (*acute*) οξύς. (*clever*) έξυπνος, ατσίδα *f.* ~ practice απάτη *f.* (*adv.*) look ~!

κουνήσου! at 10 o'clock ~ στις δέκα ακριβώς.

sharp *s.* (*mus.*) δίεση *f.*

sharpen *v.* (*whet*) ακονίζω, (*a pencil*) ξύνω. (*wits, etc.*) οξύνω. ~er *s.* ξύστρα *f.*

sharper *s.* απατεώνας *m.* card-~ χαρτοκλέφτης *m.*

sharply *adv.* (*abruptly*) απότομα, (*distinctly*) καθαρά. ~ness *s.* οξύτητα *f.*

shatter *v.* θραύω, (*fig.*) συντρίβω. ~ing *a.* συντριπτικός.

shave *v.t.* ξυρίζω. (*v.i.*) ξυρίζομαι. (*v.t.*) (*graze*) ξύνω, (*nearly graze*) περνώ ξυστά από. ~ off (*pare*) κόβω ψιλά ψιλά.

shave *s.* ξύρισμα *n.* I had a narrow ~ φτηνά τη γλύτωσα.

shaving-brush *s.* πινέλο του ξυρίσματος.

shavings *s.* ροκανίδια *n.pl.*

shawl *s.* σάλι *n.*

she *pron.* αυτή, εκείνη *f.*

sheaf *s.* (*corn*) δεμάτι *n.* (*papers*) δέσμη *f.*

shear *v.* κουρεύω. shorn κουρεμένος. shorn of (*fig.*) απογυμνωμένος από.

shears *s.* ψαλίδα *f.*

sheath *s.* θηκάρι *n.* ~e *v.* βάζω στη θήκη. (*cover*) επενδύω. ~ed in gold με χρυσή επένδυση.

shed *s.* υπόστεγο *n.*

shed *v.* (*leaves*) ρίχνω, (*tears, blood, light*) χύνω, (*clothes*) βγάζω. (*fig., get rid of*) ξεφορτώνομαι. ~ hair μαδώ. ~ leaves *or* petals φυλλορροώ.

sheen *s.* γυαλάδα *f.*

sheep *s.* πρόβατο *n.*, πρόβατα *pl.* ~'s milk πρόβιο γάλα. ~'s head (*to eat*) αρνίσιο κεφαλάκι. ~-dog *s.* τσοπανόσκυλο *n.* ~-fold *s.* στάνη *f.* ~ish *a.* αμήχανος. ~-skin *s.* προβιά *f.*

sheer *a.* (*absolute*) καθαρός. (*fine*) διαφανής. (*steep*) απόκρημνος, κατακόρυφος. by his ~ will-power με τη θέλησή του και μόνο. (*adv.*) κατακόρυφα.

sheer *v.* ~ off στρίβω, (*of vessel*) παρεκκλίνω.

sheet *s.* (*bed*) σεντόνι *n.* (*paper, metal*) φύλλο *n.* (*ice*) στρώμα *n.* (*glass*) κομμάτι *n.* (*flames*) παραπέτασμα *n.* (*rain*) in ~s με το τουλούμι. ~-iron *s.* λαμαρίνες *f.pl.*

sheik *s.* σεΐχης *m.*

shekels *s.* (*fam.*) παράδες *m.pl.*

shelf *s.* ράφι *n.*

shell *s.* (*of nut, crustacean*) κέλυφος *n.* (*of nut, egg*) τσόφλι *n.* (*of tortoise, snail*) καβούκι *n.* (*sea-~*) αχιβάδα *f.* (*of building*) σκελετός *m.* (*mil.*) οβίδα *f.* come out of one's ~ βγαίνω από το καβούκι μου.

shell *v.t.* (*open*) καθαρίζω, ξεφλουδίζω, (*mil.*) βομβαρδίζω. ~ out (*fam.*) σκάζω.

shellfish s. θαλασσινά n.pl.

shelter s. καταφύγιο n. (*also abstract*) άσυλο n. (*v.t.*) προφυλάσσω. (*v.i.*) προφυλάσσομαι. **~ed** a. προφυλαγμένος.

shelv|e v.t. (*put off*) βάζω στο χρονοντούλαπο. (*v.i., slope*) κατηφορίζω. **~ing** a. κατηφορικός.

shepherd s. βοσκός m., τσοπάνης m. (*eccl.*) ποιμήν m. (*v.t.*) οδηγώ. **~ess** s. βοσκοπούλα f.

sherbet s. σερμπέτι n.

shibboleth s. απαρχαιωμένο δόγμα.

shield s. ασπίδα f. (*in machinery, etc.*) προφυλακτήρ m. (*v.*) προστατεύω.

shift v.t. μετατοπίζω. (*v.i.*) μετατοπίζομαι. v.t. & i. (*change*) αλλάζω. ~ for oneself τα βγάζω πέρα μόνος μου. ~ one's ground αλλάζω θέση.

shift s. (*change*) αλλαγή f. (*relay*) βάρδια f. make ~ τα βολεύω.

shiftless a. ανεπρόκοπος.

shifty a. ύπουλος, ύποπτος.

shilling s. σελίνι n.

shilly-shally v. αμφιταλαντεύομαι.

shimmer v. λαμπυρίζω.

shin s. καλάμι n.

shin|e v.i. λάμπω, (*be shiny*) γυαλίζω. (*give light*) φέγγω. (*fig., excel*) διαπρέπω. ~e (a light) on φωτίζω. (*v.t.*) (*polish*) γυαλίζω. (*s.*) γυαλάδα f. **~ing** a. λαμπερός. γυαλιστερός.

shingle s. βότσαλα n.pl. (*of wood*) ταβανοσάνιδο n.

shingles s. (*med.*) έρπης ζωστήρ.

shiny a. γυαλιστερός. the trousers are ~ το παντελόνι γυαλίζει.

ship s. πλοίο n., καράβι n., σκάφος n. on ~-board επί του πλοίου. ~'s biscuit γαλέτα f.

ship v.t. (*send*) αποστέλλω, (*load*) φορτώνω. (*v.t. & i.*) (*take or go aboard*) μπαρκάρω.

shipbuild|er s. ναυπηγός m. **~ing** s. (*science*) ναυπηγική f. (*industry*) ναυπηγεία n.pl.

shipload s. φορτίο n.

shipment s. (*despatch*) αποστολή f. (*load*) φορτίο n. (*loading*) φόρτωση f.

shipowner s. εφοπλιστής m.

shipper s. (*sender*) αποστολεύς m. (*importer*) εισαγωγεύς m.

shipping s. ναυτιλία f. (*a.*) ναυτιλιακός.

shipshape a. νοικοκυρεμένος.

shipwreck s. ναυάγιο n. be ~ed ναυαγώ. ~ed sailor ναυαγός m.

shipyard s. ναυπηγείο n.

shire s. κομητεία f.

shirk v. αποφεύγω. **~er** s. φυγόπονος a.

shirt s. πουκάμισο n. in one's ~-sleeves

χωρίς σακκάκι. **~y** a. (*fam.*) θυμωμένος.

shit s. (*fam.*) σκατά n.pl. (*v.i.*) χέζω.

shiver v.t. (*break*) θραύω. (*v.i.*) θραύομαι.

shiver s. (*also ~ing*) ρίγος n. (*v.i.*) ριγώ, τουρτουρίζω. **~y** a. feel ~y αισθάνομαι ρίγος.

shoal s. (*shallows*) ρηχά n.pl. (*of fish*) κοπάδι n.

shock s. κλονισμός m. (*earthquake*) δόνηση f. (*nervous*) σοκ n. get a ~ (*surprise*) τρομάζω, (*electric*) τινάζομαι. ~absorber αμορτισέρ n.

shock v.t. (*distress*) συνταράσσω, συγκλονίζω. (*scandalize*) σοκάρω. it ~ed me (*disapproval*) μου προξένησε χειρίστη εντύπωση.

shock s. ~ of hair φουντωτά μαλλιά. **~-headed** a. αναμαλλιασμένος.

shocking a. συνταρακτικός. (*scandalous*) σκανδαλώδης. (*fam.*) φρικτός. **~ly** adv. (*fam.*) φρικτά.

shod a. see shoe, well-shod.

shoddy a. πρόστυχος, ψευτοκαμωμένος, δεύτερο πράμα.

shoe s. παπούτσι n. (*court*) γόβα f. (*patent*) λουστρίνι n. (*woman's lace-up*) σκαρπίνι n. (*wooden*) τσόκαρο n. I'd not like to be in his ~s δεν θα ήθελα να ήμουνα στη θέση του, ούτε ψύλλος στον κόρφο του. (*v.t.*) (*a horse*) πεταλώνω. be well shod φορώ καλά παπούτσια. **~black** s. λούστρος m. **~horn** s. κόκκαλο n. **~lace** s. κορδόνι n. **~maker** s. παπουτσής m. **~string** s. live on a ~string περνώ με το τίποτα.

shoo v. διώχνω.

shoot s. (*plant*) βλαστάρι n. (*v.i.*) βλαστάνω.

shoot v.i. (*move quickly*) τρέχω. ~ ahead ξεπετιέμαι μπροστά, (*surpass others*) αφήνω πίσω τους άλλους. ~ past περνώ σα βολίδα. ~ up αναπηδώ, (*in stature*) ξεπετιέμαι. ~ out πετιέμαι. go ~ing πάω κυνήγι. ~ at πυροβολώ εναντίον (*with gen.*). he shot at me μου έρριξε. (*v.t.*) (*throw, fire*) ρίχνω. (*kill with gun*) (*person*) τουφεκίζω, (*game*) χτυπώ. he shot me in the leg μου έρριξε στο πόδι. ~ down καταρρίπτω. ~ing pain σουβλιά f. ~ing star διάττων αστήρ.

shooting s. (*in war*) πυροβολισμός m. (*sport*) κυνήγι n. (*killing, execution*) τ(ο)υφεκισμός m.

shop s. μαγαζί n., κατάστημα n. (*often as derivative, e.g.*) butcher's ~ χασάπικο or κρεοπωλείο n., barber's ~ κουρείο n.

shop v.i. κάνω ψώνια. go ~ping πάω για ψώνια. go ~ping for food πάω να ψωνίσω τρόφιμα. **~per** s. αγοραστής m.

shop-assistant s. υπάλληλος m.f., πωλητής m., πωλήτρια f.

shop-keeper s. καταστηματάρχης m.

shoplifter s. κλέφτης m.

shopping s. (purchases) ψώνια n.pl. see shop v.

shop-window s. βιτρίνα f.

shore s. (sea) ακτή f. (lake) όχθη f. on ~ στην ξηρά. go on ~ αποβιβάζομαι.

shore v. ~ up αντιστυλώνω, υποστηρίζω. (s.) αντιστύλι n.

short a. (duration) μικρός, σύντομος. (height, length) κοντός. (prosody, wave) βραχύς. (curt) απότομος. for a ~ time για λίγο, για λίγη ώρα. the ~est way o συντομότερος δρόμος. in ~ με λίγα λόγια. take a ~ cut through the vineyard κόβω μέσ' από το αμπέλι. give ~ weight κλέβω στο ζύγι. we are £2 ~ μας λείπουν δύο λίρες. I am (or have run) ~ of oil έμεινα από λάδι. go ~ of στερούμαι, fall ~ of είμαι κατώτερος (both with gen.). ~ circuit βραχυκύκλωμα n. ~ sight μυωπία f. have a ~ memory έχω αδύνατη μνήμη.

short adv. (suddenly) απότομα. cut ~ διακόπτω. the oil is running ~ τελειώνει το λάδι. ~ of (except) πλην, εκτός (with gen.). it is little ~ of a miracle λίγο απέχει από το θαύμα.

shortage s. έλλειψη f.

short-change v.t. κλέβω στα ρέστα.

short-circuit v.t. (fig.) παρακάμπτω.

shortcoming s. ελάττωμα n.

short-dated a. (fin.) βραχυπρόθεσμος.

shorten v.t. (length, height) κονταίνω, (duration) συντομεύω. (v.i., of days) μικραίνω. ~ing s. κόντεμα n. συντόμευση f.

shorthand s. στενογραφία f. ~ typist στενοδακτυλογράφος m.f.

short-handed a. be ~ μου λείπουν εργατικά χέρια.

short-lived a. εφήμερος.

shortly adv. (soon) προσεχώς. ~ afterwards λίγο ύστερα. (briefly) με λίγα λόγια. (curtly) κοφτά.

short-range a. (forecast) για το άμεσο μέλλον. (gun) μικρού βεληνεκούς. (plane, etc.) μικρής αποστάσεως.

short-sighted a. (person, lit. & fig.) μύωψ, κοντόφθαλμος. (policy, etc.) μυωπικός, κοντόφθαλμος.

short-sleeved a. κοντομάνικος.

short-tempered a. ευέξαπτος.

short-term a. (temporary) προσωρινός. (fin.) βραχυπρόθεσμος.

short-winded a. be ~ έχω κοντή αναπνοή.

shot s. (discharge of gun) πυροβολισμός m. pistol ~ πιστολιά f. rifle ~ τουφεκιά f.

within rifle-~ εντός βολής τουφεκιού. small ~ σκάγια n.pl. (in games) ριξιά f. (attempt) δοκιμή f. (person) σκοπευτής m. like a ~ σαν αστραπή. **~gun** s. κυνηγετικό όπλο.

should v. I ~ go (if I had the time) θα πήγαινα, (ought to) πρέπει να πάω. why ~ I go? γιατί να πάω; (if) ~ he refuse αν (or σε περίπτωση που θα) αρνηθεί.

shoulder s. ώμος m. (fig.) straight from the ~ ορθά-κοφτά. give him the cold ~ του στρέφω τα νώτα. (v.t.) (lit. & fig.) επωμίζομαι. **~blade** s. ωμοπλάτη f. **~strap** s. (mil.) αορτήρ m. (woman's) μπρετέλλα f.

shout v. φωνάζω. ~ down προγκίζω. (s.) φωνή f. **~ing** s. φωνές f.pl.

shove v. σπρώχνω, (put) χώνω. (s.) σπρωξιά f.

shovel s. φτυάρι n. (v.) φτυαρίζω.

show s. (exhibition) έκθεση f. they are on ~ εκτίθενται. (theatre) θέατρο n. ~ business επιχειρήσεις θεαμάτων. (ostentation) επίδειξη f. a brave ~ μία εντυπωσιακή εμφάνιση. make a ~ of repentance κάνω πως μετανοώ. make a ~ κάνω φιγούρα. make a poor ~ κάνω φτωχή εντύπωση. on a ~ of hands δι' ανατάσεως των χειρών. (fam., affair, business) υπόθεση f. run the ~ διευθύνω, έχω το πρόσταγμα. steal the ~ προσελκύω το μεγαλύτερο ενθουσιασμό. good ~! μπράβο! **~case** s. βιτρίνα f. **~down** s. αναμέτρηση f. **~room** s. αίθουσα εκθέσεων.

show v.t. δείχνω, (as exhibit) εκθέτω. ~ oneself (be present) εμφανίζομαι. ~ oneself to be αποδεικνύομαι. (v.i.) (be visible) φαίνομαι.

show in v.t. οδηγώ μέσα.

show off v.t. επιδεικνύω. (v.t.) κάνω επίδειξη.

show-off s. φιγουρατζής m.

show out v.t. οδηγώ έξω.

show up v.t. αποκαλύπτω. (v.i.) διακρίνομαι. (fam., turn up) κάνω την εμφάνισή μου.

shower v.t. (make wet) καταβρέχω. (throw) ρίχνω. we were ~ed with invitations μας ήρθαν βροχή οι προσκλήσεις. (v.i.) ~ (down) πέφτω βροχή.

shower s. ελαφρή μπόρα. (fig., stones, gifts) βροχή f. **~bath** s. ντους n. **~y** a. βροχερός.

showing s. make a poor ~ κάνω φτωχή εντύπωση. on his own ~ από τα λεγόμενά του.

showman s. (fig.) ρεκλαμαδόρος m.

showy a. (appearance) θεαματικός, (behaviour) επιδεικτικός.

shred s. κουρέλι n. cut *or* tear to ~s κάνω κομμάτια, κομματιάζω, (*fig.*) κουρελιάζω. it is worn to ~s έχει γίνει κουρέλια. not a ~ ούτε ίχνος. (*v.*) κατακομματιάζω.

shrew s. (*scold*) στρίγγλα f. **~ish** a. στρίγγλα f. **~ishness** s. στριγγλιά f.

shrewd a. έξυπνος. (*well aimed*) εύστοχος. I have a ~ suspicion πολύ υποπτεύομαι. **~ly** adv. έξυπνα, εύστοχα. **~ness** s. εξυπνάδα f.

shriek v. ξεφωνίζω. (*s.*) ξεφωνητό n.

shrift s. give (*person*) short ~ (*treat severely*) μεταχειρίζομαι αυστηρά.

shrill a. διαπεραστικός. ~ cry στριγγλιά f.

shrimp s. γαρίδα f.

shrine s. (*pagan*) τέμενος n. (*Christian*) ιερός τόπος. (*wayside, etc.*) εικονοστάσι(ον) n.

shrink v.i. (*of cloth*) μπαίνω, μαζεύω. (*draw back*) (απο)τραβιέμαι, διστάζω. **~age** s. μάζεμα n.

shrive v. εξομολογώ.

shrivel v.t. & i. ζαρώνω.

shroud s. σάβανο n. (*v.*) σαβανώνω, (*fig.*) καλύπτω.

Shrove Tuesday s. Καθαρή Τρίτη.

shrub s. θάμνος m.

shrug v.t. ~ one's shoulders σηκώνω τους ώμους. ~ off αψηφώ. (*s.*) ~ of the shoulders σήκωμα των ώμων. ~ of despair κίνηση απελπισίας.

shudder v. ριγώ. (*s.*) ρίγος n.

shuffle v. ~ (one's feet) σέρνω τα πόδια μου. (*at cards*) ανακατεύω. (*s.*) σύρσιμο των ποδιών.

shun v. αποφεύγω.

shunt v. μετακινώ.

shut v.t. (*also* ~ down, in, off, up) κλείνω. ~ out εμποδίζω, κλείνω απ έξω. ~ (*person*) up (*cause to be silent*) κάνω να το βουλλώσει. (*v.i.*) κλείνω. ~ up (*fam.*) το βουλλώνω. ~ up! βούλλωσέ το! σκάσε!

shut a. (*fastened*) κλειστός, κλεισμένος. (*enclosed*) κλεισμένος.

shutter s. παραθυρόφυλλο n., παντζούρι n. drop-~ ρολό n. (*slatted*) γρίλλια f. **~ed** a. με κλειστά τα παντζούρια.

shutting s. κλείσιμο n.

shuttle s. σαΐτα f. (*v.i.*) (*fig.*) πηγαινοέρχομαι.

shy v.i. (*of horse*) σκιάζομαι. (*v.t.*) (*throw*) ρίχνω. (*s.*) (*throw*) ριξιά f. have a ~ (*attempt*) δοκιμάζω.

shy a. ντροπαλός. feel ~ ντρέπομαι. fight ~ of αποφεύγω. be ~ (*of doing sthg.*) διστάζω (να), (*of animals: lie low*) λουφάζω. **~ly** adv. ντροπαλά. **~ness** s. ντροπαλοσύνη f.

sibilant a. συριστικός.

sibling s. αμφιθαλής a.

sick a. (*ill*) άρρωστος, fall ~ αρρωσταίνω. ~-pay επίδομα ασθενείας. be ~ (*vomit*) κάνω εμετό. (*fig.*) be ~ of βαριέμαι, βαργεστίζω.

sicken v.t. αηδιάζω. (*v.i.*) αρρωσταίνω, be ~ing for κλωσώ. **~ing** a. αηδιαστικός. how ~ing! είναι αηδία.

sickle s. δρεπάνι n.

sickly a. (*delicate*) αρρωστιάρης. (*smile*) ασθενικός, (*taste, etc.*) αναγουλιαστικός.

sickness s. αρρώστια f. (*vomiting*) εμετός m.

side s. (*general*) πλευρά f. (*team*) ομάδα f. winning ~ νικητές, losing ~ χαμένοι. (*of coin, question*) όψη f. (*of cloth*) right ~ η καλή, wrong ~ η ανάποδη. he is on the tall ~ είναι μάλλον ψηλός. ~ by ~ πλάι-πλάι. by *or* at the ~ of δίπλα σε. from all ~s από παντού. on this ~ of the street στην από 'δω μεριά (*or* πλευρά) του δρόμου. take ~s μεροληπτώ. I take his ~ παίρνω το μέρος του. I take him on one ~ τον παίρνω κατά μέρος. put on one ~ βάζω κατά μέρος. (*fam.*) put on ~ κορδώνομαι.

side v. ~ with παίρνω το μέρος (*with gen.*).

side a. πλαϊνός. **~-board** s. μπουφές m. **~-effect** s. παρενέργεια f. **~-issue** s. δευτερεύον ζήτημα. **~-light** s. throw a ~light on ρίχνω λίγο φως σε. **~-line** s. πάρεργον n. **~-long** a. λοξός. **~-road** s. πάροδος f. **~-show** s. (*fig.*) πάρεργος ασχολία. **~-step** v. (*fig.*) παρακάμπτω. **~-track** v. (*fig.*) (*person*) εκτρέπω, (*matter*) παραμερίζω. **~-walk** s. πεζοδρόμιο n. **~-ways** adv. πλαγίως.

siding s. γραμμή ελιγμών.

sidle v. πλησιάζω δειλά.

siege s. πολιορκία f. lay ~ to πολιορκώ.

siesta s. μεσημεριανός ύπνος.

sieve s. κόσκινο n. (*v.*) κοσκινίζω.

sift v. κοσκινίζω. (*fig.*) εξετάζω.

sigh s. αναστεναγμός m. (*v.*) αναστενάζω, (*of wind*) μουγκρίζω. ~ for λαχταρώ, νοσταλγώ. **~ing** s. αναστεναγμοί m.pl. μουγγρητό n.

sight s. (*faculty*) όραση f. (*thing seen*) θέα f. (*spectacle*) θέαμα n. the ~s τα αξιοθέατα. catch ~ of διακρίνω. lose ~ of χάνω από τα μάτια μου. come into ~ γίνομαι ορατός. vanish from ~ χάνομαι. in ~ εν όψει. out of ~ κρυμμένος. at first ~ εκ πρώτης όψεως. love at first ~ κεραυνοβόλος έρωτας (*v.t.*) διακρίνω.

sightless a. αόμματος.

sightseeing v. go ~ επισκέπτομαι τα αξιοθέατα.

sign s. (nod, gesture) νεύμα, νόημα n. make a ~ κάνω νόημα, γνέφω. (of rain, progress, wealth, etc.) ένδειξη f. (of recognition, danger, Cross, times, life etc.) σημείο n. (trace) ίχνος n. (password) σύνθημα n. (of shop, etc.) επιγραφή f., πινακίδα f. (traffic) σήμα n. (math.) σύμβολον n.
sign v.t. (letter, etc.) υπογράφω. (v.i.) (make a ~) γνέφω. ~ on or up (v.t.) προσλαμβάνω, (v.i.) προσλαμβάνομαι, (enrol) εγγράφομαι.
signal s. σήμα n. give the ~ δίνω το σήμα, κάνω το σινιάλο. (mil.) corps of ~s σώμα διαβιβάσεων. (v.i.) κάνω σήμα. (v.t., announce) αγγέλλω. ~ize v. διακρίνω.
signal-box s. φυλάκιο σηματοδοσίας.
signalling s. σηματοδοσία f.
signator|y s. ο υπογράψας. ~ies οι συμβαλλόμενοι.
signature s. υπογραφή f.
sign-board s. επιγραφή f.
signet-ring s. δαχτυλίδι σφραγίδα.
significance s. σημασία f.
significant a. (important) σημαντικός. very ~ μεγάλης σημασίας. ~ of ενδεικτικός (with gen.).
signification s. νόημα n.
signify v.t. σημαίνω, (intimate) εκφράζω, (show) δείχνω. (v.i.) it doesn't ~ δεν έχει σημασία.
signpost s. πινακίδα f.
silence s. σιωπή, σιγή f. (v.) (a person) επιβάλλω σιωπή σε, κάνω να σωπάσει. (in argument) αποστομώνω.
silent a. σιωπηλός, (not noisy) αθόρυβος. (film) βουβός. keep or become ~ σωπαίνω. ~ly adv. σιωπηλά, αθόρυβα.
silhouette s. σιλουέττα f.
silk s. μετάξι n. (material) μεταξωτό n. (a.) μεταξωτός. ~en, ~y a. μεταξένιος. ~worm s. μεταξοσκώληξ m.
sill s. ποδιά f.
sill|y a. ανόητος, κουτός. ~iness s. ανοησία f., κουταμάρα f.
silo s. σιλό n.
silt s. ιλύς f. (v.i.) ~ up γεμίζω λάσπη.
silver s. άργυρος m., ασήμι n. (ware) ασημικά n.pl. (a.) αργυρούς, ασημένιος. ~ paper ασημόχαρτο n. ~-plated a. επάργυρος. ~y a. ασημένιος, (laugh) αργυρόηχος.
simian a. πιθηκοειδής.
similar a. παρόμοιος. ~ity s. ομοιότητα f.
simile s. παρομοίωση f.
simmer v. σιγοβράζω.
simper v. χαμογελώ κουτά.
simple s. βότανο n.
simple a. απλός. (artless) αφελής, (unadorned) απέριττος, (frugal) λιτός, (easy)

εύκολος. ~ton s. μωρόπιστος a.
simplicity s. απλότητα f.
simplif|y v. απλοποιώ. ~ication s. απλοποίηση f.
simply adv. απλά, απέριττα, λιτά. (only) μόνο, (just) απλώς, απλούστατα. (indeed) αληθινά, ασφαλώς.
simulat|e v. προσποιούμαι. (imitate) απομιμούμαι. ~ion s. προσποίηση f. απομίμηση f.
simultaneity n. ταυτόχρονο n.
simultaneous a. ταυτόχρονος. ~ly adv. ταυτοχρόνως.
sin s. αμαρτία f. (~ful act) αμάρτημα n. (v.i.) αμαρτάνω. ~ful a. αμαρτωλός. ~ner s. αμαρτωλός a.
since adv. έκτοτε, από τότε. (later on) ακολούθως. many years ~ (ago) πριν από πολλά χρόνια. (conj.) (ever) ~ I came here από τότε που ήρθα εδώ. it is a year ~ I saw him έχω να τον δω ένα χρόνο. (prep.) από (with acc.). we have lived in this house ~ 1975 ζούμε σ' αυτό το σπίτι από το 1975. (causal) αφού.
sincer|e a. ειλικρινής. ~ely adv. ειλικρινά. ~ity s. ειλικρίνεια f.
sinecure s. αργομισθία f.
sine qua non s. εκ των ων ουκ άνευ.
sinew s. νεύρο n. ~y a. νευρώδης.
sing v. άδω, τραγουδώ. (of birds) κελαηδώ. ~ praises of εξυμνώ. ~ us a song δεν μας τραγουδάς κάτι;
singe v. τσουρουφλίζω.
singer s. τραγουδιστής m., τραγουδίστρια f.
singing s. τραγούδι n. (of birds) κελάηδημα n.
single a. (not double) μονός. (sole) μόνος. μοναδικός. not a ~ book ούτε ενα βιβλίο. ~ ticket απλό εισιτήριο. (unmarried) ανύπαντρος. in ~ file εφ' ενός ζυγού. (v.) ~ out ξεχωρίζω. ~-breasted a. μονόπετος. ~-handed a. μόνος μου. ~-minded a. αφοσιωμένος στο σκοπό μου.
singleness s. ~ of purpose αφοσίωση σ' ένα σκοπό και μόνο.
singly adv. ένας-ένας, χωριστά.
singlet s. φανέλα f.
singsong s. ομαδικό τραγούδι παρέας. (a.) in a ~ manner τραγουδιστά.
singular a. (strange) παράξενος, (rare) σπάνιος. (gram.) ενικός. ~ly adv. εξαιρετικά.
sinister a. σκοτεινός, απειλητικός.
sink s. νεροχύτης m.
sink v.i. βυθίζομαι, βουλιάζω. (of ground, dregs) κατακαθίζω. (give way) κάμπτομαι. (fall) πέφτω, (into coma, decay) περιπίπτω. (of star) γέρνω, δύω. ~ in (of liquid) ποτίζω, (get stuck) κολλώ. (I told

him but) it didn't ~ in δεν το συνέλαβε.
(*v.t.*) βυθίζω, βουλιάζω. (*bury*) θάβω. (*a well*) ανοίγω, (*one's teeth*) χώνω, (*money*) ρίχνω, (*differences*) παραμερίζω. (*fam.*) I'm sunk! κάηκα! **~ing** *s.* βύθιση *f.* **~ing-fund** *s.* χρεολυτικόν ταμείον.

sinuous *a.* (*body*) φιδίσιος, (*road*) φιδωτός.

sinusitis *s.* ιγμορίτις *f.*

sip *s.* γουλιά *f.* (*v.*) πίνω γουλιά-γουλιά.

siphon *s.* σιφόνι *n.* (*v.*) μεταγγίζω διά σίφωνος.

sir *s.* (*vocative*) κύριε.

sire *s.* πατέρας *m.* (*animal*) επιβήτωρ *m.* (*vocative*) Μεγαλειότατε. (*v.*) γεννώ.

siren *s.* σειρήνα *f.*

sirloin *s.* κοντραφιλέτο *n.*

sirocco *s.* σορόκος *m.*

sissy *a. & s.* θηλυπρεπής.

sister *s.* αδελφή *f.*

sister-in-law *s.* (*brother's wife*) νύφη *f.* (*spouse's sister*) κουνιάδα *f.* (*husband's brother's wife*) συννυφάδα *f.*

sit *v.t.* καθίζω. (*v.i.*) κάθομαι. be ~ting είμαι καθισμένος, (*be in session*) συνεδριάζω. (*fam.*) be ~ting pretty την έχω καλά. (*of clothes*) it ~s well πέφτει καλά. ~ still κάθομαι ήσυχος. ~ tight δεν το κουνάω.

sit back *v.i.* (*in armchair*) ξαπλώνω. (*do nothing*) κάθομαι.

sit down *v.i.* κάθομαι (κάτω). ~ under (*insult*) καταπίνω.

sit for *v.t.* (*represent*) αντιπροσωπεύω. (*exam*) δίνω, (*portrait*) ποζάρω για.

sit in *v.i.* ~ for αντικαθιστώ. ~ on παρακολουθώ ως παρατηρητής.

sit on *v.t.* (*eggs*) κλωσσώ, (*committee*) μετέχω (*with gen.*). (*fam., squash*) (*a proposal*) απορρίπτω, (*a person*) αποπαίρνω.

sit out *v.t.* (*stay to end of*) παρακολουθώ μέχρι τέλους.

sit up *v.i.* (*at night*) ξενυχτώ. (*after lying flat*) ανακάθομαι. ~ and beg στέκομαι σούζα. (*fam.*) make (*person*) ~ ταράζω.

site *s.* τοποθεσία *f.*, θέση *f.* building ~ οικόπεδο *n.*, (*with work in progress*) γιαπί *n.* on ~ επί τόπου. (*v.*) τοποθετώ.

sitter *s.* (*artist's*) μοντέλο *f.* (*fam., easy job*) παιχνίδι *n.*

sitting *s.* (*meeting*) συνεδρίαση *f.* (*for portrait*) ποζάρισμα *n.* at a ~ (*one go*) μονορούφι. **~-room** *s.* σαλόνι *n.*, καθιστικό *n.*

situate *v.t.* τοποθετώ. ~d (*topographically*) κείμενος. be ~d βρίσκομαι. (*of person*) be awkwardly ~d βρίσκομαι σε δύσκολη θέση. whereabouts is it ~d? κατά πού πέφτει;

situation *s.* (*location*) τοποθεσία *f.*, θέση *f.* (*state of affairs*) κατάσταση *f.*, θέση *f.*

(*job*) θέση *f.*

six *num.* εξ, έξη. ~ hundred εξακόσιοι. (*fam.*) at ~es and sevens άνω κάτω. **~th** *a.* έκτος.

sixteen *num.* δεκαέξι. **~th** *a.* δέκατος έκτος.

sixt|y *num.* εξήντα. **~ieth** *a.* εξηκοστός.

size *s.* μέγεθος *n.* (*dimensions*) διαστάσεις *f.pl.* (*of clothing*) αριθμός *m.*, νούμερο *n.* (*v.*) ~ up εκτιμώ.

size *s.* (*for glazing*) κόλλα *f.*

siz(e)able *a.* μεγαλούτσικος.

sizzle *v.i.* τσιτσιρίζω.

skat|e *s.* παγοπέδιλο *n.*, πατίνι *n.* (*v.*) πατινάρω. **~ing** *s.* πατινάζ *n.*

skein *s.* κούκλα *f.*, μάτσο *n.*

skeleton *s.* σκελετός *m.* ~ key αντικλείδι *n.* (*fig.*) ~ in the cupboard κρυφή ντροπή της οικογένειας.

sketch *s.* σκαρίφημα *n.*, σκίτσο *n.* (*v.*) σκιαγραφώ, σκιτσάρω. **~y** *a.* (*rough*) πρόχειρος, (*incomplete*) ελλιπής.

skew *a.* λοξός, στραβός.

skewer *s.* σουβλάκι *n.* (*v.*) σουβλίζω.

ski *s.* σκι *n.* (*v.*) κάνω σκι. **~ing** *s.* σκι *n.*

skid *v.* ντεραπάρω. (*s.*) ντεραπάρισμα *n.*

skiff *s.* λέμβος *f.*

skilful *a.* επιδέξιος, μάστορας *m.* **~ly** *adv.* με επιδεξιότητα.

skill *s.* επιδεξιότητα *f.*, (*a* ~) τέχνη *f.* **~ed** *a.* έμπειρος. ~ed workman ειδικευμένος εργάτης.

skim *v.t.* (*milk, etc.*) ξαφρίζω. (*surface, etc.*) περνώ ξυστά από, ~ through (*book*) ξεφυλλίζω. ~med milk αποβουτυρωμένο γάλα.

skimp *v.t.* τσιγγουνεύομαι, λυπάμαι. (*v.i.*) we shall have to ~ πρέπει να σφιχτούμε.

skin *s.* πετσί *n.* (*esp. human*) δέρμα *n.* (*of fruit*) φλούδα *f.* (*crust*) κρούστα *f.* (*vessel*) ασκί *n.* thick-~ned χοντρόπετσος, thin-~ned μυγιάγγιχτος. (*worn*) next to the ~ κατάσαρκα. fear for one's ~ φοβάμαι για το πετσί (*or* το τομάρι) μου. I caught the train by the ~ of my teeth μόλις και μετά βίας πρόφτασα (*or* παρά τρίχα να χάσω) το τραίνο. **~-deep** *a.* ξώπετσος. **~-flint** *s.* τσιγγούνης *a.* **~ny** *a.* κοκκαλιάρης. **~-tight** *a.* κολλητός.

skin *v.* γδέρνω. keep your eyes ~ned τα μάτια σου δεκατέσσερα!

skip *v.i.* χοροπηδώ. (*with rope*) πηδώ σχοινάκι. (*v.i. & t.*) (*move quickly, omit*) πηδώ.

skipper *s.* καπετάνιος *m.*

skirmish *s.* αψιμαχία *f.* (*v.*) ακροβολίζομαι.

skirt *s.* φούστα *f.* (*fam.*) bit of ~ θηλυκό *n.*

skirt *v.* (*follow edge of*) πάω γύρω γύρω από. **~ing** *s.* πασαμέντο *n.*

skit s. σατιρικό νούμερο.

skittish a. ζωηρός, παιχνιδιάρης.

skittles s. τσούνια n.pl.

skulk v. (lie low) λουφάζω, (lie in wait) παραμονεύω.

skull s. κρανίο n. (as emblem) νεκροκεφαλή f. ~-cap s. καλότα f.

skunk s. (fur) σκονξ n. (fam.) βρωμόσκυλο n.

sky s. ουρανός m. praise to the skies ανεβάζω στα ουράνια. ~-blue a. ουρανής. ~lark s. κορυδαλλός m. (v.i.) παίζω. ~light s. φεγγίτης m. ~line s. ορίζοντας m. ~scraper s. ουρανοξύστης m.

slab s. πλάκα f.

slack a. (loose) χαλαρός, (sluggish) αδρανής, (lazy) τεμπέλης, (negligent) αμελής. ~ season νεκρή εποχή. business is ~ υπάρχει νέκρα (or κεσάτι) στην αγορά. ~ness s. χαλαρότητα f. τεμπελιά f.

slack v.i. τεμπελιάζω. (v.t.) ~ (off) (loosen) λασκάρω. ~er s. τεμπέλης a. ~ing s. τεμπελιά f.

slacken v.t. & i. (loosen) λασκάρω, (lessen) λιγοστεύω, κόβω. (v.i., abate) κοπάζω. ~ing s. λασκάρισμα n. λιγόστεμα n. κάμψη f.

slag s. σκουριά f.

slain a. the ~ οι σκοτωμένοι.

slake v. σβήνω.

slam v. (shut) κλείνω βιαίως. ~ down βροντώ.

slander v. συκοφαντώ, δυσφημώ. (s.) συκοφαντία f. δυσφήμηση f. ~er s. συκοφάντης m. ~ous a. συκοφαντικός.

slang s. σλανγκ n. (v.t.) βρίζω.

slant s. λόξα f. on the ~ λοξά. (point of view) άποψη f.

slant v.i. (slope) γέρνω, κλίνω. (fall obliquely) πέφτω λοξά. (v.t.) (send obliquely) ρίχνω λοξά, (news, etc.) διαστρέφω. ~ing a. λοξός.

slap s. χτύπημα n. (on face) χαστούκι n. (on head) καρπαζιά f. (v.) χτυπώ. (on face) μπατσίζω, (on head) καρπαζώνω.

slap-dash a. (person) τσαπατσούλης, (work) τσαπατσούλικος.

slap-up a. πρώτης τάξεως, σπουδαίος.

slash s. (wound) κοψιά f. (slit) σχισμή f. (with whip) καμτσικιά f.

slash v. (with blade) χαράσσω. (slit) σχίζω. (lash, lit. & fig.) μαστιγώνω. (reduce) περικόπτω. ~ed a. σχιστός. ~ing a. (of censure) δριμύς.

slat s. πήχη f. (of shutter) γρίλια f., περσίδα f.

slate s. σχιστόλιθος m. (of roof, scholar's) πλάκα f. (fig.) clean the ~ διαγράφω τα περασμένα.

slate v.t. (censure) επικρίνω.

slattern s. κατσιβέλα f.

slaughter s. σφαγή f. (v.) σφάζω. ~-house s. σφαγείο n.

Slav a. σλαβικός. (person) Σλάβος.

slave s. δούλος m., σκλάβος m. (v.i.) (also ~ away) σκοτώνομαι. ~-traffic s. δουλεμπόριο n. white ~-traffic σωματεμπορία f.

slaver v.i. σαλιάζω. (s.) σάλιο n.

slavery s. δουλεία f., σκλαβιά f.

slavish a. (condition) δουλικός, (imitation) τυφλός. ~ly adv. τυφλά.

slay v. σφάζω. ~er s. φονεύς, φονιάς m.

sleazy a. βρώμικος.

sled, ~ge s. έλκηθρο n.

sledge-hammer s. βαριά f.

sleek a. γυαλιστερός, στιλπνός. (of person's appearance) καλοζωισμένος.

sleep s. ύπνος m. go to ~ αποκοιμούμαι. drop off to ~ τον παίρνω. put to ~ κοιμίζω. (v.) κοιμούμαι, κοιμάμαι.

sleeper s. κοιμώμενος m. (of railway track) τραβέρσα f.

sleeping-car s. κλινάμαξα f.

sleeping-pill s. υπνωτικό χάπι.

sleepless a. ~ night λευκή νύχτα. ~ness s. αϋπνία f.

sleep-walk|er s. υπνοβάτης m. ~ing s. υπνοβασία f.

sleep|y a. νυσταγμένος. feel ~y νυστάζω. ~ily adv. νυσταλέως. ~iness s. νύστα f., νυσταγμός m. (inordinate desire for sleep) υπνηλία f.

sleet s. χιονόνερο n.

sleeve s. μανίκι n. (fig.) have (surprise, etc.) up one's ~ έχω εν εφεδρεία.

sleigh s. έλκηθρο n.

sleight-of-hand s. ταχυδακτυλουργία f.

slender s. λεπτός, λιγνός. (scanty) ισχνός. (hope) αμυδρός.

sleuth s. λαγωνικό n.

slew v.i. στρίβω βιαίως.

slice s. φέτα f. (v.) κόβω (σε) φέτες.

slick s. ~ of oil κηλίδα πετρελαίου.

slick a. (dextrous) επιδέξιος, (person only) επιτήδειος. (rapid) γρήγορος. ~ly adv. επιτηδείως.

slide v.t. & i. γλιστρώ, (push, glide) τσουλάω (v.t.) (slip, put) γλιστρώ, χώνω. let things ~ αφήνω τα πράγματα κι ό,τι βγει.

slide s. (act) γλίστρημα n. (icy track) τσουλήθρα f. (chute) τσουλίστρα f. (photo) διαφάνεια f. (microscope) πλάκα f. ~-rule s. λογαριθμικός κανών.

sliding a. (door) συρτός, συρόμενος. (scale) κινητός. (mech.) ολισθαίνων.

slight *a.* (*slender*) αδύνατος, (*small in degree*) ελαφρός. ~est παραμικρός, at the ~est thing με το παραμικρό. not in the ~est καθόλου. **~ly** *adv.* ελαφρώς, λίγο.

slight *v.t.* αμελώ. (*s.*) προσβολή *f.* **~ingly** *adv.* περιφρονητικά.

slim *a.* λεπτός, λιγνός. (*scanty*) ισχνός. (*hope*) ελάχιστος.

slim *v.i.* αδυνατίζω. **~ming** *s.* αδυνάτισμα *n.*

slim|e *s.* γλίτσα *f.* **~y** *a.* γλοιώδης, (*person*) σιχαμερός.

sling *s.* (*weapon*) σφενδόνη *f.* (*for hoisting*) περιλάβειον *n.*, σαμπάνιο *n.* (*for arm*) κούνια *f.* (*over shoulder*) αορτήρ *m.* (*v.t.*) (*throw*) εκσφενδονίζω, (*hang*) κρεμώ. ~ out πετώ έξω.

slink *v.* γλιστρώ. ~ off φεύγω στη ζούλα. **~y** *a.* (*gait*) φιδίσιος, (*dress*) κολλητός.

slip *s.* (*fall*) γλίστρημα *n.*, (*also fig.*) ολίσθημα *n.* (*mistake*) λάθος *n.* (*of tongue*) παραδρομή *f.* (*small piece*) κομματάκι *n.* ~ of a girl κοριτσόπουλο *n.* (*petticoat*) μεσοφόρι *n.* (*drawers*) σλιπ *n.* ~s (*theatre*) παρασκήνια *n.pl.*, (*naut.*) see slipway. give (*person*) the ~ ξεφεύγω από.

slip *v.i.* (*fall, move easily*) γλιστρώ. (*run, pop*) πετιέμαι. the knife ~ped το μαχαίρι ξέφυγε. he let it ~ out (*revealed it*) του ξέφυγε. let ~ (*lose*) χάνω. ~ up κάνω λάθος. ~ by (*of time*) κυλώ.

slip *v.t.* (*put*) γλιστρώ. (*of clothes*) ~ on φορώ, ~ off βγάζω. (*escape*) it ~ped my mind μου διέφυγε. the dog ~ped its collar ο σκύλος έβγαλε τη λαιμαριά του. **~knot** *s.* συρτοθηλιά *f.* **~way** *s.* βάζια *n.pl.*

slipper *s.* παντόφλα *f.*

slippery *a.* γλιστερός, (*also fig.*) ολισθηρός. (*fam.*) ~ customer μάρκα *f.*

slipshod *a.* τσαπατσούλικος.

slit *s.* σχισμή *f.* (*v.t. & i.*) σχίζω. (*v.t.*) (*throat*) κόβω. ~ open ανοίγω. (*a.*) σχιστός.

slither *v.* γλιστρώ.

sliver *s.* σχίζα *f.* (*of food*) φελί *n.*

slobber *v.* σαλιάζω.

slog *v.t.* κοπανάω. (*v.i.*) ~ away δουλεύω σκληρά.

slogan *s.* (*political*) σύνθημα *n.* (*catch-word*) σλόγκαν *n.*

sloop *s.* σλουπ *n.*

slop *v.t.* χύνω. (*v.i.*) ξεχειλίζω. ~ about τσαλαβουτώ.

slop *s.* (*fig., nonsense*) σάλια *n.pl.* ~s (*waste*) βρωμόνερα *n.pl.*, (*in teacup*) κατακάθια *n.pl.*, (*diet*) υγρή τροφή.

slope *s.* κλίση *f.* (*side of hill*) πλαγιά *f.* (*uphill*) ανήφορος *m.* (*downhill*) κατήφορος *m.*

slope *v.i.* (*lean*) κλίνω, γέρνω. (*go uphill*) ανηφορίζω, (*go downhill*) κατηφορίζω. (*of sun*) γέρνω. (*mil.*) ~ arms! επ' ώμου αρμ! (*fam.*) ~ off το σκάω.

sloping *a.* (*not level*) επικλινής, (*not vertical*) κεκλιμένος, λοξός. ~ handwriting πλάγια γραφή.

sloppy *a.* (*full of puddles*) με λακκούβες γεμάτες νερά. ~ food νερομπούλι *n.* (*slovenly*) τσαπατσούλικος. (*insipid*) γλυκανάλατος.

slops *s.* see slop *s.*

slosh *v.t.* (*throw*) ρίχνω. (*hit*) I ~ed him one του κοπάνησα μια. (*v.i.*) ~ about τσαλαβουτώ. (*fam.*) ~ed τύφλα στο μεθύσι. (*s.*) see slush.

slot *s.* σχισμή *f.*, χαραμάδα *f.*, τρύπα *f.* **~ted** *a.* με εγκοπές. **~machine** *s.* κερματοδέκτης *m.*

sloth *s.* νωθρότητα *f.* **~ful** *a.* νωθρός.

slouch *v.* (*walk*) σέρνομαι. ~ed in a chair σωριασμένος σε μια καρέκλα.

slough *s.* (*swamp*) βαλτοτόπι *n.* (*fig.*) ~ of despond άβυσσος της απελπισίας.

slough *s.* (*snake's*) φιδοπουκάμισο *n.* (*v.t.*) (*also* ~ off) αποβάλλω.

slovenly *a.* (*work*) τσαπατσούλικος, (*person*) τσαπατσούλης. (*in habits*) ατημέλητος.

slow *a.* αργός, βραδύς. (*~-moving*) αργοκίνητος. (*business, etc.*) αδρανής. (*tedious*) πληκτικός. be ~ αργώ, (*of clock*) πάω πίσω. in ~ motion στο ρελαντί. **~coach** *s.* αργοκίνητο καράβι. **~ness** *s.* αργοπορία *f.*, βραδύτητα *f.* **~witted** *a.* βραδύνους.

slow *adv.* αργά, βραδέως. go ~ πηγαίνω σιγά, (*delay*) καθυστερώ.

slow *v.* (*also* ~ down *or* up) (*v.t.*) επιβραδύνω. (*v.i.*) κόβω ταχύτητα.

slug *s.* γυμνοσάλιαγκος *m.*

sluggard *s.* τεμπελχανάς *m.*

sluggish *a.* νωθρός. (*river, pulse*) αργός. ~ liver ανεπάρκεια του ήπατος. **~ly** *adv.* νωθρά, αργά.

sluice *s.* υδροφράκτης *m.* **~-gate** βάνα *f.* (*v.*) (*send water over*) περιχύνω.

slum *s.* βρώμικη φτωχική συνοικία.

slumber *s.* ύπνος *m.* (*v.*) κοιμούμαι.

slump *v.i.* (*of person*) σωριάζομαι (*fin.*) πέφτω. (*s.*) κάμψη *f.*

slur *v.t.* (*speech*) δεν προφέρω καθαρά. (*mus.*) ενώνω. ~ over αντιπαρέρχομαι. (*s.*) στίγμα *n.* (*mus.*) σύζευξη *f.*

slush *s.* λασπόχιονο *n.* (*fig.*) σαχλαμάρες *f.pl.* **~y** *a.* λασπερός, (*fig.*) σαχλός.

slut *s.* τσούλα *f.*

sly *a.* πανούργος, πονηρός. on the ~ κρυφά. **~ly** *adv.* πονηρά. **~ness** *s.* πονηρία *f.*

smack *v.t.* χτυπώ, (*on face*) μπατσίζω.

(*one's lips*) πλαταγίζω. (*v.i.*) ~ of μυρίζω.
smack *s.* χτύπημα *n.* (*on face*) μπάτσος *m.*
(*on head*) καρπαζιά *f.* (*fig.*) ~ in the eye
ψυχρολουσία *f.* (*fam.*) have a ~ at δοκιμάζω (να).
smacking *s.* get a ~ τρώω ξύλο.
small *a.* μικρός. make *or* get ~er μικραίνω.
~ change ψιλά *n.pl.* a ~ eater λιγόφαγος *a.*
a ~ shopkeeper μικρέμπορος *m.* not the
~est chance ούτε η παραμικρή ελπίδα. feel
~ αισθάνομαι ταπεινωμένος. ~**holding** *s.*
μικρό αγροτεμάχιο. ~**ish** *a.* μικρούτσικος.
~**-minded** *a.* μικροπρεπής. ~**ness** *s.*
μικρότητα *f.* ~**pox** *s.* βλογιά *f.* ~**time** *a.*
(*fam.*) ασήμαντος.
small *s.* ~ of the back η μέση μου. (*fam.*) ~s
εσώρουχα *n.pl.*
smarmy *a.* γλοιώδης.
smart *v.i.* (*physically*) τσούζω, (*mentally*)
υποφέρω. I'll make him ~ for it θα τον
κάνω να πληρώσει για αυτό.
smart *a.* (*brisk*) γοργός, (*severe*) δυνατός,
(*clever*) έξυπνος. (*in appearance, dress*)
κομψός. look ~! βιάσου! ~**ly** *adv.* γοργά,
κομψά. ~**ness** *s.* εξυπνάδα *f.* κομψότητα
f.
smarten *v.t.* (*pace*) επιταχύνω. ~ up (*v.t.*)
φρεσκάρω, (*v.i.*) (*become smarter*)
κομψεύω. (*wash and brush up, etc.*)
φρεσκάρομαι.
smash *v.t.* (*break*) σπάζω, κάνω κομμάτια.
(*hit*) χτυπώ. (*v.i.*) (*break*) σπάζω, γίνομαι
κομμάτια. ~ into προσκρούω σε, πέφτω
επάνω σε. ~**ing** *a.* (*fam.*) περίφημος.
smash *s.* (*noise*) I heard a ~ άκουσα
σπασίματα. (*collision*) τρακάρισμα *n.*,
σύγκρουση *f.* (*blow*) χτύπημα *n.* (*fin.*)
κραχ *n.*, go ~ φαλίρω. ~ hit (*fam.*) μεγάλη
επιτυχία.
smattering *s.* πασάλειμμα *n.* get a ~ of
πασαλείβομαι σε.
smear *v.* πασαλείβω, (*smudge*) μουντζου-
ρώνω. (*fig., defame*) δυσφημώ. (*s.*) κηλίδα *f.* (*fig.*) συκοφαντία *f.*
smell *s.* (*sense*) όσφρηση *f.* (*odour*) οσμή *f.*,
μυρωδιά *f.* take a ~ at μυρίζομαι. there is a
~ of coffee μυρίζει καφέ. bad ~ δυσοσμία
f., βρώμα *f.*
smell *v.i.* (*also* ~ of) μυρίζω. it ~s good/of
paint μυρίζει ωραία/μπογιά. can fish ~?
έχουν τα ψάρια όσφρηση; (*v.t.*) μυρίζω.
come and ~ the roses έλα να μυρίσεις τα
τριαντάφυλλα. can you ~ gas? σου μύρισε
γκάζι; (*scent, lit. & fig.*) μυρίζομαι,
οσφραίνομαι. ~ out (*fig.*) ανακαλύπτω.
smelly *a.* δυσώδης, βρωμερός. be ~ βρωμώ.
smelt *v.* εκκαμινεύω. ~**ing** *s.* εκκαμίνευση *f.*
smile *s.* χαμόγελο *n.* μειδίαμα *n.* (*v.*)

χαμογελώ, μειδιώ. ~**ing** *a.* χαμογελαστός.
smirch *v.* κηλιδώνω.
smirk *v.* χαμογελώ κουτά.
smite *v.* (*hit*) χτυπώ, (*fig.*) πλήττω, (*of
conscience*) τύπτω. be ~ten (*with desire,
etc.*) με πιάνει, κατέχομαι από, (*with
blindness*) τυφλώνομαι, (*with love*) είμαι
τσιμπημένος.
smith *s.* σιδηρουργός *m.* ~**y** *s.* σιδηρουργείο
n.
smithereens *s.* σμπαράλια *n.pl.*
smock *s.* μπλούζα *f.*
smog *s.* νέφος *n.*
smoke *s.* καπνός *m.* (*v.*) καπνίζω. ~ out
βγάζω. ~**-bomb** *s.* καπνογόνος βόμβα.
~**-screen** *s.* προπέτασμα καπνού.
~**-stack** *s.* φουγάρο *n.*
smoker *s.* καπνιστής *m.*
smoking *s.* κάπνισμα *n.*
smoky *a.* γεμάτος καπνούς. ~ chimney
καμινάδα που καπνίζει.
smooth *v.* λειαίνω, (*stroke*) στρώνω. ~
down *or* away *or* over εξομαλύνω. ~ the
way προλειαίνω το δρόμο.
smooth *a.* (*surface*) λείος, (*ground, events*)
ομαλός, (*sea*) γαλήνιος, (*wine*) γλυκόπιοτος. (*plausible*) γαλίφης. ~**ly** *adv.* ομαλά,
κανονικά. ~**ness** *s.* ομαλότητα *f.*
smother *v.* (*cover*) σκεπάζω, (*stifle*) καταπνίγω ~ed in mud γεμάτος λάσπες.
smoulder *v.i.* σιγοκαίω. (*fig.*) υποβόσκω,
(*of person*) βράζω μέσα μου.
smudge *s.* μουντζούρα *f.* (*v.t.*) μουντζου-
ρώνω.
smug *a.* αυτάρεσκος. ~**ness** *s.* αυταρέσκεια
f.
smuggle *v.t.* περνώ λαθραίως. ~**er** *s.*
λαθρέμπορος *m.* ~**ing** *s.* λαθρεμπόριο *n.*
smut *s.* μουντζούρα *f.* (*obscenities*) βρωμιές
f.pl. ~**ty** *a.* βρώμικος.
snack *s.* κολατσιό *n.*
snag *s.* εμπόδιο *n.* that's the ~ εδώ είναι ο
κόμπος.
snail *s.* σάλιαγκας *m.*, σαλιγκάρι *n.*
snake *s.* φίδι *n.* (*fig.*) ~ in the grass δόλιος
άνθρωπος. (*v.*) ~ along πηγαίνω φιδωτά.
snaky *a.* (*road*) φιδωτός, (*hair*) φιδίσιος,
(*treacherous*) δόλιος.
snap *v.t. & i.* (*break*) σπάζω, (*close, shut*)
κλείνω. (*v.i.*) (*make ~ping noise*) κάνω
κρακ. (*speak angrily*) κακομιλώ. ~ at (*bite
at*) αρπάζω, δαγκώνω, (*fig.*) (*accept
eagerly*) αρπάζω. they were ~ped up
έγιναν ανάρπαστα. ~ one's fingers at
(*fig.*) δεν δίνω δεκάρα για.
snap *s.* (*noise*) κρακ *n.* (*with jaws*) δαγκανιά
f. (*fastener*) σούστα *f.* cold ~ ξαφνικό
κρύο. (*a., sudden*) αιφνιδιαστικός.

snappish *a.* οξύθυμος.

snappy *a.* (*quick*) γρήγορος, (*smart*) κομψός.

snapshot *s.* στιγμιότυπο *n.*

snare *s.* παγίδα *f.* (*v.*) παγιδεύω.

snarl *v.t. & i.* (*tangle*) μπλέκω. (*v.i.,* *of dog*) γρυλλίζω.

snatch *v.* αρπάζω, (*fig., steal*) κλέβω. (*s., small bit*) κομματάκι *n.* by ~es διακεκομμένα.

sneak *v.i.* (*creep*) γλιστρώ. (*v.t.*) (*steal*) βουτώ. (*s.*) μαρτυριάρης *m.* ~**ing** *a.* (*secret, furtive*) κρυφός. (*mean*) ποταπός.

sneer *v.i.* γελώ περιφρονητικά. ~ at περιφρονώ.

sneeze *v.* φτερνίζομαι. (*fam.*) not to be ~d at όχι ευκαταφρόνητος. (*s.*) φτέρνισμα *n.*

snide *a.* περιφρονητικός.

sniff *v.i.* ρουφώ τη μύτη μου. (*v.t.*) (*inhale*) εισπνέω, ρουφώ, (*smell*) μυρίζω. ~ at (*reject*) περιφρονώ.

sniff *s.* ρούφηγμα *n.* get a ~ of fresh air αναπνέω λίγο φρέσκο αέρα.

snigger *v.* κρυφογελώ. (*s.*) κρυφόγελο *n.*

snip *v.* κόβω. (*s.*) (*act*) κοψιά *f.* (*bit*) απόκομμα *n.* (*bargain*) ευκαιρία *f.*

snipe *s.* μπεκατσίνι *n.*

snipe *v.* (*also ~* at) πυροβολώ εξ ενέδρας. ~**r** *s.* ελεύθερος σκοπευτής.

snippet *s.* κομματάκι *n.*

snivel *v.* μυξοκλαίω. ~**ling** *a.* κλαψιάρης.

snob *s.,* ~**bish** *a.* σνομπ. ~**bery** *s.* σνομπισμός *m.*

snoop *v.* χώνω τη μύτη μου.

snooty *a.* ψηλομύτης.

snooze *s.* have a ~ παίρνω έναν υπνάκο.

snore *v.* ροχαλίζω. (*s.*) ροχαλητό *n.*

snort *v.* ρουθουνίζω. (*s.*) ρουθούνισμα *n.*

snot *s.* μύξα *f.*

snout *s.* ρύγχος *n.*

snow *s.* χιόνι *n.* (*v.*) it ~s χιονίζει. (*fig.*) ~ed under κατακλυσμένος. ~**drift** *s.* χιονοστιβάδα *f.* ~**fall** *s.* χιονοπτώσεις *f.pl.* ~**flake** *s.* νιφάδα *f.* ~**man** *s.* χιονάνθρωπος *m.* ~**plough** *s.* εκχιονιστήρας *m.* ~**storm** *s.* χιονοθύελλα *f.* ~**white** *a.* χιονάτος.

snowball *s.* χιονόσφαιρα, χιονιά *f.* (*v.i., fig.*) όλο και αυξάνομαι. ~**ing** *s.* χιονοπόλεμος *m.*

snowy *a.* (*covered in snow*) χιονισμένος. ~ weather χιονιά *f.*

snub *s.* προσβολή *f.* (*v.*) I ~bed him του φέρθηκα περιφρονητικά.

snub-nosed *a.* με ανασηκωμένη μύτη.

snuff *s.* ταμπάκος *m.* (*v.*) ~ out σβήνω.

snuffle *v.* ρουφώ τη μύτη μου.

snug *a.* αναπαυτικός, βολικός. (*warm*) ζεστός.

snuggle *v.i.* ~ down χώνομαι. ~ up to σφίγγομαι απάνω σε. (*v.t.*) σφίγγω.

so *adv.* (*of degree*) τόσο. ~ ... as (*comparative*) she is not ~ young as she seems δεν είναι τόσο νέα όσο φαίνεται. they're not ~ silly as to get taken in δεν είναι τόσο βλάκες ώστε (*or* που) να γελαστούν. be ~ kind as to tell me έχετε την καλοσύνη να μου πείτε; ~ much (*a.*) τόσος, τόσο πολύς, (*adv.*) τόσο πολύ. it is ~ much nonsense είναι πραγματική ανοησία. ~ much for the financial aspect of the case ως εδώ για την οικονομική άποψη της υπόθεσης. he didn't ~ much as look at me ούτε καν με κοίταξε.

so *adv.* (*of manner*) (*thus*) έτσι, (*in such a way*) κατά τέτοιο τρόπο. ~ that, ~ as to ώστε να, έτσι που να. we have arranged things ~ that someone is always at home κανονίσαμε έτσι ώστε κάποιος να είναι πάντοτε σπίτι. stand by the window ~ that I can see you να σταθείς κοντά στο παράθυρο ώστε (*or* έτσι που) να σε βλέπω. he parked his car ~ as to block the entrance παρκάρισε κατά τέτοιο τρόπο ώστε να κλείνει την είσοδο. it ~ happened that the train was late συνέβη ώστε ν' αργήσει το τραίνο. I fear ~ το φοβάμαι. I told you ~ σου το είπα. ~ do I κι εγώ επίσης. forty or ~ περίπου σαράντα. and ~ on και ούτω καθ' εξής. if ~ αν ναι. ~ to speak ούτως ειπείν. ~-and~ (*person*) ο τάδε, ο δείνα. just ~ (*yes*) ακριβώς, (*all in order*) στην εντέλεια.

so *conj.* (*that is why*) κι έτσι, επομένως, γι' αυτό. (*exclamatory*) ~ you're not coming! ώστε δεν θα 'ρθεις; ~ what! ε και;

soak *v.t. & i.* μουσκεύω. (*v.t.*) ~ up απορροφώ. (*v.i.*) ~ in *or* through διεισδύω. ~ into διαποτίζω. (*fig.*) ~ oneself in εμποτίζομαι υπό (*with gen.*). I got ~ed μούσκεψα. ~**ing** *s.* μούσκεμα *n.*

soap *s.* σαπούνι *n.* (*v.*) σαπουνίζω, ~**bubble** *s.* σαπουνόφουσκα *f.* ~ powder σαπούνι σκόνη. ~**flakes** *s.* τριμμένο σαπούνι. ~**y** *a.* από σαπούνι. (*like ~*) σαπωνοειδής. ~**y** water σαπουνάδα *f.*

soar *v.* υψώνομαι, ανέρχομαι.

sob *s.* αναφυλητό *n.*, λυγμός *m.* (*v.i.*) κλαίω με λυγμούς.

sober *a.* (*not drunk*) νηφάλιος, (*restrained*) εγκρατής, σοβαρός. (*colour*) μουντός. (*v.i.*) ~ down σοβαρεύομαι, ~ up ξεμεθώ. ~**ly** *adv.* σοβαρά.

sobriety *s.* εγκράτεια *f.*

soccer *s.* ποδόσφαιρο *n.*

sociab|le *a.* κοινωνικός. ~**ility** *s.* κοινωνικότητα *f.*

social *a.* κοινωνικός. ~ activities κοσμική κίνηση. ~ science κοινωνιολογία *f.* ~ security κοινωνική ασφάλιση. ~ worker κοινωνικός λειτουργός.

social|ism *s.* σοσιαλισμός *m.* **~ist** *a.* & *s.* (*thing, idea*) σοσιαλιστικός, (*person*) σοσιαλιστής *m.* **~ize** *v.* κοινωνικοποιώ.

society *s.* (*community*) κοινωνία *f.* (*companionship*) συντροφιά *f.* (*people of fashion*) ο καλός κόσμος. (*organization*) εταιρεία *f.*

sociology *s.* κοινωνιολογία *f.*

sock *s.* κάλτσα *f.* (*inside shoe*) πάτος *m.*

sock *v.t.* (*fam., hit*) I ~ed him (one) του έδωσα μια.

socket *s.* (*eye*) κόγχη *f.* (*electric*) πρίζα *f.*, υποδοχή *f.*

sod *s.* κομμένη λουρίδα χλόης (προς μεταφύτευση).

soda *s.* (*also* ~ water) σόδα *f.*

sodden *a.* διάβροχος, μουσκεμένος. (*dough*) λασπωμένος. (*with drink*) σουρωμένος. get ~ (*wet*) γίνομαι μούσκεμα.

sodium *s.* νάτριο *n.*

sodom|ite *s.* σοδομίτης *m.* **~y** *s.* σοδομία *f.*

sofa *s.* καναπές *m.*

soft *a.* (*not hard, gentle, lenient*) μαλακός, (*to the touch*) απλός. (*colour, light, sound, breeze*) απαλός. (*tender*) τρυφερός. (*effeminate*) μαλθακός. (*barmy*) χαζός. (*footstep*) ελαφρύς. (*job, option*) εύκολος. ~ drinks αναψυκτικά *n.pl.* have a ~ spot for έχω αδυναμία σε. he is ~ on her είναι τσιμπημένος μαζί της. **~-boiled** *a.* μελάτος. **~-pedal** *v.i.* (*fig.*) απαλύνω τη στάση μου. **~-soap** *v.t.* κολακεύω. **~-spoken** *a.* γλυκομίλητος. **~-ware** *s.* (*computer*) λογισμικό *n.*

soften *v.t.* & *i.* μαλακώνω, απαλύνω. **~ing** *s.* μαλάκωμα *n.* (*of brain*) μαλάκυνση *f.*

softly *adv.* μαλακά, απαλά, τρυφερά, ελαφρά.

softness *s.* μαλακότητα *f.* απαλότητα *f.* τρυφερότητα *f.*

soggy *a.* κάθυγρος, (*bread, etc.*) λασπωμένος.

soil *s.* (*ground*) έδαφος *n.*, γη *f.* (*earth*) χώμα *n.*

soil *v.t.* λερώνω. **~-pipe** *s.* αποχετευτικός σωλήνας αποχωρητηρίου.

sojourn *v.* διαμένω. (*s.*) διαμονή *f.*

solace *s.* παρηγοριά *f.* (*v.*) παρηγορώ.

solar *a.* ηλιακός. ~ power ηλιακή ενέργεια.

solder *s.* καλάι *n.* (*v.*) συγκολλώ.

soldier *s.* στρατιώτης *m.* foot-~ οπλίτης, φαντάρος *m.* ~ of fortune μισθοφόρος *m.* (*v.i.*) ~ on δεν το βάζω κάτω. **~ly** *a.* γενναίος.

soldiery *s.* στρατιώτες *m.pl.* (*troops*) στρατεύματα *n.pl.*

sole *s.* (*fish*) γλώσσα *f.*

sole *s.* (*of foot, shoe*) πέλμα *n.* (*of shoe*) σόλα *f.* (*v.t., a shoe*) σολιάζω.

sole *a.* μόνος, μοναδικός. (*exclusive*) αποκλειστικός. **~ly** *adv.* μόνον, αποκλειστικά.

solecism *s.* σολοικισμός *m.* (*faux pas*) γκάφα *f.*

solemn *a.* (*ritual*) επίσημος, (*grave*) σοβαρός. **~ity** *s.* επισημότητα *f.*, σοβαρότητα *f.* **~ly** *adv.* επισήμως, σοβαρώς.

solemnize *v.* τελώ.

solicit *v.* ζητώ. **~ation** *s.* παράκληση *f.* **~ing** *s.* (*by prostitute*) ψώνιο *n.*

solicitor *s.* δικηγόρος *m.f.*

solicitous *a.* be ~ ενδιαφέρομαι. **~ly** *adv.* με φροντίδα.

solicitude *s.* φροντίδα *f.*

solid *a.* στερεός. (*not hollow*) συμπαγής. (*alike all through*) ατόφιος, μασίφ. (*strong*) ισχυρός. (*unanimous*) ομόφωνος. a ~ hour μια ολόκληρη ώρα. **~ity** *s.* στερεότητα *f.* **~ly** *adv.* στερεά, ομοφώνως.

solidarity *s.* αλληλεγγύη *f.* συμπαράσταση *f.*

solidify *v.i.* στερεοποιούμαι.

soliloqu|y *s.* μονόλογος *m.* **~ize** *v.i.* μονολογώ.

solitary *a.* (*person*) μονήρης, (*thing*) μοναχικός. (*secluded*) παράμερος. not a ~ tree ούτε ένα δέντρο.

solitude *s.* μοναξιά *f.*

solo *s.* & *adv.* σόλο *n.* **~ist** *s.* σολίστ *m.f.*, σολίστας *m.*

solstice *s.* ηλιοστάσιο *n.*

solub|le *s.* διαλυτός. **~ility** *s.* διαλυτότητα *f.*

solution *s.* (*dissolving*) διάλυση *f.* (*liquid*) διάλυμα *n.* (*solving*) λύση *f.*

solve *v.* λύω, λύνω.

solv|ent *a.* διαλυτικός. (*fin.*) αξιόχρεος. (*s.*) διαλύτης *m.*, διαλυτικό *n.* **~ency** *s.* αξιόχρεον *n.*

sombre *a.* σκούρος, σκοτεινός. (*outlook, etc.*) ζοφερός.

some *a.* & *pron.* (*partitive article, often omitted*) λίγος, μερικοί. would you like ~ wine? θέλετε (λίγο) κρασί; I bought ~ apples πήρα (μερικά) μήλα. (*one unspecified*) κάποιος, κανένας. ~ friend gave it to him κάποιος φίλος του το 'δωσε. ~ day καμιά μέρα. (*several unspecified*) μερικοί, ~ people say μερικοί λένε. ~ ... others άλλοι ... άλλοι. (*a fair amount of*) αρκετός, κάμποσος. I waited for ~ time περίμενα αρκετή (*or* κάμποση) ώρα. (*approximately*) κάπου, περίπου. it lasted

~ three hours βάσταξε κάπου (or περίπου) τρεις ώρες.

some|body, ~one pron. κάποιος. see also anybody.

somehow adv. κάπως, με κάποιο τρόπο.

somersault s. τούμπα f.

something pron. κάτι. ~ of the sort κάτι τέτοιο. I gave him ~ to eat του έδωσα (κάτι) να φάει.

sometime adv. κάποτε. (formerly, former) πρώην.

sometimes adv. κάποτε, καμιά φορά. ~...~ πότε... πότε.

somewhat adv. κάπως.

somewhere adv. κάπου.

somnambul|ism s. υπνοβασία f. ~ist s. υπνοβάτης m.

somnolent a. μισοκοιμισμένος.

son s. γιος, υιός m. ~-in-law s. γαμπρός m.

sonant a. (gram.) ηχηρός.

sonata s. σονάτα f.

song s. τραγούδι n., άσμα n. (of birds) κελάδημα n. (fig.) make a ~ (and dance) φέρνω τον κατακλυσμό. for a ~ (cheap) για ένα κομμάτι ψωμί.

songster s. τραγουδιστής m.

sonic a. ηχητικός.

sonnet s. σονέτο n.

sonny s. αγοράκι μου.

sonor|ous a. ηχηρός. ~ity s. ηχηρότητα f.

soon adv. (shortly) σύντομα, σε λίγο, προσεχώς. (early) γρήγορα, νωρίς. it will ~ be three years since they left κοντεύουν τρία χρόνια από τότε που έφυγαν. ~ after three o'clock λίγο μετά τις τρεις. he came an hour too ~ ήρθε μια ώρα νωρίτερα. how ~ can it be done? πόσο γρήγορα μπορεί να γίνει; as ~ as possible το γρηγορότερο. as ~ as (or no ~er had) they got home μόλις γύρισαν σπίτι. ~er or later αργά ή γρήγορα. no ~er said than done αμ' έπος αμ' έργον. I would ~er go hungry κάλλιο (or προτιμότερο) να πεινάσω. I would (just) as ~ stay at home δεν θα μ' ένοιαζε αν έμενα σπίτι.

soot s. καπνιά f. ~y a. γεμάτος καπνιά.

sooth|e v. (pain) καταπραΰνω, (person) ησυχάζω. ~ing a. καταπραϋντικός, (emollient) μαλακτικός.

soothsayer s. μάντης m.

sop s. (fig.) εξιλαστήριο δώρο. (v.) ~ up σφουγγίζω.

sophist|ical a. σοφιστικός. ~ry s. σοφιστεία f.

sophisticated a. (person) μπασμένος στα του κόσμου. (taste) εκλεπτυσμένος, (ideas) εξελιγμένος, (machinery) υπερσύγχρονος.

soporific a. υπνωτικός.

sopping a. μουσκεμένος.

soprano s. υψίφωνος f., σοπράνο f.

sorcer|er s. μάγος m. ~ess s. μάγισσα f. ~y s. μαγεία f.

sordid a. άθλιος.

sore a. (hurting) πονεμένος, ερεθισμένος. (annoyed) χολωμένος. (great, grievous) μεγάλος, οδυνηρός. be ~ (hurting) πονώ. ~ throat πονόλαιμος m. ~ point ευαίσθητο σημείο. (s.) πληγή f. ~ly adv. σοβαρά, πολύ. ~ness s. πόνος m.

sorrow s. θλίψη f., λύπη f. to my ~ προς λύπη μου. ~s δεινά, βάσανα n.pl. ~ful a. θλιμμένος, (causing ~) θλιβερός.

sorry a. λυπημένος. be ~ (regret) λυπούμαι, λυπάμαι. (repent) μετανοώ, μετανοιώνω. I am ~ for him τον λυπάμαι. ~! (asking pardon) συγγνώμη! (deplorable) ελεεινός.

sort s. είδος n. all ~s of people όλων των ειδών (or κάθε λογής) άνθρωποι. that ~ of people τέτοιοι (or τέτοιου είδους) άνθρωποι, something of the ~ κάτι τέτοιο. nothing of the ~! καθόλου! a meal of ~s ένα είδος γεύματος ας πούμε. in a ~ of way κατά κάποιον τρόπο. he's not my ~ δεν είναι ο τύπος μου. out of ~s αδιάθετος. see also kind.

sort v. (arrange) ταξινομώ. ~ out (select) ξεδιαλέγω, (settle) κανονίζω, ξεδιαλύνω, διευθετώ.

sortie s. έξοδος f. make a ~ εξορμώ.

SOS s. ες ο ες n.

so-so adv. έτσι κι έτσι.

sot s., ~tish a. μπέκρας m.

soul s. ψυχή f. the ~ of honour η εντιμότητα προσωποποιημένη. ~ful a. παθητικός. ~less a. απάνθρωπος. ~-stirring a. συναρπαστικός.

sound s. (strait) πορθμός m.

sound s. ήχος m. within ~ of the sea σε απόσταση που ακούγεται η θάλασσα. (a.) (pertaining to ~) ηχητικός. ~-barrier s. φράγμα του ήχου. ~-proofing s. ηχομόνωση f. ~-track s. ηχητική ζώνη. ~-waves s. ηχητικά κύματα n.pl.

sound v.t. (pronounce) προφέρω. (on trumpet) σαλπίζω. (strike, ring, etc.) χτυπώ. ~ chest of ακροάζομαι. (test depth of & fig.) βολιδοσκοπώ. (v.i.) ηχώ, χτυπώ. (seem) φαίνομαι, μοιάζω. ~ings s. take ~ings βυθομετρώ.

sound a. (healthy) γερός, υγιής. (in good condition, capable, through) γερός, (right) σωστός, ορθός. (reliable) αξιόπιστος. (sleep) βαθύς. ~ly adv. γερά, καλά, βαθιά.

soup s. σούπα f. (fam.) I'm in the ~ έμπλεξα άσχημα. ~-kitchen s. συσσίτιο n.

sour *a.* ξινός. make *or* become ~ ξινίζω. ~ grapes όμφακες εισίν. **~ness** *s.* ξινίλα *f.*

source *s.* πηγή *f.*

soused *a.* (*salted*) αλίπαστος. (*drenched*) καταβρεγμένος. (*fam., drunk*) σκνίπα.

south *s.* νότος *m.*, νοτιά *f.* μεσημβρία *f.* to the ~ of Athens νοτίως των Αθηνών.

south, ~erly, ~ern *a.* νότιος, μεσημβρινός. ~ wind νοτιάς *m.*, όστρια *f.*

south-east *a.* νοτιοανατολικός. ~ wind σιρόκος *m.*

southward(s) *adv.* νοτίως.

south-west *a.* νοτιοδυτικός. ~ wind γαρμπής *m.*

souvenir *s.* ενθύμιο *n.*

sovereign *a.* κυρίαρχος, υπέρτατος. (*remedy*) ασφαλής. (*s.*) (*monarch*) μονάρχης *m.* (*pound*) λίρα *f.* **~ty** *s.* κυριαρχία *f.*

sow *s.* γουρούνα *f.*

sow *v.* σπέρνω, (*fig.*) ενσπείρω. **~er** *s.* σποριάς *m.* **~ing** *s.* σπορά *f.*, σπάρσιμο *n.*

sozzled *a.* (*fam., drunk*) σκνίπα.

spa *s.* λουτρά *n.pl.* (*town*) λουτρόπολη *f.*

space *s.* διάστημα *n.* (*void*) κενό *n.* (*room*) χώρος *m.* take up ~e πιάνω τόπο. there isn't enough ~e for it δεν χωράει. (*v.*) ~e out αραιώνω, (*events*) κλιμακώνω. **~e-man** *s.* κοσμοναύτης *m.* **~e-ship** *s.* διαστημόπλοιον *n.*

spacing *s.* (*distance*) απόσταση *f.* in single (*or* double) ~ σε μονό (*or* διπλό) διάστημα.

spacious *a.* ευρύχωρος.

spade *s.* φτυάρι *n.* (*cards*) μπαστούνι *n.*, πίκα *f.*, **~-work** *s.* προκαταρκτική εργασία.

spaghetti *s.* σπαγκέτο *n.*

span *s.* (*of hand*) σπιθαμή *f.* (*of bridge, wing*) άνοιγμα *n.* (*of time*) διάστημα *n.* (*of life*) διάρκεια *f.* (*v.*) (*in space*) περνώ, (*in time*) καλύπτω.

spangles *s.* πούλιες *f.pl.*

Spanish *a.* ισπανικός. (*person*) Ισπανός *m.*, Ισπανίδα *f.*

spank *v.t.* (*smack*) χτυπώ με την παλάμη. (*v.i.*) ~ along τρέχω.

spanking *s.* ξύλο *n.* (*a.*) (*quick*) γοργός, (*showy*) φανταχτερός, (*first rate*) θαυμάσιος.

spanner *s.* κλειδί *n.* (*adjustable*) κάβουρας *m.*, παπαγάλος *m.* (*fig.*) ~ in the works σαμποτάζ.

spar *s.* (*naut.*) αντενοκάταρτο *n.*

spar *v.i.* ερίζω.

spare *a.* (*lean*) ισχνός, (*scanty*) λιτός. (*not needed*) διαθέσιμος, (*in reserve*) ρεζέρβα. ~ parts ανταλλακτικά *n.pl.* **~ly** *adv.* λιτά.

spare *v.* (*show mercy to, be frugal with*) φείδομαι (*with gen.*). (*afford*) διαθέτω.

I cannot ~, I have none to ~ δεν μου περισσεύει. ~ (*person*) the trouble of ... απαλλάσσω από τον κόπο να. enough and to ~ περίσσεια. ~ the rod and spoil the child το ξύλο βγήκε από τον παράδεισο.

sparing *a.* φειδωλός. be ~ of, go ~ly with φείδομαι (*with gen.*). **~ly** *adv.* με οικονομία.

spark *s.* σπίθα *f.* (*fig., trace*) ίχνος *n.* (*v.*) ~ off προκαλώ. **~ing-plug** *s.* μπουζί *n.*

spark|le *v.* σπιθοβολώ, αστράφτω. (*s.*) σπιθοβόλημα *n.* **~ling** *a.* σπινθηροβόλος, αστραφτερός. (*eye*) λαμπερός, (*wine*) αφρώδης. **~ling** cleanliness απαστράπτουσα καθαριότης.

sparrow *s.* σπουργίτης *m.*

sparse *a.* αραιός. **~ly** *adv.* αραιά.

Spartan *s.* σπαρτιατικός.

spasm *s.* σπασμός *m.*

spasmodic *a.*, **~ally** *adv.* με διαλείμματα.

spastic *a.* σπαστικός.

spate *s.* πλημμύρα *f.* be in ~ πλημμυρίζω.

spatial *a.* διαστηματικός.

spatter *v.* πιτσιλίζω.

spatula *s.* σπάτουλα *f.*

spawn *s.* αβγά *n.pl.* (*also fig.*) γόνος *m.* (*v.*) ωοτοκώ.(*fig.*) γεννοβολώ.

speak *v.i.* ομιλώ, μιλώ. (*in public*) αγορεύω. ~ out *or* up μιλώ έξω από τα δόντια. ~ up μιλώ πιο δυνατά. ~ up for μιλώ υπέρ (*with gen.*). it ~s for itself μιλάει μόνο του. nothing to ~ of τίποτε το άξιον λόγου. so to ~ ούτως ειπείν. (*v.t.*) λέω, ~ English μιλώ αγγλικά. (*fig.*) ~ volumes λέω πολλά.

speaker *s.* ομιλητής *m.* S~ (*of Parliament*) Πρόεδρος *m.*

speaking *s.* το ομιλείν. public ~ το αγορεύειν. they are not on ~ terms δεν μιλιούνται. (*a.*) (*expressive*) εκφραστικός. English-~ αγγλόφωνος.

spear *s.* δόρυ *n.* (*v.*) καρφώνω. **~head** *s.* αιχμή *f.* (*v.t.*) αποτελώ την αιχμή (*with gen.*).

spec *s.* on ~ στην τύχη. **~s** *s.* γυαλιά *n.pl.*

special *a.* ειδικός, ιδιαίτερος. (*exceptional*) έκτακτος. ~ offer (*in shop*) οκαζιόν *f.* **~ly** *adv.* ειδικώς, ιδιαιτέρως.

special|ist *s.* ειδικός *a.* she is a ~ist in economics είναι ειδική στα οικονομολογικά *or* ειδικός οικονομολόγος. **~ity** *s.* ειδικότητα *f.*

specializ|e *v.* ~e in ειδικεύομαι σε. **~ation** *s.* ειδίκευση *f.*

species *s.* είδος *n.* human ~ ανθρώπινον γένος.

specific *a.* (*precise*) σαφής, συγκεκριμένος, ειδικός. (*order*) ρητός. (*s.*) ειδικό φάρμακο. ~**ally** *adv.* συγκεκριμένως.

specifications *s.* (*of work to be done*) συγγραφή υποχρεώσεων, προδιαγραφή *f.* (*of car, etc.*) χαρακτηριστικά *n.pl.*

specify *v.* καθορίζω.

specimen *s.* δείγμα *n.* (*fam.,type*) τύπος *m.*

specious *a.* εύσχημος.

speck *s.* κόκκος *m.* (*tiny distant object*) κουκκίδα *f.* not a ~ ούτε ίχνος.

speckled *a.* πιτσιλωτός.

spectacle *s.* θέαμα *n.* ~s *s.* (*glasses*) γυαλιά *n.pl.* see things through rose-coloured ~s τα βλέπω όλα ρόδινα.

spectacular *a.* θεαματικός.

spectator *s.* θεατής *m.*

spectre *s.* φάσμα *n.*

spectrum *s.* φάσμα *n.* (*fig.*) έκταση *f.*, κλίμακα *f.*

speculat|e *v.* κάνω υποθέσεις. (*fin.*) κερδοσκοπώ. ~**ion** *s.* υποθέσεις *f.pl.* (*fin.*) κερδοσκοπία *f.* ~**ive** *a.* υποθετικός, κερδοσκοπικός. ~**or** *s.* κερδοσκόπος *m.*

speech *s.* (*faculty*) λόγος *m.* (*manner, process*) ομιλία *f.* (*address*) λόγος *m.* make a ~ βγάζω λόγο. ~**less** *a.* άναυδος. ~**training** *s.* ορθοφωνία *f.*

speed *s.* ταχύτητα *f.*, γρηγοράδα *f.* at full ~ ολοταχώς ~ limit όριο ταχύτητος. (*v.i.*) σπεύδω, (*drive fast*) τρέχω. (*v.t.*) ~ up επιταχύνω, επισπεύδω. ~**y** *a.* ταχύς, γρήγορος.

speedometer *s.* κοντέρ *n.*

spell *s.* (*sorcerer's*) μάγια *n.pl.* put a ~ on κάνω μάγια σε. (*fascination*) γοητεία *f.* under the ~ of μαγεμένος από.

spell *s.* (*period*) περίοδος *f.* (*of duty*) βάρδια *f.* for a ~ για λίγο.

spell *v.* γράφω. he can ~ ξέρει ορθογραφία. ~ out συλλαβίζω, (*fig.*) επεξηγώ. ~**ing** *s.* ορθογραφία *f.* (*a.*) ορθογραφικός.

spend *v.t.* (*pay out, consume*) ξοδεύω, (*pass*) περνώ. ~**thrift** *s.* σπάταλος *a.*

spent *a.* εξαντλημένος.

sperm *s.* σπέρμα *n.* ~**atozoa** *s.* σπερματοζωάρια *n.pl.*

spew *v.* ξερνώ. ~**ing** *s.* ξέρασμα *n.*

spher|e *s.* σφαίρα *f.* (*surroundings*) κύκλος *m.*, περιβάλλον *n.* ~**ical** *a.* σφαιρικός.

sphinx *s.* σφιγξ, σφίγγα *f.*

spic|e *s.* μπαχαρικό *n.* (*v.t.*) καρυκεύω. ~**y** *a.* πικάντικος.

spick *a.* ~ and span στην τρίχα, νοικοκυρεμένος.

spider *s.* (*also* ~'s web) αράχνη *f.* ~**y** *a.* (*writing*) με μακριές λεπτές γραμμές.

spigot *s.* τάπα *f.*

spik|e *s.* μύτη *f.* (*on shoe*) καρφί *n.* (*of plant*) στάχυς *m.* (*v.t.*) καρφώνω. ~**e** (*person's*)

guns ματαιώνω τα σχέδιά του. ~**y** *a.* μυτερός.

spill *v.t.* (*liquid, powder*) χύνω, (*small objects, rider*) ρίχνω, (*overturn*) ανατρέπω. (*v.i.*) χύνομαι. (*s.*) πέσιμο *n.*

spin *v.t.* & *i.* (*turn*) στροβιλίζω, στριφογυρίζω. (*v.t.*) (*form into thread*) κλώθω, γνέθω. (*fam.*) ~ a yarn λέω μια ιστορία. ~ out παρατείνω. he spun round γύρισε απότομα. send (*thing, person*) ~ning τινάζω μακριά. (*s.*) περιδίνηση *f.* (*ride*) βόλτα *f.* (*fam.*) flat ~ πανικός *m.* ~**off** *s.* απορρέον όφελος.

spinach *s.* σπανάκι *n.*

spinal *a.* σπονδυλικός. ~ cord νωτιαίος μυελός.

spindl|e *s.* (*for spinning*) αδράχτι *n.* (*pivot*) άξονας *m.* ~**y** *a.* ~y legs πόδια οδοντογλυφίδες.

spine *s.* σπονδυλική στήλη, ραχοκοκκαλιά *f.* (*thorn*) άκανθα *f.*, αγκάθι *n.* ~**less** *a.* (*fig.*) λαπάς *m.*

spinner *s.* κλώστης *m.*, κλώστρα *f.*

spinney *s.* δασύλλιο *n.*

spinning *s.* κλωστική *f.* ~**wheel** *s.* ροδάνι *n.*

spinster *s.* ανύπαντρη γυναίκα.

spiny *a.* αγκαθωτός.

spiral *a.* ελικοειδής. (*s.*) έλιξ *f.*, σπείρα *f.* (*v.i.*) ανέρχομαι σπειροειδώς. prices are ~ling οι τιμές όλο ανέρχονται.

spire *s.* βέλος *n.*

spirit *s.* πνεύμα *n.* (*of the dead*) ψυχή *f.* (*mettle*) ψυχή *f.* (*vigour*) ζωή *f.* be in good ~s έχω κέφι. be in low ~s δεν είμαι στα κέφια μου. moving ~ πρωτεργάτης *m.* take it in the wrong ~ το παίρνω στραβά. (*liquor*) σπίρτο *n.*, ~s οινοπνευματώδη ποτά. (*v.*) be ~ed away εξαφανίζομαι. ~**lamp** *s.* καμινέτο *n.* ~**level** *s.* αλφάδι *n.*

spirited *a.* θαρραλέος, γεμάτος ψυχή. (*horse*) θυμοειδής. high-~ ζωηρός, low-~ άθυμος *n.*

spiritless *a.* ψόφιος.

spiritual *a.* ψυχικός, (*of the church*) εκκλησιαστικός. (*s.*) τραγούδι νέγρων της Αμερικής. ~**ly** *adv.* ψυχικώς.

spiritual|ism *s.* πνευματισμός *m.* ~**ist** *s.* πνευματιστής *m.*

spirituous *a.* πνευματώδης.

spit *s.* (*spike*) οβελός *m.*, σούβλα *f.* (*of land*) γλώσσα *f.* (*v.t.*) οβελίζω, σουβλίζω.

spit *v.* (*from mouth*) (*also* ~ on *or* out) φτύνω. (*s.*) φτύμα *n.* (*fam.*) the very ~ (and image) of his father φτυστός ο πατέρας του. ~**ting** *s.* φτύσιμο *n.* ~**tle** *s.* φτύμα *n.*

spite *s.* κακία *f.* (*grudge*) I have a ~ against

him του κρατώ κακία in ~ of παρά (with acc.). (v.t.) σκάω. he does it to ~ her το κάνει για να την πεισματώσει.

spiteful a. κακός. **~ly** adv. με κακία. **~ness** s. κακία f.

spiv s. (fam.) μάγκας m.

splash v.t. & i. (spatter) πιτσιλίζω. (v.t.) (scatter) σκορπίζω, ρίχνω. (v.i.) (of waves, fountain) παφλάζω. ~ about τσαλαβουτώ.

splash s. πιτσιλιά f. παφλασμός m. make a ~ (fam.) κάνω στράκες. ~ of colour κηλίδα χρώματος.

splay v. πλαταίνω. (a.) ~ feet πόδια γυρισμένα προς τα έξω.

spleen s. (organ) σπλήνα f. (fig.) κακία f.

splendid a. μεγαλοπρεπής, (excellent) υπέροχος. **~ly** adv. μεγαλοπρεπώς, (well) υπέροχα, λαμπρά.

splendour s. μεγαλοπρέπεια f. (brilliance) λαμπρότητα f.

splice v. ματίζω. (s.) μάτισμα n.

splint s. (med.) νάρθηξ m.

splinter s. θραύσμα n. (of wood) σχίζα f. (fig.) they formed a ~ group αποσχίστηκαν. (v.i.) θραύομαι, σπάζω.

split v.t. & i. σχίζω, (divide) χωρίζω. ~ hairs τα ψιλολογώ. ~ one's sides with laughter ξεκαρδίζομαι. let us ~ the difference ας μοιραστούμε τη διαφορά. I have a ~ting headache το κεφάλι μου πάει να σπάσει. ~ the atom διασπώ το άτομο. (v.i.) ~ up (become divided) διασπώμαι. ~ on (person) μαρτυρώ.

split s. σχισμή f. (rupture) ρήξη f. (disagreement) διάσπαση f. do the ~s κάνω γκραντ εκάρ.

split a. σχιστός. (in disagreement) the government was ~ (on this issue) η κυβέρνησις διεσπάσθη.

splutter v.i. μπερδεύω τα λόγια μου. (hiss) συρίζω.

spoil v.t. & i. χαλώ, χαλνώ. (v.t.) (children) κακομαθαίνω, (cosset) χαϊδεύω. (rob) ληστεύω. (v.i.) be ~ing for a fight έχω όρεξη για καβγά.

spoil, ~s λάφυρα n.pl. ~s of office αποῤέοντα οφέλη του αξιώματος. ~s system ρουσφέτια n.pl.

spoil-sport s. γρουσούζης m.

spoilt s. χαλασμένος. (child, etc.) κακομαθημένος, χαϊδεμένος.

spoke s. ακτίνα f.

spokesman s. εκπρόσωπος m.

spoliation s. λεηλασία f.

spondee s. σπονδείος m.

spong|e s. σφουγγάρι n. throw in (or up) the ~e εγκαταλείπω τον αγώνα. (v.) σφουγγίζω. (fig.) ~e on αρμέγω. **~er** s. σελέμης

m. **~y** a. σπογγώδης.

sponsor s. ανάδοχος m.f. stand ~ to (child) βαπτίζω. (v.t.) αναδέχομαι. **~ship** s. εγγύηση f.

spontane|ity s. αυθόρμητον n. **~ous** a. αυθόρμητος. **~ous** combustion αυτοανάφλεξη f. **~ously** adv. αυθόρμητα.

spoof v. γελώ. (s.) φάρσα f.

spook s. φάντασμα n. **~y** a. στοιχειωμένος.

spool s. πηνίο n., μπομπίνα f.

spoon s. κουτάλι n. (v.t.) παίρνω με κουτάλι. **~-feed** v. (fig.) τα δίνω μασημένα σε. **~ful** s. κουταλιά f.

spooning s. (fam.) γλυκοκουβεντιάσματα n.pl.

spoor s. ίχνη n.pl.

sporadic a. σποραδικός.

spore s. σπόριο n.

sport s. (amusement) παιγνίδι n. (games) αθλητισμός m., σπορ n.pl. make ~ of κοροϊδεύω. (v.i.) (play) παίζω. (v.t.) (wear) φορώ.

sporting a. (of sport) αθλητικός. (fair) τίμιος. with ~ interests φίλαθλος. he has a ~ chance of winning έχει κάποια πιθανότητα επιτυχίας.

sportive a. παιγνιδιάρης.

sportsman s. (who shoots, hunts) κυνηγός m. (who plays games) σπόρτσμαν m. (fig.) τίμιος παίκτης. **~ship** s. τίμιο παιγνίδι.

spot s. (place) τόπος m. (point) σημείο n. on the ~ (at the place in question) επί τόπου. put on the ~ (embarrass) βάζω σε δύσκολη θέση, (kill) ξεκάνω. tender ~ αδύνατο σημείο. I found his weak ~ του βρήκα το κουμπί. ~ check αιφνιδιαστικός έλεγχος.

spot v.t. & i. (stain) λεκιάζω. it is ~ting with rain ψιχαλίζει. (v.t.) (discern) διακρίνω, (guess) μαντεύω, (find) βρίσκω. **~ter** s. παρατηρητής m.

spot s. (mark, stain) λεκές m. (pimple) σπυρί n. (dot) βούλλα f. (drop) στάλα f. (fam.) a ~ of λίγος. see also speck. **~less** a. πεντακάθαρος, (morally) άσπιλος. **~ted** a. (with ~s) με βούλες, ~ted with διάστικτος με. **~ty** a. (face) με σπυριά, (object) λεκιασμένος.

spotlight s. προβολεύς m.

spouse s. σύζυγος m.f.

spout s. (of teapot, gutter) στόμιο n.

spout v.t. ξεχύνω, (verbally) απαγγέλλω. (v.i.) ξεχύνομαι, (of whale) εκτοξεύω νερό.

sprain v. στραμπουλίζω. (s.) στραμπούλισμα n.

sprat s. ψαράκι n.

sprawl v.i. (of person) ξαπλώνομαι άχαρα, (of town) απλώνομαι χωρίς σχέδιο. send

spray (*person*) ~ing τινάζω μακριά.

spray *s.* (*of flower*) κλαδάκι *n.* (*of sea*) αφρός *m.* (*vaporizer*) ψεκαστήρας *m.* (*aerosol*) αεροζόλ *n.* (*v.t.*) ραντίζω. **~ing** *s.* ράντισμα *n.*

spread *v.t.* (*open or lay out*) απλώνω. (*tar, manure, gravel, etc.*) ρίχνω. (*scatter*) σκορπίζω. (*table*) στρώνω. (*smear*) αλείβω, ~ butter on the bread αλείβω το ψωμί με βούτυρο. (*news, rumour, knowledge*) διαδίδω, (*illness*) μεταδίδω. (*extend, make go further*) επεκτείνω. ~ out (*at wider intervals*) αραιώνω, (*one's fingers*) ανοίγω.

spread *v.i.* απλώνω, ξαπλώνω, απλώνομαι, εκτείνομαι. (*go further*) επεκτείνομαι, (*be transmitted*) μεταδίδομαι. (*of news*) διαδίδομαι, (*of butter*) αλείβομαι. (*be scattered*) σκορπίζομαι. the plain ~s as far as the eye can see η πεδιάδα απλώνεται (*or* εκτείνεται) όσο βλέπει το μάτι. the branches of the apple-tree ~ right over the pavement τα κλαδιά της μηλιάς απλώνονται μέχρι το πεζοδρόμιο. the fire ~ quickly η πυρκαγιά ξάπλωσε (*or* επεκτάθηκε) γρήγορα, it ~ to the nearby school μεταδόθηκε στο διπλανό σχολείο.

spread *s.* (*expanse*) έκταση *f.* (*extension*) επέκταση *f.* (*diffusion*) διάδοση *f.* (*transmission*) μετάδοση *f.* (*fam., feast*) τσιμπούσι *n.*

spread-eagled *a.* ξαπλωμένος ανάσκελα.

spreading *s.* άπλωμα *n. see also* spread *s.*

spree *s.* go on the ~ το ρίχνω έξω.

sprig *s.* κλαδάκι *n.* (*fig.*) βλαστάρι *n.*

sprightly *a.* ζωντανός, σβέλτος.

spring *s.* (*season*) άνοιξη *f.* (*water*) πηγή *f.* (*jump*) πήδημα *n.* (*of clock, vehicle*) ελατήριο *n.* (*of seat*) σούστα *f.* **~-mattress** *s.* στρωματέξ *n.*

spring *a.* (*seasonal*) ανοιξιάτικος, εαρινός. (*of water*) πηγαίος. **~-cleaning** *s.* γενικές καθαριότητες.

spring *v.i.* (*originate*) πηγάζω, (*jump*) (ανα)πηδώ. ~ up (*jump*) πετάγομαι, (*sprout, pop up*) ξεφυτρώνω, (*of wind, etc.*) σηκώνομαι, (*develop*) δημιουργούμαι, γεννιέμαι. it ~s to the mind έρχεται στο μυαλό. be sprung from κατάγομαι από. (*v.t.*) ~ a surprise on κάνω έκπληξη σε. **~-board** *s.* βατήρας *m.*

springe *s.* θηλιά *f.*

springy *a.* ελαστικός, μαλακός.

sprink||e *v.* ραντίζω. **~-er** *s.* ψεκαστήρας *m.* **~ing** *s.* ράντισμα *n.* (*few*) λίγοι, a fair ~ing of κάμποσοι.

sprint *v.* τρέχω.

sprite *s.* νεράιδα *f.*

sprout *v.i.* βλαστάνω. (*get taller*) ρίχνω μπόλ. (*v.t.*) βγάζω. (*s.*) βλαστάρι *n.*

spruce *a.* κομψός. (*v.*) ~ oneself up καλοντύνομαι.

spry *a.* ζωντανός, σβέλτος.

spud *s.* (*fam.*) πατάτα *f.*

spume *s.* αφρός *m.*

spunk *s.* (*fam.*) θάρρος *n.*

spur *s.* σπιρούνι *n.* (*fig.*) κίνητρο *n.* (*of hill*) αντέρεισμα *n.* on the ~ of the moment αυθορμήτως. (*v.t.*) σπιρουνίζω, (*fig.*) παρακινώ.

spurious *a.* κίβδηλος, πλαστός.

spurn *v.* απορρίπτω περιφρονητικά.

sputter *v.i.* συρίζω.

sputum *s.* πτύελο *n.*

spy *s.* κατάσκοπος *m.* σπιούνος *m.* (*v.*) (*also* ~ on) κατασκοπεύω, (*clap eyes on*) ματιάζω. ~ out (*land*) εξερευνώ. **~ing** *s.* κατασκοπεία *f.*

squabble *s.* καβγάς *m.* (*v.*) καβγαδίζω.

squad *s.* ομάδα *f.*

squadron *s.* (*naut., aero.*) μοίρα *f.* (*cavalry*) ίλη *f.* (*tanks*) επιλαρχία *f.* **~-leader** *s.* (*aero.*) επισμηναγός *m.*

squal|id *a.* βρωμερός, άθλιος. **~-or** *s.* βρώμα *f.*, κακομοιριά *f.*

squall *s.* (*wind*) σπιλιάδα *f.* (*rain*) μπόρα *f.* **~-y** *a.* θυελλώδης.

squall *v.i.* στριγγλίζω.

squander *v.* σπαταλώ. **~-er** *s.* σπάταλος *a.*

square *s.* (*geom.*) τετράγωνο *n.* (*of city*) πλατεία *f.* T-square γνώμων *m.* (*fam.*) back to ~ one φτου κι απ' την αρχή.

square *v.t.* τετραγωνίζω. (*balance*) ισοσκελίζω, (*bribe*) λαδώνω. (*v.i.*) (*agree*) συμφωνώ. ~ up (*settle accounts*) λογαριάζομαι. ~ up (*to*) παίρνω επιθετική στάση.

square *a.* τετράγωνος. (*metre, root, etc.*) τετραγωνικός. ~ deal δίκαιη μεταχείριση on the ~ έντιμος. get ~ with πατσίζω με, we are all ~ είμαστε πάτσι. have a ~ meal τρώω αρκετά. (*adv.*) καθαρά, ντόμπρα.

squarely *adv.* (*honestly*) εντίμως, (*positively*) καθαρά, (*straight*) ίσια. ~ in the eye κατάματα.

squash *v.* ζουλώ, (*in crowd*) στριμώχνω. (*s.*) στρίμωγμα *n.*

squat *v.* κάθομαι ανακούρκουδα. (*occupy premises illegally*) εγκαθίσταμαι παρανόμως.

squat *a.* κοντόχοντρος. (*object*) κοντός και πλατύς.

squawk *v.* κρώζω. (*s.*) κρωγμός *m.*

squeak *v.* τσιρίζω, (*of machinery, shoe*) τρίζω. (*s.*) τσίρισμα *n.*, τρίξιμο *n.* **~-y** *a.* ~y shoes παπούτσια που τρίζουν.

squeal v. σκληρίζω. (fam., inform) ξερνάω. (s.) σκλήρισμα n.

squeamish a. σιχασιάρης. (fig.) λεπτεπίλεπτος.

squeeze v.t. (squash) ζουλώ, (clasp) σφίγγω. (mop, sponge, lemon) στύβω. (jostle) στρυμώχνω. (one's hand in door, etc.) μαγγώνω. ~ (out) (extract) βγάζω. (v.i.) ~ in or into στρυμώχνομαι, χώνομαι μέσα σε.

squeez|e, ~ing s. σφίξιμο n. στύψιμο n. στρύμωγμα n.

squelch v.i. κάνω πλατς-πλουτς.

squib s. τρακατρούκα. f. (fig.) σάτιρα f. damp ~ φιάσκο n.

squid s. καλαμάρι n.

squint v. αλληθωρίζω. (s.) στραβισμός m. (fam., glance) have a ~ at ρίχνω μια ματιά σε.

squire v. πυργοδεσπότης m. (fam., escort) καβαλιέρος m.

squirm v. συστρέφομαι. (fig.) ντροπιάζομαι.

squirrel s. σκίουρος m.

squirt v.t. εκτοξεύω. (v.i.) εκτοξεύομαι. they ~ed water all over me με περιχύσανε με νερό.

stab v. μαχαιρώνω. (s.) μαχαιριά f. (fam.) have a ~ at it κάνω μια προσπάθεια.

stability s. ευστάθεια f., σταθερότητα f.

stabiliz|e v. σταθεροποιώ. **~ation** s. σταθεροποίηση f.

stable a. σταθερός, ευσταθής.

stable s. στάβλος m.

stack s. (pile) στοίβα f., σωρός m. (hay, corn) θημωνιά f. (chimney) καπνοδόχος f. (fam.) ~s of πολλοί. (v.) στοιβάζω.

stadium s. στάδιο n.

staff s. (stick) ράβδος f. (fig. support) στήριγμα n. (flag, etc.) κοντός m., κοντάρι n. (personnel) προσωπικό n. (mil.) General S~ επιτελείον n. Chief of S~ επιτελάρχης m. (v.t.) βρίσκω προσωπικό για.

stag s. ελάφι n.

stage v.t. (a play) ανεβάζω, (also fig.) σκηνοθετώ. (organize) οργανώνω.

stage s. (of theatre) σκηνή f., the ~ (fig.) θέατρο n. go on the ~ βγαίνω στη σκηνή. (platform) εξέδρα f. (step, point reached) στάδιο n. (phase) φάση f. by ~s σταδιακώς, κλιμακωτά. **~-fright** s. τρακ n. **~-production** s. σκηνοθεσία f. **~-scenery** s. σκηνογραφία f., σκηνικά n.pl.

stagger v.i. παραπαίω, τρικλίζω. (v.t.) συγκλονίζω. (working hours, etc.) κλιμακώνω. **~ing** a. συγκλονιστικός.

stagnant a. στάσιμος. be ~ λιμνάζω.

stagnat|e v. (fig.) αποτελματώνομαι. **~ion** s. αποτελμάτωση f. (of affairs) στασιμότητα f.

staid a. σεμνός.

stain s. κηλίδα f. (dye) βαφή f. (v.t.) κηλιδώνω, (dye) βάφω. **~less** a. ακηλίδωτος, (steel) ανοξείδωτος.

stair s. σκαλοπάτι n. **~s** σκάλα f. **~case** s. κλιμακοστάσιο n.

stake s. (post) παλούκι n., πάσσαλος m. (money) μίζα f. (interest) ενδιαφέρον n. be at ~ διακυβεύομαι. play for high ~s διακυβεύω μεγάλα ποσά.

stake v. (place ~s) βάζω πασσάλους. (wager) ποντάρω, (risk) διακυβεύω, (both) παίζω.

stale a. (food) μπαγιάτικος, get ~ μπαγιατεύω, (fig., of person) έχω χάσει. (old) παλιός. **~ness** s. μπαγιατίλα f.

stalemate s. αδιέξοδο n.

stalk s. στέλεχος n., κοτσάνι n., μίσχος m.

stalk v.t. κυνηγώ, παρακολουθώ. (v.i.) βαδίζω με αργά βήματα. **~ing-horse** s. (fig.) πρόσχημα n.

stall s. (booth) μπάγκος m. (in church) στασίδι n. (for cattle) στάβλος m. ~s (theatre) πλατεία f.

stall v.i. (of engine) σταματώ, (not reply) αποφεύγω να απαντήσω. (v.t.) ~ off αποτρέπω. **~ing** s. υπεκφυγές f.pl.

stallion s. βαρβάτο άλογο.

stalwart s. παλληκάρι n.

stamina s. αντοχή f.

stammer v. τραυλίζω, **~er** s. τραυλός a.

stamp v.i. (with feet) χτυπώ τα πόδια μου. ~ on ποδοπατώ. (v.t.) (a document, letter) σφραγίζω. (imprint) εντυπώνω, (characterize) χαρακτηρίζω. ~ out (stifle) καταπνίγω, (uproot) ξεριζώνω.

stamp s. (die, mark, & fig.) σφραγίδα f. (mould) καλούπι n. (postage) γραμματόσημο n. (for ~-duty) χαρτόσημο n. **~-collecting** s. φιλοτελισμός m. **~-duty** s. χαρτόσημο n.

stampede s. (rush) ορμή f. (flight) φευγάλα f., άτακτος φυγή. (v.i.) ορμώ. (v.t.) πανικοβάλλω.

stance s. στάση f.

stanch v σταματώ.

stand s. (to ~ things on) βάση f. (plinth) υπόβαθρο n. (for hats, etc.) καλόγερος m. (with seats) εξέδρα f. (booth, kiosk, pavilion) περίπτερο n. make a ~ λαμβάνω αμυντική θέση. take one's ~ on βασίζομαι σε. he took his ~ by the window στάθηκε κοντά στο παράθυρο.

stand v.t. (set, put) ακουμπώ, στήνω, βάζω. (be resistant to) αντέχω σε, (put up with) ανέχομαι. (*treat to*) κερνώ. ~ a chance έχω ελπίδες.

stand *v.i.* στέκομαι (όρθιος). (*get up*) σηκώνομαι. (*stay*) παραμένω. (*be, be situated*) υπάρχω, βρίσκομαι, how (*or where*) do we ~? πού βρισκόμαστε; the matter ~s thus η υπόθεση έχει ως εξής. it ~s ten metres high έχει δέκα μέτρα ύψος. ~ to lose κινδυνεύω να χάσω. it ~s to reason είναι προφανές. ~ fast *or* firm κρατώ. ~ back *or* clear στέκομαι μακριά. ~ back! μην πλησιάζετε! ~ aside παραμερίζω, (*yield*) υποχωρώ.

stand by *v.t.* (*support*) υποστηρίζω, (*keep to*) εμμένω εις. (*v.i.*) (*be ready*) είμαι εν επιφυλακή. (*take no part*) μένω αμέτοχος.

stand-by *s.* αποκούμπι *n.*, στήριγμα *n.* as a ~ για ώρα ανάγκης. be on ~ είμαι εν επιφυλακή. (*a.*) εν αναμονή.

stand down *v.i.* αποσύρομαι.

stand for *v.* (*mean*) σημαίνω, (*represent*) αντιπροσωπεύω. (*put up with*) ανέχομαι, he won't ~ it δεν το σηκώνει. ~ Parliament πάω για βουλευτής.

stand in *v.i.* ~ with ενώνομαι με. ~ for αντικαθιστώ.

stand-in *s.* αντικαταστάτης *m.*

stand off *v.i.* μένω παράμερα. (*v.t., dismiss*) απολύω προσωρινά.

stand-offish *a.* be ~ κρατώ απόσταση. ~**ness** *s.* ακαταδεξία *f.*

stand out *v.i.* (*resist*) αντιστέκομαι, (*be prominent*) ξεχωρίζω.

stand over *v.i.* παραμένω εκκρεμής. let it ~ το αναβάλλω. (*v.t., supervise*) στέκομαι πάνω από το κεφάλι (*with gen.*).

stand to *v.t.* (*man*) παίρνω θέση σε (*v.i.*) (*be on alert*) παραμένω εν επιφυλακή.

stand-to *s.* συναγερμός *m.*

stand up *v.i.* σηκώνομαι (όρθιος). ~ for υποστηρίζω, παίρνω το μέρος (*with gen.*). ~ to (*resist*) αντίσταμαι εις, (*be resistant*) αντέχω σε.

standard *s.* (*flag*) σημαία *f.* (*measure*) μέτρο *n.*, κανόνας *m.* (*level*) επίπεδο *n.*, στάθμη *f.* moral ~ μέτρο της ηθικής. ~ of living βιοτικό επίπεδο. gold ~ χρυσούς κανών.

standard *a.* (*regular*) κανονικός, (*usual*) συνηθισμένος. (*work, authority*) κλασσικός, παραδεδεγμένος. (*measure*) επίσημος. ~ English τα σωστά αγγλικά. ~ lamp λαμπατέρ *n.*

standardiz|e *v.* τυποποιώ. ~**ation** *s.* τυποποίηση *f.*

standing *s.* be in good ~ έχω καλό όνομα. person of ~ σημαίνον πρόσωπον, άνθρωπος περιωπής of high social ~ ανωτέρας

κοινωνικής θέσεως. friendship of long ~ παλιά φιλία.

standing *a.* όρθιος, (*permanent*) μόνιμος.

standpoint *s.* άποψη *f.*

standstill *s.* σταμάτημα *n.* bring *or* come to a ~ σταματώ.

stanza *s.* στροφή *f.*

staple *s.* συνδετήρας *m.* (*v.t.*) συνδέω με συνδετήρα.

staple *a.* (*chief*) κύριος. ~ commodity κύριο προϊόν.

star *s.* αστήρ *m.*, άστρο, αστέρι *n.* (*of stage*) βεντέτα *f.* ~ turn κυριότερο νούμερο. S~s and Stripes Αστερόεσσα *f.* (*fam.*) see ~s βλέπω τον ουρανό σφοντύλι. (*v.i.*) (*play lead*) πρωταγωνιστώ. ~**-crossed** *a.* κακότυχος. ~**fish** *s.* αστερίας *m.* ~**-gazer** *s.* (*fig.*) ονειροπόλος *a.* ~**light** *s.* αστροφεγγιά *f.*

starboard *a.* δεξιός. to ~ δεξιά.

starch *s.* άμυλο *n.*, κόλλα *f.* (*v.*) κολλάρω. ~**y** *a.* αμυλώδης, (*prim*) τυπικός.

star|e *v.* ~e at καρφώνω τα μάτια μου επάνω σε. why are you ~ing at me so? γιατί με κοιτάς έτσι επίμονα; ~e in surprise μένω κατάπληκτος. (*of mislaid object*) it's ~ing you in the face είναι κάτω από τη μύτη σου. (*s.*) βλέμμα *n.* ~**ing** *a.* (*colour*) χτυπητός, (*eyes*) γουρλωμένος. be (*stark*) ~ing mad είμαι (τρελλός) για δέσιμο.

stark *a.* (*stiff*) άκαμπτος, (*bare*) γυμνός. (*utter*) πλήρης. (*adv.*) τελείως. ~ naked θεόγυμνος. ~ mad για δέσιμο.

starling *s.* μαυροπούλι *n.*, ψαρόνι *n.*

starry *a.* ξάστερος, έναστρος. ~ night ξαστεριά. *f.* ~**-eyed** *a.* ουτοπιστής.

start *v.i.* (*begin*) αρχίζω. (*set out, depart*) ξεκινώ, φεύγω, αναχωρώ. (*of motor*) παίρνω μπρος. (*jump*) τινάζομαι, (*come loose*) ανοίγω. his eyes ~ed out of his head γούρλωσε τα μάτια του. ~ up (*arise*) ανακύπτω, (*rise suddenly*) πετάγομαι. ~ in on, ~ out (*to*) αρχίζω. he ~ed out to learn English βάλθηκε να μάθει αγγλικά. to ~ with πρώτα πρώτα, (*at the beginning*) στην αρχή.

start *v.t.* αρχίζω. (*chat, work, friendship*) πιάνω. (*motor, scheme*) βάζω μπρος. it has ~ed me thinking με έβαλε σε σκέψεις.

start *s.* (*beginning*) αρχή *f.* (*jump*) τίναγμα *n.* give (*person*) a ~ ξαφνίζω. get the ~ of ξεπερνώ.

starter *s.* (*of race*) αφέτης *m.* (*of motor*) μίζα *f.*

starting|-point, ~-post *s.* αφετηρία *f.*

start|le *v.* τρομάζω, ξαφνιάζω. be ~led τρομάζω, ξαφνιάζομαι. ~**ling** *a.* τρομακτικός.

starvation s. πείνα f.

starv|e v.i. πεινώ. ~e to death πεθαίνω από πείνα. be ~ing for πεθαίνω για. (v.t.) they ~ed us μας τάραξανε στην πείνα.

starv|ling a., ~**eling** s. πειναλέος.

state s. (condition) κατάσταση f. (political) πολιτεία f. the S~ το κράτος. (civil power) το δημόσιον. (ceremony, pomp) επισημότητα f., in ~ επισήμως. lying-in-~ λαϊκό προσκύνημα. (fam.) be in a ~ (worried) ανησυχώ.

state a. κρατικός, (ceremonial) επίσημος.

state v.t. δηλώνω, (expound) εκθέτω. ~**ment** s. δήλωση f. έκθεση f. (by witness) κατάθεση f. (fin.) κατάσταση f.

stately a. επιβλητικός, (distinguished) αρχοντικός.

state-of-the-art a. (fam.) εκπληκτικής τελειότητας.

statesman s. πολιτικός ανήρ. ~**ship** s. πολιτική ικανότητα.

static a. στατικός.

station s. σταθμός m. (police) τμήμα n. (naval) βάση f. (place) θέση f. (rank) κοινωνική θέση. (v.t.) τοποθετώ.

stationary a. (not changing) στάσιμος, (not movable) μόνιμος, (having stopped) σταθμεύων, εν στάσει.

station|er s. χαρτοπώλης m. ~**er's** s. χαρτοπωλείο n. ~**ery** s. είδη χαρτοπωλείου, χαρτοφάκελλα n.pl.

statistical a. στατιστικός. ~**ly** adv. στατιστικώς.

statistic|s s. στατιστική. f. ~**ian** s. στατιστικός m.

statu|e s. άγαλμα n. (only of person) ανδριάς m. ~**ary** s. αγάλματα n.pl. ~**esque** a. αγαλματένιος, σαν άγαλμα. ~**ette** s. αγαλματάκι n.

stature s. ανάστημα n., μπόι n.

status s. (legal) κατάσταση f. (official) ιδιότητα f. (social) θέση f. ~ quo καθεστώς n.

statute s. θέσπισμα n., νόμος m. ~s (of corporation, etc.) καταστατικόν s.

statutory a. επιβεβλημένος, θεσπισμένος.

staunch a. πιστός. ~**ly** adv. πιστά.

staunch v. σταματώ.

stave s. (of barrel) ντούγια f. (stanza) στροφή f. (mus.) πεντάγραμμο n.

stave v. ~ off αποσοβώ. ~ in τρυπώ, σπάζω.

stay v.i. (remain) μένω, (wait) περιμένω, (sojourn) διαμένω, (continue to be) παραμένω. ~ put μένω στην ίδια θέση. ~ up (at night) κοιμάμαι αργά. (v.t.) (stop) σταματώ, (delay) αναστέλλω. ~ the course αντέχω ως το τέλος. ~ one's hunger κόβω την πείνα μου. ~-**at-home** s. σπιτόγατος

m. ~-**ing-power** s. αντοχή f.

stay s. (sojourn) διαμονή, παραμονή f. (delay) αναστολή f.

stay s. (support) στήριγμα n. ~s κορσές m. (v.t.) ~ (up) στηρίζω.

stead s. in his ~ στη θέση του. I acted in his ~ τον αντικατέστησα. it stood me in good ~ μου φάνηκε πολύ χρήσιμο.

steadfast a. σταθερός, πιστός.

stead|y a. σταθερός. (regular) κανονικός, (continuous) συνεχής. (in habits) μετρημένος. (v.t.) σταθεροποιώ. ~**ily** adv. σταθερά, (all the time) όλο.

steak s. φιλέτο n.

steal v.t. κλέβω. ~ a march on προλαβαίνω. (v.i.) (creep) γλιστρώ. ~**ing** s. κλέψιμο n., κλεψιά f.

stealth s. by ~ λαθραίως, κρυφά. ~**y** a. (glance) κρυφός, (silent) αθόρυβος.

steam (v.i.) βγάζω ατμό, αχνίζω. (sail) πλέω. (v.i.) (apply ~ to) αχνίζω, (cook) ψήνω στον αχνό. (fig.) get ~ed up εκνευρίζομαι. the window is ~ed up το παράθυρο θόλωσε.

steam s. ατμός m., αχνός m. at full ~ ολοταχώς. let off ~ ξεθυμαίνω. under one's own ~ με τις ίδιες μου τις δυνάμεις. ~-**er**, ~-**boat**, ~-**ship** s. ατμόπλοιο n. ~-**engine** s. ατμομηχανή f. ~-**navigation** s. ατμοπλοΐα f. ~-**roller** s. οδοστρωτήρας m.

steed s. άτι n.

steel s. χάλυψ m., ατσάλι n. (a.) χαλύβδινος, ατσαλένιος. (v.t.) χαλυβδ(δ)ώνω, σκληραίνω. ~ one's heart χαλυβώνω την ψυχή μου. ~**y** a. σκληρός.

steep a. απότομος, (fig.) υπερβολικός. ~**ly** adv. απότομα. ~**ness** s. απότομο n.

steep v. μουσκεύω. ~**ed** a. (fig.) εμποτισμένος, (in vice, ignorance) βουτηγμένος.

steeple s. καμπαναριό με βέλος.

steeplechase s. δρόμος μετ' εμποδίων.

steer s. βόδι n.

steer v.t. κυβερνώ, (car) οδηγώ. (v.i., in ship) κρατώ το πηδάλιο. ~ clear of αποφεύγω. ~-**ing-wheel** s. τιμόνι n. ~-**sman** s. τιμονιέρης m.

steerage s. travel ~ ταξιδεύω τρίτη θέση (or κατάστρωμα).

stellar a. αστρικός.

stem s. στέλεχος n., μίσχος m., κοτσάνι n. (of glass) πόδι n. (gram.) θέμα n. from ~ to stern από πλώρη σε πρύμνη.

stem v.t. (check) συγκρατώ, βάζω φράγμα σε. (fig.) αναχαιτίζω. (v.i.) ~ from προέρχομαι από.

stench s. μπόχα f.

stencil s. (for duplicating) μεμβράνη πολυγράφου, στένσιλ n. (for design) αχνάρι n.

stenograph|er s. στενογράφος m.f. ~y s. στενογραφία f.

stentorian a. στεντόρειος.

step s. βήμα n. (foot~) πάτημα n. (action taken) διάβημα n. (stair) σκαλί, σκαλοπάτι n. ~s (of entrance, etc.) σκάλες f.pl. pair of ~s σκάλα f. take ~s λαμβάνω μέτρα. watch one's ~ προσέχω. keep ~ with συμβαδίζω με. ~ by ~ βήμα προς βήμα.

step v.i. (walk) περπατώ, (take a ~) κάνω ένα βήμα. ~ over δρασκελίζω. ~ on πατώ, (fam.) ~ on it πατώ γκάζι, κάνω γρήγορα. ~ in (fig.) επεμβαίνω, μπαίνω στη μέση. ~ down (fig.) παραιτούμαι. ~ out ανοίγω το βήμα. (v.t.) ~ up αυξάνω, εντείνω.

step|brother s. ετεροθαλής αδελφός. ~child s. προγονός m. ~father s. πατριός m. ~mother s. μητριά f. ~sister s. ετεροθαλής αδελφή.

steppe s. στέππα f.

stereo a. στερεοφωνικός.

stereoscopic a. στερεοσκοπικός.

stereotyped a. στερεότυπος.

steril|e a. άγονος, στείρος. ~ity s. στειρότητα f. ~ize v. αποστειρώνω. ~ization s. αποστείρωση f.

sterling s. (fin.) στερλίνα f. (fig.) γνήσιος.

stern a. αυστηρός. ~ly adv. αυστηρά. ~ness s. αυστηρότητα f.

stern s. πρύμνη f.

stertorous a. ρογχώδης.

stethoscope s. στηθοσκόπιο n.

stevedore s. φορτοεκφορτωτής m.

stew v.t. & i. σιγοβράζω. (s.) εντράδα f., ραγού n. (fam.) get in a ~ γίνομαι άνω κάτω. ~ed a. βραστός. ~ed fruit κομπόστα f. ~pot s. τσουκάλι n., χύτρα f.

steward s. οικονόμος m. (at function) επιμελητής m. (naut.) καμαρότος m. ~ship s. διαχείριση f.

stick s. (bit of wood) ξύλο n. (walking) μπαστούνι n., (candle) ραβδί n. (piece) κομμάτι n. (fig.) in a cleft ~ σε δίλημμα. get hold of the wrong end of the ~ καταλαβαίνω στραβά.

stick v.t. (thrust) χώνω, μπήζω. (transfix) καρφώνω. (put) βάζω, (endure) ανέχομαι, (fix, fasten) κολλώ. ~ it out αντέχω ως το τέλος. ~ it on (overdo it) τα παρακάνω (v.i.) (also get stuck) κολλώ. ~ together παραμένουμε ενωμένοι. ~ at nothing είμαι ικανός για όλα. it ~s in my throat μου κάθεται στο λαιμό. ~ing-plaster s. λευκοπλάστης m. ~-in-the-mud s. ρουτινιέρης m.

stick out v.t. βγάζω, απλώνω. (v.i.) προεξέχω. ~ for επιμένω εις.

stick to v. (friends) μένω πιστός σε. (abide by) εμμένω εις.

stick up v.i. προεξέχω. ~ for υποστηρίζω.

stickler s. be a ~ for... είμαι σχολαστικός σε.

sticky a. it is ~ κολλάει.

stiff a. σκληρός, (with cold) ξυλιασμένος, (of muscles) πιασμένος. (formal) τυπικός, (difficult) δύσκολος, (strong) δυνατός. I have a ~ neck πιάστηκε ο λαιμός μου. ~ly adv. (of movement) αλύγιστα, σαν πιασμένος. (of manner) ψυχρά.

stiffen v.t. & i. σκληραίνω, (with cold) ξυλιάζω. (intensify) δυναμώνω. (v.i.) (of muscles) πιάνομαι.

stiffness s. σκληρότητα f. (of body) δυσκαμψία f., πιάσιμο n. (manner) ψυχρότητα f.

stifl|e v.t. πνίγω, (fig.) καταπνίγω. (v.i.) πνίγομαι. ~ing a. αποπνικτικός, πνιγηρός.

stigma s. στίγμα n. ~tize v. στιγματίζω.

still s. λαμπίκος m.

still adv. ακόμα. (but) ~ (nevertheless) όμως, ωστόσο.

still a. (unmoving) ακίνητος, (calm) ήρεμος. keep or stand ~ κάθομαι ήσυχος, δεν κουνιέμαι. ~ life νεκρά φύση. ~ waters run deep (αυτός είναι) σιγανό ποτάμι. (v.t.) καθησυχάζω. ~ness s. ηρεμία f.

still-born a. the baby was ~ το μωρό γεννήθηκε νεκρό. (fig., plan, etc.) θνησιγενής.

stilts s. ξυλοπόδαρα n.pl.

stilted a. προσποιητός, επιτηδευμένος.

stimulant s. διεγερτικό n.

stimulat|e v. διεγείρω, κεντρίζω, ερεθίζω. ~ing a. τονωτικός. ~ion s. διέγερση f.

stimulus s. κίνητρο n. (physiological) ερέθισμα n.

sting s. (organ) κεντρί n. (wound, pain) κέντημα n. (v.t.) κεντρίζω, (fig., overcharge) γδέρνω (v.i., smart) τσούζω.

sting|y a. τσιγγούνης. ~iness s. τσιγγουνιά f.

stink s. μπόχα f., βρώμα f. (v.) σκυλοβρωμάω. (also fig.) (also ~ of) βρωμάω. ~ing a. βρωμερός.

stint v.t. (food, etc.) τσιγγουνεύομαι. ~ oneself of στερούμαι (with acc. or gen.).

stint s. without ~ αφειδώς. (amount of work) ανατιθέμενη εργασία. I did my ~ έκανα το αναλογούν μου.

stipend s. μισθός m. ~iary a. έμμισθος.

stipple s. πουαντιγέ n.

stipulat|e v. ορίζω ρητώς. ~ion s. όρος m.

stir *v.i.* σαλεύω, (*bestir oneself*) κουνιέμαι. (*v.t.*) σαλεύω, κινώ. (*rouse*) εγείρω, (*excite, move*) συγκινώ. (*mix*) ανακατώνω. ~ up (*trouble, etc.*) υποκινώ. (*s.*) (*commotion*) σάλος *m.* **~ring** *a.* (*adventure*) συναρπαστικός, (*speech*) συγκινητικός.

stirrup *s.* αναβολεύς *m.*

stitch *s.* βελονιά *f.* (*knitting*) πόντος *m.* (*of wound*) ράμμα *n.* (*pain*) σουβλιά *f.* (*v.*) ράβω. **~ing** *s.* ράψιμο *n.*

stoat *s.* νυφίτσα *f.*

stock *s.* (*supply*) παρακαταθήκη *f.*, απόθεμα *n.* (*in shop*) στοκ *n.* it is out of ~ εξαντλήθηκε. take ~ of εκτιμώ. (*handle*) λαβή *f.* (*of rifle*) κοντάκι *n.* (*of anchor*) τσίπος *m.* (*beasts*) ζώα *n.pl.* (*liquor*) ζωμός *m.* (*flower*) βιόλα *f.* (*breed*) γένος *n.*, σόι *n.* he comes of English ~ κατάγεται από Άγγλους. (*fin.*) μετοχές *f.pl.*, τίτλοι *m.pl.*, αξίες *f.pl.* his ~ is going up (*fig.*) οι μετοχές του ανεβαίνουν. (*shipbuilding & fig.*) on the ~s στα σκαριά. ~s and stones άψυχα αντικείμενα.

stock *v.t.* (*equip, fill*) εφοδιάζω, γεμίζω. (*have, keep*) έχω. (*v.i.*) ~ up with εφοδιάζομαι με, αποθηκεύω.

stockade *s.* πασσαλόπηγμα *n.*

stock-breeding *s.* κτηνοτροφία *f.*

stockbroker *s.* χρηματιστής *m.*

stock-exchange, ~-market *s.* χρηματιστήριον *n.*

stocking *s.* κάλτσα *f.*

stock-in-trade *s.* σύνεργα *n.pl.* (*fig.*) παρακαταθήκη *f.*

stockpile *s.* αποθέματα *n.pl.*

stock-still *a.* τελείως ακίνητος.

stock-taking *s.* απογραφή *f.*

stocky *a.* γεροδεμένος.

stockyard *s.* μάντρα *f.*

stodgy *a.* δυσκολοχώνευτος, βαρύς.

stoic *s.*, **~al** *a.* στωικός. **~ally** *adv.* στωικά. **~ism** *s.* στωικότητα *f.*

stoke *v.* τροφοδοτώ. **~r** *s.* θερμαστής *m.*

stole *s.* (*eccl.*) πετραχήλι *n.* (*woman's*) σάρπα *f.*

stolid *a.* φλεγματικός.

stomach *s.* στομάχι *n.* (*fig.*) (*appetite*) όρεξη *f.*, (*courage*) θάρρος *n.* (*v.t.*) ανέχομαι. (*a.*) στομαχικός.

stone *v.* λιθοβολώ, πετροβολώ. (*remove ~ of*) ξεκουκουτσιάζω. **~d** *a.* με βγαλμένα τα κουκούτσια. (*fam., drunk*) σκνίπα.

stone *s.* λίθος *m.f.*, πέτρα *f.* (*of fruit*) κουκούτσι *n.* throw ~s *see* stone *v.* leave no ~ unturned βάζω λυτούς και δεμένους. a ~'s throw δύο βήματα. (*a.*) λίθινος, πέτρινος. **~breaker** *s.* λιθοκόπος *m.* **~built** *a.* λιθόκτιστος. **~cold** *a.* παγωμέ-

νος. **~deaf** *a.* θεόκουφος. **~mason** *s.* λιθοξόος *m.* **~pine** *s.* κουκουναριά *f.* **~wall** *v.i.* κωλυσιεργώ. **~work** *s.* λιθοδομή *f.*

stony *a.* (*ground*) πετρώδης. (*fig.*) σκληρός. **~broke** *a.* (*fam.*) μπατίρης.

stooge *s.* (*fam.*) υποτελής *a.*

stool *s.* σκαμνί *n.* (*faeces*) κενώσεις *f.pl.* go to ~ αποπατώ.

stoop *v.* σκύβω, (*deign*) καταδέχομαι.

stop *s.* (*stopping, stopping-place*) στάση *f.* come to a ~ σταματώ. put a ~ to θέτω τέρμα εις, τερματίζω. (*punctuation*) σημείο στίξεως. (*gram.*) ~ consonant στιγμιαίο *n.*

stop *v.t.* (*fill up, block*) κλείνω, βουλ-λώνω, φράζω. (*withhold*) κόβω. (*halt*) σταματώ, (*break off*) διακόπτω, (*prevent*) εμποδίζω. (*v.i.*) (*cease*) σταματώ, παύω. (*stand still*) σταματώ, στέκομαι, (*of bus, etc.*) κάνω στάση. (*stay, reside*) μένω. ~ off or over διακόπτω το ταξίδι μου. **~cock** *s.* γενικός διακόπτης του νερού. **~gap** *s.* προσωρινή λύση. **~press** *s.* ειδήσεις επί του πιεστηρίου. **~watch** *s.* χρονόμετρο με διακόπτη.

stoppage *s.* σταμάτημα *n.* διακοπή *f.* στάση *f.*

stopper *s.* πώμα *n.*

stopping *s.* (*of tooth*) σφράγισμα *n.*

storage *s.* (*act*) αποθήκευση *f.* (*place*) αποθήκη *f.* (*space*) χώρος για αποθήκευση. in cold ~ στην κατάψυξη.

store *s.* (*stock*) απόθεμα *n.* (*~house*) αποθήκη *f.* (*shop*) κατάστημα *n.* ~s (*supplies*) εφόδια *n.pl.* προμήθειες *f.pl.* in ~ (*stored*) αποθηκευμένος, (*to come*) what the future has in ~ το τι μας επιφυλάσσει το μέλλον. set ~ by αποδίδω σημασίαν εις. (*v.t.*) αποθηκεύω, βάζω στην αποθήκη. **~keeper** *s.* αποθηκάριος *m.*

storey *s.* όροφος *m.*, πάτωμα *n.*, two- ~ed διώροφος, δίπατος.

storied *a.* θρυλικός.

stork *s.* πελαργός *m.*, λελέκι *n.*

storm *s.* (*lit. & fig.*) θύελλα *f.* (*at sea*) φουρτούνα, τρικυμία *f.* (*uproar*) θόρυβος *m.* ~ in a teacup μεγάλο κακό για το τίποτα. (*v.i.*) λυσσομανώ. he ~ed in όρμησε μέσα αποθηριωμένος. (*v.t., capture*) καταλαμβάνω εξ εφόδου. **~tossed** *a.* be ~-tossed θαλασσοδέρνομαι. **~y** *a.* θυελλώδης, (*sea*) τρικυμιώδης.

story *s.* ιστορία *f.* (*esp. literary*) διήγημα *n.* (*fib*) παραμύθι *n.*

stout *a.* (*brave*) θαρραλέος, (*fattish*) γεμάτος, (*strongly made*) γερός, αντοχής. **~ly** *adv.* θαρραλέως, γερά. (*insistently*) επιμόνως.

stove s. (heating) σόμπα f., θερμάστρα f. (cooking) κουζίνα f. (paraffin) γκαζιέρα f. ~**pipe** s. μπουρί n.

stow v.t. ταχτοποιώ, βάζω. ~**away** s. λαθρεπιβάτης m.

straddle v.i. στέκομαι με τα πόδια ανοικτά. (v.t.) (sit astride) καβαλλικεύω, (span) περνώ, (a target) περιβάλλω.

straggl|e v.i. (spread untidily) απλώνομαι ακατάστατα, (drop behind) βραδυπορώ. ~**er** s. βραδυπορών m. ~**ing** a. σκορπισμένος.

straight a. (not crooked) ίσιος, (erect) ορθός, ίσιος. (direct) ευθύς, (honest) ευθύς, ίσιος. (unmixed) σκέτος. (in good order) ταχτοποιημένος. ~ line ευθεία f. ~ face σοβαρή έκφραση.

straight adv. (also ~ ahead) ίσια. (directly) κατ' ευθείαν. ~ away αμέσως. put ~ see straighten. keep or go ~ μπαίνω στον ίσιο δρόμο.

straighten v.t. ισιώνω. (settle, tidy) (also ~ out) ταχτοποιώ, σιάζω. (v.i.) σιάζω. ~ up τεντώνομαι.

straightforward a. (simple) απλός, (honest) ευθύς.

straightway adv. αμέσως.

strain s. (mech.) πίεση f. (nervous) υπερένταση f. (hard work, trouble) κόπος m. (burden) επιβάρυνση f. the ~ of modern life ο εντατικός ρυθμός της σύγχρονης ζωής. (tendency) τάση f., φλέβα f. (tone, style) ύφος n. ~s (of music) ήχοι m.pl. (stock, race) σόι n., ράτσα f.

strain v.t. (stretch) τεντώνω, (tire) κουράζω, (sprain) στραμπουλίζω. (filter) στραγγίζω, σουρώνω. (patience) εξαντλώ, (meaning, etc.) παρατραβώ. (v.i.) (make effort) σφίγγομαι. ~ on or at (pull) τραβώ. ~ after (seek) κυνηγώ, επιδιώκω. ~**ed** a. (manner) αφύσικος, (relations) τεταμένος, (nerves) σε υπερένταση. ~**er** s. τρυπητό, σουρωτήρι n.

strait s. (geographical) στενό n. ~s, ~ened circumstances δυσχέρειες f.pl. ~**jacket** s. ζουρλομανδύας m. ~**laced** a. σεμνότυφος.

strand v.t. (run aground) ρίχνω έξω. (fig.) be ~ed βρίσκομαι χωρίς πόρους.

strand s. (thread) νήμα n. (shore) ακτή f.

strange a. (alien, not one's own) ξένος, (unfamiliar) ασυνήθιστος. (peculiar) παράξενος, ιδιαίτερος. ~**ly** adv. περίεργα. ~**ness** s. the ~ness of his surroundings το ασυνήθιστο του περιβάλλοντός του.

stranger s. ξένος m. he is a ~ to me μου είναι άγνωστος. be no ~ to (know all about) ξέρω από.

strangle v. στραγγαλίζω, πνίγω. ~**hold** s. have a ~ hold on κρατώ από το λαιμό.

strangulation s. στραγγαλισμός m.

strap s. ιμάς m., λουρί n. (v.) δένω με λουρί.

strapping a. γεροδεμένος.

stratagem s. στρατήγημα n.

strateg|y s. στρατηγική f. ~**ic** a. στρατηγικός.

stratification s. στρωμάτωση f.

stratosphere s. στρατόσφαιρα f.

stratum s. στρώμα n.

straw s. άχυρο n. (woven) ψάθα f. (for drinking) καλαμάκι n. (a.) αχυρένιος, ψάθινος.

strawberry s. φράουλα f.

stray v. (of animals) ξεφεύγω. (lose the way) χάνω το δρόμο, (morally) παραστρατώ. (a.) (animal, bullet) αδέσποτος. (scattered) σκόρπιος. (chance) τυχαίος.

streak s. γραμμή f. (stripe) ρίγα f. (of light) αχτίδα f. (trace) φλέβα f. (v.i.) περνώ αστραπιαίως. ~**ed**, ~**y** a. ριγωτός.

stream s. (current) ρεύμα n. (river) ποτάμι n. (brook) ρυάκι n. (fig., people) κύμα n. ~ of abuse υβρεολόγιον n.

stream v.i. (pour) τρέχω. ~ in εισρέω, ~ out ξεχύνομαι. (in the wind) κυματίζω. ~**er** s. σερπαντίνα f.

streamline v. (fig.) εκσυγχρονίζω. ~**d** a. (vehicle) αεροδυναμικός.

street s. δρόμος m., οδός f. ~ musician πλανόδιος μουσικός. ~ market λαϊκή αγορά. ~wise a. μάγκας m.

strength s. δύναμη f., ισχύς f. (esp. moral, physical) σθένος n. on the ~ of (relying) βασιζόμενος σε. ~**en** v.t. & i. δυναμώνω. (v.t.) ενισχύω.

strenuous a. εντατικός, σκληρός, (effort, denial) επίμονος. ~**ly** adv. σκληρά, επιμόνως.

stress s. (pressure) πίεση f. (mech.) τάση f. (emotional) ένταση f. (emphasis) έμφαση f. (accent) τόνος m. (accentuation) τονισμός m. (v.t.) τονίζω, υπογραμμίζω.

stretch v.t. τεντώνω. ~ out (hand, etc.) εκτείνω, απλώνω. ~ one's legs (fig.) πάω να ξεμουδιάσω. ~ a point κάνω μία παραχώρηση.

stretch v.i. ~ (oneself) τεντώνομαι. ~ (oneself) out ξαπλώνομαι. (extend) απλώνομαι, εκτείνομαι. (of materials) ανοίγω.

stretch s. (land. water) έκταση f. (time) διάστημα n. (of fingers, etc.) άνοιγμα n. at a ~ (continuously) μονοκοπανιά.

stretcher s. φορείο n.

strew v. (cover) στρώνω, (scatter) σκορπίζω.

stricken a. (with grief) συντετριμμένος, (by

disease) προσβεβλημένος, (*by misfortune,
love*) χτυπημένος. ~ in years προχωρημένης ηλικίας.

strict *a.* αυστηρός, (*absolute*) απόλυτος. **~ly**
adv. αυστηρώς, ~ly speaking για να
είμαστε ακριβείς. **~ness** *s.* αυστηρότητα *f.*

stricture *s.* επίκριση *f.* pass ~s on επικρίνω.

stride *v.* βαδίζω με μεγάλα βήματα. (*s.*)
μεγάλο βήμα, δρασκελιά *f.* take it in one's
~ το κάνω χωρίς δυσκολία.

strident *a.* τραχύς.

strife *s.* διαμάχη *f.*

strike *s.* (*raid*) επιδρομή *f.* (*find*) εύρημα *n.*
(*refusal to work*) απεργία *f.* go on ~
απεργώ.

strike *v.t.* (*hit, cause to hit*) χτυπώ. (*impress*)
εντυπωσιάζω. how does it ~ you? πώς σου
φαίνεται; it ~s me (*I think*) νομίζω. (*a
blow*) καταφέρω. ~ a blow for (*fig.*) αγωνίζομαι για. (*a match*) ανάβω, (*sparks*)
βγάζω, (*coins*) κόβω. (*a flag*) κατεβάζω, (*a
tent*) ξεστήνω. ~ terror into τρομοκρατώ. ~
a bargain κλείνω συμφωνία. ~ an attitude
παίρνω πόζα.

strike *v.i.* (*attack*) επιτίθεμαι, χτυπώ. (*of
clock*) χτυπώ, βαρώ. (*of match*) ανάβω.
(*stop work*) απεργώ. ~ into *or* across
(*traverse*) διασχίζω.

strike down *v.t.* πλήττω.

strike in *v.i.* (*intervene*) πετιέμαι.

strike off *v.t.* (*cut off*) κόβω, (*remove*)
διαγράφω, (*print*) τυπώνω.

strike out *v.t.* (*erase*) σβήνω. (*v.i.*) (*give
blows*) γρονθοκοπώ. (*swim*) κολυμπώ. (*in
enterprise*) απλώνομαι. ~ for (*proceed*)
βάζω μπρος (*or* τραβώ) για.

strike up *v.t.* ~ a song αρχίζω να τραγουδώ.
~ a friendship πιάνω φιλίες.

striker *s.* απεργός *m.*

striking *a.* εντυπωσιακός. **~ly** *adv.* εκπληκτικά.

string *s.* (*twine*) σπάγγος *m.* (*row, sequence*)
σειρά *f.* (*of bow, heart, mus.*) χορδή *f.*, ~s
έγχορδα *n.pl.* (*onions, lies*) ορμαθός *m.* ~
of beads κομπολόι *n.* pull the ~s (*of
puppets & fig.*) κρατώ τα νήματα. pull ~s
(*use influence*) βάζω μέσα. without ~s
χωρίς όρους. have two ~s to one's bow το
έχω δίπορτο. **~ed** *a.* έγχορδος. **~y** *a.*
ινώδης.

string *v.t.* (*beads, etc.*) περνώ. (*hang*)
κρεμώ. ~ together (*words*) συντάσσω. ~
out αραιώνω.

stringen|t *a.* αυστηρός. **~cy** *s.* αυστηρότητα
f.

strip *s.* λουρίδα *f.*, ταινία *f.* ~ cartoon κόμικς
n. ~ lighting λαμπτήρες φθορισμού.

strip *v.t.* (*remove by stripping*) βγάζω.

(*person*) γδύνω. ~ (of) (*rob, deprive*)
απογυμνώνω (*with gen. or* από). he was
~ped of his powers τον απογύμνωσαν από
τις εξουσίες του. (*v.i.*) (*undress*) γδύνομαι.

stripe *s.* ρίγα *f.* (*mil.*) γαλόνι *n.* ~s (*of
animal*) ραβδώσεις *f.pl.* **~d** *a.* ριγωτός,
ριγέ. (*animal*) με ραβδώσεις.

stripling *s.* νεανίσκος *m.*

striv|e *v.i.* (*try*) πασχίζω. (*struggle*) παλεύω,
πολεμώ. **~e** after επιδιώκω. **~ing** *s.* πάλη
f.

stroke *s.* (*blow*) χτύπημα *n.* (*movement*)
κίνηση *f.* (*in swimming*) απλωτή *f.* (*of
clock*) χτύπημα *n.* (*of hammer*) σφυριά *f.* ~
of luck ευτύχημα *n.*, λαχείο *n.* brush-~
πινελιά *f.* I haven't done a ~ of writing δεν
έβαλα μία πεννιά. (*illness*) συμφόρηση *f.*
at one ~ με μιας.

stroke *v.* (*caress*) χαϊδεύω. (*s.*) χάδι *n.*

stroll *v.i.* σουλατσάρω. *s.* go for a ~ κάνω
μία βόλτα.

strong *a.* δυνατός, ισχυρός. (*vivid, intense*)
έντονος. (*durable*) γερός. ~ point *or* suit
φόρτε *n.* he is going ~ βαστιέται καλά.
~-box *s.* χρηματοκιβώτιο *n.* **~ly** *adv.* δυνατά, ισχυρώς, γερά.

stronghold *s.* φρούριο *n.* (*fig.*) προπύργιο
n.

strop *v.* ακονίζω στο λουρί.

structural *a.* (*alteration, etc.*) δομικός. **~ly**
adv. δομικά.

structure *s.* (*way thing is made*) κατασκευή
f. (*building*) οικοδόμημα *n.*

struggle *v.* παλεύω, πολεμώ. (*s.*) πάλη *f.*

strum *v.* (*fam.*) γρατσουνάω, κοπανάω.

strumpet *s.* πόρνη *f.*

strut *s.* (*support*) υποστήριγμα *n.*

strut *v.i.* περπατώ κορδωμένος.

stub *s.* (*counterfoil*) στέλεχος *n.* (*bit left*)
υπόλειμμα *n.* (*of cigarette*) γόπα *f.*

stub *v.t.* ~ out σβήνω. ~ one's toe σκοντάφτω.

stubble *s.* καλαμιές *f.pl.*

stubborn *a.* πεισματάρης, (*insistent*) επίμονος. **~ly** *adv.* με πείσμα, επιμόνως.
~ness *a.* πείσμα *n.*, επιμονή *f.*

stuck *a.* κολλημένος. **~-up** *a.* ψηλομύτης.

stud *s.* (*fastener*) κουμπί *n.* (*in boot, road*)
καρφί *n.* (*for breeding*) ιπποτροφείο *n.*
(*v.*) ~ded with γεμάτος (*από*).

student *s.* σπουδαστής *m.*, σπουδάστρια *f.*
(*university*) φοιτητής *m.* φοιτήτρια *f.*

studied *a.* μελετημένος, εσκεμμένος.

studio *s.* στούντιο *n.*

studious *a.* μελετηρός, (*deliberate*) μελετημένος. **~ly** *adv.* επιμελώς.

study *v.* (*learn*) σπουδάζω. (*examine, make
a ~ of*) μελετώ. (*care for*) φροντίζω για. he
is ~ing to be a doctor σπουδάζει γιατρός.

study s. μελέτη f., σπουδή f. (*room*) γραφείο n. (*written work*) μελέτη f. (*painting*) σπουδή f. (*music*) ετύντ n.

stuff s. υλικό n. (*cloth*) ύφασμα n. (*of character*) πάστα f. (*thing*) πράμα n. ~ and nonsense μπούρδες f.pl.

stuff v.t. (*cushion, fowl*) (παρα)γεμίζω. (*with food*) μπουκώνω. (*preserve*) ταριχεύω, (*thrust*) χώνω. ~ (up) (*block*) βουλλώνω. (*v.i., overeat*) τρώω τον περίδρομο. ~ed (*food*) γεμιστός.

stuffing s. γέμιση f.

stuffy a. it is ~ μυρίζει κλεισούρα. (*fam., person*) στενόμυαλος.

stultify v. αχρηστεύω.

stumb|le v. σκοντάφτω. (*in speech*) μπερδεύω τα λόγια μου. ~le on πέφτω επάνω σε. ~**ling-block** s. εμπόδιο n.

stump s. (*of tree*) κούτσουρο n. (*bit left*) υπόλειμμα n. ~**y** a. κοντόχοντρος.

stump v.i. (*walk*) περπατώ βαριά. (*v.t.*) be ~ed δεν γνωρίζω την απάντηση. ~ up πληρώνω.

stun v. ζαλίζω. ~**ning** a. καταπληκτικός.

stunt s. διαφημιστικό κόλπο.

stupef|y v. (*amaze*) καταπλήσσω, (*fuddle*) αποβλακώνω. ~**action** s. κατάπληξη f. αποβλάκωση f. ~**ied** a. κατάπληκτος, αποβλακωμένος.

stupendous a. τεράστιος, καταπληκτικός.

stupid a. κουτός, ηλίθιος. ~**ity** s. κουταμάρα f. ηλιθιότητα f., βλακεία f. ~**ly** adv. κουτά, ηλίθια.

stupor s. χαύνωση f.

sturd|y a. γερός. ~**iness** s. ρώμη f., δύναμη f.

stutter v. τραυλίζω.

sty s. (*pigs'*) χοιροστάσιο n. (*on eye*) κριθαράκι n.

style s. (*manner, way of writing, etc.*) ύφος n. (*art*) ρυθμός m. (*fashion*) στυλ n. life-~ τρόπος του ζην. live in ~ κάνω μεγάλη ζωή. (*title*) τίτλος m.

stylish a. κομψός, (*performer*) be ~ έχω στυλ. ~**ly** adv. με κομψότητα.

stylist s. be a ~ έχω στυλ. ~**ic** a. του ύφους.

stylized a. στυλιζαρισμένος.

styptic a. στυπτικός.

suav|e a. γλυκομίλητος, αβρός. ~**ity** s. αβρότητα f.

subaltern s. κατώτερος αξιωματικός.

subconscious s. υποσυνείδητο n. ~**ly** adv. υποσυνειδήτως.

subdiv|ide v. υποδιαιρώ. ~**ision** s. υποδιαίρεση f.

subdue v. υποτάσσω. ~**d** a. ήσυχος, (*sound*) χαμηλός.

sub-editor s. συντάκτης m.

subhuman a. ~ person υπάνθρωπος m.

subject s. (*citizen*) υπήκοος m.f. (*topic*) θέμα n. (*for experiment & gram.*) υποκείμενο n. (*matter under discussion, point*) προκείμενο n. (*theme for play, etc.*) υπόθεση f. (*lesson*) μάθημα n. ~-**matter** s. περιεχόμενο n.

subject a. & adv. (*not independent*) υποτελής. be ~ to υπόκειμαι εις. ~ to υποκείμενος εις, (*conditionally on*) υπό τον όρον ότι.

subject v. (*subdue*) υποτάσσω, (*cause to undergo*) υποβάλλω. ~**ion** s. (*state*) υποταγή f. ~**ive** a. υποκειμενικός.

subjoin v. προσθέτω.

sub judice a. (*law*) εκκρεμής. the case is ~ η υπόθεσις τελεί υπό διάσκεψιν.

subjugat|e v. υποδουλώνω. ~**ion** s. υποδούλωση f.

subjunctive s. (*gram.*) υποτακτική f.

sublet v. υπεκμισθώνω.

sub-lieutenant s. (*naut.*) ανθυποπλοίαρχος m.

sublimate v. (*fig.*) εξιδανικεύω.

sublime a. θείος, θεσπέσιος. S~ Porte Υψηλή Πύλη.

subliminal a. υπό το κατώφλιον της συνειδήσεως.

submarine s. υποβρύχιον n. (a.) υποβρύχιος.

submerge v.t. (*flood*) κατακλύζω, (*sink*) βυθίζω. (v.i., *dive*) καταδύομαι.

submiss|ion s. (*obedience*) υποταγή. (*law*) ισχυρισμός m. ~**ive** a. υπάκουος.

submit v.i. (*surrender*) ενδίδω. ~ to ανέχομαι. (v.t.) (*present*) υποβάλλω, (*suggest*) προτείνω.

subnormal a. κάτω του κανονικού.

subordinat|e v. υποτάσσω. (a. & s.) υφιστάμενος, κατώτερος. (*gram.*) υπο(τε)ταγμένος. ~e clause δευτερεύουσα πρόταση. ~**ion** s. (*gram.*) υπόταξη f.

suborn v. δεκάζω.

subpoena s. (*law*) κλήσις εκ μέρους του δικαστηρίου.

subscribe v. (*sign*) υπογράφω. (*to journal*) είμαι συνδρομητής, (*to fund*) συνεισφέρω. ~ to (*opinion*) εγκρίνω.

subscript a. (*gram.*) iota ~ υπογεγραμμένη f.

subscription s. συνδρομή f.

subsequent a. μεταγενέστερος. ~**ly** adv. κατόπιν.

subservient a. υποκείμενος, (*servile*) δουλοπρεπής.

subsid|e v. (*sink, settle*) βουλιάζω, παθαίνω καθίζηση. (*abate*) κοπάζω. (*of floods*) υποχωρώ. ~**ence** s. καθίζηση f.

subsidiary a. (*secondary*) δευτερεύων,

(*company*) θυγατρικός.
subsid|y s. επιχορήγηση f. **~ize** v. επιχορηγώ.
subsist v. συντηρούμαι, ζω. **~ence** s. συντήρηση f.
subsoil s. υπέδαφος n.
substance s. (*essence*) ουσία f. (*material*) ύλη f. (*wealth*) περιουσία f. (*content*) περιεχόμενο n.
sub-standard a. κατώτερος.
substantial a. ουσιαστικός. (*considerable*) σημαντικός, αξιόλογος. (*strong*) γερός. **~ly** adv. ουσιωδώς, γερά.
substantiate v. αποδεικνύω το βάσιμο (*with gen.*).
substantive a. ουσιαστικός. (s.) ουσιαστικό n.
substitut|e v.t. **~e** A for B αναπληρώνω (*or* αντικαθιστώ) το B με το A. (v.i.) **~e** for (*act for*) αναπληρώνω. (s.) (*person*) αναπληρωτής m. (*thing*) υποκατάστατο n. **~ion** s. αναπλήρωση f. αντικατάσταση f.
substratum s. υπόστρωμα n.
subterfuge s. υπεκφυγή f.
subterranean a. υπόγειος.
subt|le a. λεπτός, φίνος. (*clever*) δαιμόνιος. **~lety** s. λεπτότητα f. (*cleverness*) οξύνοια f. **~ly** adv. (*cleverly*) εντέχνως.
subtract v. αφαιρώ. **~ion** s. αφαίρεση f.
suburb s. προάστιο n. **~an** a. των προαστίων, (*pej.*) μικροαστικός.
subvention s. επιχορήγηση f.
subvers|ion s. ανατροπή f. **~ive** a. ανατρεπτικός.
subvert v. ανατρέπω.
subway s. υπόγεια διάβαση. (*USA*) υπόγειος σιδηρόδρομος.
succeed v. πετυχαίνω, επιτυγχάνω. (*come after*) διαδέχομαι. **~ to** (*title*) κληρονομώ.
success s. επιτυχία f. a **~ see** successful. **~ful** a. πετυχημένος. **~fully** adv. με επιτυχία, επιτυχώς.
succession s. διαδοχή f. (*series*) σειρά f. in **~** διαδοχικώς, απανωτά.
successive a. διαδοχικός, αλλεπάλληλος. **~ly** adv. διαδοχικώς, απανωτά.
successor s. διάδοχος m.
succinct a. περιληπτικός. **~ly** adv. περιληπτικά.
succour s. συνδρομή f. (v.) συντρέχω.
succulent a. ζουμερός, εύχυμος. (*plant*) χυμώδης.
succumb v. υποκύπτω.
such a. & pron. τέτοιος, παρόμοιος, (*with adjective*) τόσο. **~ a** hero (ένας) τέτοιος ήρωας. some **~** thing κάτι τέτοιο. I said no **~** thing δεν είπα τέτοιο πράγμα. **~** big houses τόσο μεγάλα σπίτια. we had **~ a**

nice time περάσαμε τόσο ωραία. in a city **~** as Athens σε μια πόλη σαν την Αθήνα (*or* όπως η Αθήνα). in **~** and **~ a** street στον τάδε δρόμο. I was in **~ a** hurry that I left without money τόσο πολύ βιαζόμουνα που έφυγα χωρίς λεφτά. the weather is not **~** as to prevent us going out ο καιρός δεν είναι τέτοιος που να μας εμποδίσει να βγούμε. (you can use my car) **~** as it is αν και δεν είναι της προκοπής.
suchlike a. παρόμοιος.
suck v. (*draw in*) ρουφώ, (*lick*) γλείφω. (*hold in mouth*) πιπιλίζω. (*blood, honey*) απομυζώ, (*breast*) βυζαίνω. **~ up** απορροφώ. (*fam.*) **~ up to** κωλογλείφω. (s.) ρουφηξιά f.
sucker s. (*plant*) παραφυάδα f. (*fam.*) κορόιδο n., χάπι n.
sucking-pig s. γουρουνόπουλο του γάλακτος.
suck|le v. βυζαίνω. **~ling** s. βυζανιάρικο n.
suction s. αναρρόφηση f.
sudden a. αιφνίδιος, ξαφνικός, απότομος. **~ly** adv. (*also* all of a **~**) ξαφνικά.
suds s. σαπουνάδα f.
sue v.t. (*law*) ενάγω. (v.i.) **~ for** ζητώ.
suede s. σουέντ n. (*fam.*) καστόρι n.
suet s. λίπος n.
suffer v. πάσχω, (*also* tolerate) υποφέρω, (*allow*) επιτρέπω. (*undergo*) υφίσταμαι, παθαίνω.
sufferance s. on **~** κατ' ανοχήν.
sufferer s. **~s** from rheumatism οι πάσχοντες (*or* υποφέροντες) από ρευματισμούς.
suffering s. δοκιμασία f. **~s** βάσανα, πάθη n.pl.
suffice v.i. επαρκώ.
suffici|ent a. επαρκής, αρκετός. **~ency** s. επάρκεια f. **~ently** adv. αρκετά.
suffix s. (*gram.*) επίθημα n.
suffocat|e v.t. πνίγω. (v.i.) πνίγομαι, ασφυκτιώ. **~ing** a. ασφυκτικός. **~ion** s. ασφυξία f.
suffrag|e s. (*vote*) ψήφος m. f. (*right to vote*) δικαίωμα ψήφου. **~ette** s. σουφραζέτα f.
suffuse v. βάφω, πλημμυρίζω, χύνομαι σε.
sugar s. ζάχαρη f. (*chem.*) σάκχαρο n. (v.t.) βάζω ζάχαρη σε. (*fig.*) **~ the** pill χρυσώνω το χάπι. **~-beet** s. σακχαρότευτλον n. **~-cane** s. σακχαροκάλαμον n. **~y a**. ζαχαρωμένος. (*fig.*) (*the music*) is too **~y** παραείναι γλυκιά.
suggest v. (*propose*) προτείνω. (*bring to mind*) θυμίζω, (*hint*) υπαινίσσομαι, (*indicate*) δείχνω. **~ion** s. πρόταση f. (*psychological*) υποβολή f. (*trace*) υποψία f.
suggestive a. be **~ of** φέρνω στο νου. **~ joke**

αστείο με υπονοούμενα.

suicide s. αυτοκτονία f. commit ~ αυτοκτονώ.

suicidal a. feel ~ έχω τάση προς αυτοκτονία. a ~ policy σωστή αυτοκτονία.

suit s. (men's) κοστούμι n. (women's) ταγιέρ n. ~ of armour πανοπλία f. (request) αίτηση f. (law) αγωγή f. (cards) χρώμα n. follow ~ κάνω το ίδιο.

suit v. (be satisfactory to) ικανοποιώ, (be convenient to) βολεύω, (requirements) ανταποκρίνομαι εις. the coat ~s you to παλτό σού πάει. (adapt) προσαρμόζω. ~ the action to the word αμ' έπος αμ' έργον. be ~ed (to, for) είμαι κατάλληλος. a couple who are ~ed to each other ένα ταιριαστό ζευγάρι.

suitab|le a. κατάλληλος. it isn't ~le δεν κάνει, δεν πάει, δεν ταιριάζει. ~ly adv. καταλλήλως, όπως πρέπει.

suitcase s. βαλίτσα f.

suite s. (retinue) συνοδεία f. (rooms) διαμέρισμα n. drawing-room ~ έπιπλα σαλονιού, σαλόνι n. (mus.) σουίτα f.

suitor s. μνηστήρας m.

sulk v. κάνω μούτρα. ~y a. μουτρωμένος.

sullen a. σκυθρωπός.

sully v. κηλιδώνω.

sulphate s. θειικό άλας.

sulphur s. θείον, θειάφι n. ~ic a. θειικός.

sultan s. σουλτάνος m. ~a s. σουλτάνα f.

sultana s. (fruit) σουλτανίνα f.

sultry a. ~ weather (with cloud) κουφόβρασή f., συννεφόκαμα f. (fig.) φλογερός.

sum s. (amount) ποσόν n. (total) σύνολο n. ~ total άθροισμα n. good at ~s καλός στην αριθμητική. in ~ εν συνόψει. (v.) ~ up συνοψίζω, (weigh up) ζυγίζω, αναμετρώ.

summar|ize v. συνοψίζω. ~y s. περίληψη f. (a.) συνοπτικός.

summer s. καλοκαίρι n., θέρος n. Indian ~ γαϊδουροκαλόκαιρο n. (a.) καλοκαιρινός, θερινός. ~ time (of clock) θερινή ώρα. (v.i.) παραθερίζω. ~y a. καλοκαιρινός.

summit s. κορυφή. (a.) κορυφής.

summon v. (call) καλώ, (convoke) συγκαλώ. ~ up συγκεντρώνω.

summons s. κλήση f. serve a ~ on (law) κλητεύω.

sumptuous a. πολυτελέστατος, (meal) λουκούλλειος.

sun s. ήλιος m. ~'s ηλιακός. ~-blind s. τέντα f. ~-drenched a. ηλιόλουστος. ~-dried a. λιαστός. ~-glasses s. μαύρα γυαλιά.

sunbaked a. ηλιοψημένος.

sunbath|e v. κάνω ηλιοθεραπεία. ~ing s. ηλιοθεραπεία f.

sunbeam s. ηλιαχτίδα f.

sunburnt a. ηλιοκαμένος.

Sunday s. Κυριακή f. (a.) κυριακάτικος.

sunder v. χωρίζω.

sundial s. ηλιακό ωρολόγιο.

sundown s. see sunset.

sundr|y a. διάφοροι. all and ~y όλος ο κόσμος. (s.) ~ies διάφορα n.pl.

sunflower s. ήλιος m.

sunken a. (hollow) βαθουλωμένος. (ship) βουλιαγμένος. (let in to floor, etc.) χωνευτός.

sunless a. ανήλιος.

sun|light s. ηλιακό φως. ~lit a. ηλιοφώτιστος.

sunny a. (getting the sun) ευήλιος, προσηλιακός. (disposition) χαρούμενος. a ~ day μέρα με λιακάδα.

sunrise s. ανατολή του ηλίου.

sunset s. ηλιοβασίλεμα n., δύση του ηλίου.

sunshade s. ομπρέλα f.

sunshine s. λιακάδα f. bathed in ~ ηλιόλουστος.

sunstroke s. ηλίαση f.

sup v. δειπνώ.

super s. (actor) κομπάρσος m. (a., fam.) υπέροχος.

super- υπερ-, επι-.

superabund|ant a. υπεράφθονος. ~ance s. υπεραφθονία f.

superannuat|ed a. (fam.) απαρχαιωμένος. ~ion s. συνταξιοδότηση f. ~ion fund ταμείο συντάξεων.

superb a. εξαίσιος. ~ly adv. εξαίσια.

supercilious a. (person) ψηλομύτης, (manner) υπεροπτικός.

superficial s. (on surface) επιφανειακός. (of people, ideas) επιπόλαιος. ~ity s. επιπολαιότης f.

superfine a. υπερεκλεκτός, έξτρα.

superflu|ous a. περιττός. ~ity s. περίσσεια f.

superhuman a. υπεράνθρωπος.

superimpose v. υπερθέτω, βάζω από πάνω.

superintend v. επιθεωρώ. ~ent s. επιθεωρητής m.

superior a. & s. ανώτερος (with gen.). (in office) προϊστάμενος m. (supercilious) υπεροπτικός. ~ity s. ανωτερότης f.

superlative a. εξαίσιος, (fam.) αριστούργημα m. (gram.) υπερθετικός. ~ly adv. ~ly well αριστουργηματικά.

superman s. υπεράνθρωπος m.

supermarket s. υπεραγορά f.

supernatural a. υπερφυσικός.

supernumerary a. υπεράριθμος.

superscription s. επιγραφή f.

supersede v. αντικαθιστώ.

supersonic a. υπερηχητικός.

superstit|ion s. πρόληψη f., δεισιδαιμονία f. **~ious** a. προληπτικός.

superstructure s. υπερκατασκευή f.

supervene v. επακολουθώ.

supervis|e v. επιστατώ. **~ion** s. επιστασία f. **~or** s. επιστάτης m.

supine a. (on one's back) ανάσκελα (adv.) (lazy) νωθρός.

supper s. δείπνο n. have ~ δειπνώ. **~less** a. νηστικός.

supplant v. υποσκελίζω.

supple a. εύκαμπτος, ευλύγιστος.

supplement s. συμπλήρωμα n. (of publication) παράρτημα n. (v.) συμπληρώνω. **~ary** a. συμπληρωματικός.

suppli(c)ant s. ικέτης m.

supplicat|e v. ικετεύω. **~ion** s. ικεσία f.

supplier s. προμηθευτής m.

supply v. (things) προμηθεύω, (people) εφοδιάζω. (with fuel, services) τροφοδοτώ. (provide) παρέχω. (make up for) συμπληρώνω.

suppl|y s. (stock) παρακαταθήκη f., απόθεμα n. (in shop) στοκ n. (act of ~ying) προμήθεια f., εφοδιασμός m. (of fuel, services) παροχή f. ~ies εφόδια n.pl. ~y and demand προσφορά και ζήτηση. lay in a ~y of προμηθεύομαι.

support v. (prop up) στηρίζω, (back up) υποστηρίζω. (confirm) ενισχύω. (dependants) συντηρώ.

support s. (prop) στήριγμα n., έρεισμα n. (backing) υποστήριξη f. (confirmation) ενίσχυση f. (maintenance) συντήρηση f.

suppos|e v. υποθέτω. he is ~ed to come tomorrow υποτίθεται ότι θα έρθει αύριο. we're not ~ed to smoke in here δεν επιτρέπεται το κάπνισμα εδώ μέσα. (think) θεωρώ, νομίζω. (imply) προϋποθέτω. **~e** (or ~ing) you won? πες πως κέρδισες. **~e** we have a drink? τι θα έλεγες για ένα ποτό; **supposed** a. υποτιθέμενος, δήθεν. **~ly** adv. δήθεν.

supposition s. υπόθεση f.

suppository s. υπόθετο n.

suppress v. (quell) καταπνίγω. (hush up) συγκαλύπτω, (hide) κρύβω, σκεπάζω. (stop) παύω, (ban) απαγορεύω. **~ion** s. κατάπνιξη f. συγκάλυψη f. παύση f. απαγόρευση f.

supremacy s. υπεροχή f., κυριαρχία f.

supreme a. (authority) ανώτατος. (happiness, good, Being) υπέρτατος. **~ly** adv. εις το έπακρον.

surcharge s. (extra charge) πρόσθετο τέλος. (overprint) επισήμασμα n.

sure a. (stable, steady) σταθερός. (proven, reliable) ασφαλής, (certain) βέβαιος,

σίγουρος. make ~ (of, that) βεβαιώνομαι (with για or ότι). it is ~ to rain ασφαλώς θα βρέξει. they are ~ to be late ασφαλώς θα αργήσουν. (adv.) (indeed) ασφαλώς. ~ enough πράγματι. I don't know for ~ δεν ξέρω στα σίγουρα. **~-footed** a. be ~-footed έχω σταθερό πόδι. **~ness** s. βεβαιότητα f. σταθερότητα f.

surely adv. ασφαλώς, σίγουρα, δίχως άλλο. ~ you're not leaving yet? μη μου πεις πως θα φύγεις από τώρα! ~ that isn't right βεβαίως δεν είναι σωστό.

surety s. (person) εγγυητής m. stand ~ for μπαίνω εγγυητής για.

surf s. σπάσιμο κυμάτων με αφρό.

surface s. επιφάνεια f. (v.i.) (also break ~) αναδύομαι. (v.t.) (rood, etc.) στρώνω. (a.) (also on the ~) επιφανειακός.

surfeit s. πληθώρα f. (v.) be ~ed with χορταίνω, μπουχτίζω. **~ed** χορτασμένος, μπουχτισμένος.

surge v. ξεχύνομαι, ορμώ. ~ up (of anger) πλημμυρίζω, (of blood) ανεβαίνω, (s.) (people) συρροή f. (water, anger, etc.) πλημμύρα f. (waves) φουσκοθαλασσιά f.

surgeon s. χειρουργός m., (fam.) χειρούργος m.

surg|ery s. (science) χειρουργική f. (room) ιατρείο n. **~ical** a. χειρουργικός.

surl|y a. σκαιός. **~iness** s. σκαιότητα f.

surmise v. εικάζω. (s.) εικασία f.

surmount v. υπερπηδώ. ~ed by a dome μ' έναν τρούλο από πάνω.

surname s. επώνυμο n.

surpass v. ξεπερνώ, υπερβαίνω. **~ingly** adv. ανυπέρβλητα.

surplus s. περίσσευμα n. (a.) πλεονάζων.

surprise s. έκπληξη f. (taking by ~) αιφνιδιασμός m. to my ~ προς έκπληξή μου. (a.) (unexpected) αιφνιδιαστικός.

surprise v. εκπλήσσω. (also take by ~) αιφνιδιάζω. (take aback) ξαφνιάζω. I shouldn't be ~d if he got married δεν θα παραξενευτώ αν παντρευτεί.

surprising a. εκπληκτικός.

surreal|ism s. σουρρεαλισμός m. **~ist** s. σουρρεαλιστής m.

surrender v.t. παραδίδω, (relinquish) παραχωρώ. (v.i.) παραδίδομαι, ενδίδω. (s.) παράδοση f.

surreptitious a. κρυφός. he gave me a ~ wink μου έκλεισε με τρόπο το μάτι. **~ly** adv. κρυφά.

surrogate s. see substitute.

surround v. περιβάλλω, (stand round) περιστοιχίζω, (encircle) περικυκλώνω. (s.) with a wood ~ με ξύλο γύρω γύρω.

surrounding a. the ~ mountains τα γύρω

βουνά. (s.) ~s περιβάλλον n.

surveillance s. under ~ υπό επιτήρησιν.

survey v. (view) επισκοπώ, (review) ανασκοπώ. (of surveyor: buildings) εξετάζω, (land) καταμετρώ. (s.) ανασκόπηση f. εξέταση f. καταμέτρηση f.

surveyor s. (of buildings) πολιτικός μηχανικός. (of land) τοπογράφος μηχανικός. (of weights & measures) επόπτης m.

surviv|e v. επιζώ (with gen.) ~al s. επιβίωση f. ~or s. (of species) ο επιζών, (of accident) ο επιζήσας.

susceptible a. (impressionable) αισθαντικός. ~ to (colds) επιρρεπής (or υποκείμενος) εις, (charms) ευαίσθητος εις. be ~ of επιδέχομαι.

susceptibilit|y s. ευαισθησία f. ~ies s. ευαίσθητα σημεία.

suspect v.t. & i. υποψιάζομαι, υποπτεύομαι. (a. & s.) ύποπτος.

suspend v.t. (hang) κρεμώ. (put a stop to) αναστέλλω, (defer) αναβάλλω. (from office) θέτω εις διαθεσιμότητα. be ~ed (hang) κρέμομαι.

suspenders s. ζαρτιέρες f.pl. (USA) τιράντες f.pl.

suspense s. (anxiety) αγωνία f. in ~ (unsettled) εκκρεμής.

suspension s. (springs) ανάρτηση f. (dismissal) απόλυση f. (chem.) αιώρημα n. ~-bridge s. κρεμαστή γέφυρα.

suspicion s. υποψία f.

suspicious a. (suspect) ύποπτος, (mistrustful by nature) καχύποπτος, feel ~ about υποπτεύομαι.

suspiciously adv. υπόπτως, (with mistrust) με καχυποψία. it looks ~ like... μοιάζει πολύ με.

sustain v. (hold, hold up) κρατώ, βαστώ, (effort, life, etc.) συντηρώ. (objection, etc.) κάνω δεκτόν. (undergo) υφίσταμαι. ~ing a. (food) θρεπτικός.

sustenance s. τροφή f.

suture s. ραφή f.

suzerainty s. κυριαρχία f.

svelte a. λυγερόκορμος.

swab s. (med.) βύσμα n. take a ~ λαμβάνω έκκριμα. (v.) σφουγγαρίζω.

swaddle v. φασκιώνω. ~ling-clothes s. φασκιές f. pl.

swag s. (fam.) λεία f.

swagger v. παίρνω πόζες. ~ along περπατώ κορδωμένος.

swain s. her ~ ο καλός της.

swallow s. (bird) χελιδόνι n.

swallow v.t. (lit. & fig.) καταπίνω, (fig.) χάφτω. ~ up καταβροχθίζω. ~-hole s.

καταβόθρα f.

swamp s. τέλμα n. ~y a. τελματώδης.

swamp v.t. κατακλύζω.

swan s. κύκνος m. ~-song s. κύκνειον άσμα.

swank s. επίδειξη f. (v.) κάνω επίδειξη, (boast) καυχιέμαι. ~y a. επιδεικτικός, λουξ.

swarm s. (people) τσούρμο n. (things) πλήθος n. (esp. birds, insects) σμήνος n.

swarm v.i. (of bees) ξεσμαρίζω. (fig.) συρρέω, ξεχύνομαι. ~ up (climb) αναρριχώμαι εις. be ~ing with είμαι γεμάτος από. ~ing s. (of bees) ξεσμάρισμα n.

swarthy a. μελαψός.

swash v.i. παφλάζω. ~ing noise παφλασμός m.

swashbuckler s. φανφαρόνος m.

swastika s. αγκυλωτός σταυρός.

swat v. χτυπώ.

swath s. δρεπανιά f.

swathe v. τυλίγω.

sway v.t. κουνώ, σείω. (influence) επηρεάζω. (v.i.) κουνιέμαι, σείομαι, ταλαντεύομαι. (s.) (rule) κυριαρχία f. hold ~ κυριαρχώ. (movement, also ~ing) ταλάντευση f.

swear v.i. ορκίζομαι. (curse, also ~ at) βρίζω. ~ on or by ορκίζομαι σε or στο όνομα (with gen.) (v.t.) ~ in ορκίζω. ~ing s. βλαστήμιες f.pl. ~-word s. βλαστήμια f.

sweat v. ιδρώνω. (s.) ιδρώτας m. ~ing s. ίδρωμα n. εφίδρωση f. (a.) ιδρωμένος. ~y a. ιδρωμένος. (work) κοπιαστικός.

sweater s. πουλόβερ n.

Swedish a. σουηδικός. (person) Σουηδός m., Σουηδέζα f.

sweep s. (act of ~ing) σάρωμα n., σκούπισμα n. (of arm) κίνηση f. (of weapon) χτύπημα n. (of oar) κουπιά f. (curve) καμπύλη f. (tract) έκταση f. make a clean ~ of ξεφορτώνομαι.

sweep v.i. (along, over, etc., of crowd, flood) ξεχύνομαι. ~ in καταφθάνω. (v.t.) (brush) σκουπίζω. (of wind, wave, plague) σαρώνω. (with eyes, telescope) ψάχνω. ~ along (drag) παίρνω σβάρνα. ~ up μαζεύω. ~ off or away παρασύρω. be swept off one's feet παρασύρομαι. ~ down on πέφτω επάνω σε. ~ clean or clear of καθαρίζω από. ~ the board τινάζω την μπάγκα, (fig.) κερδίζω όλα τα βραβεία. ~ all before one γίνομαι εποποιία. ~ing a. (reform, etc.) ριζικός. ~ing statement υπερβολή f. ~ings s. σκουπίδια n.pl.

sweet a. γλυκός, γλυκύς. at one's own ~ will όπως και όποτε μου καπνίσει. he is ~ on Mary πονεί το δόντι του για τη Μαρία.

have a ~ tooth είμαι γλυκατζής. **~breads**
s. γλυκάδια n. pl. **~-corn** s. καλαμπόκι n.
~heart s. αγαπημένος m., αγαπημένη f.
~meat s. ζαχαρωτό n. **~-ly** adv. γλυκά.
~-smelling a. εύοσμος.

sweet s. (pudding) γλύκισμα m. (bonbon)
καραμέλα f. ~s καραμέλες pl., (fig. pleasures) απολαύσεις f. pl.

sweeten v. γλυκαίνω, (add sugar to) βάζω
ζάχαρη σε. **~ing** s. γλυκαντικό n.

sweetish-sour a. γλυκόξινος.

sweetness s. γλύκα, γλυκάδα f., γλυκύτης
f.

swell v.t. & i. φουσκώνω, (intensify) δυναμώνω. (v.i.) (of face, leg. etc.) πρήζομαι.
~ing s. πρήξιμο n.

swell s. (of sea) φουσκοθαλασσιά f. (fam.,
big shot) μεγαλουσιάνος m. (toff) δανδής
m. (a., high-life) του καλού κόσμου.

swelter v. σκάω από τη ζέστη. ~ing heat
λάβρα f.

swerve v. παρεκκλίνω.

swift a. γρήγορος, ταχύς. **~-ly** adv. γρήγορα,
ταχέως. **~ness** s. γρηγοράδα f., ταχύτητα
f.

swig s. μεγάλη ρουφηξιά f. (v.) ρουφώ.

swill s. (slops) απόπλυμα n. (rinse) ξέπλυμα
n. (v.) (wash) ξεπλένω. (drink) ρουφώ.

swim v.i. κολυμπώ. (v.t.) ~ (across) διασχίζω κολυμπώντας. my head is ~ming το
κεφάλι μου γυρνάει. (s.) go for a ~ πάω να
κολυμπήσω, κάνω μπάνιο. (fam.) in the ~
ενήμερος. **~-mer** s. κολυμβητής m.

swimming s. κολύμπι n. **~-ly** adv. μια χαρά.
~-bath, ~-pool s. πισίνα f.

swindle v. εξαπατώ. (s.) απάτη f. **~r** s.
απατεώνας m.

swin|e s. χοίρος m. (pl. χοίροι). (pej.)
γουρούνι n. **~-eherd** s. χοιροβοσκός m.
~-ish a. γουρουνίσιος.

swing v.i. αιωρούμαι. (hang) κρεμιέμαι. ~
round στρέφομαι, γυρίζω. (of opinion,
etc.) μεταστρέφομαι, γυρίζω. ~ along
βαδίζω με μπρίο. (v.t.) κουνώ, (turn)
στρέφω. ~ public opinion κάνω την κοινή
γνώμη να μεταστραφεί. **~-ing** a. ρυθμικός.

swing s. (seat) αιώρα f., κούνια f.
(movement) αιώρηση f., μεταστροφή f.
(rhythm) ρυθμός m. it went with a ~ (of
party, etc.) είχε σουξέ. in full ~ στο φόρτε,
στο αποκορύφωμα. get into the ~ of
συνηθίζω.

swingeing a. τεράστιος.

swipe v. κοπανώ. (fam., steal) σουφρώνω.

swirl v.i. περιδινούμαι. (s.) δίνη f.

swish s. (noise) θρόισμα n. (movement)
κούνημα n.

Swiss a. ελβετικός. (person) Ελβετός m.,
Ελβετίδα f.

switch s. (rod) βέργα f., βίτσα f. (railway)
κλειδί n. (electric) διακόπτης m. (change)
αλλαγή f.

switch v.t. (shift, change) στρέφω, γυρίζω.
(pull) τραβώ. (a train, tram) μεταφέρω. ~
on (light) ανάβω, (radio) ανοίγω. ~ off
(light) σβήνω, (radio) κλείνω. (v.t. & i.) ~
over αλλάζω.

switchback s. τραινάκι σε λούνα-παρκ που
ανεβοκατεβαίνει. (fig.) ~ road δρόμος που
ανεβοκατεβαίνει.

switchboard s. ταμπλό n. (exchange)
(τηλεφωνικό) κέντρο n.

swivel v.i. περιστρέφομαι. (s.) στροφείο n.
~-chair s. περιστρεφόμενη καρέκλα.

swollen a. φουσκωμένος, (limb) πρησμένος.

swoon v. λιποθυμώ. (s.) λιποθυμία f.

swoop v.i. εφορμώ, επιπίπτω. (s.) εφόρμηση
f. at one (fell) ~ με ένα ξαφνικό χτύπημα.

swop v. ανταλλάσσω.

sword s. σπαθί n. cross ~s διαξιφίζομαι. put
to the ~ σφάζω. **~-fish** s. ξιφίας m.

sworn a. (given under oath) ένορκος.
(bound by oath) ορκισθείς, ορκωτός. ~
enemy άσπονδος εχθρός.

swot v. σπάω στο διάβασμα. (s.) σπασίκλας
m.

sybarit|e s. συβαρίτης m. **~-ic** a. συβαρι-
τικός.

sycophant s. τσανακογλείφτης m. **~-ic** a.
δουλοπρεπής.

syllab|le s. συλλαβή f. **~-ification** s. συλλα-
βισμός m.

syllabus s. πρόγραμμα n.

syllogism s. συλλογισμός m.

sylph s. συλφίς f.

sylvan a. των δασών, ειδυλλιακός.

symbol s. σύμβολο n. **~-ic(al)** a. συμβολικός.
~-ically adv. συμβολικά.

symbol|ism s. συμβολισμός m. **~-ize** v.
συμβολίζω. **~-ization** s. συμβολισμός m.

symmetr|y s. συμμετρία f. **~-ical** a. συμ-
μετρικός.

sympathetic a. (nerve, ink) συμπαθητικός.
(full of sympathy) συμπονετικός, γεμάτος
συμπάθεια. **~-ally** adv. με συμπάθεια.

sympathize v. ~ with συμπονώ, (share
views, etc.) συμμερίζομαι. (express sorrow) εκφράζω τη λύπη μου. **~r** s. οπαδός
m.

sympathy s. συμπάθεια f. (compassion)
συμπόνια f. (condolences) συλλυπητήρια
n.pl. be in ~ with ευνοώ.

symphon|y s. συμφωνία f. **~-ic** a. συμ-
φωνικός.

symposium s. (meeting) συνέδριο n.

symptom *s.* σύμπτωμα *n.*, ένδειξη *f.* **~atic** *a.* ενδεικτικός.

synagogue *s.* συναγωγή *f.*, χάβρα *f.*

synchronize *v.t.* συγχρονίζω. (*v.i.*) συμπίπτω.

syncopat|ed *a.* (*mus.*) κοντρατέμπο. **~ion** *s.* συγκοπή *f.*

syndicate *s.* συνδικάτο *n.*

syndrome *s.* σύνδρομο *n.*

synod *s.* σύνοδος *f.*

synonym *s.* συνώνυμο *n.* **~ous** *a.* συνώνυμος.

synop|sis *s.* σύνοψη *f.* **~tic** *a.* συνοπτικός.

syntax *s.* συντακτικό *n.*

synthe|sis *s.* σύνθεση *f.* **~tic** *a.* συνθετικός. **~sizer** *s.* συνθεσάιζερ *n.*

syphil|is *s.* σύφιλη *f.* **~itic** *a.* συφιλιδικός.

syphon *see* siphon.

syringe *s.* σύριγξ *f.* (*garden*) ψεκαστήρας *m.*, τρόμπα *f.* (*enema*) κλύσμα *n.* (*v.t.*) πλένω με σύριγγα, (*plants*) ραντίζω.

syrup *s.* σιρόπι *n.* **~y** *a.* σοροποειδής. ~y music (*fam.*) σιρόπια *n.pl.*

system *s.* σύστημα *n.* (*network*) δίκτυο *n.* (*bodily*) οργανισμός *m.*

systematic *a.* συστηματικός. **~ally** *adv.* συστηματικώς.

systematiz|e *v.* συστηματοποιώ. **~ation** *s.* συστηματοποίηση *f.*

T

T *s.* ~-square γωνιόμετρο *n.* it suits me to a ~ μου πάει κουτί.

tab *s.* γλωσσάκι *n.* (*with name*) ετικέτα *f.* (*loop*) θηλειά *f.* (*fam.*) keep ~s on παρακολουθώ.

tabby *s.* (*cat*) τιγράκι *n.*

tabernacle *s.* (*biblical*) Σκηνή *f.* (*church*) εκκλησία *f.*

table *s.* (*furniture*) τραπέζι *n.* (*list*) πίνακας *m.* turn the ~s παίρνω το επάνω χέρι. (*a., wine, etc.*) επιτραπέζιος. (*v., a motion*) καταθέτω. **~cloth** *s.* τραπεζομάντιλο *n.* **~land** *s.* οροπέδιο *n.* **~napkin** *s.* πετσέτα *f.* **~spoon** *s.* κουτάλι της σούπας. **~tennis** *s.* πινγκ-πονγκ *n.*

tablet *s.* πλαξ *f.* (*soap, etc.*) πλάκα *f.* (*med.*) δισκίον *n.*

tabloid *a.* (*fig.*) in ~ form συνοπτικός.

taboo *s. & a.* ταμπού *n.*

tabular *a.* in ~ form σε πίνακες.

tabulate *v.* κατατάσσω σε πίνακες.

tacit *a.* σιωπηρός. **~ly** *adv.* σιωπηρώς.

taciturn *a.* (ο) λιγόλογος.

tack *s.* (*nail*) καρφάκι *n.* (*stitch*) τρύπωμα *f.*

hard ~ γαλέτα *f.* be on the right ~ είμαι στο σωστό δρόμο. (*v.*) (*nail*) καρφώνω, (*stitch*) τρυπώνω. (*naut.*) παίρνω βόλτα. ~ on προσθέτω.

tackle *v.* (*job*) καταπιάνομαι με. (*seize*) αρπάζω, (*confront*) αντιμετωπίζω.

tackle *s.* (*gear*) εξαρτήματα *n.pl.*

tacky *a.* be ~ κολλώ.

tact *s.* τακτ *n.* **~ful** *a.*, **~fully** *adv.* με τακτ.

tactic(s) *s.* τακτική *f.*

tactical *a.* τακτικός.

tactician *s.* ειδικός στην τακτική.

tactile *a.* της αφής.

tadpole *s.* γυρίνος *m.*

taffeta *s.* ταφτάς *m.*

tag *s.* (*label*) ετικέτα *f.* (*of shoelace*) σιδεράκι *n.* (*saying*) ρητό *n.*

tag *v.* (*attach*) προσθέτω. ~ along ακολουθώ.

tail *s.* ουρά *f.* turn ~ το βάζω στα πόδια, στρίβω. keep one's ~ up δεν απο-θαρρύνομαι. heads or ~s κορόνα γράμματα **~back** *s.* ουρά *f.* **~coat** *s.* φράκο *n.* **~end** *s.* άκρον *n.*, ουρά *f.*

tail (*v.t., follow*) ακολουθώ από κοντά. (*v.i.*) ~ off (*disperse*) αραιώνω, (*fade*) σβήνω.

tailor *s.* ράφτης *m.* (*v.*) ράβω. well ~ed καλοραμμένος. **~-made** επί μέτρω.

taint *v.t.* μολύνω. become ~ed σαπίζω, χαλώ. (*s.*) (*of corruption*) στίγμα *n.* (*of bias*) ίχνος *n.*

take *v.* (*general*) παίρνω. (*convey*) πηγαίνω, (*construe*) εκλαμβάνω. (*bath, walk*) κάνω, (*room, job*) πιάνω, (*lesson, prize*) παίρνω, (*photo*) βγάζω. ~ an exam δίνω εξετάσεις. ~ a note (*or* message) κρατώ μία σημείωση. ~ your time μη βιάζεσαι, με την ησυχία σου. it ~s two men (*to do job*) χρειάζονται δύο άνθρωποι. it took us (*or* we took) an hour to get here κάναμε (*or* μας πήρε) μία ώρα να 'ρθούμε εδώ. what do you ~ me for? για ποιον με παίρνεις; I took her for the maid την πήρα για την καμαριέρα. ~ it that (*suppose*) υποθέτω ότι. ~ to pieces (*v.t.*) λύνω, (*v.i.*) λύνομαι. ~ place γίνομαι, λαμβάνω χώραν. ~ hold of πιάνω.

take aback *v.t.* ξαφνιάζω.

take after *v.t.* does he ~ his father? έχει πάρει από τον πατέρα του;

take apart *v.t.* λύνω, ξεμοντάρω. (*v.i.*) λύνομαι.

take away *v.t.* παίρνω. (*subtract, deprive person of*) αφαιρώ. (*withdraw*) αποσύρω, αποτραβώ. ~ from (*diminish*) μειώνω.

take back *v.t.* παίρνω πίσω, (*convey*) πάω πίσω.

take down *v.t.* κατεβάζω (*write*) γράφω, σημειώνω. I took him down a peg του

έκοψα το βήχα (*or* τον αέρα). (*dismantle*)
see take apart.
take in *v.t.* (*receive*) παίρνω, (*include*)
περιλαμβάνω, (*understand*) καταλα-
βαίνω. (*make smaller*) μαζεύω, it needs
taking in θέλει μάζεμα. (*trick*) γελώ. ~
lodgers νοικιάζω δωμάτια.
take off *v.t.* βγάζω. (*mimic*) μιμούμαι
χιουμοριστικά. I could not take my eyes
off her τα μάτια μου δεν ξεκολλούσαν από
πάνω της. (*v.i.*) απογειώνομαι.
take-off *s.* χιουμοριστική μίμηση. (*aero.*)
απογείωση *f.*
take on *v.t.* (*assume*) παίρνω, αποκτώ.
(*undertake*) αναλαμβάνω, (*engage*) προσλαμβάνω. (*a wager*) δέχομαι, (*at game*)
παίζω. (*v.i.*) (*become popular*) πιάνω.
(*make a fuss*) κάνω φασαρία.
take out *v.t.* βγάζω. (*insurance*) κάνω. take
it out of (*tire*) κουράζω. take it out on
ξεσπάω (*or* ξεθυμαίνω) πάνω σε.
take over *v.t.* αναλαμβάνω, (*by force*)
καταλαμβάνω. I'll take you over the house
θα σε πάω να σου δείξω το σπίτι (*v.i.*) (*in
new office*) παραλαμβάνω υπηρεσίαν.
take to *v.t.* (*start*) αρχίζω (να). (*like*)
συμπαθώ, μου αρέσει. ~ the hills (*as
outlaw*) βγαίνω στο κλαρί. ~ one's heels το
βάζω στα πόδια.
take up *v.t.* (*pick up*) παίρνω. (*start*)
αρχίζω να ασχολούμαι με. ~ the carpets
ξεστρώνω. ~ room πιάνω τόπο. it takes up
all my time μου τρώει όλο τον καιρό μου.
(*absorb*) απορροφώ, (*continue*) συνεχίζω.
take (*sthg.*) up (*complain*) κάνω παράπονα (*or* ζητώ εξηγήσεις) για. be taken
up with ενδιαφέρομαι για.
take upon *v.* ~ oneself αναλαμβάνω.
taken *a.* (*occupied*) πιασμένος. be ~ ill
αρρωσταίνω.
taking *s.* (*of decision, drug, etc.*) λήψη *f.* (*of
city, etc.*) πάρσιμο *n.* ~s εισπράξεις *f.pl.*
(*a.*) ελκυστικός.
talc *s.* ταλκ *n.*
tale *s.* ιστορία *f.* (*literary*) διήγημα *n.* tell ~s
κουτσομπολεύω.
talent *s.* (*money*) τάλαντον *n.* (*ability*)
ταλέντο *n.* ~ed με ταλέντο.
talisman *s.* φυλαχτό *n.*
talk *s.* ομιλία *f.*, κουβέντα *f.* (*discussion*)
συζήτηση *f.*, συνδιάλεξη *f.*
talk *v.* (ο)μιλώ, κουβεντιάζω. ~ it over το
συζητώ. ~ (*person*) round πείθω. ~
(*person*) into it καταφέρνω, ~ (*person*)
out of it μεταπείθω. he ~ed down to them
τους μίλησε με περιφρονητική συγκα-
τάβαση. ~**ative** *a.* ομιλητικός. ~**er** *s.*
ομιλητής *m.* (*loquacious*) φλύαρος *a.*

~ing-to *s.* (*fam.*) κατσάδα *f.*
tall *a.* ψηλός. how ~ is he? τι ύψος έχει; ~
order δύσκολο πράγμα, ~ stories παραμύθια *n.pl.*
tallow *s.* ξίγκι *n.* ~ candle στεατοκήριον *n.*
tally *v.i.* συμφωνώ. (*s.*) τσέτουλα *f.* (*token*)
μάρκα *f.* keep ~ of σημειώνω.
talon *s.* νύχι *n.*
tambourine *s.* ντέφι *n.*
tame *a.* ήμερος, (*dull*) ανούσιος. (*v.*) δαμάζω, εξημερώνω. ~**ly** *adv.* (*without spirit*)
ψόφια, (*dully*) ανούσια.
tamper *v.* ~ with πειράζω, σκαλίζω και
χαλώ, (*falsify*) παραποιώ, (*bribe*) λαδώνω.
tan *s.* (*complexion*) μαύρισμα *n.*, μπρού-
ντζινο χρώμα. ~ leather καφέ δέρμα.
tan *v.t.* (*hides & fig.*) αργάζω. (*v.t. & i.*) (*of
complexion*) μαυρίζω.
tandem *adv.* ο ένας πίσω από τον άλλον.
tang *s.* (*freshness*) δροσιά *f.* (*piquancy*)
αψάδα *f.*
tangent *s.* (*geom.*) εφαπτομένη *f.* (*fig.*) go
off at a ~ φεύγω από το θέμα.
tangerine *s.* μανταρίνι *n.*
tangib|le *a.* απτός. ~**ly** *adv.* απτώς.
tangle *v.t.* μπερδεύω. (*v.i.*) μπερδεύομαι. (*s.*)
(*state*) ανακατωσούρα *f.* (*things*) μπερ-
δεμένη μάζα. in a ~ μπερδεμένος.
tango *s.* ταγκό *n.*
tankard *s.* κύπελλο *n.*
tanker *s.* ντεπόζιτο *n.* (*w.c.*) καζανάκι *n.*
(*mil.*) άρμα μάχης, τανκ *n.*
tanker *s.* (*ship*) δεξαμενόπλοιο *n.* (*motor*)
βυτιοφόρο *n.*
tanned *a.* (*leather*) κατειργασμένος, (*face*)
ηλιοκαμένος.
tann|er *s.* βυρσοδέψης *m.* ~**ery** *s.* βυρ-
σοδεψείο *n.* ~**ing** *s.* βυρσοδεψία *f.*
tantaliz|e *v.* (*tempt*) δελεάζω, (*torment*)
βασανίζω. ~**ing** *a.* δελεαστικός, γαργαλιστικός.
tantamount *a.* it is ~ to saying είναι σα να
λες.
tantrum *s.* νευράκια *n.pl.*
tap *s.* (*knock*) ελαφρό κτύπημα. (*v.*) κτυπώ
ελαφρά.
tap *s.* (*for liquid*) βρύση *f.* (*of barrel*)
κάνουλα *f.* (*gas*) κουμπί *n.* on ~ (*beer*) του
βαρελιού, (*fig.*) έτοιμος.
tap *v.* (*open*) ανοίγω, (*draw*) τραβώ,
(*intercept*) υποκλέπτω. (*fam.*) I ~ped him
for £5 του τράκαρα πέντε λίρες.
tape *s.* ταινία *f.* red ~ γραφειοκρατία *f.* (*v.*)
(*tie*) δένω, (*record*) ηχογραφώ. I've got
him ~d τον έχω ζυγίσει.
tape-deck *s.* κασετόφωνο *n.*
tape-measure *s.* μεζούρα *f.*, μέτρο *n.*

taper *v.i.* λεπτύνομαι.
taper *s.* λεπτό κερί για άναμμα.
tape-recorder *s.* μαγνητόφωνο *n.*
tapestry *s.* γκομπλέν *n.*
tar *s.* πίσσα *f.,* κατράμι *n.* (*v.*) στρώνω με πίσσα. ~red with the same brush ίδια φάρα.
tard|y *a.* αργός, βραδύς. ~**ily** *adv.* αργά. ~**iness** *s.* καθυστέρηση *f.*
target *s.* στόχος *m.* (*fig.*) σκοπός *m.*
tariff *s.* δασμολόγιο *n.* (*price-list*) τιμοκατάλογος *m.*
tarmac *s.* άσφαλτος *f.*
tarn *s.* λιμνούλα των βουνών.
tarnish *v.t. & i.* μαυρίζω, θαμπώνω. get ~ed (*fig.*) αμαυρώνομαι.
tarpaulin *s.* μουσαμάς *m.*
tarry *v.* αργώ.
tart *a.* ξινός, (*remark*) ξερός.
tart *s.* (*pastry*) τάρτα *f.* (*woman*) τσούλα *f.* (*v.*) ~ up στολίζω φτηνά.
tartan *s.* σκωτσέζικο ύφασμα.
tartar *s.* (*in teeth*) πουρί *n.*
Tartar *s.* Τάταρος *m.* (*fig.*) αγριάνθρωπος *m.*
task *s.* καθήκον *n.,* δουλειά *f.* take to ~ μαλώνω, βάζω μπροστά. ~**force** *s.* τακτική δύναμις. ~**master** *s.* hard ~-master σκληρός εργοδότης.
tassel *s.* φούντα *f.,* θύσανος *m.*
taste *s.* (*sense, flavour*) γεύση *f.* (*aesthetic*) γούστο *n.* good ~ καλαισθησία *f.* to my ~ του γούστου μου, in bad ~ κακόγουστος, in good ~ καλόγουστος. just a ~ μια ιδέα. get a ~ of δοκιμάζω. give me a ~ of your wine δώσε μου να δοκιμάσω από το κρασί σου. have a ~ for έχω αδυναμία σε, μου αρέσει. it leaves a bad ~ (*fig.*) κάνει κακή εντύπωση.
taste *v.t.* (*sample*) δοκιμάζω, (*perceive taste of*) αισθάνομαι, can you ~ the cinnamon? αισθάνεσαι την κανέλα; (*v.i.*) it ~s good έχει ωραία γεύση. it ~s of almonds έχει γεύση αμυγδάλου.
tatter|s *s.* κουρέλια *n.pl.* be ~ed *or* in ~s έχω γίνει κουρέλι, (*of reputation*) έχω στραπατσαριστεί.
tattle *v.* κουτσομπολεύω. (*s.*) κουτσομπολιό *n.*
tattoo *s.* (*marking*) τατουάζ *n.* (*drumming*) τυμπανοκρουσία *f.* (*fig.*) beat a ~ παίζω ταμπούρλο.
tatty *a.* φθαρμένος.
taunt *v.* χλευάζω, βρίζω. (*s.*) χλευασμός *m.,* ύβρις *f.*
taut *s.* (*nerves*) τεταμένος, (*rope*) τεντωμένος. ~**en** *v.t.* τεντώνω, τεζάρω.
tautology *s.* ταυτολογία *f.*

tavern *s.* καπηλειό *n.*
tawdr|iness *s.* ψεύτικη πολυτέλεια. ~**y** *a.* ψευτοπολυτελής.
tawny *a.* κοκκινόξανθος.
tax *s.* φόρος *m.* (*burden*) βάρος *n.* free of ~ ατελής. (*v.*) φορολογώ. (*make demands on*) δοκιμάζω. ~ (*person*) with κατηγορώ για. ~**able** *a.* φορολογήσιμος. ~**ation** *s.* φορολογία *f.* ~**collector** *s.* εισπράκτωρ φόρων ~**office** *s.* εφορία *f.*
taxi, ~**cab** *s.* ταξί *n.* ~**driver** *s.* ταξιτζής *m.* ~**rank** *s.* πιάτσα *f.,* σταθμός ταξί.
taxidermy *s.* βαλσάμωμα ζώων.
tea *s.* τσάι *n.* (*fam.*) it's not my cup of ~ δεν είναι της προτιμήσεώς μου. ~**caddy** *s.* κουτί του τσαγιού. ~**cup** *s.* φλιτζάνι του τσαγιού. ~**pot** *s.* τσαγιέρα *f.* ~**spoon** *s.* κουταλάκι του τσαγιού.
teach *v.* διδάσκω, (*person only*) μαθαίνω. ~**able** *a.* (*subject*) που διδάσκεται, (*person*) επιδεκτικός μαθήσεως.
teacher *s.* (*general*) διδάσκαλος *m.* (*primary*) (δι)δάσκαλος *m.,* (δι)δασκάλισσα *f.,* δασκάλα φ. '(*secondary & higher*) καθηγητής *m.,* καθηγήτρια *f.*
teaching *s.* διδασκαλία *f.* ~s διδάγματα *n.pl.* (*a.*) διδακτικός.
team *s.* ομάδα *f.* (*beasts*) ζευγάρι *n.* ~ spirit πνεύμα συνεργασίας. ~**work** ομαδική εργασία. (*v.t.*) ζεύω. (*v.i.*) ~ up συνεργάζομαι.
tear *s.* (*drop*) δάκρυ *n.* I found her in ~s τη βρήκα να κλαίει (*or* κλαμένη).
tear *v.t.* (*rend*) (*also* ~ up) σχίζω, (*hair*) τραβώ, (*snatch*) αρπάζω, αποσπώ. ~ off βγάζω, αποσπώ. ~ up (*uproot*) ξεριζώνω. ~ to pieces καταξεσχίζω. be torn between αμφιταλαντεύομαι μεταξύ (*with gen.*). I couldn't ~myself away δεν μπορούσα να ξεκολλήσω. (*v.i.*) it ~s easily σχίζεται εύκολα. (*s.*) σχίσιμο *n.*
tear *v.i.* (*rush*) ορμώ.
tearful *a.* δακρυσμένος.
tear-gas *s.* δακρυγόνα αέρια.
tearing *a.* (*fam.*) (*row*) τρικούβερτος. in a ~ hurry τρομερά βιαστικός.
teas|e *v.* πειράζω. (*s., person*) πειραχτήρι *n.* ~**er** *s.* δύσκολο πρόβλημα. ~**ing** *s.* άκακο πείραγμα.
teat *s.* θηλή *f.*
technical *a.* τεχνικός. ~**ity** *s.* τεχνική λεπτομέρεια. ~**ly** *adv.* από τεχνικής απόψεως.
technician *s.* τεχνίτης *m.,* τεχνικός *m.*
technique *s.* τεχνική *f.*
technocrat *s.* τεχνοκράτης *m.*
technolog|ical *a.* τεχνολογικός. ~**ist** *s.* τεχνολόγος *m.* ~**y** *s.* τεχνολογία *f.*
Te Deum *s.* (*eccl.*) δοξολογία *f.*

tedi|ous *a.* ανιαρός. **~um** *s.* ανία *f.*

teem *v.* αφθονώ. ~ with βρίθω (*with gen.*). it ~ed with rain έβρεξε καταρρακτωδώς.

teenager *s.* νέος (*or* νέα *f.*) δεκατριών μέχρι δεκαεννέα ετών.

teens *s.* εφηβική ηλικία.

teeter *v.* τρικλίζω.

teething *s.* πρώτη οδοντοφυΐα. ~ troubles (*fig.*) δυσχέρειες του αρχικού σταδίου.

teetotaller *s.* ο απέχων από τα οινοπνευματώδη.

telecommunications *s.* τηλεπικοινωνία *f.*

telegram *s.* τηλεγράφημα *n.*

telegraph *s.* τηλέγραφος *m.* (*v.*) τηλεγραφώ. **~ic** *a.* τηλεγραφικός.

telepath|ic *a.* τηλεπαθητικός. **~y** *s.* τηλεπάθεια *f.*

telephone *s.* τηλέφωνο *n.*(*v.*) τηλεφωνώ.

telephon|e, ~ic *a.* τηλεφωνικός. **~e** call τηλεφώνημα *n.* **~e** operator τηλεφωνητής *m.*, τηλεφωνήτρια *f.* (*see also* phone).

teleprinter *s.* τηλέτυπο *n.*

telescop|e *s.* τηλεσκόπιο *n.* (*v.i.*) πτύσσομαι. **~ic** *a.* (*sliding*) πτυσσόμενος.

television *s.* τηλεόραση *f.* (*a.*) τηλεοπτικός.

tell *v.t. & i.* λέω, λέγω. (*relate*) αφηγούμαι. (*distinguish*) ξεχωρίζω. ~ me the news πες μου τα νέα. I told him I was coming του είπα πως θα έρθω. I told him to come του είπα να έρθει. (*v.i.*) (*know*) ξέρω, καταλαβαίνω. (*count, have effect*) δείχνω, φαίνομαι. his age is beginning to ~ on him του φάνηκαν τα χρόνια του. ~ against επιβαρύνω. ~ on (*report*) μαρτυρώ. ~ off μαλώνω. all told συνολικά.

teller *s.* (*bank*) ταμίας *m.*

telling *a.* αποτελεσματικός.

telltale *s.* μαρτυριάρης *m.* (*a., fig.*) αποκαλυπτικός, που λέει πολλά.

temerity *s.* τόλμη *f.*

temper *s.* (*humour*) ψυχική διάθεση, (*anger*) θυμός *m.* lose one's ~ χάνω την ψυχραιμία μου. get in a ~ θυμώνω, ανάβω. good-~ed μειλίχιος, bad-~ed δύστροπος.

temper *v.* (*metal*) στομώνω. (*mitigate*) μετριάζω.

temperament *s.* ιδιοσυγκρασία *f.*, ταμπεραμέντο *n.* **~al** *a.* be ~al έχω ιδιοτροπίες.

temperance *s.* εγκράτεια *f.*

temperate *a.* εγκρατής. (*climate*) εύκρατος.

temperature *s.* θερμοκρασία *f.* have a ~ έχω πυρετό. take (*person's*) ~ θερμομετρώ.

tempest *s.* θύελλα *f.* (*at sea*) φουρτούνα *f.* **~uous** *a.* θυελλώδης.

temple *s.* (*church*) ναός *m.* (*anat.*) κρόταφος *m.*

tempo *s.* ρυθμός *m.*

temporal *a.* (*of time*) χρονικός. (*worldly*)

εγκόσμιος, (*not spiritual*) κοσμικός.

temporar|y *a.* προσωρινός. **~ily** *adv.* προσωρινά.

temporize *v.* επιφυλάσσομαι να δώσω θετική απάντηση.

tempt *v.* δελεάζω, βάζω σε πειρασμό. (*lead on*) παρασύρω, (*persuade*) καταφέρνω. **~ation** *s.* πειρασμός *m.* **~ing** *a.* δελεαστικός.

ten *num.* δέκα. ~ to one (*bet*) δέκα προς ένα.

tenable *a.* (*argument*) υποστηρίξιμος, (*office*) ισχύων.

tenacious *a.* επίμονος. **~ly** *adv.* επίμονα.

tenacity *s.* επιμονή *f.*

tenant *s.* ενοικιαστής, νοικάρης *m.*

tend *v.* (*lean*) τείνω, έχω τάση. (*watch*) περιποιούμαι, φυλάω. **~ency** *s.* τάση *f.*

tendentious *a.* μεροληπτικός.

tender *a.* τρυφερός, (*sensitive*) ευαίσθητος. **~ly** *adv.* τρυφερά.

tender *v.* (*offer*) προσφέρω, (*submit*) υποβάλλω. (*for contract*) κάνω προσφορά. (*s.*) προσφορά *f.*

tender *s.* (*vessel, wagon*) εφοδιοφόρον *n.*

tendon *s.* τένων *m.*

tendril *s.* έλιξ *f.*

tenement *s.* ~-house πολυκατοικία *f.*

tenet *s.* αρχή *f.*

tennis *s.* τέννις *n.*

tenor *s.* (*mus.*) τενόρος *m.* (*drift*) νόημα *n.*

tense *s.* (*gram.*) χρόνος *m.*

tense *a.* τεταμένος. (*v.*) σφίγγω.

tension *s.* ένταση *f.* (*voltage*) τάση *f.* (*nervous*) εκνευρισμός *m.*

tent *s.* σκηνή *f.*

tentacle *s.* πλοκάμι *n.*

tentative *a.* (*trial*) δοκιμαστικός, (*provisional*) προσωρινός.

tenterhooks *s.* be on ~ κάθομαι στα καρφιά.

tenth *a.* δέκατος.

tenuous *a.* λεπτός, (*evidence, etc.*) ισχνός.

tenure *s.* κατοχή *f.*

tepid *a.* χλιαρός.

tergiversation *s.* υπεκφυγή *f.*

term *s.* (*period*) περίοδος *f.* (*duration*) διάρκεια *f.* school ~ σχολική περίοδος. (*of loan, etc.*) προθεσμία *f.* long-~ μακροπρόθεσμος, short-~ βραχυπρόθεσμος.

term *s.* (*expression, condition, limit*) όρος *m.* come to ~s with συμβιβάζομαι με. we are on good ~s τα έχουμε καλά, βρισκόμαστε σε φιλικές σχέσεις. on equal ~s επί ίσοις όροις. they are not on speaking ~s δεν μιλιούνται. in ~s of όσον αφορά. in flattering ~s με κολακευτικά λόγια.

term *v.* ονομάζω, καλώ.

terminal *a.* ακραίος, τελευταίος. (*s.*) τέρμα *n.* (*electric*) ακροδέκτης *m.*

terminate *v.t.* θέτω τέρμα εις, λύω. (*v.i.*) λήγω.

terminology *s.* ορολογία *f.*

terminus *s.* τέρμα *n.*

termite *s.* τερμίτης *m.*

terrace *s.* (*on hillside*) πεζούλι *n.* (*houses*) σειρά ομοιόμορφων σπιτιών. (*walk*) επιχωματωμένος χώρος που παρέχει θέα και περίπατο.

terracotta *s.* οπτή γη.

terra firma *s.* στερεά *f.*

terrain *s.* έδαφος *n.*

terrestrial *a.* γήινος.

terrib|le *a.* τρομερός, φρικτός. **~ly** *adv.* τρομερά.

terrific *a.* τρομακτικός, (*great*) τρομερός, (*amazing*) καταπληκτικός. **~ally** *adv.* τρομερά.

terrify *v.* κατατρομάζω. **~ing** *a.* τρομακτικός.

territorial *a.* εδαφικός. ~ army πολιτοφυλακή *f.* ~ waters χωρικά ύδατα.

territory *s.* (*land*) έδαφος *n.* (*area*) περιοχή *f.*

terror *s.* τρόμος *m.*, τρομάρα *f.*, φόβος *m.* reign of ~ τρομοκρατία *f.* have a ~ of τρέμω. (*fam.*) he's a ~ είναι τρομερός. **~struck** *a.* τρομοκρατημένος.

terror|ism *s.* τρομακρατία *f.* **~ist** *s.* τρομοκράτης *m.* **~ize** *v.* τρομοκρατώ.

terse *a.* σύντομος.

terylene *s.* τερυλέν *n.*

test *s.* (*esp. of machine*) δοκιμή *f.* (*esp. of character*) δοκιμασία *f.* (*examination*) εξέταση *f.* (*a.*) ~ case υπόθεση που δημιουργεί προηγούμενον. ~ flight δοκιμαστική πτήση. ~ tube δοκιμαστικός σωλήν. (*v.*) δοκιμάζω, εξετάζω.

testament *s.* διαθήκη *f.*

testicle *s.* όρχις *f.*

testify *v.* μαρτυρώ.

testimonial *s.* πιστοποιητικό *n.*

testimony *s.* μαρτυρία *f.*

test|iness *s.* δυστροπία *f.* **~y** *a.* δύστροπος.

tetanus *s.* τέτανος *m.*

tetchy *a.* δύστροπος.

tête-à-tête *s.* τετατέτ *n.*

tether *v.* δένω. (*s.*) σχοινί *n.* be at the end of one's ~ δεν αντέχω πια.

text *s.* κείμενο *n.* **~book** *s.* εγχειρίδιο *n.*

textile *a.* ~ industry υφαντουργία *f.* (*s.*) ~s υφαντουργικά προϊόντα.

texture *s.* υφή *f.* close ~d με πυκνή υφή.

than *conj.* από, παρά. Helen is prettier ~ Mary η Ελένη είναι πιο όμορφη από τη Μαρία. he is older ~ me είναι πιο μεγάλος από μένα (*or* μεγαλύτερός μου). it is cheaper to buy one ~ make it yourself συμφέρει καλύτερα να το αγοράσεις παρά να το φτιάξεις μόνος σου. I would rather go by train ~ fly προτιμώ να πάω σιδηροδρομικώς παρά αεροπορικώς. no sooner had I gone out ~ it started to rain μόλις βγήκα (*or* δεν πρόφτασα να βγω και) άρχισε να βρέχει. it is later ~ I thought είναι αργότερα απ' ό,τι νόμιζα.

thank *v.* ευχαριστώ. ~ God δόξα τω Θεώ.

thankful *a.* ευγνώμων. **~ly** *adv.* ευγνωμόνως. **~ness** *s.* ευγνωμοσύνη *f.*

thankless *a.* αχάριστος.

thanks *s.* ευχαριστίες *f.pl.* ~! ευχαριστώ. ~ to χάρη σε.

that *pron.* (*demonstrative*) εκείνος. ~ house εκείνο το σπίτι. (*relative*) που, the book ~ you were reading το βιβλίο που διάβαζες.

that *conj.* ότι, πως. I heard ~ he is coming tomorrow έμαθα ότι (*or* πως) φτάνει αύριο. (*after so much, etc.*) που, I arrived so late ~ I couldn't find a room έφτασα τόσο αργά που δεν μπορούσα να βρω δωμάτιο. so ~ (*result*) ώστε, in order ~ για να.

thatch *s.* αχυρένια στέγη.

thaw *v.t. & i.* λειώνω. (*s.*) λειώσιμο του χιονιού. (*fig.*) ύφεση *f.*

the *article* ο *m.*, η *f.*, το *n.*

theatr|e *s.* θέατρο *n.*, **~ical** *a.* θεατρικός.

theft *s.* κλοπή *f.*, κλεψιά *f.*

their, ~s *pron.* (δικός) τους.

the|ism *s.* θεϊσμός *m.* **~ist** *s.* θεϊστής *m.*

them *pron.* αυτούς *m.*, αυτές *f.*, αυτά *n.*

theme *s.* θέμα *n.*

themselves *see* self.

then *adv.* (*at that time, therefore*) τότε, (*next*) έπειτα, μετά. (*also*) εκτός απ' αυτό.

thence *adv.* (*from there*) από 'κει. (*for that reason*) συνεπώς, για αυτό.

thence|forth, ~forward *adv.* έκτοτε, από τότε (κι έπειτα).

theocr|acy *s.* θεοκρατία *f.* **~atic** *a.* θεοκρατικός.

theolog|ian *s.* θεολόγος *m.* **~ical** *a.* θεολογικός. **~y** *s.* θεολογία *f.*

theorem *s.* θεώρημα *n.*

theoretical *a.* θεωρητικός. **~ly** *adv.* θεωρητικά.

theor|etician, ~ist *s.* θεωρητικός *m.*

theor|ize *v.* κάνω θεωρίες. **~y** *s.* θεωρία *f.*

therap|eutic *a.* θεραπευτικός. **~y** *s.* θεραπευτική αγωγή.

there *adv.* εκεί. over ~ εκεί πέρα. ~ is έχει, υπάρχει. ~ are έχει, υπάρχουν. ~ doesn't seem to be a solution δεν φαίνεται να υπάρχει λύση. ~ was a knock at the door χτύπησε η πόρτα. ~'s Mary! να η Μαρία! ~ she is! να την. ~ are (*fine*) apples for you! να κάτι μήλα! (*fam.*) he's not all ~ είναι χαζός.

thereabouts *adv.* (*place*) εκεί γύρω, (*roughly*) πάνω κάτω.

thereafter *adv.* κατόπιν, έπειτα.

thereby *adv.* έτσι.

therefore *adv.* γι' αυτό, επομένως, άρα.

therein *adv.* εκεί (μέσα).

thereupon *adv.* οπότε, συνεπεία τούτου.

therm *s.* θερμίδα *f.* **~al** *a.* ~al springs θέρμαι *f.pl.*

thermodynamics *s.* θερμοδυναμική *f.*

thermometer *s.* θερμόμετρο *n.*

thermonuclear *a.* θερμοπυρηνικός.

thermos *s.* (*also* ~ flask) θερμός *n.*

thermostat *s.* θερμοστάτης *m.*

thesaurus *s.* λεξικό συνωνύμων και συγγενικών.

these *pron.* αυτοί, τούτοι. ~ houses αυτά (*or* τούτα) τα σπίτια.

thesis *s.* (*theme*) θέμα *n.* (*for degree*) διατριβή *f.*

they *pron.* εκείνοι, αυτοί.

thick *a.* χονδρός, χοντρός. (*dense, bushy*) πυκνός. (*of wall, layer*) παχύς, a wall 30 cm. ~ τοίχος με πάχος τριάντα εκατοστών. (*fluid*) πηχτός. he's rather ~ (*dull*) δεν είναι σπίρτο. that's a bit ~ αυτό παραείναι (*or* πάει πολύ). he stood by me through ~ and thin στάθηκε στο πλευρό μου και στις καλές και στις κακές περιστάσεις. they are very ~ (*or* as ~ as thieves) (*intimate*) είναι κώλος και βρακί. we are in the ~ of it (*activity*) είμαστε στη βράση της δουλειάς.

thicken *v.t.* κάνω πιο παχύ. (*v.t. & i.*) πυκνώνω. (*v.i.*) (*of liquid*) πήζω, δένω. (*of plot*) περιπλέκομαι.

thicket *s.* λόχμη *f.*

thick-headed *a.* ξεροκέφαλος.

thickly *adv.* (*densely*) πυκνά. (*lying*) σε πυκνό στρώμα, (*spread*) σε παχύ στρώμα.

thickness *s.* (*of book, tree-trunk, etc.*) όγκος *m.* (*of wall, layer*) πάχος *n.* (*density*) πυκνότητα *f.*

thickset *a.* πυκνός, (*sturdy*) γεροδεμένος.

thick-skinned *a.* χοντρόπετσος.

thief *s.* κλέφτης *m.*

thieve *v.* κλέβω.

thiev|ing, ~ish *a.* (*person*) κλεφταράς *m.* (*animal*) κλέφτικος.

thigh *s.* μηρός *m.*, μπούτι *n.*

thimble *s.* δαχτυλήθρα *f.* **~ful** *a.* (*fig.*) μία σταγόνα.

thin *a.* (*not thick*) λεπτός, ψιλός. (*not dense*) αραιός, (*of liquid*) αραιός. (*person*) λεπτός, λιγνός, αδύνατος. (*of poor*

quality) φτωχός. you've got ~ner (*of person*) αδυνάτισες, λέπτυνες. we had a ~ time δεν ευχαριστηθήκαμε. (*v.t. & i.*) αραιώνω.

thing *s.* πράγμα *n.*, πράμα *n.* the poor ~! ο καημένος! the very (*or* just the) ~ ό,τι χρειάζεται. the ~ is το ζήτημα είναι. for one ~ αφ' ενός, πρώτα πρώτα. it's not the ~ δεν επιτρέπεται, είναι απαράδεκτο. it's the usual (*or* quite the) ~ συνηθίζεται. the latest ~ (*fashion*) η τελευταία λέξη της μόδας. what with one ~ and another με το ένα και με το άλλο. he doesn't know a ~ about music δεν έχει ιδέα (*or* δεν σκαμπάζει *or* έχει μεσάνυχτα) από μουσική. the ~ I should like best αυτό που θα προτιμούσα.

thingummy *s.* (*fam.*) (*person*) ο τέτοιος, (*thing*) μαραφέτι *n.*

think *v.i. & i.* (*reflect*) (*also* ~ of, about) σκέπτομαι, συλλογίζομαι. (*opine*) νομίζω, μου φαίνεται ότι. (*expect, imagine*) φαντάζομαι. ~ of (*remember*) θυμάμαι. (*entertain project*) σκέφτομαι να, λέω να. (*entertain possibility of*) διανοούμαι. I should never ~ of doing such a thing δεν θα μπορούσα ποτέ να διανοηθώ (*or* δεν θα περνούσε ποτέ απ' το νου μου) να κάνω τέτοιο πράγμα. I (*get the idea of*) σκέφτομαι. who first thought of this plan? ποιος πρωτοσκέφτηκε αυτό το σχέδιο; ~ highly (*or* much) of έχω περί πολλού. ~ out μελετώ, (*a plan*) καταστρώνω. ~ over μελετώ, σκέφτομαι. ~ up επινοώ. it is thought that πιστεύεται ότι.

thinkable *a.* νοητός.

thinker *s.* στοχαστής *n.*

thinking *s.* σκέψη *f.* to my ~ κατά τη γνώμη μου. (*a.*) ~ people οι σοβαρώς σκεπτόμενοι άνθρωποι.

third *a.* τρίτος. ~ party τρίτος *m.* ~ degree, ανάκριση τρίτου βαθμού. (*s.*) τρίτον *n.* **~-rate** *a.* κακής ποιότητος.

thirst *s.* δίψα *f.* (*v.*) διψώ.

thirsty *a.* διψασμένος. feel ~ διψώ.

thirteen (*num.*) δεκατρία. (*a.*) δεκατρείς. **~th** *a.* δέκατος τρίτος.

thirt|y (*num.*) τριάντα. **~ieth** *a.* τριακοστός.

this *pron.* αυτός, τούτος. ~ house αυτό (*or* τούτο) το σπίτι. like ~ έτσι. with ~ and that με το ένα και με το άλλο.

thistle *s.* γαϊδουράγκαθο *n.*

thither *adv.* εκεί.

thong *s.* ιμάς *m.*

thorax *s.* θώραξ *m.*

thorn *s.* αγκάθι *n.* (*fig.*) ~ in one's flesh κακός μπελάς. **~y** *a.* ακανθώδης.

thorough *a.* (*treatment*) καλός, γερός.

(*worker*) καλός, ευσυνείδητος. (*know-ledge*) πλήρης. (*out-and-out*) τέλειος, πέρα για πέρα.

thoroughbred *a.* καθαρόαιμος.

thoroughfare *s.* αρτηρία *f.* no ~ απαγορεύεται η δίοδος.

thorough-going *a. see* thorough.

thoroughly *adv.* τελείως, καλά, απολύτως.

those *pron.* εκείνοι. ~ houses εκείνα τα σπίτια.

thou *pron.* εσύ.

though *conj.* μολονότι, αν και, μ' όλο που, παρ' όλο που. strange ~ it may seem όσο και να φαίνεται παράξενο. as ~ σαν να (*with indicative*), as ~ he wanted to speak σαν να ήθελε να μιλήσει. it looks as ~ they are not coming σαν να μου φαίνεται πως δε θα 'ρθουν. (*adv., yet, however*) όμως, ωστόσο.

thought *s.* σκέψη *f.* (*idea*) ιδέα *f.* (*concern*) μέριμνα *f.* take ~ for νοιάζομαι για. at the mere ~ of it και μόνο με τη σκέψη. on second ~s κατόπιν ωριμωτέρας σκέψεως.

thoughtful *a.* (*pensive*) σκεπτικός, (*kind*) ευγενικός, (*not superficial*) σοβαρός. **~ness** *s.* ευγένεια *f.* σοβαρότητα *f.*

thoughtless *a.* απερίσκεπτος. **~ness** *s.* απερισκεψία *f.*

thousand *a.* a ~ χίλιοι, two ~ δύο χιλιάδες. one in a ~ ένας στους χίλιους. **~th** *a.* χιλιοστός.

thrall *s.* (*slave*) δούλος *m.* held in ~ υποδουλωμένος. *see* enthral.

thrash *v.* μαστιγώνω, δέρνω. (*defeat*) κατατροπώνω. ~ out συζητώ λεπτομερώς.

thrashing *s.* δάρσιμο *n.* get a ~ τρώω ξύλο.

thread *s.* κλωστή *f.* (*also fig.*) νήμα *n.* (*of screw*) βόλτες *f.pl.* (*v.*) (*needle, beads, etc.*) περνώ. ~ one's way through περνώ ανάμεσα σε.

threadbare *a.* τριμμένος, λειωμένος. (*fig.*) τετριμμένος.

threat *s.* απειλή *f.*

threaten *v.* απειλώ. **~ing** *a.* απειλητικός.

three (*num.*) τρία. (*a.*) τρεις. ~ hundred τριακόσιοι. (*s.*) (*at cards*) τριάρι *n.*

three-cornered *a.* τριγωνικός. ~ hat τρίκωχο *n.*

threefold *a.* τριπλάσιος.

three-legged *a.* ~ stool τρίποδο *n.*

three-ply *a.* (*thread*) τρίκλωνος.

three-quarters *adv.* κατά τα τρία τέταρτα.

three-score *a.* εξήντα.

thresh *v.* αλωνίζω. **~ing** floor αλώνι *n.*

threshold *s.* κατώφλι *n.*

thrice *adv.* τρεις φορές, τρις.

thrift *s.* οικονομία *f.* **-less** *a.* σπάταλος. **~y** *a.* οικονόμος.

thrill *s.* (*tremor*) τρεμούλα *f.*, ανατριχίλα *f.* (*emotion*) συγκίνηση *f.* (*v.t.*) συγκινώ, συναρπάζω. (*v.i.*) συγκινούμαι. **~er** *s.* αστυνομικό *n.* **~ing** *a.* συγκινητικός, συναρπαστικός.

thrive *v.* πάω καλά, ευδοκιμώ. he ~s on work η εργασία τον τρέφει.

thriving *a.* be ~ ακμάζω, πάω περίφημα.

throat *s.* λαιμός *m.* have a sore ~ πονάει ο λαιμός μου. he rammed his ideas down our ~s μας έπρηξε με τα κηρύγματά του. it stuck in my ~ (*fig.*) δεν μπόρεσα να το χωνέψω.

throaty *a.* (*hoarse*) βραχνός, (*deep*) του λάρυγγος.

throb *v.* χτυπώ, πάλλομαι. (*fig.*) σφύζω. (*s.*) χτύπος *m.*, παλμός *m.*

throes *s.* ~ of childbirth ωδίνες *f.pl.* we are in the ~ of the election είμαστε πάνω στην αναστάτωση των εκλογών.

thrombosis *s.* θρόμβωση *f.*

throne *s.* θρόνος *m.* **~ed** (*fig.*) θρονιασμένος.

throng *s.* πλήθος *n.* (*great*) κοσμοσυρροή *f.* (*v.i.*) συρρέω. (*v.t.*) πλημμυρίζω.

throttle *v.* πνίγω (*fig.*) καταπνίγω. ~ down (*engine*) κόβω ταχύτητα.

through *prep.* (*across*) διά μέσου (*with gen.*) μέσα από. he got in ~ the window μπήκε από το παράθυρο. he was shot ~ the heart η σφαίρα διαπέρασε την καρδιά του. (*time*) all ~ the winter όλο το χειμώνα, κατά τη διάρκεια του χειμώνα. (*agency*) I got to know him ~ my brother τον γνώρισα από τον αδελφό μου. I got the job ~ my father-in-law πήρα τη θέση μέσω του πεθερού μου. (*as a result of*) ~ carelessness από αμέλεια, *see also* from.

through *adv.* go *or* get *or* let ~ περνώ. read it ~ το διαβάζω μέχρι τέλους. ~ and ~ ως το κόκκαλο, πέρα για πέρα. all the way ~ (*performance, etc.*) καθ' όλη τη διάρκεια. be ~ with (*finished*) έχω τελειώσει, (*fed up*) έχω βαρεθεί. (*without break of journey*) κατευθείαν. ~ ticket εισιτήριο συνεχείας.

throughout *prep.* (*space*) σ' όλο το διάστημα, (*time*) καθ' όλη τη διάρκεια (*both with gen.*). ~ the country σ' ολόκληρη τη χώρα. (*adv.*) (*house*) close-carpeted ~ στρωμένο όλο (*or* απ' άκρη σ' άκρη) με μοκέτα.

throw *s.* ριξιά *f.* a stone's ~ δύο βήματα.

throw *v.t.* (*general*) ρίχνω, (*esp. objects*) πετώ. ~ open (*premises*) ανοίγω. ~ a glance ρίχνω μια ματιά. ~ light on ρίχνω φως σε. ~ into confusion προκαλώ σύγχυσιν εις. ~ oneself on *or* into ρίχνομαι σε. ~ oneself on (*mercy, etc.*) καταφεύγω σε. they were ~n out of work έμειναν άνεργοι. we were ~n together by chance η τύχη μας έκανε να γνωριστούμε. get ~n about (*jolted*) τραντάζομαι.

throw away *v.t.* πετώ. this is to be thrown away αυτό είναι για πέταμα. my words were thrown away (*wasted*) τα λόγια μου πήγαν χαμένα.

throw back *v.t.* πετώ πίσω. (*repel*) απωθώ, (*shoulders*) τεντώνω. be thrown back on αναγκάζομαι να καταφύγω σε.

throw-back *s* (*biol.*) παράδειγμα προγονισμού.

throw in *v.t.* (*a remark*) πετώ. (*add*) δίνω επιπλέον. ~ one's hand παραιτούμαι.

throw off *v.t.* (*get rid of*) ξεφορτώνομαι, γλυτώνω από. (*clothes*) βγάζω, (*poem, etc.*) σκαρώνω.

throw out *v.t.* (*eject*) πετώ έξω, (*reject*) απορρίπτω. (*radiate*) εκπέμπω, (*confuse*) μπερδεύω, (*a suggestion*) πετώ.

throw over *v.t.* εγκαταλείπω.

throw up *v.t.* (*vomit*) κάνω εμετό, βγάζω. (*cast ashore*) βγάζω έξω, (*give up*) παρατώ. (*build*) σκαρώνω στα γρήγορα, (*produce*) βγάζω. (*show up*) προβάλλω. (*raise*) σηκώνω.

throwing *s.* ρίξιμο *n.* (*in athletics*) ρίψη *f.*

thrush *s.* τσίχλα *f.*

thrust *v.t.* χώνω. (*knife, etc.*) καρφώνω, μπήγω. ~ aside σπρώχνω κατά μέρος. ~ out (*heat, tongue*) βγάζω. ~ (*duty*) upon φορτώνω σε. ~ oneself forward (*fig.*) προβάλλομαι. be ~ing προβάλλομαι αναιδώς. (*s.*) σπρωξιά *f.* ώθηση *f.* (*with sword*) ξιφισμός *m.* (*attack in war*) διείσδυση *f.*

thud *s.* γδούπος *m.* (*v.*) βροντώ υπόκωφα.

thug *s.* κακούργος *m.*

thumb *v.* (*book*) φυλλομετρώ. ~ one's nose at κοροϊδεύω. ~ a lift κάνω ωτοστόπ.

thumb *s.* αντίχειρ *m.* he is under her ~ είναι υποχείριός της. by rule of ~ εμπειρικά. ~nail sketch σύντομη περιγραφή. give the ~s up εγκρίνω, give the ~s down απορρίπτω.

thump *v.* χτυπώ, βαράω. (*s.*) χτύπημα *n.*

thunder *s.* βροντή *f.* ~ and lightning αστραπόβροντο *n.* (*fig.*) steal ~ of προλαμβάνω. (*v.*) βροντώ. ~**bolt** *s.* κεραυνός *m.* ~**clap** *s.* μπουμπουνητό *n.* ~**ing** *a.* (*fam.*) δυνατός, πελώριος. ~**ous** *a.* βροντώδης. ~**struck** *a.* εμβρόντητος. ~**y** *a.* the weather looks ~y ο καιρός προμηνύει καταιγίδα.

Thursday *s.* Πέμπτη *f.*

thus *adv.* έτσι.

thwart *v.* χαλώ, φέρνω εμπόδια σε.

thy *pron.* δικός σου.

thyme *s.* θυμάρι *n.*

tiara *s.* τιάρα *f.*

tic *s.* τικ *n.*

tick *v.* (*of clock*) κάνω τικ-τακ. (*mark*) τσεκάρω. ~ (*person*) off μαλώνω. (*fam.*) what makes him ~ τι σόι άνθρωπος είναι. ~ing over στο ραλαντί.

tick *s.* (*noise*) τικ-τακ *n.* (*fam.*) in a ~ σε δύο λεπτά, on the ~ στο τσακ. (*mark*) σημάδι ελέγχου. (*credit*) I got it on ~ το πήρα βερεσέ.

ticker *s.* (*fam.*) (*heart*) καρδιά *f.* (*watch*) ρολόι *n.* ~ tape ταινία τηλετύπου.

ticket *s.* εισιτήριο *n.* (*label*) ετικέτα *f.* ~ office θυρίδα *f.*

tick|le *v.* γαργαλάω. (*amuse*) διασκεδάζω. ~**lish** *a.* be ~lish γαργαλιέμαι. (*difficult*) λεπτός, δύσκολος.

tidal *a.* που εξαρτάται από την παλίρροια. ~ wave παλιρροϊκό κύμα.

tide *s.* παλίρροια *f.* high ~ πλημμυρίδα *f.* low ~ άμπωτη *f.* the ~ is in η θάλασσα είναι έξω. the ~ is out η θάλασσα είναι μέσα. it was washed up by the ~ το έχει εκβράσει η θάλασσα. (*fig., trend*) κύμα, ρεύμα *n.* the turn of the ~ η μεταστροφή της τύχης.

tide *v.* ~ over a difficult spell ξεπερνώ μια δύσκολη περίπτωση.

tidings *s.* νέα *n.pl.*

tidy *v.* (*also* ~ up) συγυρίζω, σιάζω, τακτοποιώ. ~ oneself up σιάζομαι.

tid|y *a.* (*person*) τακτικός, (*room*) τακτοποιημένος, νοικοκυρεμένος. a ~y sum σεβαστό ποσό. a ~y way (*distance*) αρκετά μακριά. ~**iness** *s.* τάξη *f.*

tie *v.t.* (*lit & fig.*) δένω. (*come equal*) έρχομαι ισοπαλία. be ~d (down) to the house είμαι καρφωμένος στο σπίτι. ~ in with συμφωνώ με. ~ up (*parcel, ship*) δένω, (*funds*) δεσμεύω. be ~d up (*occupied*) είμαι πιασμένος *or* απασχολημένος, (*connected*) συνδέομαι.

tie *s.* (*attachment*) δεσμός *m.* (*coming equal*) ισοπαλία *f.* (*neck*) γραβάτα *f.*

tier *s.* σειρά *f.* in ~s κλιμακωτός.

tiff *s.* καβγαδάκι *n.*

tiger *s.* τίγρις *f.*

tight *a.* (*cord*) τεντωμένος, (*clothes*) στενός. (*embrace, knot, lips*) σφιχτός. my shoes are ~ με σφίγγουν τα παπούτσια μου. in a ~ corner *or* spot σε δύσκολη θέση. air-~, water-~ στεγανός. (*fam., drunk*) σουρωμένος. it was a ~ squeeze for us in the taxi ήμασταν στουμωμένοι στο ταξί. (*adv.*) σφιχτά, καλά. sit ~ δεν το κουνάω. ~-**fisted** *a.* σφιχτός. ~-**fitting** *a.* εφαρμοστός. ~**ly** *adv.* σφιχτά, καλά.

tighten *v.t. & i.* σφίγγω, τεντώνω, (*bonds*) συσφίγγω. **~ing** *s.* σφίξιμο *n.*, τέντωμα *n.*

tightrope *s.* τεντωμένο σχοινί. **~** walker σχοινοβάτης *m.*

tights *s.* καλσόν *n.*

tile *s.* (*roof*) κεραμίδι *n.* (*floor, wall*) πλακάκι *n.* (*fam.*) go out on the ~s ξενυχτώ γλεντώντας.

till *prep.* ώς, μέχρι (*with acc.*) ~ now ώς τώρα, μέχρι τούδε. (*conj.*) I shall wait ~ you come θα περιμένω ώσπου να (*or* ωσότου να *or* μέχρις ότου) έρθεις.

till *s.* ταμείο *n.*

till *v.* καλλιεργώ.

tiller *s.* (*of soil*) καλλιεργητής *m.* (*helm*) λαγουδέρα *f.*, δοιάκι *n.*

tilt *v.t. & i.* (*lean*) γέρνω. (*v.i.*) ~ at (*fig.*) επιτίθεμαι εναντίον (*with gen.*) (*s.*) (*slant*) κλίση *f.* (*combat*) κονταροχτύπημα *n.* at full ~ ορμητικά.

timber *s.* ξυλεία *f.* (*piece*) δοκάρι *n.* (*fig.*) he is of presidential ~ έχει τα προσόντα να γίνει πρόεδρος.

timbre *s.* τίμπρο *n.*, ποιόν *n.*

time *s.* χρόνος *m.*, καιρός *m.* (*of day*) ώρα *f.* (*occasion*) φορά *f.* (*date, period*) καιρός *m.*, εποχή *f.* one at a ~ ένας ένας. how many ~s? πόσες φορές; what's the ~? τι ώρα είναι; on ~ στην ώρα μου. in (*or* at the right) ~ εγκαίρως. at the same ~ συγχρόνως. at ~s, from ~ to ~ πότε πότε, κάπου κάπου. all the ~ διαρκώς, συνέχεια. half the ~ το μισό καιρό. for a ~ για κάμποσο καιρό, για ένα διάστημα. for a long ~ (*days, years*) πολύ καιρό, (*hours*) πολλή ώρα. at one ~, once upon a ~ μια φορά. some ~ κάποτε. another ~ καμμιά άλλη φορά. ~ after ~, ~ and again επανειλημμένως. for the ~ being για την ώρα. in no ~ στο πι και φι. this ~ last year πέρυσι τέτοια εποχή. at the best of ~s υπό τις καλύτερες προϋποθέσεις. lose ~ χασομερώ. waste ~ (*act in vain*) χάνω τον καιρό μου. beat ~ κρατώ το χρόνο. keep (in) ~ είμαι στο χρόνο. in old ~s τον παλιό καιρό, τα παλιά χρόνια. in my ~ στον καιρό μου. he was before my ~ δεν τον έφτασα. signs of the ~s σημεία των καιρών. be behind the ~s έχω μείνει πίσω. have a good ~ περνώ ωραία, (*as wish*) καλή διασκέδαση! it is high ~ that... είναι καιρός πια να. by the ~ we got there ώσπου να φτάσουμε εκεί. you've been a long ~ coming έκανες πολλή ώρα να 'ρθεις. we took a long ~ to get there μας πήρε πολλή ώρα να φτάσουμε εκεί. I haven't seen him for a long ~ έχω πολύ καιρό να τον δω. he takes his ~ over his work εργάζεται με το

πάσσο του (*or* με την ησυχία του). it's three ~s the size of yours είναι τρεις φορές πιο μεγάλο από το δικό σου (*or* τριπλάσιο του δικού σου).

time *v.* (*measure*) χρονομετρώ, (*arrange*) κανονίζω. ill ~d άκαιρος. it was well ~d έγινε όταν έπρεπε (*or* στην ώρα του).

time-consuming *a.* που σου τρώει χρόνο.

time-honoured *a.* καθιερωμένος από τον χρόνο.

time-lag *s.* χρονική υστέρηση.

timeless *a.* αιώνιος.

time-limit *s.* χρονικόν όριον, (*term*) προθεσμία *f.*

timely *a.* έγκαιρος.

timepiece *s.* ρολόι *n.*

time-saving *a.* που σε βοηθάει να οικονομήσεις χρόνο.

time-server *s.* καιροσκόπος *a.*

timetable *s.* (*trains*) δρομολόγιο *n.* (*work*) ωρολόγιο *n.*, χρονοδιάγραμμα *n.*

timid *a.* άτολμος, δειλός, (*shy*) ντροπαλός. (*animal*) φοβισμένος. **~ity** *s.* ατολμία *f.*, δειλία *f.* ντροπαλοσύνη *f.* **~ly** *adv.* δειλά.

timing *s.* (*of words, actions*) εκλογή του κατάλληλου χρόνου. (*regulating*) ρύθμιση *f.* (*of race*) χρονομέτρηση *f.*

timorous *see* timid.

tin *s.* (*metal*) κασσίτερος *m.* (*container*) τενεκές *m.* (*gallon-size*) μπιντόνι *n.*, (*little*) κουτί *n.* (*baking*) ταψί *n.*, νταβάς *m.* (*cake*) φόρμα *f.* (*a.*) τενεκεδένιος. (*v.*) ~ned του κουτιού. ~ned food κονσέρβα *f.*

tincture *s.* (*med.*) βάμμα *n.*

tinder *s.* προσάναμμα *n.* **~box** *s.* τσακμάκι *n.*

tine *s.* δόντι *n.*

tinfoil *s.* ασημόχαρτο *n.*

tinge *s.* απόχρωση *f.* (*trace*) ίχνος *n.* (*v.*) χρωματίζω. ~d with sadness με ελαφρά απόχρωση θλίψεως.

tingle *v.i.* τσούζω, (*have pins & needles*) μυρμηγκιάζω. **~ing** *s.* τσούξιμο *n.* μυρμήγκιασμα *n.*

tinker *s.* γανωτζής *m.*, γανωματής *m.* (*v.*) ~ with μαστορεύω.

tinkle *v.t. & i.* κουδουνίζω. (*s.*) κουδούνισμα *n.* (*fam.*) τηλεφώνημα *n.*

tinny *a.* τενεκεδένιος.

tinsel *s.* φανταχτερή φτηνή διακόσμηση.

tint *s.* (*shade*) απόχρωση *f.* (*hair dye*) βαφή *f.* (*v.*) χρωματίζω ελαφρά, (*hair*) βάφω.

tiny *a.* πολύ μικρός, μικροσκοπικός.

tip *s.* (*end*) άκρη *f.* (*point*) μύτη *f.* (*fig.*) to one's finger ~s πέρα για πέρα. (*advice*) πληροφορία *f.* (*dump*) σωρός απορριμμάτων. (*money*) φιλοδώρημα *n.*, πουρμπουάρ *n.*

tip *v.t.* (*tilt*) γέρνω, (*discharge*) αδειάζω. (*give money to*) δίνω πουρμπουάρ σε. ~ off προειδοποιώ. ~ out αδειάζω. (*v.t. & i.*) ~ over αναποθυρίζω. (*v.i., of lorry*) ~ up ανατρέπομαι.

tip-and-run *a.* (*raid*) αιφνιδιαστικός.

tipple *v.i.* μπεκρουλιάζω. **~r** *s.* μπεκρής *m.*

tipsy *a.* στο κέφι, ζαλισμένος.

tiptoe *adv.* on ~ ακροποδητί, στις μύτες των ποδιών.

tip-top *a.* πρώτης τάξεως.

tirade *s.* κατηγορητήριο *n.*

tire *v.t.* κουράζω. (*v.i.*) ~ of βαριέμαι.

tired *a.* κουρασμένος. get ~ κουράζομαι. get ~ of βαριέμαι. ~ out κατακουρασμένος, ψόφιος στην κούραση. **~ness** *s.* κόπωση *f.*

tireless *a.* ακούραστος.

tiresome *a.* ενοχλητικός.

tiro *s.* αρχάριος *m.*

tissue *s.* (*fabric*) ύφασμα *n.* (*biol.*) ιστός *m.* ~ paper τσιγαρόχαρτο *n.* ~ of nonsense κομβολόγιον ανοησιών.

tit *s.* give ~ for tat ανταποδίδω τα ίσα.

titanic *a.* τιτάνιος.

titbit *s.* μεζές *m.*

tithe *s.* δεκάτη *f.*, εκκλησιαστικός φόρος.

titillat|e *v.* γαργαλίζω. **~ing** *a.* γαργαλιστικός.

titivate *v.i.* φτιάχνομαι.

title *s.* τίτλος *m.* **~d** *a.* ευγενής.

titter *v.* γελώ πνιχτά (*s.*) πνιχτό γέλιο.

tittle *s.* not a ~ ούτε ίχνος. **~-tattle** *s.* κουτσομπολιό *n.*

to (*prep.*) (*direction*) εις, σε, (*towards*) προς, (*as far as*) μέχρι, ως (*all with acc.*). (*compared with*) μπροστά σε, εν συγκρίσει με. (*purpose*) (για) να. (*with infinitive*) να. I gave it ~ John του έδωσα στο Γιάννη (*or* του Γιάννη). I gave it ~ him του το έδωσα. the man I gave money ~ εκείνος στον οποίον έδωσα λεφτά. the man I sat next ~ εκείνος που κάθησα κοντά του. he was the first ~ volunteer ήταν ο πρώτος που προσφέρθηκε. I want you ~ help me θέλω να με βοηθήσεις. I'm ready ~ go είμαι έτοιμος να φύγω. (*as substantival infinitive*) his motto is ~ live dangerously πιστεύει εις το ζην επικινδύνως. (*various*) ~ a man όλοι ανεξαιρέτως. ~ the last man μέχρις ενός. ~ the end μέχρι τέλους. it is five ~ six (*o'clock*) είναι έξι παρά πέντε. from two ~ three (*o'clock*) δύο με τρεις. ten ~ one (*wager*) δέκα με ένα. back ~ back πλάτη με πλάτη. it is similar ~ mine μοιάζει με το δικό μου. private secretary ~ the minister ιδιαιτέρα του υπουργού. he fell ~ the ground έπεσε κατά γης. so ~ say ούτως ειπείν. a joy ~ behold χάρμα

ιδέσθαι. (*adv.*) ~ and fro πέρα δώθε. come ~ (*recover*) συνέρχομαι. push the door ~ γέρνω την πόρτα.

toad *s.* βάτραχος *m.*

toady *s.* κόλακας *m.* (*v.*) κολακεύω.

toast *s.* piece of ~ φρυγανιά *f.* (*fam.*) have (*person*) on ~ στρυμώχνω. (*v.t.*) κάνω φρυγανιά. (*fig., warm*) ψήνω.

toast *s.* (*drunk*) πρόποση *f.* the ~ of the evening (*person*) η ατραξιόν της βραδιάς. (*v.t.*) προπίνω εις, πίνω στην υγεία (*with gen.*).

tobacco *s.* καπνός *m.* **~nist** *s.* καπνοπώλης *m.* **~nist's** *s.* καπνοπωλείο *n.*

toboggan *s.* τόμπογκαν *n.*

tocsin *s.* κώδων κινδύνου.

today *adv.* σήμερα. ~ week σήμερα οχτώ. of ~, **~'s** (*a.*) σημερινός.

toddle *v.* κάνω στράτα. **~r** *s.* νήπιο *n.*

to-do *s.* (*fam.*) φασαρία *f.*

toe *s.* δάχτυλο του ποδιού. (*of shoe*) μύτη *f.* on one's ~s (*fig.*) έτοιμος προς δράσιν. (*fig.*) tread on (*person's*) ~s θίγω, πατώ τον κάλο (*with gen.*). from top to ~ από την κορφή ως τα νύχια. (*v.*) ~ the line (*fig.*) υπακούω.

toff *s.* (*fam.*) σπουδαίο πρόσωπο, (*dandy*) δανδής *m.*

toffee *s.* καραμέλα του γάλακτος.

tog *v.* (*fam.*) ντύνω. (*s.*) ~s ρούχα *n.pl.*

toga *s.* τήβεννος *f.*

together *adv.* μαζί. we put our heads ~ συσκεφθήκαμε. gather ~ (*v.t.*) μαζεύω, (*v.i.*) μαζεύομαι. tie ~ δένω.

toil *v.* μοχθώ (*s.*) μόχθος *m.*, κόπος *m.*

toilet *s.* (*dress, dressing*) τουαλέτα *f.* (*w.c.*) μέρος *n.* ~ paper χαρτί υγείας.

toils *s.* δίχτυα *n.pl.*

toilsome *a.* κουραστικός.

token *s.* (*of surrender, office*) σήμα *n.* (*of esteem*) δείγμα *n.*, ένδειξη *f.* (*counter*) μάρκα *f.* (*a.*) ~ payment συμβολική πληρωμή.

tolerab|le *a.* υποφερτός, ανεκτός. **~ly** *adv.* αρκετά.

tolerance *s.* ανοχή *f.*

tolerant *a.* ανεκτικός. **~ly** *adv.* με ανεκτικότητα.

tolerat|e *v.* ανέχομαι, υποφέρω. **~ion** *s.* ανοχή *f.*, ανεκτικότητα *f.*

toll *s.* διόδια *n.pl.* (*fig.*) take heavy ~ προξενώ σοβαρές απώλειες.

toll *v.i.* (*of bell*) χτυπώ πένθιμα.

tom *s.* (*cat*) γάτος *m.*

tomato *s.* ντομάτα *f.*

tomb *s.* τάφος *m.* (*ancient*) τύμβος *m.* **~stone** *s.* ταφόπετρα *f.*

tomboy *s.* αγοροκόριτσο *n.*

tome s. χοντρό βιβλίο.

tomfoolery s. ανόητη συμπεριφορά.

tomorrow adv. αύριο. day after ~ μεθαύριο. ~'s, of ~ (a.) αυριανός.

ton s. τόννος m. (fam.) ~s of ένα σωρό. ~s of money λεφτά με ουρά.

tonal a. τονικός. ~ity s. τονικότης f.

tone s. τόννος m. (character) ατμόσφαιρα f. ~less a. άτονος.

tone v.t. ~ down (soften) μαλακώνω, (curb) μετριάζω. ~ up τονώνω. (v.i.) ~ in with εναρμονίζομαι με.

tongs s. λαβίδα f., τσιμπίδα f.

tongue s. γλώσσα f. (of bell) γλωσσίδι n. hold one's ~ δεν μιλώ. have a ready ~ έχω λυτή τη γλώσσα μου. have one's ~ in one's cheek αστειεύομαι.

tongue-tied a. he was ~ τον έπιασε γλωσσοδέτης.

tongue-twister s. γλωσσοδέτης m.

tonic a. (of sounds) τονικός. (bracing) τονωτικός. (s.) τονωτικό n.

tonight adv. απόψε.

tonnage s. χωρητικότητα f.

tonsil s. αμυγδαλή f.

tonsure s. ξυρισμένο κεφάλι μοναχού, κουρά f.

too adv. (also) και, επίσης. me ~ κι εμένα. ~ much πλέον του δέοντος, you gave me ~ much gravy μου 'βαλες πολλή σάλτσα. it's ~ late for her to go out είναι πολύ αργά πια για να βγει έξω. we are one ~ many (to be accommodated) ένας περισσέύει.

tool s. εργαλείο n. (fig., pawn) όργανο n.

toot v.t. (fam.) (horn) κορνάρω. (s.) κορνάρισμα n.

tooth s. δόντι n., οδούς m. have a ~ out βγάζω δόντι. false teeth μασέλα f. ~ and nail αγρίως. in the teeth of ενάντια σε. have a sweet ~ μου αρέσουν τα γλυκά. ~ache s. πονόδοντος m. ~brush s. οδοντόβουρτσα f. ~paste s. οδοντόπαστα f. ~pick s. οδοντογλυφίδα f. ~some a. ορεκτικός.

top s. (toy) σβούρα f. I slept like a ~ κοιμήθηκα μονορρούφι.

top s. (summit) κορυφή f. (upper part of building, bus, etc.) επάνω μέρος. on (the) ~ of επάνω σε. at the ~ (of trunk, drawer, etc.) πάνω-πάνω. be at the ~ (of one's profession) είμαι κορυφή. at the ~ of his form (school) πρώτος στην τάξη του, (performance) στην πιο καλή του φόρμα. at the ~ of his voice με όλη τη δύναμη της φωνής του.

top a. (highest) ανώτατος, (leading) κορυφαίος. ~ floor τελευταίο πάτωμα. ~ button το επάνω κουμπί. ~ dog (fam.)

νικητής m. ~ hat ψηλό καπέλλο. out of the ~ drawer καλής κοινωνικής τάξεως. ~~coat s. παλτό n. ~~heavy a. ασταθής. ~~most a. ο πιο ψηλός.

top v.t. (plants) κορφολογώ, (surpass) ξεπερνώ. ~ the list έρχομαι πρώτος. ~ up γεμίζω.

toper s. μπεκρής m.

topic s. θέμα n. ~al a. επίκαιρος.

topography s. τοπογραφία f.

topple v.i. πέφτω (v.t.) ανατρέπω.

topsy-turvy adv. μαλλιά-κουβάρια, άνω-κάτω.

torch s. (brand) δας f., δαυλός m., πυρσός m. (electric) φακός m., ηλεκτρικό φαναράκι. ~~bearer s. μεταλαμπαδευτής m.

torment v. βασανίζω, τυραννώ. (s.) μαρτύριο n. ~er s. βασανιστής m.

tornado s. λαίλαψ f.

torpedo s. τορπίλλη f. (v.) τορπιλλίζω. ~~boat s. τορπιλλοβόλον n.

torpid a. (numbed) ναρκωμένος, (sluggish) νωθρός.

torpor s. νάρκη f. νωθρότης f.

torrent s. (also ~~bed) (lit. & fig.) χείμαρρος m., (lit.) ρέμα n. ~ial a. καταρρακτώδης.

torrid a. ~ heat or passion φωτιά και λάβρα. ~ zone διακεκαυμένη ζώνη.

torso s. κορμός m.

tortoise s. χελώνα f. ~~shell s. ταρταρούγα f.

tortuous a. (path) σκολιός. (fig.) περιπλεγμένος.

torture v. βασανίζω. (s.) (infliction) βασανισμός m., (means) βασανιστήριο n. ~er s. βασανιστής m.

tosh s. ανοησίες f.pl.

toss v.t. (throw) ρίχνω, πετώ, (the head) τινάζω. be ~ed (by waves) κλυδωνίζομαι. (v.i., of ship) σκαμπανεβάζω. ~ up παίζω κορώνα-γράμματα. ~ off (poem) σκαρώνω, (drink) πίνω μονορρούφι.

toss s. ρίξιμο n., τίναγμα n.

tot s. (child) μικρό παιδί, (dram) δαχτυλάκι n.

tot v.t. ~ up αθροίζω. (v.i.) συμποσούμαι.

total a. (entire) ολικός, συνολικός. (absolute) τέλειος, πλήρης. (s.) σύνολον n., άθροισμα n. in ~ συνολικά. (v.i.) ανέρχομαι εις. ~ly adv. εξ ολοκλήρου, τελείως, πλήρως.

total|ity s. ολότης f. ~itarian a. ολοκληρωτικός.

tote v. κουβαλώ.

totem s. τοτέμ n.

totter v. τρικλίζω, (of regime) καταρρέω.

touch s. (sense) αφή f. (act of ~ing) άγγιγμα n. (contact) επαφή f. (mus.) τουσέ n. (of

painter's brush) πινελιά *f. (small quantity)* δόση *f.*, ίχνος *n.* just a ~ μία ιδέα. be in ~ with έχω επαφή με.
touch *v.t.* αγγίζω, *(move)* συγκινώ, *(concern)* αφορώ. *(come up to, reach)* φτάνω. *(of ship, put in at)* πιάνω σε. no one can ~ him at chess δεν του βγαίνει κανείς στο σκάκι. ~ bottom βρίσκω πάτο. ~ wood χτύπα ξύλο. *(be in contact with)* εφάπτομαι *(with gen.)*, *(of land, properties)* συνορεύω με. *(v.i.)* ~ on θίγω.
touch down *v.i. (aero.)* προσγειώνομαι.
touch off *v.t. (fig.)* ανάβω, προκαλώ.
touch up *v.t.* ρετουσάρω.
touch-and-go *a.* it was ~ whether he survived παρά τρίχα να πεθάνει.
touched *a. (a bit mad)* βλαμμένος.
touching *a. (moving)* συγκινητικός. *(in contact)* they are ~ ακουμπάνε, εφάπτονται. *(prep.) (about)* σχετικά με.
touchstone *s.* Λυδία λίθος.
touch|y *a.* εύθικτος. **~iness** ευθιξία *f.*
tough *a. (hard)* σκληρός, *(strong)* γερός. *(inured)* σκληραγωγημένος. *(difficult)* δύσκολος. ~ luck! τι κρίμα! *(s.)* τραμπούκος *m.*, αλήτης *m.*
tough|en *v.* σκληραίνω, *(person)* σκληραγωγώ. **~ness** *s.* σκληρότητα *f.*
toupée *s.* περούκα *f.*
tour *s.* γύρος *m.*, περιήγηση *f. (on business)* περιοδεία *f. (v.)* ~ through περιηγούμαι, περιοδεύω.
tour de force *s.* κατόρθωμα *n.*
tour|ing *s.* τουρισμός *m.* **~ist** *s.* τουρίστας *m. (a.)* τουριστικός.
tournament *s. (medieval)* γιόστρα *f. (sports)* πρωτάθλημα *n.*
tousled *a.* with ~ hair ξεμαλλιασμένος.
tout *s.* κράχτης *m. (v.)* κυνηγώ πελατεία.
tow *s. (fibre)* στουπί *n.*
tow *v.* ρυμουλκώ. *(fig.)* have *(person)* in ~ συνοδεύομαι από.
towards *prep. (place)* προς, *(time)* κατά, περί *(all with acc.)*. England's attitude ~ America η στάση της Αγγλίας έναντι της Αμερικής. his feelings ~ me τα αισθήματά του απέναντί μου *(or* για μένα).
towel *s.* πετσέτα *f. (fig.)* throw in the ~ παραιτούμαι.
tower *s.* πύργος *m. (v.i.)* υψώνομαι. ~ above *(fig.)* υπερέχω *(with gen.)* **~ing** *a.* πανύψηλος. be in a ~ing rage μαίνομαι.
town *s.* πόλις *f. (fam.)* go to ~ *(on spree)* το ρίχνω έξω, *(act enthusiastically)* πέφτω με τα μούτρα. ~ council δημοτικόν συμβούλιον. ~ crier ντελάλης *m.* ~ hall δημαρχείον *n. (a.)* της πόλεως, *(urban)*

αστικός.
town-planning *s.* πολεοδομία *f.* ~ authority σχέδιον πόλεως.
township *s.* κωμόπολις *f.*
townsman *s.* αστός *m.*
toxic *a.* τοξικός.
toy *s.* παιχνίδι *n. (a.)* ~ train σιδηροδρομάκι *n. (v.)* ~ with παίζω με.
trace *s. (mark)* ίχνος *n. (fam.)* kick over the ~s επαναστατώ.
trac|e *v. (find)* βρίσκω, ανακαλύπτω. *(follow)* παρακολουθώ. *(draw)* σχεδιάζω, *(copy)* ξεσηκώνω. **~ing-paper** *s.* χαρτί ξεσηκώματος.
track *v.* ανιχνεύω, ψάχνω να βρω. *(hunt)* κυνηγώ. ~ down ανακαλύπτω.
track *s. (path)* μονοπάτι *n. (racing)* στίβος *m. (of wheel, comet)* τροχιά *f. (railway)* σιδηροτροχιά *f.*, γραμμές *f.pl,* single ~ μονή γραμμή. the beaten ~ η πεπατημένη. keep ~ of παρακολουθώ. be on the right (wrong) ~ ακολουθώ σωστό (εσφαλμένο) δρόμο. ~s ίχνη, αποτυπώματα *n.pl.* make ~s for κατευθύνομαι προς. ~ **suit** *s.* αθλητική φόρμα.
track *s. (of tank)* ερπύστρια *f.* ~ed vehicle ερπυστριοφόρον *n.*
tracker *s.* κυνηγός *m.* ~ dog ιχνηλάτης κύων.
tract *s. (written)* φυλλάδιο *n.*
tract *s. (area)* έκταση *f. (anat.)* οδός *f.* digestive ~ πεπτικός σωλήν.
tractable *a.* ευπειθής, *(material)* εύχρηστος.
traction *s.* έλξη *f.*
tractor *s.* τρακτέρ *n.*
trade *v.t. & i. (also* ~ in) εμπορεύομαι, *(exchange)* ανταλλάσσω. ~ in an old car δίνω παλιό αυτοκίνητο έναντι μέρους της τιμής νέου. ~ on *(exploit)* εκμεταλλεύομαι.
trade *s.* εμπόριο *n. (calling)* επάγγελμα *n. (work)* δουλειά *f. (dealings)* συναλλαγές *f.pl. (a.)* εμπορικός. **~mark** *s.* σήμα κατατεθέν.
trader *s.* εμπορευόμενος *m.*
tradesman *s.* έμπορος λιανικής πώλησης.
trade-union *s.* εργατικό συνδικάτο.
trade-wind *s.* αληγής άνεμος.
trading *s.* εμπόριο *n. (a.)* εμπορικός.
tradition *s.* παράδοση *f.*
traditional *a.* παραδοσιακός. **~ist** *s.* οπαδός της παράδοσης. **~ly** *adv.* κατά παράδοσιν.
traduce *v.* δυσφημώ.
traffic *s. (trading)* συναλλαγή *f.*, εμπόριο *n. (movement)* κίνηση *f.*, κυκλοφορία *f.* ~ police τροχαία *f. (v.)* ~ in εμπορεύομαι.
tragedian *s. (playwright)* τραγικός *m. (actor)* τραγωδός *m.*
tragedy *s.* τραγωδία *f. (sad event)* πλήγμα

π., τραγικό συμβάν.
tragic a. τραγικός. **~ally** adv. τραγικά.
tragi-comedy s. ιλαροτραγωδία f.
trail s. (path) μονοπάτι n. (marks) ίχνη n.pl. (of smoke, people) ουρά f. leave a ~ of destruction αφήνω καταστροφή στο πέρασμά μου.
trail v.t. (drag) σέρνω, (follow) παρακολουθώ. (v.i.) σέρνομαι, (of plant) έρπω.
train s. (suite) ακολουθία f. (of dress) ουρά f. (of thought) ειρμός m. (of people, objects, events) σειρά f. in their ~ (behind them) στο πέρασμά τους. (railway) τραίνο n., αμαξοστοιχία f., συρμός m. in ~ έτοιμος.
train v.t. (εκ)γυμνάζω, ασκώ. (sports) προπονώ. (direct) κατευθύνω. (v.i.) (εκ)γυμνάζομαι, ασκούμαι, προπονούμαι. **~er** s. εκγυμναστής m. (sports) προπονητής m.
training s. εκγύμναση f. προπόνηση f. in ~ (fit) σε φόρμα. **~college** s. διδασκαλείον n.
traipse v.i. σέρνομαι.
trait s. χαρακτηριστικό n.
traitor s. προδότης m. **~ous** a. (deed) προδοτικός.
trajectory s. τροχιά f.
tram s. τραμ n.
trammels s. εμπόδια n.pl.
tramp v.i. περπατώ βαριά, (on long walks) πεζοπορώ. (v.t.) διασχίζω πεζή.
tramp s. (steps) βήματα n.pl. (a walk) περίπατος m. (vagrant) αλήτης m. (steamer) τραμπ n.
trample v. ~ down or on ποδοπατώ, (fig.) θίγω, προσβάλλω.
trance s. κατάστασις υπνώσεως, (ecstasy) έκστασις f.
tranquil a. γαλήνιος, ήρεμος. **~lity** s. γαλήνη f., ηρεμία f. **~lizer** s. ηρεμιστικό n.
transact v. διεξάγω. **~ion** s. (doing) διεξαγωγή f. (thing done) συναλλαγή f., πράξη f. (of learned society) **~s** πεπραγμένα n.pl.
transatlantic a. υπερατλαντικός.
transcend v.t. υπερβαίνω τα όρια (with gen.) (excel) ξεπερνώ.
transcendent a. ύψιστος. **~al** a. υπερβατικός.
transcontinental a. διηπειρωτικός.
transcribe v. μεταγράφω.
transcript s. μεταγραφημένο αντίγραφο. **~ion** s. μεταγραφή f.
transept s. πτέρυξ f.
transfer s. (removal, conveyance) μεταφορά f. (of employee) μετάθεση f., (of property) μεταβίβαση f. (design) χαλκομανία f.

transfer v.t. μεταφέρω, (employee) μεταθέτω. (make over) μεταβιβάζω. (v.i.) μεταφέρομαι. (change conveyance) αλλάζω, κάνω αχταρμά.
transferable a. μεταβιβασμός. non- ~ αμεταβίβαστος.
transfigur|ation s. μεταμόρφωση f. **~e** v. μεταμορφώνω.
transfix v. διαπερώ, (fig., with gaze) καρφώνω. he was ~ed έμεινε μάρμαρο.
transform v. μεταμορφώνω. (convert) μετατρέπω, (alter) μεταβάλλω. **~ation** s. μεταμόρφωση f. μετατροπή f. μεταβολή f. **~er** s. μετασχηματιστής m.
transfus|e v. μεταγγίζω. **~ion** s. μετάγγιση f.
transgress v. παραβιάζω, παραβαίνω. **~ion** s. παράβαση f. (sin) αμάρτημα n. **~or** s. παραβάτης m.
transient a. παροδικός, περαστικός, εφήμερος.
transistor s. τρανζίστορ n.
transit s. μεταφορά f. in ~ υπό διαμετακόμισιν. ~ camp στρατόπεδον διερχομένων.
transition s. μετάβαση f. **~al** a. μεταβατικός.
transitive a. (gram.) μεταβατικός.
transitory a. παροδικός, εφήμερος.
translat|e v. μεταφράζω. **~ion** s. μετάφραση f. **~or** s. μεταφραστής m.
transliterate v. μεταγράφω με στοιχεία άλλου αλφαβήτου.
translucent a. ημιδιαφανής.
transmigration s. (of souls) μετεμψύχωση f.
transmission s. μετάδοση f., εκπομπή f.
transmit v. μεταδίδω, εκπέμπω. **~ter** s. (radio) πομπός m.
transmute v. μεταστοιχειώνω.
transpar|ent a. διαφανής, (obvious) καταφανής. **~ency** s. διαφάνεια f.
transpire v. it ~d that κατέστη γνωστόν ότι. (fam., happen) συμβαίνω.
transplant v. μεταφυτεύω. (med.) μεταμοσχεύω. (s.) μεταμόσχευση f.
transport v. μεταφορά f. (means of ~) συγκοινωνία f., μεταφορικά μέσα. (ship) μεταγωγικό n. (of emotion) παραφορά f.
transport v. μεταφέρω, (exile) εκτοπίζω. be ~ed with joy γίνομαι έξαλλος από χαρά. **~ation** s. (of prisoner) εκτόπιση f. **~er** s. μεταφορεύς m.
transpos|e v. αντιμεταθέτω, (mus.) μεταφέρω. **~ition** s. διαφάνεια f.
transubstantiation s. μετουσίωση f.
transverse a. εγκάρσιος.
trap s. παγίδα f. (for small game) δόκανο n. (for mice) φάκα f. (in pipe) σιφόνι n. (cart) σούστα f. set a ~ στήνω παγίδα. (v.) παγιδεύω, (retain) συγκρατώ. get ~ped (in lift, etc.) κλείνομαι. **~door** s. καταπακτή f.

trapeze s. τραπέζιον n.

trapper s. κυνηγός m.

trappings s. (fig.) σύμβολα και στολίδια αξιώματος.

traps s. (fam.) μπαγκάζια n.pl., αποσκευές f.pl.

trash s. (goods) ψευτοπράματα n.pl. (nonsense) σαχλαμάρες f.pl. (rubbish) σκουπίδια n.pl. ~y a. φτηνός, (novel, etc.) της πεντάρας.

trauma s. τραύμα n. ~**tic** a. τραυματικός.

travail s. ωδίνες f.pl.

travel s. (also ~s) ταξίδια n.pl. ~ agency γραφείον ταξιδίων. (a.) ταξιδιωτικός.

travel v.i. ταξιδεύω. (go) πηγαίνω, τρέχω, κινούμαι. ~ back (mentally) ανατρέχω. ~**led** a. ταξιδεμένος, κοσμογυρισμένος.

traveller s. ταξιδιώτης m. (for pleasure) περιηγητής m. fellow ~ συνοδοιπόρος m. ~'s cheque ταξιδιωτική επιταγή.

travelling a. (mobile) κινητός. (for travel) ταξιδιωτικός, ταξιδίου. (s.) ταξίδια n.pl.

traverse v. διασχίζω.

travesty v. κοροϊδεύω, παραποιώ. (s.) κοροϊδία f. παραποίηση f.

trawl, ~er s. τράτα f.

tray s. δίσκος m. in-~ εισερχόμενα n.pl. out-~ εξερχόμενα n.pl.

treacherous a. (person) προδότης (m.) άπιστος. (deed) προδοτικός, μπαμπέσικος, άτιμος, δόλιος. (weather) ύπουλος. ~**ly** adv. προδοτικά, δολίως.

treachery s. προδοσία f. μπαμπεσιά f., δόλος m.

treacly a. λιγωτικός.

tread v.i. (also ~ on) πατώ. (walk) περπατώ. ~ on toes of (fig.) θίγω. (s.) πάτημα n., βήμα n. (of tyre) πέλμα n.

treadmill s. (fig.) μαγγανοπήγαδο n.

treason s. (high) ~ (εσχάτη) προδοσία f. ~**able** a. προδοτικός.

treasure s. θησαυρός m. (v.) φυλάω ως κόρην οφθαλμού. (esteem) εκτιμώ. ~**d** πολύτιμος.

treasur|er s. ταμίας m. ~**y** s. θησαυροφυλάκιον n. (fig.) θησαυρός m.

treat s. απόλαυση f.

treat v.t. (behave to) μεταχειρίζομαι, φέρομαι σε. (use) μεταχειρίζομαι. (construe) θεωρώ, εκλαμβάνω. (handle topic) χειρίζομαι. (process) επεξεργάζομαι. (patients) κουράρω, (illness) θεραπεύω. (entertain) I ~ed him to a beer τον κέρασα μια μπίρα. (v.i.) ~ of πραγματεύομαι. ~ with διαπραγματεύομαι με.

treatise s. πραγματεία f.

treatment s. (use) μεταχείριση f. (of patient) κούρα f., ιατρική αγωγή. (processing) επεξεργασία f.

treaty s. συνθήκη f.

treble a. τριπλάσιος. (v.t.) τριπλασιάζω.

treble s. (mus.) πρίμο n.

tree s. δέντρο n., δένδρον n. (for boots) καλαπόδι n.

trefoil s. τριφύλλι n.

trek v. (fig.) οδοιπορώ.

trellis s. καφασωτός φράκτης για αναρριχητικά φυτά.

trembl|e v. τρέμω. ~**ing** s. τρεμούλα f.

tremendous a. (huge) τεράστιος, πελώριος. (remarkable) τρομερός. ~**ly** adv. αφάνταστα, τρομερά.

tremor s. (body) τρεμούλα f. (earth) δόνηση f.

tremulous a. τρεμουλιαστός.

trench s. (irrigation) αυλάκι n. (mil.) χαράκωμα n. ~ coat τρενσκότ n.

trenchant a. απερίφραστος, κοφτός. ~**ly** adv. ορθά-κοφτά.

trencher s. ξύλινο πιάτο. ~**man** s. good ~ man γερό πηρούνι.

trend s. κατεύθυνση f., τάση f. set the ~ λανσάρω τη μόδα. ~**y** a. υπερμοντέρνος.

trepidation s. ανησυχία f.

trespass v.i. ~ on καταπατώ, (abuse) καταχρώμαι. (s.) καταπάτηση f. (sin) αμάρτημα n. ~**er** s. καταπατητής m.

tress s. πλόκαμος m. ~**es** μαλλιά n.pl.

trestle s. στρίποδο n.

trial s. (testing) δοκιμή f. on ~ υπό δοκιμήν. (contest) αγώνας m. (painful test) δοκιμασία f. (trouble) βάσανο n., μαρτύριο n. (law) δίκη f. bring to ~ δικάζω, stand ~ δικάζομαι. (a.) δοκιμαστικός.

triang|le s. τρίγωνο n. ~**ular** a. τριγωνικός.

trib|e s. φυλή f. ~**al** a. της φυλής, φυλετικός.

tribulation s. δοκιμασία f., βάσανο n.

tribunal s. δικαστήριον n.

tribune s. (person) δημαγωγός m. (platform) βήμα n.

tributary a. υποτελής (s., river) παραπόταμος m.

tribute s. (exacted) φόρος υποτελείας. (of respect) φόρος τιμής.

trice s. (fam.) in a ~ στο πι και φι.

trick s. (device, knack) τέχνασμα n., κόλπο n. (deceit) απάτη f. (habit) συνήθεια f. (joke) φάρσα f. (at cards) χαρτωσιά f. three-card ~ παπάς m. it did the ~ ήταν αποτελεσματικό. (v.) κοροϊδεύω, εξαπατώ. ~ out στολίζω. ~**ery** s. απάτη f.

trickle v.t. & i. (drip) στάζω (v.i., roll) κυλώ.

trickster s. απατεώνας m.

tricky a. (wily) πονηρός, (difficult) δύσκολος.

tricycle s. τρίκυκλο n.

trident s. τρίαινα f.

tried a. δοκιμασμένος.

trifle s. μικρό (or ασήμαντο) πράμα. at the merest ~ με το παραμικρό. a ~ (adv.) λιγάκι (v.) παίζω. ~ away σπαταλώ.

trifling a. ασήμαντος.

trigger s. σκανδάλη f. (v.) ~ off προκαλώ.

trigonometry s. τριγωνομετρία f.

trilby s.(hat) ρεπούμπλικα f.

trill s. τρίλλια f.

trilogy s. τριλογία f.

trim a. κομψός, περιποιημένος, τακτικός (s.) (of hair) ψαλίδισμα n., (balance) ισορροπία f. in good ~ σε καλή κατάσταση.

trim v. (tree) κλαδεύω, (beard) ψαλιδίζω. (decorate) γαρνίρω, (balance) ισορροπώ. ~ one's sails to the wind πηγαίνω όπως φυσάει ο άνεμος. **~mer** s. καιροσκόπος m. **~ming** s. γαρνιτούρα f.

trinity s. τριάς, τριάδα f.

trinket s. φτηνό κόσμημα.

trio s. τριάδα f. (mus.) τρίο n.

trip s. εκδρομή f., ταξίδι n.

trip v.i. (walk) περπατώ με ελαφρό βήμα. (stumble) σκοντάφτω. (v.t.) I ~ped him (up) τον έκανα να σκοντάψει, (fig.) τον τσάκωσα.

tripartite a. τριμερής.

tripe s. πατσάς m.

triple a. τριπλούς (v.t.) τριπλασιάζω. ~t s. (child) τρίδυμον n. (mus.) τρίηχον n.

triplicate s. in ~ εις τριπλούν.

tripod s. τρίποδο n.

tripper s. εκδρομεύς m.

triptych s. τρίπτυχον n.

trite a. τετριμμένος.

triumph s. θρίαμβος m. (v.) θριαμβεύω. ~ over υπερνικώ. **~al** a. ~al arch αψίς θριάμβου.

triumphant a. θριαμβευτικός, (person) θριαμβευτής m. **~ly** adv. θριαμβευτικά.

trivet s. πυροστιά f.

trivial a. ασήμαντος, επιπόλαιος. **~ity** s. επιπολαιότητα f. (thing) ασήμαντο πράγμα.

Trojan a. τρωικός. (person) Τρως m. (pl.) Τρώες. ~ horse δούρειος ίππος. work like a ~ δουλεύω ακούραστα.

troll v. τραγουδώ.

trolley s. καροτσάκι n. (of tram) τρολλές m. **~bus** s. τρόλλεϋ n.

trollop s. τσούλα f.

trombone s. τρομπόνι n.

troop s. (people) ομάδα f. (friends) παρέα f. (mil.) ~ of cavalry, etc. ουλαμός m. ~s στρατεύματα n.pl. (v.i.) ~ in εισέρχομαι ομαδικώς. **~ship** s. οπλιταγωγόν n.

trophy s. τρόπαιο n. (prize) βραβείο n.

tropic s. ~s τροπικές χώρες. **~al** a. τρο-

πικός.

trot v.i. τροχάζω. (v.t.) ~ out παρουσιάζω. (s.) τροχασμός m. be on the ~ είμαι στο πόδι.

troth s. πίστη f.

troubadour s. τροβαδούρος m.

trouble v.t. (annoy) ενοχλώ, (inconvenience) βάζω σε κόπο. (worry) στενοχωρώ, ανησυχώ. (v.i.) (take pains) κοπιάζω. (worry, care) νοιάζομαι, σκοτίζομαι. don't ~ about that μη σε νοιάζει (or μη νοιάζεσαι) για αυτό. fish in ~d waters ψαρεύω σε θολά νερά.

trouble s. (pains) κόπος m. take ~ μπαίνω σε κόπο. (unrest) φασαρίες f.pl. (worries) έννοιες f.pl., βάσανα n.pl., σκοτούρες f.pl. (nuisance) μπελάς m. be in ~ έχω φασαρίες (or τράβαλα). get into ~ βρίσκω τον μπελά μου. be looking for ~ πάω γυρεύοντας για καβγάδες. **~maker** s. ταραχοποιός m. **~shooter** s. (in disputes) συμφιλιωτής m. (engineering) ειδικός δι' επιδιόρθωσιν μηχανικών βλαβών. **~some** a. ενοχλητικός, (difficult) δύσκολος.

trough s. σκάφη f. (of wave) κοίλον n.

trounce v. δέρνω.

troupe s. θίασος m.

trouser|s s. παντελόνι n. **~leg** s. μπατζάκι n.

trousseau s. προικιά f.

trout s. πέστροφα f.

trowel s. μυστρί n.

tru|ancy s. σκασιαρχείο n. **~ant** s. σκασιάρχης m. play ~ant το σκάω.

truce s. εκεχειρία f.

truck s. (railway) βαγόνι n. (lorry) φορτηγό n. (fam.) have no ~ with δεν έχω δοσοληψίες με.

trucker s. οδηγός φορτηγού.

truckle v.i. ~ to υποτάσσομαι σε.

truculent a. με εριστική διάθεση. **~ly** adv. βιαίως.

trudge v. βαδίζω με κόπο.

true a. αληθινός, αληθής. (actual) πραγματικός, (exact) ακριβής, (faithful) πιστός. ~ to life παρμένος από τη ζωή. come ~ επαληθεύομαι, βγαίνω αληθινός. is it ~ that...? είναι αλήθεια ότι...; (adv.) σωστά. **~blue** a. απολύτως έντιμος. **~born** a. γνήσιος.

truffle s. τρούφα f.

truism s. κοινοτοπία f.

truly adv. αληθινά, αληθώς, πραγματικά.

trump s. ατού n. (fam., person) εν τάξει άνθρωπος. turn up ~ δείχνομαι εν τάξει. (v.) παίρνω με ατού. ~ up σκαρώνω, ~ed up κατασκευασμένος.

trumpery a. της κακιάς ώρας, ευτελής.

trumpet s. σάλπιξ f. blow one's own ~

παινεύομαι. (v.) διαφημίζω. ~er s. σαλπιγκτής m.
truncate v. κουτσουρεύω.
truncheon s. γκλομπ n.
trundle v.t. & i. κυλώ.
trunk s. (tree, body) κορμός m. (box) μπαούλο n. (elephant's) προβοσκίδα f. ~s (garment) παντελονάκι n.
trunk a. ~ call υπεραστικό τηλεφώνημα. ~ road κυρία οδική αρτηρία.
truss s. (hay) δεμάτι n. (med.) κηλεπίδεσμος m. (v.) δένω.
trust v.t. εμπιστεύομαι, (believe) δίνω πίστη σε. I wouldn't ~ him with my car δεν θα του εμπιστευόμουνα το αυτοκίνητό μου. (v.i., hope) ελπίζω.
trust s. εμπιστοσύνη f. on ~ καλή τη πίστει. (responsibility) ευθύνη f. (fin.) καταπίστευμα n. (business) τραστ n.
trustee s. επίτροπος m.
trust|ful, ~ing a. έχων εμπιστοσύνη.
trust|worthy, ~y a. (person) έμπιστος, (news, etc.) αξιόπιστος.
truth s. αλήθεια f.
truthful a. (person) φιλαλήθης, (story) αληθινός. ~ly adv. με ειλικρίνεια.
try v. (attempt) προσπαθώ. (sample, test) (also ~ on, out) δοκιμάζω. ~ for πάω για. (tire) κουράζω. (law) δικάζω. ~ one's best βάζω τα δυνατά μου. (s.) προσπάθεια f.
trying a. κουραστικός, εκνευριστικός.
try-on s. απόπειρα f., δοκιμή f.
tryst s. ραντεβού n.
tsar s. τσάρος m.
T-shirt s. κοντομάνικη φανέλλα.
tub s. (barrel) βαρέλι n. (bath) μπάνιο n. (wash-~) σκάφη f. (fam., boat) σκυλοπνίχτης m.
tubby a. κοντόχοντρος.
tube s. σωλήνας m. (for medicament) σωληνάριο n.
tubercul|ar a. φυματικός. ~osis s. φυματίωση f.
tubular a. σωληνοειδής.
tuck v.t. (thrust) χώνω, (stow) κρύβω. ~ up (in bed) σκεπάζω. (v.i.) ~ in (fam.) τρώω με όρεξη.
tuck s. (fold) σούφα f.
Tuesday s. Τρίτη f.
tuft s. θύσανος m. (hair) τούφα f.
tug v. τραβώ. (s.) τράβηγμα n. (boat) ρυμουλκό n. ~ of war διελκυστίνδα f.
tuition s. διδασκαλία f. (fees) δίδακτρα n.pl.
tulip s. τουλίπα f.
tulle s. τούλι n.
tumble v.i. πέφτω, (of tumbler) κάνω τούμπες. (fam.) ~ to καταλαβαίνω. (v.t.)

ρίχνω. (s.) τούμπα f.
tumbledown a. ετοιμόρροπος.
tumbler s. (acrobat) ακροβάτης m. (glass) ποτήρι n.
tumid a. πρησμένος. (fig.) στομφώδης.
tummy s. κοιλίτσα f.
tumour s. όγκος m.
tumult s. οχλοβοή f., χάβρα f. ~uous a. θορυβώδης.
tune s. σκοπός m. sing out of ~ φαλτσάρω. (fig.) in ~ with the times προσαρμοσμένος στους καιρούς. his same old ~ το βιολί του. change one's ~ αλλάζω χαβά. ~ful a. μελωδικός.
tune v. (instrument) κουρδίζω, (engine) ρυθμίζω. ~ in to (radio) πιάνω, (become receptive) προσαρμόζομαι σε.
tuning s. κούρδισμα n. ρύθμιση f. ~ fork διαπασών n.
tunic s. χιτών m. (mil.) χιτώνιον n.
tunnel s. σήραγγα f. (v.) ανοίγω σήραγγα.
tunny s. τόννος m.
turban s. τουρμπάνι n.
turbid a. θολός.
turbine s. στρόβιλος m.
turbo-charged a. στροβιλοκίνητος.
turbot s. φούσκα m.
turbul|ence s. στροβιλισμός m. (ανα)ταραχή f. ~ent a. (air) στροβιλώδης, (crowd) θορυβώδης, (times, water) ταραγμένος.
turd s. (fam.) κουράδα f.
tureen s. (soup) σουπιέρα f. (gravy) σαλτσιέρα f.
turf s. (grass) χλόη f. (sod) κομμένη λωρίδα χλόης. (fig.) κούρσες f.pl.
turgid a. στομφώδης.
turkey s. γάλλος m, διάνος m.
Turkish a. τουρκικός. (person) Τούρκος m. Τουρκάλα f. ~ delight λουκούμι n. ~ bath χαμάμ(ι) n.
turmoil s. αναστάτωση f.
turn s. (movement, bend) στροφή f. (of circumstances) τροπή f. (change) αλλαγή f. (inclination) κλίση f. (occasion) σειρά f. (stroll) βόλτα f. take a ~ for the better πάω καλύτερα. do a good ~ to εξυπηρετώ. it will serve my ~ μου κάνει. you gave me a ~ με τρόμαξες. each in ~ ο καθένας με τη σειρά του. by ~s εκ περιτροπής, μια ο ένας μια ο άλλος. at every ~ συνεχώς. to a ~ στην εντέλεια, ό,τι πρέπει. (act, number) νούμερο n.
turn v.t. (tap, head, eyes, steering-wheel) γυρίζω, στρέφω, στρίβω. (rotate) περιστρέφω. (page, key, car, conversation, mattress, garment) γυρίζω. (attention) στρέφω. (shape, on lathe & fig.) τορνεύω. (cause to be, make) κάνω, (change,

convert) μετατρέπω (*with* σε). (*get round*) καβαντζάρω, (*outflank*) υπερκερώ. ~ inside out *or* upside down αναποδογυρίζω. ~ loose αφήνω ελεύθερο. ~ the corner, στρίβω τη γωνία. ~ one's back στρέφω τα νώτα. ~ to good use κάνω καλή χρήση (*with gen.*) ~ one's hand to αναλαμβάνω. it ~ed his head του πήρε τα μυαλά. it ~ed my stomach μου χάλασε το στομάχι. he didn't ~ a hair έμεινε ατάραχος. he's ~ed forty έκλεισε τα σαράντα. it's ~ed six o' clock είναι περασμένες έξη.

turn *v.i.* γυρίζω, στρίβω, στρέφομαι. (*rotate*) περιστρέφομαι. (*become*) γίνομαι, (*be converted*) μετατρέπομαι. (*of milk*) κόβω. ~ red γίνομαι κόκκινος, κοκκινίζω. the weather's ~ed cold ο καιρός γύρισε στο κρύο, ο καιρός κρύωσε.

turn about *v.i.* γυρίζω, κάνω μεταβολή.

turn against *v.i.* (*become hostile to*) στρέφομαι εναντίον (*with gen.*). (*v.t.*) they turned him against me τον έστρεψαν εναντίον μου.

turn aside *v.i.* αλλάζω δρόμο, εκτρέπομαι, παρεκκλίνω. (*v.t.*) εκτρέπω.

turn away *v.t.* (*one's face*) αποστρέφω, (*people*) διώχνω. (*v.i.*) γυρίζω την πλάτη.

turn back *v.i.* παίρνω το δρόμο της επιστροφής. (*v.t.*) κάνω να υποχωρήσει.

turn down *v.t.* (*fold*) διπλώνω, (*lower*) χαμηλώνω, (*reject*) απορρίπτω.

turn in *v.t.* (*give back*) επιστρέφω. (*v.i.*) (*go to bed*) πάω για ύπνο.

turn off *v.t.* (*water, radio*) κλείνω, (*light, oven*) σβήνω.

turn on *v.* (*water, radio*) ανοίγω, (*light*) ανάβω. (*attack*) στρέφομαι εναντίον. (*with gen.*) (*depend on*) εξαρτώμαι από.

turn out *v.t.* (*expel*) διώχνω, (*empty*) αδειάζω. (*light*) σβήνω, (*produce*) παράγω, βγάζω. (*v.i.*) βγαίνω. (*be revealed as*) αποδεικνύομαι, it turns out that... φαίνεται ότι. it turned out well for me μου βγήκε σε καλό.

turn-out *s.* there was a large ~ μαζεύτηκε πολύς κόσμος.

turn over *v.t.* γυρίζω, (*think about*) σκέπτομαι, (*give*) παραδίδω. (*v.i.*) γυρίζω. (*be upset*) αναποδογυρίζομαι.

turnover *s.* (*fin.*) τζίρος *m.*

turn round *v.t.* γυρίζω. (*v.i.*) γυρίζω, στρέφομαι. (*look round*) γυρίζω το κεφάλι μου.

turn to *v.* (*address oneself to*) αποτείνομαι εις, στρέφομαι προς. (*get busy*) στρώνομαι στη δουλειά.

turn up *v.t.* (*raise*) ανασηκώνω, (*find*)

ξεθάβω. (*gas, radio*) δυναμώνω. ~ one's nose at περιφρονώ. (*v.i.*) παρουσιάζομαι, εμφανίζομαι. (*of people, unexpectedly*) πλακώνω. (*be found*) βρίσκομαι.

turn-up *s.* (*of trousers*) ρεβέρ *n.*

turn upon *v.* εξαρτώμαι από.

turncoat *s.* αποστάτης *m.*

turncock *s.* νεροκράτης *m.*

turner *s.* τορνευτής *m.*

turning *s.* (*action*) γύρισμα, στρίψιμο *n.* (*bend*) στροφή *f.* (*road*) δρόμος *m.* **~-point** *s.* κρίσιμο σημείο.

turnip *s.* ρέβα *f.*

turnkey *s.* δεσμοφύλακας *m.*

turnpike *s.* δρόμος με διόδια.

turnstile *s.* περιστροφικό φράγμα εισόδου.

turpentine *s.* τερεβινθέλαιο *n.,* νέφτι *n.*

turpitude *s.* κακοήθεια *f.*

turquoise *s.* τουρκουάζ *n.*

turret *s.* πυργίσκος *m.*

turtle *s.* χελώνα *f.* turn ~ μπατάρω, τουμπάρω. **~-dove** *s.* τρυγόνι *n.*

tusk *s.* χαυλιόδους *m.*

tussle *v.* παλεύω. (*s.*) πάλη *f.*

tutelage *s.* κηδεμονία *f.*

tutelary *a.* προστατευτικός.

tutor *s.* παιδαγωγός *m.* (*university*) καθηγητής *m.* (*v.*) (*train*) διαπαιδαγωγώ, εκπαιδεύω. (*teach*) διδάσκω, (*prime*) δασκαλεύω.

TV τηλεόρασις *f.*

twaddle *s.* ανοησίες *f.pl.*

twain *s.* δύο.

twang *s.* (*of string*) ήχος παλλομένης χορδής. (*speech*) έρρινος τόνος. (*v.t.*) χτυπώ. (*v.i.*) ηχώ.

tweak *v.* τσιμπώ, τραβώ.

tweed *s.* τουήντ *n.*

tweezers *s.* τσιμπιδάκι *n.*

twelfth *a.* δωδέκατος.

twelve *num.* δώδεκα. **~month** *s.* δωδεκάμηνο *n.*

twenty *num.* είκοσι. **~ieth** *a.* εικοστός.

twice *adv.* δις, δύο φορές. ~ as much (*or* big) as διπλάσιος. (*with gen.*) he's ~ my age έχει τα διπλά μου χρόνια. think ~ σκέπτομαι καλά. *see also* **double, doubly** (*adv.*).

twiddle *v.t.* στριφογυρίζω. ~ one's thumbs παίζω (με) τα δάχτυλά μου.

twig *s.* κλαδάκι *n.* (*v., fam.*) καταλαβαίνω.

twilight *s.* (*dusk*) λυκόφως *n.* (*dawn*) λυκαυγές *n.*

twill *s.* βαμβακερό ύφασμα με ύφανση καμπαρντίνας.

twin *a.* δίδυμος (*s.*) ~s δίδυμα *n.pl.*

twine *v.t.* τυλίγω, πλέκω. (*v.i.*) τυλίγομαι. (*s.*) σπάγκος *m.*

twinge s. (pain) σουβλιά f. (conscience) τύψη f.

twink|le v. λαμπυρίζω. in the ~ling of an eye εν ριπή οφθαλμού. (s.) λαμπύρισμα n.

twirl v.t. στριφογυρίζω. (v.i.) στροβιλίζομαι.

twist v.t. (turn) στρίβω, (twine) τυλίγω, (distort) διαστρεβλώνω, (sprain) στραμπουλώ. ~ (person's) arm ζορίζω. she can ~ him round her little finger τον κάνει ό,τι θέλει. (v.i.) (coil) τυλίγομαι, (of road) έχω στροφές. (s.) (movement) στρίψιμο n. (bend) στροφή f. ~er s. κατεργάρης m.

twisted a. (crooked) στραβός, (mixed up) μπερδεμένος, (smashed) στρεβλωμένος.

twit v. δουλεύω. (s., fam.) όρνιο n. μπούφος m.

twitch v.t. (jerk) τινάζω, (pull) τραβώ, (contract) συσπώ. (v.i.) συσπώμαι. (s.) τίναγμα n. τράβηγμα n. (spasm) σπασμός m., τικ n.

twitter v. τερετίζω. (s.) τερετισμός m.

two num. δύο. (at cards) δυάρι n. fold in ~ διπλώνω στα δύο. put ~ and ~ together βγάζω τα συμπεράσματά μου. ~ hundred διακόσιοι.

two-edged a. (lit. & fig.) δίκοπος.

two-faced a. (fig.) διπρόσωπος.

twofold a. διπλάσιος.

two-legged a. δίπους.

twopenny-ha'penny a. (fam.) της πεντάρας.

two-stroke a. (engine) δίχρονος.

two-way a. (street) διπλής κατευθύνσεως. (switch) αλέ-ρετούρ, δύο θέσεων.

tycoon s. μεγιστάν m.

type s. (sort) τύπος m., είδος n. (print) τύπος m., (letters) στοιχεία n.pl. (v.) δακτυλογραφώ.

typescript s. δακτυλογραφημένο κείμενο.

typesetting s. στοιχειοθεσία f.

typewriter s. γραφομηχανή f.

typhoid a. τυφοειδής.

typhoon s. τυφών m.

typhus s. εξανθηματικός τύφος m.

typical a. χαρακτηριστικός. ~ly adv. (as usual) ως συνήθως. he is ~ly English είναι ο αντιπροσωπευτικός τύπος του Άγγλου.

typify v. αντιπροσωπεύω, είμαι ο τύπος (with gen.)

typing s. δακτυλογραφία f.

typist s. δακτυλογράφος m.f.

typograph|ical a. τυπογραφικός. ~y s. τυπογραφία f.

tyrannical a. τυραννικός. ~ly adv. τυραννικώς.

tyrann|ize v. (also ~ over) καταδυναστεύω. ~y s. τυραννία f.

tyrant s.τύραννος m.

tyre s. λάστιχο n.

tyro s. αρχάριος m.

U

ubiquitous a. πανταχού παρών.

udder s. μαστάρι n.

UFO ιπτάμενος δίσκος.

ugl|y a. άσχημος, (threatening) απειλητικός. ~iness s. ασχημία f.

UK Ηνωμένον Βασίλειον.

ulcer s. έλκος n.

ulterior a. απώτερος.

ultimate a. τελικός, (basic) βασικός. ~ly adv. τελικώς, σε τελευταία ανάλυση.

ultimatum s. τελεσίγραφον n.

ultra- υπερ-.

ultrasonic a. υπερηχητικός.

ululat|e v. ουρλιάζω. ~ion s. ούρλιασμα n.

umbilical a. ~ cord ομφάλιος λώρος.

umbrage s. take ~ προσβάλλομαι, θίγομαι.

umbrella s. ομπρέλλα f.

umpire s. διαιτητής m. (v.) διαιτητεύω.

umpteen a. (fam.) ~ times δεν ξέρω πόσες φορές. for the ~th time για πολλοστή φορά.

un- (private) α-, αν-, αντι-, ξε-.

UN Ηνωμένα Έθνη.

unabashed a. he was ~ δεν ίδρωσε το αφτί του.

unabated a. αμείωτος.

unable a. be ~ δεν μπορώ, δεν δύναμαι, αδυνατώ, δεν μου είναι δυνατόν.

unacceptable a. απαράδεκτος.

unaccountable a. ανεξήγητος, (not responsible) ανεύθυνος.

unaccounted a. he is ~ for (missing) αγνοείται η τύχη του. (in invoice, etc.) there are £10 ~ for παρουσιάζεται διαφορά δέκα λιρών.

unaccustomed a. (thing, person) ασυνήθιστος.

unaffected a. (natural) απροσποίητος, (uninfluenced) ανεπηρέαστος.

unafraid a. χωρίς φόβο.

unaided a. αβοήθητος, χωρίς βοήθεια.

unalloyed a. (happiness) ανέφελος.

unalterable a. αμετάβλητος.

unanim|ous a. ομόφωνος. ~ity s. ομοφωνία f.

unapproachable a. απλησίαστος.

unarmed a. άοπλος.

unassailable a. απρόσβλητος.

unattached a. αδέσμευτος.

unattainable a. ανέφικτος.

unattended a. (not guarded) αφύλαχτος.

(*not accompanied*) ασυνόδευτος. ~ το (*neglected*) παραμελημένος.

unavailing *a.* μάταιος.

unavoidab|le *a.* αναπόφευκτος. **~ly** *adv.* αναποφεύκτως.

unaware *a.* I was ~ of the situation ήμουν εν αγνοία της καταστάσεως, αγνοούσα την κατάσταση. **~s** *adv.* εξ απροόπτου.

unbalanced *a.* ανισόρροπος.

unbearable *a.* ανυπόφορος.

unbeaten *a.* αήττητος, (*record*) αξεπέραστος.

unbecoming *a.* (*dress, etc.*) αταίριαστος. (*unseemly*) απρεπής.

unbeknownst *adv.* (*fam.*) κρυφά.

unbeliev|er *s.*, **~ing** *a.* άπιστος.

unbend *v.i.* χαλαρώνομαι. **~ing** *a.* αλύγιστος.

unbiassed *a.* αμερόληπτος.

unbidden *a.* ακάλεστος.

unbind *v.* λύνω.

unblushing *a.* αναιδής.

unborn *a.* αγέννητος.

unbosom *v.* ~ oneself ανοίγω την καρδιά μου.

unbounded *a.* (*space*) απέραντος, (*scope*) απεριόριστος.

unbowed *a.* αλύγιστος.

unbreakable *a.* άθραυστος, (*bond*) άρρηκτος.

unbridled *a.* αχαλίνωτος.

unbroken *a.* (*continuous*) αδιάκοπος, συνεχής. (*untamed*) αδάμαστος. (*front*) άρρηκτος. (*record*) αξεπέραστος.

unburden *v.* ~ oneself (*of secret, etc.*) ανοίγω την καρδιά μου.

unbusinesslike *a.* αμεθόδευτος, χωρίς σύστημα.

unbutton *v.* ξεκουμπώνω. **~ed** ξεκούμπωτος.

uncalled-for *a.* περιττός, αδικαιολόγητος.

uncanny *a.* αφύσικος, μυστηριώδης.

uncared-for *a.* παραμελημένος.

unceasing *a.* ακατάπαυστος.

unceremonious *a.* (*abrupt*) απότομος, (*simple*) απλός. **~ly** *adv.* απότομα.

uncertain *a.* αβέβαιος. **~ty** *s.* αβεβαιότητα *f.*

unchang|eable, **~ing** *a.* αμετάβλητος.

uncharitable *a.* χωρίς ευσπλαχνία, σκληρός.

uncivil *a.* αγενής.

uncivilized *a.* απολίτιστος, (*fig.*) βάρβαρος.

uncle *s.* θείος *m.*

unclean *a.* ακάθαρτος.

uncomfortable *a.* άβολος, (*embarrassing*) ενοχλητικός. feel ~ (*ill at ease*) δεν αισθάνομαι άνετα.

uncommitted *a.* αδέσμευτος.

uncommon *a.* σπάνιος. **~ly** *adv.* πολύ, εξαιρετικά.

uncompromising *a.* (*severe*) αυστηρός, (*unyielding*) αδιάλλακτος.

unconcerned *a.* (*not involved*) αμέτοχος (*with gen. or* εις). (*not caring*) αδιάφορος.

unconditional *a.*, **~ly** *adv.*, άνευ όρων.

uncongenial *a.* ασυμπαθής. he is ~ to me δεν ταιριάζουν τα χνώτα μας.

unconnected *a.* άσχετος.

unconscionable *a.* we took an ~ time κάναμε πάρα πολλή ώρα.

unconscious *a.* (*senseless*) αναίσθητος. (*unwitting: person*) ασυνείδητος, (*action*) ασυναίσθητος. (*s.*) ασυνείδητον *n.* **~ly** *adv.* ασυναισθήτως. **~ness** *a.* αναισθησία *f.*

unconstitutional *a.* αντισυνταγματικός.

uncontrollable *a.* ασυγκράτητος.

unconventional *a.* ανορθόδοξος, αντίθετος προς τα καθιερωμένα.

uncork *v.* ξεβουλλώνω.

uncouth *a.* άξεστος.

uncover *v.t.* ξεσκεπάζω. (*v.i.*) (*take hat off*) αποκαλύπτομαι.

unct|ion *s.* (*chrism*) χρίσμα *n.* (*pej.*) (ψεύτικο) κατανυκτικό ύφος. **~uous** *a.* γλοιώδης.

uncultivated *a.* ακαλλιέργητος.

uncut *a.* (*stone*) ακατέργαστος. (*page, tree*) άκοπος, άκοφτος. (*entire*) άνευ περικοπών.

undaunted *a.* απτόητος.

undeceive *v.* βγάζω από την πλάνη.

undecided *a.* (*person*) αναποφάσιστος, (*issue*) εκκρεμής.

undecipherable *a.* δυσανάγνωστος.

undeclared *a.* (*war*) ακήρυκτος. (*goods*) αδήλωτος.

undemonstrative *a.* όχι εκδηλωτικός.

undeniable *a.* αναμφισβήτητος.

under *prep.* 1. (*place*) κάτω από, υπό (*with acc.*) ~ the tree κάτω από το δέντρο. the cat came out from ~ the table η γάτα βγήκε κάτω από το τραπέζι. nothing new ~ the sun ουδέν καινόν υπό τον ήλιον. 2. (*grade, rank*) κάτω (*with gen.*) ~ 10 years old κάτω των δέκα ετών. ~ the rank of admiral κάτω του βαθμού του ναυάρχου. 3. (*condition*) ~ the yoke υπό τον ζυγόν. ~ arms υπό τα όπλα. ~ observation υπό παρατήρησιν. 4. (*in accordance with*) κατά, συμφώνως προς (*both with acc.*) ~ article ten κατά το άρθρον δέκα. ~ his will συμφώνως προς την διαθήκην του. 5. (*various*) ~ way εν κινήσει. ~ age ανήλικος. look it up ~ the letter B να κοιτάξεις στο βήτα. in ~ two days σε λιγότερο από

δύο ημέρες. come ~ (*belong to*) υπάγομαι εις.
under *adv.* (από) κάτω. go ~ (*sink*) βυθίζομαι, (*fig.*) αποτυγχάνω. keep ~ (*v.t.*) (*oppress*) καταπιέζω, (*stifle*) καταπνίγω.
underbid *v.i.* μειοδοτώ.
undercloth|es, ~ing *s.* εσώρουχα *n.pl.*
undercoat *s.* αστάρι *n.*
undercover *a.* μυστικός.
undercurrent *s.* υποβρύχιο ρεύμα. (*fig.*) συγκεκαλυμμένο ρεύμα.
underdeveloped *a.* υπανάπτυκτος.
underdog *s.* ο αδικημένος από τη ζωή.
underdone *a.* όχι πολύ ψημένος.
underestimate *v.* υποτιμώ.
underfoot *adv.* it is wet ~ το έδαφος είναι υγρό. trample ~ ποδοπατώ.
undergo *v.* υφίσταμαι, υποφέρω.
undergraduate *s.* φοιτητής *m.*, φοιτήτρια *f.*
underground *a.* υπόγειος, (*secret*) κρυφός. (*adv.*) κάτω από τη γη. go ~ (*fig.*) περνώ στην παρανομία. (*s.*) (*resistance movement*) αντίσταση *f.* (*railway*) υπόγειος *m.*
undergrowth *s.* χαμηλή βλάστηση.
underhand *a.* πονηρός.
underlie *v.t.* αποτελώ τη βάση (*with gen.*) what ~s all this? τι κρύβεται κάτω από όλα αυτά;
underline *v.* υπογραμμίζω.
underling *s.* (*pej.*) άνθρωπος *m.*, υπαλληλίσκος *m.*, τσιράκι *m.*
undermanned *a.* ανεπαρκώς επανδρωμένος.
undermine *v.* υπονομεύω.
underneath *adv.* από κάτω. (*prep.*) κάτω από (*with acc.*)
undernourishment *s.* υποσιτισμός *m.*
underpass *s.* υπόγεια διάβαση.
underpin *v.* υποστυλώνω.
underprivileged *a.* the ~ οι απόκληροι της τύχης.
underrate *v.* υποτιμώ.
undersigned *a.* the ~ ο υπογεγραμμένος.
understaffed *a.* με ανεπαρκές προσωπικό.
understand *v.* καταλαβαίνω, αντιλαμβάνομαι, εννοώ. (*learn*) μαθαίνω. it is understood (*that*) εννοείται, εξυπακούεται. make oneself understood γίνομαι αντιληπτός. am I to ~ that...? εννόησα καλά ότι...;
understand|able *a.* (*book, etc.*) ευκολονόητος. it is ~able (*that*) καταλαβαίνει κανείς (ότι). **~ably** *adv.* ως είναι ευνόητον.
understanding *s.* (*intelligence, grasp*) αντίληψη *f.* (*sympathetic*) κατανόηση *f.* (*agreement*) συνεννόηση *f.* reach an ~ συννενοούμαι. on the ~ that υπό τον όρον ότι. (*a.*) με κατανόηση.

understate *v.* υποτιμώ. ~ one's age κρύβω τα χρόνια μου. **~ment** *s.* that's an ~ment! δεν λες τίποτα!
understudy *s.* αντικαταστάτης *m.* (*v.*) αντικαθιστώ.
undertak|e *v.* αναλαμβάνω. (*enter upon*) επιχειρώ. (*promise*) εγγυώμαι (ότι). **~er**'s *s.* γραφείον κηδειών. **~ing** *s.* επιχείρηση *f.* εγγύηση *f.*
undertone *s.* in an ~ χαμηλοφώνως (*fig.*) ~ of discontent υπολανθάνουσα δυσφορία.
undertow *s.* αντιμάμαλο *n.*
undervalue *v.* υποτιμώ.
underwater *a.* υποβρύχιος.
underwear *s.* εσώρουχα *n.pl.*
underweight *a.* (*goods*) λιποβαρής, ξίκικος. (*person*) έχων βάρος κάτω του κανονικού.
underworld *s.* άδης *m.*, ο κάτω κόσμος. (*social*) υπόκοσμος *m.*
underwrite *v.* συνυπογράφω, εγγυώμαι. **~r** *s.* εγγυητής *m.* (*marine*) ναυτασφαλιστής *m.*
undeserved *a.* (*blame*) άδικος, (*praise*) παρ' αξίαν. **~ly** *adv.* αδίκως, παρ' αξίαν.
undesirable *a.* ανεπιθύμητος.
undeveloped *a.* (*land, etc.*) αναξιοποίητος, (*person*) ανεξέλικτος.
undeviating *a.* απαρέγκλιτος.
undischarged *a.* (*debt*) ανεξόφλητος, (*task*) ανεκτέλεστος.
undisciplined *a.* απειθάρχητος.
undiscovered *a.* (*not known of*) άγνωστος.
undisguised *a.* απροκάλυπτος.
undistinguished *a.* μέτριος.
undisturbed *a.* (*person*) ατάραχος, (*objects*) όπως ήταν.
undo *v.* (*open*) ανοίγω, ξετυλίγω. (*knot*) λύνω, (*stitches*) ξηλώνω, (*hair*) λύνω, χαλώ. (*unbutton*) ξεκουμπώνω. (*destroy*) χαλώ. come ~ne λύνομαι, ξεκουμπώνομαι. it can't be ~ne δεν ξεγίνεται.
undoing *s.* (*fig.*) καταστροφή *f.*
undone *a.* (*task*) ακάμωτος, απραγματοποίητος. (*rope, hair*) λυτός, λυμένος. (*lost*) χαμένος. *see* undo.
undoubtedly *adv.* αναμφισβητήτως.
undreamed *a.* ~ of που δεν το ονειρεύτηκε κανείς, αφάνταστος.
undress *v.t.* γδύνω, (*v.i.*) γδύνομαι. **~ed** *a.* γδυτός. (*salad*) σκέτος, (*stone*) ακατέργαστος.
undrinkable *a.* it is ~ δεν πίνεται.
undue *a.* υπέρμετρος.
undulat|e *v.* (*wave*) κυματίζω. **~ing** country κυματιστό τοπείο, όλο λοφάκια και πεδιάδες. **~ory** *a.* κυματοειδής.
undying *a.* αιώνιος.
unearth *v.* ξεθάβω, ανακαλύπτω.

unearthly *a.* υπερφυσικός, (*frightening*) τρομακτικός, (*preposterous*) απίθανος.

uneas|y *a.* ανήσυχος. ~iness *s.* ανησυχία *f.*

uneat|able *a.* που δεν τρώγεται. ~en *a.* αφάγωτος.

unedifying *a.* εξευτελιστικός.

uneducated *a.* αμόρφωτος.

unemotional *a.* απαθής.

unemploy|ed *a.* άνεργος. ~ment *s.* ανεργία *f.*

unending *a.* ατέλειωτος. ~ly *adv.* συνεχώς, όλο.

unenlightened *a.* (*not informed*) απληροφόρητος. (*backward*) μη φωτισμένος.

unenterprising *a.* be ~ δεν έχω επιχειρηματικό δαιμόνιο.

unequal *a.* άνισος. he is ~ to the task δεν είναι ικανός για τη δουλειά. ~led *a.* άφθαστος. ~ly *adv.* ανίσως.

unequivocal *a.* κατηγορηματικός.

unerring *a.* ασφαλής, αλάθητος.

UNESCO Ουνέσκο *n.*

uneven *a.* (*irregular*) ακανόνιστος, (*not the same*) ανομοιογενής, (*unequal*) άνισος, (*not smooth*) ανώμαλος. ~ly *adv.* άνισα.

unexceptionable *a.* άψογος.

unexpected *a.* απροσδόκητος, απρόοπτος, αναπάντεχος. ~ly *adv.* εξ απροόπτου, αναπάντεχα.

unexploited *a.* ανεκμετάλλευτος.

unfailing *a.* ανεξάντλητος, αμείωτος. ~ly *adv.* διαρκώς.

unfair *a.* άδικος, όχι σωστός. ~ly *adv.* άδικα. ~ness *s.* αδικία *f.*

unfaithful *a.* άπιστος. ~ness *s.* απιστία *f.*

unfamiliar *a.* (*not known*) άγνωστος, ασυνήθιστος. I am ~ with his works τα έργα του δεν μου είναι γνωστά.

unfathomable *a.* ανεξιχνίαστος.

unfavourable *a.* δυσμενής.

unfeeling *a.* αναίσθητος, άπονος.

unfinished *a.* ημιτελής, ατελείωτος.

unfit *a.* ακατάλληλος, (*person only*) ανίκανος. (*ill*) ανήμπορος.

unflagging *a.* ακούραστος.

unfledged *a.* απουπούλιαστος. (*fig.*) άπειρος.

unflinching *a.* απτόητος.

unfold *v.t.* ξεδιπλώνω, ξετυλίγω. (*make known*) αναπτύσσω. (*v.i.*) (*of events*) εκτυλίσσομαι, (*of landscape*) απλώνομαι.

unforgettable *a.* αξέχαστος, αλησμόνητος.

unforgivable *a.* ασυγχώρητος.

unfortunate *a.* (*luckless*) άτυχος, (*poor*) δύστυχος, κακόμοιρος. (*circumstance*) ατυχής, λυπηρός. ~ly *adv.* δυστυχώς.

unfounded *a.* αβάσιμος.

unfrequented *a.* ασύχναστος.

unfriendly *a.* μη φιλικός, εχθρικός. (*atmosphere*) δυσμενής.

unfruitful *a.* άκαρπος.

unfurl *v.* ανοίγω.

unfurnished *a.* χωρίς έπιπλα.

ungainly *a.* άχαρος, άγαρμπος.

ungentlemanly *a.* που δεν αρμόζει σε κύριο.

unget-at-able *a.* απρόσιτος.

ungodly *a.* αθεόφοβος (*fam.*) απίθανος.

ungovernable *a.* (*ship, country*) ακυβέρνητος. (*passion*) ασυγκράτητος. (*child*) σκληρός.

ungrateful *a.* αγνώμων. (*of task*) άχαρος. ~ness *s.* αγνωμοσύνη *f.*, αχαριστία *f.*

unguarded *a.* (*prisoner, camp*) αφρούρητος, (*words*) αδιάκριτος. (*careless*) απρόσεκτος.

unguent *s.* αλοιφή *f.*

unhandy *a.* (*person*) αδέξιος, (*thing*) άβολος.

unhapp|y *a.* (*person*) δυστυχής, δυστυχισμένος. (*event, remark*) ατυχής. ~ily *adv.* δυστυχώς, (*ill*) ατυχώς. ~iness *s.* δυστυχία *f.*

unhealthy *a.* (*bad for health*) ανθυγιεινός. (*having bad health*) αρρωστιάρης.

unheard *a.* ~ of ανήκουστος.

unhinge *v.* (*derange*) διαταράσσω. his mind is ~d έπαθε το μυαλό του.

unholy *a.* ανόσιος. (*fam.*) τρομερός.

unhoped *a.* ~for ανέλπιστος.

unhurt *a.* αβλαβής.

Uniat(e) *s.* Ουνίτης *m.*

uniform *s.* στολή *f.* (*a.*) ομοιόμορφος ~ity *s.* ομοιομορφία *f.*

unif|y *v.* ενοποιώ. ~ication *s.* ενοποίηση *f.*

unilateral *a.* μονομερής.

unimpeachable *a.* (*blameless*) άμεμπτος, (*sure*) ασφαλής.

unimportant *a.* επουσιώδης.

uninhabited *a.* ακατοίκητος.

uninhibited *a.* αδιάφορος για τους τύπους. be ~ δεν έχω ψυχολογικές αναστολές.

uninitiated *a.* αμύητος.

unintelligible *a.* ακατανόητος.

unintentional *a.* ακούσιος, αθέλητος. ~ly *adv.* άθελα, όχι εκ προθέσεως.

uninterrupted *a.* αδιάκοπος.

uninvit|ing *a.* που δεν σε ελκύει. ~ed *a.* απρόσκλητος.

union *s.* ένωση *f.* (*worker's*) συνδικάτο *n.*, σωματείο *n.* U~ Jack *s.* βρεττανική σημαία.

unique *a.* μοναδικός.

unisex *a.* για άνδρες και γυναίκες.

unison *s.* in ~ ομοφώνως, από συμφώνου.

unit *s.* μονάδα *f.*

unit|e *v.t.* ενώνω. (*v.i.*) ενώνομαι. U~ed Kingdom Ηνωμένον Βασίλειον. ~y *s.* ενότης *f.*

universal *a.* γενικός, παγκόσμιος. (*suffrage*) καθολικός. ~ly *adv.* γενικώς, παγκοσμίως.

universe *s.* σύμπαν *n.*, υφήλιος *f.*

university *s.* πανεπιστήμιο *n.* (*a.*) (*matters*) πανεπιστημιακός, (*persons*) του πανεπιστημίου.

unjust *a.* άδικος. ~ly *adv.* άδικα.

unkempt *a.* ατημέλητος, (*hair*) αχτένιστος.

unkind *a.* σκληρός. be ~ to (*ill-treat*) κακομεταχειρίζομαι, (*not show compassion*) δεν δείχνω συμπάθεια σε. ~ly *adv.* σκληρά, χωρίς συμπάθεια. take it ~ly μου κακοφαίνεται. ~ness *s.* σκληρότητα *f.*

unknowing *a.* αγνοών. ~ly *adv.* όντας εν αγνοία, εν αγνοία μου.

unknown *a.* άγνωστος.

unlawful *a.* παράνομος.

unleash *v.* αμολάω.

unleavened *a.* άζυμος.

unless *conj.* αν δεν, εκτός αν. ~ I am mistaken αν δεν κάνω λάθος. I'll get something ready to eat ~ you'd rather we ate out θα ετοιμάσω κάτι να φάμε εκτός αν προτιμάς να φάμε έξω.

unlettered *a.* αγράμματος.

unlike *a.* ανόμοιος, διαφορετικός. (*different from*) διαφορετικός από. (*prep.*) όχι όπως, διαφορετικά από.

unlikeable *a.* αντιπαθητικός.

unlikely *a.* απίθανος.

unlimited *a.* απεριόριστος. (*ad lib.*) κατά βούλησιν.

unload *v.t.* (*ship, mule, cargo*) ξεφορτώνω. (*get rid of*) ξεφορτώνομαι. ~ing *s.* εκφόρτωση *f.*

unlock *v.* ξεκλειδώνω, (*fig.*) ανοίγω. ~ed *a.* ξεκλείδωτος.

unloose *v.* αμολάω.

unluck|y *a.* άτυχος, (*not successful*) ατυχής. (*bringing bad luck: person*) γουρσούζης, (*thing*) γουρσούζικος. I think it ~y (*to do sthg.*) το έχω σε κακό. ~ily *adv.* δυστυχώς.

unmade *a.* άφτιαστος.

unman *v.* (*shake, upset*) κλονίζω το ηθικό (*with gen.*). ~ly *a.* άνανδρος. ~ned *a.* (*without crew*) μη επανδρωμένος.

unmanageable *a.* (*object*) δύσχρηστος, (*child*) σκληρός, (*house, situation, etc.*) δύσκολος.

unmannerly *a.* αγενής, ανάγωγος.

unmask *v.t.* αποκαλύπτω.

unmatch|able, ~ed *a.* απαράμιλλος.

unmentionable *a.* ακατονόμαστος.

unmerciful *a.* ανηλεής, άσπλαχνος.

unmerited *a.* (*blame*) άδικος, (*praise*) παρ' αξίαν.

unmindful *a.* be ~ of (*forget*) ξεχνώ, (*not heed*) αδιαφορώ για.

unmistakable *a.* σίγουρος, προφανής.

unmitigated *a.* πέρα για πέρα.

unmoved *a.* ασυγκίνητος.

unnatural *a.* αφύσικος.

unnecessar|y *a.* περιττός, άσκοπος. it was ~y δεν χρειαζόταν. ~ily *adv.* ασκόπως, άδικα. ~ily large μεγαλύτερος απ' ό,τι χρειάζεται.

unnerve *v. see* unman.

unnoticed *a.* απαρατήρητος.

unobtrusive *a.* (*person*) διακριτικός, (*object*) μη εμφανής.

unoccupied *a.* ελεύθερος.

unorganized *a.* ανοργάνωτος.

unorthodox *a.* ανορθόδοξος.

unpack *v.t.* (*trunk*) αδειάζω, ανοίγω. (*contents*) βγάζω.

unparalleled *a.* άνευ προηγουμένου, (*matchless*) απαράμιλλος.

unpardonable *a.* ασυγχώρητος.

unpick *v.* ξηλώνω.

unpleasant *a.* δυσάρεστος. ~ness *s.* (*trouble*) δυσαρέσκειες *f.pl.*

unpopular *a.* (*politically*) αντιδημοτικός. become ~ (*of person*) παύω να είμαι δημοφιλής. such ideas are ~ τέτοιες ιδέες δεν εγκρίνονται. this line (*of goods*) is ~ today αυτό το είδος δεν είναι ζητημένο σήμερα. ~ity *s.* αντιδημοτικότης *f.*

unprecedented *a.* πρωτοφανής, άνευ προηγουμένου.

unprejudiced *a.* αμερόληπτος.

unpremeditated *a.* απρομελέτητος.

unprepared *a.* απροετοίμαστος.

unprincipled *a.* χωρίς ηθικές αρχές, ασυνείδητος.

unprofessional *a.* στερούμενος επαγγελματικής συνειδήσεως.

unprofitable *a.* ασύμφορος, μη επικερδής.

unpromising *a.* που δεν υπόσχεται πολλά.

unpropitious *a.* δυσοίωνος.

unprotected *a.* (*orphan, etc.*) απροστάτευτος, (*exposed*) εκτεθειμένος.

unproved *a.* (*case*) αναπόδεικτος, (*untested*) αδοκίμαστος.

unprovided *a.* ~ with χωρίς, μη έχων, στερούμενος (*with gen.*). ~ for (*person*) χωρίς πόρους, (*contingency*) που δεν έχει προβλεφθεί.

unprovoked *a.* απρόκλητος.

unpublished *a.* ανέκδοτος.

unqualified *a.* (*without reserve*) πλήρης, ανεπιφύλακτος. (*for job, purpose*) ακατάλληλος. (*without competent authority*)

αναρμόδιος.
unquestion|able *a.* αναμφισβήτητος. **~ing** *a.* τυφλός.
unquiet *a.* ταραγμένος.
unquote *v. imper.* κλείσατε τα εισαγωγικά.
unravel *v.t.* (*disentangle*) ξετυλίγω, (*knitting*) ξηλώνω. (*mystery*) λύω, ξεδιαλύνω. (*v.i., get frayed*) ξεφτώ.
unready *a.* ανέτοιμος.
unreal *a.* φανταστικός, εκτός πραγματικότητος. **~istic** *a.* απροσγείωστος, μακράν της πραγματικότητος.
unreason|able *a.* παράλογος. he is ~able δεν είναι λογικός. **~ing** *a.* τυφλός.
unrelenting *a.* (*inflexible*) αλύγιστος, (*incessant*) αδιάλειπτος.
unreliable *a.* που δεν μπορείς να βασιστείς. (*news, witness*) αναξιόπιστος.
unrelieved *a.* (*absolute*) απόλυτος, (*unvaried*) χωρίς ποικιλία. (*of plain dress*) σκέτος, αγαρνίριστος.
unremitting *a.* (*care*) αμείωτος, (*person*) ακούραστος.
unrequited *a.* ανανταπόδοτος.
unreservedly *adv.* ανεπιφυλάκτως.
unresolved *a.* (*problem*) άλυτος, (*person*) αναποφάσιστος.
unrest *s.* αναβρασμός *m.* δυσφορία *f.*
unresrtain|ed, ~able *a.* ασυγκράτητος.
unrestricted *a.* απεριόριστος, ελεύθερος.
unrighteous *a.* ανίερος.
unripe *a.* άγουρος, αγίνωτος. (*also fig.*) ανώριμος.
unrivalled *a.* απαράμιλλος.
unroll *v.t.* ξετυλίγω, (*spread*) στρώνω. (*v.i.*) (*of events*) εκτυλίσσομαι.
unruffled *a.* ατάραχος.
unruly *a.* άτακτος.
unsafe *a.* ανασφαλής.
unsatisfactory *a.* μη ικανοποιητικός.
unsavoury *a.* αηδής, βρωμερός.
unsay *v.* ξελέω, ανακαλώ.
unscathed *a.* σώος.
unscrew *v.t. & i.* ξεβιδώνω. come ~ed ξεβιδώνομαι.
unscrupulous *a.* ασυνείδητος. **~ness** *s.* ασυνειδησία *f.*
unseasonable *a.* (*untimely*) άκαιρος. (*weather*) ασυνήθης για την εποχή.
unseasoned *a.* (*food*) ακαρύκευτος, (*person*) άπειρος. the wood was ~ τα ξύλα δεν είχαν τραβήξει.
unseat *v.* ανατρέπω, ρίχνω.
unseaworthy *a.* μη πλόιμος. ~ boat (*fam.*) σκυλοπνίχτης *m.*
unseeml|y *a.* απρεπής. **~iness** *s.* απρέπεια *f.*
unseen *a.* (*hidden*) αθέατος, (*unnoticed*)

απαρατήρητος.
unsettle *v.* αναστατώνω. **~d** *a.* (*troubled*) ταραγμένος. (*weather*) ακατάστατος, (*matter*) εκκρεμής, (*bill*) απλήρωτος.
unshakeable *a.* ακλόνητος.
unshaven *a.* αξύριστος.
unshod *a.* ανυπόδητος, (*horse*) απετάλωτος.
unsightly *a.* άσχημος, αντιαισθητικός.
unsigned *a.* ανυπόγραφος.
unskilled *a.* (*worker*) ανειδίκευτος. be ~ in δεν ξέρω από.
unsociable *a.* ακοινώνητος.
unsolicited *a.* μη ζητηθείς.
unsolved *a.* άλυτος.
unsophisticated *a.* απλός, αφελής, he is ~ δεν είναι μπασμένος στα του κόσμου.
unsound *a.* (*erroneous*) εσφαλμένος, (*not safe*) επισφαλής. be of ~ mind έχω χάσει τα λογικά μου.
unsparing *a.* (*lavish*) αφειδής, σπάταλος. (*demanding*) ανηλεής.
unspeakable *a.* ανείπωτος, (*awful*) φρικτός.
unspecified *a.* απροσδιόριστος.
unspoilt *a.* απείραχτος, αχάλαστος. (*child*) μη κακομαθημένος.
unstable *a.* ασταθής.
unsteady *a.* ασταθής.
unstick *v.t.* ξεκολλώ. *see* unstuck.
unstinting *a.* αφειδής.
unstop *v.* ξεβουλλώνω.
unstuck *a.* come ~ ξεκολλώ, (*fig.*) ναυαγώ.
unstudied *a.* ανεπιτήδευτος.
unsubstantiated *a.* αναπόδεικτος.
unsuccessful *a.* ανεπιτυχής, αποτυχημένος. **~ly** *adv.* ανεπιτυχώς.
unsuit|ed, ~able *a.* ακατάλληλος, (*incompatible*) αταίριαστος.
unsung *a.* μη δοξασθείς.
unsupported *a.* (*wall, opinion*) αστήρικτος, (*person*) ανυποστήρικτος.
unsure *a.* αβέβαιος. I am ~ whether... δεν είμαι βέβαιος αν.
unsurpass|ed, ~able *a.* άφθαστος.
unsuspected *a.* κρυμμένος.
unsuspecting *a.* ανυποψίαστος.
unsuspicious *a.* ανυποψίαστος.
unswerving *a.* σταθερός.
unsympathetic *a.* (*cold*) ψυχρός, αδιάφορος. (*unlikeable*) ασυμπάθιστος.
untainted *a.* αμόλυντος.
untamed *a.* αδάμαστος.
untapped *a.* (*fig., resources*) αναξιοποίητος.
untaught *a.* αδίδακτος.
untaxed *a.* αφορολόγητος.
unteachable *a.* (*person*) ανεπίδεκτος μα-

θήσεως, (subject) που δεν διδάσκεται.
untenable a. αστήρικτος.
unthink|able a. αδιανόητος. **~ing** a. απερίσκεπτος.
untid|y a. ακατάστατος. **~iness** s. ακαταστασία f.
untie v.t. λύνω, come ~d λύνομαι.
until prep. & conj. see till.
untimely a. άκαιρος, (premature) πρόωρος.
untiring a. ακούραστος.
unto prep. see to.
untold a. (quantity) αφάνταστος, που δεν λέγεται. (joy, sorrow) ανείπωτος. for ~ centuries δεν ξέρω επί πόσους αιώνες.
untouchable s. παρίας m.
untoward a. δυσάρεστος.
untrained a. αγύμναστος. (worker) ανειδίκευτος.
untried a. αδοκίμαστος.
untrodden a. απάτητος.
untrue a. αναληθής.
untrustworthy a. αναξιόπιστος.
untruth s. ψεύδος n. **~ful** a. (person) ψεύτης. m. **~fulness** a. αναλήθεια f. (lying) το ψεύδεσθαι.
untutored a. αδίδακτος. (not primed) αδασκάλευτος.
untwist v.t. ξεστρίβω.
unusable a. άχρηστος, ακατάλληλος.
unused a. (not in use) αχρησιμοποίητος. (new) αμεταχείριστος. (not accustomed) I am ~ to the heat είμαι ασυνήθιστος (or δεν είμαι συνηθισμένος) στη ζέστη.
unusual a. ασυνήθης, ασυνήθιστος. it is ~ for him to be late δεν έχει τη συνήθεια (or δεν συνηθίζει) να αργεί, συνήθως δεν αργεί. **~ly** adv. ασυνήθως.
unutterab|le a. απερίγραπτος. **~ly** adv. που δεν λέγεται.
unvarnished a. (fig.) απλός. ~ truth ωμή αλήθεια.
unveil v.t. αποκαλύπτω. **~ed** a. (fig.) ασυγκάλυπτος. **~ing** s. αποκαλυπτήρια n.pl.
unversed a. be ~ in δεν ξέρω από.
unviolated a. απαραβίαστος.
unvoiced a. μη εκφρασθείς.
unwarrant|ed, ~able s. αδικαιολόγητος.
unwary a. απρόσεκτος.
unwashed a. άπλυτος.
unwavering a. αχλόνητος.
unwed a. ανύπανδρος.
unwelcome a. μη ευπρόσδεκτος.
unwell a. αδιάθετος.
unwept a. άκλαυτος.
unwholesome a. ανθυγιεινός.
unwieldy a. μπατάλικος, βαρύς και δυσκί-

νητος, δυσκολομεταχείριστος.
unwillingly adv. παρά τη θέλησή μου.
unwind v.t. ξετυλίγω. (v.i.) (fam., relax) ησυχάζω.
unwise a. ασύνετος.
unwitting a. ασυνείδητος. **~ly** adv. άθελα, ασυναισθήτως.
unwomanly a. που δεν αρμόζει σε γυναίκα.
unwonted a. ασυνήθης.
unworkable a. (plan) ανεφάρμοστος.
unworldly a. απλός, αφελής.
unworn a. (clothes) αφόρετος.
unworth|y a. ανάξιος. **~iness** s. αναξιότητα f.
unwound a. ξεκουρδισμένος.
unwrap v. ξετυλίγω.
unwritten a. άγραφος.
unwrought a. ακατέργαστος.
unyielding a. ανένδοτος.
unyoke v. ξεζεύω.
unzip v.t. ~ it (dress, etc.) ανοίγω το φερμουάρ.
up adv. & prep. (α)πάνω, επάνω. ~ and down επάνω κάτω, walk ~ ˙ and down πηγαινοέρχομαι. ~s and downs τα πάνω και τα κάτω. be ~ (out of bed) είμαι σηκωμένος. time's ~ πέρασε η ώρα. it's all ~ δεν υπάρχει πια ελπίδα. what's ~? τι τρέχει; what's ~ with you? τι έχεις; ~ in arms ξεσηκωμένος. be well ~ in (a subject) κατέχω καλά. he is one ~ on me με έχει ξεπεράσει. I was ~ all night ξενύχτησα. be ~ against αντιμετωπίζω. ~ to (as far as) μέχρι(ς), ώς (with acc.) ~ to now ώς τώρα. it's ~ to you από σένα εξαρτάται. he's ~ to something κάτι μαγειρεύει. what will he be ~ to next? τι θα μας παρουσιάσει ακόμα; I don't feel ~ to it δεν έχω τα κότσια. his latest book isn't ~ to his first one το τελευταίο βιβλίο του δεν φτάνει το πρώτο. ~ a tree πάνω σ' ένα δέντρο. just ~ the street στο δρόμο πιο πάνω.
up-and-coming a. που υπόσχεται πολλά.
upbraid v. μαλώνω.
upbringing s. ανατροφή f.
up-country adv. στο εσωτερικό.
update v. (accounts, etc.) ενημερώνω, (modernize) εκσυγχρονίζω.
upend v. στήνω όρθιον.
upgrade v. προβιβάζω. (s.) be on the ~ προοδεύω, βρίσκομαι σε άνοδο.
upheaval s. αναστάτωση f.
uphill a. ανηφορικός, (task) δύσκολος. ~ slope ανήφορος m. (adv.) go ~ ανηφορίζω.
uphold v. υποστηρίζω, (maintain) κρατώ, (confirm) επικυρώνω.
upholster v. ταπετσάρω, ντύνω. **~er** s. ταπετσιέρης m. **~y** s. ταπετσαρία f.

upkeep s. συντήρηση f.
upland s. υψίπεδο n.
uplift v. εξυψώνω. (s.) ηθική εξύψωση.
up-market a. ανωτέρας κατηγορίας.
upon prep. see on. ~ my word! μα την αλήθεια!
upper s. (of shoe) ψίδι n. (fam.) on one's ~s απέντσρος.
upper a. ανώτερος, άνω, (ε)πάνω. ~ class ανωτέρα τάξις. U~ House Άνω Βουλή. U~ Egypt Άνω Αίγυπτος. ~ floor το επάνω πάτωμα. get the ~ hand παίρνω το πάνω χέρι. ~ crust (fam.) αφρόκρεμα της κοινωνίας. ~most a. ανώτατος, ψηλότερος, ο πιο πάνω. (adv.) πάνω - πάνω, προς τα πάνω.
uppish a. get ~ κάνω τον σπουδαίο.
uppity a. θρασύς.
upright a. όρθιος. (morally) ακέραιος, ίσιος.
uprising s. ξεσήκωμα.
uproar s. οχλαγωγία. f. in an ~ ανάστατος. ~ious a. θορυβώδης.
uproot v. εκριζώ, ξερριζώνω.
upset v.t. (trouble) στενοχωρώ, (disorganize) αναστατώνω, (overturn) αναποδογυρίζω. the food ~ me το φαΐ μού χάλασε το στομάχι. she ~ the tea έχυσε το τσάι. (s.) (shock) κλονισμός m. (disturbance) αναστάτωση f. (of vehicle) ανατροπή f. ~ting a. ενοχλητικός.
upshot s. αποτέλεσμα n.
upside-down adv. ανάποδα. turn ~ (v.t. & i.) αναποδογυρίζω., (v.t., fig.) κάνω άνω κάτω.
upstage v.t. (fig.) φέρνω σε μειονεκτική θέση.
upstairs adv. & a. επάνω.
upstanding a. ευθυτενής. ~ fellow λεβέντης m.
upstart s. πρόσωπον σκοτεινής καταγωγής που έχει αναρριχηθεί σε εξουσία ή πλούτη.
upstream adv. προς τα ανάντη του ποταμού.
upsurge s. σφοδρά εκδήλωσις.
uptake s. he is quick in the ~ είναι σπίρτο.
up-to-date a. (modern) σύγχρονος, μοντέρνος. (knowing the latest particulars) ενημερωμένος. bring ~ ενημερώνω, (modernize) εκσυγχρονίζω.
upturn s. στροφή προς τα άνω (or πάνω). ~ed a. (head, etc.) γυρισμένος προς τα πάνω. (upside-down) αναποδογυρισμένος.
upward a. (trend, etc.) ανοδικός.
upward, ~s adv. προς τα πάνω. ~s of a thousand άνω των χιλίων.
uranium s. ουράνιον n.

urban a. αστικός, των πόλεων. ~ization s. μετατροπή αγροτικής περιοχής εις αστικήν.
urban|e a. ευγενής, αβρός. ~ity s. ευγένεια f. αβροσύνη f.
urchin s. ζωηρό παιδί.
urge v. παροτρύνω, πιέζω. (advice) συμβουλεύω, (recommend) συνιστώ. (s.) παρόρμηση f., επιθυμία f.
urgen|t a. επείγων. ~cy s. επείγουσα ανάγκη. ~tly adv. επειγόντως.
urin|e s. ούρα n.pl. ~al s. ουρητήριον n. ~ary a. ουρητικός. ~ate v. ουρώ.
urn s. υδρία f.
us pron. εμάς, μάς, μας.
USA Ηνωμένες Πολιτείες της Αμερικής.
usage s. χρήση f.
use s. χρήση f. make ~ of κάνω χρήση (with gen.) χρησιμοποιώ. for the ~ of προς χρήσιν (with gen.) in ~ εν χρήσει. of ~ χρήσιμος. what ~ is it? σε τι χρησιμεύει; it's no ~ our waiting longer άδικα περιμένουμε. it's no ~ protesting δεν ωφελούν οι διαμαρτυρίες. fall out of ~ περιπίπτω εις αχρηστίαν, αχρηστεύομαι.
use v. (make use of) χρησιμοποιώ, μεταχειρίζομαι. (treat) μεταχειρίζομαι. ~ (up) (consume) καταναλίσκω, ξοδεύω, (exhaust) εξαντλώ.
used v.i. I ~ to go every Sunday πήγαινα (or συνήθιζα να πηγαίνω) κάθε Κυριακή.
used a. (not new) μεταχειρισμένος. (accustomed) συνηθισμένος. get ~ to συνηθίζω (with να, σε or acc.). I am ~ to it το έχω συνηθίσει.
useful a. χρήσιμος. I found it ~ μου φάνηκε χρήσιμο. ~ness s. χρησιμότης f.
useless a. άχρηστος, (vain) μάταιος. ~ly adv. ματαίως.
user s. ο μεταχειριζόμενος. (consumer) καταναλωτής m.
usher v. οδηγώ ~ in εισάγω, (fig.) εγκαινιάζω. (s.) κλητήρας m. ~ette s. ταξιθέτρια f.
usual a. συνήθης, συνηθισμένος. as ~ ως συνήθως. it is ~ (or the ~ thing) συνηθίζεται. ~ly adv. συνήθως, κανονικά. he ~ly goes to bed early συνηθίζει να κοιμάται νωρίς.
usufruct s. (law) επικαρπία f.
usur|er s. τοκογλύφος m. ~y s. τοκογλυφία f.
usurp v. σφετερίζομαι. ~ation s. σφετερισμός m. ~er s. σφετεριστής m.
utensil s. σκεύος n.
uter|us s. μήτρα f. ~ine a. μητρικός.
utilitarian a. ωφελιμιστικός (practical) πρακτικός. (s.) ωφελιμιστής m. ~ism s.

ωφελιμισμός *m.*

utility *s.* χρησιμότητα *f.* public ~ επιχείρησις κοινής ωφελείας.

utilize *v.* χρησιμοποιώ.

utmost *a.* with the ~ care με την πιο μεγάλη προσοχή. of the ~ importance υψίστης σημασίας. (*s.*) fifty at the ~ πενήντα το πιο πολύ. do one's ~ βάζω τα δυνατά μου.

Utopia *s.* ουτοπία *f.* ~**n** *a.* ουτοπικός, (*person*) ουτοπιστής *m.*

utter *a.* πλήρης, τέλειος. ~**ly** *adv.* πλήρως, τελείως, πέρα για πέρα.

utter *v.* βγάζω, προφέρω. (*issue*) θέτω εις κυκλοφορίαν. ~**ance** *s.* (*speaking*) ομιλία *f.* (*thing said*) κουβέντα *f.* give ~ance to εκφράζω.

uttermost *a. see* utmost. to the ~ ends of the earth στα απώτερα άκρα της γης.

uvula *s.* σταφυλή *f.*

uxorious *a.* υπερβολικά αφοσιωμένος στη γυναίκα μου.

V

vac *s.* (*fam.*) διακοπές πανεπιστημίου.

vacancy *s.* (*space*) κενό *n.* (*position*) χηρεύουσα θέση. (*lodging*) διαθέσιμο δωμάτιο (*of mind*) κενότης *f.*

vacant *a.* (*space*) κενός, άδειος, (*seat, etc.*) ελεύθερος, μη κατειλημμένος. (*look, etc.*) απλανής, ανέκφραστος. become ~ (*of throne, post*) χηρεύω, (*of house*) ξενοικιάζομαι. ~ plot άχτιστο οικόπεδο. ~**ly** *adv.* με ανέκφραστο βλέμμα.

vacate *v.* αφήνω, αδειάζω.

vacation *s.* διακοπές *f.pl.* (*v.i.*) περνώ τις διακοπές.

vaccinat|e *v.* εμβολιάζω. get ~ed κάνω εμβόλιο. ~**ion** *s.* εμβολιασμός *m.*

vaccine *s.* εμβόλιο, μπόλι *n.*, βατσίνα *f.*

vacillat|e *v.* αμφιταλαντεύομαι. ~**ion** *s.* αμφιταλάντευση *f.*

vacu|ous *a.* ανέκφραστος. ~**ity** *s.* χαζομάρα *f.*

vacuum *s.* κενό *n.* ~ cleaner ηλεκτρική σκούπα. ~ flask θερμός *n.*

vade-mecum *s.* εγκόλπιον *n.*

vagabond *a.* πλανόδιος, περιπλανώμενος. (*s.*) αλήτης *m.*, μπαγαπόντης *m.*

vagary *s.* ιδιοτροπία *f.*

vagina *s.* κόλπος *m.*

vagr|ant *a.* πλανόδιος, περιπλανώμενος. (*s.*) αλήτης *m.* ~**ancy** *s.* αλητεία *f.*

vague *a.* (*undefined*) αόριστος, ασαφής, ακαθόριστος. (*dim*) αμυδρός. be ~ (*of person*) δεν είμαι θετικός. his answer was

~ η απάντησή του δεν ήταν καθαρή. not have the ~st idea δεν έχω την παραμικρή ιδέα. ~**ly** *adv.* αόριστα, ασαφώς, αμυδρώς. ~**ness** *s.* αοριστία *f.*

vain *a.* (*fruitless*) άκαρπος, κενός, μάταιος. in ~ άδικα, του κάκου, εις μάτην. labour in ~ ματαιοπονώ. (*conceited*) ματαιόδοξος. be ~ of είμαι υπερήφανος για. ~**ly** *adv.* ματαίως, άδικα.

vainglor|y *s.* ματαιοδοξία *f.* ~**ious** *a.* ματαιόδοξος.

valance *s.* φραμπαλάς *m.*

vale *s.* κοιλάδα *f.*

valedict|ion *s.* αποχαιρετισμός *m.* ~**ory** *a.* αποχαιρετιστήριος.

valency *s.* (*chem.*) σθένος *n.*

valentine *s.* ανώνυμο ερωτικό γράμμα που στέλνεται στη γιορτή του Αγίου Βαλεντίνου (14 Φεβρουαρίου).

valet *s.* υπηρέτης κυρίου.

valetudinarian *a. & s.* υποχονδριακός.

valiant *a.* ανδρείος, γενναίος. ~**ly** *adv.* ανδρείως.

valid *a.* έγκυρος, ισχύων. it is not ~ δεν ισχύει. ~**ate** *v.* κυρώνω. ~**ity** *s.* κύρος *n.*, ισχύς *f.* have ~ity ισχύω.

valise *s.* βαλίτσα *f.*

valley *s.* κοιλάδα *f.*

val|our *s.* ανδρεία *f.* ~**orous** *a.* ανδρείος, γενναίος.

valuable *a.* πολύτιμος, αξίας. (*s.*) ~s αντικείμενα αξίας.

valuation *s.* εκτίμηση *f.*

value *s.* αξία *f.* be of ~ έχω αξία. it is good ~ for money αξίζει τα λεφτά του. (*math.*) τιμή *f.* (*chem.*) δείχτης *m.* ~s (*moral*) αξίες *f.pl.* (*v.*) εκτιμώ. I ~ your help η βοήθειά σας μου είναι πολύτιμη. ~**less** *a.* χωρίς αξία.

valve *s.* βαλβίδα *f.* (*radio*) λάμπα *f.*, λυχνία *f.*

vamoose *v.* (*fam.*) το κόβω λάσπη.

vamp *s.* (*of shoe*) ψίδι *n.* (*woman*) ξελογιάστρα *f.* (*v.*) (*fam.*) ~ up σκαρώνω.

vampire *s.* βρυκόλακας *m.*

van *s.* (*vehicle*) φορτηγό αυτοκίνητο. (*front*) *see* vanguard.

vandal *s.* βάνδαλος *m.* ~**ism** *s.* βανδαλισμός *m.*

vane *s.* ανεμοδείκτης *m.* (*also fig.*) ανεμοδούρα *f.* (*blade of mill, etc.*) πτερύγιον *n.*

vanguard *s.* (*mil.*) εμπροσθοφυλακή *f.* (*fig.*) πρωτοπορία *f.*

vanilla *s.* βανίλια *f.* ~ ice παγωτό κρέμα.

vanish *v.i.* εξαφανίζομαι, χάνομαι.

vanity *s.* (*futility*) ματαιότητα *f.* (*conceit*) ματαιοδοξία *f.*

vanquish *v.* κατανικώ, (*fears*) υπερνικώ.

vantage s. point or coign of ~ παρατηρητήριο n., (fig.) πλεονεκτική θέση.

vapid a. ανούσιος, σαχλός. ~**ity** s. σάχλα f.

vaporiz|e v.t. εξαερώνω. ~**ation** s. εξαέρωση f.

vaporous a. αχνώδης.

vapour s. υδρατμός m. ~ bath ατμόλουτρο n. have the ~s έχω νεύρα. ~**ing(s)** s. αερολογία f.

vari|able a. μεταβλητός, ευμετάβλητος, άστατος. (s.) μεταβλητόν n. ~**ability** s. ευμετάβλητον n.

variance s. we are at ~ έχουμε διαφορές. set at ~ βάζω στα αίματα.

variant a. διαφορετικός. (s.) παραλλαγή f.

variation s. (variant, also mus.) παραλλαγή f. (change) μεταβολή f. (deviation) απόκλιση f. (fluctuation) διακύμανση f.

varicose a. ~ vein κιρσός m.

varied a. ποικίλος.

variegated a. ποικίλος, ποικιλόχρωμος.

variety s. ποικιλία f. a ~ of things διάφορα πράματα. (sort, kind) είδος n. (theatre) βαριετέ n.

various a. ποικίλοι, διάφοροι. ~**ly** adv. ~ly known as γνωστός υπό τα ποικίλα ονόματα...

varnish s. βερνίκι n. (v.) βερνικώνω.

vary v.t. & i. ποικίλλω. (v.t.) αλλάζω. (v.i.) διαφέρω.

vase s. βάζο n.

vaseline s. βαζελίνη f.

vassal a. & s. υποτελής.

vast a. (amount, bulk) τεράστιος, (space) απέραντος, αχανής. ~**ly** adv. πάρα πολύ. ~**ness** s. απεραντοσύνη f.

VAT ΦΠΑ.

vat s. κάδος m., δεξαμενή f. (cauldron) καζάνι n.

vault s. (dome) θόλος m. (tomb) τάφος m. (cellar) υπόγειο n. ~**ed** a. θολωτός.

vault v. (also ~ over) πηδώ. ~**ing** horse εφαλτήριο n.

vaunt v.t. & i. καυχιέμαι. (v.i.) κοκορεύομαι. ~**ingly** adv. κομπαστικά.

VDU οθόνη f.

veal s. μοσχάρι n. (a.) μοσχαρίσιος.

veer v. γυρίζω.

vegan s. αυστηρός χορτοφάγος.

vegetable s. λαχανικό n., χορταρικό n., ζαρζαβάτι n. (a.) φυτικός.

vegetarian a. & s. χορτοφάγος, (diet, etc.) χορτοφάγων. ~**ism** s. χορτοφαγία f.

vegetat|e v. διάγω μονότονη ζωή χωρίς πνευματικά ενδιαφέροντα, φυτοζωώ. ~**ion** s. βλάστηση f., φυτεία f.

vehem|ent a. ορμητικός, (desire) σφοδρός, (protest) έντονος, βίαιος. ~**ently** adv. με

σφοδρότητα, εντόνως. ~**ence** s. σφοδρότητα, βιαιότητα f.

vehicle s. (conveyance) όχημα n. (fig., means) μέσον n. (chem.) φορεύς m.

veil s. πέπλος m., βέλο n. (Moslem) φερετζές m. (fig.) draw a ~ over αποσιωπώ. (v.t.) κρύβω. ~**ed** a. πεπλοφόρος. (fig.) συγκεκαλυμμένος.

vein s. φλεψ, φλέβα f. (mood) in the same ~ με το ίδιο πνεύμα.

vellum s. βιδέλο n.

velocity s. ταχύτητα f.

velvet s. βελούδο n. (a.) (also ~y) βελούδινος, βελουδένιος.

venal a. εξαγοραζόμενος. ~**ity** s. το ότι εξαγοράζεται κανείς.

vend v. πουλώ. ~**ing** machine αυτόματος πωλητής. ~**or** s. πωλητής m.

vendetta s. βεντέτα f.

veneer s. καπλαμάς m. (fig.) βερνίκι n.

venerable a. σεβάσμιος.

venerat|e v. σέβομαι. ~**ed** σεβαστός. ~**ion** s. σεβασμός m.

venereal a. αφροδίσιος. V.D. specialist αφροδισιολόγος m.

Venetian a. ενετικός, βενετσιάνικος. (person) Βενετός m. ~ blind στόρι με γρίλλιες.

vengeance s. εκδίκηση f. take ~ on εκδικούμαι. (fam.) it's raining with a ~ ! βροχή μια φορά!

vengeful a. εκδικητικός.

venial a. ελαφρός.

venison s. κρέας ελαφιού.

venom s. (poison) ιός m. (fig.) κακία f., μοχθηρία f. ~**ous** a. δηλητηριώδης, (fig.) μοχθηρός. ~ous person φαρμακομύτης m. ~**ously** adv. με κακία.

vent s. (hole) οπή f. (outlet) διέξοδος f. (in jacket) σχισμή f. give ~ to αφήνω να μου ξεφύγει (v.) ~ one's feelings ξεθυμαίνω.

ventilat|e v. εξαερίζω. (make known) φέρω στη δημοσιότητα. ~**ion** s. εξαερισμός m. ~**or** s. εξαεριστήρας m.

ventriloqu|ism s. εγγαστριμυθία f. ~**ist** s. εγγαστρίμυθος m.

venture v.t. (put at risk) διακινδυνεύω. (v.i., dare) τολμώ. ~ to suggest παίρνω το θάρρος να προτείνω. ~ out of the house τολμώ να βγω έξω.

venture s. εγχείρημα n. (business) επιχείρηση f. ~**some** a. τολμηρός.

venue s. τόπος συναντήσεως.

verac|ious a. φιλαλήθης. ~**ity** s. φιλαλήθεια f.

veranda(h) s. σκεπασμένη βεράντα του ισογείου.

verb s. ρήμα n. ~**al** a. (of verbs) ρηματικός,

(*spoken*) προφορικός. ~**ally** *adv*. προφορικώς.
verbatim *adv*. κατά λέξιν.
verbiage *s*. περιττολογία *f*.
verbos|e *a*. περιττολόγος. ~**ity** *s*. περιττολογία *f*.
verdant *a*. χλοερός.
verdict *s*. (*of court*) ετυμηγορία *f*. (*opinion*) γνώμη *f*.
verdigris *s*. γανίλα στα χαλκώματα.
verdure *s*. πρασινάδα *f*., χλόη *f*.
verge *s*. (*edge*) άκρη *f*. (*fig., brink*) χείλος *n*. (*of winter, bankruptcy*) πρόθυρα *n.pl*. be on the ~ of sixty κοντεύω τα εξήντα. on the ~ of tears έτοιμος να ξεσπάσω σε κλάματα. (*v*.) ~ on πλησιάζω, εγγίζω.
verger *s*. νεωκόρος *m*.
verif|y *v*. εξακριβώνω, επαληθεύω. ~**ication** *s*. επαλήθευση *f*.
verisimilitude *s*. αληθοφάνεια *f*., πιθανότητα *f*.
veritab|le *a*. πραγματικός. ~**ly** *adv*. πραγματικώς.
verity *s*. αλήθεια *f*.
vermilion *s*. ζωηρό κόκκινο χρώμα.
vermin *s*. (*animals*) επιβλαβή ζώα. (*insects*) παράσιτα *n.pl*. (*fig., of person*) κάθαρμα *n*. ~**ous** *a*. ψειριασμένος.
vermouth *s*. βερμούτ *n*.
vernacular *s*. καθομιλουμένη γλώσσα.
vernal *a*. εαρινός.
versatil|e *a*. (*mind, talent*) πολύπλευρος, (*thing*) με πολλαπλές χρήσεις. ~**ity** *s*. (*of person*) ικανότητα *f*.
verse *s*. (*line*) στίχος *m*. (*stanza*) στροφή *f*. (*poetry*) έμμετρος λόγος, ποίηση *f*.
versed *a*. εντριβής. be well ~ in ξέρω πολλά από, κατέχω καλά.
versification *s*. μετρική *f*.
version *s*. (*account*) άποψη *f*., εκδοχή *f*. (*translation*) μετάφραση *f*.
versus *prep*. (*sport*) εναντίον, (*law*) κατά (*both with gen.*).
vertebr|a *s*. σπόνδυλος *m*. ~**al** *a*. σπονδυλικός. ~**ate** *a*. σπονδυλωτός.
vertical *a*. κάθετος, κατακόρυφος. ~**ly** *adv*. κάθετα, κατακόρυφα.
vertig|o *s*. ίλιγγος *m*. ~**inous** *a*. ιλιγγιώδης.
verve *s*. ζωντάνια *f*.
very *adv*. πολύ. ~ much πάρα πολύ. at the ~ most το πολύ πολύ. he's not so ~ young δεν είναι και τόσο (πολύ) νέος. it is my ~ own είναι κατιδικό μου. the ~ first ο πρώτος πρώτος. (*also by prefix*) ~ busy πολυάσχολος, ~ expensive πανάκριβος, ~ white κάτασπρος.
very *a*. the ~ idea frightens me και μόνον η ιδέα με τρομάζει. at that ~ moment εκείνη

ακριβώς τη στιγμή. it's the ~ thing I want είναι ακριβώς αυτό που θέλω. at (*or* to) the ~ edge άκρη άκρη. her ~ words τα ίδια της τα λόγια. the ~ image of his father φτυστός (*or* ο ίδιος) ο πατέρας του. the veriest babe knows that ακόμα κι ένα μωρό παιδί το ξέρει.
vespers *s*. εσπερινός *m*.
vessel *s*. (*receptacle*) δοχείο *n*. (*also* blood-~) αγγείο *n*. (*ship*) σκάφος *n*. πλοίο *n*.
vest *s*. φανέλλα *f*. (*waistcoat*) γιλέκο *n*.
vest *v.t*. περιβάλλω. have a ~ed interest έχω ίδιον συμφέρον. ~ed interests κεκτημένα δικαιώματα. be ~ed in (*belong to*) ανήκω σε.
vestal *a*. ~ virgin εστιάς *f*.
vestibule *s*. είσοδος *f*.
vestig|e *s*. ίχνος *n*. ~**ial** *a*. υποτυπώδης.
vestment *s*. άμφιον *n*.
vestry *s*. (*sacristy*) σκευοφυλάκιον *n*. (*meeting*) ενοριακή επιτροπή.
vet *s*. κτηνίατρος *m*. (*v*.) (*fam*.) εξετάζω. (*fig*.) ελέγχω.
veteran *a*. παλαίμαχος. (*s*.) βετεράνος *m*.
veterinary *a*. κτηνιατρικός. ~ surgeon κτηνίατρος *m*.
veto *s*. αρνησικυρία *f*., βέτο *n*. (*v*.) απαγορεύω.
vex *v*. πειράζω, ενοχλώ, εκνευρίζω. ~ed question επίμαχο θέμα. ~**ation** *s*. ενόχληση *f*., εκνευρισμός *m*. ~**atious** *a*. ενοχλητικός, εκνευριστικός.
via *prep*. μέσω (*with gen*.).
viable *a*. βιώσιμος.
viaduct *s*. οδογέφυρα *f*.
viand *s*. έδεσμα *n*.
vibrant *a*. (*resonant*) ηχηρός. (*fig*.) be ~ with (*health, etc*.) σφύζω από.
vibrat|e *v.i*. δονούμαι, σείομαι. (*of voice*) πάλλομαι. (*resound*) αντηχώ. (*of engine*) κραδαίνομαι. (*v.t*.) δονώ, σείω. ~**ion** *s*. δόνηση *f*., σείσιμο *n*. κραδασμός. *m*.
vicar *s*. εφημέριος *m*. ~**age** *s*. κατοικία εφημερίου.
vicarious *a*. (*authority*) δοτός. (*action*) γινόμενος υπό άλλου. ~**ly** *adv*. δι' αντιπροσώπου.
vice *s*. (*bad habit*) βίτσιο *n*., ελάττωμα *n*. (*depravity*) φαυλότητα *f*., κακία *f*., ανηθικότητα *f*.
vice *s*. (*tool*) μέγγενη *f*.
vice-consul *s*. υποπρόξενος *m*.
vice-president *s*. αντιπρόεδρος *m*.
viceroy *s*. αντιβασιλεύς *m*.
vice versa *adv*. αντιστρόφως, τανάπαλιν.
vicinity *s*. γειτονιά *f*. in the ~ (*of*) κοντά (σε).
vicious *a*. φαύλος, ανήθικος, (*spiteful*) κακός. (*living a depraved life*) φαυλόβιος.

~ circle φαύλος κύκλος. ~ly adv. με κακία.
~ness s. κακία f.
vicissitudes s. περιπέτειες f.pl.
victim s. θύμα n., ο παθών. ~ of earthquake σεισμόπληκτος m. ~ize v. κατατρέχω.
~ization s. κατατρεγμός m.
victor s. νικητής m., νικήτρια f.
Victorian a. βικτωριανός. early ~ (art, etc.) που ανήκει στα πρώτα χρόνια της βικτωριανής εποχής.
victor|y s. νίκη f. ~y celebrations νικητήρια n.pl. ~ious a. νικηφόρος. be ~ious νικώ.
victual v.t. ανεφοδιάζω (v.i.) ανεφοδιάζομαι. (s.) ~s τρόφιμα n.pl.
vide v. imper. ίδε, βλέπε.
video s.a. βίντεο n.
vie v. ~ (with) αμιλλώμαι, συναγωνίζομαι.
view s. (what is seen) θέα f. (picture) εικόνα f. (beholding) όψις f. at first ~ εκ πρώτης όψεως. point of ~ άποψη f. (opinion) γνώμη f., ιδέα f., αντιλήψεις f.pl. be put on ~ εκτίθεμαι. come into ~ εμφανίζομαι, φαίνομαι. come in ~ of αντικρύζω, βλέπω μπροστά μου. in full ~ of the audience μπρος στα μάτια όλου του ακροατηρίου. in ~ (prospective) εν όψει. have in ~ (be interested in) κοιτάζω, (intend) σκοπεύω. keep in ~ έχω υπ' όψιν. in ~ of what happened yesterday έχοντας υπ' όψιν το τι συνέβη χθες (or τα χθεσινά συμβάντα). with a ~ to με προοπτική να.
view v. βλέπω, κοιτάζω. (examine) εξετάζω. ~er s. θεατής m. ~point s. άποψη f.
vigil s. αγρυπνία f. keep ~ αγρυπνώ. ~ance s. επαγρύπνηση f. ~ant a. άγρυπνος. be ~ant γρηγορώ. ~antly adv. αγρύπνως.
vignette s. βινιέτα f. (description) σκιαγράφηση f.
vigorous a. δυνατός, ρωμαλέος, γερός, ενεργητικός. (forcible) έντονος. ~ly adv. δυνατά, γερά, έντονα.
vigour s. δύναμη f. ρώμη f. (youthful) σφρίγος n. (vitality) ζωτικότητα f. (activeness) ενεργητικότητα f.
Viking s. Βίκινγχ m.
vile a. αχρείος, βδελυρός. (fam., awful) απαίσιος, ελεεινός, άθλιος. ~ly adv. απαίσια. ~ness s. αχρειότητα f. the ~ness of the weather ο απαίσιος καιρός.
vilif|y v. δυσφημώ. ~ication s. δυσφήμηση f.
villa s. έπαυλη f., βίλλα f.
village s. χωριό n. (a.) χωρικός. ~r s. χωρικός, χωριανός m.
villain s. παλιάνθρωπος m. κακοήθης, αχρείος a. (joc.) κατεργάρης m. (of drama) ο κακός. ~ous a. κακοήθης, αχρείος. (fam., awful) απαίσιος, ελεεινός. ~y s. κακοήθεια f.

vim s. (fam.) ενεργητικότητα f.
vindicat|e v. δικαιώνω. ~ion s. δικαίωση f.
vindictive a. εκδικητικός. ~ness s. εκδικητικότης f.
vine s. κλήμα n., άμπελος m. ~-arbour s. κληματαριά f. ~-leaf s. κληματόφυλλο n. ~yard s. αμπέλι n. αμπελώνας m.
vinegar s. ξίδι n., όξος n. ~y a. (taste) ξιδάτος, (temper) ξινισμένος.
vintage s. τρύγος m. ~ wine κρασί καλής χρονιάς. (fig.) ~ model παλαιού τύπου.
vintner s. οινοπώλης m.
viola s. (mus.) βιόλα f.
violat|e v. (infringe) παραβιάζω, (rape) βιάζω, (desecrate) βεβηλώνω. ~ion s. παραβίαση f. βιασμός m. βεβήλωση f.
violence s. βία f., βιαιότητα f. σφοδρότητα f. do ~ to παραβιάζω.
violent a. βίαιος (storm, etc.) σφοδρός, (feeling) έντονος. become ~ εξαγριώνομαι. ~ly adv. βιαίως, σφοδρώς, έντονα, με βία.
violet s. βιολέττα f., μενεξές m. (a.) ιόχρους, ~ιο ιόχρους. ~ι εξεδένιος.
violin s. βιολί n. ~ist s. βιολιστής m.
violoncello s. βιολοντσέλλο n.
VIP προσωπικότης f.
viper s. οχιά f., έχιδνα f.
virago s. μέγαιρα f.
virgin s. παρθένος, παρθένα f. (a.) (modesty, etc.) παρθενικός, (soil, forest, snow) παρθένος. ~al a. παρθενικός. ~ity s. παρθενία f.
viril|e a. ανδρικός, (style) αρρενωπός. ~ity s. ανδρισμός m.
virtual a. ουσιαστικός. ~ly adv. στην ουσία, κατ' ουσίαν.
virtue s. αρετή f. (morality) ηθικότητα f. (power, property) ιδιότητα f. (advantage) it has the ~ (of being) έχει το πλεονέκτημα (or το προσόν or το καλό) (with va, ότι, που). of easy ~ ελαφρών ηθών. make a ~ of necessity κάνω την ανάγκη φιλοτιμία. by ~ of δυνάμει, βάσει (with gen.).
virtuos|o s. δεξιοτέχνης m. ~ity s. δεξιοτεχνία f.
virtuous a. ενάρετος, ηθικός. (smug) αυτάρεσκος. ~ly adv. (morally) ηθικά, (smugly) με αυταρέσκεια.
virul|ent a. (toxic) τοξικός. (bitter) δριμύς, κακεντρεχής. ~ence s. τοξικότητα f. δριμύτητα f., κακεντρέχεια f.
virus s. ιός m.
visa s. βίζα f., θεώρηση f. (v.t.) θεωρώ.
visage s. πρόσωπον n.
vis-a-vis adv. ο ένας απέναντι στον άλλον. (prep.) σχετικώς με, έναντι (with gen.)
viscera s. σπλάχνα n.pl.

viscount s. υποκόμης m. **~ess** s. υποκόμησσα f.

visc|ous a. γλοιώδης **~osity** s. το γλοιώδες.

visib|le a. ορατός, θεατός. (evident) καταφανής. **~ly** adv. καταφανώς. **~ility** s. ορατότητα f.

vision s. (faculty) όραση f. (apparition) όραμα n., οπτασία f. (clear-sightedness) διορατικότητα f. **~ary** a. (unreal) χιμαιρικός. (s.) οραματιστής m.

visit s. επίσκεψη f., βίζιτα f. pay a ~ κάνω επίσκεψη. (v.t.) επισκέπτομαι. ~ing card επισκεπτήριο n. **~or** s. επισκέπτης m. we have ~ors έχουμε επισκέψεις.

visitant s. επισκέπτης m.

visitation s. επίσημη επίσκεψη. (divine) θέλημα Θεού. (scourge) πληγή f. V~ (eccl. feast) Ασπασμός m.

visor s. (of helmet) προσωπίς f. (of cap) γείσο n.

vista s. (view) θέα f. (fig.) προοπτική f.

visual a. οπτικός. ~ aids οπτικά βοηθήματα. ~ display unit οθόνη f. **~ize** v. φαντάζομαι, βλέπω με το νου μου **~ly** adv. οπτικώς.

vital a. ζωτικός. ~ importance ζωτικότητα f. ~ spot καίριο σημείο, ~ statistics δημογραφική στατιστική, (joc.) σωματικές αναλογίες γυναίκας. (s.) **~s** ζωτικά όργανα.

vitality s. ζωτικότητα f.

vitalize v. τονώνω, ζωογονώ.

vitally adv. ζωτικώς, ~ important ζωτικής (or υψίστης) σημασίας.

vitamin s. βιταμίνη f.

vitiate v. (spoil) χαλώ, μολύνω. (an argument) καταστρέφω. (law) ακυρώ.

vitreous s. υαλώδης.

vitriol s. βιτριόλι n. **~ic** a. φαρμακερός.

vituperat|e v. βρίζω. **~ion** s. βρισιές f.pl. **~ive** a. υβριστικός.

viva v.t. εξετάζω προφορικώς. (s.) προφορική εξέταση. (adv.) ~ voce προφορικώς.

vivac|ious a. ζωηρός, ζωντανός. **~ity** s. ζωηρότητα, ζωντάνια f.

vivid a. (colour, memories) ζωηρός (picture, description) ζωντανός. (flash) εκθαμβωτικός. **~ly** adv. ζωηρά, (description) με περιγραφικότητα.

vivisection s. ζωοτομία f.

vixen s. θηλυκιά αλεπού. (fig., also ~ish) στρίγλα f.

viz adv. δηλαδή, τουτέστι.

vizier s. βεζίρης m.

vocabulary s. λεξιλόγιο n.

vocal a. (of the voice) φωνητικός. be ~ (about sthg.) εκφράζομαι εντόνως. **~ly** adv. από φωνητική άποψη. **~ic** a. (of

vowels) φωνήεντος. **~ist** s. τραγουδιστής m.

vocation s. (call) αποστολή f. (calling) επάγγελμα n. (inclination) κλίση f. **~al** a. επαγγελματικός.

vocative s. (gram.) κλητική f.

vocifer|ate v. κραυγάζω. **~ous** a. κραυγαλέος.

vogue s. μόδα. f. in ~ της μόδας.

voice s. φωνή f. give ~ to εκφράζω. I lost my ~ πιάστηκε η φωνή μου. he has no ~ in the matter δεν του πέφτει λόγος. (v.) εκφράζω. **~d** a. (phonetics) ηχηρός. **~less** a. άφωνος.

void a. (empty) κενός, (not valid) άκυρος. ~ of εστερημένος, άνευ (both with gen.). (s.) κενό n. (v.t.) (invalidate) ακυρώ, (discharge) βγάζω.

volatile a. (chem.) αιθέριος, πτητικός. (person) ασταθής.

volcan|o s. ηφαίστειο n. **~ic** a. ηφαιστειογενής (fig.) ηφαιστειώδης.

vole s. αρουραίος m.

volition s. βούληση f., θέληση f. of one's own ~ θεληματικά.

volley s. (missiles, applause) ομοβροντία f. (stones, abuse) βροχή f. **~ball** s. βόλλεϋ n.

voltage s. τάση f., βολτάζ n.

volte-face s. (fig.) μεταστροφή f.

volub|le a. φλύαρος. **~ility** s. φλυαρία f.

volume s. (mass) όγκος m. (capacity) χωρητικότητα f. (loudness) ένταση f. (of tone) πλάτος n. (book) τόμος m. speak ~s for λέω πολλά για.

voluminous a. (work) ογκώδης, (writings) πολύτομος, (writer) πολυγράφος, (clothing) πολύπτυχος.

voluntar|y a. εθελοντικός, εκούσιος, θεληματικός. **~ily** adv. εθελοντικώς, εκουσίως, θεληματικά.

volunteer s. εθελοντής m. (v.t.) προσφέρομαι εθελοντικώς. (v.t.) he ~ed information προσφέρθη να δώσει πληροφορίες.

voluptu|ary s. ηδυπαθής, φιλήδονος a. **~ous** a. αισθησιακός, γεμάτος ηδονές.

vomit v.t. & i. ξερνώ, κάνω εμετό. **~ing** s. εμετός m., ξέρασμα n.

vorac|ious a. αδηφάγος, αχόρταγος. **~ity** s. αδηφαγία f., βουλιμία f.

vortex s. δίνη f.

votary s. θιασώτης m.

vote s. ψήφος m.f. put to the ~ θέτω εις ψηφοφορίαν. solicit ~s ψηφοθηρώ. win equal number of ~s ισοψηφώ. ~ of censure πρόταση μομφής. (v.) ψηφίζω, (propose) προτείνω, (declare) αναγνωρίζω. ~ against or down καταψηφίζω. ~ for

(υπερ)ψηφίζω. **~er** *s.* ψηφοφόρος *m.*
voting *s.* ψηφοφορία *f.*
votive *a.* ~ offering τάμα *n.,* ανάθημα *n.*
vouch *v.* ~ for εγγυώμαι για.
voucher *s.* κουπόνι *n.,* δελτίο *n.*
vouchsafe *v.i.* καταδέχομαι, ευδοκώ.
vow *s.* όρκος *m.* (*as act of piety*) τάμα *n.* (*v.i.*) ορκίζομαι, τάζομαι. I have ~ed to make a pilgrimage το έχω τάμα να πάω προσκύνημα. (*v.t.*) (*an offering*) τάζω.
vowel *s.* φωνήεν *n.*
voyage *s.* (θαλασσινό) ταξίδι *n.* (*v.*) ταξιδεύω. **~r** *s.* θαλασσοπόρος *m.*
voyeur *s.* ηδονοβλεψίας *m.*
vulcanize *v.* βουλκανιζάρω.
vulgar *s.* χυδαίος, πρόστυχος. (*tongue, fraction, error*) κοινός. **~ian** *s.* χυδαίος, πρόστυχος *a.* **~ism** *s.* χυδαϊσμός *m.* **~ity** *s.* χυδαιότητα *f.,* προστυχιά *f.*
vulgariz|e *v.* εκχυδαΐζω **~ation** *s.* εκχυδαϊσμός *m.*
vulnerab|le *a.* τρωτός, ευπρόσβλητος. **~ility** *s.* τρωτόν *n.*
vulture *s.* γύψ *m.* (*fig., person*) αρπακτικό όρνιο.
vying *s.* συναγωνισμός *m.* see vie.

W

wad *s.* (*paper, notes*) μάτσο *n.* (*soft material*) ταμπόν(ι) *n.* **~ding** *s.* βάτα *f.*
waddle *v.* περπατώ κουνιστά (*or* σαν πάπια).
wade *v.i.* ~ across the stream διασχίζω το ρυάκι περπατώντας. (*fig.*) ~ through (*book, etc.*) τελειώνω με κόπο. ~ into (*opponent*) επιτίθεμαι κατά (*with gen.*). (*task*) ρίχνομαι σε, πέφτω με τα μούτρα σε.
wafer *s.* γκοφρέτα *f.* (*thin piece*) πολύ λεπτό κομμάτι. (*eccl.*) όστια *f.*
waffle *s.* (*nonsense*) σαλιάρισμα *n.* (*v.i.*) σαλιαρίζω.
waft *v.* be ~ed σκορπίζομαι απαλά από το αεράκι.
wag *v.t.* κουνώ (*v.i.*) (*fig.*) tongues are ~ging το κουτσομπολιό παίρνει και δίνει.
wag *s.* αστειολόγος *m.* **~gish** *a.* αστείος.
wage *v.t.* διεξάγω, κάνω.
wage,~s *s.* μισθός *m.* a day's ~ μεροκάματο *n.,* ημερομίσθιον *n.* ~ claim απαιτήσεις για αύξηση μισθών. **~-packet** *s.* βδομαδιάτικο *n.* **~-earner** *s.* μισθοσυντήρητος, μεροκαματιάρης *n.*
wager *s.* στοίχημα *n.* (*v.*) στοιχηματίζω.
waggle *v.t.* σείω, κουνώ.
waggon *s.* τετράτροχο κάρρο. (*railway*)

βαγόνι *n.* water ~ βυτίο *n.* (*joc.*) go on the ~ απέχω από οινοπνευματώδη.
waif *s.* πεντάρφανος *a.*
wail *v.i.* ολοφύρομαι, οδύρομαι. (*of baby, wind, etc.*) ουρλιάζω. (*s.*) (*also* ~ing) οδυρμός *m.* ουρλιαχτό *n.*
wainscot *s.* ξύλινη επένδυση κάτω μέρους τοίχου.
waist *s.* μέση *f.* **~band** *s.* ζωνάρι *n.* **~coat** *s.* γιλέκο *n.*
wait *v.* (*also* ~ for) περιμένω. ~ on (*serve*) υπηρετώ. ~ upon (*visit*) επισκέπτομαι. ~ at table σερβίρω. we must ~ and see πρέπει να περιμένουμε να δούμε. I kept him ~ing two hours τον είχα και περίμενε δύο ώρες.
wait *s.* we had a long ~ περιμέναμε πολύ. lie in ~ παραμονεύω, παραφυλάω. they lay in ~ for the coach παραφύλαξαν ώσπου να φανεί η άμαξα.
waiting *s.* αναμονή *f.* lady-in- ~ κυρία επί των τιμών. ~ list κατάλογος προτεραιότητος. ~ room αίθουσα αναμονής.
wait|er *s.* γκαρσόνι *n.,* σερβιτόρος *m.* **~ress** *s.* σερβιτόρα *f.*
waive *v.* (*forgo*) παραιτούμαι από. (*set aside*) βάζω κατά μέρος.
wake *s.* (*for dead*) αγρυπνία παρά το πλευρόν νεκρού.
wake *s.* (*of ship*) απόνερα *n.pl.* (*fig.*) in the ~ of πίσω από.
wake,~n *v.t. & i.* (*also* ~ up) ξυπνώ. ~ up to (*realize*) καταλαβαίνω. (*v.t.*) (*rouse*) εγείρω. ~ memories of φέρνω στο νου, ενθυμίζω.
wakeful *a.* άγρυπνος, ~ night νύχτα αϋπνίας. **~ness** *s.* αγρυπνία *f.*
waken *v.* see wake. **~ing** *s.* ξύπνημα *n.*
waking *a.* ξύπνιος.
walk *v.i.* περπατώ. (*go on foot*) πάω με τα πόδια. ~ away with (*contest*) κερδίζω εύκολα. ~ off with (*take*) βουτώ, σηκώνω. ~ into (*meet, collide with*) πέφτω επάνω σε, (*a trap*) πέφτω σε. he ~ed out σηκώθηκε κι έφυγε. ~ out on παρατώ. ~ up to πλησιάζω. (*v.t.*) (*cover on foot*) περπατώ. ~ the streets γυρίζω στους δρόμους. they ~ed him off to the police station τον πήγαν στο τμήμα. (*s.*) περίπατος *m.,* βόλτα *f.* (*gait*) βάδισμα *n.* people from all ~s of life κάθε είδους άνθρωποι. ten minutes' ~ δέκα λεπτά με τα πόδια.
walk|er *s.* πεζοπόρος *m.* **~ing** *s.* πεζοπορία *f.,* περπάτημα *n.*
walkie-talkie *s.* φορητός πομποδέκτης.
walk-over *s.* εύκολη νίκη.
wall *s.* (*of building*) τοίχος *m.,* ντουβάρι *n.* (*of city, fortress*) τείχος *n.* (*of garden, etc.*) μάντρα *f.* (*inner surface of vessel or*

organ) τοίχωμα *n.* (*fig.*) go to the ~ παραμερίζομαι. he has his back to the ~ τον κόλλησαν στον τοίχο. (*v.t.*) περιτοιχίζω. ~ up (*close*) κλείνω, (*a victim*) χτίζω. **~paper** *s.* ταπετσαρία *f.*

wallet *s.* πορτοφόλι *n.*

wallflower *s.* (*fam., at dance*) κορίτσι που δεν το ζητούν να χορέψει.

wallop *v.* ξυλοκοπώ. (*s.*) ξυλοκόπημα *n.*

wallow *v.* κυλιέμαι. (*fig.*) be ~ing in money κολυμπάω στα λεφτά.

walnut *s.* καρύδι *n.* (*tree, wood*) καρυδιά *f.*

walrus *s.* θαλάσσιος ίππος.

waltz *s.* βαλς *n.* (*v.*) βαλσάρω.

wan *a.* ωχρός.

wand *s.* ράβδος *f.*

wander *v.* περιπλανώμαι, περιφέρομαι, γυρίζω. ~ from the subject φεύγω από το θέμα. (*have one's thoughts elsewhere*) αφαιρούμαι, τα έχω χαμένα. **-er** *s.* περιπλανώμενος *a.* **-ings** *s.* περιπλανήσεις *f.pl.*

wan|e *v.* ελαττώνομαι, (*of glory*) δύω. the moon is ~ing (*or* on the ~e) το φεγγάρι είναι στη χάση του.

wangle *v.* (*fam., get*) καταφέρνω να πάρω, πετυχαίνω. (*fix*) μαγειρεύω.

want *s.* (*lack*) έλλειψη *f.*, στέρηση *f.* (*need*) ανάγκη *f.* (*penury*) ένδεια *f.* (*desire*) επιθυμία *f.* for ~ of money ελλείψει χρημάτων. I am in ~ of money μου λείπουν χρήματα, στερούμαι χρημάτων.

want *v.* (*desire*) θέλω, επιθυμώ. (*require, ought to have*) χρειάζομαι. (*lack, not have*) she will not ~ for anything δεν θα της λείψει τίποτα. *see also* need.

wanting *a.* be ~ (*missing*) λείπω. he is ~ in tact του λείπει το τακτ. be found ~ αποδεικνύομαι υστερών.

wanton *a.* (*licentious*) ακόλαστος. (*extravagant*) εξωφρενικός. ~ act αδικαιολόγητως κακόβουλη πράξη.

war *s.* πόλεμος *m.* state of ~ εμπόλεμος κατάστασις. make ~ on (*fig.*) πολεμώ. (*a.*) πολεμικός. (*v.*) πολεμώ. ~ring states εμπόλεμα κράτη. ~ criminal εγκληματίας πολέμου. **--cry** *s.* πολεμική κραυγή, (*fig.*) σύνθημα *n.* **~fare** *s.* πόλεμος *m.* **~head** *s.* κώνος *m.* **--horse** *s.* (*fig.*) παλαίμαχος πολεμιστής. **--like** *a.* αρειμάνιος. **~monger** *s.* πολεμοκάπηλος *m.* **--path** *s.* be on the **--path** έχω λυμένο το ζωνάρι μου για καβγά. **~ship** *s.* πολεμικόν *n.*

warble *v.* κελαϊδώ.

ward *s.* (*person under guardian*) κηδεμονευόμενος *m.* (*guardianship*) κηδεμονία *f.* (*of hospital*) θάλαμος *m.* (*district*

περιοχή *f.*

ward *v.* ~ off (*avert*) αποτρέπω, (*an attack*) αποκρούω, (*illness*) προλαμβάνω.

warden *s.* (*principal*) διευθυντής *m.* (*guard*) φύλακας *m.*

warder *s.* δεσμοφύλακας *m.*

wardrobe *s.* (*cupboard*) ιματιοθήκη *f.* (*clothes*) ρούχα *n. pl.* (*also theatrical*) γκαρνταρόμπα *f.*

ware *s.* (*in compounds*) silver ~ ασημικά *n.pl.* glass ~ γυαλικά *n.pl.* **~s** *s.* εμπορεύματα *n.pl.*

warehouse *s.* αποθήκη *f.*

warm *a.* ζεστός. the weather is ~ έχουμε ευχάριστα ζεστό καιρό. I feel ~ αισθάνομαι ζεστά. (*of feelings*) θερμός. (*v.t.*) (*also* ~ up) ζεσταίνω. (*v.i., fig.*) ~ up ζεσταίνομαι. **--blooded** *a.* θερμόαιμος. **--hearted** *a.* καλόκαρδος. **-ly** *adv.* (*dressed*) ζεστά, (*feeling*) θερμά.

warmth *s.* ζεστασιά *f.* (*feeling*) θερμότητα *f.*

warn *v.* προειδοποιώ. **-ing** *s.* προειδοποίηση *f.* (*lesson*) μάθημα *n.* (*a.*) προειδοποιητικός.

warp *v.t. & i.* (*bend*) σκεβρώνω. (*v.t., mentally*) διαστρεβλώνω. (*s.*) σκέβρωμα *n.* (*in weaving*) στημόνι *n.*

warrant *s.* (*order*) ένταλμα *n.* (*justification*) δικαιολογία *f.* (*guarantee*) εγγύηση *f.* (*v.*) εγγυώμαι. **--officer** *s.* (*naut.*) αρχικελευστής *m.* (*mil.*) ανθυπασπιστής *m.* (*aero.*) αρχισμηνίας *m.*

warren *s.* τόπος γεμάτος κουνελοφωλιές. (*fig.*) λαβύρινθος μικρών δρόμων (*or* διαδρόμων).

warrior *s.* πολεμιστής *m.*

wart *s.* κρεατοελιά *f.*

war|y *a.* προσεκτικός. be **~y** προσέχω. **-ily** *adv.* προσεκτικά. **-iness** *s.* επιφυλακτικότητα *f.*

was *v.* I ~ ήμουν, he ~ ήταν, I ~ to have gone ήταν να πάω.

wash *v.t.* πλένω. (*v.i.*) πλένομαι. ~ one's hair λούζω τα μαλλιά μου, λούζομαι. (*of sea ~ing shores*) (περι)βρέχω. it won't ~ δεν πλένεται, (*fig.*) δεν στέκεται. ~ one's hands of it νίπτω τας χείρας μου. ~ one's dirty linen in public βγάζω τα άπλυτά μου στη φόρα. ~ off *or* out *or* away (*remove*) καθαρίζω, βγάζω, (*come out*) καθαρίζω, βγαίνω. be ~ed away (*by waves*) παρασύρομαι. ~ed out (*person*) χλωρός. ~ down (*food*) κατεβάζω, (*car*) πλένω. ~ up (*cast ashore*) εκβράζω, (*v.i.*) πλένω τα πιάτα. ~ed up (*fig., person*) αποτυχημένος.

wash *s.* (*act of washing*) πλύσιμο *n.* have a

~ πλένομαι. give (sthg.) a ~ δίνω ένα πλύσιμο σε. my clothes are at (or in) the ~ τα ρούχα μου πλένονται. (of ship) κύματα προξενούμενα από διερχόμενο πλοίο.
washable a. be ~ πλένομαι.
washer s. (for screw, etc.) ροδέλλα f., λαστιχάκι n. **~-woman** s. πλύστρα f.
wash-house s. πλυσταριό n.
washing s. (act) πλύσιμο n. (linen for ~) ρούχα για πλύσιμο. (~ of linen) μπουγάδα f. it's ~ day έχουμε μπουγάδα. **~-machine** s. πλυντήριο ρούχων. **~-up** s. πλύσιμο πιάτων.
wash-out s. (fam.) φιάσκο n.
wash-tub s. σκάφη f.
washy a. νερουλός, (fig.) ανούσιος.
wasp s. σφήκα f. ~ waist πολύ λεπτή μέση. **~-ish** a. (person) τσούχτρα (f.), (remark) τσουχτερός.
wastage s. απώλεια f.
waste s. σπατάλη f., απώλεια f. ~ of time χάσιμο χρόνου. go to ~ πάω χαμένος. (refuse) σκουπίδια, απορρίμματα. n.pl. (barren region) έρημος f. ~ pipe σωλήν αποχετεύσεως.
waste a. (barren) έρημος, ~ land χέρσο έδαφος. (useless) άχρηστος, για πέταμα. lay ~ ερημώνω. **~-paper-basket** κάλαθος αχρήστων, (fam.) καλάθι n.
wast|e v.t. σπαταλώ, (time, opportunity) χάνω **~-e** one's breath χάνω τα λόγια μου. it's **~ed** on him πάει χαράμι. (v.i.) πάω χαμένος. **~-e** away φθίνω, λειώνω. **~-eful** a. σπάταλος.
wast|er, ~rel s. ανεπρόκοπος a.
watch s. (timepiece) ρολόι n. **~-maker** ρολογάς m.
watch s. (guard) βάρδια f. keep ~ κάνω βάρδια, αγρυπνώ. be on the ~ προσέχω, έχω τα μάτια μου τέσσερα.
watch v.t. (look at) κοιτάζω, (mind) προσέχω, (follow, observe) παρακολουθώ. (v.i.) ~ over φυλάω. ~ out προσέχω, έχω το νου μου. **~-dog** s. (fig.) κέρβερος m. **~-ful** a. άγρυπνος, προσεκτικός. **~-man** s. φύλακας m. (night) νυχτοφύλακας m. **~-tower** s. βίγλα f. **~-word** s. σύνθημα n.
water s. νερό n., ύδωρ n. in Greek ~s στα ελληνικά ύδατα. make ~ κάνω το νερό μου, (of ship) κάνω νερά. of the first ~ αρίστης ποιότητος, it won't hold ~ (fig.) δεν στέκεται. be in low ~ έχω αναπαραδιές. get into deep ~(s) τα βρίσκω μπαστούνια. get into hot ~ βρίσκω τον μπελά μου. pour cold ~ on υποδέχομαι χωρίς ενθουσιασμό.
water v.t. ποτίζω, (wine, milk, etc.) νερώνω. ~ down μετριάζω. (v.i.) (of eyes) δακρύζω.

my mouth ~s τρέχουν τα σάλια μου, (of ship, etc.) παίρνω νερό. **~-ed** a. (silk) μουαρέ, (milk) νερωμένος.
water-bottle s. (soldier's) παγούρι n.
water-cart s. (sprinkler) καταβρεχτήρας m. (for drinking) υδροφόρο βυτίο.
water-closet s. αποχωρητήριο n.
water-colour s. νερομπογιά f. (picture) ακουαρέλλα f.
water-conduit s. υδραγωγείο n.
water-cooled a. υδρόψυκτος.
watercourse s. (brook) ρυάκι n. (torrent) ρέμα n.
water-driven a. υδροκίνητος.
waterfall s. καταρράκτης m.
waterfront s. παραλία f.
water-ice s. γρανίτα f.
watering s. πότισμα n. **~-can** s. ποτιστήρι n. **~-place** s. (spa) λουτρόπολη f. (seaside) παραθαλάσσιο θέρετρο. (for beasts) ποτίστρα f.
water-key s. υδρονομεύς m.
waterless a. άνυδρος.
water-level s. υδροστάθμη f. ·
waterlily s. νούφαρο n.
waterline s. ίσαλος f.
waterlogged s. πλημμυρισμένος.
water-main s. κύριος υδραγωγός.
watermark s. φιλιγκράν, υδατόσημο n.
watermelon s. καρπούζι n.
water-mill s. νερόμυλος m.
water-nymph s. νηρηίς f.
waterproof a. αδιάβροχος.
water-pump s. υδραντλία f.
watershed s. υδροκρίτης m. (fig.) διαχωριστική γραμμή.
waterspout s. (at sea) σίφουνας m.
water-supply s. ύδρευση f.
watertight a. υδατοστεγής, στεγανός. (fig.) in ~ compartments χωριστά.
waterway s. πλωτή διόδος.
waterworks s. μηχανοστάσιον υδρεύσεως.
watery a. νερουλός, (eyes) που τρέχουν, (moon, etc.) που προμηνύει βροχή. ~ pap νερομπούλι n.
watt s. βατ n.
wattle s. πλέγμα από κλαδιά. ~ hut τσαρδάκι n.
wave s. κύμα n. (in hair) σκάλα f. heat ~ κύμα καύσωνος. (of hand) κούνημα n. **~-band** s. ζώνη συχνοτήτων. **~-length** s. μήκος κύματος.
wave v.t. κουνώ, ανεμίζω. (brandish) κραδαίνω. (v.i.) κυματίζω, ανεμίζω. (give sign) κάνω νόημα, γνέφω. he **~d** to us μας χαιρέτησε κουνώντας το χέρι.
waver v. (hesitate) αμφιταλαντεύομαι, (of courage) κλονίζομαι. (of flame) τρε-

μουλιάζω. ~er s. αναποφάσιστος a.

wavy v. κυματοειδής, (hair) σκαλωτός.

wax s. κερί n. (a.) κέρινος. (v.t.) κερώνω.

wax v.i. (grow) the moon is ~ing το φεγγάρι είναι στη γέμισή του. (become) γίνομαι.

way s. 1. (road, route) δρόμος m. ~ in είσοδος f. ~ out έξοδος f. this ~ από 'δω. a long ~ μακριά. half ~ στα μισά του δρόμου. on the ~ καθ' οδόν, (coming now) he's on the ~ έρχεται. all (along) the ~ καθ' όλη τη διαδρομή. go the right ~ πάω σωστά. by ~ of Athens μέσω Αθηνών. lead the ~ (lit. & fig.) προπορεύομαι. make one's ~ προχωρώ. out-of-the-~ (remote) απόμερος, (rare) σπάνιος. parting of the ~s σταυροδρόμι n. go one's own ~ κάνω του κεφαλιού μου. by the ~ (int.) αλήθεια. 2. (manner) τρόπος m. this ~ έτσι. the right ~ σωστά, καλά. the wrong ~ όχι σωστά, λάθος. in such a ~ as to έτσι που να. I did it the ~ you said το έκανα όπως μου είπες. the wrong ~ up (or round) ανάποδα. the right ~ up όρθιος. be in a bad ~ πάω άσχημα. get one's own ~ μου περνάει το δικό μου. have it your own ~ κάνε ό,τι θέλεις. 3. (habit) συνήθεια f. his little ~s οι ιδιοτροπίες του. mend one's ~s διορθώνομαι. 4. (point of view) in many ~s από πολλές απόψεις. 5. (impetus) gather ~ παίρνω φόρα. under ~ (ship) εν κινήσει, (work) υπό εκτέλεσιν. 6. (free passage) make ~ κάνω τόπο. give ~ (collapse, yield) υποχωρώ. get out of the ~ φεύγω από τη μέση. it's in the ~ εμποδίζει. right of ~ (passage) δουλεία οδού, (priority) προτεραιότητα f.

wayfarer s. οδοιπόρος m.

waylay v. στήνω καρτέρι σε.

wayside s. by the ~ κάθω στο δρόμο.

wayward a. πεισματάρης.

WC αποχωρητήριο n.

we pron. εμείς.

weak a. αδύνατος, (powerless) ανίσχυρος. (memory) ασθενής, (tea, coffee) ελαφρός, (argument) φτωχός. ~er sex ασθενές φύλο. (fig.) (person's) ~ point σφυγμός m. ~ly adv. (of will) αναποφάσιστα, (of voice) με αδύνατη φωνή. ~ness a. αδυναμία f. have a ~ness for έχω αδυναμία σε.

weaken v.t. εξασθενίζω. (v.i.) εξασθενώ, (in intensity) πέφτω. ~ing s. εξασθένιση f. πέσιμο n.

weak-kneed a. (fig.) δειλός.

weakling s. αδύναμος a. (fam., contemptuous) ψοφίμι n.

weal s. (mark) σημάδι από μαστίγωμα. (good) the general ~ το κοινό καλό.

wealth s. πλούτος m., πλούτη n.pl. ~y a. πλούσιος, πάμπλουτος.

wean v. αποκόβω.

weapon s. όπλο n.

wear 1. (v.t.) (clothes) φορώ. she ~s her hair in a bun κάνει (or έχει) τα μαλλιά της κότσο. (impair, also ~ away, down, out) τρώω, χαλώ, φθείρω, (resistance) κάμπτω. (make disappear) σβήνω, εξαλείφω. ~ out (tire) εξαντλώ. 2. (v.i.) (last) ~ well (of person) κρατιέμαι καλά, (of material) αντέχω. (become impaired, also ~ away, down, out) τρώγομαι, χαλώ, φθείρομαι, (of resistance) κάμπτομαι, (disappear) σβήνω, εξαλείφομαι. (continue) time ~s on τα χρόνια κυλάνε. (pass away) the pain wore off ο πόνος πέρασε. ~ing a. κουραστικός.

wear s. (clothing) winter ~ χειμωνιάτικα ρούχα. men's ~ ανδρικά είδη. (resistance) there's still some ~ left in this suit αυτό το κοστούμι φοριέται ακόμα. (use) the carpet gets a lot of hard ~ το χαλί πατιέται πολύ. (damage, also ~ and tear) φθορά. f. the worse for ~ στραπατσαρισμένος.

wear|y v.t. κουράζω. (v.i.) κουράζομαι, αποκάνω. (a.) κουρασμένος, αποκαμωμένος. ~iness s. κούραση f. ~isome a. κουραστικός.

weasel s. νυφίτσα f.

weather s. καιρός m. good ~ καλοκαιρία f. bad ~ κακοκαιρία f. in cold ~ όταν κάνει κρύο. under the ~ αδιάθετος. make heavy ~ of κάνω αδικαιολόγητη φασαρία για. (a.) ~ conditions καιρικές συνθήκες. ~ forecast πρόγνωση καιρού. ~ satellite μετεωρολογικός δορυφόρος. ~beaten a. ψημένος από τον ήλιο και τον αέρα. ~cock, ~vane s. ανεμοδείκτης m. ~proof a. στεγανός.

weather v.t. (come through) καβαντζάρω, ξεπερνώ. ~ed a. (worn) φαγωμένος, (faded) ξεθωριασμένος. (of wood) τραβηγμένος.

weav|e v.t. υφαίνω. (garland, basket, romance) πλέκω, (plot) εξυφαίνω. (v.i.) ~e in and out of ελίσσομαι μέσα από. (s.) ύφανση f. ~er s. υφαντής m. ~ing s. ύφανση f.

web s. ιστός m. (fig., network) δίκτυο n. ~-footed a. στεγανόπους.

webbing s. ενισχυτική ταινία για ταπετσαρία επίπλων και ζώνες.

wed v. t. (of priest) παντρεύω, (of spouse) παντρεύομαι. (fig., unite) ενώνω. be ~ded to (idea, etc.) είμαι προσηλωμένος (or προσκολλημένος) σε.

wedding s. γάμος m. silver ~ αργυροί

γάμοι. (a.) (present, ceremony) γαμήλιος.
~ ring βέρα f. ~ dress νυφικό φόρεμα.
wedge s. σφήνα f. (slice) φέτα f. (fig.) thin
end of the ~ απαρχή απαιτήσεων. (v.t.)
σφηνώνω.
wedlock s. γάμος m. born out of ~ εξώγαμος.
Wednesday s. Τετάρτη f. Ash ~ Καθαρά
Τετάρτη.
wee a. μικροσκοπικός. ~ bit λιγουλάκι.
weed s. αγριόχορτο n. ζιζάνιο n. (fam.,
tobacco) καπνός m. (person) see weedy.
(v.) ξεχορταριάζω. (fig.) ~ out ξεχωρίζω,
βγάζω. ~**y** a. (person) ψηλός και αδύνατος.
weeds s. widow's ~ πένθιμα ρούχα χηρείας.
week s. εβδομάδα f. today ~ σήμερα οκτώ.
on Monday ~ την άλλη Δευτέρα. ~**day** s.
καθημερινή f. ~**end** s. σαββατοκύριακο n.
~**ly** a. εβδομαδιαίος.
weep v. (also ~ for) κλαίω. ~**ing** s. κλάψιμο
n., κλάματα n.pl.
weigh v.t. (also ~ up, out) ζυγίζω. ~ down
(oppress) βαραίνω. ~ anchor σαλπάρω.
(v.i.) ζυγίζω. ~ on βαραίνω. it ~s with me
έχει σημασία για μένα. ~ in (at contest)
ζυγίζομαι, (in debate) ρίχνομαι στη
συζήτηση. ~**ing-machine** s. ζυγαριά f.
weight s. βάρος n. (gravity) βαρύτητα f.
(authority) κύρος n. under ~ (goods)
λιποβαρής, λειψός, (person) κάτω του
κανονικού βάρους. carry ~ βαρύνω. throw
one's ~ about κάνω τον καμπόσο. by ~ με
το ζύγι. ~s (for weighing) σταθμά, ζύγια
n.pl. (of clock, net, etc.) βαρίδια n.pl.
~**lessness** s. έλλειψις βαρύτητος.
weighty a. βαρύς, (momentous) βαρυσήμαντος, (authoritative) βαρύνων.
weir s. φράγμα n.
weird a. παράξενος, αλλόκοτος.
welcome s. καλωσόρισμα n. (reception)
υποδοχή f. (v.) καλωσορίζω. (receive)
υποδέχομαι. ~! καλώς ορίσατε! I ~ the
opportunity χαίρομαι που μου δίνεται η
ευκαιρία.
welcome a. ευπρόσδεκτος. you are ~ to use
my car έχεις το ελεύθερο (or μπορείς
ελεύθερα) να χρησιμοποιήσεις το αυτοκίνητό μου. (iron.) you are ~ to it να το
χαίρεσαι.
weld v. συγκολλώ. (fig.) συνδέω. ~**ing** s.
οξυγονοκόλλησις f.
welfare s. (good) καλόν n., συμφέρον n.
(social work) πρόνοια f. ~ state κράτος
προνοίας.
well s. (water, oil) πηγάδι n., (also of lift)
φρέαρ n. light-~ φωταγωγός m.
well v.i. αναβλύζω.

well adv. καλά, καλώς. pretty ~ (nearly)
σχεδόν. very ~ (agreement) καλά, σύμφωνοι, εντάξει. I am ~ είμαι καλά. all is ~
όλα είναι εντάξει. all being ~ αν όλα πάνε
βολικά. that's all very ~ (but...) καλά και
άγια όλα αυτά. do ~ πάω καλά, (make
good) προκόβω. do oneself ~ ζω καλά,
καλοπερνώ. do ~ out of κερδίζω από. do ~
by φέρομαι καλά σε. be ~ up in (know)
κατέχω καλά, είμαι μπασμένος σε. ~
before nightfall αρκετά πριν νυχτώσει. he
is ~ over forty είναι αρκετά πάνω από τα
σαράντα. as ~ (too) επίσης. as ~ as (and
also) καθώς και. he's keen on music as ~
as painting έχει πάθος με τη μουσική
καθώς και με τη ζωγραφική. it would be as
~ to... καλό θα ήταν να. it may ~ be that he
was away πιθανόν να έλειπε. it was just as
~ I didn't go καλά που δεν πήγα. you did ~
to come καλά έκανες και ήρθες. I couldn't
very ~ refuse δεν ήταν εύκολο να αρνηθώ.
we might (or may) as ~ stay here
μπορούμε (or θα μπορούσαμε) να μείνουμε εδώ. let ~ alone μη θίγετε τα καλώς κείμενα! ~ on (advanced) προχωρημένος.
well int. (resumptive, introductory) λοιπόν.
(surprised) ~! τι λες; ~ I never (did)! για
φαντάσου, μη μου πεις. (resigned, concessive) ε. (yes but) ναι, αλλά...
well-appointed a. καλοβαλμένος.
well-balanced a. ισορροπημένος.
well-behaved a. με καλούς τρόπους,
(animal) ήσυχος.
well-being s. (good) καλό n. sense of ~
ευεξία f.
well-born a. από σπίτι.
well-bred a. καλοαναθρεμμένος.
well-connected a. που ανήκει στον καλό
κόσμο.
well-disposed a. καλώς διατεθειμένος.
well-dressed a. καλοντυμένος.
well-educated a. μορφωμένος.
well-founded a. βάσιμος.
well-groomed a. κομψός, καλοβαλμένος.
well-heeled a. (fam.) παραλής.
well-informed a. μπασμένος, καλώς πληροφορημένος.
wellingtons s. ψηλές μπότες από καουτσούκ.
well-known a. πασίγνωστος.
well-made a. καλοφτιαγμένος.
well-marked a. σαφής, έντονος.
well-matched a. ταιριαστός.
well|-meaning, ~-meant a. καλοπρο-
αίρετος.
well-nigh adv. σχεδόν.
well-off a. (rich) εύπορος, he is ~ κρατιέται

καλά. (*fortunate*) τυχερός.
well-preserved *a.* καλοδιατηρημένος.
well-rounded *a.* τορνευτός.
well-shod *a.* καλοποδεμένος.
well-spoken *a.* ευγενικός στην ομιλία. ~ of με καλό όνομα.
well-timed *a.* επίκαιρος.
well-to-do *a.* ευκατάστατος.
well-tried *a.* δοκιμασμένος.
well-turned *a.* τορνευτός.
well-wisher *s.* υποστηρικτής *m.*, οπαδός *m.*
well-worn *a.* πολυχρησιμοποιημένος.
Welsh *a.* ουαλλικός. (*person*) Ουαλλός *m.*
welsh *v.i.*το σκάω χωρίς να πληρώσω. I ~ed on him του έβαλα φέσι.
Weltanschauung *s.* κοσμοθεωρία *f.*
welter *s.* σύμφυρμα *n.* (*v.*) κυλιέμαι.
wench *s.* κοπέλλα *f.* ~**ing** *s.* κοριτσοκυνήγι *n.*
wend *v.* ~ one's way τραβώ.
were *v.* you ~ (*sing.*) ήσουν, (*pl.*) ήσαστε. we ~ ήμασταν, ήμαστε. they ~ ήταν. ~ you to go αν πήγαινες.
west *s.* δύση *f.*, δυσμαί *f.pl.* to the ~ of Athens δυτικώς των Αθηνών.
west, ~erly, ~ern *a.* δυτικός. ~ wind πονέντες *m.*
west, ~ward(s) *adv.* δυτικώς. he's gone ~ (*fam.*) πάει αυτός.
westerniz|e *v.* εκδυτικοποιώ. ~**ation** *s.* εκδυτικοποίηση *f.*
wet *a.* (*liquid, moist*) υγρός, (*covered in water*) βρεμμένος, (*rainy*) βροχερός. get ~ βρέχομαι, μουσκεύω. (*fam., person*) ψόφιος. ~ blanket κάποιος ανάποδος που χαλάει τη συντροφιά. (*s.*) the ~ βροχή *f.* (*v.t.*) βρέχω, μουσκεύω. ~**nurse** *s.* παραμάννα *f.* ~**ting** *s.* βρέξιμο *n.*, μούσκεμα *n.*
whack *v.* κοπανάω. ~*ed* (*exhausted*) εξαντλημένος. (*s.*) (*also* ~ing) κοπάνισμα *n.* (*share*) μερίδιο *n.* get a ~ing τρώω ξύλο. have a ~ at δοκιμάζω.
whal|e *s.* φάλαινα *f.* (*fam.*) have a ~e of a time περνώ περίφημα. ~**ebone** *s.* μπανέλα *f.* ~**ing** *s.* φαλαινοθηρία *f.*
wharf *s.* αποβάθρα *f.* (*v.i.*) πλευρίζω.
what *a., pron. & adv.* 1. (*int.*) ~! τι; πώς; ~ a pity τι κρίμα. ~ a nice day τι ωραία μέρα. 2. (*interrogative*) τι; πώς; ~ news? τι νέα; ~ 's the matter?) τι έχεις; ~ for? γιατί; so ~? και; ~ of it? τι μ' αυτό; ~'s it like? πώς είναι; σαν τι μοιάζει; ~ book did you get? τι βιβλίο πήρες; ~ about a drink? τι θα λέγατε για ένα ποτό; ~ about your exams? (*how did you fare?*) πώς τα πήγες (*or* τι έγινε) με τις εξετάσεις σου; ~ does it matter? και τι πειράζει; ~ if (*supposing*) he didn't come? πες πως δεν ήρθε. ~ on

earth is that? τι είναι αυτό για το Θεό; I know ~ 's ~ ξέρω τι μου γίνεται. 3. (*relative*) αυτό που, ό,τι. ~ he told me αυτό που μου είπε. come ~ may ό,τι και να συμβεί. say ~ you will λέγε ό,τι θέλεις. ~ 's more επι πλέον, από πάνω. ~ with one thing and another με το ένα και το άλλο.
whatever *pron.* (*interrogative*) τι; ~ next! μη χειρότερα! ~ will he say next? να δούμε τι άλλο ακόμα θα μας πει. (*relative*) ό,τι. ~ you may say ό,τι και να πεις. ~ time you like ό,τι ώρα θέλεις. there is no doubt ~ δεν υπάρχει καμιά αμφιβολία. of ~ kind οιουδήποτε είδους.
whatsoever *pron.* οτιδήποτε. no person ~ κανένας απολύτως.
wheat *s.* σιτάρι *n.* ~**en** *a.* σιταρένιος.
wheedle *v.* καλοπιάνω.
wheel *s.* τροχός *m.*, ρόδα *f.* (*steering*) τιμόνι *n.* (*v.t.*) κυλώ, σπρώχνω. (*v.i.*) (*turn*) στρέφω, (*circle*) διαγράφω κύκλους. ~**ed** *a.* τροχοφόρος.
wheel|barrow, ~chair *s.* καρροτσάκι *n.*
wheeze *v.* ασθμαίνω. (*s.*) (*fam.*) κόλπο *n.* ~**y** *a.* ασθματικός.
whelp *s.* κουτάβι *n.*
when *adv.* (*interrogative*) πότε; ~ do they leave? πότε θα φύγουν; (*conj.*) σαν, όταν. ~ we left σαν (*or* όταν) φύγαμε. (*relative*) που. one day ~ I'm not busy μια μέρα που δεν είμαι απασχολημένος.
whence *adv.* απ' όπου, όθεν.
whenever *conj.* όποτε, κάθε φορά που.
where *adv.* πού; (*relative*) (εκεί) που. ~ are you going? πού πάς; leave it ~ I told you άφησέ το εκεί που σου είπα. is this ~ you live? εδώ μένετε; the district ~ I live η γειτονιά που μένω.
whereabouts *adv.* πού. (*s.*) do you know his ~? ξέρεις πού βρίσκεται;
whereas *conj.* ενώ.
whereat *adv.* επί του οποίου, οπότε.
wherefore *conj.* διατί. whys and ~s τα διατί και τα διότι.
wherein *adv.* εις τι; (*relative*) εις το οποίον.
whereof *adv.* περί του οποίου.
whereon *adv.* επί του οποίου.
whereupon *adv.* κατόπιν του οποίου, οπότε.
wherever *adv.* όπουδήποτε. ~ you (may) go όπου και να πας.
wherewithal *s.* τα μέσα, τα αναγκαία.
whet *v.* ακονίζω, (*appetite*) ανοίγω. ~**stone** *s.* ακόνι *n.*
whether *conj.* (*if*) αν. let me know ~ you're coming (or not) πες μου αν θα 'ρθεις (ή όχι). (*in alternatives*) ~ ... or... είτε... είτε. he's got to go ~ he likes it or not είτε το

θέλει είτε όχι (*or* θέλει δε θέλει) πρέπει να πάει. ~ it rains or not, I'm leaving tomorrow βρέχει ξεβρέχει φεύγω αύριο.

whey *s.* ορρός *m.*

which *pron.* (*interrogative*) ποιος; ~ book do you want? ποιο βιβλίο θέλεις; ~ church did you go to? σε ποια εκκλησία πήγατε; (*relative*) που, ο οποίος. the letter ~ got lost το γράμμα που χάθηκε. the ship in ~ we travelled το βαπόρι με το οποίο ταξιδέψαμε. I can't tell ~ is ~ δεν μπορώ να τα ξεχωρίσω.

whichever *pron.* όποιος. take ~ books you want πάρε όποια βιβλία θέλεις.

whiff *s.* (*smell*) μυρωδιά. *f.* (*at pipe, etc.*) ρουφηξιά. *f.* get a ~ of air παίρνω λίγο αέρα. I got a ~ of drains μύρισε βόθρος. (*v.*) ρουφώ.

while *conj.* (*for as long as*) όσον καιρό, καθ' όλο το διάστημα που. (*at the time when*) καθώς, ενώ, την ώρα που. (*whereas*) ενώ. let us settle it now ~ we have the chance ας το κανονίσουμε τώρα που έχουμε την ευκαιρία.

while *s.* καιρός *m.* for a ~ για κάμποσο καιρό. after a ~ σε λίγο. all the ~ όλο τον καιρό, όλη την ώρα. once in a ~ κάπου κάπου. it is not worth ~ δεν αξίζει τον κόπο. (*v.*) ~ away the time σκοτώνω την ώρα.

whim *s.* καπρίτσιο *n.*

whimper *v.* κλαψουρίζω. (*s.*) (*also* ~ing) κλαψούρα *f.*

whims|y *s.* ιδιορρυθμία. *f.* ~**ical** *a.* ιδιόρ-ρυθμος.

whin|e *v.* κλαψουρίζω. ~**ing** *s.* κλαψούρα *f.*, γκρίνια *f.* (*a.*) γκρινιάρης.

whinny *v.* χλιμιντρίζω.

whip *s.* μαστίγιο *n.*, καμ(ου)τσίκι *n.* (*fig.*) I have the ~ hand over him τον έχω του χεριού μου.

whip *v.* μαστιγώνω, δέρνω. (*defeat*) συν-τρίβω. (*eggs, cream*) χτυπώ. (*fam., steal*) βουτώ. ~ up (*arouse*) εξάπτω. ~ off, out, away βγάζω απότομα.

whipper-snapper *s.* (*fam.*) αναιδόμουτρο *n.*

whippet *s.* μικρό λαγωνικό.

whipping *s.* μαστίγωμα *n.* get a ~ τρώω ξύλο.

whirl *v.i.* (*rush*) τρέχω, ορμώ. (*v.t.*) παρα-σύρω. ~ round (*v.i.*) στροβιλίζομαι, στρι-φογυρίζω. (*s.*) a ~ of social events στρόβιλος κοινωνικών απασχολήσεων. my mind is in a ~ το κεφάλι μου γυρίζει.

whirligig *s.* (*top*) στρόβιλος *m.* (*at fair*) αλογάκια *n.pl.* ~ of time «του κύκλου τα γυρίσματα».

whirlpool *s.* ρουφήχτρα *f.*

whirlwind *s.* ανεμοστρόβιλος *m.*

whirr *s.* (*of machinery*) βούισμα *n.* (*of wings*) φτεροκόπημα *n.* (*v.i.*) βουίζω, περνώ φτεροκοπώντας.

whisk *v.* (*tail*) κουνώ, (*eggs*) χτυπώ. (*convey*) κουβαλώ αρον-άρον. (*s.*) (*move-ment*) κούνημα *n.* (*for eggs, cream*) χτυπητήρι *n.* (*for flies*) ξεμυγιαστήρι *n.*

whiskers *s.* (*men's*) φαβορίτες *f.pl.* (*beast's*) μουστακάκια *n.pl.*

whisk|y, ~ey *s.* ουίσκι *n.*

whisper *v.* ψιθυρίζω. (*s.*) ψίθυρος *m.* in a ~ ψιθυριστά. ~**ing** *s.* ψιθυρισμοί *m.pl.*

whistl|e *s.* σφυρίζω. (*fam.*) he can ~e for it (*money*) δεν θα πάρει φράγκο. (*s.*) (*instrument*) σφυρίχτρα *f.* (*vocal, also* ~ing) σφύριγμα *n.* wet one's ~e βρέχω το λαρύγγι μου.

whit *s.* not a ~ ούτε ίχνος, καθόλου.

white *a.* άσπρος. (*race, flag, corpuscle, night, House, Sea*) λευκός. (*pale*) χλωμός. turn ~ ασπρίζω, (*pale*) χλωμιάζω. all *or* very ~ κάτασπρος. ~ elephant (*fig.*) ογκώδες και άχρηστο απόκτημα. ~ heat λευκοπύρωση *f.* ~ horses (*waves*) αρνάκια *n.pl.* ~ lie αθώο ψέμα. W~ Paper λευκή βίβλος. ~ slave trade σωματεμπορία *f.* show the ~ feather (*fig.*) δειλιάζω. they bled him ~ (*fig.*) τον άφησαν πανί με πανί.

white *s.* (*person*) λευκός *m.* (*of eye, egg*) ασπράδι. *n.* she was dressed in ~ φορούσε άσπρα. ~**n** *v.t.* & *i.* ασπρίζω. ~**ness** *s.* ασπρίλα *f.*

whitebait *s.* μαρίδα *f.*

white-collar *a.* ~ workers υπαλληλική τάξη.

white-hot *a.* λευκοπυρωμένος.

whitewash *s.* ασβέστης *m.*, υδρόχρωμα *n.* (*v.*) ασβεστώνω, ασπρίζω. (*fig.*) he was ~ed τον εμφάνισαν ως αθώο καλύπτοντας την ενοχή του.

whither *adv.* πού; (*relative*) που.

whitlow *s.* παρωνυχίς *f.*

Whitsun *s.* Πεντηκοστή *f.*

whittle *v.* σκαλίζω. (*fig.*) ~ down περι-κόπτω.

whiz *v.* (*rush*) ορμώ, (*whistle*) σφυρίζω. ~ past περνώ σαν αστραπή. (*s.*) (*noise*) σφύριγμα *n.*

who *pron.* (*interrogative*) ποιος; (*relative*) ο οποίος, που.

whoever *pron.* (*interrogative*) ~ heard of such a thing? πού ακούστηκε ποτέ τέτοιο πράμα; (*relative*) όποιος, οποιος.

whole *a.* (*undamaged*) ακέραιος, γερός. (*full, complete*) ολόκληρος. (*attributive*) ολόκληρος, όλος. for a ~ week επί μία ολόκληρη εβδομάδα. he talked the ~ time μιλούσε όλη την ώρα. throughout the ~ house σ' όλο το σπίτι.

whole *s.* όλον *n.*, σύνολον *n.* the ~ of his fortune το σύνολον της περιουσίας του. as a ~ (*in one lot*) συνολικώς. on the ~ γενικά. **~heartedly** *adv.* ολοψύχως.

wholefood *a.* ολικής αλέσεως.

wholemeal *a.* από ατόφιο αγνό αλεύρι.

wholesale *adv.* χονδρικώς. (*fig.*) (*widely*) δεξιά και αριστερά. (*a.*) γενικός, μαζικός. **~r** *s.* χονδρέμπορος *m.*

wholesome *a.* υγιεινός, (*morally*) υγιής.

wholly *adv.* τελείως, απολύτως.

whom *pron.* (*interrogative*) ποιον; (*relative*) τον οποίον, που.

whoop *s.* κραυγή *f.* (*v.*) κραυγάζω. **~ing-cough** *s.* κοκκύτης *m.*

whoopee *s.* make ~ γλεντώ.

whop *v.* χτυπώ. (*s.*) χτύπημα *n.* **~per** *s.* (*fam.*) κάτι το πελώριο. **~ping** *a.* πελώριος.

whor|e *s.* πόρνη *f.* **~ing** *s.* πορνεία *f.*

whose *pron.* (*interrogative*) ποιανού; τίνος; (*relative*) του οποίου.

whosoever *pron.* οποιοσδήποτε.

why *adv.* γιατί; the reason ~ I didn't come ο λόγος που (*or* για τον οποίο) δεν ήρθα. (*int.*) έ!

wick *s.* φυτίλι *n.*

wicked *a.* κακός, κακοήθης. (*a sinner*) αμαρτωλός. **~ness** *s.* κακοήθεια *f.* (*sin*) αμαρτία *f.*

wicker *a.* ψάθινος.

wicket *s.* παραπόρτι *n.*

wide *a.* φαρδύς, πλατύς, (*esp. range, experience*) ευρύς. how ~ is it? τι φάρδος (*or* πλάτος) έχει; it is ten metres ~ έχει φάρδος δέκα μέτρων. (*great*) μεγάλος.

wide *adv.* far and ~ παντού. ~ open ορθάνοιχτος. ~ awake ξυπνητός, ξύπνιος, (*fig.*) ανοιχτομάτης. be ~ of the mark πέρτω έξω.

widely *adv.* ευρέως, (*greatly*) πολύ. ~ read (*newspaper*) με ευρεία κυκλοφορία, (*author*) που διαβάζεται πολύ, (*scholar*) πολύ διαβασμένος. ~ scattered houses αραιά σκορπισμένα σπίτια.

widen *v.t. & i.* φαρδαίνω, πλαταίνω. (*v.t.*) ευρύνω. **~ing** *s.* (*of garment*) φάρδεμα *n.* (*of road*) διεύρυνση *f.*

widespread *a.* διαδεδομένος, γενικός.

widow *s.* χήρα *f.* grass ~ γυναίκα της οποίας ο σύζυγος λείπει. (*v.*) be ~ed χηρεύω. **~er** *s.* χήρος *m.* **~hood** *s.* χηρεία *f.*

width *s.* φάρδος *n.*, πλάτος *n.*

wield *v.* (*pen, sword*) χειρίζομαι, (*sceptre*) κρατώ, (*authority*) ασκώ.

wife *s.* γυναίκα *f.* σύζυγος *f.*

wig *s.* περρούκα *f.*, φενάκη *f.*

wigging *s.* (*fam.*) κατσάδα *f.*, λούσιμο *n.*

wigwam *s.* σκηνή ερυθροδέρμων.

wild *a.* άγριος, (*land*) χέρσος. (*angry*) έξω φρενών. (*mad, preposterous*) εξωφρενικός, τρελλός. make *or* become ~ (*fierce*) αγριεύω. (*fam.*) be ~ about τρελλαίνομαι για. run ~ αποχαλινούμαι, (*of plants*) οργιάζω. **~ly** *adv.* (*fiercely*) άγρια, (*madly*) σαν τρελλός, (*at random*) στην τύχη, (*absolutely*) τελείως.

wildfire *s.* like ~ αστραπιαίως.

wild-goose *s.* go on a ~ chase κάνω άδικο κόπο επιδιώκοντας κάτι.

wilderness *s.* ερημιά *f.*

wilds *s.* άγριοι τόποι.

wiles *s.* τεχνάσματα *n.pl.* (*flattery*) γαλιφιές *f.pl.*

wilful *a.* (*intentional*) εκ προθέσεως, (*stubborn*) πεισματάρης.

will *s.* (*also* ~-power) θέλησις *f.* (*volition*) βούλησις *f.* (*of God, fate*) θέλημα *n.* free ~ ελευθέρα βούληση. good ~ αγαθές προθέσεις. ill ~ έχθρα *f.* at ~ κατά βούλησιν. against one's ~ με το ζόρι. of one's own free ~ με τη θέλησή μου, αυτοβούλως. (*do sthg.*) with a ~ με την καρδιά μου. strong ~ed ισχυράς θελήσεως. weak ~ed άβουλος.

will *s.* (*testament*) διαθήκη *f.*

will *v.* (*desire, ordain*) θέλω. (*auxiliary*) he ~ come θα έρθει. it won't be ready δεν θα είναι έτοιμο. the door won't shut η πόρτα δεν κλείνει. they ~ have arrived by now τώρα θα έχουν φτάσει πια. ~ you kindly tell me? μου λέτε παρακαλώ; (*insist*) he ~ have it that coffee is bad for you επιμένει ότι ο καφές σε βλάπτει. he ~ keep whistling όλο σφυρίζει. he won't give up smoking δεν εννοεί να κόψει το τσιγάρο.

willing *a.* πρόθυμος, (*disposed*) διατεθειμένος. **~ly** *adv.* πρόθυμα, (*without coercion*) αυτοβούλως. **~ness** *s.* προθυμία *f.*

will-o'-the-wisp *s.* χίμαιρα *f.*

willow *s.* ιτέα, ιτιά *f.* **~y** *a.* λυγερός.

willy-nilly *adv.* εκών άκων.

wilt *v.t.* μαραίνω. (*v.i.*) μαραίνομαι.

wily *a.* πανούργος, πονηρός.

win *v.i.* κερδίζω, νικώ. (*v.t.*) κερδίζω, (*prize*) παίρνω, (*glory, fame*) αποκτώ, (*hearts*) κατακτώ. ~ back ανακτώ. ~ over παίρνω με το μέρος μου. ~ through υπερνικώ τις δυσκολίες μου. (*s.*) νίκη *f.*, επιτυχία *f.*

wince *v.* συσπώ το πρόσωπό μου.

winch *s.* βαρούλκο *n.*, βίντσι *n.*

wind *s.* άνεμος *m.*, αέρας *m.* (*med.*) αέρια *n.pl.* break ~ αφήνω αέρια. get ~ of παίρνω χαμπάρι, μυρίζομαι. get the ~ up (*fam.*)

τα χρειάζομαι. put the ~ up (*person*)
τρομάζω. raise the ~ βρίσκω τα απαιτούμενα λεφτά. what's in the ~? τι μα-
γειρεύεται; get one's second ~ ξαναβρίσκω την αναπνοή μου. sail too close to
the ~ παρά λίγο να παραβώ τα όρια του
πρέποντος. (*a.*) ~ instrument πνευστό
όργανο. **~bag** *s.* πολυλογάς *m.* **~break** *s.*
ανεμοφράκτης *n.* **~fall** *s.* (*fig.*) απρο-
οδόκητο κέρδος. **~mill** *s.* ανεμόμυλος *m.*
~pipe *s.* τραχεία *f.* **~screen** *s.* παρμπρίζ
n. ~screen-wiper καθαριστήρας του τζα-
μιού. **~swept** *a.* ανεμόδαρτος. **~ward** *a.*
προσήνεμος. to ~ward προσνήεμως. **~y** *a.*
it is ~y φυσάει. (*person*) φλύαρος.

wind *v.t.* (*deprive of breath*) κόβω την
αναπνοή (*with gen.*). (*detect*) μυρίζομαι.

wind *v.t.* (*twist, wrap*) τυλίγω, (*into a ball*)
κουβαριάζω. (*turn*) γυρίζω. (*clock*)
κουρδίζω. ~ one's arms around σφίγγω
στην αγκαλιά μου. ~ one's way (*into*)
χώνομαι, γλιστρώ. ~ up (*clock*) κουρδίζω,
(*hoist*) σηκώνω, (*finish*) τερματίζω,
κλείνω, (*liquidate*) διαλύω. (*v.i.*) (*of road,
river, procession*) ελίσσομαι, (*of plant,
snake*) τυλίγομαι. ~ up (*end up*)
καταλήγω. get wound up κουρδίζομαι.
~ing *a.* φιδωτός. ~ing sheet σάβανο *n.*

windlass *s.* βαρούλκο *n.*

window *s.* παράθυρο *n.* (*of shop*) βιτρίνα *f.*
~-dressing *s.* (*fig.*) βιτρίνα *f.* **~-pane** *s.*
τζάμι *n.* **~-sill** *s.* περβάζι *n.*

wine *s.* κρασί *n.*, οίνος *m.* (*v.t.*) ~ and dine
περιποιούμαι. **~-cellar** *s.* κάβα *f.* **~-press**
s. ληνός *m.*, πατητήρι *n.* **~-skin** *s.* ασκί *n.*

wing *s.* (*of bird, car*) φτερό *n.* (*of plane,
army, building, party*) πτέρυξ *f.* **~s** (*theatre
& fig.*) παρασκήνια *n.pl.* on the ~ εν
πτήσει. under one's ~ υπό την προστασίαν
μου. take ~ κάνω φτερά. lend ~s to δίνω
φτερά σε. (*v.t.*) (*injure*) τραυματίζω
ελαφρά. ~ one's way πετώ. **~ed** *a.*
πτερωτός. **~less** *a.* άπτερος.

wink *v.i.* κλείνω (*or* κάνω) το μάτι. (*of
distant lights*) τρεμοσβήνω. ~ at it (*ignore
it*) κλείνω τα μάτια, κάνω στραβά μάτια.
(*fam.*) like ~ing στο πι και φι.

wink *s.* tip the ~ to προειδοποιώ. have forty
~s παίρνω έναν υπνάκο. I didn't sleep a ~
δεν έκλεισα μάτι.

winkle *v.t.* ~ out βγάζω, (*worm out*)
εκμαιεύω.

winner *s.* ο κερδισμένος, (*victory*) νικητής
m. (*in lottery*) ο τυχερός. (*horse*) γκανιάν
n.

winning *a.* (*engaging*) γοητευτικός. ~
number κερδίζων αριθμός. ~ team νική-
τρια ομάδα. **~-post** *s.* τέρμα *n.* **~s** *s.* κέρδη

n.pl.

winnow *v.* λιχνίζω. (*fig.*) ξεχωρίζω. **~ing** *s.*
λίχνισμα *n.* **~ings** *s.* απολιχνίδια *n.pl.*

winsome *a.* χαριτωμένος.

winter *s.* χειμώνας *m.* spend the ~ ξεχει-
μωνιάζω. (*a.*) χειμερινός, χειμωνιάτικος.
(*v.i.*) παραχειμάζω.

wintry *a.* χειμωνιάτικος, (*fig.*) ψυχρός.

wipe *v.* σκουπίζω, (*sponge*) σφουγγίζω. ~
away (*tears*) σκουπίζω. ~ off (*marks*)
βγάζω. ~ out (*efface*) σβήνω, εξαλείφω,
(*settle, kill*) καθαρίζω, (*army, etc.*)
εξολοθρεύω. ~ up καθαρίζω, (*dishes*)
σκουπίζω. (*s.*) σκούπισμα, σφούγγισμα *n.*

wire *s.* σύρμα *n.* (*telegram*) τηλεγράφημα *n.*
~ netting, barbed ~ συρματόπλεγμα *n.* live
~ (*fig.*) δυναμικόν πρόσωπον. pull the ~s
ενεργώ παρασκηνιακώς. ~ tapping υποκλοπή τηλεφωνικών συνδιαλέξεων. (*a.*)
συρμάτινος, από σύρμα.

wir|e *v.* (*install circuit in*) κάνω τη
συρμάτωση σε. (*fix.*) στερεώνω με σύρμα.
(*telegraph*) τηλεγραφώ. **~ing** *s.* συρμάτωση *f.*

wireless *s.* ασύρματος *m.* (*set*) ραδιόφωνο
n. (*a.*) ραδιοφωνικός.

wiry *a.* (*body*) νευρώδης.

wisdom *s.* σοφία *f.* (*good sense*) γνώση *f.*
~ **-tooth** φρονιμίτης *m.*

wise *s.* σοφός, (*sensible*) γνωστικός. I was
none the ~r δεν διαφωτίστηκα περισσότερο. get ~ to μαθαίνω. put (*person*) ~
(to) ενημερώνω (για). **~ly** *adv.* με σοφία,
σοφά.

wise *s.* (*way*) in this ~ έτσι. in no ~ καθόλου.

wise *adv.* (*suffix of manner, respect*) cross ~
σταυροειδώς, crab ~ σαν τον κάβουρα,
clock ~ δεξιόστροφα, money ~ από άποψη
χρημάτων.

wisecrack *s.* εξυπνάδα *f.*

wish *s.* (*expression of desire, prayer*) ευχή *f.*
(*want, desire*) επιθυμία *f.*

wish *v.* εύχομαι, (*want*) θέλω. ~ for θέλω,
επιθυμώ. I ~ I had seem him μακάρι (*or* θα
ήθελα) να τον είχα δει. I ~ I could play the
piano θα ήθελα να έπαιζα (*or* να μπορούσα να παίζω) πιάνο. I ~ you were not
so gullible θα επιθυμούσα να μην είσαι
τόσο εύπιστος. I ~ you success σου
εύχομαι καλή επιτυχία.

wishful *a.* be ~ θέλω. indulge in ~ thinking
τα βλέπω όπως τα θέλω.

wisp *s.* (*hair*) τσουλούφι *n.* (*smoke*) τολύπη
f. (*straw*) φούχτα *f.* ~ of a girl κοριτσούδι
n.

wistaria *s.* γλυσίνα *f.*

wistful *a.* μελαγχολικός.

wit *s.* (*sense*) μυαλό *n.* (*quickness of mind*)

ετοιμότητα *f.*, οξύνοια *f.* (*liveliness of spirit*) ευφυΐα *f.*, πνεύμα *n.* out of one's ~s τρελλός. at one's ~s' end εις απόγνωσιν. live by one's ~s ζω με κομπίνες. keep one's ~s about one έχω τα μάτια μου τέσσερα.

witch *s.* μάγισσα *f.* (*fig. charmer*) γόησσα *f.* **~craft, ~ery** *s.* μαγεία *f.* **~-doctor** *s.* μάγος *m.* **~-hunt** *s.* κακόβουλη δίωξη αντιφρονούντων.

with *prep.* με (*with acc.*), μετά (*with gen.*). (*in spite of*) παρά (*with acc.*). I went ~ him πήγα μαζί του. ~ difficulty μετά βίας. ~ pleasure μετά χαράς. ~ all his faults παρ' όλα τα ελαττώματά του. it rests ~ you εξαρτάται από σας. to cry out ~ pain ξεφωνίζω από πόνο. (*a room*) filled ~ people γεμάτο κόσμο. off ~ you! δίνε του! ~ that (*thereupon*) μετά το οποίο, οπότε. ~ child έγκυος.

withdraw *v.t.* αποσύρω, (*revoke*) ανακαλώ, (*take back*) παίρνω πίσω. (*v.i.*) αποσύρομαι, (απο)τραβιέμαι. **~al** *s.* (*of troops*) αποχώρηση *f.* (*of money*) ανάληψη *f.* **~n** *a.* αποτραβηγμένος.

wither *v.t.* μαραίνω. (*with scorn*) ξεραίνω, κατακεραυνώνω. (*v.i.*) μαραίνομαι, ξεραίνομαι. **~ed** *a.* μαραμένος, ξεραμένος. **~ing** *a.* (*glance, etc.*) κεραυνοβόλος.

withhold *v.* (*conceal*) κρύβω, αποκρύπτω. (*not grant*) αρνούμαι, (*not pay*) κρατώ. (*retain unlawfully*) κατακρατώ.

within *adv.* εντός, μέσα. (*prep.*) εντός (*with gen.*), μέσα σε. ~ one's income εντός των ορίων του εισοδήματός μου. ~ call εις απόστασιν φωνής.

without *adv.* έξω (*prep.*) (*not having*) χωρίς, δίχως (*with acc.*), άνευ (*with gen.*) (*outside*) εκτός (*with gen.*) do *or* go ~ κάνω χωρίς. ~ fail χωρίς (*or* δίχως) άλλο. ~ precedent άνευ προηγουμένου. ~ the walls εκτός των τειχών. ~ doubt αναμφιβόλως. ~ number αναρίθμητος. he left ~ saying a word έφυγε χωρίς να πει λέξη.

withstand *v.* αντέχω σε.

witless *a.* ανόητος.

witness *s.* (*person*) μάρτυς *m.* (*testimony*) μαρτυρία *f.* bear ~ to μαρτυρώ. bear false ~ ψευδομαρτυρώ.

witness *v.t.* (*show*) μαρτυρώ, (*see*) βλέπω. (*be present at*) παρίσταμαι εις, παρακολουθώ. (*a signature*) βεβαιώ. (*v.i.*) (*give evidence*) καταθέτω.

witticism *s.* ευφυολόγημα *n.*

wittingly *adv.* συνειδητώς.

witt|y *a.* ευφυής, πνευματώδης. **~ily** *adv.* ευφυώς, με πνεύμα.

wizard *s.* μάγος *m.* **~ry** *s.* μαγεία *f.*

wizened *a.* ζαρωμένος.

woad *s.* λουλάκι *n.*

wobbl|e *v.i.* κουνιέμαι. (*of voice, jelly*) τρεμουλιάζω. **~y** *a.* be ~y κουνιέμαι, τρέμω, δεν στηρίζομαι καλά.

woe *s.* θλίψη *f.*, δυστυχία *f.* ~s βάσανα *n.pl.* tale of ~ κατάλογος των βασάνων. ~ betide you! αλίμονό σου! **~begone** *a.* θλιμμένος.

woeful *a.* (*person*) θλιμμένος (*tale, ignorance*) θλιβερός. **~ly** *adv.* θλιβερά.

wold *s.* υψίπεδο *n.*

wolf *s.* λύκος *m.* she- ~ λύκαινα *f.* (*fig.*) keep the ~ from the door εξασφαλίζω τα προς το ζην. cry ~ σημαίνω κίνδυνο αδικαιολόγητος.

woman *s.* γυναίκα *f.* old ~ γριά *f.* old-~ish ιδιότροπος. ~ doctor γιατρίνα *f.* ~ driver οδηγός. *f.* women and children γυναικόπαιδα *n.pl.* women's γυναικείος. Women's Lib φεμινιστικό κίνημα. **~hood** *s.* she has reached ~hood έγινε πια γυναίκα. **~ish** *a.* (*man*) θηλυπρεπής. **~izer** *s.* γυναικάς *m.* **~ly** *a.* γυναικείος.

womb *s.* μήτρα *f.* (*ailment*) of the ~ μητρικός.

wonder *s.* (*feeling*) θαυμασμός *m.* (*marvel*) θαύμα *n.* no ~ he was annoyed δεν είναι περίεργο που ενοχλήθηκε. it's a ~ he wasn't injured είναι θαύμα πώς γλύτωσε. work ~s κάνω θαύματα. ~s will never cease! κάποιος φούρνος γκρέμισε!

wonder *v.* (*ask oneself*) αναρωτιέμαι, ας ήξερα. I ~ what he is up to αναρωτιέμαι (*or* ας ήξερα) τι μαγειρεύει. I ~ what we're having for supper θα ήθελα να ήξερα τι θα φάμε το βράδυ. ~ at (*be puzzled by*) θαυμάζω, απορώ με. I ~ at his indifference θαυμάζω (*or* απορώ με) την αδιαφορία του. can you ~ at it? είναι να σου φαίνεται περίεργο; I ~ he wasn't injured είναι άξιον απορίας πώς δεν τραυματίστηκε. I shouldn't ~ if he got married δεν θα μου φανεί περίεργο (*or* δεν θα παραξενευτώ) αν παντρευτεί.

wonderful *a.* θαυμάσιος. **~ly** *adv.* θαυμάσια.

wondering *a.*, **~ly** *adv.* με έκπληξη.

wonderland *s.* χώρα των θαυμάτων.

wonderment *s.* θαυμασμός *m.*, έκπληξη *f.*

wondrous *a.* θαυμάσιος.

wont *s.* συνήθεια *f.*

won't *v. see* will *v.*

woo *v.* κορτάρω (*fig.*) κυνηγώ. **~er** *s.* μνηστήρας *m.* **~ing** *s.* κόρτε *n.*

wood *s.* (*substance*) ξύλο *n.* (*timber*) ξυλεία *f.* (*pieces of ~*) ξύλα *n.pl.* (*tree-covered ground*) δάσος *n.* (*wine*) from the ~ του βαρελιού. (*fig.*) out of the ~ εκτός κινδύνου. touch ~! χτύπα ξύλο! (*a.*) ξύλινος, από ξύλο. **~bine** *s.* αγιόκλημα *n.* **~-carver** *s.* ξυλογλύπτης *m.* **~cock** *s.* μπεκάτσα *f.* **~cut** *s.* ξυλογράφημα *n.* **~cutter** *s.* ξυλοκόπος *m.* **~ed** *a.* δασωμένος. **~pecker** *s.* δρυοκολάπτης *m.* **~work** *s.* ξυλεία *f.* (*joinery*) ξυλουργική *f.*

wooden *a.* ξύλινος, από ξύλο. (*fig.*) δύσκαμπτος, αδέξιος. **~-headed** *a.* χοντροκέφαλος.

wool *s.* μαλλί *n.* (*fig.*) dyed in the ~ βαμμένος. pull the ~ over (*person's*) eyes κοροϊδεύω, ξεγελώ. be ~-gathering είμαι αφηρημένος. (*a.*) (*also* ~len) μάλλινος. ~lens μάλλινα *n.pl.*

woolly *a.* (*like wool*) σαν μαλλί, (*woollen*) μάλλινος. (*fig., unclear*) ασαφής.

word *s.* (*singly*) λέξη *f.* **~s** (*in connected speech*) λόγια *n.pl.* (*promise*) λόγος *m.* ~ of command διαταγή *f.* send ~ μηνώ. ~ for ~ λέξιν προς λέξιν. in a ~ με λίγα λόγια. in other ~s με άλλα λόγια. by ~ of mouth διά ζώσης. in the ~s of Plato κατά τον Πλάτωνα. they had ~s τα χόντρυναν. I want a ~ with you θέλω να σου πω δυο λόγια. I take your ~ for it σας πιστεύω. I take him at his ~ βασίζομαι στο λόγο του. be as good as one's ~ κρατώ το λόγο μου. ~s fail me δεν βρίσκω λόγια. he is too comic for ~s είναι τόσο κωμικός που δεν περιγράφεται. his ~ is law είναι νόμος η λέξη του και δεσμείν. he didn't say a ~ δεν έβγαλε λέξη. my ~! βρε βρε βρε! (*warning*) not a ~! μιλιά! μόκο! τσιμουδιά!

word *v.* διατυπώνω. **~ing** *s.* διατύπωση *f.*

word-perfect *a.* be ~ (*in role, lesson*) το ξέρω απ' έξω και ανακατωτά.

word-processor *s.* επεξεργαστής κειμένου.

wordy *a.* (*person*) περιττολόγος, (*text*) μακροσκελής.

work *s.* (*employment, what is done*) δουλειά *f.*, εργασία *f.* (*substantial achievement*) έργον *n.* (*things to be done*) δουλειές *f.pl.* it is the ~ of criminals είναι δουλειά εγκληματιών. (*book, painting, etc.*) έργον *n.* ~ of art έργον τέχνης. complete ~s of Solomos τα άπαντα του Σολωμού. ~ permit άδεια εργασίας. out of ~ άνεργος. make short ~ of καθαρίζω, κανονίζω. I had my ~ cut out to find it είδα κι έπαθα για να το βρω. he had set to ~ to... είχε βαλθεί να. *see also* works.

work *v.i.* (*do work*) δουλεύω, εργάζομαι. (*function, go*) πάω, δουλεύω, λειτουργώ. (*ferment*) βράζω. (*take effect, come off*) πιάνω. ~ together συνεργάζομαι. (*v.t.*) (*perform*) κάνω, (*operate*) χειρίζομαι,

δουλεύω. (*exploit*) εκμεταλλεύομαι. (*embroider*) κεντώ. (*fashion, knead*) δουλεύω. I ~ed it (*pulled it off*) το πέτυχα. ~ one's passage πληρώνω τα ναύλα μου δουλεύοντας. ~ oneself up (*get roused*) εξάπτομαι. he ~ed his way up from nothing κατάφερε να ανέβει από το τίποτα. ~ through (*penetrate*) διαπερνώ. ~ loose (*v.t. & i.*) λασκάρω.

work at *v.* καταπιάνομαι με.

work in (into) *v.t.* (*introduce*) εισάγω. (*v.i.*) (*get in*) μπαίνω, εισχωρώ. ~ with (*plans, etc.*) συνδυάζομαι.

work off *v.t.* (*get rid of*) απαλλάσσομαι από, ξεφορτώνομαι (*v.i., get detached*) βγαίνω.

work on *v.t.* (*influence*) επηρεάζω. ~ the assumption that βασίζομαι στο ότι...

work out *v.t.* (*details, plans*) επεξεργάζομαι. (*solve*) λύνω, (*find, think of*) βρίσκω, (*reckon*) υπολογίζω. (*v.i.*) (*turn out*) εξελίσσομαι, καταλήγω. it worked out well for me μου βγήκε σε καλό.

work-out *s.* (*test*) δοκιμή *f.* (*practice*) άσκηση *f.*

work up *v.t.* (*create*) δημιουργώ, (*arouse*) διεγείρω. (*prepare, fashion*) δουλεύω. get worked up εξάπτομαι. (*v.i.*) we're working up to an election πάμε για εκλογές. what is he working up to? πού θέλει να καταλήξει;

workable *a.* (*scheme*) πρακτικός, εφαρμόσιμος.

workaday *a.* καθημερινός.

worker *s.* εργάτης *m.* **~'s** εργατικός.

workhouse *s.* πτωχοκομείον *n.*

working *a.* (*person*) εργαζόμενος. ~ day εργάσιμος ημέρα, ~ hours ώρες εργασίας. ~ clothes ρούχα της δουλειάς. ~ party επιτροπή πραγματογνωμόνων. hard-~ εργατικός. ~ class εργατική τάξη. ~-class district εργατική συνοικία. ~-class mentality νοοτροπία ανθρώπου της εργατικής τάξεως. it is in ~ order λειτουργεί καλά.

working *s.* λειτουργία *f.* ~ out (*of plan, etc.*) επεξεργασία *f.*

workman *s.* εργάτης *m.* **~like** *a.* μαστορικός, καλός. **~ship** *s.* δουλειά *f.* (*of higher quality*) εργασία *f.*

works *s.* (*mechanism*) μηχανισμός *m.* (*factory*) εργοστάσιο *n.* (*public, construction, artistic, etc.*) έργα *n.pl.*

workshop *s.* εργαστήριο *n.*

work-shy *a.* φυγόπονος, τεμπέλης.

world *s.* κόσμος *m.* end of the ~ συντέλεια του κόσμου. on top of the ~ (*fig.*) κεφάτος, πανευτυχής. it did me a ~ of good με ωφέλησε αφάνταστα. what in the ~ does he want? τι στην οργή (*or* στην ευχή) θέλει; I wouldn't do that for the ~ για τίποτα στον κόσμο δεν θα το 'κανα αυτό. she thinks the ~ of him τον έχει σε πολύ μεγάλη υπόληψη. (*a.*) παγκόσμιος. ~ power μεγάλη δύναμις. **~wide** *a.* παγκόσμιος.

worldly *a.* (*not spiritual*) εγκόσμιος, του κόσμου. (*person*) ενδιαφερόμενος για τα του κόσμου. ~ interest ενδιαφέρον για τα υλικά αγαθά. ~ goods κοσμικά αγαθά.

worm *s.* σκουλήκι *n.* (*v.*) ~ one's way σέρνομαι, χώνομαι, τρυπώνω. ~ out (*secret, etc.*) εκμαιεύω. **~eaten** *a.* σκωληκόβρωτος, σκουληκιασμένος.

wormwood *s.* (*fig.*) φαρμάκι *n.*

worn *a.* (*worse for wear*) φθαρμένος, φαγωμένος. (*garment*) τριμμένος, (*face*) τραβηγμένος. ~out *a.* (*thing*) λειωμένος, (*idea*) ξεπερασμένος, (*person*) εξαντλημένος.

worr|y *s.* ανησυχία *f.*, έννοια *f.*, στενοχώρια, *f.* ~**ies** βάσανα *n.pl.*

worr|y *v.t.* ανησυχώ, στενοχωρώ. (*bother*) ενοχλώ. (*v.i.*) ανησυχώ, στενοχωριέμαι. ~**ied** *a.* ανήσυχος, στενοχωρημένος. ~**ying** *a.* ανησυχητικός, (*tiresome*) ενοχλητικός.

worse *a.* χειρότερος. get ~ χειροτερεύω, επιδεινώνομαι. the ~ for wear *see* worn. (*adv.*) χειρότερα. he is none the ~ for it δεν έπαθε τίποτα. be ~ off βρίσκομαι σε χειρότερη κατάσταση.

worsen *v.t. & i.* χειροτερεύω. ~**ing** *s.* χειροτέρεμα *n.*, επιδείνωση *f.*

worship *v.* λατρεύω. (*s.*) λατρεία *f.* ~**per** *s.* λάτρης *m.*

worst *a.* ο χειρότερος, χείριστος. if the ~ comes to the ~ στη χειρότερη περίπτωση. get the ~ of it ηττώμαι. do one's ~ κάνω ό,τι χειρότερο μπορώ. (*v.*) νικώ.

worsted *s.* κασμίρι από στριμμένο μαλλί.

worth *a.* be ~ αξίζω. it is ~ reading αξίζει να το διαβάσεις κανείς. it's ~ a lot έχει μεγάλη αξία. how much is he ~? τι περιουσία έχει; he pulled for all he was ~ τράβηξε με όλη του τη δύναμη.

worth *s.* αξία *f.* thirty pound's ~ of books βιβλία αξίας τριάντα λιρών. it wasn't a good money's ~ δεν άξιζε τα χρήματα που έδωσα. ~**less** *a.* άνευ αξίας, (*person*) ανάξιος.

worthwhile *a.* που αξίζει τον κόπο.

worth|y *a.* άξιος (*equal in* ~) αντάξιος. ~y of note αξιόλογος, αξιοσημείωτος. (*s.*) (*personage*) προεστώς *m.* ~**ily** *adv.* επαξίως. ~**iness** *s.* αξία *f.*

would *v.* ~ you like some more? θέλετε (*or* θα θέλατε) λίγο ακόμα; they ~ like to live

in the country θα τους άρεσε να ζουν στην εξοχή. I ~ rather not go out yet θα προτιμούσα να μη βγω ακόμα. he said he ~ come tomorrow είπε πως θα έρθει αύριο. he ~ insist επέμενε σώνει και καλά (*with* να, ότι). he ~ have a nap every day after lunch κάθε μέρα έπαιρνε έναν υπνάκο μετά το μεσημεριανό φαγητό. ~ that it were possible! μακάρι να γινόταν!

would-be *a.* υποψήφιος.

wound *s.* πληγή *f.* (*also fig.*) τραύμα *n.* (*v.*) πληγώνω, τραυματίζω. ~**ed** person τραυματίας *m.*

wraith *s.* φάντασμα *n.*

wrangle *v.* λογομαχώ. (*s.*) λογομαχία *f.*

wrap *v.* (*also* ~ up) τυλίγω, περιτυλίγω. (*cover*) σκεπάζω. ~ oneself up κουκουλώνομαι. ~ped up in (*absorbed*) αποροφημένος σε. (*s.*) (*rug, etc.*) σκέπασμα *n.* (*shawl*) σάλι *n.* ~**per** *s.* (*of book*) κάλυμμα *n.* (*of newspaper*) ταινία *f.* ~**ping** *s.* (*act*) τύλιγμα *n.* (*stuff*) περιτύλιγμα *n.*

wrath *s.* οργή *f.* ~**ful** *a.* οργισμένος.

wreak *v.* ~ havoc on καταστρέφω. ~ vengeance on εκδικούμαι. ~ one's anger on ξεσπάω το θυμό μου σε.

wreath *s.* στέφανος *m.* (*marriage*) στεφάνι *n.* (*pl.* στέφανα). (*funeral*) στεφάνι *n.* (*pl.* στεφάνια). (*of smoke*) τολύπη *f.*

wreathe *v.t.* (*wrap*) τυλίγω, (*cover*) σκεπάζω. ~ oneself (*round sthg.*) περιτυλίγομαι. ~**d** (*covered*) σκεπασμένος. (*with flowers*) στολισμένος. (*face*) ~d in smiles φωτισμένο με χαμόγελο. (*v.i., of smoke*) ανεβαίνω σε τολύπες.

wreck *s.* (*destruction, ~ed ship*) ναυάγιο *n.* (*vehicle, building, person*) ερείπιο *n.* (*the car*) was a complete ~ καταστράφηκε εντελώς. I feel an absolute ~ αισθάνομαι ράκος. (*v.*) ρημάζω, καταστρέφω. be ~ed (*of ship, enterprise*) ναυαγώ. ~**age** *s.* συντρίμματα *n.pl.* (*of building*) ερείπια *n.pl.*

wrench *v.* αποσπώ βιαίως, (*sprain*) στραμπουλίζω. (*s.*) απότομο τράβηγμα, στραμπούλισμα *n.* (*tool*) κάβουρας *m.* ~ of parting πόνος (*or* σπαραγμός) του χωρισμού.

wrest *v.* αρπάζω, (*also fig.*) αποσπώ.

wrestl|e *v.* παλεύω. ~**ler** *s.* παλαιστής *m.* ~**ling** *s.* πάλη *f.*

wretch *s.* (*scoundrel*) παλιάνθρωπος *m.* (*pitiful*) κακομοίρης *a.* (*fam.*) poor ~ φουκαράς *m.*

wretched *a.* (*unhappy*) δυστυχισμένος, (*to be pitied*) κακομοίρης. (*mean, poor*) άθλιος, ελεεινός, της κακιάς ώρας. be in a ~ state είμαι χάλια. (*fam., as expletive*)

(*person*) ευλογημένος, (*thing*) παλιο-. **~ly**
adv. δυστυχισμένα, ελεεινά. **~ness** *s.*
δυστυχία *f.* αθλιότητα *f.*

wriggle *v.t.* κουνώ, παίζω, (*v.i.*) (*of fish*)
σπαρταρώ, (*of worm*) συστρέφομαι, (*of
person*) κουνιέμαι. (*slip*) γλιστρώ, (*crawl*)
σέρνομαι. ~ out of ξεγλιστρώ από.

wrinkle *s.* ρυτίδα *f.*, ζάρα *f.* (*v.*) (*also get
~d*) ρυτιδώνω, ζαρώνω.

wring *v.* σφίγγω, (*extract*) αποσπώ. I'll ~ his
neck θα του στρίψω το λαρύγγι. ~ out
(*clothes*) στύβω. ~ing wet μουσκεμένος.
(*s.*) σφίξιμο *n.* στύψιμο *n.*

wrist *s.* καρπός *m.* **~band** *s.* μανικέτι *n.*
~watch *s.* ρολόι του χεριού.

writ *s.* ένταλμα *n.* Holy W~ Αγία Γραφή.

write *v.* γράφω, (*a learned work*) συγγράφω.
~r *s.* συγγραφεύς *m.*

write down *v.t.* (*note*) σημειώνω. (*fam.*,
consider) ~ as θεωρώ, παίρνω για.

write off *v.t.* ξεγράφω.

write-off *s.* (*fam.*) it is a ~ είναι για πέταμα.

write out *v.t.* γράφω, (*copy*) αντιγράφω. he
has written himself out εξόφλησε.

write up *v.t.* (*update*) ενημερώνω. (*praise*)
γράφω ευνοϊκή κριτική για. (*describe*)
κάνω γραπτή περιγραφή (*with gen.*).

write-up *s.* it got a good ~ είχε ευνοϊκή
κριτική.

writhe *v.i.* (*in coils*) συστρέφομαι, (*in pain*)
σπαράζω.

writing *s.* (*action*) γράψιμο *n.* (*form*)
γραφή. *f.* (*art, profession*) το συγγράφειν.
(*hand~*) γράψιμο *n.*, γραφή *f.* in ~ (*a.*)
γραπτός, (*adv.*) γραπτώς. this paper's got
~ on it αυτό το χαρτί είναι γραμμένο. ~ on
the wall (*fig.*) προειδοποίηση μελλο-
ντικής καταστροφής. ~s γραπτά *n.pl.*

written *a.* γραπτός, γραμμένος, έγγραφος
~ by hand χειρόγραφος.

wrong *a. & adv.* 1. (*out of order*) go ~ χαλώ.
there is something ~ with my watch κάτι
έχει το ρολόι μου. there's nothing ~ with
her (*in health*) δεν έχει (*or* δεν της
συμβαίνει) τίποτα. 2. (*morally*) κακός,
(*unjust*) άδικος. be ~ (*of person*) έχω
άδικο. it is ~ to tell lies είναι κακό (*or* δεν
είναι σωστό) να λέει ψέματα κανείς. 3.
(*erroneous*) λανθασμένος. be ~ (*of person*)
δεν έχω δίκιο, έχω λάθος. go ~ (*of plans*)
πηγαίνω στραβά, αποτυχαίνω. the ~ way
round ανάποδα. it's the ~ colour δεν είναι
το χρώμα που πρέπει. I got the ~ number
(*on phone*) πήρα λάθος. I took the ~ bus
πήρα λάθος λεωφορείο. it's the ~ address
δεν είναι η σωστή διεύθυνση. it's in the ~
place είναι σε λανθασμένη θέση. it has
come at the ~ time δεν ήρθε όταν έπρεπε.

that is where we went ~ εκεί κάναμε λάθος
(*or* πήγαμε στραβά). get hold of the ~ end
of the stick το παίρνω στραβά. get caught
on the ~ foot καταλαμβάνομαι ανέτοιμος.
~ side (*of cloth*) ανάποδη *f.*

wrong *s.* κακό *n.*, άδικο *n.* (*injustice*) αδικία
f. do (*person*) ~ αδικώ. be in the ~
σφάλλω, βρίσκομαι εν αδίκω. (*v.t.*) αδικώ.

wrong-do|er *s.* παραβάτης του νόμου
(*legally*) *or* των ηθικών αρχών (*morally*).
~ing *s.* παράβαση *etc.* (*as above*).

wrongful *a.* παράνομος. **~ly** *adv.* παρανόμως.

wrong-headed *a.* ανάποδος.

wrongly *adv.* κακά, στραβά. (*unjustly*)
άδικα. (*mistakenly*) εσφαλμένως, (*fam.*)
λάθος.

wroth *a.* εξοργισμένος.

wrought *a.* (*metal*) σφυρήλατος. ~ up
αναστατωμένος.

wry *a.* (*crooked*) στραβός, (*ironic*) ειρωνικός. make a ~ face στραβομου-
τσουνιάζω, στραβώνω τα μούτρα μου. ~
smile βεβιασμένο χαμόγελο. **~ly** *adv.*
ειρωνικά.

X

xenophob|ia *s.* ξενοφοβία *f.* **~ic** *a.* be ~ic
έχω ξενοφοβία.

xerox *v.t.* βγάζω φωτοαντίγραφο από.

X-ray|s *s.* ακτίνες X. ~ photograph(y)
ακτινογραφία *f.* (*v.*) ~ ακτινοσκοπώ. get
~ed κάνω ακτινοσκόπηση.

xylophone *s.* ξυλόφωνο *n.*

Y

yacht *s.* θαλαμηγός *f.*, γιωτ *n.* **~club**
ναυτικός όμιλος. **~ing** *s.* ιστιοπλοΐα *f.*

yahoo *s.* κτηνώδης άνθρωπος.

yank *v.t.* τραβώ απότομα.

yap *v.* γαυγίζω, (*fig.*) φλυαρώ. **~ping** *s.*
γαύγισμα *n.* φλυαρία *f.*

yard *s.* (*measure*) γυάρδα *f.* (*naut.*) κεραία *f.*
~stick *s.* μέτρο *n.*

yard *s.* (*of building*) αυλή *f.* (*storage,
builder's*) μάντρα *f.* (*USA*) κήπος *m.*

yarn *s.* (*thread*) νήμα *n.* (*story*) ιστορία *f.* (*v.*)
λέω ιστορίες.

yashmak *s.* γιασμάκι *n.*

yawn *v.* χασμουριέμαι, (*fig., gape open*)
χαίνω, χάσκω. (*s.*) (*also* ~ing) χασμού-
ρημα, χασμουρητό *n.*

ye *pron.* εσείς.

yea *adv.* ναι. ~s and nays οι υπέρ και οι κατά.

year *s.* χρόνος *m.* (*pl.* χρόνια *n.*), έτος *n.* this ~ φέτος, this ~'s φετεινός. last ~ πέρ(υ)σι, last ~'s περσινός. next ~ του χρόνου, next ~'s του ερχόμενου χρόνου. the ~ before last πρόπερσι. ~s ago προ χρόνων. ten ~s old δέκα χρονών (*or* ετών). ~ in ~ out χρόνος μπαίνει χρόνος βγαίνει. the school ~ σχολικόν έτος. a ~'s pay χρονιάτικο *n.* New Y~'s Day Πρωτοχρονιά. *f.* this was a good ~ for peaches η φετινή ήταν μια καλή χρονιά για τα ροδάκινα. the best pupil of his ~ ο καλύτερος μαθητής της χρονιάς του. ~**book** *s.* επετηρίδα *f.* ~**ling** *s.* ~ling colt πουλάρι ενός έτους.

yearly *a.* (*also* lasting one year) ετήσιος, ενιαύσιος. (*adv.*) ετησίως, κάθε χρόνο.

yearn *v.* (*also* ~ for) λαχταρώ. ~**ing** *s.* λαχτάρα *f.* (*regret*) καημός *m.*, μεράκι *n.*

yeast *s.* μαγιά *f.*

yell *v.t. & i.* φωνάζω. (*v.i.*) ξεφωνίζω. (*s.*) ξεφωνητό *n.*

yellow *a.* κίτρινος, (*cowardly*) δειλός. turn ~ κιτρινίζω. ~**ish** *a.* κιτρινωπός.

yelp *v.* ουρλιάζω. (*s.*) ούρλιασμα *n.*

yeoman *s.* αγρότης μικροκτηματίας. (*fig.*) ~ service πολύτιμη βοήθεια σε ώρα ανάγκης.

yes *adv.* ναι, μάλιστα. (*strongly affirmative*) αμέ, αμ' πώς! (*esp. as response to neg. question*) πώς.

yesterday *adv.* χθες, χτες, ~ morning χθες το πρωί ~ week σα χθες οκτώ. ~'s χθεσινός.

yet *adv.* (*still*) ακόμα, not ~ όχι ακόμα. have you been paid ~? πληρώθηκες ή ακόμα; as ~ (*up to now*) ως (*or* μέχρι) τώρα, για την ώρα. (*up to then*) ως (*or* μέχρι) τότε. and ~ (*nevertheless*) κι όμως. nor ~ ούτε και. (*conj.*) αλλά.

yew *s.* τάξος *f.*

YHA ξενώνες νέων.

yield *v.t.* (*produce*) παράγω, αποδίδω, βγάζω. (*surrender*) παραδίδω. (*v.i.*) παραδίδομαι. (*give way*) υποχωρώ, ενδίδω. (*s.*) απόδοση *f.* ~**ing** *a.* (*person*) υποχωρητικός, (*substance*) μαλακός, εύκαμπτος.

yob *s.* κτηνώδης νέος.

yoga *s.* γιόγκα *f.*

yog(h)urt *s.* γιαούρτι *n.*

yoke *s.* ζυγός *m.* (*of oxen*) ζευγάρι *n.* (*in dressmaking*) νωμίτης *m.* (*v.t.*) ζεύω. ~**d** ζεμένος.

yokel *s.* βλάχος *m.*

yolk *s.* κρόκος *m.*

yon, ~der *adv.* εκεί πέρα. (*a.*) εκείνος.

yore *s.* in days of ~ τον παλιό καιρό.

you *pron.* (*subject*) (ε)σύ *sing.* (ε)σείς *pl.* (*object*) (ε)σένα, (ε)σάς. (*unemphatic object*) σε, σας, (*indirect*) σου, σας.

young *a.* νέος, νεαρός, μικρός ~ man έφηβος, νέος *m.*, παλληκάρι *n.* ~ woman νεαρή γυναίκα. ~ lady δεσποινίδα *f.* ~ people νέοι *m.pl.*, η νεολαία. my ~er brother ο μικρός μου αδελφός. (*s.*) (*offspring*) μικρά *n.pl.* with ~ έγκυος. ~**ster** *s.* παιδί *n.*

your *pron.* ~ house το σπίτι σου (*sing.*), σας (*pl.*). ~ own, ~s δικός σου/σας ~**self** *see* self.

youth *s.* (*state*) νεότης *f.* νιάτα *n.pl.* (*young people*) νεολαία *f.*, νιάτα *n.pl.* (*lad*) νεανίας, νέος *m.*

youthful *a.* (*person*) νέος, (*appearance*) νεανικός. ~**ness** *s.* νεανικότητα *f.*

yule, ~tide *s.* Χριστούγεννα *n.pl.*

Z

zany *a.* παλαβός.

zeal *s.* ζήλος *m.* ~**ot** *s.* φανατικός *a.*

zealous *a.* ένθερμος, γεμάτος ζήλο. ~**ly** *adv.* με ζήλο.

zebra *s.* ζέβρα *f.* ~ crossing διάβαση πεζών.

zenith *s.* ζενίθ *n.*

zephyr *s.* ζέφυρος *m.*

zero *s.* μηδέν *n.* below ~ κάτω του μηδενός. ~ hour ώρα μηδέν.

zest *s.* ενθουσιασμός *m.* (*gusto*) κέφι *n.* (*piquancy*) νοστιμάδα *f.* (*appetite*) όρεξη *f.*

zeugma *s.* ζεύγμα *n.*

zig-zag *s.* ζιγκ-ζαγκ *n.* (*v.*) κάνω ζιγκ-ζαγκ ~ down κατεβαίνω με ζιγκ-ζαγκ.

zinc *s.* ψευδάργυρος *m.*, τσίγκος *m.*

zionism *s.* σιωνισμός *m.*

zip *s.* (*noise*) σφύριγμα *n.* (*energy*) ενεργητικότητα *f.* (*v.t.*) ~ it up (*dress, etc.*) κλείνω με φερμουάρ. (*v.i.*) ~ past περνώ σαν σφαίρα. ~**-code** *s.* ταχυδρομικός τομεύς. ~**-fastener** *s.* φερμουάρ, εκλέρ *n.*

zither *s.* είδος σαντουριού.

zodiac *s.* ζωδιακός κύκλος. sign of the ~ ζώδιον *n.*

zone *s.* ζώνη *f.* (*v.*) χωρίζω σε ζώνες.

zoo *s.* ζωολογικός κήπος.

zoolog|y *s.* ζωολογία *f.* ~**ical** *a.* ζωολογικός. ~**ist** *s.* ζωολόγος *m.*

zoom *v.i.* (*rise*) υψώνομαι απότομα. (*buzz*) βουίζω.

Mars Άρης	**Priam** Πρίαμος
Martha Μάρθα	**Prometheus** Προμηθεύς
Mary Μαρία, (*fam.*) Μάρω, Μαριγώ	**Psyche** Ψυχή
Matilda Ματθίλδη	**Ptolemy** Πτολεμαίος
Matthew Ματθαίος	**Pyrrhus** Πύρρος
Medea Μήδεια	**Pythagoras** Πυθαγόρας
Menelaus Μενέλαος	
Mercury Ερμής	**Rachel** Ραχήλ
Michael Μιχαήλ, Μιχάλης (*fam.*) Μίχος	**Racine** Ρακίνας
Midas Μίδας	**Raphael** Ραφαήλ
Minerva Αθηνά	**Rebecca** Ρεβέκκα
Minos Μίνως	**Richard** Ριχάρδος
Moses Μωυσής	**Robert** Ροβέρτος
	Roland Ρολάνδος
Nadia Νάντια	**Rousseau** Ρουσσώ
Napoleon Ναπολέων	**Rudolph** Ροδόλφος
Narcissus Νάρκισσος	**Ruth** Ρουθ
Natalie Ναταλία	
Nathaniel Ναθαναήλ	**Samson** Σαμψών
Nausicaa Ναυσικά	**Samuel** Σαμουήλ
Neptune Ποσειδών	**Sappho** Σαπφώ
Nero Νέρων	**Sarah** Σάρρα
Nestor Νέστωρ	**Saturn** Ουρανός
Newton Νεύτων	**Saul** Σαούλ
Nicholas Νικόλαος, (*fam.*) Νίκος	**Scipio** Σκιπίων
Nimrod Νεμρώδ	**Scylla** Σκύλλα
Niobe Νιόβη	**Sebastian** Σεβαστιανός
Noah Νώε	**Sergius** Σέργιος
	Shakespeare Σαίξπηρ
Octavius Οκτάβιος	**Shaw** Σω
Odysseus Οδυσσεύς	**Simon** Σίμων, Σίμος
Oedipus Οιδίπους	**Socrates** Σωκράτης
Olga Όλγα	**Solomon** Σολομών
Orpheus Ορφεύς	**Sophia** Σοφία
Otto Όθων	**Sophocles** Σοφοκλής
Ovid Οβίδιος	**Spyridion** Σπυρίδων
	Stella Στέλλα
Pallas (*Athene*) Παλλάς	**Stephen** Στέφανος
Pan Παν	**Sulla** Σύλλας
Patroclus Πάτροκλος	**Susanna** Σουζάννα
Paul Παύλος	
Pausanias Παυσανίας	**Tantalus** Τάνταλος
Penelope Πηνελόπη	**Telemachus** Τηλέμαχος
Pericles Περικλής	**Thales** Θαλής
Persephone Περσεφόνη	**Themistocles** Θεμιστοκλής
Perseus Περσεύς	**Theocritus** Θεόκριτος
Peter Πέτρος	**Theodora** Θεοδώρα
Pharaoh Φαραώ	**Theodore** Θεόδωρος
Phidias Φειδίας	**Theresa** Θηρεσία
Philip Φίλιππος	**Theseus** Θησεύς
Phoebe Φοίβη	**Thetis** Θέτις
Phoebus Φοίβος	**Thomas** Θωμάς
Pilate Πιλάτος	**Thucydides** Θουκυδίδης
Plato Πλάτων	
Pluto Πλούτων, Άδης	**Ulysses** Οδυσσεύς
Polyphemus Πολύφημος	**Uranus** Ουρανός
Poseidon Ποσειδών	
Praxiteles Πραξιτέλης	**Venus** Αφροδίτη

Vespasian Βεσπασιανός
Vesta Εστία
Victor Βίκτωρ
Victoria Βικτωρία
Vincent Βικέντιος
Virgil Βιργίλιος
Voltaire Βολταίρος
Vulcan Ἥφαιστος

William Γουλιέλμος

Xenophon Ξενοφών
Xerxes Ξέρξης

Zeus Ζευς, Δίας
Zoe Ζωή

APPENDIX III

SELECT LIST OF PRINCIPLE PARTS OF VERBS

(I) = Imperfect (P) = Present

Present	Aorist active	Aorist passive	Past participle
αγαπίζω	αγάπισα		
αγαπώ	αγάπησα	αγαπήθηκα	αγαπημένος
αγγέλλω	άγγειλα	αγγέλθηκα	αγγελμένος
αγγίζω	άγγιξα	αγγίχτηκα	αγγιγμένος
αγνοώ	αγνόησα	αγνοήθηκα	αγνοημένος
αγοράζω	αγόρασα	αγοράστηκα	αγορασμένος
αγορεύω	αγόρευσα		
άγω	ήγαγον	ήχθην	
αγωνίζομαι		αγωνίστηκα	
αγωνιώ	αγωνιούσα (I)		
αδημονώ	αδημονούσα (I)		
αδυνατίζω	αδυνάτισα		αδυνατισμένος
αδυνατώ	αδυνατούσα (I)		
αηδιάζω	αηδίασα		αηδιασμένος
αιμάσσω	ήμαξα		ηματωμένος
αίρω	ήρα	ήρθην	ηρμένος
αισθάνομαι		αισθάνθηκα	
ακούω	άκουσα	ακούστηκα	ακουσμένος
αλέθω	άλεσα	αλέστηκα	αλεσμένος
αλείβω	άλειψα	αλείφτηκα	αλειμμένος
αλλάζω	άλλαξα	αλλάχτηκα	αλλαγμένος
αμείβω	άμειψα	αμείφτηκα	
αμφιβάλλω	αμφέβαλλα (I)		
αναβάλλω	ανέβαλα	αναβλήθηκα	αναβεβλημένος
ανάβω	άναψα		αναμμένος
αναγαλλιάζω	αναγάλλιασα		αναγαλλιασμένος
αναδίνω	ανάδωσα		
αναδύομαι		ανεδύθην	
αναθέτω	ανέθεσα	ανετέθην	
αναιρώ	αναίρεσα	αναιρέθηκα	αναιρεμένος
ανακατεύω	ανακάτεψα	ανακατεύθηκα	ανακατεμένος
ανακατώνω	ανακάτωσα	ανακατώθηκα	ανακατωμένος
ανακοινώ	ανακοίνωσα	ανακοινώθηκα	
ανακρίνω	ανέκρινα	ανακρίθηκα	
ανακτώ	ανέκτησα	ανακτήθηκα	ανακτημένος
ανακύπτω	ανέκυψα		
αναμιγνύω	ανέμιξα	ανεμίχθην	αναμεμιγμένος
αναξαίνω	ανέξανα		αναξασμένος
αναπνέω	ανέπνευσα		
αναπτύσσω	ανέπτυξα		ανεπτυγμένος
αναρριχώμαι		αναρριχήθηκα	
αναρωτιέμαι		αναρωτήθηκα	
ανασαίνω	ανάσανα		
ανατέλλω	ανέτειλα		
ανεβάζω	ανέβασα		ανεβασμένος

Present	Aorist active	Aorist passive	Past participle
ανεβαίνω	ανέβηκα		
ανελκύω	ανέλκυσα		
ανέρχομαι		ανήλθον	
ανέχομαι		ανέχτηκα	
ανήκω	ανήκα (I)		
ανθίσταμαι		αντέστην	
ανοίγω	άνοιξα	ανοίχτηκα	ανοιγμένος
αντέχω	άντεχα/άνθεξα		
αντιδρώ	αντέδρασα		
αντιθέτω	αντέθεσα	αντετέθην	
αντιλαμβάνομαι		αντιλήφθηκα	
αντιτείνω	αντέτεινα		
αντιτίθεμαι		αντετέθην	
απελαύνω	απήλασα	απηλάθην	
απεχθάνομαι		απεχθανόμουν (I)	
αποβάλλω	απέβαλα	αποβλήθηκα	αποβλημένος
απογίνομαι		απέγινα	
αποθαρρύνω	αποθάρρυνσα	αποθαρρύνθηκα	αποθαρρημένος
αποκάνω	απέκαμα		αποκαμωμένος
αποκρίνομαι		αποκρίθηκα	
αποκτώ	απέκτησα	αποχτήθηκα	αποκτηθείς
απολαμβάνω	απέλαβον		
απολαύω	απόλαψα		
απομένω	απόμεινα		
απονέμω	απένειμα	απενεμήθην	
αποπλέω	απέπλευσα		
απορρέω	απέρρευσα		
απορρίπτω	απέρριψα	απερρίφθην	
απορρίχνω	απόρριξα		
αποσυντίθεμαι		αποσυντέθηκα	αποσυντεθειμένος
αποσύρω	απόσυρα	αποσύρθηκα	αποσυρμένος
αποτελώ	αποτέλεσα	αποτελέστηκα	αποτελούμενος (P)
αποτίω	απέτισα		
αποτυχαίνω	απέτυχα		αποτυχημένος
αποφέρω	απέφερα		
αρέσω	άρεσα		
αρκώ	ήρκεσα	ηρκέσθην	
αρμόζω	άρμοζα (I)		
αρπάζω	άρπαξα	αρπάχτηκα	αρπαγμένος
αρρωσταίνω	αρρώστησα		αρρωστημένος
αρχίζω	άρχισα		αρχινισμένος
αυξάνω	αύξησα	αυξήθηκα	αυξημένος
αφαιρώ	αφαίρεσα	αφαιρέθηκα	αφαιρεμένος
αφήνω	άφησα/άφηκα	αφέθηκα	αφημένος
αφίσταμαι		απέστην	
αφοσιούμαι		αφοσιώθηκα	αφοσιωμένος
βάζω	έβαλα	βάλθηκα	βαλμένος
βαίνω	έβην		
βάλλω	έβαλα	εβλήθην	βεβλημένος
βαραίνω	βάρυνα		βαρεμένος
βαριέμαι		βαρέθηκα	βαριεστισμένος
βαρύνω	εβάρυνα	εβαρύνθην	βεβαρημένος
βαρώ	βάρεσα	βαρέθηκα	βαρεμένος
βαστώ	βάσταξα	βαστάχτηκα	βασταγμένος

Present	Aorist active	Aorist passive	Past participle
βάφω	έβαψα	βάφ(τ)ηκα	βαμμένος
βγάζω	έβγαλα	βγάλθηκα	βγαλμένος
βγαίνω	βγήκα		βγαλμένος
βλάπτω	έβλαψα	βλάφτηκα	βλαμμένος
βλέπω	είδα	ειδώθηκα	ιδωμένος
βουτώ	βούτηξα	βουτήχτηκα	βουτηγμένος
βρέχω	έβρεξα	βράχηκα	βρεγμένος
βρίσκω	βρήκα	βρέθηκα	βρισκόμενος (P)
βρυχώμαι		βρυχήθηκα	
γδέρνω	έγδαρα	γδάρθηκα	γδαρμένος
γδύνω	έγδυσα	γδύθηκα	γδυμένος
γελώ	γέλασα	γελάστηκα	γελασμένος
γέρνω	έγειρα		γερμένος
γερνώ	γέρασα		γερασμένος
γίνομαι		έγινα	γινωμένος
γράφω	έγραψα	γράφ(τ)ηκα	γραμμένος
δεικνύω	έδειξα	εδείχθην	δεδειγμένος
δείχνω	έδειξα	δείχτηκα	δειγμένος
δέρνω	έδειρα	δάρθηκα	δαρμένος
δέχομαι		εδέχθην/δέχτηκα	δεδεγμένος
διαιρώ	διαίρεσα	διαιρέθηκα	διαιρεμένος
διακρίνω	διέκρινα	διακρίθηκα	διακεκριμένος
διανέμω	διένειμα	διανεμήθηκα	διανεμημένος
διαρρέω	διέρρευσα		
διαρρηγνύω	διέρρηξα		
διδάσκω	δίδαξα	διδάχτηκα	διδαγμένος
δίδω	έδωσα	εδόθην	δεδομένος
διευκολύνω	διευκόλυνα	διευκολύνθηκα	
διευθύνω	διεύθυνα	διευθύνθηκα	
δίνω	έδωσα	δόθηκα	δοσμένος
διψώ	δίψασα		διψασμένος
διώχνω	έδιωξα	διώχτηκα	διωγμένος
δρω	έδρασα		
δύναμαι		ηδυνήθην	
εγγράφω	ενέγραψα	εγγράφηκα	εγγεγραμμένος
εγγυώμαι		εγγυήθηκα	εγγυημένος
εγείρω	ήγειρα	ηγέρθην	εγερμένος
εγκαταλείπω	εγκατέλειψα	εγκαταλείφθηκα	εγκατα(λε)λειμμένος
εκδίδω	εξέδωσα	εκδόθηκα	εκδομένος
εκθέτω	εξέθεσα	εκτέθηκα	εκτεθειμένος
εκλέγω	εξέλεξα	εξελέγην	εκλεγμένος
εκπέμπω	εξέπεμψα	εξεπέμφθην	
εκρήγνυμαι		εξερράγην	
εκτείνω	εξέτεινα	εκτάθηκα	εκτεταμένος
εκτίω	εξέτισα		
ελέγχω	έλεγξα	ελέγχθηκα	ηλεγμένος
ελκύω	έλκυσα	ελκύστηκα	ελκυσμένος
εμμένω	ενέμεινα		
εμπνέω	ενέπνευσα		
ενδιαφέρω	ενδιέφερα	ενδιαφέρθηκα	ενδιαφερόμενος (P)
ενθαρρύνω	ενθάρρυνα	ενθαρρύνθηκα	
εννοώ	εννόησα	εννοήθηκα	

Present	Aorist active	Aorist passive	Past participle
εξαίρω	εξήρα	εξήρθην	εξηρμένος
εξαιρώ	εξαίρεσα	εξαιρέθηκα	εξαιρεμένος
επεκτείνω	επέκτεινα	επεκτάθηκα	επεκταμένος
επεμβαίνω	επενέβην		
επιβάλλω	επέβαλα	επιβλήθηκα	επιβεβλημένος
επιθυμώ	πεθύμησα		
επιμένω	επέμεινα		
επιστρέφω	επέστρεψα	επιστράφηκα	
επιτείνω	επέτεινα	επετάθην	επιτεταμένος
επιτίθεμαι		επετέθην	
επιτρέπω	επέτρεψα	επιτράπηκα	επιτετραμμένος
επιτυγχάνω	επέτυχα		επιτυχημένος
εργάζομαι		εργάστηκα	εργασμένος
έρχομαι		ήρθα	
ερωτεύομαι		ερωτεύθηκα	ερωτευμένος
ερωτώ	ηρώτησα	ηρωτήθην	
ευρίσκω	εύρον	ευρέθην	
εύχομαι		ευχήθηκα	
εφευρίσκω	εφεύρα	εφευρέθηκα	εφευρεμένος
εφιστώ	επέστησα		
έχω	είχα		
ζεσταίνω	ζέστανα	ζεστάθηκα	ζεσταμένος
ζεύω	έζεψα	ζεύτηκα	ζεμένος
ζητώ	ζήτησα	ζητήθηκα	ζητημένος
ζω	έζησα		
ηγούμαι		ηγήθην	
ηττώμαι		ηττήθην	ηττημένος
θαρρώ	θάρρεψα		
θέλω	θέλησα		
θέτω	έθεσα	τέθηκα	θεμένος
θίγω	έθιξα	εθίγην	
θυμίζω	θύμισα		θυμισμένος
θυμούμαι		θυμήθηκα	
θυμώνω	θύμωσα		θυμωμένος
ιδρύω	ίδρυσα	ιδρύθηκα	ιδρυμένος
ιδρώνω	ίδρωσα		ιδρωμένος
ίσταμαι		έστην/εστάθην	
ισχύω	ίσχυσα		
καθίζω	κάθισα		
καθιστώ	κατέστησα	κατέστην	κατεστημένος
κάθομαι		κάθισα	καθισμένος
καίω	έκαψα	κάηκα	καμένος
καλώ	κάλεσα	καλέστηκα	καλεσμένος
κά(μ)νω	έκαμα/έκανα		καμωμένος
κατάγομαι		κατήχθην	κατηγμένος
καταθέτω	κατέθεσα	κατετέθην	κατατεθειμένος
καταλαβαίνω	κατάλαβα		
καταλαμβάνω	κατέλαβα	κατελήφθην	κατειλημμένος
καταναλίσκω	κατανάλωσα	καταναλώθηκα	καταναλωμένος
κατανέμω	κατένειμα	κατανεμήθηκα	κατανεμημένος

Present	Aorist active	Aorist passive	Past participle
καταπίνω	κατάπια	καταπιώθηκα	καταπιωμένος
καταπλήσσω	κατέπληξα	κατεπλάγην	
καταριέμαι		καταράστηκα	καταραμένος
καταρρέω	κατέρρευσα		
καταρρίπτω	κατέρριψα	κατερρίφθην	
κατασταίνω	κατάστησα	καταστάθηκα	καταστημένος
κατάσχω	κατάσχεσα	κατασχέθηκα	κατασχεμένος
καταφέρνω	κατάφερα	καταφέρθηκα	καταφερμένος
καταφέρω	κατέφερα	κατεφέρθην	
κατεβάζω	κατέβασα		κατεβασμένος
κατεβαίνω	κατέβηκα		
κατέχω	κατείχα		
κέκτημαι		εκτήθην	κεκτημένος
κερνώ	κέρασα	κεράστηκα	κερασμένος
κηρύσσω	εκήρυξα	εκηρύχθην	(κε)κηρυγμένος
κλαίω	έκλαψα	κλάφτηκα	κλαμένος
κλέβω	έκλεψα	κλέφτηκα	κλεμμένος
κλεί(ν)ω	έκλεισα	κλείστηκα	(κε)κλεισμένος
κλέπτω	έκλεψα	εκλάπην	(κε)κλεμμένος
κλίνω	έκλινα	κλίθηκα	(κε)κλιμένος
κόβω	έκοψα	κόπηκα	κομμένος
κοιμούμαι		κοιμήθηκα	κοιμισμένος
κοιτάζω	κοίταξα	κοιτάχτηκα	κοιταγμένος
κοντεύω	κόντεψα		
κόπτω	έκοψα	εκόπην	(κε)κομμένος
κρεμ(ν)ώ	κρέμασα	κρεμάστηκα	κρεμασμένος
κρίνω	έκρινα	κρίθηκα	κεκριμένος
κρύβω	έκρυψα	κρύφτηκα	κρυμμένος
λαβαίνω	έλαβα	λήφθηκα	
λαμβάνω	έλαβον	ελήφθην	ειλημμένος
λανθάνω	έλαθον		λανθασμένος
λαχαίνω	έλαχα		
λέ(γ)ω	είπα	ειπώθηκα	ειπωμένος
λείβομαι		λείφτηκα	
λείπω	έλειψα	ελείφθην	
λήγω	έληξα		
λύνω	έλυσα	λύθηκα	λυμένος
λυπούμαι		λυπήθηκα	λυπημένος
λύω	έλυσα	ελύθην	λελυμένος
μαζεύω	μάζεψα	μαζεύτηκα	μαζεμένος
μαθαίνω	έμαθα	μαθεύτηκα	μαθημένος
μαίνομαι		εμαινόμην (I)	
μεθώ	μέθυσα		μεθυσμένος
μένω	έμεινα		
μεταχειρίζομαι		μεταχειρίστηκα	μεταχειρισμένος
μετέχω	μετείχα		
μηνύω	εμήνυσα	εμηνύθην	μηνυθείς
μηνώ	μήνυσα		
μιλώ	μίλησα	μιλήθηκα	μιλημένος
μιμούμαι		μιμήθηκα	
μοιάζω	έμοιασα		
μπαίνω	μπήκα		μπασμένος
μπορώ	μπόρεσα		

Present	Aorist active	Aorist passive	Past participle
νοιάζομαι		νοιάστηκα	
νοικιάζω	νοίκιασα	νοικιάστηκα	νοικιασμένος
νοιώθω	ένοιωσα		
ντρέπομαι		ντράπηκα	
ντύνω	έντυσα	ντύθηκα	ντυμένος
ξανοίγω	ξάνοιξα	ξανοίχτηκα	ξανοιγμένος
ξαπλώνω	ξάπλωσα	ξαπλώθηκα	ξαπλωμένος
ξαποστέλνω	ξαπόστειλα	ξαποστάλθηκα	ξαποσταλμένος
ξεμένω	ξέμεινα		
ξενοιάζω	ξένοιασα		ξενοιασμένος
ξέρω	ήξερα		
ξεχνώ	ξέχασα	ξεχάστηκα	ξεχασμένος
ξυπνώ	ξύπνησα		
οικειοποιούμαι		οικειοποιήθηκα	
ονειρεύομαι		ονειρεύτηκα	ονειρεμένος
οφείλω	όφειλα		οφειλόμενος (P)
παθαίνω	έπαθα		
παίζω	έπαιξα	παίχτηκα	παιγμένος
παίρνω	πήρα	πάρθηκα	παρμένος
παραγίνομαι		παραέγινα	παραγινωμένος
παραείμαι	παραήμουν		
παραπονιέμαι		παραπονέθηκα	παραπονεμένος
παραστέχομαι		παραστάθηκα	
παρατώ	παράτησα	παρατήθηκα	παρατημένος
παρεμβαίνω	παρενέβηκα		
παρέρχομαι		παρήλθα	
παρευρίσκομαι		παρευρέθηκα	
παρέχω	παρέσχον		
παρίσταμαι		παρέστην	
παριστάνω	παρέστησα	παρεστάθην	
πάσχω	έπαθα		
παύω	έπαψα	παύτηκα	παυμένος
πάω	πήγα		
πεθαίνω	πέθανα		πεθαμένος
πείθω	έπεισα	πείστηκα	πεισμένος
πεινώ	πείνασα		πεινασμένος
πέμπω	έπεμψα	επέμφθην	
περηφανεύομαι		περηφανεύτηκα	
περιέρχομαι		περιήλθα	
περιέχω	περιείχα		
περιλαμβάνω	περιέλαβα	περιελήφθην	
περιμένω	περίμενα		
περισπώ	περιέσπασα	περιεσπάσθην	
περνώ	πέρασα	περάστηκα	περασμένος
πετυχαίνω	πέτυχα		πετυχημένος
πετώ	πέταξα	πετάχτηκα	πεταμένος
πέφτω	έπεσα		πεσμένος
πηγαίνω	πήγα		
πιάνω	έπιασα	πιάστηκα	πιασμένος
πιέζω	πίεσα	πιέστηκα	(πε)πιεσμένος
πίνω	ήπια	πιώθηκα	πιωμένος
πλάθω	έπλασα	πλάστηκα	πλασμένος

Present	Aorist active	Aorist passive	Past participle
πλένω/πλύνω	έπλυνα	πλύθηκα	πλυμένος
πλέω	έπλευσα		
πλησιάζω	πλησίασα		
πλήττω	έπληξα		
πνέω	έπνευσα		
πνίγω	έπνιξα	πνίγηκα	πνιγμένος
ποικίλλω	εποίκιλα		ποικιλμένος
ποιούμαι		ποιήθηκα	ποιημένος
πράττω	έπραξα	επράχθην	πεπραγμένος
πρέπει	έπρεπε		
πρήζω	έπρηξα	πρήστηκα	πρησμένος
προαναφέρω	προανέφερα	προανεφέρθην	προαναφερθείς
προβάλλω (v.t.)	προέβαλα		
προγράφω	προέγραψα	προγράφ(τ)ηκα	προγραμμένος
προδίδω	πρόδωσα	προδόθηκα	προδομένος
προηγούμαι		προηγήθηκα	προηγούμενος (P)
προΐσταμαι		προέστην	προϊστάμενος (P)
προκαλώ	προκάλεσα	προκλήθηκα	
προκαταλαμβάνω	προκατέλαβον		προκατειλημμένος
πρόκειται		επρόκειτο (I)	
προλαβαίνω	πρόλαβα	προλήφθηκα	
προσαρμόζω	προσάρμοσα	προσαρμόστηκα	προσαρμοσμένος
προσέχω	πρόσεξα		
προστυχαίνω	προστύχεψα		
προσφέρω	προσέφερα	προσεφέρθην	
προτείνω	πρότεινα	προτάθηκα	προτεταμένος
προϋποθέτω	προϋπέθεσα		
προϋπολογίζω	προϋπελόγισα	προϋπελογίσθην	
προφταίνω	πρόφτασα		
ράβω	έραψα	ράφτηκα	ραμμένος
ρέω	έρρευσα		
ρημάζω	ρήμαξα		ρημαγμένος
ριγώ	ρίγησα		
ριγώνω	ρίγωσα		ριγωμένος
ριζώνω	ρίζωσα	ριζώθηκα	ριζωμένος
ρίχνω	έρριξα	ρίχτηκα	ριγμένος
ρουφώ	ρούφηξα		ρουφηγμένος
σβήνω	έσβησα	σβήστηκα	σβησμένος
σέβομαι		σεβάστηκα	
σέρνω	έσυρα	σύρθηκα	συρμένος
σημαίνω	σήμανα		σεσημασμένος
σημειώνω	σημείωσα	σημειώθηκα	σημειωμένος
σκέπτομαι		σκέφτηκα	εσκεμμένος
σπάζω/σπω	έσπασα	σπάστηκα	σπασμένος
σπείρω	έσπειρα	σπάρθηκα	σπαρμένος
σπεύδω	έσπευσα		εσπευσμένος
στάζω	έσταξα		
στέκω/-ομαι		στάθηκα	
στέλνω/στέλλω	έστειλα	στάλθηκα	σταλμένος
στενοχωρώ	στενοχώρησα	στενοχωρήθηκα	στενοχωρημένος
στηρίζω	στήριξα	στηρίχτηκα	στηριγμένος
στρέφω	έστρεψα	στράφηκα	στραμμένος
στρυμώχνω	στρύμωξα	στρυμώχτηκα	στρυμωγμένος

Present	Aorist active	Aorist passive	Past participle
στρώνω	έστρωσα	στρώθηκα	στρωμένος
συγγράφω	συνέγραψα	συγγράφηκα	συγγραμμένος
συγκαλύπτω	συνεκάλυψα	συνεκαλύφθην	συγκεκαλυμμένος
συγκρίνω	συνέκρινα	συνεκρίθην	συγκεκριμένος
συγχαίρω		συγχάρηκα (v.t.)	
συγχωρώ	συγχώρεσα	συγχωρέθηκα	συγχωρεμένος
συλλαμβάνω	συνέλαβα	συνελήφθην	
συμβαίνει	συνέβηκε		
συμβουλεύω	συμβούλεψα	συμβουλεύτηκα	
συμπεριφέρομαι		συμπεριφέρθηκα	
συμπίπτω	συνέπεσα		
συμπτύσσω	συνέπτυξα	συνεπτύχθην	συνεπτυγμένος
συμφέρει	συνέφερε		
συνάπτω	συνήψα	συνήφθην	συνημμένος
συνδέω	σύνδεσα	συνδέθηκα	συνδεδεμένος
συνεισφέρω	συνεισέφερα		
συνεπαίρνω	συνεπήρα		
συνέρχομαι		συνήλθα	
συνεφέρνω	συνέφερα		
συνιστώ	συνέστησα	συνεστήθην	συστημένος
συρρέω	συνέρρευσα		
σύρω	έσυρα	εσύρθην	(σε)συρμένος
συσσωρεύω	συσσώρευσα	συσσωρεύθηκα	συσσωρευμένος
συστέλλω	συνέστειλα	συνεστάλην	συνεσταλμένος
συστήνω	σύστησα	συστήθηκα	συστημένος
σφάζω	έσφαξα	σφάχτηκα	σφαγμένος
σφάλλω	έσφαλα	εσφάλην	εσφαλμένος
σφίγγω	έσφιξα	σφίχτηκα	σφιγμένος
σχολάζω	σχόλασα		
σχολιάζω	σχολίασα	σχολιάστηκα	
σώζω/σώνω	έσωσα	σώθηκα	σωσμένος
σωπαίνω	σώπασα		
τάζω	έταξα	τάχτηκα	τα(γ)μένος
ταΐζω	τάισα	ταΐστηκα	ταϊσμένος
ταξιδεύω	ταξίδεψα		ταξιδεμένος
τάσσω	έταξα	ετάχθην	τεταγμένος
τείνω	έτεινα	ετάθην	τεταμένος
τελώ	ετέλεσα	ετελέσθην	τετελεσμένος
τέμνω	έταμον	ετμήθην	τετμημένος
τέρπω	έτερψα		
τήκω	έτηξα		τετηγμένος
τίθεμαι		ετέθην	τεθειμένος
τίκτω	έτεκον	ετέχθην	
τινάζω	τίναξα	τινάχτηκα	τιναγμένος
τραβώ	τράβηξα	τραβήχτηκα	τραβηγμένος
τρατάρω/-έρνω	τράταρα		
τρελλαίνω	τρέλλανα	τρελλάθηκα	
τρέπω	έτρεψα	τράπηκα	τραμμένος
τρέφω	έθρεψα	τράφηκα	θρεμμένος
τρέχω	έτρεξα		
τρίβω	έτριψα	τρίφτηκα	τριμμένος
τρίζω	έτριξα		
τρομάζω	τρόμαξα		τρομαγμένος
τρώω	έφαγα	φαγώθηκα	φαγωμένος

Present	Aorist active	Aorist passive	Past participle
τυγχάνω	έτυχον		
τυλίγω	τύλιξα	τυλίχτηκα	τυλιγμένος
τυχαίνω	έτυχα		
υδρεύω	ύδρευσα	υδρεύθην	υδρευμένος
υιοθετώ	υιοθέτησα	υιοθετήθηκα	υιοθετημένος
υπάρχω	υπήρξα		
υπεκφεύγω'	υπεξέφυγα		
υπερηφανεύομαι			
see περη-			
υπερτερώ	υπερτέρησα		
υπηρετώ	υπηρέτησα	υπηρετήθηκα	υπηρετημένος
υποκύπτω	υπέκυψα		
υπονοώ	υπονόησα	υπονοήθηκα	
υποπτεύομαι		υποπτεύθηκα	
υπόσχομαι		υποσχέθηκα	
υποφέρω	υπέφερα		
υφαίνω	ύφανα	υφάνθηκα	υφασμένος
υφίσταμαι		υπέστην	
φαίνομαι		φάνηκα	
φαντάζομαι		φαντάστηκα	φαντασμένος
φαντάζω	φάνταξα		
φέγγω	έφεξα		
φείδομαι		εφείσθην	
φέρ(ν)ω	έφερα	φέρθηκα	φερμένος
φεύγω	έφυγα		
φθάνω	έφθασα		φθασμένος
φθείρω	έφθειρα	εφθάρην	φθαρμένος
φοβούμαι		φοβήθηκα	φοβισμένος
φορώ	φόρεσα	φορέθηκα	φορεμένος
φράζω	έφραξα	φράχτηκα	φραγμένος
φρεσκάρω	φρεσκάρισα	φρεσκαρίστηκα	φρεσκαρισμένος
φταίω	έφταιξα		
φτάνω	έφτασα		φτασμένος
φτιάνω	έφτιαξα	φτιάχτηκα	φτιαγμένος
φυλακίζω	φυλάκισα	φυλακίστηκα	φυλακισμένος
φυλάω	φύλαξα	φυλάχτηκα	φυλαγμένος
φυσώ	φύσηξα		
φωνάζω	φώναξα		
χαίρω/-ομαι		χάρηκα	
χαλώ	χάλασα	χαλάστηκα	χαλασμένος
χάνω	έχασα	χάθηκα	χαμένος
χορταίνω	χόρτασα		χορτασμένος
χρειάζομαι		χρειάστηκα	
χρεοκοπώ	χρεοκόπησα		χρεοκοπημένος
χρεώνω	χρέωσα	χρεώθηκα	χρεωμένος
χρ(ε)ωστώ	χρ(ε)ωστούσα (I)		
χρησιμεύω	χρησίμεψα		
χρησιμοποιώ	χρησιμοποίησα	χρησιμοποιήθηκα	χρησιμοποιημένος
χτίζω	έχτισα	χτίστηκα	χτισμένος
χυμώ	χύμηξα		
χύνω	έχυσα	χύθηκα	χυμένος
χωρίζω	χώρισα	χωρίστηκα	χωρισμένος

Present	Aorist active	Aorist passive	Past participle
χωρώ	χώρεσα		
ψάχνω	έψαξα	ψάχτηκα	
ψέλνω	έψαλα		ψαλμένος
ψεύδομαι		εψεύσθην	
ωφελώ	ωφέλησα	ωφελήθηκα	ωφελημένος
ωχριώ	ωχρίασα		